The chemistry of

Organocopper Compounds

Patai Series: The Chemistry of Functional Groups

A series of advanced treatises founded by Professor Saul Patai and under the general editorship of Professor Zvi Rappoport

The **Patai Series** publishes comprehensive reviews on all aspects of specific functional groups. Each volume contains outstanding surveys on theoretical and computational aspects, NMR, MS, other spectroscopic methods and analytical chemistry, structural aspects, thermochemistry, photochemistry, synthetic approaches and strategies, synthetic uses and applications in chemical and pharmaceutical industries, biological, biochemical and environmental aspects.
To date, over 120 volumes have been published in the series.

Recently Published Titles

The chemistry of the Cyclopropyl Group (Volume 2)
The chemistry of the Hydrazo, Azo and Azoxy Groups (Volume 2, 2 parts)
The chemistry of Double-Bonded Functional Groups (Volume 3, 2 parts)
The chemistry of Organophosphorus Compounds (Volume 4)
The chemistry of Halides, Pseudo-Halides and Azides (Volume 2, 2 parts)
The chemistry of the Amino, Nitro and Nitroso Groups (2 volumes, 2 parts)
The chemistry of Dienes and Polyenes (2 volumes)
The chemistry of Organic Derivatives of Gold and Silver
The chemistry of Organic Silicon Compounds (2 volumes, 4 parts)
The chemistry of Organic Germanium, Tin and Lead Compounds (Volume 2, 2 parts)
The chemistry of Phenols (2 parts)
The chemistry of Organolithium Compounds (2 volumes, 3 parts)
The chemistry of Cyclobutanes (2 parts)
The chemistry of Peroxides (Volume 2, 2 parts)
The chemistry of Organozinc Compounds (2 parts)
The chemistry of Anilines (2 parts)
The chemistry of Organomagnesium Compounds (2 parts)
The chemistry of Hydroxylamines, Oximes and Hydroxamic Acids (Volume 1, 2 parts)
The chemistry of Metal Enolates (2 parts)

Forthcoming Titles

The chemistry of Hydroxylamines, Oximes and Hydroxamic Acids (Volume 2)
The chemistry of Organomanganese Compounds
The chemistry of Selenium and Tellurium

The Patai Series Online

Starting in 2003 the **Patai Series** is available in electronic format on Wiley InterScience. All new titles will be published as online books and a growing list of older titles will be added every year. It is the ultimate goal that all titles published in the **Patai Series** will be available in electronic format.
For more information see under **Online Books** on:

www.interscience.wiley.com

R — Cu

Dedicated to my wife, Cecile

Ilan

Contributing authors

Özge Aksin-Artok	Dortmund University of Technology, Organic Chemistry, Otto-Hahn-Strasse 6, D-44227 Dortmund, Germany Fax: +49-231-7553884; e-mail: oezge.aksin@tu-dortmund.de
Alexandre Alexakis	University of Geneva, Department of Organic Chemistry, 30, Quai Ernest Ansermet 1211 Genève 4, Switzerland Fax: +41-22-3793215; e-mail: alexandre.alexakis@unige.ch
Sarah J. Aves	Department of Chemistry, University of Cambridge, Lensfield Road, Cambridge, CB2 1EW, UK Fax: +44-1223-336362; e-mail: sja41@cam.ac.uk
Asuncion Barbero	Departamento de Química Orgánica, Facultad de Ciencias, Universidad de Valladolid, 47011 Valladolid, Spain Fax: +34-983-423013; e-mail: barbero@qo.uva.es
Donald J. Burton	Department of Chemistry, University of Iowa, Iowa City, IA 52242, USA Fax: +1-319-335-1270; e-mail: donald-burton@uiowa.edu
Fabrice Chemla	Institut Parisien de Chimic Moléculaire, UPMC-Univ Paris 6, CNRS UMR 7201, Tour 44–45, 2ème Etage, Case 183, 4 Place Jussieu, 75252 Paris Cedex 05, France Fax: +33-1-44-27-75-67; e-mail: fabrice.chemla@upmc.fr
Charles R. Davis	Sigma-Aldrich, 6000 N. Teutonia Avenue, Milwaukee, WI 53209, USA Fax: +1-414-438-4224; e-mail: charles.davis@sial.com
R. Karl Dieter	Howard L. Hunter Laboratory, Department of Chemistry, Clemson University, Clemson, South Carolina 29634-0973, USA Fax: +1-864-656-6613; e-mail: dieterr@clemson.edu
Franck Ferreira	Institut Parisien de Chimie Moléculaire, UPMC-Univ Paris 6, CNRS UMR 7201, Tour 44–45, 2ème Etage, Case 183, 4 Place Jussieu, 75252 Paris Cedex 05, France Fax: +33-1-44-27-75-67; e-mail: franck.ferreira@upmc.fr
Tobias Gärtner	Institut für Organische Chemie, Universität Regensburg, Universitätsstraße 31, D-93040 Regensburg, Germany Fax: +49-941-943-4617; e-mail: tobias.gaertner@chemie.uni-regensburg.de

Dmitri Gelman	Institute of Chemistry, The Hebrew University of Jerusalem, Edmund Safra Campus, Givat Ram, 91904 Jerusalem, Israel Fax: +972-2-6585279; e-mail: dgelman@chem.ch.huji.ac.il
Saul Gómez-Manzo	Facultad de Química, Universidad Nacional Autónoma de México, Ciudad Universitaria, Coyoacán, 04510, D.F. México, México Fax: +52-5556162010; e-mail: sgomez@ifc.unam.mx
Ruth M. Gschwind	Institut für Organische Chemie, Universität Regensburg, Universitätsstraße 31, D-93040 Regensburg, Germany Fax: +49-941-943-4617; e-mail: ruth.gschwind@chemie.uni-regensburg.de
Johann T. B. H. Jastrzebski	Chemical Biology & Organic Chemistry, Debye Institute for Nanomaterials Science, Faculty of Science, Utrecht University, Padualaan 8, 3584 CH Utrecht, The Netherlands Fax: +31-30-2523615; e-mail: j.t.b.h.jastrzebski@uu.nl
Jan S. Jaworski	Faculty of Chemistry, University of Warsaw, 02-093 Warszawa, Poland Fax: +48-22-822-5996; e-mail: jaworski@chem.uw.edu.pl
Norbert Krause	Dortmund University of Technology, Organic Chemistry, Otto-Hahn-Strasse 6, D-44227 Dortmund, Germany Fax: +49-231-7553884; e-mail: norbert.krause@tu-dortmund.de
Peter M. H. Kroneck	Fachbereich Biologie, Universität Konstanz, 78457 Konstanz, Germany Fax: +49-7531882966, e-mail; peter.kroneck@uni-konstanz.de
Al Mokhtar Lamsabhi	Departamento de Química, Universidad Autónoma de Madrid, Cantoblanco, 28049-Madrid, Spain Fax: +34-91-4975238; e-mail: mokhtar.lamsabhi@uam.es
Joel F. Liebman	Department of Chemistry and Biochemistry, University of Maryland, Baltimore County, 1000 Hilltop Circle, Baltimore, Maryland 21250, USA Fax: +1-410-455-2608; e-mail: jliebman@umbc.edu
Eiichi Nakamura	Department of Chemistry, The University of Tokyo, Hongo, Bunkyo-ku, Tokyo 113-0033, Japan Fax: +81-3-5800-6889; e-mail: nakamura@chem.s.u-tokyo.ac.jp
Liza Penn	Institute of Chemistry, The Hebrew University of Jerusalem, Edmund Safra Campus, Givat Ram, 91904 Jerusalem, Israel Fax: +972-2-6585279; e-mail: elizabet.penn@mail.huji.ac.il

Damien Polet — University of Geneva, Department of Organic Chemistry, 30, Quai Ernest Ansermet 1211 Genève 4, Switzerland
Fax: +41-22-3793215; e-mail: damien.polet@addexpharma.com

Francisco J. Pulido — Departamento de Química Orgánica, Facultad de Ciencias, Universidad de Valladolid, 47011 Valladolid, Spain
Fax: +34-983-423013; e-mail: pulido@qo.uva.es

Olivier Riant — Unité de Chimie Organique et Médicinale, Département de Chimie, Place Louis Pasteur 1, Université Catholique de Louvain, 1348 Louvain la Neuve, Belgium
Fax : +32-010474168; e-mail: olivier.riant@uclouvain.be

Christopher J. Rosenker — Department of Chemistry, Parkman Ave. CSC 1301, University of Pittsburgh, Pittsburgh, PA 15260, USA
Fax: +1-412-624-0787; e-mail: cjr34@pitt.edu

Jean-Yves Salpin — CNRS—Université d'Evry Val d'Essonne—Laboratoire Analyse et Modélisation pour la Biologie et l'Environnement—UMR 8587—Bâtiment Maupertuis, Boulevard François Mitterrand, 91025 Evry, France
Fax: +33-1 69 47 76 55; e-mail: jean-yves.salpin@univ-evry.fr

Juan Pablo Saucedo-Vázquez — Facultad de Química, Universidad Nacional Autónoma de México, Ciudad Universitaria, Coyoacán, 04510, D.F. México, México
Fax: +52-5556162010; e-mail: jpablo_sa@correo.unam.mx

Mathias O. Senge — School of Chemistry, SFI Tetrapyrrole Laboratory, Trinity College Dublin, Dublin 2, Ireland
Fax: +353-1-608-8536; e-mail: sengem@tcd.ie
Institute of Molecular Medicine, Medicinal Chemistry, Trinity Centre for Health Sciences, Trinity College Dublin, St. James's Hospital, Dublin 8, Ireland

Natalia N. Sergeeva — School of Chemistry, SFI Tetrapyrrole Laboratory, Trinity College Dublin, Dublin 2, Ireland
Fax: +353-1-608-8536; e-mail: sergeevn@tcd.ie

Suzanne W. Slayden — Department of Chemistry, George Mason University, 4400 University Drive, Fairfax, Virginia 22030, USA
Fax: +1-703-993-1055; e-mail: sslayden@gmu.edu

Martha E. Sosa-Torres — Facultad de Química, Universidad Nacional Autónoma de México, Ciudad Universitaria, Coyoacán, 04510, D.F. México, México
Fax: +52-5556162010; e-mail: mest@servidor.unam.mx

Claude Spino — Université de Sherbrooke, Département de Chimie, Sherbrooke, QC, J1K 2R1, Canada
Fax: +1-819-821-8017; e-mail: claude.spino@usherbrooke.ca

David R. Spring

Department of Chemistry, University of Cambridge, Lensfield Road, Cambridge, CB2 1EW, UK
Fax: +44-1223-336362; e-mail: drspring@ch.cam.ac.uk

Jeanine Tortajada

CNRS—Université d'Evry Val d'Essonne—Laboratoire Analyse et Modélisation pour la Biologie et l'Environnement—UMR 8587—Bâtiment Maupertuis, Boulevard François Mitterrand, 91025 Evry, France
Fax: +33-1 69 47 76 55; e-mail: jeanine.tortajada@chimie.univ-evry.fr

Gerard van Koten

Chemical Biology & Organic Chemistry, Debye Institute for Nanomaterials Science, Faculty of Science, Utrecht University, Padualaan 8, 3584 CH Utrecht, The Netherlands
Fax: +31-30-2523615; e-mail: g.vankoten@uu.nl

Rhett T. Watson

Division of Natural Sciences and Engineering, University of South Carolina Upstate, 800 University Way, Spartanburg, South Carolina 29303, USA
Fax: +1-864-503-5366; e-mail: rwatson@uscupstate.edu

Peter Wipf

Department of Chemistry, Parkman Ave. CSC 1301, University of Pittsburgh, Pittsburgh, PA 15260, USA
Fax: +1-412-624-0787; e-mail: pwipf@pitt.edu

Manuel Yáñez

Departamento de Química, Universidad Autónoma de Madrid, Cantoblanco, 28049-Madrid, Spain
Fax: +34-91-4975238; e-mail: manuel.yanez@uam.es

Naohiko Yoshikai

Division of Chemistry & Biological Chemistry, School of Physical & Mathematical Sciences, Nanyang Technological University, SPMS-04-01, 21 Nanyang Link, Singapore 637371
Fax: +65-6791-1961; e-mail: nyoshikai@ntu.edu.sg

Foreword

The present book, *The Chemistry of Organocopper Compounds,* belongs to the sub-group of volumes dealing with organometallic derivatives within 'The Chemistry of Functional Groups' series. Previous related recent volumes are the three parts of *The Chemistry of Organolithium Compounds* (2004, 2006), *The Chemistry of Organozinc Compounds* (2006) and *The Chemistry of Organomagnesium Compounds* (2008), all edited by the editors of the present volume. The use of organocopper compounds in organic synthesis is increasing in recent decades, especially in selective reactions, new classes of reagents are available, and mechanistic understanding has been enhanced by using more sophisticated methods.

The two parts of the present volume consist of the most extensive review of organocopper compounds since N. Krause's *Modern Organocopper Chemistry* (Wiley-VCH 2002) and they follow the tradition of the series by dealing with a wide range of aspects of organocopper chemistry. The 19 chapters written by experts from 15 countries include theoretical aspects, structural analysis, thermochemistry, NMR spectra, mass spectrometric behavior, electrochemistry, photochemistry of selected sub-groups and biochemistry. Several chapters are devoted to synthetic aspects including the preparation of various kinds of organocopper reagents such as α-aminoalkylcuprates, fluorinated or silyl and stannyl derivatives and the synthetic uses of organocopper reagents in transmetallation reactions, carbocupration reactions, copper mediated reactions on extended multiple bond systems, asymmetric allylic alkylations, reductive reactions and cross-coupling reactions, copper catalyzed enantioselective conjugate additions, as well as oxidation of organocopper reagents. We gratefully acknowledge the contributions of all the authors of the book.

Unfortunately, the planned chapter on 'Copper mediated radical reactions' was not delivered.

The literature coverage is mostly up to the end of 2008, but includes few references in early 2009.

We will be grateful to readers who draw our attention to mistakes in the present volume and to omissions of important chapters related to organocopper chemistry.

ZVI RAPPOPORT, Jerusalem
ILAN MAREK, Haifa
June 2009

The Chemistry of Functional Groups
Preface to the series

The series 'The Chemistry of Functional Groups' was originally planned to cover in each volume all aspects of the chemistry of one of the important functional groups in organic chemistry. The emphasis is laid on the preparation, properties and reactions of the functional group treated and on the effects which it exerts both in the immediate vicinity of the group in question and in the whole molecule.

A voluntary restriction on the treatment of the various functional groups in these volumes is that material included in easily and generally available secondary or tertiary sources, such as Chemical Reviews, Quarterly Reviews, Organic Reactions, various 'Advances' and 'Progress' series and in textbooks (i.e. in books which are usually found in the chemical libraries of most universities and research institutes), should not, as a rule, be repeated in detail, unless it is necessary for the balanced treatment of the topic. Therefore each of the authors is asked not to give an encyclopaedic coverage of his subject, but to concentrate on the most important recent developments and mainly on material that has not been adequately covered by reviews or other secondary sources by the time of writing of the chapter, and to address himself to a reader who is assumed to be at a fairly advanced postgraduate level.

It is realized that no plan can be devised for a volume that would give a complete coverage of the field with no overlap between chapters, while at the same time preserving the readability of the text. The Editors set themselves the goal of attaining reasonable coverage with moderate overlap, with a minimum of cross-references between the chapters. In this manner, sufficient freedom is given to the authors to produce readable quasi-monographic chapters.

The general plan of each volume includes the following main sections:

(a) An introductory chapter deals with the general and theoretical aspects of the group.

(b) Chapters discuss the characterization and characteristics of the functional groups, i.e. qualitative and quantitative methods of determination including chemical and physical methods, MS, UV, IR, NMR, ESR and PES—as well as activating and directive effects exerted by the group, and its basicity, acidity and complex-forming ability.

(c) One or more chapters deal with the formation of the functional group in question, either from other groups already present in the molecule or by introducing the new group directly or indirectly. This is usually followed by a description of the synthetic uses of the group, including its reactions, transformations and rearrangements.

(d) Additional chapters deal with special topics such as electrochemistry, photochemistry, radiation chemistry, thermochemistry, syntheses and uses of isotopically labeled compounds, as well as with biochemistry, pharmacology and toxicology. Whenever applicable, unique chapters relevant only to single functional groups are also included (e.g. 'Polyethers', 'Tetraaminoethylenes' or 'Siloxanes').

This plan entails that the breadth, depth and thought-provoking nature of each chapter will differ with the views and inclinations of the authors and the presentation will necessarily be somewhat uneven. Moreover, a serious problem is caused by authors who deliver their manuscript late or not at all. In order to overcome this problem at least to some extent, some volumes may be published without giving consideration to the originally planned logical order of the chapters.

Since the beginning of the Series in 1964, two main developments have occurred. The first of these is the publication of supplementary volumes which contain material relating to several kindred functional groups (Supplements A, B, C, D, E, F and S). The second ramification is the publication of a series of 'Updates', which contain in each volume selected and related chapters, reprinted in the original form in which they were published, together with an extensive updating of the subjects, if possible, by the authors of the original chapters. Unfortunately, the publication of the 'Updates' has been discontinued for economic reasons.

Advice or criticism regarding the plan and execution of this series will be welcomed by the Editors.

The publication of this series would never have been started, let alone continued, without the support of many persons in Israel and overseas, including colleagues, friends and family. The efficient and patient co-operation of staff-members of the Publisher also rendered us invaluable aid. Our sincere thanks are due to all of them.

The Hebrew University
Jerusalem, Israel

SAUL PATAI
ZVI RAPPOPORT

Sadly, Saul Patai who founded 'The Chemistry of Functional Groups' series died in 1998, just after we started to work on the 100th volume of the series. As a long-term collaborator and co-editor of many volumes of the series, I undertook the editorship and I plan to continue editing the series along the same lines that served for the preceeding volumes. I hope that the continuing series will be a living memorial to its founder.

The Hebrew University
Jerusalem, Israel
May 2000

ZVI RAPPOPORT

Contents

1. Theory of organocopper-mediated reactions 1
   Eiichi Nakamura and Naohiko Yoshikai

2. Structural organocopper chemistry 23
   Gerard van Koten and Johann T. B. H. Jastrzebski

3. Thermochemistry of organocopper compounds 145
   Joel F. Liebman and Suzanne W. Slayden

4. NMR of organocopper compounds 163
   Tobias Gärtner and Ruth M. Gschwind

5. Photochemical transformations involving copper porphyrins and phthalocyanines 217
   Natalia N. Sergeeva and Mathias O. Senge

6. Electrochemistry of organocopper compounds 247
   Jan S. Jaworski

7. Gas-phase chemistry of organocopper compounds 279
   Al Mokhtar Lamsabhi, Manuel Yáñez, Jean-Yves Salpin and Jeanine Tortajada

8. Biochemistry of organocopper compounds 347
   Martha E. Sosa-Torres, Juan P. Saucedo-Vázquez, Saul Gómez-Manzo and Peter M. H. Kroneck

9. *N*-Functionalized organocuprates 365
   R. Karl Dieter and Rhett T. Watson

10. Transmetalation reactions producing organocopper compounds 443
    Christopher J. Rosenker and Peter Wipf

11. Carbocupration of alkynes 527
    Fabrice Chemla and Franck Ferreira

12. Oxidation of organocopper compounds 585
    Sarah J. Aves and David R. Spring

13. Copper-mediated asymmetric allylic alkylations 603
    Claude Spino

14. Copper-catalyzed enantioselective conjugate addition 693
Damien Polet and Alexandre Alexakis

15. Copper(I) hydride reagents and catalysts 731
Olivier Riant

16. Silyl and stannyl derivatives of organocopper compounds 775
Francisco J. Pulido and Asunción Barbero

17. Copper-mediated and copper-catalyzed addition and substitution reactions of extended multiple bond systems 857
Norbert Krause and Özge Aksin-Artok

18. Copper-mediated cross-coupling reactions 881
Liza Penn and Dmitri Gelman

19. Fluorinated organocopper reagents 991
Charles R. Davis and Donald J. Burton

Author index 1033

Subject index 1097

List of abbreviations used

Ac	acetyl (MeCO)
acac	acetylacetone
Ad	adamantyl
AIBN	azoisobutyronitrile
Alk	alkyl
All	allyl
An	anisyl
Ar	aryl
Bn	benzyl
Bu	butyl (C_4H_9)
Bz	benzoyl (C_6H_5CO)
c-	cyclo
CD	circular dichroism
CI	chemical ionization
CIDNP	chemically induced dynamic nuclear polarization
CNDO	complete neglect of differential overlap
Cp	η^5-cyclopentadienyl
Cp*	η^5-pentamethylcyclopentadienyl
DABCO	1,4-diazabicyclo[2.2.2]octane
DBN	1,5-diazabicyclo[4.3.0]non-5-ene
DBU	1,8-diazabicyclo[5.4.0]undec-7-ene
DIBAH	diisobutylaluminium hydride
DME	1,2-dimethoxyethane
DMF	*N*,*N*-dimethylformamide
DMSO	dimethyl sulphoxide
E-	entgegen
ee	enantiomeric excess
EI	electron impact
ESCA	electron spectroscopy for chemical analysis
ESR	electron spin resonance
Et	ethyl
eV	electron volt

Fc	ferrocenyl
FD	field desorption
FI	field ionization
FT	Fourier transform
Fu	furyl(OC_4H_3)
GLC	gas liquid chromatography
Hex	hexyl(C_6H_{13})
c-Hex	cyclohexyl(c-C_6H_{11})
HMPA	hexamethylphosphortriamide
HOMO	highest occupied molecular orbital
HPLC	high performance liquid chromatography
i-	iso
ICR	ion cyclotron resonance
Ip	ionization potential
IR	infrared
LAH	lithium aluminium hydride
LCAO	linear combination of atomic orbitals
LDA	lithium diisopropylamide
LUMO	lowest unoccupied molecular orbital
M	metal
M	parent molecule
MCPBA	*m*-chloroperbenzoic acid
Me	methyl
MNDO	modified neglect of diatomic overlap
MS	mass spectrum
n-	normal
Naph	naphthyl
NBS	*N*-bromosuccinimide
NCS	*N*-chlorosuccinimide
NMR	nuclear magnetic resonance
Pen	pentyl(C_5H_{11})
Ph	phenyl
Pip	piperidyl($C_5H_{10}N$)
ppm	parts per million
Pr	propyl (C_3H_7)
PTC	phase transfer catalysis or phase transfer conditions
Py	pyridine (C_5H_5N)
Pyr	pyridyl (C_5H_4N)
R	any radical
RT	room temperature

s-	secondary
SET	single electron transfer
SOMO	singly occupied molecular orbital
t-	tertiary
TCNE	tetracyanoethylene
TFA	trifluoroacetic acid
TFE	2,2,2-trifluoroethanol
THF	tetrahydrofuran
Thi	thienyl(SC_4H_3)
TLC	thin layer chromatography
TMEDA	tetramethylethylene diamine
TMS	trimethylsilyl or tetramethylsilane
Tol	tolyl(MeC_6H_4)
Tos or Ts	tosyl(*p*-toluenesulphonyl)
Trityl	triphenylmethyl(Ph_3C)
Vi	vinyl
XRD	X-ray diffraction
Xyl	xylyl($Me_2C_6H_3$)
Z-	zusammen

In addition, entries in the 'List of Radical Names' in *IUPAC Nomenclature of Organic Chemistry*, 1979 Edition, Pergamon Press, Oxford, 1979, p. 305–322, will also be used in their unabbreviated forms, both in the text and in formulae instead of explicitly drawn structures.

CHAPTER **1**

Theory of organocopper-mediated reactions

EIICHI NAKAMURA and NAOHIKO YOSHIKAI

Department of Chemistry, The University of Tokyo, Hongo, Bunkyo-ku, Tokyo 113-0033, Japan
Fax: +81-3-5800-6889; e-mail: nakamura@chem.s.u-tokyo.ac.jp, yoshikai@chem.s.u-tokyo.ac.jp

I. INTRODUCTION 1
II. STRUCTURE AND REACTIVITY OF ORGANOCOPPER COMPOUNDS 3
III. CONJUGATE ADDITION VERSUS S_N2 ALKYLATION REACTIONS . . 5
IV. ALLYLATION, ALKENYLATION AND ACYLATION REACTIONS . . . 10
V. EFFECT OF LEWIS ACID AND LEWIS BASE 15
A. BF_3 Effect 15
B. Me_3SiCl Effect 15
C. Dummy Ligand Effect 16
D. Cyano-Gilman Cuprates 17
VI. CONCLUDING REMARKS 18
VII. REFERENCES 19

I. INTRODUCTION

Organocopper compounds in organic chemistry appear most frequently in the form of nucleophilic organocopper(I) reagents, which are used either as stoichiometric reagents or as catalytic species generated *in situ* from a small amount of a copper(I) salt and a large amount of organomagnesium or zinc reagent. Generally formulated as R_2CuM bearing a variety of metal atoms and organic groups, metal organocuprates and related species are uniquely effective synthetic reagents for nucleophilic delivery of hard anionic nucleophiles such as alkyl, vinyl and aryl anions[1–7]. A wide variety of transformations including conjugate addition, carbocupration, alkylation, allylation, alkenylation and acylation reactions can be achieved readily with organocuprate reagents but not with other organometallics. The chemistry of organocopper reagents has been reviewed many times with emphasis

The chemistry of organocopper compounds
Edited by Z. Rappoport and I. Marek

on synthetic utility and on structural properties, but rarely on reaction mechanisms[8,9]. Organocopper chemistry has indeed developed to the current level of sophistication on the basis of working hypotheses that had rather weak experimental and theoretical support.

The history of nucleophilic organocopper chemistry dates back to 1941, when Kharasch and Tawney reported the 1,4-addition reaction of a Grignard reagent to an α,β-unsaturated ketone in the presence of a catalytic amount of a CuI salt[10]. Gilman and coworkers reported in 1952 that addition of one equivalent of MeLi to CuI results in the formation of a yellow precipitate, which then afford a colorless solution upon addition of another equivalent of MeLi[11]. In 1966, Costa and coworkers isolated a complex between phenylcopper(I) and magnesium, as well as crystals of a lithium diphenylcuprate(I) complex[12]. Since then, a large number of structures of lithium organocuprates(I) have been determined[13–15].

The synthetic chemistry of organocuprates started to develop quickly in 1966, when House and coworkers showed that the reactive species of conjugate addition is the lithium diorganocuprate called the Gilman reagent[16]. The foundation for the subsequent development was laid by Corey and Posner. Many important transformations such as substitution reactions of alkyl, alkenyl and allyl electrophiles[17–21] and carbocupration of acetylene[22] had been discovered by the mid-1970s.

Organocopper chemistry is still rapidly expanding its synthetic scope. Landmark achievements in the last decade are catalytic asymmetric conjugate addition and allylic substitution reactions by combinations of chiral copper catalysts and a variety of organometallic reagents such as Grignard, organozinc and organoaluminum reagents[23,24]. Even chiral quaternary carbon centers can be constructed in a highly enantioselective manner. Copper hydride chemistry and directing group strategies have also been developed[25,26].

Though the dominant part of synthetic organocopper chemistry has focused on the synthesis of biologically active compounds, some interesting developments in material science have been reported. Quantitative fivefold arylation of C_{60} and threefold arylation of C_{70} using magnesium-based organocopper reagents are good examples, providing a new class of cyclopentadienyl and indenyl structural motifs[27].

Despite synthetic utility of organocopper reagents, their reaction mechanisms had been poorly understood for a long time. This was primarily due to the structural complexity of organocopper reagents in solution[28]. Even the nature of the standard 'Gilman reagents' needs careful definition. Although numerous reports described the Gilman reagents as R_2CuLi, most of them are now known to participate in the reaction as an equilibrating mixture of various aggregate species such as $R_2CuLi{\bullet}LiX$, which is an inevitable side product of the preparation of 'R_2CuLi' from RLi with CuX. Even small differences in solvent and the accompanying salt affect the composition of the reagent and hence the reactivity[29]. The equilibrium nature of $R_2CuLi{\bullet}LiX$ has made it difficult to conduct mechanistic studies such as kinetic studies. Because of this complexity, it is now customary to indicate all ingredients used when describing a reagent. Understanding of the aggregation state in solution is fundamental for discussion of the reaction mechanism. In general, Gilman reagents largely exist as contact ion pairs (R_2CuLi dimers) in weakly coordinating solvents such as diethyl ether, while they prefer to take solvent-separated ion pair structures in well-solvating solvents such as THF. These species are in equilibrium with each other, which is closely associated with the reactivity of the reagent.

Numerous mechanistic investigations have been carried out to obtain crystallographic and spectroscopic information on reactants and products, as well as to understand the nature of reactive intermediates through kinetic and NMR spectroscopic studies. Nevertheless, the information gained until the mid-1990s was rather fragmentary and incomplete to provide a comprehensive mechanistic picture of organocopper reactions. A breakthrough was provided in the mid-1990s by the development of *ab initio* and density functional theories (DFT) that can quickly carry out electron-correlated calculations. With this computation/theoretical capacity, it became possible for the first time to understand the behavior of the multi-metallic cuprate systems and the role of the 3d orbitals of the copper atom. The theoretical studies were strongly supported by the comparison of the theoretical results with new sets of experimental results such as high-precision kinetic isotope effects (KIEs) and new NMR data such as those obtained by diffusion experiments, new pulse sequences and low-temperature techniques.

In this chapter, the current status of mechanistic understanding of organocopper-mediated reactions is summarized by focusing on lithium dialkylcuprate(I) clusters. Section II describes the fundamental relationship of the structure, molecular orbital and reactivity of an organocuprate (R_2Cu^-) molecule. Sections III and IV describe the mechanisms of basic organocuprate reactions with a focus on the frontier orbital interaction between the cuprate and each electrophilic substrate. Section V describes the theoretical aspects behind effects of Lewis acids or bases on the reactivity and selectivity of organocuprate reactions. Section VI describes the summary and the future outlook.

II. STRUCTURE AND REACTIVITY OF ORGANOCOPPER COMPOUNDS

Structures of diorganocuprate(I) species determined in crystals or by theory have invariably indicated a linear C−Cu−C coordination geometry as the most stable cuprate geometry. The highest occupied molecular orbital (HOMO) of a linear R_2Cu^- molecule mainly consists of the copper $3d_{z^2}$ orbital[30,31]. The $3d_{z^2}$ orbital is highest in energy among other copper 3d orbitals due to its strong anti-bonding interaction with the ligand 2p orbitals (Figure 1, left). Bending of the C−Cu−C geometry results in not only destabilization of the R_2Cu^- molecule (*ca* 20 kcal mol^{-1} increase in energy at 113°), but also mixing of the copper $3d_{xz}$ orbital with the ligand 2p orbitals (Figure 1, right). The energy level of the $3d_{xz}$ orbital is raised by this orbital mixing, and becomes the HOMO of a bent R_2Cu^- at $\theta < 150°$.

The above analysis demonstrates the strong correlation between the geometry and the nucleophilic reactivities of an organocuprate anion. For a linear R_2Cu^- molecule, the high-lying cuprate orbital will have symmetry suitable for interaction with C−X (X = halogen or heteroatom) σ^* orbital (Figure 2a). Alternatively, a bent R_2Cu^- species is now suitable for interaction with the π^* orbital of C=C double or C≡C triple bond (Figure 2b). Such an interaction is a typical case of Dewar–Chatt–Duncanson d−π^* back-donation[32], which largely compensates for the energy loss associated with the bending.

With the aid of these back-donation interactions, an organocuprate(I) reagent undergoes oxidative addition reaction with various electrophiles such as α,β-unsaturated carbonyl compounds and organic halides to form an organocopper(III) intermediate. The role of such Cu(III) species in the cuprate reactions has long been unclear, but now is accepted as the ubiquitous intermediates in the organocopper-mediated C−C bond formation reactions. Theory and experiments demonstrated that a T-shaped trialkylcopper(III) species is kinetically very unstable and undergoes reductive elimination without energy barrier, but can be stabilized by a fourth donative ligand through formation of a square-planar $R_3Cu{\bullet}L$ complex (Scheme 1)[33–39]. In real synthetic transformations, L can be either a neutral ligand such as ethereal solvents or heteroatom ligands or anionic ligands such as halides, cyanide or an intramolecular enolate anion (cf. Section III). The C−C bond formation reactions of organocuprate reagents commonly takes place through reductive elimination of such organocopper(III) species.

The above analysis provides the basis for understanding the mechanisms of the organocopper-mediated addition and substitution reactions, which will be detailed in the following sections. Similar analysis can be applied to the element in the same group such as gold, which however forms much more stable C−Au(I) bonds and therefore is unreactive[40,41].

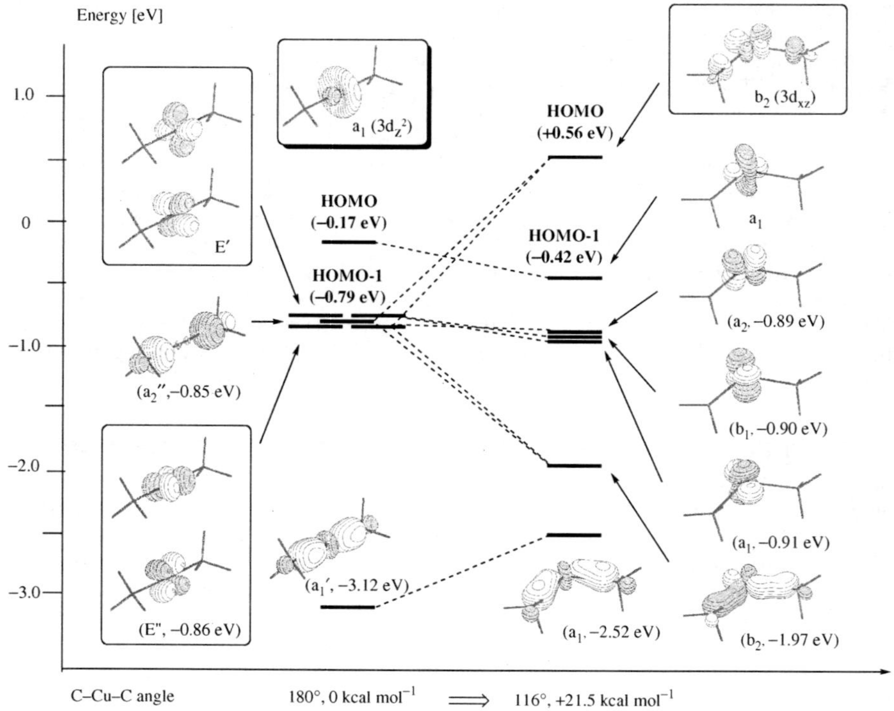

FIGURE 1. Correlation of the molecular orbital and the C−Cu−C angle of a $(CH_3)_2Cu$−molecule (B3LYP/Ahlrichs DZP for Cu, 6-311+G(d,p) for C, H//B3LYP/Ahlrichs SVP for Cu, 6-31G(d) for C, H)

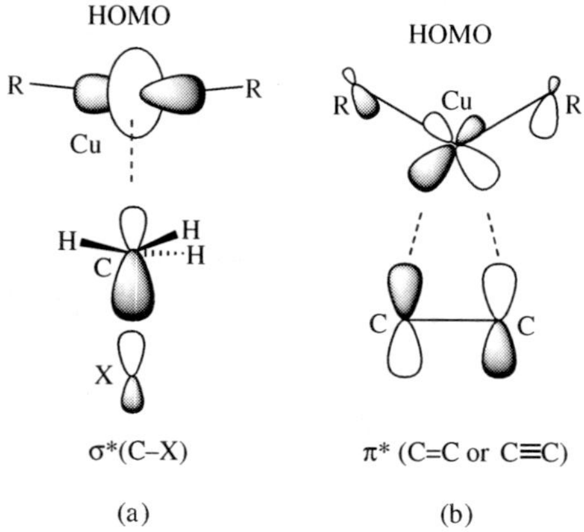

FIGURE 2. Schematic representations of orbital interactions between R_2Cu^- and electrophiles. (a) Interaction with an alkyl halide. (b) Interaction with an olefin or an acetylene

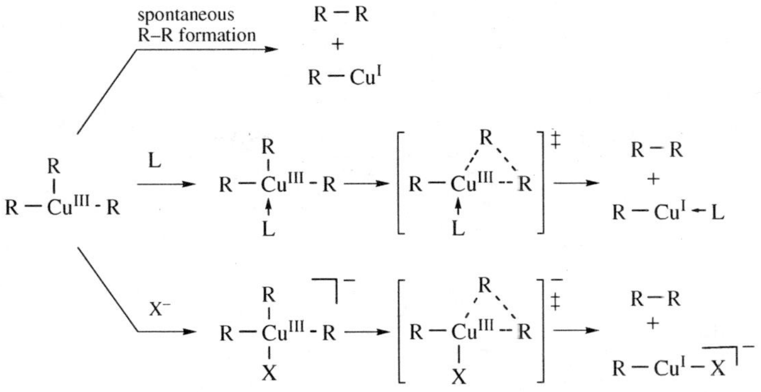

SCHEME 1. Reaction pathways of triorganocopper(III) complex (L = neutral ligand, X = anionic ligand)

On the other hand, the d-orbitals of zinc(II), a main group neighbor, are too low-lying to be nucleophilic and, as a result, organozinc compounds behave more like a Grignard reagent[42,43].

III. CONJUGATE ADDITION VERSUS S_N2 ALKYLATION REACTIONS

The conjugate addition to an α,β-unsaturated carbonyl compound and the S_N2 reaction with an alkyl halide or pseudohalide are illustrative examples that show the relationship of the geometry, molecular orbital and reactivity of an organocuprate reagent. The lowest unoccupied molecular orbitals (LUMOs) of these electrophilic substrates are the C=C π^* and C−X σ^* orbitals, respectively, which are entirely different from each other with respect to the orbital symmetry. An organocuprate reagent, by flexibly changing its geometry, can undergo an efficient frontier orbital interaction with each type of LUMO, which allows smooth reactions of the entirely different classes of electrophiles. In this section, reaction mechanisms and molecular orbital interactions involved in these and related reactions will be overviewed.

The conjugate addition reaction of organocuprates to α,β-unsaturated carbonyl compounds and related compounds is undoubtedly the most important organocopper reaction, but the mechanism has remained unclear for a long time. High-level quantum mechanical analyses coupled with experimental studies provide a solid mechanistic picture of the reaction. Thus, the reaction pathways of addition of $[Me_2CuLi]_2$ and $Me_2CuLi{\cdot}LiCl$ were studied first for acrolein[44] and later for cyclohexenone and its derivatives[45,46]. The conjugate addition of an organocuprate to an enone involves complicated multi-step equilibrium processes such as structural reorganization of the Cu/Li cluster and complexation of the Cu and Li atoms to the olefinic and carbonyl moieties, respectively. The most important intermediate of the reaction is an organocopper(III) intermediate that may be denoted as a β-cuprio(III) enolate, which is formed by the oxidative addition of the enone to the cuprate(I). This species undergoes rate-determining reductive elimination to afford the conjugate adduct (Scheme 2). The reductive elimination step is therefore the stage at which the stereochemistry of the conjugate adduct is determined. Theoretical analysis of this process reproduced a series of experimental data such as an activation energy[47], ^{13}C NMR chemical shift values[48] and $^{12}C/^{13}C$ KIEs[49].

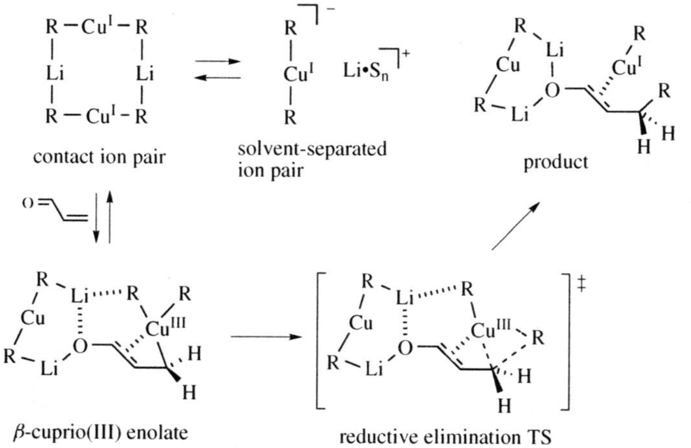

SCHEME 2. General reaction pathway of organocuprate conjugate addition (S = ethereal solvent). Coordinating solvents on lithium atoms are omitted for clarity

From a molecular orbital point of view, the organocopper conjugate addition represents a typical transition metal/olefin reaction. One can draw a pair of donation and back-donation schemes for Me_2Cu^- and ethylene (Figure 3a). This reaction, however, cannot be experimentally observed because ethylene is not electrophilic enough to accept electrons from the cuprate. When ethylene is substituted with a carbonyl group (e.g. acrolein), it becomes electrophilic enough to react with the cuprate. The donation/back-donation scheme is now unsymmetrical, as shown in Figure 3b. The complex can be viewed as a T-shaped Cu(III) complex bearing an enolate anion as the fourth ligand in a square-planar coordination sphere. For the reductive elimination to take place, the Cu(III) center needs to recover electrons from the Cu−C bond, since the two electrons localized in this bond have largely originated from the copper atom (Figure 3c).

Besides the d−π^* interaction as a fundamental driving force of conjugate addition, the lithium atom acts as a Lewis acid to activate the carbonyl oxygen and hence to assist the oxidative addition of the cuprate. Both experiments and theory demonstrated that the Lewis acidity of the lithium atom, which is attenuated by solvent coordination, is a critical factor of conjugate additions. For instance, addition of a crown ether (i.e. 12-crown-4) is known to retard most of the organocuprate reactions including conjugate addition. Evaluation of solvation of lithium atoms is therefore essential for theoretical prediction of activation energies (Figure 4).

The β-cuprio(III) enolate, a key intermediate in the conjugate addition, can undergo migration of the Cu(III) center when the β-position is substituted with an unsaturated group (Scheme 3)[50–52]. Reductive elimination of the resulting organocopper(III) species furnishes a remote conjugate addition product. The remote conjugate addition can be performed most successfully when the conjugation is terminated by an alkynyl group[53], because an allenylcopper(III) species involved is kinetically unstable to undergo facile reductive elimination.

The carbocupration of acetylene provides a reliable stereoselective route to *cis*-alkenylcopper(I) species, which can undergo further C−C bond formation as an alkenyl nucleophile. This reaction used to be considered to take place through a four-centered mechanism, and hence to be mechanistically different from conjugate addition. However, theoretical analysis of the reaction of a lithium organocuprate cluster ($[Me_2CuLi]_2$ or

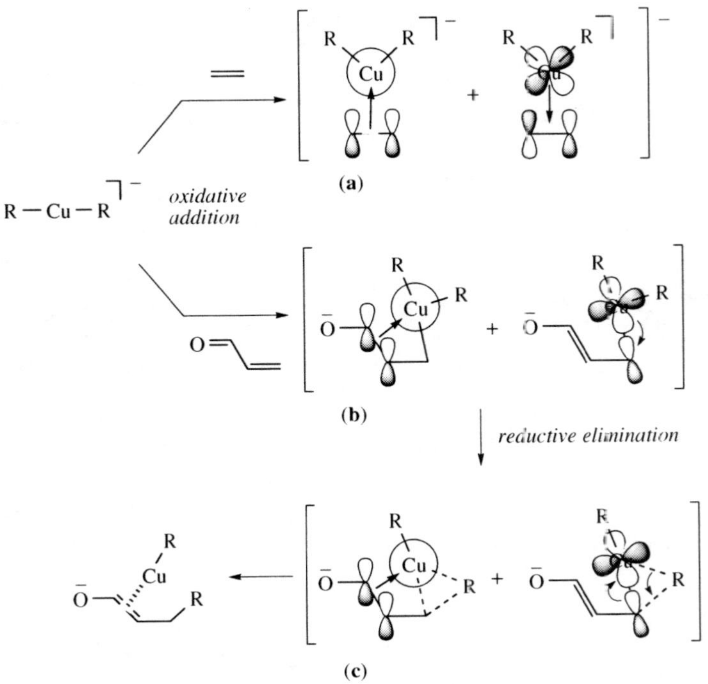

FIGURE 3. Schematic representations of frontier orbital interactions in (a) cuprate/olefin complex, (b) cuprate/enone complex, and (c) its reductive elimination TS

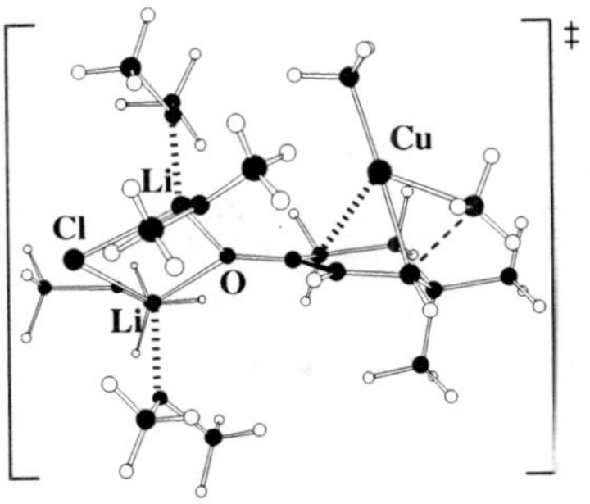

FIGURE 4. Transition state of the reaction of 4,4-dimethylcyclohex-2-en-1-one with $Me_2CuLi{\bullet}LiCl$ solvated with four molecules of dimethyl ether

$Me_2CuLi{\bullet}LiCl$) with acetylene revealed kinship of the two reactions (Scheme 4)[54]. While not substituted with a carbonyl group as an enone, intrinsically electron-deficient sp carbon atoms of acetylene allow facile back-donation interaction with an organocuprate (step i in Scheme 4; see also Figure 5)[55]. The resulting cuprio(III)cyclopropane intermediate, assisted by coordination of the lithium cation to the olefinic π-bond, undergoes

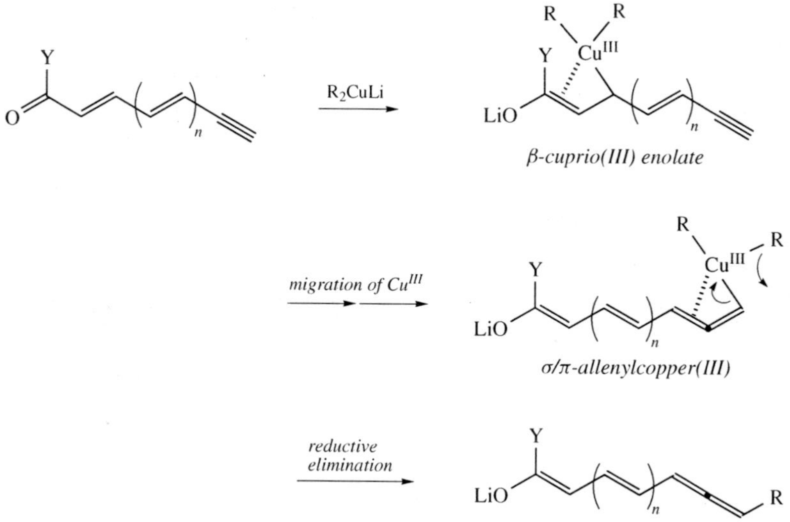

SCHEME 3. Reaction pathway of the remote conjugate addition of organocuprate

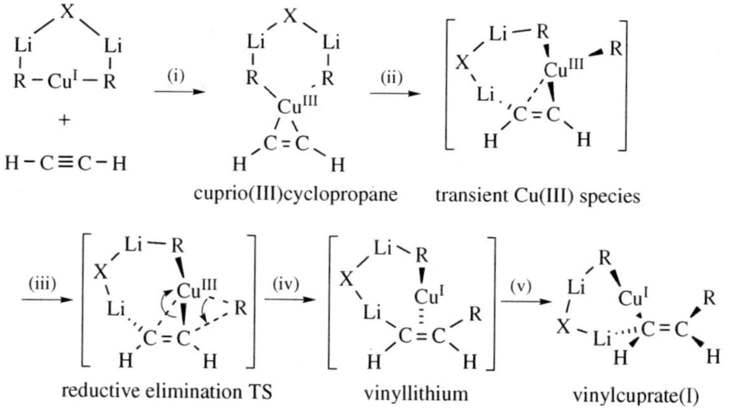

SCHEME 4. Reaction pathway for the carbocupration of acetylene

reductive elimination to form a C−C bond (step ii to iv). The initial product of carbocupration (i.e. alkenyllithium) undergoes Li/Cu transmetalation and affords an alkenylcuprate species (step v). Thus, the conjugate addition and the carbocupration involve in common an inner-sphere electron-transfer process that converts the stable $C{-}Cu^{I}$ bond into an unstable $C{-}Cu^{III}$ bond and the subsequent reductive elimination forms a new C−C bond.

The mechanistic bridge between the conjugate addition and the carbocupration was clearly demonstrated through theoretical analysis of cuprate additions to electron-deficient alkynes (i.e. ynones and ynoates)[56]. These reactions also involve the oxidative addition (d/π^* back-donation)/reductive elimination processes as the parent carbocupration reaction[57,58]. However, the final experimentally detectable products depend on the stability of the newly formed copper−carbon bond (Scheme 5). If the copper−carbon bond

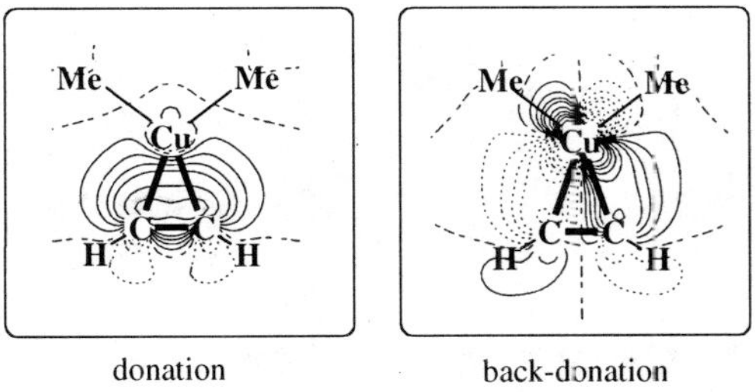

FIGURE 5. Localized Kohn–Sham orbitals of the complex between $Me_2CuLi{\bullet}LiCl$ and acetylene. Positions of the lithium and chlorine atoms are not indicated for clarity

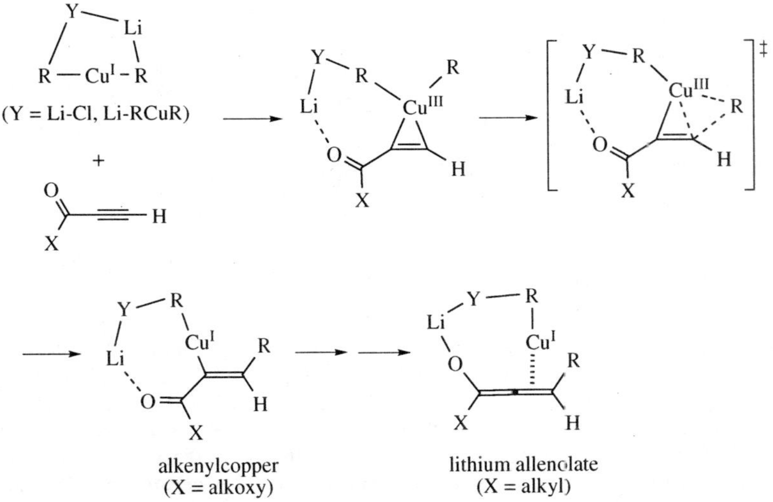

SCHEME 5. Reaction pathways of organocuprate addition to alkynyl carbonyl compounds

is stable enough (i.e. the case of ynoate), the product is an alkenylcopper species. If the bond is covalently weak (i.e. the case of ynone), the copper–carbon bond is replaced by a lithium–oxygen bond to give a lithium allenolate.

All of the addition reactions discussed above feature inner-sphere charge transfer from $3d_{xz}$ orbital of bent R_2Cu^- species to π^* orbitals of the electrophiles. Another typical reaction of an organocuprate reagent, which features a linear R_2Cu^- geometry as the reactive conformation, is S_N2–alkylation of alkyl halides, tosylates, epoxides or aziridines. The reaction takes place with inversion of stereochemistry at the electrophilic carbon[59], and the reaction rate is first-order both to the R_2CuLi dimer and to the electrophile[60]. Kinetic isotope effects (KIEs) for the reaction of $Me_2CuLi{\bullet}LiI{\bullet}PBu_3$ and CH_3I suggested that the rate-determining step of the reaction is the displacement of the leaving group[61].

Theoretical analyses of reactions of alkyl halides (CH_3I and CH_3Br) and epoxides (ethylene oxide and cyclohexene oxide) with lithium organocuprate clusters revealed a mechanistic aspect of this class of organocopper reactions that is entirely different from addition reactions (Scheme 6)[62,63]. DFT calculations showed that the rate-determining step of the alkylation reaction is the displacement of the leaving group (i.e. oxidative

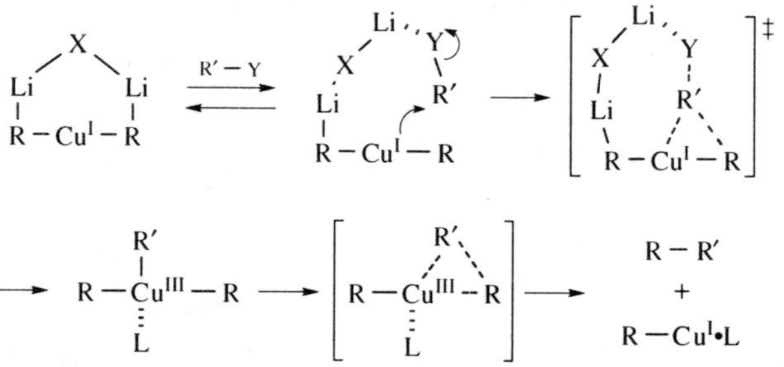

SCHEME 6. Reaction of $R_2CuLi{\bullet}LiX$ and an alkylating agent R'Y. The ligand L can be solvent, halide or metal salt

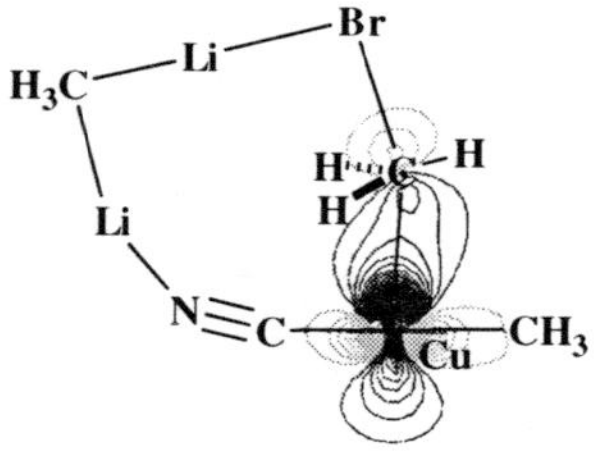

FIGURE 6. Localized Kohn–Sham orbital in the S_N2 reaction between a cyanocuprate (MeCu(CN) Li•LiMe) and MeBr. For the cyanocuprate, see Section V.D

addition) by a nucleophilic copper(I) atom, which involves efficient overlap of the copper $3d_{z^2}$ orbital of the linear R_2Cu^- species and the C–Y σ^* orbital (Figure 6). The presence of the lithium atom in the cuprate is critical, as it assists the C–Y bond cleavage by coordinating to the leaving Y atom. Such a 'push–pull' mode of substrate activation is a common feature of organocopper reactions, whereas the key copper d-orbital is different between the addition and S_N2 reactions. The calculated and experimental KIE values for the reaction of methyl iodide showed good agreement with each other.

The rate-determining displacement of the leaving group leads to formation of a T-shaped trialkylcopper(III) intermediate, which accepts coordination of a fourth ligand (halide, solvent etc.) to form a square-planar structure. The copper(III) complex features the *trans* relationship of the two R groups and the *cis* relationship of the R and R′ groups (Scheme 6), which result from the linear geometry of the R_2Cu^- moiety in the oxidative addition step. Hence, reductive elimination of the copper(III) complex exclusively gives the cross-coupling product R–R'.

A similar reaction pathway was found for the ring-opening alkylation of an epoxide with an organocuprate cluster. In contrast to the CH_3Br reaction, the stereochemistry of the electrophilic carbon is already inverted in the transition state, providing the reason for the preferred 'trans-diaxial opening' of cyclohexene oxides widely observed in synthetic studies. BF_3, a highly oxophilic Lewis acid, accelerates the epoxide opening reaction by participating in the C–O bond cleavage step (cf. Section V.A.)[64].

IV. ALLYLATION, ALKENYLATION AND ACYLATION REACTIONS

In this section, mechanisms of organocopper-mediated substitution reactions of allyl, alkenyl and acyl electrophiles will be overviewed. The common characteristics of these electrophiles is that they have low-lying C=C (or C=O) π^* and C−X σ^* orbitals, which are energetically close to each other and hence can mix with each other in an appropriate geometry. The key processes involved in the reaction of an organocuprate reagent with such an electrophile are (1) bending of the R_2Cu^- fragment, which raises the energy level of the $3d_{xz}$ orbital, and (2) structural change of the electrophile, which results in mixing of the π^* and σ^* orbitals. With these structural changes, the organocuprate reagent can undergo effective inner-sphere electron transfer to the electrophile, which eventually leads to the C−X bond cleavage[65]. Some details of each reaction system are discussed below.

The substitution reaction of an organocopper reagent with an allylic electrophile such as halides and esters provides an invaluable tool in organic synthesis. The reactivity profile is much more complex than the substitution reaction of an alkyl halide, because the C−C bond formation can take place either at the position α or γ to the position of the leaving group, and on the face *anti* or *syn*[66,67] to the side of the leaving group. An illustrative example of this complexity is shown in equation 1. Thus, the reaction of a dialkylcuprate is anti-stereoselective but non-regioselective[68,69], while a heterocuprate MeCu(CN)Li undergoes the reaction in an anti- and γ-selective manner[70]. While an allylcopper(III) species has long been an accepted intermediate in such reactions[71], it was only recently that theory shed light on the mechanism of its formation and reaction.

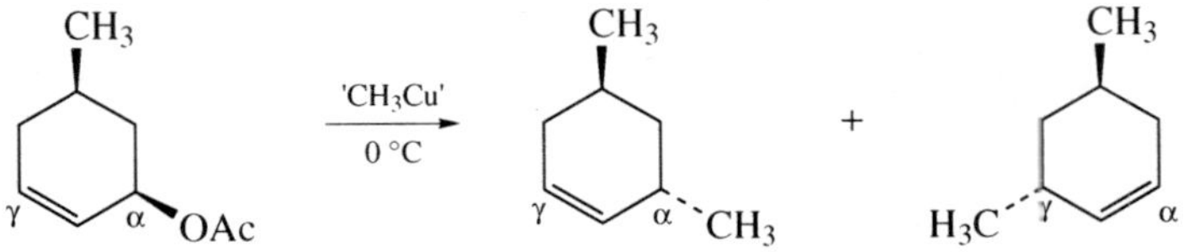

(1)

$CH_3Cu(CN)Li$	4	:	96
$(CH_3)_2CuLi$	50	:	50

The reaction pathway of the substitution reaction of a homocuprate and an allyl acetate is illustrated in Scheme 7[72]. The homocuprate reversibly forms a square-planar olefin π-complex, which then irreversibly releases an acetate anion in an *anti* fashion, leading to a symmetrical π-allylcopper(III) complex. The *anti* pathway is much more favorable than the *syn* pathway due to more effective overlap of the copper $3d_{xz}$ orbital and the C=C π^*/C−O σ^* mixed orbital of the allyl acetate. The cleavage of the C−OAc bond is intramolecularly assisted by a lithium cation. The π-allylcopper(III) is entirely symmetric in its geometry and electronic structure, and therefore undergoes reductive elimination either at the α or γ carbon atom. The π-allylcopper(III) can equilibriate with a less stable σ-allylcopper(III), which is however much less responsible for the C−C bond formation (Scheme 8).

If the starting material is substituted differently at the α and γ atoms, reductive elimination of the resulting π-allylcopper(III) intermediate takes place at different rates at the α and γ carbon atoms[73]. Analysis of a series of unsymmetrically substituted π-allylcopper(III) complexes indicated that the regioselectivity of the reductive elimination of such intermediates is mainly controlled by the electronic effect, and correlated well to the Hammett $\sigma_p{}^+$ constant[74]. In summary, a homocuprate undergoes *anti*-displacement of the leaving group to give a π-allylcopper(III) intermediate, and the regiochemistry of C−C bond formation from this intermediate is governed by electronic and steric effects of the substituent.

In contrast to the homocuprate reaction, the *trans* effect in the oxidative addition step is a critical factor for the γ-regioselective allylic substitution of a heterocuprate (see also Section V.C for heterocuprates). Such heterocuprates a priori offer two diastereomeric

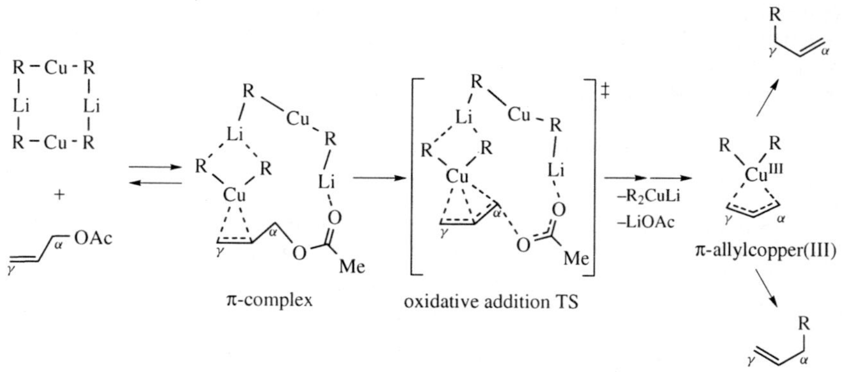

SCHEME 7. Reaction of R_2CuLi dimer and allyl acetate

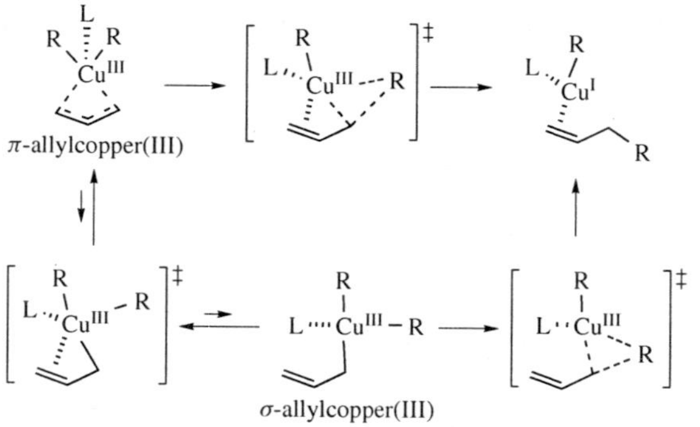

SCHEME 8. Isomerization and reductive elimination pathways of allylcopper(III) species

transition states of oxidative addition. For the case of the cyanocuprate, the alkyl ligand has higher σ-donor ability than the cyano ligand and thus causes stronger *trans* effect. This and the disymmetry of the C=C π*/C–O σ* mixed orbital make a transition state with the alkyl ligand on the γ side enjoy better FMO interaction than the opposite isomeric TS (Scheme 9). The *trans* effect is also reflected in the resulting allylcopper(III) intermediate, which takes an enyl $[\sigma + \pi]$ geometry. Due to the configurational stability of a tetracoordinated organocopper(III) species, once formed, the allylcopper(III) complex **A** undergoes reductive elimination exclusively at the γ position. Overall, the regioselectivity of allylic substitution of heterocuprate is controlled by the orbital interaction in the oxidative addition step.

Ligand effect on the rate of the oxidative addition/reductive elimination steps should be noted. Thus, less σ-donating ligands such as cyanide reduce the nucleophilicity of the organocuprate, and decelerate the oxidative addition. On the other hand, such ligands accelerate the reductive elimination because the allylcopper(III) species bearing such ligands are less stabilized.

Nucleophilic substitution on an alkenyl or acyl sp^2-carbon atom is another important class of organocopper-mediated reactions. In particular, substitution of an alkenyl halide initially reported in 1967 changed the accepted wisdom that a nucleophilic substitution on an unactivated sp^2 carbon is synthetically impracticable. While the original organocuprate

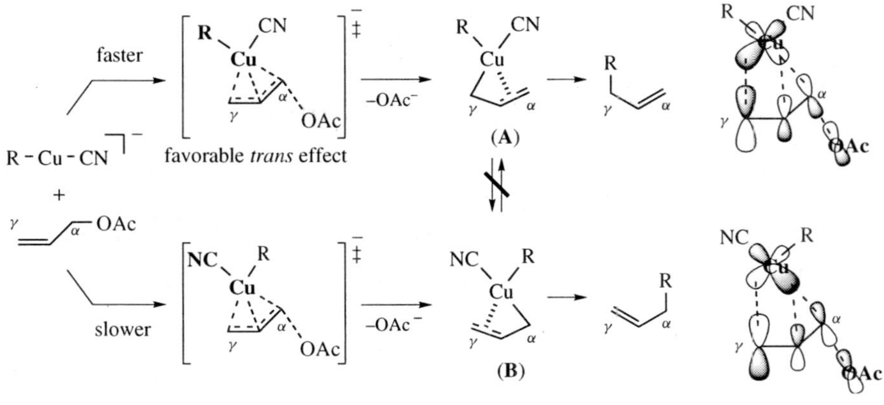

SCHEME 9. Reaction pathways of substitution of allyl acetate with a cyanocuprate. Schematic representations of the frontier orbital interaction during oxidative addition are shown on the right

reaction is now not as popular as conjugate addition and allylic substitution reactions, it can be regarded as the prototype of modern catalytic cross-coupling reactions of alkenyl and aryl electrophiles utilizing nickel(0) and palladium(0) complexes[75].

The substitution reaction has been considered to involve insertion of the Cu(I) atom into the sp^2 carbon–halogen bond (oxidative addition) and reductive elimination of the resulting organocopper(III) intermediate[76]. However, theoretical study offered a significant mechanistic modification as outlined in Scheme 10[77]. First, the cuprate forms a π-complex with the alkenyl bromide as it does with enones, acetylenes and allylic electrophiles (*vide supra*). The subsequent C–Br bond cleavage may go through a three-centered or an eliminative pathway. In the three-centered pathway, the leaving bromide migrates to the Cu(III) center, while the eliminative pathway is characterized by the trapping of the eliminated bromide by a Li cation. Comparison of experimental and calculated KIE data indicated that the latter pathway is more likely.

The common structural features of the three-centered and eliminative TSs are bending of the R_2Cu^- moiety and sp^2-to-sp^3 hybridization of the α-carbon atom. These structural changes of the reagent and substrate allow interaction of the Cu $3d_{xz}$ orbital (HOMO) with the C=C π^*/C–Br σ^* mixed orbital (LUMO), which is an important driving force of the C–Br bond cleavage (Figure 7). Note that this charge-transfer interaction is taking place already in the π-complex.

The theoretical study on the cuprate reaction revealed several important mechanistic issues in the oxidative addition of alkenyl and aryl electrophiles: (1) The formation of a π-complex prior to C–X bond cleavage[78], (2) intrinsic mechanistic dichotomy (three-centered vs. eliminative) of C–X bond cleavage, (3) assistance of C–X bond cleavage by a Lewis acid. This mechanistic framework offers a useful guideline for the design of a catalytic process involving oxidative addition as a critical step[79].

The mechanistic dichotomy of the three-centered and eliminative oxidative addition is also found in the substitution of an acyl electrophile by organocuprate reagents, which offers a reliable method for the synthesis of ketones[80,81]. The reaction pathway depends on the nature of the leaving group[82]. For instance, substitution of a thioester takes place through a three-centered insertion of the Cu atom into the C–S bond due to high affinity of thiolate to the Cu atom. On the other hand, an acid chloride prefers an eliminative pathway (Scheme 11).

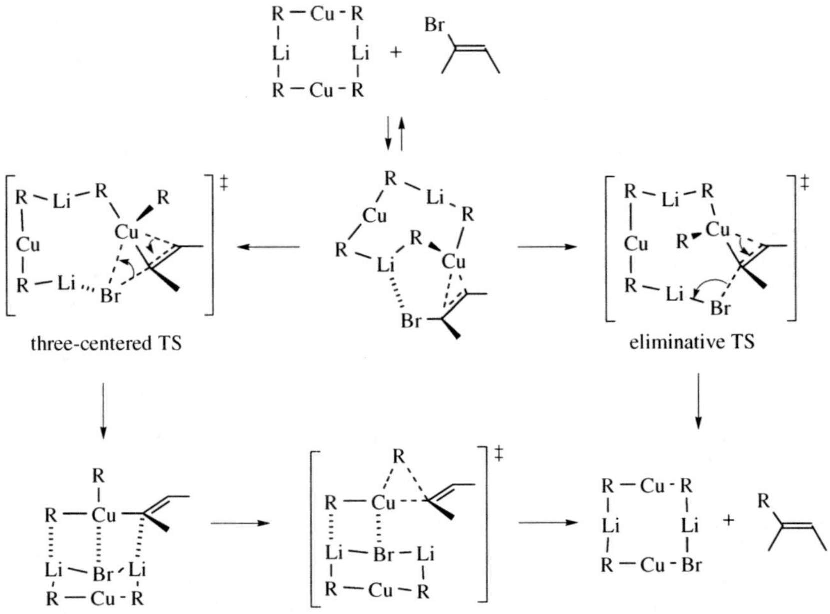

SCHEME 10. Reaction of R_2CuLi dimer and an alkenyl bromide

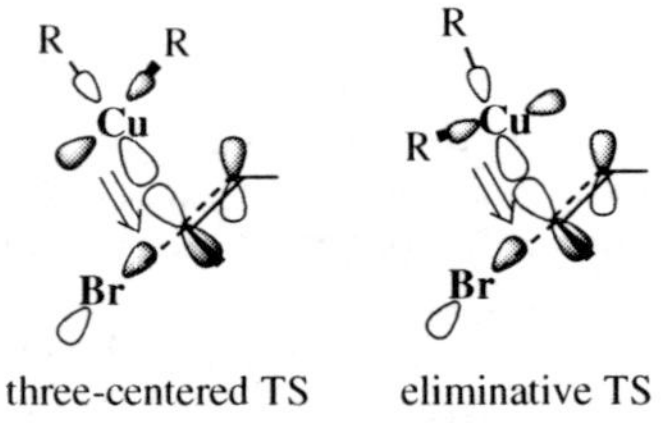

FIGURE 7. Schematic representations of orbital interaction during the oxidative addition of an alkenyl bromide to R_2Cu^-

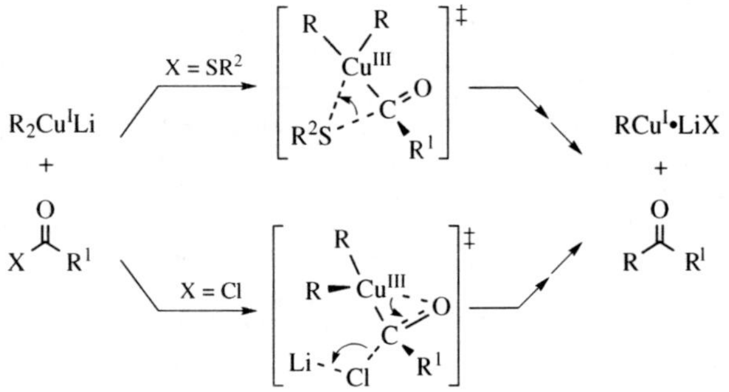

SCHEME 11. Reaction of an organocuprate and an acyl electrophile

V. EFFECT OF LEWIS ACID AND LEWIS BASE

A wide variety of modified organocopper reagents have been developed during the half-century history of organocopper chemistry. The modifiers, typically Lewis acidic or basic additives, are employed mainly for acceleration and regio-, stereo- and chemoselectivity control of the reaction. In this section, the theoretical basis behind some representative cases including BF_3 effect, Me_3SiCl effect, dummy ligand effect and cyano-Gilman cuprate will be discussed.

A. BF_3 Effect

It has been amply demonstrated that the rate and selectivity of organocopper reactions are significantly affected by countercations or Lewis acidic additives (e.g. BF_3[83], Zn(II)[84], Ti(IV)[85], Al(III)[86]). A representative example of such cases is that a Lewis acid such as BF_3 dramatically increases the reaction rate and changes the selectivity of conjugate addition of an organocopper reagent to an unsaturated carbonyl compound[87]. The Lewis acid activation of the carbonyl group was a working hypothesis of the use of BF_3. However, this idea contradicts the mechanism of the conjugate addition, that is, reductive elimination of organocopper(III) species is the rate-determining step (Section III). Alternatively, theoretical studies indicate that removal of one of the two alkyl groups by complexation of BF_3 on triorganocopper(III) accelerates the reductive elimination (Scheme 12).

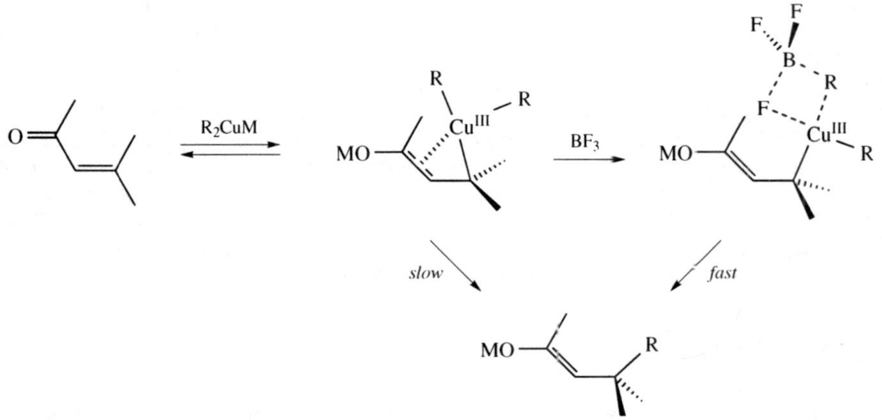

SCHEME 12. Proposed mechanism of the acceleration of the cuprate conjugate addition by BF_3

The origin of the acceleration of epoxide alkylation by BF_3 has also been examined theoretically. A plausible pathway for BF_3 participation in the reaction is shown in Scheme 13. The cooperative interaction of BF_3 fluorine and boron atoms with the cuprate and epoxide is responsible for the acceleration of the C–O bond cleavage.

B. Me_3SiCl Effect

Since the initial discovery by Nakamura and Kuwajima in 1984, chlorotrimethylsilane (Me_3SiCl) has been a standard reagent for acceleration of conjugate addition reactions. The effect was first reported for copper-catalyzed conjugate additions of zinc homoenolates[88], and followed by applications to Grignard-based catalytic organocopper

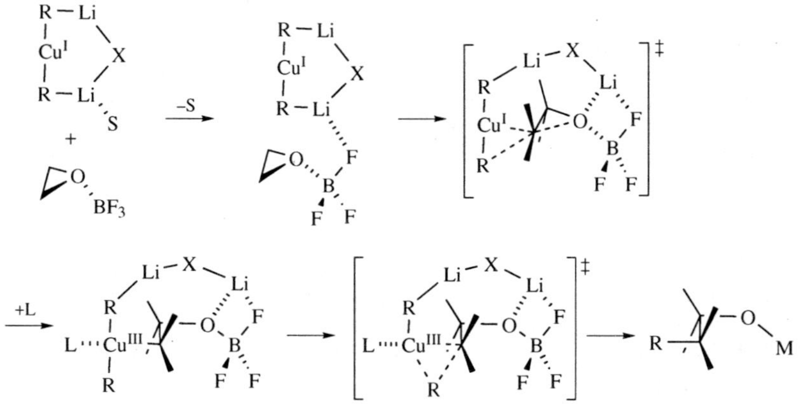

SCHEME 13. Mechanism of the acceleration of an epoxide alkylation reaction by BF_3 (S = ethereal solvent, L = ligand such as solvent or halide, M = metal cluster)

reagents and stoichiometric lithium diorganocuprates[89]. Acceleration of conjugate additions and modification of their selectivities by means of silylating agents are now well established.

The mechanism has become the subject of considerable discussion. The key steps in representative mechanistic proposals include silylation of a π-complex of an enone and a cuprate[90,91], Lewis acid activation of the enone substrate with Me_3SiCl[92], chloride coordination to the lithium cation[93] and chloride coordination to the copper atom[94], while none of them has thus far gained strong experimental and theoretical support. While these proposals failed to provide a direct answer to the mechanism of Me_3SiCl acceleration, the positive correlation between the silylating ability of the reagent and the magnitude of rate acceleration[95] strongly suggests that the rate-determining step of the reaction is the silylation step rather than the C−C bond forming step. Recent studies of kinetic isotope effects supported this conjecture[96]. Further mechanistic analysis including reaction rate, stereochemistry and theoretical analysis are still awaited.

C. Dummy Ligand Effect

A synthetic problem associated with the use of homocuprates R_2Cu^- is that the reagent can transfer only one of the two R ligands to the target electrophiles, with one R ligand being lost as an unreactive RCu species. The introduction of mixed organocuprates, in which the X group acts as a nontransferable dummy ligand, provided the first general solution to this problem. Typical dummy ligands include alkynyl[97], cyano[98], phenylthio[99], dialkylamino[100], phosphino[100] and trimethylsilylmethyl[101] groups. It has been accepted that the ligand transfer selectivity of a mixed organocuprate depends on the Cu−X bond strength. While this hypothesis has been successful for the design of dummy ligands, theoretical analysis of reductive elimination of organocopper(III) complex bearing alkyl and dummy ligands showed that the selectivity is accounted for by two factors, thermodynamic stability and kinetic reactivity of the Cu(III) intermediate[102,103].

For the typical dummy ligands, the *trans* effect and the strong Li−X affinity are the reason why these ligands stay on the copper atom (Figure 8a and 8b). In contrast, the trimethylsilylmethyl group acts as a dummy ligand due to a high activation energy required

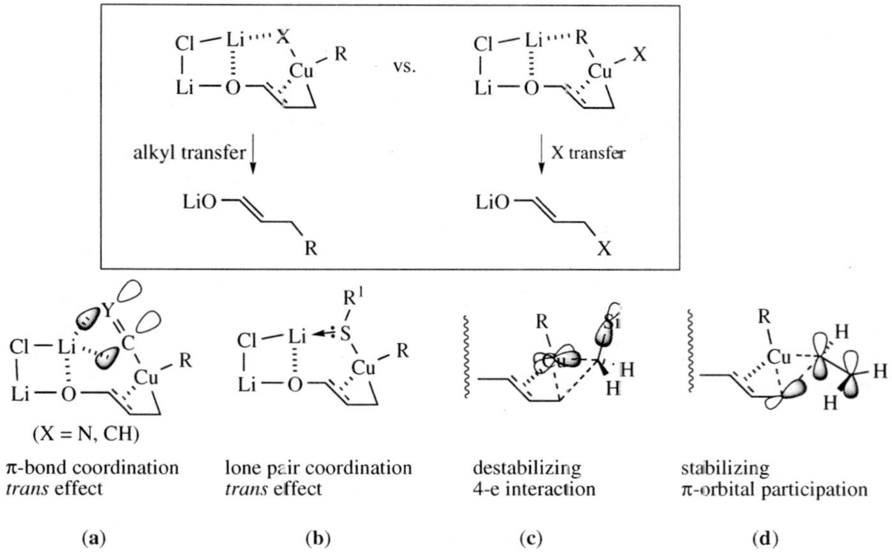

FIGURE 8. Ligand transfer selectivity in conjugate addition and its rationale

for its participation in C–C bond forming reductive elimination (Figure 8c). Transfer of an alkenyl group is preferred due to participation of C=C π-orbital in the C–C bond formation (Figure 8d). As described for allylic substitution reaction in Section IV, dummy ligands play decisive roles not only in the group transfer selectivity but also in the regio- and stereoselectivity of the reaction.

D. Cyano-Gilman Cuprates

Lipshutz and coworkers reported in 1981 that reagents formed by addition of two equivalents of RLi to CuCN give higher yields than the corresponding Gilman cuprates (R_2CuLi) or mixed cyanocuprates (RCu(CN)Li), and described them as higher-order cyanocuprates, assuming a triply coordinated structure[104]. The higher-order structure of this reagent has been a subject of controversy since the report of Bertz[105]. The current consensus based on a large number of structural studies is that the cyanide anion is not attached to copper and the cuprate must be regarded as a cyano-Gilman cuprate[106]. Despite this conclusion, the Lipshutz cuprate still remains the one of choice in many synthetic transformations. Furthermore, it shows some unusual reactivity, which suggests that a minute amount of alkyllithium is generated in equilibrium with the major R_2Cu^- species.

Theoretical studies on the kinetic reactivity of the Lipshutz reagent in S_N2 alkylation and conjugate addition reactions have been reported[107–109]. Exploration of the reaction pathways of organocuprate clusters of the $Me_2Cu(CN)Li_2$ composition showed that, while less stable than the cyano-Gilman structure **C**, the reactant **D** undergoes S_N2 alkylation and conjugate addition as smoothly as **C** (Scheme 14). Thus, such a minor and undetectable cuprate species may be responsible for the distinctly higher reactivity of the Lipshutz cuprate in some cases. The origin of the high reactivity of the minor isomer **D** is ascribed to an anisotropic coordination behavior of the cyanide anion.

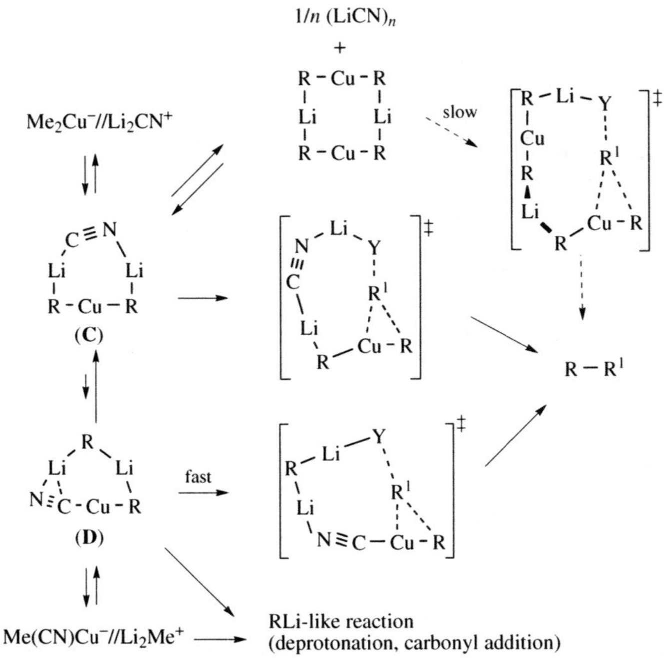

SCHEME 14. Proposed reaction pathway of the Lipshutz cuprate

VI. CONCLUDING REMARKS

Extensive mechanistic studies in the past 15 years as guided by theory have established a unified mechanism of the nucleophilic organocopper chemistry. The reactivity and the high synthetic versatility of organocopper reagents originate from two important factors, namely (1) the polarizable 3d orbitals of the copper(I) atom, the energy level of which is suitable for efficient inner-sphere electron transfer to a variety of electrophiles, and (2) their ability to form cluster structures bearing countercations, Lewis acids and bases, and ligands. The energy levels of the copper(I) 3d orbitals are much higher than those in zinc(II), and become even higher upon mixing with the 2p orbital of the alkyl ligand through R_2Cu^- formation. Organonickel and silver species are less stable and hence less synthetically useful than organocopper(I) reagents, while organogold(I) complexes are too stable for nucleophilic delivery of alkyl groups. The C−Cu−C angle is intimately related to the orbital symmetry and the reactivity of a diorganocuprate. The Cu(I)/Cu(III) redox sequence is assisted by cooperative interactions of an electrophilic substrate with a polymetallic organocopper cluster, which forms under appropriate conditions (solvent, ligand, countercation etc.).

The design of organocopper reagents and catalysts can now be considered as an issue of supramolecular chemistry, which requires elaborate control of cooperative transition state structures through organotransition metal chemistry, Lewis acid chemistry, coordination chemistry and so on. It is certain that the reactive species in catalytic organocopper chemistry is a cuprate, either R_2Cu^- or $R(X)Cu^-$, but details need to be the subject of systematic study[110–118]. Nonetheless, the information obtained from the stoichiometric cuprate has proven to be useful for designing the reactivity of catalytic copper reagent[119].

VII. REFERENCES

1. J. F. Normant, *Synthesis*, 63 (1972).
2. Y. Yamamoto, *Angew. Chem., Int. Ed. Engl.*, **25**, 947 (1986).
3. E. Nakamura, *Synlett*, 539 (1991).
4. B. H. Lipshutz and S. Sengupta, *Org. React.*, **41**, 135–631 (1992).
5. R. J. K. Taylor (Ed.), *Organocopper Reagents, A Practical Approach*, Oxford University Press, Oxford, 1994.
6. N. Krause and A. Gerold, *Angew. Chem., Int. Ed. Engl.*, **36**, 186 (1997).
7. N. Krause (Ed.), *Modern Organocopper Chemistry*, Wiley-VCH, Weinheim, 2002.
8. E. Nakamura and S. Mori, *Angew. Chem., Int. Ed.*, **39**, 3750 (2000).
9. S. Mori and E. Nakamura, in *Modern Organocopper Chemistry* (Ed. N. Krause), Wiley-VCH, Weinheim, 2002, pp. 315–346.
10. M. S. Kharasch and P. O. Tawney, *J. Am. Chem. Soc.*, **63**, 2308 (1941).
11. H. Gilman, R. G. Jones and L. A. Woods, *J. Org. Chem.*, **17**, 1630 (1952).
12. G. Costa, A. Camus, L. Gatti and N. Marsich, *J. Organomet. Chem.*, **5**, 568 (1966).
13. P. P. Power, *Prog. Inorg. Chem.*, **39**, 75 (1991).
14. G. van Koten, S. L. James and J. T. B. H. Jastrzebski, in *Comprehensive Organometallic Chemistry II* (Eds E. W. Abel, F. G. A. Stone and G. Wilkinson), Vol. 3, Pergamon, Oxford, 1995, pp. 57–133.
15. J. T. B. H. Jastrzebski and G. van Koten, in *Modern Organocopper Chemistry* (Ed. N. Krause), Wiley-VCH, Weinheim, 2002, pp. 1–44.
16. H. O. House, W. L. Respess and G. M. Whitesides, *J. Org. Chem.*, **31**, 3128 (1966).
17. E. J. Corey and G. H. Posner, *J. Am. Chem. Soc.*, **89**, 3911 (1967).
18. E. J. Corey and G. H. Posner, *J. Am. Chem. Soc.*, **90**, 5615 (1968).
19. G. M. Whitesides, W. F. Fischer Jr., J. San Filippo Jr., R. W. Bashe and H. O. House, *J. Am. Chem. Soc.*, **91**, 4871 (1969).
20. P. Rona, L. Tökes, J. Tremble and P. Crabbé, *J. Chem. Soc., Chem. Commun.*, 43 (1969).
21. H. L. Goering and S. Kantner, *J. Org. Chem.*, **46**, 2144 (1981).
22. A. Alexakis, J. Normant and J. Villiéras, *Tetrahedron Lett.*, **17**, 3461 (1976).
23. A. Alexakis, J.-E. Bäckvall, N. Krause, O. Pàmies and M. Diéguez, *Chem. Rev.*, **108**, 2796 (2008).
24. S. R. Harutyunyan, T. D. Hartog, K. Geurts, A. J. Minnaard and B. L. Feringa, *Chem. Rev.*, **108**, 2824 (2008).
25. C. Deutsch, N. Krause and B. H. Lipshutz, *Chem. Rev.*, **108**, 2916 (2008).
26. B. Breit and Y. Schmidt, *Chem. Rev.*, **108**, 2928 (2008).
27. Y. Matsuo and E. Nakamura, *Chem. Rev.*, **108**, 3016 (2008).
28. R. M. Gschwind, *Chem. Rev.*, **108**, 3029 (2008).
29. M. John, C. Auel, C. Behrens, M. Marsch, K. Harms, F. Bosold, R. M. Gschwind, P. R. Rajamohanan and G. Boche, *Chem. Eur. J.*, **6**, 3060 (2000).
30. S. Mori and E. Nakamura, *Tetrahedron Lett.*, **40**, 5319 (1999).
31. S. Mori, A. Hirai, M. Nakamura and E. Nakamura, *Tetrahedron*, **56**, 2805 (2000).
32. J. P. Collman, L. S. Hegedus, J. R. Norton and R. G. Finke, *Principles and Applications of Organotransition Metal Chemistry*, 2nd edn, University Science Books, Mill Valley, CA, 1987.
33. A. E. Dorigo, J. Wanner and P. v. R. Schleyer, *Angew. Chem., Int. Ed. Engl.*, **34**, 476 (1995).
34. J. P. Snyder, *J. Am. Chem. Soc.*, **117**, 11025 (1995).
35. E. Nakamura, M. Yamanaka and S. Mori, *J. Am. Chem. Soc.*, **122**, 1826 (2000).
36. S. H. Bertz, S. Cope, M. Murphy, C. A. Ogle and B. J. Taylor, *J. Am. Chem. Soc.*, **129**, 7208 (2007).
37. S. H. Bertz, S. Cope, D. Dorton, M. Murphy and C. A. Ogle, *Angew. Chem., Int. Ed.*, **46**, 7082 (2007).
38. T. Gärtner, W. Henze and R. M. Gschwind, *J. Am. Chem. Soc.*, **129**, 11362 (2007).
39. E. R. Bartholomew, S. H. Bertz, S. Cope, M. Murphy and C. A. Ogle, *J. Am. Chem. Soc.*, **130**, 11244 (2008).
40. W. Nakanishi, M. Yamanaka and E. Nakamura, *J. Am. Chem. Soc.*, **127**, 1446 (2005).
41. S. Komiya, T. A. Albright, R. Hoffmann and J. K. Kochi, *J. Am. Chem. Soc.*, **98**, 7255 (1976).
42. M. Uchiyama, S. Nakamura, T. Furuyama, E. Nakamura and K. Morokuma, *J. Am. Chem. Soc.*, **129**, 13360 (2007).

43. M. Uchiyama, S. Nakamura, T. Ohwada, M. Nakamura and E. Nakamura, *J. Am. Chem. Soc.*, **126**, 10897 (2004).
44. E. Nakamura, S. Mori and K. Morokuma, *J. Am. Chem. Soc.*, **119**, 4900 (1997).
45. S. Mori and E. Nakamura, *Chem. Eur. J.*, **5**, 1534 (1999).
46. M. Yamanaka and E. Nakamura, *Organometallics*, **20**, 5675 (2001).
47. J. Canisius, A. Gerold and N. Krause, *Angew. Chem., Int. Ed.*, **38**, 1644 (1999).
48. S. H. Bertz and R. A. J. Smith, *J. Am. Chem. Soc.*, **111**, 8276 (1989).
49. D. E. Frantz, D. A. Singleton and J. P. Snyder, *J. Am. Chem. Soc.*, **119**, 3383 (1997).
50. N. Yoshikai, T. Yamashita and E. Nakamura, *Angew. Chem., Int. Ed.*, **44**, 4721 (2005).
51. N. Yoshikai, T. Yamashita and E. Nakamura, *Chem. Asian. J.*, **1**, 322 (2006).
52. S. Mori, M. Uerdingen, N. Krause and K. Morokuma, *Angew. Chem., Int. Ed.*, **44**, 4715 (2005).
53. N. Krause and S. Thorand, *Inorg. Chim. Acta*, **296**, 1 (1999).
54. E. Nakamura, S. Mori, M. Nakamura and K. Morokuma, *J. Am. Chem. Soc.*, **119**, 4887 (1997).
55. S. Mori and E. Nakamura, *J. Mol. Struct. (THEOCHEM)*, **461–462**, 167 (1999).
56. S. Mori, E. Nakamura and K. Morokuma, *Organometallics*, **23**, 1081 (2004).
57. K. Nilsson, C. Ullenius and N. Krause, *J. Am. Chem. Soc.*, **118**, 4194 (1996).
58. K. Nilsson, T. Andersson, C. Ullenius, A. Gerold and N. Krause, *Chem. Eur. J.*, **4**, 2051 (1998).
59. C. R. Johnson and G. A. Dutra, *J. Am. Chem. Soc.*, **95**, 7783 (1973).
60. R. G. Pearson and C. D. Gregory, *J. Am. Chem. Soc.*, **98**, 4098 (1976).
61. C.-Y. Guo, M. L. Brownawell and J. San Filippo Jr., *J. Am. Chem. Soc.*, **107**, 6028 (1985).
62. E. Nakamura, S. Mori and K. Morokuma, *J. Am. Chem. Soc.*, **120**, 8273 (1998).
63. S. Mori, E. Nakamura and K. Morokuma, *J. Am. Chem. Soc.*, **122**, 7294 (2000).
64. A. Ghribi, A. Alexakis and J. F. Normant, *Tetrahedron Lett.*, **25**, 3075 (1984).
65. E. J. Corey and N. W. Boaz, *Tetrahedron Lett.*, **25**, 3063 (1984).
66. C. Gallina and P. G. Ciattini, *J. Am. Chem. Soc.*, **101**, 1035 (1979).
67. B. Breit and P. Demel, *Adv. Synth. Catal.*, **343**, 429 (2001).
68. H. L. Goering and V. D. Singleton, *J. Am. Chem. Soc.*, **98**, 7854 (1976).
69. H. L. Goering and V. D. Singleton, *J. Org. Chem.*, **48**, 1531 (1983).
70. H. L. Goering and S. S. Kantner, *J. Org. Chem.*, **49**, 422 (1984).
71. A. S. E. Karlström and J.-E. Bäckvall, *Chem. Eur. J.*, **7**, 1981 (2001).
72. N. Yoshikai, S.-L. Zhang and E. Nakamura, *J. Am. Chem. Soc.*, **130**, 12862 (2008).
73. M. Yamanaka, S. Kato and E. Nakamura, *J. Am. Chem. Soc.*, **126**, 6287 (2004).
74. C. Hansch, A. Leo and R. W. Taft, *Chem. Rev.*, **91**, 165 (1991).
75. A. de Meijere and F. Diederich (Eds.), *Metal-Catalyzed Cross-Coupling Reactions*, Wiley-VCH, New York, 2004.
76. R. Bruckner, *Advanced Organic Chemistry: Reaction Mechanisms*, Chap. 13., Academic Press, New York, 2002.
77. N. Yoshikai and E. Nakamura, *J. Am. Chem. Soc.*, **126**, 12264 (2004).
78. N. Yoshikai, H. Matsuda and E. Nakamura, *J. Am. Chem. Soc.*, **130**, 15258 (2008).
79. N. Yoshikai, H. Mashima and E. Nakamura, *J. Am. Chem. Soc.*, **127**, 17978 (2005).
80. G. H. Posner and C. E. Whitten, *Tetrahedron Lett.*, 4647 (1970).
81. R. J. Anderson, C. A. Henrick and L. D. Rosenblum, *J. Am. Chem. Soc.*, **96**, 3654 (1974).
82. N. Yoshikai, R. Iida and E. Nakamura, *Adv. Synth. Catal.*, **350**, 1063 (2008).
83. K. Maruyama and Y. Yamamoto, *J. Am. Chem. Soc.*, **99**, 8068 (1977).
84. M. Arai, T. Kawasuji and E. Nakamura, *J. Org. Chem.*, **58**, 5121 (1993).
85. M. Arai, E. Nakamura and B. H. Lipshutz, *J. Org. Chem.*, **56**, 5489 (1991).
86. S. Flemming, J. Kabbara, K. Nickisch, J. Westermann and J. Mohr, *Synlett*, 183 (1995).
87. A. B. Smith, III and P. J. Jerris, *J. Am. Chem. Soc.*, **103**, 194 (1981).
88. E. Nakamura and I. Kuwajima, *J. Am. Chem. Soc.*, **106**, 3368 (1984).
89. E. Nakamura, in *Organocopper Reagents* (Ed. R. J. K. Taylor), Oxford University Press, Oxford, 1994, pp. 129–142.
90. E. J. Corey and N. W. Boaz, *Tetrahedron Lett.*, **26**, 6015 (1985).
91. E. J. Corey and N. W. Boaz, *Tetrahedron Lett.*, **26**, 6019 (1985).
92. Y. Horiguchi, M. Komatsu and I. Kuwajima, *Tetrahedron Lett.*, **30**, 7087 (1989).
93. B. H. Lipshutz, S. H. Dimock and B. James, *J. Am. Chem. Soc.*, **115**, 9276 (1993).

94. S. H. Bertz, G. B. Miao, B. E. Rossiter and J. P. Snyder, *J. Am. Chem. Soc.*, **117**, 11023 (1995).
95. M. Eriksson, A. Johansson, M. Nilsson and T. Olsson, *J. Am. Chem. Soc.*, **118**, 10904 (1996).
96. D. E. Frantz and D. A. Singleton, *J. Am. Chem. Soc.*, **122**, 3288 (2000).
97. E. J. Corey and D. J. Beames, *J. Am. Chem. Soc.*, **94**, 7210 (1972).
98. J. P. Gorlier, L. Hamon, J. Levisalles and J. Wagnon, *J. Chem. Soc., Chem. Commun.*, 88 (1973).
99. G. H. Posner, C. E. Whitten and J. J. Sterling, *J. Am. Chem. Soc.*, **95**, 7788 (1973).
100. S. H. Bertz, G. Dabbagh and G. M. Villacorta, *J. Am. Chem. Soc.*, **104**, 5824 (1982).
101. S. H. Bertz, M. Eriksson, G. B. Miao and J. P. Snyder, *J. Am. Chem. Soc.*, **118**, 10906 (1996).
102. E. Nakamura and M. Yamanaka, *J. Am. Chem. Soc.*, **121**, 8941 (1999).
103. M. Yamanaka and E. Nakamura, *J. Am. Chem. Soc.*, **127**, 4697 (2005).
104. B. H. Lipshutz, R. S. Wilhelm and D. M. Floyd, *J. Am. Chem. Soc.*, **103**, 7672 (1981).
105. S. H. Bertz, *J. Am. Chem. Soc.*, **112**, 4031 (1990).
106. N. Krause, *Angew. Chem., Int. Ed.*, **38**, 79 (1999).
107. E. Nakamura and N. Yoshikai, *Bull. Chem. Soc. Jpn.*, **77**, 1 (2004).
108. E. Nakamura, M. Yamanaka, N. Yoshikai and S. Mori, *Angew. Chem., Int. Ed.*, **40**, 1935 (2001).
109. M. Yamanaka, S. Mori and E. Nakamura, *Bull. Chem. Soc. Jpn.*, **75**, 1815 (2002).
110. M. Kitamura, T. Miki, K. Nakano and R. Noyori, *Bull. Chem. Soc. Jpn.*, **73**, 999 (2000).
111. K. Nakano, Y. Bessho and M. Kitamura, *Chem. Lett.*, **32**, 224 (2003).
112. H. Zhang and R. M. Gschwind, *Angew. Chem., Int. Ed.*, **45**, 6391 (2006).
113. H. Zhang and R. M. Gschwind, *Chem. Eur. J.*, **13**, 6691 (2007).
114. K. Schober, H. Zhang and R. M. Gschwind, *J. Am. Chem. Soc.*, **130**, 12310 (2008).
115. T. Pfretzschner, L. Kleemann, B. Janza, K. Harms and T. Schrader, *Chem. Eur. J.*, **10**, 6048 (2004).
116. E. Gallo, F. Ragaini, L. Bilello, S. Cenini, C. Gennari and U. Piarulli, *J. Organomet. Chem.*, **689**, 2169 (2004).
117. A. Pichota, P. S. Pregosin, M. Valentini, M. Worle and D. Seebach, *Angew. Chem., Int. Ed.*, **39**, 153 (2000).
118. S. R. Harutyunyan, F. Lopez, W. R. Browne, A. Correa, D. Pena, R. Badorrey, A. Meetsma, A. J. Minnaard and B. L. Feringa, *J. Am. Chem. Soc.*, **128**, 9103 (2006).
119. A. Hajra, N. Yoshikai and E. Nakamura, *Org. Lett.*, **8**, 4153 (2006).

CHAPTER **2**

Structural organocopper chemistry

GERARD VAN KOTEN and JOHANN T. B. H. JASTRZEBSKI

Chemical Biology & Organic Chemistry, Debye Institute for Nanomaterials Science, Faculty of Science, Utrecht University, Padualaan 8, 3584 CH Utrecht, The Netherlands
Fax: +31-30-2523615; e-mail: j.t.b.h.jastrzebski@uu.nl

I. INTRODUCTION 23
II. ALKYL- AND ARYLCOPPER(I) COMPOUNDS 27
A. Donor-base-free Organocopper Compounds, RCu 27
B. Organocopper Compounds Containing Intramolecularly Coordinating Substituents, (LR)Cu 34
C. Donor–Acceptor Complexes of Organocopper Compounds (RCu)-L . . 39
III. ORGANOCOPPER(I) COPPER(I)-X AGGREGATES (X = MONOANIONIC GROUP) 51
IV. CYCLOPENTADIENYLCOPPER(I) COMPOUNDS 63
V. COPPER CARBORANE COMPOUNDS 68
VI. COPPER(I) ACETYLIDES 73
VII. ORGANOCUPRATES 90
A. Neutral Organocuprates 90
B. Neutral Heterocuprates 101
C. Ionic Structures 104
D. Cyanocuprates 108
VIII. ORGANOCOPPER(II) AND ORGANOCOPPER(III) COMPOUNDS . . . 112
IX. COPPER CARBENE COMPLEXES 119
X. CONCLUSIONS 133
XI. REFERENCES 135

I. INTRODUCTION

The history of organocopper chemistry dates back to 1859 when Böttger reported the preparation of the *highly explosive* copper(I) acetylide Cu_2C_2, a compound most probably involving copper-to-carbon bonds[1]. In the same year the reaction of diethylzinc with CuCl was reported by Buckton[2]. Instead of the anticipated 'ethylcopper' only metallic copper mirrors were obtained, from which he concluded that it was impossible to bind an organic

The chemistry of organocopper compounds
Edited by Z. Rappoport and I. Marek

group to copper. However, about sixty years later the isolation of phenylcopper from the reaction of a phenyl Grignard reagent with CuI was reported by Reich[3]. In 1952 Gilman and coworkers reported the formation of methylcopper as a yellow solid, obtained from the reaction of two equivalents of MeLi with $CuCl_2$ in diethyl ether as a solvent[4]. The first step of this reaction involved reduction of Cu(II) to Cu(I) by which the first equivalent of MeLi acts as a one-electron reductor (equation 1) followed in the second step by the transfer of the methyl group from the second equivalent of MeLi from lithium to copper (equation 2). Methylcopper is a yellow solid, stable as a suspension in diethyl ether at 0 °C, but which decomposes *explosively* when made dry.

$$2\,\mathrm{MeLi} + 2\,\mathrm{CuCl_2} \longrightarrow \mathrm{Cu_2Cl_2} + 2\,\mathrm{LiCl} + \mathrm{C_2H_6} \qquad (1)$$

$$2\,\mathrm{MeLi} + \mathrm{Cu_2Cl_2} \longrightarrow 2\,\mathrm{MeCu} + 2\,\mathrm{LiCl} \qquad (2)$$

In 1936 the applicability of organocopper reagents in synthetic organic chemistry was demonstrated by the pioneering work of Gilman and Straley[5]. Somewhat later, it appeared that catalytic amounts of copper halides favored the 1,4-addition over the more usually observed 1,2-addition mode in the reaction of Grignard reagents with α,β-unsaturated ketones (equation 3)[6] and α,β-unsaturated esters[7], a discovery that appeared to be of crucial importance for the further development of organocopper reagents.

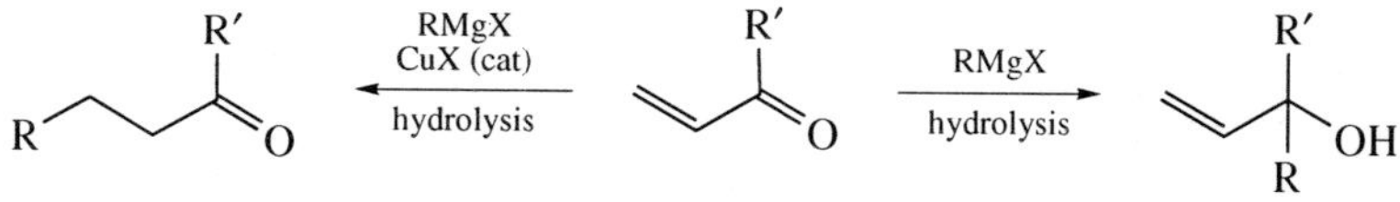

(3)

The synthesis of the Gilman cuprate Me_2CuLi[4,8–10] and related R_2CuLi reagents, and the demonstration of their synthetic potential by House and coworkers[8,9] and Corey and Posner[10], resulted in a major breakthrough in the field of copper-mediated synthetic organic chemistry. An obvious disadvantage of this type of cuprates, which are used in stoichiometric amounts, is the fact that only one equivalent of the (precious) organic groups is used in the reaction. This was already recognized at an early stage of these investigations and later was addressed by Posner and coworkers[11], who introduced the concept of cuprate reagents containing one non-transferable (non-precious) group and one transferable grouping. An example of such a non-transferable group is the PhS^- anion, introduced in the reaction mixture as CuSPh. Much later this idea was applied in the enantioselective 1,4-addition reaction of Grignard reagents to α,β-unsaturated ketones in the presence of catalytic amounts of a chiral copper(I) 2-aminoarenethiolate in which the chiral 2-aminoarenethiolato anion not only acts as a non-transferable group but also induces enantioselective transfer of the organic group to the prochiral substrate[12,13].

Another class of compounds that much contributed to the applicability of organocopper reagents in organic synthesis are the so-called higher order cyanocuprates $R_2Cu(CN)Li_2$, initially developed by Lipshutz and coworkers[14–16].

Recently, the mechanistic insights in both reactions involving cuprates and other organocopper compounds and in the nature of these reagents itself have been reviewed[17–19]. Moreover, these insights are corroborated by modern computational studies[20,21].

Systematic studies towards the synthesis, isolation and characterization of organocopper compounds started in the late 1960s and early 1970s. Important contributors to these early studies were: (i) Camus and coworkers, on phenylcopper, as well as 2-, 3- and 4-tolylcopper[22,23], (ii) Cairncross and coworkers, on pentafluorophenylcopper and $3\text{-}CF_3C_6H_4Cu$[24,25], (iii) Lappert and Pearce, on (trimethylsilyl)methylcopper[26] and

(iv) van Koten and coworkers, who introduced the concept of stabilization by intramolecular coordination as, e.g., in 2-$Me_2NCH_2C_6H_4Cu$ and other functionally substituted arylcopper compounds[27,28]. Already at an early stage of these studies it became clear that organocopper compounds have a great tendency to form aggregates with all kinds of metal salts, including copper(I) salts. It appeared that materials, earlier regarded as impure organocopper compounds, in fact were pure materials in which the copper halide is an integral part of the formed polynuclear organocopper compound, as, e.g., in (2-$Me_2NC_6H_4)_4Cu_6Br_2$ (*vide infra*)[29,30].

Copper is a late transition element and may occur in formal oxidation states ranging from 0 to 4+ and of which the ions readily form complexes to yield a wide variety of coordination compounds. Oxidation states 1+, 2+ and 3+ are the most common ones, while oxidation states 0 and 4+ in copper compounds are extremely rare. In inorganic and coordination chemistry the 2+ oxidation state is the most abundant one, and is regarded to be more stable than the 1+ oxidation state. However, the organometallic chemistry of copper, in terms of isolable compounds, is strongly dominated by compounds in which copper is in the 1+ oxidation state.

The intrinsic instability of hypothetical diorganocopper(II) compounds R_2Cu is most probably associated with the redox properties of copper and the stability of the R radical. Decomposition of such compounds can occur via two different routes: (i) the formation of an organocopper compound with copper in its 1+ oxidation state and a radical $R^\bullet$, which is formally a one-electron reduction process (equation 4), and (ii) direct formation of R–R and Cu(0), which is formally a two-electron reduction process (reductive elimination) (equation 5).

$$R{-}Cu{-}R \longrightarrow R{-}Cu + R^\bullet \quad (4)$$

$$R{-}Cu{-}R \longrightarrow R{-}R + Cu \quad (5)$$

Recently, a few organocopper(II) compounds in which copper is in its 2+ oxidation state have been isolated. In these compounds the copper atom is 'caught' in a very rigid ligand environment, e.g. as in *N*-confused porphyrins (*vide infra*) that provides a rigid tetradentate, square planar coordination environment to the copper(II) center.

Organocopper compounds with copper in its 3+ oxidation state are also rare, but recently a few have been isolated in which either the organic group which is σ-bonded to the copper(III) center contains highly electronegative groups, as, e.g., in $(CF_3)_2Cu$-S_2CNEt_2[31], or in which the carbon anion is part of a multidentate ligand system that provides a stabilizing ligand environment for the copper(III) center.

A variety of coordination numbers and geometries has been observed for structures of copper(I) compounds in the solid state, especially for its inorganic complexes (Figure 1)[32]. However, in organometallic chemistry mainly coordination geometries such as linear, trigonal planar and trigonal planar with a strong distortion towards T-shaped geometries have been encountered.

Already at the early stages of the development of organocopper chemistry it became clear that organocopper(I) compounds are not simple monomeric species RCu. Molecular weight determinations by ebulliometry and cryoscopy showed that in solution organocopper compounds can exist as dimeric, tetrameric or even octameric species. Definite proof for the existence of such associated species in the solid state was obtained from the earliest X-ray crystal structure determinations of organocopper(I) compounds. In the early days such compounds were denoted 'organocopper clusters'[33] because in many of these structures short Cu···Cu distances could be observed. However, as theoretical studies later confirmed direct metal–metal bonding is not involved in these organocopper structures and therefore a better name for this type of compounds is 'organocopper aggregates'.

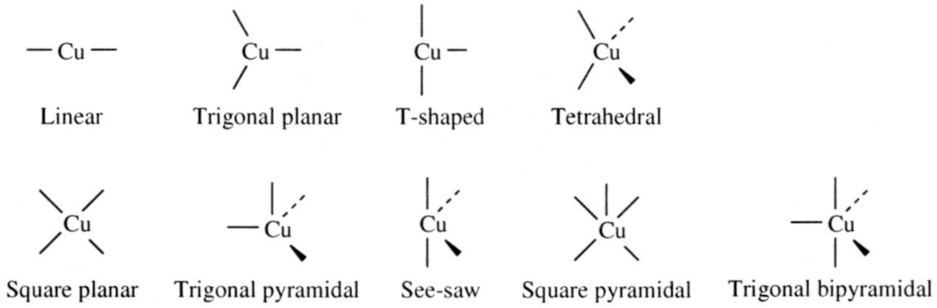

FIGURE 1. The observed coordination geometries for the copper(I) center in copper(I) compounds

It must be noted that the recently reported structures of organocopper-(II) and -(III) compounds all have $[R_xCuL_n]Anion_{z-x}$ stoichiometry (in which z is either 2 or 3 and x can be either 1, 2 or 3). These compounds are mononuclear and all have a direct two-electron two-center C−Cu bond.

Simple σ-bonding between copper and a carbon ligand has only been observed in a few organocopper(I) complexes with RCu(L) stoichiometry in which L is a neutral two-electron donor ligand. In most organocopper(I) compounds RCu aggregation occurs as the result of electron-deficient two-electron three-center bonding between an anionic carbon atom (of the R-anion) and two copper(I) centers. A simplified model for such a bonding between a phenyl anion and two copper atoms is shown schematically (Figure 2). The lowest molecular orbital is a bonding combination of the sp^2-carbon orbital with bonding orbitals on the two copper atoms (**A**), while a molecular orbital of higher energy is the result of overlap of a π-orbital at carbon with an antibonding combination of copper orbitals (**B**). In addition, back-donation from the copper atoms to the bridging carbon atom occurs via overlap of filled antibonding orbitals at copper to a π^*-carbon orbital. It should be noted that the second molecular orbital increases the electron density at the bridging carbon atom, and thus the kinetic stability of the copper−carbon bond. It is obvious that an orientation of the plane of the aryl group perpendicular to the Cu−Cu vector is the energetically most favorable situation. Such a situation is often observed in the solid state structures of most arylcopper compounds (*vide infra*) and is enhanced when *ortho*-substituents are present.

Finally, it is interesting to note that initially the structural characterization of organocopper compounds in the solid state by X-ray crystallography was hampered by several factors which are: (i) thermal instability, (ii) extreme sensitivity towards oxygen and moisture and (iii) limited solubility in suitable solvents to grow single crystals.

Nowadays, it has been well established that the thermal stability of organocopper compounds increases in the order: alkyl $<$ aryl $\approx$ alkenyl $<$ alkynyl[34]. It is therefore

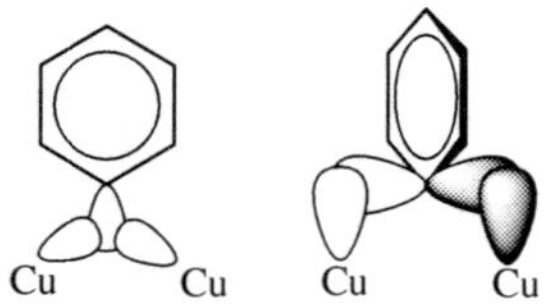

(**A**) (**B**)

FIGURE 2. Simplified model for the phenyl-to-copper bonding in an electron-deficient two-electron three-center bond

not surprising that the first organocopper compound characterized by X-ray diffraction was an alkynylcopper complex, i.e. $PhC{\equiv}CCu(PMe_3)$[35], which appeared to exist as a tetrameric aggregate in the solid state. The first alkyl- and arylcopper compounds that were structurally characterized by X-ray diffraction were Me_3SiCH_2Cu[26] and 2-(Me_2NCH_2)-5-(Me)C_6H_3Cu[36,37], respectively. Both compounds exist as tetrameric aggregates in the solid state.

When modern X-ray crystallographic techniques became more available to synthetic chemists, a renaissance towards the elucidation of the solid state structures of organocopper compounds started. Nowadays, a few hundred structures of organocopper compounds are known. In fact, in the August 2008 version of the CSD database[38], 351 structures containing a σ-copper-to-carbon bond (excluding inorganic carbon monoxide-, isocyanide- and π-complexes) could be retrieved. Several of these structures have been reviewed (in part) earlier[34,39–41]. The present chapter gives an overview of the structural investigations of all organocopper compounds known so far. Depending on the particular type of aggregate the structures have been divided into several classes of compounds and are discussed in corresponding sub-chapters. Although copper–carbene complexes formally are not 'real' organocopper compounds, their structures have been included in this chapter because of the nature of the direct Cu–C interaction in these compounds.

II. ALKYL- AND ARYLCOPPER(I) COMPOUNDS

A. Donor-base-free Organocopper Compounds, RCu

The intrinsic thermal lability of alkylcopper compounds (*vide supra*, many decompose violently when dry), together with their very low solubility in both polar and apolar organic solvents, prevented the growth of single crystals suitable for X-ray crystal structure determination and therefore hampered the structural characterization of simple alkylcopper compounds.

One exception is (trimethylsilyl)methylcopper, which has been obtained from the reaction of the corresponding organolithium compound with (insoluble) CuI in diethyl ether. Me_3SiCH_2Cu showed a remarkably improved thermal stability (mp 79 °C) and, moreover, appeared to be soluble in hydrocarbon solvents[26]. A tetrameric aggregate structure $(Me_3SiCH_2)_4Cu_4$ (**1**) was proposed for this compound based on the observed mass spectra. However, unequivocal evidence for this tetrameric structure had to await the results of an X-ray crystal structure determination[42]. The molecular geometry of **1** comprises four copper atoms arranged in a square plane (Figure 3). Each of the four trimethylsilylmethyl groups are electron deficient, i.e. bridge bonded between neighboring copper atoms along the edges of the square in a symmetric way [Cu(1)–C(1) 1.98(1) and Cu(2)–C(1) 2.00(1) Å]. The four bridging carbon atoms lie roughly in the Cu_4-plane while the coordination geometry at the copper atoms is close to a linear one with C(1)–Cu–C(1a) amounting to 163.1(1)°. It should be noted that **1** represents the first and so far only example of a pure, uncomplexed alkylcopper compound of which the structure in the solid state was characterized by X-ray crystallography.

Although already in the early days of organocopper chemistry simple arylcopper compounds like $PhCu$[22], 2-, 3- and 4-MeC_6H_4Cu[23], C_6F_5Cu[24], 2-$CF_3C_6H_4Cu$[24] and 3-CF_3-C_6H_4Cu[25] had been synthesized and isolated as pure compounds, knowledge about their actual structures remained speculative. For insoluble PhCu and some of the alkyl-substituted ones, a polymeric structure was proposed (Figure 4). However, molecular weight determinations in solution by ebulliometry and cryoscopy of some soluble compounds indicated that they exist as discrete aggregates in solution. For example, these measurements showed that C_6F_5Cu and $3\text{-}CF_3C_6H_4Cu$ in solution most likely have tetrameric and octameric aggregated structures, respectively[24,25]. In later studies, based on cryoscopic

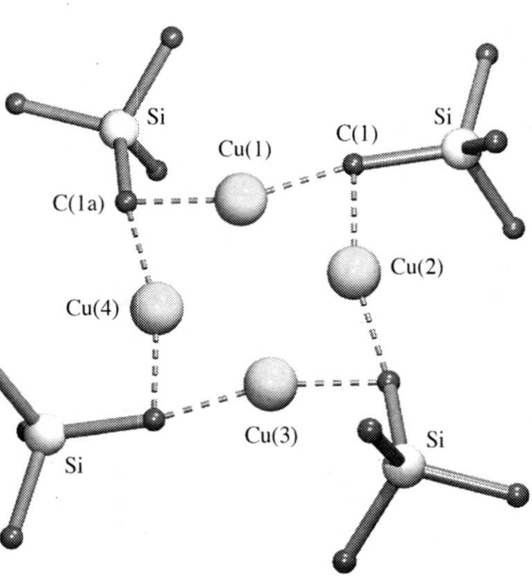

FIGURE 3. Molecular geometry of $(Me_3SiCH_2)_4Cu_4$ (**1**) in the solid state

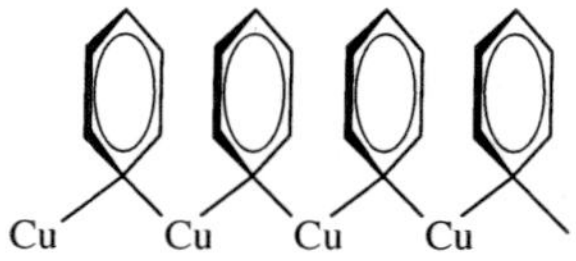

FIGURE 4. Proposed polymeric structure for insoluble PhCu in the solid state

measurements in benzene, a tetrameric aggregated structure was observed for 4-Me-C_6H_4Cu[43].

When modern X-ray crystallography became more readily available to the organocopper chemist, structures of various alkyl-substituted arylcopper compounds could be determined more routinely. It must be noted, however, that as the application of X-ray techniques often appeared crucial for gathering information on structural features of organocopper compounds, the growth of single crystals often has been a limiting (and frustrating) factor.

The molecular geometries of $[Me_5C_6]_4Cu_4$ (**2**)[44], $[2,4,6\text{-}Et_3C_6H_2]_4Cu_4$ (**3**)[45], $[2,6\text{-}(i\text{-}Pr)_2C_6H_3]_4Cu_4$ (**4**)[46], $[2,4,6\text{-}(i\text{-}Pr)_3C_6H_2]_4Cu_4$ (**5**)[47] and $[2\text{-(vinyl)}C_6H_4]_4Cu_4$ (**6**)[48] all exhibit a tetrameric, aggregate structure as a common structural motif. As an example the structure of pentamethylphenylcopper is shown (Figure 5).

The molecular geometry of $[Me_5C_6]_4Cu_4$ (**2**) comprises four copper atoms in a perfect square-planar arrangement [Cu-Cu 2.413(2) Å]. The four aryl groups are each two-electron three-center bridge bonded via C_{ipso} between two copper atoms along the edges of the Cu_4-plane in a highly symmetrical manner [Cu−C 1.99(1) Å] with the planes of the aryl groups orientated perpendicularly to the Cu_4-plane.

The structures of $[2,4,6\text{-}Et_3C_6H_2]_4Cu_4$ (**3**), $[2,6\text{-}(i\text{-}Pr)_2C_6H_3]_4Cu_4$ (**4**) and $[2\text{-(vinyl)}C_6H_4]_4Cu_4$ (**6**) are closely related to the structure of **2**, while the structure of $[2,4,6\text{-}(i\text{-}Pr)_3C_6H_2]_4Cu_4$ (**5**) deviates considerably, which is particularly expressed by the asymmetric bonding of the bridging aryl groups (Figure 5). Although the Cu−C bond distances in the bridge only slightly deviate [Cu(1)−C(1) and Cu(1a)−C(1) are 1.958(7)

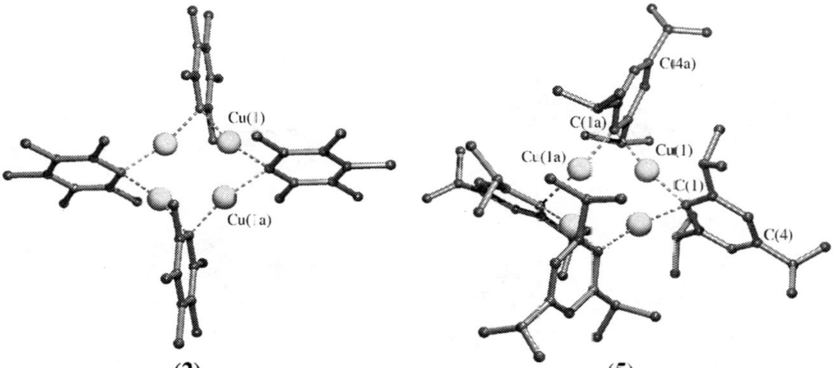

FIGURE 5. Molecular geometry of $[Me_5C_6]_4Cu_4$ (**2**) and $[2,4,6\text{-}(i\text{-}Pr)_3C_6H_2]_4Cu_4$ (**5**) in the solid state

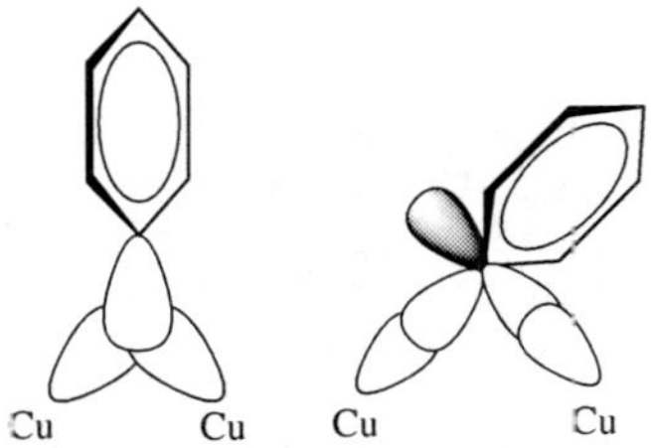

FIGURE 6. The different bridge-bonding situations in **2**, **3**, **4** and **6** versus the one encountered in **5**

and 2.018(7) Å, respectively], the angle of the vector C(4)–C(1) with that of the Cu(1)–C(1) bond is close to linear (165.4°) while the angle of the C(4)–C(1) vector and the Cu(1a)–C(1) bond is 118.6°. This asymmetric bonding has been explained in terms of different orbital overlap in the bridge bond (Figure 6). In **2, 3, 4** and **6** aggregate formation is the result of pure two-electron three-center bonding (bonding situation **A**) while in **5** formation of the tetramer is described as the result of association of four monomeric units, in which the copper atom is bonded via a two-center two-electron bond, and aggregation is provided by an interaction of the electron-rich π-orbital of C_{ipso}, perpendicular to the plane of the aryl ring, with an empty orbital of the neighboring copper atom (bonding situation **B**). A similar type of bonding has been observed in neutral aryl cuprates[49] and is even more expressed in the corresponding aurates[50].

An X-ray crystal structure determination of 2-thienylcopper revealed a molecular geometry consisting of a tetrameric aggregate $(C_4H_4S)_4Cu_4$ (**7**)[51] with structural features closely related to those of **2**–**5**. The 2-thienyl groups are each spanning the edges of a Cu_4-square binding two copper atoms via two-electron three-center bridge bonds. The planes of the 2-thienyl groups are orientated perpendicularly to the Cu_4-plane, but in such a way that two adjacent ones have the sulfur atoms above and the other two below the Cu_4-plane.

More than 35 years after its discovery, *vide supra*, the solid state structure of pentafluorophenylcopper was unambiguously established by an X-ray crystal structure determination. Recrystallization from a 1,2-dichloroethane/cyclohexane mixture afforded crystalline

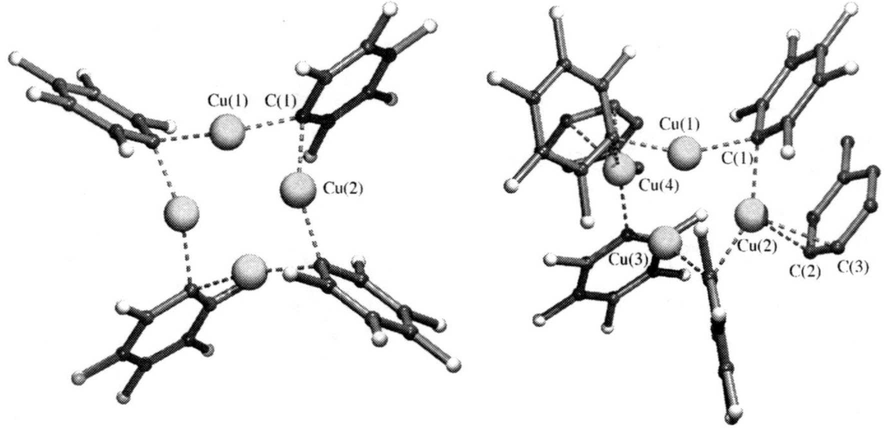

(**8**) (**9**)

FIGURE 7. Molecular geometry of $(C_6F_5)_4Cu_4$ (**8**) and $(C_6F_5)_4Cu_4(C_6H_5Me)_2$ (**9**) in the solid state

$(C_6F_5)_4Cu_4$ (**8**) whereas the crystalline material obtained after recrystallization from toluene appeared to be $(C_6F_5)_4Cu_4(C_6H_5Me)_2$ (**9**)[52,53]. The molecular geometry of **8** is similar to that of other tetrameric aggregates (*vide supra*) with the four copper atoms in an almost perfect square-planar arrangement and the pentafluorophenyl groups symmetrically bridge bonded [Cu(1)–C(1) 1.998(2) and Cu(2)–C(1) 2.002(2) Å] between the copper atoms (Figure 7).

A striking difference is observed in the solid state structure of **9** which contains, in addition to the four pentafluorophenyl groups, two π-coordinate bonded toluene groups. The four copper atoms do not lie in one plane but form a butterfly structure with the angle between the Cu(1)–Cu(2)–Cu(3) and Cu(1)–Cu(3)–Cu(4) planes amounting to 37.5° (Figure 7). The pentafluorophenyl groups are each rather symmetrically bridge bonded [Cu(1)–C(1) 1.957(2) and Cu(2)–C(1) 2.006(2) Å] via C_{ipso} to two copper atoms. The two toluene (solvent) molecules are each η^2-bonded [Cu(2)–C(2) and Cu(2)–C(3) 2.339(2) and 2.298(2) Å, respectively] to opposite copper atoms. Consequently, two of the copper atoms [Cu(1) and Cu(3)] are two-coordinate, whereas the other two [Cu(2) and Cu(4)] have a distorted trigonal-planar coordination geometry.

A remarkable structural diversity was observed for the structures of mesitylcopper in the solid state. Single crystals, obtained by adding diethyl ether to a THF solution of mesitylcopper, afforded crystalline Mes_4Cu_4 (**10**)[54] with a symmetric square-planar Cu_4-geometry (Figure 8) similar to those found in other tetrameric organocopper aggregates (*vide supra*). However, recrystallization of the same material from toluene yielded a cyclic pentameric aggregate Mes_5Cu_5 (**11**)[55,56]. The molecular geometry of **11** comprises a ten-membered ring of alternating copper and bridging carbon atoms (Figure 8). The ring is puckered with a total puckering amplitude[57] of 1.47 Å. The dihedral angle of the plane of the mesityl groups with the least-squares plane of the five copper atoms ranges from 91° to 105°.

It appeared that from a THF/diethyl ether solution of **10**, which was stored at −80 °C, after prolonged time a so far unknown polymorph crystallized. An X-ray crystal structure determination, using low-temperature handling techniques, showed this solid phase to be that of a pentameric Mes_5Cu_5 aggregate with non-coordinating disordered THF molecules in the crystal lattice. Prolonged standing of this material at −20 °C in THF/diethyl ether

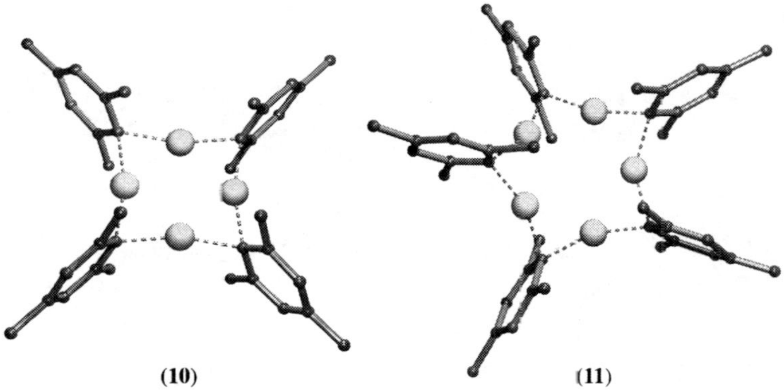

(10) (11)

FIGURE 8. Molecular geometry of tetrameric Mes_4Cu_4 (**10**) and pentameric Mes_5Cu_5 (**11**) in the solid state

solution resulted in complete conversion to crystalline tetrameric **10**. These observations indicate that in solution an equilibrium exists between tetrameric and pentameric aggregates that was also supported by NMR studies[56] and DFT calculations[58], which indicated that the tetramer and pentamer are very close in energy.

The X-ray crystal structure determination of 2,6-di(phenyl)phenylcopper showed that this compound exists as a trimeric aggregate $[2,6\text{-}Ph_2C_6H_3]_3Cu_3$ (**12**)[59] in the solid state with a triangular arrangement [Cu(1)–Cu(2) 2.3758(7), Cu(1)–Cu(3) 2.4224(10) and Cu(2)–Cu(3) 2.9136(11) Å] of the three copper atoms (Figure 9). Two of the monoanionic aryl groups are two-electron three-center bridge bonded between the copper atoms [Cu(1)–C(1) 1.989(3), Cu(2)–C(1) 2.085(3), Cu(1)–C(2) 2.047(3) and Cu(3)–C(2) 1.989(4) Å]. The third aryl anion is η^1-bonded to copper via C_{ipso} [Cu(2)–C(3) 1.924(3) Å] while one of the *ortho*-phenyl groups is η^2-bonded (π-coordination) to Cu(3) [Cu(3)–C(4) 2.158(3) and Cu(3)–C(5) 2.232(4) Å].

In the solid state 2,6-di(mesityl)phenylcopper exists as a symmetric dimeric aggregate $(2,6\text{-}Mes_2C_6H_3)_2Cu_2$ (**13**)[59] (Figure 9). The aryl group is η^1-bonded to copper via C_{ipso} [Cu–C(1) 1.927(5) Å] while dimerization occurs via π-coordination of one of the mesityl groups in a η^2-fashion [Cu–C(3) 2.295(4) and Cu–C(4) 2.123(4) Å] to copper in the other half of the dimer.

The X-ray crystal structure determination of the product obtained from the reaction of two equivalents of ferrocenyltrimethyltin, $FcMe_3Sn$, with $(C_6F_5)_4Cu_4$ showed the formation of heteroleptic $Fc_2(C_6F_5)_2Cu_4$ (**14**)[60] (Figure 10).

In **14**, the copper atoms are arranged in a parallelogram geometry with each of the ferrocenyl and pentafluorophenyl groups two-electron three-center bridge bonded to two copper atoms [Cu(1)–C(1) 1.971(5), Cu(2)–C(1) 1.969(5), Cu(2)–C(2) 2.018(5) and Cu(3)–C(2) 2.031(5) Å]. The two pentafluorophenyl groups, and consequently the two ferrocenyl groups, are in *trans*-position.

A similar reaction of 1,2-bis(trimethyltin)ferrocene with $(C_6F_5)_4Cu_4$ yielded the heteroleptic aggregate $Fc_4(C_6H_5)_8Cu_{16}$ (**15**)[60], as shown by an X-ray crystal structure determination (Figure 10). This aggregate, which consists of four 1,2-$FcCu_2$ and eight C_6F_5Cu units, making a total of sixteen copper atoms, represents the largest organocopper aggregate known to date. In **15** each ferrocene moiety is bound to four copper atoms, of which two are located above the Cp plane and the other two bound from the *endo*-side. Of the four copper atoms attached to each ferrocene moiety, three are involved in bridging with

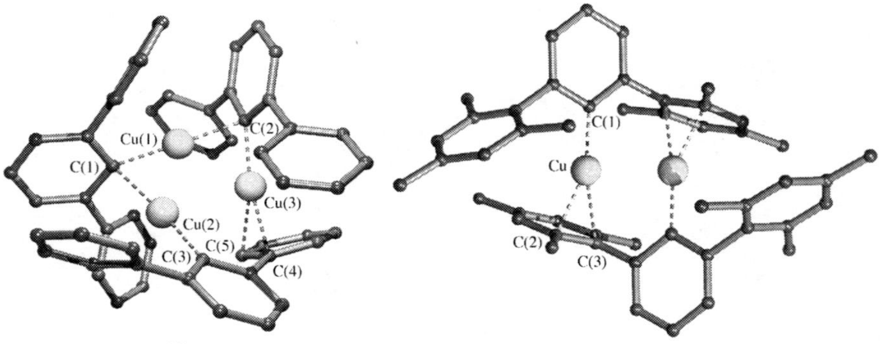

(12) (13)

FIGURE 9. Molecular geometry of trimeric $[2,6\text{-}Ph_2C_6H_3]_3Cu_3$ (**12**) and dimeric $[2,6\text{-}Mes_2C_6H_3]_2Cu_2$ (**13**) in the solid state

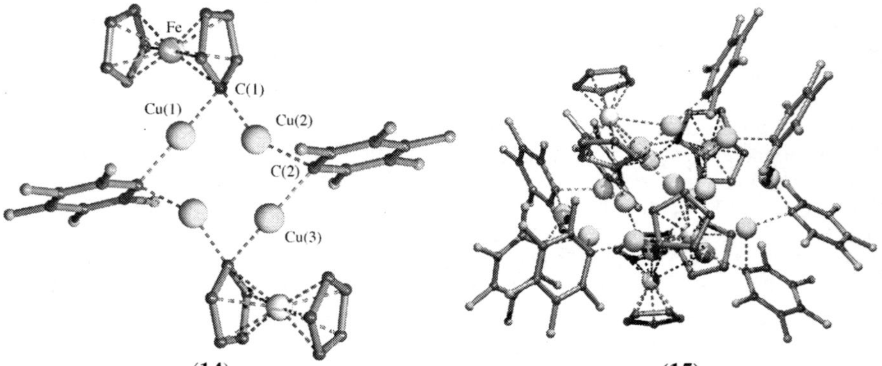

(14) (15)

FIGURE 10. Molecular geometry of the heteroleptic aggregates $Fc_2(C_6F_5)_2Cu_4$ (**14**) and $Fc_4(C_6H_5)_8Cu_{16}$ (**15**) in the solid state

other ferrocene units. This arrangement leads to a core of four ferrocene units and six copper atoms. The fourth copper atom is only involved in bridging C_6F_5 groups, resulting in the formation of two interlocking crown-like sub-structures, each consisting of four C_6F_5Cu units (Figure 10).

The reaction of Cu_2Br_2(S-Alkyne)(DMS) with PhLi afforded the heteroleptic aggregate (**16**) in low yield (equation 6)[61]. The formation of **16** is most likely the result of the addition of PhLi over the alkyne bond of the ligand and subsequent transmetallation. The structure of **16** in the solid state was established by an X-ray crystal structure determination and is shown schematically in equation 6. The molecular geometry of this heteroleptic aryl alkenyl copper aggregate comprises four copper atoms in a square-planar arrangement. The two phenyl groups are two-electron three-center bridge bonded between the copper atoms [Cu–C 1.989(4) and 2.005(4) Å] and *trans*-oriented with respect to each other in the aggregate. Likewise, the two alkenylic groups are two-electron three-center bridge bonded between the copper atoms [Cu–C 2.030(4) and 2.029(4) Å].

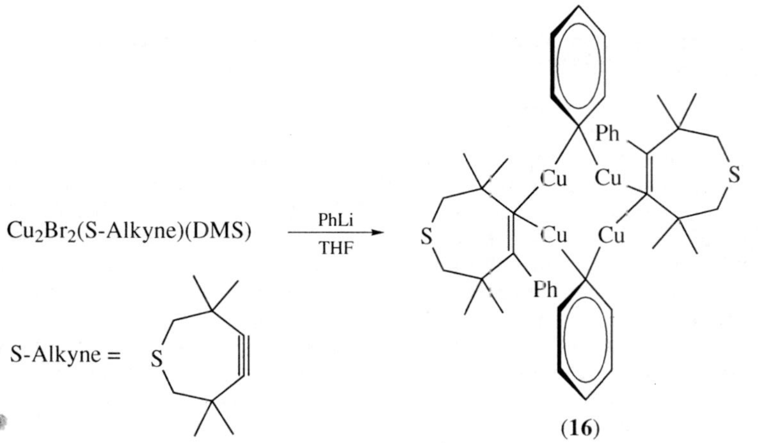

(16)

(6)

A remarkable observation is made when the structural features of 2-[(dimethylamino)-methyl]ferrocenylcopper (**17**)[62] are compared with those of the 2-[(dimethylamino)-methyl]phenylcopper compound (*vide infra*). The ferrocenylcopper compound (**17**) appeared to be a tetrameric aggregate with a square-planar Cu_4-structural motif with each of the ferrocenyl groups two-electron three-center bridge bonded [Cu(1a)–C(1) 2.05(4) and Cu(1b)–C(1) 2.05(4) Å] to two copper atoms (Figure 11). Although the 2-[(dimethylamino)methyl]ferrocenyl ligand is a potentially *C*,*N*-bidentate, chelating ligand, in **17** the nitrogen atoms are not involved in coordination to copper (shortest Cu–N distance exceeds 3 Å). This is in striking contrast to the organocopper derivatives, containing

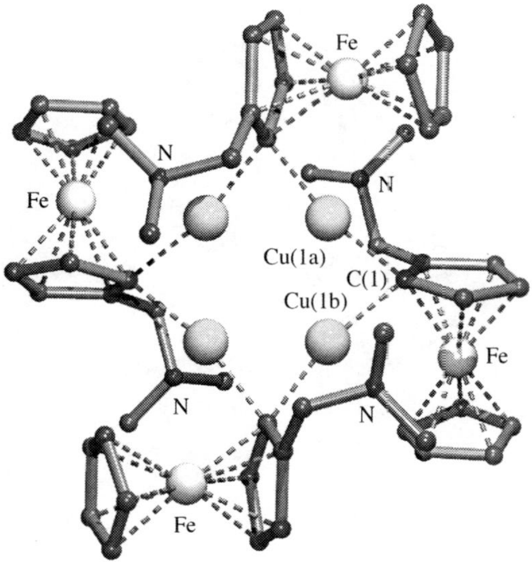

FIGURE 11. Molecular geometry of $[2\text{-}Me_2NCH_2Fc]_4Cu_4$ (**17**) in the solid state

the very similar monoanionic, *C,N*-bidentate 2-[(dimethylamino)methyl]phenyl ligand, in which intramolecular Cu–N coordination does occur; cf. **18**, *vide infra.*

B. Organocopper Compounds Containing Intramolecularly Coordinating Substituents, (LR)Cu

In the early days of organocopper chemistry the concept of 'stabilization by intramolecular coordination' was introduced by van Koten. Such intramolecular coordination enhances the thermal stability of organocopper compounds. This concept involves the use of carbon-anionic ligands that, in addition to the carbon anion, contain a heteroatom functionalized substituent, capable of additional, intramolecular coordination to copper. Making use of this concept a series of arylcopper compounds with *ortho*-substituents, e.g. (dimethylamino)methyl, dimethylamino, methoxymethyl, methoxy, diphenylphosphino and dimethylsulfamoyl, were prepared, which all showed enhanced thermal stability[27,28].

For one of these compounds, i.e. 2-[(dimethylamino)methyl]-5-(methyl)-phenylcopper (**18**), the structure in the solid state could be determined by an X-ray crystal structure determination and actually **18** represents the first organocopper compound for which the structure was confirmed by X-ray crystallography[36,37]. This compound exists in the solid state as a tetrameric aggregate of which the molecular geometry comprises four copper atoms in a butterfly arrangement with each of the aryl groups two-electron three-center bridge bonded [Cu(1)–C(1) 2.16 and Cu(2)–C(1) 1.97 Å] between two copper atoms (Figure 12). The planes of the aryl groups are orientated perpendicular to the Cu–Cu vector of the respective bonded copper atoms. To each of the copper atoms a nitrogen atom of the (dimethylamino)methyl substituent is intramolecularly coordinate bonded [Cu(1)–N 2.19 Å] resulting in a trigonal planar coordination geometry at each of the copper atoms.

The molecular geometry of the parent organocopper compound lacking the 5-methyl substituent present in **18**, i.e. 2-[(dimethylamino)methyl]phenylcopper (**19**), was also established by an X-ray crystal structure determination and shows a structural motif identical to that of **18**[63].

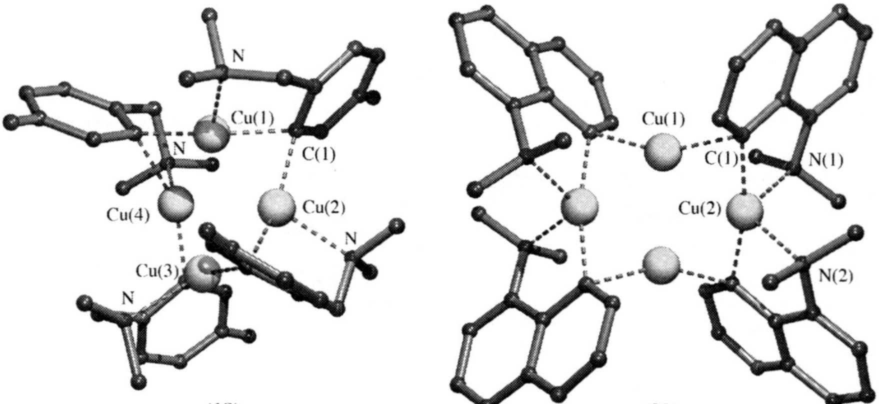

(**18**) (**20**)

FIGURE 12. Molecular geometry of $[2\text{-}Me_2NCH_2\text{-}5\text{-}MeC_6H_3]_4Cu_4$ (**18**) and $[8\text{-}Me_2NC_{10}H_6\text{-}1]_4Cu_4$ (**20**) in the solid state

8-(Dimethylamino)-1-naphthylcopper (**20**) also exists in the solid state as a tetrameric aggregate[64]. Its molecular geometry comprises four copper atoms arranged in a parallelogram with the naphthyl groups two-electron three-center bridge bonded via C_{ipso} [Cu(1)–C(1) 2.055(8) and Cu(2)–C(1) 2.019(9) Å] between two copper atoms (Figure 12). The nitrogen atoms of the dimethylamino substituents are pairwise coordinate bonded to two copper atoms [Cu(2)–N(1) 2.243(8) and Cu(2)–N(2) 2.269(7) Å] resulting in two copper atoms with a linear [Cu(1) and Cu(1a)] and two copper atoms [Cu(2) and Cu(2a)] with a distorted tetragonal coordination geometry.

One of the few examples of a heteroleptic organocopper aggregate is $(Vin)_2(2\text{-}Me_2NC_6H_4)_2Cu_4$ (**21**) (Vin = *cis*-$(2\text{-}Me_2NC_6H_4)C{=}C(Me)(C_6H_4Me\text{-}4)$) in which alkenyl and aryl groups are combined in one aggregate. This compound is accessible via two different synthetic approaches: (i) reaction of $(Vin)_2Cu_4Br_2$ (*vide infra*) with two equivalents of $Me_2NC_6H_4Li$, and (ii) reaction of $(Me_2NC_6H_4)_4Cu_6Br_2$ (*vide infra*) with two equivalents of VinLi (equation 7)[65]. The structure of **21** in the solid state was established by X-ray crystallography and is shown schematically in equation 7[66]. The four copper atoms in **21** are arranged as a rhombus with the alkenyl groups two-electron three-center bridge bonded [Cu–C 1.98(3) and 2.04(3) Å] over two adjacent sides of the rhombus. The nitrogen atom of the 2-*N*,*N*-dimethylaniline substituents of each alkenyl group are coordinate bonded to opposing copper atoms [Cu–N 2.26(4) and 2.33(4) Å]. The other two sides of the rhombus are each occupied by two-electron three-center bonded 2-dimethylaminophenyl groups [Cu–C 2.02(5) and 2.07(4) Å] of which the nitrogen atoms are not involved in coordination to copper. Consequently, the copper atoms in **21** are alternating two- and three-coordinate.

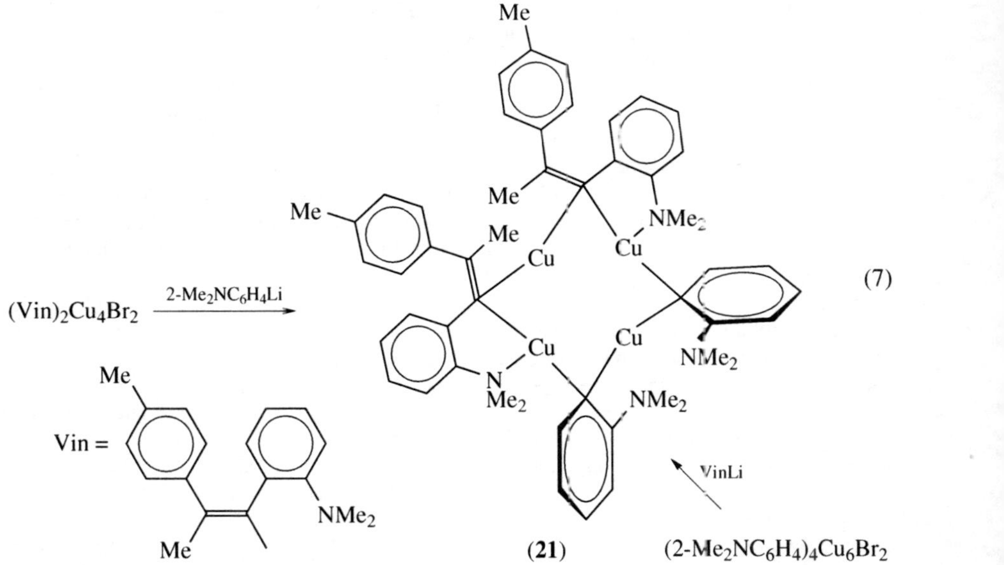

(7)

(**21**)

Substitution of the two bromine atoms in hexanuclear $(2\text{-}Me_2NC_6H_4)_4Cu_6Br_2$ via reaction with a monosubstituted lithium acetylide (LiC≡CR) leads to the formation of heteroleptic hexanuclear $(2\text{-}Me_2NC_6H_4)_4Cu_6(C{\equiv}CR)_2$ aggregates[67,68]. For one of these, i.e. $(2\text{-}Me_2NC_6H_4)_4Cu_6(C{\equiv}CC_6H_4Me\text{-}4)_2$ (**22**), the structure in the solid state was established

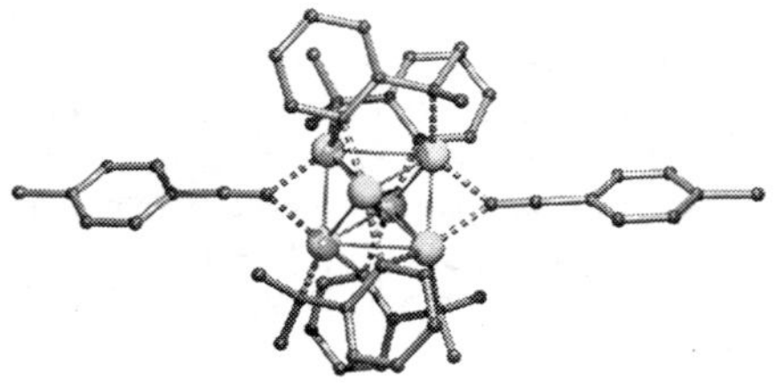

(A)

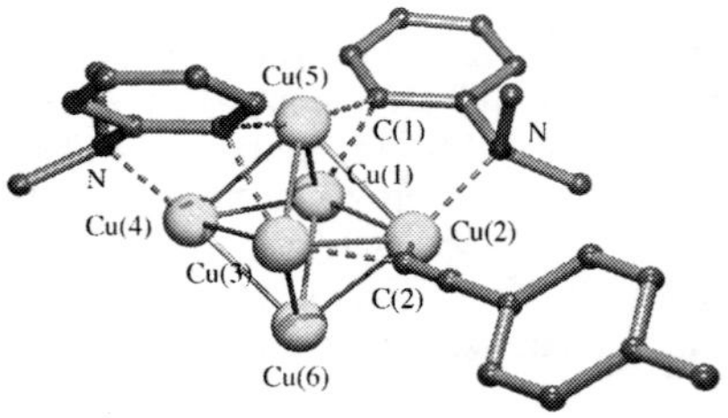

(B)

FIGURE 13. (**A**) Overall molecular geometry of $(2\text{-}Me_2NC_6H_4)_4Cu_6(C{\equiv}CC_6H_4Me\text{-}4)_2$ (**22**) (view along the $Cu_{apical}-Cu_{apical}$ axis. (**B**) Detailed view of the bonding at the Cu_6-octahedron (two of the $2\text{-}Me_2NC_6H_4$ groups and one acetylide group are omitted for clarity)

by an X-ray crystal structure determination[69]. In **22**, the six copper atoms are arranged as an octahedron (Figure 13). Two of the $2\text{-}Me_2NC_6H_4$ groups are two-electron three-center bridge bonded between the top, apical copper atom, Cu(5), and two opposing equatorial copper atoms, Cu(1) and Cu(3), respectively, while the nitrogen atoms of the 2-dimethylamino substituents are coordinate bonded to the other equatorial copper atoms, Cu(2) and Cu(4) (Figure 13, **B**). For the other two $2\text{-}Me_2NC_6H_4$ groups the bonding situation is reversed, i.e. two-electron three-center bridge bonded between the bottom, apical copper atom and Cu(2) and Cu(4), respectively, and the nitrogen atoms coordinate bonded to Cu(1) and Cu(3). The two acetylide groups are μ_2-bonded between two copper atoms at opposing sides in the equatorial plane. Thermolysis of **22** at 80 °C in benzene affords exclusively the cross-coupling product $2\text{-}Me_2NC_6H_4C{\equiv}CC_6H_4Me\text{-}4$[67]. Based on this observation a synthetic protocol was developed for the high-yield synthesis of asymmetrically substituted acetylenes $2\text{-}Me_2NC_6H_4C{\equiv}CR$ from $2\text{-}Me_2NC_6H_4Cu$ and the corresponding copper acetylide[70].

An X-ray crystal structure determination of 2-anisylcopper showed this compound to exist as an octameric aggregate $(2\text{-}MeOC_6H_4)_8Cu_8$ (**23**) in the solid state[71]. In **23**, the eight copper atoms form a slightly distorted square antiprism (Figure 14). The asymmetric unit contains four copper atoms which results in two cystallographically independent square anti- prismatic Cu_8 arrangements as a consequence of the symmetry in space group *Fddd*. As these arrangements are chemically identical and only differ slightly with respect to bonding features, details for only one of them are given. The phenyl groups are two-electron three-center bridge bonded via C_{ipso} over the edges of the Cu_4-square faces [Cu(1)–C(1) 2.03(2), Cu(2)–C(1) 2.05(2), Cu(5)–C(2) 2.01(2) and Cu(8)–C(2) 2.02(2) Å], while the oxygen atom of the methoxy substituent is coordinate bonded to the opposite apex [Cu(1)–O(2) 2.31(2) and Cu(5)–O(1) 2.43(2) Å] (Figure 14, **B**). As a consequence of this bonding mode, each of the copper atoms is coordinating to two carbon and one oxygen atom in a distorted pyramidal geometry. The planes of the aryl groups are orientated perpendicularly to the Cu–Cu vector of the C-bonded copper atoms (Figure 14, **A**). The bridge-bonded carbon atoms are displaced above the upper square face of the antiprism and below the lower square face of 0.68 Å (mean) and of 0.62 Å (mean), respectively.

In the organocopper aggregates discussed so far, aggregation primarily involved multicenter (electron-deficient) bonding of an anionic carbon atom with two copper atoms while intramolecular coordination of the heteroatom containing substituent occurred to either one of the C-bonded copper atoms or another copper atom in the aggregate. Another bonding mode resulting in aggregation occurs, however, via a direct σ-carbon

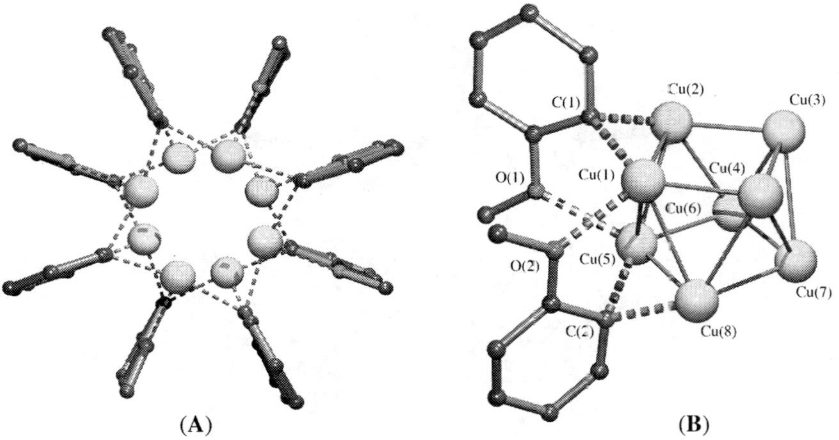

FIGURE 14. (**A**) Overal molecular geometry of $(2\text{-}MeOC_6H_4)_8Cu_8$ (**23**) in the solid state (viewed perpendicular to the square faces of the square antiprism). (**B**) Details of the Cu_8-square antiprism and bonding of the two crystallographically independent aryl groups

anion-to-copper bond, while aggregation involves coordination of the heteroatom containing substituent to the copper atom of a neighboring molecule. An example of the latter mode of aggregation is demonstrated by the structure in the solid state of (2-pyridyl)(trimethylsilyl)methylcopper. The X-ray crystal structure determination showed that this compound exists as a cyclic, tetrameric aggregate $[(2\text{-}C_5H_4N)(Me_3Si)CH]_4Cu_4$ (**24**) in the solid state[72]. In **24**, the copper atoms are arranged in a square while the monoanionic $(2\text{-}C_5H_4N)(Me_3Si)CH$ ligands are *C*,*N*-bridge bonded [Cu(1)–C(1) 1.934(5) and Cu(1)–N(4) 1.935(4) Å] over the edges of the Cu_4-square, coming alternatingly from above and from below the Cu_4-plane (Figure 15). As a result of this bonding mode the four copper atoms in **24** have a close-to-linear [C–Cu–N 165.5(2)°] coordination geometry.

X-ray structure determinations of a whole series of organocopper compounds derived from monoanionic, potentially *C*,*N*-bidentate bridging ligands exhibit a dimeric structural motif in the solid state. The compounds of this series have in common that dimerization occurs via two monoanionic, bidentate *C*,*N*-bridge bonded groups between two copper atoms which, as a result, have a linear coordination geometry. These compounds are: $[(2\text{-}C_5H_4N)(Me_3Si)_2C]_2Cu_2$ (**25**)[73], $\{[2\text{-}[6\text{-}MeC_5H_3N](Me_3Si)_2C]\}_2Cu_2$ (**26**)[74], $\{[2\text{-}[6\text{-}(Me_3SiCH_2)C_5H_3N](Me_3Si)_2C]\}_2Cu_2$ (**27**)[74], $\{[2\text{-}C_5H_4N](Me_3Si)_2C(Me_3Si)_2C\}_2Cu_2$ (**28**)[75], $[2\text{-}(Me_3Si)N{=}P(Ph)_2C_6H_4]_2Cu_2$ (**29**)[76], $[Me_3SiN{=}(t\text{-}Bu)CC(H)(SiMe_3)]_2Cu_2$ (**30**)[77] and $(OXL)_2Cu_2$ (**31**)[78] (OXL = 2-(4,4-dimethyl-2-oxazolinyl)phenyl). The structures of **25–30** are shown schematically (Figure 16).

As an example of this dimeric structural motif, details for the structure of 2-(4,4-dimethyl-2-oxazolinyl)phenylcopper (**31**) are given[78]. The molecular geometry of **31** comprises a centrosymmetric dimer as the result of two *C*,*N*-bidentate bridge bonded (oxazolinyl)aryl ligands (Figure 17). The aryl group is σ-bonded via C_{ipso} to one of the copper atoms (Cu(1)) [Cu(1)–C(1) 1.899(5) Å] while the nitrogen atom of the oxazolinyl group forms a coordination bond to the other copper atom (Cu(1a)) [Cu(1a)–N(1) 1.902(4) Å]. For the second (oxazolinyl)aryl ligand this bonding is reversed, i.e. σ-bonded to Cu(1a) via C_{ipso} and with its nitrogen atom coordinating to Cu(1), thus forming two close-to-linear C–Cu–N arrangements [C(1)–Cu(1)–N(1a) 177.8(2)°].

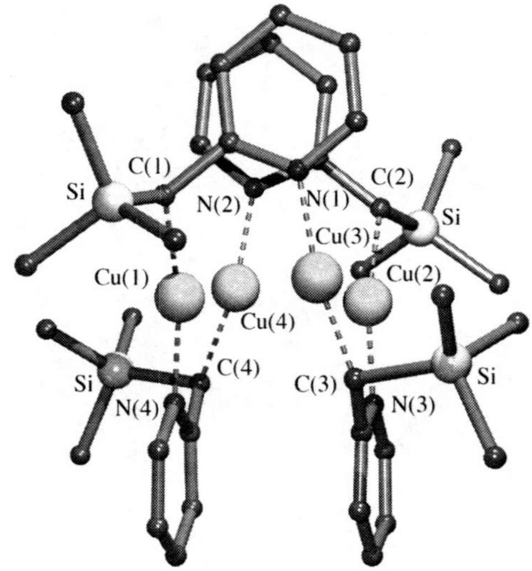

FIGURE 15. Molecular geometry of $[(2\text{-}C_5H_4N)(Me_3Si)CH]_4Cu_4$ (**24**) in the solid state

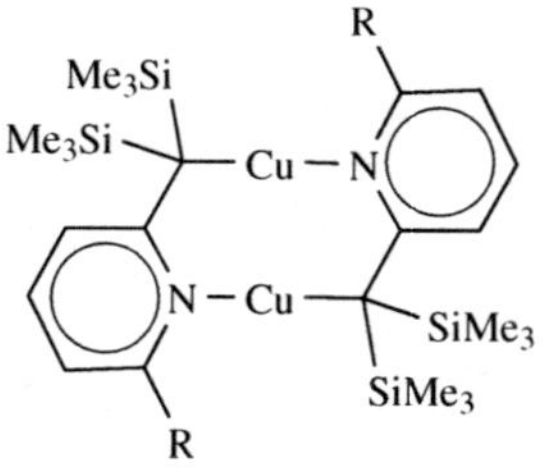

(**25**) R = H
(**26**) R = Me
(**27**) R = CH_2SiMe_3

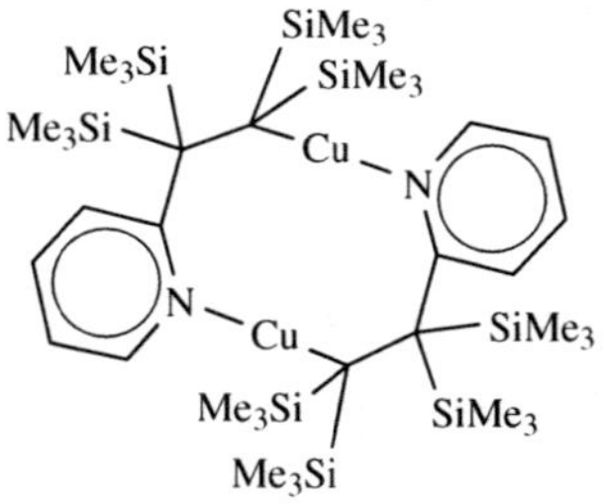

(**28**)

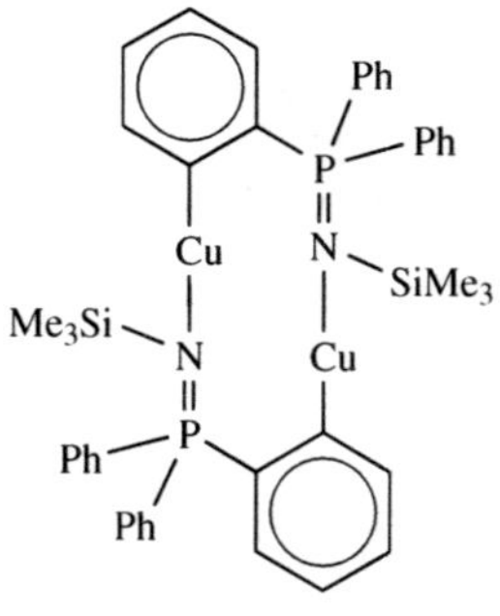

(**29**)

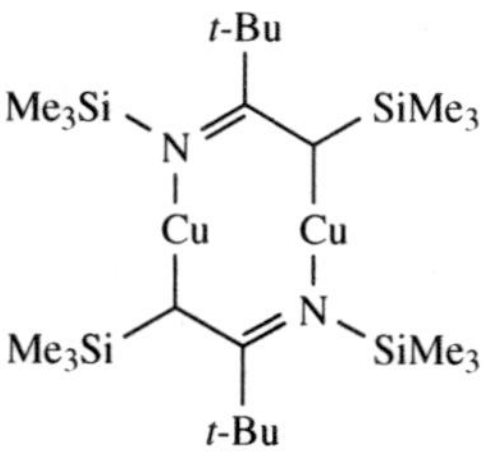

(**30**)

FIGURE 16. Schematic representation of the structures of dimeric organocopper aggregates **25–30** in the solid state

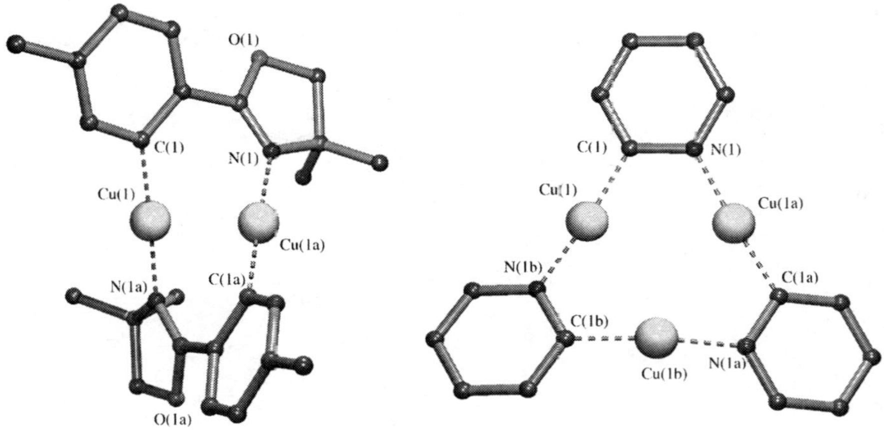

(**31**) (**32**)

FIGURE 17. Molecular geometry of dimeric **31** and trimeric $(2\text{-}C_5H_5N)_3Cu_3$ (**32**) in the solid state

The X-ray crystal structure determination of 2-pyridylcopper showed this compound to exist as a trimeric aggregate $(2\text{-}C_5H_5N)_3Cu_3$ (**32**) in the solid state[79] comprising a triangular arrangement of copper atoms. Each of the 2-pyridyl groups is σ-bonded via C_{ipso} to one of the copper atoms and forms a nitrogen coordination bond to a neighboring copper atom (Figure 17). Due to crystallographic disorder (twofold crystallographic axis through Cu(1b) and the center of the C(1)−N(1) bond) it is not possible to assign particular Cu−C or Cu−N bonds. The average Cu−C and thus C−N bond length is 1.90 Å while the C−Cu−N arrangements are close to linear.

C. Donor–Acceptor Complexes of Organocopper Compounds (RCu)-L

An obvious way to arrive at organocopper compounds RCu with improved stability was to convert them into their corresponding coordination complexes (RCu)-L using phosphorus or nitrogen containing additional ligands L. In the early days of organocopper chemistry, several attempts were undertaken to react methylcopper with triphenylphosphine which afforded the isolable $MeCu(PPh_3)_3$[80,81] and $MeCu(PPh_3)_2$[81] complexes indeed. Also, the synthesis and isolation of 2,2′-bipyridine MeCu(bipy)[82] and dimethylformamide MeCu(DMF)[82] complexes have been reported, but all are thermally unstable.

The interaction of arylcopper compounds such as [PhCu], $[2\text{-}MeC_6H_4Cu]$, $[3\text{-}MeC_6H_4\text{-}Cu]$ and $[4\text{-}MeC_6H_4Cu]$ with phosphorus containing ligands like PPh_3, 1,2-bis(diphenylphosphino)ethane (diphos) and 1,1-bis(diphenylphosphino)methane (DPM) and nitrogen containing ligands like 2,2′-bipyridine, 4,4′-bipyridine and 1,10-phenanthroline were likewise studied[83,84]. Ultimately it was concluded that complex formation does occur, however detailed information concerning the structures of these complexes could not be gathered.

An exception was the detailed study towards the interaction of $(2\text{-}Me_2NCH_2C_6H_4)_4Cu_4$ (one of the few well-characterized organocopper compounds at that time) with phosphorus containing ligands. Making use of a microwave titration technique it was shown that $(2\text{-}Me_2NCH_2C_6H_4)_4Cu_4$ (**19**) does not interact with triphenylphosphine, whereas with diphos it reacts to afford a 1:1 complex $2\text{-}Me_2NCH_2C_6H_4Cu$(diphos) (**33**) (equation 8)[85]. Addition of excess diphos results in an unprecedented C−P bond cleavage reaction with the formation of a diphenylphosphidocopper complex $(Ph_2P)_2Cu_2(\text{diphos})_2$ (**34**), diphenylvinylphosphine and *N,N*-dimethylbenzylamine (equation 8). The structure of $(Ph_2P)_2Cu_2(dppe)_2$ in the solid state was established by an X-ray crystal structure determination[86].

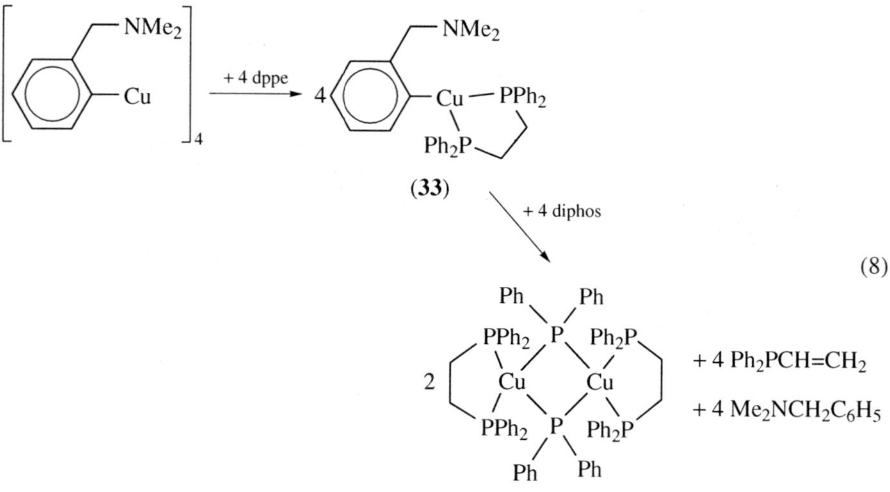

(34)

(8)

In more recent studies, complexes between organocopper compounds [RCu] and Lewis bases L have been isolated and were structurally characterized by X-ray crystallography. Depending on the particular organocopper compound and the particular ligand involved, deaggregation of the parent organocopper aggregate to a monomeric or lower aggregated organocopper-ligand adduct (RCu)-L may occur, but also ligand-to-copper coordination with retention of the initial organocopper aggregate structure has been observed.

Tris(triphenylphosphine)methylcopper $MeCu(PPh_3)_3$ (**35**) has been isolated at $-20\,^{\circ}C$ as a crystalline material and its structure in the solid state was established by an X-ray crystal structure determination at $-150\,^{\circ}C$[87]. Its molecular geometry comprises a distinct monomeric species with the methyl group σ-bonded to copper [Cu–C(1) 2.043(12) Å] and the three triphenylphosphine ligands *P*-coordinate bonded to copper [Cu–P(1) 2.311(3), Cu–P(2) 2.313(3) and Cu–P(3) 2.313(3) Å] (Figure 18). As is indicated by the bond angles around copper, this copper atom exhibits a close-to-perfect tetrahedral coordination geometry [C(1)–Cu–P(1) 106.9(4)°, C(1)–Cu–P(2) 106.7(4)°, C(1)–Cu–P(3) 106.9(4)°, P(1)–Cu–P(2) 113.1(1)°, P(1)–Cu–P(3) 109.6(1)° and P(2)–Cu–P(3) 113.1(1)°].

As mentioned above, triphenylphosphine and the bidentate phosphines diphos and DPM do form complexes with phenylcopper; the actual structures of these complexes are still unknown. It appeared, however, that the tridentate phosphine triphos (triphos = [1,1,1-tris(diphenylphosphino)methyl]ethane) effectively deaggregates polymeric phenylcopper into a monomeric $PhCu[(Ph_2PCH_2)_3CMe]$ complex (**36**)[88]. An X-ray crystal structure determination of crystalline **36** revealed a distinct monomeric molecular geometry in the solid state (Figure 18) with the phenyl group σ-bonded via C_{ipso} [Cu–C(1) 2.020(4) Å] and the three phosphorus atoms of the triphos ligand forming Cu–P coordination bonds [Cu–P(1) 2.276(2), Cu–P(2) 2.295(2) and Cu–P(3) 2.342(2) Å], resulting in a distorted tetrahedral coordination geometry at the copper atom. The C–Cu–P angles range from 119.4(1)° to 126.9(1)° while the P–Cu–P angles range from 90.0(1)° to 93.3(1)°, values far from the ideal tetrahedral value, but which are a logical consequence of the strain in the chelating tripodal ligand.

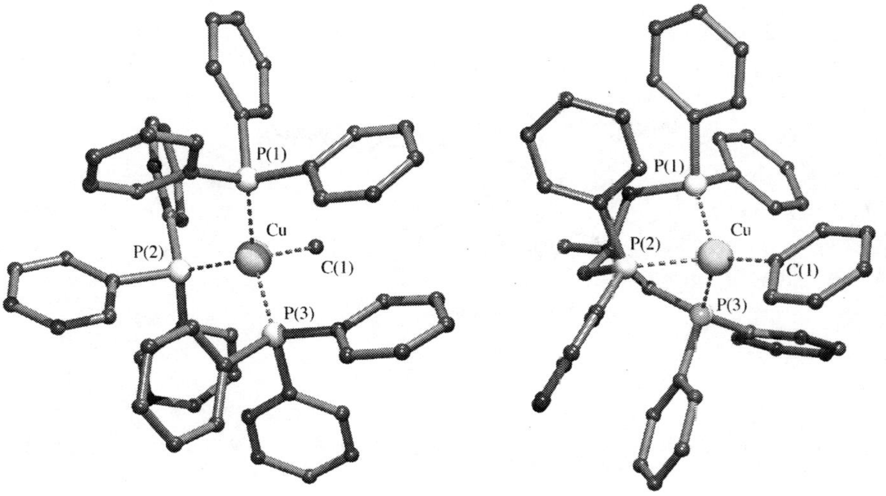

(35) (36)

FIGURE 18. Molecular geometry of monomeric $MeCu(PPh_3)_3$ (**35**) and $PhCu[(Ph_2PCH_2)_3CMe]$ (**36**) in the solid state

Various organocopper–monodentate ligand complexes with a 1:1 RCu-to-ligand molar ratio have been prepared and were characterized structurally by X-ray crystallography. The copper atom in these complexes has adopted a linear coordination geometry with a σ-C–Cu bond in a monomeric structure. Ligands employed included phosphorus-containing ligands as in MeCuP(Hex-*c*)$_3$ (**37**)[89] and 2,6-$Mes_2C_6H_3CuPPh_3$ (**38**)[90], sulfur-containing ligands as in 2,6-[2,6-(*i*-Pr)$_2C_6H_3]_2C_6H_3CuSMe_2$ (**39**)[91] and 2,4,6-(*t*-Bu)$_3C_6H_2CuSMe_2$ (**40**)[92] and nitrogen-containing ligands as in $C_6F_5CuNC_5H_5$ (**41**)[93] and $C_6F_5CuNC_4H_4N$-CuC_6F_5 (**42**)[93]. The structures of these complexes are shown schematically (Figure 19).

The C–Cu–heteroatom bond angle in all these compounds is very close to the ideal linear value of 180°. It is interesting to note that both the pyridine complex **41** and the pyrazine complex **42** are perfectly planar molecules with the pentafluorophenyl ring and the ring of the ligand in one plane. These molecules form supramolecular structures in the crystal lattice as the result of π-stacking interactions.

A special tweezer-type ligand $(Me_3SiC_5H_4)_2Ti(C{\equiv}CSiMe_3)_2$, containing two η^5-bonded (trimethylsilyl)cyclopentadienyl groups and two σ-bonded (trimethylsilyl)ethynyl groups to titanium, appeared to be extremely effective in stabilizing alkyl-, alkenyl- and aryl-copper compounds in the form of their monomeric complexes $RCu[(Me_3SiC_5H_4)_2Ti(C{\equiv}CSiMe_3)_2]$ (R = Me, Et, *n*-Bu, *i*-Pr, *c*-C_5H_9, CH_2SiMe_3, vinyl, Ph, 4-MeC_6H_4, 4-$MeOC_6H_4$, Mes, 1-$C_{10}H_7$ and 2,4,6-$Ph_3C_6H_2$)[94–97]. These compounds could be isolated as crystalline materials of which for some, i.e. $MeCu[(Me_3SiC_5H_4)_2Ti(C{\equiv}CSiMe_3)_2]$ (**43**)[95], *n*-$BuCu[(Me_3SiC_5H_4)_2Ti(C{\equiv}CSiMe_3)_2]$ (**44**)[97], $MesCu[(Me_3SiC_5H_4)_2Ti(C{\equiv}CSiMe_3)_2]$ (**45**)[94] and 2,4,6-$Ph_3C_6H_2Cu[(Me_3SiC_5H_4)_2Ti(C{\equiv}CSiMe_3)_2]$ (**46**)[96], the structure in the solid state was established by X-ray crystallography (Figure 20). These complexes have comparable structural features with the organic group σ-bonded to copper and both acetylenic functionalities of the Ti-ligand in a tweezer-type fashion bonded to the same copper atom. As an example the molecular geometry of **44** is shown (Figure 20). In **44**, the *n*-butyl group is σ-bonded to copper [1.981(6) Å] and the two acetylenic groups

Me — Cu — P(Hex-*c*)$_3$

(**37**)

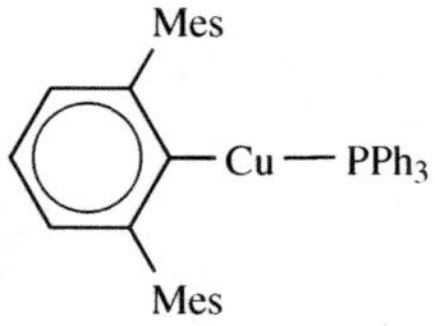

(**38**)

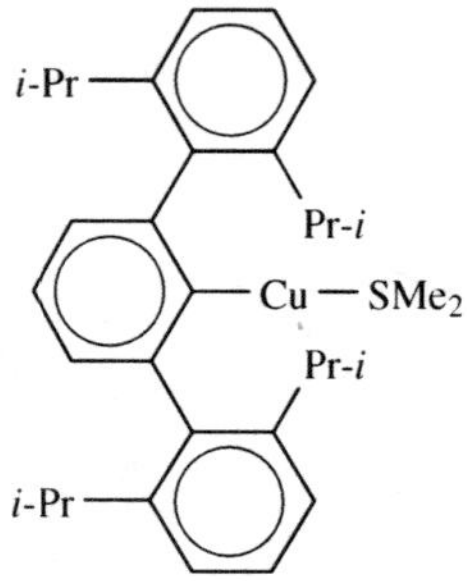

(**39**)

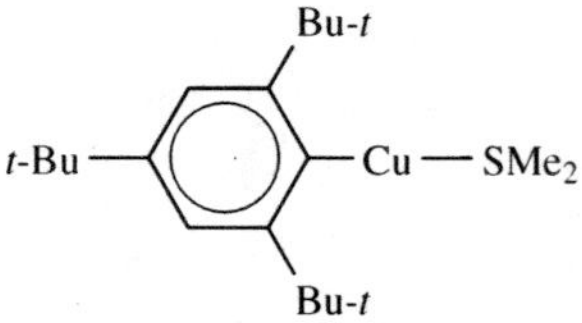

(**40**)

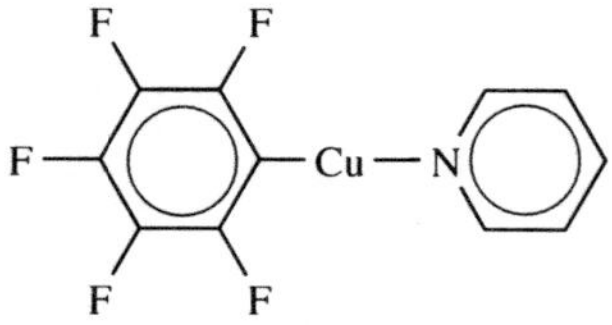

(**41**)

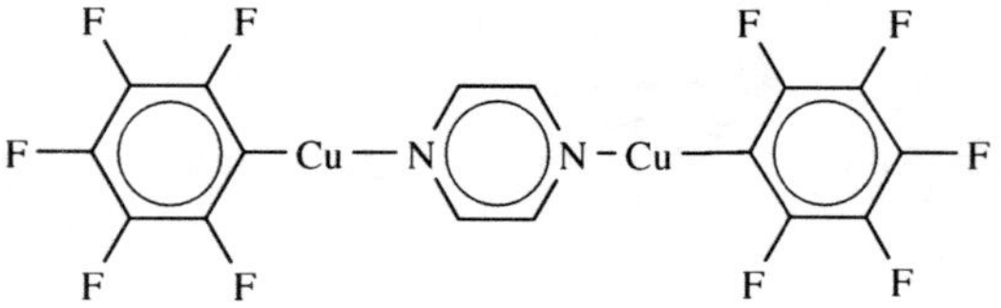

(**42**)

FIGURE 19. Schematic representation of the molecular geometry of complexes **37–42** in the solid state

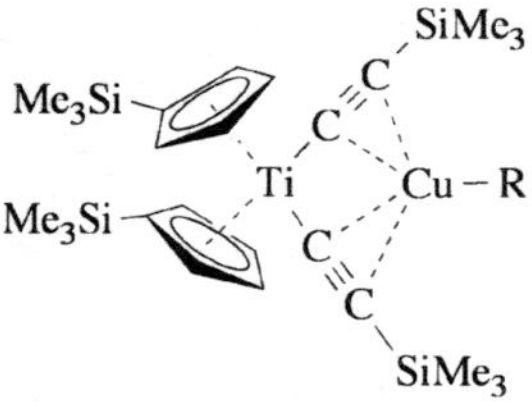

(**43**) R = Me
(**44**) R = *n*-Bu
(**45**) R = Mes
(**46**) R = 2,4,6-Ph$_2$C$_6$H$_2$

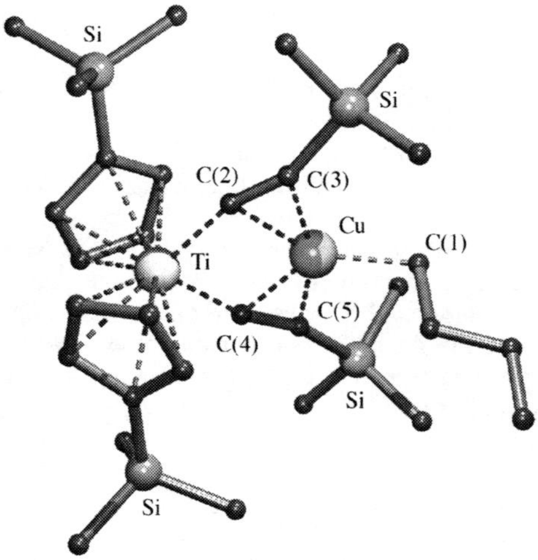

(**44**)

FIGURE 20. Monomeric organocopper complexes **43–46** and molecular geometry of *n*-BuCu-[(Me$_3$SiC$_5$H$_4$)$_2$Ti(C≡CSiMe$_3$)$_2$] (**44**) in the solid state

are η^2-bonded to copper [Cu–C(2) 2.067(6), Cu–C(3) 2.076(6), Cu–C(4) 2.053(3) and Cu–C(5) 2.077(6) Å] resulting in a pseudo-trigonal coordination geometry at the copper atom.

During these studies a rather unexpected rearrangement was observed. When the organocopper copper bromide aggregate $[2,6\text{-}(Me_2NCH_2)_2C_6H_3]_2Cu_4Br_2$ (*vide infra*) was reacted with $(Me_3SiC_5H_4)_2Ti(C{\equiv}CSiMe_3)_2$, its presumed monomeric organocopper complex $2,6\text{-}(Me_2NCH_2)_2C_6H_3Cu[(Me_3SiC_5H_4)_2Ti(C{\equiv}CSiMe_3)_2]$ (**47**) was most likely formed first, but then immediately rearranged to compound **48** via addition of the σ-carbon–copper bond over the C≡C bond of one of the acetylenic moieties of the Ti-ligand by which the organic group R migrates to the β-carbon atom of one of the coordinated acetylenic groups, thus forming a 1,2-dimetalla alkenyl unit (equation 9)[98].

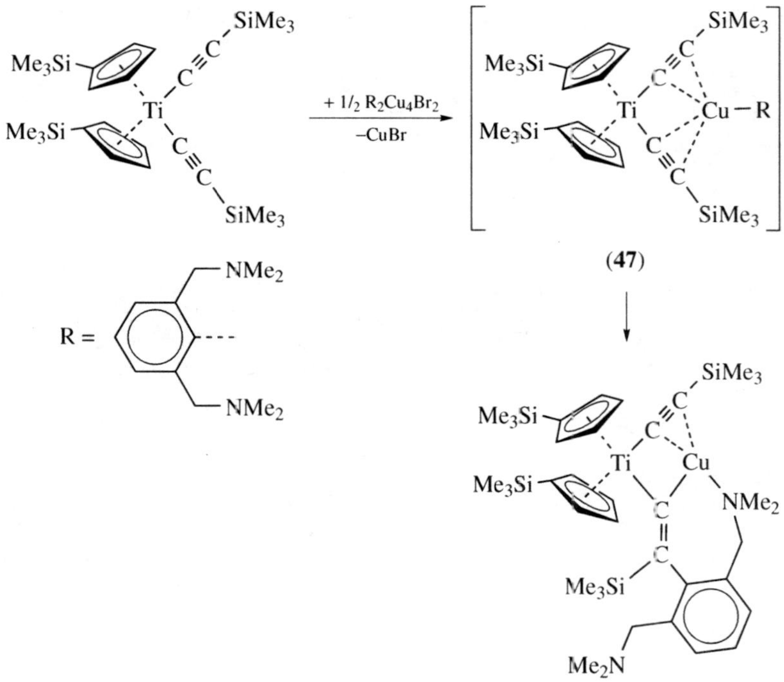

(47)

(48)

(9)

The molecular geometry of **48** in the solid state was established by an X-ray crystal structure determination (Figure 21). In **48**, the two (trimethylsilyl)cyclopentadienyl groups are η^5-bonded to titanium while the (trimethylsilyl)ethynyl group is σ-bonded via C(2) to titanium. The alkenyl group resulting from the rearrangement reaction is both σ-bonded to copper [Cu-C(1) 2.013(14)] and to titanium [Ti-C(1) 2.044(14) Å] via its dianionic C_{ipso} center, thus forming a 1,1-dimetalla alkenyl grouping. The unaffected (trimethylsilyl)ethynyl grouping is η^2-bonded to copper [Cu–C(2) 1.972(16) and Cu–C(3) 2.236(18) Å]. A trigonal planar coordination geometry at the copper atom is finally attained by coordination of the nitrogen atom of one of the (dimethylamino)methyl substituents of the bis-*ortho*-substituted aryl group to copper [Cu–N(1) 2.027(12) Å].

A particular type of ligands are the *N*-heterocyclic carbenes (NHC). These ligands are not only capable of forming complexes with copper salts (*vide infra* Section IX), but

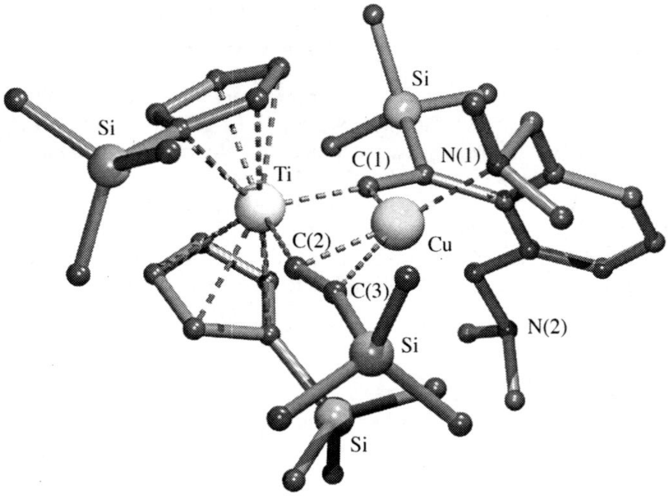

FIGURE 21. Molecular geometry of **48** in the solid state

also form complexes with organocopper compounds. The structures in the solid state of some of these, i.e. $F_3CCu(NHC)$ (**49**)[99], $F_3CCu(NHC)$ (**50**)[99], MeCu(NHC) (**51**)[100], [3-$C_2H_5C{=}C(H)C_2H_5$]Cu(NHC) (**52**)[101], 2,6-$Mes_2C_6H_3Cu(NHC)$ (**53**)[90] and 2,4,6-$Me_3C_6H_2$ Cu(NHC) (**54**)[90], were established by X-ray crystal structure determinations. Their molecular geometries are shown schematically (Figure 22).

Also, these complexes are distinct monomers with the organic group σ-bonded to copper and the carbene carbon atom of the NHC, forming a coordination bond to copper in a linear $C_{ipso}-Cu-C_{carbene}$ arrangement.

The reaction of the borylcopper NHC complex (**55**) with styrene affords selectively the β-boraalkylcopper–NHC complex (**56**) that could be isolated in high yield (91%) as a crystalline solid (equation 10)[102]. This compound represents one of the few examples of a benzylic organocopper compound which has sufficient thermal stability to be isolable.

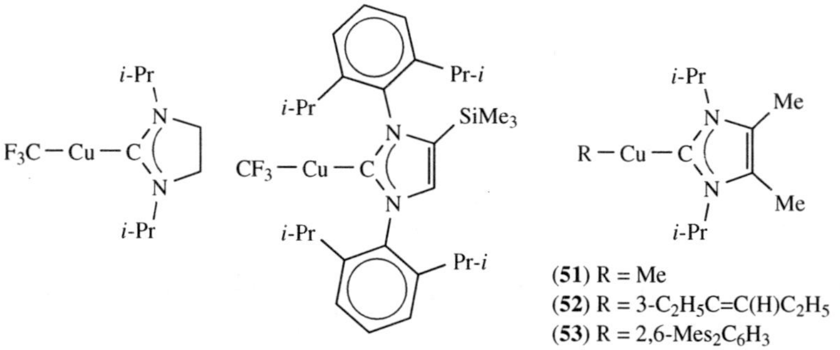

(**51**) R = Me
(**52**) R = 3-$C_2H_5C{=}C(H)C_2H_5$
(**53**) R = 2,6-$Mes_2C_6H_3$
(**54**) R = 2,4,6-$Me_3C_6H_2$

(**49**) (**50**)

FIGURE 22. Schematic representation of the molecular geometry of NHC complexes **49–54** in the solid state

Prolonged heating of **56** did not result in noticeable decomposition, but instead a rearrangement reaction occurred resulting in the formation of the α-boraalkylcopper–NHC complex (**57**) (equation 10). It was proposed that this rearrangement occurs via a β-hydride elimination reaction resulting in the formation of a copper hydride NHC complex and the corresponding borylalkene followed by a reinsertion of the borylalkene into the copper–hydride bond. Also, **57** could be isolated as a crystalline material. For both **56** and **57** the structures in the solid state were established by X-ray crystal structure determinations.

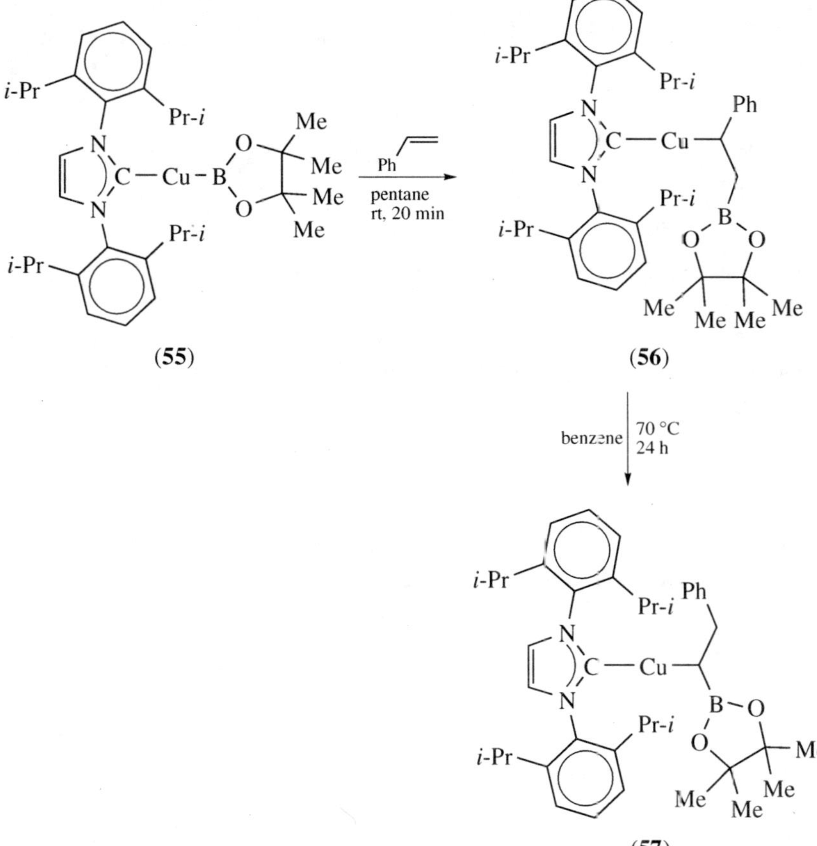

(**55**) (**56**) (**57**) (10)

The molecular geometry of **56** comprises a linearly coordinated copper atom [C(1)–Cu–C(2) 175.07(16)°] with the boraalkyl group σ-bonded to copper [C(1)–Cu 1.948(3) Å] via its β-carbon atom and the carbene carbon atom coordinate bonded to copper [Cu–C(2) 1.898(4) Å] (Figure 23). The molecular geometry of **57** is quite similar, now with the α-carbon of the boraalkyl group σ-bonded to copper [Cu-C(1) 1.959(3) Å] and, like in **56**, the carbene carbon atom coordinate bonded [Cu–C(2) 1.895(3) Å] (Figure 23). The C(1)–Cu–C(2) bond angle 169.51(13)° deviates slightly more from the ideal linear value compared to this bond angle in **56**.

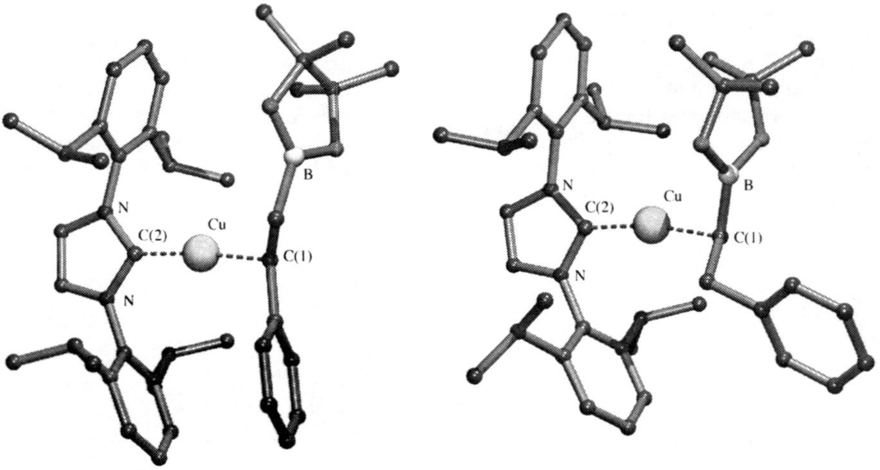

(**56**) (**57**)

FIGURE 23. Molecular geometries of **56** and **57** in the solid state

A similar reaction of the borylcopper–NHC complex **55** with 2,4,6-trimethylbenzaldehyde as substrate gives rise to the exclusive formation of the 1,2-insertion product (**58**) (equation 11) and was isolated as a crystalline material[103].

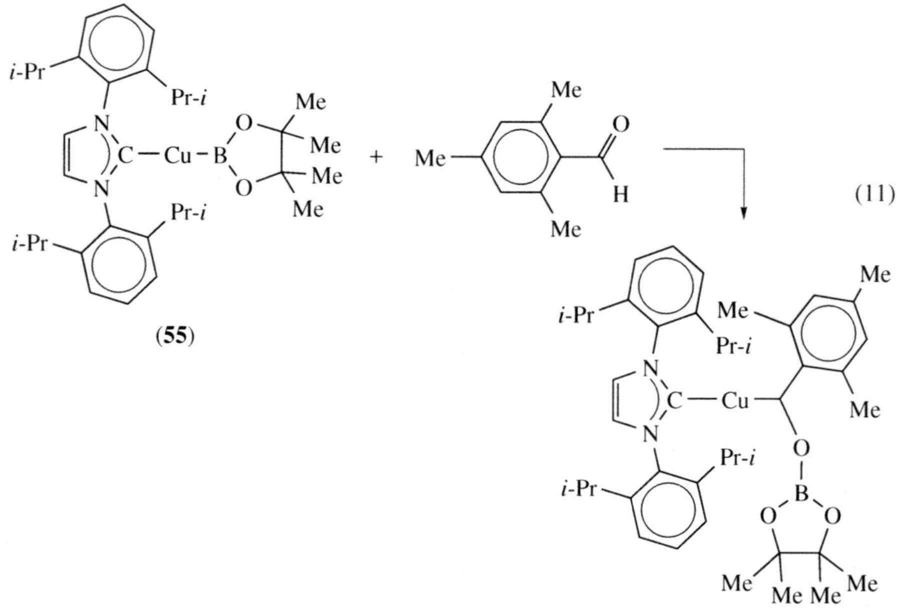

(**55**)

(11)

(**58**)

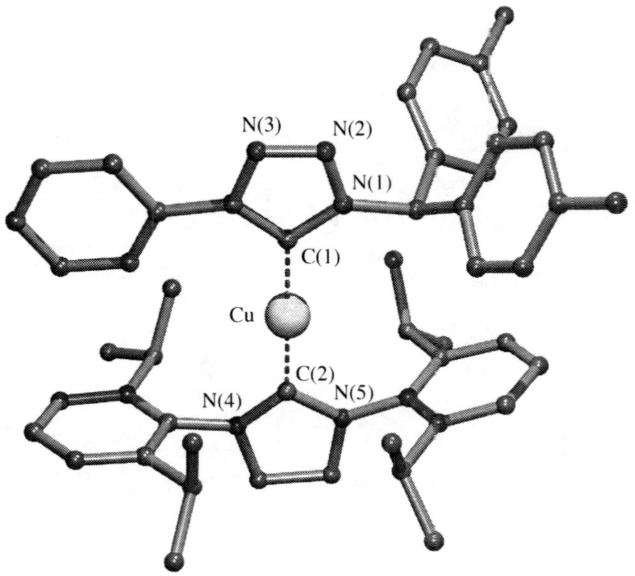

FIGURE 24. Molecular geometry of **59** in the solid state

An X-ray crystal structure determination revealed its molecular geometry consisting of a σ-bonded boroxyalkylcopper group [Cu–C 1.947(1) Å] and a coordinate bonded carbene carbon atom [Cu–C 1.898(2) Å] in a linear arrangement [C–Cu–C 175.46(8)°].

The copper triazolide NHC copper complex (**59**) (Figure 24) was obtained from the reaction of the corresponding PhC≡CCu(NHC) complex with $N_3CH(C_6H_4Me\text{-}4)_2$[104]. This complex represents an intermediate of the 'Click' reaction[105]. The molecular geometry of **59** in the solid state was established by an X-ray crystal structure determination (Figure 24). In **59**, the triazolide group is σ-bonded via C(1) to copper [Cu–C(1) 1.904(2) Å] and the *N*-heterocyclic carbene is coordinate bonded to copper via its carbene carbon atom [Cu–C(2) 1.909(5) Å] affording an almost perfect linear coordination geometry [C(1)–Cu–C(1) 178.0(3)°] at the copper center.

The organocopper donor ligand adducts discussed so far are all distinctly monomeric species, either formed by deaggregation of the parent organocopper aggregate or prepared as such. However, some cases have been reported in which deaggregation did not occur, but adduct formation did. In these compounds the donor molecule became an integral part of the aggregate.

When phenylcopper, for which a polymeric structure in the solid state has been proposed (*vide supra*), was prepared from PhLi and CuBr in dimethyl sulfide (DMS) as solvent, a crystalline material was obtained for which a structural analysis by X-ray crystallography revealed a tetranuclear aggregate structure $Ph_4Cu_4(DMS)_2$ (**60**)[106]. In **60**, the four copper atoms are arranged in a rhombus with almost equal Cu–Cu sides [2.444(2) and 2.475(1) Å] but largely different Cu–Cu–Cu angles [Cu(1)–Cu(2)–Cu(1a) 67.0(1)° and Cu(2)–Cu(1a)–Cu(2a) 113.0(1)°] (Figure 25). The four phenyl groups are two-electron three-center bridge bonded via C_{ipso} to the Cu–Cu edges of the rhombus [Cu(1)–C(1) 1.997(8) and Cu(2)–C(1) 2.070(6) Å] and are orientated perpendicularly to the Cu_4-plane.

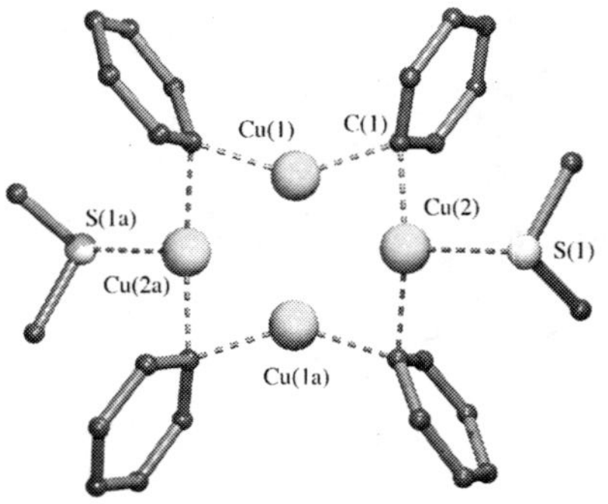

(60)

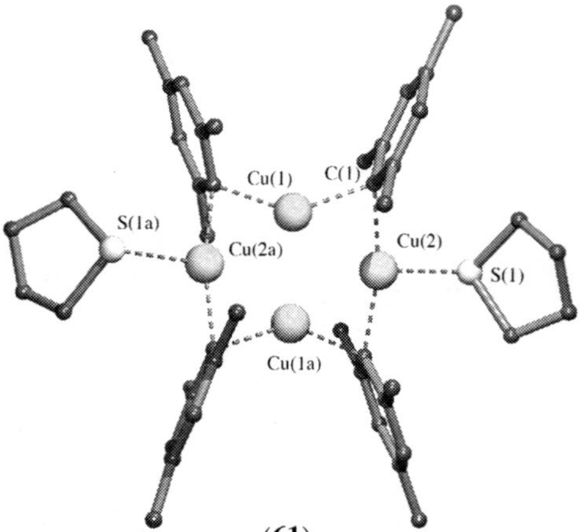

(61)

FIGURE 25. Molecular geometry of $Ph_4Cu_4(DMS)_2$ (**60**) and $Mes_4Cu_4(THT)_2$ (**61**) in the solid state

The C_{ipso} atoms which have either a digonal or a trigonal coordination geometry are not coplanar with the Cu_4-array but are disposed in adjacent pairs above and below this plane. Two dimethyl sulfide ligands are coordinate bonded to copper [Cu(2)–S(1) 2.383(2) Å] at opposing sides of the rhombus.

Depending on the solvent and the temperature, mesitylcopper crystallizes as either a cyclic pentameric or a tetrameric aggregate (*vide supra*). However, addition of THT (THT = tetrahydrothiophene) to a THF solution of the pentameric aggregate resulted in the crystallization of the tetranuclear aggregate $Mes_4Cu_4(THT)_2$ (**61**) containing *S*-coordinate bonded THT[56]. Its solid state structure was determined by an X-ray crystal structure determination (Figure 25) showing a structure which is closely related to that of **60**, i.e a structure with two-electron three-center bridge bonded mesityl groups to the Cu–Cu edges of a Cu_4-rhombus [Cu(1)–C(1) 2.055(8) and Cu(2)–Cu(1) 2.054(7) Å] and THT ligands *S*-coordinate bonded to copper [Cu(2)–S(1) 2.369(4) Å] at opposing sides of the rhombus.

A similar overall structural motif was found for the dimethyl sulfide adduct of *o*-tolylcopper $(2\text{-}MeC_6H_4)_4Cu_4(DMS)_2$ (**62**)[107]. An X-ray crystal structure determination of **62** showed that two of the two-electron three-center bridge bonded *o*-tolyl groups are disordered. A careful analysis of this disorder showed the presence of two geometrically different conformers in the crystal lattice, an α-conformer with all four 2-Me substituents pointing to one side of the Cu_4-plane (Figure 26, **A**) and a β-conformer with the 2-Me substituents alternating above and below the Cu_4-plane (Figure 26, **B**). These α- and β-conformers have site occupancies of 75% and 25%, respectively.

In an attempt to prepare heteroleptic cuprates containing both an aryl group and a chiral aminoarenethiolato group from the reaction of $Mes_2Mg(THF)_2$ with $[2\text{-}(Me_2N(Me)CH)\text{-}C_6H_4S]_3Cu_3$, a rather unexpected product, $Mes_4Cu_4[Mg(SC_6H_4CH(Me)NMe_2)_2]_2$ (**63**), was obtained[108].

The stoichiometry of the product indicates that a quantitative transmetallation reaction had occurred to yield neutral [MesCu] and $[2\text{-}(Me_2N(Me)CH)C_6H_4S]_2Mg$ groupings which are both present in the final aggregate. An X-ray crystal structure determination of **63** showed this compound to consist of a central Mes_4Cu_4 structural motif identical

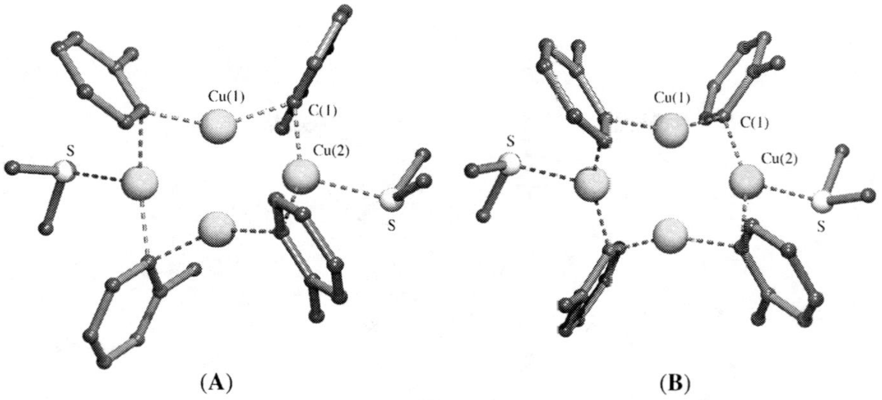

FIGURE 26. Molecular geometry of the α-conformer (**A**) and β-conformer (**B**) of $(2\text{-MeC}_6H_4)_4Cu_4(DMS)_2$ (**62**) in the solid state

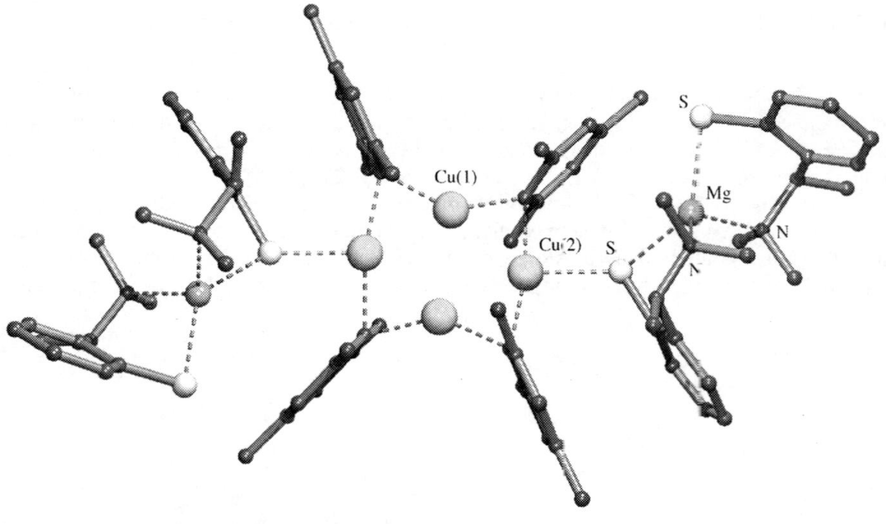

FIGURE 27. Molecular geometry of $Mes_4Cu_4[Mg(SC_6H_4CH(Me)NMe_2)_2]_2$ (**63**) in the solid state

to that in $Mes_4Cu_4(THT)_2$ (cf. **61**, *vide supra*) in which the THT ligands have been replaced by the thiolate-S atoms of $[2\text{-}(Me_2N(Me)CH)C_6H_4S]_2Mg$ groupings, i.e. they are *S*-coordinate bonded to copper at opposing sides of the Cu_4-rhombus (Figure 27). In fact, an alternate interpretation of the binding in **63** is that each of these thiolato groups are monoanions which are bridge bonded between copper and magnesium [Cu(2)–S 2.389(2) and Mg–S 2.390(2) Å]. Formally, in the latter interpretation **63** may be regarded as being a heterocuprate $[Mg(SC_6H_4CH(Me)NMe_2]_2[Mes_4Cu_4(SC_6H_4CH(Me)NMe_2)_2]$.

An asymmetric dimeric structure in the solid state was found for the dimethyl sulfide adduct of 2,4,6-triphenylphenylcopper $[2,4,6\text{-}Ph_3C_6H_2]_2Cu_2(DMS)_2$ (**64**)[92]. In **64**, one of the aryl groups is two-electron three-center bridge bonded between the two copper

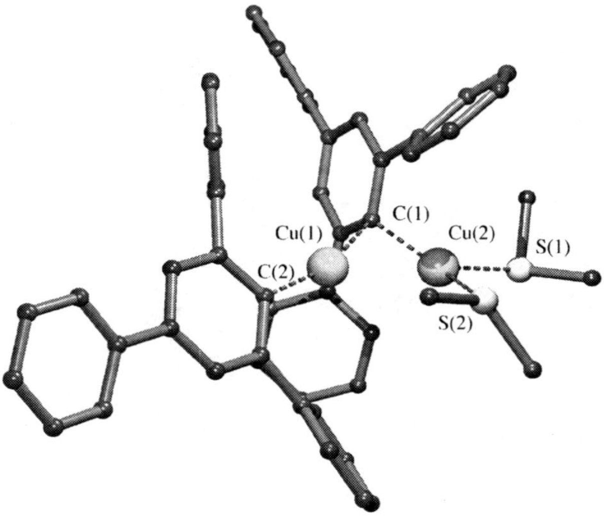

FIGURE 28. Molecular geometry of $[2,4,6\text{-}Ph_3C_6H_2]_2Cu_2(DMS)_2$ (**64**) in the solid state

atoms [Cu(1)–C(1) 2.017(5) and Cu(2)–C(1) 2.027(6) Å] whereas the other aryl group is σ-bonded to C(1) [Cu(1)–C(2) 1.935(5) Å] (Figure 28). A trigonal planar coordination geometry at Cu(2) is attained by the additional coordination of two DMS ligands to this copper atom [Cu(2)–S(1) 2.340(2) and Cu(2)–S(2) 2.273(2) Å].

When $Ph_4Cu_4(DMS)_2$ in dimethyl sulfide solution is treated with excess pmdta (pmdta = N,N,N',N'',N''-pentamethylethylenetriamine) and the resulting solution kept at $-20\,^\circ$C, a crystalline material slowly crystallizes which has $Ph_8Cu_8(pmdta)_2$ stoichiometry (**65**)[109]. An X-ray crystal structure determination of this material showed the presence

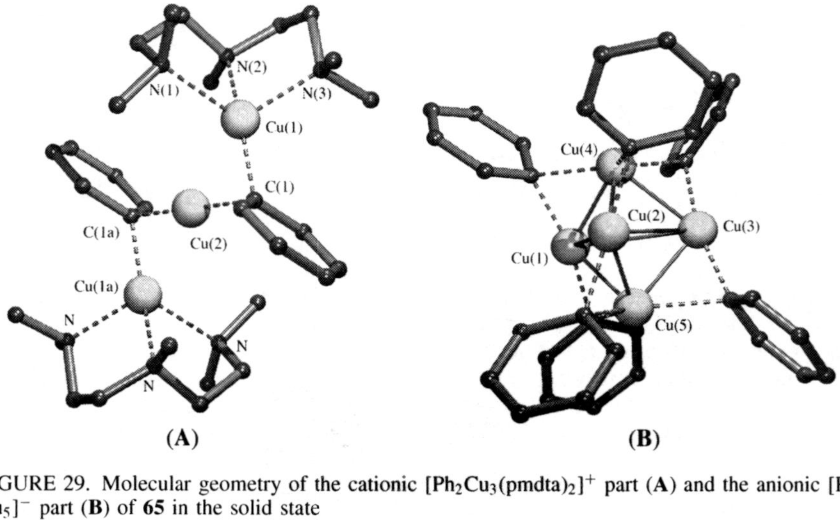

FIGURE 29. Molecular geometry of the cationic $[Ph_2Cu_3(pmdta)_2]^+$ part (**A**) and the anionic $[Ph_6Cu_5]^-$ part (**B**) of **65** in the solid state

of isolated $[Ph_2Cu_3(pmdta)_2]^+$ cations and $[Ph_6Cu_5]^-$ anions in the crystal lattice. The cation consists of a linear array of three copper atoms with each of the two phenyl groups two-electron three-center bridge bonded between the central copper atom and one of the terminal copper atoms [Cu(1)–C(1) 1.989(5) and Cu(2)–C(1) 2.006(7) Å] (Figure 29, **A**). The terminal copper atoms are each capped by a pmdta ligand of which all three nitrogen atoms are involved in coordination to copper [Cu(1)–N(1) 2.467(5), Cu(1)–N(2) 2.087(3) and Cu(1)–N(3) 2.206(5) Å]. As a result, the central copper atom has a linear coordination geometry while the terminal copper atoms have a distorted tetrahedral coordination geometry. The anion consists of five copper atoms arranged in a flattened trigonal bipyramidal arrangement, with the six phenyl groups spanning the equatorial-axial edges of this trigonal bipyramid via two-electron three-center bonds (Figure 29, **B**). A similar structural arrangement has been found in the anions of cuprate-like compounds (*vide supra*, Section VII).

III. ORGANOCOPPER(I) COPPER(I)-X AGGREGATES (X = MONOANIONIC GROUP)

During the early studies towards the synthesis and isolation of pure organocopper compounds, it became clear that organocopper compounds were difficult to purify from metal salts used (e.g. copper halides) or set free during the synthesis (e.g. magnesium or lithium salts). Later studies by van Koten and coworkers showed that organocopper compounds can form distinct aggregates with metal halides including copper halides. These observations have been of importance not only for the development of protocols that allowed the synthesis of pure organocopper compounds, but also established that the different products obtained, i.e. the pure organocopper and the organocopper metal halide compounds, often have distinctly different physical properties, e.g. different thermal stabilities or, more importantly, even distinctly different chemical reactivities, for example as group transfer reagents in organic synthesis. An example is outlined below.

The reaction of $2\text{-}Me_2NC_6H_4Li$ with CuBr in a 1:1 molar ratio in diethyl ether afforded, irrespective of the order of addition of the reagents, the organocopper copper bromide aggregate $(2\text{-}Me_2NC_6H_4)_4Cu_6Br_2$ (**66**) (equation 12)[30]. It is obvious that at this stage of the reaction unreacted organolithium starting material still was present in the reaction mixture. Even prolonged reaction times or heating the mixture to the boiling point of diethyl ether did not result in further reaction. Only after removal of the diethyl ether by evaporation and addition of benzene as a solvent did a smooth further reaction occur to ultimately yield the pure, insoluble organocopper compound $2\text{-}Me_2NC_6H_4Cu$ (equation 12). The molecular geometry of **66** was determined by an X-ray crystal structure determination (*vide infra*) and had been one of the first examples of an organocopper compound for which the structure in the solid state was established unambiguously.

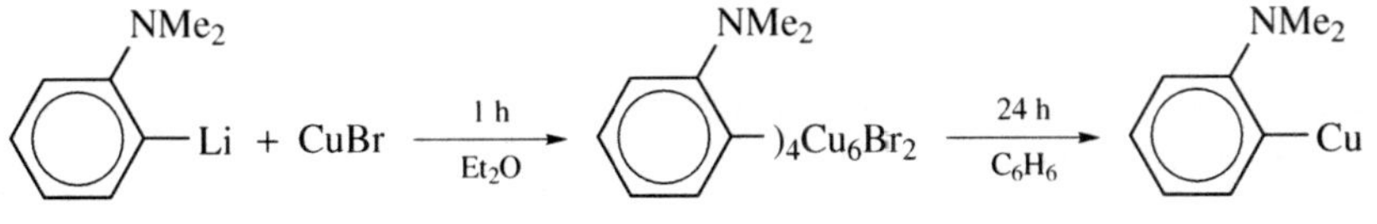

(12)

(**66**)

Another example, showing the importance of the order of addition of the reagents, is the reaction of closely related $2\text{-}Me_2NCH_2C_6H_4Li$ with CuBr. When the organolithium compound was slowly added to a suspension of CuBr in diethyl ether, invariably the organocopper copper bromide adduct $2\text{-}Me_2NCH_2C_6H_4Cu_2Br$ (**67**) was formed (equation 13)[110]. It appeared impossible to convert **67** into the pure organocopper compound, even with excess of the organolithium reagent, most probably as a consequence of the insolubility of **67** in common organic solvents. Moreover, it appeared that **67** has a lower thermal stability as compared to the parent organocopper compound.

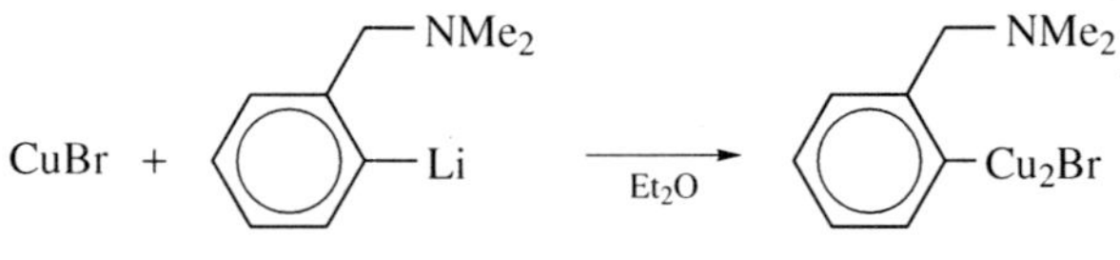

(**67**)

(13)

Applying the alternate order of addition of the reagents, i.e. when solid CuBr was gradually added to a suspension of $2\text{-}Me_2NCH_2C_6H_4Li$ in diethyl ether, half way during the addition of CuBr a clear solution was obtained due to the formation of the cuprate $(2\text{-}Me_2NCH_2C_6H_4)_2CuLi$ (equation 14), discussed in Section VII. Addition of the second half of CuBr to the initially formed $(2\text{-}Me_2NCH_2C_6H_4)_2CuLi$ resulted in the formation of pure $2\text{-}Me_2NCH_2C_6H_4Cu$[111] discussed in the previous section.

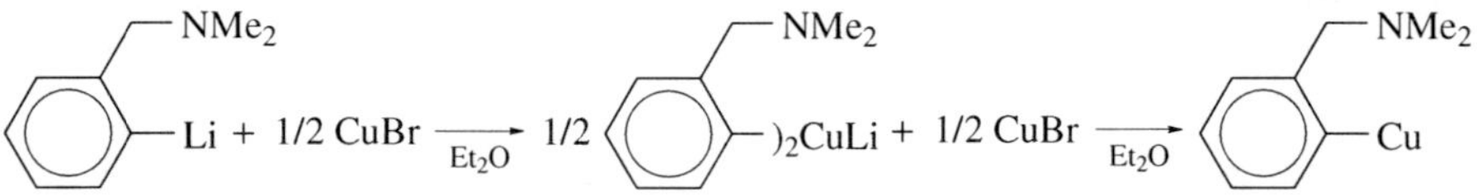

(14)

Later, it was unambiguously established that the nature of the anionic group X in organocopper copper-X aggregates (X = monoanionic group) largely influences the thermal stability of the formed aggregate. For example, treatment of $2\text{-}Me_2NC_6H_4Cu$ with CuX (X = Cl, Br or I) resulted in the formation of the corresponding $(2\text{-}Me_2NC_6H_4)_4Cu_6X_2$ aggregates which have extremely high thermal stability (>200 °C)[112]. However, when $2\text{-}Me_2NC_6H_4Cu$ was treated with CuOTf (OTf = trifluoromethanesulfonate) in an exact 2:1 molar ratio, $(2\text{-}Me_2NC_6H_4)_4Cu_6(OTf)_2$ (**68**) was formed (equation 15), which not only is thermally less stable (explosive decomposition at 120 °C) than the corresponding copper halide aggregates but, in the presence of a slight excess of CuOTf, decomposes quantitatively to form the symmetric biaryl and metallic copper (equation 15)[113]. Based on spectroscopic data, a structural motif has been proposed for **68**, similar to the one that was proposed for **66** (*vide infra*). When pure $4\text{-}MeC_6H_4Cu$ was treated with half an equivalent of CuOTf, no aggregates containing CuOTf were formed, but instead direct and quantitative formation of 4,4′-bitolyl was observed. It appeared that even catalytic amounts of CuOTf induced decomposition of $4\text{-}MeC_6H_4Cu$ to 4,4′-bitolyl and metallic copper (equation 16)[43].

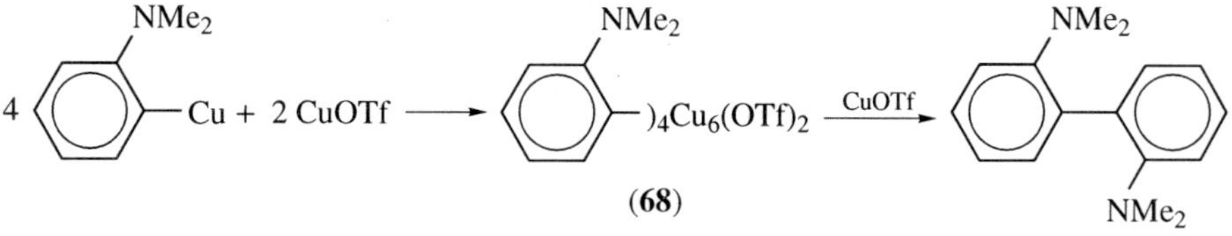

(**68**)

(15)

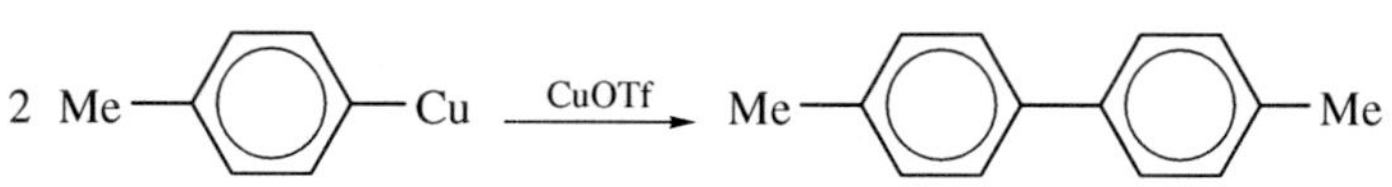

(16)

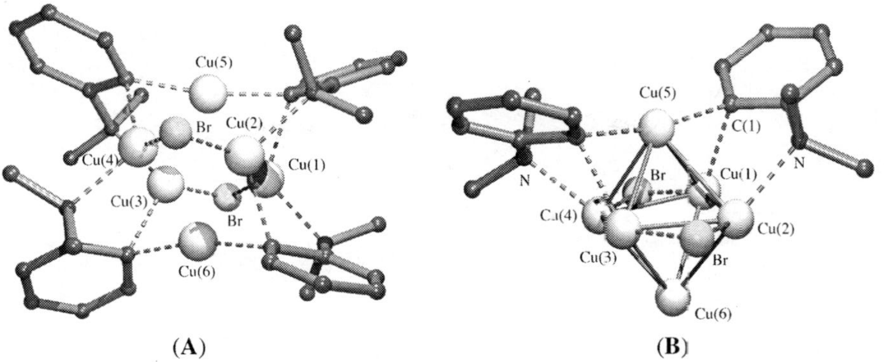

FIGURE 30. (**A**) Molecular geometry of (2-$Me_2NC_6H_4)_4Cu_6Br_2$ (**66**). (**B**) Detailed view of the bonding at the Cu_6-octahedron (two of the 2-$Me_2NC_6H_4$ groups are omitted for clarity)

As mentioned above, the structure of the organocopper copper bromide aggregate (2-$Me_2NC_6H_4)_4Cu_6Br_2$ (**66**) in the solid state was established by X-ray crystallography[29]. Its molecular geometry consists of six copper atoms arranged in an octahedron (Figure 30). Two of the 2-$Me_2NC_6H_4$ groups are two-electron three-center bridge bonded via C_{ipso} between the top, apical copper atom (Cu(5)) and two opposing equatorial copper atoms (Cu(1) and Cu(3), respectively) while the nitrogen atoms of the 2-dimethylamino substituents are coordinate bonded to the other equatorial copper atoms (Cu(2) and Cu(4), respectively). Consequently, each 2-$Me_2NC_6H_4$ anion is capping a face of the Cu_6-octahedron (Figure 30, **B**). For the other two 2-$Me_2NC_6H_4$ groups the bonding situation is reversed, i.e. two-electron three-center bridge bonded via C_{ipso} between the bottom, apical copper atom and Cu(2) and Cu(4), respectively, while the nitrogen atoms are coordinate bonded to Cu(1) and Cu(3). The two bromide groups are μ_2-bridge bonded between the copper atoms at opposing sides of the equatorial plane.

When 2-(4,4-dimethyl-2-oxazolinyl)-4-methylphenylcopper was treated with half an equivalent of CuBr, an aggregate with overall stoichiometry $(Oxl)_4Cu_6Br_2$ (**69**) was obtained (Oxl = 2-(4,4-dimethyl-2-oxazolinyl)-4-methylphenyl)[114]. The X-ray crystal structure determination of **69** revealed a molecular geometry with the same structural motif as was found for **66**, with the nitrogen atoms of the 2-oxazolinyl substituent coordinate bonded to the equatorial copper atoms of the Cu_6-octahedron.

From several studies, it appeared that the structural motifs of organocopper copper bromide aggregates derived from monoanionic groups containing potentially coordinating amine substituents are largely influenced by: (i) the nature and the number of potentially coordinating substituents, and (ii) the spatial orientation of these substituents. The molecular geometries of such copper halide aggregates containing either monoanionic, bidentate (Vin-N), tridentate (Ph-CNN, 2-Naph-CNN and Ph-NCN) or pentadentate (Ph-NNCNN) substituents (Figure 31) were established by X-ray crystal structure determinations.

The vinylic anion, Vin-N, forms an organocopper copper bromide aggregate with stoichiometry $(Vin\text{-}N)_2Cu_4Br_2$ (**70**) as established from an X-ray crystallographic study[65, 115]. Its molecular geometry comprises four copper atoms in a planar rhombus-type arrangement (Figure 32). The aggregate has twofold axial symmetry about the axis through Cu(2) and Cu(3). The vinylic groups are two-electron three-center bridge bonded via C(1) between Cu(1) and Cu(2), and Cu(2) and Cu(1a), respectively [Cu(1)–C(1) 2.00(2) and Cu(2)–C(1) 2.00(2) Å]. The nitrogen atoms of the dimethylamino substituents form intramolecular coordinate bonds to Cu(1) and Cu(1a) [Cu(1)–N(1) 2.15(2), Cu(1a)–N(1a)

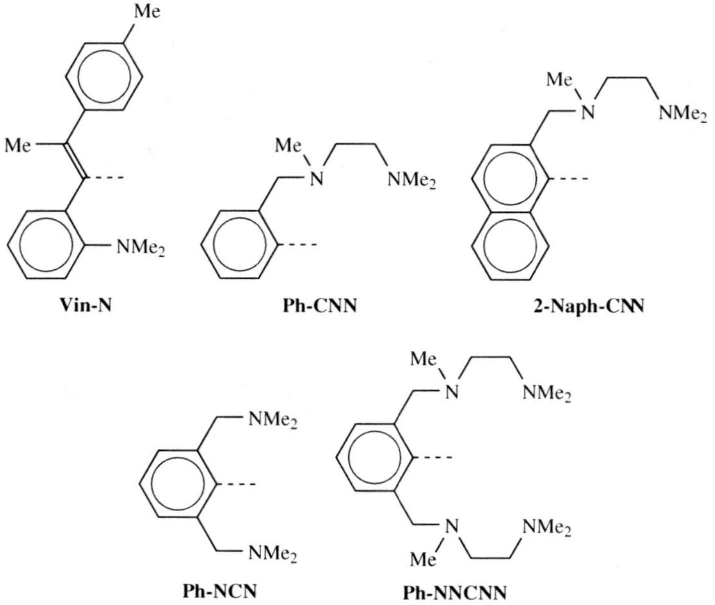

FIGURE 31. The monoanionic bidentate-, tridentate- and pentadentate ligands

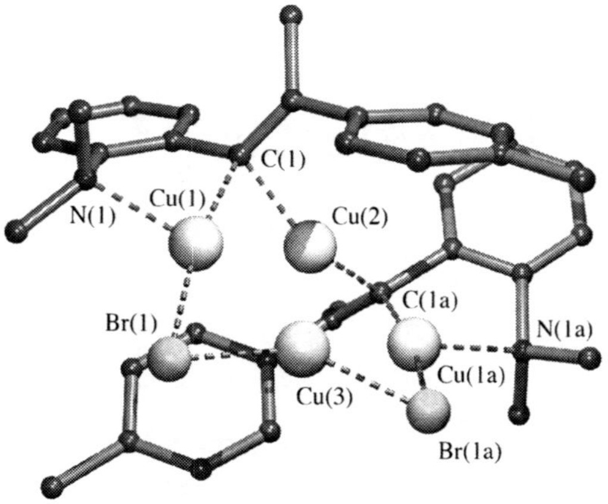

FIGURE 32. Molecular geometry of (Vin-N)$_2$Cu$_4$Br$_2$ (**70**) in the solid state

2.15(2) Å], respectively. The two bromine atoms are μ_2-bonded between the copper atoms at the other edges of the rhombus [Cu(1)–Br(1) 2.323(3) and Cu(3)–Br(1) 2.323(3) Å], but are displaced 0.617(7) Å from the central Cu$_4$-plane. As a result of the various bonding modes two of the copper atoms (Cu(2) and Cu(3)) are two-coordinate, but strongly distorted from linear [C(1)–Cu(2)–C(1a) 145.5(5)° and Br(1)–Cu–Br(1a) 135.8(1)°], while Cu(1) and Cu(1a) have a distorted trigonal planar coordination geometry.

From reaction of (Ph-CNN)Li with 1.5 equivalents of CuBr, an organocopper copper bromide aggregate having the stoichiometry $(Ph\text{-}CNN)_2Cu_3Br$ (**71**) was isolated in high yield of which the structure in the solid state was establihed by an X-ray crystal structure determination[63]. When two or more equivalents of CuBr were used, a product with stoichiometry $(Ph\text{-}CNN)Cu_2Br$ was isolated. The latter compound most probably has a dimeric structure, as was found for the corresponding 2-naphthyl compound (*vide infra*). It should be noted that from the same reaction with exactly one or less than one equivalent of CuBr, only **71** could be isolated, though in lower yield. From this observation it was concluded that the parent organocopper compound (Ph-CNN)Cu does not exist.

In the solid state **71** consists of three copper atoms in a triangular arrangement (Figure 33). The two aryl groups each span one of the edges of the Cu_3-triangle via two-electron three-center bonds [Cu(1)–C(1) 1.956(6) and Cu(2)–C(1) 2.108(5) Å] while the bromine atom resides at the third Cu–Cu edge by forming a μ_2-bridge bond between two copper atoms [Cu(2)–Br 2.432(1) and Cu(2a)–Br 2.417(1) Å]. Both nitrogen atoms of the two Ph-CNN ligands are involved in coordination to copper [Cu(2)–N(1) 2.224(5) and Cu(2)–N(2) 2.158(5) Å] resulting in a distorted tetrahedral coordination geometry at Cu(2) and Cu(2a) and a digonal one at Cu(1).

Irrespective of the number of equivalents of CuBr used, the reaction of the naphthyllithium compound (2-Naph-CNN)Li with CuBr always led to the formation of $(2\text{-Naph-CNN})_2Cu_4Br_2$ (**72**) as the only isolable compound. This indicates that the hypothetical parent organocopper compound (2-Naph-CNN)Cu and the aggregate with stoichiometry $(2\text{-Naph-CNN})_2Cu_3Br$ are the thermodynamically less favorable species[63]. The molecular geometry of **72** in the solid state was established by an X-ray crystal structure determination and shows similarities with that of **70**. Like in **70**, the four copper atoms form a rhombus (Figure 33). The two naphthyl groups are two-electron three-center bridge bonded via C(1) between the copper atoms [Cu(1)–C(1) 2.032(8) and Cu(2)–C(1) 1.986(8) Å] at two adjacent sides of the rhombus. The two bromine atoms are μ_2-bridge bonded, each between two copper atoms [Cu(1)–Br(1) 2.458(1) and Cu(3)–Br(1) 2.289(1) Å] at the remaining sides of the rhombus. Like in **71**, both nitrogen atoms of the two 2-Naph-CNN ligands are involved in coordination to copper [Cu(1)–N(1) 2.235(8) and Cu(1)–N(2) 2.140(8) Å], resulting in a distorted tetrahedral coordination geometry at Cu(1) and Cu(1a). Consequently, the copper atoms Cu(2) and Cu(3) are

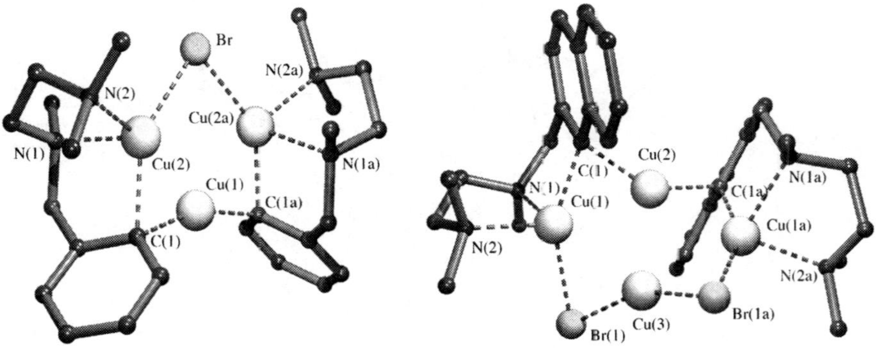

(71) **(72)**

FIGURE 33. Molecular geometry of $(Ph\text{-}CNN)_2Cu_3Br$ (**71**) and $(2\text{-Naph-CNN})_2Cu_4Br_2$ (**72**) in the solid state

two-coordinate, but deviate considerably from linear as is indicated by the respective bond angles [C(1)–Cu(1)–C(1a) 142.4(3)° and Br(1)–Cu(3)–Br(1a) 155.32(6)°].

During the attempted synthesis of pure 2,6-$(Me_2NCH_2)_2C_6H_3Cu$ ((Ph-NCN)Cu) from the reaction of 2,6-$(Me_2NCH_2)_2C_6H_3Li$ with CuBr in a 1:1 ratio, it appeared that the pure organocopper compound could never be isolated but instead, according to elemental analysis, products were formed containing LiBr, which most likely is an integral part of an aggregated species containing organocopper fragments and LiBr. However, when two or more equivalents of CuBr were used, a distinct oranocopper copper bromide aggregate [2,6-$(Me_2NCH_2)_2C_6H_3]_2Cu_4Br_2$ (**73**) could be isolated and was characterized structurally by X-ray crystallography[116].

In **73**, the four copper atoms are positioned in a butterfly arrangement of which two adjacent Cu–Cu edges are bridged by the two two-electron three-center bonded aryl groups [Cu(1)–C(1) 2.172(8) and Cu(2)–C(1) 2.029(9) Å] (Figure 34). At the two other Cu–Cu edges, the bromine atoms are μ_2-bridge bonded between the respective copper atoms [Cu(1)–Br(1) 2.317(2) and Cu(3)–Br(1) 2.414(2) Å]. The two nitrogen atoms of the two *ortho*-(dimethylamino)methyl substituents of each of the Ph-NCN ligands are coordinate bonded to adjacent copper atoms [Cu(1)–N(1) 2.247(8) and Cu(2)–N(2) 2.188(8) Å]. As a result all four copper atoms are three-coordinate, but three of them have a different environment. The copper atoms Cu(1) and Cu(1a) are three-coordinate as the result of the bonding of one carbon, one nitrogen and one bromine atom, Cu(2) is three-coordinate due to two bonding carbon atoms and one nitrogen atom and Cu(3) is three-coordinate as the result of the bonding of two bromine and one nitrogen atom.

The potentially pentadentate, monoanionic Ph-NNCNN ligand (Figure 31) combines the binding and chelating properties of the Ph-CNN and Ph-NCN ligands. The reaction of the corresponding organolithium compound (Ph-NNCNN)Li with CuBr afforded an organocopper copper bromide aggregate with the formulation $(Ph\text{-}NNCNN)_2Cu_5Br_3$ (**74**). The molecular geometry of **74** in the solid state was established by an X-ray crystal structure determination (Figure 35)[117].

Formally, the molecular geometry of **74** can be regarded as consisting of two identical $[(Ph\text{-}NNCNN)Cu_2]^+$ cationic units linked together via a central $[CuBr_3]^{2-}$ dianion (Figure 35). In the cation, the Ph-NNCNN anionic ligand is two-electron three-center

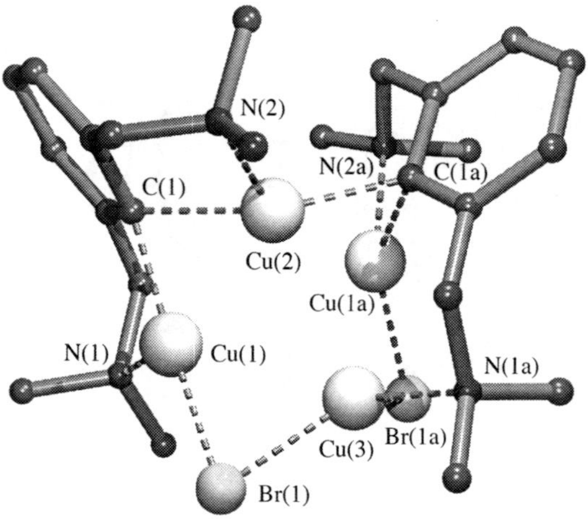

FIGURE 34. Molecular geometry of [2,6-$(Me_2NCH_2)_2C_6H_3]_2Cu_4Br_2$ (**73**) in the solid state

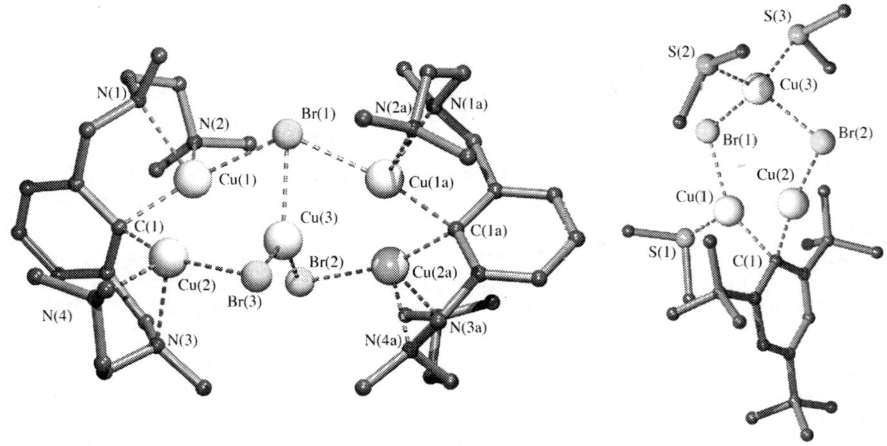

(74) **(75)**

FIGURE 35. Molecular geometry of $(Ph\text{-}NNCNN)_2Cu_5Br_3$ (**74**) and of 2,4,6-$(t\text{-}Bu)_3C_6H_2Cu_3Br_2(DMS)_3$ (**75**) in the solid state

bonded between two copper atoms via C_{ipso} [Cu(1)–C(1) 2.020(6) and Cu(2)–C(1) 2.076(6) Å]. Each of the $CH_2N(Me)CH_2CH_2NMe_2$ *ortho*-substituents is *N*,*N'*-chelate bonded to a copper atom [Cu(1)–N(1) 2.316(6), Cu(1)–N(2) 2.288(6), Cu(2)–N(3) 2.275(6) and Cu(2)–N(4) 2.203(6) Å]. In the central $[CuBr_3]^{2-}$ dianion the sum of the three Br–Cu–Br bond angles is 360° within experimental error, pointing to a trigonal planar coordination geometry for this copper atom. Linkage of the two $[(Ph\text{-}NNCNN)Cu_2]^+$ cations with the central $[CuBr_3]^{2-}$ anion is achieved via one μ_3-bonded bromine atom bridging between Cu(1), Cu(1a) and Cu(3) [Cu(1)–Br(1) 2.398(1), Cu(3)–Br(1) 2.459(1) and Cu(1a)–Br(1) 2.494(1) Å], one μ_2-bonded bromine atom bridging between Cu(2) and Cu(3) [Cu(2)–Br(3) 2.448(1) and Cu(3)–Br(3) 2.398(1) Å] and one μ_2-bonded bromine atom bridging between Cu(2a) and Cu(3) [Cu(2a)–Br(2) 2.403(1) and Cu(3)–Br(2) 2.389(1) Å].

The reaction of 2,4,6-$(t\text{-}Bu)_3C_6H_2Li$ with three equivalents of CuBr afforded, after the addition of dimethyl sulfide (DMS), a crystalline material which appeared to be the organocopper copper bromide aggregate 2,4,6-$(t\text{-}Bu)_3C_6H_2Cu_3Br_2(DMS)_3$ (**75**) of which the molecular geometry in the solid state was established by an X-ray crystal structure determination[118]. In **75**, the organic group is two-electron three-center bonded via C_{ipso} to two copper atoms [Cu(1)–C(1) 2.09(2) and Cu(2)–C(1) 1.97(1) Å] (Figure 35). To each of these copper atoms a bromine atom is bonded, each having a bonding interaction with a third copper atom [Cu(1)–Br(1) 2.420(4), Cu(3)–Br(1) 2.510(3), Cu(2)–Br(2) 2.295(3) and Cu(3)–Br(2) 2.558(4) Å] resulting in a C(1)–Cu(1)–Br(1)–Cu(3)–Br(2)–Cu(2) six-membered ring which is essentially flat (average deviation = 0.009 Å). The aryl group has a perpendicular orientation (88.2°) with respect to that plane. To one of the C_{ipso} bonded copper atoms a DMS ligand is *S*-coordinate bonded [Cu(1)–S(1) 2.306(6) Å], resulting in a trigonal planar coordination geometry at this copper atom. To the third copper atom two additional DMS ligands are coordinated [Cu(3)–S(2) 2.278(5) and Cu(3)–S(3) 2.267(5) Å], completing a distorted tetrahedral coordination geometry at this copper atom. In fact this structure can be interpreted as consisting of a [2,4,6-$(t\text{-}Bu)_3C_6H_2Cu_2]^+$ cation and a $[CuBr_2(DMS)_3]^-$ anion which make contact via μ_2-bonded bromine atoms.

The reaction of the Grignard reagent 2-vinylphenylMgBr (ViPhMgBr) with CuCl resulted in the formation of various differently aggregated species of which two could be isolated and appeared to be $[Mg(THF)_6][ViPh_2Cu_5Br_4]_2$ (**76**) and $[MgCl(THF)_5]$-$[ViPh_4Cu_5Br_2]$ (**77**) according to their respective X-ray crystal structure determinations[48]. The anions of both **76** and **77** are anionic organocopper copper bromide aggregates, although formally they also could be regarded as 'cuprate' species. The molecular geometry of the monoanion of **76** consists of five copper atoms in a square-pyramidal arrangement. The four bromine atoms are μ_2-bridge bonded between the basal copper atoms of the square pyramid (Figure 36). The two aryl groups are each two-electron three-center bonded via C_{ipso} between the apical copper atom and opposing copper atoms, Cu(3) and Cu(5), at the basal plane [Cu(1)–C(1) 1.95(4) and Cu(5)–C(1) 2.10(4) Å], respectively. The *ortho*-vinyl substituents are intramolecular η^2-bonded [Cu(2)–C(2) 2.17(4) and Cu(2)–C(3) 2.04(4) Å] to the basal copper atoms that are not bonded to the respective C_{ipso} atoms, i.e. Cu(2) and Cu(4). As a result of the various bonding modes, the apical copper atom has a linear coordination geometry [C(1)–Cu(1)–C(4) 169.8(11)°] while the basal copper atoms have distorted trigonal planar geometries.

In the solid state structure of the dianion of **77**, four of the five copper atoms form a square-planar arrangement while three of the four aryl groups are two-electron three-center bridge bonded via C_{ipso} over three edges of the Cu_4-square (Figure 36). At the fourth edge of the square, the fourth aryl group is bonded via C_{ipso} to one of the copper atoms while the vinyl substituent of this aryl group is η^2-bonded to the other copper atom. This aryl group is also bonded via C_{ipso} to a fifth copper atom, located above the Cu_4-square, to which also a terminal bromine atom is bonded. Furthermore, at the fourth edge of the square a bromine atom is μ_2-bridge bonded between the copper atoms.

During investigations on the mechanism and possible intermediates in the Hurtley reaction, an elegant synthetic procedure was developed towards the synthesis of (substituted) copper(I) benzoates. This procedure involves the controlled protonolysis of mesitylcopper with (substituted) benzoic acid (equation 17)[119,120]. Sometimes, a side product was formed in low yield and appeared to contain both mesityl and benzoate groups. These

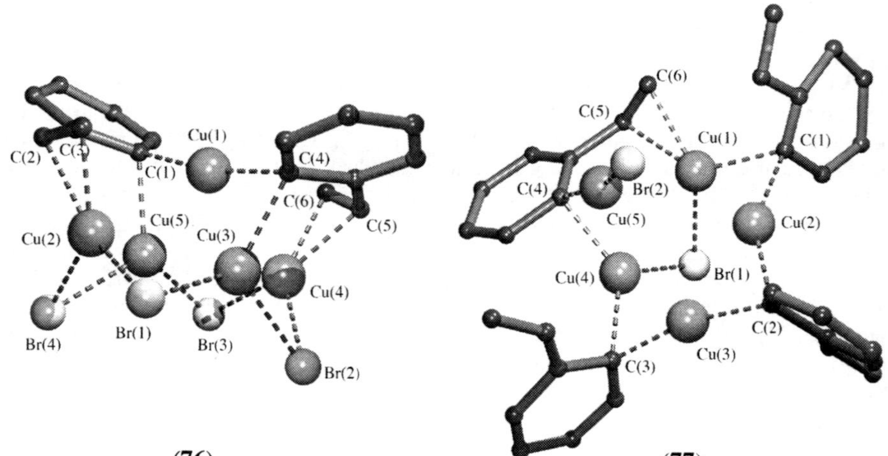

(76) **(77)**

FIGURE 36. Molecular geometry of the monoanion of $[Mg(THF)_6ViPh_2Cu_5Br_4]_2$ (**76**) and the dianion of $[MgCl(THF)_5][ViPh_4Cu_5Br_2]$ (**77**) in the solid state

side products could be prepared independently in high yield via a redistribution reaction of pure mesitylcopper and the corresponding copper benzoate (equation 18).

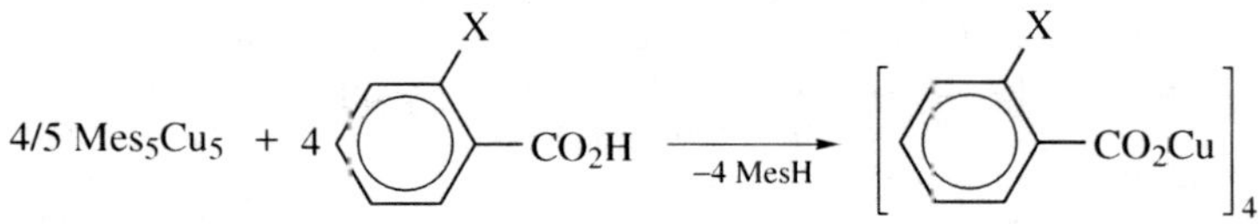

(17)

X = H, Cl or Br

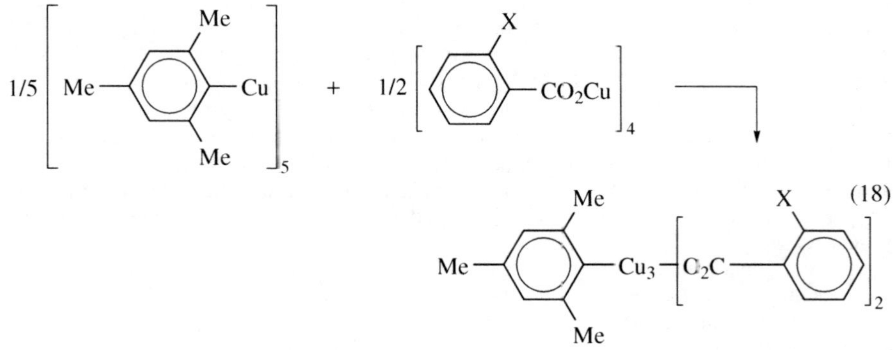

(18)

(**78**) X = H

According to an X-ray crystal structure determination of the product obtained from the reaction of mesitylcopper with copper(I) benzoate, this product appeared to be the mesitylcopper copper benzoate aggregate $MesCu_3(O_2CPh)_2$ (**78**) consisting of three copper atoms, one mesityl group and two benzoate groups (Figure 37).

The molecular geometry of **78** comprises three copper atoms arranged in a triangle with the mesityl group two-electron three-center bridge bonded between two of the copper

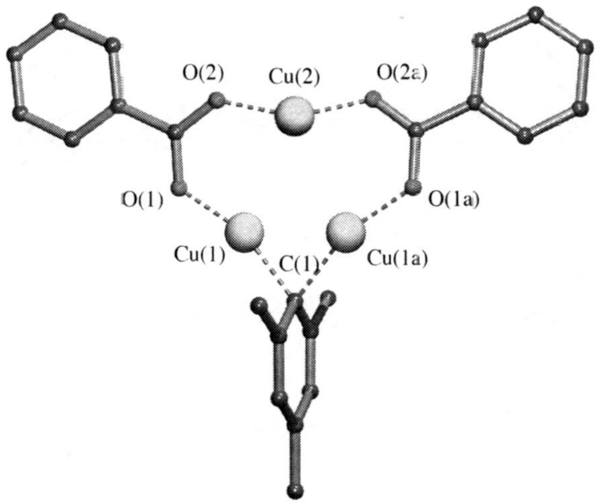

FIGURE 37. Molecular geometry of $MesCu_3(O_2CPh)_2$ (**78**) in the solid state

atoms via C_{ipso} [Cu(1)–C(1) and Cu(1a)–C(1) 1.957(6) Å]. The two benzoate groups are spanning the other two sides of the Cu_3-triangle via an *O,O′*-bridge bonding mode [Cu(1)–O(1) 1.868(5) and Cu(2)–O(2) 1.855(5) Å]. It should be noted that aggregates with structural motifs as found for **78** are also possible intermediates that can be formed during either the insertion reaction of CO_2 into Cu–C bonds of organocopper compounds, or the decarboxylation reaction (i.e. elimination of CO_2) of copper carboxylates to yield organocopper compounds[121, 122].

A rather unexpected aggregate consisting of mesitylcopper and bis(trimethylsilyl) amidocopper was obtained from the reaction of Mes_5Cu_5 and the germylene $[(Me_3Si)_2N]_2$-Ge. According to an X-ray crystal structure determination this aggregate appeared to be $Mes_2Cu_4[N(SiMe_3)_2]_2$ (**79**)[123]. In **79**, the four copper atoms are arranged in a Cu_4-square while the two mesityl groups are two-electron three-center bridge bonded [Cu(1)–C(1) 1.982(3) and Cu(4)–C(1) 1.998(3) Å] via C_{ipso} at opposing sides of the Cu_4-square (Figure 38). At the other sides of the square, two $(Me_3Si)_2N$ groups are μ_2-bonded between the respective copper atoms [Cu(1)–N(1) and Cu(2)–N(1) 1.935(2) Å]. The $C_2Cu_4N_2$ eight-membered ring is slightly puckered with C(1) and C(2) above (0.215 Å) and N(1) and N(2) below (0.125 Å) the Cu_4-plane. The coordination geometry at all four copper atoms is close to linear, as is indicated by the C–Cu–N bond angles (average C–Cu–N 170.0°).

A side product containing copper, 2,6-$Mes_2C_6H_3$ groups, a *tert*-butoxy group and iodide was obtained from the reaction of 2,6-$Mes_2C_6H_3Li$ with *t*-BuOCu in the presence of LiI. The molecular geometry of this aggregate (2,6-$Mes_2C_6H_3)_2Cu_4$(OBu-*t*)I (**80**) was established by an X-ray crystal structure determination (Figure 38)[124]. Like in **79**, the four copper atoms are arranged in a square and the two 2,6-$Mes_2C_6H_3$ groups are two-electron three-center bonded via C_{ipso} to two copper atoms at opposing edges of the Cu_4-square [Cu(1)–C(1) 1.946(4) and Cu(2)–C(1) 2.031(4) Å]. At one of the remaining edges a *tert*-butoxy group is μ_2-bonded between the copper atoms [Cu(1)–O 1.856(3) and Cu(3)–O 1.861(3) Å] and at the opposing edge an iodide is μ_2-bonded between the copper atoms [Cu(2)–I 2.4457(6) and Cu(4)–I 2.4499(6) Å]. The thus formed C_2Cu_4OI eight-membered ring is essentially flat.

Arylcopper arenethiolate copper aggregates $Mes_2Cu_4(SAr)_2$ were obtained from the interaggregate exchange reaction of the parent organocopper compound Mes_5Cu_5 and

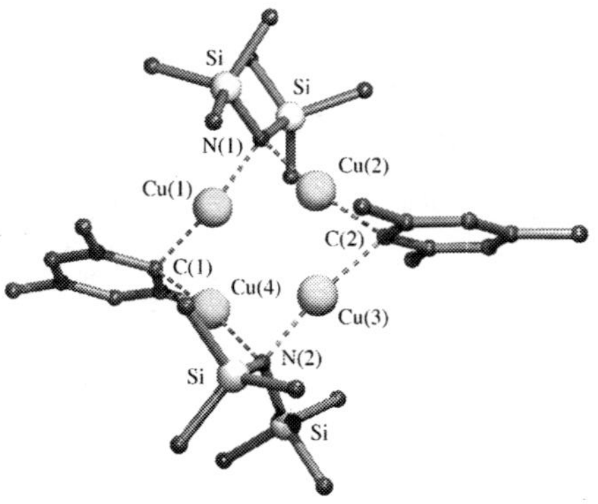

(**79**)

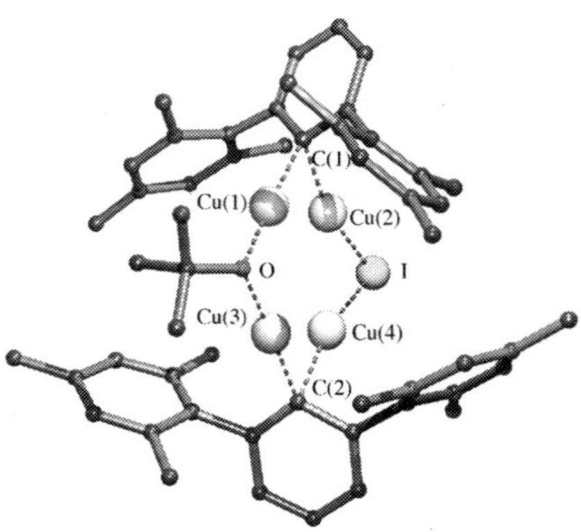

(**80**)

FIGURE 38. Molecular geometry of $Mes_2Cu_4[N(SiMe_3)_2]_2$ (**79**) and $[2,6\text{-}Mes_2C_6H_3]_2Cu_4$(OBu-*t*)I (**80**) in the solid state

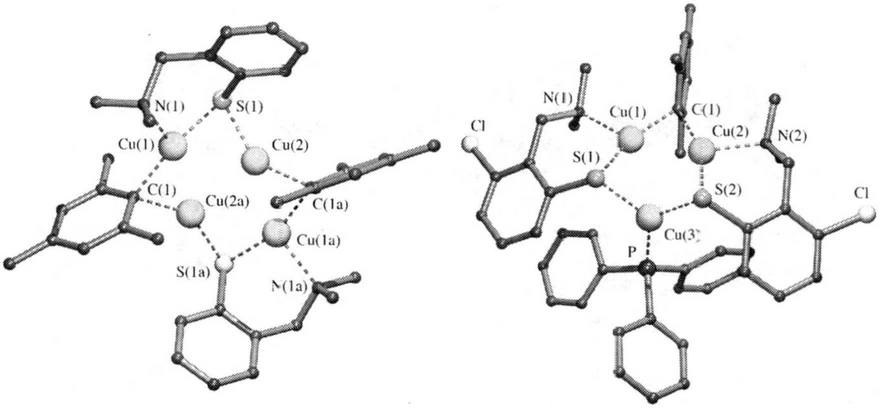

(81) **(82)**

FIGURE 39. Molecular geometry of $Mes_2Cu_4(SC_6H_4CH_2NMe_2\text{-}2)_2$ (**81**) and $MesCu_3(SC_6H_4CH_2NMe_2\text{-}2\text{-}Cl\text{-}3)_2(PPh_3)$ (**82**) in the solid state

the copper arenethiolate $(ArS)_3Cu_3$. The molecular structure in the solid state of one of these aggregates, i.e. of $Mes_2Cu_4(SC_6H_4CH_2NMe_2\text{-}2)_2$ (**81**), was established by an X-ray crystal structure determination[13]. Its structure comprises four copper atoms in a butterfly arrangement with a torsion angle Cu(1)–Cu(2a)–Cu(2)–Cu(1a) of 159.61(2)° (Figure 39). The aggregate contains a non-crystallographic C_2 axis, perpendicular to the Cu(1)–Cu(1a) and Cu(2)–Cu(2a) vectors. The mesityl groups are two-electron three-center bonded via C_{ipso} to two copper atoms at opposite edges [Cu(1)–C(1) 2.005(2) and Cu(2a)–C(1) 2.019(2) Å]. Likewise, the thiolate groups are μ_2-bonded between the respective copper atoms at the other two opposite edges [Cu(1)–S(1) 2.2439(9) and Cu(2)–S(1) 2.1928(8) Å], while the nitrogen atoms of the 2-amino substituents form intramolecular coordination bonds to two opposite positioned copper atoms [Cu(1)–N(1) 2.143(2) Å]. As a result of these bonding features two of the copper atoms [Cu(2) and Cu(2a)] are two-coordinate, but the respective angles deviate considerably from linear [C(1)–Cu(2a)–S(1a) 141.28(7)°], while the other two [Cu(1) and Cu(1a)] have a distorted trigonal planar coordination geometry.

When a Lewis base like PPh_3 was added to $Mes_2Cu_4(SAr)_2$, a deaggregation and rearrangement reaction occurs leading to the quantitative formation of an aggregate $MesCu_3(SAr)_2(PPh_3)$, and as a requirement of the reaction stoichiometry also Mes_5Cu_5 was reformed. The structure in the solid state of one of these products, i.e. of $MesCu_3(SC_6H_4CH_2NMe_2\text{-}2\text{-}Cl\text{-}3)_2(PPh_3)$ (**82**), was established by an X-ray crystal structure determination[13]. In **82** the three copper atoms are arranged in a triangle, with the mesityl group two-electron three-center bonded via C_{ipso} [Cu(1)–C(1) 2.021(4) and Cu(2)–C(1) 2.040(4) Å] and the two thiolate groups μ_2-bonded [Cu(1)–S(1) 2.234(1) and Cu(3)–S(1) 2.280(1) Å] over the Cu–Cu edges of the Cu_3-triangle (Figure 39). The nitrogen atoms of the *ortho*-(dimethylamino)methyl substituents form intramolecular coordination bonds to Cu(1) and Cu(2), respectively [Cu(1)–N(1) 2.150(4) Å]. Finally, a PPh_3 molecule is coordinate bonded to the third copper atom [Cu(3)–P 2.214(1) Å].

A rather unique aggregate with the stoichiometry $Mes_6Cu_{10}O_2$ (**83**) was isolated from the partial oxidation of mesitylcopper dissolved in hexanes with dioxygen. The molecular geometry of **83** in the solid state was determined by an X-ray crystal structure determination (Figure 40)[125].

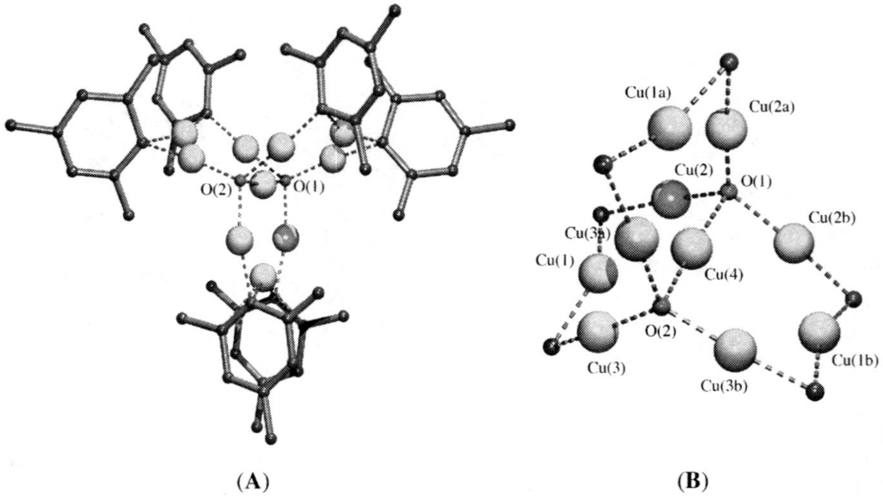

FIGURE 40. (**A**) Molecular geometry of $Mes_6Cu_{10}O_2$ (**83**) in the solid state. (**B**) Detailed view of the $Cu_{10}O_2$ core of the aggregate including the C_{ipso} carbons of the mesityl groups

The structural geometry of **83** can be described as consisting of a central O–Cu–O trianionic unit to which three Mes_2Cu_3 cationic units are bonded via their peripheral copper atoms, each to an oxygen atom of the central O–Cu–O unit, in such a way that the aggregate has approximate D_{3h}-symmetry with, as its approximate threefold axis, the O(1)–Cu(4)–O(2) vector of the central unit (Figure 40, **B**). In the Mes_2Cu_3 unit the copper atoms are arranged in a triangle [Cu(1)–Cu(2) 2.414(6), Cu(1)–Cu(3) 2.337(7) and

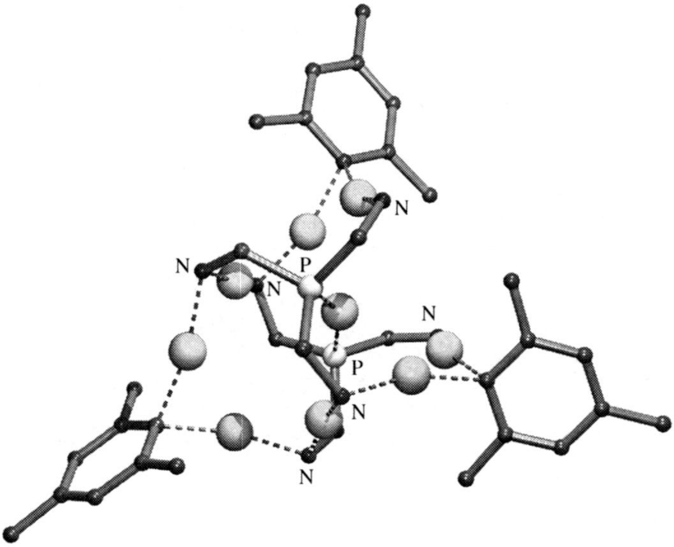

FIGURE 41. Molecular geometry of $Mes_3Cu_9\{[3,5\text{-}(CF_3)_2C_6H_3NCH_2]_3P\}_2$ (**84**) in the solid state

Cu(2)–Cu(3) 3.816(6) Å]. The two mesityl groups are two-electron three-center bridge bonded via C_{ipso} to the two copper atoms at the short edges of the triangle [Cu(1)–C(1) 1.88(4), Cu(3)–C(1) 2.02(4), Cu(1)–C(2) 1.99(5) and Cu(2)–C(2) 2.02(5) Å]. The two terminal copper atoms [Cu(2) and Cu(3)] are each bonded to an oxygen atom of the central O–Cu–O unit [Cu(2)–O(1) 1.80(3) and Cu(3)–O(2) 1.88(3) Å]. As a result of the bonding of three Mes_2Cu_3 units to the oxygen atoms of the central O–Cu–O unit, both oxygen atoms gain a rather unusual trigonal-pyramidal geometry, one with Cu(2), Cu(2a) and Cu(2b) at the base and the other one with Cu(3), Cu(3a) and Cu(3b) at the base and Cu(4) as the common apex.

From the controlled protonolysis reaction of mesitylcopper with the tripodal ligand $[3,5\text{-}(CF_3)_2C_6H_3N(H)CH_2]_3P$, a mesitylcopper copper amido aggregate with the formula $Mes_3Cu_9\{[3,5\text{-}(CF_3)_2C_6H_3NCH_2]_3P\}_2$ (**84**) was isolated and characterized structurally by X-ray crystallography[126]. The aggregate has a rather irregular geometry (Figure 41), but may be regarded as consisting of a central $[Cu_3[(3,5\text{-}(CF_3)_2C_6H_3NCH_2)_3P]_2]^{3-}$ trianion to which three $[MesCu_2]^+$ cations are bonded via Cu–N(amido) bonds. Within the $[MesCu_2]^+$ cations the mesityl group is bonded via C_{ipso} in its usual bonding mode, i.e. two-electron three-center bonded to two copper atoms.

IV. CYCLOPENTADIENYLCOPPER(I) COMPOUNDS

With the discovery and structural characterization of ferrocene[127–129] and the subsequent developement of other cyclopentadienyl metal compounds, a new era in organometallic chemistry started. In the early days most studies concerned the development of synthetic pathways to new organometallic compounds covering almost the whole periodic system of the elements and the structural characterization of these new compounds. Since then thousands of this type of compounds are known, but till today the number of well-characterized organocopper compounds of this type remains limited.

The synthesis and characterization of the triethylphosphine complex of cyclopentadienylcopper, $CpCu(PEt_3)$ (**85**), was reported as early as 1956 by Wilkinson and Piper[130]. Based on physical and chemical evidence it was suggested that the cyclopentadienyl group should be σ- or η^1-bonded to copper. Later, a careful reexamination of the infrared spectra of $CpCu(PEt_3)$ and $CpCu(PPh_3)$ (**86**) by Cotton and Marks let them conclude that both in the crystalline state and in solution the cyclopentadienyl group in both compounds is η^5-bonded to copper[131].

Unequivocal evidence for η^5-bonding of the cyclopentadienyl groups in both **85** and **86**, at least in the solid state, was obtained from X-ray crystal structure determinations (Figure 42)[132,133].

In **85**, the cyclopentadienyl group is bound to copper with almost equal Cu–C distances [Cu–C 2.25(3), 2.23(2), 2.24(3), 2.25(3) and 2.24(3) Å] while the triethylphosphine ligand is coordinate bonded [Cu–P 2.136(9) Å] to copper. The P–Cu vector is pointing to the center of the cyclopentadienyl group. Due to space-group symmetry the ethyl groups of the triethylphosphine moiety are crystallographically disordered.

A closely related structure in the solid state was found for **86**. Like in **85**, the cyclopentadienyl group is η^5-bonded to copper [Cu–C 2.200(3), 2.229(3), 2.232(2), 2.205(2) and 2.190(3) Å]. Also in this compound the P–Cu vector of the coordinating PPh_3 molecule [Cu–P 1.135(1) Å] is pointing to the center of the cyclopentadienyl group. A very similar molecular geometry in the solid state was observed for (methylcyclopentadienyl)(triphenylphosphine)copper $MeC_5H_4Cu(PPh_3)$ (**87**)[134], the only difference compared to **86** being the presence of one methyl group at the cyclopentadienyl ligand.

(Pentaphenylcyclopentadienyl)(triphenylphosphine)copper, $(Ph_5C_5)Cu(PPh_3)$ (**88**), was prepared from the reaction of the corresponding cyclopentadienylsodium derivative with

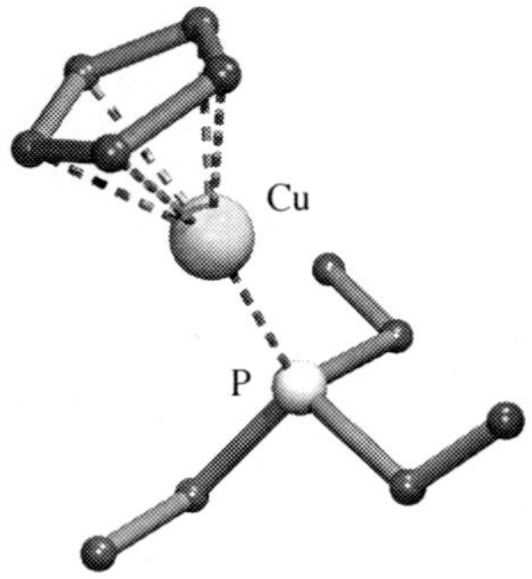

(85)

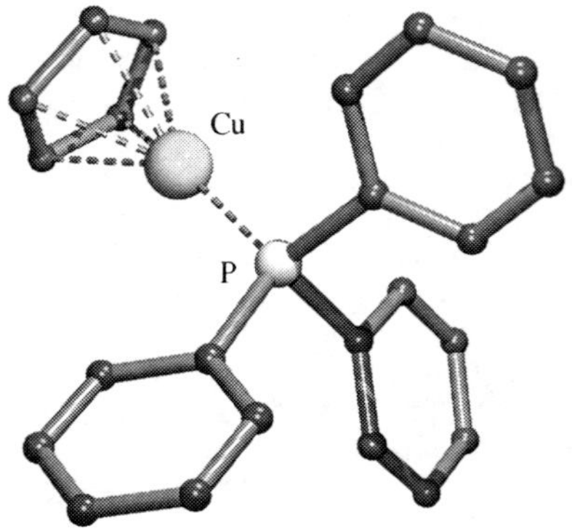

(86)

FIGURE 42. Molecular geometry of $CpCu(PEt_3)$ (**85**) (only one of the PEt_3 disorder components is shown) and $CpCu(PPh_3)$ (**86**) in the solid state

$CuCl(PPh_3)$. The molecular geometry of **88** was established by an X-ray crystal structure determination and comprises an η^5-bonded cyclopentadienyl group [Cu–C 2.247(8), 2.244(10), 2.235(9), 2.223(8) and 2.228(9) Å] and a coordinate bonded triphenylphosphine ligand [C–P 2.126(2) Å] (Figure 43)[135]. The angle between the phosphorus atom, copper and the centroid of the cyclopentadienyl group is almost linear (179.1°). It is interesting to note that **88** crystallizes in the acentric space group $P2_12_12_1$. As a consequence of the propeller like arrangement of the five phenyl groups (the tilt angles between the plane of the cyclopentadienyl ring and the planes of the phenyl rings are 43.1(5)°, 50.0(5)°, 69.6(5)°, 47.2(5)° and 55.9(5)°, respectively), the molecule is chiral.

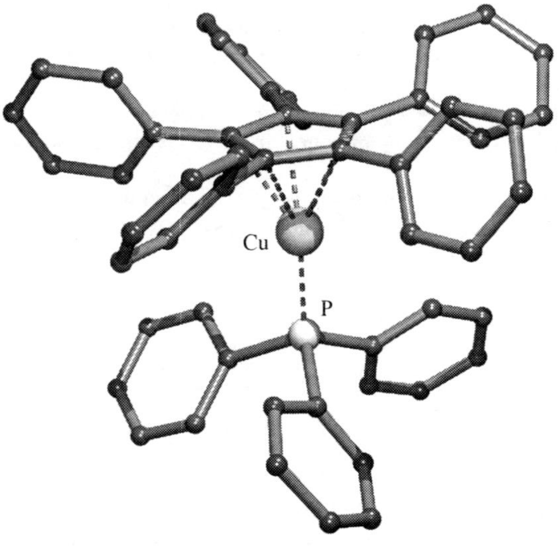

FIGURE 43. Molecular geometry of $(Ph_5C_5)Cu(PPh_3)$ (**88**) in the solid state

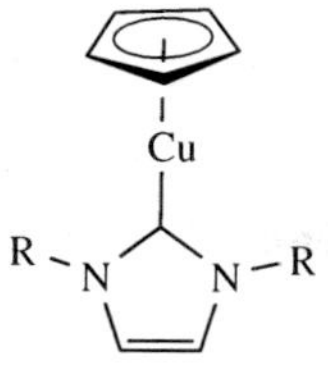

(**89**) R = *t*-Bu
(**90**) R = *i*-Pr
(**91**) R = 2,4,6-$Me_3C_6H_2$
(**92**) R = 2,6-*i*-$Pr_2C_6H_3$

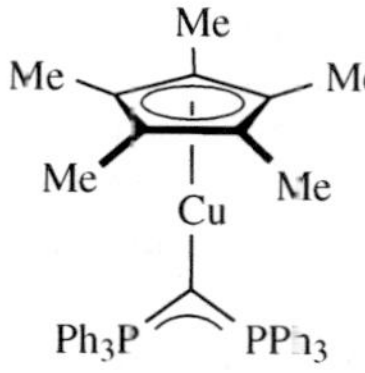

(**93**)

FIGURE 44. Schematic structures of the carbene cyclopentadienylcopper complexes **89–93**

A series of cyclopentadienylcopper complexes (**89–92**), containing an *N*-heterocyclic carbene as an additional ligand, has been prepared and characterized structurally in the solid state by X-ray crystallography (Figure 44)[136]. The structures of these compounds are closely related. In all compounds the cyclopentadienyl group is η^5-bonded to copper and the *N*-heterocyclic carbene is coordinate bonded to copper via its carbene carbon atom [Cu–C 1.886(3), 1.878(3), 1.864(4) and 1.861(5) Å in **89–92**, respectively].

The copper chloride complex of the double phosphorus ylide hexaphenyldiphosphorane was successfully converted into the corresponding pentamethylcyclopentadienyl (Cp*) derivative Cp*Cu[C(PPh$_3$)$_2$ (**93**) via reaction with pentamethylcyclopentadienylpotassium. The structure of **93** in the solid state was established by an X-ray crystal structure determination and is shown schematically (Figure 44)[137]. Like in all other cyclopentadienylcopper compounds, the pentamethylcyclopentadienyl group is η^5-bonded to copper [Cu–C 2.213(6), 2.240(5), 2.328(6), 2.344(7) and 2.271(6) Å]. The bent [P–C–P 136.0(4)°] hexaphenyldiphosphorane ligand is coordinate bonded via its ylidic carbon atom to copper [Cu–C 1.992(6) Å].

From the reaction of two equivalents of CpLi with one equivalent of CuCl in the presence of [Ph_4P]Cl, a crystalline product was isolated with stoichiometry [Cp_2Cu][PPh_4] (**94**)[138]. An X-ray crystal structure determination of this material showed the presence of isolated [Cp_2Cu]$^-$ anions and [Ph_4P]$^+$ cations in the crystal lattice. It appeared that the anion exhibits a 'slipped-sandwich' structure in which two cyclopentadienyl groups are each η^2-bonded to copper [Cu–C(1) 2.15(3), Cu–C(2) 2.22(2), Cu–C(3) 2.15(3) and Cu–C(4) 2.20(3) Å] (Figure 45).

That the cation plays an important role whose particular compound crystallizes from solution became evident from the reaction of two equivalents of CpLi with one equivalent of CuCl in the presence of 12-crown-4 (12-C-4). From this reaction mixture selective

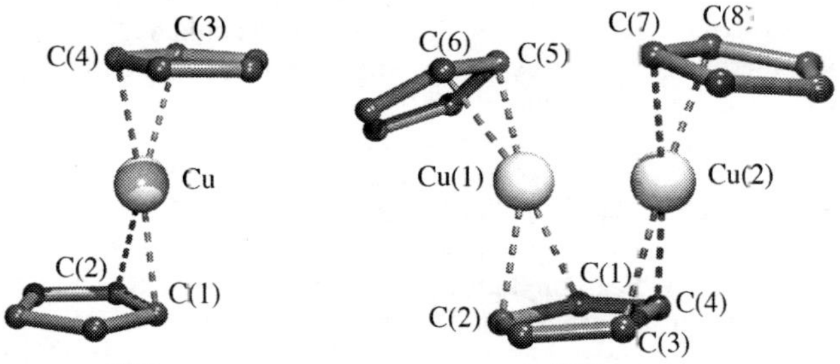

(**94**) (**95**)

FIGURE 45. Molecular geometry of the anions of **94** and **95** in the solid state

crystallization of a compound with stoichiometry $[Cp_3Cu][Li(12\text{-}C\text{-}4)_2]$ (**95**) occurred[138]. Its X-ray crystal structure determination showed the presence of isolated $[Cp_3Cu_2]^-$ anions and $[Li(14\text{-}C\text{-}4)_2]^+$ cations in the crystal lattice. In the anion, one of the cyclopentadienyl groups acts as a bridge between two copper atoms in a *cis*-μ^2-(η^2 : η^2) bonding mode [Cu−C 2.055(7), 2.243(7), 2.075(7) and 2.274(11) Å] (Figure 45). To each of the copper atoms, one terminal cyclopentadienyl group is η^2-bonded [Cu−C(5) 2.087(7), Cu−C(6) 2.180(7), Cu−C(7) 2.075(7) and Cu−C(8) 2.274(11) Å] resulting in a 'slipped-sandwich' structure for the anion.

The reaction of 1-Me_3Sn-3,5-di-*tert*-butyl-1,2,4-triphosphole with $CuCl(PPh_3)$ affords the corresponding copper complex (3,5-di-*ortho*-*tert*-butyl-1,2,4-triphospholyl)$Cu(PPh_3)$ (**96**) (equation 19) as an orange microcrystalline solid[139].

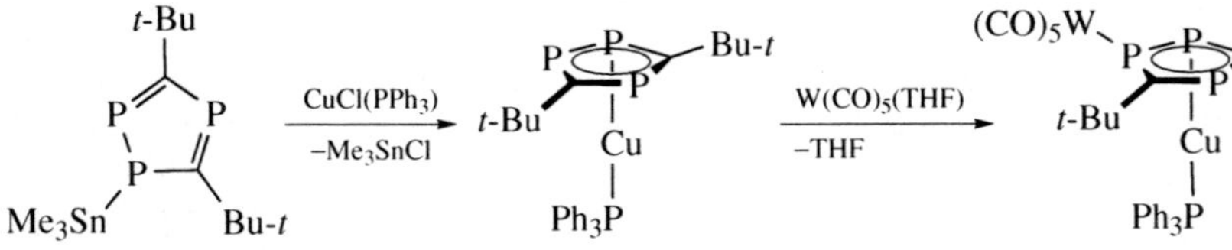

(**96**) (**97**) (19)

Unfortunately, all attempts to grow single crystals of **96** to establish its solid state structure by X-ray crystallography failed. Based on variable temperature ^1H, ^{13}C and ^{31}P NMR studies, it was suggested that in solution **96** exists as a complicated mixture of interchanging isomers in which the triphospholyl ligand is both η^1- and η^5-bonded to copper. Compared to the cyclopentadienyl ligand, in the triphospholyl ligand it is not only the delocalized π-system but also the phosphorus lone pairs that can play an important role in the bonding features of this ligand. The highly fluxional behavior of **96** could be blocked by using the lone pair of one of the phosphorus atoms of the triphospholyl ligand for an additional coordination bond to a second metal. Compound **96** was converted quantitatively into the corresponding tungsten complex η^5-(*P*-$W(CO)_5$-3,5-di-*ortho*-*tert*-butyl-1,2,4-triphospholyl)$Cu(PPh_3)$ (**97**) (equation 19). The molecular geometry of **97** in the solid state was determined by an X-ray crystal structure determination (Figure 46)[139]. In **97**, the 1,2,4-triphospholyl anionic ligand is η^5-bonded to copper via its two carbon atoms [Cu−C 2.274(8) and 2.277(8) Å] and three phosphorus atoms [Cu−P(1) 2.438(3), Cu−P(2) 2.473(3) and Cu−P(3) 2.417(3) Å]. To reach coordination saturation at the copper atom the triphenylphosphine ligand is coordinate bonded via its phosphorus atom to copper [Cu−P(4) 2.180(3) Å]. The $W(CO)_5$ moiety is bonded to the 1,2,4-triphospholyl skeleton via the lone pair of phosphorus atom P(1) [W−P(1) 2.532(2) Å].

The trimetallic cluster compounds $Cp^*{}_3Rh_2Cu(CO)_2$ (**98**) and $Cp^*CpCuPtW(CC_6H_4Me\text{-}4)(PMe_3)_2(CO)_2$ (**99**) have in common that they both contain a pentamethylcyclopentadienylcopper moiety. For both clusters the structure in the solid state was established by X-ray crystal structure determinations[140,141].

The molecular geometry of **98** in the solid state comprises a central core of two rhodium atoms and one copper atom arranged in a triangle with almost equal metal−metal distances [Cu−Rh(1) 2.579(12), Cu−Rh(2) 2.529(19), Rh(1)−Rh(2) 2.571(12) Å] while to each of the rhodium atoms and the copper atom a pentamethylcyclopentadienyl group is η^5-bonded (Figure 47). The two carbonyl groups are each μ^3-bridge bonded between all three metals at opposite sites of the triangular M_3-face. One of the carbonyl groups is bonded to the metals with one relatively short carbon−rhodium bond [Rh−C 1.82(5) Å], one relatively short carbon−copper bond [2.05(6) Å] and one relatively long carbon−rhodium bond [Rh−C 2.23(6) Å]. The other carbonyl group is bonded to the three metals with

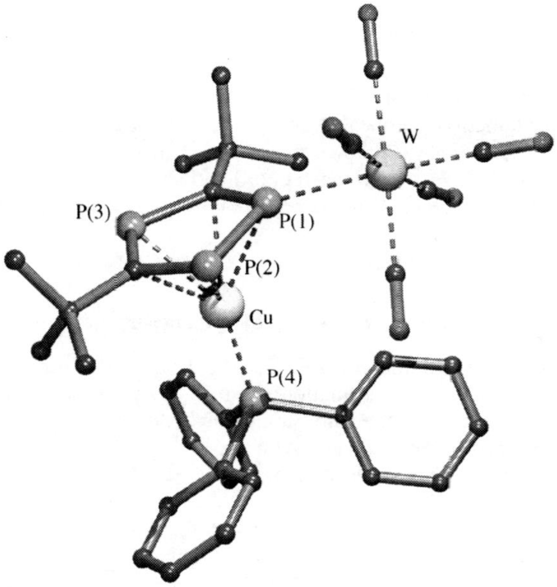

FIGURE 46. Molecular geometry of **97** in the solid state

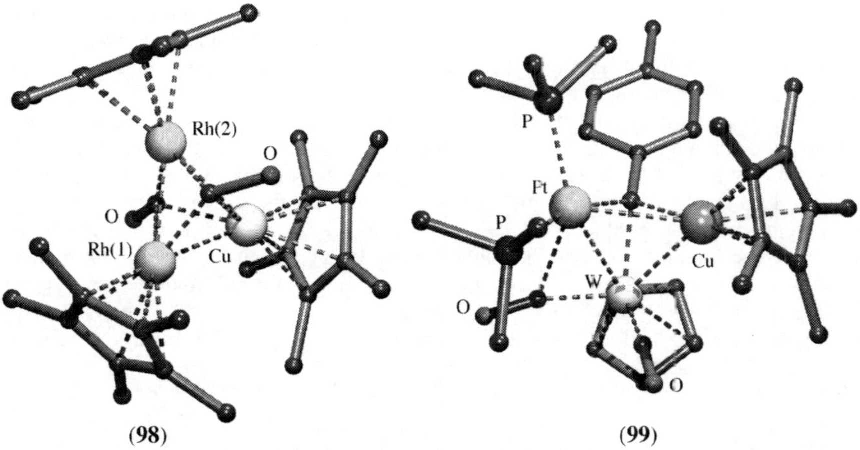

(**98**) (**99**)

FIGURE 47. Molecular geometry of the trimetallic clusters $Cp^*{}_3Rh_2Cu(CO)_2$ (**98**) and Cp*CpCu $PtW(CC_6H_4Me\text{-}4)(PMe_3)_2(CO)_2$ (**99**) in the solid state

two relatively short carbon–rhodium bonds [Rh–C 1.92(5) and 1.93(5) Å] and a longer carbon–copper bond [Cu–C 2.26(3) Å].

The X-ray crystal structure determination of **99** also showed the presence of the three metals in a triangular arrangement [Pt–W 2.779(2), Cu–W 2.648(3) Å and Cu–Pt 2.807(3) Å] (Figure 47). Above one of the triangular faces the carbyne carbon atom of the CC_6H_4Me-4 group is symmetrically μ^3-bridge bonded to all three metals [Cu–C 1.96(2), Pt–C 2.00(2) and W–C 2.03(2) Å]. One of the CO ligands is rather unsymmetrically μ^2-bridge bonded between the tungsten and platinum atom [C–W 1.92(3) and C–Pt 2.45(3) Å] while the second CO ligand is end-on coordinated to tungsten [C–W 1.97(3) Å]. The pentamethylcyclopentadienyl group is bonded to copper in a η^5-bonding mode [mean Cu–C 2.27(3) Å]. Likewise, the cyclopentadienyl group is η^5-bonded to tungsten [mean W–C 2.42(4) Å]. To reach coordination saturation at the platinum center, two trimethylphosphine molecules are coordinate bonded to platinum [Pt–P 2.285(6) and 2.341(7) Å].

V. COPPER CARBORANE COMPOUNDS

In the mid-sixties of the previous century the fields of boron hydride and carborane chemistry and transition metal coordination chemistry merged, when it was recognized that in particular carborane anions exhibit a similar bonding behavior towards transition metals like the well-known cyclopentadienyl anion[142].

Icosahedral 1,2-dicarba-*closo*-dodecaborane, 1,2-$C_2B_{10}H_{12}$ can be converted with a base into the anion $[C_2B_9H_{12}]^-$ via the formal extraction of a $[BH]^{2+}$ vertex from the icosahedron followed by proton addition to the resulting $[C_2B_9H_{11}]^{2-}$ anion. It was expected that the twelfth hydrogen atom became bonded to the open pentagonal face of the icosahedral fragment, bridging between two boron atoms. It was found that $[C_2B_9H_{12}]^-$ could be easily deprotonated to form the new bis-anionic ligand $[C_2B_9H_{11}]^{2-}$. The discovery of a ferrocene-type analog $[(C_2B_9H_{11})_2Fe]^{2-}$ initiated studies into a new class of extremely stable organometallic complexes[143,144].

Lipscomb, who was awarded the Nobel Prize in 1976 for his fundamental studies on the bonding principles in borane and carborane cluster compounds, has described with coworkers the bonding in the hypothetical $[B_{11}H_{11}]^{4-}$, a description which is also applicable to the understanding of the bonding in the isoelectronic $[C_2B_9H_{11}]^{2-}$ anion[145]. Accordingly, the bonding in this type of metal carborane complexes can be described as follows. The applied molecular orbital scheme places 6 electrons in the apical region of the pentagonal pyramidal $[C_2B_9H_{11}]^{2-}$ anion, 14 electrons between the two five-membered rings, and 6 electrons in five sp^3 orbitals directed toward the vacant apex position associated with the open pentagonal face. These last five atomic orbitals may be used to generate three bonding and two anti-bonding molecular orbitals. These would include a strongly bonding a_1 orbital, two degenerate and bonding e_1 orbitals and two strongly antibonding e_2 orbitals. Based on these considerations it is expected that the $[C_2B_9H_{11}]^{2-}$ anion resembles the well-known $[C_5H_5]^-$ anion in its reaction with transition metal ions and type of bonding to these metals.

The copper carboranes $[PPN][(1,2\text{-}C_2B_9H_{11})Cu^{I}(PPh_3)]$ (**100**)[146,147] (PPN = bis-(triphenylphosphoranylidene)ammonium), $[Et_4N]_2[(1,2\text{-}C_2B_9H_{11})_2Cu^{II}]$ (**101**)[148] and $[Ph_3MeP][(1,2\text{-}C_2B_9H_{11})_2Cu^{III}]$ (**102**)[149] have been prepared and were characterized structurally in the solid state by X-ray crystallography. It should be noted that in **100** the copper atom is in its 1+ oxidation state, in its 2+ oxidation state in **101**, while in **102** it is present in the very rare 3+ oxidation state.

The solid state structures of these compounds comprise separate cationic and anionic moieties in their respective unit cells of which the anions are shown (Figure 48). To compare the bonding features of the $[C_2B_9H_{11}]^{2-}$ anion in these compounds with each other, the relevant bond distances are compiled in Table 1. It should be noted that in both **101** and **102** the two $[C_2B_9H_{11}]^{2-}$ anions bonded to copper are symmetry related due to space group symmetry.

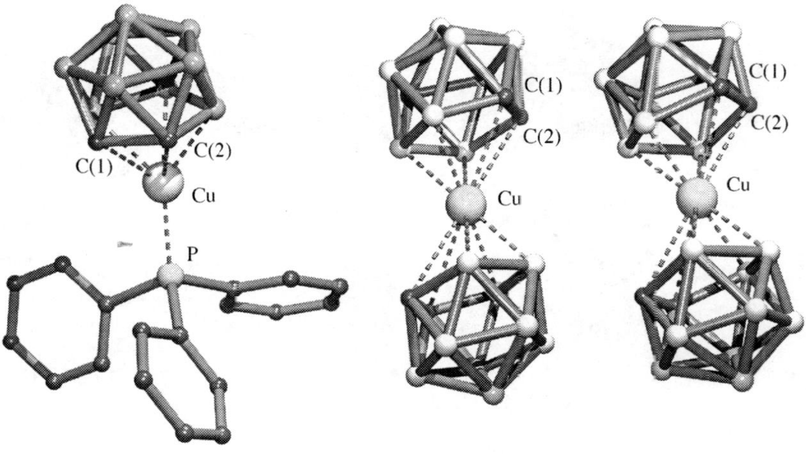

(**100**) (**101**) (**102**)

FIGURE 48. Molecular geometry of the anions $[(1,2\text{-}C_2B_9H_{11})Cu^I(PPh_3)]^-$ (**100**), $[(1,2\text{-}C_2B_9H_{11})_2\text{-}Cu^{II}]^{2-}$ (**101**) and $[(1,2\text{-}C_2B_9H_{11})_2Cu^{III}]^-$ (**102**), respectively

In complex **100**, the bonding of the $[C_2B_9H_{11}]^{2-}$ anion to copper can be described as being η^5 via two carbon and three boron atoms resulting in an overall *closo*-metallacarborane structural motif. The copper atom lies 1.679 Å from the mean plane of the C_2B_3 pentagonal face, which is significantly shorter than the one observed in isolobal $CpCu(PPh_3)$ (1.862 Å). The Cu–P distance (2.147(2) Å) is comparable to the corresponding value observed in $CpCu(PPh_3)$ (2.135 Å). The phosphorus atom is slightly displaced towards the carbon–carbon edge of the C_2B_3 face.

In the $[(1,2\text{-}C_2B_9H_{11})_2Cu^{II}]^{2-}$ anion of **101** the Cu–C distances are significantly elongated compared to those of **100** (Table 1). As a consequence the copper atom is displaced 0.6 Å from the pseudo-fivefold axis through the C_2B_9 cage resulting in a 'slipped-sandwich' structure. The bonding to copper therefore can be best described as being distorted towards η^3 via three boron atoms.

The overall structural geometry of the $[(1,2\text{-}C_2B_9H_{11})_2Cu^{III}]^-$ anion in **102** is closely related to that of $[(1,2\text{-}C_2B_9H_{11})_2Cu^{II}]^{2-}$ in **101**, the only chemical difference being the oxidation state of copper: 2+ in **101** and 3+ in **102**, respectively. In **102**, however, the distortion towards η^3-bonded $[C_2B_9H_{11}]^{2-}$ anions is even more pronounced and also expressed in a significantly shorter C(1)–C(2) bond distance (Table 1).

TABLE 1. Relevant bond distances (Å) of the $[C_2B_9H_{11}]^{2-}$ anion in carborane copper complexes **100–102**

Complex	**100**	**101**	**102**
Cu–C(1)	2.316(6)	2.583	2.499(7)
Cu–C(2)	2.317(6)	2.569	2.506(6)
Cu–B(1)	2.182(7)	2.211	2.173(8)
Cu–B(2)	2.166(7)	2.132	2.152(7)
Cu–B(3)	2.115(7)	2.245	2.053(7)
C(1)–C(2)	1.586(8)	1.534	1.450(9)

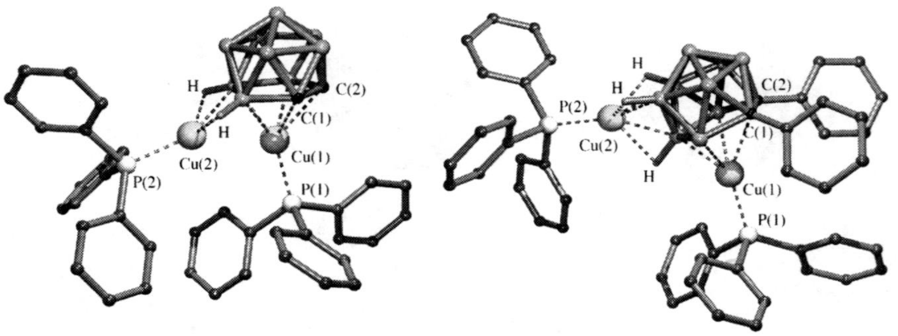

(**103**) (**104**)

FIGURE 49. Molecular geometry of *closo*-[*exo*-4,8-(μ-H)$_2$Cu^I(PPh$_3$)-3-(PPh$_3$)-3,1,2-Cu^{I}C$_2$B$_9$H$_9$] (**103**) and [Cu(PPh$_3$)]$_2$[Ph$_2$C$_2$B$_9$H$_9$] (**104**) in the solid state

Reaction of [Tl]$_2$[C$_2$B$_9$H$_{11}$] with two equivalents of CuCl(PPh$_3$) in the absence of other cations afforded neutral *closo*-[*exo*-4,8-(μ-H)$_2$Cu^I(PPh$_3$)-3-(PPh$_3$)-3,1,2-Cu^{I}C$_2$B$_9$H$_9$] (**103**)[146, 147] of which the structure in the solid state was established by an X-ray crystal structure determination. The molecular geometry of **103** comprises a [1,2-C$_2$B$_9$H$_{11}$]$^{2-}$ dianion to which two [Cu(PPh$_3$)]$^+$ cations are bonded (Figure 49). One of the [Cu(PPh$_3$)]$^+$ cations [Cu(1)–P(1) 2.164(1) Å] is η^5-bonded to the two carbon and three boron atoms of the C$_2$B$_3$ pentagonal face [Cu(1)–C(1) 2.280(4), Cu(1)–C(2) 2.331(4), Cu(1)–B 2.164(5), 2.226(5) and 2.140(5) Å]. The second Cu(PPh$_3$) moiety is bonded to two boron atoms of the C$_2$B$_3$ pentagonal face [Cu(2)–B 2.173(5) and 2.210(4) Å] and two *exo*-polyhedral boron-hydride-copper three-center two electron linkages [Cu–H 1.91(4) and 2.03(4) Å].

The introduction of phenyl substituents at each of the carbon atoms of the [1,2-C$_2$B$_9$H$_{11}$]$^{2-}$ framework in **103** results in the complex [Cu(PPh$_3$)]$_2$[Ph$_2$C$_2$B$_9$H$_9$] (**104**)[150]. An X-ray crystal structure determination of **104** revealed entirely different bonding modes for the two copper atoms (Figure 49). In **104**, one of the [Cu(PPh$_3$)]$^+$ cations [Cu(1)–P(1) 2.183(10) Å] is located above the C$_2$B$_3$ pentagonal face of the [Ph$_2$C$_2$B$_9$H$_9$] dianion, but is slipped by 0.624 Å towards the boron atoms. The observed distances between the copper atom and the atoms of the C$_2$B$_3$ pentagonal face [Cu(1)–C(1) 2.56(3), Cu(1)–C(2) 2.67(3), Cu(1)–B 2.07(4), 2.20(4) and 2.32(4) Å] suggest rather an η^3-bonding mode to three boron atoms than η^5-bonding. The other [Cu(PPh$_3$)]$^+$ moiety [Cu(2)–P(2) 2.128(10) Å] is bonded to three boron atoms of the B$_3$ trigonal face [Cu(2)–B 2.27(3), 2.18(4) and 2.24(4) Å], adjacent to the C$_2$B$_3$ pentagonal face, and three additional *exo*-polyhedral B-H-Cu three-center two-electron linkages [Cu–H 2.01, 2.05 and 2.07 Å].

A structure very closely related to the structure of **103** was found for [Cu(PPh$_3$)]$_2$-[C$_2$B$_{10}$H$_{12}$] (**105**) in which one of the [Cu(PPh$_3$)]$^+$ cations is η^6-bonded to two carbon and four boron atoms of the C$_2$B$_4$ hexagonal face of the carborane skeleton while the other [Cu(PPh$_3$)]$^+$ cation is bonded to two boron atoms of the C$_2$B$_4$ hexagonal face and two *exo*-polyhedral boron-hydride-copper bridges[151].

Replacement of a terminal hydride on the boron atom adjacent to one of the carbon atoms at the pentagonal C$_2$B$_3$ face of the [C$_2$B$_9$H$_{11}$]$^{2-}$ anion by a Lewis base like methyl isonicotinate or dimethyl sulfide results in the zwitter-monoanions, [C$_2$B$_9$H$_{10}$(NC$_5$H$_4$CO$_2$-Me-4)]$^-$ and [C$_2$B$_9$H$_{10}$(SMe$_2$)]$^-$, respectively. With Cu$^+$ in the presence of PPh$_3$ these anions form the neutral complexes [Cu(PPh$_3$)][C$_2$B$_9$H$_{10}$(NC$_5$H$_4$CO$_2$Me-4)] (**106**) and [Cu(PPh$_3$)][C$_2$B$_9$H$_{10}$(SMe$_2$)] (**107**), respectively. For both compounds the structures in the solid state were established by X-ray crystal structure determinations (Figure 50)[147, 152].

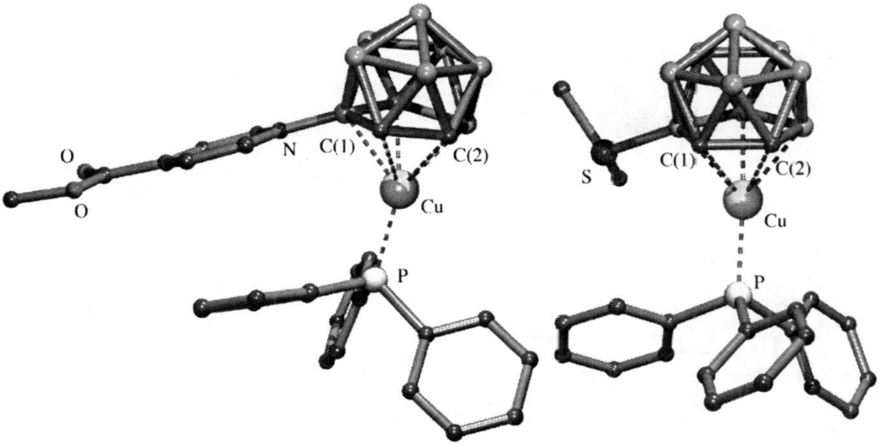

(**106**) (**107**)

FIGURE 50. Molecular geometry of the neutral complexes $[Cu(PPh_3)][C_2B_9H_{10}(NC_5H_4CO_2Me\text{-}4)]$ (**106**) and $[Cu(PPh_3)][C_2B_9H_{10}(SMe_2)]$ (**107**) in the solid state

In **106**, the $[Cu(PPh_3)]^+$ moiety [Cu−P 2.163(3) Å] is located with its Cu atom above the C_2B_3 pentagonal face of the carborane cage. The observed bond distances [Cu−C(1) 2.383(9), Cu−C(2) 2.480(9), Cu−B 2.254(4), 2.150(12) and 2.116(11) Å] suggest a strong distortion towards an η^3-boroallylic bonding rather than an η^5-type of bonding to the C_2B_3 face. A similar type of bonding of the $[Cu(PPh_3)]^+$ unit to the pentagonal face of the $[C_2B_9H_{10}(SMe_2)]^-$ anion was observed in **107**[152].

When the $[C_2B_9H_{10}(NC_5H_4CO_2Me\text{-}4)]^-$ anion is treated with Cu^+ in the absence of PPh_3, trimeric $[CuC_2B_9H_{10}(NC_5H_4CO_2Me\text{-}4)]_3$ (**108**) is formed. An X-ray crystal structure determination of **108** showed that this trimer consists of three $[CuC_2B_9H_{10}(NC_5H_4\text{-}CO_2Me\text{-}4)]$ units symmetry related by a threefold crystallographic axis resulting in a 'clustered cluster' with a 'pinwheel' ligand array (Figure 51)[147,153]. Each of the copper atoms is located above the C_2B_3 pentagonal face of a carborane unit. However, the observed bond distances [Cu−C(1) 2.634(9), Cu−C(2) 2.635(9), Cu−B 2.265(8), 2.116(8) and 2.307(8) Å] indicate a distortion towards η^3-boroallylic bonding. The three $[CuC_2B_9H_{10}(NC_5H_4CO_2Me\text{-}4)]$ units are linked via a Cu−B bond [Cu−B 2.120(8) Å] to an adjacent $[CuC_2B_9H_{10}(NC_5H_4CO_2Me\text{-}4)]$ unit and an additional B-H-Cu three-center two-electron linkage [Cu−H 1.608 Å].

Reaction of the pyrazolyl anion with two equivalents of the carborane $C_2B_9H_{11}$ affords the monoanionic bis-carborane substituted pyrazole $[C_3H_3N_2(C_2B_9H_{11})_2]^-$. Double deprotonation results in the formation of the trianion $[C_3H_3N_2(C_2B_9H_{10})_2]^{3-}$, which has been successfully applied in the synthesis of a series of transition metal complexes including a copper(III) derivative $Cu^{III}[C_3H_3N_2(C_2B_9H_{10})_2]$ (**109**)[154]. An X-ray crystal structure determination of **109** revealed an *ansa*-bridged sandwich-type structure in which the pentagonal C_2B_3 faces of both carborane units are η^5-bonded to the copper atom (Figure 51).

From the reaction of two equivalents of the dilithium salt $Li_2[(C_2B_{10}H_{10})_2]$, derived from bis(*o*-carborane), with $CuCl_2$, after exchange of the cations with Et_4N^+ an ionic compound $[(Et_4N)_2][Cu^{II}\{(C_2B_{10}H_{10})_2\}_2]$ (**110**) was isolated[155,156]. An X-ray crystal structure determination of **110** revealed separated Et_4N^+ cations and $[Cu^{II}\{(C_2B_{10}H_{10})_2\}_2]^{2-}$ anions in the crystal lattice[157]. The anion consists of two chelating bis(*o*-carborane) dianions

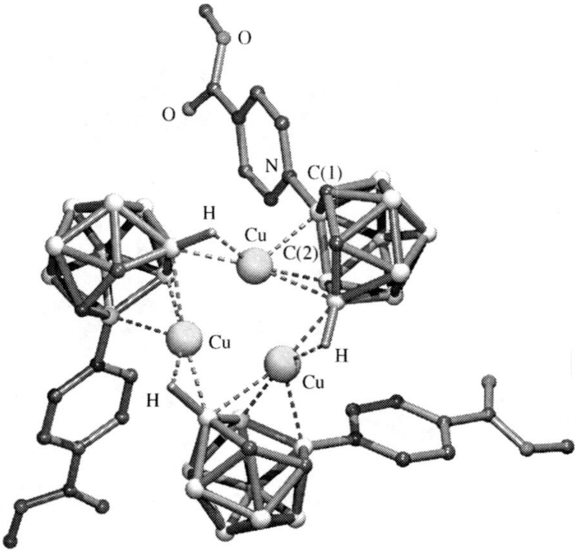

(**108**)

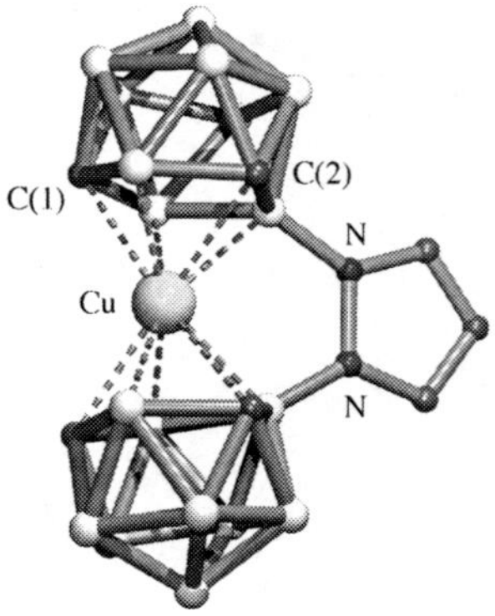

(**109**)

FIGURE 51. Molecular geometry of $[CuC_2B_9H_{10}(NC_5H_4CO_2Me\text{-}4)]_3$ (**108**) and $Cu^{III}[C_3H_3N_2\text{-}(C_2B_9H_{10})_2]$ (**109**) in the solid state

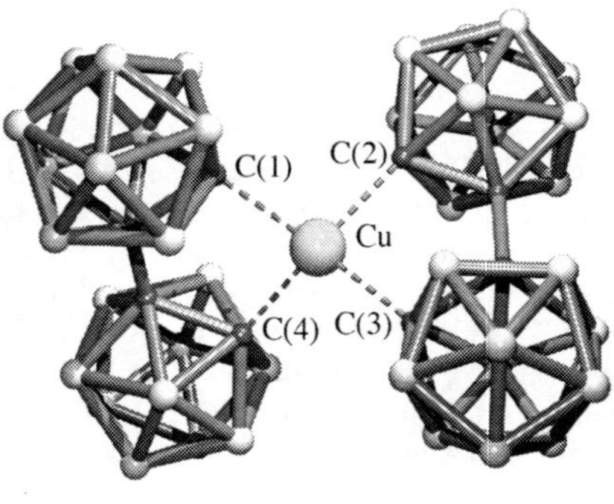

(**110**)

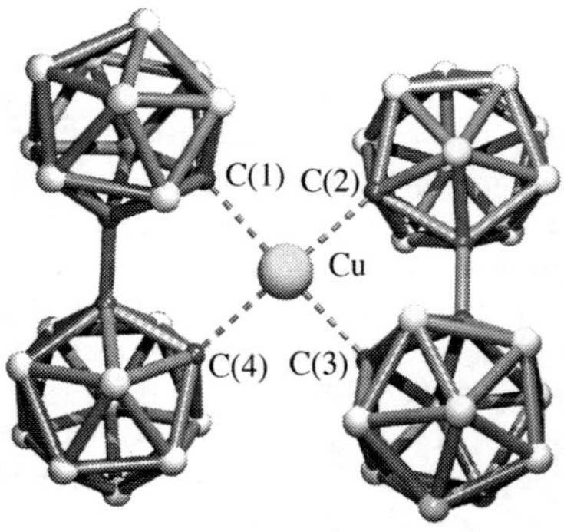

(**111**)

FIGURE 52. Molecular geometry of the anions $[Cu^{II}\{(C_2B_{10}H_{10})\}_2]_2]^{2-}$ (**110**) and $[Cu^{III}\text{-}\{(C_2B_{10}H_{10})_2\}_2]^{-}$ (**111**) in the solid state

σ-bonded via the carbon atoms [Cu−C(1) 2.070(10), Cu−C(2) 2.072(10), Cu−C(3) 2.072(10) and Cu−C(4) 2.070(10) Å] to divalent copper (Figure 52). The coordination geometry at copper is rather irregular, as is indicated by the various bond angles around copper [C(1)−Cu−C(2) 102.3(4)°, C(1)−Cu−C(3) 142.4(4)°, C(1)−Cu−C(4) 90.0(4)°, C(2)−Cu−C(3) 89.4(4)°, C(2)−Cu−C(4) 142.4(4)° and C(3)−Cu−C(4) 102.3(4)°]. The angle formed by the normals to the planes of the two bidentate ligands is 54°, while the ideal angle for a square-planar situation is 0° and a tetrahedral one is 90°.

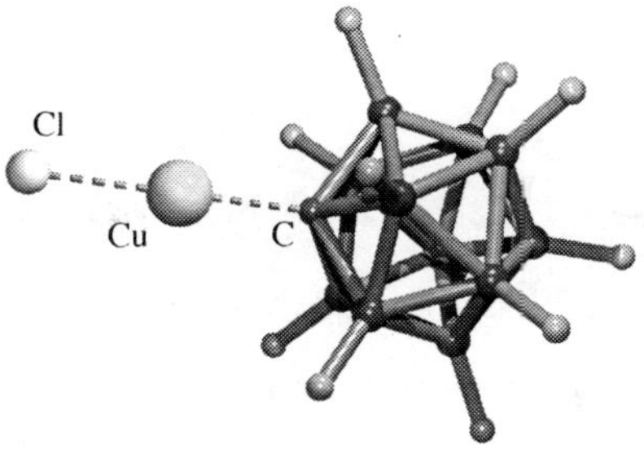

FIGURE 53. Molecular geometry of the $[CuCl(CB_{11}F_{11})]^{2-}$ dianion of **112** in the solid state

Single electron oxidation of $Li_2[Cu\{(C_2B_{10}H_{10})_2\}_2]$ using $CuCl_2$ as the oxidant affords, after exchange of the cations with Et_4N^+, the complex $[Et_4N][Cu^{III}\{(C_2B_{10}H_{10})_2\}_2]$ (**111**) with copper in its trivalent oxidation state[156]. An X-ray crystal structure determination of **111** showed, as for **110**, separated Et_4N cations and $[Cu^{III}\{(C_2B_{10}H_{10})_2\}_2]^-$ anions in the crystal lattice[157]. The anion has the same structural motif as the one in **110**. The Cu–C bond distances [Cu–C(1) 2.030(6), Cu–C(2) 2.005(6), Cu–C(3) 2.029(6) and Cu–C(4) 2.015(6) Å] are slightly shorter compared to those in **110**. The various bond angles around copper [C(1)–Cu–C(2) 92.3(2)°, C(1)–Cu–C(3) 162.4(2)°, C(1)–Cu–C(4) 90.5(2)°, C(2)–Cu–C(3) 90.5(2), C(2)–Cu–C(4) 161.2(2)° and C(3)–Cu–C(4) 92.5(2)°] point to an irregular coordination geometry at the copper atom.

The dodecahedral $[HCB_{11}F_{11}]^-$ monocarborane monoanion represents an example of the so-called 'highly fluorinated weakly coordinating anions'. This monoanion can be easily deprotonated to the corresponding dianion $[CB_{11}F_{11}]^{2-}$. Its adduct with CuCl was isolated as the $[n\text{-}Bu_4N]_2[CuCl(CB_{11}F_{11})]$ complex (**112**) of which the structure in the solid state was established by X-ray crystallography[158]. It solid state structure comprises isolated $[n\text{-}Bu_4N]^+$ cations and $[CuCl(CB_{11}F_{11})]^{2-}$ dianions in the crystal lattice. In the dianion the copper atom is σ-bonded [Cu–C 1.917(5) Å] to the carbon atom of the dodecahedral $CB_{11}F_{11}$ cluster and the chlorine atom [Cu–Cl 2.136(1) Å] (Figure 53). The carbon to copper to chlorine bond angle is close to linear [C–Cu–Cl 176.0(2)°].

VI. COPPER(I) ACETYLIDES

The potential of copper acetylides CuC≡CR as a synthetic tool in organic chemistry was recognized in the early sixties of the previous century[159–161]. Compared to alkyl- and arylcopper compounds, copper acetylides have a better thermal stability and are less sensitive towards oxidation and hydrolysis. The formation of the most simple copper acetylide, i.e. copper acetylide (Cu_2C_2) itself, obtained by passing acetylene gas through an ammoniacal solution of a Cu(I) salt, was reported as early as 1897[162]. It is a red solid material, highly *explosive* in a *dry* state! The monosubstituted copper acetylides are easily obtained in high yield from the reaction of the parent acetylene with an ammoniacal solution of a Cu^I salt (equation 20)[159].

$$HC{\equiv}CR + Cu^+ \xrightarrow{H_2O/NH_3} CuC{\equiv}CR + NH_4^+ \qquad (20)$$

Due to its electronic properties a monosubstitued, monoanionic acetylide ligand can act as two; four- or even six-electron donor and therefore can easily form bridge bonds

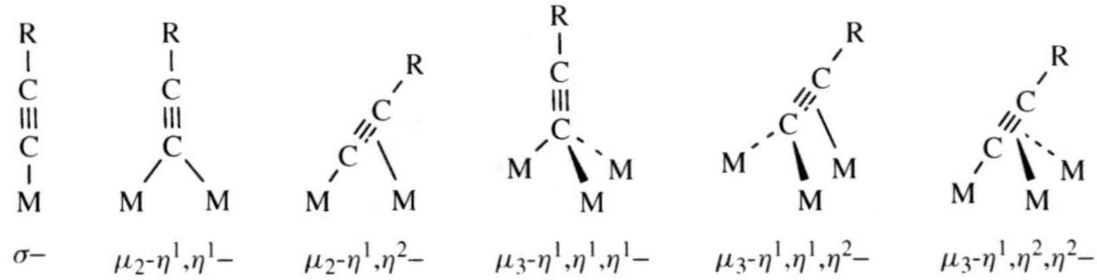

FIGURE 54. The various bonding modes of the monoanionic acetylide ligand

between two, three or, in exceptional cases, even four metal centers via a variety of bonding modes (Figure 54).

Due to these versatile bonding modes copper acetylides have a strong tendency to form highly aggregated species, both in the solid state and in solution. Although synthetic procedures and their application in synthetic organic chemistry have been reported for a large variety of pure, monofunctionalized copper acetylides, CuC≡CR, only for a few of these could the structures in the solid state be established by X-ray crystallography. The major problem is the insolubility of the greater part of these compounds in common organic solvents which prevents growing of single crystals suitable for X-ray structure determinations.

The structure of the parent copper acetylide, i.e. CuC≡CH, is unknown, but the solid state structure of a rather complicated inorganic material, obtained from an electrochemical synthetic procedure, formulated as $Rb_{11}[Cu^{I}_{15}Cl_{16}Br_6(Cu^{II}Cl_6)CuC{\equiv}CH]$ (**113**) containing the CuC≡CH fragment, has been determined[163]. The crystal lattice is built up of anionic $[Cu^{I}(Cu^{I}_{15}Cl_{16}Br_6)]^{7-}{}_n$ layers, perpendicular to the [201] direction, and contains octahedral macrocavities. Each macrocavity is occupied by a $[Cu^{II}Cl_6]^{4-}$ octahedron surrounded by eight Rb cations. Space between the anions is occupied by other Rb cations and neutral CuC≡CH molecules. The short Cu−C σ-bond [1.84(5) Å] in the CuC≡CH fragment is notable. However, the considerable deviation from linearity [Cu−C−C 142(5)°] suggests a significant π-interaction. The only other structures containing CuC≡CH fragments are $[Ca(NH_3)_6][Cu(C{\equiv}CH)_3]$ (**114**) and $Rb_2[Cu(C{\equiv}CH)_3]{\cdot}NH_3$ (**115**)[164]. The solid state structures of **114** and **115** were determined from powder diffraction data. The unit cell of **114** contains isolated $[Cu(C{\equiv}CH)_3]$ dianions and $[Ca(NH_3)_6]$ dications (Figure 55). In the $[Cu(C{\equiv}CH)_3]$ dianion the three C≡CH groups are σ-bonded to copper [Cu−C 2.031(7) Å] with equal bond distances, as the result of space group symmetry. The C−Cu−C bond angles around copper are 120° within experimental error, indicating a perfect trigonal planar coordination geometry for this copper atom. The solid state structure of **115** is similar, with isolated $[Cu(C{\equiv}CH)_3]$ dianions and [Rb] cations in the unit cell. Formally, these two compounds belong to the class of ionic cuprates.

Pure copper *tert*-butylacetylide has been prepared from the reaction of CuBr(DMS) with LiC≡CBu-*t* in a mixture of diethyl ether/hexane as a solvent. According to an X-ray crystal structure determination this material appeared to be $[(CuC{\equiv}CBu\text{-}t)_{24}]{\bullet}2C_6H_{14}$ (**116**)[165]. Within the $(CuC{\equiv}CBu\text{-}t)_{24}$ aggregate (Figure 56), the acetylide groups adopt a variety of μ_2- and μ_3-bridge bonding modes with Cu−C distances varying from 1.939 to 2.541 Å.

When the same compound is crystallized from benzene, an X-ray crystal structure determination revealed the formation of an entirely different aggregate $[(CuC{\equiv}CBu\text{-}t)_{20}]{\bullet}C_6H_6$ (**117**) (Figure 57)[166].

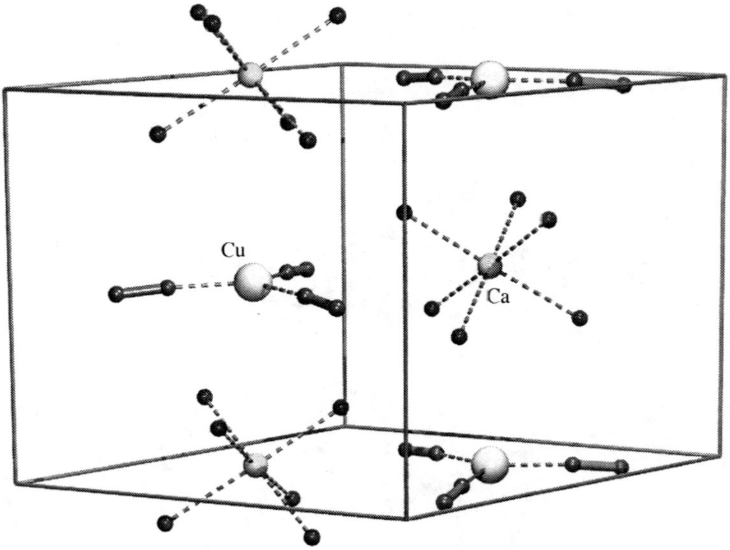

FIGURE 55. Unit cell content of $[Ca(NH_3)_6][Cu(C{\equiv}CH)_3]$ (**114**) in space group $P6_3mc$

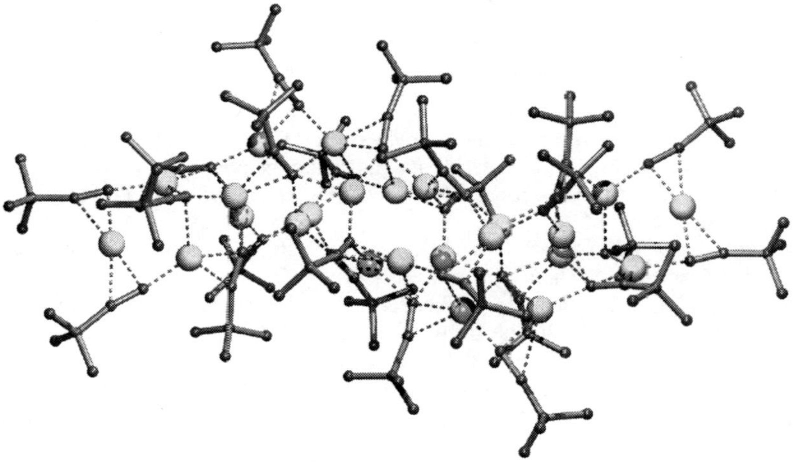

FIGURE 56. Molecular geometry of $[(CuC{\equiv}CBu\text{-}t)_{24}]{\bullet}2C_6H_{14}$ (**116**) in the solid state

Also in this aggregate, the acetylide groups adopt a variety of bonding modes to copper with Cu−C distances varying from 1.862(6) to 2.471(7) Å. On a macromolecular level, the overall structural geometry of the aggregate can be seen to consist of a central Cu_8 ring interlocking with two puckered Cu_6 rings resulting in a catenane-type of architecture (Figure 58).

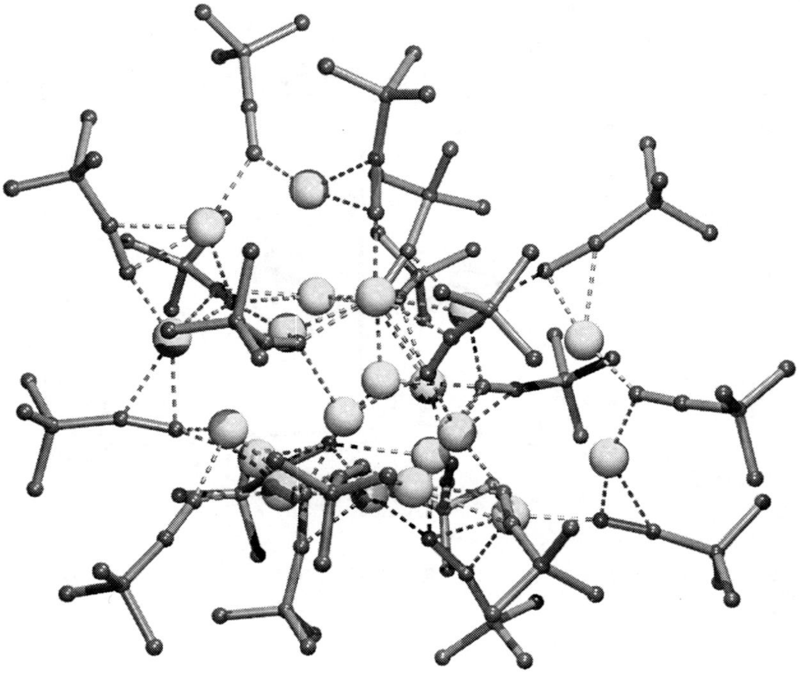

FIGURE 57. Overall structural geometry of the $(CuC{\equiv}CBu\text{-}t)_{20}$ aggregate **117**

The solid state structure of copper phenylacetylide CuC≡CPh (**118**) was determined from powder X-ray diffraction data[166]. Its molecular geometry comprises linear infinite chains via symmetrically μ_2-η^1,η^1-bonded (3c-2e) [Cu−C 1.93(1) and 1.95(1) Å] acetylide groups (Figure 59).

The copper acetylide CuC≡CPr-*n* (**119**) derived from 1-pentyne forms infinite polymeric sheets in the crystal lattice[166]. These sheets are formed by μ_2-η^1,η^2-bridging and μ_3-η^1,η^1,η^2-bridging acetylide groups between two and three copper atoms, respectively, with Cu−C bond distances ranging from 1.96(1) to 2.23(1) Å.

The structures in the solid state of the cationic, mixed copper−silver aggregate $[Cu_6Ag_8(Cl)(C{\equiv}CFc)_{12}]^+$ (**120**)[167] (Fc = ferrocenyl) and the anionic, mixed copper−silver aggregate $[Cu_7Ag_6(C{\equiv}CPh)_{14}]^-$ (**121**)[168,169] have been established by X-ray crystal structure determinations. The six copper and eight silver atoms of **120** are arranged in such a way that they form a rhombic dodecahedron consisting of fourteen Cu_2Ag_2 quadrangles (Figure 60). The eight silver atoms are arranged at the eight apices of a cube while the six copper atoms are located at the apices of an octahedron. A chloride anion is located at the center of the metal cage. To each of the copper atoms are η^1-bonded two acetylide groups while each of these acetylides is η^2,η^2-bonded to two silver atoms.

The overall structural geometry of the metal framework in **121** can be described as consisting of three square pyramids which are interconnected via the common copper atom at the apex (Figure 60). Each of the square pyramids consists of two copper atoms and two silver atoms at the base and a copper atom at the apex. To the central copper atom, two acetylide groups are σ-bonded in an almost linear fashion [C−Cu−C 178.5(9)°] and have no further interaction with other metals. To each of the peripheral copper atoms,

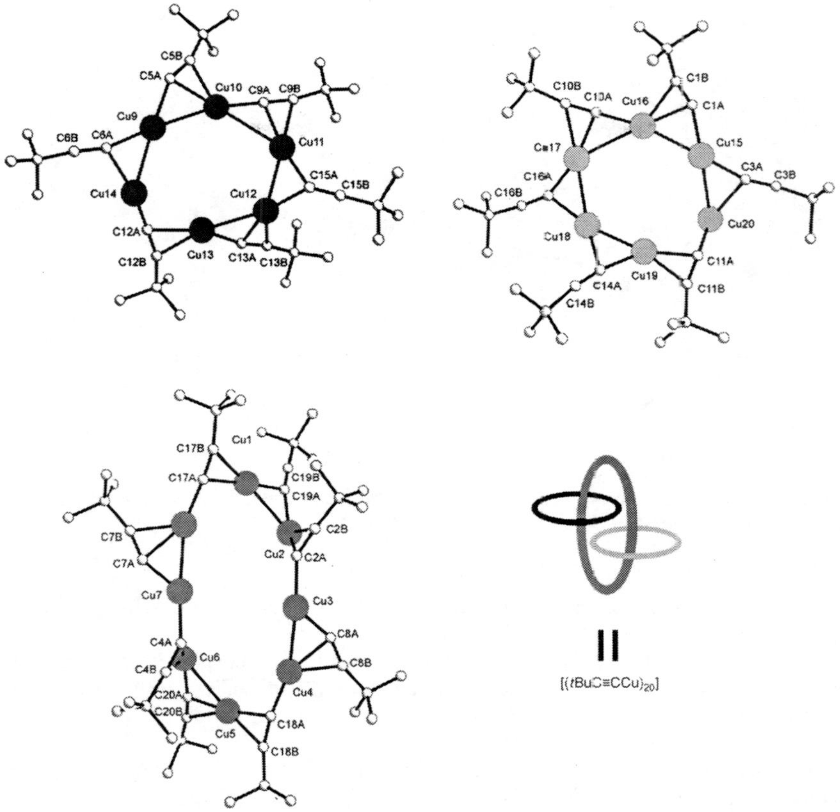

FIGURE 58. Schematic representation showing the two six-membered Cu_6 units and the central Cu_8 moiety and the assembly to the catenane-type architecture of **117**. From Reference 166. © Wiley-VCH Verlag GmbH & Co. KGaA. Reproduced with permission

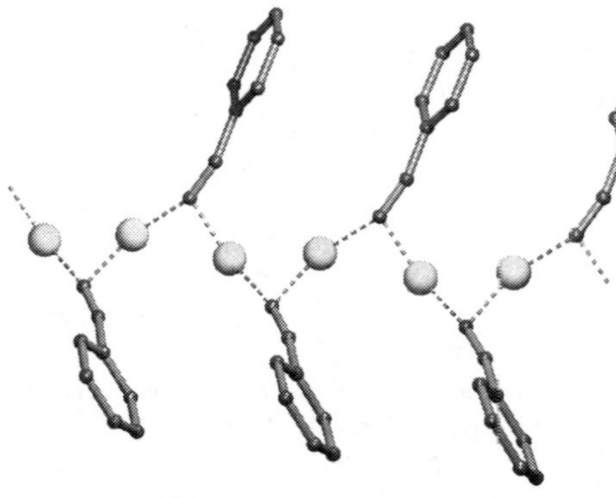

FIGURE 59. Part of the infinite polymeric chain of CuC≡CPh (**118**) in the solid state

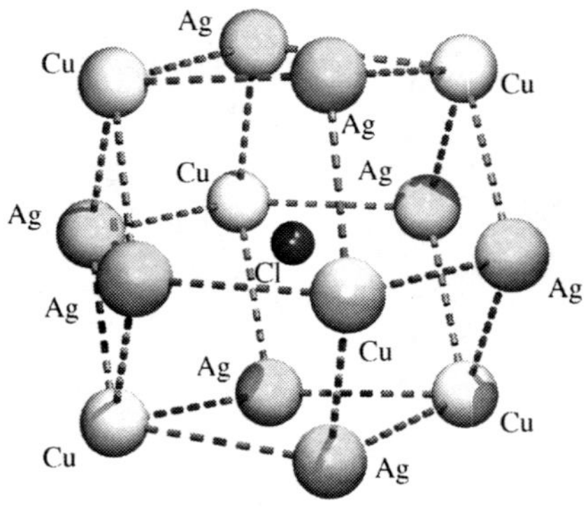

(120)

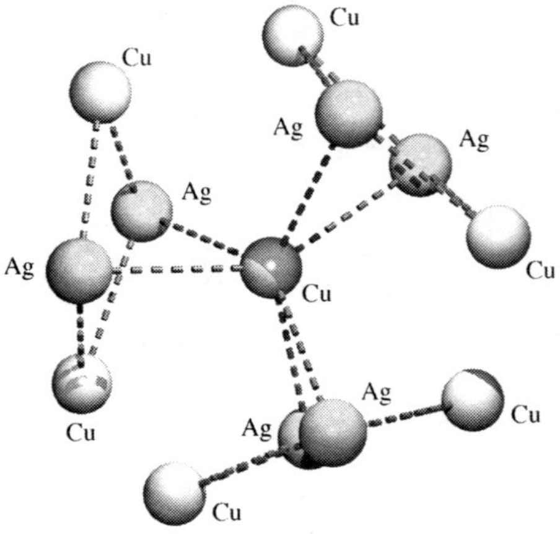

(121)

FIGURE 60. Structural geometry of the metal frameworks of cationic $[Cu_6Ag_8(Cl)(C{\equiv}CFc)_{12}]^+$ (**120**) and anionic $[Cu_7Ag_6(C{\equiv}CPh)_{14}]^-$ (**121**), respectively

two acetylide groups are η^1-bonded while each of these acetylide groups are η^2,η^2-bonded to two silver atoms.

It is fascinating that addition of simple Lewis bases L like phosphines or amines do not deaggregate polymeric or oligomeric copper acetylides $[CuC{\equiv}CR]_n$ to the expected simple, monomeric adducts LCuC≡CR but instead cause the formation of dimeric, trimeric or tetrameric complexes (*vide infra*).

It has been found that only a special bidentate 'tweezer'-like ligand is capable of monomerizing copper acetylides. This ligand $(\eta^5\text{-}C_5H_4SiMe_3)_2Ti(C{\equiv}CR^1)_2$ (R^1 = *t*-Bu or $SiMe_3$) (Figure 61) binds to the copper atom of a monomeric copper acetylide unit in a bidentate fashion via two η^2-bonded acetylenic functionalities. Four of these monomeric complexes, **122a**[170] (R^1 = *t*-Bu, R^2 = *t*-Bu), **122b**[171] (R^1 = *t*-Bu, R^2 = C≡CEt), **122c**[172] (R^1 = *t*-Bu, $R^2 = C_6H_4NO_2$-4) and **122d**[173] ($R^1 = SiMe_3$, $R^2 = SiMe_3$), have been prepared and were characterized structurally in the solid state by X-ray crystallography.

The molecular geometry of these four complexes in the solid state are closely related. The structure of **122a** is shown as an example (Figure 61). In **122a**, the acetylide moiety is σ-bonded to copper [Cu−C(1) 1.903(4) Å] in an almost linear fashion [Cu−C(1)−C(2) 173.1(3)°]. The two acetylenic substituents are both η^2-bonded to copper [Cu−C(3) 2.068(4), Cu−C(4) 2.157(4) Å and Cu−C(5) 2.066(3) and Cu-C(6) 2.146(4) Å], affording a pseudo-trigonal planar coordination geometry at the copper atom.

Reaction of polymeric copper phenylacetylide with methyldiphenylphosphine afforded a dimeric complex $[PhC{\equiv}CCu(PPh_2Me)_2]$ (**123**) of which the structure in the solid state was established by X-ray crystallography[174]. The molecular geometry of **123** is shown schematically (Figure 62) and comprises a perfectly planar Cu−C−Cu−C four-membered ring as a result of two μ_2-η^1-bonded acetylide groups between the two copper atoms. The bridge bond is slightly asymmetric [Cu−C 2.011(3) and 2.209(4) Å]. The Cu−C−Cu bond angles are rather acute (70.95(11)°), but compensated by a larger C−Cu−C bond angle (109.05(14)°). To each of the copper atoms, perpendicular to the central Cu−C−Cu−C plane, two PPh_2Me molecules are coordinate bonded via their respective phosphorus atoms to copper [Cu−P 2.2801(12) and 2.2822(12) Å], one from above and one from below that

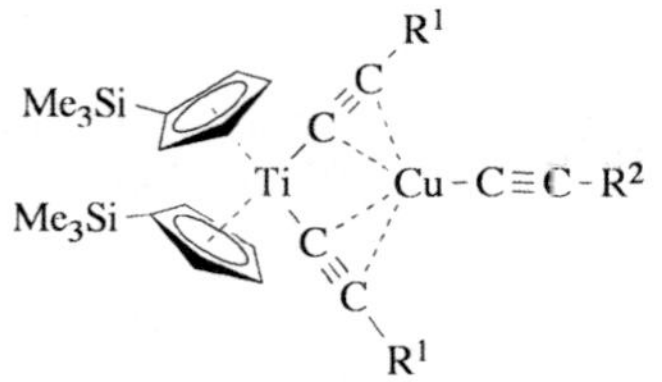

(**122a**) R^1 = *t*-Bu, R^2 = *t*-Bu
(**122b**) R^1 = *t*-Bu, R^2 = C≡CEt
(**122c**) R^1 = *t*-Bu, R^2 = $C_6H_4NO_2$-4
(**122d**) R^1 = $SiMe_3$, R^2 = $SiMe_3$

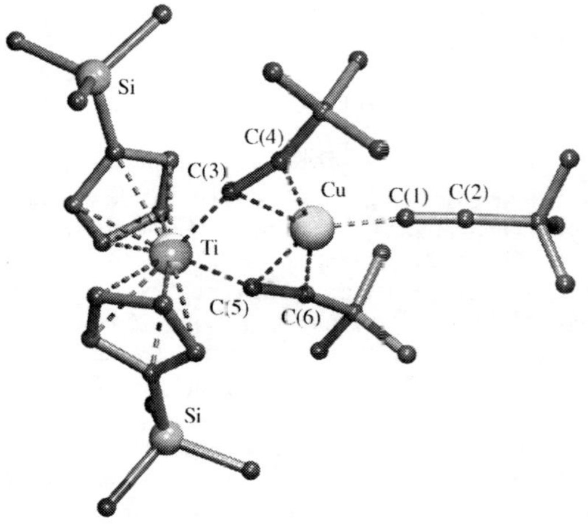

(**122a**)

FIGURE 61. Monomeric copper acetylide units in 'tweezers' **122a–122d** and molecular geometry of **122a** in the solid state

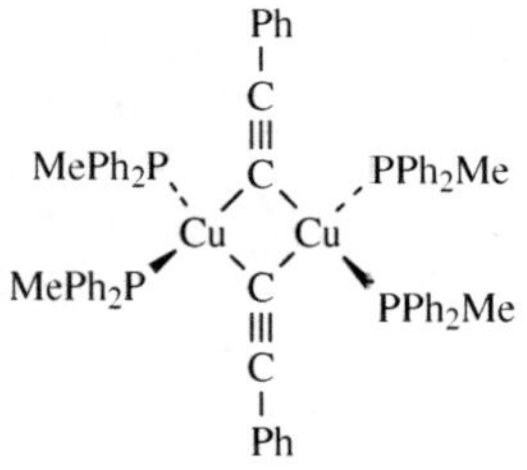

(**123**)

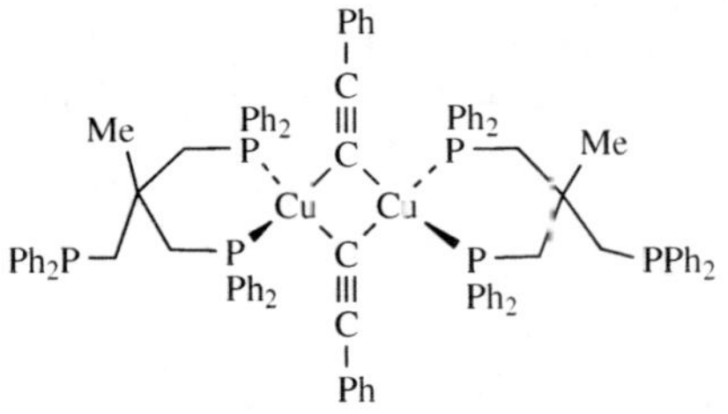

(**124**)

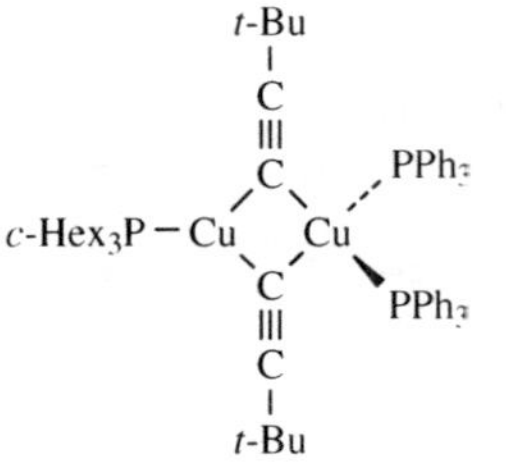

(**125**)

FIGURE 62. Schematic representation of the molecular geometry of **123–125** in the solidstate

plane. The coordination geometry of the copper atoms is almost perfectly tetrahedral, as is indicated by the P–Cu–P bond angle (109.75(4)°).

The potentially tridentate ligand triphos is also capable of depolymerizing copper phenylacetylide. An X-ray crystal structure determination of the obtained product surprisingly revealed a dimeric complex [PhC≡CCu(triphos)] (**124**)[175] instead of the expected monomeric structure with a tridentate bonded triphos ligand. Instead, the overall structural features of **124** with respect to the central Cu–C–Cu–C plane are closely related to those of **123** (Figure 62). In **124**, coordination saturation at copper is reached by a bidentate *P*,*P*′-chelate bonded triphos ligand at each of the copper atoms.

The reaction of $CpCu(PPh_3)$ with LiC≡CBu-*t* in the presence of one equivalent of tricyclohexylphosphine affords the complex $(t\text{-BuC≡C})_2Cu_2(Pc\text{-Hex}_3)(PPh_3)_2$ (**125**)[176] of which the structure in the solid state was determined by an X-ray crystal structure determination. The molecular geometry of **125** is shown schematically (Figure 62) and comprises, like **123** and **124**, a central flat Cu–C–Cu–C four-membered ring as the result of two μ_2-η^1-bridge bonded acetylide groups between two copper atoms in a slightly asymmetric way [Cu–C 2.054(9) and 2.143(9) Å]. One of the copper atoms has a tetrahedral coordination geometry as the result of two additional coordinate bonded triphenylphosphine ligands [Cu–P 2.272(3) and 2.273(3) Å]. One tricyclohexylphosphine ligand is coordinate bonded to the other copper atom [Cu–P 2.224(3) Å], resulting in a trigonal planar coordination geometry for this copper atom.

The copper acetylide dppf complex $(4\text{-TolC≡C})_2Cu_2(dppf)_2$ (dppf = 1, 1′-bis(diphenylphosphino)ferrocene) (**126**) is also dimeric in nature. Its structure in the solid state was established by an X-ray crystal structure determination[177]. The molecular geometry (Figure 63) of **126** comprises a similar planar Cu–C–Cu–C arrangement as the result of two μ_2-η^1-bridge bonded acetylide groups [Cu–C 2.06(2) and 2.136(2) Å] as was found for **123–125**. The two dppf ligands adopt a *P*,*P*′-bridging bonding mode, this bridging occurring between the two copper atoms [Cu–P 2.310(6) and 2.317(7) Å] of the central Cu_2C_2 motif.

The only example of a dimeric copper acetylide complex in which the additional ligand is not a phosphine is the tmtch complex of copper phenylacetylide $(PhC≡C)_2Cu_2(tmtch)_2$

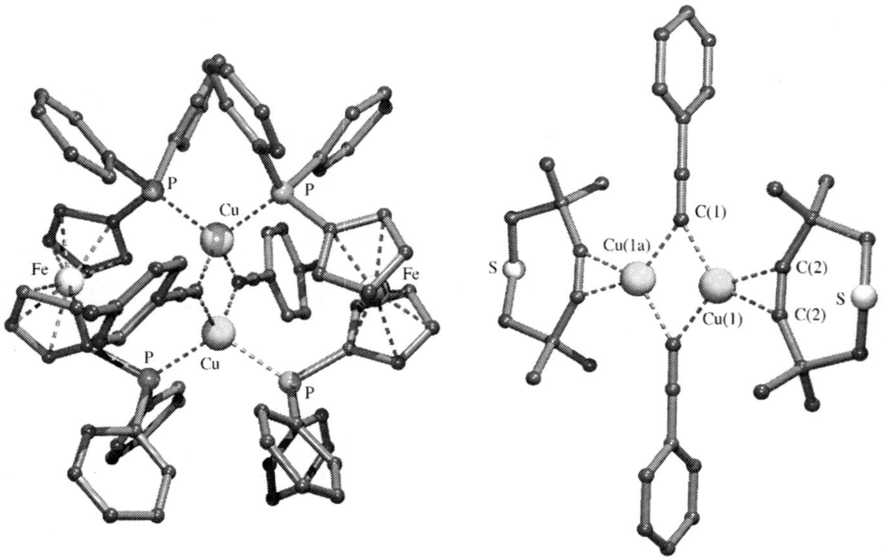

(**126**) (**127**)

FIGURE 63. Molecular geometry of $(4\text{-TolC≡C})_2Cu_2(dppf)_2$ (**126**) and $(PhC≡C)_2Cu_2(tmtch)_2$ (**127**) in the solid state

(**127**) (tmtch = 3,3,6,6-tetramethyl-1-thia-4-cycloheptyne)[178]. The molecular geometry of **127** in the solid state was established by an X-ray crystal structure determination (Figure 63) and comprises two symmetrically μ_2-η^1-bridge-bonded acetylide groups between the two copper atoms [Cu(1)–C(1) 2.014(3) and Cu(1a)–C(1) 2.022(3) Å]. To each of the copper atoms a cyclic acetylene is η^2-bonded via the acetylenic carbon atoms [Cu(1)–C(2) 1.940(4) and Cu(1)–C(3) 1.951(4) Å].

Copper acetylide phosphine complexes often exist as tetrameric aggregates having a central heterocubane structural motif. This has been observed for a variety of complexes, i.e. $(RC{\equiv}C)_4Cu_4(P)_4$ [R = Ph; P = PPh_3 (**128**)[179], R = Ph; P = $P(4\text{-}MeC_6H_4)_3$ (**129**)[180], R = Ph; P = $PPh_2(2\text{-}Py)$ (**130**)[181], R = 4-$MeOC_6H_4$; P = PPh_3 (**131**)[182], R = Me_3Si; P = PPh_3 (**132**)[183], R = C≡CPh; P = PPh_3 (**133**)[184]] (Figure 64). In these complexes the acetylide groups are μ_3-η^1-bridge bonded between three copper atoms in such a way that the four bridging carbon atoms and the four copper atoms form a slightly distorted cube. To each of the copper atoms an additional phosphine is coordinate bonded via its phosphorus atom, thus giving each copper atom a distorted tetrahedral coordination geometry.

The structures of the ionic copper acetylide complexes $[(4\text{-}MeC_6H_4C{\equiv}C)_3Cu_4(P(4\text{-}MeC_6H_4)_3)_4][PF_6]$ (**134**)[185] and $[(4\text{-}MeOC_6H_4C{\equiv}C)_3Cu_4(PPh_3)_4][PF_6]$ (**135**)[186] in the solid state have been established by X-ray crystal structure determinations. The molecular geometry of the cations is shown schematically in Figure 64. They exhibit an 'open' heterocubane structural motif, i.e. a structure lacking one carbon atom in the cube. Consequently, one of the copper atoms has a distorted tetrahedral coordination geometry whereas the other three copper atoms are three-coordinate. These cationic complexes have interesting luminescent properties[185, 186].

The trimethylphosphine complex of copper phenylacetylide also exists as a tetrameric aggregate, $(PhC{\equiv}C)_4Cu_4(PMe_3)_4$ (**136**)[35], not with a central heterocubane structural motif but with an open structure in which the four copper atoms are coplanar and arranged in a zigzag chain.

The tetrameric aggregate is assembled via two μ_2-η^1-bridge-bonded acetylide groups, each between two copper atoms [Cu(1), Cu(3) and Cu(2), Cu(4), respectively], and two μ_3-η^1,η^2-bridge-bonded acetylide groups, each between three copper atoms (Figure 65). The bridge bonding of the η^1-bonded acetylide groups is rather asymmetric [Cu(1)–C 2.223(17) and Cu(3)–C(1) 1.957(16) Å], in contrast to the μ_3-η^1,η^2-bridge-bonded acetylide group [Cu(1)–C(2) 2.073(15), Cu(3)–C(2) 2.073(15), Cu(2)–C(2) 2.061(16) and

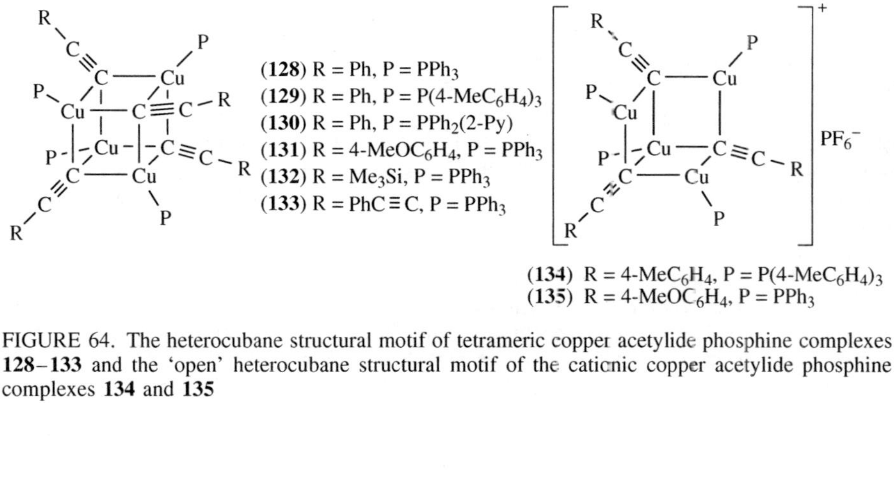

FIGURE 64. The heterocubane structural motif of tetrameric copper acetylide phosphine complexes **128–133** and the 'open' heterocubane structural motif of the cationic copper acetylide phosphine complexes **134** and **135**

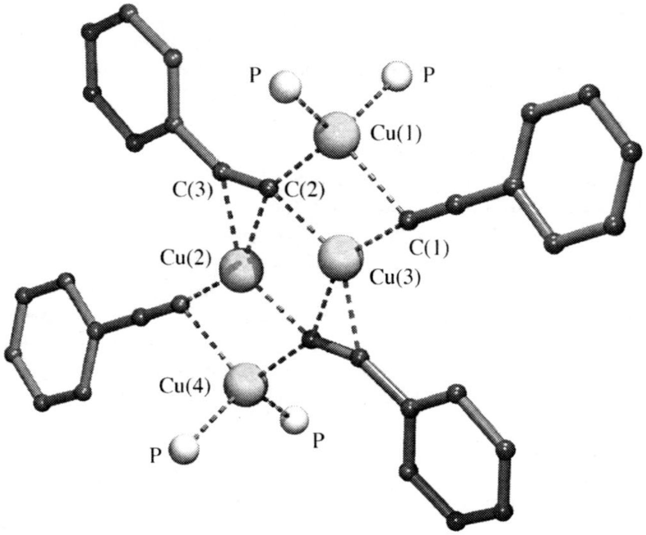

FIGURE 65. Molecular geometry of $(PhC{\equiv}C)_4Cu_4(PMe_3)_4$ (**136**) in the solid state; the methyl groups of the trimethylphosphine ligands are omitted for clarity

Cu(2)–C(3) 2.085(14) Å]. A tetrahedral coordination geometry at the peripheral copper atoms Cu(1) and Cu(4) is reached by additional coordination of two triphenylphosphine ligands to each of these copper atoms [Cu–P 2.230(5) Å average].

Copper phenylacetylide and 1,8-bis(diphenylphosphino)-3,6-dioxaoctane (bdpoo) form a complex $(PhC{\equiv}C)_4Cu_4(bdpoo)_2$ (**137**)[187], which is also tetrameric in nature. An X-ray crystal structure determination of **137** revealed a structural motif that, with respect to the bonding of the four acetylide groups to the four copper atoms, is closely related to that of **136**, i.e. two μ_2-η^1-bridge-bonded and two μ_3-η^1,η^2-bridge-bonded acetylide groups (Figure 66). The two bidentate phosphine ligands are both *P*,*P*-bridge bonded spanning between the two peripheral copper atoms Cu(1) and Cu(4). Complex **137** exhibits luminescent behavior with extreme high quantum yields.

The nature of the acetylide group can play an important role in the formation of particular aggregates which became evident from the different structures in the solid state of the complexes formed between tmtch (tmtch = 3,3,6,6-tetramethyl-1-thia-4-cycloheptyne) and copper *tert*-butyl- and phenylacetylide, respectively. The copper *tert*-butylacetylide complex appeared to be tetrameric $(t\text{-}BuC{\equiv}C)_4Cu_4(tmtch)_2$ (**138**)[178], whereas the corresponding copper phenylacetylide complex $(PhC{\equiv}C)_2Cu_2(tmtch)_2$ has a dimeric structure (*vide supra*). The molecular geometry of **138** comprises four copper atoms in a planar cyclic arrangement. All four acetylide groups are μ_2-η^1,η^2-bridge bonded between two copper atoms (Figure 66). Cu(1) and Cu(3) experience η^1-bonding of the acetylide groups exclusively while Cu(2) and Cu(4) exclusively experience η^2-bonding. Coordination saturation at Cu(1) and Cu(3) is reached by the additional η^2-bonded cyclic acetylene.

As to their interesting photophysical properties, a series of cationic tricopper acetylide complexes having general formula $[(R^1C{\equiv}C)(R^2C{\equiv}C)Cu_3(dppm)_3][Y]$ (Figure 67) were prepared in order to study their structural features in detail. For several of these complexes the structures in the solid state could be established by X-ray crystal structure determinations, namely **139**[188], $R^1 = R^2 = Ph$, $Y = BF_4^-$; **140**[189], $R^1 = R^2 = C{\equiv}CH$, $Y = PF_6^-$;

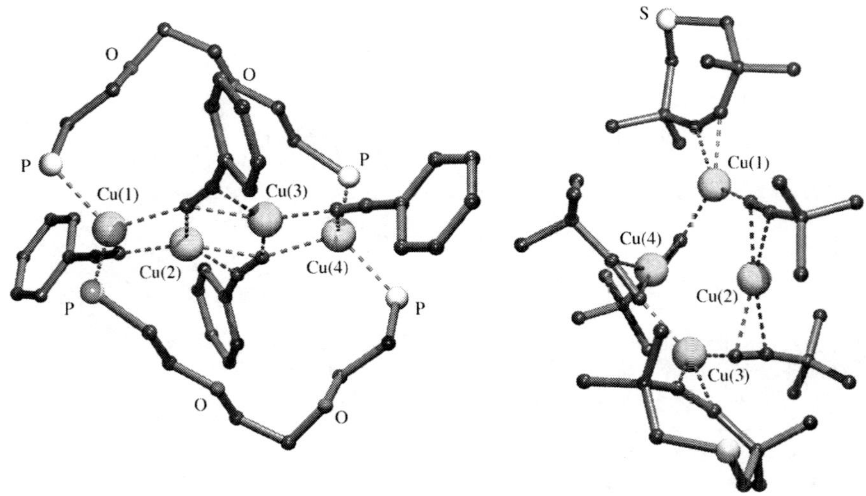

(**137**) (**138**)

FIGURE 66. Molecular geometry of $(PhC{\equiv}C)_4Cu_4(bdpoo)_2$ (**137**) (P-bonded phenyl groups are omitted for clarity) and $(t\text{-}BuC{\equiv}C)_4Cu_4(tmtch)_2$ (**138**) in the solid state

141[189], $R^1 = R^2 = C{\equiv}CPh$, $Y = PF_6^-$; **142**[190], $R^1 = R^2 = C_5H_4FeC_5H_5$, $Y = PF_6^-$; **143**[191], $R^1 = R^2 =$ benzo-15-crown-5, $Y = PF_6^-$; **144**[192], $R^1 = R^2 = C{\equiv}CRe(CO)_3$(4-Me-bipy), $Y = PF_6^-$; **145**[193], $R^1 = R^2 = C{\equiv}CC_6H_4C{\equiv}CRe(CO)_5$(bipy), $Y = PF_6^-$; **146**[194], $R^1 = 4\text{-}MeOC_6H_4$, $R^2 = 4\text{-}EtOC_6H_4$, $Y = PF_6^-$ and **147**[194], $R^1 = 4\text{-}MeOC_6H_4$, $R^2 = 4\text{-}NO_2C_6H_4$, $Y = PF_6^-$.

These complexes have as a common structural motif a regular trigonal plane of three copper atoms linked via three *P,P*-bridge-bonded dppm ligands (dppm = 1,1-bis(diphenylphosphino)methane). Each of the two acetylide groups are μ_3-η^1-bonded to the three copper atoms, one approaching the trigonal plane from above and one from below. As a representative example the molecular geometry of **139** is shown (Figure 68). In **139**, the copper–copper distances in the triangular plane are almost equal [Cu(1)–Cu(2) 2.570(3), Cu(2)–Cu(3) 2.615(3) and Cu(1)–Cu(3) 2.598(3) Å] while the six-coordinate bonded phosphorus atoms of the three dppm ligands lie roughly in the same plane. The μ_3-η^1-bonding of the acetylide groups to the three copper atoms is rather asymmetric [Cu(1)–C(1) 2.12(2), Cu(2)–C(1) 2.17(2) and Cu(3)–C(1) 2.34(2) Å].

A closely related type of cationic tricopper complexes has general formula $[(RC{\equiv}C)Cu_3X(dppm)_3][Y]$ (Figure 67) of which three, i.e. **148**[195], R = Ph, X = Cl, $Y = BF_4^-$; **149**[196], R = *t*-Bu, X = Cl, $Y = PF_6^-$ and **150**[197], $R = C{\equiv}CAuC{\equiv}CC{\equiv}CH$, X = I, have been characterized structurally in the solid state by X-ray crystallography. It should be noted that the latter compound is a zwitterionic, neutral complex due to the formal negative charge on the gold atom.

The molecular geometry of **148** is shown as an example (Figure 68). The structural features of the central Cu_3P_6 core of **148** are closely related to that of **139**. Like in **139**, the acetylide group is μ_3-η^1-bonded in an asymmetric way to the three copper atoms [Cu(1)–C(1) 1.98(2), Cu(2)–C(1) 2.18(2) and Cu(3)–C(1) 2.34(2) Å]. The opposite site of the triangular plane of copper atoms is capped by a μ_3-η^1-bonded chlorine atom [Cu(1)–Cl 2.568(4), Cu(2)–Cl 2.586(4) and Cu(3)–Cl 2.535(4) Å].

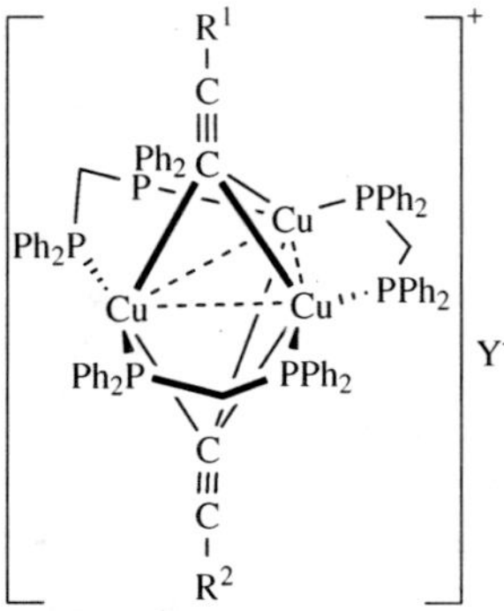

(**139**) $R^1 = R^2 = Ph$, $Y = BF_4^-$
(**140**) $R^1 = R^2 = C{\equiv}CH$, $Y = PF_6^-$
(**141**) $R^1 = R^2 = C{\equiv}CPh$, $Y = PF_6^-$
(**142**) $R^1 = R^2 = C_5H_4FeC_5H_5$, $Y = PF_6^-$
(**143**) $R^1 = R^2$ = benzo-15-crown-5, $Y = PF_6^-$
(**144**) $R^1 = R^2 = C{\equiv}CRe(CO)_3$(4-Me-bipy), $Y = PF_6^-$
(**145**) $R^1 = R^2 = C{\equiv}CC_6H_4C{\equiv}CRe(CO)_3$(bipy), $Y = PF_6^-$
(**146**) R^1 = 4-$MeOC_6H_4$, R^2 = 4-$EtOC_6H_4$, $Y = PF_6^-$
(**147**) R^1 = 4-$MeOC_6H_4$, R^2 = 4-$NO_2C_6H_4$, $Y = PF_6^-$

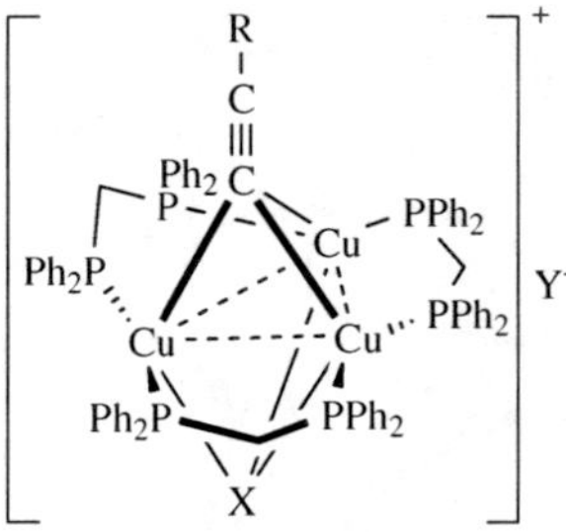

(**148**) R = Ph, X = Cl, $Y = BF_4^-$
(**149**) R = *t*-Bu, X = Cl, $Y = PF_6^-$
(**150**) $R = C{\equiv}CAuC{\equiv}CC{\equiv}CH$, X = I

FIGURE 67. The cationic tricopper complexes $[(R^1C{\equiv}C)(R^2C{\equiv}C)Cu_3(dppm)_3][Y]$ (**139**–**147**) and $[(RC{\equiv}C)Cu_3X(dppm)_3][Y]$ (**148**–**150**)

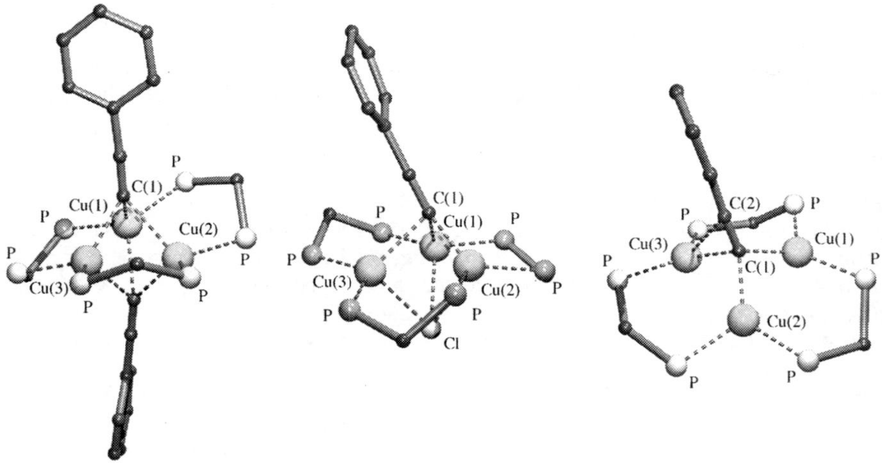

(**139**) (**148**) (**151**)

FIGURE 68. Molecular geometry of the cations of $[(PhC{\equiv}C)_2Cu_3(dppm)_3][BF_4]$ (**139**), $[(PhC{\equiv}C)Cu_3Cl(dppm)_3][BF_4]$ (**148**) and $[(PhC{\equiv}C)Cu_3(dppm)_3][BF_4]_2$ (**151**) in the solid state; the phenyl groups of the dppm ligands are omitted for clarity

A third class of complexes is the one lacking the second μ_3-η^1-bonded group, i.e. $[(PhC{\equiv}C)Cu_3(dppm)_3][BF_4]_2$ (**151**)[198] and $[(t\text{-}BuC{\equiv}C)Cu_3(dppm)_3][PF_6]_2$ (**152**)[199], and consequently is dicationic in nature. These two compounds have closely related structures, of which that of **151** is shown (Figure 68). The overall geometry of the central Cu_3P_6 core is similar to those of **139** and **148**, but the Cu_3 triangle is less regular as is indicated by the various Cu–Cu distances [Cu(1)–Cu(2) 2.813(2), Cu(2)–Cu(3) 2.904(3) and Cu(1)–Cu(3) 3.274(3) Å]. The bonding of the acetylide group to the three copper atoms is rather symmetric [Cu(1)–C(1) 1.96(1), Cu(2)–C(1) 2.04(1) and Cu(3)–C(1) 2.08(1) Å], but the relatively short Cu(3)–C(2) distance, 2.57(1) Å, suggests rather a μ_3-η^1,η^1,η^2- than a μ_3-η^1,η^1,η^1-bonding to the three copper atoms.

A complex that contains twice the structural motif of **151** is $[(dppm)_3Cu_3C{\equiv}CC_6H_4C{\equiv}CCu_3(dppm)_3][BF_4]_4$ (**153**)[200], containing a 1,4-diacetylidephenylene unit. The molecular geometry of the tetracationic part of **153** is shown (Figure 69). The structural features of the two $(dppm)_3Cu_3$ moieties are closely related to that of **151**. Each of these moieties is μ_3-η^1,η^1,η^2-bonded to an acetylide functionality, like in **151**.

Ionic $[(C{\equiv}C)Cu_4(dppm)_4][BF_4]_2$ (**154**) was obtained as an air-stable, yellow crystalline solid when $[Cu_2(dppm)_2(MeCN)_2][BF_4]_2$ was added to a solution of trimethylsilylacetylene and n-BuLi in THF at room temperature. An X-ray crystal structure determination of **154** showed the presence of isolated $[(C{\equiv}C)Cu_4(dppm)_4]^{2+}$ dications and BF_4^- anions in the crystal lattice[201]. The molecular geometry of the dicationic part involves a square of four copper atoms linked via P,P-bridge-bonded dppm ligands [Cu–P 2.253(4) Å mean value] (Figure 70). The acetylide dianion is end-on, η^1-bonded to Cu(1) and Cu(1a) [Cu–C 1.91(1) Å] and side-on η^2-bonded to Cu(2) and Cu(2a) [Cu–C 2.12(1) and 2.13(1) Å].

An X-ray crystal structure determination of $[(FcC{\equiv}C)_4Cu_6(dppm)_2][ClO_4]_2$ (**155**) (Fc = ferrocenyl) showed that this complex exists in the solid state as a hexanuclear aggregate with the copper atoms in an octahedral arrangement (Figure 71)[202].

The four copper atoms Cu(1)–Cu(4) reside at the equatorial plane while Cu(5) and Cu(6) occupy the axial positions of a Cu_6-octahedron. The two dppm ligands are each P,P-bridge bonded between two adjacent copper atoms in the equatorial plane, Cu(2) and Cu(3), and Cu(1) and Cu(4), respectively [Cu–P 2.215(3) Å mean value]. The anionic acetylide carbon atoms are located above opposing triangular faces of the octahedron, and thus μ_3-η^1,η^1,η^1-bonded to two equatorial and one axial copper atom. The

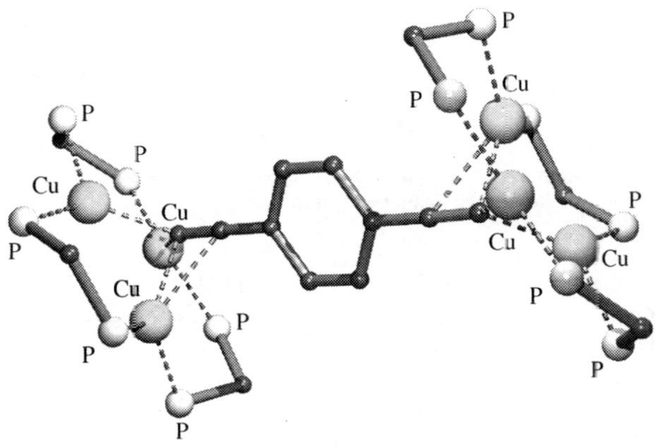

FIGURE 69. Molecular geometry of the tetracation of $[(dppm)_3Cu_3C{\equiv}CC_6H_4C{\equiv}CCu_3(dppm)_3]$ $[BF_4]_4$ (**153**) in the solid state; the phenyl groups of the dppm ligands are omitted for clarity

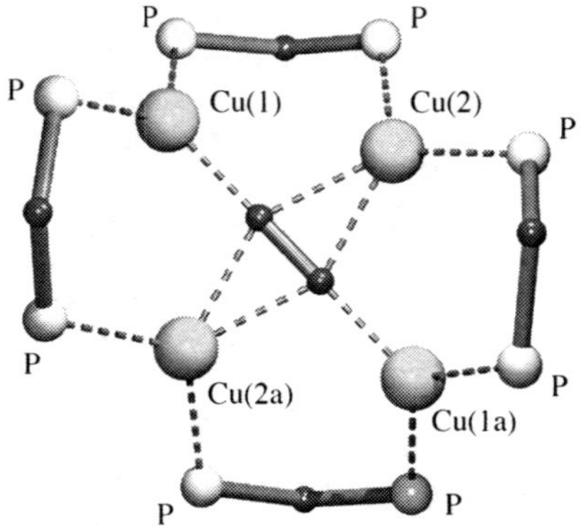

FIGURE 70. Molecular geometry of the dication of $[(C{\equiv}C)Cu_4(dppm)_4][BF_4]_2$ (**154**) in the solid state; the phenyl groups of the dppm ligands are omitted for clarity

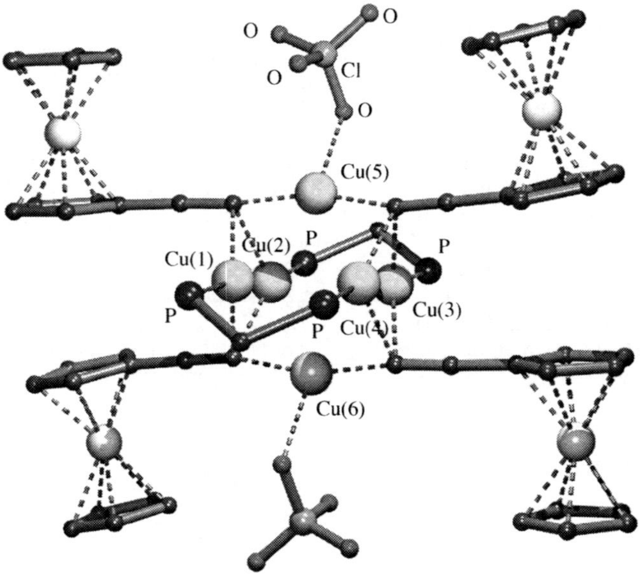

FIGURE 71. Molecular geometry of hexanuclear $[(FcC{\equiv}C)_4Cu_6(dppm)_2][ClO_4]_2$ (**155**) in the solid state; the phenyl groups of the dppm ligands are omitted for clarity

carbon-to-axial-copper bond distance is relatively short [Cu(5)−C 1.897(10) Å] compared to the carbon-to-equatorial-copper bond distances [Cu(1)−C 2.116(10) and Cu(2)−C 2.127(10) Å]. The perchlorate anions are within bonding distance of the axial copper atoms [Cu(5)−O 2.353(11) Å].

A particular type of copper acetylide aggregates are the heteroleptic ones, i.e. those which, apart from the acetylide group, contain other anionic groups. The solid state structures of three aggregates containing acetylide and 2-aminoarenethiolato anionic groups have

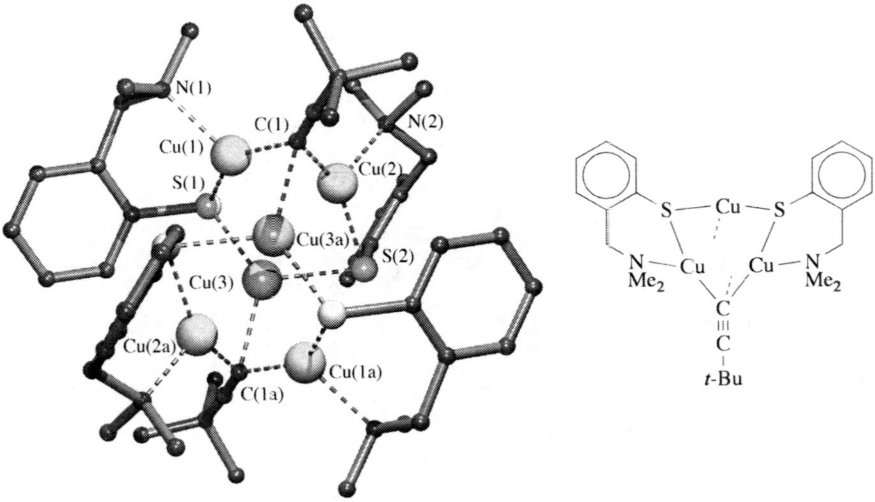

(**A**) (**B**)

FIGURE 72. (**A**) Molecular geometry of $(t\text{-BuC}{\equiv}\text{C})_2\text{Cu}_6(\text{SC}_6\text{H}_4\text{CH}_2\text{NMe}_2\text{-}2)_4$ (**156**) in the solid state. (**B**) One of the trinuclear units of **156**

been established by X-ray crystal structure determinations. These compounds $(t\text{-BuC}{\equiv}\text{C})_2\text{Cu}_6(\text{SC}_6\text{H}_4\text{CH}_2\text{NMe}_2\text{-}2)_4$ (**156**)[12, 203], $(t\text{-BuC}{\equiv}\text{C})_2\text{Cu}_6(\text{SC}_6\text{H}_4\text{CH(Me)NMe}_2\text{-}2)_4$ (**157**)[203] and $(t\text{-BuC}{\equiv}\text{C})_2\text{Cu}_6(1\text{-SC}_{10}\text{H}_6\text{NMe}_2\text{-}8)_4$ (**158**)[204] have comparable structures in the solid state. As an example the molecular geometry of **156** is shown (Figure 72, **A**).

The molecular geometry of **156** can be described as consisting of two symmetry-related trinuclear units $t\text{-BuC}{\equiv}\text{CCu}_3(\text{SC}_6\text{H}_4\text{CH}_2\text{NMe}_2\text{-}2)_2$. This trinuclear unit contains Cu(1), Cu(2) and Cu(3) (Figure 72, **B**), the two thiolate groups which are μ_2-bridge bonded via their sulfur atoms between Cu(1) and Cu(3) [Cu(1)–S(1) 2.258(3) and Cu(3)–S(1) 2.245(3) Å] and Cu(2) and Cu(3) [Cu(2)–S(2) 2.327(3) and Cu(3)–S(2) 2.229(3) Å], respectively. The nitrogen-containing *ortho*-substituents of the arenethiolato groups are intramolecularly coordinating to Cu(1) [Cu(1)–N(1) 2.10(1) Å] and Cu(2) [Cu(2)–N(2) 2.07(1) Å]. The two trinuclear units are linked via two μ_3-η^1,η^1,η^1-bonded acetylides, one between Cu(1), Cu(2) and Cu(3a) [Cu(1)–C(1) 1.96(1), Cu(2)–C(1) 1.96(1) and Cu(3a)–C(1) 2.13(1) Å] and the other between Cu(1a), Cu(2a) and Cu(3).

Aggregates containing acetylides and hfac (hfac = hexafluoroacetylacetonate) represent another type of heteroleptic copper acetylide aggregates. A variety of these compounds, i.e. $(n\text{-PrC}{\equiv}\text{C})_{15}\text{Cu}_{26}(\text{hfac})_{11}$ (**159**)[205], $(t\text{-BuC}{\equiv}\text{C})_8\text{Cu}_{16}(\text{hfac})_8$ (**160**)[206], $(\text{PhCH}_2\text{C}{\equiv}\text{C})_{12}\text{Cu}_{20}(\text{hfac})_8$ (**161**)[206], $(n\text{-BuC}{\equiv}\text{C})_{14}\text{Cu}_{26}(\text{hfac})_{12}$ (**162**)[207], $(n\text{-PenC}{\equiv}\text{C})_{14}\text{Cu}_{26}(\text{hfac})_{12}$ (**163**)[207] and $(n\text{-HexC}{\equiv}\text{C})_{14}\text{Cu}_{26}(\text{hfac})_{12}$ (**164**)[207], have been characterized structurally by X-ray crystallography. These are all highly aggregated species with rather complicated structures. A common structural feature is the location of the hfac groups at the periphery of the aggregate and their *O*,*O*-chelate bonding to one copper atom. Consequently, the hfac groupings are not involved in aggregation (e.g. by bridging between two copper centers). Aggregation occurs via multicenter bonding of the acetylide groups with various bonding modes, μ_2-η^1,η^1, μ_2-η^1,η^2, μ_3-η^1,η^1,η^1, μ_3-η^1,η^1,η^2 and μ_3-η^1,η^2,η^2. The molecular geometry of **161** is shown as an example (Figure 73).

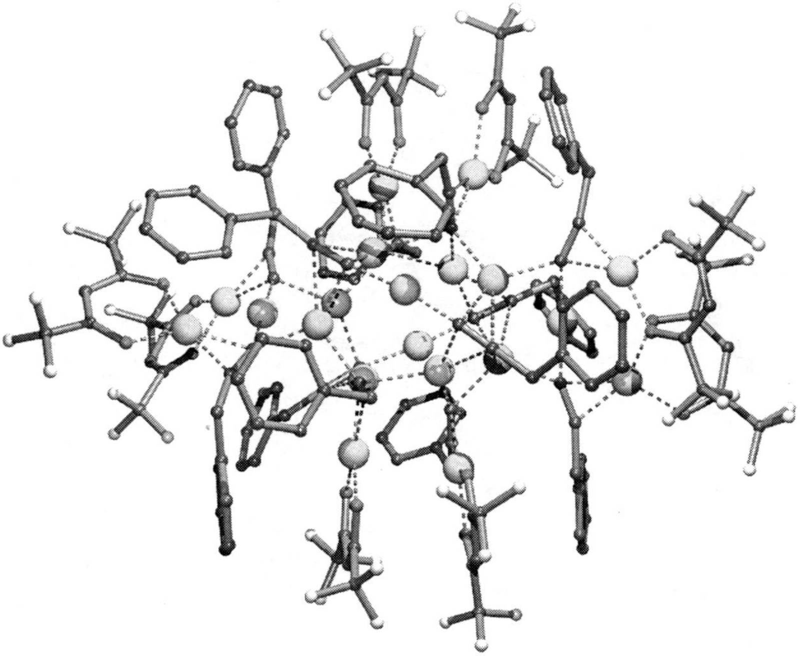

FIGURE 73. Molecular geometry of $(PhCH_2C{\equiv}C)_{12}Cu_{20}(hfac)_8$ (**161**) in the solid state

Apart from the mixed copper–silver acetylide aggregates (*vide supra*), a few compounds have been characterized structurally in which the acetylide group is involved in bridging between copper and another transition metal.

The symmetric trimeric aggregate $(HC{\equiv}C)_3Cu_3[Co_2(CO)_6]_3$ (**165**)[208] represents an example of a so-called 'cluster of clusters'. The molecular geometry of **165** involves three copper atoms arranged in an almost perfect triangle [Cu(1)–Cu(2) 2.494(2), Cu(2)–Cu(3) 2.508(2) and Cu(2)–Cu(3) 2.499(2) Å] (Figure 74). The acetylide groups are symetrically μ_2-η^1-bridge bonded between the copper atoms [Cu(1)–C(1) 1.962(11) and Cu(3)–C(1) 1.939(11) Å]. Each of the acetylide groups binds a $Co_2(CO)_6$ moiety via a μ_2-η^2,η^2-bonding mode [Co(1)–C(1) 2.142(11), Co(1)–C(2) 2.004(14), Co(2)–C(1) 2.087(11) and Co(2)–C(2) 1.989(12) Å]. Additionally, two of the carbonyl groups of each $Co_2(CO)_6$ moiety forms an unsymmetric bridge bond between cobalt and the adjacent copper atom with short Co–C bond distances [Co(1)–C 1.808(16) and Co(2)–C 1.813(17) Å] and long Cu–C bond distances [Cu(1)–C 2.408(16) and 2.545(17) Å].

An X-ray crystal structure determination of $(PhC{\equiv}C)_6Cu_2Yb_2(THF)_4$ (**166**)[209] shows the presence of two different types of bridging acetylide groups (Figure 75). Four of these are μ_2-η^1-bridge bonded between copper and ytterbium [Yb(1)–C(1) 2.685(6), Cu(1)–C(1) 1.930(6), Yb(2)–C(1a) 2.800(6) and Cu(1)–C(1a) 1.941(6) Å] while the remaining two are μ_3-η^1-bridge bonded between two ytterbium and one copper atom [Yb(1)–C(2) 2.694(6), Yb(2)–C(2) 2.699(6) and Cu(1)–C(2) 1.969(7) Å]. To complete coordination saturation at each of the ytterbium atoms, two THF molecules are coordinate bonded to each of them.

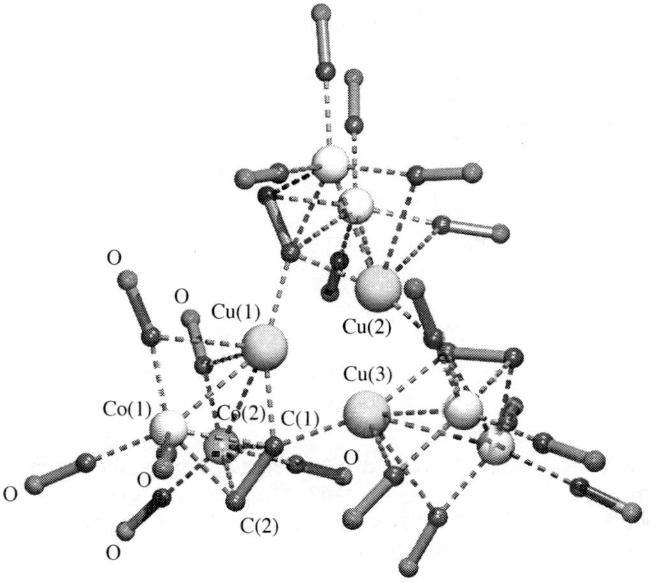

FIGURE 74. Molecular geometry of $(HC{\equiv}C)_3Cu_3[Co_2(CO)_6]_3$ (**165**) in the solid state

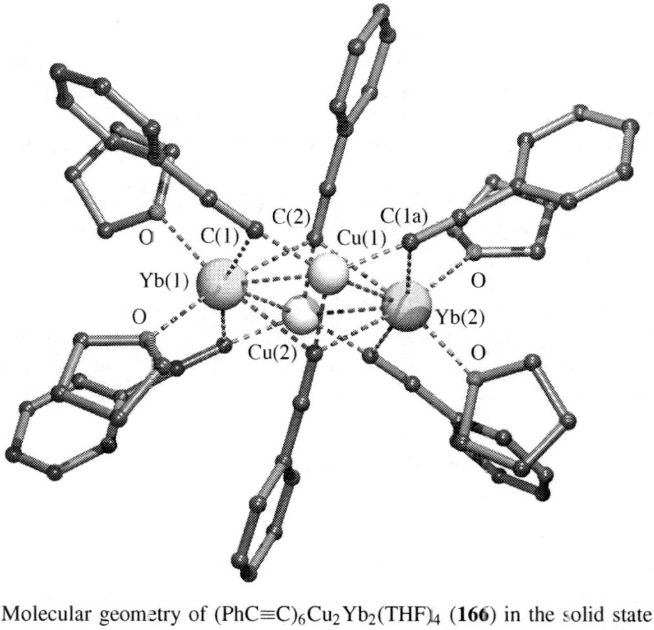

FIGURE 75. Molecular geometry of $(PhC{\equiv}C)_6Cu_2Yb_2(THF)_4$ (**166**) in the solid state

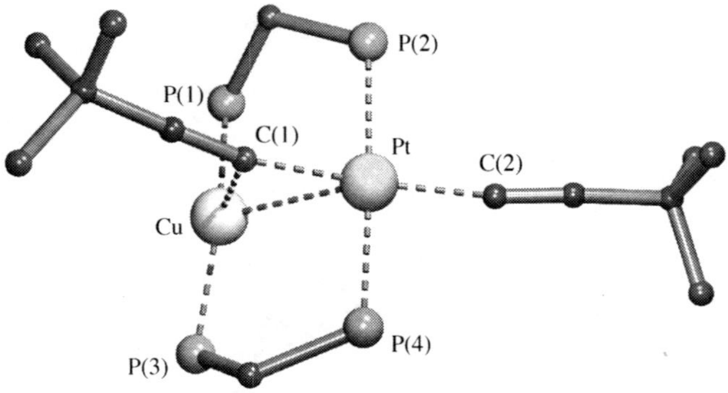

FIGURE 76. Molecular geometry of the cation of $[(t\text{-BuC}\equiv\text{C})_2\text{CuPt(dppm)}_2][\text{SbF}_6]$ (**171**) in the solid state; the phenyl groups of the dppm ligands are omitted for clarity

With respect to the bonding modes of the acetylide groups, the europium-containing aggregate $(\text{PhC}\equiv\text{C})_6\text{Cu}_2\text{Eu}_2(\text{THF})_4(\text{py})_2$ (**167**)[209] has a solid state structure comparable to that of **166**, the only difference being the presence of an additional coordinate-bonded pyridine ligand at each of the europium atoms.

The mixed copper–platinum acetylide complexes $[(\text{RC}\equiv\text{C})_2\text{CuPt(dppm)}_2][\text{X}]$ ($\text{X} = \text{ClO}_4^-$ or SbF_6^-) were studied in detail[210] because of their interesting photophysical properties. These complexes exhibit photoluminescence with lifetimes in the microsecond range, indicating spin—forbidden triplet excited states. For some of these complexes, i.e. $[(\text{PhC}\equiv\text{C})_2\text{CuPt(dppm)}_2][\text{ClO}_4]$ (**168**), $[(4\text{-MeC}_6\text{H}_4\text{C}\equiv\text{C})_2\text{CuPt(dppm)}_2][\text{ClO}_4]$ (**169**), $[(4\text{-MeOC}_6\text{H}_4\text{C}\equiv\text{C})_2\text{CuPt(dppm)}_2][\text{ClO}_4]$ (**170**) and $[(t\text{-BuC}\equiv\text{C})_2\text{CuPt(dppm)}_2][\text{SbF}_6]$ (**171**), the structures in the solid state were established by X-ray crystal structure determinations. All compounds are ionic, as is indicated by the presence of separate $[(\text{RC}\equiv\text{C})_2\text{CuPt(dppm)}_2]^+$ cations and X^- anions in the unit cell. The cations of these compounds have comparable structures and, as an example, the molecular geometry of **171** is shown (Figure 76).

In the cation of **171**, the short Pt–Cu distance [Pt–Cu 2.767(2) Å] points to a strong Pt–Cu bonding interaction[211]. The two dppm ligands are both bridging between the platinum and copper atom [Cu–P(1) 2.226(4), Cu–P(3) 2.234(4), Pt–P(2) 2.308(4) and Pt–P(4) 2.311(4) Å]. One of the acetylide groups is η^1-bonded to platinum [Pt–C(2) 1.980(14) Å] while the other one is μ_2-η^1,η^1-bridge bonded between copper and platinum [Cu–C(1) 2.153(16) and Pt–C(1) 2.011(12) Å]. The coordination geometry at the platinum center is square planar with *trans*-orientated phosphorus atoms while the copper atom exhibits a T-shaped coordination geometry.

VII. ORGANOCUPRATES

A. Neutral Organocuprates

The discovery of the Gilman reagent Me_2CuLi[4], representing the first example of an organocuprate, and its application in synthetic organic chemistry caused major breakthroughs in the development of organocopper chemistry. Since then, the actual constitution of Me_2CuLi and related species had been heavily debated. Three possible structures were proposed (Figure 77): (i) **A**, an ionic structure consisting of $[\text{Me}_2\text{Cu}]^-$ anions and (solvated) lithium cations, (ii) **B**, a tetrahedral structure comparable to the structure of methyllithium in which two of the lithium atoms in the Li_4 tetrahedron were replaced by

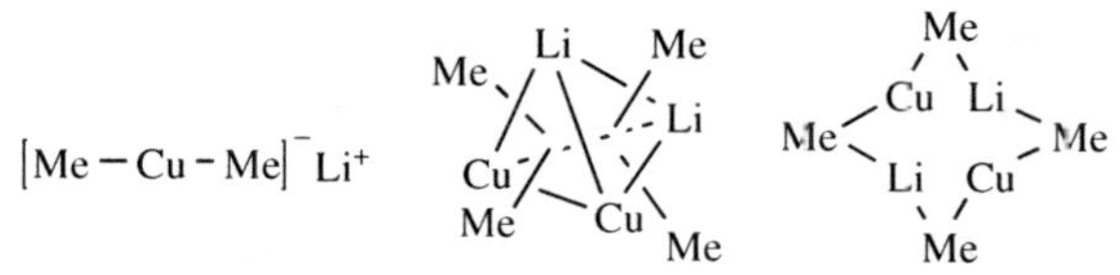

(A) (B) (C)

FIGURE 77. Proposed structures for the Gilman reagent in the seventies

two copper atoms[212], and (iii) **C**, a planar cyclic structure in which the methyl groups were two-electron three-center bridge bonded via C_{ipso} between a lithium and a copper atom at the edges of a Cu_2Li_2 square. Based on molecular weight determinations using vapor pressure depression, NMR spectroscopic studies and solution X-ray scattering techniques, structure **C** for the Gilman reagent and related species seemed to be the most likely one[213].

At the same time the synthesis of a diarylcuprate, i.e. of $(2\text{-}Me_2NCH_2C_6H_4)CuLi$, which could be isolated as a pure, crystalline material, was reported[214]. Molecular weight determinations by cryoscopy and ebulliometry in benzene solution pointed to a dimeric aggregate $(2\text{-}Me_2NCH_2C_6H_4)_4Cu_2Li_2$ (**172**). Based on ^1H and ^{13}C NMR spectroscopic data, a structure was proposed as shown schematically in Figure 78, **A**. Conclusive evidence supporting the existence of the proposed structure in solution came from a detailed ^{13}C NMR spectroscopic investigation of the corresponding diarylargentate $(2\text{-}Me_2NCH_2C_6H_4)_4Ag_2Li_2$ (**173**)[215]. Due to the presence of magnetically active metal nuclei (^7Li, ^{107}Ag and ^{109}Ag), detailed structural information could be gathered from the observed ^1H and ^{13}C NMR spectra. From these data the following conclusions could be drawn: (i) because in both the ^1H and ^{13}C NMR spectrum one resonance pattern was observed for the organic moieties, all four groups are (magnetically) equivalent; (ii) the observation of two resonances for the NMe_2 group and an AB pattern for the benzylic CH_2 group indicated that these groups are diastereotopic and thus nitrogen-to-metal coordination (most likely to lithium) is inert on the NMR time scale; and (iii) the ^{13}C resonance for C_{ipso} shows scalar coupling with one ^7Li nucleus and one Ag nucleus (Figure 78, **B**) indicating that this C_{ipso} is bonded to one lithium and one silver atom. Based on these conclusions the structure as proposed and shown schematically in Figure 78, **A** was most likely.

When techniques to handle extremely air-sensitive crystals became available, the molecular structures in the solid state of **172** (M = Cu)[49] and of the corresponding aurate $(2\text{-}Me_2NCH_2C_6H_4)_4Au_2Li_2$ (**174**)[50] (M = Au in Figure 78, **A**) could be established by X-ray crystal structure determinations. It appeared that the molecular geometry of **172** in the solid state was fully in agreement with the structure in solution that was proposed twelve years earlier based on spectroscopic data. In **172**, the two copper and two lithium atoms are arranged in an almost perfect planar arrangement with the aryl groups bridge-bonded via C_{ipso} between copper and lithium (Figure 79). The aggregate has twofold symmetry as a consequence of a crystallographic C_2-axis perpendicular to both the Cu–Cu(a) and Li–Li(a) vector. The lithium atoms adopt a distorted tetrahedral coordination geometry as a result of intramolecular coordination [Li–N(1) 2.139(6) and Li–N(2a) 2.123(6) Å] of the nitrogen atoms of each of the two *ortho*-(dimethylamino)methyl substituents to the same lithium atom. The observed bridge-bonding mode is rather asymmetric as is obvious both from the different bonding distances between C_{ipso} and copper and lithium, respectively [Cu–C(1) 1.936(3) and Li–C(1) 2.374(7) Å], and from the orientation of the aryl group, i.e. the angle between the C(4)–C(1) vector and the C(1)–Cu bond which is close to linear (165°) while the angle between the C(4)–C(1) vector and the C(1)–Li

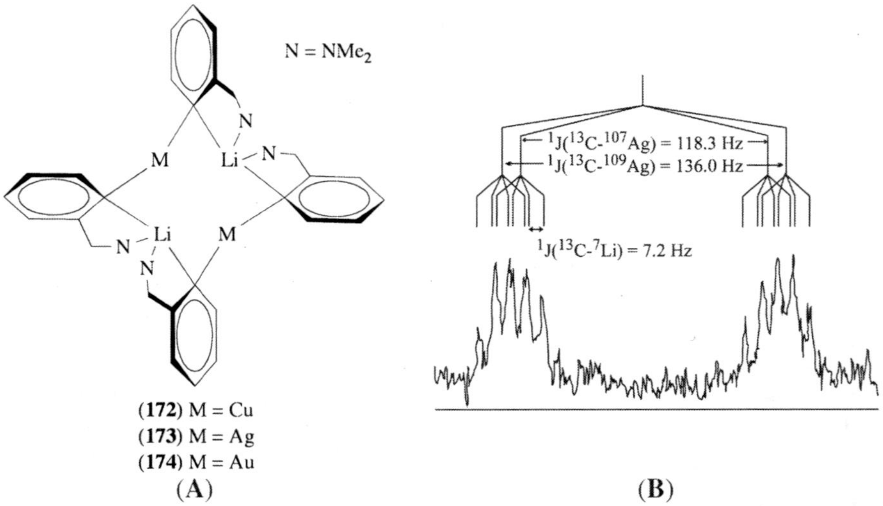

FIGURE 78. (**A**) Proposed structures for diarylcuprate (M = Cu, **172**) and the corresponding argentate (M = Ag, **173**). (**B**) Observed ^{13}C NMR resonance pattern for C_{ipso} of **173**

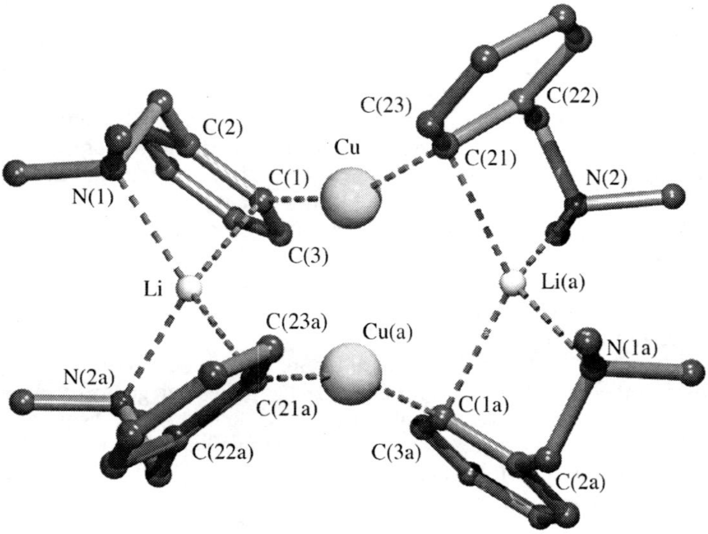

FIGURE 79. Molecular geometry of $(2\text{-}Me_2NCH_2C_6H_4)_4Cu_2Li_2$ (**172**) in the solid state

bond is 115°. The latter values point to a considerable tilting of the aryl plane towards lithium, resulting in a Li−C_{ipso} bond almost perpendicular to the plane of the aryl ring.

Comparison, in a simplified model (Figure 80), of the molecular orbitals involved in a symmetric two-electron three-center bond (**A**) with those describing the asymmetric bonding as observed in **172** (**B**) shows that in the latter situation the overlap between the

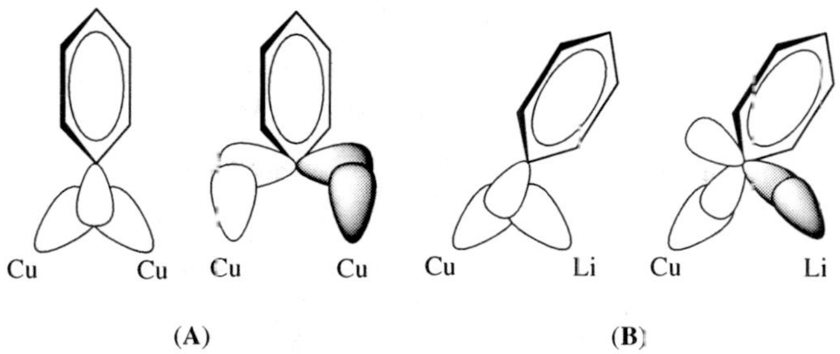

FIGURE 80. Molecular orbitals involved in (**A**) symmetric two-electron three-center bonding and (**B**) asymmetric bridge bonding

aryl-sp^2 orbital and copper orbitals is enhanced, i.e. the bonding is shifted towards a two-electron two-center bonding situation[216,217]. In its extreme, the solid state structure of **172** could be described in terms of consisting of two monoanionic $[(2\text{-}Me_2NCH_2C_6H_4)_2Cu]^-$ fragments linked by two lithium cations via two Li–N coordination bonds and electrostatic interactions with the *ipso*-carbon atoms. However, the observation of a scalar coupling of ^{7}Li with ^{13}C in the ^{13}C NMR spectrum of **172** [$^1J(^7\text{Li}-^{13}\text{C}) = 7.0$ Hz] indicates that, at least in solution, considerable s-electron density is present between the *ipso*-carbon atom and lithium. In this respect it is interesting to note that in an X-ray crystal structure determination of the corresponding aurate $(2\text{-}Me_2NCH_2C_6H_4)_4Au_2Li_2$ this asymmetric bonding is even more expressed[50].

A structural feature that was recognized in early stages of these studies, is the fact that, due to the asymmetric substitution of the aryl group (Me_2NCH_2 substituent at the 2-position) and the bonding of C_{ipso} to two different metals, the *ipso*-carbon atom is a chiral center[218,219]. In the solid state structure of **172** the configuration of each of the four *ipso*-carbon atoms is identical, but as a consequence of the centro-symmetric space group (C2/c) both enantiomers of the diastereoisomer in which the *ipso*-carbon atoms have either all four of the (*R*) or all four of the (*S*) configuration are present in the crystal lattice. It should be noted that the configurations of the *ipso*-carbon atoms bound to the same lithium atom, i.e. C(1) and C(21a) (Figure 79), are controlled by each other and should be either both (*R*) or both (*S*). This is a logical consequence of the fact that one of the nitrogen atoms approaches the lithium atom from above the Cu_2Li_2-plane while the other nitrogen atom must approach the lithium atom from below that plane. The reversed situation implies inversion of configuration of both *ipso*-carbon atoms. Such a process indeed was observed by NMR spectroscopy at higher temperatures and implies dissociation of the Li–N coordination bond and concomitant rotation of the aryl group by 180° around the C(1)–C(4) vector as shown schematically (Figure 81).

At this point it is noteworthy that the configuration of the *ipso*-carbon atoms at the two different lithium atoms are independent and therefore, in principle, two diastereoisomers of the aggregate are possible: One with all four *ipso*-carbon atoms having the same configuration (as observed in the solid state) and a second one with the *ipso*-carbon atoms bonded to one lithium with the (*R*)-configuration and with (*S*)-configuration at the *ipso*-carbon atoms bonded to the other lithium atom. However, the ^{1}H and ^{13}C NMR spectra of **172** show only one distinct resonance pattern, indicating the presence of only one diastereoisomer in solution, most likely the same one as in the solid state.

A much more complicated situation arises when the organic group also contains an additional (rigid) chiral center as, e.g., in the 2-[1-(dimethylamino)ethyl]phenyl group. This topic has been studied in detail by ^{1}H and ^{13}C NMR spectroscopy for the cuprates

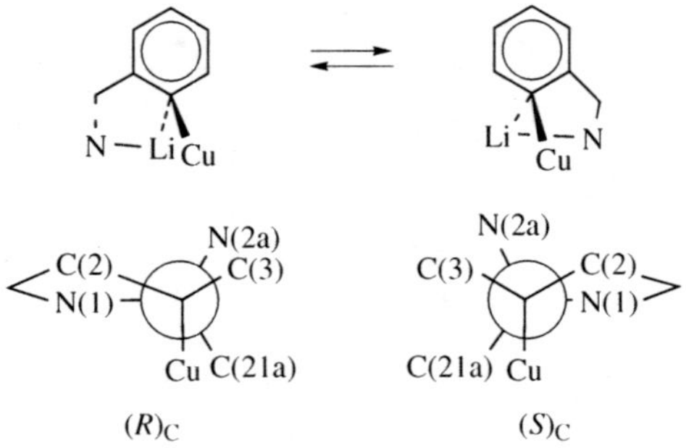

FIGURE 81. Process describing the inversion of configuration at the chiral *ipso*-carbon atom (Newman projection along the C_{ipso}–lithium bond)

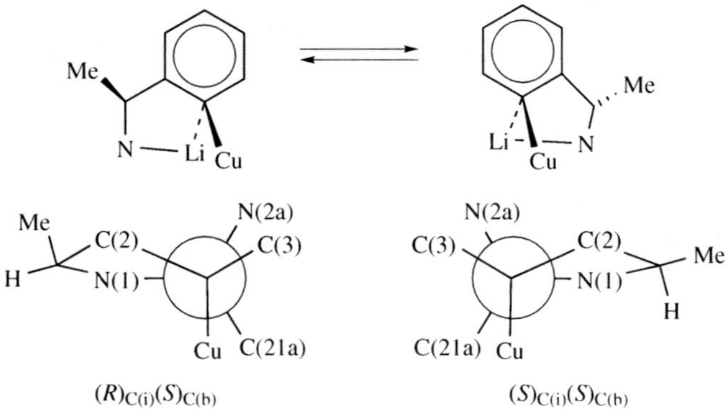

FIGURE 82. Interconversion of the diastereoisomeric ligand in **175** (Newman projection along the *ipso*-C–lithium bond)

$[2\text{-}Me_2NCH(Me)C_6H_4]_4Cu_2Li_2$ containing both the enantiopure and the racemic ligand, (*S*)-**175** and (rac)-**175**, respectively[216–219]. In these aggregates, within one ligand, two chiral centers are present: (i) the *ipso*-carbon atom C(i) and (ii) the benzylic carbon atom C(b). If inversion of configuration at the *ipso*-carbon atom occurs via a similar process as described above for **172**, this results in interconversion of one diastereoisomeric form of the ligand into another diastereoisomer (Figure 82). Also in these aggregates, for reasons outlined above, the configuration of the *ipso*-carbon atoms bonded to one lithium atom are controlled by each other, but are independent of the ones bonded to the other lithium atom. Thus, for (*S*)-**175** in principle three diastereoisomeric aggregates are possible: (i) one with the configurations $(R)_{C(i)}(S)_{C(b)}(R)_{C(i)}(S)_{C(b)}/(R)_{C(i)}(S)_{C(b)}(R)_{C(i)}(S)_{C(b)}$, (ii) another one with configurations $(S)_{C(i)}(S)_{C(b)}(S)_{C(i)}(S)_{C(b)}/(S)_{C(i)}(S)_{C(b)}(S)_{C(i)}(S)_{C(b)}$, and (iii) a third one with the configurations $(R)_{C(i)}(S)_{C(b)}(R)_{C(i)}(S)_{C(b)}/(S)_{C(i)}(S)_{C(b)}(S)_{C(i)}(S)_{C(b)}$. For the racemic material (rac)-**175**, many more combinations of configurations of the chiral centers are possible. However, in both the ^1H and ^{13}C NMR spectra of (*S*)-**175** and (rac)-**175** only one resonance pattern is observed for the organic ligand, indicating that only one of all possible diastereoisomers is present in solution, which is a nice illustration of diastereoselective self-assembly. In this respect it is interesting to note that the organocopper

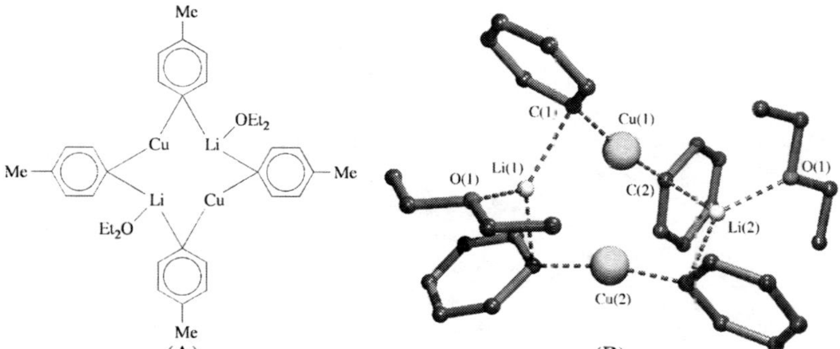

(A) (B)

FIGURE 83. (**A**) Proposed structure for $(4\text{-}MeC_6H_4)_4Cu_2Li_2(OEt_2)_2$. (**B**) Molecular geometry of **176** in the solid state

compound $[2\text{-}Me_2NCH(Me)C_6H_4]_4Cu_4$ derived from the racemic ligand in solution exists as a mixture of various diastereoisomeric aggregates[217].

Based on molecular weight determinations and ^{1}H NMR spectroscopic studies, already in 1977, a structure for the simple neutral diarylcuprates $(4\text{-}MeC_6H_4)_4Cu_2Li_2(OEt_2)_2$ and $Ph_4Cu_2Li_2(OEt_2)_2$ (**176**) had been proposed as is shown schematically (Figure 83, **A**)[220].

About thirteen years later an X-ray crystal structure determination of **176** confirmed this proposal (Figure 83)[221]. The molecular geometry of **176** comprises a four-membered Cu–Li–Cu–Li ring that is slightly folded (154.3°) along the Cu–Cu vector. The aryl groups are two-electron three-center bridge bonded via C_{ipso} in an asymmetric way [Cu(1)–C(1) 1.922(7) and Li(1)–C(1) 2.254(10) Å]. Each of the lithium atoms is three-coordinate as a result of an additional coordinating diethyl ether molecule [Li(1)–O(1) 1.926(9) Å]. Like for **172**, the overall molecular geometry of **176** can be described as consisting of two almost linear $[Aryl_2Cu]^-$ anions [C(1)–Cu(1)–C(2) 167.7(3)°] linked by two $[Li(OEt_2)]^+$ cations via electrostatic interactions.

When diphenylcopperlithium was prepared in dimethyl sulfide (DMS) as solvent and the resulting reaction product was recrystallized from DMS, a crystalline material was obtained with formula $Ph_4Cu_2Li_2(DMS)_3$ (**177**)[106]. An X-ray crystal structure determination revealed a molecular geometry (Figure 84) with structural features related to those of **176**. Two almost linear (average C–Cu–C 162.8°) $[Ph_2Cu]^-$ anions are linked by two lithium cations (average Li–C_{ipso} 2.284(5) Å). One of the lithium cations is solvated with one coordinate bonded DMS molecule [Li(1)–S(1) 2.524(5) Å] whereas the other lithium cation is solvated by two DMS molecules [Li(2)–S(2) 2.552(5) and Li(2)–S(3) 2.526(5) Å].

The only examples of neutral dialkylcuprate compounds that have been structurally characterized in the solid state by X-ray crystallography have Me_3SiCH_2 groups as the alkyl groupings. Recrystallization of the corresponding dialkylcuprate from diethyl ether afforded an aggregate with the composition $(Me_3SiCH_2)_4Cu_2Li_2(OEt_2)_3$ (**178**)[222] while recrystallization from DMS gave a crystalline material with the composition $(Me_3SiCH_2)_4Cu_2Li_2(SMe_2)_2$ (**179**)[223]. An X-ray crystal structure determination of **178** showed a structural motif closely related to that of **177**. Two $[(Me_3SiCH_2)_2Cu]^-$ anions [average Cu–C 1.955(7) Å] which have a close to linear arrangement (average C–Cu–C 171.8°) are bridged by one $[Li(OEt)]^+$ cation at one side and a $[Li(OEt_2)_2]^+$ cation at the other side of the aggregate (Figure 85). The bond distances to the three-coordinate lithium atom are significantly shorter [Li(1)–C(1) 2.199(14), Li(1)–C(2) 2.264(14) and Li(1)–O(1)

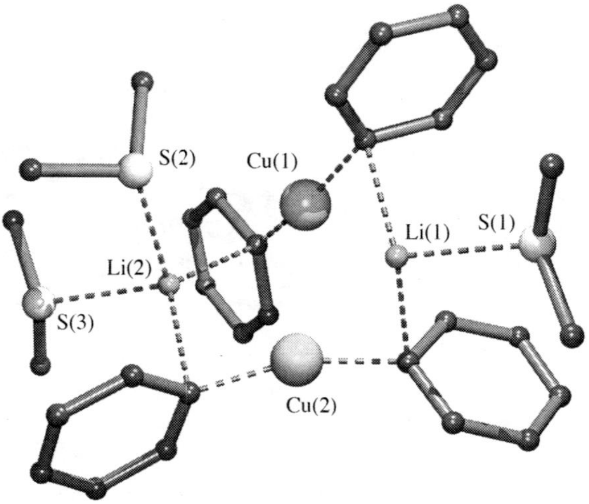

FIGURE 84. Molecular geometry of $Ph_4Cu_2Li_2(DMS)_3$ (**177**) in the solid state

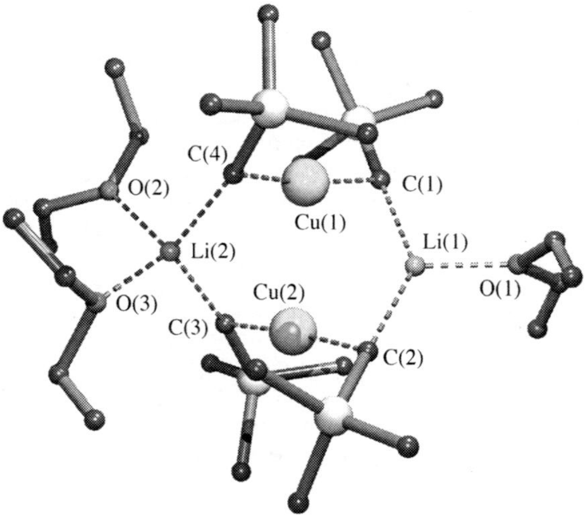

FIGURE 85. Molecular geometry of $(Me_3SiCH_2)_4Cu_2Li_2(OEt_2)_3$ (**178**) in the solid state

1.958(14) Å] as compared to the corresponding distances at the four-coordinate lithium atom [Li(2)–C(3) 2.321(14), Li(2)–C(4) 2.304(14), Li(2)–O(2) 2.022(13) and Li(2)–O(3) 2.022(13) Å]. This is not unexpected due to the difference in coordination numbers of the lithium atoms, i.e. three- vs. four-coordination.

An X-ray crystal structure determination of **179** showed this compound to exist as an infinite polymeric chain in the solid state (Figure 86). The infinite chain is built-up of $[(Me_3SiCH_2)_4Cu_2Li_2]$ units with structural features almost identical to that of **178**. The $[(Me_3SiCH_2)_4Cu_2Li_2]$ units are linked to a chain via two μ^2-*S*-bridge-bonded DMS

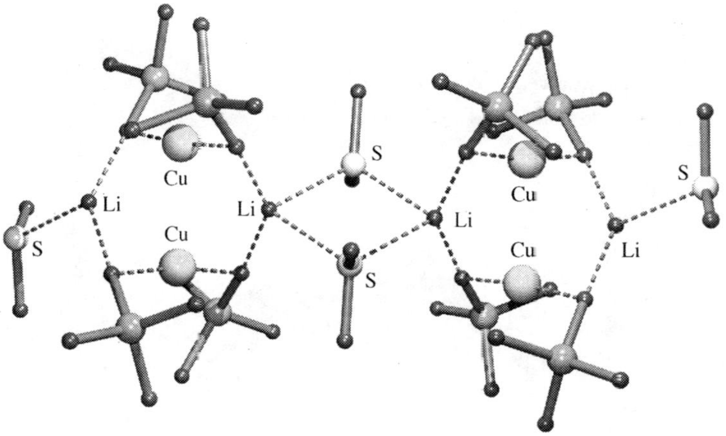

FIGURE 86. Part of the infinite polymeric chain of $(Me_3SiCH_2)_4Cu_2Li_2(SMe_2)_2$ (**179**) in the solid state

molecules between two lithium atoms of adjacent $[(Me_3SiCH_2)_4Cu_2Li_2]$ units. The relatively long Li–S bond distances [average Li–S 2.680(9) Å] point to a weak solvation of the lithium cations and results in slightly shorter Li–C distances as compared to the corresponding Li–C distances in **178**.

Reaction of the intramolecularly chelated aryllithium compound $[(CNN)Li]_2$ with two equivalents of a copper arylacetylide afforded the unique, neutral heteroleptic organocuprates **180** and **181** (equation 21)[224]. The structures in the solid state of both **180** and **181** were established by X-ray crystal structure determinations and appeared to be isostructural, and therefore only details of the molecular geometry of **180** are given below.

CNN = 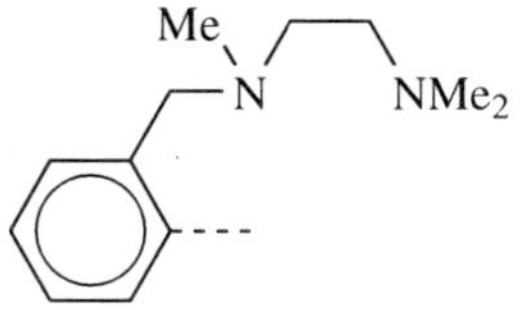

(21)

$[(CNN)Li]_2$ + 2 $4\text{-}RC_6H_4C{\equiv}CCu$ ⟶ $(4\text{-}RC_6H_4C{\equiv}C)_2(CNN)_2Cu_2Li_2$

(**180**) R = Me
(**181**) R = Me_3Si

The overall structural motif of **180** is identical to that found in the neutral organocuprates discussed so far (*vide supra*), i.e. two $[R_2Cu]^-$ monoanionic units linked by two lithium cations (Figure 87). The molecular geometry of **180** consists of two different monoanionic units, however: one $[(CNN)_2Cu]^-$ anion with relative short $Cu-C_{ipso}$ bond distances [Cu(1)–C(1) 1.932(2) and Cu(1)–C(2) 1.933(2) Å] and a second one, a $[(4\text{-}MeC_6H_4C{\equiv}C)_2Cu]^-$ anion with even shorter Cu–C bond distances [Cu(2)–C(3) 1.868(2) and Cu(2)–C(4) 1.871(2) Å]. These two fragments are linked to each other by two lithium cations that have an interaction with the *ipso*-carbon atoms of the first unit [Li(1)–C(1) 2.513(4) and Li(2)–C(2) 2.513(4) Å] and the α-acetylenic carbon atoms of the $[(4\text{-}MeC_6H_4C{\equiv}C)_2Cu]$ unit [Li(1)–C(4) 2.219(4) and Li(2)–C(3) 2.233(4) Å]. Additionally,

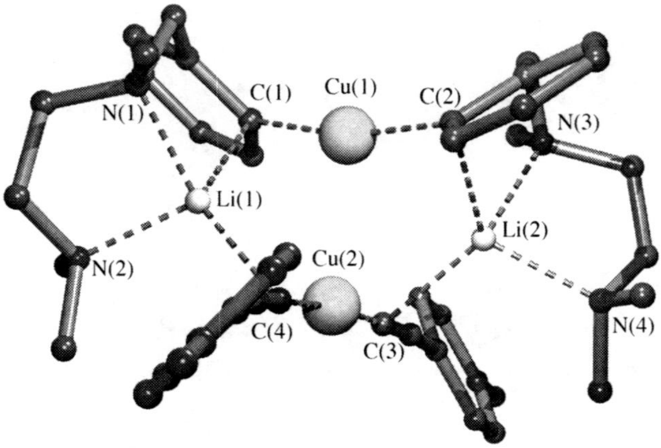

FIGURE 87. Molecular geometry of heteroleptic cuprate **180** in the solid state

the 2-$Me_2NCH_2CH_2N(Me)CH_2$ substituents are intramolecular *N*,*N*′-chelate bonded to the respective lithium atoms [Li(1)–N(1) 2.045(4), Li(1)–N(2) 2.056(4), Li(2)–N(3) 2.094(4) and Li(2)–N(4) 2.096(4) Å].

From the reaction of mesitylcopper with mesityllithium in a 1:1 molar ratio in toluene as the solvent, two products were isolated with the compositions $Mes_4Cu_2Li_2$ (**182**) and Mes_4Cu_3Li (**183**), respectively[225]. The structures in the solid state of both aggregates were established by X-ray crystal structure determinations.

The molecular geometry of **182** comprises two symmetry-related $[Mes_2Cu]^-$ anions that have an almost perfect linear arrangement [C(1)–Cu(1)–Cu(2) 178.34(7)°] and are linked by two lithium cations (Figure 88). The two $[Mes_2Cu]^-$ anions are arranged in such a way that the lithium atoms have a bonding interaction with one of *ipso*-carbon atoms in one $[Mes_2Cu]^-$ anion [Li(1)–C(1) 2.129(4) Å] and a η^6-bonding interaction with the aryl ring in the other $[Mes_2Cu]^-$ unit with Li–C bond distances ranging from 2.271(4) to 2.365(4) Å.

The molecular geometry of **183** contains a structural element that closely resembles that of Mes_4Cu_4 in the solid state. The three copper atoms are arranged in a triangle with two relatively short edges [Cu(1)–Cu(3) 2.441 and Cu(2)–Cu(3) 2.415 Å] and one long edge [Cu(1)–Cu(2) 3.599 Å]. At the short edges two of the mesityl groups are two-electron three-center bonded via C_{ipso} between the copper atoms via C_{ipso} [Cu(1)–C(1)

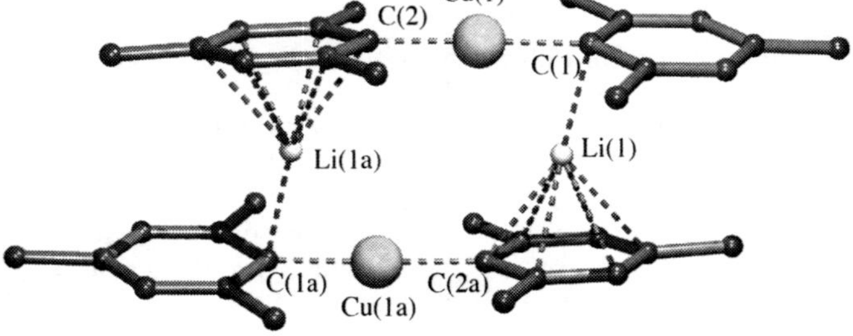

FIGURE 88. Molecular geometry of $Mes_4Cu_2Li_2$ (**182**) in the solid state

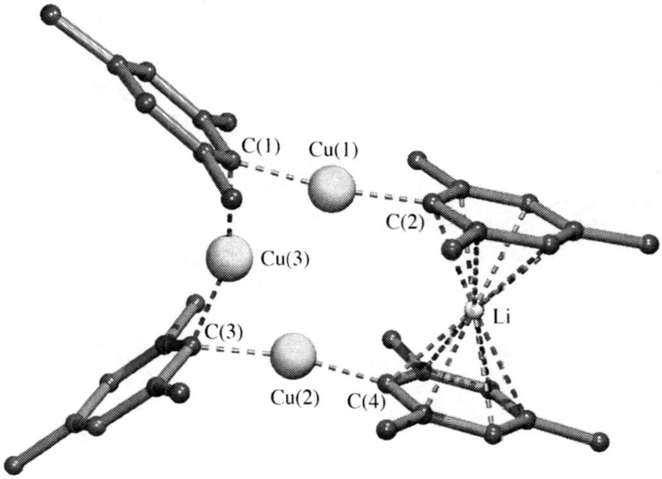

FIGURE 89. Molecular geometry of Mes_4Cu_3Li (**183**) in the solid state

2.010(4), Cu(3)–C(1) 2.016(4), Cu(2)–C(3) 2.012(4) and Cu(3)–C(3) 1.999(4) Å] and are orientated perpendicular to the respective Cu–Cu vectors (Figure 89). Another mesityl group is bonded to both Cu(1) and Cu(2), but in η^1-fashion [Cu(1)–C(2) 1.951(4) and Cu(2)–C(4) 1.932(4) Å]. These latter two mesityl groups are perfectly coplanar, arranged with a lithium cation that is sandwiched in between both η^1-bonded mesityl groups via η^6-bonding interactions with bond distances ranging from 2.304(7) to 2.531(8) Å.

From the reaction of CuBr with three equivalents of PhLi in dimethyl sulfide, a so-called higher-order organocuprate with formula $Ph_5Cu_2Li_3(DMS)_4$ (**184**) was isolated. Its structure in the solid state was characterized by X-ray crystallography[226]. The molecular geometry of **184** is rather irregular but may be described as consisting of $[Ph_2Cu]^-$ and $[Ph_3Cu_2]^-$ anions which are linked together by three (DMS solvated) lithium cations (Figure 90). In **184**, two of the phenyl groups are two-electron three-center bridge bonded via C_{ipso} between copper and lithium [C(1) bonded to Li(1) and Cu(1), and C(5) bonded to Cu(1) and Li(3)]. The other three phenyl groups are two-electron four-center bonded between two lithium atoms and one copper atom via C_{ipso} [C(2) bonded to Li(1), Li(2) and Cu(2), C(3) bonded to Li(1), Li(3) and Cu(2), and C(4) bonded to Li(2), Li(3) and Cu(2)]. To attain a distorted tetrahedral coordination geometry at the respective lithium atoms, one DMS molecule is coordinate bonded to Li(1) and Li(3), respectively, while to Li(2) two DMS molecules are bonded.

The only example of a neutral cuprate that has been characterized structurally in the solid state and that has a metal other than lithium is the aggregate $Ph_6Cu_4Mg(OEt_2)$ (**185**)[227]. The structure of **185** consists of a trigonal bipyramidal array with three copper atoms at the equatorial positions, one copper atom at the top, axial position and one magnesium atom at the bottom, axial position (Figure 90). Three two-electron three-center bonded phenyl groups are spanning via C_{ipso} the three Cu–Cu axial-equatorial edges of the trigonal bipyramid. The C_{ipso}–Cu_{axial} bond distances [average C_{ipso}–Cu(4) 2.098(12) Å] are slightly longer than the C_{ipso}–$Cu_{equatorial}$ bond distances [average C_{ipso}–$Cu_{equatorial}$ 1.974(12) Å]. The other three phenyl groups are bonded in a two-electron three-center bonding mode spanning the Mg–Cu axial equatorial edges with relatively short C_{ipso}–$Cu_{equatorial}$ bond distances [average C_{ipso}–$Cu_{equatorial}$ 1.945(12) Å] and relatively long

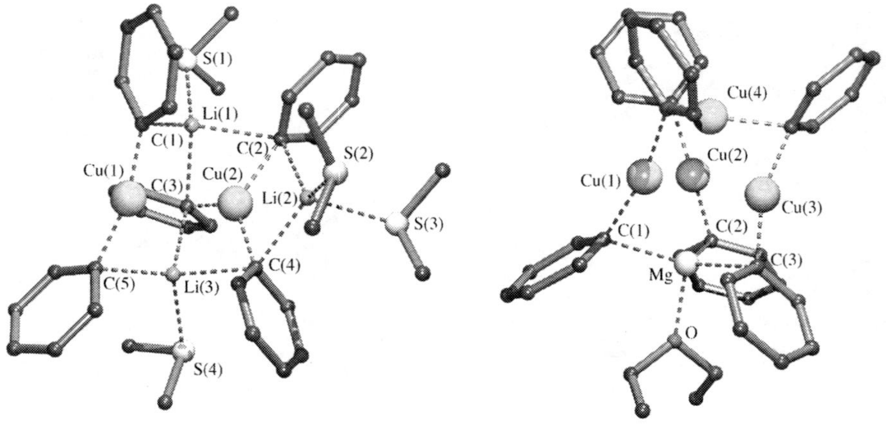

(**184**) (**185**)

FIGURE 90. Molecular geometry of $Ph_5Cu_2Li_3(DMS)_4$ (**184**) and $Ph_6Cu_4Mg(OEt_2)$ (**185**) in the solid state

C_{ipso}–Mg bond distances [average C_{ipso}–Mg 2.353(11) Å]. The fourth coordination site at the magnesium atom is occupied by a coordinate-bonded diethyl ether molecule [Mg–O 2.045(9) Å].

Treatment of *N*-(trimethylsilyl)triethylphosphinimine with *n*-butyllithium resulted in α-lithiation of one of the ethyl groups bound to phosphorus. A subsequent transmetallation reaction with CuI afforded the metallacyclic diorganocuprate (**186**) of which the structure in the solid state was established by an X-ray crystal structure determination (equation 22)[228]. In **186**, the two organic moieties are σ-bonded via their α-carbon atoms to copper [Cu–C(1) 1.951(2) and Cu–C(2) 1.956(2) Å] in an almost linear fashion [C(1)–Cu–C(2) 167.20(11)°]. The two nitrogen atoms are each coordinate bonded to the lithium cation [Li–N(1) 1.940(5) and Li–N(2) 1.944(5) Å], in this way forming an eight-membered $C_2P_2N_2CuLi$ ring. The unusual linear, two-coordinate geometry of the lithium cation is notable [N(1)–Li–N(2) 169.8(3)°].

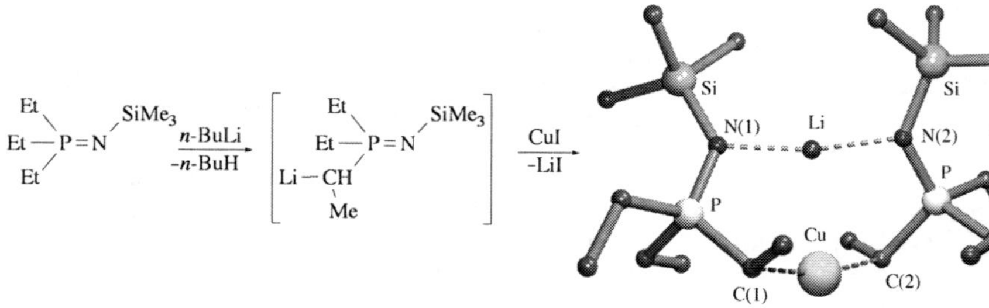

(22)

(**186**)

A remarkable dimetalladiphospha-cyclic organocopper compound (**187**) was obtained from the reaction of trimethylmethylenephosphorane and CuCl and for which, based on spectroscopic data, a structure was proposed, as shown schematically in equation 23[229].

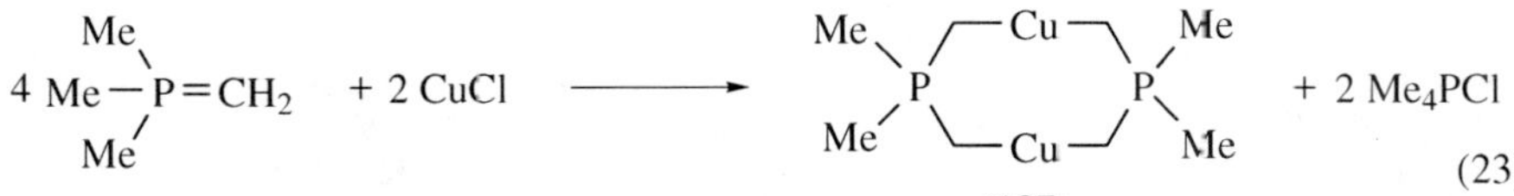

(187)

(23)

An X-ray crystal structure determination of **187** confirmed this proposal[230]. The copper and C_{ipso} carbon atoms of the $C_4Cu_2P_2$ dimetalladiphospha-cyclic ring lie in one plane while the respective phosphorus atoms are positioned either 0.79 Å above or below this plane, respectively. The copper–carbon bond distances are relatively short [Cu–C 1.96(1) and 1.95(1) Å]. The coordination geometry at the copper atoms is close to linear [C–Cu–C 175.8(8)°]. This compound may be regarded as a zwitterionic cuprate with formal negative charges on both copper atoms and positive charges at the phosphorus atoms.

B. Neutral Heterocuprates

The neutral organocuprates discussed so far all contained monoanicnic, carbon-bonded organic groups as the only anionic moieties present in the cuprate agggregates. However, in this section neutral organocuprates are described that contain, apart from the monoanionic carbon-bonded organic group, also other monoanions, like halogen, amido, phosphido or cyano groups. These cuprates are denoted 'neutral heterocuprates'. A particular type of these heterocuprate aggregates, the so-called 'cyanocuprates', will be discussed separately because of their particular importance as a synthetic tool in organic chemistry.

The reaction of the organolithium compounds $[CNN]_2Li_2$ with CuBr in a 1:1 molar ratio afforded the heterocuprates $[CNN]_2CuLi_2Br$ (**188**–**192**) in quantitative yield (equation 24)[231,232].

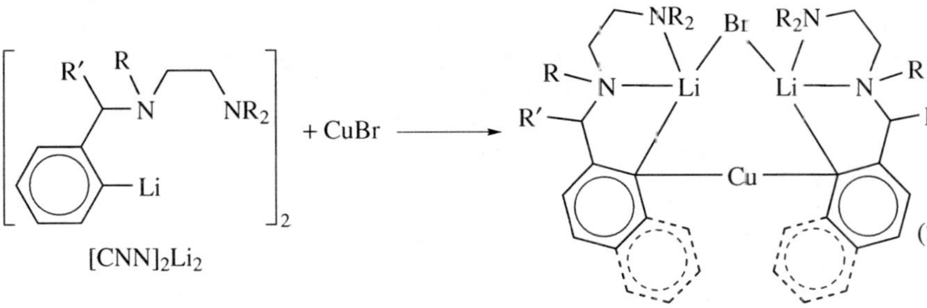

$[CNN]_2Li_2$

(24)

(**188**) Aryl, R = Me, R′ = H
(**189**) Aryl, R = Me, R′ =Me, (*R*)
(**190**) Aryl, R = Me, R′ =Me, (rac)
(**191**) Aryl, R = Et, R′ = H
(**192**) Naphthyl, R = Me, R′ = H

The structures in the solid state of three of these, i.e. **188, 191** and **192**, were established by X-ray crystal structure determinations. These three aggregates are isostructural and have a structural motif closely related to that of the organocopper copper bromide aggregate $[CNN]_2Cu_3Br$ (*vide supra*, **71**). As an example the molecular geometry of **188** is shown (Figure 91). In **188**, the copper and two lithium atoms are arranged in a

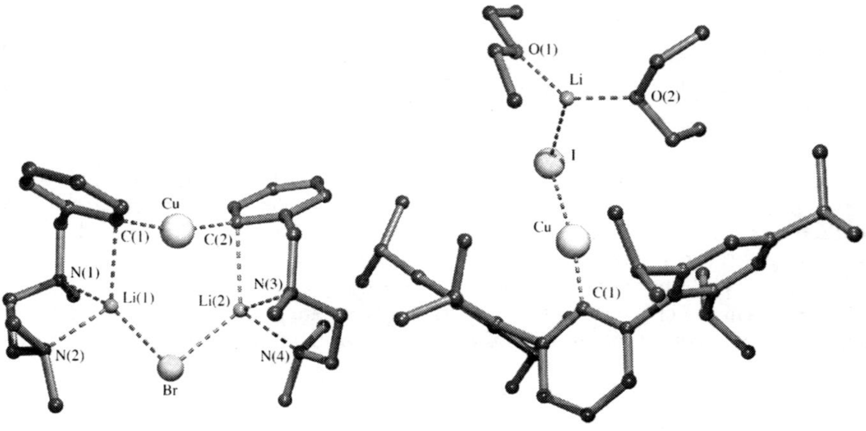

(188) **(193)**

FIGURE 91. Molecular geometry of heterocuprates $[CNN]_2CuLi_2Br$ (**188**) and 2,6-$Tip_2C_6H_3CuLiI(OEt_2)_2$ (**193**) in the solid state

triangle with the aryl groups two-electron three center bonded via C_{ipso} between the copper and lithium atoms at the Cu–Li edges of the triangle [Cu–C(1) 1.929(1), Li(1)–C(1) 2.305(2), Cu–C(2)1.927(2) and Li(2)–C(2) 2.393(3) Å]. At the Li–Li edge of the triangle a bromine atom is μ_2-bridge bonded between the lithium atoms [Li(1)–Br 2.411(2) and Li(2)–Br 2.430(2) Å]. The *ortho*-$CH_2N(Me)CH_2CH_2NMe_2$ substituents are each N,N'-chelate bonded to one lithium atom [Li(1)–N(1) 2.139(2), Li(1)–N(2) 2.189(2), Li(2)–N(3) 2.113(3) and Li(2)–N(4) 2.149(3) Å]. The aryl groups are orientated in such a way that the benzylic-N atom of one of the *ortho*-$CH_2N(Me)CH_2CH_2NMe_2$ substituents approaches the lithium atom from below the $CuLi_2$ plane while the benzylic-N atom from the other substituent coordinates to lithium from its opposite side. It should be noted that careful analysis of the X-ray data showed that the positions of the lithium atoms are slightly disordered with copper (site population <5%), indicating that lithium might be substituted by copper with retention of the triangular structural motif, and thus points to the existence of three further aggregates, i.e. $[CNN]_2CuLi_2Br$, $[CNN]_2Cu_2LiBr$ and $[CNN]_2Cu_3Br$ (which actually has been characterized, cf. **71**).

The reaction of 2,6-$Tip_2C_6H_3Li$ (Tip = 2,4,6-$(i\text{-}Pr)_3C_6H_2$) with CuI in a 1:1 molar ratio in diethyl ether as the solvent afforded a product which, after recrystallization from apolar solvents like hexanes, still contained LiI[233]. An X-ray crystal structure determination revealed the formation of a heterocuprate 2,6-$Tip_2C_6H_3CuLiI(OEt_2)_2$ (**193**) instead of the anticipated pure organocopper compound 2,6-$Tip_2C_6H_3Cu$. The molecular geometry of **193** in the solid state comprises a 2,6-$Tip_2C_6H_3$ group that is σ-bonded via C_{ipso} to copper [Cu–C(1) 1.902(5) Å] while the iodide anion bridges between copper and lithium [Cu–I 2.4512(8) and Li–I 2.690(11) Å] (Figure 91). The carbon–copper–iodine arrangement is close to linear [C(1)–Cu–I 171.4(2)°], whereas the Cu–I–Li bond angle is rather acute, 95.5(2)°. Two additional diethyl ether molecules are coordinate bonded to lithium [Li–O(1) 1.905(11) and Li–O(2) 1.893(11) Å], resulting in trigonal planar coordination geometry at lithium.

From a solution containing a 1:1 mixture of mesitylcopper and $LiN(CH_2Ph)_2$ in toluene, a crystalline material was obtained that, according to an X-ray crystal structure determination, appeared to be the dimeric mesitylamidocuprate $Mes_2Cu_2Li_2[N(CH_2Ph)_2]_2$ (**194**)[234].

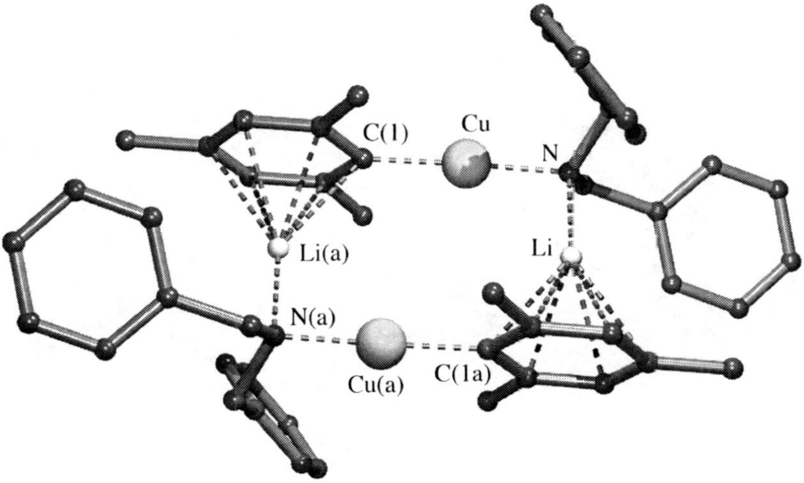

FIGURE 92. Molecular geometry of the dimeric mesitylamidocuprate $Mes_2Cu_2Li_2[N(CH_2Ph)_2]_2$ (**194**) in the solid state

The asymmetric unit of the centrosymmetric dimer consists of a mesityl group σ-bonded via C_{ipso} to copper [Cu–C(1) 1.913(5) Å] and a dibenzylamido group μ_2-bridge bonded between copper and lithium [Cu–N 1.753(2) Å and Li–N 1.921(4) Å]. The C(1)–Cu–N bond angle is close to linear [C(1)–Cu–N 175.3(2)°]. Dimerization occurs via an η^6-interaction of the mesityl group of the adjacent monomeric unit with lithium (Figure 92) with Li–C bond distances ranging from 2.334(10) to 2.445(10) Å.

The arylamidocuprate 2-$(Ph_3P)C_6H_4CuNHMes$ (**195**) was obtained in 55% yield from a reaction mixture containing $Cu(SCN)_2$, LiNHMes and Ph_4PCl in 1,2-dimethoxyethane as the solvent, which was kept under reflux conditions for 2 hours. The molecular geometry of **195** in the solid state was established by an X-ray crystal structure determination (Figure 93)[235]. In **195**, the 2-(triphenylphosphonium)phenyl unit is σ-bonded via C_{ipso} to copper [Cu–C(1) 1.895(6) Å] while the amido anion MesNH is σ-bonded via N to copper [Cu–N 1.849(5) Å] in an almost linear fashion [C(1)–Cu–N 175.1(3)°]. This compound has a formal negative charge on copper (which, in fact, resides partly on the C_{ipso} and the amido-N centers) and a formal positive charge at the phosphorus atom. Interestingly, the cationic phosphorus center in this monometallic structure can be considered as assuming the function that the lithium cation has in common heterocuprate structures.

It should be noted that the mechanistic aspects of the formation of **195** from the abovementioned reaction mixture are not understood yet. It is obvious, however, that in one of the reaction steps a reduction of copper to its monovalent oxidation state must have taken place. Most likely, *ortho*-lithiation of one of the phenyl groups of the tetraphenylphosphonium cation prior to transmetallation is the primary step in the formation of **195**.

From the reaction of MeLi with t-Bu_2PCu in THF as the solvent, the monomeric methylphosphidocuprate $MeCuLiP(Bu\text{-}t)_2(THF)_3$ (**196**) was isolated as a crystalline material[236]. This compound represents a unique and early example of a neutral heterocuprate. The X-ray crystal structure determination of **196** revealed its monomeric nature with the methyl group σ-bonded to copper [Cu–C 1.940(4) Å] and the phosphorus atom of the phosphido group μ_2-bridge bonded between copper and lithium [Cu–P 2.217(2) and Li–P 2.54(1) Å] (Figure 93). The coordination geometry at copper is close to perfectly linear,

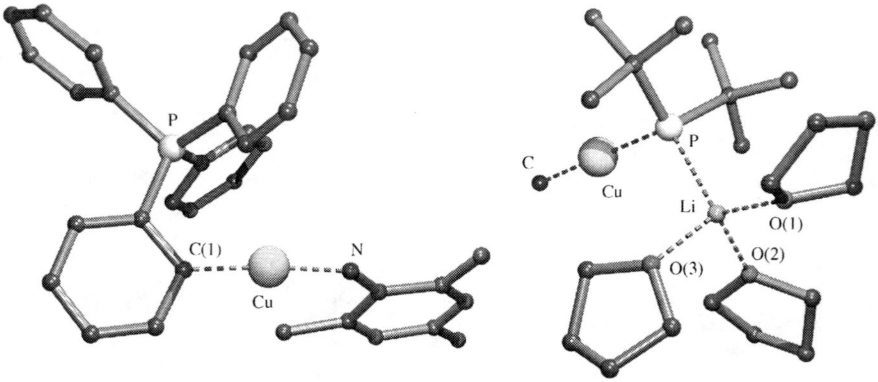

(**195**) (**196**)

FIGURE 93. Molecular geometry of 2-(Ph_3P)C_6H_4CuNHMes (**195**) and MeCuLiP(Bu-*t*)$_2$(THF)$_3$ (**196**) in the solid state

as is indicated by the C–Cu–P bond angle of 179.0(3)°. As a result of three additional coordinate-bonded THF molecules, the lithium atom has a tetrahedral coordination geometry.

C. Ionic Structures

For a better understanding of the mechanisms which give organocuprates their special reactivity properties, knowledge about the structures of such species in solution is of prime importance. It should be noted that the various structures in the solid state of the neutral organocuprates discussed above do not necessarily represent their actual structure when dissolved. It has now been well established that in solution organocuprates usually exist as neutral aggregates or contact ion pairs (CIP) in equilibrium with solvent-separated ion pairs (SSIP), as shown schematically (Figure 94)[222]. It is obvious that in strongly coordinating solvents like THF, this equilibrium will be in favor of SSIP structures whereas in apolar or weakly coordinating solvents like diethyl ether neutral or CIP structures are favored. Also, the addition of ligands that have a strong affinity to complex the lithium cation, such as [12]crown-4, will shift the equilibrium in favor of the SSIP structure. From various studies, in paticular NMR spectroscopic and computational studies (see References 19, 20 and 222 and references cited therein), it became clear that CIP structures are the active species in organic reactions with organocuprate reagents. For example, the reaction of Me_2CuLi with 4-methylcyclohexenone in diethyl ether as the solvent afforded the expected 1,4-addition product in 90% yield. When the same reaction was carried out

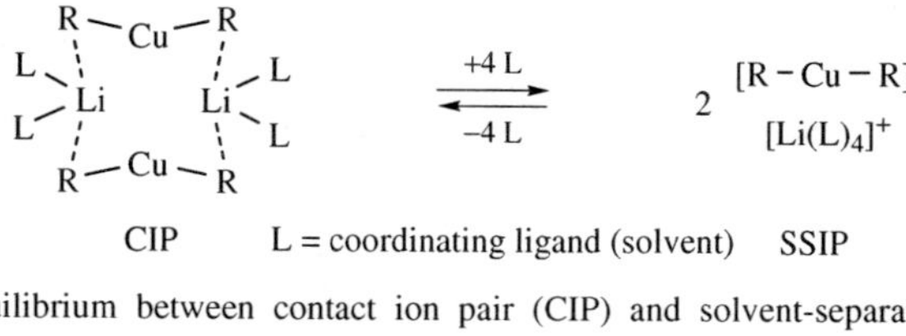

FIGURE 94. Equilibrium between contact ion pair (CIP) and solvent-separated ion pair (SSIP) structures

in the presence of 2.2 equivalents of [12]crown-4, which shifts the CIP/SSIP equilibrium completely to the side of the SSIP structures, no 1,4-addition product was formed at all[237]. This structure–activity relationship was furthermore supported by systematic studies of logarithmic reactivity profiles of reactions of organocuprates with 2-cyclohexenone and iodocyclohexane[238–241].

Ionic diorganocuprates exhibiting SSIP structures in the solid state were isolated by recrystallization from strongly polar solvents or recrystallized in the presence of coordinating ligands. The following compounds were characterized structurally by X-ray crystallography: $[Li(DME)_3][Me_2Cu]$ (**197**)[222] (DME = 1,2-dimethoxyethane), $[Li([12]crown\text{-}4)_2][Me_2Cu]$ (**198**)[242], $[Cu(PMe_3)_4][Me_2Cu]$ (**199**)[243], $[Li([12]crown\text{-}4)_2][Ph_2Cu]$ (**200**)[242], $[Ph_3Ca_2(THF)_6][Ph_2Cu]$ (**201**)[244], $[Li(DME)_3][((Me_3Si)_2CH)_2Cu]$ (**202**)[222],

TABLE 2. Relevant structural data of SSIP structures **197**–**205**

Compound	Anion	Cation	Cu–C (Å)	C–Cu–C (°)
197	$[Me_2Cu]$	$[Li(DME)_3]$	1.929(3)	179.2(1)
198	$[Me_2Cu]$	$[Li([12]crown\text{-}4)_2]$	1.935(8)	180
199	$[Me_2Cu]$	$[Cu(PMe_3)_4]$	1.93(1)	180
200	$[Ph_2Cu]$	$[Li([12]crown\text{-}4)_2]$	1.925(10)	178.5(4)
201	$[Ph_2Cu]$	$[Ph_3Ca_2(THF)_6]$	1.910(3)	180.0(2)
202	$[((Me_3Si)_2CH)_2Cu]$	$[Li(DME)_3]$	1.935(5)	178.9(2)
203	$[((Me_3Si)_3C)_2Cu]$	$[Li(THF)_4]$	2.027(7)	180
204	$[Mes_2Cu]$	$[Cu(DPPE)_2]$	1.915(9)	180.0(7)
205	$[((Me_3Si)_2CH)CuBr]^a$	$[Li([12]crown\text{-}4)_2]$	1.920(6)	178.7(2)

[a] Cu–Br 2.267(2) Å.

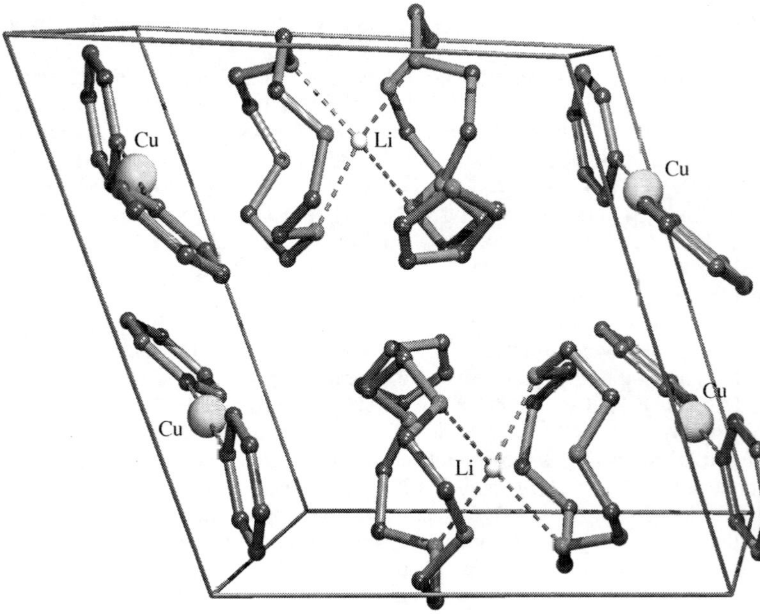

FIGURE 95. Unit cell of crystalline $[Li([12]crown\text{-}4)_2][Ph_2Cu]$ (**200**)

$[Li(THF)_4][((Me_3Si)_3C)_2Cu]$ (**203**)[245], $[Cu(DPPE)_2][Mes_2Cu]$ (**204**)[246] (DPPE = 1,2-bis (diphenylphosphino)ethane) and $[Li([12]crown\text{-}4)_2][(Me_3Si)_2CHCuBr]$ (**205**)[242]. The relevant structural parameters are compiled in Table 2; the solid state structure of **200** is shown as an example (Figure 95).

All compounds have in common that their structure in the solid state consists of isolated $[R\text{-}Cu\text{-}R]^-$ anions and isolated cations, usually solvated lithium cations, in the crystal lattice. In the anions the copper atoms have a linear digonal coordination geometry with relatively short (<2.00 Å) copper–carbon bond distances. In this respect it should be noted, however, that the calculated Cu–C bond distance in **203** is not very reliable due to crystallographic disorder of the copper atom[245].

Cyclopentadienylcopper triphenylphosphine undergoes a nucleophilic substitution reaction when treated with two equivalents of fluorenyllithium. This reaction resulted in the formation of a unique ionic cuprate $[Li(THF)_4][(fluorenyl)_2Cu(PPh_3)]$ (**206**) (equation 25)[176].

$$CpCu(PPh_3) + 2\,(fluorenyl)Li \xrightarrow[THF]{} \underset{\mathbf{(206)}}{[Li(THF)_4][(fluorenyl)_2Cu(PPh_3)]} + CpLi \qquad (25)$$

An X-ray crystal structure determination of **206** revealed its ionic nature by the presence of isolated $[(fluorenyl)_2Cu(PPh_3)]^-$ anions and $[Li(THF)_4]^+$ cations in the crystal lattice. The copper atom in the anion has a trigonal-planar coordination geometry as the result of two σ-bonded fluorenyl groups [Cu–C(1) 2.079(9) and Cu–C(2) 2.170(9) Å] and a coordinated triphenylphosphine molecule [Cu–P 2.223(3) Å] (Figure 96). To our knowledge the latter coordination geometry is unprecedented for ionic organocuprate structures.

A remarkable structure was found for the compound obtained from the reaction of $2,6\text{-}Mes_2C_6H_3Li$ with excess CuI in THF as the solvent. The product $[Li(THF)_4][2,6\text{-}Mes_2C_6H_3Cu_2I_2]$ (**207**) combines an organocopper copper iodide aggregate with an ionic 'ate' structural motif[233].

An X-ray crystal structure determination showed the presence of distinct, isolated $[2,6\text{-}Mes_2C_6H_3Cu_2I_2]^-$ anions and $[Li(THF)_4]^+$ cations in the crystal lattice. In the anion of **207** the aryl group is two-electron three-center bonded via C_{ipso} between the two copper atoms in a slightly asymmetric manner [Cu(1)–C(1) 1.967(12) and Cu(2)–C(1) 1.974(15) Å] (Figure 96). To each of the copper atoms an iodine atom is end-on bonded [Cu(1)–I(1) 2.414(2) and Cu(2)–I(2) 2.423(2) Å]. The C(1)–Cu–I bond angles deviate from linear [C(1)–Cu(1)–I(1) 164.1(4)° and C(1)–Cu(2)–I(2) 166.8(4)°]. As expected for two-electron three-center bonded aryl groups, the aryl group in **207** is orientated perpendicular to the Cu(1)–Cu(2) vector.

From the reaction of PhLi with copper salts under various conditions, several aggregated ionic cuprates were isolated. Their structures in the solid state were characterized by X-ray crystallography. These ionic cuprates are $[Li(THF)_4][Ph_6Cu_5]$ (**208a**)[247], $[Li(PMDTA)(THF)][Ph_6Cu_5]$ (**208b**)[247], $[Li(OEt_2)_4][Ph_6Cu_4Li]$ (**209**)[227] and $[Li_4Cl_2(Et_2O)_{10}][Ph_6Cu_3Li_2]_2$ (**210**)[248]. The anionic organocuprate moieties in these aggregates have a structural motif that is closely related to that of the neutral organocopper magnesium cuprate Ph_6Cu_4Mg[227] (*vide supra,* **185**). This structural motif comprises a trigonal bipyramidal arrangement of the five metals, i.e. five copper atoms in **208**, three copper atoms at the equatorial positions and one copper atom and one lithium atom residing at the axial sites of a trigonal bipyramid in **209**, and three copper atoms at the equatorial positions and both lithium atoms at the axial sites in **210** (Figure 97, **A**). In these anionic aggregates, all six phenyl groups are two-electron three-center bonded via C_{ipso} between two metals over the six equatorial–axial edges of the trigonal bipyramid. The

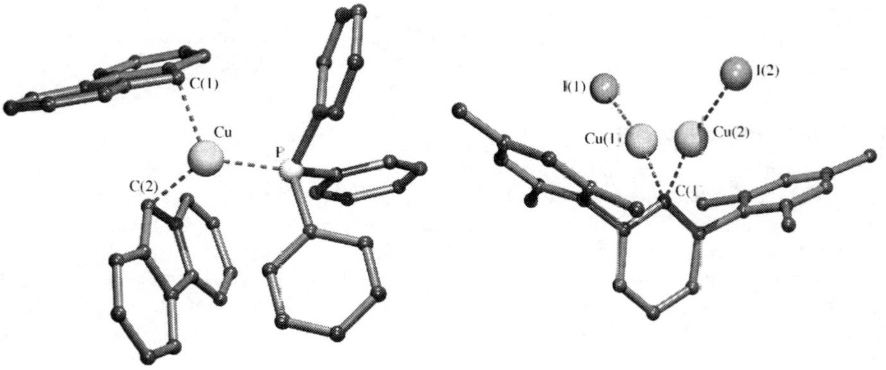

(206) **(207)**

FIGURE 96. Molecular geometry of the $[(fluorenyl)_2Cu(PPh_3)]^-$ anion of $[Li(THF)_4][(fluorenyl)_2Cu(PPh_3)]$ (**206**) and $[2,6\text{-}Mes_2C_6H_3Cu_2I_2]^-$ anion of $[Li(THF)_4][2,6\text{-}Mes_2C_6H_3Cu_2I_2]$ (**207**) in the solid state

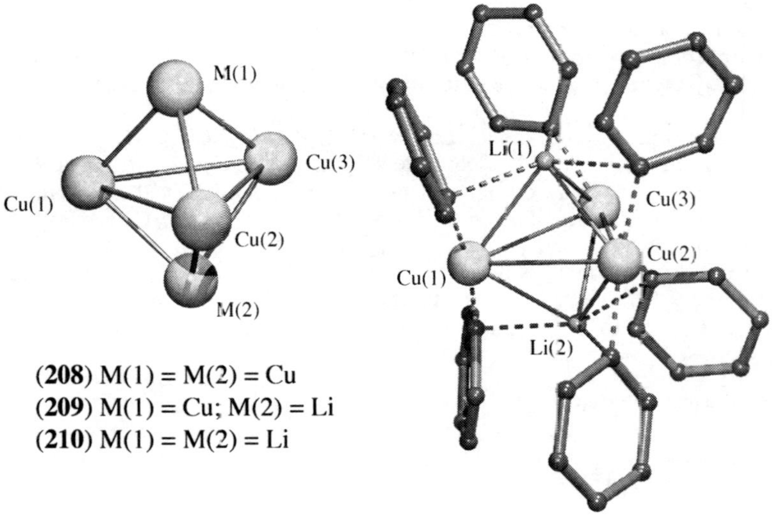

(**A**) (**B**) (**210**)

FIGURE 97. (**A**) The trigonal bipyramidal array of metal atoms in **208**–**210**. (**B**) Molecular geometry of **210** in the solid state

bond distances between the respective C_{ipso} atoms and the equatorial copper atoms [average $Cu_{equatorial}$–C: in **208a** 1.99(2) Å, in **208b** 1.96(4) Å, in **209** 1.95(1) Å and in **210** 1.923(6) Å] are systematically shorter than the distance between the C_{ipso} atom and the axial metal atom [average Cu_{axial}–C: in **208a** 2.19(2) Å, in **208b** 2.14(3) Å, in **209** 2.33(3) and Li_{axial}–C 2.16(4) Å, in **210** Li_{axial}–C 2.257(13) Å]. As a result of the bonding of the six phenyl groups to the trigonal bipyramidal array of five metal atoms, the equatorial copper atoms have a linear C_{ipso}–Cu–C_{ipso} coordination geometry while the two axially positioned metal atoms have a trigonal planar $M(C_{ipso})_2$ coordination geometry (Figure 97, **B**) (however, note that each of the metal atoms are six-coordinate).

D. Cyanocuprates

The addition of either one or two equivalents of an organolithium reagent to CuCN results in the formation of the so-called 'lower-order' or 'higher-order' cyanocuprates, respectively (Figure 98)[249], of which the importance as a synthetic tool in organic chemistry has been well established[16,250–252]. Soon after its discovery, especially for the 'higher-order' cyanocuprates, a debate started concerning the actual constitution of this reagent in solution. Two models to describe its structure have been put forward: (i) a bis-anionic species in which the two organic groups and the cyanide anion are bound to copper while the two lithium cations are separated from this dianion (**A** in Figure 98), and (ii) a cyano-Gilman cuprate in which the two organic groups are bound to copper and the cyanide is bridge bonded to two lithium cations (**B** in Figure 98). In this respect it should be noted that it was already demonstrated in an early stage by ^{13}C labeling and ^{13}C NMR studies that in the 'lower-order' cyanocuprates the organic and the cyanide group are bound to the same copper atom[253].

^{1}H and ^{13}C NMR studies[254–256], EXAFS and XANES investigations[257–260] and computational studies[259,261,262] strongly indicated that in the 'higher-order' cyanocuprates the cyanide is not bound to copper and a structure like **B** (Figure 98) would be more realistic. This controversy came to an end in 1999[263] in favor of model **B** when the results of the first X-ray crystal structure determinations of two 'higher-order' cyanocuprates were reported.

The 'higher-order' cyanocuprate $[Li_2CN(THF)_4][(2\text{-}Me_2NCH_2C_6H_4)_2Cu]$ (**211**) was obtained from the reaction of two equivalents $2\text{-}Me_2NCH_2C_6H_4Li$ with CuCN in THF as the solvent[264]. An X-ray crystal structure determination showed this compound to exist as a linear (zigzag) polymeric chain in which $[(2\text{-}Me_2NCH_2C_6H_4)_2Cu]^-$ Gilman cuprate anionic moieties and $[Li_2CN(THF)_4]^+$ cations are distinguishable (Figure 99). The observed structural motif is fully in agreement with that of the earlier proposed structure **B** (Figure 98). In the anionic moiety, the two aryl groups are σ-bonded via C_{ipso} to copper [Cu–C(1) 1.918(2) Å] in a perfect linear arrangement [C(1)–Cu–C(1a) 180°]. In

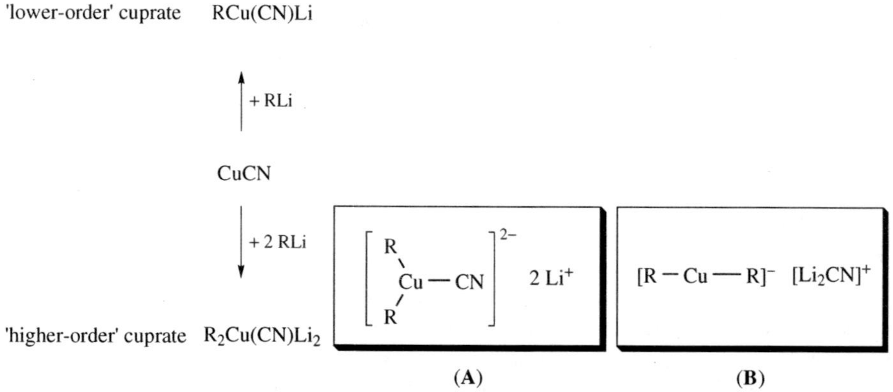

FIGURE 98. Formation of 'higher-order' and 'lower-order' cyanocuprates and proposed structures for the 'higher-order' cyanocuprates

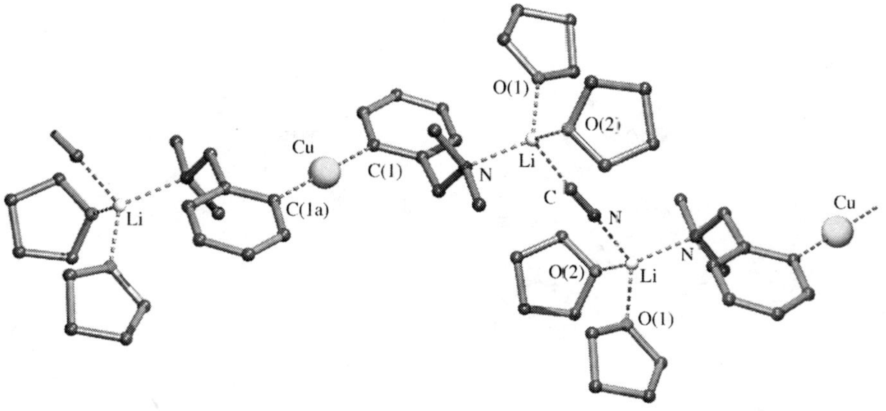

FIGURE 99. Part of the polymeric chain of $[Li_2CN(THF)_4][(2-Me_2NCH_2C_6H_4)_2Cu]$ (**211**) in the solid state. Note that the positions of the C and N in the cyanide are chosen arbitrarily (see text)

the cationic $[Li_2CN(THF)_4]$ fragment the cyanide group is *C*,*N*-bridge bonded between two lithium atoms. Due to a crystallographic inversion center located at the center of the C≡N bond, the nitrogen and carbon atoms are indistinguishable and thus the C−Li and N−Li bond distances are found equal, 2.090(4) Å. To each of the lithium atoms two THF molecules are coordinate bonded [Li−O(1) 1.925(4) and Li−O(2) 2.002(4) Å]. The polymeric (zigzag) chain is formed due to coordination of the nitrogen atom of the *ortho*-Me_2NCH_2 substituent to a lithium atom in the neighboring cationic $[Li_2CN(THF)_4]$ fragment [Li−N 2.090(4) Å]. Aggregation state studies by cryoscopy in THF and conductivity measurements in THF pointed to the presence of solvent-separated ions in THF solution (SSIP structure). However, addition of benzene to a solution of **211** in THF resulted in the formation of a precipitate (LiCN) and the neutral organocuprate $(2-Me_2NCH_2C_6H_4)_4Cu_2Li_2$ having a CIP structure (*vide supra*, see **172**).

The only other example of a 'higher-order' cyanocuprate that was characterized structurally in the solid state by X-ray crystallography is $[Li_2CN(PMDTA)(THF)][t\text{-}Bu_2Cu]$ (**212**)[265]. That this cyanocuprate exists in the solid state as a real SSIP structure is indicated by the presence of isolated $[t\text{-}Bu_2Cu]^-$ anions and $[Li_2CN(PMDTA)(THF)]^+$ cations in the crystal lattice (Figure 100). In the $[t\text{-}Bu_2Cu]^-$ anion the two *tert*-butyl groups are σ-bonded via C_{ipso} to copper [Cu−C(1) 1.957(4) Å] and, like in **211**, the coordination geometry at the copper atom is perfectly linear [C(1)−Cu−C(1a) 180°]. In the $[Li_2CN(PMDTA)(THF)]^+$ cation the cyanide group is *C*,*N*-bridge bonded between the two lithium atoms. But also here the center of the C≡N bond is located on a crystallographic inversion center making the carbon and nitrogen atoms of the cyanide group indistinguishable, and consequently the C−Li and N−Li bond distances are found equal, 2.105(7) Å. To each of the lithium atoms an additional THF molecule is coordinate bonded via its oxygen atom (2.024(8) Å), and all three nitrogen atoms of the PMDTA ligand are involved in coordination to lithium [Li−N(1) 2.104(8), Li−N(2) 2.270(9) and Li−N(3) 2.356(7) Å]. The coordination geometry at lithium can be described as trigonal bipyramidal, with the oxygen atom and N(1) and N(3) at the equatorial sites and the cyanide group and N(2) at the axial positions.

For the structures of the 'lower-order' cyanocuprates $(Me_2PhSi)_3CCu(CN)Li(THF)_3$ (**213**)[266] and $2,6\text{-}Mes_2C_6H_3Cu(CN)Li(THF)_3$ (**214**)[267] in the solid state, X-ray crystal structure determinations revealed distinct monomeric structures with the organic moiety

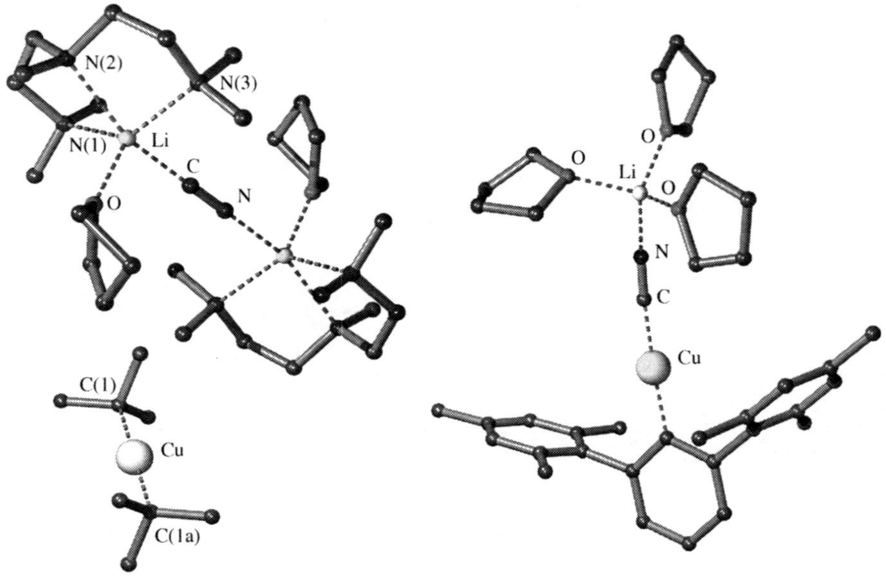

(**212**) (**214**)

FIGURE 100. The isolated $[t\text{-Bu}_2\text{Cu}]^-$ anion and $[\text{Li}_2\text{CN(PMDTA)(THF)}]^+$ cation in the unit cell of $[\text{Li}_2\text{CN(PMDTA)(THF)}][t\text{-Bu}_2\text{Cu}]$ (**212**) and molecular geometry of 2,6-$\text{Mes}_2\text{C}_6\text{H}_3\text{Cu(CN)Li(THF)}_3$ (**214**) in the solid state

σ-bonded via C_{ipso} to copper and the cyanide group bonded via its carbon atom to copper in a linear fashion [C−Cu−C; in **213** 175.2(6)° and in **214** 173.5(2)°] (Figure 100). In both structures the nitrogen atom of the cyanide group forms a bond to the lithium atom of a $[\text{Li(THF)}_3]$ moiety. Actually, in contrast to the ionic structures found for the 'higher-order cuprates' the 'lower-order cuprates' seem to have neutral structures in the solid state.

Neutral dimeric structures, as shown schematically (Figure 101), were found by X-ray crystallography for the 'lower-order' cyananocuprates: $[t\text{-BuCu(CN)Li(OEt}_2)_2]_2$ (**215**)[265], $[\text{Me}_3\text{SiCH}_2\text{Cu(CN)Li(OEt}_2)_2]_2$ (**216**)[268], $[(\text{Me}_3\text{Si})_3\text{CCu(CN)Li(THF)}_2]_2$ (**217**)[269], $[\text{Me}_2\text{N(Me)}_2\text{SiC(SiMe}_3)_2\text{Cu(CN)Li(THF)}_2]_2$ (**218**)[269], $[(\text{PhMe}_2\text{Si})_2(\text{Me}_3\text{Si})\text{CCu(CN)-Li(THF)}_2]_2$ (**219**)[266] and $[2,6\text{-Tip}_2\text{C}_6\text{H}_3\text{Cu(CN)Li(THF)}_2]_2$ (**220**)[270] (Tip = 2,4,6-$(i\text{-Pr})_3$-C_6H_2). They all exhibit the same structural motif consisting of two linear, monoanionic [R-Cu-CN] units linked via the nitrogen atom of the cyanide group by bridging between the two $[\text{LiL}_2]$ cations, thus forming a dimeric structure. As an example, the molecular geometry of **215** is shown (Figure 101). In the monoanionic [t-Bu-Cu-CN] fragments of **215** the *tert*-butyl group is σ-bonded via C_{ipso} to copper [Cu−C(1) 1.969(7) Å] and the cyanide group is bonded to copper via its carbon atom [Cu−C(2) 1.878(8) Å] in a linear arrangement [C(1)−Cu−C(2) 170.0(3)°]. The nitrogen atoms of two cyanide groups are each μ_2-bridge bonded between two lithium atoms forming a central flat N_2Li_2 four-membered ring. The bonding of nitrogen to the two lithium atoms is slightly asymmetric [N−Li(1) 2.149(13) and N−Li(2) 2.050(13) Å]. To complete a tetrahedral coordination geometry at each lithium atom two diethyl ether molecules are coordinate bonded [Li−O(1) 1.970(14) and Li−O(2) 1.964(13) Å].

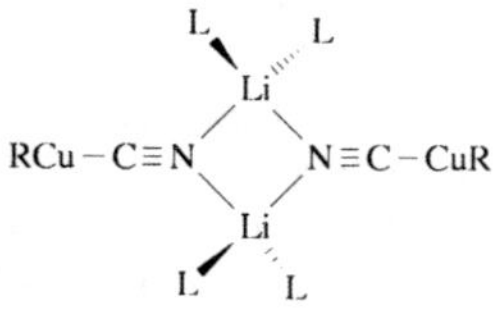

(215) R = *t*-Bu; L = Et_2O
(216) R = Me_3SiCH_2; L = Et_2O
(217) R = $(Me_3Si)_3C$; L = THF
(218) R = $Me_2N(Me)_2SiC(SiMe_3)_2$; L = THF
(219) R = $(PhMe_2Si)_2(Me_3Si)C$; L = THF
(220) R = 2,6-$Tip_2C_6H_3$; L = THF

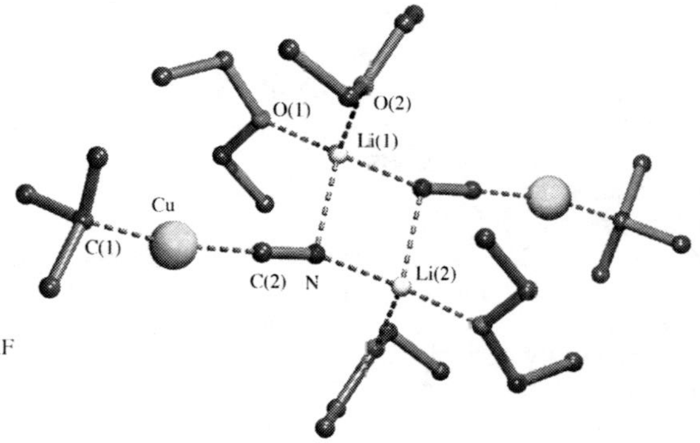

(215)

FIGURE 101. The 'lower-order' cyanocuprates **215–220** and molecular geometry of **215** in the solid state

The potassium cyanocuprate [$(PhMe_2Si)_2(Me_3Si)CCu(CN)K$] (**221**)[266] has an entirely different structure than the corresponding lithium derivative **219**. An X-ray crystal structure determination of **221** showed this compound to exist as a tetrameric aggregate with a heterocubane structural motif, as shown schematically (Figure 102). The tetramer is built up of four linear, monoanionic [$(PhMe_2Si)_2(Me_3Si)C)Cu(CN)$] units with the nitrogen atoms of the cyanides and the four potassium atoms alternating at the corners of the N_4K_4-cube.

A crystalline material with formula [$Cu_2Li_4(CN)_4(PPh_3)_4(THF)_{10}$][$Ph_2Cu$] (**222**) was obtained from the reaction of PhLi with CuCN in the presence of one equivalent of PPh_3 in THF as the solvent. An X-ray crystal structure determination of this material showed the presence of isolated $[Ph_2Cu]^-$ anions with a linear structure and $[Cu_2Li_4(CN)_4(PPh_3)_4(THF)_{10}]^{2+}$ cations in the crystal lattice[271]. The solid state structure of **222** is shown schematically (Figure 102). This cyanocuprate has an ionic structure with a cation that exhibits a strikingly complex coordination network structure containing both copper and lithium and a seemingly simple cuprate anion.

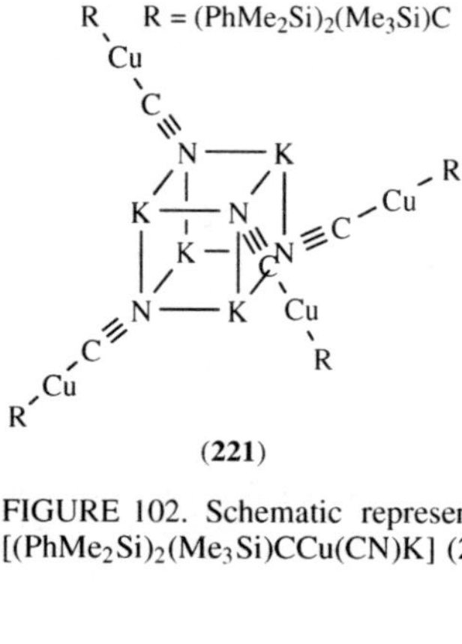

(221)

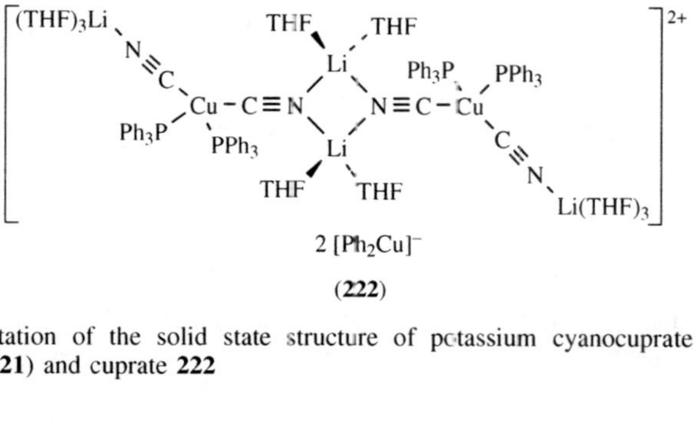

(222)

FIGURE 102. Schematic representation of the solid state structure of potassium cyanocuprate [$(PhMe_2Si)_2(Me_3Si)CCu(CN)K$] (**221**) and cuprate **222**

VIII. ORGANOCOPPER(II) AND ORGANOCOPPER(III) COMPOUNDS

The inorganic and coordination chemistry of copper covers various oxidation states, Cu(I) (d^{10}), Cu(II) (d^9), Cu(III) (d^8) and Cu(IV) (d^7), of which the last one is rarely encountered in complexes. The greater part of the compounds have copper in the formal Cu(II) oxidation state. The organometallic chemistry of copper is essentially dominated by organocopper compounds with copper in the formal Cu(I) oxidation state; see other chapters. It is only thanks to the use of special organic groups and/or ligand environments that so far also a few examples of organocopper compounds could be isolated and characterized structurally with copper in an oxidation state higher than +1.

In the early days of organocopper chemistry organocopper(II) compounds were regarded as intrinsically unstable. However, the existence and key role of organocopper(III) species as transient species or intermediates in various reactions has been frequently proposed, e.g. in cross-coupling reactions of copper(I) carboxylates with organic halides[272], in conjugate addition reactions of cuprates to enones[273], in copper(I)-induced Ullmann coupling reactions[274] and in reactions of dialkyl- and diarylcuprates with organic halides[275]. That organocopper(III) intermediates, formed during such reactions, could be relatively stable indeed was supported by computational studies[21, 276–278].

It is only recently that the presence of such species in solution (at low temperature) could be established by NMR spectroscopy. It must be noted that both Cu(I) (d^{10}) and Cu(III) (d^8) are diamagnetic species and are NMR observable. By making use of rapid injection NMR techniques, the organocuprates (**223**) (X = I, CN, SCN or SPh) (Figure 103) with copper in its trivalent oxidation state were identified in the reaction mixture of a Gilman reagent Me_2Cu^ILiLiX with EtI at low temperature (−100 °C)[279]. It is exactly these species that were proposed before as intermediates in the reaction of the Gilman reagent with alkyl halides[275]. Moreover, the neutral triorganocopper(III) complexes $EtMe_2Cu^{III}L$ (**224**) were observed and characterized when this reaction was carried out in the presence of neutral donor ligands like tertiary amines, phosphines or phosphites[280].

Using ^{13}C-labeled CN-lithium salts, the formation of $Li[Me_3Cu^{III}CN]$ (**225**) (Figure 103), which was prepared from MeLi, $Cu^{13}CN$ and MeI, could be followed while the structure of **225** was studied in detail by NMR spectroscopy at low temperatures[281]. The intermediacy of trialkylcopper(III) species **226** (Figure 103) was observed making use of the rapid injection NMR technique during the reaction of $Me_2Cu^ILiLi^{13}CN$ with 2-cyclohexenone in the presence of Me_3SiCl at −100 °C[282]. Accordingly, species **226** has been proposed as intermediate, formed prior to the reductive elimination step, in this conjugate addition[276].

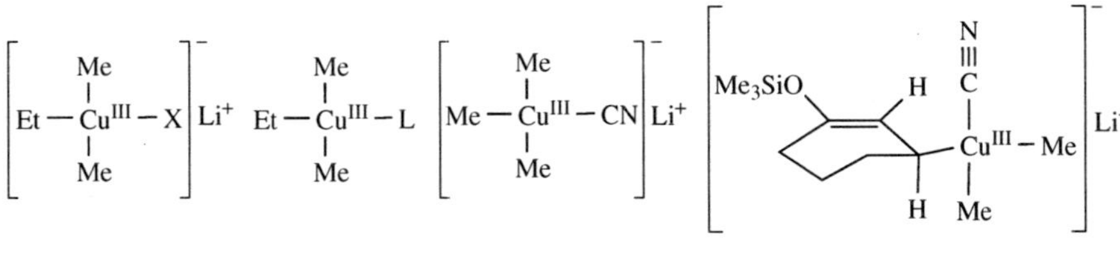

(**223**) (**224**) (**225**) (**226**)

FIGURE 103. The alkylcopper(III) species **223**–**226** identified and studied in solution at low temperature

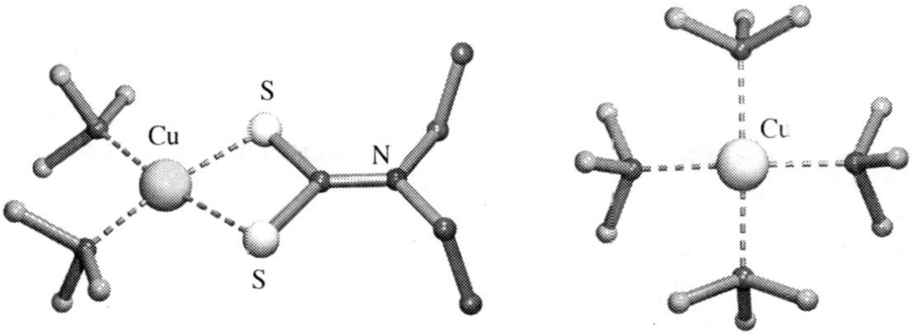

(228) (230)

FIGURE 104. Molecular geometry of $(CF_3)_2Cu^{III}S_2CNEt_2$ (**228**) and the anion $[(CF_3)_4Cu^{III}]^-$ of **230** in the solid state

In the early days of organocopper chemistry it was recognized that perfluoroalkylcopper(I) compounds have a much enhanced thermal stability as compared to their non-fluorinated hydrocarbon analogs. As new routes for the introduction of fluorine-containing functional groups are important for a range of applications, the reactivity of these perfluoroalkylcopper(I) reagents were studied extensively, especially their use in synthetic organic chemistry[283,284]. It appeared that in contrast to the CF_3Cu^IL (L = metal halide) system, which is relatively stable towards oxidation, the cuprate $[CdI][(CF_3)_2Cu^I]$ (**227**) is readily oxidized by a variety of oxidants to stable organocopper(III) compounds. Oxidation of **227** by *N,N*-diethylthiuram disulfide afforded $(CF_3)_2Cu^{III}S_2CNEt_2$ (**228**), which represents the first example of an isolated organocopper(III). The structure of **228** was unambiguously established by an X-ray crystal structure determination (Figure 104)[31,285]. The molecular geometry of **228** comprises a copper atom to which two CF_3 groups are σ-bonded [Cu–C both 1.907(5) Å]. The monoanionic *N,N*-diethyldithiocarbamate group is *S,S*-chelate bonded to the copper atom [Cu–S both 2.205(1) Å]. The coordination geometry at copper can be described as slightly distorted square-planar with the two CF_3 groups in *cis*-position. The sum of the bond angles around copper is 360° within experimental error, consistent with a square-planar coordination geometry of the copper atom. However, the individual bond angles slightly deviate [S–Cu–C both 93.2(1)°, C–Cu–C 94.5(2)° and S–Cu–S 79.32(5)°] from their ideal values (90°), but it is a logical consequence of the small bite angle of the chelating dithiocarbamate group.

A very similar structure was observed in the solid state for the perfluoroethyl analog $(C_2F_5)_2Cu^{III}S_2CNEt_2$ (**229**)[286].

The formation of $[(CF_3)_4Cu^{III}]^-$ anions, in which the copper atom has the 3+ oxidation state, are further examples of the oxidation reaction of an organocuprate(I), i.e. $[CdI][(CF_3)_2Cu^I]$ with various oxidants like O_2, XeF_2, I_2, Br_2, Cl_2 or ICl[31,287]. Addition of bulky cations like n-Bu_4N^+, Ph_4P^+ or PNP^+ (PNP = $[Ph_3PNPPh_3]^+$) to such reaction mixtures afforded the corresponding organocuprate(III) salts as crystalline materials[287]. The structure of $[PNP][(CF_3)_4Cu^{III}]$ (**230**) in the solid state was established by an X-ray crystal structure determination. Its solid state structure comprises isolated [PNP] cations and $[(CF_3)_4Cu^{III}]^-$ anions in the crystal lattice[287]. The anions consist of a copper atom to which four CF_3 groups are σ-bonded [Cu–C 1.946(18), 1.984(14), 1.955(18) and 1.981(18) Å] via its carbon atoms (Figure 104). The C–Cu–C bond angles around copper are all close to 90° [ranging from 89.1(8)° to 91.4(6)°], pointing to a slightly distorted square-planar coordination geometry.

A crystalline material (BEDT-TTF)$_2$Cu^{III}(CF$_3$)$_4$(TCE) (**231**) (BEDT-TTF = bis(ethylenedithio)tetrathiafulvalene, TCE = 1,1,2-trichloroethane) exhibiting superconducting properties below 4 K has been prepared and was characterized structurally in the solid state by X-ray crystallography[288, 289]. The crystal lattice contains [BEDT-TTF]$_2$ radical cations and [(CF$_3$)$_4$Cu^{III}] anions. The structural features of the latter are closely related to that of **230**.

The paramagnetic Cu(II) complexes (**232**) derived from triazamacrocyclic ligands underwent a remarkable reaction in MeCN solution at room temperature (equation 26). From this reaction two diamagnetic products, present in a 1:1 molar ratio, could be identified: (i) the Cu(I) complex of the protonated triazamacrocyclic ligand and (ii) unique organocopper(III) complexes (**233**)[290, 291]. This reaction involves a C–H activation step, which is unprecedented in organocopper chemistry.

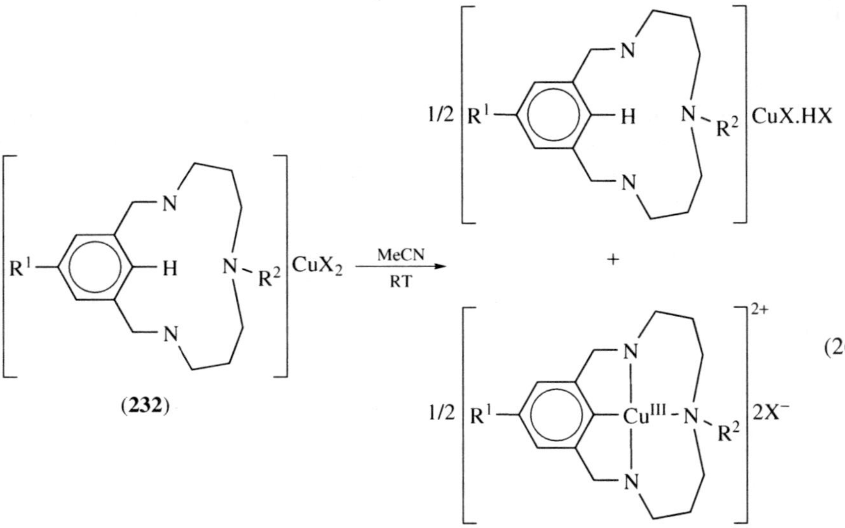

(**232**)

(26)

(**233a**) R^1 = H, R^2 = Me, X = ClO_4^-
(**233b**) R^1 = H, R^2 = H, X = $CF_3SO_3^-$
(**233c**) R^1 = Me, R^2 = H, X = $CF_3SO_3^-$
(**233d**) R^1 = Me, R^2 = Me, X = $CF_3SO_3^-$
(**233e**) R^1 = H, R^2 = H, X = $CF_3SO_3^-$/Cl$^-$

The molecular geometries of **233a**–**233e** in the solid state were established by X-ray crystal structure determinations[291]. These complexes have comparable structural features and therefore only details for **233a** are given below. In **233a** (Figure 105) the copper atom is σ-bonded to C(1) of the aryl group [Cu–C(1) 1.905(3) Å], while the three nitrogen atoms of the triazamacrocyclic ligand are all involved in coordination to copper [Cu–N(1) 1.905(3), Cu–N(2) 2.031(3) and Cu–N(3) 1.961(3) Å]. The copper atom, C(1), and the three nitrogen atoms lie in one plane, as is indicated by the sum of the bond angles around copper, which is 360° within experimental error. The two perchlorate anions are located above and below this plane with shortest Cu···O interactions with the oxygen atoms O(1) and O(2) of 2.428(3) and 2.584(3) Å, respectively. These relatively long Cu–O distances point to weak interactions and therefore the coordination geometry at copper could be best described as distorted square-planar.

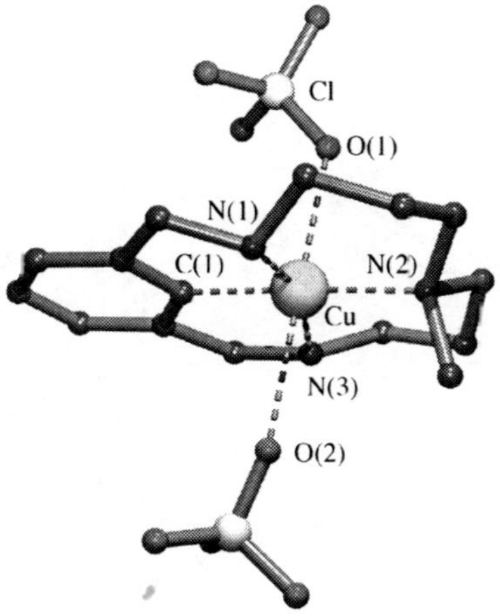

FIGURE 105. Molecular geometry of **233a** in the solid state

Reaction of tris(2-pyridylthio)methane (tptmH) with [Cu^{I}(MeCN)$_4$][PF$_6$] afforded the unique organocopper(II) complex [(tptm)Cu^{II}(MeCN)][PF$_6$] (**234**), again via C−H activation (equation 27)[292]. Addition of KF or n-Bu$_4$NX (X = Cl, Br or I) to a solution of **234** in MeCN gave the neutral complexes (tptm)Cu^{II}F (**235a**), (tptm)Cu^{II}Cl (**235b**), (tptm)Cu^{II}Br (**235c**) and (tptm)Cu^{II}I (**235d**), respectively (equation 27)[293].

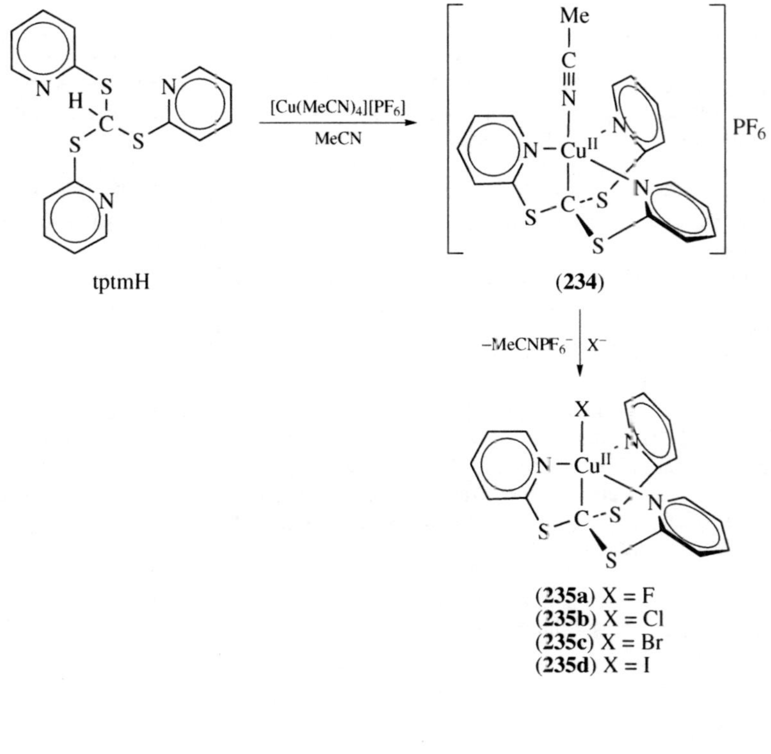

(27)

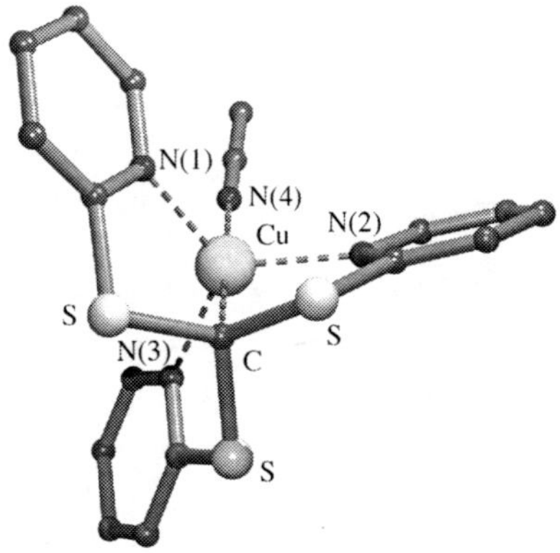

(**234**)

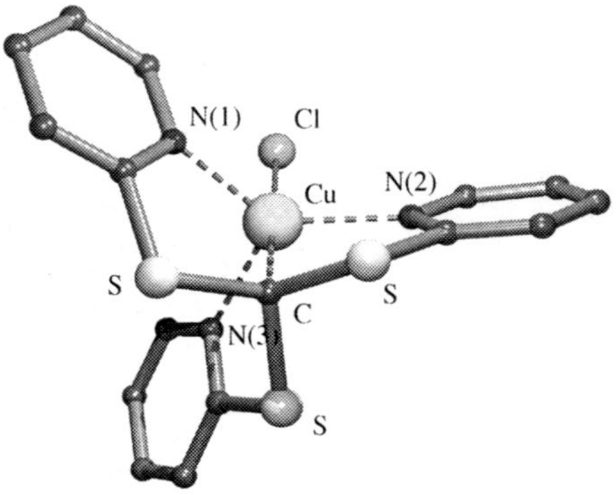

(**235b**)

FIGURE 106. Molecular geometry of the $[(tptm)Cu^{II}(MeCN)]^+$ cation of **234** and neutral (tptm) $Cu^{II}Cl$ (**235b**) in the solid state

The molecular geometry of **234** in the solid state was established by an X-ray crystal structure determination and shows isolated $[(tptm)Cu^{II}(MeCN)]^+$ cations and $[PF_6]^-$ anions in the crystal lattice[292]. The molecular geometry of the $[(tptm)Cu^{II}(MeCN)]^+$ cation comprises a copper atom with a trigonal-bipyramidal coordination geometry (Figure 106). The σ-bonded carbon atom [Cu−C 2.004(3) Å] and the coordinating nitrogen atom of the acetonitrile molecule [Cu−N(4) 2.028(3) Å] reside at the axial positions. The equatorial positions are occupied by the three coordinating nitrogen atoms [Cu−N(1) 2.074(3), Cu−N(2) 2.077(3) and Cu−N(3) 2.084(3) Å] of the three pyridyl groups.

The molecular geometries in the solid state of the neutral complexes $(tptm)Cu^{II}F$[293] (**235a**), $(tptm)Cu^{II}Cl$ (**235b**) and $(tptm)Cu^{II}I$[293] (**235d**) were also established by an X-ray crystal structure determination. These three complexes have similar overall structural features and are closely related to those of the $[(tptm)Cu^{II}(MeCN)]^+$ cation (Figure 106). In **235a, 235b** and **235d** the σ-bonded carbon atom and the halogen atom are at the axial positions of a trigonal-bipyramidal arrangement. The only significant difference between these three structures are the different copper−halogen bond lengths [Cu−F 1.874(7) Å in **235a**, Cu−Cl 2.321(3) Å in **235b** and Cu−I 2.717(1) Å in **235d**]. The observed structural parameters are in agreement with the values obtained from DFT calculation for these three complexes[293]. Moreover, these DFT calculations indicate that the spin density is located along the z-axis of the trigonal bipyramid, i.e. on the σ-bonded carbon atom, copper and the halogen atom, which is consistent with the observed EPR spectra.

Chemical oxidation with Ce(IV) or electrochemical oxidation of $(tptm)Cu^{II}Cl$ in the presence of PF_6^- afforded $[(tptm)Cu^{III}Cl][PF_6]$ (**236**), in which copper is oxidized to its trivalent oxidation state. An X-ray crystal structure determination of **236** showed the presence of isolated $[(tptm)Cu^{III}Cl]^+$ cations and $[PF_6]^-$ anions in the crystal lattice[294]. The molecular geometry of the $[(tptm)Cu^{III}Cl]^+$ cation is almost identical to that of neutral $(tptm)Cu^{II}Cl$. Only the copper−nitrogen bond distances in the equatorial plane are slightly shorter, which is most likely a consequence of a smaller ionic radius for Cu(III) versus Cu(II).

When $(tptm)Cu^{II}F$ was crystallized from dichloromethane/cyclohexane, containing trace amounts of water, a crystalline material with the composition $4[(tptm)Cu^{II}F].12(H_2O)$ was obtained. An X-ray crystal structure determination of this material showed it to be a two-dimensional clathrate hydrate sandwiched by planar arrays of the copper compound[295].

The *N*-confused porphyrin C_6F_5-NCP (equation 28) reacted with $Cu(OAc)_2$ to form the corresponding Cu(II) derivative $Cu^{II}(C_6F_5\text{-NCP})$ (**237**)[296]. Its structure in the solid state was established by an X-ray crystal structure determination. In **237**, the copper has a square-planar coordination geometry as the result of one σ-bonded carbon atom, one σ-bonded nitrogen atom and two coordinating nitrogen atoms. Due to crystallographic disorder as the result of space group symmetry, the position of the σ-bonded carbon atom could not be located and therefore only the average bond lengths between Cu(II) and the ligand atoms can be given [1.980(9) and 2.018(9) Å].

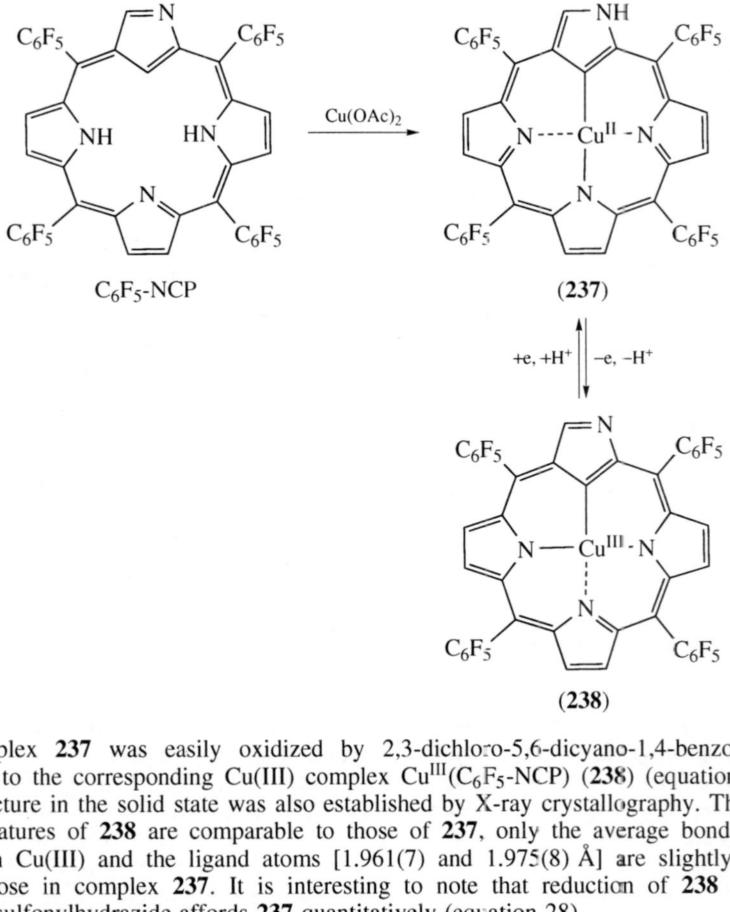

C_6F_5-NCP (**237**) (**238**) (28)

Complex **237** was easily oxidized by 2,3-dichloro-5,6-dicyano-1,4-benzoquinone (DDQ) to the corresponding Cu(III) complex $Cu^{III}(C_6F_5\text{-NCP})$ (**238**) (equation 28)[297]. Its structure in the solid state was also established by X-ray crystallography. The structural features of **238** are comparable to those of **237**, only the average bond lengths between Cu(III) and the ligand atoms [1.961(7) and 1.975(8) Å] are slightly shorter than those in complex **237**. It is interesting to note that reduction of **238** with *p*-toluenesulfonylhydrazide affords **237** quantitatively (equation 28).

The doubly *N*-confused porphyrins C_6F_5-*trans*-N_2CP and C_6F_5-*cis*-N_2CP (Figure 107) reacted with $Cu(OAc)_2$ to form diamagnetic complexes $Cu^{III}(C_6F_5\text{-}trans\text{-}N_2CP)$ (**239**) and $Cu^{III}(C_6F_5\text{-}cis\text{-}N_2CP)$ (**240**), respectively. The observed ^1H NMR spectra showed the absence of the inner CH and NH resonances, suggesting the presence of copper in its trivalent oxidation state in the core[298,299].

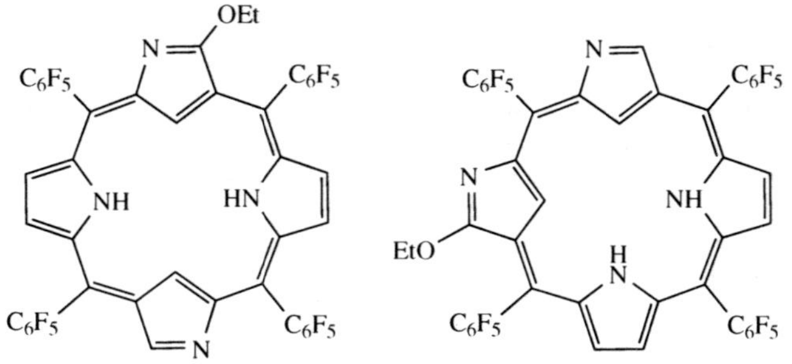

FIGURE 107. The doubly *N*-confused porphyrins C_6F_5-*trans*-N_2CP and C_6F_5-*cis*-N_2CP

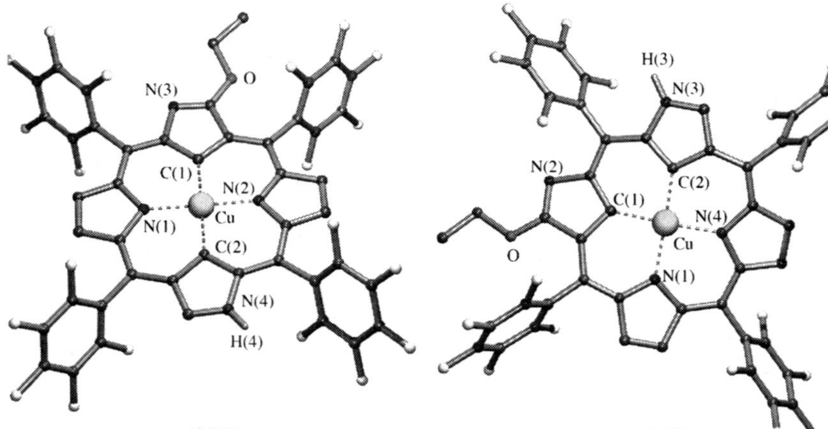

FIGURE 108. Molecular geometry of the organocopper(III) complexes Cu^{III}(C_6F_5-*trans*-N_2CP) (**239**) and Cu^{III}(C_6F_5-*cis*-N_2CP) (**240**)

For both **239** and **240** the structures in the solid state were established by X-ray crystal structure determinations (Figure 108)[298, 299]. In **239** the copper atom exhibits square-planar coordination geometry [Cu−C 1.940(7) and 1.944(7) Å, and Cu−N 1.965(6) and 1.967(6) Å]. The molecule is essentially planar with only a small deviation (0.06 Å) from the mean plane consisting of 25 core atoms. The molecules of **239** form polymeric linear chains in the crystal lattice via intramolecular hydrogen bonding between H(4) and N(3) of neighboring molecules.

For complex **240** a similar structure was found with square-planar coordinated copper [Cu−C 1.939(3) and 1.934(3) Å, and Cu−N 1.969(3) and 1.954(4) Å]. Also, this complex forms polymeric chains in the crystal lattice via intramolecular H(3) to N(2) hydrogen bridges of a neighboring molecule, but in this complex the chain has a zigzag arrangement.

The copper(II) complex **241** (Figure 109) was prepared from the reaction of the corresponding calic[4]phyrin derivative with $Cu(OAc)_2$. An ESR spectroscopic study of **241** points to the formation of an organocopper(II) compound with copper in its divalent

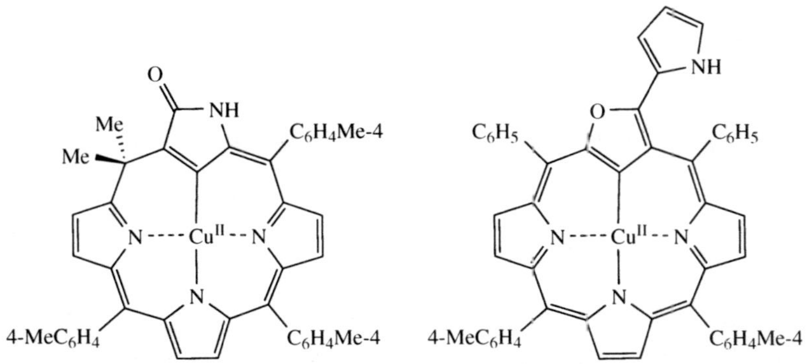

(**241**) (**242**)

FIGURE 109. Schematic representation of the molecular geometry of **241** and **242** in the solid state

oxidation state. The molecular geometry of **241** in the solid state was established by an X-ray crystal structure determination and is shown schematically (Figure 109)[300]. The copper atom in **241** has a square-planar coordination geometry as a result of one σ-bonded carbon [Cu–C 2.007(4) Å], one σ-bonded nitrogen atom [Cu–N 2.010(3) Å] and two coordinating nitrogen atoms [Cu–N 2.076(4) and 2.056(4) Å].

In a similar reaction, organocopper(II) complex **242** (Figure 109) was obtained from the reaction of the corresponding pyrrole-appended, *O*-confused oxaporphyrin with $Cu(OAc)_2$[301]. Its structure in the solid state was established by an X-ray crystal structure determination and is shown schematically (Figure 109). Like in **241** the copper atom has a square-planar coordination geometry as a result of one σ-bonded carbon atom [Cu–C 1.939(4) Å], one σ-bonded nitrogen atom [Cu–N 2.020(3) Å] and two coordinating nitrogen atoms [Cu–N 1.977(3) and 2.065(3) Å].

Finally, it has to be recalled that also various Cu^{II}–and Cu^{III}–carborane complexes have been reported; see Section V, **101** (Cu^{II}) and **102, 103** and **109** (Cu^{III}).

IX. COPPER CARBENE COMPLEXES

Formally, copper carbene complexes do not belong to the class of organocopper compounds. Although the carbon-to-copper distance observed in copper carbene complexes is of the same order as this distance in true organocopper compounds, it is the type of bonding that makes the difference. In organocopper compounds the carbon atom formally contributes one electron to the carbon–copper bond while in the copper carbene complexes the carbene carbon atom contributes two electrons to the carbon–copper interaction.

Because of its growing potential and applications as a catalyst in many organic transformations, the structures of copper carbene complexes in the solid state known so far will be discussed briefly. These transformations include copper-hydride-mediated reactions like carbonyl reductions and hydrosilylation reactions, carbene transfer reactions, cyclopropanation and insertion reactions, [3 + 2] cycloaddition reactions of azides and alkynes and 'click chemistry', topics that have been reviewed recently[302].

Most of the copper carbene complexes are derived from *N*-heterocyclic carbenes (NHC), discovered in the early 1960s[303]. In the early days NHCs were considered as phosphine mimics in organometallic chemistry[304,305], but experimental data show that NHC-transition metal catalysts can surpass their phosphine-based analogs in both activity and scope[306–308]. NHC–copper complexes can be easily prepared by the *in situ* deprotonation of an azolinium salt or alternatively by reaction of free, stable NHC[309,310] with a copper salt (equation 29).

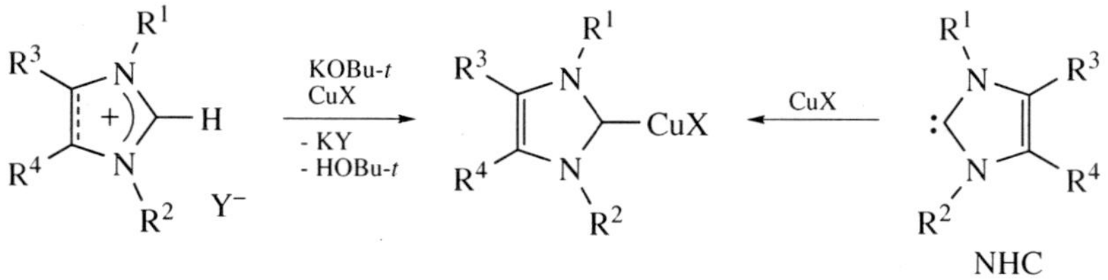

(29)

A series of NHC–copper complexes, NHC–CuX (**243–258**), derived from NHCs bearing sterically demanding groups at the nitrogen atoms (Figure 110) were prepared, characterized by X-ray crystallography and applied as catalyst in various organic transformations. In the solid state these complexes are discrete monomers having comparable structural features. The relevant structural data of these compounds are compiled in Table 3.

In complexes **243–258** the carbene carbon-to-copper bond length only varies slightly and the bond angle (carbene) carbon-to-copper-to-heteroatom (anion) is in all cases very close to 180°, rendering a linear coordination geometry at the copper atoms. The overall structural geometry of all complexes is closely related and only differs in the anion bound to copper. The structure of **243** is shown as a representative example (Figure 111).

It is notable that in complexes containing a potentially bidentate anion, i.e. acetate in **250–252** and triflate in **253**, only one oxygen atom is bonded to copper, as can be concluded from the (preferred) linear coordination geometry at the copper atom. This is different in NHC–copper complexes containing a β-diketonate or β-diketiminate as the anion (Figure 112).

In (IMes)Cu(DBM) (**259**)[316] (DBM = dibenzoylmethanato), both oxygen atoms are symmetrically bonded to copper [Cu–O(1) 1.973(3) and Cu–O(2) 1.986(3) Å] while the carbene carbon atom occupies the third coordination site [Cu–C 1.861(4) Å]. The C–Cu–O bond angles are rather large [C–Cu–O(1) 136.0(2)° and C–Cu–O(2) 132.5(2)°], but is compensated by an acute O(1)–Cu–O(2) bond angle of 91.3(1)°, most likely as a

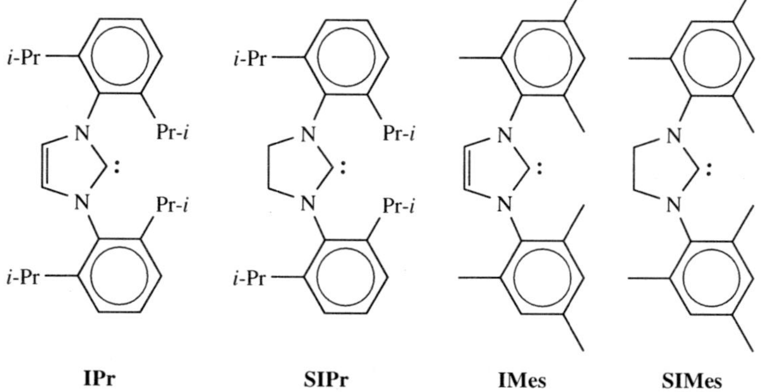

FIGURE 110. Sterically demanding NHCs

TABLE 3. Relevant structural data for the NHC–copper complexes **243–258**

Compound	Anion (X)	Cu–C (Å)	Cu–X (Å)	C–Cu–X (°)	Reference
(IPr)CuX (**243**)	Cl	1.953(8)	2.089(3)	180.0(2)	311
(IPr)CuX (**244**)	Cl	1.881(7)	2.106(2)	176.7(2)	100
(SIMes)CuX (**245**)	Cl	1.882(4)	2.099(11)	178.48	312
(IPr)CuX (**246**)	OBu-*t*	1.864(2)	1.810(1)	179.05(7)	101
(SIPr)CuX (**247**)	OEt	1.861(8)	1.793(7)	175.9(4)	313
(SIPr)CuX (**248**)	OPh	1.881(4)	1.831(3)	178.3(2)	313
(IMes)CuX (**249**)	OPh	1.864(2)	1.833(1)	175.53(6)	313
(IPr)CuX (**250**)	O_2CMe	1.854(4)	1.850(3)	177.2(2)	100
(SIPr)CuX (**251**)	O_2CMe	1.886(3)	1.838(2)	172.5(2)	314
(IMes)CuX (**252**)	O_2CMe	1.859(3)	1.867(3)	171.1(1)	314
(IPr)CuX (**253**)	O_3SCF_3	1.863(3)	1.875(2)	175.6(1)	314
(IPr)CuX (**254**)	SPh	1.895(2)	2.139(1)	178.3(1)	315
(IPr)CuX (**255**)	SCH_2Ph	1.898(2)	2.127(1)	171.5(1)	315
(SIPr)CuX (**256**)	SPh	1.896(3)	2.145(1)	177.5(1)	315
(SIPr)CuX (**257**)	SCH_2Ph	1.897(3)	2.121(1)	169.5(1)	315
(SIPr)CuX (**258**)	NHPh	1.876(2)	1.846(2)	175.21(7)	313

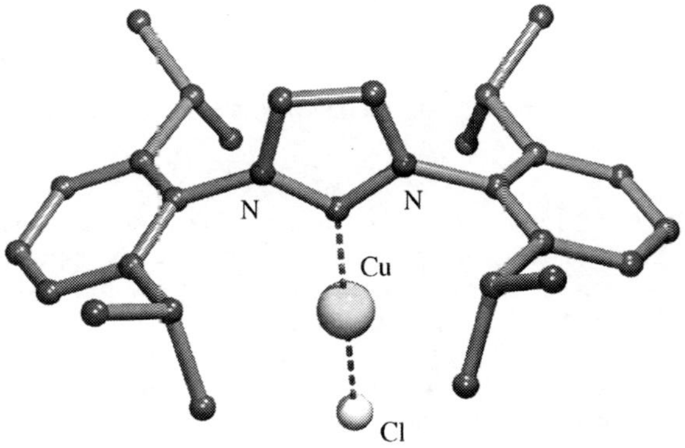

FIGURE 111. Molecular geometry of (IPr)CuCl (**243**) in the solid state

consequence of the small bite angle of the DBM anion. However, the sum of the bond angles around copper is 360° within experimental error, indicating a perfect trigonal planar geometry.

A closely related structure was found for the β-diketiminato complex (IMes)Cu (NacNac) (**260**) (NacNac = $CH[C(Me)N(C_6H_3Me_2\text{-}2,6)]_2^-$) (Figure 112)[317]. The β-diketiminate is chelate bonded to copper [Cu–N(1) 1.995(2) and Cu–N(2) 2.002(2) Å] in a symmetric way. The carbene carbon-to-copper distance 1.918(2) Å is slightly elongated compared to this distance observed in linear copper carbene complexes (cf. Table 3). The sum of the bond angles around copper [C–Cu–N(1) 135.38(8)°, C–Cu–N(2) 131.05(8)° and N(1)–Cu–N(2) 93.57(6)°)] is 360°, indicating that also in this complex the coordination geometry at copper is trigonal planar. It is notable that the N(3)–C–N(4) plane of the carbene is orientated perpendicular to the N(1)–Cu–N(2) plane, which was considered to arise from steric interactions and, furthermore, due to the fact that NHCs are weak π-acceptor ligands for Cu(I)[318].

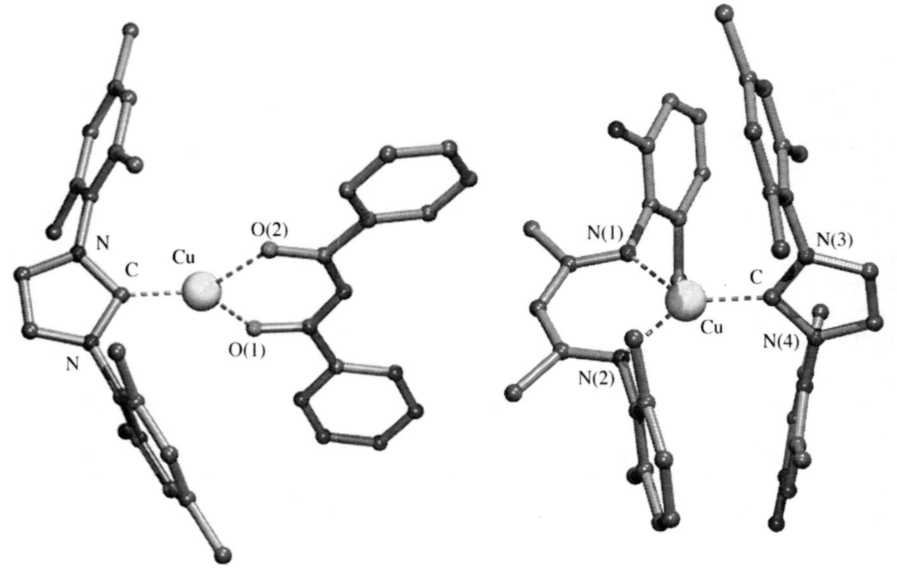

(**259**) (**260**)

FIGURE 112. Molecular geometry of (IMes)Cu(DBM) (**259**) and (IMes)Cu(NacNac) (**260**) in the solid state

Reaction of (IPr)Cu(OBu-*t*) (**246**) with bis(pinacolato)diboron (pinB-Bpin) afforded the copper boryl complex (IPr)Cu(Bpin) (**261**). This complex appeared to be capable of abstracting an oxygen atom from carbon dioxide, resulting in the formation of the borate complex (IPr)Cu(OBpin) (**262**) and carbon monoxide (equation 30)[319]. For both **261** and **262** the structures in the solid state were established by X-ray crystal structure determinations. These structures show large similarities with the structures observed for other NHC–copper complexes in which the copper atom is linear dicoordinate (*vide supra*).

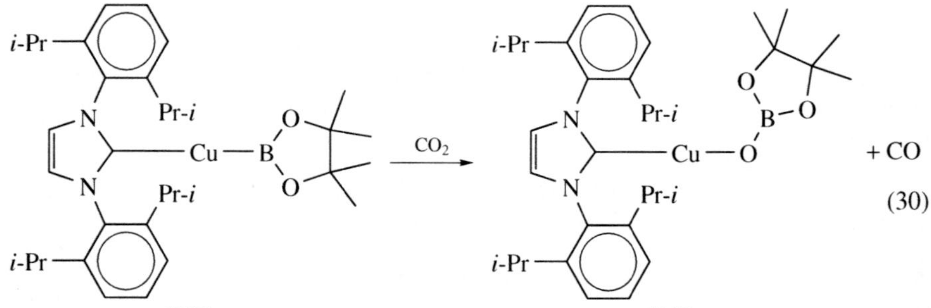

(**261**) (**262**) (30)

Based on NMR spectroscopic studies, the reaction of (IPr)Cu(OBu-*t*) (**246**) with triethoxysilane afforded a carbene copper hydride complex (IPr)CuH (**263**) (equation 31).

The observation of an absorbance at 638 cm^{-1} in its IR spectrum is indicative for bridging copper hydrides[320].

$$\underset{\textbf{(246)}}{\text{(IPr)Cu(OBu-}t\text{)}} \xrightarrow[\text{-(EtO)}_3\text{SiOBu-}t]{\text{(EtO)}_3\text{SiH}} \underset{\textbf{(263)}}{\text{(IPr)CuH}} \qquad (31)$$

The structure of **263** in the solid state was established by an X-ray crystal structure determination[101]. Unfortunately, due to the limited quality of the dataset, the hydrogen atoms could not be located. The structure of **263** comprises two (IPr)Cu units [C–Cu 1.878(6) Å] most likely linked via two bridging hydrogen atoms between the copper atoms (Figure 113). The observed copper–copper distance [Cu–Cu 2.3059(11) Å] is in agreement with the distance observed in other copper-hydride dimers with bridging hydrides[320].

The cationic copper carbene complex $[(IMes)_2Cu][CF_3SO_3]$ (**264**) was isolated from the reaction of copper triflate with two equivalents of IMes in THF. Its structural characterization by X-ray crystallography was hampered by a very poor dataset. From these data it could only be concluded that its overall structural features are identical to those of the corresponding silver complex, i.e. the two coordinating carbene carbon atoms and the metal are in a linear arrangement[321].

The cationic metalla crown ether (**265**) was prepared via a transmetallation reaction of the corresponding silver compound with CuI (equation 32). Its structure in the solid state was established by an X-ray crystal structure determination[322]. The 13-membered ring comprises two NHCs linked via ethylene bridges to the oxygen atoms of 1,2-dihydroxybenzene. Ring closure occurs by coordination of the carbene carbon atoms

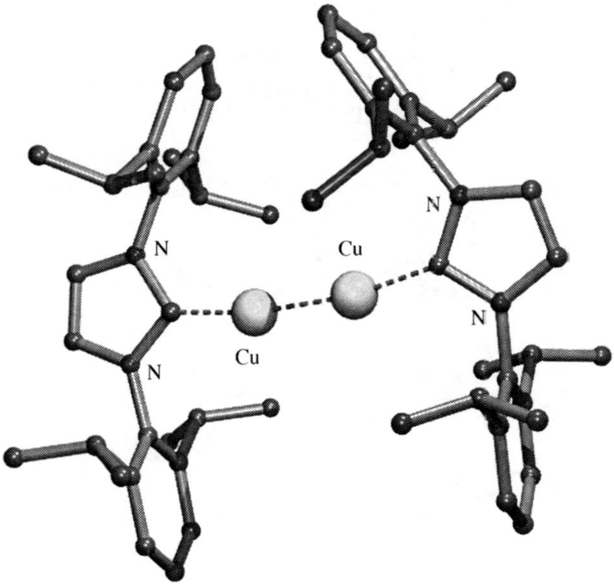

FIGURE 113. Molecular geometry of hydride-bridged (IPr)CuH dimer **263** in the solid state

of the two NHCs to a copper cation [Cu–C 1.960(6) and 1.940(6) Å] with an almost linear coordination geometry [C–Cu–C 174.2(2)°] at copper.

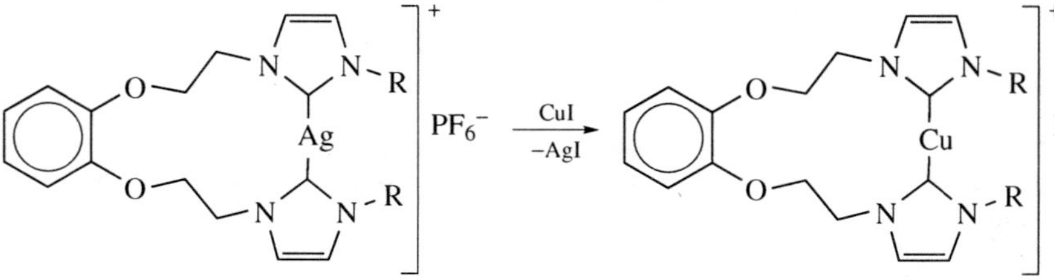

(32)

(**265**)

A macrocyclic ligand, consisting of four NHCs linked via propane-1,3-diyl bridges between the nitrogen atoms of the NHCs, has been prepared and was converted to the dicopper complex $[-CH_2CH_2CH_2(NHC)-]_4Cu_2I_2$ (**266**). An X-ray crystal structure determination of **266** confirmed the isolation of dications, consisting of two copper atoms encapsulated in the macrocyclic ligand, and separate iodide anions arranged in the crystal lattice[323]. In the dication (Figure 114) two carbene carbon atoms of oppositely positioned NHCs are bonded to each of the two copper cations [Cu(1)–C(1) 1.920(9), Cu(1)–C(2) 1.948(11), Cu(2)–C(3) 1.923(10) and Cu(2)–C(4) 1.915(10) Å]. Both copper atoms have an almost linear coordination geometry [C(1)–Cu(1)–C(2) 173.9(4)° and C(3)–Cu(2)–C(4) 174.6(5)°]. Interestingly, the Cu···Cu distance is 2.553(2) Å.

The copper complexes derived from the NHC-containing tripodal ligands TIMEN-Me, TIMEN-Bn, TIMEN-*t*-Bu and TIME-Me (Figure 115) were prepared and characterized structurally by X-ray diffraction.

An X-ray crystal structure determination of (TIMEN-Bn)CuBr (**267**)[324] showed that its solid state structure comprises isolated (TIMEN-Bn)Cu cations and bromide anions in the crystal lattice. The asymmetric unit contains three independent molecules that are chemically identical and differ only slightly with respect to bond angles and bond distances. Details are given for one of these molecules (Figure 116).

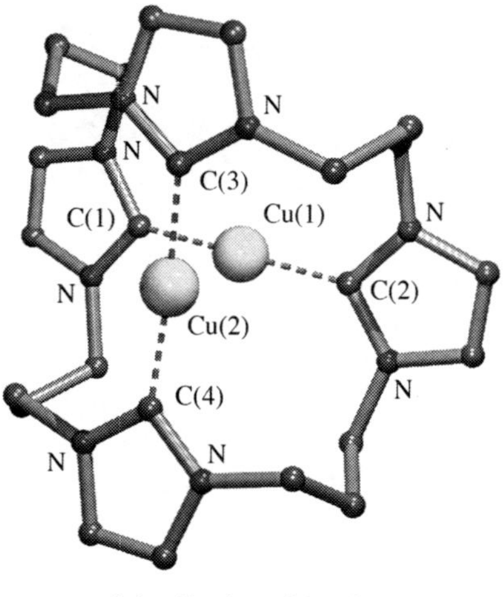

FIGURE 114. Molecular geometry of the dication of **266** in the solid state

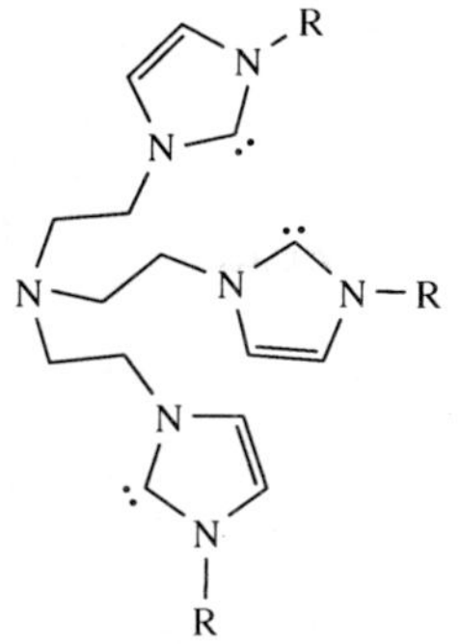

TIMEN-Me; R = Me
TIMEN-Bn; R = Bn
TIMEN-*t*-Bu; R = *t*-Bu

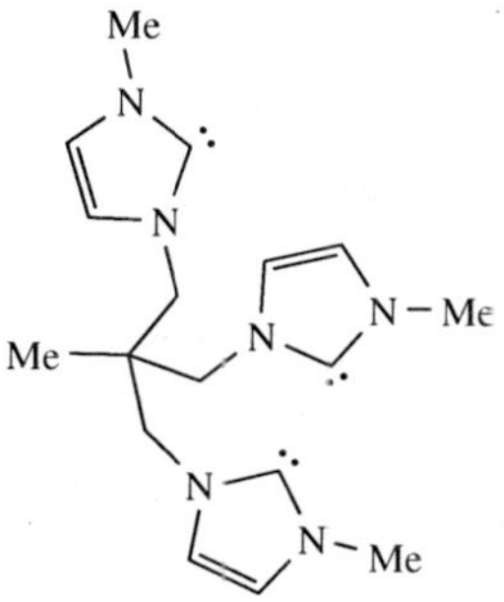

TIME-Me

FIGURE 115. NHC-containing tripodal ligands

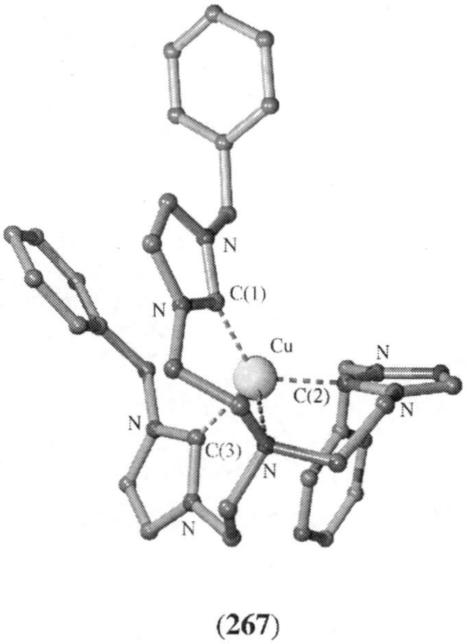

(267)

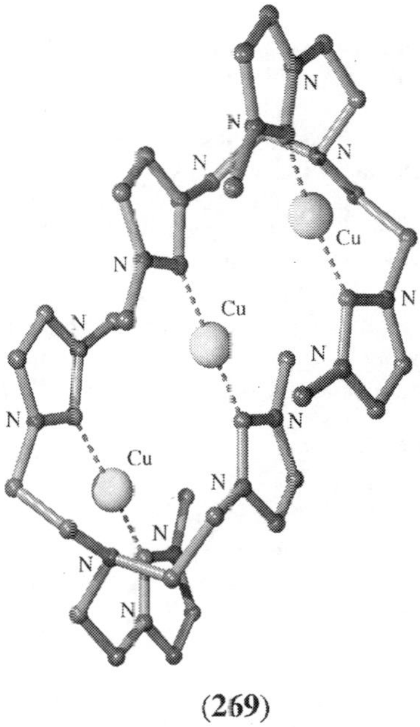

(269)

FIGURE 116. Molecular geometry of the cations $[(TIMEN\text{-}Bn)Cu]^+$ (**267**) and $[(TIMEN\text{-}Me)_2Cu_3]^{3+}$ (**269**)

In the (TIMEN-Bn)Cu cation the three carbene carbon atoms of the tripodal ligand are bonded to copper [Cu−C(1) 1.949(6), Cu−C(2) 1.948(6) and Cu−C(3) Å]. The relatively long Cu−N(amine) distance, 2.365(4) Å, suggests that the copper-to-nitrogen interaction is weak. The sum of the C−Cu−C bond angles is 358.7° [C(1)−Cu−C(2) 115.3(3), C(1)−Cu−C(3) 123.3(3)° and C(2)−Cu−C(3) 120.1(1)°]. These values are close to the ideal values for a trigonal planar coordination geometry at copper. However, taking into account the copper-to-amine nitrogen interaction, the coordination geometry at copper could be best described as slightly distorted towards a trigonal pyramidal one.

The carbene copper complex [(TIMEN-*t*-Bu)Cu]PF_6 (**268**) also has been prepared and was characterized structurally by X-ray crystallography. The overall structural features of **268** in the solid state are very close to those of **267** and therefore will not be discussed in detail.

When the sterically less demanding TIMEN-Me tripodal ligand was used, a copper complex with an entirely different and surprising structure was obtained. Instead of the formation of discrete monomers, a novel structure was obtained in which two TIMEN-Me ligands associate with three copper cations to form [(TIMEN-Me)$_2$Cu$_3$](PF_6)$_3$ (**269**)[324]. An X-ray crystal structure determination showed the presence of isolated [(TIMEN-Me)$_2$Cu$_3$]$^{3+}$ cations (Figure 116) and PF_6^- anions in the crystal lattice. The cation may be regarded as consisting of two [(TIMEN-Me)Cu]$^+$ cations in which two carbene carbon atoms are bonded to the copper cation in an almost linear fashion [C–Cu–C 172.2(2)°]. Like in **267** the amine nitrogen-to-copper distance is relatively long [Cu–N 2.365(4) Å]. When this latter interaction is considered to be a nitrogen-to-copper interaction, the coordination geometry at copper should be described as T-shaped. Two of these cations are linked via the third carbene carbon atom of the tripodal ligand to an additional copper cation [Cu–C 1.909(4) and 1.910(4) Å]. The coordination geometry of this latter copper atom is close to perfectly linear [C–Cu–C 178.7(2)°].

The tripodal ligand TIME-Me (Figure 115), which has a (Me)C instead of nitrogen atom as a bridgehead group, formed the complex [(TIME-Me)$_2$Cu$_3$](PF_6)$_3$ (**270**) with three copper cations[325]. An X-ray crystal structure determination revealed a solid state structure with overall structural features very close to those of related **269**.

The copper complexes of a series of 2-pyridyl- and oxazolinyl-substituted NHCs (Figure 117) were prepared and their structures in the solid state were characterized by X-ray crystallography.

In the solid state the copper carbene complex (IPr-2-Py)CuBr (**271**) exists as discrete monomers (Figure 118)[326]. The bonding carbene carbon atom [Cu–C 1.880(6) Å], the copper atom and the bonding bromine atom [Cu–Br 2.208(1) Å] form an almost linear arrangement [C–Cu–Br 170.3(2)°]. The pyridyl nitrogen atom only weakly coordinates

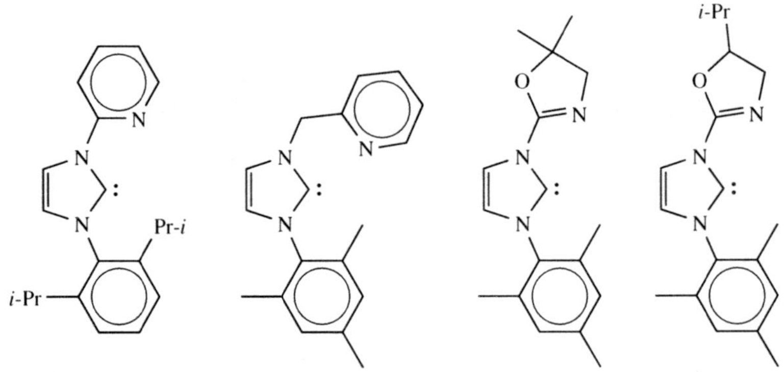

IPr-2-Py **Mes-2-CH$_2$Py** **MesOx** **MesOxIPr**

FIGURE 117. 2-Pyridyl- and oxazolinyl-substituted NHCs

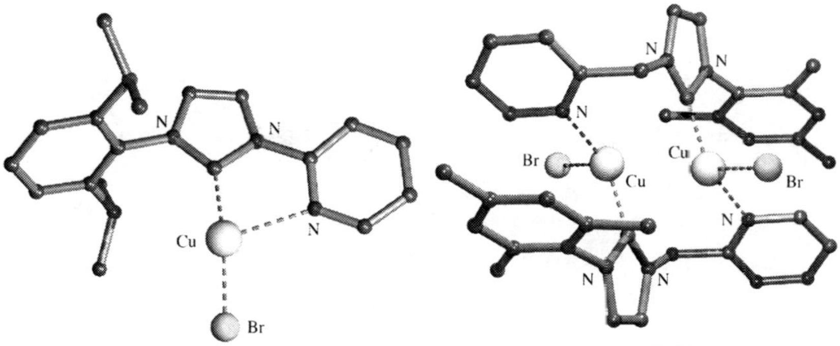

(271) **(272a)**

FIGURE 118. Molecular geometry of (IPr-2-Py)CuBr (**271**) and [(Mes-2-CH_2Py)CuBr]$_2$ (**272a**) in the solid state

intramolecularly to copper, as is indicated by a relatively long Cu–N bonding distance [Cu–N 2.454(5) Å]. Taking this latter interaction into account as a real bond, the coordination geometry at copper should be described as T-shaped.

The structure in the solid state of the carbene copper complex (Mes-2-CH_2Py)CuBr (**272**) depends on the solvent of crystallization.

When **272** was crystallized by slow cooling of a dichloromethane/diethyl ether solution, the solid state structure of the crystalline material comprised discrete dimers [(Mes-2-CH_2Py)CuBr]$_2$ (**272a**) (Figure 118). This dimer consists of two (Mes-2-CH_2Py)CuBr units in which the bonding carbene carbon–copper distance is in the expected range [Cu–C 1.931(2) Å], but the bromine–copper distance is rather long [Cu–Br 2.5147(4) Å] compared to this distance observed in **271**. Dimer formation occurs via the pairwise coordination of the pyridyl nitrogen atoms to a copper atom of an adjacent (Mes-2-CH_2Py)CuBr unit. The bond angles around copper [C–Cu–Br 106.14(7)°, C–Cu–N 133.88(8)° and Br–Cu–N 98.08(5)°] point to a rather irregular three-coordination at the copper atoms.

However, when **272** was crystallized from a concentrated chloroform solution, the solid state structure of the crystals comprised an infinite polymeric chain [(Mes-2-CH_2Py) CuBr]$_n$ (**272b**) (Figure 119)[326]. This chain consists of (Mes-2-CH_2Py)CuBr units [Cu–C 1.914(4) and Cu–Br 2,3758(7) Å] linked via a coordination bond [Cu–N 2.044(3) Å] of the pyridyl nitrogen atom to a copper atom of a neighboring (Mes-2-CH_2Py)CuBr unit. The bond angles around copper [C–Cu–Br 130.7(1)°, C–Cu–N 125.2(2)° and Br–Cu–N 103.4(1)°], and especially the sum of these bond angles being 360° within experimental error, points to a trigonal planar coordination geometry for each of the copper atoms.

An X-ray crystal structure determination of the copper complex containing the MesOx ligand (Figure 117) revealed a dimeric structure for [(MesOx)CuBr]$_2$ (**273**)[327] in the solid state with structural features closely related to that of **272a**. Like **272b**, the solid state structure of ligand MesOxIPr (Figure 117) with copper comprises an infinite polymeric chain of (MesOxIPr)CuBr units (**274**) with closely related structural features.

A chelating ligand that combines a hard alkoxy functionality with two *N*-heterocyclic carbenes has been designed and was subsequently converted into its copper complex (equation 33)[328]. The first step involved reaction of two equivalents of *N*-*t*-butylimidazole with 2-chloromethyl-2-phenyloxirane to afford the zwitterionic diimidazolinium salt (**275**). Treatment of this diimidazolinium salt with Ag_2O afforded the alkoxy silver carbene

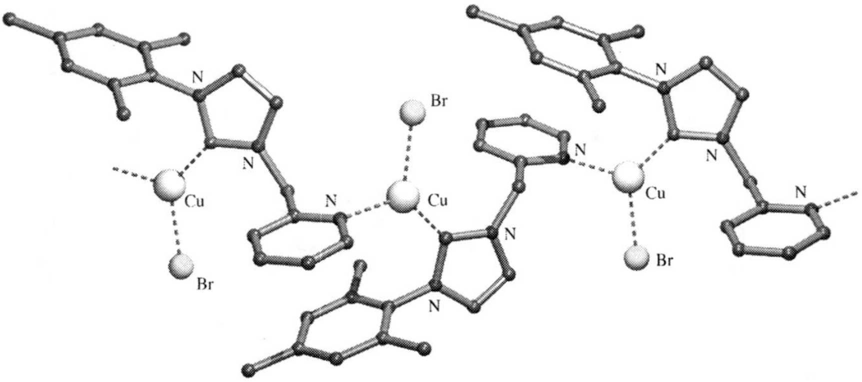

FIGURE 119. Part of the polymeric chain of **272b** in the solid state

complex (**276**). Subsequently, this silver complex was converted into the alkoxy copper carbene complex (**277**) via a transmetallation reaction with CuI.

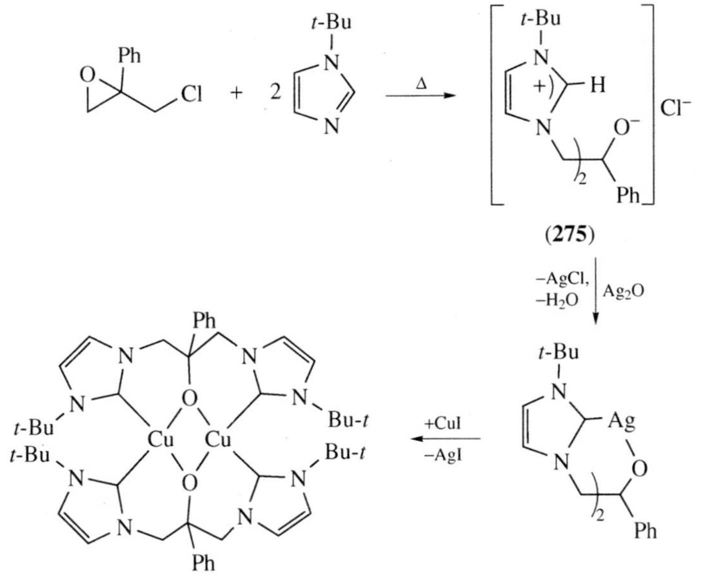

(33)

The structure of **277** in the solid state was established by an X-ray crystal structure determination and is shown schematically (equation 33). The structure is a dimeric one in which two oxygen atoms are symmetrically bridge bonded to two copper atoms [Cu–O 1.941(3) and 1.959(2) Å]. Two carbene carbon atoms are bonded to each of the copper atoms [Cu–C 1.960(5) and 1.966(4) Å]. The angle between the CuO_2 and CuC_2 planes is

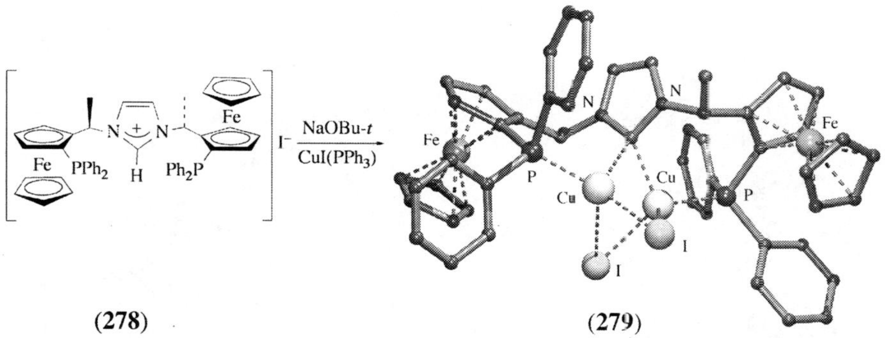

(278) **(279)**

FIGURE 120. Molecular geometry of **279** in the solid state

only 14.0(2)°, indicating that the coordination geometry at copper is nearly square-planar, which is rather unusual for Cu(I).

The chiral imidazolinium salt (**278**)[329], a precursor for a potentially tridentate PCP ligand in which C is a *N*-heterocyclic carbene, has been prepared and was successfully converted into its dicopper complex (**279**)[330] (Figure 120). The central core of the complex consists of two copper atoms bridged by two iodine atoms in a rather asymmetric way [Cu–I 2.559(1) and 3.110(1) Å]. The tridentate PCP ligand is bridge bonded via its carbene carbon atom to both copper atoms [Cu–C 2.113(5) and 2.174(5) Å]. One phosphorus atom of the ferrocenylphosphine moieties is bonded to each of the copper atoms [Cu–P 2.181(1) and 2.206(1) Å]. As a result both copper atoms adopt a distorted tetrahedral coordination geometry.

Also, copper complexes containing 1,3-substituted 3,4,5,6-tetrahydropyrimidine-2-ylidenes as the NHC have been prepared and were characterized structurally[331]. The structure of 1,3-diisopropyl-3,4,5,6-tetrahydropyrimidine-2-ylidene copper chloride (**280**) in the solid state is straightforward (Figure 121). The copper atom adopts a linear coordination geometry [C–Cu–Cl 171.18(6)°] as the result of bonding of the carbene carbon atom [Cu–C 1.915(2) Å] and the chlorine atom [Cu–Cl 2.1106(6) Å] to copper.

An unexpected structure was found for 1,3-dimesityl-3,4,5,6-tetrahydropyrimidine-2-ylidene copper bromide (**281**)[331]. Instead of the anticipated monomeric structure as found for **280**, an X-ray crystal structure determination showed the presence of separated bis (carbene) copper cations (Figure 121) and $[CuBr_2]^-$ anions. In the cation the Cu–C bond distances are identical [Cu–C 1.934(2) Å] as the consequence of crystallographic symmetry. A linear coordination geometry at copper is reflected by the C–Cu–C bond angle, 179.29(4)°.

The thiazolinylidene copper chloride complex (**282**) was prepared starting from the corresponding lithiated thiazole (equation 34). An X-ray crystal structure determination showed that this compound exists in the solid state as a discrete monomer of which the structure is shown schematically (equation 34)[332]. The copper atom adopts a linear coordination geometry [C–Cu–Cl 178.6(1)°] as the result of bonding of the carbene carbon atom [Cu–C 1868(6) Å] and the chlorine atom [Cu–Cl 2.099(2) Å] to copper.

Reaction of the copper β-diketiminate $[(MesNC(Me))_2CH]_2Cu_2(MeC_6H_5)$ (**283**) with one equivalent of diphenyldiazomethane in diethyl ether as the solvent afforded the dicopper diphenylcarbene complex $[(MesN(Me)C)_2CH]_2Cu_2(CPh_2)$ (**284**) (equation 35)[318,333].

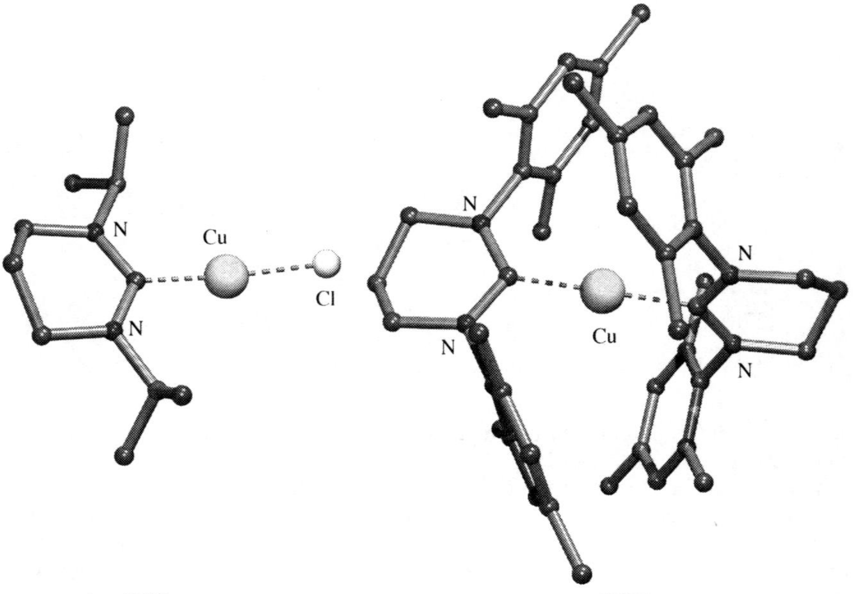

(**280**) (**281**)

FIGURE 121. Molecular geometry of **280** and the cation of **281** in the solid state

It appeared that in aromatic solvents the dicopper carbene complex **284** is in equilibrium with the monocopper carbene complex [(MesN(Me)C)$_2$CH]Cu(CPh$_2$) (**285**) and the copper β-diketiminate **283** (equation 35)[318,333].

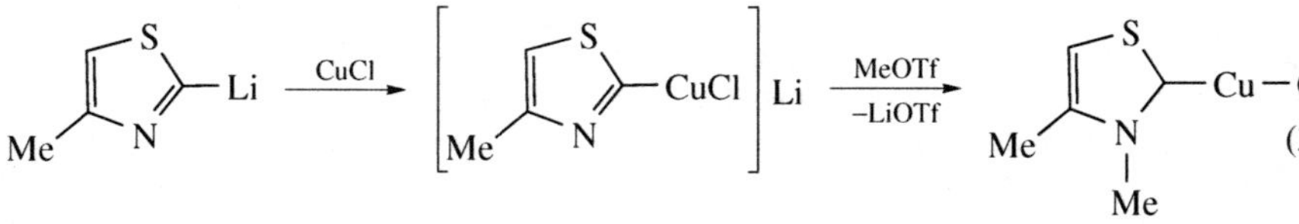

(34)

(**282**)

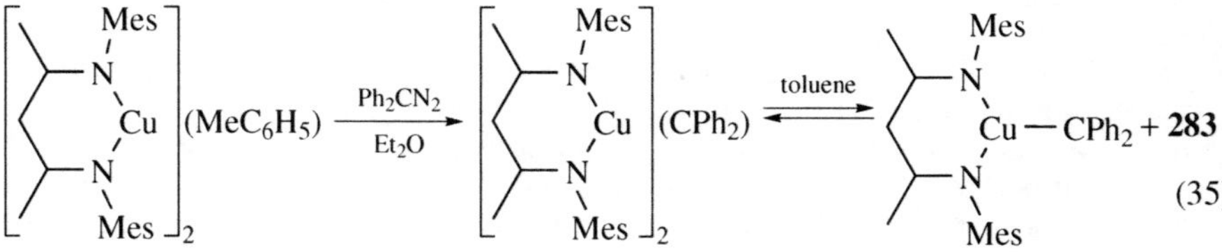

(35)

(**283**) (**284**) (**285**)

For both **284** and **285** the structure in the solid state was established by X-ray crystal structure determinations[318,333]. In the dicopper carbene complex **284** the diphenylcarbene is μ-bridge bonded with its carbene carbon atom between the two copper atoms [Cu(1)–C 1.933(5), Cu(2)–C 1.931(5) Å] of two copper β-diketiminate units (Figure 122). In each

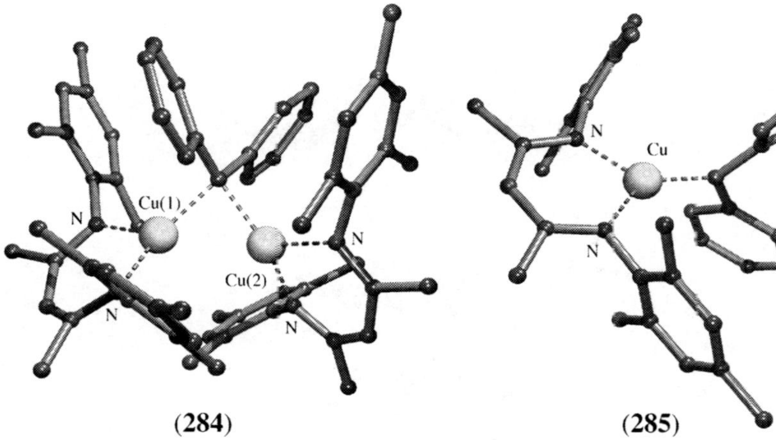

FIGURE 122. Molecular geometry of **284** and **285** in the solid state

of the copper β-diketiminate moieties the copper atom is chelate bonded symmetrically to two nitrogen atoms [Cu−N 1.971(4) and 1.963(4) Å].

As expected, in the monocopper carbene complex **285** (Figure 122) the bond distance between the copper atom and the carbene carbon atom [Cu−C 1.834(3) Å] is significantly shorter compared to the copper−carbon distances in **284**. Like in **284**, the copper atom in **285** is bonded symmetrically [Cu−N 1.906(3) and 1.922(3) Å] to the two nitrogen atoms of the copper β-diketiminate unit. The sum of the bond angles around copper in **285** [C−Cu−N(1) 132.3(1)°, C−Cu−N(2) 132.5(1)° and N(1)−Cu−N(2) 96.1(1)°] is 360° within experimental error, indicating a perfect trigonal planar coordination geometry at copper.

So far, only copper carbene complexes have been discussed with copper in its monovalent oxidation state. Copper carbene complexes with formally divalent copper are very rare and only three examples of well-characterized copper(II) carbene complexes are known.

Reaction of the free NHC ligand IPr (Figure 110) with $Cu(OAc)_2$ in toluene afforded the copper(II) carbene complex $(IPr)Cu^{II}(OAc)_2$ (**286**)[334]. Its structure in the solid state was established by an X-ray crystal structure determination (Figure 123).

In **286**, the carbene carbon-to-copper distance [Cu−C 1.942(4) Å] does not significantly differ from the distances observed in copper(I) carbene complexes (*vide supra*). The two acetate groups are crystallographically symmetry related and are *O*,*O*-chelate bonded to copper with one relatively long [Cu−O(1) 2.259(4) Å] and one relatively short [Cu−O(2) 1.943(3) Å] bond. The coordination geometry around copper can be best described in terms of a distorted trigonal bipyramid with the carbene carbon atom, O(1) and O(3) in the equatorial plane and O(2) and O(4) at the axial positions. The sum of the bond angles around copper in the equatorial plane is 360°, pointing to a trigonal bipyramidal geometry. However, the C−Cu−O(2) (99.82(9)°) and O(1)−Cu−O(2) (60.5(1)°) bond angles, and consequently the O(2)−Cu−O(4) (160.4(2)°) bond angle, deviate considerably from the ideal values (90°, 90° and 180°, respectively) for a trigonal bipyramid, which is a logical consequence of the small bite angle of the chelate-bonded acetate group. It is noteworthy that **286** has been successfully applied as an alternative for Stryker's reagent in hydrosilylation reactions[334].

In copper(II) carbene complex (**287**)[335] (Figure 124), containing divalent copper, two oxygen-anionic functionalized *N*-heterocyclic carbene ligands are *C*,*O*-chelate bonded

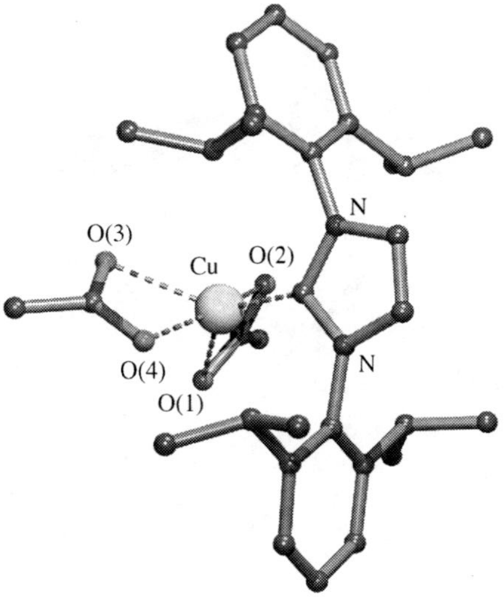

FIGURE 123. Molecular geometry of (IPr)Cu^{II}(OAc)$_2$ (**286**) in the solid state

to copper. Its structure in the solid state was established by an X-ray crystal structure determination and is shown schematically in Figure 124. The bonding of two carbene carbon atoms [Cu–C 1.946(4) and 1.940(4) Å] and two anionic oxygen atoms [Cu–O 1.950(3) and 1.960(3) Å] results in a distorted square planar coordination geometry at copper.

A monoanionic ligand precursor that combines a chiral binaphthyl core, a monoanionic oxygen atom at the 2-position of one of the naphthyl groups and a *N*-heterocyclic carbene at the 2-position of the other naphthyl group has been developed[336]. The enantiopure ligand precursor was successfully converted into the aryloxycopper(II) chloride carbene complex (**288**) (Figure 124). The structure of this complex in the solid state was established by an X-ray crystal structure determination[337]. It appeared to be a dimer with

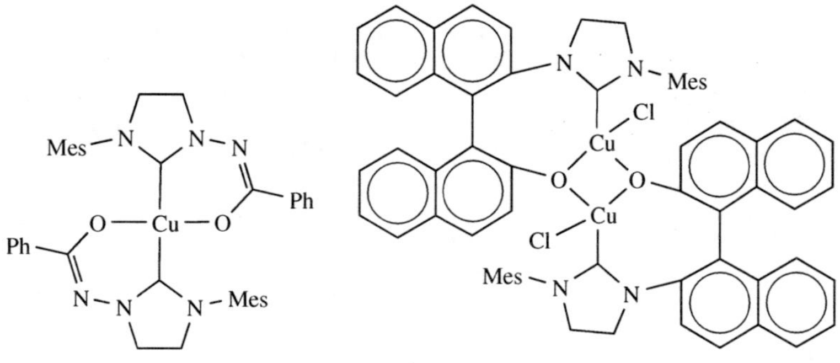

(**287**) (**288**)

FIGURE 124. Schematic structures of the copper(II) carbene complexes **287** and **288** in the solid state

the two monoanionic *O,C*-chelating ligands bridge-bonded via its oxygen atom between two copper atoms [Cu–O 1.986(6) and 1.950(6) Å] while each of the carbene carbon atoms forms a coordination bond to a copper atom [Cu–C 1.926(8) and 1.964(8) Å]. One chlorine atom is bonded to each of the copper atoms [Cu–Cl 2.226(3) and 2.222(3) Å], resulting in a slightly distorted square-planar coordination geometry at each of the copper atoms. This enantiopure copper complex has been successfully applied as a catalyst in enantioselective allylic allylation reactions with ee values up to 98%[337].

X. CONCLUSIONS

Today's major interest in organocopper compounds is connected to their use in organic synthesis. Although the first examples date back from the first half of the last century comprising Ullmann-type cross-coupling reactions and the use of Gilman-type reagents in conjugate addition chemistry, contemporary interest is primarily in the development of coupling reactions of copper reagents, such as cuprates and heterocuprates, as well as of copper-catalyzed stereoselective transformations.

Much attention has currently been paid to circumvent the limitations of the conventional and present synthetic routes for the generation of organocopper reagents. Initially, organocopper compounds were prepared from organolithium and Grignard reagents which excluded the use of substituents and reactive groups that are incompatible with the reactivity of the respective metal–carbon bonds in these reagents. However, with the new synthetic approaches the *in situ* synthesis of a broad variety of organocopper reagents with a wide choice of substituents became available. It must be noted that these reagents comprise the organocopper entity RCu and the metal salts present in the transmetallating reagent as well as the salts that are formed during the formation of RCu. Often, it is the unique composition of salts and organocopper entities in this mixture that provides the special reactivity of the *in situ* formed organocopper reagent. The various unique RCu-MX structures encountered in the solid state presented in this overview (see, e.g., Section III) as well as those emerging from computational studies substantiate the reasons for this special reactivity. It provides insight in structure and reactivity properties of the key intermediates and transient species present in the reaction mixture of copper-mediated organic synthesis. These aspects have been highlighted in a number of recent reviews[20, 21, 277].

The structural work as well as the reactivity studies indicated that, in order to synthesize and isolate *pure* organocopper compounds, special precautions have to be taken, i.e. sometimes the order of addition of the reagents to the reaction solution appears to be of crucial importance, the purity of the transmetallating reagents (no metal salts present) has to be high, the nature of the metal salts formed during the synthesis has to be controlled in order to prevent undesired organocopper–metal salt aggregate formation and often special reaction conditions have to be applied, such as low reaction temperatures, severe exclusion of oxygen and water because of the extreme sensitivity of most of the known organocopper compounds and derivatives thereof.

Moreover, many pure organocopper compounds, e.g. methylcopper, decompose violently below room temperature. Finally, also the choice of the polarity of the solvents used for both the reaction and the subsequent purification and crystallization of the organocopper compound often appeared of crucial importance owing to the tendency of organocopper species to exist either as SSIP (Solvent Separated Ion Pair) or CIP (Contact Ion Pair)/neutral aggregates in solution and in the solid state, respectively[19, 222].

The identification and structural characterization of organocopper compounds is often difficult and rely in many cases on the use of X-ray diffractions techniques. The present review reveals the great structural variety that organocopper compounds can exhibit. Of course, from the elemental analytical data of purified material the atomic composition can be derived, while the NMR data can reveal the connectivity of the building blocks, but in most cases the nature of the interactions of the copper and eventually other metal centers with their respective anions as well as the aggregation state of the species remained obscure. In most of the cases the structural information gathered with X-ray diffraction techniques appeared indispensable.

In former days, molecular weight determinations (cryoscopic, osmometric or ebulliometric measurements) were used to shed light on the aggregation state of the organocopper compounds in solution. However, even at the beginning of organocopper chemistry it was the results of the first X-ray structure determinations that showed the true bonding characteristics in the solid state of organocopper compounds, i.e. the presence of dimeric, trimeric, tetrameric etc. aggregate formation, the (electron-deficient) bridge bonding of the anions to the copper (metal) cations, and the way organocopper and metal-anion entities assemble to form aggregates with unique structural features.

It should be noted that the structures found in the solid state do not necessarily reflect those present in a particular solvent. Moreover, in many cases the single crystal used for the X-ray structure determination appeared after all not to belong to the major product, but represented only the structural features of a minor product present in the material isolated from the reaction mixture. The latter point is of special interest when the structural information of an organocopper compound is used to interprete its reactivity as an organocopper reagent.

The wealth of structural information emerging from X-ray structural studies has been inspirational for the computational chemist. In particular, the discovery of the many organocopper metal-anion aggregated (including cuprates) structures and their versatile bonding patterns were leading the computational chemist in the selection process of possible reaction intermediates and reaction pathways of organocopper-mediated synthetic transformations.

The focus of this review was on the structural features of organocopper compounds in the solid state. One aspect that remained underscored, however, but that has been discussed intensively immediately after the structural elucidation of the first organocopper compounds, are the sometimes very short Cu···Cu distances [e.g. in **19** of 2.372(8) Å[63] vs. 2.48 Å in metallic Cu] and acute Cu–C_{ipso}–Cu bond angles (71.35(8)°) encountered in the majority of these structures. A qualitative explanation is provided in Figure 6 (see also discussion in Reference 63), which shows that the electron density in the electron-deficient (two-electron three-center) bonding does not coincide with the vectors between the atoms involved but is directed more to the center of the triangle. Consequently, this leads to more linear digonal C_{ipso}–Cu–C_{ipso} bond angles; see Figure 30, **B**[29] for example. Furthermore, it leads to the acute bond angles (Cu–C_{ipso}–Cu) in these aggregates and to short Cu–Cu distances[63].

Another aspect is the observation that the organocopper and the corresponding organolithium compounds both have a strong tendency to form electron-deficient C_{ipso}–Cu bonds leading to the formation of aggregated structures. In a number of cases the copper atoms in an aggregated structure could even be replaced by lithium atoms with retention of the structural features of the aggregate, or even appeared to be disordered in the structure (cf. **188** in Figure 91). In a number of cases this notion was effectively used for the clean and selective synthesis of a given mixed aggregate (cf. the synthesis of the diarylcuprate **172** from the the reaction of the arylcopper **18** with the corresponding aryllithium in a 1:1 molar ratio in an apolar solvent)[49]. This example also demonstrates that the formation of heteroaggregates, such as $R_nX_mCu_{n+m}$, are the result of selective self-assembly of the homoaggregates, RCu and CuX, which apparently is the thermodynamically most stable species. In this respect organocopper and organolithium chemistry have much in common[338].

Finally, the structural information on the CIP and SSIP structures of cuprates could be connected to their different reactivity patterns in C–X coupling reactions: it is the cuprate in its CIP structure that exists in solvents with low polarity and coordinating properties which is the reactive intermediate, whereas the SSIP structure of the cuprate in polar solvents is unreactive. This indicates that the nature of the metal ion next to the copper ion in the CIP structure of the cuprate plays an important role in the reaction with the substrate, i.e. it is often an anchoring point for the substrate (e.g. the keto-O of an α,β-unsaturated ketone) while the copper center is involved in the group transfer step (to the alkenic grouping of the α,β-unsaturated ketone).

A recent and exciting development in organocopper chemistry is the synthesis and structural characterization of organocopper(II) and -(III) compounds. The presented structures show the important role the ancillary multidentate ligand systems play in these compounds to arrive at a stable σ-bond between the C-anion and the copper(II) or -(III) center.

It is to be expected that because of the growing importance of copper-mediated and catalyzed reactions in organic synthesis, there will be further and increasing interest in the synthesis and characterization of organocopper compounds. So far, novel aggregates and compounds often have shown new and unexpected molecular geometries and bonding patterns. In many cases these findings appeared instrumental for increasing our understanding of the mechanistic details of copper-mediated synthetic routes. Besides this aspect it is the synthesis and manipulation of organocopper compounds that remain a great challenge for the organometallic synthetic chemist. Since 1970, the beauty and fascinating structural versatility of these compounds have been a tremendous reward for the authors of this review.

XI. REFERENCES

1. R. C. Böttger, *Liebigs Ann. Chem.*, **109**, 351 (1859).
2. G. Buckton, *Liebigs Ann. Chem.*, **109**, 225 (1859).
3. M. R. Reich, *Compt. Rend. Acad. Sci.*, **177**, 322 (1923).
4. H. Gilman, R. G. Jones and A. L. Woods, *J. Org. Chem.*, **17**, 1630 (1952).
5. H. Gilman and J. M. Straley, *Recl. Trav. Chim. Pays-Bas*, **55**, 821 (1936).
6. M. S. Kharasch and P. O. Tawney, *J. Am. Chem. Soc.*, **63**, 2308 (1941).
7. J. Munch-Petersen, *J. Org. Chem.*, **22**, 170 (1957).
8. H. O. House, W. L. Respess and G. M. Whitesides, *J. Org. Chem.*, **31**, 3128 (1966).
9. G. M. Whitesides, W. F. Fisher, J. San Filippo, R. W. Bashe and H. O. House, *J. Am. Chem. Soc.*, **91**, 4871 (1969).
10. E. J. Corey and G. H. Posner, *J. Am. Chem. Soc.*, **89**, 3911 (1967).
11. G. H. Posner, C. E. Whitten and J. J. Sterling, *J. Am. Chem. Soc.*, **95**, 7788 (1973).
12. D. M. Knotter, A. L. Spek and G. van Koten, *J. Chem. Soc., Chem. Commun.*, 1738 (1989).
13. D. M. Knotter, D. M. Grove, W. J. J. Smeets, A. L. Spek and G. van Koten, *J. Am. Chem. Soc.*, **114**, 3400 (1992).
14. B. H. Lipshutz, R. S. Wilhelm and J. A. Koslowski, *Tetrahedron*, **40**, 5005 (1984).
15. B. H. Lipshutz, S. Sharma and E. L. Ellisworth, *J. Am. Chem. Soc.*, **112**, 4032 (1990).
16. B. H. Lipshutz and S. Sengupta, *Org. React.*, **41**, 135 (1992).
17. A. Alexakis, J. E. Bäckvall, N. Krause, O. Pàmies and M. Diéguez, *Chem. Rev.*, **108**, 2796 (2008).
18. B. Breit and Y. Schmidt, *Chem. Rev.*, **108**, 2928 (2008).
19. R. M. Gschwind, *Chem. Rev.*, **108**, 3029 (2008).
20. E. Nakamura and S. Mori, *Angew. Chem., Int. Ed.*, **39**, 3750 (2000).
21. N. Yoshikai, R. Iida and E. Nakamura, *Adv. Synth. Catal.*, **350**, 1063 (2008).
22. G. Costa, A. Camus, L. Gatti and N. Marsich, *J. Organomet. Chem.*, **5**, 568 (1966).
23. A. Camus and N. Marsich, *J. Organomet. Chem.*, **14**, 441 (1968).
24. A. Cairncross, H. Omura and W. A. Sheppard, *J. Am. Chem. Soc.*, **93**, 248 (1971).

25. A. Cairncross and W. A. Sheppard, *J. Am. Chem. Soc.*, **93**, 247 (1971).
26. M. F. Lappert and R. Pearce, *J. Chem. Soc., Chem. Commun.*, 24 (1973).
27. G. van Koten, A. J. Leusink and J. G. Noltes, *J. Chem. Soc., Chem. Commun.*, 1107 (1970).
28. G. van Koten, A. J. Leusink and J. G. Noltes, *Inorg. Nucl. Chem. Lett.*, **7**, 227 (1971).
29. J. M. Guss, R. Mason, K. M. Thomas, G. van Koten and J. G. Noltes, *J. Organomet. Chem.*, **40**, C79 (1972).
30. G. van Koten, A. J. Leusink and J. G. Noltes, *J. Organomet. Chem.*, **85**, 105 (1975).
31. M. A. Willert-Porada, D. J. Burton and N. C. Baenziger, *J. Chem. Soc., Chem. Commun.*, 1633 (1989).
32. B. J. Hathaway, in *Comprehensive Coordination Chemistry* (Ed. G. Wilkinson), Vol. 5, Pergamon, Oxford, 1987, pp. 533–774, and references cited therein.
33. J. G. Noltes, *Phil. Trans. R. Soc. Lond.*, **A308**, 35 (1982).
34. G. van Koten, J. T. B. H. Jastrzebski and S. L. James, in *Comprehensive Organometallic Chemistry II* (Eds E. W. Abel, F. G. A. Stone and G. Williamson), Vol. 3, Pergamon, Oxford, 1995, pp. 57–133.
35. P. W. R. Corfield and H. M. M. Shearer, *Acta Crystallogr.*, **21**, 957 (1966).
36. J. M. Guss, R. Mason, I. Søtofte, G. van Koten and J. G. Noltes, *J. Chem. Soc., Chem. Commun.*, 446 (1972).
37. G. van Koten and J. G. Noltes, *J. Organomet. Chem.*, **84**, 129 (1975).
38. Cambridge Structural Database, release 5.29, August 2008.
39. J. T. B. H. Jastrzebski and G. van Koten, in *Modern Organocopper Chemistry* (Ed. N. Krause), Wiley-VCH, Weinheim, 2002, pp. 1–44.
40. G. van Koten and J. G. Noltes, in *Comprehensive Organometallic Chemistry* (Eds G. Wilkinson, F. G. A. Stone and E. W. Abel), Vol. 2, Pergamon, Oxford, 1982, pp. 709–763.
41. P. J. Pérez and M. M. Diaz-Requejo, in *Comprehensive Organometallic Chemistry III* (Eds R. H. Crabtree, D. M. P. Mingos and K. Meyer), Vol. 2, Elsevier, Amsterdam, 2007, pp. 153–195.
42. J. A. J. Jarvis, R. Pearce and M. F. Lappert, *J. Chem. Soc., Dalton Trans.*, 999 (1977).
43. G. van Koten, J. T. B. H. Jastrzebski and J. G. Noltes, *J. Org. Chem.*, **42**, 2047 (1977).
44. H. Eriksson, M. Håkansson and S. Jagner, *Inorg. Chim. Acta*, **277**, 233 (1998).
45. M. Håkansson, H. Eriksson and S. Jagner, *Inorg. Chim. Acta*, **359**, 2519 (2006).
46. R. T. Boere, J. D. Masuda and P. Tran, *J. Organomet. Chem.*, **691**, 5585 (2006).
47. D. Nobel, G. van Koten and A. L. Spek, *Angew. Chem., Int. Ed. Engl.*, **28**, 208 (1989).
48. H. Eriksson, M. Örtendahl and M. Håkansson, *Organometallics*, **15**, 4823 (1996).
49. G. van Koten, J. T. B. H. Jastrzebski, F. Muller and C. H. Stam, *J. Am. Chem. Soc.*, **107**, 697 (1985).
50. G. van Koten, J. T. B. H. Jastrzebski, C. H. Stam and N. C. Niemann, *J. Am. Chem. Soc.*, **106**, 1880 (1984).
51. M. Håkansson, H. Eriksson, A. B. Åhman and S. Jagner, *J. Organomet. Chem.*, **595**, 102 (2000).
52. A. Sundararaman, R. A. Lalancette, L. N. Zakharov, A. L. Rheingold and F. Jäkle, *Organometallics*, **22**, 3526 (2003).
53. F. Jäkle, *Dalton Trans.*, 2851 (2007).
54. H. Eriksson and M. Håkansson, *Organometallics*, **16**, 4243 (1997).
55. S. Gambarotta, C. Floriani, A. Chiesi-Villa and C. Guastini, *J. Chem. Soc., Chem. Commun.*, 1156 (1983).
56. E. M. Meyer, S. Gambarotta, C. Floriani, A. Chiesi-Villa and C. Guastini, *Organometallics*, **8**, 1067 (1989).
57. D. Cremer and J. A. Pople, *J. Am. Chem. Soc.*, **97**, 1354 (1975).
58. P. Belanzoni, M. Rosi, A. Sgamellotti, E. J. Baerends and C. Floriani, *Chem. Phys. Lett.*, **257**, 41 (1996).
59. M. Niemeyer, *Organometallics*, **17**, 4649 (1998).
60. K. Venkatasubbaiah, A. G. DiPasquale, M. Bolte, A. L. Rheingold and F. Jäkle, *Angew. Chem., Int. Ed.*, **45**, 6838 (2006).
61. P. Schulte, U. Behrens and F. Olbrich, *Z. Anorg. Allg. Chem.*, **626**, 1692 (2000).
62. A. N. Nesmeyanov, Yu. T. Struchkov, N. N. Sedova, V. G. Andrianov, Yu. V. Volgin and V. A. Sazonova, *J. Organomet. Chem.*, **137**, 217 (1977).

63. M. D. Janssen, M. A. Corsten, A. L. Spek, D. M. Grove and G. van Koten, *Organometallics*, **15**, 2810 (1996).
64. E. Wehman, G. van Koten, M. Knotter, H. Spelten, D. Heijdenrijk, A. N. S. Mak and C. H. Stam, *J. Organomet. Chem.*, **325**, 293 (1987).
65. R. W. M. ten Hoedt, G. van Koten and J. G. Noltes, *J. Organomet. Chem.*, **179**, 227 (1979).
66. J. G. Noltes, R. W. M. ten Hoedt, G. van Koten, A. L. Spek and J. C. Schoone, *J. Organomet. Chem.*, **225**, 365 (1982).
67. G. van Koten and J. G. Noltes, *J. Chem. Soc., Chem. Commun.*, 575 (1974).
68. R. W. M. ten Hoedt, G. van Koten and J. G. Noltes, *J. Organomet. Chem.*, **133**, 113 (1977).
69. R. W. M. ten Hoedt, J. G. Noltes, G. van Koten and A. L. Spek, *J. Chem. Soc., Dalton Trans.*, 1800 (1978).
70. G. van Koten, R. W. M. ten Hoedt and J. G. Noltes, *J. Org. Chem.*, **42**, 2705 (1977).
71. A. Camus, N. Marsich, G. Nardin and L. Randaccio, *J. Organomet. Chem.*, **174**, 121 (1979).
72. R. I. Papasergio, C. L. Raston and A. H. White, *J. Chem. Soc., Dalton Trans.*, 3085 (1987).
73. R. I. Papasergio, C. L. Raston and A. H. White, *J. Chem. Soc., Chem. Commun.*, 1419 (1983).
74. T. R. van den Ancker, S. K. Bhargava, F. Mohr, S. Papadopoulos, C. L. Raston, B. W. Skelton and A. H. White, *J. Chem. Soc., Dalton Trans.*, 3069 (2001).
75. C. Eaborn, M. S. Hill, P. B. Hitchcock and J. D. Smith, *J. Chem. Soc., Dalton Trans.*, 2467 (2002).
76. S. Wingerter, H. Gornitzka, G. Bertrand and D. Stalke, *Eur. J. Inorg. Chem.*, 173 (1999).
77. P. B. Hitchcock, M. F. Lappert and M. Layh, *J. Chem. Soc., Dalton Trans.*, 1619 (1998).
78. E. Wehman, G. van Koten, J. T. B. H. Jastrzebski, M. A. Rotteveel and C. H. Stam, *Organometallics*, **7**, 1477 (1988).
79. F. Garcia, A. D. Hopkins, R. A. Kowenicki, M. McPartlin, M. C. Rogers and D. S. Wright, *Organometallics*, **23**, 3884 (2004).
80. G. Costa, G. Pellizer and F. Rubessa, *J. Inorg. Nucl. Chem.*, **26**, 961 (1964).
81. A. Yamamoto, A. Miyashita, T. Yamamoto and S. Ikeda, *Bull. Chem. Soc. Jpn.*, **45**, 1583 (1972).
82. K.-H. Thiele and J. Köhler, *J. Organomet. Chem.*, **12**, 225 (1968).
83. G. Costa, A. Camus, N. Marsich and L. Gatti, *J. Organomet. Chem.*, **8**, 339 (1967).
84. A. Camus and N. Marsich, *J. Organomet. Chem.*, **21**, 249 (1970).
85. G. van Koten and J. G. Noltes, *J. Chem. Soc., Chem. Commun.*, 452 (1972).
86. G. van Koten, J. G. Noltes and A. L. Spek, *J. Organomet. Chem.*, **159**, 441 (1978).
87. P. S. Coan, K. Folting, J. C. Huffmann and K. G. Caulton, *Organometallics*, **8**, 2724 (1989).
88. S. Gambarotta, S. Strologo, C. Floriani, A. Chiesi-Villa and C. Guastini, *Organometallics*, **3**, 1444 (1984).
89. F. Schaper, S. R. Foley and R. F. Jordan, *J. Am. Chem. Soc.*, **126**, 2114 (2004).
90. M. Niemeyer, *Z. Anorg. Allg. Chem.*, **629**, 1535 (2003).
91. B. Schiemenz and P. P. Power, *Organometallics*, **15**, 958 (1996).
92. X. He, M. M. Olmstead and P. P. Power, *J. Am. Chem. Soc.*, **114**, 9668 (1992).
93. A. Sundararaman, L. N. Zakharov, A. L. Rheingold and F. Jäkle, *Chem. Commun.*, 1708 (2005).
94. M. D. Janssen, M. Herres, A. L. Spek, D. M. Grove, H. Lang and G. van Koten, *J. Chem. Soc., Chem. Commun.*, 925 (1995).
95. M. D. Janssen, K. Kohler, M. Herres, A. Dedieu, W. J. J. Smeets, A. L. Spek, D. M. Grove, H. Lang and G. van Koten, *J. Am. Chem. Soc.*, **118**, 4817 (1996).
96. H. Lang, K. Kohler, G. Rheinwald, L. Zsolnai, M. Buchner, A. Driess, G. Huttner and J. Strahle, *Organometallics*, **18**, 598 (1999).
97. T. Stein and H. Lang, *J. Organomet. Chem.*, **664**, 142 (2002).
98. M. D. Janssen, W. J. J. Smeets, A. L. Spek, D. M. Grove, H. Lang and G. van Koten, *J. Organomet. Chem.*, **505**, 123 (1995).
99. G. G. Dubinia, H. Furutachi and D. A. Vicic, *J. Am. Chem. Soc.*, **130**, 8600 (2008).
100. N. P. Mankad, T. G. Gray, D. S. Laitar and J. P. Sadighi, *Organometallics*, **23**, 1191 (2004).
101. N. P. Mankad, D. S. Laitar and J. P. Sadighi, *Organometallics*, **23**, 3369 (2004).
102. D. S. Laitar, E. Y. Tsui and J. P. Sadighi, *Organometallics*, **25**, 2405 (2006).
103. D. S. Laitar, E. Y. Tsui and J. P. Sadighi, *J. Am. Chem. Soc.*, **128**, 11036 (2006).
104. C. Nolte, P. Mayer and B. F. Straub, *Angew. Chem., Int. Ed.*, **46**, 2101 (2007).
105. K. B. Sharpless and M. G. Finn, *Angew. Chem., Int. Ed.*, **40**, 2004 (2001).

106. M. M. Olmstead and P. P. Power, *J. Am. Chem. Soc.*, **112**, 8008 (1990).
107. B. Lenders, D. M. Grove, W. J. J. Smeets, P. van der Sluis, A. L. Spek and G. van Koten, *Organometallics*, **10**, 786 (1991).
108. D. M. Knotter, W. J. J. Smeets, A. L. Spek and G. van Koten, *J. Am. Chem. Soc.*, **112**, 5895 (1990).
109. X. He, K. Ruhlandt-Senge and P. P. Power, *J. Am. Chem. Soc.*, **116**, 6963 (1994).
110. G. van Koten and J. G. Noltes, *J. Organomet. Chem.*, **84**, 419 (1975).
111. G. van Koten, A. J. Leusink and J. G. Noltes, *J. Organomet. Chem.*, **84**, 117 (1975).
112. G. van Koten and J. G. Noltes, *J. Organomet. Chem.*, **102**, 551 (1975).
113. G. van Koten, J. T. B. H. Jastrzebski and J. G. Noltes, *Inorg. Chem.*, **16**, 1782 (1977).
114. E. Wehman, G. van Koten and J. T. B. H. Jastrzebski, *J. Organomet. Chem.*, **302**, C35 (1986).
115. W. J. J. Smeets and A. L. Spek, *Acta Crystallogr., Sect. C*, **43**, 870 (1987).
116. E. Wehman, G. van Koten, C. J. M. Erkamp, D. M. Knotter, J. T. B. H. Jastrzebski and C. H. Stam, *Organometallics*, **8**, 94 (1989).
117. G. M. Kapteijn, I. C. M. Wehman-Ooyevaar, D. M. Grove, W. J. J. Smeets, A. L. Spek and G. van Koten, *Angew. Chem., Int. Ed. Engl.*, **32**, 72 (1993).
118. C.-S. Hwang, M. M. Olmstead, X. He and P. P. Power, *J. Chem. Soc., Dalton Trans.*, 2599 (1998).
119. H. L. Aalten, G. van Koten, K. Goubitz and C. H. Stam, *J. Chem. Soc., Chem. Commun.*, 1252 (1985).
120. H. L. Aalten, G. van Koten, K. Goubitz and C. H. Stam, *Organometallics*, **8**, 2293 (1989).
121. A. Cairncross, J. R. Roland, R. M. Henderson and W. A. Sheppard, *J. Am. Chem. Soc.*, **92**, 3187 (1970).
122. T. Tsuda, T. Nakatsuka, T. Hirayama and T. Deagusa, *J. Chem. Soc., Chem. Cummun.*, 557 (1974).
123. M. Niemeyer, *Acta Crystallogr., Sect. E: Struct. Rep. Online*, **57**, m491 (2001).
124. M. Niemeyer, *Acta Crystallogr., Sect. E: Struct. Rep. Online*, **57**, m416 (2001).
125. M. Håkansson, M. Ürtendahl, S. Jagner, M. P. Sigalas and O. Eisenstein, *Inorg. Chem.*, **32**, 2018 (1993).
126. A. L. Keen, M. Doster, H. Han and A. A. Johnson, *Chem. Commun.*, 1221 (2006).
127. T. J. Kealy and P. L. Pauson, *Nature*, **168**, 1039 (1951).
128. S. A. Miller, J. A. Tebboth and J. F. Tremaine, *J. Chem. Soc.*, 632 (1952).
129. P. Laszlo and R. Hoffmann, *Angew. Chem., Int. Ed.*, **39**, 123 (2000).
130. G. Wilkinson and T. S. Piper, *J. Inorg. Nucl. Chem.*, **2**, 32 (1956).
131. F. A. Cotton and T. J. Marks, *J. Am. Chem. Soc.*, **91**, 7281 (1969).
132. L. T. J. Delbaere, D. W. McBride and R. B. Ferguson, *Acta Crystallogr., Sect. B*, **26**, 515 (1970).
133. F. A. Cotton and J. Takats, *J. Am. Chem. Soc.*, **92**, 2353 (1970).
134. T. P. Hanusa, T. A. Ulibarri and W. J. Evans, *Acta Crystallogr., Sect. C*, **41**, 1036 (1985).
135. Q. T. Anderson, E. Erkizia and R. R. Conry, *Organometallics*, **17**, 4917 (1998).
136. H. Ren, X. Zhao, S. Xu, H. Song and B. Wang, *J. Organomet. Chem.*, **691**, 4109 (2006).
137. C. Zybill and G. Muller, *Organometallics*, **6**, 2489 (1987).
138. P. Jutzi, W. Wieland, B. Neumann and H.-G. Stammler, *J. Organomet. Chem.*, **501**, 369 (1995).
139. F. W. Heinemann, M. Zeller and U. Zenneck, *Organometallics*, **23**, 1689 (2004).
140. G. A. Carriedo, J. A. K. Howard and F. G. A. Stone, *J. Organomet. Chem.*, **250**, C28 (1983).
141. G. A. Carriedo, J. A. K. Howard and F. G. A. Stone, *J. Chem. Soc., Dalton Trans.*, 1555 (1984).
142. M. F. Hawthorne, D. C. Young and P. A. Wegner, *J. Am. Chem. Soc.*, **87**, 1818 (1965).
143. M. F. Hawthorne, D. C. Young, T. D. Andrews, D. V. Howe, R. L. Pilling, A. D. Pitts, M. Reintjes, L. F. Warren Jr. and P. A. Wegner, *J. Am. Chem. Soc.*, **90**, 879 (1968).
144. M. F. Howthorne and G. B. Dunks, *Science*, **178**, 462 (1972).
145. E. B. Moore, L. L. Lohr and W. N. Lipscomb, *J. Chem. Phys.*, **35**, 1329 (1961).
146. Y. Do, H. C. Kang, C. B. Knobler and M. F. Hawthorne, *Inorg. Chem.*, **26**, 2348 (1987).
147. H. C. Kang, Y. Do, C. B. Knobler and M. F. Hawthorne, *Inorg. Chem.*, **27**, 1716 (1988).
148. R. M. Wing, *J. Am. Chem. Soc.*, **89**, 5599 (1967).
149. R. M. Wing, *J. Am. Chem. Soc.*, **90**, 4828 (1968).

150. K. J. Adams, J. Cowie, S. G. D. Henderson, G. J. McCormick and A. J. Welch, *J. Organomet. Chem.*, **481**, C9 (1994).
151. B. E. Hodson, T. D. McGrath and F. G. A. Stone, *Organometallics*, **24**, 3386 (2005).
152. E. J. M. Hamilton and A. J. Welch, *Polyhedron*, **10**, 471 (1991).
153. H. C. Kang, Y. Do, C. B. Knobler and M. F. Hawthorne, *J. Am. Chem. Soc.*, **109**, 6530 (1987).
154. A. Varadarajan, S. E. Johnson, F. A. Gomez, S. Chakrabarti, C. B. Knobler and M. F. Hawthorne, *J. Am. Chem. Soc.*, **114**, 9003 (1992).
155. D. A. Owen and M. F. Hawthorne, *J. Am. Chem. Soc.*, **92**, 3194 (1970).
156. D. A. Owen and M. F. Hawthorne, *J. Am. Chem. Soc.*, **93**, 873 (1971).
157. D. E. Harwell, J. McMillan, C. B. Knobler and M. F. Hawthorne, *Inorg. Chem.*, **36**, 5951 (1997).
158. S. V. Ivanov, J. J. Rockwell, O. G. Polyakov, C. M. Gaudinski, O. P. Anderson, K. A. Solntsev and S. H. Strauss, *J. Am. Chem. Soc.*, **120**, 4224 (1998).
159. R. D. Stephens and C. E. Castro, *J. Org. Chem.*, **28**, 3313 (1963).
160. J. Burdon, P. L. Coe, C. R. Marsh and J. C. Tatlow, *J. Chem. Soc., Chem. Commun.*, 1259 (1967).
161. C. E. Castro, R. Havlin, V. K. Honwad, A. Malte and S. Moje, *J. Am. Chem. Soc.*, **91**, 6464 (1969).
162. H. G. Soderbaum, *Chem. Ber.*, **30**, 760 (1897).
163. B. M. Mykhalichko, T. Glovyak and M. G. Mys'kiv, *Russ. J. Inorg. Chem.*, **40**, 733 (1995).
164. U. Cremer, S. Disch and U. Ruschewitz, *Z. Anorg. Allg. Chem.*, **630**, 2304 (2004).
165. F. Olbrich, J. Kopf and E. Weiss, *Angew. Chem., Int. Ed. Engl.*, **32**, 1077 (1993).
166. S. S. Y. Chui, M. F. Y. Ng and C.-M. Che, *Chem. Eur. J.*, **11**, 1739 (2005).
167. Q. H. Wei, G. Q. Yin, L. Y. Zhang and Z. N. Chen, *Organometallics*, **25**, 4941 (2006).
168. O. M. Abu-Salah, M. S. Hussain and E. O. Schlemper, *J. Chem. Soc., Chem. Commun.*, 212 (1988).
169. M. S. Hussain and O. M. Abu-Salah, *J. Organomet. Chem.*, **445**, 295 (1993).
170. K. Kohler, H. Pritzkow and H. Lang, *J. Organomet. Chem.*, **553**, 31 (1998).
171. W. Frosch, S. Back, H. Muller, K. Kohler, A. Driess, B. Schiemenz, G. Huttner and H. Lang, *J. Organomet. Chem.*, **619**, 99 (2001).
172. W. Frosch, S. Back, G. Rheinwald, K. Kohler, H. Pritzkow and H. Lang, *Organometallics*, **19**, 4016 (2000).
173. M. D. Janssen, M. Herres, L. Zsolnai, D. M. Grove, A. L. Spek, H. Lang and G. van Koten, *Organometallics*, **14**, 1098 (1995).
174. V. W.-W. Yam, W.-K. Lee, K. K. Cheung, H.-K. Lee and W.-P. Leung, *J. Chem. Soc., Dalton Trans.*, 2889 (1996).
175. M. I. Bruce, N. N. Zaitseva, B. W. Skelton, N. Somers and A. H. White, *Inorg. Chim. Acta*, **360**, 681 (2007).
176. A. J. Edwards, M. A. Paver, P. R. Raithby, M.-A. Rennie, C. A. Russell and D. S. Wright, *Organometallics*, **13**, 4967 (1994).
177. J. Diez, M. P. Gamasa, J. Gimeno, A. Aguirre, S. Garcia-Granda, J. Holubova and L. R. Falvello, *Organometallics*, **18**, 662 (1999).
178. F. Olbrich, U. Behrens and E. Weiss, *J. Organomet. Chem.*, **472**, 365 (1994).
179. L. Naldini, F. Demartin, M. Manassero, M. Sansoni, G. Rassu and M. A. Zoroddu, *J. Organomet. Chem.*, **279**, C42 (1985).
180. V. W.-W. Yam, W.-K. Lee and K.-K. Cheung, *J. Chem. Soc., Dalton Trans.*, 2335 (1996).
181. M. P. Gamasa, J. Gimeno, E. Lastra and X. Solans, *J. Organomet. Chem.*, **346**, 277 (1988).
182. V. W.-W. Yam, W. K.-M. Fung and K.-K. Cheung, *J. Cluster Sci.*, **10**, 37 (1999).
183. K. Osakada, T. Takizawa and T. Yamamoto, *Organometallics*, **14**, 3531 (1995).
184. V. W.-W. Yam, C.-H. Lam and N. Zhu, *Inorg. Chim. Acta*, **331**, 239 (2002).
185. C.-L. Chan, K.-K. Cheung, W. H. Lam, E. C.-C. Cheng, N. Zhu, S. W.-K. Choi and V. W.-W. Yam, *Chem. Asian J.*, **1**, 273 (2006).
186. V. W.-W. Yam, S. W.-K. Choi, C.-L. Chan and K.-K. Cheung, *Chem. Commun.*, 2067 (1996).
187. W.-H. Chan, Z.-Z. Zhang, T. C. W. Mak and C.-M. Che, *J. Organomet. Chem.*, **556**, 169 (1998).
188. J. Diez, M. P. Gamasa, J. Gimeno, A. Aguirre and S. Garcia-Granda, *Organometallics*, **10**, 380 (1991).

189. W.-Y. Lo, C.-H. Lam, V. W.-W. Yam, N. Zhu, K.-K. Cheung, S. Fathallah, S. Messaoudi, B. Le Guennic, S. Kahlal and J.-F. Halet, *J. Am. Chem. Soc.*, **126**, 7300 (2004).
190. J. H. K. Yip, J. Wu, K.-Y. Wong, K.-W. Yeung and J. J. Vittal, *Organometallics*, **21**, 1612 (2002).
191. V. W.-W. Yam, C.-H. Lam and K.-K. Cheung, *Inorg. Chim. Acta*, **316**, 19 (2001).
192. V. W.-W. Yam, W.-Y. Lo, C.-H. Lam, W. K.-M. Fung, K. M.-C. Wong, V. C.-Y. Lau and N. Zhu, *Coord. Chem. Rev.*, **245**, 39 (2003).
193. V. W.-W. Yam, W. K.-M. Fung, K. M.-C. Wong, V. C.-Y. Lau and K.-K. Cheung, *Chem. Commun.*, 777 (1998).
194. V. W.-W. Yam, W. K.-M. Fung and K.-K. Cheung, *Organometallics*, **17**, 3293 (1998).
195. J. Diez, M. P. Gamasa, J. Gimeno, E. Lastra, A. Aguirre and S. Garcia-Granda, *Organometallics*, **12**, 2213 (1993).
196. V. W. Yam, W.-K. Lee and T.-F. Lai, *Organometallics*, **12**, 2383 (1993).
197. M. I. Bruce, B. C. Hall, B. W. Skelton, M. E. Smith and A. H. White, *J. Chem. Soc., Dalton Trans.*, 995 (2002).
198. M. P. Gamasa, J. Gimeno, E. Lastra, A. Aguirre and S. Garcia-Granda, *J. Organomet. Chem.*, **378**, C11 (1989).
199. V. W.-W. Yam, W.-K. Lee, K.-K. Cheung, B. Crystall and D. Phillips, *J. Chem. Soc., Dalton Trans.*, 3283 (1996).
200. V. W.-W. Yam, W. K.-M. Fung and K.-K. Cheung, *Chem. Commun.*, 963 (1997).
201. V. W.-W. Yam, W. K.-M. Fung and K.-K. Cheung, *Angew. Chem., Int. Ed. Engl.*, **35**, 1100 (1996).
202. Q.-H. Wei, L.-Y. Zhang, L.-X. Shi and Z.-N. Chen, *Inorg. Chem. Commun.*, **7**, 286 (2004).
203. D. M. Knotter, A. L. Spek, D. M. Grove and G. van Koten, *Organometallics*, **11**, 4083 (1992).
204. M. D. Janssen, J. G. Donkervoort, S. B. van Berlekom, A. L. Spek, D. M. Grove and G. van Koten, *Inorg. Chem.*, **35**, 4752 (1996).
205. T. C. Higgs, P. J. Bailey, S. Parsons and P. A. Tasker, *Angew. Chem., Int. Ed.*, **41**, 3038 (2002).
206. C. W. Baxter, T. C. Higgs, A. C. Jones, S. Parsons, P. J. Bailey and P. A. Tasker, *J. Chem. Soc., Dalton Trans.*, 4395 (2002).
207. T. C. Higgs, S. Parsons, P. J. Bailey, A. C. Jones, F. McLachlan, A. Parkin, A. Dawson and P. A. Tasker, *Organometallics*, **21**, 5692 (2002).
208. A. Vega, V. Calvo, E. Spodine, A. Zarate, V. Fuenzalida and J.-Y. Saillard, *Inorg. Chem.*, **41**, 3389 (2002).
209. L. N. Bochkarev, O. N. Druzhkova, S. F. Zhiltsov, L. N. Zakharov, G. K. Fukin, S. Ya. Khorshev, A. I. Yanovsky and Y. T. Struchkov, *Organometallics*, **16**, 500 (1997).
210. G.-Q. Yin, Q.-H. Wei, L.-Y. Zhang and Z.-N. Chen, *Organometallics*, **25**, 580 (2006).
211. P. Pyykkö, *Chem. Rev.*, **97**, 597 (1997).
212. H. O. House and M. J. Umen, *J. Am. Chem. Soc.*, **94**, 5495 (1972).
213. R. G. Pearson and C. D. Gregory, *J. Am. Chem. Soc.*, **98**, 4098 (1976).
214. G. van Koten and J. G. Noltes, *J. Chem. Soc., Chem. Commun.*, 940 (1972).
215. A. J. Leusink, G. van Koten, J. W. Marsman and J. G. Noltes, *J. Organomet. Chem.*, **55**, 419 (1973).
216. G. van Koten, J. T. B. H. Jastrzebski, C. H. Stam and C. Brevard, in *Biological & Inorganic Copper Chemistry* (Eds. K. D. Karlin and J. Zubieta), Vol. II, Adenine Press, New York, 1986, pp. 267–285.
217. G. van Koten and J. T. B. H. Jastrzebski, *Tetrahedron*, **45**, 569 (1989).
218. G. van Koten and J. G. Noltes, *J. Organomet. Chem.*, **171**, C39 (1979).
219. G. van Koten and J. G. Noltes, *J. Am. Chem. Soc.*, **101**, 6539 (1979).
220. G. van Koten, J. T. B. H. Jastrzebski and J. G. Noltes, *J. Organomet. Chem.*, **140**, C23 (1977).
221. N. P. Lorenzen and E. Weiss, *Angew. Chem., Int. Ed. Engl.*, **29**, 300 (1990).
222. M. John, C. Auel, C. Behrens, M. Marsch, K. Harms, F. Bosold, R. M. Gschwind, P. R. Rajamohanan and G. Boche, *Chem. Eur. J.*, **6**, 3060 (2000).
223. M. M. Olmstead and P. P. Power, *Organometallics*, **9**, 1720 (1990).
224. C. M. P. Kronenburg, J. T. B. H. Jastrzebski, M. Lutz, A. L. Spek and G. van Koten, *Organometallics*, **22**, 2312 (2003).
225. R. P. Davies, S. Hornauer and A. J. P. White, *Chem. Commun.*, 304 (2007).
226. M. M. Olmstead and P. P. Power, *J. Am. Chem. Soc.*, **111**, 4135 (1989).

227. S. I. Khan, P. G. Edwards, H. S. H. Yuan and R. Bau, *J. Am. Chem. Soc.*, **107**, 1682 (1985).
228. A. Müller, B. Neumüller and K. Dehnicke, *Angew. Chem., Int. Ed. Engl.*, **36**, 2350 (1997).
229. H. Schmidbaur, J. Adlkofer and W. Buchner, *Angew. Chem., Int. Ed. Engl.*, **12**, 415 (1973).
230. G. Nardin, L. Randaccio and E. Zangrando, *J. Organomet. Chem.*, **74**, C23 (1974).
231. C. M. P. Kronenburg, J. T. B. H. Jastrzebski, J. Boersma, M. Lutz, A. L. Spek and G. van Koten, *J. Am. Chem. Soc.*, **124**, 11675 (2002).
232. C. M. P. Kronenburg, C. H. M. Amijs, J. T. B. H. Jastrzebski, M. Lutz, A. L. Spek and G. van Koten, *Organometallics*, **21**, 4662 (2002).
233. C.-S. Hwang and P. P. Power, *Organometallics*, **18**, 697 (1999).
234. R. P. Davies, S. Hornauer and P. B. Hitchcock, *Angew. Chem., Int. Ed.*, **46**, 5191 (2007).
235. P. Reiss and D. Fenske, *Z. Anorg. Allg. Chem.*, **626**, 1317 (2000).
236. S. F. Martin, J. R. Fishpaugh, J. M. Power, D. M. Giolando, R. A. Jones, C. M. Nunn and A. H. Cowley, *J. Am. Chem. Soc.*, **110**, 7226 (1988).
237. C. Quannes, G. Dressaire and Y. Langlois, *Tetrahedron Lett.*, 815 (1977).
238. S. H. Bertz, G. Miao, B. E. Rossiter and J. P. Snyder, *J. Am. Chem. Soc.*, **117**, 11023 (1995).
239. S. H. Bertz, G. Miao and M. Eriksson, *Chem. Commun.*, 815 (1996).
240. S. H. Bertz, M. Eriksson, G. Miao and J. P. Snyder, *J. Am. Chem. Soc.*, **118**, 10906 (1996).
241. S. H. Bertz, A. Chopra, M. Eriksson, C. A. Ogle and P. Seagle, *Chem. Eur. J.*, **5**, 2680 (1999).
242. H. Hope, M. M. Olmstead, P. P. Power, J. Sandell and X. Xu, *J. Am. Chem. Soc.*, **107**, 4337 (1985).
243. D. F. Dempsey and G. S. Girolami, *Organometallics*, **7**, 1208 (1988).
244. R. Fischer, H. Gorls and M. Westerhausen, *Organometallics*, **26**, 3269 (2007).
245. C. Eaborn, P. B. Hitchcock, J. D. Smith and A. C. Sullivan, *J. Organomet. Chem.*, **263**, C23 (1984).
246. P. Leoni, M. Pasquali and C. A. Ghilardi, *J. Chem. Soc., Chem. Commun.*, 240 (1983).
247. P. G. Edwards, R. W. Gellert, M. W. Marks and R. Bau, *J. Am. Chem. Soc.*, **104**, 2072 (1982).
248. H. Hope, D. Oram and P. P. Power, *J. Am. Chem. Soc.*, **106**, 1149 (1984).
249. B. H. Lipshutz, R. S. Wilhelm and D. M. Floyd, *J. Am. Chem. Soc.*, **103**, 7672 (1981).
250. B. H. Lipshutz, *Synthesis*, 325 (1987).
251. B. H. Lipshutz, in *Organometallics in Synthesis* (Ed. M. Schlosser), John Wiley & Sons, Ltd, Chichester, 1994, pp. 283–382.
252. B. H. Lipshutz, in *Advances in Metal-Organic Chemistry*, Vol. 4 (Ed. L. S. Liebeskind), JAI Press, Greenwich, CT, 1995, pp. 1–64.
253. S. H. Bertz, *J. Am. Chem. Soc.*, **113**, 5470 (1991).
254. S. H. Bertz, *J. Am. Chem. Soc.*, **112**, 4031 (1990).
255. S. H. Bertz, K. Nilsson, Ö. Davidsson and J. P. Snyder, *Angew. Chem., Int. Ed.*, **37**, 314 (1998).
256. T. A. Mobley, F. Müller and S. Berger, *J. Am. Chem. Soc.*, **120**, 1333 (1998).
257. T. Stemmler, J. E. Penner-Hahn and P. Kochel, *J. Am. Chem. Soc.*, **115**, 348 (1993).
258. T. M. Barnhart, H. Huang and J. E. Penner-Hahn, *J. Org. Chem.*, **60**, 4310 (1995).
259. T. L. Stemmler, T. M. Barnhart, J. E. Penner-Hahn, C. E. Tucker, P. Knochel, M. Böhme and G. Frenking, *J. Am. Chem. Soc.*, **117**, 12489 (1995).
260. H. Huang, C. H. Liang and J. E. Penner-Hahn, *Angew. Chem., Int. Ed.*, **37**, 1564 (1998).
261. J. P. Snyder, D. P. Spangler, J. R. Behling and B. E. Rossiter, *J. Org. Chem.*, **59**, 2665 (1994).
262. J. P. Snyder and S. H. Bertz, *J. Org. Chem.*, **60**, 4312 (1995).
263. N. Krause, *Angew. Chem., Int. Ed.*, **38**, 79 (1999).
264. C. M. P. Kronenburg, J. T. B. H. Jastrzebski, A. L. Spek and G. van Koten, *J. Am. Chem. Soc.*, **120**, 9688 (1998).
265. G. Boche, F. Bosold, M. Marsch and K. Harms, *Angew. Chem., Int. Ed.*, **37**, 1684 (1998).
266. C. Eaborn, M. S. Hill, P. B. Hitchcock and J. D. Smith, *Organometallics*, **19**, 5780 (2000).
267. C.-S. Hwang and P. P. Power, *Bull. Korean Chem. Soc.*, **24**, 605 (2003).
268. F. Bosold, M. Marsch, K. Harms and G. Boche, *Z. Kristallogr. New Cryst. Struct.*, **216**, 143 (2001).
269. C. Eaborn, S. M. El-Hamruni, M. S. Hill, P. B. Hitchcock and J. D. Smith, *J. Chem. Soc., Dalton Trans.*, 3975 (2002).
270. C.-S. Hwang and P. P. Power, *J. Am. Chem. Soc.*, **120**, 5409 (1998).
271. R. P. Davies and S. Hornauer, *Eur. J. Inorg. Chem.*, 51 (2005).
272. A. H. Lewin and N. L. Goldberg, *Tetrahedron Lett.*, 491 (1972).

273. H. O. House, *Acc. Chem. Res.*, **9**, 59 (1976).
274. T. Cohen and I. Cristea, *J. Am. Chem. Soc.*, **98**, 748 (1976).
275. G. M. Whitesides, W. F. Fischer Jr., J. S. Filippo Jr., R. W. Bashe and H. O. House, *J. Am. Chem. Soc.*, **91**, 4871 (1969).
276. J. P. Snyder, *J. Am. Chem. Soc.*, **117**, 11025 (1995).
277. M. Yamanaka, A. Inagaki and E. Nakamura, *J. Comput. Chem.*, **24**, 1401 (2002).
278. H. Hu and J. P. Snyder, *J. Am. Chem. Soc.*, **129**, 7210 (2007).
279. S. H. Bertz, S. Cope, D. Dorton, M. Murphy and C. Ogle, *Angew. Chem., Int. Ed.*, **46**, 7082 (2007).
280. E. R. Bartholomew, S. H. Bertz, S. Cope, D. C. Dorton, M. Murphy and C. A. Ogle, *Chem. Commun.*, 1176 (2008).
281. T. Gärtner, W. Henze and R. M. Gschwind, *J. Am. Chem. Soc.*, **129**, 11362 (2007).
282. S. H. Bertz, S. Cope, M. Murphy, C. A. Ogle and B. J. Taylor, *J. Am. Chem. Soc.*, **129**, 7208 (2007).
283. V. C. R. McLoughlin and J. Thrower, *Tetrahedron*, **25**, 5921 (1969).
284. D. M. Wiemers and D. J. Burton, *J. Am. Chem. Soc.*, **108**, 832 (1986).
285. R. E. Marsch, *Acta Crystallogr.*, **B53**, 317 (1997).
286. D. Naumann, T. Roy, B. Caeners, D. Hütten, K.-F. Tebbe and T. Gilles, *Z. Anorg. Allg. Chem.*, **626**, 999 (2000).
287. D. Naumann, T. Roy, K.-F. Tebbe and W. Crump, *Angew. Chem., Int. Ed. Engl.*, **32**, 1482 (1993).
288. J. A. Schlueter, U. Geiser, J. M. Williams, H. H. Wang, W.-K. Kwok, J. A. Fendrich, K. D. Carlson, C. A. Achenbach, J. D. Dudek, D. Naumann, T. Roy, J. E. Schrirber and W. R. Bayless, *J. Chem. Soc., Chem. Commun.*, 1599 (1994).
289. U. Geiser, J. A. Schlueter, J. M. Williams, D. Naumann and T. Roy, *Acta Crystallogr.*, **B51**, 789 (1985).
290. X. Ribas, D. A. Jackson, B. Donnadieu, J. Mahia, T. Parella, R. Xifra, B. Hedman, K. O. Hodgson, A. Llobet and T. D. P. Stack, *Angew. Chem., Int. Ed.*, **41**, 2991 (2002).
291. R. Xifra, X. Ribas, A. Llobet, A. Poater, M. Duran, M. Sola, T. D. P. Stack, J. Benet-Buchholz, B. Donnadieu, J. Mahia and T. Parella, *Chem. -Eur. J.*, **11**, 5146 (2005).
292. I. Kinoshita, L. J. Wright, S. Kubo, K. Kimura, A. Sakata, T. Yano, R. Miyamoto, T. Nishioka and K. Isobe, *Dalton Trans.*, 1993 (2003).
293. R. Miyamoto, R. Santo, T. Matsushita, T. Nishioka, A. Ichimura, Y. Teki and I. Kinoshita, *Dalton Trans.*, 3179 (2005).
294. R. Santo, R. Miyamoto, R. Tanaka, T. Nishioka, K. Sato, K. Toyota, M. Obata, S. Yano, I. Kinoshita, A. Ichimura and T. Takui, *Angew. Chem., Int. Ed.*, **45**, 7611 (2006).
295. R. Miyamoto, R. T. Hamazawa, M. Hirotsu, T. Nishioka, I. Kinoshita and L. J. Wright, *Chem. Commun.*, 4047 (2005).
296. H. Maeda, A. Osuka, Y. Ishikawa, I. Aritome, Y. Hisaeda and H. Furuta, *Org. Lett.*, **5**, 1293 (2003).
297. H. Maeda, Y. Ishikawa, T. Matsuda, A. Osuka and H. Furuta, *J. Am. Chem. Soc.*, **125**, 11822 (2003).
298. H. Maeda, A. Osuka and H. Furuta, *J. Am. Chem. Soc.*, **125**, 15690 (2003).
299. H. Furuta, H. Maeda and A. Osuka, *J. Am. Chem. Soc.*, **122**, 803 (2000).
300. H. Furuta, T. Ishizuka, A. Osuka, Y. Uwatoko and Y. Ishikawa, *Angew. Chem., Int. Ed.*, **40**, 2323 (2001).
301. M. Pawlicki, I. Kańska and L Latos-Grażyński, *Inorg. Chem.*, **46**, 6575 (2007).
302. S. Diez-González and S. P. Nolan, *Synlett*, **14**, 2158 (2007).
303. H.-W. Wanzlick, *Angew. Chem., Int. Ed. Engl.*, **1**, 75 (1962).
304. K. Öffle, *J. Organomet. Chem.*, **12**, P42 (1968).
305. J. C. Green, R. G. Scur, P. L. Arnold and G. L. Cloke, *Chem. Commun.*, 1963 (1997).
306. E. Peris and R. H. Crabtree, *Coord. Chem. Rev.*, **248**, 2239 (2004).
307. V. César, S. Bellemin-Laponnaz and L. H. Gade, *Chem. Soc. Rev.*, **33**, 619 (2004).
308. F. Glorius (Ed.), *N-Heterocyclic Carbenes in Transition Metal Catalysis, Topics in Organometallic Chemistry*, Vol. 21, Springer, Berlin, 2007.
309. A. J. Arduengo III, R. L. Harlow and M. Kline, *J. Am. Chem. Soc.*, **113**, 361 (1991).
310. A. J. Arduengo III, *Acc. Chem. Res.*, **32**, 913 (1999).
311. H. Kaur, F. K. Zinn, E. D. Stevens and S. P. Nolan, *Organometallics*, **23**, 1157 (2004).

312. S. Díez-González, H. Kaur, F. K. Zinn, E. D. Stevens and S. P. Nolan, *J. Org. Chem.*, **70**, 4784 (2005).
313. L. A. Goj, E. D. Blue, S. A. Delp, T. B. Gunnoe, T. R. Cundari, A. W. Pierpont, J. L. Petersen and P. D. Boyle, *Inorg. Chem.*, **45**, 9032 (2006).
314. L. A. Goj, E. D. Blue, S. A. Delp, T. B. Gunnoe, T. R. Cundari and J. L. Petersen, *Organometallics*, **25**, 4097 (2006).
315. S. A. Delp, C. Munro-Leighton, L. A. Goj, M. A. Ramirez, T. B. Gunnoe, J. L. Petersen and P. D. Boyle, *Inorg. Chem.*, **46**, 2365 (2007).
316. A. Welle, S. Díez-González, B. Tinant, S. P. Nolan and O. Riant, *Org. Lett*, **8**, 6059 (2006).
317. J. T. York, V. G. Young Jr. and W. B. Tolman, *Inorg. Chem.*, **45**, 4191 (2006).
318. X. Dai and T. H. Warren, *J. Am. Chem. Soc.*, **126**, 10085 (2004).
319. D. S. Laitar, P. Muller and J. P. Sadighi, *J. Am. Chem. Soc.*, **127**, 17196 (2005).
320. G. V. Goeden, J. C. Huffman and K. C. Caulton, *Inorg. Chem.*, **25**, 2484 (1986).
321. A. J. Arduengo III, H. V. R. Dias, J. C. Calabrese and F. Davidson, *Organometallics*, **12**, 3405 (1993).
322. X.-Jian Wan, F.-B. Xu, Q.-S. Li, H.-B. Song and Z.-Z. Zhang, *Inorg. Chem. Commun.*, **8**, 1053 (2005).
323. R. McKie, J. A. Murphy, S. R. Park, M. D. Spicer and S.-Z. Zhou, *Angew. Chem., Int. Ed.*, **46**, 6525 (2007).
324. X. Hu, I. Castro-Rodriguez and K. Meyer, *J. Am. Chem. Soc.*, **125**, 12237 (2003).
325. X. Hu, I. Castro-Rodriguez, K. Olsen and K. Meyer, *Organometallics*, **23**, 755 (2004).
326. A. A. D. Tulloch, A. A. Danopoulos, S. Kleinhenz, M. E. Light, M. B. Hursthouse and G. Eastham, *Organometallics*, **20**, 2027 (2001).
327. N. Schneider, V. Cesar, S. Bellemin-Laponnaz and L. H. Gade, *J. Organomet. Chem.*, **690**, 5556 (2005).
328. P. L. Arnold, A. C. Scarisbrick, A. J. Blake and C. Wilson, *Chem. Commun.*, 2340 (2001).
329. S. Gischig and A. Togni, *Organometallics*, **23**, 2479 (2004).
330. S. Gischig and A. Togni, *Organometallics*, **24**, 203 (2005).
331. B. Bantu, D. Wang, K. Wurst and M. R. Buchmeiser, *Tetrahedron*, **61**, 12145 (2005).
332. H. G. Raubenheimer, S. Cronje and P. J. Olivier, *J. Chem. Soc., Dalton Trans.*, 313 (1995).
333. Y. M. Badiei and T. H. Warren, *J. Organomet. Chem.*, **690**, 5989 (2005).
334. J. Yun, D. Kim and H. Yun, *Chem. Commun.*, 5181 (2005).
335. C. Y. Legault, C. Kendall and A. B. Charette, *Chem. Commun.*, 3826 (2005).
336. J. J. Van Veldhuizen, S. B. Garber, J. S. Kingsbury and A. H. Hoveyda, *J. Am. Chem. Soc.*, **124**, 4954 (2002).
337. A. O. Larsen, W. Leu, C. N. Oberhuber, J. E. Campbell and A. H. Hoveyda, *J. Am. Chem. Soc.*, **126**, 11130 (2004).
338. R. A. Gossage, J. T. B. H. Jastrzebski and G. van Koten, *Angew. Chem., Int. Ed.*, **44**, 1448 (2005).

CHAPTER **3**

Thermochemistry of organocopper compounds

JOEL F. LIEBMAN

Department of Chemistry and Biochemistry, University of Maryland, Baltimore County, 1000 Hilltop Circle, Baltimore, Maryland 21250, USA
Fax: (+1)410-455-2608; e-mail: jliebman@umbc.edu

and

SUZANNE W. SLAYDEN

Department of Chemistry, George Mason University, 4400 University Drive, Fairfax, Virginia 22030, USA
Fax: (+1)703-993-1055; e-mail: sslayden@gmu.edu

I. INTRODUCTION . . . 146
II. HYDROCARBYL DERIVATIVES OF COPPER . . . 146
A. Methyl Copper and its Derivatives . . . 146
B. Copper Acetylides . . . 147
C. Alkyl Radicals on the Surface of Elemental Copper . . . 148
D. Conjectures on the Energetics of Dialkyl Cuprates . . . 149
E. Anionic Copper . . . 149
F. Copper Bonded to Methylene . . . 149
G. Cyclopropanes, Other Cycloalkanes and Copper-containing Species . . . 150
H. Vinyl and Other Alkenylcoppers . . . 150
I. Phenyl and Arylcopper Derivatives . . . 151
J. Polycuprioalkanes (Polycopperalkanes) . . . 153
K. Bonds with Higher Oxidation State Copper . . . 153
III. THE THERMOCHEMISTRY OF BINARY COPPER–CARBON SPECIES, Cu_xC_y . . . 155
IV. COPPER COMPLEXES WITH CARBON MONOXIDE AND CARBON DIOXIDE . . . 156
A. Homoleptic Copper Carbonyls and Related Species . . . 156
B. Copper Halide Carbonyl Complexes . . . 157

The chemistry of organocopper compounds
Edited by Z. Rappoport and I. Marek

C. Cu(I) vs Cu(II) Binding of CO 158
D. Copper Carbon Dioxide Complexes 158
V. REFERENCES 158

I. INTRODUCTION

Copper has played an illustrious role in organic reaction chemistry both as reagent and catalyst, and there is an extensive chemical literature in consequence. Diverse coordination compounds of copper with a wide variety of ligands fill the textbooks. However, there are few organocopper species for which knowledge of their energetics accompanies that of their structural and reaction chemistry. These few copper–carbon species include conventionally bonded R–Cu compounds, gas-phase metal cation complexes and organic compounds chemisorbed on clean metal surfaces of the bulk element. Copper cyanide and its derivatives (such as anionic and isocyanide complexes, and the *N*-oxide, the metal fulminate) are not discussed in this chapter despite Cu–C bonding. Likewise, π-bonded complexes of copper with olefins, acetylenes and arenes are not considered here. Although we would prefer to consider only enthalpies of formation and of reaction, or even Gibbs energies and equilibrium constants, there are very little such data. What there is in that regard are included, along with some reports of bond energies and binding energies. As a transition metal, copper has a variety of oxidation states: +1 and +2 are commonplace, and +3 is important—even +4 is increasingly well-known, although seemingly not yet to the organic chemist. Most of the available data are for the +1 and +2 states, with a section on the higher oxidation state.

II. HYDROCARBYL DERIVATIVES OF COPPER

The hydrocarbyl derivatives of copper are classical σ-bonded copper-containing species. They are formally analogous to other organometallic compounds of lithium, magnesium, zinc, cadmium and mercury, aluminum, tin and lead. Most commonly they contain Cu(I) and have the generic formula 'RCu'. Other species with formal R–Cu σ bonds are formed from reactions of organic species on metallic copper surfaces. RCu is also found in the anionic 'ate' complexes, $[R_2Cu]^-$.

A. Methyl Copper and its Derivatives

Methyl copper is the simplest hydrocarbyl derivative of Cu(I). From the analysis of the energetics of gas-phase Cu^+/alkane reactions, the 0 K bond energy for CH_3Cu is found to be 243 ± 8 kJ mol^{-1}[1]. This value is in fine agreement with an earlier lower bound of 249 kJ mol^{-1} found from related cation/alcohol reactions[2]. From the sublimation enthalpy of Cu (337 kJ mol^{-1})[3], reaction 1 is exothermic by *ca* 94 kJ mol^{-1}. This is much less exothermic than for the corresponding reactions of Ag and Au which are −150 and −159 kJ mol^{-1}, respectively[4]. The bond energy value does not reflect particularly weak metal–carbon bonding because there is no additional stabilization due to agostic interactions between the closed-shell Cu and the C–H bonds. No new information about any C–Cu bond energies was found for any other saturated hydrocarbon.

$$\text{MeCu(g)} \longrightarrow \text{Me}^\bullet\text{(g)} + \text{Cu(s)} \quad (1)$$

Approximating the enthalpy of dissociation by the dissociation energy, the equality in equation 2 can be rewritten as equation 3.

$$D(\text{X–Y}) = \Delta H_f(\text{X}) + \Delta H_f(\text{Y}) - \Delta H_f(\text{XY}) \quad (2)$$

$$\Delta H_f(\text{MeCu}) = \Delta H_f(\text{Me}^\bullet) + \Delta H_f(\text{Cu}) - D(\text{Me–Cu}) \quad (3)$$

Using the 0 K enthalpies of formation of CH_3• and Cu(g), the 0 K enthalpy of formation of gaseous MeCu is 243 kJ mol^{-1}. Equating the 0 K–298 K enthalpy difference for MeCu with that of MeI (copper is a rather non-electropositive metal and iodine is a rather non-electronegative nonmetal) of 10 kJ mol^{-1}, or with MeH (methane) of 8 kJ mol^{-1} [3], gives an estimated enthalpy of formation at 298 K for MeCu of *ca* 234 kJ mol^{-1}.

The RCu species is simply understood as containing Cu(I). The related Cu(II) species, MeCuH and MeCuF, were formed by the reactions of atomic copper with CH_4[5] and CH_3F[6] in a matrix environment. However, because the temperature and electronic state of the atomic copper are ambiguous (the copper is formed by laser excitation of either a metal-containing matrix or bulk metal), thermochemical results cannot be derived from these data. However, the structural and stoichiometric assignments are secure because of the careful spectroscopic analyses of the organocopper species as well as alternative syntheses of the relevant species.

B. Copper Acetylides

Lebedev and Kulagina published a review of thermodynamic characteristics of copper, silver, gold and mercury acetylides from calorimetric measurements[7]. Enthalpies of formation of the copper-containing acetylides of the type RC≡CCu in the solid state are reported to be: R = Vi, 296.0 ± 2.8; *n*-Bu, 73.2 ± 3.5; *n*-Hx, −100.4 ± 3.3; and Ph, 296.0 ± 2.1 kJ mol^{-1}. Of related interest is the enthalpy of formation of silver phenylacetylide, −346.0 ± 1.7 kJ mol^{-1}. There is an apparent anomaly in the data. It is well-known that there is a linear relationship between the enthalpies of formation and number of carbons in a homologous series of compounds, $CH_3(CH_2)_nZ$, in both the liquid and gas phases[8]. There are very few series for which solid state enthalpies of formation are available, but for two of them the methylene increments are −31.5 (*n*-alkanols) and −27.0 (*n*-alkanamides) kJ mol^{-1}. The enthalpy of formation difference for the two copper acetylides R = *n*-Bu and R = *n*-Hex is −86.8 kJ mol^{-1} per methylene group which is too large to be credible. Assuming that the enthalpy of hydrogenation in the solid state is not too different from that in the liquid or gaseous state (*ca* −130 kJ mol^{-1} for hydrogenation of 1-buten-3-yne to 1-butyne), then the enthalpy of formation of copper ethylacetylide derived from hydrogenation of copper vinylacetylide is *ca* 166 kJ mol^{-1}. From the derived enthalpy of formation of copper ethylacetylide and the average of the two known methylene increments, the enthalpy of formation for the butyl derivative would be about 108 and for the hexyl derivative about 50 kJ mol^{-1}.

Vinyl- and phenyl-substituted compounds, Vi-Z and Ph-Z, are observed to have approximately the same enthalpies of formation[9] and this holds true for the measured ViC≡CCu and PhC≡CCu species as well. In an earlier chapter[4] we reported that the endothermicities of reaction 4 are essentially the same for M = Ag and Cu and so reaction 5 is essentially thermoneutral.

$$\text{MCN(s)} + \text{PhC≡CH(lq)} \longrightarrow \text{PhC≡CM(s)} + \text{HCN(lq)} \quad (4)$$

$$\text{CuCN(s)} + \text{PhC≡CAg(s)} \longrightarrow \text{PhC≡CCu(s)} + \text{AgCN(s)} \quad (5)$$

The measured enthalpies of combustion of the copper acetylides are problematic even though the error bars are relatively small and the analysis of the compounds suggest high purity samples. The combustion process is written with the explicit assumption that solid CuO, in the presence of water and carbon dioxide, is the product. However, the solid CuO, the lower valence solid Cu_2O and the hydrated copper oxides $Cu(OH)_2$ and CuOH are comparably stable. To further confound the issue, there are also well-known copper hydroxide carbonates (such as the minerals malachite and azurite) that could be formed.

Small differences in combustion conditions, or the identity of the copper acetylide, could result in different amounts of these species. As such, careful measurements of enthalpy of combustion are not immediately related to well-defined values of enthalpy of formation and so the reported enthalpies of formation and even their differences must regrettably be viewed with skepticism.

There is an electrochemical measurement[10] for the enthalpy of formation of the parent monocopper acetylide, CuC≡CH, in aqueous solution. There are no additional data to help interrelate this aqueous-phase data with those from the solid.

C. Alkyl Radicals on the Surface of Elemental Copper

These solid-phase experiments deal with organometallics containing R−Cu bonds[11,12] but in a different phase from those in which discrete molecular species are found, or at least those compounds with well-defined stoichiometries. Perhaps had there been ample thermochemical data for more 'normal' compounds, these findings from the solid state might not have been included. In what follows, the crystal face that is being considered is generally disregarded, despite careful experiments that allowed for this distinction[13]. Summarizing numerous studies from the solid state, the bond energies are: methyl−Cu, 121 ± 9 kJ mol^{-1} [12]; ethyl−Cu, 138 ± 7 kJ mol^{-1} [14]; primary alkyl−Cu, 136 ± 25 kJ mol^{-1} [12]; and allyl−Cu, 109 ± 17 kJ mol^{-1} [15]. A comparison could be made between methyl and primary alkyl in terms of the methyl deviation from the 'universal methylene increment' and the anomalies of methyl substitution[8]. The weakened allyl−Cu bond could be explained simply in terms of resonance stabilization of the allyl radical[16] formed on breaking this bond modulated by an interaction of the metal with the π system in the organometallic. However, additional factors complicate the explanations. For example, the double bond in ethylene has been suggested to be tilted away from the Cu(110) surface[17] but bonded to the Cu(100) and Cu(111) surfaces[11], i.e. the π-metal interaction is seemingly not always stabilizing. C−H bonds seem to favorably interact with the Cu surface[18] as shown by the increase by 5 ± 1 kJ mol^{-1} for the bonding of *n*-alkyl halides for each additional methylene group[19]. α-Hydride elimination from organocopper compounds is at least seven orders of magnitude less favorable than β-hydride elimination[20], making methyl-bonded copper considerably more stable kinetically than the higher *n*-alkyl copper species despite a meaningfully weaker C−Cu bond in the former.

All secondary alkyl coppers seem to have comparable bond strengths, or at least have comparable activation energies for decomposition[21]. This is the case whether the secondary alkyl group is acyclic, monocyclic or bicyclic. However, their reactivity with regard to decomposition by β-hydride elimination exhibits variation of some six powers of ten when the rates are corrected to a common temperature of 200 K. The variation was ascribed as largely due to free-energy differences needed to orient the hydrocarbyl group with a 0° dihedral angle between the β-H and the surface atom bearing the Cu−C bond. No interconversion of primary and secondary alkyl groups on Cu is known to the authors. No evidence for any special, i.e. non-classical, behavior was found as for the 2-norbornyl carbocation. No mention was made about tertiary alkyl coppers and so no comparison is possible between 1- and 2-norbornyl copper derivatives and interaction of copper with the anti-Bredt olefin 1-norbornene as opposed to the more mildly strained 2-norbornene that are formed upon decomposition of the corresponding bicycloalkyl. The 'discrete' molecular species 1-norbornylcopper exhibits high thermal stability[22] as it is this 1-norbornene that would be produced by the 'natural' β-elimination decomposition pathway.

D. Conjectures on the Energetics of Dialkyl Cuprates

The enthalpy of formation of any dialkyl cuprate, $[RR^1Cu]^-$, remains unmeasured for any R and R^1. Likewise, any enthalpy of reaction of such species is unknown. However, there is reason for optimism about such determinations and what follows are outlines of some suggested experiments. Such anions are readily formed in the gas phase by sequential decomposition of copper(I) biscarboxylate anions[23]. Both homoleptic organocuprates (R_2Cu^-, where R = Me and Et) and diverse heteroleptic organocuprate anions (RR^1Cu^-, where R = Me and R^1 = Et, *n*-Pr and *i*-Pr, *t*-Bu, allyl and Bn) have been prepared, as well as monocarboxylate complexes, as in reaction 6. It would appear that competition reactions involving loss of the various hydrocarbyl groups in the $RCuR^1$ by 'kinetic method' measurements[24] should provide information on relative carbon–copper bond strengths.

$$(RCOO)(R^1COO)Cu^- \longrightarrow (RCOOCuR^1)^- \text{ or } (RCuOOCR^1)^- \longrightarrow (RCuR^1)^- \quad (6)$$

In the gas phase[25], dimethylcuprate has been shown to react with methyl iodide to form $CH_3CuI^- + C_2H_6$. From this, the bounds for the relative dissociation energies of other R–Cu and Cu–halogen bonds are $D(\text{Cu–I}) - D(\text{C–I}) > D(\text{Cu–C}) + D(\text{C–C})$.

Finally, cupric acetate reacts with bis(hydrocarbyl) magnesium R_2Mg in the presence of tertiary phosphines, $\boldsymbol{P}$[26], to give organocopper species, both the ionic salt $[Cu\boldsymbol{P}_4]^+[CuR_2]^-$ and/or the covalent ligated $RCu\boldsymbol{P}_3$ depending on the identity of the R group. Evaluation of the equilibrium constant for the ligand exchange reaction 7 would provide useful information on the R–Cu bond strength in neutral and ionic organocopper compounds.

$$2\ RCu\boldsymbol{P}_3 \rightleftharpoons [Cu\boldsymbol{P}_4]^+[CuR_2]^- + 2\boldsymbol{P} \quad (7)$$

E. Anionic Copper

Anionic copper is the monoatomic species Cu^- for which the energetics are well known. Simplistically, Cu^- may be expected to behave like a halide ion because it is monatomic, singly charged and has no unpaired electrons. However, the electron affinity of Cu corresponding to formation of the ^1S anion is 119 kJ mol^{-1} [27–29] as compared to >300 kJ mol^{-1} for the halogens F, Cl, Br and I. Some comparisons of the hydrogen bonding behavior of this anion with halides and with its congeners Ag^- and Au^- have been made[30].

Elemental copper[31] and Cu(I) ate complexes[32] have been suggested as synthetic intermediates from the naphthalene radical anion reduction of Cu(I) complexes. It was proposed that Cu^- is formed by electron transfer to the metal, which then undergoes a displacement reaction with the organic compound. We await more classical chemistry comparisons involving Cu^- as the nucleophile[33] analogous to S_N2 reactions of alkyl halides for which considerable understanding of gas-phase chemical energetics has been deduced.

F. Copper Bonded to Methylene

There are at least three meanings to 'copper bonded to methylene': the four-atom species $CuCH_2$; a CH_2 bonded to a metallic copper surface; and the binding of CH_2 as found in an *n*-alkane or derivative thereof to Cu. With regard to the first meaning, and related to the aforementioned gas-phase ion–molecule reactions of Cu^+ with diverse alkanes[1], are those with ethylene oxide and cyclopropane that have been shown to form CH_2Cu^+ [34]. While the CH_3–Cu^+ bond energy is 240 ± 7 kJ mol^{-1} (this is *ca* twice that for the neutral methylcopper), the CH_2-Cu^+ bond energy is 267 ± 6 kJ mol^{-1}. Although discrete, neutral tetraatomic CH_2Cu has been observed and spectroscopically characterized subsequent to laser synthesis from atomic Cu and either cyclopropane[35] or diazomethane[36], again it is impossible to derive thermochemical information without knowing the temperature or electronic state of Cu. No measurement for its ionization energy has been reported and so the CH_2Cu bond energy cannot be derived to compare with that of CH_3Cu.

The C−Cu bond energy for discrete molecular species and that for a surface bound organic group are significantly different. For surface bound CH_2 on metallic copper, a bond energy of 314 ± 42 kJ mol^{-1} has been determined[37], more than twice the corresponding binding energy of CH_3 on the surface, 138 ± 33 kJ mol^{-1}—it remains unknown whether this is a methylidene ($=CH_2$) or μ^2-methylene ($-CH_2-$) complex. The last meaning of 'CH_2Cu bond energy', namely that from the interaction of a methylene group as found in an alkane or other alkyl chain with the Cu metal surface, has a value of *ca* 5 kJ mol^{-1} [19].

G. Cyclopropanes, Other Cycloalkanes and Copper-containing Species

Earlier we cited the gas-phase reaction[34] of Cu^+ with cyclopropane to form $CuCH_2{}^+$. Now consider the reaction of cyclopropane with metallic copper. The simple adsorption process on Cu(110) and Cu(111) surfaces is experimentally shown[38] to be exothermic by *ca* 40 kJ mol^{-1}. Irradiation of cyclopropane on copper surfaces with 10 eV electrons (*ca* 900 kJ mol^{-1}) result in both C−C and C−H bond cleavage showing the presence of both cupracyclobutane and cyclopropyl-copper species in an approximately 1:4 ratio. That the cupracyclobutane is the product and not the 'pentanuclear' 1,2-dicupracyclopentane was shown by IR spectroscopy[39,40]. That the cyclopropyl-copper maintains the 3-membered ring from the parent cyclopropane was demonstrated and comparison was made with other cyclopropyl and 'linear alkyl' (e.g. η^1-allyl) derivatives. All of these results are consistent with the spectroscopy of other organometallics, although 1,2-dimetallocyclopentanes are notably scarce in the literature, e.g. both 1,2- and 1,3-dibromopropane form propene on aluminum surfaces, as does 1,3-diiodopropane[41]. 1,3-Diiodopropane reacts[42] with α,β-unsaturated esters (i.e. acrylates and the isomeric fumarates and maleates), copper and isocyanides to form the corresponding substituted cyclopentyl isocyanides (RNC)•CuI complex. The authors suggest the intermediacy of 3-iodopropyl copper derivatives rather than either cupracyclobutanes or 1,2-dicupracyclopentanes, although none of these intermediates are precluded by the data. 1,4-Diiodobutane likewise forms the substituted cyclohexane—there is no evidence for or against cupracyclopentanes or 1,2-dicupracyclohexanes.

H. Vinyl and Other Alkenylcoppers

Surprisingly little is known about the energetics of vinylcopper species. Vinyl radical bonds to metallic copper surfaces[43]. The dissociation of vinyl copper is more reminiscent of methyl than other alkyl coppers in that butadiene is formed by its dissociation rather than β-hydrogen elimination to form acetylene, whereas surface bound ethyl forms ethylene and not butane. The lability of surface bound alkyl groups on Cu affirms the exoergicity (at least at temperatures exceeding 200 K) of the 'generic' alkyl/alkene reaction 8.

$$\text{Cu(surface)}-\text{C}-\text{C}-\text{H} \longrightarrow \text{Cu(surface)}-\text{H} + {>}\text{C}{=}\text{C}{<}\text{(g)} \tag{8}$$

However, not enough is known to deduce the energetics of the unsaturated counterpart process for the vinyl/acetylene reaction 9.

$$\text{Cu(surface)}-\text{C}{=}\text{C}-\text{H} \longrightarrow \text{Cu(surface)}-\text{H} + -\text{C}{\equiv}\text{C}-\text{(g)} \tag{9}$$

Indeed, we cannot thermochemically compare the gas, liquid or solution phase reaction 10 with reaction 11.

$$\text{Cu--C--C--H} \longrightarrow \text{CuH} + {>}\text{C=C}{<} \quad (10)$$

$$\text{Cu--C=C--H} \longrightarrow \text{CuH} + -\text{C}{\equiv}\text{C}- \quad (11)$$

The hydrocupration reaction[44], the reverse of equation 11, of 3-hexyne with dimeric 1,3-bis(2,6-diisopropylphenyl)imidazol-2-ylidene copper hydride forms the isolable *cis*-3-hexenylcopper as a complex, reaction 12.

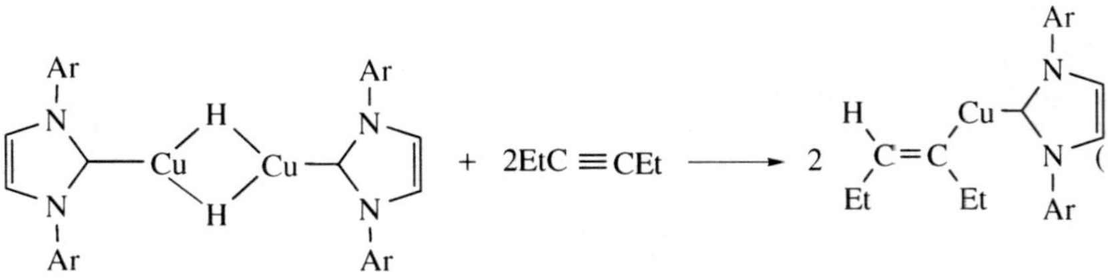

(12)

Ar = 2,6-diisopropylphenyl

In contrast, CuH is formed by *in situ* reduction of CuCl, and then reacts[45] with 1-hexyne to form 5,7-dodecadiene via the sequential reaction 13.

$$\begin{aligned} 2\text{CuH} + 2\text{BuC}{\equiv}\text{CH} &\longrightarrow 2\text{BuCH=CHCu} \\ &\longrightarrow (E,E)\text{-BuCH=CHCH=CHBu} + 2\text{Cu(s)} \end{aligned} \quad (13)$$

It is too difficult to disentangle the simultaneously large steric and electronic effects of the imidazolylidene ligand on copper and the corresponding ones in the two isomeric hexenyl radicals and dimeric diene to derive bounds on Cu−C bond strengths despite whatever knowledge we have about the thermochemistry of the diene products[46]. (*Z*)- and (*E*)-1-propenylcopper, as their respective tributylphosphine complexes, thermally decompose to form (*Z,Z*)- and (*E,E*)-2,4-hexadiene, respectively, with stereochemical integrity[47]. 2-Butenyl copper derivatives likewise decompose[47] with stereochemical integrity. In both cases, the (*Z,Z*)-diene is less stable than its (*E,E*)-counterpart. The difference is 8.0 kJ mol^{-1} for the propenyl dimer[48] and 1.6 kJ mol^{-1} for the 2-butenyl dimer[49].

I. Phenyl and Arylcopper Derivatives

Little is known about the energetics of arylcopper species. Cu^+ binds to benzene with a 0 K dissociation energy[50,51] of *ca* 217 ± 10 kJ mol^{-1}. An unrealized challenge is to find a base that will effect deprotonation (reaction 14) but not demetallation (reaction 15). It is generally agreed that $C_6H_6Cu^+$ is a π complex (η^2 and η^6) in which case considerable geometric reorganization is needed to accompany the deprotonation to form the σ-bonded phenyl copper.

$$C_6H_6Cu^+ + B \longrightarrow PhCu + BH^+ \quad (14)$$

$$C_6H_6Cu^+ + B \longrightarrow C_6H_6 + BCu^+ \quad (15)$$

A successful reaction would allow the determination of at least an upper bound to the enthalpy of formation of PhCu and thereby a lower bound to the Ph−Cu bond energy.

The same question applies for the variety of other aromatic–Cu^+ complexes, although in the case of the species with phenol it has recently been suggested that this π-complex and a η^1-carbon bonded complex are nearly isoenergetic, with the latter being somewhat more stable[52]. The bond energy of PhOH•Cu^+ is *ca* 215 kJ mol^{-1} [51,52].

Earlier in this chapter we mentioned the anions $[MeCuR]^-$ for a variety of alkyl groups (R = Me and R^1 = Et, *n*-Pr and *i*-Pr, *t*-Bu, allyl and Bn) and their gas-phase synthesis[23] from carboxylate complexes. In the same study R^1 = Ph was also investigated and all subsequent discussion for these dialkylcuprates applies to methyl(phenyl)cuprate here as well. Indeed, diphenylcuprate(−1) is known as a discrete anion in complexed lithium and calcium cation salts $Li(12\text{-crown-}4)_2\ CuPh_2^-$ [53] and $[(THF)_3Ca(\mu\text{-Ph})_3Ca(THF)_3]^+$ $CuPh_2^-$ [54].

Phenyl copper is an oligomer in the condensed phase. In dimethyl sulfide solution it has been shown to be a mixture of trimer and tetramer[55]; a variety of oligomerizations has been reported for aryl copper depending on the aryl group and the solvent. An enthalpy of aggregation of -15 ± 4 kJ mol^{-1} was reported for the oligomer shuffling reaction 16.

$$3(PhCu)_4 \longrightarrow 4(PhCu)_3 \qquad (16)$$

The tetramer is known as the dimethyl sulfide disolvate in the solid but solvation data are absent[56]. 2,6-Diphenylphenylcopper (2-cuprioterphenyl) (**1**) is a trimer in the solid phase[57]. In benzene solution, it has an aggregation number of almost 2, suggesting the dimer dominates in solution. 2,6-Dimesitylphenyl copper is a dimer in the solid. In benzene solution, it has an aggregation number of 1.2 suggesting somewhat greater stability of the monomer over the dimer or higher oligomers.

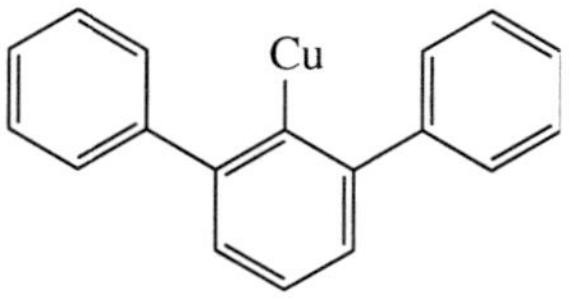

(**1**)

Benzene reacts with gaseous homoatomic copper anion clusters to form $PhCu_m^-$ for $m = 1$, 2 and 3[58,59]. A combination of the photoionization energy of these species[59] and the enthalpies of formation of the polymetal anions[60,61] provides at least a bound for the enthalpy of formation of the neutral phenyl copper species. Phenyl-copper clusters also exist in the condensed phase, albeit generally with rather exotic counterions and affixed ligands. For example, the $[Cu_5Ph_6]^-$, $[Cu_4LiPh_6]^-$ and $[Cu_3Li_2Ph_6]^-$ anionic clusters with D_{3h} symmetry have been crystallographically characterized[62–64]. This 'equivalence' of Cu and Li as evidenced by the previous series of three anions is not unprecedented, even with no assumption of similar geometries or even structures[65]. Neutral and cationic clusters have also been seen and characterized, e.g. Cu_4MgPh_6 (isoelectronic to the previous '5-atom cluster'[63]) and $[Cu_3Ph_2]^+$ [66]. The last ion, its linear CuCuCu chain doubly complexed with *N,N,N′,N″,N″*-pentamethyldiethylenetriamine (PMDTA), was formed as its $[Cu_5Ph_6]^-$ salt. Unfortunately, the energetics—or even relative energetics—of these species are essentially unknown. We can say reaction 17 is energetically favorable but little more.

$$\text{'CuPh'} + \text{PMTDA} \longrightarrow [Cu_3Ph_2\text{•}2PMDTA]^+[Cu_5Ph_6]^- \qquad (17)$$

As for the phenyl copper derivatives, very little is quantitatively known about the energetics of other classes of cluster species such as metal carbonyls and carboranes, at least in comparison with classical organic compounds, inorganic and organometallic compounds.

J. Polycuprioalkanes (Polycopperalkanes)

In previous sections, we discussed alkyl copper compounds, also known as cuprioalkanes. We also discussed, however briefly, dicuprioalkanes. In this section we return to this latter class of species and generalize them with the title 'polycuprioalkanes' to allow for the possibility of multiple copper atoms being present. We start with dicupriomethane. There is no evidence for any discrete molecular species with the formula CH_2Cu_2. The closest to this species is CH_2 absorbed on metallic copper[37] but, as acknowledged before, there is no information about how many coppers the CH_2 is bonded to. Tri- and tetracupriomethane have not been studied—there are thermochemical data for carbon presumably bonded to two coppers but not four, while for lithium there are thermochemical data on CLi_4 but not CH_2Li_2.

A species with the formula $C_2H_4Cu_2$ is known from gas-phase experiments[67]. From modeling the observed data, the Cu_2/C_2H_4 binding energy for the gas-phase species has been suggested to be at least 83 $kJ\,mol^{-1}$. Should this species be understood as an ethylene dicopper cluster[67], a 1,2-disubstituted ethane[68], or as a perpendicular C_2H_4•Cu–Cu complex[69]? There is also a spectroscopically observed and characterized[70–74] species with this formula, accompanied by the likewise characterized $C_4H_8Cu_2$ and $C_6H_{12}Cu_2$, C_2H_4Cu, C_4H_8Cu and $C_6H_{12}Cu$. These last five species, like $C_2H_4Cu_2$, were described as ethylene-copper complexes but structural proof is likewise absent.

Continuing now with species with the formula $C_nH_{2n}Cu_2$, the copper reaction chemistry of α,ω-diiodoalkanes with suitable olefins was briefly discussed in the previous section (we just mentioned $C_nH_{2n}Cu_2$ for $n = 2$, 4 and 6). The study of these $I(CH_2)_nI$ with activated copper, but without olefin, has also been reported[75]. The yield of cycloalkane is not monotonic with increasing n: $n = 4$, 14%; 5, 40%; 6, 18%; 8, 0%. The yield of more conventional cyclization reactions reflects entropy and conformational freedom as well as enthalpic considerations; e.g. formation of cyclopropanes and epoxides is rather easy despite their high strain.

In principle, other alkane derivatives with even more coppers may be suggested, such as the theoretical suggestion of $C_2H_2Cu_4$, alternatively described[68] as 1,1,2,2-tetracupricethane and as a 1:2 acetylene dicopper complex. No experimental data, structural or energetic, are yet available, however.

K. Bonds with Higher Oxidation State Copper

Copper may be considered as having an oxidation state of +1 in the simple alkyl coppers, copper acetylides or the dialkylcuprates, or even in the surface bound alkyl group containing species. The copper in the $MeCu^+$ ion is +2 by assuming the Me is carbanionic, as we did in the simple alkyl coppers and in magnesium species such as RMg^+, RMgX and R_2Mg. We know of no thermochemical data for any neutral alkyl Cu(II) species, neither R_2Cu nor even RCuX complex to parallel the corresponding organomagnesium species, R_2Mg and RMgX[76]. However, photoionization of the aforementioned R_2Cu^- (and likewise hetero-organocuprates RR^1Cu^- and $RCuX^-$) would seemingly provide us with this information.

Rather much more can be inferred thermochemically about organo-Cu(III) species. Cu(II) oxidizes alkyl radicals to the corresponding carbocations and a copper–carbon bonded intermediate was suggested, as in reaction 18[77,78].

$$R^{\bullet} + Cu(II) \longrightarrow [R{-}Cu]^{+2} \longrightarrow Cu(I) + R^{+} \longrightarrow \text{carbocation derived products} \quad (18)$$

The putative $[R{-}Cu]^{+2}$ intermediate is Cu(III) by assuming the R is carbanionic. Using this oxidation state definition, $[R_2Cu]^{+}$ species is also considered as Cu(III). One example is the cuprocyclobutane cation suggested[79] as an intermediate in the gas-phase reaction of quadricyclane with Cu^{+} to form $[C_5H_6CuC_2H_2]^{+}$ by way of a norbornadiene–copper complex which fragments but leaves all of the pieces attached to the metal. Likewise, an intermediate mixed π-complexed propargyl-alkyl copper cation has been suggested for the fragmentation of octynes into mixed 'lower' alkyne-alkene–Cu(I) complexes[80].

Neutral trihydrocarbyl copper species have long been discussed. However, only recently have Me_2EtCu with affixed anionic and neutral ligands and $[Me_3EtCu]^{-}$ been observed and well-characterized by low temperature 'rapid injection NMR'[81–83] in organocuprate reactions such as reaction 19.

$$Me_2CuLi{\bullet}LiX(X = I, CN, SCN, SPh) + EtI \longrightarrow [Me_2EtCuX]^{-} \quad (19)$$

Additional reaction then proceeds to form $[Me_3EtCu]^{-}$. Alternatively, if the CuI complex is first reacted with suitable electron-donating Lewis bases such as phosphines, phosphates, isocyanides and pyridines, then neutral Me_2EtCuL are formed. These were found to be highly fragile suggestive of weak C–Cu bonds. Not surprisingly, the stability depends on the affixed X and L. Thermal decomposition reactions at the 'hot' temperature of $-80\,^{\circ}C$ result in propane. However, these studies cannot be used to deduce relative Me–Cu and Et–Cu bond strengths because the relatively small methyl groups are situated *trans* from each other to allow the two larger groups to be *trans* in the square-plane Cu(III) complex.

The anionic Cu(III) species, $(CF_3)_4Cu^{-}$, is isolable as a 'surprisingly stable' salt[84]. It would appear that this Cu(III) species is particularly stable in that its synthesis from $CuCF_3$, formed *in situ*, can be achieved by oxidation with the rather strong oxidant XeF_2 but also with any of the three elemental halogens Cl_2, Br_2 and I_2 (the last two rather weak) when accompanied by the large cations Bu_4N^{+}, Ph_4P^{+} or PNP^{+} ($PNP = Ph_3PNPPh_3$). To quantify this assumption of stability, it might be possible to 'volatilize' this anion from its salts and sequentially chip away Cu–C bonds, determining their energy by use of collision-induced dissociation measurements, as in reaction 20.

$$(CF_3)_4Cu^{-} + M \longrightarrow (CF_3)_nCu^{-} + (4 - n)CF_3^{\bullet} + M \quad (20)$$

Perhaps this species and the less trifluoromethylated anions can be photoionized to form the neutral formal Cu(IV) species using the same methodology that was used for Cu^{-} in an earlier section of this review, as in reaction 21.

$$(CF_3)_nCu^{-} \longrightarrow (CF_3)_nCu + e^{-} (n = 1\text{–}4) \quad (21)$$

The superconducting bis(ethylenedithio)tetrathiafulvalene salt of $(CF_3)_4Cu^{-}$ has the stoichiometry $(BEDT\text{-}TTF)_2Cu(CF_3)_4$ with cocrystallized 1,1,2-trichloroethane solvent[85]. It most likely contains Cu(III) rather than Cu(II) despite the long-known existence of $[Cu(II)Cl_4]^{2-}$ and other cuprate salts. The same study reports the existence of the superconducting $SF_5CH_2CF_2SO_3$ salt with the same formal 2:1 cation/anion ratio, as well as cites numerous other salts of this cation with well-defined anions of charge -1. Earlier ESR experiments[86] did not show Cu(II) in this $(CF_3)_4Cu^{-}$ salt and noted that superconductivity is lost in the absence of the halocarbon solvent in the solid.

We now note the remarkably stable Cu(III) complex[87], $(CF_3)_2CuS_2CNEt_2$. The phrase 'remarkably stable' is used because this species, like $(CF_3)_4Cu^-$, can be isolated and characterized, unlike most other Cu(III) species and unlike most other species with Cu−C bonds. Comparison of its Cu−S bond strength with copper(II) dithiocarbamates[88] and its Cu−C bond with the above copper(III)-containing anion $(CF_3)_4Cu^-$ should prove instructive.

III. THE THERMOCHEMISTRY OF BINARY COPPER–CARBON SPECIES, Cu_xC_y

There are apparently no direct measurements, whether it be a dissociation energy for the CuC diatomic or of bulk solid Cu_2C or CuC_2, to mimic the long-known binary carbides of aluminum or calcium, Al_4C_3 or CaC_2, respectively. Analogous to the carbide chemistry of other metals, these latter species are 'methanides' and 'acetylides' (also called 'acetylenides' and 'ethynediides'). For the latter class of species there are some fragmentary data. There is the long-known species with the formal formula Cu_2C_2, cuprous (or copper(I)) acetylide. This species has usually been considered as highly explosive[89–92], although recently it has been 'tamed'[93]. There is also Cu_2C_4, the copper butadiyne or diacetylene derivative[94], as well as a polyyne copper derivative of ambiguous stoichiometry[95]. However, while structure and composition are becoming better understood, thermochemical information is still generally lacking.

It has been observed[96] that silver acetylide is a more energetic explosive than its copper counterpart. Assume the simplest decomposition reaction for M = Ag and Cu is equation 22.

$$M_2C_2(s) \longrightarrow 2M(s) + 2C(s) \tag{22}$$

If it is also assumed that the carbonaceous products are the same for both metals[97], then the reaction for M = Cu is less exothermic than that for M = Ag. The enthalpy of formation of $Ag_2C_2(s)$ has been given[3] as 350.6 kJ mol^{-1}, and so that value is an upper bound for the enthalpy of formation of $Cu_2C_2(s)$. However, it was also shown[98] that the transmetallation reaction 23 proceeds rapidly and quantitatively.

$$Ag_2C_2 + 2Cu^+(aq) \longrightarrow Cu_2C_2 + 2Ag^+(aq) \tag{23}$$

Assuming the entropy change is negligible and accepting the stoichiometry of both metal acetylides as written, an even better (i.e. smaller or tighter) upper bound for the enthalpy of formation of $Cu_2C_2(s)$ is 282 kJ mol^{-1}. This correctly confirms that solid Cu_2C_2 is highly unstable. $Na_2C_2(s)$ has an enthalpy of formation of 17 kJ mol^{-1} and so is likewise unstable relative to the elements, but much less so than the other two metals. All of the carbides are relatively unstable because the C−C bond strength in graphite is so much stronger than the covalent bonds in most other metal compounds with nonmetals. Nitrides, likewise, are generally quite unstable. Since sodium is more electropositive than either copper or silver, the sodium salt is expected to be most stable.

Solid Cu_2S has an enthalpy of formation of −80 kJ mol^{-1}, while Ag_2S (derived as the average of two crystalline forms) has the value of *ca* −30 kJ mol^{-1}. Na_2S has an enthalpy of formation of −365 kJ mol^{-1} and so the order of stability for the sulfides is the same as for acetylides. Consider now the formal solid state reaction 24. As written, this reaction is at least 29 kJ mol^{-1} exothermic.

$$Na_2C_2 + Cu_2S \longrightarrow Na_2S + Cu_2C_2 \tag{24}$$

Recently, it has been found[99] that the 'undercoordinated' (410) surface of solid Cu reacts with ethylene to form chemisorbed C_2. By use of 'a simple thermodynamic calculation' involving the strength of C−H and Cu−H bonds, the authors deduced that the Cu−C bond energy is 200 ± 30 kJ mol^{-1}. However, as do the original authors, we forego attempting a precise description on the molecular level of the Cu(s)−C_2 surface species of interest.

Gas-phase copper–carbon cluster ions have been investigated with rather nonclassical stoichiometries even when it is acknowledged that it is C_2, not C, which is the fundamental carbonaceous building block. Some species have been ascribed as having 'magic numbers for stability', e.g. $Cu_{11}C_{10}^{+}$ [100], and others as sandwich structures, e.g. $Cu_3C_4^{-}$ [101] in which a triangular Cu_3^{3+} is sandwiched between two C_2^{2-} ions. Both qualitative and quantitative data have been derived for the energetics of these ions, but none of these findings result in an enthalpy of formation for any neutral species containing only copper and carbon, or, for that matter, containing copper, carbon and anything else such as hydrogen as would be formed by protonating an anion.

IV. COPPER COMPLEXES WITH CARBON MONOXIDE AND CARBON DIOXIDE

In this section, copper complexes with carbon monoxide and carbon dioxide qualify as organometallics because they have Cu−C bonds. For the synthetic chemist, copper has long been associated with carbon monoxide. The classic Gattermann synthesis of aromatic aldehydes employs CO, along with some catalyst, as the formylating agent of the arene. Copper salts have often served as the catalyst. The Koch carbonylation of aliphatics (e.g. alkenes, alcohols, alkanes) with CO to form the acid with one more carbon (or some derivative thereof) has long been known. Again copper-containing species have shown catalytic activity. However, these processes have not been studied from a thermochemical vantage point, much less from a calorimetric perspective. Other examples are the binding of CO by copper in the active oxygen binding sites of hemocyanins of crustaceans and some other invertebrates, and by copper-containing artificial silicate and aluminosilicate minerals. There are no values for biologically relevant Cu−CO bonding energies. We will ignore the binding of CO to Cu as found in cage-structured metal aluminosilicates. Most commonly these are zeolites, and the copper may be found in the oxidation state(s) Cu(0), Cu(I) and/or Cu(II).

A. Homoleptic Copper Carbonyls and Related Species

Like other transition metals, copper forms a variety of carbonyls—neutral, cationic and anionic species defined by the general stoichiometry $Cu_x(CO)_y$. Whether these species be charged or not, there are no other affixed groups or other ligands. These are most generally not like organic carbonyl compounds. The $x = 2$, $y = 1$ species, $Cu_2(CO)$, should not be assumed to resemble formaldehyde, H_2CO, even though both copper and hydrogen are often found in their univalent oxidation state. The $x = 1$, $y = 2$ species, $Cu(CO)_2$, should not be considered an inorganic analog of carbon suboxide, C_3O_2 (O=C=C=C=O). Neutral copper carbonyl species are all unstable, i.e. none of them are isolable under conventional laboratory conditions at or near STP, and, as a consequence, there are comparatively few measurements of their bond energies, much less enthalpies of formation. This instability also suggests that the bond strengths in these species are rather low compared to the carbonyls of many other metals such as copper's neighbors in the periodic table, iron, cobalt and nickel. For the neutral copper carbonyls, the bond energies are only known[102] for $x = 1$ with $y = 1$ and 2 to be 25 ± 5 and >50 kJ mol^{-1}, respectively. This may suggest that $Cu(CO)_2$ is a viable gas-phase species under ambient conditions and that gives optimism for the existence of $Cu(CO)_3$. However, we know of no evidence for this or any higher copper carbonyl under ambient conditions. That is, one must consider the low temperature and high dilutions that characterize cryogenic matrices[103–105]. As adsorbed on the surface of metallic copper, 250 mV (24 kJ mol^{-1}) was adequate to remove[106] a CO from $Cu(CO)_2$ to form CuCO, but 500 mV (48 kJ mol^{-1}) was needed to remove and then exothermically transfer the remaining CO to form Cu. For bulk metallic copper, it has been shown that the binding energy of CO starts at 70 kJ mol^{-1} for pristine surfaces and then decreases to *ca* 53 kJ mol^{-1} for coverages between 0.15 and 0.35 of a monolayer[107]. Alloys are complicated and effects are usually nonadditive with regard to their metallic components. For example, the binding of CO to copper–platinum alloys is stronger than to either of these metals alone[108].

Sequential Cu–CO bond energies have been measured[109] for $Cu(CO)_y{}^+$ ions, $y = 1, 2,$ 3 and 4, which are respectively 148 ± 9, 172 ± 3, 75 ± 4 and 53 ± 3 kJ mol^{-1}. These ions are also known in the solid phase[110] and are well-characterized in solution as well[111,112]. It is perhaps surprising that $Cu(CO)_4{}^+$ is not found in a solid salt. After all, the isoelectronic $Co(CO)_4{}^-$ and $Fe(CO)_4{}^{2-}$ are well-known in condensed-phase media and the isoelectronic neutral $Ni(CO)_4$ is well-studied in the gas phase as well. Perhaps as a tetrahedral cation, $Cu(CO)_4{}^+$ lacks a coordination site that would allow it to 'snuggle' with its counterion. It thereby derives adequate additional stabilization that would allow for its isolation in the solid as part of an ionic salt.

The electron affinities of some $Cu_x(CO)_y$ species have been measured in the gas phase[113]. For $x = 1$, $y = 2, 3, 4$ and $x = 2$, $y = 4, 5$ these values are 92, 98, 100, 138 and 115 kJ mol^{-1}, respectively. The energetics of some 'copper-rich' ($x > y$) anionic carbonyl clusters have also been measured. For $Cu_x(CO)^-$ the recommended[114] $Cu_x{}^-$–CO binding energies are respectively: $x = 3$, 77 ± 29; $x = 4$, 62 ± 14; $x = 5$, 119 ± 21; $x = 6$, 84 ± 24; and $x = 7$, 59 ± 16 kJ mol^{-1}.

Gas-phase $Cu_x(CO)^+$ clusters with $x = 7$ and 17 show particular stability compared to those with nearby values of x[115]. Gas-phase CO-rich cationic clusters $Cu_x(CO)_y{}^+$ ($x \leqslant 14$, y generally $\geqslant x$) have also been studied[116] but while rate constants are reported, bond energies and stoichiometries are generally absent. Of these species only for the $x = 1$, $y = 2$ case are there thermochemical data for the cation[109], neutral[102] and anion[113].

For CS that is isoelectronic to CO, the Cu^+–CS bond energy has been determined[117] to be 238 ± 12 kJ mol^{-1} in $Cu(CS)^+$. It is interesting to note that the first synthesis of CS involved passing carbon disulfide over copper metal[118]. Revisiting this experiment with contemporary spectroscopic instrumentation should prove interesting. While some neutral copper thiocarbonyl species such as $Cu(CS)_2$ and Cu_2CS[119] were observed in cryogenic matrices a century later, thermochemical data are absent for these species as well as for any anionic copper–carbon monosulfide complex. All that can be said is that metal CS complexes are usually more strongly bound than the isoelectronic CO complexes.

B. Copper Halide Carbonyl Complexes

As evidenced by reports from the mid and late 19th and early 20th centuries, copper halide complexes with CO have a long history[120–123]. More recently, CO was cocondensed in a matrix with CuCl and $CuCl_2$ and their respective monocarbonyls, CuCl•CO[124–127] and $CuCl_2$•CO[127], were characterized structurally as discrete molecules. The thermochemistry of CuCl/CO complexation has been studied[128,129]. The enthalpy of the carbonylation reaction 25 is exothermic by 46.6 ± 0.1 kJ mol^{-1} with an associated Gibbs energy of -1.374 ± 0.006 kJ mol^{-1} [128].

$$CuCl(s) + CO(g) \longrightarrow CuCl{\bullet}CO(s) \qquad (25)$$

By contrast, $CuAlCl_4$ in toluene[130] reversibly binds CO by 64.5 kJ mol^{-1}, as found by taking the difference between the activation energies for the forward and reverse directions for CO complexation. However, the binding enthalpy is only *ca* 30 kJ mol^{-1} as found from enthalpy of solution measurements involving the explicit formation of copper carbonyl $AlCl_4$ complexes[131]. On the assumption that CuCl forms $Cu(NH_3)_3Cl$ in water in the presence of NH_3, CO complexation of 'ammoniacal cuprous chloride' is suggested to be exothermic by 39.7 kJ mol^{-1} [132].

C. Cu(I) vs Cu(II) Binding of CO

Although there is a relatively rich Cu(I)–carbon monoxide chemistry, there is comparatively little for Cu(0) and Cu(II). One of the few direct comparisons of Cu(I) and Cu(II) is for their *t*-butoxides. While copper(I) *t*-butoxide forms a monocarbonyl complex stable enough[133] that it may be sublimed (60 °C at 1 mm), this same study reports that copper(II) *t*-butoxide reacts with CO at 90 °C to form this copper (I) alkoxide carbonyl and di-*t*-butyl carbonate. The enthalpy of formation of copper(II) *t*-butoxide is unknown[134], as are that of other copper and transition metal alkoxides. Accordingly, the presumed exothermicity of this reaction cannot be used to provide an estimate for the enthalpy of formation of Cu(O-Bu-*t*)•CO. Combined electrochemical and spectroscopic measurements[135] have shown that Cu(I) prefers binding CO over acetonitrile, while the reverse is true for Cu(II).

D. Copper Carbon Dioxide Complexes

CO_2 may bind copper by one or both oxygens to form a Lewis base/Lewis acid complex, or by the carbon to produce a formate-like species. We concern ourselves here only with carbon-bonded species. The study of the binding of CO_2 to copper is related to that of CO. The water gas shift reaction that interconverts CO and H_2O, and CO_2 and H_2, is catalyzed by metallic copper. Not only have C-bonded Cu–CO complexes been suggested, but more recently so have C-bonded $Cu{-}CO_2$ complexes[136]. This same $CuCO_2$ species has been suggested[137] as part of the radiation chemistry of CO_2 in the presence of copper ions and hence metallic copper. This is the reaction chemistry of solvated electrons and the resultant formation of a variety of transient species: the one-carbon containing $CuCO_2$, its anion $CuCO_2^-$, the protonated form $CuCOOH^+$ (not to be confused with a form of Cu(II) formate), CuCO and the cation $CuCO^+$. Two-carbon transients Cu(CO)COOH, $Cu(CO)(CO_2)$ and $Cu(CO)_2^+$ are also observed. There are other possible structural motifs for $CuCO_2$ and $CuCO_2^-$. There is the copper carbonyl OCuCO formed by the reaction[138] of atomic copper and CO_2 and those reactions involving Cu^- or CO_2. Of all these species, there are thermochemical data only for the Cu–CO bond energies in the $CuCO^+$ and corresponding dicarbonyl cation complexes.

V. REFERENCES

1. R. Georgiadis, E. R. Fisher and P. B. Armentrout, *J. Am. Chem. Soc.*, **111**, 4251 (1989).
2. A. Weil and C. L. Wilkins, *J. Am. Chem. Soc.*, **107**, 7315 (1985).
3. D. D. Wagman, W. H. Evans, V. B. Parker, R. H. Schumm, I. Halow, S. M. Bailey, K. L. Churney and R. L. Nuttall, *The NBS tables of chemical thermodynamic properties: Selected values for inorganic and C1 and C2 organic substances in SI units, J. Phys. Chem. Ref. Data*,

11 (Supplement No. 2) (1982). Unless otherwise said, all thermochemical data of inorganic compounds are from this source.
4. J. F. Liebman, J. A. Martinho Simões and S. W. Slayden, 'The Thermochemistry of the organometallic compounds of silver and gold', in *The Chemistry of Organic Derivatives of Gold and Silver* (Eds. S. Patai and Z. Rappoport), John Wiley & Sons, Ltd, Chichester, 1999, p. 51.
5. J. M. Parnis, S. A. Mitchell, J. Garcia-Prieto and G. A. Ozin, *J. Am. Chem. Soc.*, **107**, 8169 (1985).
6. L. B. Knight, Jr., S. T. Cobranchi, B. W. Gregory and G. C. Jones, Jr., *J. Chem. Phys.*, **88**, 524 (1988).
7. B. V. Lebedev and T. G. Kulagina, *Russ. Chem. Bull.*, **50**, 1121 (2001).
8. J. D. Cox and G. Pilcher, *Thermochemistry of Organic and Organometallic Compounds*, Academic Press: London, 1970.
9. (a) J. F. Liebman, *Mol. Struct. Energ.*, **3**, 267 (1986).
(b) P. George, C. W. Bock and M. Trachtman, *Mol. Struct. Energ.*, **4**, 163 (1987).
(c) E. P. Hunter, S. G. Lias, C. M. Rooney, J. L. Winstead and J. F. Liebman, *Int. J. Mass Spectrom. Ion Proc.*, **179/180**, 261 (1998).
10. G. S. Natarajan and K. A. Venkatachalam, *J. Appl. Chem. Biotech.*, **22**, 1019 (1972).
11. F. Zaera, *Chem. Rev.*, **95**, 2651 (1995).
12. B. E. Bent, *Chem. Rev.*, **96**, 1361 (1996).
13. C. J. Jenks, B. E. Bent and F. Zaera, *J. Phys. Chem. B*, **104**, 3017 (2000).
14. C. J. Jenks, M. Xi, M. X. Yang and B. E. Bent, *J. Phys, Chem.*, **98**, 2152 (1994).
15. A. B. Gurevich, A. V. Teplyakov, M. X. Yang, B. E. Bent, M. T. Holbrook and S. R. Bare, *Langmuir*, **14**, 1419 (1998).
16. G. B. Ellison, G. E. Davico, V. M. Bierbaum and C. H. DePuy, *Int. J. Mass Spectrom. Ion Proc.*, **156**, 109 (1996).
17. C. J. Jenks, B. E. Bent, N. Bernstein and F. Zaera, *Surf. Sci.*, **277**, L89 (1992).
18. C. J. Jenks, B. E. Bent, N. Bernstein and F. Zaera, *J. Phys. Chem. B*, **104**, 3008 (2000).
19. J. L. Lin and B. E. Bent, *J. Phys. Chem*, **96**, 8529 (1992).
20. C. J. Jenks, C. M. Chiang and B. E. Bent, *J. Am. Chem. Soc.*, **113**, 6308 (1991).
21. A. V. Teplyakov and B. E. Bent, *J. Am. Chem. Soc.*, **117**, 10076 (1995).
22. V. Dimitrov and K. H. Thiele, *Z. Anorg. Allg. Chem.*, **510**, 7 (1984).
23. N. Rijs, G. N. Khairallah, T. Waters and R. A. J. O'Hair, *J. Am. Chem. Soc.*, **130**, 1069 (2008).
24. G. Chen and R. G. Cooks, *J. Mass Spectrom.*, **32**, 1258 (1997).
25. P. F. James and R. A. J. O'Hair, *Org. Lett.*, **6**, 2761 (2004).
26. D. F. Dempsey and G. S. Girolami, *Organometallics*, **7**, 1208 (1988).
27. H. Hotop, R. A. Bennett and W. C. Lineberger, *J. Chem. Phys.*, **58**, 2373 (1973).
28. R. C. Bilodeau, M. Scheer and H. K. Haugen, *J. Phys. B: At. Mol. Opt. Phys.*, **31**, 3885 (1998).
29. G. J. Rathbone, T. Sanford, D. Andrews and W. C. Lineberger, *Chem. Phys. Lett.*, **401**, 570 (2005).
30. H. Schneider, A. D. Boese and J. M. Weber, *J. Chem. Phys.*, **123**, art. No. 084307 (2005).
31. D. E. Stack, W. R. Klein and R. D. Rieke, *Tetrahedron Lett.*, **34**, 3063 (1993).
32. R. D. Rieke, B. T. Dawson, D. E. Stack and D. E. Stinn, *Synth. Commun.*, **20**, 2711 (1990).
33. Unpublished discussions of J. M. Weber and J. F. Liebman.
34. (a) E. R. Fisher and P. B. Armentrout, *J. Phys. Chem.*, **94**, 1674 (1990).
(b) E. R. Fisher and P. B. Armentrout, *J. Phys. Chem.*, **98**, 8260 (1994).
35. L. B. Knight, Jr., S. T. Cobranchi, J. Petty and D. P. Cobranchi, *J. Chem. Phys.*, **91**, 4587 (1989).
36. (a) S. C. Chang, Z. H. Kafafi, R. H. Hauge, W. E. Billups and J. L. Margrave, *J. Am. Chem. Soc.*, **109**, 4508 (1987).
(b) S. C. Chang, Z. H. Kafafi, R. H. Hauge, J. L. Margrave and W. E. Billups, *Tetrahedron Lett.*, **28**, 1733 (1987).
37. C. M. Chiang, T. H. Wentzlaff and B. E. Bent, *J. Phys. Chem.*, **96**, 1836 (1992).
38. R. Martel and P. H. McBreen, *J. Chem. Phys.*, **107**, 8619 (1997).
39. R. Martel, A. Rochefort and P. H. McBreen, *J. Am. Chem. Soc.*, **116**, 5965 (1994).
40. R. Martel, A. Rochefort and P. H. McBreen, *J. Am. Chem. Soc.*, **119**, 7881 (1997).

41. B. E. Bent, R. G. Nuzzo, B. R. Zegarski and L. H. Dubois, *J. Am. Chem. Soc.*, **113**, 1143 (1991).
42. Y. Ito, K. Nakayama, K. Yonezawa and T. Saegusa, *J. Org. Chem.*, **39**, 3273 (1974).
43. M. X. Yang, J. Eng, Jr., P. W. Kash, G. W. Flynn, B. E. Bent, M. T. Holbrook, S. R. Bare, J. L. Gland and D. A. Fischer, *J. Phys. Chem.*, **100**, 12431 (1996).
44. N. P. Mankad, D. S. Laitar and J. P. Sadighi, *Organometallics*, **23**, 3369 (2004).
45. S. A. Rao and M. Periasamy, *J. Chem. Soc., Chem. Commun.*, 495 (1987).
46. J. F. Liebman, in *The Chemistry of Functional Groups Supplement A2: The Chemistry of Dienes and Polyenes*, Vol. 1 (Ed. Z. Rappoport), John Wiley & Sons, Ltd, Chichester, 1997, p. 67.
47. G. M. Whitesides, C. P. Casey and J. K. Krieger, *J. Am. Chem. Soc.*, **93**, 1379 (1971).
48. W. Fang and D. W. Rogers, *J. Org. Chem.*, **57**, 2294 (1992).
49. W. R. Roth, H. W. Lennartz, W. v. E. Doering, W. R. Dolbier, Jr. and J. C. Schmidhauser, *J. Am. Chem. Soc.*, **110**, 1883 (1988).
50. F. Meyer, F. A. Khan and P. B. Armentrout, *J. Am. Chem. Soc.*, **117**, 9740 (1995).
51. C. Ruan, Z. Yang and M. T. Rodgers, *Phys. Chem. Chem. Phys.*, **9**, 5902 (2007).
52. P. Milko, J. Roithova, D. Schröder, J. Lemaire, H. Schwarz and M. C. Holthausen, *Eur. J. Chem.*, **14**, 4318 (2008).
53. H. Hope, M. M. Olmstead, P. P. Power, J. Sandell and X. Xu, *J. Am. Chem. Soc.*, **107**, 4337 (1985).
54. R. Fischer, H. Görls and M. Westerhausen, *Organometallics*, **26**, 3269 (2007).
55. S. H. Bertz, G. Dabbagh, X. He and P. P. Power, *J. Am. Chem. Soc.*, **115**, 11640 (1993).
56. M. M. Olmstead and P. P. Power, *J. Am. Chem. Soc.*, **112**, 8008 (1990).
57. M. Niemeyer, *Organometallics*, **17**, 4649 (1998).
58. X. Xing, H. Liu and Z. Tang, *Phys. Chem. Commun.*, 32 (2003).
59. X.-J. Liu, X. Zhang, K.-L. Han, X.-P. Xing, S.-T. Sun and Z.-C. Tang, *J. Phys. Chem. A*, **111**, 3248 (2007).
60. J. Ho, K. M. Ervin and W. C. Lineberger, *J. Chem. Phys.*, **93**, 6987 (1990).
61. K. J. Taylor, C. L. Pettiette-Hall, O. Cheshnovsky and R. E. Smalley, *J. Chem. Phys.*, **96**, 3319 (1992).
62. P. G. Edwards, R. W. Gellert, M. W. Marks and R. Bau, *J. Am. Chem. Soc.*, **104**, 2072 (1982).
63. S. I. Khan, P. G. Edwards, H. S. H. Yuan and R. Bau, *J. Am. Chem. Soc.*, **107**, 1682 (1985).
64. H. Hope, D. Oram and P. P. Power, *J. Am. Chem. Soc.*, **106**, 1149 (1984).
65. X. He, M. M. Olmstead and P. P. Power, *J. Am. Chem. Soc.*, **114**, 9668 (1992).
66. X. He, K. Ruhlandt-Senge, P. P. Power and S. H. Bertz, *J. Am. Chem. Soc.*, **116**, 6963 (1994).
67. L. Lian, F. Akhtar, P. A. Hackett and D. M. Rayner, *Chem. Phys. Lett.*, **205**, 487 (1993).
68. M. Boehme, T. Wagener and G. Frenking, *J. Organomet. Chem.*, **520**, 31 (1996).
69. S. Roszak and K. Balasubramanian, *Chem. Phys. Lett.*, **231**, 18 (1994).
70. H. Huber, D. McIntosh and G. A. Ozin, *J. Organomet. Chem.*, **112**, C50 (1976).
71. G. A. Ozin, H. Huber and D. McIntosh, *Inorg. Chem.*, **16**, 3070 (1977).
72. P. H. Kasai, D. McLeod, Jr. and T. Watanabe, *J. Am. Chem. Soc.*, **102**, 179 (1980).
73. T. Merle-Mejean, S. Bouchareb and M. Tranquille, *J. Phys. Chem.*, **93**, 1197 (1989).
74. J. A. Howard, H. A. Joly and B. Mile, *J. Phys. Chem.*, **94**, 1275 (1990).
75. F. O. Ginah, T. A. Donovan, Jr., S. D. Suchan, D. R. Pfennig and G. W. Ebert, *J. Org. Chem.*, **55**, 584 (1990).
76. J. F. Liebman, T. Holm and S. W. Slayden, in *The Chemistry of Organomagnesium Compounds* (Eds. Z. Rappoport and I. Marek), John Wiley & Sons, Ltd, Chichester, 2008, p. 101.
77. J. K. Kochi and A. Bemis, *J. Am. Chem. Soc.*, **90**, 4038 (1968).
78. J. K. Kochi and J. D. Bacha, *J. Org. Chem.*, **33**, 2746 (1968).
79. D. K. MacMillan, R. N. Hayes, D. A. Peake and M. L. Gross, *J. Am. Chem. Soc.*, **114**, 7801 (1992).
80. D. K. MacMillan, M. L. C. Schulze and H. Schwarz, *Organometallics*, **11**, 2079 (1992).
81. S. H. Bertz, S. Cope, M. Murphy, C. A. Ogle and B. J. Taylor, *J. Am. Chem. Soc.*, **129**, 7208 (2007).
82. S. H. Bertz, S. Cope, D. Dorton, M. Murphy and C. A. Ogle, *Angew. Chem., Int. Ed.*, **46**, 7082 (2007).
83. E. R. Bartholomew, S. H. Bertz, S. Cope. D. C. Dorton, M. Murphy and C. A. Ogle, *Chem. Commun.*, 1176 (2008).

84. D. Naumann, T. Roy, K. F. Tebbe and W. Crump, *Angew. Chem., Int. Ed. Engl.*, **32**, 1482 (1993).
85. H. H. Wang, M. L. VanZile, U. Geiser, J. A. Schlueter, J. M. Williams, A. M. Kini, P. P. Sche, P. G. Nixon, R. W. Winter, G. L. Gard, D. Naumann and T. Roy, *Synth. Met.*, **85**, 1533 (1997).
86. H. H. Wang, J. A. Schlueter, U. Geiser, J. M. Williams, D. Naumann and T. Roy, *Inorg. Chem.*, **34**, 5552 (1995).
87. M. A. Willert-Porada, D. J. Burton and N. C. Baenziger, *J. Chem. Soc., Chem. Commun.*, 1633 (1989).
88. M. A. V. Ribeiro da Silva and A. M. M. V. Reis, *J. Chem. Thermodyn.*, **21**, 167 (1989).
89. I. M. Dolgopol'skii, I. M. Dobromil'skaya, K. S. Moiseeva and F. B. Nankina, *Zh. Prikl. Khim.*, **19**, 1281 (1946); *Chem. Abstr.*, **41**, 33565 (1947).
90. A. Chambionnat, *Bull. Soc. Sci. Nat. Maroc.*, **28**, 77 (1948).
91. J. Eggert, *Proc. Roy. Soc. London A*, **246**, 240 (1958).
92. A. N. Sukachev, *Dopov. Akad. Nauk Ukr. RSR, Ser. B, Geol. Geof. Him. Biol.* (11), **21** (1985); *Chem. Abstr.*, **104**, 36438 (1986).
93. B. Balamurugan, B. R. Mehta and S. M. Shivaprasad, *Appl. Phys. Lett.* **82**, 115 (2003).
94. F. Cataldo, *Eur. J. Solid State Inorg. Chem.*, **35**, 281 (1998).
95. F. Cataldo, G. Compagnini, A. Scandurra and G. Strazzulla, *Fuller. Nanotub. Car. N*, **16**, 126 (2008).
96. J. Koehn, *Amts Mitteilungsbl. Bundesanst. Materialpruef. Berl.*, **8**, 57 (1978).
97. This assumption is plausible, at least within a 'few' kJ mol^{-1} as evidenced by the near constancy of the enthalpy of formation of carbon produced by the thermochemically relevant reaction of elemental sodium with per- and polyfluorinated organic compounds, V. P. Kolesov, O. G. Talakin and S. M. Skuratov, *Zh. Fiz. Khim.*, **42**, 2307 (1968); *Chem. Abstr.*, **70**, 14804 (1968). However, it should not be assumed that it is the same elemental carbon formed in the explosive decomposition of the metal acetylides as formed from this highly exothermic reduction reaction—from these organometallics we assume it has more carbyne character, i.e. it would be more reminiscent of the 1-dimensional carbon allotrope −CCCCCC−.
98. A. K. Babko and M. I. Grebel'skaya, *Zh. Obshch. Khim.*, **22**, 66 (1952); *Chem. Abstr.*, **46**, 38234 (1952).
99. T. Kravchuk, L. Vattuone, L. Burkholder, W. T. Tysoe and M. Rocca, *J. Am. Chem. Soc.*, **130**, 12552 (2008).
100. Y. Yamada and A. W. Castleman, Jr., *Chem. Phys. Lett.*, **204**, 133 (1993).
101. A. N. Alexandrova, A. I. Boldyrev, H.-J. Zhai and L. S. Wang, *J. Phys. Chem. A*, **109**, 562 (2005).
102. M. A. Blitz, S. A. Mitchell and P. A. Hackett, *J. Phys. Chem.*, **95**, 8719 (1991).
103. H. Huber, E. P. Kuendig, M. Moskovits and G. A. Ozin, *J. Am. Chem. Soc.*, **97**, 2097 (1975).
104. J. H. B. Chenier, C. A. Hampson, J. A. Howard and B. Mile, *J. Phys. Chem.*, **93**, 114 (1989).
105. M. F. Zhou and L. Andrews, *J. Chem. Phys.*, **111**, 4548 (1999).
106. H. J. Lee and W. Ho, *Phys. Rev. B*, **61**, R16347 (2000).
107. C. M. Truong, J. A. Rodriguez and D. W. Goodman, *Surf. Sci.*, **271**, L385 (1992).
108. M. J. Gladys, O. R. Inderwildi, S. Karakatsani, V. Fiorin and G. Held, *J. Phys. Chem.*, **112**, 6422 (2008).
109. F. Meyer, Y.-M. Chen and P. B. Armentrout, *J. Am. Chem. Soc.*, **117**, 4071 (1995).
110. J. J. Rack, J. D. Webb and S. H. Strauss, *Inorg. Chem.*, **35**, 277 (1996).
111. Y. Souma, J. Iyod and H. Sano, *Inorg. Chem.*, **15**, 968 (1976).
112. N. Tsumori, Q. Xu, M. Hirahara, S. Tanihata, Y. Souma, Y. Nishimura, N. Kuriyama and S. Tsubota, *Bull. Chem. Soc. Jpn.*, **75**, 2257 (2002).
113. J. Stanzel, E. F. Aziz, M. Neeb and W. Eberhardt, *Collect. Czech. Chem. Commun.*, **72**, 1 (2007).
114. V. A. Spasov, T.-H. Le and K. M. Ervin, *J. Chem. Phys.*, **112**, 1713 (2000).
115. M. A. Nygren, P. E. M. Siegbahn, C. Jin, T. Guo and R. E. Smalley, *J. Chem. Phys.*, **95**, 6181 (1991).
116. R. E. Leuchtner, A. C. Harms and A. W. Castleman, *J. Chem. Phys.*, **92**, 6527 (1990).
117. C. Rue, P. B. Armentrout, I. Kretzschmar, D. Schroeder and H. Schwarz, *J. Phys. Chem. A*, **106**, 9788 (2002).
118. J. Thomsen, *Zt. Anorg. Chem.*, **34**, 187 (1903) as found in *J. Chem. Soc. Abstr.* **II**, 188 (1903).

119. Q. Kong, A. Zeng, M. Chen, M. Zhou and Q. Xu, *J. Chem. Phys.*, **118**, 7267 (2003).
120. F. LeBlanc, *C. R. Acad. Sci.*, 483 (1850).
121. W. A. Noyes and J. W. Shepherd, *J. Am. Chem. Soc.*, **20**, 343 (1898).
122. M. P. E. Berthelot, *Ann. Chim. Phys.*, **23**(vii), 32 (1901).
123. W. Manchot and J. N. Friend, *Justus Liebigs Ann. Chem.*, **359**, 100 (1908).
124. M. Haakansson and S. Jagner, *Inorg. Chem.*, **29**, 5241 (1990).
125. M. Haakansson, S. Jagner and S. F. A. Kettle, *Spectrochim. Acta*, **48A**, 1149 (1992).
126. H. S. Plitt, M. R. Baer, R. Ahlrichs and H. Schnoeckel, *Inorg. Chem.*, **31**, 463 (1992).
127. L. Shao, L. Zhang, M. Zhou and O. Qin, *Organometallics*, **20**, 1137 (2001).
128. W. Backen and R. Vestin, *Acta Chem. Scand.*, **A33**, 85 (1979).
129. (a) L. Ciavatta and M. Iuliano, *Ann. Chim. (Rome)*, **87**, 583 (1997).
(b) L. Ciavatta and M. Iuliano, *Ann. Chim.(Rome)*, **88**, 743 (1998).
130. R. V. Gholap and R. V. Chaudhari, *Can. J. Chem. Eng.*, **70**, 505 (1992).
131. S. V. Korbutov, L. S. Tyurina and Yu. G. Karpova, *Khim. Prom.*, 178 (1988); *Chem. Abstr.*, **108**, 227569 (1988).
132. R. V. Gholap and R. V. Chaudhari, *Ind. Eng. Chem. Res.*, **27**, 2105 (1988).
133. T. Tsuda, H. Habu, S. Horiguchi and T. Saegusa, *J. Am. Chem. Soc.*, **96**, 5930 (1974).
134. There are measurements to determine the enthalpy of formation values of some Mg(II) and Ca(II) alkoxides in T. Barreira and J. P. Leal, *Eur. J. Inorg. Chem.*, 987 (2000) as well as estimates for some other alkaline earth alkoxides. We hesitate to use the analysis for such ionic species for the more covalent Cu(II) alkoxides. Likewise, we opt not to use the related study on alkali metal alkoxides and phenoxides, P. Nunes, J. P. Leal, V. Cachata, H. Raminhos and M. E. Minas da Piedade, *Eur. J. Chem.*, **9**, 2095 (2003) for deriving enthalpies of formation of uncomplexed Cu(I) alkoxides.
135. R. M. Hernandez, L. Aiken, P. K. Baker and M. Kalaji, *J. Electroanal. Chem.*, **520**, 53 (2002).
136. A. A. Gokhale, J. A. Dumesic and M. Mavrikakis, *J. Am. Chem. Soc.*, **130**, 1402 (2008).
137. J. Grodkowski and P. Neta, *J. Phys. Chem. B*, **105**, 4967 (2001).
138. M. Zhou, B. Liang and L. Andrews, *J. Phys. Chem. A*, **103**, 2013 (1999).

CHAPTER **4**

NMR of organocopper compounds

TOBIAS GÄRTNER and RUTH M. GSCHWIND

Institut für Organische Chemie, Universität Regensburg, Universitätsstraße 31, D-93040 Regensburg, Germany
Fax: +49 941 943 4617; e-mail: ruth.gschwind@chemie.uni-regensburg.de

I. INTRODUCTION 163
A. General Aspects of NMR of Organocopper Compounds 164
B. NMR Techniques Applied to Organocopper Compounds 166
II. NMR STRUCTURE DETERMINATION OF ORGANOCOPPER REAGENTS 169
A. Stoichiometric Organocopper Reagents, An Introduction 169
B. Diorganocuprates—The Free Reagent 171
1. Monomer structure 171
2. Solvent separated ion pairs (SSIPs) vs. contact ion pairs (CIPs) 173
C. Supramolecular Aggregation 178
III. NMR SPECTROSCOPY OF INTERMEDIATE COMPLEXES OF ORGANOCUPRATES 184
A. Cu(I) Organocuprate Intermediates 184
B. Cu(III) Organocuprate Intermediates 194
IV. NMR STRUCTURE ELUCIDATION IN Cu(I)-CATALYZED REACTIONS 200
A. Catalytic Copper Complexes with Thiol-TADDOL Ligands 200
B. Catalytic Copper Complexes with Phosphoramidite Ligands 201
V. CONCLUSION 209
VI. ACKNOWLEDGMENT 210
VII. REFERENCES 210

I. INTRODUCTION

The importance of Nuclear Magnetic Resonance (NMR) spectroscopy for the structure elucidation of inorganic materials, organometallic complexes and metal-containing biological systems is well documented by numerous recent publications[1–9]. In these studies, the direct NMR observation of metal resonances provides valuable information about the

The chemistry of organocopper compounds
Edited by Z. Rappoport and I. Marek

TABLE 1. Nuclear properties (spin quantum number I, natural abundance N.A., gyromagnetic ratio γ, quadrupole moment Q and receptivity relative to ^{13}C $R_{N.A.}$) of isotopes used for NMR investigations on organocopper compounds

Isotope	I	N.A. (%)	γ (10^7 rad s^{-1} T^{-1})	Q (10^{-30} m^2)	$R_{N.A.}$
^1H	1/2	99.985	26.7522205	—	5.87×10^3
^6Li	1	7.59	3.937127	−0.0808	3.79×10^0
^7Li	3/2	92.41	10.397704	−4.01	1.59×10^3
^{13}C	1/2	1.108	6.728286	—	1.00×10^0
^{15}N	1/2	0.37	−2.7126188	—	2.23×10^{-2}
^{31}P	1/2	100	10.8394	—	3.91×10^2
^{63}Cu	3/2	69.09	7.111791	−22.0	3.82×10^2
^{65}Cu	3/2	30.91	7.6043	−20.4	2.08×10^2

physical and chemical environment of the metal atom and additional information is gained from investigations of the ligand resonances. However, for copper compounds there was little information available about NMR of copper substances in the early volume, *The Chemistry of the Metal–Carbon Bond*, of the Patai series in 1982. Neither direct copper-detected nor ligand-detected structural information was available, which was interpreted as 'possibly indicating a lack of interest in Cu(I) chemistry'[10]. Since then, the relevance of organocopper complexes has grown dramatically. This is documented by a series of recent reviews, which describe the wide applicability of organocopper compounds in catalytic and stoichiometric organic reactions and the actual interest in their reaction mechanisms[11–23]. The major relevance in organic synthesis also necessitates a better understanding of the structure and dynamics of organocopper compounds, in order to enable faster reaction optimization processes and, to some extent, a rational control of the reactivity.

However, the NMR properties of the two copper isotopes (Table 1) allow the direct NMR detection of copper resonances mainly in highly symmetric structural arrangements due to their high quadrupole moments. Therefore, for most of the copper complexes, the NMR spectroscopic approach relies on different NMR active nuclei available in the ligands. In Table 1, the NMR properties[24] of selected isotopes, which have been used successfully in structure elucidation of various organocopper compounds, are given.

Nowadays, NMR is the most powerful method for structure analysis in solution, but it is an indirect method and not a direct one, as e.g. X-ray analysis. Therefore, for an accurate structure elucidation via NMR, a sufficient number of structure parameters have to be spectroscopically available. Due to the fact that organocopper compounds and copper complexes often form highly symmetrical supramolecular structures, the available NMR parameters, such as chemical shifts δ, scalar couplings J, dipolar interactions and diffusion coefficients, sometimes do not reveal sufficient information for a complete and independent structure determination by NMR. Therefore, in structure elucidation of copper complexes in solution, very often information from X-ray structures, theoretical calculations and further spectroscopic methods are combined with NMR spectroscopic results to reveal structural aspects.

A. General Aspects of NMR of Organocopper Compounds

In solution and solid-state NMR spectroscopy, the direct detection of Cu resonances shows some limitations, which are typical for nuclei with large quadrupole moments. Copper possesses two NMR active natural isotopes, ^{63}Cu and ^{65}Cu, with a natural abundance of 69% and 31%, respectively. Both have gyromagnetic ratios similar to that of

^{13}C and their receptivities show very acceptable values, with the slightly better one for ^{63}Cu (Table 1). However, the most restricting parameter for NMR of copper isotopes is the large quadrupole moment (Q) of both copper isotopes. Quadrupole moments arise in every nuclei with a spin quantum number $I \geqslant 1$ and Table 1 shows that the NMR spectroscopically favorable isotopes ^{6}Li and ^{7}Li also possess quadrupole moments. This seeming contradiction is caused by the fact that the principal NMR accessibility of an isotope depends on the absolute value of the quadrupole moment and of the electric field gradient (EFG) across the nucleus, with large values being detrimental in case of both parameters. In the following, the influence of quadrupole moment and EFG is briefly explained to understand the limitations in copper NMR.

The quadrupole moment is a measure of the deviations from the spherical symmetric charge distribution of the nucleus, which can be either prolate (lengthened) or oblate (flattened). Consequentially, a quadrupole nucleus owns degenerate energy levels due to different orientations of the quadrupole, in addition to the nuclear spin orientations, and further transitions between these energy levels are possible, which are used in Nuclear Quadrupole Resonance (NQR). These energy levels are quantized and split according to an electronic field gradient (EFG), which results from an asymmetric charge distribution around the nucleus due to an anisotropic arrangement of neighboring electrons and atoms. In solution, the quadrupole coupling cannot be detected due to the isotropic tumbling of the molecules, but acts as an effective relaxation source and can lead to enormous line broadening of the signals. The effective amount of quadrupolar relaxation is directly correlated with the magnitude of Q^2 and the local EFG[24–28]. As a result, complexes with a high symmetry, i.e. small EFGs and/or small quadrupole moments, show advantageous NMR properties.

In the case of $^{63/65}Cu$ NMR spectroscopy, it could be shown that in highly symmetric tetrahedral $Cu(I)L_4$ or octahedral $Cu(I)L_6$ complexes the EFGs are sufficiently minimized to enable the detection of copper resonances[26]. Contrarily, already an exchange of only one ligand at the copper site to a $Cu(I)L_3L'$-type structure usually leads to extremely broad line widths or even to an undetectable copper signal. Even for CuL_4 complexes, disturbances of the symmetry induced by solvent, temperature, concentration or chemical composition are observed as line broadening of the signals[29–32]. Recently, it was also reported that in the case of $Cu(I)L_3L'$-type complexes the right choice of the ligands reduces significantly the line width by matching exactly the EFG[31]. Nevertheless, despite the large restrictions in NMR spectroscopy of Cu(I) complexes in solution, quite a large amount of $^{63/65}Cu$ spectra of highly symmetrical complexes with varying ligands, e.g. phosphites[27,30,33–37], phosphines[35,38,39], diphosphines[38,40–45], nitriles[29,30,32,36,37,46–55] and carbonyl compounds[26] are reported in the literature and some reviews have been published[25,26,28,56]. For copper complexes with reduced symmetry and not detectable copper resonances, only the NMR active nuclei of the ligands can be used for structure elucidation.

In addition, the line widths of all NMR signals are very sensitive to the presence of paramagnetic compounds. Therefore, it is of great importance to avoid paramagnetic nuclei in high-resolution NMR. Considering organocopper compounds, the Cu oxidation states +I and +II are by far the most common and only the diamagnetic Cu(I) is observable via NMR. For paramagnetic Cu(II) compounds, electron spin resonance (ESR) is the method of choice. Therefore, for the application of high-resolution NMR spectroscopy the absence of Cu(II) ions is very important, because otherwise the line-broadening effects are tremendous[30]. In recent studies, the copper oxidation state +III has become more and more important[57–59]. In this case the ligand field theory predicts for the d^8 electron configuration of Cu(III) a structure-dependent situation where square-planar complexes are diamagnetic and tetrahedral ones are paramagnetic. For square-planar Cu(III) complexes, this could be experimentally confirmed and consequently tetrahedral and square-planar complexes should be distinguishable by different line widths. Hence, NMR spectroscopy of organocopper reagents is rather restricted to Cu(I) and square planar Cu(III) compounds.

B. NMR Techniques Applied to Organocopper Compounds

High-resolution NMR investigations are in general based on the determination of the fundamental NMR parameters, chemical shift and scalar coupling. The chemical shifts of $^{63/65}$Cu resonances in different copper complexes can reach the higher positive or negative three-digit area (*ca* −400 ppm to 800 ppm)[26,60]. Originally, CuCl or $K_3[Cu(CN)_4]$ in D_2O were used as standards for $^{63/65}$Cu, but nowadays a solution of the tetrakis(acetonitrile) complex $[Cu(CH_3CN)_4]^+$ is commonly accepted and acts as chemical shift reference of 0 ppm[26]. Because in organocopper chemistry severe line-broadening effects very often lead to undetectable $^{63/65}$Cu resonances, the chemical shift values and the coupling patterns of other nuclei, such as ^{1}H, ^{13}C, $^{6/7}$Li and ^{31}P, are used for structure elucidation.

Besides the chemical shift, the information from scalar coupling constants, i.e. the multiplicity of the signals, is the second important classical parameter in high-resolution NMR spectroscopy. Due to the unfavorable nuclear properties of $^{63/65}$Cu, direct couplings to copper are only detected in highly symmetrical complexes. In complexes with reduced symmetry sometimes valuable scalar couplings of the NMR active nuclei across copper are reported. For example, in temperature-dependent studies of copper–phosphoramidite complexes[61] and in organocuprate Cu(I) or Cu(III) π-intermediates[57–59,62–65] either direct observations of scalar coupling constants across copper or magnetization transfers via scalar couplings across copper were possible. These studies show that not only the absolute electronegativities and the resulting EFGs are the critical factors for the detection of scalar couplings between ligand nuclei across copper, but also the exchange rate, i.e. the lability of the ligands.

Especially for structural studies of lithium organocopper compounds and intermediates, fully and partially ^{13}C labelled compounds were synthesised with much effort in order to observe reliable $J_{H,C}$ and $J_{C,C}$ coupling constants. In achiral Cu(I) π-complexes, INADEQUATE and HMBC experiments were successfully applied for the determination of $J_{C,C}$ and magnetization transfers in partially labelled complexes. This spectroscopic approach without any scalar couplings to Li is in contrast to that of other organometallic compounds like, e.g., organolithium reagents, where direct scalar couplings to Li are commonly used to determine structures and aggregation levels in solution[66–74]. When Li ions are part of organocopper reagents, the applicability of $J_{Li,X}$ scalar couplings depends on the individual binding properties. For lithium amidocuprates, the existence of $J_{Li,N}$ couplings facilitates the data interpretation, whereas in the case of lithium dialkylcuprates the covalent character of the organocopper–lithium bonds is not sufficient for a detection of scalar couplings.

In addition to chemical shifts and scalar couplings, qualitative dipolar ^{1}H,^{1}H homonuclear and ^{1}H,X heteronuclear (X = $^{6/7}$Li, ^{13}C, ^{15}N) interactions can provide further structural details[75–86] and the next step of structural refinements is the quantitative determination of NOEs or HOEs, which provide distances between different nuclei. In most of the organic molecules and organometallic complexes with several dipolar interactions, it is possible to assign one NOE/HOE to a known distance, which then serves as a distance reference for the other NOEs and HOEs. In the case of organocuprates, which form highly symmetric species by supramolecular assembling (Section II.B.2), quantitative NOEs or HOEs for distance measurements are difficult to access, because sometimes only one resonance signal exists. Even for these systems a quantitative NOE/HOE determination is possible, but for that the reintroduction and determination of correlation times τ_C and the measurement of build-up curves is necessary[87]. For example, the estimation

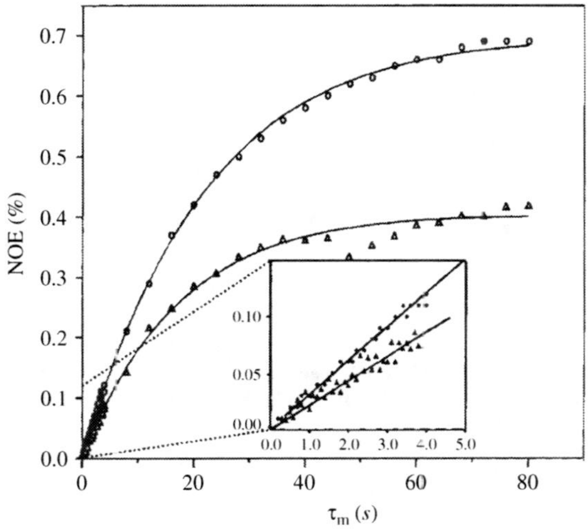

FIGURE 1. $^1H,^6Li$ build-up curves of $Me_2CuLi{\bullet}LiCN$ (Δ, **1**•LiCN) and Me_2CuLi (O, **1**), both 0.72 M in diethyl ether at 239 K. The initial build-up region is enlarged. Reprinted with permission from Reference 87. Copyright 2001 American Chemical Society

of τ_C of organocuprates was made via the maximum HOE enhancement η_{max} of the $^1H-^6Li$ HOE and with the help of the Solomon equations[87,88]. The initial build-up rate $\sigma_{^1H,^6Li}$ of the HOE (Figure 1) then provides the H–Li distance. Because in the cross-relaxation rate of the $^1H-^6Li$ HOE the distance r_{H-Li} is the only unknown parameter, in case the correlation time the isotope specific constants, the gyromagnetic ratio γ and the resonance frequency ω are known (equation 1)

$$\sigma_{^1H,^6Li} = \frac{4}{15}\left(\frac{\mu_0}{4\pi}\right)^2\left(\frac{h}{2\pi}\right)^2\frac{\gamma_H^2\gamma_{Li}^2}{r_{H-Li}^6}\tau_C \times\left[\frac{6}{1+(\omega_H+\omega_{Li})^2\tau_C^2}-\frac{1}{1+(\omega_H-\omega_{Li})^2\tau_C^2}\right] \quad (1)$$

The determination of homonuclear $^1H-^1H$ NOE build-up curves in highly symmetric molecules sometimes requires a determination of NOEs between chemically equivalent groups. In these structures the symmetry problem can be solved by using the two different isotopomers $^1H-^{13}C$ and $^1H-^{12}C$ (Figure 2a). The basic experiments for this purpose are the HMQC–ROESY[89] and HSQC–NOESY pulse sequences[90]. However, in the case of long interproton distances, even with a 20% ^{13}C labelling the sensitivity of these two methods is too low, because mixing times up to 1 s have to be used, which lead to an extreme diffusion-like signal attenuation caused by the applied pulsed-field gradients. To circumvent this obstacle, a NOESY–HSQC pulse sequence was developed in which the gradients for coherence selection are separated only by short refocusing delays and diffusion effects are minimized (Figure 2b). This approach was successfully applied to organocuprates[87,91].

The pulse sequences used in exchange spectroscopy (EXSY) are closely related to the basic NOESY experiments. Both pulse sequences are identical and only the length of the mixing time is varied. The EXSY experiments can be used to detect and quantify exchange processes, which are slow on the NMR time scale, without applying

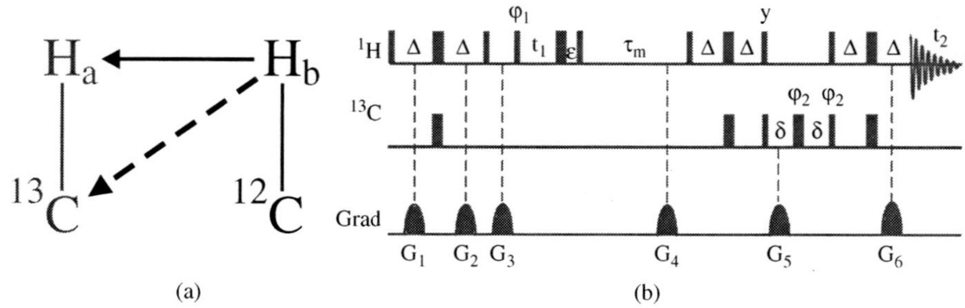

FIGURE 2. (a) Schematic description of the two different isotopomers, which are used in (b) the NOESY–HSQC to determine $^1H,^1H$ NOEs between chemically equivalent groups. From Reference 91. Copyright Wiley-VCH Verlag GmbH & Co. KGaA. Reproduced with permission

temperature-dependent NMR, which would be disadvantageous in case of temperature-sensitive compounds.

A NMR spectroscopic method for the determination of the size of supramolecular assemblies is the diffusion-ordered spectroscopy (DOSY)[84, 86, 92–96]. In DOSY experiments the spatial molecular motion in solution by virtue of thermal energy is used for the determination of self-diffusion coefficients. In the 1960s, Stejskal and Tanner[97] carried out the first PFG-SE-Experiment (Pulsed-Field-Gradient Spin-Echo). Due to the change of the spatial position within a distinct time interval between two pulsed gradients, an attenuation of the signal is observed which can be used to calculate the self-diffusion coefficient. The obtained diffusion coefficient D is inversely correlated to the hydrodynamic radius r_H, which is a measure of the size of supramolecular assemblies. For an accurate calculation of the hydrodynamic radius from the experimental diffusion coefficients, a modified Stokes–Einstein equation (equation 2) has to be applied which considers the relative solvent/solute size (c) and the shape of the molecules (f_S)[93, 98].

$$D = \frac{kT}{c(r_{solv}, r_H) f_S \pi \eta r_H} \qquad (2)$$

In equation 2, k represents the Boltzmann constant, T the temperature and η the viscosity of the solvent. For reliable, reproducible and quantitative DOSY measurements, variations in the viscosity and possible contributions of thermal convection have to be especially considered. Viscosity changes, e.g. due to variable sample composition or concentration, have to be eliminated via viscosity standards[93, 99]. Convection in the NMR tube can falsify the diffusion value dramatically, because of the principal translational character of the self-diffusion coefficient. Convection effects are significantly present in high- or low-temperature measurements and strengthen with increasing difference from room temperature. To compensate contributions from ideal convection, a convection compensating pulse sequence, developed by Jerschow and Müller, is a reliable method[100]. Later on, attempts were made to circumvent the low sensitivity of this method by shorter and more sensitive pulse sequences[101].

For the stabilization of reaction intermediates, rapid injection NMR (RI-NMR) is a very promising technical approach which was developed in the last twenty years[102, 103]. An insert inside the NMR spectrometer allows one to inject substances directly into the NMR tube, while the tube remains in the probe ready for the next experiment. This technique affords minimal dead times between injection and NMR detection and is therefore ideal for the observation of reaction intermediates with short life times.

II. NMR STRUCTURE DETERMINATION OF ORGANOCOPPER REAGENTS

A. Stoichiometric Organocopper Reagents, An Introduction

The chemistry of stoichiometric organocopper(I) compounds is mostly covered by the chemistry of organocuprates. Since the first observations of Gilman and Straley[104], who found soluble organocopper reagents after treatment of copper(I)salts with two equivalents of organolithium reagents, organocuprates have become a widely used organometallic reagent in organic synthesis. The general synthesis of homoleptic organocuprates is given in Scheme 1a. The reaction of 1 equivalent of Cu(I) salt and 2 equivalents of organolithium compound yields the desired Gilman-type cuprate[104]. Using Grignard or organozinc reagents, instead of alkyllithium, Normant-type[105,106] or Knochel-type[107] cuprates are derived, respectively.

$$\text{(a) } 1\ \mathrm{CuX} + 2\ \mathrm{RLi} \longrightarrow \mathrm{R_2CuLi{\bullet}LiX}$$

$$\text{(b) } 1\ \mathrm{CuCN} + \mathrm{RLi} \longrightarrow \mathrm{RCu(CN)Li}$$

$$\text{(c) } \mathrm{R^1MgCl} + \mathrm{LiNR_2} + \mathrm{CuCl} \longrightarrow \mathrm{R^1Cu(NR_2)Li}$$

SCHEME 1. Schematic description of the synthesis of (a) homoleptic Gilman cuprates, (b) heteroleptic cyanocuprates and (c) heteroleptic amidocuprates

In case only one equivalent of alkylation agent is used, heteroleptic organocuprates (Scheme 1b) or amidocuprates (Scheme 1c) are obtained, which are sometimes of higher synthetic importance due to the non-transferable ligand[108]. Especially, the amidocuprates provide the introduction of chiral information via substituted chiral amido ligands[109]. Considering the three equations in Scheme 1, it is obvious that the exact ratio of copper(I) salt to the alkylation agent is crucial when the structures of free organocuprate reagents are discussed. In comparison to the synthetically highly valued heteroleptic cuprates, the Gilman-type dimethyl cuprates Me_2CuLi (**1**) have become a generally accepted model for mechanistic and structural studies on copper-mediated reactions. The structure elucidation of these Gilman cuprates caused the famous and long-standing scientific discussion about 'higher order' and 'lower order' organocuprates[110], which could be finalized in favor of the Gilman cuprates[110], and continued with numerous theoretical and spectroscopic studies about the structures and reaction intermediates of dimethyl cuprates[19,111,112].

Synthetically, it was early recognized that dialkylcuprates (Scheme 1a) are able to form highly chemo- and diastereoselectively C−C bonds and this property is used throughout organic synthesis[108,113–115]. Scheme 2 shows schematically the three standard reaction types of organocuprates, addition reactions to unsaturated carbonyl compounds (Scheme 2a), S_N2-like substitution reactions (Scheme 2b) and S_N2' allylic substitutions (Scheme 2c). Amidocuprates are also frequently used reagents in synthesis[108,116–119]. If additional redox agents, such as chloranil, are used, even coupling reactions between the alkyl and the amido substituents are possible and therefore amidocuprates provide access to tertiary amines[116–118].

The famous discussion about 'higher order' organocuprates started, because Lipshutz and coworkers had reported higher reactivities of cyanocuprates than of iodocuprates[120,121]. Also later on, strong salt and solvent dependencies were found in synthetic studies of various organocuprate reactions[108,122]. Even the only two detailed studies with experimental setups enabling a direct comparison of the reactivities of cyano- versus iodocuprates show deviating results. In a study using logarithmic reactivity profiles, similar reactivities were reported for iodo- and cyanocuprates[123] whereas a combined kinetic and spectroscopic study showed higher reactivities of the cyanocuprate in pure diethyl ether[124].

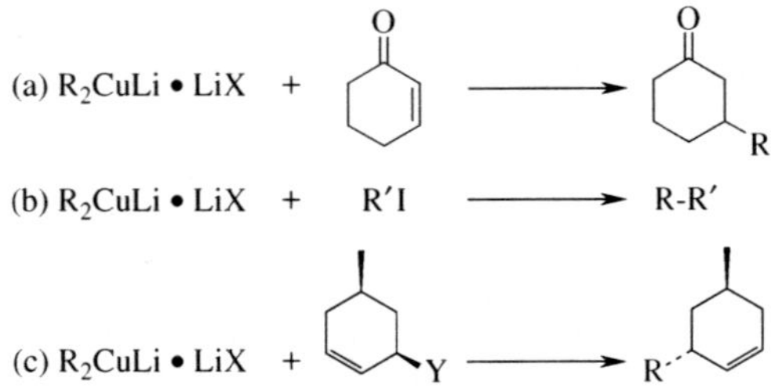

SCHEME 2. Schematic description of (a) the 1,4-addition to α,β-unsaturated Michael acceptors, (b) S_N2-like substitution reactions and (c) S_N2' allylic substitution reactions of Gilman cuprates (X = CN, I; Y = halide, OAc)

The identification of the Gilman-type cuprates as the dominating monomer structure in solution for all dialkylcuprates[110] shifted the focus to possibly different supramolecular cluster structures of cyano- and iodocuprates as the reason for the deviating reactivities. Especially in diethyl ether, colligative measurements[125–127], broad line widths in ^{13}C and ^{15}N spectra[128], crystal structures[129, 130] and mass spectrometric investigations[131] consistently indicated supramolecular aggregation to be present. In 2005, the deviating reactivity of cyanocuprates and iodocuprates (Figure 3) as well as salt-free cuprates were explained by different supramolecular structures in solution by using combined kinetic and NMR spectroscopic studies of 1,4-addition reactions (Scheme 2a)[132]. This study revealed that the variations in the reaction rates of **1**•LiI (Figure 3a) and **1**•LiCN (Figure 3b) in diethyl ether upon addition of THF correlate with a disaggregation of the supramolecular structure or solvent-induced changes in the supramolecular cluster structures (see Section II.C).

Also in the stabilization and structure elucidation of organocuprate intermediates, impressive progress has been made during the last decade. Investigations on reaction intermediates of addition reactions revealed Cu(I) π-complexes as important intermediate structures[62, 63, 133–135] and in the past few years even the detection of decisive Cu(III) intermediates in addition, as well as in S_N2-like/S_N2' substitutions, had been successful[57–59, 64, 65].

In the course of structure determination of the supramolecular complexes and the intermediates of organocuprates in solution, NMR spectroscopy turned out to be a very

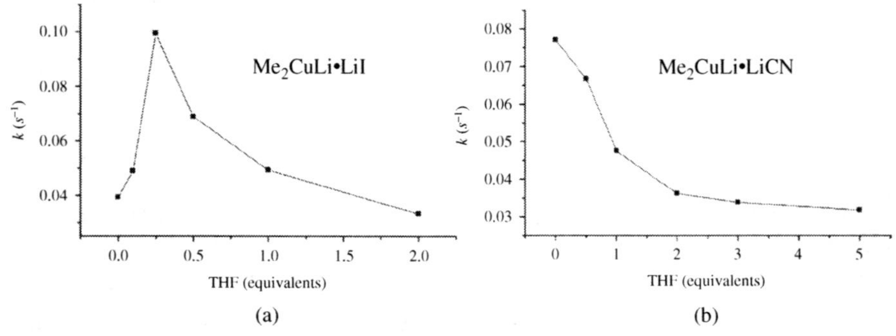

FIGURE 3. Rate constants k (s^{-1}) of the 1,4-addition reaction of (a) $Me_2CuLi \bullet LiI$ and (b) $Me_2CuLi \bullet LiCN$ to 4,4-dimethylcyclohex-2-enone in diethyl ether upon addition of THF. Reprinted with permission from Reference 124. Copyright 2005 American Chemical Society

powerful method even for complicated and highly symmetric aggregate structures. Especially, a step by step NMR analysis of small structural aspects in combination with results from theoretical calculations and X-ray analyses allowed to solve structural details of organocuprates and their intermediates, knowledge which is crucial for further developments in organocopper chemistry.

B. Diorganocuprates — The Free Reagent

1. Monomer structure

The reliable determination of the monomer structure was the basis for the structure elucidation of the free organocuprate reagent and its supramolecular structures in solution. At first, δ chemical shift values served as a source for structure information. However, with the chemical shifts as sole structural parameters, the differentiation of homoleptic and heteroleptic organocuprates was difficult and the influence of solvent, aggregation and temperature on organocuprates could not be explained for decades. Hence, the discussion about 'higher order' ($R_2Cu(CN)Li_2$) and 'lower order' ($R_2CuLi{\bullet}LiCN$) cuprates had not been finalized for a long time[110].

'Higher order' cuprates were proposed to have three ligands attached to one Cu(I) center in contrast to the 'lower order' cuprates, in which two ligands are bound to Cu(I). To detect these differences in the coordination sphere of copper, the measurement of $^2J_{C,C}$ coupling constants across copper is a powerful method. The existence of scalar couplings directly reveals the connectivity in the complexes and the number and arrangements of the substituents are evident from the multiplicity pattern and the absolute coupling constant value of the signals. For this purpose $^2J_{C,C}$ coupling constants were determined in samples with and without cyanide-containing cuprates to provide evidence for either 'higher order' or 'lower order' cuprates. First, $^2J_{C,C}$ coupling constants in 1D ^{13}C spectra were observed in heteroleptic RCu(CN)Li cuprates in THF, with phenyl, ethyl and methyl groups as substituents (Table 2)[136]. The fact that one cyanide and one alkyl substituent are bound to the same Cu center was proven upon ^{13}C labelling of the cyanide, which caused a doublet splitting of the alkyl group. Examples of temperature-dependent ^{13}C chemical shifts and $^2J_{C,C}$ of heteroleptic MeCu(CN)Li (**2**) and EtCu(CN)Li (**3**) are listed in Table 2.

Interestingly, the coupling constants in Table 2 show strong temperature dependencies, that is, starting from a minimum value of 12.3 Hz, the absolute values increase with decreasing temperature, indicating similar structures at low temperature and a partial

TABLE 2. ^{13}C NMR chemical shift values and $^2J_{C,C}$ coupling constants of selected heteroleptic cuprates at different temperatures in THF or diethyl ether as solvent. Reprinted with permission from Reference 136. Copyright 1991 American Chemical Society

Cuprates	Solvent	T (°C)	**C1** (ppm)	$^2J_{C,C}$ (Hz)	CN (ppm)
$^{13}CH_3Cu(^{13}CN)Li$ (**2***)	THF-d_8	−78	−12.85		149.34
	THF-d_8	−100	−12.60		149.13
	THF-d_8	−110	−12.46	20.8	148.97
$CH_3Cu(^{13}CN)Li$ (**2**)	ether-d_{10}	−78	−12.58		151.01
	ether-d_{10}	−100	−12.25		150.20
	ether-d_{10}	−110	−12.10	12.3	149.95
	ether-d_{10}	−120	−11.93	21.6	149.78
$CH_3CH_2Cu(^{13}CN)Li$ (**3**)	THF-d_8	−78	1.64	21.6	149.11
	THF-d_8	−100	1.74	22	148.96
	ether-d_{10}	−78	1.85		150.86
	ether-d_{10}	−100	1.89	20.8	150.10

decoupling at higher temperatures due to exchange processes. For example, at −110 °C, the $^{2}J_{C,C}$ coupling constant of **2** in diethyl ether is significantly smaller than that of **2*** in THF. But a temperature reduction of a sample of **2** to −120 °C causes a $^{2}J_{C,C}$ coupling constant even slightly larger than that of **2*** at −110 °C. A comparison of $^{2}J_{C,C}$ of **3** in THF at −78 °C (21.6 Hz) and −100 °C (22 Hz) suggests a maximum of the experimental coupling constant at the range of $^{2}J_{C,C} = 20.8–24.2$ Hz. These results showed that it is principally possible to determine the number and kind of organic substituents on copper by measuring scalar couplings across copper. Therefore, this approach was ideal to prove or disprove the existence of 'higher order' or cyano-Gilman cuprates in solution. For this purpose, the scalar coupling patterns of ^{13}C-labelled Me_2CuLi (**1**) and $Me_2CuLi•LiCN$ (**1•**LiCN) were measured in THF (Figure 4)[137].

The salt-free cuprate Me_2CuLi (**1**) was used to provide the coupling constants of the basic Gilman dimethylcuprate unit (Figure 4a) and, interestingly, for both salt-containing cuprates $Me_2CuLi•LiCN$ (**1•**LiCN) and $Me_2CuLi•LiI$ (**1•**LiI) an identical multiplicity pattern compared to **1** (Figure 4c) was detected. A comparison with simulated spectra (Figure 4b) showed clearly the existence of an $A_3XX'A_3'$ spin system, which reveals identical 'lower order' cuprate structures for **1•**LiCN and **1•**LiI[137]. In addition, the simulation provided the scalar coupling constants of $^{1}J_{C,H} = 109.5$ Hz, $^{2}J_{C,C} = 21.0$ Hz and $^{3}J_{C,H} = -0.8$ Hz. A comparison of the $^{1}J_{H,C}$ scalar coupling with that of MeLi ($^{1}J_{H,C} = 98$ Hz) reveals the metal-bound character of the methyl group and the value of $^{2}J_{C,C} = 21$ Hz is in accordance with the maximum $^{2}J_{C,C}$ values of the heteroleptic organocuprates in Table 2. From these results, a linear structure with either two

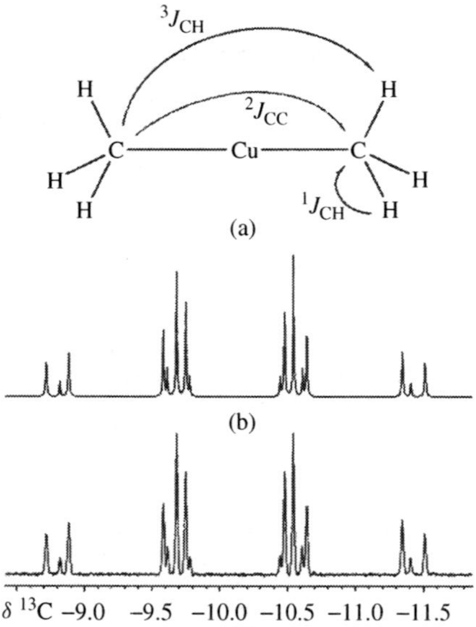

FIGURE 4. (a) Monomeric cuprate unit with the observed scalar couplings indicated by arrows; (b) simulated and (c) experimental ^{13}C spectrum of Me_2CuLi (**1**) in THF. The detection of identical $^{1}J_{C,H}$, $^{2}J_{C,C}$ and $^{3}J_{C,H}$ scalar coupling constants in 1D ^{13}C spectra of $Me_2CuLi•LiCN$ showed that the Gilman cuprate is the general structure for all dialkylcuprates. Reprinted with permission from Reference 137. Copyright 1998 American Chemical Society

alkyl substituents or alkyl/cyanide (1:1) can be concluded for homoleptic cuprates and cyanide-containing heteroleptic cuprates, which was also confirmed by various other theoretical and spectroscopic results[110].

2. Solvent separated ion pairs (SSIPs) vs. contact ion pairs (CIPs)

Numerous synthetic studies revealed a strong solvent dependence of reactions with organocuprates, which hinted at the existence of supramolecular structures in solution being relevant for their reactivity[122,123,138–140]. The first investigations of the aggregation level of organocuprates started with colligative measurements in diethyl ether[125,126,141] followed by mass spectrometric investigations[131], NMR spectroscopic measurements[129,142,143] and theoretical calculations[111,144–146]. Especially, theoretical calculations proposed a dimer as a minimal cluster, necessary for conjugate addition reactions of organocuprates[145]. In NMR spectroscopic investigations, Li coordinating agents, such as HMPA and crown ethers, influenced the $^2J_{C,C}$ coupling constants in heteroleptic cuprates across copper and this effect was attributed to the complexation of the Li cation[136]. Another obvious NMR spectroscopic hint of aggregation was the observation of broad line widths in ^{13}C and ^{15}N spectra of organocuprates in diethyl ether[128]. In addition, a study on phenyl- and diphenylcopper(I) species with variable-temperature ^{13}C NMR spectra revealed some details about aggregation. An examination of δ(*ipso*-C) showed that for differently aggregated PhLi and Ph_2CuLi complexes the chemical shift of the *ipso*-C decreases with an increasing number of metal atoms bound to it (Figure 5)[143], an effect which can be attributed to the paramagnetic shielding term[147].

Figure 5 shows that for PhLi a chemical shift decrease of approximately 25 ppm is observed upon aggregation to $(PhLi)_4$, and the aggregation from Ph_2CuLi to $(Ph_2CuLi)_2$ causes a decrease of approximately 15 ppm[143]. But this useful correlation seems to be only valid for diphenylcuprates, because the homoleptic and heteroleptic alkylcuprates **1**, **1**•LiI, **1**•LiCN, **2** and **3** (Table 2) show only small and even increasing chemical shift differences switching from THF (monomers) to diethyl ether (supramolecular aggregates).

Another NMR spectroscopic approach was initiated by the observation of different aggregation levels in crystal structures. Polar solvents like THF and Li coordinating agents force the cuprate to form solvent separated ion pairs (SSIPs, Figure 6b), while diethyl ether, which is a less coordinating solvent supports the formation of contact ion pairs (CIPs), in which the Li atom is a part of the supramolecular assembly (Figure 6a).

In general, a transfer of structure information from crystal structures to the situation in solution has to be done with great care. In studies of organolithium compounds, it was shown that completely different structures can be present either in solution or in the solid state[148–151]. But with selected NMR measurements, structural aspects of crystal structures can be verified in solution. Traditionally, aggregation studies on Li-containing complexes are performed by determination of scalar couplings between Li and the heteroatom, as is done for lithium amidocuprates (see later in this section). However, in the case of homoleptic organocuprates, $J_{Li,C}$ scalar couplings have not been detected up to now. Therefore, in solution, aggregation trends and supramolecular structures of organocuprates can only be derived via the measurement of diffusion coefficients and various dipolar interactions. Using Heteronuclear Overhauser Spectroscopy (HOESY), the quite good spectroscopic properties of ^6Li and ^7Li allow determining qualitative and sometimes even quantitative distances in solution. From crystallographic[129,130] and theoretical studies[111,144,146,152,153] it was known that in organocuprate CIPs the distances between the Li ions and the alkyl substituents are less than 250 pm, i.e. quite intense HOE cross-peaks can be detected. In contrast, in SSIPs the Li atom and the organocuprate units are separated more than 500 pm, which is beyond the cut-off limit of HOEs. Therefore, no HOE cross-peaks can be detected in SSIPs, if alternative magnetization transfers via solvent molecules, chemical

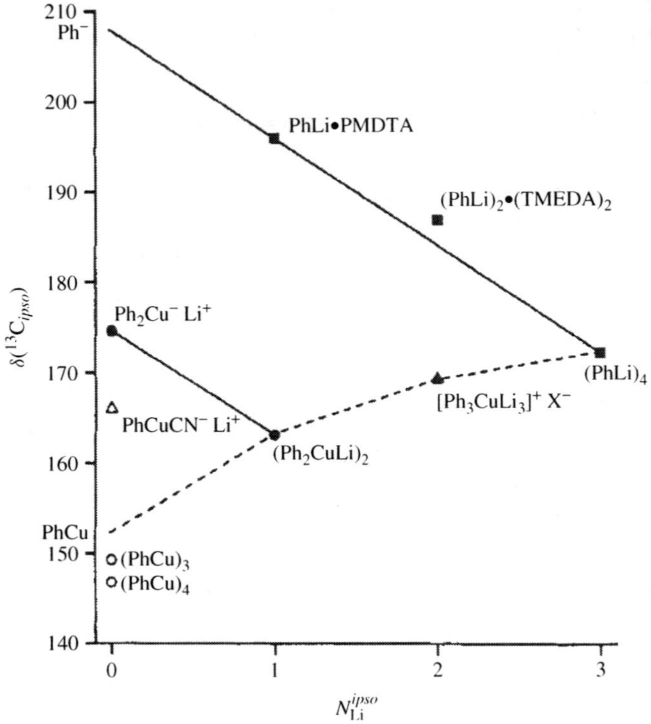

FIGURE 5. Plots of $\delta(^{13}C_{ipso})$ vs. N_{ipso}(Li), the number of Li atoms per *ipso*-C. Note that the $(Ph_2CuLi)_n$ line (●) is parallel to the $(PhLi)_n$ line (■). Reprinted with permission from Reference 143. Copyright 1993 American Chemical Society

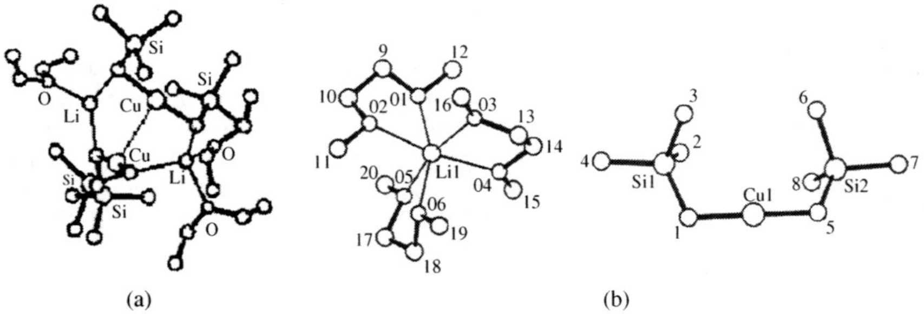

FIGURE 6. Two examples showing the principle structure of (a) CIPs in $[Li_2Cu_2(CH_2SiMe_3)_4(Et_2O)_3$ and (b) SSIPs in $[Li(dme)_3]^+[(Me_3SiCH_2)_2Cu]$. From Reference 129. Copyright Wiley-VCH Verlag GmbH & Co. KGaA. Reproduced with permission

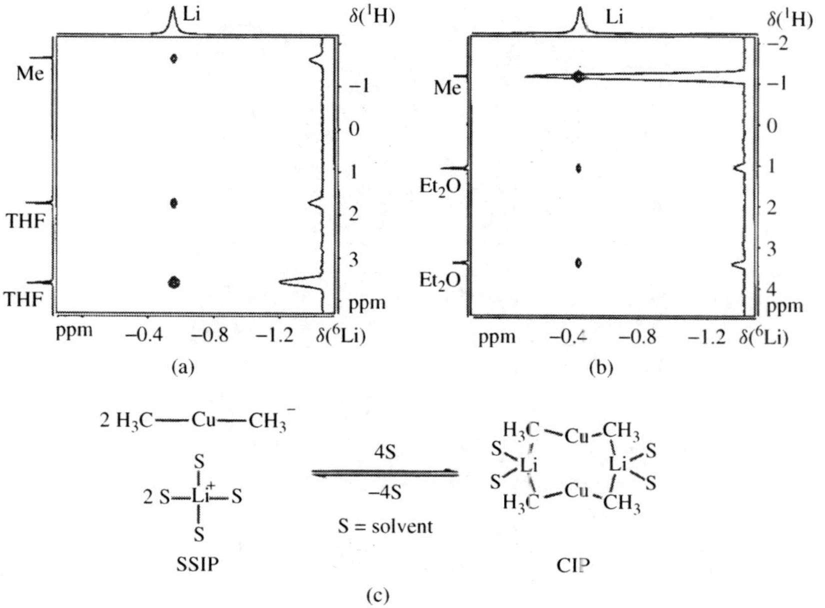

FIGURE 7. $^{1}H,^{6}Li$ HOESY spectra of **1** in (a) THF and (b) diethyl ether, and (c) the corresponding equilibrium of solvent separated ion pairs (SSIPs) and contact ion pairs (CIPs); the Me/Li cross-peak intensity in (a) indicates only small amounts of CIPs in THF, whereas in diethyl ether (b) mainly CIPs exist. Reprinted with permission from Reference 142. Copyright 2000 American Chemical Society

exchange or concentration-dependent background signals can be excluded as accomplished for organocuprates[142]. Consequently, qualitative HOE measurements of organocuprates can be used to reveal the amount of SSIPs and CIPs in different samples, as was shown for the model reagent Me_2CuLi (**1**) in THF and diethyl ether (Figure 7)[129].

In THF, a weak interaction between Li and dimethylcuprate and strong cross-signals between Li and THF are detected (Figure 7a). In contrast, in diethyl ether the interaction between Li and dimethylcuprate is strong and that between Li and diethyl ether reduced (Figure 7b). To visualise these intensity differences, the 1D projections of the cross-peaks are additionally given on the right side of the spectra in Figure 7. These $^{1}H,^{6}Li$ HOESY data clearly indicate that in THF only a small amount of CIPs exist, whereas in diethyl ether the formation of CIPs is preferred. Thus, for organocuprates a solvent-dependent equilibrium between SSIPs and CIPs was established in solution. This equilibrium could be correlated with the reactivity of organocuprates in 1,4-addition reactions and, in accordance with theoretical calculations[111], the CIPs were identified as the reactive species[129].

In order to identify the structure of these synthetically so important CIPs in solution, quantitative $^{1}H,^{7}Li$ HOEs and $^{1}H,^{1}H$ NOEs of dimethylcuprates were measured in diethyl ether[87]. Salt-free Me_2CuLi was used as archetype of organocuprate homodimers and cyanide-containing Me_2CuLi•LiCN was used as model for the heterodimer structures, which were proposed in several theoretical calculations[111, 154–160]. Based on crystal structures and theoretical calculations, the $^{1}H,^{1}H$ NOE and $^{1}H,^{6}Li$ HOE ratios between

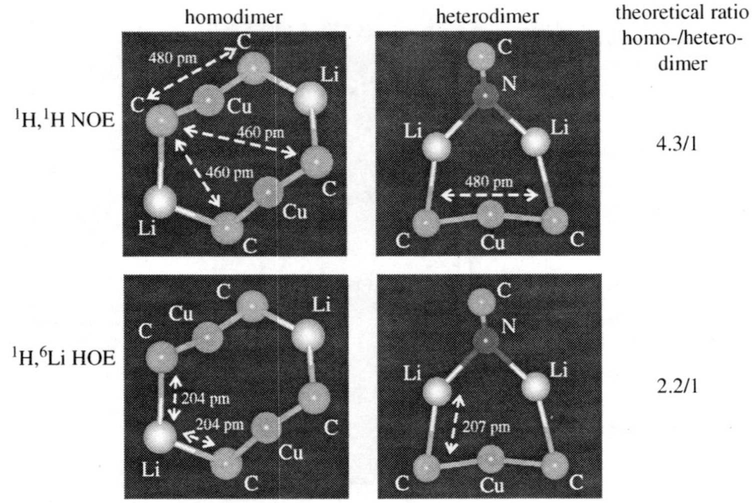

FIGURE 8. Homodimer, $(Me_2CuLi)_2$, and heterodimer structures ($Me_2CuLi{\bullet}LiCN$) of organocuprates with the characteristic distances resulting in differently strong $^1H,^1H$ NOEs and $^1H,^6Li$ HOEs. Reprinted with permission from Reference 87. Copyright 2001 American Chemical Society

homo- and heterodimers were calculated (Figure 8) and the pronouncedly different values, especially for the $^1H,^1H$ NOE, show that a structure differentiation is possible if these NMR parameters can be observed.

As evident from Figure 8, the symmetric structures of organocuprates only allow for a detection of one $^1H,^6Li$ HOE, both in homodimers and in heterodimers. This means that no reference distance is available. As a consequence, the correlation time (τ_C) had to be measured and the Solomon equations[88] were used to quantify the 1D HOE build-up rates[87]. In the case of dimethylcuprates, the maximum $^1H,^6Li$ HOE was used for the determination of τ_C as the most appropriate method[87]. The subsequent analysis of the $^1H,^6Li$ HOE build-up curves revealed similar NOE intensities for both cuprates and H–Li distances of 243 ± 3 pm and 242 ± 9 pm for **1** and **1**•LiCN, respectively. This indicates very similar homodimer structures of both **1** and **1**•LiCN in diethyl ether. To confirm this conclusion, additionally ^1H, ^1H NOE measurements were performed. In the case of **1** and **1**•LiCN, this means that NOEs between chemically equivalent protons must be detected. Therefore, solutions of 20% ^{13}C-labelled cuprates were prepared to differentiate the chemically equivalent groups by means of the different isotopomers $^1H-^{12}C$ and $^1H-^{13}C$ (Figure 9a). This allows one to measure NOE build-up curves from the central ^1H signal ($^1H-^{12}C$) to the ^{13}C satellites ($^1H-^{13}C$ isotopomer) with a sensitivity-improved 1D NOESY–HSQC pulse sequence[87]. The results for **1** and **1**•LiCN in diethyl ether are displayed in Figure 9b. The build-up curves of **1** and **1**•LiCN show a similar curve progression, which corroborates a homodimer structure of both **1** and **1**•LiCN.

In contrast to homoleptic alkylcuprates with lithium exchange rates being fast on the NMR time scale, in the case of lithium amidocuprates slow chemical exchange rates of Li are observed. This enables the detection of different Li signals as well as separated proton signals in amidocuprates with a reduced symmetry and facilitates the structure elucidation of lithium amidocuprates, because a more classical NMR spectroscopic approach can be applied. As a result, the structure elucidation of amidocuprates is primarily based

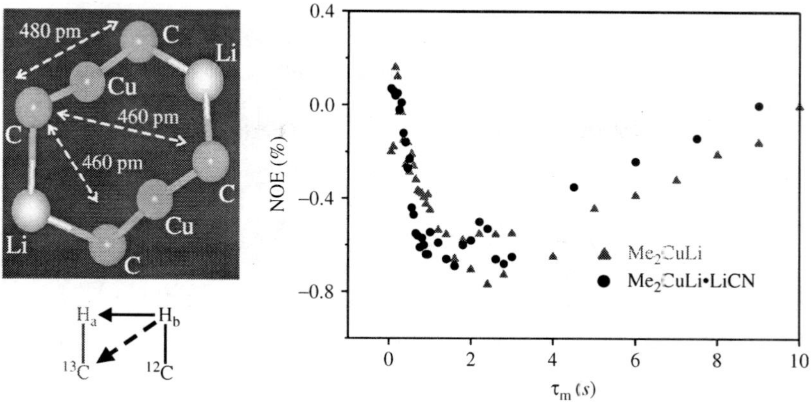

(a) (b)

FIGURE 9. (a) ^{1}H,^{1}H NOEs between chemically equivalent groups can be detected using the different isotopomers ^{1}H–^{13}C and ^{1}H–^{12}C; (b) ^{1}H,^{1}H NOE–HSQC build-up curves of Me_2CuLi (▲) and $Me_2CuLi{\cdot}LiCN$ (●) in diethyl ether show a similar structure of both compounds. Reprinted with permission from Reference 87. Copyright 2001 American Chemical Society

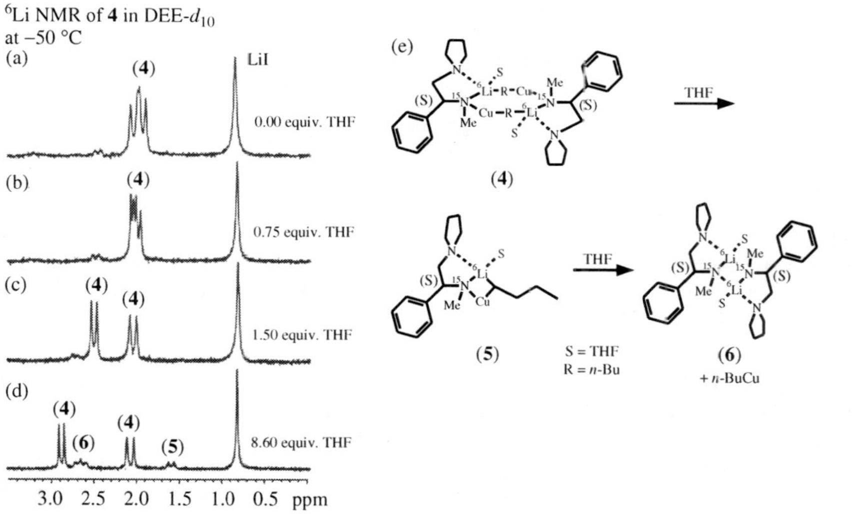

FIGURE 10. ^{6}Li spectra of **4** in (a) diethyl ether (DEE) and with additional (b) 0.75 equiv, (c) 1.5 equiv and (d) 8.60 equiv THF; (e) schematic disaggregation process of the dimer upon addition of THF. Reprinted with permission from Reference 109. Copyright 2000 American Chemical Society

on different $^{6/7}$Li signals, which allow a detailed interpretation of $J_{Li,N}$ scalar coupling constants and multiplicity patterns (Figure 10a–d)[161–163], and of ^{1}H,$^{6/7}$Li HOESY spectra (Figure 11a) in the classical manner. As an example the ^{6}Li spectra of the amidocuprate **4** are shown in Figure 10, for which $J_{Li,N}$ values and multiplicity patterns in combination with 1D and 2D NMR spectroscopy suggest a dimer structure in diethyl ether, which

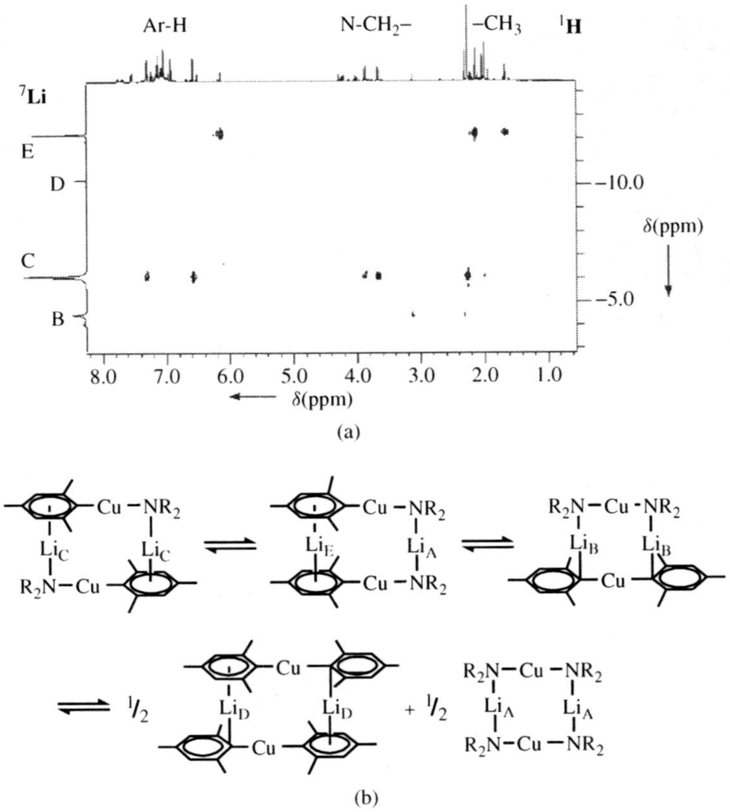

FIGURE 11. (a) ^{1}H,^{7}Li HOESY spectrum of **7**, showing different species in toluene, which are in accordance with (b) a Schlenk-like equilibrium of **7**. The signal of Li_A (*ca* 1 ppm, not shown) does not show HOE signals, due to broad line width. From Reference 167. Copyright Wiley-VCH Verlag GmbH & Co. KGaA. Reproduced with permission

is disaggregated upon addition of THF[109]. In Figure 10e the proposed disaggregation is shown from the dimer **4** to the monomer **5** and finally to **6**, which consists of separated Li amide and *n*-BuCu compounds.

In further studies on $[Cu_2Li_2Mes_2(N(CH_2Ph)_2)_2]$ (**7**), indirectly detected ^{1}H,^{7}Li HOESY spectra[164, 165] revealed several species in toluene, which are obvious from different Li signals (Figure 11a). With the aid of lithium chemical shift data[166–168a] the different species were assigned to the Schlenk-like equilibrium shown in Figure 11b. Recently, similar NMR studies were performed to investigate the influence of THF on the structures and reactivities of these amidocuprates[168b].

C. Supramolecular Aggregation

After the homodimeric core structure was elucidated as the main structural motif of dialkylcuprates in diethyl ether and the CIPs were identified as the reactive species in 1,4-addition reactions to enones, the question arose whether there possibly exist even higher

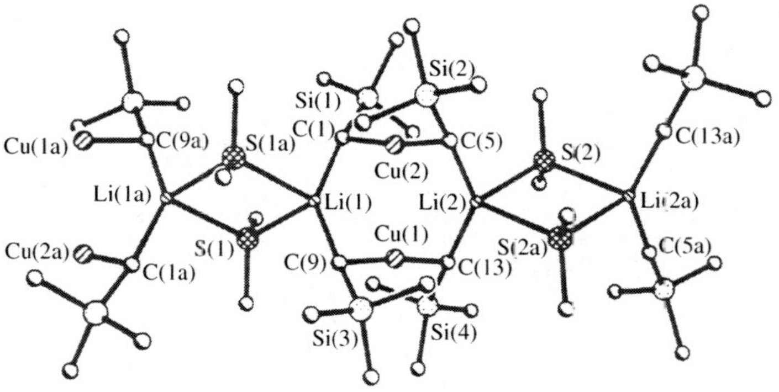

FIGURE 12. Solid state structure of $[Li_2Cu_2(CH_2SiMe_3)_4(SMe_2)_2]_\infty$ ($[\mathbf{8}_2{\bullet}(SMe_2)_2]_\infty$). Reprinted with permission from Reference 130. Copyright 1990 American Chemical Society

supramolecular assemblies with impact on the reactivity of these reagents. In the case of the homoleptic dimethylcuprates, **1** and **1**•LiCN, the negative sign of the $^1H,^1H$ NOE build-up curves (Figure 9b) indicated larger assemblies than homodimers in solution[87] and polymeric structures were found in crystal structures, e.g. that of $[Li_2Cu_2(CH_2SiMe_3)_4(SMe_2)_2]_\infty$ (Figure 12)[130].

Aggregation tendencies beyond the formation of homodimers were additionally indicated by mass spectrometric investigations[131] and broad line width of ^{13}C and ^{15}N signals of organocuprate reagents in diethyl ether[128]. In synthetic studies, an influence of different copper salts, concentrations and varying alkyl substituents on the reactivity and selectivity of organocuprates was observed[108]. As discussed in detail in Section I.B, pulsed field gradient (PFG) DOSY experiments can be used to measure the diffusion coefficient D of supramolecular aggregates in solution, which can be correlated to the hydrodynamic radii and the aggregation level of these assemblies. One great advantage of DOSY measurements is that no special sample preparation is necessary, but correctly applied DOSY experiments (see Section I.B and references therein) can be used to monitor the influence of different concentrations, temperatures and alkyl substituents on the aggregation level.

The tendency of organocuprates to form supramolecular structures in diethyl ether is shown in Table 3 by experimental and theoretical diffusion coefficients. Depending on the steric hindrance of the alkyl residues and the presence and kind of copper salts, aggregation levels between dimers and oligomers are found. For $(Me_3SiCH_2)_2CuLi$ (**8**), an example for sterically hindered cuprates, a slight trend towards higher diffusion values D, i.e. smaller aggregates, is observed. The diffusion data of cuprates with the same alkyl substituent, but different or no Li salt units attached, show that salt-free **1** and **8** and iodide-containing **1**•LiI and **8**•LiI have similar diffusion values, while **1**•LiCN and **8**•LiCN reveal much lower diffusion coefficients, which indicate larger assemblies.

For an accurate quantitative interpretation of the diffusion values in terms of aggregation numbers, presumptions and/or measurements of the solvent shell, the chemical composition and, especially in organometallic chemistry, possible exchange contributions have to be made. In addition, for non-spherical molecules such as organocuprate oligomers (see Figures 12 and 13), shape correction factors are necessary for a quantitative interpretation of diffusion coefficients (Section I.B and equation 2). Therefore, the models shown in Figure 13 were used for the interpretation of the diffusion values in Table 3 and their hydrodynamic radii and cylindrical shape factors were derived from crystal structures[130], theoretical calculations[111, 152, 153] and hard-sphere increments[170, 171].

TABLE 3. Diffusion coefficients D (10^{-9} m^2 s^{-1}), molecular radii r_C (10^{-10} m)^a, length indices n and $n_{mf}{}^b$, solvation indices n_{solv} and theoretical solvation indices n_{solv}(t) of different organocuprates in diethyl ether. Reprinted with permission from Reference 169. Copyright 2003 American Chemical Society

Complex		$r_c{}^{a,c}$	D	$n_{mf}{}^b$	n^c	n_{solv}	$n_{solv}(t)^c$
$(Me_3SiCH_2)_2CuLi$	(**8**)	5.39	0.59	1.3	1.7	4.8	3.2
$(Me_3SiCH_2)_2CuLi{\bullet}LiI$	(**8**•LiI)	6.05	0.54	1.1	1.4	7.5	5.4
		(6.39)			(1.3)		(7.5)
$(Me_3SiCH_2)_2CuLi{\bullet}LiCN$	(**8**•LiCN)	6.01	0.35	4.5	3.6	6.9	4.6
		(6.35)			(3.2)		(6.6)
Me_2CuLi	(**1**)	4.22	0.53	4.4	3.1	2.4	2.6
$Me_2CuLi{\bullet}LiI$	(**1**•LiI)	5.20	0.51	2.2	2.3	6.3	4.9
		(5.64)			(1.9)		(7.0)
$Me_2CuLi{\bullet}LiCN$	(**1**•LiCN)	5.14	0.33	9.0	5.2	5.1	4.4
		(5.58)			(4.5)		(6.4)

$^a r_c$ = radius of the core units calculated by molecular hard-sphere volume increments.
$^b n_{mf}$ is the aggregation number calculated by a model-free approach (see text for details).
^{c}For salt-containing complexes, two sets of values are given: those obtained from model (c) (without brackets) and from model (d) (in brackets) of Figure 13.

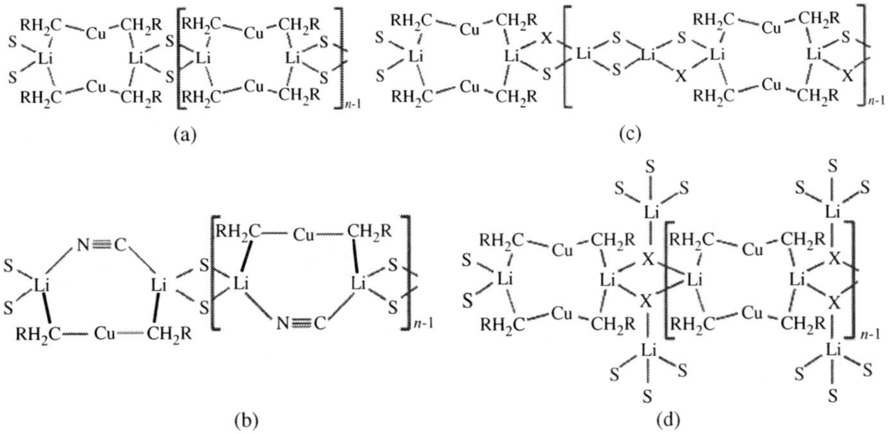

R = H, Me_3Si; S = solvent molecule; X= I, CN.

FIGURE 13. Structure models of dialkylcuprate aggregates beyond dimers; salt-free homodimers (a), salt-containing heterodimers (b), different salt-containing homodimers (c) and (d). Reprinted with permission from Reference 169. Copyright 2003 American Chemical Society

In organometallic compounds, the properties of the solvent are often decisive for their structures in solution. In addition, the solvent shell usually has a significant size and is sometimes even larger than the organometallic compound itself. Therefore, it is crucial for the interpretation of DOSY data to determine and include the number of solvent molecules attached to the complex, i.e. the solvation index n_{solv}. In principle, the solvation of organometallic complexes can be calculated from the normalized diffusion constant of the pure solvent D_{free} and that of the solvent in the reagent sample D_{obs} according to equations 3 and 4 below (D_{cup} represents the diffusion coefficient of the cuprate and α

the percentage of coordinated solvent out of the total amount of solvent n_{tot}).

$$D_{obs} = \alpha D_{cup} + (1 - \alpha) D_{free} \tag{3}$$

$$n_{solv} = \alpha n_{tot} \tag{4}$$

Equation 3 shows that the diffusion coefficient of the solvent in the cuprate samples is averaged between free and complexed solvent molecules. Considering the usual error range of 2–5% in DOSY measurements, the determination of solvation is only possible in the case of large oligomers or highly concentrated samples. Applying the models of Figure 13 including the amount of solvent molecules attached, aggregation numbers (length indices) n can be calculated (Table 3)[169, 172]. To evaluate the influence of the shape factors, which were derived from linear polymeric chains in crystal structures, also the aggregation indices based on spherical shapes, i.e. without any model (n_{mf}), are given in Table 3. These n_{mf} values have similar relative aggregation trends, but different absolute values and highly increased oligomerization numbers for **1**•LiCN and **8**•LiCN. These data show that for an absolute quantification of the oligomerization, reliable shape factors are necessary, but that, independent of the model used, the presence of LiCN leads to significantly larger oligomers.

DOSY measurements combined with kinetic investigations can also be used to test whether the degree or oligomerization of organocuprates is correlated with their reactivity in 1,4-addition reactions to enones[124]. For this purpose, the oligomers were stepwise disaggregated by using different solvent mixtures of diethyl ether and THF and parallel kinetic measurements were performed (see Figure 3 for kinetic and Figure 14 for diffusion results). A disaggregation of **1**•LiCN upon increasing equivalents of THF was indeed detected by normalized diffusion coefficients (Figure 14b), whereas in **1**•LiI samples no disaggregation effect was observed within an experimental error range of 5% (Figure 14a). The parallel kinetic data of **1**•LiCN showed significantly reduced rate constants upon addition of THF and, thus, the supramolecular structures of **1**•LiCN were found to be essential for its reactivity in 1,4-additions.

The kinetic data in Figure 3 clearly show a pronounced effect of THF on the reactivity of **1**•LiI, which is not detectable by DOSY experiments. Therefore, $^{1}H,^{7}Li$ HOE and $^{1}H,^{1}H$ NOE experiments were applied, because dipolar interactions are more sensitive towards small structural changes due to the r^{-6} dependence of the NOE/HOE and the maximum range of approximately 5 Å[124]. From a NMR spectroscopic point of view, it is difficult for these highly symmetrical and flexible oligomers to find reliable reference

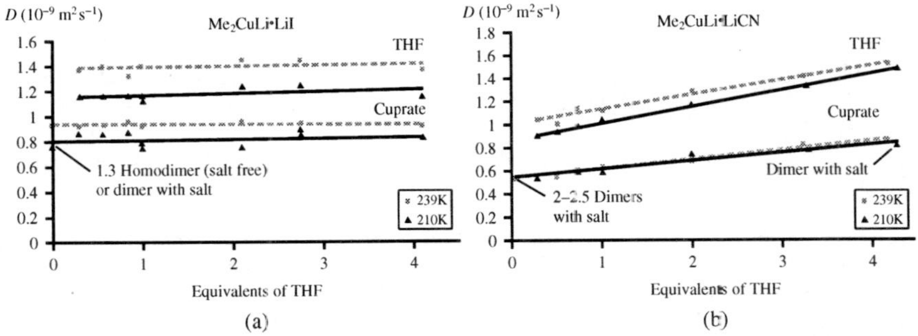

FIGURE 14. Diffusion coefficients of (a) Me_2CuLi•LiI and (b) Me_2CuLi•LiCN in different solvent mixtures of diethyl ether and THF. Reprinted with permission from Reference 124. Copyright 2005 American Chemical Society

distances to interpret the observed cross-peak intensities of a number of HOE/NOE signals originating from different samples. Based on the result that homodimeric core structures exist in diethyl ether (Section II.B.2), all $^{1}H,^{7}Li$ cross-signals could be calibrated relative to the known cuprate $^{1}H,^{7}Li$ HOE cross-signal. With this method, the effect of increasing amounts of THF on the structures of **1**•LiCN and **1**•LiI was elucidated. In the case of **1**•LiCN (Figure 15c), the HOE between Li and diethyl ether is decreasing in the same manner as the HOE between Li and THF is increasing upon addition of increasing amounts of THF. In samples of **1**•LiI, the HOE to THF increases dramatically, while

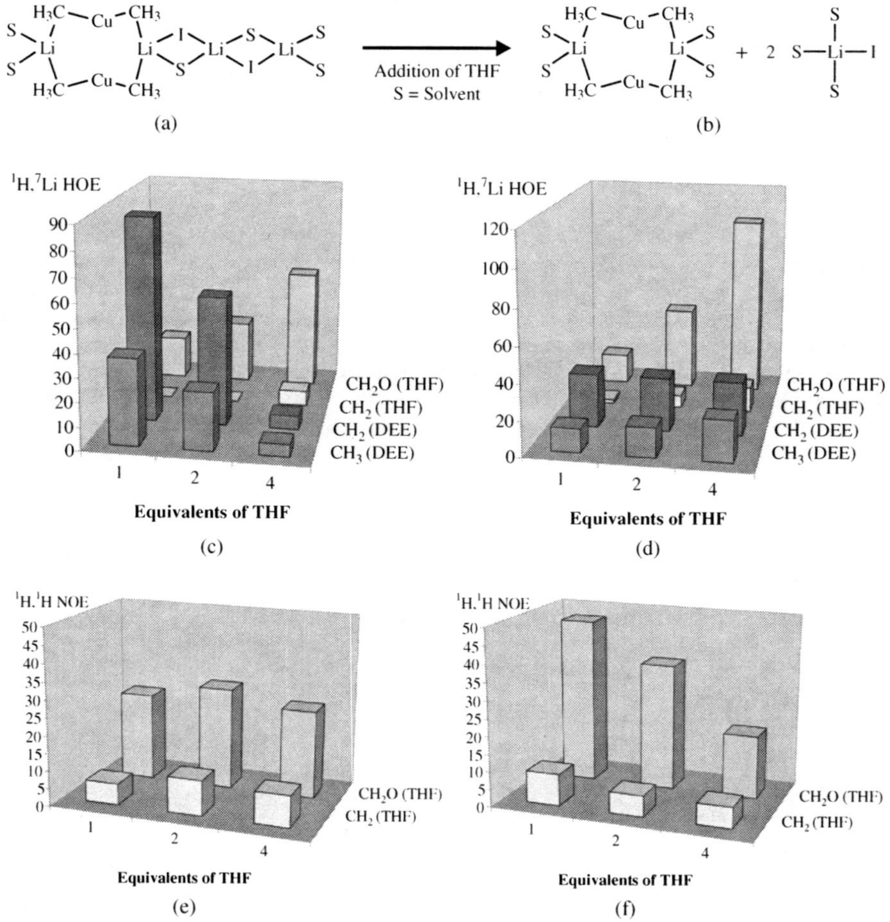

FIGURE 15. (a) Postulated aggregate structure and (b) disaggregation in the case of **1**•LiI. In addition, bar charts are displayed summarizing the $^{1}H,^{7}Li$ HOE volume integrals of the cross-peaks between lithium and the protons of diethyl ether (DEE) and THF for (c) Me_2CuLi•LiCN and (d) Me_2CuLi•LiI and bar charts summarizing the $^{1}H,^{1}H$ NOEs between the methyl groups of the cuprate and the CH_2 groups of THF for (e) Me_2CuLi•LiCN and (f) Me_2CuLi•LiI. Reprinted with permission from Reference 124. Copyright 2005 American Chemical Society

the HOE to diethyl ether remains constant (Figure 15d). These HOE patterns indicate that in **1•**LiCN solvent molecules are exchanged from diethyl ether to THF, while the general supramolecular structure of **1•**LiCN remains and is disaggregated as a whole. In contrast, the addition of THF to **1•**LiI causes additional coordination sites for solvent molecules at Li, which can be interpreted as dissociation of salt units from the homodimer, which is shown schematically in Figures 15a and 15b. Both structural interpretations were confirmed by ^{1}H,^{1}H NOE experiments, in which the distance between the two CH_2 groups of THF was chosen as reference distance, after normalization of the increasing amounts of THF. In **1•**LiCN, the ^{1}H,^{1}H NOE between the methyl groups of the cuprate and THF remains constant (Figure 15e), as expected for an unmodified core structure. In contrast, in **1•**LiI, the ^{1}H,^{1}H NOEs between cuprates and THF decrease upon addition of THF (Figure 15f), which is in accordance with THF solvated Li ions dissociating from the cuprate unit.

The previously discussed studies show that certain combinations of ^{1}H,^{1}H NOE and ^{1}H,^{7}Li HOE measurements are a sensitive method to elucidate even the structural changes of disaggregation processes in supramolecular aggregates. However, these studies do not reveal the absolute position of the anion, either iodide or cyanide, in the supramolecular core structure. For this purpose, ^{13}C-labelled Cu^{13}CN was used to elucidate the position of ^{13}CN by ^{1}H,^{13}C HOEs. In a sample in which the exchange between cuprate coordinated THF (THF* in Figure 16a) and bulk THF was slow on the NMR time scale, it was possible to detect ^{1}H,^{13}C HOEs between ^{13}CN and the cuprate-bound THF molecules, but none to the cuprate itself (Figure 16a). This surprising result was interpreted as an orientation of the ^{13}C away from the cuprate moiety (Figure 16b).

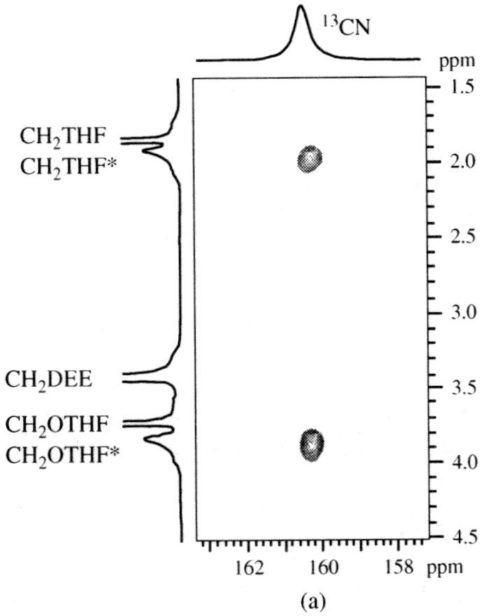

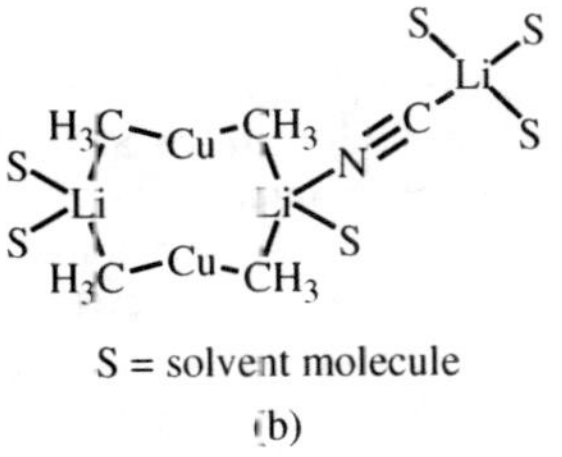

FIGURE 16. (a) ^{1}H,^{13}C HOESY spectrum of Me_2CuLi•LiCN with 12 equivalents of THF. Two sets of signals are observed for THF: THF in the solvent bulk and THF* bound to the cuprate aggregate; the observed ^{1}H,^{13}C HOE cross-peaks indicate the orientation of CN shown in (b). Reprinted with permission from Reference 124. Copyright 2005 American Chemical Society

III. NMR SPECTROSCOPY OF INTERMEDIATE COMPLEXES OF ORGANOCUPRATES

The results presented for the free organocuprate reagents in the previous section show impressively that NMR spectroscopy is a powerful method for the structure determination of organometallic compounds in solution, even in the case of flexible and oligomeric aggregates. NMR is also the method of choice for the structure elucidation of reaction intermediates. However, the basic prerequisite for any NMR investigation of transient species is to stabilize sufficient amounts of it for a certain time period, because NMR is a very insensitive and slow method. Consequently, the NMR methods applicable to reaction intermediates are limited by the existing life time and amount of the intermediate and isotope labelling is often used to increase sensitivity.

A. Cu(I) Organocuprate Intermediates

In conjugate addition reactions of organocuprates to Michael acceptors, π-complexes between cuprates and Michael acceptors were proposed theoretically as first reaction intermediates[111, 112, 173–179] and were confirmed experimentally (e.g. Figure 17)[62, 134, 135, 180–189]. Furthermore, in copper-mediated click reactions of copper acetylides with azides, a π-complex formation between Cu and the acetylene is reported as the initial step, too[190]. In these intermediate π-complexes, the π-bond carbons are expected to experience the highest chemical shift variations and can be used as sensors for the formation of π-intermediates. In one of the first literature-available NMR studies of organocuprate intermediates[134], an organocuprate π-complex was stabilized by using low-temperature NMR in combination with cinnamic ester as quite unreactive Michael acceptor. In the ^{13}C spectra of cinnamic ester and its organocuprate π-complex, upfield shifts of the π-bond carbons of $\Delta\delta = -67.2$ ppm and -82.6 ppm were detected. In addition, a small downfield shift of the carbonyl carbon indicates a Li coordination at the carbonyl oxygen. Later on, also in further studies of organocuprate π-complexes, these characteristic ^{13}C chemical shift differences were detected and used as evidence for π-complexation in organocuprate intermediates (for a typical example, see Figure 17)[62, 180–183, 185, 186].

Similar to the chemical shifts, scalar couplings as second fundamental NMR parameter can also be used for the detection and structure elucidation of intermediate complexes. However, especially in organocopper complexes, line broadening due to quadrupolar relaxation often hampers the detection of scalar couplings. In addition, exchange processes may lead to a reduction of the detected scalar coupling constant, as already mentioned for heteroleptic cuprates in Table 2 (Section II.B.1). Dealing with this problem, elaborate intermediate stabilization strategies, low-temperature NMR and specific isotope labelling rendered not only magnetization transfers via scalar couplings, but also the quantitative determination of scalar coupling constants possible in various intermediate species. Information from both methods allowed impressive insights into bonding orders and structures of organocuprate intermediates.

The first $J_{C,C}$ coupling constants in organocuprate π-complexes were detected in the cuprate ynoate complex **11** (Figure 18)[135]. To enable the detection of $^{1}J_{C,C}$ scalar couplings, compound **9** was ^{13}C-labelled at C-2, C-3 and C-5. The $^{1}J_{C,C}$ coupling constants of free **9** were determined with the aid of the INADEQUATE technique. After addition of the sterically demanding t-$Bu_2CuLi{\bullet}LiCN$, the π-complex **11** and the corresponding $^{1}J_{C,C}$ coupling constants were detected (Figure 18).

Comparing the coupling constants in **9** and **11**, it is evident that the most varying $^{1}J_{C,C}$ coupling constant is $^{1}J_{C,C}$ between C-2 and C-3, which decreases from 74 Hz to 51 Hz and indicates the interaction of the cuprate with the π-bond. As $^{1}J_{C,C}$ scalar couplings imply information about hybridization and bond orders, the significant decrease of the $^{1}J_{C,C}$

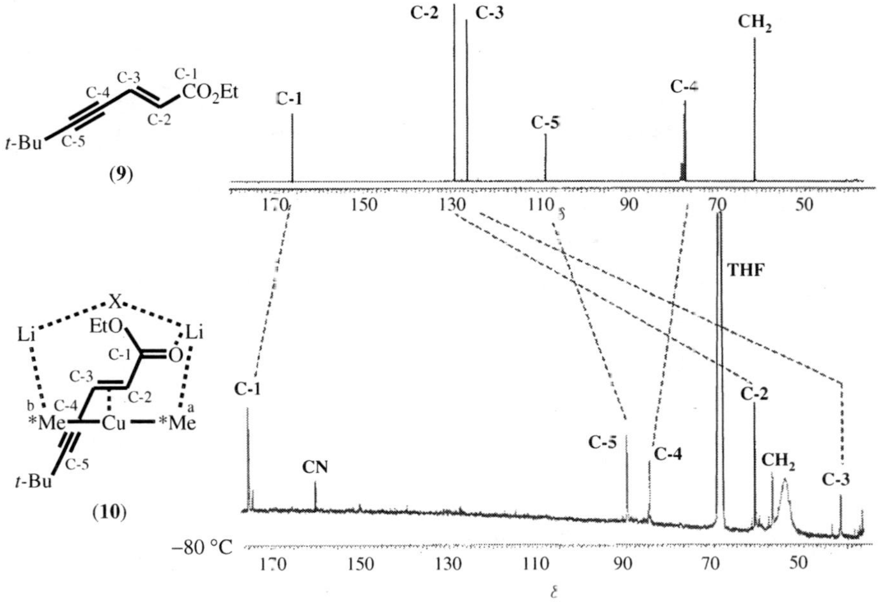

FIGURE 17. The ^{13}C NMR spectra of the ethyl 2-en-4-ynoate **9** and its cuprate-enyne π-complex **10** show typical ^{13}C chemical shift changes upon carbonyl complexation and π-complex formation. From Reference 62. Copyright Wiley-VCH Verlag GmbH & Co. KGaA. Reproduced with permission

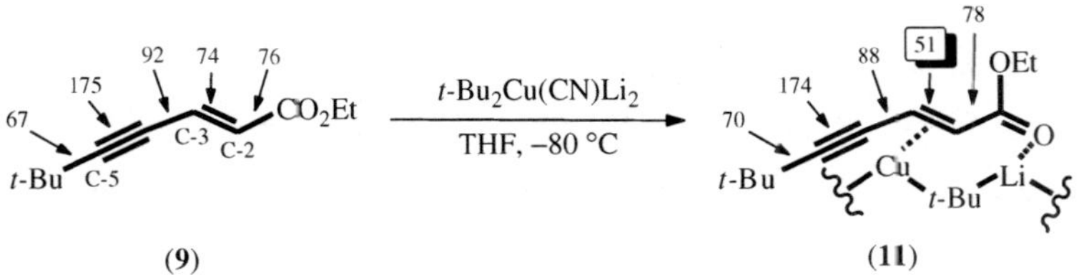

FIGURE 18. Comparison of the $^{1}J_{C,C}$ coupling constants of the ynoate **9** before and after formation of the π-complex **11** show the exclusive coordination of the cuprate to the former double bond. Reprinted with permission from Reference 135. Copyright 1994 American Chemical Society

shows that the hybridization and bond order of C-2 and C-3 in the π-complex is similar to sp^2-carbons, which are connected via a single bond. For comparison, in 1,3-butadiene the $^{1}J_{C,C}$ coupling constant of the single bond, which is connecting the two sp^2-carbons, is 53.7 Hz[135]. Later studies showed that in organocuprate intermediates scalar couplings between the cuprate and the enone moiety can also be detected. For this purpose, samples with specifically ^{13}C-labelled **9** and completely ^{13}C-labelled **1•LiCN** were prepared and the ^{13}C spectra of either C-2* or C-3* labelled intermediates (Figure 19b and 19c) were compared with that of completely unlabelled **9** in the π-complex **10** (Figure 19a). Introducing a ^{13}C label in the C3 position, the methyl group *Me^a at −6.9 ppm is split into a doublet with a coupling constant of 12 Hz. Consequently, the ^{13}C signal of C3 is also a doublet with 12 Hz[62]. In contrast, the labelling in position C2 produces no observable

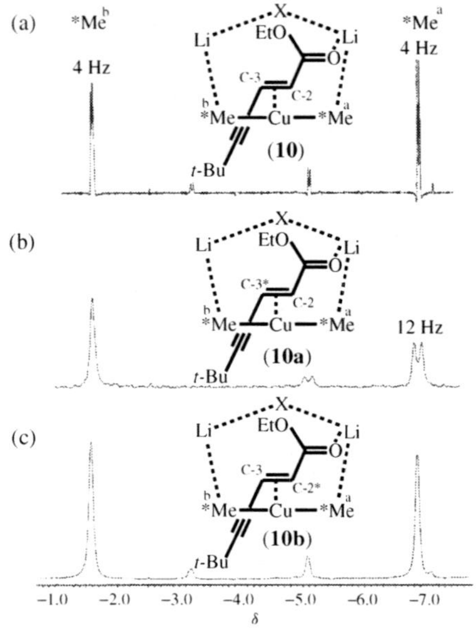

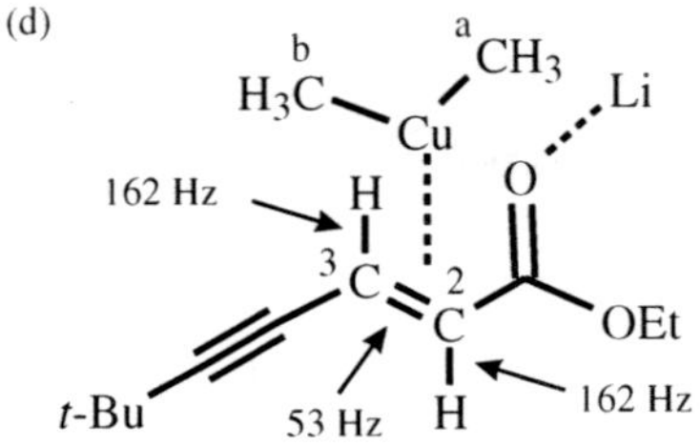

C,C Coupling constants	
$J_{2,3}$	53 Hz
$J_{a,b}$	≤ 4 Hz
$J_{2,a}$	0 Hz
$J_{2,b}$	*ca* 0 Hz
$J_{3,a}$	12 Hz
$J_{3,b}$	0 Hz

FIGURE 19. ^{13}C NMR spectra of the π-complexes (a) **10**, (b) **10a**, (c) **10b** with completely labelled **1**•LiCN (methyl groups termed a and b) and selectively labelled Michael acceptors. Labels are marked with asterisks. (d) The detected coupling constants of the π-complex demonstrate the partly covalent connection of cuprate and enone and the bent structure of the cuprate moiety. From Reference 62. Copyright Wiley-VCH Verlag GmbH & Co. KGaA. Reproduced with permission

coupling pattern (Figure 19c). This difference in the scalar coupling constants within the π-complex is in accordance with a bent structure of the cuprate moiety in the intermediate (Figure 19d).

The orientation of the two methyl groups $^*Me^a$ and $^*Me^b$ in these intermediates was confirmed by NOESY cross-signals to the vinyl protons H-2 and H-3 (Figure 20). The cross-peak intensities show that the methyl group $^*Me^b$ at −0.6 ppm is directed towards the *t*-butyl group and the methyl group $^*Me^a$ at *ca* −1.1 ppm towards the carbonyl function (Figure 20).

The described general characteristic of organocuprates in THF to form π-complexes as first detectable intermediate in 1,4-addition reactions was also confirmed in intermediate studies with diethyl ether as solvent. From the studies of the organocuprate reagents it was known that oligomeric supramolecular assemblies exist in diethyl ether, which could be correlated to their different reactivity (see Section II.A). Consequently, the question arose whether these supramolecular assemblies persist in the π-intermediates. Extremely broad line widths and the gel-like textures of concentrated π-complexes in diethyl ether indicated high supramolecular structures, but did not allow any detailed NMR investigations[63, 132]. Therefore, distinct amounts of THF were used to disaggregate the supramolecular assemblies until spectroscopically acceptable line widths were observed, a strategy which was based on the studies of oligomeric aggregates of the free organocuprates (Section II.C). In order to slow down the reaction rates compared to unsubstituted 2-cyclohexenone and in an attempt to stabilize the π-intermediates, additionally different substitution patterns were used in the cyclic enones **12**, **13**, **14** and **15** (Scheme 3)[63, 132, 191].

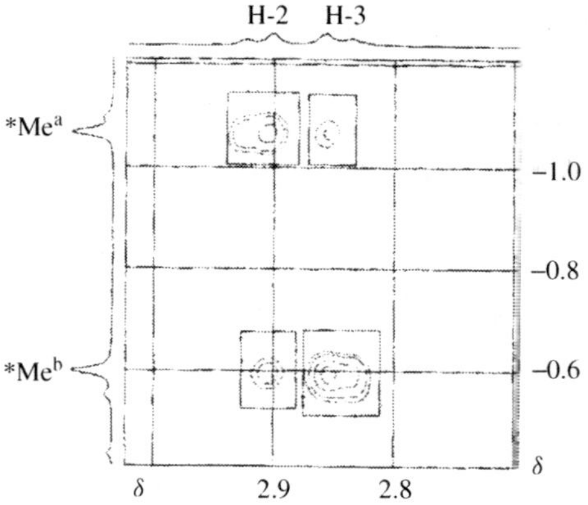

FIGURE 20. $^{1}H,^{1}H$ NOESY section of the vinyl protons and the cuprate methyl groups of the cuprate intermediate **10**, which confirms the orientation of the methyl groups shown in Figure 19d. From Reference 62. Copyright Wiley-VCH Verlag GmbH & Co. KGaA. Reproduced with permission

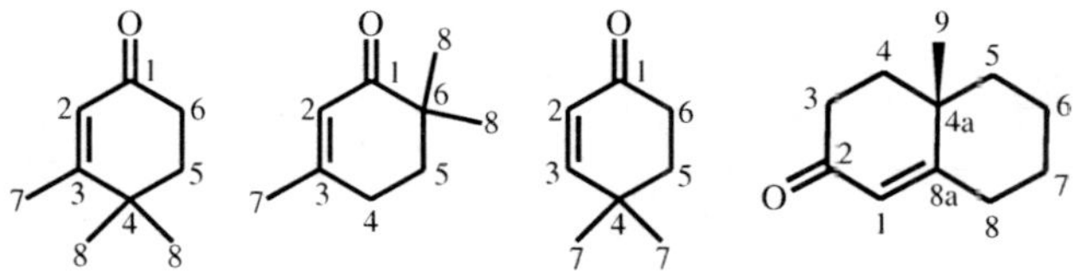

(**12**) (**13**) (**14**) (**15**)

SCHEME 3. Selected sterically demanding achiral enones **12**, **13** and **14** and chiral enone **15**, which build NMR observable enantiomeric (**12**, **13**, **14**) and diastereomeric (**15**) π-complexes in diethyl ether. Reprinted with permission from Reference 63. Copyright 2008 American Chemical Society

In investigations of the π-complexes of 4,4a,5,6,7,8-hexahydro-4a-methylnaphthalen-2(3H)-one (**15**), it could be shown that diastereomeric complexes are formed due to an α- and β-face coordination of the cuprate (Figure 21a). This significantly complicates the NMR spectra of the resulting π-intermediates, because two sets of signals exist in varying amounts due to the two diastereomeric complexes (Figure 21c). The resulting problems of signal overlap and sensitivity were solved by using achiral enones, in which α- and β-face coordination of the cuprate leads to enantiomeric complexes producing only one set of signals (Figures 21b and 21d)[63].

Using disaggregation with THF and achiral enones, the NMR spectroscopic foundations were laid to demonstrate that the π-complexes described in THF, including their bended structure, are a general structural motif also in diethyl ether as solvent. In the example of **14** with 2 equivalents of **1**•LiI, the characteristic ^{13}C and ^{1}H chemical shift differences before (Figure 22a,b) and after (Figure 22c,d) π-complexation are shown in diethyl ether.

Interestingly, in diethyl ether the carbonyl carbon C1 experiences a small upfield shift upon π-complexation (Figure 22a,c), whereas in THF a downfield shift was observed, which was assigned to a Li coordination (Figure 17)[62]. This observation indicates that in diethyl ether more complex species than a single Li ion are responsible for the carbonyl complexation[63].

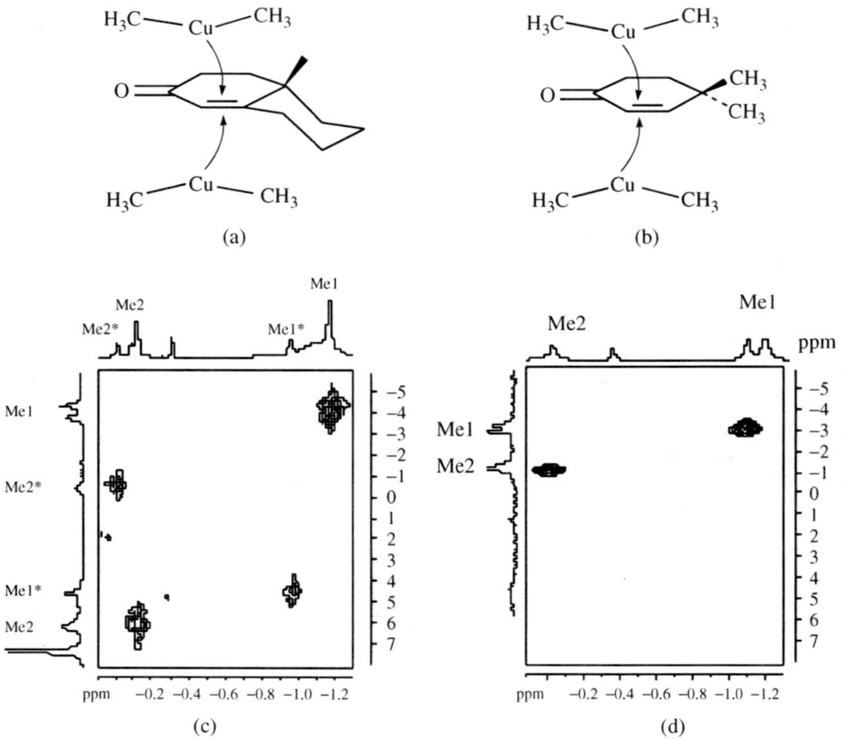

FIGURE 21. Schematic representation of the α- and β-face complexation of (a) chiral **15** and (b) achiral **14** and π-complex cuprate sections of the corresponding ^{1}H,^{13}C HMQC spectra in diethyl ether at 180 K. Diastereomeric π-complexes with chiral enones show (c) two sets of signals, while (d) enantiomeric π-complexes produce only one set of signals. Reprinted with permission from Reference 63. Copyright 2008 American Chemical Society

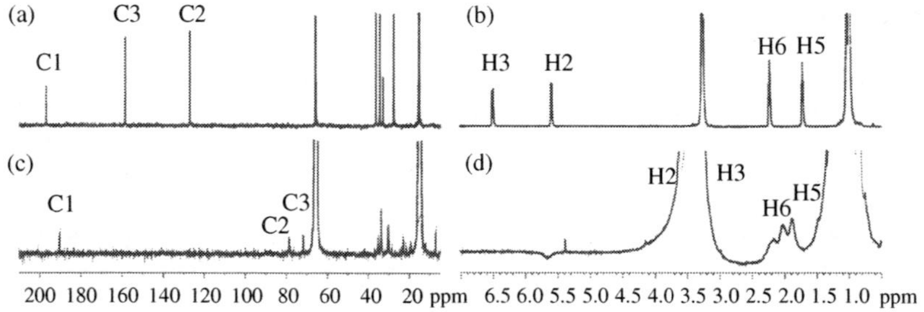

FIGURE 22. Typical change of ^{13}C spectra (a, c) and ^{1}H spectra (b, d) of free enone **14** (a, b) and upon π-complexation (c, d) with 2 equivalents of **1**•LiI in diethyl ether at 170 K. Reprinted with permission from Reference 63. Copyright 2008 American Chemical Society

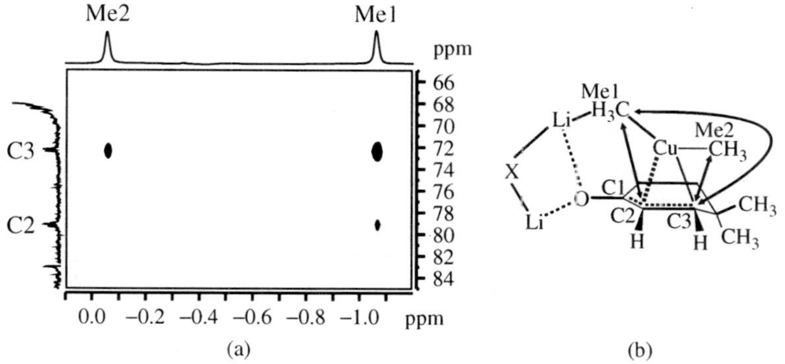

FIGURE 23. (a) $^1H,^{13}C$ HMBC spectrum of a π-complex of **1**•LiCN and **14** showing the cuprate section in the ^1H dimension and the coordinated double-bond section in the ^{13}C dimension. The cross-signals indicate scalar couplings, which are illustrated by arrows in the schematic π-complex (b). Reprinted with permission from Reference 63. Copyright 2008 American Chemical Society

The most impressive result of the described optimization of the experimental conditions was the detection of scalar couplings in π-complexes even without ^{13}C labelling. By applying a HMBC magnetization transfer across copper from the methyl groups of the cuprate to the enone, it was possible without specific ^{13}C labelling of the enone to adduce direct evidence for π-complexation and the bended cuprate structure in these intermediates also in diethyl ether (Figure 23).

Previously, it was discussed for π-complexes in THF that different scalar coupling constants were used to deduce the bent structure of the cuprate moiety (Figure 19). Despite the described extensive improvements of the experimental conditions, in diethyl ether a similar approach was not applicable, because the scalar couplings were smaller than the line widths. In such cases an indirect approach can be used, because in HMBC spectra the integrals of the cross-peaks are correlated qualitatively with the coupling constants[192–195]. Therefore, in principle qualitative coupling constants can be derived from HMBC spectra and, e.g., the cross-peak pattern shown in Figure 23a confirms the bent structure of the cuprate unit also for organocuprate π-intermediates in diethyl ether (Figure 23b). As additional structural feature in diethyl ether compared to THF, DOSY experiments revealed that the π-complexes also form supramolecular assemblies and that the oligomeric cuprate aggregates even increase after addition of enones[63].

In synthetic studies about conjugate addition reactions of cuprates, high and sometimes unexpected diastereoselectivities were obtained for a variety of chiral cyclic enones[108]. For example, the chiral enone **15** (Scheme 4a) yields almost exclusively the β-methyloctalone, which means that surprisingly a *cis*-selective 1,4-addition reaction takes place.

In order to test whether this unexpected and high diastereoselectivity is caused by a conformational preference of the π-intermediate, π-complexes of **15** and methyl cuprate were prepared in diethyl ether and revealed two sets of signals due to diastereomeric α- and β-face complexes (see above)[63]. The ^1H and ^{13}C chemical shifts were assigned via a combination of $^1H,^{13}C$ HSQC, $^1H,^{13}C$ HMBC, $^1H,^1H$ NOESY and $^1H,^{13}C$ INEPT INADEQUATE experiments, including a diastereotopic assignment of the CH_2 groups (Scheme 4b). With the aid of INEPT INADEQUATE experiments[196,197], in both π-intermediate conformations scalar couplings between the cuprate Me1 and C8a in **15** were detected even without ^{13}C labelling of the enone, indicating the bent cuprate structure in both intermediate conformations (Figure 24).

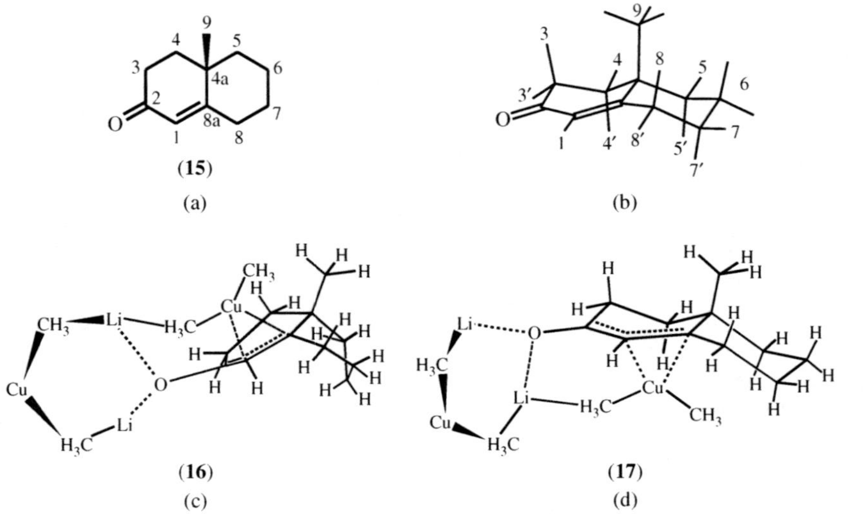

SCHEME 4. (a) 4,4*a*,5,6,7,8-Hexahydro-4*a*-methylnaphthalen-2(3*H*)-one (**15**), (b) the differentiation of the protons within the CH_2 groups, (c) the major conformation or β-face complexation (**16**) and (d) the minor conformation or α-face complexation (**17**) of the π-complex composed of **15** and Me_2CuLi and $Me_2CuLi{\bullet}LiI$. Reprinted with permission from Reference 63. Copyright 2008 American Chemical Society

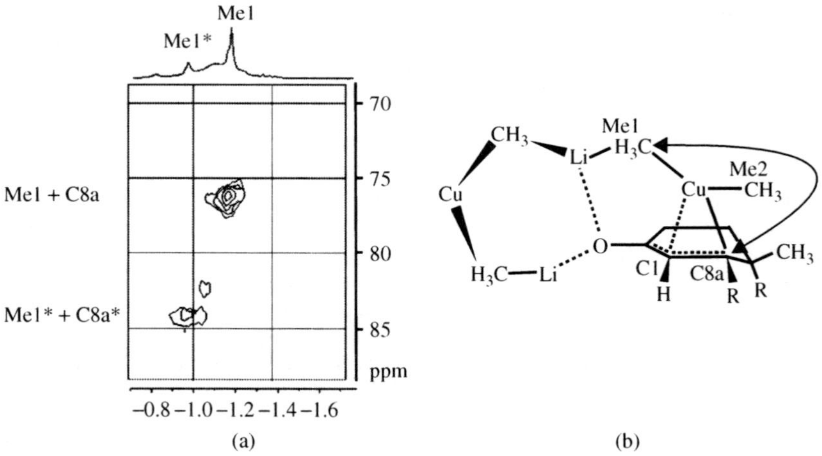

FIGURE 24. (a) Section of a $^{1}H,^{13}C$ INEPT INADEQUATE spectrum of the π-intermediates composed of **15** (natural abundance) and ^{13}C-labelled **1•**LiI in diethyl ether at 180 K. The cross-signals appearing at ^{13}C chemical shifts equal to $\delta^{13}C_{Me1(*)} + \delta^{13}C_{C8a(*)}$ are the result of $^{2}J_{C,C}$ scalar couplings across copper as indicated by the arrow in (b). Reprinted with permission from Reference 63. Copyright 2008 American Chemical Society

From the schematic drawings of the α- and β-face π-complexes of **15** in Scheme 4, it is evident that the β-face π-complex is the precursor of the detected *cis*-addition and the α-face π-complex is the precursor of a possible *anti*-addition. Therefore, the further structural features of the two π-complexes of **15**, i.e. identification of α- and β-face complex and determination of the enone conformations in both complexes, were performed with NOESY experiments. Based on the qualitative interpretation of the $^1H,^1H$ NOESY spectrum shown in Figure 25 and the theoretically calculated conformations of pure **15**[198], the major π-complex could be identified as the β-face π-complex shown in Scheme 4c and the minor π-complex was assigned to the α-face complex shown in Scheme 4d. These results indicate impressively that the conformational preferences of the enone change significantly upon cuprate complexation. In addition, the β-face complex is already the major π-intermediate species. This means that the nearly exclusive *cis*-selective formation of β-methyloctalone in the product is to some extent already preformed in the α-/β-face ratio of the π-intermediates (Scheme 4, **16** and **17**) and further enhanced by the subsequent reaction pathway via the Cu(III) intermediates.

The second principal technique used for the stabilization of organocuprate intermediates is the RI-NMR technique[102, 103], which reduces the dead time before the first NMR scan to a minimum and allows 1D NMR spectra within the first seconds of a reaction. After the very first scan, the starting time for the second experiment is limited by the relaxation properties of the sample and typical repetition times are between 1–2 s. For classical 2D NMR experiments, e.g. COSY or NOESY, also in RI-equipped NMR spectrometers a stable equilibrium state is necessarily induced by low temperatures or with slowly reacting compounds.

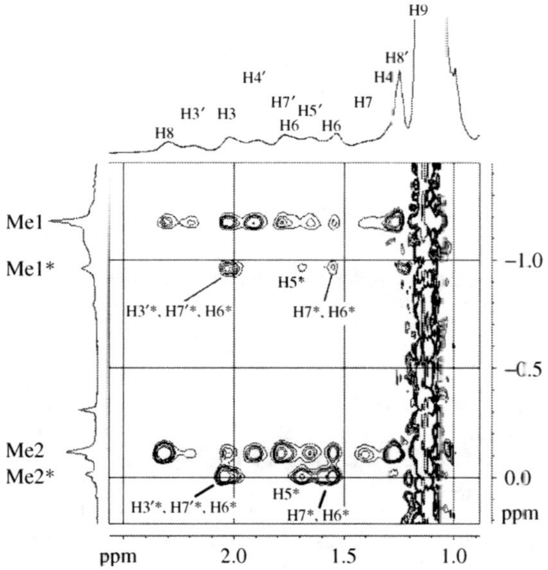

FIGURE 25. Section of a $^1H,^1H$ NOESY spectrum of the π-complexes composed of **15** and **1**•LiI in diethyl ether at 180 K. The different patterns of the cross-signals indicate the different structures presented in Scheme 4c,d. The minor intermediate is labelled with an asterisk; for a visualization of the numbers see Scheme 4b. Reprinted with permission from Reference 63. Copyright 2008 American Chemical Society

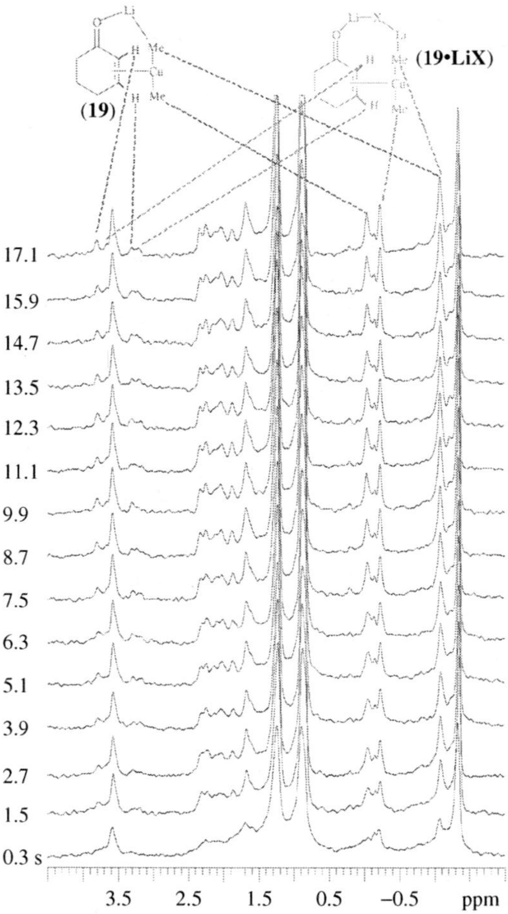

FIGURE 26. Stacked plot of rapid-injection ^{1}H spectra of the reaction of **1**•LiI with 2-cyclohexenone at $-100\,^{\circ}C$ in THF. Reprinted with permission from Reference 133. Copyright 2002 American Chemical Society

A typical series of spectra, performed with RI-NMR, is shown in Figure 26 for the reaction of **1**•LiI and 2-cyclohexenone in THF at $-100\,^{\circ}C$[133]. The assignment of the different detected species was based on 2D COSY and NOESY experiments after equilibrium was reached.

From the integrals of the signals presented in Figure 26, the rate constants for the formation of the individual π-complexes **19** and **19**•LiX can be determined. Such a measurement of reaction rates is a typical and powerful application of RI and standard NMR on reacting systems. In the case of **19** and **19**•LiX, the observed rate constants in combination with EXSY measurements were used to propose the exchange equilibria of organocuprate reagents and intermediate species shown in Figure 27[133].

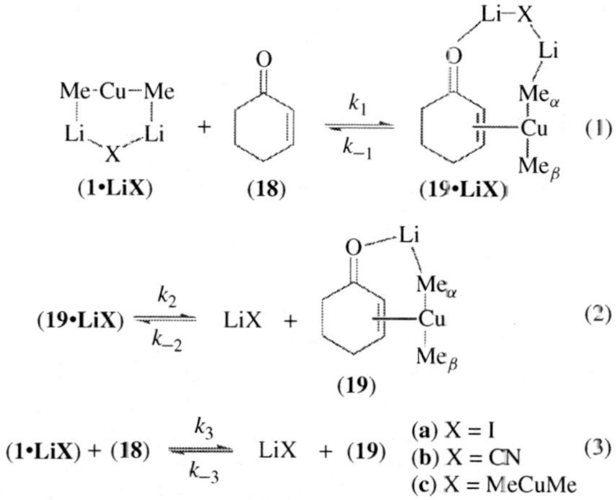

FIGURE 27. Equilibria in the reaction of **1**•LiI with 2-cyclohexenone in THF at $-100\,^\circ$C based on experimental reactions rates and exchange equilibria. Reprinted with permission from Reference 133. Copyright 2002 American Chemical Society

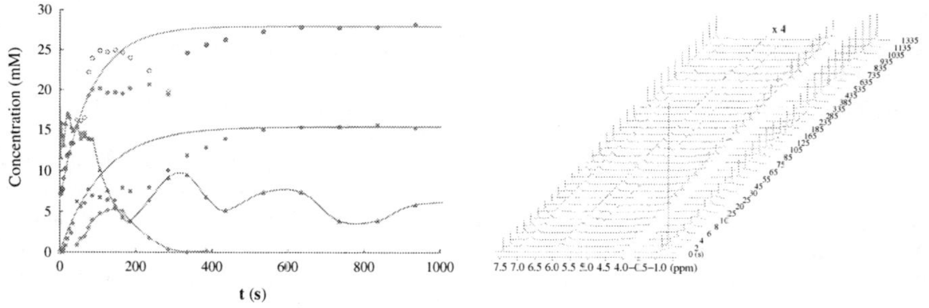

(a) (b)

FIGURE 28. (a) Concentration vs. time plots for the reaction of cyano-Gilman reagent Me_2CuLi• LiCN (♦) with 2-cyclohexenone (▲) at $-70\,^\circ$C. Additionally, the enolate product (■) and residual copper species (●) are displayed. (b) Stacked plots of ^{1}H NMR spectra for the addition of cyclohexenone to Me_2CuLi•LiCN at $-70\,^\circ$C. The first spectrum shows the cuprate solution before injection. Reproduced by permission of the Royal Society of Chemistry from Reference 184

The importance of low-temperature stabilization even in RI-NMR is impressively demonstrated in the following example. For the RI ^{1}H spectra shown in Figure 26 and the intensities derived from RI ^{1}H spectra shown in Figure 28, similar experimental setups were used (2-cyclohexenone, THF) and the temperature was raised from $-100\,^\circ$C to $-70\,^\circ$C. At $-70\,^\circ$C (see Figure 28), the amount of π-complex is too low for detection but the enolate product is observed[184]. Additionally, an interesting ocillatory process becomes obvious. While directly after the injection of cyclohexenone (▲) no free Me_2CuLi•LiCN (♦) is detected, later on its concentration rises again and reaches a maximum at 145 s. The curve of cyclohexenone (▲) in turn shows an unexpected oscillating behavior during the whole measurement.

B. Cu(III) Organocuprate Intermediates

Numerous theoretical calculations predicted Cu(III) species as second essential intermediate in the prototypical reactions of organocuprates, such as conjugate additions to α,β-unsaturated carbonyls, S_N2-like cross-couplings and S_N2' allylic substitutions[111, 173–176, 199–202]. In contrast to this extensive theoretical work, experimental evidence of these elusive Cu(III) species has been missing for decades. Therefore, it was a real breakthrough that recently the first Cu(III) intermediates were detected by NMR spectroscopy[57–59, 64, 65]. Later on, additional Cu(III)-containing organometallic complexes were reported[203].

In 2007, Bertz and Ogle succeeded in the very first experimental detection of a Cu(III) intermediate in organocuprate reactions. In these experiments the Cu(III) intermediate of a 1,4-addition reaction was detected with the aid of rapid-injection NMR and TMSCN as stabilizing agent for the Cu(III) enolate species (Figure 29)[58]. The intermediate Cu(III) enolate was trapped by the formation of a stable silyl enol ether (compound **20** in Figure 29) and this stabilization trick allowed extensive NMR investigations of the Cu(III) intermediate. With a double application of the rapid-injection technique it was even possible to show experimentally that neither the kind of starting cuprate (**1**•LiI or **1**•LiCN) nor the injection sequence influences the formation of the Cu(III) compound. Route A first allows the detection of π-complexes **19** and **19**•LiI, whereas the further injection of TMSCN leads to the formation of the Cu(III) σ-complex (**20** in Figure 29). In route B the injection sequence starts with TMSCl and the identical Cu(III) σ-complex is observed immediately after the injection of the cyclohexenone.

In Table 4 the observed chemical shifts for the different compounds of Figure 29 are listed. The chemical shifts of the cuprates (**1**•LiI or **1**•LiCN) and the Cu(III) species **20** can be distinguished by an appreciable downfield shift of the methyl groups in **20**. While **1**•LiI (−9.12 ppm/−1.40 ppm for $^{13}C/^{1}H$) and **1**•LiCN (−9.04 ppm/−1.35 pm) show common cuprate chemical shifts, **20** exhibits a striking chemical shift combination of the ^{13}C and ^{1}H methyl signals (12.43 ppm/0.05 ppm for Me^t and 25.31 ppm/0.53 ppm for Me^c). These numbers denote that the ^{13}C signals of the Cu(III) species shift dramatically downfield in

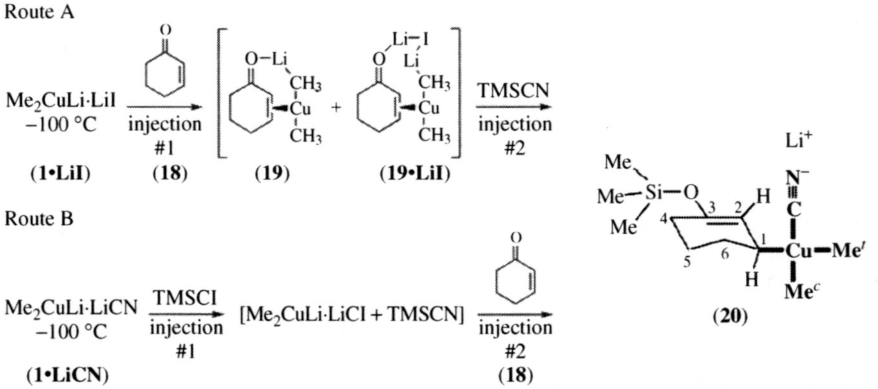

FIGURE 29. Two routes to generate the Cu(III) σ-complex **20** of the 1,4-addition to 2-cyclohexenone in THF. Reprinted with permission from Reference 58. Copyright 2007 American Chemical Society

TABLE 4. Comparison of ^{13}C and ^{1}H (parentheses) chemical shifts[a] for organocuprate Cu(I) π-complexes and Cu(III) σ-complexes. Reprinted with permission from Reference 58. Copyright 2007 American Chemical Society

Group	18	1•LiI	1•LiCN	19	19•LiI	19•LiCN	20
$CH_3(CH_3{}^t)$[b]		**−9.12** (−1.40)	**−9.04** (−1.35)	**−5.02** (−1.12)	**−5.56** (−1.16)	**−5.76** (−1.15)	**12.43** (0.05)
CH_3 $(CH_3{}^t)$[b]		**−9.12** (−1.40)	**−9.04** (−1.35)	**−0.57** (−0.10)	**−1.85** (−0.24)	**−2.14** (−0.21)	**25.31** (0.53)
CN			158.89			159.20	**153.78**
C_1 (C_3)[b]	198.65			194.75	193.34	193[c]	144.73
C_2-H(C_2-H)[b]	130.12 (5.90)			**77.45** (3.77)	**75.82** (3.68)	**75.27** (3.71)	116.28 (5.02)
C_3-H(C_1-H)[b]	151.65 (7.08)			**61.50** (3.26)	**61.50** (3.19)	**61.51** (3.17)	**39.68** (2.74)

[a]Parts per million from TMS. Values for C atoms attached to Cu are in boldface.
[b]Labelling for **20**. Note that C_1 of **18** becomes C_3 of **20** and C_3 of **18** becomes C_1 of **20**.
[c]Shift could not be measured accurately, owing to broadening.

the range of 20 to 35 ppm compared to cuprates. The observation of two distinguishable methyl resonances Me^t and Me^c in compound **20** (*trans* and *cis* to the ring) is caused by the asymmetric chemical environment in the Cu(III) complex and was confirmed by NOE spectra.

The connectivity in **20** was directly proven using $^2J_{C,C}$ coupling patterns across copper. For this purpose, a sample with ^{13}C-labelled methyl groups and labelled $Cu^{13}CN$ was prepared and 1D ^{13}C measurements were performed (Figure 30). In the resulting spectra, all $^2J_{C,C}$ scalar couplings across copper were detected, which were expected for the connectivity in **20** ($^2J_{C,C}$ ring methine carbon (C_1), Me^t = 38.1 Hz, $^2J_{C,C}$ cyano substituent, Me^c = 35.4 Hz). Additionally, Me^t is coupled to the cyano group with $^2J_{C,C}$ = 5.4 Hz and to Me^c with $^2J_{C,C}$ = 2.9 Hz. The ^{13}C chemical shifts and $^2J_{C,C}$ coupling constants

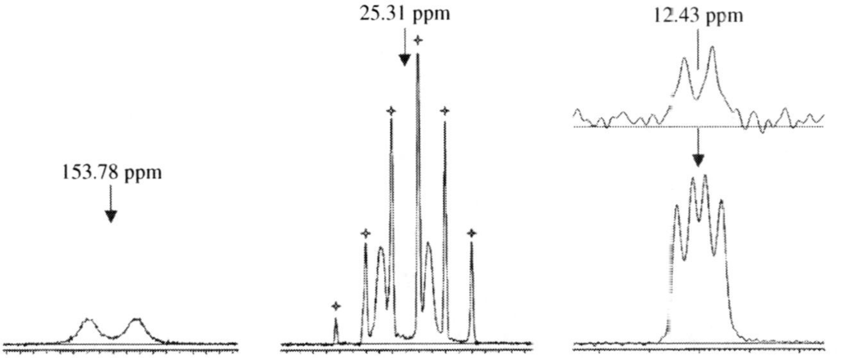

FIGURE 30. ^{13}C NMR sections for labelled **20** with solvent (+). In the upper spectrum ^{13}CN is ^{13}C-labelled and in the lower spectra both $^{13}CH_3$ and ^{13}CN are ^{13}C-labelled. Reprinted with permission from Reference 58. Copyright 2007 American Chemical Society

in **20** were consistent with theoretical calculations of **20**, which confirmed the proposed square-planar structure of the Cu(III) σ-complex **20**[199].

Shortly after the detection of the first Cu(III) σ-complex in 1,4-additions, the preparation and NMR spectroscopic detection of the first Cu(III) σ-complexes in cross-coupling reactions was also successful[59, 65].

During the investigation of organocuprate π-complexes conventional low-temperature NMR, ^{13}C labelling of the cyanide and diethyl ether as solvent were standard experimental tools for stabilizing the Cu(I) intermediates[63]. Surprisingly, with this experimental setup an additional Cu(III) species was also detected in various ^{1}H,^{13}C HMBC spectra, which later turned out to be the intermediate of cross-coupling reactions[59]. This species showed cross-signals of two chemically non-equivalent methyl groups and one cyanide group attached to the same copper. Initially, the amount of this species was so low that in the corresponding proton spectra no signal could be detected even at a high number of scans and only the $^{3}J_{H,C}$ coupling between the methyl groups and the ^{13}C-labelled cyanide led to a signal enhancement allowing its detection in the ^{1}H,^{13}C HMBC spectra (Figure 31c). By variations in the ratio of $Cu^{13}CN$ to MeLi and the presence of small amounts of MeI, it was possible to increase the concentration of this Cu(III) species to such an extent that not only the detection of ^{1}H signals (Figure 31a), but also extensive HMBC (Figure 31b, c)

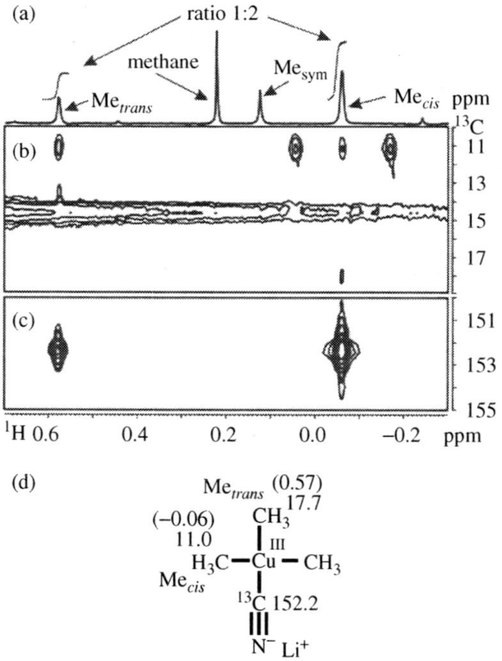

(**21**)

FIGURE 31. Selected high-field section of (a) a 1D ^{1}H and (b), (c) a ^{1}H,^{13}C HMBC spectrum, which prove the existence of (d) a square-planar lithium trimethylcyanocuprate(III) **21** due to cross-signals between (b) the different methyl signals (^{1}H chemical shifts in parentheses in d) and (c) cross-signals between the methyl signals Me_{trans}/Me_{cis} and the cyanide Me_{sym} is a separate, symmetrical species. Reprinted with permission from Reference 59. Copyright 2007 American Chemical Society

and NOESY NMR measurements were possible. The two methyl signals Me_{trans} ($\delta(^1H)$ = 0.57 ppm) and Me_{cis} ($\delta(^1H) = -0.06$ ppm) in the ^1H spectrum, their integral ratio of 1:2, respectively, and the HMBC coupling pattern shown in Figure 31b,c directly indicate the formation of the square-planar $Me_3Cu(III)(CN)Li$ complex (**21**) presented in Figure 31d. In contrast, in a tetrahedral Cu(III) complex only one signal would be observable for all three methyl groups in the ^1H spectrum. Considering the supramolecular assemblies as a general feature of organocuprates and Cu(I) intermediates in diethyl ether, it is interesting that up to now $^1H,^1H$ NOESY spectra have not revealed any hint for supramolecular aggregates of the $Me_3Cu(III)(CN)Li$ intermediate in cross-coupling reactions in diethyl ether.

Simultaneously to the described investigation of **20**, further Cu(III) intermediates of organocuprate substitution reactions were reported by applying rapid-injection NMR in THF as solvent[65]. In Figure 32, the time-dependent intensity developments of the ^1H signals of the cyano Cu(III) and the iodo Cu(III) intermediates are shown as examples. These Cu(III) intermediates were formed immediately after the injection of ethyl iodide into a cuprate/THF solution at −100 °C and could be assigned via 1D ^1H, ^{13}C, 2D NOESY and HMQC spectra. This experiment was repeated with various cuprates, $Me_2CuLi{\bullet}LiI$ (**1**•LiI), $Me_2CuLi{\bullet}LiCN$ (**1**•LiCN), $Me_2CuLi{\bullet}LiSCN$ (**1**•LiSCN) and $Me_2CuLi{\bullet}LiSPh$ (**1**•LiSPh), and yielded quite a number of different Cu(III) σ-complexes (Table 5)[65]. Throughout all of these Cu(III) species, $^2J_{C,C}$ coupling constants were used to confirm the connectivity. With ^{13}C-labelled cuprates and $CH_3{}^{13}CH_2I$, several coupling constants were determined for these compounds (Table 5), with the coupling constants across copper being consistent with the stereochemistry ($^2J_{trans} \gg {}^2J_{cis}$)[65,204,205]. Again the carbon chemical shifts of the Cu(III) species show very deshielded values in the range of 13 ppm to 20 ppm for the CH_3 substituents and 28 ppm to 39 ppm for the CH_2 group of the ethyl substituent. The proton chemical shifts reveal values in the range of −0.5 ppm to 0.8 ppm

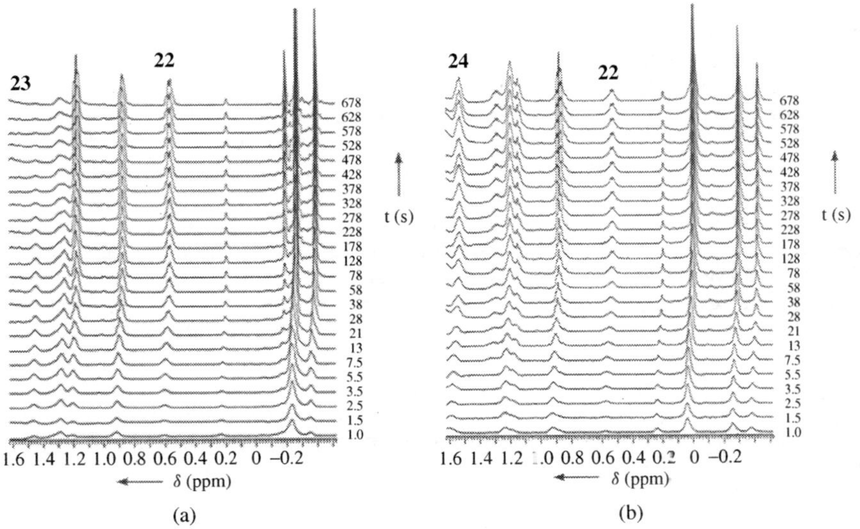

FIGURE 32. Time-dependent ^1H NMR spectra from the rapid-injection treatment of (a) **1**•LiI and (b) **1**•LiCN with ethyl iodide. In (a) Me_2EtCuI (**23**) and in (b) $Me_2EtCu(CN)$ (**24**) are observed, and both samples show signals assigned to Me_3EtCu (**22**). From Reference 65. Copyright Wiley-VCH Verlag GmbH & Co. KGaA. Reproduced with permission

TABLE 5. ^{13}C NMR (^{1}H NMR) chemical shifts[a] for tetracoordinate, square-planar Cu(III) σ-complexes in THF. From Reference 65. Copyright Wiley-VCH Verlag GmbH & Co. KGaA. Reproduced with permission

Complex	δ Me	δ Et (CH_2, CH_3)	δ Cuprate[b]
Me_3EtCu[c] **(22)**	13.26, 16.03[d] (−0.42, −0.29[d])	28.17, 16.13 (0.54, 1.15)	*e*
Me_2EtCuI **(23)**	16.03 (−0.07)	33.67, ND (1.48, 1.28)	−9.07 (−1.39)
$Me_2EtCuCN$[f] **(24)**	14.06 (0.00)	34.66, 16.03 (1.54, 1.20)	−9.11 (−1.41)
$Me_2EtCuSCN$ **(25)**	17.42 (−0.21)	39.07, ND (1.36, 1.26)	−9.08 (−1.31)
trans-$Me_2EtCuSPh$ **(26a)**	18.06 (−0.31)	34.21, 17.53 (1.83, 1.31)	−9.50 (−1.31)
cis-$Me_2EtCuSPh$ **(26b)**	14.83, 20.40 (−0.50, 0.83)	30.04, 15.60 (0.53, 1.14)	−9.50 (−1.31)

[a]Chemical shifts δ (ppm) vs. TMS. ND = not determined.
[b]Shifts of starting cuprate, **1**•LiX (X = I, CN, SCN, SPh), before injection.
[c]Shifts for **22** from **24**, see [e].
[d]2 × area of other Me peak.
[e]Prepared from **1**•LiX. The shifts of the ^{13}C atom bonded to copper in **22** vary with Cu salt by ±0.1 ppm.
[f]Cyanide ^{13}C shifts are δ = 152.66 ppm for **24** and δ = 158.95 ppm for **1**•LiCN.

and 0.5 ppm to 1.8 ppm, respectively (Table 5). In contrast, in cuprates the ^{13}C chemical shifts vary only between −9.07 ppm and −9.50 ppm and those of ^{1}H between −1.31 ppm and 1.41 ppm (Table 5). These deviating chemical shift differences in Cu(I) and Cu(III) complexes reflect the different binding properties between the copper compound and the anions of the previous copper salts (e.g. CN^-, I^-, SCN^-). In the Cu(I) complexes mainly ionic interactions exist between these species, whereas in the Cu(III) complexes the salt anions are covalently bound.

In addition to the trialkyl Cu(III) species with iodide (**23**), cyanide (**24**), thiocyanate (**25**) or thiophenolate (**26a**, **26b**) attached, quite often also the tetraalkyl species Me_3EtCu (**22**) was detected in this study (Table 5, Figure 32)[65]. Interestingly, the tetraalkyl species **22** is exceedingly stable, because after warming up the sample to −10 °C and re-cooling to −100 °C, the tetraalkyl species predominated in solution[65]. In the case of an injection of ethyl iodide to **1**•LiSPh, an isomerization process between *trans*-$Me_2EtCuSPh$ (**26a**) and *cis*-$Me_2EtCuSPh$ (**26b**) was also detected. In further studies, in which Gilman cuprates were treated with different stabilizing agents, such as pyridine or PBu_3, even neutral copper complexes were detected in THF[64]. In addition, the described Cu(III) investigations demonstrate impressively the influence of the solvent on the reaction rates of organocuprates even on the level of Cu(III) intermediates. In THF, the reaction rate of the cross-coupling with methyl iodide as alkyl halide is too fast to observe any intermediate signals even with rapid-injection NMR. In contrast, in diethyl ether a long-term stabilization of the Cu(III) intermediate **21**, which is produced from MeI, is possible.

For S_N2' substitution reactions of organocuprates and allylic substrates, a reaction mechanism including Cu(III) intermediates similar to that of classical substitution reactions was proposed[111,173,174] and recently theoretical calculations on the origin of the regio- and stereoselectivity in S_N2' reactions were published[206]. With rapid-injection

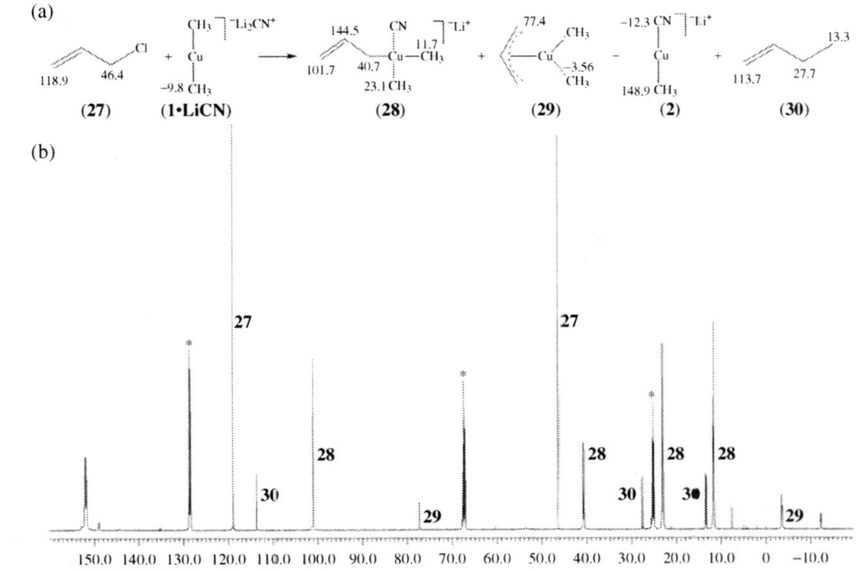

FIGURE 33. (a) Reaction of allyl chloride **27** (allyl-1,3-^{13}C chloride, 50 atom% at each position) with ($^{13}CH_3)_2CuLi$•$Li^{13}CN$ (**1**•LiCN) and (b) ^{13}C NMR spectrum of products **28** (major) and **29** (minor) in THF-d_8 at $-100\,^{\circ}$C. Asterisks denote solvent peaks. Reprinted with permission from Reference 57. Copyright 2008 American Chemical Society

NMR, the reaction of allyl chloride with the Gilman cuprates Me_2CuLi•LiX (X = I, CN) was investigated and Cu(III) species were found also for these allylic substitution reactions[57]. Interestingly, after injection of 1,3-^{13}C-labelled allyl chloride (**27**) into a solution of **1**•$Li^{13}CN$ in THF (Figure 33a), the ^{13}C chemical shifts of the allyl part indicate two different Cu(III) complexes being present in solution, a Cu(III) σ-complex and a Cu(III) π-allyl complex (Figure 33b).

For compound **28** (Figure 33a), the methyl ^{13}C chemical shifts of 11.7 ppm and 23.1 ppm indicate the presence of a Cu(III) σ-complex in agreement with the previous Cu(III) NMR studies. For the π-allyl copper complex, identical ^{13}C chemical shifts ($\delta(^{13}C) =$ 77.4 ppm) of C1 and C3 of the allyl ligand (**29**, Figure 33a) and one chemical shift for the methyl groups bound to copper ($\delta(^{13}C) = -3.56$ ppm) indicate an η_3 π-allyl complex. Interestingly, the cuprate ^{13}C chemical shifts and $^2J_{C,C}$ coupling constants in this π-allyl complex are quite similar to those observed in Cu(I) π-complexes of cuprates and α,β-unsaturated Michael acceptors ($^2J_{C,C} = 12$ Hz, see Figure 19). This is in accordance with theoretical calculations, which show that the formal Cu(III) π-allyl complex has a binding situation similar to that of Cu(I) π-complexes.[57]

Rapid-injection NMR spectra of these Cu(III) σ-allyl and π-allyl complexes show the time-dependent interconversion of the two species (Figure 34b). The σ-allyl complex is formed directly after the injection of **27** (Figure 34a) and then its amount decreases in favor of the π-allyl complex. This suggests a thermodynamically higher stability of the π-allyl copper complex **29**.

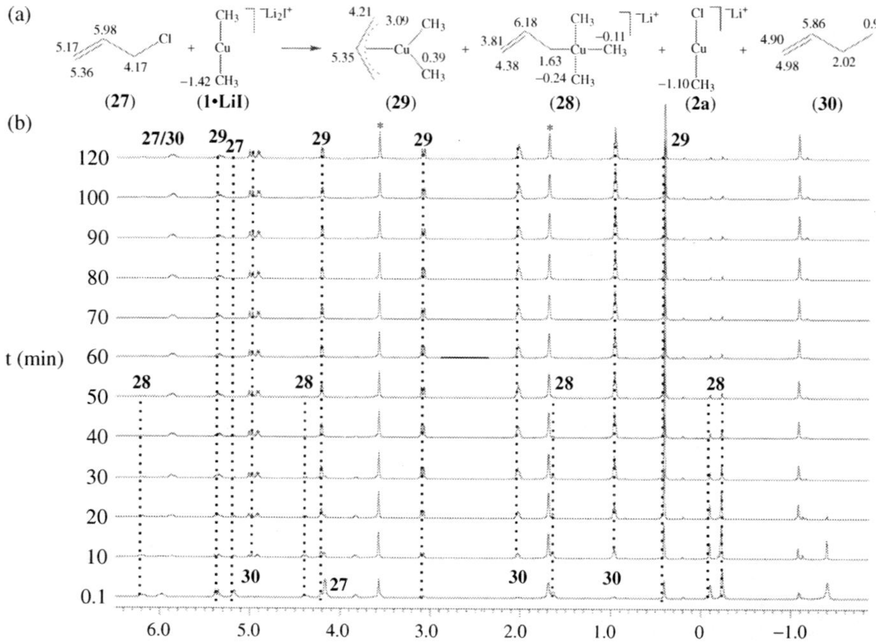

FIGURE 34. ¹H NMR (a) spectra of the products **29** and **28** at regular time intervals after injection of **27** into **1•**LiI (b). Reprinted with permission from Reference 57. Copyright 2008 American Chemical Society

IV. NMR STRUCTURE ELUCIDATION IN Cu(I)-CATALYZED REACTIONS

A. Catalytic Copper Complexes with Thiol-TADDOL Ligands

Nowadays, a multitude of catalytic metal/ligand combinations is accessible in organic synthesis. For copper and its catalytic properties, a recent series of reviews and the edition of this book describe the synthetic potential of copper-catalyzed reactions in detail[11–15, 18, 21, 23]. However, structure elucidation reports on catalytic copper systems are very rare and the amount not at all comparable to the numerous studies with, for instance, Pt, Pd or Rh as central transition metals. For example, in copper-catalyzed 1,4-addition reactions only a few crystal structures of precatalytic copper phosphoramidite complexes are published[207, 208]. All of these crystal structures show a tetrahedral coordination on copper, which does not explain the ligand accelerated catalysis observed in these reactions. Only in the last few years, NMR spectroscopic investigations revealed some structural details about precatalytic copper complexes in solution[19]. In the following, two examples will be discussed in detail, although diimine copper complexes[209–215] and ferrocene-derived diphosphine copper complexes[216–218] have also been investigated.

A very remarkable study compares the structures of Cu(I) complexes with thiol-TADDOL ligands in the solid state and in solution[219]. In copper-catalyzed enantioselective 1,4-additions of Grignard reagents, high *er* values are obtained in the presence of thiol-TADDOL ligands (Scheme 5). In this reaction, also an inversion of the *er* ratio can be achieved by applying the different derivatives **31, 32** and **33**. Ligand **31** (Scheme 5a) catalyzes the formation of (−)-(*S*)-3-butylcycloheptanone, whereas ligands **32** and **33**

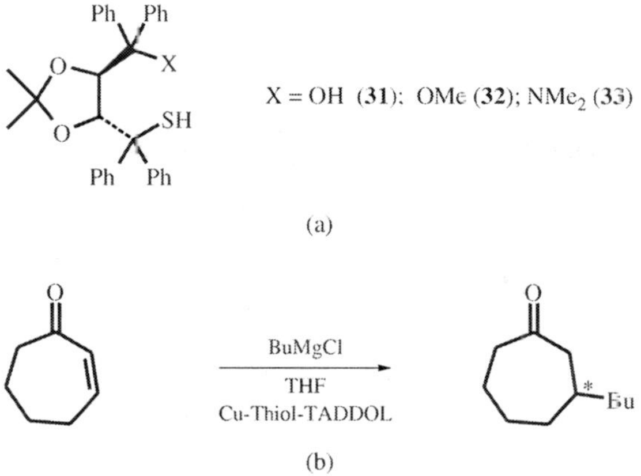

SCHEME 5. (a) Thiol-TADDOL ligands **31**–**33**. (b) Copper-catalyzed conjugate addition: $er = 92:8$ (with **31**) and $er = 8:92$ (with **32** or **33**). From Reference 219. Copyright Wiley-VCH Verlag GmbH & Co. KGaA. Reproduced with permission

(Scheme 5a) form (+)-(*R*)-3-butylcycloheptanone[220]. Additionally, for ligands **31** and **32** a small non-linear effect was found, which suggests the participation of more than one ligand in the catalytically active complex[219].

For X-ray studies, crystals could be obtained by treating **31**, **32** or **33** with butyl lithium and adding afterwards CuCl. For each complex, the solid state structure shows a Cu_4S_4 unit in which each sulfur atom is coordinated to two copper atoms. Interestingly, no interaction between the $-$OH, $-$OMe or $-NMe_2$ groups of the ligand and the metal center is observed, which means that the thiol-TADDOL acts as a monodentate ligand[219].

Subsequent NMR diffusion measurements of the pure ligands **31** and **32** (Scheme 5a) and the corresponding Cu_4L_4 complexes revealed that the Cu_4S_4-core unit remains intact in solution. Now the question arose as to whether the Cu_4L_4 complex remains stable upon transmetalation. To model the transmetalation process and simultaneously to produce NMR suitable samples, isocyanide was used as additional ligand and diffusion measurements indicated also Cu_4L_4 complexes for the isocyanide derivatives **34**–**36** (Figure 35).

In addition, $^1H,^1H$ NOESY spectra of **34** (Figure 36a) and **35** (Figure 36b) were recorded in order to gain information as to whether the observed inversion of enantioselectivity for **31** compared to **32/33** can be correlated with the 3-dimensional structures of their isocyanide complexes. Indeed, the NOESY cross-peak sections of the isocyanide protons and the aryl protons of **34** and **35** show different signal intensities. For **34**, the methyl protons of the butyl group exhibit strong cross-peaks to both sets of phenyl protons (Figure 36a), whereas in **35** the butyl protons have only a very weak interaction with the phenyl protons adjacent to the methoxy functional group (Figure 36b). This indicates a conformational change via rotation around the C$-$S bond (indicated in Figure 35 by an arrow), derivatizing the free alcohol (**34**) into the methoxy substituent (**35**), and explains the observed inversion of enantioselectivity.

B. Catalytic Copper Complexes with Phosphoramidite Ligands

In the past few years, interest in chiral monodentate ligands has grown enormously[221,222]. In particular, the biphenol- or binaphthol-based phosphoramidite ligands[223–226] are reported to yield high *ee* values and tolerate a wide range of different reaction types and

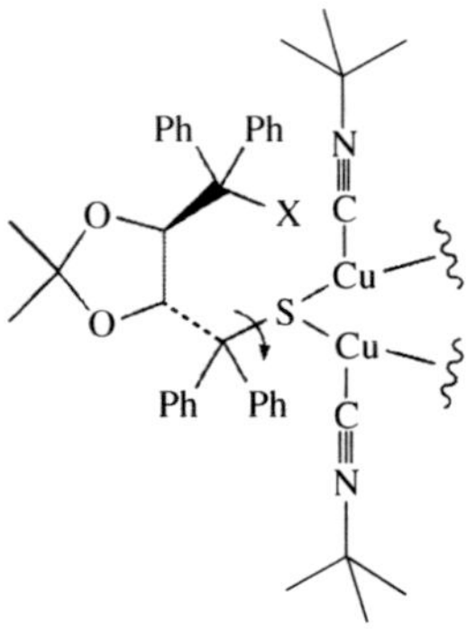

X = OH (**34**); OMe (**35**); NMe_2 (**36**)

FIGURE 35. Part of the Cu_4 (thiol-TADDOL)$_4$ structure, adopting a different conformation for **34** in contrast to **35** and **36**. This conformational change is indicated by an arrow. From Reference 219. Copyright Wiley-VCH Verlag GmbH KGaA. Reproduced with permission

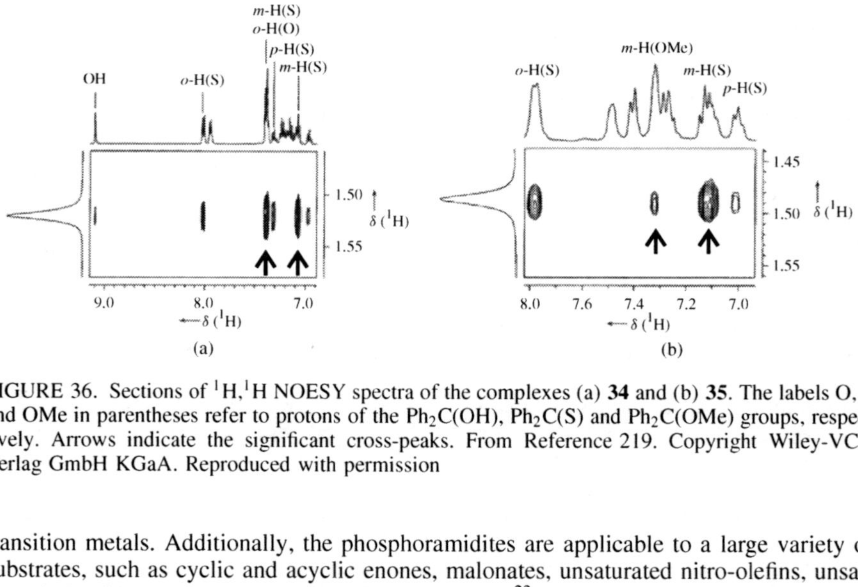

FIGURE 36. Sections of $^1H,^1H$ NOESY spectra of the complexes (a) **34** and (b) **35**. The labels O, S and OMe in parentheses refer to protons of the $Ph_2C(OH)$, $Ph_2C(S)$ and $Ph_2C(OMe)$ groups, respectively. Arrows indicate the significant cross-peaks. From Reference 219. Copyright Wiley-VCH Verlag GmbH KGaA. Reproduced with permission

transition metals. Additionally, the phosphoramidites are applicable to a large variety of substrates, such as cyclic and acyclic enones, malonates, unsaturated nitro-olefins, unsaturated piperidones and unsaturated imines or amines[23]. This broad range of applications suggests the existence of so-called privileged ligand structures in the case of phosphoramidite ligands. In 1,4-addition reactions of dialkylzinc reagents, catalytic amounts of copper(I) salts (Scheme 6) in the presence of chiral monodentate phosphoramidites (Figure 37) yield excellent *ee* values[18, 23, 108, 219, 227–229]. Two promising ligands, which combine atropisomerism with two stereogenic centers, are the binaphthol-based (**37**) and biphenol-based (**38**) phosphoramidites shown in Figure 37.

In 2002 Alexakis reported that highly enantioselective copper-catalyzed 1,4-addition reactions of diethylzinc to cyclohexenone in the presence of **38** do not necessarily need toluene as solvent but are also possible in several other organic solvents and with a couple

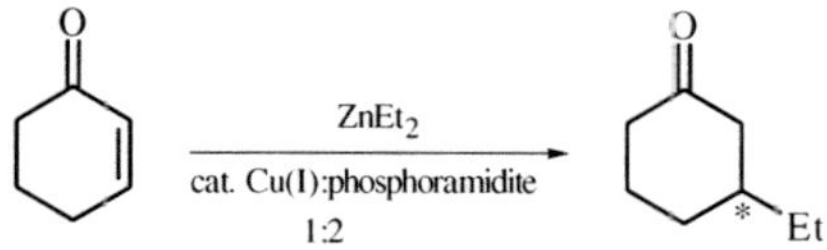

SCHEME 6. Schematic description of the Cu(I)-catalyzed asymmetric 1,4-addition of non-stabilized carbon nucleophiles. From Reference 231. Copyright Wiley-VCH Verlag GmbH KGaA. Reproduced with permission

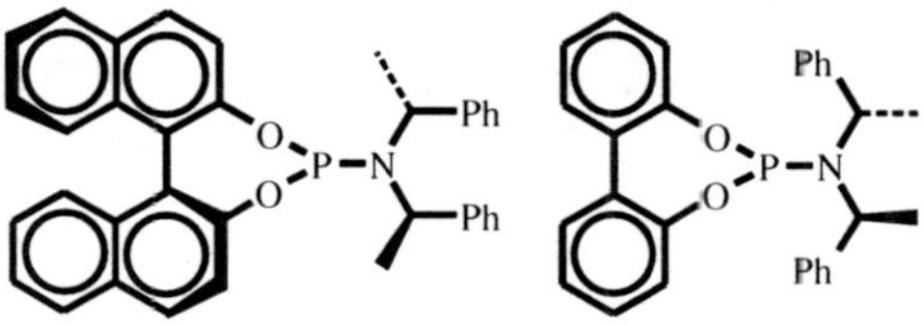

(37) (38)

FIGURE 37. Two representative phosphoramidite ligands derived from binaphthol and biphenol[228,230]. From Reference 231. Copyright Wiley-VCH Verlag GmbH KGaA. Reproduced with permission

of different copper salts[230]. These synthetic results laid the basis for subsequent NMR spectroscopic investigations of precatalytic phosphoramidite copper complexes, because it was now possible to use several combinations of solvents and copper salts for the optimization of the spectroscopic properties (aggregation, line widths, number of compounds, relaxation properties) without losing relevance for the interpretation of synthetic results.

Under experimental conditions close to those used in synthetic 1,4-addition reactions, the ^{31}P spectra of 1:2 mixtures of different copper salts and **37** or **38** show the coexistence of free ligand and at least one copper phosphoramidite complex in solution (see Figure 38a). In the corresponding ^{1}H spectra the signals of the free ligand and the copper complexes show no separate signals but nearly completely overlapping chemical shifts. Therefore, 1D ^{31}P spectra are the key to distinguish different species of phosphoramidite Cu complexes in solution (for spectroscopic properties, see Table 1) and were used to identify the solvent dependence of the complex species (Figure 38a)[231]. In THF and toluene, broad signals of several complex species are observed, whereas in $CDCl_3$ and CD_2Cl_2 a single complex signal is detected (denoted as **C2**) besides the free ligand. By reducing the ligand-to-copper ratio from 2:1 to 1:1, one of the other complexes in Figure 38a could be assigned to the 1:1 phosphoramidite copper complex **C1** (Figure 38b). Based on these results, $CDCl_3$ and CD_2Cl_2 were chosen because these solvents allow a separation of the two species **C1** and **C2** and simultaneously provide high *ee* values in 1,4-addition reactions.

From the synthetically optimized conditions it was proposed that a L_2Cu complex is the catalytically active species. Therefore, it was surprising that a 2:1 ratio of ligand to copper salt produced signals of **C2** plus free ligand. At this point, ^{31}P spectra with varying ratios of copper salt to ligand revealed the stoichiometry of **C2** (Figure 39). A 1:1 ratio exclusively produces **C1** ($\delta = 121.7$ ppm). The addition of more ligand leads to the formation of **C2** ($\delta = 126.6$ ppm) and to a decrease of **C1**. At 1.5:1 mainly **C2** is present besides small amounts of free ligand and **C1** and at higher ratios **C2** and increasing amounts of free ligand are detected. These spectra show that only in the case of a 1:1 ratio a single species (**C1**) exists in solution, which can be characterized directly with diffusion experiments. For all other ratios and especially for the diffusion characterization

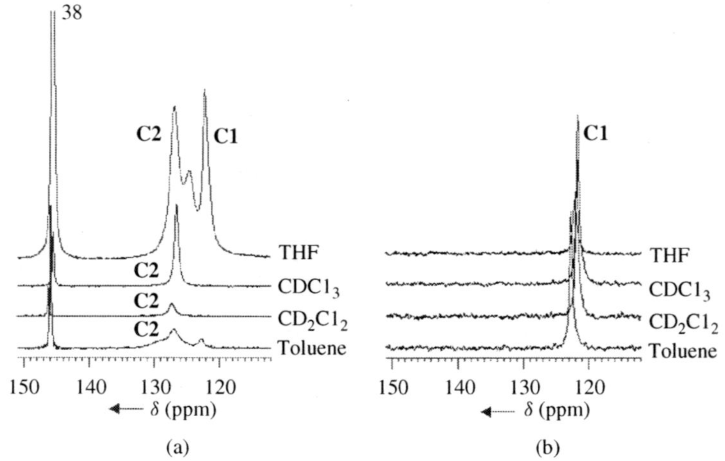

FIGURE 38. ^{31}P spectra of ligand **38** and CuCl in varying solvents and at a ratio of ligand:CuCl of (a) 2:1 and (b) 1:1 at 220 K. From Reference 231. Copyright Wiley-VCH Verlag GmbH KGaA. Reproduced with permission

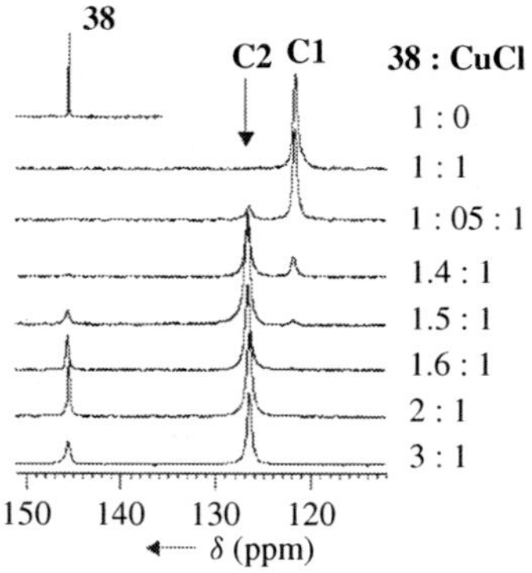

FIGURE 39. ^{31}P spectra of CuCl and ligand **38** at varying ratios in $CDCl_3$ at 220 K. From Reference 231. Copyright Wiley-VCH Verlag GmbH KGaA. Reproduced with permission

of **C2**, separated signals for free and coordinated ligands would be necessary, which do not exist in the ^{1}H spectra due to severe signal overlap. Also, the well-separated ^{31}P signals were not suitable for DOSY experiments, because quadrupole relaxation and exchange phenomena cause too short transversal relaxation times visible in extremely broad line widths. At this point serendipity helped to cut this Gordian knot. The internal dynamic of the phosphoramidite ligands is considerably influenced by its kind of complexation. As a result, the methine proton in **38** and **37** features different line widths in the free ligand, **C1** and **C2**, with the line width of **C2** being fortunately by far the smallest one at 220 K. This allows to use the quite long convection compensating DOSY pulse sequence of Jerschow and Müller[100] as complex selective T_2 filter for the exclusive detection of the DOSY attenuation of **C2**[231].

TABLE 6. Diffusion constants D (10^{-10} m^2s^{-1}) of the free ligands **37** and **38**, and the complexes **C1** and **C2** consisting of CuCl and ligands **37** or **38**. From Reference 231. Copyright Wiley-VCH Verlag GmbH KGaA. Reproduced with permission

Ligand	*D*(ligand)	*D*(**C1**)	*D*(**C2**)
37	2.30	1.60	1.62
38	2.68	1.81	1.83

The experimental diffusion constants of **C1** and **C2** with the identical ligand are very similar, while the free ligands, **37** and **38**, and the corresponding complexes with different ligands reveal well-separated diffusion constants according to their size (Table 6). To interpret these experimental diffusion data, structural models were derived from crystal structures with ligand (L) to copper salt ratios between 1:1 and 3:1 and a maximum number of four ligands in the complex (Figure 40).

From experimentally determined ligand volumes and hard-sphere increments of the copper salts, the theoretical volumes of these models were derived and compared to the experimentally detected ones. With this method it could be shown that three phosphoramidite ligands exist in **C1** and **C2**. The volume of different amounts of copper salts in the two complexes is within the experimental error of the DOSY measurements. Therefore, this information was taken from the ^{31}P spectra shown in Figure 39 and **C1** could be proposed to be structure **41** and **C2** correlated with structure **42** (Figure 40). Due to the fact that in synthetic protocols the 2:1 ratio was reported to give the highest *ee* values, and at a 2:1 ratio only **C2** is present besides free ligand (Figure 39), complex **C2** can be proposed to be the precatalytic complex.

A further NMR spectroscopic screening with three phosphoramidite ligands and four Cu(I)X salts (X = Cl, Br, I, thiophene-2-carboxylate) confirmed the mixed trigonal/tetrahedral Cu coordination in **42** as basic structural motif of precatalytic phosphoramidite copper complexes with a free coordination site for transmetalation[232].

Due to the fact that the above-described structure elucidation process of **C2** does not conform to a classical NMR structure determination, additional low-temperature ^{31}P spectra of phosphoramidite copper complexes were recorded. The resolved low-temperature spectra of CuCl/**37** and CuI/**38** are shown in Figure 41, representing the two principal

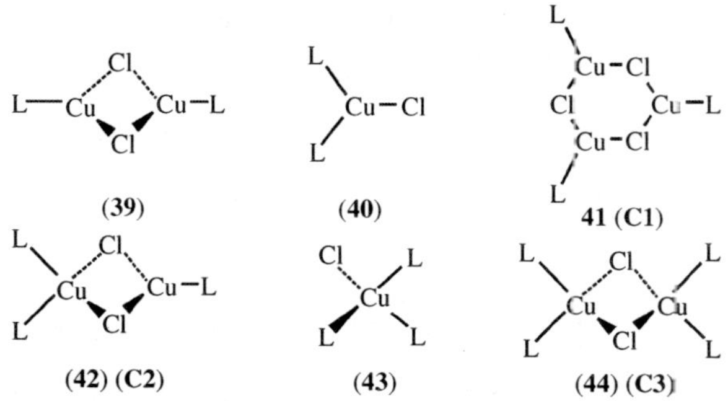

FIGURE 40. Schematic models of copper(I) complexes with one, two or three metal atoms based on crystal structures of phosphoramidite and phosphine Cu complexes. From Reference 231. Copyright Wiley-VCH Verlag GmbH KGaA. Reproduced with permission

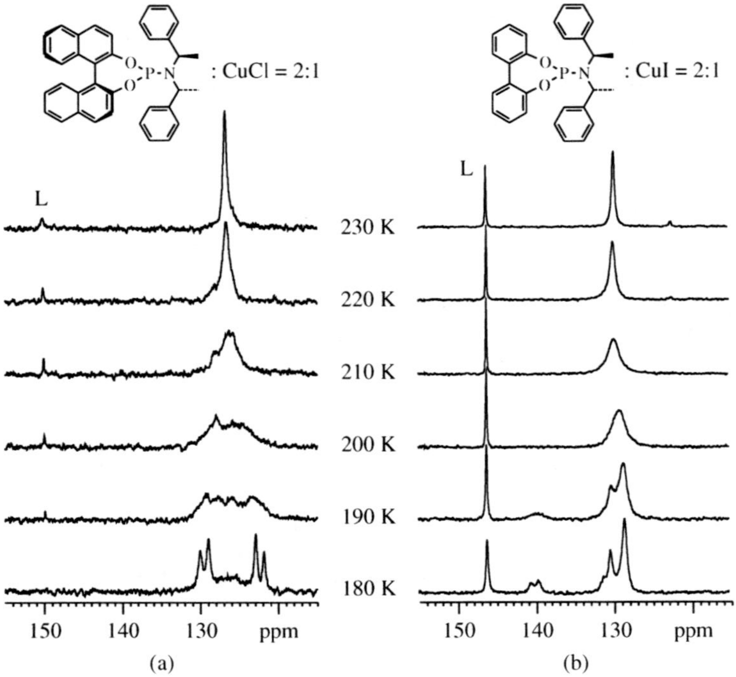

FIGURE 41. ^{31}P NMR spectra of the combinations (a) ligand **37**/CuCl and (b) ligand **38**/CuI at a ratio of 2:1 at different temperatures in CD_2Cl_2. L indicates the free ligand. Reprinted with permission from Reference 61. Copyright 2008 American Chemical Society

signal patterns found in the low-temperature spectra of various phosphoramidite copper complexes[61].

The CuCl-containing sample (Figure 41a and 42a) is the first example of phosphoramidite copper complexes, in which a resolved AA′BB′ signal pattern with a $^{2}J_{P,P}$ coupling of 260 Hz is detected[61]. This and cross-signals in the low-temperature ^{31}P,^{31}P COSY spectra indicate CuLL′ subunits, in which the two ligands have different three-dimensional orientations with chemically non-equivalent phosphorous atoms. Diffusion measurements of closely related complexes showed that these CuLL′ subunits are part of a $L_2L_2'Cu_2Cl_2$ complex (Figure 42) as already found in a phosphoramidite crystal structure[208]. Furthermore, dynamic NMR simulations indicate high ligand exchange rates in these complexes (Figure 42b)[61].

A comparison of the spectrum of the $L_2L_2'Cu_2Cl_2$ complex of **37**/CuCl (Figure 41a) with that of **38**/CuI (Figure 41b) suggests for **38**/CuI a combination of two complex species with one part being the already identified type $L_2L_2'Cu_2X_2$. An intensity-adapted simulation of the spectra reveals the second complex species to have two signals with an intensity ratio of 1:2, which fits perfectly to the previously proposed mixed trigonal/tetrahedral precatalyst. Hence, a separated simulation of the ^{31}P spectra of $L_2L_2'Cu_2I_2$ (Figure 43a) and of $LL_2'Cu_2I_2$ (Figure 43b) was the basis for the interpretation of the low-temperature species of **38**/CuI at 180 K. An intensity-adapted superposition (Figure 43c) of the simulated spectra of $L_2L_2'Cu_2I_2$ (Figure 43a) and of $LL_2'Cu_2I_2$ (Figure 43b) shows a nearly perfect agreement with the experimental spectrum in Figure 43d. Thus, the

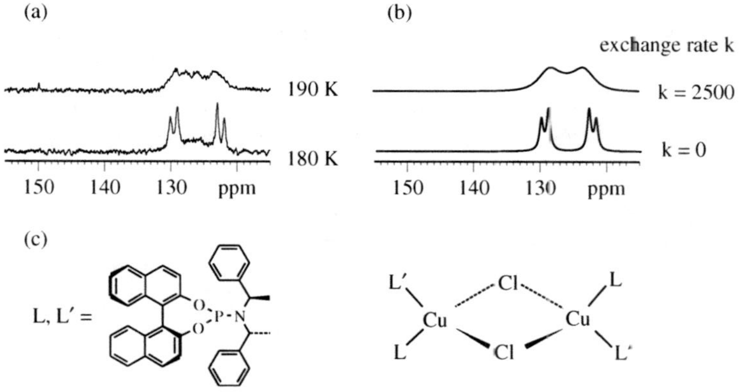

FIGURE 42. (a) Experimental and (b) simulated spectra of the low-temperature complex composed of **37** and CuCl. (c) Schematic structures of the ligand and the $L_2L_2'Cu_2X_2$ complex are given. L and L′ represent the identical ligand in different sterical arrangements resulting in separated ^{31}P signals. Reprinted with permission from Reference 61. Copyright 2008 American Chemical Society

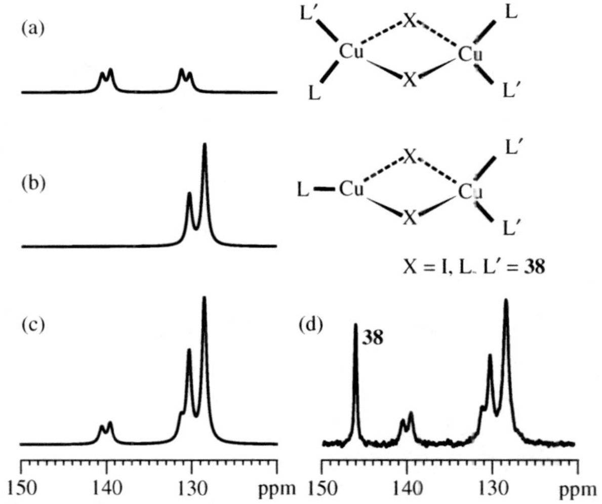

FIGURE 43. Simulated ^{31}P NMR spectra of (a) the $L_2L_2'Cu_2X_2$ complex, (b) the $LL_2'Cu_2X_2$ complex and (c) an intensity-adapted superposition of (a) and (b). The superposition (c) shows an excellent agreement with (d) the experimental ^{31}P spectrum of a 2:1 ratio of **38** to CuI at 180 K. L and L′ represent an identical ligand in different sterical arrangements resulting in separated ^{31}P signals. Reprinted with permission from Reference 61. Copyright 2008 American Chemical Society

existence of the mixed trigonal/tetrahedral complex **C2** was also proven by classical NMR spectroscopic methods.

The combined interpretation of low-temperature ^{31}P spectra and temperature-dependent DOSY information of various phosphoramidite copper complexes hinted at a temperature-dependent interconversion of different copper complex species in solution. For example, in DOSY measurements, CuL_2 complexes were detected at higher temperatures (300 K) and these copper complexes aggregate with decreasing temperature up to $L_4Cu_2X_2$ complexes at 180 K. A temperature-dependent interconversion of different catalytic species in solution is of extreme interest for synthetic applications and would explain the observed temperature sensitivity of these reactions. However, from the low spectral resolution and the averaged signals of the temperature-dependent ^{31}P spectra shown in Figure 44 it is

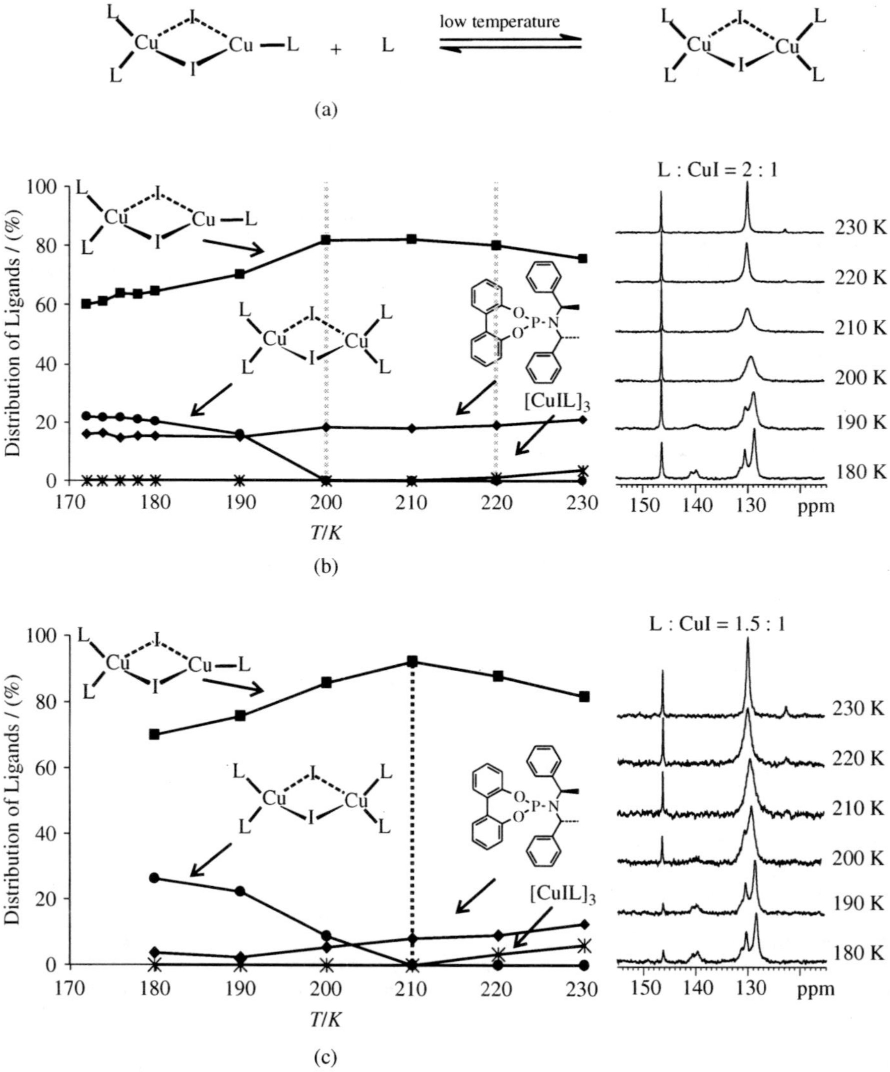

FIGURE 44. (a) Schematic representation of the temperature-dependent interconversion of **C2** into **C3**. Relative ^{31}P integral values and experimental ^{31}P spectra of a **38**/CuI ratio of (b) 2:1 and (c) 1.5:1 at temperatures between 170 and 230 K; free ligand (♦), **C1** (x), **C2** (■) and **C3** (●). Reprinted with permission from Reference 61. Copyright 2008 American Chemical Society

obvious that a classical NMR quantification of the different complex species, for which well-separated and well-resolved signals are necessary, is not possible. Therefore, the well-resolved and sharp ^{31}P signal of the free ligand was used as indicator for the existing copper species in solution. From the investigations described, the possibly different coexisting copper complexes were identified as **C1**, **C2** and **C3** (for schematic models, see Figure 40), possessing the different ligand to copper salt ratios of 1:1, 1.5:1 and 2:1, respectively. That means, in the case of a temperature-dependent interconversion of **C1** into **C2** and then into **C3**, that the free ligand is stepwise consumed and reflects the interconversion step as shown by the example in Figure 44a. The temperature-dependent amounts of **C1**, **C2**, **C3** and free ligand, which are derived from the ^{31}P integrals of a 2:1 ratio of **38**/CuI, are displayed in Figure 44b. These graphs show that above 210 K small amounts of **C1** coexist beside the main complex **C2**. At 210 and 200 K, **C2** exists exclusively in solution and it partially interconverts into **C3** at temperatures below 200 K. Interestingly, at a 1.5:1 ratio of ligand to copper, i.e. at the stoichiometry of the precatalytic complex **C2** ($L_3Cu_2X_2$), the stability of **C2** is reduced. This is obvious from Figure 44c showing a significantly reduced temperature range, in which **C2** exists exclusively. With this method, the amount of the different copper complexes can be detected even in temperature regions in which spectroscopically unresolved complex signals exist, because the free ligand is used as a spy for the interconversion of stoichiometrically deviating complexes.

V. CONCLUSION

For decades, the disadvantageous properties of the two Cu isotopes, the presence of dynamic equilibria and the complex supramolecular structures hampered the structure elucidation of organocopper complexes in solution. With continuous developments in NMR spectroscopic techniques, the switch to the NMR active nuclei in the ligands of organocopper complexes and step-by-step structural approaches, it was recently possible to elucidate the supramolecular structures of stoichiometric and catalytically active copper reagents in solution. In these complexes, the high symmetry of the aggregates often hampers the application of classical NMR spectroscopic approaches. However, elaborate DOSY, HOE and NOE NMR spectroscopic measurements, which are tailored to the specific structure problems, allow to gain insights into the structures and sizes of these supramolecular assemblies, which is still a challenging task.

For organocuprates, it could be shown that the linear cuprate units form homodimer core structures in diethyl ether solution, which tend to aggregate in a chain-like manner bridged by salt and solvent molecules. These supramolecular aggregates of organocuprates could be correlated with their reactivities in addition reactions to α,β-unsaturated enones. Furthermore, the two approaches of intermediate stabilization, i.e. the rapid-injection NMR at low temperatures and the preparative stabilization in conventional low-temperature NMR, in combination with isotopic labelling, allowed the detection of elusive intermediate structures of copper-mediated reactions. Thus, it was proven experimentally via NMR spectroscopy that both π- and σ-complexes of the Cu(I)/Cu(III) redox system are the decisive intermediate complexes in copper-mediated conjugate addition and S_N2/S_N2' reactions.

In the case of catalytic copper complexes, the continuous spectroscopic improvement allowed for a structure elucidation revealing multinuclear complexes. Again, via DOSY and NOE measurements, for both TADDOL-like as well as phosphoramidite ligands, detailed structural information was derived from NMR in solution. Especially, for phosphoramidites and their precatalytic complexes, it was possible to obtain important information about the temperature-dependent stability of the precatalytic copper complexes, which may be very helpful for a further design of catalytically active complexes.

Thus, the recent progress in structure elucidation of organocopper compounds in solution by NMR spectroscopy shows impressively the capability of this method, which should encourage further research in the field of supramolecular assemblies throughout organocopper chemistry.

VI. ACKNOWLEDGMENT

We would like to thank all other members of the Gschwind working group for continuous support during the preparation of this manuscript and the Deutsche Forschungsgemeinschaft for financial support of our copper projects.

VII. REFERENCES

1. S. J. Berners-Price, L. Ronconi and P. J. Sadler, *Prog. Magn. Reson. Spectrosc.*, **49**, 65 (2006).
2. C. Dybowski and G. Neue, *Prog. Magn. Reson. Spectrosc.*, **41**, 153 (2002).
3. J. M. Ernsting, S. Gämers and C. J. Elsevier, *Magn. Reson. Chem.*, **42**, 721 (2004).
4. D. Johnels and H. Guenther, in *The Chemistry of Organolithium Compounds* (Eds Z. Rappoport and I. Marek), Vol. 1, Chap. 4, John Wiley & Sons, Ltd, Chichester, 2004, pp. 137–203.
5. G. H. Penner and X. Liu, *Prog. Magn. Reson. Spectrosc.*, **49**, 151 (2006).
6. J. R. L. Priqueler, I. S. Butler and F. D. Rochon, *Appl. Spectrosc. Rev.*, **41**, 185 (2006).
7. S. Ramaprasad, *Prog. Magn. Reson. Spectrosc.*, **47**, 111 (2005).
8. B. Wrackmeyer, *Mod. Magn. Reson.*, **1**, 457 (2006).
9. B. M. Still, P. G. A. Kumar, J. R. Aldrich-Wright and W. S. Price, *Chem. Soc. Rev.*, **36**, 665 (2007).
10. J. A. Davies, in *The Chemistry of the Metal–Carbon Bond* (Eds F. R. Hartley and S. Patai), Vol. 1, Chap. 21, Wiley, Chichester, 1982, pp. 813–918.
11. K.-i. Yamada and K. Tomioka, *Chem. Rev.*, **108**, 2874 (2008).
12. L. M. Stanley and M. P. Sibi, *Chem. Rev.*, **108**, 2887 (2008).
13. M. Shibasaki and M. Kanai, *Chem. Rev.*, **108**, 2853 (2008).
14. T. B. Poulsen and K. A. Jørgensen, *Chem. Rev.*, **108**, 2903 (2008).
15. M. Meldal and C. W. Tornøe, *Chem. Rev.*, **108**, 2952 (2008).
16. Y. Matsuo and E. Nakamura, *Chem. Rev.*, **108**, 3016 (2008).
17. B. H. Lipshutz and Y. Yamamoto, *Chem. Rev.*, **108**, 2793 (2008).
18. S. R. Harutyunyan, T. den Hartog, K. Geurts, A. J. Minnaard and B. L. Feringa, *Chem. Rev.*, **108**, 2824 (2008).
19. R. M. Gschwind, *Chem. Rev.*, **108**, 3029 (2008).
20. G. Evano, N. Blanchard and M. Toumi, *Chem. Rev.*, **108**, 3054 (2008).
21. C. Deutsch, N. Krause and B. H. Lipshutz, *Chem. Rev.*, **108**, 2916 (2008).
22. B. Breit and Y. Schmidt, *Chem. Rev.*, **108**, 2928 (2008).
23. A. Alexakis, J. E. Bäckvall, N. Krause, O. Pamies and M. Dieguez, *Chem. Rev.*, **108**, 2796 (2008).
24. J. Mason, *Multinuclear NMR*, Plenum Press, New York, 1987.
25. P. Granger, *Transition Metal Nuclear Magnetic Resonace*, Elsevier, Amsterdam, 1991.
26. J. Malito, *J. Annu. Rep. NMR Spectrosc.*, **38**, 265 (1999).
27. A. Marker and M. J. Gunter, *J. Magn. Reson.*, **47**, 118 (1982).
28. I. Szymanska, *Polish J. Chem.*, **80**, 1095 (2006).
29. J. K. Irangu and R. B. Jordan, *Inorg. Chem.*, **42**, 3934 (2003).
30. P. Kroneck, J. Kodweiss, O. Lutz, A. Nolle and D. Zepf, *Z. Naturforsch.*, **37A**, 186 (1982).
31. M. Kujime, T. Kurahashi, M. Tomura and H. Fujii, *Inorg. Chem.*, **46**, 541 (2007).
32. U. Ochsenbein and C. W. Schläpfer, *Helv. Chim. Acta*, **63**, 1926 (1980).
33. S. Kitagawa, M. Munakata and M. Sasaki, *Inorg. Chim. Acta*, **120**, 77 (1986).
34. P. Kroneck, O. Lutz, A. Nolle and H. Oehler, *Z. Naturforsch.*, **35A**, 221 (1980).
35. O. Lutz, H. Oehler and P. Kroneck, *Z. Naturforsch.*, **33A**, 1021 (1978).
36. O. Lutz, H. Oehler and P. Kroneck, *Z. Physik A*, **288**, 17 (1978).
37. E. Szlyk and I. Szymanska, *Polyhedron*, **18**, 2941 (1999).

38. J. R. Black, W. Levason, M. D. Spicer and M. Webster, *J. Chem. Soc., Dalton Trans.*, 3129 (1993).
39. D. J. Fife, W. M. Moore and K. W. Morse, *Inorg. Chem.*, **23**, 1684 (1984).
40. S. J. Berners Price, C. Brevard, A. Pagelot and P. J. Sadler, *Inorg. Chem.*, **25**, 596 (1986).
41. C. L. Doel, A. M. Gibson, G. Reid and C. Frampton, *Polyhedron*, **14**, 3139 (1995).
42. B. Mohr, E. E. Brooks, N. Rath and E. Deutsch, *Inorg. Chem.*, **30**, 4541 (1991).
43. E. Szlyk, R. Kucharek and I. Szymanska, *Polish J. Chem.*, **75**, 337 (2001).
44. E. Szlyk, R. Kucharek and I. Szymanska, *J. Coord. Chem.*, **53**, 55 (2001).
45. E. Szlyk, R. Kucharek, I. Szymanska and L. Pazderski, *Polyhedron*, **22**, 3389 (2003).
46. J. A. Connor and R. J. Kennedy, *Polyhedron*, **7**, 161 (1988).
47. K. Endo, K. Yamamoto, K. Deguchi and K. Matsushita, *Bull. Chem. Soc. Jpn.*, **60**, 2803 (1987).
48. R. L. Geerts, J. C. Huffman, K. Folting, T. H. Lemmen and K. G. Caulton, *J. Am. Chem. Soc.*, **105**, 3503 (1983).
49. D. S. Gill, L. Byrne and T. I. Quickenden, *Z. Naturforsch.*, **53a**, 1004 (1998).
50. D. S. Gill, U. Kamp, A. Doelle and M. D. Zeidler, *Indian J. Chem.*, **40A**, 693 (2001).
51. D. S. Gill, L. Rodehüser and J. J. Delpuech, *J. Chem. Soc., Faraday Trans.*, **86**, 2847 (1990).
52. D. S. Gill, L. Rodehüser, P. Rubini and J. J. Delpuech, *J. Chem. Soc., Faraday Trans.*, **91**, 2307 (1995).
53. D. S. Gill, J. Singh, R. Singh, T. Zamir and T. I. Quickenden, *Indian J. Chem.*, **38A**, 913 (1999).
54. S. Kitagawa and M. Munakata, *Inorg. Chem.*, **23**, 4388 (1984).
55. K. B. Nilsson and I. Persson, *Dalton Trans.*, 1312 (2004).
56. R. Goodfellow, in *Multinuclear NMR* (Ed. J. Mason), Plenum Press, New York, 1987.
57. E. R. Bartholomew, S. H. Bertz, S. Cope, M. Murphy and C. A. Ogle, *J. Am. Chem. Soc.*, **130**, 11244 (2008).
58. S. H. Bertz, S. Cope, M. Murphy, C. A. Ogle and B. J. Taylor, *J. Am. Chem. Soc.*, **129**, 7208 (2007).
59. T. Gärtner, W. Henze and R. M. Gschwind, *J. Am. Chem. Soc.*, **129**, 11362 (2007).
60. J. A. Tang, B. D. Ellis, T. H. Warren, J. V. Hanna, C. L. B. Macdonald and R. W. Schurko, *J. Am. Chem. Soc.*, **129**, 13049 (2007).
61. K. Schober, H. Zhang and R. M. Gschwind, *J. Am. Chem. Soc.*, **130**, 12310 (2008).
62. J. Canisius, T. A. Mobley, S. Berger and N. Krause, *Chem. Eur. J.*, **7**, 2671 (2001).
63. W. Henze, T. Gärtner and R. M. Gschwind, *J. Am. Chem. Soc.*, **130**, 13718 (2008).
64. E. R. Bartholomew, S. H. Bertz, S. Cope, D. C. Dorton, M. Murphy and C. A. Ogle, *Chem. Commun.*, 1176 (2008).
65. S. H. Bertz, S. Cope, D. Dorton, M. Murphy and C. A. Ogle, *Angew. Chem,. Int. Ed.*, **46**, 7082 (2007).
66. D. B. Collum, *Acc. Chem. Res.*, **26**, 227 (1993).
67. A. Corruble, D. Davoust, S. Desjardins, C. Fressigne, C. Giessner-Prettre, A. Harrison-Marchand, H. Houte, M.-C. Lasne, J. Maddaluno, H. Oulyadi and J.-Y. Valnot, *J. Am. Chem. Soc.*, **124**, 15267 (2002).
68. G. Fränkel, A. M. Fränkel, M. J. Geckle and F. Schloss, *J. Am. Chem. Soc.*, **101**, 4745 (1979).
69. K. Gregory, P. v. R. Schleyer and R. Snaith, *Adv. Inorg. Chem.*, **37**, 47 (1991).
70. B. L. Lucht and D. B. Collum, *Acc. Chem. Res.*, **32**, 1035 (1999).
71. R. E. Mulvey, *Chem. Soc. Rev.*, **20**, 167 (1991).
72. A.-M. Saspe and P. v. R. Schleyer, *Lithium Chemistry: A Theoretical and Experimental Overview*, John Wiley & Sons, Inc., New York, 1995.
73. D. Seebach, R. Hässig and J. Gabriel, *Helv. Chim. Acta*, **66**, 308 (1983).
74. R. Sott, J. Granander and G. Hilmersson, *J. Am. Chem. Soc.*, **126**, 6798 (2004).
75. A. G. Avent, C. Eaborn, M. N. A. El-Kehli, M. E. Molla, J. D. Smith and A. C. Sullivan, *J. Am. Chem. Soc.*, **108**, 3854 (1986).
76. W. Bauer, T. Clark and P. v. R. Schleyer, *J. Am. Chem. Soc.*, **109**, 970 (1987).
77. W. Bauer, P. A. A. Klusener, S. Harder, J. A. Kanters, A. J. M. Duisenburg, L. Brandsma and P. v. R. Schleyer, *Organometallics*, **7**, 552 (1988).
78. W. Bauer, G. Müller and P. v. R. Schleyer, *Angew. Chem., Int. Ed.*, **25**, 1103 (1986).
79. W. Bauer and P. v. R. Schleyer, *Magn. Reson. Chem*, **26**, 827 (1988).
80. W. Bauer and P. v. R. Schleyer, *Adv. Carbanion Chem.*, **1**, 89 (1992).

81. T. Brand, E. J. Cabrita and S. Berger, *Prog. Magn. Reson. Spectrosc.*, **46**, 159 (2005).
82. H. Günther, D. Moskau and D. Schmalz, *Angew. Chem.*, **99**, 1242 (1987).
83. G. Hilmersson, P. I. Arvidsson, O. Davidson and M. J. Hakansson, *J. Am. Chem. Soc.*, **120**, 8143 (1998).
84. A. Macchioni, *Chem. Rev.*, **105**, 2039 (2005).
85. H. Mo and T. C. Pochapsky, *Prog. Magn. Reson. Spectrosc.*, **30**, 1 (1997).
86. P. S. Pregosin, A. P. G. Kumar and I. Fernandez, *Chem. Rev.*, **105**, 2977 (2005).
87. R. M. Gschwind, X. Xie, P. R. Rajamohanan, C. Auel and G. Boche, *J. Am. Chem. Soc.*, **123**, 7299 (2001).
88. D. Neuhaus and M. Williamson, *The Nuclear Overhauser Effect in Structural and Conformational Analysis*, VCH, Weinheim, 1989.
89. J. Kawabata, E. Fukushi and J. Mizutani, *J. Am. Chem. Soc.*, **114**, 1115 (1992).
90. R. Wagner and S. Berger, *Magn. Reson. Chem.*, **35**, 199 (1997).
91. R. M. Gschwind, X. Xie and P. R. Rajamohanan, *Magn. Reson. Chem.*, **42**, 308 (2004).
92. A. Dehner and H. Kessler, *ChemBioChem*, **6**, 1550 (2005).
93. A. Macchioni, G. Ciancaleoni, C. Zuccaccia and D. Zuccaccia, *Chem. Soc. Rev.*, **37**, 479 (2008).
94. P. S. Pregosin, *Prog. Magn. Reson. Spectrosc.*, **49**, 265 (2006).
95. P. Stilbs, *Prog. Magn. Reson. Spectrosc.*, **19**, 1 (1987).
96. Y. Cohen, L. Avram and L. Frish, *Angew. Chem., Int. Ed.*, **44**, 520 (2005).
97. E. O. Stejskal and J. E. Tanner, *J. Chem. Phys.*, **42**, 288 (1965).
98. D. Zuccaccia and A. Macchioni, *Organometallics*, **24**, 3476 (2005).
99. E. J. Cabrita and S. Berger, *Magn. Reson. Chem.*, **39**, S142 (2001).
100. A. Jerschow and N. Müller, *J. Magn. Reson.*, **125**, 372 (1997).
101. G. H. Sørland, J. G. Seland, J. Krane and H. W. Anthonsen, *J. Magn. Reson.*, **142**, 323 (2000).
102. J. F. McGarrity, C. A. Ogle, Z. Brich and H. R. Loosli, *J. Am. Chem. Soc.*, **107**, 1810 (1985).
103. J. F. McGarrity and J. Prodolliet, *J. Org. Chem.*, **49**, 4465 (1984).
104. H. Gilman and J. M. Straley, *Recl. Trav. Chim. Pays-Bas*, **55**, 821 (1936).
105. J. F. Normant, *Synthesis*, **2**, 63 (1972).
106. E. C. Ashby, A. B. Goel and R. S. Smith, *J. Organomet. Chem.*, **212**, C47 (1981).
107. M. C. P. Yeh, P. Knochel and L. E. Santa, *Tetrahedron Lett.*, **29**, 3887 (1988).
108. N. Krause, *Modern Organocopper Chemistry*, Wiley-VCH, Weinheim, 2002.
109. J. Eriksson, P. I. Arvidsson and O. Davidsson, *J. Am. Chem. Soc.*, **122**, 9310 (2000).
110. N. Krause, *Angew. Chem., Int. Ed.*, **38**, 79 (1999).
111. E. Nakamura and S. Mori, *Angew. Chem., Int. Ed.*, **39**, 3750 (2000).
112. E. Nakamura and N. Yoshikai, *Bull. Chem. Soc. Jpn.*, **77**, 1 (2004).
113. N. Krause, *Metallorganische Chemie*, Spectrum Akademischer Verlag, Heidelberg, 1996.
114. B. H. Lipshutz, in *Organometallics in Synthesis* (Ed. M. Schlosser), John Wiley & Sons, Ltd, Chichester, 1994.
115. R. J. K. Taylor, *Organocopper Reagents: A Practical Approach*, Oxford University Press, Oxford, 1994.
116. N. Boudet, S. R. Dubbaka and P. Knochel, *Org. Lett.*, **10**, 1715 (2008).
117. S. R. Dubbaka, M. Kienle, H. Mayr and P. Knochel, *Angew. Chem., Int. Ed.*, **46**, 9093 (2007).
118. M. Kienle, S. R. Dubbaka, K. Brade and P. Knochel, *Eur. J. Org. Chem.*, 2007, 4166 (2007).
119. S. H. Bertz, C. A. Ogle and A. Rastogi, *J. Am. Chem. Soc.*, **127**, 1372 (2005).
120. B. H. Lipshutz, R. S. Wilhelm and D. M. Floyd, *J. Am. Chem. Soc.*, **103**, 7672 (1981).
121. B. H. Lipshutz, J. A. Kozlowski and R. S. Wilhelm, *J. Org. Chem*, **49**, 3943 (1984).
122. G. Hallnemo and C. Ullenius, *Tetrahedron*, **39**, 1621 (1983).
123. S. H. Bertz, A. Chopra, M. Eriksson, C. A. Ogle and P. Seagle, *Chem. Eur. J.*, **5**, 2680 (1999).
124. W. Henze, A. Vyater, N. Krause and R. M. Gschwind, *J. Am. Chem. Soc.*, **127**, 17335 (2005).
125. E. C. Ashby and J. J. Watkins, *J. Am. Chem. Soc.*, **99**, 5312 (1977).
126. A. Gerold, J. T. B. H. Jastrzebski, C. M. P. Kronenburg, N. Krause and G. van Koten, *Angew. Chem., Int. Ed.*, **36**, 755 (1997).
127. C. D. Gregory and R. G. Pearson, *J. Am. Chem. Soc.*, **98**, 4098 (1976).
128. S. H. Bertz, K. Nilsson, Ö. Davidsson and J. P. Snyder, *Angew. Chem., Int. Ed.*, **37**, 314 (1998).
129. M. John, C. Auel, C. Behrens, M. Marsch, K. Harms, F. Bosold, R. M. Gschwind, P. R. Rajamohanan and G. Boche, *Chem. Eur. J.*, **6**, 3060 (2000).

130. M. M. Olmstead and P. P. Power, *Organometallics*, **9**, 1720 (1990).
131. B. H. Lipshutz, J. Keith and D. J. Buzard, *Organometallics*, **18**, 1571 (1999).
132. W. Henze, PhD Thesis, Rheinische Friedrich-Wilhelms-Universität, Bonn, 2005.
133. S. H. Bertz, C. M. Carlin, D. A. Deadwyler, M. D. Murphy, C. A. Ogle and P. Seagle, *J. Am. Chem. Soc.*, **124**, 13650 (2002).
134. G. Hallnemo, T. Olsson and C. Ullenius, *J. Organomet. Chem.*, **282**, 133 (1985).
135. N. Krause, R. Wagner and A. Gerold, *J. Am. Chem. Soc.*, **116**, 381 (1994).
136. S. H. Bertz, *J. Am. Chem. Soc.*, **113**, 5470 (1991).
137. T. A. Mobley, F. Müller and S. Berger, *J. Am. Chem. Soc.*, **120**, 1333 (1998).
138. S. H. Bertz and G. Dabbagh, *Tetrahedron*, **45**, 425 (1989).
139. C. Ouannes, G. Dressaire and Y. Langlois, *Tetrahedron Lett.*, 815 (1977).
140. G. M. Whitesides, W. F. Fischer, J. SanFilippo, R. W. Bashe and H. O. House, *J. Am. Chem. Soc.*, **91**, 4871 (1969).
141. R. G. Pearson and C. D. Gregory, *J. Am. Chem. Soc.*, **98**, 4098 (1976).
142. R. M. Gschwind, P. R. Rajamohanan, M. John and G. Boche, *Organometallics*, **19**, 2868 (2000).
143. S. H. Bertz, G. Dabbagh, X. He and P. P. Power, *J. Am. Chem. Soc.*, **115**, 11640 (1993).
144. S. Mori and E. Nakamura, *Chem. Eur. J.*, **5**, 1534 (1999).
145. E. Nakamura, S. Mori and K. Morokuma, *J. Am. Chem. Soc.*, **119**, 4900 (1997).
146. E. Nakamura, S. Mori, M. Nakamura and K. Morokuma, *J. Am. Chem. Soc.*, **119**, 4887 (1997).
147. M. Karplus and J. A. Pople, *J. Chem. Phys.*, **38**, 2803 (1963).
148. D. B. Collum, D. Kahne, S. A. Gut, R. T. DePue, F. Mohamadi, R. A. Wanat, J. Clardy and G. Van Duyne, *J. Am. Chem. Soc.*, **106**, 4865 (1984).
149. M. J. Kaufman and A. Streitwieser, Jr., *J. Am. Chem. Soc.*, **109**, 6092 (1987).
150. L. R. Liou, A. J. McNeil, A. Ramirez, G. E. S. Toombes, J. M. Gruver and D. B. Collum, *J. Am. Chem. Soc.*, **130**, 4859 (2008).
151. F. Xu, R. A. Reamer, R. Tillyer, J. M. Cummins, E. J. J. Grabowski, P. J. Reider, D. B. Collum and J. C. Huffman, *J. Am. Chem. Soc.*, **122**, 11212 (2000).
152. S. H. Bertz, A. S. Vellekoop, R. A. J. Smith and J. P. Snyder, *Organometallics*, **14**, 1213 (1995).
153. M. Böhme, G. Frenking and M. T. Reetz, *Organometallics*, **13**, 4237 (1994).
154. H. Huang, K. Alveraz, Q. Liu, T. M. Barnhart, J. P. Snyder and J. E. Penner-Hahn, *J. Am. Chem. Soc.*, **118**, 8808 and 12252 (correction) (1996).
155. H. Huang, C. H. Liang and J. E. Penner-Hahn, *Angew. Chem., Int. Ed.*, **37**, 1564 (1998).
156. C. M. P. Kronenburg, C. H. M. Amijs, J. T. B. H. Jastrzebski, M. Lutz, A. L. Spek and G. van Koten, *Organometallics*, **21**, 4662 (2002).
157. C. M. P. Kronenburg, J. T. B. H. Jastrzebski, J. Boersma, M. Lutz, A. L. Spek and G. van Koten, *J. Am. Chem. Soc.*, **124**, 11675 (2002).
158. J. P. Snyder and S. H. Bertz, *J. Org. Chem.*, **60**, 4312 (1995).
159. J. P. Snyder, D. P. Spangler, J. R. Behling and B. E. Rossiter, *J. Org. Chem.*, **59**, 2665 (1994).
160. T. L. Stemmler, T. M. Barnhart, J. E. Penner-Hahn, C. E. Tucker, P. Knochel, M. Böhme and G. Frenking, *J. Am. Chem. Soc.*, **117**, 12489 (1995).
161. D. R. Armstrong, K. W. Henderson, A. R. Kennedy, W. J. Kerr, F. S. Mair, J. H. Moir, P. H. Moran and R. Snaith, *J. Chem. Soc., Dalton Trans.*, 4063 (1999).
162. K. B. Aubrecht, B. L. Lucht and D. B. Collum, *Organometallics*, **18**, 2981 (1999).
163. T. Koizumi, K. Morihashi and O. Kikuchi, *Bull. Chem. Soc. Jpn.*, **69**, 305 (1996).
164. T. M. Alam, D. M. Pedrotty and T. J. Boyle, *Magn. Reson. Chem*, **40**, 361 (2002).
165. W. Bauer, *Magn. Reson. Chem.*, **34**, 532 (1996).
166. W. Bauer, in *Lithium Chemistry: A Theoretical and Experimental Overview* (Eds A.-M. Sapse and P. v. R. Schleyer), Wiley, New York, 1995, pp. 125–172.
167. (a) R. P. Davies, S. Hornauer and P. B. Hitchcock, *Angew. Chem., Int. Ed.*, **46**, 5191 (2007).
(b) R. K. Dieter, T. W. Hanks and B. Lagu, *Organometallics*, **11**, 3549 (1992).
168. (a) H. Günther, in *Advanced Applications of NMR to Organometallic Chemistry* (Eds M. Gielen, R. Willem and B. Wrackmeyer), Wiley, Chichester, 1996.
(b) R. Bomparola, R. P. Davies, S. Hornauer and A. J. P. White, *Dalton Trans.*, 1104 (2009).
169. X. Xie, C. Auel, W. Henze and R. M. Gschwind, *J. Am. Chem. Soc.*, **125**, 1595 (2003).
170. A. Bondi, *J. Phys. Chem.*, **68**, 441 (1964).
171. Y. Marcus and G. Hefter, *Chem. Rev.*, **104**, 3405 (2004).

172. V. A. Bloomfield, 'Survey of Biomolecular Hydrodynamics', in *On-Line Biophysics Textbook* (Ed. T. M. Schuster), Internet Edition, 2000.
173. J. Norinder, J.-E. Bäckvall, N. Yoshikai and E. Nakamura, *Organometallics*, **25**, 2129 (2006).
174. M. Yamanaka, S. Kato and E. Nakamura, *J. Am. Chem. Soc.*, **126**, 6287 (2004).
175. M. Yamanaka and E. Nakamura, *Organometallics*, **20**, 5675 (2001).
176. N. Yoshikai, T. Yamashita and E. Nakamura, *Angew. Chem., Int. Ed.*, **44**, 4721 (2005).
177. S. Mori, E. Nakamura and K. Morokuma, *Organometallics*, **23**, 1081 (2004).
178. S. Mori, M. Uerdingen, N. Krause and K. Morokuma, *Angew. Chem., Int. Ed.*, **44**, 4715 (2005).
179. M. Yamanaka and E. Nakamura, *J. Am. Chem. Soc.*, **127**, 4697 (2005).
180. S. H. Bertz and R. A. Smith, *J. Am. Chem. Soc.*, **111**, 8276 (1989).
181. B. Christenson, T. Olsson and C. Ullenius, *Tetrahedron*, **45**, 523 (1989).
182. N. Krause, *J. Org. Chem.*, **57**, 3509 (1992).
183. E. L. Lindstedt, M. Nilsson and T. Olsson, *J. Organomet. Chem.*, **334**, 255 (1987).
184. M. D. Murphy, C. A. Ogle and S. H. Bertz, *Chem. Commun.*, 854 (2005).
185. S. Sharma and A. C. Oehlschläger, *Tetrahedron*, **47**, 1177 (1991).
186. C. Ullenius and B. Christenson, *Pure Appl. Chem.*, **60**, 57 (1988).
187. J. Eriksson and O. Davidsson, *Organometallics*, **20**, 4763 (2001).
188. K. Nilsson, C. Ullenius and N. Krause, *J. Am. Chem. Soc.*, **118**, 4194 (1996).
189. A. S. Vellekoop and R. A. Smith, *J. Am. Chem. Soc.*, **116**, 2902 (1994).
190. C. Nolte, P. Mayer and B. F. Straub, *Angew. Chem.*, **119**, 2147 (2007).
191. K. Nilsson, T. Andersson, C. Ullenius, A. Gerold and N. Krause, *Chem. Eur. J.*, **4**, 2051 (1998).
192. A. Bax and S. Pochapsky, *J. Magn. Reson.*, **99**, 638 (1992).
193. A. Bax and M. F. Summers, *J. Am. Chem. Soc.*, **108**, 2093 (1986).
194. J. Ruiz-Cabello, G. W. Vuister, C. T. W. Moonen, P. van Gelderen, J. S. Cohen and P. C. M. van Zijl, *J. Magn. Reson.*, **100**, 282 (1992).
195. W. Willker, D. Leibfritz, R. Kerssebaum and W. Bermel, *Magn. Reson. Chem.*, **31**, 287 (1993).
196. B. Reif, M. Köck, R. Kerssebaum, H. Kang, W. Fenical and C. Griesinger, *J. Magn. Res., Series A*, **118**, 282 (1996).
197. J. Weigelt and G. Otting, *J. Magn. Res., Series A*, **113**, 128 (1995).
198. A. Aamouche, F. J. Devlin and P. J. Stephens, *J. Am. Chem. Soc.*, **122**, 7358 (2000).
199. H. Hu and J. P. Snyder, *J. Am. Chem. Soc.*, **129**, 7210 (2007).
200. S. Mori and E. Nakamura, in *Modern Organocopper chemistry* (Ed. N. Krause), Wiley-VCH, Weinheim, 2002, pp. 315–346.
201. S. Mori, E. Nakamura and K. Morokuma, *J. Am. Chem. Soc.*, **122**, 7294 (2000).
202. J. P. Snyder, *J. Am. Chem. Soc.*, **117**, 11025 (1995).
203. L. M. Huffman and S. S. Stahl, *J. Am. Chem. Soc.*, **130**, 9196 (2008).
204. P. S. Pregosin and R. W. Kunz, in *NMR—Basic Principles and Progress*, **16** (Eds P. Diehl, E. Fluck and R. Kosfeld), Springer, Berlin, 1979.
205. S. J. B. Price, M. J. DiMartino, D. T. Hill, R. Kuroda, M. A. Mazid and P. J. Sadler, *Inorg. Chem.*, **24**, 3425 (1985).
206. N. Yoshikai, S.-L. Zhang and E. Nakamura, *J. Am. Chem. Soc.*, **130**, 12862 (2008).
207. A. H. M. de Vries, A. Meetsma and B. L. Feringa, *Angew. Chem.*, **108**, 2526 (1996).
208. W.-J. Shi, L.-X. Wang, Y. Fu, S.-F. Zhu and Q.-L. Zhou, *Tetrahedron: Asymmetry*, **14**, 3867 (2003).
209. R. R. Conry and W. S. Striejewske, *Organometallics*, **17**, 3146 (1998).
210. V. Desvergnes-Breuil, V. Hebbe, C. Dietrich-Buchecker, J.-P. Sauvage and J. Lacour, *Inorg. Chem.*, **42**, 255 (2003).
211. K. Kunz, U. Scholz and D. Ganzer, *Synlett*, 2428 (2003).
212. S. V. Ley and A. W. Thomas, *Angew. Chem., Int. Ed.*, **42**, 5400 (2003).
213. A. Ouali, M. Taillefer, J.-F. Spindler and A. Jutand, *Organometallics*, **26**, 65 (2007).
214. I. Pianet and J.-M. Vincent, *Inorg. Chem.*, **43**, 2947 (2004).
215. T. Posset and J. Bluemel, *J. Am. Chem. Soc.*, **128**, 8394 (2006).
216. S. R. Harutyunyan, F. Lopez, W. R. Browne, A. Correa, D. Pena, R. Badorrey, A. Meetsma, A. Minnaard and B. L. Feringa, *J. Am. Chem. Soc.*, **128**, 9103 (2006).

217. F. Lopez, S. R. Harutyunyan, A. Meetsma, A. J. Minnaard and B. L. Feringa, *Angew. Chem., Int. Ed.*, **44**, 2752 (2005).
218. F. Lopez, A. J. Minnaard and B. L. Feringa, *Acc. Chem. Res.*, **40**, 179 (2007).
219. A. Pichota, P. S. Pregosin, M. Valentini, M. Worle and D. Seebach, *Angew. Chem., Int. Ed.*, **39**, 153 (2000).
220. D. Seebach, G. Jäschke, A. Pichota and L. Audergon, *Helv. Chim. Acta*, **80**, 2515 (1997).
221. I. V. Komarov and A. Börner, *Angew. Chem., Int. Ed.*, **40**, 1197 (2001).
222. W. Tang and X. Zhang, *Chem. Rev.*, **103**, 3029 (2003).
223. A. Alexakis, S. Rosset, J. Allamand, S. March, F. Guillen and C. Benhaim, *Synlett*, 1375 (2001).
224. A. E. Arnold, PhD Thesis, Rijksuniversiteit, Groningen, 2002.
225. I. S. Mikhel, G. Bernardinelli and A. Alexakis, *Inorg. Chim. Acta*, 1825 (2006).
226. M. van den Berg, A. J. Minnaard, E. P. Schudde, J. van Esch, A. H. M. de Vries, J. G. de Vries and B. L. Feringa, *J. Am. Chem. Soc.*, **122**, 11539 (2000).
227. A. Alexakis and C. Benhaim, *Eur. J. Org. Chem.*, 3221 (2002).
228. L. A. Arnold, R. Imbos, A. Mandoli, A. H. M. De Vries, R. Naasz and B. L. Feringa, *Tetrahedron*, **56**, 2865 (2000).
229. K. Li and A. Alexakis, *Tetrahedron Lett.*, **46**, 8019 (2005).
230. A. Alexakis, C. Benhaim, S. Rosset and M. Humam, *J. Am. Chem. Soc.*, **124**, 5262 (2002).
231. H. Zhang and R. M. Gschwind, *Angew. Chem., Int. Ed.*, **45**, 6391 (2006).
232. H. Zhang and R. M. Gschwind, *Chem. Eur. J.*, **13**, 6691 (2007).

CHAPTER 5

Photochemical transformations involving copper porphyrins and phthalocyanines

NATALIA N. SERGEEVA[a] and MATHIAS O. SENGE[a,b]

[a]*School of Chemistry, SFI Tetrapyrrole Laboratory, Trinity College Dublin, Dublin 2, Ireland*
and
[b]*Institute of Molecular Medicine, Medicinal Chemistry, Trinity Centre for Health Sciences, Trinity College Dublin, St. James's Hospital, Dublin 8, Ireland*
Fax: +353-1-608-8536; e-mail: sengem@tcd.ie

I. INTRODUCTION 218
A. Abbreviations 218
B. General Introduction 218
II. BASIC PHOTOCHEMISTRY OF PORPHYRINS AND PHTHALOCYANINES 218
A. General Concepts and Theoretical Background 218
B. Photophysics of Copper Tetrapyrroles 219
III. ELECTRON TRANSFER SYSTEMS AND PHOTOCHEMICAL REACTIONS 220
A. Introduction 220
B. Donor–Acceptor Electron Transfer Compounds 220
C. Heteroligand Systems 222
D. Photochemical Reactions 224
IV. PHOTOINDUCED ELECTRON TRANSFER IN APPLIED PHOTOCHEMISTRY 225
A. Nanomaterials—Molecular Electronic Devices 225
B. Solar Energy 228
C. Hydrogen Production 230
D. Dye Industry 230
V. GREEN CHEMISTRY OF COPPER-BASED DYES: PHOTODEGRADATION, STABILIZATION AND PHOTOCATALYSIS 232

The chemistry of organocopper compounds
Edited by Z. Rappoport and I. Marek

A. Photodegradation of CuP and CuPc Dyes 232
B. CuP and CuPc Photocatalysts Based on Titanium Dioxide 235
C. Other Hybrid CuP and CuPc Photocatalysts 237
VI. PHOTOBIOCHEMISTRY 237
A. Photodynamic Therapy and Singlet Oxygen Production 237
VII. ACKNOWLEDGEMENTS 241
VIII. REFERENCES 241

I. INTRODUCTION

A. Abbreviations

D-A	donor–acceptor systems	PDT	photodynamic therapy
C_{60}	[60]fullerene	PET	photoinduced electron transfer
ET	electron transfer	S	singlet state
P	porphyrin	T	triplet state
Pc	phthalocyanines		

B. General Introduction

Copper tetrapyrroles have found industrial uses for oil desulfurization, as photoconducting agents in photocopiers, deodorants, germicides, optical computer disks, semiconductor devices, photovoltaic cells, optical and electrochemical sensing, and as molecular electronic materials. Nevertheless, even in this area the body of available literature is large and we only use selected examples to highlight the state of the art of this field. A description of syntheses, methodology or electron transfer reactions is outside the purview of this work and the present work can only give a broad overview and selected examples of studies in this area.

II. BASIC PHOTOCHEMISTRY OF PORPHYRINS AND PHTHALOCYANINES

A. General Concepts and Theoretical Background

Tetrapyrroles are macrocyclic heteroaromatic compounds and the aromatic character of the underlying tetrapyrrole moiety, the central metal and the reactivity of the functional groups in the side chains govern their chemistry. The tetrapyrrole ligand is capable of coordinating almost any known metal with the core nitrogen atoms. Together with the conformational flexibility of the macrocycle and the variability of its side chains, this accounts for their unique role in photosynthesis, medicine, biochemistry and applications[1–3].

Porphyrins and phthalocyanines **1–3** are well-known representatives of the tetrapyrrole class of heteroaromatic compounds and closely related. Tetraazaporphyrin (or porphyrazine) **2** is the first member of the phthalocyanine class. In phthalocyanines, carbon atoms of methine bridges (positions: 5, 10, 15 and 20 in **1**) have been replaced by four nitrogen atoms and generally referred to the structure **3** of a tetrabenzotetraazoporphyrin.

They contain an extended π-conjugated system which is responsible for their use in many applications ranging from technical (pigments, catalysts, photoconductors) to medicinal (photodynamic therapy) uses. The electronic absorption spectra are governed by the aromatic 18 π-electron system and typically consist of two main bands. In phthalocyanines, the Q band around 660–680 nm is the most intense one accompanied by a weaker Soret band near 340 nm[4]. In porphyrins, the situation is reversed with an intense Soret band around 380–410 nm and weaker Q bands in the 550–650 nm region. The position and intensity of the absorption bands are affected by the central metal,

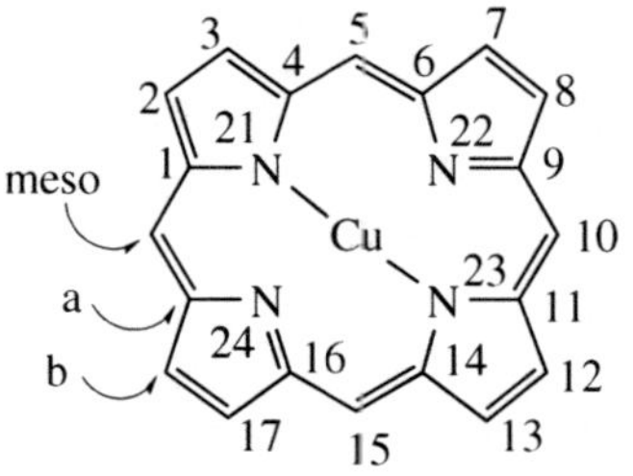

(1)

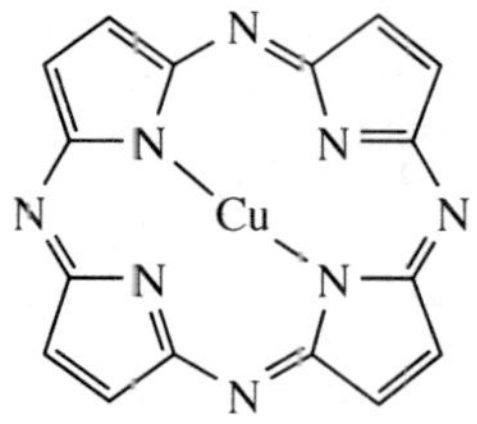

(2)

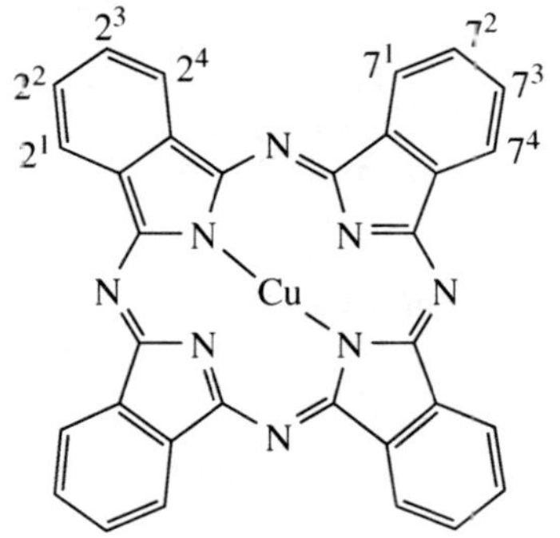

(3)

axial ligands, solvation, substituents and their regiochemical arrangement, and aggregation. The theoretical background has been widely reviewed and established in pioneering works by Gouterman[5] and Mack and Stillman[6]. The spectral characteristics depend strongly on the substituent pattern. By now, almost all possible combinations of electron-donating, electron-withdrawing or sterically demanding groups have been prepared. Copper tetrapyrroles **1–3** behave like most other organic chromophores. The absorption of light results in the rapid formation of the lowest excited singlet state via promotion of an electron from the HOMO to the LUMO. The excited state can then either relax to the ground state via radiative (fluorescence) or non-radiative processes (internal conversion of vibrational relaxation). Another possibility is inter-system crossing to form a triplet state which again can relax either via radiative (phosphorescence) or non-radiative processes. In our context, both types of excited states can take part in photochemical reactions and, in the presence of donor or acceptor units, energy or electron transfer between the chromophores can compete with these processes[7]. In addition, metallo(II) porphyrins and phthalocyanines may form ions upon illumination. These are either anion or π-cation radicals and undergo further photochemical reactions[8,9].

B. Photophysics of Copper Tetrapyrroles

Cu-ions are known to be excited-state quenchers with the partially filled d orbitals. Copper(II) ions are capable of fluorescence quenching by electron or energy transfer. Furthermore, Cu(II), with its d_9 valence electron configuration, is paramagnetic. These special features of the copper have a strong influence on the subsequently discussed photochemical tetrapyrrole transformations. Many spectroscopic studies provide details on magnetic interactions and spin dynamics of states with different multiplicities, such as doublets, triplets and charge-transfer states. The communication between these states strongly depends on the temperature and the solvent, and the spectroscopic studies established the existence of radical species deduced through ps optical experiments and the corresponding theoretical calculations[10–13].

Copper(II) porphyrins are interesting complexes from both a theoretical and experimental point of view. The unpaired electron in the $d_{x^2-y^2}$ orbital couples with the normal porphyrin (π, π^*) excited states to form the singdoublet [^{2}S(π, π^*)], tripdoublet [^{2}T (π, π^*)] and quartet [^{4}T (π, π^*)] states.[14] These complexes do not exhibit the typical fluorescence of closed-shell metalloporphyrins, but rather show moderately strong phosphorescence from the tripdoublet–quartet manifold[15–19].

The relaxation processes in excited Cu-porphyrins have been extensively studied by picosecond transient absorption spectroscopy[20–22]. It was found that photoexcitation of a CuP in the $^2S_0 \rightarrow {}^2S_n$ channel was followed by an extremely fast (1 ps) inter-system crossing to the excited $^2T_1$ state, thus suppressing porphyrin fluorescence. The equilibrium formation between the $^2T_1$ and $^4T_1$ states proceeds within hundreds of picoseconds. Note that splitting between the excited $^2T_1$ and $^4T_1$ states for different Cu-porphyrins varies, depending on the porphyrin macrocycle structure. A variety of studies were undertaken to investigate the effects of porphyrin structure[23–25] and axial ligands[26] on the photophysical and photochemical processes of copper porphyrins.

Similar to metalloporphyrins, the central metal in metallophthalocyanines has a significant effect on the nature and the lifetime of the excited states. An interesting feature of CuPc's[27,28] and other metallophthalocyanines is their ability to participate in stacked assemblies with other CuPc's or with metalloporphyrins. Investigations of these assemblies have shown that metallophthalocyanines with transition metals often display an excited-state relaxation pathway associated with the interactions of the macrocyclic π-system and d-orbitals of the central metal[29,30].

III. ELECTRON TRANSFER SYSTEMS AND PHOTOCHEMICAL REACTIONS

A. Introduction

Studies on photoinduced energy and electron transfer in supramolecular assemblies have witnessed a rapid growth in the past decade. Most of these studies focused on the mechanistic details of light-induced chemical processes. Researchers attempt to generate systems with ultra-fast charge transfer and charge recombination applicable as light-induced switches or with long-lived charge-separated states for solar-energy conversion[31]. These endeavours have produced an expanding body of information on porphyrin/phthalocyanine dyads, their design, and energy, exciton and charge transfer properties. Incorporation of these systems into larger architectures now offers the possibility for applications in molecular photonics, electronics, solar-energy conversion, nanomaterials and quantum optics.

B. Donor–Acceptor Electron Transfer Compounds

CuP and CuPc play a very important role in electron transfer processes. The simplest covalently linked systems consist of a tetrapyrrole unit linked to an electron acceptor or donor moiety with appropriate redox properties.

Classic D–A systems. Biomimetic systems comprised of porphyrins and quinones have been studied extensively with regard to their electron transfer and charge transfer properties as they are related to photosynthesis and solar-energy conversion. In many cases, the magnitude of the reduction potentials of quinones covalently linked to porphyrins suggests light-induced radical-pair generation. However, use of quinones as acceptors decreases the energy that can be stored in the photoinduced radical pair, thus diminishing the likelihood of a mechanism via radical-pair generation[32]. For mechanistic studies, it is essential to have acceptor moieties other than quinones with systematically variable reduction potentials so that the energy of $P^{+}-A^{-}$ falls between the energies of the excited singlet and triplet states and beyond. One example for alternatives to porphyrin–quinone model systems are the copper picryl porphyrins **4**[33]. Here, the interaction between the porphyrin and trinitroaryl group is relatively stronger than in porphyrin–quinone systems[34].

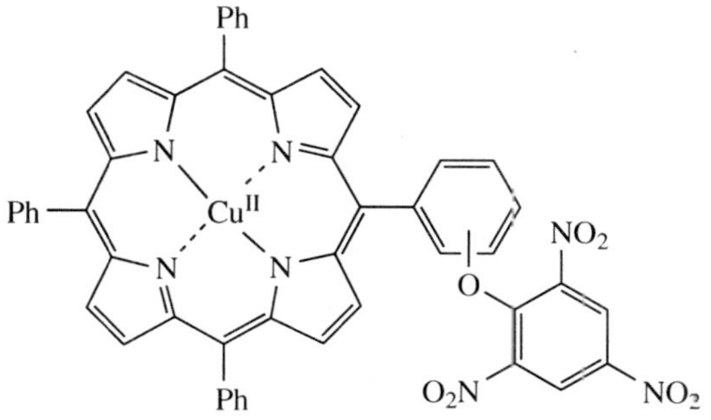

(**4**)

Fullerenes. Applications of fullerenes in photoinduced electron-transfer studies derive from the fact that back electron transfer is inhibited in donor–acceptor systems incorporating [60]fullerene (C_{60}). Fullerenes can accept up to six electrons and exhibit small reorganization energies while the photoinduced charge-separation is accelerated and charge recombination is slowed. The effect of the introduction of C_{60} as an electron acceptor into tetrapyrrole molecules should enhance our knowledge of the dependence of photoinduced electron-transfer dynamics on molecular topology. Several self-assembled donor–acceptor systems containing fullerenes as three-dimensional electron acceptors, and porphyrins as electron donors have been described. Likewise, different types of non-covalently and covalently linked copper porphyrin–fullerene dyads have been synthesized and intensively studied[35–37].

A series of fulleropyrrolidinophthalocyanines containing free base, Zn and Cu, were investigated as electron-donating building blocks in fullerene dyes of type **5**[38]. These experiments support the view that the $C_{60}/C_{60}^{\bullet-}$ couple exhibits the strongest electron affinity, while the different $Pc/Pc^{\bullet+}$ couples play the role of electron donors.

Further investigations have shown that the donor ability increases in the following order: CuPc < H_2Pc < ZnPc. In other words, starting from the free base (H_2Pc), addition of either Cu (CuPc) or Zn (ZnPc) increases or decreases the electron density on the phthalocyanine, respectively. Similarly, the energy for the radical ion pair $CuPc^{\bullet-}-C_{60}^{\bullet+}$ formed in a photoinduced electron-transfer reaction (*vide infra*) decreases in the following order: (1.40 eV)> H_2Pc-C_{60} (1.36 eV)> $ZnPc-C_{60}$ (1.23 eV).

Although H_2Pc and ZnPc are both strongly fluorescent probes in room-temperature emission experiments, neither fluorescence nor phosphorescence was observed for CuPc. Implicit in the fluorescence silence is an ultra-fast deactivation of the initially excited state of this paramagnetic compound with a d_9-metal centre. A similar mechanism has been proposed for the analogous copper porphyrin (CuP), which showed, however, activation of moderately strong and long-lived phosphorescence as a result of thermally equilibrated

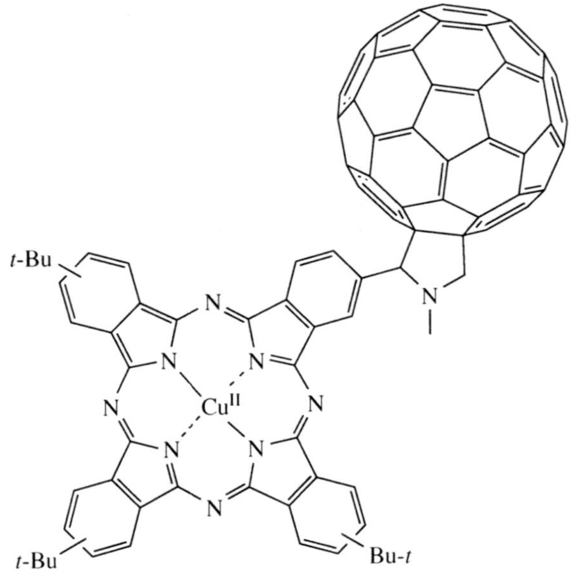

(5)

triplet–doublet/triplet–quartet states[39,40]. Due to the comparable nature of the porphyrin and phthalocyanine macrocycles and binding of the transition-metal centre (i.e. Cu–N), it is likely that a similar ultra-fast deactivation pattern governs the photophysics of CuPc.

C. Heteroligand Systems

Photoinduced electron transfer (PET) between metalloporphyrins and free bases in dimeric, trimeric and oligomeric porphyrin systems has been studied extensively. Depending on the choice of the donor and acceptor unit, electron transfer from either singlet or triplet states can be observed. Electron transfer studies in systems based on heterodimers with covalent or electrostatic bonds is of particular interest as it relates directly to the natural photosynthesis process.

ET from CuP to its free-base partner was observed in a CuP–H_2P hybrid dimer **6**[41]. The advantage of a hybrid system is that the energy donor and an acceptor are distinct. The lowest excited singlet and triplet states of the free-base porphyrin monomer are lower than those of the copper porphyrin. Based on T–T transient absorption spectra, the triplet yield of the free-base porphyrin was found to increase in the hybrid dimer compared to a monomeric free-base porphyrin. According to a time-resolved ESR study, the intermolecular energy transfer occurs between the triplets. The lowest excited triplet state of the CuP–H_2P dimer is generated both by energy transfer processes from the triplet manifolds in the copper porphyrin moiety and by inter-system crossing from the lowest excited singlet state of the free-base moiety. As a result of the paramagnetic perturbation, inter-system crossing in CuP is remarkably accelerated and it seems to prevent singlet–singlet energy transfer to another porphyrin moiety.

Electrostatically linked dimers can be easily produced by a coupling between two monomers with oppositely charged substituents in the liquid phase. An example of such

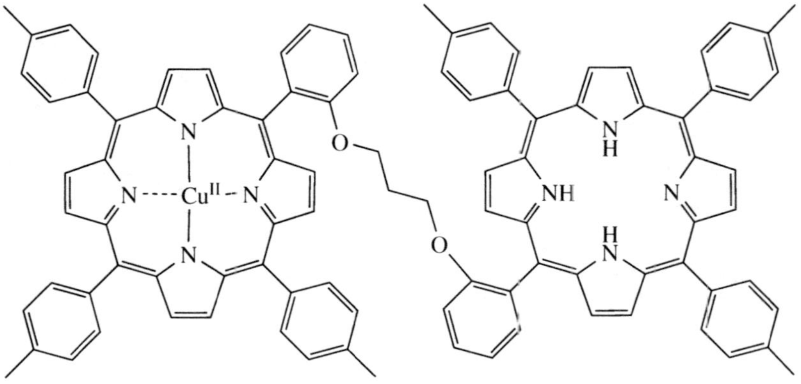

(6)

systems has been illustrated by Tran-Thi and coworkers[42], where Zn-porphyrin units were linked with Cu-phthalocyanines. Drastic changes in the ground-state absorption of the electrostatically linked dimer compared to the corresponding monomers indicate a strong interaction between the two chromophores.

According to femto- and nano-second absorption spectroscopy, very efficient inter-system conversion takes place in the excited ZnP–CuPc, leading to the final 'triplet' state. In contrast, excitation of the ZnP–AlPc dimer led to an electron transfer from the porphyrin to the phthalocyanine moiety. The different behaviour in ET can be explained by the peculiar properties of the paramagnetic CuPc.

The whole mechanistic scheme can be summarized as follows for ZnP–CuPc and ZnP–AlPc, respectively:

Cu–Zn

$${}^{S_0}(\text{Zn–Cu}) + h\nu \rightarrow {}^{2S}(\text{Zn–Cu})^*$$
$${}^{2S}(\text{Zn–Cu}) \rightarrow {}^{S_0}(\text{Zn})\text{–}{}^{2T-4T}(\text{Cu})^*$$
$$(\Phi \geqslant 0.75;\ k_{isc} = 5 \cdot 10^{11})$$
$${}^{S_0}(\text{Zn})\text{–}{}^{2T-4T}(\text{Cu})^* \rightarrow {}^{S_0}(\text{Zn–Cu})$$
$$(k_T = 3.2 \cdot 10^{7})$$
$${}^{S_0}(\text{Zn}) \rightarrow {}^{1S}(\text{Zn})^* \rightarrow {}^{1T}(\text{Zn})^*$$
$$(\Phi = 0.85;\ k_{isc} = 6.7 \cdot 10^{8})$$
$${}^{1T}(\text{Zn})^* \rightarrow {}^{S_0}(\text{Zn})$$
$$(k_T = 2 \cdot 10^{4})$$

Zn–Al

$$\text{Zn–Al} + h\nu \rightarrow {}^{1S}(\text{Zn–Al})^*$$
$${}^{1S}(\text{Zn–Al})^* \rightarrow {}^{2S}(\text{Zn}^{\bullet+}\text{–Al}^{\bullet-})$$
$$({}^{1}k_{et} = 3.8 \cdot 10^{11})$$
$${}^{1S}(\text{Zn–Al})^* \leftarrow {}^{1T}(\text{Zn–Al})^*$$
$$(k_{isc} = 9.5 \cdot 10^{9})$$
$${}^{2S}(\text{Zn}^{\bullet+}\text{–Al}^{\bullet-}) \rightarrow {}^{S_0}(\text{Zn–Al})$$
$$({}^{1}k_{-et} = 1.4 \cdot 10^{10})$$
$${}^{1T}(\text{Zn–Al})^* \rightarrow {}^{2T}(\text{Zn}^{\bullet+}\text{–Al}^{\bullet-})$$
$$({}^{1T}k_{et} = 2.45 \cdot 10^{4})$$
$${}^{2T}(\text{Zn}^{\bullet+}\text{–Al}^{\bullet-}) \rightarrow {}^{S_0}(\text{Zn–Al})$$
$$({}^{1T}k_{-et} = 3.8 \cdot 10^{3})$$

Here, k_{et} (s^{-1}) and k_{-et} (s^{-1}) are the forward and backward ET rate constants, respectively, k_T is the rate constant of the conversion from triplet (T) to ground state and k_{isc} (s^{-1}) is the rate constant for inter-system conversion from singlet to triplet.

For systems such as the copper-copper dibenzofuran-bridged co-facial bisporphyrins **7** and **8**, EPR spectroscopy has been shown to be an essential analytical tool that provides information not available from optical studies. It also complements crystallographic studies by probing intra-molecular metal–metal arrangements in frozen solution[43].

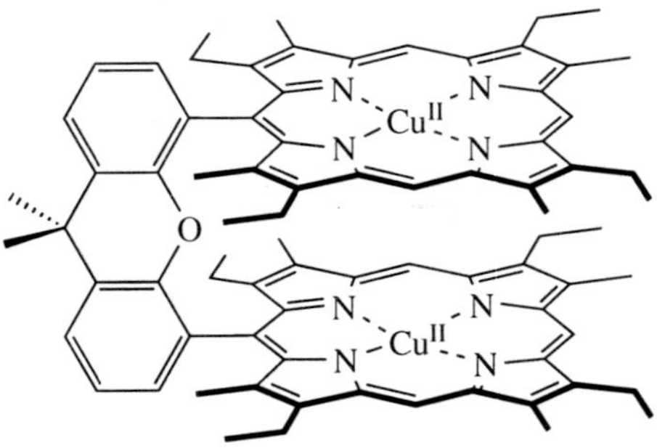

(7)

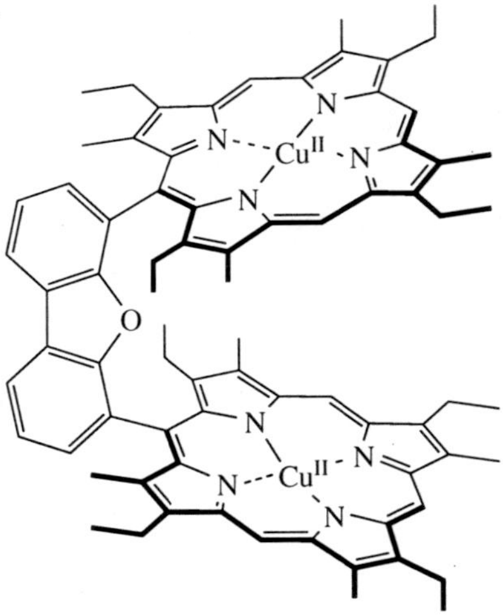

(8)

D. Photochemical Reactions

In the early 1970s photoinduced metallation of porphyrin series was reported[44]. It was found that etioporphyrin I, tetraphenylporphyrin, phyllo-type petroporphyrins (porphyrins derived from chlorophyll phyllo-type) and certain Cu(II) 1,3-diketonates underwent a quantitative conversion into Cu(II) porphyrins in a short period of time (time frame varied with light intensity) upon light irradiation (*ca* 400 nm). Experiments in the dark required several hours or days to achieve a comparable conversion. Upon illumination in the dark experiments, reaction was found to occur at the photoinduced rate. The length of this period was proportional to the length of the prior irradiation time. In addition, no spectrophotometric evidence was found for any porphyrin-containing compounds other than a free base and a Cu(II) porphyrin.

$$\begin{aligned} H_2P &+ h\nu &&\rightarrow H_2P^* \\ H_2P^* &+ Cu(II)L_2 &&\rightarrow Cu(I)L + L^{\bullet} + H_2P \\ Cu(I)L &+ H_2P &&\rightarrow Cu(I)HP + HL \\ Cu(I)HP &+ L^{\bullet} &&\rightarrow Cu(II)P + HL \end{aligned}$$

where L is a ligand and H_2P is a free-base porphyrin.

Based upon the observation that the photochemical reaction occurs upon irradiation by light (*ca* 400 nm) in the region of the Soret band, the formation of a porphyrin excited form would be logical.

Other interesting examples for photochemical transformations were observed in copper(II) porphyrins **9** and **10**, which bear a fused methylenepropano-ring. They were shown to undergo unusual self-sensitized photo-oxygenation reactions[45]. These porphyrins were perfectly stable in the dark under inert gas, however once exposed to both light and air, they were rapidly transformed into oxygenated adducts.

Compound **10** appeared to be more unstable in the presence of both light and air and its major oxygenation product was identified as the α,β-unsaturated ketone **11** and obtained in 52% yield. Differently, the co-facial bis-porphyrin **12** has been isolated from the reaction of **9** in the presence of DBU in very good yield (60%) and its structure was confirmed by X-ray analysis.

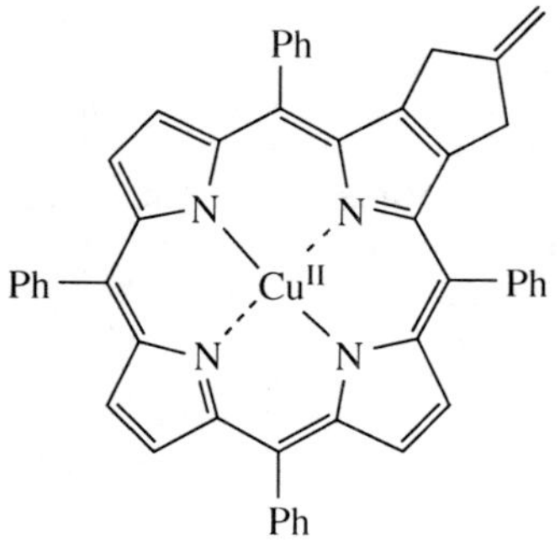

(**9**)

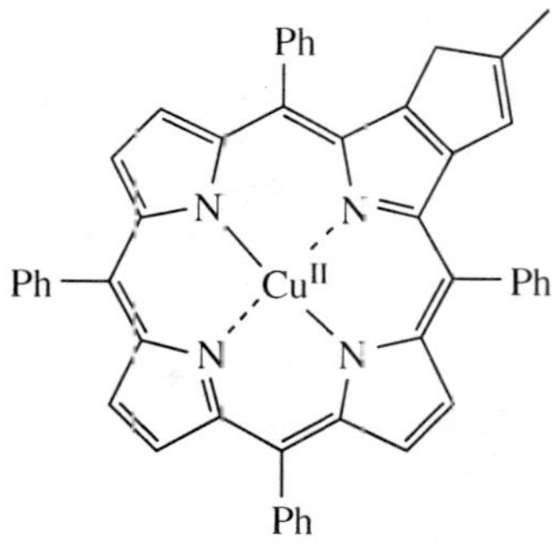

(**10**)

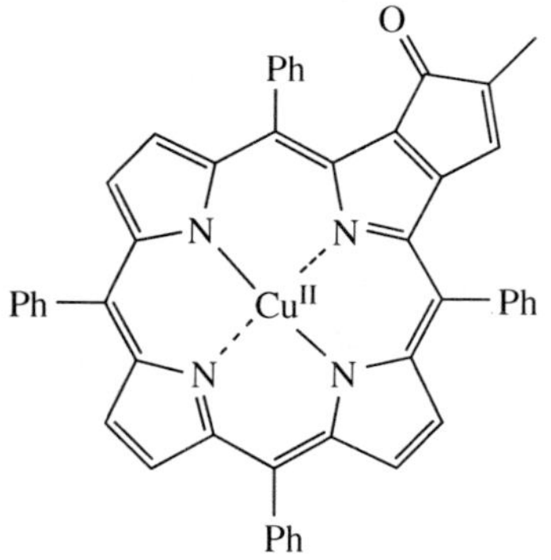

(**11**)

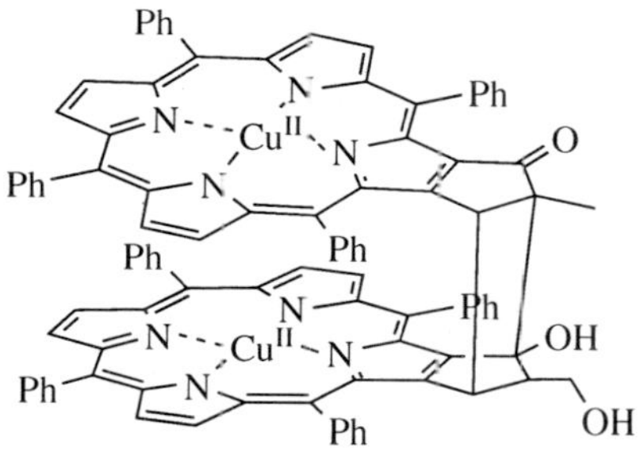

(**12**)

IV. PHOTOINDUCED ELECTRON TRANSFER IN APPLIED PHOTOCHEMISTRY

Solar-energy conversion, electrophotography, hydrogen production and photocatalysis are the most popular topics involving copper-based tetrapyrroles. CuPc-based photoenergy conversion systems have been designed using donor–acceptor molecules, polymers and carbon nanotubes. The mechanism of photoinduced electron transfer (PET) is extensively studied and a variety of the artificial systems applying PET is examined for use as molecular devices.

A. Nanomaterials — Molecular Electronic Devices

The discovery of the charge transfer across the donor–acceptor systems opened an effective way to improve the photo-optical response of the organic materials. Development and use of organic ultra-thin monolayers as potential models for optoelectronic devices and catalysts in solid photochemistry has attracted great attention. Some studies focused on the fabrication of CuPc and CuP thin films using photoinduced deposition methodologies[46–48] and investigated their optical[49], optical memory and photoswitching[50] and xerographic[51] properties.

Many memory and switching phenomena in organic solid films have been examined, as prototypical candidates for molecular electronic devices and photogeneration in these organic superlattices under an applied electric field may play a fundamental role. Some work has been reported on PET and back electron transfer (charge recombination) of hetero-Langmuir–Blodgett film devices with CuPc as a sensitizer. A transient electric current was observed upon the selective photoexcitation of CuPc and the resulting charge-separated state in the device survived for a few minutes. Applying electric biases and/or lowering temperatures caused the charge-separated state to last much longer[52].

Disk-shaped like monomers can form rod-like aggregates (discotic mesophase materials) and they exhibit long-range order, large electrical anisotropies. The high interest in these materials stems from their promise as organic electronic devices requiring high charge mobilities and dense integration (e.g. organic field effect transistors, organic light-emitting diodes and photovoltaic cells). CuPc's, especially with flexible hydrocarbon chains, exhibit a remarkable feature to form the discotic mesophased architectures[53].

Syntheses and thin-film formation have been performed for octa-substituted phthalocyanines **13–15** with and without polymerizable styryl side chains: $CuPc(OC_2H_4OCH{=}CHPh)_8$ (**13**) with styryl groups at the termini on the side chains, or with one alkoxy group removed, $CuPc(OC_2H_4CH{=}CHPh)_8$[54] (**15**), and $CuPc(OCH_2)_2OBz)_8$ (**14**). The latter compound (**14**) has been shown to form highly coherent rod-like aggregates in Langmuir–Blodgett films with excellent control of rod orientation[55,56].

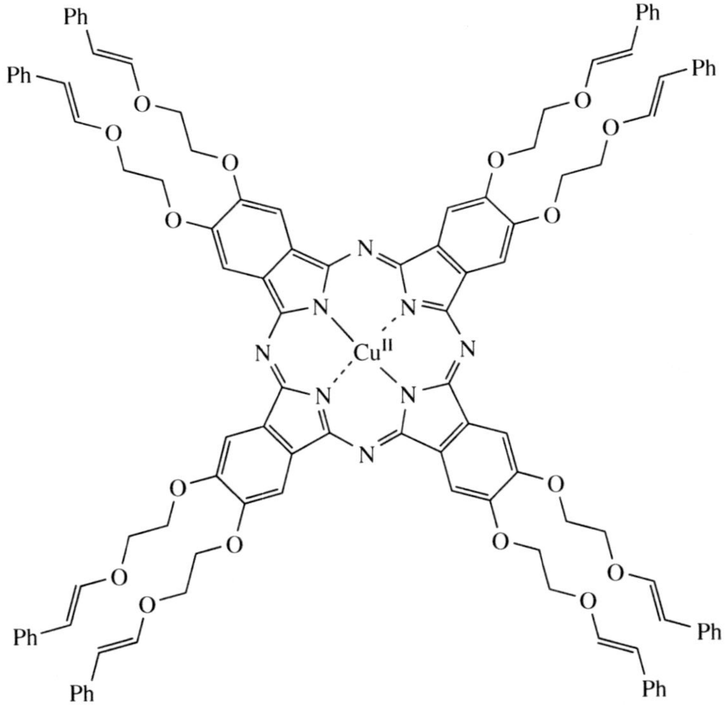

(**13**)

Irradiation of the styryl $\pi-\pi^*$ absorbance bands with λ *ca* 254 nm for horizontally transferred LB films of **13** and **15** resulted in a stabilization of their rod-like aggregates, through formation of cyclobutane links between adjacent side chains (Scheme 1). In the case of compound **13**, styryl group polymerization was achieved up to 75% versus *ca* 55% for **15**[57].

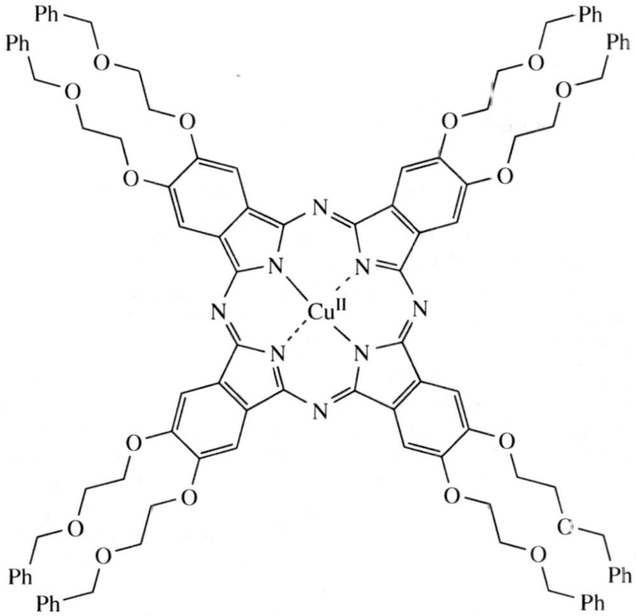

(**14**)

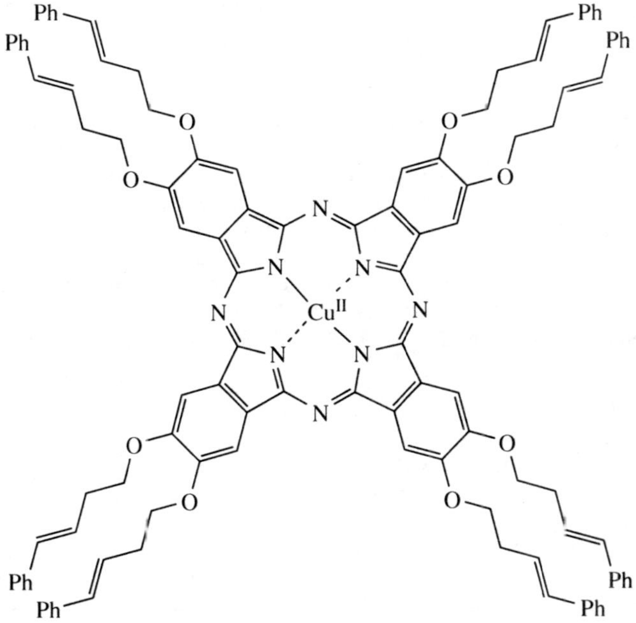

(**15**)

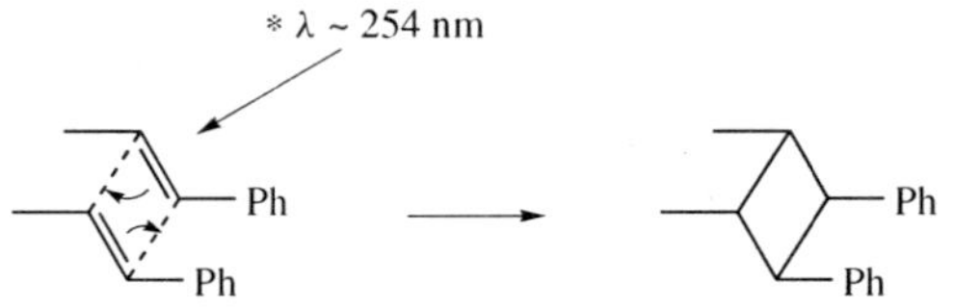

SCHEME 1. Photoinduced cyclobutane formation within adjacent side chains of **13** and **15**

Polymerized thin films have shown a long-range order that was confirmed by atomic force microscopy and X-ray reflectometry. The differences in the orientation of individual Pc's, between films of **13** and **14**, were determined by transmission and reflectance. Higher dark photoconductivities and electrical anisotropies were observed in films of **13** after annealing and polymerization, compared with those found for films of **14**.

Photoelectrodes. Several studies were aimed at the fabrication of molecular-based photoelectrodes and special attention has been given to the efficient uptake of visible light and conversion into output of the photosystems. Typical examples are the construction, determination of photoelectric characteristics and an evaluation of the photoelectrodes[58,59]. For example, photoelectrode devices constructed from CuPc as a p-type semiconductor in combination with *N*,*N*′-diphenylglyoxaline-3,4,9,10-perylenetetracarboxylic acid bisbenzimidazole as n-type semiconductor in the water phase were investigated in terms of kinetics. Each film of the p/n bilayer presents a photoanode, where the photoinduced oxidation of thiol occurs. The holes originate on account of the photophysical events in the p/n interior, involving the charge separation of excitons at the p/n interface[60] (Figure 1).

Gas-sensitive electrodes can be fabricated by use of CuPc incorporation into a polypyrrole backbone. The photoresponse in electrochemical mode increases due to the presence of phthalocyanine in the films, and enhanced sensitivity towards nitrogen gas as compared to pure polypyrrole film was observed.[61]

Junctions. Molecular conductance junction[62] is another important utilization of phthalocyanines. Here, a molecule or a small cluster of molecules conduct electrical current between two electrodes. Copper phthalocyanine (CuPc) has been used in silicon-based molecular nanotechnology[63] to integrate molecular electronic function with silicon surfaces. In these applications the knowledge about mechanisms of charge transfer and the role of the outer phenyl ring is important to understand the interaction with the surrounding substrate when the CuPs molecule is bound to the surface via its central copper atom[64]. One example is the intercalation of CuPc pyridinium complex and iron phthalocyanine into two layered titanates from $Na_2Ti_3O_7$ to study the microstructure control[65].

B. Solar Energy

One of the main targets in solar-energy conversion is reaching 10% conversion efficiency. This is within reach and has caused an intensive development of organic and hybrid thin-film photovoltaics (third-generation solar cells). Substantial improvements have been made resulting in new device concepts and improved understanding of energy-conversion processes. Films of tetrapyrrole semiconductors are promising candidates for stable and low-cost electrode materials that can be used for solar-energy conversion. Various copper phthalocyanines have been investigated as thin-film electrodes in photoelectrochemical cells[66,67].

The behaviour of the phthalocyanines as p-type materials results under illumination in cathodic photocurrents with reduction of a suitable acceptor in solution. These thin-film

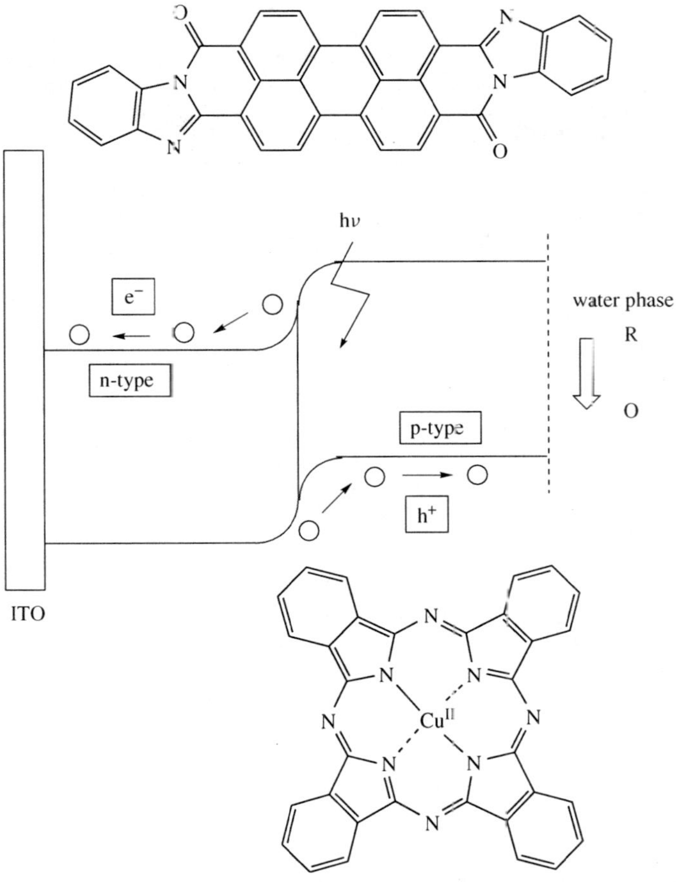

FIGURE 1. Schematic illustration of the photoanodic reaction at a CuPc/water interface

materials are robust and absorbing at *ca* 650–750 nm. n-Type semiconduction in organic materials has been reported for porphyrins. For example, Yamashita and coworkers[68] described the semiconducting behaviour of metal-free and copper tetraphenylporphyrins to be switched from p-type into n-type in a photoelectrochemical cell when more than three phenyl groups were replaced by pyridyl residues.

Organic–inorganic hybrid cells also demonstrated enhanced performance by use of nanoporous Si as n-type and CuPc. The CuPc device shows a conversion efficiency up to 2% under white-light illumination (20–30 $mW\,cm^{-1}$)[69]. Theoretical studies on a microkinetical model of a solar cell included electron–hole pair recombination between donor and acceptor sites. A simulation has been performed on CuPc and 3,4,9,10-perylenetetracarboxylic acid dianhydride (PTCDA) and interpreted in a simple kinetic picture of an electron–hole pair generation step at the CuPc–PTCDA interface and subsequent transport in the CuPc and PTCDA domains[70].

Other studies reported the construction and analysis of photovoltaic cells fabricated with CuPc and TiO_2[71–73] or ZnS[74] nanoparticles. The dye-sensitized TiO_2 solar cell showed conversion efficiencies of 7–10% under standard solar conditions. The high efficiencies are associated with ultra-fast charge transfer from the dye to TiO_2, the high internal surface area of the TiO_2 films, the broad absorption of the dye and the efficient separation of opposite charges between the domains. Recently, some attention has been given to composites of metal phthalocyanines and carbon nanotubes because of their high quantum efficiency facilitated by charge transfer between them and the complementary properties of the composites. These are regarded as promising candidates for the fabrication of donor–acceptor heterojunction diodes and photovoltaic devices. An example of such constructions has been reported recently for a copper phthalocyanine with long dodecyl chains covalently attached to modified multi-walled carbon nanotubes[75].

C. Hydrogen Production

Photochemical processes capable of the evolution of molecular hydrogen are one of the most important areas of the current research. Considerable interest has followed after Fujishima and Honda reported the photoelectrochemical water splitting using a UV-responsive TiO_2 electrode[76]. Nowadays, dihydrogen (H_2) energy is attracting much attention as a 'clean energy' source with water photolysis being one of the most promising procedures to produce H_2. Some of the strategies are based on the development of an efficient solid/water interface to achieve the (photo)chemical energy conversion under solar irradiation.

Being p-type of semiconductors, CuPc's can be used to construct p/n type of organic bilayers as photoelectrodes and used in water phase[77]. In aqueous solution, copper tetrapyrroles can be used to reduce methyl viologen (MV) from MV^{2+} to $MV^{\bullet +}$ that is capable of decomposing water in the presence of a catalyst. A hydrogen-producing three-component system [chromophore/donor/MV] was analysed using CuPc as a chromophore[78]. However, sulfonated phthalocyanines compare unfavourably to metalloporphyrins[79]. Some studies were carried out on the photodynamics of these systems for future applications[80].

D. Dye Industry

Improvements in the lightfastness of disperse dyes is one of the major goals in the dye industry. The relationship between chemical structure and lightfastness of disperse dyes and complete determinations of the reaction mechanisms that characterize the photofading of disperse dyes on polyamide and PET substrates is of great interest. Many studies have been focused on investigating the light-induced reduction, photooxidation by singlet oxygen and wavelength-dependence of photoreduction in various organic solvents, to name a few.

Himeno and coworkers[81] reported that reductive fading of monoazo disperse dyes combined with vinylsulfonyl copper-phthalocyanine on nylon or polyester substrates was decreased compared with the rate of reductive fading of disperse dyes without the CuPc dye. Singlet oxygen generated by the photosensitization of CuPc dye suppressed the reductive fading of disperse dyes. Nylon had a greater tendency to give light-induced reduction and oxidation of disperse dyes than polyester, depending upon the properties of the adsorbed dye.

Some investigations were carried out purely on CuPc dyes; examples are the photosensitized oxidation of copper phthalocyanine (CuPc) reactive dye on cellulose under different conditions[82–84] and additives[85], and photochemical reduction of the tetrasodium salt of copper tetrasulfonatophthalocyanines with amines[86,87].

Dyes are utilized in polymer materials to enhance their colour-changing properties. However, these additives can considerably affect the polymer stability towards degradation. Interestingly, CuPc can play the role of a polymer stabilizer against degradation as

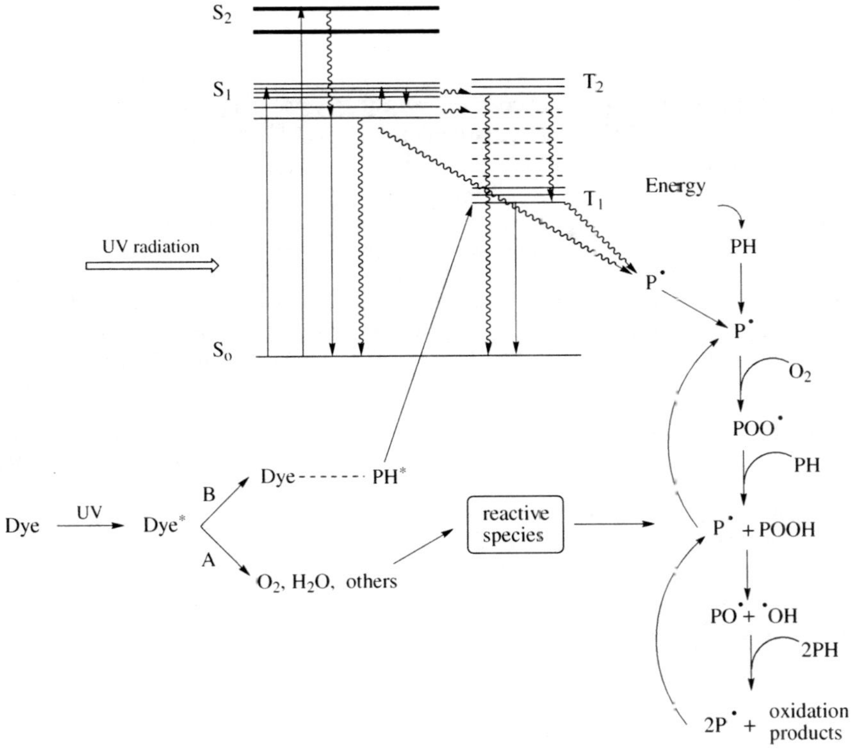

FIGURE 2. Schematic illustration of dye participation in the photodegradation mechanism of polymers (P)

well as in some cases to be its trigger. The interaction mechanism of dyes can occur via two pathways[88–90] (Figure 2). In pathway A, after the light absorption in oxygen-containing media, the dye produces chemical species such as singlet oxygen and superoxide anions. The reactive chemical species thus produced accelerate polymer degradation. In pathway B, the dye reaches excited singlet (S_1, S_2) and triplet (T_1, T_2) states. The energy absorbed by the dye is then transferred to the chemical groups in the polymeric chain via intermolecular energy transference. The energetic excitation of the polymer by the colourant increases the potential formation of free radicals ($P^{\bullet}$).

For example, CuPc incorporated into polycarbonate resulted in accelerated degradation under UV-radiation. Being a strong sensitizer, CuPc very likely enhances the formation of reactive species in polycarbonate. Excited states of CuPc may intercept hydrogen atoms from methyl groups in polycarbonate, increasing the formation of free radicals, which is the beginning for the sequential photo-oxidation processes leading to the degradation of a polycarbonate. Alternatively, photodegradation of polycarbonate triggered by CuPc may go via electron transfer sensitization[91]. In contrast, the three-component system involving hindered piperidine (known to be effective photostabilizers for some polymers), antioxidant and CuPc have shown synergistic effects. Interestingly, CuPc alone offered little photoprotection and some enhanced photoprotection was observed for antioxidant CuPc. These results indicate that CuPc and a hindered piperidine exhibit a highly favourable interaction for photostabilization[92]. Finally, CuPc dyes can also be used as heterogeneous sensitizers to induce the photopolymerization of trimethylolpropane triacrylate[93].

V. GREEN CHEMISTRY OF COPPER-BASED DYES: PHOTODEGRADATION, STABILIZATION AND PHOTOCATALYSIS

High levels of environmental contamination have been associated with dye wastewater contents from the textile dyeing and finishing industry. Various types of synthetic dyes are used in the textile industry including azo, anthraquinone, triarylmethane and phthalocyanines designed to be highly robust. Moreover, due to the complexity and variety of the dyes used for the different purposes, as well as their non-biodegradability, it is rather difficult to find unique treatment methods that would result in effective degradation of all types of dyes. Azo dyes are the largest class of commercially used colourants, while phthalocyanines are mainly utilized as blue and green dyes. Copper phthalocyanines (**16**–**20** are several examples of the commercially used CuPc dyes) are the most important derivatives of the phthalocyanine pigment class[94]. Their commercial success is based on three features: beautiful bright blue to green shades and strength, remarkable chemical stability and excellent fastness to light. Due to their low tendency to migrate in materials, CuPc's are used in inks and printing, coatings, some plastics, textile and leather. However, environmentally, these advantages as robust and efficient colourants decrease the efficiency of convenient treatment technologies. Additional contamination associates with metals such as copper released during the process from the metallized dyes.

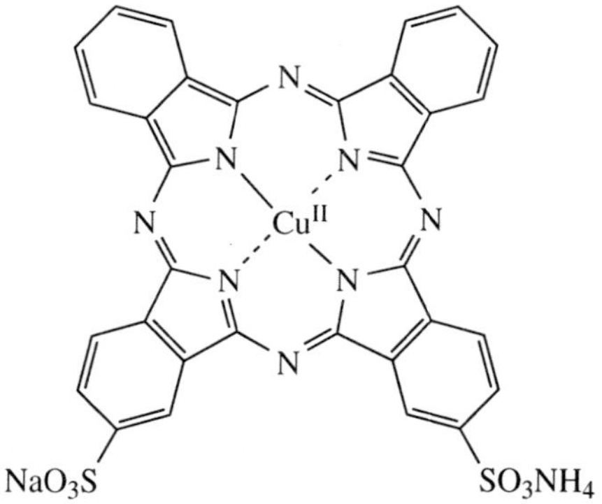

C.I. Direct Blue 199

(16)

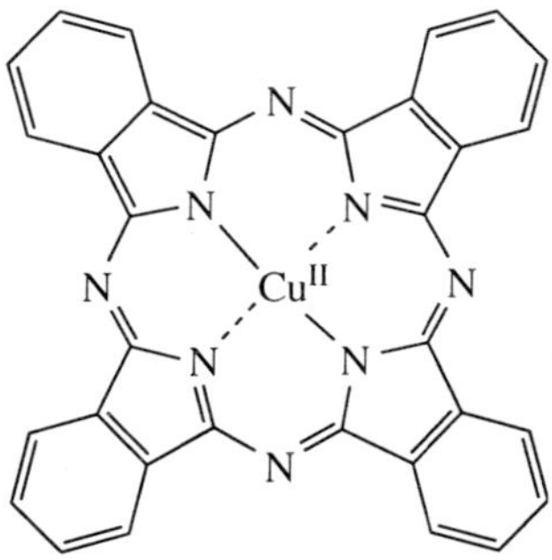

Phthalocyanine Blue BN

(17)

A. Photodegradation of CuP and CuPc Dyes

Although porphyrins and especially phthalocyanines are stable compounds, both will undergo photo-oxidative degradation or photoexcited ET reactions. The phthalocyanine systems are very stable even in advanced catalytic processes such as photocatalytic oxidation. Photoelectrocatalytic oxidation attracts considerable attention as a way to increase the photocatalytic efficiency in degradation of organic pollutants and studies have been carried out to apply this methodology for decolourization of Pc systems.

The decolourization of commercially relevant reactive CuPc dyes, such as *Remazol Turquoise Blue 133*, was studied comparatively in aqueous solution with regards to

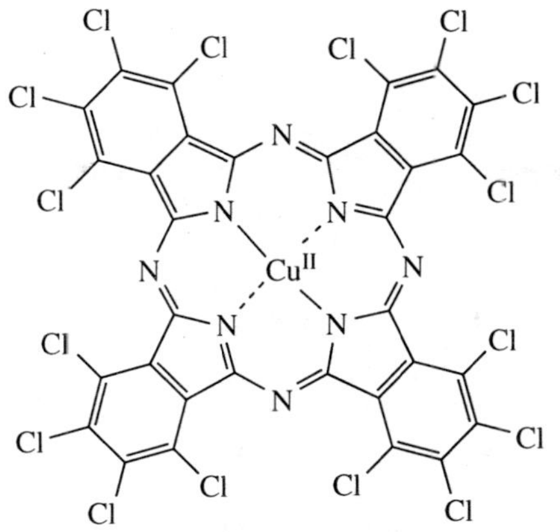

Phthalocyanine Green G

(**18**)

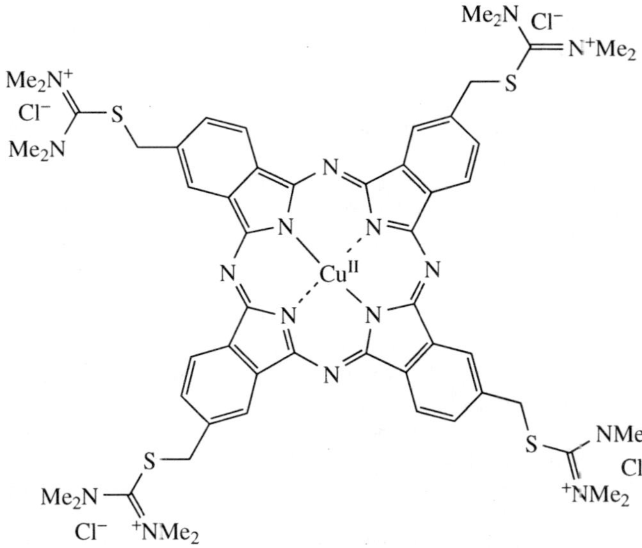

Alcian Blue 8 GX

(**19**)

TiO_2-mediated photocatalytic and photoinduced and dark Fenton/Fenton-like reactions[95]. However, the authors reported that none of those advanced oxidation systems was effective in the degradation of the CuPc dye. Further investigations were performed on Remazol Turquoise dyes. For example, *Remazol Turquoise Blue 15* (**20**) was investigated by direct cathodic reduction on platinum electrode and photoelectrocatalytic oxidation on Ti/TiO_2 as nanostructured semiconductor thin-film electrodes with higher efficiency for the combined process of photoelectrocatalytic oxidation and electrochemical reduction[96].

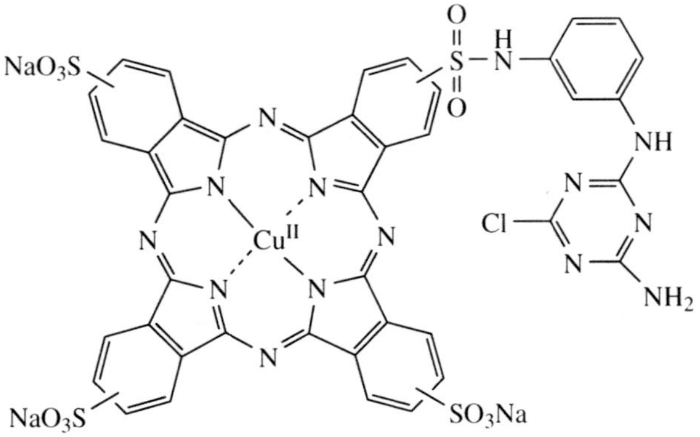

Remazol Turquoise Blue 15

(**20**)

The direct cathodic reduction was not successful for dye mineralization, but almost complete mineralization of the dye was reached using photoelectrocatalytic oxidation (thin-film Ti/TiO_2 photoanodes) and UV irradiation. Maximum photoelectrocatalytic degradation was achieved at $E = +1.2$ V and 0.1 mol l^{-1} Na_2SO_4. At pH *ca* 2 or pH *ca* 8 100% colour removal of the Cu-dye solution, almost complete mineralization values of 95% (pH *ca* 2) and 85% (pH *ca* 8.0) were recorded[97]. However, this method is limited to dyes at high concentrations.

Some studies on photostable CuPc sulfonate sodium salts were carried out under UV ($\lambda > 320$ nm) or visible light ($\lambda > 450$ nm) in the presence of a TiO_2 semiconductor in an aqueous medium. They underwent notably photobleaching in the presence of TiO_2. The spectral analysis showed that the dye photobleaching led to complete destruction of the phthalocyanine ring. In addition, the stability of the dye towards visible light was greatly affected by the physical properties of TiO_2 semiconductor, and the dye photostability could be improved through addition of electron sacrificers such as 4-chlorophenol[98].

Another example of photodegradation catalysed by titanium dioxide is illustrated for the CuPc dye **19** (*Alcian Blue 8 GX*). This dye is one of the so-called biological stains, such as *Eosin Y, Auramine O, Hematoxylin* or *Rose Bengal* etc., which are widely used in biomedical research laboratories and for diagnostic purposes. Some of them are known to be toxic or mutagenic for humans and animals, highly resistant to micro-organisms such that biological wastewater treatment processes are very inefficient in treating these dyes.

The photocatalytic degradation of *Alcian Blue 8 GX* was carried out in aqueous suspensions containing the commercial catalyst TiO_2 P-25. According to the Langmuir–Hinshelwood model, photodegradation followed approximately a pseudo-first kinetic order. It was shown that the photocatalytic degradation reaction can be described mathematically as a function of parameters such as pH, H_2O_2 concentration and irradiation time using the response surface methodology[99].

Decolourization and mineralization of phthalocyanine dye **16** (*C.I. Direct Blue 199*) has been studied using an advanced oxidation process. The authors varied experimental conditions such as hydrogen peroxide and UV dosage, dye concentration and the pH of the medium. The operating conditions for 90% decolourization of the dye and 74% removal of total organic carbon were found to be an initial dye concentration of 20 mg l^{-1}, hydrogen peroxide dosage of 116.32 mM and a UV dosage of 560 W at pH *ca* 9 in 30 min.

Interestingly, decolourization of *C.I. Direct Blue 199* was found to be more difficult than that of an azo dye (*C.I. Acid Black 1*) under the same operating conditions[100].

B. CuP and CuPc Photocatalysts Based on Titanium Dioxide

Great attention has been drawn to other applications of porphyrins and phthalocyanines as very promising (photo)catalysts. Semiconductor photocatalysis is an advanced oxidation process and can be used to eliminate organic pollution. TiO_2-based photocatalysts are being used in photocatalytic technologies for the degradation of organic compounds in water. In past decades, great interest has focused on the development of economically and environmentally friendly new catalytic systems with enhanced activity of TiO_2-based catalysts for use in the oxidative degradation of various organic pollutants.

Titanium dioxide (TiO_2) is one of the most popular materials to be used as a photocatalyst for air treatment and degradation of organic pollutants in water. Having a large band gap (3.2 eV with $\lambda \leqslant 385$ nm), TiO_2 can consume only about 5% (with energy above 3.0 eV, $\lambda \leqslant 410$ nm) of solar light reaching the earth's surface. Thus, to increase the efficiency of solar-energy utilization through TiO_2, it is often doped with transition metal ions or organic dyes to shift the photoresponse of the catalyst into the visible region. In view of its low cost and feasibility, dye sensitization is considered to be an effective method to modify the photochemical properties of TiO_2 particles. Mechanistically, the photosensitization of TiO_2 proceeds via initial excitation of the dye. A photoinduced electron transfer from excited dye to TiO_2 and the surface reaction of TiO_2 conductive band electron were suggested as the key processes.

It was shown that TiO_2 impregnated with Cu-based sensitizers is beneficial for the photoactivation of TiO_2, and CuP's[101–103] and CuPc's have been shown to improve the photocatalytical activities of these photocatalysts.

Comparative studies show that TiO_2 samples doped with CuP and FeP are more photoactive than simple TiO_2 (anatase), whereas samples including MnP showed less photoreactivity[104]. Free-base porphyrins have shown slightly higher photoactivity than TiO_2, suggesting that the porphyrin macrocycle is photocatalytic as well. However, samples impregnated with CuP exhibited the highest photoactivity. These results, related to the photodegradation of 4-nitrophenol (4-NP) in an aqueous heterogeneous environment, suggest that the Cu(II)–Cu(I) photocatalytic redox cycle plays the main role in the process.

Cu(II) could be reduced to Cu(I) by electrons of the conduction band of TiO_2 where additional electrons are injected, due to the presence of the sensitizer:

$$\begin{aligned}
TiO_2[Cu(II)P] + h\nu &\rightarrow TiO_2[Cu(II)P]^* \\
TiO_2[Cu(II)P]^* &\rightarrow TiO_2[Cu(I)P]^+ + e^-{}_{CB} \\
TiO_2[Cu(II)P] + e^-{}_{CB} &\rightarrow TiO_2[Cu(I)P] \\
TiO_2[Cu(I)P] + {}^3O_2 &\rightarrow TiO_2[Cu(II)P] + O_2^{\bullet -} \\
TiO_2[Cu(I)P] + {}^1O_2({}^1\Delta) &\rightarrow TiO_2[Cu(II)P] + O_2^{\bullet -} \\
TiO_2[Cu(I)P] + H_2O_2 &\rightarrow TiO_2[Cu(II)P] + {}^{\bullet}OH + OH^-
\end{aligned}$$

Systematic investigations demonstrated the existence of an optimum photoreactivity depending on the amount of impregnated porphyrin. Several reports have shown that the copper porphyrins-TiO_2 photocatalysts are (photo)stable under irradiation conditions and during photocatalytic experiments and retain those catalytic activities even after several cycles.

Similar results were shown for a series of TiO_2-CuPc's systems where different Cu(II)Pc (TiO_2–CuPc) were examined for their photocatalytic degradation of 4-NP[105]. Here, the presence of modified CuPc was beneficial for the photoactivation of TiO_2 (anatase), while only in few cases a slightly enhanced photoactivity for TiO_2 (rutile) was observed.

Recent studies of the photocatalytic activity of polycrystalline TiO_2 samples impregnated with rare-earth-metal diphthalocyanines in the 4-NP photocatalytic degradation show better photoactivity for the complexes of lanthanide metals, such as Nd, Sm and Ho[106]. Improved photocatalytic activity was observed in the decomposition rates of 4-NP for TiO_2-lanthanide diphthalocyanines over those impregnated with Cu(II)P's.

Other examples of photocatalysis illustrate the utilization of hybrid sol-gel TiO_2/organosilica films doped with CuPc as photocatalysts for the oxidative degradation of organics under visible light irradiation[107]. This process has been successfully demonstrated for the aliphatic 2-propanol using copper and iron phthalocyanines, methylsilica and amorphous titania.

Plastic used widely all over the world and its waste as the so-called 'white pollution' has been recognized as one of the central environmental problems. Heterogeneous photocatalytic oxidation can occur under moderate conditions, such as room temperature, atmospheric pressure and molecular oxygen as the only oxidant. Recently, studies of the solid-phase photodegradation of TiO_2-embedded plastic upon irradiating the composite film for 300 h under air have shown reduction of its average molecular weight by two-thirds and weight by 27%[108]. Comparative studies have been carried out on polystyrene (PS) incorporated in (TiO_2/CuPc) and TiO_2 alone. Faster photocatalytic degradation of polystyrene was observed in PS-TiO_2/CuPc than in PS-TiO_2. During photocatalytic degradation, activation of polystyrene by the reactive oxygen species attacking neighbouring polymer chains leads to chain cleavage and production of new reactive radicals. The higher charge-separation efficiency of TiO_2/CuPc photocatalyst results in more effective generation of reactive oxygen species both on surface and inside thin film, polystyrene undergoes faster and more complete mineralization over TiO_2/CuPc than over TiO_2 photocatalyst[109].

Another serious environmental problem has been caused by azo dyes from textile industries. Here, methyl orange is often selected as a reliable model to investigate decolourization and decomposition of azo dyes using photocatalysts. It has been shown that photostable copper 2,9,16,23-tetracarboxyl phthalocyanine (CuTcPc) doped with amorphous TiO_2 (am-TiO_2) hybrid photocatalyst exhibits excellent photocatalytic activity towards methyl orange under visible irradiation ($\lambda > 550$ nm)[110] (Figure 3). In contrast, no photobleaching of methyl orange was observed with bare am-TiO_2 at $\lambda > 550$ nm. Besides the active oxygen species, $TcPc^{+\bullet}$ radical cation produced as well also reacts with methyl orange and induces the photodegradation of methyl orange. Since no valence band hole

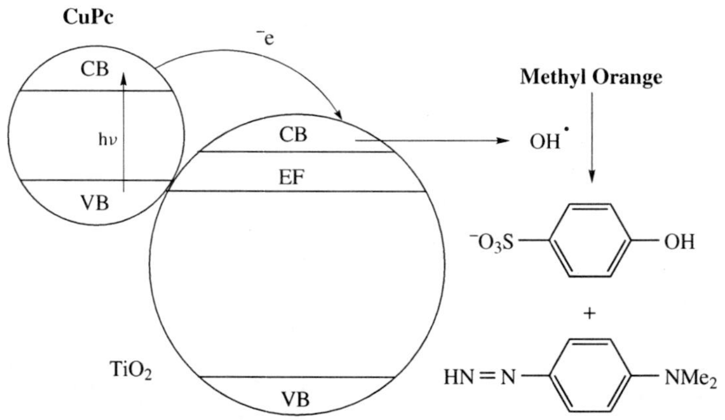

FIGURE 3. Visible-light-induced reductive degradation of methyl orange on CuPc/TiO_2

is produced in am-TiO_2, the interior charge recombination is avoided in dye sensitization photocatalysis.

Similar results were reported for hybrid photocatalysts *in situ* hydrothermally synthesized from TiO_2 and water-soluble copper phthalocyanine tetrasulfonate[111]. Its efficiency with TiO_2 was tested on methyl orange. However, methyl orange was not completely degraded by this catalyst under visible light.

C. Other Hybrid CuP and CuPc Photocatalysts

Several studies have been carried out to investigate the influence of different semiconductor types and phthalocyanine complexes on the photoactivity. For instance, copper phthalocyanine complexes supported on Al_2O_3, TiO_2 or WO_3 were prepared and the photooxidation of Na_2S and $Na_2S_2O_3$ studied. It was found that the photocatalytic activity of the phthalocyanine complex, supported on TiO_2 or WO_3, was much higher compared to dielectric Al_2O_3. This high photocatalytic activity was explained by an electron transfer from the conduction band of the excited Pc to the conduction band of a semiconductor. The increase in the quantum yield of the redox process and the higher degree of oxidation of the substrate resulted in the additional formation of superoxide radicals on the TiO_2 or WO_3 conduction band[112].

Zn/Al hydrotalcite-like compounds (general formula for these materials is $\{[M^{II}{}_{1-x}M^{III}{}_x(OH)_2]^{x+}\bullet(A_{x/n})^{n-}\bullet mH_2O\}$, where M^{II} is a divalent, M^{III} is a trivalent cation and A^{n-} is the anion) with various Zn/Al molar ratios were prepared through co-precipitation and used as supports for the immobilization of copper tetrasulfonated phthalocyanine. The photocatalytic bleaching of methylene blue was studied under different conditions upon solar radiation. Interestingly, the Zn/Al molar ratio has almost no effect, but an increase in pH enhances the photodecolourization of methylene blue. A maximum of 75% dye decolourization was achieved in 4 h solar irradiation in contrast to 33% in dark control[113].

Photocatalysts prepared *in situ* by CuPc immobilized on Al and encapsulated inside zeolite-X show a remarkable affinity towards CN^-. Photo-oxidation is based on transferring electrons from the zeolite to cyanide ions in solution. It was found that CuPc affects zeolite cages through distortions, and accommodation inside them results in strong interaction with the complex and can enhance activity of the whole system[114].

Mesoporous catalysts were prepared through reaction of 3-aminopropyltrimethoxysilane and (hexadecafluorophthalocyaninato)copper(II), followed by co-condensation of tetraethylorthosilicate around a micelle formed by *n*-dodecylamine. The surfactant was removed from the pores via continuous extraction with ethanol, yielding the Si–$CuF_{16}Pc$ catalyst. SEM images confirmed catalyst formation as nanoaggregates with a diameter of 100 nm. This novel material shows an excellent photocatalytic activity, degrading almost 90% of 2,4-dichlorophenoxyacetic acid in up to 30 min, while only approximately 40% of photodegradation was obtained in its absence. This catalyst demonstrates efficacy for photodegradation of organochloride compounds, industrial dyes and pesticides with unproblematic recuperation from the reaction medium[115].

CuP's are not very often employed as photocatalysts for organic molecule degradation compared with CuPc; an example has been given by the photocatalytic degradation of 2,4,6-trinitrotoluene by copper porphyrins under light irradiation[116].

VI. PHOTOBIOCHEMISTRY

A. Photodynamic Therapy and Singlet Oxygen Production

The interaction of tetrapyrrole derivatives with oxygen under the influence of light, and their photostability, has been a matter of great interest due to its medicinal relevance. Photodynamic therapy (PDT) presents the only one clearly established medicinal application of tetrapyrroles and relies on the selective accumulation of a tetrapyrrole photosensitizer in target tissue where it can be activated with light to produce toxic singlet oxygen resulting in, e.g., tumour necrosis. Several porphyrin-based compounds have been approved for medicinal applications and others are in Phase-2 trials[117–120].

Some studies have been carried out to investigate copper tetrapyrroles with regards to singlet oxygen production for use in PDT. For example, an iminium salt of octaethylbenzochlorin copper complex **21** was tested for its tumouricidal effects on the AY-27 N-[4-(5-nitro-2-furyl)-2-thiazolyl] formamide tumour line and was found to be an effective photosensitizer *in vivo* in combination with non-coherent light sources.

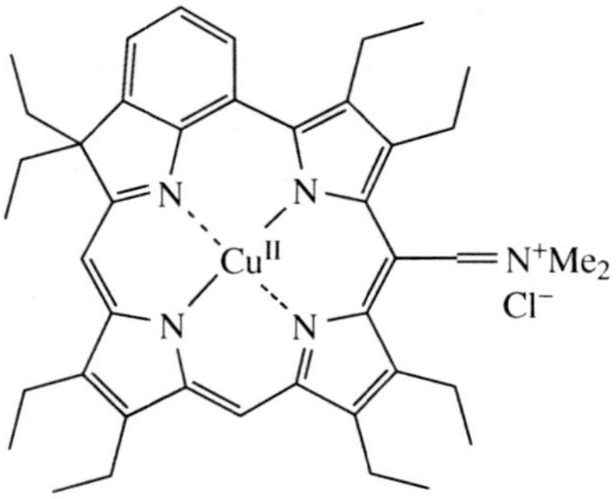

(**21**)

Skin photosensitization was found to be minimal when drug-injected mice were illuminated in a solar simulator[121]. The photodynamic ability of this complex required the presence of molecular oxygen in order to generate reactive species upon light activation[122, 123].

Sometimes, the efficacy of one-photon PDT is limited by hypoxia, which can prevent the production of the cytotoxic singlet oxygen species, leading to tumour resistance for PDT. To solve this problem, two-photon excitation of the photosensitizer can be employed. Excitation of the CuPc triplet state leads to an upper excited triplet state with distinct photochemical properties, which could inflict biological damage independently from the presence of molecular oxygen[124]. The potential of a two-photon excitation process was investigated on Jurkat cells incubated with copper phthalocyanine tetrasulfonate[125].

Other studies have shown that copper-based tetrapyrroles can also exhibit photonucleolytic activity and may be used for the DNA cleavage. Many copper complexes have been shown to exhibit nuclease activity in the presence of external co-oxidants[126–131]. Very few reports are available where copper(II) complexes promote DNA cleavage on their own[132, 133]. For example, hexaporphyrin dyads were synthesized from *n*-butyl stannonic acid and 5-(4-carboxyphenyl)-10,15,20-tritolylporphyrin ($H_2TTP–CO_2H$) followed by copper insertion as artificial nucleases[134]. The Cu-hexaporphyrin **22** showed high nuclease activity towards DNA supercoil.

Interestingly, DNA cleavage did not occur in the presence of the free-base hexaporphyrin alone and a monomeric CuP complex also failed to cleave it. Considerable inhibition of DNA cleavage in the presence of singlet oxygen quencher NaN_3 suggests the involvement of reactive oxygen species for cleavage.

Another example of potentially useful biomolecules is presented by the porphyrin **23**[135] bearing the enediyne structural fragments. Enediynes[136, 137] themselves are known to undergo rearrangements to form diradicals upon photoactivation leading to H-abstraction from DNA and cell death[138].

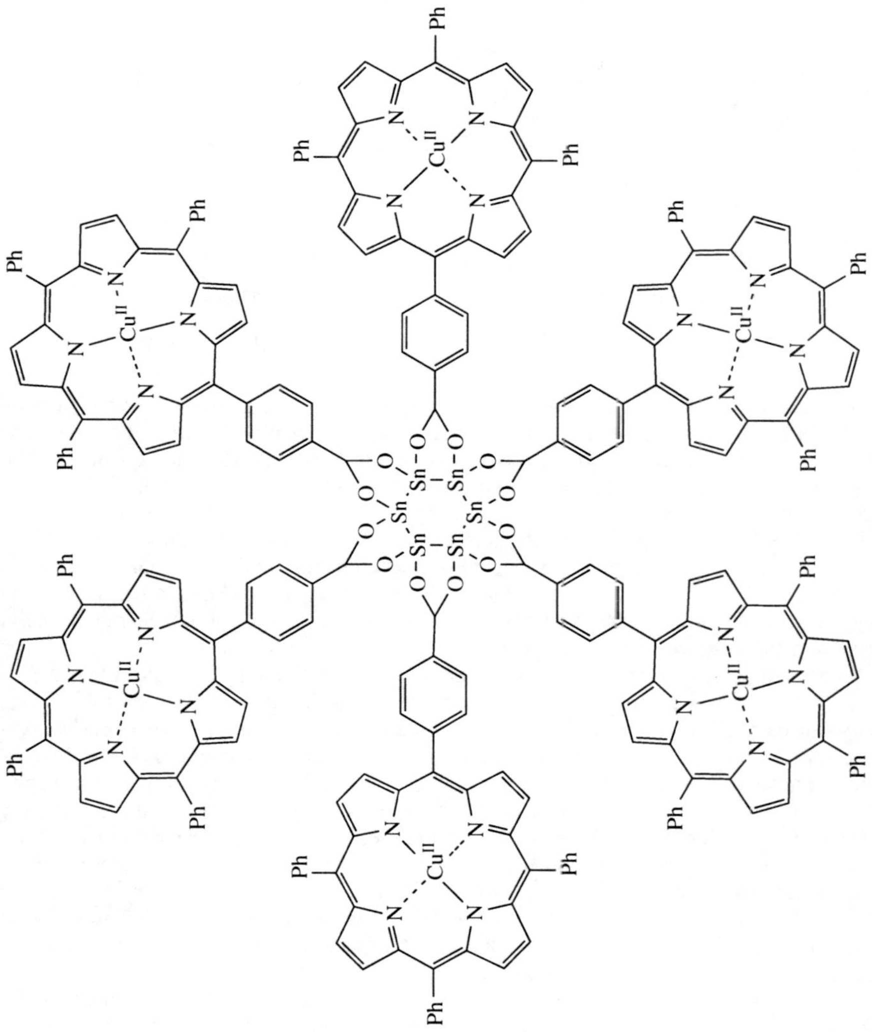

(**22**)

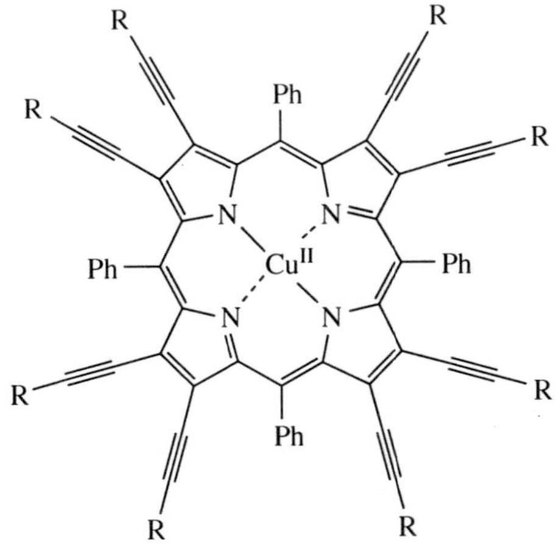

R = Ph, $SiMe_3$

(**23**)

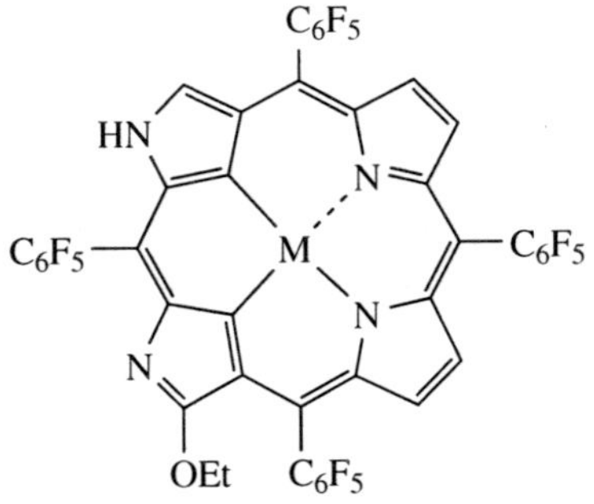

(**24**) M = 2H

(**25**) M = Cu(III)

It is interesting to note that 'doubly *N*-confused' copper porphyrins have also been synthesized and investigated in terms of their singlet oxygen production. The free-base 'doubly *N*-confused' porphyrin complex **24** exhibits a low fluorescence quantum yield of 9.1×10^{-3}, which is completely suppressed in the Cu(III)P complex **25**. However, a high singlet oxygen quantum yield ($\phi\Delta = 0.66$) was observed for the Cu(III).

These results were explained in terms of enhanced inter-system crossing rates, induced by a heavy atom effect, rather than a redox quenching involving the metal ions in an unusually high oxidation state. The 'confused porphyrins' are photochemically stable in oxygen-free solution but are bleached more or less rapidly by photogenerated singlet oxygen in the absence of other quenchers[139].

Phthalocyanines can act as photosensitizers for the production of singlet oxygen in a similar manner to the porphyrins. Besides their catalytic properties, as discussed above, TiO_2 can be used to kill bacteria under UV irradiation. For instance, *Lactobacillus acidophilus, Saccharomyces cerevisiae* and *Escherichia coli* can be completely sterilized when incubated with platinum-loaded TiO_2. TiO_2 photocatalysis is known to generate various active oxygen species, such as hydroxyl radicals and peroxide, and superoxide radical anions under UV irradiation. These properties can be used to develop disinfection materials and self-disinfecting thin films particularly attractive in places such as hospitals. So far, dye-sensitized TiO_2 films prepared with CuPc as sensitizers have shown bactericidal activities. It was found that the films can kill *E. coli* DH5α bacteria under visible-light irradiation ($\lambda > 420$ nm) and that the bactericidal activity is related to the adsorption amount of the dye on film[140].

Dyes can also be attached to polymers soluble in water or/and in polar solvents, or to the insoluble polymer supports for recovery from the reaction medium for further re-use. CuPc can be easily functionalized for preparation of polymer hybrid thin films[141]. For example, CuPc was linked to an amino group of *Amberlite IRA-93* and is strongly fluorescent in the solid state and in DMF suspension. Singlet oxygen quantum yields were determined in suspension by monitoring the photo-oxidation of diphenylisobenzofuran. Fluorescence and singlet oxygen production are environmentally dependent, e.g., on dye aggregation[142].

The degree of aggregation plays an important role in the photochemical mechanisms of photo-oxidation. Singlet oxygen production in these systems is expected to be different for monomers[143].

CuP and CuPc have found applications in different areas of current research targeting highly interesting topics of organic and medicinal chemistry, material and nanoscience. They have been shown to be effective photocatalysts for water-waste and organic-pollutant degradation in green chemistry. On the other hand, they can also be used as efficient stabilizing agents for polymers and dyes. The search for alternative energy sources is one of the important aims of the present century and has attracted much attention recently. We have also highlighted their future potential as molecular devices in emerging areas of nanoscience. Copper tetrapyrroles, especially copper phthalocyanines, have found utilization as prospective candidates in hydrogen production as (photo)catalysts and in solar energy as photovoltaic cells and junctions.

VII. ACKNOWLEDGEMENTS

Writing of this article was made possible by generous funding from Science Foundation Ireland (Research Professorship—SFI 04/RP1/B482) and the Health Research Board (translational research award 2007 TRA/2007/11).

VIII. REFERENCES

1. M. O. Senge, *J. Photochem. Photobiol. B*, **16**, 3 (1992).
2. K. M. Kadish, K. M. Smith and R. Guilard, *The Porphyrin Handbook*, Vol. 1–10, Academic Press, San Diego, 2000.
3. K. M. Kadish, K. M. Smith and R. Guilard, *The Porphyrin Handbook*, Vol. 10–20, Academic Press, San Diego, 2003.
4. T. Nyokong and H. Isago, *J. Porphyrins Phthalocyanines*, **8**, 1083 (2004).
5. M. Gouterman, in *The Porphyrins* (Ed. D. Dolphin), Vol. 3, Academic Press, New York, 1978, p. 1.
6. J. Mack and M. J. Stillman, in *The Porphyrin Handbook* (Eds K. M. Kadish, K. M. Smith and R. Guilard), Vol. 16, Academic Press, San Diego, 2003, p. 43.
7. D. Gust and T. A. Moore, in *The Porphyrin Handbook* (Eds K. M. Kadish, K. M. Smith and R. Guilard), Vol. 8, Academic Press, San Diego, 2000, p. 153.
8. M. O. Senge and N. N. Sergeeva, in *The Chemistry of Organozinc Compounds* (Eds Z. Rappoport and I. Marek), Chap. 9, John Wiley & Sons, Ltd, Chichester, 2006, p. 395.
9. N. N. Sergeeva and M. O. Senge, in *The Chemistry of Organomagnesium Compounds* (Eds Z. Rappoport and I. Marek), Chap. 5, John Wiley & Sons, Ltd, Chichester, 2008, p. 189.
10. N. Armaroli, G. Accorsi, F. Cardinali and A. Listorti, *Top. Curr. Chem.*, **280**, 69 (2007).
11. J. Sykora, *Coord. Chem. Rev.*, **159**, 95 (1997).
12. G. Ferraudi, *Adv. Chem. Ser.*, **238**, 83 (1993).
13. D. R. McMillin, J. R. Kirchhoff and K. V. Goodwin, *Coord. Chem. Rev.*, **64**, 83 (1985).
14. D. Kim, D. Holten and M. Gouterman, *J. Am. Chem. Soc.*, **106**, 2793 (1984).
15. B. E. Smith and M. Gouterman, *Chem. Phys. Lett.*, **2**, 517 (1968).
16. M. Gouterman, R. A. Mathies, B. E. Smith and W. S. Caughey, *J. Chem. Phys.*, **52**, 3795 (1970).
17. D. Eastwood and M. Gouteman, *J. Mol. Spectrosc.*, **30**, 437 (1969).
18. J. Bohandy and B. F. Kim, *J. Chem. Phys.*, **78**, 4331 (1983).
19. N. Van Dijk, M. Noort and J. H. van der Waals, *Mol. Phys.*, **44**, 913 (1981).
20. S. G. Kruglik, P. A. Apanasevich, V. S. Chirvony, V. V. Kvach and V. A. Orlovich, *J. Phys. Chem.*, **99**, 2978 (1995).
21. E. F. Hilinski, K. D. Straub and P. M. Rentzepis, *Chem. Phys. Lett.*, **111**, 333 (1984).
22. N. Serpone, H. Ledon and T. Netzel, *Inorg. Chem.*, **23**, 454 (1984).

23. K. Heinze and A. Reinhart, *Dalton Trans.*, 469 (2008).
24. M. W. Renner, K. M. Barkigia, Y. Zhang, C. J. Medforth, K. M. Smith and J. Fajer, *J. Am. Chem. Soc.*, **116**, 8582 (1994).
25. A. C. Jones, M. J. Dale, G. A. Keenan and P. R. R. Langridge-Smith, *Chem. Phys. Lett.*, **219**, 174 (1994).
26. S. C. Jeoung, D. Kim and D. W. Cho, *J. Raman Spectrosc.*, **31**, 319 (2000).
27. A. V. Nikolaitchik, O. Korth and M. A. J. Rodgers, *J. Phys. Chem. A*, **103**, 7587 (1999).
28. D. R. Prasad and G. Ferraudi, *Inorg. Chem.*, **21**, 2967 (1982).
29. A. Antipas and M. Gouterman, *J. Am. Chem. Soc.*, **105**, 4896 (1983).
30. C. D. Tait, D. Holten and M. Gouterman, *Chem. Phys. Lett.*, **100**, 268 (1983).
31. Y. Iseki and S. Inoue, *J. Chem. Soc., Chem. Commun.*, 2577 (1994).
32. J. S. Connolly, in *Photochemical Conversion and Storage of Solar Energy—1982* (Ed. J. Rabani), Part A, The Weizmann Science Press of Israel, Jerusalem, 1982, p. 175.
33. G. B. Maiya and V. Krishnan, *J. Phys. Chem.*, **89**, 5225 (1985).
34. S. Yamada, T. Sata, K. Kano and T. Ogawa, *Photochem. Photobiol.*, **37**, 257 (1983).
35. K. Li, D. I. Schuster, D. M. Guldi, M. A. Herranz and L. Echegoyen, *J. Am. Chem. Soc.*, **126**, 3388 (2004).
36. J. F. Nierengarten, J. F. Eckert, D. Felder, J. F. Nicoud, N. Armaroli, G. Marconi, V. Vicinelli, C. Boudon, J. P. Gisselbrecht, M. Gross, G. Hadziioannou, V. Krasnikov, L. Ouali, L. Echegoyen and S. G. Liu, *Carbon*, **38**, 1587 (2000).
37. M. E. El-Khouly, Y. Araki, M. Fujitsuka and O. Ito, *Phys. Chem. Chem. Phys.*, **4**, 3322 (2002).
38. D. M. Guldi, I. Zilbermann, A. Gouloumis, P. Vázquez and T. Torres, *J. Phys. Chem. B*, **108**, 18485 (2004).
39. M. Asano-Someda, S. I. Sato, K. Aoyagi and T. Kitagawa, *J. Phys. Chem.*, **99**, 13800 (1995).
40. M. Asano-Someda and Y. Kaizu, *J. Photochem. Photobiol. A*, **87**, 23 (1995).
41. M. Asano-Someda, T. Ichino and Y. Kaizu, *Coord. Chem. Rev.*, **132**, 243 (1994).
42. T. H. Tran-Thi, J. F. Lipskier, D. Houde, C. Pepin, E. Keszei and J. P. Jay-Gerin, *J. Chem. Soc., Faraday Trans.*, **88**, 2129 (1992).
43. C. J. Chang, E. A. Baker, B. J. Pistorio, Y. Deng, Z. H. Loh, S. E. Miller, S. D. Carpenter and G. Nocera, *Inorg. Chem.*, **41**, 3102 (2002).
44. A. Bluestein and J. M. Sugihara, *J. Inorg. Nucl. Chem.*, **35**, 1050 (1973).
45. L. Jiao, B. H. Courtney, F. R. Fronczek and K. M. Smith, *Tetrahedron Lett.*, **47**, 501 (2006).
46. H. Yamanouchi and T. Saji, *Chem. Lett.*, 531 (1996).
47. K. Hoshino, K. Kurasako, T. Inayama and H. Kokado, *J. Electroanal. Chem.*, **406**, 175 (1996).
48. J. P. Bearinger, G. Stone, A. T. Christian, L. Dugan, A. L. Hiddessen, K. J. J. Wu, L. Wu, J. Hamilton, C. Stockton and J. A. Hubbell, *Langmuir*, **24**, 5179 (2008).
49. S. Yamada, K. Kuwata, H. Yonemura and T. Matsuo, *J. Photochem. Photobiol. A*, **87**, 115 (1995).
50. S. Liu, M. Fujihira and T. Saji, *Chem. Lett.*, 1855 (1994).
51. M. S. Xu, J. B. Xu, M. Wang and D. L. Que, *J. Appl. Phys.*, **91**, 748 (2002).
52. K. Naito and A. Miura, *J. Am. Chem. Soc.*, **115**, 5185 (1993).
53. L. K. Chau, E. J. Osburn, N. R. Armstrong, D. F. O'Brien and B. A. Parkinson, *Langmuir*, **10**, 351 (1994).
54. A. S. Drager, R. A. P. Zangmeister, N. R. Armstrong and D. F. O'Brien, *J. Am. Chem. Soc.*, **123**, 3595 (2001).
55. P. Smolenyak, R. Peterson, K. Nebesny, M. Törker, D. F. O'Brien and N. R. Armstrong, *J. Am. Chem. Soc.*, **121**, 8628 (1999).
56. R. A. P. Zangmeister, P. E. Smolenyak, A. S. Drager, D. F. O'Brien and N. R. Armstrong, *Langmuir*,**17**, 7071 (2001).
57. C. L. Donley, W. Xia, B. A. Minch, R. A. P. Zangmeister, A. S. Drager, K. Nebesny, D. F. O'Brien and N. R. Armstrong, *Langmuir*,**19**, 6512 (2003).
58. S. Meshitsuka and K. Tamaru, *J. Chem. Soc., Faraday Trans. 1*, 760 (1977).
59. M. Ichikawa, H. Fukumura and H. Masuhara, *J. Phys. Chem.*, **98**, 12211 (1994).
60. T. Abe, S. Miyakushi, K. Nagai and T. Norimatsu, *Phys. Chem. Chem. Phys.*, **10**, 1562 (2008).
61. S. Radhakrishnan and S. D. Deshpande, *Mater. Lett.*, **48**, 144 (2001).
62. A. Nitzan and M. A. Ratner, *Science*, **300**, 1384 (2003).
63. M. C. Hersan, N. P. Guighges and J. W. Lyding, *Nanotechnology*, **11**, 70 (2000).

64. H. Abramczyk, B. Brożek-Płuska, K. Kurczewski, M. Kurczewska, I. Szymczyk, P. Krzyczmonik, T. Bslashlaszczyk, H. Scholl and W. Czajkowski, *J. Phys. Chem. A*, **110**, 8627 (2006).
65. R. Kaito, N. Miyamoto, K. Kuroda and M. Ogawa, *J. Mater. Chem.*, **12**, 3463 (2002).
66. H. Yanagi, K. Tsukatani, H. Yamaguchi, M. Ashida, D. Schlettwein and D. Wöhrle, *J. Electrochem. Soc.*, **140**, 1942 (1993).
67. Y. Osada and A. Mizumoto, *J. Appl. Phys.*, **59**, 1776 (1986).
68. K. Yamashita, Y. Harima and T. Matsubayashi, *J. Phys. Chem.*, **93**, 5311 (1989).
69. I. A. Levitsky, W. B. Euler, N. Tokranova, B. Xu and J. Castracane, *Appl. Phys. Lett.*, **85**, 6245 (2004).
70. K. O. Sylvester-Hvid, *J. Phys. Chem. B*,**110**, 2618 (2006).
71. H. Ding, X. Zhang, M. K. Ram and C. Nicolini, *J. Colloid Interface Sci.*, **290**, 166 (2005).
72. C. Arbour, D. K. Sharma and C. H. Langford, *J. Phys. Chem.*, **94**, 331 (1990).
73. M. Gratzel and M. Halmann, *Solar Energy Mater.*, **20**, 177 (1990).
74. B. Zhang, J. Mu and D. Wang, *J. Dispers. Sci. Tech.*, **26**, 521 (2005).
75. Y. Wang, H. Z. Chen, H. Y. Li and M. Wang, *Mater. Sci. Eng. B*, **117**, 296 (2005).
76. A. Fujishima and K. Honda, *Nature*, **238**, 37 (1972).
77. T. Abe and K. Nagai, *Organic Electronics*, **8**, 262 (2007).
78. H. Yarnaguchi, R. Fujiwara and K. Kusuda, *Makromol. Chem., Rapid Commun.*, **7**, 225 (1986).
79. A. Harrimana, M. C. Reichoux, *J. Chem. Soc., Faraday Trans. 2*, **76**, 1618 (1980).
80. M. Ichikawa, H. Fukumura and H. Masuhara, *J. Phys. Chem.*, **99**, 12072 (1995).
81. K. Himeno, Y. Okada and Z. Morita, *Dyes & Pigments*, **45**, 109 (2000).
82. Y. Okada, *Dyes & Pigments*, **19**, 203 (1992).
83. Y. Okada, M. Hirose, T. Kato, H. Motomura and Z. Morita, *Dyes & Pigments*, **14**, 113 (1990).
84. Y. Okada, M. Hirose, T. Kato, H. Motomura and Z. Morita, *Dyes & Pigments*, **14**, 265 (1990).
85. T. Hihara, Y. Okada and Z. Morita, *Dyes & Pigments*, **50**, 185 (2001).
86. Y. Kaneko, Y. Nishimura, T. Arai, H. Sakuragi, K. Tokumaru and D. Matsunaga, *J. Photochem. Photobiol. A*, **89**, 37 (1995).
87. Y. Kaneko, T. Arai, H. Sakuragi, K. Tokumaru and C. Pac, *J. Photochem. Photobiol. A*, **97**, 155 (1996).
88. E. I. Sánchez, M. Calderón and M. I. Gutiérrez, *Polym. Bull.*, **51**, 271 (2004).
89. N. S. Allen, *Polym. Degrad. Stab.*, **44**, 357 (1994).
90. C. H. Bamford and M. J. S. Dewar, *Nature*, **163**, 214 (1949).
91. C. Saron, F. Zulli, M. Giordano and M. I. Felisberti, *Polym. Degrad. Stab.*, **91**, 3301 (2006).
92. N. S. Allen and A. Parkinson, *Polym. Degrad. Stab.*, **5**, 189 (1983).
93. K. Rosche, C. Decker, G. Israel and J. P. Fouassier, *Eur. Polym. J.*, **33**, 849 (1997).
94. P. F. Gordon and P. Gregory, *Organic Chemistry in Colour*, Springer-Verlag, New York, 1983.
95. I. Arslan and I. A. Balciodlu, *Dyes & Pigments*, **43**, 95 (1999).
96. M. E. Osugi, G. A. Umbuzeiro, F. J. V. De Castro and M. B. Zanoni, *J. Hazard. Mater. B*, **137**, 871 (2006).
97. M. E. Osugi, G. A. Umbuzeiro, M. A. Anderson and M. V. B. Zanoni, *Electrochim. Acta*, **50**, 5261 (2005).
98. A. Sun, G. Zhang and Y. Xu, *Mater. Lett.*, **59**, 4016 (2005).
99. A. F. Caliman, C. Cojocaru, A. Antoniadis and I. Poulios, *J. Hazard. Mater. B*, **144**, 265 (2007).
100. H. Y. Shu and M. C. Chang, *J. Hazard. Mater. B*, **125**, 96 (2005).
101. G. Mele, R. Del Sole, G. Vasapollo, G. Marci, E. García-López, L. Palmisano, J. M. Coronado, M. D. Hernandez-Alonso, C. Malitesta and M. R. Guascito, *J. Phys. Chem. B*, **109**, 12347 (2005).
102. G. Mele, R. Del Sole, G. Vasapollo, E. García-López, L. Palmisano and M. Schiavello, *J. Catalysis*, **217**, 334 (2003).
103. C. Wang, J. Li, G. Mele, G.-M. Yang, F. X. Zhang, L. Palmisano and G. Vasapollo, *Appl. Catalysis B*, **76**, 218 (2007).
104. G. Mele, R. Del Sole, G. Vasapollo, E. García-López, L. Palmisano, L. Jun, R. Słota and G. Dyrda, *Res. Chem. Intermed.*, **33**, 433 (2007).
105. G. Mele, G. Ciccarella, G. Vasapollo, E. García-López, L. Palmisano and M. Schiavello, *Appl. Catal. B*, **38**, 309 (2002).

106. G. Mele, E. García-López, L. Palmisano, G. Dyrda and R. Słota, *J. Phys. Chem. C*, **111**, 6581 (2007).
107. G. Palmisano, M. Concepción Gutierréz, M. L. Ferrer, M. D. Gil-Luna, V. Augugliaro, S. Yurdakal and M. Pagliaro, *J. Phys. Chem. C*, **112**, 2667 (2008).
108. S. Cho and W. Choi, *J. Photochem. Photobiol. A*, **143**, 221 (2001).
109. J. Shang, M. Chai and Y. Zhu, *Environ. Sci. Technol.*, **37**, 4494 (2003).
110. F. Chen, Z. Deng, X. Li, J. Zhang and J. Zhao, *Chem. Phys. Lett.*, **415**, 85 (2005).
111. W. Zhiyu, C. Haifeng, T. Peisong, M. Weiping, Z. Fuan, Q, Guodong and F. Xianping, *Colloids & Surfaces A: Physicochem. Eng. Aspects*, **289**, 207 (2006).
112. V. Iliev, D. Tomova, L. Bilyarska, L. Prahov and L. Petrov, *J. Photochem. Photobiol. A*, **159**, 281 (2003).
113. K. M. Parida, N. Baliarsingh, B. S. Patra and J. Das, *J. Mol. Catalysis A: Chem.*, **267**, 202 (2007).
114. R. M. Mohamed and M. M. Mohamed, *Appl. Catalysis A: Gen.*, **340**, 16 (2008).
115. E. DeOliveira, C. R. Neri, A. O. Ribeiro, V. S. Garcia, L. L. Costa, A. O. Moura, A. G. S. Prado, O. A. Serra and Y. Iamamoto, *J. Colloid Interface Sci.*, **323**, 98 (2008).
116. H. J. Harmon, *Chemosphere*, **63**, 1094 (2006).
117. R. Bonnett, *Chemical Aspects of Photodynamic Therapy*, Gordon and Breach Sci. Publ., Amsterdam, 2000.
118. A. Wiehe, Y. M. Shaker, J. C. Brandt, S. Mebs and M. O. Senge, *Tetrahedron*, **61**, 5535 (2005).
119. T. J. Dougherty, C. J. Gomer, B. W. Henderson, G. Jori, D. Kessel, M. Korbelik, J. Moan and Q. Peng, *J. Natl. Cancer Inst.*, **90**, 889 (1998).
120. D. Kessel and T. J. Dougherty, *Reviews in Contemporary Pharmacotherapy*, **10**, 19 (1999).
121. S. H. Selman, J. A. Hampton, A. R. Morgan, R. W. Keck, A. D. Balkany and D. Skalkos, *Photochem. Photobiol.*, **57**, 681 (1993).
122. J. A. Hampton, D. Skalkos, P. M. Taylor and S. H. Selman, *Photochem. Photobiol.*, **58**, 100 (1993).
123. J. A. S. Cavaleiro, H. Görner, P. S. S. Lacerda, J. G. MacDonald, G. Mark, M. G. P. M. S. Neves, R. S. Nohr, H. P. Schuchmann, C. von Sonntag and A. C. Tomé, *J. Photochem. Photobiol. A*, **144**, 131 (2001).
124. M. Fournier, C. Pépin, D. Houde, R. Ouellet and J. E. van Lier, *Photochem. Photobiol. Sci.*, **3**, 120 (2004).
125. Y. Mir, J. E. van Lier, B. Paquette and D. Houde, *Photochem. Photobiol.*, **84**, 1182 (2008).
126. S. Borah, M. S. Melvin, N. Lindquist and R. A. Manderville, *J. Am. Chem. Soc.*, **120**, 4557 (1998).
127. E. L. Hegg and J. N. Burstyn, *Inorg. Chem.*, **35**, 7474 (1996).
128. O. Baudoin, M. P. T. Fichou, J. P. Vigneron and J. M. Lehn, *J. Chem. Soc., Chem. Commun.*, 2349 (1998).
129. T. Itoh, H. Hisada, T. Sumiya, M. Hosono, Y. Usui and Y. Fajii, *J. Chem. Soc., Chem. Commun.*, 677 (1997).
130. S. Dhar and A. R. Chakravarty, *Inorg. Chem.*, **42**, 2483 (2003).
131. M. S. Melvin, J. T. Tomlinson, G. R. Saluta, G. L. Kucera, N. Lindquist and R. A. Manderville, *J. Am. Chem. Soc.*, **122**, 6333 (2000).
132. P. A. N. Reddy, M. Nethaji and A. R. Chakravarthy, *Eur. J. Inorg. Chem.*, **7**, 1440 (2004).
133. S. Verma, S. G. Srivatsan and C. Madhavaiah, *Nucleic Acids Mol. Biol.*, **13**, 129 (2004).
134. V. Chandrasekhar, S. Nagendran, R. Azhakar, M. R. Kumar, A. Srinivasan, K. Ray, T. K. Chandrashekar, C. Madhavaiah, S. Verma, U. D. Priyakumar and G. N. Sastry, *J. Am. Chem. Soc.*, **127**, 2410 (2005).
135. T. Chandra, B. J. Kraft, J. C. Huffman and J. M. Zaleski, *Inorg. Chem.*, **42**, 5158 (2003).
136. J. Golik, G. Dubay, G. Groenewold, H. Kawaguchi, M. Konishi, B. Krishnan, H. Ohkuma, K. Saitoh and T. W. Doyle, *J. Am. Chem. Soc.*, **109**, 3462 (1987).
137. M. D. Lee, T. S. Dunne, C. C. Chang, G. A. Ellestad, M. M. Siegel, G. O. Morton, W. J. McGahren and D. B. Borders, *J. Am. Chem. Soc.*, **109**, 3466 (1987).
138. A. L. Smith and K. C. Nicolaou, *J. Med. Chem.*, **39**, 2103 (1996).
139. F. M. Engelmann, I. Mayer, K. Araki, H. E. Toma, M. S. Baptista, H. Maeda, A. Osuka and H. Furuta, *J. Photochem. Photobiol. A*, **163**, 403 (2004).

140. J. C. Yu, Y. Xie, H. Y. Tang, L. Zhang, H. C. Chan and J. Zhao, *J. Photochem. Photobiol. A*, **156**, 235 (2003).
141. Y. Choe, T. Kim and W. Kim, *Colloids & Surfaces B*, **38**, 155 (2004).
142. J. L. Bourdelande, M. Karzazi, L. E. Dicelio, M. I. Litter, G. M. Tura, E. San Román and V. Vinent, *J. Photochem. Photobiol. A*, **108**, 273 (1997).
143. H. Abramczyk, I. Szymczyk, G. Waliszewska and A. Lebioda, *J. Phys. Chem. A*, **108**, 264 (2004).

CHAPTER 6

Electrochemistry of organocopper compounds

JAN S. JAWORSKI

Faculty of Chemistry, University of Warsaw, 02-093 Warszawa, Poland
Fax: +48-22-822-5996; e-mail: jaworski@chem.uw.edu.pl

I. INTRODUCTION 247
II. ELECTROCHEMICAL SYNTHESIS OF ORGANOCOPPER COMPOUNDS 248
III. ELECTROCHEMICAL STABILITY OF ORGANOCOPPER COMPOUNDS 255
A. Mononuclear Compounds 255
B. Binuclear Compounds 257
1. Mixed compounds with Cu and Pt or Ti atoms 257
2. $\{Cu(\mu\text{-}L)\}_2$ compounds 259
C. Polynuclear Compounds 261
IV. CONCLUDING REMARKS 276
V. REFERENCES 276

I. INTRODUCTION

Electrochemical investigations into organocopper compounds are very scarce. In a number of chapters and reviews dealing with electrochemistry of organometallic compounds published in the last decades only a few examples of such studies are mentioned[1–4], often as a part of broader investigations including copper complexes with organic ligands. This is mainly caused by the fact that electrochemical reactions of organocopper compounds are dominated by changes of the oxidation state of the metal and hence the most important problem to be investigated was the effect of different organic moieties on the stability of reactants, in particular of the copper(I) oxidation state. On the other hand, the extreme versatility of oragnocopper reagents in organic synthesis and the great interest in them have no counterpart in the field of electrochemical reactivity. One more drawback should be mentioned here: a large number of organocopper compounds are either insoluble or unstable in solvents that are usually used in electrochemical measurements, even though they are quite stable in the crystalline state.

The chemistry of organocopper compounds
Edited by Z. Rappoport and I. Marek

First of all, Section II of this Chapter presents a review of studies on the synthesis of organocopper compounds including mainly the use of direct electrolysis based on the simultaneous cathodic reduction of an organic compound and the dissolution of a sacrificial copper anode. These methods appeared suitable for the preparation of copper(I) π-complexes with the coordination of organic cations by the metal atom through the C=C bond, in particular for the direct formation of such compounds in the form of crystals.

The discussion of electrochemical behavior of the title compounds is focused mainly on their stability at different oxidation states of copper atoms. Compounds with only one copper atom (Section III.A) and binuclear compounds with copper and platinum or titanium atoms as well as with two copper atoms in the molecule (Section III.B) have been successively reviewed. Finally, Section III.C presents electrochemical data on polynuclear compounds including copper clusters (very often with three copper atoms) bounded to organic moieties in homonuclear as well as heteronuclear compounds. The synthesis of such compounds and investigations of their structures and photophysical behavior have received a lot of attention in recent years because of their potential technological applications as precursors of nonlinear optical materials and rigid-rod molecular wires[5]. A number of papers were also devoted to the electrochemical study of electronic communication between the redox centers, like ferrocene or titanocene, at the ends of heteronuclear organometallic molecules. Such investigations (see references cited in Reference 6), in general directed toward the search for new materials of interest for molecular electronic devices, also dealt with organocopper compounds, of which some examples are mentioned in Section III.C.

Finally, the Chapter ends with a suggestion concerning further use of electrochemical methods in investigations into organocopper compounds.

It should be explained at the beginning that the use of the terms ‘reversibility’ and ‘irreversibility’ in this Chapter follows the same sense as used in the original papers cited and does not correspond exactly to the definition of electrochemical reversibility. Namely, in the cyclic voltammetry method the process is called irreversible if reversing the potential scan after a cathodic peak gives no anodic peak, and vice versa. And conversely, the process is called reversible if both cathodic and anodic peaks are observed, even if the difference between potentials of a cathodic (E_{pc}) and anodic (E_{pa}) peak, ΔE_p, is greater than the theoretical value. In such a case the formal potential of a given redox couple is approximated as the midpoint between both peaks, $E_o = \frac{1}{2}(E_{pc} + E_{pa})$.

The following common abbreviations are used in this Chapter, beside the general abbreviations of the book: CV, cyclic voltammetry; EDX, energy dispersive X-ray spectra; ESI-MS, electrospray ionization mass spectrometry; FAB-MS, positive-ion fast atom bombardment mass spectroscopy; cod, 1,5-cyclooctadiene; dppm, bis(diphenylphosphino) methane; dppf, 1,1′-bis(diphenylphosphino)ferrocene; bipy, 2,2′-bipyridyl; TBAP, tetrabutylammonium perchlorate; $TBAPF_6$, tetrabutylammonium hexafluorophosphate; $TBABF_4$, tetrabutylammonium tetrafluoroborate; GC electrode, glassy carbon electrode; F, Faraday constant; and M, mole dm^{-3}.

The quoted potentials are originally measured mainly versus an aqueous saturated calomel electrode (SCE) or versus the Ag/Ag^+ couple in acetonitrile. Sometimes they were converted to the standard hydrogen electrode (SHE) scale. However, in many reports the ferricinium ferrocene redox couple (Fc^+/Fc) was additionally used as the internal standard and potentials are often expressed (sometimes after recalculations) in this scale.

II. ELECTROCHEMICAL SYNTHESIS OF ORGANOCOPPER COMPOUNDS

Olefin copper(I) π-complexes can be easily prepared in the form of crystals using electrochemical techniques. Out of a great number of cyclic olefins which can be bonded to copper(I), a stable compound **1** with 1,5-cyclooctadiene (cod) was prepared by Manahan[7] employing the direct current electrolysis (1–2 V and 20 mA) of a solution of copper(II) perchlorate and cod in absolute methanol under nitrogen atmosphere with the use of copper electrodes. The remarkably stable crystals of **1** formed at both electrodes in cathodic (reaction 1a) and anodic (reaction 1b) processes are moderately soluble in acetone and their voltammetric properties will be discussed in Section III. $Cu(BF_4)_2$ can also be used as a starting reactant[8] instead of $Cu(ClO_4)_2$.

$$Cu^{2+} + 2cod + ClO_4^- + e \longrightarrow \underset{(\mathbf{1})}{Cu(cod)_2ClO_4}\downarrow \qquad (1a)$$

$$Cu + 2cod + ClO_4^- - e \longrightarrow Cu(cod)_2ClO_4\downarrow \qquad (1b)$$

Reactions 1a and 1b clearly show that the main idea of electrochemical synthesis is the use of copper electrodes and a solution of copper(II) compound in order to produce copper(I) species in the redox reaction: $Cu^0 + Cu^{2+} \rightarrow 2Cu^+$. Then, in the presence of a π-electron ligand in the solution, the crystals of the final copper(I) π-complex are formed. Thus, the original method of Manahan[7] was improved and converted to an alternating-current electrolysis[9] which was found to be very powerful for direct preparation of high-quality single crystals of copper(I) compounds. The growth of crystals can be conveniently controlled by changing the voltage and current density, and moreover, for unstable copper(I) π-complexes the difficult process of recrystallization is omitted. A great number of Cu(I) compounds (**2–28**) with the metal atom coordinated to C=C bonds in allyl groups of different ligands were prepared using the ac electrochemical method by Goreshnik and coworkers[10–32] and crystal structures of the products were determined. However, in some cases σ-complexes without copper–carbon bonds were also produced by this method. The synthesized organocopper compounds as well as the applied organic reactants containing allyl groups are listed in Table 1. The second reactant was $CuBr_2$ or $CuCl_2{\bullet}2H_2O$ and ethanol was mainly used as solvent. The alternating current of a frequency of 50 Hz was applied to copper electrodes and the potential difference was equal to 0.30 V. Crystals were formed on the electrodes after time ranging from four hours (for **3**), ten hours (for **2**), 48 hours (for **15**, **16**, **19** and **20**) to several days (for **13**, **23** and **24**) and even twenty days for **21**, but in order to prepare **25** after two days of electrolysis the solution was placed in a refrigerator at $-1\,^{\circ}C$ for a month. Moreover, it should be added that a change from chlorine to bromine atoms in the starting reactant can change the bonding to the copper atom and the final structure of the compound; for example, instead of copper(I) π-complex **16** the use of $CuBr_2$ and the same ligand yields $[C_4H_8N_2(C_3H_5)_4]^{2+}$ $[CuBr_3]^{2-}$, which is not an organocopper compound[20].

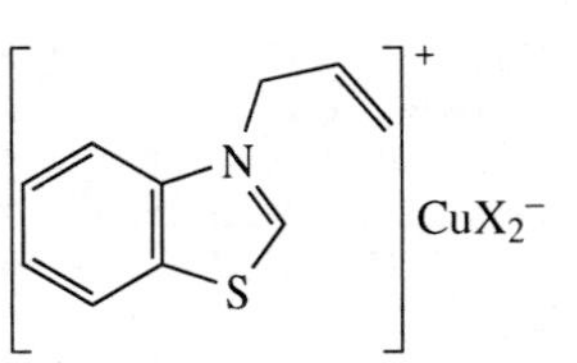

(**2**) X = Cl
(**3**) X = Cl/Br

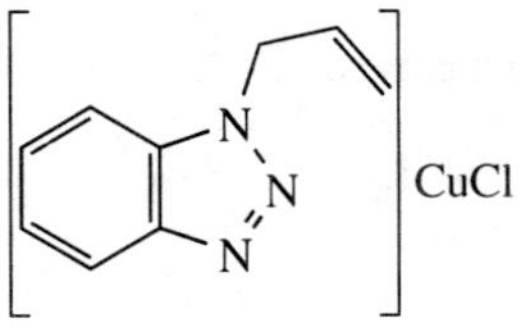

(**4**)

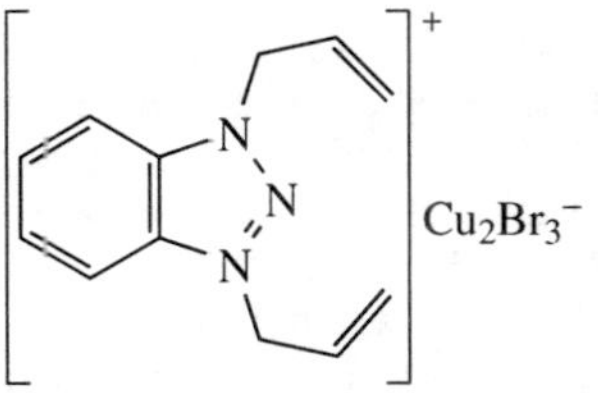

(**6**)

The determined crystal structures allowed the author to establish the bonding of copper atoms (including a coordination with C=C bonds as well as with nitrogen atoms and bridging halogen atoms) and to discuss the effectiveness of metal–olefin interactions. For

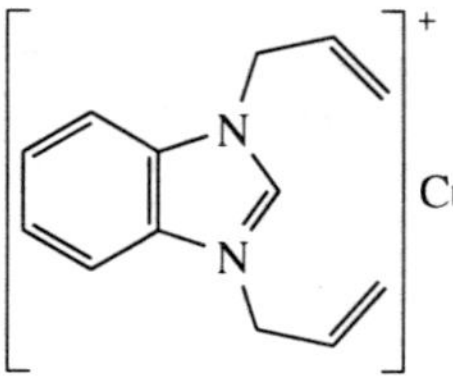

(**7**) X = Cl
(**8**) X = Cl/Br

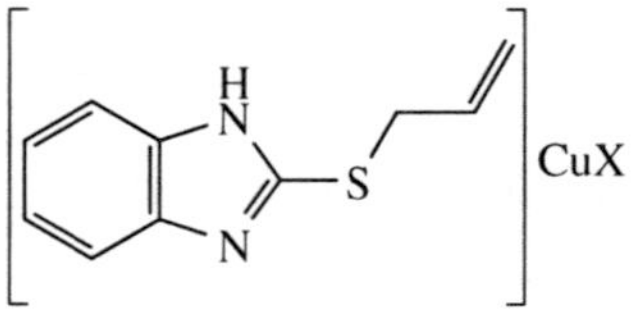

(**9**) X = Cl or Br

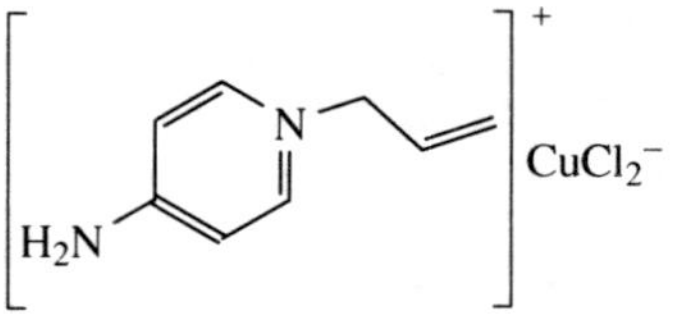

(**12**)

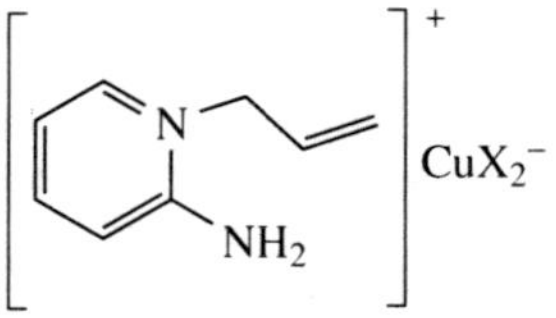

(**13**) X = Cl
(**14**) X = Cl/Br

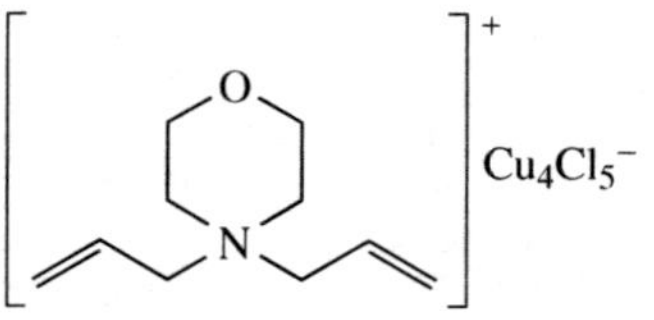

(**15**)

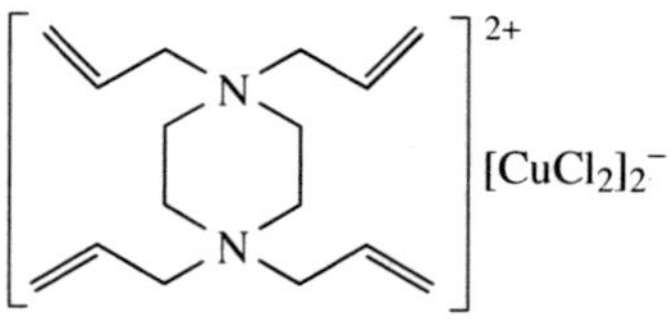

(**16**)

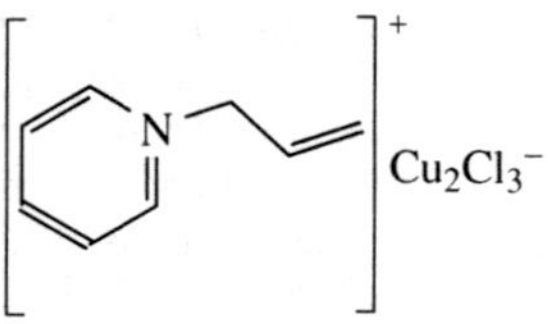

(**17**)

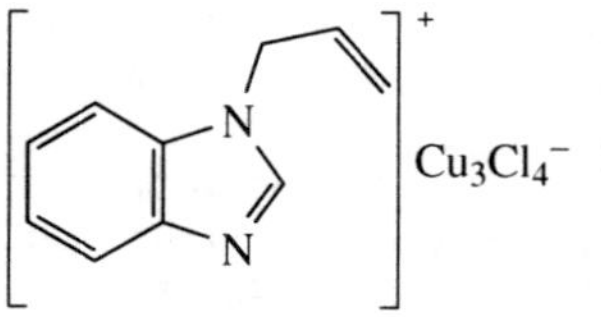

(**18**)

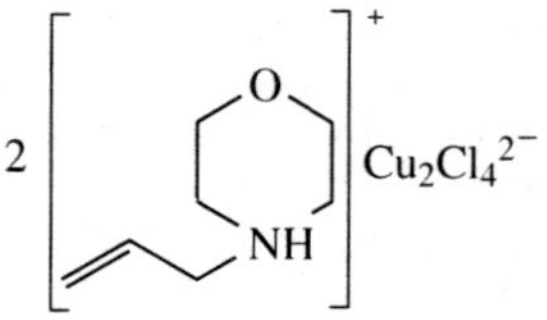

(**19**)

example, it was found[27] that two copper atoms in **21** have trigonal–pyramidal arrangements with a varying degree of tetrahedral distortion, indicating different strengths of metal C=C interactions for each Cu atom. Thus, a better view of the structure can be represented by dimers $Cu_2[C_6H_4N_3(OC_3H_5)]_2Br_4$ (**21a**). In a similar manner **2** and **13** can be represented by the formulas $[C_7H_5NS(C_3H_5)]\ [CuCl_2)]$ (**2**)[4] and $[H_2NC_5H_4N(C_3H_5)][CuCl_2]$ (**13**)[18] shown in Table 1 or by the dimeric formulas $[C_7H_5NS(C_3H_5)]_2[Cu_2Cl_4]$ (**2a**) and $[H_2NC_5H_4N(C_3H_5)]_2[Cu_2Cl_4]$ (**13a**), respectively, which not only means that there are interactions between two organic cations and the dinegative $(Cu_2Cl_4)^{2-}$ ion, but also allows one to show properly all σ- and π-bonds of copper(I), as is shown below. Dimeric fragments **10a** and **11a** with a weakly bonded C=C group to each copper atom were also established[17] for crystals of **10** and **11** and the dimeric formula of **19** is shown as **19a**. However, the details of structures and bonding are beyond the scope of this chapter and will not be discussed here.

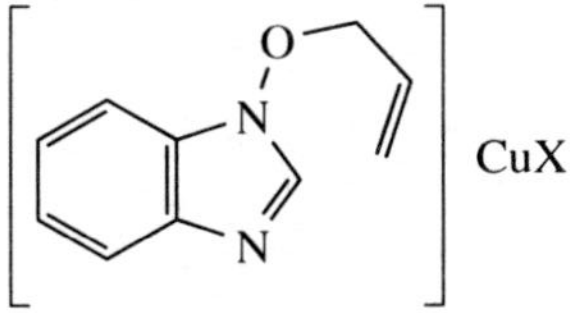

(**20**) X = Cl
(**21**) X = Br

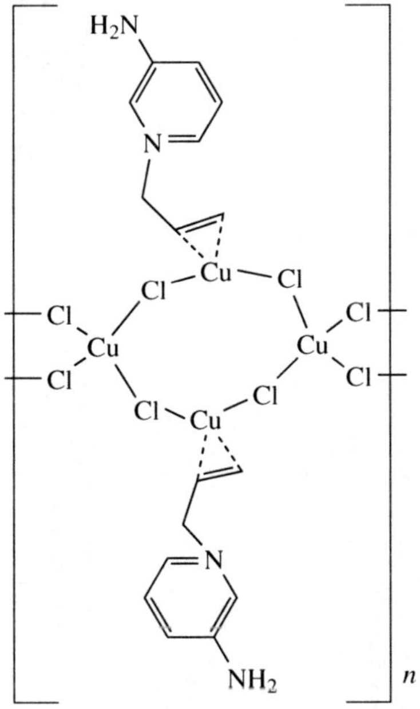

(**23**)

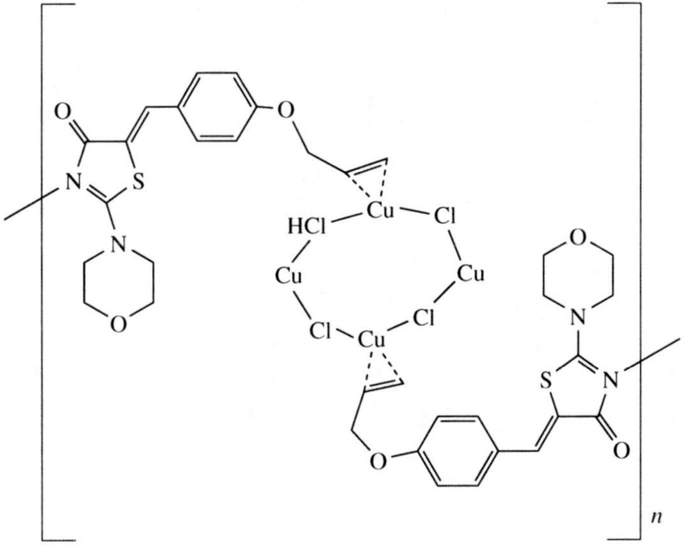

(**24**)

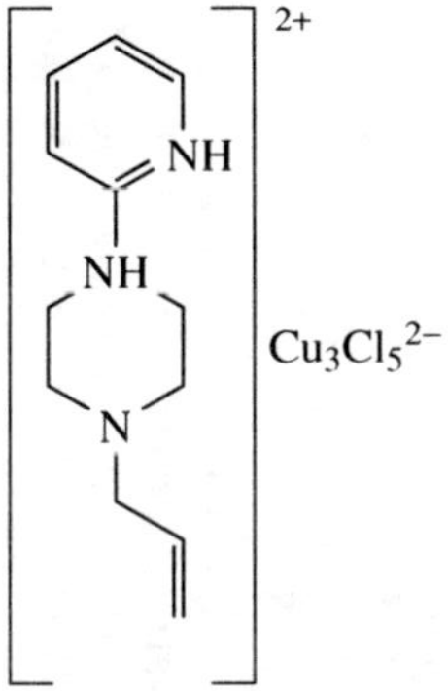

(**25**)

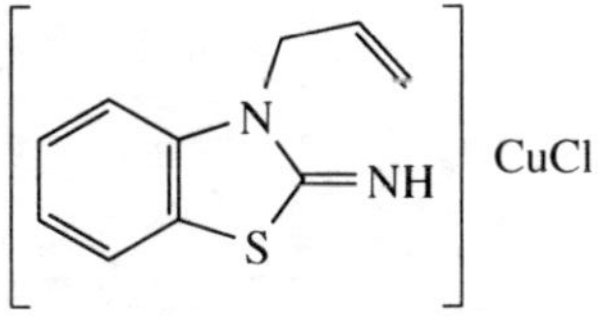

(**26**)

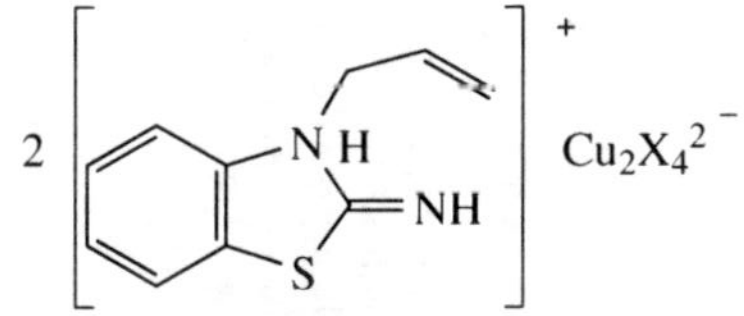

(27) X = Cl
(28) X = Br

TABLE 1. Electrochemical syntheses of copper(I) π-complexes with bonding by allyl groups

Reactant with the π-ligand	Product Cu(I) compound	Reference
N-Allylbenzothiazolium halide	$[C_7H_5NS(C_3H_5)]^+(CuCl_2)^- \cdot H_2O$ **2**[a]	10
	$[C_7H_5NS(C_3H_5)]^+ (CuCl_{1.06}Br_{0.94})^- \cdot H_2O$ **3**[a]	
1-Allylbenzotriazole	$[C_6H_4N_3(C_3H_5)] \cdot CuCl$ **4**[a]	11
Diallyl sulfide	$[(C_3H_5)_2S] \cdot 2CuCl$ **5**	12
1,3-Diallylbenzotriazolium bromide	$[C_6H_4N_3(C_3H_5)_2]^+ (Cu_2Br_3)^-$ **6**	13
1,3-Diallylbenzimidazolium chloride	$[C_7H_5N_2(C_3H_5)_2]^+ (Cu_2Cl_3)^-$ **7**	14
1,3-Diallylbenzimidazolium halide	$[C_7H_5N_2(C_3H_5)_2]^+ [Cu_2(Cl_{0.67}Br_{2.33})]^-$ **8**	15
(2-Allylthio)benzimidazole	$[C_7H_5N_2S(C_3H_5)] \cdot CuX$ (X = Cl or Br) **9**	16
1-Allyl-4-aminopyridinium halide	$\{Cu[H_2NC_5H_4N(C_3H_5)]Br_2\} \cdot H_2O$ **10**	17
	$\{Cu[H_2NC_5H_4N(C_3H_5)]Br_{0.65}Cl_{1.35}\} \cdot H_2O$ **11**	
	$\{Cu[H_2NC_5H_4N(C_3H_5)]Cl_2\}$ **12**	
1-Allyl-2-aminopyridinium chloride	$[H_2NC_5H_4N(C_3H_5)]^+ [CuCl_2]^-$ **13**	18
1-Allyl-2-aminopyridinium bromide	$[H_2NC_5H_4N(C_3H_5)]^+ [CuBr_{1.10}Cl_{0.90}]^-$ **14**	
N,N'-Diallylmorpholinium chloride	$[C_4H_8ON(C_3H_5)_2]^+ [Cu_4Cl_5]^-$ **15**	19
N,N,N',N'-Tetraallylpiperazinium chloride	$[C_4H_8N_2(C_3H_5)_4]^{2+} [CuCl_2]_2^-$ **16**	20
1-Allylpyridinium chloride	$[C_5H_5N(C_3H_5)]^+ [Cu_2Cl_3]^-$ **17**	22
1-Allylbenzimidazolium chloride	$[C_7H_5N_2H(C_3H_5)]^+ [Cu_3Cl_4]^-$ **18**	23
N-Allylmorpholine hydrochloride	$\{[C_4H_8ONH(C_3H_5)]^+\}_2 [Cu_2Cl_4]^{2-}$ **19**	24
1-Allyloxybenzotriazole	$C_6H_4N_3(OC_3H_5) \cdot CuCl$ **20**	25
1-Allyloxybenzotriazole	$C_6H_4N_3(OC_3H_5) \cdot CuBr$ **21**	27
	$C_6H_4N_3(OC_3H_5) \cdot 2CuCl$ **22**	
1-Allyl-3-aminopyridinium chloride	$\{[H_2NC_5H_4N(C_3H_5)]_2 \cdot Cu_4Cl_6\}_n$ **23**	28
3-[(2-Morpholino-4-oxo-4,5-dihydro-1,3 thiazol-5-ylidene)methylphenoxy] propene	$[(C_{17}H_{18}N_2O_3S)_2 \cdot Cu_4Cl_4]_n$ **24**[b]	29
1-(2-Pyridyl)-4-allylpiperazinium dichloride	$[(C_5H_4NH)NC_4H_8NH(C_3H_5)]^{2+} [Cu_3Cl_5]^{2-}$ **25**	30
Mixture of 2-imino-3-allyl-benzothiazole with 2-imino-3-allyl-benzothiazolium bromide	$[C_7H_5N_2S(C_3H_5) \cdot (CuCl)]$ **26**[a]	31
	$\{[C_7H_6N_2S(C_3H_5)]^+\}_2 [Cu_2Cl_4]^{2-}$ **27**	
	$\{[C_7H_6N_2S(C_3H_5)]^+\}_2 [Cu_2Br_4]^{2-}$ **28**	

[a] Electrochemical synthesis was performed in acetonitrile solution.
[b] Electrochemical synthesis was performed at 323 K because of low solubility of the organic reactant.

Heterovalent copper(I,II) π,σ-complexes were also synthesized by the same method using 2,4,6-triallyloxy-1,3,5-triazine (**29**) as the ligand[21,26,32]. In compound $Cu_7Br_{6.48}Cl_{1.52} \cdot 2C_3N_3(OC_3H_5)_3$ (**30**) each ligand **29** is coordinated to one copper(II) atom through the nitrogen atom and to two copper(I) atoms through the C=C bonds[26]. Higher efficiency of the π-interaction between Cu(I) and C=C in a similar coordination was found in $Cu_7Cl_8 \cdot 2C_3N_3(OC_3H_5)_3$ (**31**), which is a mixed-valence Cu(I),Cu(II) π,σ-complex obtained[21] by electrochemical reduction of the copper(II) complex $[CuCl_2 \cdot 2C_3N_3(OC_3H_5)_3]$. Further reduction yields[21] $[Cu_8Cl_8 \cdot 2C_3N_3(OC_3H_5)_3] \cdot 2C_2H_5OH$, with all three allylic groups and a nitrogen atom of **29** coordinated to four copper atoms. Another mixed-valence copper π-complex, $[Cu_3Cl_4 \cdot 2C_3N_3(OC_3H_5)_3]$ (**32**), was obtained[32] after a two-hour electrolysis of **29** and $CuCl_2 \cdot 2H_2O$. In **32** the copper(I) atom coordinates two bridging chlorine atoms and the C=C bond of the allyl group which is in the *trans* position to the nitrogen atom of the triazine ring coordinated to the copper(II) atom. This is in sharp contrast to the *cis* arrangement found earlier[21] in **31**. It should also be added

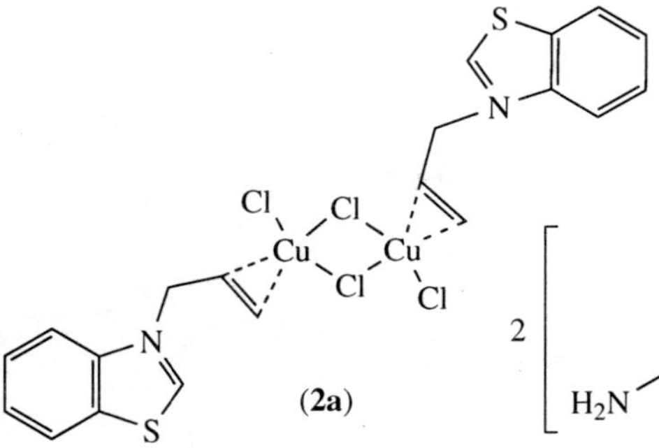

(**2a**)

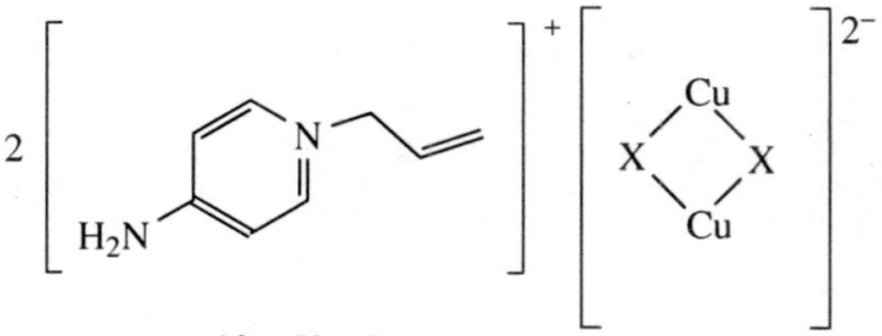

(**10a**) X = Br
(**11a**) X = Cl/Br

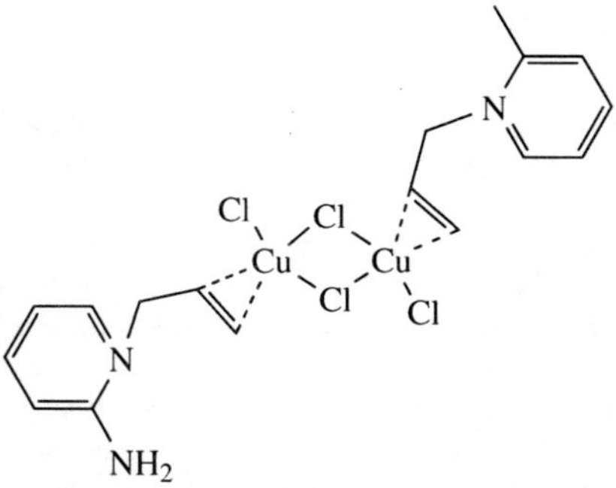

(**13a**)

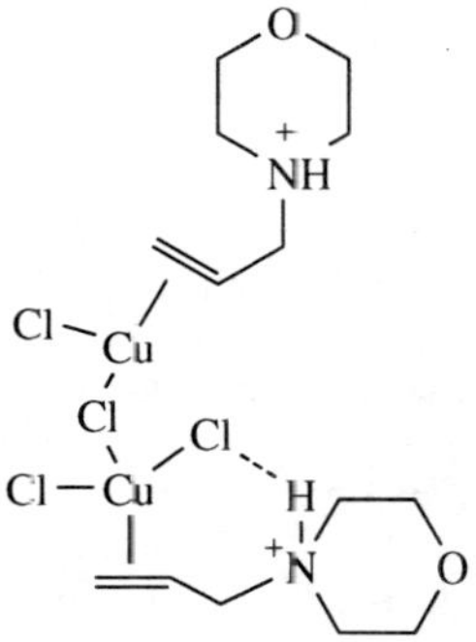

(**19a**)

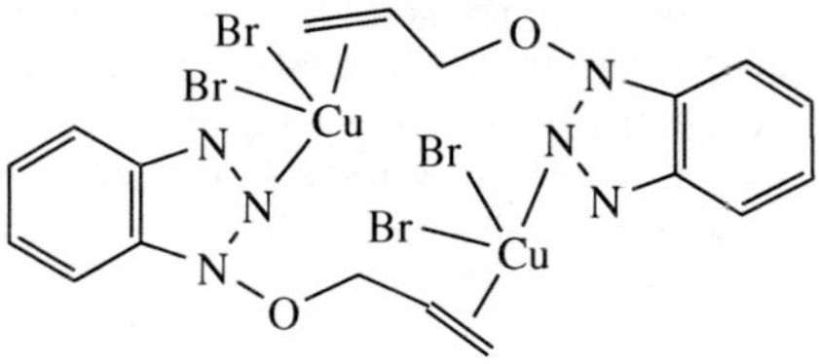

(**21a**)

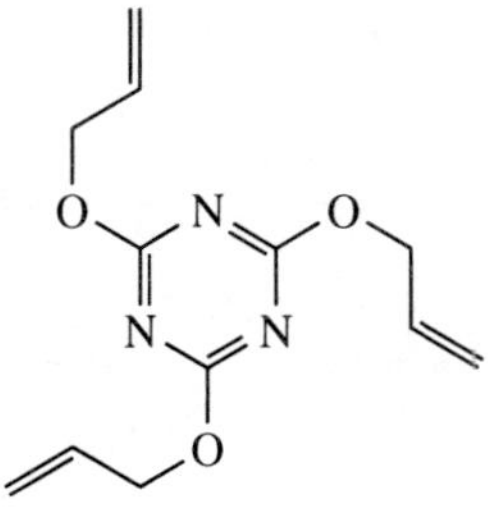

(**29**)

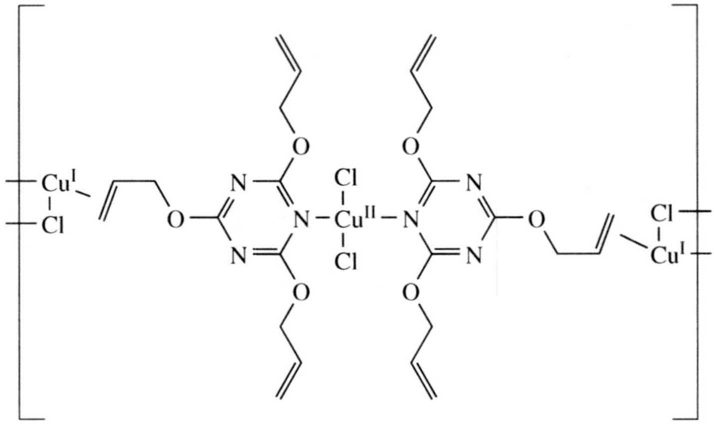

(32)

that in the case of some ligands, during the electrochemical synthesis some intermediate complexes of copper(II) coordinated only by nitrogen atoms were obtained and their structures were determined[33].

The synthesis of cyano copper(I) complexes in one-step reactions was achieved[34–36] by electrochemical dissolution of a sacrificial copper anode and a simultaneous cathodic reduction of nitriles in acetonitrile solutions. A similar method was next used for preparation of two organocopper compounds[36]. Fluorenylbis(triphenylphosphane)copper(I)–diacetonitrile $[Cu(C_{13}H_9)(PPh_3)_2]\bullet[2MeCN]$ (**33**) was synthesized[36] by electrolysis at 0 °C in an undivided cell, equipped with a copper anode and a platinum cathode, of the acetonitrile solution containing fluorene, triphenylphosphane (**34**) and $TBABF_4$ as a supporting electrolyte. A constant current of 50 mA was applied and, after passing the charge of 1F per mole, a white precipitate of **33** (quite soluble in DMSO) was formed. The proposed structure of **33** was supported by elemental analysis, EDX and ESI-MS (most abundant ions are $[Cu(C_{13}H_9)(PPh_3)_2]^+$ and $[Cu(C_{13}H_9)(PPh_3)(PPh)]^+$), as well as ^1H, ^{13}C, ^{31}P NMR and IR spectroscopy. Similar electrolysis of an acetonitrile solution containing triphenylmethane (**35**), **34** and $TBABF_4$ under the same conditions yielded triphenylmethyl(triphenylphosphane)copper(I)–diacetonitrile $[Cu(CPh_3)(PPh_3)]\bullet[2MeCN]$ (**36**) as a greenish precitipate. The formula of **36** was proposed on the basis of elemental analysis, EDX and FAB-MS results. Thus, the following cathodic (2a), anodic (2b) and final (2c) reactions were suggested:

$$\underset{(\mathbf{35})}{Ph_3C\text{-}H} + e \longrightarrow Ph_3C^- + {}^1\!/\!{}_2H_2\uparrow \qquad (2a)$$

$$Cu + \underset{(\mathbf{34})}{PPh_3} + MeCN - e \longrightarrow [Cu(MeCN)(PPh_3)]^+ \qquad (2b)$$

$$[Cu(MeCN)(PPh_3)]^+ + Ph_3C^- + MeCN \longrightarrow \underset{(\mathbf{36})}{[Cu(CPh_3)(PPh_3)]\bullet[2MeCN]\downarrow} \qquad (2c)$$

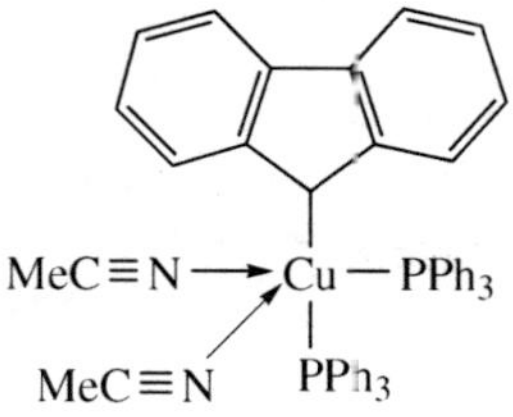

(33)

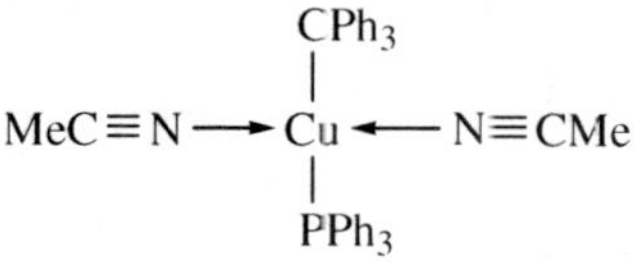

(36)

III. ELECTROCHEMICAL STABILITY OF ORGANOCOPPER COMPOUNDS

Electrochemical measurements can be very useful when evaluating the stability of copper compounds in different oxidation states. This method was often used to compare the effects of the structure and ligand nature on the stability of copper complexes. Some examples of this approach to organocopper compounds, including a determination of equilibrium constants or only a report of redox potentials, are given in this Section.

A. Mononuclear Compounds

Voltammetric examinations of **1** in acetone solutions at the rotating platinum electrode showed[7] the shifts of half-wave potentials for reversible processes of Cu(II)/Cu(I) and Cu(I)/Cu(0) couples with the cod concentration as shown in Figure 1. The slopes of the linear relationships shown support the 1:1 stoichiometry of the $Cu(cod)^+$ complex. The overall formation constant for this complex obtained from Cu(II)/Cu(I) and Cu(I)/Cu(0) data (Figure 1) was equal to $\log \beta = 4.4 \pm 0.1$ and $\log \beta = 4.6$, respectively.

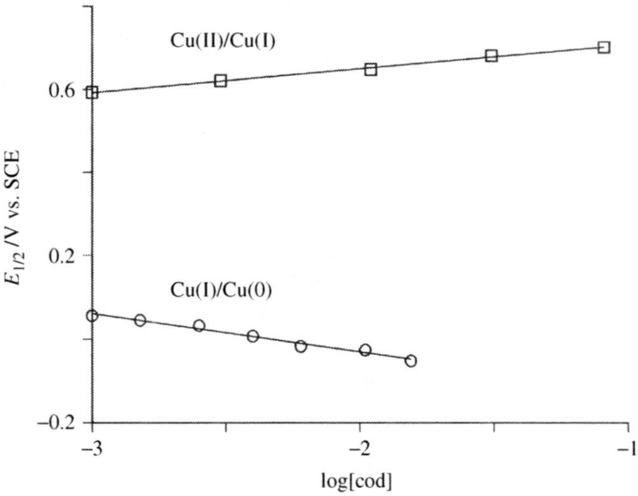

FIGURE 1. Dependence of half-wave potentials for the Cu(II)/Cu(I) and Cu(I)/Cu(0) couples in **1** on the log of cod concentration in acetone. Data from Reference 7

In a similar manner stability constants for the formation of copper(I) π-complexes with maleic acid H_2L (**37**) (equation 3) and its deprotonated anions, HL^- and L^{2-}, were determined[37]. For this purpose reversible potentials of two consecutive one-electron steps for the reduction of Cu(II) in aqueous solutions of different pH and the variable concentrations of **37** were measured at the hanging mercury drop electrode using cyclic and square-wave voltammetry. The observed increase in the stability of the complexes with deprotonated **37** in the order of log K (M^{-1}) = 3.36, 4.08 and 4.45 for CuH_2L^+, CuHL and CuL^-, respectively evidently points to additional electrostatic attractions between copper ion and the negatively charged oxygen atoms[37].

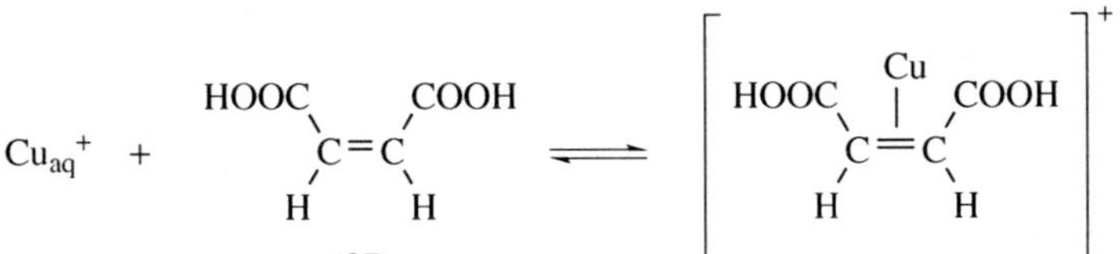

(**37**) (3)

Sandwich complexes of copper with the (3)-1,2-dicarbollide ligand, $B_9C_2H_{11}^{2-}$, involving Cu(II) d^9 paramagnetic ion, $Cu(B_9C_2H_{11})_2^{2-}$ (**38**$^{2-}$), and Cu(III) d^8 diamagnetic ion, $Cu(B_9C_2H_{11})_2^-$ (**38**$^-$), were prepared and their spectral and electrochemical properties were reported[38]. They are formally analogous to the bis(cyclopentadienyl)metallocenes, yet isostructural anions of crystal salts $[(C_2H_5)_4N]_2Cu^{II}(B_9C_2H_{11})_2$ (**38a**) and $[(C_6H_5)_3PCH_3Cu^{III}(B_9C_2H_{11})_2$ (**38b**) do not form symmetrical sandwich structures, but are distorted by a slippage of the carborane moieties parallel to one another[39,40], as shown in Figure 2. It is evident from this Figure that three boron atoms of each carborane face are closer to the copper atom than the two carbon atoms. Cyclic voltammograms of both **38a** and **38b** recorded at a platinum electrode in MeCN solutions containing 0.1 M TEAP or TBAP showed a reversible one-electron process corresponding to the **38b/38a** redox couple with the cathodic peak potential equal to −0.35 V vs. SCE. The second, quasi-reversible process corresponding to Cu^{II}/Cu^{I} couple with a cathodic peak potential varying

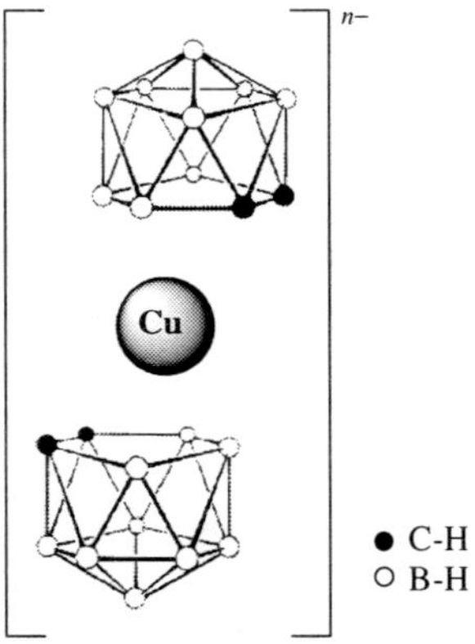

(**38**)

FIGURE 2. The structure of **38**$^{2-}$ and **38**$^-$ ions. Reprinted with permission from Reference 38. Copyright 1968 American Chemical Society

around −1 V vs. SCE suggests the existence of Cu(I) compound but its stability was not investigated. The obtained results indicate that the Cu^{III} compound **38b** exhibits the most stable oxidation state. On the other hand, **38a** is a strong reducing agent, as supported by gradual oxidation of its solutions in the presence of air, producing **38b**. The properties of **38a** and **38b** were compared[38] with those of similar compounds containing other transition metals, in particular gold, and the most probable metal–ligand orbital interactions were suggested.

Electrochemical and spectroelectrochemical behavior of **39** in which unusually high oxidation state of copper(III) is stabilized by two copper–carbon bonds in the central core of *cis*-doubly *N*-confused porphyrin ligand was investigated in CH_2Cl_2 solutions containing 0.1 M $TBAPF_6$ electrolyte[41]. Apart from a number of CV peaks similar to those observed for the ligand alone, but also connected with the residual water, the additional reduction peak found around the potential of −0.9 V vs. SHE was assigned to the Cu^{III}/Cu^{II} couple.

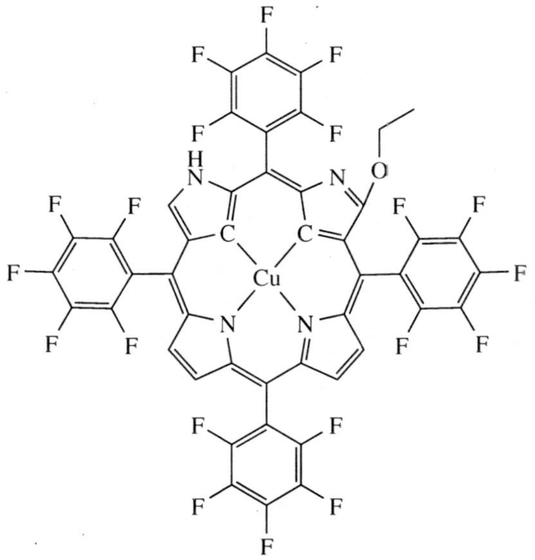

(39)

B. Binuclear Compounds

1. Mixed compounds with Cu and Pt or Ti atoms

Low oxidation states of transition metals in mixed-metal-alkynyl complexes, interesting as precursors of organometallic polymers with potential electronic applications, can be stabilized by di-imine ancillary ligands[42]. Two such complexes containing 1:1 Pt:Cu metals, namely [(*t*-Bu_2bipy)Pt(C≡CR)$_2$Cu(SCN)], with R = *p*-MeC_6H_4 (**40a**) and R = $SiMe_3$ (**40b**), were prepared[42] by reacting (4,4′-bis-*tert*-butyl-2,2′-bipyridyl) platinum bis-alkynyl with Cu(SCN) in MeCN. Their X-ray analysis showed the trigonal planar coordination of copper with η^2 bonding to two C≡C bonds. Electrochemical measurements of **40** in CH_2Cl_2 solutions containing 0.5 M $TBABF_4$ using a platinum electrode revealed a reversible one-electron reduction of bipyridyl ligand at a potential around $E_{red} = -1.3$ V vs. Ag/AgCl, close to a value of $E_{red} = -1.29$ V vs. Ag/AgCl observed[42,43] under the same conditions for the mononuclear platinum complex [Pt(Me$_2$bipy)(C≡CPh)$_2$]. The more interesting one-electron oxidation processes of **40** assigned to Cu^{I}/Cu^{II} redox couple ($E_{ox} > 1$ V vs. Ag/AgCl) are completely irreversible because of decomplexation of unstable Cu^{II} compounds formed in the electron transfer. A similar irreversible oxidation of analogue platinum–silver complex, [(*t*-Bu$_2$bipy)Pt(C≡CSiMe$_3$)$_2$Ag(SCN)] (**41**), was found at lower potential ($E_{ox} = 0.85$ V vs. Ag/AgCl), indicating much lower stability of the Ag^{II} than Cu^{II} compound[42].

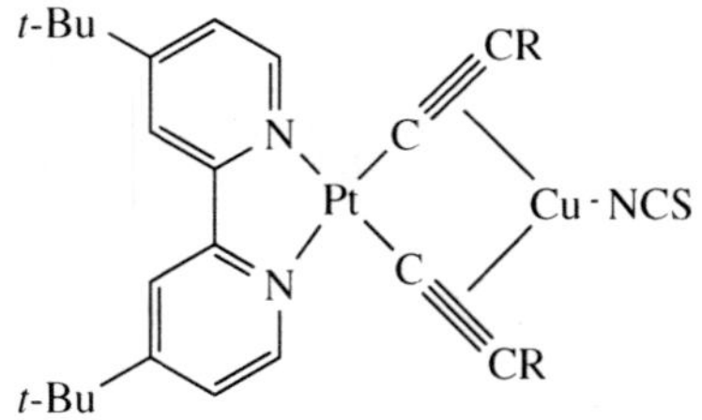

	E_{red} (V)	E_{ox} (V)
40a	-1.36	1.07
40b	-1.29	1.25

(**40a**) R = *p*-MeC$_6$H$_4$
(**40b**) R = SiMe$_3$

By applying chelating ligands, the so-called π-tweezers like bis(alkynyl) titanocene (η^5-C$_5$H$_4$SiMe$_3$)$_2$Ti(C≡CSiMe$_3$)$_2$, it was possible to synthesize several heterobimetallic compounds including early transition metal in a high oxidation state and late transition metal in a low oxidation state. Electronic interactions between both metal centers in such compounds were investigated, often using CV measurements. The discussion of the above problem and comparisons for compounds with different metals are beyond the scope of this Chapter. However, considering only Ti(IV)–Cu(I) compounds the reduction and oxidation potentials obtained[44–46] for {(η^5-C$_5$H$_4$SiMe$_3$)$_2$Ti(μ-σ,π-C≡CR)$_2$}CuX (**42**) and thiol end-capped titanium–copper complexes {(η^5-C$_5$H$_4$SiMe$_3$)$_2$Ti(μ-σ,π-C≡CR)$_2$}CuSC$_6$H$_4$R′-4 (**43**) are given in Table 2.

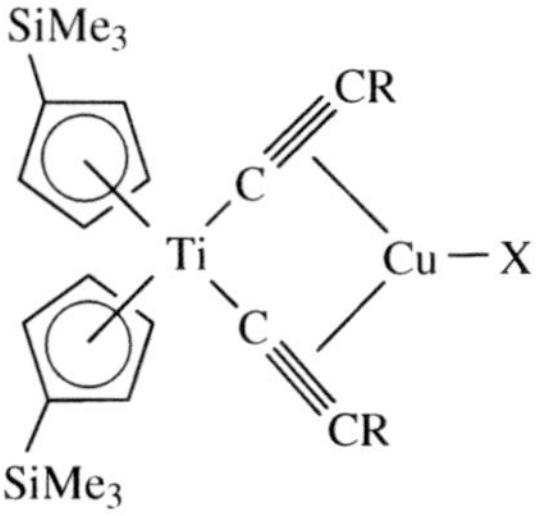

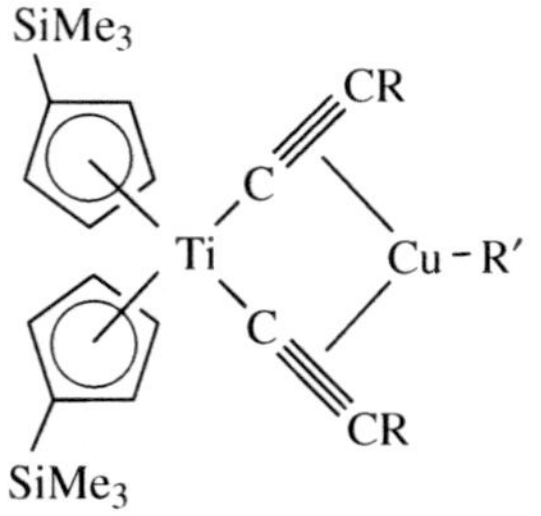

(**42a**) R = Ph; X = Br
(**42b**) R = SiMe$_3$; X = Cl
(**42c**) R = SiMe$_3$; X = Br
(**42d**) R = SiMe$_3$; X = I

(**43a**) R = SiMe$_3$; R′ = 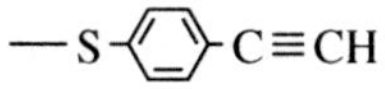

(**43b**) R = SiMe$_3$; R′ = 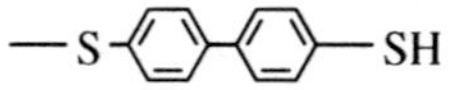

(**43c**) R = *t*-Bu; R′ = 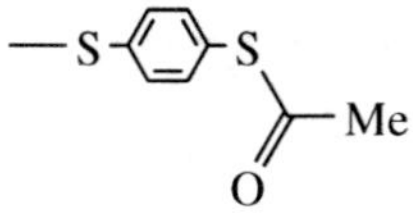

TABLE 2. Cathodic and anodic potentials for the reduction and oxidation at the platinum electrode of compounds **42** and **43** in THF solutions[a] at 298 K

Compound	Reduction		Oxidation	Reference
	$-E_o$ (V vs. Fc^+/Fc)		E_{pa} (V vs. Fc^+/Fc)	
	Cu^I/Cu^0	Ti^{IV}/Ti^{III}		
42a	1.55[b]	1.74[b]	0.17[c]	44
42b	1.73	1.84		45[g]
42c	1.75	1.89		45[g]
42d	1.72	1.85[b]		45[g]
43a	1.51[b]	1.96[b]	0.86	46
43b[d]	1.78[b,e]	2.74[b]	0.91, 1.02, 1.36	46
43c	1.56[b]	1.97[b,f]	—	46

[a] Supporting electrolyte: 0.1M $TBAPF_6$; $\upsilon = 0.2$ or 0.1 V s^{-1}.
[b] E_{pc} value is given for the irreversible peak.
[c] E_o value is given for the reversible process; $\Delta E_p = 100$ mV.
[d] CH_2Cl_2 solution was used.
[e] Small reoxidation peak was observed at $E_{pa} = -1.60$ V vs. Fc^+/Fc.
[f] Small reoxidation peak was observed at $E_{pa} = -1.7$ V vs. Fc^+/Fc.
[g] Data from Reference 45 with permission from Elsevier.

The reduction of copper(I) is in most cases completely irreversible, i.e. no reoxidation peak is observed, and presumably this process results in the disintegration of the complex with the formation of metallic particles[45]. However, the last reaction is probably slower for **43b** and thus a small reoxidation peak was observed[46]. Reversible oxidation of copper(I) to copper(II) was reported only for **42a**. On the other hand, the irreversible anodic peak for **43a** and the first peak for **43b** (Table 2) were assigned[46] most probably to the oxidation of the copper–thiolate unit.

2. $\{Cu(\mu\text{-}L)\}_2$ compounds

Binuclear copper compound $[\{Cu[\mu\text{-}C(SiMe_3)_2(C_5H_4N)]\}_2]$ (**44**), stable to the thermal and oxidative decomposition, was investigated[47] using the cyclic voltammetry in THF solutions containing 0.2 M $TBABF_4$. The reversible reduction process observed at $E_o = -2.47$ V vs. SCE results in an unstable product (its half-life time determined by the CV method is $t_{1/2} < 1$ s). On the other hand, the oxidation product corresponding to $E_o = 0.71$ V vs. SCE exhibits much higher stability ($t_{1/2} > 100$ s) and the oxidation process (equation 4) involves two electrons as shown by the controlled-potential coulometry at 0.64 V. Thus, the formally copper(II) alkyl species $\mathbf{44}^{2+}$ are produced of unexpected

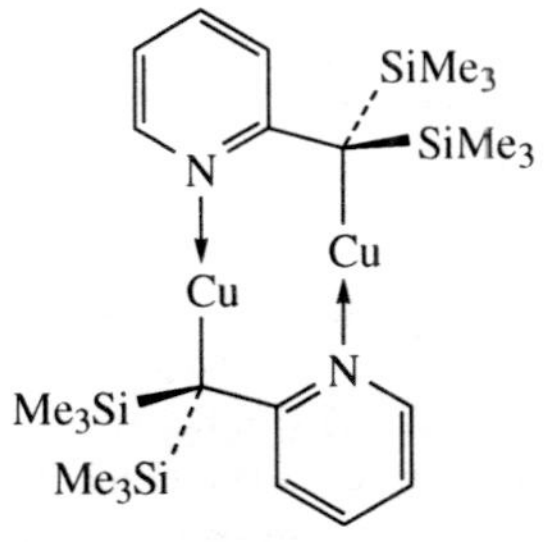

(**44**)

TABLE 3. Cathodic and anodic peak potentials for the irreversible reduction and oxidation[a] of compounds **45–48** in THF at 298 K[50]

Potentials (V vs. Fc^+/Fc)	Step	Compounds			
		45	**46**	**47**	**48**
E_{pc}	1	−3.19	−2.92	−3.20	−2.97
	2	—	−3.09	—	−3.13
E_{pa}	1	0.61	0.47	0.81	0.66
	2	—	0.81	—	1.36

[a] Glassy carbon working electrode; supporting electrolyte: 0.1M $TBAPF_6$, $\upsilon = 0.1\ V\,s^{-1}$.

stability in comparison with earlier reports on unstable copper(II) alkyls[48]. It should be added that copper hydrocarbyl species in the cationic form are also rare.

$$\mathbf{44} - 2e \longrightarrow \mathbf{44}^{2+} \tag{4}$$

Among a number of copper(I) complexes with 1-azaallyl ligand investigated by Hitchcock and coworkers[49,50] there are two dimeric organocopper compounds, $[\{Cu[\mu\text{-}N(SiMe_3)C(t\text{-}Bu)C(H)SiMe_3]\}_2]$ (**45**) and $[\{Cu[\mu\text{-}N(SiMe_3)C(Ph)C(SiMe_3)_2]\}_2]$ (**46**), for which some interactions between both copper(I) centers were discussed. Their examination by several electroanalytical methods revealed that in THF solutions at 298 K, **45** gives one cathodic and one anodic peak, but **46** gives two separate cathodic and two anodic peaks. All the peaks are irreversible and their potentials are collected in Table 3 together with potentials of irreversible peaks obtained for the copper complex **47** and compound **48** which is the precursor of **46**. A comparison of reduction potentials given in Table 3 indicates[50] that the electron is transferred to an orbital mainly centered on the azaallyl ligand. Thus, the electron-withdrawing phenyl substituents facilitate the electroreduction shifting the potential of **46** to less negative values than those observed for **45**, but **45** and **47** have the same E_{pc} value. It was also shown that the following rapid reaction was responsible for the lack of reoxidation peaks, but this could be obtained by lowering the temperature. Then for **46** at 268 K the formal potentials and peak separations are equal to $E_o = -2.91$ V and $\Delta E_p = 96$ mV and $E_o = -3.06$ V and $\Delta E_p = 199$ mV, for the first and the second reduction steps, respectively. In addition, the third, irreversible reduction peak appeared then at −3.55 V.

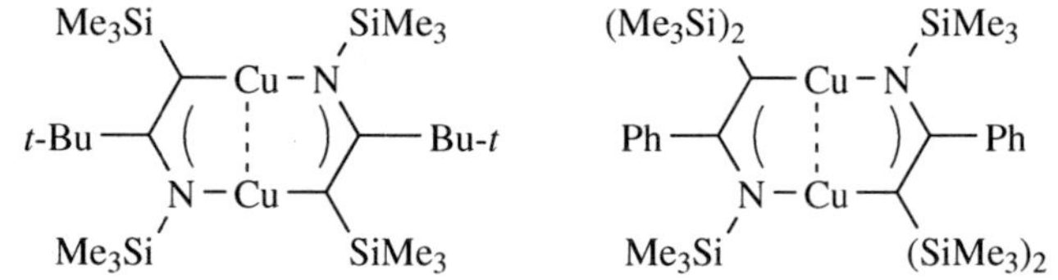

(**45**) (**46**)

At 268 K, a first oxidation peak of **46** also shows a cathodic response giving $E_o =$ 0.45 V and $\Delta E_p = 112$ mV. Moreover, in CH_2Cl_2 solution at 298 K the first oxidation step of **46** is reversible ($E_o = 0.27$ V, $\Delta E_p = 84$ mV) but the second one is not ($E_{pa} =$ 1.23 V). However, the most interesting result is that the first oxidation process corresponds to a one-electron step, in contrast to the behavior of **44** described previously. This result was confirmed for **46** by the square-wave voltammetry and for an analogue compound

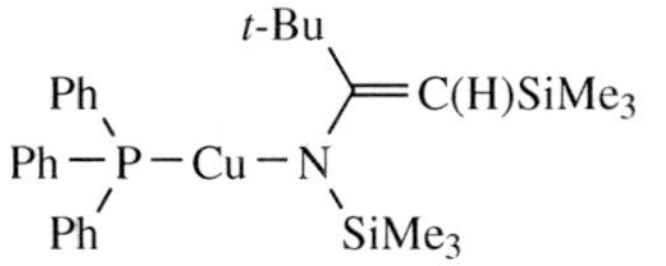

(**47**)

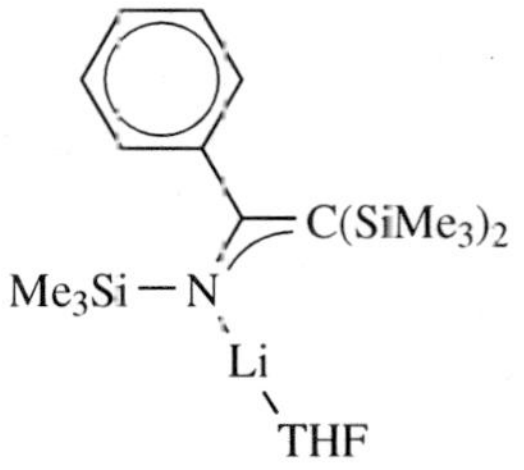

(**48**)

of gold using also chronocoulometry and chronoamperometry. Moreover, the formation at the electrode of a paramagnetic radical cation of gold analogue of **47** was confirmed by an ESR signal obtained in spectroelectrochemical experiments at 3.3 K. In conclusion, one-electron oxidation processes take place at the metal centers, leading to mixed-valent Cu^{+}/Cu^{2+} species[50], which are more stable for compound **46** than for **45**. The contrast to independent oxidation of both Cu(I) centers at the same potential observed[47] for **44** was noticed but not explained[49]. In this context it seems interesting to point out the problem of metal–metal interactions (in crystals the Cu–Cu distances are 2.499(2) Å for **45**[49] and 2.412(1) Å for **44**[47]) and the problem of electronic communication (no substantial communication is expected for saturated ligands, similarly as observed for **44**).

C. Polynuclear Compounds

A great number of polynuclear alkynylcopper(I) clusters, in particular trinuclear copper(I) acetylides, were synthesized by Yam and coworkers[51–58], and their structures as well as remarkable photophysical and photochemical properties were investigated because of their potential applications as nonlinear optical materials. Electrochemical methods were used regularly in their research to compare the stability of different oxidation states. However, attempts to use electrochemical techniques in order to prepare mixed-valence Cu(I)Cu(II) compounds were not successful because in most cases the one-electron oxidation was not reversible, indicating some instability of the mixed-valence complex. Thus, faster spectroscopic techniques had to be used[52].

Cyclic voltammetry at a GC electrode of $[Cu_3(\mu\text{-}Ph_2PCH_2PPh_2)_3(\mu_3\text{-}\eta^1\text{-}C{\equiv}CR)_2]PF_6$ (**49**) in acetonitrile solutions shows the one-electron quasi-reversible oxidation of the copper(I) center[51,57], with the exception of *p*-nitrophenylacetylide (**49c**) which oxidizes irreversibly. The oxidation potentials are given in Table 4; for the quasi-reversible behavior the E_o values were obtained from peak potentials, E_{pa} and E_{pc}. A similar oxidation of the copper(I) center was also observed for a parallel series of monocapped acetylides $[Cu_3(\mu\text{-}Ph_2PCH_2PPh_2)_3(\mu_3\text{-}\eta^1\text{-}C{\equiv}CR)][BF_4]_2$ (**50**) and the oxidation potentials obtained[51,57] are given in Table 4; note the different anion for $Cu_3(\mu\text{-}Ph_2PCH_2PPh_2)_3(\mu_3\text{-}\eta^1\text{-}C{\equiv}CBu\text{-}t)][PF_6]_2$ (**50b**). Some trends in oxidation potentials can be easily understood[51,57]. A greater σ-donating ability of alkyl substituents over aromatic groups, i.e. *tert*-butylacetylide and *n*-hexylacetylide over phenylacetylide, makes the oxidation of **49b** and **49g** easier than the oxidation of **49a** owing to higher stabilization of the product copper(II) center in the first two compounds[51,57]. A similar trend was also observed for parallel compounds in a series of monocapped acetylides **50**. For aryl substituents, the E_o values clearly decrease with increasing electron-donating abilities of the substituents[57] and a rough correlation with the Hammett substituent constants[59] can be noticed (for five compounds **50a**, **50c–50f** the correlation coefficient is 0.907). For both compounds containing a strong electron-withdrawing *p*-nitrophenylacetylide group, **49c** and **50c**, the oxidation

TABLE 4. Potentials for the quasi-reversible (or irreversible) oxidation at a GC electrode of bicapped (**49**) and monocapped (**50**) trinuclear copper(I) acetylides in MeCN solutions at 298 K[a]

Compound	E_o (V vs. Fc^+/Fc)	Reference	Compound	E_o (V vs. Fc^+/Fc)	Reference
49a	0.41	51	**50a**	0.38	53
49b	0.37	51	**50b**	0.36	53
49c	0.49[b]	57[c]	**50c**	0.48[b]	57[c]
49d	0.37	57[c]	**50d**	0.36	57[c]
49e	0.33	57[c]	**50e**	0.36	57[c]
49f	0.34	57[c]	**50f**	0.15	57[c]
49g	0.27	57[c]	**50g**	0.31	57[c]

[a] Supporting electrolyte: 0.1M $TBAPF_6$; scan rate 0.1 V s^{-1}.
[b] Irreversible peak; E_{pa} is given instead of E_o.
[c] Data from Reference 57 with kind permission of Springer Science and Business Media.

process is irreversible, indicating most probably[57] the decomposition of the unstable copper(II) product. Only these two compounds exhibit the reduction process (at −1.40 V vs. Fc^+/Fc) which certainly corresponds to the electroactivity of the nitrophenyl group. It should also be added that at more anodic potentials further oxidation processes were observed[51,57] with completely irreversible CV peaks suggesting the decomposition of clusters.

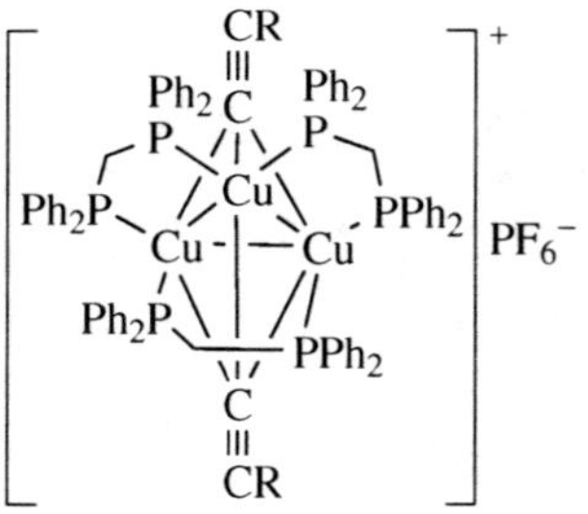

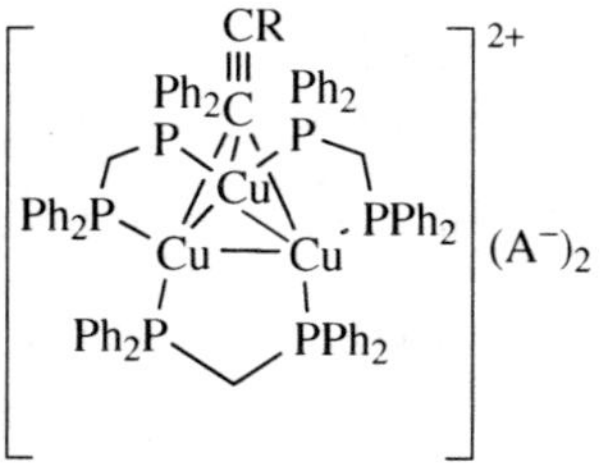

(**49a**) R = Ph
(**49b**) R = *t*-Bu
(**49c**) R = *p*-$NO_2C_6H_4$
(**49d**) R = *p*-PhC_6H_4
(**49e**) R = *p*-$MeOC_6H_4$
(**49f**) R = *p*-$NH_2C_6H_4$
(**49g**) R = *n*-Hex

(**50a**) R = Ph; A = BF_4
(**50b**) R = *t*-Bu; A = PF_6
(**50c**) R = *p*-$NO_2C_6H_4$; A = BF_4
(**50d**) R = *p*-PhC_6H_4; A = BF_4
(**50e**) R = *p*-$MeOC_6H_4$; A = BF_4
(**50f**) R = *p*-$NH_2C_6H_4$; A = BF_4
(**50g**) R = *n*-Hex; A = BF_4

Oxidation potentials for a series of trinuclear copper(I) acetylides (**51**), exhibiting photoluminescent behavior as **49** and **50**, but built with bis(diphenylphosphino)alkylamines and—arylamines bridging ligands (**52**) instead of dppm (i.e. $Ph_2PCH_2PPh_2$ as in **49** and **50**), are given in Table 5. Further oxidation of **51** at more anodic potentials probably results in the decomposition of clusters showing irreversible processes[54].

The inspection of potentials given in Table 5 shows that the oxidation becomes easier with increasing electron-donating ability of the acetylide ligands (for the same R = *n*-Pr the E_o values decrease for R′ in the order **51d** ⩾ **51e** > **51b** > **51c** > **51a** ⩾ **51f** > **51g** and for R = Ph in the order **51j** > **51h** > **51k**) as expected[54] due to preferential

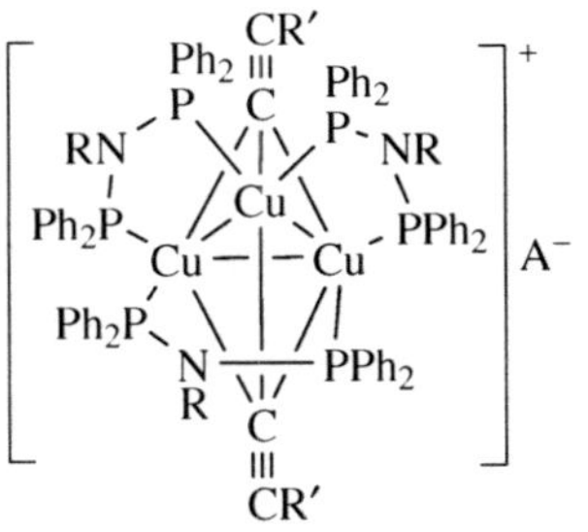

(**51**)

51	R	R′	A
(**a**)	*n*-Pr	*p*-EtOC$_6$H$_4$	BF$_4$
(**b**)	*n*-Pr	*p*-PhC$_6$H$_4$	BF$_4$
(**c**)	*n*-Pr	Ph	BF$_4$
(**d**)	*n*-Pr	*p*-NO$_2$C$_6$H$_4$	BF$_4$
(**e**)	*n*-Pr	*p*-py	PF$_6$
(**f**)	*n*-Pr	*t*-Bu	PF$_6$
(**g**)	*n*-Pr	*n*-C$_{10}$H$_{21}$	PF$_6$
(**h**)	Ph	*p*-EtOC$_6$H$_4$	PF$_6$
(**i**)	Ph	*p*-PhC$_6$H$_4$	PF$_6$
(**j**)	Ph	Ph	PF$_6$
(**k**)	Ph	*n*-Hex	PF$_6$
(**l**)	*p*-MeC$_6$H$_4$	*p*-EtOC$_6$H$_4$	PF$_6$
(**m**)	*p*-MeC$_6$H$_4$	*p*-PhC$_6$H$_4$	PF$_6$
(**n**)	*p*-MeC$_6$H$_4$	Ph	PF$_6$
(**o**)	*p*-FC$_6$H$_4$	*p*-EtOC$_6$H$_4$	PF$_6$
(**p**)	*p*-FC$_6$H$_4$	*p*-PhC$_6$H$_4$	PF$_6$

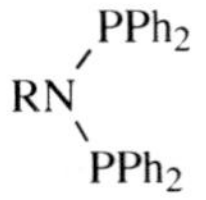

(**52**)

(**52a**) R = *n*-Pr
(**52b**) R = Ph
(**52c**) R = *p*-MeC$_6$H$_4$
(**52d**) R = *p*-FC$_6$H$_4$

TABLE 5. Potentials for the quasi-reversible oxidation at a GC electrode of trinuclear copper(I) acetylide compounds in MeCN solutions[a] at 298 K. Reprinted with permission from Reference 54. Copyright 1997 American Chemical Society

Compound	E_o V vs. Fc$^+$/Fc)	Compound	E_o (V vs. Fc$^+$/Fc)	Compound	E_o (V vs. Fc$^+$/Fc)
51a	0.30	**51g**	0.23	**51m**	0.37
51b	0.34	**51h**	0.38	**51n**	0.38
51c	0.32	**51i**	0.37	**51o**	0.40
51d	0.45	**51j**	0.41	**51p**	0.44
51e	0.44	**51k**	0.31		
51f	0.29	**51l**	0.35		

[a] Supporting electrolyte: 0.1M TBAPF$_6$; scan rate 0.1 V s^{-1}.

stabilization of the higher oxidation state of copper(II). The same trend was observed with increasing the electron richness of ligand **52**; for example, for a constant R′ = *p*-EtOC$_6$H$_4$ the E_o values increase in the order of **51a** > **51l** > **51h** > **51o**[54]. The results obtained[54] for **51** with aryl R and R′ substituents (with the exception of **51i**) can be described in a more quantitative way using the Hammett equation with substituent constants[59] for both ligands, $\sigma_p{}^R$ and $\sigma_p{}^{R'}$, in a two-parameter equation (equation 5):

$$E_o = 0.24(\pm 0.08)\sigma_p{}^R + 0.13(\pm 0.07)\sigma_p{}^{R'} + 0.42(\pm 0.01) \quad (5)$$

In equation 5, 95% errors of regression coefficients are given in parentheses and this equation holds for the seven compounds shown in Figure 3 with the correlation coefficient of $r = 0.976$ and the Snedecor parameter $F = 40.8$, indicating that the relationship is statistically important with the probability of 99.8%; moreover, the addition of the second parameter is statistically significant with a probability of 92.8%. Evidently, the more

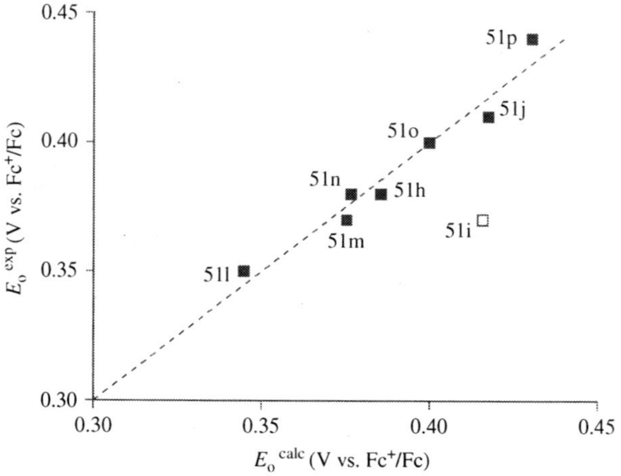

FIGURE 3. Relationship between the experimental[54] oxidation potentials of **51** (Table 5) and the calculated values from equation 5. The theoretical line with unit slope is shown. The open square denotes the most deviating point not included in the regression

negative σ_p^R and $\sigma_p^{R'}$ values correspond to a higher electron-donating ability, which stabilizes the copper(II) state, resulting in lower oxidation potentials.

A similar effect of an electron-donating ability of ligands on the stabilization of copper(II) oxidation state which decreases the oxidation potentials was reported[55] for copper(I)–rhenium(I) mixed-metal acetylide compounds (**53**); cations **53**$^+$ are shown in the formula drawn below, the corresponding anions being PF_6^- or BF_4^-. Thus, the quasi-reversible oxidation process corresponding to Cu(I)/Cu(II) couple occurs more easily for **53b** with the 2,5-dimethyl-1,4-diethynylbenzene ligand than for **53a** with the 1,4-diethynylbenzene ligand and more easily for **53c** with t-Bu_2bipy than for **53a** with bipyridine ligand (cf. the potentials given in Table 6). However, the discussed effect is rather weak and oxidation potentials for all compounds **53** have similar values. Moreover, the effect of the second metalloligand, rhenium(I) acetylide backbone, is also small, as is evident from a comparison with the results discussed previously: **53d** with **51c** ($E_o = 0.34$ and 0.32, respectively) and **53a** with **49a** ($E_o = 0.39$ and 0.41, respectively).

TABLE 6. Potentials for the quasi-reversible oxidation and reduction at a GC electrode of trinuclear copper(I) acetylide compounds in MeCN solutions[a] at 298 K

Compound	Oxidation E_o (V vs. Fc+/Fc)	Reduction $-E_o$ (V vs. Fc+/Fc)	References
53a	0.39	1.82	55
53b	0.31	1.83	55
53c	0.33	1.91	55
53d	0.34	1.90	55
53e	0.30	1.93	55
54a	0.33	—	56
54b	[b]	1.39	56

[a] Supporting electrolyte: 0.1M $TBAPF_6$; scan rate 0.1 V s^{-1}.
[b] Irreversible process.

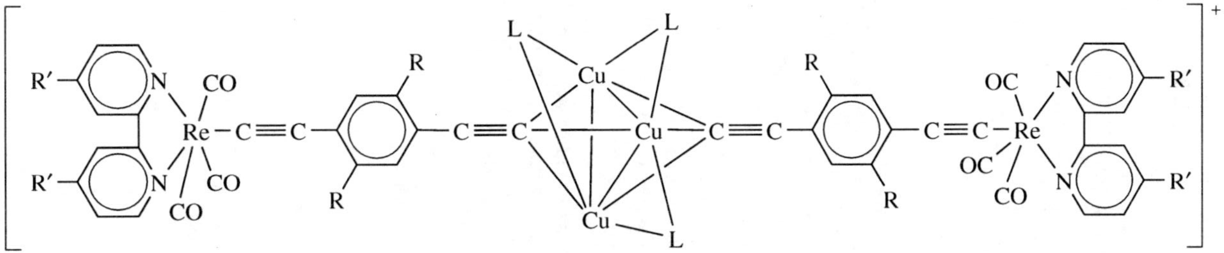

(**53**$^+$)

53	R	R′	L
(**a**)	H	H	dppm
(**b**)	Me	H	dppm
(**c**)	H	*t*-Bu	dppm
(**d**)	H	H	*n*-Pr-*N*(PPh_2)$_2$
(**e**)	Me	H	*n*-Pr-*N*(PPh_2)$_2$

The quasi-reversible reduction process of **53** was assigned[55] to the one-electron transfer from an electrode to bpy ligands on the basis of electrode potentials for similar reactants described in the literature. The E_o values measured[55] are also given in Table 6. It is evident that for the same ligand L = dppm coordinated to the Cu_3 cluster the reduction is more difficult for **53c** having t-Bu_2bipy ligand coordinated to the rhenium atom than for **53a** with unsubstituted bipyridine of a greater π-acceptor ability. The observed behavior is fully conceivable, i.e. the opposite effect was discussed above for the oxidation process.

Quasi-reversible oxidation corresponding to the Cu(I) → Cu(II) process was observed[56] under similar conditions as for **53** (Table 6) for mix-capped trinuclear copper(I) compound $[Cu_3(\mu\text{-}Ph_2PCH_2PPh_2)_3(\mu_3\text{-}\eta^1\text{-}C{\equiv}CC_6H_4OMe\text{-}p)(\mu_3\text{-}\eta^1\text{-}C{\equiv}CC_6H_4OEt\text{-}p)]PF_6$ (**54a**), but the process was completely irreversible for $[Cu_3(\mu\text{-}Ph_2PCH_2PPh_2)_3(\mu_3\text{-}\eta^1\text{-}C{\equiv}CC_6H_4OMe\text{-}p)(\mu_2\text{-}\eta^1\text{-}C{\equiv}CC_6H_4NO_2\text{-}p)]PF_6$ (**54b**). It was suggested that irreversibility is caused by the decomposition of the oxidized form of **54b**, which is unstable due to strong electron-withdrawing properties of the nitro group. On the other hand, for **54b** a quasi-reversible one-electron reduction was found at potential of *ca* −1.39 V, close to the reduction potential of 4-nitrophenylacetylene. Thus, the acetylide ligand itself is most probably reduced, similarly as the ligand reduction observed for **53** (Table 6). The recorded cyclic voltammograms were shown in the original paper[56].

Recently, Yam and coworkers[58] extended their studies of photoluminescence and electrochemical properties of monoynyl trinuclear copper(I) complexes to a series of diynyl compounds $[Cu_3(\mu\text{-}Ph_2PCH_2PPh_2)_3(\mu_3\text{-}\eta^1\text{-}C{\equiv}CC{\equiv}CR)_2]PF_6$ (**55**). Oxidation potentials obtained[59] for the quasi-reversible one-electron processes are given in Table 7, after recalculating from SCE to Fc^+/Fc scale. The trinuclear silver(I) diynyl complexes, analogous to **55a** and **55c**, have higher oxidation potentials, around 0.72–0.81 V vs. Fc^+/Fc, and therefore Yam and coworkers suggested[58] that the oxidation of **55** 'has substantial metal-centered character but with mixing of some diynyl ligand centered contributions'. The last suggestion was supported by theoretical calculations indicating that some HOMO orbitals are delocalized over the organic ligand–metal triangle backbone. The substantial participation of the oxidation of Cu(I) to Cu(II) can explain the observed shift of the oxidation potentials, namely the compounds with more electron-rich diynyl groups are more easily oxidized[58] (Table 7). It was also noticed[58] that diynyl complexes have somewhat more

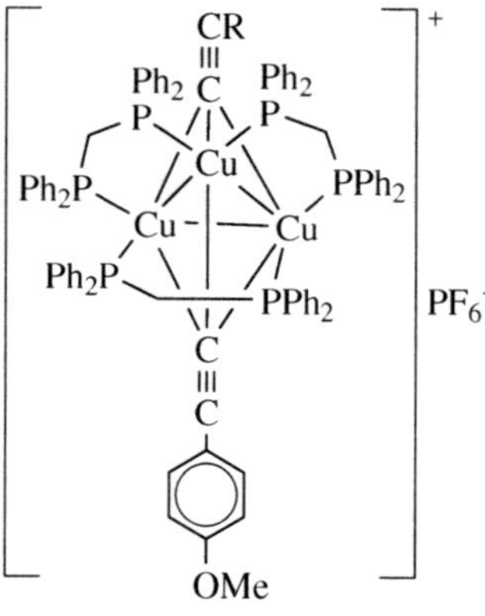

(**54a**) R = p-$EtOC_6H_4$
(**54b**) R = p-$NO_2C_6H_4$

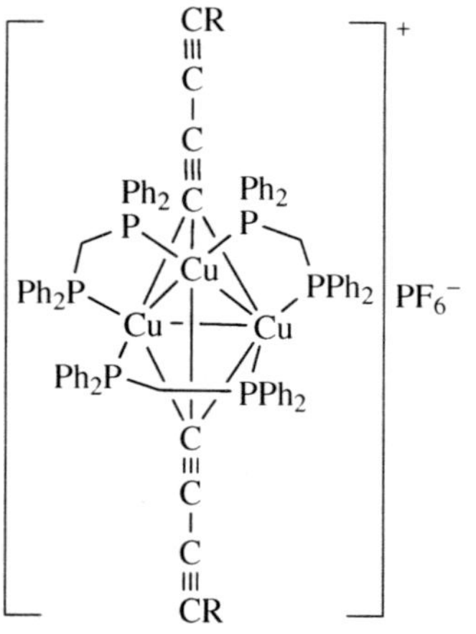

(**55**)

55	R
(**a**)	Ph
(**b**)	p-MeC_6H_4
(**c**)	p-$MeOC_6H_4$
(**d**)	n-Hex
(**e**)	H

TABLE 7. Potentials for the quasi-reversible oxidation at a GC electrode of trinuclear copper(I) diynyl compounds **55** in MeCN solutions[a] at 298 K. Reprinted with permission from Reference 58. Copyright 2004 American Chemical Society

Compound	**55a**	**55b**	**55c**	**55d**	**55e**
E_o (V vs. Fc^+/Fc)	0.43	0.37	0.34	0.37	0.50

[a] Supporting electrolyte: 0.1M $TBAPF_6$, $\upsilon = 0.1$ V s^{-1}.

positive potentials than analogous monoynyl compounds, as is evident from a comparison of E_o values of **55a**, **55c** and **55d** given in Table 7 with E_o values of **49a**, **49e** and **49g** (Table 4), respectively.

Trinuclear copper(I) cluster acetylide acting as a bridge between two ferrocene molecules $[Cu_3(\mu\text{-}Ph_2PCH_2PPh_2)_3(\mu_3\text{-}\eta^1\text{-}C{\equiv}CFc)_2]PF_6$ (**56**) was used[60] as a model compound for the investigation of electronic interactions between metal centers and the effect of spacer nature on electron transfer processes between the electron donor and acceptor. These problems look interesting in view of the continuous search for molecular wires. The CV measurements at 298 K for MeCN solutions of **56** containing 0.1 M $TBAPF_6$ as a supporting electrolyte show on the platinum disc anode two almost merging oxidation peaks of similar area at the potential around 0.15 V vs. Ag/0.1 M $AgNO_3$, which corresponds[60] to 0.09 V vs. Fc^+/Fc. They are assigned to the successive oxidation of two ferrocenyl groups on the basis of the lack of oxidation peaks at similar potentials for **49a**, which has the same formula except that there is Ph instead of Fc. Moreover, equal values of the anodic and cathodic peak currents and the difference of 60 mV between the anodic and cathodic peak potentials showed clearly[60] the reversibility of both processes. Thus, they can be schematically described (omitting dppm ligands) by equations 6a and 6b.

$$[Fc^{II}-C{\equiv}C-Cu_3-C{\equiv}C-Fc^{II}]^+ - e \rightleftharpoons [Fc^{II}-C{\equiv}C-Cu_3-C{\equiv}C-Fc^{III}]^{2+} \quad (6a)$$

(**56**$^+$) (**56**$^{2+}$)

$$[Fc^{II}-C{\equiv}C-Cu_3-C{\equiv}C-Fc^{III}]^{2+} - e \rightleftharpoons [Fc^{III}-C{\equiv}C-Cu_3-C{\equiv}C-Fc^{III}]^{3+} \quad (6b)$$

(**56**$^{2+}$) (**56**$^{3+}$)

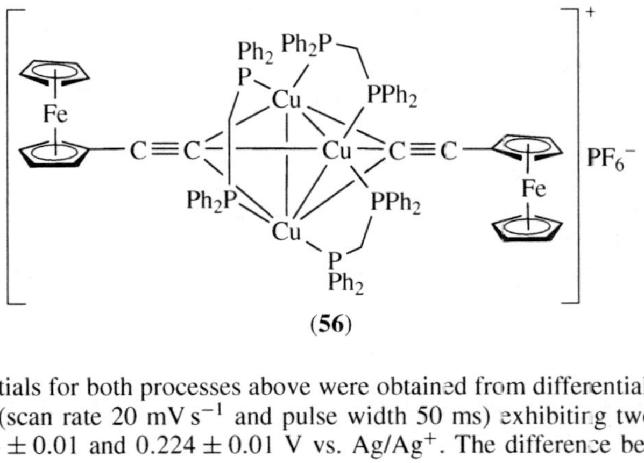

(**56**)

Exact potentials for both processes above were obtained from differential pulse voltammetric curves (scan rate 20 mV s^{-1} and pulse width 50 ms) exhibiting two almost equal peaks at 0.114 ± 0.01 and 0.224 ± 0.01 V vs. Ag/Ag^+. The difference between them of

0.110 ± 0.014 V corresponds to the equilibrium constant $K_c = 77 \pm 30$ for the comproportionation reaction (equation 7).

$$\mathbf{56}^{+} + \mathbf{56}^{3+} \rightleftharpoons 2\,\mathbf{56}^{2+} \quad (7)$$

Relatively weak electronic communication across the $-C{\equiv}C-Cu_3-C{\equiv}C-$bridge in $\mathbf{56}^{2+}$ was discussed[60] in comparison with the behavior of other organometallic compounds. It was concluded that the stability of the mixed-valence cation $\mathbf{56}^{2+}$ does not really depend on the electron delocalization. In fact, it depends mainly on the reduction of electrostatic repulsion between the oxidized Fc^+ and the positively charged Cu_3 cluster as well as on statistical distribution. The CV curves of **56** show a further irreversible oxidation peak at the potential 1.046 ± 0.01 vs. Ag/Ag^+ and a small cathodic peak at 0.509 ± 0.01 V vs. Ag/Ag^+, with the peak current only around 6% of the above-mentioned anodic peak. Finally, those peaks were assigned[60] to the oxidation and the reduction of the tricopper(I) core. After a comparison with the irreversible copper(I) oxidation of **49c** which has the electron-deficient *p*-nitrophenylacetylide group, it was suggested[60] that the oxidation of ferrocenylacetylide ligands decreases the stabilization of the copper(II) state, which explains the irreversibility of the copper(I) oxidation in **56**. On the other hand, the reduction of organic ligand in **56** was not observed in the available potential window.

Very recently, a number of heteromultimetallic transition metal complexes, some of them with organocopper bonds, were synthesized by Lang and coworkers[61,62] and investigated by spectroscopic and electrochemical methods. Among these reactants there were compounds with copper and titanium, or copper and platinum brought in close proximity to each other by organic π-conjugated bridging units, but the molecules involved also iron or ruthenium atoms. In four compounds, Fc-C≡C-bipy{[Pt(μ-σ,π-C≡CSiMe$_3$)$_2$]CuFBF$_3$} (**57**), [Fc-C≡C-bipy{(η^5-C$_5$H$_4$SiMe$_3$)$_2$Ti(μ-σ,π-C≡CSiMe$_3$)$_2$}Cu]PF$_6$ (**58**), [(η^5-Cp)(PPh$_3$)$_2$Ru-C≡C-bipy{(η^5-C$_5$H$_4$SiMe$_3$)$_2$Ti(μ-σ,π-C≡CSiMe$_3$)$_2$}Cu]BF$_4$ (**59a**) and [(η^5-Cp)(dppf)Ru-C≡C-bipy{(η^5-C$_5$H$_4$SiMe$_3$)$_2$Ti(μ-σ,π-C≡CSiMe$_3$)$_2$}Cu]PF$_6$ (**59b**), the copper(I) is η^2-coordinated by two C≡C bonds in the chelating ligand and in **58** and **59** additionally Cu(I) is chelate-bonded to 2,2′-bipyridine through nitrogen atoms. The CV curves obtained in THF or CH_2Cl_2 solutions of **58** and **59** exhibit quasi-reversible peaks corresponding to the Cu^I/Cu^0 redox couple and some additional peaks corresponding to the oxidation of iron(II) and/or ruthenium(II), which are beyond the scope of this Chapter. However, all E_o and ΔE_p values reported are given in Table 8 for comparison. For **57** the reduction of copper(I) is irreversible, suggesting[61] that this process takes place initially, and it results in fragmentation of the reactant. Similar irreversibility was in fact observed for **58** unless the potential scan was reversed just after the reduction peak. The behavior above supports the conclusion about some chemical transformation of the copper(0) reduction product. It should also be noted that only for **57** was the two-step reduction of bipy observed. The reduction of copper(I) center in **59** is more difficult than that in **58** (cf. the E_o values in Table 8) as a consequence of the increase in electron density. The suggestion above is in line with the observed positive shift of E_o potential for Ru^{II}/Ru^{III} couple after coordination of bipy to a bis(alkynyl)titanocene–copper(I) fragment in **59**, as well as with IR data[62].

A series of heterobimetallic compounds **42** and **43** with bis(η^2-alkynyl) titanocene (η^5-C$_5$H$_4$SiMe$_3$)$_2$Ti(C≡CSiMe$_3$)$_2$ as the π-tweezer (discussed in Section III.B.1) was extended[63] to tetranuclear complexes after the addition of two ferrocenyl groups to the π-tweezer. One of the resulting organocopper compounds {(η^5-C$_5$H$_4$SiMe$_3$)$_2$Ti(μ-σ,π-C≡CFc)$_2$}CuBr (**60**) was investigated at 298 K by the CV method in THF solutions containing 0.1 M TBAPF$_6$. The reversible pair of cathodic and anodic peaks was observed at $E_o = 0.01$ V vs. Fc^+/Fc ($\Delta E_p = 150$ mV) which corresponded to the transfer of a

TABLE 8. Potentials E_o and anodic–cathodic peak separations ΔE_p from quasi-reversible CV curves recorded at a platinum electrode for solutions of **57**, **58** and **59** at 298 K[a]

Compound	Solvent	Cu^I/Cu^0	bipy/bipy$^{-\bullet}$	bipy$^{-\bullet}$/bipy^{2-}	Fe^{II}/Fe^{III}	Ru^{II}/Ru^{III}	Reference
		$-E_o$ (V) ΔE_p (mV)	$-E_o$ (V) ΔE_p (mV)	$-E_o$ (V) ΔE_p (mV)	E_o (V) ΔE_p (mV)	E_o (V) ΔE_p (mV)	
57	THF	1.82[b]	1.60	2.23	0.16	—	61[c]
		—	130	120	120		
58	THF	1.39	2.67[b]	—	0.12	—	61[c]
		120	—		150		
59a	CH_2Cl_2	1.50	—	—	—	0.215	62
		110				100	
59b	CH_2Cl_2	1.52	—	—	0.545	0.20	62
		95			165	115	

[a] Potentials E_o and E_{pc} given vs. Fc^+/Fc scale; supporting electrolyte: 0.1M $TBAPF_6$; $\upsilon = 0.1$ V s^{-1}.
[b] E_{pc} value for the irreversible reduction peak.
[c] Reprinted with permission from Reference 61. Copyright 2006 American Chemical Society.

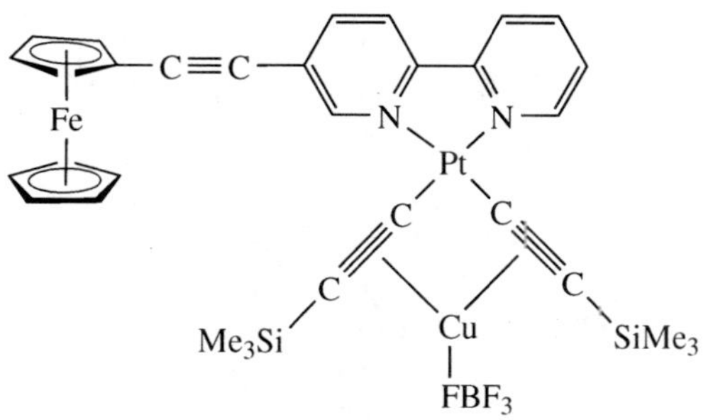

(**57**)

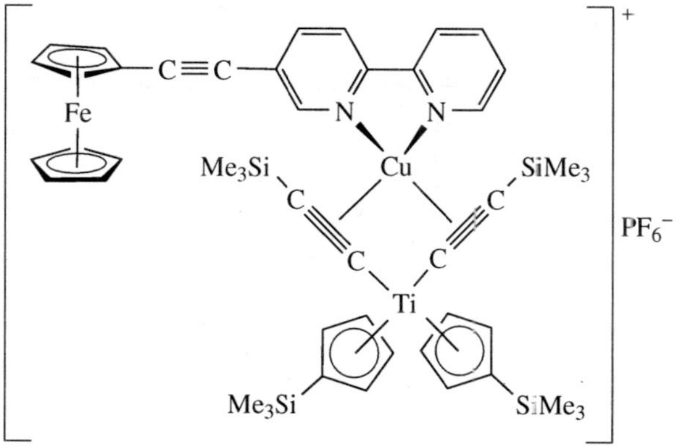

(**58**)

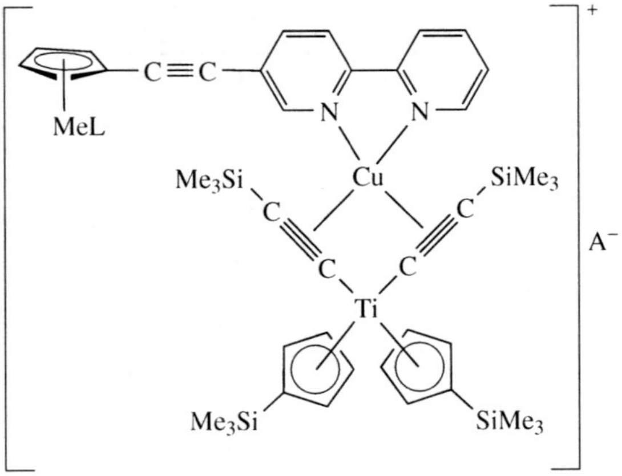

(**59**)

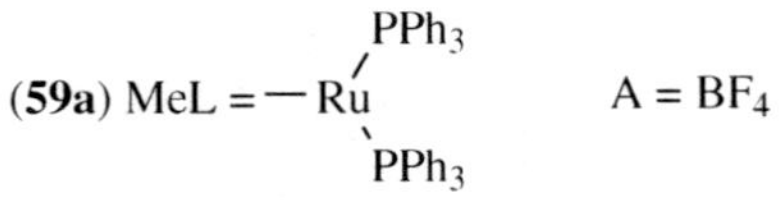

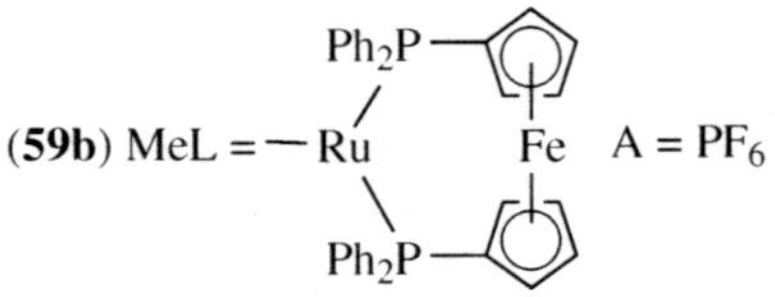

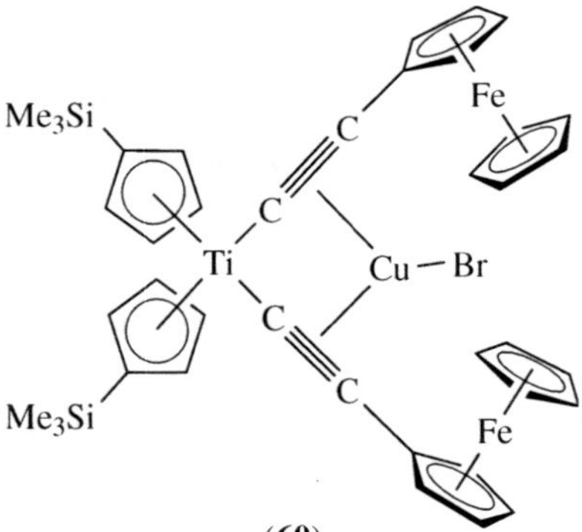

(**60**)

total of three electrons. A comparison with the CV behavior of **42** and of the π-tweezer itself[64,65] allowed the authors to assign the observed peaks to three synchronous steps: one for Cu^{I}/Cu^{II} couple and two for Fe^{II}/Fe^{III} couples. The reversibility of each step found in repeated multicyclic experiments[63] evidently demonstrated the stability of σ-bonds between titanium and alkynyl units after oxidation of the ferrocenyl groups. The last behavior is in sharp contrast to those reported[64,65] for the π-tweezer alone, namely the oxidation of the remote ferrocenyl units in $[(\eta^5\text{-}C_5H_4SiMe_3)_2Ti(\mu\text{-}\sigma,\pi\text{-}C{\equiv}CFc)_2]$ immediately induced the bond cleavage. Two irreversible reduction peaks also observed for **60** at potentials −1.71 and −1.96 V vs. Fc^+/Fc were assigned[63] to Cu^I/Cu^0 and Ti^{IV}/Ti^{III}, respectively.

In **61**, two titanium-tweezer fragments are coordinated to copper(I) atom forming the molecule which looks interesting for a comparative study of electronic communication between the remote titanium(IV) ions[66]. Cyclic voltammograms recorded at 298 K in THF solutions (0.1 M $TBAPF_6$, $\upsilon = 0.2\ V\,s^{-1}$) exhibited a pair of peaks at $E_o = 0.10$ V vs. Fc^+/Fc ($\Delta E_p = 130$ mV) that corresponded to the Cu^I/Cu^{II} couple, which was unexpected in light of unsuccessful attempts at a synthesis of copper(II) compounds with similar π-tweezers. The irreversible reduction of copper(I) was observed at $E_{pc} = -1.64$ V vs. Fc^+/Fc, typical of similar compounds[67], and the precipitation of elemental copper was then observed. However, the chemical nature of other products was unknown and thus it was impossible to draw reasonable conclusions as concerns the further irreversible reduction process observed at $E_{pc} = -2.26$ V vs. Fc^+/Fc, which probably corresponds to the Ti^{IV}/Ti^{III} couple[66].

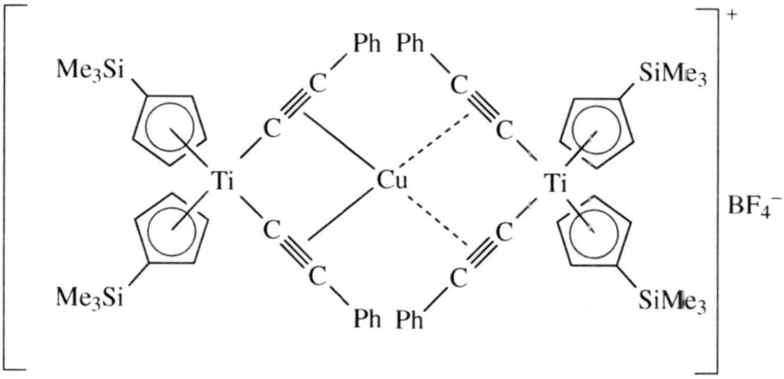

(**61**)

Tetrametallic bis(alkynyl) titanocene complexes **62** with two heterobimetallic Ti(IV)–Cu(I) or Ti(IV)–Ag(I) building blocks (similar to **42**) coupled together by a single halide or pseudo-halide bridge looked more promising for the investigation of electronic communication between both titanium centers[45,67]. However, electrochemical investigations[45,67] did not support this expectation, indicating that halide-bridged copper(I) or silver(I) centers act as an impedance rather than as a transmitter. Peak potentials obtained[45] for the reduction of **62** are given in Table 9; the values for organosilver compounds are also included for a clear demonstration of conclusions concerning the electrode mechanism proposed for compounds with a −Cu−X−Ag− bridge. The electroreduction of copper(I) and silver(I) centers is completely irreversible, resulting in fragmentation of the reactants, similar to what was observed[45] for the corresponding heterobinuclear compounds **42**. However, potentials for Cu^I/Cu^0 couples in **62a–62c** are more positive than for the corresponding **42b–42d** (cf. Table 2). The observed effect was explained[45] taking into account a decrease of the charge density at copper(I) ions in **62** due to the effect of the bridging halide ions. Moreover, numerical simulations of the CV curves supported the suggestion that cathodic peaks correspond to two one-electron reduction steps of tetranuclear and binuclear compounds, respectively[45]. An interesting conclusion was drawn from the fact that for **62d–62f**, peak potentials for the Ag^I/Ag^0 couple are less negative than for **62g–62i**

TABLE 9. Peak potentials E_{pc} for the reduction of **62** at a platinum electrode in THF solutions[a] at 298 K. Reprinted from Reference 45 with permission from Elsevier

Compound	$-E_{pc}$ (V vs. Fc^+/Fc)		Compound	$-E_{pc}$ (V vs. Fc^+/Fc)	
	Cu^I/Cu^0	Ti^{IV}/Ti^{III}		Ag^I/Ag^0	Ti^{IV}/Ti^{III}
62a	1.61	2.04	**62g**	1.42	1.69
62b	1.70	2.01	**62h**	1.64	2.05
62c	1.67	1.96	**62i**	1.54	2.03

Compound	$-E_{pc}$ (V vs. Fc^+/Fc)		
	Cu^I/Cu^0	Ti^{IV} Ti^{III}	Ag^I/Ag^0
62d	1.79	2.20	1.26
62e	1.74	1.99	1.28
62f	1.73	2.08	1.32

[a] Supporting electrolyte: 0.1 M $TBAPF_6$; $v = 0.2$ V s^{-1}.

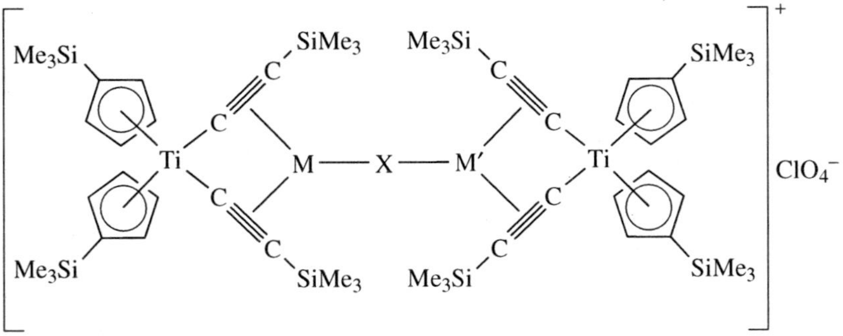

(62)

62	M	M′	X
(a)	Cu	Cu	Cl
(b)	Cu	Cu	Br
(c)	Cu	Cu	I
(d)	Cu	Ag	Cl
(e)	Cu	Ag	Br
(f)	Cu	Ag	I
(g)	Ag	Ag	Cl
(h)	Ag	Ag	Br
(i)	Ag	Ag	I
(j)	Cu	Cu	CN
(k)	Cu	Cu	SCN
(l)	Cu	Ag	SCN
(m)	Ag	Ag	SCN

with the same halide, but peak potentials for the Cu^I/Cu^0 couple are more negative than for the corresponding **62a–62c**. This behavior implies[43] that the reduction of **62d–62f** starts from the silver(I) center and results in the fragmentation into Ag^0 and **42b–42d**, respectively, which are further reduced to Cu^0 and bis(alkynyl) titanocene; other products of both processes were not identified. The proposed fragmentation is in line with the observed independence of potentials for the Ti^{IV}/Ti^{III} couple which were around −2 V (with the exception of **62g**), i.e. similar to those observed for the free π-tweezer ligand. A similar mechanism was also proposed[67] for thiocyanato-bridged compounds **62k–62m**.

Tetranuclear **63** exhibits[46] a similar electrochemical behavior as binuclear **43b**, discussed in Section III.B.1. Irreversible oxidation peaks observed in CH_2Cl_2 solutions (with 0.1 M $TBAPF_6$) at $E_{pa} = 0.95$ V and 1.10 V vs. Fc^+/Fc correspond most probably to the oxidation of the copper–thiolate system and the further peak at $E_{pa} = 1.47$ V vs. Fc^+/Fc to the oxidation of disulfide formed in an accompanying chemical reaction. Two reduction peaks with close potentials $E_{pc} = -1.74$ V and −1.88 V vs. Fc^+/Fc correspond to the reduction of copper(I) and a small difference between them points to the existence of weak electronic interactions[46].

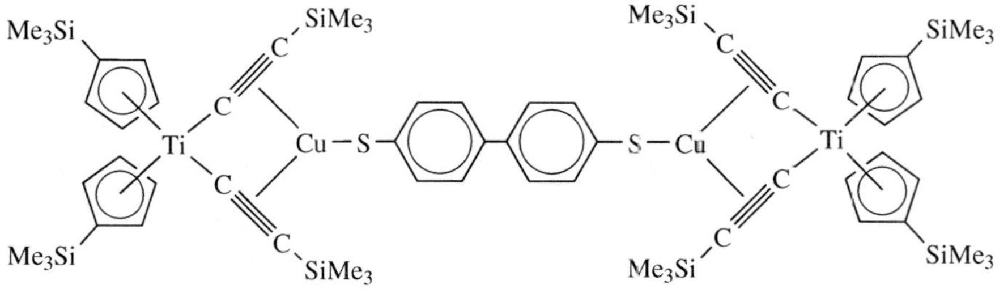

(**63**)

Trimetallic Ti(IV)–Cu(I)–Fe(II) and pentametallic Ti(IV)–Cu(I)–Fe(II)–Cu(I)–Ti(IV) compounds **64**, **65** and **66** were prepared[68,69] in the same laboratory using bis(alkynyl) titanocene and ferrocenecarboxylate as building blocks. Their electrochemical behavior was examined[68,69] as potentially interesting from the standpoint of electronic communication and the cooperative effect in catalysis. The electrode potentials obtained[68,69] for the reduction and oxidation processes are given in Table 10; for the irreversible reduction cathodic peak potentials are given, for reversible or quasi-reversible processes E_o values obtained from the CV curves and the corresponding ΔE_p values are shown. The first reduction process corresponds to Cu^I/Cu^0 and its irreversible character is similar to that observed for binuclear Cu–Ti compounds. The second reduction process, occurring at more negative potentials and sometimes showing the corresponding reoxidation peak, was assigned to the Ti^{IV}/Ti^{III} couple. It was emphasized[68] that the last process does not depend strongly on the chemical identity of the central part of molecules. This supports the suggested fragmentation of the product of the first reduction step. For oxidation processes after reversing the potential scan the cathodic peaks were always visible. Sometimes only one oxidation peak was found. However, if two anodic peaks were observed, one of them corresponded to the Cu^I/Cu^{II} couple and the other to the Fc/Fc^+ couple. Unfortunately, the identities of these processes were suggested only on the basis of a comparison with the oxidation of the parent ferrocenecarboxylic acids and were not persuasively proved. Nevertheless, the observation of only one pair of copper(I) redox processes for bicopper compounds **66** indicated[69] that electronic communication between Cu centers is not favored. In the final conclusion[68,69] the bridging carboxylate units were not recognized as transmitters.

Very recently, two heterotetranuclear Fe(II)–Au(I)–Cu(I)–Ti(IV) complexes **67** and **68** were investigated[70] by the CV methods and electrode potentials obtained are given

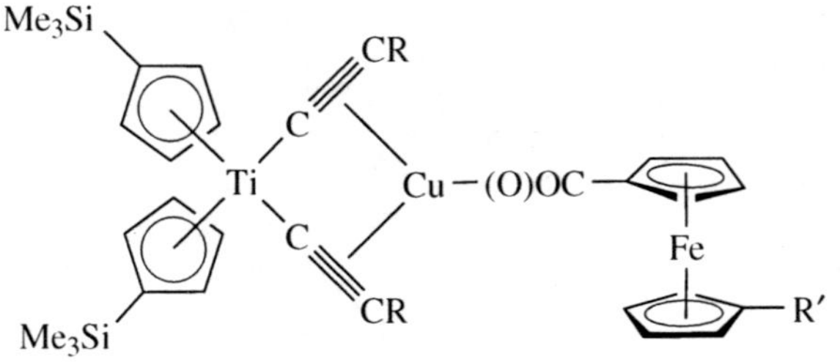

(**64a**) R = $SiMe_3$, R′ = H
(**64b**) R = *t*-Bu, R′ = COOH

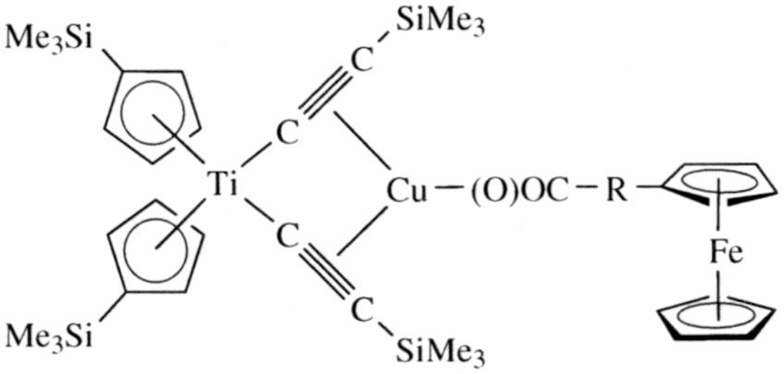

(**65a**) R = *trans*-CH=CH
(**65b**) R = CH_2CH_2

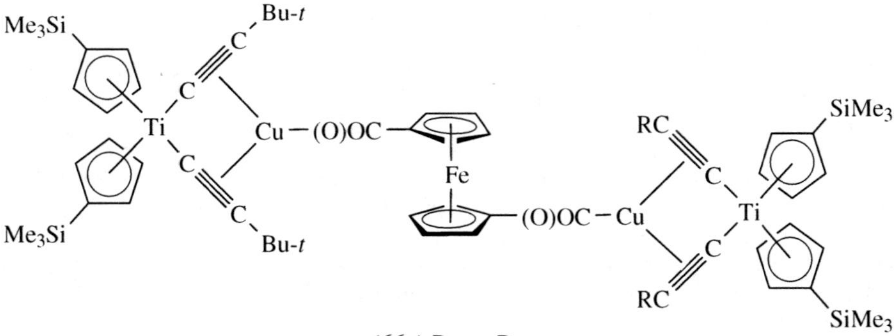

(**66a**) R = *t*-Bu
(**66b**) R = $SiMe_3$

in Table 11. The irreversible reduction of copper(I) in **67** results in fragmentation of the reactant and the π-tweezer $[(\eta^5\text{-}C_5H_4SiMe_3)_2Ti(C{\equiv}CSiMe_3)_2]$ formed is further reduced at a more negative potential, similar to what was previously observed[45] for other copper complexes with bis(alkynyl) titanocene. The conclusion above was supported[70] by multicyclic experiments, which showed an increase of the peak corresponding to the Ti^{IV}/Ti^{III} couple while the reduction peak of Cu^{I}/Cu^{0} disappeared[70].

TABLE 10. Potentials for the reduction and oxidation at a platinum electrode of **64**, **65** and **66** at 298 K[a]

Compound	Solvent	E_{pc} (V vs. Fc^+/Fc) or E_o (V vs. Fc^+/Fc) [ΔE_p (mV)]				Reference
		Reduction		Oxidation		
64a	CH_2Cl_2	−1.69	−1.92	−0.02 [95]	0.25 [105]	68
64b	THF	−1.79	−2.01 [100]	0.11 [70]	0.40 [70]	69[b]
65a	CH_2Cl_2	−1.73	−1.92	0.05 [90]	—	68
65b	CH_2Cl_2	−1.70	−1.92	−0.03 [100]	—	68
66a	THF	−1.84	−2.00 [120]	0.09 [80]	0.39 [90]	69[b]
66b	THF	−1.81	−1.99 [110]	0.11 [60]	0.41 [60]	69[b]

[a] Supporting electrolyte: 0.1M $TBAPF_6$; $\upsilon = 0.1$ V s^{-1}.
[b] Reprinted with permission from Reference 69. Copyright 1999 American Chemical Society.

TABLE 11. Reduction and oxidation potentials at a platinum electrode for THF solutions[a] of **67** and **68** at 298 K[70]

Compound	E_{pc} (V vs. Fc^+/Fc) or E_o (V vs. Fc^+/Fc) [ΔE_p (mV)]			
	Reduction Cu^I/Cu^0	Ti^{IV}/Ti^{III}	bipy/$bipy^{\bullet -}$	Oxidation Fe^{II}/Fe^{III}
67	−1.30	−1.80 [180]	—	0.28 [115]
68	−1.41 [115]	−1.635 [180] −1.76 [180]	−2.72 [160]	0.27 [115]

[a] Supporting electrolyte: 0.1M $TBAPF_6$; $\upsilon = 0.1$ V s^{-1}.

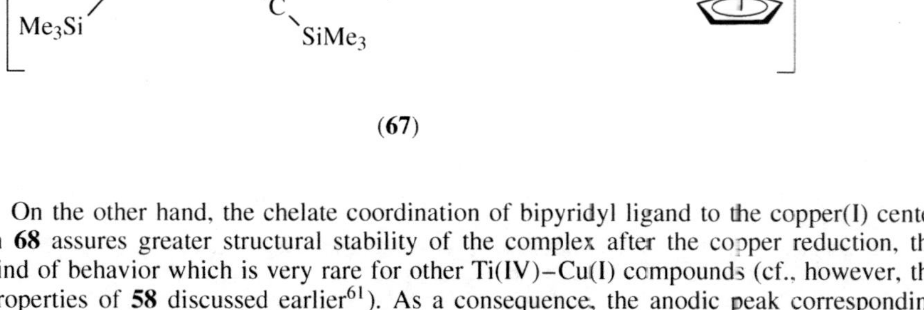

(**67**)

On the other hand, the chelate coordination of bipyridyl ligand to the copper(I) center in **68** assures greater structural stability of the complex after the copper reduction, the kind of behavior which is very rare for other Ti(IV)–Cu(I) compounds (cf., however, the properties of **58** discussed earlier[61]). As a consequence, the anodic peak corresponding to partial reoxidation of the copper(0) was observed[70] as well as two pairs of peaks at $E_o = -1.635$ V and -1.76 V for the Ti^{IV}/Ti^{III} reduction in **68** and in the free π-tweezer molecule, respectively. The reversible one-electron reduction of bipy was also observed (Table 11).

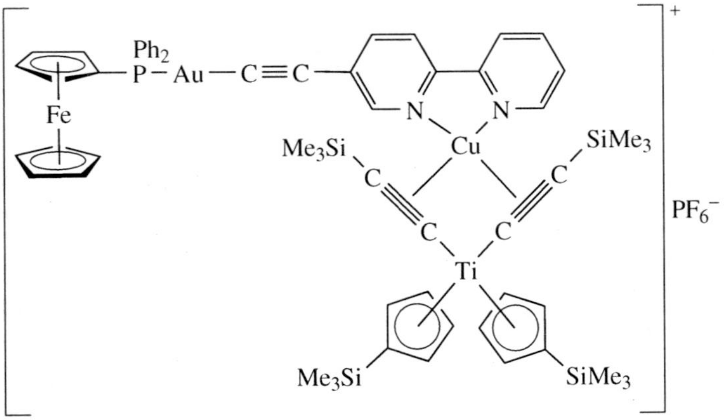

(**68**)

IV. CONCLUDING REMARKS

It is evident from this review that electrochemical methods were mostly used to investigate the stability of new synthesized organocopper compounds at different oxidation states as well as to study the problem of electronic communication between redox centers in molecules. For that purpose the most desirable electrochemical data were reversible formal potentials for a given redox couple. The irreversibility observed in many cases was only valuable to indicate that the products of electrode reactions are unstable. On the other hand, the mechanism of electrode processes was beyond the authors' interest and its examination is still open for electrochemists' activity. Contemporary CV measurements are able to prove whether the overall irreversibility is caused by the following chemical reaction, by the slow electron transfer itself or whether the process should rather be described as the concerted electron transfer and bond breaking. Thus, modern synthetic routes leading to multinuclear metalloorganic complexes with fascinating structures make it a challenge to use advanced electrochemical investigations in this area in order to recognize the details of electrode mechanisms. In view of recent achievements in the synthesis of novel organocopper compounds one can suppose as well that their ability to undergo electron transfer processes, not accessible for direct investigations in the case of unstable compounds, like methylcopper reagents for which attempts to measure oxidation potentials failed[71], will be available for more stable analogous compounds[72, 73].

V. REFERENCES

1. C. K. Mann and K. Barnes, *Electrochemical Reactions in Nonaqueous Systems*, Chap. 13.6, Dekker, New York, 1970.
2. H. Lehmkuhl, in *Organic Electrochemistry. An Introduction and a Guide* (Ed. M. M. Baizer), Chap. 18, Dekker, New York, 1973, pp. 621–678.
3. D. A. White, in *Organic Electrochemistry. An Introduction and a Guide* (Eds M. M. Baizer and H. Lund), 2nd ed., Chap. 19, Dekker, New York, 1983, pp. 591–636.
4. A. P. Tomilov, I. N. Chernyh and Yu. M. Kargin, *Elektrokhimiya elementoorganicheskih soiedinienij*, Nauka, Moscow, 1985.
5. H. B. Fyfe, M. Mlekuz, G. Stringer, N. J. Taylor and T. B. Marder, in *Inorganic and Organometallic Polymers with Special Properties* (Ed. R. M. Laine), NATO ASI Series, Vol. 206, Kluwer Academic Publishers, Dordrecht, 1992, pp. 331–344.
6. D. Astruc, *Acc. Chem. Res.*, **30**, 383 (1997).

7. S. E. Manahan, *Inorg. Chem.*, **5**, 2063 (1966).
8. A. P. Tomilov, I. N. Chernyh and Yu. M. Kargin, *Elektrokhimiya elementoorganicheskih soiedinienij*, Nauka, Moscow, 1985, p. 23.
9. B. M. Mykhalichko and M. G. Mys'kiv, Ukraine Patent UA25450A (1998), Bull. N.6; cited in References 23–33.
10. E. A. Goreshnik and M. G. Mys'kiv, *Pol. J. Chem.*, **73**, 1245 (1999).
11. E. A. Goreshnik, *Pol. J. Chem.*, **73**, 1253 (1999).
12. E. A. Goreshnik and M. G. Mys'kiv, *Russ. J. Coord. Chem.*, **25**, 137 (1999).
13. E. A. Goreshnik, A. V. Pavlyuk, D. Schollmeyer and M. G. Mys'kiv, *Russ. J. Coord. Chem.*, **25**, 653 (1999).
14. E. A. Goreshnik, V. N. Davydov, A. V. Pavlyuk and M. G. Mys'kiv, *Russ. J. Coord. Chem.*, **25**, 732 (1999).
15. E. A. Goreshnik, D. Schollmeyer, M. G. Mys'kiv and O. V. Pavl'uk, *Z. Anorg. Allg. Chem.*, **626**, 1016 (2000).
16. E. A. Goreshnik, D. Schollmeyer and M. G. Mys'kiv, *Z. Anorg. Allg. Chem.*, **628**, 2118 (2002).
17. E. A. Goreshnik, D. Schollmeyer and M. G. Mys'kiv, *Z. Anorg. Allg. Chem.*, **629**, 2040 (2003).
18. E. Goreshnik, D. Schollmeier and M. Mys'kiv, *Acta Crystallogr., Sect. C*, **59**, m478 (2003).
19. E. A. Goreshnik and M. G. Mys'kiv, *Russ. J. Coord. Chem.*, **29**, 505 (2003).
20. E. A. Goreshnik and M. G. Mys'kiv, *Russ. J. Coord. Chem.*, **29**, 871 (2003).
21. E. A. Goreshnik, L. Z. Ciunik, Y. K. Gorelenko and M. G. Mys'kiv, *Z. Anorg. Allg. Chem.*, **630**, 2743 (2004).
22. O. V. Pavlyuk, E. A. Goreshnik, Z. Ciunik and M. G. Mys'kiv, *Z. Anorg. Allg. Chem.*, **631**, 793 (2005).
23. E. A. Goreshnik, D. Schollmeyer and M. G. Mys'kiv, *Russ. J. Coord. Chem.*, **31**, 185 (2005).
24. E. A. Goreshnik, V. N. Davydov and M. G. Mys'kiv, *Russ. J. Coord. Chem.*, **31**, 286 (2005).
25. E. A. Goreshnik and M. G. Mys'kiv, *Russ. J. Coord. Chem.*, **31**, 341 (2005).
26. E. A. Goreshnik, D. Schollmeyer and M. G. Mys'kiv, *Russ. J. Coord. Chem.*, **31**, 429 (2005).
27. E. A. Goreshnik, D. Schollmeyer and M. G. Mys'kiv, *J. Chem. Crystallogr.*, **35**, 565 (2005).
28. E. Goreshnik, D. Schollmeyer and M. Mys'kiv, *Acta Crystallogr., Sect. C*, **61**, m127 (2005).
29. E. Goreshnik, V. Karpyak and M. Mys'kiv, *Acta Crystallogr., Sect. C*, **61**, m390 (2005).
30. E. A. Goreshnik and M. G. Mys'kiv, *Russ. J. Coord. Chem.*, **32**, 25 (2006).
31. E. A. Goreshnik and M. G. Mys'kiv, *Z. Anorg. Allg. Chem.*, **633**, 1723 (2007).
32. E. A. Goreshnik and M. G. Mys'kiv, *Russ. J. Coord. Chem.*, **34**, 443 (2008).
33. E. A. Goreshnik, D. Schollmeyer and M. G. Mys'kiv, *Z. Anorg. Allg. Chem.*, **631**, 835 (2005).
34. M. F. Kamte, U. Baumeister and W. Schäfer, *Z. Anorg. Allg. Chem.*, **629**, 1919 (2003).
35. M. F. Kamte, C. Wagner and W. Schaefer, *J. Coord. Chem.*, **57**, 55 (2004).
36. M. F. Kamte, *Electrosynthesis and Mechanism of Copper(I) Nitrile Complexes*, PhD Thesis, Martin-Luther-Universität, Halle-Wittenberg, 2004; http://nbn-resolving.de/urn/resolver.pl?urn=nbn%3Ade%Agbv%3A3-000007437.
37. N. Navon, A. Masarawa, H. Cohen and D. Meyerstein, *Inorg. Chim. Acta*, **261**, 29 (1997).
38. L. F. Warren, Jr. and M. F. Hawthorne, *J. Am. Chem. Soc.*, **90**, 4823 (1968).
39. R. M. Wing, *J. Am. Chem. Soc.*, **89**, 5599 (1967).
40. R. M. Wing, *J. Am. Chem. Soc.*, **90**, 4828 (1968).
41. K. Araki, H. Winnischofer, H. E. Toma, H. Maeda, A. Osuka and H. Furuta, *Inorg. Chem.*, **40**, 2020 (2001).
42. C. J. Adams and P. R. Raithby, *J. Organometal. Chem.*, **578**, 178 (1999).
43. S. L. James, M. Younus, P. R. Raithby and J. Lewis, *J. Organometal. Chem.*, **543**, 233 (1997).
44. S. Back, T. Stein, W. Frosch, I-Y. Wu, J. Kralik, M. Büchner, G. Huttner, G. Rheinwald and H. Lang, *Inorg. Chim. Acta*, **325**, 94 (2001).
45. T. Stein, H. Lang and R. Holze, *J. Electroanal. Chem.*, **520**, 163 (2002).
46. H. Lang, K. Rößler, D. Taher, R. Holze and B. Walfort, *Inorg. Chim. Acta*, **361**, 1659 (2008).
47. R. I. Papasergio, C. L. Raston and A. H. White, *J. Chem. Soc., Dalton Trans.*, 3085 (1987).
48. K. Wada, M. Tamura and J. Kochi, *J. Am. Chem. Soc.*, **92**, 6656 (1970).
49. P. B. Hitchcock, M. F. Lappert and M. Layh, *J. Chem. Soc., Dalton Trans.*, 1619 (1998).
50. P. B. Hitchcock, M. F. Lappert, M. Layh and A. Klein, *J. Chem. Soc., Dalton Trans.*, 1455 (1999).
51. V. W. W. Yam, W. K. Lee and T. F. Lai, *Organometallics*, **12**, 2383 (1993).
52. V. W. W. Yam, W. K. Lee and P. K. Y. Yeung, *J. Phys. Chem.*, **98**, 7545 (1994).

53. V. W. W. Yam, W. K. Lee, K. K. Cheung, B. Crystall and D. Phillips, *J. Chem. Soc., Dalton Trans.*, 3283 (1996).
54. V. W. W. Yam, W. K. M. Fung and M. T. Wong, *Organometallics*, **16**, 1772 (1997).
55. V. W. W. Yam, W. K. M. Fung, K. M. C. Wong, V. C. Y. Lau and K. K. Cheung, *J. Chem. Soc., Chem. Commun.*, 777 (1998).
56. V. W. W. Yam, W. K. M. Fung and K. K. Cheung, *Organometallics*, **17**, 3293 (1998).
57. V. W. W. Yam, W. K. M. Fung and K. K. Cheung, *J. Cluster Sci.*, **10**, 37 (1999).
58. W. Y. Lo, C. H. Lam, V. W. W. Yam, N. Zhu, K. K. Cheung, S. Fathallah, S. Messaoudi, B. Le Guennic, S. Kahlal and J. F. Halet, *J. Am. Chem. Soc.*, **126**, 7300 (2004).
59. C. Hansch, A. Leo and R. W. Taft, *Chem. Rev.*, **91**, 165 (1991).
60. J. H. K. Yip, J. Wu, K. Y. Wong, K. W. Yeung and J. J. Vittal, *Organometallics*, **21**, 1612 (2002).
61. R. Packheiser, B. Walfort and H. Lang, *Organometallics*, **25**, 4579 (2006).
62. R. Packheiser, P. Ecorchard, B. Walfort and H. Lang, *J. Organometal. Chem.*, **693**, 933 (2008).
63. S. Back, G. Rheinwald and H. Lang, *J. Organometal. Chem.*, **601**, 93 (2000).
64. Y. Hayashi, M. Osawa and Y. Wakatsuki, *J. Organometal. Chem.*, **542**, 241 (1997).
65. S. Back, H. Pritzkow and H. Lang, *Organometallics*, **17**, 41 (1998).
66. H. Lang, T. Stein, S. Back and G. Rheinwald, *J. Organometal. Chem.*, **689**, 2690 (2004).
67. H. Lang and Th. Stein, *J. Organometal. Chem.*, **641**, 41 (2002).
68. J. Kühnert, M. Lamač, T. Rüffer, B. Walfort, P. Štěpnička and H. Lang, *J. Organometal. Chem.*, **692**, 4303 (2007).
69. W. Frosch, S. Back and H. Lang, *Organometallics*, **18**, 5725 (1999).
70. R. Packheiser, A. Jakob, P. Ecorchard, B. Walfort and H. Lang, *Organometallics*, **27**, 1214 (2008).
71. Y. Chounan, T. Ibuka and Y. Yamamoto, *J. Chem. Soc., Chem. Commun.*, 2003 (1994).
72. W. Frosch, S. Back, G. Rheinwald, K. Köhler, H. Pritzkow and H. Lang, *Organometallics*, **19**, 4016 (2000).
73. M. D. Janssen, K. Köhler, M. Herres, A. Dedieu, W. J. J. Smeets, A. L. Spek, D. M. Grove, H. Lang and G. van Koten, *J. Am. Chem. Soc.*, **118**, 4817 (1996) and references cited therein.

CHAPTER **7**

Gas-phase chemistry of organocopper compounds

AL MOKHTAR LAMSABHI and MANUEL YÁÑEZ

Departamento de Química, Universidad Autónoma de Madrid, Cantoblanco, 28049-Madrid, Spain
Fax: +34-91-4975238; e-mail: manuel.yanez@uam.es

and

JEAN-YVES SALPIN and JEANINE TORTAJADA

Université d'Evry Val d'Essonne — Laboratoire d'Analyse et Modélisation pour la Biologie et l'Environnement-CNRS — UMR 8587 — Bâtiment Maupertuis, Boulevard François Mitterrand, 91025 Evry, France
Fax: +33-1 69 47 76 55; e-mail: jean-yves.salpin@univ-evry.fr

I. INTRODUCTION 280
II. PERFORMANCE OF DIFFERENT THEORETICAL MODELS 281
III. LIGAND-FIELD SPECTROSCOPY AND PHOTOFRAGMENTATION PROCESSES 285
IV. Cu^+ AND Cu^{2+} BINDING ENERGIES 289
A. Role of Agostic-type Interactions on Cu^+ Binding Energies 296
B. Binding Energies of Solvated Cu^+ and Cu^{2+} Ions 298
V. COORDINATION 300
A. Coordination in Complexes with One Ligand 301
B. Coordination in Complexes with More than One Ligand 303
C. Coordination of Cu(II) in Blue Proteins 307
VI. BONDING 308
A. Bonding in Cu^+ Complexes 308
B. Bonding in Cu^{2+} Complexes 309
VII. Cu^+/Cu^{2+} REACTIONS WITH SMALL ORGANIC BASES 310
VIII. GAS-PHASE REACTIVITY OF COPPER(I) AND COPPER(II) IONS TOWARD MOLECULES OF BIOLOGICAL RELEVANCE 318

The chemistry of organocopper compounds
Edited by Z. Rappoport and I. Marek

A. Reactivity with Amino Acids and Peptides 318
1. Binary mixtures 319
2. Ternary complexes 326
3. Interaction with proteins 330
B. Reactivity with Saccharides and Derivatives 332
C. Reactivity with Nucleic Acid Building Blocks 337
IX. REFERENCES 340

I. INTRODUCTION

Organocopper compounds play very important roles in chemistry and biochemistry, and they have received a great deal of attention as useful reagents or because they are basic components of proteins and enzymes. Among the former, the so-called blue copper proteins are particularly relevant because they exhibit high reduction potentials. Actually, for some proteins the presence of Cu(II) is required to interact with the appropriate peptides or other proteins to carry out their specific activity. Copper forms also part of oxidation enzymes such as catechol oxidase[1], peptidylglycine α-hydroxylating monooxygenase[2] or indophenoloxidase as well as of many other biomolecules like tyrosinase[3], superoxide dismutase[4] or copper amine oxidase[5] whose biochemical activity is directly related to the presence of this transition metal in different oxidation states. Not surprisingly, many efforts have been devoted to unravel the mechanisms associated with the biochemical activity of these systems, to determine which is the active site and what is the role played by the metal. Different approaches can be found in the literature with this common objective, but only two subsets can be considered gas-phase investigations: those based on the use of mass spectrometry techniques, and those based on computational models of non-solvated systems. Of course, most of the studies carried out on the proteins or enzymes themselves do not fall in this category, and neither do many other studies, based on the use of appropriate chemistry model systems in the condensed phase. However, a great deal of effort was devoted to understand the intrinsic behavior of these complicated systems, by investigating the interactions of some of its basic components with Cu(I) and Cu(II). In this respect it is important to mention that there is more and more evidence that, more often than expected, the behavior of biochemical systems in the physiological medium is closer to that shown in the gas phase than to that shown in solution, reflecting the low polarity of the physiological medium. Good examples are provided by the acidities of uracil and uracil analogues[6,7]. The enhanced intrinsic acidity of the N1 site is consistent with the stability of anionic uracil in the active site of uracil-DNA glycosylase, which is, conversely, contrary to the expectation based on solution acidities[8,9]. Also, it has been found, for instance, that blue copper proteins show structures which are very close (within 1.7 kcal mol^{-1}) to their optimal gas-phase geometries[10], and that the effects of base stacking and hydrogen bonding on DNA duplexes are similar in the gas phase and in solution[11], although the latter dominate in the gas phase.

This research activity was carried out either through the use of mass spectrometry techniques, through the use of computational models or by a wise combination of both. In the sections which follow we shall present an overview of the reactivity of Cu(I) and Cu(II), the two oxidation states which are more relevant from the chemical viewpoint, with molecules of different size, and different characteristics. Since both oxidation states are present in biochemical media, a great deal of attention was paid, mainly through the use of different mass spectrometry techniques, to the interactions of these two metal ions with small or medium-size systems, which are basic components of larger biochemical compounds or which constitute reasonable model compounds of larger and more complicated systems, in an effort to understand their mode of action. A review dealing with copper complexes of a variety of organic and bioorganic molecules generated experimentally by different ionization methods has been recently published, mainly focused on ternary complexes of Cu(II)[12]. Hence, we shall not discuss in detail in this chapter this kind of complexes, and we will focus a little bit more on the use of ternary complexes to generate odd-electron ions of biomolecules. On the other hand, in order to keep the number of citations within reasonable limits we will refer exclusively to papers published in the last ten years, although papers published before 1998 might be quoted if necessary.

Although gas-phase experimental investigations involving species containing Cu(I) and Cu(II) are numerous, gas-phase experimental results on neutral organocopper compounds are, on the contrary, very rare. The very few studies reported in the literature are reduced to infrared or photochemical investigations of some small species[13,14]. A section will be devoted to overview the spectroscopic and photochemical aspects of organocopper compounds and to the photofragmentations produced.

As mentioned above, a great deal of the information on bonding, structure, binding energies, catalytic effects etc. comes from computational approaches, so we thought it convenient to include in this chapter a section devoted to analyzing the performance of the different theoretical models available for the treatment of Cu-containing systems as well as the problems which arise depending on the oxidation state of the metal.

Particular attention will be also paid to the coordination of Cu as a function of its oxidation state, because this coordination sometimes dictates the behavior of many Cu-containing proteins or enzymes.

It is also common that metal cations such as Cu(II) when dissolved in polar liquids show Jahn–Teller distortions, which of course have important effects on its specific interaction with different kind of ligands. Hence, some attention will be also paid to this question.

II. PERFORMANCE OF DIFFERENT THEORETICAL MODELS

Since many of the results on structure, bonding, coordination and other properties of organocopper compounds to be discussed in forthcoming sections are based on or use computational techniques, we have considered it pertinent to begin this chapter by presenting a brief overview on the performance of *ab initio* and density functional theory (DFT) calculations normally used to offer a rationale to the properties of these compounds. For more general and complete analyses of the performance of different computational techniques when dealing with molecular ions in the gas phase, the reader is referred to two general reviews published recently[15,16], where *ab initio* molecular orbital theory, DFT, quantum Monte Carlo theory and the methods to calculate the rate of complex chemical reactions involving ionic species in the gas phase are analyzed.

Nowadays, it is well established that accounting for electron correlation effects is practically mandatory in any computational method aiming at a correct description of a chemical system. In the framework of the molecular orbital theory this is usually done by using the Møller–Plesset (MPn) perturbation theory, at different orders ($n = 2, 3, 4, \ldots$) or coupled cluster (CCSD(T)) approaches. Correlation effects are particularly important when dealing with transition metal ions in general and with Cu(I) and Cu(II) in particular, because they can be considered as highly-correlated systems with significant electron clustering. Therefore, models based on Hartree–Fock (HF) calculations are usually very unreliable. It has been unambiguously shown, for instance, that the HF method yields wrong coordination numbers. This is indeed the case for different mixed complexes between Cu(I) and water and ammonia, where the three-coordination is systematically preferred at the HF level[17], whereas the inclusion of correlation effects reduces in all cases the level of coordination to two. Similarly, the fact that the HF approach exaggerates the electrostatic and polarization contributions to the stabilization energies results in wrong coordination numbers in $Cu(H_2O)_n^+$ ($n = 3$–6) complexes. Whereas at the HF level the three-coordinated complexes were the most stable minima for $n = 3–5$ and the four-coordinated one for $n = 6$, the inclusion of electron correlation effects reduces the optimum coordination number to two in all cases[18]. For smaller clusters, in particular for the dehydrated system, the RHF method predicts the D_{2d} structure to be the only stable form of the complex, while MP2 and CCSD(T) methods predict a lower symmetry (C_2) for its ground state. It has also been found that in molecular dynamic simulations based on QM/MM approaches, the inclusion of correlation effects leads to a significant improvement in the characterization of the Jahn–Teller effects observed[19] in the solvation of Cu^{2+}. Correlation effects are also important when evaluating relative binding energies of Cu^+ to guanine and adenine where, in general, the HF values are too small because they do not account for the stabilizing intersystem correlation energy and do not describe properly the covalent character of the bonds between the metal and the base[20]. Similar findings have been reported for the interactions between Cu^+ and DNA base pairs[21].

Much less demanding than the MP2 or CCSD(T) methods to account for dynamical correlation effects are the different methods based on the DFT. Among the different methods based on the local density approximation (LDA) or the gradient generalized approximation (GGA), the most commonly used is the B3LYP hybrid functional which combines the Becke's three-parameter nonlocal hybrid exchange potential[22] with the nonlocal correlation functional of Lee, Yang and Parr[23]. This approach has been shown to yield reliable geometries for a wide variety of systems[24–28] and its performance to describe Cu^+ complexes has been also assessed in different studies. Luna and coworkers[29] showed that for Cu^+–nitrogen base complexes this DFT approach was a very good alternative, as far as binding energies, vibrational frequencies, rotational constants and electron densities were concerned, to high-level *ab initio* methods such as G2 theory or CCSD(T) calculations. Almost simultaneously, in their study of the complexes of α-amino acids with Cu^+, Hoyau and Ohanessian[30] found that the agreement between B3LYP and MP2 optimized geometries was excellent, although the DFT approach places the Cu^+ ion systematically closer to the basic site of the amino acid. Later, Sodupe and coworkers[31] also found that the B3LYP method was very reliable for the study of Cu^+ and Cu^{2+} complexes with glycine. Also importantly, Yáñez and coworkers[32] found that the B3LYP method was superior to G2 and other high-level *ab initio* approaches when trying to reproduce the Cu^+ affinities of a wide set of neutral bases. Similarly, the B3LYP method was found to be superior to MP2 in reproducing the sequential binding energies of $[Cu(NH_3)_n]^+$ ($n = 1$, 2, 3, 4) complexes[33] and the relative Cu^+ ion affinities of glycine and alanine[34], although the performance is slightly worse when dealing with absolute values[35]. Klippenstein and Yang[36] also found reasonably good agreement between CCSD(T) and B3LYP Cu^+ binding energies, although the latter are almost systematically about 4 kcal mol^{-1} higher than the former. As we shall discuss later in a little more detail, Sodupe and coworkers found for different Cu^{2+} complexes[31,37,38] that functionals with a larger amount of exact exchange than B3LYP, such as BHLYP, yielded results in close agreement with CCSD(T) values.

It is important to emphasize, however, that not always do DFT and MP2 approaches yield the same optimized geometries for Cu^+ complexes. This is the case, for instance, for some aromatic compounds including benzene. DFT methods favor for benzene–Cu^+ complex the less symmetric (C_s) η^2 structure, whereas the MP2 method yields as the ground state a C_{6v} η^6 conventional π-complex[39]. Similar behavior was reported for phenylsilane–Cu^+ and phenylgermane–Cu^+ complexes[40]. Also, while the MP2 method predicts a η^6 conventional π-structure for the complexes of aniline and phenol with Cu^+, the B3LYP approach yields as global minima η^1 complexes in which the metal ion is bonded to the *para* carbon atom[41,42]. More importantly, when correlation corrections are included beyond second order by carrying out quadratic configuration interaction calculations (QCISD), the structures predicted by the DFT approach are clearly stabilized with respect to MP2 ones. As a matter of fact, for benzene, QCISD/6-311G(d,p) calculations predict both the η^6 conventional π-complex found at the MP2 level, and the η^2 structure found when the B3LYP approach is used, to be local minima of the potential energy surface (PES), the former being only slightly more stable. For phenol and aniline, both η^1 and η^6 structures are also found to be practically degenerate[41]. The origin of the stabilization of these lower symmetry structures in which the metal cation interacts specifically with a CC bond will be discussed later on, when analyzing the behavior of other unsaturated systems.

An added difficulty inherent to the theoretical treatment of transition metal ions is the existence of different states close in energy, so nondynamical correlation effects may also be important, in which case single-reference methods, as are most of the standard *ab initio* and density functional approaches, become unreliable. For the particular case of Cu(I) this situation is aggravated by the fact that, very often, more than one solution is obtained in the HF calculations, which is the origin of dramatic failures when the G2 approach is used to evaluate Cu(I) binding energies. As a matter of fact, when more than one HF solution exist none is a good representation of the wave function, and even high-correlated methods based on them become unreliable[43]. This is already the case when dealing with CuH^+, the simplest system containing Cu^+, where two different HF solutions can be found, leading to wrong descriptions of the ion even at the CCSD(T) level of theory[43]. The situation is even worse for CuO^+ where up to three different HF solutions can be found[32]. The most dramatic consequence is that when the standard G2 procedure, as implemented in the Gaussian series of programs[44], is used, both CuH^+ and CuO^+ are predicted to be unbound. For the particular case of CuO^+, this unphysical result is due to the fact that the G2 standard procedure starts from an unstable HF wave function. In principle, such solutions can be discarded 'a priori' by analyzing the stability of the HF wave function or by obtaining the right orbitals in the initial guess of the HF calculations[45]. Nevertheless, as we have mentioned above, when more than one HF solutions are found none of them is a good zeroth-order wave function and, as a consequence, even when the right orbitals are chosen, the G2 and CCSD(T) estimated binding energies are still off with respect to the experimental value by more than 15.5 and 14.3 kcal mol^{-1}, respectively[45].

It must be also taken into account that in Gn ($n = 1, 2, 3$) theories, the correlation energy is estimated by assuming additivity of partial contributions evaluated at different levels, by means of the MPn perturbation series which, for Cu(I)-containing systems, converges very slowly. As illustrated in Figure 1, already for the bare Cu^+ ion the MPn series up to fifth order presents significant oscillations, and these oscillations increase significantly when the basis set includes diffuse components and high angular momentum polarization functions[32]. This pathological behavior is typically found in systems which present a significant electron clustering[46] and is not exclusive of Cu(I)-containing systems, but it has been also found in other transition metal ions, such as Ni^+. For instance, $NiOH^+$ or $Ni(OH)_2{}^+$ presents two or four different HF solutions within a 2.5 kcal mol^{-1} energy gap, respectively[47]. $Ni^+(^2D)$ presents also similar strong oscillations of the MPn perturbation terms[48].

Interestingly, the behavior of different density functional approaches is much more regular, and for those cases in which several HF solutions exist, only a unique solution is found when the density functional theory is used[43,47]. Furthermore, it has been shown that the B3LYP/6-311+G(2df,2p) calculations on B3LYP/6-311G(d,p) optimized geometries yielded Cu(I) binding energies in very good agreement with the known experimental values[32]. This is probably one of the reasons why most of the computational studies reported on Cu-containing systems have been performed by using this DFT approach.

The situation is, however, slightly different as far as the next oxidation state of Cu is considered. On the one hand, and to the best of our knowledge, no pathological behavior

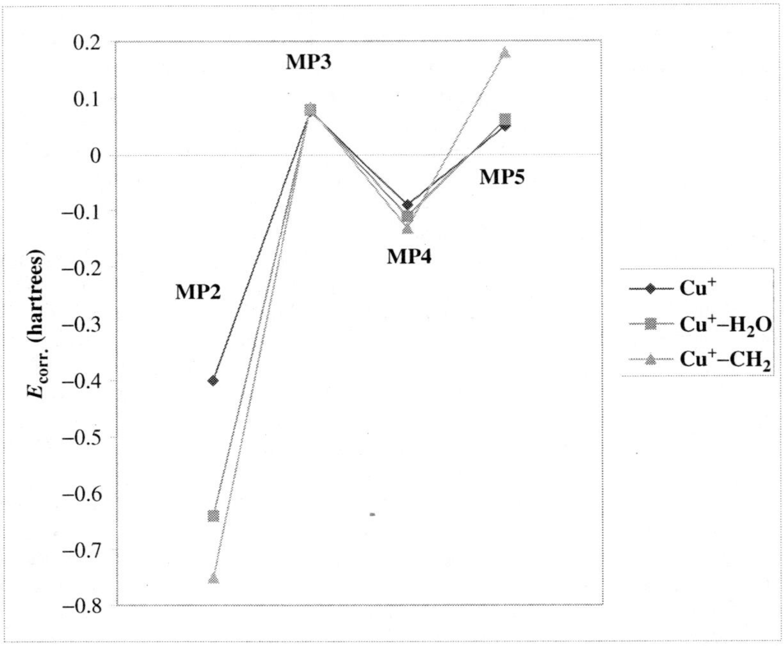

FIGURE 1. Electron correlation energies ($E_{corr.}$) evaluated using the MPn ($n = 2, 3, 4, 5$) perturbation series for Cu^+ and its complexes with H_2O and CH_2, showing that this perturbative series converges very slowly for Cu^+-containing systems. Values taken from Reference 32

has been described regarding *ab initio* calculations on Cu(II) complexes, in spite of the fact that Cu(II) is an open-shell species. On the other hand, there are some interesting differences with respect to Cu(I) concerning the performance of different DFT approaches. It has been shown[38] using $Cu(H_2O)^{2+}$ as a suitable model system that a correct description of its ground state depends significantly on the degree of mixing of the exact Hartree–Fock and DFT exchange functionals. MP2 and CCSD(T) calculations predicted for this complex a C_{2v} symmetry associated to a $^2A_1$ state. This is not an unexpected result taking into account that the $3d_z{}^2$ orbital is the one with larger overlap with H_2O, and the larger this overlap the larger the metal–ligand repulsion is and therefore the less stable becomes the corresponding electronic state. This is nicely reflected in the Cu–O distance which follows the same trend as the electronic states stability: $^2A_1 < {}^2B_1 < {}^2B_2 < {}^2A_2$. In contrast, most of the different hybrid and nonhybrid functionals found as global minimum a C_s structure associated to a $^2A'$ ground state, and only the BHLYP approach, where equal amounts of Hartree–Fock and DFT exchange are used, reproduced the CCSD(T) results. In fact, it was shown that within the C_{2v} symmetry the relative stabilities of the different electronic states are rather sensitive to the proportion of exact exchange introduced in the functional. The origin of this discrepancy is that both LDA and GGA functionals usually overestimate the stability of delocalized situations as those associated with C_s structures in contrast with C_{2v} ones where the electron hole is more localized at the metal ion. This problem is significantly reduced by increasing the contribution of the exact exchange, explaining the good behavior of the BHLYP method. For some systems larger than $Cu(H_2O)^{2+}$ it was found that this effect is not very significant[49], but it cannot be concluded that this will be, in general, the behavior to be expected. Hence, when dealing with Cu(II) compounds, it is convenient to verify whether the results obtained are sensitive to the amount of exact exchange included in the functional.

It is important to note that, as indicated above, Cu(II) is an open-shell system and the possible spin contamination in the corresponding unrestricted treatment can, in some specific cases, be important and it may render DFT calculations completely unreliable[50].

III. LIGAND-FIELD SPECTROSCOPY AND PHOTOFRAGMENTATION PROCESSES

Ligand-field spectroscopy has been considered traditionally a very useful technique in organometallic chemistry, but experiments are usually undertaken in the condensed phase. This implies that the solvent and the presence of a counterion significantly limit the knowledge one can gain on the structure of the chromophore from the spectral transitions. However, significant advances have been made in the last decade in spectroscopy and photodissociation studies of size-selected metal ion complexes in the gas phase. Using a laser vaporization technique combined with a supersonic beam expansion, Cu^+–furan complexes were generated and analyzed by means of time-of-flight mass spectrometry techniques[51]. A ground-state binding energy of 37 kcal mol^{-1} for this complex[51] could be estimated from the photofragment spectra. Similarly, neutral Cu–benzene complexes, $Cu_n(benzene)_m$, were produced in the gas phase by using the laser vaporization method and characterized by mass spectrometry, photoionization spectroscopy and chemical probe experiments[13]. Although only the complexes with $n = 1$ and $m = 1, 2$ were generated, their benzene–Cu binding energies could be measured. The values obtained, 0.18 and 0.22 eV, respectively, are small and lower than those measured for early transition metals like Sc or Ti[13]. This is due to the fact that Cu has the *d* shell completely full, so no d–π interactions but repulsive interactions between the 4*s* orbital and the benzene ligand should be expected.

Photodissociation experiments have also been reported for Cu^+–pyridine complexes[52]. These complexes were obtained by combining the laser ablation technique with a supersonic molecular beam. Although when this technique is used, up to five pyridine molecules are coupled to Cu^+, only the photofragmentation of the monomer Cu^+–pyridine complex was investigated by irradiating with UV light in the 202–284.1 nm wavelength range. Even though one can reasonably expect the positive charge of the complex to be localized on the metal, since Cu has a lower ionization energy (7.72 eV) than pyridine (9.25 eV)[53], pyridine$^+$ was the only photoproduct detected in the experiments carried out. This means that the photodissociative process induced intramolecular ligand-to-metal charge transfer.

Similar experiments on Cu^+–ketone (acetone, 2-butanone, 3-methyl-2-butanone and 2-pentanone) complexes showed[54] a significantly different behavior to that outlined above for Cu^+–pyridine complexes. In this case the photofragmentation was produced by using UV radiation at 266 nm, which should correspond to ligand excitation. The first important finding is that no charged L^+ fragments are observed for any of the ketones investigated, in clear contrast also with the behavior of $[Ag(acetone)]^+$ complex[55]. The same behavior was observed when the number of ligands is greater than 2, in which case the loss of neutral ligands to produce very stable $[CuL_2]^+$ ions clearly dominates over bond cleavage processes within the ligand. Conversely, when the number of ligands is one or two, there are significant differences between the different ketones as well as in respect to Cu^+–pyridine complexes. For instance, $[Cu(acetone)]^+$ does not show any photodissociation, whereas 2-butanone and 2-pentanone lose both CH_3 and the complementary alkyl group simultaneously or sequentially, indicating that 266-nm photons have enough energy to cleave two α C–C bonds in the ligand. For 3-methyl-2-butanone the dominant dissociation process corresponds to the loss of CH_3. When the number of ligands is 2, with the only exception of 3-methyl-2-butanone, the dominant photofragment is $[CuLCO]^+$, which once more involves the cleavage of two α C−C bonds. In $[Cu(3\text{-methyl-2-butanone})]^+$ complexes only one metal–ligand bond is broken and the dominant product is $[CuL]^+$. Very recently, a photofragmentation and theoretical study of $[Cu\text{-}(PhOH)(PhO)]^+$ complex[42] showed that the phenolate anion interacts with the metal preferentially through the oxygen atom, which is accompanied by an electron transfer which locates the unpaired electron on the aromatic moiety. Furthermore, the bond strength between copper and the oxygen atom of the phenoxy radical is weakened by the presence of neutral phenol.

Spectroscopic information of complexes involving neutral Cu and organic ligands is almost nonexistent, and we are only aware of the ZEKE (zero kinetic energy) photoelectron study of pyridine–Cu complex[14]. In this study, it was found that the ionization energy of the Cu–pyridine complex was 2.308 eV smaller than that of the free Cu atom, indicating significant stabilization upon ionization. Using these data and the known binding energy for the Cu^+–pyridine complex, it was possible to obtain for the binding energy of the Cu–pyridine complex a value of 5.5 ± 2.4 kcal mol^{-1}. However, this value was estimated using the first reported collision induced dissociation (CID) binding energy for the Cu^+–pyridine complex[56], which is too low as compared with the one measured by means of photodissociation techniques[52] as well as the new reported CID value[57]. If this new CID value is used instead, the Cu–pyridine binding energy amounts to 9.5 ± 2.4 kcal mol^{-1}, which is in reasonably good agreement with the B3LYP/6-311+G** calculated value (12.1 kcal mol^{-1})[58]. Also interestingly, the shift in the ionization energy observed for Cu–pyridine complex with respect to Cu[14] is larger than that measured using similar techniques[59] for $CuNH_3$, which means that the Cu^+–pyridine binding energy is larger than the Cu^+–NH_3 value.

One of the difficulties associated with the spectroscopic investigation of Cu^{2+} complexes is that the preparation in the gas phase of these doubly charged species is a challenge, because many of them are unstable with respect to charge transfer or to proton loss. Nevertheless, even in these cases the ligand-field photochemistry of the resulting monocations offers important insights on their structure and reactivity. A paradigmatic example is offered by the study of Spence and coworkers[60] on the $[Cu(II)(bpy)(serine\text{-}H)]^{+\bullet}$ complex **1** in Figure 2, generated by means of electrospray ionization. In this study, the photodissociation of the serine moiety triggered by the excitation of a d–d transition is compared with the collision activated dissociation (CAD) of this ion complex, previously investigated by Tureček and coworkers[61] who proposed the scheme shown in Figure 2 to account for the four peaks observed on the CAD spectra. Ligand-field excitation of the aforementioned complex at 575 nm triggers photofragmentation reactions yielding three of the four products observed in CAD and metastable decomposition of $[Cu(II)(bpy)(serine\text{-}H)]^{+\bullet}$ complex, dissociation product **2** being absent. Although a detailed mechanism for the observed photofragmentations could not be established, the formation of products **3** and **4** follows the loss of H_2CO, whereas the formation of **5** follows the bond cleavage between Cu and the amino nitrogen of serine. Interestingly, the sum of the branching ratios for the formation of **3** and **4** (0.46) is comparable to that observed for CAD and metastable decomposition (0.44), which was taken as an indication that the barriers to elimination of formaldehyde and transfer of OH followed by the loss of $H_2CCH(NH_2)CO_2$ must be rather similar and much lower than the photon energy (2.16 eV).

The advantage of being able to investigate transition metal ion complexes in which the metal is in a high oxidation state is that very often, as for Cu, this is the state in which they are found in solution or in the physiological medium. Hence, significant efforts have been devoted to the generation of Cu^{2+} complexes in the gas phase. Although the number of attempts that succeeded to do so are not many, it is important to discuss here the pioneering work of Stace and coworkers[54,62–64] who were able to produce $[Cu(pyridine)_n]^{2+}$

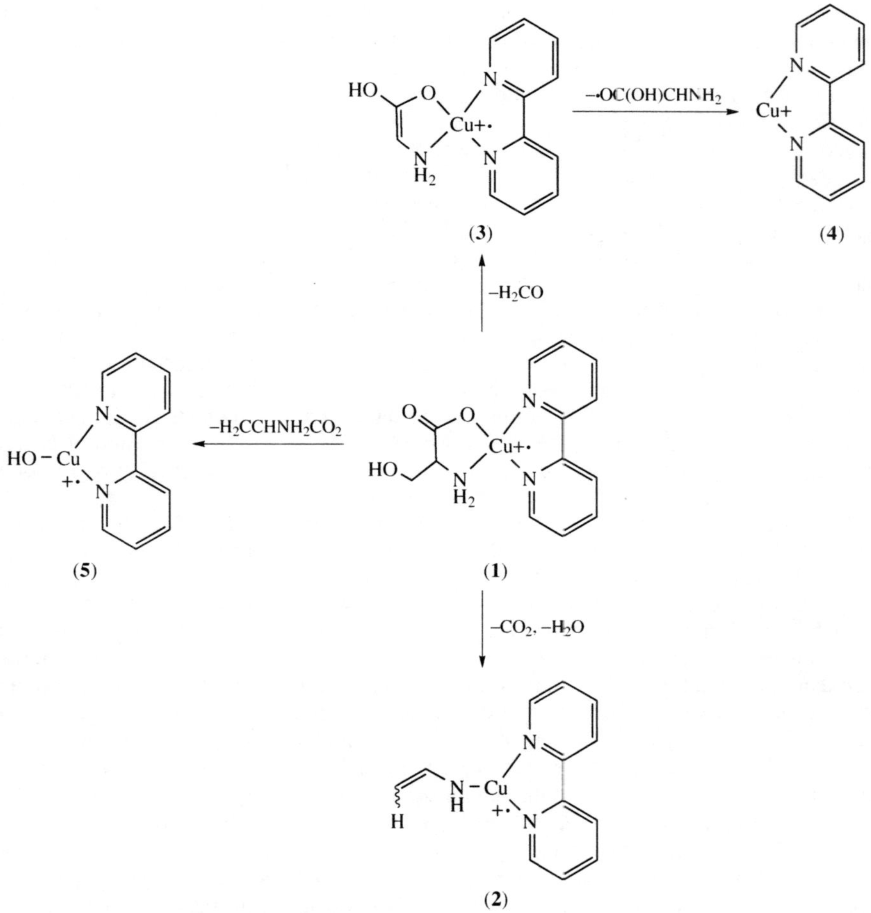

FIGURE 2. Reaction scheme mechanism proposed by Tureček and coworkers[6] for CAD of [Cu(II)(bpy)(serine-H)]$^{+\bullet}$ complex in the gas phase

(n = 4–7) and Cu[(acetone)$_n$]$^{2+}$ (n = 4–8) complexes in the gas phase in order to study their photofragmentation. A detailed description of the modern experimental setup used by this group to study the photofragmentation of doubly charged species in the gas phase can be found elsewhere[65,66]. The goal of these investigations, as those carried out on singly charged species, was to determine whether d–d excitations within the metal ion can lead to metal-to-ligand charge transfer or whether the excitation of the ligand leads to a photofragmentation of the complex. [Cu(pyridine)$_4$]$^{2+}$ when UV-irradiated at about 280 nm, which in principle can excite pyridine to either the $S_1(n\pi^*)$ or the $S_2(\pi\pi^*)$ state[67], loses a neutral pyridine molecule, with practically no evidence of dissociative charge transfer. This is in clear contrast with the behavior observed[54] in the case of [Cu(acetone)$_4$]$^{2+}$ when light of 26 nm is employed. In this case, after excitation, the complex loses one neutral and one charged acetone molecule, the product of the reaction being a singly charged [Cu(acetone)$_2$]$^+$ species. This can be taken as an indication[54] that two acetone molecules are enough to stabilize a Cu^{2+} ion in the gas phase, while three molecules of pyridine are needed to produce a similar effect. The fragmentation of larger $[Cu(pyridine)_n]^{2+}$ complexes exhibit a pattern with very abrupt transitions yielding $[Cu(pyridine)_4]^{2+}$ as the dominant product, likely indicating a very large stability for this complex. Again, the behavior of $[Cu(acetone)_n]^+$ systems when $n \geqslant 5$ is different. For $n = 5$ two acetone molecules are lost yielding a tri-coordinated $[Cu(acetone)_3]^+$ complex[54], while no photodissociations were observed after excitation of this doubly charged species. It is also worth mentioning that the fragmentations undergone by $[Cu(acetone)_4]^{2+}$ under collision-induced dissociation (CID) conditions[68] differ from those observed upon UV irradiation, since after losing one neutral acetone molecule, a charge transfer process leads to two stable singly charged species, namely $[Cu(acetone)]^+$ and $[Cu(acetone)_2]^+$, pointing to a quite different mechanism behind both kinds of processes. When the ligand was 2-butanone, 2-pentanone and 3-methyl-2-butanone, the reactivity pattern was also different. For the latter, no photofragmentations were detected under any circumstances. However, for the other two ketones, when the number of ligands around the metal ion was 5, one and two ligands were lost, the former process being, for both 2-butanone and 2-pentanone, dominant. Also, for the particular case of 2-butanone, a very small proportion of charge transfer leading to $[Cu(2\text{-butanone})_3]^+$ singly charged ions was detected. For $n = 6$, only the loss of two neutral ketone molecules was observed.

The nature of the photofragments when $[Cu(pyridine)_n]^{2+}$ ($n = 4$–7) are irradiated with visible light in the range of 450–1000 nm does not change with respect to that observed when UV light was used[62], and a loss of neutral pyridine is observed, although the mechanism is certainly different. Using $Ar_3{}^+$ (Ar = argon) as reference, the molar extinction coefficient for the transitions recorded around 600 nm, which corresponds to spin-allowed d–d transitions in Cu(II)[69], was 100 L mol^{-1} cm^{-1}. This value is lower than that measured for Ag(II) complexes (500 L mol^{-1} cm^{-1}) for which the charge transfer mechanism clearly dominates. Furthermore, the extinction coefficient estimated for $[Cu(pyridine)_4]^{2+}$ is in good agreement with the values reported for spin-allowed d–d transitions in Cu(II) complexes in aqueous solutions[69–71]. They correspond to transitions from the d_{xz} and d_{yz} into the antibonding $d_{x^2-y^2}$ and are very likely also responsible for the observed absorption spectra in the gas phase. For the smallest $[Cu(pyridine)_3]^{2+}$ complex investigated, it was found[64] that when irradiated in the same region, a weak signal associated with the loss of one charged pyridine unit was detected, leaving a stable $[Cu(pyridine)_2]^+$ ion. Interestingly, this finding is consistent with the results observed in CAD experiments carried out by the same group[65]. The same behavior has been observed for $[Cu(picoline)_n]^{2+}$ ($n = 3$–6) complexes[64], where $[Cu(picoline)_3]^{2+}$ photodissociates by losing a $(4\text{-picoline})^+$ ion, whereas larger complexes just lose a neutral picolone molecule. However, the corresponding absorption bands appear slightly blue-shifted with respect to those of the corresponding pyridine complexes.

Although in recent years the number of studies on metal cation complexes using infrared multiphoton dissociation techniques (IRMPD) has grown dramatically, mainly due to the availability of free-electron laser (FEL) facilities, very few of these efforts were focused on doubly charged complexes, and even less on Cu^{2+}-containing systems. As a matter of fact, we are only aware of the paper published by Stace and coworkers[66] in which the infrared multiphoton spectrum for $[Cu(pyridine)_4]^{2+}$ complexes was recorded. This technique is indeed an interesting complementary tool of visible or UV photofragmentations, even for systems with high binding energies. Although the energy required to dissociate a $[Cu(pyridine)_4]^{2+}$ complex into $[Cu(pyridine)_3]^{2+}$+ pyridine is as large as 59.8 kcal mol^{-1}, which is equivalent to the energy of about 21 IR photons, the available experimental setups can trigger the sequential absorption of up to 40 photons. In the IRMPD study of the $[Cu(pyridine)_4]^{2+}$ complex, a line-tunable CO_2 laser over the

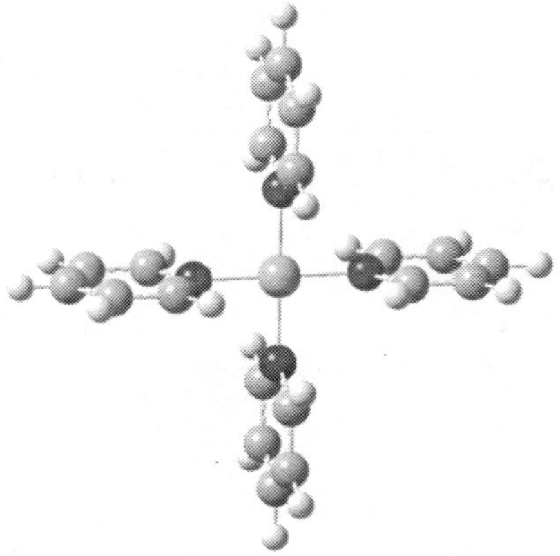

FIGURE 3. D_{4h} structure of the $[Cu(pyridine)_4]^{2+}$ complex

range of 910–1090 cm^{-1} was used[66]. The assignment of the signals observed was guided through the use of DFT calculations, which showed that the only structure that was compatible with a band at 1042 cm^{-1}, accompanied by a much weaker one which is only detected when the laser power is increased, was a D_{4h} square-planar one[66] (Figure 3).

These results are in line with previous DFT calculations which showed that indeed the D_{4h} structure is the ground state for $[Cu(pyridine)_4]^{2+}$ complexes, the D_{2h} and D_{2d} conformers being at least 14.3 kcal mol^{-1} higher in energy[64].

New IRMPD studies carried out by the same group on $[^{63}Cu(pyridine)_4]^{2+}$ at a laser power of 550mW introduced some variations with respect to previous photodissociations using UV and visible light[66]. In these IR experiments there is a charge transfer, the dominant product being $C_5H_5N^+$. However, the fact that there is no complementary $[^{63}Cu(pyridine)_3]^+$ was interpreted[66] as the reaction pathway being sequential, so that the first step is the loss of neutral pyridine from $[^{63}Cu(pyridine)_4]^{2+}$, the same reaction path observed in UV and visible photofragmentations, which is followed by the photoexcitation of $[^{63}Cu(pyridine)_3]^+$ to yield $C_5H_5N^+$ and $[^{63}Cu(pyridine)_2]^+$.

Finally, it should be mentioned that some absorption spectra of Cu(II) complexes with bacteriochlorin[72] and texaphyrin[73] have been studied theoretically in an attempt to investigate the potential behavior of these molecules as photosensitizers in photodynamic therapy.

IV. Cu^+ AND Cu^{2+} BINDING ENERGIES

The number of experimental Cu^+ binding energies is really scarce, and even scarcer if one excludes inorganic compounds such as NH_3, H_2O, H_2S etc. A good compilation was published in 1996 by Freiser[74], but after this compilation not many new values have been reported in the literature dealing with organic compounds, and most of the Cu^+ binding energies known to date come from *ab initio* or DFT calculations. This is indeed the case as far as Cu^{2+} binding energies are concerned due to the difficulty in generating long-lived Cu^{2+} complexes in the gas phase.

Since the development of what are usually called high-level *ab initio* methods, it was possible to obtain thermodynamic magnitudes within chemical accuracy (± 1 kcal mol^{-1}). Hence, very accurate *ab initio* calculations could be used to anchor gas-phase ion affinity scales, or to detect some anomalous experimental values. Cu^+ affinities present nice

TABLE 1. Cu^+ and Cu^{2+} (values within brackets) affinities (in kcal mol^{-1}) of α-amino acids[a] and some glycine oligomers

Amino acid	Cu^+ affinity	Amino acid	Cu^+ affinity
Glycine	64.3 68.1[b] 68.6[c] 67.1[d] [214.8][b] [205.2][d] [242.7][e]	Tyrosine	72.5
Alanine	65.9 68.2[d] [220.7][d]	Cysteine	72.9 [268.1][g]
Serine	67.4	Glutamine	74.0
Valine	67.9	Methionine	74.6
Leucine	68.4	Tryptophan	75.7
Isoleucine	68.6	Histidine	77.6
Threonine	68.8	Lysine	>77.6
Proline	69.1	Arginine	>77.6
Aspartic acid	69.3	GG[f]	87.5[e] 80.3[c] [256.4][e]
Asparagine	70.9	GGG[f]	97.5[e] 94.8[c] [301.7][e]
Glutamic acid	71.5	GGGG[f]	104.7[e] [338.1][e]
Phenylalanine	72.2		

[a]The values reported, except otherwise stated, were taken from Reference 30 and were obtained by combining an accurate theoretical estimate for the Cu^+ affinity of glycine with the relative experimental scale reported in Reference 76.
[b]Value obtained at the CCSD(T) level using B3LYP geometries taken from Reference 81.
[c]Values obtained at the MP2(fc)/6-311++G(2df,2p)//B3LYP/DZVP level taken from Reference 77.
[d]BHandHLYP/6-311++G** calculated values taken from Reference 35.
[e]B3LYP calculated values taken from Reference 31.
[f]GG, GGG and GGGG stand for diglycine, triglycine and tetraglycine, respectively.
[g]B3LYP calculated value taken from Reference 78.

examples of both situations. In some other cases, theory was ahead of experiment predicting Cu^+ binding energies before they were experimentally known. Quite interestingly, in many cases these predictions were very accurate. This is the case, for instance, for a series of different aromatic compounds, whose experimental Cu^+ binding energies published in 2007[75] agreed very well with previous calculated values reported in 2004[40] and 2006[41].

An absolute Cu^+ affinity scale for all amino acids (Table 1) was built up by combining a rather accurate CCSD(T) Cu^+ affinity value for glycine with the experimental relative scale reported by Cerda and Wesdemiotis[76] for all the amino acids. This scale was expanded some years later by incorporating theoretical estimates for both the Cu^+ and Cu^{2+} affinities of (glycyl)$_n$glycine ($n = 1-3$) oligomers (Table 1) obtained by means of B3LYP calculations[31]. These values differ slightly from those obtained previously by Hopkinson and coworkers[77] using the same method but a smaller basis set. For both oxidation states of the metal, due to larger electrostatic interactions, the binding energies increase as the peptide chain becomes longer[31]. The gap between Cu^+ and Ag^+ binding energies in glycine (glycyl)$_n$glycine series was also found to increase as the size of the ligand increases[77].

Later on, new theoretical estimates of Cu^+ affinities of some α-amino acids were reported by the group of Russo[34,35,78] but these values did not differ substantially from those obtained by Hoyau and Ohanessian[30]. It is worth mentioning that in some of these studies Cu^+ was found to bind neutral bases systematically stronger than Ag^+[33,58,79]. Also, in some of these publications Cu^+ affinities were found to be significantly larger than Li^+ affinities[34,40,80]. Although in some cases this was attributed to simple differences in the electrostatic interactions between the metal cation and the base[34], in forthcoming sections we will have the opportunity to show that the enhanced Cu^+ affinities reflect a non-negligible covalent character in the Cu^+–ligand interactions.

Cu^+ affinities for biochemical systems other than α-amino acids or their oligomers have also been published, although most of them correspond to theoretical estimates. As a matter of fact and to the best of our knowledge, only the experimental Cu^+ affinity of adenine has been reported[82]. B3LYP/6-311+G(2df,2p) Cu^+ binding energies for uracil, thymine, cytosine, guanine and adenine were published by Russo and coworkers[83] (Table 2). Those for uracil and all the thiouracil derivatives, at the same level, were also published by Yáñez and coworkers[84]. Agreement between calculated[83] and experimental values[82] for adenine is fairly good.

Not much experimental information is available either on Cu^+ affinities to alkanes. Nevertheless, a compilation of B3LYP/6-31G* calculated values for the first twelve members of the series, excluding heptane and nonane, was reported in 2001[80] showing a clear increase along the series which becomes clearly attenuated as the length of the chain increases (Figure 4). Similar behavior was found for complexes with polyethylene glycols (PEG)[80], although the absolute binding energies are much larger. For both series of compounds the calculated Cu^+ affinities are significantly larger than Na^+ affinities, pointing to quite different bonding patterns, as we shall discuss in forthcoming sections.

B3LYP/6-311+G(2df,2p) Cu^+ affinities of small saturated and unsaturated bases containing N, P and As as active sites[85–87] show that unsaturated systems are systematically less basic than their saturated analogues. For methylenimine ($H_2C{=}NH$) only the nitrogen-attached species was found to be stable, whereas for methylenphosphine ($H_2C{=}PH$) and methylenarsine ($H_2C{=}AsH$) both C- and heteroatom-attached species are local minima of the PES. For $H_2C{=}AsH$ the carbon attached complex is the global minimum. For methylidynephosphine (HCP) and methylidynearsine (HCAs) only the carbon attached species have been found[85]. One year later, an experimental value for the Cu^+ affinity of acetonitrile was reported[88]. As expected, this value was larger than that calculated for the unsubstituted parent compound (HCN)[85] and only slightly smaller than that predicted for methanimine[85]. It is noteworthy that, although the same trend was found for their proton affinities, the gap is significantly larger (17.2 kcal mol^{-1})[53] when the reference acid is a H^+ than when it is a Cu^+ ion (2.3 kcal mol^{-1}) (Table 2).

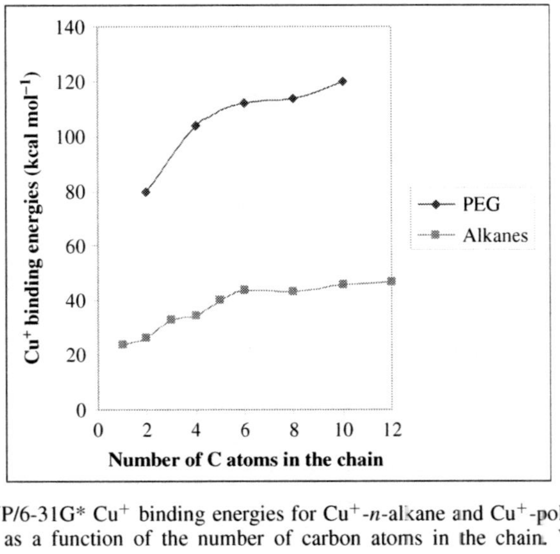

FIGURE 4. B3LYP/6-31G* Cu^+ binding energies for Cu^+-n-alkane and Cu^+-poly(ethylene glycol) (PEG) complexes as a function of the number of carbon atoms in the chain. Values taken from Reference 80

TABLE 2. Cu^+ affinity (in kcal mol^{-1}) of different organic compounds

Compound	Cu^+ affinity	Compound	Cu^+ affinity
Uracil	59.9[a]; 54.8[b]	Ethynylphosphine	55.3[p]
2-Thiouracil	63.2[b]	Ethynylarsine	51.2[p]
4-Thiouracil	65.9[b]	Methylidynephosphine	47.0[n]
2,4-Dithiouracil	67.1[b]	Methylarsine	58.5[n]
Thymine	60.0[a]	Methylenarsine	51.3[n]
Cytosine	80.2[a]	Methylidynearsine	51.8[n]
Guanine	88.0[a]	Dimethyl ether (DME)	44.3 ± 2.8[q]
Adenine	70.3 ± 2.5[c]; 69.0[a]	Dimethoxyethane (DXE)	63.1 ± 1.8[r]
Methane	23.6[d]	Benzene	52 ± 5[s]; 50.6[t]; 51.9 ± 1.4[u]; 50.3[v]
Ethane	26.0[d]	Toluene	52.7[s]; 52.3 ± 2.2[u]
Propane	32.9[d]; 28.3[e]	Phenylsilane	51.8[w]
Butane	34.5[d]	Phenylgermane	52.9[w]
Pentane	40.4[d]	Phenol	52.1[x]; 50.3 ± 2.8[u]
Hexane	43.6[d]	Aniline	59.8[x]; 60.2 ± 2.5[u]
Octane	43.2[d]	Benzaldehyde	52.2[x]
Decane	46.0[d]	Benzoic acid	55.9[x]
Dodecane	47.0[d]	Trifluoromethylbenzene	41.7[x]
Ethylene glycol	79.6[d]	Pyridine	58.7 ± 2.4[y]; 65.5[z]; 57.8[aa]; 65.7[bb]; 62.7[cc]
Ethylene	44-50[f]; 45.4[g]; 42.2[h]; 46.5[e]; 43.4[i]; 42.0 ± 3[j]	4,4′-Bipyridine	63.0[dd]
Acetylene	40.3[h]; 43.9[e]	2,2′-Bipyridine	88.5[ee]
Ethylsilane	40.8[i]	1,10-Phenanthroline	94.4[ff]
Ethylgermane	42.9[i]	Pyrimidine	59.6 ± 2.3[gg]
Prop-1-ene	47.8[i]	Imidazole	68.7 ± 1.7[hh]
Vinylsilane	46.0[i]	Furan	37[ii]; 37.6[jj]
Vinylgermane	47.6[i]	Fluorobenzene	43.8 ± 1.1[u]
Prop-1-yne	48.9[k]	Chlorobenzene	46.5 ± 1.1[u]
Ethynylsilane	47.2[k]	Bromobenzene	50.7 ± 1.8[u]
Ethynylgermane	50.3[k]	Iodobenzene	52.8 ± 1.8[u]
Methylamine	59.8[l]	Anisole	56.5 ± 2.1[u]
Methanimine	59.3[l]	*N*-Methylaniline	63.8 ± 2.5[u]
Hydrogen cyanide	50.0[l]	*N*,*N*-Dimethylaniline	55.6 ± 2.8[u]
Acetonitrile	57.0[m]	Pyrrole	57.2 ± 3.0[u]
Methylphosphine	63.4[n]	*N*-Methylpyrrole	63.9 ± 1.8[u]
Methylenphosphine	52.9[n]	Indole	60.6 ± 2.1[u]
Ethylamine	62.3[o]	Naphthalene	55.3 ± 2.1[u]
Ethylphosphine	65.7[o]	Coronene	59.7[v]
Ethylarsine	61.0[o]	Cyclododeca-1,5,9-trien-3,7,11-triyne	92.4[v]

TABLE 2. (*continued*)

Compound	Cu^+ affinity	Compound	Cu^+ affinity
Vinylamine	62.8[o]	Tribenzocyclyne	107.5[v]
Vinylphosphine	62.2[o]	bis(2,2′-bipyridine)	38.8[kk]
Vinylarsine	53.9[o]		

[a]B3LYP/6-311+G(2df,2p) calculated value at 0 K taken from Reference 83.
[b]B3LYP/6-311+G(2df,2p) calculated value at 298 K taken from Reference 84.
[c]Experimental value taken from Reference 82.
[d]B3LYP calculated value taken from Reference 80.
[e]B3LYP calculated value taken from Reference 36.
[f]Experimental estimate taken from Reference 90.
[g]CCSD(T) calculated value taken from Reference 90.
[h]CCSD(T) calculated value taken from Reference 36.
[i]B3LYP/6-311+G(2df,2p) calculated value taken from Reference 104.
[j]Experimental value taken from Reference 92.
[k]B3LYP/6-311+G(2df,2p) calculated value taken from Reference 105.
[l] B3LYP/6-311+G(2df,2p) calculated value taken from Reference 29.
[m]Experimental value taken from Reference 88. The values for $Cu^+(CH_3CN)_2$, $Cu^+(CH_3CN)_3$, $Cu^+(CH_3CN)_4$, $Cu^+(CH_3CN)_5$ are: 56.6, 16.6, 11.2 and 5.0 kcal mol^{-1}, respectively.
[n]B3LYP/6-311+G(2df,2p) calculated value taken from Reference 85.
[o]B3LYP/6-311+G(2df,2p) calculated value taken from Reference 86.
[p] B3LYP/6-311+G(2df,2p) calculated value taken from Reference 87.
[q]Experimental value taken from Reference 100. The values for $Cu^+(DME)_2$, $Cu^+(DME)_3$, $Cu^+(DME)_4$ are: 57.2, 16.4, and 18.0 kcal mol^{-1}, respectively.
[r]Experimental value taken from Reference 99. The value for Cu^+ $(DXE)_2$ is 43.1 kcal mol^{-1}.
[s]Experimental value taken from Reference 97.
[t]Value estimated using the variable reaction coordinate transition state theory taken from Reference 94.
[u]Experimental value taken from Reference 75.
[v]B3LYP calculated value taken from Reference 36.
[w]B3LYP/6-311+G(2df,2p) calculated value taken from Reference 40.
[x]B3LYP/6-311+G(3df,2p) calculated value taken from Reference 41.
[y]Experimental value taken from Reference 56.
[z]Experimental value taken from Reference 52.
[aa]G2 calculated value taken from Reference 52.
[bb]B3LYP calculated value taken from Reference 58. In the same reference a Cu binding energy of 12.1 kcal mol^{-1} is also reported.
[cc]Experimental value taken from Reference 57. The values for Cu^+(pyridine)$_2$, Cu^+(pyridine)$_3$, Cu^+(pyridine)$_4$ are: 56.5 ± 2.1, 19.7 ± 0.5 and 14.4 ± 0.6 kcal mol^{-1}, respectively.
[dd]Experimental value taken from ref. 57. The values for Cu^+(4,4′-bipyridine)$_2$, Cu^+(4,4′-bipyridine)$_3$ are: 55.6 ± 2.5 and 15.1 ± 0.4 kcal mol^{-1}, respectively.
[ee]Experimental value taken from Reference 57. The value for Cu^+(2,2′-bipyridine)$_2$ is: 56.8 ± 2.3 kcal mol^{-1}.
[ff]Experimental value taken from Reference 57. The value for Cu^+(1,10-phenanthroline)$_2$ is: 55.7 ± 3.0 kcal mol^{-1}.
[gg]Experimental value taken from Reference 98.
[hh]Experimental value taken from Reference 106. The values for Cu^+(imidazole)$_2$, Cu^+(imidazole)$_3$, Cu^+(imidazole)$_4$ are: 61.5 ± 2.2, 19.1 ± 0.6 and 15.2 ± 0.5 kcal mol^{-1}, respectively.
[ii]Experimental value taken from Reference 51.
[jj]CCSD(T) calculated value taken from Reference 51.
[kk]B3LYP calculated value taken from Reference 33.

In spite of the fact that the interactions between ethylene and transition metal ions received much attention because this compound is a good prototype to understand the behavior of unsaturated organic ligands in the bond-activation process and catalysis, its experimental Cu^+ affinity is still not known with precision. In 1990, Fisher and Armentrout[89] published a lower bound for the Cu^+–ethylene binding energy of 23 kcal mol^{-1} which was about half the value estimated by using reasonably accurate quantum chemistry methods[90]. This motivated a revision of the aforementioned experimental value using mass spectrometry techniques in which an equilibrium between $(H_2O)Cu^+$

or $(NH_3)_nCu^+$ ($n = 1, 2$) and $(C_2H_4)Cu^+$ was tried to be established. Combining the calculated NH_3-Cu^+ binding energy[91] with the fact that according to the aforementioned experiments the bond dissociation energy of NH_3-Cu^+ exceeds that of C_2H_4-Cu^+ at least by 6 kcal mol^{-1}, a more realistic experimental value for the ethylene Cu^+ affinity of 44–50 kcal mol^{-1} was proposed[90]. Later on, using guided-ion beam mass spectrometry techniques, the mono- and bis-ethylene complexes with Cu^+ were studied by collisionally dissociating the complex ions with xenon[92]. The new Cu^+ affinity so obtained (42 ± 3 kcal mol^{-1}) was not far from the previous experimental range. Posterior theoretical estimates obtained at the CCSD(T) and B3LYP levels[36] are in reasonably good agreement with this new experimental range (Table 2). Also surprisingly, and to the best of our knowledge, no precise experimental Cu^+ affinities have been reported for acetylene. However, the CCSD(T) and B3LYP estimates are reasonably close and within the range 40–44 kcal mol^{-1} [36], although an older MP2 value is slightly lower (36.2 kcal mol^{-1})[93]. The new B3LYP estimated value for the Cu^+ affinity for benzene reported in the aforementioned publication[36] was in very good agreement with a previous value obtained using the variable reaction coordinate transition state theory[94]. Also, estimates for the Cu^+ affinity of coronene, cyclododeca-1,5,9-trien-3,7,11-triyne and tribenzocyclyne (Chart 1), which are interesting ligands in organometallic chemistry, were reported[94]. The Cu^+ affinity of coronene was predicted to be 5 kcal mol^{-1} larger than that of benzene, reflecting the larger polarizability of the former. However, tribenzocyclyne binds Cu^+ almost twice as strongly as benzene, because the corresponding Cu^+ complex is planar, with the metal cation inserted directly into the ligand cavity, a fact that had been postulated before by Dunbar and coworkers[95,96].

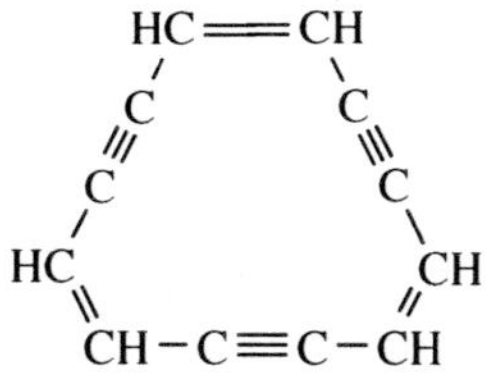

cyclododeca-1,5,9-trien-3,7,11-triyne

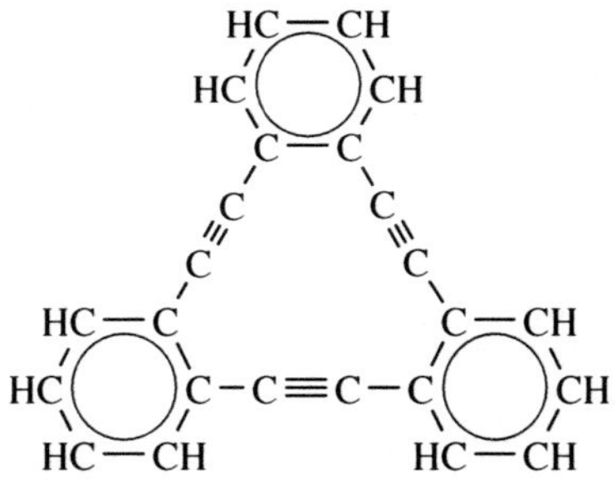

tribenzocyclyne

CHART 1

As expected, the Cu^+ affinities measured for pyridine and pyrimidine are larger than that of benzene[75,97] (Table 2), but for the former the reported values differ significantly depending on the experimental procedure used to obtain them, the binding energy measured in photodissociation experiments (65.5 kcal mol^{-1})[52] being much higher than the one obtained by means of threshold collision-induced dissociation experiments (58.7 ± 2.4 kcal mol^{-1})[56]. The reported calculated values do not solve completely this dichotomy because the MP2 estimates are closer to the CID experiments (57.8 kcal mol^{-1}), whereas the B3LYP estimates (65.5 kcal mol^{-1})[58] agree better with the values obtained in photodissociation experiments. More recent CID measurements[57] yield a value (62.7 kcal mol^{-1}) which seems to support both the value obtained in photodissociation experiments and by B3LYP calculations. For imidazole, agreement between the experimental values obtained in collision-induced dissociation experiments and the B3LYP/6-311+G(2d,2p) calculated values is very good. Interestingly, in parallel with their behavior in protonation processes, imidazole was found to be more basic than pyridine when Cu^+ is the reference acid (Table 2). However, the gap between the Cu^+ affinities is more than three times larger than the gap between the corresponding proton affinities. Conversely, pyrimidine, which is a weaker base than pyridine in protonation process[53], was predicted to be slightly more basic[98] than pyridine when the reference acid is Cu^+, if for the pyridine Cu^+ affinity the CID value is adopted[56]. However, if either the photodissociation value[52] or the new CID[57] are used, pyridine would still be more basic than pyrimidine with respect to Cu^+, which seems to support these last two estimates. Yáñez and coworkers have shown[85] for different series of organic bases the existence of a rough correlation between Cu^+ affinities and proton affinities; however, these correlations depend significantly on the nature of the basic site. Hence, the correlations found for P- and As-containing bases present similar slopes, but very different from that found for the corresponding N-containing analogues. This seems to indicate significant dissimilarities between proton and Cu^+ association processes, even if in the latter the covalent character, which is very important in the former, is not negligible.

Recently, a compilation of the Cu^+ affinities of a large series of aromatic compounds obtained by means of threshold collision-induced dissociation techniques has been published[75]. It is worth mentioning that several of these experimental values, namely those of phenol, toluene and aniline, were in very good agreement with calculated values obtained at the B3LYP level and reported in the literature several years earlier[40,41]. As indicated in Section III, the Cu^+ binding energy to furan was measured by means of photodissociation techniques and the experimental value is very well reproduced by CCSD(T) calculations[51].

Of particular interest are comparisons of the Cu^+ affinity of dimethoxyethane (DXE)[99] and dimethyl ether (DME)[100] and those of pyridine, 4,4′-bipyridine, 2,2′-bipyridine and 1,10-phenanthroline (Chart 2), which illustrate the chelating effect when the metal ion interacts with bidentate ligands. Since DXE is bidentate, it binds more strongly to Cu^+ than DME. However, while the second DME molecule binds to Cu^+ slightly more strongly than the first one, the second DXE is bound much more weakly than the first one. As we shall discuss later on, this is related to the fact that for Cu^+ di-coordination is strongly favored through the participation of *sd* hybrids, which reduces metal–ligand repulsion. This is also consistent with the fact that Cu^+ binds a second DXE molecule much more strongly than the third and fourth DME ligands.

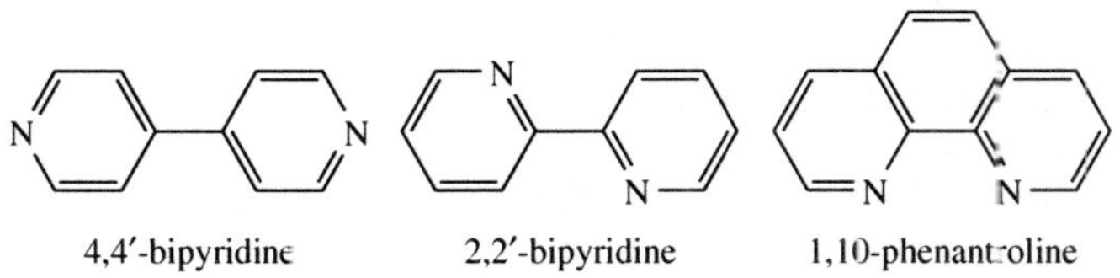

CHART 2

2,2′-Bipyridine and 1,10-phenanthroline exhibit very strong binding as compared to pyridine or 4,4′-bipyridine, because the latter behave as monodentate ligands while the former yield chelated complexes in which Cu^+ is simultaneously bound to the two ring nitrogens. This means that although the *trans* isomer is by far the most stable structure of 2,2′-bipyridine, the interaction with Cu^+ triggers an internal rotation around the single bond connecting both aromatic rings in order to facilitate the interaction of the metal ion with both N atoms. This is also reflected in the incremental binding energies of these systems, as we shall discuss at the end of this section.

A compilation of DFT calculated interaction energies between Cu^{2+} and typical bidentate ligands, namely ethyl xanthate, ethyl trithiocarbonate, dithiobutyric acid, ethyl dithiocarbamate, diethyl dithiocarbamate, diethylphosphinecarbodithioic acid and diethoxyphosphinecarbodithioic acid, showed[101] that the stronger interactions occur for dithiocarbamates, in agreement with their ability as powerful collectors, i.e. as chemicals able to separate valuable minerals from their ores[102]. The Cu^{2+} binding energies to dimethyl dithiocarbamate were also calculated at the DFT level[103].

A. Role of Agostic-type Interactions on Cu^+ Binding Energies

Particularly interesting are the enhanced Cu^+ affinities of silicon and germanium, saturated and unsaturated compounds. The calculated Cu^+ affinity of propane is in line with the values reported for methane but, surprisingly, those calculated for ethylsilane and ethylgermane are almost twice as large[104] (Table 2). Also, some unexpected Cu^+ affinities were reported for the corresponding unsaturated compounds. The Cu^+ affinity for propene is, as expected, only slightly larger than that estimated for ethylene, but surprisingly the Cu^+ affinities for vinylsilane and vinylgermane are similar to that of propene. Analogously, ethynylsilane and ethynylgermane exhibit Cu^+ affinities rather close or even slightly higher than propyne[105] (Table 2). The behavior of the analogous aromatic derivatives is similar and phenylsilane and phenylgermane were predicted to be as basic as toluene with respect to Cu^+ [40].

The enhanced Cu^+ affinity exhibited by ethylsilane and ethylgermane is due to the fact that the metal cation prefers to bond the terminal SiH_3 and GeH_3 groups rather than the terminal methyl group. The global minimum of the propane–Cu^+ complex corresponds to a structure in which Cu^+ interacts simultaneously with both terminal CH_3 groups[104, 107] (Figure 5). Conversely, although for ethylsilane and ethylgermane the analogous structures in which Cu^+ interacts with the CH_3 and XH_3 groups simultaneously are also local minima of the potential energy surface, the global minima correspond to complexes in which the metal cation interacts exclusively with the XH_3 group (X = Si, Ge) (Figure 5). The important finding is, however, that the latter structures are significantly more stable than the former. An analysis of the bonding, using second-order orbital analysis within the natural bond orbital (NBO) formalism[108], shows the existence of a charge donation from the σ_{X-H} (X = C, Si, Ge) bonding orbital toward the empty 4*s* orbital of the metal and a back-donation from the occupied *d* orbitals of the metal into the σ^*_{X-H} (X = C, Si, Ge) antibonding orbital of the organic moiety. These interactions are particularly strong when X is not very electronegative, so that a $X^{+\delta}-H^{-\delta}$ polarity is expected. This kind of interaction resembles closely the agostic bonds defined by Brookhart and Green[109] to account for the effects of transition metals on the C–H bond of organic ligands, and therefore they have been named as agostic-type interactions.

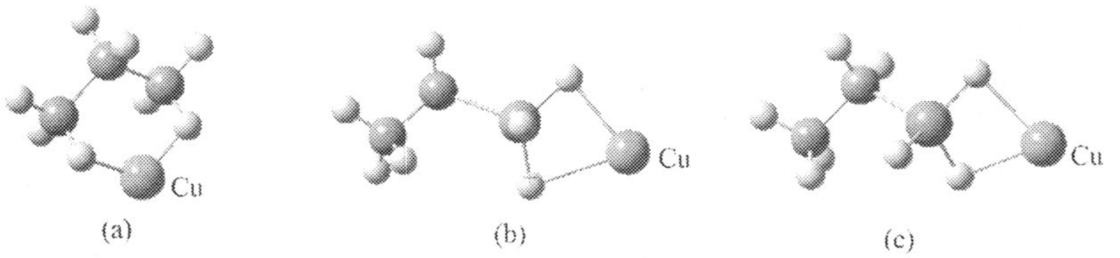

FIGURE 5. B3LYP/6-311G(d,p) optimized geometries corresponding to the global minima of the (a) propane–Cu^+, (b) ethylsilane–Cu^+ and (c) ethylgermane–Cu^+ complexes, respectively, taken from Reference 104

The unexpected high Cu^- affinities exhibited by the unsaturated and aromatic analogues, namely vinylsilane, vinylgermane, ethynylsilane and ethynylgermane, and the phenyl derivatives, are also a direct consequence of the stabilizing effect of the aforementioned agostic-type interactions. For the unsaturated and aromatic hydrocarbons, propene, propyne and toluene, the most stable conformation of the Cu^+ complexes in the gas phase correspond to conventional π-complexes (Figure 6), in which the bonding is usually described in terms of the Dewar–Chatt–Duncanson model[110,111], which was analyzed quantitatively for the case of Cu^+–ethylene complexes by Ziegler and Rauk[112]. According to this model the metal ion interacts with the π-system, through a donation of charge from π_{CC} bonding orbitals of the organic moiety toward the empty 4s orbital of the metal and a back-donation from the occupied d orbitals of the metal toward the π^*_{CC} antibonding orbitals of the organic moiety, as illustrated in Figure 7.

Conversely, for Si- and Ge-containing compounds, the most stable structure corresponds to a nonconventional complex in which a typical three-center–two-electron bond is formed through the interaction of a p orbital of the C atom, a d orbital of the metal and the s orbital of the H atoms of the XH_3 (X = Si, Ge) group (Figure 8).

The existence of these agostic-type interactions leads to a significant activation of the X–H (X = Si, Ge) bond which participates in the interaction, because the depopulation

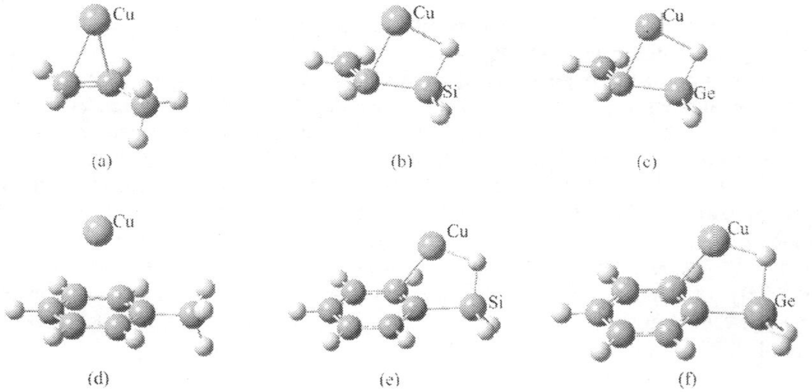

FIGURE 6. B3LYP/6-311G(d,p) optimized geometries corresponding to the global minima of the (a) propene–, (b) vinylsilane–, (c) vinylgermane–, (d) toluene–, (e) phenylsilane–and (f) phenylgermane–Cu^+ complexes, respectively, taken from References 104 and 40

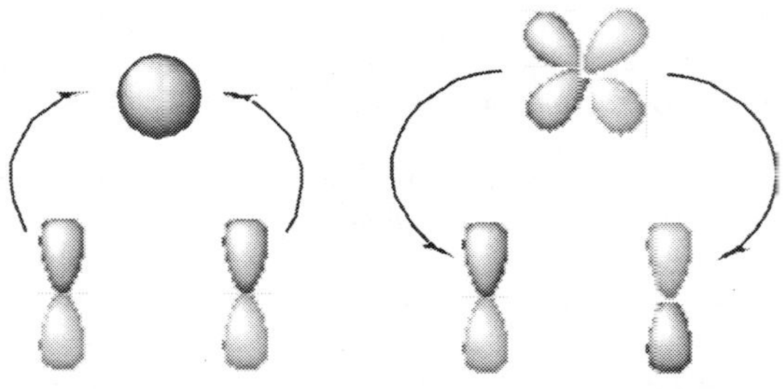

FIGURE 7. Orbital interactions associated with conventional π-complexes between Cu^+ and unsaturated carbon compounds

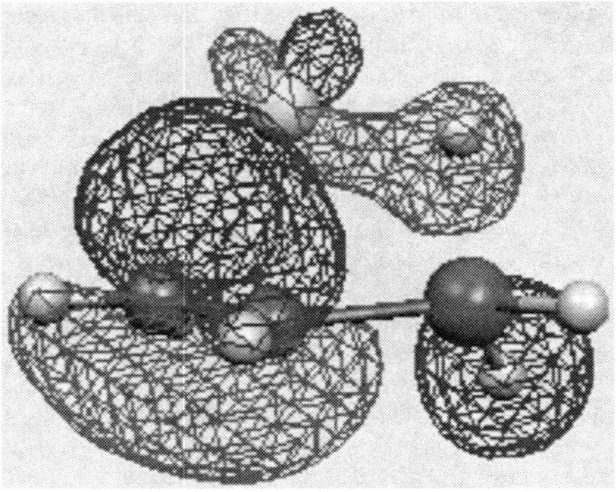

FIGURE 8. Molecular orbital of the vinylsilane–Cu^+ complex showing the overlap between a *d* orbital in Cu with the π_{CC} orbital and *s* orbital of one of the H of the SiH_3 group

of the σ_{XH} bonding orbital and the population of the σ^*_{XH} antibonding orbital leads necessarily to a significant weakening of this linkage. This is clearly reflected in a significant lengthening of the bond which becomes 0.11–0.15 Å longer than the X–H bonds that do not interact with the metal. This weakening has also a dramatic effect on the infrared spectra of these species. These differences are clearly seen when the spectrum of the conventional π-complex of vinylsilane, taken as a suitable example, is compared with that of the corresponding nonconventional π-complex. As illustrated in Figure 9, the infrared spectrum of the nonconventional π-complex presents two strong absorption bands at about 1700 cm^{-1} and 1200 cm^{-1}, respectively, which are not observed for the conventional π-complex. They correspond to the Si–H stretching and Si–H bending modes, respectively, of the SiH bond directly interacting with the metal. Due to the weakening triggered by the agostic interaction, the former appears red-shifted by about 600 cm^{-1} and the latter blue-shifted by about 400 cm^{-1}.

B. Binding Energies of Solvated Cu^+ and Cu^{2+} Ions

To finish this section it is convenient to indicate that some effort has also been addressed to investigate the effects that the solvation of Cu^+ and Cu^{2+} may have on their binding energies, as well as to measuring incremental binding enthalpies with different organic ligands[88,106,113]. For obvious reasons, most of this effort is concentrated on the interactions with systems of biochemical relevance. For $[Cu^+(H_2O)_5Guanine]^+$ complexes, the most stable conformers always correspond to bisligated Cu ions bonded to one water molecule and to N7 of guanine, the binding energies being in the range 75–83 kcal mol^{-1}[114]. This implies a decrease with respect to the binding energy of the bare metal ion of about 5–13 kcal mol^{-1}. When the metal cation is in its higher Cu(II) oxidation state, a clear attenuation of the binding energy as a function of the number of water molecules around the metal is also observed[115], as illustrated in Figure 10. Hence, while the binding energy of guanine to a bare Cu^{2+} ion amounts to 302 kcal mol^{-1}, addition of one water molecule leads to a reduction of *ca* 70 kcal mol^{-1}. When two and three water molecules are added the calculated decreases are 143 and 180 kcal mol^{-1}, respectively. When the maximum number of coordinated water molecules (5) is reached, the overall decrease

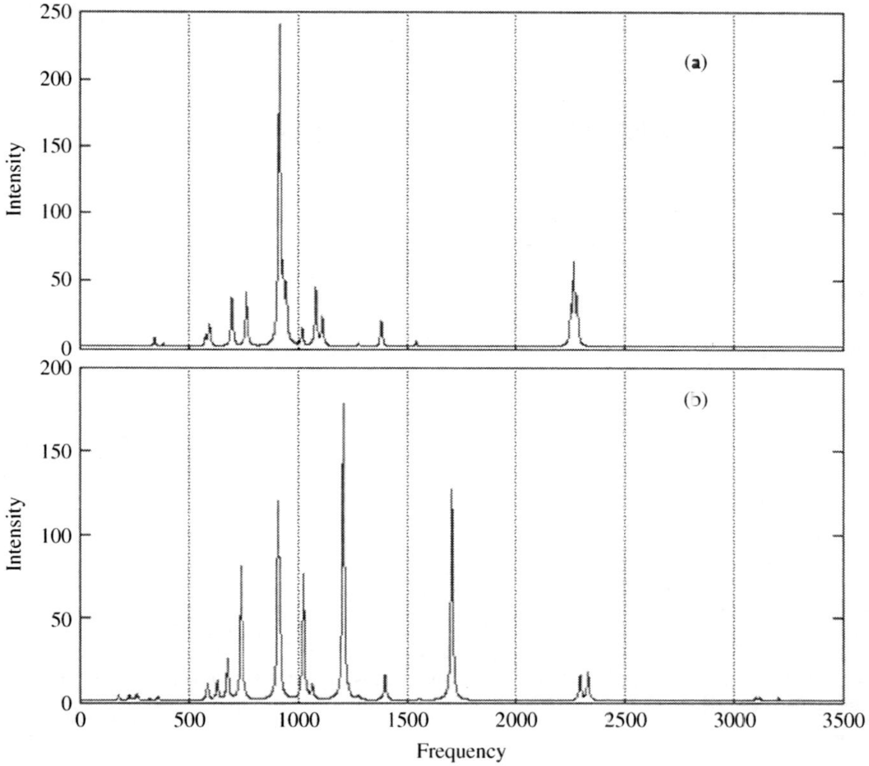

FIGURE 9. Infrared spectra for the (a) conventional π-complex and (b) nonconventional Cu^+–vinylsilane complex, calculated at the B3LYP/6-311G(d,p) level. Frequencies are in cm^{-1} and intensities in KM mol^{-1}

is *ca* 212 kcal mol^{-1} [115]. Similar findings have been reported for complexes involving $[Cu(H_2O)_3]^{2+}$ and guanine or GGG triplets[116]. The replacement of water by guanine leads to a stabilization of the complex of about 24.3 kcal mol^{-1}. This stabilization amounts to 35–45 kcal mol^{-1} when water is replaced by TGGG or CGGG triplets. The interaction of hydrated Cu^{2+} ions with guanine and an anionic phosphate group was also studied through the use of DFT methods[117].

One typical signature as far as Cu^+ sequential binding enthalpies is concerned is that the values measured or calculated for the second ligand are nearly equal or even slightly larger than those obtained for the first ligand. This is indeed the case when dealing with Cu^+(dimethyl ether)$_n$, Cu^+(methanol)$_n$ and Cu^+(acetone)$_n$ ($n = 1$–4) complexes, where the Cu^+ bond dissociation energy for $n = 2$ is about 2.0 kcal mol^{-1} larger than that measured for $n = 1$[100, 118, 119]. This bonding enhancement is well reproduced by CCSD(T)[113] and DFT calculations[118, 119], and is a consequence of the hybridization of the d_{z^2} and 4*s* orbitals of the metal which leads to a significant reduction of the Pauli repulsion between the metal cation and the ligand in the corresponding linear arrangement. Consistently with this explanation, for the three kinds of complexes the binding of the third ligand implies a dramatic decrease (about 40 kcal mol^{-1} for dimethyl ether, 28 kcal mol^{-1} for methanol

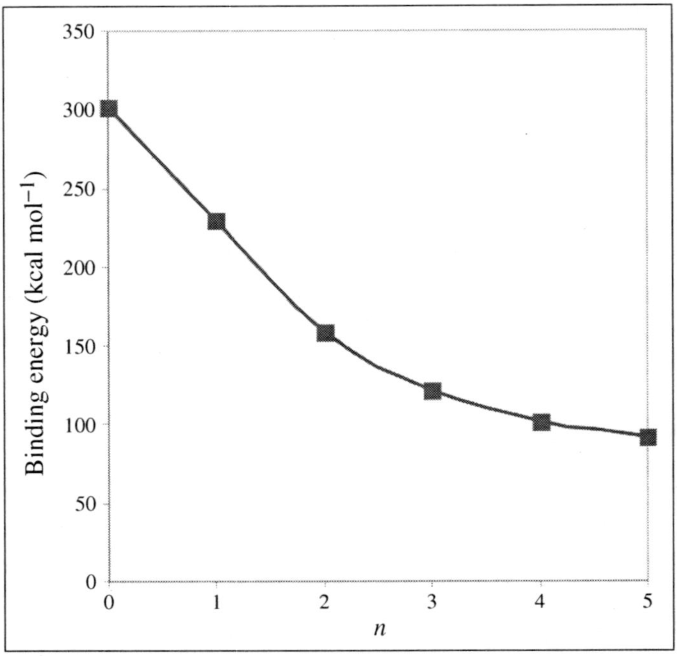

FIGURE 10. Binding energy of $[Cu(H_2O)_n(guanine)]^{2+}$ complexes as a function of the number of water molecules (n)

and 34 kcal mol^{-1} for acetone) of the interaction energy. The situation is similar when the ligands are acetonitrile[88], imidazole[106] or pyridine[57], the only difference being that the binding energy of the second ligand is, in these cases, slightly smaller than that of the first ligand (Table 2). A particular case is represented by 2,2′-bipyridine and 1,10-phenanthroline, where the change in the Cu^+ binding energies on going from complexes with one ligand to complexes with two ligands is significantly large (31 and 39 kcal mol^{-1}, respectively)[57], reflecting the fact that in the complex with only one ligand the metal ion is already bisligated. Accordingly, the electrostatic contributions to the binding decrease sharply, because the chelating ligands provide two donor interactions such that the charge retained by Cu^+ decreases quite significantly. On the other hand, the addition of the second ligand yields a structure in which the coordination number of the metal ion is 4, which is accompanied by a significant increase in the repulsion between the two ligands.

V. COORDINATION

Different evidence seems to be consistent with the idea that copper coordination preferences in Cu-containing enzymes could play a significant role in their biological functions, and therefore the preferred coordination of copper and its dependence on its oxidation state was the subject of many studies. There is also evidence of the effect of the coordination geometry on the gas-phase reactivity of tetra-coordinated Cu(II) ion complexes. In the gas-phase reactions of ammonia with $Cu(EN\text{-}(py)_2)^{2+}$, $Cu((en)\text{-}(phen)_2)^{2+}$, $Cu((phen)_2)^{2+}$ (EN = 1,6-bis(2-pyridyl)-2,5-triazahexane; en = ethylenediamine; phen = 1,10-phenanthroline) the reactivity of the tetra-coordinate complexes depends strongly on the nature of the ligand. When EN-$(py)_2$ is the ligand, the rate constant is about half that measured when (en)$(phen)_2^+$ is the ligand. Even more significant is the change observed when two phen ligands are bound to the metal ion, because in this case no significant addition of NH_3 was observed[120].

A. Coordination in Complexes with One Ligand

It is interesting to discuss in the first place what should be the expected coordination when the metal ion interacts with only one ligand. We have already discussed in the section devoted to analyzing the performance of various theoretical models the different coordination patterns exhibited by Cu^+ when the base is a benzene derivative. Of particular interest are the situations which arise when the ligand presents two or more basic sites. We have already discussed several cases in which the metal forms chelated structures when the base presents more than one basic site; however, it is important to emphasize that not always is a chelated structure found to be the global minimum of the PES of bidentate or polydentate bases. Glycine and urea are two suitable examples. Both bases present as possible basic sites a carbonyl group, an amino group and, in the case of glycine, in addition a hydroxyl group. However, whereas glycine leads to a bisligated complex[30] in which Cu^+ interacts with the carbonyl oxygen and with the amino *N* (Figure 11), in urea this conformer lies 12.9 kcal mol^{-1} higher in energy than the global minimum[121] in which Cu^+ interacts exclusively with the carbonyl oxygen. This can be understood if one takes into account that in urea the amino nitrogen lone-pair conjugates with the carbonyl group and is not available to interact with the metal. It is worth noting, however, that the attachment of Cu^+ to the carbonyl group is not linear, as is the case for alkali ions, pointing to different bonding patterns, as we shall discuss later. The same behavior is found for uracil[83,84] and thymine complexes[83]. Also for 2-, 4- and 2,4-thiouracil, the Cu^+ adducts are mono-coordinated Cu^+ complexes in which the metal ion is attached preferentially to the sulfur atom. It is also interesting to indicate that any attempt to find a tri-coordinated glycine–Cu^+ complex in which the metal interacts with both oxygen atoms and with the nitrogen atom failed[30], as all these structures collapsed to the bisligated global minimum. The situation is different, however, as far as the complexes with serine and cysteine are concerned. In these two cases the global minimum corresponds to conformers in which Cu^+ interacts with the amino nitrogen, the carbonyl oxygen and the alcohol oxygen in serine, or the thiol sulfur in cysteine (Figure 11).

This propensity to yield highly coordinated complexes reflects the importance of polarization interactions and depends on the flexibility of the base to adequately orient its lone pairs toward the metal. We have already mentioned the case of 2,2′-bipyridine, where the association with the metal is preceded by an internal rotation around the single bond connecting both aromatic rings in order to favor the interaction of both ring nitrogens

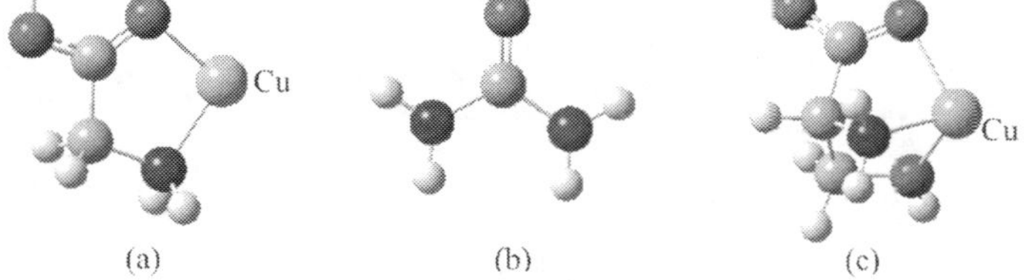

FIGURE 11. Optimized geometries of (a) glycine–, (b) urea– and (c) serine–Cu^+ complexes

with the metal ion. Also in the case of serine and cysteine, the neutral base undergoes a rearrangement to favor the interaction of the three basic sites with the metal ion, which implies little energy. Conversely, the concomitant interaction with the OH group of the carboxylic group would require a strong distortion and breaking of the intramolecular hydrogen bond between this group and the carbonyl oxygen, and therefore this interaction does not occur. This is corroborated by the fact that the complexes between diglycine and triglycine with Cu^+ always correspond to bisligated structures in which the metal ion is chelated to the terminal amino group and the carbonyl oxygen, and all attempts to produce tri-coordinate complexes were not successful[31,77]. Conversely, for Ag^+, tri- (in the case of diglycine) and tetra-coordinate (in di- and triglycine) structures are the most stable[77]. It should also be emphasized that for diglycine, triglycine and tetraglycine, the complexes with a structure similar to that shown in Figure 11 for glycine–Cu^+ are not the global minimum of the potential energy surface because Cu^+ prefers a linear coordination environment rather than an angular one. As already explained in Section IV.B, associated with the linear arrangement is a significant reduction of the Pauli repulsion between the metal cation and the ligand through the formation of *sd* hybrids[122–124] (Figure 12). Conversely, in an angular arrangement the hybridization involves necessarily, by symmetry reasons, the $4p$ and d_{xz} orbital, and due to the higher energy of the $4p$ orbitals the hybridization is less effective and the repulsion decrease is smaller[31]. We will find again a similar conformation in many complexes in which Cu^+ interacts with two or more ligands.

The structure of the glycine complexes does change dramatically when Cu^+ is replaced by Cu^{2+} since, although the global minimum still corresponds to a bisligated structure, in this case the metal ion interacts with the CO_2^- terminus of the zwitterionic form of glycine[81] (Figure 13). Similar behavior has been found when the structure of α-alanine–Cu^+ complex is compared with that of α-alanine–Cu^{2+}[35]. Changes are apparently less significant when dealing with urea and uracil[49,125], where the replacement of

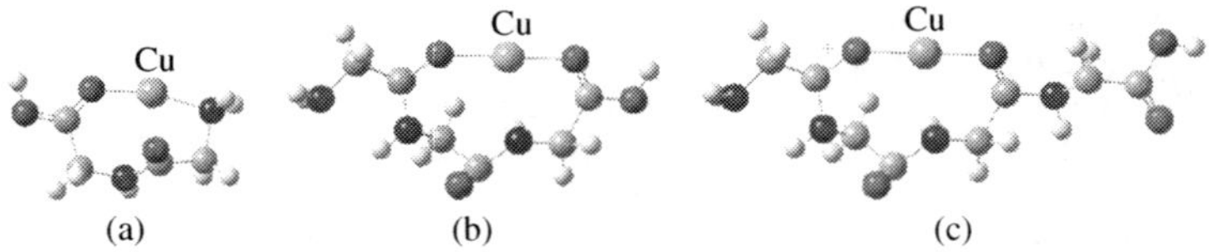

FIGURE 12. B3LYP optimized geometries for the global minima of the (a) diglycine–, (b) triglycine– and (c) tetraglycine–Cu^+ potential energy surfaces taken from Reference 31

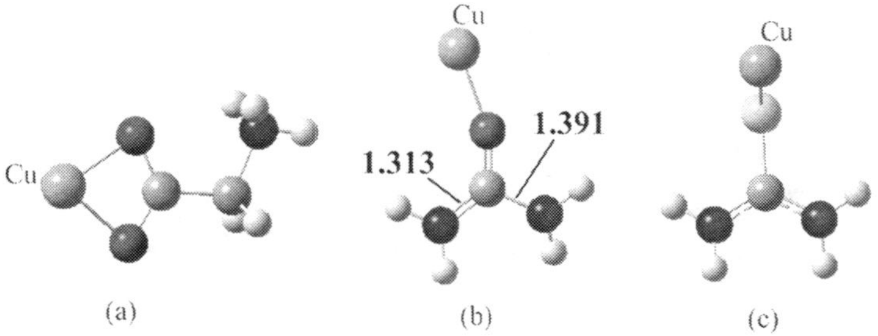

FIGURE 13. B3LYP optimized geometries of the Cu^{2+} complexes with (a) glycine, taken from Reference 10, (b) urea and (c) thiourea, taken from Reference 49

Cu^+ by Cu^{2+} triggers a noticeable opening of the C−O−Cu angle, which for uracil becomes very close to 180°. More importantly, Cu^{2+} association leads to a significant symmetry change in the urea moiety, where one of the C−N bonds becomes sizably shorter than the other one[49] (Figure 13). Conversely, attachment of Cu^{2+} to thiourea yields a nearly symmetric structure in which Cu^{2+} binds to the sulfur atom practically in the plane which bisects the NCN angle[49], the CSCu angle being 109.7° (Figure 13). Similar CSCu bending arrangements are also found for thiouracil derivatives[125].

These significant dissimilarities between Cu^+ and Cu^{2+} complexes when interacting with one ligand reflect completely different bonding patterns[81,125]. As pointed out by Bertrán and coworkers[81], the ionization potential of Cu^- (20.8 eV) is much larger than that of most of the aforementioned organic bases, and certainly larger than that of glycine (9.38 eV), urea (10.27 eV), thiourea (8.5 eV) or uracil (9.2 eV)[53], so one may expect Cu^{2+} to be able to oxidize the base so that the Cu^{2+} complexes would behave more like Cu^+-base$^{+\bullet}$ than like Cu^{2+}-base systems. This is mirrored in the corresponding electron densities. $[Cu{-}uracil]^{2+}$ complexes can be actually viewed as Cu^+–uracil$^+$ complexes, not only because the uracil moiety bears a net positive charge very close to unity, but because the spin density, which in a Cu^{2+} complex should be located on the metal (with a d^9 configuration), is located on the organic moiety. The same electron distribution was observed in $[Cu{-}thiouracils]^{2+}$ [125] and $[Cu{-}urea]^{2+}$ complexes[49]. As we shall discuss in detail in forthcoming sections, the oxidative capacity of Cu^{2+} is one of its specific signatures and dictates many of the properties of Cu(II)-containing systems. Actually, the fact that the global minimum of the $[glycine{-}Cu]^{2+}$ complex corresponds to a η^2-*O,O* structure of the zwitterionic isomer indicates that Cu^{2+} association strongly favors a proton transfer from the carbonylic group toward the amino group[126]. This was also observed in cysteine–Cu^{2+} interactions[78] where the association with Cu^{2+} triggers a proton transfer from the S−H group toward the amino group, the global minimum of the $[cysteine{-}Cu]^{2+}$ complex being a bisligated structure in which Cu interacts with the carbonyl oxygen and the sulfur atom of this zwitterionic form[78]. It is important to mention that the charge transfer from the base toward the metal ion seems to be strongly favored in monodentate complexes, whereas for bidentate structures the spin density is more delocalized, as found for glyoxylic acid oxime–Cu^{2+} interactions[37].

B. Coordination in Complexes with More than One Ligand

When Cu(I) or Cu(II) interact with more than one ligand, it seems well established that Cu(I) prefers lower coordination numbers than Cu(II)[4]. Actually, different studies confirmed a preference for Cu(I) to be bisligated, whereas Cu(II) is usually found in tetra-coordinated planar structures. So, even though it is usually assumed that d^{10} metal ions should form tetrahedral complexes[127], this is not always the case when dealing with Cu^+. We have already indicated the preference of Cu(I) to yield complexes in which Cu is linear or nearly bound to two active sites of a polydentate base, whenever the base is flexible enough to favor these linear arrangements. When two or more ligands interact with the metal ion this is indeed the kind of coordination usually found. Obviously, this is the arrangement in $Cu^+(pyridine)_2$ complexes[128] as well as in the neutral counterpart, $Cu(pyridine)_2$, although in the latter the Cu−N distances are larger because of the increase of the Pauli electron repulsion between the *N* lone-pair and the 4*s* electron of the metal. This preference of Cu(I) for di-coordination is nicely reflected in the solvation patterns exhibited by Cu^+ in $Cu^+(H_2O)_n$ ($n = 2, 4, 6$) clusters[129]. At the MP2 and CCSD(T) levels of theory, the $Cu^+(H_2O)_2$ cluster is predicted to be a C_2 structure (Figure 14)[130]. When a third water molecule is added, the lowest energy structure corresponds to that depicted in Figure 14, in which the third water molecule forms hydrogen bonds with one in the first

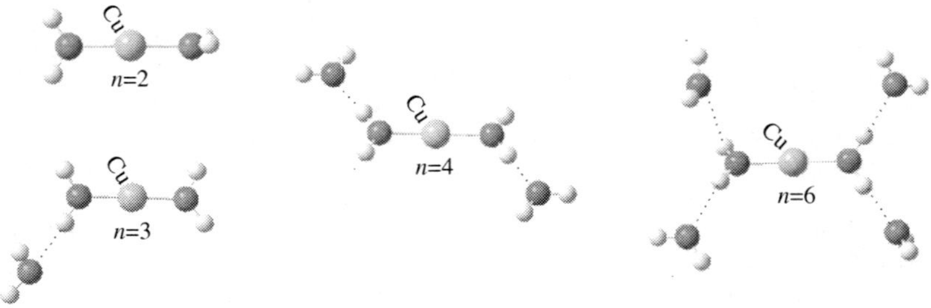

FIGURE 14. Structure of the global minima for $Cu^+(H_2O)_n$ ($n = 2, 3, 4, 6$) taken from References 130 and 129 showing the preference of Cu^+ to form linear L–Cu–L′ arrangements when interacting with several ligands

shell. Analogously, the most stable structure for Cu^+ $(H_2O)_4$ clusters can be viewed as the result of the solvation of the C_2 dimer. That is, two water molecules are covalently bonded to the metal, while the other two interact with the first-shell water molecules through the formation of hydrogen bonds, acting as proton acceptors (Figure 14). Also consistently, the global minimum of the $Cu^+(H_2O)_6$ has only two water molecules in the first shell linearly bound to the metal, while the other four solvate these two, acting as proton acceptors (Figure 14).

This bonding pattern is also found in many complexes between Cu^+ and organic ligands. For instance, in the Cu^+(imidazole)$_2$ complexes Cu is linearly coordinated to the two bases. Furthermore, when a third imidazole molecule is added it interacts through hydrogen bonds with one of the imidazoles of the first shell, whereas the local minima in which the three ligands are in the first solvation shell is predicted to be 4.4 kcal mol^{-1} higher in energy[106]. Similarly, in the global minimum for the Cu^+(imidazole)$_4$ complex only two imidazole molecules are directly connected to the metal, while the structures with three and four ligands in the first solvation shell are predicted to be 6.3 and 11.5 kcal mol^{-1} less stable, respectively. In Cu^+(dimethyl ether)$_2$, again the ground state corresponds to a nearly linear bisligated complex[113]. In this case, when a third ligand is added, it does not solvate one of the ligands of the first shell because in this case there is no possibility of forming hydrogen bonds, but the structure of the complex corresponds to a distorted T-type arrangement, in which the third ligand is much weakly bound than the other two (Figure 15).

A similar bonding pattern was obtained at the B3LYP/DZVP level for complexes formed by the interaction of glycine–Cu^+ ions with CO, water and ammonia[131].

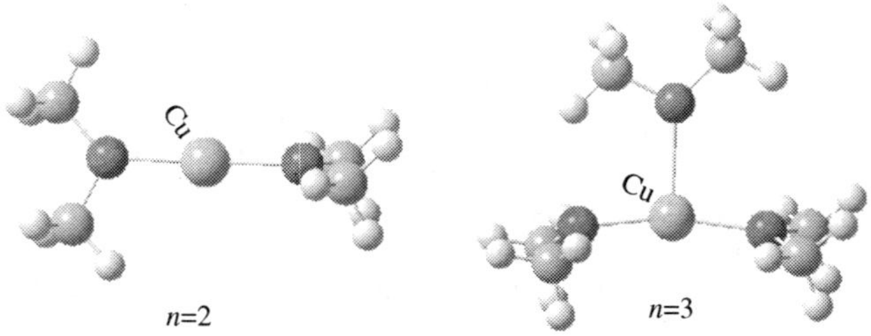

FIGURE 15. CCSD(T) optimized geometries for the Cu^+(dimethyl ether)$_n$ ($n = 2, 3$) complexes taken from Reference 113

Whereas Cu(I) prefers in general di-coordinated structures, Cu(II) yields quite stable tetra-coordinated complexes, and when the coordination number is six, there is a significant Jahn–Teller effect which forces two of the ligands to be much less tightly bound than the other four. This happens, for example, in complexes between Cu(II) and acetone[132]. These complexes were generated in the gas phase by a technique able to produce complexes of a controlled size. Each complex is first produced as a neutral Cu-solvent species, which is subsequently ionized by electron impact. The data obtained for $[Cu(acetone)_n]^{2+}$ complexes show a maximum abundance when the number of ligands is four, accompanied by a rapid decline in the abundance when n increases. On the other hand, the results show no evidence of a stable octahedral unit when $n = 6$, due to a strong Jahn–Teller effect leading to a distorted octahedral structure in which two ligands are weakly bound. Very interestingly, however, when the ligand is ethylene, the maximum intensity is observed for $n = 3$[132]. This was interpreted in terms of donations and back-donations between the CC π-system and the metal, because the optimum number of π-bonded ligands would be three, through the interaction of d_{xy}, d_{xz} or d_{yz} orbitals with the π^*_{CC} antibonding orbital[132]. The same experimental technique was used to investigate the coordination of Cu(II) to different alcohols, namely methanol, ethanol, 1-propanol, 2-propanol, different ketones (acetone, butanone, 2-pentanone, 2,4-pentanedione) and other organic bases such as pyridine, pyrazine, tetrahydrofuran, dioxane, benzene, benzonitrile and ethylenediamine[65]. The intensity distributions recorded for the interactions with acetone, pyridine, tetrahydrofuran and acetonitrile are clearly dominated by a very intense $[CuL_4]^{2+}$ unit, followed by a rapid decline in the intensity measured for complexes of larger size. Furthermore, in $[CuL_4]^{2+}$ complexes containing acetone, 2-butanone or 2-pentanone, the charge transfer reaction leads to the loss of two ligands to form $[CuL_2]^+$ complexes. The situation is different for alcohol ligands. For 1- and 2-propanol, both Cu(II) complexes exhibit intensity maxima at $n = 4$ as for the case of the aprotic solvents mentioned above. However, for ethanol and methanol, up to eight ligands can be gathered around the metal because of the ability to form hydrogen bonds between the ligands in the first shell and those in the second one. In fact, as shown by Ziegler and coworkers[133], in the case of methanol, four ligands are located in the first solvation shell in a square-planar distribution, with another four connected to the first ones via hydrogen bonds. On going to ethanol there is a downward shift in maximum intensity to $n = 6$, with $[CuL_4]^{2+}$ becoming more prominent. The association with benzene constitutes a special case since $[Cu(C_6H_6)_2]^{2+}$ is the only ion observed, suggesting that the structure very likely corresponds to a sandwich-like arrangement in which the metal dication is centrally located between both aromatic rings.

In $[Cu(pyridine)_n]^{2+}$ ($n = 4$, 6) complexes[64], when $n = 4$ the ground state corresponds to a D_{4h} structure, but the ground state of the hexamer can be viewed as the result of associating to the D_{4h} tetramer two additional ligands in an overall octahedral arrangement. However, while the Cu–N distance for four ligands is 1.95 Å, for two of them it is 3.72 Å, showing a clear Jahn–Teller distortion, similar to the one found in clusters in which Cu^{2+} is solvated by water[19, 134–137], ammonia[135, 138] or water and ammonia molecules simultaneously[19, 135, 139, 140]. This preference for tetra-coordination in the case of Cu(II) complexes was also observed in clusters between GGG (G = glycine) and hydrated Cu^{2+} ions[116]. Even in complexes between glycine and $Cu^{2+}(H_2O)_4$, where the coordination around the metal ion is forced to be five, one of the water ligands is always far away from it, or moves to the second solvation shell[141]. Very recently, the structures of complexes between guanine and hydrated Cu(II) were reported[115]. The B3PW91/6-31+G(d) optimized geometries showed that the global minimum of the diaqua system corresponds to a tetra-coordinated complex, in which the metal forms a chelated structure with the base and interacts simultaneously with the two water molecules (Chart 3). Importantly however, when two more water molecules are added, the coordination of the metal ion

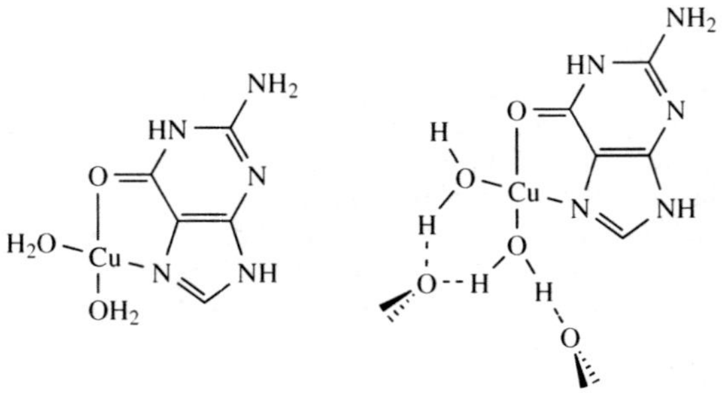

CHART 3

does not change, and as illustrated in Chart 3, the two new water ligands form hydrogen bonds with the water molecules of the first shell.[115]

Even in the case of copper-bound proteins, where the rigidity imposed by the protein backbone may prevent ligand exchange or ligand loss, Cu(I) shows a clear preference to yield di-coordinated species[4,142,143]. Two thorough theoretical studies of the Cu(I) and Cu(II) binding sites for a His–His peptide model[4,142] using 3-(1*H*-imidazol-5-yl)-*N*-[2-(1*H*-imidazol-5-yl)ethyl] propanamide (**6**) as a suitable model system (Chart 4) showed a preference of Cu(II) to yield tetra-coordinated complexes, forming in general distorted square-planar structures. Since this modeling was carried out by taking into account the specific solvation of the metal cation by water molecules, it was found that Cu(II) releases water molecules when the histidines and the backbone get involved in the coordination sphere, in order to keep the coordination number equal to four[142]. A more recent study using the same model compound[4] showed that a reduction of the metal was accompanied by a decrease in the coordination number to three or two. As a matter of fact, it was found that lowering the coordination number of Cu(I) from four to three or two was favored irrespective of the identity of the ligand lost. Actually, in the most stable complex between Cu(I) and compound **6**, the metal ion forms two linear bonds with the N atoms of the imidazole rings. A third, very weak Cu(I)–O interaction in the stem position leads to a distorted T geometry.

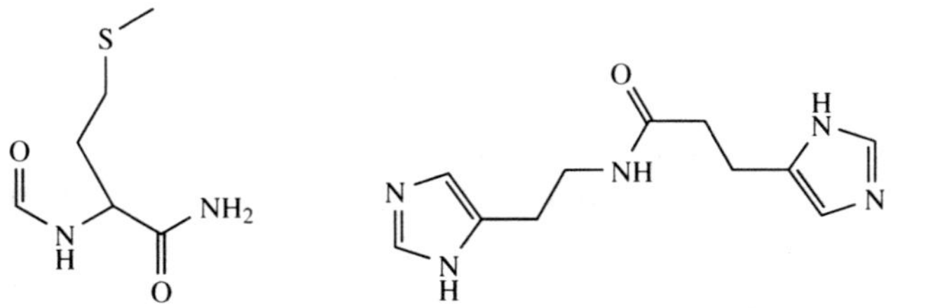

N-formylmethioninamide   3-(1*H*-imidazol-5-yl)-*N*-[2-(1*H*-imidazol-5-yl)ethyl] propanamide (**6**)

CHART 4

In a systematic investigation of the bonding of Cu(II) and Cu(I) to *N*-formylmethioninamide (NFMET) (Chart 4)[143] as a suitable methionine model peptide, analogous results were found. Hence, in general, tetra-coordinated Cu(II)–NFMET complexes were found to be more stable than those with higher coordination numbers. However, for Cu(I)–NFMET the tri-coordinate complexes were more stable than the tetra-coordinated ones.

C. Coordination of Cu(II) in Blue Proteins

The coordination of Cu(II) in blue proteins has received particular attention, because it seems well established that coordination dictates their oxidation capacity, i.e. their role as electron-transfer proteins as well as other of their properties. As a matter of fact, mononuclear copper proteins have been usually classified into two types, which differ clearly in the coordination of Cu(II). In the so-called 'normal' proteins, Cu(II) exhibits two different arrangements depending on the coordination number, tetragonal or elongated octahedral due to a typical Jahn–Teller effect. In contrast, in blue proteins Cu(II) is usually surrounded by a trigonal coordination sphere involving two histidine nitrogen atoms and a cysteine thiolate group. Our knowledge of the coordination of Cu(II) in blue proteins comes essentially from theoretical modeling using small systems. A quite simple model, in which Cu(II) interacts with two or three ammonia molecules and with a SH^-, SH_2 or $S(CH_3)_2$ ligand, permitted to gain some insight into the origin of these preferences for a trigonal coordination[144]. This model clearly reproduced the unusually short Cu–cysteine and the unusually long Cu–methionine bond lengths. Methionine occupies an axial position in the protein and leads to an unfavorable interaction with the doubly occupied *d* orbital of the metal ion. At the same time, the model clearly showed the stabilizing role played by the thiolate ligand because of its soft character that gives covalent Cu–S bonds, stabilizing the trigonal arrangement around the metal. In a posterior analysis[10], it was concluded that for small and hard ligands the tetragonal structure around Cu(II) was the most stable. However, for large, soft and polarizable ligands, trigonal and tetragonal coordinations are very close in energy, although in complexes like $Cu(NH_3)_2(SH)(SH_2)^+$ the trigonal structure is more stable. This was taken as an indication of the important role played by methionine in this kind of protein, as a consequence of the large amount of charge transferred from the ligand to the copper ion, whose charge becomes rather close to +1.

More sophisticated models, in which $S(CH_3)_2$ and CH_3CONH_2 ligands simulate the effects of methionine and glutamine residues, respectively, and $CH_3CONH(CH_2)_2$-imidazole simulates the effect of the backbone amide group, were used[145] to gain insight into the coordination of a series of blue copper proteins such as azurine or stellacyanin (Figure 16). The main conclusion of this analysis was that the axial ligands had a rather small influence on the reduction potentials of these proteins, which seems to depend essentially on the solvent accessibility to the copper site and on the orientation of the protein dipoles around the metal.

QM/MM calculations[146] were also reported for azurin, plastocyanin and stellacyanin in which the QM region, which contains complexes similar to those described in the previous

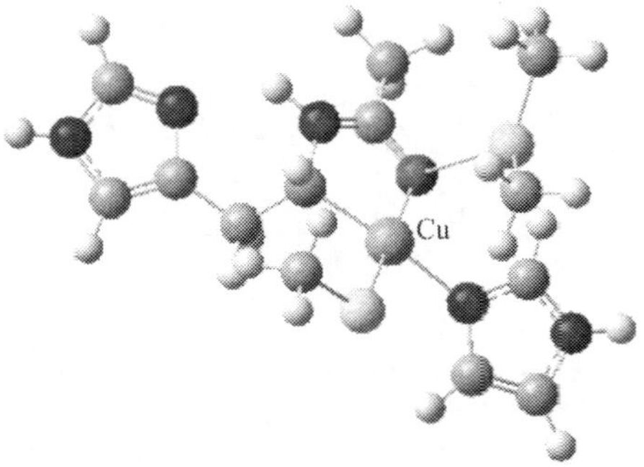

FIGURE 16. Complex used in Reference 145 to model blue copper proteins

paragraph, was treated at the DFT level of theory, whereas the rest of the protein was described using molecular mechanic approaches. This model reproduces the trends in the relative values of the redox potentials for plastocyanin and stellacyanin, suggesting that the much smaller value found for stellacyanin is associated with a much shorter Cu–O distance in the oxidized form.

VI. BONDING

A. Bonding in Cu^+ Complexes

It has been usually assumed that the interaction of Cu^+ with neutral systems is essentially electrostatic, but more and more evidence indicates that covalent contributions are not at all negligible. This covalency is reflected both in the binding energies and in the structures of the Cu^+ complexes, through the contribution of dative bonds from the ligand to the 4*s* empty orbital of the metal and back-donations from occupied *d* orbitals of the metal into empty antibonding orbitals of the base[39–41, 90, 104, 105, 112, 147]. We have already discussed in preceding sections that these interactions were the origin of the enhanced stability of nonconventional π-complexes between Cu^+ and unsaturated or aromatic derivatives of silicon and germanium. We will emphasize here the significant differences between Cu^+ and alkali metal ions, where the interactions are dominantly electrostatic. The differences in bonding are clearly illustrated in Figure 17, which shows the energy density contour maps for the conventional π-complexes of Li^+ and Cu^+ with phenylgermane[40]. It should be recalled here that negative values of the energy density[148] indicate that the potential energy density dominates over the kinetic energy density as in covalent bonds, and vice versa, a positive value of the energy density indicates that the kinetic energy density term dominates as in typical ionic bonds.

Figure 17 shows that in the phenylgermane–Li^+ and phenylgermane–Cu^+ conventional π-complexes the metal ion sits on the electronic hole of the aromatic ring, but whereas the energy density is clearly positive between the metal cation and the base for the Li^+ complex, for Cu^+ it is negative, providing evidence that this interaction has a non-negligible covalent character. This is coherent with the interaction energies between the σ_{C1C6}, σ_{C2C3} and σ_{C4C5} occupied orbitals of the base toward the 4*s* empty orbital of Cu (15, 14 and 13 kcal mol^{-1}, respectively) calculated using a second order NBO perturbation analysis and those from filled *d* orbitals of the metal and the corresponding antibonding

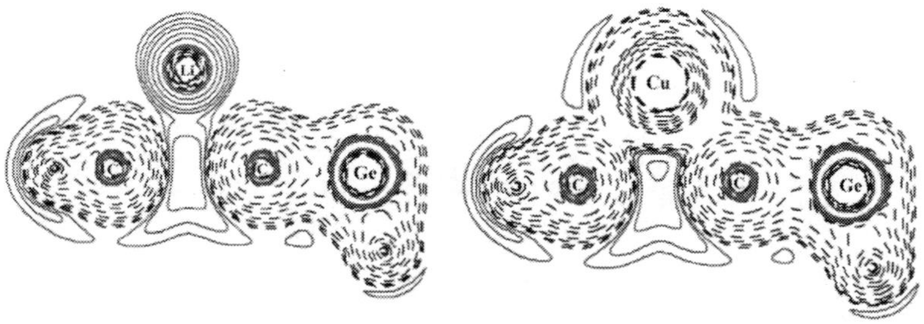

(a) (b)

FIGURE 17. Energy density contour maps for the phenylgermane–Li^+ π-complex (a) and phenylgermane–Cu^+ conventional π-complex (b). The energy density has been plotted in a plane perpendicular to the aromatic ring and containing the substituted and *para* carbon atoms, as well as the metal ion. Dashed lines and solid lines correspond to negative and positive values of the energy density, respectively

orbitals of the base (7, 6 and 5 kcal mol^{-1}, respectively)[40]. The depopulation of the σ_{C1C6}, σ_{C2C3} and σ_{C4C5} bonding orbitals and the concomitant population of the σ^*_{C1C6}, σ^*_{C2C3} and σ^*_{C4C5} antibonding ones results in a lengthening (0.018 Å, in average) of all the CC bonds of the aromatic ring, whose stretching frequencies appear shifted to the red by *ca* 36–51 cm^{-1} [40]. Of course, these kinds of interactions are not possible in Li^+ complexes where the first empty 2*s* orbital lies very high in energy. The situation is not different when dealing with saturated compounds[147]. As shown in Figure 18 for complexes of Li^+ and Cu^+ with germane, again the energy density is positive between the metal ion and the base in the case of Li^+, but clearly negative when the metal ion is Cu^+.

The non-negligible covalent character of Cu^+ complexes is mirrored in binding affinities which are, in average, 1.3 times larger than the corresponding Li^+ affinities[40]. Similar behavior was reported for guanine and adenine Cu^+ complexes when compared with the corresponding Li^+ ones[20]. Finally, it is worth mentioning that besides electrostatic terms, the energy employed in the deformation of the ligand is crucial to rationalize the stability trends of different complexes[31,81].

B. Bonding in Cu^{2+} Complexes

The most significant characteristic of the bonding in Cu(II) complexes is the oxidation undergone by the ligand with the result that both the spin density and the electron deficiency is concentrated in it[4,31,42,81,101,115,116,125,149–151]. A similar oxidation effect has been reported for Cu(III) complexes[93] where the metal does not have a d^8 electronic configuration explaining why, quite unexpectedly, the Cu–L distances are uniformly longer in Cu(III) complexes than in their Cu(I) analogues[93]. As could be anticipated, the more coordinated the metal is, the less oxidized are the ligands[115]. For instance, while in Cu^{2+}–guanine complexes reduction of Cu(II) to Cu(I) takes place and the spin density on the metal is almost zero, when the number of water molecules solvating the metal increases, this effect decreases dramatically and the spin density on Cu (0.72) is again closer to unity[115]. Interestingly, the same conclusion was reached when dealing with complexes in which only one (polydentate) ligand interacts with the metal[31,37].

The aforementioned oxidation of the ligand enhances its intrinsic acidity which facilitates its deprotonation[150,152]. This explains why in general $[Cu{-}L{-}H]^+$ monocations are readily detected in the gas phase instead of the corresponding $[Cu{-}L]^{2+}$ doubly charged species[151]. Cu(II) has also a clear tendency to oxidize amino acids and peptides. For instance, $[Cu(II)(2,2'\text{-bypiridine})(AA\text{-}H)]^+$ complexes (where AA is an amino

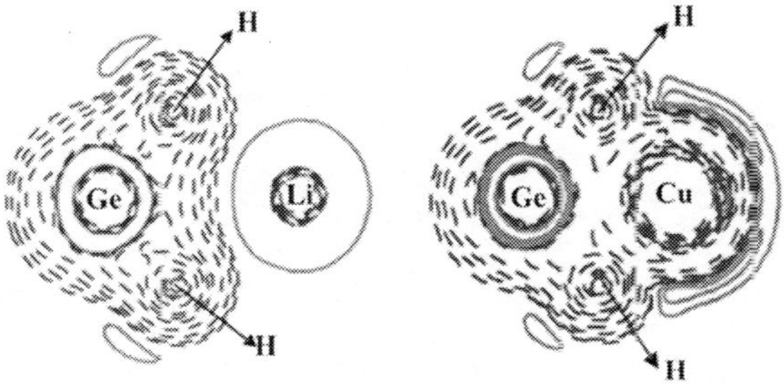

(a) (b)

FIGURE 18. Energy density contour maps for the germane–Li^+ (a) and germane–Cu^+ (b) complexes. The energy density has been plotted in a plane containing the silicon atom, as well as the two hydrogen atoms interacting with the metal ion. Same conventions as in Figure 17

acid) lead to the loss of CO_2 followed by further radical-driven fragmentation of the resulting Cu(I) complex[61,153,154]. Similarly, CID experiments carried out on $[Cu(2,2'\text{-bypiridine})(Arg)]^{2+}$ and $[Cu(His)_2]^{2+}$ complexes lead to the formation of $[Arg\text{-}CO_2]^{+\bullet}$ and $[His\text{-}CO_2]^{+\bullet}$ radical cations, which was taken as evidence of the formation of arginine and histidine radical cations as transient species[153,155]. Similar results were reported for complexes between Cu(II) and histidine-containing peptides, as we will see in the section dealing with the reactivity of copper ions with biomolecules. On the other hand, the reduction of Cu(II) to Cu(I) is the basic mechanism behind the electron-transfer role of blue copper proteins[10,144], whereas the oxidation capacity of Cu(II) is also the driving force of proton-transfer reactions within base pairs[156,157]. Experiments conducted in the gas phase[158] showed significant differences in the bonding of Cu(II) with respect to Ag(II) and Au(II).

VII. Cu^+/Cu^{2+} REACTIONS WITH SMALL ORGANIC BASES

Over the past decades, a great body of studies on gas-phase reactions of bare Cu^+/Cu^{2+} cations with small molecules containing prototypical bonds (e.g. N−H, C−H, C−C, O−H) have been performed[159–168]. In this section we will present an overview of the most significant results reported in the period of time covered in this chapter.

The interaction of Cu^+ with carbonyl compounds has been widely studied in the literature. The binding energy has been measured or calculated for a large set of compounds, as has been discussed in preceding sections. For Cu^{2+}, the deprotonation usually accompanying the association of this doubly charged metal to organic compounds makes the measurement of binding energies practically impossible, since the species experimentally accessible is the singly-charged deprotonated complex. As a matter of fact, as we have already noted when discussing the bonding, it appears that the most significant difference in reactivity between both copper oxidation states is the oxidation capacity of Cu(II). The interaction of Cu^+/Cu^{2+} with thiouracil derivatives (Chart 5) provides a good example for these differences[84,125,150]. While Cu^+ interaction occurs preferentially with one oxygen or a sulfur lone pair, Cu^{2+} interacts preferentially with the two oxygen lone pairs simultaneously. In the first case, the attachment of the metal occurs at the most basic site, which is the heteroatom at C4. For Cu^{2+} the interaction is preferred at the oxygen atom, independently of its position. This was explained by a greater affinity of Cu^{2+} toward carbonyl than toward the thiocarbonyl group[125], even though in general thiocarbonyl derivatives exhibit a gas-phase basicity larger than their carbonyl analogues[169,170].

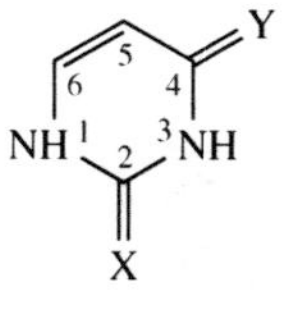

X = Y = O uracil
X = S, Y = O 2-thiouracil
X = O, Y = S 4-thiouracil
X = Y = S 2,4-dithiouracil

CHART 5

To the best of our knowledge there are no studies dealing with the reactivity of uracil and thiouracil derivatives toward Cu^+. However, as these compounds can be viewed as formed by two different groups, formamide and urea or their thio derivatives, the reactivity patterns of uracils and thiouracils might be considered similar to those of formamide and urea or thiourea, if one excludes ring effects which are only present in the former and which will be discussed in subsequent sections. Tortajada and coworkers[171] have reported a nice experimental and theoretical study on the fragmentation of formamide–Cu^+ complexes in the gas phase. In a preliminary study, the isomerization of formamide using different levels of theory was analyzed in order to estimate its affinity toward Cu^+ and to locate the most basic centers[172]. Subsequently, by means of Chemical Ionization Fast-Atom Bombardment (CI-FAB) mass spectrometry techniques, the dissociation products of the formamide–Cu^+ complex were investigated, the main precursors for these fragmentations being the oxygen- and nitrogen-attached species. The exploration of the PES at the B3LYP/6-311+G(2df,2p)//B3LYP/6-311G(d,p) level of theory revealed that $OCCu^+$ and $CuNH_3^+$, which were the most abundant fragments observed by Mass Analyzed Ion Kinetic (MIKE)-CAD spectra, may be produced through three different mechanisms. If the *O*-attached complex is assumed to be the precursor, the reaction should overcome, as a first step, an activation barrier of about 93.4 kcal mol^{-1} associated with a 1,2-hydrogen shift from carbon to nitrogen (Figure 19). A similar energetic barrier, *ca* 95.7 kcal mol^{-1}, corresponding to a 1,2-hydrogen transfer from carbon to oxygen is the first step when the same fragments are produced if the *N*-attached complex is considered to be the precursor for the unimolecular fragmentation. The calculated activation barrier is higher (about 104.9 kcal mol^{-1}, Figure 20) if a direct hydrogen transfer from the CH group to the amino group is envisaged.

In the case of urea, two amino groups are available for the interaction with the metal ion. The effects of this second amino group are reflected in both the interaction patterns of copper and the nature of the fragments obtained in gas-phase unimolecular decompositions, because new fragments can be generated when dealing with urea, which are not possible when the base is formamide. For instance, while the loss of ammonia leads to $COCu^+$ in the case of formamide–Cu^+ reactions, in urea–Cu^+ reactions the same process produces a $HNCOCu^+$ ion, which is the most abundant one[121,173]. The DFT calculated PESs showed a quite different mechanism in both cases. While in urea–Cu^+ reaction ammonia is formed through a 1,3-hydrogen transfer between both amino groups, for formamide–Cu^+ reactions, this is a three-step process, the former being a 1,2-H shift from the C–H group toward the carbonyl oxygen, which is subsequently transferred to the amino group by means of a 1,3-H shift. Furthermore, the $HNCOCu^+$ fragment is only produced from the *O*-attached adduct as precursor. In both reactions a high intensity peak corresponding to the $CuNH_3^+$ ion is observed with origin in the *N*-attached complex. In both cases the barriers are lower than the entrance channel, which indicates that these mechanisms are thermodynamically favored (Figure 21).

In the reaction of formamide with copper, it should be mentioned that the loss of water was found to be a quite favorable process whereas such a process was not observed for urea. A theoretical survey of the corresponding PES showed that for formamide a double hydrogen transfer toward the carbonyl oxygen atom is feasible, a first one from the nearest carbon and a second one from the amino group[171]. However, for urea–Cu^+ these hydrogen transfers to oxygen do not compete with those taking place between the two amino groups which ultimately lead to the loss of NH_4^+[121]. Interestingly, this is not the case when the carbonyl group is replaced by an isoelectronic NH group as in the case of guanidine–Cu^+ complexation[174] where the loss of NH_4^+ was not observed in the mass spectra. It is worth noting, however, that this is the only significant difference between guanidine–Cu^+ and urea–Cu^+ fragmentations. For instance, if we take into account that the difference between the urea and guanidine molecule is the replacement of the oxygen

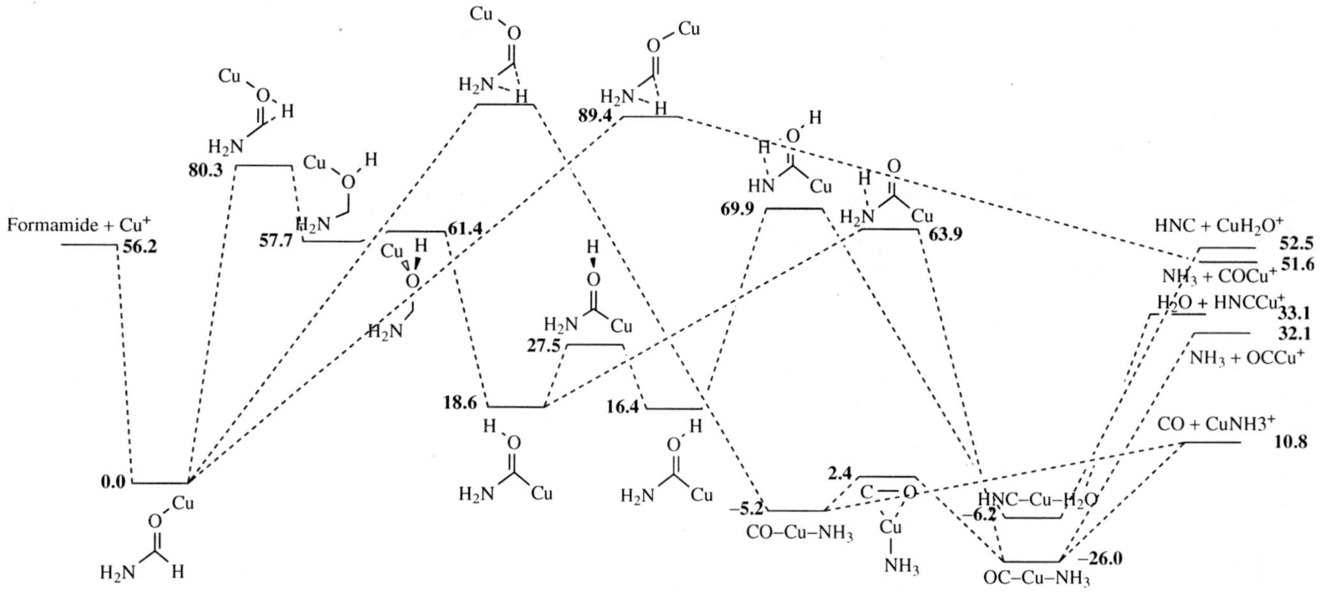

FIGURE 19. Energy profile of the unimolecular reactivity of the oxygen-attached formamide–Cu^+ complex. Energy values are in kcal mol^{-1}

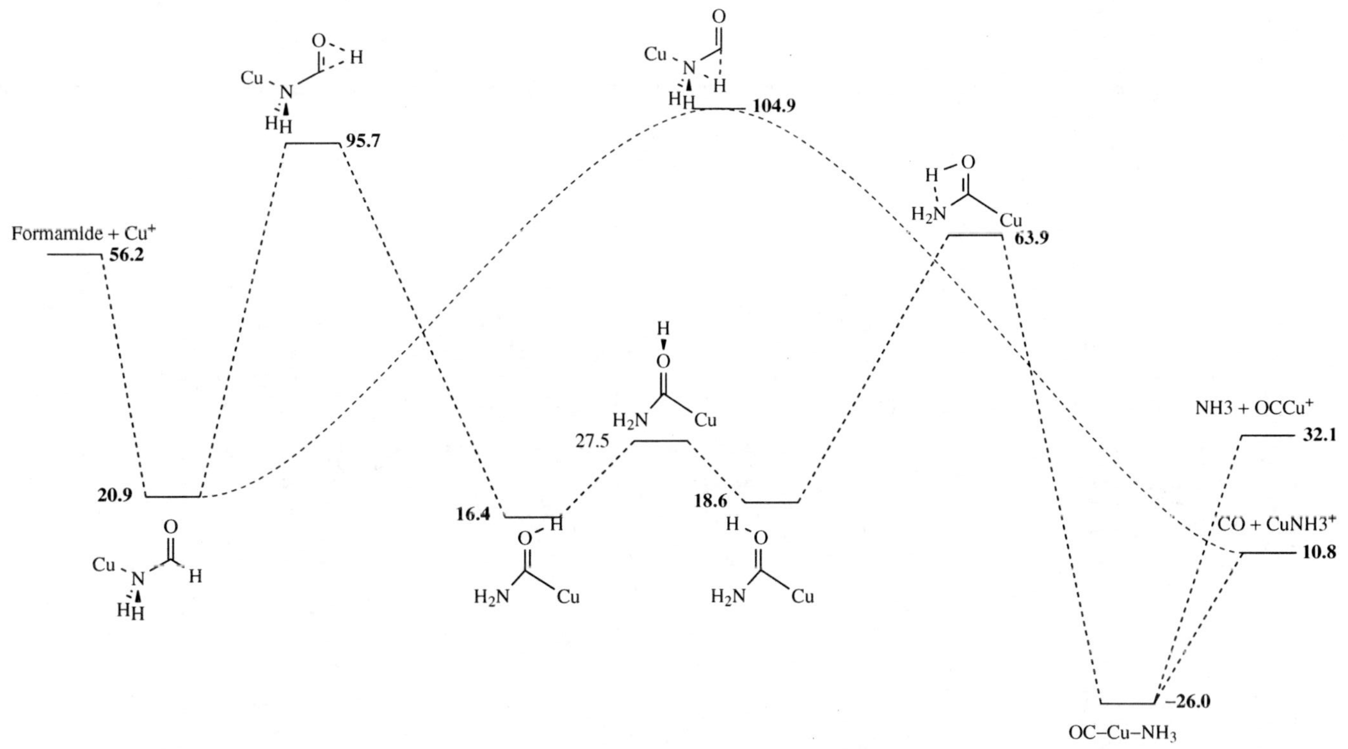

FIGURE 20. Energy profile of the unimolecular reactivity of the nitrogen-attached formamide–Cu^+ complex. Energy values are in kcal mol^{-1}

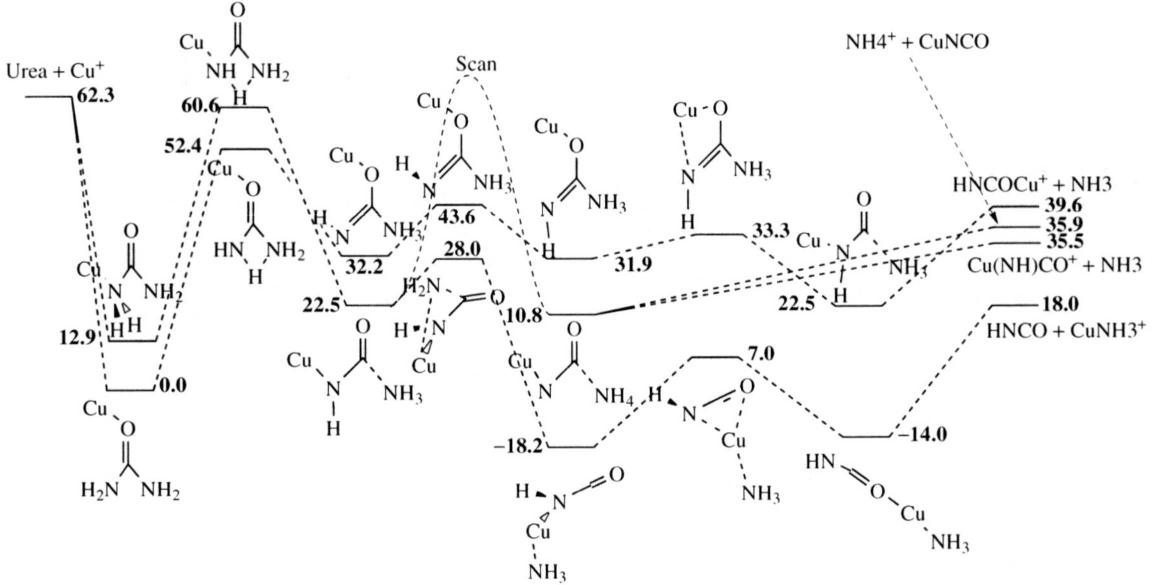

FIGURE 21. Energy profile associated with the unimolecular reactivity of the urea–Cu^+ system. Energy values are in kcal mol^{-1}

by a NH group, the loss of HNCO observed in urea–Cu^- reactions[121] appears as the loss of HNCNH in guanidine–Cu^+ reactions[174]. In both cases the dominant product of the reaction was ammonia.

Another aspect to be emphasized concerning the Cu^+ reactivity is that the metal cation can act as a hydrogen carrier for the loss of H_2. For urea– and formamide–Cu^+ systems, the experimental results do not provide any indication of such a loss. However, in 2-propanol–Cu^+ [175, 176], ethylenediamine–Cu^+ [177] and alkene–Cu^+ reactions[178], the loss of a H_2 molecule was systematically observed with different intensity. All the theoretical studies which aimed at explaining the dissociation of H_2 in these complexes pointed to the role of copper as a carrier. Concerning the alkene–Cu^+ complexes, the attachment of the metal cation to the π-system forces a coiling of the alkyl chain, which favors the formation of a hexa-coordinated intermediate (form **7** in Scheme 1) that eventually evolves to yield form **8**, which could be a good precursor for the loss of H_2.

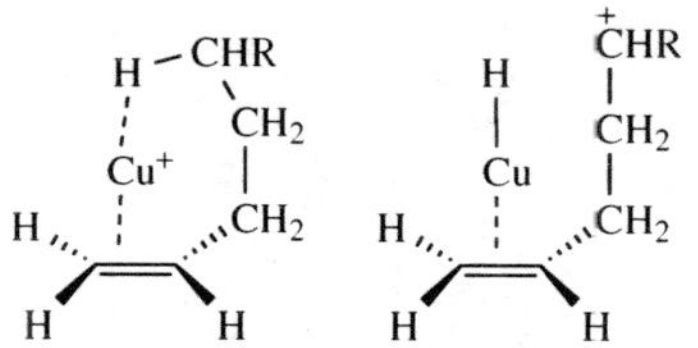

(7) (8)

SCHEME 1

However, much more stable complexes, such as species **9** (Scheme 2), can be formed when the metal ion interacts simultaneously with two C–H bonds of the alkyl chain, through typical agostic-type interactions, already discussed in detail in Section IV.A. The activation of both C–H bonds lead to the formation of structure **10** (Scheme 2), which finally dissociates by losing H_2.

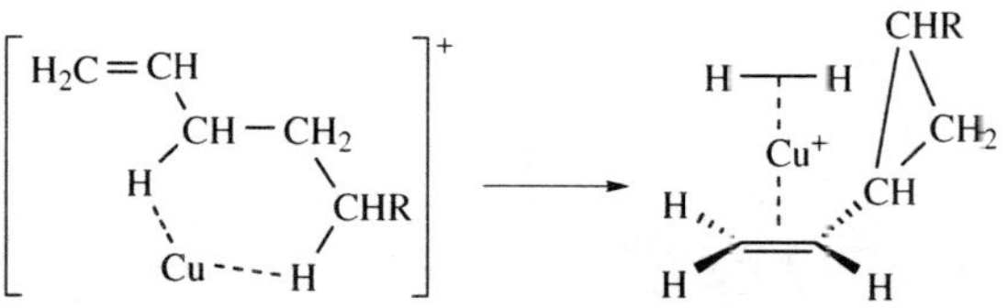

(9) (10)

SCHEME 2

The most favorable process in the reactions between Cu^+ and alkenes corresponds, however, to the loss of one olefin molecule. This happens, as proposed by Morizur and coworkers[179], through a pseudo-insertion mechanism, in which form **7**, where the alkyl chain coils up to enhance its interaction with the metal cation, plays an important role. From **7**, and for the particular case of 1-pentene, two different unimolecular dissociation processes leading to $C_2H_4 + [H_3C{-}CH{=}CH_2{--}Cu]^+$ and to $C_2H_6 + [H_2C{=}CH_2{--}Cu]^+$ are possible, the former being more favorable due to the larger Cu^+ binding energy to $H_3C{-}CH{=}CH_2$.

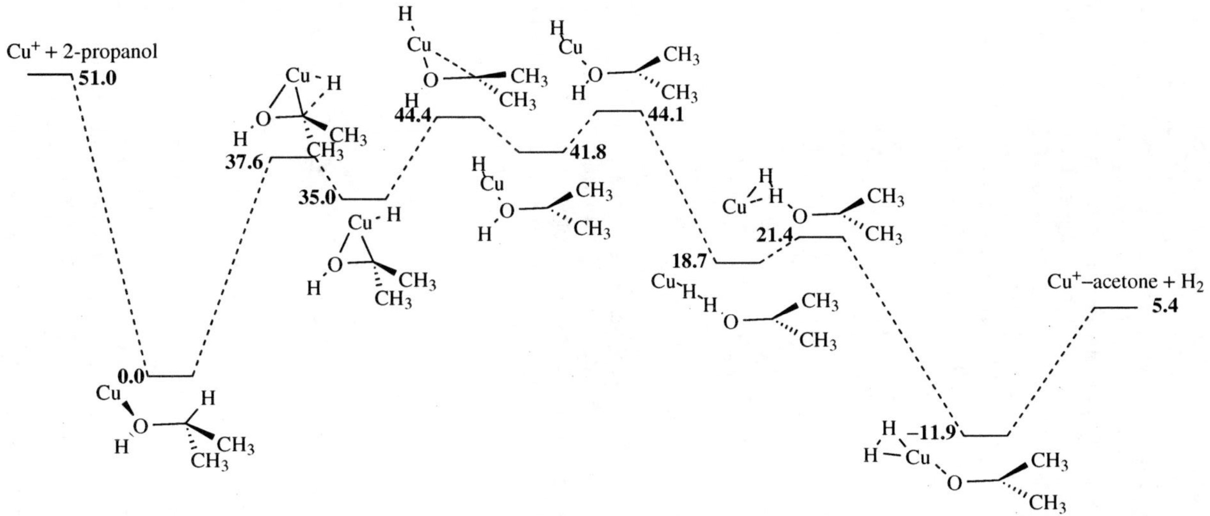

FIGURE 22. Energy profile associated with the unimolecular decomposition of the 2-propanol–Cu^+ complex. Energy values are in kcal mol^{-1}

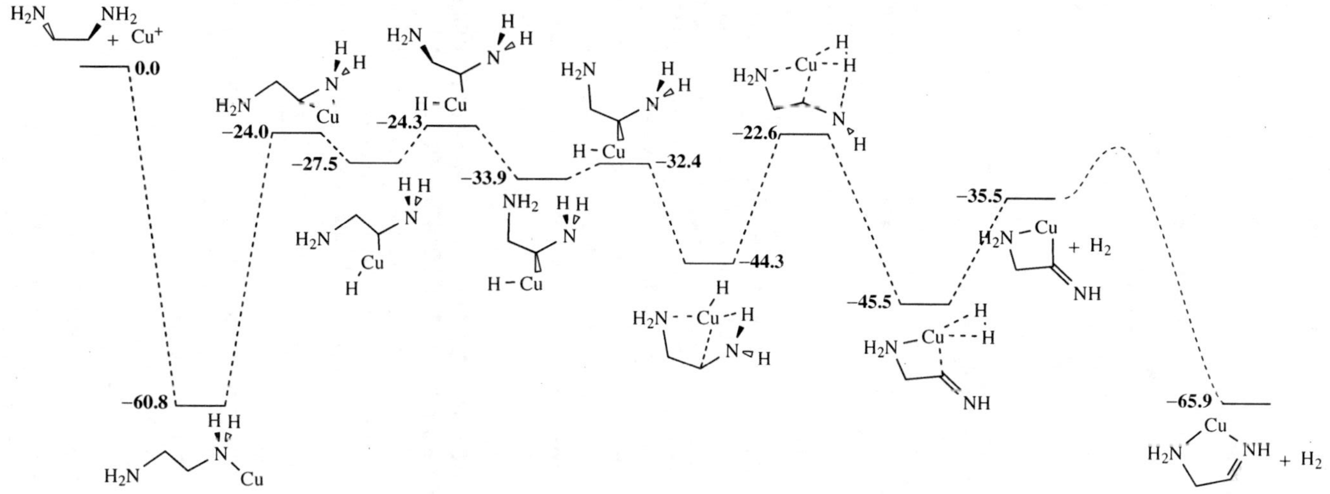

FIGURE 23. Energy profile associated with the unimolecular decomposition of the ethylenediamine Cu^+ complex. Energy values are in kcal mol^{-1}

The loss of H_2 in Cu^+–2-propanol reactions was investigated by Cheng and Hu[176] by means of B3LYP/6-311+G(d,p) calculations. Two different mechanisms would be compatible with the experimental observations of Huang and coworkers[175] where the most abundant fragment is $[Cu\text{-}(3C,6H,O)]^+$. One of them would yield 2-propenol–Cu^+ while the other would produce acetone–Cu^+. In both cases, the 2-propanol–Cu^+ complex loses a hydrogen molecule. The most favorable mechanism was the one schematized in Figure 22, where acetone–Cu^+ appears as the product ion. The exit channel is reached via a C–OH metal insertion followed by a hydrogen shift. It can be seen that, also in this case, copper acts as carrier of the hydrogen molecule to yield the acetone–Cu^+ fragment and the energy barriers involved in this mechanism are below the entrance channel. Note that the C–H insertion mechanism was found energetically unfavorable and consequently was discarded.

For ethylenediamine–Cu^+ reaction, Yáñez and coworkers[177] also showed the active role played by copper in the H_2 loss. In this case, the activation energies also remain below the entrance channel all along the route from the initial Cu^+ adduct to the products (Figure 23).

Cu^{2+} reactivity is dominated by its oxidation capacity, already mentioned when discussing its bonding. Accordingly, and unlike inorganic molecules[180], the doubly charged complexes resulting from the association of Cu^{2+} to any typical organic base have not been experimentally observed by the experimental techniques available to date, since an immediate deprotonation of the system takes place and only the singly charged deprotonated copper complex is detected. Hence, Cu^{2+} reactivity is rather similar in many aspects to that of Cu^+, since what is usually observed is the fragmentation of the singly charged species produced upon deprotonation of the doubly charged complexes[125,150,181]. Hence, not surprisingly the fragmentation of uracil–Cu^{2+} complexes, to be discussed in a forthcoming section, follows similar patterns to those observed, for instance, in the fragmentation of formamide–Cu^+ and urea–Cu^+ complexes.

The most significant difference between formamide–Cu^+ and urea–Cu^+ with respect to uracil–Cu^{2+} reactions is the loss of $NCO^\bullet$ radical, which is only observed for the latter, reflecting, as we shall discuss later, the important role played in this case by π-type complexes.

VIII. GAS-PHASE REACTIVITY OF COPPER(I) AND COPPER(II) IONS TOWARD MOLECULES OF BIOLOGICAL RELEVANCE

As already mentioned in the introduction, copper plays a significant role in different biochemical processes. For example, in the mechanism of action of several enzymes, it is the cofactor of redox reactions involving molecular oxygen, and it is involved in antibody production.

Gas-phase studies on the interaction of copper with small model molecules is of great interest because knowledge of their intrinsic properties can provide important clues to understand the behavior of more complicated systems of biological importance. These studies have received much more attention since the introduction of fast atom bombardment (FAB), laser desorption and electrospray ionization (ESI) as methods of forming such complexes in the gas phase.

A. Reactivity with Amino Acids and Peptides

A typical example is the study of the reactivity of copper ions toward amino acids and peptides, and numerous studies involving either of the aforementioned ionization methods have been published during the last decade. These studies can be divided into

TABLE 3. Code names of the various amino acids

Name	Code name	
	3 letters	1 letter
Alanine	Ala	A
Arginine	Arg	R
Asparagine	Asn	N
Aspartic acid	Asp	D
Cysteine	Cys	C
Glutamic acid	Glu	E
Glutamine	Gln	Q
Glycine	Gly	G
Histidine	His	H
Isoleucine	Ile	I
Leucine	Leu	L
Lysine	Lys	K
Methionine	Met	M
Phenylalanine	Phe	F
Proline	Pro	P
Serine	Ser	S
Threonine	Thr	T
Tryptophan	Trp	W
Tyrosine	Tyr	Y
Valine	Val	V

two categories. The first one corresponds to the direct examination of 'binary' mixtures of copper salts and amino acids or peptides, and will be described first. The second category deals with the generation of ternary complexes by using an auxiliary organic ligand which occupies some coordination sites in the copper ion, and thus directs the coordination of the biomolecule. For the sake of simplicity in the following sections the different amino acids will be identified by the abbreviations shown in Table 3.

1. Binary mixtures

Gas-phase interaction between Cu(I) ions and the simplest amino acid, namely glycine, were first studied by Plasma Desorption Mass Spectrometry[182]. Complexes of interest were generated in a plasma desorption ion source by bombarding a mixture of glycine and cupric salts with fission fragments of ^{252}Cf. Regardless of the copper salt used, the resulting adducts are of the type MCu^+. Copper in its cuprous oxidation state [Cu(I)] is indeed exclusively observed, not only for glycine but also for the other nine amino acids considered in this study (Val, Phe, Tyr, Trp, His, Asp, Asn, Glu, Gln). Under metastable conditions, the unimolecular decomposition of the $[Cu{-}glycine]^+$ complex is characterized by the predominant elimination of a 46u $[H_2,C,O_2]$ fragment. It is noteworthy that this fragmentation is commonly observed for both cationized and protonated amino acids. This elimination may correspond either to the loss of formic acid, HCOOH, or to the consecutive loss of H_2 and CO_2, or H_2O and CO. Use of deuteriated amino acids suggested that the most probable mechanism involves the consecutive elimination of water and carbon monoxide, likely leading to the formation of a metal-cationized imine.

The Cu^+/glycine system has been recently reinvestigated by combining electrospray ionization tandem mass spectrometry and DFT calculations[183]. Collisional activation experiments were carried out on $[^{63}Cu{-}glycine]^+$ ions (*m/z* 138). Again, the most favorable dissociation process corresponds to the formation of $[Cu{\cdots}HN{=}CH_2]^+$ concomitant with the elimination of [H_2,C,O_2]. Total H/D exchange of the acidic hydrogens showed that the hydrogen atoms eliminated in the 46u are labile hydrogens. In order to rationalize these experimental findings, the authors explored the potential energy surfaces associated with this system. As mentioned in Section VII and confirming previous theoretical studies, the most stable form of copper–glycine involves metal chelation between the nitrogen atom and the oxygen atom of the carbonyl group. Several mechanisms were considered to account for the fragmentations observed, involving either metal insertion into covalent bonds (C–C, C–O) or dissociative attachment whereby the metal ion catalyzes the fragmentation by its distant electronic influence. It was shown that the formation of [Cu$\cdots$HN=CH_2]$^+$ is exclusively associated with the loss of (H_2O+CO) as elimination of other neutral species such as HCOOH, $C(OH)_2$ and H_2+CO_2 implies energy barriers located higher in energy than the decationization energy and therefore are not competitive. The dissociation proceeds through mechanisms involving copper insertion into the C–C and C–OH bonds. It may start either with insertion into the C–C bond, followed by insertion into the C–OH bond (Figure 24), or these two elementary steps may occur in the reverse order.

This elimination of 46u is in fact commonly observed for various amino acids, and notably for the four aromatic amino acids histidine, tryptophan, tyrosine and phenylalanine[155, 182, 184]. However, the reactivities of glycine and of these amino acids also differ. For example, elimination of CO_2+NH_3 is systematically observed and is the main fragmentation for the [Cu–His]$^+$ complex[184]. An additional characteristic fragmentation detected at high collision energies corresponds to the loss of a 137u consistent with a [Cu,C_2,O_2,N,H_4] group, and leading to the ionized side chain R^+. The observed losses are found to be very similar for all the aromatic systems and only differ in their relative intensities. They do not involve cleavage within the amino acid side chain, in agreement with other previous metal cation–amino acid studies[77, 185]. Given these similarities, Sodupe and coworkers[184] have only investigated theoretically the [Cu–Phe]$^+$ system. The

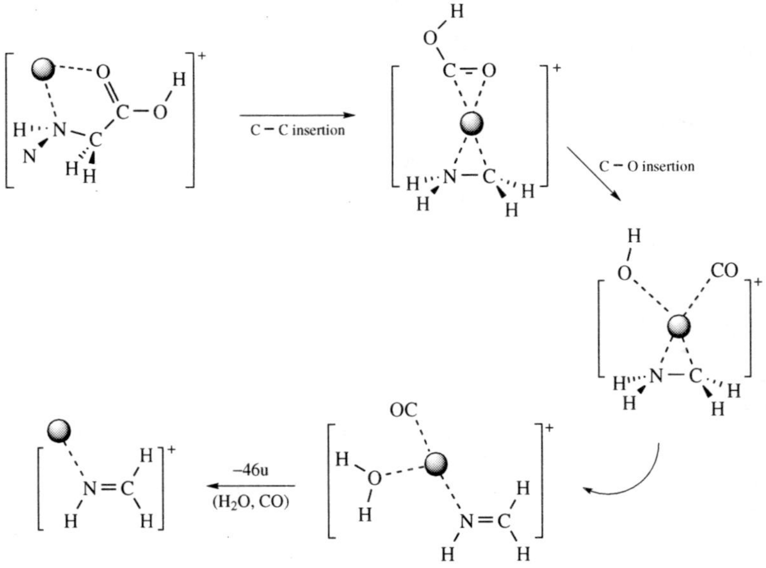

FIGURE 24. Most favorable mechanism proposed for the elimination of 46u from the [Cu–glycine]$^+$ complex

most stable structure corresponds to the metal cation interacting with the aromatic ring, the amine nitrogen and the carbonyl oxygen. Exploration of the various potential energy surfaces showed that the observed eliminations are produced from insertion of the metal cation either into the backbone C−C bond of Phe (losses of 46u) or into the C−R bond (losses of 137u). Concerning the latter process, it has been demonstrated that the direct elimination of a HOOC−CH−NH_2Cu moiety is favored over the consecutive loss of CuCOOH and CH_2=NH.

In addition to the $[Cu-AA]^+$ (AA = amino acids) species, complexes of general formula $[Cu(AA)-H]^+$ can also be generated by decreasing the cone voltage of an electrospray interface. This was observed for histidine[155]. Expectedly, the gas-phase reactivities of $[Cu-His]^+$ and $[Cu(His)-H]^+$ complexes are different. Upon collision, the latter species not only expels a $[C_2,O_2,H_3,N]$ entity, but also yields the immonium I_{His} as a fragment ion. Based on thermochemical estimates, the formation of these two particular ions has been interpreted as elimination of CO_2, immediately followed by loss of formaldimine and the neutral metal, respectively. Note that for zinc and copper the decarboxylated complexes appeared unstable (not observed) while they are detected in significant abundance for other metal ions such as Ni^{2+}, Co^{2+} and Fe^{2+}. In the literature, the decarboxylation is usually interpreted as a radical-like mechanism involving reduction of the metal (Figure 25). Note that the elimination of formaldimine implies the insertion of the metal within the amino acid side-chain.

Adopting mild source/interface conditions also allows the formation of multiply charged and solvated ions by electrospray. As a matter of fact, with such conditions and by using water as solvent, Seto and Stone[186] managed to generate stable and relatively abundant $[Cu(gly)(H_2O)]^{2+}$, $[Cu(gly)(H_2O)_2]^{2+}$ and $[Cu(gly)_2]^{2+}$ complexes in the gas phase, even

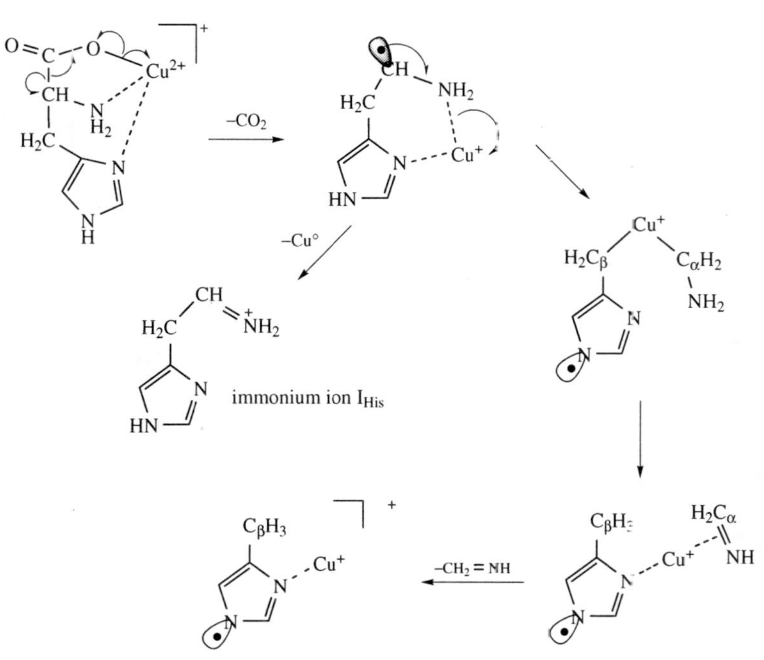

FIGURE 25. Mechanism proposed for the characteristic fragmentations of the $[Cu(His)-H]^+$ complex

if the major ion in the spectrum is that of protonated glycine, $GlyH^+$. On the other hand, these authors did not manage to produce in the source the complex in which the copper dication interacts with a single glycine molecule, but this behavior under ESI is typical of this metal, as mentioned before. It is worth noting that a doubly charged species isobaric to the $[Cu(gly)]^{2+}$ ion was detected in the CAD spectra of $[Cu(gly)(H_2O)]^{2+}$ and $[Cu(gly)(H_2O)_2]^{2+}$, strongly suggesting that the $[Cu(gly)]^{2+}$ is indeed stable thermodynamically.

The unimolecular reactivity of the doubly charged complexes is characterized by two kinds of product ions, depending on whether a charge reduction occurs during the fragmentation. If one examines the CID spectra of the $[Cu(gly)(H_2O)]^{2+}$ species, it can be seen that there is a preference for loss of neutral water rather than of neutral glycine, in agreement with the fact that glycine should have the higher enthalpy of association.

An interesting fragment ion in the MS/MS spectrum of $[Cu(gly)_2]^{2+}$ is the loss of water. The postulated mechanism for the formation of this ion is the elimination of a molecule of water in an intracomplex condensation reaction, leading to the formation of a peptide bond between the two glycine residues, the dipeptide remaining attached to the metal. Even if there is no experimental evidence in the gas phase supporting this assumption, it is worth mentioning that such a reaction occurs in aqueous solutions containing a Cu(II) salt and glycine[187, 188].

Chiral assignments are of great interest because the chemical or biological activity of a substance often depends on its stereochemistry. While condensed-phase techniques such as NMR, circular dichroism and chromatography are widely used to study enantioselective intermolecular interactions, increasing attention is being paid to gas-phase techniques, in particular mass spectrometry. Chiral recognition by mass spectrometry has been made possible through the use of the well-known kinetic method[189, 190], which was initially developed by Cooks and coworkers for the determination of thermochemical values such as gas-phase basicities, metal ion affinities or ionization energies of organic compounds[191]. Recently, the kinetic method has been successfully applied to the chiral recognition of *D*- and *L*-amino acids by analysis of the kinetics of competitive fragmentations of trimeric copper(II)-bound complexes[192]. Singly charged cluster ions of general formula $([Cu(II)(AA)_n - H]^+$ ($n = 2, 3, 4$) can be generated easily in the gas phase by electrospray ionization of Cu(II)–amino acid mixtures. CID experiments showed that dimeric cluster ions ($n = 2$) dissociate by losing CO_2. In contrast, heterotrimeric cluster ions appeared to have two singly bound ligands (Figure 26), as attested by the MS/MS spectra of $[Cu(AA1)_2(AA2)-H]^+$, resulting solely in the formation of dimeric cluster ions by elimination of either AA1 or AA2.

Six amino acids (AA) were studied, i.e. Tyr, Leu, Met, Phe, Thr, Asp. By using *L*-proline as reference and by assuming that only the monoligated amino acids can be lost, the singly charged trimeric cluster ions $[Cu(II)(L\text{-proline})_2(AA)-H]^+$ were mass selected

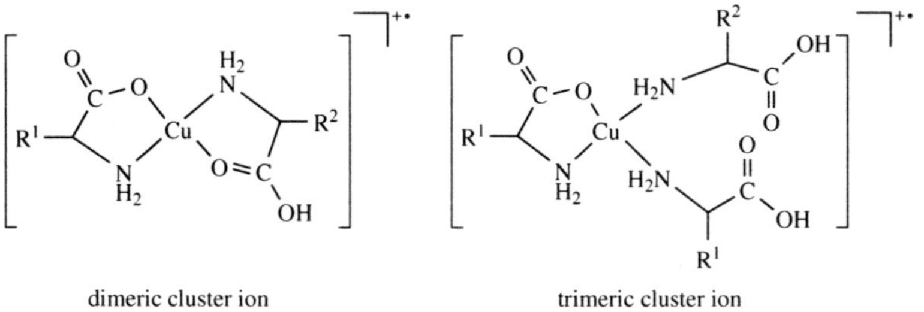

FIGURE 26. Coordination scheme of copper in binary and ternary cluster ions

and collisionally excited in a quadrupole ion trap, where they are dissociated competitively to form the dimeric complexes [Cu(II)(*L*-proline)(AA)−H]$^+$ and [Cu(II)(*L*-proline)$_2$−H]$^+$ by the loss of the neutral reference, or the amino acids AA studied, respectively.

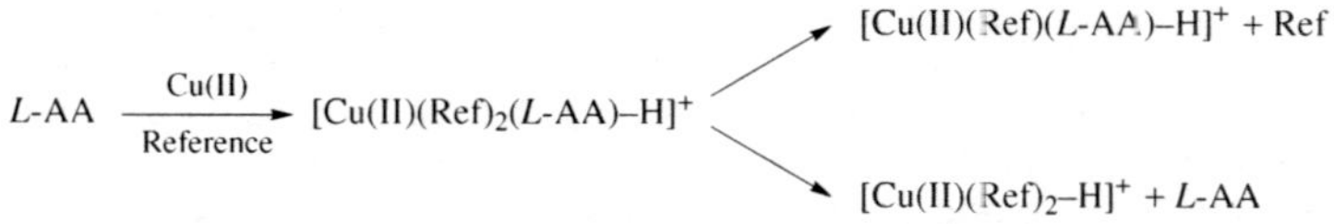

with $r_L = I(\text{Cu(II)}(L\text{-proline})(L\text{-AA})\text{–H}]^+)/I([\text{Cu(II)}(L\text{-proline})_2\text{–H}]^+)$

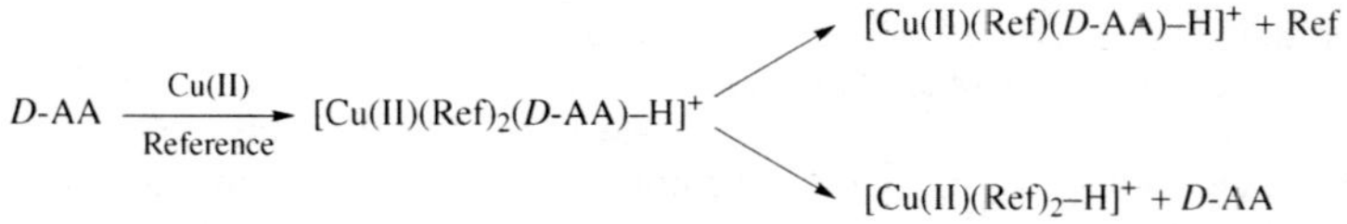

and $r_D = I(\text{Cu(II)}(L\text{-proline})(D\text{-AA})\text{–H}]^+)/I([\text{Cu(II)}(L\text{-proline})_2\text{–H}]^+)$

The difference in stability of the fragment ions [Cu(II)(*L*-proline)(AA)−H]$^+$, due to the two enantiomeric forms of the AA, results in differences in the corresponding product ion abundances, measured relative to the abundance of [Cu(II)(ref)$_2$−H]$^+$. The two abundance ratios r_L and r_D allow evaluating the chiral resolution factor R, equal to r_D/r_L. The data obtained revealed that amino acids with an aromatic side chain show a very high chiral effect ($R \gg 1$), while aliphatic amino acids are characterized by R values very close to 1.

Additional experiments demonstrated for aromatic amino acids the quantitative nature of the chiral distinction achieved by the kinetic method. Proline mixtures with various optical purities (fraction of *D* isomer = 0, 0.25, 0.5, 0.75 and 1) were used as the reference and *D*- or *L*-tyrosine was used as the analyte. As expected, the largest chiral discrimination was observed when pure *D*- or *L*-proline was used as the reference. There was no chiral distinction for *D*- and *L*-tyrosine when proline racemate was used as the reference. These first results showed that the kinetic method could be applicable to measurement of enantiomeric excess.

During the last decade, tandem mass spectrometry has become an invaluable tool for identification of peptides and proteins. In this context, the most widely applied strategy is to perform low-energy CID experiments on protonated peptides ([M+nH]$^{n+}$ ions). Not only these ions can be easily formed by several ionization methods, and especially matrix-assisted laser desorption ionization (MALDI) and electrospray ionization (ESI), but they also provide primary sequence information upon activation by cleavage of the amide linkages to yield complementary b_n/y_n series of ions (Figure 27).

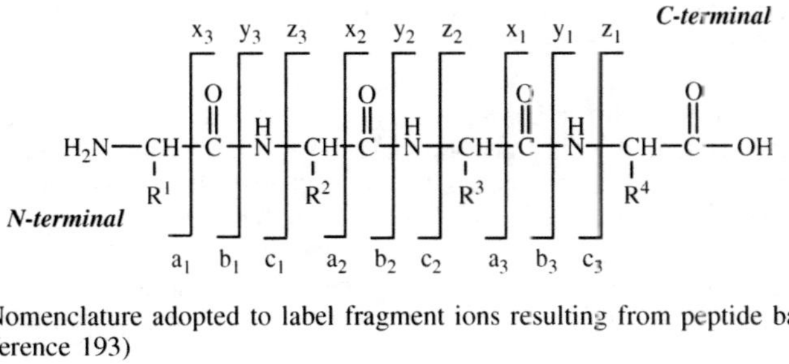

FIGURE 27. Nomenclature adopted to label fragment ions resulting from peptide backbone cleavages (from Reference 193)

Although fragmentation of protonated peptides is highly useful, this approach suffers from several limitations. The sequence information can be limited because protonated peptides may not fragment at every amide bond, leading to incomplete sequence information. Products from only one or a few specific cleavages often dominate product ion spectra, depending upon the charge state and the primary sequence of the peptide. Finally, the location of post-translational modifications (PTMs) cannot be determined. For these reasons, alternative means for deriving peptide structural information are desirable. In this context, metal-specific fragmentations can provide a useful complement toward full structural characterization of peptides. Consequently, besides studies describing the mechanisms of interaction of copper with peptides, the reactivity of Cu(I) and Cu(II) ions has been applied for sequencing purpose, and also to distinguish isomeric peptides.

As mentioned earlier, the elimination of 46u from $[M+Cu]^+$ ions is commonly observed for amino acids. On the other hand, this fragmentation is not observed for peptides. Instead, Shields and coworkers[194] have shown that the metastable ion spectra of $[M+Cu]^+$ complexes obtained with small peptides containing a *N*-terminal arginine are characterized by loss of neutral fragments such as water, formaldehyde or acetaldehyde when peptides include a serine, threonine or glutamic acid residue, and ammonia or guanidine for peptides containing basic residues such as lysine or arginine. Unlike the $[M+H]^+$ ions, immoniums ions and internal fragments were practically not detected. Interestingly, series of ions of general formula $[a_n+Cu\text{-}H]^+$, $[b_n+Cu\text{-}H]^+$; and $[c_n+Cu+H]^+$, resulting from the cleavage of the peptidic chain, are also particularly abundant. Such ions have also been observed with alkali metals[195] and Ag^+ [196], and provide interesting sequence information. Authors also noted that the dissociation processes of $[M+Cu]^+$ ions were governed by the presence and position of the arginine residue. All the fragment ions of peptides containing a *N*-terminal arginine contained the arginine residue and Cu^+.

$[b_n+Cu\text{-}H]^+$ ions are formed by cleavage of peptide backbone amide bonds associated with the migration of hydrogen to the neutral leaving group. Several mechanisms have been proposed in the literature in order to explain their formation, one involving the transfer of a hydrogen from the C_α-carbon to the neutral leaving group[197] and two involving instead a hydrogen from the amide NH group[198, 199]. Data obtained with glycine-containing peptides suggested that the proton transferred to the leaving group does not originate from the side chain as glycine does not have a side chain. Furthermore, results deduced from H/D exchange experiments indicated that the hydrogen transferred is an amide nitrogen from the *N*-terminal side of the cleavage. Consequently, to interpret the difference of reactivity between $[M+H]^+$ and $[M+Cu]^+$ ions, the authors proposed that the Cu^+ ion is anchored at the *N*-terminal arginine, and that $[b_n+Cu\text{-}H]^+$ fragment ions are formed via a mobile proton transferred from the *N*-terminus to other amide nitrogens along the peptide backbone. The results obtained during this particular study also showed that the $[a_n+Cu\text{-}H]^+$ ion were either generated directly from the precursor ions, or by loss of CO from the corresponding $[b_n+Cu\text{-}H]^+$ species. A second study[200] confirmed that the side chains of the basic amino acids arginine and, to a lesser extent, histidine and lysine were the Cu^+ binding sites within peptides. B3LYP calculations have also been performed in order to establish a relative copper affinity scale for monodentate and bidentate interactions. Relative binding energies of the model monodentate ligand–Cu^+ systems were as follows: Arg>His>Lys>Cys>Ser. A slightly different order has been found for bidentate Cu^+ binding energies (Arg>Lys>His>Gln>Asn>Glu>Asp), indicating the importance of multidentate Cu^+ coordination. In a MALDI study published several years ago[201], it was deduced from the analysis of the interactions with sixteen synthetic peptides that the compounds containing no histidine did not form stable metal complexes under MS conditions, whereas all histidine-containing sequences formed metal complexes. This confirms histidine as an important binding site of copper in proteins.

The use of metallation by divalent metal ions for investigating peptide structures has not been widely explored under negative ESI conditions so far. One reason could be the relatively low yield of negative-ion complexes generated under ESI conditions in comparison with those achieved in positive-ion mode. However, the group of Tabet has demonstrated that the copper reactivity can be successfully applied for the differentiation of isomeric peptides[202, 203]. By mixing various modified enkephalins (YGGFX, X = I, L, M, Q, K) and copper chloride, pseudo-molecular ions such as $[(M\text{-}3H)+Cu(II)]^-$, $[M\text{-}H]^-$ and $[M\text{-}2H+CuCl]^-$ can be generated in the gas phase by electrospray. The cationization yield strongly depends on the metal/ligand ratio and, by using a 10/1 ratio, the $[(M\text{-}3H)+Cu(II)]^-$ complex can be produced in significant abundance.

The major pathway observed in the low-energy CID spectra of $[(M\text{-}3H)+Cu(II)]^-$ corresponds to the loss of carbon dioxide. This differs strongly from the reactivity observed for binary complexes involving either of the other first-row transition metal ions, as with Mn^{2+}, Fe^{2+}, Co^{2+}, Ni^{2+} or Zn^{2+} there is no loss of CO_2 (or it is very weak) and H_2O elimination appears to be preferred. Formation of the $[(M-3H)+Cu(II)]^-$ complexes requires removal of three labile protons, e.g. from the amide and/or the carboxylic groups as well as the phenol group of the tyrosine side chain. The elimination of CO_2 strongly suggests that the terminal carboxylic site is deprotonated and does not coordinate the metallic center.

The distinction of the *C*-terminal YGGFL and YGGFI (I and L at the *C*-terminal position) isomers by MS/MS experiments is neither possible from $[M\text{-}H]^-$ ions nor from $[M+H]^+$ species. On the other hand, these two peptides can be unambiguously distinguished by fragmentation of the $[(M-3H)+Cu(II)]^-$ complexes (Figure 28).

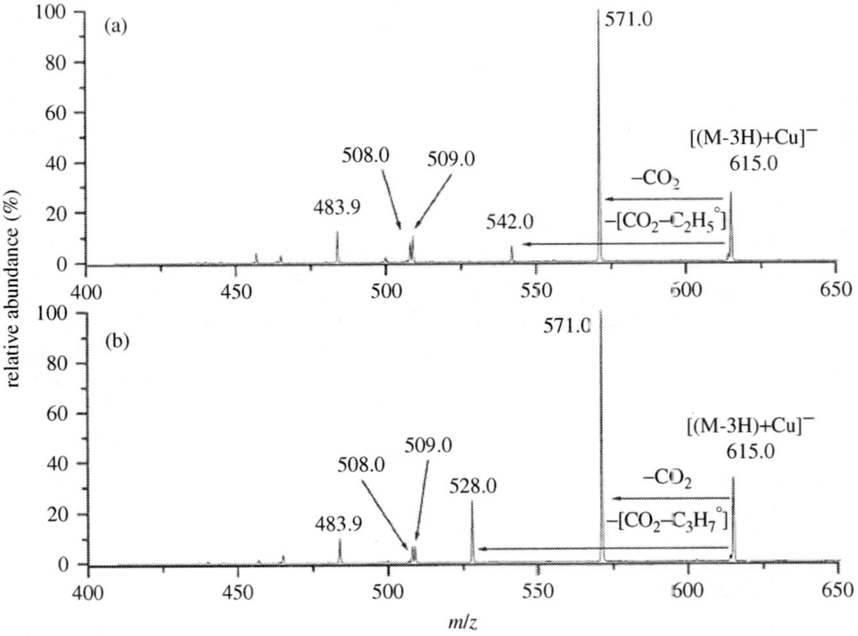

FIGURE 28. Low-energy CID spectrum of $[M\text{-}3H+Cu(II)]^-$ complexes obtained for (a) YGGFI and (b) YGGFL. Reproduced by permission of John Wiley & Sons, Ltd from Reference 202

Following decarboxylation, the loss of a $C_2H_5^{\bullet}$ radical is indeed characteristic of YGGFI loss while elimination of $C_3H_7^{\bullet}$ is specific to YGGFL. Other transition metal ions (e.g. Mn^{2+}, Fe^{2+}, Co^{2+}, Ni^{2+} or Zn^{2+}) did not lead to such diagnostic ions. Such radical losses (produced by $\beta-\gamma$ bond cleavages) were observed only with Cu^{2+} ions and indicate that the reduction of copper occurred during the decomposition process. Note, however, that such a distinction is not achieved when the I and L residues occupy internal position along the peptidic chain. It should also be pointed out that under high-energy CID, radical losses have been observed from anionic binary complexes containing alkali-earth metals[204,205]. However, $\alpha-\beta$ bond cleavages take place rather than $\beta-\gamma$ bond cleavages. Such $\alpha-\beta$ bond cleavages did not allow the Leu/Ile isomers to be distinguished.

The two isobaric peptides, namely YGGFK and YGGFQ, were also studied[203]. Both spectra exhibit losses of CO_2 and $CH_2{=}C_6H_4{=}O$. Interestingly, YGGFQ exhibits an abundant additional product ion attributed to the loss of H_2NCOCH_2 (Gln side-chain radical), again leading to the distinction of *C*-terminal Gln/Lys residues.

Isomeric peptide distinction can also be achieved through the use of the kinetic method, by a strategy similar to that presented for the distinction of enantiomeric amino acids (*vide supra*). More precisely, Cooks and coworkers evaluated the capacity of the kinetic method for the differentiation of isomeric dipeptides and for the quantitation of mixtures of isomers[206]. Six pairs of isomeric dipeptides (X-Y/Y-X) were successfully distinguished, by using an appropriate reference, and the relative amount of each partner in binary mixtures could be determined. This approach is particularly powerful because the spectra are simple and therefore easy to interpret. Although the information content is low, the fragmention abundances are dependent on stereochemical interactions that are very sensitive to isomeric form. More recently, the kinetic method has been applied to differentiate isomeric tripeptides[207] and was extended further to determine compositions of ternary mixtures of the isomers Gly-Gly-Ala (GGA), Ala-Gly-Gly (AGG) and Gly-Ala-Gly (GAG). Different metals ions were tested and among the first-row transition metal ions, Cu(II) yields remarkably effective isomeric differentiation for both the isobaric tripeptides, GGI/GGL using GAG as the reference ligand, and the positional isomers GAG/GGA using GGI as the reference ligand. The procedure also allowed one to perform chiral quantification of a ternary mixture of optical isomers.

Finally, complexation by Fe^{2+} and Cu^{2+} metal ions was used successfully to differentiate diastereomeric YAGFL, $Y^{(D)}$AGFL and $Y^{(D)}AGF^{(D)}$L pentapeptides[208]. Binary $[Metal(II)(M\text{-}H)]^+$ metal–peptide complexes were generated by ESI in the positive-ion mode. Their fragmentations were studied under low-energy collision conditions in an ion trap mass spectrometer. Again, with copper, the decarboxylation process is common to the three systems, but a careful examination of the CID spectra showed interesting differences. Indeed, elimination of $CH_2{=}C_6H_4{=}O$ is characteristic of the YAGFL peptide, thereby allowing its distinction from its $Y^{(D)}$AGFL isomer. Concerning $Y^{(D)}$AGFL and $Y^{(D)}$AGFL pentapeptides, their distinction is less straightforward and is based on the intensity of the fragment ion resulting from the decarboxylation. These distinctions were attributed to the stereochemical effects due to the $^{(D)}$Leu/$^{(L)}$Leu and $^{(D)}$Ala/$^{(L)}$Ala residues yielding various steric interactions which direct relative dissociation rate constants of the binary $[M{-}H{+}Metal(II)]^+$ complexes.

2. *Ternary complexes*

As biomolecules such as amino acids and peptides are practically always polydentate ligands, metal ions may interact with a variety of binding sites and coordination numbers, thereby resulting in a multitude of possible structures. In addition, numerous computational studies of transition metal ion complexes have shown that these structures are often very close in energy, and that the most stable forms in the gas phase do not necessarily correspond to the most stable forms in solution. Consequently, the actual structures generated in the gas phase by simply mixing the metal ion and the ligand of interest are often unknown.

A way to circumvent this problem is to use an auxiliary organic ligand that occupies some coordination sites in the metal ion ligand sphere and thus direct the coordination of the biomolecule to involve only one or two of its functional groups. In solution, copper(II) ions form stable complexes with multidentate polar molecules, such as diimines, nitrogen heterocycles, amino acids, peptides or nucleosides. Very stable complexes self-assemble in solution when Cu(II) salts are mixed with an equivalent of the analyte and an aromatic diimine, for example 2,2′-bipyridine (bpy) or 1,10-phenanthroline (phen) (Chart 6), and the pioneering work of the Tureček's group has demonstrated that these solutions result in the production by electrospray of very abundant gaseous $[Cu(II)(ligand)(Analyte-H)]^{+\bullet}$ ternary complexes.

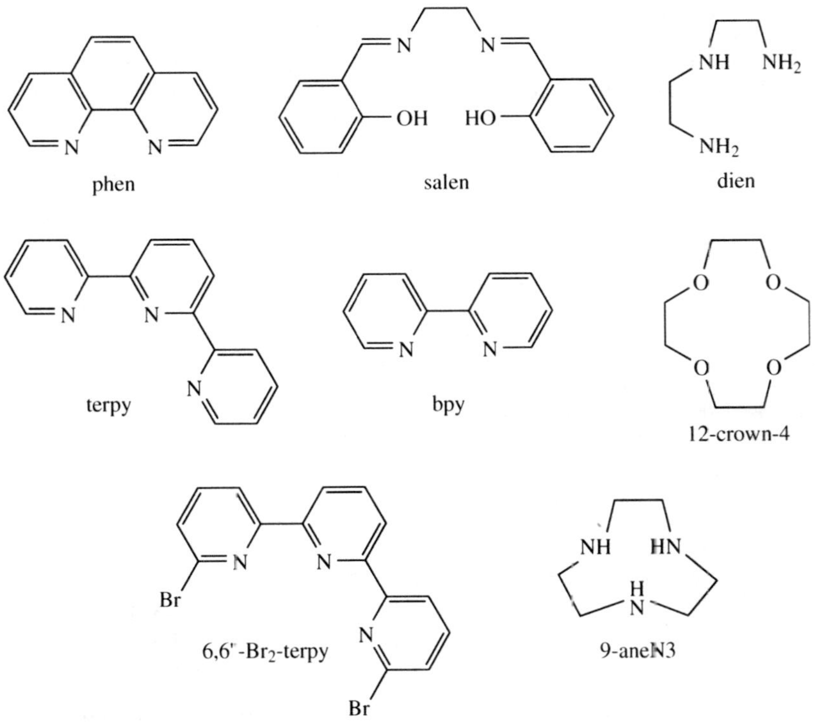

CHART 6

Amino acids were the first biomolecules studied by this group[61,153,154] and the structure of these ternary complexes has been characterized by combining mass spectrometry and theoretical calculations (Figure 29).

Experiments have shown that amino acids bind to Cu(II) by deprotonated carboxylate acid groups, blocking the carboxylic group as a methyl ester practically prevents

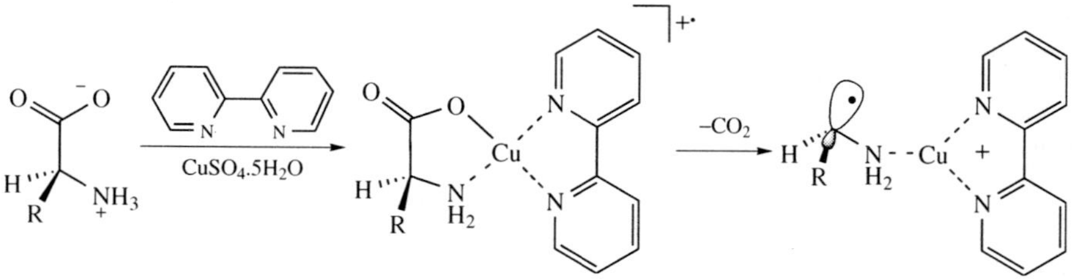

FIGURE 29. Structure of ternary Cu(II) complexes obtained with amino acids and the auxiliary ligand bipyridine

the complexes from being formed by electrospray[61,209]. The ternary complexes are odd-electron species showing a tetra-coordinated Cu(II) with a square-planar geometry. The bpy and amino acid amine nitrogens form regular Cu−N bonds with lengths in the 1.994 to 2.042 Å range. Analysis of the electron density reveals that most of the positive charge is located on the auxiliary ligand.

Upon collision, these singly charged ternary complexes undergo facile decarboxylation (Figure 29) yielding an amino acid C_α ion radical. In turn, this ion undergoes a cleavage of the $C_\beta-C_\gamma$ bond resulting in the loss of a radical from the side chain. Consequently, these ternary complexes allow one to notably distinguish leucine from isoleucine, and lysine from glutamine[210]. Isomeric amino acids leucine and isoleucine could also be quantified in 90:10 to 10:90 binary mixtures; Tureček and coworkers also studied a series of tripeptides (GGA, LGG, GGL, GGI, FGG, GGF, LGF, GLF, GFL, GYA and GAY)[211]. CID spectra of ternary complexes showed fragments that were indicative of the amino acid sequence in the peptide. The CID spectra of GGL and GGI complexes exhibit diagnostic peaks allowing *C*-terminal leucine and isoleucine to be distinguished. Interestingly, the formation of the $[Cu(bpy)peptide]^{\bullet 2+}$ ternary complexes is also observed when adopting mild electrospray ionization conditions. Their MS/MS spectra are clearly distinct for peptide isomers. Their fragmentation proceeds from both the *N*- and *C*-termini. Dissociations starting from the *N*-terminus are analogous to those in protonated peptides and produce b- and a-type ions. The complementary Cu-containing singly charged ions are analogous to ternary complexes of dipeptides or amino acids and are useful for structure elucidation of the *C*-termini in the peptides. The reader will find a complete description of these studies in a review published in 2007 by Tureček[12].

A new type of redox reaction involving Cu(II) ions was discovered by Siu and coworkers for ternary complexes containing peptides and diethylenetriamine (dien) as auxiliary ligand (L)[212]. By mixing in methanol the $Cu(II)dien(NO_3)_2$ complex and a peptide M, $[Cu(II)(dien)(M)]^{2+}$ ions could be produced by electrospray. Upon collision, this complex dissociates by a charge transfer process to give rise to the radical cation of the peptide, $M^{+\bullet}$. The product ion spectra of $M^{+\bullet}$ appeared very different from those of the corresponding $(M+H)^+$ and display a rich fragmentation chemistry, providing complementary structural information. This new reaction has attracted considerable attention because of the unique fragmentation behavior of such radical cations. Indeed, while the fragmentation of even-electron peptide ions is dominated by the cleavage of amide bonds, radical cations are also characterized by cleavages of N−Cα bonds, leading to c and z product ions (Figure 27). Consequently, many studies have been published over the last eight years and nowadays the use of ternary complexes is a well-established approach, besides electron capture dissociation (ECD) and electron transfer dissociation (ETD), to produce gaseous radical cations of amino acids and peptides. However, the charge transfer reaction is not the only dissociation observed. By using enkephalin derivatives[213], the following dissociation scheme could be established:

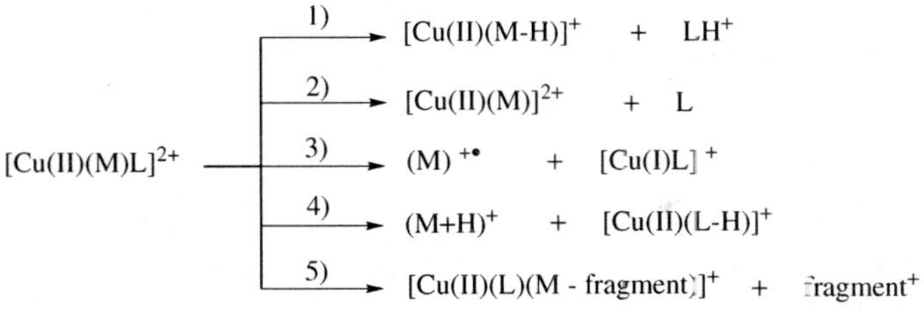

In order to favor the charge transfer process, much effort has been devoted to the choice of the auxiliary ligand (Figure 29). The first experiments were carried out by using dien as auxiliary ligand[212,213]. However, with dien, this particular methodology was only applicable to peptides containing tyrosyl or tryptophanyl residues. Attempts to produce radical cations of peptides including basic amino acids failed. NH-containing ligands such as dien indeed promote proton transfer reactions to the peptide (channel 4). Several studies have extended this approach to the use of terpy (terpy = 2,2′:6′,2″-terpyridine) for generating peptide radical cations[214–216]. In addition to peptides with tyrosyl and tryptophanyl residues, the $[Cu(II)(terpy)(M)]^{\bullet 2+}$ complex was also found capable of generating radical cations of oligopeptides containing basic amino acid residues such as arginine, lysine and histidine. It was also established that the abundance of the peptide radical cation could depend on the sequence of the peptide. By using simple di- and tripeptides GX, GGX, GXG, XG and XGG, the influence of the position of the basic residue, X (X=R, K and H), on the formation of peptide radical cations was probed[217], and it was found that $M^{\bullet +}$ is formed with greatest abundance when the basic residue is at the *C*-terminus. In order to gain some insight into the binding modes of these peptides to $[Cu(II)(terpy)]^{\bullet 2+}$, the formation and fragmentation of copper(II) complexes of tripeptides protected as their carboxymethyl/ethyl esters ($M-OCH_3$ or $M-OC_2H_5$) were also studied. For lysine and arginine, no ternary complexes were observed, suggesting that arginine- and histidine-containing peptides bind to $[Cu(II)(terpy)]^{\bullet 2+}$ as zwitterions.

By using terpy as auxiliary ligand, it was still not possible to generate radical cations containing only aliphatic amino acid residues, which is a very challenging problem due to the relatively high ionization energies (IEs) of aliphatic amino acids. Recently, Chu and coworkers[218–220] reported the benefits of sterically encumbered auxiliary ligands. They demonstrated that it was indeed possible to generate molecular radical cations of a series of aliphatic tripeptides GGX (W=G, A, P, I, K, L and V), by using either 12-crown-4, 6,6″-Br_2-terpy or 9-aneN3 ligands (Chart 6). 12-Crown-4 is a cyclic polyether ligand that has no exchangeable protons, and therefore competitive proton transfer can be avoided during the formation of radical cations of M. The enhancement in the formation of peptide radical cations observed with Cu(II)(9-aneN3) is at first glance surprising, because both 9-aneN3 and dien contain three amino nitrogen atoms and acidic hydrogen (NH) atoms. The facile formation of GGK^+ ions from $[Cu(II)(9\text{-aneN3})GGK]^{2+}$ complex ions presumably occurs because of inefficient proton transfer between the peptide and the sterically constrained 9-aneN3 ligand. For the whole series of GGX tripeptides, use of 12-crown-4 ligand resulted in more abundant peptide radical cations. The reasons for the enhanced peptide radical cation formation when using sterically encumbered ligands, relative to their open-chain analogs, are not obvious. It was suggested that the presence of a constrained macrocyclic ligand could weaken metal–peptide chelation through steric repulsion between the ligand and the peptide, a situation which may lead to more favorable peptide radical cation formation.

Finally, Chu and Lam have recently shown that the formation of deprotonated radical cations of peptides $(M+nH)^{(n+1)+\bullet}$ by MS/MS spectra of ternary complexes was also possible[221]. This had been presently achieved from triply charged $[Cu(II)(terpy)(M)]^{\bullet 3+}$ complexes. $(M+H)^{2+\bullet}$ ions have indeed been formed in significant abundance, especially for relatively long oligopeptides that present a basic amino acid residue at either the *C*- or *N*-terminus. Dissociation of $[M+H]^{2+\bullet}$ dications through MS^3 experiments leads to the formation of fragment ions $a_n{}^+$, $y_n{}^+$, $[z_n+H]^+$ and $w_n{}^+$. This fragmentation is similar to that of the dissociation of odd-electron peptide radical cations $M^{+\bullet}$, but is significantly different from those of the corresponding $[M+2H]^{2+}$ and $[M+H]^+$ ions. A very striking feature is that the differentiation between internal isomeric leucine and isoleucine residues in polypeptides is possible through the analysis of the secondary fragmentation products of the $[z_n+H]^+$, which generates even-electron $w_n{}^+$ ions. This is illustrated in Figure 30.

This distinction is demonstrated by the presence of peaks associated with the loss of diagnostic neutral radicals, $CH(CH_3)_2$ (43u) and CH_2CH_3 (29u), for leucine and isoleucine, respectively, by inspecting the consecutive mass differences between the $[z_6+H]^+$ and $w_6{}^+$ ions. This approach, although not universal, therefore appears particularly promising not only because these hydrogen-deficient radical cations are now accessible to rather cheap instruments (ion traps), but also because these ions provide particularly structural information sensibly different from that obtained from even-electron protonated ions[222, 223].

3. *Interaction with proteins*

To conclude this section about peptides and proteins, it is worth noting that several studies dealt with the interaction of copper ions with large proteins[224–230] and notably metallothioneins (MTs)[224, 230]. MTs are small cysteine-rich proteins that typically comprise some 60 amino acid residues, many of which are serine and threonine. Many mono- and divalent metals have been observed to complex with MTs, including Cu(I), Ag(I), Zn(II), Cd(II), Au(I) and Hg(II)[231–233]. Siu and coworkers[230] studied the Rabbit metallothionein MT2A and combined electrospray ionization to ion-mobility experiments to determine their binding properties toward Cd(II), Zn(II), Ag(I), Hg(II) and Cu(I) ions. They found that the number of metal ions bound by MTs depends on the pH of the solution. This number decreases as the pH is lowered. Observation of $[Cu_4MT2A+4H]^{4+}$, $[Cu_4MT2A+5H]^{5+}$ and $[Cu_4MT2A+6H]^{6+}$ ions suggests a 4:1 stoichiometry. Ion-mobility measurement showed that the conformations of the ions obtained with Cu(I) are similar to those observed for Ag(I) ions and that in all cases metal binding to MTs induces a conformational rigidity to the metal–MT complexes. The same stoichiometry has been observed by Jensen and coworkers during their study of the metallothionein isoform 3 (MT3)[224]. They also demonstrated that Cu(I) appears to bind preferentially to the β domain of MTs. As a matter of fact, addition of 2 molar equivalents of Cu(I) to the βMT3 domain peptide resulted in a prominent tetracopper species, while mixing Cu(I)–acetonitrile with αMT3 did not generate a higher molecular weight species in the electrospray spectrum.

Finally, one may also cite the study of Baldwin and coworkers[225] about the interaction of copper ions with the Prion protein (PrP). The *N*-terminal side of the Syrian hamster (SHa) PrP, spanning from the 57th till the 91st residue, is highly conserved as it consists in $WCQ(PHGGGWGQ)_4$. The repetition of the eight amino acid sequence PHGGGWGQ, each copy of which is referred to as one octarepeat, was identified as a potential Cu^{2+} binding motif, which might induce secondary structure and protect the *N*-terminus against

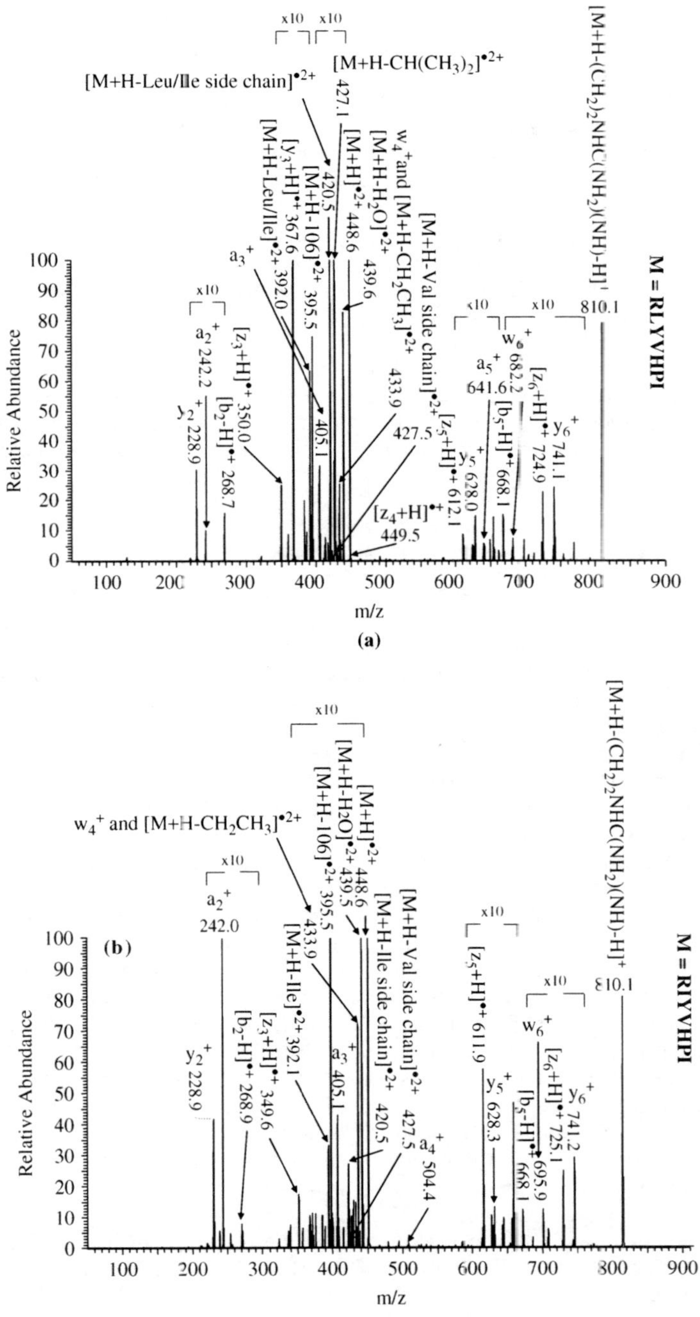

FIGURE 30. MS^3 spectrum of $[M+H]^{2+}$ ions with (a) M = RLYVHPI and (b) M = RIYVHPI. Reproduced by permission of Elsevier, Copyright 2005 from Reference 221

proteolysis. Baldwin and coworkers[225] used ESI-MS techniques in order to evaluate the binding of Cu^{2+} ions to the mature Prion protein (PrP), by studying synthetic peptides corresponding to sections of the sequence of the mature prion protein (PrP). This ESI-MS study demonstrated that Cu^{2+} is unique among divalent metal ions in binding to PrP and defines the location of the major Cu^{2+} binding site as the octarepeat region in the *N*-terminal domain. Furthermore, it appeared that two adjacent octarepeats were sufficient to bind one Cu^{2+} ion, but also that the histidines were essential ligands for this. The stoichiometries of the complexes measured directly by ESI-MS were found to depend on the pH: a peptide containing four octarepeats chelates two Cu^{2+} ions at pH 6, and four at pH 7.4. Dissociation constants (k_D) for each Cu^{2+} ion binding to the octarepeat peptides were determined by electrospray. This is possible by assuming that the total signal response for each individual ion is proportional to the concentration of that species in the gas phase and, by extension, in solution. This also assumes that the free peptide and the peptide with metal bound give the same signal response. Measured k_D values lie mostly in the low micromolar range. Circular dichroism measurements also reveal a pH-dependent structural change. At pH 6, only a modest spectral change is induced by binding of Cu^{2+} ions, while Cu^{2+} binding at pH 7.4 induces a major conformational change.

B. Reactivity with Saccharides and Derivatives

Carbohydrates are the most abundant biomolecules in nature. They are mainly found as polysaccharides that play a crucial role in both animal and vegetal life. Monosaccharides not only bind together to form polysaccharides, but also with purines and pyrimidines (ribose and 2-deoxyribose) to give nucleosides, and subsequently RNA and DNA. Moreover, most naturally occurring proteins and lipids are glycosylated. Carbohydrates are thus involved in many biological functions, such as cell–cell recognition, cell–cell adhesion, and act also as antigens and as blood group substances. Such a variety makes carbohydrate analysis a challenging task for mass spectrometry. A complete structural description of carbohydrates implies notably exact mass measurements, determination of the sequence, sites and anomeric configuration of the glycosidic linkages and, when possible, stereochemical characterization of the different asymmetric centers of the sugar ring(s). Many studies have been carried out on those topics for more than three decades, using different ionization techniques. For many years, derivatization (methylation, permethylation and peracetylation) in conjunction with electron impact or positive-ion chemical ionization have been the primary method used. Analysis of underivatized oligosaccharides has also been performed by negative-ion chemical ionization or FAB. With the advent of soft ionization methods such as FAB, or more recently ESI, the particular reactivity of the metal ions in the gas phase has been used in order to distinguish structural isomers of mono- or polysaccharides. Among the metal ions considered, several studies published during the last ten years implied Cu(I) or Cu(II) ions, even if interactions with sugars were much less studied than interactions with amino acids or peptides.

In this context, our group compared the analytical potential of three metal ions Ag^+, Cu^+ and Pb^{2+}, by studying their interactions with *D*-glucose, *D*-galactose and *D*-fructose, *O*-methyl-α-*D*-glucose and *O*-methyl-β-*D*-glucose[234]. Mixing copper chloride and the monosaccharides in a glycerol matrix resulted in the formation of intense $[Cu(I)(monosaccharide)]^+$ complexes. MIKE spectra of the $[Cu(I)(monosaccharide)]^+$ ions are significantly different from those obtained for the silver-glycoside adducts. Similar losses are observed (dehydrogenation, dehydration), but copper cationization induces additional species. Fragmentation of the various complexes produced exclusively Cu^+-containing fragment ions. Metastable-ion fragmentation pathways common to all monosaccharides are losses of H_2, 16u and H_2O. Interestingly, the main fragmentation of the three hexoses is different for each copper complex, namely dehydration for *D*-glucose, dehydrogenation for *D*-galactose and finally loss of 16 mass units for *D*-fructose. Consequently, interactions with Cu(I) allows these three monosaccharides to be easily distinguished, experiments carried out with lead being also successful. As methylation of the anomeric hydroxyl blocks the $\alpha \rightleftarrows \beta$ anomerization reaction, we tried to characterize the absolute configuration of the anomeric carbon by considering the two methyl-glycosides α-1-*O*-methyl-*D*-glucose and β-1-*O*-methyl-*D*-glucose. The distinction of these two molecules was not as straightforward as with silver cations, but the combined elimination of water and methanol appeared characteristic of the β anomer.

DFT calculations have been carried out to propose reliable structures for the complexes observed with *D*-glucose[235]. Cramer and Truhlar[236] have estimated at nearly 3000 the number of possible conformers of glucopyranose. This number becomes even larger when the interaction with Cu^+ is considered. In order to restrict the survey of the possible adducts to a reasonable number, Tortajada and coworkers[235] have considered only those structures in which the Cu^+ is attached to at least two different oxygens of glucose, assuming that the sugar moiety retains its cyclic structure.

The calculations performed demonstrated that the most stable cyclic forms obtained for both anomers imply interaction of the metal with the hydroxymethyl group, the anomeric hydroxyl and the endocyclic oxygen (O5) (Figure 31).

In the case of β-glucose, it was possible to obtain a conformation in which the Cu^+ is interacting with four different hydroxyl groups. Another important finding of this survey is that Cu^+ association produces such important distortions of the conformation of the sugar that we can no longer talk of boat or chair conformations.

The AIM analysis showed that activation of the C1−O5 bond is favored in di-coordinated complexes, while the activation of the C5−O5 bond is favored in tri-coordinated ones. The C1−O5 fission should be preceded by a 1,3-H shift from C5 to C1, resulting in **OG1** opened forms (Figure 31). Alternatively, a 1,3-H shift from C1 toward C5 would lead to the C5−O5 bond fission, and to the production of **OG2** opened structures. It is worth mentioning that the activation barriers associated with these hydrogen shifts and leading ultimately to the cleavage of the six-membered ring, although high (around 61.9 kcal mol^{-1} at the 6-31G* level), are still lower than the glucose−Cu^+ binding energy, which is estimated to be 66.9 kcal mol^{-1} at the same level of theory. Consequently, the complex formed by a direct attachment of Cu^+ to glucose has enough internal energy to overpass these activation barriers leading to opened-structures **OG1** and **OG2**. The second important quantitative result is that these opened structures are

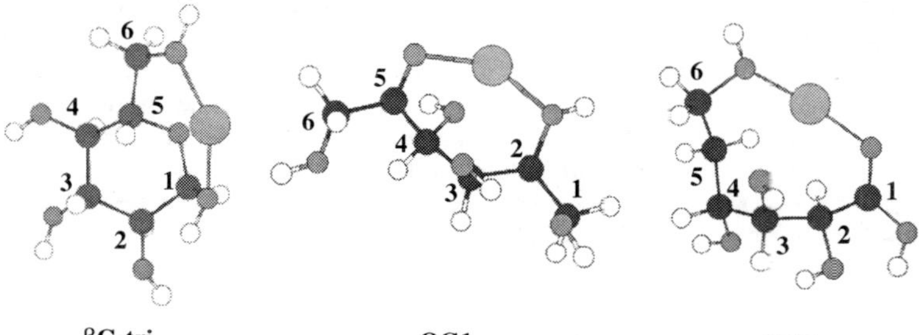

FIGURE 31. Some of the most stable cyclic and opened forms for the [Cu(I)(*D*-glucose)]$^+$ complex

also much more stable than the cyclic ones. The enhanced stability of the opened structures is clearly associated with a more efficient bonding between Cu^+ and the oxygen atoms of the sugar moiety. In the opened structures the O–Cu–O fragment exhibits a practically linear arrangement, while in cyclic structures this possibility is hindered by the rigidity of the ring. The potential energy surface associated with the loss of water (*m/z* 225, main fragmentation) was also explored and it was concluded that the most stable final product ions have their origin in opened structures. The most stable *m/z* 225 product ions are those formed by a spontaneous fragmentation of the most stable opened structure **OG2**, while the less stable structures are those produced by the unimolecular dissociation of cyclic complexes.

As metastable decomposition studies showed that Ag^+, Cu^+ and Pb^{2+} were of potential interest in the structural characterization of isomeric glycosides, we reconsidered these systems by using electrospray ionization a couple of years later[223]. With the same metallic salt, ESI mass spectra showed the presence of abundant protonated and coppered species. The electrospray mass spectra of $CuCl_2$/monosaccharide mixtures are characterized by complexes in which the formal oxidation state of copper remained Cu(II), such as $[Cu(II)(monosaccharide)-H]^+$, metallic complexes of reduced Cu(I)-adduct ions like $[Cu(I)(monosaccharide)]^+$ (at *m/z* 243/245) and $[Cu(I)(monosaccharide)_2]^+$ (at *m/z* 423/425). Doubly-charged species were never observed. The ratio $[Cu(II)(monosaccharide)-H]^+/[Cu(I)(monosaccharide)]^+$ was found to depend on nozzle-skimmer voltage variations. At low cone voltage, metallic complexes involving Cu(II) are predominant. On the contrary, high cone voltage leads to an increase of the $[Cu(I)(monosaccharide)]^+$ abundance. The formation under electrospray conditions of a mixture of organometallic species containing copper ions in two different oxidation states, already known for amino acids and peptides[12], is therefore also observed with carbohydrates. Globally, positive-ion ESI spectra of coppered complexes do not really allow the aldoses and the ketose to be differentiated.

The low-energy CID spectra of $[^{63}Cu(I)(monosaccharide)]^+$ adduct ions recorded at a collision energy of 20 eV show an amazing reactivity. The copper cation was able to induce activation of practically all C–C and C–O bonds resulting in many fragmentation pathways. Consequently, unlike in FAB conditions, the use of copper salt does not allow the distinction between *D*-glucose, *D*-galactose and *D*-fructose, neither from the positive-ion ESI spectra nor from the ESI-MS/MS spectra. Globally, lead(II) ions exhibited the greatest potential for characterizing isomeric saccharides under electrospray conditions[222,237].

The ternary complex approach has also been employed in order to discriminate glucose from its aldose isomers (talose, mannose and galactose)[238]. Ternary complexes were generated by using nickel, copper and zinc. In addition, several auxiliary ligands were considered in order to address the effect of the size, number and coordination number of the ligands in the complex. By using a quadrupole ion trap mass spectrometer, tandem mass spectrometric experiments were performed on the electrospray-generated metal *N*-glycoside complexes. Diaminopropane (dap) and ethylenediamine (en) were used to generate tri-coordinate $[Cu(L/hexose)-H)]^+$ complexes[238]. The product ion spectra of all tri-coordinate Cu/*dap* complexes were very simple, possessing a single prominent product ion at *m/z* 178 (elimination of $C_4H_8O_4$ by cross-ring cleavage). Such simple spectra did not allow any stereochemical differentiation. On the other hand, the $[Cu(en/monosaccharide)-H]^+$ MS/MS spectra obtained under identical experimental conditions from all four diastereomeric Cu/*en* complexes are characterized by various cross-ring cleavages (elimination of $C_nH_{2n}O_n$ moieties, $n = 1-3$) and/or loss of water, and revealed unique product ion spectra for each of these precursor ions. Whereas a full differentiation of the four diastereomeric monosaccharides was achieved with the Cu/*en* complexes, such a distinction appeared not possible with zinc. Finally, unlike the tri-coordinate complexes, tetra- and penta-coordinate ions did not lead to a successful distinction of the four isomers.

The gas phase reactivity of copper ions has also been applied to the structural analysis of carbohydrate derivatives, and notably flavonoids[239]. Flavonoids are a class of phytochemicals sharing a common chemical structure, based on a C_{15} skeleton with a chromane ring bearing a second aromatic ring B in position 2, 3 or 4 (presently 2 in that particular study, Chart 7). Their ubiquitous presence in plants makes them an integral part of the human diet. The high degree of variety in flavonoid structure makes accurate identification a difficult task.

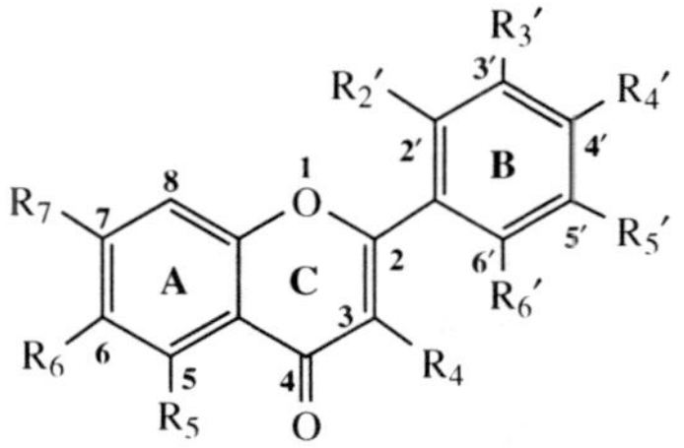

CHART 7

ESI, APCI and MALDI are used to analyze flavonoids and some significant progress has been made toward systematic structural characterization of flavonoids by mass spectrometry. However, mass spectrometric analysis is still not able to achieve de novo identification of flavonoids. Only tentative identifications, generally based on collision-induced dissociations of protonated or deprotonated pseudo-molecular ions, can be made, particularly in terms of saccharide location and identity, unless complementary analytical methods, such as UV, FTIR or NMR spectroscopy, are employed. In order to develop a simple and robust method, Davis and Brodbelt[239] chose to form flavonoid glucoside/metal complexes in order to get a broader array of fragments for structural analysis. Five metals, Ca(II), Mg(II), Co(II), Ni(II) and Cu(II), were evaluated for their ability to form such complexes. The complexes were produced from methanolic solutions containing 1:1 flavonoid glycoside/metal salt. No pH adjustment was performed on the analyte solutions. Under these conditions, the flavonoid glycosides formed 1:1 and 2:1 analyte/metal complexes of the type $[M(II)(L-H)]^+$ and $[M(II)(L)(L-H)]^+$. The 2:1 complexes were found more abundant, and gave simple CID spectra with easily-assigned fragments and a variety of dissociation pathways for structural determination, and were therefore chosen for an analytical purpose. Despite the efficient formation of transition metal complexes of the type $[M(II)(L-H)(L)]^+$ where M = Co, Ni or Cu, none of these complexes permitted complete identification of all glycosylation sites studied (C6, C8, O3, O7 and O4′). Neither cobalt nor nickel complexation provide sufficient differentiation of 3-*O*- and 4′-*O*-glucosyl flavonoids. The copper complexes were even less useful for locating the glycosylation sites. It turned out that the Mg(II) complexes offered the best and most complete identification and differentiation of all five categories of flavonoid glucosides studied.

Numerous mass spectrometry studies on the fragmentation of polysaccharides have shown that the main fragmentation pathways consist of glycosidic cleavages (ions of the type B, C, Y or Z; Figure 32) that involve single bond rupture between the sugar rings, and cross-ring cleavages of the rings themselves (A and X ions). The former predominate and provide sequence information whereas the latter yield additional information on the linkage position of one residue to the next. In the positive-ion mode, $[M+H]^+$, $[M+Li]^+$ and $[M+Na]^+$ ions are often used but they mainly dissociate by glycosidic bond cleavage.

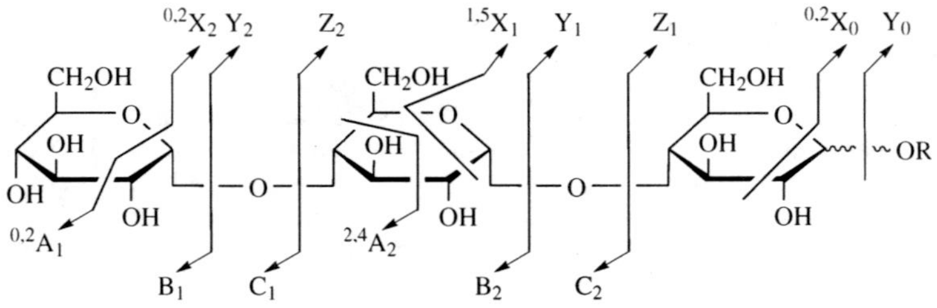

FIGURE 32. Nomenclature of the fragmentation of carbohydrates (from Reference 240)

Several groups have also considered other metal ions. Recently, Harvey reported the use of Mg^{2+}, Ca^{2+}, Mn^{2+}, Co^{2+} and Cu^{2+} cations in order to analyze Maltoheptaose (linear $\beta 1 \rightarrow$ 4-linked glucose) and several *N*-linked glycans[241]. Again, electrospray ionization was used and the dominant species obtained with all salts were $[M+metal]^{2+}$ complexes. Fragmentation of the doubly charged ions became prominent as the cone voltage was increased, except those ions formed with copper. In-source fragmentation involved dissociation of the $[M+metal]^{2+}$ ions into singly charged $[M+metal]^{+}$ ions with loss of a monosaccharide residue (mannose 162 mass units). In the case of copper, the $[M+Cu]^{+}$ ion was formed without apparent loss of carbohydrate, indicating that its oxidation state must have changed to Cu(I).

MS/MS of the $[M+metal]^{2+}$ ions produced both singly and doubly charged ions with the relative abundance of doubly charged ions decreasing in the order Ca>Mg>Mn>Co>Cu. Whatever the polysaccharides considered, fragmentation of $[M+Cu]^{2+}$ or $[M+Cu]^{+}$ complexes mostly corresponded to singly charged ions arising from glycosidic bond cleavages, therefore yielding useful sequence information. In the particular case of maltoheptaose, the MS/MS spectrum of the $[M+Cu]^{+}$ ion was strikingly similar to that of $[M+Na]^{+}$ complexes. Globally, MS/MS spectra of copper-cationized polysaccharides were devoid of cross-ring fragments, hence failing in providing the position of the glycosidic linkages.

The interactions between copper and polysaccharides arising from the degradation of cellulose, chitin and chitosan were also investigated[242]. Chitosan is a linear homopolymer of β-1,4-linked *D*-glucosamine residues possessing high selectivities for transition metal ions. Chitin refers to the acetylated form of chitosan, consisting of β-1,4-linked *N*-acetyl-*D*-glucosamine. Owing to their abundance in the marine environment, chitosan and chitin can control metal ion equilibria in their surroundings. Cellulose was chosen as a benchmark because it does not exhibit any amino or *N*-acetylamino groups and therefore was not supposed to bind strongly copper ions. These systems were investigated by electrospray ionization and the goal of these studies was to understand the degree to which electrospray mass spectra can reflect the solution chemistry of copper–oligosaccharide complexes[242]. To this end, mass spectrometry data were compared with potentiometric studies. Sugar–copper complexes of general formula $[M_2+Cu]^{2+}$, $[M_3+Cu]^{2+}$ and $[M-H+Cu]^{+}$ have been observed in the gas phase with the chito- and cello-tetrasaccharides. It is worth noting that $[M+Cu]^{2+}$ were also detected, suggesting that tetrasaccharides are polarizable enough to stabilize the double charge of copper under electrospray conditions. On the other hand, the potentiometric data revealed no copper binding for mono- and tetrasaccharides of cellulose and chitin. The complexes observed on mass spectra were therefore attributed to charge-attachment phenomena during desolvation from the highly charged electrospray droplets. In contrast, the chitosan tetrasaccharide due to its available amino groups was expected to bind copper effectively and to give intense electrospray spectra. However, no complexes were detected by mass spectrometry. The authors suggested that neutral complexes could be formed in solution. In order to check this assumption, they measured the $[M+2H]^{2+}$ abundance as a function of the copper concentration. Its intensity declined as the $[Cu^{2+}]$ concentration in solution was increased, suggesting removal of the free tetrasaccharide by complexation to Cu^{2+} ions. This experiment was repeated with a sodium salt but no decline in the $[M+2H]^{2+}$ response was observed, confirming the formation of neutral and/or negatively charged Cu(II)/chitosan complexes, hence not detectable in the positive-ion mode. This result was in agreement with the potentiometric studies indicating that at pH 7, Cu(II) ions and chitosan form strong complexes which are essentially neutral or negatively charged, explaining the absence of metal species in the positive-ion ESI spectrum.

C. Reactivity with Nucleic Acid Building Blocks

Metal cations can both stabilize and destabilize DNA[243]. The interaction of divalent cations with nucleic acids plays an important role in promoting and maintaining their functionalities[243–245]. As illustrated throughout this chapter, the open-shell Cu^{2+} cation has a rich redox chemistry and some studies have emphasized that the exposure of DNA to Cu^{2+} can have different consequences, such as single and double strand cleavage, base modification and formation of basic sites[246,247].

The interactions and coordination chemistry of metal cations with nucleic acid building blocks have been studied quite extensively in the condensed phase. In contrast, data about the intrinsic gas-phase behavior of such interactions are rather scarce. A good knowledge of the mechanisms at the molecular level is still lacking in most cases. One reasonable way of approaching this chemistry is to adopt a 'bottom-up' strategy based on the gradual increase of the size and complexity of the nucleic acid building blocks. In this context, we began our gas-phase investigations with uracil, one of the five nucleobases. As mentioned in Section VI, we first studied in the last few years the interactions (binding, energetics) of uracil and its thio-derivatives toward proton[169], copper(I)[84] and copper(II)[125]. Then, we went a step further by exploring the experimental unimolecular reactivity of the complexes produced in the gas phase by electrospray ionization[151].

Under electrospray conditions, interaction between copper(II) ions and uracil gave rise to different types of ions, the structure of which, as expected, clearly depended on the cone voltage applied. At low cone voltage, a pair of peaks associated to copper/uracil interaction, namely $[Cu(uracil)_n-H]^+$ complexes ($n = 1, 2$), were observed. Increasing the cone voltage resulted in the additional apparition of $[Cu(uracil)]^+$ ions through a reduction process. Neither $[Cu(uracil)]^{2+}$ ions nor higher homologues were observed, suggesting that the source conditions were likely not mild enough to allow the production of multiply charged complexes.

The $[^{63}Cu(uracil-H)]^+$ species was selected and allowed to dissociate upon collision with nitrogen. Two main fragment ions were detected at m/z 130.9 and 131.9, corresponding to the loss of [H,N,C,O] and $[N,C,O]^\bullet$, respectively. A weak fragment ion at m/z 145.9 attributed to the elimination of carbon monoxide was also observed. By using appropriate labeled uracils, it could be unambiguously concluded that the loss of both $NCO^\bullet$ and HNCO involves specifically C2 and N3, whereas the loss of carbon monoxide involves only C4.

In order to rationalize these experimental findings, DFT calculations were performed on the $[Cu(uracil-H)]^+$ system. The most stable complex corresponds to a form in which Cu^+ bridges between N1 and the oxygen atom of the C(2)O carbonyl group (Figure 33). Conventional π-complexes in which the metal cation is above the plane of the ring also correspond to local minima of the potential energy surface, although they were found to be significantly less stable than σ-complexes.

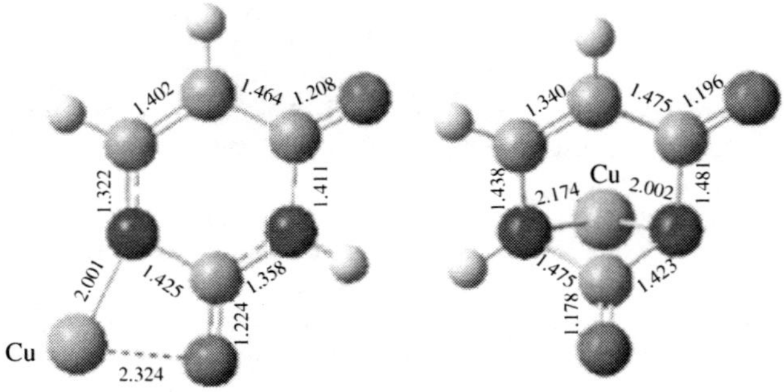

FIGURE 33. The most stable σ- and π-forms optimized for the [Cu(uracil–H)]$^+$ complex. Bond lengths are in Å

Considering the C–N bond distances as well as the charge densities at the bond critical point deduced from an AIM analysis, the N3–C4 (Chart 5) fission seemed to be the most favorable process. A survey of the PES (Figure 34) indicates that the loss of HNCO has its origin in the global minimum (σ-complex). Interestingly, the most favorable mechanism for the elimination of NCO$^\bullet$ involves π-type complexes. The out-of-plane copper interaction leads to specific bond activations within the ring, which result in bond cleavages associated with the NCO$^\bullet$ loss process. Consequently, π-complexes were found to play an important role in the gas-phase reactivity of [Cu(uracil–H)]$^+$ complexes. The fact that the HNCO and NCO$^\bullet$ losses follow complete different pathways allowed one to explain why in both cases the same atoms, namely N3 and C2, are involved.

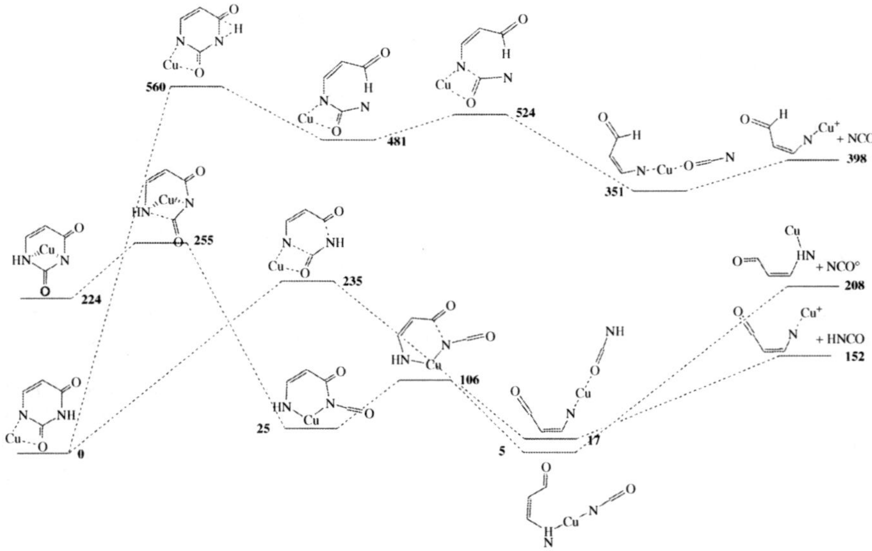

FIGURE 34. Energy profile corresponding to the loss of HNCO and NCO$^\bullet$ from [(uracil–H)Cu]$^+$ complexes. Energy values are in kcal mol^{-1}

Recently, Sodupe and coworkers[184] also showed the important role of π-complexes in the interaction of Cu^+ with phenylalanine, which are the precursors for the C−C insertion which triggers the fragmentation of phenylalanine−Cu^+ complexes.

During the last decade, electrospray ionization has also been employed to examine the formation and reactivity of $[Cu(L)(N)]^{2+}$ ternary complexes involving nucleic acid building blocks (N). In their study, Wee and coworkers considered both Pt(II) and Cu(II) cations that they mixed with two possible auxiliary ligands (dien and terpy) and nucleobases, nucleosides or nucleotides[248]. Ternary complexes could be obtained with all the nucleic acid buildings blocks and the data obtained showed that all three components played a role in the formation of the ternary complex. More of these complexes could be formed for Cu(II) than for Pt(II).

Like for peptides, the MS/MS spectra of the ternary complexes are characterized by different dissociation processes: a redox reaction which results in the formation of the radical cation of the nucleic acid constituent, $N^{+\bullet}$; the loss of the nucleic acid constituent in its protonated form; and the fragmentation of the nucleic acid constituent. $N^{+\bullet}$ radical cation were only observed with copper, confirming that the metal must have suitable redox properties to promote the formation of radical cations. Again, using terpy in place of dien allowed limiting the efficiency of the proton transfer process. In fact, N has the biggest influence on the fragmentations observed. Nucleobase radical cations have been obtained solely with copper and terpy ligand. The relative yield of the radical cations of each of the nucleobases from the copper ternary complexes exactly followed their relative vertical ionization potentials (IPs) G<A<C<T. Consequently, nucleobases with the lower IPs are the most easily ionized and form the greater yield of their radical cation upon dissociation of the metal complex. As the nucleobase was changed for the nucleosides, the redox reaction no longer occurred. Finally, changing the nucleoside to the nucleotide yields a new type of product ion in which the metal remained bound to the phosphate and the nucleobase was lost in its protonated form. In general, all of the Cu(II) complexes of the nucleotides fragmented via loss of the protonated nucleobase, suggesting that the copper was bound to the phosphate. This is consistent with the known binding of copper to the phosphate moiety in nucleotides in the condensed phase[249].

In a recent study of Cheng and Bohme[250], an electrosprayed water/methanol solution of guanosine and copper nitrate resulted in the formation of a gas-phase copper complex of $[CuL_n]^{2+\bullet}$, $[CuL(MeOH)_n]^{2+\bullet}$ and $[CuG_n(NO_3)]^{+\bullet}$, as well as the ions $[L]^{+\bullet}$, $[L+H]^+$, $[G]^{+\bullet}$ and $[G+H]^+$ (L = guanosine, G = guanine). The observation under mild electrospray conditions of an abundant $[L]^{+\bullet}$ radical cation was quite unexpected and constitutes one of the very first examples of formation of radical cations of guanosine from binary mixtures. MS/MS of the $[Cu(L)_3]^{2+\bullet}$ ions suggested that the formation of $[L]^{+\bullet}$ can be achieved through the transfer of an electron from L to Cu^{2+} within the complex before Coulomb repulsion dissociates the complex into a singly charged $[Cu(L_2)]^+$ complex and $[L]^{+\bullet}$. The formation of the $[G]^{+\bullet}$ ions observed on the MS/MS spectrum proceeds through the dissociation of the *N*-glycoside as confirmed by the CID spectrum of $[L]^{+\bullet}$. Adding or removing one L unit resulted in very different CID spectra. The electron transfer channel disappears for $[Cu(L)_2]^{\bullet 2+}$, while a $[L_2]^{+\bullet}$ is generated for the $[Cu(L)_4]^{\bullet 2+}$ species.

Finally, bigger oligonucleotides have also been considered and Hettich published several years ago a MALDI/FTICR (Fourier Transform Ion Cyclotron Resonance) study of the interactions of transition metal ions with a series of single-stranded dinucleotides[251]. While metal ions such as Na^+, K^+, Mg^{2+} and Ca^{2+} exhibit a strong affinity for phosphate groups, the transition metals are more likely to form covalent bonds with the nucleobases themselves. In general, the most common binding sites for heavy metal ions appear to be the N7 atoms in adenine and guanine.

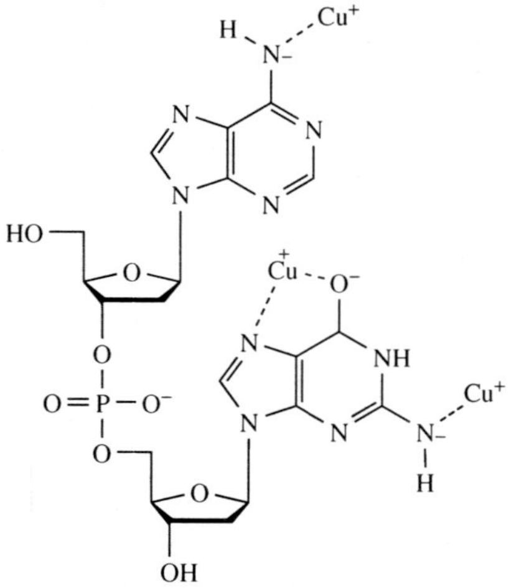

FIGURE 35. Proposed structure for the [*d*AG+3Cu-4H]$^-$ complex observed by MALDI

Various dinucleotides were considered during this study, namely *d*TG, *d*CA, *d*CT, *d*GG and *d*AG. Most of the results concerned the latter dinucleotide because *d*AG provides a variety of acidic protons and possible binding sites for metal ions. In the particular case of copper ions, up to three copper ions could be attached to the *d*AG dinucleotide, as attested by the presence of [M+Cu-2H]$^-$, [M+2Cu-3H]$^-$ and [M+3Cu-4H]$^-$ ions on the MALDI mass spectrum. High-resolution measurements revealed that copper is present only at the Cu(I) oxidation state. The copper ion thus appears to act similarly to the alkali metal ions, and simply displaces acidic protons of the dinucleotide.

Collisional dissociation of the [*d*AG+Cu-2H]$^-$ ion revealed loss of adenine and deoxyadenosine as the primary fragment products. Collisional dissociation of the ion [*d*AG+2Cu-3H]$^-$ also revealed loss of deoxyadenosine, verifying the presence of both copper ions in the remaining fragment ion. From these results, it could be deduced that at least two copper ions were associated strongly with the deoxyguanosine, replacing the 3′-hydroxyl proton and/or the nucleobase protons. In order to examine this metal ion binding site with more detail, the interaction of copper with dinucleotides of varying sequences has been investigated. Up to 2, 2, 3, 3 and 4 protons could be replaced by copper for *d*CA, *d*CT, *d*TG, *d*AG and *d*GG, respectively. Because copper is present in all of these complexes as Cu^+, it does not require multisite attachment like the other multivalent metal ions examined in this article, and thus behaves like an alkali metal, and it simply replaces available acidic hydrogens in the dinucleotides. Consequently, copper allowed the estimate of the maximum number of dinucleotide protons that can be replaced by a singly charged metal ion. As the results obtained for the series of dinucleotides suggest that the copper coordination depended primarily on the identity of the nucleobases, the author concluded that copper ions primarily replace acidic hydrogens on the nucleobases of the dinucleotides and that the involvement of the deoxyribose hydroxyl groups in metal ion binding seemed to be minimal. Based on these results, a structure (Figure 35) could be proposed for the [*d*AG+3Cu-4H]$^+$ complex.

IX. REFERENCES

1. M. Güell and P. E. M. Siegbahn, *J. Biol. Inorg. Chem.*, **12**, 1251 (2007).

2. S. T. Prigge, A. S. Kolhekar, B. A. Eipper, R. E. Mains and L. M. Amzel, *Science*, **278**, 1300 (1997).
3. A. Granata, E. Monzani, L. Bubacco and L. Casella, *Chem. Eur. J.*, **12**, 2504 (2006).
4. D. F. Raffa, G. A. Rickard and A. Rauk, *J. Biol. Inorg. Chem.*, **12**, 147 (2007).
5. R. Prabhakar and P. E. M. Siegbahn, *J. Phys. Chem. B*, **107**, 3944 (2003).
6. M. A. Kurinovich and J. K. Lee, *J. Am. Soc. Mass Spectrom.*, **13**, 985 (2002).
7. J. K. Lee, *Int. J. Mass Spectrom.*, **240**, 261 (2005).
8. A. C. Drohat and J. T. Stivers, *J. Am. Chem. Soc.*, **122**, 1840 (2000).
9. S. Pan, X. J. Sun and J. K. Lee, *J. Am. Soc. Mass Spectrom.*, **17**, 1383 (2006).
10. U. Ryde, M. H. M. Olsson, B. O. Roos, J. O. D. Kerpel and K. Pierloot, *J. Biol. Inorg. Chem.*, **5**, 565 (2000).
11. S. Pan, X. J. Sun and J. K. Lee, *Int. J. Mass Spectrom.*, **253**, 238 (2006).
12. F. Tureček, *Mass Spectrom. Rev.*, **26**, 563 (2007).
13. T. Kurikawa, H. Takeda, M. Hirano, K. Judai, T. Arita, S. Nagao, A. Nakajima and K. Kaya, *Organometallics*, **18**, 1430 (1999).
14. J. Miyawaki and K. Sugawara, *Chem. Phys. Lett.*, **386**, 196 (2004).
15. M. Alcamí, O. Mó and M. Yáñez, *Mass Spectrom. Rev.*, **20**, 195 (2001).
16. J. M. Mercero, J. M. Matxain, X. Lopez, D. M. York, A. Largo, L. A. Eriksson and J. M. Ugalde, *Int. J. Mass Spectrom.*, **240**, 37 (2005).
17. M. Pavelka and J. V. Burda, *Chem. Phys.*, **312**, 193 (2005).
18. J. V. Burda, M. Pavelka and M. Simanek, *J. Mol. Struct. (Theochem)*, **683**, 183 (2004).
19. C. F. Schwenk and B. M. Rode, *J. Am. Chem. Soc.*, **126**, 12786 (2004).
20. J. V. Burda, J. Sponer and P. Hobza, *J. Phys. Chem.*, **100**, 7250 (1996).
21. J. V. Burda, J. Sponer, J. Leszczynski and P. Hobza, *J. Phys. Chem. B*, **101**, 9670 (1997).
22. A. D. Becke, *J. Chem. Phys.*, **98**, 1372 (1993).
23. C. Lee, W. Yang and R. G. Parr, *Phys. Rev. B*, **37**, 785 (1988).
24. A. M. Mebel, K. Morokuma and M. C. Lin, *J. Chem. Phys.*, **103**, 7414 (1995).
25. A. Ricca and C. W. Bauschlicher, *Theor. Chim. Acta*, **92**, 123 (1995).
26. A. L. Llamas-Saiz, C. Foces-Foces, O. Mó, M. Yáñez, E. Elguero and J. Elguero, *J. Comput. Chem.*, **16**, 263 (1995).
27. J. A. Montgomery Jr., M. J. Frisch, J. Ochterski and G. A. Petersson, *J. Chem. Phys.*, **110**, 2822 (1999).
28. L. A. Curtiss, P. C. Redfern, K. Raghavachari and J. A. Pople, *J. Chem. Phys.*, **114**, 108 (2001).
29. A. Luna, B. Amekraz and J. Tortajada, *Chem. Phys. Lett.*, **266**, 31 (1997).
30. S. Hoyau and G. Ohanessian, *J. Am. Chem. Soc.*, **119**, 2016 (1997).
31. A. Rimola, E. Constantino, L. Rodríguez-Santiago and M. Sodupe, *J. Phys. Chem. A*, **112**, 3444 (2008).
32. A. Luna, M. Alcamí, O. Mó and M. Yáñez, *Chem. Phys. Lett.*, **320**, 129 (2000).
33. G. Corongiu and P. Nava, *Int. J. Quantum Chem.*, **93**, 395 (2003).
34. T. Marino, N. Russo and M. Toscano, *J. Inorg. Biochem.*, **79**, 179 (2000).
35. T. Marino, N. Russo and M. Toscano, *J. Mass Spectrom.*, **37**, 786 (2002).
36. S. J. Klippenstein and C. N. Yang, *Int. J. Mass Spectrom.*, **201**, 253 (2000).
37. I. Georgieva, N. Trendafilova, L. Rodríguez-Santiago and M. Sodupe, *J. Phys. Chem. A*, **109**, 5668 (2005).
38. J. Poater, M. Solà, A. Rimola, L. Rodríguez-Santiago and M. Sodupe, *J. Phys. Chem. A*, **108**, 6072 (2004).
39. T. K. Dargel, R. H. Hertwig and W. Koch, *Mol. Phys.*, **96**, 583 (1999).
40. I. Corral, O. Mó and M. Yáñez, *Eur. J. Mass Spectrom.*, **10**, 921 (2004).
41. I. Corral, O. Mó and M. Yáñez, *Int. J. Mass Spectrom.*, **255–256**, 20 (2006).
42. P. Milko, J. Roithová, D. S. Schröder, J. Lemaire, H. Schwarz and M. C. Holthausen, *Chem. Eur. J*, **14**, 10 (2008).
43. T. K. Ghanty and E. R. Davidson, *Int. J. Quantum Chem.*, **77**, 291 (2000).
44. M. J. Frisch, G. W. Trucks, H. B. Schlegel, G. E. Scuseria, M. A. Robb, J. R. Cheeseman, V. G. Zakrzewski, J. J. A. Montgomery, T. Vreven, K. N. Kudin, J. C. Burant, J. M. Millam, S. S. Iyengar, J. Tomasi, V. Barone, B. Mennucci, M. Cossi, G. Scalmani, N. Rega, G. A. Petersson, H. Nakatsuji, M. Hada, M. Ehara, K. Toyota, R. Fukuda, J. Hasegawa, M. Ishida, T. Nakajima, Y. Honda, O. Kitao, C. Adamo, J. Jaramillo, R. Gomperts, R. E. Stratmann,

O. Yazyev, J. Austin, R. Cammi, C. Pomelli, J. Ochterski, P. Y. Ayala, K. Morokuma, G. A. Voth, P. Salvador, J. J. Dannenberg, V. G. Zakrzewski, S. Dapprich, A. D. Daniels, M. C. Strain, O. Farkas, D. K. Malick, A. D. Rabuck, K. Raghavachari, J. B. Foresman, J. V. Ortiz, Q. Cui, A. G. Baboul, S. Clifford, J. Cioslowski, B. B. Stefanov, G. Liu, A. Liashenko, P. Piskorz, I. Komaromi, R. L. Martin, D. J. Fox, T. Keith, M. A. Al-Laham, C. Y. Peng, A. Nanayakkara, M. Challacombe, P. M. W. Gill, B. Johnson, W. Chen, M. W. Wong, C. Gonzalez and J. A. Pople, 'Gaussian03, Revision E.01' (Gaussian, Inc., Wallingford CT, 2003).
45. B. J. Lynch and D. G. Truhlar, *Chem. Phys. Lett.*, **361**, 251 (2002).
46. D. Cremer and Z. He, *J. Phys. Chem.*, **100**, 6173 (1996).
47. A. Ricca and C. W. Bauschlicher, Jr., *J. Phys. Chem. A*, **101**, 8949 (1997).
48. A. Luna, M. Alcamí, O. Mó, M. Yáñez and J. Tortajada, *Int. J. Mass Spectrom.*, **217**, 119 (2002).
49. C. Trujillo, A. M. Lamsabhi, O. Mó and M. Yáñez, *Phys. Chem. Chem. Phys.*, **10**, 3229 (2008).
50. M. Alcamí, O. Mó, M. Yáñez and I. L. Cooper, *J. Chem. Phys.*, **112**, 6131 (2000).
51. P. Su, F. Lin and C. S. Yeh, *J. Phys. Chem. A*, **105**, 9643 (2001).
52. Y. S. Yang, W. Y. Hsu, H. F. Lee, Y. C. Huang, C. S. Yeh and C. H. Hu, *J. Phys. Chem. A*, **103**, 11287 (1999).
53. *NIST Chemistry Webbook. Standard Reference Database Number 69* (Eds P. J. Linstrom and W. G. Mallard), Release June 2005, National Institute of Standards and Technology, Gaithersburg MD, 20899 (http://webbook.nst.gov) (2005).
54. L. Puskar and A. J. Stace, *Mol. Phys.*, **103**, 1829 (2005).
55. K. F. Willey, P. Y. Cheng, M. B. Bishop and M. A. Ducan, *J. Am. Chem. Soc.*, **113**, 4721 (1991).
56. M. T. Rodgers, J. R. Stanley and R. Amunugama, *J. Am. Chem. Soc.*, **122**, 10969 (2000).
57. N. S. Rannulu and M. T. Rodgers, *J. Phys. Chem. A*, **111**, 3465 (2007).
58. D. Wu, B. Ren, Y. Jiang, X. Xu and Z. Tian, *J. Phys. Chem. A*, **106**, 9042 (2002).
59. J. Miyawaki and K. Sugawara, *J. Chem. Phys.*, **119**, 6539 (2003).
60. T. G. Spence, B. T. Trotter and L. A. Posey, *Int. J. Mass Spectrom.*, **177**, 187 (1998).
61. C. L. Gatlin, F. Tureček and T. Vaisar, *J. Mass Spectrom.*, **30**, 1617 (1995).
62. L. Puskar, P. E. Barran, R. R. Wright, D. A. Kirkwood and A. J. Stace, *J. Chem. Phys.*, **112**, 7751 (2000).
63. L. Puskar and A. J. Stace, *J. Chem. Phys.*, **114**, 6499 (2001).
64. L. Puskar, H. Cox, A. Goren, G. Aitken and A. J. Stace, *Faraday Disc.*, **124**, 259 (2003).
65. R. R. Wright, N. R. Walker, S. Firth and A. J. Stace, *J. Phys. Chem. A*, **105**, 54 (2001).
66. G. Wu, J. Guan, G. Aitken, H. Cox and A. J. Stace, *J. Chem. Phys.*, **124**, 201103 (2006).
67. M. Chachisvilis and A. H. Zewail, *J. Phys. Chem. A*, **103**, 7408 (1999).
68. A. J. Stace, *Phys. Chem. Chem. Phys.*, **3**, 1935 (2001).
69. A. B. P. Lever, *Inorganic Electronic Spectroscopy*, 2nd edn, Elsevier, Amsterdam, 1984.
70. D. L. Leussing and R. C. Hansen, *J. Am. Chem. Soc.*, **79**, 4270 (1957).
71. K. Ozutsumi and T. Kawashima, *Polyhedron*, **11**, 169 (1992).
72. L. Petit, C. Adamo and N. Russo, *J. Phys. Chem. B*, **109**, 12214 (2005).
73. A. D. Quartarolo, N. Russo, E. Sicilia and F. Lelj, *J. Chem. Theory Comput.*, **3**, 860 (2007).
74. B. S. Freiser, *Organometallic Ion Chemistry*, Kluwer Academic Publishers, Dordrecht, 1996.
75. C. H. Ruan, Z. B. Yang and M. T. Rodgers, *Phys. Chem. Chem. Phys.*, **9**, 5902 (2007).
76. B. A. Cerda and C. Wesdemiotis, *J. Am. Chem. Soc.*, **117**, 9734 (1995).
77. T. Shoeib, C. Rodriquez, K. W. M. Siu and A. C. Hopkinson, *Phys. Chem. Chem. Phys.*, **3**, 853 (2001).
78. M. Belcastro, T. Marino, N. Russo and M. Toscano, *J. Mass Spectrom.*, **40**, 300 (2005).
79. D. Schröder, R. Wesendrup, R. H. Hertwig, T. K. Dargel, H. Grauel, W. Koch, B. R. Bender and H. Schwarz, *Organometallics*, **19**, 2608 (2000).
80. A. W. Ehlers, C. G. d. Koster, R. J. Meier and K. Lammertsma, *J. Phys. Chem. A*, **105**, 8691 (2001).
81. J. Bertrán, L. Rodríguez-Santiago and M. Sodupe, *J. Phys. Chem. B*, **103**, 2310 (1999).
82. M. T. Rodgers and P. B. Armentrout, *J. Am. Chem. Soc.*, **124**, 2678 (2002).
83. N. Russo, M. Toscano and A. Grand, *J. Mass Spectrom.*, **38**, 265 (2003).
84. A. M. Lamsabhi, M. Alcamí, O. Mó and M. Yáñez, *ChemPhysChem*, **4**, 1011 (2003).

85. A. Luna, M. Alcamí, O. Mó and M. Yáñez, *Int. J. Mass Spectrom.*, **201**, 215 (2000).
86. L. Galiano, M. Alcamí, O. Mó and M. Yáñez, *J. Phys. Chem. A*, **106**, 9306 (2002).
87. L. Galiano, M. Alcamí, O. Mó and M. Yáñez, *ChemPhysChem*, **4**, 72 (2003).
88. G. Vitale, A. B. Valina, H. Huang, R. Amunugama and M. T. Rodgers, *J. Phys. Chem. A*, **105**, 11353 (2001).
89. E. R. Fisher and P. B. Armentrout, *J. Phys. Chem.*, **94**, 1674 (1990).
90. R. H. Hertwig, W. Koch, D. Schroder, H. Schwarz, J. Hrusak and P. Schwerdtfeger, *J. Phys. Chem.*, **100**, 12253 (1996).
91. C. W. Bauschlicher, Jr., S. R. Langhoff and H. Partridge, *J. Chem. Phys.*, **94**, 2068 (1991).
92. M. R. Sievers, L. M. Jarvis and P. B. Armentrout, *J. Am. Chem. Soc.*, **120**, 1891 (1998).
93. J. Bera, A. Samuelson and J. Chandrasekhar, *Organometallics*, **17**, 4136 (1998).
94. C. N. Yang and S. J. Klippenstein, *J. Phys. Chem. A*, **103**, 1094 (1999).
95. R. C. Dunbar, D. Solooki, C. A. Tessier, W. J. Youngs and B. Asamoto, *Organometallics*, **10**, 52 (1991).
96. R. C. Dunbar, G. T. Uechi, D. Solooki, C. A. Tessier, W. Youngs and B. Asamoto, *J. Am. Chem. Soc.*, **115**, 12477 (1993).
97. F. Meyer, F. A. Khan and P. B. Armentrout, *J. Am. Chem. Soc.*, **117**, 9740 (1995).
98. R. Amunugama and M. T. Rodgers, *J. Phys. Chem. A*, **105**, 9883 (2001).
99. H. Koizumi and P. B. Armentrout, *J. Am. Soc. Mass Spectrom.*, **12**, 480 (2001).
100. H. Koizumi, X. Zhang and P. B. Armentrout, *J. Phys. Chem. A*, **105**, 2444 (2001).
101. M. Porento and P. Hirva, *Theor. Chem. Acc.*, **107**, 200 (2002).
102. D. W. Fuerstenau, R. Herrera-Urbina and D. W. McGlashan, *Int. J. Miner. Process.*, **58**, 15 (2000).
103. I. Georgieva and N. Trendafilova, *J. Phys. Chem. A*, **111**, 13075 (2007).
104. I. Corral, O. Mó and M. Yáñez, *Int. J. Mass Spectrom.*, **107**, 1370 (2003).
105. I. Corral, O. Mó and M. Yáñez, *J. Phys. Chem. A*, **107**, 1370 (2003).
106. N. S. Rannulu and M. T. Rodgers, *Phys. Chem. Chem. Phys.*, **7**, 1014 (2005).
107. Y. D. Hill, B. S. Freiser and C. W. Bauschlicher, *J. Am. Chem. Soc.*, **113**, 1507 (1991).
108. E. D. Glendening, A. E. Reed and F. Weinhold, 'NBO version 3.1'.
109. M. Brookhart and M. L. H. Green, *J. Organomet. Chem.*, **250**, 395 (1983).
110. M. J. S. Dewar, *Bull. Soc. Chim. Fr.*, **C79**, 18 (1951).
111. J. Chatt and L. A. Duncanson, *J. Chem. Soc.*, 2939 (1953).
112. T. Ziegler and A. Rauk, *Inorg. Chem.*, **18**, 1558 (1979).
113. D. Feller and D. A. Dixon, *J. Phys. Chem. A*, **106**, 5136 (2002).
114. J. V. Burda, M. K. Shukla and J. Leszczynski, *J. Mol. Model.*, **11**, 362 (2005).
115. M. Pavelka, M. K. Shukla, J. Leszczynski and J. V. Burda, *J. Phys. Chem. A*, **112**, 256 (2008).
116. Y. Yoshioka, H. Kawai, T. Sato, K. Yamaguchi and I. Saito, *J. Am. Chem. Soc.*, **125**, 1968 (2003).
117. L. Rulisek and J. Sponer, *J. Phys. Chem. B*, **107**, 1913 (2003).
118. Z. Yang, N. S. Rannulu, Y. Chu and M. T. Rodgers, *J. Phys. Chem. A*, **112**, 388 (2008).
119. Y. Chu, Z. Yang and M. T. Rodgers, *J. Am. Soc. Mass Spectrom.*, **13**, 453 (2002).
120. M. Y. Combariza and R. W. Vachet, *J. Phys. Chem. A*, **108**, 1757 (2004).
121. A. Luna, B. Amekraz, J. P. Morizur, J. Tortajada, O. Mó and M. Yáñez, *J. Phys. Chem. A*, **104**, 3132 (2000).
122. M. Rosi and C. W. Bauschlicher, Jr., *J. Chem. Phys.*, **90**, 7264 (1989).
123. M. Rosi and C. W. Bauschlicher, Jr., *J. Chem. Phys.*, **92**, 1876 (1990).
124. C. W. Bauschlicher, S. R. Langhoff and H. Partridge, *J. Chem. Phys.*, **94**, 2068 (1991).
125. A. M. Lamsabhi, M. Alcamí, O. Mó, M. Yáñez and J. Tortajada, *ChemPhysChem*, **5**, 1871 (2004).
126. S. Pulkkinen, M. Noguera, L. Rodríguez-Santiago, M. Sodupe and J. Bertran, *Chem. Eur. J*, **6**, 4393 (2000).
127. C. Sivasankar, N. Sadhukhan, J. Bera and A. Samuelson, *New J. Chem.*, **31**, 385 (2007).
128. X. Wang, B. Sohnlein, S. Li, J. Fuller and D. Yang, *Can. J. Chem.*, **85**, 714 (2007).
129. B. Ni, J. R. Kramer and N. H. Werstiuk, *J. Phys. Chem. A*, **109**, 1548 (2005).
130. D. Feller, E. D. Glendening and W. A. d. Jong, *J. Chem. Phys.*, **110**, 1475 (1999).
131. D. Caraiman, T. Shoeib, K. M. Siu, A. C. Hopkinson and D. K. Bohme, *Int. J. Mass Spectrom.*, **228**, 629 (2003).
132. N. R. Walker, S. Firth and A. J. Stace, *Chem. Phys. Lett.*, **292**, 125 (1998).

133. A. Bérces, T. Nukada, P. Margl and T. Ziegler, *J. Phys. Chem. A*, **103**, 9693 (1999).
134. G. W. Marini, K. R. Liedl and B. M. Rode, *J. Phys. Chem. A*, **103**, 11387 (1999).
135. H. D. Pranowo, A. H. B. Setiaji and B. M. Rode, *J. Phys. Chem. A*, **103**, 11115 (1999).
136. C. F. Schwenk and B. M. Rode, *J. Chem. Phys.*, **119**, 9523 (2003).
137. C. F. Schwenk and B. M. Rode, *ChemPhysChem*, **4**, 931 (2003).
138. C. F. Schwenk and B. M. Rode, *ChemPhysChem*, **5**, 342 (2004).
139. H. D. Pranowo and B. M. Rode, *J. Phys. Chem. A*, **103**, 4298 (1999).
140. C. F. Schwenk and B. M. Rode, *Chem. Commun.*, 1286 (2003).
141. T. Marino, M. Toscano, N. Russo and A. Grand, *J. Phys. Chem. B*, **110**, 24666 (2006).
142. D. F. Raffa, R. Gomez-Balderas, P. Brunelle, G. A. Rickard and A. Rauk, *J. Biol. Inorg. Chem.*, **10**, 887 (2005).
143. R. Gómez-Balderas, D. F. Raffa, G. A. Rickard, P. Brunelle and A. Rauk, *J. Phys. Chem. A*, **109**, 5498 (2005).
144. M. H. M. Olsson, U. Ryde, B. O. Roos and K. Pierloot, *J. Biol. Inorg. Chem.*, **3**, 109 (1998).
145. M. H. M. Olsson and U. Ryde, *J. Biol. Inorg. Chem.*, **4**, 654 (1999).
146. K. Paraskevopoulos, M. Sundararajan, R. Surendran, M. Hough, R. Eady, I. Hillier and S. Hasnain, *Dalton Trans.*, 3067 (2006).
147. I. Corral, O. Mó and M. Yáñez, in *Encyclopedia of Computational Chemistry* (Eds P. v. R. Schleyer, N. L. Allinger, T. Clark, J. Gasteiger, P. A. Kollman, H. F. Schaefer III and P. R. Schreiner), John Wiley & Sons, Ltd, Chichester, 2004; *URL: http://mrw.interscience.wiley.com/emrw/9780470845011/ecc/article/cn0062/current/abstract?hd=All,Corral*.
148. D. Cremer and E. Kraka, *Angew. Chem.*, **96**, 612 (1984).
149. J. Zhang, V. Frankevich, R. Knochenmuss, S. D. Friess and R. Zenobi, *J. Am. Soc. Mass Spectrom.*, **14**, 42 (2003).
150. A. M. Lamsabhi, M. Alcamí, O. Mó, M. Yáñez and J. Tortajada, *J. Phys. Chem. A*, **110**, 1943 (2006).
151. A. M. Lamsabhi, M. Alcamí, O. Mó, M. Yáñez, J. Tortajada and J. Y. Salpin, *ChemPhys Chem*, **8**, 181 (2007).
152. D. Xing, X. Tan, X. Chen and Y. Bu, *J. Phys. Chem. A*, **112**, 7418 (2008).
153. C. L. Gatlin, F. Tureček and T. Vaisar, *J. Mass Spectrom.*, **30**, 1605 (1995).
154. C. L. Gatlin, F. Tureček and T. Vaisar, *J. Am. Chem. Soc.*, **117**, 3637 (1995).
155. H. Lavanant, E. Hecquet and Y. Hoppilliard, *Int. J. Mass Spectrom.*, **185/186/187**, 11 (1999).
156. M. Noguera, J. Bertran and M. Sodupe, *J. Phys. Chem. A*, **108**, 333 (2004).
157. M. Noguera, J. Bertran and M. Sodupe, *J. Phys. Chem. B*, **112**, 4817 (2008).
158. N. R. Walker, R. R. Wright, P. E. Barran, J. N. Murrell and A. J. Stace, *J. Am. Chem. Soc.*, **123**, 4223 (2001).
159. J. Allison, R. B. Freas and D. P. Ridge, *J. Am. Chem. Soc.*, **101**, 1332 (1979).
160. P. B. Armentrout and J. L. Beauchamp, *Acc. Chem. Res.*, **22**, 315 (1989).
161. P. B. Armentrout, *Ann. Rev. Phys. Chem*, **41**, 313 (1990).
162. K. Eller and H. Schwarz, *Chem. Rev.*, **91**, 1121 (1991).
163. J. C. Weisshaar, *Adv. Chem. Phys.*, **82**, 213 (1992).
164. P. B. Armentrout, in *Selective Hydrocarbons Activation: Principles and Progress* (Eds J. A. Davis, P. L. Watson, A. Greenberg and J. F. Liebman), VCH, New York, 1990, pp. 467–533.
165. J. C. Weisshaar, *Acc. Chem. Res.*, **26**, 213 (1993).
166. G. A. Somorjai, *Introduction to Surface Chemistry and Catalysis*, John Wiley & Sons, Inc., New York, 1994.
167. P. B. Armentrout and B. L. Kickel, in *Organometallic Ion Chemistry* (Eds B. S. Freiser), Kluwer, Dordrecht, 1996, pp. 1–45.
168. P. B. Armentrout, in *Topics in Organometallics Ion Chemistry, Vol. 4. Gas-Phase Organometallic Chemistry* (Eds J. Brown and P. Hofman), Springer-Verlag, Berlin, 1999, pp. 1–45.
169. A. M. Lamsabhi, M. Alcamí, O. Mó, W. Bouab, M. Esseffar, J. L. Abboud and M. Yáñez, *J. Phys. Chem. A*, **104**, 5122 (2000).
170. M. Esseffar, W. Bouab, A. M. Lamsabhi, J. L. Abboud, R. Notario and M. Yáñez, *J. Am. Chem. Soc.*, **122**, 2300 (2000).
171. A. Luna, B. Amekraz, J. Tortajada, J. P. Morizur, M. Alcamí, O. Mó and M. Yáñez, *J. Am. Chem. Soc.*, **120**, 5411 (1998).
172. A. Luna, J. P. Morizur, J. Tortajada, M. Alcamí, O. Mó and M. Yáñez, *J. Phys. Chem. A*, **102**, 4652 (1998).

173. D. Schröder, T. Weiske and H. Schwarz, *Int. J. Mass Spectrom.*, **219**, 729 (2002).
174. A. Luna, B. Amekraz, J. Morizur, J. Tortajada, O. Mó and M. Yáñez, *J. Phys. Chem. A*, **101**, 5931 (1997).
175. Y. Huang, P. Su and C. S. Yeh, *Bull. Chem. Soc. Jpn.*, **74**, 677 (2001).
176. M. Cheng and C. H. Hu, *J. Phys. Chem. A*, **106**, 11570 (2002).
177. M. Alcamí, A. Luna, O. Mó, M. Yáñez, J. Tortajada and B. Amekraz, *Chem. Eur. J.*, **10**, 2927 (2004).
178. A. Luna, O. Mó, M. Yáñez, J. P. Morizur, E. Leclerc, B. Desmazières, V. Haldys, J. Chamot-Rooke and J. Tortajada, *Int. J. Mass Spectrom.*, **228**, 359 (2003).
179. P. J. Fordham, J. Chamot-Rooke, E. Giudice, J. Tortajada and J. P. Morizur, *J. Mass Spectrom.*, **34**, 1007 (1999).
180. D. S. Schroder, H. Schwarz, J. Wu and C. Wesdemiotis, *Chem. Phys. Lett.*, **343**, 258 (2001).
181. J. A. Stone, T. Su and D. Vukomanovic, *Can. J. Chem.*, **83**, 1921 (2005).
182. H. Lavanant and Y. Hoppilliard, *J. Mass Spectrom.*, **32**, 1037 (1997).
183. Y. Hoppilliard, G. Ohanessian and S. Bourcier, *J. Phys. Chem. A*, **108**, 9687 (2004).
184. A. Rimola, M. Sodupe, J. Tortajada and L. Rodríguez-Santiago, *Int. J. Mass Spectrom.*, **257**, 60 (2006).
185. E. R. Talaty, B. A. Perera, A. L. Gallardo, J. M. Barr and M. J. Van Stipdonk, *J. Phys. Chem. A*, **105**, 8059 (2001).
186. C. Seto and J. A. Stone, *Int. J. Mass Spectrom.*, **192**, 289 (1999).
187. R. Tauler and B. M. Rode, *Inorg. Chim. Acta*, **173**, 93 (1990).
188. K. R. Liedl and B. M. Rode, *Chem. Phys. Lett.*, **197**, 181 (1992).
189. S. A. McLuckey, D. Cameron and R. G. Cooks, *J. Am. Chem. Soc.*, **103**, 1313 (1981).
190. W. A. Tao, D. Zhang, E. N. Nikolaev and R. G. Cooks, *J. Am. Chem. Soc.*, **122**, 10598 (2000).
191. R. G. Cooks, J. S. Patrick, T. Kotiaho and S. A. McLuckey, *Mass Spectrom. Rev.*, **13**, 287 (1994).
192. W. A. Tao, D. Zhang, F. Wang, P. D. Thomas and R. G. Cooks, *Anal. Chem.*, **71**, 4427 (1999).
193. P. Roepstorff and J. Fohlman, *Biomed. Mass Spectrom.*, **11**, 601 (1984).
194. S. J. Shields, B. K. Bluhm and D. H. Russell, *J. Am. Soc. Mass Spectrom.*, **11**, 626 (2000).
195. T. Lin, A. H. Payne and G. L. Glish, *J. Am. Soc. Mass Spectrom.*, **12**, 497 (2001).
196. I. K. Chu, X. Guo, T. C. Lau and K. W. M. Siu, *Anal. Chem.*, **71**, 2364 (1999).
197. L. M. Teesch and J. Adams, *J. Am. Chem. Soc.*, **113**, 812 (1991).
198. T. Yalcin, C. Khouw, I. G. Csizmadia, M. R. Peterson and A. G. Harrison, *J. Am. Soc. Mass Spectrom.*, **6**, 1165 (1995).
199. T. Yalcin, I. G. Csizmadia, M. R. Peterson and A. G. Harrison, *J. Am. Soc. Mass Spectrom.*, **7**, 233 (1996).
200. B. K. Bluhm, S. J. Shields, C. A. Bayse, M. B. Hall and D. H. Russell, *Int. J. Mass Spectrom.*, **204**, 31 (2001).
201. S. Kuenzel, D. Pretzel, J. Andert, K. Beck and S. Reissmann, *J. Peptide Sci.*, **9**, 502 (2003).
202. A. Bossée, F. Fournier, O. Tasseau, B. Bellier and J. Tabet, *Rapid Commun. Mass Spectrom.*, **17**, 1229 (2003).
203. A. Bossée, C. Afonso, F. Fournier, O. Tasseau, C. Pepe, B. Bellier and J. Tabet, *J. Mass Spectrom.*, **39**, 903 (2004).
204. P. Hu and M. L. Gross, *J. Am. Chem. Soc.*, **114**, 9153 (1992).
205. P. Hu and M. L. Gross, *J. Am. Chem. Soc.*, **114**, 9161 (1992).
206. W. A. Tao, L. Wu and R. G. Cooks, *J. Am. Soc. Mass Spectrom.*, **12**, 490 (2001).
207. L. Wu, K. Lemr, T. Aggerholm and R. G. Cooks, *J. Am. Soc. Mass Spectrom.*, **14**, 152 (2003).
208. M. Lagarrigue, A. Bossée, C. Afonso, F. Fournier, B. Bellier and J. Tabet, *J. Mass Spectrom.*, **41**, 1073 (2006).
209. T. Vaisar, J. Heinecke, J. Seymour and F. Tureček, *J. Mass Spectrom.*, **40**, 608 (2005).
210. J. Seymour and F. Tureček, *J. Mass Spectrom.*, **35**, 566 (2000).
211. T. Vaisar, C. L. Gatlin, R. D. Rao, J. Seymour and F. Tureček, *J. Mass Spectrom.*, **36**, 306 (2001).
212. I. K. Chu, C. Rodriquez, T. Lau, A. C. Hopkinson and K. W. M. Siu, *J. Phys. Chem. B*, **104**, 3393 (2000).
213. I. K. Chu, C. F. Rodriguez, F. Rodriguez, A. C. Hopkinson and K. W. M. Siu, *J. Am. Soc. Mass Spectrom.*, **12**, 1114 (2001).

214. S. Wee, R. A. J. O'Hair and W. D. McFadyen, *Rapid Comm. Mass Spectrom.*, **16**, 884 (2002).
215. S. Wee, R. A. J. O'Hair and W. D. Mcfadyen, *Int. J. Mass Spectrom.*, **234**, 101 (2004).
216. E. Bagheri-Majdi, Y. Ke, G. Orlova, I. K. Chu, A. C. Hopkinson and K. W. M. Siu, *J. Phys. Chem. B*, **108**, 11170 (2004).
217. S. Wee, R. A. J. O'Hair and W. D. Mcfadyen, *Int. J. Mass Spectrom.*, **249–250**, 171 (2006).
218. I. K. Chu, S. Siu, C. Lam, J. Chan and C. Rodriquez, *Rapid Commun. Mass Spectrom.*, **18**, 1798 (2004).
219. I. K. Chu, C. N. W. Lam and S. O. Siu, *J. Am. Soc. Mass Spectrom.*, **16**, 763 (2005).
220. C. Lam, S. Siu, G. Orlova and I. K. Chu, *Rapid Commun. Mass Spectrom.*, **20**, 790 (2006).
221. I. K. Chu and C. N. W. Lam, *J. Am. Soc. Mass Spectrom.*, **16**, 1795 (2005).
222. J. Y. Salpin and J. Tortajada, *J. Mass Spectrom.*, **37**, 379 (2002).
223. L. Boutreau, E. Leon, J. Y. Salpin, B. Amekraz, C. Moulin and J. Tortajada, *Eur. J. Mass Spectrom.*, **9**, 377 (2003).
224. L. T. Jensen, J. M. Peltier and D. R. Winge, *J. Biol. Inorg. Chem.*, **3**, 627 (1998).
225. R. M. Whittal, H. L. Ball, F. E. Cohen, A. L. Burlingame, S. B. Prusiner and M. A. Baldwin, *Protein Sci.*, **9**, 332 (2000).
226. T. Kurahashi, A. Miyazaki, S. Suwan and M. Isobe, *J. Am. Chem. Soc.*, **123**, 9268 (2001).
227. A. Urvoas, B. Amekraz, C. Moulin, L. Le Clainche, R. Stöcklin and M. Moutiez, *Rapid Commun. Mass Spectrom.*, **17**, 1889 (2003).
228. R. Nagane, T. Koshigoe and M. Chikira, *J. Inorg. Biochem.*, **93**, 204 (2003).
229. A. Sinz, A. J. Jin and O. Zschornig, *J. Mass Spectrom.*, **38**, 1150 (2003).
230. Y. Guo, Y. Ling, B. A. Thomson and K. W. M. Siu, *J. Am. Soc. Mass Spectrom.*, **16**, 1787 (2005).
231. A. Arseniev, P. Schultze, E. Worgotter, W. Braun, G. Wagner, M. Vasak, J. H. R. Kagi and K. Wuthrich, *J. Mol. Biol.*, **201**, 637 (1988).
232. H. Li and J. D. Otvos, *Biochemistry*, **35**, 13937 (1996).
233. M. J. Stillman, *Metal-Based Drugs*, **6**, 277 (1999).
234. J. Y. Salpin, L. Boutreau, V. Haldys and J. Tortajada, *Eur. J. Mass Spectrom.*, **7**, 321 (2001).
235. M. Alcamí, A. Luna, O. Mó, M. Yáñez, L. Boutreau and J. Tortajada, *J. Phys. Chem. A*, **106**, 2641 (2002).
236. C. J. Cramer and D. G. Truhlar, *J. Am. Chem. Soc.*, **115**, 5745 (1993).
237. A. El Firdoussi, M. Lafitte, J. Tortajada, O. Kone and J.-Y. Salpin, *J. Mass Spectrom.*, **42**, 999 (2007).
238. G. Smith, A. Kaffashan and J. A. Leary, *Int. J. Mass Spectrom.*, **183**, 299 (1999).
239. B. D. Davis and J. S. Brodbelt, *J. Am. Soc. Mass Spectrom.*, **15**, 1287 (2004).
240. B. Domon and C. E. Costello, *Glycoconjugate J.*, **5**, 397 (1988).
241. D. J. Harvey, *J. Am. Soc. Mass Spectrom.*, **12**, 926 (2001).
242. M. Shahgholi, J. H. Callahan, B. J. Rappoli and D. A. Rowley, *J. Mass Spectrom.*, **32**, 1080 (1997).
243. B. Lippert, *Coord. Chem. Rev.*, **200**, 487 (2000).
244. C. Harford and B. Sarkar, *Acc. Chem. Res.*, **30**, 123 (1997).
245. C. J. Burrows and J. G. Muller, *Chem. Rev.*, **98**, 1109 (1998).
246. B. Halliwell and O. I. Aruoma, *FEBS Lett.*, **281**, 9 (1991).
247. R. Drouin, H. Rodriguez, S. W. Gao, Z. Gebreyes, T. R. Oconnor, G. P. Holmquist and S. A. Akman, *Free Radical Biol. Med.*, **21**, 261 (1996).
248. S. Wee, R. A. J. O'Hair and W. D. Mcfadyen, *Rapid Commun. Mass Spectrom.*, **19**, 1797 (2005).
249. R. Cini and C. Pifferi, *J. Chem. Soc., Dalton Trans.*, 699 (1999).
250. P. Cheng and D. K. Bohme, *J. Phys. Chem. B*, **111**, 11075 (2007).
251. R. L. Hettich, *Int. J. Mass Spectrom.*, **204**, 55 (2001).

CHAPTER **8**

Biochemistry of organocopper compounds

MARTHA E. SOSA-TORRES, JUAN P. SAUCEDO-VÁZQUEZ and SAUL GÓMEZ-MANZO

Facultad de Química, Universidad Nacional Autónoma de México, Ciudad Universitaria, Coyoacán, 04510, D.F. México, México
Fax: 0052-55-56162010, e-mail: mest@servidor.unam.mx

and

PETER M. H. KRONECK

Fachbereich Biologie, Universität Konstanz, 78457 Konstanz, Germany
Fax: 0049-7531-882966, e-mail: peter.kroneck@uni-konstanz.de

I. INTRODUCTION . . . 348
II. TYPES OF BIOLOGICAL COPPER CENTERS . . . 348
A. Type 1 Cu . . . 349
B. Type 2 Cu . . . 349
C. Type 3 Cu . . . 350
D. Novel Cu Centers . . . 352
III. COPPER PROTEINS WITH MONONUCLEAR COPPER SITES . . . 352
A. Plastocyanin . . . 352
B. Galactose Oxidase . . . 352
C. Cu,Zn Superoxide Dismutase . . . 354
IV. COPPER PROTEINS WITH MULTINUCLEAR COPPER SITES . . . 355
A. Catechol Oxidase . . . 355
B. Ascorbate Oxidase . . . 356
C. Nitrous Oxide Reductase . . . 357
D. Carbon Monoxide Dehydrogenase . . . 359
V. OUTLOOK . . . 360
VI. ACKNOWLEDGMENTS . . . 361
VII. REFERENCES . . . 361

The chemistry of organocopper compounds
Edited by Z. Rappoport and I. Marek

I. INTRODUCTION

Many of the active sites of currently known enzymes harbor transition metals, such as Mo, Mn, Fe, Co, Ni, Cu and Zn, thus turning these biomolecules into potent catalysts. The choice of the transition metal usually depends on its ionic radius, the type of reaction, the required redox-potential, the nature of available binding sites and, perhaps most important, on the bioavailability of the metal itself. In the early reducing atmosphere, copper was practically unavailable and locked in its univalent Cu(I) form as insoluble Cu_2S[1,2]. This extremely stable form of copper kept it from use in biological systems, in contrast to iron which existed as soluble Fe(II) under these conditions. The arrival of dioxygen in the atmosphere, as a result of the evolution of photosynthetic organisms, increased the stability and hence the concentration of oxidized transition metal ions. Oxidation of Cu(I) lead to soluble Cu(II), whereas soluble Fe(II) was converted into Fe(III) which, in water, occurred mainly as insoluble $Fe(OH)_3$. The occurrence, functions and reactions catalyzed by copper enzymes indicate that copper only became biologically relevant after dioxygen began to accumulate in the atmosphere.

Today, most living organisms need copper to serve as a cofactor in crucial redox proteins and enzymes[3,4], such as plastocyanin, a blue type 1 Cu electron transfer protein used in the photosynthetic apparatus of green plants, superoxide dismutase, a type 2 Cu, Zn enzyme which detoxifies superoxide anion ($O_2^{\bullet -}$) to dioxygen and hydrogen peroxide, or cytochrome *c* oxidase, the redox-driven mitochondrial proton pump converting dioxygen to water. Yet paradoxically, larger doses of copper are cytotoxic, and nature has evolved highly sophisticated mechanisms to handle and detoxify copper ions. Therefore, it is not too surprising that any imbalance of copper in humans will lead to diseases that can seriously impair the quality of life. These include Menkes syndrome and Wilson's disease, involved in severe neurological disorders, and characterized by the inability to distribute correctly copper to all cells and tissues[5].

In this chapter we summarize structural and functional aspects of selected copper proteins and enzymes, and discuss recent advances in the field of novel copper enzymes, such as nitrous oxide reductase and carbon monoxide dehydrogenase. Physiological, structural and mechanistic aspects of these proteins and their active sites will be described. Many of them carry complex multi-metal catalytic sites with unique chemical and electronic properties which have attracted the interest of chemists, biologists and spectroscopists. In view of the enormous number of publications accumulating in the field, we recommend as primary references the two Handbooks on Metalloproteins[6,7], and the special issues of Chemical Reviews on Biomimetic Chemistry[8,9]. The following comprehensive reviews are also recommended for further reading[10–18].

II. TYPES OF BIOLOGICAL COPPER CENTERS

In biological systems, copper occurs as part of the active site of proteins or enzymes. The various manners in which copper is bound to these centers are of special interest. In most cases, the metal ion participates directly in the catalytic cycle. However, in some enzymes, copper generates an intrinsic organic cofactor without participating in catalysis. Usually, the copper ions change their redox-states during catalysis, electrons are transferred either between the metal ion and its substrate, or between the metal ion and its partner with the individual redox-potentials determining the electron flux. In biological systems, relevant oxidation states are confined to diamagnetic Cu(I) ($S = 0$; $3d^{10}4s^0$) and paramagnetic Cu(II) ($S = 1/2$; $3d^94s^0$). In agreement with the HSAB-concept of Pearson[19], the 'soft' Cu(I) will be preferentially complexed by amino acids such as cysteine (Cys), histidine (His) and methionine (Met), while the 'harder' Cu(II) can also bind 'harder' ligands, such as H_2O, or amino acids tyrosine (Tyr) and histidine (His). The different oxidation states

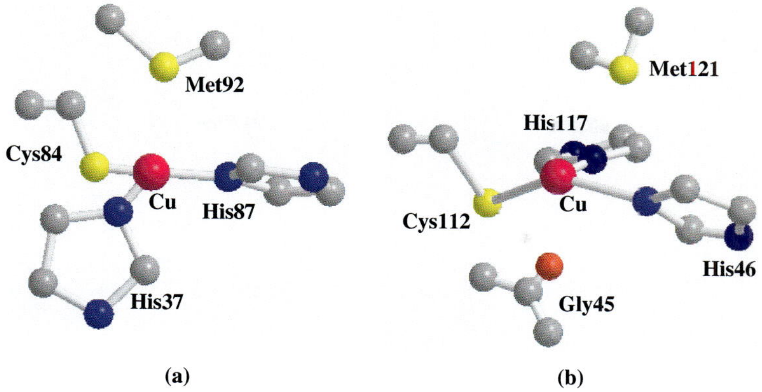

PLATE 1. Schematic representation of the type 1 copper sites in (A) plastocyanin[25] and (B) azurin[26]

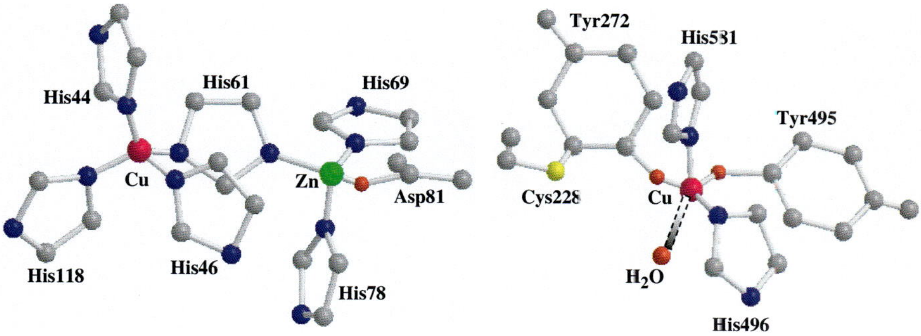

PLATE 2. Schematic representation of the type 2 Cu sites in Cu,Zn superoxide dismutase[31] (left) and galactose oxidase[32] (right)

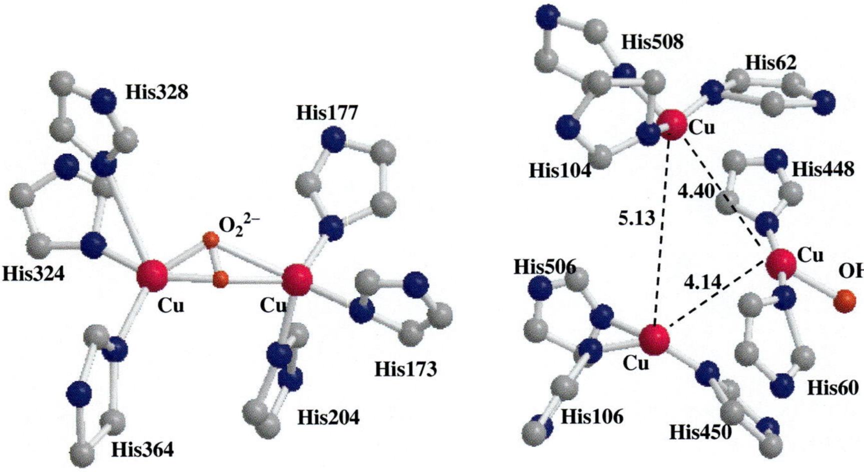

PLATE 3. Schematic representation of the type 3 copper site in the oxygen-bound form hemocyanin[33] (left), and of the trinuclear copper site in the reduced state of ascorbate oxidase[34] (right)

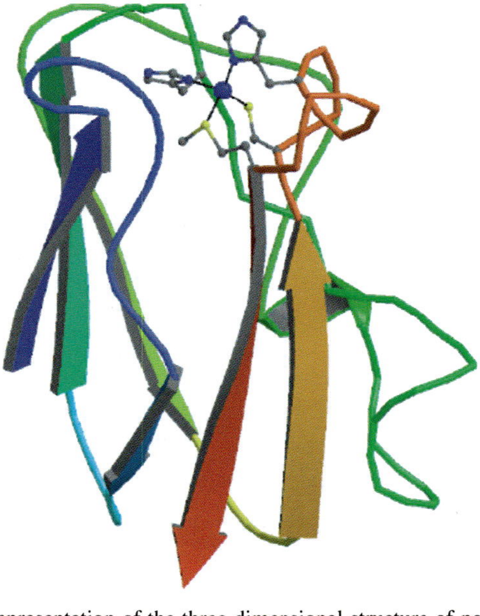

PLATE 4. Schematic representation of the three-dimensional structure of poplar plastocyanin (PDB code 1PLC)

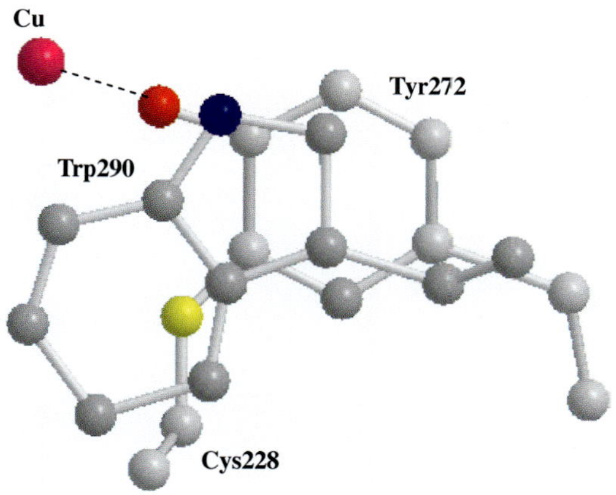

PLATE 5. Stacking of tryptophan290 and the tyrosine272-cysteine228 thioether moiety in galactose oxidase[44, 45]

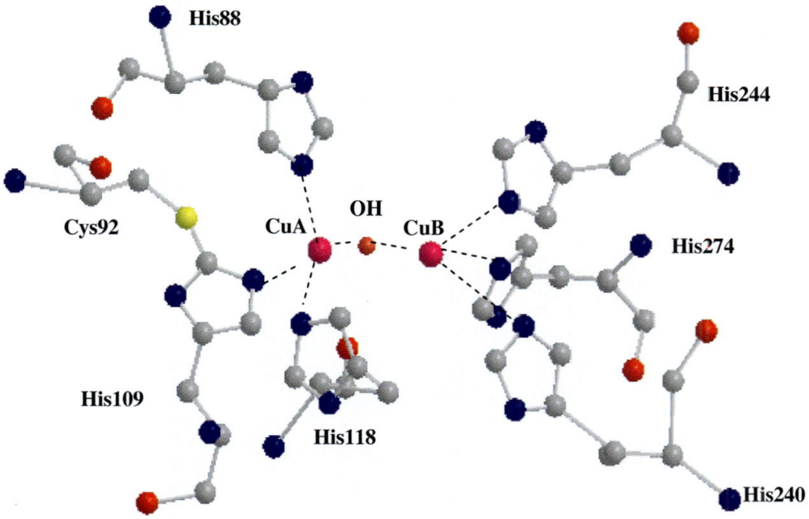

PLATE 6. Schematic representation of the dinuclear copper site of catechol oxidase[53]

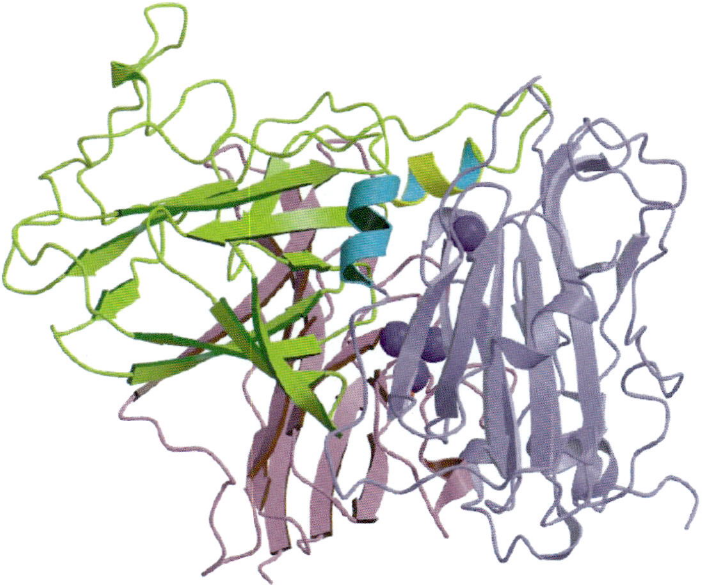

PLATE 7. Schematic representation of the three-dimensional structure of ascorbate oxidase, showing the monomer with the mononuclear type 1 Cu and the trinuclear copper site (PDB code 1AOZ)

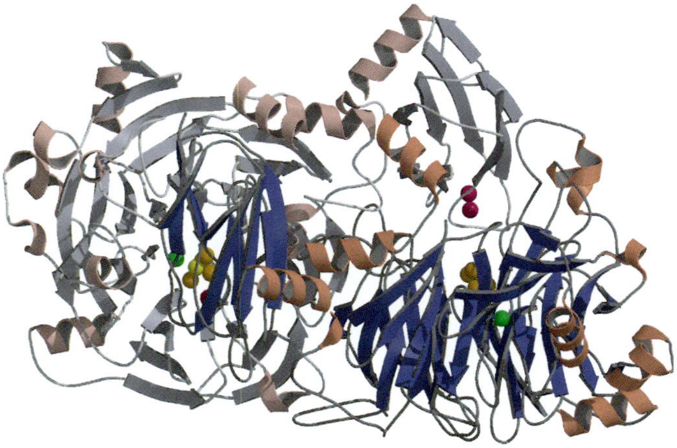

PLATE 8. Schematic representation of the three-dimensional structure of nitrous oxide reductase from *Paracoccus denitrificans*, showing the head-to-tail dimer with the mixed-valent Cu_A (purple) and the tetranuclear Cu_Z (yellow) sites (PDB code 1FWX)

of copper not only favor distinct ligands but also distinct geometries. Sterical hindrances excluded, Cu(I) (coordination number 4) favors a tetrahedral conformation in contrast to Cu(II), which favors a square-planar conformation. In copper proteins, ligand sites and geometries may differ significantly from those observed in regular copper complexes. A redox-active copper center within a protein must switch between the oxidation states Cu(I) and Cu(II), with a minimum of energy for structural reorganization. This goal is usually achieved in two ways: first, by ligands (CySH and His) which favor both Cu(I) and Cu(II), and second, by employing the proper coordination geometries through the protein scaffold which will allow rapid electron transfer. The choice of ligands and geometry will strongly influence the redox potential of the central copper ion. In aqueous solution, the redox potential of Cu^{2+}/Cu^{+} is $E^{\circ} = +153$ mV, *vs* a range of potentials found in copper proteins and enzymes from +183 mV in halocyanin to +785 mV of the type 1 Cu center of laccase[20,21]. This wide range of relatively high redox potentials was achieved by the evolution of various copper-binding centers. In these centers, the coordinating amino acid residues are fixed within the protein backbone, forcing the metal ion into a stereochemically rigid environment. In all cases for efficient electron transfer, there is minimal change in geometry with redox (i.e. a low reorganizational energy) and efficient electronic coupling through the protein[14,22]. Furthermore, the high redox potentials of copper proteins make them suitable for reactions with dioxygen[2].

The large pool of potential coordinating ligands for copper in proteins, and its proximity to neighboring copper ions, or other metals centers, allowed for three distinct classes of copper metal centers within proteins which were denoted types 1, 2 and 3[23]. Malkin and Malmström classified the copper centers according to their spectroscopic properties, mainly UV/Vis and electron paramagnetic resonance (EPR), as briefly discussed in the following section.

A. Type 1 Cu

A remarkable and unique spectroscopic feature of numerous copper proteins is a blue color as a result of the electronic properties of the type 1 copper center, with a characteristic intense absorption maximum around 600 nm ($\varepsilon > 3000$ M^{-1} cm^{-1}) and a small (*ca* 180–240 MHz) value for the hyperfine coupling constant $A_{||}$ in the EPR spectra. By comparison, the corresponding coefficient of a 'normal' Cu(II) complex, such as square-planar D_{4h}-$CuCl_4{}^{2-}$, comes to $\varepsilon = 40$ M^{-1} cm^{-1}, and the magnitude for A_{II} is significantly larger[14]. A greek key β-barrel structural motif is typical for type 1 Cu proteins denoting the cupredoxin fold[24]. Basically, type 1 copper is coordinated by two histidines (2N), one cysteine (1S) and one methionine (1SMet)[25,26] (Figure 1).

Depending on the individual protein, the geometry of the metal center is between a distorted tetrahedron (e.g. plastocyanin) and a distorted trigonal bipyramide (e.g. azurin)[27]. In both cases, the methionine residue is the ligand most distant from the copper center. In addition, azurin carries a fifth ligand, usually a carbonyl oxygen atom opposite to methionine. In some type 1 Cu proteins, the methionine ligand was exchanged against glutamine[28], or has been omitted completely[29]. Type 1 copper centers have not been only observed in small copper proteins, such as azurin or plastocyanin; they also act as electron transfer sites in multi-copper enzymes, e.g. laccase, ascorbate oxidase or nitrite reductase.

B. Type 2 Cu

The spectroscopic properties of the non-blue type 2 copper center, by comparison to the blue type 1 center, differ less from those of regular Cu(II) complexes[2,23]. A copper site was classified as type 2 copper if its absorption bands in the visible region were

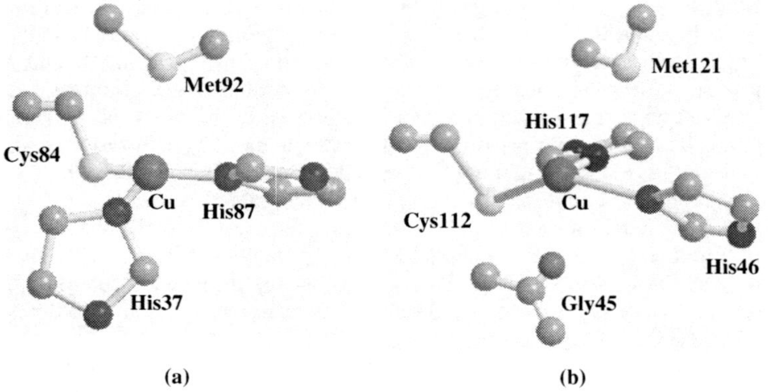

FIGURE 1. Schematic representation of the type 1 copper sites in (A) plastocyanin[25] and (B) azurin[26]

relatively weak ($\varepsilon \leqslant 500$ M^{-1} cm^{-1}) and the magnitude of A_{II} in the EPR spectrum was large (*ca* 450–550 MHz). Type 2 copper has been identified in numerous dioxygen activating enzymes (e.g. oxidases and oxygenases), nitrite reductase and in Cu,Zn superoxide dismutase[30–32]. While type 1 copper centers are restricted to electron transport, type 2 copper centers are directly involved in the metabolism of oxygen and nitrogen. In amine oxidases, the type 2 Cu is involved in the biosynthesis of an intrinsic organic cofactor. In these oxidases, dioxygen is reduced to hydrogen peroxide by a two-electron-transfer process. On the other hand, copper nitrite reductase, a constituent of the bacterial denitrification pathway, converts nitrite to nitric oxide (NO). Cu,Zn superoxide dismutase catalyzes the disproportionation reaction of the superoxide anion radical ($O_2^{-\bullet}$) to hydrogen peroxide and dioxygen:

$$O_2^{\bullet-} + O_2^{\bullet-} + 2\,H^+ \longrightarrow H_2O_2 + O_2$$

Type 2 copper centers are quite flexible with regard to their ligands and structural geometries. Clearly, the favored ligand of type 2 Cu is the imidazole nitrogen of histidine, but tyrosine also plays an important role as will be discussed for galactose oxidase[31,32] (Figure 2). One common feature is, however, that in those enzymes where the type 2 Cu actively participates in catalysis, one coordination site is always accessible for the substrate. Peptidylglycine-α-hydroxylating monooxygenase harbors two type 2 copper centers per functional unit denoted Cu_A and Cu_B, which do not directly interact as they are separated by a solvent-filled cleft about 11 Å wide. In this case, Cu_A is coordinated by the N_δ atoms of three histidines, and Cu_B is tetrahedrally coordinated by the N_ε atoms of two histidines, the sulfur of one methionine and an additional ligand. In the oxidized enzyme, this fourth ligand is either H_2O or OH^-[30]. Last but not least, the type 2 copper ion in Cu,Zn superoxide dismutase is coordinated by four histidine residues, one of which connects the copper ion to the zinc ion, the second metal ion bound to the active site of the enzyme[31] (Figure 2).

C. Type 3 Cu

A type 3 copper center consists of a pair of copper ions which lacks an EPR signal in the oxidized state because of strong antiferromagnetic coupling between the metal

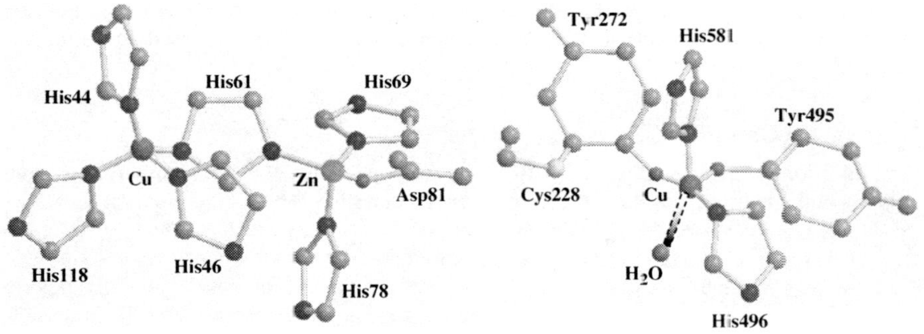

FIGURE 2. Schematic representation of the type 2 Cu sites in Cu,Zn superoxide dismutase[31] (left) and galactose oxidase[32] (right)

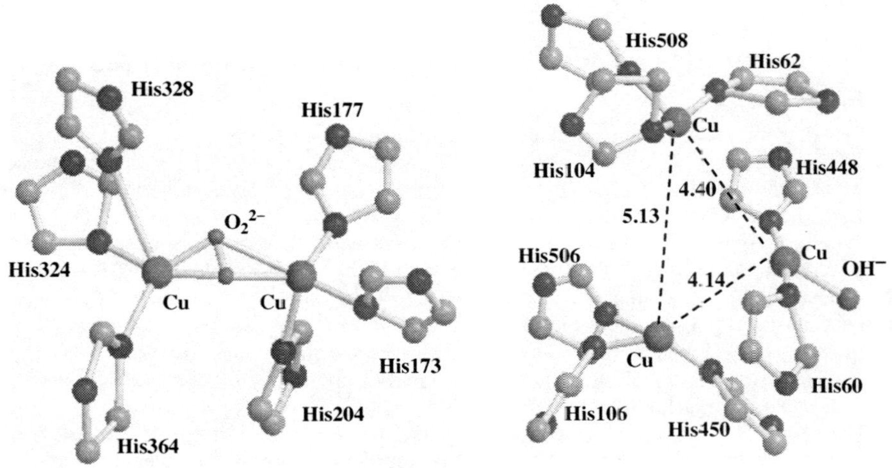

FIGURE 3. Schematic representation of the type 3 copper site in the oxygen-bound form hemocyanin[33] (left), and of the trinuclear copper site in the reduced state of ascorbate oxidase[34] (right)

centers of the Cu(II)–Cu(II) pair[23]. Type 3 copper sites occur in hemocyanin, in tyrosinase and in catechol oxidase (Figure 3). Hemocyanin binds O_2 reversibly (the *blue blood* of arthropods and molluscs)[33], catechol oxidase converts catechols to the corresponding *o*-quinones and tyrosinase hydroxylates monophenols (e.g. tyrosine) converts catechols to quinones. Usually, each of the two copper ions is coordinated by three histidines. While reduced, the binuclear copper site in hemocyanin is colorless but turns blue upon binding of dioxygen. The Cu–Cu distances in catechol oxidase range from 2.9 Å for the hydroxo-bridged Cu(II),Cu(II) form to 4.5 Å for the Cu(I),Cu(I) deoxy form. Interestingly, both catechol oxidase and hemocyanin contain a covalent thioether cross-link between a histidine ligand and a nearby cysteine residue. Multi-copper oxidases, also denoted blue oxidases (e.g. laccase, ascorbate oxidase and ceruloplasmin), contain trinuclear copper sites, formally built by a type 2 and a type 3 copper center (Figure 3). Both type 3 copper ions are coordinated by three histidines, whereas the third copper center is ligated by two histidines and a third ligand (H_2O or OH^-)[34]. These trinuclear copper centers represent the catalytic site for reduction of dioxygen to water as will be discussed for ascorbate oxidase.

D. Novel Cu Centers

In addition to the canonical type 1, 2 and 3 copper sites found in numerous copper proteins and enzymes, novel copper centers have been discovered and structurally characterized over the past two decades. These include the mixed-valent [Cu(1.5+)...Cu(1.5+)] Cu_A electron transfer center in nitrous oxide reductase and cytochrome *c* oxidase[35], the tetranuclear μ-sulfido [Cu_4S] cluster in nitrous oxide reductase[36] and the hetero-dinuclear Mo-μ-sulfido-Cu site in carbon monoxide dehydrogenase isolated from aerobic microorganisms[37]. In addition, two important hetero-dinuclear metal centers have to be mentioned, the Cu,Zn site in superoxide dismutase (Figure 2) and the heme a_3-Cu site in cytochrome *c* oxidase[38]. These centers are involved in the conversion of oxygen species, N_2O as well as CO. They will be introduced when discussing the structural and functional aspects of the corresponding enzymes.

III. COPPER PROTEINS WITH MONONUCLEAR COPPER SITES

A. Plastocyanin

Plastocyanin consists of a single peptide and carries a single type 1 Cu center. Depending on the organism, its amino acid chain length ranges from 97 to 105, with a molecular mass of approximately 10 kD[39]. Plastocyanins are found in photosynthetic bacteria, in algae and in plant chloroplasts where they are loosely bound to the surface of the inner thylakoid membrane. They are part of the photosynthetic electron-transport chain where they connect the cytochrome b_6-f-complex and P700 of photosystem I. When copper became available following the oxygenation of the primordial atmosphere, the copper protein plastocyanin was increasingly replaced by the iron protein cytochrome c_6^2.

In 1978, Hans Freeman and colleagues[39] determined the crystal structure of the first blue copper site, that of plastocyanin, which couples photosystem I with photosystem II through electron transfer (Figure 4).

The geometric structure of the copper site was a distorted tetrahedron, not the tetragonal geometry normally observed for Cu(II). This raised the issue of the role of protein in determining the geometric structure of this active site[40]. There were also two chemically interesting ligands: a short thiolate S–Cu bond of Cys (*ca* 2.1 Å) and a long thioether S–Cu bond of Met (*ca* 2.9 Å). The coordination environment was completed by two His N–Cu bonds at *ca* 2.05 Å. Associated with this unusual geometry and ligation are the unique spectroscopic features of the blue copper site. In contrast to the weak d $\rightarrow$ d transitions of normal tetragonal Cu(II) complexes with ε *ca* 40 M^{-1} cm^{-1} at *ca* 600 nm, the blue copper site exhibited an intense absorption maximum at *ca* 600 nm (ε *ca* 5000 M^{-1} cm^{-1}). In the EPR spectrum, the hyperfine coupling (A_{II}) of the electron spin to the nuclear spin on copper was reduced by more than a factor of 2 for the blue type 1 copper relative to normal copper complexes. For details on the ground- and excited-state properties of the copper center in plastocyanin and their functional importance, refer to the article by Solomon[14], and for structural details to the article by Freeman and Guss[41].

B. Galactose Oxidase

A rather unusual copper center is present in galactose oxidase in which a Cu(II) ion is coupled to a modifed tyrosine radical (Figure 2). The fungal enzyme (*Dactylium*

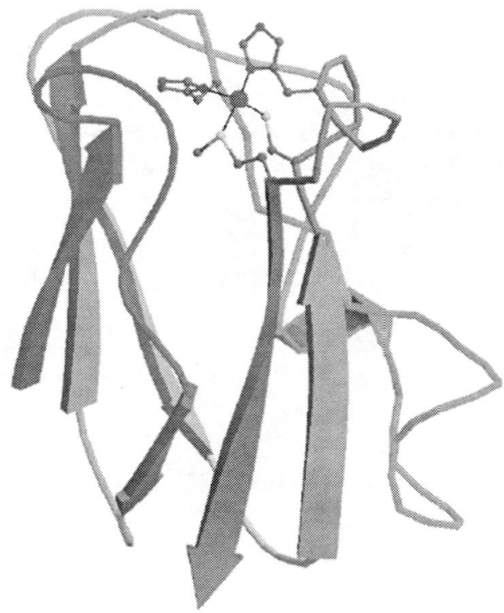

FIGURE 4. Schematic representation of the three-dimensional structure of poplar plastocyanin (PDB code 1PLC)

dendroides) consists of a single peptide chain which, in contrast to the type 2 Cu dependent amine oxidases, does not contain the typical organic cofactor TOPA (2,4,5-trihydroxyphenylalanine). Its chemistry and spectroscopy was recently reviewed[42,43]. Galactose oxidase oxidizes primary alcohols to aldehydes, with dioxygen being converted to hydrogen peroxide.

$$R - CH_2OH + O_2 \longrightarrow R - CHO + H_2O_2$$

Its substrate specificity is remarkably low. A wide range of compounds from small alcohols, such as propane-1,2-diol, to polysaccharides will be accepted. Note that galactose oxidase has a very low substrate specificity but a high stereospecificity: D-glucose and L-galactose will not be oxidized[44].

The three-dimensional structure of galactose oxidase consists almost entirely of β-strands; only a single, short α-helix (residues 328–331) occurs. The D-galactose binding site within domain 1 is perhaps the site of attachment of the enzyme to the cell walls of trees—the natural habitat of *Dactylium dendroides*. Galactose oxidase performs a two-electron reduction, although it only contains a single type 2 copper center. However, in the active state of the enzyme, the copper center is coordinated to a tyrosine (Tyr272, Figure 2) which forms a tyrosine radical producing an antiferromagnetically coupled Cu(II)-Tyr radical site. The copper center sits on the surface of domain 2, close to its sevenfold axis, and surrounded by a multitude of aromatic amino acid residues, which might help to generate and stabilize the tyrosyl radical[44]. In the crystal structure, the copper ion is coordinated by four internal ligands and one external ligand: the O from Tyr272, N from His496, His581, and a solvent molecule (H_2O) almost form a square around the metal ion; the O from Tyr495 is the fifth, axial ligand (Figure 2). The tyrosyl radical is bound to Cys228 via a thioether bond, and Trp290, through stacking, stabilizes the tyrosyl radical[45] (Figure 5).

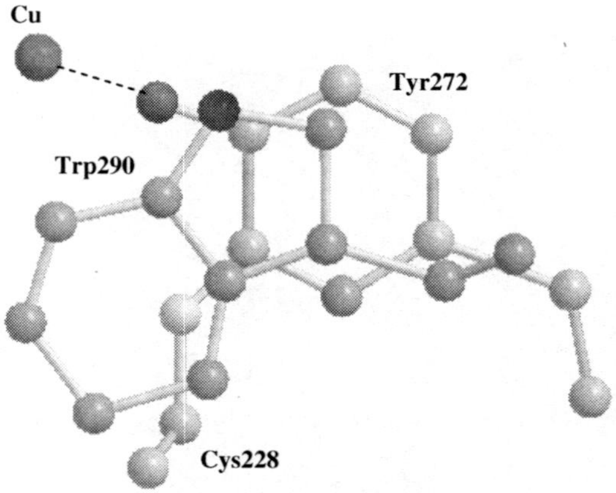

FIGURE 5. Stacking of tryptophan290 and the tyrosine272-cysteine228 thioether moiety in galactose oxidase[44, 45]

The Tyr-Cys-Trp triad participates in the generation and stabilization of the tyrosyl radical which exhibits a redox potential around + 0.40 V[46], in contrast to + 0.94 V for free tyrosine[47], probably the result of the thioether bond. The incoming substrate, R-CH_2-OH, replaces the water molecule at the active Cu(II)-Tyr radical and gets converted to its aldehyde R-CHO. In parallel, the formed Cu(I)-Tyr center undergoes reoxidation by dioxygen to produce hydrogen peroxide. For further details on galactose oxidase and its mechanism of action, refer to the article by Wilmot and colleagues[45].

C. Cu,Zn Superoxide Dismutase

Cu,Zn superoxide dismutase (CuZnSOD) has been found in all eukaryotic species. The intracellular form of CuZnSOD is a homodimer (2 × 16 KDa); each subunit carries an active center consisting of a Cu(II) and a Zn(II) ion[31] (Figure 2). The postulated physiological function of CuZnSOD is the disproportionation of the superoxide radical anion, $O_2^{\bullet -}$, to O_2 and H_2O_2. The value of the redox potential of the copper center in CuZnSOD allows it to shuttle between reduction and oxidation of $O_2^{\bullet -}$ [48].

$$\mathrm{SOD} - \mathrm{Cu}^{2+} + \mathrm{O_2}^{\bullet -} \longrightarrow \mathrm{SOD} - \mathrm{Cu}^{+} + \mathrm{O_2}$$

$$\mathrm{SOD} - \mathrm{Cu}^{+} + \mathrm{O_2}^{\bullet -} + 2\mathrm{H}^{+} \longrightarrow \mathrm{SOD} - \mathrm{Cu}^{2+} + \mathrm{H_2O_2}$$

The active center of CuZnSOD is located at the end of a channel consisting of highly conserved, charged residues directing the negatively charged substrate to the active copper site[49]. In fact, because of the electrostatic forces exerted by the protein, the substrate diffusion towards the active site is faster than its diffusion in bulk water. The active center of CuZnSOD does not only contain a type 2 copper ion but a further metal ion, Zn(II), as well (Figure 2). The Cu(II) ion is coordinated in a distorted, square-planar conformation by His44 ($N_{\delta 1}$), His46 ($N_{\varepsilon 2}$), His61 ($N_{\varepsilon 2}$) and His118 ($N_{\varepsilon 2}$). The conformation surrounding the Zn(II) ion is a distorted tetrahedron built by three histidines and an aspartate residue. The bridging imidazolate anion (His61) acts as a straight bridge between the metal ions, which are separated by 6 Å. During catalytic turnover, copper is the active redox partner of the superoxide radical; the Zn(II) ion does not appear to be directly involved in catalysis. For further details on the properties of the heterometallic copper protein superoxide dismutase and its mechanism of action, refer to the article by Bordo and colleagues[50].

IV. COPPER PROTEINS WITH MULTINUCLEAR COPPER SITES

The reduction of dioxygen is a major step in many important biological processes including respiration and ligand oxidation. Enzymes containing either iron or copper, or both elements, are the key players in this process and are the target of intensive biochemical, spectroscopic and structural studies[51,52].

A. Catechol Oxidase

Like type 3 copper proteins hemocyanin and tyrosinase, catechol oxidase carries a dinuclear copper center. The dinuclear copper centers in these proteins are strongly coupled through a bridging ligand which provides a direct mechanism for the two-electron reduction of dioxygen[52] (Figure 6). Catechol oxidase has been isolated from several plant sources; its physiological function might be related to the protection of the damaged plant from pathogens and insects. The dinuclear copper site of catechol oxidase is found in the center of a hydrophobic pocket of the enzyme. In the oxidized Cu(II)–Cu(II) form of the enzyme the two metal ions are 2.9 Å apart. In addition to six histidine ligands, a bridging ligand (most likely OH^-) appears to be present, completing the four-coordinate trigonal pyramidal coordination sphere for both Cu(II) ions, denoted CuA and CuB. An interesting feature of the dinuclear metal center in catechol oxidase is a covalent thioether bond formed between one of the histidine ligands (CuA, His109) and an adjacent cysteine sulfur atom (Cys92) (Figure 6). Such a covalent linkage had been also described for the mononuclear copper site of galactose oxidase, which had been proposed to stabilize a tyrosyl radical (Figure 5). Most recently, a catalytic mechanism based on biochemical, spectroscopic and X-ray structural data has been postulated. Overall, two molecules of

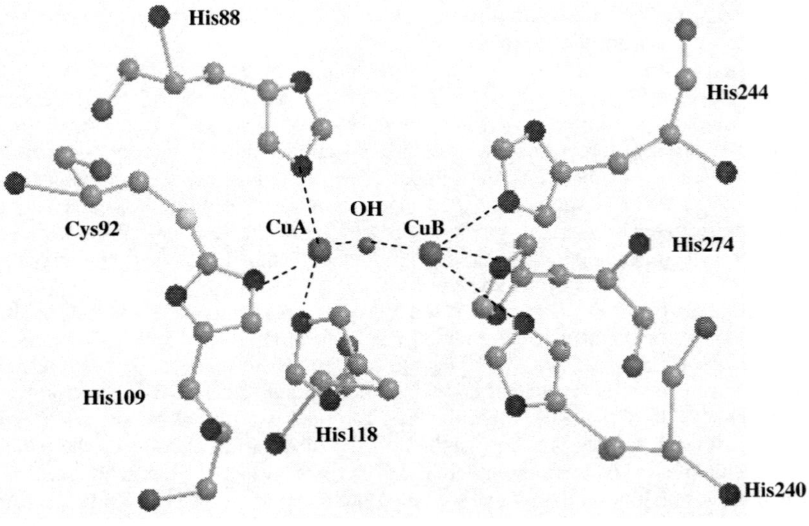

FIGURE 6. Schematic representation of the dinuclear copper site of catechol oxidase[53]

catechol are converted to the corresponding quinones, with dioxygen being reduced to water. Based on the structure of the enzyme with its inhibitor phenylurea, a monodentate binding of the catechol substrate to the CuB center seems most likely to reduce the Cu(II)–OH–Cu(II) center to its Cu(I)–Cu(I) form, which had been successfully crystallized and structurally characterized. In the next step, after release of the first quinone, O_2 and a second catechol molecule react with the Cu(I)–Cu(I) state of the enzyme forming the catechol peroxide–Cu(II)–Cu(II) adduct, which releases the second water molecule and the oxidized substrate.

For further details on structural and functional aspects of catechol oxidase, refer to the articles by Eicken, Gerdemann and Krebs[53], and Reedijk and coworkers[54].

B. Ascorbate Oxidase

Multi-copper enzymes that react with O_2 include the well-studied multi-copper oxidases. These enzymes couple the reduction of dioxygen to water with substrate oxidation, and contain a total of four copper ions arranged as a type 1 blue copper site, a type 2 copper site and a type 3 copper site similar to that in tyrosinase and catechol oxidase. The type 2 and type 3 copper ions form a trinuclear cluster for O_2 binding and activation. Among the crystallographically characterized members of this class are laccase, ascorbate oxidase and ceruloplasmin from eukaryotes, and CueO and CotA laccase from prokaryotes[13]. Recently, two new multi-copper oxidase structures have been determined, phenoxazinone synthase and ferroxidase Fet3. Phenoxazinone synthase catalyzes the last step in the biosynthesis of the antibiotic actinomycin D. Fet3 is the yeast homolog of ceruloplasmin, which catalyzes the oxidation of Fe(II) to Fe(III). These ferroxidases are a subfamily of the multi-copper oxidases that contain residues supporting a specific reactivity towards Fe^{2+} and play a vital role in iron metabolism in bacteria, algae, fungi and mammals. Additional CotA structures have also been described recently and mechanistic hypotheses related to these structures have been reviewed[51,55].

The functional unit of the classical blue oxidases, e.g. laccase, ascorbate oxidase and ceruloplasmin, contains one or more type 1 copper centers and a trinuclear center which, formally, consists of a type 2 and two type 3 copper atoms. The type 1 center usually serves as the electron transfer site, while the trinuclear copper site functions as the catalytically active center, converting dioxygen to water.

Ascorbate oxidase occurs mainly in higher plants. Its physiological function is still not fully understood; most likely it plays a role in ripening, growth control or protection[56,57]. Ascorbate oxidase is a glycosylated homodimeric enzyme (2×70 kDa). As in laccase, the subunits consist of three domains and harbor 4 copper ions. The type 1 copper center is located in domain 3; the trinuclear copper center is bound between domains 1 and 3 (Figures 3 and 7). The subunits are glycosylated on their asparagine 92 residues. Ascorbate, the physiological reductant, binds on the surface of the enzyme, close to the type 1 copper center. The trinuclear catalytic copper center is embedded inside the enzyme and accessible by two channels of different sizes.

The type 2 and type 3 centers form the trinuclear copper cluster, the active site for the reduction of dioxygen to water, which has been strongly supported by X-ray crystallography for fungal laccase, ceruloplasmin and ascorbate oxidase. The distant type 1 Cu center is linked to the trinuclear copper cluster through the Cys-His electron-transfer pathway. The T1Hg derivative, where the type 1 copper is replaced by the redox inactive Hg^{2+}, has allowed one to gain an insight into the electronic structure of the trinuclear copper-cluster site and its interaction with dioxygen. For further structural and mechanistic information on reaction intermediates of the reduction of O_2 to H_2O, refer to the reviews by Rosenzweig and Sazinsky[13], Lindley and coworkers[51], Solomon and coworkers[52] and Messerschmidt[58].

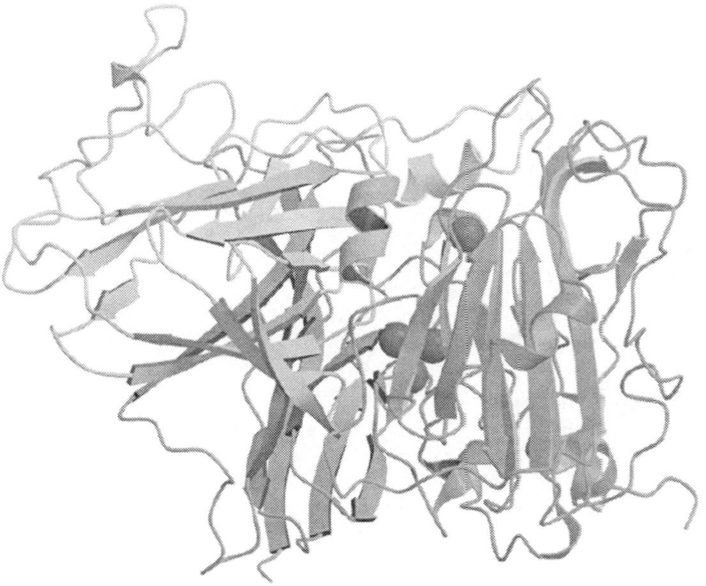

FIGURE 7. Schematic representation of the three-dimensional structure of ascorbate oxidase, showing the monomer with the mononuclear type 1 Cu and the trinuclear copper site (PDB code 1AOZ)

C. Nitrous Oxide Reductase

Nitrous oxide (N_2O) is a potent greenhouse gas that has been steadily on the rise since the beginning of industrialization. It is an obligatory metabolite of denitrifying bacteria, and some production of N_2O is also found in nitrifying and methanotrophic bacteria. In denitrification, N_2O is converted to dinitrogen and water, catalyzed by the multi-copper enzyme nitrous oxide reductase. This respiratory process provides the cell with an electron sink for anaerobic growth of bacteria such as *Pseudomonas stutzeri* or *Paracoccus denitrificans*:

$$N_2O + 2H^+ + 2e^- \longrightarrow N_2 + H_2O \quad E^{\circ\prime} = +1.35\ \text{V};$$
$$\Delta G^{\circ\prime} = -339.5\ \text{KJ mol}^{-1};\ (\text{pH } 7.0)$$

The Cu centers in nitrous oxide reductase are both novel structures among the Cu proteins: the mixed-valent dinuclear Cu_A and the tetranuclear Cu_Z center. A modified nitrous oxide reductase exists in *Wolinella succinogenes* which carries a covalently attached *c*-type cytochrome in addition to the two copper centers[59].

The nitrous oxide reductases isolated so far are homodimers that contain two multinuclear Cu centers per subunit: (i) the Cu_A mixed-valent electron transfer site with two bridging cysteine ligands forming a highly spin-delocalized Cu_2S_2 rhomb, and (ii) the catalytic tetranuclear Cu_Z site, which represents the first biologically active $Cu{-}S^{2-}$ cluster known. In Cu_Z, the Cu atoms are coordinated solely to imidazole nitrogen from seven histidine residues (Figure 8). As Cu usually adopts the oxidation states +1 and +2 in biological systems, in principle three different oxidation and resulting spin states can arise for Cu_A, and five different states for Cu_Z (Figure 8). For the Cu_A center only the $S = 1/2$ and $S = 0$ spin states have been observed, whereas for Cu_Z the states $[Cu_4S]^{4+}$, $S = 0$, 1

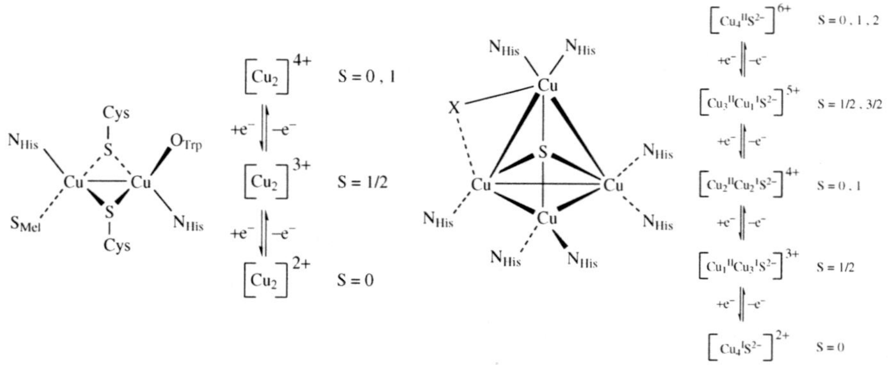

FIGURE 8. Schematic representation of the mixed-valent Cu_A center (left) and Cu_Z center (right) in nitrous oxide reductase

and $[Cu_4S]^{3+}$, $S = 1/2$, and the colorless super-reduced state $[Cu_4S]^{2+}$, $S = 0$, have been characterized by various spectroscopic techniques including multifrequency electron paramagnetic resonance (EPR), circular dichroism (CD), magnetic circular dichroism (MCD) and extended X-ray absorption fine structure (EXAFS) spectroscopy[60–65].

Almost two decades after the identification of nitrous oxide reductase as a copper enzyme[66], the first X-ray structure was reported[67]. Whereas the Cu_A center was anticipated accurately from spectroscopic studies, the structure of the Cu_Z site was unexpected. It comprises four Cu ions arranged in a novel 'butterfly' cluster solely coordinated by histidine residues. The structure of nitrous oxide reductase therefore marks the discovery of a new type of biological center[68]. In the first crystal structure of the enzyme from *Ma. hydrocarbonoclasticus*, in addition to the seven histidine residues bound to the Cu_Z site, three exogenous ligands were assigned as being either water or a hydroxide anion. At that time, a sulfur-containing ligand was not considered, which made it difficult to reconcile the structure of the Cu_Z center with the spectroscopic data that strongly suggested sulfur ligation. However, this picture changed rapidly. Elemental analysis and resonance Raman spectroscopy of isotopically labelled nitrous oxide reductase conclusively demonstrated the presence of one acid-labile sulfur ligand[62, 69]. Furthermore, the structure of the enzyme from *Pa. denitrificans* (1.6 Å resolution) became available[70] and confirmed the existence of one sulfide at the Cu_Z site. Each nitrous oxide reductase monomer is composed of two distinct domains: the *C*-terminal domain with a classical cupredoxin fold (Figure 9), carrying the mixed-valent $[Cu_A(1.5+)\ldots Cu_A(1.5+)]$ electron transfer center, and the *N*-terminal domain, a β-propeller structure, which hosts the catalytic, μ_4-sulfide bridged Cu_Z site. The *N*-terminal domain adopts a seven-bladed β-propeller homologous to other enzymes, such as methylamine dehydrogenase or galactose oxidase. The shortest Cu_A–Cu_Z distance within a monomer is *ca* 40 Å, which is too long for an efficient electron transfer between the two sites. On the other hand, the corresponding intermonomeric distance of 10 Å is well-suited for fast electron transfer within a protein[71] (Figure 9).

The Cu_Z site comprises four copper ions, Cu_I to Cu_{IV}, arranged in a tetranuclear μ_4-sulfide bridged Cu cluster solely coordinated by histidine residues (Figure 8). It appears that Cu_I is the predominantly oxidized Cu center, which is consistent with Cu_I having a four-coordinate structure and the other three Cu atoms lower coordination numbers. Thus, the Cu_I center sits roughly above a plane built by the three Cu centers Cu_{II}, Cu_{III} and

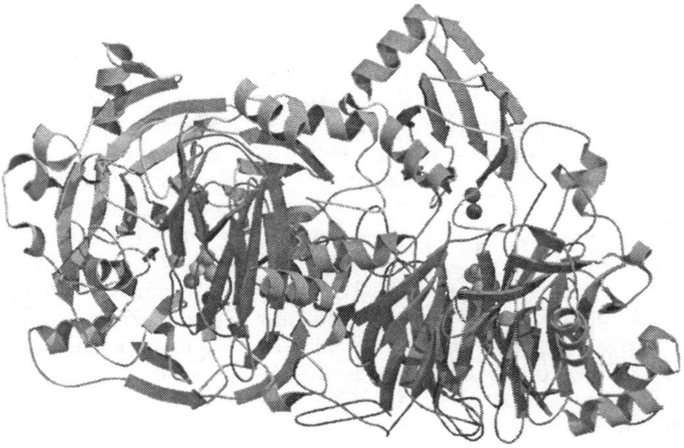

FIGURE 9. Schematic representation of the three-dimensional structure of nitrous oxide reductase from *Paracoccus denitrificans*, showing the head-to-tail dimer with the mixed-valent Cu_A (purple) and the tetranuclear Cu_Z (yellow) sites (PDB code 1FWX)

Cu_{IV}, and the inorganic sulfur[64]. Biomimetic complexes, featuring this structural unit with Cu in higher oxidation levels, have not been reported yet[15].

The enzyme from *Pseudomonas stutzeri* could be activated by base treatment[60], or incubation with CO[61]. One might speculate that at higher pH, or with CO, the Cu_I-Cu_{IV} edge of the cluster will get altered; the edge is thought to bind the substrate (Figure 8). Furthermore, nitrous oxide reductase sources will be activated *in vitro* by reduction with the redox dyes methyl viologen or benzyl viologen[59], indicating that the catalytically relevant form of Cu_Z is the fully reduced $[Cu_4S]^{2+}$, $S = 0$ state. Incubation of the reductively activated nitrous oxide reductase from *Achromobacter cycloclastes* with N_2O-saturated buffer led to a partially oxidized enzyme according to optical and EPR spectra. The activated form converts N_2O to N_2 and thus represents the first report of substrate reduction *in vitro*[72]. The results of computational investigations suggest that in the lowest energy structure of the Cu_Z-N_2O complex, N_2O binds at the Cu_I-Cu_{IV} edge in a bent, μ-1,3 bridging mode with the terminal N atom coordinating to Cu_I[63] (Figure 8). By $Cu \rightarrow N_2O$ π^* backbonding interaction the N$-$O bond will be weakened significantly, thus facilitating the direct N$-$O bond cleavage via simultaneous transfer of two electrons from Cu_Z to the μ-1,3 bridged N_2O.

D. Carbon Monoxide Dehydrogenase

Carbon monoxide dehydrogenases of aerobic or anaerobic microorganisms, which represent the essential catalyst in the global biogeochemical cycle of atmospheric carbon monoxide (CO), catalyze the oxidation of CO to carbon dioxide, or the reverse reaction:

$$CO + H_2O \longrightarrow CO_2 + 2H^+ + 2e^-$$

The annual removal of CO from the lower atmosphere and earth by these microorganisms has been estimated to be *ca* 10^8 tons[37]. The CO dehydrogenase isolated from the aerobe *Oligotropha carboxidovorans* is a complex multi-site Mo- and Cu-containing iron$-$sulfur flavoprotein. It carries a dinuclear heterometal [CuSMo(=O)OH] center at the active site. The cluster is coordinated through interactions of the Mo with the dithiolate pyran ring of molybdopterin cytosine dinucleotide and of the Cu with the thiolate sulfur of Cys388. Earlier, the presence of a SeH-group in place of copper had been reported for the active-site structure which could not be confirmed[73]. The three-dimensional structure of CO dehydrogenase with the inhibitor *n*-butyl isocyanide bound led to a model for the catalytic mechanism of CO oxidation which involves a thiocarbonate-like intermediate state. The dinuclear [CuSMo(=O)OH] cluster of CO dehydrogenase establishes a previously uncharacterized class of dinuclear molybdoenzymes containing the pterin cofactor. The μ-sulfido bridge between molybdenum and copper as well as the copper are cyanolyzable and react with cyanide, yielding thiocyanate (NCS^-) and, most likely, a Cu cyanide complex[37]. More recently, a CO dehydrogenase from an anaerobic microorganism, *Carboxydothermus hydrogenoformans*, has been structurally characterized in different states[74]. In this case, catalysis is performed by a complex Ni-, Fe- and S-containing metal center called cluster C. In a reduced state, exogenous CO_2 is bound and reductively activated by cluster C. In the intermediate structure, CO_2 acts as a bridging ligand between Ni and the asymmetrically coordinated Fe, where it completes the square-planar coordination of the Ni ion.

V. OUTLOOK

Copper along with iron active sites clearly dominate the field of biological oxygen chemistry and play important roles in homogeneous and heterogeneous catalysis[52]. Recently, a role was assigned to copper in the biosynthesis of the molybdopterin cofactor which binds both molybdenum and tungsten via its dithiolene sulfur atoms[75]. Note that for the reduction of nitrite (NO_2^-) to nitric oxide (NO), a key step within the important denitrification pathway as part of the global nitrogen cycle, microorganisms can either use the copper-dependent nitrite reductase (one type 1 and one type 2 copper), or the iron-dependent nitrite reductase (one heme *c* and one heme *d*)[76]. Two molecules of NO will be transformed to nitrous oxide (N_2O) by NO reductase, a complex membrane-bound iron enzyme, followed by the conversion of N_2O to dinitrogen by the copper enzyme nitrous oxide reductase with the mixed-valent Cu_A electron transfer center and the tetranuclear Cu_Z sulfide cluster[59] (Figure 8).

Over the past decade, a complicated network of metal-specific trafficking proteins has been identified. These proteins exhibit metal binding motifs with unprecedented coordination chemistry tailored to their function[77]. For example, the CysXXCys sequence, found in cytosolic copper chaperones and trafficking proteins, provides for facile Cu(I) transfer via low-coordination-number anionic intermediates[78]. Extracellular methionine-rich motifs seem to play an important role in copper trafficking factors, such as the CusF protein. CusF is a component of the copper and silver efflux machinery used by *Escherichia coli* to tolerate elevated and otherwise toxic concentrations of these metal ions. This protein uses a new metal recognition site wherein Cu(I) is tetragonally displaced from a Met2His ligand plane toward a conserved tryptophan residue[79]. Spectroscopic studies demonstrate that both thioether ligation and strong cation-π interactions with tryptophan stabilize metal binding. Molecular recognition in this methionine-rich site is quite different from the Cu(I) coordination chemistry in other copper chaperones and metalloregulatory proteins. Thus, this methionine-rich site carries no negatively charged ligands, and Cu(I) binding leads to a cationic coordination site, in contrast to the CysXXCys recognition sequence, which confer an overall anionic character to the coordination site.

In many methanotrophic bacteria—these are aerobically living microorganisms that oxidize CH_4 to methanol for carbon and energy and play a major role in the global carbon cycle—copper requirements can be up to fourfold higher than iron requirements. Thus, copper plays a central role in metabolism, regulating expression of two types of methane monooxygenases, which convert methane to methanol, a soluble iron-dependent methane monooxygenase and a copper-dependent particulate methane monooxygenase. Recently, evidence was presented for a small fluorescent chromopeptide called methanobactin[80]. Methanobactin is a copper transport molecule analogous to the siderophores that mediate iron transport into many cells. It is composed of a tetrapeptide, a tripeptide, and several unusual moieties, including two 4-thionyl-5-hydroxyimidazole chromophores that coordinate the copper, a pyrrolidine that confers a bend in the overall chain and an amino-terminal isopropyl ester group. The copper coordination environment includes a dual nitrogen- and sulfur-donating system derived from the thionyl imidazolate moieties. Obviously, structural elucidation of methanobactin has broad implications in terms of organocopper chemistry, biological methane oxidation and global carbon cycling[80]. In this context, the crystal structure of the integral membrane copper protein methane monoxygenase from *Methylococcus capsulatus* had been determined (2.8 Å), and both copper and zinc sites had been structurally assigned[81]. However, despite the efforts of several research groups, definitive identification of the active site of particulate methane monooxygenase has been elusive, and the involvement of a putative di-iron active site, as identified earlier in soluble methane monooxygenase, cannot be excluded[82].

VI. ACKNOWLEDGMENTS

This work was supported by DGAPA-UNAM (R.P. 1N210108, M.E.S.T.) and Deutsche Forschungsgemeinschaft (DFG Kr 451, P.M.H.K).

VII. REFERENCES

1. R. J. P. Williams and J. J. R. da Silva Fraústo, *The Natural Selection of the Chemical Elements*, Oxford University Press, 1996.
2. B. Abolmaali, H. V. Taylor and U. Weser, *Struct. Bond.*, **91**, 91 (1998).
3. W. Kaim and J. Rall, *Angew. Chem., Int. Ed. Engl.*, **35**, 43 (1996).
4. S. Puig and D. J. Thiele, *Curr. Opin. Chem. Biol.*, **6**, 171 (2002).
5. M. J. Sutcliffe, *Structure*, **12**, 517 (2004).
6. I. Bertini, A. Sigel and H. Sigel, *Handbook on Metalloproteins*, M. Dekker Inc., New York, 2001.
7. A. Messerschmidt, R. Huber, T. Poulos and K. Wieghardt, *Handbook of Metalloproteins*, Wiley-VCH Verlag GmbH, Weinheim, 2001.
8. R. H. Holm, P. Kennepohl and E. I. Solomon, *Chem. Rev.*, **96**, 2239 (1996).
9. R. H. Holm and E. I. Solomon, *Chem. Rev.*, **104**, 347 (2004).
10. K. A. Koch, M. M. O Peña and D. J. Thiele, *Chem. Biol.*, **4**, 549 (1997).
11. P. Chen and E. I. Solomon, *Proc. Natl. Acad. Sci. U.S.A.*, **101**, 13105 (2004).
12. C. Dennison, *Coord. Chem. Rev.*, **249**, 3025 (2005).
13. A. C. Rosenzweig and M. H. Sazinsky, *Curr. Opin. Struct. Biol.*, **16**, 729 (2006).
14. E. I. Solomon, *Inorg. Chem.*, **45**, 8012 (2006).
15. W. B. Tolman, *J. Biol. Inorg. Chem.*, **11**, 261 (2006).
16. E. I. Solomon, R. Sarangi, J. S. Woertink, A. J. Augustine, J. Yoon and S. A. Ghosh, *Acc. Chem. Res.*, **40**, 581 (2007).
17. L. Quintanar, C. Stoj, A. B. Taylor, P. J. Hart, D. J. Kosman and E. I. Solomon, *Acc. Chem. Res.*, **40**, 445 (2007).
18. B.-E. Kim, T. Nevitt and D. J. Thiele, *Nature Chem. Biol.*, **4**, 176 (2008).
19. R. G. Pearson, *J. Am. Chem. Soc.*, **85**, 3533 (1963).
20. M. Brischwein, B. Scharf, M. Engelhard and W. Mäntele, *Biochemistry*, **32**, 13710 (1993).
21. B. Reinhammar, *Biochim. Biophys. Acta*, **275**, 245 (1972).
22. R. Marcus and N. Sutin, *Biochim. Biophys. Acta*, **811**, 265 (1985).
23. R. Malkin and B. G. Malmström, *Adv. Enzymol.*, **33**, 177 (1970).
24. E. T. Adman, *Adv. Prot. Chem.*, **42**, 145 (1991).
25. C. A. Collyer, J. M. Guss and H. C. Freeman, *J. Mol. Biol.*, **211**, 617 (1990).

26. H. Nar, A. Messerschmidt, R. Huber, M. Van de Kamp and G. W. Canters, *J. Mol. Biol.*, **221**, 765 (1991).
27. E. N. Baker, *J. Mol. Biol.*, **203**, 1071 (1988).
28. A. M. Nersissian, P. J. Hart, Z. B. Mehrabian, R. M. Nalbandyan, J. Peisach, R. G. Herrmann, D. Eisenberg and J. S. Valentine, *J. Inorg. Biochem.*, **64**, 93 (1996).
29. I. Zaitseva, A. Zaitsev, G. Card, K. Moshkov, B. Bax, A. Ralph and P. Lindley, *J. Biol. Inorg. Chem.*, **1**, 15 (1996).
30. S. T. Prigge, B. A. Eipper, R. E. Mains and L. M. Amzel, *Science*, **304**, 864 (2004).
31. J. A. Tainer, E. D. Getzoff, K. M. Beem, J. S. Richardson and D. C. Richardson, *J. Mol. Biol.*, **160**, 181 (1982).
32. N. Ito, S. E. V. Phillips, C. Stevens, Z. B. Ogel, J. N. McPherson, K. P. Keen, K. K. S. Yadav and P. F. Knowles, *Nature*, **350**, 87 (1991).
33. K. A. Magnus, B. Hazes, H. Tonthat, C. Bonaventura, J. Bonaventura and W. G. Hol, *Protein Struct. Funct. Genet.*, **19**, 302 (1994).
34. A. Messerschmidt, A. Rossi, R. Ladenstein, R. Huber, M. Bolognesi, G. Gatti, A. Marchesini, R. Petruzelli and A. Finazzi-Agrò, *J. Mol. Biol.*, **206**, 513 (1989).
35. P. M. H. Kroneck, W. A. Antholine, J. Riester and W. G. Zumft, *FEBS Lett.*, **242**, 70, (1988).
36. K. Brown, K. Djinovic-Carugo, T. Haltia, I. Cabrito, M. Saraste, J. J. G. Moura, I. Moura, M. Tegoni and C. Cambillau, *J. Biol. Chem.*, **275**, 41133 (2000).
37. H. Dobbek, L. Gremer, R. Kiefersauer, R. Huber and O. Meyer, *Proc. Natl. Acad. Sci. U.S.A.*, **99**, 15971 (2002).
38. T. Tsukihara, H. Aoyama, E. Yamashita, T. Tomizaki, H. Yamaguchi, K. Shinzawa-Itoh, R. Nakashima, R. Yaono and S. Yoshikawa, *Science*, **269**, 1069 (1995).
39. P. M. Colman, H. C. Freeman, J. M. Guss, M. Murata, V. A. Norris, J. A. M. Ramshaw and M. P. Venkatappa, *Nature*, **272**, 319 (1978).
40. B. G. Malmström, *Eur. J. Biochem.*, **223**, 711 (1994).
41. H. C. Freeman and J. M. Guss, in *Handbook of Metalloproteins* (Eds A. Messerschmidt, R. Huber, T. Poulos and K. Wieghardt), Wiley-VCH Verlag GmbH, Weinheim, 2001, pp. 1153–1168.
42. J. W. Whittaker, *Met. Ions Biol. Syst.*, **30**, 315 (1994).
43. I. W. C. E. Arends, P. Gamez and R. A. Sheldon, *Adv. Inorg. Chem.*, **58**, 235 (2006).
44. N. Ito, S. E. V. Phillips, K. S. S. Yadav and P. F. Knowles, *J. Mol. Biol.*, **238**, 794 (1994).
45. M. J. McPherson, M. R. Parsons, R. K. Spooner and C. M. Wilmot, in *Handbook of Metalloproteins* (Eds. A. Messerschmidt, R. Huber, T. Poulos and K. Wieghardt), Wiley-VCH Verlag GmbH, Weinheim, 2001, pp. 1272–1283.
46. G. A. Hamilton, P. K. Adolf, J. de Jersey, G. C. DuBois, G. R. Dyrkacz and R. D. Libby, *J. Am. Chem. Soc.*, **100**, 1899 (1978).
47. M. R. DeFelippis, C. P. Murthy, M. Faraggi and M. H. Klapper, *Biochemistry*, **28**, 4847 (1989).
48. H. A. Azab, L. Banci, M. Borsari, C. Luchinat, M. Sola and M. S. Viezzoli, *Inorg. Chem.*, **31**, 4649 (1992).
49. E. D. Getzoff, J. A. Tainer, P. K. Weiner, P. A. Kollmann, J. S. Richardson and D. C. Richardson, *Nature*, **306**, 287 (1983).
50. D. Bordo, A. Pesce, M. Bolognesi, M. E. Stroppolo, M. Falconi and M. Dessideri, in *Handbook of Metalloproteins* (Eds A. Messerschmidt, R. Huber, T. Poulos and K. Wieghardt), Wiley-VCH Verlag GmbH, Weinheim, 2001, pp. 1284–1300.
51. I. Bento, M. A. Carrondo and P. F. Lindley, *J. Biol. Inorg. Chem.*, **11**, 539 (2006).
52. E. I. Solomon, P. Chen, M. Metz, S.-K. Lee and A. E. Palmer, *Angew. Chem., Int. Ed.*, **40**, 4570 (2001).
53. C. Eicken, C. Gerdemann and B. Krebs, in *Handbook of Metalloproteins* (Eds A. Messerschmidt, R. Huber, T. Poulos and K. Wieghardt), Wiley-VCH Verlag GmbH, Weinheim, 2001, pp. 1319–1329.
54. I. A. Koval, P. Gamez, C. Belle, K. Selmeczib and J. Reedijk, *Chem. Soc. Rev.*, **35**, 814 (2006).
55. A. J. Terzulli and D. J. Kosman, *J. Biol. Inorg. Chem.*, **14**, 315 (2009).
56. A. Marchesini, P. Cappalletti, L. Canonica, B. Danieli and S. Tollari, *Biochim. Biophys. Acta*, **484**, 290 (1977).

57. P. M. H. Kroneck, F. A. Armstrong, H. Merkle and A. Marchesini, in *Ascorbic Acid: Chemistry, Metabolism, and Uses* (Eds. P. A. Seib and B. M. Tolbert), Advances in Chemistry Series, 200, American Chemical Society, Washington, DC, 1982, pp. 223–248.
58. A. Messerschmidt, in *Handbook of Metalloproteins* (Eds A. Messerschmidt, R. Huber, T. Poulos and K. Wieghardt), Wiley-VCH Verlag GmbH, Weinheim, 2001, pp. 1345–1358.
59. W. G. Zumft and P. M. H. Kroneck, *Adv. Microb. Physiol.*, **52**, 107 (2007).
60. C. L. Coyle, W. G. Zumft, P. M. H. Kroneck, H. Körner and W. Jakob, *Eur. J. Biochem.*, **153**, 459 (1985).
61. J. Riester, W. G. Zumft and P. M. H. Kroneck, *Eur. J. Biochem.*, **178**, 751 (1989).
62. T. Rasmussen, B. C. Berks, J. Sanders-Loehr, D. M. Dooley, W. G. Zumft and A. J. Thomson, *Biochemistry*, **39**, 12753 (2000).
63. P. Chen, S. I. Gorelsky, S. Gosh and E. I. Solomon, *Angew. Chem., Int. Ed.*, **43**, 4132 (2004).
64. V. S. Oganesyan, T. Rasmussen, S. Fairhurst and A. J. Thomson, *Dalton Trans.*, 996 (2004).
65. P. Wunsch, H. Körner, F. Neese, R. J. M. van Spanning, P. M. H. Kroneck and W. G. Zumft, *FEBS Lett.*, **579**, 4605 (2005).
66. W. G. Zumft and T. Matsubara, *FEBS Lett.*, **148**, 107 (1992).
67. K. Brown, M. Tegoni, M. Prudêncio, A. S. Pereira, S. Besson, J. J. Moura, I. Moura and C. Cambillau, *Nature Struct. Biol.*, **7**, 191 (2000).
68. A. C. Rosenzweig, *Nature Struct. Biol.*, **7**, 169 (2000).
69. M. L. Alvarez, J. Y. Ai, W. G. Zumft, J. Sanders-Loehr and D. M. Dooley, *J. Am. Chem. Soc.*, **123**, 576 (2001).
70. T. Haltia, K. Brown, M. Tegoni, C. Cambillau, M. Saraste, K. Mattila and K. Djinovic-Carugo, *Biochem. J.*, **369**, 77 (2003).
71. C. C. Page, C. C. Moser, X. Chen and P. L. Dutton, *Nature*, **402**, 47 (1999).
72. J. M. Chan, J. A. Bollinger, C. L. Grewell and D. M. Dooley, *J. Am. Chem. Soc.*, **126**, 3030 (2004).
73. H. Dobbek, L. Gremer, M. Meyer and R. Huber, *Proc. Natl. Acad. Sci. U.S.A.*, **96**, 8884 (1999).
74. J.-H. Jeoung and H. Dobbek, *Science*, **318**, 1461 (2007).
75. G. Schwarz, *Cell. Mol. Life Sci.*, **62**, 2792 (2005).
76. M. Rudolf and P. M. H. Kroneck, *Met. Ions Biol. Syst.*, **43**, 75 (2005).
77. A. C. Rosenzweig and T. V. O'Halloran, *Curr. Opin. Chem. Biol.*, **4**, 140 (2000).
78. K. J. Franz, *Nature Chem. Biol.*, **4**, 85 (2008).
79. Y. Xue, A. V. Davis, G. Balakrishnan, J. P. Stasser, B. M. Staehlin, P. Focia, T. G. Spiro, J. E. Penner-Hahn and T. V. O'Halloran, *Nature Chem. Biol.*, **4**, 107 (2008).
80. H. J. Kim, D. W. Graham, A. A. DiSpirito, M. A. Alterman, N. Galeva, C. K. Larive, D. Asunskis and P. M. A. Sherwood, *Science*, **305**, 1612 (2004).
81. A. S. Hakemian and A. C. Rosenzweig, *Annu. Rev. Biochem.*, **76**, 223 (2007).
82. A. C. Rosenzweig, *Biochem. Soc. Trans.*, **36**, 1134 (2008).

CHAPTER **9**

N-Functionalized organocuprates

R. KARL DIETER

Howard L. Hunter Laboratory, Department of Chemistry, Clemson University, Clemson, South Carolina 29634-0973, USA
Fax: +1 864 656 6613; e-mail: dieterr@clemson.edu

and

RHETT T. WATSON

Division of Natural Sciences and Engineering, University of South Carolina Upstate, 800 University Way, Spartanburg, South Carolina 29303, USA
Fax: +1 864 503 5366; e-mail: rwatson@uscupstate.edu

I. INTRODUCTION . . . 366
II. ORGANOCUPRATES WITH A N–Cu BOND . . . 366
A. Amidocuprates . . . 366
1. Non-transferable ligands and cuprate structure . . . 366
2. Chiral non-transferable ligands . . . 367
3. Transferable *N*-ligands . . . 373
B. *N*-Carbamoyl(R_T)cuprates . . . 379
III. ORGANOCUPRATES WITH AN α-*N*-SUBSTITUENT . . . 383
A. Carbamoyl, α-Imidomethyl, α-Iminomethyl, α-Formamidinyl- and α-Azoalkylcuprates . . . 383
B. α-(*N*-Carbamoyl)alkylcuprates . . . 387
C. *N*-Heterocyclic Carbene–Copper (NHC–Cu) Complexes . . . 400
D. α-*N*-Heteroarylcuprates . . . 402
IV. ORGANOCUPRATES WITH A β-*N*-SUBSTITUENT . . . 406
A. β-(*N*-Carbonyl)alkylcuprates . . . 406
1. β-Cuprio-*N*-carbamoyl reagents . . . 410
2. β-Cuprioamide reagents . . . 411
B. β-Cuprio-α-aminocarboxylates . . . 416
C. Aryl and Heteroarylcuprates with a β-*N*-Atom . . . 419
1. β-*N*-Substituted arylcuprates . . . 421
2. β-*N*-Heteroarylcuprates . . . 423

The chemistry of organocopper compounds
Edited by Z. Rappoport and I. Marek

D. Cuprates of Nitrogen Enolates 426
1. α-Cuprionitriles 426
2. α-Cuprioxazolidines 428
3. α-Cuprioazaenolates 428
V. SUMMARY 432
VI. REFERENCES 435

I. INTRODUCTION

The early development of organocopper chemistry focused upon lithium dialkylcuprates (Gilman reagents) and copper-catalyzed 1,4-conjugate additions of Grignard reagents[1]. The observation that stoichiometric quantities of Gilman reagents only transferred one of the two ligands on copper prompted a search for non-transferable ligands, and amines were found to be suitable residual ligands both for transferable ligand economy and asymmetric induction. More recently, procedures for copper-mediated amine, amide, and imide alkylation and arylation have been developed (see Chapter 18 by Penn and Gelman). The importance of alkaloids and *N*-containing compounds in natural products and medicinal chemistry led to protocols for metallation adjacent to a protected nitrogen atom and from these beginnings transmetallations from the lithium compounds to copper were investigated as a potentially effective strategy for C−C bond formation. This chapter will focus on *N*-functionalized organocuprates where the *N*-atom is directly bound to Cu or located at positions α or β to the metal. The latter reagents provide a tactical approach to forming C−C bonds at positions α or β to the nitrogen atom.

II. ORGANOCUPRATES WITH A N−Cu BOND

A. Amidocuprates

1. Non-transferable ligands and cuprate structure

Amido(organo)cuprates (i.e. R_2NCuR_TLi where R_T = a transferable ligand) were first examined as potentially useful reagents containing a non-transferable heteroatom ligand but found to be unstable[2]. Subsequent investigations revealed that amidocuprates are not formed below −50 °C, but once formed are more thermally stable than monoalkylcyanocuprates, acetylenic and phenylthio[2] mixed cuprates, or Gilman reagents prepared from a variety of Cu(I) salts[3]. Thus, conditions necessary for complete cuprate formation should always be a focus of attention. These cuprates can be prepared by either addition of R_TLi to R_2NCu or by addition of R_2NLi to R_TCu, and although the same reagents are produced the sequence may impact the reaction outcomes. Although the mixed amidocuprates are generally less thermally stable than the mixed phosphidocuprates, steric hindrance in the amine increases thermal stability[3a] as does the more coordinating solvent THF. By solvent coordination to sites on copper, β-hydride elimination is suppressed. The phosphido(R_T)cuprates gave higher yields than the amido(R_T)cuprates upon reaction with 2-cyclohexenone, 1-iodooctane and cyclohexene oxide, and both reagents displayed improvements of 10–20% upon addition of LiBr[4]. Both reagents afford comparable yields in coupling reactions with allylic halides, enones, acid chlorides, epoxides and alkyl halides, although sterically hindered enones such as isophorone give significantly higher yields with the phosphido(R_T)cuprates[3b]. Mixed amido(R_T)cuprates prepared from lithiated imidazole or pyrrole and CuCN are reasonably effective in reactions with epoxides, alkyl halides, enones and vinyl triflates, although the mixed cuprate appears to be in dynamic equilibrium with RCu(CN)Li and the lithiated heterocycle[5]. The lithium hexamethyldisilazido(*n*-Bu)cuprate displays excellent reactivity (greater than

2-thienylCu(n-BuLi)•LiCN but less than n-Bu_2CuLi•LiCN) in Et_2O, but not in THF, for coupling with 2-cyclohexenone[6]. These reagents have not received widespread use in synthesis.

Computational studies reveal that the amido ligands do not competitively transfer from copper (i.e. vinyl > Me ≫ alkynyl, cyano, heteroatom, CH_2SiMe_3) due to the greater *trans*-effect of the heteroatom ligand, and the greater affinity of the *N*-atom for the Li-cation in the cuprate cluster in the transition state[7]. While the C−N bond strength may play a minor role[7b], the cuprate cluster dominates in regulating ligand selectivity and reactivity.

Although an early NMR study hinted at multiple species in MeLi/Et_2NCu solutions, little structural information was elucidated[8]. The preference of the lithium ions in the cuprate cluster for the *N*-atoms was confirmed by NMR solution studies and X-ray structure determinations[9,10]. In Et_2O, the amidocuprates exist largely as head-to-tail dimers in equilibrium with the monomer. In THF, the dimer collapses to a monomer and then ultimately to the lithium amide dimer and RCu[9a]. The alkylcopper decomposes via β-hydride elimination if β-hydrogen atoms are available. In toluene, amidocuprates prepared from mesitylcopper undergo a Schlenk-type equilibrium to afford three mixed amido(aryl)cuprate dimers **1**–**3**, the homo arylcuprate dimer **4** and the homo amidocuprate dimer **5** (Figure 1)[10]. These structural studies provide crucial insight into enantioselectivites obtained from enantioenriched amido(alkyl)cuprates prepared with chiral amide ligands (*vide infra*).

2. Chiral non-transferable ligands

Enantioselective cuprate-mediated reactions can be effected by the use of substrates containing a chiral auxiliary[11], complexation of a dialkyl or diarylcuprate to a neutral chiral ligand[12] or by utilization of heteroatomcuprate reagents where the chirality resides in the non-transferable heteroatom ligand[11,13]. The use of (−)-sparteine by Kretchmer in the copper-catalyzed conjugate addition of Grignard reagents to enones was the first reported attempt at enantioselective organocopper reactions[14]. Early work on the use of chiral amidocuprates to effect asymmetric conjugate addition reactions has been reviewed[11]. Although high enantiomeric excesses (ee) were obtained (e.g. as high as 88%), in some instances, the ligands did not prove general for a range of substrates and transferable ligands[15]. The use of bidentate and tridentate ligands containing additional heteroatoms for coordination to the cuprate cluster was frequently examined.

A large number of enantioenriched amido ligands were examined by Bertz[16], Dieter[17], Rossiter[18] and their coworkers for asymmetric conjugate addition to enones. The studies largely involved simple cyclic ketones and ligand screening was generally performed with 2-cyclohexenone. Initial examination of proline-derived amidocuprates gave promising results[17a] that later work found difficult to replicate[17b]. Bertz and coworkers[16] examined nearly fifty chiral amines, amino alcohols and alcohols for transfer of the phenyl group, and found the vast majority to give very low to no enantioselectivity. Alkoxyamines proved superior to hydroxyamines, which were converted to the *N,O*-dianions for cuprate formation. Poor reproducibility was achieved with (−)-ephedrine, which interestingly gave higher ees in THF (50%) than in Et_2O (20%). Enantioselectivities were slightly affected by the cation (e.g. Li, Mg), added salt, solvent and temperature, and the best conditions involved use of a lithium reagent, CuI, diethyl ether and reaction at −78 °C. In some instances, reversal of the (*R*/*S*) selectivity was observed between Et_2O and THF. The best amines were (*R*)- or (*S*)-α-methylbenzylamine (**6**), (*R*)- or (*S*)-α-1-naphthylethylamine (**7**) and (4*S*,5*S*)-(+)-5-amino-2,2-dimethyl-4-phenyl-1,3-dioxane (**8**) (Figure 2, A; 30–50% ee).

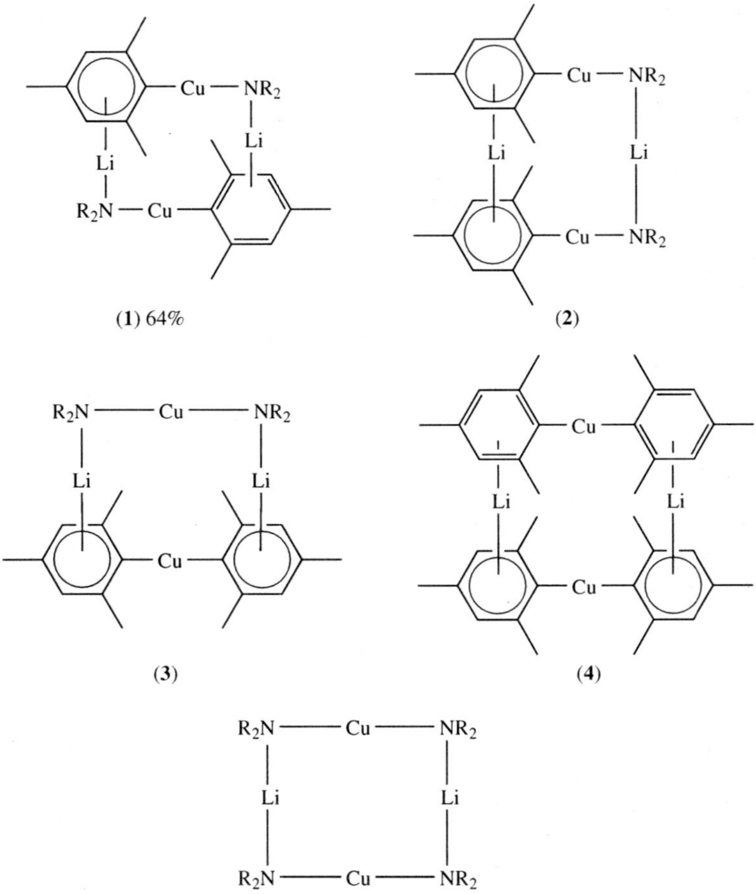

FIGURE 1. ^{1}H–^{7}LI HOESY NMR determined Schlenk equilibrium species for $[Cu_2Li_2Mes_2(N(CH_2Ph)_2)_2]$ in PhMe ($R = PhCH_2$)[10]. Reprinted with permission from reference[10] (Structural studies on a Lithium Organo-Amidocuprates in the Solid State and in Solution) Copyright Wiley-VCH Verlag GmbH & Co.

Dieter and coworkers examined several ephedrine and pseudoephedrine derived diamines and triamines[17b,c], while Rossiter and coworkers[18] explored a series of chiral diamines and triamines prepared from phenylglycine. These series of chiral amine ligands (Figure 2, B; **9–21**) all display a similar structural motif containing a phenyl group attached to the stereogenic center bearing the cuprate appended *N*-atom with one or two additional *N*-atoms present in the side chain for intramolecular coordination in the cuprate structure. Several interesting patterns emerged from these studies. Cuprates prepared from CuCN gave low to no enantioselectivity[16,18d], while CuBr generally gave lower chemical yields but higher ees than CuI[17b,c]. For the (*S*)-*N*-methyl-1-phenyl-2-(1-piperidinyl)ethanamine[(*S*)-MAPP, **9**] enantioselectivity increased with ring size (cyclopentenone < cyclohexenone < cycloheptenone > cyclooctenone: % ee, 45, 83, 96, 86) reflecting additional steric bias in the larger rings[18b].

Although no asymmetric induction was observed in THF[7b,c,18d], both bidentate and tridentate amidocuprates gave comparable enantioselectivity in Et_2O or dimethyl sulfide (DMS) (Figure 2, B; **9**, **11–14** vs. **18**, **19**, **21**). Cuprates containing ephedrine-derived tridentate ligands (i.e. **19**) afforded good optical yields, which were suppressed by N-methylation (i.e. **20**), while the reverse occurred with the phenylglycine-derived ligands (i.e. **17** vs. **18**). The sole difference in these two sets of tridentate ligands, which display complementary ee trends between the primary and secondary amides, is the presence (ephedrine series) or absence (phenyl glycine series) of a methyl group on the carbon framework. In addition, the two sets of bidentate ligands reveal a subtle but significant interplay of substitution patterns on the Cu-bound N-atom and the ligand backbone. In the ephedrine or pseudoephedrine series, either N-methylation or introduction of a piperidine ring alone increased the observed ees (compare **13–16**) and the effect is cumulative (i.e. **14**) and independent (**14** vs. **16** and **19** vs. **21**) of the relative stereochemistry in the ephedrine and pseudoephedrine ligands. The phenylglycine-derived ligands required both N-methylation and piperidine ring introduction in order to achieve effective asymmetric induction. Finally, N-methylation in the ephedrine and pseudoephedrine bidentate series inverts the sense of asymmetric induction (i.e. S for **13** and **15** and R for **14** and **16**) but not for the tridentate ligands in either series (**19–21**). Although a single quadradentate ligand (i.e. **22**) was found that gave modest ees, this strategy did not prove superior to the di- and tri-dentate ligands first examined[18f].

A series of chiral amidocuprates derived from β-aminothioethers gave poor enantioselectivity, but confirmed the change in facial selectivity for N-heterocuprates upon changing the solvent from Et_2O to toluene perhaps reflecting significant changes in cuprate solvation and aggregation[19].

While amidocuprates derived from the dianion of norephedrine or the monoanion of O-methylnorephedrine gave poor enantioselectivity in n-butylcuprate additions to chalcone, the N-silyl-O-methylnorephedrine-derived cuprates gave excellent yields and enantioselectivities (Figure 2, **23**)[20]. The β-silyl substituent greatly increased both the chemical yields and enantioselectivities in these reactions. Again no to low enantioselectivity was observed in THF.

Although the precise origin of these Cu(I) salt and ligand structural effects upon the sense and extent of asymmetric induction are difficult to interpret, a model involving a *trans* mixed chiral amido(alkyl)cuprate dimer invoked by Dieter and coworkers[17a], and elaborated on by Rossiter and coworkers[18d], is consistent with recent NMR[9] studies of chiral amido(n-Bu)cuprates suggesting that these species are dimers in Et_2O that rapidly decompose to lithium amide dimers and alkylcopper species in THF solutions[9a]. In addition, the NMR studies suggest that the two lithium cations in the cuprate cluster are in different environments[9] that should be sensitive to structural variations in the chiral ligands. Titration of a chiral amido(n-Bu)cuprate with 2-cyclohexenone reveals cleavage of the dimer to a monomer upon Cu–alkene complexation, suggesting that enantioselectivity is determined in the complexation event between the cuprate dimer and cyclohexenone[9b]. Thus dimer–enone complexation affords a strongly bound enone–cuprate monomer that undergoes reductive elimination to afford enantioenriched 2-n-butylcyclohexanone. Since tridentate alkoxy(alkyl)cuprates give good ees in THF[21] and no enantioselectivity in ether, the implication arises that alkoxy(alkyl)cuprates display different solution dynamics[17c]. It is also interesting that the cuprates derived from the bidentate alkoxides gave no enantioselectivity in THF[17c].

Grignard and organozinc reagents are generally required for reactions catalytic in copper due to the reactivity and basicity of organolithium reagents. The asymmetric 1,4-addition

A. Transfer of Ph to Cyclohexenone from Scalemic Amidocuprates:

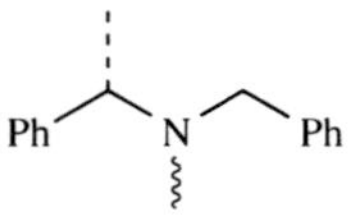

(**6**)

Et_2O 23% (40% ee) *R*

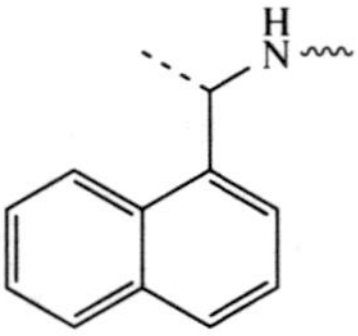

(**7**)

Et_2O 74% (30% ee) *S*

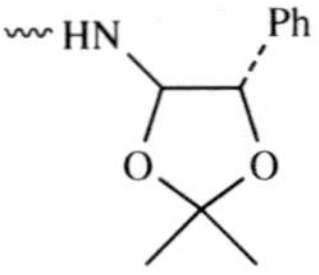

(**8**)

Et_2O 62% (50% ee) *R*
THF 26% (0% ee) -
Hexane 51% (30% ee) *S*

B. Transfer of *n*-Bu to Cyclohexenone (or Cycloalkanones Noted) from Scalemic Amidocuprates:

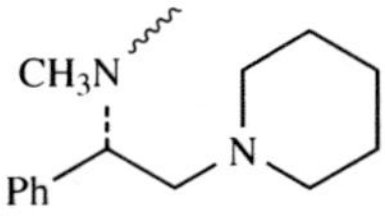

(**9**)

DMS 92% (81% ee) *S*
Et_2O 71% (88% ee) *S*

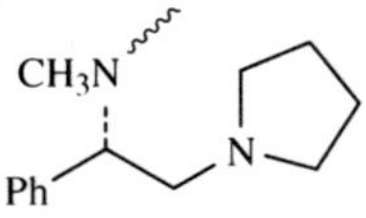

(**10**)

Et_2O 75% (37% ee) *S*

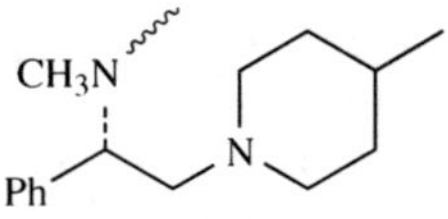

(**11**)

Et_2O 69% (84% ee) *S*

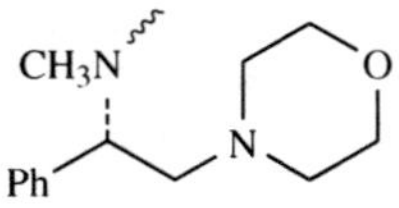

(**12**)

Et_2O 35% (83% ee) *S*
for 2-cycloheptenone

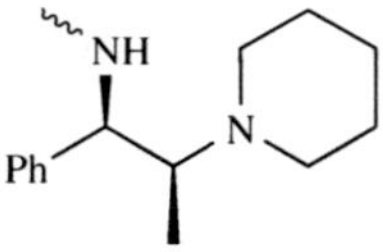

(**13**)

Et_2O 86% (78% ee) *S*

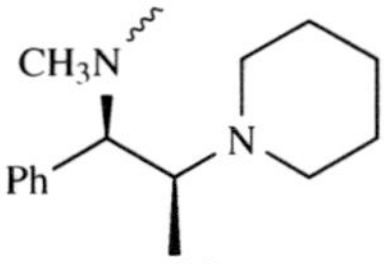

(**14**)

Et_2O 96% (82% ee) *R*

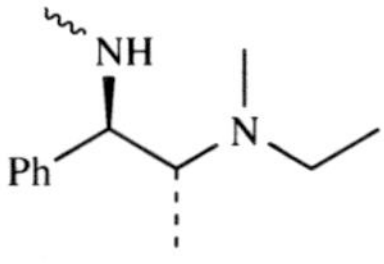

(**15**)

Et_2O 60% (30% ee) *S*

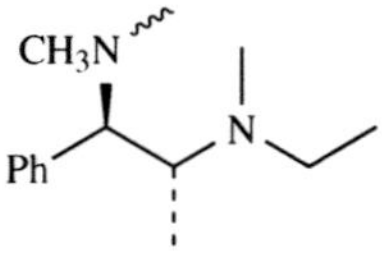

(**16**)

Et_2O 50% (53% ee) *R*

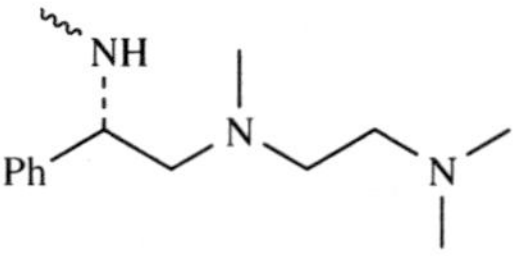

(**17**)

DMS 61% (4% ee) *S*

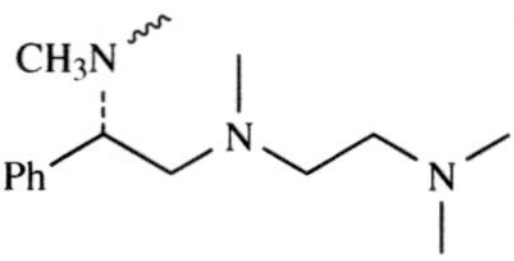

(**18**)

DMS 92% (81% ee) *S*
Et_2O 75% (55% ee) *S*

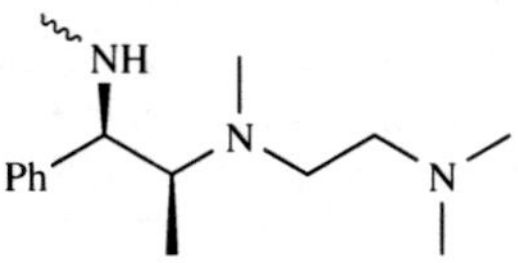

(**19**)

Et_2O 79% (70% ee) *R*

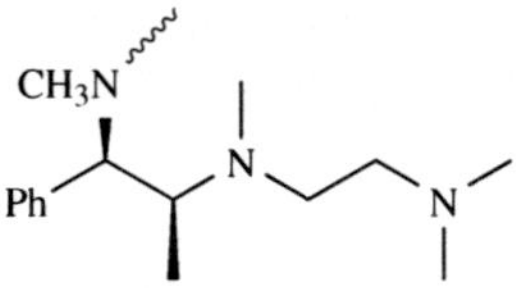

(**20**)

Et_2O 56% (10% ee) *R*

FIGURE 2. (*continued*)

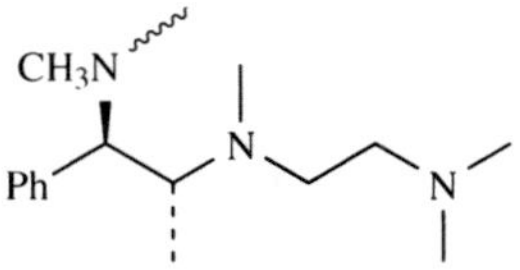

(**21**)

Et_2O 70% (66% ee) *R*

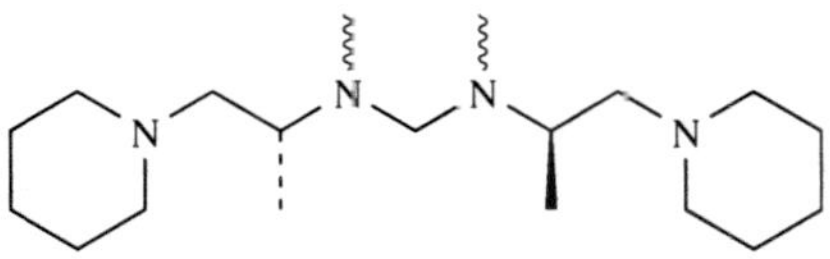

(**22**)

PhH/Et_2O 64% (77% ee) *R*

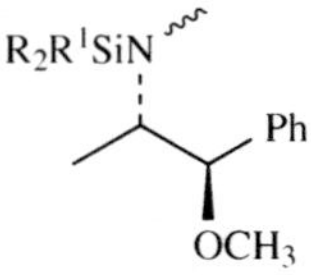

(**23**)

DMS 92% (96% ee) *S*
Et_2O 99% (99.6% ee) *S*

FIGURE 2. Chemical yields and enantioselectivities in the reaction of chiral amido(*n*-Bu)CuLi with 2-cyclohexenone (unless otherwise noted)[16, 17b,c, 18d,f, 20]

of *n*-BuMgCl to 2-cyclohexenone is catalyzed by an aminotropone iminate[22]. Low enantioselectivities are obtained in THF (20%) and optimal conditions (74% ee) require the addition of HMPA and TMSCl, which favor more rapid silylation of the carbonyl oxygen in the rate-determining step involving reductive elimination[22b]. This is consistent with kinetic isotope measurements[23]. This catalyst gives poor ees with other Grignard reagents (i.e. RMgX, R = Me, Et, Ph, vinyl) and/or enones. Chiral amidocuprates derived from diphenyl- or dicyclohexylphosphino-substituted pyrrolidinones have been used for the catalytic asymmetric conjugate addition of butylmagnesium chloride and diethylzinc to 2-cyclohexenone affording good chemical yields, but poor to modest enantioselectivities in Et_2O[24]. Low ees are obtained for the Grignard reagent in other solvents (e.g. hexane, PhMe, CH_2Cl_2, THF or MeCN) and when the cuprate is prepared from Cu(I) salts [e.g. CuCl, CuCN, $Cu(OTf)_2$] other than CuI. The zinc reagents offered no improvements under these conditions.

The highest enantioselectivities observed for a copper-catalyzed 1,4-addition of diethylzinc to acyclic enones (83–98% ee) were obtained with a 2-amino-2′-hydroxy-1,1′-binaphthyl derivative and $Cu(OTf)_2$ in non-polar solvents such as PhMe or PhMe/chloroalkanes[25]. Much lower ees are obtained in more polar solvents such as THF. Given the use of $Cu(OTf)_2$ and the reaction conditions (i.e. no prior deprotonation of the amide) it is unlikely that an amidocuprate reagent is generated, and the rigidity imposed by the amide linkage is likely a function of complexation between the amide of the binaphthyl ligand, the $Cu(OTf)_2$ and the zinc reagent.

Major advances in the asymmetric conjugate addition of dialkylzinc reagents to enones have been made using catalytic quantities of $Cu(OTf)_2$ and sulfonamides[26] or peptidic amides[27]. Although the sulfonamides and peptidic amides contain acidic N−H hydrogen atoms that could in principle lead to amido cuprates, the emerging picture is that covalent N−Cu complexes are not formed. Rather the sulfonamides and peptidic amides serve as templates to coordinate both the zinc reagent and the copper catalyst with both metals ultimately binding to the enone substrates in a pre-organization leading to eventual transfer of the ligand from zinc to copper followed by reductive elimination to afford the coupled product.

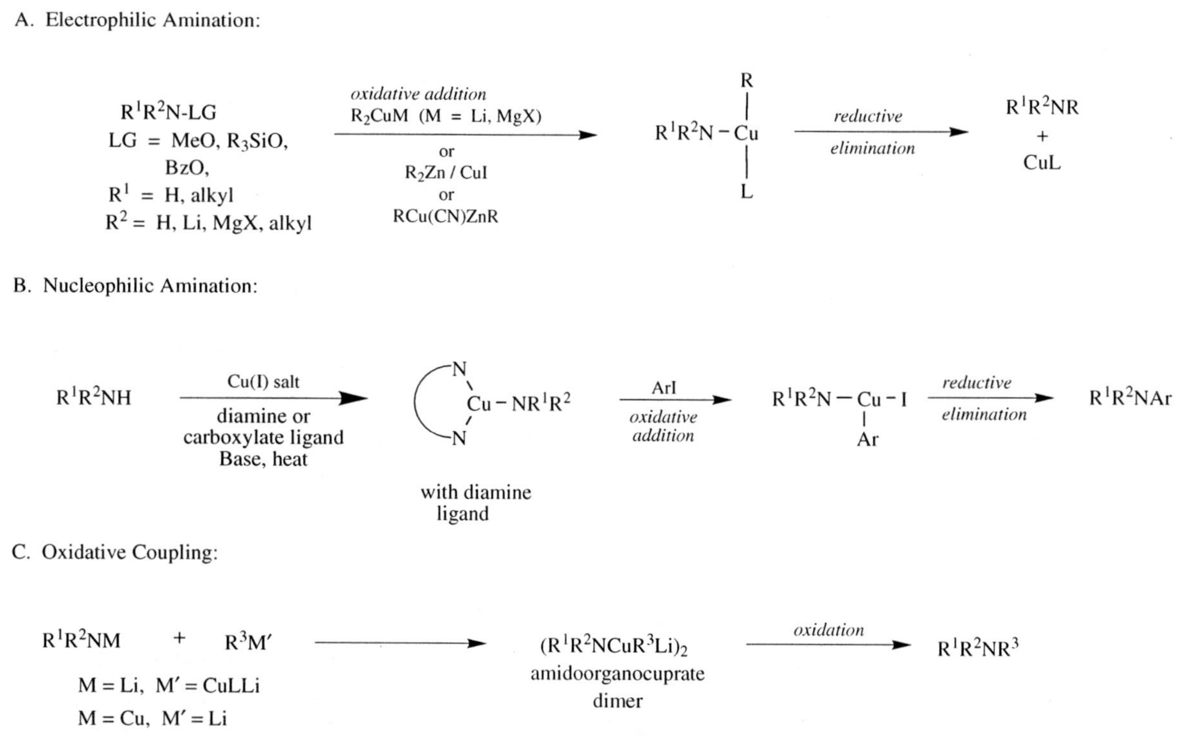

SCHEME 1. General scheme for copper-mediated amination reactions: A = Nitrogen electrophile, B = Carbon electrophile, C = Oxidative coupling

3. Transferable N-ligands

Although *N*-heteroatoms generally function as non-transferable ligands, reaction conditions have been developed for copper-mediated amine alkylation and arylation. These copper-mediated amination[28–30] reactions generally involve the intermediacy of mixed *N*-heteroatom(R)copper(III) species, and can be classified as electrophilic[28] or nucleophilic[29] depending upon whether the leaving group is on the *N*- or *C*-centered substrate, respectively (Scheme 1). Electrophilic aminations involve reaction of an organocopper reagent with an electrophilic *N*-center to afford a mixed amido(R)copper (III) complex, which affords the alkylated amine upon reductive elimination (A, Scheme 1). Alternatively, reaction of a nucleophilic R_2NCu species (e.g. amido, amidate, imidoyl etc.) with an aryl halide affords the mixed $R_2NCuArL$ Cu(III) species that gives the arylated amine upon reductive elimination (B, Scheme 1). The copper-mediated coupling of an aryl amine and an aryl halide is known as the Goldberg reaction[31] and is discussed by Penn and Gelman in this volume. A third mechanistic pathway involves oxidative coupling of amidocuprates (C, Scheme 1)[30].

Alkyllithium reagents react with lithiated alkoxyamines to afford alkylated primary or secondary amines[32a] isolated as the amides, and the chemical yields in this amine synthesis decrease along the organometallic series RLi > R_2CuLi > RMgBr > R_3ZnLi (equation 1)[32b]. The endocyclic restriction test (i.e. use of a substrate with potential intramolecular pathways) coupled with cross-over experiments revealed that the reaction followed an S_N2-mechanistic pathway involving displacement of the alkoxy leaving group by the organometallic reagent. For copper reagents, this substitution step is followed by a reductive elimination. Reaction of lithium *N-tert*-butyl-*N*-tosyloxycarbamate and copper enolates provided for the preparation of *N*-Boc protected α-amino imides and α-amino phosphonic esters (equation 2)[33a], while utilization of lithium allyl-*N*-tosyloxycarbamate affords allyloxycarbonyl protected α-aminophosphonic esters[33b] or allyloxycarbonyl protected aryl and heteroarylamines[33b]. These reactions involve either α-cuprioimides, α-cupriophosphonic esters or arylcopper reagents and are likely to proceed via a more reactive halocuprate reagent (e.g. $ArCuBrLi{\bullet}SMe_2$). Cuprates derived from enantioenriched amides containing a *cis*-aminoindanolyl chiral auxiliary (equation 3) afforded, with excellent diastereoselectivity (96.3 to >99% de), the *N*-Boc protected α-amino amides, which upon hydrolysis with aqueous acid afforded scalemic (i.e. enantioenriched) α-amino acids of high optical purity (89.4–98.2% ee)[33b].

$$\text{RM} \xrightarrow[\text{2. PhCOCl}]{\text{1. LiNHOMe}} \text{RNHCOPh} \qquad (1)$$

	% Yield	
M	R = *s*-Bu	R = Ph
Li	68	91
MgBr	19	54
RCuLi	58	83
Me_2ZnLi	18	-

$$RM \xrightarrow{p\text{-}MeC_6H_4SO_3N(Li)Boc} RNHBoc \qquad (2)$$

RM		% Yield
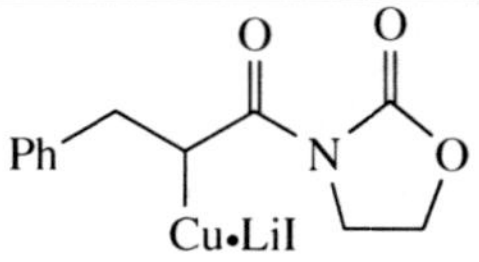		55
$R^1CHPO(OEt)_2$ (Cu•LiI)	R^1 = Me	80
	R^1 = Ph	50

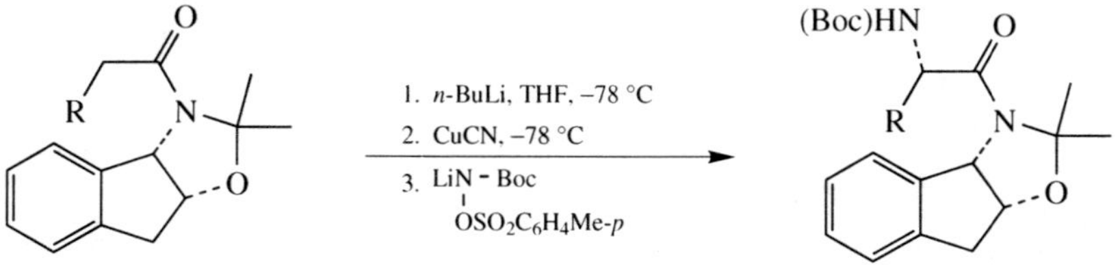

(3)

R = Me, *n*-Bu, *i*-Pr, $PhCH_2$, Ph   51–77% (de = 99%)

The methodology was extended to *N*,*O*-bis(trimethylsilyl)hydroxylamine[34a], and to *N*-alkyl(silyloxy)amines[34b] in combination with Gilman and $R_2CuLi{\bullet}LiCN$ reagents [R = aryl (79–90% and 45–88% yields, respectively), heteroaryl (58–72% and 60–70% yields, respectively), alkyl (48–80% yields)] where one ligand is presumably sacrificed to deprotonate the amine. The formation of alcohol by-products (5–18%) in the former reaction is explained by a silyl migration from oxygen to the lithiated *N*-atom where bis(trimethylsilyl)amide becomes the leaving group. Speculatively, a concerted amine deprotonation and ligand substitution was suggested[34a]. Phenylzinc reagents in the presence of catalytic quantities of Cu(I) salts react with *O*-methylhydroxylamine to afford aniline with increasing yields along the series PhZnCl < Ph_2Zn < Ph_3ZnLi[35a,b]. Lower yields were obtained with the stoichiometric zinc cuprates $Ph_2CuZnCl$ and PhCu(CN)ZnCl. A kinetic study and linear Hammett plots for the reaction of $R_2CuMgBr$ and RCu(CN)ZnR (prepared from R_2Zn and CuCN) with *O*-methylhydroxylamine are consistent with S_N2-type aminations[35c]. The reaction with $Me_3SiNHOSiMe_3$ has been applied to the synthesis of ferrocenyl amines[36]. Although the use of mixed 2-thienyl(cyano)- or *tert*-butyl(cyano) cuprates gave modest yields (39–56%), the reaction could not be scaled up to 50 mmol. The bis(ferrocenyl)cuprates derived from CuCN gave lower yields (36%), but excellent reproducibility and scale-up potential.

Nevertheless, both *N*-monoalkyl- and *N*,*N*-dialkylbenzoyloxy amines (equation 4) undergo copper-catalyzed substitution reactions with diorganozinc reagents to afford secondary and tertiary amines, respectively[37]. Grignard reagents can also be employed, but deprotonate the *N*-monoalkylbenzoyloxy amines failing to give secondary amine products[37d]. The reaction can be catalyzed by a variety of Cu(II) and Cu(I) salts (e.g. CuI, $CuCl_2$, $[CuOTf]_2C_6H_6$), and shown by the endocyclic restriction test and cross-coupling experiments to be consistent with an S_N2-mechanistic pathway[37d]. Thus, it is interesting that the reaction appears insensitive to steric hindrance about the electrophilic nitrogen center (equation 4)[37c]. While the endocyclic restriction test and cross-over experiments rule out an intramolecular oxidative addition, the insensitivity to steric hindrance appears inconsistent with a simple S_N2 pathway.

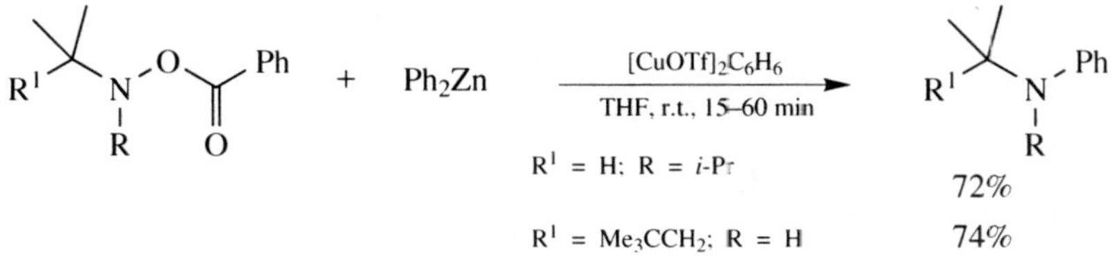

(4)

The reaction of amidocopper reagents with ICH_2ZnI affords α-aminomethylcopper intermediates, which react with allylic halides to afford homoallylic amines (equation 5)[38]. The proposed mechanism invokes an amidocopper–ICH_2ZnI complex **24** within which *N*-ligand migration with loss of the iodide leaving group affords the α-aminocopper species, which can undergo further reaction.

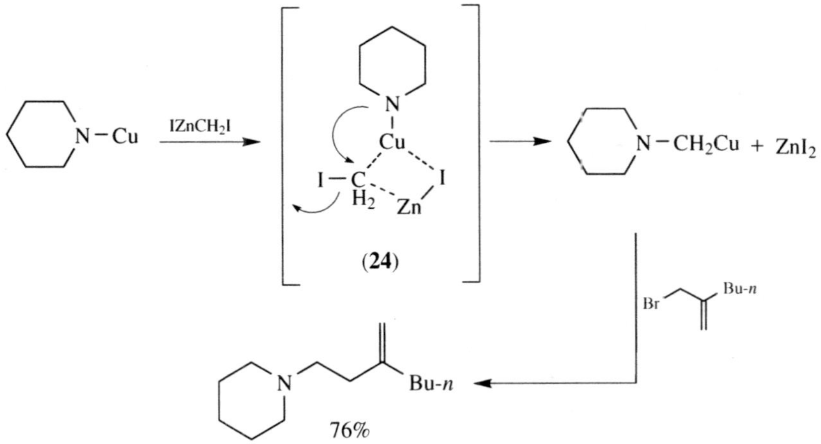

(5)

Primary amines can be prepared by reaction of alkylcopper species generated from either alkyllithium or alkyl Grignard reagents with 4,4′-bis(trifluoromethyl)benzophenone-*o*-methylsulfonyloxime, followed by hydrolysis of the resultant *N*-alkyldiarylimine[39] **26** (Scheme 2). Chemical yields are improved by addition of polar additives such as HMPA or TMEDA. Grignard or lithium reagents in the absence of copper fail to give substitution products in the polar solvents. Lithium dialkylcuprates afford the unsubstituted imine presumably arising via hydrolysis of a more stable (i.e. **27** vs. **25**) *N*-diarylmethanimino(dialkyl)copper(III) intermediate **27**. Oxidation of this imino(dialkyl) copper(III) intermediate with molecular oxygen affords the alkylated imine in low yield (39%) along with the 4,4′-bis(trifluoromethyl)benzophenone azine dimer (21%). It is speculated that alkylcopper reagents afford an $Ar_2C{=}NCuR(OSO_2OMe)$ copper(III) intermediate **25** more prone to reductive elimination to give $Ar_2C{=}NR$[39b]. It is therefore noteworthy that Grignard reagents in the presence of catalytic quantities of CuCN•2LiCl, where the reactive agent is most likely R_2CuMgX, afford good yields of primary amines. The reaction gave very low yields for arylcopper compounds and primary arylamines were prepared from aryl Grignard reagents in an Et_2O/PhMe solvent mixture[39b].

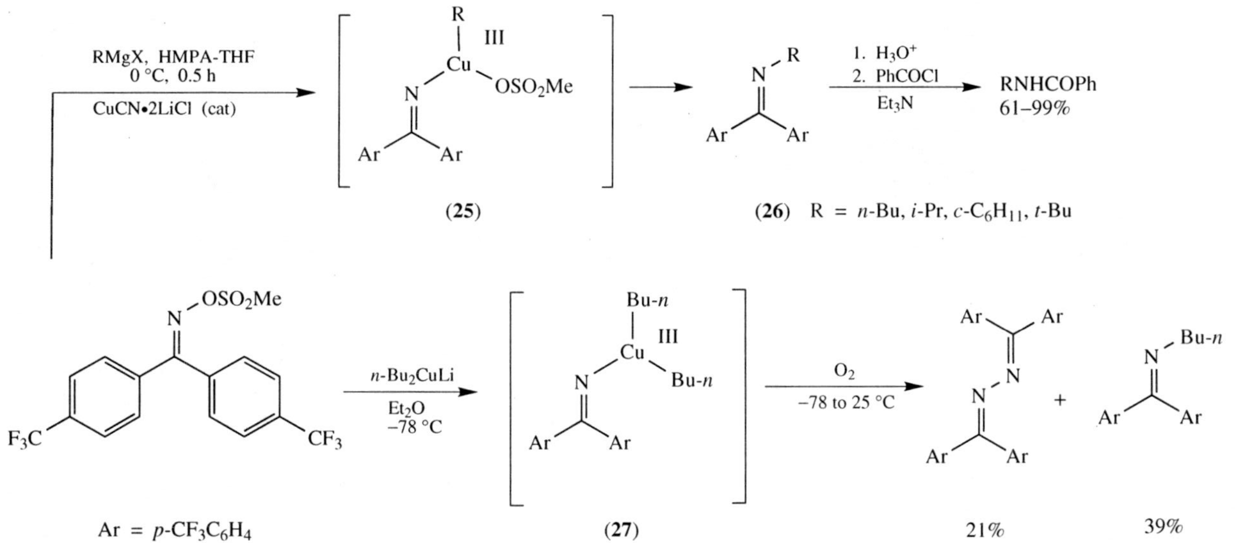

SCHEME 2. Reaction of a sulfonyl oxime with Grignard reagents/CuCN (cat) and with cuprates[39]

Acetone *O*-(2,4,6-trimethylphenylsulfonyl)oxime reacts with arylzinc reagents in the presence of catalytic amounts of CuCN to afford modest yields of *N*-arylated imines[35a]. Yields increase along the series Ph_2Zn < PhZnCl < Ph_3ZnLi and 20 mol% of CuCN affords higher yields. Yields could be increased by use of two equivalents of 1,3-dimethyl-3,4,5,6-tetrahydro-2(1*H*)-pyrimidinone (DMPU)[40]. Lower yields of product are obtained with the stoichiometric cuprate reagents. Low yields were obtained with PhLi and poor to modest yields with Gilman cuprates[35b]. Modest yields could be achieved with Grignard reagents, magnesium cuprates, zinc cuprates and lithium aryl(cyano)cuprates with the latter giving the highest yields (76%). Although modest yields could be achieved with benzylzinc reagents (56–62%), alkylzinc reagents gave low yields of *N*-alkyl imines. Competitive kinetic studies and Hammett plots gave reaction constants for the Grignard ($\rho = -2.94$; $r = 0.977$) and zinc cyanocuprate ($\rho = -0.84$; $r = 0.936$) substitution reactions on the *N*-atom[41], consistent with a decrease of charge on the aryl organometallic reagent in the transition state for substitution. The greater ρ-value for the Grignard reagents reflects the difference between the carbon-centered nucleophile in the Grignard vs. the Cu-centered nucelophile in the cuprate reagents. The authors posit an S_N2 pathway for the Grignard reagents and an oxidative–addition reductive–elimination pathway for the copper reagents.

This *N*-imination strategy has been extended to the copper-catalyzed reaction of oxime *O*-carboxylates with boronic acids and organostannanes[42] (equation 6). The reaction is catalyzed by a variety of Cu(I) and Cu(II) salts and suggested to proceed via copper oxidative addition to the N–O bond, followed by transmetallation of a Cu(III) intermediate (i.e. **28**) and subsequent reductive elimination.

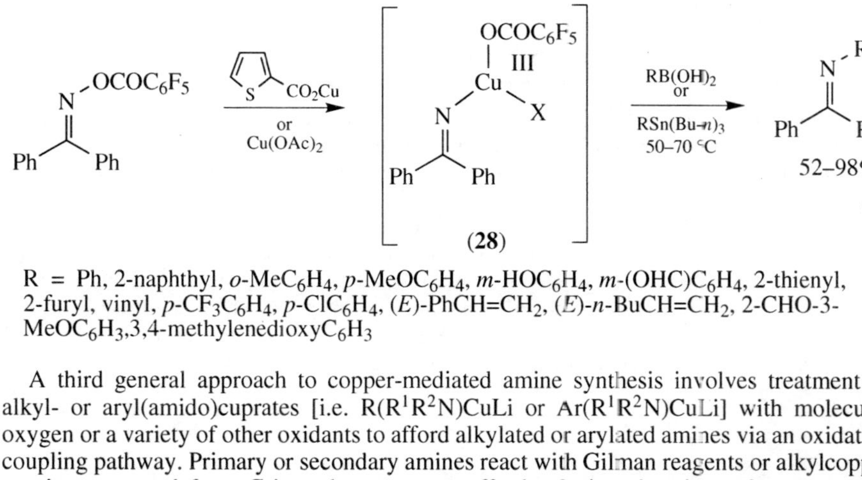

(6)

R = Ph, 2-naphthyl, *o*-MeC_6H_4, *p*-$MeOC_6H_4$, *m*-HOC_6H_4, *m*-$(OHC)C_6H_4$, 2-thienyl, 2-furyl, vinyl, *p*-$CF_3C_6H_4$, *p*-ClC_6H_4, (*E*)-PhCH=CH_2, (*E*)-*n*-BuCH=CH_2, 2-CHO-3-$MeOC_6H_3$,3,4-methylenedioxyC_6H_3

A third general approach to copper-mediated amine synthesis involves treatment of alkyl- or aryl(amido)cuprates [i.e. $R(R^1R^2N)CuLi$ or $Ar(R^1R^2N)CuLi$] with molecular oxygen or a variety of other oxidants to afford alkylated or arylated amines via an oxidative coupling pathway. Primary or secondary amines react with Gilman reagents or alkylcopper species generated from Grignard reagents to afford substituted amines after quenching with molecular oxygen[43]. Higher yields are obtained in polar solvents and with an excess of the organocopper reagent. The alternative sequence involving reaction of butylcopper with lithium butylheptylamide gave similar yields of product suggesting the intermediacy of lithium alkyl(amido)cuprates. The formation of *N*-(*o*,*o*′-diethoxyphenyl)piperidine in 51% yield when copper piperidide and the aryllithium reagent are heated at reflux in THF–hexane illustrates the robust stability of aryl(amido)cuprate reagents.

Reaction of *o*-lithiated tertiary benzamides with anilinocuprate reagents (e.g. ArRN-CuXLi: R = H, Me and X = Cl, CN, 5.0 equivalents) afforded *N*-arylanthranilamides (23–63%), which could be parlayed into a synthesis of acridones (25–95%)[44]. Alkyl-, aryl- and alkenyl(amido)cuprates prepared by either addition of lithium amides to RCu(CN)Li, amines to cyano Gilman cuprates (i.e. $R_2CuLi{\bullet}LiCN$) or by addition of cuprioamides to organolithium derivatives all gave similar yields (45–62%) of *N*-alkylated, arylated or vinylated amines upon treatment with molecular oxygen (equation 7). Variable amounts of C–C homocoupling products were observed[45]. The procedure could be adapted to lithiated hydrazines to afford *N*-alkylated or arylated hydrazines (20–60%). EPR observation of diisopropyl, diethyl or diphenyl nitroxide signals upon addition of dioxygen to lithium amidocuprate solutions suggested the formation of aminyl radicals. Oxygen-quenched solutions also displayed EPR signals for Cu(II) complexes upon warming to room temperature. Improved yields were achieved by utilization of zinc cyanocuprates and lithium amides coupled with careful selection of oxidizing agents[45b]. An oxygen/*o*-dinitrobenzene (20 mol%) combination proved effective for the zinc cuprates, while a $Cu(NO_3)_2/O_2$ oxidizing system proved more effective for the lithium cuprates[45b].

$$i\text{-}Pr_2NLi + RCu(CN)M' \xrightarrow{\text{oxidant}} i\text{-}Pr_2NR \qquad (7)$$

		% Yield		
Cuprate	Oxidant	R = *n*-Bu	PhCH=CH	Ph
RCu(CN)Li	O_2	50	45	60
RCu(CN)ZnCl·LiCl	O_2	85	-	30
RCu(CN)ZnCl·LiCl	$O_2/o\text{-}C_6H_4(NO_2)_2$ (cat)	-	60	70

Good yields of monoaryl-, diaryl- or triarylamines as well as heteroaryl amines can be prepared by oxidation of lithium(amido)cuprates with chloranil[46]. Halogen metal exchange of aryl iodides with *i*-PrMgCl (equation 8), of aryl bromides with *i*-$PrMgCl{\bullet}2LiCl$ or Directed Ortho-Metallation (DOM, equation 9) with 2,2,6,6-tetramethylpiperidin-1-ylmagnesium chloride–lithium chloride complex ($TMPMgCl{\bullet}LiCl$) generates the aryl magnesium reagents, which afford the arylcopper reagents upon addition of $CuCN{\bullet}2LiCl$ and one equivalent of $(Me_2NCH_2CH_2)_2O$. Addition of 2–3 equivalents of a lithium amide followed by chloranil oxidation affords the amines. The procedure is compatible with a range of functional groups[46a,b] including nitriles, esters, aryl iodides, aryl bromides, aryl chlorides, ethers, carbonates and a variety of heteroaryl[46c] ring systems.

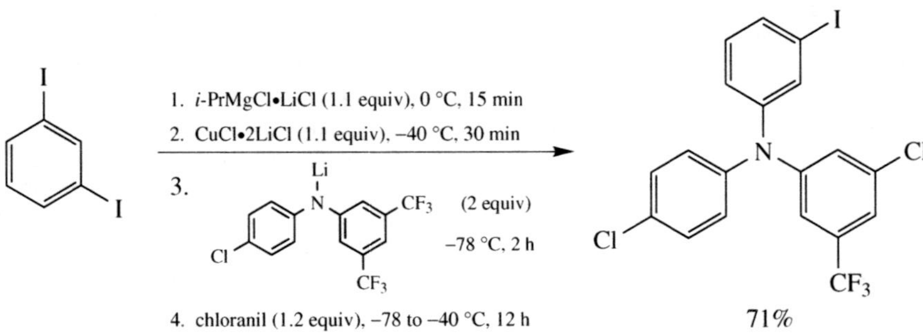

(8)

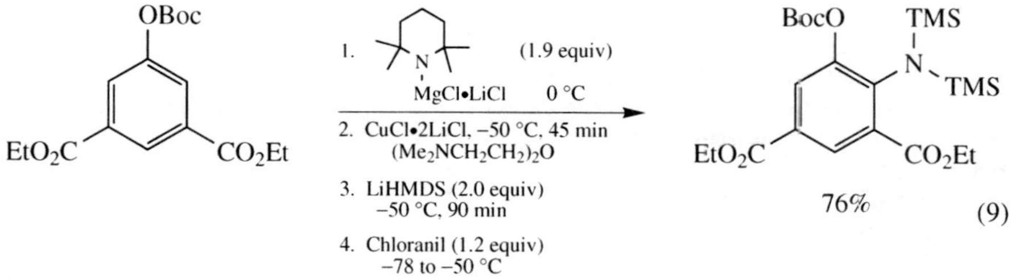

(9)

B. N-Carbamoyl(R_T)cuprates

Gallina and coworkers introduced allylic carbamates containing an acidic hydrogen on the N-atom as a directing group for *syn*-selective S_N2'-substitution reactions[47]. Although the initial procedure[47a] gave complete selectivity with three equivalents of Me_2CuLi in diethyl ether, later work[47b] revealed substantial sensitivity of the regio- and stereoselectivity to the solvent and cuprate stoichiometry. Subsequent mechanistic studies by Goering and coworkers elucidated the origin of these effects[48]. Slow addition of Me_2CuLi (3.0 equivalents) to Et_2O solutions of carbamate gave excellent *syn*-S_N2' substitution (i.e. **29**) since this allowed cuprate deprotonation of the carbamate (with formation of MeCu), N-heteroatom(Me)cuprate formation and subsequent intramolecular allylic substitution (equation 10). Rapid addition of three equivalents of Me_2CuLi in Et_2O resulted in intermolecular *anti*-S_N2'-substitution that was competitive with formation of the N-heteroatom(methyl)cuprate, where the competing pathways led to reduced selectivities. Conducting the reaction in THF accelerated carbamate deprotonation, and N-heteroatomcuprate formation and excellent *syn*-S_N2' substitution was again achieved. Alternatively, the carbamate can be deprotonated with MeLi, followed by sequential addition of CuI and then an alkyllithium to introduce the alkyl group to be transferred. The latter procedure is preferred, as it requires only one equivalent of the alkyl ligand to be introduced. Stoichiometric amounts of copper are required in order to form the amidocopper intermediate, and use of catalytic copper is likely to lead to dialkylcuprates which will react intermolecularly with *anti*-S_N2' or S_N2 substitution. This *syn*-S_N2' substitution procedure is sensitive to steric hindrance where decomposition of the cuprate is competitive with allylic substitution. A sterically hindered *trans*-α-methyl-γ-mesitylallyl system gave exclusive γ-substitution as a 97:3 mixture of E/Z alkenes in 25% yield after several days at 0 °C.

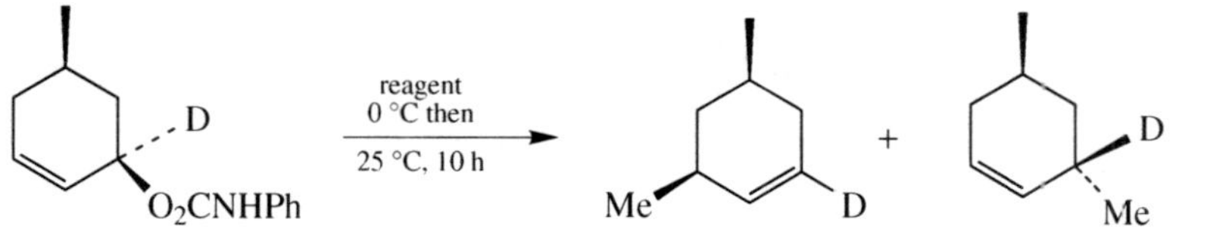

(29) (30) (10)

Reagent	Conditions	**29**:**30**	*cis*:*trans*	% Yield
Me_2CuLi (3.0 equiv)	Et_2O, add slowly	98:2	96:4	79
Me_2CuLi (3.0 equiv)	Et_2O, add rapidly	52:48	13:87	88
Me_2CuLi (3.0 equiv)	THF	>99:<1	>99:<1	66
i. MeLi. ii. CuI. iii. MeLi	THF	>99:<1	>99:<1	70

The procedure has been employed in a number of synthetic applications[49] and works well in the absence of steric congestion[49a].

This intramolecular allylic substitution protocol has been applied to the synthesis of allylsilanes[50a], and extended to the preparation of β-carbonyl siloxanes via an intramolecular conjugate addition reaction[50b]. While carbamates derived from simple acyclic primary alcohols give very modest γ-selectivity [e.g. *E*-2-pentenol (65:35), 3-methyl-2-pentenol (56:44)], S_N2-substitution can predominate [e.g. *Z*-2-pentenol (29:71, S_N2':S_N2)][50a]. Carbamates derived from simple acyclic secondary and tertiary alcohols generally give clean γ-substitution (i.e. S_N2') and *E*-olefin geometries, but can afford *E*/*Z* mixtures of double bond isomers **32** and **33** (equation 11). The *Z*-carbamates are often far more *E*-selective than the *E*-carbamates, particularly when amido(silyl)cuprates are employed (equation 12). Preference for the *E*-alkene product arises from a smaller $A^{1,3}$-strain in the transition states **34** and **36** than in the transition states **35** and **37** leading to the Z-alkene product (Scheme 3). The transition state energy differences (e.g. **36** vs. **37**) are greater in the *Z*-allylic carbamates than in the *E*-isomers (e.g. **34** vs. **35**) leading to a greater stereoselectivity.

(**31**) (**32**) (**33**)

Carbamate	R	% Yield	(**32**:**33**)
E-**31**	Me	40	90:10
Z-**31**	Me	74	100:0
E-**31**	Ph	59	93:7
Z-**31**	Ph	87	100:0

(11)

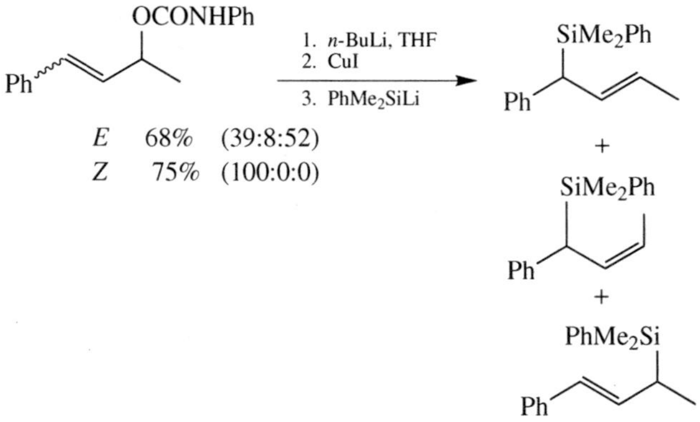

(12)

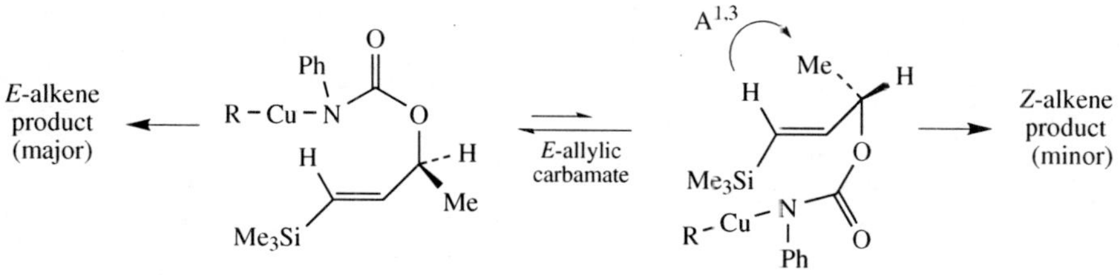

(**34**) (**35**)

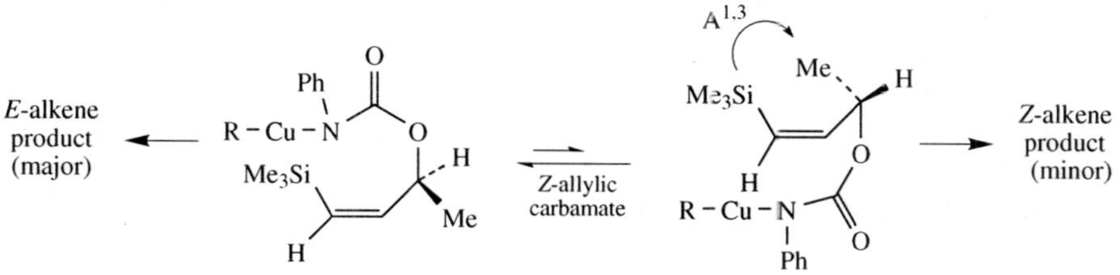

(**36**) (**37**)

SCHEME 3. A[1,3]-strain responsible for (E)-selectivity in carbamate mediated *syn*-S_N2'-allylic substitutions

Steric hindrance can significantly impact chemical yields, regiochemistry and alkene geometry (equations 13 and 14). Carbamate-directed intramolecular conjugate addition of silylcuprates in δ-carbamoyl-α,β-enonates affords β-diphenyl(hydroxy)silyl-substituted esters with modest diastereoselectivity (85:15)[50b].

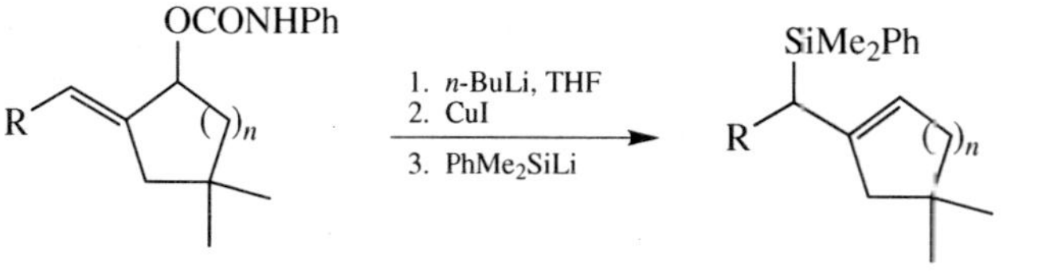

(13)

R = Me, Ph, *i*-Pr
n = 1,2

20–45% (51–92%)
(based on recovered
starting material)

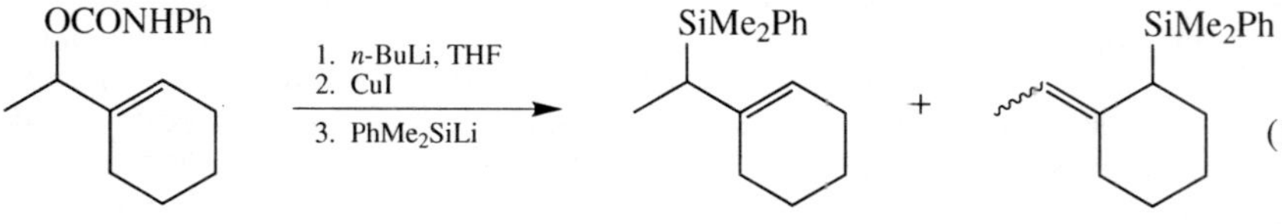

(14)

16% 34% (E:Z 1:1)

Magnesium amido(silyl)cuprates are Z-selective with E-allylic carbamates and E-selective with Z-allylic carbamates[51]. CuI, $CuBr{\bullet}SMe_2$ and $CuI{\bullet}2LiCl$ all give comparable Z-selectivity, while $CuCN{\bullet}2LiCl$ gives significant decrease in Z-selectivity. This reversal of selectivity with the magnesium cuprates is attributed to aggregation phenomena and a detailed explanation awaits further experimentation. Amido(alkyl)cuprates derived from lithium, zinc and titanium reagents are E-selective with E-carbamates, while aluminum reagents afford ligand-dependent E/Z-selectivity (equation 15)[51a]. Z-Allylic carbamates are highly E-selective (equation 16).

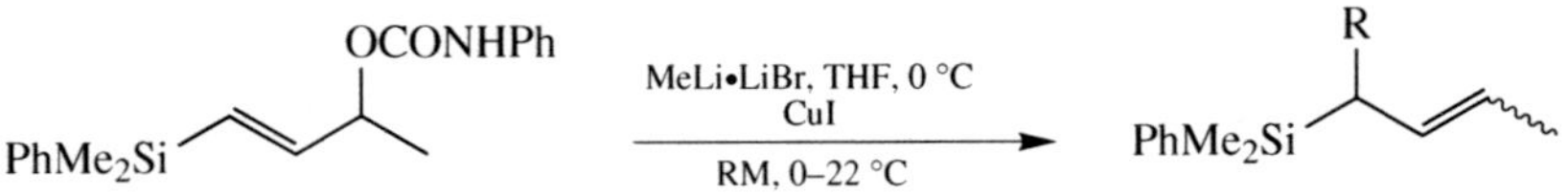

RM	E:Z	γ:α	% Yield
Me_3SiCH_2Li	60:40	-	52
Me_3SiCH_2MgCl	5:95	-	82
$Me_2Ti(OPr\text{-}i)_3Li$	89:11	99:1	82
Et_2Zn	90:10	>99:<1	77
$AlMe_3$	76:24	>99:<1	76
MeLi	92:8	94:6	72

(15)

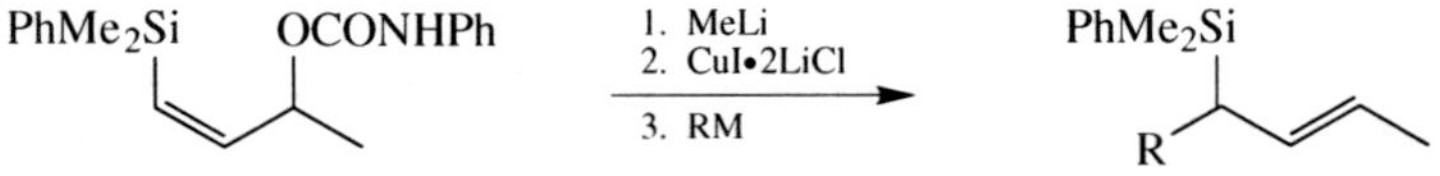

RM	E:Z	γ:α	% Yield
i-BuMgCl	93:7	98:2	93
i-BuLi	>99:<1	100:0	93
PhLi	>99:<1	>99:<1	55
Me_3SiCH_2MgCl	93:7	>99:<1	78

(16)

Reaction of carbamates with bis(triorganosilyl)zinc reagents in the presence of catalytic amounts of CuI (5 mol%) afford γ-substitution with *anti*-selectivity (i.e. *anti*-S_N2'), and this selectivity was observed even when the carbamate was deprotonated prior to addition of the zinc reagent and copper catalyst[52]. High levels of asymmetric induction can be achieved intramolecularly when the substrate functionality and heteroatom ligand are contained in the same molecule. Chiral N-carbamoyl(alkyl)cuprates derived from allylic carbamates [i.e. ($RCH{=}CHCH_2OC(O)NR^*$) CuR^1] undergo an intramolecular allylic rearrangement with excellent enantioselectivities (R^1 = Me, n-Bu, Ph; 82–95% ee), although yields are modest (42–70%) due to unreacted starting carbamate[53].

This protocol for the stereoselective preparation of allylic silanes has been utilized in synthetic applications[54]. Nevertheless, the methodology can fail when steric hindrance in the substrate becomes a problem[55].

III. ORGANOCUPRATES WITH AN α-*N*-SUBSTITUENT

Metallation adjacent to a *N*-heteroatom generally involves lithiation followed by transmetallation to copper. These α-*N*-heteroatom-substituted alkyl or heteroarylcuprates generally undergo typical cuprate reactions, but with modification of cuprate reactivity. The only exceptions to the transmetallation protocol appears to be the insertion of CO or zinc carbenoids into a Cu—N bond.

A. Carbamoyl, α-Imidomethyl, α-Iminomethyl, α-Formamidinyl- and α-Azoalkylcuprates

Lithium bis(*N*,*N*-diethylamido)cuprate absorbs two equivalents of carbon monoxide at ordinary pressure and ambient temperature in THF:HMPA (4:1) to afford lithium bis(*N*,*N*-diethylcarbamoyl)cuprate [i.e. $(Et_2NCO)_2CuLi$], which reacts with PhI, allyl bromide and RCOX (R = Me, Ph, EtO) to give amides and α-oxoamides (or oxamic esters for R = EtO), respectively, in variable yields[56a]. Higher, but modest, yields (49–61%) were obtained when the reaction was conducted between 60–80 °C and at elevated CO pressures (50 kg cm^{-2}). Reaction with methyl vinyl ketone at −78 °C gave $Et_2NCOCH_2CH_2COMe$ in 78% yield. Preparation of the carbamoylcuprate with CuI instead of CuCl extended the method to *N*-methylanilide derivatives[56b]. Reaction of carbamoylcuprates **38** prepared from CuI in less polar solvents (e.g. Et_2O) with mineral acids, Br_2/NaOMe, carbon dioxide or PhI/Pd^0 afforded formamides **39**, carbamates **41**, oxamic acids **42** and amides **43**, respectively, under mild conditions (Scheme 4)[57]. Oxidation of the carbamoylcuprate with Br_2 or $TiCl_4$ effected oxidative dimerization to oxamides **40**.

As already noted, zinc carbenoids (e.g. ICH_2ZnI) insert into the Cu—N bond to afford α-aminoalkyl cuprates, which can be alkylated with very reactive allyl halides, but are unreactive toward less reactive electrophiles[38]. Similarly, the zinc phthalimidomethylcuprate reacted with 3-iodo-2-cyclohexenone to give the substitution product in good yield (72%), but was unreactive with other electrophiles[58], and did not react with nitroalkenes in conjugate addition reactions.

α-Imidocopper reagents (i.e. copper(I) aldimines) generated by addition of organolithium reagents to *tert*-butyl isocyanide followed by treatment with one equivalent of a Cu(I) salt (e.g. CuI, $CuBr{\bullet}SMe_2$) are unstable at room temperature and decompose to form amides and imine dimers[59]. In the presence of $BF_3{\bullet}Et_2O$, *t*-BuN=C(Cu)Bu-*n* underwent conjugate addition to cyclohexenone (71%), 3-penten-2-one (95%), methyl vinyl ketone (39%) and hexenal (88%) in modest to excellent yields, although with enals the imine product was hydrolyzed during work-up. This acyl anion protocol has not been utilized in synthetic applications.

Glycine templates have been developed employing bislactim ethers for the asymmetric synthesis of non-proteinogenic α-amino acids[60] (equation 17). Although these organometallic species may be viewed as α-metallo imines or metallo aza enolates (*vide infra*) depending upon which nitrogen atom is the focus of attention, the latter description best reflects their reactivity profile. The lithium aza enolate can be converted into the organocuprate and the latter reagent undergoes conjugate addition reactions with α,β-unsaturated carbonyl compounds to afford conjugate adducts **44**. Small amounts of 1,2-addition products (e.g. **45**) arise from enals and β,β-dialkyl-substituted enones. Extended 1,6-conjugate addition occurs in preference (2:1) to the 1,4-addition when both opportunities exist[61]. Excellent stereoselectivity is achieved at the ring carbon atom of the bislactim ether, while stereocontrol at the newly created stereogenic center β to the

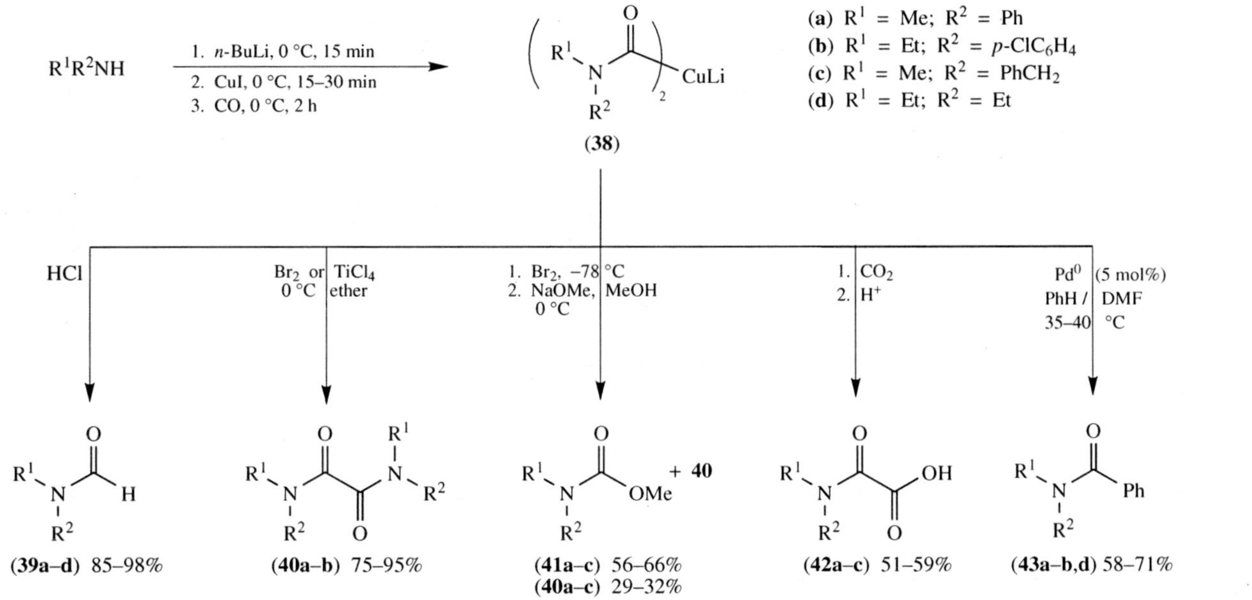

SCHEME 4. Preparation of formamides **39**, oxamides **40**, carbamates **41**, oxamic acids **42**, and amides **43** from lithium bis(carbamoyl)cuprates **38**[57]

carbonyl of the original enone or enal varies widely (i.e., 1′ in **44**). The lithium bislactim ethers can be effectively alkylated with simple alkyl halides, but the copper reagents have been employed for benzyl halide derivatives bearing a nitro group[62] or when the lithium reagent failed with alkyl halides[63].

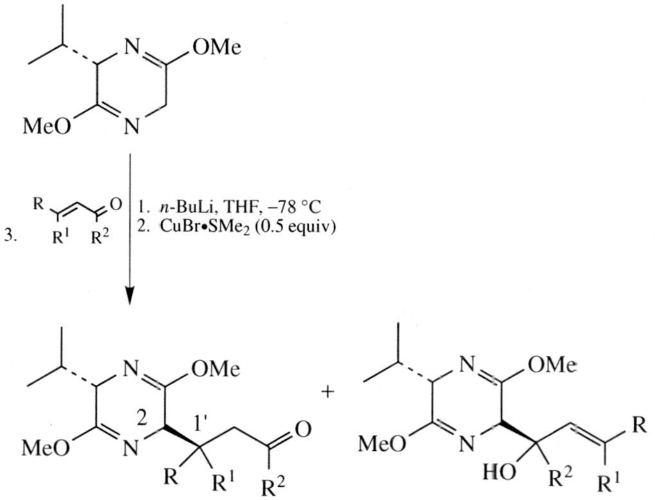

(**44**) (**45**)

(17)

R	R[1]	R[2]	**44** (2*R*,1′*R*):(2*R*,1′*S*)	**44**:**45**	% Yield
H	−$(CH_2)_2$−		98:2	>99:<1	71
H	−$(CH_2)_3$−		94:6	97:3	71
H	−$(CH_2)_4$−		89:11	>99:<1	66
Me	−$(CH_2)_2$−		50:50	84:16	52
Ph	H	Me	45:55	94:6	62
2-Furyl	H	Me	61:39	82:11	60
H	H	Me	-	95:5	39

N-Methyldiphenylmethanimine (i.e. $Ph_2C{=}NMe$) can be lithiated with LDA or *n*-BuLi, and converted into an organocopper reagent with one equivalent of $CuBr{\bullet}SMe_2$ that effects an S_N2'-allylic substitution with propargyl mesylates to afford α-allenic amines in low yields (14–54%)[64]. Gilman reagents and organocopper reagents prepared with other Cu(I) salts give lower yields. Imine hydrolysis could be achieved with aqueous oxalic acid, and the amines isolated as the oxalate salts or free amines upon treatment with NaOH. The procedure has been utilized for the preparation of scalemic α-amino allenes from enantioenriched propargyl mesylates utilizing CuCN•2LiCl to generate the organocopper species (i.e. RCu)[65].

Vinylborates derived from 1-(N-*tert*-butylimidoyl)-2-pyrrolinyllithium (i.e. 2-Li-*c*-C_4H_5N-CH=NBu-*t*) undergo transmetallation with CuCN (1.46 equiv) to afford organocopper or cuprate species that can be alkylated with allyl and benzyl halides (50–60%) and vinyliodonium salts (61%), although the reaction fails with heteroaryl- and alkynyliodonium salts[66]. Treatment of the 2-pyrrolinyllithium reagent with pentynylcopper affords a mixed homocuprate that can be allylated with allyl bromide in 87% yield[67].

Reaction of N-sulfonyl-2-pyrrolinyllithium with Gilman reagents results in coupling with concomitant ring opening to afford stereoselectively a vinyl organometallic species that can be trapped with alkyl and allyl halides, I_2, and chlorotrimethylstannanes to afford homoallylic amines with stereodefined olefin geometry (equation 18)[68].

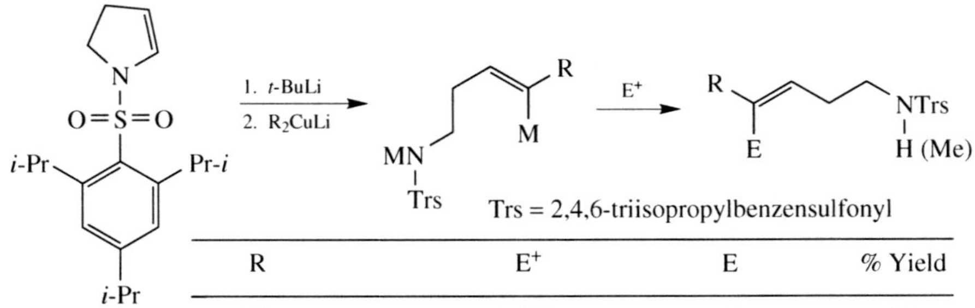

R	E^+	E	% Yield
Me	H_2O	H	71
n-Bu	MeI	Me	92
t-Bu	CH_2=$CHCH_2Br$	allyl	88
t-Bu	I_2	I	92
Me_3Sn	MeI	Me	37
Me	Me_3SnCl	Me_3Sn	75
Me_2C=$CH(CH_2)_2$	MeI/HMPA	Me	73[a]

(18)

[a] N-Methylation also occurred.

As the above examples illustrate, the early success of cuprate reagents containing an α-nitrogen atom consisted of sp^2-centered ligands[56,59,66–68], although aza allylcuprates also gave good yields in conjugate addition reactions[60–63] and modest yields with propargyl substrates[64,65].

The development of procedures for lithiation adjacent to the N-atom in amine derivatives provided opportunities for cuprate chemistry via transmetallation[69]. Meyers and coworkers[70] as well as Gawley and coworkers[71] prepared cuprate reagents from α-lithio-*tert*-butylformamidines in order to avoid SET reactions observed in the alkylation of the lithium reagents with alkyl halides. Dieter and coworkers extended the reactions of formamidinylcuprates to 1,4-conjugate additions with α,β-enones[72] and also examined cuprates prepared from N-lithiated hydrazones[73]. The latter reagents underwent conjugate addition reactions with cyclic enones, enoates and alkylidene malonates with poor diastereoselectivity. In both cases, difficulty in removing the *tert*-butylformamidine and N-*tert*-butylazo protecting groups in the presence of the carbonyl functionality precluded successful synthetic applications of these methods.

B. α-(*N*-Carbamoyl)alkylcuprates

The easier removal of *N*-Boc (i.e. *tert*-butoxycarbonyl) protecting groups prompted an examination of Beak's α-lithiated carbamates for the preparation of α-*N*-functionalized alkylcuprate reagents[74]. In initial studies, cuprates prepared by deprotonation of *N*-Boc-pyrrolidine **46**, with *s*-BuLi followed by addition of CuCN, gave low and capricious yields of 1,4-adducts with 2-cyclohexenone (equation 19)[74a]. In marked contrast, utilization of the *N*-Boc-2-pyrrolidinylstannane **47** to effect a transmetallation cascade from Sn to Li to Cu afforded cuprate solutions that gave nearly quantitative yields in the conjugate addition reaction with 2-cyclohexenone[74a]. Although the conjugate addition yields appeared to depend upon the diamine (and its purity) employed to assist carbamate deprotonation, a subsequent study explored the effect of Cu(I) salt, *s*-butyllithium quality, solvent and temperature upon α-(*N*-carbamoyl)alkylcuprate chemistry[75]. Good yields of conjugate adducts could be achieved when the cuprate was prepared from solid CuCN for extended periods at low temperatures (e.g. −78 to −50 °C for 40–80 min) or shorter periods at higher temperatures (e.g. 25 °C, 15 min). A more reliable procedure resulted when the cuprate was prepared from THF soluble CuCN•2LiCl at low temperature (i.e. −78 °C). A thermal stability study indicated that both the *N*-Boc-2-pyrrolidinyl- and *N*-Boc-2-piperidinylcuprates displayed good stability at 20 °C for two hours. Modest yields of 1,4-addition products could also be obtained when poor quality *s*-BuLi containing LiH (solid suspension) and lithium alkoxides was employed in the deprotonation step. These impurities appeared to increase the rate of cuprate decomposition as excellent yields were obtained with high quality *s*-BuLi freshly prepared. These results suggest that the α-(*N*-carbamoyl)alkylcuprates are significantly more stable than the lithium reagents and that slow cuprate formation (e.g. solid THF insoluble CuCN, −78 °C) results in competitive decomposition of the lithium reagent and diminished yields in the copper-mediated reactions. Utilization of THF soluble CuCN•2LiCl allows cuprate formation at −78 °C where the lithiated carbamates are more stable.

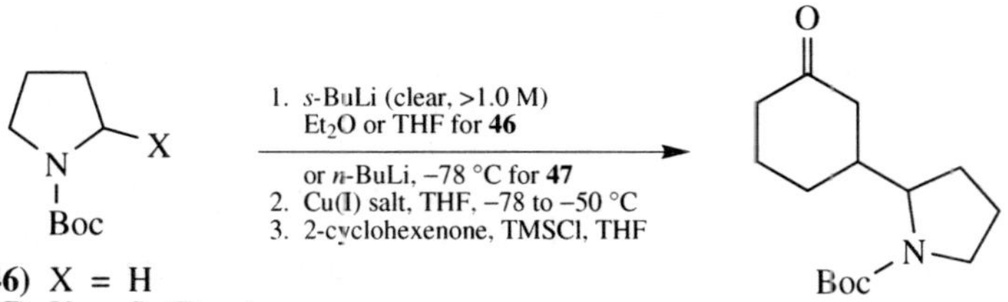

(**46**) X = H
(**47**) X = $Sn(Bu\text{-}n)_3$

cpd	Diamine	CuX[a]	% Yield
46	TMEDA	CuCN[b]	<12–41
47	-	CuCN	98
46	(−)-Sparteine	CuCN	71–73
46	TMEDA (distilled)	CuCN	71–73
46	(−)-Sparteine (distilled)	CuCN	95
46	(−)-Sparteine	CuCN•2LiCl[b]	90

(19)

[a] Cuprate formation at −55 to −40 °C unless noted. [b] Cuprate formation at −78 °C.

Utilizing the THF soluble CuCN•2LiCl protocol, α-(N-carbamoyl)alkylcuprates undergo 1,4-conjugate addition to α,β-enones[74b], α,β-enals[74b], α,β-unsaturated carboxylic acid derivatives (e.g. esters, thiolesters, imides)[76], α,β-ynones[77], α,β-ynoates[76b,78], and α-allenyl esters[78,79] (Scheme 5). Conjugate addition to α,β-γ,δ-allenyl esters occurred stereoselectively *anti* to the substituent at the γ-carbon atom to afford (E)-3-carbamoylalkyl-β,γ-unsaturated esters. These conjugate addition reactions require the use of TMSCl to accelerate the reactions and can be accomplished with both cyano Gilman reagents [i.e. R_2CuLi•LiCN, R = α-(N-carbamoyl)alkyl] and alkyl(cyano)cuprates [i.e. RCu(CN)Li, R = α-(N-carbamoyl)alkyl] where the most effective reagent is substrate-dependent. Alkyl (cyano)cuprate reagents give lower yields with α,β-enoates and higher yields with α,β-ynoates or ynones than the Gilman reagent. Reaction of cuprates with α,β-alkynyl ketones and esters may well proceed via a 1,2-addition or carbocupration process[80] and the use of TMSCl and CuCN•2LiCl in the α-(N-carbamoyl)alkylcuprate reactions facilitates E:Z isomerization of the intermediate α-cuprioalkenoate via the ketone or ester lithium allenolate. Low to modest yields of conjugate adducts were generally obtained with α,β-unsaturated nitriles and sulfoxides[76b], while very modest diastereoselectivity was observed in the ynone and ynoate conjugate adducts. This protocol could be extended to α,β-unsaturated nitro compounds if dialkylzinc reagents and copper catalysts were employed[81] where the procedure was used in the preparation of triplex DNA-specific intercalators.

α-(N-Carbamoyl)alkylcuprates also participate in substitution reactions (Scheme 6) with acyl halides[82], vinyl iodides[78,83], vinyl triflates[84], allylic substrates (halides[85], sulfides[85], phosphates[85,86]) and propargl substrates[87] (mesylates, halides and phosphates). For the 2-pyrrolidinylcuprate, vinylation with (E)-1-iodohexene proceeded in good yields with both the Gilman (80–88%), RCu(Cl)Li (77–83%, R = 2-pyrrolidinyl) and RCu(CN) Li (69–73%) reagents. Higher yields were generally achieved with Gilman and RCu(Cl)Li reagents in the acylation reactions. Use of the RCu(CN)Li or RCu(Cl)Li reagents is efficient in α-(N-carbamoyl)alkyl ligand, although yields are slightly lower than those obtained with R_2CuLi•LiCN. Alkyl(cyano)cuprates gave lower yields and excellent enantioselectivities in reactions with enantioenriched allylic phosphates, while the Gilman reagents gave higher chemical yields but lower enantioselectivities[86]. Propargyl mesylates generally afforded comparable yields of α-amino allenes upon reaction with both Gilman and alkyl(cyano)cuprates[87c], but the latter were employed in reactions with enantioenriched propargyl mesylates[87d].

Substitution reactions with cinnamyl allyl halides or phosphates afforded mixtures of rearranged (S_N2') and unrearranged (S_N2) products and little regioselective control could be achieved[85]. These results are consistent with initial formation of an olefin–copper π-complex followed by allylic inversion (i.e. S_N2' with generally *anti* stereoselectivity) to give a σ-alkylcopper complex. This σ-allyl complex can undergo reductive elimination to afford the S_N2' substitution product or isomerize via a π-allyl complex to a rearranged σ-allyl complex, which affords the S_N2 substitution product upon reductive elimination. Alkylation of α-(N-carbamoyl)alkylcuprate reagents with allylic sulfides prepared from benzothiazole-2-thiol gave regiospecific S_N2 substitution in modest to good yields (31–80%). Excellent regiocontrol could also be achieved with propargyl sulfonates and epoxides resulting in exclusive S_N2' substitution for most systems[87]. Propargyl acetates were unreactive. Substitution without allylic rearrangement was only observed when severe steric crowding was present in the α-(N-carbamoyl)alkyl ligand or in the propargyl substrate.

These conjugate addition and substitution reactions of α-(N-carbamoyl)alkylcuprates were exploited in the synthesis of nitrogen heterocycles[78]. Substitution reactions of α-(N-carbamoyl)alkylcuprates with (E)-6-chloro-3-iodo-2-hexenoate (**48**) gave good to

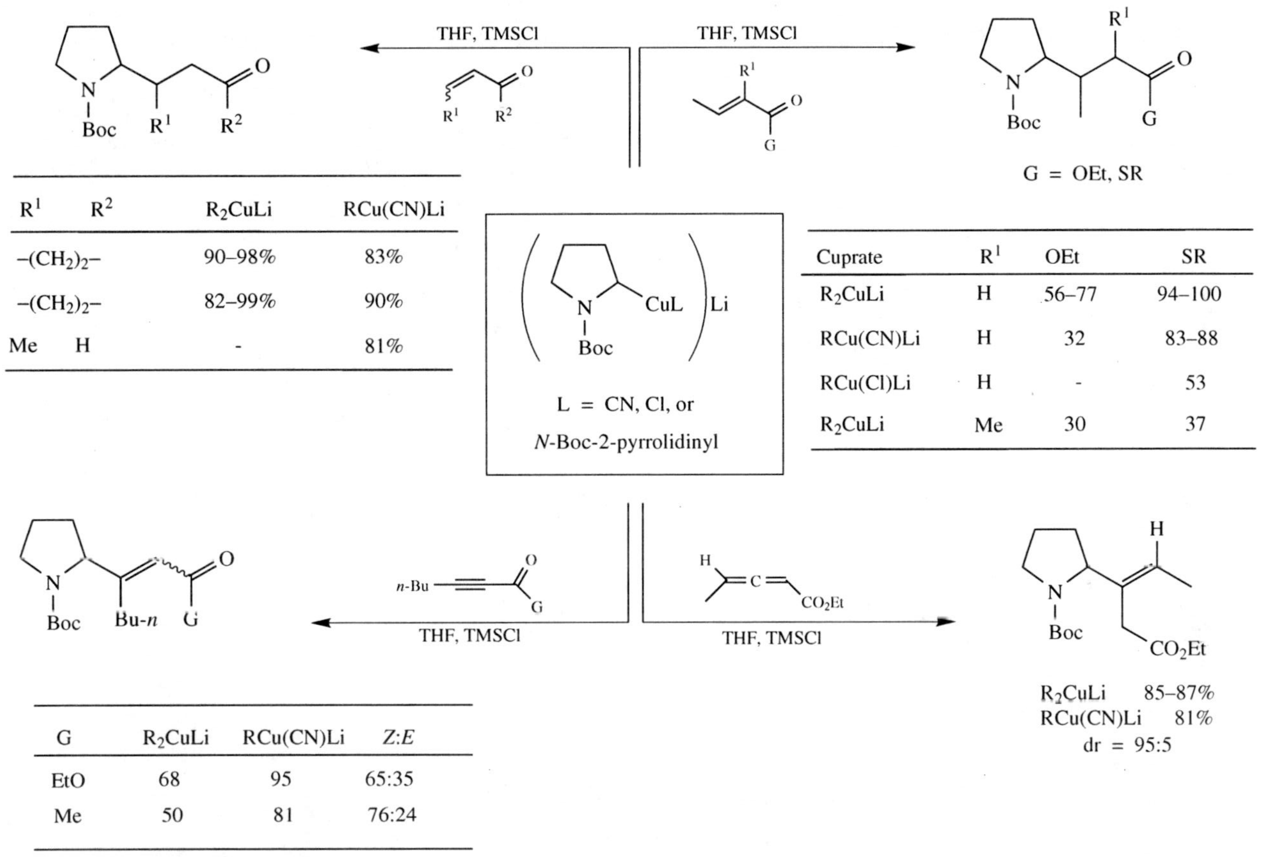

R^1	R^2	R_2CuLi	RCu(CN)Li
$-(CH_2)_2-$		90–98%	83%
$-(CH_2)_2-$		82–99%	90%
Me	H	-	81%

Cuprate	R^1	OEt	SR
R_2CuLi	H	56–77	94–100
RCu(CN)Li	H	32	83–88
RCu(Cl)Li	H	-	53
R_2CuLi	Me	30	37

G	R_2CuLi	RCu(CN)Li	*Z*:*E*
EtO	68	95	65:35
Me	50	81	76:24

SCHEME 5. Conjugate addition reactions of *N*-Boc-2-pyrrolidinylcuprates with enones[74b], enals[74b], enoates[76], thioesters[76], α-allenyl esters[78, 79], ynones[77], and ynoates[76, 78]

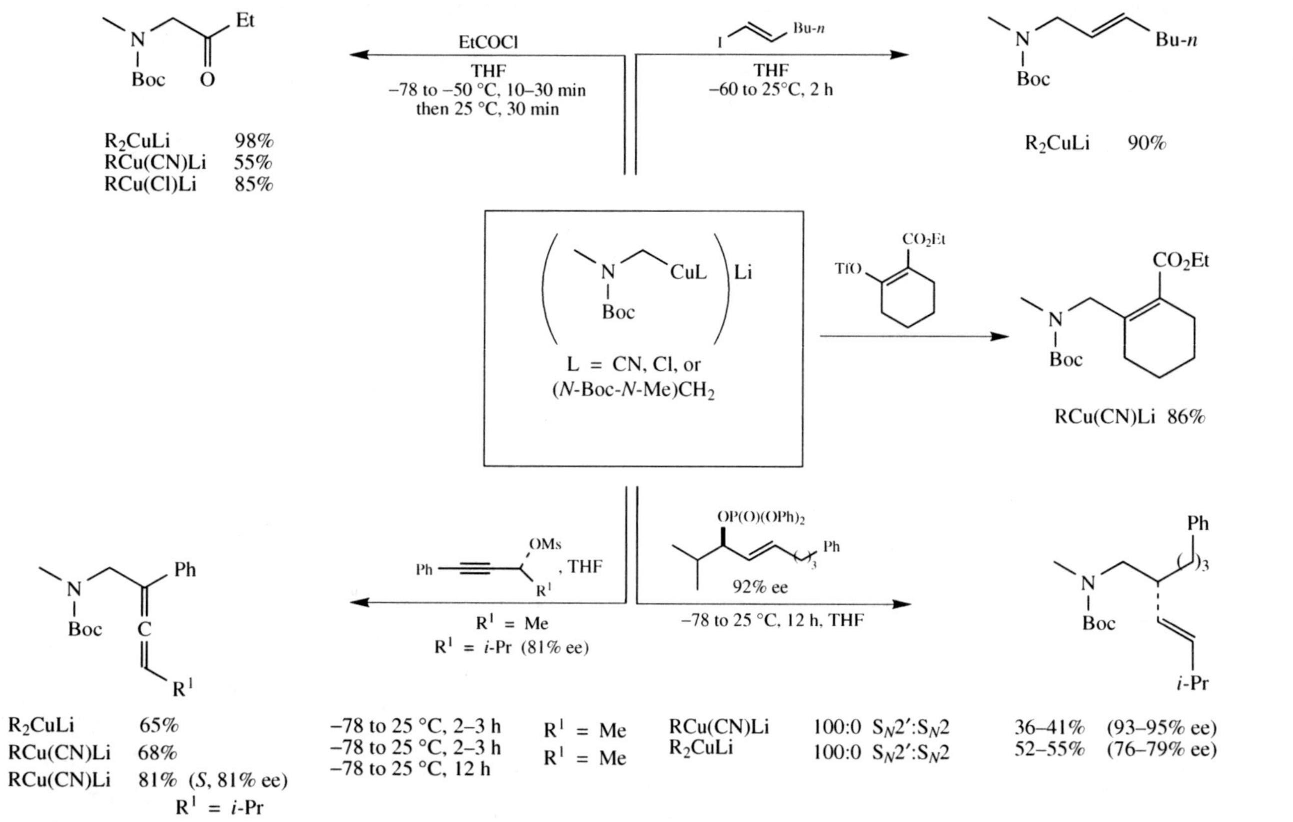

SCHEME 6. Substitution reactions of (*N*-Boc-*N*-methyl)methylcuprates with acid chlorides[82], vinyl iodides[83], vinyl triflates[84b], allylic phosphates[86] and propargyl mesylates[87c,d]

excellent yields of piperidines **49a**, indolizidines **49b** and quinolizidines **49c** containing a carboalkoxyalkylidene moiety on the carbon framework (Scheme 7) after *N*-Boc deprotection and cyclization, while reactions with the (*Z*)-isomer **50** followed by *N*-Boc cleavage and cyclization led to pyrrolin-2-ones **51a** and pyrrolizidin-2-ones **51b** in good yields and indolizidin-2-ones **51c** in poor yields. This route using the 3-iodoenoates is preferable to the use of alkynyl esters (see Scheme 5), which afford mixtures of (*E*)- and (*Z*)-γ-carbamoylenoates with poor diastereoselectivity. The conjugate addition reactions α-(*N*-carbamoyl)alkylcuprates to α-allenyl esters **52** provide a diastereselective route to 4-alkylidene pyrrolidinones **53** and pyrrolizidinones[78, 79], while conjugate addition to α,β-alkynylketones provided a synthesis of simple (e.g. **54**) and fused pyrroles[77] (Scheme 8). The procedure is versatile permitting the introduction of substituents at three of the four carbon atoms of the pyrrole ring system. Reaction of α-(*N*-carbamoyl)alkylcuprates with propargyl mesylates provides a synthetic route to 3-pyrrolines[87c] and utilization of enantioenriched propargyl mesylates **55** affords enantioenriched 3-pyrrolines **56** wherein the ee of the propargyl mesylate is preserved in the final product (Scheme 8)[87b,d]. Arylpalladium(II) species (e.g. $PhPdL_2I$) can mediate nucleophilic cyclization of the intermediate α-allenyl amines to afford, after reductive elimination, substituted 3-phenyl-3-pyrrolines or 3-phenylpyrroles[87b]. Utilization of $Ru_3(CO)_2$ to effect amino allene cyclization gave either 3-pyrrolines, pyrroles or 3-pyrrolidin-2-ones[87c] depending upon reaction conditions.

Control of absolute stereochemistry in bond formations adjacent to the *N*-carbamoyl moiety was first achieved using chiral auxiliary methodology (equation 20). Transmetallation of diastereomeric *N*-(α-stannylalkyl) lactams **57** epimeric at the alkylstannane stereocenter affords an epimeric mixture of organolithium reagents that rapidly equilibrates to the more stable epimer **58**. Treatment of the lithium reagent with CuCN (1.0 or 0.5 equivalents) affords enantiopure α-aminoalkylcuprates that gave a single diastereomer upon reaction with acrolein[88]. Conjugate addition to α,β-enones gave mixtures of diastereomers epimeric at the β-carbon of the original enone. Diastereoselectivities are poor with acyclic enones (56:44 dr) and modest to excellent with cyclic enones. The poor diastereoselectivity at the β-carbon of cyclic enones arises from poor facial selectivity during cuprate addition. Acyclic enones may also give poor diastereoselectivity at the β-carbon center via *E*:*Z* isomerization arising from an equilibrium between a cuprate–enone d–π^* complex and starting materials.

The asymmetric deprotonation of carbamates developed by Beak and coworkers[89] provided opportunities for the preparation of enantioenriched α-(*N*-carbamoyl)alkylcuprates provided conditions could be found to maintain configurational stability during cuprate preparation and subsequent reaction. Asymmetric deprotonation of *N*-Boc pyrrolidine could be achieved with *s*-BuLi/(−)-sparteine at low temperatures in Et_2O, but rapid racemization occurred when conducted in THF[89]. This posed a problem since the α-(*N*-carbamoyl)alkylcuprate conjugate addition and substitution reactions were generally conducted in THF. Generation of the enantioenriched *N*-Boc-2-lithiopyrrolidine in Et_2O followed by addition of a THF solution of CuCN•2LiCl allowed rapid cuprate formation at −78 °C with the latter reagent undergoing racemization at a significantly slower rate than the lithium derivative[90]. Utilizing this protocol, good to excellent enantioselectivities could be achieved in the vinylation, allylation, propargyl substitution reactions and the conjugate addition reaction with methyl vinyl ketone (Figure 3). Less reactive substrates afforded lower enantioselectivities and a low 20% ee could be achieved with benzyl acrylate using a TMSCl/HMPA additive mixture that greatly accelerated the conjugate addition reaction. Reactive cuprate/electrophile combinations that reacted rapidly at low temperatures generally gave good enantioselectivities, while reagent/substrate combinations that

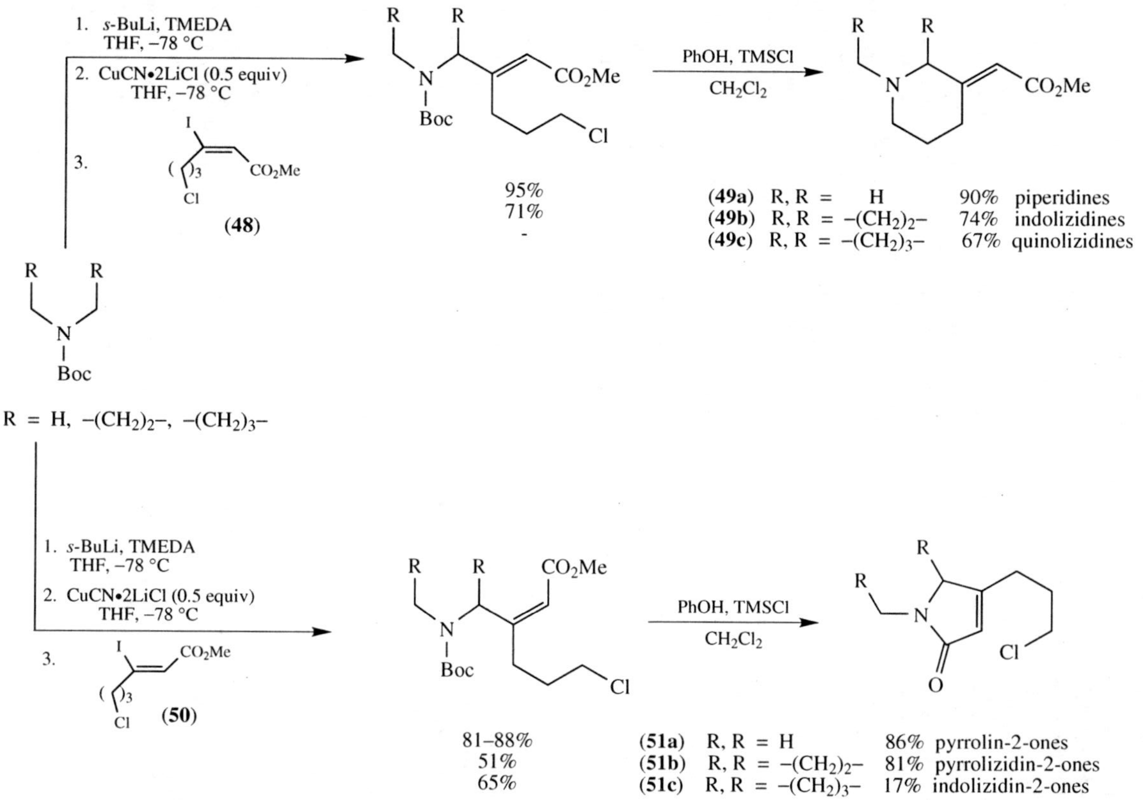

SCHEME 7. Synthesis of mono and bicyclic *N*-heterocycles via vinylation of lithium α-(*N*-carbamoyl)alkyl(cyano)cuprate reagents with methyl (*E*)- and (*Z*)-6-chloro-3-iodo-2-hexenoate[78]

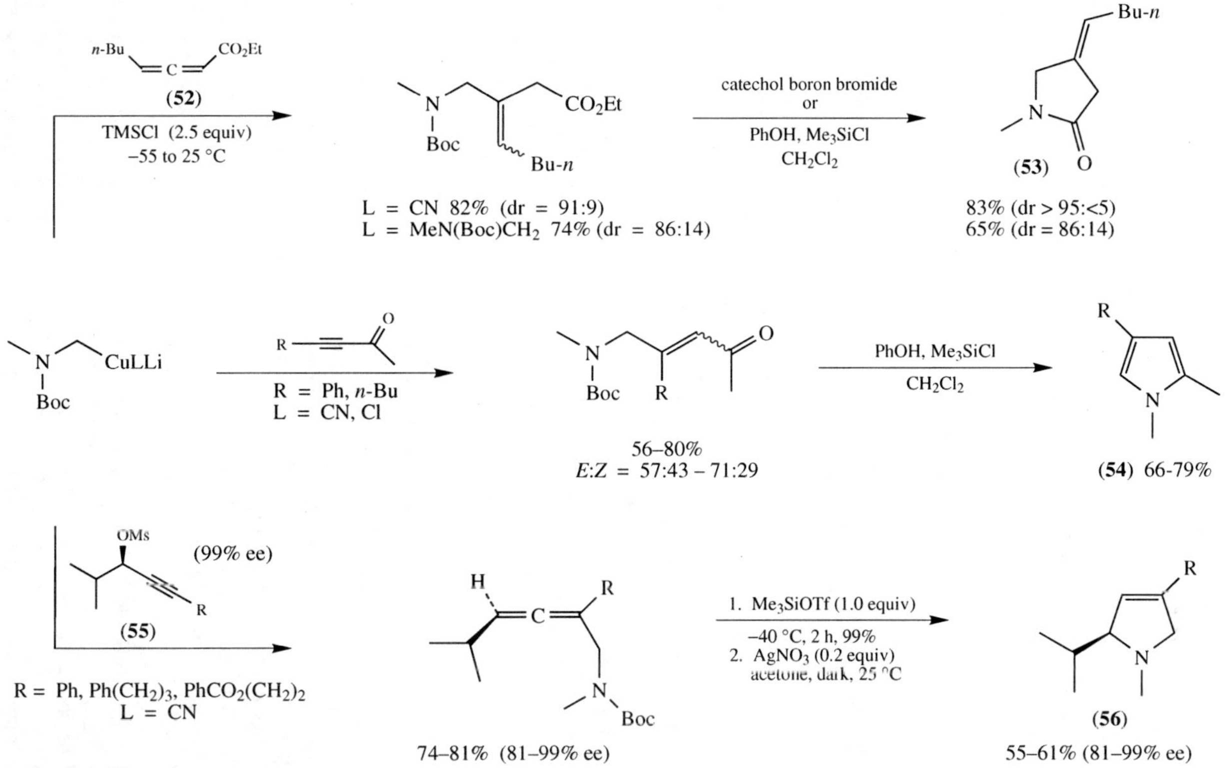

SCHEME 8. Reaction of α-(*N*-carbamoyl)alkylcuprates with α-allenyl esters[78,79], α,β-alkynyl esters[77] and propargyl mesylates[87d] in synthetic routes to 4-alkylidene-2-pyrrolidinones, substituted pyrroles and 3-pyrrolines, respectively

were less reactive and required higher temperatures and/or longer reaction times generally gave lower enantioselectivities.

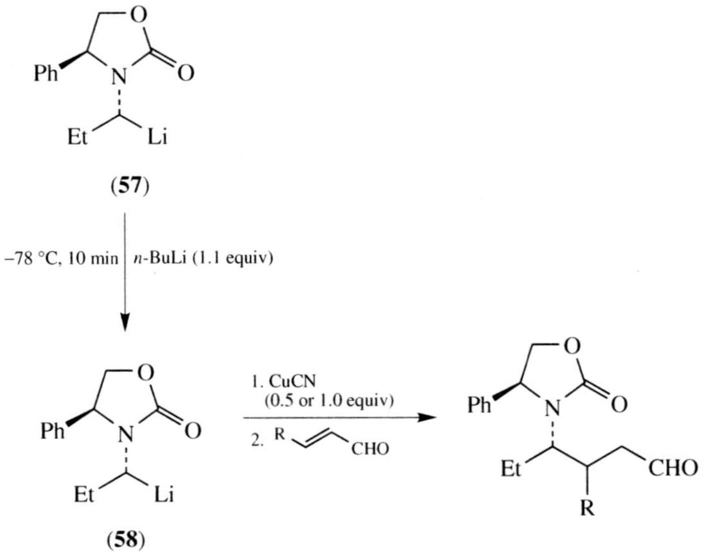

(20)

R	CuCN	% Yield	% dr
H	1.0	78	100:0
H	0.5	96	100:0
Me	1.0	85	100:0 (C4)
			71:29 (C3)

Preparation of enantioenriched acyclic α-(N-carbamoyl)alkylcuprates proved more problematic since direct deprotonation could not be achieved. An enantioenriched stannane **59** was prepared and converted into scalemic cuprates **60** via a Sn to Li to Cu transmetallation cascade (Figure 4)[91]. Gilman and zinc alkyl(cyano)cuprates gave modest enantioselectivities in THF and no ee in Et_2O, and enantioselectivities were again electrophile-sensitive. The solvent effect is interesting and suggests that the acyclic analogs are more configurationally stable than the cyclic derivatives. It is also interesting to note that the α-(alkyloxy)alkylcuprate analogs gave higher ees in Et_2O than in THF.

These asymmetric reactions were applied to the synthesis of several pyrrolizidine and indolizidine alkaloids[92–94]. Alkylation of the cuprates derived from (S)-N-Boc-2-lithiopyrrolidine with 2-iodo-5-chloro-1-pentene (**61a**) afforded a substitution product **62a** that could be converted into (+)-*ent*-heliotridane or (+)-isoretronecanol, while alkylation of the cuprates with 2-iodo-6-chloro-1-hexene (**61b**) provided a synthetic route via **62b** to (+)-tashiromine contaminated with its diastereomer (+)-5-epitashiromine generated in the hydroboration reaction (Scheme 9)[92]. The enantiopurity of the alkaloids matched the enantioselectivity of the cuprate alkylation reaction, which was higher for Gilman reagents than for the alkyl(cyano)cuprates. The diastereoselectivities reflect the hydrogenation and

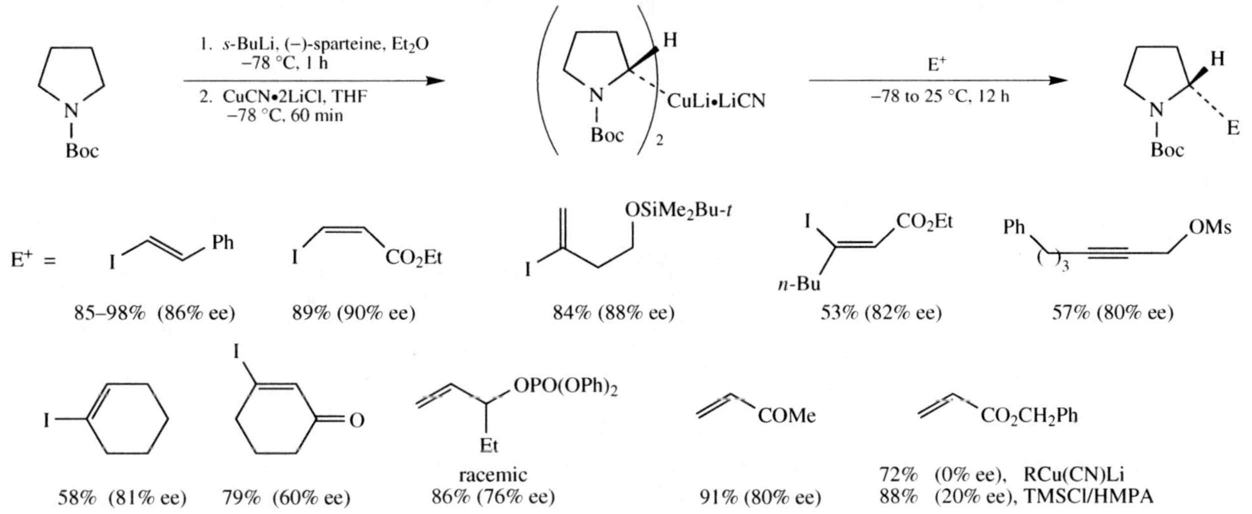

FIGURE 3. Enantiomeric excesses for reaction of bis(*N*-Boc-2-pyrrolidinyl)cuprate with various electrophiles[90]

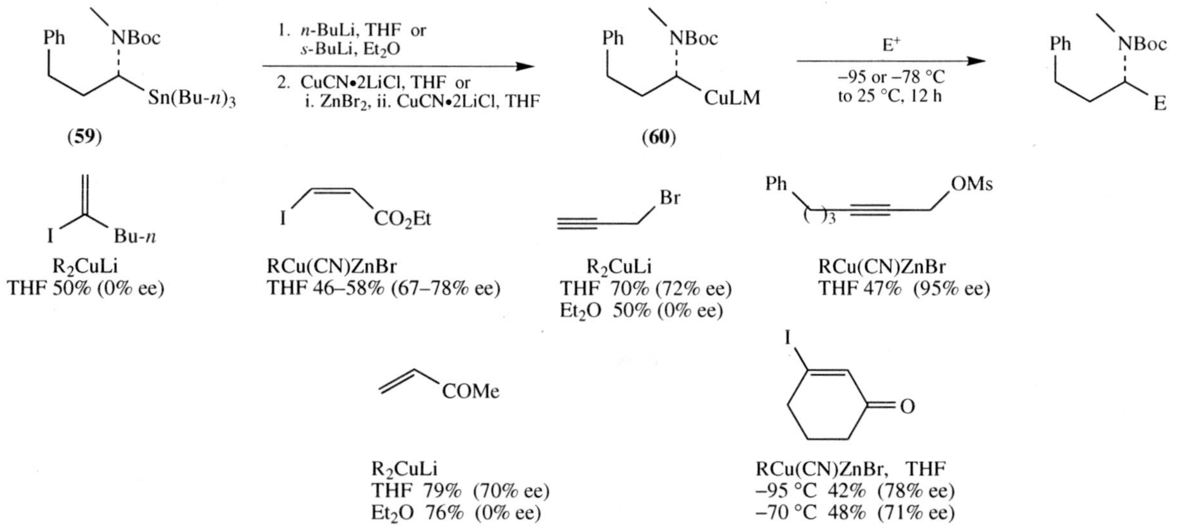

FIGURE 4. Enantiomeric excesses for reaction of a scalemic acyclic α-(*N*-carbamoyl)alkylcuprate with various electrophiles[91]

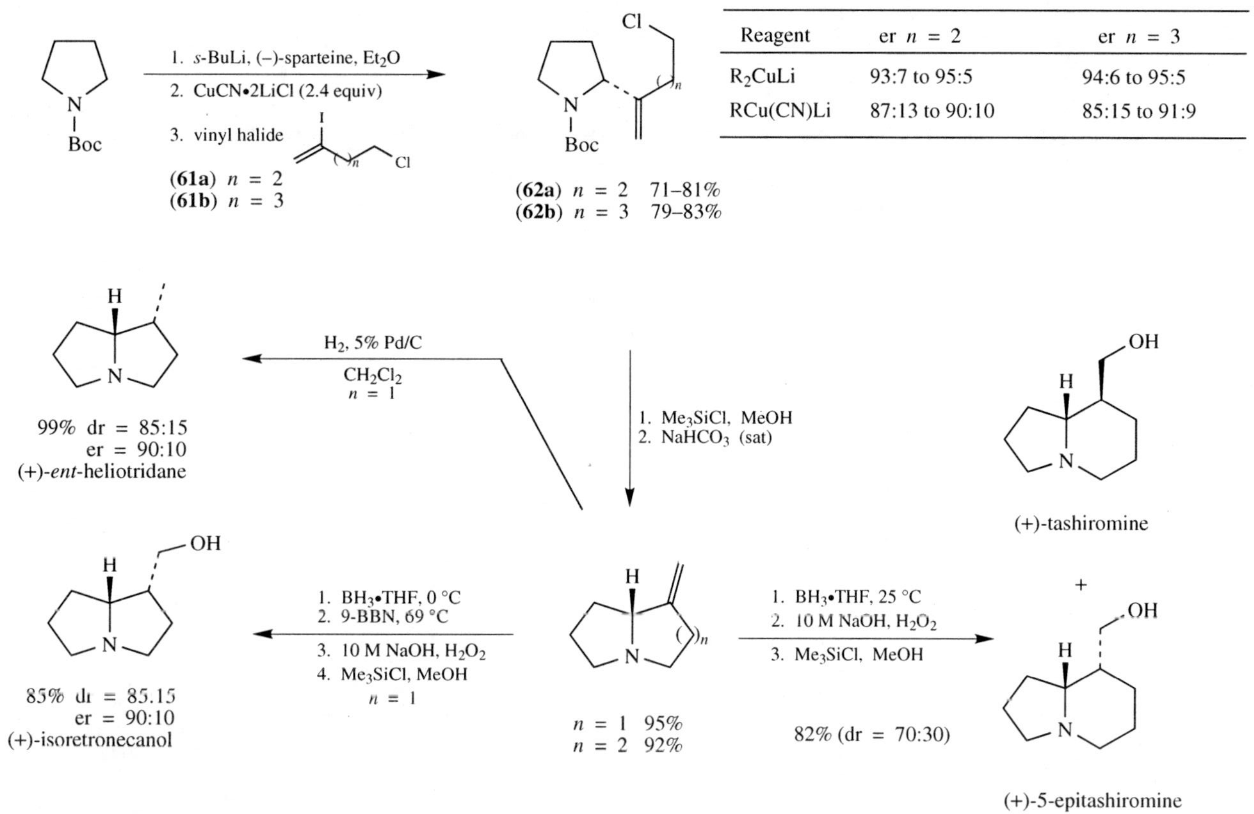

Reagent	er $n = 2$	er $n = 3$
R_2CuLi	93:7 to 95:5	94:6 to 95:5
RCu(CN)Li	87:13 to 90:10	85:15 to 91:9

SCHEME 9. Asymmetric synthesis of (+)-*ent*-heliotridane, (+)-isoretonecanol, (+)-tashiromine and (+)-5-epitashiromine via vinylation of enantioenriched lithium di-2-(*N-tert*-butoxycarbonylpyrrolidinyl)cuprate[92b]

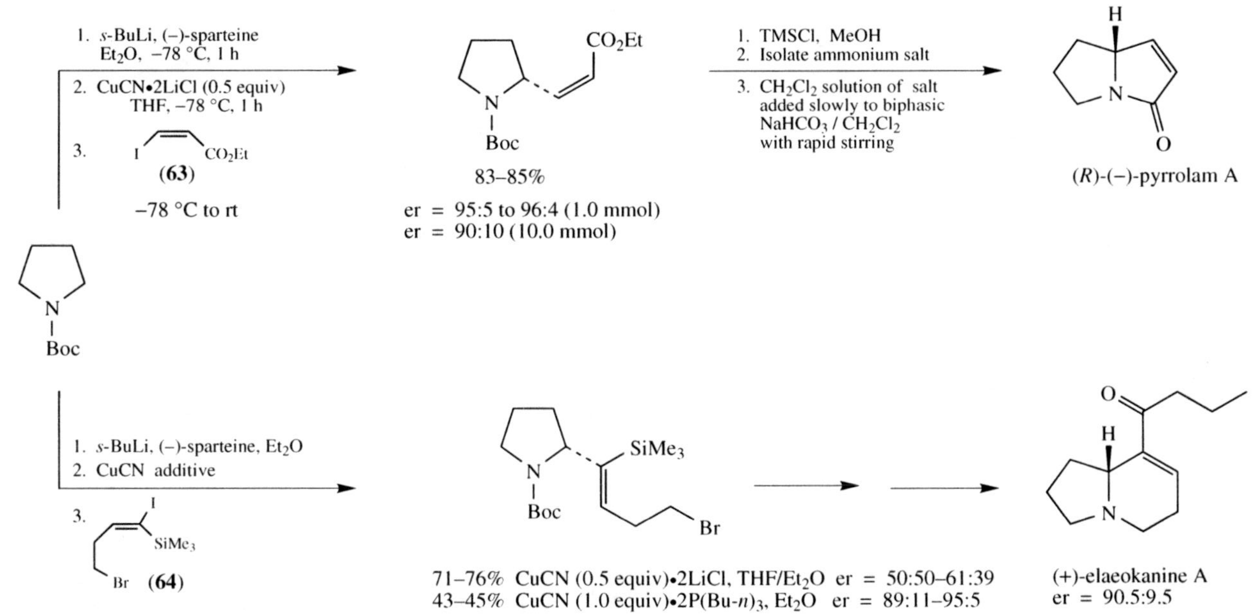

SCHEME 10. Asymmetric synthesis of (+) pyrrolam A[93] and (+)-elaeokanine A[94] via vinylation of enantioenriched lithium 2-(*N-tert*-butoxycarbonylpyrrolidinyl)cuprates

hydroboration stereoselectivities and optimization studies could examine more selective reagents for these reactions.

The asymmetric vinylation reactions were also employed in synthetic routes to (*R*)-(−)-pyrrolam A[93] and (+)-elaeokanine A[94] (Scheme 10). Enantioselectivities were particularly sensitive to cuprate reagent, substrate structure and reaction conditions in the (+)-elaeokanine synthesis. The Gilman reagent gave excellent ees when vinylated with ethyl (Z)-3-iodopropenoate (**63**), while the alkyl(cyano)cuprate in the presence of n-Bu_3P was more effective for vinylation with (*E*)-1-iodo-1-trimethylsilyl-4-bromo-1-butene (**64**). Optimal conditions required vinylation at −40 °C to enhance the rate of vinylation against the competitive rates of cuprate decomposition and racemization.

α-(*N*-Carbamoyl)alkylcuprate chemistry has been extended to the allylation of 3,7-diazabicyclo[3.3.1]nonanes (e.g. *N*-bis-bispidine, equation 21)[95], allylation of piperazines (equation 22)[96], diastereoselective 1,2-additions of α-(*N*-carbamoyl)propargylcuprates to aldehydes (equation 23)[97] and the allylation of 2-indolylcuprates with π-allyl Mo complexes (equation 24)[98]. The α-lithiocarbamate can be generated by reduction of the α-cyanocarbamate and the resultant cuprate undergoes alkylation and conjugate addition reactions (equation 25)[99]. Dieter and coworkers have shown that two different carbamates can be coupled by oxidation of the mixed α-(*N*-carbamoyl)alkylcuprate with molecular iodine (equation 26)[100].

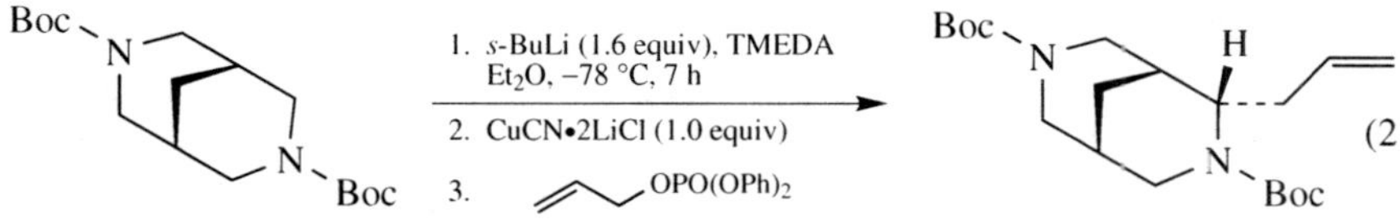

(21)

N-Boc-bispidine

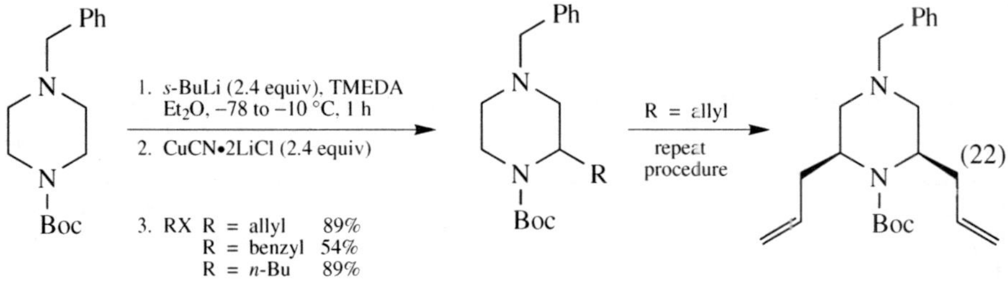

(22)

83%

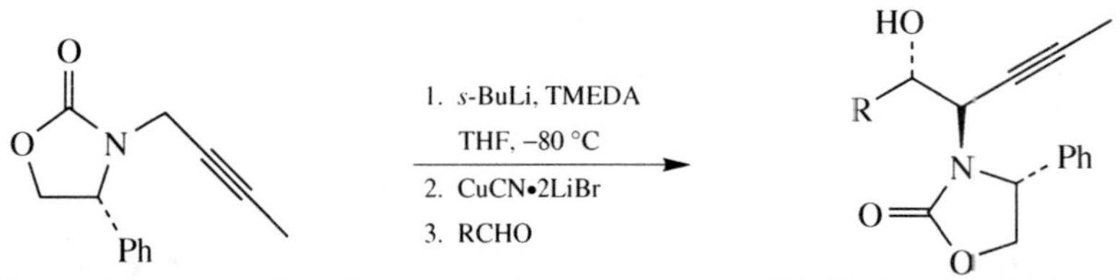

(23)

R = *i*-Pr, *c*-Hex, *n*-Pr, *t*-Bu, propenyl, pentynyl 75–88% (dr = 86:14 to >95:<5)

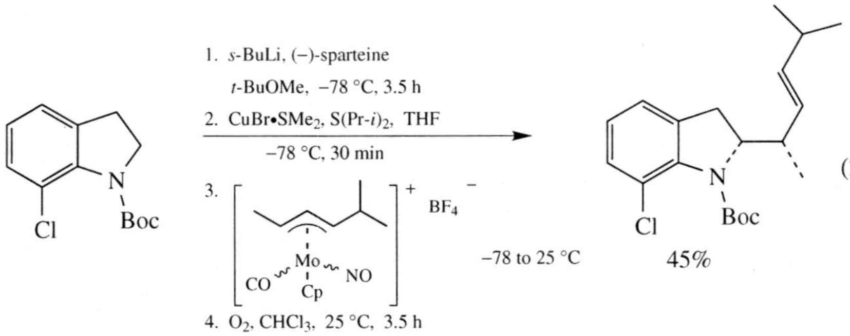

(24)

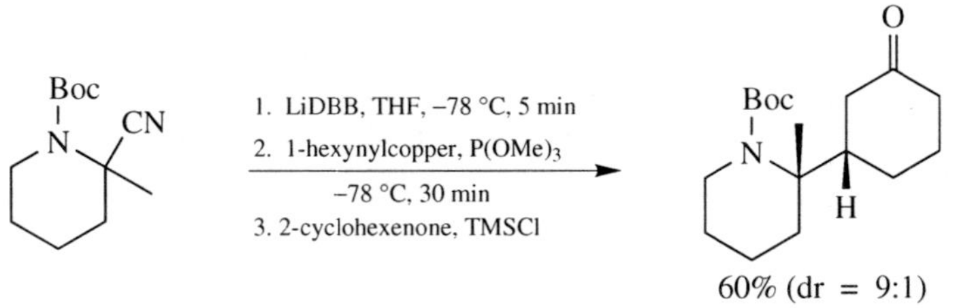

(25)

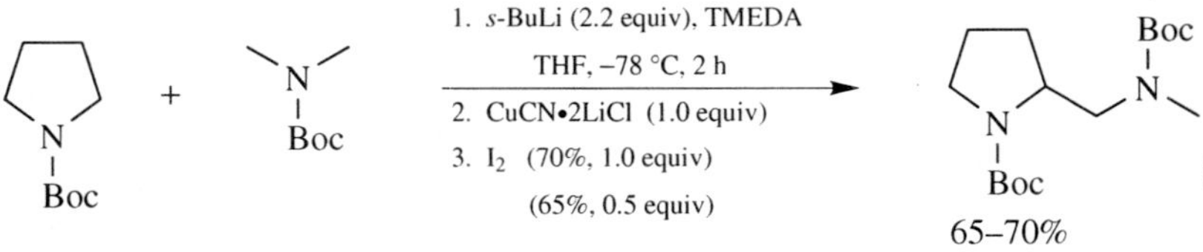

(26)

C. *N*-Heterocyclic Carbene–Copper (NHC–Cu) Complexes

N-Heterocyclic carbenes (NHC), containing two nitrogen heteroatoms alpha to the carbene metal ligation, are an important emerging class of ligands for transition metal mediated reactions[101]. They are effective ligands for a variety of transition metals and provide significant ligand accelerated catalysis (LAC) in a number of transformations. The *N*-heterocyclic carbenes can be readily generated from imidazolium and dihydroimidazolium salts via deprotonation with a variety of bases. Application of these ligands to organocopper chemistry is relatively new, and developments have been outlined in two recent reviews[102]. In many instances, the NHC–Cu complexes appear to be copper(I) species with relatively few examples involving the cuprate or ate complexes.

NHC–Cu complexes were first reported as effective agents for copper-catalyzed conjugate addition reactions of dialkylzinc reagents[103] and asymmetric conjugate additions employing scalemic NHC–Cu complexes soon followed[104]. The LAC of these ligands is attributed to their ability to stabilize the often-invoked Cu(III) intermediate via σ-donor stabilization[103]. High enantioselectivities were achieved with lithium alkoxy carbene ligands suggesting the possible intermediacy of alkoxycuprate reagents (equation 27)[104d]. $Cu(OTf)_2$ is often used and is presumably reduced to a Cu(I) species prior to formation of the active copper reagent promoting the conjugate addition reaction[12b,105]. The superior performance of $Cu(OTf)_2$ appears to be related to its higher solubility in organic solvents. Scalemic NHC−Cu complexes have also been developed that permit the asymmetric conjugate addition of Grignard reagents to enones for the generation of quaternary centers (equation 28)[106]. Restriction of conformational mobility about the N−C bond of the *N*-substituent can be achieved by chelation (e.g. **65**)[106a] or by steric biasing via substituents on the imidazole ring (e.g. **66**)[106b]. The Grignard reactions appear to involve ate complexes (e.g. $NHCCuEt_2MgX$)[106a] or higher-order cuprates if the metal alkoxide is intramolecularly bound to copper. In contrast, the organozinc reactions are more likely to involve organocopper(I) reagents in the absence of intramolecular coordination of a metal alkoxide or amide[104f] heteroatom center.

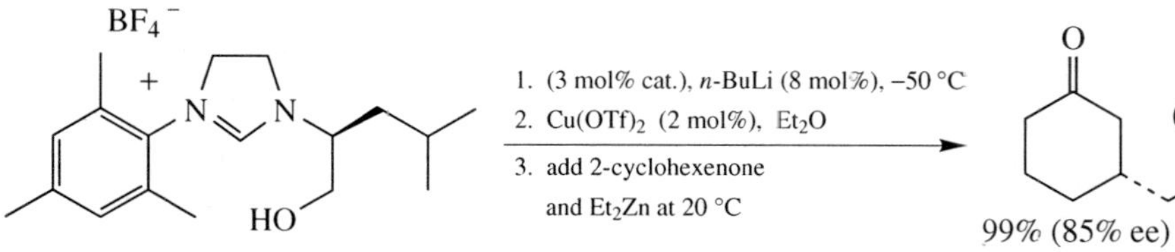

(27)

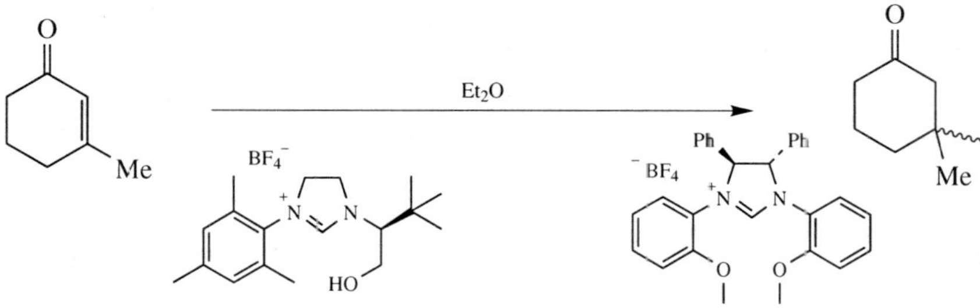

(28)

(65)		(66)
(1.2 equiv)	EtMgBr	(2.0 equiv)
(3 mol%)	$Cu(OTf)_2$	(6 mol%)
(4 mol%)	ImH^+	(3 mol%)
0 or 30 °C, 30 min	*T*, time	−60 °C, 1.5 h
99% conversion (81% isolated) (80% ee) *S*		99% conversion (80% ee) *R*

The use of Grignard reagents and NHC−Cu complexes favor S_N2' (**67**) over S_N2 (**68**) substitution pathways in allylic substitution reactions (equation 29)[107] and cuprates (i.e. $NHC{-}CuR_2^-\ {}^+MgX$) were invoked as the active agents[107b]. In cyclohexenol derivatives, *syn*-S_N2' substitutions are favored over the *anti*-S_N2' pathways[107b]. Scalemic dimeric NHC−Ag(I) complexes in the presence of Cu(II) salts catalyze asymmetric S_N2' substitutions of allyl phosphates with dialkyl-[108a], diallyl-[108b] and diarylzinc[108b] reagents. Utilization of these NHC−Ag(I) complexes permits the use of air-stable $CuCl_2{\bullet}2H_2O$ in place of the moisture-sensitive $Cu(OTf)_2$[108a]. The dimeric NHC−Cu(II) complexes were also isolated and shown to catalyze the enantioselective alkylation reaction, although a copper(I) species (i.e. NHC−CuR) is most likely the catalytic active agent[105]. These reagents were used for the preparation of allylsilanes containing either tertiary or quaternary silicon-substituted carbon centers[108b]. They have also been utilized for the enantioselective (73–95% ee) generation of quaternary centers by catalyzing the conjugate addition of dialkylzinc reagents to 3-carboalkoxy-2-enones[109]. Similarly, both a monomeric copper(II) alkoxy–NHC complex and a dimeric copper(II)–bisalkoxy-NHC complex serve as pre-catalysts for the conjugate addition of Et_2Zn to 2-cyclohexenone[110].

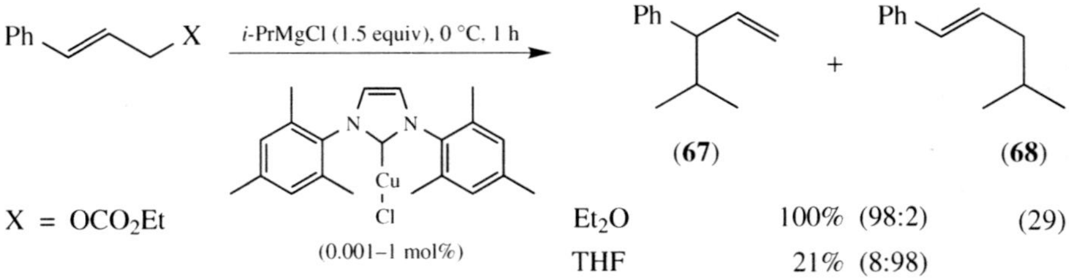

(**67**) (**68**)

X = OCO_2Et	Et_2O	100% (98:2)	(29)
	THF	21% (8:98)	
X = Cl	Et_2O	100% (96:4)	
	THF	100% (89:11)	

The NHC–Cu complexes (e.g. **69a–c**) catalyze a number of copper-mediated transformations (Scheme 11). They can be used as Stryker reagents for the hydrosilylation of ketones[111] and aldehydes[111c,d], hydrocupration of alkynes[112] and for conjugate reductions[111d,113] of α,β-enones and enoates. A tandem conjugate reduction–aldol reaction has been reported with modest diastereoselectivity[114]. These NHC–Cu complexes also catalyze the 1,2-addition of dialkylzinc reagents to *N*-tosylimines[115], the addition of arylboranes[116] or alkyl groups[117] to carbon dioxide and the 1,4-conjugate addition of amines[118b,c], alcohols[118b,c] and thiols[118d] to enones, acrylate esters and α,β-unsaturated nitriles. A NHC–Cu complex of pinacol borane adds to alkenes providing vicinal Cu–B substitution potentially allowing for subsequent transformations at the Cu and B substituted centers (equation 30)[119]. The pinacol borane NHC–Cu complex also effects the 1,2-diboration of aldehydes[120].

D. α-*N*-Heteroarylcuprates

α-Heteroarylzinc cuprates prepared from 2-iodo-imidazoles[121], thiazoles[121], pyridines[122], indoles[123] or quinolines[121] react with allylic halides, 1-iodo-1-alkynes and 3-iodo-2-cyclohexen-1-one to afford coupled products in good yields. These results illustrate the facility with which sp^2-centers bound to copper participate in ligand transfer.

While exploring the reactions of metallated heteroaryls, Knochel and coworkers found the α-*N*-heteroarylzinc reagents to be generally unreactive with most organic electrophiles. However, in the presence of stoichiometric or substoichiometric quantities of CuCN•2LiCl efficient coupling with a variety of substrates (allylic, vinyl and alkynyl halides) was observed[121]. The organozinc precursors were obtained by direct insertion of zinc metal into the carbon–iodine bond of 2-iodo-1-methylimidazole, 2-iodoquinoline or 2-iodo-1,3-thiazole in THF or *N,N*-dimethylacetamide at 25 °C for one hour (zinc insertion into some brominated analogues failed even when heated to reflux). Upon generation of the organozinc species (e.g. **70–72**), 5–10 mol% of CuCN•2LiCl in THF was employed for cuprate formation before reaction with the electrophiles. The α-*N*-heteroarylcuprates gave

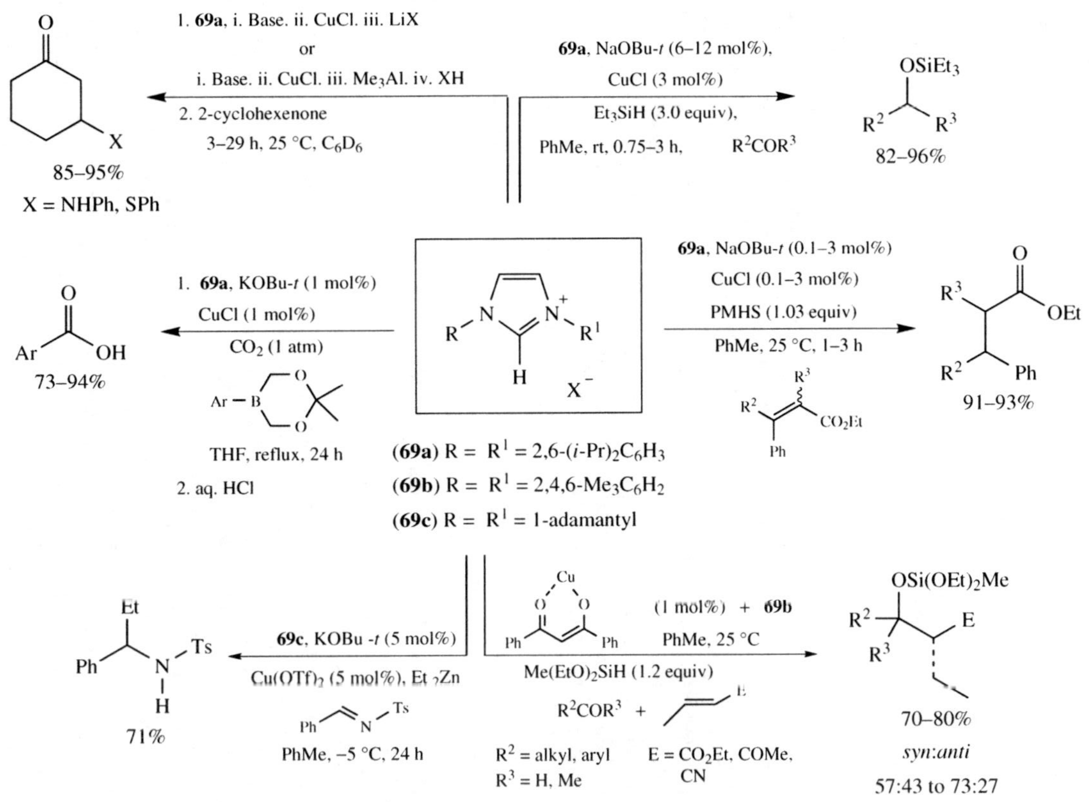

SCHEME 11. Hydrosilylation[111a], conjugate reduction[113], conjugate reduction–aldol[114], 1,2-additions to imines[115], arylborate carboxylation[116] and heteroatom conjugated additions[118b,c] mediated by *N*-heterocyclic carbene–copper complexes

good to excellent yields (68–91%, Scheme 12) of the substituted imidazoles (**73a** and **b**), thiazols (**74**) and quinolines (**75a** and **b**). Smooth reaction with 3-iodo-2-cyclohexen-1-one and 1-iodohexyne required stoichiometric amounts of CuCN•2LiCl, but resulted in good chemical yields (Scheme 12, 70 and 68%).

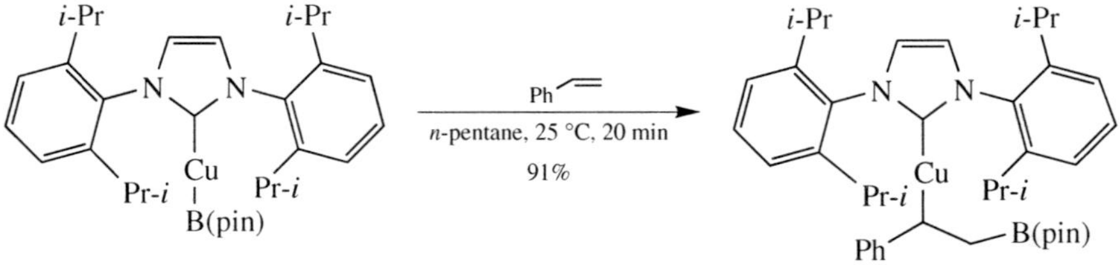

(30)

Later, Knochel and coworkers employed cuprate reagents derived from magnesiated indoles (equation 31) and imidazoles (equation 32) for use in allylic substitution reactions[123]. These Grignard reagents were prepared from the corresponding iodoindoles (e.g. **76**) or iodo- or bromoimidazoles (e.g. **78**) by halogen–magnesium exchange using *i*-PrMgBr in THF[124]. Transmetallation from magnesium to copper was effected by treatment with THF soluble CuCN•LiCl at −30 to −40 °C for 2 to 3 hours[125, 126]. Addition of allyl bromide afforded good to excellent yields of the coupled products **77** and **79** (57–84%). Perhaps the most striking feature of this work is the extent to which other functional groups are tolerated. In these examples, selective halogen–magnesium exchange was achieved on polyhalogenated five-membered heterocycles (e.g. **76** and **78**). Subsequent halogen–metal exchange is exceedingly slow because of the electron-rich nature of the heterocyclic Grignard reagent. Control of regiochemistry is also achieved in the case of tribromoimidazole (**78**, equation 32). Halogen–metal exchange is directed to the 2-position via chelation of the Grignard reagent to the ether oxygen atom. The vigorous reaction conditions (−80 to +100 °C, 1 h, equation 32) illustrate the stability of the cuprate reagent. This methodology has also been applied in the preparation of substituted, polyfunctionalized pyridines (equation 33)[122].

α-*N*-Heteroarylcuprates have also been used in the construction of novel, substituted allyl purines[127]. Purines lithiated at the 6-position suffer from extensive rearrangement to give the more stable 8-lithiated derivatives even at −80 °C. The corresponding purinylzinc reagents have been prepared, but only Pd-mediated cross-coupling with aryl halides has been reported[121]. Allylation at the 6-position was accomplished only with cuprates derived from the corresponding Grignard reagents. The Grignard was prepared from 6-iodopurine by halogen–magnesium exchange (*i*-PrMgBr, THF, −78 °C), and transmetallation to copper was achieved using a variety of Cu(I) salts (e.g. CuCN, CuI). To determine optimal conditions, several protocols, both stoichiometric and catalytic in copper, were examined. The authors found that CuI (20 mol%) was superior to CuCN (10–100 mol%). The Cu(I) salts were not solubilized with LiCl, and all experiments using CuCN were conducted from 0 °C to room temperature presumably to insure cuprate formation (i.e. R_2CuMgI, **80**), whereas reactions with CuI were allowed to warm to room temperature from −80 °C. Using the optimal conditions (CuI, 20 mol%, THF, −80 °C to rt), seven allylic halides were chosen to probe the scope of the reaction (Scheme 13). Yields were poor to good (7–54%), although poor regioselectivity was observed with 1-chloro-2-butene, which gave a mixture of regioisomers **81** and **82** (Scheme 13, S_N2: S_N2' = 59:41).

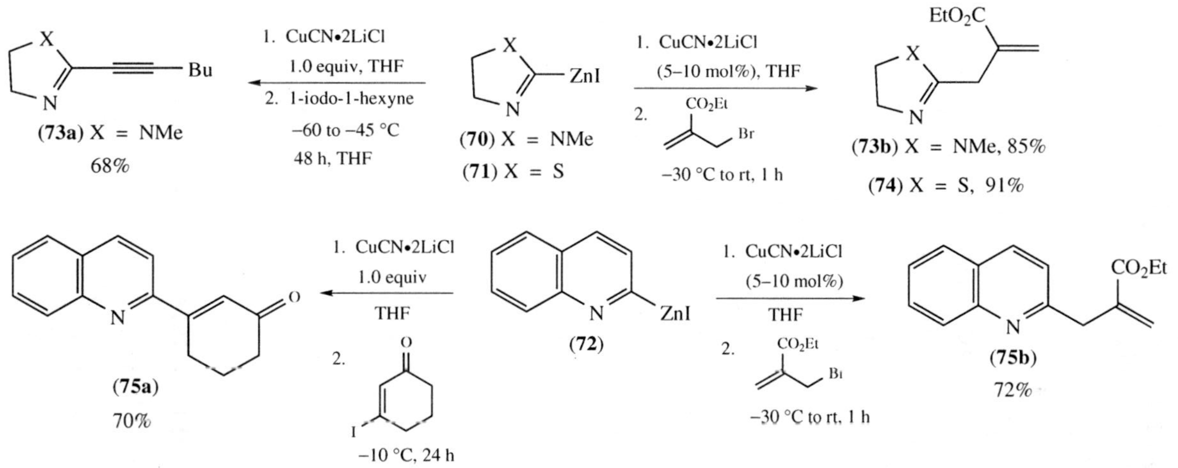

SCHEME 12. Reactions of α-*N*-heteroarylzinc cuprates with allylic, vinyl and alkynyl halides[121]

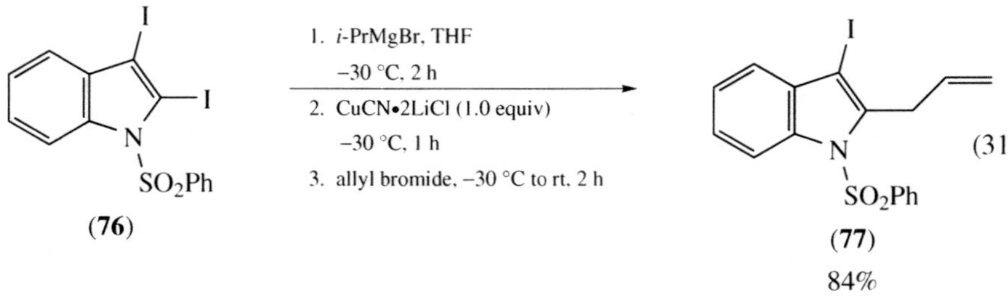

(76) (77) (31)

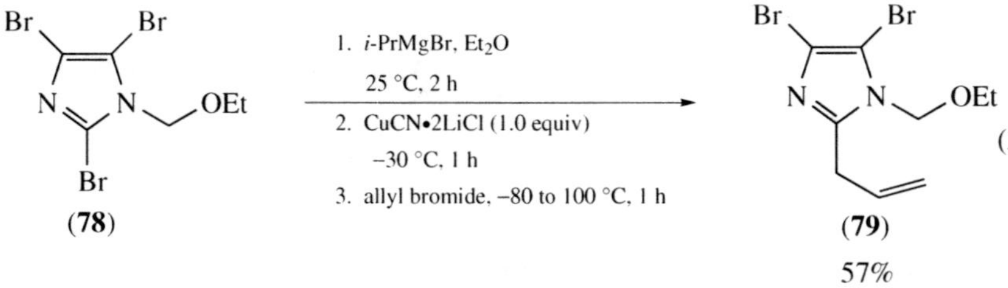

(78) (79) (32)

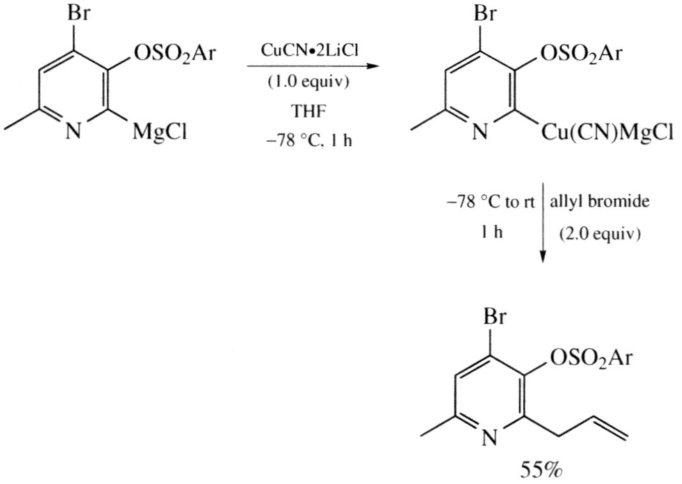

(33)

IV. ORGANOCUPRATES WITH A β-*N*-SUBSTITUENT

A. β-(*N*-Carbonyl)alkylcuprates

There are two β-aminocuprate derivatives where the *N*-atom is protected with a carbonyl functionality. These include the carbamate and amide derivatives and involve the formation and utilization of zinc cuprate reagents.

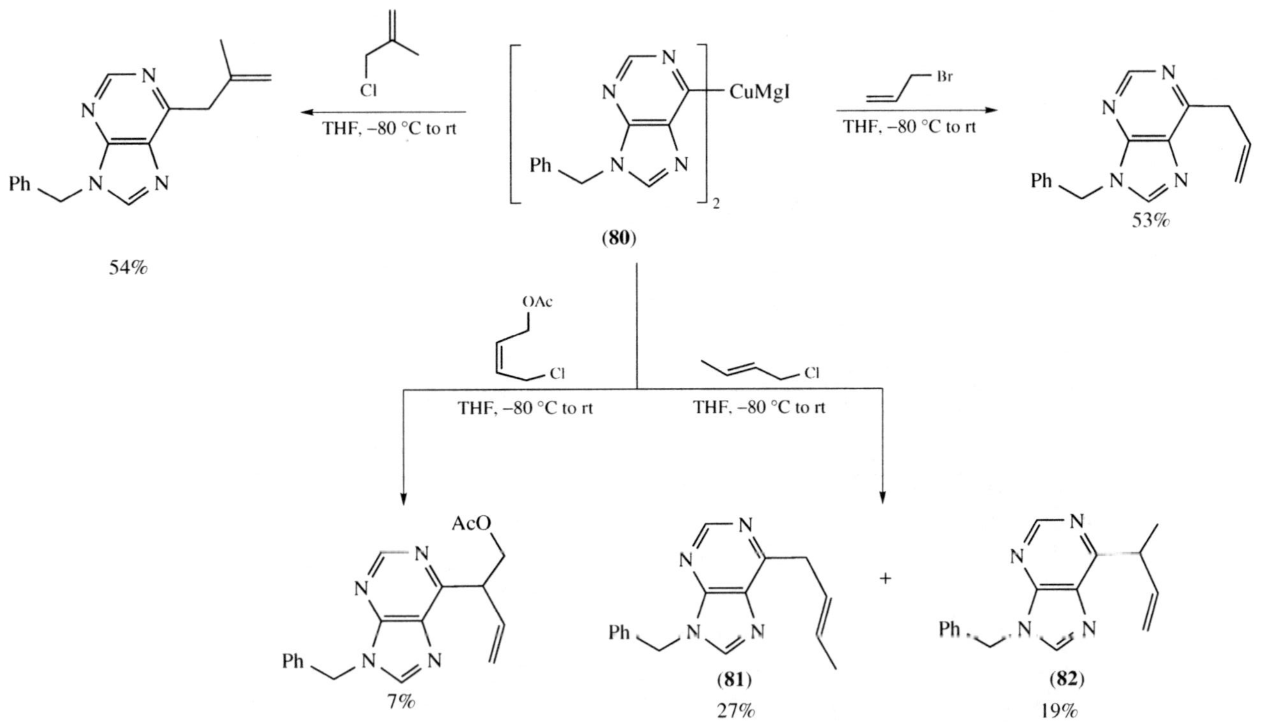

SCHEME 13. Preparation of 6-allylpurine derivatives using α-*N*-heteroarylcuprates[127]

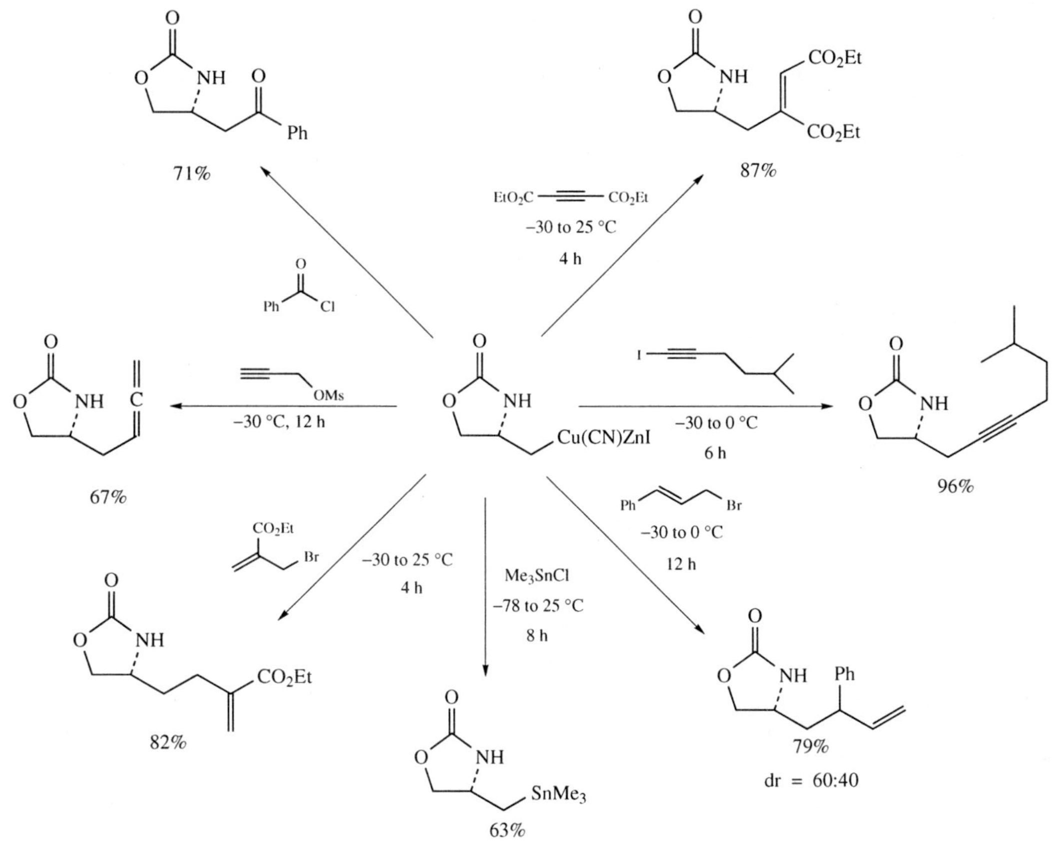

SCHEME 14. Reactions of β-cuprio-*N*-carbamoyl reagents with ynoates, 1-haloalkynes, allylic halides, chlorostannanes, propargyl mesylates and acyl halides in THF[128]

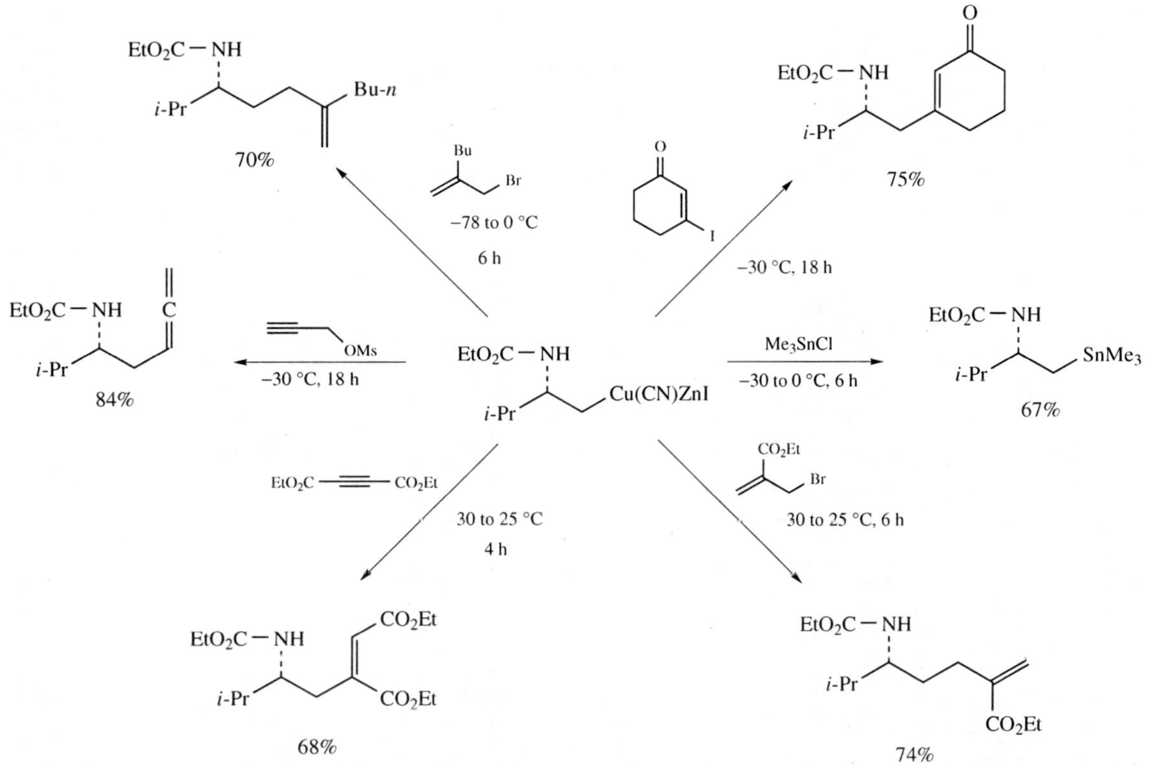

SCHEME 15. Reactions of β-cuprio-*N*-carbamoyl reagents with vinyl halides, propargyl mesylates and ynoates in THF[128]

1. β-Cuprio-N-carbamoyl reagents

Knochel and coworkers developed a protocol by which chiral, amino acid derived, β-*N*-carbamoyl cuprates may be prepared from the corresponding organozinc reagents[128]. These reagents are of particular interest because of the intact stereogenic center β to the copper atom. Once generated, the organozinc reagents undergo transmetallation to copper (CuCN•2LiCl, 1.0 equivalents, 0 °C, 5 min) and the cuprate reagents react with several classes of electrophiles to afford polyfunctional amines protected as carbamates. These β-(*N*-carbamoyl)alkylcuprates react with a wide array of substrates including allyl and vinyl halides, 1-iodoalkynes, acid chlorides, propargyl mesylates, trialkylchlorostannanes and α,β-alkynyl esters. Reactions of cyclic β-(*N*-carbamoyl)alkylcuprates afford good to excellent yields (Scheme 14, 63–96%) of coupled products. Reactions of chiral acyclic ethyl carbamates were also reported to give good results (Scheme 15, 67–84%).

Other carbamates that have been used in cuprate-mediated allylation reactions include the *N*-Boc-azetidines introduced by Billotte[129]. In this procedure, 3-iodo-*N*-Boc-azetidine is converted to the organozinc reagent by treatment with Zn dust. Transmetallation to copper (CuCN•2LiCl, 1.0 equivalents, 0 °C) followed by addition of allyl bromide gives the coupled product in moderate yield (equation 34, 58%).

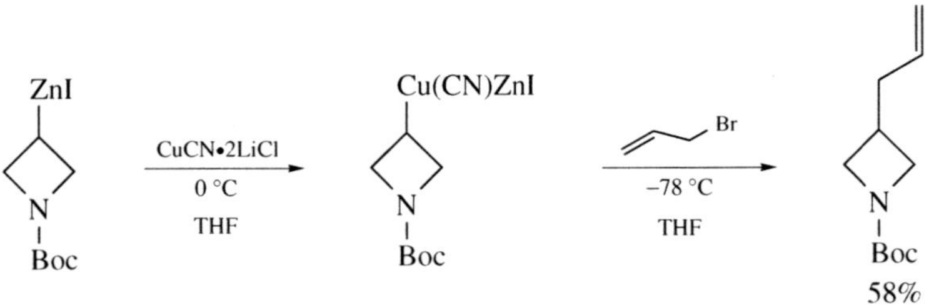

(34)

Hiemstra and coworkers used β-(*N*-carbamoyl)alkylcuprate reagents in reactions with propargyl tosylates (or mesylates) to prepare allene precursors for use in synthetic routes to bicyclic carbamates and lactams[130]. Only two examples with carbamates are noted, and yields are modest to moderate (equation 35, 39–49%). The cuprates were prepared from the corresponding organozinc species using a catalytic amount of $CuBr•SMe_2$ in DMF, which was preferred over using stoichiometric amounts of CuCN•2LiCl.

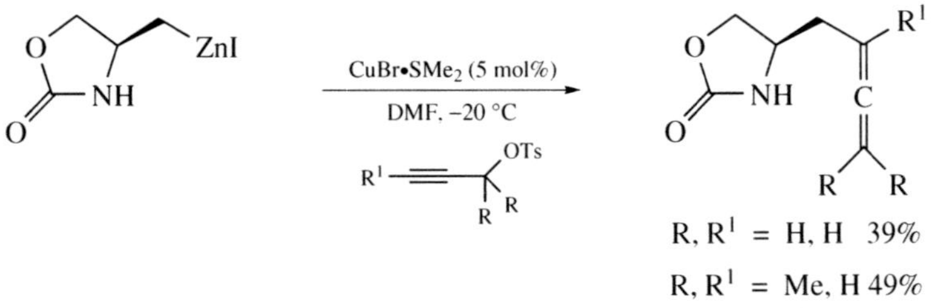

(35)

Reactions of β-(*N*-carbamoyl)alkylcuprates with aromatic, aliphatic and alkenyl acid chlorides have also been reported[131]. The cuprates were prepared from the corresponding organozinc reagents in THF and *N*,*N*-dimethylacetamide (THF:DMA 1:1) at −25 °C with stoichiometric amounts of CuCN•2LiCl. Yields were generally good (equation 36, 65–78%), with the exception of propenoyl chloride that failed completely. An important consequence of these reactions is that a *N*-substituted stereogenic center is installed *β* to the newly generated ketone carbonyl.

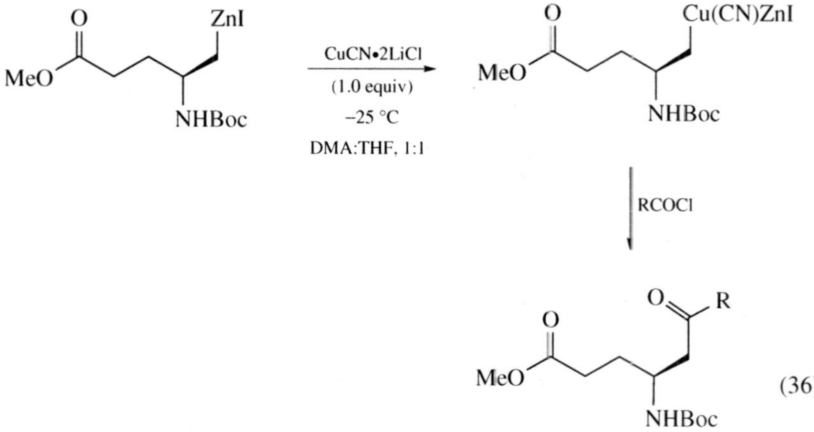

(36)

R	% Yield[131]
(*E*)-C(=$CHOCH_3$)CH_2CH_3	68
Ph	75
CH=CH_2	0
CH_2OAc	65
2-furyl	78
$(CH_2)_4CH_3$	77

2. *β*-Cuprioamide reagents

Cuprate reagents, which possess an amide functional group *β* to the copper atom, have also been reported to be effective conduits for installation of a nitrogen functionality into organic frameworks. Knochel and coworkers explored these reagents while examining the corresponding *β*-amidozinc reagents from which the cuprates were prepared[128]. Cuprates prepared from stoichiometric amounts of CuCN•2LiCl (1.0 equivalents) reacted with allylic bromides, acid chlorides, trialkylchlorostannanes and *α*,*β*-alkynyl esters to give the coupled products in moderate to excellent yields (Scheme 16, 34–93%). For reactions involving secondary *β*-cuprioamide reagents the chemical yields are good to excellent (76–93%), but diastereoselectivity remains uncontrolled (Scheme 17).

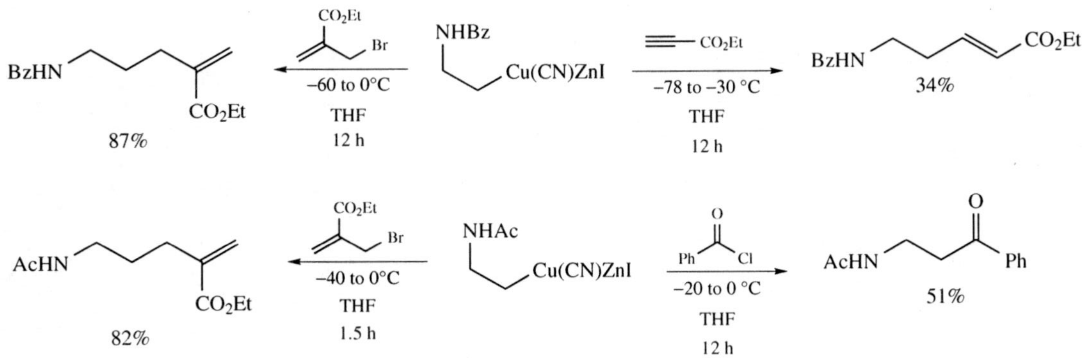

SCHEME 16. Reactions of β-cuprioamide reagents with allyl halides, ynoates and acyl halides[128]

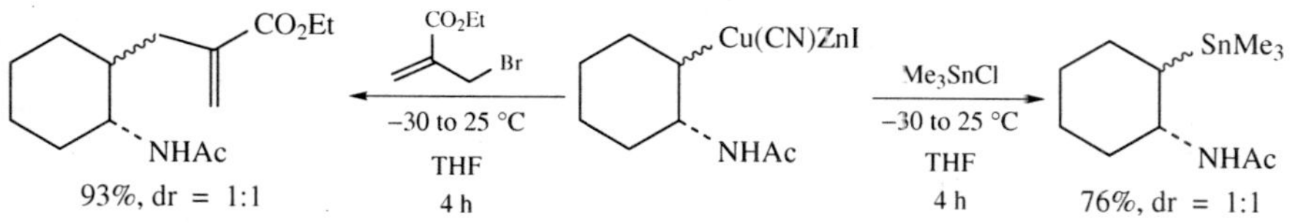

SCHEME 17. Reactions of β-cuprioamide reagents with allylic halides and chlorostannanes[128]

Hiemstra and coworkers also investigated β-cuprioamide reagents in reactions with propargyl tosylates and mesylates[130]. The cuprates were prepared from the corresponding organozinc reagents using a catalytic amount of $CuBr{\bullet}SMe_2$ in DMF. Chemical yields are moderate (equation 37, 51–59%), but this methodology does allow incorporation of oxygen-containing substituents and aromatic groups in the propargyl substrate.

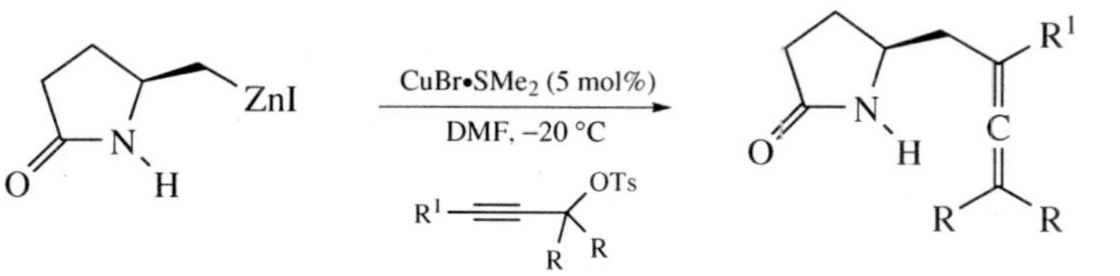

R^1	R	% Yield[130]
H	H	58
Me	H	59
H	Me	54
Ph	H	56
CH_2OTBS[a]	H	51
CH_2OAr	H	54

(37)

[a] TBS = *tert*-Butyldimethylsilyl

With designs on preparing bicyclic amines and lactams, Hiemstra and coworkers later extended this work to include reactions with 1-iodoalkynes and allenyl bromides. Although earlier work employed catalytic amounts of copper ($CuBr{\bullet}SMe_2$), stoichiometric quantities of copper ($CuCN{\bullet}2LiCl$, 1.0 equivalents) were employed in these reactions[132]. Results were moderate with yields ranging from 49–52% (Scheme 18).

Hielmgaard and Tanner reported reactions of β-cuprioamide reagents with a range of aroyl, alkanoyl and alkenoyl chlorides[131]. The cuprates were prepared from the corresponding organozinc reagents in a mixture of THF and dimethylacetamide (THF:DMA 1:1) with one equivalent of $CuCN{\bullet}2LiCl$. Coupling of these reagents with acid chlorides appears to be capricious with isolated yields ranging from moderate (44–51%) to complete failure of the reaction (equation 38).

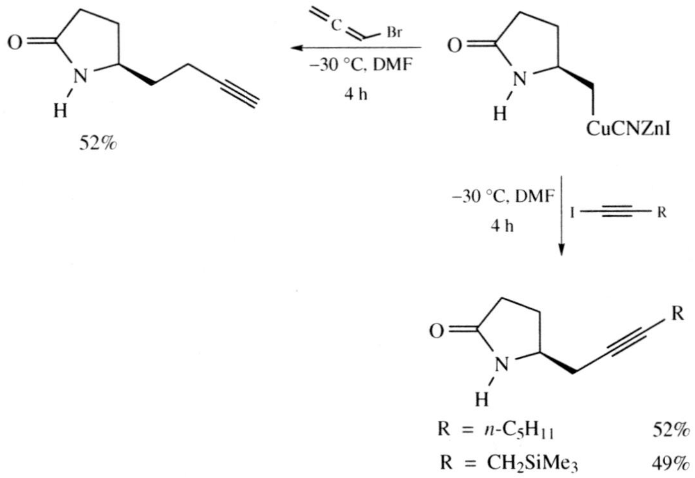

SCHEME 18. Reactions of β-cuprioamide reagents with allenyl and alkynyl halides[132]

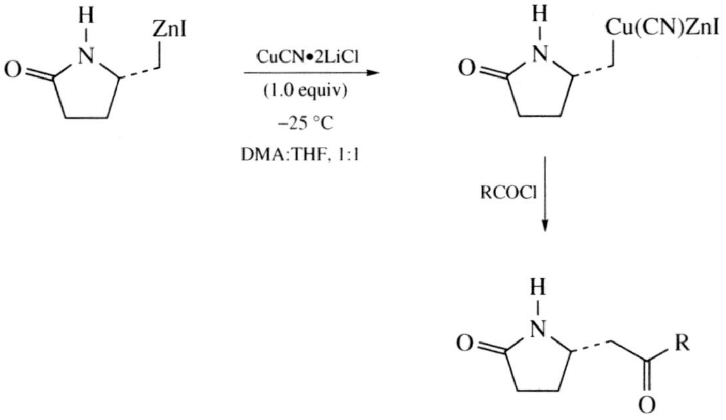

(38)

R	% Yield[131]
(*E*)-C(=CHOCH$_3$)CH$_2$CH$_3$	51
Ph	52
CH=CH$_2$	0
CH$_2$OAc	0
2-furyl	44
(CH$_2$)$_4$CH$_3$	0

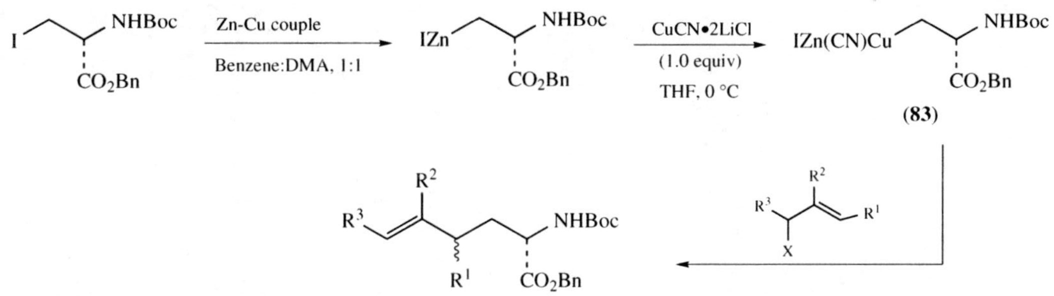

R^1	R^2	R^3	X	% Yield
H	H	H	Cl	65
CH_2Br	H	H	Br	48
Ph	H	H	Cl	48
H	Me	H	OTs	56
H	CO_2Me	H	Br	51
H	H	H	Br	32

SCHEME 19. Reactions of zinc–copper reagent **83** with allylic substrates[133]

B. β-Cuprio-α-aminocarboxylates

β-Cuprio-α-aminocarboxylates such as **83** (Scheme 19) have been prepared by Jackson and Dunn using the corresponding alkylzinc reagents derived from serine[133]. The goal of this work was to develop a novel, convenient route to new unsaturated *N*-protected amino acids. Initial work was largely confined to allylic substitution reactions employing simple allylic halides and tosylates. Allylic chlorides outperformed bromides generally giving higher chemical yields. Only one example of a propargyl system was examined (propargyl bromide, 55%). The reactions proceed in an S_N2' fashion with modest to good chemical yields (32–65%), but with poor diastereoselectivity. The authors suggest that internal alkenes are less reactive than terminal ones, however the yields hover around 40–60% irrespective of alkene substitution (Scheme 19).

In later work, the authors expanded this methodology to include reactions with acid chlorides, α,β-enones, 1-bromo- or 1-iodoalkynes and propargyl halides[134]. Reactions with propargyl tosylates resulted in selective formation of allenes (no acetylenes) in moderate to good yields (51–81%) and with impressive diastereoselectivity (equation 39, de 88–89%). Cuprate formation was achieved by treating the corresponding alkylzinc reagent with THF soluble CuCN•2LiCl according to Knochel's protocol[135]. The stereochemical integrity of the stereocenter β to copper remained intact with no observable epimerization. Clive and coworkers used this methodology to prepare the angiotensin-converting enzyme inhibitors (−) A58365A and (−) A58365B[136].

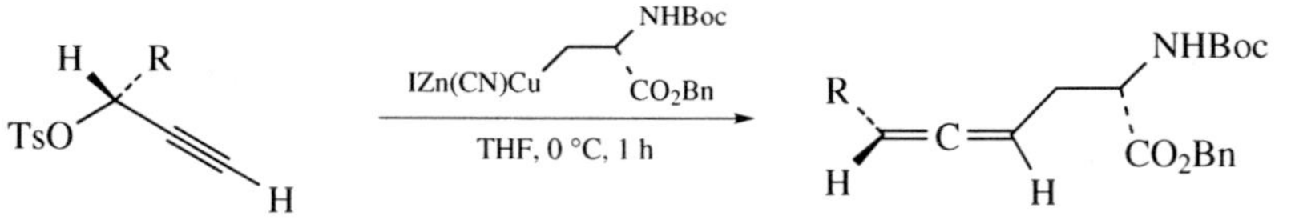

(39)

R = Me 88% ee
R = Ph 85% ee

R = Me 81% dr = 94.5:5.5
R = Ph 51% dr = 94:6

These serine-derived zinc alkyl(cyano)cuprate reagents have also been used with α,β-alkynyl iminium triflates in DMF to give α,β-unsaturated ketones after aqueous workup[137]. Utilization of these reagents in conjugate addition reactions has been reported to give poor results[128,138]. However, switching to a more polar aprotic solvent such as DMF resulted in modest yields (28–33%) of the conjugate adducts and in some reactions eliminated the need for additives such as chlorotrimethylsilane. However, in reactions with allyl chloride, the results were better with yields ranging from 73–82% (equation 40)[139]. Formation of the cuprate reagent was achieved by warming the reaction from −55 to 0 °C. Upon cuprate formation, the reaction was recooled to −78 °C before the electrophiles were added.

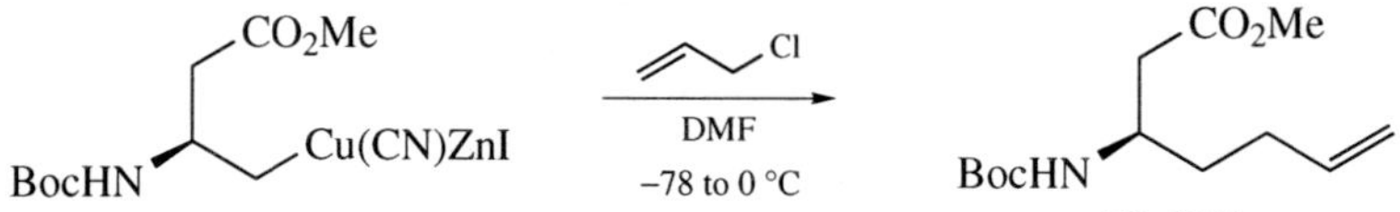

(40)

73–82%

More recently, Jackson and coworkers developed procedures catalytic in copper (CuBr•DMS) as a means of exerting regiocontrol in reactions with substituted allyl halides (e.g. 1-bromo-2,3-dimethyl-2-butene)[140]. However, while the use of catalytic amounts of copper did simplify the workup procedure, no enhancement of regiocontrol was observed

TABLE 1. Chemical yields as a function of CuX (X = CN, Br) stoichiometry in the reactions of β-cuprio-α-aminocarboxylates with allyl and propargyl bromide[141]

Zn reagent	Electrophile	Product	% Yield (stoichiometric Cu)	% Yield (catalytic Cu)
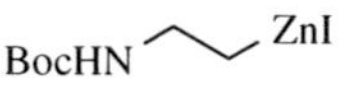	allyl bromide	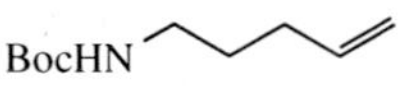	33	51
	allyl bromide	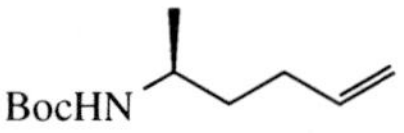	36	55
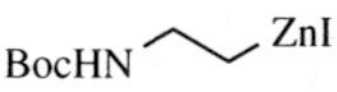	propargyl bromide	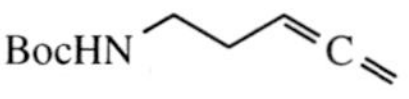	30	63
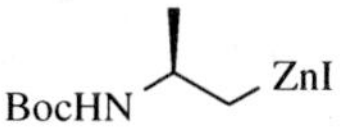	propargyl bromide	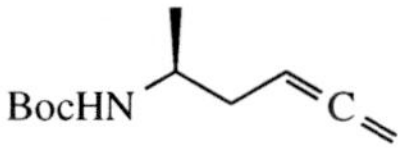	29	60

(equation 41, S_N2': S_N2 = 49:51). Furthermore, diastereoselection was modest with the tosylate derived from (E)-2-methyl-2-buten-1-ol (59%, dr 63:37).

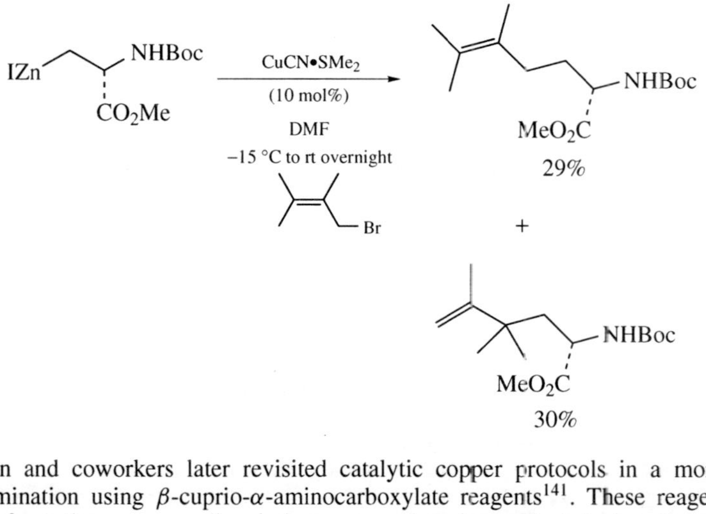

(41)

Jackson and coworkers later revisited catalytic copper protocols in a more extensive examination using β-cuprio-α-aminocarboxylate reagents[141]. These reagents were prepared from the corresponding iodo compounds using activated zinc dust in DMF. Transmetallation to copper was accomplished using a catalytic amount of copper bromide (CuBr•DMS, 5 mol%) at −55 °C. To insure cuprate formation, the solutions were allowed to warm briefly to 0 °C, and then re-cooled to −55 °C before electrophile addition. Allylic and propargylic halides were investigated along with one example of a vinyl halide (3-iodocyclohex-2-enone, Table 1). Results were generally better when the catalytic protocol was employed with chemical yields ranging from 47–78%. This was a marked improvement over the reactions stoichiometric in copper where yields were more modest (29–59%). The R_2CuZnI species is most likely the active reagent in the catalytic procedure.

This methodology has also been used to prepare 9-fluorenylmethoxycarbonyl (Fmoc)-protected amino acids for use in automated solid-phase peptide synthesis[142]. Jackson and coworkers employed a stoichiometric quantity of soluble CuCN•LiCl to facilitate cuprate formation from the organozinc species in THF. Subsequent reaction with allyl chloride provided the coupled product in good chemical yield (60%).

Primary and secondary β-iodo amino acids have also been coupled with acid chlorides of aryl and alkyl phosphinous acids in moderate to good yields (equation 42)[143]. The β-iodo amino acid is converted to the corresponding zinc iodide and subsequently treated with stoichiometric amounts of CuCN•LiCl followed by chlorodiphenylphosphine. The initially formed phosphine is protected as the phosphinothione, which may later be unmasked with Raney nickel. Yields range from poor (equation 42, 32%) with secondary iodides to good (75%) with primary iodides. Reaction of the cuprate reagent with bis(*tert*-butyl)chlorophosphine fails to give the coupled product in any proportion. The authors found that while secondary iodides do participate in these reactions, they do so with complete loss of stereochemistry at the carbon bearing the iodine. The authors cite several articles related to this problem[144]. Gilbertson and coworkers later exploited this work in the synthesis of a bisphosphine ligand system for the palladium-catalyzed addition to cyclic allyl acetates[145].

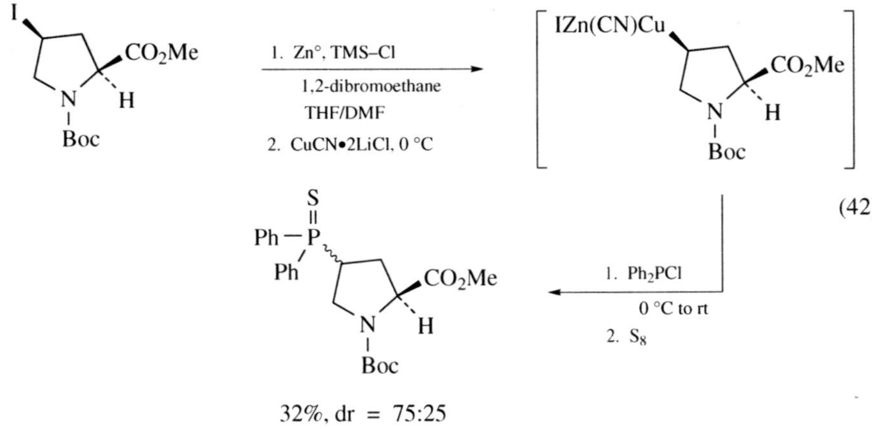

(42)

Isobe and coworkers employed serine-derived zinc–copper reagents in reactions with alkynyl iodides for the preparation of α-*C*-glycosylamino acids[146]. The reaction, which proceeds in modest yield (43%), tolerates oxygen-rich functionality well as evidenced by the sugar moiety present on the acetylenic halide[147] (equation 43). Seeking to explore the issue of stereocontrol in these coupling reactions, Wilson and Jackson used the secondary L-threonine-derived zinc reagents as precursors in copper-catalyzed (CuBr•DMS) couplings with three allylic halides[148]. The chemical yields were modest (34–38%), although no diastereoselectivity was observed.

Alkenyl phosphates have also been coupled to these serine-derived zinc cuprate reagents to prepare β-cycloalkylalanine derivatives[149]. The reactions were effective for six- and seven-membered racemic endocyclic allylic phosphates as modest to good yields were achieved (40–55%), but diastereoselectivities were poor ($n = 1$, 53:47; $n = 2$, 60:40, equation 44) as little kinetic resolution occurred. It has been suggested that the decomposition of β-*N*-carbamoylzinc reagents (cuprate precursors) by β-elimination is a unimolecular *syn*-process and is solvent-dependent[150]. The nature of the protecting group on nitrogen also seems to play an important role. The authors speculate that coordination of zinc by the protecting group facilitates the elimination event (equation 45). This problem can be circumvented by judicial choice of reaction solvent. It was demonstrated that elimination occurred four times faster in THF than in the more polar DMF, and this is attributed to the more robust coordination of zinc to DMF rather than to the carbamate, thereby slowing decomposition.

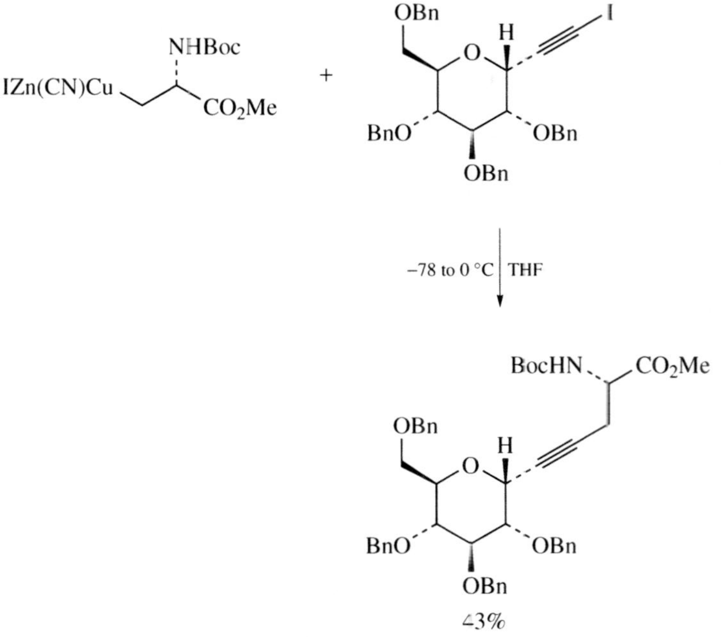

(43)

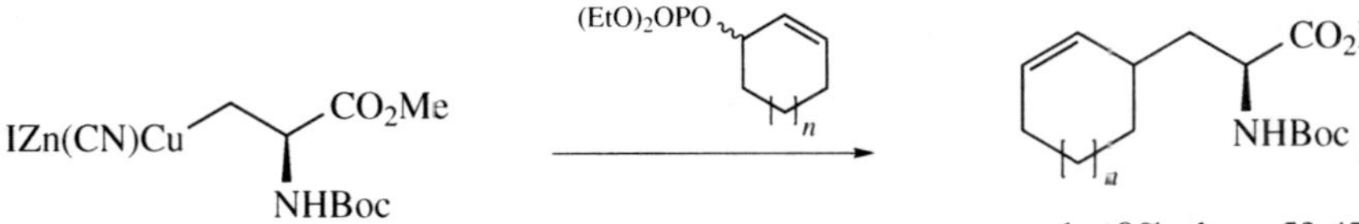

(44)

$n = 1$, 40%, dr = 53:47

$n = 2$, 55%, dr = 60:40

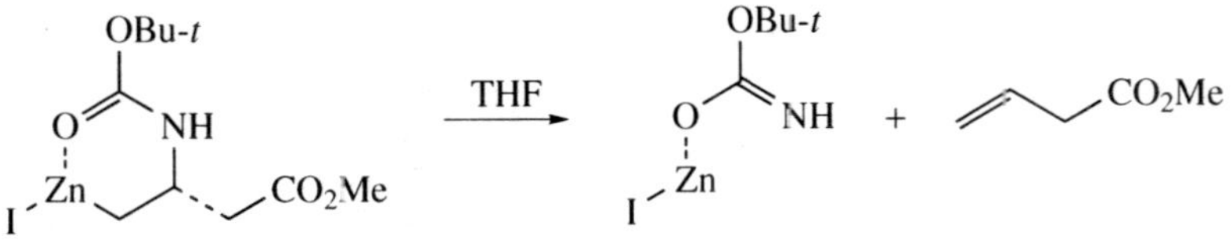

(45)

C. Aryl and Heteroarylcuprates with a β-*N*-Atom

Aromatic cuprates containing a β-*N*-atom consist of aromatic compounds with a *N*-substitutent *ortho* to the carbon–metal bond or *N*-heterocyclic compounds with metallation occurring β to the nitrogen atom. Initial metallation of the aromatic ring can be achieved by Directed Ortho Metallation (DOM) protocols or by halogen metal exchange.

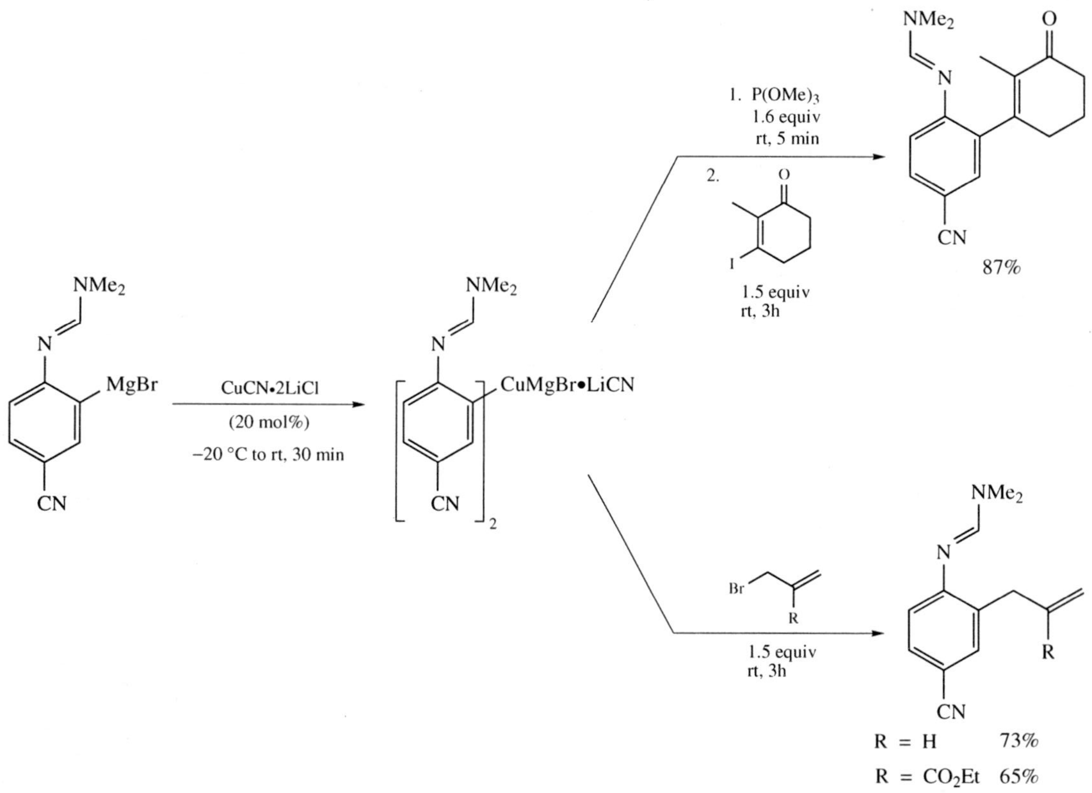

SCHEME 20. Reactions of β-N-formamidinylarylcuprates with allylic and vinylic halides[151]

1. β-N-Substituted arylcuprates

Much of the impetus for the development of zinc cuprates arose from the incompatibility of the corresponding alkyllithium or Grignard reagents toward many important functional groups. Furthermore, *N*-substituted aryl Grignard reagents often require masking of the nitrogen functionality in the form of a protecting group. Knochel and coworkers found the formamidine moiety[70] to be an excellent protecting group for the generation of *N*-substituted aryl Grignard reagents. These Grignard reagents could then be added directly to various electrophiles, or alternately, transmetallated with Cu(I) salts to form cuprate reagents[151,152]. These *β*-*N*-substituted arylcuprates performed admirably in both allylic and vinylic substitution reactions. The Grignard reagent was prepared from the corresponding aryl bromide by halogen–magnesium exchange using *i*-PrMgBr in THF and then treated with THF soluble CuCN•LiCl (20 mol%) at −20 °C to effect cuprate formation. Addition of allyl bromide or ethyl-2-(bromomethyl)prop-2-enoate afforded good chemical yields of the coupled products (73 and 65%, respectively, Scheme 20). This same protocol was employed with 3-iodo-2-methylcyclohex-2-en-1-one in the presence of trimethylphosphite to give the vinylation product in excellent yield (87%, Scheme 20).

Knochel and coworkers later extended this methodology to include reactions between functionalized, primary alkyl iodides and *β*-*N*-substituted arylcuprates. It was demonstrated that the formamidine moiety was not crucial to the success of the reaction and so it was replaced with either simple alkyl groups (e.g. *N*,*N*-dimethyl), or more easily removed allyl groups (equation 46)[153]. The cuprate reagents RCu(CN)MgI from CuCN•2LiCl, 1.0 equivalents) were again prepared from the corresponding Grignard reagents that were obtained by magnesium–halogen exchange from the iodoaniline derivatives. The chemical yields were uniformly good with ethyl 4-iodobutanoate ranging from 65% (*N*,*N*-bis allyl) to 73% (*N*,*N*-dimethyl). Likewise, reaction with ethyl-2-(bromomethyl)acrylate afforded the coupled product in good chemical yield (81%) demonstrating the broad utility of these cyanocuprate reagents (equation 46)[154]. The use of primary anilines was also explored[155]. Iodoanilines were treated with two equivalents of *i*-PrMgCl to effect metal–halogen exchange as well as deprotonation of the aniline nitrogen. Conversion to the corresponding cuprate was accomplished using one equivalent of CuCN•2LiCl in THF at −30 °C for one hour. Subsequent coupling reactions afforded good to excellent yields with allylic bromides (70–91%), propargyl bromides (82–89%) as well as *α*,*β*-alkynyl esters (69–71%). Interestingly, reactions with propargyl bromide show a preference for S_N2 rather than S_N2' substitution (i.e. no allenes formed).

After the initial success of this protocol, the emphasis was redirected toward sequential allylic substitutions on polyfunctional aniline derivatives[156,157]. This work showed that diiodoanilines, protected as formamidines, will tolerate the sequence of reactions involving halogen–magnesium exchange with *i*-PrMgBr to prepare the Grignard reagent, transmetallation to copper with THF soluble CuCN•2LiCl and subsequent coupling to allylic halides, while leaving the second iodine available for a second alkyl or allylic substitution. Isolated yields for the first reaction were good ranging from 68–75%. Likewise, subsequent reactions with primary alkyl halides (e.g. ethyl 4-iodobutanoate) afforded good results (53%). The results for allylic substitutions were even more impressive (88%, Scheme 21). This work showcases the power of *α*-*N*-substituted arylcuprate reagents as efficient synthetic tools for constructing carbon–carbon bonds in aromatic compounds with diverse and sensitive functional groups. It also underscores the importance of these reagents to organic synthesis in general as is illustrated by the preparation of polyfunctional indoles[123,152,156], quinolines[156,157] and quinazolinones[156].

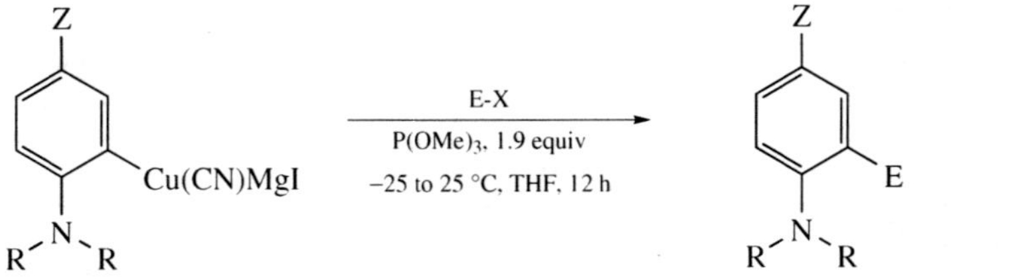

(46)

R	Z	E	X	% Yield[153,154]
$CH_2CH{=}CH_2$	CO_2Et	$CH_2(CH_2)_3Cl$	I	69
$CH_2CH{=}CH_2$	CO_2Et	$CH_2C(CO_2Et){=}CH_2$	Br	81
$CH_2CH{=}CH_2$	CO_2Et	$CH_2(CH_2)_2CO_2Et$	I	65
CH_3	H	$CH_2(CH_2)_2CO_2Et$	I	73

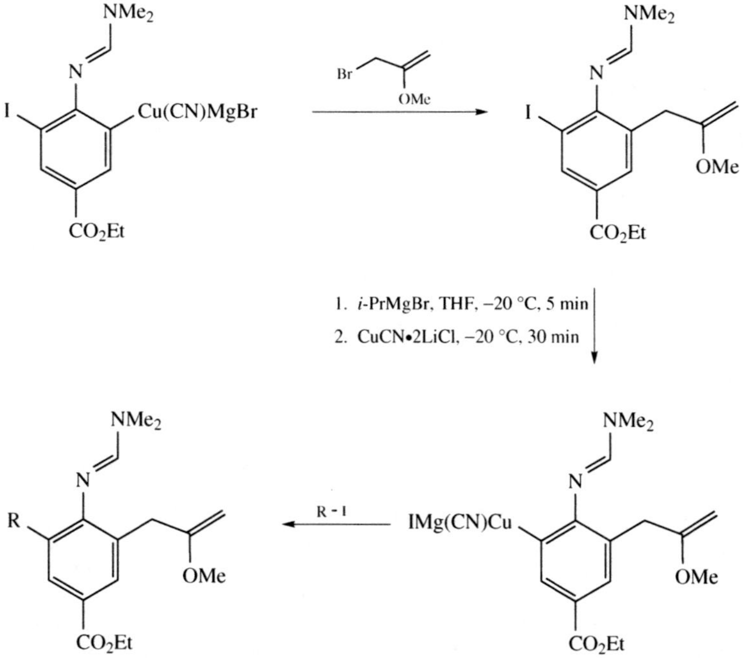

R = $CH_2CH{=}CH_2$, −20 °C to rt, 30 min 88%

R = $(CH_2)_3CO_2Et$, −20 °C to rt, 16 h 53%

SCHEME 21. Sequential coupling reactions of β-N-formamidinylarylcuprates[156]

TABLE 2. Reactions of β-nitroarylcuprates with allylic and acyl halides[158]

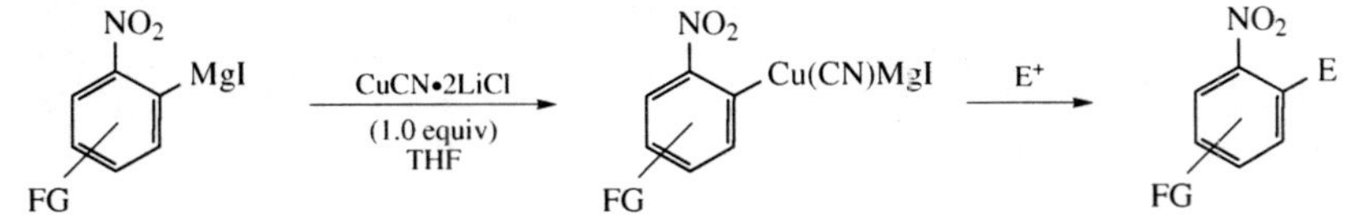

E	FG	Product	% Yield
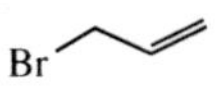	H	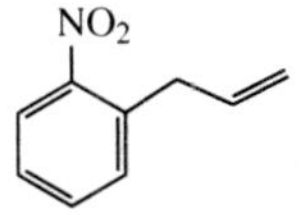	75
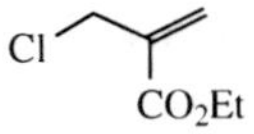	OMe	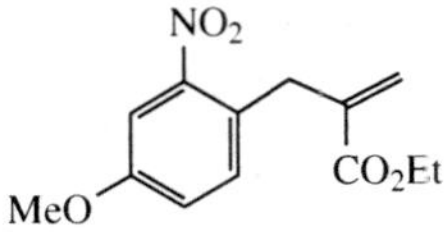	72
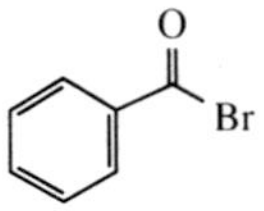	CO_2Et	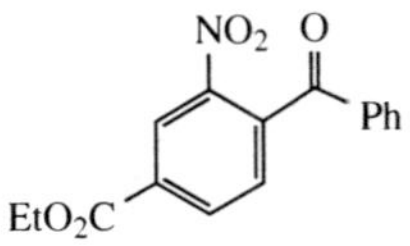	76
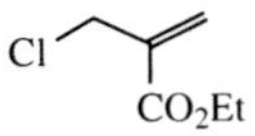	I	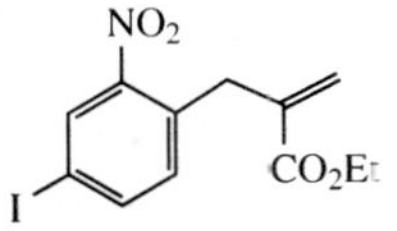	74

The effects of a nitro group on the arylcuprates have also been explored[158, 159]. Sapountzis and Knochel[158] exploited the magnesium–halogen exchange protocol (PhMgCl) to prepare β-nitroaryl Grignard reagents from the corresponding functionalized iodonitrobenzenes. Transmetallation to copper was accomplished with stoichiometric amounts of CuCN•2LiCl in THF and the resulting cuprate reagents were treated with allylic halides or acyl bromides. Isolated yields ranged from 72–76% (4 examples) indicating that no reduction of the nitro group by PhMgCl takes place. Reduction of nitro groups by Grignard reagents has been previously reported[158, 160, 161]. Presumably, with allylic substrates a preference for S_N2' would be predicted; however, this issue is not addressed and remains ambiguous with the choice of allylic halides (Table 2). This work was later extended to include reactions with vinyl halides (e.g. 3-iodo-2-methylcyclopentenone) as well as acyl halides[162]. Reaction with benzoyl chloride afforded the acylation product in modest yield (45%), while the more reactive benzoyl bromide and aliphatic propionyl chloride gave better results (76% and 61%, respectively).

2. *β-N-Heteroarylcuprates*

While exploring the reactions of metallated heteroaromatic compounds, Knochel and coworkers employed β-*N*-heteroarylcuprate reagents as tools for constructing substituted quinolines and uracils through allylic substitution reactions[121]. Organozinc precursors were used rather than Grignard reagents for cuprate preparation and were obtained by insertion of zinc metal into the carbon–iodine bond of 3-iodoquinoline (*N,N*-dimethylacetamide,

70 °C, 1 hour) or 5-iodo-1,3-dimethyl-1,3,5,6-tetrahydropyrimidine-2,4-dione (THF, 70 °C, 3 hours). Upon generation of the organozinc species, 5–10 mol% of CuCN•2LiCl in THF was employed for cuprate formation before treatment with ethyl (2-bromomethyl) acrylate resulting in good yields (equations 47 and 48, 71–74%). The cuprate species in these reactions is presumably of the type R_2CuZnI.

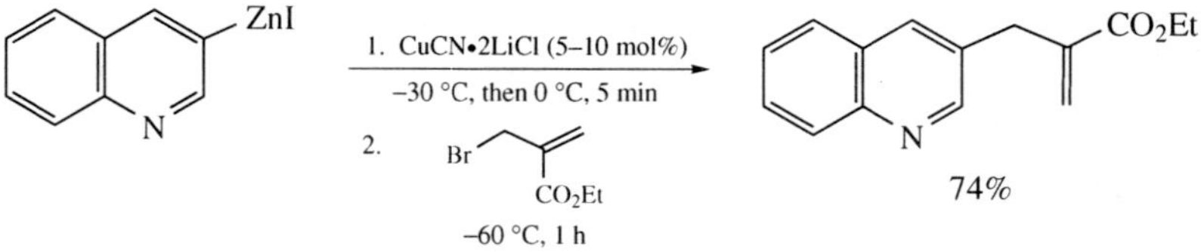

(47)

Knochel and coworkers later extended this work to include reactions with allyl and propargyl bromide, and also in conjugate addition reactions to α,β-alkynyl esters[123, 157]. Cuprate reagents were prepared from the corresponding Grignard reagents. Yields of allylic substitutions are good (70–77%), while the yields for conjugate additions are more modest (41%). Reaction with propargyl bromide results in formation of a small quantity (10%) of allene, indicating a strong but incomplete preference for an S_N2-reaction pathway that leads to the acetylene as the major product (Table 3).

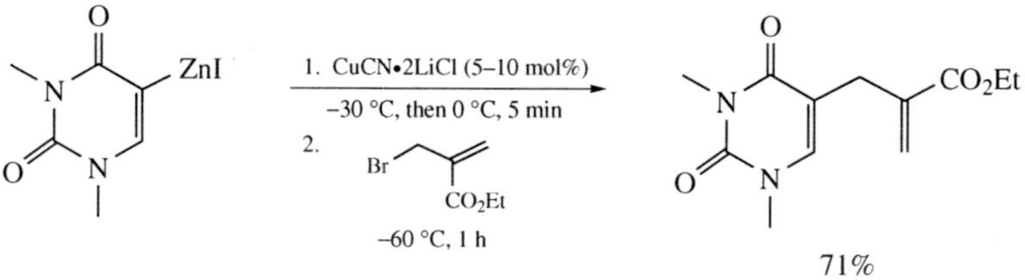

(48)

TABLE 3. Reactions of β-*N*-heteroarylcuprates with allyl bromides, propargyl bromides and *tert*-butyl prop-2-ynoate[121]

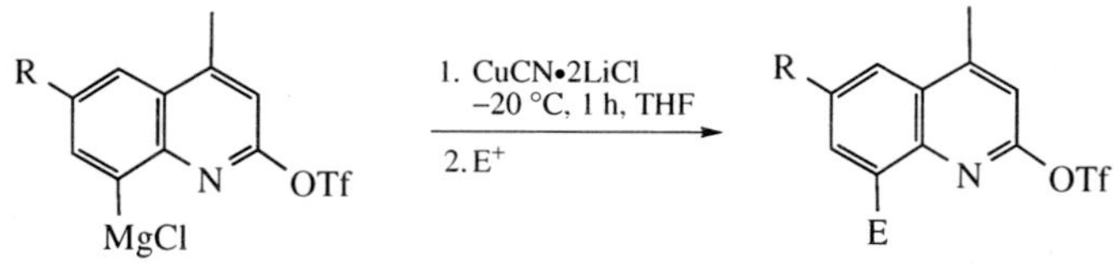

R	E^+	E	% Yield
CF_3	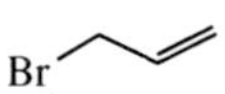	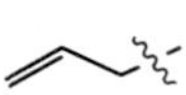	70
CO_2Et	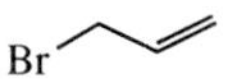	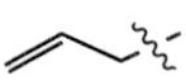	77
CO_2Et	≡—CO_2Bu-*t*	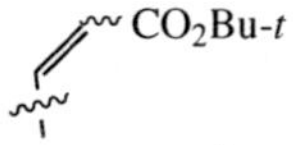	41
CO_2Et	≡—CH_2Br		70

β-N-Heteroarylcuprate reagents have also been used in allylic substitution reactions to prepare polyfunctionalized pyridine derivatives[154]. Chelate-directed halogen–magnesium exchange allows selective metallation at the 3-position by coordination of the formamidine nitrogen, giving rise to the Grignard reagent. This is then converted to the corresponding cuprate reagent ($R_2CuMgBr$) by treatment with catalytic amounts of CuCN•2LiCl in THF. Subsequent reaction with allyl bromide affords the 3-allylated pyridine in good yield (equation 49, 78%).

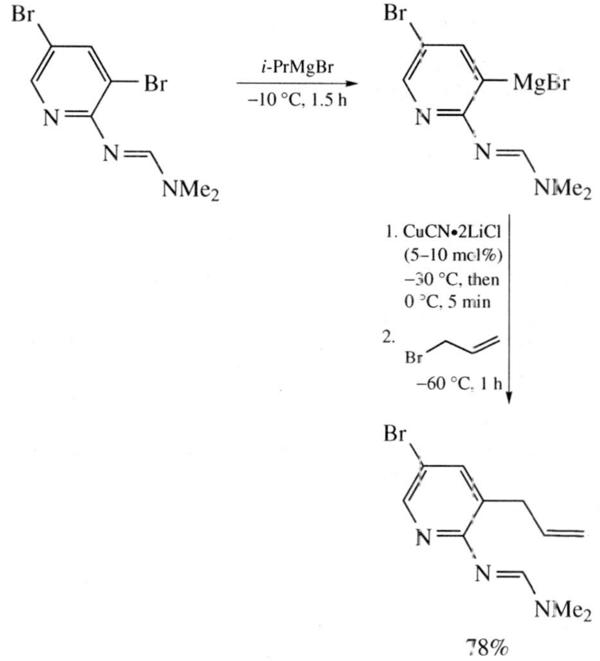

(49)

Iron-catalyzed aryl cross-coupling with magnesium-derived cuprates has also been used to prepare substituted pyridines and indoles[163]. β-Iodoindoles or β-bromopyridines are converted to the corresponding Grignard reagents, which are then transmetallated to copper (CuCN•2LiCl, 1.0 equivalents). These cuprate reagents undergo smooth coupling with aryl iodides in the presence of an iron catalyst [Fe(acac)$_3$]. Yields of substituted indoles are good (85%), while the pyridine variants are more modest (equations 50 and 51, 57%).

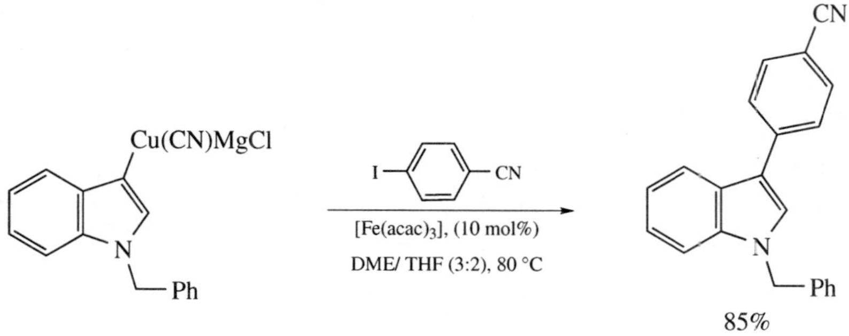

(50)

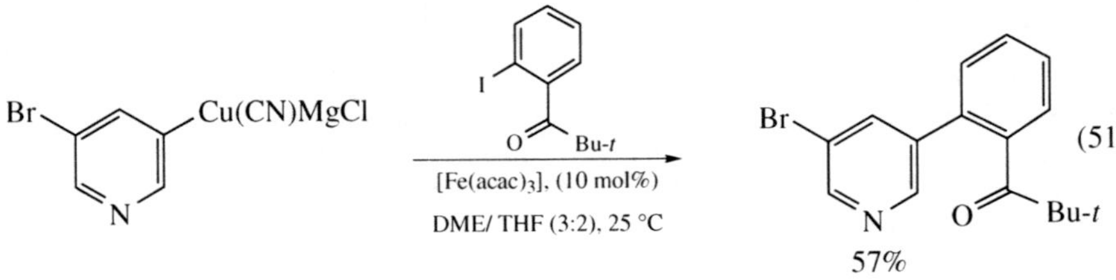

(51)

D. Cuprates of Nitrogen Enolates

In contrast to carbonyl compounds, enolates generated from compounds containing a nitrogen–carbon double or triple bond form cuprate reagents upon treatment with a Cu(I) salt. Cuprate reagents can be generated from nitriles, oxazolidines, hydrazones, oximes and imines.

1. α-Cuprionitriles

α-Metallated nitriles are powerful nucleophiles, which possess a range of reactivity depending on what metal is chosen. α-Cuprionitriles have emerged as useful tools for carbon–carbon bond formation in allylic, propargylic and acyl substitution reactions[164]. The preparation of these reagents is perhaps as interesting as their synthetic utility. In contrast to most organocopper species, which are prepared by transmetallation from the corresponding alkyllithium, alkylzinc or Grignard reagents, α-cuprionitriles can be prepared by addition of the α-bromo- or α-chloronitriles directly to a solution of the Gilman reagent, lithium dimethylcuprate (Me_2CuLi). This leads initially to formation of an *ate* complex (**84**, equation 52) which fragments to give the α-cuprionitrile **85**[165]. Access to these α-cuprionitriles can also be gained by conventional transmetallation from lithium nitrile enolates. Once formed, these reagents undergo substitution reactions with allylic and propargylic halides in moderate to good yields (54–85%, equation 53). Likewise, acylations with cyanoformates afford excellent yields of the corresponding α-cyanoesters (87%).

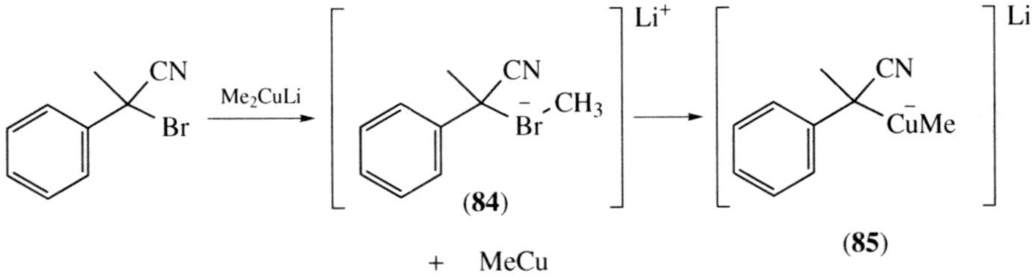

(52)

Another interesting observation made by Fleming and coworkers in this outstanding work is that high diastereoselectivities can be achieved by internal 1,2-asymmetric induction[166]. Thus, when bromonitrile (**88**) was treated with Me_2CuLi followed by methylcyanoformate, a single diastereomer is formed as a result of conformational bias (59%, equation 54).

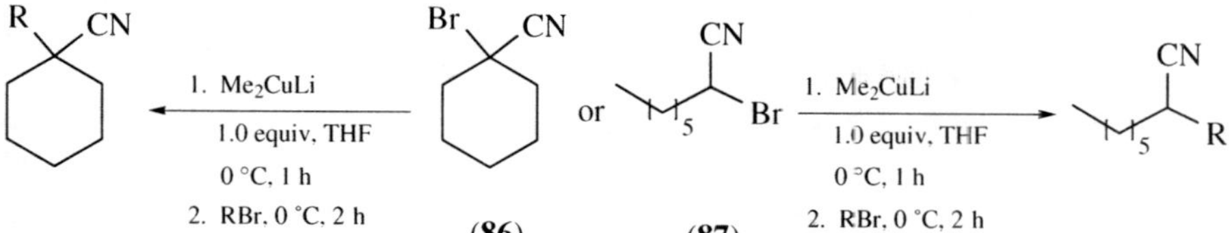

(86) (87)

α-Bromonitrile	R	% Yield[164]
86	$CH_2CH{=}CH_2$	69
86	$CH_2CH{=}CH(Ph)$	54
86	$CH_2C(CO_2Et){=}CH_2$	70
86	$CH{=}C{=}CH_2$	73
87	$CH_2CH{=}CH_2$	85
87	$CH{=}C{=}CH_2$	77
87	$CH_2C(CO_2Et){=}CH_2$	78

(53)

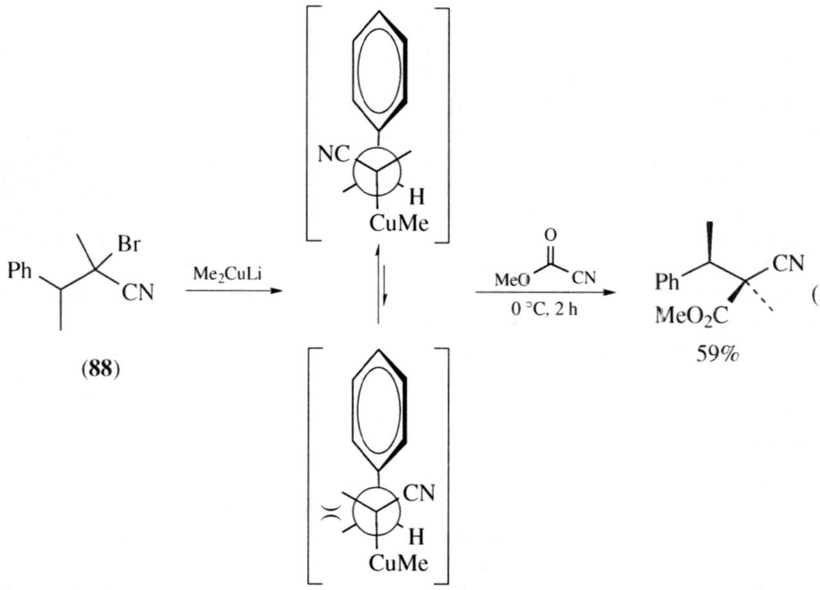

(88)

(54)

Some α-metallated nitriles also display a clear preference for S_N2 vs. S_N2' pathways in reactions with propargylic halides[167]. When cyclohexanecarbonitrile is deprotonated (LDA, THF) and the resulting α-lithionitrile treated with propargyl bromide, exclusive formation of the acetylene by an S_N2 pathway results (72%, equation 55). However, the preference shifts to an S_N2' pathway affording allenes when the α-lithionitrile is treated with methylcopper (MeCu) before addition of the propargyl bromide (equation 55, 70%).

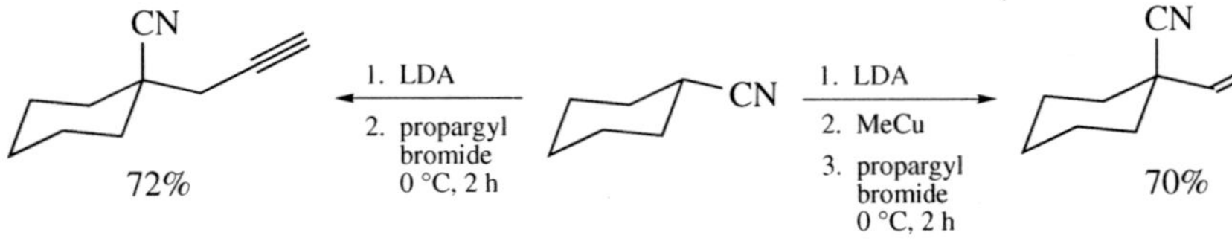

(55)

2. α-Cuprioxazolidines

α-Metallated oxazolidines are nucleophilic carboxymethyl synthons which can be used to construct new carbon–carbon bonds while simultaneously adding useful functionality to the product molecule. Not surprisingly, α-cuprioxazolidines have been employed in reactions that are traditional mainstays of organocuprate chemistry (e.g. conjugate additions, allylic and vinyl substitutions). Simonelli and coworkers explored the utility of these α-cuprioxazolidines in conjugate addition reactions to nitroalkenes[168]. The cuprates were prepared by transmetallation from the corresponding alkyllithium reagents, which were in turn generated by direct deprotonation of 2,2,4-trimethyl-2-oxazoline with *n*-butyllithium at $-78\,^{\circ}C$. The stoichiometrically prepared dialkylcuprates (R_2CuLi) were then reacted with a series of nitroalkenes. Reactions were run at $-78\,^{\circ}C$ for two hours before quenching, and no additives such as chlorotrimethylsilane, HMPA or $BF_3 \bullet Et_2O$ were used. The results are promising with yields ranging from moderate to excellent (67–96%, Scheme 22), allowing quaternary centers to be generated.

Reactions of these α-cuprioxazolidines in 1,4-conjugate additions to α,β-unsaturated ketones have also been reported by Simonelli and coworkers[169]. The cuprates were again prepared (stoichiometrically) from the corresponding alkyllithiums to give, presumably, the lithium dialkylcuprate species (R_2CuLi). The results, which are summarized in Table 4, illustrate the capricious nature of these reagents and their sensitivity to the reaction conditions. In some examples, yield of the 1,4-adduct is good (71%, cyclopentenone). With other substrates, the conjugate adduct is not formed at all, while 1,2-addition prevails (e.g. 3,5,5-trimethylcyclohexenone). Additives such as chlorotrimethylsilane/HMPA or $BF_3 \bullet Et_2O$ were investigated but were found to be ineffective in assisting the conjugate addition. In several cases, the addition of such additives favored formation of the 1,2-adduct.

3. α-Cuprioazaenolates

α-Cuprioazaenolates were first prepared by Corey and Enders for the preparation of 1,5-dicarbonyl compounds from α,β-unsaturated ketones and esters[170]. The sequence begins with deprotonation (LDA) of *N*,*N*-dimethylhydrazones derived from acetone or acetaldehyde in THF. Subsequent transmetallation to copper is achieved using half an equivalent of copper cyanide complexed with isopropyl sulfide ($CuCN \bullet S(Pr\text{-}n)_2$) that presumably leads to the lithium dialkylcuprate (R_2CuLi). Upon cuprate formation, the enones or enoates are added and the reactions are allowed to warm to $0\,^{\circ}C$ over 12 hours. Good yields ranging from 75–90% (Scheme 23) of the 1,5-dicarbonyl compounds are obtained after hydrolysis of the hydrazone moiety. An important feature of this work is that α-lithiated *N*,*N*-dimethylhydrazones circumvent some of the limitations encountered with lithium enolates.

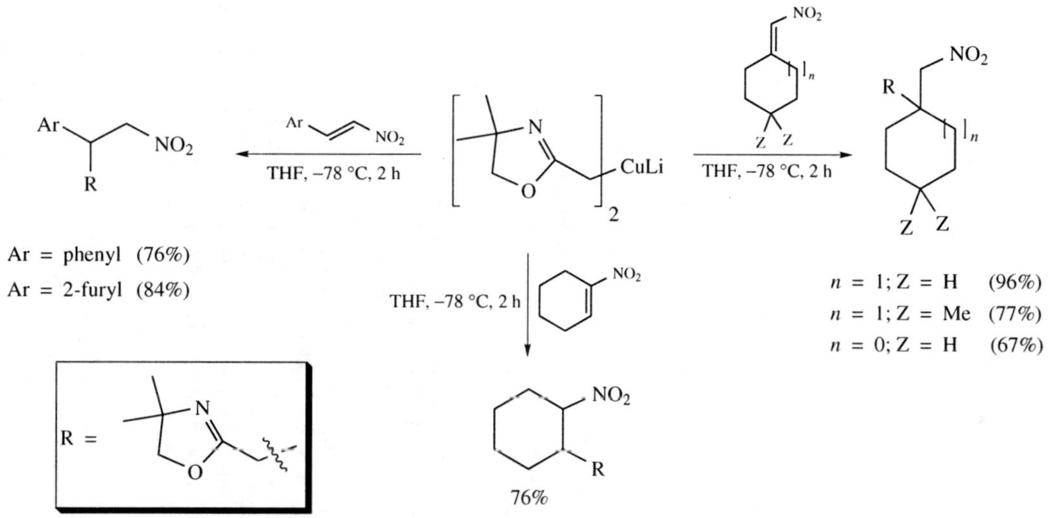

SCHEME 22. Reactions of α-cuprioxazolidines with nitroalkenes[168]

TABLE 4. Conjugate addition reactions of α-cuprioöxazolidine[169]

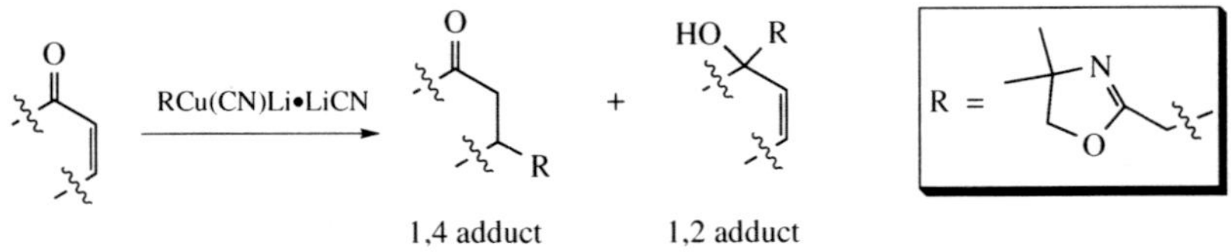

Enone	% Yield ; 1,4 adduct	% Yield ; 1,2 adduct
Cyclohex-2-en-1-one	70	22
Cyclopent-2-en-1-one	71	7
5,5-Dimethylcyclohex-2-en-1-one	63	17
2,6-Dimethylhepta-2,5-dien-4-one	0	40
(5*R*)-2-Methyl-5-(1-methylvinyl)cyclohex-2-en-1-one	26	49
3-Methylcyclopent-2-en-1-one	62	30
4,6,6-Trimethylbicyclo[3.1.1]hept-3-en-2-one	21	40
3,5,5-Trimethylcyclohex-2-en-1-one	0	75
6-Methylcyclohex-2-en-1-one	45	40
3-Ethoxy-5,5-dimethylcyclohex-2-en-1-one	0	93
(2*E*)-1,3-Diphenylprop-2-en-1-one	0	84

Gawley and coworkers later expanded the work of Corey and Enders by including the reactions of oxime ethers and by trapping the resulting enolates with various electrophiles such as TMSCl, acetyl chloride or acetic anhydride[171]. Lithium dialkylcuprates (R_2CuLi) were prepared from the organolithium precursors using half an equivalent of copper bromide complexed with dimethyl sulfide ($CuBr{\bullet}Me_2S$ in THF). The isolated yields for the trapped enolates were good to excellent (equation 56, 78–98%).

Conjugate addition to prochiral α,β-unsaturated ketones and esters leads to formation of a new stereogenic center and, as such, development of an enantiofacial-discriminating reaction is of significant synthetic importance. Previous attempts have been made to effect conjugate addition of simple alkyl groups in an enantioselective fashion with the aid of chiral ligands on copper[11–22, 24–27, 172]. Yamamoto and coworkers prepared several enantioenriched ketimines that afford scalemic α-cuprioazaenolates after deprotonation (*n*-BuLi) and transmetallation to copper (R_2CuLi, CuI, 0.5 equivalents, THF)[173]. Subsequent reaction with 2-cyclohexenone or 2-cyclopentenone followed by removal of the chiral amine gives the corresponding optically active 1,5-dicarbonyl compounds. Chemical yields were moderate to good (41–89%), while enantiomeric excesses ranged from modest to good (17–85% ee, Scheme 24). This work was later employed in an asymmetric approach to *trans*-dihydrindandione[174].

Seeking to boost enantioselectivities, Yamamoto and coworkers prepared another set of chiral ketimines containing two chiral centers (Scheme 25)[175]. After deprotonation with *n*-BuLi, treatment with a copper acetylide gives a mixed cuprate employing the acetylide as a non-transferable ligand. After reaction with 2-cyclohexenone or 2-cyclopentenone and subsequent hydrolysis of the imine, the scalemic 1,5-dicarbonyl compounds are obtained

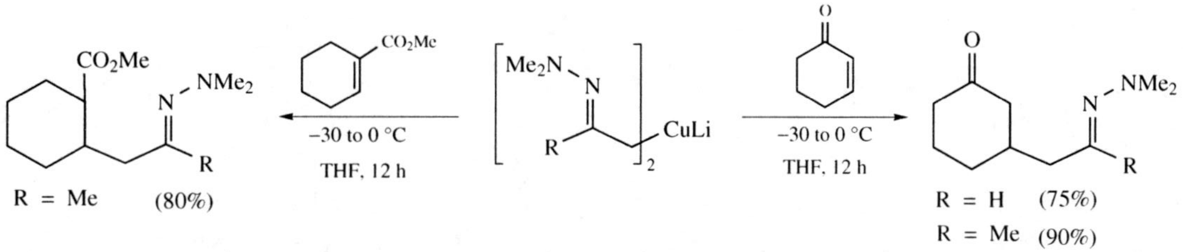

SCHEME 23. Conjugate addition reactions of α-cuprioazaenolates with α,β-unsaturated ketones and esters[170]

in moderate to good yields approximating earlier work[173], but with dramatically improved enantioselectivity (71–82% ee).

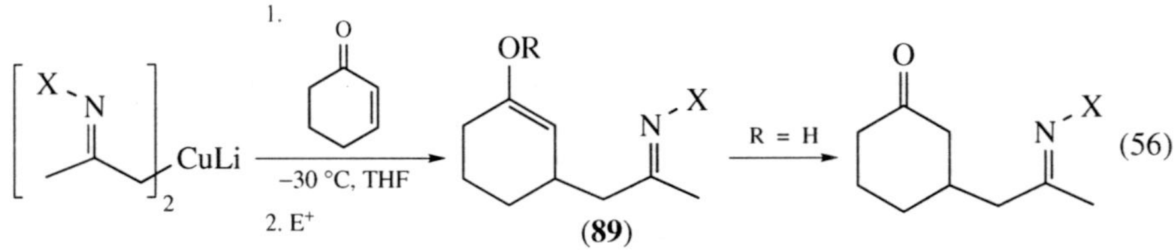

X	E^+	R	% Yield (**89**)
NMe_2	NH_4Cl	H	95
NMe_2	AcCl	Ac	88
NMe_2	Me_3SiCl	Me_3Si	92
OMe	NH_4Cl	H	98
OMe	AcCl	Ac	95
OMe	Me_3SiCl	Me_3Si	78

α-Cuprioazaenolates were also used in conjugate addition reactions by Movassaghi and Chen to prepare tricyclic imino and amino alcohols from α,β-enones[176]. The readily available cyclic iminium chlorides were treated with n-butyllithium in the presence of $CuBr{\bullet}SMe_2$ (5–10 mol% at −78 °C in THF), and the enone was added. The 1,4-adducts of the five-membered-ring iminium chlorides were isolated in good yield (equation 57, 67–88%). The procedure is also successful for six-membered iminium ring analogues; however, the 1,4-adducts undergo spontaneous tautomerization and addition to the carbonyl to give the tricyclic imino alcohol. Diastereoselectivity in imino alcohol formation is excellent for both five- and six-membered-ring iminium chlorides as all reactions resulted in formation of a single diastereomer.

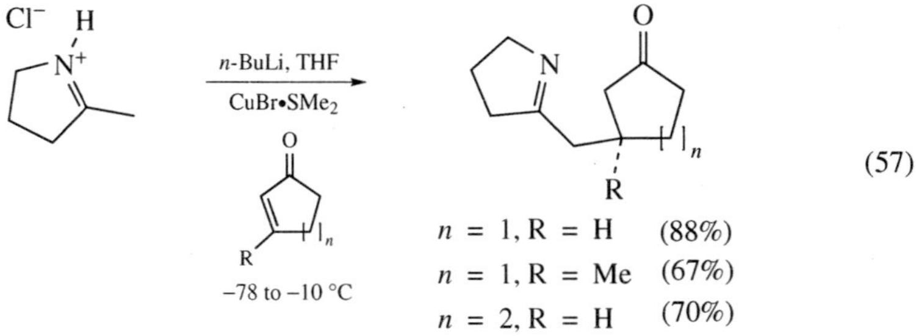

V. SUMMARY

Lithiated amines were first employed as non-transferable ligands in mixed amido(alkyl) cuprates to improve the efficient utilization of the transferable ligand and to afford cuprate reagents of increased stability. Subsequent utilization of chiral amines afforded enantioenriched amidocuprates that effected transfer of alkyl and aryl ligands to cyclic enones with modest to good enantioselectivities that could be enhanced by the use of silylated amine

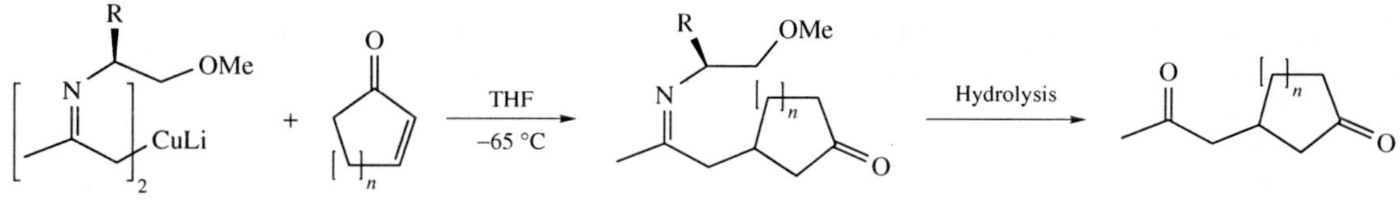

n	R	% Yield	% Optical yield
2	CH_2Ph	41	27.5 (*R*)
2	CH_2Ph	21	28.6 (*R*)
2	*i*-Pr	46	22.5 (*R*)
2	*t*-Bu	30	44.2 (*S*)
2	*t*-Bu	31	43.6 (*R*)
1	CH_2Ph	54	16.5 (*R*)
1	*i*-Pr	75	26.9 (*S*)
1	*t*-Bu	89	75.4 (*R*)

SCHEME 24. Conjugate addition reactions of enantioenriched α-cuprioazaenolates with α,β-unsaturated ketones[173]

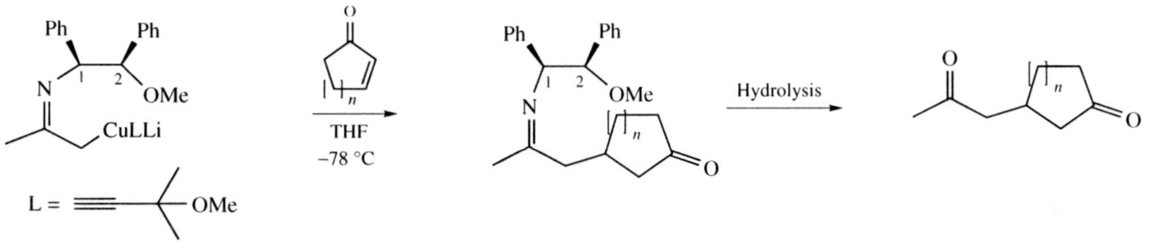

n	auxiliary	solvent	% Yield	% Optical yield
1	1*S*,2*R*	THF	78	78 (*R*)
1	1*R*,2*S*	Et_2O	41	80 (*S*)
1	1*R*,2*S*	Et_2O	56	82 (*S*)
2	1*R*,2*S*	THF	78	71 (*S*)

SCHEME 25. Conjugate addition reactions of scalemic α-cuprioazaenolates with α,β-unsaturated ketones[175]

ligands. These procedures employed stoichiometric amounts of enantioenriched amines and have been supplanted by methodologies catalytic in copper. Zinc cuprates derived from β-iodo serine derivatives provide useful synthetic routes to non-natural α-amino acids. The development of α-(N-carbamoyl)alkylcuprates provides versatile synthetic routes to a rich variety of heterocyclic compounds, and enantioselective synthetic routes to N-heterocycles containing a simple or fused pyrrolidine ring. More recent developments revolve around the use of N-heterocyclic carbene ligands (NHC) in copper-catalyzed reactions as well as the generation of cuprate reagents from N-heterocyclic Grignard reagents prepared via halogen–magnesium exchange reactions. Application and development of NHC–Cu complexes in cuprate-mediated reactions is a promising area of research. These developments in N-functionalized organocuprate chemistry provide useful modifications of cuprate reactivity and selectivity profiles, and synthetic methodologies for the preparation of N-functionalized compounds and N-heterocycles. Much work remains to be done in the development of enantioselective methodologies and of procedures catalytic in copper.

VI. REFERENCES

1. For leading reviews see:
(a) H. Heaney and S. Christie, in *Science of Synthesis: Organometallics, Compounds of Groups 12 and 11 (Zn, Cd, Hg, Cu, Ag, Au)* (Ed. S. V. Ley), Vol. 3, Chap. 4, Georg Thieme Verlag, Stuttgart, 2004, p. 305.
(b) N. Krause, (Ed.), *Modern Organocopper Chemistry*, Wiley-VCH, Weinheim, 2002.
(c) B. H. Lipshutz and S. Sengupta, *Org. React.*, **41**, 135 (1992).
2. G. H. Posner, C. E. Whitten and J. J. Sterling, *J. Am. Chem. Soc.*, **95**, 7788 (1973).
3. (a) S. H. Bertz and G. Dabbagh, *J. Chem. Soc., Chem. Commun.*, 1030 (1982).
(b) S. H. Bertz, G. Dabbagh and G. M. Villacorta, *J. Am. Chem. Soc.*, **104**, 5824 (1982).
4. S. H. Bertz and G. Dabbagh, *J. Org. Chem.*, **49**, 1119 (1984).
5. B. H. Lipshutz, P. Fatheree, W. Hagen and K. L. Stevens, *Tetrahedron Lett.*, **33**, 1041 (1992).
6. S. H. Bertz, M. Eriksson, G. Miao and J. P. Snyder, *J. Am. Chem. Soc.*, **118**, 10906 (1996).
7. (a) M. Yamanaka and E. Nakamura, *J. Am. Chem. Soc.*, **127**, 4697 (2005).
(b) E. Nakamura and M. Yamanaka, *J. Am. Chem. Soc.*, **121**, 8941 (1999).
8. R. K. Dieter, T. W. Hanks and B. Lagu, *Organometallics*, **11**, 3549 (1992).
9. (a) J. Eriksson, P. I. Arvidsson and O. Davidsson, *J. Am. Chem. Soc.*, **122**, 9310 (2000).
(b) J. Eriksson and O. Davidsson, *Organometallics*, **20**, 4763 (2001).
10. R. P. Davies, S. Hornauer and P. B. Hitchcock, *Angew. Chem., Int. Ed.*, **46**, 5191 (2007).
11. B. E. Rossiter and N. M. Swingle, *Chem. Rev.*, **92**, 771 (1992).
12. For reviews on copper catalysis involving neutral chiral heteroatom ligands see:
(a) A. Alexakis, *Chimia*, **54**, 55 (2000).
(b) B. L. Feringa, *Acc. Chem. Res.*, **33**, 346 (2000).
13. For reviews on catalysis involving chiral heteroatom cuprates prepared from anionic ligands see:
(a) N. Krause and A. Gerold, *Angew. Chem., Int. Ed. Engl.*, **36**, 186 (1997).
(b) N. Krause, *Angew. Chem., Int. Ed.*, **37**, 283 (1998).
14. R. A. Kretchmer, *J. Org. Chem.*, **37**, 2744 (1972).
15. (a) F. Leyendecker, F. Jesser and D. Laucher, *Tetrahedron Lett.*, **24**, 3513 (1983).
(b) T. Imamoto and T. Mukaiyama, *Chem. Lett.*, 45 (1980).
16. S. H. Bertz, G. Dabbagh and G. Sundararajan, *J. Org. Chem.*, **51**, 4953 (1986).
17. (a) R. K. Dieter and M. Tokles, *J. Am. Chem. Soc.*, **109**, 2040 (1987).
(b) R. K. Dieter, B. Lagu, N. Deo and J. W. Dieter, *Tetrahedron Lett.*, **31**, 4105 (1990).
(c) R. K. Dieter, S.-Y. Lin, N. Deo, B. Lagu and J. W. Dieter, unpublished results.
18. (a) B. E. Rossiter and M. Eguchi, *Tetrahedron Lett.*, **31**, 965 (1990).
(b) B. E. Rossiter, M. Eguchi, A. E. Hernandez, D. Vickers, J. Medich, J. Marr and D. Heinis, *Tetrahedron Lett.*, **32**, 3973 (1991).
(c) B. E. Rossiter, G. Miao, N. M. Swingle, M. Eguchi, A. E. Hernández and R. G. Patterson, *Tetrahedron: Asymmetry*, **3**, 231 (1992).

(d) B. E. Rossiter, M. Eguchi, G. Miao, N. M. Swingle, A. E. Hernández, D. Vickers, E. Fluckiger, R. G. Patterson and K. V. Reddy, *Tetrahedron*, **49**, 965 (1993).
(e) N. M. Swingle, K. V. Reddy and B. E. Rossiter, *Tetrahedron*, **50**, 4455 (1994).
(f) G. Miao and B. E. Rossiter, *J. Org. Chem.*, **60**, 8424 (1995).
19. G. A. Cran, C. L. Gibson, S. Handa and A. R. Kennedy, *Tetrahedron: Asymmetry*, **7**, 2511 (1996).
20. S. H. Bertz, C. A. Ogle and A. Rastogi, *J. Am. Chem. Soc.*, **127**, 1372 (2005).
21. E. J. Corey, R. Naef and F. J. Hannon, *J. Am. Chem. Soc.*, **108**, 7114 (1986).
22. (a) G. M. Villacorta, C. P. Rao and S. J. Lippard *J. Am. Chem. Soc.*, **110**, 3175 (1988).
(b) K. H. Ahn, R. B. Klassen and S. J. Lippard, *Organometallics*, **9**, 3178 (1990).
23. D. E. Frantz and D. A. Singleton, *J. Am. Chem. Soc.*, **122**, 3288 (2000).
24. Y. Nakagawa, K. Matsumoto and K. Tomioka, *Tetrahedron*, **56**, 2857 (2000).
25. X. Hu, H. Chen and X. Zhang, *Angew. Chem., Int. Ed.*, **38**, 3518 (1999).
26. M. Kitamura, T. Miki, K. Nakano and R. Noyori, *Bull. Chem. Soc. Jpn.*, **73**, 999 (2000).
27. (a) T. Seota, K. Selim, M. Kuriyama and K. Tomioka, *Adv. Synth. Catal.*, **349**, 629 (2007).
(b) K. Selim, T. Soeta, K. Yamada and K. Tomioka, *Chem. Asian J.*, **3**, 342 (2008).
(c) M. K. Brown, S. J. Degrado and A. H. Hoveyda, *Angew. Chem., Int. Ed.*, **44**, 5306 (2005).
(d) M. K. Brown and A. H. Hoveyda, *J. Am. Chem. Soc.*, **130**, 12904 (2008).
28. For a short review see: P. Dembech, G. Seconi and A. Ricci, *Chem. Eur. J.*, **6**, 1281 (2000).
29. (a) E. R. Strieter, D. G. Blackmond and S. L. Buchwald, *J. Am. Chem. Soc.*, **127**, 4120 (2005) and references cited therein.
(b) For a review see: S. R. Chemler and P. H. Fuller, *Chem. Soc. Rev.*, **36**, 1153 (2007).
30. For a review see: D. S. Surry and D. R. Spring, *Chem. Soc. Rev.*, **35**, 218 (2006).
31. I. Goldberg, *Ber. Dtsch. Chem. Ges.*, **39**, 1691 (1906).
32. (a) P. Beak, A. Basha, B. Kokko and D. Loo, *J. Am. Chem. Soc.*, **108**, 6016 (1986).
(b) P. Beak and G. W. Selling, *J. Org. Chem.*, **54**, 5574 (1989).
33. (a) J. P. Genet, S. Mallart, C. Greck and E. Piveteau, *Tetrahedron Lett.*, **32**, 2359 (1991).
(b) C. Greck, L. Bischoff, F. Ferreira and J. P. Genet, *J. Org. Chem.*, **60**, 7010 (1995).
(c) N. Zheng, J. D. Armstrong, III, J. C. McWilliams and R. P. Volante, *Tetrahedron Lett.*, **38**, 2817 (1997).
34. (a) A. Casarini, P. Dembech, D. Lazzari, E. Marini, G. Reginato, A. Ricci and G. Seconi, *J. Org. Chem.*, **58**, 5620 (1993).
(b) P. Bernardi, P. Dembech, G. Fabbri, A. Ricci and G. Seconi, *J. Org. Chem.*, **64**, 641 (1999).
35. (a) E. Erdik and T. Daskapan, *Synth. Commun.*, **29**, 3989 (1999).
(b) E. Erdik and T. Daskapan, *J. Chem. Soc., Perkin Trans. 1*, 3139 (1999).
(c) E. Erdik, F. Eroglu and D. Kahya, *J. Phys. Org. Chem.*, **18**, 950 (2005).
36. O. Riant, O. Samuel, T. Flessner, S. Taudien and H. B. Kagan, *J. Org. Chem.*, **62**, 6733 (1997).
37. (a) A. M. Berman and J. S. Johnson, *J. Am. Chem. Soc.*, **126**, 5680 (2004).
(b) A. M. Berman and J. S. Johnson, *J. Org. Chem.*, **70**, 364 (2005).
(c) A. M. Berman and J. S. Johnson, *J. Org. Chem.*, **71**, 219 (2006).
(d) M. J. Campbell and J. S. Johnson, *Org. Lett.*, **9**, 1521 (2007).
38. P. Knochel, N. Jeong, M. J. Rozema and M. C. P. Yeh, *J. Am. Chem. Soc.*, **111**, 6474 (1989).
39. (a) H. Tsutsui, Y. Hayakashi and K. Narasaka, *Chem. Lett.*, 317 (1997).
(b) H. Tsutsui, T. Ichikawa and K. Narasaka, *Bull. Chem. Soc. Jpn.*, **72**, 1869 (1999).
(c) K. Narasaka, *Pure Appl. Chem.*, **74**, 143 (2002).
40. T. Daskapan, *Tetrahedron Lett.*, **47**, 2879 (2006).
41. E. Erdik and O. Omur, *Appl. Organometal. Chem.*, **19**, 887 (2005).
42. S. Liu, Y. Yu and L. S. Liebeskind, *Org. Lett.*, **9**, 1947 (2007).
43. H. Yamamoto and K. Maruoka, *J. Org. Chem.*, **45**, 2739 (1980).
44. M. Iwao, J. N. Reed and V. Snieckus, *J. Am. Chem. Soc.*, **104**, 5531 (1982).
45. (a) A. Alberti, F. Canè, P. Dembech, D. Lazzari, A. Ricci and G. Seconi, *J. Org. Chem.*, **61**, 1677 (1996).
(b) F. Canè, D. Brancaleoni, P. Dembech, A. Ricci and G. Seconi, *Synthesis*, 545 (1997).
46. (a) V. del Amo, S. R. Dubbaka, A. Krasovskiy and P. Knochel, *Angew. Chem., Int. Ed.*, **45**, 7838 (2006).
(b) M. Kienle, S. R. Dubbaka, V. del Amo and P. Knochel, *Synthesis*, 1272 (2007).

(c) N. Boudet, S. R. Dubbaka and P. Knochel, *Org. Lett.*, **10**, 1715 (2008).

47. (a) C. Gallina and P. G. Ciattini, *J. Am. Chem. Soc.*, **101**, 1035 (1979).
(b) C. Gallina, *Tetrahedron Lett.*, **23**, 3093 (1982).
48. H. L. Goering, S. S. Kantner and C. C. Tseng, *J. Org. Chem.*, **48**, 715 (1983).
49. (a) N. R. Schmuff and B. M. Trost, *J. Org. Chem.*, **48**, 1404 (1983).
(b) M. Angeles Rey, J. A. Martinez-Perez, A. Fernandez-Gacio, K. Halkes, Y. Fall, J. Granja and A. Mourino, *J. Org. Chem.*, **64**, 3196 (1999).
(c) T. J. Brocksom, F. Coelho, J.-P. Depres, A. E. Greene, M. E. Freire de Lima, O. Hamelin, B. Hartmann, A. M. Kanazawa and Y. Wang, *J. Am. Chem. Soc.*, **124**, 15313 (2002).
50. (a) I. Fleming, D. Higgins, N. J. Lawrence and A. P. Thomas, *J. Chem. Soc., Perkin Trans. 1*, 3331 (1992).
(b) M. R. Hale and A. H. Hoveyda, *J. Org. Chem.*, **59**, 4370 (1994).
51. (a) J. H. Smitrovich and K. A. Woerpel, *J. Am. Chem. Soc.*, **120**, 12998 (1998).
(b) J. H. Smitrovich and K. A. Woerpel, *J. Org. Chem.*, **65**, 1601 (2000).
52. M. Oestreich and G. Auer, *Adv. Synth. Catal,*, **347**, 637 (2005).
53. S. E. Denmark and L. K. Marble, *J. Org. Chem.*, **55**, 1984 (1990).
54. (a) B. Pedram, A. van Oeveren, D. E. Mais, K. B. Marschke, P. M. Verbost, M. B. Groen and L. Zhi, *J. Med. Chem.*, **51**, 3696 (2008).
(b) Z.-H. Peng and K. A. Woerpel, *Org. Lett.*, **3**, 675 (2001).
(c) M. Ahmar, C. Duyck and I. Fleming, *J. Chem. Soc., Perkin Trans. 1*, 2721 (1998).
(d) I. Fleming and S. B. D. Winter, *J. Chem. Soc., Perkin Trans. 1*, 2687 (1998).
(e) K. Tamao, A. Kawachi, Y. Tanaka, H. Ohtani and Y. Ito, *Tetrahedron*, **52**, 5765 (1996).
55. (a) S. C. Archibald, D. J. Barden, J. F. Y. Bazin, I. Fleming, C. F. Foster, A. K. Mandal, A. K. Mandal, D. Parker, K. Takaki, A. C. Ware, A. R. B. Williams and A. B. Zwicky, *Org. Biomol. Chem.*, **2**, 1051 (2004).
(b) I. Fleming and D. Higgins, *J. Chem. Soc., Perkin Trans. 1*, 2673 (1998).
56. (a) T. Tsuda, M. Miwa and T. Saegusa, *J. Org. Chem.*, **44**, 3734 (1979).
(b) Y. Wakita, T. Kobayashi, M. Maeda and M. Kojima, *Chem. Pharm. Bull.*, **30**, 3395 (1982).
57. Y. Wakita, S. Noma, M. Maeda and M. Kojima, *J. Organomet. Chem.*, **297**, 379 (1985).
58. (a) P. Knochel, T. S. Chou, C. Jubert and D. Rajagopal, *J. Org. Chem.*, **58**, 588 (1993).
(b) C. Jubert and P. Knochel, *J. Org. Chem.*, **57**, 5431 (1992).
59. Y. Ito, H. Imai, T. Matsuura and T. Saegusa, *Tetrahedron Lett.*, **25**, 3091 (1984).
60. U. Schollkopf, D. Pettig, E. Schulze, M. Klinge, E. Egert, B. Benecke and M. Noltemeyer, *Angew. Chem., Int. Ed. Engl.*, **27**, 1194 (1988).
61. H. Wild and L. Born, *Angew. Chem., Int. Ed. Engl.*, **30**, 1685 (1991).
62. (a) R. Beugelmans, A. Bigot and J. Zhu, *Tetrahedron Lett.*, **35**, 7391 (1994).
(b) R. Beugelmans, A. Bigot, M. Bois-Choussy and J. Zhu, *J. Org. Chem.*, **61**, 771 (1996).
(c) M. Bois-Choussy, L. Neuville, R. Beugelmans and J. Zhu, *J. Org. Chem.*, **61**, 9309 (1996).
63. J. E. Baldwin, R. M. Adlington and M. B. Mitchell, *J. Chem. Soc., Chem. Commun.*, 1332 (1993).
64. A. Claesson and C. Sahlberg, *Tetrahedron*, **38**, 363 (1982).
65. A. Horvath, J. Benner and J.-E. Bäckvall, *Eur. J. Org. Chem.*, 3240 (2004).
66. M. Ishikura and M. Terashima, *Heterocycles*, **27**, 2641 (1998).
67. A. I. Meyers, P. D. Edwards, T. R. Bailey and G. E. Jagdmann, *J. Org. Chem.*, **50**, 1019 (1985).
68. C. E. Neipp, J. M. Humphrey and S. F. Martin, *J. Org. Chem.*, **66**, 531 (2001).
69. For a review of α-lithiocarbamate chemistry see:
(a) P. Beak, A. Basu, D. J. Gallagher, Y. S. Park and S. Thayumanavan, *Acc. Chem. Res.*, **29**, 552 (1996).
(b) P. Beak, T. A. Johnson, D. D. Kim and S. H. Lim, *Topics Organomet. Chem.*, **5**, 139 (2003).
70. (a) A. I. Meyers, P. D. Edwards, W. F. Rieker and T. R. Bailey, *J. Am. Chem. Soc.*, **106**, 3270 (1984).
(b) P. D. Edwards and A. I. Meyers, *Tetrahedron Lett.*, **25**, 939 (1984).
71. R. E. Gawley, G. C. Hart and L. J. Bartolotti, *J. Org. Chem.*, **54**, 175 (1989).
72. R. K. Dieter and C. W. Alexander, *Tetrahedron Lett.*, **33**, 5693 (1992).
73. C. W. Alexander, S.-Y. Lin and R. K. Dieter, *J. Organomet. Chem.*, **503**, 213 (1995).
74. (a) R. K. Dieter and C. W. Alexander, *Synlett*, 407 (1993).

(b) R. K. Dieter, C. W. Alexander and L. E. Nice, *Tetrahedron*, **56**, 2767 (2000).
75. R. K. Dieter, C. M. Topping and L. E. Nice, *J. Org. Chem.*, **66**, 2302 (2001)
76. (a) R. K. Dieter and S. E. Velu, *J. Org. Chem.*, **62**, 3798 (1997).
(b) R. K. Dieter, K. Lu and S. E. Velu, *J. Org. Chem.*, **65**, 8715 (2000).
77. R. K. Dieter and H. Yu, *Org. Lett.*, **2**, 2283 (2000).
78. R. K. Dieter and K. Lu, *J. Org. Chem.*, **67**, 847 (2002).
79. R. K. Dieter and K. Lu, *Tetrahedron Lett.*, **40**, 4011 (1999).
80. K. Nilsson, T. Andersson, C. Ullenius, A. Gerold and N. Krause, *Chem. Eur. J.*, **4**, 2051 (1998).
81. L. Strekowski, Y. Gulevich, K. Van Aken, D. W. Wilson and K. R. Fox, *Tetrahedron Lett.*, **36**, 225 (1995).
82. (a) R. K. Dieter, R. R. Sharma and W. Ryan, *Tetrahedron Lett.*, **38**, 783 (1997).
(b) R. K. Dieter, R. R. Sharma, H. Yu and V. K. Gore, *Tetrahedron*, **59**, 1083 (2003).
83. R. K. Dieter and R. R. Sharma, *Tetrahedron Lett.*, **38**, 5937 (1997).
84. (a) R. K. Dieter, J. W. Dieter, C. W. Alexander and N. S. Bhinderwala, *J. Org. Chem.*, **61**, 2930 (1996).
(b) S. Li and R. K. Dieter, *J. Org. Chem.*, **68**, 969 (2003).
85. R. K. Dieter, S. E. Velu and L. E. Nice, *Synlett*, 1114 (1997).
86. R. K. Dieter, V. K. Gore and N. Chen, *Org. Lett.*, **6**, 763 (2004).
87. (a) R. K. Dieter and L. E. Nice, *Tetrahedron Lett.*, **40**, 4293 (1999).
(b) R. K. Dieter and H. Yu, *Org. Lett.*, **3**, 3855 (2001).
(c) R. K. Dieter, N. Chen, H. Yu, L. E. Nice and V. Gore, *J. Org. Chem.*, **70**, 2109 (2005).
(d) R. K. Dieter, N. Chen and V. K. Gore, *J. Org. Chem.*, **71**, 8755 (2006).
88. T. Tomoyasu, K. Tomooka and T. Nakai, *Tetrahedron Lett.*, **41**, 345 (2000).
89. P. Beak, S. T. Kerrick, S. Wu and J. Chu, *J. Am. Chem. Soc.*, **116**, 3231 (1994).
90. (a) R. K. Dieter, C. M. Topping, K. R. Chandupatla and K. Lu, *J. Am. Chem. Soc.*, **123**, 5132 (2001).
(b) R. K. Dieter, G. Oba, K. R. Chandupatla, C. M. Topping, K. Lu and R. T. Watson, *J. Org. Chem.*, **69**, 3076 (2004).
91. R. K. Dieter, R. T. Watson and R. Goswami, *Org. Lett.*, **6**, 253 (2004).
92. (a) R. K. Dieter, N. Chen and R. T. Watson, *Tetrahedron*, **61**, 3221 (2005).
(b) For a preliminary account in the racemic series see: R. K. Dieter and R. Watson, *Tetrahedron Lett.*, **43**, 7725 (2002).
93. R. T. Watson, V. K. Gore, K. R. Chandupatla, R. K. Dieter and J. P. Snyder, *J. Org. Chem.*, **69**, 6105 (2004).
94. R. K. Dieter and N. Chen, *J. Org. Chem.*, **71**, 5674 (2006).
95. D. Stead, P. O'Brien and A. J. Sanderson, *Org. Lett.*, **7**, 4459 (2005).
96. M. Berkheij, L. S. van der Sluis, C. Sewing, D. J. den Boer, J. W. Terpstra, H. Hiemstra, W. I. I. Bakker, A. van den Hoogenband and J. H. van Maarseveen, *Tetrahedron Lett.*, **46**, 2369 (2005).
97. N. Alouane, F. Bernaud, J. Marrot, E. Vrancken and P. Mangeney, *Org. Lett.*, **7**, 5797 (2005).
98. P. J. Kocienski, J. A. Christopher, R. Bell and B. Otto, *Synthesis*, 75 (2005).
99. S. A. Wolckenhauer and S. D. Rychnovsky, *Org. Lett.*, **6**, 2745 (2004).
100. S. Li and R. K. Dieter, *J. Org. Chem.*, **68**, 969 (2003).
101. For reviews see:
(a) V. Cesar, S. Bellemin-Loponnaz and L. H. Gade, *Chem. Soc. Rev.*, **33**, 619 (2004).
(b) S. P. Nolan (Ed.), *N-Heterocyclic Carbenes in Synthesis*, Wiley-VCH, Weinheim, 2006, p. 55.
(c) F. Glorius (Ed.), *N-Heterocyclic Carbenes in Transition Metal Catalysis, Vol. 21, Topics in Organometallic Series*, Springer, Berlin, 2007.
102. For reviews see:
(a) S. Diez-Gonzalez and S. P. Nolan, *Synlett*, 2158 (2007).
(b) S. Diez-Gonzalez and S. P. Nolan, *Aldrichimica Acta*, **41**, 43 (2008).
103. P. K. Fraser and S. Woodward, *Tetrahedron Lett.*, **42**, 2747 (2001).
104. (a) F. Guillen, C. L. Winn and A. Alexakis, *Tetrahedron: Asymmetry*, **12**, 2083 (2001).
(b) C. L. Winn, F. Guillen, J. Pytkowicz, S. Roland, P. Mangeney and A. Alexakis, *J. Organomet. Chem.*, **690**, 5672 (2005).

(c) H. Clavier, L. Coutable, J.-C. Guillemin and M. Mauduit, *Tetrahedron: Asymmetry*, **16**, 921 (2005).
(d) H. Clavier, L. Coutable, L. Toupet, J.-C. Guillemin and M. Mauduit, *J. Organomet. Chem.*, **690**, 5237 (2005).
(e) H. Clavier, J.-C. Guillemin and M. Mauduit, *Chirality*, **19**, 471 (2007).
(f) T. Moore, M. Merzouk and N. Williams, *Synlett*, 21 (2008).

105. B. L. Feringa, R. Naasz, R. Imbos and L. A. Arnold, in *Modern Organocopper Chemistry* (Ed. N. Krause), Chap. 7, Wiley-VCH, Weinheim, 2002, p. 224 (see also p. 231).
106. (a) D. Martin, S. Kehrli, M. D'Augustin, H. Clavier, M. Mauduit and A. Alexakis, *J. Am. Chem. Soc.*, **128**, 8416 (2006).
(b) Y. Matsumoto, K. Yamada and K. Tomioka, *J. Org. Chem.*, **73**, 4578 (2008).
107. (a) S. Tominaga, Y. Oi, T. Kato, D. K. An and S. Okamoto, *Tetrahedron Lett.*, **45**, 5585 (2004).
(b) S. Okamoto, S. Tominaga, N. Saino, K. Kase and K. Shimoda, *J. Organomet. Chem.*, **690**, 6001 (2005).
108. (a) A. O. Larsen, W. Leu, C. N. Oberhuber, J. E. Campbell and A. H. Hoveyda, *J. Am. Chem. Soc.*, **126**, 11130 (2004).
(b) M. A. Kacprzynski, T. L. May, S. A. Kazane and A. H. Hoveyda, *Angew. Chem., Int. Ed.*, **46**, 4554 (2007).
109. M. K. Brown, T. L. May, C. A. Baxter and A. H. Hoveyda, *Angew. Chem., Int., Ed.*, **46**, 1097 (2007).
110. P. L. Arnold, M. Rodden, K. M. Davis, A. C. Scarisbrick, A. J. Blake and C. Wilson, *J. Chem. Soc., Chem. Commun.*, 1612 (2004).
111. (a) H. Kaur, F. K. Zinn, E. D. Sevens and S. P. Nolan, *Organometallics*, **23**, 1157 (2004).
(b) S. Diez-Gonzalez, H. Kaur, F. K. Zinn, E. D. Stevens and S. P. Nolan, *J. Org. Chem.*, **70**, 4784 (2005).
(c) S. Diez-Gonzalez, N. M. Scott and S. P. Nolan, *Organometallics*, **25**, 2355 (2006).
(d) J. Yun, D. Kim and H. Yun, *J. Chem. Soc., Chem. Commun.*, 5181 (2005).
112. N. P. Mankad, D. S. Laitar and J. P. Sadighi, *Organometallics*, **23**, 3369 (2004).
113. V. Jurkauskas, J. P. Sadighi and S. L. Buchwald, *Org. Lett.*, **5**, 2417 (2003).
114. A. Welle, S. Diez-Gonzalez, B. Tinant, S. P. Nolan and O. Riant, *Org. Lett.*, **8**, 6059 (2006).
115. A. B. Md, Y. Suzuki and M. Sato, *Chem., Pharm. Bull.*, **56**, 57 (2008).
116. T. Ohishi, M. Nishiura and Z. Hou, *Angew. Chem., Int. Ed.*, **47**, 5792 (2008).
117. N. P. Mankad, T. G. Gray, D. S. Laitar and J. P. Sadighi, *Organometallics*, **23**, 1191 (2004).
118. (a) L. A. Goj, E. D. Blue, S. A. Delp, T. B. Gunnoe, T. R. Cundari, A. W. Pierpont, J. L. Petersen and P. D. Boyle, *Inorg. Chem.*, **45**, 9032 (2006) for synthesis of the complexes.
(b) C. Munro-Leighton, E. D. Blue and T. B. Gunnoe, *J. Am. Chem. Soc.*, **128**, 1446 (2006).
(c) C. Munro-Leighton, S. A. Delp, E. D. Blue and T. B. Gunnoe, *Organometallics*, **26**, 1483 (2007).
(d) S. A. Delp, C. Munro-Leighton, L. A. Goj, M. A. Ramirez, T. B. Gunnoe, J. L. Petersen and P. D. Boyle, *Inorg. Chem.*, **46**, 2365 (2007).
119. D. S. Laitar, E. Y. Tsui and J. P. Sadighi, *Organometallics*, **25**, 2405 (2006).
120. D. S. Laitar, E. Y. Tsui and J. P. Sadighi, *J. Am. Chem. Soc.*, **128**, 11036 (2006).
121. A. S. B. Prasad, T. M. Stevenson, J. R. Citineni, V. Nyzam and P. Knochel, *Tetrahedron*, **53**, 7237 (1997).
122. W. Lin, L. Chen and P. Knochel, *Tetrahedron*, **63**, 2787 (2007).
123. H. Ila, O. Baron, A. J. Wagner and P. Knochel, *J. Chem. Soc., Commun. Soc.*, 583 (2006).
124. P. Knochel, W. Dohle, N. Gommermann, F. F. Kneisel, F. Kopp, T. Korn, I. Sapountzis and V. A. Vu, *Angew. Chem., Int. Ed.*, **42**, 4302 (2003).
125. M. Abarbri, J. Thibonnet, L. Berillon, F. Dehmel, M. Rottlander and P. Knochel, *J. Org. Chem.*, **65**, 4618 (2000).
126. F. Dehmel, M. Abarbri and P. Knochel, *Synlett*, 345 (2000).
127. M. Klecka, T. Tobrman and D. Dvorak, *Collect. Czech. Chem. Commun.*, **71**, 1221 (2006).
128. R. Duddu, M. Eckhardt, M. Furlong, H. P. Knoess, S. Berger and P. Knochel, *Tetrahedron*, **50**, 2415 (1994).
129. S. Billotte, *Synlett*, 379 (1998).
130. W. F. J. Karstens, M. Stol, F. P. J. T. Rutjes and H. Hiemstra, *Synlett*, 1126 (1998).
131. T. Hjelmgaard and D. Tanner, *Org. Biomol. Chem.*, **4**, 1796 (2006).

132. W. F. J. Karstens, M. J. Moolenaar, F. P. J. T. Rutjes, U. Grabowska, W. M. Speckamp and H. Hiemstra, *Tetrahedron Lett.*, **40**, 8629 (1999).
133. M. J. Dunn and R. F. W. Jackson, *J. Chem. Soc., Chem. Commun.*, 319 (1992).
134. M. J. Dunn, R. F. W. Jackson, J. Pietruszka and D. Turner, *J. Org. Chem.*, **60**, 2210 (1995).
135. H. P. Knoess, M. T. Furlong, M. J. Rozema and P. Knochel, *J. Org. Chem.*, **56**, 5974 (1991).
136. D. L. J. Clive, D. M. Coltart and Y. Zhou, *J. Org. Chem.*, **64**, 1447 (1999).
137. M. Reisser and G. Maas, *Synthesis*, 1129 (1998).
138. E. Nakamura, S. Aoki, K. Sekiya, H. Oshino and I. Kuwajima, *J. Am. Chem.* **109**, 8056 (1987).
139. C. S. Dexter and R. F. W. Jackson, *J. Org. Chem.*, **64**, 7579 (1999).
140. H. J. C. Deboves, U. Grabowska, A. Rizzo and R. F. W. Jackson, *J. Chem. Soc., Perkin Trans. 1*, 4284 (2000).
141. C. Hunter, R. F. W. Jackson and H. K. Rami, *J. Chem. Soc., Perkin Trans. 1*, 1349 (2001).
142. H. J. C. Deboves, C. A. G. N. Montalbetti and R. F. W. Jackson, *J. Chem. Soc., Perkin Trans. 1*, 1876 (2001).
143. S. J. Greenfield and S. R. Gilbertson, *Synthesis*, 2337 (2001).
144. (a) L. Jobran and G. Hummel, *Org. Lett.*, **2**, 2265 (2000).
(b) K. K. Schumacher, J. Jiang and M. M. Joullie, *Tetrahedron: Asymmetry*, **9**, 47 (1998).
(c) A. Boudier, C. Darcel, F. Flachsman, L. Micouin, M. Oestreich and P. Knochel, *Chem. Eur. J.*, **6**, 2748 (2000).
145. S. J. Greenfield, A. Agarkov and S. R. Gilbertson, *Org. Lett.*, **5**, 3069 (2003).
146. T. Nishikawa, K. Wada and M. Isobe, *Biosci. Biotechnol. Biochem.*, **66**, 2273 (2002).
147. M. C. Yeh and P. Knochel, *Tetrahedron Lett.*, **30**, 4799 (1989).
148. I. Wilson and R. F. W. Jackson, *J. Chem. Soc., Perkin Trans. 1*, 2845 (2002).
149. T. Carrillo-Marquez, L. Caggiano, R. F. W. Jackson, U. Grabowska, A. Rae and M. J. Tozer, *Org. Biomol. Chem.*, **3**, 4117 (2005).
150. C. S. Dexter, C. Hunter and R. F. W. Jackson, *J. Org. Chem.*, **65**, 7417 (2000); R. F. W. Jackson, I. Rilatt and P. J. Murray, *Chem. Commun.*, 1242 (2003).
151. G. Varchi, A. E. Jensen, W. Dohle, A. Ricci, G. Cahiez and P. Knochel, *Synlett*, 477 (2001).
152. P. Knochel, E. Hupe, W. Dohle, D. M. Lindsay, V. Bonnet, G. Queguiner, A. Boudier, F. Kopp, S. Demay, N. Seidel, I. Calaza, V. A. Vu, I. Sapountzis and T. Bunlaksananusorn, *Pure Appl. Chem.*, **74**, 11 (2002).
153. W. Dohle, D. M. Lindsay and P. Knochel, *Org. Lett.*, **3**, 2871 (2001).
154. A. E. Jensen, W. Dohle, I. Sapountzis, D. M. Lindsay, V. A. Vu and P. Knochel, *Synthesis*, 565 (2002).
155. G. Varchi, C. Kofink, D. M. Lindsay, A. Ricci and P. Knochel, *Chem. Commun.*, 396 (2003).
156. D. M. Lindsay, W. Dohle, A. E. Jensen, F. Kopp and P. Knochel, *Org. Lett.*, **4**, 1819 (2002).
157. A. Staubitz, W. Dohle and P. Knochel, *Synthesis*, 233 (2003).
158. I. Sapountzis and P. Knochel, *Angew. Chem., Int. Ed.*, **41**, 1610 (2002).
159. W. Dohle, A. Staubitz and P. Knochel, *Chem. Eur. J.*, **9**, 5323 (2003).
160. G. Koebrich and P. Buck, *Chem. Ber.*, **103**, 1412 (1970).
161. (a) C. E. Tucker, T. N. Majid and P. Knochel, *J. Am. Chem. Soc.*, **114**, 3983 (1992).
(b) J. F. Cameron and J. M. J. Frechet, *J. Am. Chem. Soc.*, **113**, 4303 (1991).
162. I. Sapountzis, H. Dube, R. Lewis, N. Gommermann and P. Knochel, *J. Org. Chem.*, **70**, 2445 (2005).
163. I. Sapountzis, W. Lin, C. C. Kofink, C. Despotopoulou and P. Knochel, *Angew. Chem., Int. Ed.*, **44**, 1654 (2005).
164. F. F. Fleming, Z. Zhang, W. Liu and P. Knochel, *J. Org. Chem.*, **70**, 2200 (2005).
165. (a) R. W. Hoffmann, M. Bronstrup and M. Muller, *Org. Lett.*, **5**, 313 (2003).
(b) V. Schulze, M. Bronstrup, V. P. W. Bohm, P. Schwerdtfeger, M. Schimeczek and R. W. Hoffmann, *Angew. Chem., Int. Ed.*, **37**, 824 (1998).
166. F. F. Fleming, W. Liu, S. Ghosh and O. W. Steward, *J. Org. Chem.*, **73**, 2803 (2008).
167. F. F. Fleming, Y. Wei, W. Liu and Z. Zhang, *Tetrahedron*, **64**, 7477 (2008).
168. F. Simonelli, G. C. Clososki, A. A. dos Santos, A. R. M. de Oliveira, F. de A. Marques and P. H. G. Zarbin, *Tetrahedron Lett.*, **42**, 7375 (2001).
169. A. A. dos Santos, A. R. M. de Oliveira, F. Simonelli, F. de A. Marques, G. C. Clososki and P. H. G. Zarbin, *Synlett*, 975 (2003).
170. E. J. Corey and D. Enders, *Tetrahedron Lett.*, 11 (1976).

171. R. E. Gawley, E. J. Termine and J. Aube, *Tetrahedron Lett.*, **21**, 3115 (1980).
172. (a) J. S. Zweig, J. L. Luche, E. Barreiro and P. Crabbe, *Tetrahedron Lett.*, 2355 (1975).
(b) W. Langer and D. Seebach, *Angew. Chem., Int. Ed. Engl.*, **14**, 634 (1975).
(c) T. Imamoto and T. Mukaiyama, *Chem. Lett.*, 45 (1980).
(d) M. Huche, J. Berlan, G. Pourcelot and P. Cresson, *Tetrahedron Lett.*, **22**, 1329 (1981).
173. K. Yamamoto, M. IIjima and Y. Ogimura, *Tetrahedron Lett.*, **23**, 3711 (1982).
174. K. Yamamoto, M. IIjima, Y. Ogimura and J. Tsuji, *Tetrahedron Lett.*, **25**, 2813 (1984).
175. K. Yamamoto, M. Kanoh, N. Yamamoto and J. Tsuji, *Tetrahedron Lett.*, **28**, 6347 (1987).
176. M. Movassaghi and B. Chen, *Angew. Chem., Int. Ed.*, **46**, 565 (2007).

CHAPTER **10**

Transmetalation reactions producing organocopper compounds

CHRISTOPHER J. ROSENKER and PETER WIPF

Department of Chemistry, University of Pittsburgh, Pittsburgh, PA 15260, USA
Fax: +1 412-624-0787; e-mail: pwipf@pitt.edu

I. INTRODUCTION 444
II. TRANSMETALATIONS IN THE FIRST TRANSITION METAL SERIES 444
A. Transmetalation from Organotitanium Compounds 444
B. Transmetalation from Organomanganese Compounds 447
C. Transmetalation of Chromium Fischer Carbene Complexes 450
III. TRANSMETALATIONS IN THE SECOND TRANSITION METAL SERIES 452
A. Transmetalation from Organozirconium Compounds 452
1. From alkyl and alkenyl organozirconium compounds 452
2. From zirconacycles 462
IV. TRANSMETALATIONS IN THE LANTHANIDE SERIES 471
A. Transmetalation from Organosamarium Compounds 471
V. TRANSMETALATIONS IN GROUP IIB 471
A. Transmetalation from Organozinc Compounds 471
B. Transmetalation from Organomercury Compounds 493
VI. TRANSMETALATIONS IN GROUP IIIA 496
A. Transmetalation from Organoboron Compounds 496
B. Transmetalation from Organoaluminum Compounds 502
VII. TRANSMETALATIONS IN GROUP IVA 509
A. Transmetalation from Organotin Compounds 509
B. Transmetalation from Organosilicon Compounds 513
VIII. TRANSMETALATIONS IN GROUP VIA 514
A. Transmetalation from Organotellurium Compounds 514
IX. APPLICATIONS TO NATURAL PRODUCT SYNTHESIS 516

The chemistry of organocopper compounds
Edited by Z. Rappoport and I. Marek

X. CONCLUSION 518
XI. REFERENCES 519

I. INTRODUCTION

Organocopper compounds are among the most important organometallic reagents used in contemporary organic synthesis[1–3]. They are used in key carbon–carbon bond formations such as conjugate additions to α,β-unsaturated carbonyl compounds, additions to Michael acceptors, nucleophilic displacement of activated and non-activated leaving groups, epoxide openings, additions to acetylenes and a range of cross-coupling reactions. More recently, the development of copper-catalyzed enantioselective conjugate addition and allylic displacement reactions has greatly enhanced the utility of organocopper reagents in the synthesis of complex molecules. The general method for the preparation of organocopper compounds has been the transmetalation from the corresponding lithium or magnesium organometallics; however, due to the lack of functional group compatibility, this method of preparation is often unable to generate highly functionalized organocopper compounds. Alternatively, transmetalation reactions can be used to generate organocopper intermediates from a variety of organometallic compounds, including organoaluminum, -boron, -chromium, -manganese, -mercury, -samarium, -silicon, -tin, -tellurium, -titanium, -zinc and -zirconium reagents. In general, the transmetalation route provides access to a wide range of functionalized organocopper compounds, which inherit reactivity characteristics from the residual metal salts often closely associated with the organocopper reagent or the substrate. To a first approximation, the relative electonegativity of an organometallic precursor can be used to predict the equilibrium of the ligand transfer process, which generally is shifted toward the more electronegative metal complex.

The scope of this chapter covers transmetalation reactions producing organocopper compounds from various organometallic precursors as well as their subsequent reactions with electrophiles. The preparation of organocopper compounds via transmetalation has previously been reviewed[4,5], and this chapter therefore focuses on more recent developments where appropriate.

II. TRANSMETALATIONS IN THE FIRST TRANSITION METAL SERIES

A. Transmetalation from Organotitanium Compounds

Organotitanium compounds were first transmetalated to form organocopper compounds over 15 years ago. Despite ongoing use and investigation of titanium in organic synthesis[6–9], new methodologies utilizing transmetalation to organocopper compounds remain rare. Nakamura and coworkers developed the first titanium to copper transmetalation methodology utilizing an organotitanium ate complex to perform alkylation of allylic chlorides and phosphates under copper catalysis (Scheme 1)[10,11].

Alkyl titanates are prepared by the addition of 2 equivalents of an alkyllithium to a solution of chlorotriisopropoxytitanium (**1**) in THF at $-70\,^{\circ}$C. The reaction with butyltitanate **2** and allylic phosphate **3** under catalytic copper(I) conditions gives the carvone derivative **4** in 92% yield with excellent S_N2' regioselectivity and stereoselectivity (inversion with respect to the phosphate). The copper-catalyzed alkylation of the chlorinated cyclohexenone derivative **5** shows exclusive chemoselectivity toward S_N2'-addition to the allylic chloride over the conjugate addition to the α,β-unsaturated ketone to give **6** in excellent yield. It should be noted that by adding TMSCl, the chemoselectivity could be changed to favor 1,4-addition. The exact mechanism of these additions remains unknown; however, based on the lack of reactivity of the organotitanium reagents alone and the lack

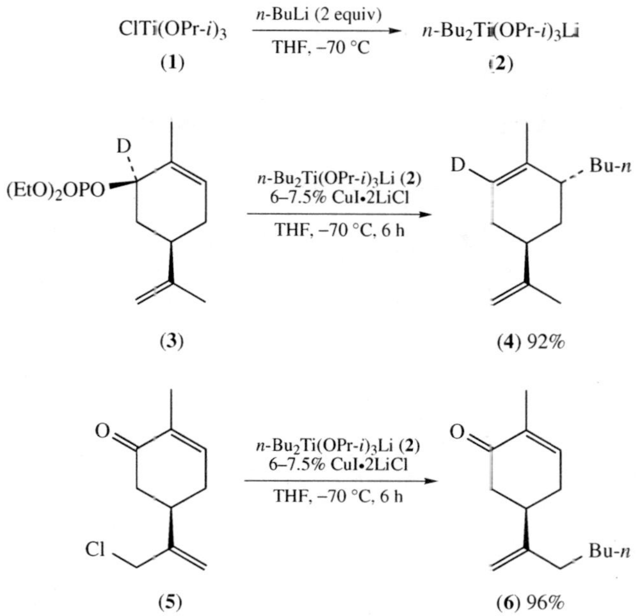

SCHEME 1. Alkylation of allylic chlorides and phosphates

of regioselectivity of alkylcopper, lithium dialkyl cuprate and copper(I)-catalyzed organolithium additions to enones, it was argued that both titanium and copper are necessary for these reactions to proceed in high yields and selectivities. One explanation for the regioselectivity favoring the S_N2' pathway could be due to the presence of an organocopper–titanium Lewis acid complex (similar to a Yamamoto-type $RCu{\bullet}BF_3$ complex) formed by a transmetalation process[4, 12–14].

More recently, Sato and Urabe have used a copper(I)-mediated reaction to perform selective mono-allylation of titanacycles (Scheme 2)[15]. Enyne **7** reacts with (η^2-propene) $Ti(OPr\text{-}i)_2$ (**8**), generated *in situ* from *i*-PrMgCl and $Ti(OPr\text{-}i)_4$, to yield the intermediate titanacyclopentene **9**, which upon treatment with stoichiometric $Li_2Cu(CN)Cl_2$ and allyl bromide gives the mono-allylation product **10** in good yield (95%) and regioselectivity (>95:<5; with respect to the other allylation regioisomer). Enyne **11** and diyne **13** can also be mono-allylated to give diene **12** and triene regioisomers **14** and **15**, respectively. This reaction is analogous to that of Cp_2-zirconacyclopentenes and Cp_2-zirconacyclopentadienes discussed later (see Section III.A.2).

Furthermore, Sato and coworkers have performed a one-pot, two-step hydrotitanation of 1-trimethylsilyl-1-octyne **16** (Scheme 3) to give a (β-silylalkenyl)titanium species **17** by selective C-protonation of the intermediate titanocyclopropene **18** with *sec*-butanol[16]. It should be noted that other hydrometalation reactions of 1-trimethylsilyl-1-octyne produce the regioisomeric α-silylalkenyl-metal species[17–23]. The alkenyltitanium species **17** underwent a copper(I)-mediated addition to alkylidene malonic ester **19**, cyclohexenone and allylic bromide **22** to provide alkenylsilanes **20**, **21** and **23**, respectively (Scheme 4). The hydrotitanation of 1-tributylstannyl-1-octyne **24** gives the β-titanated alkenylstannane lynchpin **25** that can be selectively functionalized across the olefin. For example, the β-titanated alkenylstannane **25** can be added in a conjugate fashion under copper catalysis to alkylidene malonic ester **19**, yielding a vinylstannane that is then iodinated to give

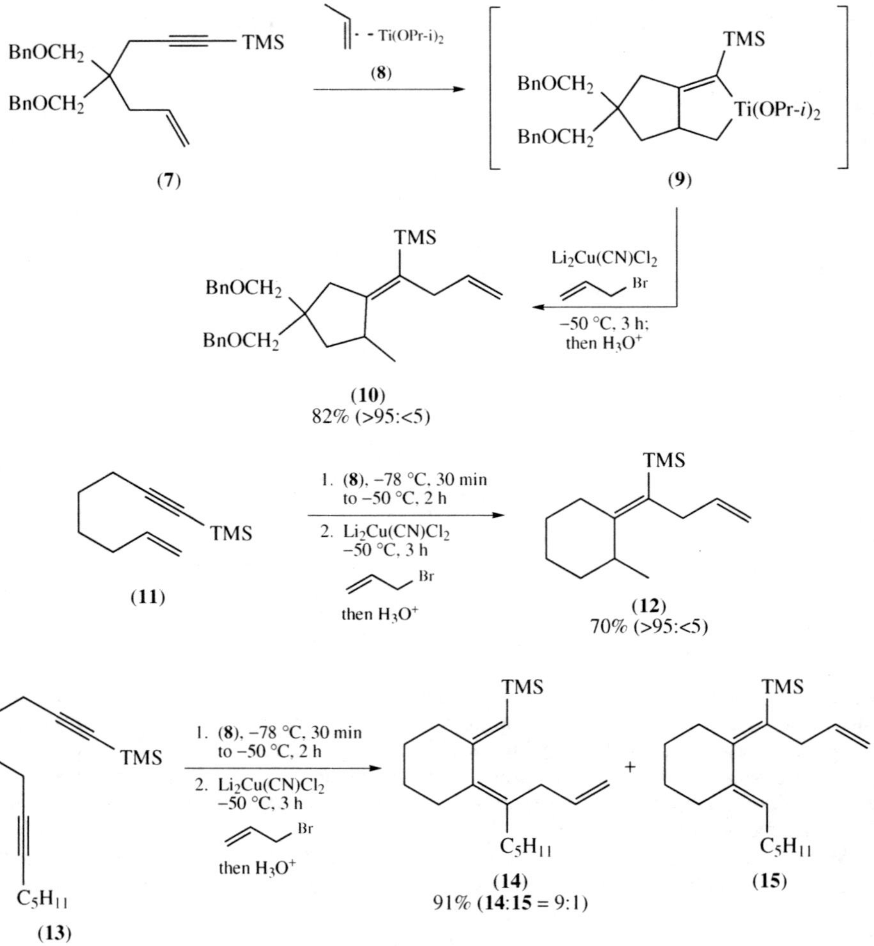

SCHEME 2. Copper-mediated allylation of titanacycles derived from enynes and diynes

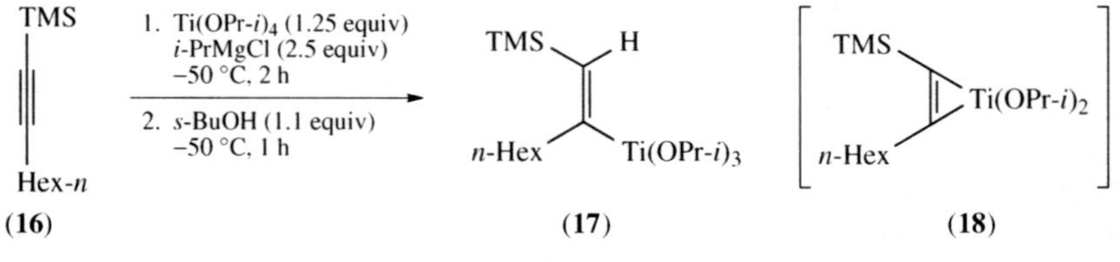

SCHEME 3. Hydrotitanation of 1-trimethylsilyl-1-octyne

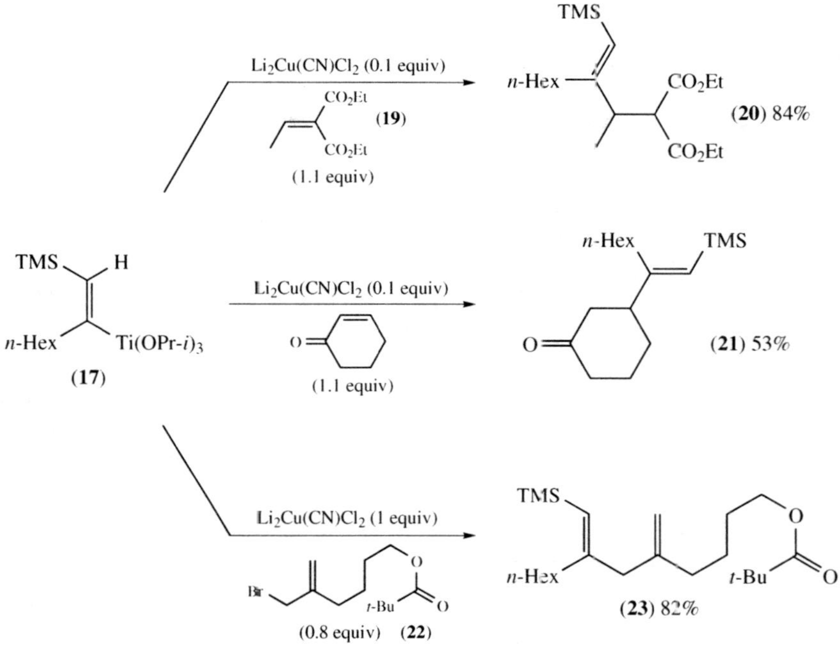

SCHEME 4. Hydrotitanation and copper-mediated reactions with electrophiles

vinyl iodide **26**. The vinyl iodide **26** can be subjected to a Sonogashira coupling to give the ene-yne **27** (Scheme 5).

B. Transmetalation from Organomanganese Compounds

The chemistry of organomanganese reagents has recently been reviewed[24–26]. Organomanganese reagents can be prepared by a transmetalation reaction of manganese halides and lithium/magnesium organometallics. Alternatively, organomanganese compounds can be prepared by the insertion of activated manganese into organic halides[25,26]. Activated manganese can be prepared by reducing manganese halides with lithium aluminum hydride[27], potassium/graphite[28], lithium and naphthalene[29,30], or lithium and 2-phenylpyridine[31]. Preparation of organomanganese reagents from activated manganese is attractive because the preparation of these reagents from lithium/magnesium organometallics can be limited by functional group compatibilities. The transmetalation of organomanganese to organocopper compounds is appealing as it can lead to improved yields and chemoselectivity, usually favoring 1,4- instead of 1,2-addition, which is preferred in standard organomanganese chemistry.

Cahiez and coworkers have investigated the coupling of organomanganese reagents to acid chlorides with and without catalytic copper(I) salts[32–34]. In the presence of catalytic amounts of CuCl, improvements in yield were observed for the additions of methyl-, aryl-, alkenyl-, and secondary and tertiary alkylmanganese chlorides to acid chloride **28** (Table 1). In many cases, the addition of 1% CuCl provided a substantial increase in yield. Addition of *t*-BuMnCl to acid chloride **28** in the absence of CuCl failed to form the desired product; however, upon the addition of 1% CuCl, the reaction proceeded

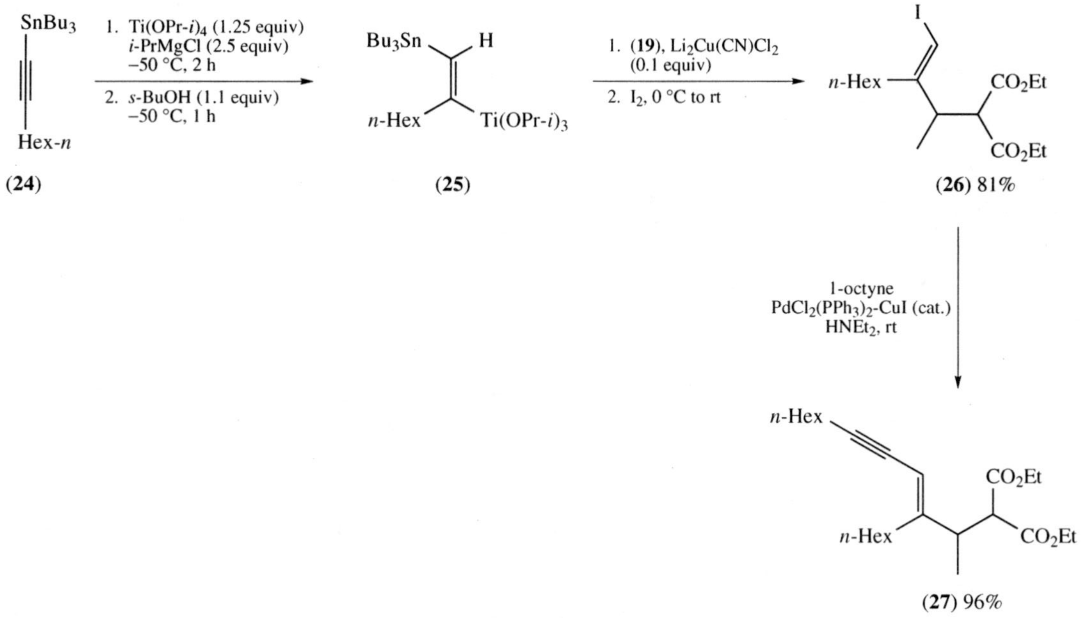

SCHEME 5. Hydrotitanation of 1-tributylstannyl-1-octyne **24** and selective functionalization of olefin **25**

TABLE 1. Copper-catalyzed addition to acid chloride

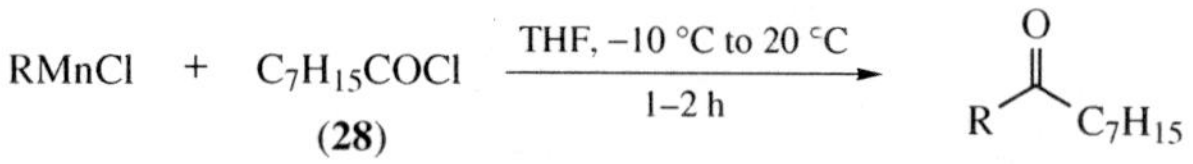

(28)

R	no CuCl (% yield)	1% CuCl (% yield)
Ph	75	92
i-Pr	69	93
t-Bu	0	96[a]
Me	40	91
$(CH_3)_2C{=}CH$	80	87

[a] −50 °C.

in excellent 96% yield at −50 °C. Addition of the MeMnCl to **28** proceeded in much higher yields under the CuCl-catalyzed conditions. Acid chlorides can also be alkylated by Grignard reagents in the presence of catalytic amounts of $MnCl_4Li_2$ and CuCl, albeit in slightly lower yields[33].

Cahiez and Alami have also developed copper-catalyzed conditions for the addition of organomanganese reagents to enones and vinylogous esters[35]. In the presence of 0.1 mol% of CuCl, butylmanganese chloride was added to cyclohexenone to afford the β-substituted ketone **29** in excellent yield. Copper-catalyzed conditions achieved higher yields than the more conventional conjugate addition of the butyl Grignard reagent, organocopper and higher order cuprate, possibly due to the cooperative effects in the active organometallic composite of copper and manganese salts (Table 2). Addition of butylmanganese chloride to cyclohexeneone gave yields of less than 10% in the absence of copper, illustrating the advantage of the transmetalation process to achieve high yields in conjugate additions to reactive enones. Cahiez and coworkers have also developed conditions for the conjugate addition of organomagnesium chloride reagents in the presence of catalytic $MnCl_2$ and CuCl to achieve similar yields and selectivity compared to the CuCl-catalyzed conjugate addition of organomanganese reagents to enones[36]. Presumably, both processes involve a very similar alkyl-Mn/Cu intermediate. Conjugate additions to vinylogous ester **30** were performed by the addition of methylmanganese chloride in the presence of 5% CuCl and TMSCl to give enone **32** after quenching of the silyl enol ether **31** (Scheme 6)[35]. It should be noted that for this substrate other organometallic reagents, such as methylcopper or

TABLE 2. Conditions for conjugate addition to cyclohexenone

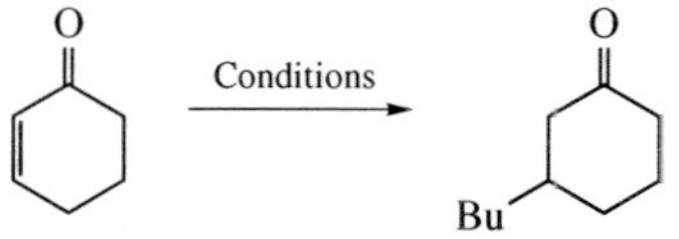

(29)

Conditions	Yield
BuMnCl, 0.1% CuCl, THF, 0 °C, 30 min	95%
BuMgCl, 5% CuCl, THF, 0 °C	54%
BuCu, Bu_3P (2.2 equiv), Et_2O, −78 to −40 °C	82%
$Bu_2Cu(CN)Li_2$, Et_2O, −78 °C, 5 h	86%

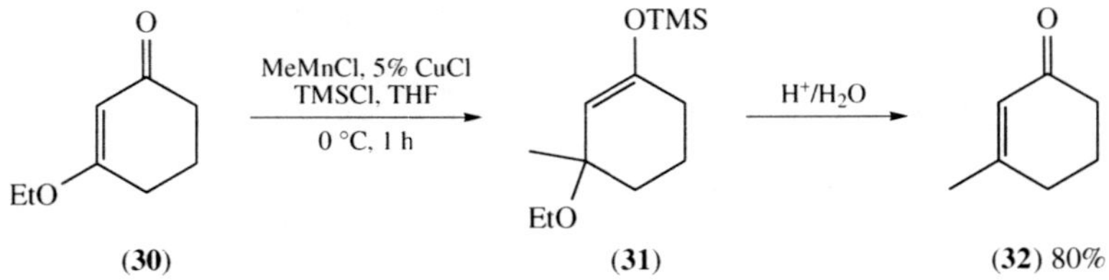

SCHEME 6. Copper-mediated conjugate addition of methylmanganese chloride to vinylogous ester **30**

lithium dimethyl cuprate, fail to react, while lithium dimethyl cuprate in the presence of TMSCl results in a mixture of 1,4- and 1,2-addition, favoring the 1,2-addition product[37].

Conjugate addition to α,β-unsaturated esters and alkylidene malonic esters were accomplished using organomanganese reagents under CuCl catalysis (Scheme 7). Copper(I) chloride-catalyzed addition of *i*-PrMnCl to the α,β-unsaturated ester **33** gave ester **34** in excellent yield[38]. The corresponding conjugate addition of alkyl Grignard reagents to **33** under identical conditions proceeded in a slightly decreased yield of 86%[39]. The beneficial effect of CuCl is remarkable in the conjugate addition of vinylmanganese chloride **36** to alkylidene malonic ester **35**, which gave the alkylated product **37** in 46% yield without CuCl and in 92% yield in the presence of 5% CuCl[40].

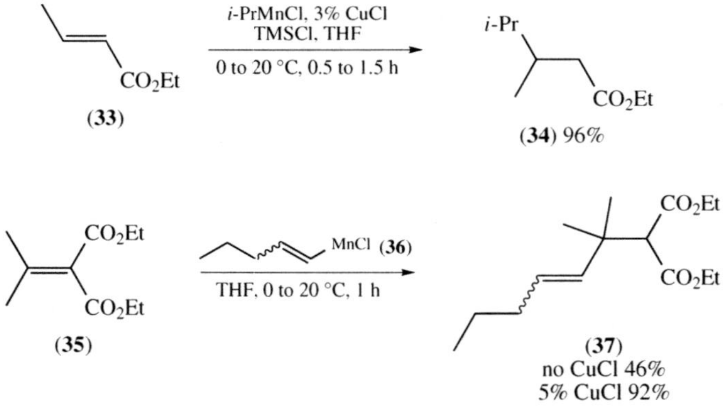

SCHEME 7. Conjugate addition to α,β-unsaturated ester and alkylidene malonic ester substrates

Catalytic quantities of CuCl in the addition of organomanganese chlorides to α,β-unsaturated aldehydes provided a dramatic change in the chemoselectivity (Scheme 8)[41]. Addition of butylmanganese chloride to aldehyde **38** resulted in the selective formation of the 1,2-addition product **39** in good yield; however, upon the addition of 5% CuCl, the 1,4-addition product **40** predominates in a comparable yield. The conjugate addition products from alkyl organomanganese compounds (R = Ph and *n*-Bu) and aldehyde **41** were trapped by TMSCl to provide the corresponding silyl enol ethers **42a** and **42b** in good yields (Scheme 9).

C. Transmetalation of Chromium Fischer Carbene Complexes

Catalytic transmetalation reactions of chromium carbene complexes have recently been reviewed and shown to be useful in expanding the synthetic utility of Fischer carbenes[42].

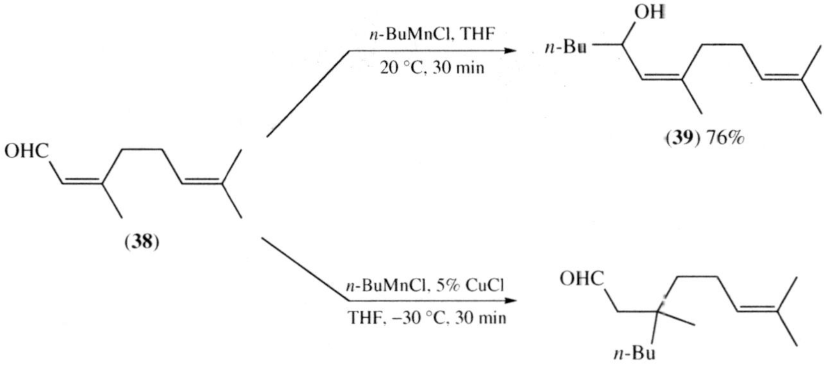

(38) (39) 76% (40) 78%

SCHEME 8. Controlling chemoselectivity of organomanganese compounds with copper(I) chloride

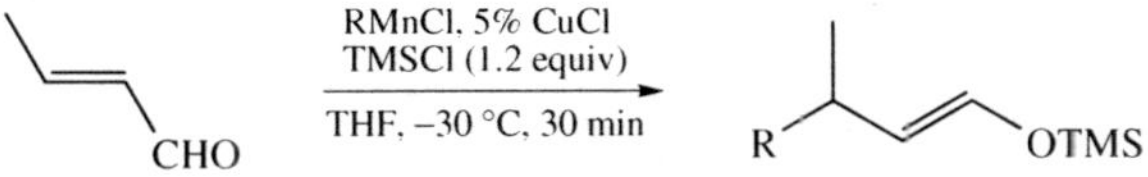

(41)

R = Ph (**42a**) 82%
R = n-Bu (**42b**) 76%

SCHEME 9. Trapping of conjugate addition products gives silyl enol ethers

The first observation of a copper(I) centered carbene formed by carbene ligand transfer from a Fischer carbene complex was recently reported by Aumann and coworkers[43]. After this initial report, Barluenga and coworkers showed that Fischer chromium carbene complexes **43a–d** could be coupled with ethyl diazoacetate (**44**) under copper(I) catalysis to form the vinylogous carbonates **45a–d** in high yield (Scheme 10)[44]. Apparently, a copper(I)-bis-carbene species led to the observed products, regenerating the active catalyst.

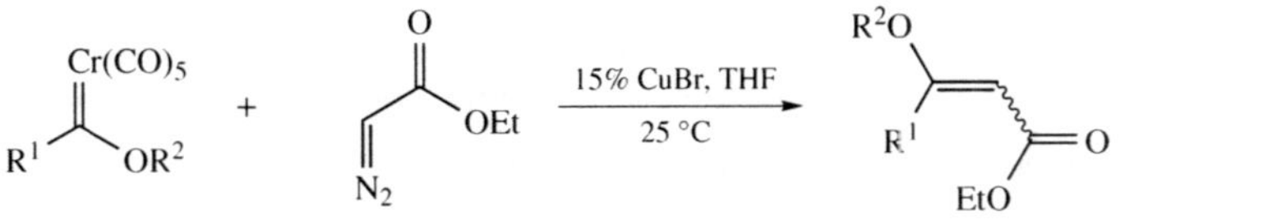

(**43a–d**) (**44**) (**45a–d**) 85–95%

R^2 = Me; E:Z = 1:1
R^2 = menthyl; E:Z = >30:<1

(**a**) R^1 = Bu; R^2 = Me
(**b**) R^1 = Ph; R^2 = Me
(**c**) R^1 = (E)-CH=CH-2-furyl; R^2 = Me
(**d**) R^1 = (E)-CH=CH-2-furyl; R^2 = menthyl

SCHEME 10. Copper(I)-catalyzed coupling of diazoester **44** and Fischer carbene **43a–d**

While investigating the dimerization of Fischer chromium carbene complexes, Barluenga and coworkers observed an increase in the E:Z selectivity and yield when changing the copper(I) source from CuBr to $[Cu(MeCN)_4]BF_4$[45]. This advancement led to the ability to generate a number of dienes and trienes in good stereoselectivity. For example, dimerization of Fischer chromium carbene **43c** provided conjugated triene **46** in excellent yield and high selectivity (Scheme 11). Trienes were subsequently cyclized under Nazarov

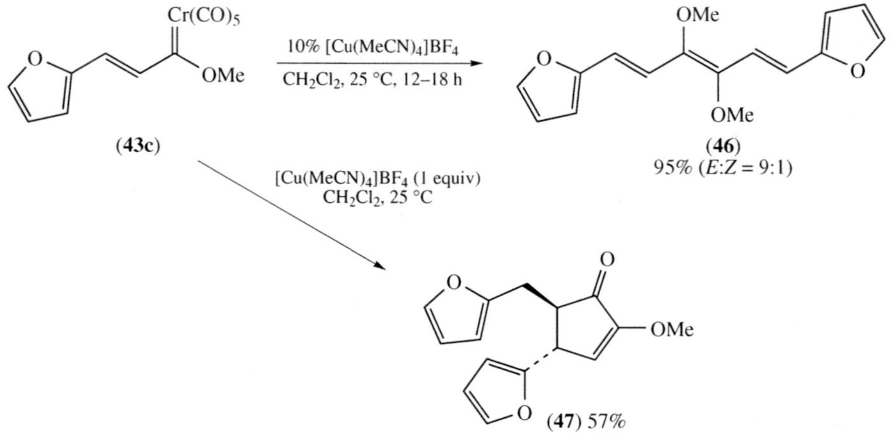

SCHEME 11. Dimerization of Fischer carbene **43c** and one-pot dimerization/Nazarov cyclization producing **47**

conditions to generate α-ether-cyclopentenones. Interestingly, Fischer chromium carbene **43c** was chemo- and stereoselectively dimerized and cyclized in one-pot by treatment with stoichiometric $[Cu(MeCN)_4]BF_4$ to give α-ether-cyclopentenone **47**.

III. TRANSMETALATIONS IN THE SECOND TRANSITION METAL SERIES

According to the Pauling electronegativity scale, zirconium (1.33), niobium (1.60), cadmium (1.69), hafnium (1.30) and tantalum (1.50) are prone to spontaneous transmetalation reactions with copper (1.90). However, synthetically relevant examples for transmetalations to copper(I) salts from the second and third transition metal series are currently still limited to zirconium.

A. Transmetalation from Organozirconium Compounds

1. From alkyl and alkenyl organozirconium compounds

Organozirconium reagents[6,46] can be easily prepared by hydrozirconation[47–51] of alkenes and alkynes using Schwartz reagent (Cp_2ZrHCl)[52]. However, the lack of reactivity of the resulting sterically shielded alkyl[53,54] and alkenyl[55] zirconium species with organic electrophiles necessitates transmetalation in order to gain utility for organic synthesis. Hydrozirconation has considerable functional group compatibility, greatly exceeding that of lithium and magnesium, and can therefore provide access to various functionalized organocopper compounds. After the pioneering work of Wailes and coworkers[56], Schwartz and coworkers were the first to recognize the potential gained by transmetalation reactions of organozirconium compounds to form organocopper compounds[57]. Thermal decomposition of the alkenyl copper(I) complex **49** derived from alkenyl zirconocene **48** resulted in formation of a copper mirror (Cu^0) and diene **50** with retention of olefin configuration (Scheme 12). The transmetalation of alkenyl zirconocene **48** to the alkenyl copper(I) intermediate and subsequent conversion to the corresponding ate complexes in the presence of lithium iodide allowed for the 1,4-addition to an α,β-unsaturated ketone to give **51**.

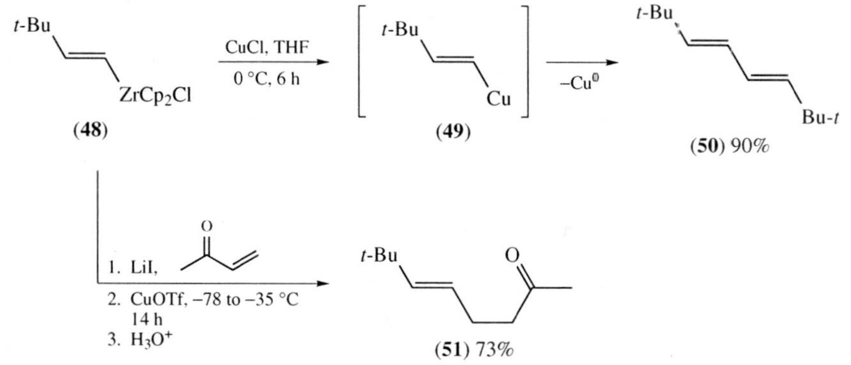

SCHEME 12. The first transmetalation reaction of organozirconium compounds to form products derived from transient organocopper compounds

However, this particular method is limited by the use of stoichiometric copper and the very slow rate at which alkyl groups transmetalate from zirconium to copper[58].

Lipshutz and Ellsworth[59] and Wipf and Smitrovich[60] reported modifications that expanded the scope and utility of Schwartz's original process. In the work performed by Lipshutz and Ellsworth, treatment of alkenyl zirconocene **53**, obtained by hydrozirconation of alkyne **52**, with one equivalent of methyl lithium followed by addition of a higher order cyanocuprate, led to transmetalation and formation of organocopper compound **54**. The resulting alkenyl cyanocuprate **54** was added in a conjugate fashion to enone **55** to afford ketone **56** (Scheme 13).

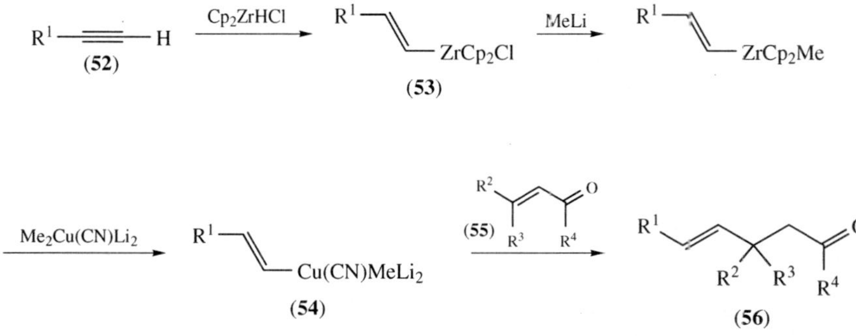

SCHEME 13. General method of transmetalation developed by Lipshutz and Ellsworth

During the transmetalation sequence, the addition of $MeLi/Me_2Cu(CN)Li_2$ is analogous to adding three equivalents of MeLi and one equivalent of LiCl-solubilized CuCN. Thus, the hydrozirconation of alkyne **57** and treatment with MeLi (3 equiv) and CuCN•2LiCl provided intermediate cyanocuprate **58**. Subsequent 1,4-conjugate addition to sterically hindered enone **59** works well (Scheme 14)[59].

The methodology developed by Lipshutz and Ellsworth has also been successfully applied to the large-scale synthesis of the protected prostaglandin misoprostol (**63**)[61]. Alkyne **60** was hydrozirconated and transmetalated to form cyanocuprate **61**, which was added to enone **62** to give the protected prostaglandin misoprostol (**63**) in good yield (Scheme 15).

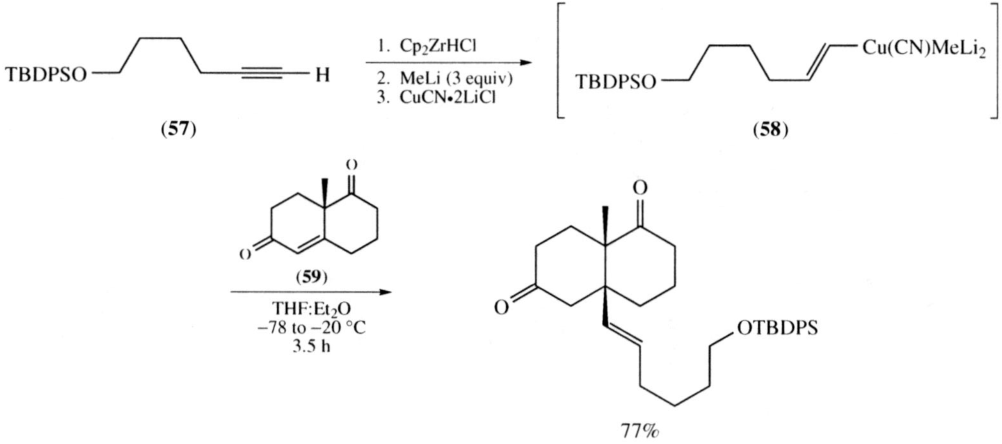

SCHEME 14. 1,4-Conjugate addition of alkenylcuprate **58** to enone **59**

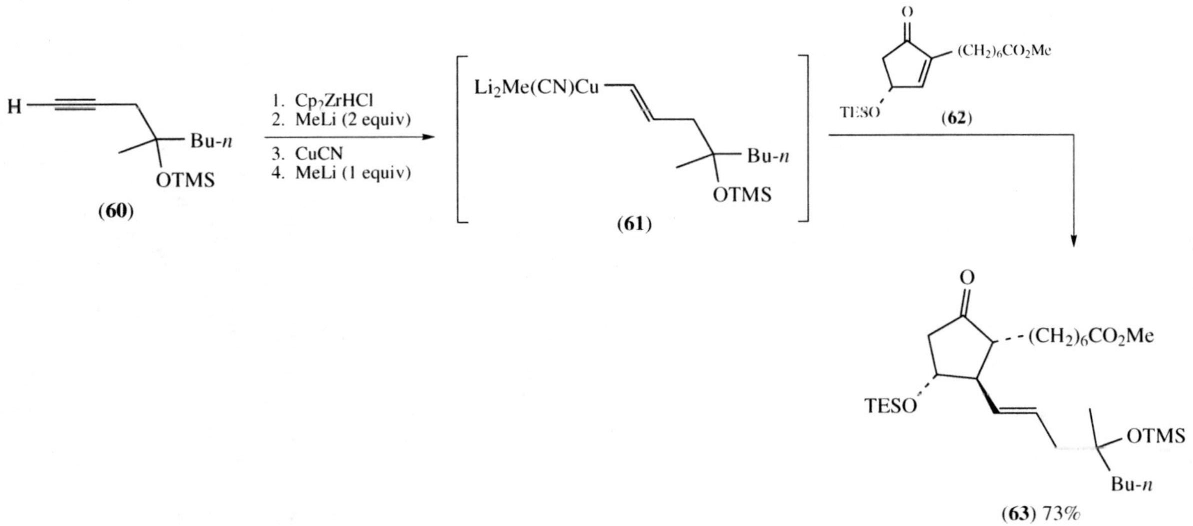

SCHEME 15. Application of the hydrozirconation/transmetalation protocol to yield protected misoprostol (**63**)

Congruently, Wipf and Smitrovich expanded upon Schwartz's original zirconium/copper transmetalation methodology by utilizing copper(I) salts catalytically[60]. Hydrozirconation of 1-hexene generated the alkylzirconium species **64**, which was added to cyclohexenone in the presence of 5–10% CuBr•Me_2S to give ketone **65** in good yield (Scheme 16). This methodology was the first to generate alkyl cuprates *in situ* from alkenes.

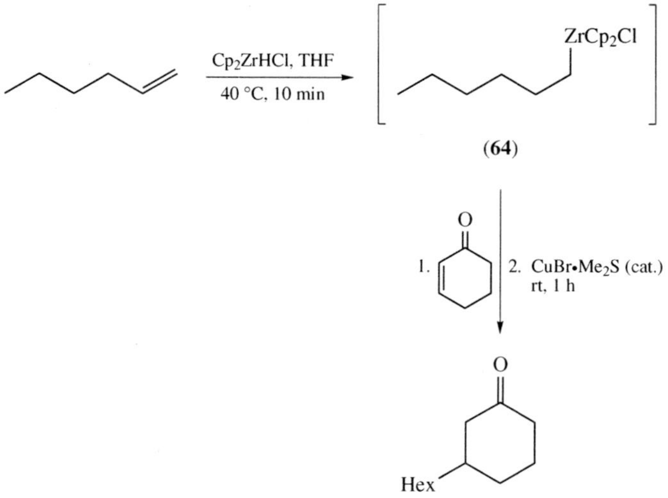

SCHEME 16. Transmetalation of alkylzirconium compound **64** and 1,4-addition to cyclohexenone in the presence of catalytic copper(I) salt

Independently, Lipshutz and coworkers utilized catalytic copper(I) conditions to transmetalate from zirconium to copper and investigated the regioselectivity of alkyl cuprate addition to allylic halides and phosphates[62]. Reactions of alkyl cuprates and allylic halides proceeded in good yields and selectivity for the S_N2'-addition, illustrating a remarkable reactivity similarity to Yamamoto-type Lewis acid complexes (RCu•BF_3)[63,64]. The comparison to the regioselectivity of addition of dialkyl cuprates to allylic halides is also interesting (Table 3)[65]. The chemoselectivity of S_N2'-displacement vs. 1,4-conjugate addition would most likely favor S_N2' based on the work done by Nakamura and coworkers with titanium/copper complexes (see Section II.A)[11].

Another feature of utilizing the hydrozirconation process is the tendency for alkyl zirconium species to migrate to a less sterically hindered position through an oxidative addition–reductive elimination cycle[66–68]. This property can be used to obtain primary alkyl zirconocenes from internal olefins, but it also detracts from the opportunity to use internal and secondary chiral organozirconocenes. The hydrozirconation of internal olefin **66** and subsequent conjugate addition to enone **67** in the presence of a catalytic amount of copper yields the straight-chain ketone **68** (Scheme 17)[60]. Interestingly, the hydrozirconation of 3,4-dihydro-2*H*-pyran (**69**) with two equivalents of Schwartz reagent, subsequent transmetalation with copper(I) and addition to cyclohexenone provided a straight-chain alcohol in the product **72**. The mechanistic hypothesis involves a tandem hydrozirconation–β-elimination–hydrozirconation sequence to give an acyclic bis-zirconocene **70**

TABLE 3. Ratio of α- and γ-substitution of allylic halides using various cuprates

Substrate		$n\text{-}Bu_2CuLi$	$n\text{-}BuCu\bullet BF_3$	$Ph(CH_2)_4ZrCp_2Cl$ CuCN (yield)
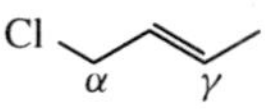	α	96	6	0 (90%)
	γ	4	94	100
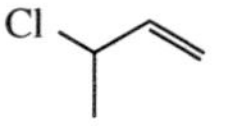	α	22	4	<2 (98%)
	γ	78	96	>98
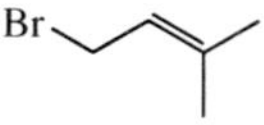	α	100	10	11 (89%)
	γ	0	90	89

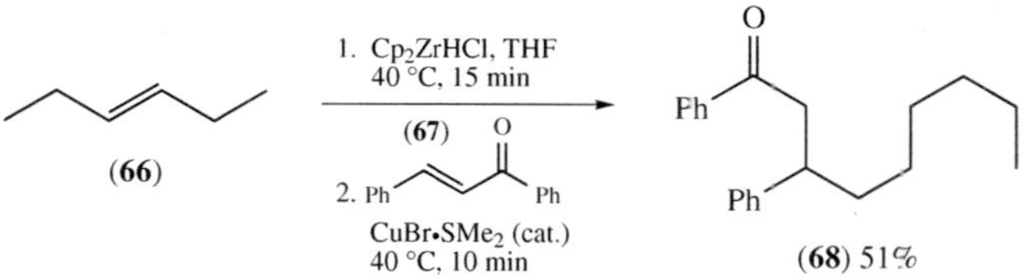

SCHEME 17. Hydrozirconation followed by spontaneous migration of the alkylzirconocene, transmetalation and 1,4-conjugate addition to enone **67**

(path a) or cyclic zirconocene oxide **71** (path b) (Scheme 18)[4,60]. This approach represents a strategy for a protective group-free Michael addition of unsaturated alcohol chains.

The scope of the methodology using hydrozirconation of alkenes and alkynes and further transmetalation to alkyl and alkenyl cuprates was further expanded to the synthesis of ketones from acid chlorides[69]. Alkyl zirconium compounds will add to acid chlorides, albeit slowly; however, alkenyl zirconium compounds do not[70]. Alkene **73** was hydrozirconated and added to a sterically demanding trimethyl acetyl chloride to give ketone **74** in good yield. Hydrozirconation of alkyne **75** and addition to benzoyl chloride furnished the ketone **76**, while showing tolerance to the homopropargylic iodide (Scheme 19). This methodology is attractive, considering the alternative of using Grignard reagents, which notoriously gives alcohol side products and are limited in their tolerance to other functional groups[71–73].

Mixed cuprates containing methyl or other alkyl groups, such as **54** and **61**, are not generally useful for substitution reactions, because alkyl group transfer to the electrophile competes with vinyl group transfer. However, the use of a non-transferable ligand in the mixed cuprate solves this problem and allows for a significant increase in the substrate scope[74]. The ligand exchange reaction of mixed cyanocuprate **78** and alkenyl zirconocene **77** provides the mixed cyanocuprate **79**, which efficiently added to epoxide **80**, allylic chloride **82** and enol triflate **84** to afford homoallylic alcohol **81**, skipped diene **83** and unsaturated ester **85**, respectively (Scheme 20).

Lipshutz and Wood modified their zirconium/copper transmetalation methodology and applied it to the three-component coupling protocol[75]. Vinylzirconocene **86**, obtained via hydrozirconation of the corresponding alkyne, was treated with methyl lithium, methyl zincate and a catalytic quantity of dimethyl cyanocuprate, leading to the vinylcyanocuprate

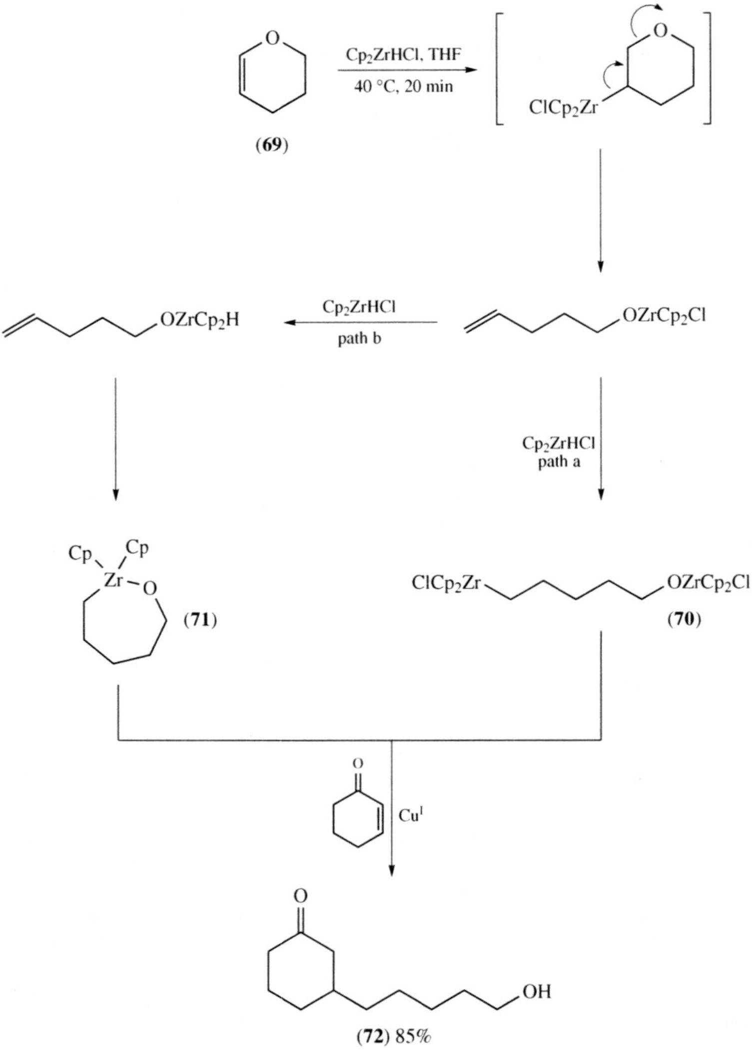

SCHEME 18. Mechanism of ring opening of pyran **69** and subsequent conjugate addition to give hydroxy ketone **72**

via transmetalation. Subsequent 1,4-addition to cyclopentenone provided the intermediate copper enolate, which was transmetalated by the organozincate, leading to regeneration of the cyanocuprate catalyst. The resulting zinc enolate **87** was trapped with aldehyde **88** to provide the three-component coupling, prostaglandin-like product **89** (Scheme 21).

Wipf and Takahashi were able to apply their hydrozirconation/copper-catalyzed conjugate addition sequence to *N*-acyl oxazolidinones[76]. The hydrozirconation of 1-hexene followed by copper-catalyzed conjugate addition to *N*-acyl oxazolidinone **90** in the presence of Lewis acid provided the *β*-substituted addition product **91** in good yield and

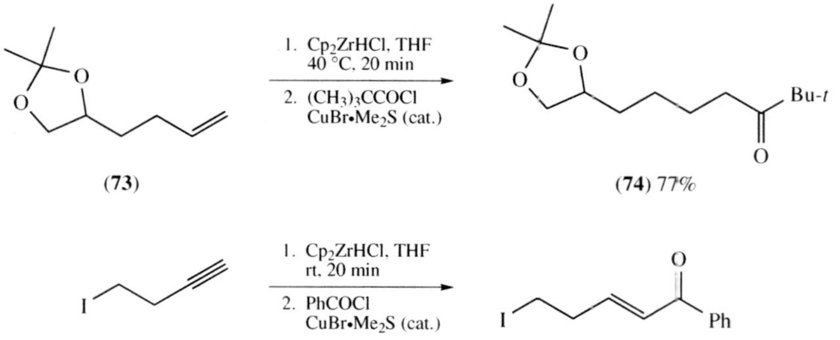

(**73**) (**74**) 77%

(**75**) (**76**) 79%

SCHEME 19. Hydrozirconation of alkenes or alkynes followed by transmetalation to alkyl or alkenyl cuprates and addition to acid chlorides to generate ketones

diastereoselectivity (Scheme 22). Interestingly, the intermediate enolate was trapped using benzaldehyde to give the α,β-substituted *syn*-aldol product **92** in good yield and excellent diastereoselectivty. In the latter one-pot reaction, three contiguous stereocenters are set. The exact role of $BF_3•Et_2O$ is not known, but it was hypothesized that it assists in the formation of chelated cationic zirconium species **93** via reversible chloride abstraction. The facial selectivity of the enolate addition to the aldehyde was rationalized using the 6-membered Zimmerman–Traxler transition state model **94**.

Furthermore, Wipf and coworkers closely investigated the hydrozirconation/transmetalation to copper and addition to a variety of enones, with the goal of gaining more insight into the reaction[77]. They observed that the mode of preparation of Schwartz reagent[52] had dramatic effects on the rate of 1,4-addition. *In situ* generation by reduction of zirconocene dichloride with $LiEt_3BH$[78] and *t*-BuMgCl[79,80] resulted in the formation of LiCl or $MgCl_2$ as byproducts, which significantly reduced the rate of conjugate addition and the chemical yield. Schwartz reagent prepared by reduction of zirconocene dichloride using $LiAlH(OBu\text{-}t)_3$[81] or $LiAlH_4$[82,83] and purified by precipitation works well to obtain the 1,4-addition product. Wipf and coworkers also postulated a mechanism for the hydrozirconation and subsequent copper-catalyzed transmetalation–conjugate addition reaction (Scheme 23). Direct decomposition of the alkyl zirconocene **95** to give alkane **97** in the absence of Cu(I) salts is extremely slow. It is likely that copper forms a complex with the alkyl zirconocene **95**, resulting in the short-lived alkylcopper-zirconium species **96**, which upon decomposition provides a copper mirror and alkane **97**. The presence of an enone facilitates chelation from the carbonyl to the alkylzirconium species and subsequently Cu(I) complexation followed by transmetalation of the alkyl group, giving Cu(III) intermediate **98**. Reductive elimination leads to zirconium enolate **99** and regeneration of Cu(I). It is unclear if the transmetalation proceeds through the formation of a Cu(III)[84–88] intermediate or if a SET mechanism is involved.

More recently, Bergdahl and coworkers performed conjugate additions of alkenyl zirconium compounds to α,β-unsaturated ketones and aldehydes in the presence of catalytic $CuI•0.75\ Me_2S$[89,90]. The latter complex was prepared from $CuI•Me_2S$ which, under vacuum, loses Me_2S to form the more stable stoichiometry of $CuI•0.75\ Me_2S$[91–93]. The use of 10% $CuI•0.75\ Me_2S$ gave significantly improved yields compared to other Cu(I) salts (i.e. $CuBr•Me_2S$ and CuCN). The formation of diene and a copper mirror, as seen in the initial work done by Schwartz and coworkers (Scheme 12)[57], was less effective using the $CuI•0.75\ Me_2S$ complex. Conjugate addition of vinylzirconium species **100** in the

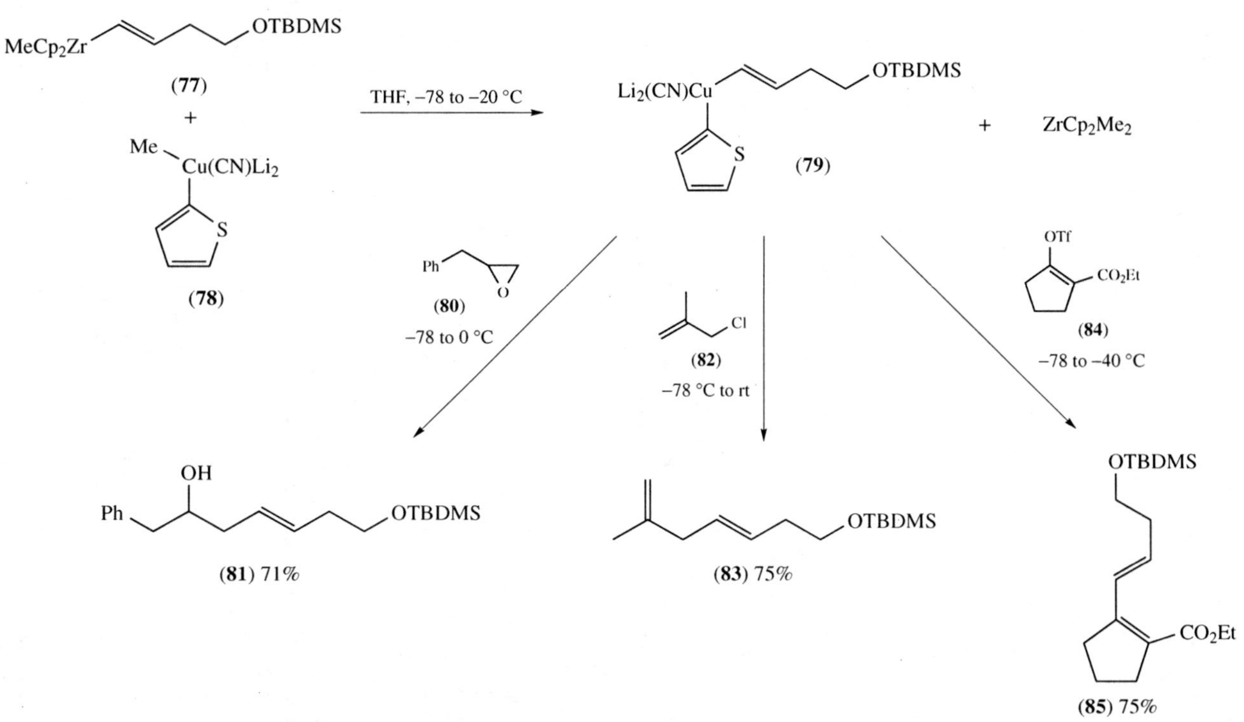

SCHEME 20. Substitution reaction of mixed cuprate **79** with various electrophiles

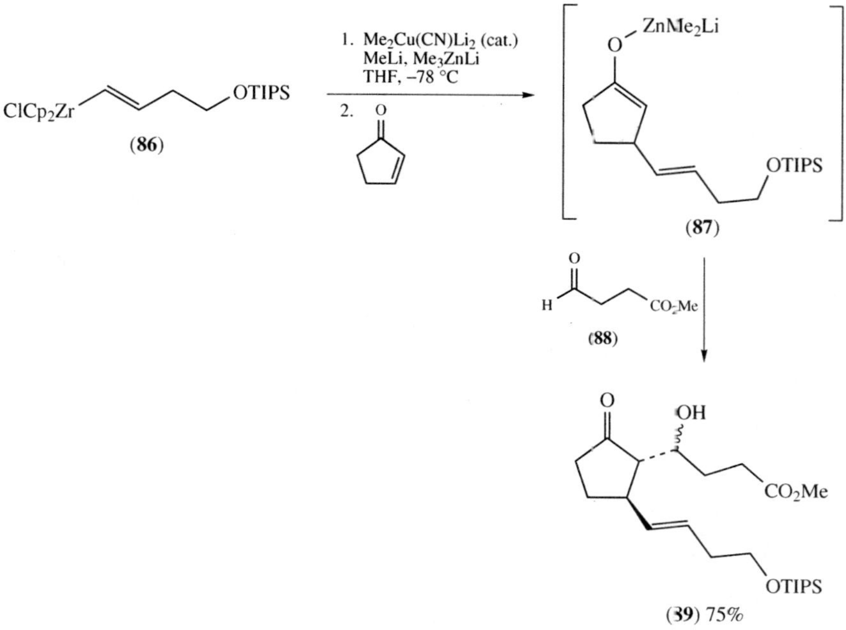

SCHEME 21. Three-component coupling methodology

presence of catalytic CuI•0.75 Me_2S to enone **101** and, more remarkably, crotonaldehyde, gave products **102** and **103**, respectively, in excellent yield (Scheme 24). It should also be noted that performing the addition of alkyl zirconocene to cyclohexenone, as in Scheme 16[77], with catalytic CuI•0.75 Me_2S, did not yield the conjugate addition product.

In an interesting extension to heterocylic chemistry, Hanzawa and coworkers utilized zirconium/copper transmetalation to perform asymmetric 1,2-additions to 3,4-dihydroisoquinoline (**104**)[94]. Treatment of **104** with vinylzirconocene **105** in the presence of an acylating agent (Boc_2O), copper(I) salt and chiral amine **106** yielded the alkylated product **107** in good yield and enantioselectivity (Scheme 25). The yield was increased to 80% in the non-asymmetric version by utilizing ethyl chloroformate as the electrophile.

The acylzirconocene[95] species is isoelectronic to an acyl anion and can be prepared by the hydrozirconation of alkenes or alkynes and subsequent insertion into CO. In the presence of a catalytic amount of copper iodide, acylzirconocenes **108** and **109** were added to allylic bromide to provide β,γ-unsaturated ketones **110** and **111**, which presumably occurred through an intermediate acyl copper compound (Scheme 26)[96]. Hanzawa and coworkers further expanded the use of the acylzirconocene **112** and performed a conjugate addition to α,β-unsaturated ketone **113** to provide 1,4-dicarbonyl compound **114** (Scheme 27)[97]. The reaction was sluggish in the absence of BF_3•Et_2O, and would not proceed without Cu(I).

Sato and coworkers were able to generate dialkoxy allylic zirconium species **115** by treatment of triethyl orthoacrylate with the Negishi reagent[98,99], and upon addition of CuCN they observed S_N2'-additions to allylic and propargylic phosphates[100] and 1,2-additions to imines[101]. The dialkoxy allylic zirconium species functions as a homoenolate

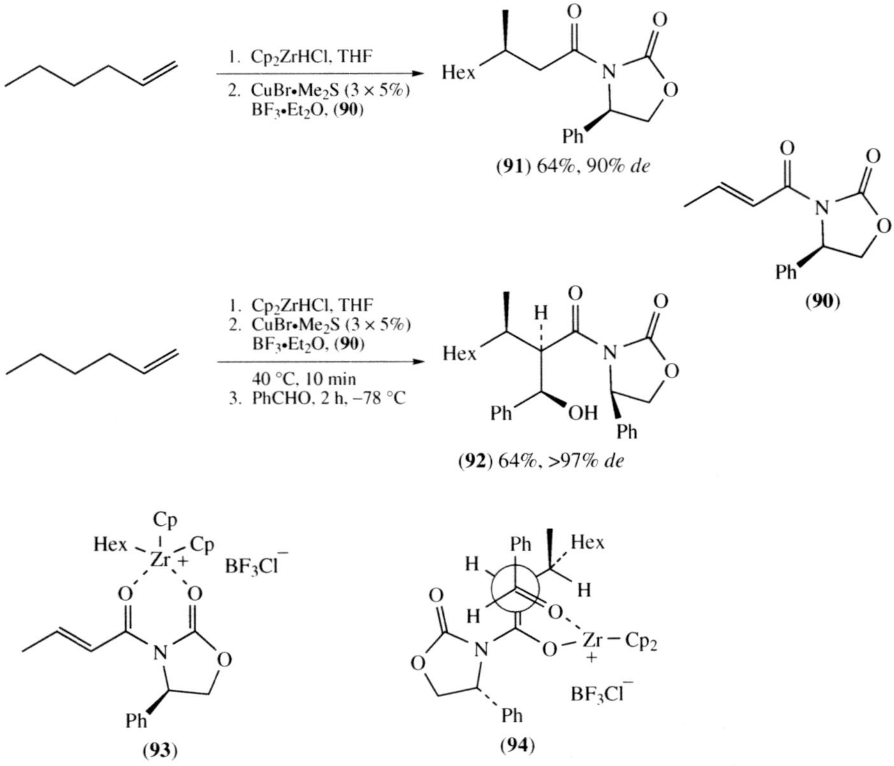

SCHEME 22. Hydrozirconation/copper-catalyzed conjugate addition to *N*-acyl oxazolidinones

equivalent. Sato and coworkers proposed that a dialkoxy allylic copper intermediate was formed via transmetalation. Zirconium species **115** was added in S_N2' fashion to allylic phosphate **116** to give ester **117**, which was deprotected and cyclized to form lactone **118**. Alternatively, **115** was added to imine **119**, to give ketal **120**, which was subsequently deprotected to give α-amino ketone **121** (Scheme 28).

2. *From zirconacycles*

Takahashi and coworkers reported the first example of a copper transmetalation using zirconacyclopentadienes[102]; in the presence of catalytic copper chloride and allyl chloride, zirconacyclopentadiene **122** was di-allylated[103]. Treatment of the resulting tetraene **123** with the Negishi reagent and carbon monoxide led to tricycle **124** (Scheme 29). This zirconium/copper allylation methodology was further applied to functionalized allylic chlorides to generate six- and seven-membered carbocycles[104]. Six-membered rings were formed by treating trimethylsilyl-substituted zirconacyclopentadiene **125** with two equivalents of copper chloride and allylic chloride **126** to give the acyclic triene intermediate **127**, formed via transmetalation from zirconium to copper and subsequent S_N2'-substitution. Presumably, a second transmetalation from zirconium to copper occurred to give an alkenyl copper(I) complex followed by an intramolecular S_N2'-displacement of the allylic chloride to provide cyclohexadiene **128**. Alternatively, the seven-membered **131** was obtained from the

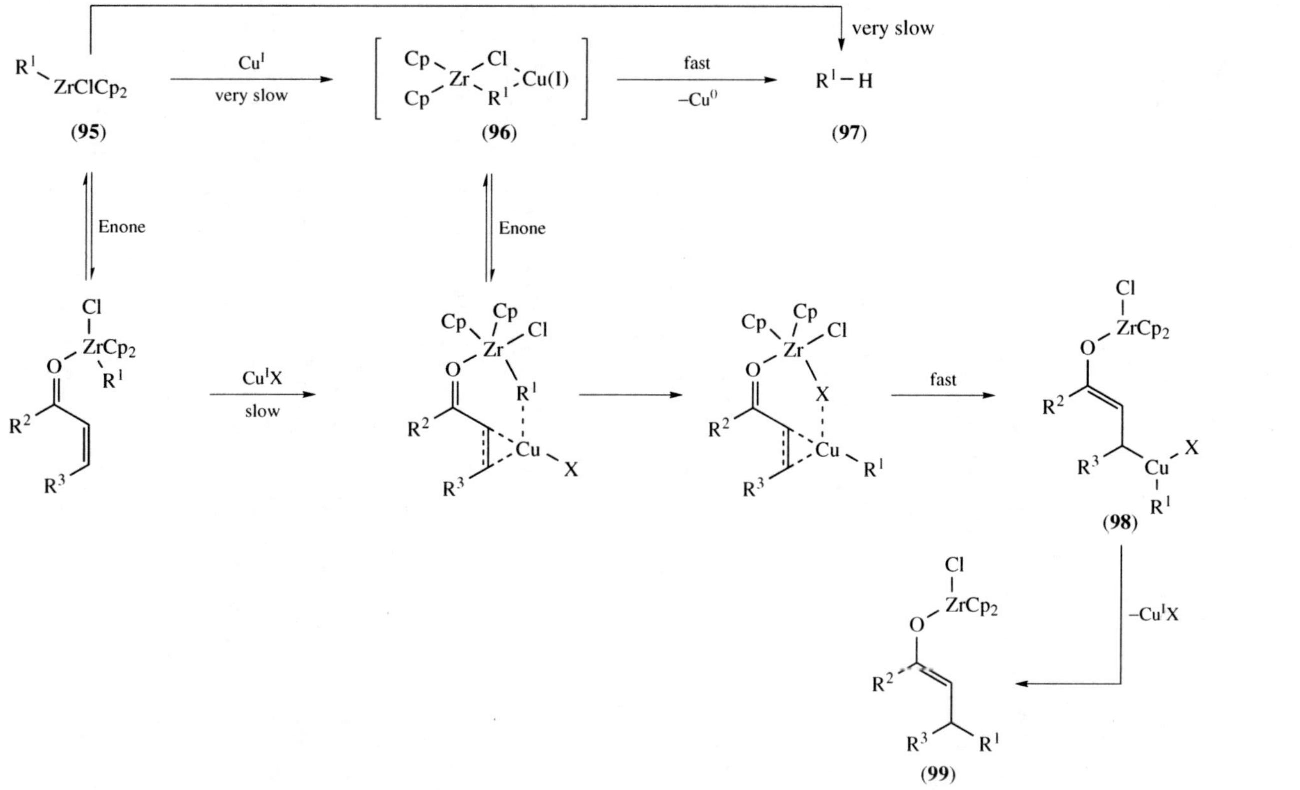

SCHEME 23. Proposed mechanism for copper-catalyzed 1,4-addition of an alkyl zirconium species

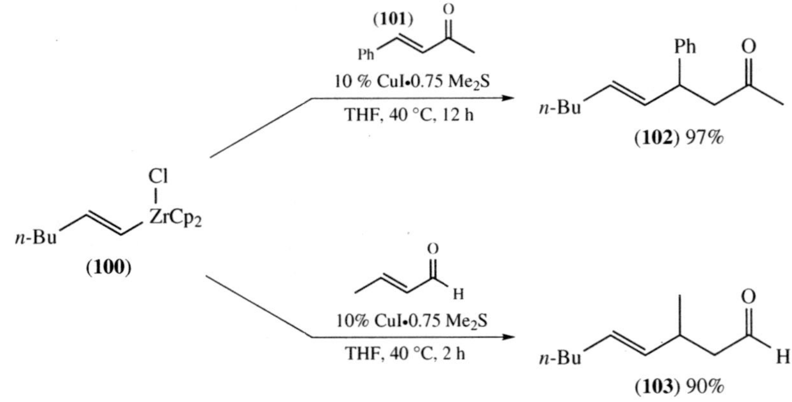

SCHEME 24. CuI•0.75 Me_2S-catalyzed conjugate addition reaction of alkenyl zirconocene to α,β-unsaturated ketone **101** and crotonaldehyde

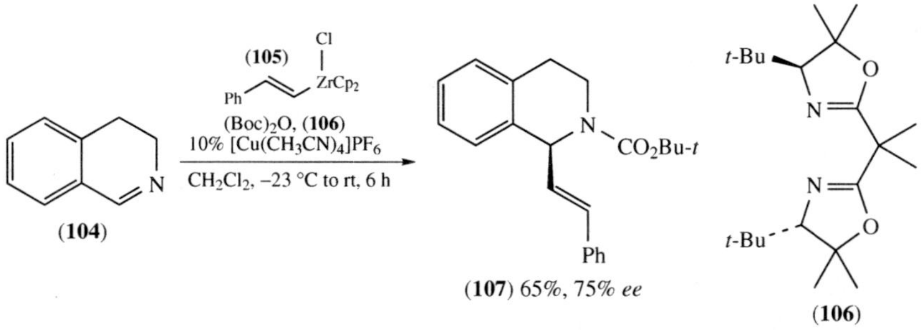

SCHEME 25. Copper(I)-catalyzed asymmetric addition of vinylzirconocene **105** to dihydroisoquinoline **104**

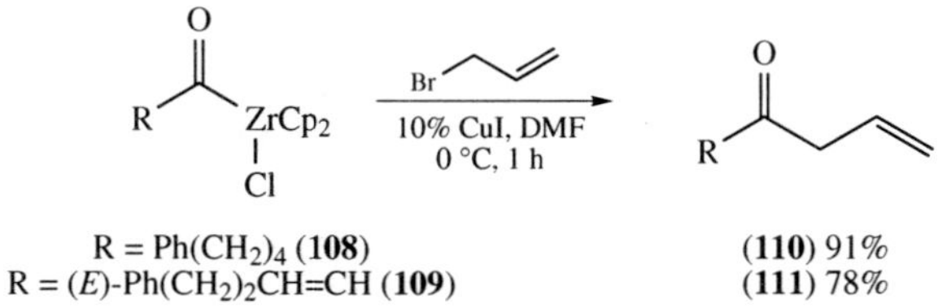

SCHEME 26. Addition of acyl zirconium compounds to allyl bromide

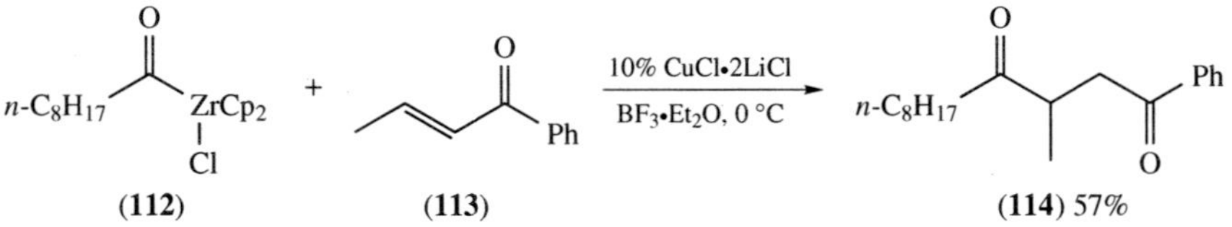

SCHEME 27. Conjugate addition of acyl zirconium compound **112** to α,β-unsaturated ketone **113**

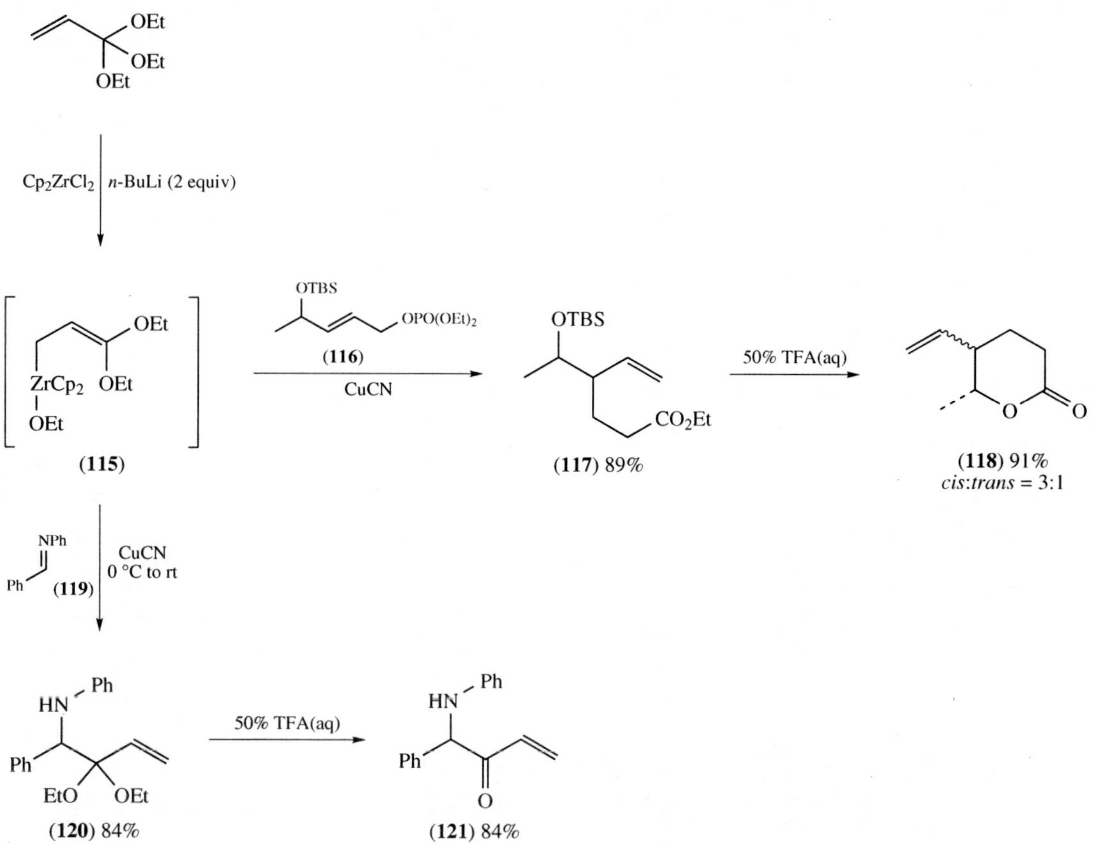

SCHEME 28. Addition of dialkoxy allylic zirconium species **115** to allylic phosphate **116** and imine **119**

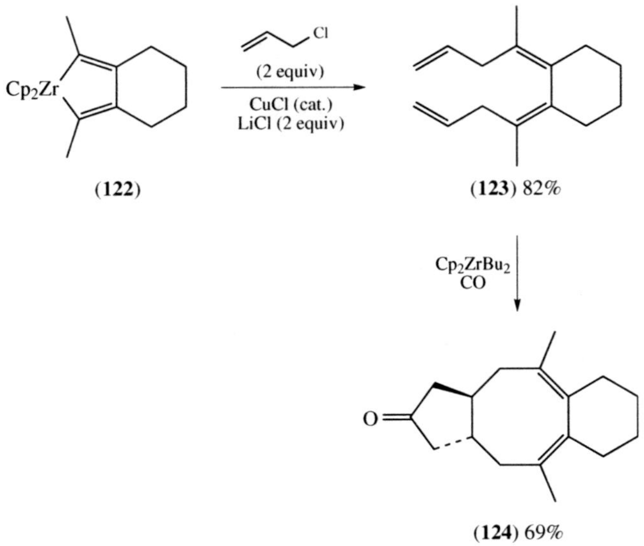

SCHEME 29. Copper-catalyzed transmetalation/allylation of zirconacyclopentadiene **122** and subsequent CO insertion and cyclization

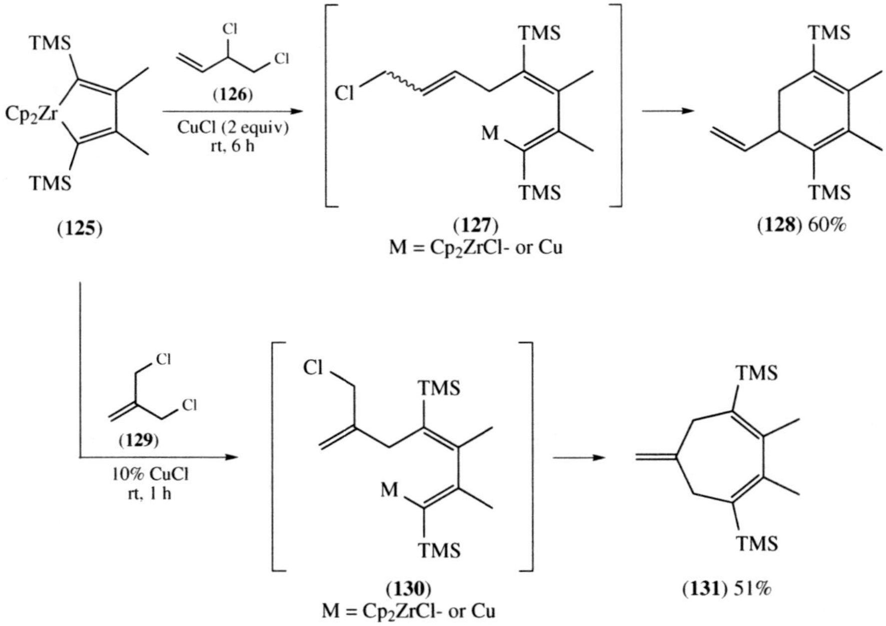

SCHEME 30. Further application of the zirconacyclopentadiene allylation methodology to form 6- and 7-membered carbocycles

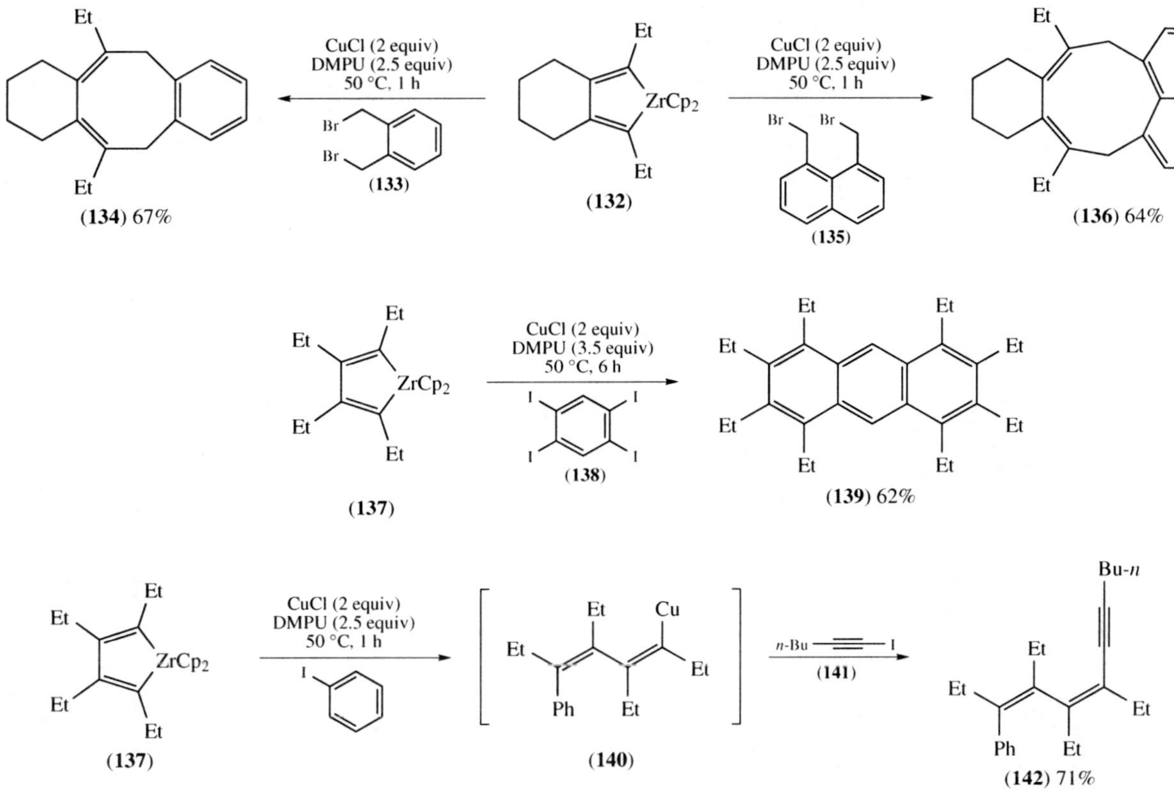

SCHEME 31. Various copper-mediated transformations of zirconacyclopentadienes

zirconacyclopentadiene **125** in the presence of catalytic copper chloride and bis-allylic chloride **129**. The formation of **131** occurred through intermediate **130**, via transmetalation from zirconium to copper and either S_N2- or S_N2'-substitution (Scheme 30). Other substituted zirconacyclopentadienes can be used; however, the relationship between yield and the amount of copper used would need to be optimized experimentally on an individual basis.

Zirconacyclopentadienes have also been coupled with benzylic halides, alkynyl halides and aryl halides in the presence of copper(I) salts[105–107]. Zirconacyclopentadiene **132**, in the presence of two equivalents of copper(I) chloride and DMPU, forms 8- and 9-membered rings, **134** and **136**, by coupling with bis(halomethyl) arenes **133** and **135**, respectively (Scheme 31)[106]. These reactions proceeded in the presence of catalytic amounts of CuCl (10 mol%); however, the yields of the polycyclic products were typically lower. Under similar conditions, **137** was coupled with tetraiodobenzene **138** to afford anthracene derivative **139**[105]. Zirconacyclopentadienes also reacted with diiodobenzenes in the presence of copper to give naphthalene derivatives. Iodobenzene was efficiently coupled with **137**, and the intermediate alkenyl copper **140** was quenched with alkynyl iodide **141** to give dienyne **142**[107].

Takahashi and coworkers also used the transmetalation reaction between copper and zirconacyclopentadienes to synthesize substituted arene **144** from three different alkynes in a one-pot annulation (Scheme 32)[108]. An intermediate zirconacyclopentadiene **143** was synthesized using the Negishi reagent and sequential treatment with two alkynes. The substituted arene was formed by the addition of a third alkyne in the presence of two equivalents of copper chloride. This method was used to synthesize a variety of aromatic esters, including **145**, **146** and **147**, in acceptable yields.

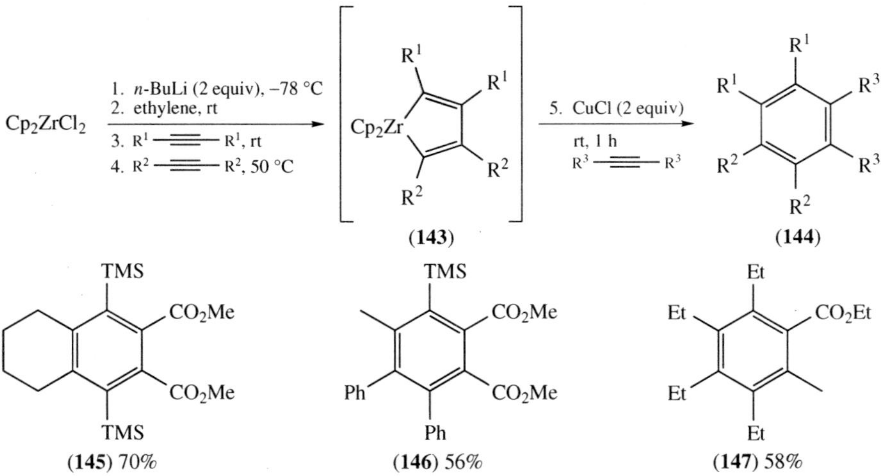

SCHEME 32. One-pot synthesis of highly substituted arenes via transmetalation of zirconacyclopentadienes

The mechanism of this reaction was postulated to proceed by one of three possible pathways from zirconacyclopentadiene **143** (Scheme 33). After the initial transmetalation of **143** to form the bimetallic species **148**, a second transmetalation occurs to form **150**, which can undergo alkyne insertion to form bis(vinylcopper) species **151**. Alternatively, **148** can perform the alkyne insertion first, providing **149**, and subsequent transmetalation gives **151**. Takahashi and coworkers proposed that **144** originated from **151** by either a ring-closing coupling reaction of the two alkenyl copper moieties or an intramolecular Michael addition and subsequent reductive elimination. They based this hypothesis on the isolation of quenched triene products from the hydrolysis of **149** or **151**. A concerted Diels–Alder-type cyclization of **150** was also not ruled out; however, it does not appear to be favored based on the isolation of quenched triene intermediates[108].

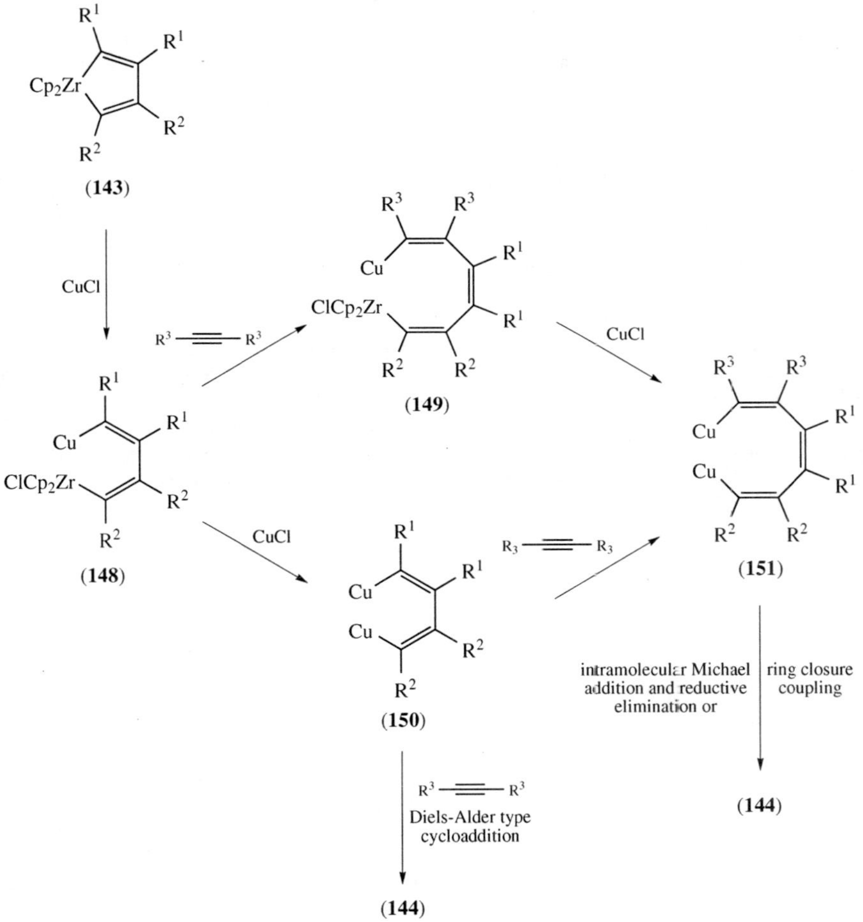

SCHEME 33. Possible mechanisms of substituted arene formation from zirconacyclopentadienes

Transmetalation reactions of zirconacyclopentanes and subsequent carbometalations with unactivated alkynes have been shown to provide carbocycles[109]. Zirconacyclopentane **152** reacted with phenylacetylene in the presence of two equivalents of copper(I) chloride to give carbocycle **153**. Olefin **155** was obtained from **152** via transmetalation, cross-coupling with an alkynyl bromide **154** and carbometalation (Scheme 34). In a noteworthy attempt to support the occurrence of the transmetalated copper species, the intermediate alkenyl metal was successfully quenched with an allylic halide and an acid chloride to give a functionalized carbocycle.

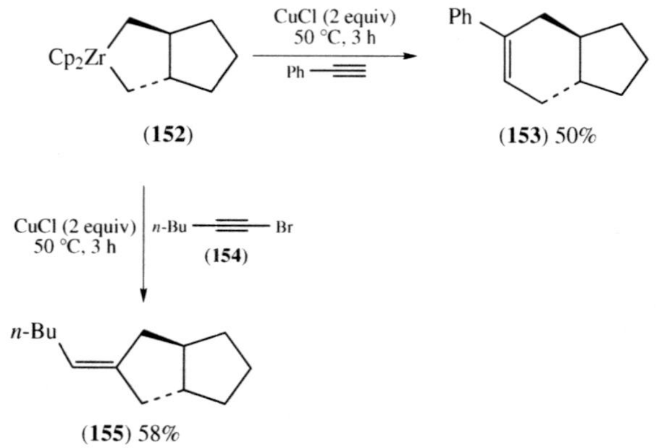

SCHEME 34. Copper-mediated carbometalation of alkynes to achieve carbocycles

Takahashi and coworkers further expanded the transmetalation of zirconacycles to the synthesis of spirocycles[110]. Spirocycles **158**, **161** and **163** were obtained by treatment of zirconacyclopentene **156**, zirconacyclopentane **159** and zirconacyclopentadiene **162**, respectively, with two equivalents of copper(I) chloride and 3-iodocycloenones **157** or **160** (Scheme 35). It was hypothesized that the initial event in the reaction was the cross-coupling of an organocopper species, obtained by transmetalation from zirconium, followed by a 1,4-conjugate addition of a second organocopper species.

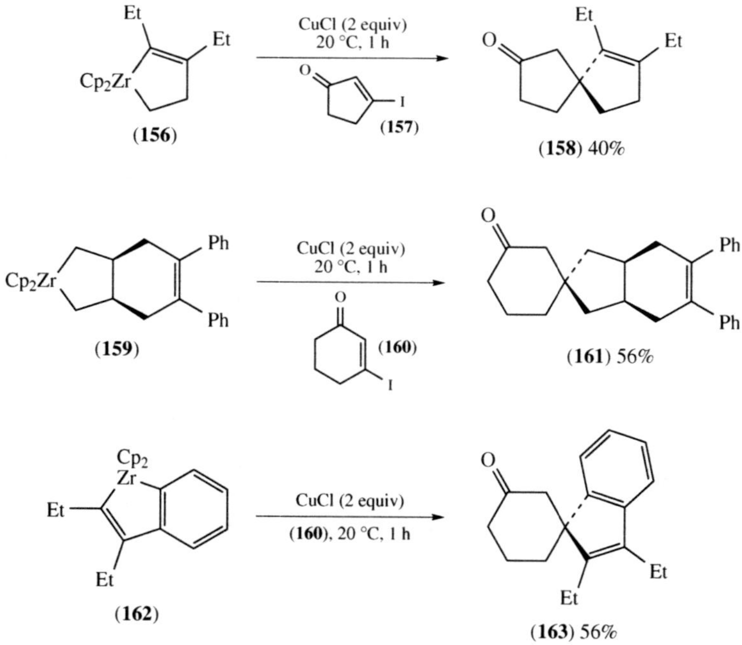

SCHEME 35. Copper-mediated conjugate addition of zirconacycles to cyclic 3-iodoenones

IV. TRANSMETALATIONS IN THE LANTHANIDE SERIES

Although electronegativities suggest that transmetalation of organolanthanide compounds to copper occurs readily, investigations in this area are limited. Presumably, the lanthanide–copper transmetalation appears of low synthetic utility since most organolanthanide compounds are prepared from the corresponding organomagnesium or organolithium species. However, the reactivity of the resulting cuprate will be strongly influenced by the presence of an oxophilic lanthanide salt, thus providing mixed organometallic species with inherently different reactivity. The relatively tolerant nature of both of these derivatives toward various functional groups is also attractive.

A. Transmetalation from Organosamarium Compounds

Wipf and coworkers were first to investigate the samarium–copper transmetalation of organosamarium compounds prepared by the reduction of alkyl iodides with samarium(II) iodide[111,112]. Initial investigations used a solution-stable organosamarium species (i.e. **166**), and subsequent transmetalation to a copper(I) iodide–triethyl phosphite complex ($CuI{\bullet}P(OEt)_3$). Organosamarium species **166** was prepared by the reduction of alkyl iodide **165**, or alternatively, aryl iodide **164**, via an intramolecular 5-*exo*-trig cyclization. Transmetalation using $CuI{\bullet}P(OEt)_3$ to form the bisalkyl Gilman-type copper complex **167** followed by a 1,4-conjugate addition to cyclohexenone provided ketone **168** (Scheme 36). In general, the use of 2 equivalents of halide, 4 equivalents of SmI_2 and 1 equivalent of both $CuI{\bullet}P(OEt)_3$ and enone are necessary to obtain the conjugate addition product for a variety of halides and enones.

The initial methodology developed by Wipf and coworkers was later improved upon by adding a slight excess of halide to the enone in the presence of TMSCl and catalytic amounts of copper(I) salts[112]. Reduction of alkyl iodide **169**, followed by transmetalation with a catalytic amount of $CuBr{\bullet}Me_2S$ and addition to functionalized enone **170** activated with TMSCl gave silyl enol ether **171** (Scheme 37). The corresponding ketone **172** was obtained in good overall yield by silyl deprotection of **171**. This improved methodology resulted in higher yields compared to the alternative approach using stoichiometric copper.

Berkowitz and Wu employed similar conditions to perform the sp^3–sp^3 cross-coupling reaction of simple alkyl iodides and bromides[113]. This cross-coupling methodology minimized homo-coupling and was applied to the synthesis of an archaebacterial lipid[114]. For the synthesis of a fragment of this archaebacterial lipid, alkyl iodide **173** was reduced with SmI_2 and coupled to iodo ester **174** in the presence of catalytic copper(I) bromide in good yield (Scheme 38). Berkowitz and Wu proposed a dialkylsamarium copper complex as an intermediate, as suggested by Wipf and coworkers.

V. TRANSMETALATIONS IN GROUP IIB

A. Transmetalation from Organozinc Compounds

The chemistry of organozinc compounds, including transmetalation reactions to organocopper compounds, has been covered extensively[5,115–117]. As early as 1859, the first experiments involving the preparation of alkylcopper species from copper halides and organozinc derivatives were performed. These initial findings paved the way for significant advances, which have been reviewed previously by Wipf[4] and Knochel[5,117].

The ease of preparation, versatility of application and functional group compatibility contribute to the frequent use of organozinc compounds in organometallic chemistry. Organozinc reagents are classified into three groups: organozinc halides (RZnX), diorganozincs (R_2Zn) and triorganozincates (R_3ZnLi). Among these, triorganozincates

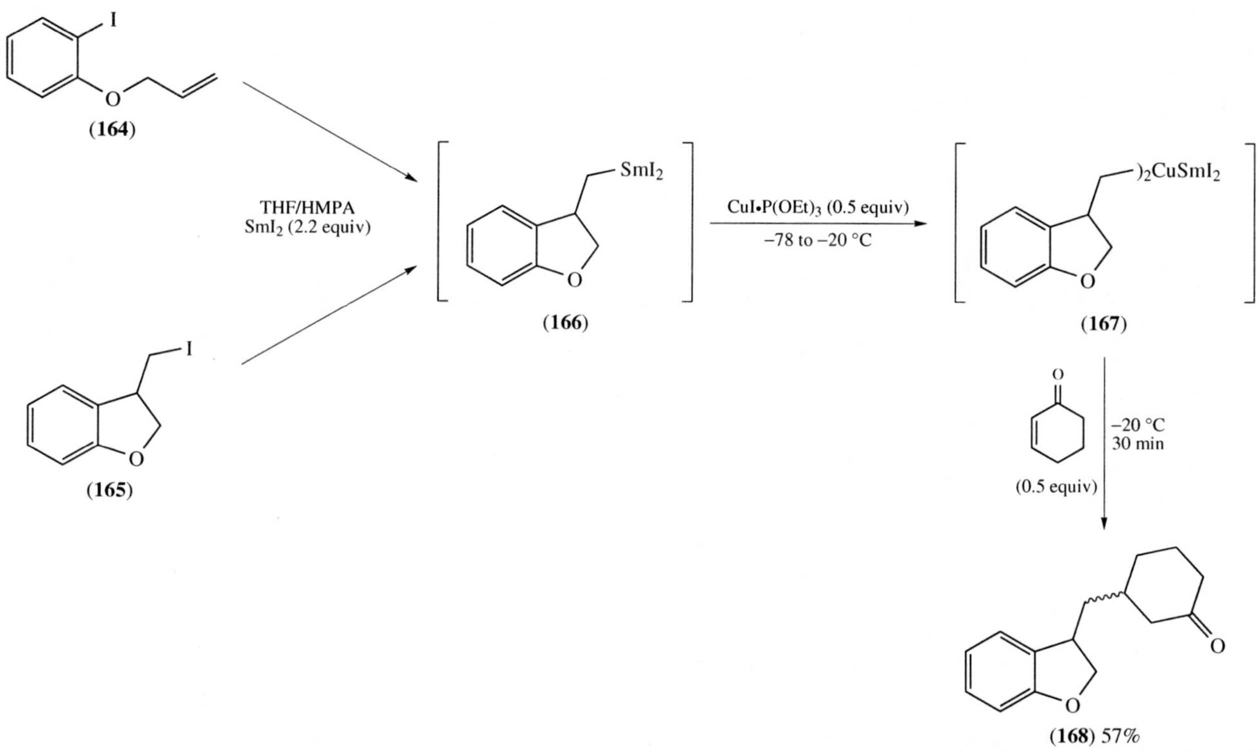

SCHEME 36. Transmetalation of alkyl samarium **166** and conjugate addition to cyclohexenone

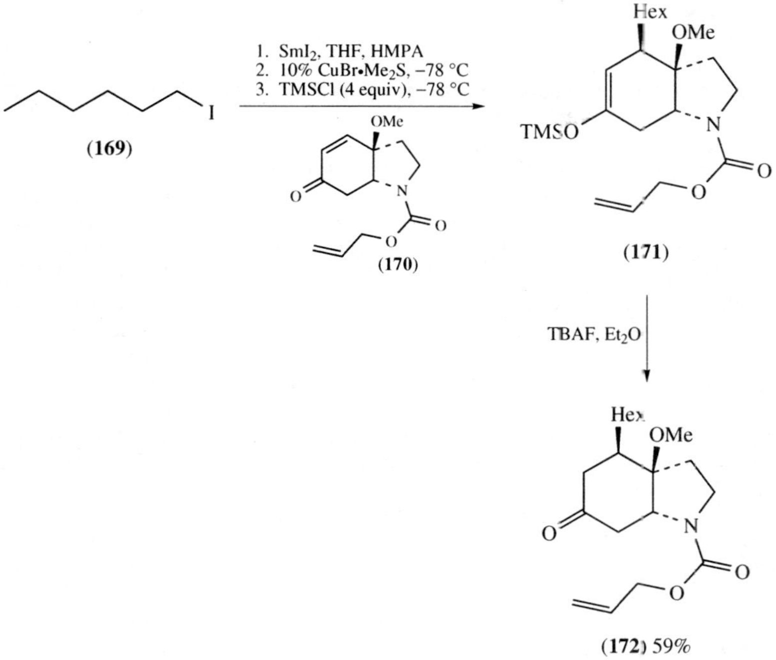

SCHEME 37. Copper-catalyzed conjugate addition of alkyl samarium to **170**

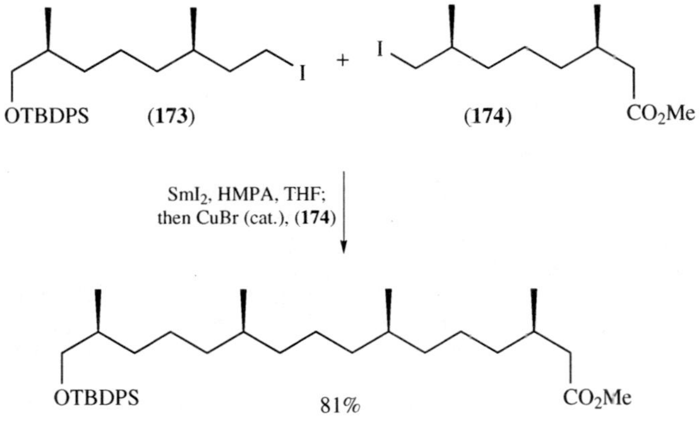

SCHEME 38. Copper-catalyzed cross-coupling of alkyl iodides

exhibit the highest reactivity, followed by diorganozincs and organozinc halides. The latter are usually prepared by direct insertion of zinc dust into organohalides. Rieke zinc, prepared by the reduction of zinc halides, is more active than zinc metal and can insert more easily into organohalides[118,119]. Alternatively, a palladium(0)- or nickel(0)-catalyzed halogen–zinc exchange can be used to synthesize organozinc halides through a radical mechanism[120,121]. Although less general, organozinc halides can also be prepared by the transmetalation of organometallic compounds with zinc halides (e.g. transmetalation from organolithium compounds)[122,123]. The preparation of diorganozinc reagents can generally be accomplished through an iodine–zinc[124,125] or boron–zinc exchange[126,127]. It is not possible to prepare enantiomerically pure diorganozinc reagents by iodine–zinc exchange due to the radical nature of the reaction; however, the boron–zinc exchange has been shown to occur with retention of stereochemical configuration[126,127].

Tanner and Hjelmgaard studied the convergent synthesis of ketones similar to **178** and **181** using the copper-mediated cross-coupling of an alkyl iodide and an acid chloride[128]. Insertion of zinc into alkyl iodide **175**, readily synthesized from *L*-pyroglutamic acid, provided alkyl zinc iodide **176**. Transmetalation to copper led to the corresponding copper–zinc species **177**, which was added to benzoyl chloride. Addition to aliphatic acid chlorides did not result in the desired products. Additionally, the methodology was applied to alkyl iodide **179**, synthesized from *L*-glutamic acid, to afford ketone **181** in good yield (Scheme 39), bypassing the usual β-elimination problem. Functionalized alkylzinc–copper species **180** was efficiently added to a wide range of acid chlorides, including those with aryl and alkyl substituents. This methodology complements the Pd(0)-catalyzed addition of organozinc halides to acid chlorides, and in cases where low or no yields are obtained, the zinc to copper transmetalation methodology may be useful in gaining access to the desired product[129].

In an interesting study, the relative abilities of *n*-butyl and phenyl groups in a mixed diorganozinc reagent **182** to transfer selectively in copper(I)-catalyzed benzoylations were investigated[130]. Benzophenone (**184**) or aryl alkyl ketone **183** were obtained selectively, depending on the solvent mixture used in the reaction (Scheme 40). Erdik and Pekel speculate that the coordinating ligands (n-Bu_3P or TMEDA) favor the formation of the reactive intermediate **185** to provide **183**, or of **186** to provide **184**. The exact cause of the selectivity remains under investigation.

Enantioselective 1,4-conjugate additions to Michael acceptors catalyzed by copper and other metals have recently been an area of intense investigation and several reviews have been published[131–135]. Herein, we limit the discussion to the enantioselective addition of organozinc compounds via transmetalation to copper.

During the investigation of palladium-catalyzed conjugate additions of diorganozinc reagents to enals, Marshall and coworkers performed comparison studies using copper-catalyzed conditions[136]. Generally, conjugate additions of lithium cyanocuprates and magnesium cuprates to enals provide mixtures of 1,2- and 1,4-addition products as well as various side products[137]. Marshall and coworkers expanded upon the conditions developed by Lipshutz and coworkers[138] for copper-catalyzed addition of diorganozinc reagents to enones, and successfully achieved the conjugate addition of diorganozincs to enals. Copper-catalyzed conjugate addition of diethylzinc to aldehyde **187** in the presence of TMSCl and subsequent deprotection of the silyl enol ether provided **188** (Scheme 41). Interestingly, when the reaction was performed in the presence of both enone **189** and **187**, the selective conjugate addition to the enal proceeded in identical yield, and unreacted **189** was isolated from the reaction mixture. Conjugate addition to the chiral enal **190** proceeded to give a β,γ-substituted aldehyde **191** in quantitative yield with excellent diastereoselectivity. The copper-catalyzed reaction is an excellent complement to the palladium-catalyzed addition of diorganozincs previously investigated by Marshall and coworkers.

Recently, Feringa and coworkers have expanded their work on enantioselective Michael additions to enones to include stereoselective conjugate additions to α,β-unsaturated lactams[139]. The conjugate addition of diethylzinc to unsaturated lactam **192** proceeded smoothly in the presence of catalytic quantities of copper(II) triflate and BINOL-based phosphoramidite (*R*,*S*,*S*)-**193** to provide the alkylated lactam **195** in high enantioselectivity. In a multi-component reaction application, zinc enolate **194** was quenched with

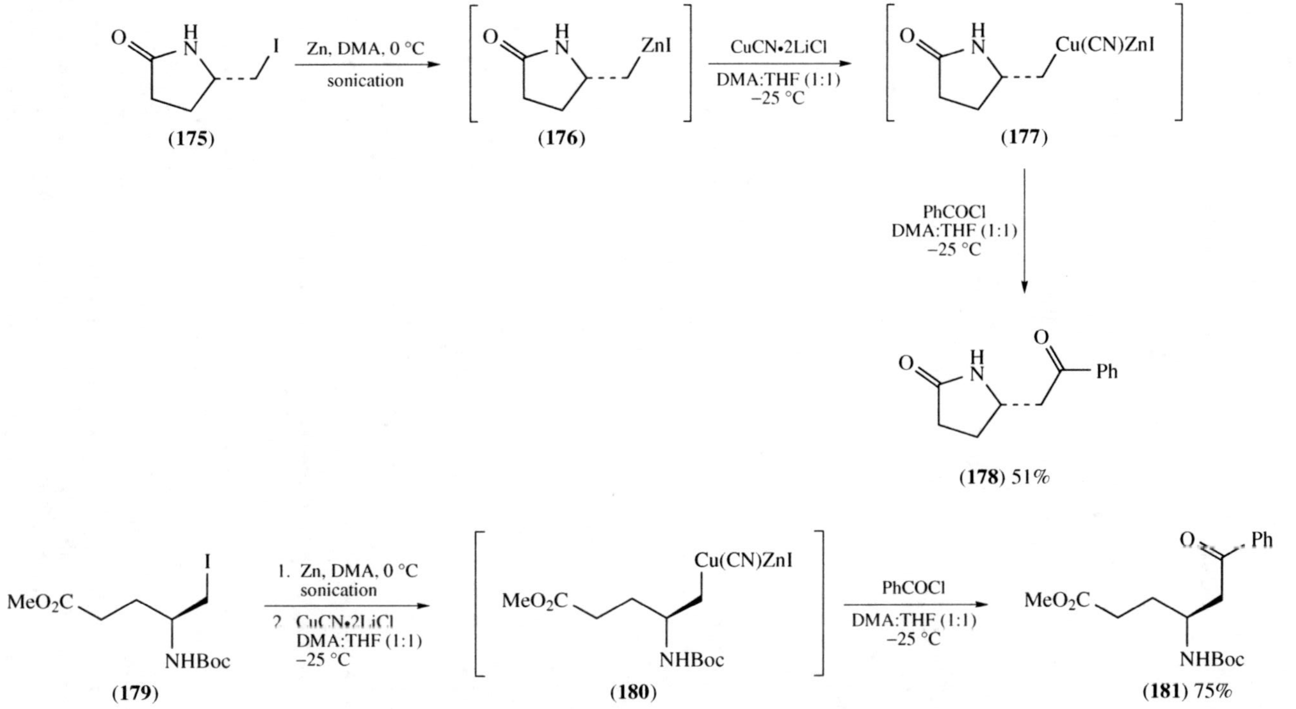

SCHEME 39. Copper-mediated addition of organozinc halides to acid chlorides

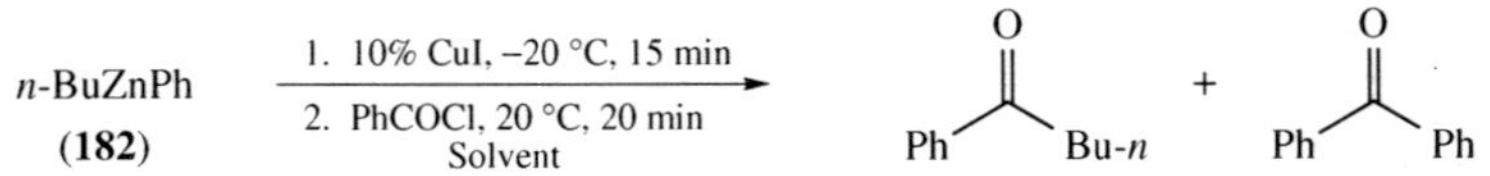

(182) (183) (184)

Solvent	Yield	183:184
THF:n-Bu_3P (1 equiv)	82%	100:0
THF:TMEDA (2:1)	67%	8:92

n-BuCu^{I}ZnPh (L) (185)   PhCu^{I}Zn(Bu-n) (L) (186)

SCHEME 40. Solvent-dependent selective benzoylation of mixed diorganozinc reagents

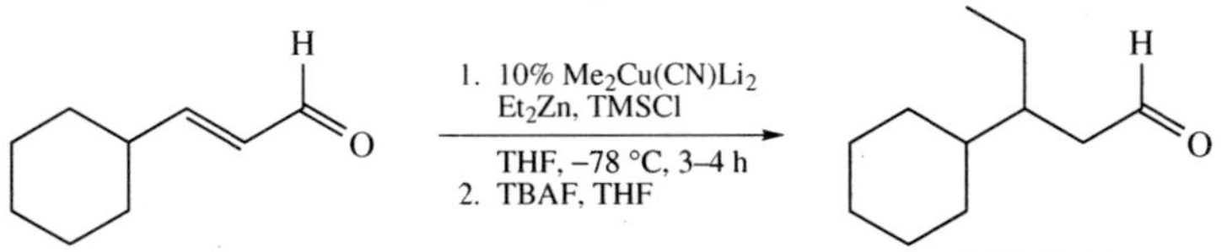

(187) (188) 82%

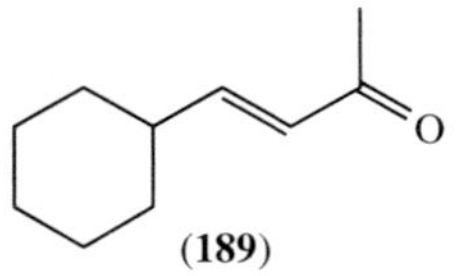

(189)

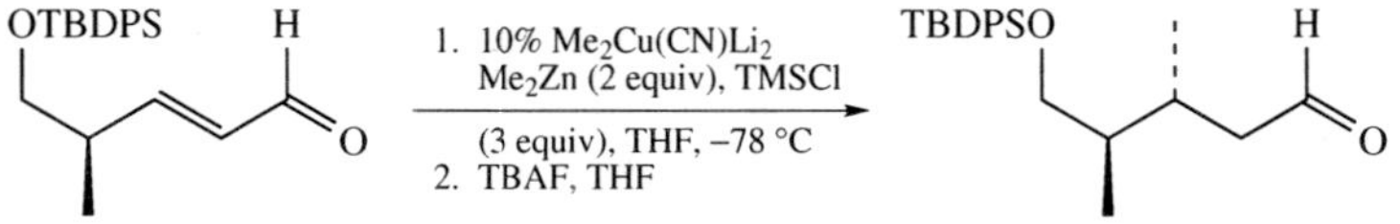

(190) (191) 100%, >20:<1 *dr*

SCHEME 41. Copper-catalyzed conjugate addition of diorganozincs to enals **187** and **190**

acetaldehyde to produce alcohol **196** as a mixture of alcohol epimers (Scheme 42). In this reaction sequence, it was observed that an electron-withdrawing group was necessary on nitrogen; otherwise, the reactivity of the enone is too low for the conjugate addition to take place. The conjugate addition of organozinc reagents to pyridone **197** was also successful[140]. The copper-catalyzed 1,4-addition of n-Bu_3ZnLi in the presence of TMSCl provided **198**; similarly, the treatment of **197** with diethylzinc, Cu(I) salt and TMSCl resulted in an analogous alkylation to give pyridone **199**. The corresponding enantioselective conjugate addition to **197** was accomplished using the chiral phosphoramidite (S,R,R)-**193** and copper(I) thiophenecarboxylate (CuTC) to deliver **200** (Scheme 43).

Feringa and coworkers also applied their copper phosphoramidite-catalyzed enantioselective conjugate addition methodology toward the formation of synthetically useful

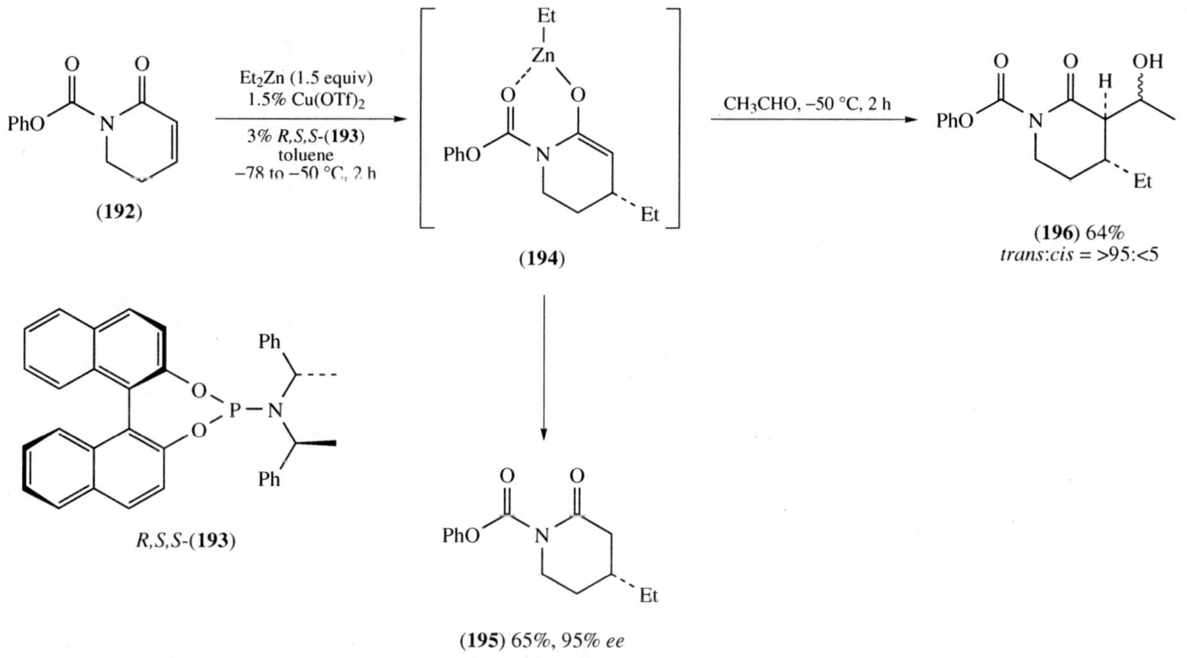

SCHEME 42. Enantioselective Michael addition to an α,β-unsaturated lactam

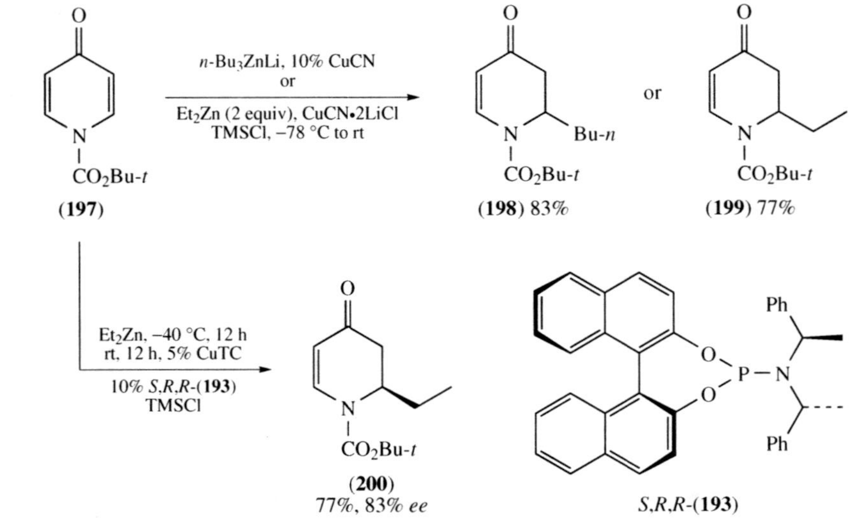

SCHEME 43. Conjugate addition of organozinc reagents to pyridone **197**

vinyl triflates[141]. A Michael addition to cyclohexenone provided the intermediate zinc enolate **201** and, upon treatment of the zinc alkoxide with triflic anhydride, vinyl triflate **202** was obtained in excellent enantioselectivity. Subsequently, the enol triflate was subjected to palladium cross-coupling conditions (Negishi reaction), and chiral styrene **203** was isolated. Interestingly, they also applied this reaction sequence to a one-pot protocol to provide **203** in a lower yield but with retention of configuration (Scheme 44). Cycloheptenones could also be used as substrates.

Further investigations into the enantioselective copper-catalyzed conjugate addition reaction to enones have led to the development of other chiral ligands that control the stereochemical outcome of the initial C−C bond formation. Nakamura and coworkers developed the aminohydroxyphosphine ligand **205**, and investigated its efficacy as a chiral ligand for the conjugate addition of dialkylzincs[142]. They found that diethylzinc in the presence of catalytic quantities of copper(II) triflate can be added in high enantioselectivity to acyclic enones **204** and **206** with this ligand (Scheme 45). Based on theoretical calculations and previous work in nickel-catalyzed cross-coupling reactions, the authors propose that the phosphine ligand has the ability to bridge the bimetallic transition state **207**[143, 144]. This results in virtually complete control of the facial selectivity during the addition of the alkyl group.

The synthesis of all-carbon quaternary stereocenters is an important and challenging process that remains the subject of intense interest in organic chemistry[145]. Several copper-catalyzed methodologies have been developed that perform enantioselective 1,4-conjugate additions to β-disubstituted enones resulting in the desired quaternary center[146–148]. In the presence of catalytic amounts of the dipeptide-based aniline ligand **209** and copper(I) cyanide, conjugate additions of dialkylzincs to β-disubstituted enone **208** occur in high yields and enantioselectivites[147]. This methodology was also suitable for conjugate additions to similarly substituted cyclopentenones to form quaternary stereocenters in 5-membered carbocycles. Elaboration of the resulting ketones **210** and **211** was demonstrated to provide access to the chiral ketone **212** and the α,β-unsaturated ester **213** (Scheme 46).

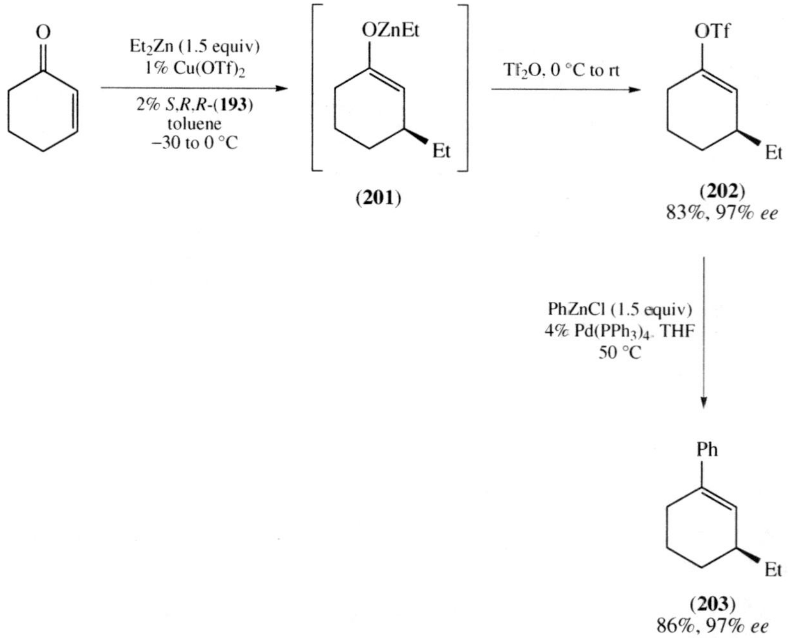

SCHEME 44. Enantioselective conjugate addition to cyclohexenone followed by enol triflate formation and Pd-catalyzed cross-coupling

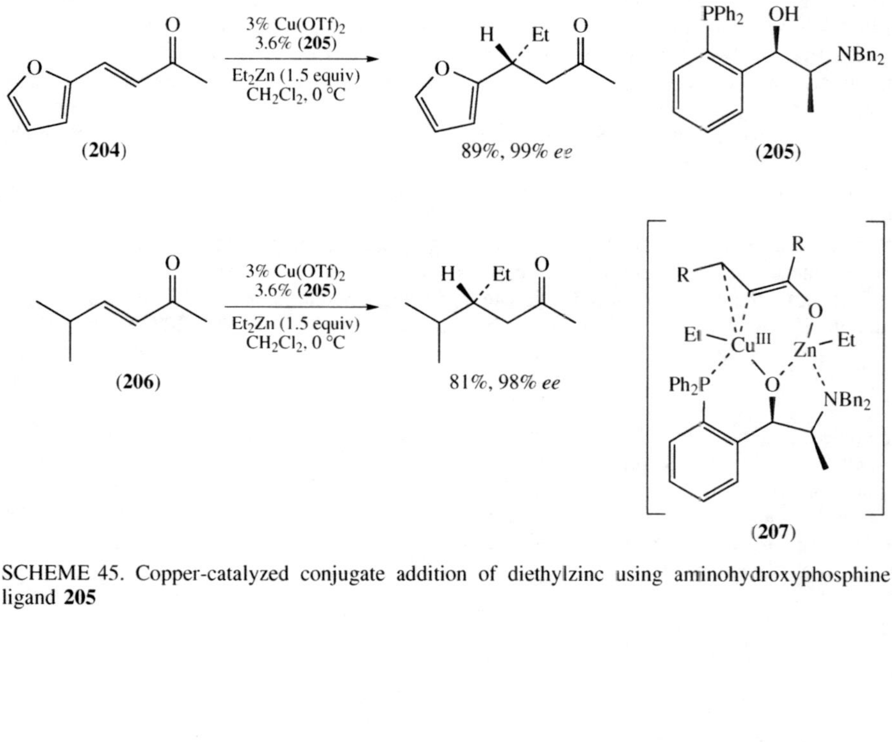

SCHEME 45. Copper-catalyzed conjugate addition of diethylzinc using aminohydroxyphosphine ligand **205**

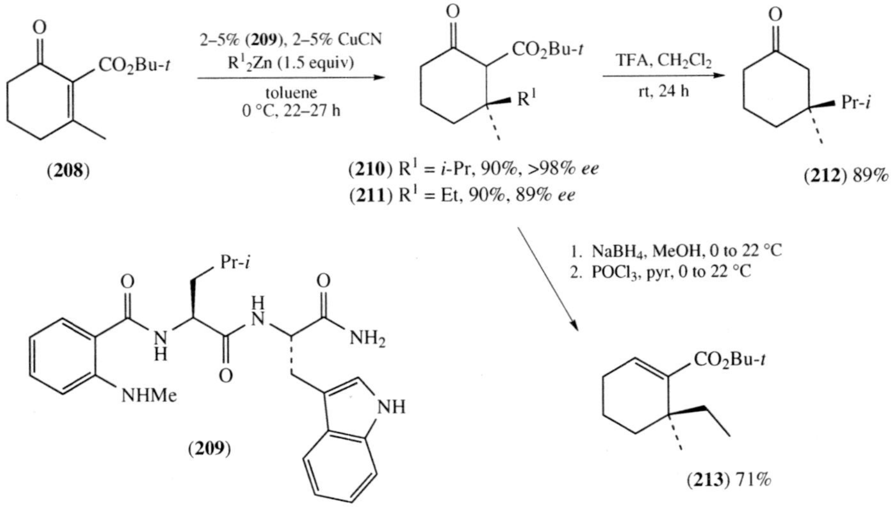

SCHEME 46. Enantioselective conjugate additions to tetrasubstituted enone **208** and further transformations to access **212** and **213**

Subsequent investigation in the enantioselective conjugate addition of dialkylzincs led Hoveyda and coworkers to use air-stable chiral bidentate Ag(I) *N*-heterocyclic carbenes (NHC)[146, 148]. At the same time, similar results were obtained for the enantioselective conjugate addition of Grignard reagents utilizing monodentate NHC complexes[149]. Diaryl- and dialkylzincs were successfully added to enone **214** in high yield and enantioselectivity in the presence of catalytic quantities of NHC–Ag(I) complex **215** and a copper(I) triflate–benzene complex provides β-disubstituted ketones **217** and **219** (Scheme 47). The active catalyst (Cu(I)-NHC) is generated *in situ* during the reaction[150, 151]. It should be noted that the facial selectivity changes for diarylzinc and dialkylzinc additions. The authors propose transitions states **216** and **218**, which illustrate the change in orientation of the enone to minimize steric interactions. The substrate scope is limited to relatively small diorganozinc reagents (i-Pr_2Zn results in low conversion) and small substituents at the β-position; however, the reaction can be applied similarly to β-substituted cycloheptenones in good yields and enantioselectivities. The conjugate additions of dialkylzincs to γ-keto ester **220** using NHC–Ag(I) complex **221** gave products in good yield and enantiomeric excess[146]; the addition of a methyl group provides keto-ester **223**. Selective reduction of the ester functionality is accomplished by treatment of the intermediate zinc enolate **222** with lithium aluminum hydride to give the chiral alcohol **224** (Scheme 47).

Copper-catalyzed asymmetric allylic substitution reactions are well precedented[117, 152, 153], and a review by Oshima and Yorimitsu in 2005 summarizes the progress in this area[154]. At the time protocols by Hoveyda and coworkers utilizing bidentate NHC complexes **226**[150] and Alexakis and coworkers using chiral phosphoramidites **228**[155] emerged as the most promising methods to accomplish the enantioselctive substitution reactions on allylic phosphates **225** and allylic bromides **227** (Scheme 48). More recently, methodologies have been developed to achieve higher enantioselectivites with a broader range in substrate scope that allows access to a variety of useful organic synthons.

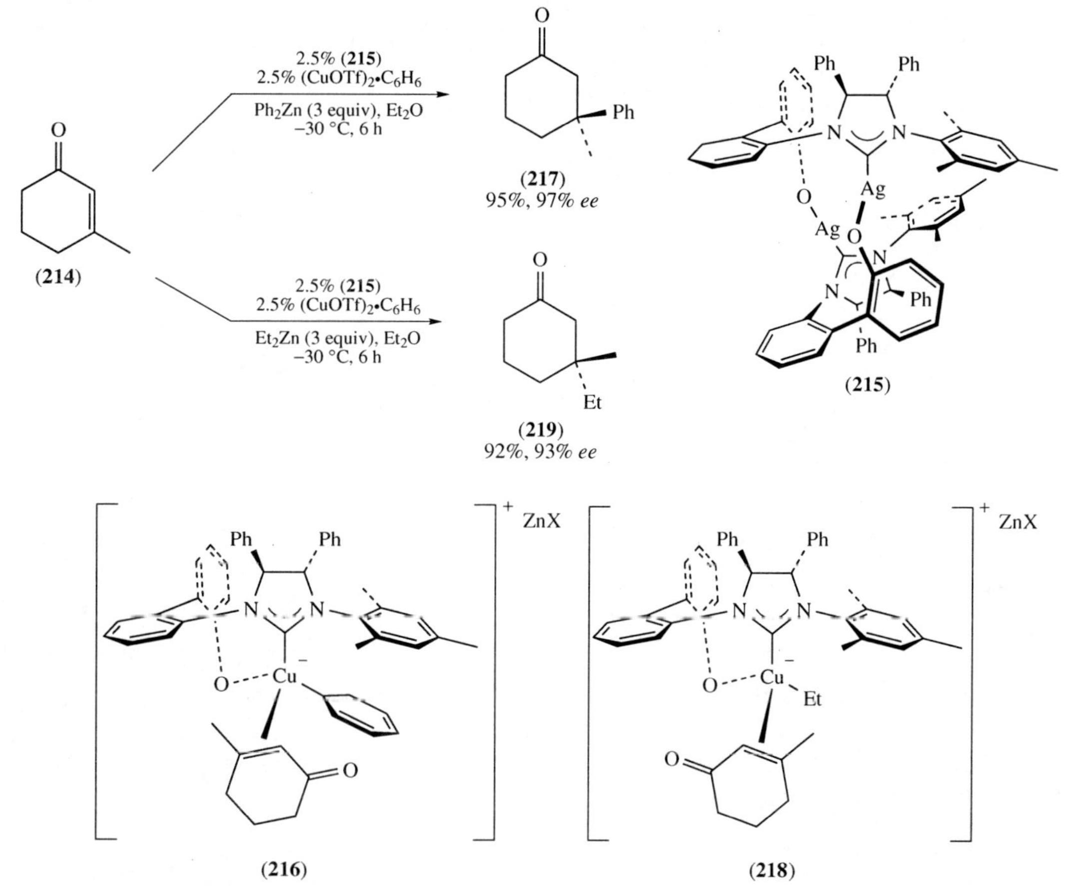

SCHEME 47. Copper-catalyzed enantioselective conjugate additions of diorganozinc to enones **214** and **220** using NHC complexes **215** and **221**

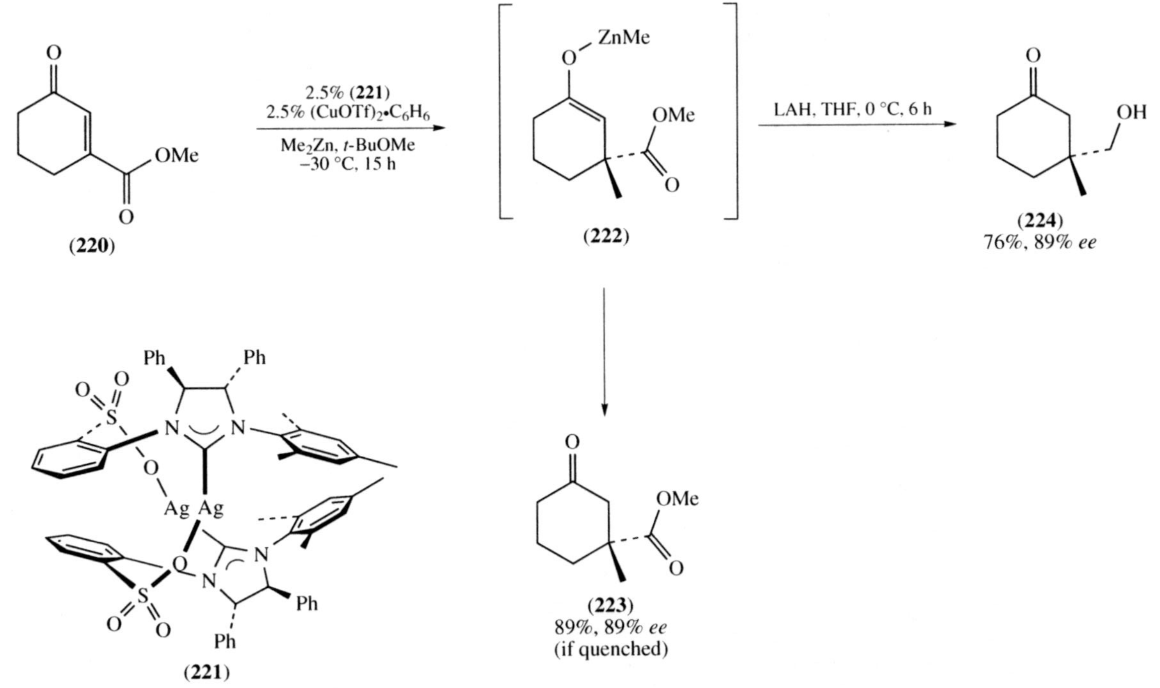

SCHEME 47. (*continued*)

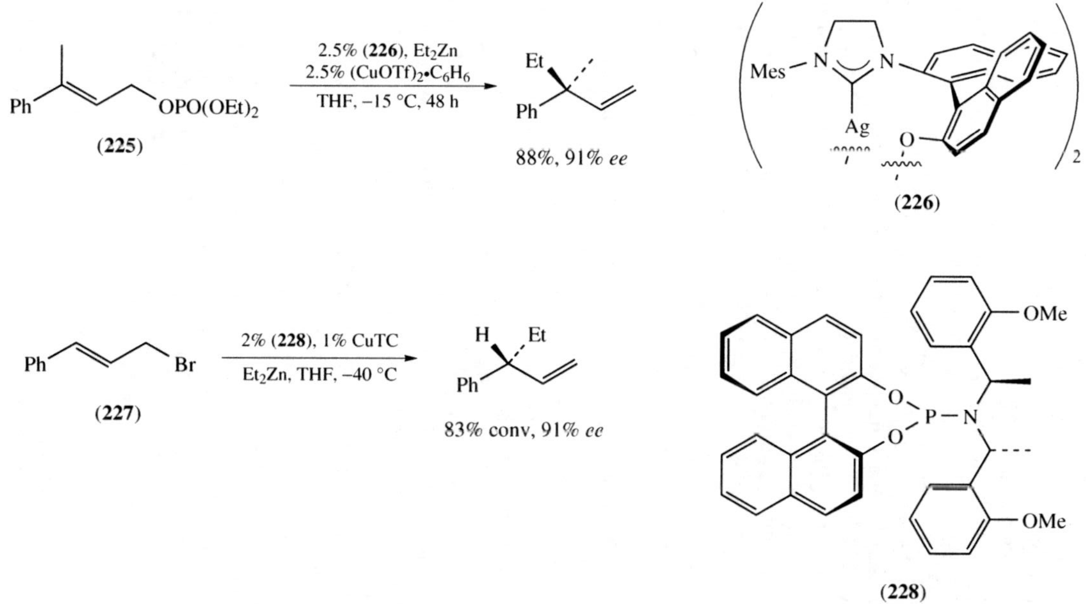

SCHEME 48. Copper-catalyzed asymmetric allylic substitution reactions using chiral ligands **226** and **228**

Knochel and coworkers have taken advantage of the high level of chirality transfer in S_N2'-additions of diorganozincs to enantiomerically pure allylic pentafluorobenzoates[156–158]. Using this methodology, enantiomerically pure tertiary alcohols, amines and isocyanates were synthesized[157]. Chiral allylic pentafluorobenzoates **229** were obtained as virtually single enantiomers, and upon S_N2'-addition of a dialkylzinc in the presence of copper, chiral olefin **231** was obtained in good yield with almost complete transfer of chirality in generating the all-carbon quaternary center. The chirality transfer is due to the minimization of the allylic 1,3-strain in transition state **230**. The olefin **231** was then converted to the tertiary alcohol **232** by oxidative olefin cleavage and Baeyer–Villiger rearrangment with retention of configuration. Alternatively, olefin **231** was subjected to oxidative olefin cleavage and Curtius rearrangement to provide the corresponding amine **233** or isonitrile **234** with no loss in enantiopurity (Scheme 49). The same methodology was applied to enantioenriched allylic pentafluorobenzoate **235** to produce chiral alkenylsilanes that were used for the synthesis of chiral α,β-unsaturated ketones **237** or chiral styrene derivatives **239**[158]. Dialkyl- and diarylzincs were added to **235** in good yield with excellent transfer of chirality to give alkenylsilane **236**, which was transmetalated to aluminum and added to various acid chlorides to deliver α,β-unsaturated ketones **237** with a γ-stereocenter. Alternatively, alkenylsilane **236** was transformed into pinacol borane **238** and in turn engaged in a Suzuki coupling to provide the enantioenriched styrene derivative **239** (Scheme 50).

Breit and coworkers have investigated a similar approach for generating quaternary stereocenters in a stereospecific manner by using a directed/non-directing leaving group strategy to control the absolute stereochemistry[159–161]. Treatment of enantiomerically pure allylic *ortho*-diphenylphosphanylbenzoate (*o*-DPPB) **240** with alkyl Grignard reagents in the presence of a catalytic quantity of copper resulted in a neighboring group directed S_N2'-addition. The reaction proceeded through the chelated transition state **241** to give (−)-**242** with nearly complete chirality transfer. Interestingly, treatment of allylic phosphine oxide **243**, obtained from the oxidation of **240**, with a dialkylzinc reagent and a copper(I) salt proceeded to give the opposite S_N2' product (+)-**242** with no erosion of enantiomeric excess. Presumably, nucleophilic addition to oxidized **243** occurred through the non-chelated transition state **244**. Thus, as an interesting caveat to this methodology, either enantiomer of **242** can be accessed from a single enantiomer of the starting material **240** by controlling the oxidation state of the phosphine (Scheme 51).

In an interesting application for the synthesis of (*E*)-alkene dipeptide isosteres (EADIs)[162–164], Fujii and coworkers incorporated unnatural side chains at the α-position of EADIs using organozinc–copper reagents[165]. Selective S_N2'-addition of naphthylmethylzinc bromide **246** to aziridine **245** furnished EADI **247**. The opposite naphthylmethyl diastereomer could be accessed by treatment of aziridine **245** with methanesulfonic acid to form mesylate **248**, which was eliminated via an anti-selective S_N2'-addition of **246** in the presence of copper(I) and $BF_3{\bullet}Et_2O$ to give EADI **249** in good yield over the two steps. Thus, they were able to obtain the desired compounds **247** and **249**, epimeric at the α-position, from the common intermediate **245** (Scheme 52).

Expanding on their previous work using chiral bidentate NHC–copper complexes, Hoveyda and coworkers were able to synthesize allylic silanes **251** containing tertiary and quaternary carbon stereocenters in high enantioselectivies[166]. Dialkylzincs were efficiently added to di- and tri-substituted alkenylsilanes **250** in the presence of NHC complex **226** to obtain allylic silanes **251** with high enantiomeric excess. Alternatively, diarylzincs were added to **250** in slightly reduced yield and enantioselectivity. Di- and tri-substituted alkenes required the use of NHC complexes **215** and **221**, respectively. The resulting allylic silane **252** was coupled to the unsaturated ester **253** under metathesis conditions using the Hoveyda–Grubbs catalyst to provide ester **254** with excellent *E*-olefin selectivity and no loss in enantiomeric excess at the γ-carbon. Diol **256** was obtained with no

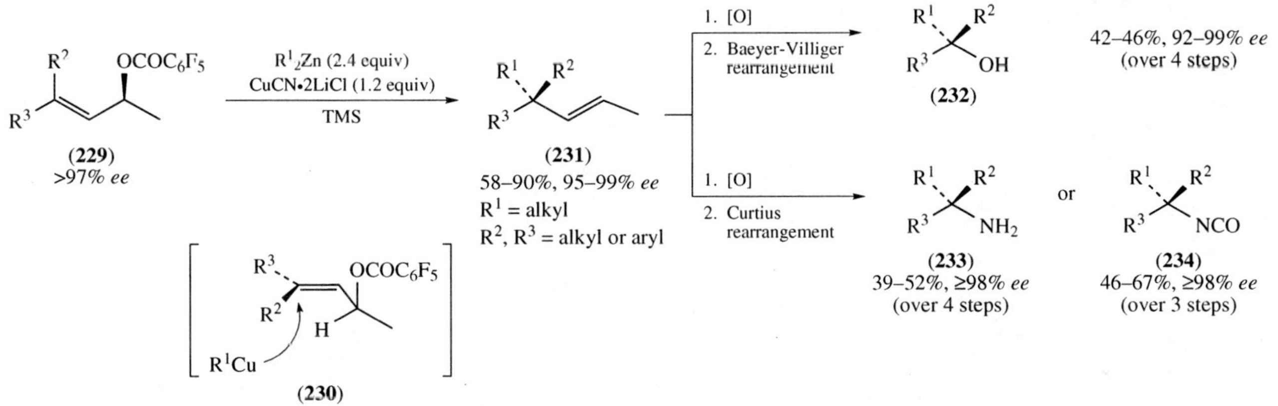

SCHEME 49. Diasteoselective allylic S_N2'-substitution by copper-mediated addition of dialkylzincs and further conversion to alcohols, amines, and isocyanates

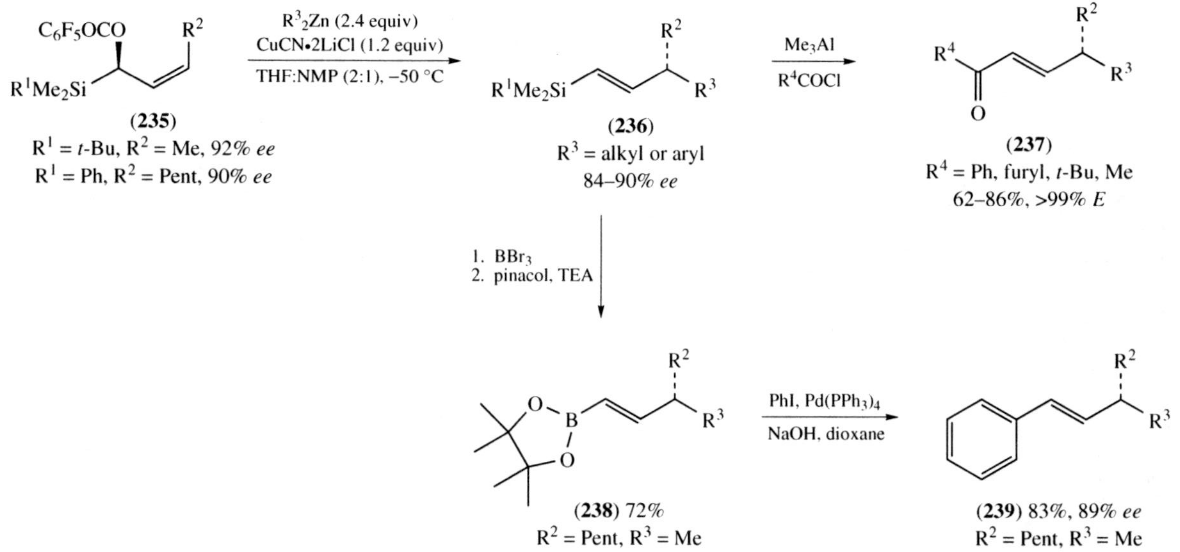

SCHEME 50. S_N2'-addition of diorganozinc reagents to allylic pentafluorobenzoate **235** to obtain chiral alkenylsilanes **236** for further synthetic manipulation

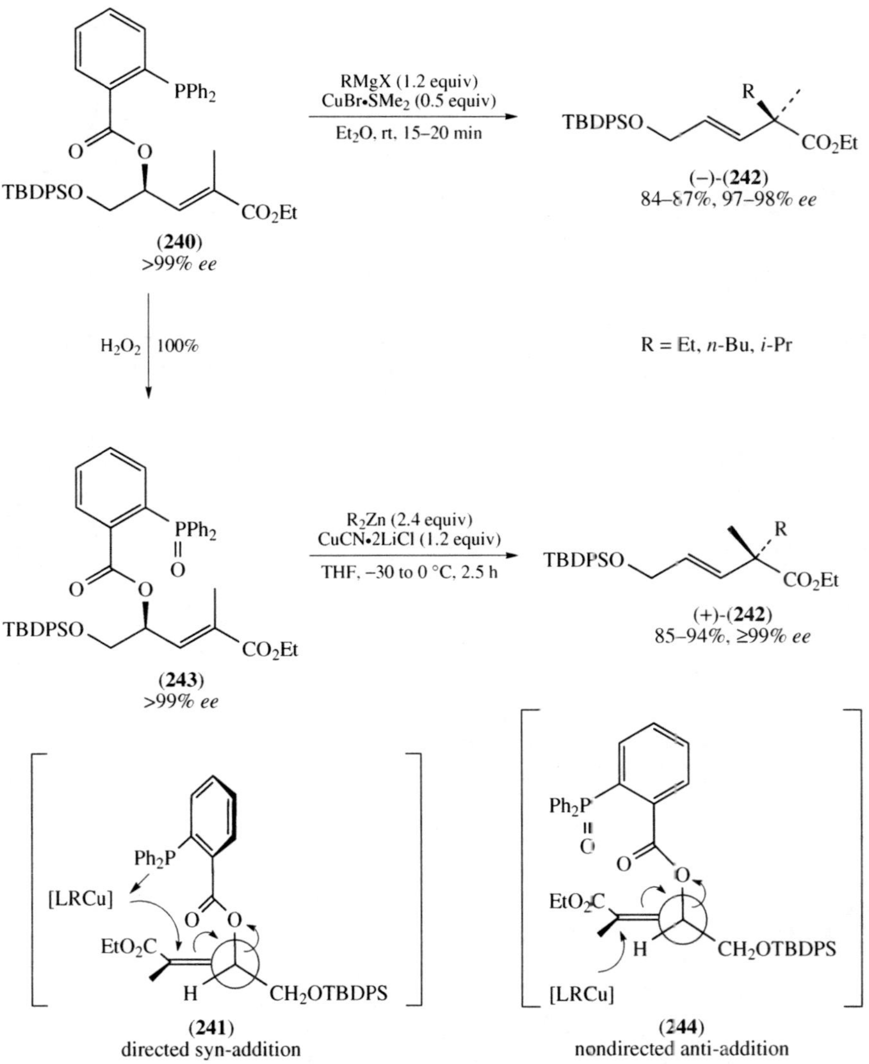

SCHEME 51. Directed and non-directed S_N2'-displacement of organocopper reagents on a tunable allylic *ortho*-diphenylphosphanylbenzoate (*o*-DPPB)

deterioration of enantiopurity by a hydroboration–oxidation/Fleming–Tamao oxidation sequence from silane **255** (Scheme 53).

In a useful application of the copper-catalyzed enantioselective conjugate addition of dialkylzincs to nitro olefins **257** and **261**, alkyl-substituted β-amino acids were synthesized[167–169]. Sewald and Rimkus performed a copper-catalyzed conjugate addition of diethylzinc to nitro olefin **257** in the presence of chiral phosphoramidite **258** to prepare ester **259** in high yield and enantioselectivity[168]. Reduction of the nitro group, protection of the amine and hydrolysis of the ester afforded the β-amino acid **260** in 71% yield

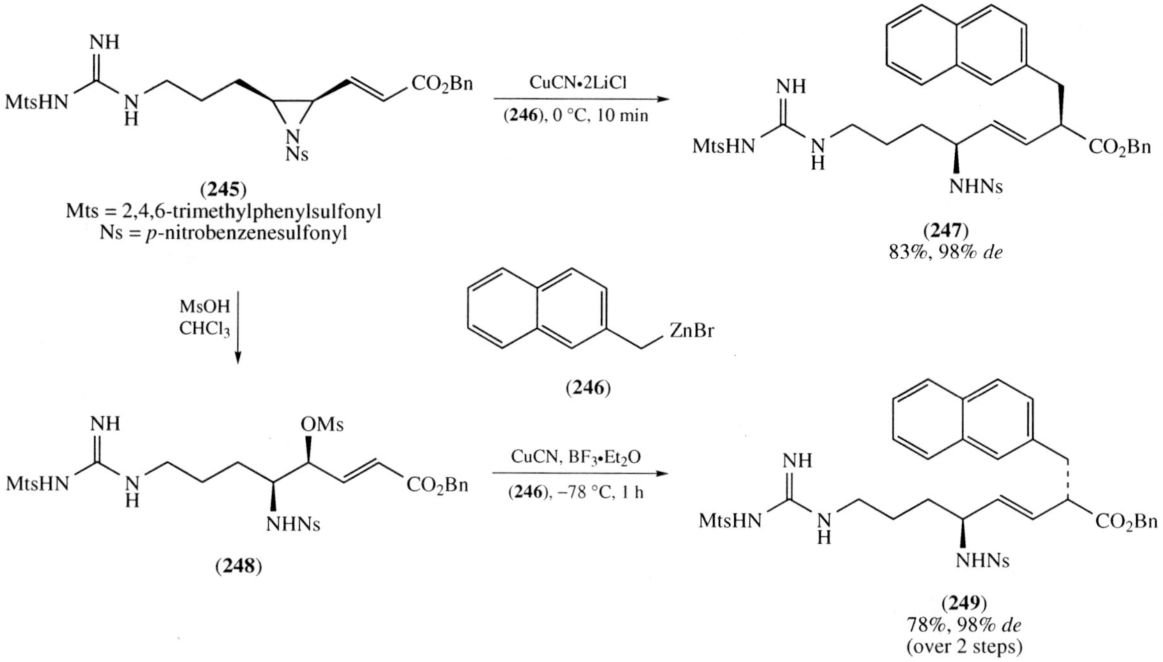

SCHEME 52. Copper-mediated S_N2'-addition of organozinc bromide to synthesize precursors to (*E*)-alkene dipeptide isosteres (EADIs)

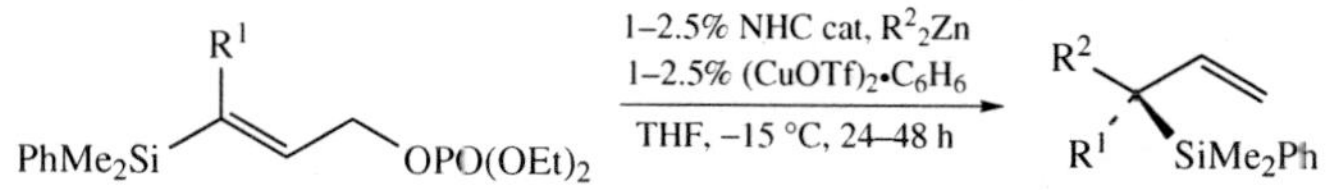

(250) R^1 = H or Me

(251) R^2 = alkyl or aryl

R^1	R^2	NHC cat	Result
CH_3 or H	alkyl	**226**	72–94%, 91–98% *ee*
H	aryl	**215**	82–88%, 90–91% *ee*
CH_3	aryl	**221**	72–80%, 76–85% *ee*

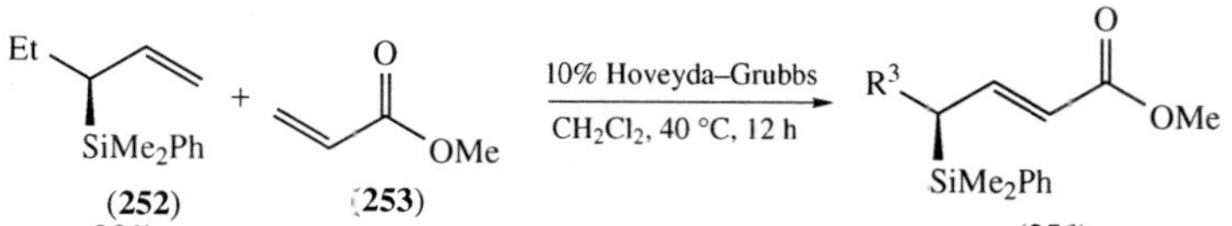

(252) 98% *ee*

(253)

(254) 84%, >98% *E*, 98% *ee*

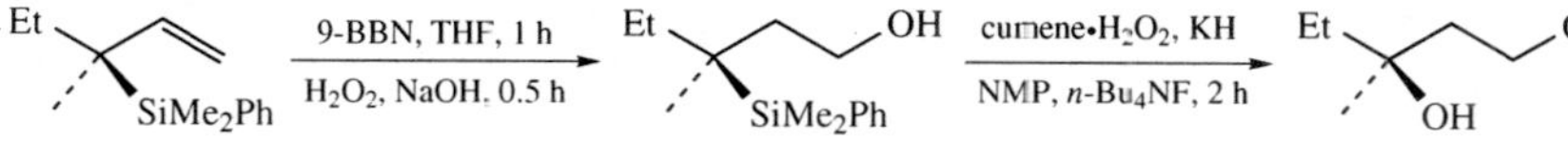

(255) 91% *ee*

(256) 80%, 91% *ee*

SCHEME 53. Copper-catalyzed asymmetric allylic S_N2'-substitution reaction to yield chiral allylic silanes **251** and subsequent transformations to provide **254** and **256**

from **257**. Alternatively, Feringa and coworkers performed enantioselective conjugate additions of several dialkylzincs to nitro olefin **261** to obtain ketals **263** in high yield and enantiopurity[167]. Hydrogenolysis of the nitro group, Boc-protection and subsequent hydrolysis and oxidation of the ketal **264** provided β-amino acid **265** with no deterioration of enantiomeric excess. Ojima and coworkers achieved slightly higher enantioselectivites using the chiral phosphoramidite **262** in the conjugate addition of diethylzinc to nitro olefins (Scheme 54)[169].

Organozinc reagents in the presence of copper salts have been added enantioselectively to aldimines and ketimines to generate amines. Several recent reviews have been published covering this topic[170,171], and only some pertinent examples will be covered here. Previous studies on the enantioselective addition of diorganozincs to *N*-diphenylphosphinoylimines of type **266** have been performed by Charette and coworkers, providing amides **268** in high yield and enantioselctivities[172,173]. Charette and Lauzon also exploited the similar reactivities of aldimine **266** and trifluoromethyl ketimine **271**, and were able to perform copper-catalyzed enantioselective additions of diethyl- and dimethylzinc[174]. Synthesizing **271** directly proceeded in very low yields; however, treatment of ketone **269** with titanium(IV) ethoxide and diphenylphosphinamide provided hemiaminal **270** as a colorless, air-stable solid. The corresponding imine **271** could be generated *in situ* upon addition of dialkylzinc. In the presence of catalytic copper(II) triflate and (*R*,*R*)-BozPHOS (**267**), the desired addition products **272** could be isolated in very high enantiopurities. The diphenylphosphinoyl protecting group required slightly harsher conditions for removal; nonetheless, the amine hydrochloride **273** could be accessed in a reasonable yield (Scheme 55).

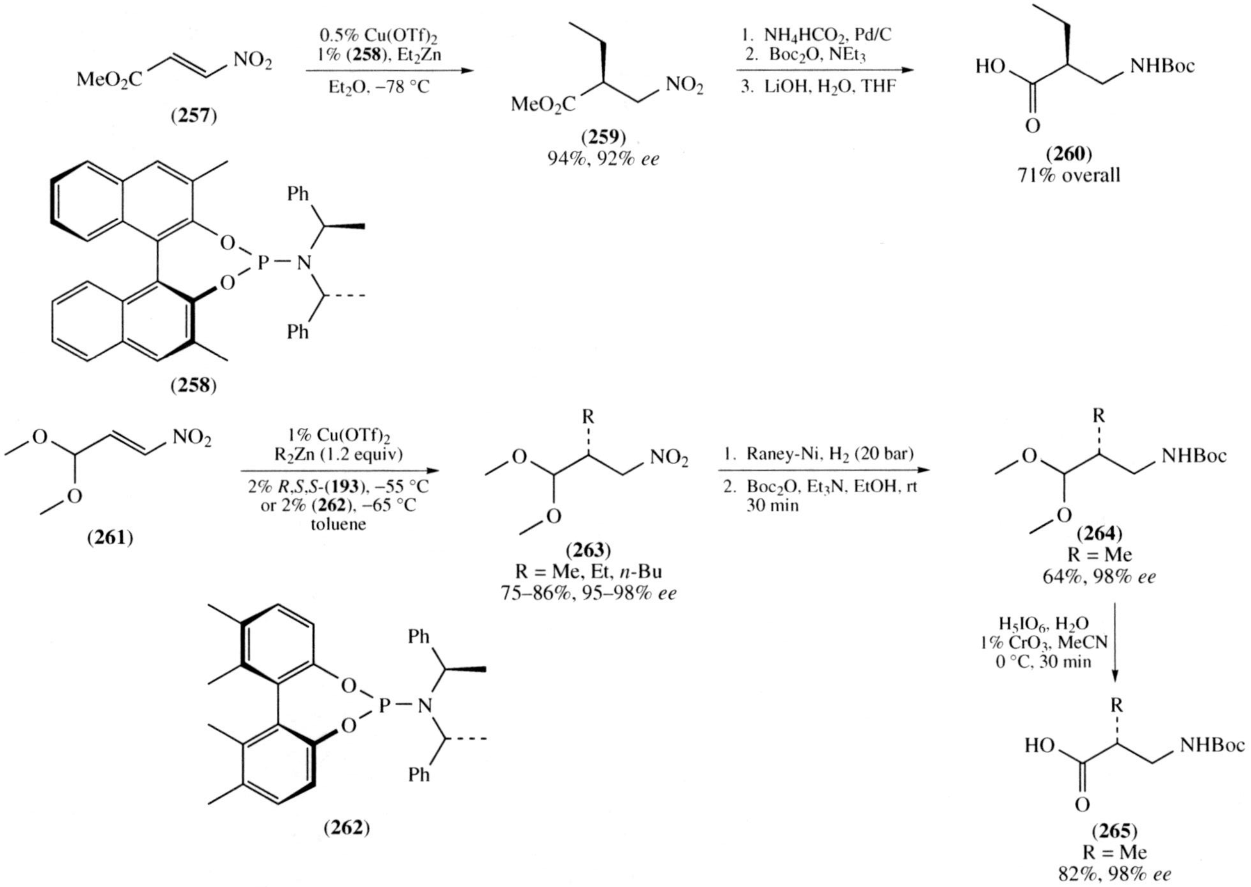

SCHEME 54. Synthesis of alkyl-substituted β-amino acid derivatives by enantioselective conjugate addition to nitro olefins

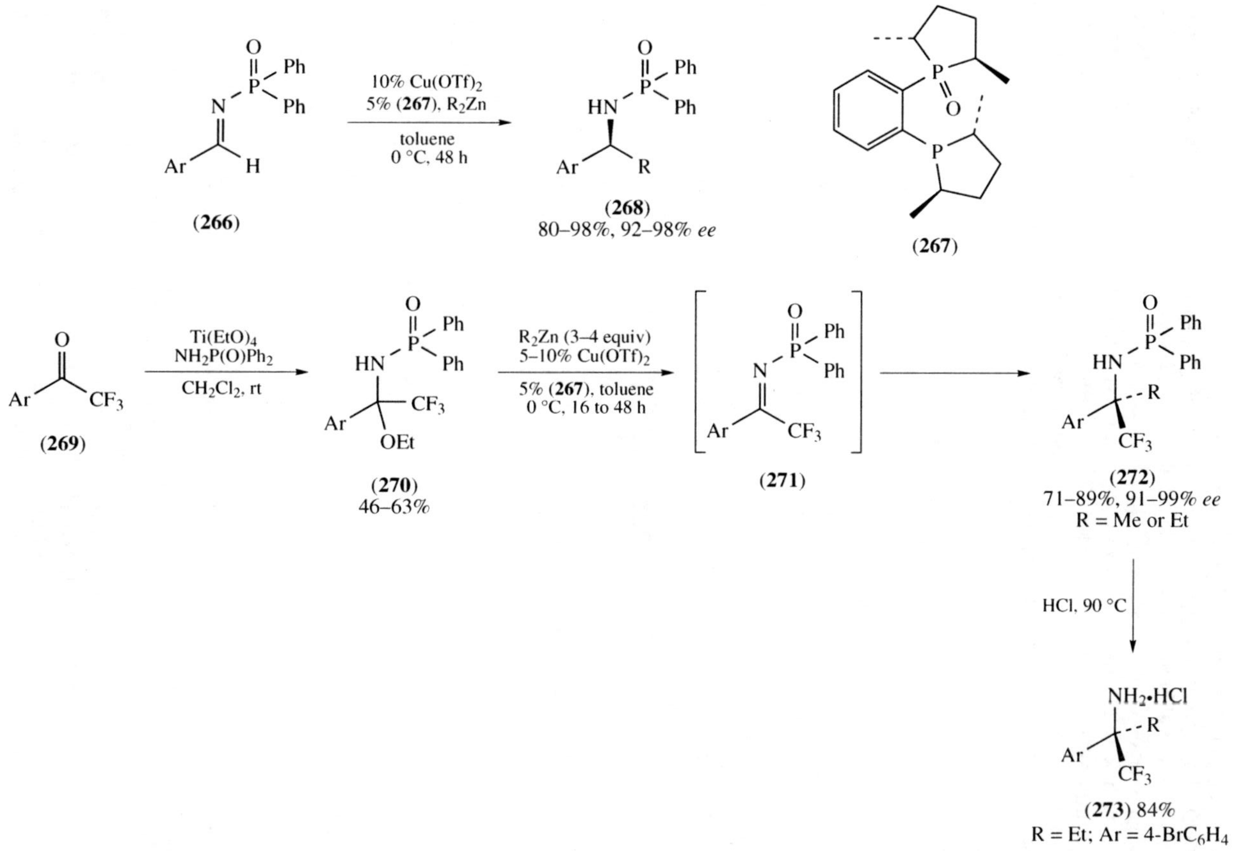

SCHEME 55. Copper-catalyzed enantioselective addition of alkylzinc reagents to *N*-diphenylphosphinoylimines

The copper-catalyzed addition of alkylzinc halides to *N*-(heteroarylsulfonyl)aldimines was investigated by Carretero and coworkers[175]. In the presence of 5% copper(II) triflate, various alkylzinc halides were added efficiently to aldimine **274**, providing various pyridylsulfonyl amines **275**. Presumably, the Lewis base donor ability of the *N*-pyridylsulfonyl group localizes the less nucleophilic alkylzinc–copper species in close proximity to the aldimine to facilitate addition. The resulting pyridylsulfonyl amines **275** were deprotected in the presence of Mg^0. In an application to the synthesis of pyridones, alkylzinc halide **277** was added to pyridylsulfonyl imine **276** and subsequently deprotected, affording the cyclized pyridone **278** (Scheme 56). The broader functional group tolerance of alkylzinc halides compared to more reactive organometallics could be useful if this methodology were to be expanded to an asymmetric variant.

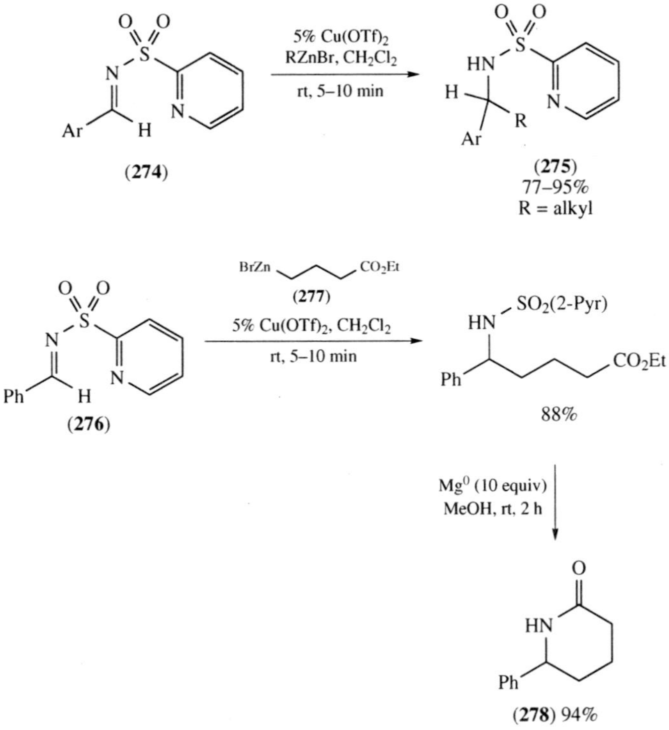

SCHEME 56. Addition of alkylzinc halides to *N*-pyridylsulfonyl imines **274** under copper catalysis

Expanding again on their use of BINOL-derived phosphoramidites, Feringa and coworkers have successfully used chiral phosphoramidites to control the stereochemical outcome of the copper-catalyzed diorganozinc imine addition[176]. Treatment of imine precursor **279** with catalytic copper(II) triflate, 4% phosphoramidite ligand (*S*,*R*,*R*)-**193** and diethyl-, diisopropyl- or dibutylzinc resulted in the formation of amide **280** in high enantiomeric excess. Alternatively, the addition of dimethylzinc under identical conditions proceeded with very low enantioselectivity. However, treatment of **279** with a copper source, phosphoramidite ligand and trimethylaluminum resulted in the formation of the opposite enantiomer **281** in good enantioselectivity (Scheme 57). The stereochemical outcome was

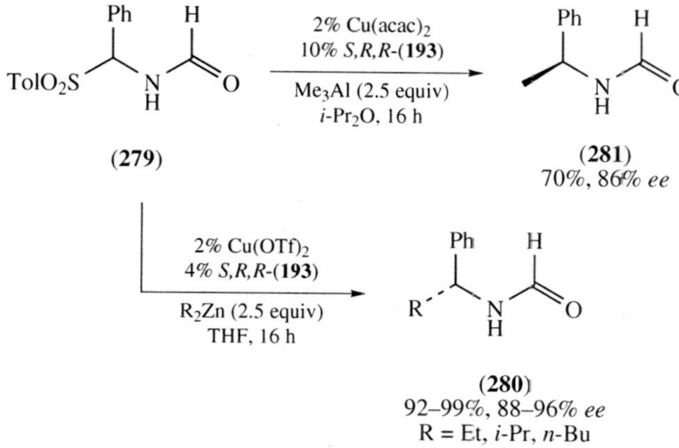

SCHEME 57. Copper-catalyzed enantioselective addition to *in situ* generated imines from **279** using chiral phosphoramidite ligand (*S*,*R*,*R*)-**193**

rationalized by the presence of two active catalysts in the aluminum/copper system, as demonstrated earlier by Feringa and coworkers[177].

Johnson and Berman have investigated the synthesis of secondary and tertiary amines by a copper-catalyzed electrophilic amination of diaryl- and dialkylzinc reagents[178, 179]. Treatment of various primary and secondary amines **282** with benzoyl peroxide (**283**) resulted in the formation of electrophilic amination reagents **284**. In the presence of catalytic quantities of copper reagent ($CuCl_2$ or $(CuOTf)_2$•C_6H_6) and diorganozinc reagents, the secondary and tertiary amines **285** were formed in moderate to excellent yields with tolerance to aryl triflates, nitriles and acetates. Several examples (**286**–**289**) of the product scope can be seen (Scheme 58). The formation of the very hindered amine **286** is particularly noteworthy. The mechanism of the reaction was investigated by the endocyclic restriction test[180], and it was proposed that the diorganozinc reagent transmetalates to an intermediate copper–zinc species and performs an S_N2-reaction on the electrophilic amine[181].

B. Transmetalation from Organomercury Compounds

The toxicity of organomercury compounds prohibits their widespread use in organic synthesis; however, these reagents have been used in transmetalation reactions with copper to achieve some interesting results. The electronegativity of mercury (2.00) is similar to copper, leading to a sluggish transmetalation and an equilibrium favoring the organomercury reagent. Initial investigations by Bergbreiter and Whitesides showed that alkylmercury reagents could be transmetalated to copper and quenched with an electrophile[182]. The process occurred through a reactive ate complex containing copper(I), mercury(II) and lithium(I). Several years later, Šrogl and Kočovský utilized the transmetalation of an alkylmercury reagent to copper in the formation of a cyclobutane-containing steroid derivative[183, 184]. Treatment of cyclopropane **290** with mercury(II) resulted in the rearranged alkylmercury **291**. Subsequent transmetalation from mercury to copper and intramolecular 1,2-addition to the pendant aldehyde provided cyclobutanol **292** in high yield. The authors also performed an intramolecular 1,4-conjugate addition to the α,β-unsaturated ester **293**, formed by Horner–Emmons olefination of aldehyde **291**, by transmetalation to copper to provide cyclobutane **294** (Scheme 59). The nature of the

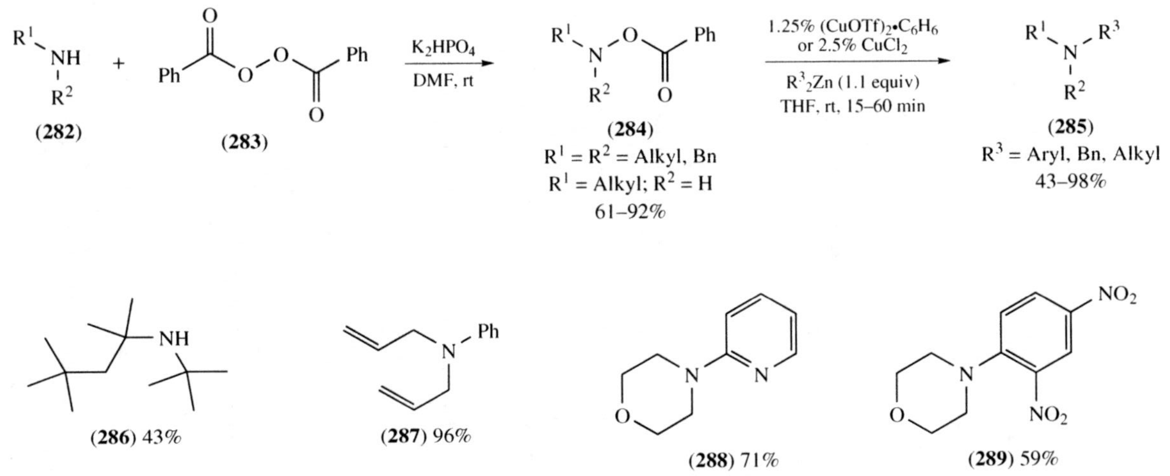

SCHEME 58. Electrophilic amination of diorganozinc reagents using copper catalysis

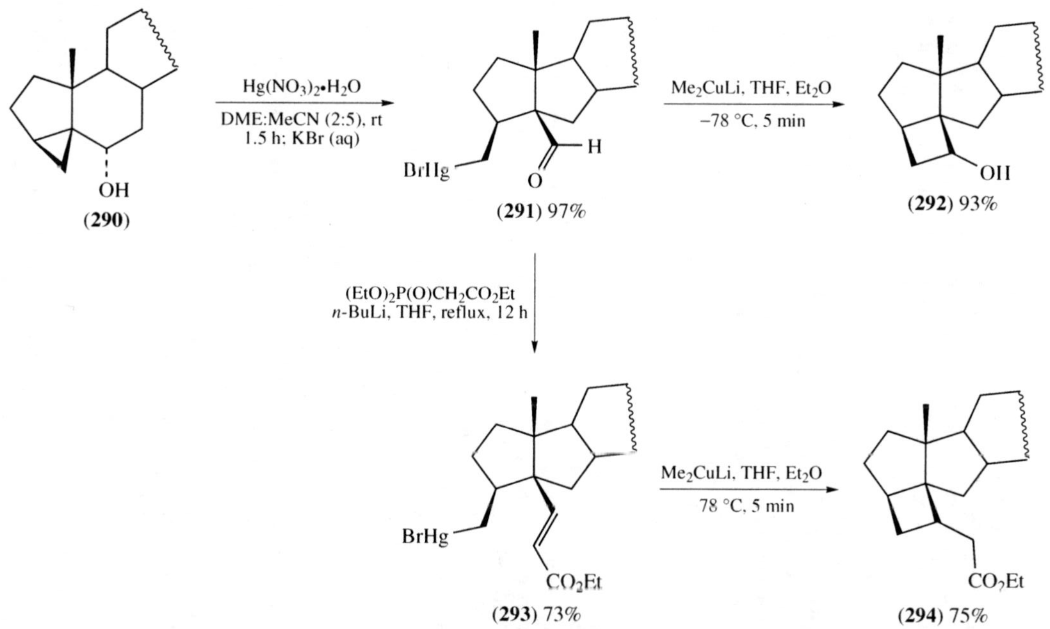

SCHEME 59. Transmetalation of organomercury compounds to copper in the synthesis of cyclobutane-containing steroid derivatives

copper species in these reactions remains to be elucidated; however, previous work suggested the presence of a bimetallic organomercury–copper reagent.

VI. TRANSMETALATIONS IN GROUP IIIA

A. Transmetalation from Organoboron Compounds

Boron (2.04) is more electronegative than copper(I) (1.90) and almost equal to copper(II) (2.00), which points toward an unfavorable equilibrium in a transmetalation from boron to copper(I). However, formation of boron ate complexes can facilitate ligand transfer from boron to less electron-rich copper salts[185–189]. Formation of the boron ate complex occurs by treatment with alkyl lithium or sodium methoxide, thus limiting the methodology to the preparation of unfunctionalized alkenyl or alkyl copper reagents. It should be noted that although the transmetalation reactions from boron to copper are not general, a transmetalation scheme from boron to copper through an intermediate organozinc has been used to generate highly functionalized organocopper reagents[126, 127, 190].

Treatment of dialkenylchloroborane **295** with three equivalents of methyl copper resulted in the formation of the alkenyl copper species **296**, which afforded the homo-coupled product **297**. Alternatively, in the presence of crotyl chloride, alkenyl copper **296** reacted to yield a 73:27 mixture of the S_N2'- (**298**) and the S_N2-products (**299**) in a combined 50% yield, as well as the homo-coupled product **297** (Scheme 60)[188, 191, 192]. A higher ratio of S_N2'- to S_N2-products was achieved by using alkenyl-9-BBN **300**, proceeding through boron ate complex **301** to result in a 88:12 mixture of S_N2'- (**302**) and S_N2-products (**303**) (Scheme 60)[188].

The preparation of alkyl aryl ketone **307** was accomplished by treatment of organoborane **304** with methyl lithium, to form lithium alkylboron ate complex **305**, followed by addition of copper(I) chloride–cyclooctadiene complex to generate the copper alkylboron complex **306** and benzoyl chloride (Scheme 61)[192]. In an application of this ketone formation methodology, Ichikawa and coworkers used a unique one-pot fluoride-induced selective transmetalation. Tosylate **308** was treated with two equivalents of *n*-BuLi followed by tributylborane to provide difluorovinylborane **309**. In the presence of lithium fluoride, the boron ate complex **310** is formed and upon transmetalation to generate vinylcopper **311**, cross-coupling with benzoyl chloride affords ketone **312** (Scheme 62)[193, 194]. More recently, Ichikawa has used this methodology to perform cross-coupling reactions of intermediate vinylcopper **311** with various electrophiles[195].

Furthermore, ligand transfer reactions can be achieved by exposing the requisite alkenylborane to sodium methoxide, forming an active boron ate complex. The homo-coupling product **297** was obtained by the reaction of borane **313** with sodium methoxide, followed by transmetalation from boron to copper, forming vinylcopper **296** (Scheme 63)[189]. Transmetalation of the activated boron ate complex **314** and addition of allyl bromide provided the skipped diene **315**[196]. Brown and Molander applied this methodology toward the synthesis of 1,3-enyne **316**[197], whereas Suzuki and coworkers have used a similar methodology to form 1,5-enynes[198].

Miyaura and coworkers[199, 200] and Hosomi and coworkers[201] concurrently reported copper-mediated borylation of Michael acceptors. Miyaura and coworkers have performed the borylation reaction on α,β-unsaturated ester **317** and acrylonitrile (**321**) to obtain alkyl boranes **320** and **322**, respectively. Hosomi and coworkers utilized conditions catalytic in copper to achieve the borylation of (*E*)-chalcone in high yield followed

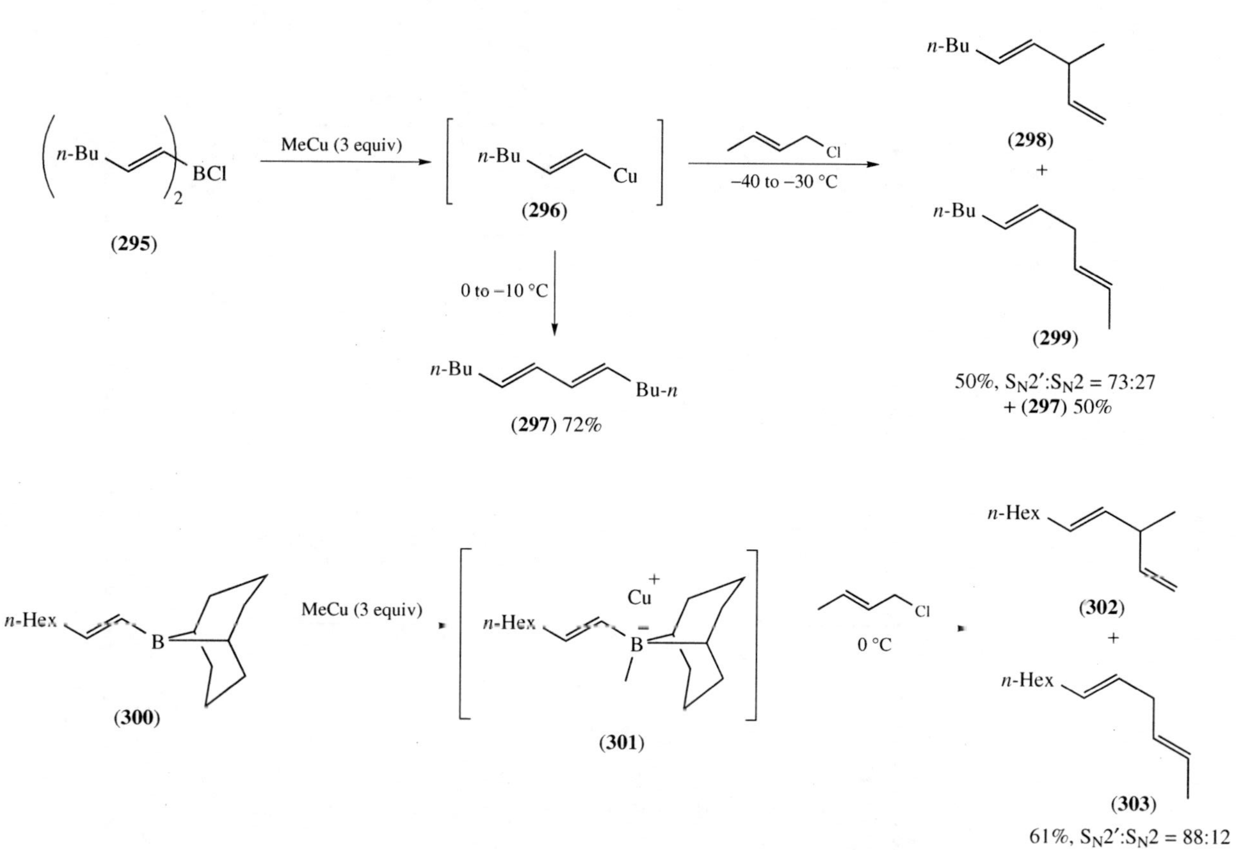

SCHEME 60. Transmetalation reactions from boron to copper and substitution reactions with crotyl chloride

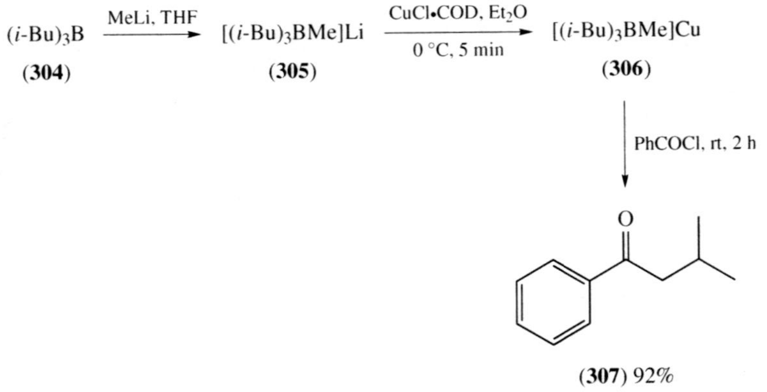

SCHEME 61. Synthesis of aryl alkyl ketones from organoboranes by coupling of copper alkylboron ate complex **306**

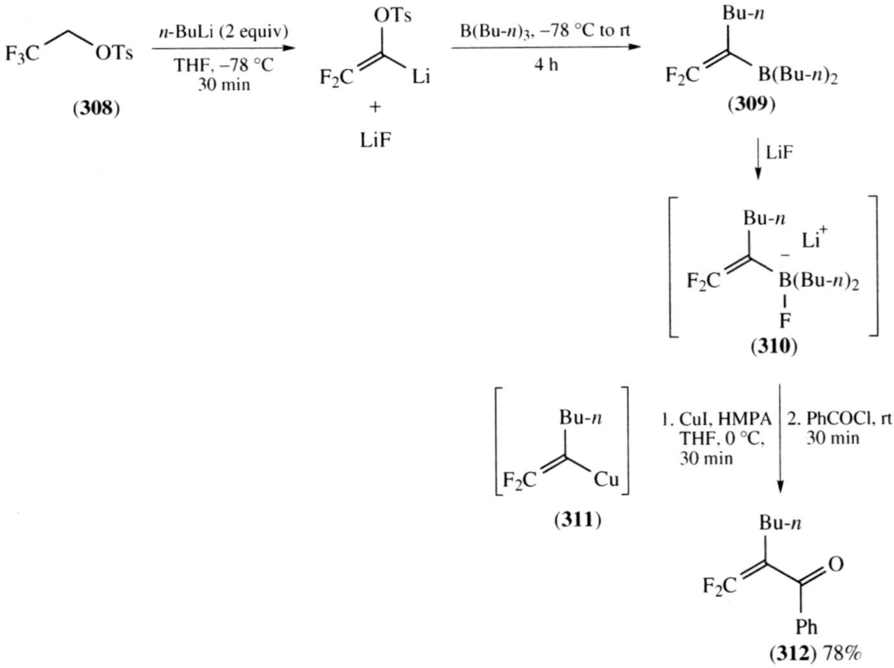

SCHEME 62. One-pot ketone synthesis featuring a fluoride-induced transmetalation from boron to copper

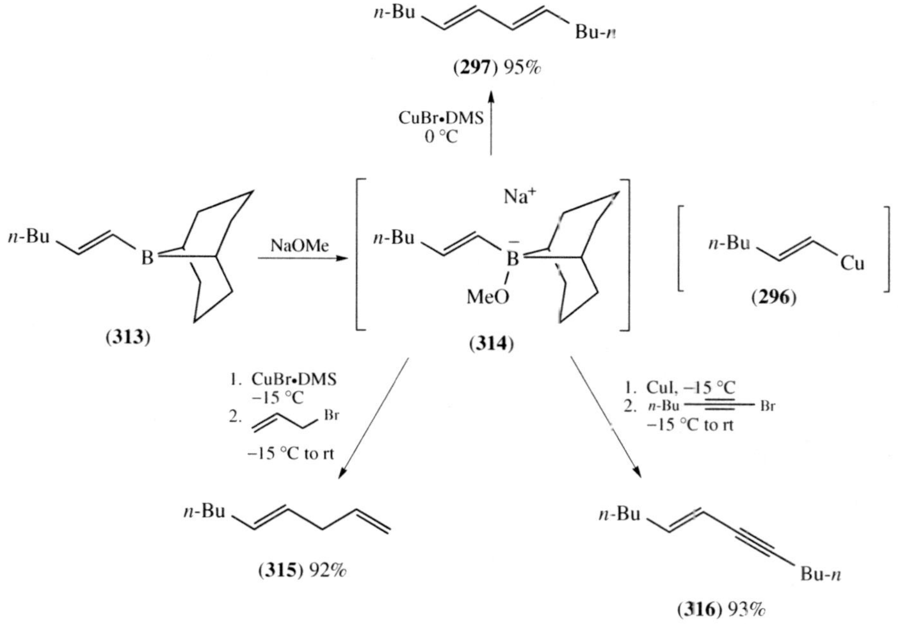

SCHEME 63. Activation of borane **313** using sodium methoxide followed by copper-mediated formation of dienes and enyne

by an oxidation of the borate **323** to obtain the β-hydroxy ketone **324**, quantitatively (Scheme 64). The transmetalation between diboron **318** and copper(I) salt is believed to generated boron–copper complex **319**, a key intermediate in conjugate addition reactions. Ramachandran and coworkers utilized their vinylalumination methodology to synthesize the typical Baylis–Hillman product **325**, which upon treatment with **318** under Miyaura's conditions produced allylboronate **326** in high yield[202]. Subsequent crotylboration reactions catalyzed by Lewis acids have been performed to obtain γ-lactones[203].

Recently, Ito and coworkers have studied the copper-catalyzed addition of diboron **318** to allylic and propargylic carbonates[204–207]. Addition of **318**, via boron–copper species **319**, to racemic allylic carbonate **327** in the presence of catalytic quantities of copper(I) alkoxide and Xantphos ligand provided alkyl boronate **328**. Extending these conditions for the addition of **318** to enantiomerically pure allylic carbonates **329** and **331** resulted in the desired allylic boronates **330** and **332**, respectively, in high regio- and stereoselectivity (Scheme 65)[204]. In an analogous reaction, Ito and coworkers successfully added diboron compound **318** to propargylic carbonates, affording allenes. Diboron **318** was added to the enantioenriched propargylic carbonate **333**, yielding allenylboronate **334**; subsequent treatment with Lewis acid and i-PrCHO provided a mixture of homopropargylic alcohols **335** and **336** with little or no erosion of enantiopurity (Scheme 66)[207].

Ito and coworkers have applied their methodology to the enantioselective synthesis of cyclopropanes. The addition of diboron reagent **318** to α-silyl allylic carbonate **337** in the presence of the (R)-segphos complex provided the optically active B-Si bifunctionalized cyclopropane **338** (Scheme 67). Successive stereoselective Suzuki–Miyaura coupling and Tamao–Fleming oxidation provided cyclopropanol **339** with no loss in enantiopurity[206].

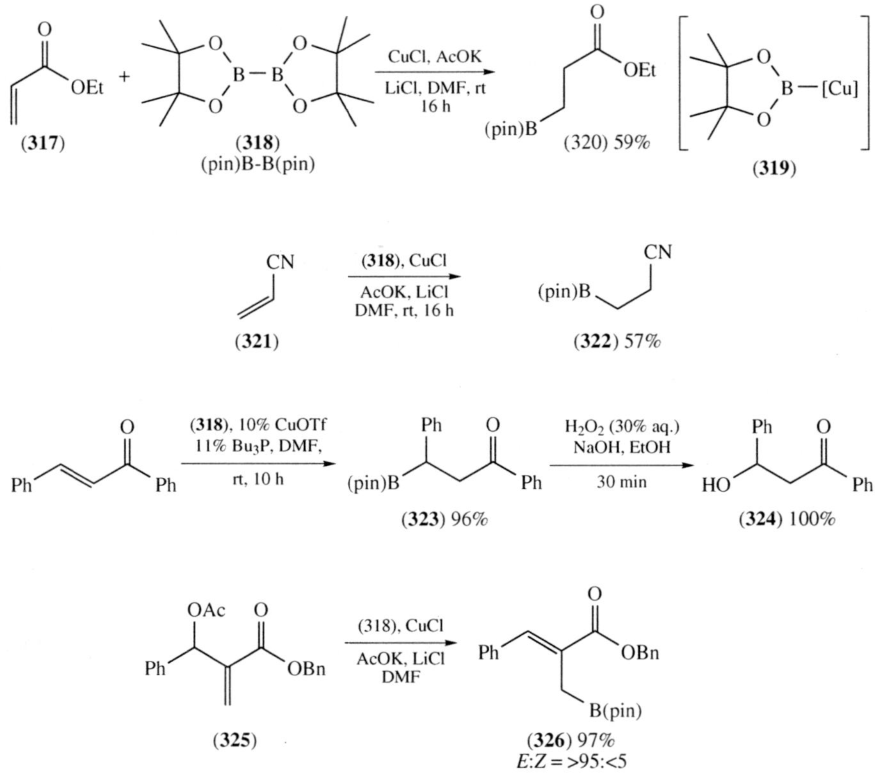

SCHEME 64. Transmetalation from boron to copper and subsequent Michael addition

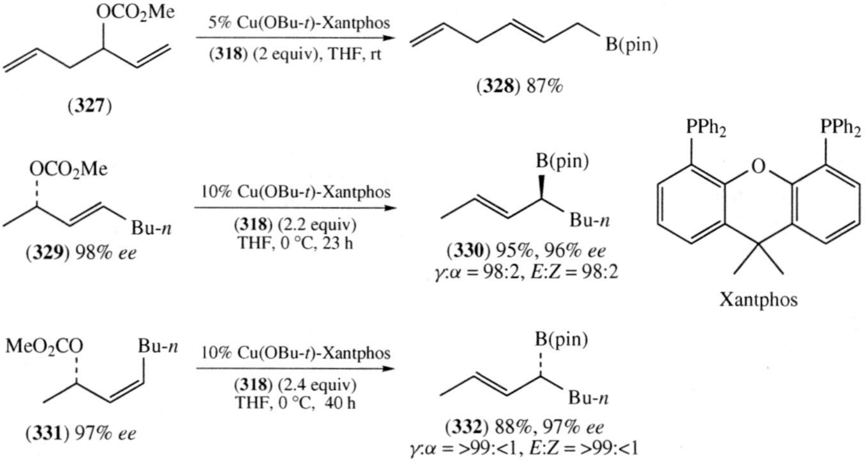

SCHEME 65. Copper-catalyzed S_N2'-addition of diboron species **318** to allylic carbonates

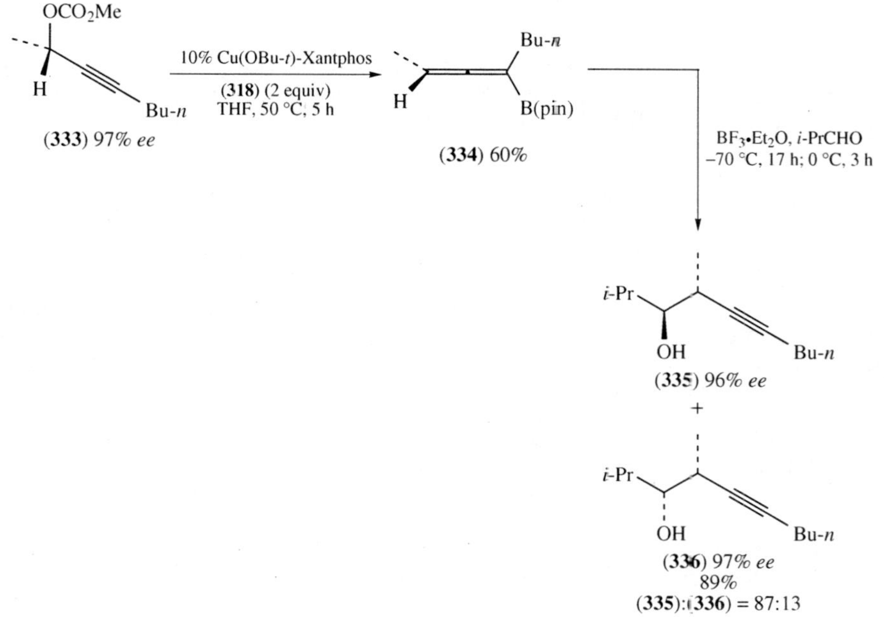

SCHEME 66. Copper-catalyzed substitution of enantiomerically pure propargylic carbonate with diboron reagent **318** and subsequent Lewis acid-mediated addition to an aldehyde

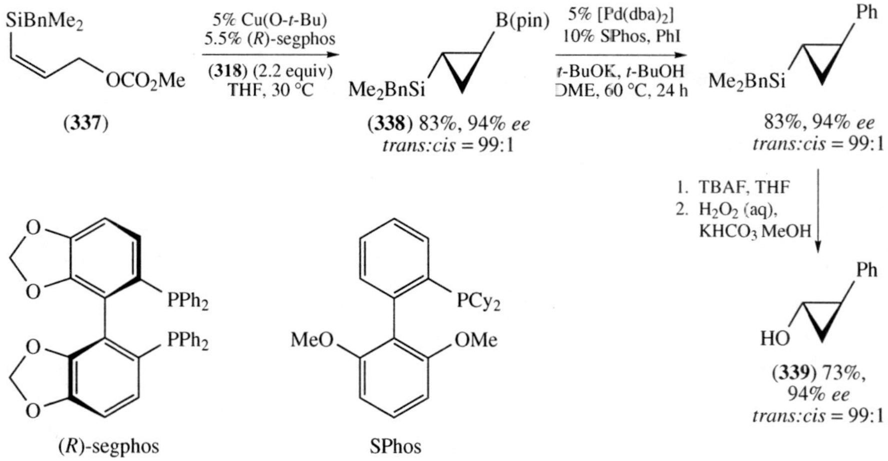

SCHEME 67. Enantioselective addition of diboron reagent **318** to α-silyl allylic carbonate to form cyclopropanol

B. Transmetalation from Organoaluminum Compounds

Compared to boron, the electronegativity of aluminum (1.61) facilitates transmetalation reactions with copper more readily. Early reports of copper(I)-mediated cross-coupling of vinylalanes[208,209] and the coupling between acid chlorides and organoaluminates in the presence of copper(I)[73,210] and copper(II)[211] salts alluded to the viability of an aluminum–copper exchange, although these reactions may proceed through a radical mechanism[73].

Ireland and Wipf[212] and Wipf and coworkers[213] investigated conditions to perform carboalumination of terminal alkynes followed by transmetalation to copper and conjugate addition to enones. Initially, transmetalation from aluminum to copper was achieved by a stepwise process[214,215] involving iodination, lithium–halogen exchange and transmetalation to copper. Subsequently, Ireland and Wipf were able to perform the transmetalation of vinylalanes directly. Vinylalane **341** was obtained by the carboalumination of terminal alkyne **340**. *In situ* transmetalation to higher order cyanocuprate or dimethyl cuprate followed by addition of cyclohexenone resulted primarily in methyl addition; however, using the bisalkynyl cyanocuprate **342**, with two non-transferable ligands, the desired olefin **343** was obtained in high yield (Scheme 68). Control experiments with the vinylalane **341** and the corresponding aluminum ate complex in both the presence or absence of $Ni(acac)_2$ resulted in only trace amounts of product. Carbocupration of the alkyne using MeCuMgBr[216], to arrive at the analogous vinylcopper species, failed to show signs of the product, even after 5 days. The control experiments clearly demonstrated the utility of the carboalumination/transmetalation/conjugate addition methodology. The carboalumination

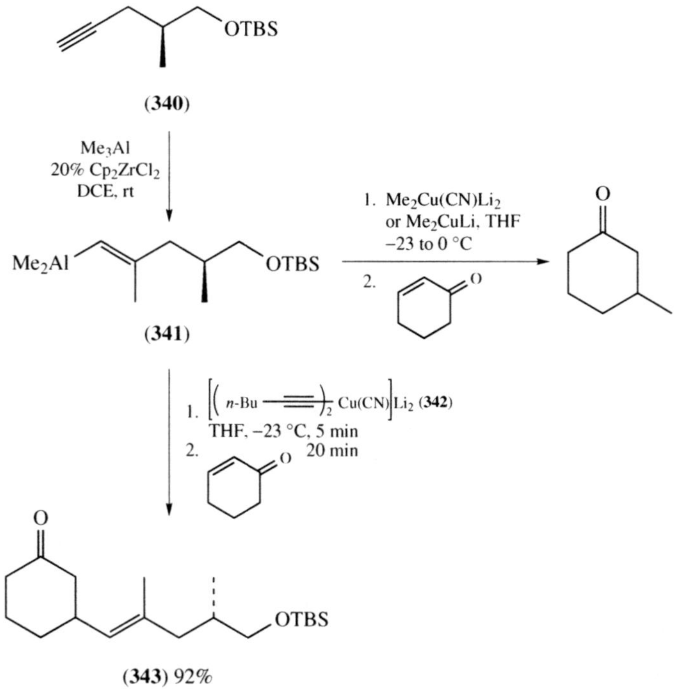

SCHEME 68. Carboalumination of alkyne **340**, transmetalation to copper and conjugate addition to cyclohexenone

of an internal alkyne resulted in the vinylalane **344**, which was transmetalated to copper and added to an enone, providing the tetrasubstituted olefin **345** in modest yield. The addition of vinylalane **341** to the functionalized enone **346** provided the FK-506 segment **347** (Scheme 69)[212,213].

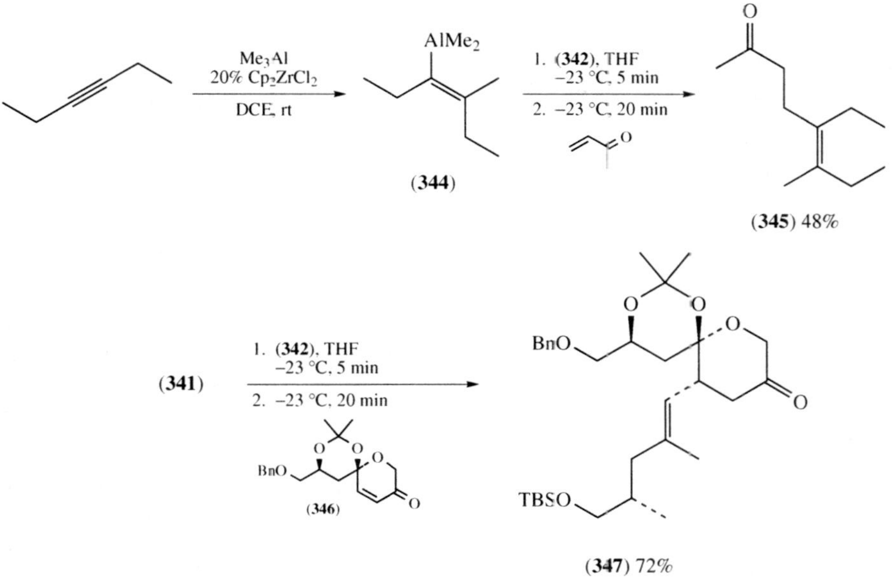

SCHEME 69. Conjugate addition of internal vinylalane **344** and conjugate addition to functionalized enone **346**

Wipf and coworkers proposed that the reaction mechanism proceeded through an initial ligand transfer from the copper ate complex **342** to the vinylalane **348**, followed by a second ligand transfer between aluminum ate complex **349** and copper salt **350**. The ligand exchange processes established an equilibrium that led to the formation of vinylic cyanocuprate **351** and/or mixed cuprate **352**, which irreversibly adds to enone **353** to provide ketone **354**. The open coordination site on vinylalane **348** was essential in establishing the equilibrium with copper ate complex **342**, and thus the addition of strongly coordinating additives or Lewis acids significantly reduced the yield and rate of the process. If the rate of conjugate addition is fast compared to that of various side reactions, the ligand transfer reactions do not need to be thermodynamically favorable (Scheme 70)[213].

Lipshutz and Dimock investigated a copper-catalyzed approach for the conjugate addition of vinylalanes and found that the use of 10% CuCN•2LiCl was sufficient to produce the 1,4-addition product[217]. They applied the copper-catalyzed methodology toward the synthesis of the prostaglandin derivative **355** by carboalumination of **60** and copper-catalyzed conjugate addition to enone **62** (Scheme 71).

In 1993, a group from Schering AG explored the copper-catalyzed 1,4-addition reaction of trialkylalane reagents to enones[218–221] and enals[221,222]. Addition of trialkylanes to enones **356a** or **356b** in the presence of copper(I) bromide proceeded in high yields to give the desired 1,4-addition products[220]. The methyl addition to enals **356c** and **356d** resulted in a mixture of 1,2- and 1,4-addition products; although a high yield and preference for conjugate addition was observed (Scheme 72)[222].

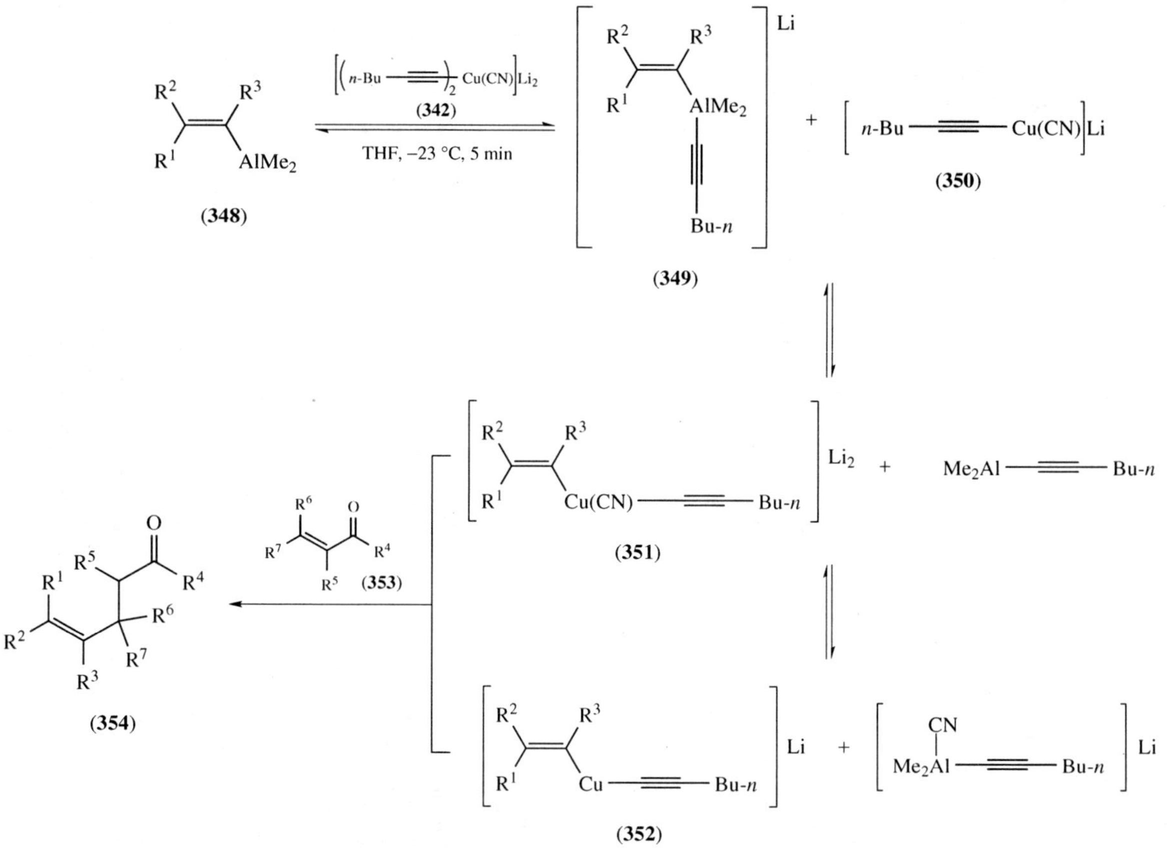

SCHEME 70. Mechanistic proposal for the copper-mediated addition of vinylalanes to enones

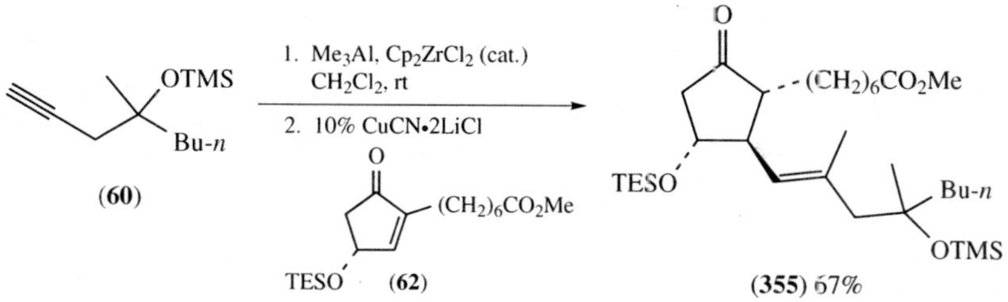

SCHEME 71. Application of the copper-catalyzed addition of vinylalane to form prostaglandin derivative **355**

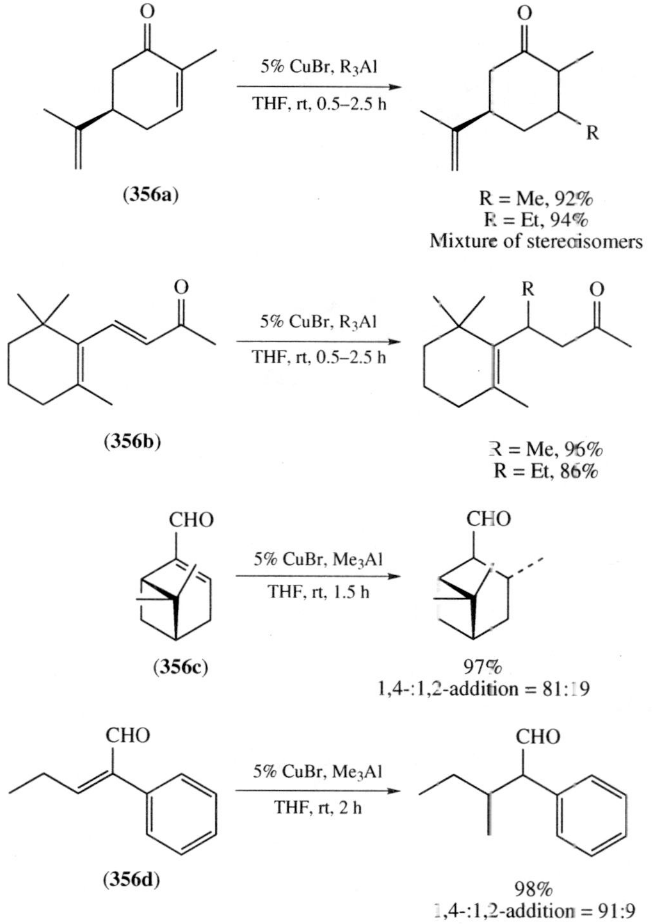

SCHEME 72. Copper-catalyzed conjugate addition of trialkylalanes reagents to enones and enals

More recently, Alexakis and coworkers[223] and Hoveyda and coworkers[224] have expanded their methodologies of copper-catalyzed asymmetric conjugate additions of organozinc reagents (see Section V.A) to the asymmetric conjugate addition of alanes in the presence of copper salts[135]. Alexakis and coworkers were able to perform an asymmetric conjugate addition of trimethylalane to the β-substituted enone **357** in the presence of 10% CuTC and chiral phosphoramidite **358** to achieve the β-disubstituted ketone **359** in high enantioselectivity. Treatment of **359** with acid resulted in a hydrolysis of the acetal and cyclization to provide bicyclic ketone **360**[225]. Alexakis and Vuagnoux-d'Augustin were able to apply this methodology to the asymmetric conjugate addition of vinylalane **362**, obtained by hydroalumination of the corresponding alkyne **361**, in the presence of phosphoramidite **358** and 30% $[Cu(MeCN)_4]BF_4$ to yield ketone **363** in moderate yield and enantioselectivity. The use of CuTC to catalyze the conjugate addition of **362** led to the 1,2-addition/dehydration product (Scheme 73)[226].

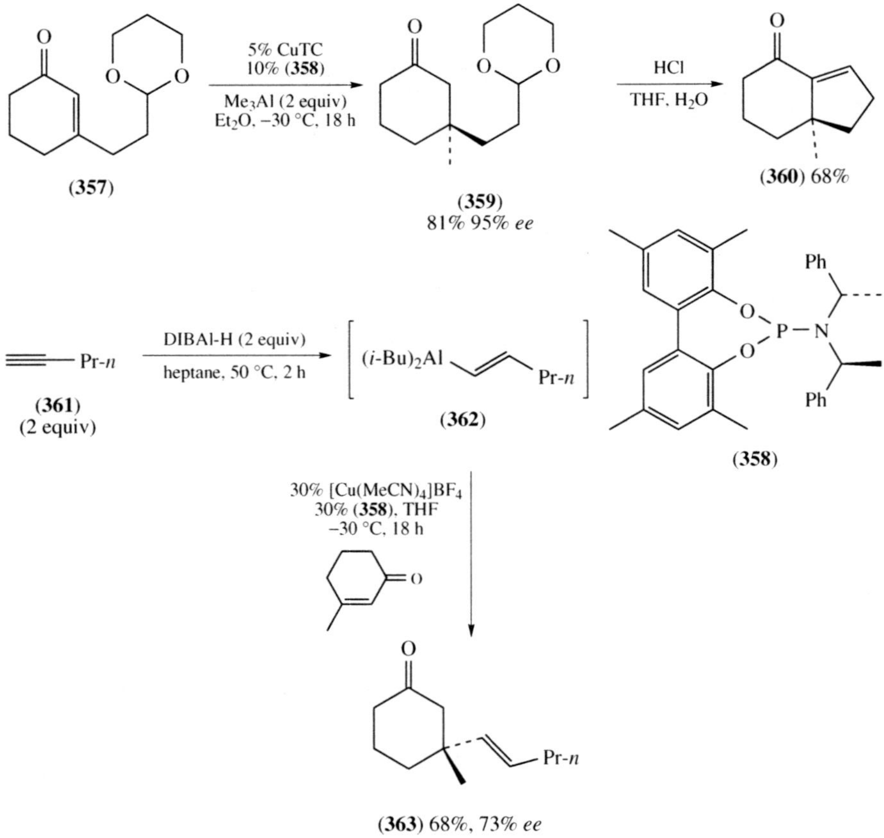

SCHEME 73. Copper-catalyzed enantioselective conjugate additions of trimethylalane and vinylalane **362** to β-substituted enones

The asymmetric addition of aryl groups to α,β-unsaturated ketones is a challenging synthetic task; however, the copper-catalyzed enantioselective addition of dialkylarylalanes has been accomplished by Alexakis and coworkers[227]. Dialkylarylalanes can be generated

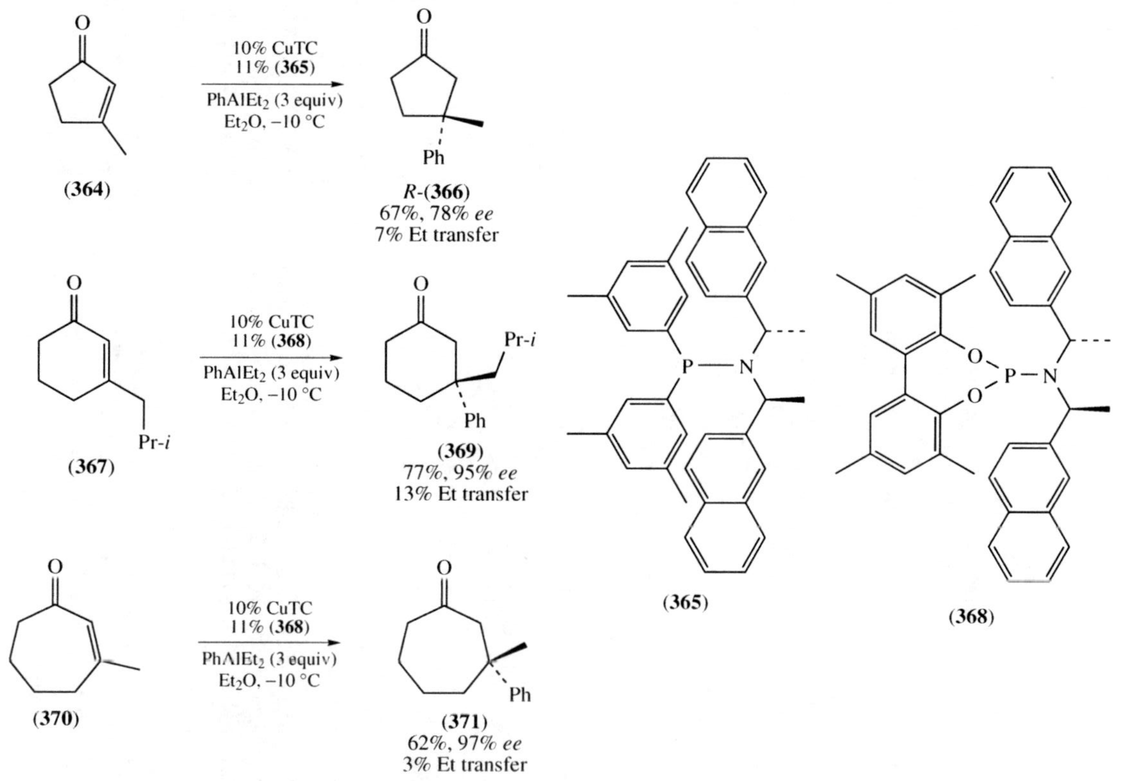

SCHEME 74. Asymmetric conjugate addition of arylalanes in the presence of catalytic quantities of copper(I) and phosphoramidite ligands

from their corresponding aryl iodides by lithium–halogen exchange followed by quenching with diethylaluminum chloride. The 1,4-addition of dialkylarylalanes was performed on 5-, 6- and 7-membered-β-substituted enones (**364**, **367** and **370**, respectively), using catalytic amounts of copper(I) and chiral phosphoramidite ligands **365** and **368** to access enantioenriched ketones *R*-**366**, **369** and **371** (Scheme 74)[227].

Hoveyda and coworkers applied their chiral bidentate silver *N*-heterocyclic carbene (NHC) complexes in the presence of catalytic amounts of copper(II) salts, generating their active copper(I)–NHC complex *in situ*[150,151], to perform enantioselective conjugate addition reactions of trialkylalanes to form all-carbon quaternary centers[224]. Triethylalane addition to β-phenyl enone **372** upon treatment with 2.5% of Ag(I)–NHC complex **373** and 5% copper(II) triflate provided **374** in high yield and enantioselectivity. Alternatively, the diethylarylalane can be applied analogously in the addition to enone **364** to give ketone *S*-**366** in decreased enantioselectivity (Scheme 75). The methodology can be applied to a variety of enones and generally results in good enantioselectivities[224].

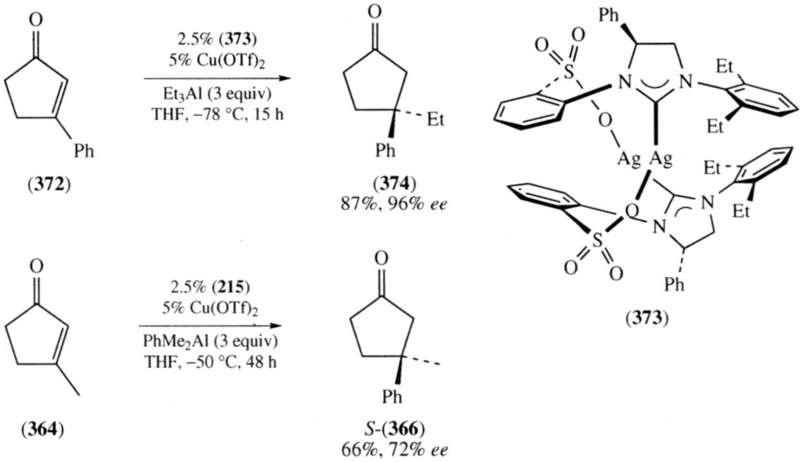

SCHEME 75. Enantioselective 1,4-conjugate addition of aryl- and alkylalanes to cyclopentenones using Ag–NHC complexes

Hoveyda and coworkers also utilized the *in situ* generated copper(I)–NHC complexes to perform enantioselective S_N2'-additions to allylic phosphates[228]. Hydroalumination of alkyne **375** followed by addition of the vinylalane to allylic phosphonate **376** in the

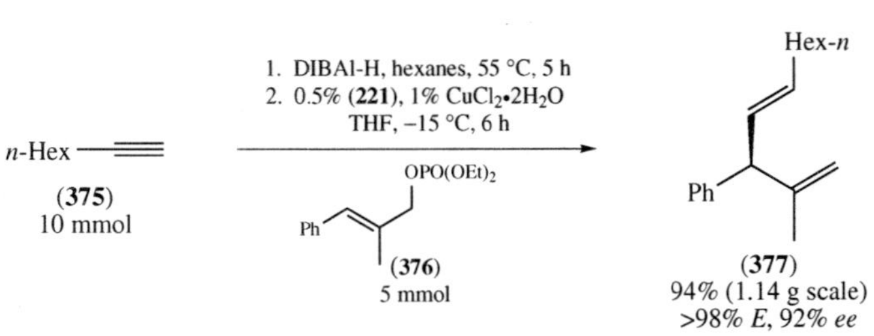

SCHEME 76. Enantioselective S_N2'-addition of vinylalane to allylic phosphate **376** catalyzed by an *in situ* generated copper–NHC complex

presence of 0.5% NHC complex **221** and copper(II) chloride dihydrate resulted in clean S_N2'-addition, providing diene **377** in high yield and enantioselectivity (Scheme 76). Additionally, this methodology has been proven to be scalable up to 10 mmol without a decrease in yield or enanticselectivity[228].

VII. TRANSMETALATIONS IN GROUP IVA

Silicon (1.90), tin [Sn(II): 1.80; Sn(IV): 1.96], and lead [Pb(II): 1.87; Pb(IV): 2.33] have similar electronegativities compared to copper(I) and have been used in ligand exchange reactions with copper(I). The transmetalation chemistry from lead to copper is limited and has been previously addressed[4].

A. Transmetalation from Organotin Compounds

Corey and Wollenberg synthesized vinylcuprates *in situ* from the corresponding vinylstannane by sequential transmetalation from tin to lithium to copper[229,230]. The direct transmetalation from vinylstannanes to copper ate complexes was first achieved by Campbell, Lipshutz and coworkers[231,232]. Treatment of vinylstannane **378** with a mixed cyanocuprate led to the formation of vinylcuprate **379**, which upon exposure to enone **380** yielded the conjugate addition product **381**. Bis-stannane **382** was transmetalated to the vinylcuprate **383** and provided vinylstannane **385** in a conjugate addition reaction with enone **384** (Scheme 77). Further application of their methodology led to the synthesis of highly reactive allylic organocuprates[233]. Allylic cuprate **387** was obtained by treating the

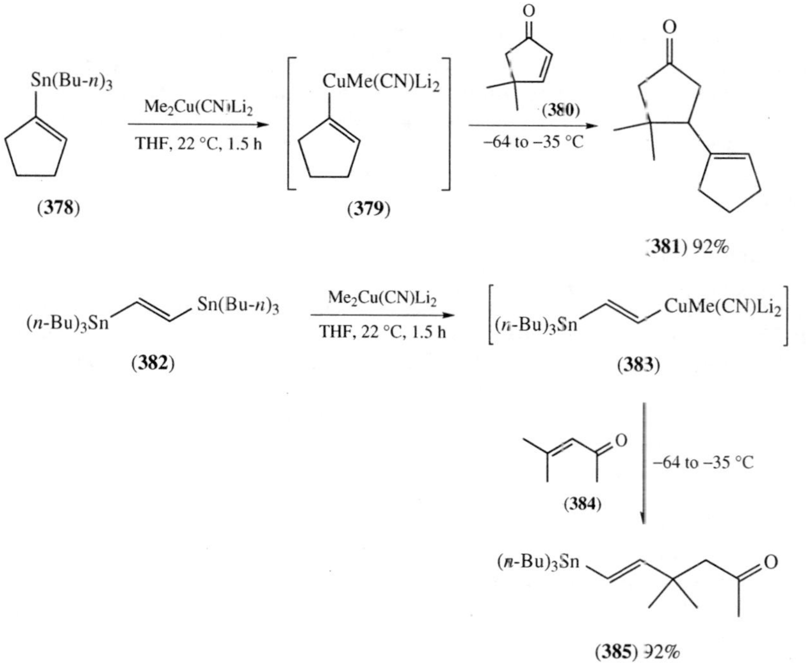

SCHEME 77. Transmetalation of vinylstannanes to vinylcuprates and subsequent conjugate addition reactions with α,β-unsaturated ketones

corresponding allylstannane **386** with 0.5 equivalents of dimethylcyanocuprate. Cuprate **387** was reactive enough to perform a substitution on the non-activated primary alkyl chloride **388**, resulting in **389**, and to open epoxide **390**, providing alcohol **391** (Scheme 78).

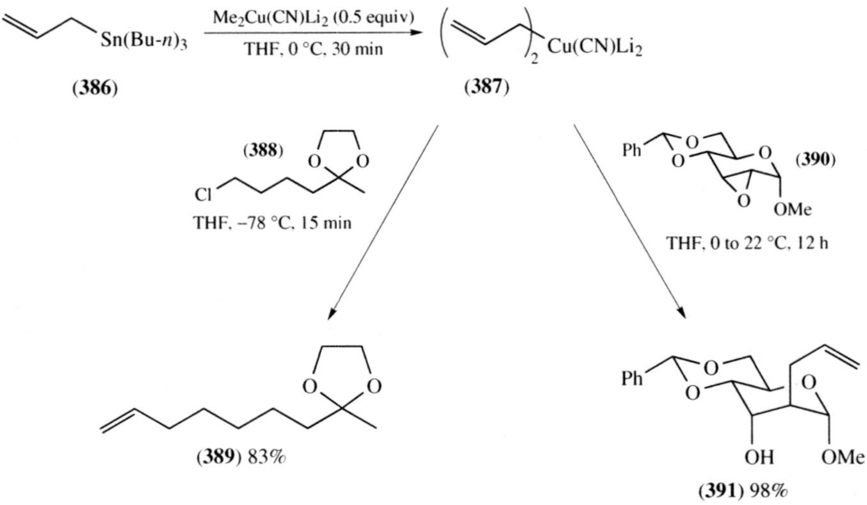

SCHEME 78. Substitution reactions of highly reactive diallylcuprate **387**

The transmetalation reaction of mixed silylstannanes upon exposure to higher order cuprates resulted in the formation of either a tin–copper species or a silicon–copper species; however, the outcome is strongly dependent on the bulk of the substituents on the silylstannane[234,235]. Mixed silylstannane **392** resulted in a transmetalation to form stannyl-cuprate **393**, while mixed silylstannane **396** provided silyl-cuprate **397**. The intermediate cuprates were added to enone **394** to give β-stannylketone **395** and β-silylketone **398** (Scheme 79).

Piers and Wong have demonstrated the use of an intramolecular copper(I)-mediated cross-coupling reaction to construct diene-carbocycles containing 4-, 5- and 6-membered rings[236]. Treatment of vinylstannanes **399** and **401** with copper(I) chloride in DMF at elevated temperatures resulted in the cyclized products **400** and **402**, presumably through a vinylcopper intermediate obtained by transmetalation of the vinylstannane (Scheme 80). The yield of the cross-coupling reaction decreases with the use of other copper(I) sources and/or different solvents. This methodology was applied further to the intramolecular conjugate addition of vinylstannanes to enones[237,238]. Addition of vinylstannane **403** to 2.5 equivalents of copper(I) chloride in DMF at room temperature resulted in the formation of an intermediate vinylcopper species leading to the bicyclic ketone **404** via an intramolecular Michael addition. Analogously, the 5-membered ring enone **405** was treated with CuCl to give ketone **406** (Scheme 81).

In an application to heterocyclic chemistry, Bennasar and coworkers were able to synthesize the functionalized indole **409** by exposing indole **407** to a higher order cuprate, effecting a transmetalation followed by a 1,4-addition to the unsaturated lactam **408** (Scheme 82)[239]

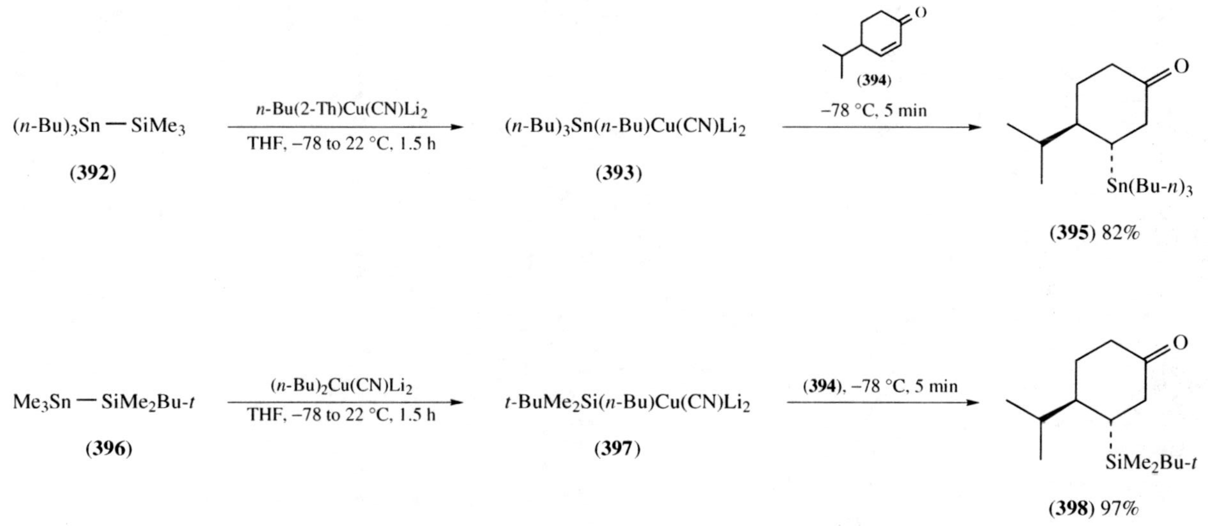

SCHEME 79. Transmetalation of mixed silylstannanes **392** and **396** followed by conjugate additions to enone **394**

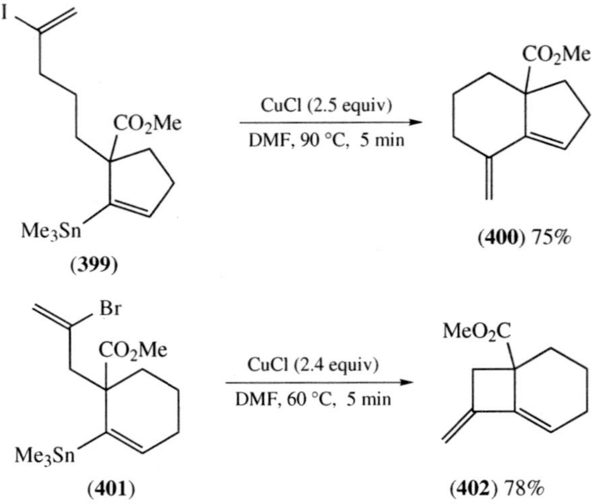

SCHEME 80. Intramolecular copper-mediated cross-coupling reaction of vinylstannanes and vinylhalides

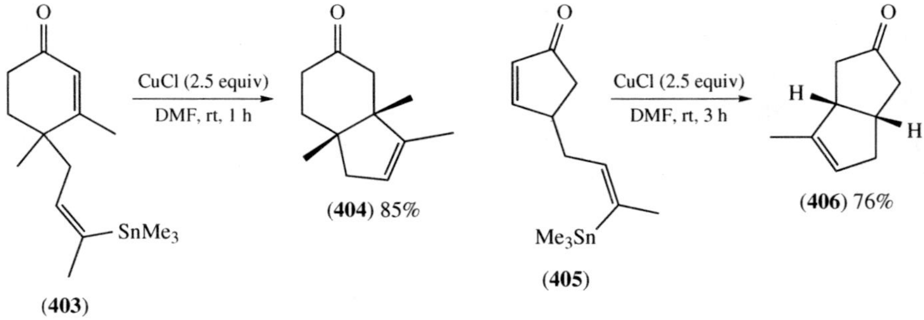

SCHEME 81. Copper-mediated intramolecular conjugate addition reactions of vinylstannanes and enones

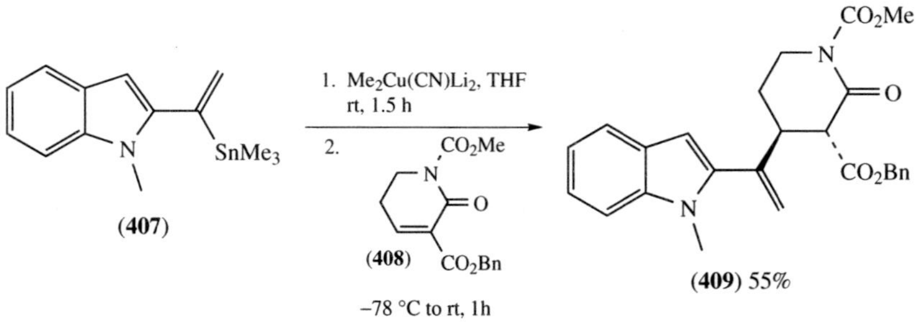

SCHEME 82. Conjugate addition of functionalized vinylstannane **407** to the α,β-unsaturated lactam **408**

B. Transmetalation from Organosilicon Compounds

Hosomi and coworkers were first to isolate an organocopper species after the transmetalation from an organosilicon compound[240]. Transmetalation was achieved by exposing silyl alkyne **410** to 20% CuCl in 1,3-dimethyl-2-imidazolidinone (DMI), and the alkynyl copper **411** could be isolated in 81% yield. Alternatively, in the presence of benzoyl chloride, a coupling occurred to give ketone **412** (Scheme 83)[240, 241].

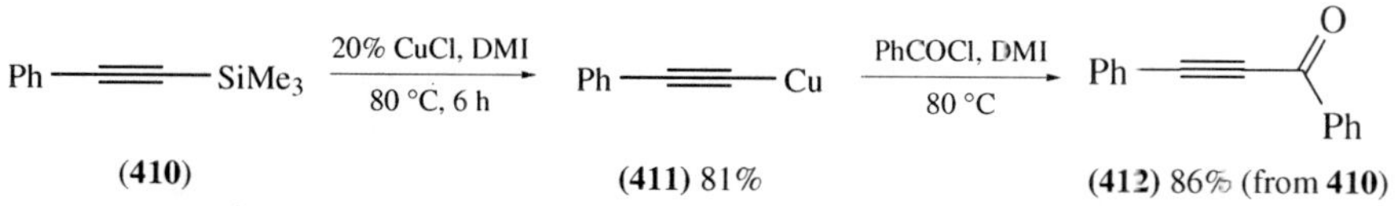

SCHEME 83. Transmetalation from silicon to copper and subsequent acid chloride coupling

Similar to diboron compounds (see Section VI.A), disilane **413** has been transmetalated in the presence of copper to generate a silyl-copper species **415** that was added to Michael acceptors[242]. Disilane **413** was added to cyclohexenone and enal **416** by treatment of catalytic amounts of copper(I) triflate-benzene complex and tributylphosphine in DMF to provide β-silyl-carbonyl compounds **414** and **417**, respectively, in high yield (Scheme 84). The authors propose that tributylphosphine plays an important role in the stabilization of the intermediate silyl-copper species **415**.

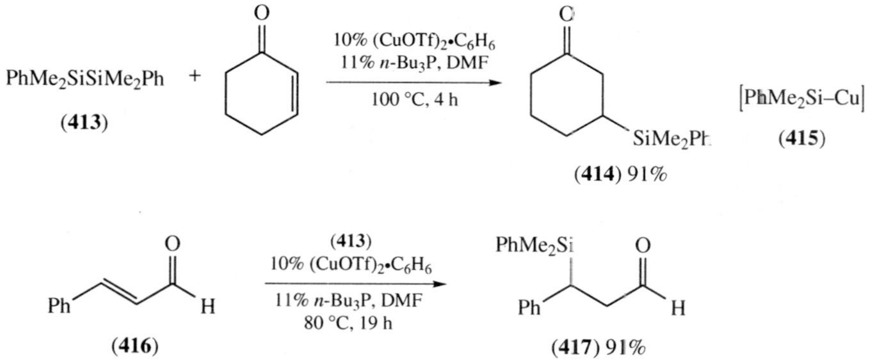

SCHEME 84. Copper-catalyzed conjugate addition of disilane **413** to cyclohexenone and enal **416**

Very recently, Hoveyda and Lee have used the chiral NHC ligand **418** in the presence of catalytic quantities of copper(I) bromide to achieve the enantioselective conjugate addition of alkenyl- and aryl-trifluorosilanes to enones[243]. The copper–NHC complex **419** was generated *in situ* from NHC ligand **418**, copper(I) bromide and NaOBu-*t*. The alkyl- and/or alkenyl-trifluorsilanes react with the fluoride source, TASF, generating a hypervalent silicate, which subsequently undergoes a ligand exchange reaction from silicon to copper, forming an organocuprate intermediate. The alkyl or alkenyl group is added in a conjugate fashion to cyclohexenone to provide the corresponding enantioenriched β-substituted ketones **420** and **421** (Scheme 85).

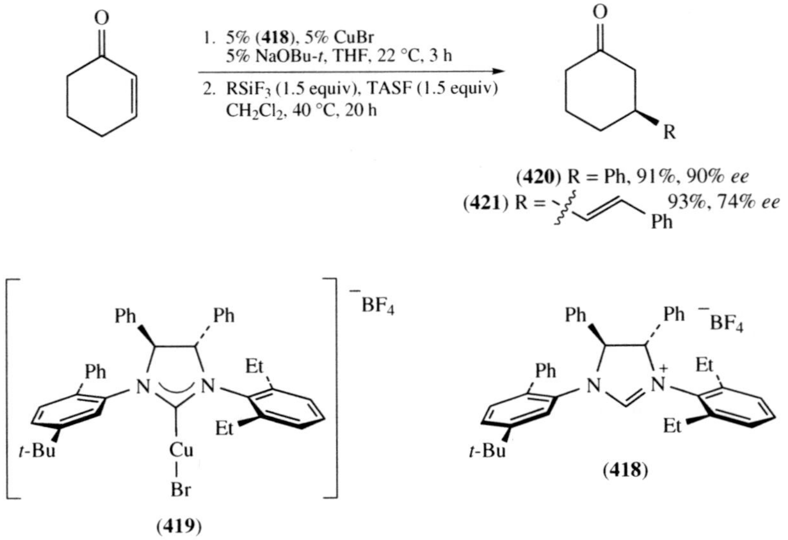

SCHEME 85. Copper-catalyzed enantioselective conjugate addition of aryl- and alkenyl-trifluorosilanes

VIII. TRANSMETALATIONS IN GROUP VIA

A. Transmetalation from Organotellurium Compounds

Transmetalation from the relatively electropositive tellurides (2.10) presumably occurs through a tellurium ate complex. Vinyltelluride **423** was obtained by *syn*-reduction of alkynyltelluride **422**. Subsequent transmetalation with a higher order cuprate resulted in the *Z*-vinylcuprate **424** and a conjugate addition to enone **384** provided ketone **425** with retention of olefin configuration (Scheme 86)[244].

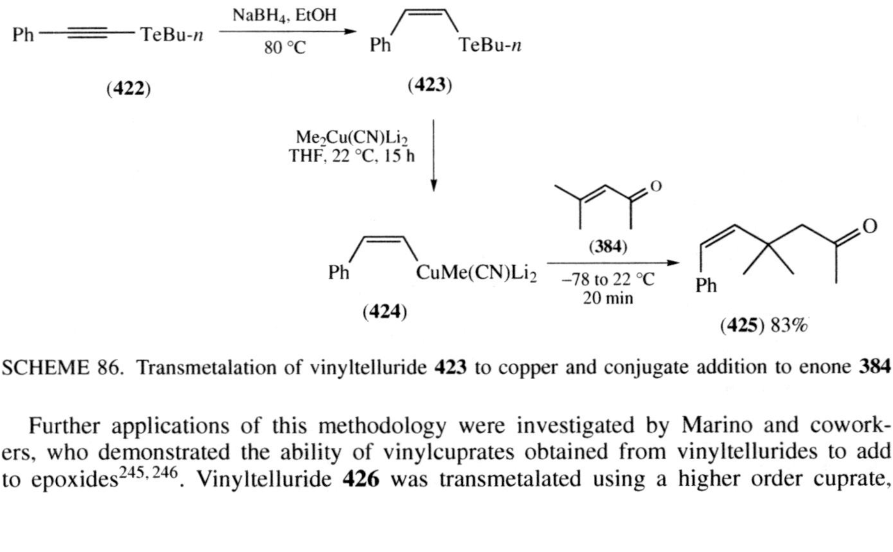

SCHEME 86. Transmetalation of vinyltelluride **423** to copper and conjugate addition to enone **384**

Further applications of this methodology were investigated by Marino and coworkers, who demonstrated the ability of vinylcuprates obtained from vinyltellurides to add to epoxides[245,246]. Vinyltelluride **426** was transmetalated using a higher order cuprate,

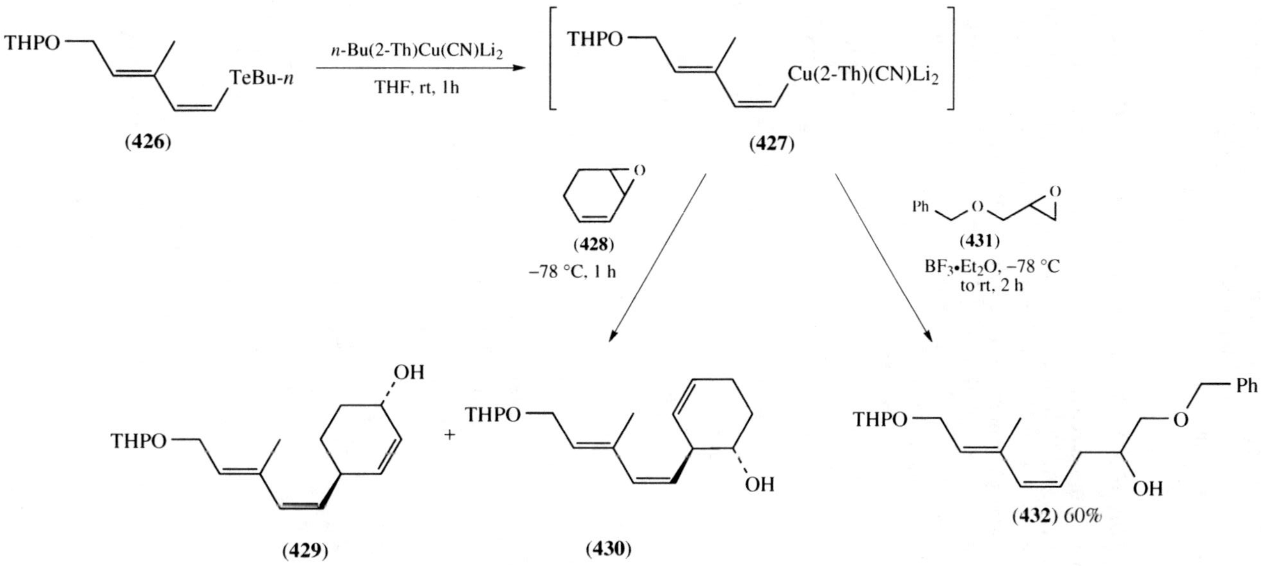

90%, (**429**).(**430**) – 5.2:1

SCHEME 87. Transmetalation of vinyltelluride **426** and epoxide opening reactions generating E-vinylcuprate **427**, which was added into allylic epoxide **428** to give a mixture of allylic alcohol **429** and homoallylic alcohol **430**. The vinylcuprate **427** could also be added into epoxide **431** in the presence of $BF_3 \cdot Et_2O$ to give the homoallylic alcohol **432** (Scheme 87). The obvious utility of organotelluride transmetalation chemistry is the ability to easily access Z-vinylcuprates.

IX. APPLICATIONS TO NATURAL PRODUCT SYNTHESIS

As a result of the high chemoselectivity, functional group tolerance and stereospecific or enantioselective nature of some of the organocopper methodologies mentioned previously, these methods have been useful tools in the construction of various natural products[247]. Examples of natural product syntheses that utilize transmetalation reactions to form organocopper compounds are bongkrekic acid[248], cortisone[249, 250], adrenosterone[250], FK-506[251–253], methyl epijasmonate[254], β-cedrene[255], (−)-mesembrine[256], iso[7]-levuglandin D_2[257], kendomycin[258] and baconipyrone C[259].

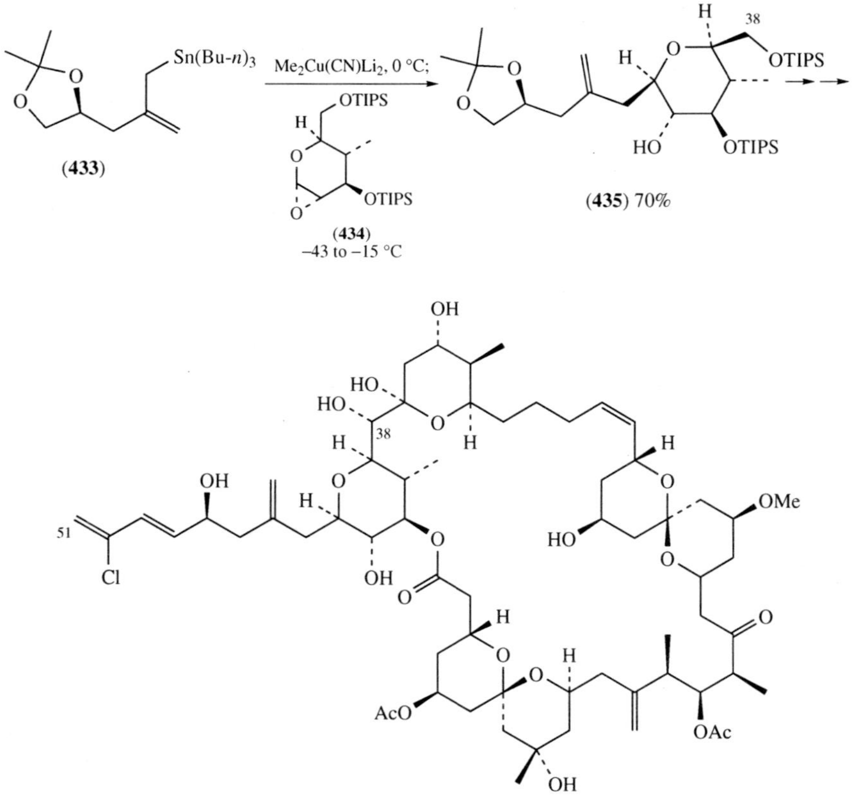

altohyrtin A
(**436**)

SCHEME 88. Application of a Sn−Cu transmetalation to achieve the C31−C58 sequence of altohyrtin A

During the total synthesis of altohyrtin A (**436**)[260,261], Kishi and coworkers applied the tin to copper transmetalation methodology developed by Lipshutz and coworkers[233]. Allylic stannane **433** was transmetalated to the corresponding allylic cuprate, which was used to open epoxide **434** to provide alcohol **435**[261]. It should be noted that an excess of the stannane was required to achieve a good yield; however, the excess reagent could be completely recovered. Fragment **435** was further functionalized to form the C38 to C51 sequence of altohyrtin A (**436**) (Scheme 88).

Hoveyda and Brown have recently applied transmetalations from zinc and aluminum to copper in the total synthesis of clavirolide C (**444**)[262]. The copper-catalyzed enantioselective conjugate addition of dimethylzinc to the α,β-unsaturated ester **437** in the presence of **438** proceeded in high yield and enantioselectivity; however, it was necessary to trap

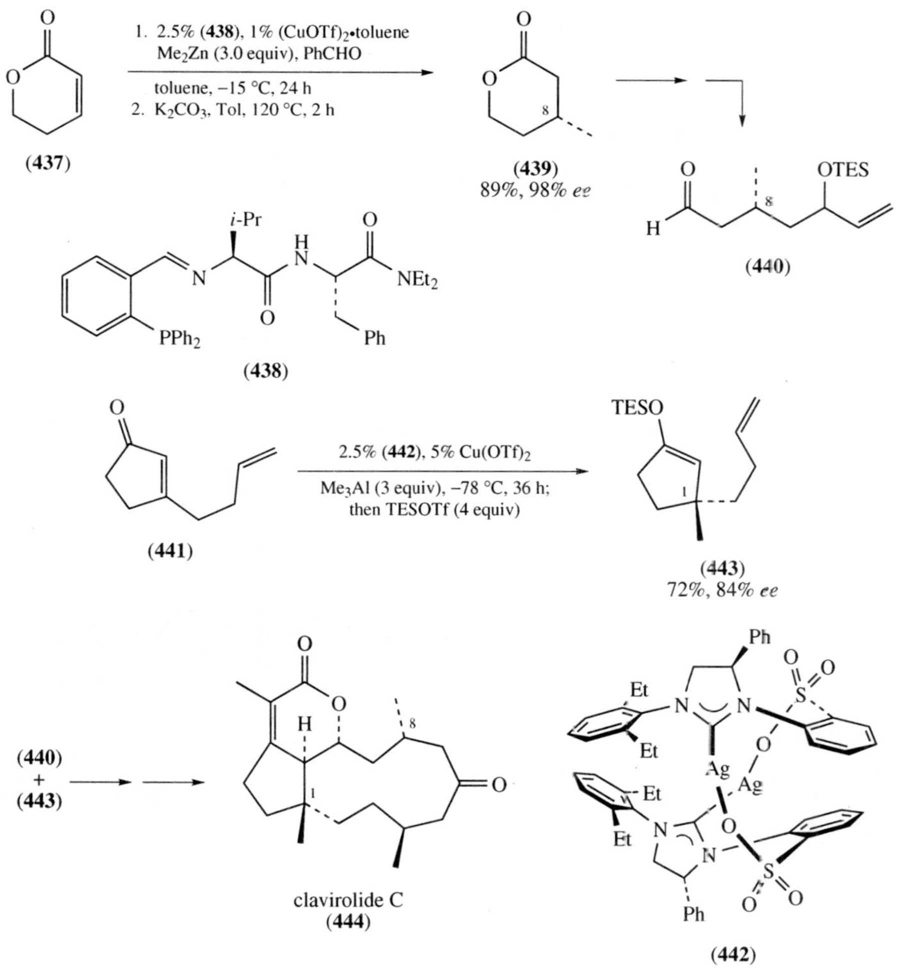

SCHEME 89. Application of the copper-catalyzed enantioselective conjugate addition of dimethylzinc and trimethylalane in the total synthesis of clavirolide C

out the intermediate zinc enolate with benzaldehyde to achieve the high yield. Treatment with base provided ester **439** with the requisite stereocenter at C8. The absolute stereochemistry at C1 was set by a copper–NHC-catalyzed enantioselective conjugate addition to enone **441** using **442** and subsequent trapping with TESOTf provided the desired silyl enol ether **443**. Fragments **440** and **443** were joined and further functionalized to arrive at clavirolide C (**444**) (Scheme 89).

Charette and Barbe completed the total synthesis of (+)-lepadin B, using a transmetalation of zirconium to copper to install a pendant diene moiety in a late stage transformation[263]. Alkyne **446** was hydrozirconated, transmetalated to copper(I) and added to enone **445** in a conjugate fashion, from the less hindered face, leading to diene **447**[59,89]. Subsequent reduction of the ketone and deprotection steps provided (+)-lepadin B (**448**) (Scheme 90).

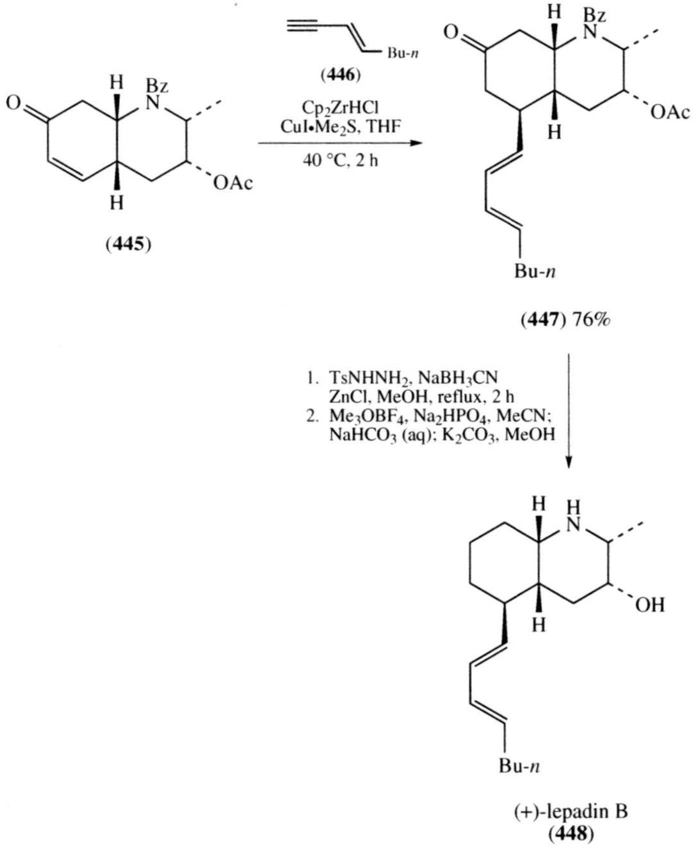

SCHEME 90. Transmetalation from zirconium to copper in the total synthesis of (+)-lepadin B

X. CONCLUSION

The functional group compatibility of the organometallic reagents shown to transmetalate to copper salts provides access to a wide range of organocopper intermediates. This strategy successfully integrates reactivity patterns characteristic for copper as well as the original organometallic precursor species, while oftentimes enhancing overall chemoselectivity. The transmetalation scheme therefore considerably widens the scope of traditional organocopper chemistry, not only as a consequence of the presence of the residual metal salts but also due to the nature of precursor processes that are accessible through a different metal manifold. The emergence of copper-catalyzed enantioselective and stereospecific reactions to generate chiral building blocks further enhances the utility of this strategy in synthetic organic chemistry. The computational and experimental investigation of mechanistic features and/or the identification of active metallic species is complicated by the various species present in solution, but further investigations are likely to deepen our understanding of organometallic processes and enhance the rational design of novel transformations.

XI. REFERENCES

1. B. H. Lipshutz, in *Organometallics in Synthesis. A Manual*, 2nd edn., (Ed. M. Schlosser), Chap. 6, John Wiley & Sons, Ltd, Chichester, 2002, pp. 665–815.
2. N. Krause (Ed.), *Modern Organocopper Compounds*, Wiley-VCH, Weinheim, 2002.
3. P. Knochel, X. Yang and N. Gommermann, in *Handbook of Functionalized Organometallics. Applications in Synthesis*, Vol. 2 (Ed. P. Knochel), Chap. 9, Wiley-VCH, Weinheim, 2005, pp. 379–398.
4. P. Wipf, *Synthesis*, 537 (1993).
5. P. Knochel and B. Betzemeier, in *Modern Organocopper Chemistry* (Ed. N. Krause), Chap. 3, Wiley-VCH, Weinheim, 2002, pp. 45–78.
6. I. Marek and H. Chechik-Lankin, in *Handbook of Functionalized Organometallics. Applications in Synthesis*, Vol. 2 (Ed. P. Knochel), Chap. 12, Wiley-VCH, Weinheim, 2005, pp. 503–540.
7. M. T. Reetz, in *Organometallics in Synthesis. A Manual*, 2nd edn. (Ed. M. Schlosser), Chap. 7, Wiley, Chichester, 2002, pp. 817–923.
8. F. Sato and H. Urabe, in *Titanium and Zirconium in Organic Synthesis* (Ed. I. Marek), Chap. 9, Wiley-VCH, Weinheim, 2002, pp. 319–354.
9. F. Sato, H. Urabe and S. Okamoto, *Chem. Rev.*, **100**, 2835 (2000).
10. M. Arai, B. H. Lipshutz and E. Nakamura, *Tetrahedron*, **48**, 5709 (1992).
11. M. Arai, E. Nakamura and B. H. Lipshutz, *J. Org. Chem.*, **56**, 5489 (1991).
12. J. Kang, W. Cho and W. K. Lee, *J. Org. Chem.*, **49**, 1838 (1984).
13. R. W. Rickards and H. Rönneberg, *J. Org. Chem.*, **49**, 572 (1984).
14. J. Kallmerten, *Tetrahedron Lett.*, **25**, 2843 (1984).
15. H. Urabe and F. Sato, *J. Am. Chem. Soc.*, **121**, 1245 (1999).
16. H. Urabe, T. Hamada and F. Sato, *J. Am. Chem. Soc.*, **121**, 2931 (1999).
17. Y. Gao and F. Sato, *J. Chem. Soc., Chem. Commun.*, 659 (1995).
18. J. J. Eisch and M. W. Foxton, *J. Org. Chem.*, **36**, 3520 (1971).
19. Y. Gao, K. Harada, T. Hata, H. Urabe and F. Sato, *J. Org. Chem.*, **60**, 290 (1995).
20. A. Hassner and J. A. Soderquist, *J. Organomet. Chem.*, **131**, C1 (1977).
21. F. Sato, *J. Organomet. Chem.*, **285**, 53 (1985).
22. B. B. Snider, M. Karras and R. S. E. Conn, *J. Am. Chem. Soc.*, **100**, 4624 (1978).
23. I. Hyla-Kryspin, R. Gleiter, C. Krüeger, R. Zwettler and G. Erker, *Organometallics*, **9**, 517 (1990).
24. K. Oshima, *J. Organomet. Chem.*, **575**, 1 (1999).
25. G. Cahiez and F. Mahuteau-Betzer, in *Handbook of Functionalized Organometallics. Applications in Synthesis*, Vol. 2 (Ed. P. Knochel), Chap. 13, Wiley-VCH, Weinheim, 2005, pp. 541–568.
26. G. Cahiez, C. Duplais and J. Buendia, *Chem. Rev.*, **109**, 1434 (2009).
27. T. Hiyama, M. Obayashi and A. Nakamura, *Organometallics*, **1**, 1249 (1982).
28. A. Fürstner and H. Brunner, *Tetrahedron Lett.*, **37**, 7009 (1996).
29. S.-H. Kim and R. D. Rieke, *J. Org. Chem.*, **63**, 6766 (1998).
30. S.-H. Kim, M. V. Hanson and R. D. Rieke, *Tetrahedron Lett.*, **37**, 2197 (1996).

31. G. Cahiez, A. Martin and T. Delacroix, *Tetrahedron Lett.*, **40**, 6407 (1999).
32. G. Cahiez and B. Laboue, *Tetrahedron Lett.*, **30**, 7369 (1989).
33. G. Cahiez and B. Laboue, *Tetrahedron Lett.*, **33**, 4439 (1992).
34. G. Cahiez, P.-Y. Chavant and E. Metais, *Tetrahedron Lett.*, **33**, 5245 (1992).
35. G. Cahiez and M. Alami, *Tetrahedron Lett.*, **30**, 3541 (1989).
36. S. Marquais, M. Alami and G. Cahiez, *Org. Synth.*, **72**, 135 (1995).
37. A. Alexakis, J. Berlan and Y. Besace, *Tetrahedron Lett.*, **27**, 1047 (1986).
38. G. Cahiez and M. Alami, *Tetrahedron Lett.*, **31**, 7423 (1990).
39. G. Cahiez and M. Alami, *Tetrahedron Lett.*, **31**, 7425 (1990).
40. G. Cahiez and M. Alami, *Tetrahedron*, **45**, 4163 (1989).
41. G. Cahiez and M. Alami, *Tetrahedron Lett.*, **30**, 7365 (1989).
42. M. Gómez-Gallego, M. J. Mancheño and M. A. Sierra, *Acc. Chem. Res.*, **38**, 44 (2005).
43. I. Göttker-Schnetmann, R. Aumann and K. Bergander, *Organometallics*, **20**, 3574 (2001).
44. J. Barluenga, L. A. López, O. Löber, M. Tomás, S. García-Granda, C. Alvarez-Rúa and J. Borge, *Angew. Chem., Int. Ed.*, **40**, 3392 (2001).
45. J. Barluenga, P. Barrio, R. Vicente, L. A. López and M. Tomás, *J. Organomet. Chem.*, **689**, 3793 (2004).
46. E. Negishi, in *Organometallics in Synthesis. A Manual*, 2nd edn. (Ed. M. Schlosser), Chap. 8, Wiley, Chichester, 2002, pp. 925–1002.
47. P. Wipf and H. Jahn, *Tetrahedron*, **52**, 12853 (1996).
48. P. Wipf, W. Xu, H. Takahashi, H. Jahn and P. D. G. Coish, *Pure Appl. Chem.*, **69**, 639 (1997).
49. P. Wipf, H. Takahashi and N. Zhuang, *Pure Appl. Chem.*, **70**, 1077 (1998).
50. B. H. Lipshutz, S. S. Pfeiffer, K. Noson and T. Tomioka, in *Titanium and Zirconium in Organic Synthesis* (Ed. I. Marek), Chap. 4, Wiley-VCH, Weinheim, 2002, pp. 110–148.
51. P. Wipf and C. Kendall, *Top. Organomet. Chem.*, **8**, 1 (2004).
52. J. Schwartz and J. A. Labinger, *Angew. Chem., Int. Ed. Engl.*, **15**, 333 (1976).
53. U. Annby, S. Karlsson, S. Gronowitz, A. Hallberg, J. Alvhäll and R. Svenson, *Acta Chem. Scand.*, **47**, 425 (1993).
54. P. J. Chirik, M. W. Day, J. A. Labinger and J. E. Bercaw, *J. Am. Chem. Soc.*, **121**, 10308 (1999).
55. P. Wipf and R. L. Nunes, *Tetrahedron*, **60**, 1269 (2004).
56. P. C. Wailes, H. Weigold and A. P. Bell, *J. Organomet. Chem.*, **27**, 373 (1971).
57. M. Yoshifuji, M. J. Loots and J. Schwartz, *Tetrahedron Lett.*, 1303 (1977).
58. J. Schwartz, M. J. Loots and H. Kosugi, *J. Am. Chem. Soc.*, **102**, 1333 (1980).
59. B. H. Lipshutz and E. L. Ellsworth, *J. Am. Chem. Soc.*, **112**, 7440 (1990).
60. P. Wipf and J. H. Smitrovich, *J. Org. Chem.*, **56**, 6494 (1991).
61. K. A. Babiak, J. R. Behling, J. H. Dygos, K. T. McLaughlin, J. S. Ng, V. J. Kalish, S. W. Kramer and R. L. Stone, *J. Am. Chem. Soc.*, **112**, 7441 (1990).
62. L. M. Venanzi, R. Lehmann, R. Keil and B. H. Lipshutz, *Tetrahedron Lett.*, **33**, 5857 (1992).
63. B. H. Lipshutz, E. L. Ellsworth and S. H. Dimock, *J. Am. Chem. Soc.*, **112**, 5869 (1990).
64. Y. Yamamoto and K. Maruyama, *J. Am. Chem. Soc.*, **100**, 3240 (1978).
65. Y. Yamamoto, S. Yamamoto, H. Yatagai and K. Maruyama, *J. Am. Chem. Soc.*, **102**, 2318 (1980).
66. T. Gibson, *Organometallics*, **6**, 918 (1987).
67. S. Karlsson, A. Hallberg and S. Gronowitz, *J. Organomet. Chem.*, **403**, 133 (1991).
68. I. Marek, N. Chinkov and A. Levin, *Synlett*, 501 (2006).
69. P. Wipf and W. Xu, *Synlett*, 718 (1992).
70. D. B. Carr and J. Schwartz, *J. Am. Chem. Soc.*, **101**, 3521 (1979).
71. T. Fujisawa, T. Mori, K. Higuchi and T. Sato, *Chem. Lett.*, 1791 (1983).
72. F. Sato, M. Inoue, K. Oguro and M. Sato, *Tetrahedron Lett.*, 4303 (1979).
73. F. Sato, H. Kodama, Y. Tomuro and M. Sato, *Chem. Lett.*, 623 (1979).
74. B. H. Lipshutz and K. Kato, *Tetrahedron Lett.*, **32**, 5657 (1991).
75. B. H. Lipshutz and M. R. Wood, *J. Am. Chem. Soc.*, **116**, 11689 (1994).
76. P. Wipf and H. Takahashi, *Chem. Commun.*, 2675 (1996).
77. P. Wipf, W. Xu, J. H. Smitrovich, R. Lehmann and L. M. Venanzi, *Tetrahedron*, **50**, 1935 (1994).
78. B. H. Lipshutz, R. Keil and E. L. Ellsworth, *Tetrahedron Lett.*, **31**, 7257 (1990).
79. D. R. Swanson, T. Nguyen, Y. Noda and E. Negishi, *J. Org. Chem.*, **56**, 2590 (1991).

80. E. Negishi, J. A. Miller and T. Yoshida, *Tetrahedron Lett.*, **25**, 3407 (1984).
81. P. C. Wailes and H. Weigold, *Inorg. Synth.*, **19**, 223 (1979).
82. S. L. Buchwald, S. J. LaMaire, R. B. Nielsen, B. T. Watson and S. M. King, *Org. Synth.*, **71**, 77 (1993).
83. S. L. Buchwald, S. J. LaMaire, R. B. Nielsen, B. T. Watson and S. M. King, *Tetrahedron Lett.*, **28**, 3895 (1987).
84. D. Naumann, T. Roy, K.-F. Tebbe and W. Crump, *Angew. Chem., Int. Ed. Engl.*, **32**, 1482 (1993).
85. R. Eujen, B. Hoge and D. J. Brauer, *J. Organomet. Chem.*, **519**, 7 (1996).
86. M. A. Willert-Porada, D. J. Burton and N. C. Baenziger, *J. Chem. Soc., Chem. Commun.*, 1633 (1989).
87. E. Nakamura, S. Mori, M. Nakamura and K. Morokuma, *J. Am. Chem. Soc.*, **119**, 4887 (1997).
88. M. Yamanaka, S. Kato and E. Nakamura, *J. Am. Chem. Soc.*, **126**, 6287 (2004).
89. A. El-Batta, T. R. Hage, S. Plotkin and M. Bergdahl, *Org. Lett.*, **6**, 107 (2004).
90. A. El-Batta and M. Bergdahl, *Org. Synth.*, **84**, 192 (2007).
91. M. Eriksson, T. Iliefski, M. Nilsson and T. Olsson, *J. Org. Chem.*, **62**, 182 (1997).
92. H. O. House, C.-Y. Chu, J. M. Wilkins and M. J. Umen, *J. Org. Chem.*, **40**, 1460 (1975).
93. M. Eriksson, A. Hjelmencrantz, M. Nilsson and T. Olsson, *Tetrahedron*, **51**, 12631 (1995).
94. A. Saito, K. Iimura, M. Hayashi and Y. Hanzawa, *Tetrahedron Lett.*, **50**, 587 (2009).
95. Y. Hanzawa, in *Titanium and Zirconium in Organic Synthesis* (Ed. I. Marek), Chap. 5, Wiley-VCH, Weinheim, 2002, pp. 149–179.
96. Y. Hanzawa, K. Narita and T. Taguchi, *Tetrahedron Lett.*, **41**, 109 (2000).
97. Y. Hanzawa, K. Narita, M. Yabe and T. Taguchi, *Tetrahedron*, **58**, 10429 (2002).
98. E. Negishi and T. Takahashi, *Acc. Chem. Res.*, **27**, 124 (1994).
99. E. Negishi, F. E. Cederbaum and T. Takahashi, *Tetrahedron Lett.*, **27**, 2829 (1986).
100. A. Sato, H. Ito and T. Taguchi, *J. Org. Chem.*, **70**, 709 (2005).
101. A. Sato, H. Ito, M. Okada, Y. Nakamura and T. Taguchi, *Tetrahedron Lett.*, **46**, 8381 (2005).
102. T. Takahashi and Y. Li, in *Titanium and Zirconium in Organic Synthesis* (Ed. I. Marek), Chap. 2, Wiley-VCH, Weinheim, 2002, pp. 50–85.
103. T. Takahashi, M. Kotora, K. Kasai, N. Suzuki and K. Nakajima, *Organometallics*, **13**, 4183 (1994).
104. M. Kotora, C. Umeda, T. Ishida and T. Takahashi, *Tetrahedron Lett.*, **38**, 8355 (1997).
105. T. Takahashi, R. Hara, Y. Nishihara and M. Kotora, *J. Am. Chem. Soc.*, **118**, 5154 (1996).
106. T. Takahashi, W.-H. Sun, Y. Liu, K. Nakajima and M. Kotora, *Organometallics*, **17**, 3841 (1998).
107. T. Takahashi, W.-H. Sun, C. Xi, H. Ubayama and Z. Xi, *Tetrahedron*, **54**, 715 (1998).
108. T. Takahashi, Z. Xi, A. Yamazaki, Y. Liu, K. Nakajima and M. Kotora, *J. Am. Chem. Soc.*, **120**, 1672 (1998).
109. Y. Liu, B. Shen, M. Kotora and T. Takahashi, *Angew. Chem., Int. Ed.*, **38**, 949 (1999).
110. C. Xi, M. Kotora, K. Nakajima and T. Takahashi, *J. Org. Chem.*, **65**, 945 (2000).
111. M. J. Totleben, D. P. Curran and P. Wipf, *J. Org. Chem.*, **57**, 1740 (1992).
112. P. Wipf and S. Venkatraman, *J. Org. Chem.*, **58**, 3455 (1993).
113. W. F. Berkowitz and Y. Wu, *Tetrahedron Lett.*, **38**, 3171 (1997).
114. W. F. Berkowitz and Y. Wu, *Tetrahedron Lett.*, **38**, 8141 (1997).
115. E. Nakamura, in *Organometallics in Synthesis. A Manual*, 2nd edn. (Ed. M. Schlosser), Chap. 5, Wiley, Chichester, 2002, pp. 579–664.
116. P. Knochel, H. Leuser, L. Gong, S. Perrone and F. F. Kneisel, in *Handbook of Functionalized Organometallics. Applications in Synthesis*, Vol. 1 (Ed. P. Knochel), Chap. 7, Wiley-VCH, Weinheim, 2005, pp. 251–346.
117. P. Knochel, H. Leuser, L. Gong, S. Perrone and F. F. Kneisel, in *The Chemistry of Organozinc Compounds*, Part 1 (Eds. Z. Rappoport and I. Marek), Chap. 8, John Wiley & Sons, Ltd, Chichester, 2006, pp. 287–393.
118. P. Knochel and R. D. Singer, *Chem. Rev.*, **93**, 2117 (1993).
119. E. Erdik, *Tetrahedron*, **43**, 2203 (1987).
120. A. Vaupel and P. Knochel, *Tetrahedron Lett.*, **35**, 8349 (1994).
121. H. Stadtmüller, R. Lentz, C. E. Tucker, T. Stüdemann, W. Dörner and P. Knochel, *J. Am. Chem. Soc.*, **115**, 7027 (1993).
122. J. P. Gillet, R. Sauvêtre and J. F. Normant, *Synthesis*, 538 (1986).

123. C. E. Tucker, T. N. Majid and P. Knochel, *J. Am. Chem. Soc.*, **114**, 3983 (1992).
124. M. J. Rozema, A. R. Sidduri and P. Knochel, *J. Org. Chem.*, **57**, 1956 (1992).
125. J. Furukawa, N. Kawabata and J. Nishimura, *Tetrahedron Lett.*, 3353 (1966).
126. F. Langer, J. Waas and P. Knochel, *Tetrahedron Lett.*, **34**, 5261 (1993).
127. F. Langer, L. Schwink, A. Devasagayaraj, P.-Y. Chavant and P. Knochel, *J. Org. Chem.*, **61**, 8229 (1996).
128. T. Hjelmgaard and D. Tanner, *Org. Biomol. Chem.*, **4**, 1796 (2006).
129. C. S. Dexter, C. Hunter, R. F. W. Jackson and J. Elliott, *J. Org. Chem.*, **65**, 7417 (2000).
130. E. Erdik and Ö. Ö. Pekel, *J. Organomet. Chem.*, **693**, 338 (2008).
131. T. Jerphagnon, M. G. Pizzuti, A. J. Minnaard and B. L. Feringa, *Chem. Soc. Rev.*, **38**, 1039 (2009).
132. J. Christoffers, G. Koripelly, A. Rosiak and M. Rössle, *Synthesis*, 1279 (2007).
133. B. L. Feringa, R. Naasz, R. Imbos and L. A. Arnold, in *Modern Organocopper Chemistry* (Ed. N. Krause), Chap. 7, Wiley-VCH, Weinheim, 2002, pp. 224–258.
134. F. López, A. J. Minnaard and B. L. Feringa, *Acc. Chem. Res.*, **40**, 179 (2007).
135. T. Thaler and P. Knochel, *Angew. Chem., Int. Ed.*, **48**, 645 (2009).
136. J. A. Marshall, M. Herold, H. S. Eidam and P. Eidam, *Org. Lett.*, **8**, 5505 (2006).
137. R. J. Linderman and J. R. McKenzie, *J. Organomet. Chem.*, **361**, 31 (1989).
138. B. H. Lipshutz, M. R. Wood and R. Tirado, *J. Am. Chem. Soc.*, **117**, 6126 (1995).
139. M. Pineschi, F. Del Moro, F. Gini, A. J. Minnaard and B. L. Feringa, *Chem. Commun.*, 1244 (2004).
140. R. K. Dieter and F. Guo, *J. Org. Chem.*, **74**, 3843 (2009).
141. R. M. Suárez, D. Peña, A. J. Minnaard and B. L. Feringa, *Org. Biomol. Chem.*, **3**, 729 (2005).
142. A. Hajra, N. Yoshikai and E. Nakamura, *Org. Lett.*, **8**, 4153 (2006).
143. E. Nakamura and S. Mori, *Angew. Chem., Int. Ed.*, **39**, 3751 (2000).
144. N. Yoshikai, H. Mashima and E. Nakamura, *J. Am. Chem. Soc.*, **127**, 17978 (2005).
145. C. J. Douglas and L. E. Overman, *Proc. Natl. Acad. Sci. U. S. A.*, **101**, 5363 (2004).
146. M. K. Brown, T. L. May, C. A. Baxter and A. H. Hoveyda, *Angew. Chem., Int. Ed.*, **46**, 1097 (2007).
147. A. W. Hird and A. H. Hoveyda, *J. Am. Chem. Soc.*, **127**, 14988 (2005).
148. K.-S. Lee, M. K. Brown, A. W. Hird and A. H. Hoveyda, *J. Am. Chem. Soc.*, **128**, 7182 (2006).
149. D. Martin, S. Kehrli, M. d'Augustin, H. Clavier, M. Mauduit and A. Alexakis, *J. Am. Chem. Soc.*, **128**, 8416 (2006).
150. A. O. Larsen, W. Leu, C. N. Oberhuber, J. E. Campbell and A. H. Hoveyda, *J. Am. Chem. Soc.*, **126**, 11130 (2004).
151. J. J. Van Veldhuizen, J. E. Campbell, R. E. Giudici and A. H. Hoveyda, *J. Am. Chem. Soc.*, **127**, 6877 (2005).
152. B. Breit and P. Demel, in *Modern Organocopper Chemistry* (Ed. N. Krause), Chap. 6, Wiley-VCH, Weinheim, 2002, pp. 188–223.
153. A. S. E. Karlström and J. Bäckvall, in *Modern Organocopper Chemistry* (Ed. N. Krause), Chap. 8, Wiley-VCH, Weinheim, 2002, pp. 259–288.
154. H. Yorimitsu and K. Oshima, *Angew. Chem., Int. Ed.*, **44**, 4435 (2005).
155. K. Tissot-Croset, D. Polet and A. Alexakis, *Angew. Chem., Int. Ed.*, **43**, 2426 (2004).
156. N. Harrington-Frost, H. Leuser, M. I. Calaza, F. F. Kneisel and P. Knochel, *Org. Lett.*, **5**, 2111 (2003).
157. H. Leuser, S. Perrone, F. Liron, F. F. Kneisel and P. Knochel, *Angew. Chem., Int. Ed.*, **44**, 4627 (2005).
158. S. Perrone and P. Knochel, *Org. Lett.*, **9**, 1041 (2007).
159. B. Breit, P. Demel and C. Studte, *Angew. Chem., Int. Ed.*, **43**, 3786 (2004).
160. B. Breit, P. Demel, D. Grauer and C. Studte, *Chem. Asian J.*, **1**, 586 (2006).
161. B. Breit and Y. Schmidt, *Chem. Rev.*, **108**, 2928 (2008).
162. P. Wipf and P. C. Fritch, *J. Org. Chem.*, **59**, 4875 (1994).
163. P. Wipf and J. Xiao, *Org. Lett.*, **7**, 103 (2005).
164. J. Xiao, B. Weisblum and P. Wipf, *Org. Lett.*, **8**, 4731 (2006).
165. H. Tamamura, K. Hiramatsu, S. Ueda, Z. Wang, S. Kusano, S. Terakubo, J. O. Trent, S. C. Peiper, N. Yamamoto, H. Nakashima, A. Otaka and N. Fujii, *J. Med. Chem.*, **48**, 380 (2005).

166. M. A. Kacprzynski, T. L. May, S. A. Kazane and A. H. Hoveyda, *Angew. Chem., Int. Ed.*, **46**, 4554 (2007).
167. A. D. Duursma, A. J. Minnaard and B. L. Feringa, *J. Am. Chem. Soc.*, **125**, 3700 (2003).
168. A. Rimkus and N. Sewald, *Org. Lett.*, **5**, 79 (2003).
169. H. Choi, Z. Hua and I. Ojima, *Org. Lett.*, **6**, 2689 (2004).
170. G. K. Friestad and A. K. Mathies, *Tetrahedron*, **63**, 2541 (2007).
171. K.-i. Yamada and K. Tomioka, *Chem. Rev.*, **108**, 2874 (2008).
172. A. A. Boezio and A. B. Charette, *J. Am. Chem. Soc.*, **125**, 1692 (2003)
173. A. A. Boezio, J. Pytkowicz, A. Côté and A. B. Charette, *J. Am. Chem. Soc.*, **125**, 14260 (2003).
174. C. Lauzon and A. B. Charette, *Org. Lett.*, **8**, 2743 (2006).
175. J. Esquivias, R. G. Arrayás and J. C. Carretero, *Angew. Chem., Int. Ed.*, **46**, 9257 (2007).
176. M. G. Pizzuti, A. J. Minnaard and B. L. Feringa, *J. Org. Chem.*, **73**, 940 (2008).
177. S. R. Harutyunyan, F. López, W. R. Browne, A. Correa. D. Peña, R. Badorrey, A. Meetsma, A. J. Minnaard and B. L. Feringa, *J. Am. Chem. Soc.*, **128**, 9103 (2006).
178. A. M. Berman and J. S. Johnson, *J. Org. Chem.*, **71**, 219 (2006).
179. A. M. Berman and J. S. Johnson, *Org. Synth.*, **83**, 31 (2006).
180. P. Beak, *Acc. Chem. Res.*, **25**, 215 (1992).
181. M. J. Campbell and J. S. Johnson, *Org. Lett.*, **9**, 1521 (2007).
182. D. E. Bergbreiter and G. M. Whitesides, *J. Am. Chem. Soc.*, **96**, 4937 (1974).
183. P. Kočovský and J. Šrogl, *J. Org. Chem.*, **57**, 4565 (1992).
184. P. Kočovský, J. Šrogl, M. Pour and A. Gogoll, *J. Am. Chem. Soc.*, **116**, 186 (1994).
185. N. Miyaura, M. Itoh and A. Suzuki, *Tetrahedron Lett.*, 255 (1976).
186. N. Miyaura, N. Sasaki, M. Itoh and A. Suzuki, *Tetrahedron Lett.*, 173 (1977).
187. N. Miyaura, N. Sasaki, M. Itoh and A. Suzuki, *Tetrahedron Lett.*, 3369 (1977).
188. H. Yatagai, *J. Org. Chem.*, **45**, 1640 (1980).
189. J. B. Campbell and H. C. Brown, *J. Org. Chem.*, **45**, 549 (1980).
190. P. Knochel, *Synlett*, 393 (1995).
191. Y. Yamamoto, H. Yatagai, A. Sonoda and S.-I. Murahashi, *J. Chem. Soc., Chem. Commun.*, 452 (1976).
192. Y. Yamamoto, H. Yatagai, K. Maruyama, A. Sonoda and S.-I. Murahashi, *J. Am. Chem. Soc.*, **99**, 5652 (1977).
193. J. Ichikawa, S. Hamada, T. Sonoda and H. Kobayashi, *Tetrahedron Lett.*, **33**, 337 (1992).
194. J. Ichikawa, T. Minami, T. Sonoda and H. Kobayashi, *Tetrahedron Lett.*, **33**, 3779 (1992).
195. J. Ichikawa, *J. Fluorine Chem.*, **105**, 257 (2000).
196. H. C. Brown and J. B. Campbell, *J. Org. Chem.*, **45**, 550 (1980).
197. H. C. Brown and G. A. Molander, *J. Org. Chem.*, **46**, 645 (1981).
198. S. Hara, Y. Satoh and A. Suzuki, *Chem. Lett.*, 1289 (1982).
199. K. Takahashi, T. Ishiyama and N. Miyaura, *Chem. Lett.*, 982 (2000).
200. K. Takahashi, T. Ishiyama and N. Miyaura, *J. Organomet. Chem.*, **625**, 47 (2001).
201. H. Ito, H. Yamanaka, J.-i. Tateiwa and A. Hosomi, *Tetrahedron Lett.*, **41**, 6821 (2000).
202. P. V. Ramachandran, D. Pratihar, D. Biswas, A. Srivastava and M. V. R. Reddy, *Org. Lett.*, **6**, 481 (2004).
203. P. V. Ramachandran and D. Pratihar, *Org. Lett.*, **9**, 2087 (2007).
204. H. Ito, C. Kawakami and M. Sawamura, *J. Am. Chem. Soc.*, **127**, 16034 (2005).
205. H. Ito, S. Ito, Y. Sasaki, K. Matsuura and M. Sawamura, *Pure Appl. Chem.*, **80**, 1039 (2008).
206. H. Ito, Y. Kosaka, K. Nonoyama, Y. Sasaki and M. Sawamura, *Angew. Chem., Int. Ed.*, **47**, 7424 (2008).
207. H. Ito, Y. Sasaki and M. Sawamura, *J. Am. Chem. Soc.*, **130**, 15774 (2008).
208. G. Zweifel and R. L. Miller, *J. Am. Chem. Soc.*, **92**, 6678 (1970).
209. R. A. Lynd and G. Zweifel, *Synthesis*, 658 (1974).
210. M. Nakada, S.-i. Nakamura, S. Kobayashi and M. Ohno, *Tetrahedron Lett.*, **32**, 4929 (1991).
211. K. Takai, K. Oshima and H. Nozaki, *Bull. Chem. Soc. Jpn.*, **54**, 1281 (1981).
212. R. E. Ireland and P. Wipf, *J. Org. Chem.*, **55**, 1425 (1990).
213. P. Wipf, J. H. Smitrovich and C. Moon, *J. Org. Chem.*, **57**, 3178 (1992).
214. R. Baker and J. L. Castro, *J. Chem. Soc., Chem. Commun.*, 378 (1989).

215. S. V. Ley, N. J. Anthony, A. Armstrong, M. G. Brasca, T. Clarke, D. Culshaw, C. Greck, P. Grice, A. B. Jones, B. Lygo, A. Madin, R. N. Sheppard, A. M. Z. Slawin and D. J. Williams, *Tetrahedron*, **45**, 7161 (1989).
216. A. Marfat, P. R. McGuirk and P. Helquist, *J. Org. Chem.*, **44**, 3888 (1979).
217. B. H. Lipshutz and S. H. Dimock, *J. Org. Chem.*, **56**, 5761 (1991).
218. J. Westermann and K. Nickisch, *Angew. Chem., Int. Ed. Engl.*, **32**, 1368 (1993).
219. J. Kabbara, S. Flemming, K. Nickisch, H. Neh and J. Westermann, *Tetrahedron Lett.*, **35**, 8591 (1994).
220. J. Kabbara, S. Flemming, K. Nickisch, H. Neh and J. Westermann, *Chem. Ber.*, **127**, 1489 (1994).
221. J. Kabbara, S. Flemming, K. Nickisch, H. Neh and J. Westermann, *Tetrahedron*, **51**, 743 (1995).
222. J. Kabbara, S. Flemming, K. Nickisch, H. Neh and J. Westermann, *Synlett*, 679 (1994).
223. A. Alexakis, V. Albrow, K. Biswas, M. d'Augustin, O. Prieto and S. Woodward, *Chem. Commun.*, 2843 (2005).
224. T. L. May, M. K. Brown and A. H. Hoveyda, *Angew. Chem., Int. Ed.*, **47**, 7358 (2008).
225. M. d'Augustin, L. Palais and A. Alexakis, *Angew. Chem., Int. Ed.*, **44**, 1376 (2005).
226. M. Vuagnoux-d'Augustin and A. Alexakis, *Chem. Eur. J.*, **13**, 9647 (2007).
227. C. Hawner, K. Li, V. Cirriez and A. Alexakis, *Angew. Chem., Int. Ed.*, **47**, 8211 (2008).
228. Y. Lee, K. Akiyama, D. G. Gillingham, M. K. Brown and A. H. Hoveyda, *J. Am. Chem. Soc.*, **130**, 446 (2008).
229. E. J. Corey and R. H. Wollenberg, *J. Am. Chem. Soc.*, **96**, 5581 (1974).
230. E. J. Corey and R. H. Wollenberg, *J. Org. Chem.*, **40**, 2265 (1975).
231. J. R. Behling, K. A. Babiak, J. S. Ng, A. Campbell, R. Moretti, M. Koerner and B. H. Lipshutz, *J. Am. Chem. Soc.*, **110**, 2641 (1988).
232. B. H. Lipshutz and J. I. Lee, *Tetrahedron Lett.*, **32**, 7211 (1991).
233. B. H. Lipshutz, R. Crow, S. H. Dimock, E. L. Ellsworth, R. A. J. Smith and J. R. Behling, *J. Am. Chem. Soc.*, **112**, 4063 (1990).
234. B. H. Lipshutz, D. C. Reuter and E. L. Ellsworth, *J. Org. Chem.*, **54**, 4975 (1989).
235. B. H. Lipshutz, S. Sharma and D. C. Reuter, *Tetrahedron Lett.*, **31**, 7253 (1990).
236. E. Piers and T. Wong, *J. Org. Chem.*, **58**, 3609 (1993).
237. E. Piers, E. J. McEachern and P. A. Burns, *J. Org. Chem.*, **60**, 2322 (1995).
238. E. Piers, E. J. McEachern and P. A. Burns, *Tetrahedron*, **56**, 2753 (2000).
239. M.-L. Bennasar, C. Juan, T. Roca, M. Monerris and J. Bosch, *Tetrahedron*, **57**, 10125 (2001).
240. H. Ito, K. Arimoto, H.-o. Sensui and A. Hosomi, *Tetrahedron Lett.*, **38**, 3977 (1997).
241. Y. Nishihara, M. Takemura, A. Mori and K. Osakada, *J. Organomet. Chem.*, **620**, 282 (2001).
242. H. Ito, T. Ishizuka, J.-i. Tateiwa, M. Sonoda and A. Hosomi, *J. Am. Chem. Soc.*, **120**, 11196 (1998).
243. K.-s. Lee and A. H. Hoveyda, *J. Org. Chem.*, **74**, 4455 (2009).
244. J. V. Comasseto and J. N. Berriel, *Synth. Commun.*, **20**, 1681 (1990).
245. J. P. Marino, F. Tucci and J. V. Comasseto, *Synlett*, 761 (1993).
246. F. C. Tucci, A. Chieffi, J. V. Comasseto and J. P. Marino, *J. Org. Chem.*, **61**, 4975 (1996).
247. Y. Chounan and Y. Yamamoto, in *Modern Organocopper Chemistry* (Ed. N. Krause), Chap. 9, Wiley-VCH, Weinheim, 2002, pp. 289–314.
248. E. J. Corey and A. Tramontano, *J. Am. Chem. Soc.*, **106**, 462 (1984).
249. Y. Horiguchi, E. Nakamura and I. Kuwajima, *J. Org. Chem.*, **51**, 4323 (1986).
250. Y. Horiguchi, E. Nakamura and I. Kuwajima, *J. Am. Chem. Soc.*, **111**, 6257 (1989).
251. R. E. Ireland, P. Wipf and T. D. Roper, *J. Org. Chem.*, **55**, 2284 (1990).
252. R. E. Ireland, J. L. Gleason, L. D. Gegnas and T. K. Highsmith, *J. Org. Chem.*, **61**, 6856 (1996).
253. R. E. Ireland, L. Liu and T. D. Roper, *Tetrahedron*, **53**, 13221 (1997).
254. H. Stadtmüller and P. Knochel, *Synlett*, 463 (1995).
255. J. H. Rigby and M. Kirova-Snover, *Tetrahedron Lett.*, **38**, 8153 (1997).
256. S. E. Denmark and L. R. Marcin, *J. Org. Chem.*, **62**, 1675 (1997).
257. S. C. Roy, L. Nagarajan and R. G. Salomon, *J. Org. Chem.*, **64**, 1218 (1999).
258. D. R. Williams and K. Shamim, *Org. Lett.*, **7**, 4161 (2005).
259. D. G. Gillingham and A. H. Hoveyda, *Angew. Chem., Int. Ed.*, **46**, 3860 (2007).

260. J. Guo, K. J. Duffy, K. L. Stevens, P. I. Dalko, R. M. Roth, M. M. Hayward and Y. Kishi, *Angew. Chem., Int. Ed.*, **37**, 187 (1998).
261. M. M. Hayward, R. M. Roth, K. J. Duffy, P. I. Dalko, K. L. Stevens, J. Guo and Y. Kishi, *Angew. Chem., Int. Ed.*, **37**, 192 (1998).
262. M. K. Brown and A. H. Hoveyda, *J. Am. Chem. Soc.*, **130**, 12904 (2008).
263. G. Barbe and A. Charette, *J. Am. Chem. Soc.*, **130**, 13873 (2008).

CHAPTER **11**

Carbocupration of alkynes

FABRICE CHEMLA and FRANCK FERREIRA

Institut Parisien de Chimie Moléculaire, UPMC-Univ Paris 6, CNRS UMR 7201, Tour 44–45, 2ème Etage, Case 183, 4 Place Jussieu, 75252 Paris Cedex 05, France Fax: +33 1 44 27 75 67; e-mail: fabrice.chemla@upmc.fr, franck.ferreira@upmc.fr

I. GENERAL INTRODUCTION . . . 528
II. CARBOCUPRATION OF ACETYLENE . . . 529
III. CARBOCUPRATION OF SUBSTITUTED ALKYNES . . . 539
A. Alkyl-substituted Alkynes . . . 539
1. Non-functionalized alkynes . . . 539
2. Functionalized alkynes . . . 543
a. Propargyl amines, ethers and thioethers . . . 543
b. Alkynyl epoxides . . . 545
c. Alkynyl acetals . . . 545
d. Alkynyl orthoesters . . . 547
e. Fluoroalkylated alkynes . . . 548
B. Carbocupration and Copper-catalyzed *Carbomagnesiation* Reaction of Propargyl Alcohols . . . 550
C. Enynes and Arylacetylenes . . . 555
D. Heteroatom-substituted Alkynes . . . 557
1. Alkynyl ethers and *O*-alkynyl carbamates . . . 557
a. Alkynyl ethers . . . 557
b. *O*-Alkynyl carbamates . . . 560
2. Alkynyl thioethers, sulfoxides, sulfones and sulfoximines . . . 560
a. Alkynyl thioethers . . . 560
b. Alkynyl sulfoxides, sulfones and sulfoximines . . . 563
3. Alkynyl amines, amides and *N*-alkynyl carbamates . . . 567
a. Alkynyl amines . . . 567
b. Alkynyl amides and *N*-alkynyl carbamates . . . 567
4. Alkynyl phosphines, phosphonates and phosphine oxides . . . 569
a. Alkynyl phosphines . . . 569
b. Alkynyl phosphonates and phosphine oxides . . . 569
5. Alkynyl silanes . . . 571
E. Intramolecular Carbocupration Reactions . . . 572

The chemistry of organocopper compounds
Edited by Z. Rappoport and I. Marek

1. Synthesis of carbocycles 572
a. Non-stabilized organocopper reagents 572
b. Stabilized organocopper reagents 574
2. Synthesis of heterocycles 576
IV. CONCLUSION 579
V. REFERENCES 581

I. GENERAL INTRODUCTION

The addition of organocopper reagents to alkynes (i.e. 'carbocupration') is one of the most useful carbometallation reactions, as it allows the straightforward synthesis of stereodefined alkenyl copper reagents. Therefore, it represents a powerful tool for the stereoselective preparation of di-, tri- and tetra-substituted olefins. The reaction can be conducted with various organocopper reagents, including organocopper species from Grignard reagents $RCu{\bullet}MgBr_2$, Gilman-type cuprates $R_2CuLi{\bullet}LiX$ and organocopper–zinc mixed species RCu(CN)ZnX. The carbocupration reaction occurs in most cases in a strict *syn*-addition pathway (Scheme 1). Throughout this review, the [Cu] notation will be used when no information concerning the reactive copper(I)-based species is available.

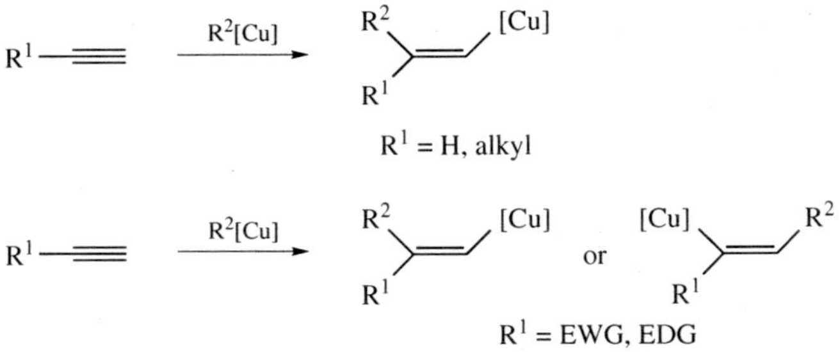

SCHEME 1

The regioselectivity is related to the substitution pattern of the alkyne partner. With acetylene and simple monosubstituted alkylalkynes, the addition occurs to lead to the less substituted (or branched) alkenyl copper species. However, the presence of a donor or acceptor group on the acetylenic moiety or close to it can interfere with this general pattern and can lead to a different regioselectivity.

The case of the acetylene carbocupration will be examined first, including the large applications of this reaction in the field of total synthesis. The carbocupration reaction of alkyl-substituted acetylenes will then be summarized. Reactions involving alkynes substituted with a heteroatom and intramolecular carbocuprations will be examined separately. The special case of the copper-catalyzed *carbomagnesiation* of propargyl alcohols will also be examined, although this reaction is not strictly a carbocupration reaction.

The carbocupration reaction of acetylenes has been reviewed one time specifically[1], and the most important features of this reaction have appeared as a part of an important review concerning the reactivity of the organocopper species[2]. An overview of this reaction, amongst other carbometallation reactions, has appeared several times[3,4] and some aspects of this approach have also been overviewed in more specific reviews[5–7].

II. CARBOCUPRATION OF ACETYLENE

The addition of organocopper reagents onto acetylene is one of the most efficient ways to prepare (*Z*)-alkenyl organometallic species and has thus found a widespread audience in the field of organic synthesis. Early works have shown that organocopper reagents RCu•LiI[8], prepared from organolithiums, or $RCu \cdot MgBr_2$[9,10], prepared from Grignard reagents, add to acetylene in low polarity solvents (in THF or HMPA, deprotonation of acetylene is observed[9]). However, the best reagents to achieve the carbocupration reaction of acetylene[11] are Gilman organocuprates R_2CuLi derived from transmetallation reaction of two equivalents of organolithium reagents with one equivalent of CuI in Et_2O (Scheme 2). The carbometallation reaction occurs smoothly at low temperature and both alkyl (or R) groups are transferred to give the corresponding alkenyl cuprate reagents **1a**. More recently, it was found[12] that some mixed zinc–copper organometallic species can also undergo the carbometallation reaction of acetylene to give the corresponding alkenyl mixed species **1b**.

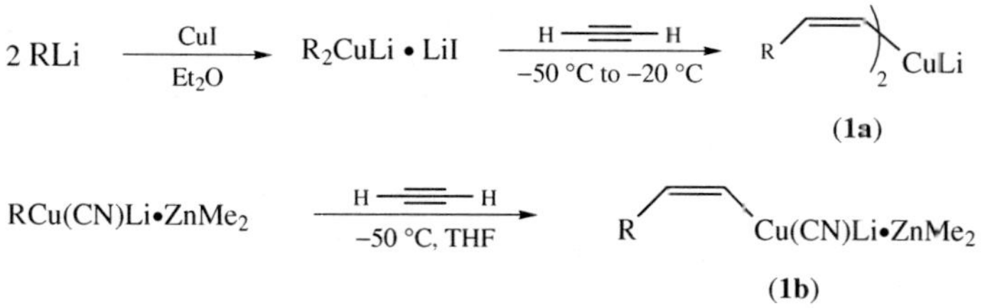

(**1a**)

(**1b**)

SCHEME 2

Mixed organocuprate reagents (obtained from transmetallation of organolithium reagents with PhSCu or copper acetylides) could also be used[8].

A large array of alkyl groups (mainly primary) can be introduced by this reaction. Several functionalized organolithium reagents such as **2** can be successfully used in this carbocupration reaction, thus providing a general access to functionalized alkenyl copper reagents (Scheme 3)[13,14].

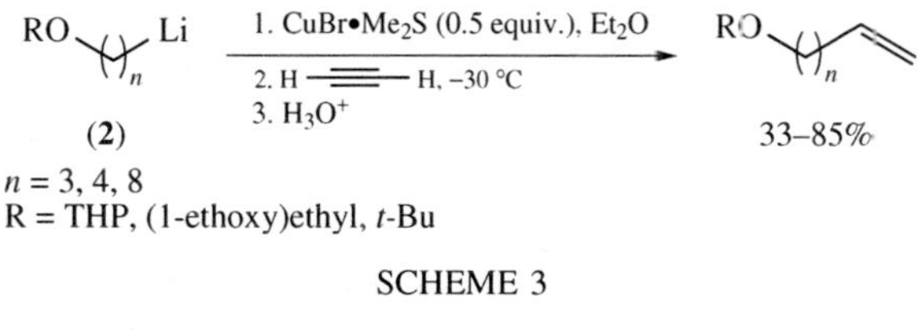

SCHEME 3

Methyl-, phenyl-, allyl- and alkynyl-organocopper reagents fail to react with acetylene. 2-Substituted alkenyl copper reagents **1a** (Scheme 4) react slowly[15], allowing the carbocupration of acetylene to stop after the first addition. The same behavior is observed starting from alkyl copper reagents RCu[10]. However, the double addition can be achieved in good yields by warming the reaction mixture to 0 °C and by adding acetylene in excess[16,17]. Using these modified conditions, the corresponding (*Z,Z*)-*dienyl* cuprate **3** can be obtained and further functionalized (Scheme 4). 1-Substituted alkenyl cuprates and 1,2-disubstituted alkenyl cuprates **4** have also been shown to react with acetylene, thus providing an excellent and stereospecific preparation of conjugated dienes.

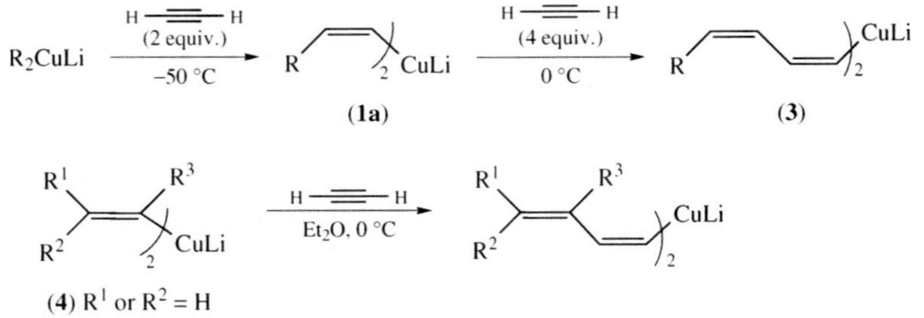

SCHEME 4

The alkenyl organocopper species resulting from the carbocupration of acetylene can be further functionalized with various electrophiles[1]. *Iodolysis* leads to the corresponding alkenyl iodides with complete retention of the configuration, affording a totally stereocontrolled access to monosubstituted (Z)-alkenyl iodides **5** (Scheme 5)[11]. Functionalized organolithiums such as **6a** can be used[14], as well as some functionalized organocopper–zinc mixed species such as **6b**[12]. In the same way, bromination and chlorination can be achieved using NBS and NCS, respectively[18,19].

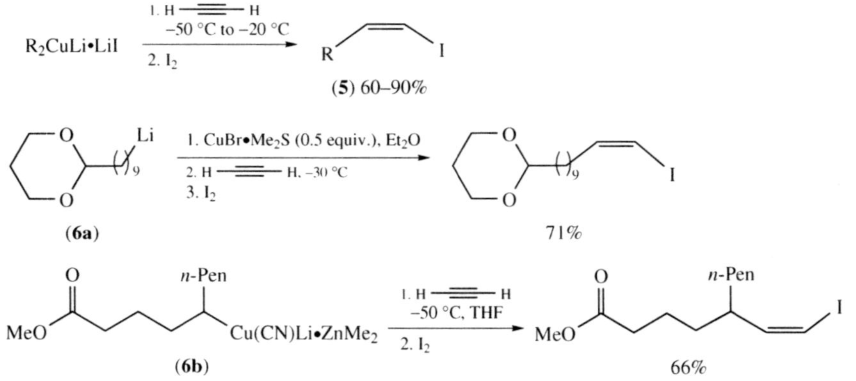

SCHEME 5

2-Substituted alkenyl cuprates derived from carbocupration of acetylene can also be functionalized through aminomethylation[20], thiomethylation[21], carboxylation[13,22,23] or Pd-catalyzed acylation[24] reactions. Several representative examples are listed in Scheme 6.

Alkenyl cuprate species derived from the carbocupration of acetylene have also been shown to react with *epoxides*. However, under standard conditions, only one alkenyl group is transferred, due to the lower reactivity of the heteroalkenyl cuprate species resulting from the first reaction (Scheme 7). The addition of hex-1-ynyl lithium **(7)** results in the formation of a new and more reactive cuprate species, allowing better yields in this alkylation reaction.

The same type of moderate reactivity is found in the reaction with *aldehydes*. Reaction of alkenyl organocuprates with formaldehyde results in the formation of the corresponding allylic alcohol **8** (Scheme 8)[22]. However, in the case of primary aliphatic aldehyde **10**, the addition reaction occurs in good yield only by using the mixed organocuprate **9** in the presence of $BF_3\bullet Et_2O$[25].

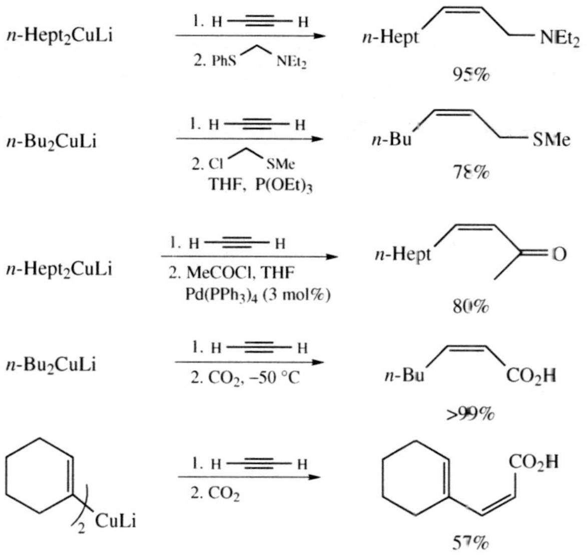

SCHEME 6

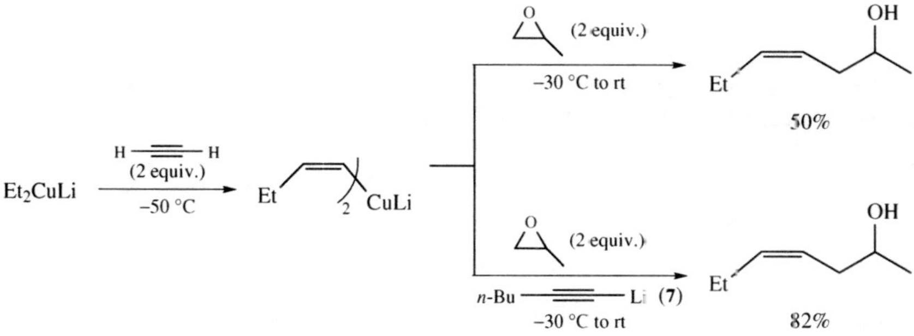

SCHEME 7

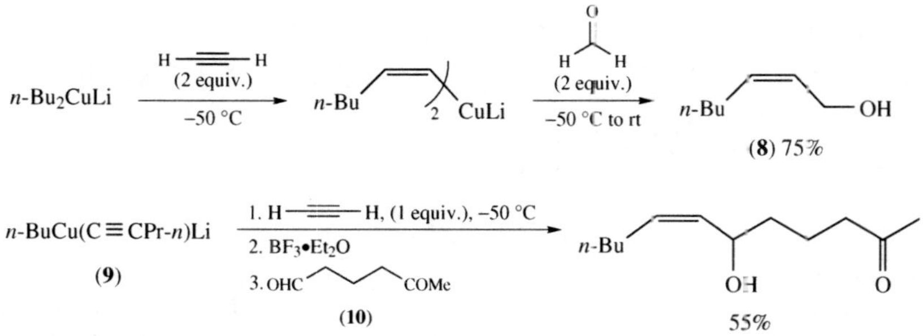

SCHEME 8

The reactivity in alkylation reactions of alkenyl cuprates derived from the carbocupration of acetylene reaction has also been studied. Lithium di-alkenyl cuprates can react twice, but the second alkylation reaction is much slower than the first one[26]. To obtain good yields for the second alkylation step, the addition of HMPA and $P(OEt)_3$ is necessary for simple alkyl bromides and iodides. Alkyl chlorides do not react at all (Scheme 9). *Dienyl* cuprates can also react[16], as well as functionalized alkenyl cuprates[14].

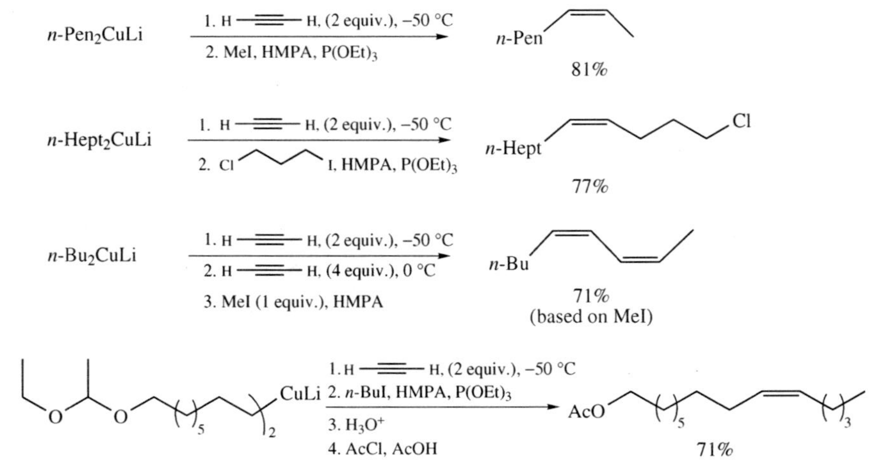

SCHEME 9

In the case of more reactive halides (allyl halides, benzyl halides), the addition of phosphite is not necessary to obtain high yields for the alkylation products (Scheme 10)[16,17,26].

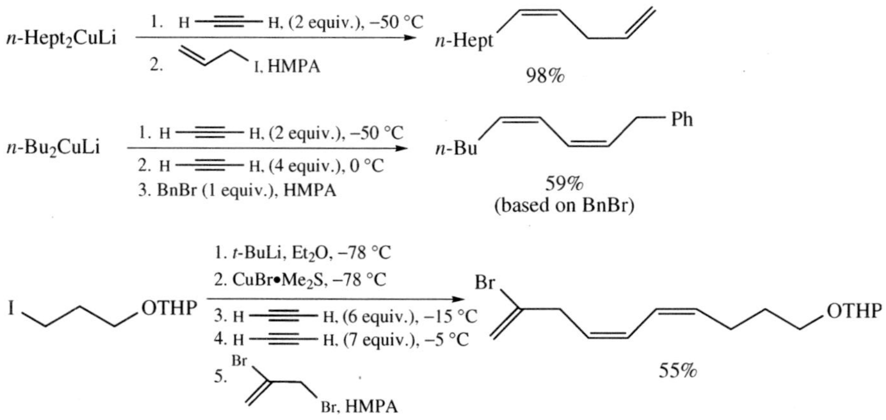

SCHEME 10

Another synthetically interesting reaction of vinyl copper species with electrophiles is the *coupling* reaction with alkenyl iodides. The reaction has been shown to occur smoothly and stereospecifically under Pd(0) catalysis, after the addition of zinc salts (Scheme 11)[13,14]. The *coupling* reaction with alkynyl halides occurs in moderate yields,

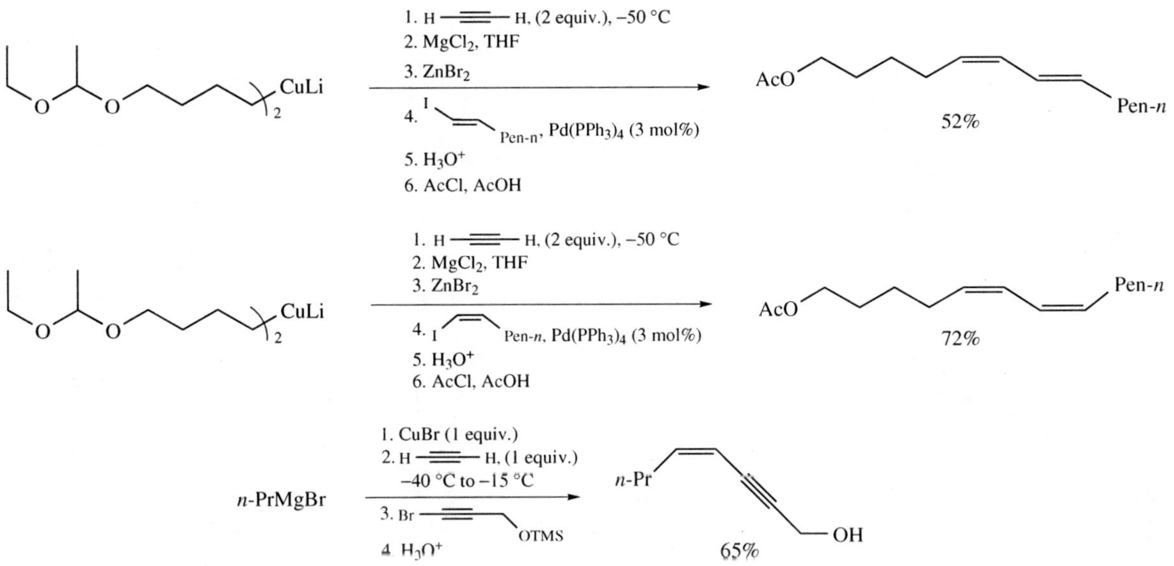

CuLi
2
1. H ─≡─ H, (2 equiv.), −50 °C
2. MgCl2, THF
3. ZnBr2
4. I, Pen-n, Pd(PPh3)4 (3 mol%)
5. H3O+
6. AcCl, AcOH
AcO
Pen-n
52%
CuLi
2
1. H ─≡─ H, (2 equiv.), −50 °C
2. MgCl2, THF
3. ZnBr2
4. I, Pen-n, Pd(PPh3)4 (3 mol%)
5. H3O+
6. AcCl, AcOH
AcO
Pen-n
72%
n-PrMgBr
1. CuBr (1 equiv.)
2. H ─≡─ H, (1 equiv.)
−40 °C to −15 °C
3. Br, OTMS
4. H3O+
n-Pr
OH
65%

SCHEME 11

due to competitive halogen–metal exchange, especially in the case of lithium dialkenyl cuprates[27,28]. If the carbocupration of acetylene is run from the Grignard reagent, the corresponding (*Z*)-enyne is obtained in fair yield.

Partly due to the lack of information concerning the real structure of organocopper clusters and the difficulty to illustrate any intermediate, the mechanism of the carbocupration of acetylene is still under debate. However, a number of theoretical *ab initio* calculations have been conducted by Nakamura and coworkers[29–33].

Carbocupration of acetylene has been studied systematically through calculations for five model species: MeCu, $[Me_2Cu]$, Me_2CuLi, $Me_2CuLi{\bullet}LiCl$ and $(Me_2CuLi)_2$. All these complexes have been invoked in the discussions concerning organocuprate mechanisms. Some general conclusions have been obtained for the reactivity of these reagents with π-acceptors. MeCu can only undergo addition by a four-centered mechanism, but this path requires high energy as the covalent Me–Cu bond (13.16 kJ mol^{-1}) must be cleaved. The neutral RCu species is therefore not a reactive nucleophile. Lithium organocuprates such as $(R_2CuLi)_2$ bind tightly to acetylene through two-electron donation from the copper atom. The proposed mechanism involves the intermediacy of such a Cu(III) cluster **11** formed from the reaction between lithium organocuprate and acetylene. After reductive elimination, the alkenyl lithium–Cu(I) complex **12** is obtained and undergoes transmetallation to the carbocupration product **13** (Scheme 12).

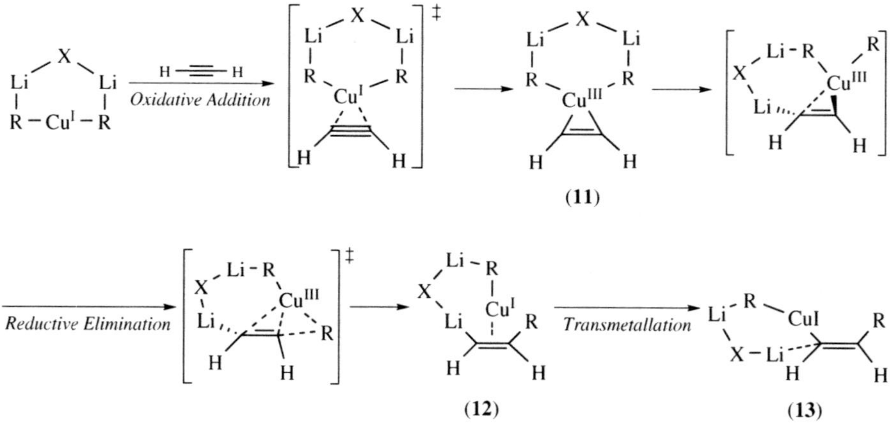

SCHEME 12

The overall pathway can be viewed as a Cu(I)-assisted carbolithiation followed by an intracluster transmetallation reaction. These calculations have been corroborated by the spectroscopic observation of related ynoates–Cu(I)[34,35] and enone–Cu(I) complexes[36] as well as by the recent isolation and spectroscopic characterization of organocopper(III) complexes[37–40].

The synthetic applications of the carbocupration reaction of acetylene have taken advantage of its high stereoselectivity in forming (*Z*)-alkenyl metals and of the good reactivity of the resulting alkenyl copper species with various electrophiles. Not surprisingly, the main field of application is related to the insect *pheromone* total synthesis[41], as numerous *pheromones* present one or more disubstitued alkene functionalities with a specific (*E*)- or (*Z*)-configuration.

The easy stereoselective formation of 1,2-disubstituted (z)-alkenes was applied early in the total synthesis of *pheromone* of *Cossus Cossus* **14** (Scheme 13)[42,43], *Eupoecilia Ambiguella* **15** and of the Douglas-fir tussock moth *Orgyia Pseudotsugata* **16**, as well as in an advanced intermediate towards *exo*-brevicomin. In all these syntheses, a carbocupration–alkylation sequence is applied, leading in a few steps to the *pheromones*. One main advantage of the carbocupration reaction is its ability to be performed on a large scale (180 mmol).

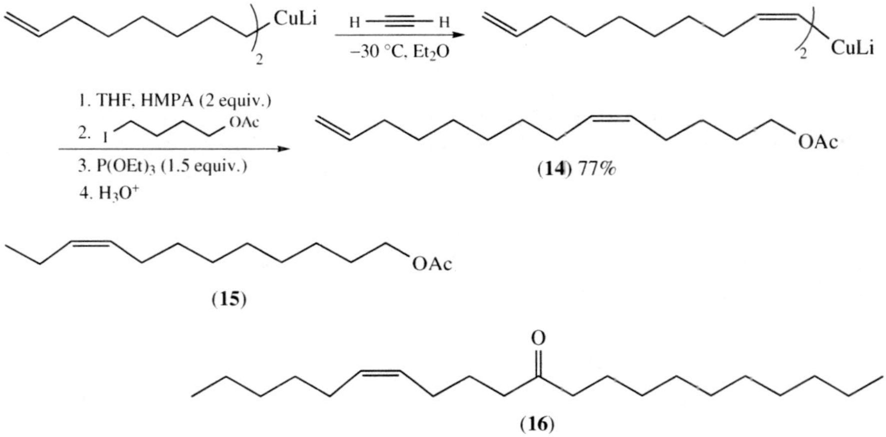

SCHEME 13

The acetylene carbocupration–1,4-addition sequence has been applied to the *pheromone* of peach fruit moth *Carposia Niponensis* **17** in high overall yield (Scheme 14)[44].

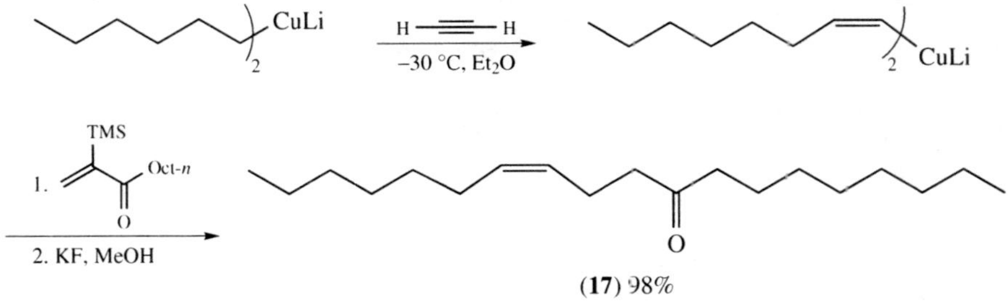

SCHEME 14

A stereoselective access to the sex *pheromone* of *Phtorimaea Operculella* **18**, presenting a (Z,E)-1,4-diene functionality, has been disclosed following an acetylene carbocupration-*epoxide* opening (Scheme 15)[45].

The Pd(0)-catalyzed *coupling* reaction of alkenyl cuprates prepared through the carbocupration of acetylene with (E)-alkenyl halides is a highly efficient method to prepare stereospecifically (Z,E)-1,3-dienes. This approach has been also applied several times for the synthesis of insect *pheromones*. For example, the *pheromone* from *Lobesia Botrana*

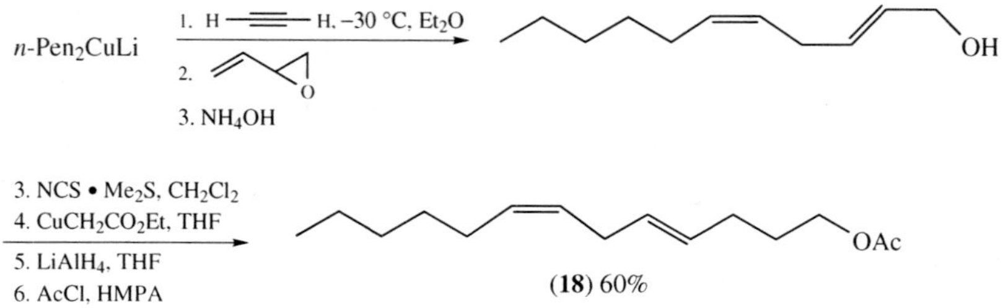

SCHEME 15

19, as well as those from *Bombyx Mori* **20**, *Malascoma Disstria* **21** and the sex *pheromone* **22** from the egyptian cotton leafworm *Spodoptera Littoralis* and from *Stenoma Cecropia*, have been prepared following this approach (Scheme 16)[46].

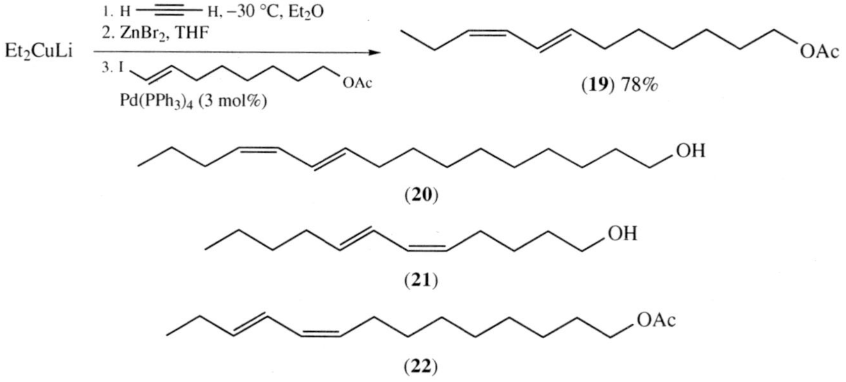

SCHEME 16

The same approach using Ni catalysis has been used to prepare $(1,3E,5Z)$-undecatriene **23**, which is the main odoriferant component of the essential oils of *Galbanum* (Scheme 17)[47].

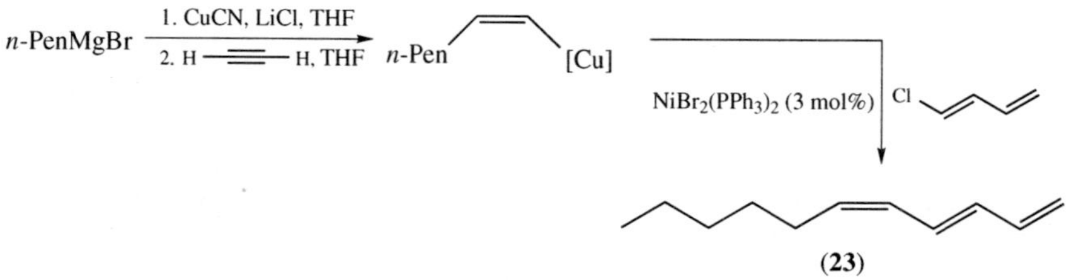

SCHEME 17

Alternative approaches to the (E,Z)-1,3-diene system involve the reduction of the enynol **24** prepared through a carbocupration reaction of acetylene followed by a *coupling* reaction with an alkynyl bromide. This was used in an alternate approach to bombykol **20** (Scheme 18)[27,28].

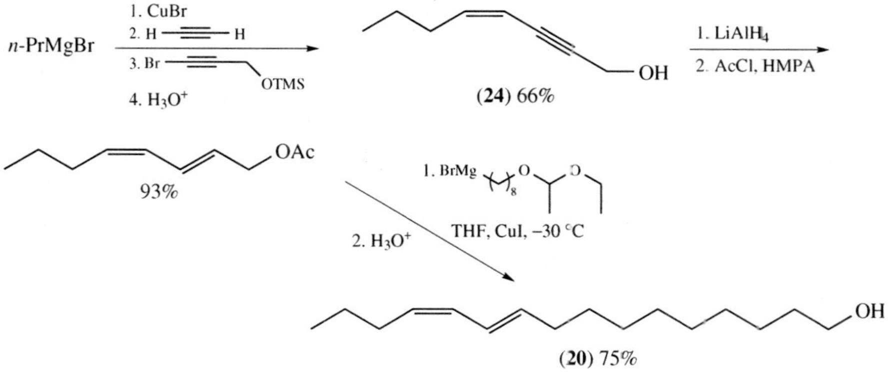

SCHEME 18

Another efficient possibility is to perform the addition of the alkenyl copper reagents, derived from the carbocupration reaction of acetylene, onto phenylthioacetylene to obtain the corresponding (Z,E)-dienyl phenyl sulfide. This approach has been applied to the synthesis of *pheromones* **19–22**[48]. An example of this methodology is depicted in the case of the synthesis of *pheromone* **19** (Scheme 19).

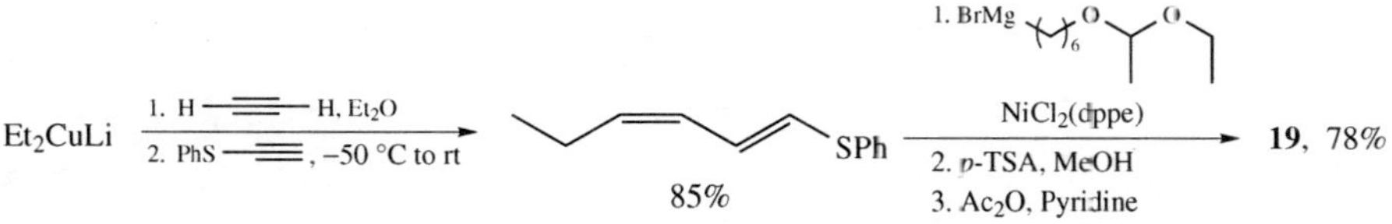

SCHEME 19

Although (Z,Z)-1,3-dienes can also be prepared through *coupling* of (Z)-alkenyl cuprates with (Z)-alkenyl halides, partial isomerization reactions or low yields limit the synthetic utility of this approach[46]. Better results are obtained through double carbocupration of acetylene, as described above. This methodology has been applied to the synthesis of navel orangeworm *pheromone* **25**[16,25] and of the thermally unstable sex *pheromone* **26** from the stink bug *Thyanta Pallidovirens* (Scheme 20)[49]. It has also been applied to a synthetic approach of Ajudazol A[17].

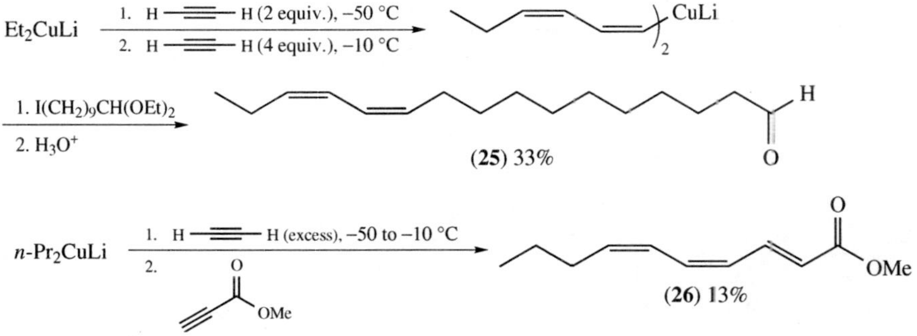

SCHEME 20

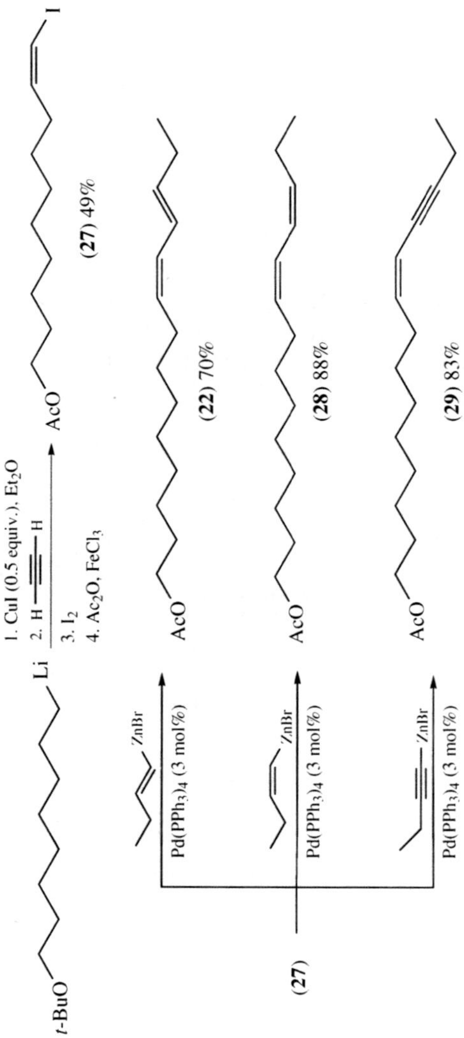
t-BuO
Li
1. CuI (0.5 equiv.), Et2O
2. H — H
3. I2
4. Ac2O, FeCl3
AcO
I
(27) 49%
(27)
ZnBr
Pd(PPh3)4 (3 mol%)
AcO
(22) 70%
AcO
(28) 88%
AcO
(29) 83%

SCHEME 21

Finally, it should be noted that alkenyl cuprates derived from the carbocupration of acetylene can be easily quenched with I_2 to give the corresponding (Z)-alkenyl iodides. These compounds can serve as stereodefined building blocks in many *coupling* reactions. For example, this approach has been used to prepare the versatile alkenyl iodide **27** (Scheme 21)[50].

This methodology has been applied to the preparation of *pheromone* **22**, which is a common intermediate to the egyptian cotton leafworm *Spodoptera Littoralis* and to the oil palm tree defoliator Lepidotere *Stenoma Cecropia*. The analogues **28** and **29** have also been prepared from **27** using this methodology.

The same stereoselective *coupling* step has been used in the total synthesis of the sex *pheromone* **30** from both *Zeuzera Pyrina* and *Vitacea Polistiformis* (Scheme 22)[51].

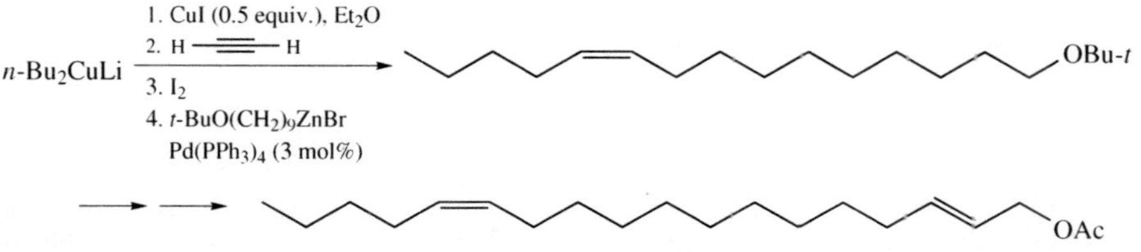

(30)

SCHEME 22

III. CARBOCUPRATION OF SUBSTITUTED ALKYNES

A. Alkyl-substituted Alkynes

1. Non-functionalized alkynes

The carbocupration of non-functionalized terminal alkynes is known to be highly regio- and stereoselective leading to branched alkenyl coppers as major isomers through a *syn*-addition (Scheme 23)[1]. This regioselectivity is accurately explained by the inductive donation of the alkyl substituent, which polarizes the carbon–carbon triple bond. By contrast, internal alkynes generally undergo poorly regioselective carbocupration due to the opposite inductive influence of the two alkyl substituents, leading to mixtures of the two isomeric branched and linear alkenyl coppers (Scheme 23).

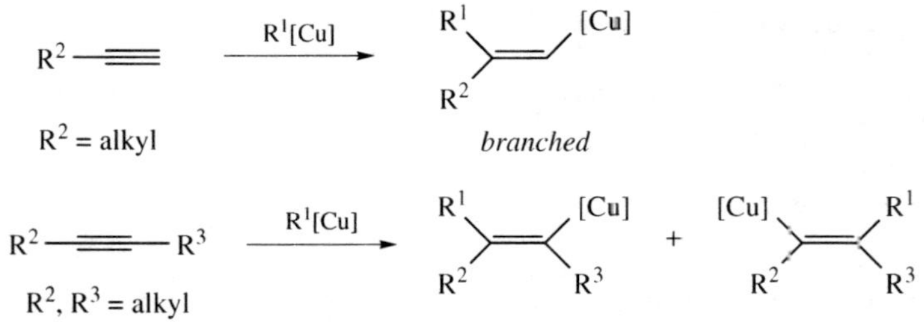

SCHEME 23

Primary alkyl organocopper species R_2CuMgX or R_2CuLi can readily deliver their primary alkyl R group to terminal alkynes giving the corresponding alkenes[52–54] whereas, in the case of secondary or tertiary alkyl groups, only acetylenic proton abstraction is observed[9,55]. Moreover, the addition of secondary or tertiary alkyl groups from Grignard-derived organocopper reagents $RCu{\bullet}MgX_2$ to terminal alkynes is not successful[8,10,55]. Interestingly, $MgBr_2$ has been demonstrated to play a crucial role in the latter reaction (Scheme 24)[56]. Indeed, when using a stoichiometric amount of $MgBr_2$ in THF, secondary (*i*-butyl and *i*-propyl) as well as tertiary (*t*-butyl and *t*-amyl) alkyl groups can be transferred regio- and stereoselectively from $RCu{\bullet}MgX_2$ to terminal alkynes with modest to good conversion. Under these conditions, the corresponding branched alkenes are obtained as single isomers in good isolated yields upon acidic hydrolysis.

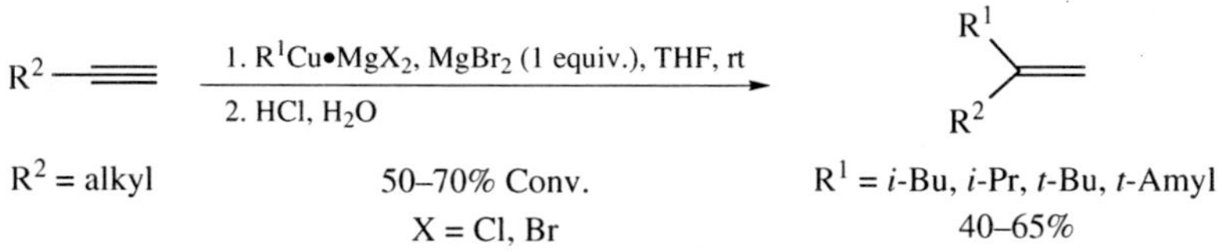

SCHEME 24

A related reaction is reported for the carbocupration of propargyltrimethylsilane with n-$BuCu{\bullet}MgBr_2$ in the presence of either LiI or $P(OEt)_3$[63]. Both ligands play a crucial role by stabilizing the unstable starting organocopper species and have a pronounced accelerating effect. Under these conditions, the regioselectivity is total, giving the branched adduct in good yields (Scheme 25).

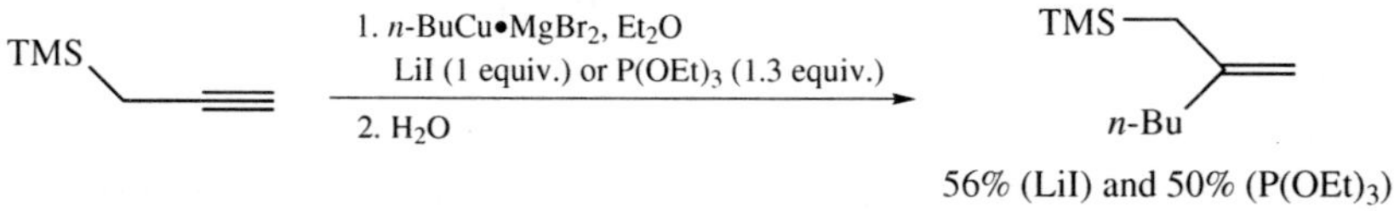

SCHEME 25

In the presence of a stoichiometric amount of LiBr, known to suppress proton abstraction from terminal alkynes[57,58], the carbocupration of terminal alkynes **31** in THF with cuprate $Et_2CuMgBr$ is reported to lead regioselectively to the corresponding branched (Z)-alkenes with high conversions (>95%) and good isolated yields along with variable amounts of dimers, presumably derived from oxidative dimerization of the intermediary alkenyl copper derivative (Scheme 26)[59].

On the other hand, when using hindered cuprate t-$Bu_2CuMgBr$ under the same conditions, 89% of conversion is attained. The desired branched (Z)-isomer is obtained in good yield accompanied by a significant amount of the linear (Z)-alkene (Scheme 26). Here, the partial loss of regioselectivity is assumed to be due to steric hindrance exerted by the bulky *t*-butyl group. In addition, while strict *syn*-addition applies for $Et_2CuMgBr$, for the more basic and reactive t-$Bu_2CuMgBr$ some *anti*-adduct is also observed (*ca* 5%), certainly as the consequence of steric effects.

It is noteworthy that all attempts to transfer a methyl group from various cuprates ($Me_2CuMgCl{\bullet}LiBr$, Me_2CuLi_2, $Me_3Cu_2MgCl{\bullet}(LiBr)_2$ or $MeCu{\bullet}MgBrCl$) to **31** has failed and only resulted in the abstraction of the acetylenic hydrogen atom.

The tandem carbocupration–*oxygenation* of terminal alkynes has been reported to give access to α-branched *aldehydes* or stereodefined trisubstituted enol esters or silyl ethers

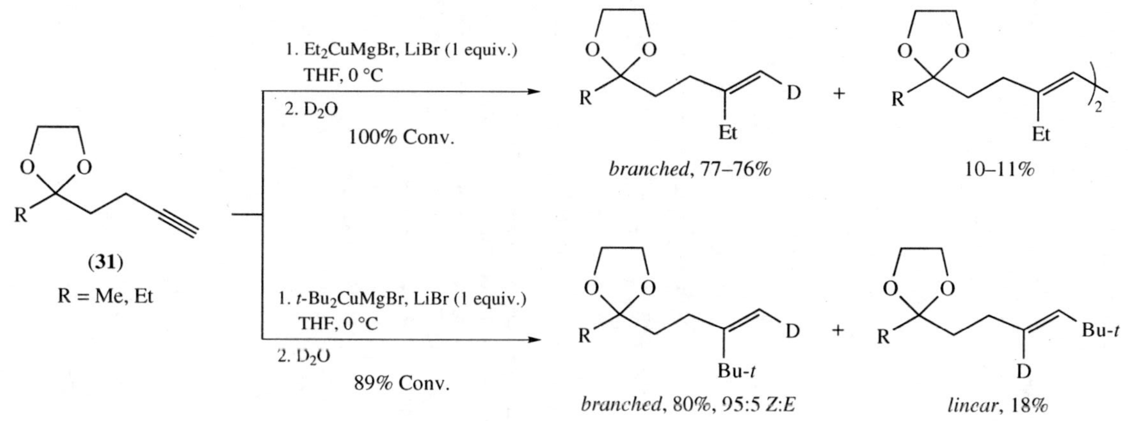
(31)
R = Me, Et
1. Et2CuMgBr, LiBr (1 equiv.)
THF, 0 °C
2. D2O
100% Conv.
branched, 77–76%
10–11%
1. t-Bu2CuMgBr, LiBr (1 equiv.)
THF, 0 °C
2. D2O
89% Conv.
branched, 80%, 95:5 Z:E
linear, 18%

SCHEME 26

(Scheme 27)[60]. The overall transformation consists in the carbocupration of terminal alkynes, subsequent *oxygenation* of the branched alkenyl copper intermediates with LiOOBu-*t* and then reaction with various electrophiles. This tandem sequence tolerates alkyl- and silyl-ethers, esters and tertiary amines and affords the corresponding oxidized adducts in good yields. Under the best conditions, the carbocupration step is performed with RMgX or RLi species in the presence of CuBr•Me_2S in variable relative amounts.

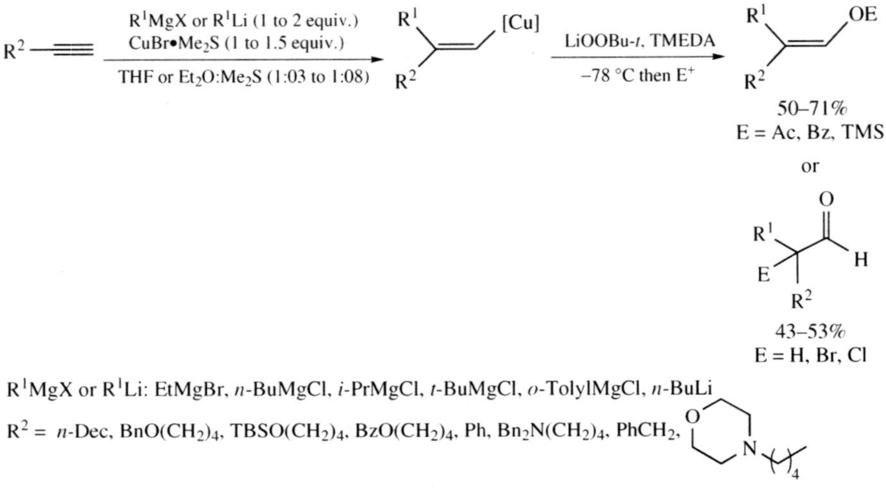

SCHEME 27

Diversely functionalized organocopper reagents can be utilized in the carbocupration of terminal alkynes. Thus, the organocopper reagent derived from *zirconacyclopentane* **32** has recently been shown to add regio- and *syn*-stereoselectively on hex-1-yne[61]. Upon warming the reaction mixture, the formed linear alkenyl copper **33** undergoes reductive *coupling* into bicyclic compound **34** in good yield (Scheme 28).

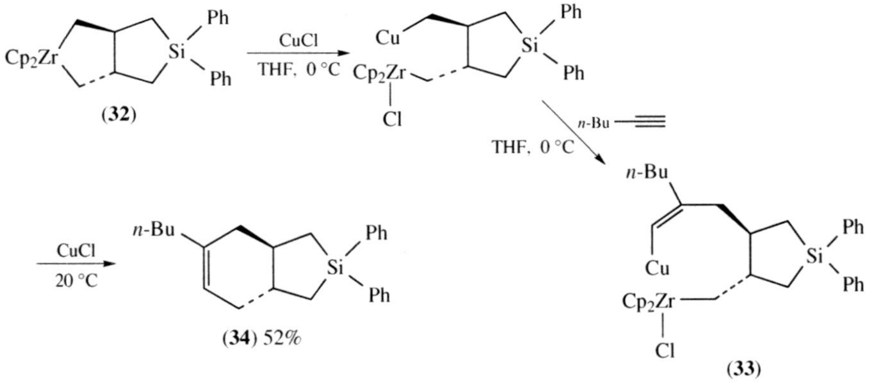

SCHEME 28

α-Silylated organocopper reagents can also react with hex-1-yne. The best results are obtained with organocopper TMSCH_2Cu•$MgBr_2$ in Et_2O in the presence of LiI or $P(OEt)_3$

which speed up the reaction. Under these conditions, the branched alkene is isolated in good yield and total regioselectivity (Scheme 29)[62–64]. By contrast, addition of LiBr leads to the corresponding symmetrical 1,4-diene, presumably by destabilizing the alkenyl copper, whereas Me_2S slows down the reaction rate. In all cases, magnesium salts are required for the reaction to occur, while cuprates $(TMSCH_2)_2CuMgX$ and $(TMSCH_2)_2CuLi$ both fail to react.

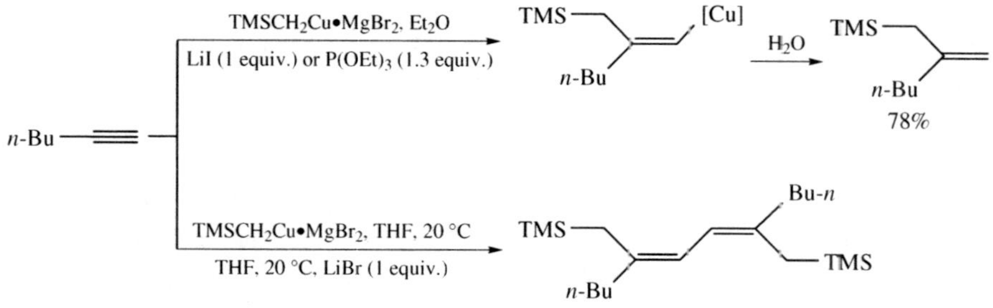

SCHEME 29

Internal non-functionalized alkynes are recognized to be poor substrates for the addition of organometallic species, and thus the carbocupration reaction involving such substrates is generally sluggish. However, a recent arylmagnesiation of 4-octyne catalyzed by $Fe(acac)_3$ (5 mol%) together with CuBr (10 mol%) has been disclosed. This allows the corresponding trisubstituted alkenes to be obtained in high yields and E:Z selectivities as the consequence of a clean *syn*-addition (Scheme 30)[65].

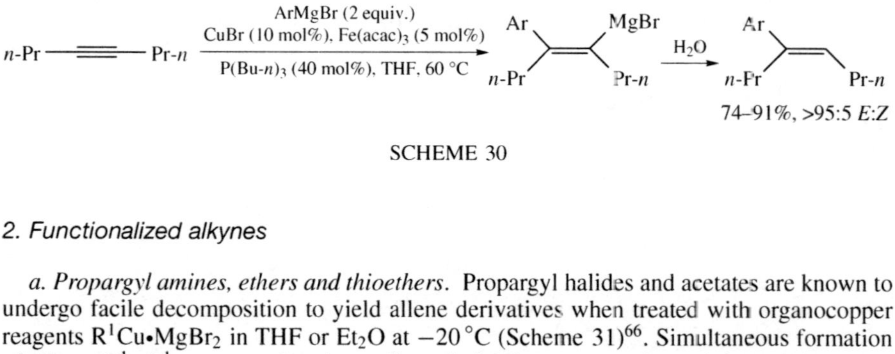

SCHEME 30

2. *Functionalized alkynes*

a. Propargyl amines, ethers and thioethers. Propargyl halides and acetates are known to undergo facile decomposition to yield allene derivatives when treated with organocopper reagents $R^1Cu{\cdot}MgBr_2$ in THF or Et_2O at −20 °C (Scheme 31)[66]. Simultaneous formation of alkane $R^1{-}R^1$ suggests that these allene derivatives are not formed through branched alkenyl copper intermediates, so that the reaction certainly does not proceed through carbometallation of the alkynes.

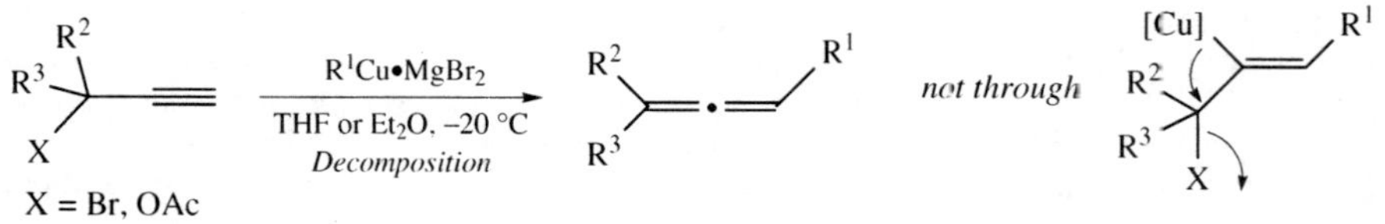

SCHEME 31

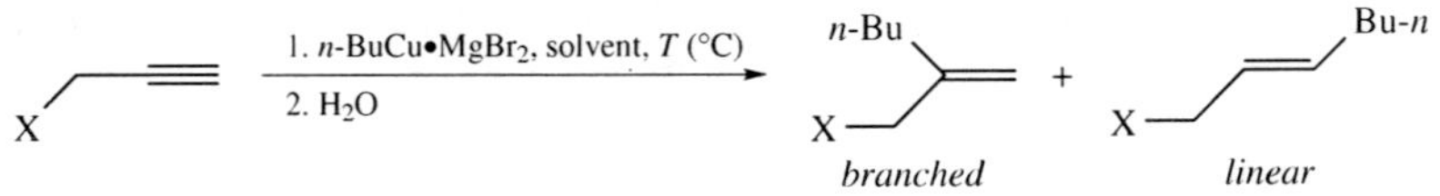

X	Solvent	T (°C)	*Branched*	*Linear*	Yield (%)
NMe_2	THF	−25	86	14	47
	Et_2O	−15	3	97	82
NEt_2	Et_2O	−15	40	60	65
OMe	THF	−25	86	14	42
	Et_2O	−55	3	97	65
OBu-*t*	Et_2O	−55	10	90	not determined
OTMS	THF	−50	85	15	53
	Et_2O	−55	0	100	72
SEt	THF	−25	97	3	32
	Et_2O	−15	97	3	44

SCHEME 32

By contrast, propargyl amines, ethers and thioethers undergo facile carbocupration with organocopper reagent n-BuCu•$MgBr_2$, in a highly *syn*-stereoselective fashion, to afford functionalized regioisomeric branched and/or linear alkenes upon hydrolysis (Scheme 32)[66].

The regioselectivity of the addition reaction is highly dependent upon the solvent. In THF, the expected branched isomer is formed regioselectively. The regioselectivity is controlled by the polarization of the acetylene moiety as a consequence of the inductive effect of the alkyl subtituent (Scheme 33).

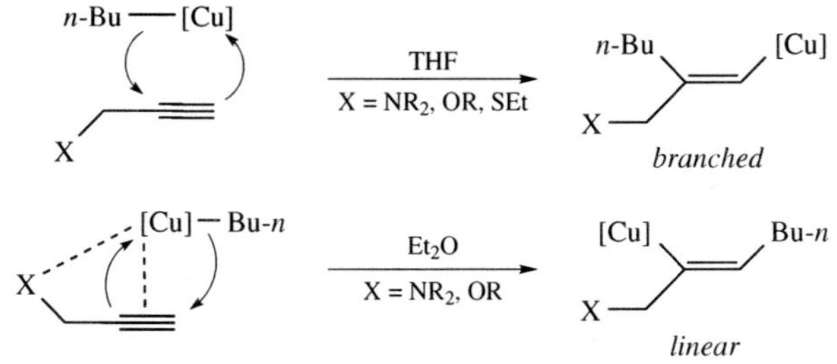

SCHEME 33

By contrast, in Et_2O, which is a less polar and less basic solvent than THF, the regioselectivity is interpreted by the formation of both π(C=C)−Cu and heteroatom−Cu complexes favoring the transfer of the n-butyl group to the acetylenic carbon distal to the alkyl substituent. The linear isomer is then formed as the major product in Et_2O (Scheme 33). One should note that virtually no linear isomer is observed from reaction involving propargyl thioether, since in this case sulfur−copper interaction would result in the formation of a highly strained ring. As evidence of this heteroatom-directed addition, steric crowding on the heteroatom hinders its complexation with the copper species and thus enhances the amount of the branched isomer.

In all cases, significant amounts of symmetrical 1,4-dienes **35** and trisubstituted alkenes **36** are obtained. The former are formed through the self-*coupling* reaction of alkenyl copper **34**, while the latter are formed through the cross-*coupling* reaction of **34** with unreacted copper reagent (Scheme 34).

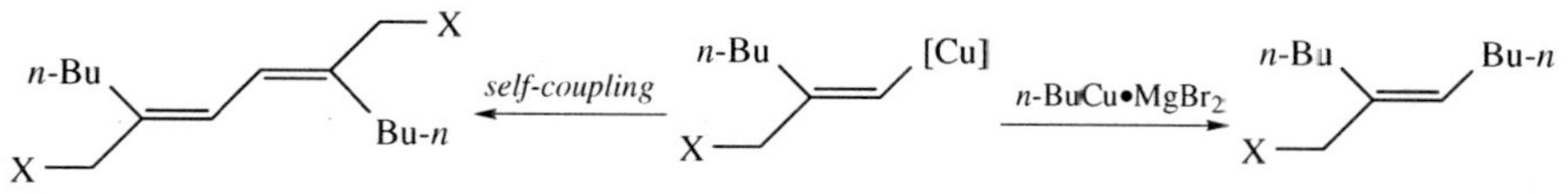

(**35**) 2–27% (**34**) (**36**) 2–15%

X = NR_2, OR, SEt

SCHEME 34

b. Alkynyl epoxides. The reaction of alkynyl *epoxides* with organocopper reagents has been recently reviewed[67]. The reaction occurs in a neat S_N2' overall process to give the corresponding allenyl alcohols. The observed stereoselectivity[68] is in most cases consistent with an overall *anti*-substitution reaction. However, some variations are observed[69] depending on the nature of the organocopper reagent. The reaction can also be conducted[70] with Grignard reagents under Cu(I) catalysis. However, the intermediacy of an alkenyl copper species could not be ascertained, and both a carbometallation–β-elimination process or copper(III)-mediated substitution process can be considered (Scheme 35).

c. Alkynyl acetals. The carbocupration of alkynyl acetals can give access to regioisomeric branched and linear alkenyl copper derivatives provided the temperature is maintained below −20 °C to prevent the decomposition of the latter (Scheme 36)[71,72]. The level of regioselectivity of the reaction is mainly dependent on both the nature of the copper reagent and the solvent.

As a general rule, mono-organocoppers RCu•LiX and RCu•MgX_2 afford mixtures of both linear and branched isomers with slight variation according to the solvent[73]. On the other hand, either branched or linear isomers can be prepared selectively with cuprates. Using R_2CuMgX in THF leads to the linear isomer as the major adduct, whereas using R_2CuLi in Et_2O leads almost exclusively to the branched isomer.

In THF, the carbocupration leads to mixtures of major branched alkenyl copper and minor allenes arising from β-elimination on the linear alkenyl copper leading to copper-alkoxide. In this solvent, the regioselectivity is controlled by the polarization of the acetylene moiety due to the inductive effect of the substituent (Scheme 36).

On the other hand, the regioselectivity observed in Et_2O, giving almost exclusively the linear isomer, is explained by the complexation of the copper salt by the oxygen atoms of the acetal. Generally, in Et_2O the reaction temperature must be maintained low enough to prevent the β-elimination on the formed linear alkenyl copper, typically below −20 °C and −50 °C for dialkoxy acetals and dioxolanes, respectively. In some cases, these organocopper intermediates exhibit a remarkable thermal stability, as for **37**, which does not β-eliminate into allene derivatives even in refluxed Et_2O (Scheme 37)[74].

Thus, various primary and secondary dialkyl-, dialkenyl- and diphenyl-lithium cuprates can be added to terminal alkynyl acetals (two equivalents) in Et_2O[62,71]. Upon hydrolysis or *iodolysis*, the corresponding branched alkenes are isolated in good to excellent yields with high regioselectivity (>90:10) and total *syn*-stereoselectivity (Scheme 38). It is noteworthy that in the case of *t*-Bu_2CuLi, the reaction is non-regioselective and non-stereoselective, since the branched isomer is obtained as well as both the (*E*)- and (*Z*)-linear isomers

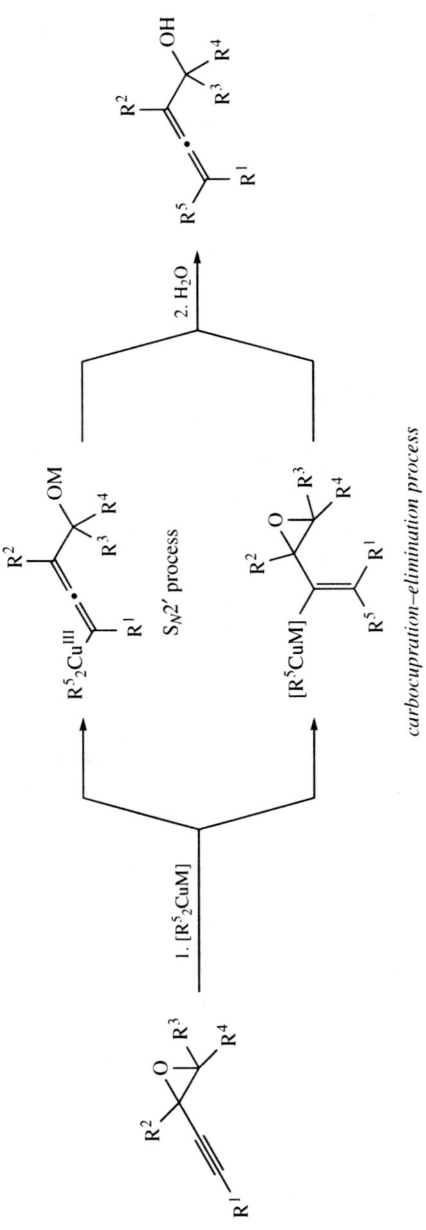
1. [R^5_2CuM]
2. H_2O
S_N2' process
carbocupration–elimination process

SCHEME 35

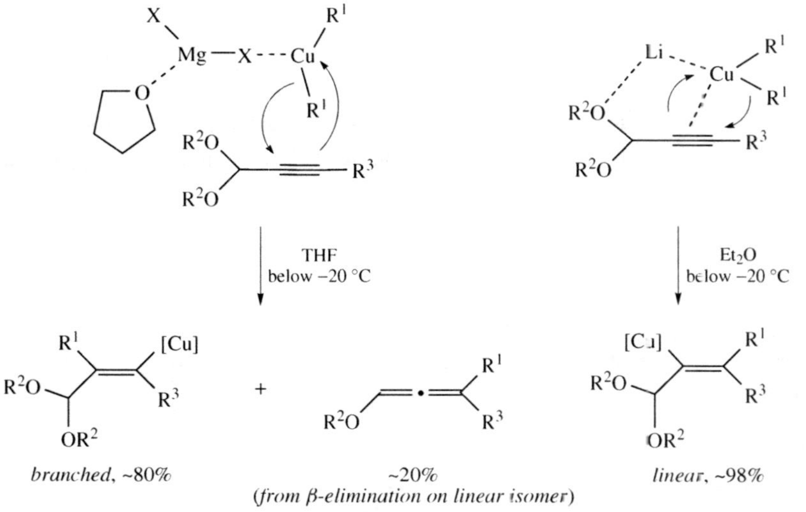

SCHEME 36

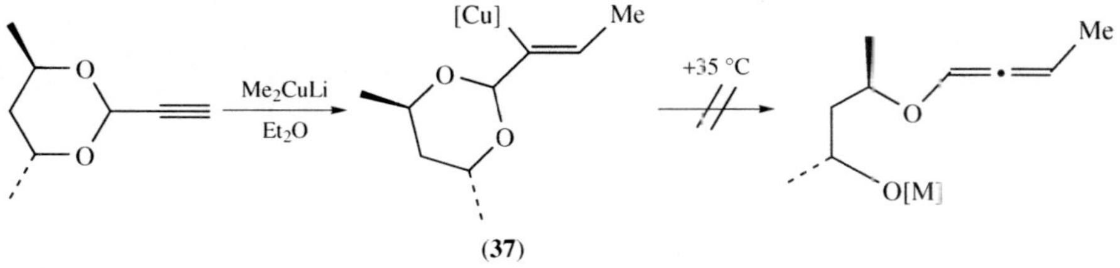

SCHEME 37

even at −50 °C in Et_2O. This result is attributed to the steric bulkiness of the *t*-butyl group. In addition, lithium dimethyl cuprate Me_2CuLi does not give reproducible results.

The carbocupration can also occur starting from the less reactive internal alkynyl acetals (two equivalents) bearing diverse substituents at the acetylenic carbon atom, such as alkyl, aryl, alkenyl and phenyl groups[71]. It should be noted that even *t*-Bu_2CuLi can be used, although a lower yield is observed in this case. On the other hand, the trimethylsilyl group impedes the addition, either by its bulkiness or by its inductive effect, which is opposite to that of the alkyl substituent. Except in the latter case, the corresponding trisubstituted alkenes are obtained in good to high yields with total regio- and stereoselectivity (Scheme 38).

Similarly, both terminal and internal alkynyl ketals can be efficiently reacted with *n*-Bu_2CuLi in Et_2O in good yields (Scheme 38).

d. Alkynyl orthoesters. The carbocupration of 1,1,1-triethoxypropyne with the deuterium-labeled mono-organocopper reagent $CD_3Cu{\bullet}MgI_2$ is reported to lead to the corresponding branched adduct in modest yield with the same regiocontrol as that observed with acetals (Scheme 39)[75].

This reaction has been used in the synthesis of deuterium-labeled methacrylates, tiglates and senecioates for biosynthetic studies of carboxylic acids in carabid beetles.

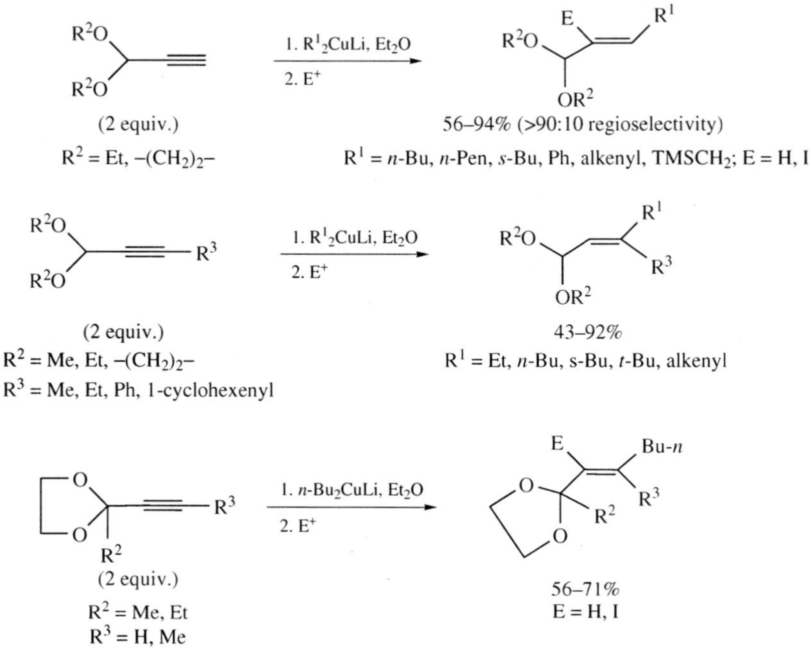

SCHEME 38

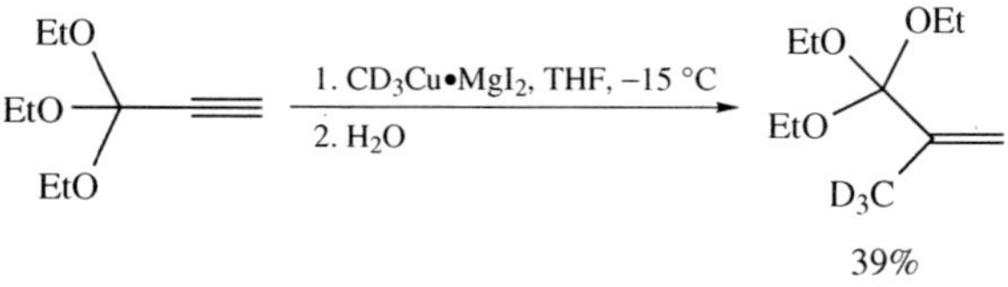

SCHEME 39

e. Fluoroalkylated alkynes. As described previously, internal alkynes are generally poorly reactive towards organocopper species. However, polyfluoroaryl copper reagents are described to undergo facile addition to highly reactive bis(trifluoromethyl)acetylene in DMF, giving fluoroalkylated trisubstituted alkenes **38** in good yields with a high preference for the (E)-isomers (>90:10 E:Z ratio) (Scheme 40)[76].

In this reaction, it is important that the stoichiometry is reasonably accurate and that only a slight excess of bis(trifluoromethyl)acetylene is added. Indeed, if two or more equivalents of the alkyne are used, the alkenyl copper initially produced may undergo a second addition to the alkyne to afford the conjugated *dienyl* copper **39**. Moreover, other copper reagents such as C_6F_5Cu, $C_6F_5Cu(CN)CdX$, $(C_6F_5)_2Cu(CN)(CdX)_2$ and $C_6F_5Cu(CN)Li_2$, only lead to complex mixtures. This methodology has been successfully applied to the total synthesis of antiestrogenic drug panomifene[78].

A similar reaction is possible with (dialkoxyphosphinyl)difluoromethyl copper reagent $(RO)_2P(O)CF_2Cu$, which allows the preparation of fluoroalkylated tetrasubstituted alkenes **40** in variable yields upon addition of alkyl-, allyl- or aryl- iodides to the reaction mixture (Scheme 40)[77].

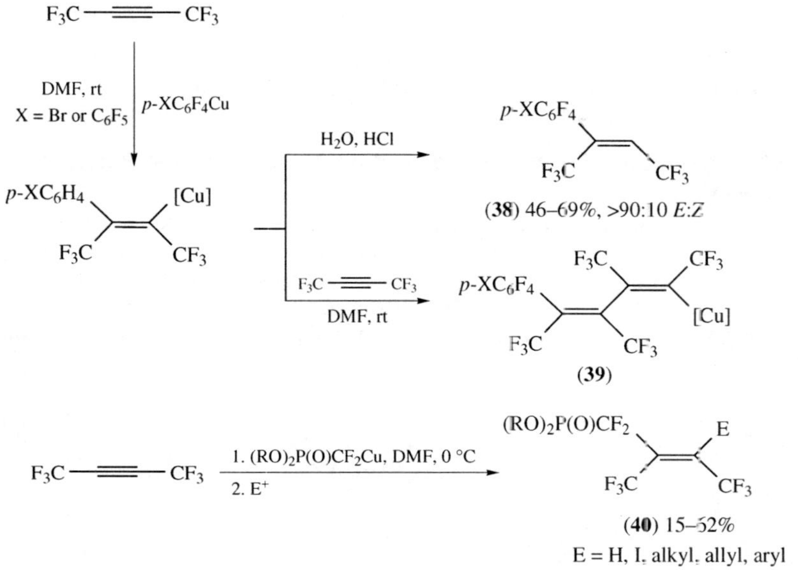

SCHEME 40

Organocopper reagents can also add to fluorine-containing internal aryl alkynes. Under the optimized conditions, i.e. with Grignard-derived cuprates $R_2CuMgBr$ and Lipshutz cuprates $R_2Cu(CN)Li_2$, internal aryl alkynes bearing a trifluoromethyl substituent afford trisubstituted fluorinated alkenes in high yields and with high regio- and stereoselectivity (Scheme 41)[78,79]. By contrast, performing the reaction with HCF_2- or $HCF_2(CF_2)_2$-substituted internal alkynes results in the formation of alkenes in fair to good yields, albeit with high regio- and stereoselectivity.

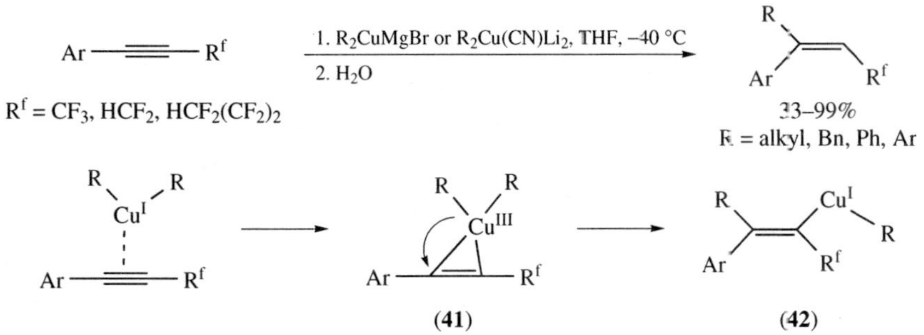

SCHEME 41

In every case, the high regioselective *syn*-carbocupration can be explained through the coordination of the Cu(I) reagent to the carbon–carbon triple bond followed by oxidative addition of the alkyne to Cu(I) to form the intermediate Cu(III) species **41**. At this stage, due to the strong electron-withdrawing effect exerted by the fluoroalkyl group R^f, the R^fC–Cu(III) bond may be stronger than Ar–Cu(III). Accordingly, alkyl substituent R on

Cu(III) is transferred to the olefinic carbon distal to the fluoroalkyl group, alkenyl copper **42** being produced preferentially (Scheme 41).

The carbocupration reaction of internal trifluoromethylated alkynyl phosphonates has also been reported and is discussed in detail in Section III.D.4.b.

B. Carbocupration and Copper-catalyzed *Carbomagnesiation* Reaction of Propargyl Alcohols

Normant and coworkers have examined the addition of organocopper reagents onto propargyl alcohols[66,80]. On propargyl alcohol itself (used as its Li or Na salt), the regioselectivity is found to be dependent on the solvent polarity, although the branched allylic alcohol is always formed as the major isomer (Scheme 42). However, in the case of internal propargyl alcohols, this regioselectivity is variable depending upon the substitution degree of the starting material.

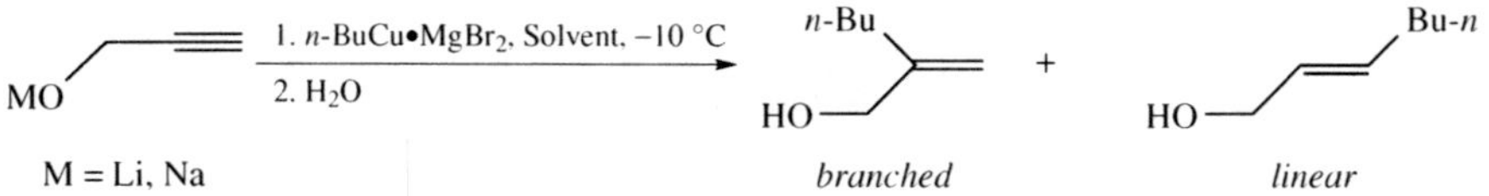

Solvent	*Branched*	*Linear*
Et_2O	60	40
THF	79	21
DME	87	13
HMPA	91	9

SCHEME 42

In the case of 3-phenylprop-2-yn-1-ol, the regioselectivity of the carbocupration is this time excellent[81]. More surprising is the observed stereoselectivity, as the *trans*-trisubstituted allylic alcohol is obtained upon addition, which has been attributed to an *anti*-addition pathway (Scheme 43). Further isomerization of the resulting alkenyl metal species is also conceivable.

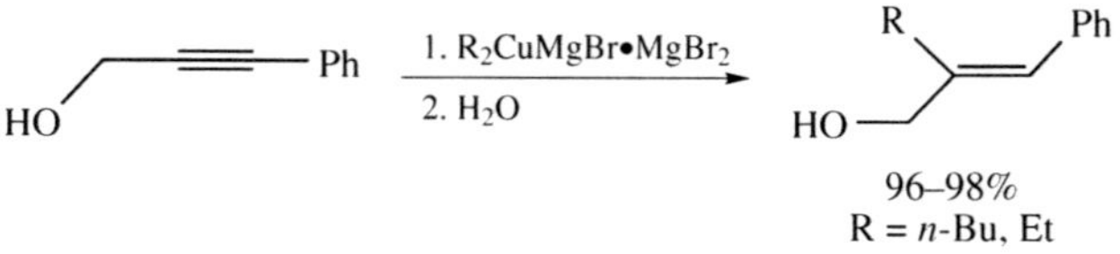

SCHEME 43

The observed regioselectivities are very variable and precluded the application of the carbocupration of propargyl alcohols in synthesis. However, some years later, Jousseaume and Duboudin found that Cu(I) salts catalyze the *carbomagnesiation* of propargyl alcohols. The regioselectivity is excellent in the case of primary propargyl alcohols, and the reaction can be applied to a wide range of substrates and reagents (Scheme 44)[63,82].

This reaction is not strictly within the scope of this review, as it has not been proved that this reaction is a 'true' carbocupration process. Indeed the fact that the observed regioselectivity of the addition is not the same as that previously observed in the case of

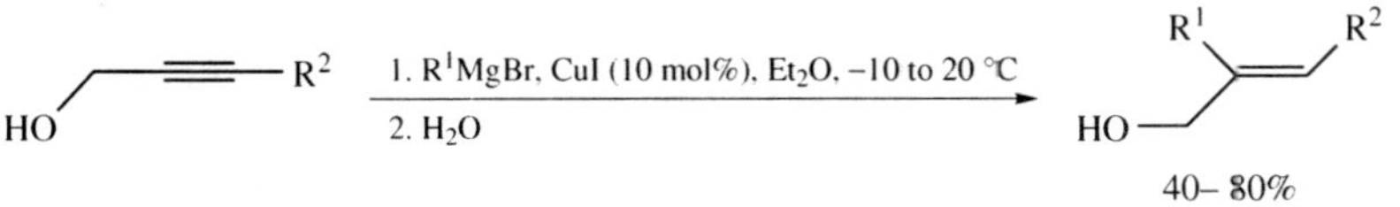

R^1 = Me, Et, *i*-Pr, *t*-Bu, Ph, allyl, crotyl, Bn, $TMSCH_2$

R^2 = H, Me, *n*-Bu, Bn, Ph, methallyl, TMS, $C_5H_{11}C{\equiv}C$, PhC≡C

SCHEME 44

the addition of organocopper reagents to propargyl alcoholates seems to indicate a different mechanism or species. Moreover, the resulting organometallic species **44** derived from propargyl alcohol **43** can be functionalized by *iodolysis*, or by reaction with allyl bromide (Scheme 45)[83]. The observed stereoselectivity proves that the addition process occurs in a *trans*-fashion, which is totally unusual for the carbocupration reaction. The intermediacy of the cyclic organometallic species **45** is also proposed[84].

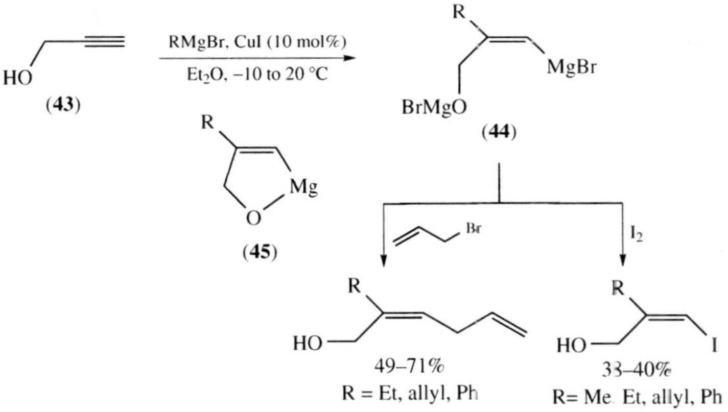

SCHEME 45

Interestingly, it was shown later that the same reaction can occur without the need of copper salt[85–92]. However, the conditions needed in this case (refluxing THF, di(isopropyl) ether or cyclohexane) are largely different from those involved in the presence of CuI catalyst (Et_2O or THF at $-10\,^{\circ}C$). Nevertheless, the extensive use of this reaction in synthesis justifies an overview here.

It should be noted that under the reaction conditions, no metallation of the acetylenic position of the propargyl alcohol occurs. Actually, a different course is followed when bimetallated propargyl alcohol **46** is used[93]. The alkenyl bimetallic species **47** is obtained and can be illustrated through deuteriation (Scheme 46). A surprising 4-*endo-trig* cyclization to bimetallic **48** occurs when the initial Grignard nucleophile is AllylMgBr. Here again, the cyclic bimetallic species **48** can be illustrated through deuteriation or reaction with allyl bromide.

The Cu(I)-catalyzed *carbomagnesiation* of propargyl alcohol **43** has been used as an excellent source of branched allylic alcohols; the reaction of the alkenyl Grignard reagent resulting from the reaction of **43** with electrophiles includes deuteriation[94], *iodolysis*[95], addition to *aldehydes*[96] and to borates[97] (Scheme 47).

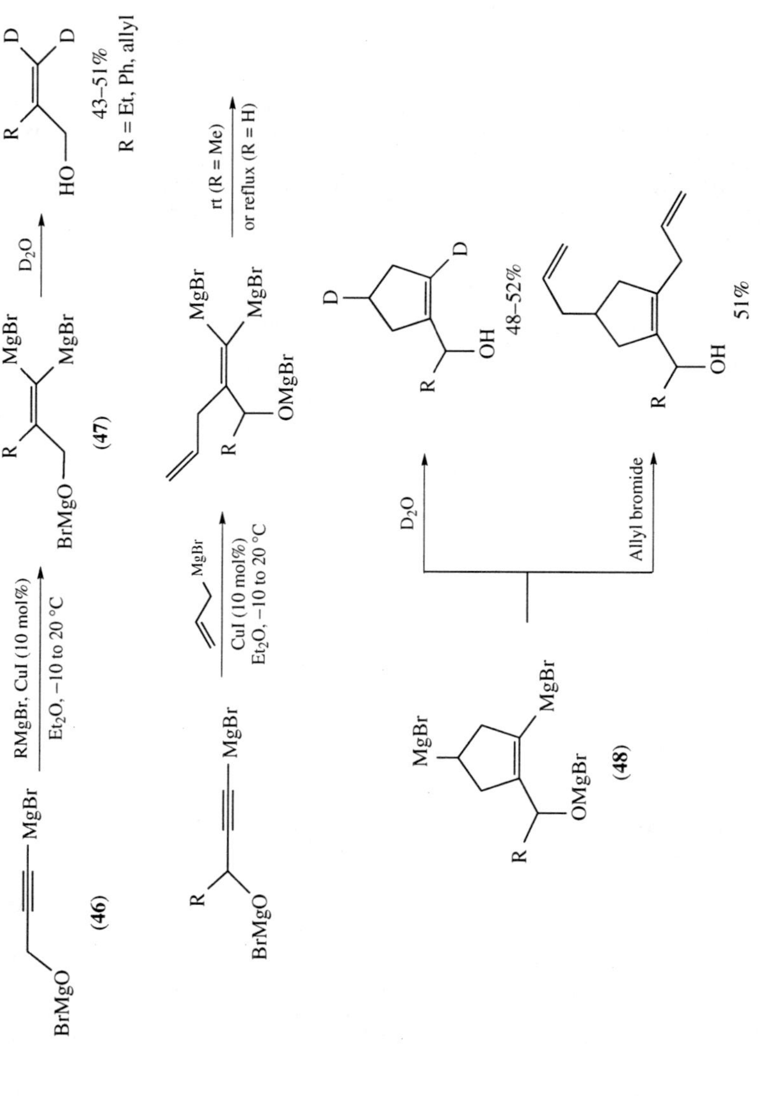

SCHEME 46

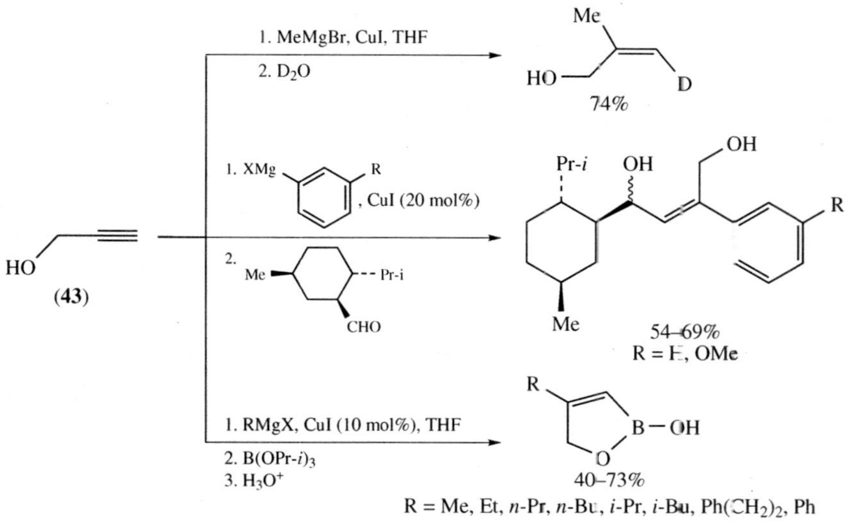

(**43**)

SCHEME 47

The Cu(I)-catalyzed *carbomagnesiation* of propargyl alcohols bearing a substituent on the acetylenic position represents also a valuable tool for the stereoselective synthesis of trisubstituted allylic alcohols. Some representative examples are listed in Scheme 48. The substituent can be a TMS group[96], an alkenyl moiety[98] or an alkyl chain[99].

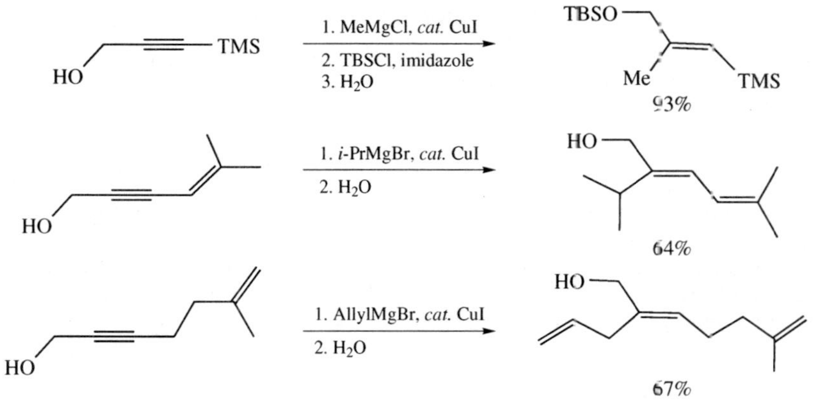

SCHEME 48

The resulting alkenyl Grignard reagent can react through *iodolysis*, leading stereoselectively to tetrasubstituted (*Z*)-iodo alkenols (Scheme 49)[95, 100].

The case of secondary propargyl alcohols was examined earlier by Jousseaume and Duboudin[82]. They found that the regioselectivity is variable under their standard conditions (Et_2O as the solvent), depending on the substrate and the nucleophile (Scheme 50). Regardless of the regioselectivity, yields are low and the reaction sluggish.

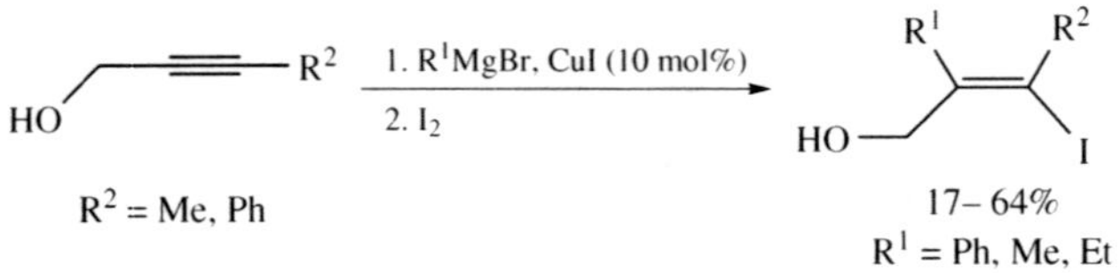

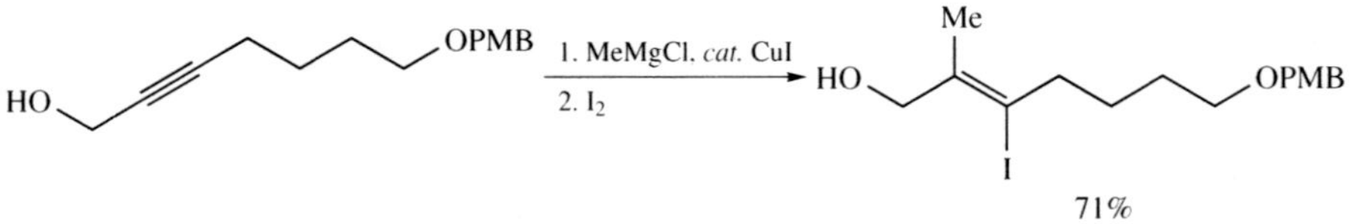

SCHEME 49

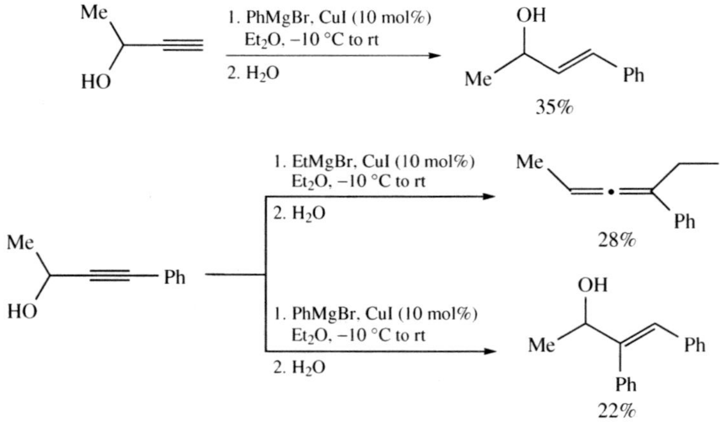

SCHEME 50

Very recently, this methodology was applied to the reaction of propargyl alcohol **49** (Scheme 51)[101]. A systematic study of the influence of solvent, amount of copper salt, nature and amount of Grignard reagent and counterion on both the regioselectivity and conversion of the reaction was conducted. The best reaction conditions allow the preparation of the allylic alcohol **50** in high isolated yield.

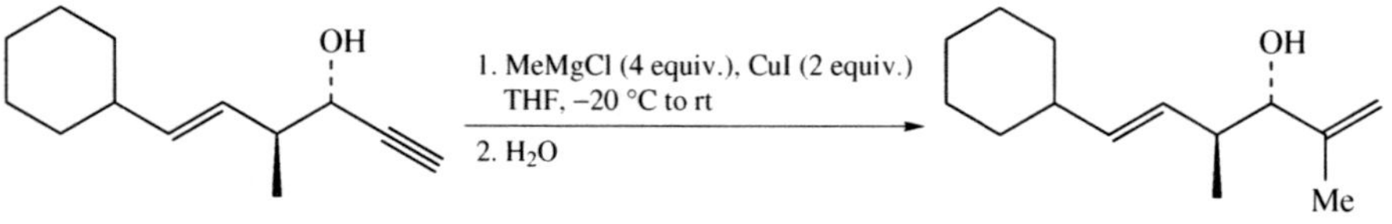

(**49**) (**50**) 81%

SCHEME 51

This reaction was recently re-examined by Lu and Ma. They found that the regioselectivity is actually dependent on the solvent and obtained excellent results in THF at low temperature, favoring the 'classical' *trans*-addition[102]. The cyclic Grignard reagent resulting from propargylic alcohol **51** can be functionalized with I_2 or engaged in a Pd(0)-catalyzed *coupling* reaction (Scheme 52).

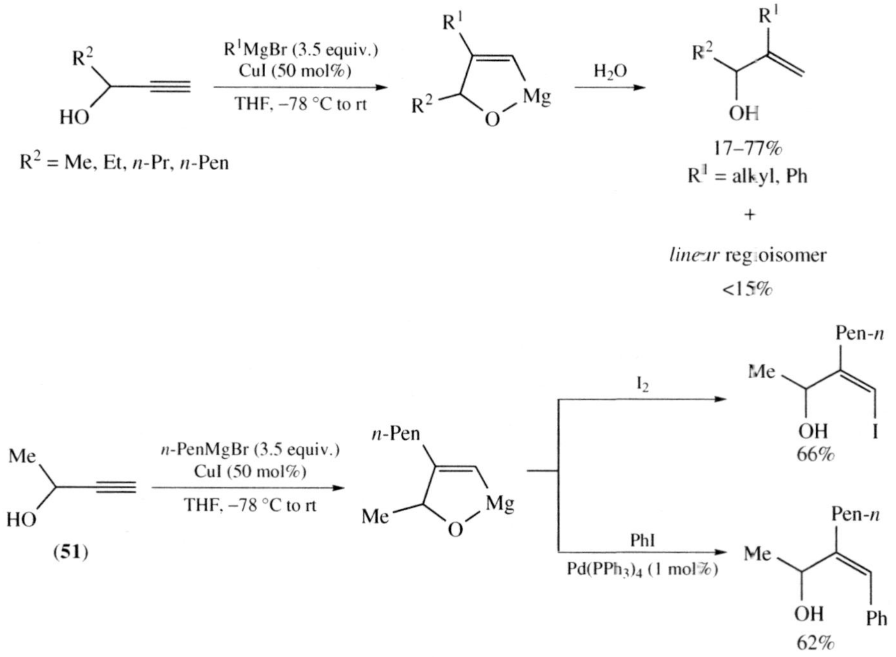

SCHEME 52

By contrast, using toluene as solvent, the regio- and stereoselectivity of the reaction are reversed, the linear product being formed as the major product in a *syn*-addition, as shown by *iodolysis* (Scheme 53)[103].

The intermediate alkenyl metal obtained from **51**, regardless of its nature, can also be functionalized with allyl bromide or engaged in a *coupling* reaction under Pd(0) catalysis.

Both reactions can be applied to enantiopure propargylic alcohol **51** to afford, depending upon the conditions, the corresponding branched or linear enantiopure alkenyl iodides (Scheme 54).

C. Enynes and Arylacetylenes

The carbocupration of enynes represents a valuable method for the preparation of 1,3-dienyl patterns. For instance, cuprate EtCu adds efficiently to isoprenylacetylene in the presence of two equivalents of LiBr in THF. The reaction is highly regio- and stereoselective, only the linear *dienyl* copper intermediate being produced as a consequence of the mesomeric electron donation from the vinyl moiety (Scheme 55)[104]. Subsequent addition of (trimethylsilyl)ethynyl iodide then leads to the corresponding (*E*)-dienyne in high yield.

The same regioselectivity has been observed by Knochel and coworkers during the carbocupration of phenylacetylene with functionalized copper–zinc reagents[12]. This reaction

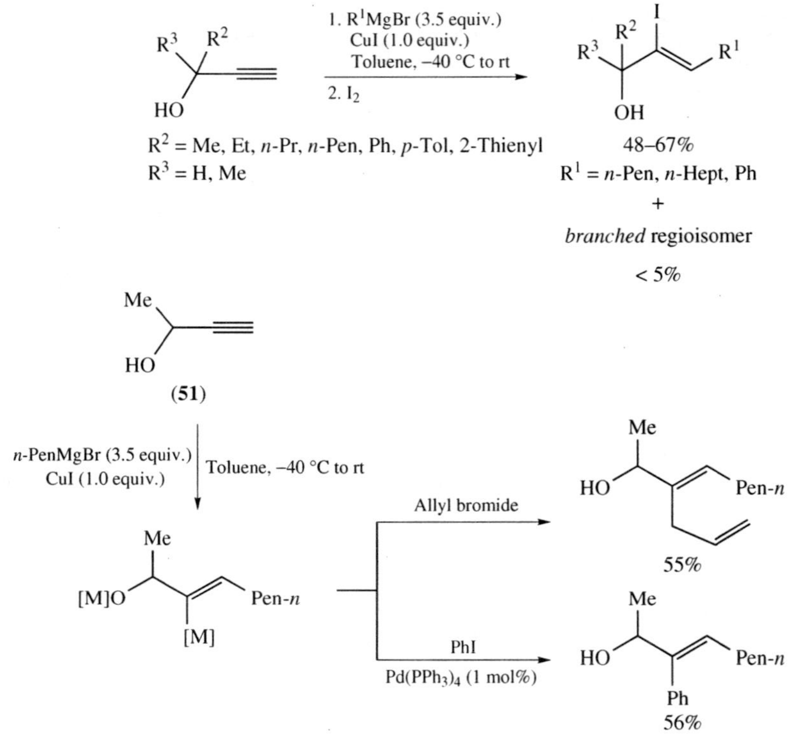

SCHEME 53

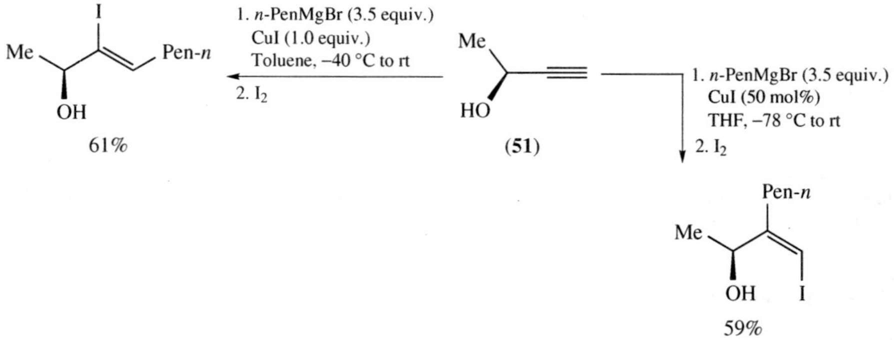

SCHEME 54

thus allows 1,5-dienes to be formed in good isolated yields and high *syn*-stereoselectivity upon treatment of the reaction mixture with allyl bromide (Scheme 56).

On the other hand, the reaction of enynes with secondary and tertiary alkyl Grignard reagents in the presence of a catalytic amount of Li_2CuCl_4 (1 mol%) affords allenyl-propargyl copper species as a consequence of the addition of the initially produced organocopper reagent to the vinyl moiety[105].

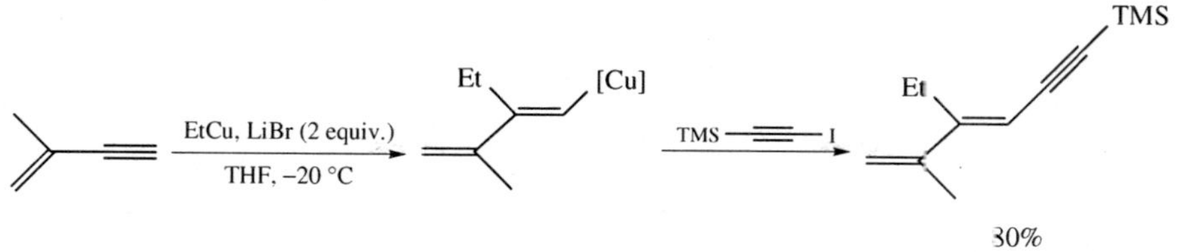

SCHEME 55

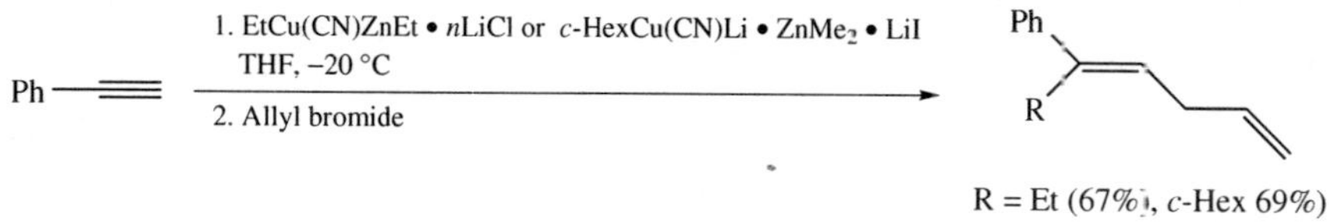

SCHEME 56

The carbocupration of alkynes bearing alkenyl or aryl groups has been applied to the synthesis of bicyclic compounds **55**. In this synthesis, both phenylacetylene and (1-cyclohexenyl)acetylene undergo a carbocupration reaction with **53**, derived from *zirconacyclopentane* **52**[61]. The reaction then affords regio- and stereoselectively the branched alkenyl coppers **54** which, upon warming the reaction mixture to room temperature, undergo reductive *coupling* to give bicyclic compounds **55** in good yields (Scheme 57).

The reversal of the regioselectivity is observed during the carbocupration of 5-ethynyl-2′-deoxyuridine. Indeed, the reaction carried out with an excess of organocopper reagent EtCuMgBr affords the corresponding linear adduct in good yield as the result of a clean regioselective *syn*-carbocupration reaction (Scheme 58)[106]. The high regiocontrol, as opposed to that 'normally' observed with enynes, is rationalized through a chelated intermediate (with the oxygen of the uracil ring). However, the major product is contaminated by a small amount of the tetrasubstituted alkene resulting from the alkenyl–alkyl *coupling* side reaction between the initially formed alkenyl copper and unreacted EtCuMgBr.

D. Heteroatom-substituted Alkynes

1. Alkynyl ethers and O-alkynyl carbamates

a. Alkynyl ethers. The carbocupration of terminal alkynyl ethers (ynol ethers) allows a straightforward access to alkenyl ethers (enol ethers). Seminal works by Normant and coworkers have shown that the carbocupration of ethoxyacetylene occurs with organocopper reagent $RCu{\bullet}MgBr_2$ in THF or Et_2O in a *syn*-fashion preferentially at the carbon β to the oxygen to afford branched (E)-2-alkoxyalkenyl coppers **56**. These organometallic species can further be trapped with various electrophiles to give the corresponding alkenyl ethers in high yields (Scheme 59)[107]. Alkenyl coppers **56** are generally unstable above $-20\,^{\circ}C$, at which temperature they undergo *trans*-β-elimination with the copper alkoxide to afford the corresponding terminal alkynes.

The observed regioselectivity, leading to the branched adduct as single isomer, is in accordance with *ab initio* theoretical calculations conducted by Nakamura's group on the carbocupration of ethynol with MeCu[30]. As shown by this study, the carbocupration of ethynol proceeds through a transition state **57** leading to the branched isomer. This was accurately rationalized by the polarization of the acetylene moiety due to the mesomeric donation from the oxygen.

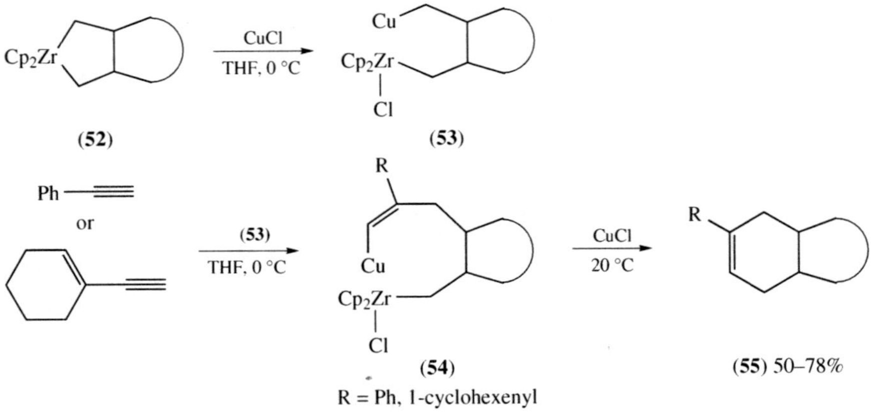

Zirconacycle (**52**)	R	Product (**55**)	Yield (%)
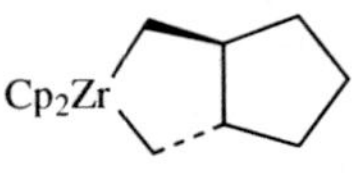	Ph	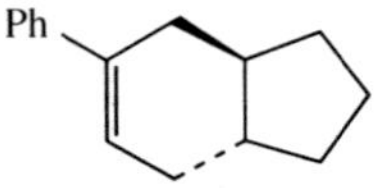	50
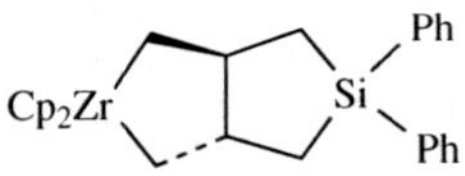	Ph	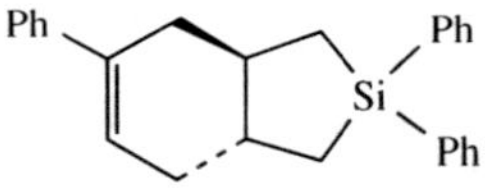	61
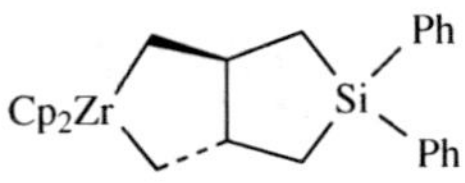	1-Cyclohexenyl	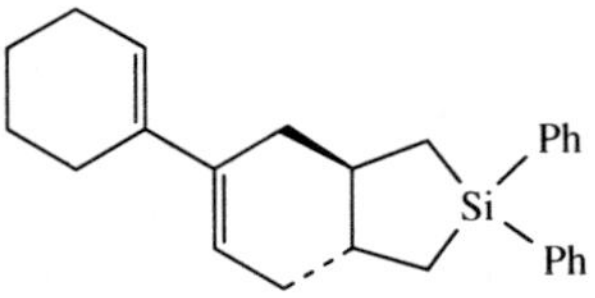	78
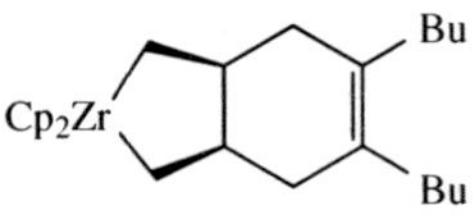	Ph	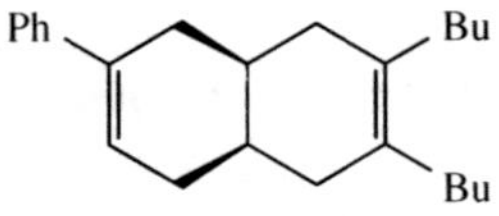	56

SCHEME 57

On the other hand, the carbocupration of internal alkynyl ethers usually gives mixtures of regioisomers, as illustrated by the reaction of 1-ethoxyhex-1-yne with the organocopper reagent $EtCu{\bullet}MgBr_2$ (Scheme 60)[30].

In this case, the low regioselectivity is the consequence of the inductive influence of the *n*-butyl substituent, which is opposite to the mesomeric effect from the oxygen. Similarly, isomeric alkenyl ethers **58** are obtained in substantial amounts in the Cu(I)-induced 1,3-substitution of 1-iodo-1-methoxypropadiene through the carbocupration of the initially formed alkenyl ethers **57** with unreacted copper reagent[108].

The carbocupration of terminal alkynyl ethers has been applied in synthesis. For instance, secondary α-silylated copper reagents have been added onto ethoxyacetylene

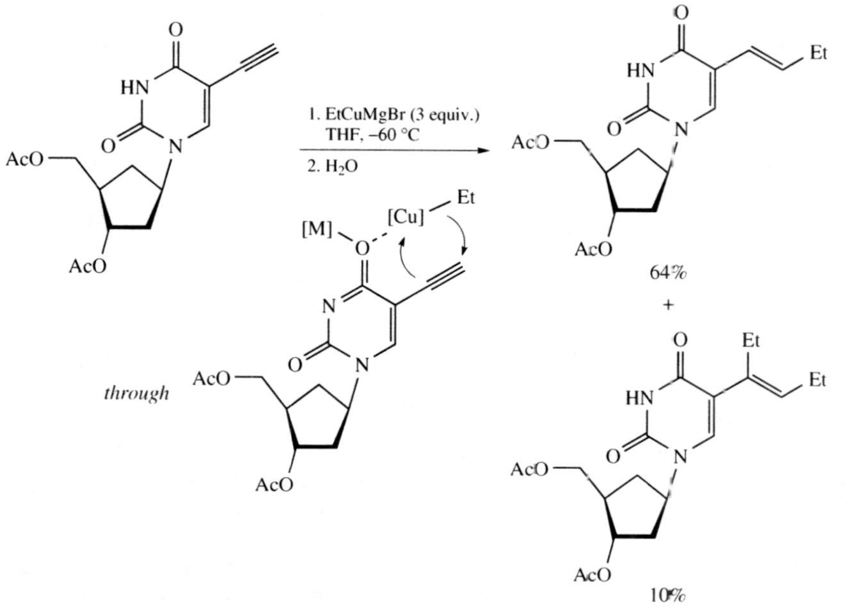

SCHEME 58

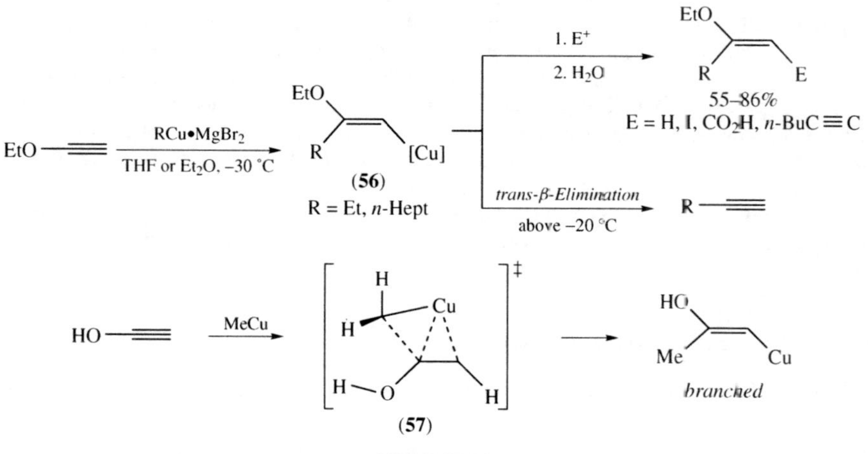

SCHEME 59

to give selectively the corresponding branched α-silylated alkenyl ethers in good isolated yields[62,64]. Recently, the *syn*-carbocupration of ethoxyacetylene with *in situ* prepared phenylcopper PhCu•MgBr$_2$ in THF followed by Pd-catalyzed cross-*coupling* reaction with iodo arenes has been used in the preparation of trisubstituted alkenyl ethers in high yields (Scheme 61)[109].

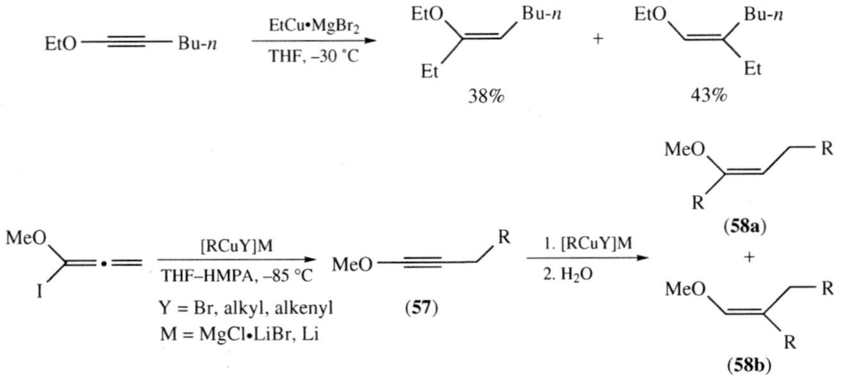

SCHEME 60

Similarly, two consecutive high-yielding regioselective *syn*-carbocuprations of alkynyl ethers involving functionalized organocopper reagents have been applied to the synthesis of the F–J fragment of the gambieric acids (Scheme 61)[110].

b. O-Alkynyl carbamates. The stereoselective carbocupration of *O*-alkynyl carbamates with various copper reagents $RCu{\bullet}MgBr_2$ has been recently described as a straightforward method for the preparation of linear *O*-alkenyl carbamates. In Et_2O as the solvent, single linear (*E*)-isomers are isolated in good to excellent yields (40–86%) (Scheme 62)[111,112]. The introduction of the copper atom in the α position of the oxygen with a high level of regioselectivity results from the coordination of the copper atom to the carbamoyl group. The Cu(I)-catalyzed *carbomagnesiation* of *O*-alkynyl carbamates is also reported in Et_2O.

Interestingly, reversal of the regioselectivity is observed when the reaction is performed in THF. This solvent, which is a stronger Lewis base than Et_2O, impedes the coordination of the copper species to the carbamoyl group, so that the regioselectivity is now mainly controlled by the polarization of the acetylene moiety. This polarization is induced by the mesomeric donation from the oxygen atom, which overrides the inductive influence of the alkyl substituent R^2. Under these conditions, carbocupration of *O*-alkynyl carbamates then affords branched (*E*)-*O*-alkenyl carbamates as the major adducts, albeit in decreased regioselectivity (*ca* 80% regioselectivity) (Scheme 62).

2. Alkynyl thioethers, sulfoxides, sulfones and sulfoximines

a. Alkynyl thioethers. The carbocupration of alkynyl thioethers was studied earlier by Normant and coworkers[107]. For instance, the carbocupration of ethyl(ethynyl)sulfane with copper reagents $RCu{\bullet}MgBr_2$ in THF has been demonstrated to take place in a *syn*-fashion preferentially at the carbon α to the oxygen. This affords the linear (*E*)-alkynyl thioethers as single isomers upon hydrolysis (Scheme 63). The observed regioselectivity is in accordance with theoretical *ab initio* calculations conducted by Nakamura and coworkers[30] on the carbocupration of ethynethiol with MeCu. Indeed, they have shown that this reaction proceeds through transition state **59** as a consequence of the electron-withdrawing inductive effect of the sulfur atom, which polarizes the acetylene moiety. One should note that performing the reaction in Et_2O instead of THF results in a slight decrease in regioselectivity, since small amounts of the branched isomers are observed.

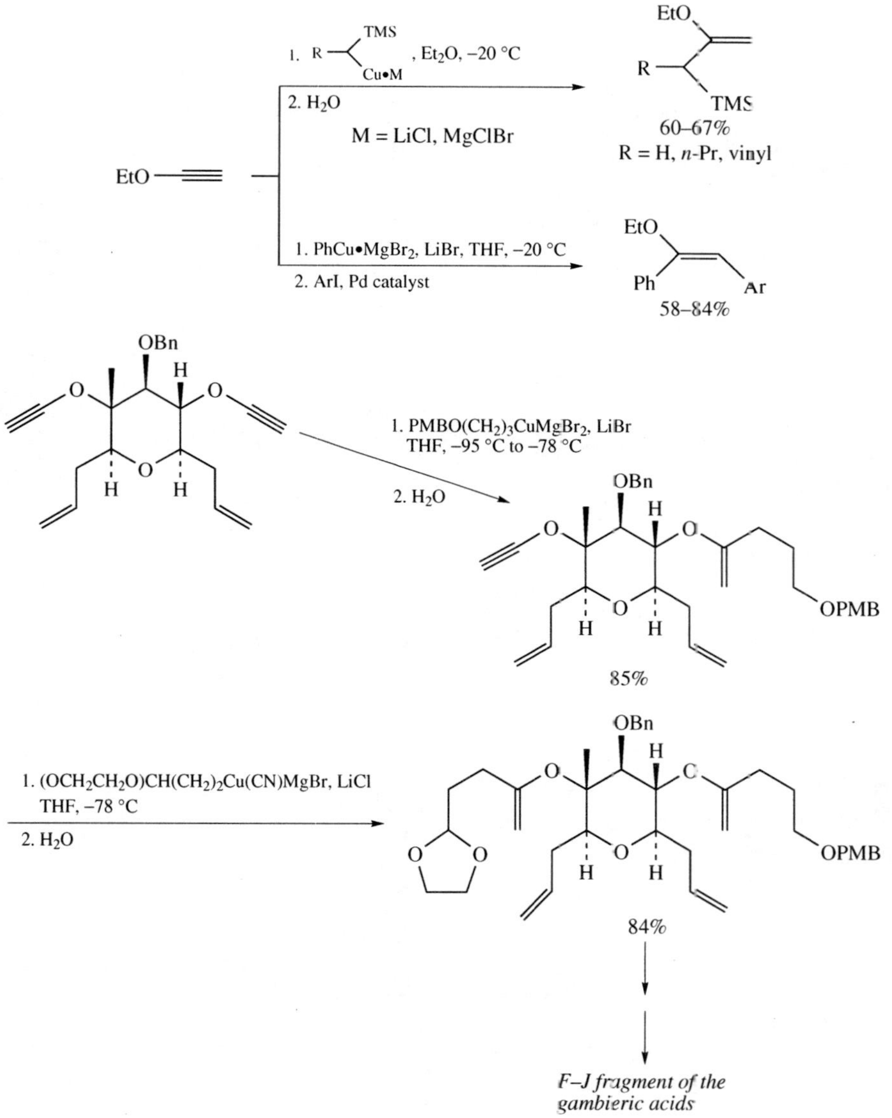

SCHEME 61

Unlike alkynyl ethers, a high level of regioselectivity is also observed with internal alkynyl thioethers **60** wherein both the alkyl substituent R^3 and the thioether moiety polarize the acetylene moiety in the same sense through their inductive effect. Thus, the corresponding alkenyl thioethers are isolated in high yields as single isomers (Scheme 63). In all cases, intermediate thioalkenyl coppers exhibit a remarkable thermal stability even in refluxed THF.

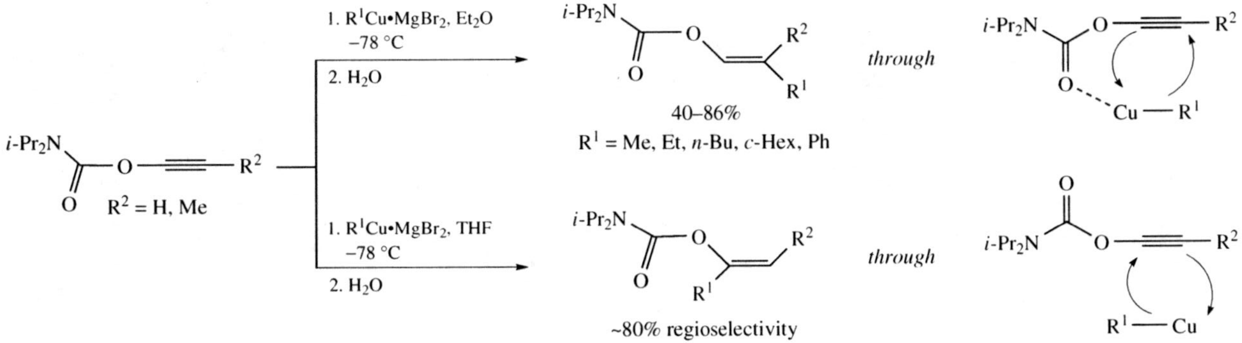

i-Pr2N
O
R2 = H, Me
1. R1Cu•MgBr2, Et2O
−78 °C
2. H2O
1. R1Cu•MgBr2, THF
−78 °C
2. H2O
40–86%
R1 = Me, Et, n-Bu, c-Hex, Ph
~80% regioselectivity
through
Cu
R1
R2

SCHEME 62

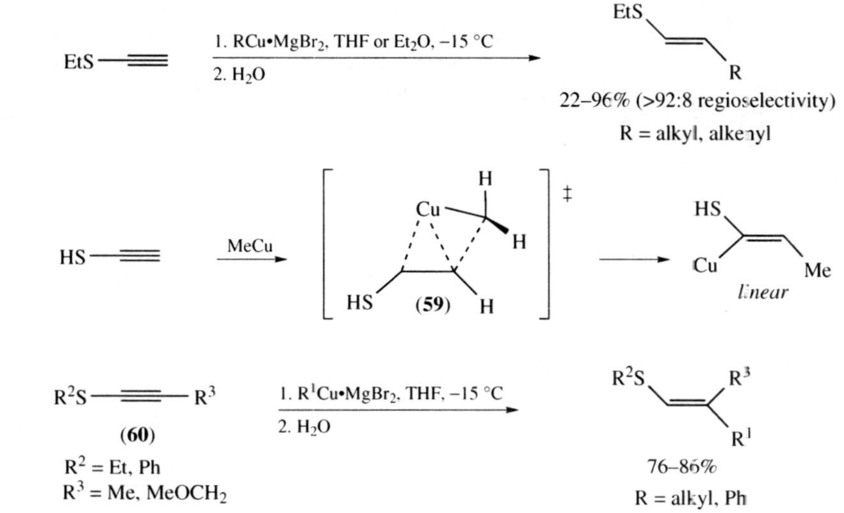

SCHEME 63

Since this early work by Normant and coworkers, the carbocupration of alkynyl thioethers has been applied to the preparation of diversely substituted alkenyl thioethers. As an example, the *syn*-carbocupration of thiophenylacetylene has been used as a key step in the synthesis of insect *pheromone* by giving a regio- and stereoselective access to linear (E,Z)-dienes (Scheme 64, see also Scheme 19)[48]. Similarly, Knochel's group has prepared tetra-substituted alkynyl thioethers through the carbocupration of (hex-1-ynyl)(methyl)sulfane with polyfunctional copper–zinc reagents (Scheme 64)[12].

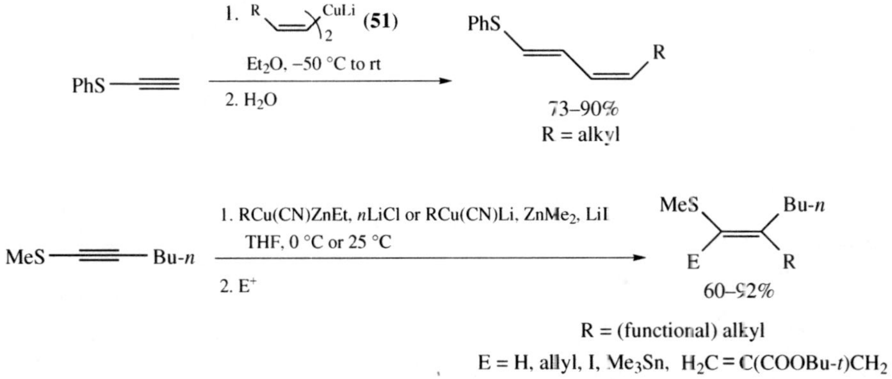

SCHEME 64

b. Alkynyl sulfoxides, sulfones and sulfoximines. Early works on the carbocupration of alkynyl sulfoxides have shown that the addition of organocopper reagents R^1Cu to such substrates allows the corresponding sulfur-substituted alkenes to be formed as single isomers (Scheme 65)[113–115]. By contrast, cuprates $R^1{}_2CuM$ are reported to lead to the formation of sulfoxides arising from the nucleophilic attack of the organometallic species to the sulfur atom rather than the carbometallation reaction. The regioselectivity of the carbocupration reaction may be explained by the chelating character of the sulfoxide moiety which forces the copper atom to add in the α position to the sulfur atom. On the other hand, from alkynyl sulfones mixtures of *syn*- and *anti*-stereomers in variable ratios are generally obtained[116,117].

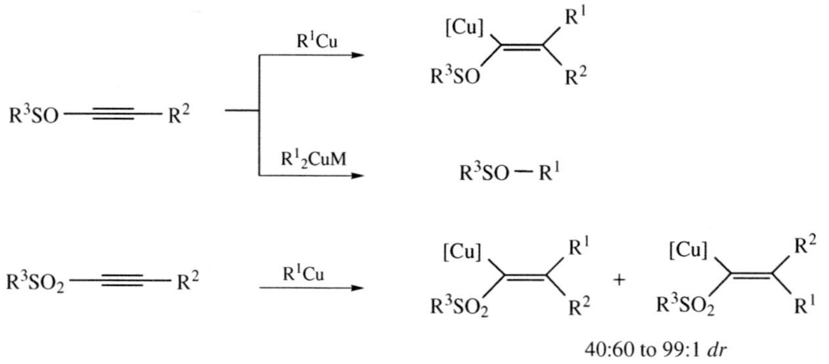

SCHEME 65

More recently, the carbocupration of *p*-tolyl alkynyl sulfoxide **61** with copper reagents RCuMgBr in THF gives a regio- and stereoselective access to the corresponding polysubstituted alkenyl sulfoxides in good to high yields after subsequent reactions of the alkenyl copper species with various electrophiles (Scheme 66)[118].

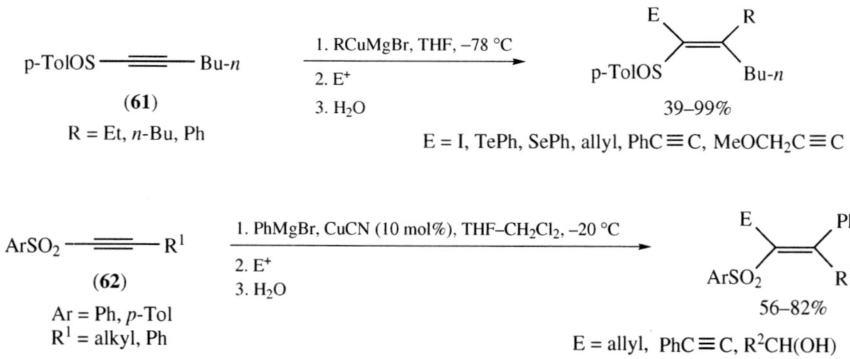

SCHEME 66

The Cu(I)-catalyzed *carbomagnesiation* of alkynyl sulfones **62** has also been described. Conducted with a stoichiometric amount of Grignard reagent PhMgBr in the presence of a catalytic amount of CuCN (10 mol%), the reaction allows tetrasubstituted alkenyl sulfones to be obtained in good yields as single isomers (Scheme 66). This methodology has been successfully applied to the preparation of arylsulfonyl allylic alcohols when employing *aldehydes* as electrophiles[119].

This reaction has been extended to the asymmetric preparation of polysubstituted propadienes by Marek, Knochel and coworkers[120,121]. The overall process involves the carbocupration of *p*-tolyl alkynyl sulfoxides and sulfones with organocopper reagents RCuMgBr followed by the zinc carbenoid homologation and subsequent *β*-elimination of the resulting organometallic species. Depending on the conditions, from the common alkenyl copper intermediates **63** either di- or trisubstituted allenes are isolated in good yields (Scheme 67).

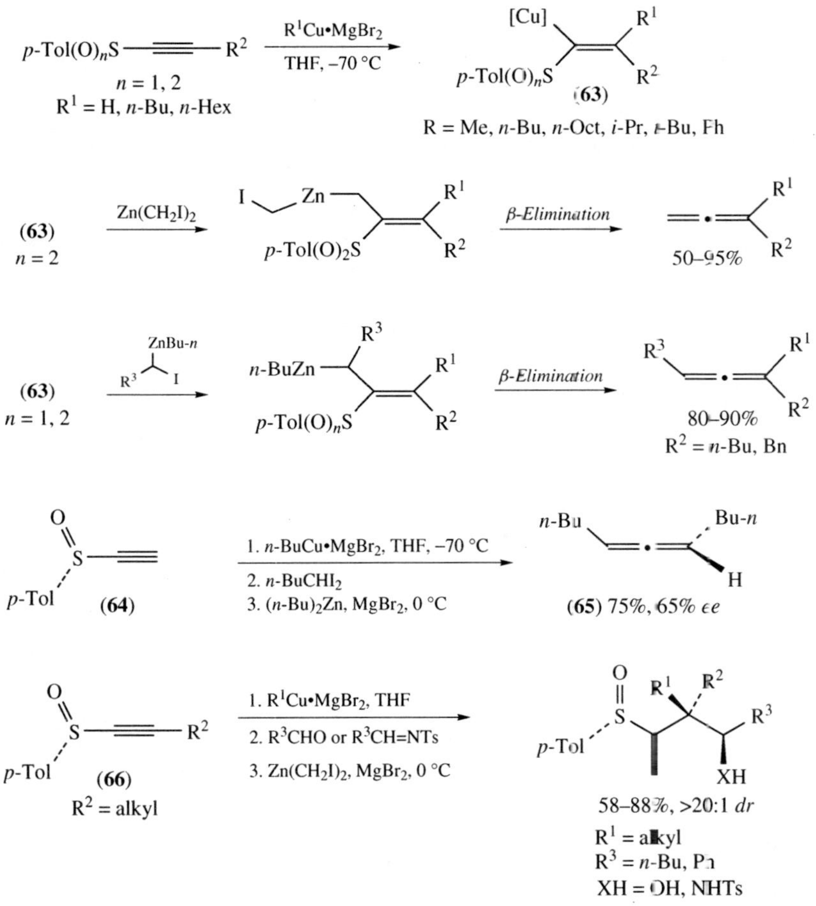

SCHEME 67

Interestingly, the enantiopure chiral alkynyl sulfoxide **64** allows the preparation of the enantioenriched allene **65**. A related reaction has been described for the highly diastereoselective synthesis of quaternary centers from internal *p*-tolyl alkynyl sulfoxides **66** (Scheme 67)[122–125].

The carbocupration reaction has been extended to alkynyl sulfoximine **67**, which can react with ethylcopper reagents to give trisubstituted alkenyl sulfoximines. In all cases, the yields are good with a total regioselectivity (Scheme 68)[126]. On the other hand, the stereoselectivity is variable, the best being attained in THF at −40 °C with EtCu as the copper

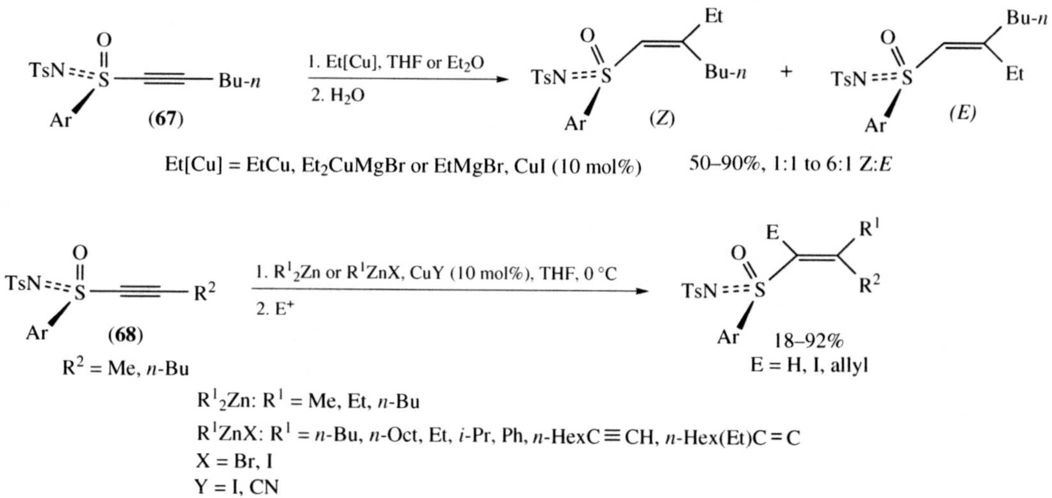
TsN
O
S
Ar
Bu-n
(67)
1. Et[Cu], THF or Et2O
2. H2O
Et
Bu-n
(Z)
+
(E)
Et[Cu] = EtCu, Et2CuMgBr or EtMgBr, CuI (10 mol%)
50–90%, 1:1 to 6:1 Z:E
R2
(68)
R2 = Me, n-Bu
1. R12Zn or R1ZnX, CuY (10 mol%), THF, 0 °C
2. E+
E
R1
18–92%
E = H, I, allyl
R12Zn: R1 = Me, Et, n-Bu
R1ZnX: R1 = n-Bu, n-Oct, Et, i-Pr, Ph, n-HexC≡CH, n-Hex(Et)C=C
X = Br, I
Y = I, CN

SCHEME 68

reagent. Interestingly, although the Cu(I)-catalyzed *carbomagnesiation* of **67** gives an equimolar mixture of (Z)- and (E)-isomers, the Cu(I)-catalyzed carbozincation of alkynyl sulfoximines **68** results in the clean formation of tetrasubstituted alkenyl sulfoximines in good yields as single isomers. Under the best conditions, the reaction is performed with a stoichiometric amount of R_2Zn or RZnX in the presence of a catalytic amount of CuI or CuCN (10 mol%) in THF (Scheme 68).

3. Alkynyl amines, amides and N-alkynyl carbamates

a. Alkynyl amines. Pioneering studies on the carbocupration of alkynyl amines (ynamines) undertaken by Normant's group have demonstrated that N-diphenyl-N-ethynylamine can react regio- and stereoselectively with ethylcopper reagent $EtCu{\bullet}MgBr_2$ in THF or Et_2O to give only the corresponding branched alkenyl amine (enamine) in high yields (Scheme 69)[107]. In this reaction, the sense of the addition is driven by the electron donation from the nitrogen polarizing the acetylene moiety.

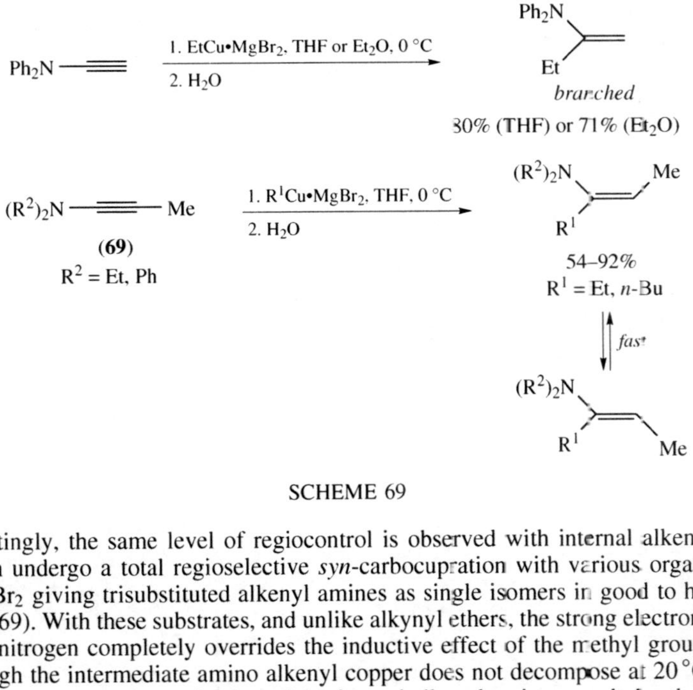

SCHEME 69

Interestingly, the same level of regiocontrol is observed with internal alkenyl amines **69**, which undergo a total regioselective *syn*-carbocupration with various organocoppers $RCu{\bullet}MgBr_2$ giving trisubstituted alkenyl amines as single isomers in good to high yields (Scheme 69). With these substrates, and unlike alkynyl ethers, the strong electron donation from the nitrogen completely overrides the inductive effect of the methyl group.

Although the intermediate amino alkenyl copper does not decompose at 20 °C in THF, the rapid double-bond isomerization of the formed alkenyl amines precludes the synthetic utility of this reaction.

b. Alkynyl amides and N-alkynyl carbamates. Both the carbocupration and Cu(I)-catalyzed *carbomagnesiation* reactions on diversely substituted N-alkynyl carbamates are described[127]. Thus, N-alkynyl carbamate **70** can react regio- and stereoselectively with various organocoppers $RCu{\bullet}MgBr_2$ in Et_2O. In all cases, upon addition of electrophiles to the reaction mixture, the corresponding N-alkenyl carbamates are produced as single isomers in good to excellent isolated yields (Scheme 70). Here again, the coordination of the copper reagent by the amide moiety is assumed to be responsible for the observed

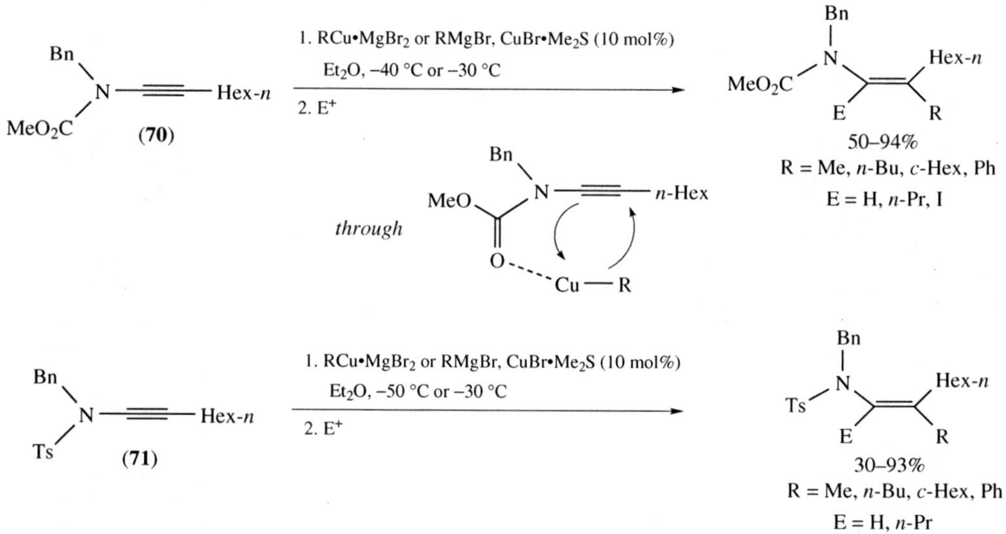
Bn
N
MeO2C
Hex-n
(70)
1. RCu•MgBr2 or RMgBr, CuBr•Me2S (10 mol%)
Et2O, −40 °C or −30 °C
2. E+
E
R
50–94%
R = Me, n-Bu, c-Hex, Ph
E = H, n-Pr, I
through
MeO
O
Cu
n-Hex
Ts
(71)
Et2O, −50 °C or −30 °C
30–93%
E = H, n-Pr

SCHEME 70

regioselectivity. Similar results are obtained during the *carbomagnesiation* of **70** with stoichiometric amounts of Grignard reagents RMgBr in the presence of a catalytic amount of $CuBr{\bullet}Me_2S$ (10 mol%).

The methodology has been successfully applied to sulfonyl-substituted alkynyl amides **71** to prepare the corresponding sulfonyl alkynyl amides in variable yields (Scheme 70)[127].

4. Alkynyl phosphines, phosphonates and phosphine oxides

a. Alkynyl phosphines. Since the late 1970s, the carbocupration of alkynyl phosphines is recognized to be highly regio- and stereoselective. For instance, the addition of ethylcopper reagent $EtCu{\bullet}MgBr_2$ onto (ethynyl)diphenylphosphine is reported by Normant and coworkers to occur in THF or Et_2O at the carbon α to the phosphorus atom to give the corresponding linear (E)-2-phosphinyl-1-alkene in high yields (Scheme 71)[107]. The total observed regioselectivity is due to the electron-withdrawing inductive influence of the phosphorous atom polarizing the acetylene moiety. A high level of regioselectivity is also observed with internal alkynyl phosphines as diphenyl(prop-1-ynyl)phosphine, which leads to adduct **72** as a single isomer in high yield. In this latter case, the regiocontrol is a consequence of the methyl substituent and the phosphorous atom which polarize the acetylene moiety in the same sense, through their inductive effects (Scheme 71).

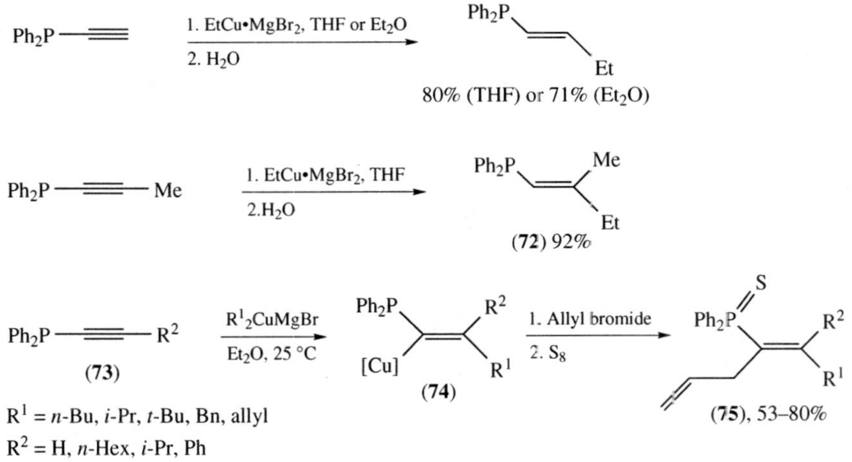

SCHEME 71

Recently, the reaction of various alkynyl phosphines **73** with diverse Grignard-derived cuprates $R_2CuMgBr$ has been used in the stereoselective synthesis of phosphorous-substituted alkenes **75** upon treatment of copper intermediates **74** with allyl bromide followed by the reaction of the phosphorus atom with S_8. Except in the case of the hindered internal alkynyl phosphine substituted by the t-butyl group for which no reaction occurs, adducts **75** are isolated in good yields as single isomers (Scheme 71)[128]. Other electrophiles such as H_2O, acyl chlorides and Ph_2PCl have also been utilized for the synthesis of diversely functionalized alkene derivatives.

b. Alkynyl phosphonates and phosphine oxides. Both alkynyl phosphonates and phosphine oxides are good substrates towards the carbocupration reaction. Thus, diethyl alkenyl

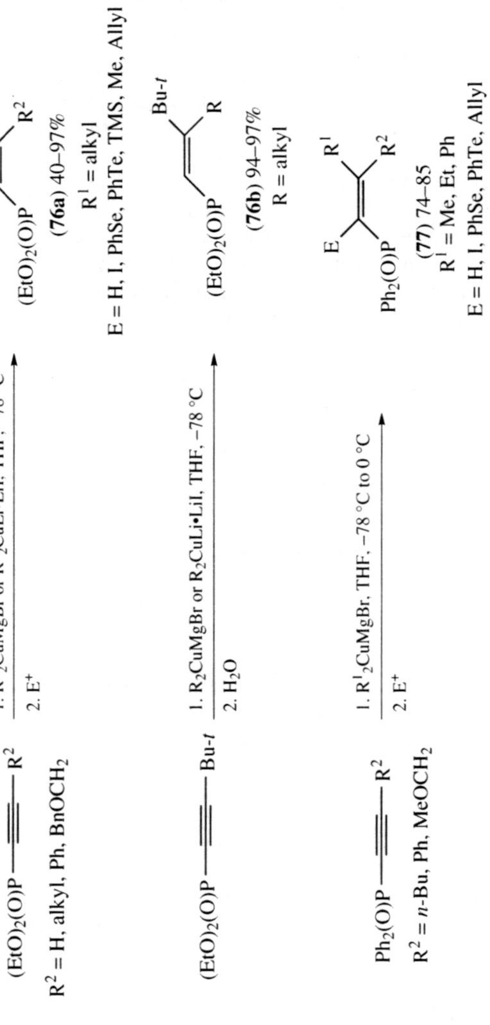
$(EtO)_2(O)P$—≡—R^2
R^2 = H, alkyl, Ph, $BnOCH_2$
1. $R^1{}_2CuMgBr$ or $R^1{}_2CuLi{\bullet}LiI$, THF, −78 °C
2. E^+
$(EtO)_2(O)P$
E
R^1
R^2
(**76a**) 40–97%
R^1 = alkyl
E = H, I, PhSe, PhTe, TMS, Me, Allyl
$(EtO)_2(O)P$—≡—Bu-*t*
1. $R_2CuMgBr$ or $R_2CuLi{\bullet}LiI$, THF, −78 °C
2. H_2O
$(EtO)_2(O)P$
Bu-*t*
R
(**76b**) 94–97%
R = alkyl
$Ph_2(O)P$—≡—R^2
R^2 = *n*-Bu, Ph, $MeOCH_2$
1. $R^1{}_2CuMgBr$, THF, −78 °C to 0 °C
2. E^+
$Ph_2(O)P$
E
R^1
R^2
(**77**) 74–85
R^1 = Me, Et, Ph
E = H, I, PhSe, PhTe, Allyl

SCHEME 72

phosphonates can be obtained through the addition of organocuprates R_2CuLi and $R_2CuMgBr$ to the parent alkynyl phosphonates (Scheme 72)[129]. In all cases, the latter undergo a smooth regioselective carbocupration in THF. The observed regioselectivity may be the consequence of both the mesomeric electron-withdrawing influence from the phosphonate group and the inductive influence of the alkyl substituent R^2, which polarize the carbon–carbon triple bond in the same sense. A phosphonate-directed carbocupration could also be invoked.

When the alkyne is substituted by a hydrogen, a primary alkyl or a phenyl group, only isomers **76a** are obtained in good isolated yields as the result of a complete *syn*-carbocupration. Surprisingly, with the alkynyl phosphonate bearing a *t*-butyl substituent, only isomers **76b** are obtained in high isolated yields as the consequence of a clean *anti*-carbocupration.

Alkynyl phosphine oxides can also react with Grignard-derived cuprates $R_2CuMgBr$ in the same highly regio- and *syn*-stereoselective manner to produce the corresponding alkenyl phosphine oxides **77** in high yields (Scheme 72)[130].

Recently, the carbocupration of trifluoromethylated alkynyl phosphonate **78** has been reported. Various organolithium species and Grignard reagents in the presence of a stoichiometric amount of CuCN allow trifluoromethylated (Z)-alkenyl phosphonates to be prepared as the major isomers in good yields and generally high regioselectivities (Scheme 73)[131].

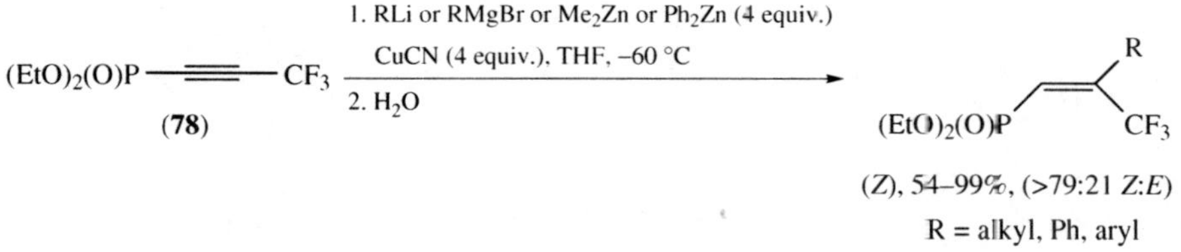

SCHEME 73

Under the same conditions, only dimethyl- and diphenylzinc reagents Me_2Zn and Ph_2Zn in the presence of CuCN lead to high regio- and stereoselectivity.

5. Alkynyl silanes

Terminal alkynyl silanes can react with various copper reagents. Both triphenylsilyl- and trimethylsilylacetylenes can react regio- and stereoselectively with primary, secondary or tertiary alkyl organocopper reagents. Except in the case of MeCu for which no reaction occurs, the corresponding linear alkenyl silanes are isolated in excellent yields upon hydrolysis (Scheme 74)[132].

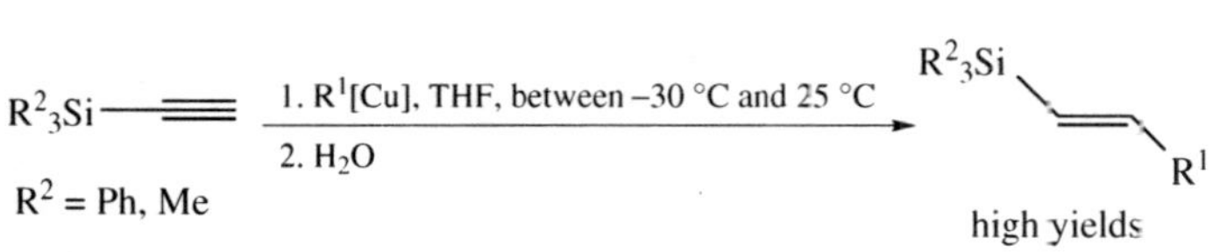

R^1Cu: R^1 = Et, *i*-Pr, *t*-Bu, *c*-Hex

$R^1{}_2CuMgCl$ or $R^1{}_3Cu_2MgCl$: R^1 = Me, Et, *i*-Pr, *t*-Bu, *c*-Hex

SCHEME 74

In this reaction, the sense of the regioselectivity is only driven by the inductive effect of the silicon atom. This is in accordance with Nakamura's *ab initio* theoretical calculations on the carbocupration of ethynyltrimethylsilane with MeCu[30]. Similar results are obtained with Grignard-derived cuprates $R_2CuMgCl$ and higher-order cuprates R_3Cu_2MgCl which can transfer, at an acceptable rate, one R group, even in the case where R is a methyl group (Scheme 74). In all the above reactions, only little dimerization of the intermediate silylated alkenyl copper is observed as a consequence of steric hindrance from the silyl group.

The carbocupration of alkynyl silanes has recently been applied to the synthesis of tetrasubstituted alkenylsilanes. Thus, the Cu(I)-catalyzed *carbomagnesiation* of alkynyl 2-pyridylsilane **79**, performed with a stoichiometric amount of Grignard reagents Ar^1MgI in the presence of a catalytic amount of CuI (30 mol%), allows alkenyl coppers **80** to be produced with a high regio- and stereocontrol. A Kumada–Tamao–Corriu-type cross-*coupling* run on **80** then affords the desired products in good to high yields (Scheme 75)[133, 134]. The high level of regiocontrol in this carbocupration is the consequence of the strong directing effect of the 2-pyridyl group on the silicon. As evidence of this complex-induced effect, the reaction does not take place at all with the corresponding 3-pyridyl-, 4-pyridyl- and phenylsilanes.

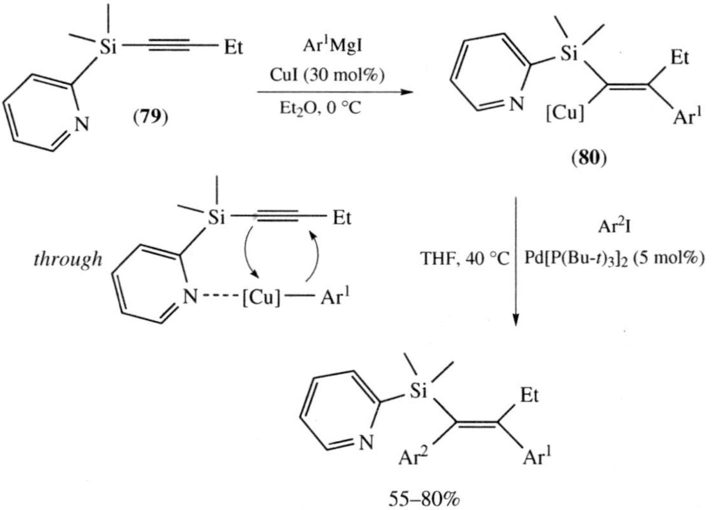

SCHEME 75

E. Intramolecular Carbocupration Reactions

1. Synthesis of carbocycles

a. Non-stabilized organocopper reagents. The intramolecular carbocupration reaction of internal alkynes with various functional non-stabilized copper–zinc reagents has been described by Knochel and coworkers. This reaction leads to functionalized five-membered carbocycles through a clean *syn*-addition, upon trapping the initially formed cyclic organocoppers **81** with electrophiles (Scheme 76)[12].

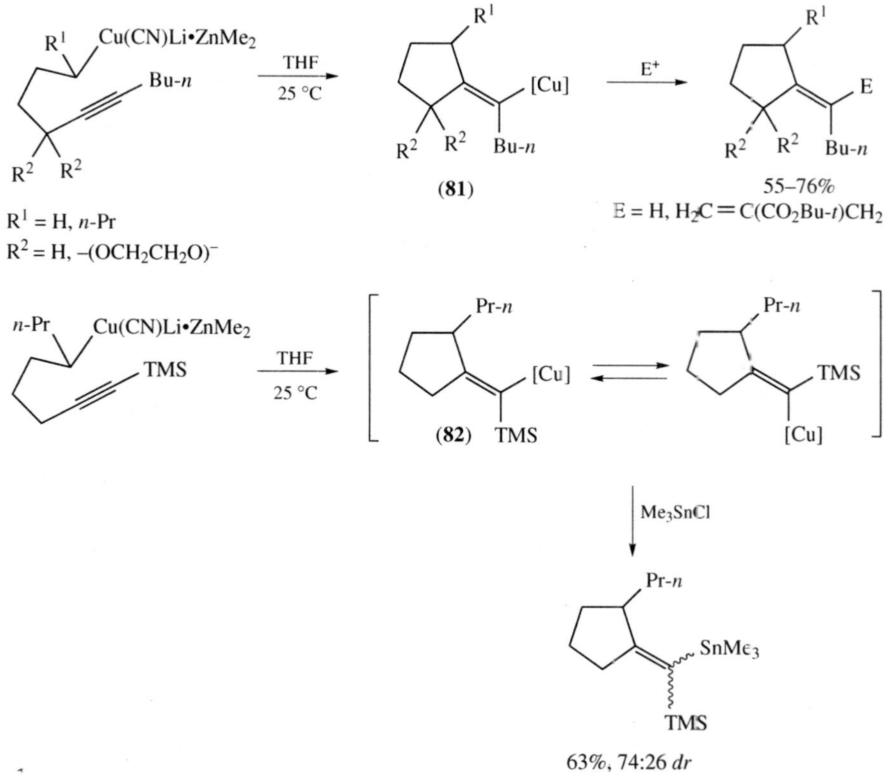

SCHEME 76

Here, the regioselectivity is controlled by the 5-*exo-dig* cyclization being more favorable than the 6-*endo-dig* cyclization. All attemps to extend this methodology to the synthesis of six-membered carbocycles failed.

When the acetylene moiety is substituted by the trimethylsilyl group, the reaction affords the copper intermediate **82** through a clean *syn*-addition[12]. However, the latter is not configurationally stable and thus is in equilibrium with its isomeric form. Upon treatment with Me_3SnCl, the corresponding five-membered carbocycle is then isolated in good yield as a mixture of two isomers (Scheme 76).

An analogous reaction is reported using organocopper reagents derived from *zirconacyclopentanes* **52** (Scheme 77)[61].

Upon successive treatments with CuCl and substituted alkynyl bromides, *zirconacyclopentanes* **52** give the intermediate copper species **83**. The latter then undergo an intramolecular highly regio- and stereoselective *syn*-carbocupration to lead to the formation of bicyclic compounds **84** in good isolated yields via a 5-*exo-dig* cyclization.

From *zirconacyclopentane* **85** and diyne **86**, tricyclic product **87** is formed in good isolated yield. Its formation is explained by two consecutive highly regio- and *syn*-stereoselective intramolecular carbocuprations: a *syn*-5-*exo-dig* cyclization followed by a *syn*-6-*exo-dig* cyclization (Scheme 78)[61].

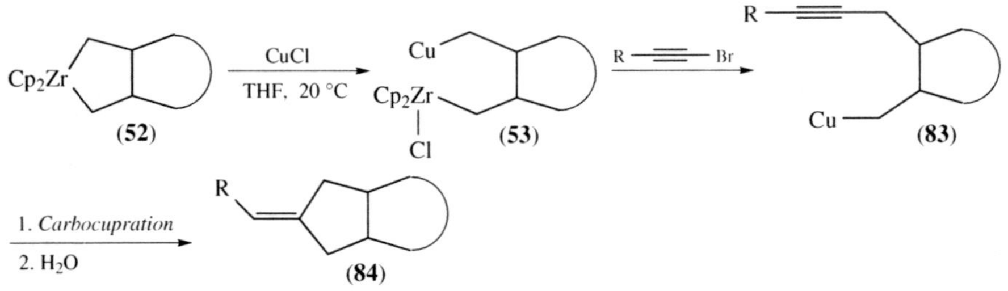

Zirconacycle (**52**)	R	Product (**84**)	Yield (%)
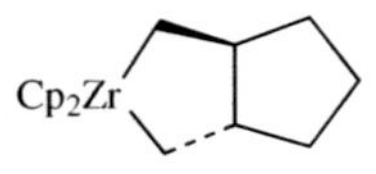	*n*-Bu	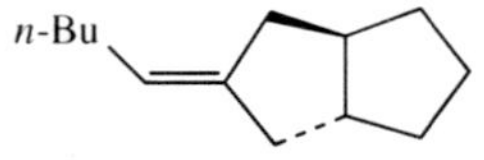	58
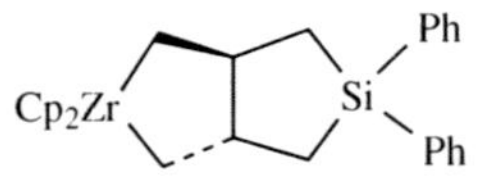	Ph	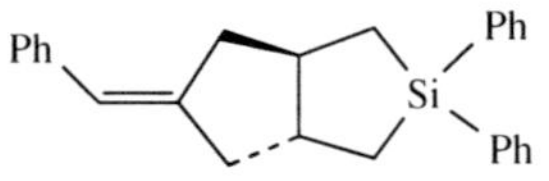	56
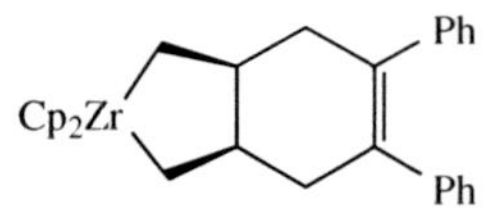	TMS	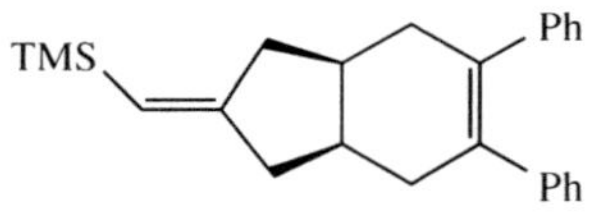	63

SCHEME 77

b. Stabilized organocopper reagents. Stabilized enolates **88**, generated by the deprotonation of the corresponding precursors with KH, are able to undergo intramolecular addition onto terminal alkynes in the presence of a catalytic amount of CuI (10 mol%)[135]. This allows (*Z*)-alkenyl coppers **89** to be formed through a 5-*exo-dig* carbocyclization in a highly regio- and *anti*-stereoselective fashion (Scheme 79). The 'unusual' *anti*-stereoselectivity of the addition has been illustrated through the Pd(0) cross-*coupling* reaction of copper species **89** with aryl halides giving (*Z*)-*exo*-methylene cyclopentanes **90** in good yields and thus supporting a clean *anti*-addition of stabilized enolates onto the terminal alkyne.

The observed *anti*-stereoselectivity supports a reaction mechanism involving an attack of the enolate nucleophile onto the alkyne activated by CuI (Scheme 79). Indeed, the formation of a copper enolate would lead to the opposite selectivity through a *syn*-addition.

The presence of Cu(I) salt in this transformation is crucial to prevent polymerization side-reactions observed if only Pd(0) is used. Variations on this methodology have further been developed, such as the use of a catalytic amount of KOBu-*t* (15 mol%)[136, 137] for the deprotonation step or multi-component Cu(I)-mediated Michael addition–cyclizations[138] (Scheme 80).

Very recently, this reaction was successfully applied to the stereoselective synthesis of 7,7-dialkynyl-5-hydroxymethyl-6-oxabicyclo[3.2.1]octane-1-carboxylic acids on a gram scale without erosion of the yield[139].

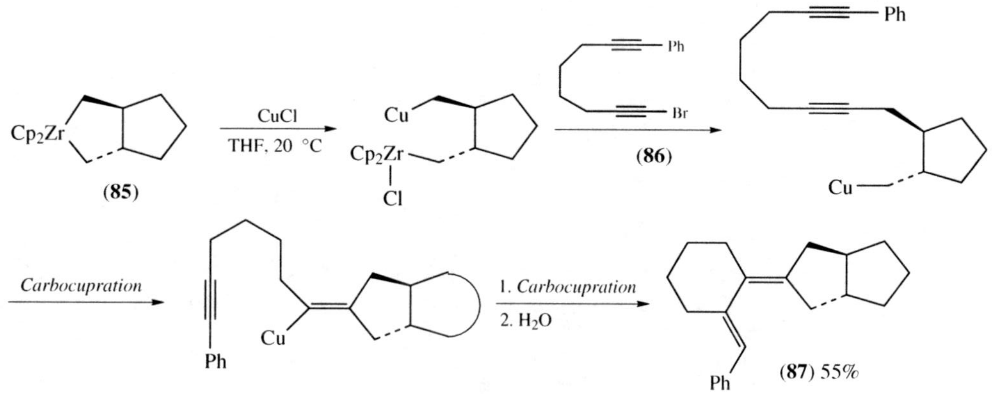
Cp2Zr
(85)
CuCl
THF, 20 °C
Cu
Cp2Zr
Cl
Ph
Br
(86)
Ph
Cu
Carbocupration
Ph
Cu
1. Carbocupration
2. H2O
Ph
(87) 55%

SCHEME 78

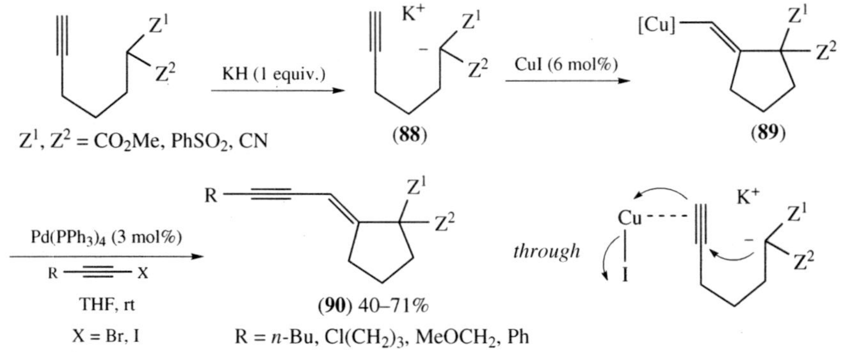

SCHEME 79

A related reaction involving Michael addition of potassium enolate derived from malonic enolate to α,β-ethylenic ester **91** has been described. Proton transfer followed by the addition of CuI (10 mol%) to the reaction mixture allows the annulation process to take place. Subsequent hydrolysis results in the formation of the desired highly functionalized *exo*-methylene cyclopentane in good isolated yield (Scheme 81).

Cu(I)-mediated 5-*exo-dig* cyclization onto internal alkynes is generally sluggish. However, the cyclization of stabilized enolate **92** can be achieved within 2 h when refluxing THF in the presence of a stoichiometric amount of CuI (Scheme 82)[137].

In the case of enolate **93**, the reaction needs a longer reaction time than with **92**, which results in the isomerization of the (Z)-alkenyl copper intermediate into its more stable (E)-isomer. In this case, a moderate selectivity in favor of the (Z)-adduct is then obtained.

Cu(I)-mediated 5-*endo-dig* carbocyclization of stabilized enolates derived from **94** onto internal aryl alkynes has been reported[140]. Treated with a catalytic amount of KOBu-*t* (5 mol%) and CuI (2 mol%), the desired cyclized indenes are isolated in moderate to excellent yields (Scheme 83). The annulation reaction tolerates a wide range of functionalities on the acetylene moiety. However, in the case of the sterically demanding trimethylsilylethynyl group, no reaction occurs even using stoichiometric amounts of KOBu-*t* and CuI. A mechanistic rationale involving an *anti* 5-*endo-dig* carbocupration analogous to the one presented for the Cu(I)-mediated 5-*exo-dig* cyclizations (see Scheme 79) has been proposed.

2. Synthesis of heterocycles

Cu(I)-promoted cyclizations of stabilized enolates give access to pyrrolidines. Addition of lithium propargylamides, generated by the deprotonation of amines with a catalytic amount of *n*-BuLi (10 mol%), to activated benzylidene derivatives **95** followed by the cyclization of the resulting anion in the presence of a catalytic amount of CuI (3 mol%) leads to *exo*-methylene pyrrolidines in high yields (73–89%) upon hydrolysis[141].

Under the same conditions, alkylidene derivatives **96** give lower yields due to the undesired deprotonation of the substrates. As in the case of the synthesis of carbocycles, the carbocupration reaction has been demonstrated to proceed in an *anti*-fashion. The mechanism involves an attack of the enolate nucleophile onto the acetylene moiety activated by the *in situ* generated propargylamine copper salt (Scheme 84).

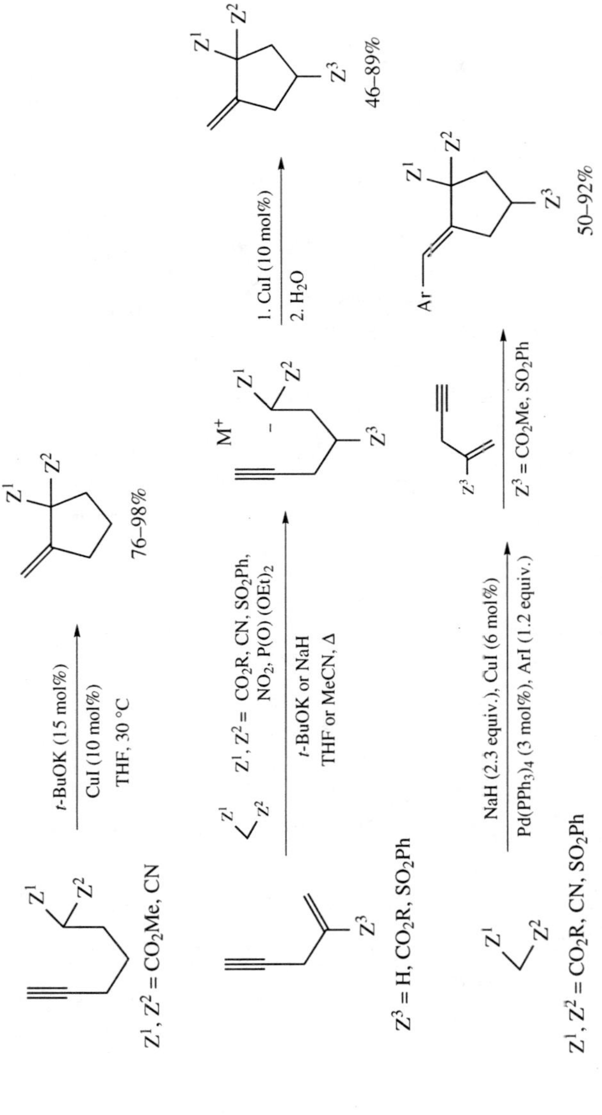

t-BuOK (15 mol%)
CuI (10 mol%)
THF, 30 °C
Z^1, Z^2 = CO_2Me, CN
76–98%
Z^1, Z^2 = CO_2R, CN, SO_2Ph, NO_2, $P(O)(OEt)_2$
t-BuOK or NaH
THF or MeCN, Δ
Z^3 = H, CO_2R, SO_2Ph
M^+
1. CuI (10 mol%)
2. H_2O
46–89%
NaH (2.3 equiv.), CuI (6 mol%)
$Pd(PPh_3)_4$ (3 mol%), ArI (1.2 equiv.)
Z^3 = CO_2Me, SO_2Ph
Z^1, Z^2 = CO_2R, CN, SO_2Ph
Ar
50–92%

SCHEME 80

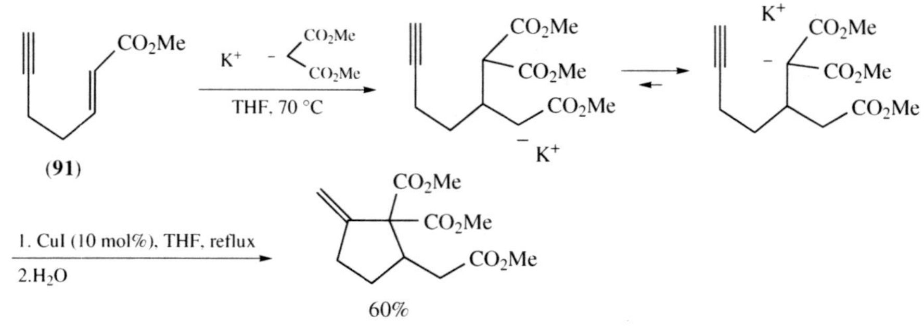

SCHEME 81

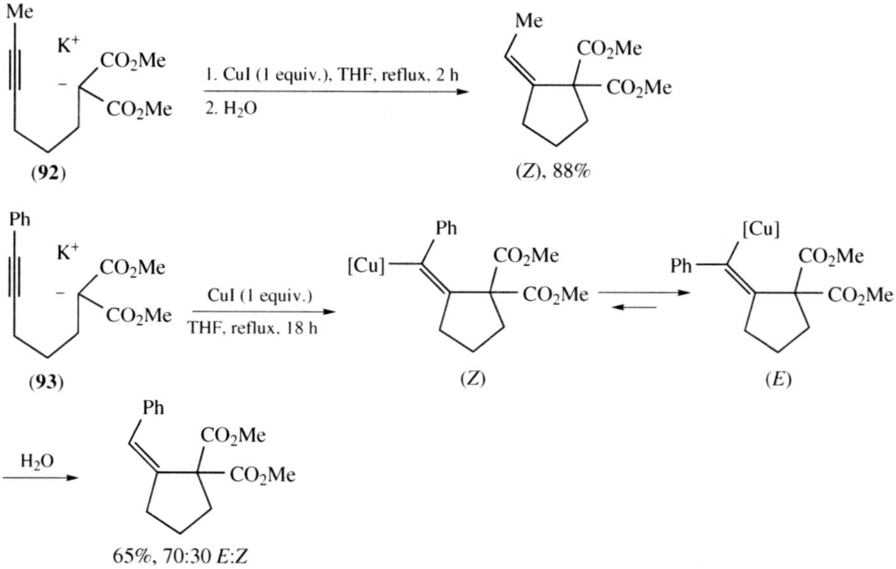

SCHEME 82

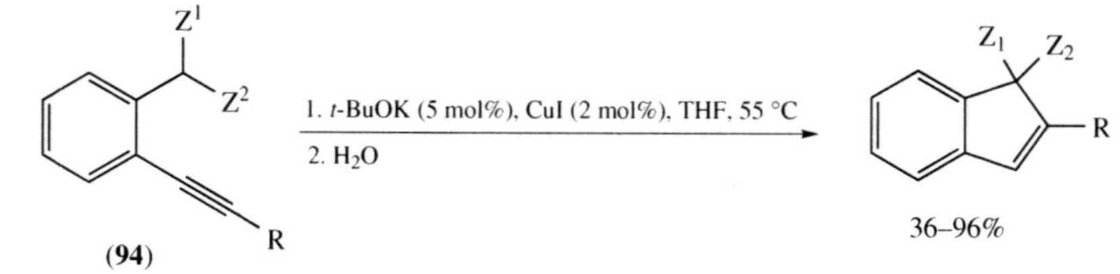

Z^1, Z^2 = CO_2Et, CN, SO_2Ph

R = (functional) alkyl, alkenyl, aryl

SCHEME 83

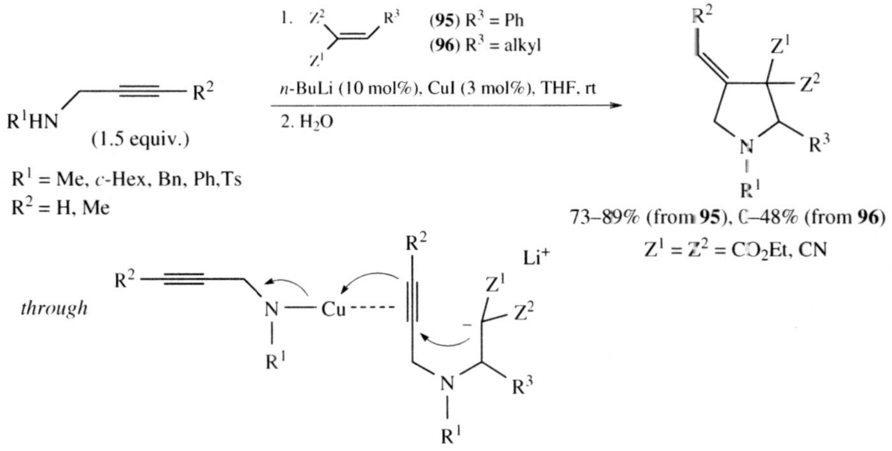

SCHEME 84

The reaction has been extended to other electron-deficient olefins **97** to give the corresponding *exo*-methylene pyrrolidines in good yields and selectivities in the presence of a catalytic amount of CuI or of the more soluble CuI•[P(Ph$_3$)]$_3$ complex (3 mol%) (Scheme 85)[141,142].

A one-pot *coupling* of nitroalkenes **98** and vinylsulfones **99** with NaOAr is also described[142]. When treated with a catalytic amount of Pd(PPh$_3$)$_4$ (4 mol%), the initially formed alkenyl coppers undergo an allylic substitution giving pyrrolidine derivatives (Scheme 85). From nitroalkenes **98**, the corresponding functionalized pyrrolidines are obtained in modest to good yields as single isomers. On the other hand, from vinylsulfones **99** inseparable mixtures of adducts are obtained with variable selectivities.

Similarly, in the presence of catalytic amounts of NaH (50 mol%) and CuI•[P(Ph$_3$)]$_3$ complex (30 mol%), propargyl alcohol **43** adds to ethyl 2-phenylsulfonylcinnamate to give the corresponding *exo*-methylene tetrahydrofuran in good yield as a single isomer (Scheme 86)[143]. The methodology has been applied to the reaction of a variety of secondary as well as tertiary alcohols **100** with benzylidene- and alkylidenemalonates[144]. In the presence of catalytic amounts of *n*-BuLi (25 mol%) and CuI (20 mol%), polysubstituted *exo*-methylene tetrahydrofurans are then obtained in moderate to high yields and high purity. Best results are obtained with a mono- or disubstitution at the propargyl position (Scheme 86).

IV. CONCLUSION

Since its discovery at the beginning of the 1970s, the carbocupration of alkynes is a reaction that has proven its utility and versatility in the stereoselective synthesis of di-, tri- and tetrasubstituted alkenes. This reaction has been applied to a number of differently substituted alkynes, and a good understanding of the mechanistic background of this reaction has been acquired over the last thirty years. All these results and studies, as well as all its successful applications to total synthesis of important natural fragments, have made the carbocupration reaction one of the most important 'classical' examples in the field of carbometallation-type reactions.

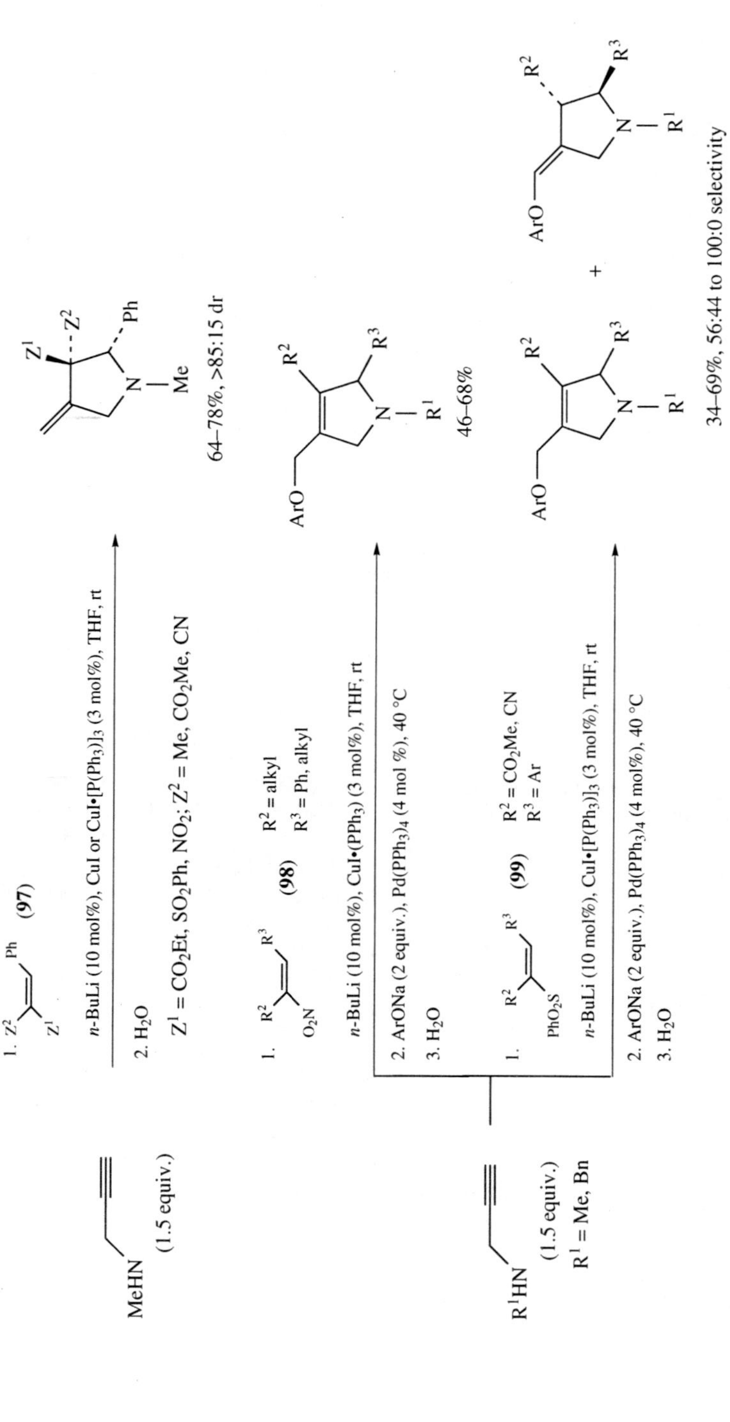
MeHN
(1.5 equiv.)
1. (97)
n-BuLi (10 mol%), CuI or CuI•[P(Ph3)]3 (3 mol%), THF, rt
2. H2O
Z1 = CO2Et, SO2Ph, NO2; Z2 = Me, CO2Me, CN
64–78%, >85:15 dr
R1HN
(1.5 equiv.)
R1 = Me, Bn
1. (98) R2 = alkyl R3 = Ph, alkyl
n-BuLi (10 mol%), CuI•(PPh3) (3 mol%), THF, rt
2. ArONa (2 equiv.), Pd(PPh3)4 (4 mol %), 40 °C
3. H2O
46–68%
1. (99) R2 = CO2Me, CN R3 = Ar
n-BuLi (10 mol%), CuI•[P(Ph3)]3 (3 mol%), THF, rt
2. ArONa (2 equiv.), Pd(PPh3)4 (4 mol%), 40 °C
3. H2O
34–69%, 56:44 to 100:0 selectivity

SCHEME 85

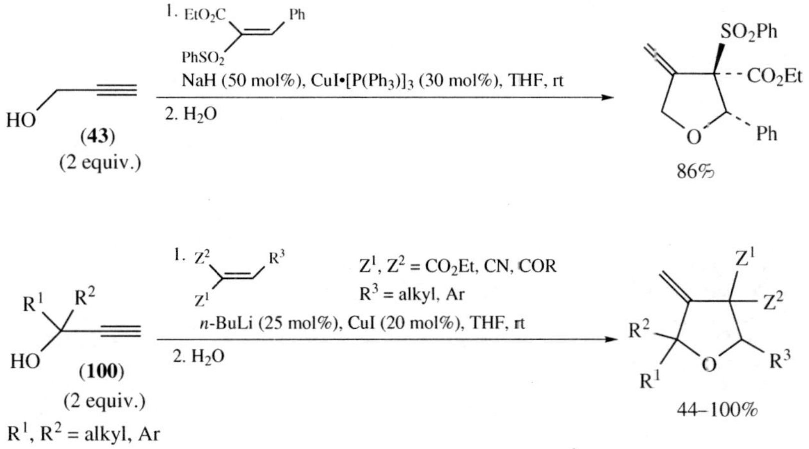

SCHEME 86

V. REFERENCES

1. J. F. Normant, Alkyne Carbocupration and Polyene Synthesis, in *Organocopper Reagents—A Practical Approach* (Ed. R. J. K. Taylor), Chap. 11, Oxford University Press, New York, 1994, pp. 237–256.
2. B. H. Lipshutz and S. Sengupta, *Org. React.*, **41**, 135 (1992).
3. J. F. Normant and A. Alexakis, *Synthesis*, 841 (1981).
4. A. G. Fallis and P. Forgione, *Tetrahedron*, **57**, 5899 (2001).
5. S. Casson and P. Kocienski, *Contemp. Org. Synth.*, **2**, 19 (1995).
6. A. B. Flynn and W. W. Ogilvie, *Chem. Rev.*, **107**, 4698 (2007).
7. B. Breit and Y. Schmidt, *Chem. Rev.*, **108**, 2928 (2008).
8. A. Alexakis, J. F. Normant and J. Villiéras, *Tetrahedron Lett.*, 3461 (1976).
9. J. F. Normant and M. Bourgain, *Tetrahedron Lett.*, 2583 (1971).
10. J. F. Normant, G. Cahiez, M. Bourgain, C. Chuit and J. Villiéras, *Bull. Soc. Chim. Fr.*, 1656 (1974).
11. A. Alexakis, G. Cahiez and J. F. Normant, *J. Organomet. Chem.*, **177**, 293 (1979).
12. S. Achyutha Rao and P. Knochel, *J. Am. Chem. Soc.*, **113**, 5735 (1991).
13. M. Gardette, A. Alexakis and J. F. Normant, *Tetrahedron Lett.*, **23**, 5155 (1982).
14. M. Gardette, A. Alexakis and J. F. Normant, *Tetrahedron*, **41**, 5887 (1985).
15. A. Alexakis and J. F. Normant, *Tetrahedron Lett.*, **23**, 5151 (1982).
16. M. Furber, R. J. K. Taylor and S. C. Burford, *J. Chem. Soc., Perkin Trans. 1*, 1809 (1986).
17. O. Krebs and R. J. K. Taylor, *Org. Lett.*, **7**, 1063 (2005).
18. H. Westmijze, J. Meijer and P. Vermeer, *Recl. Trav. Chim. Pays-Bas*, **96**, 168 (1977).
19. A. B. Levy, P. Talley and J. A. Dunford, *Tetrahedron Lett.*, 3545 (1977).
20. C. Germon, A. Alexakis and J. F. Normant, *Tetrahedron Lett.*, **21**, 3763 (1980).
21. C. Germon, A. Alexakis and J. F. Normant, *Synthesis*, 43 (1984).
22. A. Alexakis, G. Cahiez and J. F. Normant, *Tetrahedron*, **36**, 1961 (1980).
23. M. Vuagnoux-d'Augustin and A. Alexakis, *Eur. J. Org. Chem.*, 5852 (2007).
24. N. Jabri, A. Alexakis and J. F. Normant, *Tetrahedron Lett.*, **24**, 5081 (1983).
25. M. Furber, R. J. K. Taylor and S. C. Burford, *Tetrahedron Lett.*, **26**, 2731 (1985).
26. A. Alexakis, G. Cahiez and J. F. Normant, *Synthesis*, 826 (1979).
27. J. F. Normant, A. Commerçon and J. Villiéras, *Tetrahedron Lett.*, 1465 (1975).
28. J. F. Normant, A. Commerçon and J. Villiéras, *Tetrahedron*, **36**, 1215 (1980).
29. E. Nakamura and S. Mori, *Angew. Chem., Int. Ed.*, **39**, 3750 (2000).
30. E. Nakamura, Y. Miyachi, N. Koga and K. Morokuma, *J. Am. Chem. Soc.*, **114**, 6686 (1992).
31. E. Nakamura, S. Mori, M. Nakamura and K. Morokuma, *J. Am. Chem. Soc.*, **119**, 4887 (1997).
32. S. Mori and E. Nakamura, *Tetrahedron Lett.*, **40**, 5319 (1999).

33. S. Mori and E. Nakamura, *J. Mol. Struct. (Theochem)*, **461–462**, 167 (1999).
34. K. Nilsson, T. Andersson and C. Ullenius, *J. Organomet. Chem.*, **545–546**, 591 (1997).
35. K. Nilsson, T. Andersson, C. Ullenius, A. Gerold and N. Krause, *Chem. Eur. J.*, **4**, 2051 (1998).
36. S. H. Bertz, C. M. Carlin, D. A. Deadwyler, M. D. Murphy, C. A. Ogle and P. H. Seagle, *J. Am. Chem. Soc.*, **124**, 13650 (2002).
37. S. H. Bertz, S. Cope, D. C. Dorton, M. D. Murphy and C. A. Ogle, *Angew. Chem., Int. Ed.*, **46**, 7082 (2007).
38. S. H. Bertz, S. Cope, M. D. Murphy, C. A. Ogle and B. J. Taylor, *J. Am. Chem. Soc.*, **129**, 7208 (2007).
39. E. R. Bartholomew, S. H. Bertz, S. Cope, M. D. Murphy and C. A. Ogle, *J. Am. Chem. Soc.*, **130**, 11244 (2008).
40. E. R. Bartholomew, S. H. Bertz, S. Cope, D. C. Dorton, M. D. Murphy and C. A. Ogle, *Chem. Commun.*, 1176 (2008).
41. V. N. Odinokov, *Chem. Nat. Compounds*, **36**, 11 (2000); *Chem. Abstr.*, **134**, 310998 (2001).
42. G. Cahiez, A. Alexakis and J. F. Normant, *Tetrahedron Lett.*, **21**, 1433 (1980).
43. J. B. Ousset, C. Mioskowski, Y.-L. Yang and J. R. Falck, *Tetrahedron Lett.*, **25**, 5903 (1984).
44. A. F. Sviridov, M. S. Ermolenko, D. V. Yashunsky and N. K. Kotchetkov, *Tetrahedron Lett.*, **24**, 4359 (1983).
45. A. Alexakis, G. Cahiez and J. F. Normant, *Tetrahedron Lett.*, 2027 (1978).
46. M. Gardette, N. Jabri, A. Alexakis and J. F. Normant, *Tetrahedron*, **40**, 2741 (1984).
47. A. Alexakis, A.-M. Barthel, J. F. Normant, C. Fugier and M. Leroux, *Synth. Commun.*, **22**, 1839 (1992).
48. V. Fiandanese, G. Marchese, F. Naso, L. Ronzini and D. Rotunno, *Tetrahedron Lett.*, **30**, 243 (1989).
49. J. G. Millar, *Tetrahedron Lett.*, **38**, 7971 (1997).
50. F. Ramiandrasoa and F. Tellier, *Synth. Commun.*, **20**, 333 (1990).
51. F. Ramiandrasoa and C. Descoins, *Synth. Commun.*, **19**, 2703 (1989).
52. J. F. Normant and A. Alexakis, *Synthesis*, 841 (1981).
53. C. Germon, A. Alexakis and J. F. Normant, *Synthesis*, 40 (1984).
54. C. Germon, A. Alexakis and J. F. Normant, *Synthesis*, 43 (1984).
55. H. Westmije, H. Kleijn and P. Vermeer, *Tetrahedron Lett.*, 2023 (1977).
56. S. A. A. Rao and M. Periasamy, *Tetrahedron Lett.*, **29**, 4313 (1988).
57. H. Westmijze, J. Meijer, H. J. T. Bos and P. Vermeer, *Recl. Trav. Chim. Pays-Bas*, **95**, 299 (1976).
58. H. Westmijze, J. Meijer, H. J. T. Bos and P. Vermeer, *Recl. Trav. Chim. Pays-Bas*, **95**, 304 (1976).
59. W. Adam, C. Sahin and M. Schneider, *J. Am. Chem. Soc.*, **117**, 1695 (1995).
60. D. Zhang and J. M. Ready, *Org. Lett.*, **7**, 5681 (2005).
61. Y. Liu, B. Shen, M. Kotora and T. Takahashi, *Angew. Chem., Int. Ed.*, **38**, 949 (1999).
62. M. Bourgain-Commerçon, J. P. Foulon and J. F. Normant, *Tetrahedron Lett.*, **24**, 5077 (1983).
63. J. P. Foulon, M. Bourgain-Commerçon and J. F. Normant, *Tetrahedron*, **42**, 1389 (1986).
64. J. P. Foulon, M. Bourgain-Commerçon and J. F. Normant, *Tetrahedron*, **42**, 1399 (1986).
65. E. Shirakawa, T. Yamagami, T. Kimura, S. Yamaguchi and T. Hayashi, *J. Am. Chem. Soc.*, **127**, 17164 (2005).
66. A. Alexakis, J. F. Normant and J. Villiéras, *J. Mol. Catal.*, **1**, 43 (1975/76).
67. F. Chemla and F. Ferreira, *Curr. Org. Chem.*, **6**, 539 (2002).
68. J. A. Marshall and K. G. Pinney, *J. Org. Chem.*, **58**, 7180 (1993).
69. A. C. Oehlschlager and E. Czyzewska, *Tetrahedron Lett.*, **24**, 5587 (1983).
70. A. Alexakis, I. Marek, P. Mangeney and J. F. Normant, *Tetrahedron*, **47**, 1677 (1991).
71. A. Alexakis, A. Commerçon, C. Coulentianos and J. F. Normant, *Tetrahedron*, **40**, 715 (1984).
72. I. Marek, A. Alexakis, P. Mangeney and J. F. Normant, *Bull. Soc. Chim. Fr.*, **129**, 171 (1992).
73. A. Alexakis, A. Commerçon, C. Coulentianos and J. F. Normant, *Pure Appl. Chem.*, **55**, 1759 (1983).
74. A. Alexakis, P. Mangeney and J. F. Normant, *Tetrahedron Lett.*, **26**, 4197 (1985).
75. J. C. Shattuck, A. Svatos, C. M. Blazy and J. Meinwald, *Tetrahedron Lett.*, **38**, 6803 (1997).
76. K. J. MacNeil and D. J. Burton, *J. Org. Chem.*, **58**, 4411 (1993).
77. R. D. Guneratne and D. J. Burton, *J. Fluorine Chem.*, **98**, 11 (1999).

78. T. Konno, T. Daitoh, A. Noiri, J. Chae, T. Ishihara and H. Yamanaka, *Org. Lett.*, **6**, 933 (2004).
79. T. Konno, T. Daitoh, A. Noiri, J. Chae, T. Ishihara and H. Yamanaka, *Tetrahedron*, **61**, 9391 (2005).
80. J. F. Normant, A. Alexakis and J. Villiéras, *J. Organomet. Chem.*, **57**, C99 (1973).
81. J. K. Crandall and F. Collonges, *J. Org. Chem.*, **41**, 4089 (1976).
82. J. G. Duboudin and B. Jousseaume, *J. Organomet. Chem.*, **168**, 1 (1979).
83. J. G. Duboudin, B. Jousseaume and A. Bonakdar, *J. Organomet. Chem.*, **168**, 227 (1979).
84. A. G. Fallis, *Synlett*, 2249 (2004).
85. T. Wong, M. W. Tjepkema, H. Audrain, P. D. Wilson and A. G. Fallis, *Tetrahedron Lett.*, **37**, 755 (1996).
86. P. Forgione and A. G. Fallis, *Tetrahedron Lett.*, **41**, 11 (2000).
87. P. Forgione, P. D. Wilson and A. G. Fallis, *Tetrahedron Lett.*, **41**, 17 (2000).
88. P. Forgione, P. D. Wilson, G. P. A. Yap and A. G. Fallis, *Synthesis*, 921 (2000).
89. N. P. Villava-Servin, A. Laurent and A. G. Fallis, *Synlett*, 1251 (2003).
90. P. J. Tessier, A. J. Penwell, F. E. S. Souza and A. G. Fallis, *Org. Lett.*, **5**, 2989 (2003).
91. A. G. Fallis, P. Forgione, S. Woo, S. Legoupy, S. Py, C. Harwig and T. Rietveld, *Polyhedron*, **19**, 533 (2000).
92. P. Bury, G. Hareau, P. Kocienski and D. Dhanak, *Tetrahedron*, **50**, 8793 (1994).
93. J. G. Duboudin and B. Jousseaume, *Synth. Commun.*, **9**, 53 (1979).
94. J. E. Baldwin, R. M. Adlington, D. G. Marquess, A. R. Pitt, M. J. Porter and A. T. Russell, *Tetrahedron*, **52**, 2515 (1996).
95. R. C. Larock, M. J. Doty and X. Han, *J. Org. Chem.*, **64**, 8770 (1999).
96. C. Spino and C. Gobdout, *J. Am. Chem. Soc.*, **125**, 12106 (2003).
97. G.-H. Fang, Z.-J. Yan, J. Yang and M.-Z. Deng, *Synthesis*, 1148 (2006).
98. J. G. Millar, *Tetrahedron Lett.*, **49**, 315 (2008).
99. B. B. Snider, J. Y. Kiselgof and B. M. Foxman, *J. Org. Chem.*, **63**, 7945 (1998).
100. A. S. Kende, J. I. Martin Hernando and J. B. J. Milbank, *Tetrahedron*, **58**, 61 (2002).
101. K. A. Parker and Q. Xie, *Org. Lett.*, **10**, 1349 (2008).
102. Z. Lu and S. Ma, *J. Org. Chem.*, **71**, 2655 (2006).
103. S. Ma and Z. Lu, *Adv. Synth. Catal.*, **348**, 1894 (2006).
104. H. Kleijn and P. Vermeer, *J. Organomet. Chem.*, **292**, 437 (1985).
105. H. Todo, J. Terao, H. Watanabe, H. Kuniyasu and N. Kambe, *Chem. Commun.*, 1332 (2008).
106. S. Czernecki, A. Hoang and J.-M. Valéry, *Tetrahedron Lett.*, **35**, 3539 (1994).
107. A. Alexakis, G. Cahiez and J. F. Normant, *Bull. Soc. Chim. Fr.*, **99**, 693 (1977).
108. J. M. Oostveen, H. Westmijze and P. Vermeer, *J. Org. Chem.*, **45**, 1158 (1980).
109. N. Kato and N. Miyaura, *Tetrahedron*, **52**, 13347 (1996).
110. J. S. Clark, M. C. Kimber, J. Robertson, C. S. P. McErlean and C. Wilson, *Angew. Chem., Int. Ed.*, **44**, 6157 (2005).
111. H. Chechik-Lankin and I. Marek, *Org. Lett.*, **5**, 5087 (2003).
112. H. Chechik-Lankin and I. Marek, *Synthesis*, 3311 (2005).
113. W. E. Truce and M. J. Lusch, *J. Org. Chem.*, **39**, 3174 (1974).
114. P. Vermeer, J. Meijer and C. Eylander, *Recl. Trav. Chim. Pays-Bas*, **93**, 240 (1974).
115. W. E. Truce and M. J. Lusch, *J. Org. Chem.*, **43**, 2252 (1978).
116. P. Vermeer and J. Meijer, *Recl. Trav. Chim. Pays-Bas*, **94**, 15 (1975).
117. V. Fiandanese, G. Marchese and F. Naso, *Tetrahedron Lett.*, 5131 (1978).
118. Q. Xu and X. Huang, *Tetrahedron Lett.*, **45**, 5657 (2004).
119. M. Xie, L. Liu, J. Wang and S. Wang, *J. Organomet. Chem.*, **690**, 4058 (2005).
120. J. P. Varghese, P. Knochel and I. Marek, *Org. Lett.*, **2**, 2849 (2000).
121. J. P. Varghese, I. Zouev, L. Aufauvre, P. Knochel and I. Marek, *Eur. J. Org. Chem.*, 4151 (2002).
122. G. Sklute, D. Amsallem, A. Shabli, J. P. Varghese and I. Marek, *J. Am. Chem. Soc.*, **125**, 11776 (2003).
123. G. Sklute and I. Marek, *J. Am. Chem. Soc.*, **128**, 4642 (2006).
124. G. Kolodney, G. Sklute, S. Perrone, P. Knochel and I. Marek, *Angew. Chem., Int. Ed.*, **46**, 9291 (2007).
125. I. Marek and G. Sklute, *Chem. Commun.*, 1683 (2007).
126. G. Sklute, C. Bolm and I. Marek, *Org. Lett.*, **9**, 1259 (2007).

127. H. Chechik-Lankin, S. Livshin and I. Marek, *Synlett*, 2098 (2005).
128. S. Kanemura, A. Kondoh, H. Yorimitsu and K. Oshima, *Org. Lett.*, **9**, 2031 (2007).
129. J. Mo and D. Y. Oh, *J. Org. Chem.*, **64**, 2950 (1999).
130. X. Huang and Z. Wu, *Synthesis*, 2445 (2004).
131. T. Konno, A. Morigaki, K. Ninomiya, T. Miyabe and T. Ishihara, *Synthesis*, 564 (2008).
132. H. Westmijze, H. Kleijn and P. Vermeer, *J. Organomet. Chem.*, **276**, 317 (1984).
133. K. Itami, T. Kamei and J.-I. Yoshida, *J. Am. Chem. Soc.*, **125**, 14670 (2003).
134. T. Kamei, K. Itami and J.-I. Yoshida, *Adv. Synth. Catal.*, **346**, 1824 (2004).
135. D. Bouyssi, G. Balme and J. Gore, *Tetrahedron Lett.*, **32**, 6541 (1991).
136. N. Monteiro, G. Balme and J. Gore, *Synlett*, 227 (1992).
137. D. D. Bouyssi, N. Monteiro and G. Balme, *Tetrahedron Lett.*, **40**, 1297 (1999).
138. N. Coia, D. Bouyssi and G. Balme, *Eur. J. Org. Chem.*, 3158 (2007).
139. N. Pérez-Hernández, M. Febles, C. Pérez, R. Pérez, M. L. Rodríguez, C. Foces-Foces and J. D. Martín, *J. Org. Chem.*, **71**, 1139 (2006).
140. D. Zhang, Z. Liu, E. K. Yum and R. C. Larock, *J. Org. Chem.*, **72**, 251 (2007).
141. B. Clique, N. Monteiro and G. Balme, *Tetrahedron Lett.*, **40**, 1301 (1999).
142. B. Clique, S. Vassiliou, N. Monteiro and G. Balme, *Eur. J. Org. Chem.*, 1493 (2002).
143. M. Cavicchioli, X. Marat, N. Monteiro, B. Hartmann and G. Balme, *Tetrahedron Lett.*, **43**, 2609 (2002).
144. M. Woods, N. Monteiro and G. Balme, *Eur. J. Org. Chem.*, 1711 (2000).

The chemistry of

Organocopper Compounds

Dedicated to my wife, Cecile

Ilan

Contributing authors

Özge Aksin-Artok	Dortmund University of Technology, Organic Chemistry, Otto-Hahn-Strasse 6, D-44227 Dortmund, Germany Fax: +49-231-7553884; e-mail: oezge.aksin@tu-dortmund.de
Alexandre Alexakis	University of Geneva, Department of Organic Chemistry, 30, Quai Ernest Ansermet 1211 Genève 4, Switzerland Fax: +41-22-3793215; e-mail: alexandre.alexakis@unige.ch
Sarah J. Aves	Department of Chemistry, University of Cambridge, Lensfield Road, Cambridge, CB2 1EW, UK Fax: +44-1223-336362; e-mail: sja41@cam.ac.uk
Asuncion Barbero	Departamento de Química Orgánica, Facultad de Ciencias, Universidad de Valladolid, 47011 Valladolid, Spain Fax: +34-983-423013; e-mail: barbero@qo.uva.es
Donald J. Burton	Department of Chemistry, University of Iowa, Iowa City, IA 52242, USA Fax: +1-319-335-1270; e-mail: donald-burton@uiowa.edu
Fabrice Chemla	Institut Parisien de Chimie Moléculaire, UPMC-Univ Paris 6, CNRS UMR 7201, Tour 44–45, 2ème Etage, Case 183, 4 Place Jussieu, 75252 Paris Cedex 05, France Fax: +33-1-44-27-75-67; e-mail: fabrice.chemla@upmc.fr
Charles R. Davis	Sigma-Aldrich, 6000 N. Teutonia Avenue, Milwaukee, WI 53209, USA Fax: +1-414-438-4224; e-mail: charles.davis@sial.com
R. Karl Dieter	Howard L. Hunter Laboratory, Department of Chemistry, Clemson University, Clemson, South Carolina 29634-0973, USA Fax: +1-864-656-6613; e-mail: dieterr@clemson.edu
Franck Ferreira	Institut Parisien de Chimie Moléculaire, UPMC-Univ Paris 6, CNRS UMR 7201, Tour 44–45, 2ème Etage, Case 183, 4 Place Jussieu, 75252 Paris Cedex 05, France Fax: +33-1-44-27-75-67; e-mail: franck.ferreira@upmc.fr
Tobias Gärtner	Institut für Organische Chemie, Universität Regensburg, Universitätsstraße 31, D-93040 Regensburg, Germany Fax: +49-941-943-4617; e-mail: tobias.gaertner@chemie.uni-regensburg.de

Dmitri Gelman	Institute of Chemistry, The Hebrew University of Jerusalem, Edmund Safra Campus, Givat Ram, 91904 Jerusalem, Israel Fax: +972-2-6585279; e-mail: dgelman@chem.ch.huji.ac.il
Saul Gómez-Manzo	Facultad de Química, Universidad Nacional Autónoma de México, Ciudad Universitaria, Coyoacán, 04510, D.F. México, México Fax: +52-5556162010; e-mail: sgomez@ifc.unam.mx
Ruth M. Gschwind	Institut für Organische Chemie, Universität Regensburg, Universitätsstraße 31, D-93040 Regensburg, Germany Fax: +49-941-943-4617; e-mail: ruth.gschwind@chemie.uni-regensburg.de
Johann T. B. H. Jastrzebski	Chemical Biology & Organic Chemistry, Debye Institute for Nanomaterials Science, Faculty of Science, Utrecht University, Padualaan 8, 3584 CH Utrecht, The Netherlands Fax: +31-30-2523615; e-mail: j.t.b.h.jastrzebski@uu.nl
Jan S. Jaworski	Faculty of Chemistry, University of Warsaw, 02-093 Warszawa, Poland Fax: +48-22-822-5996; e-mail: jaworski@chem.uw.edu.pl
Norbert Krause	Dortmund University of Technology, Organic Chemistry, Otto-Hahn-Strasse 6, D-44227 Dortmund, Germany Fax: +49-231-7553884; e-mail: norbert.krause@tu-dortmund.de
Peter M. H. Kroneck	Fachbereich Biologie, Universität Konstanz, 78457 Konstanz, Germany Fax: +49-7531882966, e-mail; peter.kroneck@uni-konstanz.de
Al Mokhtar Lamsabhi	Departamento de Química, Universidad Autónoma de Madrid, Cantoblanco, 28049-Madrid, Spain Fax: +34-91-4975238; e-mail: mokhtar.lamsabhi@uam.es
Joel F. Liebman	Department of Chemistry and Biochemistry, University of Maryland, Baltimore County, 1000 Hilltop Circle, Baltimore, Maryland 21250, USA Fax: +1-410-455-2608; e-mail: jliebman@umbc.edu
Eiichi Nakamura	Department of Chemistry, The University of Tokyo, Hongo, Bunkyo-ku, Tokyo 113-0033, Japan Fax: +81-3-5800-6889; e-mail: nakamura@chem.s.u-tokyo.ac.jp
Liza Penn	Institute of Chemistry, The Hebrew University of Jerusalem, Edmund Safra Campus, Givat Ram, 91904 Jerusalem, Israel Fax: +972-2-6585279; e-mail: elizabet.penn@mail.huji.ac.il

Damien Polet	University of Geneva, Department of Organic Chemistry, 30, Quai Ernest Ansermet 1211 Genève 4, Switzerland Fax: +41-22-3793215; e-mail: damien.polet@addexpharma.com
Francisco J. Pulido	Departamento de Química Orgánica, Facultad de Ciencias, Universidad de Valladolid, 47011 Valladolid, Spain Fax: +34-983-423013; e-mail: pulido@qo.uva.es
Olivier Riant	Unité de Chimie Organique et Médicinale, Département de Chimie, Place Louis Pasteur 1, Université Catholique de Louvain, 1348 Louvain la Neuve, Belgium Fax : +32-010474168; e-mail: olivier.riant@uclouvain.be
Christopher J. Rosenker	Department of Chemistry, Parkman Ave. CSC 1301, University of Pittsburgh, Pittsburgh, PA 15260, USA Fax: +1-412-624-0787; e-mail: cjr34@pitt.edu
Jean-Yves Salpin	CNRS—Université d'Evry Val d'Essonne—Laboratoire Analyse et Modélisation pour la Biologie et l'Environnement—UMR 8587—Bâtiment Maupertuis, Boulevard François Mitterrand, 91025 Evry, France Fax: +33-1 69 47 76 55; e-mail: jean-yves.salpin@univ-evry.fr
Juan Pablo Saucedo-Vázquez	Facultad de Química, Universidad Nacional Autónoma de México, Ciudad Universitaria, Coyoacán, 04510, D.F. México, México Fax: +52-5556162010; e-mail: jpablo_sa@correo.unam.mx
Mathias O. Senge	School of Chemistry, SFI Tetrapyrrole Laboratory, Trinity College Dublin, Dublin 2, Ireland Fax: +353-1-608-8536; e-mail: sengem@tcd.ie Institute of Molecular Medicine, Medicinal Chemistry, Trinity Centre for Health Sciences, Trinity College Dublin, St. James's Hospital, Dublin 8, Ireland
Natalia N. Sergeeva	School of Chemistry, SFI Tetrapyrrole Laboratory, Trinity College Dublin, Dublin 2, Ireland Fax: +353-1-608-8536; e-mail: sergeevn@tcd.ie
Suzanne W. Slayden	Department of Chemistry, George Mason University, 4400 University Drive, Fairfax, Virginia 22030, USA Fax: +1-703-993-1055; e-mail: sslayden@gmu.edu
Martha E. Sosa-Torres	Facultad de Química, Universidad Nacional Autónoma de México, Ciudad Universitaria, Coyoacán, 04510, D.F. México, México Fax: +52-5556162010; e-mail: mest@servidor.unam.mx
Claude Spino	Université de Sherbrooke, Département de Chimie, Sherbrooke, QC, J1K 2R1, Canada Fax: +1-819-821-8017; e-mail: claude.spino@usherbrooke.ca

David R. Spring
Department of Chemistry, University of Cambridge, Lensfield Road, Cambridge, CB2 1EW, UK
Fax: +44-1223-336362; e-mail: drspring@ch.cam.ac.uk

Jeanine Tortajada
CNRS—Université d'Evry Val d'Essonne—Laboratoire Analyse et Modélisation pour la Biologie et l'Environnement—UMR 8587—Bâtiment Maupertuis, Boulevard François Mitterrand, 91025 Evry, France
Fax: +33-1 69 47 76 55; e-mail: jeanine.tortajada@chimie.univ-evry.fr

Gerard van Koten
Chemical Biology & Organic Chemistry, Debye Institute for Nanomaterials Science, Faculty of Science, Utrecht University, Padualaan 8, 3584 CH Utrecht, The Netherlands
Fax: +31-30-2523615; e-mail: g.vankoten@uu.nl

Rhett T. Watson
Division of Natural Sciences and Engineering, University of South Carolina Upstate, 800 University Way, Spartanburg, South Carolina 29303, USA
Fax: +1-864-503-5366; e-mail: rwatson@uscupstate.edu

Peter Wipf
Department of Chemistry, Parkman Ave. CSC 1301, University of Pittsburgh, Pittsburgh, PA 15260, USA
Fax: +1-412-624-0787; e-mail: pwipf@pitt.edu

Manuel Yáñez
Departamento de Química, Universidad Autónoma de Madrid, Cantoblanco, 28049-Madrid, Spain
Fax: +34-91-4975238; e-mail: manuel.yanez@uam.es

Naohiko Yoshikai
Division of Chemistry & Biological Chemistry, School of Physical & Mathematical Sciences, Nanyang Technological University, SPMS-04-01, 21 Nanyang Link, Singapore 637371
Fax: +65-6791-1961; e-mail: nyoshikai@ntu.edu.sg

Foreword

The present book, *The Chemistry of Organocopper Compounds,* belongs to the sub-group of volumes dealing with organometallic derivatives within 'The Chemistry of Functional Groups' series. Previous related recent volumes are the three parts of *The Chemistry of Organolithium Compounds* (2004, 2006), *The Chemistry of Organozinc Compounds* (2006) and *The Chemistry of Organomagnesium Compounds* (2008), all edited by the editors of the present volume. The use of organocopper compounds in organic synthesis is increasing in recent decades, especially in selective reactions, new classes of reagents are available, and mechanistic understanding has been enhanced by using more sophisticated methods.

The two parts of the present volume consist of the most extensive review of organocopper compounds since N. Krause's *Modern Organocopper Chemistry* (Wiley-VCH 2002) and they follow the tradition of the series by dealing with a wide range of aspects of organocopper chemistry. The 19 chapters written by experts from 15 countries include theoretical aspects, structural analysis, thermochemistry, NMR spectra, mass spectrometric behavior, electrochemistry, photochemistry of selected sub-groups and biochemistry. Several chapters are devoted to synthetic aspects including the preparation of various kinds of organocopper reagents such as α-aminoalkylcuprates, fluorinated or silyl and stannyl derivatives and the synthetic uses of organocopper reagents in transmetallation reactions, carbocupration reactions, copper mediated reactions on extended multiple bond systems, asymmetric allylic alkylations, reductive reactions and cross-coupling reactions, copper catalyzed enantioselective conjugate additions, as well as oxidation of organocopper reagents. We gratefully acknowledge the contributions of all the authors of the book.

Unfortunately, the planned chapter on 'Copper mediated radical reactions' was not delivered.

The literature coverage is mostly up to the end of 2008, but includes few references in early 2009.

We will be grateful to readers who draw our attention to mistakes in the present volume and to omissions of important chapters related to organocopper chemistry.

ZVI RAPPOPORT, Jerusalem
ILAN MAREK, Haifa
June 2009

The Chemistry of Functional Groups
Preface to the series

The series 'The Chemistry of Functional Groups' was originally planned to cover in each volume all aspects of the chemistry of one of the important functional groups in organic chemistry. The emphasis is laid on the preparation, properties and reactions of the functional group treated and on the effects which it exerts both in the immediate vicinity of the group in question and in the whole molecule.

A voluntary restriction on the treatment of the various functional groups in these volumes is that material included in easily and generally available secondary or tertiary sources, such as Chemical Reviews, Quarterly Reviews, Organic Reactions, various 'Advances' and 'Progress' series and in textbooks (i.e. in books which are usually found in the chemical libraries of most universities and research institutes), should not, as a rule, be repeated in detail, unless it is necessary for the balanced treatment of the topic. Therefore each of the authors is asked not to give an encyclopaedic coverage of his subject, but to concentrate on the most important recent developments and mainly on material that has not been adequately covered by reviews or other secondary sources by the time of writing of the chapter, and to address himself to a reader who is assumed to be at a fairly advanced postgraduate level.

It is realized that no plan can be devised for a volume that would give a complete coverage of the field with no overlap between chapters, while at the same time preserving the readability of the text. The Editors set themselves the goal of attaining reasonable coverage with moderate overlap, with a minimum of cross-references between the chapters. In this manner, sufficient freedom is given to the authors to produce readable quasi-monographic chapters.

The general plan of each volume includes the following main sections:

(a) An introductory chapter deals with the general and theoretical aspects of the group.

(b) Chapters discuss the characterization and characteristics of the functional groups, i.e. qualitative and quantitative methods of determination including chemical and physical methods, MS, UV, IR, NMR, ESR and PES—as well as activating and directive effects exerted by the group, and its basicity, acidity and complex-forming ability.

(c) One or more chapters deal with the formation of the functional group in question, either from other groups already present in the molecule or by introducing the new group directly or indirectly. This is usually followed by a description of the synthetic uses of the group, including its reactions, transformations and rearrangements.

(d) Additional chapters deal with special topics such as electrochemistry, photochemistry, radiation chemistry, thermochemistry, syntheses and uses of isotopically labeled compounds, as well as with biochemistry, pharmacology and toxicology. Whenever applicable, unique chapters relevant only to single functional groups are also included (e.g. 'Polyethers', 'Tetraaminoethylenes' or 'Siloxanes').

This plan entails that the breadth, depth and thought-provoking nature of each chapter will differ with the views and inclinations of the authors and the presentation will necessarily be somewhat uneven. Moreover, a serious problem is caused by authors who deliver their manuscript late or not at all. In order to overcome this problem at least to some extent, some volumes may be published without giving consideration to the originally planned logical order of the chapters.

Since the beginning of the Series in 1964, two main developments have occurred. The first of these is the publication of supplementary volumes which contain material relating to several kindred functional groups (Supplements A, B, C, D, E, F and S). The second ramification is the publication of a series of 'Updates', which contain in each volume selected and related chapters, reprinted in the original form in which they were published, together with an extensive updating of the subjects, if possible, by the authors of the original chapters. Unfortunately, the publication of the 'Updates' has been discontinued for economic reasons.

Advice or criticism regarding the plan and execution of this series will be welcomed by the Editors.

The publication of this series would never have been started, let alone continued, without the support of many persons in Israel and overseas, including colleagues, friends and family. The efficient and patient co-operation of staff-members of the Publisher also rendered us invaluable aid. Our sincere thanks are due to all of them.

The Hebrew University SAUL PATAI
Jerusalem, Israel ZVI RAPPOPORT

Sadly, Saul Patai who founded 'The Chemistry of Functional Groups' series died in 1998, just after we started to work on the 100th volume of the series. As a long-term collaborator and co-editor of many volumes of the series, I undertook the editorship and I plan to continue editing the series along the same lines that served for the preceeding volumes. I hope that the continuing series will be a living memorial to its founder.

The Hebrew University ZVI RAPPOPORT
Jerusalem, Israel
May 2000

Contents

1. Theory of organocopper-mediated reactions — 1
   Eiichi Nakamura and Naohiko Yoshikai

2. Structural organocopper chemistry — 23
   Gerard van Koten and Johann T. B. H. Jastrzebski

3. Thermochemistry of organocopper compounds — 145
   Joel F. Liebman and Suzanne W. Slayden

4. NMR of organocopper compounds — 163
   Tobias Gärtner and Ruth M. Gschwind

5. Photochemical transformations involving copper porphyrins and phthalocyanines — 217
   Natalia N. Sergeeva and Mathias O. Senge

6. Electrochemistry of organocopper compounds — 247
   Jan S. Jaworski

7. Gas-phase chemistry of organocopper compounds — 279
   Al Mokhtar Lamsabhi, Manuel Yáñez, Jean-Yves Salpin and Jeanine Tortajada

8. Biochemistry of organocopper compounds — 347
   Martha E. Sosa-Torres, Juan P. Saucedo-Vázquez, Saul Gómez-Manzo and Peter M. H. Kroneck

9. *N*-Functionalized organocuprates — 365
   R. Karl Dieter and Rhett T. Watson

10. Transmetalation reactions producing organocopper compounds — 443
    Christopher J. Rosenker and Peter Wipf

11. Carbocupration of alkynes — 527
    Fabrice Chemla and Franck Ferreira

12. Oxidation of organocopper compounds — 585
    Sarah J. Aves and David R. Spring

13. Copper-mediated asymmetric allylic alkylations — 603
    Claude Spino

14. Copper-catalyzed enantioselective conjugate addition 693
Damien Polet and Alexandre Alexakis

15. Copper(I) hydride reagents and catalysts 731
Olivier Riant

16. Silyl and stannyl derivatives of organocopper compounds 775
Francisco J. Pulido and Asunción Barbero

17. Copper-mediated and copper-catalyzed addition and substitution reactions of extended multiple bond systems 857
Norbert Krause and Özge Aksin-Artok

18. Copper-mediated cross-coupling reactions 881
Liza Penn and Dmitri Gelman

19. Fluorinated organocopper reagents 991
Charles R. Davis and Donald J. Burton

Author index 1033

Subject index 1097

List of abbreviations used

Ac	acetyl (MeCO)
acac	acetylacetone
Ad	adamantyl
AIBN	azoisobutyronitrile
Alk	alkyl
All	allyl
An	anisyl
Ar	aryl
Bn	benzyl
Bu	butyl (C_4H_9)
Bz	benzoyl (C_6H_5CO)
c-	cyclo
CD	circular dichroism
CI	chemical ionization
CIDNP	chemically induced dynamic nuclear polarization
CNDO	complete neglect of differential overlap
Cp	η^5-cyclopentadienyl
Cp*	η^5-pentamethylcyclopentadienyl
DABCO	1,4-diazabicyclo[2.2.2]octane
DBN	1,5-diazabicyclo[4.3.0]non-5-ene
DBU	1,8-diazabicyclo[5.4.0]undec-7-ene
DIBAH	diisobutylaluminium hydride
DME	1,2-dimethoxyethane
DMF	*N*,*N*-dimethylformamide
DMSO	dimethyl sulphoxide
E-	entgegen
ee	enantiomeric excess
EI	electron impact
ESCA	electron spectroscopy for chemical analysis
ESR	electron spin resonance
Et	ethyl
eV	electron volt

Fc	ferrocenyl
FD	field desorption
FI	field ionization
FT	Fourier transform
Fu	furyl(OC_4H_3)
GLC	gas liquid chromatography
Hex	hexyl(C_6H_{13})
c-Hex	cyclohexyl(*c*-C_6H_{11})
HMPA	hexamethylphosphortriamide
HOMO	highest occupied molecular orbital
HPLC	high performance liquid chromatography
i-	iso
ICR	ion cyclotron resonance
Ip	ionization potential
IR	infrared
LAH	lithium aluminium hydride
LCAO	linear combination of atomic orbitals
LDA	lithium diisopropylamide
LUMO	lowest unoccupied molecular orbital
M	metal
M	parent molecule
MCPBA	*m*-chloroperbenzoic acid
Me	methyl
MNDO	modified neglect of diatomic overlap
MS	mass spectrum
n-	normal
Naph	naphthyl
NBS	*N*-bromosuccinimide
NCS	*N*-chlorosuccinimide
NMR	nuclear magnetic resonance
Pen	pentyl(C_5H_{11})
Ph	phenyl
Pip	piperidyl($C_5H_{10}N$)
ppm	parts per million
Pr	propyl (C_3H_7)
PTC	phase transfer catalysis or phase transfer conditions
Py	pyridine (C_5H_5N)
Pyr	pyridyl (C_5H_4N)
R	any radical
RT	room temperature

s-	secondary
SET	single electron transfer
SOMO	singly occupied molecular orbital
t-	tertiary
TCNE	tetracyanoethylene
TFA	trifluoroacetic acid
TFE	2,2,2-trifluoroethanol
THF	tetrahydrofuran
Thi	thienyl(SC_4H_3)
TLC	thin layer chromatography
TMEDA	tetramethylethylene diamine
TMS	trimethylsilyl or tetramethylsilane
Tol	tolyl(MeC_6H_4)
Tos or Ts	tosyl(*p*-toluenesulphonyl)
Trityl	triphenylmethyl(Ph_3C)
Vi	vinyl
XRD	X-ray diffraction
Xyl	xylyl($Me_2C_6H_3$)
Z-	zusammen

In addition, entries in the 'List of Radical Names' in *IUPAC Nomenclature of Organic Chemistry*, 1979 Edition, Pergamon Press, Oxford, 1979, p. 305–322, will also be used in their unabbreviated forms, both in the text and in formulae instead of explicitly drawn structures.

CHAPTER **12**

Oxidation of organocopper compounds

SARAH J. AVES and DAVID R. SPRING

Department of Chemistry, University of Cambridge, Lensfield Road, Cambridge CB2 1EW, United Kingdom
Fax: +44 (0) 1223 336362; e-mail: drspring@ch.cam.ac.uk

I. INTRODUCTION 585
II. FORMATION OF C−C BONDS 586
A. Initial Studies 586
B. Cross-coupling 586
C. Biaryl Formation 590
D. Intramolecular Bond Formation 592
E. Dimerizations of Heteroaromatics, Alkenyl and Alkyl Groups and Macrocycle Formation 595
III. FORMATION OF C−N BONDS 598
A. Initial Studies 598
B. Further Developments 599
IV. CONCLUSION 601
V. REFERENCES 601

I. INTRODUCTION

Although organocopper reagents have been synthesized since the 1800's[1], it is only with the advent of Gilman cuprates in 1952[2] that their synthetic utility has been realized. This trend is also reflected in the oxidation chemistry of organocopper reagents, as wide-ranging procedures for the oxidative coupling of mono-organocopper reagents have been in place for over one hundred years[3] (the Glaser alkyne coupling[4] for example) whereas organocuprate oxidations have only begun to be fully exploited in the last couple of decades[5].

Theoretical studies by Nakamura and coworkers[6] have shown that the 3d orbitals on copper are around 7 eV higher in dimethyl cuprate than in methylcopper. As these orbitals constitute the HOMO, the oxidation of organocuprates is an attractive reaction as the

The chemistry of organocopper compounds
Edited by Z. Rappoport and I. Marek

high-lying nature of the orbitals means that the oxidation can be performed with relative ease, avoiding the harsher reaction conditions of classical methods. Numerous reagents can be used to effect the oxidation, including dinitroarenes, inorganic salts, oxygen and quinones.

This chemistry is beginning to find its niche in biaryl synthesis, where steric considerations may make other methods difficult, and electrophilic amination[7] where the oxidation of amidocuprates provides a complementary reaction to palladium-catalysed methods.

II. FORMATION OF C–C BONDS

A. Initial Studies

The oxidation of organocuprates was first explored systematically by Whitesides and coworkers, who examined the dimerization of a number of cuprates formed from readily accessible Grignard reagents and organolithiums[8]. The organocuprate (**2**) was formed by transmetallation of the organometallic (**1**) with copper(I) iodide/tri-*n*-butylphosphine complex and then oxidized (equation 1). Nitrobenzene, benzophenone and copper(II) chloride–TMEDA complex were all reported to be effective oxidants, but use of molecular oxygen led to the best overall yields.

$$\underset{(\mathbf{1})}{\mathrm{RM}} \xrightarrow{\mathrm{CuI.PBu_3}} \underset{(\mathbf{2})}{\mathrm{R_2CuM}} \xrightarrow{\mathrm{O_2}} \mathrm{R{-}R} \qquad (1)$$

Primary and secondary alkyl, vinyl, alkynyl and aryl centres could be coupled in good yields (67–88%) but tertiary alkyl species proved challenging. The counterion present was also shown to have an effect on the reaction, as in the case of the *n*-butyl derivative the lithium reagent gave higher yields of dimer than the Grignard reagent.

Analysis of the product mixture obtained upon oxidation of lithium di-*n*-butylcuprate showed the formation of but-1-ene (14%) and *n*-butanol (5%) as well as the desired *n*-octane (84%) and from this a tentative mechanism was proposed. Butylcopper(I) was found to be present in incompletely oxidized reaction mixtures and it is oxidation of this species which could explain the formation of both but-1-ene and *n*-butanol.

A radical mechanism was ruled out on the basis that formation of butyl radicals in a solution saturated with oxygen would lead to the formation of much larger amounts of *n*-butanol and thus it was proposed that lithium di-*n*-butylcuprate (**3**) is initially oxidized to di-*n*-butylcopper(II) (**4**), which can disproportionate to octane (**6**) and *n*-butylcopper(I) (**7**). The presence of a short-lived Cu(III) species **5** which can undergo reductive elimination to form the products is also a possibility (equation 2)[5].

$$\underset{(\mathbf{3})}{2\,\mathrm{Bu_2CuLi}} \xrightarrow{[\mathrm{O}]} \underset{(\mathbf{4})}{2\,\mathrm{Bu_2Cu^{II}}} \xrightarrow[-\,\underset{(\mathbf{7})}{\mathrm{BuCu^{I}}}]{} \underset{(\mathbf{5})}{[\mathrm{Bu_3Cu^{III}}]} \xrightarrow{\text{Reductive elimination}} \underset{(\mathbf{6})}{\mathrm{Bu{-}Bu}} + \underset{(\mathbf{7})}{\mathrm{BuCu^{I}}} \qquad (2)$$

B. Cross-coupling

In order to perform a cross-coupling procedure via organocuprate oxidation, one must first form a mixed organocuprate (i.e. the two ligands on copper are different organic species). Research in this area has concentrated on the formation of the desired cross-coupled product and the avoidance of homo-coupled products, which lead to deleterious yields.

It was Mandeville and Whitesides who first looked at this issue as part of a wider study of selectivity of organic group transfer in mixed organocuprates[9]. They found that upon oxidation the expected 1:2:1 statistical mixture of homo-coupled and cross-coupled products did not form; instead it was found that some groups had a higher propensity to couple than others. Alkynyl groups were found to be reluctant to participate in any form of coupling, whether it be homo- or cross-, despite the precedent of the Glaser reaction. The sterically hindered *t*-butyl group did not undergo homo-coupling, and so its use as one of the ligands in the mixed cuprate allowed moderate yields of cross-coupled product with less sterically demanding groups (equation 3).

$$n\text{-BuLi} \xrightarrow[\text{iii) PhNO}_2]{\text{i) CuI.PBu}_3\text{; ii) } t\text{-BuLi}} \underset{36\%}{n\text{-Bu}-\text{Bu-}n} + \underset{60\%}{n\text{-Bu}-\text{Bu-}t} + \underset{<1\%}{t\text{-Bu}-\text{Bu-}t} \qquad (3)$$

The copper salt used to form the mixed cuprate was also shown to have an influence on the amount of cross-coupling by Bertz and Gibson[10]. Electronically and sterically similar alkyl groups were used so that the dependence on the copper salt could be investigated. While lower-order cuprates (i.e. those of the form R_2CuLi) prepared from CuI were found to give statistical mixtures of homo- and cross-coupled products on oxidation with *ortho*-dinitrobenzene, the higher-order cuprates (i.e. those which have more than two organic ligands present in the cuprate structure, in this case $R_2Cu(CN)Li_2$) prepared from CuCN gave a ratio of homo- to cross-coupled products closer to 1:4:1. These contrasting results were taken to imply a difference in the ligand exchange rate (although the oxidation of a mixture of the two preformed higher-order homocuprates gave mainly homo-coupling, suggesting that ligand exchange does not occur) or differences in the solution structure of the two cuprates.

The use of α-functionalized alkyl cuprates allowed higher yields of cross-coupled product to be prepared provided that both ligands contained a heteroatom[11]. The functionalization of the cuprate also allowed iodine to be used as the oxidant which had previously been shown to be ineffective for unfunctionalized substrates. *N*-Boc-2-lithiopyrrolidone (**8**) could be coupled with *N*-Boc-*N*-methyl-1-lithiomethylamine (**9**) in 70% yield (equation 4); similar yields were obtained on coupling with *n*-BuLi. They fell off rapidly with increasing steric hindrance on the alkyl substrate, with *t*-BuLi giving poor yields. Interestingly, attempted cross-coupling of **8** with ligands containing an α-oxygen rather than nitrogen only took place with a large excess of oxidant.

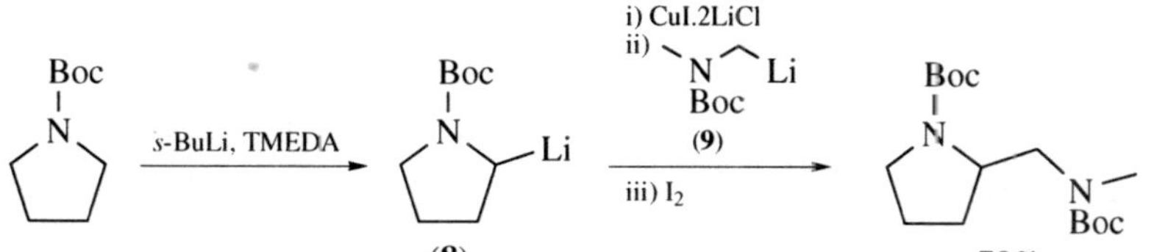

(**8**) 70% (4)

High yields of cross-coupled biaryls were achieved by the oxidation of 'kinetic' mixed organocuprates by molecular oxygen[12]. These 'kinetic' conditions developed by Lipshutz and coworkers involve the formation of a lower-order cyanocuprate (solubilized CuI was found to be ineffective) and then addition of the second aryl lithium reagent at $-125\,^{\circ}$C, with the oxidation also carried out at this low temperature. In a comparative study these conditions gave a ratio of 1:26:1 of products (equation 5) as opposed to the statistical mixture of 1:2:1 when the oxidation was carried out at $-78\,^{\circ}$C.

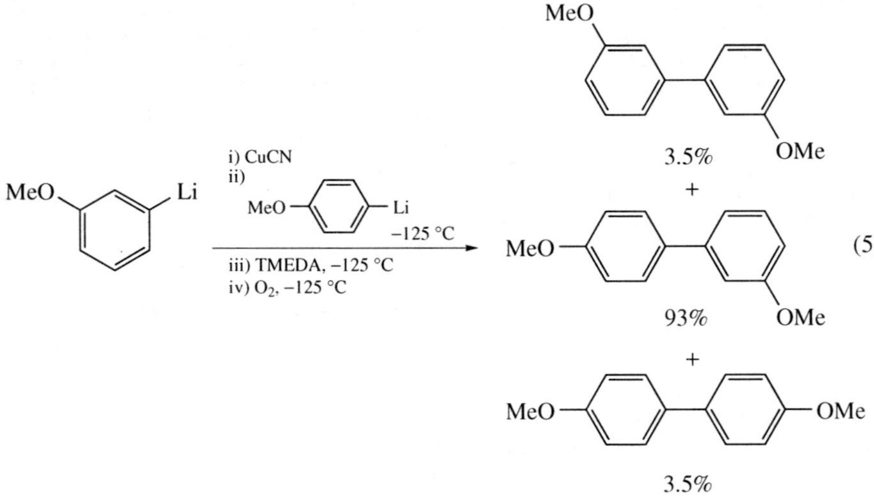

(5)

TMEDA was found to be a necessary additive to achieve high yields, and it was suggested that it activated the cuprate cluster towards decomposition.

If the cuprate was warmed to −78 °C and then re-cooled before oxidation, a statistical mixture of products was produced suggesting that a kinetically controlled species was formed initially. However, when the cuprate was studied spectroscopically, the spectra of the re-cooled cuprate was identical to that of the initial cluster rather than that of the one at −78 °C, suggesting that these are not truly kinetically formed species.

Use of the Lipshutz protocol has been undertaken by Coleman and coworkers for the formation of biaryls, most recently in the elegant synthesis of eupomatilones 4 and 6 (equation 6)[13].

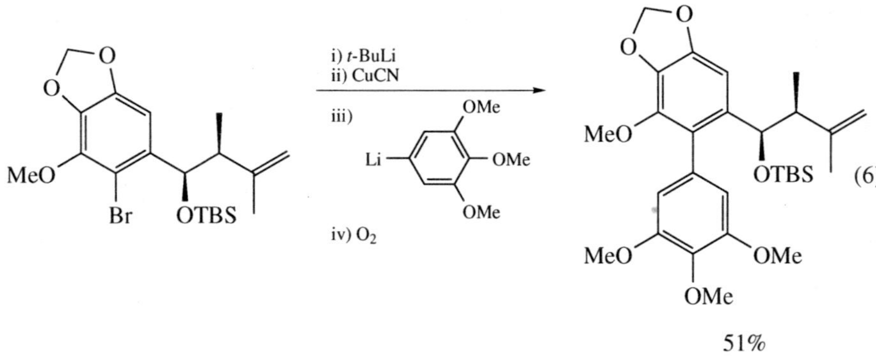

(6)

An exciting development in cross-coupling has been the advent of directed *ortho*-cupration methodology[14]. Here, a Lipshutz amidocuprate is used first as a base to effect deprotonation and secondly as a copper source to complete the metallation reaction. It was found that in order for the reaction to be high yielding and chemoselective, the amine ligand on copper must be 2,2,6,6-tetramethylpiperidine (TMP). However, a range of organic groups could be tolerated as the second group on copper.

Oxidations were carried out as part of the studies into the reactivity of the synthesized *ortho*-cuprates, and it was found that the organic ligand of the intermediate cuprate could be coupled to the aryl unit by the action of nitrobenzene (equation 7). If both groups on the cuprate used to perform the metallation were TMP, then the homo-coupled biaryl **10** resulted in quantitative yield upon oxidation (equation 8).

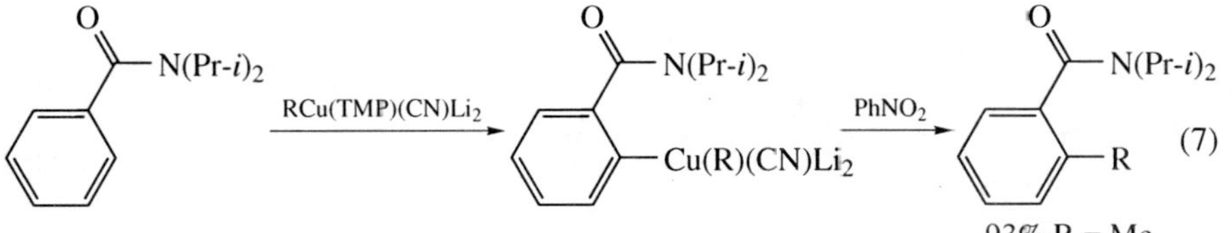

(7)

(**11**) 93% R = Me
91% R = Ph

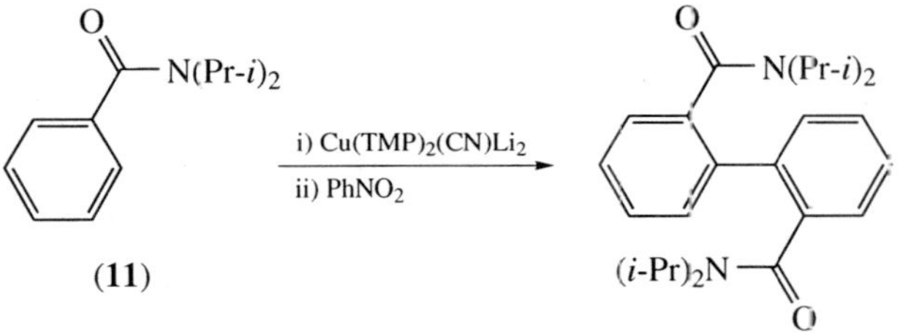

(8)

(**11**)

(**10**) 100%

It should be noted that this reaction was only carried out on substrate **11** and so the generality of this reaction is not known, however these results point towards another solution to the cross-coupling problem.

An alterative approach to forming cross-coupled products has been to use one of the coupling partners in a large excess, thus avoiding the production of statistical mixtures. Whitesides and coworkers showed that this approach could be successfully used with hindered alkyl groups (equation 9)[15].

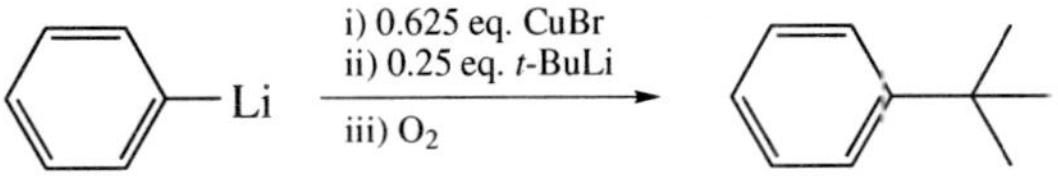

(9)

73% (based on *t*-BuLi)

Recently, Knochel and coworkers have developed a procedure for the preparation of polyfunctional alkynes via oxidation of an aryl(alkynyl)organocuprate[16]. This uses only 2 equivalents of the lithiated alkyne to form the cuprate and allows the use of sterically hindered arenes and selective mono-coupling of dihaloarenes, the latter of which can be difficult to obtain via a Sonogashira coupling.

The mixed organocuprate is formed by reacting the lithiated alkyne with an aryl copper species formed by transmetallation from the Grignard reagent. The organocuprate is then oxidized with the quinone chloranil (**12**) to give the cross-coupled product. Use of the group's directed magnesiation methodology[17] allowed the coupling of doubly *ortho*-substituted arenes in good yields. Also, 3,5-dibromopyridine (**13**) could be selectively mono-coupled with alkynes bearing a range of substituents in 68–83% yield (equation 10).

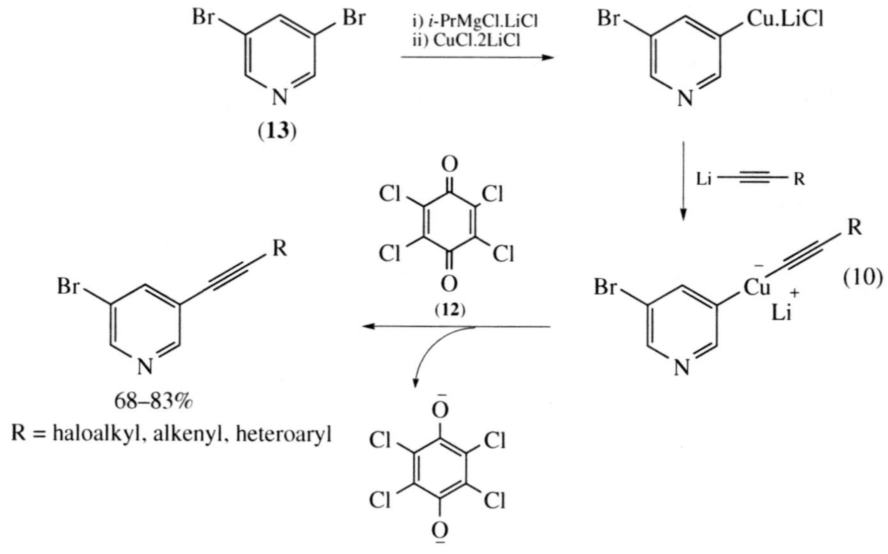

After selective mono-coupling, the remaining halogen of the dihaloarene substrates could be used in further reactions; in this case annulated pyridine formations were undertaken. Lithiation of the bromoalkyne **14** and reaction with *p*-tolylnitrile gave lithiated intermediate **15**, which cyclized on addition of iodine to give the iodinated pyridine. Removal of the iodine by *n*-BuLi gave the annulated pyridine **16** in 44% overall yield (equation 11).

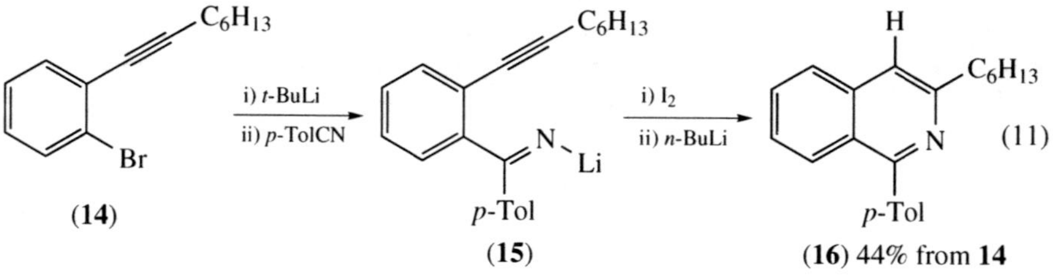

C. Biaryl Formation

The formation of sterically hindered biaryls is one area where organocuprate oxidations have proved useful as palladium- and nickel-catalysed transformations can perform poorly in this context. Work by Spring and coworkers has focussed around the development of methodology to forge biaryl bonds in substrates with multiple *ortho* substituents, using transmetallation protocols that are compatible with a wide range of functional groups. This was part of a wider study into medium ring synthesis (*vide infra*).

Starting with transmetallation from aryl lithium reagents, it was recognized that the usual protocol of lithium–halogen exchange requires the presence of a halogen moiety which can add extra synthetic steps to install. Use was therefore made of an *ortho*-lithiation strategy to form the aryl lithium species prior to transmetallation. Methoxy, alkoxy ether, sulfonamide and carboxylate groups all proved to be effective directing groups and heterocycles were also compatible (equation 12)[18]. Use of a chiral oxazoline as a directing group allowed the formation of **17** as a single diastereoisomer (equation 13).

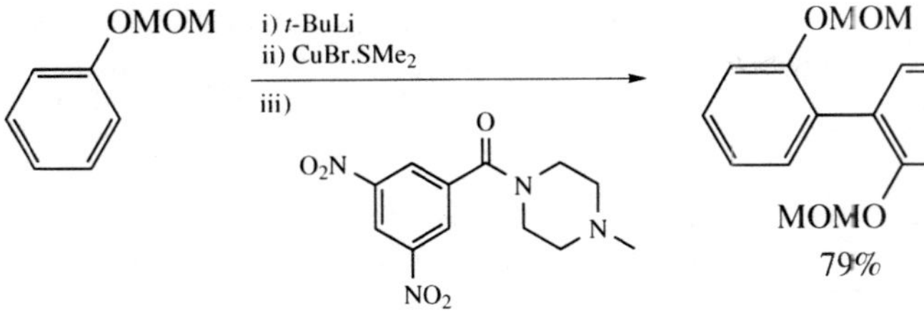

(**18**)

(12)

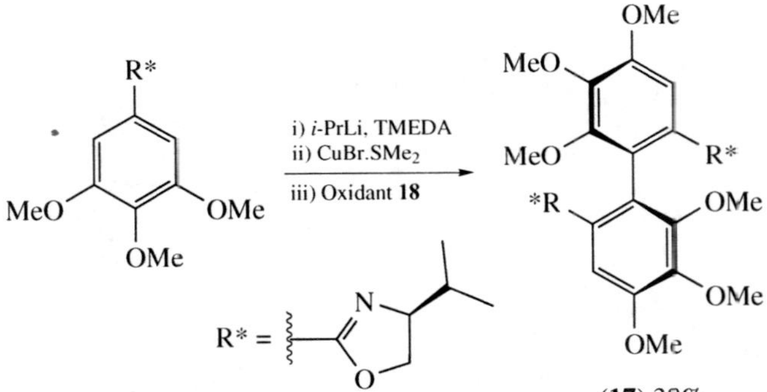

(**17**) 38%

(13)

Dinitroarene oxidant **18** was also developed, as it was recognized that dinitrobenzenes were the best oxidants in this context. However, the post-oxidation by-products were difficult to remove from the product formed. The inclusion of the piperazine unit in **18** allows by-products to be easily removed via an acid wash or filtration through silica on work-up.

Use of Knochel's Mg/I exchange methodology[19] to form aryl Grignards allowed the formation of iodinated biaryls and as the exchange could be performed regioselectively oligomerization could be avoided (equation 14). Biaryls with four *ortho* substituents could also be synthesized by this methodology (equation 15)[20].

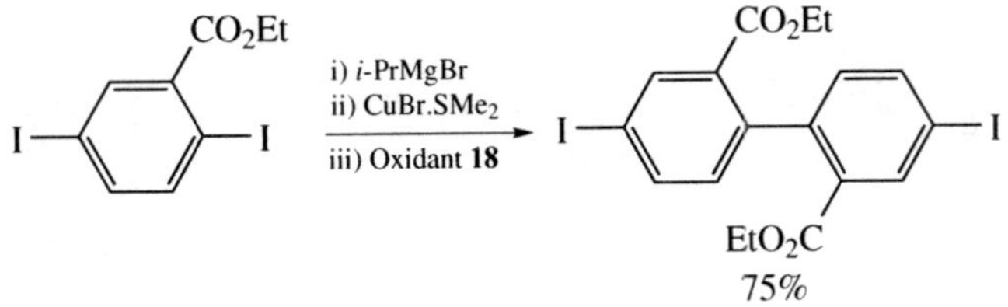

(14)

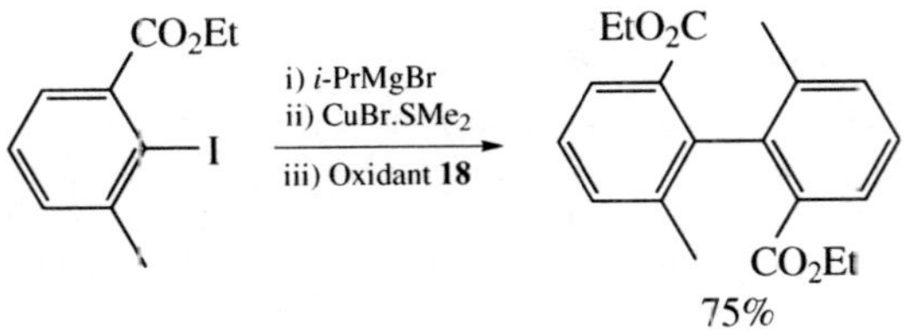

(15)

In order to further improve functional group tolerance, transmetallation from zinc was also investigated. This is notable as it is the first example of the oxidation of zinc organocuprates. (Oxidation of zinc amidocuprates is also known—see Section III.B). Use of aryl zinc reagents was advantageous as it allowed the use of more readily available aryl bromides and the aryl metals could be synthesized under milder conditions, thus allowing functional groups such as ketones to be present[21]. Use of highly active Rieke zinc (Zn*) proved useful in synthesizing the required organozinc reagents.

Studies showed that it was possible to perform the coupling using a catalytic amount of copper salt and oxidant **18** in conjunction with molecular oxygen. This proved to be useful not only in biaryl formation, as various vinyl and benzyl bromides could also be used as starting materials (equations 16 and 17).

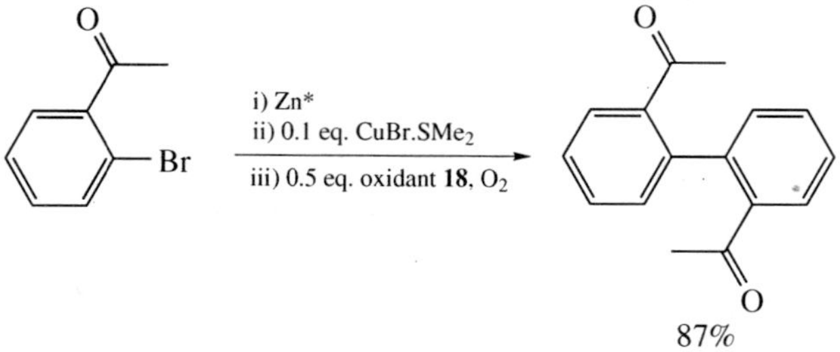

(16)

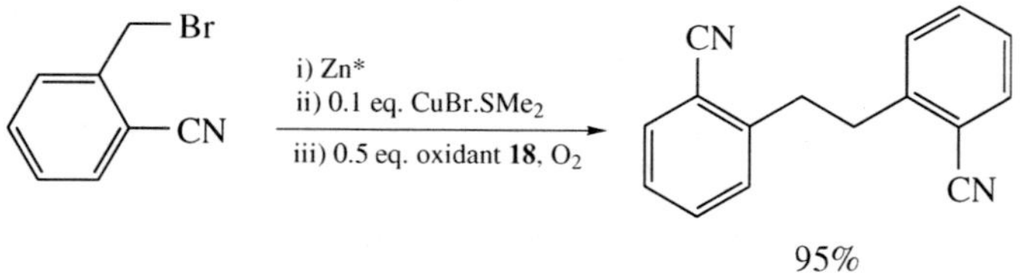

(17)

Other recent work in this area includes studies by Iyoda and coworkers on the oxidation of Lipshutz cuprates, utilizing quinone oxidants, in order to develop methodology to form macrocyclic cyclophanes[22]. The coupling reactions were in general high yielding, although the use of *t*-BuLi to form the aryl lithium for transmetallation necessarily reduces the functional group compatibility (equation 18). Methyl-, methoxy- and halogen-substituted aromatics are given as examples along with thienyl.

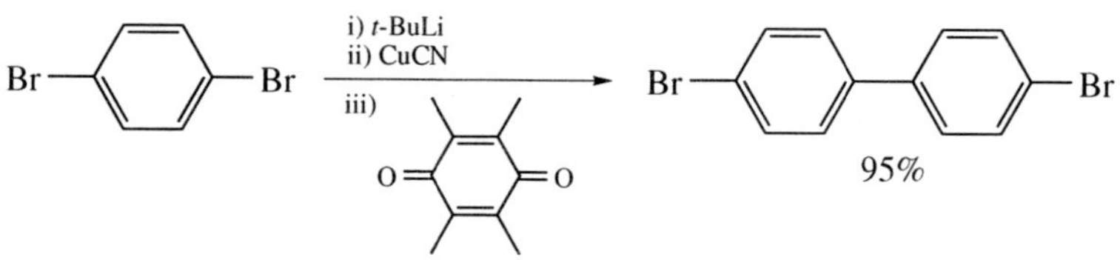

(18)

D. Intramolecular Bond Formation

The idea of intramolecular biaryl bond formation using a tether was first introduced by Lipshutz and coworkers as a way of aiding troublesome heteroaromatic cross-couplings[23]. This was quickly extended to the use of chiral tethers as a means of forming non-racemic biaryls[24]. Compound **19** was used in the synthesis of (+)-*O*-permethyltellimagrandin II on removal of the tether (equation 19).

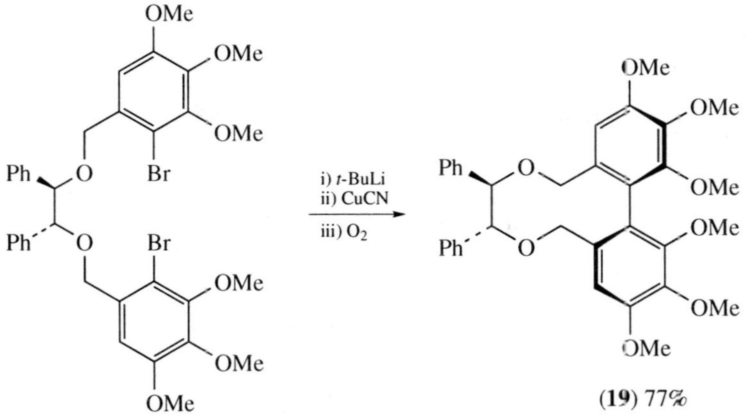

(19)

Intramolecular biaryl formation was developed further by Schreiber and coworkers as a means of accessing medium-ring biaryl compounds. The strained structure of a medium ring due to the combination of transannular, torsional and large angle strain means that these compounds can be difficult to access via palladium- or nickel-mediated processes; it was found that the use of copper allowed the formation of the desired biaryl compounds with good atropdiastereoselectivity by utilizing a chiral amino alcohol tether.

Use of polymer-supported synthesis allowed a library to be built of nine-, ten- and eleven-membered rings with a variety of biaryl substituents. Since lithium/halogen exchange was not compatible with the polymer beads, the initial aryl metal species was generated via exchange with i-PrBu$_2$MgLi[25]. A later study[26] generated a further library of nine-membered ring containing compounds via solid-phase synthesis and also an interesting dimeric biaryl **20** in the solution phase (equations 20 and 21).

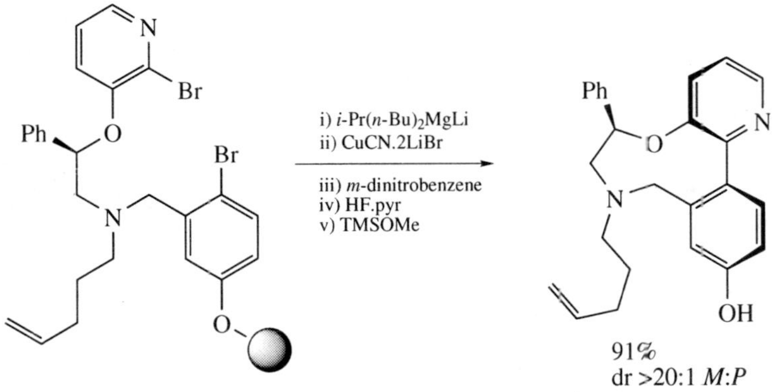

(20)

The transmetallation strategies for forming biaryls (*vide supra*) have also been used by Spring and coworkers for the formation of medium rings. *Ortho*-lithiation and Mg/I exchange both allowed the formation of examples of 10- and 11-membered rings after transmetallation to copper (equation 22)[18,20]. Cross-coupling of biaryls could be performed using a tethering strategy, and most importantly use of the magnesium organocuprate allowed the formation of **21**, a model for the highly strained core of the natural product sanguiin H-5, with complete diastereoselectivity (equation 23).

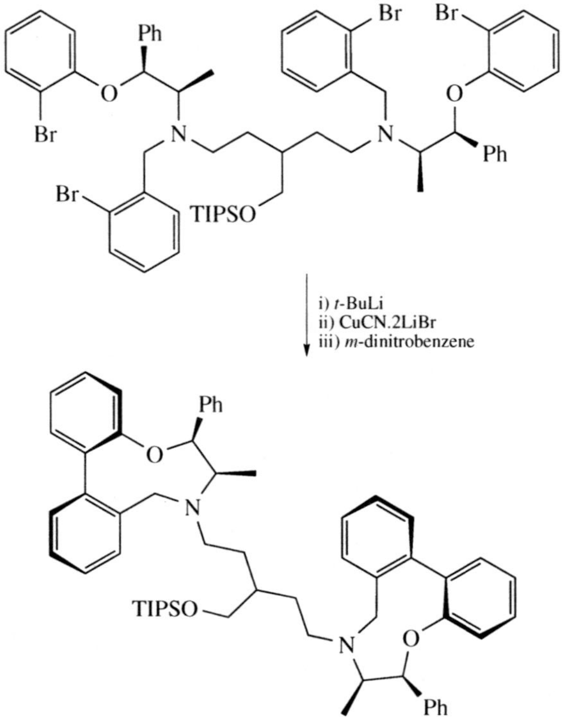

(21)

(**20**) 59%
dr >20:1 *M,M:P,P*

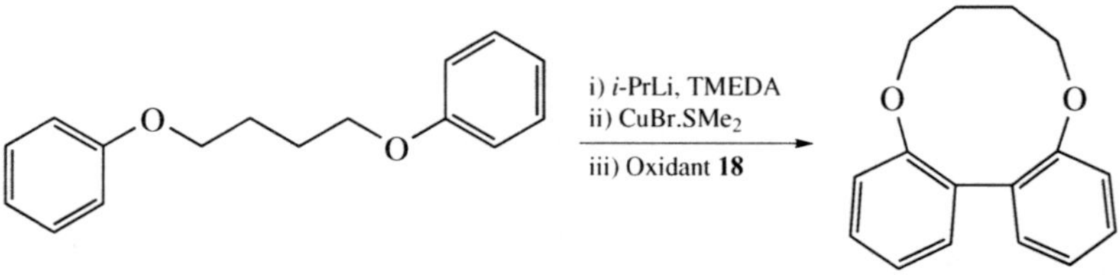

(22)

57%

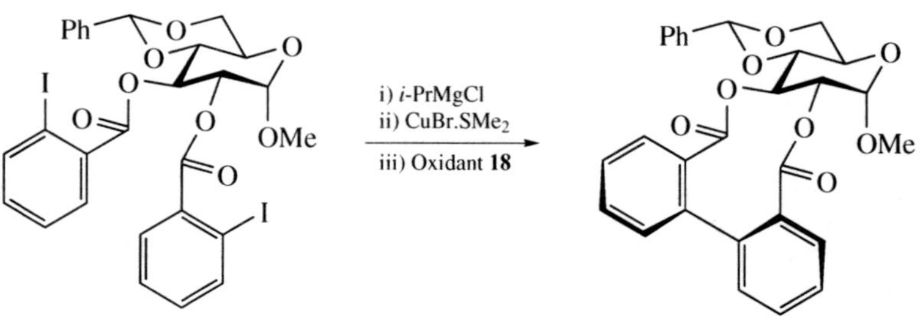

(23)

(**21**) 67%, dr >50:1

Insertion of Rieke zinc into aryl bromides allowed zinc organocuprate oxidation methodology to be used in the synthesis of buflavine (**22**) (equation 24)[21]. This represented the first synthesis of the natural product where the medium ring and the biaryl bond are formed simultaneously.

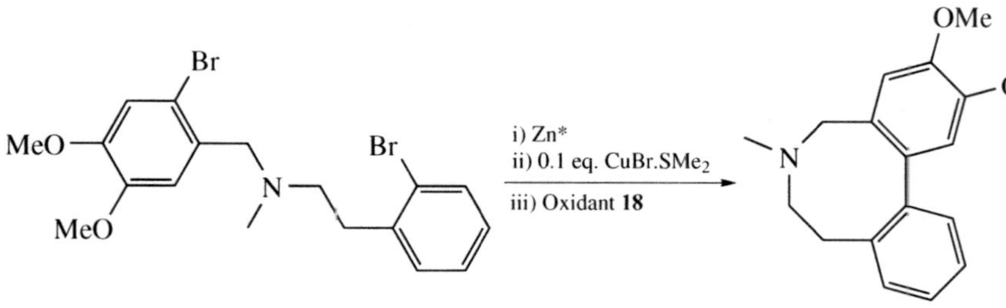

(24)

(**22**) 80%

With proof of the utility of these methods in medium-ring formation in hand, attention was turned to completing the synthesis of sanguiin H-5. While the aryl bromide/Zn* route from **23** was slightly higher yielding, both this and the use of I/Mg exchange on **24** provided the key medium-ring compound **25** in good yields. Global deprotection via hydrogenation provided the natural product (**26**) in quantitative yield (equation 25)[27].

E. Dimerizations of Heteroaromatics, Alkenyl and Alkyl Groups and Macrocycle Formation

Organocuprate oxidation to form C−C bonds has also seen use outside of biaryl couplings and medium-ring formation. Iyoda has performed oxidations of Lipshutz cuprates in order to synthesize thiophene-fused cyclophanes and nona- and dodecaphenylenes; the optioelectronic and nanostructural properties of the latter were investigated[22,28]. The ten-membered cyclophane structure was not accessible via palladium couplings and it is suggested that the linear C−Cu−C bond in the organocuprate intermediate favours the intermolecular over intramolecular coupling (equation 26).

In the second study, the ratio of nona- to dodecaphenylene (**27**:**28**) formed upon oxidation was dependent on the substituents present on the vertex aryl group, although the nona-system **27** could be formed exclusively in 46% yield when this was hydrogen (equation 27).

Sherburn and coworkers' desire to investigate the chemistry of [4]dendralene (**29**) led to the development of a synthesis using oxidation of an alkenyl cuprate. The reaction could be performed on multi-gram scale to provide a usable yield of the desired compound (equation 28)[29].

Other recent examples of the coupling of alkenyl copper species include Xi and coworkers' developments in the synthesis of cyclooctatetraene derivatives. Transmetallation from a dilithio or zirconaindene species provides the alkenyl copper intermediate for oxidation with a quinone[30]. This intramolecular coupling forms a cyclobutadiene (**30**), which then dimerizes to form the tricycle **31**. A retro-[2 + 2] reaction furnishes the cyclooctatetraene **32** (equation 29). This methodology also provides an approach to benzocyclobutadiene derivatives[31].

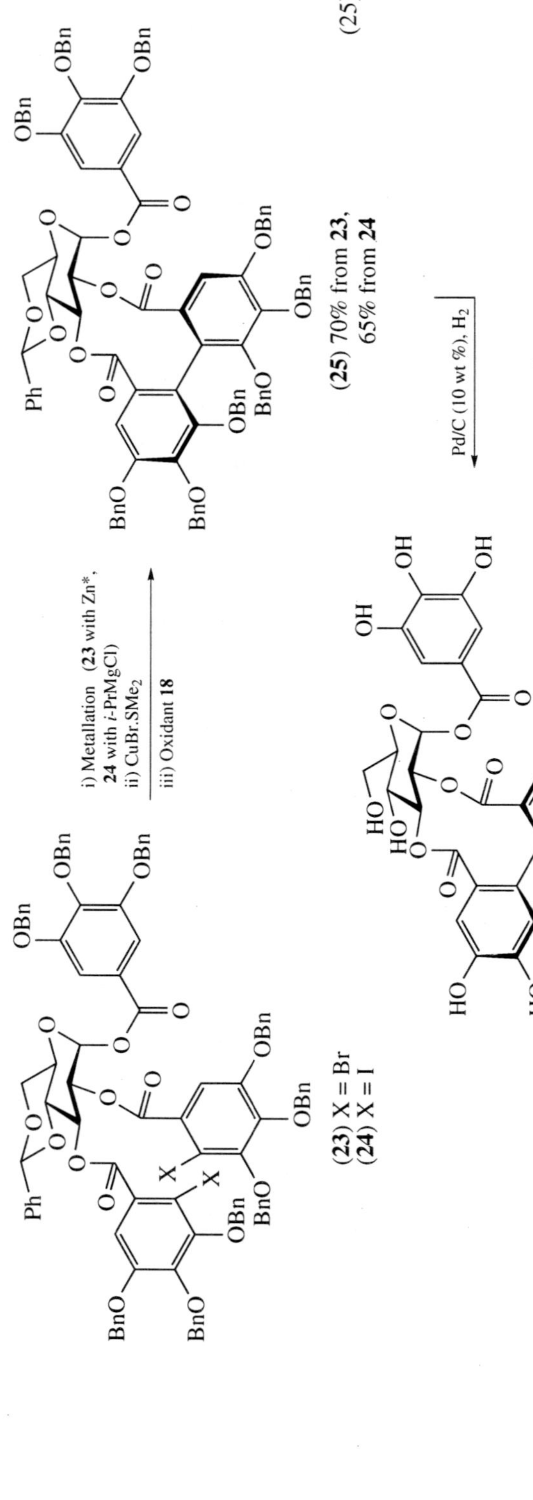
i) Metallation (23 with Zn*, 24 with i-PrMgCl)
ii) CuBr.SMe2
iii) Oxidant 18
Pd/C (10 wt %), H2
(23) X = Br
(24) X = I
(25) 70% from 23, 65% from 24
(26) 99%

(25)

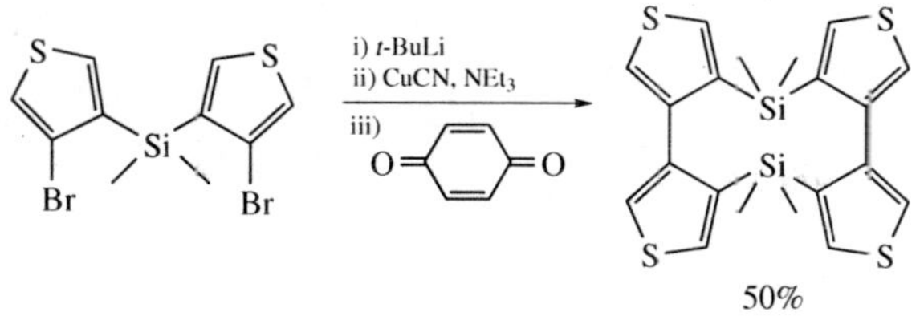

(26)

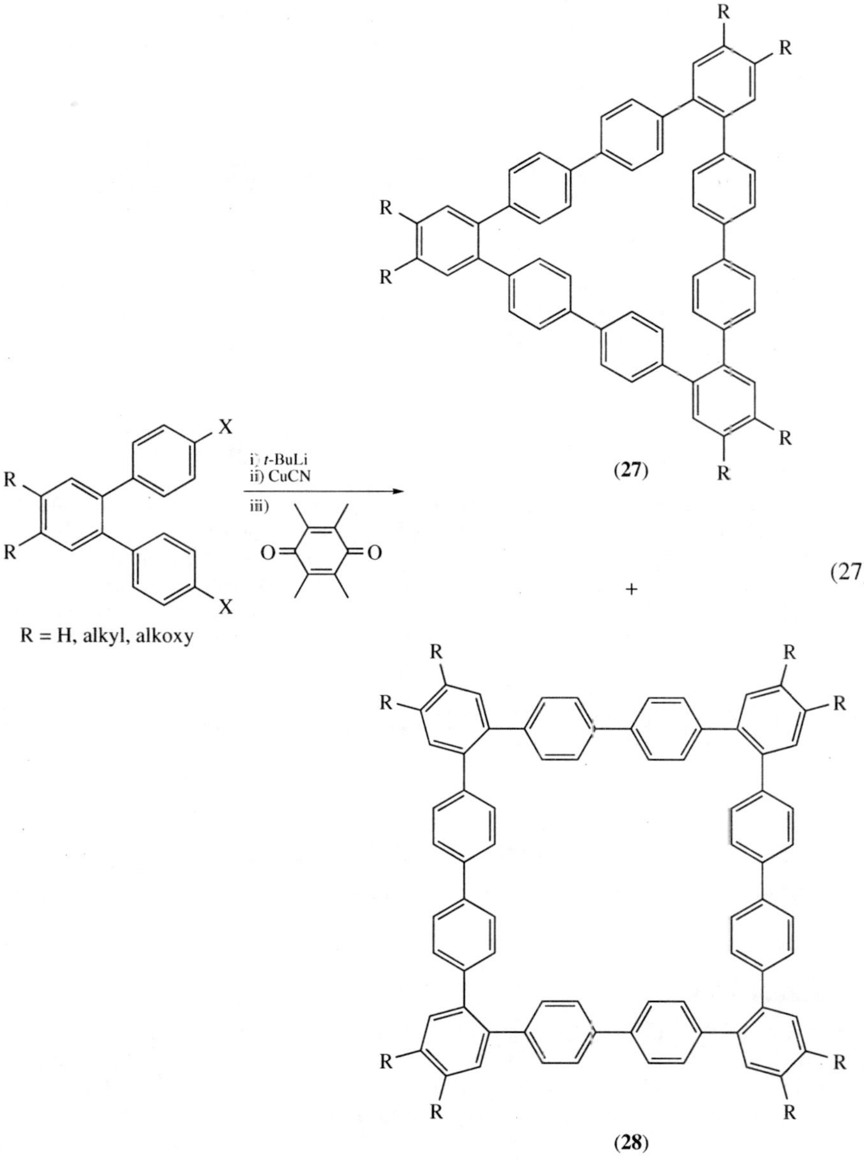

R = H, alkyl, alkoxy

(27)

(28)

(27)

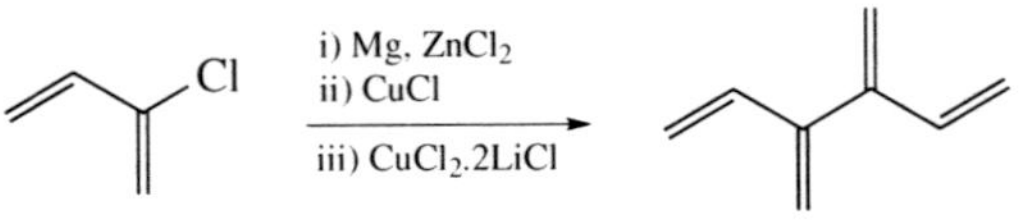

(29) 26%

(28)

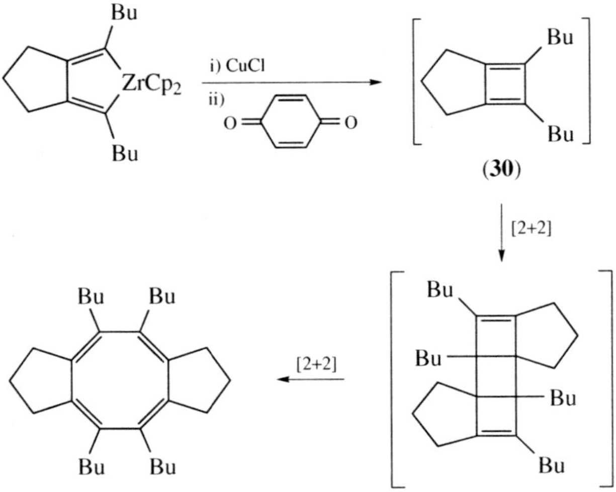

(30)

(29)

(32) 43% (31)

Organocuprate oxidation provides a solution to the formation of bonds at bridgehead positions where nucleophilic substitution reactions are impracticable. When Michl and coworkers attempted to couple bicyclo[1.1.1]pentanes using Pd- or Ni-mediated processes, products were formed in low yields (20–40%). However, the use of copper chemistry allowed the coupling to take place in reproducibly good yields (50–70%), and the methodology was used to synthesize a variety of staffane derivatives (equation 30)[32].

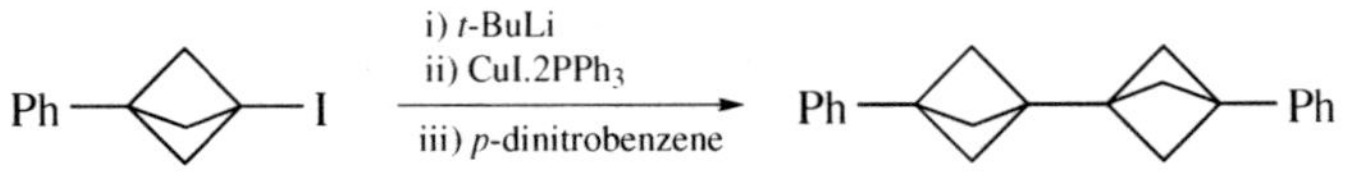

50–70%

(30)

III. FORMATION OF C–N BONDS

The oxidation of amidocuprates allows the formation of a new carbon–nitrogen bond between the ligands on copper.

A. Initial Studies

The first work in this area was carried out by Yamamoto and Maruoka[33], who used the oxidation reaction to alkylate a variety of primary and secondary amines and anilines. An excess of primary, tertiary or aryl organocuprate was added to the amine and the reaction mixture was oxidized with molecular oxygen (equation 31).

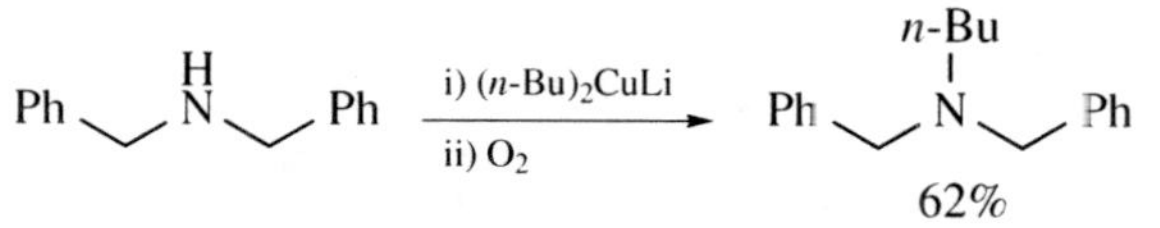
(31)

The resulting yield of amine was variable, with the lowest resulting from the transfer of tertiary alkyl groups. Interestingly, the yields of *N*-methylation products were higher when methyl magnesium chloride rather than methyl lithium was used to form dimethyl cuprate, showing that the other atoms present in the solution aggregation structure of the cuprate can also affect reactivity in amidocuprates as well as organocuprates.

Snieckus and coworkers further developed the methodology to synthesize anthranilamides as precursors to acridones[34]. Here, an excess of anilidocuprate was added to lithiobenzamides (prepared by a directed *ortho*-lithiation) and the oxidation performed by molecular oxygen (equation 32). The use of copper(I) cyanide was reported to give cleaner, higher-yielding reactions than that of copper(I) chloride; however, stated yields remained moderate despite the use of excess cuprate.

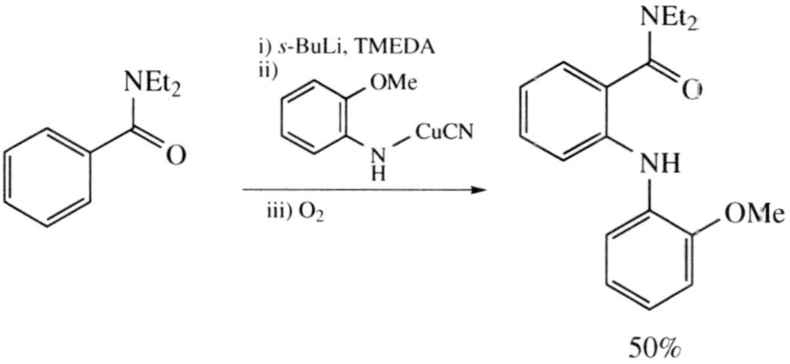
(32)

B. Further Developments

While the above two procedures involve adding an excess of cuprate to an organolithium, further work by Dembech, Ricci and coworkers showed that a 1:1 ratio of cuprioamide and organolithium could produce coupled products in useful yields[35]. Molecular oxygen was once again used as the oxidant, as other oxidizing agents were found to give lower yields. This study is noteworthy as heteroaromatics were coupled for the first time. The methodology also allows the functionalization of hydrazines, albeit in low yields due to the formation of azo compounds as a by-product. Further studies[36] showed that yields of *N*-arylation products and hydrazines could be improved by the addition of zinc chloride to the cyanocuprates before amidocuprate formation, coupled with the use of catalytic *ortho*-dinitrobenzene as a co-oxidant alongside molecular oxygen (equations 33 and 34).

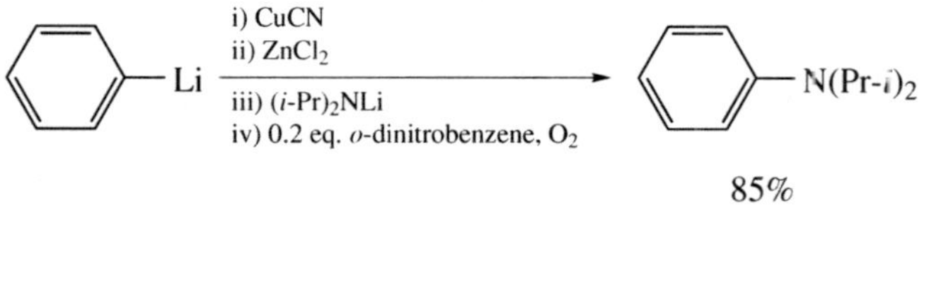
(33)

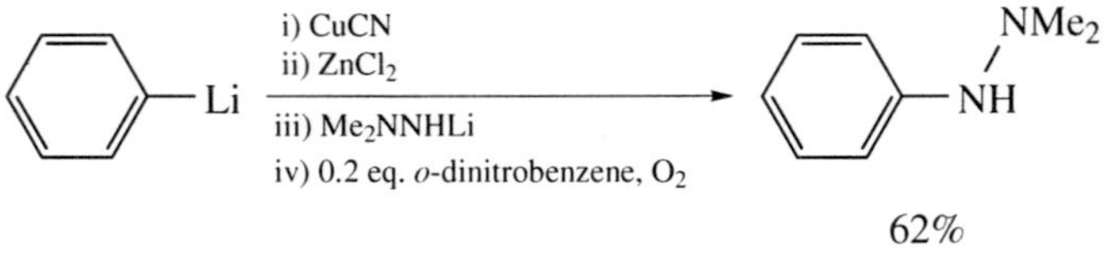

(34)

More recently, the Knochel group have used cuprate oxidation methodology to prepare primary, secondary and tertiary aromatic and heteroaromatic amines with the use of the quinone chloranil (**12**) as an oxidant[37,38]. Here, an organocopper reagent is prepared from a functionalized Grignard reagent formed by an Mg/halogen exchange[19,39] or a directed magnesiation[17] and combined with a lithiated amine to form the amidocuprate for oxidation. Various electron withdrawing and donating groups on the aromatic moieties can be tolerated, including esters and halogens.

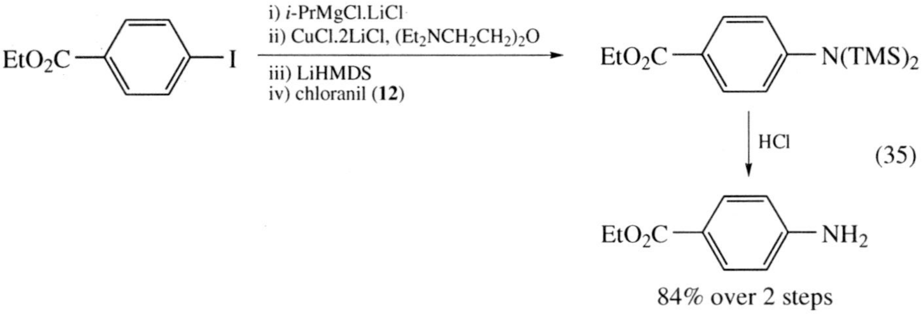

(35)

Primary amines are accessed via the use of lithium hexamethyldisilazide and subsequent desilylation with TBAF or HCl to reveal the amine (equation 35). The methodology also allows the formation of tertiary aryl and sterically hindered amines, which can be difficult to form by other methods, in yields of 55% and above (equation 36).

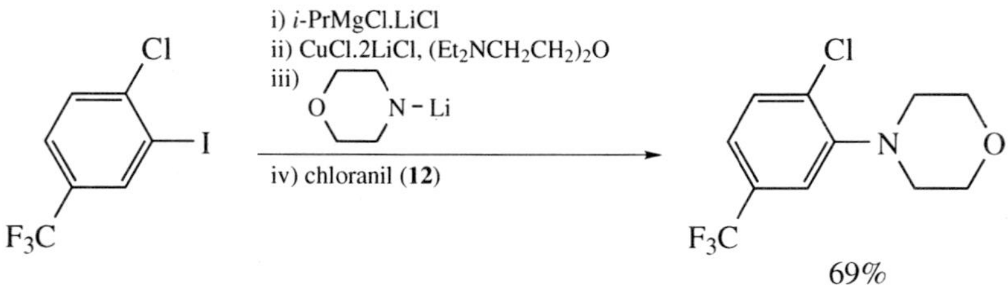

(36)

This methodology has been further extended to the amination of purine and pyrimidine derivatives[40]. Again, the starting material could be cuprated regioselectively by the appropriate choice of magnesiation method (*vide supra*). This allowed access to uracil, thymine and adenine derivatives in good yields under mild conditions, and the CDK inhibitor purvalanol A (**33**) could be synthesized in 46% overall yield (equation 37).

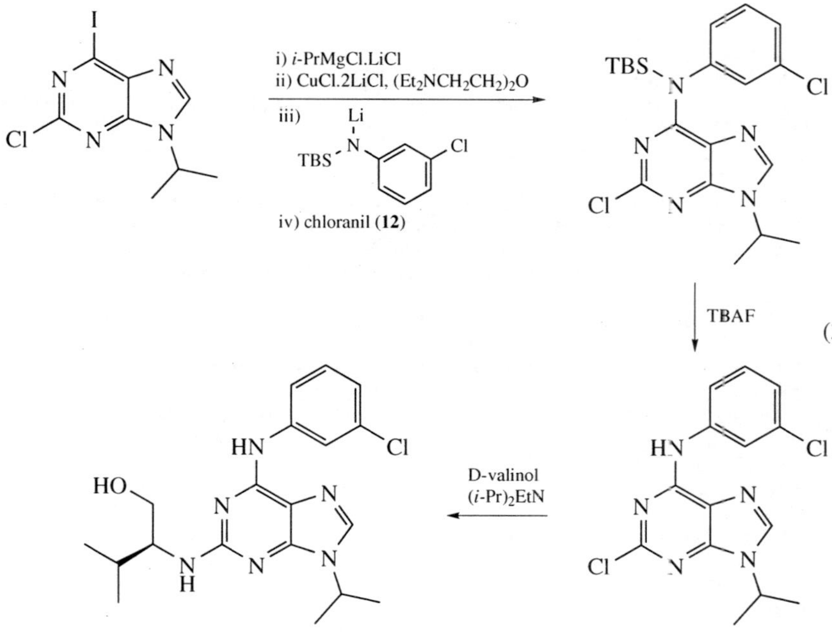

(37)

(**33**) 65%

71% over 2 steps

IV. CONCLUSION

Organocuprate oxidation has proved itself to be a useful reaction in the forty years since its development. However, it is only recently that the full potential of the reaction has begun to be exploited, with new transmetallation procedures allowing milder reaction conditions and greater functional group tolerance. Hopefully these new advances, such as directed cuprations and the catalytic use of copper, will push forward the development of oxidation methodologies especially in the cross-coupling and C−N bond-forming arenas where the reaction remains underexplored.

V. REFERENCES

1. R. Boettger, *Liebigs Ann. Chem.*, **109**, 351 (1859).
2. H. Gilman, R. G. Jones and L. A. Woods, *J. Org. Chem.*, **17**, 1630 (1952).
3. T. Kauffman, *Angew. Chem., Int. Ed. Engl.*, **13**, 291 (1974).
4. C. Glaser, *Liebigs Ann. Chem.*, **154**, 137 (1870).
5. D. S. Surry and D. R. Spring, *Chem. Soc. Rev.*, **35**, 218 (2006).
6. S. Mori, A. Hirai, M. Nakamura and E. Nakamura, *Tetrahedron*, **56**, 2805 (2000).
7. M. Kienle, S. R. Dubbaka, K. Brade and P. Knochel, *Eur. J. Org. Chem.*, 4166 (2007).
8. G. M. Whitesides, J. San Filippo, C. P. Casey and E. J. Panek, *J. Am. Chem. Soc.*, **89**, 5302 (1967).
9. W. H. Mandeville and G. M. Whitesides, *J. Org. Chem.*, **39**, 400 (1974).

10. S. H. Bertz and C. P. Gibson, *J. Am. Chem. Soc.*, **108**, 8286 (1986).
11. R. K. Dieter, S. J. Li and N. Y. Chen, *J. Org. Chem.*, **69**, 2867 (2004).
12. B. H. Lipshutz, K. Siegmann, E. Garcia and F. Kayser, *J. Am. Chem. Soc.*, **115**, 9276 (1993).
13. S. Mitra, S. R. Gurrala and R. S. Coleman, *J. Org. Chem.*, **72**, 8724 (2007).
14. S. Usui, Y. Hashimoto, J. V. Morey, A. E. H. Wheatley and M. Uchiyama, *J. Am. Chem. Soc.*, **129**, 15102 (2007).
15. G. M. Whitesides, W. F. Fischer, J. San Filippo, R. W. Bashe and H. O. House, *J. Am. Chem. Soc.*, **91**, 4871 (1969).
16. S. R. Dubbaka, M. Kienle, H. Mayr and P. Knochel, *Angew. Chem., Int. Ed.*, **46**, 9093 (2007).
17. A. Krasovskiy, V. Krasovskaya and P. Knochel, *Angew. Chem., Int. Ed.*, **45**, 2958 (2006).
18. D. S. Surry, D. J. Fox, S. J. F. Macdonald and D. R. Spring, *Chem. Commun.*, 2589 (2005).
19. A. E. Jensen, W. Dohle, I. Sapountzis, D. M. Lindsay, V. A. Vu and P. Knochel, *Synthesis*, 565 (2002).
20. D. S. Surry, X. B. Su, D. J. Fox, V. Franckevicius, S. J. F. Macdonald and D. R. Spring, *Angew. Chem., Int. Ed.*, **44**, 1870 (2005).
21. X. B. Su, D. J. Fox, D. T. Blackwell, K. Tanaka and D. R. Spring, *Chem. Commun.*, 3883 (2006).
22. Y. Miyake, M. Wu, M. J. Rahman, Y. Kuwatani and M. Iyoda, *J. Org. Chem.*, **71**, 6110 (2006).
23. B. H. Lipshutz, F. Kayser and N. Maullin, *Tetrahedron Lett.*, **35**, 815 (1994).
24. B. H. Lipshutz, Z. P. Liu and F. Kayser, *Tetrahedron Lett.*, **35**, 5567 (1994).
25. D. R. Spring, S. Krishnan, H. E. Blackwell and S. L. Schreiber, *J. Am. Chem. Soc.*, **124**, 1354 (2002).
26. S. Krishnan and S. L. Schreiber, *Org. Lett.*, **6**, 4021 (2004).
27. X. B. Su, D. S. Surry, R. J. Spandl and D. R. Spring, *Org. Lett.*, **10**, 2593 (2008).
28. M. J. Rahman, J. Yamakawa, A. Matsumoto, H. Enozawa, T. Nishinaga, K. Kamada and M. Iyoda, *J. Org. Chem.*, **73**, 5542 (2008).
29. A. D. Payne, A. C. Willis and M. S. Sherburn, *J. Am. Chem. Soc.*, **127**, 12188 (2005).
30. C. Chen, C. J. Xi, C. B. Lai, R. J. Wang and X. Y. Hong, *Eur. J. Org. Chem.*, 647 (2004).
31. C. Chen, C. J. Xi, Y. Y. Liu and X. Y. Hong, *J. Org. Chem.*, **71**, 5373 (2006).
32. C. Mazal, A. J. Paraskos and J. Michl, *J. Org. Chem.*, **63**, 2116 (1998).
33. H. Yamamoto and K. Maruoka, *J. Org. Chem.*, **45**, 2739 (1980).
34. M. Iwao, J. N. Reed and V. Snieckus, *J. Am. Chem. Soc.*, **104**, 5531 (1982).
35. A. Alberti, F. Cane, P. Dembech, D. Lazzari, A. Ricci and G. Seconi, *J. Org. Chem.*, **61**, 1677 (1996).
36. F. Cane, D. Brancaleoni, P. Dembech, A. Ricci and G. Seconi, *Synthesis*, 545 (1997).
37. V. del Amo, S. R. Dubbaka, A. Krasovskiy and P. Knochel, *Angew. Chem., Int. Ed.*, **45**, 7838 (2006).
38. M. Kienle, S. R. Dubbaka, V. del Amo and P. Knochel, *Synthesis*, 1272 (2007).
39. A. Krasovskiy and P. Knochel, *Angew. Chem., Int. Ed.*, **43**, 3333 (2004).
40. N. Boudet, S. R. Dubbaka and P. Knochel, *Org. Lett.*, **10**, 1715 (2008).

CHAPTER **13**

Copper-mediated asymmetric allylic alkylations

CLAUDE SPINO

Université de Sherbrooke, Département de Chimie, Sherbrooke, QC, Canada, J1K 2R1
Fax: (819)-821-8017; e-mail: Claude.Spino@USherbrooke.ca

I. INTRODUCTION 604
A. Preamble and a Brief History 604
B. General Considerations on the Structure of the Product 606
1. Chemoselectivity 606
a. S_N versus other reactions 606
b. Two leaving groups on the allylic moiety 607
2. Regioselectivity 608
a. Effect of substrate structure 608
b. Nature of the cuprate reagent 609
c. Effect of solvent and reaction conditions 609
d. Effect of the leaving group 609
3. Stereochemistry 610
II. DIASTEREOSPECIFIC METHODS 610
A. General Considerations 610
1. Substrate features and stereospecificity 610
2. Effect of the leaving group on the stereospecificity 612
B. Acyclic Allylic Systems 613
1. Diastereomerically pure precursors prepared from chiral non-racemic acyclic allylic alcohols 614
a. Substrates taken from the chiral pool or obtained by resolution or asymmetric reduction of ketones 614
b. Use of a chiral auxiliary 616
c. Applications of diastereospecific methodologies in synthesis 622
2. Opening of allylic epoxides and aziridines 622
3. Displacement of propargylic leaving groups 629
4. Displacement of allenic bromides 638

The chemistry of organocopper compounds
Edited by Z. Rappoport and I. Marek

C. Cyclic Systems . . . 638
1. Common cyclic systems . . . 638
2. Cyclic systems of particular synthetic interest . . . 643
a. Diastereospecific ring-opening of cyclopentenediol derivatives . . . 643
b. Diastereospecific opening of cyclopentadiene monoepoxides . . . 644
c. Diastereospecific opening of cyclopenta[*b*]furanones . . . 646
d. Pot-pourri of applications in total synthesis . . . 647
D. Miscellaneous Allylic Systems . . . 649
III. DIASTEREOSELECTIVE METHODS . . . 653
A. General Considerations . . . 653
B. Chirality δ to the Leaving Group . . . 654
1. Chiral center bearing no heteroatom . . . 654
a. Acyclic substrates . . . 654
b. Cyclic substrates . . . 654
2. Chiral center bearing a heteroatom . . . 658
a. Oxygen atom . . . 658
b. Other heteroatoms . . . 659
3. Use of a chiral auxiliary . . . 660
C. Chirality at the Homoallylic Position . . . 660
D. Chirality on the Leaving Group . . . 661
IV. ENANTIOSELECTIVE METHODS . . . 663
A. Chirality on the Leaving Group . . . 663
B. Chiral Ligands on Copper . . . 666
1. Prochiral substrates . . . 669
a. Di- and trisubstituted substrates . . . 669
b. Substrates possessing a β-substituent . . . 672
c. Miscellaneous substrates . . . 675
2. Resolutions and desymmetrizations . . . 676
a. Allylic and propargylic epoxides . . . 676
b. Miscellaneous substrates . . . 679
C. Dynamic Kinetic Asymmetric Transformation . . . 680
V. CONCLUSIONS AND OUTLOOK . . . 682
VI. REFERENCES AND NOTES . . . 682

I. INTRODUCTION

A. Preamble and a Brief History

The copper-catalyzed or copper-promoted allylic substitution, also called allylic alkylation, is arguably one of the best methods to bond an electrophilic allylic group **1** with an aryl or alkyl unit in such nucleophiles as organolithium, organomagnesium, organozinc or organotitanium reagents ('hard nucleophiles') as shown in equation 1. Leaving groups (X) include esters, carbamates, sulfonates, oxiranes, phosphates, alcohols, ethers, acetals and halides. The term 'allylic alkylation', in this chapter, will always mean that the nucleophile is being 'alkylated' by the allylic electrophile.

There are two substitution pathways available to the organocopper reagent, namely the S_N2' pathway, or γ-allylic alkylation, to give product **2**, and the S_N2 pathway, or α-allylic alkylation, to give product **3**. The issues of chemo-, regio- and stereoselectivity can be challenging and few methods can boast to control them all. A number of reviews have been published on various aspects of the asymmetric allylic alkylation of organocopper reagents[1,2]. This chapter will provide an overview of what has been achieved to date in the realm of copper-promoted asymmetric allylic substitution, focusing on the most recent developments. The chapter is divided into sections treating the various aspects of stereochemical control in copper-promoted allylic substitution and is ordered according to the source of the chiral element or its position on the allylic electrophile. Questions of chemo- and regioselectivity will be discussed wherever necessary.

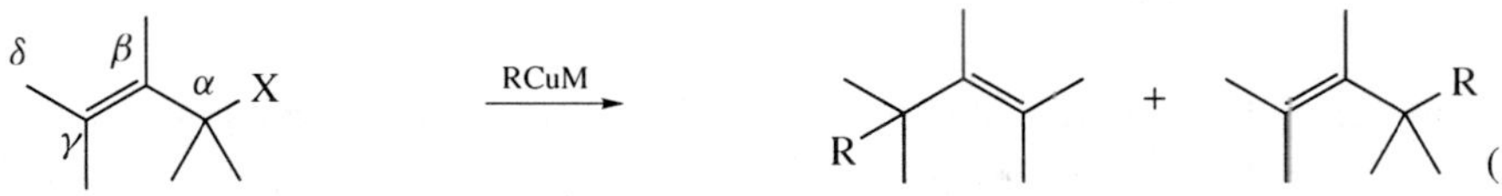

(1)

(1) **(2)** **(3)**

R = alkyl, aryl, allyl...
M = Li, MgX, ZnX, TiX$_3$...

γ-product α-product

The first report of an allylic substitution using organocopper reagents was that of Rona and Crabbé who, surprisingly, performed the reaction using propargylic acetates (as opposed to allylic acetates) to afford allenes[3]. The following year, they described the γ-substitution (S_N2') of steroidal allylic acetate **4** to give exocyclic alkene **5** using Gilman-type cuprate reagents (equation 2)[4]. It was recognized early on as a method to prepare geometrically pure alkenes[4,5] and has been used for that purpose ever since[6]. Rickborn and Staroscik noticed that the opening of 1,3-cyclohexadiene monoepoxide by lithium dimethylcuprate proceeded with *anti* stereospecificity[7]. Goering and Singleton later reported on the stereospecificity of the reaction wherein they demonstrated experimentally that lithium dimethylcuprate displaced cyclic allylic acetate **6** exclusively *anti* with respect to the leaving group to give products **7** and **8** (equation 3)[8].

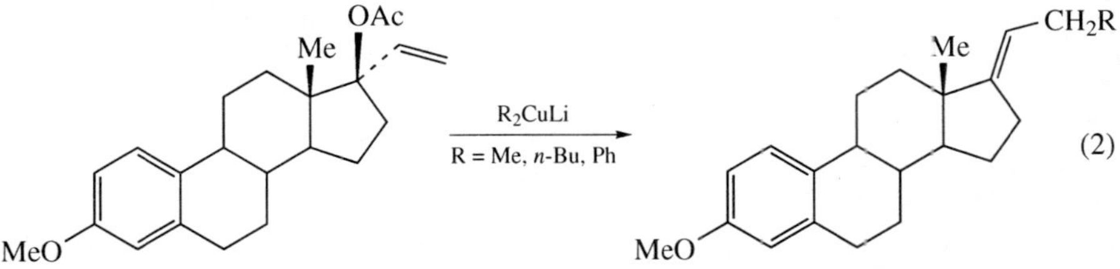

(2)

(4) **(5)**

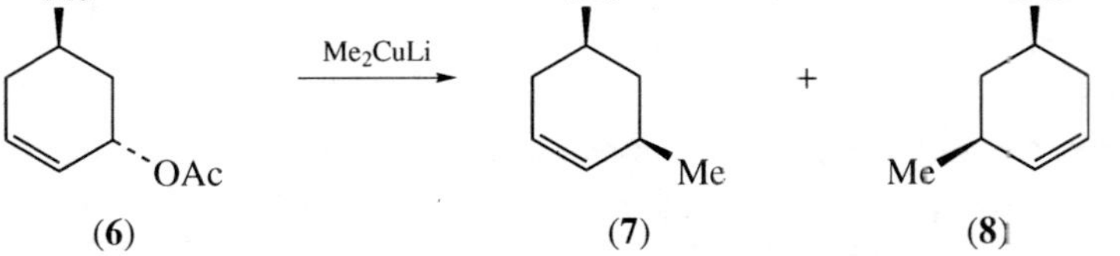

(3)

(6) **(7)** **(8)**

Further experimentation and high level calculations were subsequently performed by the research groups of Goering[8,9], Bäckvall[10], Nakamura[11] and others[12], in order to elucidate the mechanism and elements of stereo- and regiocontrol of the reaction. Soon, it became clear that, given certain reaction conditions, the copper-mediated allylic alkylation proceeds preferentially through a stereospecific *anti* addition pathway. Although there cannot be a single mechanistic model to explain all the observations detailed in the literature, there is a growing consensus around the mechanistic interpretation shown in Scheme 1[10a]. A Cu(I) π-complex **I** undergoes a stereospecific *anti* and regiospecific oxidative γ-addition of copper onto the allylic system to give a Cu(III) σ-complex **II**.

Depending on the rate of reductive coupling of **II** to **2** vs the equilibration of the σ-complexes **II** and **IV** via the π-allyl complex **III**, the overall reaction can be more or less regiospecific. Electron-withdrawing ligands Z, such as cyano, usually lead to a fast reductive coupling (-ZCu) while with Z = alkyl, equilibration becomes competitive. Nakamura and coworkers have proposed, based on a theoretical analysis, that the asymmetry of alkylcopper cyanides (as compared to dialkyl cuprates) leads to the preferential formation of a σ-complex at the γ-position of the allyl electrophile[11d].

However, collapse of the π-allyl complex **III** directly to products **2** or **3** can be regioselective and is influenced primarily by the electronics and, to a lesser extent, the sterics of the substituents. Nakamura and coworkers have proposed, based on DFT calculations, that the copper π-complex **III** undergoes reductive coupling much faster than the corresponding σ-complexes **II** or **IV** and they have proposed that the product forms from the π-complex[11b].

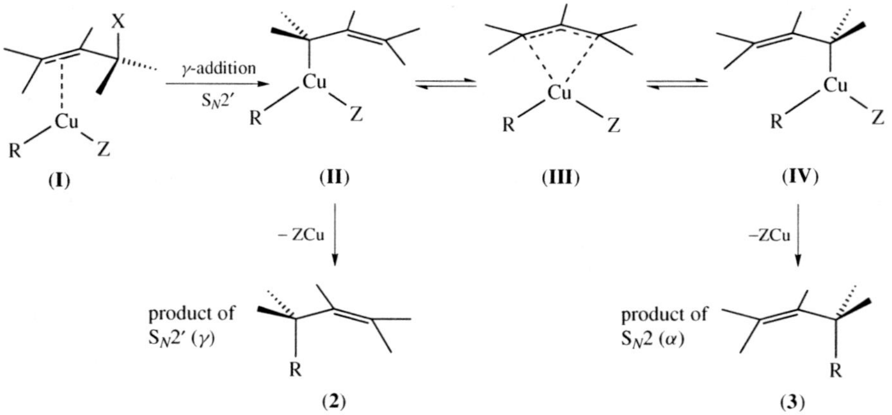

SCHEME 1

The first example of diastereoselection (induction from nearby chiral centers when the leaving group is on an achiral carbon) in allylic substitution of organocopper reagents was reported in 1989[13] and the first example of enantioselection with a chiral organocopper complex was published in 1995[14]. Since then, a large body of work and new advances have been reported.

B. General Considerations on the Structure of the Product

Before discussing in some detail the different methods and reagents that have been used or developed to effect asymmetric allylic alkylation of hard nucleophiles, it is worthwhile to describe some generalities and trends regarding the structural features of the product as a direct consequence of the reaction conditions.

1. Chemoselectivity

a. S_N versus other reactions. The reaction of organocuprates on an allylic electrophile will tolerate the presence of several *spectator* functional groups on the substrate: these include ketones (for cuprate reagents made from dialkylzinc)[15a], esters[15b], lactones[15b], ethers[15c], acetals[15c,h], alcohols[15d,i], amides[15e], nitriles[15a,f], nitros[15g], azides[15h] and dithianes[15i]. Of course, attack at the carbonyl carbon of the leaving group itself may occur with some esters and lactones and the pivalate or mesitoate esters are sometimes used to prevent this occurrence[9a,h,i].

Allylic substitutions can be faster than 1,4-addition reactions. For example, E- or Z-γ-mesyloxy-α,β-enoate **9** gave exclusively the products of allylic substitution **11a–d** when mono- or dialkylcyanocuprate reagents (R^2 = Me, Et, n-Pr, n-Bu) were used (equation 4)[16]. The nature of the leaving group in **9** influenced the reactivity of the cuprate reagent; when allylic acetates were used, for instance, much reduction side products were obtained (R^2 = H in **11** or **12**). Nonetheless, the chemoselectivity remained highly in favor of the allylic substitution (vs 1,4-addition) regardless of the leaving group, of the stereochemistry of the allylic and homoallylic carbons, or of any steric hindrance α to the ester group. The latter is exemplified by the reaction of Z-γ-mesyloxy-α,β-enoates **10** to create a quaternary chiral carbon center in **12b–d**. However, the chemoselectivity was found to be dependent upon the reaction conditions and the solvent used, THF being the preferred solvent[17].

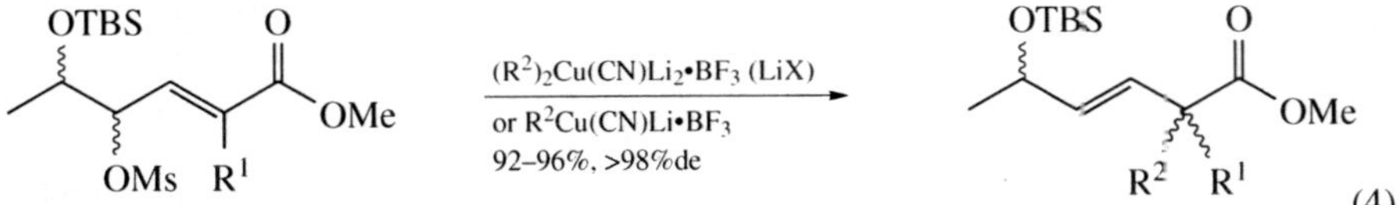

(4)

(**9**) R^1 = H
(**10**) R^1 = Me

(**11**) R^1 = H
(**12**) R^1 = Me

(**a**) R^2 = Me
(**b**) R^2 = Et
(**c**) R^2 = n-Pr
(**d**) R^2 = n-Bu

The addition of TMSCl may reverse this tendency and favor the 1,4-addition process, as shown eloquently by the experiment depicted in equation 5[18]. Without the additive, the alkyltitanium-derived copper species reacts principally with the phosphate[18,19b] **13a** (76% **15a**, 9% cyclohexanone **16**, and 79% recovered **14**) whereas when 1 equivalent of TMSCl was present, the phosphate **13a** was recovered in 96% yield (2% of **15a** was detected) and 47% of the silyl enol ether of **16** was isolated.

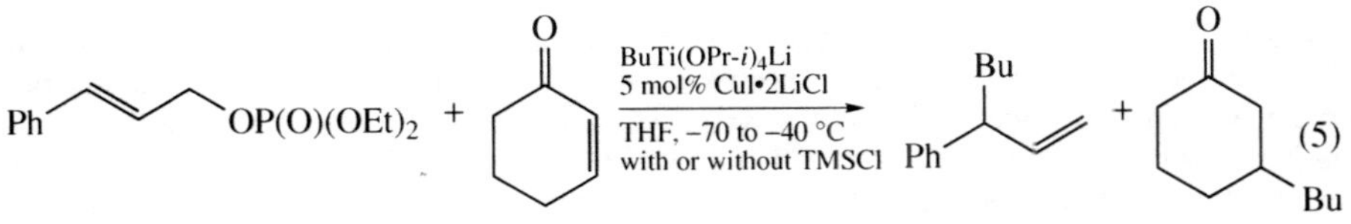

(5)

(**13a**) (**14**) (**15a**) (**16**)

b. Two leaving groups on the allylic moiety. There are few examples of allylic systems flanked by one leaving group on each side. In such cases, chemoselectivity usually depends on the leaving ability of each group. For example, a chloride (**17a**) reacted faster than an acetate to give a mixture of **18a** and **19a** (equation 6)[10d,19]. An acetate reacted faster than a phenol (**17b**) to give a 96:4 mixture of **18b** and **19b** and it reacted also faster than a thiophenol (**17c**) giving a 75:25 mixture of **18c** and **19c** (equation 7)[10d]. However, leaving groups capable of coordination with the cuprate reagent, such as a S-benzothiazolyl (**20**), seem to acquire a reactivity advantage over leaving groups that do not coordinate the metal, such as a pivalate ester, as shown in the exclusive formation of product **21** (equation 8)[20].

None of regioisomeric product **22** was observed, undoubtedly due to coordination and delivery of the cuprate reagents to the γ-position by the *S*-benzothiazole moiety.

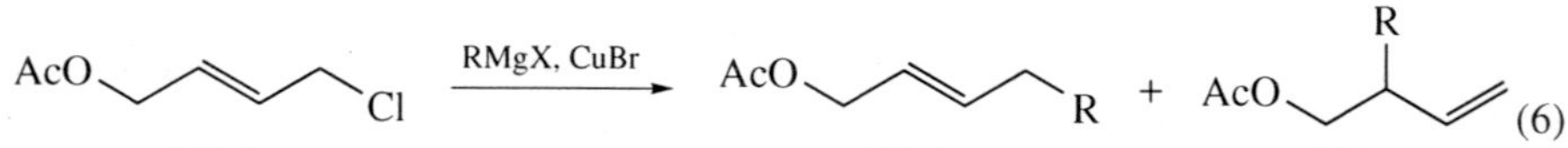

(**17a**) (**18a**) (**19a**) (6)

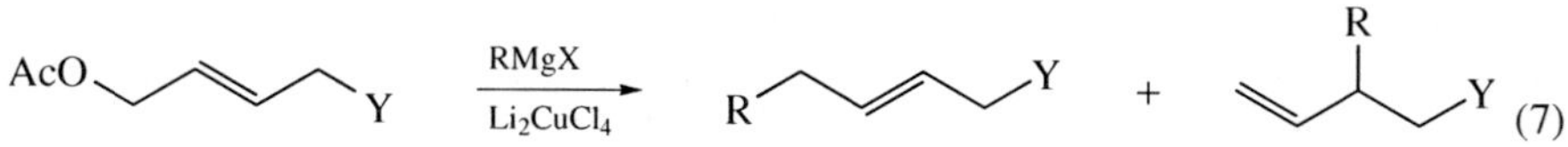

(**17b**) Y = OPh
(**17c**) Y = SPh

(**18b**) Y = SPh
(**18c**) Y = OPh

(**19b**) Y = SPh
(**19c**) Y = OPh (7)

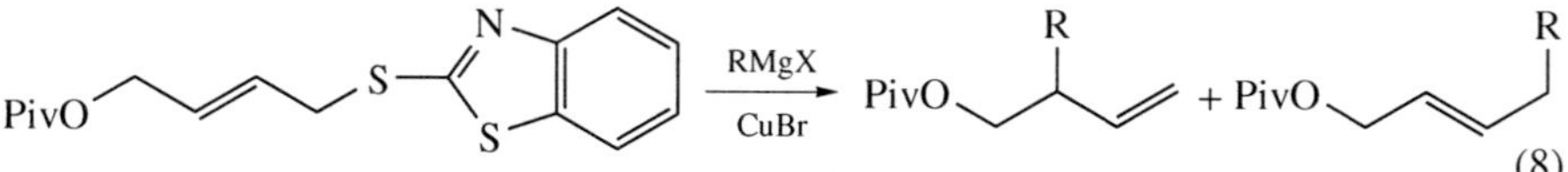

(**20**) (**21**) (**22**) (8)

2. Regioselectivity

The question of S_N2 vs S_N2' displacement (also called α- and γ-substitution, respectively) is perhaps the most challenging to answer when considering a copper-promoted allylic alkylation. Everything from substrate structure, solvent, additives, up to the nature of the organocopper reagent itself may affect the regiochemical outcome of the reaction. However, some generalities may be drawn.

a. Effect of substrate structure. Before the advent of organocyanocuprate reagents, perhaps the strongest influence on the regioselectivity was exercised by the substrate itself. This was observed very early on, using Gilman-type reagents (R_2CuLi), where it was found that the degree of substitution at the γ- and α-positions of the allylic electrophile greatly affects the regioselectivity of the substitution. For example, *trans*-methyl styryl acetate **23** and phenyl *E*-crotyl acetate **25** both gave the same major product **24** from a regioselective but not regiospecific reaction (Scheme 2)[21b]. This means that the organocopper does not force a mechanism-based regiospecificity on the reaction but rather that the substrate **23** favors the reductive coupling to one of its two reactive sites. Other substrates tend to give predominantly the more substituted double bond, which was sometimes interpreted as an attack of the organocopper onto the least substituted end of the allylic system[21a,c]. Preservation of double bond geometry in the starting material and the possible loss of double bond configuration during the reaction was also probed[9a]. No general trend could be drawn but reaction conditions, including the solvent, could affect the outcome.

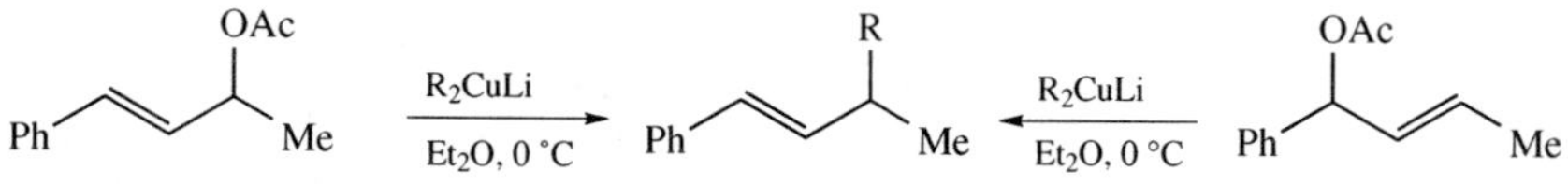

(**23**) (**24**) (**25**)

SCHEME 2

b. Nature of the cuprate reagent. The influence of the substrate structure on the regiochemical outcome of the allylic alkylation can be superseded by the influence of the cuprate reagent. Two types of organocopper reagents are most frequently used in allylic alkylation and they have been described as 'Gilman-type' organocuprates of general formula R_2CuM or RCuXM (also referred to as 'lower-order' organocuprates) and 'higher-order' organocuprates of general formula R_nCuM_{n-1} ($n > 2$) or R_nCuXM_n ($n > 2$)[1]. In addition, monoalkylcopper reagents are sometimes used, often in conjunction with a Lewis acid. The nature of the organocopper reagent has a great influence on the chemo-, regio- and stereochemical outcome of the allylic substitution. The nature of the species in solution, in turn, depends on the source of Cu(I) salt, the starting organometallic species and the solvent. Traditional sources of Cu(I) are: CuI, $CuBr{\bullet}SMe_2$, CuCN. Also, Kochi's salt[22], Li_2CuCl_4, has been used many times as a practical replacement for more traditional sources. Often the copper reagent exists as aggregates with other copper(I) complex as well as with metal salts (e.g. $R_2CuLi{\bullet}LiI$).

Due to these numerous factors, the proper choice of organocopper reagent for any particular substrate may have to be determined by trial and error. Maruyama, Yamamoto and coworkers first reported that monoalkylcopper complexed with BF_3 ($RCu{\bullet}BF_3$) gave very high γ-regioselectivity with simple allylic halides[23,24]. As stated in the preceding Section I.B.2.a, Gilman-type cuprates tend to give mixture of regioisomers with most allyl electrophiles, especially those reagents made from organolithiums. However, alkylcyanocuprate reagents[25] have earned a special place in the realm of allylic substitution, mostly because of their highly γ-regiospecific reaction, especially when prepared from organomagnesiums[9e,21a,26]. Yamamoto and coworkers developed a higher-order cuprate made from alkyllithium reagents, $ZnCl_2$ and CuCN of general formula $R_2CuCN(ZnCl)_2$ that undergo highly regiospecific allylic alkylation[27]. Organocopper species made from alkyltrialkoxytitanium[18] and organozinc[19a,28] have a tendency to give γ-alkylated products. Lithium dialkylcuprate$\bullet SMe_2$ complexes are, on the other hand, α-selective in their allylic alkylation with acyclic halides[29]. Since its inception[30], the copper-catalyzed allylic substitution of Grignard reagents has been widely used. These reaction conditions lead mostly to a regioselective α-attack of the allylic electrophile presumably involving higher-order organocopper species.

c. Effect of solvent and reaction conditions. The regioselectivity may be controlled by a change in reaction conditions. For example, adding tetrahydrofuran to a reaction done in diethyl ether results in a vastly different outcome in the product, as shown in a dramatic reversal of regiochemistry involving the conversion of allylic thiobenzothiazole **13b** into mixtures of **26** and **15a** (equation 9)[31]. In addition to the solvent, the temperature, the speed of addition of the organometallic precursor and the amount of catalyst may each affect the regioisomeric ratio of product obtained[32]. The rate of addition of the organometallic precursor presumably affects the nature of the copper species present in solution. Apart from the fact that the solvent and reaction conditions do affect the nature of the copper species and its aggregation state, there is no other satisfactory explanation for the high sensitivity of the regiochemistry of allylic alkylation to a solvent change. The conditions usually have to be optimized by trial and error.

d. Effect of the leaving group. The nature of the leaving group may have a profound effect on the regiochemistry of the allylic alkylation process. In general, coordinating leaving groups such as Gallina's carbamate (cf. Section II.A.2) and Caló's benzothiazole[31] compel the nucleophile to add with high γ-regioselectivity. Such leaving groups have a strong stereodirecting effect as well and are thus discussed in more detail in the following sections. Non-coordinating groups have much less influence on the regiochemistry of allylic alkylation.

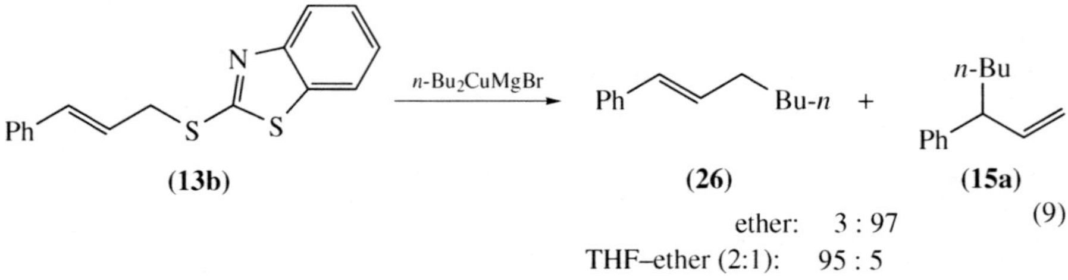

(9)

3. Stereochemistry

The subject of stereochemistry in the products of asymmetric allylic alkylation is central to this chapter and will be discussed where appropriate throughout the following sections. The different stereochemical aspects of the reaction to be considered are: a) the diastereospecificity of the reaction, which will dictate, along with conformational biases in acyclic cases, the stereochemistry of the newly created chiral center with respect to the original stereochemistry at the carbon bearing the leaving group (when that carbon is chiral); b) chiral induction exerted by chiral centers or appendages elsewhere on the substrate (diastereoselection); c) chiral induction from asymmetric ligands on copper (enantioselection); d) the geometry of the final alkene.

II. DIASTEREOSPECIFIC METHODS

A. General Considerations

1. Substrate features and stereospecificity

This section covers what is, without a doubt, the most frequently utilized type of asymmetric allylic alkylation. Cuprate reagents generally react stereospecifically *anti* to the leaving group. Staroscik and Rickborn first noticed the stereospecificity in the opening of cyclic allylic epoxides[7]. Goering and coworkers demonstrated this same stereospecificity for cyclic allylic acetates (cf. equation 3). Lithium dimethylcuprate added almost exclusively *anti* to the leaving acetate, though there was no regioselectivity in this case[8].

Corey and Boaz proposed a model engaging the copper's d orbitals with the substrate's π^* and C$-$X σ^* orbitals (Figure 1)[33]. Taken together with conformational considerations, the stereospecificity allows the stereochemistry at the newly formed chiral center in acyclic products to be predicted with confidence. For example, Ibuka, Yamamoto and coworkers have used allylic strain to attain a high level of control on the transfer of chirality at the distal S_N2' position on γ-mesyloxy-α,β-enoates (4*S*,5*R*)-(*E*)-**9** and (4*S*,5*R*)-(*Z*)-**9**

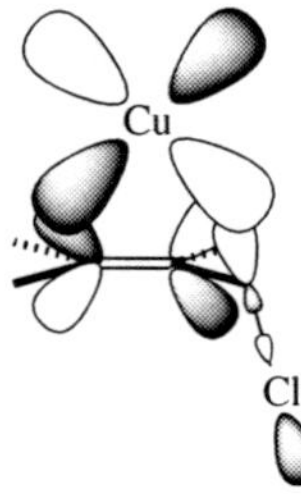

FIGURE 1. Corey–Boaz' molecular orbital model explaining the preferred *anti* displacement of cuprate reagents to allylic systems

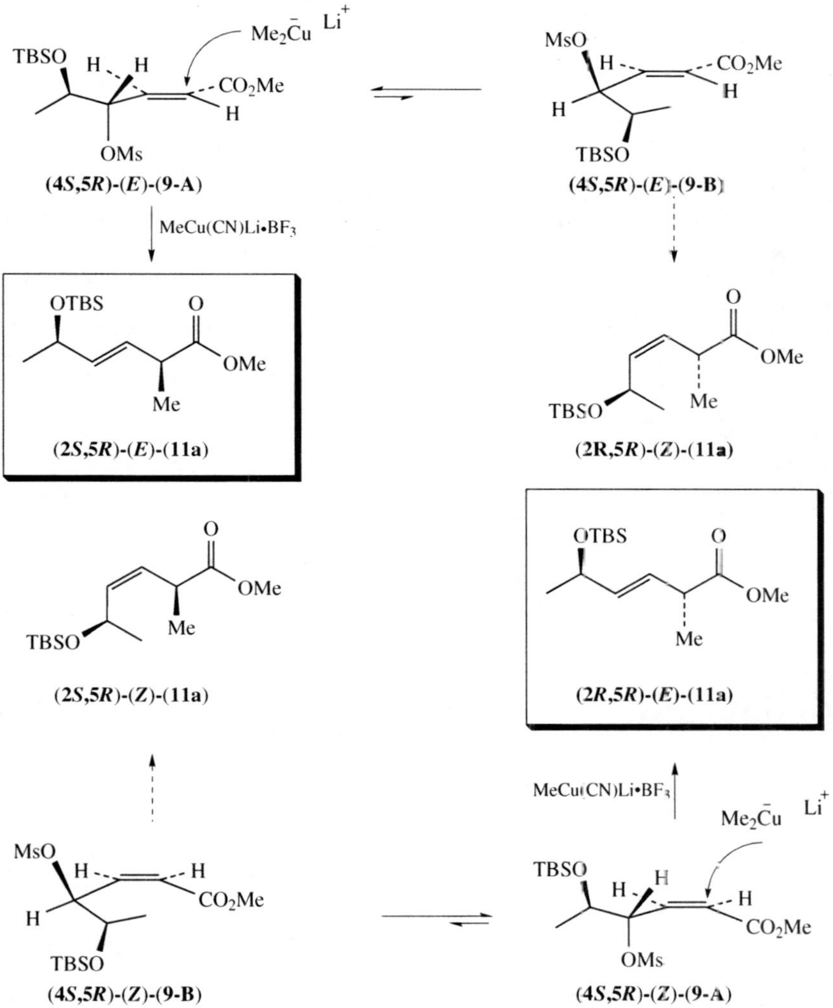

SCHEME 3

(each shown in two conformations **A** and **B**, see Scheme 3)[17b]. Allylic strain is a strong stereocontrolling element in that one conformer (**A**), where $A^{1,3}$-strain is minimized, is usually much lower in energy than the other conformer (**B**), especially for starting allylic substrates possessing a double bond of *Z* geometry. It is important to note, however, that the minor diastereomeric products (2*S*,5*R*)-(*Z*)-**11a** and (2*R*,5*R*)-(*Z*)-**11a** originating from the higher-energy conformers (4*S*,5*R*)-(*E*)-**9-B** and (4*S*,5*R*)-(*Z*)-**9-B**, respectively, possess double bonds of *Z* geometry. Thus, allylic strain in the transition state also controls the geometry of the double bond in the final product. Importantly, if the transfer of chirality is not complete, it becomes possible to uncover the source of the problem by determining the geometry of the alkene in the final product. Failure to detect allylic alkylation products possessing a *Z* alkene necessarily vindicates the occurrence of low stereoselectivity ($A^{1,3}$-strain) as a possible source of the other stereoisomer and leaves a loss in stereospecificity as the sole culprit[34]. A loss in stereospecificity may occur by way of radical or ionic intermediates and would not necessarily lead to products with a Z alkene. The reader will note that both major diastereomers (2S,5R)-(E)-**11a** and (2R,5R)-(E)-**11a** can be prepared starting from compounds with the same stereochemistry at the carbon bearing the leaving group but with different double bond geometries in the starting material.

If allylic strain controls the stereochemical outcome of the newly formed chiral center, it also controls the predominant formation of the E double bond in the product. One rare exception to this rule has been reported by Smitrovich and Woerpel[35a,b]. They described the selective formation of Z-alkenes **28a–c** from the reaction of allylic carbamates **27a** or **27b** with Grignard reagents using CuCl•2LiCl as promoter (Scheme 4). The fact that the corresponding organocuprates made from alkyllithium reagents gave the expected products with a E alkene, commensurate with the minimization of allylic strain, led the author to propose a different mechanism whereby a copper-bound carbamate **27-I** would not bring about the expected *syn* addition of the cuprate reagent but instead a regioselective *anti* addition of the free Grignard reagent to give intermediate **27-II.** This step would then be followed by a stereospecific *anti* elimination of the metallocarbamate, thereby forcing the formation of (Z)-**28**. Allylsilanes are useful synthons and were used in [3 + 2]-annulations and for the total synthesis of (+)-blastmycinone[35c,d].

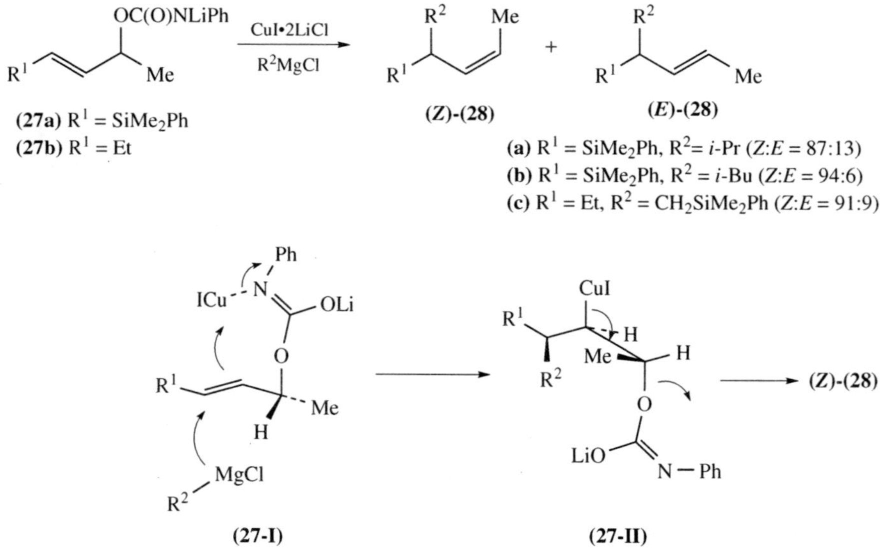

SCHEME 4

2. Effect of the leaving group on the stereospecificity

Certain leaving groups have the ability to coordinate to the copper reagent such that the stereospecificity of the allylic substitution is reversed in favor of *syn* addition products. One such group is the carbamate first introduced by Gallina and Ciattini[36]. It was shown that carbamate *syn*-**29**, under treatment with Gilman's lithium dimethylcuprate, gave exclusively the adduct *syn*-**30** (equation 10), while carbamate *anti*-**29** afforded solely the adduct *anti*-**30** (equation 11). The deuterium atom was installed to also demonstrate the regiospecificity of the process. It seems that the coordination of the cuprate reagent by the carbamate prevents its delivery to the α position.

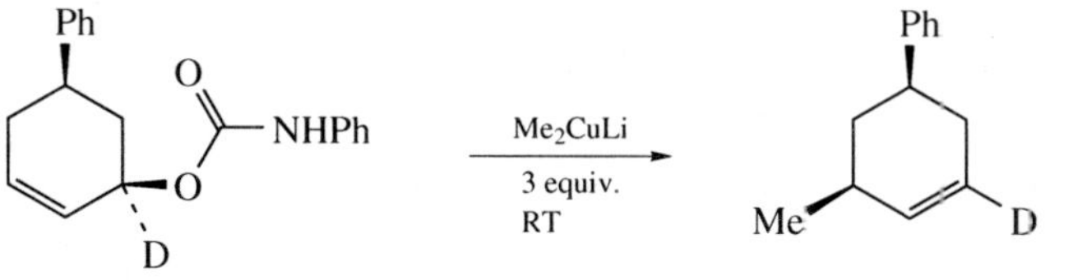

syn-(**29**) *syn*-(**30**) (10)

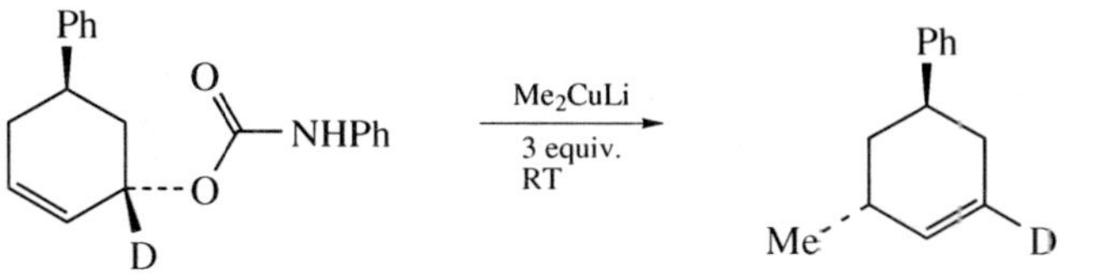

anti-(**29**) *anti*-(**30**) (11)

Caló's benzothiazoles (cf. equation 9) also effect *syn* allylations[31] while Yamamoto's monoalkylcopper BF_3 complexes can displace free hydroxyl groups by coordinating to them and delivering the alkyl *syn* to the hydroxyl moiety (not shown)[23b]. Both provide highly regioselective allylations and monoalkylcopper BF_3 complexes react with allylic acetate with the usual *anti* stereoselectivity. It is thus obvious that the stereochemical complementarity of coordinating and non-coordinating leaving groups offer the chance for stereochemical convergence (i.e. making one and the same product stereoisomer (or mixture of stereoisomers) starting from two enantiomeric or epimeric alcohols, so long as the alcohols are separable or are made separately)[37].

Breit and coworkers have designed a leaving group consisting of an aromatic ester carrying a phosphine in the *ortho* position (Scheme 5). The phosphine in **31a** may coordinate to the copper reagent in Et_2O and deliver the alkyl *syn* to the leaving group, via the $A^{1,3}$-strain minimized conformation of **32**, to give allylic alkylation product **33**. Phosphine oxide **31b** does not, however, coordinate to the cuprate reagent in THF. Thus, a dialkylzinc-derived cuprate reagent now attacks compound **31b** *anti* to the leaving group to give *ent*-**33**[38].

B. Acyclic Allylic Systems

In this section, the different methods and strategies used to obtain a new tertiary or quaternary chiral center from an acyclic allylic system will be discussed, with an emphasis on more recent literature. The sub-sections were categorized according to substrate structure, but all substrates in this section have in common the fact that the bond between the carbon bearing the leaving group and the alkene can rotate freely. Otherwise, they are covered in Section II.C. Everywhere possible and appropriate, the source of chirality and applications or the intended end-use of the product will be stated.

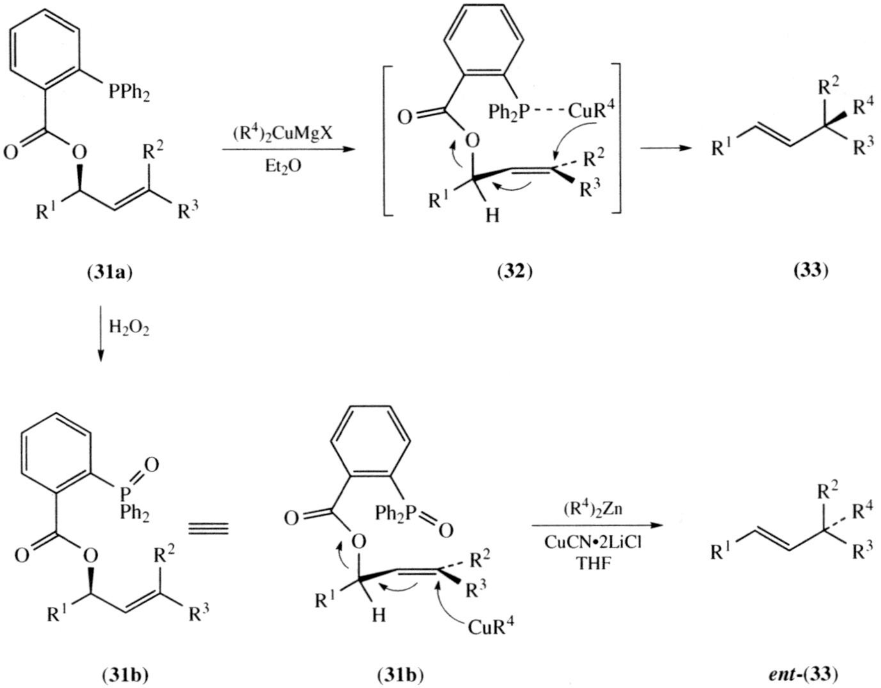

SCHEME 5

1. Diastereomerically pure precursors prepared from chiral non-racemic acyclic allylic alcohols

a. Substrates taken from the chiral pool or obtained by resolution or asymmetric reduction of ketones. A number of research groups have used pre-made chiral allylic alcohols, converting the hydroxyl group into a leaving group, and transferred chirality to the γ-position by effecting a regioselective S_N2' displacement with a cuprate reagent. Trost and Klun opened racemic 5- and 6-membered lactones, substituted with a vinyl group, via a regio- and stereoselective allylic alkylation of monoalkylcyano cuprates (not shown)[39]. Ibuka, Yamamoto and coworkers prepared chiral non-racemic δ-mesyloxy-α,β-enoates such as **9** starting from L-threonine **34**, and other δ-mesyloxy-α,β-enoates from sorbic acid or D-xylose, and then effected their regio- and stereoselective displacement with R_2CuLi, $R_2CuLi{\bullet}BF_3$ or $R_2Cu(CN)Li_2{\bullet}BF_3$ and other cuprate reagents (Scheme 6)[17b,40]. They obtained γ-allylation products like **11a** but the regioselectivity was high even for starting esters possessing a trisubstituted double bond, which provided access to the formation of all-carbon quaternary stereocenters (not shown)[17a]. Surprisingly, the method gave rise to reduction of the allylic electrophile, instead of its alkylation, when the structurally similar δ-acetoxy-α,β-enoates (as opposed to δ-mesyloxy-α,β-enoates) were used as substrates. The absence of BF_3 or the use of divinylcuprate reagents increased the occurrence of reduction[16b,41]. They then extended the chemo-, regio- and stereoselectivity of their methodology to access peptide isosteres in the form of δ-hetero-β,γ-enoates like **37** (Scheme 6)[16b]. The starting mesylate **36** was synthesized from amino acid **35** while the other mesylates **38–41** were made from other amino acids or carbohydrates. As stated

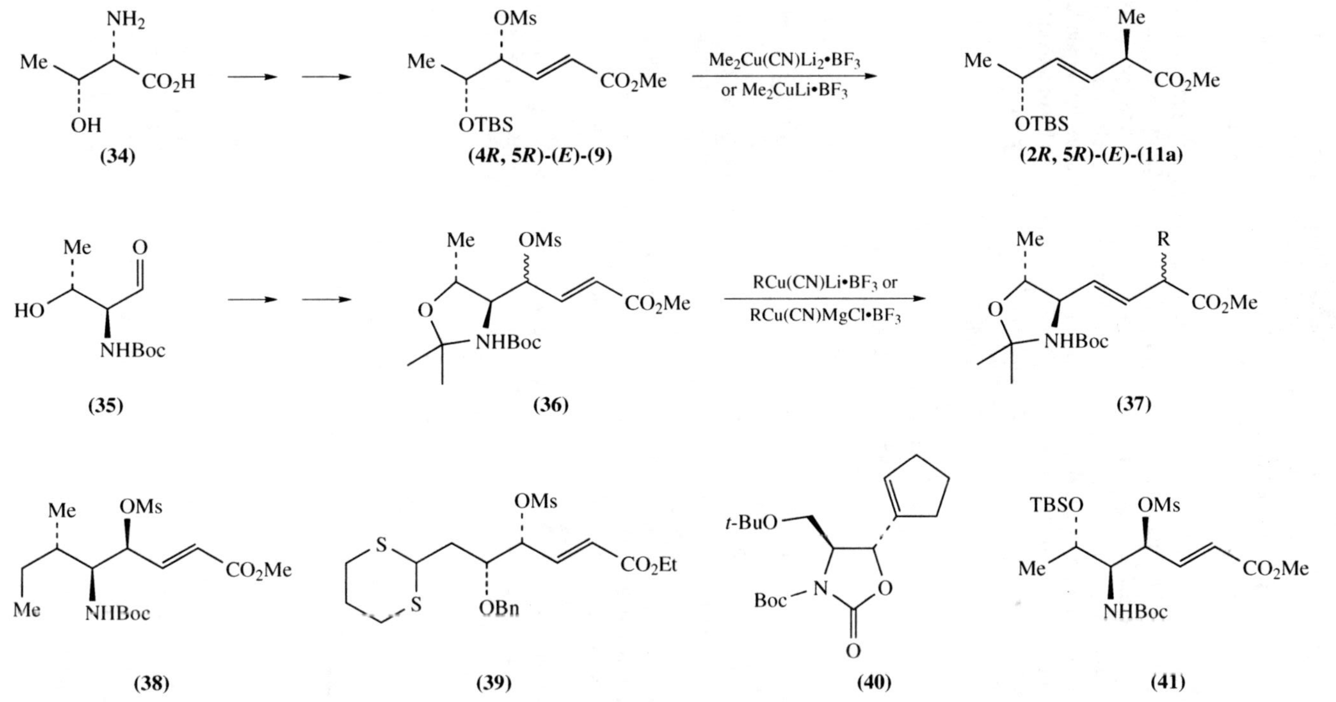
NH2
Me
CO2H
OH
(34)
OMs
Me
CO2Me
OTBS
(4R, 5R)-(E)-(9)
Me2Cu(CN)Li2•BF3
or Me2CuLi•BF3
Me
Me
CO2Me
OTBS
(2R, 5R)-(E)-(11a)
Me
O
HO
NHBoc
(35)
Me
OMs
O
CO2Me
NHBoc
(36)
RCu(CN)Li•BF3 or
RCu(CN)MgCl•BF3
Me
R
O
CO2Me
NHBoc
(37)
Me
OMs
CO2Me
Me
NHBoc
(38)
OMs
S
S
CO2Et
OBn
(39)
t-BuO
Boc
N
O
O
(40)
TBSO
OMs
Me
CO2Me
NHBoc
(41)

SCHEME 6

before (see Section I.B.1.a), the chemoselectivity (alkylation vs 1,4-addition) is dependent on reaction conditions. All substrates (**36**, **38–41**), including cyclopentene derivative **40**[42], displayed excellent chirality transfer[16b, 27b, 40, 42, 43].

Belelie and Chong have found that allylic phosphates are efficiently displaced, principally in an S_N2' fashion, using $R_2Cu(CN)Li_2$ reagents[44]. They used a kinetic resolution through a Sharpless asymmetric epoxidation[45] (SAE-KR) to efficiently separate the enantiomers of various allylic alcohols (±)-**42**, from which they prepared the allylic phosphates and submitted them to displacement with higher-order cyanocuprate reagents (equation 12). Some loss of stereochemical integrity was observed in the products (*R*)-**44** and this was attributed to a *syn* addition to the more stable $A^{1,3}$-strain-minimized conformation of (*R*)-**43**. The regioselectivity was good (usually >90:<10) but dependent upon the steric volume of the allylic substitutent R^1. The alkene was cleaved oxidatively to obtain a chiral non-racemic α-substituted acid or aldehyde (not shown).

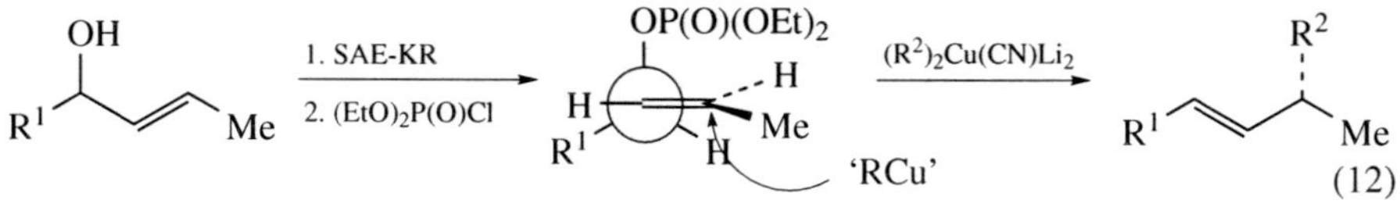

(±)-(**42**) (*R*)-(**43**) (*R*)-(**44**) (12)

Pentafluorobenzoates like **(*Z*)-(*R*)-45** and **(*E*)-(*S*)-45** were effectively used to enable the allylic alkylation of the less reactive dialkylzinc reagents with excellent regio- and stereoselectivity (Scheme 7). Functionalized diorganozinc reagent **46** or $(n\text{-Pen})_2Zn$ reacted well to give **47** and **49**, respectively, both possessing a quaternary stereogenic carbon[46]. The non-racemic alcohols, precursors of **45**, were prepared by kinetic resolution or asymmetric reduction of the ketone, as shown for ketone **48**. The method was used as a general strategy to prepare chiral alcohols and amines like **50** and **51**, respectively[46b], as well as α-chiral (*E*)-vinylsilanes (not shown)[6d]. A short synthesis of (+)-ibuprofen® using this methodology was also realized (not shown)[46a].

The picolinic ester proved to be a useful leaving group, enabling highly *anti* and S_N2' selective allylic alkylations when installed on substrates possessing a *Z*-alkene (**52**) (equation 13). Only arylmagnesium bromide-derived copper species were investigated as nucleophiles with this system to give aryl-substituted products **53**. It was suggested that $MgBr_2$, formed *in situ* in the solution, actually activates the leaving group. Chirality transfer was dependent upon temperature but was generally good[47].

Finally, a chiral non-racemic quaternary carbon containing a CF_3 unit was prepared from the copper-catalyzed allylic alkylation of ethyl Grignard with a trifluoromethyl-substituted allylic acetate (not shown)[48]. The chirality in the starting acetate originated in a stereoselective reduction of the corresponding ketone with (*R*)-Binal-H.

b. Use of a chiral auxiliary. Optically pure alcohols are efficiently prepared by the addition of nucleophiles (H^- or **59**) onto prochiral carbonyls such as **54** or **58** (Scheme 8). In that regard, a number of chiral auxiliaries (Aux*) can be used to generate the chiral alcohol (**55**, R′ = H) precursor to the corresponding allylic electrophile (**55**, R′ = LG). Most of the time, the auxiliary's role ends with the generation of the carbinol's stereochemistry though it may affect the regiochemical outcome of the allylic alkylation reaction to give **56** or its regioisomer. The difference between the various methods thus centers around how the auxiliary is later cleaved or somehow used to convert the final allylated product into a useful synthetic intermediate such as **57** or **60**.

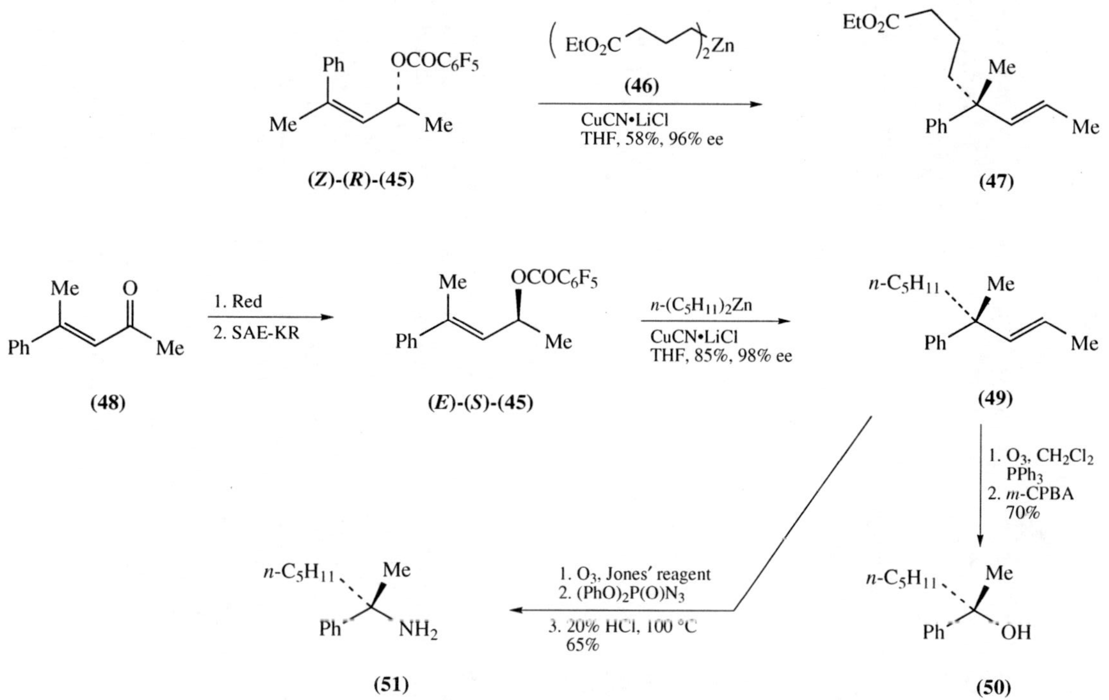
OCOC6F5
Ph
Me
Me
(Z)-(R)-(45)
EtO2C
Zn
(46)
CuCN•LiCl
THF, 58%, 96% ee
(47)
(48)
1. Red
2. SAE-KR
(E)-(S)-(45)
n-(C5H11)2Zn
THF, 85%, 98% ee
n-C5H11
(49)
1. O3, CH2Cl2
PPh3
2. m-CPBA
70%
OH
(50)
1. O3, Jones' reagent
2. (PhO)2P(O)N3
3. 20% HCl, 100 °C
65%
NH2
(51)

SCHEME 7

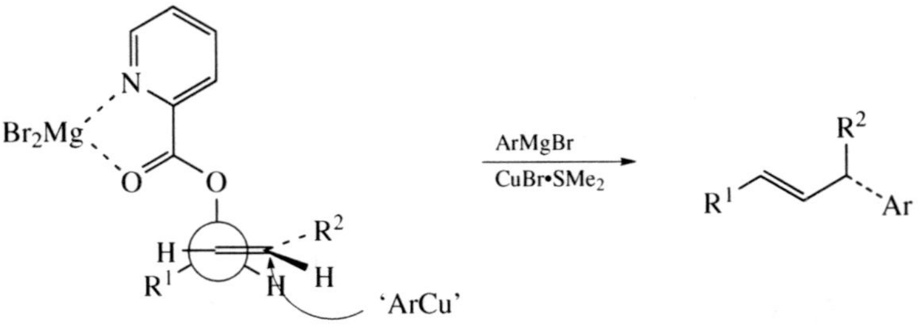

(52) (53) (13)

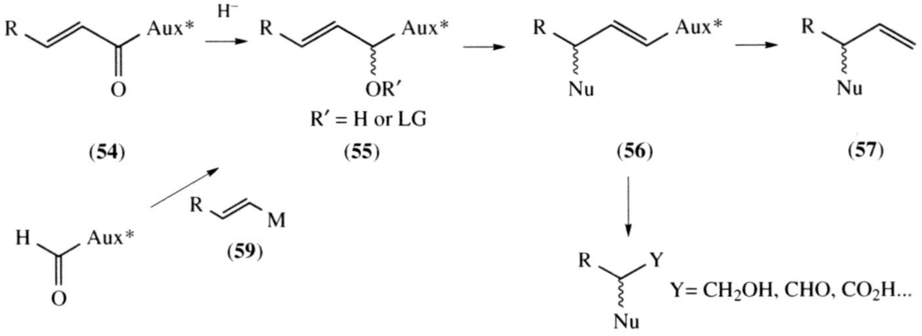

(54) (55) (56) (57) (58) (59) (60)

SCHEME 8

Chiral allylic alcohols can be accessed via an aldol-type process with chiral sulfoxides **61** acting as chiral auxiliaries (Scheme 9). The diastereoselectivity of this process is low, giving mixtures of alcohols **62a** and **65a** that were easily separated by normal silica gel chromatography and derivatized into the corresponding mesylates **62b** and **65b**. Then, displacement of each mesylate by organocopper reagents afforded (*Z*)-**63** and (*Z*)-**64** as the major product, respectively, resulting from an *anti* addition onto $A^{1,3}$-strain-minimized conformers **62b-A** and **65b-B** (Figure 2). Diastereoselectivity from **62b** was high (>91:<9 **(*Z*)-63:(*E*)-64**), but the level of diastereoselectivity from **65b** varied depending on the copper species. In that case, products **(*Z*)-64** and **(*E*)-63** were formed in ratios ranging from 70:30 up to 94:6. The reason behind the difference between **62b** and **65b** is likely due to the chiral sulfoxide's own configuration, one of them somewhat destabilizing conformer **65b-B**. We could say that **62b** represents a 'matched' pair between the chirality at the carbon bearing the mesylate and the chirality at sulfur. The chiral sulfoxide auxiliary in **63** or **64** was removed by reduction and was thus not recoverable[49].

When the starting sulfoxides were reduced to the corresponding sulfides, the subsequent allylic substitution followed more or less the same course as per the sulfoxides. However, if the sulfoxides were oxidized to the corresponding sulfone, the reaction favored the formation of the *E*-product **64**, possibly because of the added bulk around the sulfur atom favoring conformer **62b-B**.

Menthone (**67**) and its derivative *p*-menthane-3-carboxaldehyde **70** are two recyclable chiral auxiliaries that can be cleaved by oxidation or ring-closing metathesis (RCM) after the sequence of reactions involving an asymmetric allylic substitution. The addition

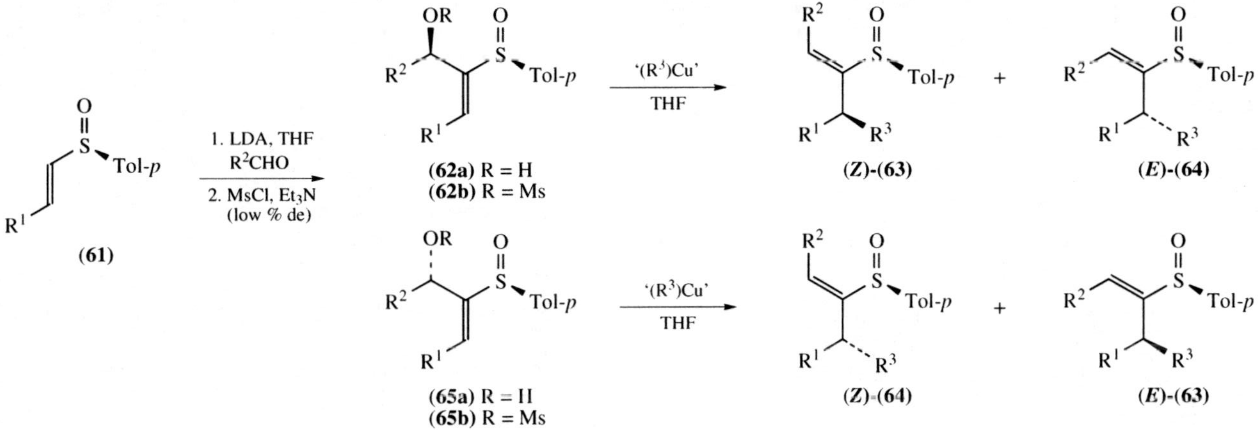
Tol-p
(61)
1. LDA, THF
R2CHO
2. MsCl, Et3N
(low % de)
(62a) R = H
(62b) R = Ms
'(R3)Cu'
THF
(Z)-(63)
(E)-(64)
(65a) R = H
(65b) R = Ms
(Z)-(64)
(E)-(63)

SCHEME 9

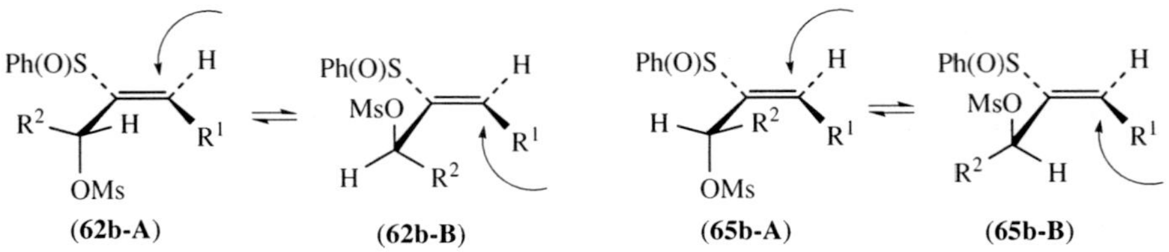

FIGURE 2

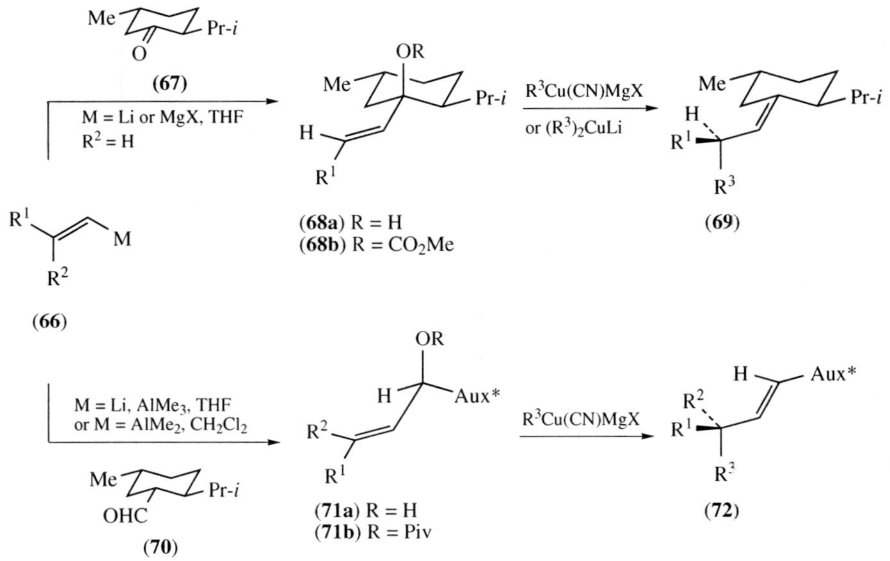

SCHEME 10

of various vinylmetal nucleophiles **66** to menthone (**67**) always gives a single alcohol **68a** (Scheme 10). The addition of vinylalanes or vinyllithiums to **70**, in the presence of trimethylaluminum, is highly diastereoselective and affords diastereomerically pure alcohols **71a** after separation by normal silica gel chromatography[50]. Activation and stereospecific allylic alkylation proceeds with high regioselectivity, probably aided by the bulk of either chiral auxiliaries. Reaction of carbonate **68b** is insensitive to the nature of the cuprate reagents[51] but pivalate **71b**, on the other hand, gave good regioselectivity only with monoalkylcyanocuprates made from alkylmagnesium halides[52]. Strong allylic strain in either case ensures that only one conformer reacts and that the transfer of chirality is generally complete. Only aryl substituents (R^1 = Ar) caused noticeably lower diastereoselectivities when bulky cuprate reagents were used. Auxiliary **70** may be used to effect the synthesis of quaternary stereocenters (**72**, R^1, R^2, R^3 = alkyl, aryl).

Oxidative cleavage of **69** or **72** with ozone affords α-chiral carboxylic acids, aldehydes or hydroxymethylenes, depending upon the work-up conditions[51,52]. An iterative approach (**66** → **73** → **74**, Schemes 10 and 11) allows for the rapid synthesis of polypropionate units such as the fragment A (**75**) of ionomycin (Scheme 11)[51d]. Both enantiomers of amino acids, including α,α-dialkylated amino acids like **76** and **78** (*ent*-**76**), are available from **77** by a stereodivergent strategy by way of a Curtius rearrangement[53]. The

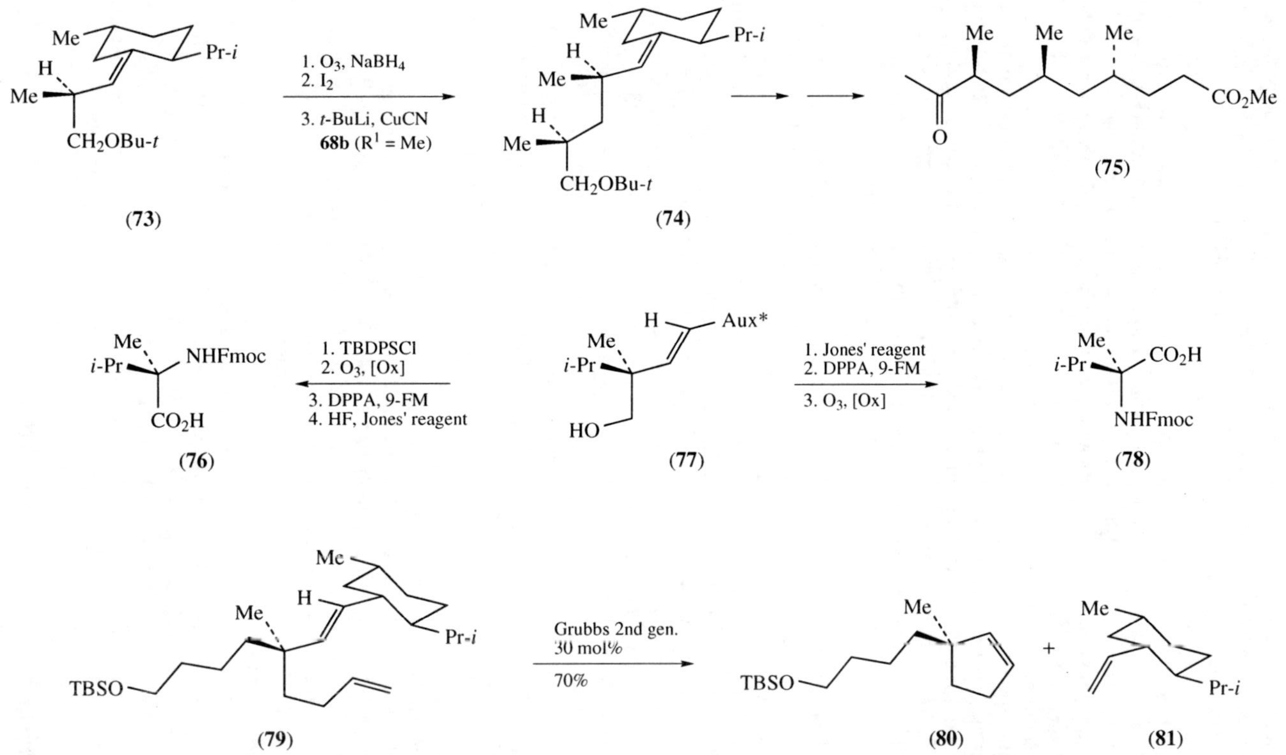
Me
Pr-i
H
Me
CH2OBu-t
(73)
1. O3, NaBH4
2. I2
3. t-BuLi, CuCN
68b (R1 = Me)
(74)
Me
Me
Me
O
CO2Me
(75)
Me
i-Pr
NHFmoc
CO2H
(76)
1. TBDPSCl
2. O3, [Ox]
3. DPPA, 9-FM
4. HF, Jones' reagent
H
Aux*
HO
(77)
1. Jones' reagent
2. DPPA, 9-FM
3. O3, [Ox]
(78)
TBSO
(79)
Grubbs 2nd gen.
30 mol%
70%
(80)
(81)

SCHEME 11

p-menthane-3-carboxaldehyde auxiliary can also be cleaved by RCM, as shown for intermediate **79**, to effect the direct synthesis of chiral carbocyclic compounds such as **80**[54]. The recovered chiral fragment **81** is very volatile and is easily removed by evaporation from the crude products mixture. It can be subsequently converted back to aldehyde **70** by ozonolysis.

A resin-bound version of this auxiliary achieves overall yields of 4–18% for the 4-step sequence. Ozonolysis releases the final products, which have a purity >80% for alcohols and aldehydes and >70% for carboxylic acids. The resin-bound auxiliary is regenerated in the process and is ready for a second iteration of the reaction sequence[55].

Couty and coworkers used a similar strategy to obtain chiral allenyl or vinyl- oxazolidines **84** or **86** (Scheme 12). The α-keto oxazoline **82** was reduced selectively to the corresponding alcohol with zinc borohydride and mesylation supplied **83**. Reaction with lithium dimethylcuprate gave the desired allenyl compound **84** with complete transfer of chirality. Alternatively, the hydroxy-alkyne can be reduced with Red-Al and the allylic alcohol mesylated to give, for example, (*E*)-**85**, or hydrogenated and mesylated to give (*Z*)-**85**. After stereospecific allylic alkylation with lithium dimethylcuprate, α,β-unsaturated oxazolidine **86a** or **86b**, respectively, are obtained[56]. The chiral auxiliary was removed in three steps by *N*-deprotection, hydrolysis and reduction to give **87a** and **87b**.

c. Applications of diastereospecific methodologies in synthesis. Using their recently developed tunable leaving group, Breit and coworkers synthesized the C_3–C_{11} fragment of the natural CDK and angiogenesis inhibitor borrelidin (**94**)[38,57]. They resolved racemic allylic alcohol (±)-**42a** by enzymatic methods and derived (*R*)-**42a** into ester **88**. Grignard reagent **90**, made from the corresponding known bromide **89**, in the presence of Cu(I), reacted with **88** in a highly regioselective and stereospecific fashion to give compound **92** (Scheme 13). In three iterations, one iteration being the transformation of bromide **89** via **9e** into the iodide **91**, they arrived at the desired C3–C11 fragment **93**.

A series of vitamin D analogs **97** were synthesized thanks to the high regioselectivity and stereospecificity of Yamamoto's cuprate reagent (Scheme 14). Phosphate **95** was displaced with nearly complete transposition of the double bond despite an obvious steric impediment to the reaction site to give **96** in >99% de[58]. However, with aryl or vinylcopper reagents, a loss in regioselectivity was observed.

Controlling the C-15 carbinol stereochemistry of prostaglandins has been a daunting challenge[59]. A stereoconvergent approach takes advantage of the possibility of achieving *syn* or *anti* stereochemistry in asymmetric allylic alkylation by converting each diastereomeric alcohol to either a coordinating or a non-coordinating leaving group. Indeed, α-alcohol **98a** was converted to its benzoate ester **99** while the β-alcohol **98b** was converted to carbamate **100** (Scheme 15). Each were separately converted to the same β-silane **101** using a silacuprate reagent and the allylsilane **101** was then oxidized to the corresponding allyl alcohol **102**. There was, therefore, no need to prepare stereoselectively alcohol **98a** or **98b**[60].

In their approach to cladiell-11-ene-3,6,7-triol **106a**, a member of a family of cembranoid diterpenes that includes the potent anti-leukemic sclerophytin A **106b**, McIntosh and coworkers utilized the strong bias offered by $A^{1,3}$-strain to effect a highly stereoselective (**103B** more stable than **103A**) and stereospecific (*anti* addition) displacement of cyclic lactone **103** to give adduct **104** (Scheme 16)[61]. The latter was converted to cladiell-11-ene-3,6,7-triol **106a** by way of the synthetic intermediate **105**.

2. Opening of allylic epoxides and aziridines

Allylic epoxides and aziridines are somewhat different than other allylic electrophiles, if only that they are usually prepared in enantiomerically enriched form by distinct synthetic

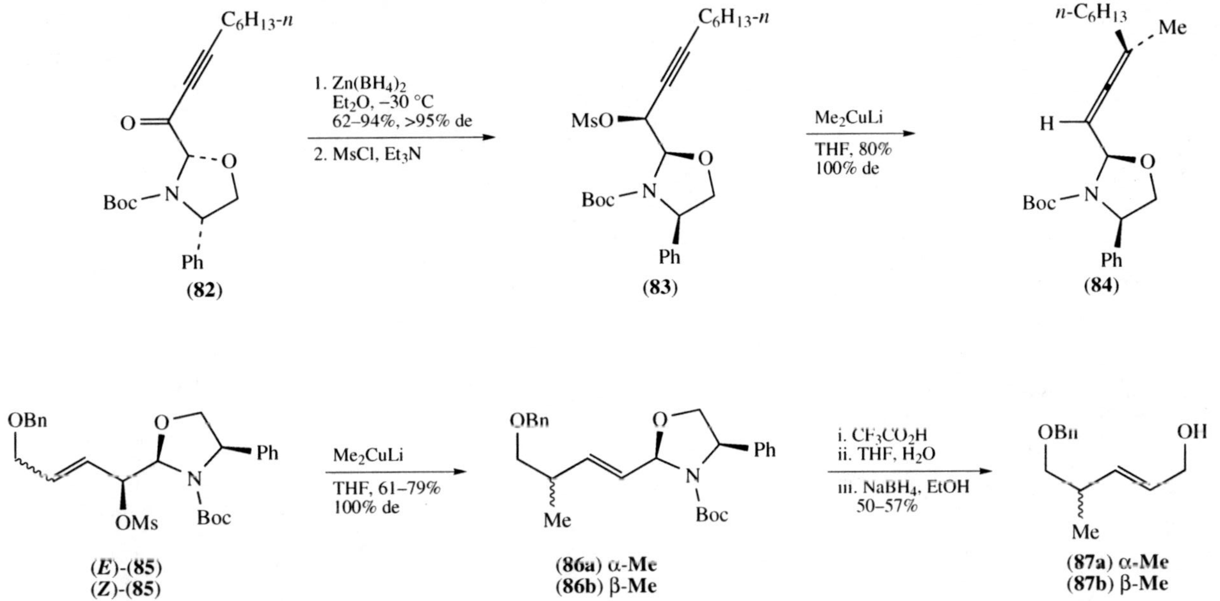
C6H13-n
O
Boc
N
O
Ph
(82)
1. Zn(BH4)2
Et2O, −30 °C
62–94%, >95% de
2. MsCl, Et3N
C6H13-n
MsO
O
Boc
N
Ph
(83)
Me2CuLi
THF, 80%
100% de
n-C6H13
Me
H
O
Boc
N
Ph
(84)
OBn
O
Ph
N
OMs
Boc
(E)-(85)
(Z)-(85)
Me2CuLi
THF, 61–79%
100% de
OBn
O
Ph
N
Me
Boc
(86a) α-Me
(86b) β-Me
i. CF3CO2H
ii. THF, H2O
iii. NaBH4, EtOH
50–57%
OBn
OH
Me
(87a) α-Me
(87b) β-Me

SCHEME 12

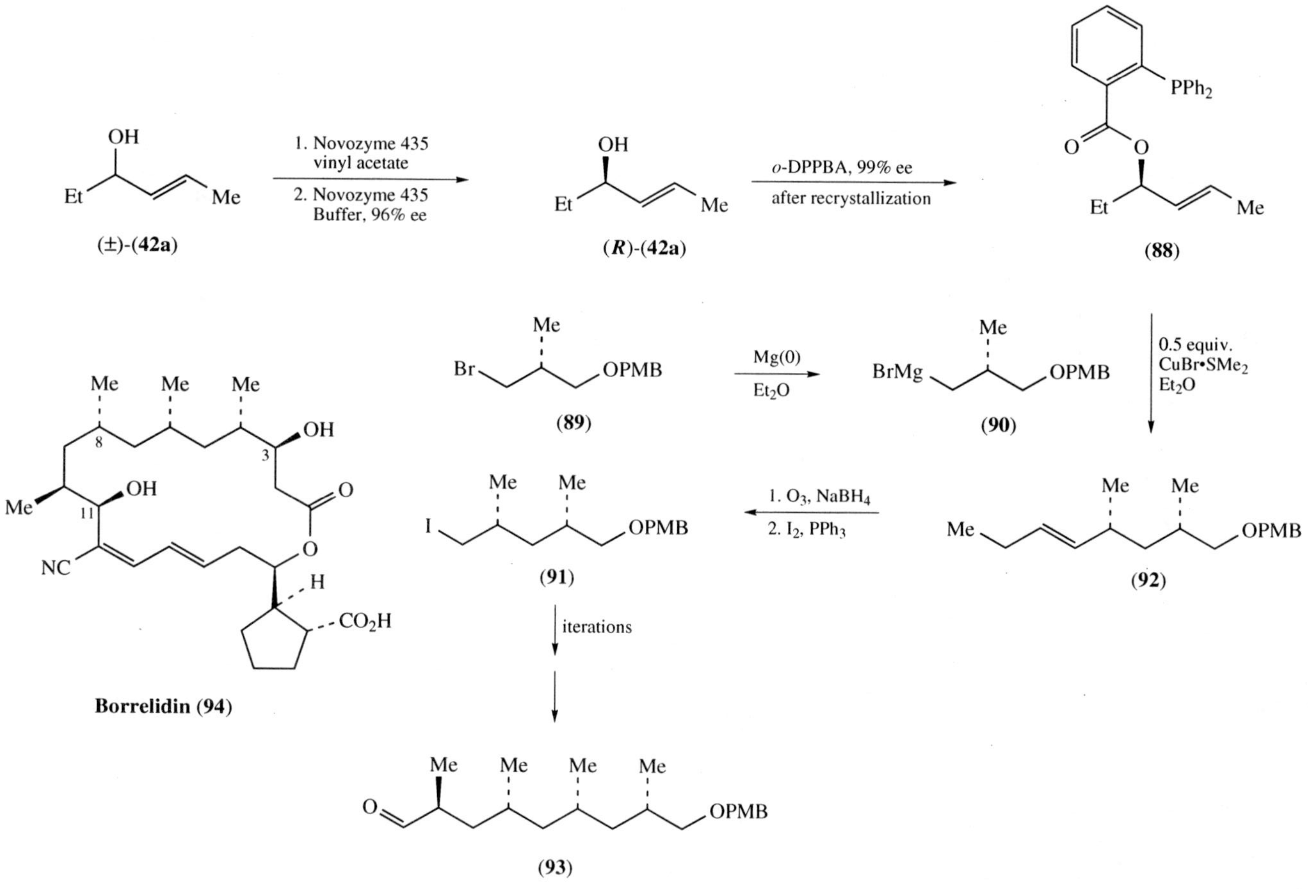
(±)-(42a)
1. Novozyme 435 vinyl acetate
2. Novozyme 435 Buffer, 96% ee
(R)-(42a)
o-DPPBA, 99% ee
after recrystallization
(88)
0.5 equiv. CuBr•SMe2 Et2O
(89)
Mg(0)
Et2O
(90)
(92)
1. O3, NaBH4
2. I2, PPh3
(91)
iterations
(93)
Borrelidin (94)

SCHEME 13

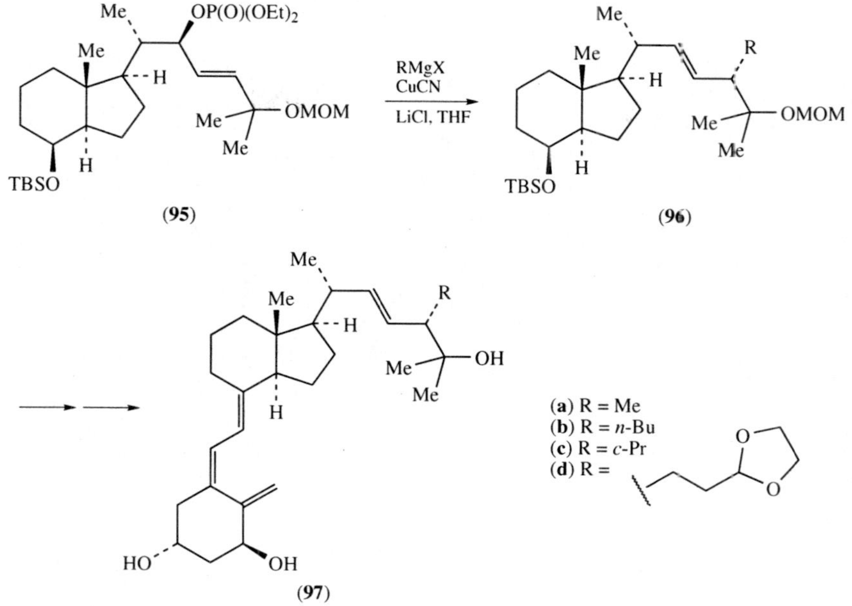

SCHEME 14

means and that the leaving oxygen or nitrogen resides, by definition, on a secondary or tertiary chiral carbon. In addition, the alkylation reaction creates an allylic alcohol or amine that may, in principle, allow a second allylic alkylation to take place.

The first example of the opening of an allylic epoxide by a cuprate reagent was reported simultaneously by Anderson[5], and by Herr and Johnson[62a]. Initial studies showed that regio- and stereoselection was normally good, with $RCu(CN)Li_2$ giving overall the best results[63,64]. Marshall and coworkers undertook an in-depth study of how the substrate's structure and double bond geometry, as well as the nature of the copper reagent, affected the stereochemistry[64a,65]. A detailed discussion of possible reaction pathways is included in Marshall's 1989 review[63]. The predicted *anti* and S_N2' selectivity is indeed the norm, though the double bond geometry in the starting material and the nature of the organocopper may affect the outcome[16b]. A copper-catalyzed version was developed in 1997 by Lipshutz and coworkers. The catalyst used was MeCu(CN)Li, which undergoes ligand exchange with lithium trialkylzincates[66].

The strategy is a good starting point towards the synthesis of homochiral natural products containing polypropionate units because the starting epoxide can be made non-racemic by a number of methods, including the Sharpless asymmetric epoxidation[45] (SAE) (for example **107** to **108**, Scheme 17). The addition lithium methylcyanocuprate to allylic epoxide **109** sets up a synthetically useful allylic alcohol, like the one found in **110**, that can be used to further elaborate the polypropionate fragment into **111** as shown in Scheme 17[67].

Aziridines like **112**, **114** or **116**, as allylic electrophiles, lead to the formation of useful amino esters that can be incorporated into peptide isosteres[40b,42b,68]. The aziridines were prepared from L-threonine or D-allothreonine in enantiomerically pure form or they were acquired through enzymatic resolution. The addition of MeCu(CN)Li or the addition of Me_2Zn or Me_3ZnLi, catalyzed by 3–30 mol% CuCN, to aziridine **112** gave excellent

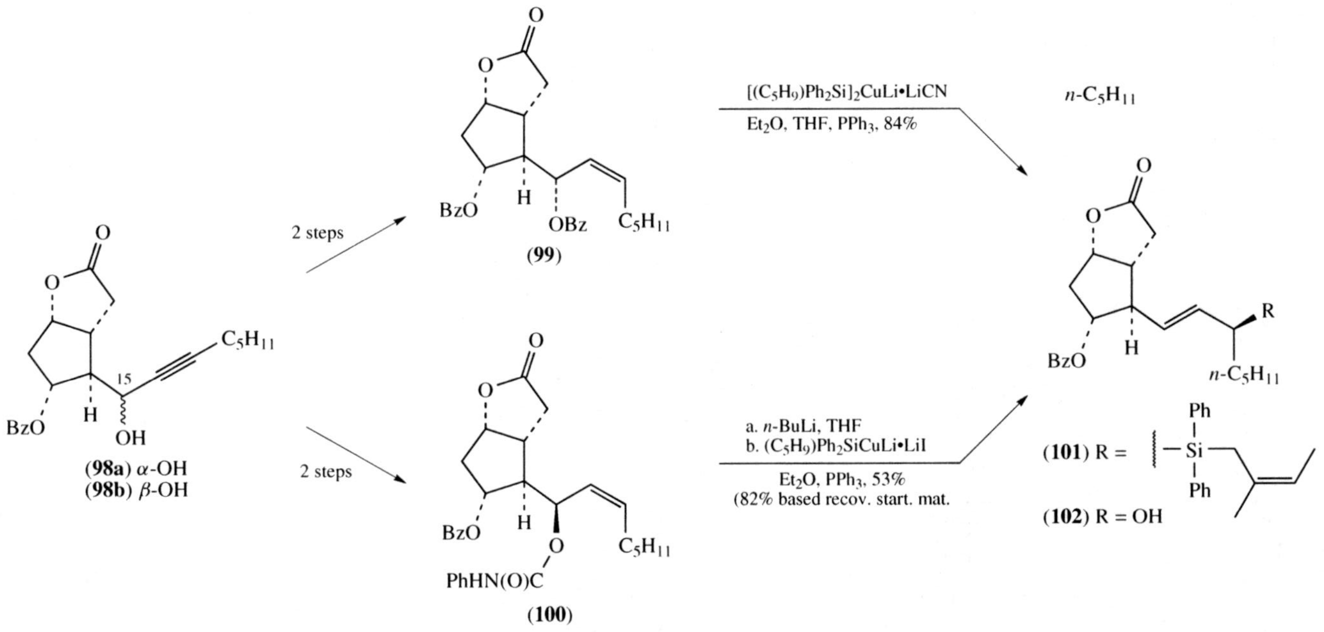

O
BzO
H
15
OH
C_5H_{11}
(98a) α-OH
(98b) β-OH
2 steps
2 steps
OBz
(99)
PhHN(O)C
(100)
$[(C_5H_9)Ph_2Si]_2CuLi•LiCN$
Et_2O, THF, PPh_3, 84%
a. n-BuLi, THF
b. $(C_5H_9)Ph_2SiCuLi•LiI$
Et_2O, PPh_3, 53%
(82% based recov. start. mat.
n-C_5H_{11}
R
(101) R =
Ph
Si
(102) R = OH

SCHEME 15

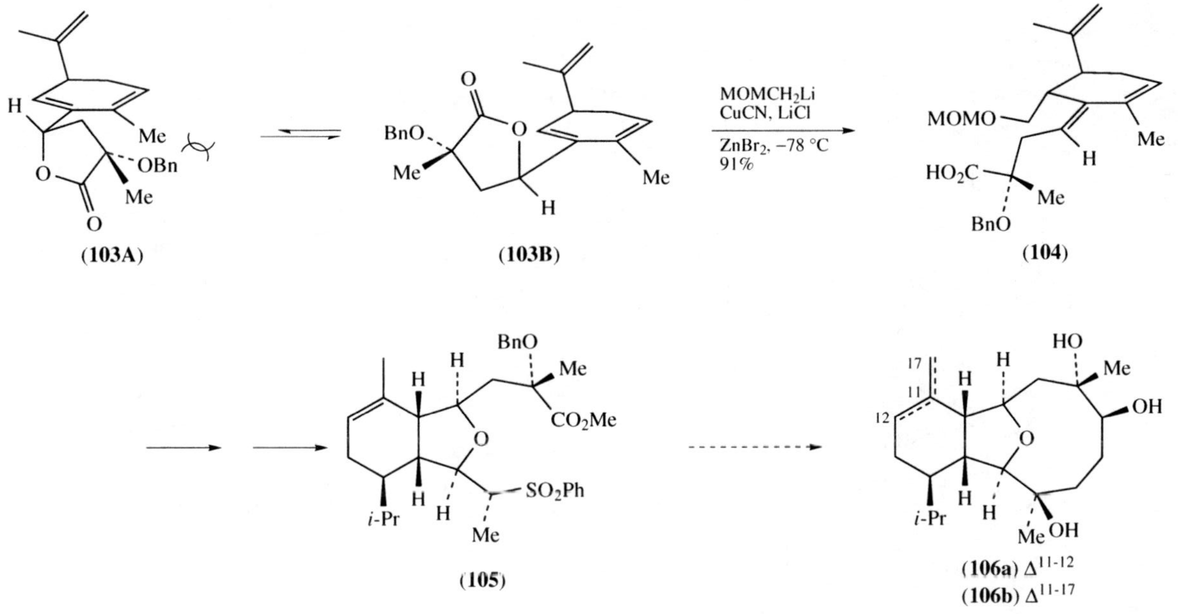
H
Me
OBn
Me
O
O
(103A)
BnO
Me
O
O
H
Me
(103B)
MOMCH$_2$Li
CuCN, LiCl
ZnBr$_2$, −78 °C
91%
MOMO
Me
H
HO_2C
Me
BnO
(104)
H
H
BnO
Me
CO_2Me
O
SO_2Ph
i-Pr
H
H
Me
(105)
HO
Me
OH
17
11
12
H
H
O
i-Pr
H
H
Me
OH
(106a) $\Delta^{11\text{-}12}$
(106b) $\Delta^{11\text{-}17}$

SCHEME 16

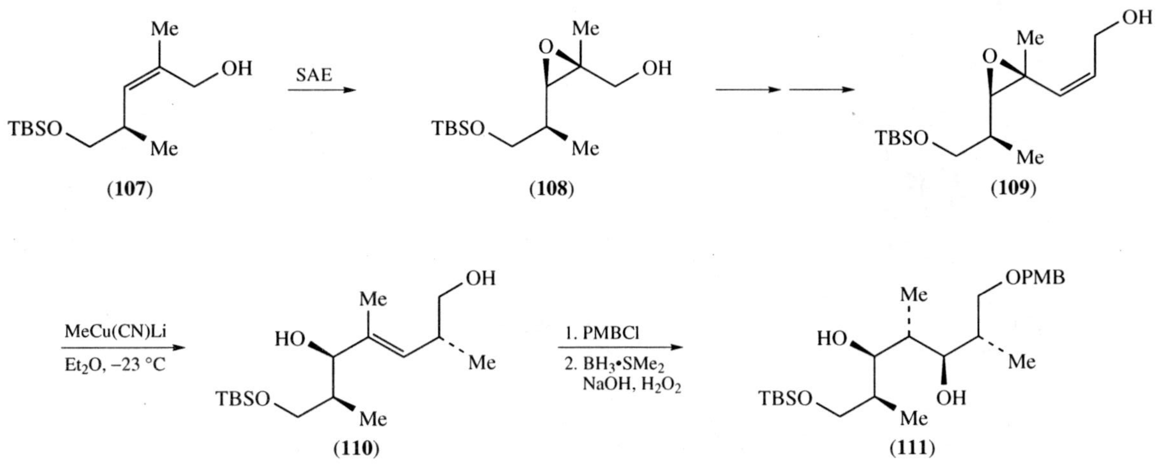

SCHEME 17

yields of the β,γ-enoates **113** (equation 14). Other copper reagents derived from organomagnesium compounds gave similar results starting with aziridines **114** or **116** and yielded products **115** and **117**, respectively (equations 15 and 16, respectively). The usual allylic strain and *anti*-selective addition arguments explain the observed results. As in the case of the δ-acetate-α,β-enoates (see Section II.B.1.a), the use of Gilman's type or higher-order cuprates led mostly to reduced products rather than the substitution product. In addition, the analogous 4- and 5-membered N-heterocycles do not undergo this ring-opening reaction[68b]. N-Acyl allylic aziridines, prepared from homochiral epoxides, were also converted by addition of cuprates to peptide isosteres[69].

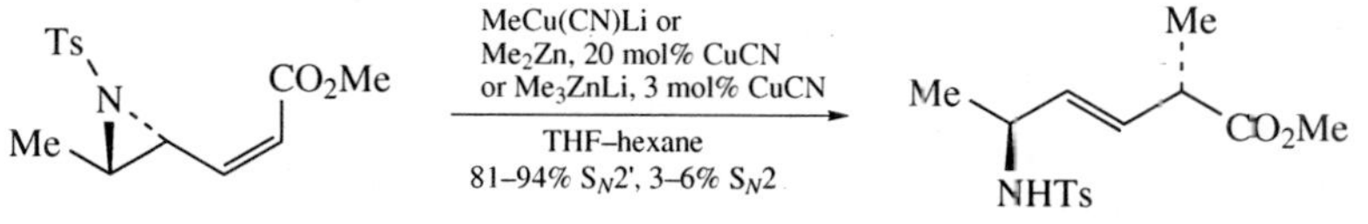

(112) (113) (14)

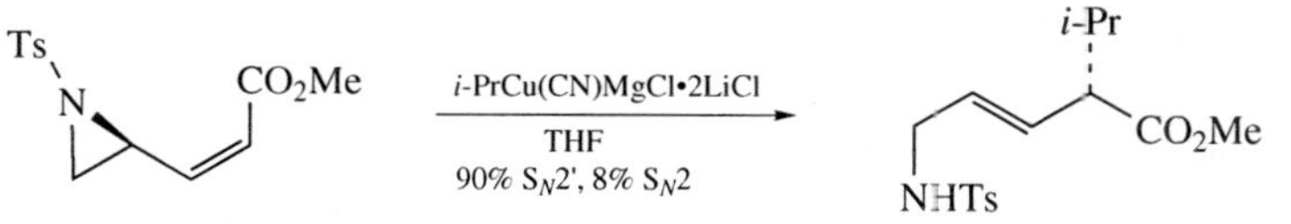

(114) (115) (15)

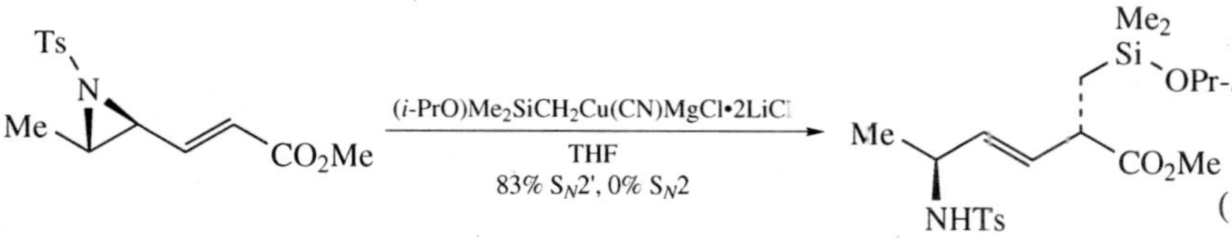

(116) (117) (16)

As was the case with the displacement of chiral α-mesyloxy vinyl sulfoxides (see Section II.B.1.b), the reaction of chiral α-epoxy vinyl sulfoxides gave a regioselective and stereospecific reaction with monoalkylcyanocuprates[70]. The epoxides were prepared by cyclization of the corresponding 1,2-dihydroxy mesylates, themselves prepared via an aldol-type condensation of the deprotonated vinyl sulfoxide onto the appropriate chiral α-alkoxy aldehyde. The aldol proceeded, again, with modest diastereoselectivity.

3. *Displacement of propargylic leaving groups*

Although not an allylic alkylation per se, the displacement of propargylic leaving groups is a directely related reaction[71]. In fact, Rona and Crabbé's original reports dealt with the reaction of lithium dimethylcuprate with propargyl acetates[3], before the report on the allylic alkylation of the same reagent[4a].

When the reaction is γ-selective, it can provide access to chiral non-racemic allenes[72,73]. Enantioenriched propargyl alcohols are accessible by a number of methods[74,75], which renders the preparation of chiral non-racemic allenes using a cuprate-mediated asymmetric allylic alkylation very efficient. Schemes 18 and 19 display two recent examples taken from the literature. Mesylate **120** was prepared via Carreira's asymmetric alkynylation using alkyne **118** and aldehyde **119** (Scheme 18)[76a]. The reaction of mesylate **120** with lithium methylcyanocuprate led to the formation of allene **121** as a single diastereomer. The chiral allene was then cyclized using silver nitrate to afford dihydrofuran **122**,

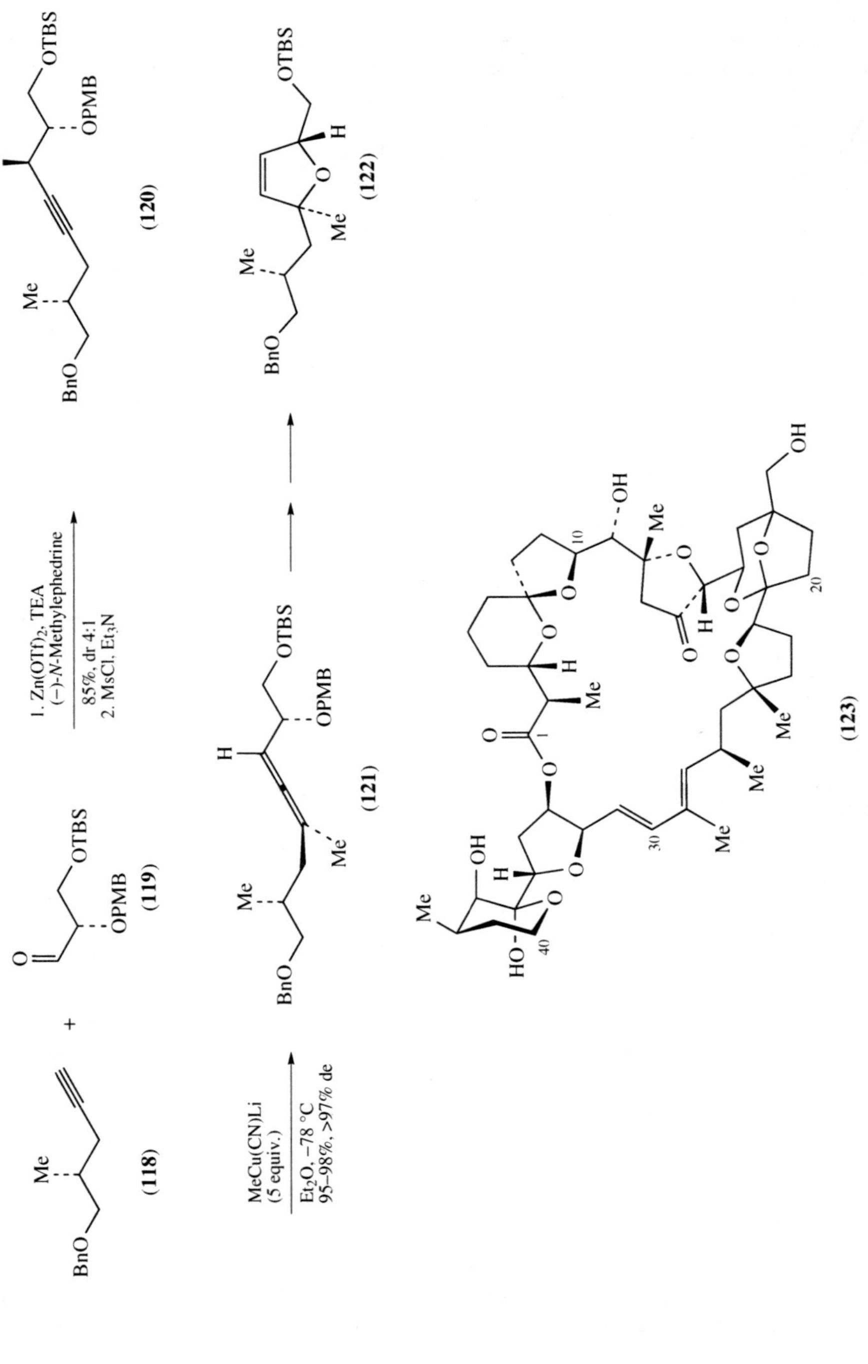
BnO
Me
(118)
+
OPMB
OTBS
(119)
1. Zn(OTf)$_2$, TEA
(−)-N-Methylephedrine
85%, dr 4:1
2. MsCl, Et$_3$N
OMs
(120)
MeCu(CN)Li
(5 equiv.)
Et$_2$O, −78 °C
95–98%, >97% de
(121)
(122)
(123)
OH
HO
Me
H
O
1
10
20
30
40

SCHEME 18

which is the C21–C28 fragment of the natural macrolide pectenotoxin-4 (**123**). Isolated from dinoflagellates residing inside shellfish, it possesses a range of interesting biological activities, including depolymerization and tumor cell repression properties[76a].

Alternatively, the related mesylate **125** was prepared using Noyori's asymmetric hydrogenation of ketone **124** (Scheme 19). The allene **126**, resulting from the diastereospecific propargylic alkylation, was oxidized to the unusual spiro diepoxide **127**[76b,c,d]. Several other substrates underwent the same sequence successfully and the spiro diepoxides could be transformed into a number of useful structural motifs. This strategy was used to prepare a C1–C15 fragment of pectenotoxin-4 (**123**, Scheme 18)[76b,e] and a formal synthesis of the anti-cancer psymberin was also reported[76f].

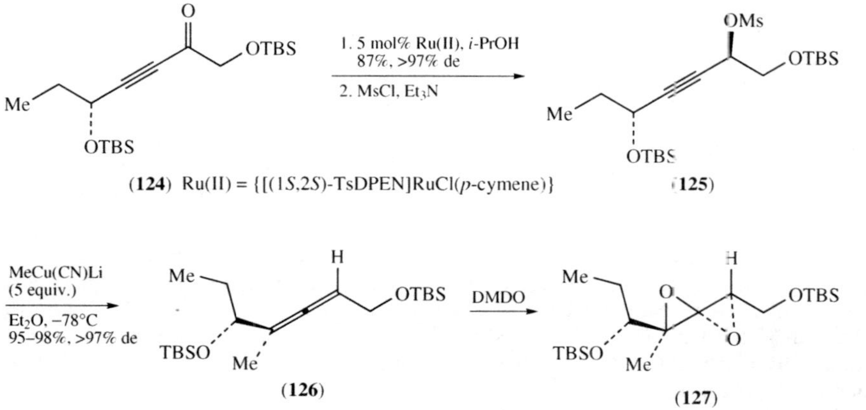

SCHEME 19

α-Hydroxyallenes have become extremely valuable synthetic intermediates, in particular since it was demonstrated that they can be smoothly converted to dihydrofuran derivatives[77,78]. Dihydrofurans are invaluable building blocks for myriad polyether natural products. Propargylic epoxides are ideal precursors because their reactions with cuprate reagents lead directly to α-hydroxyallenes. Enynes like **127a** and **130** can be epoxidized chemoselectively, as shown in equations 17 and 18[78], as well as stereoselectively by Shi and coworkers'[79] or Sharpless and coworkers' asymmetric epoxidation[45] (SAE) protocols as shown for example in equation 18[80]. The attack of alkylcopper reagents is usually highly regio- and *anti*-selective with propargyl epoxides[31] and this is what was observed for propargylic epoxide *rac*-**128b** and enantiomerically enriched epoxide **131**. α-Hydroxyallenes *rac*-**129** was cyclized to a dihydrofuran fragment of the mycotoxin verrucosidine[78] while **132** was converted to a vinylallene to be used as diene in a stereoselective Diels–Alder reaction for the synthesis of anti-cancer quassinoids[80].

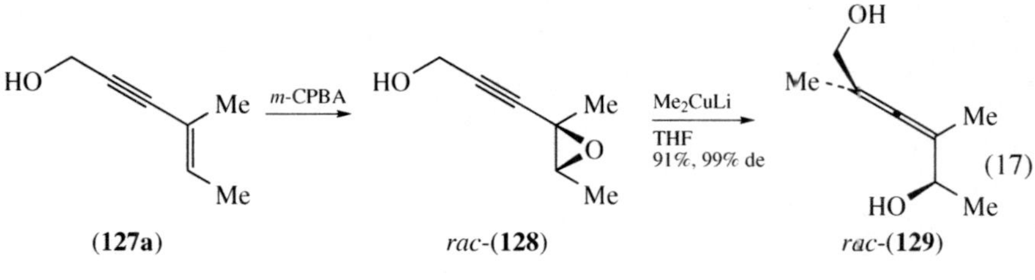

(17)

(127a) *rac*-**(128)** *rac*-**(129)**

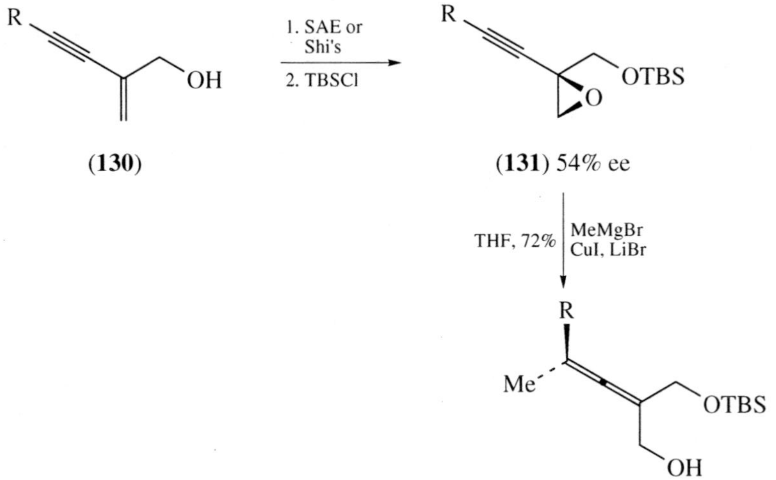

(**130**) (**131**) 54% ee (**132**) 54% ee

(18)

This method was used to generate a chiral non-racemic trisubstituted allene **135**, which was cyclized with a gold catalyst to dihydrofuran **136** (Scheme 20). The product **136** was then taken to natural alkaloid (−)-isochrysotricine **137**, which was isolated from the Rubiaceae plant *Hedyotis capitellata*[82]. The starting racemic alcohol *rac*-**133** was kinetically resolved to >98% ee using a Sharpless asymmetric epoxidation[45] (SAE-KR) and then epoxidized to give **134**.

Homochiral carbonate **138**, as well as other analogous carbonates and sulfites, underwent a regio- and stereoselective ring opening with complete transposition of the double bond to give chiral allene **139** (equation 19)[83]. It is not unusual, however, to see a good deal of racemization (or epimerization) of the allenes, apparently caused by single electron transfer by copper(0) or other copper species[84]. Phosphine or phosphite ligands often solve this problem[84]. In addition, fast exchange between copper and magnesium may provoke a *syn*-elimination, at least when propargyl ethers, including epoxides, are the leaving groups[85]. Complexation between the Mg−X (especially with X = Cl) and the leaving group would be at the origin of this behavior. The addition of phosphines favors *anti*-elimination. Using this protocol, arylcopper species, prepared by a metal−halogen exchange of aryl iodides like **140**, were added to a series of propargyl epoxides *rac*-**141** to give regioselectively and stereospecifically α-hydroxyallenes *rac*-**142** (equation 20)[86].

Krause and coworkers have shown that α-hydroxyallenes *rac*-**144** and *rac*-**147** are accessible from the copper hydride reduction of propargyl acetates *rac*-**143** and *rac*-**146**, respectively (equations 21 and 22). The copper hydride species was generated from copper chloride and polymethylhydridosiloxane (PMHS) in the presence of the NHC ligand generated from **148** as stabilizing agent. The presence of alcohols, halides and several other functional groups on the substrate is tolerated[87]. α-Hydroxyallenes are useful synthetic intermediates being, for instance, precursors for gold-catalyzed cyclizations to dihydrofurans such as *rac*-**145** (equation 21).

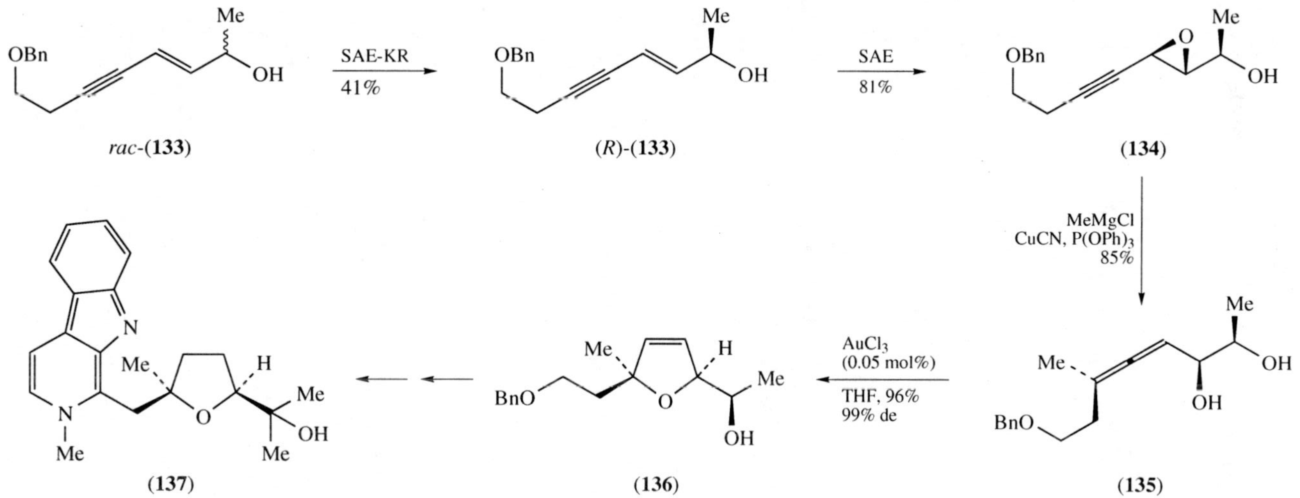
OBn
Me
OH
rac-(133)
SAE-KR
41%
(R)-(133)
SAE
81%
O
(134)
MeMgCl
CuCN, P(OPh)3
85%
Me
BnO
(135)
AuCl3
(0.05 mol%)
THF, 96%
99% de
H
(136)
N
(137)

SCHEME 20

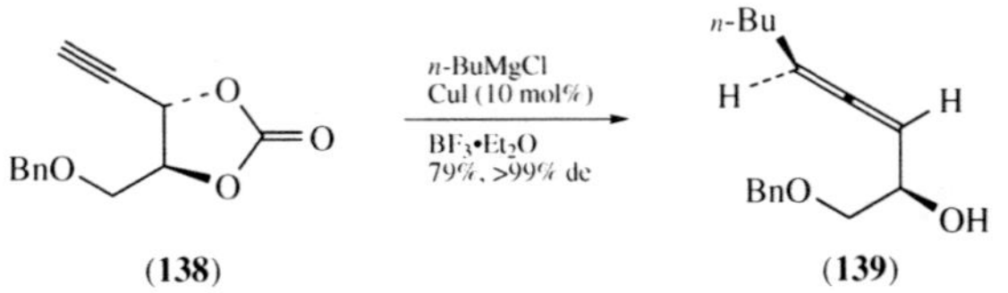

n-BuMgCl
CuI (10 mol%)
BF3•Et2O
79%, >99% de
BnO
O
O
O
n-Bu
H
H
BnO
OH
(138)
(139)

(19)

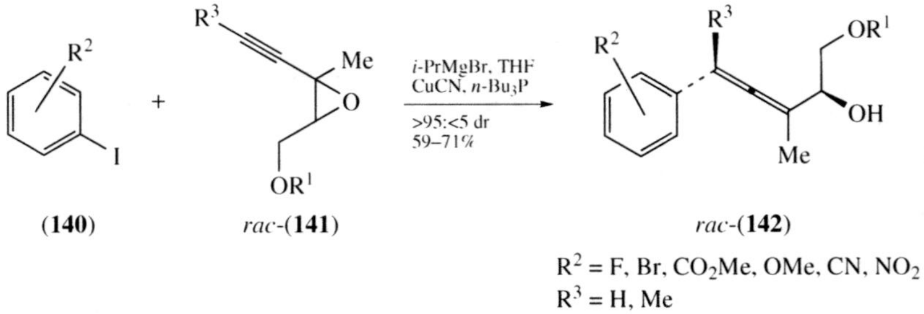

R2
I
+
R3
Me
O
OR1
i-PrMgBr, THF
CuCN, n-Bu3P
>95:<5 dr
59–71%
R3
OR1
R2
OH
Me
(140)
rac-(141)
rac-(142)
R2 = F, Br, CO2Me, OMe, CN, NO2
R3 = H, Me

(20)

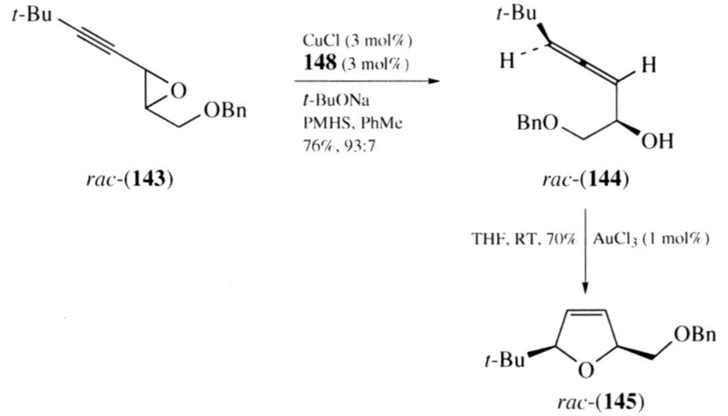

t-Bu
O
OBn
CuCl (3 mol%)
148 (3 mol%)
t-BuONa
PMHS, PhMe
76%, 93:7
t-Bu
H
H
BnO
OH
rac-(143)
rac-(144)
THF, RT, 70%
AuCl3 (1 mol%)
t-Bu
O
OBn
rac-(145)

(21)

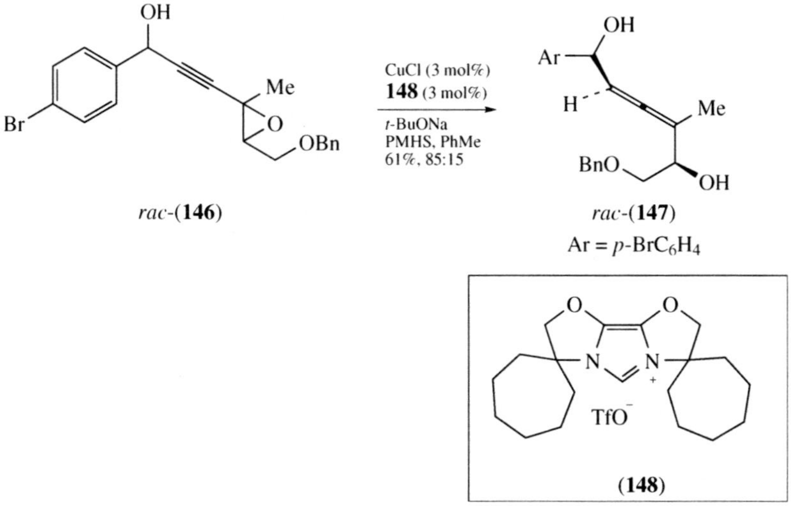

OH
Br
Me
O
OBn
CuCl (3 mol%)
148 (3 mol%)
t-BuONa
PMHS, PhMe
61%, 85:15
OH
Ar
H
Me
BnO
OH
rac-(146)
rac-(147)
Ar = p-BrC6H4
O
O
N
N
+
TfO−
(148)

(22)

Akin to α-hydroxyallenes, it is possible to form α-aminoallenes **150** from the copper-mediated opening of propargylic aziridine **149** or analogs thereof (equation 23)[88]. The former are equally important synthetic intermediates, being precursors to dihydropyrroles such as **151a** and other N-heterocycles[77c] including pyrrolines and pyrroles[89], as well as allylic aziridines and azetidines[90].

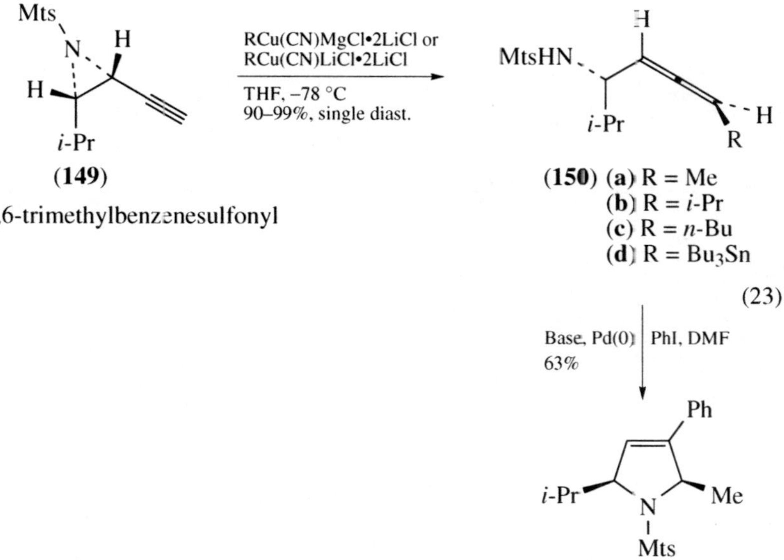

(149)

Mts = 2,4,6-trimethylbenzenesulfonyl

(150) (**a**) R = Me
(**b**) R = *i*-Pr
(**c**) R = *n*-Bu
(**d**) R = Bu_3Sn

(23)

(151a)

Terminal alkynes such as **152** are good electrophiles for the allylic alkylation of the cyanocopper reagent derived from benzhydrylidene(methyl)amine (equation 24). A single stereo- and regioisomer **153** was isolated after hydrolysis of the imine product in the reaction work-up. The alcohol precursor to mesylate **152** was prepared in five steps from (+)-diethyl tartrate[91].

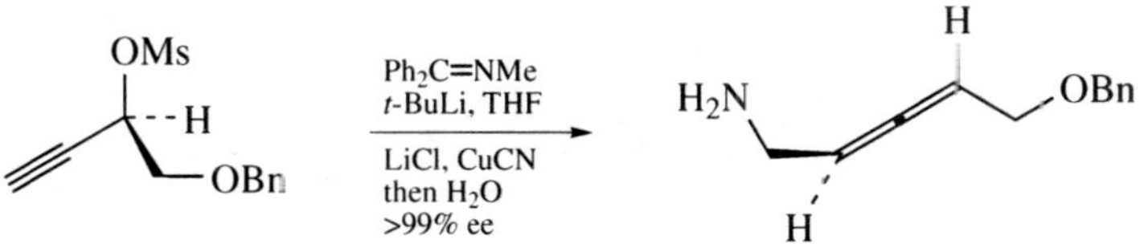

(24)

(152) **(153)**

Similarly, α-aminoallene **159** was prepared by regio- and stereoselective addition of the cyanocuprate reagent made from N-Boc dimethylamine onto propargyl mesylate **158** in good yield (Scheme 21). The alcohol **157**, precursor to mesylate **158**, was obtained by enantioselective alkynylation of isobutyraldehyde (**155**) with alkyne **154** in the presence of a catalytic amount (22 mol%) of ligand **156**[89,92]. Like their α-hydroxyallene homologues, α-aminoallenes can be cyclized, using Lewis acids, to useful chiral heterocycles such as dihydropyrrole **160**.

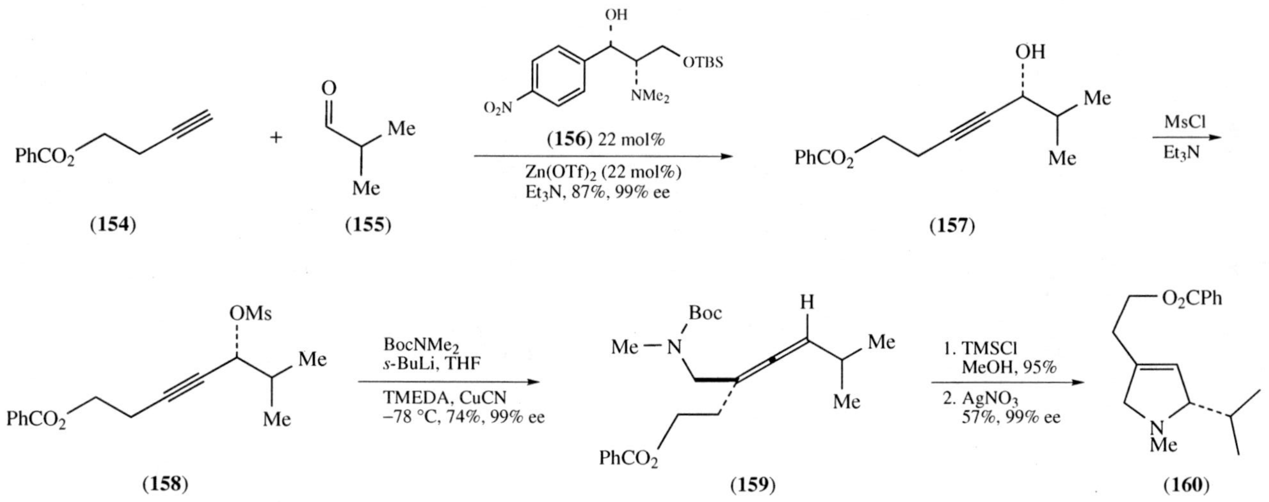
PhCO2
(154)
O
Me
Me
(155)
OH
O2N
OTBS
NMe2
(156) 22 mol%
Zn(OTf)2 (22 mol%)
Et3N, 87%, 99% ee
OH
PhCO2
Me
Me
(157)
MsCl
Et3N
OMs
PhCO2
Me
Me
(158)
BocNMe2
s-BuLi, THF
TMEDA, CuCN
−78 °C, 74%, 99% ee
Boc
Me — N
H
Me
Me
PhCO2
(159)
1. TMSCl
MeOH, 95%
2. AgNO3
57%, 99% ee
O2CPh
N
Me
(160)

SCHEME 21

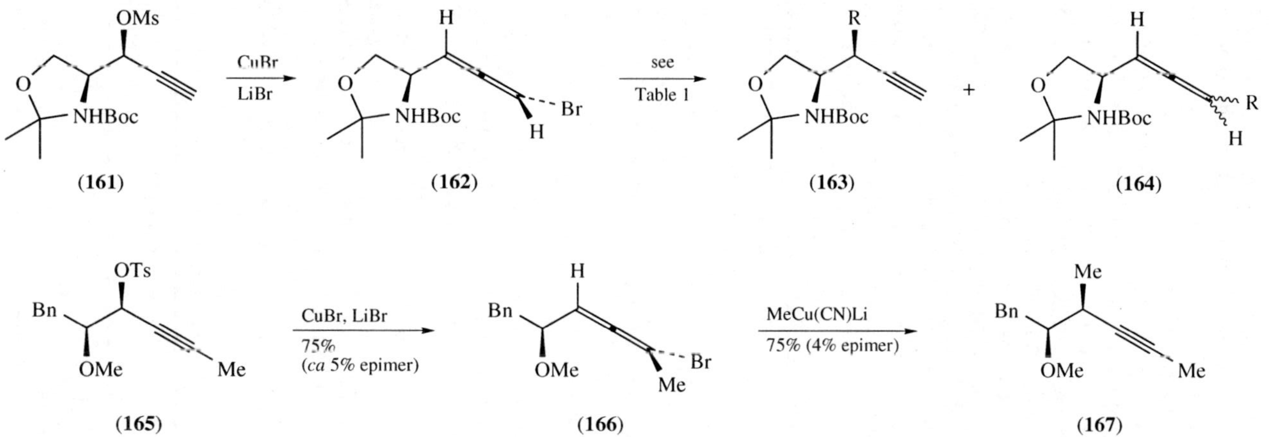
OMs
CuBr
LiBr
O
NHBoc
(161)
H
Br
H
(162)
see
Table 1
R
(163)
+
(164)
OTs
Bn
OMe
Me
(165)
CuBr, LiBr
75%
(ca 5% epimer)
(166)
MeCu(CN)Li
75% (4% epimer)
(167)

SCHEME 22

4. Displacement of allenic bromides

In an infrequently used methodology, which is nonetheless closely related to the allylic alkylation, it is possible to effect a 'propargylic alkylation' of alkylcopper reagents. For this purpose, one uses an allenic bromide like **162** or other allenic leaving groups (Scheme 22). After the report by Corey and Boaz[93], additional methods were described[94]. In the first example, the ratio of **163:164** was dependent upon the nature of the cuprate reagent used, the choice of which was based on the alkyl group that was being transferred. With the right reagent, good yields and ratios of alkynes **163** were obtained (Scheme 22, Table 1)[94a]. The method was utilized to achieve the stereocontrolled synthesis of unnatural amino acids[94]. With bromoallene **166**, made from the S_N2' displacement of mesylate **165** with copper bromide, the ratio of S_N2':S_N2-products was also satisfactory when monoalkylcyano cuprate reagents were used, giving nearly exclusively adduct **167**. Mesylated alcohols like **161** are relatively easy to obtain in enantiomerically pure form, and this type of 'relay' strategy to make α-chiral alkynes amounts to a S_N with net retention of stereochemistry at the carbon bearing the original departing sulfonate. It is of note that the opposite stereochemical result could be obtained from an α-selective (S_N2) displacement of the propargylic mesylate by a nucleophilic alkyl[95].

With silylated bromoallene **169**, the course of the reaction was quite different. A S_N2' displacement by the copper atom occurred with inversion of configuration with respect to the leaving bromide to give organocopper intermediate **170** (Scheme 23). A *syn*-stereospecific and thermodynamically-driven rearrangement then transforms the propargylic cuprate **170** into allenic cuprate **173**. 1,2-Reductive coupling finally gives the product **172**, resulting from a net S_N2 displacement of the bromide in **169**. It is possible, however, that a direct 1,4-reductive coupling from **170** occurs to give **172** directly. Direct S_N2' displacement of the mesylate **168** by $RCuMgX_2$•LiBr gave the diastereomeric allenylsilane product **171** of opposite configuration at the allene carbon[96].

C. Cyclic Systems

1. Common cyclic systems

This section encompasses all substrates incapable of rotation about their allylic bond (the one bearing the leaving group). It was on such cyclic systems that the stereospecificity of copper-mediated allylic alkylation was first observed and studied[7,8,9f,63]. Because there is no possible rotation around the allylic bond, the stereoselectivity aspect of the addition, through changes in conformations, vanishes and only the stereospecificity aspect of the reaction remains. Another difference between acyclic and cyclic substrates is that the leaving group in the latter is necessarily on a secondary carbon.

The regioselectivity of the addition can be influenced by steric factors. For example, acetates **174** and **176** gave the same product **175** with the use of lithium dimethylcuprate

TABLE 1. Copper-mediated propargylation of allenic bromide **162**

Entry	'RCu'	Yield (%)	S_N2':S_N2 (**163:164**)	%de **163**
1	$MeCuMgBr_2$•LiBr	90	90:10	>98
2	$Bu_2Cu(CN)Li_2$	76	95:5	>98
3	$Ph_2Cu(I)Li_2$	72	95:5	>98

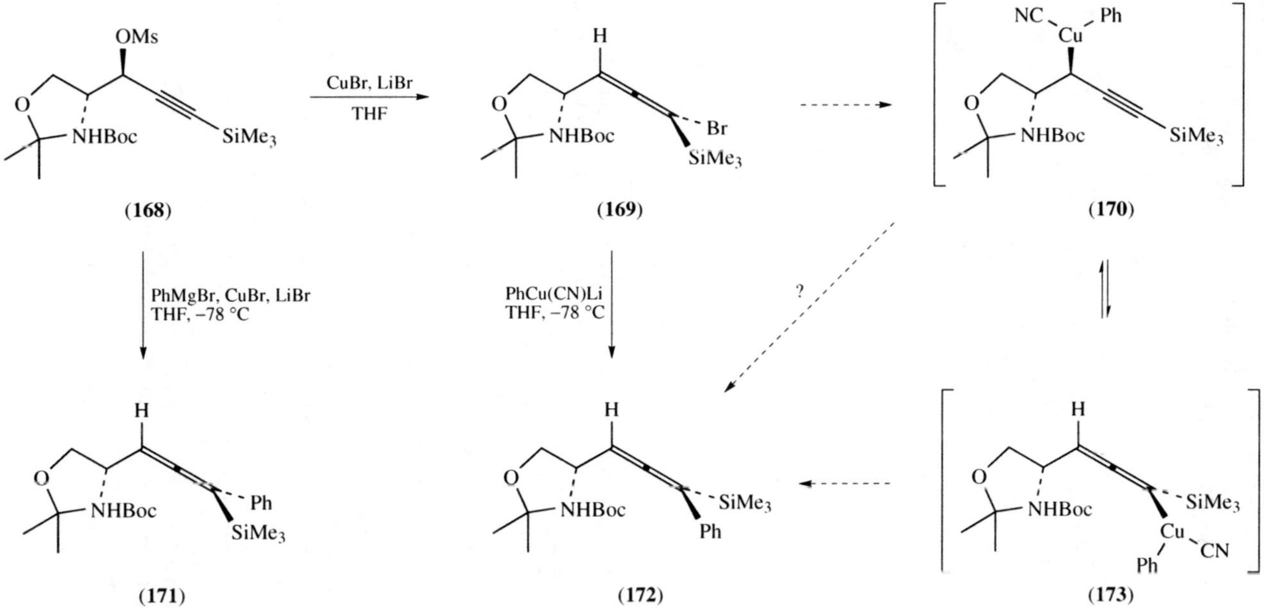
OMs
NHBoc
SiMe3
(168)
CuBr, LiBr
THF
H
NHBoc
Br
SiMe3
(169)
NC
Cu
Ph
NHBoc
SiMe3
(170)
PhMgBr, CuBr, LiBr
THF, −78 °C
PhCu(CN)Li
THF, −78 °C
?
H
NHBoc
Ph
SiMe3
(171)
H
NHBoc
SiMe3
Ph
(172)
H
NHBoc
SiMe3
Cu
CN
Ph
(173)

SCHEME 23

(Scheme 24). Of course, this reagent is known to be poorly regioselective and thus more prone to steric influence from the substrate. Other reagents may be more γ-selective, but the use of a coordinating leaving group usually ensures complete S_N2' selectivity despite severe steric impediment. This is clearly demonstrated with carbamates **177** and **180**, which gave regioisomers **178** and **179**, respectively, as the sole observable products[36b, 97].

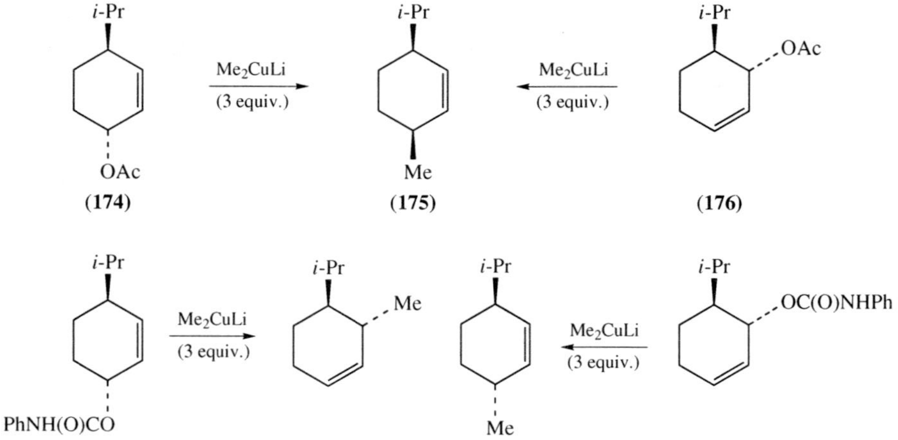

(174) (175) (176) (177) (178) (179) (180)

SCHEME 24

As stated above, some copper reagents are quite γ-selective, even with non-coordinating leaving groups. This is the case for monoalkylcopper reagents made from copper cyanide and alkyllithiums. Scheme 25 exemplifies this selectivity using cycloheptadiene monoepoxide **181**[98]. Gilman-type reagents gave a 70:30 mixture of S_N2':S_N2 products **182** and **183**, respectively. It reacted with similarly low regioselectivity with cyclohexadiene monoepoxide **186**. The lower-order alkylcyanocuprates gave good-to-nearly perfect γ-selectivity with **181**.

In absence of copper, silyllithium reagent **185** reacted with racemic cyclohexadiene monoepoxide **186** to give solely the product of S_N2 attack (not shown). However, with a stoichiometric amount of copper cyanide, the S_N2' displacement product **184** was the only one observed (Scheme 25, bottom left)[99]. By contrast, lithium dimethylcuprate gave a near-equal mixture of **187a** and **188a**. Cyclohexadiene oxide **186** was the subject of more allylic alkylation studies[100] and its enantiomerically pure form was the starting material for a short synthesis of the sesquiterpenoid (4*S*)-β-elemenone[101].

With polycyclic compounds of greater complexity, the situation is not so easily predictable. While allylic epoxide **189** gave exclusively the S_N2' product **191**, despite the presence of the axial C13 methyl group, the analogous epoxide **192** underwent the S_N2 attack of the cuprate reagent to give ultimately **195** (Scheme 26). The authors propose that initial attack of the copper atom occurs at the allylic position but, because of steric and conformational effects, the intermediate adduct **193** rearranges to **194** prior to reductive coupling to **195**, while the organocopper **190** does not[102].

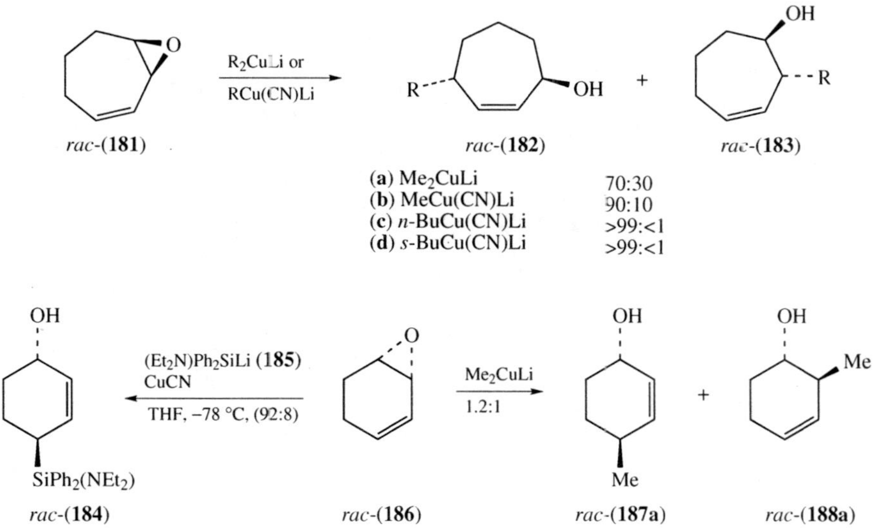

SCHEME 25

More recently, cycloalkenol-derived phosphonates like **196a** and **196b** were shown to undergo a highly regioselective and stereospecific copper-promoted addition of organozinc reagents, some of which were functionalized (R = CH_2CO_2Me, CN, CH_2OR), to adducts **197a** and **197b** as shown in equation 25[103]. The alcohols, from which the phosphonates **196a** and **196b** were made, were prepared by asymmetric reduction of the corresponding ketones[103c, 104]. Compound **200**, an intermediate in the formal synthesis of the female sex hormone (+)-estrone, was prepared using this reaction as a key step. The iodide **198** was used to allylate the dialkylzinc reagent **199** in excellent diastereospecificity and regioselectivity for the creation of the quaternary chiral carbon in **200** (equation 26)[104].

Diketopiperazine **201** is a cyclic dipeptide unit often encountered in biologically active compounds. Diketopiperazine mimetics such as **203a–f** were formed regioselectively and stereospecifically from the S_N2' addition of organocopper reagents to allylic phosphates **202** (Scheme 27)[105]. The regioselectivity was very sensitive to the nature of the organocopper species and to the presence or absence of salts. Several organocopper species were tested including Me_3CuLi_2•LiI•3LiBr, MeCu•LiI•LiBr, MeCuI•MgCl•2LiCl, MeCu(CN)•MgCl•2LiCl•BF_3 among many others. Magnesium salts led to a mixture of S_N2 and S_N2' adducts but lithium salts directed the reagent to the γ-position presumably due to their propensity, as proposed, to form clusters with the cuprate reagent. This led the authors to propose a mechanism in which a bridging of the amide carbonyl and the organocopper(III) species by LiCl molecules occurs, as seen in **204**. This bridging would explain the observed regioselectivity. High-level DFT calculations support this working hypothesis and suggest that the transfer of the methyl group occurs via a copper(II) intermediate **205** to give complex **206**. The method was used to synthesize a potent CXCR4-chemokine receptor antagonist **207**[105a].

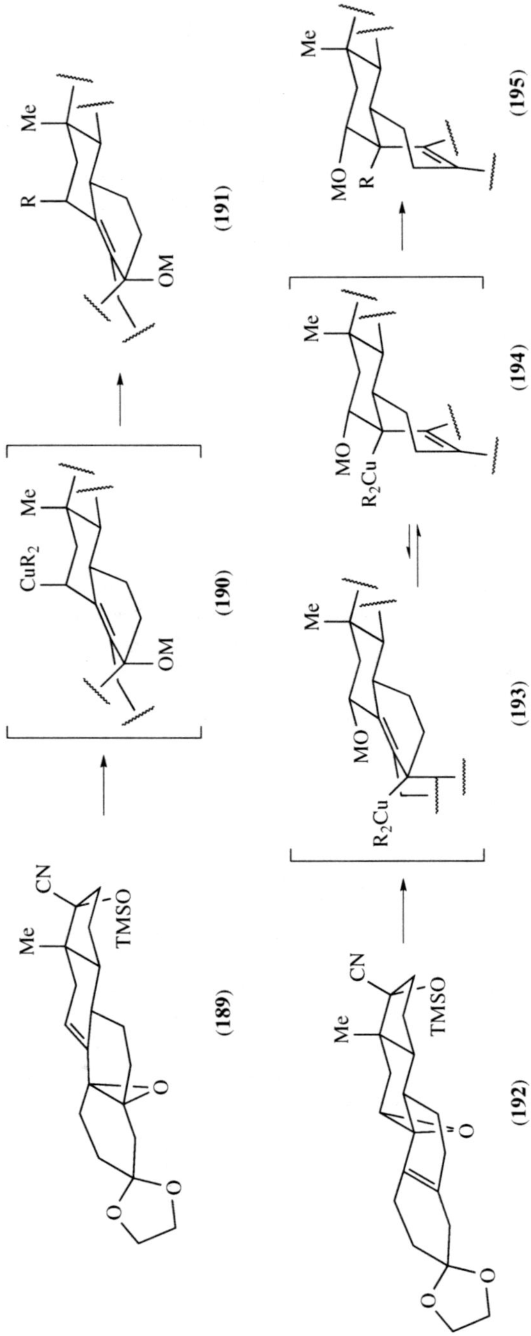

SCHEME 26

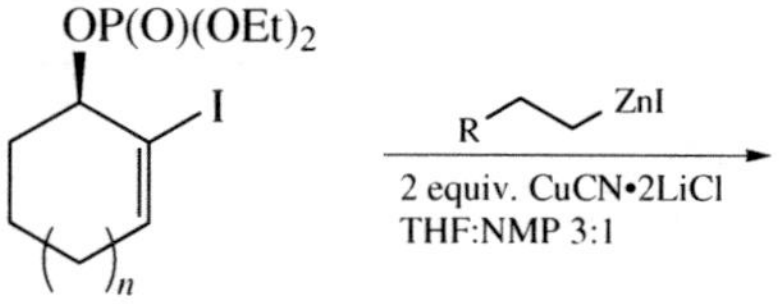

(25)

(**196a**) $n = 0$
(**196b**) $n = 1$

(**197a**) $n = 0$
(**197b**) $n = 1$

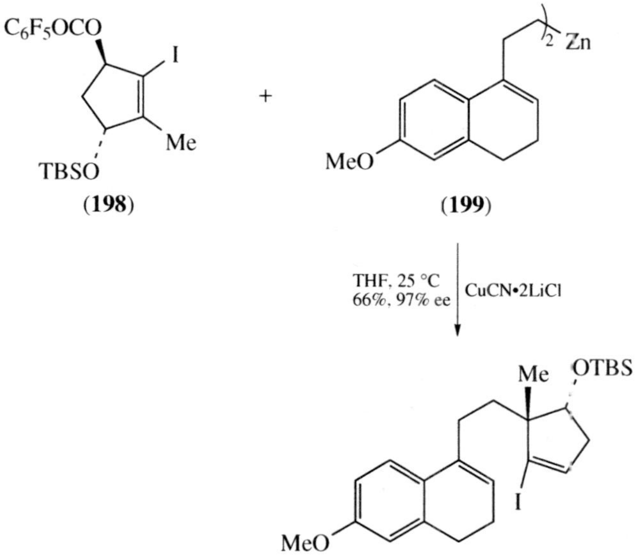

(**198**) (**199**) (**200**)

(26)

2. Cyclic systems of particular synthetic interest

a. Diastereospecific ring-opening of cyclopentenediol derivatives. *meso*-Cyclopent-2-ene-1,4-diol is one cyclic allylic system that has been studied extensively. Its monoacetate can be obtained enantiomerically pure by enzymatic resolution, among other methods, and is even available from commercial sources[106]. Achieving high regioselectivity in the copper-mediated displacement of such systems with organometallics has been a challenge, but now a number of reaction conditions and reagents allow for a high S_N2 or S_N2' regioselectivity.

Kobayashi and his coworkers have made a number of contributions to this area[107]. They surveyed a large number of reaction conditions, and in particular examined the effect the ratio of copper cyanide to organomagnesium had on the regioselectivity. They were able to control the regioselectivity of the displacements of cyclopentenediol monoacetate and obtain either the S_N2' or S_N2 product. They later found that the α-methoxyacetate leaving group, as shown in **208**, is among the best to effect the allylic alkylation of aryl- and

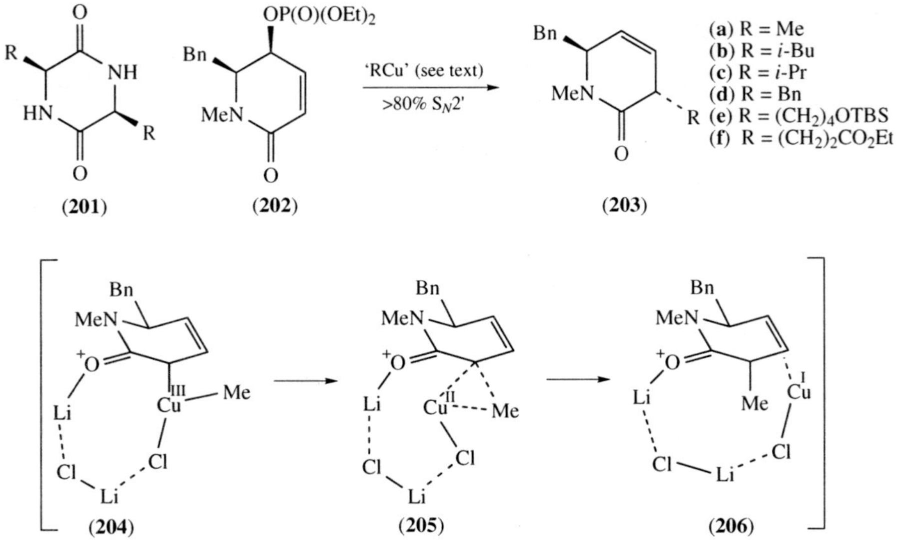

(201) (202) (203)

(204) (205) (206)

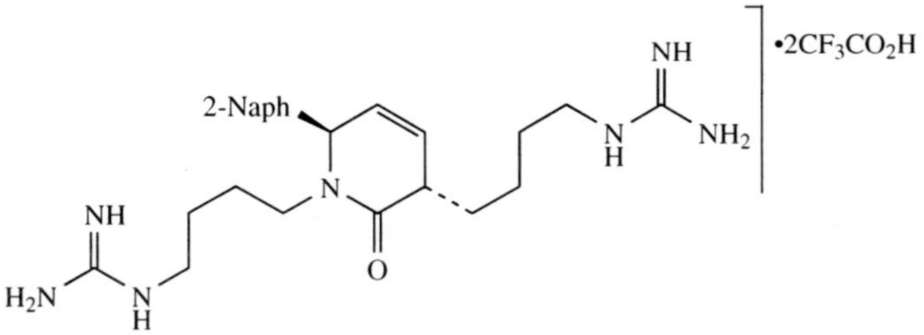

(207)

SCHEME 27

alkenylzinc reagents (equation 27)[107g]. High ratios of S_N2':S_N2 products **209:210** were obtained systematically.

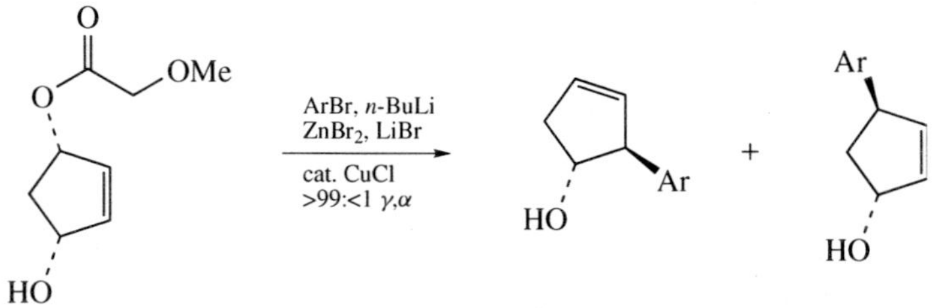

(27)

(208) (209) (210)

b. Diastereospecific opening of cyclopentadiene monoepoxides. The reader will note that cyclic alcohols like **209** and **210**, or the analogs possessing alkyl groups **212** and

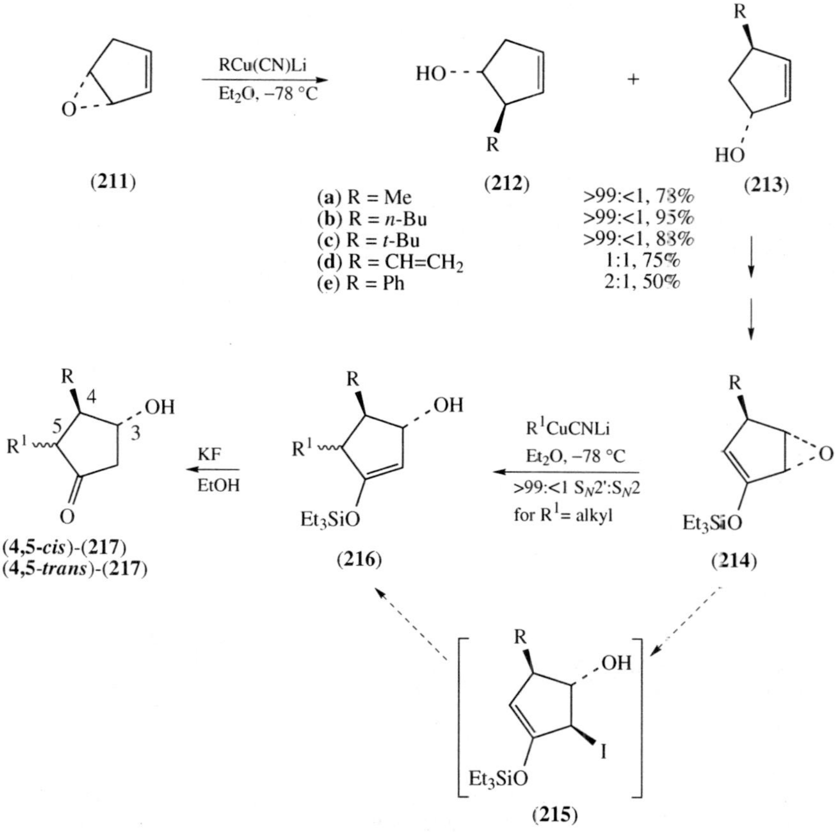

SCHEME 28

213, can also be obtained by the regioselective opening of cyclopentadiene oxide **211** (Scheme 28). In an approach to prostaglandins, cyclopentadiene oxide **211** was reacted with several cyanocopper reagents to give predominantly the S_N2' product **213a–c**. Aryl- and vinylcopper species gave, however, a mixture of the two adducts **212d–e** and **213d–e**. Then, allylic substitution products **213** were converted to their silyl enol ethers **214** in preparation for a second allylic substitution. The same copper reagents as before, under identical reaction conditions, gave prostaglandin-like adducts **217**, after hydrolysis of the intermediate silyl enol ether **216**. Again, reaction with alkylcyanocuprates afforded only the product of S_N2' attack while the use of sp^2-hybridized carbon nucleophiles resulted in a poorly regioselective reaction. This time, though, mixtures of stereoisomers 4,5-*cis* and 4,5-*trans* cyclopentanones **217** were obtained and it was suggested that the mechanism of addition may differ depending on steric encumbrance on the substrate and whether a commercial alkyllithium or one made from a metal–iodide exchange was used. When a commercial alkyllithium was used, a direct *anti* displacement gave mostly the *cis*-cyclopentenone (4,5-*cis*)-**217**, after hydrolysis of the silyl enol ether. In the case of a freshly prepared alkyllithium, the S_N2 opening of the epoxide at the allylic position by the iodide present in the solution afforded compound **215**, which preceded the iodide's own S_N2' displacement by the cuprate reagent, giving predominantly, after hydrolysis of the silyl enol ether, (4,5-*trans*)-**217**. The latter is the product of a double stereochemical inversion[108].

c. Diastereospecific opening of cyclopenta[b]furanones. The bicyclic cyclopenta[*b*] furanones **218a–c** are reactive towards various copper reagents to give 3,5-substituted cyclopentenes **219** and **220** (equation 28). It was found that lithium dialkylcuprate gave predominantly the product resulting from attack on the least hindered carbon (Table 2, entries 1–3), while copper species of general formula 'RCuM' led mostly to product **219** via a S_N2' attack (Table 2, entries 4–7)[109]. Such S_N2'-selective ring-opening of this type of bicyclic lactones has been used extensively as a key transformation for a number of natural product syntheses, in particular the triquinane sesquiterpenes[110].

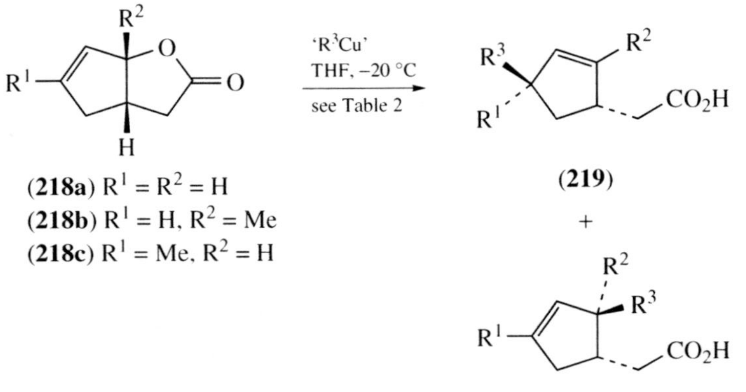

(**218a**) $R^1 = R^2 = H$
(**218b**) $R^1 = H, R^2 = Me$
(**218c**) $R^1 = Me, R^2 = H$

(**219**) + (**220**) (28)

More recently, alkylated enantiomerically enriched bicyclic lactones such as (*S,S*)-**221** have become available from the palladium-catalyzed nucleophilic opening of the cyclopentadiene oxide **211** (cf. Scheme 28). The copper-promoted addition of Grignard

TABLE 2. Copper-mediated S_N2' opening of bicyclic lactones **218a–c**

Entry	Lactone	'RCu'	Ratio of **219/220**
1	**218a**	Me_2CuLi	62:38
2	**218b**	Me_2CuLi	98:2
3	**218c**	Me_2CuLi	7:93
4	**218a**	$MeCu(CN)Li$	75:25
5	**218a**	$H_2C{=}CHCH_2MgBr$, $CuBr{\bullet}SMe_2$ (1 equiv.)	>98:<2
6	**218b**	$(O(CH_2)_3O)CH(CH_2)_2MgBr$, $CuBr{\bullet}SMe_2$ (1 equiv.)	>98:<2
7	**218c**	$MeMgBr$, $CuBr{\bullet}SMe_2$ (1 equiv.)	95:5

reagent (*R*)-**222** to (*S*,*S*)-**221** (giving **223**) has been applied to the total synthesis of cyclopentane-containing membrane lipids of *archea* bacteria (equation 29)[111].

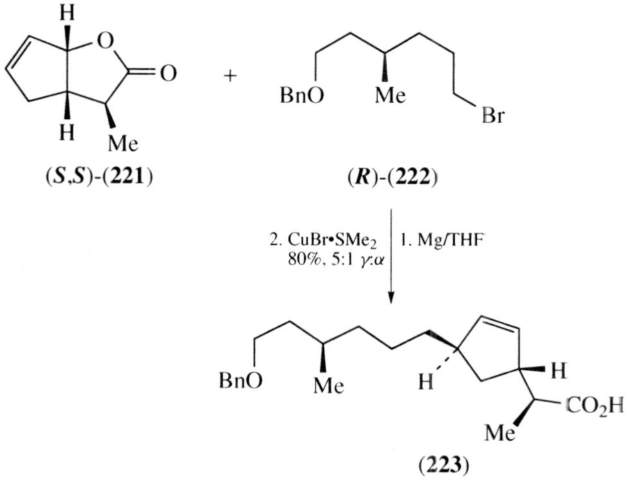

(29)

d. Pot-pourri of applications in total synthesis. There are numerous applications of S_N2' displacements with cyclic systems in the literature. As early as 1984, Rickards and Rönnenberg used a Lewis acid to promote the reaction between diarylcuprate (**225**) and acetate **224** (Scheme 29). The acetate leaving group sits on a quaternary carbon, which helped in securing a good regioselectivity. The isoprenyl group, on the other hand, assisted the stereospecificity of the reaction[112]. Adduct **226** was then taken to (−)-Δ^9-6*a*,10*a*-*trans*-tetrahydrocannabinol **227**. A total synthesis of cytotoxic bakkenolide-A (**230**) made use of the regio- and stereodirecting ability of the carbamate group on the chiral cyclohexene derivative **228** to give adduct **229**[113].

Steroidal side chains were introduced by the action of lithium alkylcyanocuprates onto a trisubstituted alkene positioned *exo* to cyclopentene oxides **231a** and **231b** (equation 30). The *anti* diastereoselectivity was complete in both cases, but the regioselectivity was higher when lithium isohexanylcyanocuprate added onto **231a** (100:0, **232a:233a**) than when lithium methylcyanocuprate added onto **231b** (1:1, **232b:233b**)[114]. Bulkier cuprates generally afford higher stereoselectivity than smaller ones, though rarely as drastically as in the example shown in equation 30[63]. The side chain of vitamin D analogs was introduced in an analogous manner from the allylic pivalate, phosphate or carbamate[115].

In constrast to the usual γ-selectivity of the cyanocuprates, the higher-order cyanocuprate **235** reacted exclusively in a S_N2 fashion on cyclic tosylaziridine **234** to give tosylamine **236** (Scheme 30)[116]. The product was used to complete the synthesis of (+)-pancratistatin (**239**). Surprisingly, the Gilman-type cuprate **237** gave 75% yield of the S_N2' product **238**, a product of *syn*-addition[116c]. This unusual stereoselectivity may be the consequence of the type of organocuprate chosen as Gilman-type cuprates are known to sometimes give rise to lower levels of regio- and stereospecificity[9a,21a,c]. The oxirane analog of **234** (NTs = O) gave mostly S_N2 displacement products with various cuprate reagents[116c].

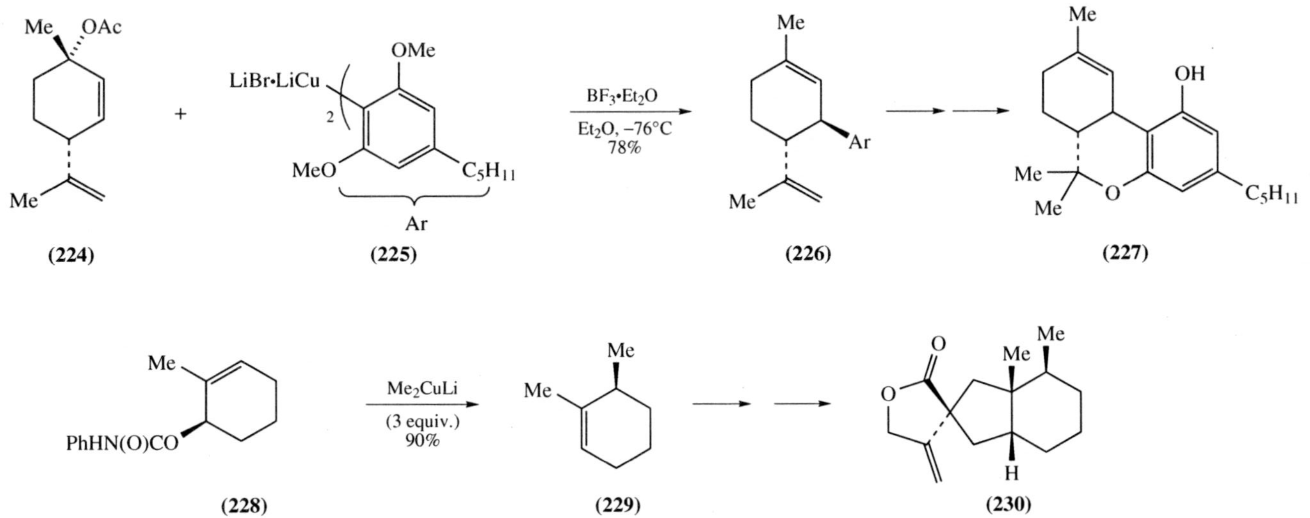

Me
OAc
Me
(224)
+
LiBr•LiCu
2
OMe
MeO
C5H11
Ar
(225)
BF3•Et2O
Et2O, –76°C
78%
Me
Ar
Me
(226)
Me
OH
Me
Me
O
C5H11
(227)
Me
PhHN(O)CO
(228)
Me2CuLi
(3 equiv.)
90%
Me
Me
(229)
O
O
Me
Me
H
(230)

SCHEME 29

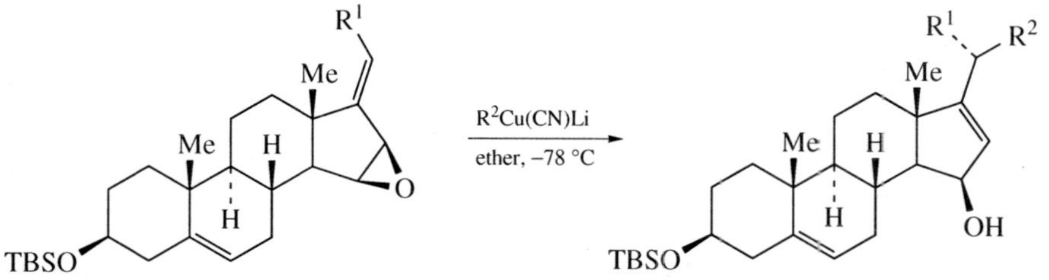

(**231a**) R^1 = Me
(**231b**) R^1 = i-C_6H_{13} = 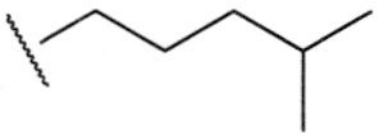

(**232a**) R^1 = Me, R^2 = i-C_6H_{13}
(**232b**) R^1 = i-C_6H_{13}, R^2 = Me

+ (30)

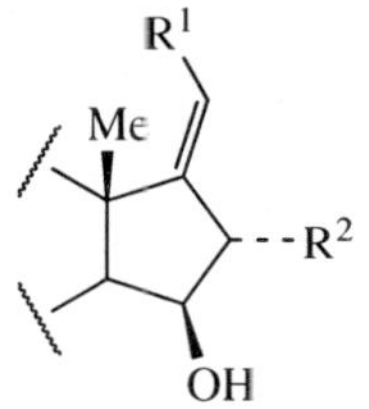

(**233a**)
(**233b**)

The S_N2 selectivity of aziridine **234** could be the result of steric hindrance by the acetonide group since, in a related approach to (+)-pancratistatin (**239**), the allylic acetate **240** reacted with arylcyanocuprate **241** to give a single product **242** resulting from an *anti*-selective S_N2' reaction (equation 31)[117].

Ernst and Helmchen took advantage of the great tolerance of organozinc reagents for many functional groups to introduce the ester side chain of 12-oxophytodienoic acid (**244**), the biosynthetic precursor of many natural jasmonoids (Scheme 31). Coupled with CuCN, the often-favored source of copper(I) for highly γ-selective reactions, the organozinc **246** reacted with allylic bromide **245** to afford a single diastereomer **247** in 97% yield[118a]. The asymmetry in **245** was created by a palladium-catalyzed enantioselective allylic alkylation of malonate by 2-cyclopentenyl chloride to give **243** in 95% ee. The same approach led to the synthesis of the iridoids isoiridomyrmecin and α-skytanthine[118b].

D. Miscellaneous Allylic Systems

The stereospecificity of allylations with enyne acetates **248** was astonishingly high considering the distance between the departing group and the reaction site (Scheme 32). Initial experiments did not include additives like phosphines or phosphates and gave mixed results: compound E-**249a** being isolated in only 20% ee, whereas Z-**249a** was isolated in 74% ee, though the sense of asymmetric induction was not specified (Table 3, entry 1)[119]. The observed low stereospecificity was apparently the result of racemization of the product allene by the cuprate reagent[85a]. The authors were therefore able to suppress this racemization by the addition of n-Bu_3P or $(EtO)_3P$[85a] and demonstrate that the actual

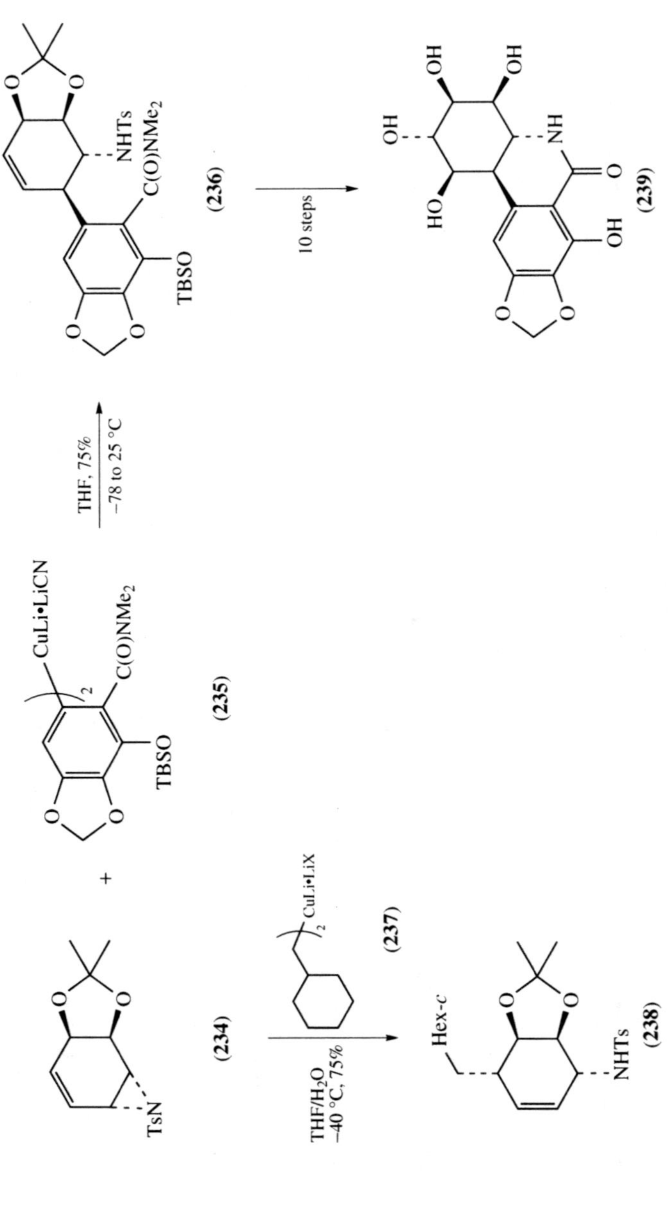
TsN
(234)
+
CuLi•LiCN
C(O)NMe2
TBSO
(235)
THF, 75%
−78 to 25 °C
NHTs
C(O)NMe2
TBSO
(236)
10 steps
OH
HO
NH
O
OH
(239)
CuLi•LiX
(237)
THF/H2O
−40 °C, 75%
Hex-c
NHTs
(238)

SCHEME 30

allylic substitution took place in a highly stereospecific manner (Table 3, entries 2–5)[120]. The stereoselectivity (i.e. **248-A** vs **248-B**), however, remained low with most substrates giving from 40:60 to 30:70 mixtures of *E* and *Z* vinylallenes **249**. The sense of asymmetric induction has not yet been established and it is premature at this stage to assume the more common *anti* stereoselectivity. Nonetheless, the predominant formation of the *Z* alkene, with the exception of **248e** (entry 5), is surprising as one would have predicted a predominance of attack onto conformer **248-A** from $A^{1,3}$-strain consideration. However, it appears that conformer **248-B** is either more reactive or predominant in solution.

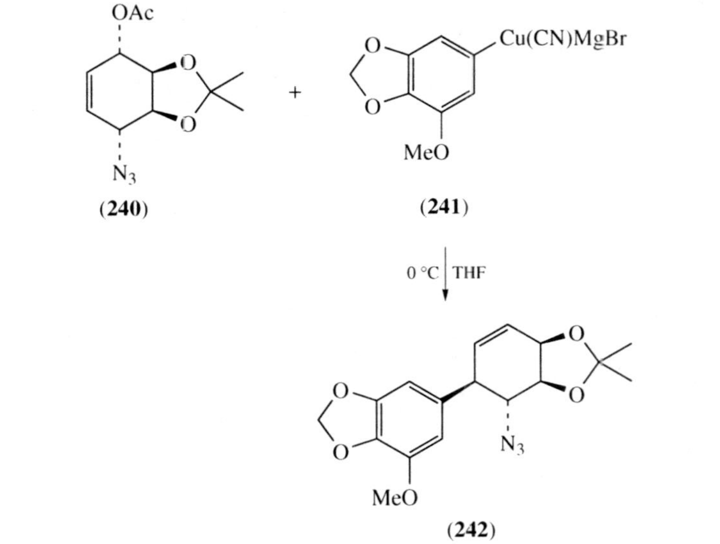

(31)

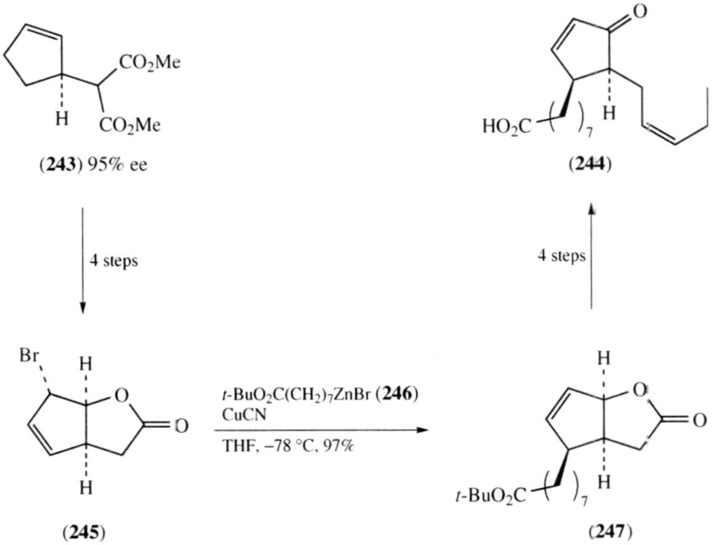

SCHEME 31

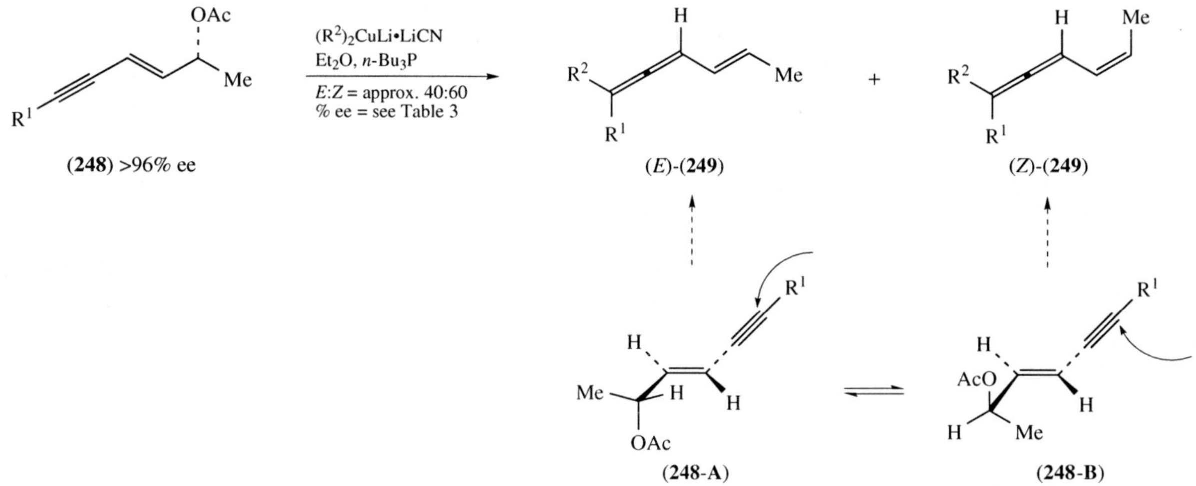
(248) >96% ee
$(R^2)_2$CuLi•LiCN
Et_2O, n-Bu_3P
E:Z = approx. 40:60
% ee = see Table 3
(E)-(249)
(Z)-(249)
(248-A)
(248-B)

SCHEME 32

TABLE 3. Long-range stereospecificity of allenylation of cuprate reagents by **248**

Entry	**248**	R^1	Additive	**249**	R^2	$E:Z$	% ee of (E)-**249**[a]	% ee of (Z)-**249**[a]
1	**a**	Me	None	**a**	t-Bu	25:75	20	74
2	**b**	Me	Bu_3P	**b**	t-Bu	25:75	92	93
3	**c**	n-Bu	Bu_3P	**c**	t-Bu	33:67	94	96
4	**d**	Ph	Bu_3P	**d**	t-Bu	40:60	92	95
5	**e**	$SiMe_3$	Bu_3P	**e**	n-Bu	60:40	99	99

[a]The sense of asymmetric induction was not specified.

Macrocycles often have several conformations of similar energies that may hamper stereoselectivity in some reactions. The 9- and 10-membered endocyclic allenes **251** and **253** were obtained from the stereospecific γ-selective displacement of the corresponding macrocyclic propargyl acetates **250** (equation 32) and **252** (equation 33)[121]. The transfer of chirality from the stereocenter (**250** and **252**) to the axis of chirality (**251** and **253**) was complete, though the stereochemistry of the 9-membered macrocyclic allenes was inferred from the results obtained with the 10-membered allenes. Both propargyl acetates were prepared enantiomerically pure by enzymatic resolution of the racemate.

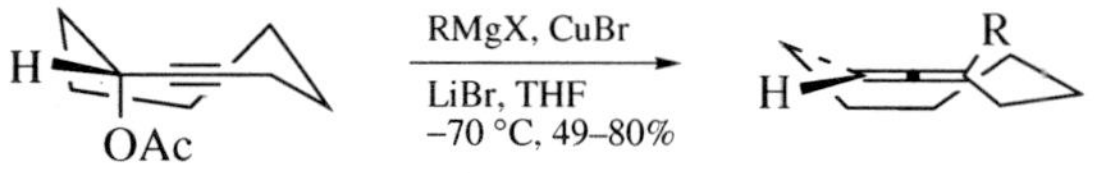

(**250**) 91% ee (**251**) 91% ee (32)

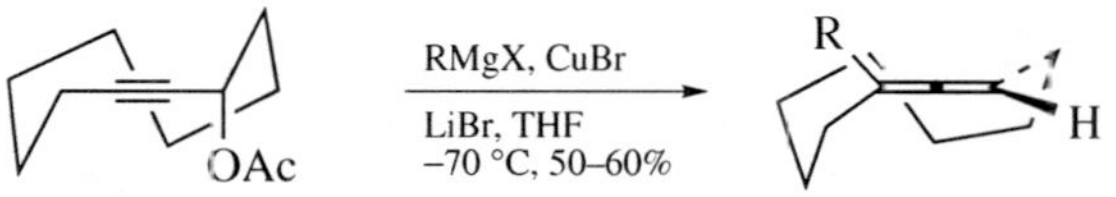

(**252**) 60% ee (**253**) 60% ee (33)

III. DIASTEREOSELECTIVE METHODS

A. General Considerations

Asymmetric induction from a nearby chiral center only applies to substrates that bear the leaving group on an achiral carbon. That is because the stereospecificity of the copper-mediated allylic alkylation reaction is rarely, if ever, supplanted by the asymmetric induction of spectator chiral centers. So when the carbon bearing the leaving group is chiral, stereospecific *anti*-addition on a $A^{1,3}$-strain-minimized conformation of the substrate will take place, whether there are nearby chiral centers or not. For example, the substituant R in **214** (cf. Scheme 28, Section II.C.2.b) does not force the organocopper reagent to attack *anti* to it, but rather the reagent will attack *anti* to the leaving group, regardless of the stereochemistry of the carbon bearing the substituent R. In the same vein, if the alkyl in the copper species is chiral, its stereochemistry will not influence the outcome of the asymmetric allylic alkylation. For example, the chiral cuprate reagents obtained from **90** by the reaction sequence shown in Scheme 13 (Section II.B.1.b) could not undo the stereospecificity of the allylic alkylation reaction that afforded **92**.

Diastereoselective allylic alkylations, as defined above, are far less numerous than diastereospecific ones, probably because the stereochemical outcome is less certain and not always easily predictable. However, because the carbon bearing the leaving group is, by definition, achiral in these cases, it so happens that there are no issues of stereospecificity or product's alkene geometry.

B. Chirality δ to the Leaving Group

1. Chiral center bearing no heteroatom

a. Acyclic substrates. Nakamura and coworkers studied extensively the 1,2-asymmetric induction in copper-mediated S_N2' displacement of primary chlorides[13,122]. The mixing of $ZnCl_2$ or $TiCl(Pr\text{-}i)_3$ together with Gilman-type cuprate reagents led to highly S_N2'-selective reactions (Scheme 33, top). The diastereoselection was excellent (approx. 95:5) across substrates **254a–c** (Scheme 33, center). In the case of trisubstituted (*Z*)-allylic chloride **257**, the diastereoselectivity was complete but the regioselectivity suffered from the added steric volume around the γ-position such that mixtures of **258** and **259** were obtained (Scheme 33, bottom).

The sense of stereoinduction for substrates like **254** may be analyzed in terms of the Felkin–Ahn model of addition (Figure 3), which predicts the predominance of adducts **255** or **256**[123]. Therefore, the preferred conformations **I** or **II** do not have a minimized allylic strain, but rather a minimized energy for the interaction between the substrate and the large metal complex. In that sense, transition states **I** or **II** are both conceivable. In accordance with this analysis is the fact that *Z* allylic chloride (*Z*)-**254a**, which should experience increased allylic strain in the transition states **I** or **II**, gave a lower stereoselectivity compared to (*E*)-**254a** (Scheme 33).

Impressive results were obtained from the copper-catalyzed allylic substitution of alkyltitanium reagents onto allylic chloride **254a** (equation 34)[18,124]. These reaction conditions consistently gave highly γ-selective allylic alkylation on various substrates and the diastereoselectivity was very high indeed for the case shown in equation 34.

A dramatic case of high stereo- and regioselectivity was recently reported using the reaction conditions reported in Scheme 33. Compound **260**, possessing an all-carbon quaternary chiral center at the δ-position, underwent a complete stereo- and regioselective copper-mediated addition in the presence of zinc chloride (equation 35)[125]. Contrary to the author's statement, product **261** was formed from the addition of the cuprate reagent onto a $A^{1,3}$-minimized conformation **260-TS**, which affords the anti-Felkin product **261**. The latter was transformed to analogs of vitamin D_3.

As part of a mechanistic study aimed at rationalizing the chiral induction of a series of ligands, allylic phosphate **262** was reacted with copper triflate and dimethylzinc to give product **263** in 80% de (equation 36)[126]. No rationale was offered for the diastereoselectivity, but the attack of the copper reagent occurred on the face of the allylic phosphate **262** as predicted from the Felkin–Ahn model (**262-TS**)[123].

b. Cyclic substrates. There are scant reports of cases of diastereoselective allylic alkylation with cyclic substrates, mostly because few cyclic substrates bear the leaving group on an achiral carbon. Allylic phosphate **264**, made from pyrraldehyde, reacted with trialkylaluminums in the presence of 10% CuCN to give 91% of a 6:1 mixture of *anti* and *syn* products **265a** and **265b** (equation 37)[127]. Also, bicyclo[3.3.0] systems bearing chiral sulfoximines were shown to add organocopper reagents on the expected convex face of the ring system (not shown)[128].

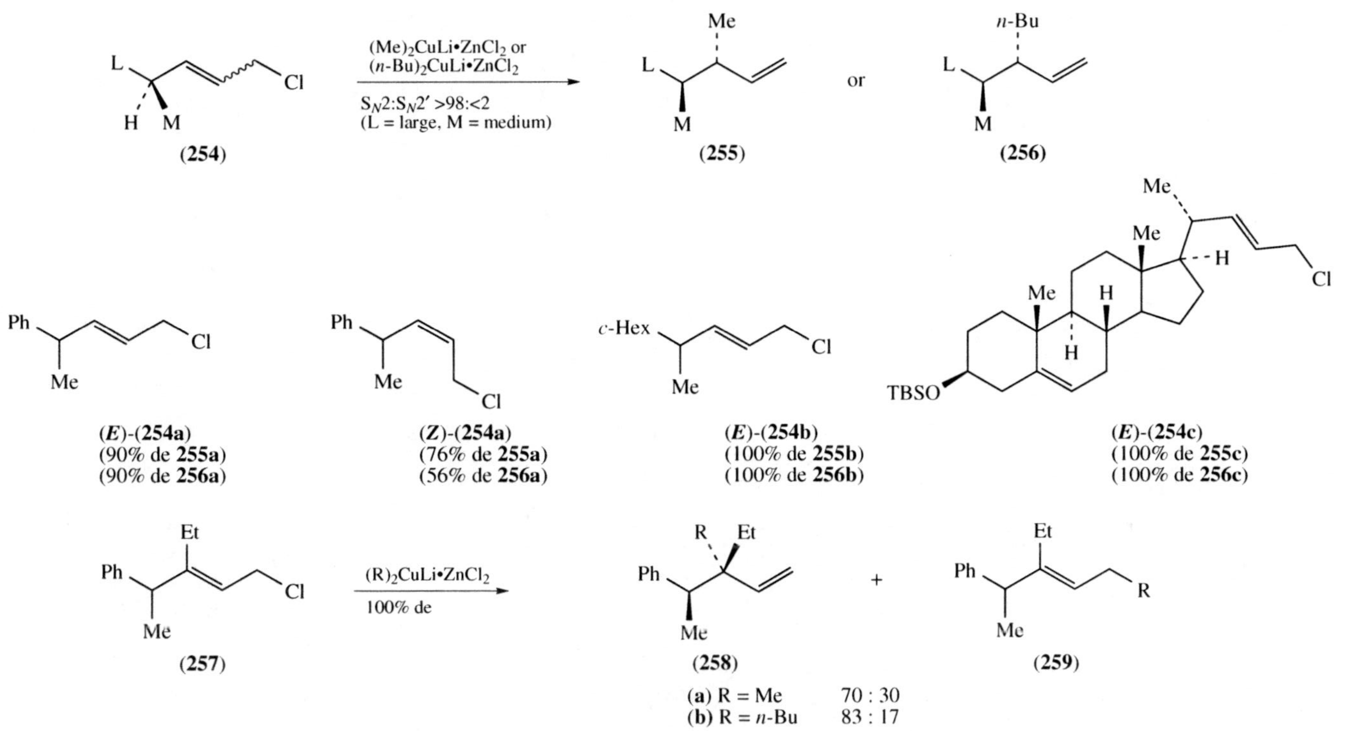
(Me)$_2$CuLi•ZnCl$_2$ or
(n-Bu)$_2$CuLi•ZnCl$_2$
S$_N$2:S$_N$2′ >98:<2
(L = large, M = medium)
or
(**254**)
(**255**)
(**256**)
(***E***)-(**254a**)
(90% de **255a**)
(90% de **256a**)
(***Z***)-(**254a**)
(76% de **255a**)
(56% de **256a**)
(***E***)-(**254b**)
(100% de **255b**)
(100% de **256b**)
(***E***)-(**254c**)
(100% de **255c**)
(100% de **256c**)
(R)$_2$CuLi•ZnCl$_2$
100% de
+
(**257**)
(**258**)
(**259**)
(**a**) R = Me 70 : 30
(**b**) R = n-Bu 83 : 17

SCHEME 33

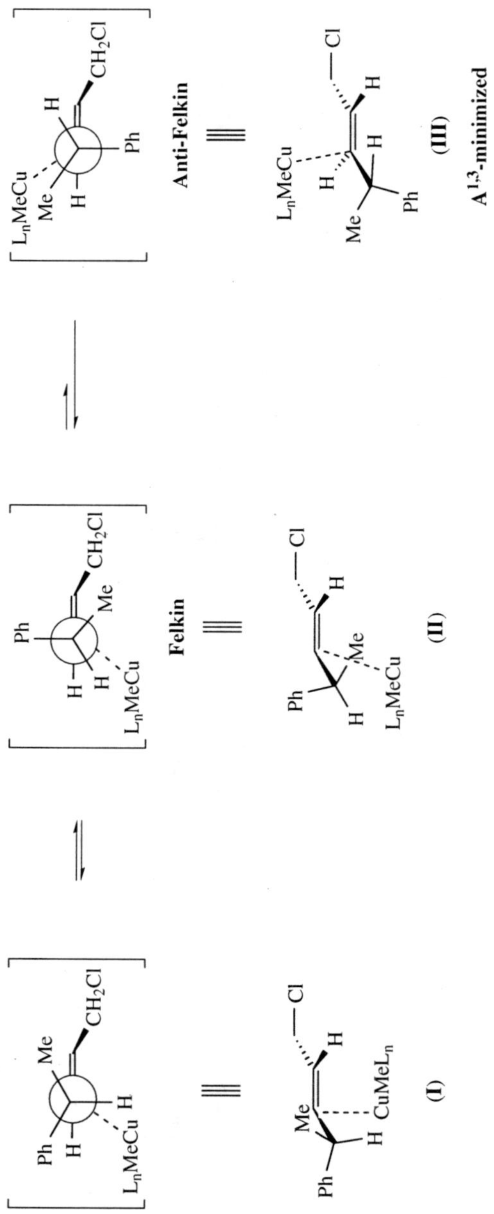

Me
CH2Cl
Ph
H
LnMeCu
H
Cl
H
CuMeLn
H
(I)
Ph
CH2Cl
Me
H
H
LnMeCu
Felkin
Cl
H
Me
LnMeCu
Ph
H
(II)
LnMeCu
H
CH2Cl
Me
H
Ph
Anti-Felkin
Cl
H
H
H
Me
Ph
(III)
A1,3-minimized

FIGURE 3

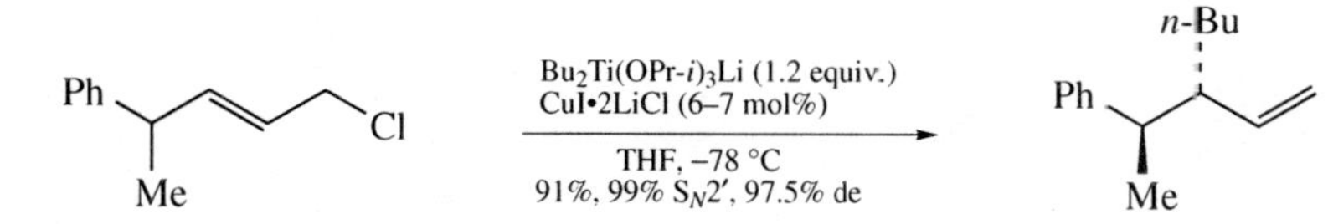

(254a) (256a)

(34)

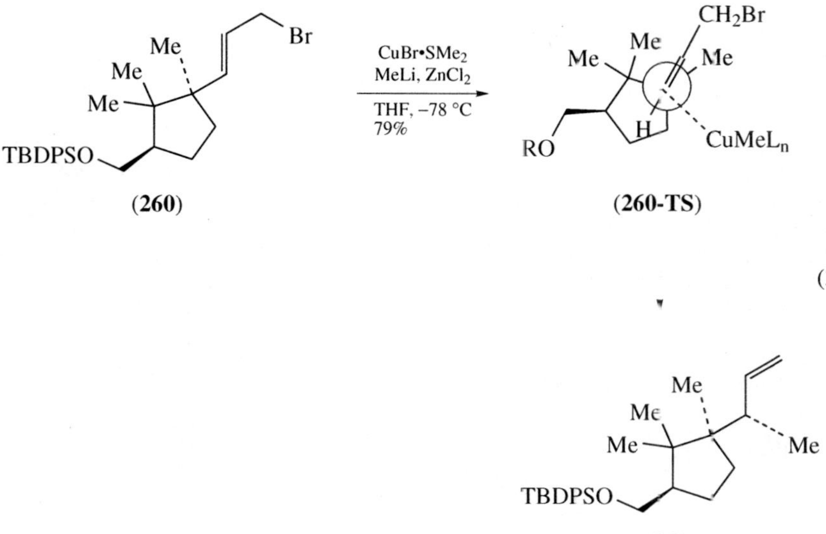

(260) (260-TS)

(35)

(261)

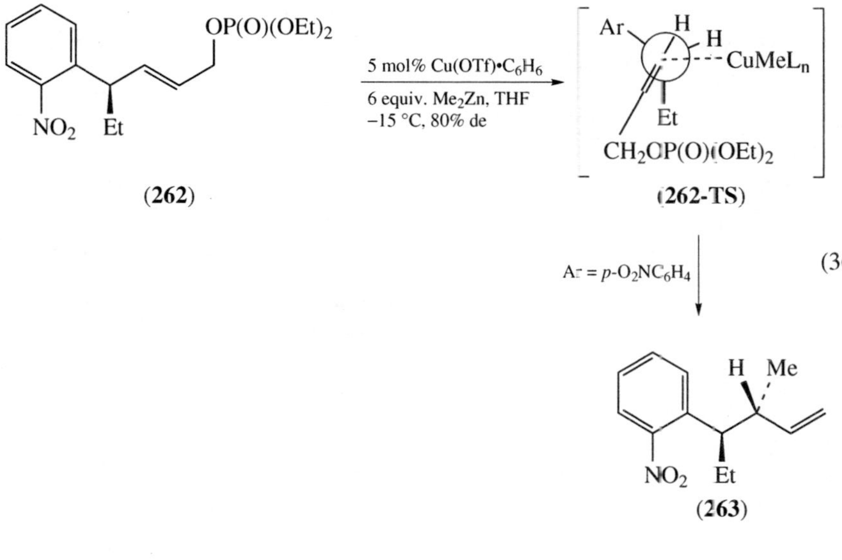

(262) (262-TS)

(36)

(263)

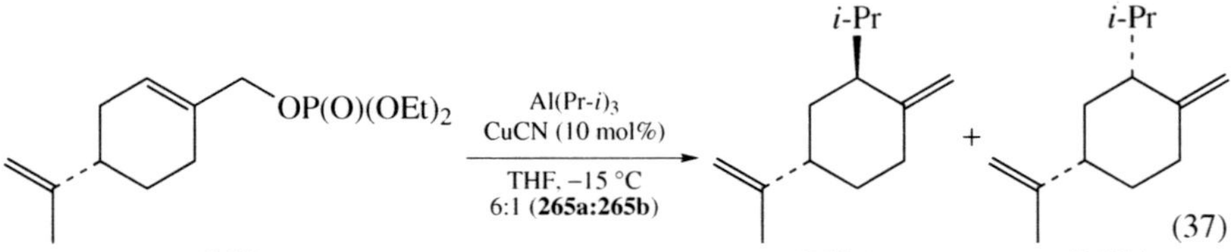

(**264**) (**265a**) (**265b**) (37)

2. Chiral center bearing a heteroatom

a. Oxygen atom. Alkoxy groups were very effective at transferring chirality over to the adjacent carbon in the regioselective displacement of allylic chloride **266** using a Gilman-type cuprate reagent in the presence of zinc chloride. The *anti*-adducts **267a** and **267b** were obtained predominantly and with less than 2% of the regioisomers **268a** and **268b** being formed (equation 38)[13, 122]. Several other substrates were investigated with similar results. Trisubstituted allylic chloride **269** underwent a less regioselective but more diastereoselective alkylation to give products **270a** and **270b**, and **271a** and **271b**, the former bearing an all-carbon quaternary stereogenic center (equation 39). The sense of asymmetric induction for δ-alkoxy allylic chlorides like **266** is as predicted from the 'inside alkoxy' effect model **272** (Figure 4)[123].

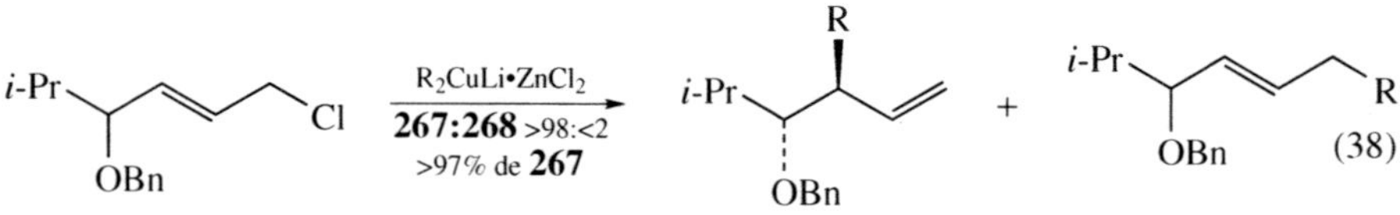

(**266**) (**267a**) R = Me (**268a**) R = Me (38)
(**267b**) R = *n*-Bu (**268b**) R = *n*-Bu

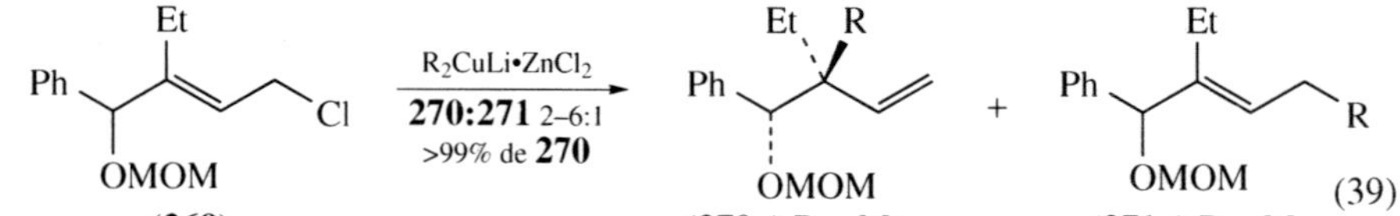

(**269**) (**270a**) R = Me (**271a**) R = Me (39)
(**270b**) R = *n*-Bu (**271b**) R = *n*-Bu

As was the case for allylic chloride **254a** (Scheme 33), allylic alkylation of alkyltitanium reagents with allylic chloride **266** worked equally well to give **267b** with complete

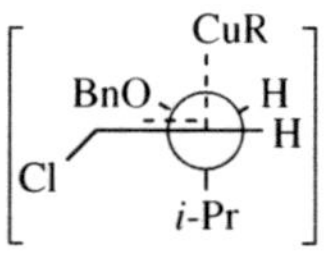

(**272**)
'inside alkoxy'

FIGURE 4

control over the regio- and stereochemistry (equation 40)[18,124].

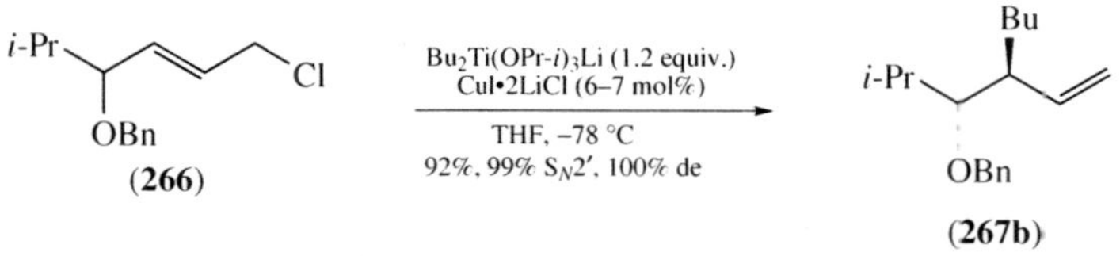

(40)

Belelie and Chong reported a follow-up on the study by Nakamura concerning the chiral induction exerted by an adjacent alkoxy group[129]. The allylic structure was rigged with a tertiary acetate designed to sterically bias the allylic system toward S_N2' attack of the cuprate reagent. They were able to exercise control over the *anti* or *syn* selectivity by choosing to protect or not the hydroxyl group at the δ-position. An example in given in equation 41. A free hydroxyl as in **273a** gave good-to-excellent diastereoselectivity in favor of *syn*-**274a**. Protected hydroxyls **273b–d** gave predominantly the *anti*-products **275b–d**, regardless of the nature of the protecting group. In either case, substrates with groups larger than *n*-Bu gave a better selectivity while larger alkylcuprate reagents led to a lower selectivity. Oxidative cleavage of the isopropylidene **275a–d** gave aldol-type products **276a–d**.

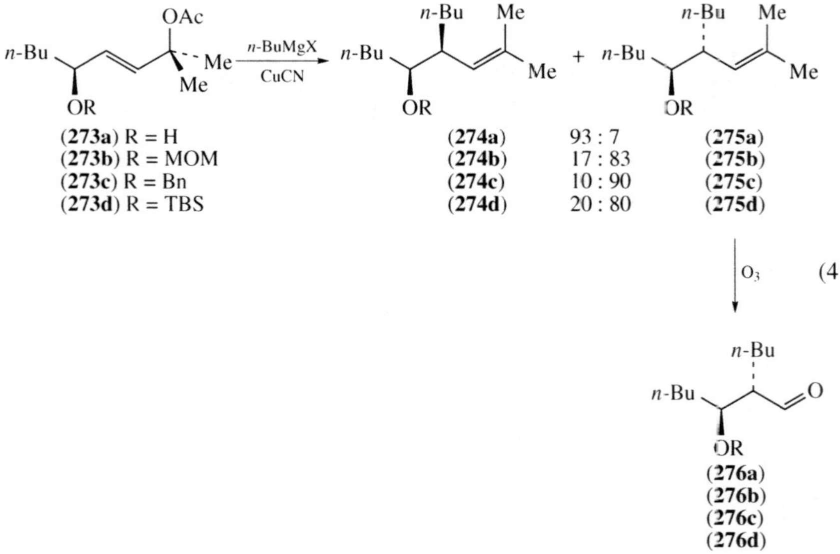

(41)

γ,γ-Dialkoxyzirconocene **277a** was obtained from orthoester **277** and transmetallated to the corresponding copper species with copper cyanide. It was then added to δ-alkoxy allylic phosphates **278a** and **278b** with modest stereoselectivity (Scheme 34)[130]. The stereochemistry assigned to each product **279a** and **279b** corresponds to an *anti* addition on conformer **278-TS**, in line with the 'inside alkoxy effect'[122,123].

b. Other heteroatoms. Breit, Mann and coworkers have recently disclosed a method to make β-branched α-amino acids. They started from Garner's aldehyde that were converted into esters **280**, which is affixed with their special tunable leaving group: the *o*-diphenylphosphanyl benzoate (equation 42)[131]. Then Grignard reagents in the presence of various

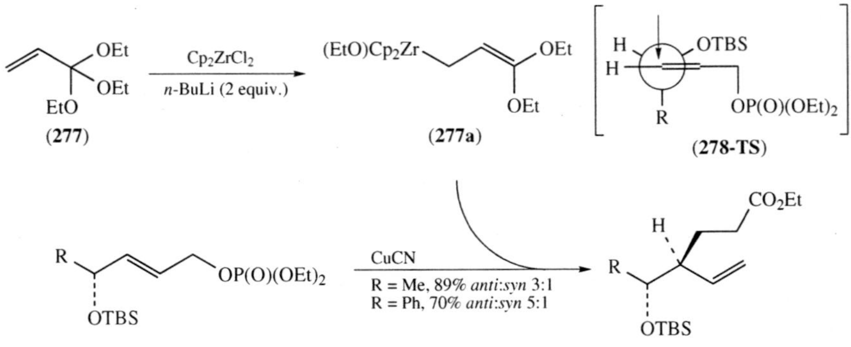

(**278a**) R = Me
(**278b**) R = Ph

(**279a**) R = Me
(**279b**) R = Ph

SCHEME 34

copper(I) sources were reacted with **280**. The geometry of the double bond in the starting ester **280** controlled the relative stereochemistry in the final adduct **281** as (*E*)-**280** led to **281a** and (*Z*)-**280** led to **281b**, both in a highly diastereoselective fashion.

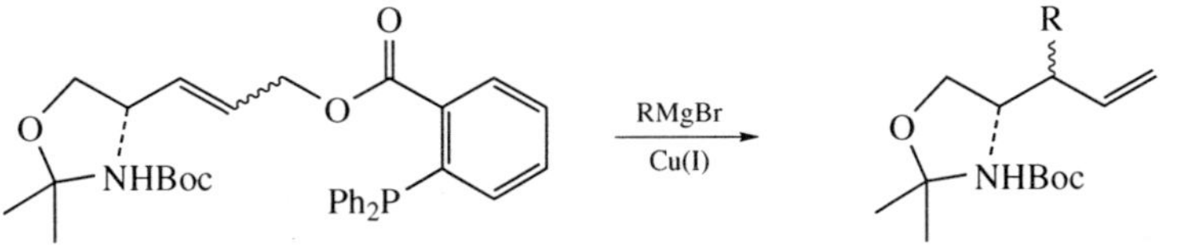

(42)

(*E*)-(**280**)
(*Z*)-(**280**)

(**281a**) *β*-R (from (*E*)-**280**)
(**281b**) *α*-R (from (*Z*)-**280**)

3. Use of a chiral auxiliary

A chiral sultam was used as a chiral auxiliary to induce asymmetry *α* to the carbonyl carbon with excellent diastereoselectivity (equation 43)[132]. This asymmetric allylic alkylation gave the same products **283a–d** that one would expect from the alkylation of the metal dienolate of **282**. The allylic alkylation's nucleophile, however, would be an electrophile in the classical enolate alkylation reaction. Note that no 1,4-addition product was detected.

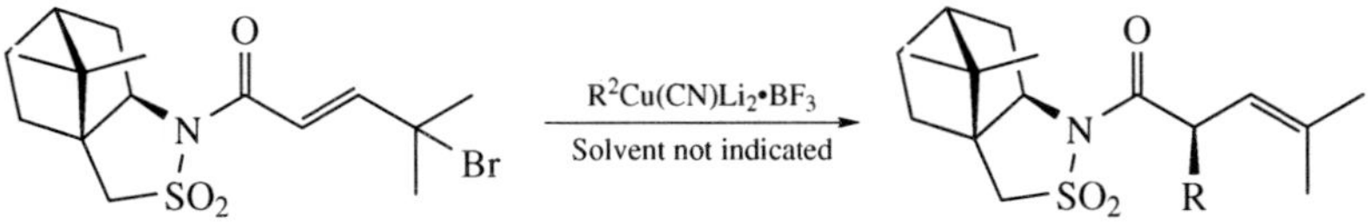

(43)

(**282**)

(**283a**) R = Me (98% de)
(**283b**) R = *n*-Bu (98% de)
(**283c**) R = $CH{=}CH_2$ (96% de)
(**283d**) R = Ph (95% de)

C. Chirality at the Homoallylic Position

A unique and interesting example of asymmetric allylic alkylation with a chiral carbon at the homoallylic position was reported by Taguchi and coworkers. Trialkylalanes could be made to displace allylic fluorides in the presence of Cu(I)[133]. It was found that copper(I) and a free alcohol at C5 in **284** were both necessary for the reaction to proceed (Scheme 35). The authors suggest activation of the departing fluoride by an alkoxydialkylaluminum on the more stable complex **287** (vs **286**), where the R^1 group and the allylic moiety are *trans* to one another. Internal delivery of the monoalkylcopper (made by transmetallation between $(R^2)_3Al$ and Cu(I)) accounts for the observed stereochemistry in the product **285**. The geometric analog, *Z*-**284**, underwent the reaction with a lower selectivity than *E*-**284**.

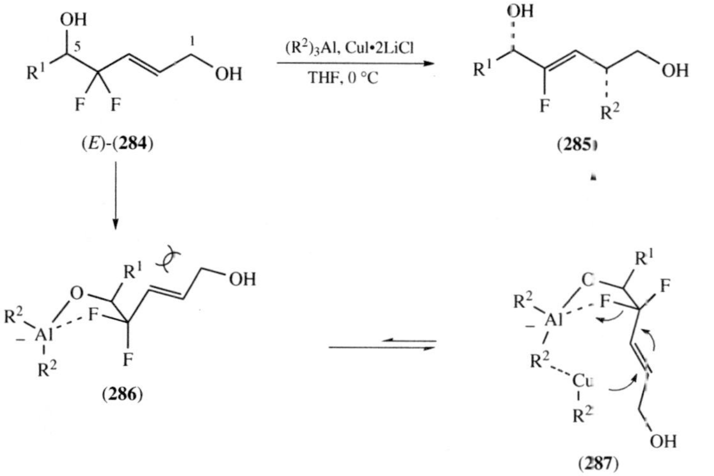

SCHEME 35

D. Chirality on the Leaving Group

In order to be classified as 'diastereoselective' (Section III), a copper-promoted displacement reaction must favor one of two diastereomeric products, not enantiomeric products. The latter cases are dealt with in Section IV (enantioselective methods). Chiral acetals produce diastereomeric products upon reaction with organocopper species, though they are subsequently hydrolyzed to enantiomeric ketones in the reaction work-up. Ethers and acetals are not usually good enough leaving groups to participate in a copper-promoted allylic alkylation reaction. However, when the cuprate reagent is associated with a Lewis acid such as BF_3, ethers do react[134a]. The first example, involving chiral ketal **288**, was reported by Alexakis, Normant and coworkers but did not seem promising as ketone **290** was isolated in only 26% ee after hydrolysis of the intermediate enol ether **289** (equation 44)[134b].

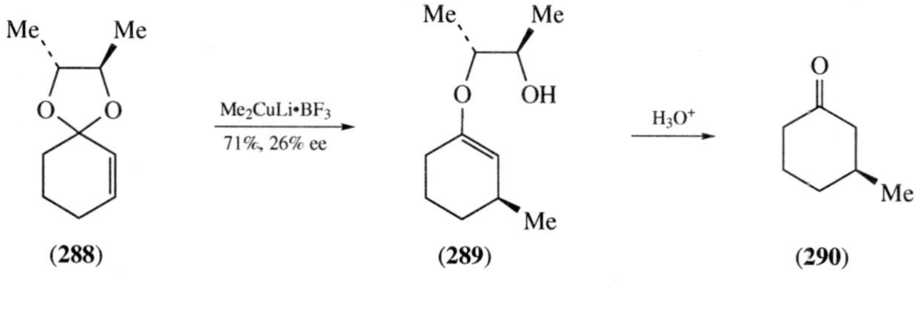

(44)

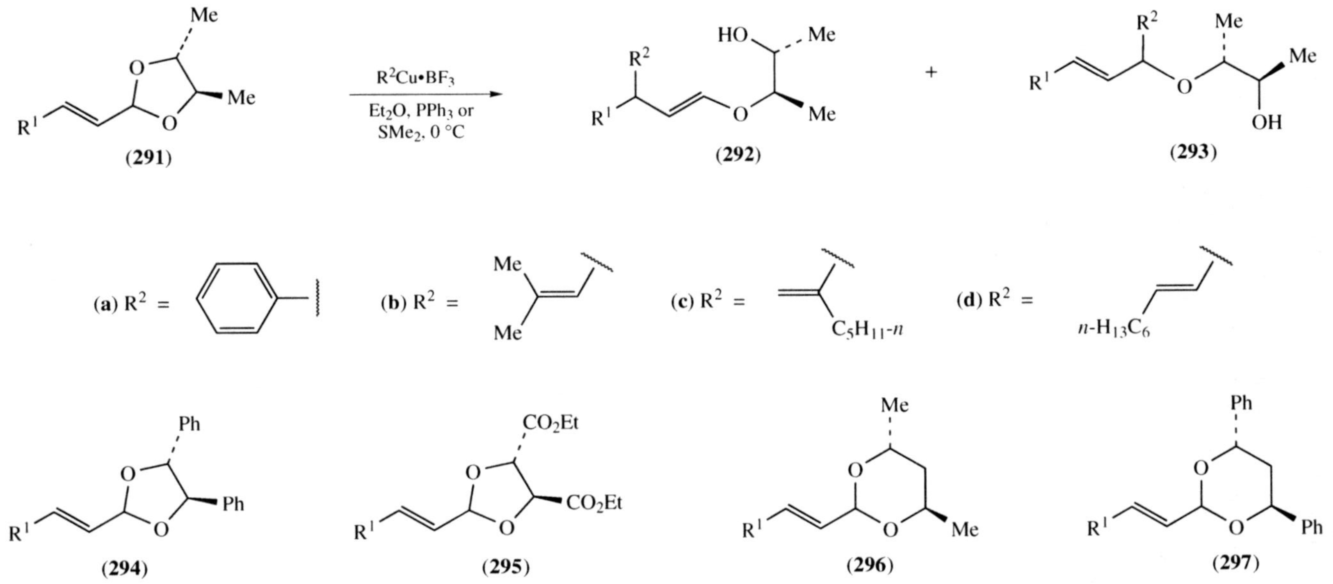

R^{2}Cu•BF$_3$
Et$_2$O, PPh$_3$ or SMe$_2$, 0 °C
(291)
(292)
(293)
(a) R^2 =
(b) R^2 =
(c) R^2 =
(d) R^2 =
C_5H_{11}-n
n-$H_{13}C_6$
(294)
(295)
(296)
(297)

SCHEME 36

While cyclic ketals gave γ-selective allylic alkylation, acyclic acetals such as **291** gave mixtures of S_N2' (**292**) and S_N2 (**293**) products when alkylcopper (R^2Cu) were used. Aryl- and vinylcopper proved to be highly γ-selective reagents for such acyclic acetals, giving addition products **292a–d** with diastereomeric excesses in the 80% range (Scheme 36)[134]. It was found that the addition of tributylphosphine or dimethyl sulfide improved the diastereoselectivity up to 95:5[134]. Other chiral acetals (**294–297**) were investigated but none was better than the one derived from 2,3-butanediol (as in **291**)[134d]. Chiral diene-acetals (**291**, replace R^1 by $R^1{-}HC{=}CH_2$) also partake in allylic substitution to give mixtures of S_N2'- and S_N2''-displacement products[135].

Propargylic acetals proved to be good substrates as well and a series of chiral non-racemic allenes was prepared in this way with enantiomeric excesses ranging from 40–100% (not shown)[85c, 136].

IV. ENANTIOSELECTIVE METHODS

Enantioselective methods are defined here as those that leave only the newly created chiral center as the source of asymmetry in the product. However, if other chiral centers present in the starting material are too far from the reacting site or otherwise incapable of influencing the course of the allylic alkylation, then these cases too will be discussed in this section. Enantioselective methods using a catalytic amount of a chiral source are of course more desirable and it is safe to say that much progress has been made in this particular area in the last decade. We start this section, however, by describing methods using stoichiometric chirality located on the leaving group.

A. Chirality on the Leaving Group

Chiral leaving groups have been used with moderate success to effect the regio- and enantioselective allylic substitution of alkylcopper species. One of the main challenges with this strategy rests with the distance between the γ-position of the allylic moiety (usually the preferred position of attack) and the chirality source on the leaving group. Denmark and Marble were the first to report a case of 1,7-asymmetric induction in the asymmetric allylic alkylation of monoalkylcuprate reagents with chiral carbamate **298** affording γ-products **299a–c** (equation 45)[137]. Although only one substrate was tested, this is a rare case of long-distance asymmetric induction and the rationale for such a phenomenon is not straightforward.

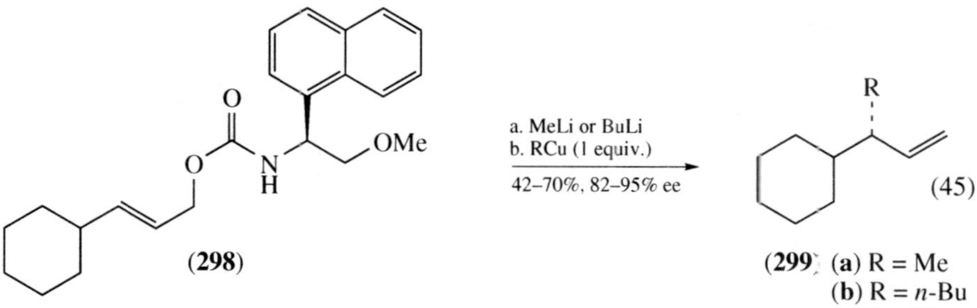

(45)

(**298**) (**299**) (**a**) R = Me
(**b**) R = *n*-Bu
(**c**) R = Ph

After trying out allylic alkylations with camphor-derived benzimidazole **300** but getting products **301** with low %ee[20], Caló and coworkers explored the copper-mediated γ-displacement of a series of chiral dihydroxazole and dihydrothiazoles (**302a–d**) with success (Scheme 37)[138]. It was crucial to put an excess of copper(I) salt with respect to

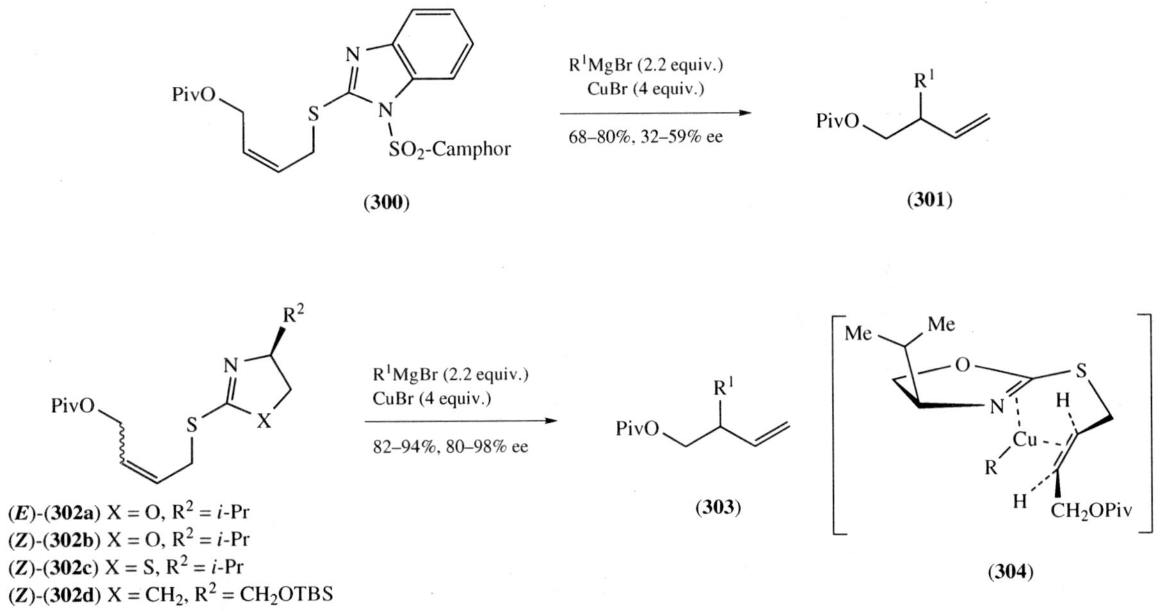
PivO
S
N
N
SO2-Camphor
(300)
R1MgBr (2.2 equiv.)
CuBr (4 equiv.)
68–80%, 32–59% ee
R1
PivO
(301)
R2
N
PivO
S
X
(E)-(302a) X = O, R2 = i-Pr
(Z)-(302b) X = O, R2 = i-Pr
(Z)-(302c) X = S, R2 = i-Pr
(Z)-(302d) X = CH2, R2 = CH2OTBS
R1MgBr (2.2 equiv.)
CuBr (4 equiv.)
82–94%, 80–98% ee
R1
PivO
(303)
Me
Me
O
S
N
H
Cu
R
H
CH2OPiv
(304)

SCHEME 37

the Grignard reagent to get products **303** with high regioselectivity and enantioselectivity. This ratio of copper(I) salt to Grignard reagent favors the formation of RCu, which is less basic and may form complex **304** for exclusive γ-delivery of the alkyl group. The alkene would wrap on the face of the oxazoline ring opposite the isopropyl group. These reactions are also examples of long-range asymmetric induction.

(Diphenylphosphanyl)ferrocene carboxylate **305**, a chiral version of Breit's *o*-DPPB leaving group (see Scheme 5, Section II.A.2), was used to effect the copper-mediated enantioselective allylic alkylation of Grignard reagents to give adducts **307** (equation 46)[139]. Delivery of the alkylcopper species by the chiral leaving group onto a lower-energy conformation of intermediate **306**, as shown in equation 46, was proposed to account for the sense of enantioselection. The reaction parameters, including temperature, solvent and reagent addition times, were thoroughly investigated. All but phenylmagnesium bromide gave good results (Table 4, entries 1–4) but the presence of an aromatic substituent on the double bond somewhat lowered the selectivity (entries 5 and 6). It was possible to generate a quaternary stereocenter starting from geraniol or nerol using this method, albeit in modest enantioselectivity (49 and 59% ee, respectively) but excellent regioselectivity.

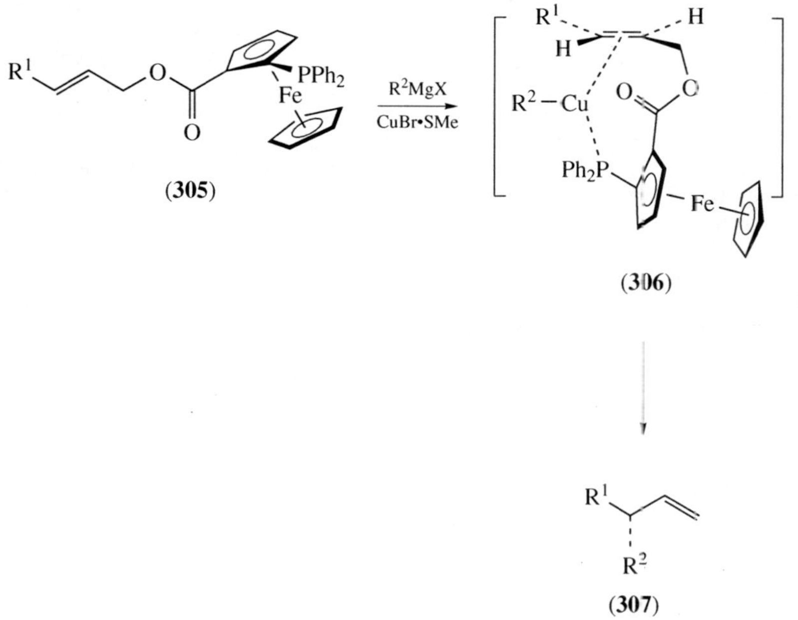

(46)

TABLE 4. Copper-mediated allylic substitution of Grignard reagents with 0.5 equiv of $CuBr{\bullet}SMe_2$ and substrates **305**

Entry	R^1	R^2MgX	S_N2: S_N2'	% ee (**307**)	Yield (%)
1	*c*-Hex	MeMgI	93:7	82	56
2	*c*-Hex	*n*-BuMgBr	97:3	95	77
3	*c*-Hex	*i*-PrMgBr	98:2	81	82
4	*c*-Hex	PhMgBr	75:25	28	nd
5	Ph	*n*-BuMgBr	87:13	78	86
6	*p*-$MeOC_6H_4$	MeMgI	84:16	65	60

TABLE 5. Copper-mediated allylic substitution of Grignard reagents with 0.5 equiv of $CuBr{\bullet}SMe_2$ and substrate **308**

Entry	n in **308**	R	S_N2: S_N2'	% ee (**309**)	Yield (%)
1	0	**a**	98:2	27	40
2	1	**a**	99:1	72	89
3	1	**b**	2:98	—	95
4	1	**c**	95:5	76	93
5	1	**d**	85:15	66	60
6	1	**e**	98:2	71	90
7	2	**a**	99:1	60	83
8	2	**b**	5:95	—	93
9	2	**c**	97:3	61	91
10	2	**e**	91:9	60	91
11	3	**a**	98:2	60	90

A chiral sulfoximine appendage induced moderate levels of asymmetry in the allylic alkylation of monoalkylcopper species in the presence of LiI and BF_3, as shown in the conversion of **308** ($n = 0$–3) into **309a–e** and **310a–e** (equation 47 and Table 5). The solvent THF was crucial as regioselectivity dropped dramatically or even reversed in Et_2O as the solvent. The case of the benzylcopper reagent is an exception and neither solvents nor additives much affected the ratio in favor of the α-product (entries 3 and 8). Otherwise, the level of asymmetric induction, though unexceptional, seemed consistent across a range of substrates, including a seven-membered ring sulfoximine (entry 11). Modifying the N-substitution had an effect on the reactivity of the electrophile, but no improvement on the selectivity was found[128, 140].

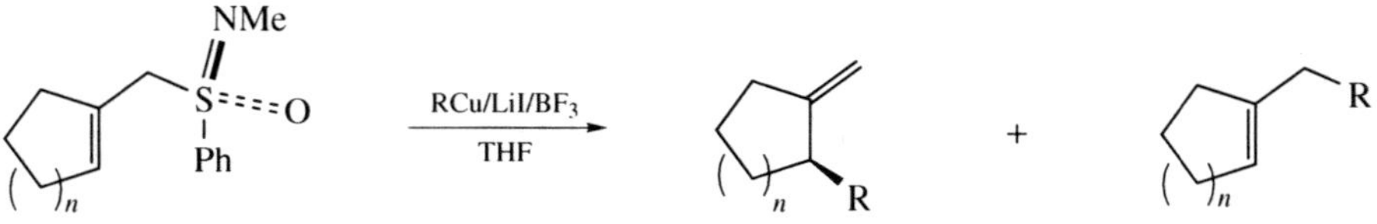

(**308**) (**309**) (**310**)

(47)

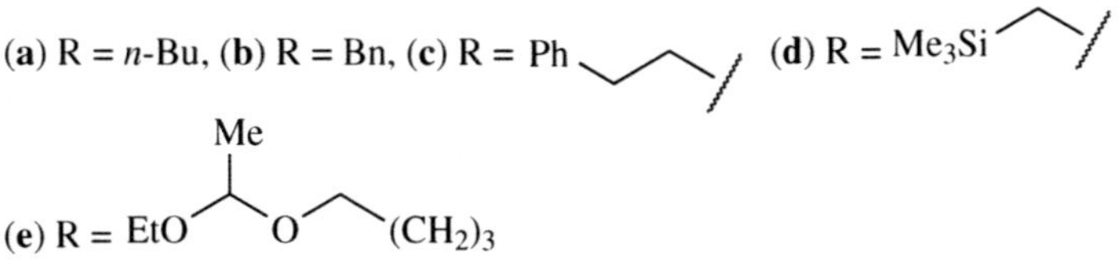

1,4-Additions followed by elimination constitute a formal S_N2' displacement. For example, α,β-unsaturated ketones **311a–c** underwent such a sequence to give enantiomerically enriched *exo*-methylene cyclohexenones **313a–c** via the Michael addition intermediates **312a–c** (Scheme 38). The enantiomeric excess of the products was very sensitive to the temperature. A transition state **314** for the conjugate addition was proposed to explain the selectivity[141].

B. Chiral Ligands on Copper

Since the first report of an asymmetric allylic substitution using chiral copper arylthiolate (**315**) in 1995[14], several research groups have described their own chiral ligands

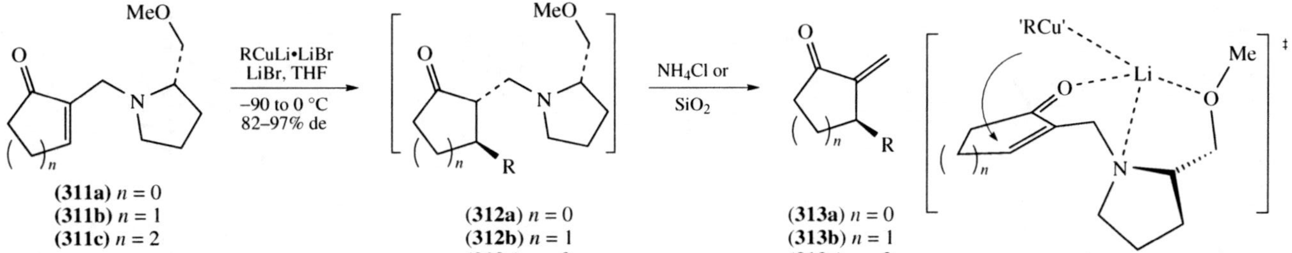

(311a) $n = 0$
(311b) $n = 1$
(311c) $n = 2$

(312a) $n = 0$
(312b) $n = 1$
(312c) $n = 2$

(313a) $n = 0$
(313b) $n = 1$
(313c) $n = 2$

(314)

SCHEME 38

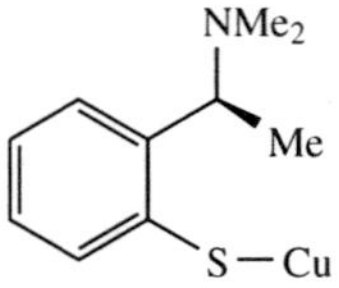

(**315**)

Bäckvall

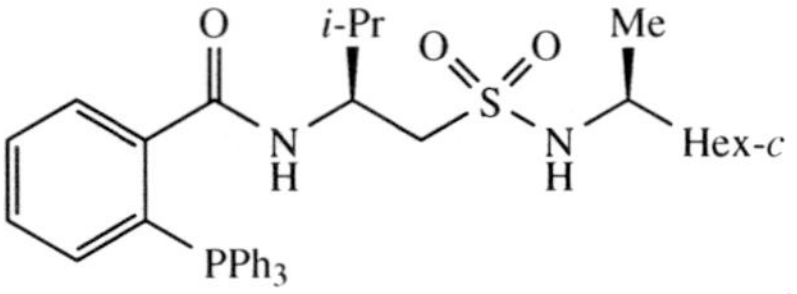

(**316**)

Gennari

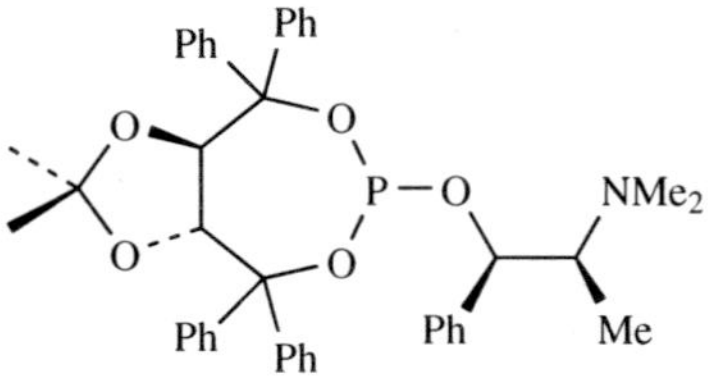

(+)-(**317**)

Alexakis

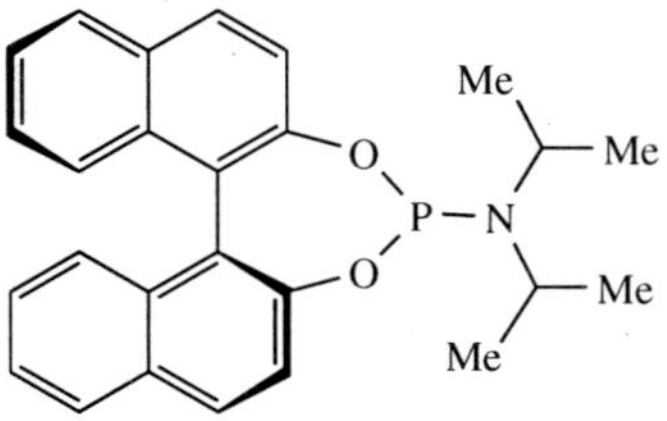

(+)-(**318**)

Feringa

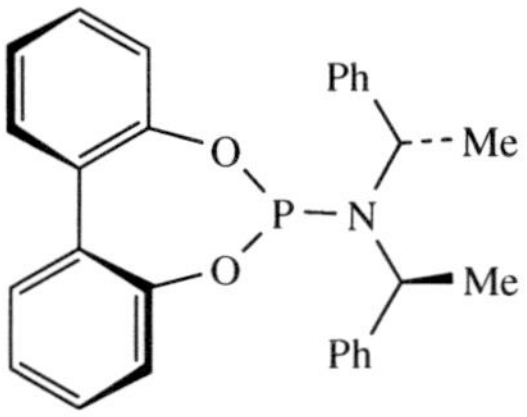

(**319**)

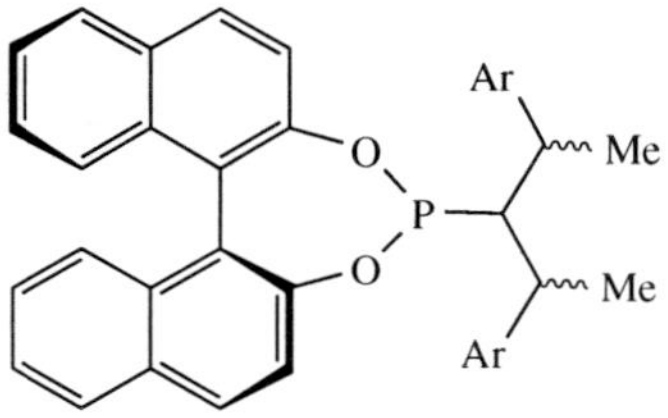

(**320a**) Ar = Ph (*S*,S,*S*)
(**320b**) Ar = Ph (*S*,*R*,*R*)
(**320c**) Ar = *o*-MeOC_6H_4 (*S*,*R*,*R*)

Feringa–Alexakis

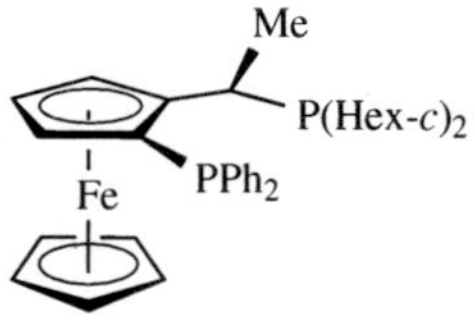

(**321**)

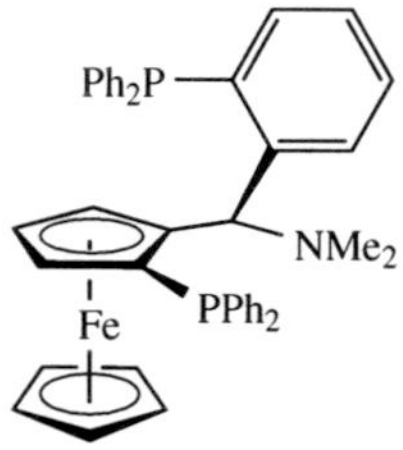

(**322**)

Feringa

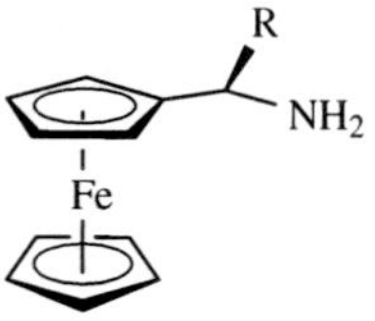

(**323a**) R = 2-naphthyl
(**323b**) R = 3,5-di-*t*-butylphenyl

Knochel

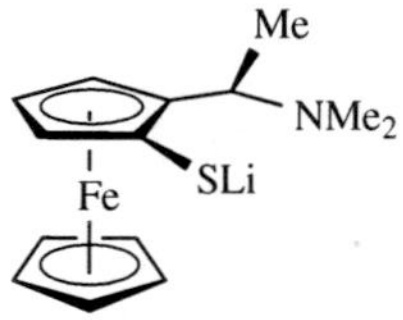

(**324**) (*R*,S_p)

Bäckvall

FIGURE 5

to effect this transformation. Figure 5 lists several ligands that were successfully used as chiral appendages to promote the copper-catalyzed asymmetric allylic alkylation of various organometallics, mostly organomagnesium and organozinc. The following sections are placed according to the structural features of the allylic substrates.

1. Prochiral substrates

a. Di- and trisubstituted substrates. The arenethiolate **315**, first developed by van Koten and coworkers[14,142], and several other arenethiolates were used as chiral ligands on copper to allylate Grignard reagents with allylic acetates **325** to give a single product **326** in modest enantioselectivity (10-53% ee) (equation 48 and Table 6, entry 1). The authors proposed that a complexation between the alkene and copper gave **328**, in which the dimethylamino group and sulfur atom are coordinated to magnesium, and accounted for the observed stereochemistry (Figure 6)[143].

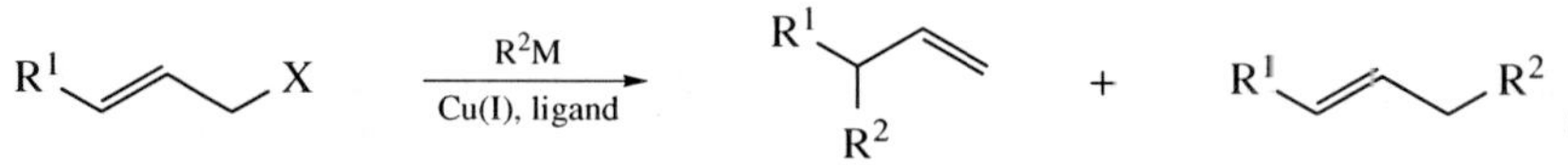

(48)

(**325**) (**326**) (**327**)

A later design of copper complexes of Cp-ferrocene-arenethiolate ligands, like **324**, gave somewhat improved results (Table 6, entry 2). A whole series of ligands analogous to **324**, including a ruthenocene analog, were prepared but with no improvement over the original ligand **324**. Thiolate **324** is neither easily purified nor stored, so a method was developed by which **324** is stored as its disulfide. The magnesium thiolate analog of **324** (with MgI instead of Li) was thus generated *in situ* by the Grignard reagent itself. However, this led to the formation of 1 equivalent of the alkyl sulfide corresponding to the *O*-alkylation of **324** that unfortunately lessens the enantioselectivity of the reaction[144].

The displacement of a substituted primary allylic halide was tackled successfully by each of the research groups of Knochel, Feringa and Alexakis. Knochel and coworkers

TABLE 6. Typical copper-catalyzed allylic substitution of primary substrates (see equation 48) with various ligands and organometallic nucleophiles

Entry	Ligand (mol%)	Cu(I) or Cu(II)	R^1	R^2M	X	**326**:**327**	%ee **326** (%Yield)
1	**315** (14)	CuI	*c*-Hex	*n*-BuMgBr	OAc	100:0	42 (100)
2	**324** (30)	CuI	*c*-Hex	*n*-BuMgBr	OAc	97:3	64 (98)
3	**323b** (10)	$CuBr{\bullet}SMe_2$	Ph	*n*-Pen_2Zn	Cl	98:2	96 (82)
4	**323b** (10)	$CuBr{\bullet}SMe_2$	Ph	R_2Zn^a	Cl	100:0	50 (75)
5	**320b** (2)	$CuBr{\bullet}SMe_2$	Ph	Et_2Zn	Br	84:16	77 (54)
6	**316** (10)	$Cu(OTf)_2{\bullet}C_6H_6$	Ph	Et_2Zn	OP^b	90:10	40 (93)
7	**317** (1)	$CuTc^c$	Ph	EtMgBr	Cl	96:4	82 (97)
8	**320c** (1)	$CuTc^c$	*c*-Hex	EtMgBr	Cl	99:1	91 (82)
9	**320c** (1)	$CuTc^c$	Ph	Et_2Zn	Br	96:4	91 (83)
10	**321** (6)	$CuBr{\bullet}SMe_2$	Ph	MeMgBr	Br	85:15	85 (98)
11	**322** (1)	$CuBr{\bullet}SMe_2$	$BnOCH_2$	*n*-BuMgBr	Br	100:0	94 (93)
12	**331b** (10)	CuCN	*o*-$O_2NC_6H_4$	Et_2Zn	OP^b	—	87 (85)
13	**330a** (10)	$Cu(OTf)_2{\bullet}C_6H_6$	$C_5H_{11}C{\equiv}C$	Et_2Zn	OP^b	82:18	96 (76)

[a] $R_2Zn = [EtO_2C(CH_2)_4]_2Zn$.
[b] OP = $OP(O)(OEt)_2$.
[c] TC = thiophene-2-carboxylate.

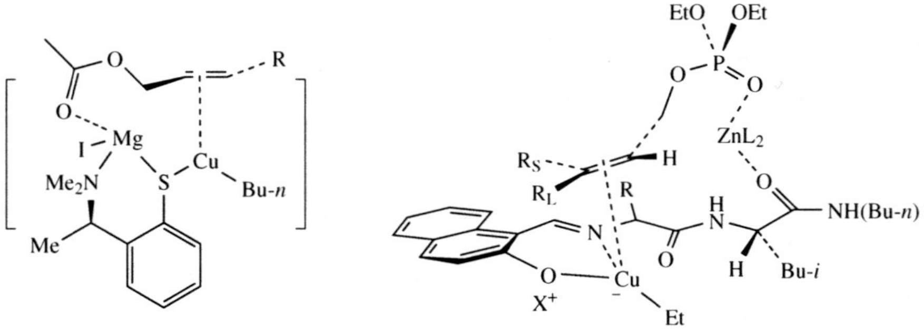

FIGURE 6

utilized allylic chlorides, dialkylzincs, copper bromide and ligand **323a** to effect a highly regioselective but moderately enantioselective (42–83% ee) allylic alkylation. Only bulky dialkylzinc reagents gave satisfactory results[145a]. However, ligand **323b** gave much better results (80–98% ee) and with a larger spectrum of dialkylzinc reagents, including functionalized ones (Table 6, entries 3 and 4)[145b].

Dialkylzinc reagents underwent moderately regioselective and enantioselective (up to 77% ee) copper-catalyzed allylic substitution of cinnamyl bromide when ligand **320b** was used (Table 6, entry 5)[146]. Cinnamyl phosphate was used in conjunction with a copper complex made with ligand **316** to produce **326** in low enantiomeric excess (up to 40%) along with 10–30% of the S_N2 displacement product **327** (Table 6, entry 6)[147]. The more readily available Grignard reagents were added, with good selectivity, to primary chlorides using ligand **320a**, or several analogs, as well as ligand **317** (Table 6, entry 7)[148]. Of all the ligands surveyed, ligand **320c** was superior for the asymmetric allylic alkylation of either Grignard-type nucleophiles or dialkylzinc reagents (Table 6, entries 8 and 9)[149]. Other phosphoramidite and phosphate ligands, containing spirocyclic moieties, met with moderate success on the same kind of allylic halides[150].

The research group of Feringa, who had achieved the copper-catalyzed enantioselective conjugate addition of Grignard reagents onto α,β-unsaturated systems with bis-phosphine ligands **321** and **322**, successfully applied the same ligands to effect the copper-catalyzed enantioselective allylic substitution of Grignard reagents (Table 6, entries 10 and 11)[151]. Taniaphos ligand (**322**) was particularly effective when the solvent was changed to dichloromethane. Its copper complex effected the allylic substitution of Grignard reagents onto a series of allylic bromides (entry 11). The regioselectivity and the enantiomeric excesses were excellent. A detailed mechanistic picture for this system is still not available for the allylic substitution but it is for the 1,4-addition to α,β-unsaturated systems[152].

Hoveyda and coworkers have developed several ligands **330a**, **330b** and **331a–c** for the copper-catalyzed asymmetric allylic alkylation of different organometallics with **325** (Figure 7). Ligands **331a–c**, out of a series of similar ligands, were used to prepare tertiary and quaternary allylic chiral centers from allylic phosphates with good enantioselectivity (Table 6, entry 12)[153]. Surprisingly, while ligands **331b** and **331c** gave products **335a–e** of *S* configuration from the allylic alkylation of diethylzinc with phosphonates **334a–e** (equation 49), the same ligand **331c** gave adduct **337** of *R* configuration with phosphonate **334d** and dialkylzinc **336** (equation 50). Hydrolysis of the adduct resulted in the synthesis

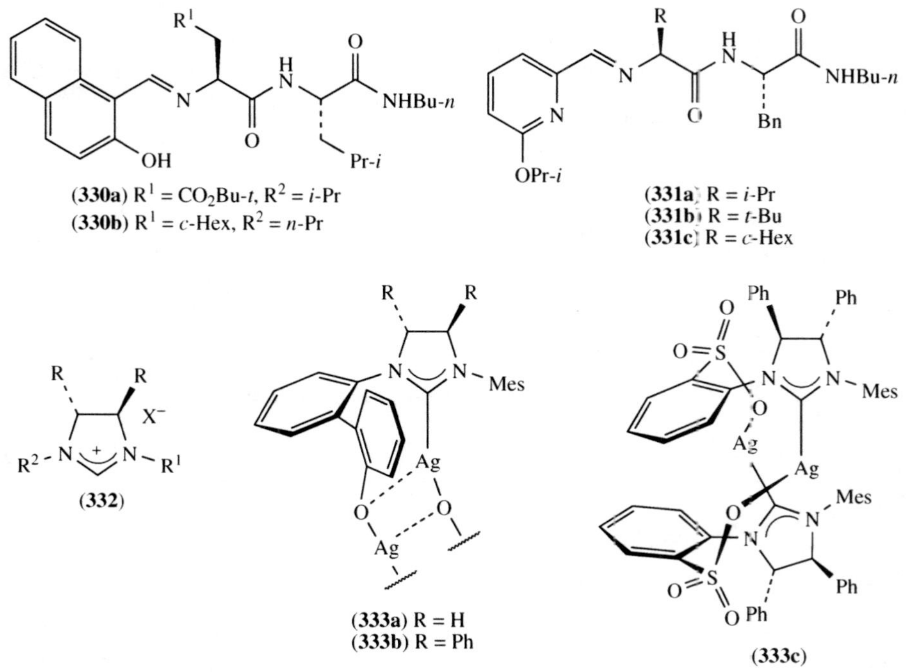

FIGURE 7

of (*R*)-(−)-sporochnol (**337**), a fish deterrent. The saturated analog of **336** added also with the same sense of asymmetry (*R*) as did **336** itself on other allylic substrates.

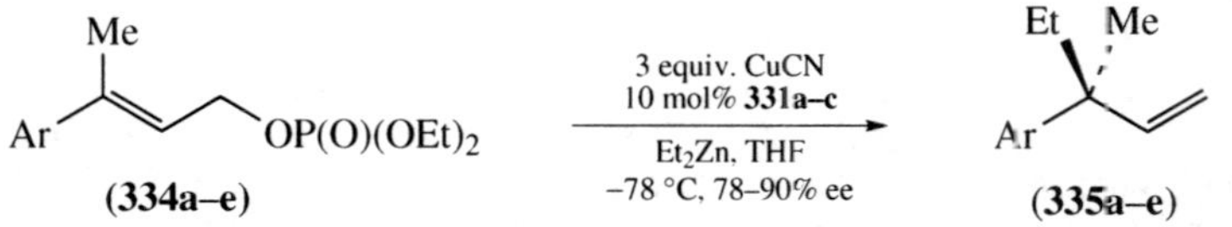

(49)

Ar = XC_6H_4 where X = (**a**) Ph, (**b**) *o*-OMe, (**c**) *p*-NO_2, (**d**) *p*-OTs, (**e**) *p*-CF_3

Ligands **330a** proved superior for a host of di- and trisubstituted phosphates with %ee values in the 78–96% range. The example shown in entry 13 of Table 6 contrasts with the work of Krause and coworkers, who could obtain high S_N2'' selectivity (1,6-addition) with enyne acetates to give ene-allenes (see Section II.D). After a structure–activity relationship study on the ligand, a working hypothesis was put forth to explain the sense and level of asymmetric induction in the reaction. A coordination between the leaving phosphate and an amide carbonyl oxygen as in **329** was implicated and the larger group R_L points away from the naphthyl moiety (Figure 6)[126].

Other members of this series of peptidic ligands, especially ligand **330b**, were capable of inducing enantioselectivity in a copper-catalyzed allylic substitution of dialkylzinc reagents on a γ-phosphonyloxy-α,β-unsaturated ester **338** (Scheme 39). Compound **339** was synthesized by this method using dimethylzinc and, after a cross metathesis of **339** with **340** to give **341**, hydrolysis completed the synthesis of natural (−)-elenic acid (**342**)[154].

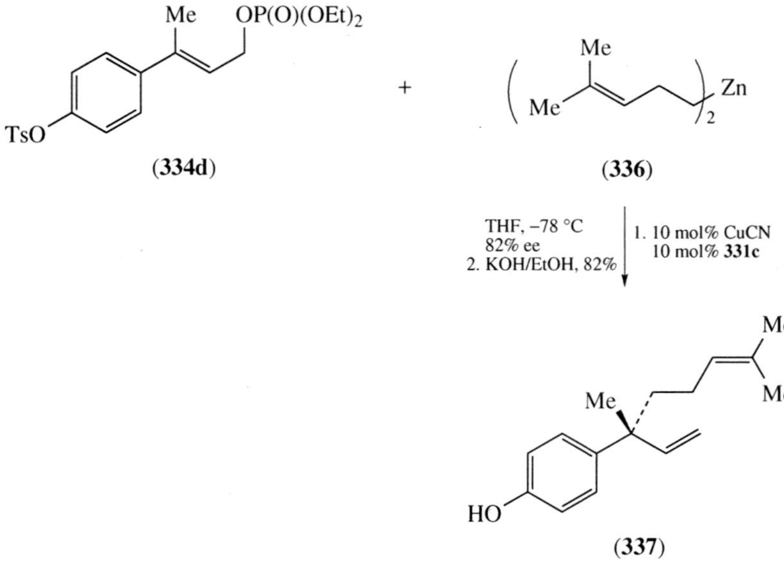

(50)

A series of NHC ligands **332** were used to effect the same transformations of di- and trisubstituted allylic phosphates with %ee values ranging from 71–98 (Figure 7)[155]. Other NHC ligands were reported to effect high S_N2'-selective allylic alkylation of one particular substrate[156]. Silver complexes **333a** and **333b** later allowed extension of the methodology to include 3,3-diaryl-1-phosphonyloxy-2-propenes[157].

b. Substrates possessing a β-substituent. A β-substituent should be detrimental to the diastereoselectivity of any allylic alkylation based on minimization of $A^{1,3}$-strain, i.e. for substrates where the leaving group sits on a secondary carbon (see Section II). That's because the β-substituent's volume approaches that of the vinyl fragment thereby diminishing the energy gap between the two reactive conformations. However, in substrates where the leaving group resides on a primary carbon, this is no longer the case as the stereoselectivity no longer depends on minimization of the allylic strain. β-Substitution allows for the formation of 1,1-disubstituted terminal olefins that should complement, in a synthetic sense, the other products so far discussed.

Enol esters, such as **343**, are compatible substrates with Feringa's reaction conditions (involving ligands **321** and **322**). His research group applied this methodology for the synthesis of chiral non-racemic allylic esters **344** (equation 51)[158]. Regioisomer **345** was only seen in traces in some cases and its formation was generally not problematic. Chiral allylic alcohols and their derivatives such as **344** are very useful building blocks. They can also be made from the enantioselective reduction of the corresponding ketone as well as by a number of different methods[159].

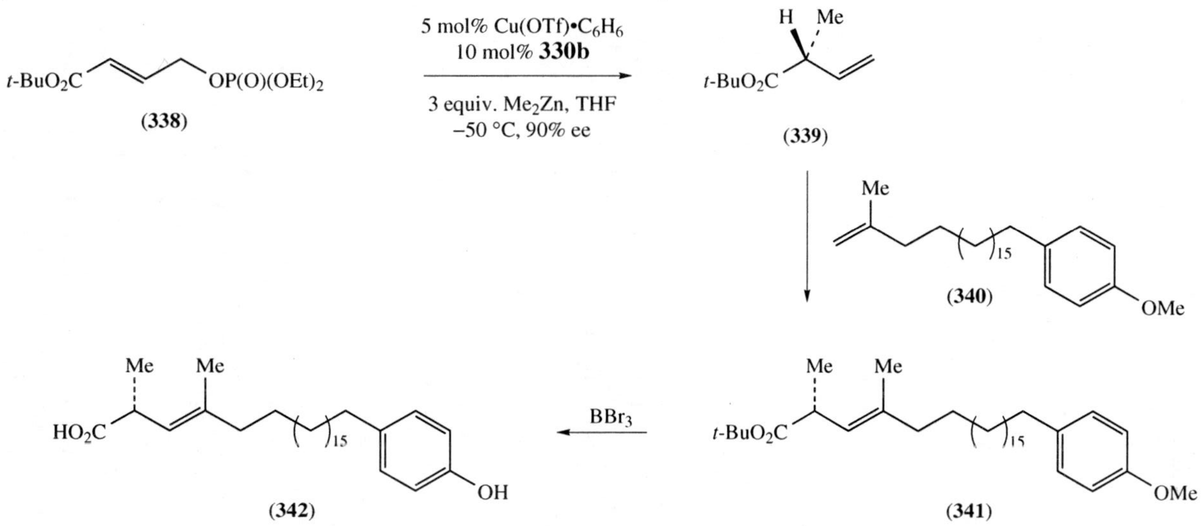

t-BuO2C
OP(O)(OEt)2
(338)
5 mol% Cu(OTf)•C6H6
10 mol% 330b
3 equiv. Me2Zn, THF
−50 °C, 90% ee
H
Me
t-BuO2C
(339)
Me
15
(340)
OMe
Me
Me
t-BuO2C
15
OMe
(341)
BBr3
Me
Me
HO2C
15
OH
(342)

SCHEME 39

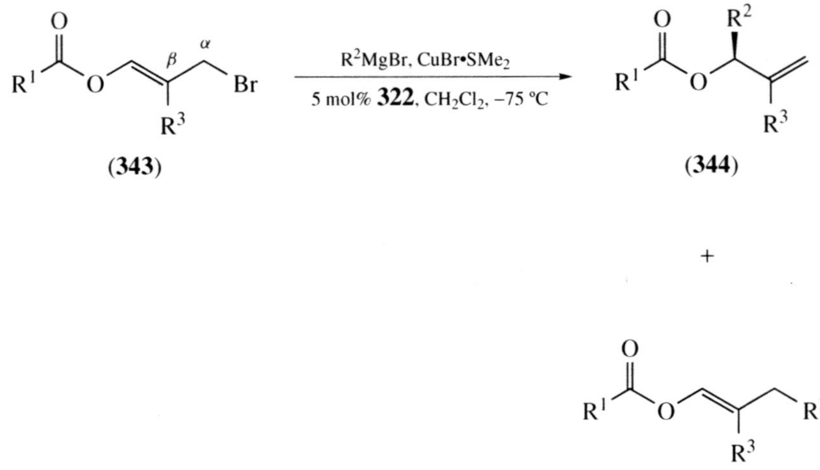

(**343**) (**344**) (**345**) (51)

Only two examples in Feringa's study involved **343** where $R^3 \neq H$. A study of wider scope on substrates that bear a β-substituent, such as **346a–c**, was reported by Alexakis and coworkers using ligands **319** and **320**[160]. After optimization, it was found that 3 mol% of ligand **320a** gave the best results. In equation 52, the ratios of γ:α attack (**347:348**) varied from 83:17 to 92:8 and the enantioselectivity from 96:4 to 98:2. Endocyclic cyclopentenyl- and cyclohexenylmethyl chloride **349** ($n = 1,2$) also gave satisfactory results, the two regioisomeric adducts **350** and **351** being obtained with regioselectivities and enantioselectivities similar to the acyclic cases (equation 53).

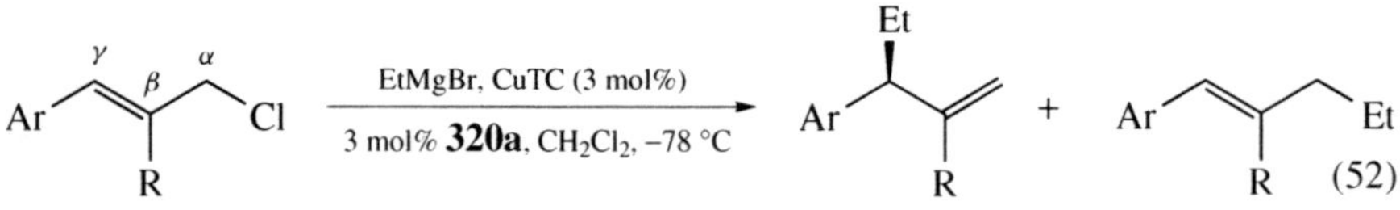

(52)

(**346a**) Ar = p-ClC_6H_4, R = Me
(**346b**) Ar = p-MeC_6H_4, R = Me
(**346c**) Ar = Ph, R = Et

(**347**) (**348**)

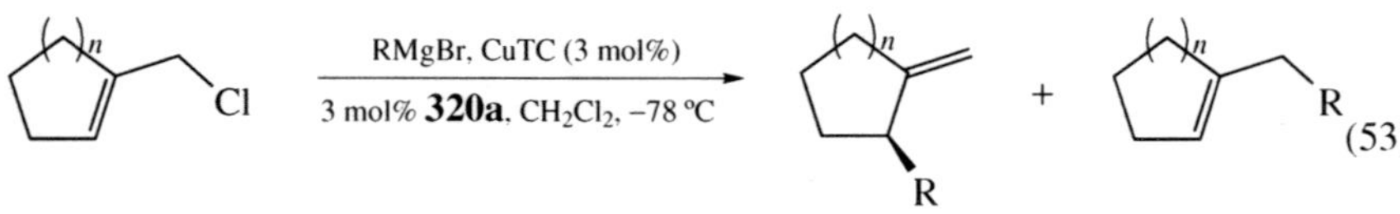

(53)

(**349**) (**350**) (**351**)

Very recently, the copper-catalyzed asymmetric allylic substitution involving a vinylmetal, in the presence of a vinylaluminum, was disclosed[161]. A series of di- or trisubstituted primary allylic phosphates **352** were efficiently converted to the corresponding 1,4-dienes **353** with vinylaluminum reagents made from the hydroalumination of alkynes (equation 54, Table 7). Best results were achieved using a copper complex **333c**

TABLE 7. Copper-catalyzed asymmetric allylic substitution of vinylalu-minum reagents using **352**

Entry	R^1	R^2	R^3	Yield (%)	% ee
1	Ph	Me	*n*-Hex	84	92
2	Ph	Me	$PhCH_2$	85	91
3	$Ph(CH_2)_2$	Me	*n*-Hex	88	77
4	*c*-Hex	H	*n*-Hex	92	86
5	$PhMe_2Si$	H	*n*-Hex	91	93

(Figure 7). Noteworthy is the high enantioselectivity despite the presence of a 2-substituent (R^2) (entries 1–3)

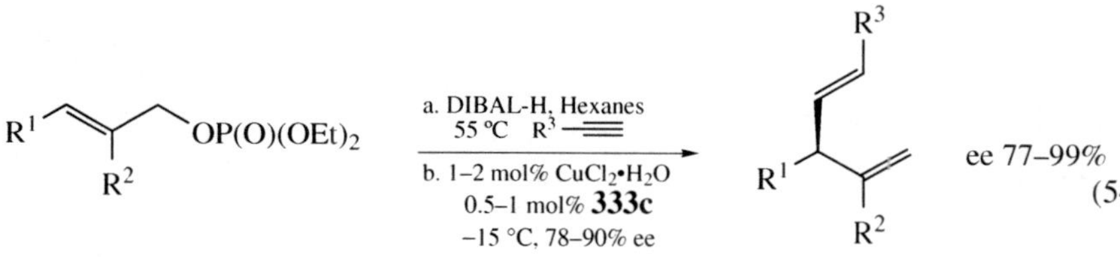

(54)

(352) **(353)**

Allylic substrates possessing an electron-withdrawing group at the β-position undergo exclusively S_N2' displacement because of their electronic bias. In that way, diethylzinc was successfully added to methyl 3-aryl-2-(chloromethyl)acrylates **354** to give **355** (equation 55). Ligand **356** was found to be more effective than several other secondary amines[162]. The formation of EtZnCl was shown to dramatically lower the enantioselectivity of the reaction. Therefore, the use of methylaluminoxane as a scavenger led to a substantial increase in the enantiomeric excess of the products. Earlier experiments used a less effective chiral ligand, a methylthiol derivative of binaphthol[162b].

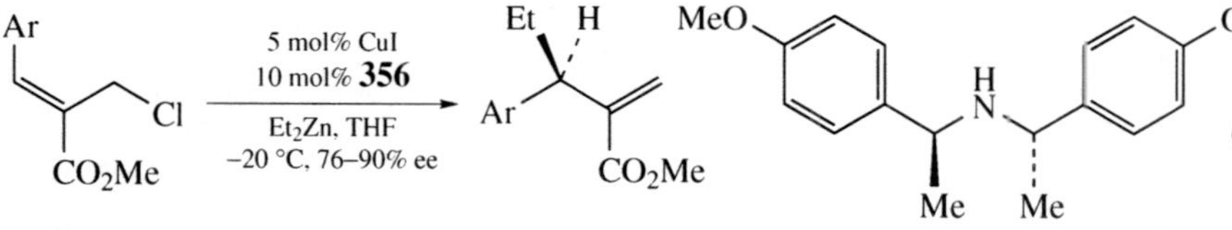

(55)

(354) **(355)** **(356)**

c. Miscellaneous substrates. Notably, bisfunctionalized alkene **357** was made to participate in the copper-catalyzed asymmetric allylic alkylation with phosphoramidate ligand **320a** and its analogs (equation 56)[163]. The corresponding dichloride gave comparable results. Remarkably, the S_N2'-selectivity was complete in every alkylation forming **358** from **357** and the stereoselectivity was generally good. The advantage of using such a difunctionalized substrate is obvious in that the allylic alkylation product is itself a useful electrophilic building block that can be used for further elaboration.

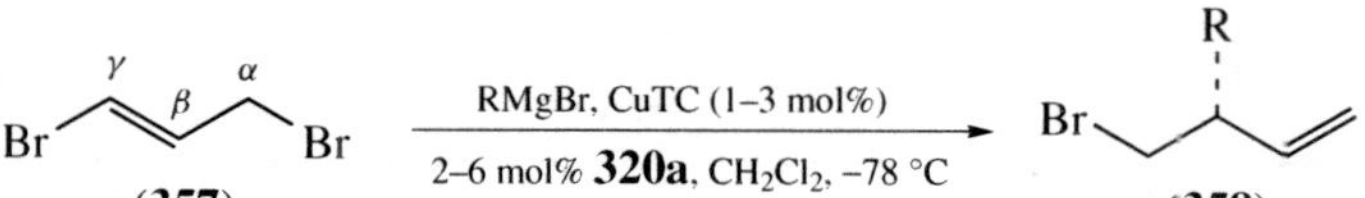

(56)

(357) **(358)**

2. Resolutions and desymmetrizations

a. Allylic and propargylic epoxides. Phosphoramidites (+)-**318** and (+)-**320a** (cf. Figure 5) were used by Feringa and coworkers to effect the kinetic resolution of racemic cyclic allylic epoxides (±)-**181**, (±)-**186** and (±)-**211** with dialkylzinc reagents (equation 57)[164a,b]. Regioselectivity was excellent in most cases, favoring the S_N2' products **213a,f**, **187a,b** and **182a,d** over their S_N2 counterparts **212a,f**, **188a,b** and **183a,d**. Moderate-to-high enantioselectivity was achieved (measured for the S_N2' products), the magnitude of which depended on the ring size of the starting epoxide and on the ligand used (Table 8). Ligand **320a** (entries 2, 3, 6 and 7) gave superior results than ligand **318** (entries 1, 4, 5 and 8). Ligand **320a**, as well as a close derivative, showed limited ability to resolve a propargylic epoxide[164c].

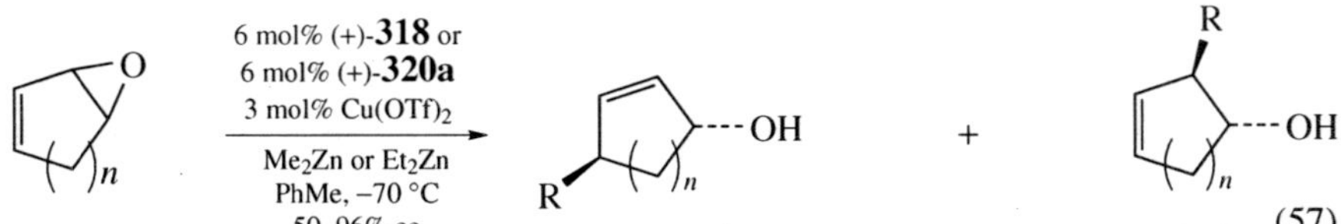

(57)

(±)-(**211**) (n = 1)
(±)-(**186**) (n = 2)
(±)-(**181**) (n = 3)

(−)-(**213a**) (n = 1, R = Me) (−)-(**212a**)
(−)-(**213f**) (n = 1, R = Et) (−)-(**212f**)
(−)-(**187a**) (n = 2, R = Me) (−)-(**188a**)
(−)-(**187b**) (n = 2, R = Me) (−)-(**188b**)
(−)-(**182a**) (n = 3, R = Me) (−)-(**183a**)
(−)-(**182d**) (n = 3, R = Et) (−)-(**183d**)

Equey and Alexakis showed that trialkylaluminums were effective in resolving epoxide **186** using copper triflate and ligand **320a** (equation 58)[165]. Cyclopentadiene oxide (**211**) and cyclooctadiene oxide (not shown) were resolved with excellent enantioselectivities (87 and 93%, respectively) but poor S_N2':S_N2 selectivity. Later, racemic 1,3-cyclohexadiene monoepoxide **186** was resolved using Grignard reagents and a whole slew of ligands, including **319**, **320c**, **321** and **322**, and many of their derivatives. The regioselectivity was excellent and the enantioselectivity was moderate-to-good[166].

Better enantiomeric excesses were obtained for the kinetic resolution of *exo*-methylene cyclohexene oxides (±)-**359** using ligands (S,S,S)-**320a** or (S,R,R)-**320b** (equation 59)[167]. Good enantioselectivities (88% ee of **360a**) was obtained for 50% (±5%) conversions of (±)-**359**. Since only 50% of the desired products (−)-**360a** could be acquired from a

TABLE 8. Resolution of cycloalkadiene monoepoxides **181**, **186**, **211** by copper-catalyzed asymmetric allylic alkylation of dialkyl zinc reagents

Entry	Epoxide	R_2Zn (0.5 equiv.)	Cat	Yield (%)^a, S_N2':S_N2	% ee of S_N2' (**182**, **187**, **213**)
1	**211**	Me	**318**	10, 4:1	46
2	**211**	Me	**320a**	12, 3:1	50
3	**211**	Et	**320a**	8, 12:1	54
4	**186**	Me	**318**	30, 13:1	62
5	**186**	Et	**318**	18, >20:<1	86
6	**186**	Me	**320a**	33, 13:1	92
7	**186**	Et	**320a**	32, 59:1	91
8	**181**	Me	**318**	32, 10:1	60
9	**181**	Me	**320a**	38, 16:1	96

^{a}Maximum yield = 50%.

kinetic resolution, the desymmetrization of bis-*exo*-methylene cycloalkene oxides **362** was also explored using the same ligands (*S*,*S*,*S*)-**320a** or (*S*,*R*,*R*)-**320b** (equation 60) with good success, giving the desired adduct (−)-**363** in 97% ee and only 2% of the regioisomer (−)-**364**[167].

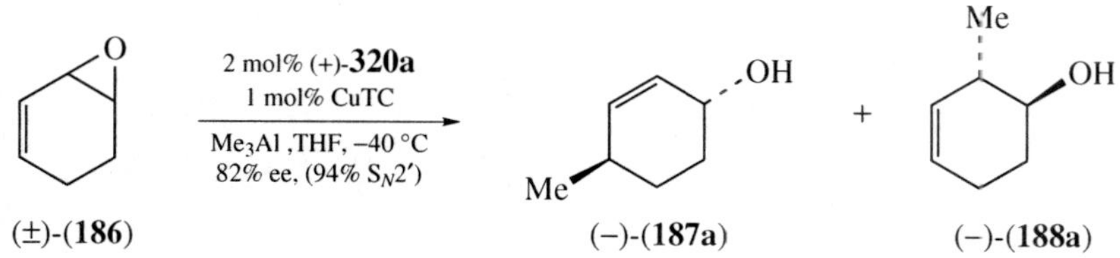

(58)

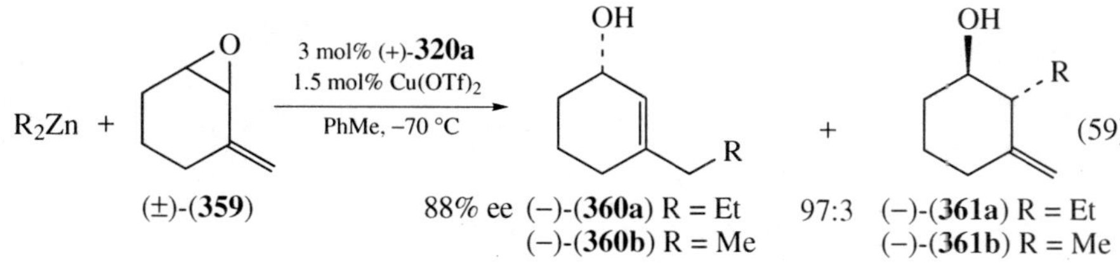

(59)

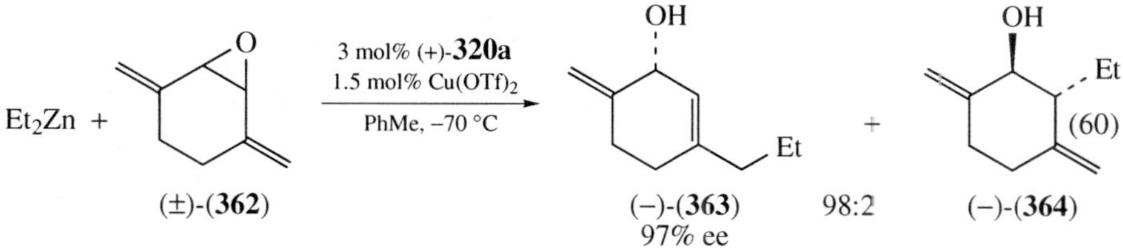

(60)

Later, it was found that one enantiomer of epoxide **359** reacted faster to give the major S_N2' displacement product (−)-**360** while the other enantiomer of the same epoxide reacted faster to give the S_N2 product (−)-**361** with the same copper–ligand complex[168]. For example, (±)-**359** reacted with 1.5 equivalent of Me_2Zn under the same reaction conditions as shown in equation 59 (except that *ent*-**320a** was used) to give a 55:45 ratio of *ent*-**360b** (96% ee) and *ent*-**361b** (92% ee). So, an excess of reagent converted 100% of the starting epoxide into two different and useful enantiomerically pure products. Epoxide (±)-**362** suffered the same regiodivergent kinetic resolution with slightly less success.

This is an interesting phenomenon, mechanistically speaking. That the chiral recognition process dictated the regiochemical outcome was demonstrated by reacting (±)-**359** in the presence of the racemic catalyst made of (±)-**320a**. Under these conditions the S_N2' product was predominant (98:2 ratio of **360:361**). The authors propose that the initially generated π-complexes **365** and **366** transform into the corresponding σ-complexes **368** and **369** (Scheme 40). In the case of a matched pair between epoxide and ligand enantiomers (e.g. **368**), reductive coupling to give **371**, and then **360**, would be faster than the equilibrium between **368** and **367** while for the other mismatched pair (e.g. **369**), reductive coupling to **372** would be slower than the equilibrium between **369** and **370**. Reductive coupling of **370** would lead to the S_N2 product **361**. However, another possibility is that the enantiodifferentiating step be the formation of the σ-complex **368** from the π-complex **365** and σ-complex **370** from π-complex **366**. In other words, that $k_1 > k_3 > k_2$ (Scheme 40).

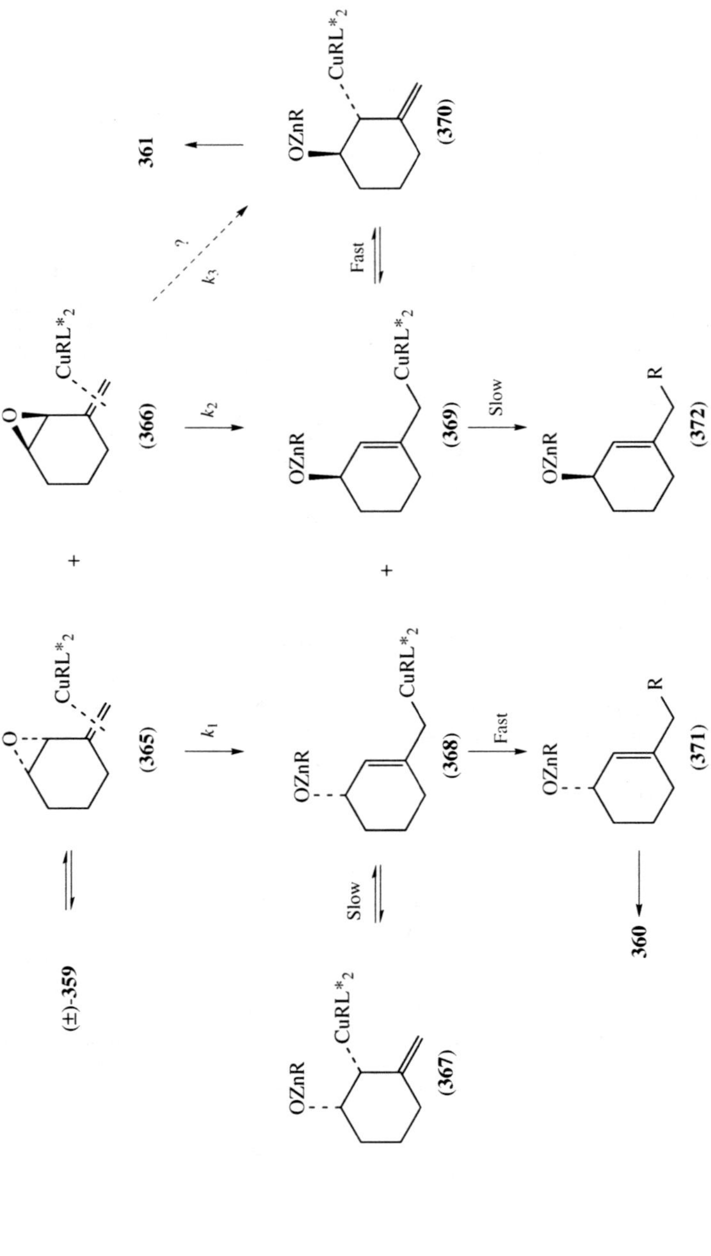

SCHEME 40

b. Miscellaneous substrates. The phosphoramidite ligand **320b**, and one other analog, were useful in desymmetrizing oxabicyclo compounds like **373** (equation 61)[169]. The ligand was responsible for the enantioselective production of (−)-**374** and also for the diastereoselectivity (**374** vs **375**) as the copper-catalyzed reaction of Et_2Zn (R = Et) *without the ligand* led predominantly to racemic *syn* product **375**. Also, the more reactive $(i\text{-}Pr)_2Zn$ led to a higher proportion of (±)-**375** (R = *i*-Pr) than Et_2Zn, presumably from the $Cu(OTf)_2$- or $Zn(OTf)_2$-catalyzed reaction.

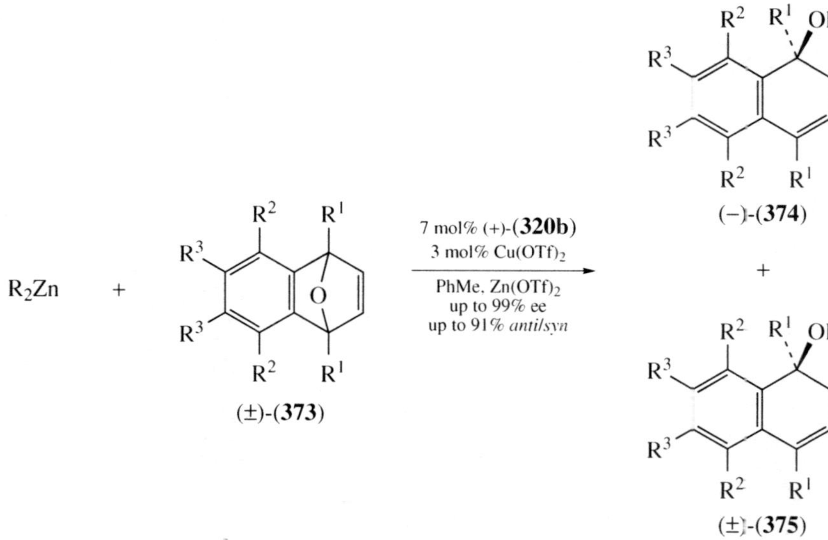

(61)

Gennari and coworkers screened a number of chiral ligands of the type **316** to achieve the copper-catalyzed desymmetrization of cyclic bis-phosphates **376** (equation 62, n = 1–3) with diethylzinc with some success (up to 88% ee)[170]. A joint effort by the research groups of Gennari and Feringa culminated in a highly selective method to desymmetrize cyclic bis-phosphates **376** using the previously described ligand **320b** as well as a series of analogous ligands. The loading of the ligand **320b** was significant, being at 20%, but the enantiomeric excesses and diastereomeric purity of the products **377a–c** were high (up to 99% ee) while the diastereoselectivity (**377a–c** vs **378a–c**) was complete for **376a** and **376c** but moderate in the case of **376b**[171].

Symmetrical diazines and other hetero-norbornene derivatives **379** are readily available from the cycloaddition of cyclopentadiene with various dienophiles. They were desymmetrized using phosphoramidate *ent*-**320a**[172a,b], trialkylaluminums and a catalytic amount of copper triflate (equation 63). However, the mechanistic picture is complicated by the fact that the ligand was shown to react with the trialkylaluminum reagent to give a different complex[172c]. This casts doubt as to the exact nature of the catalyst and, in addition, it causes the loss of some of the ligand. The cyclic amines **380** were nonetheless isolated in good yield if only in moderate enantiomeric purity (equation 63).

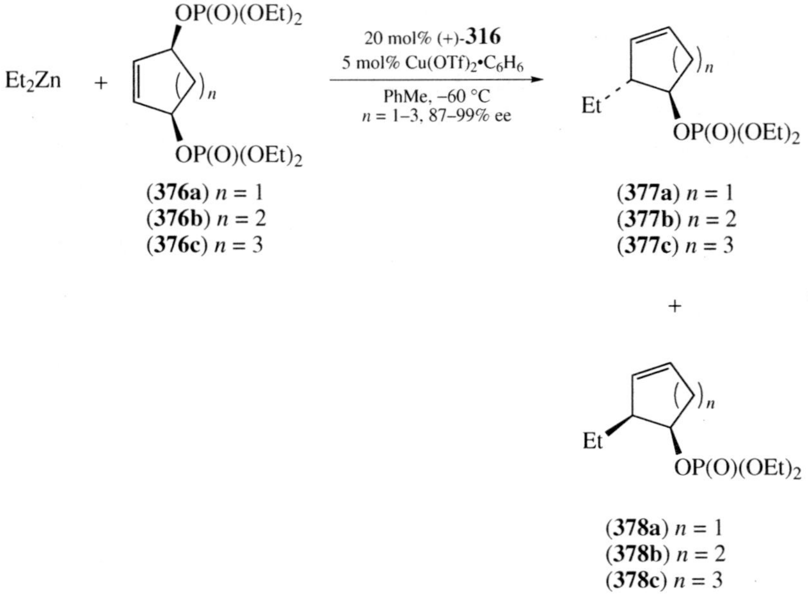

(62)

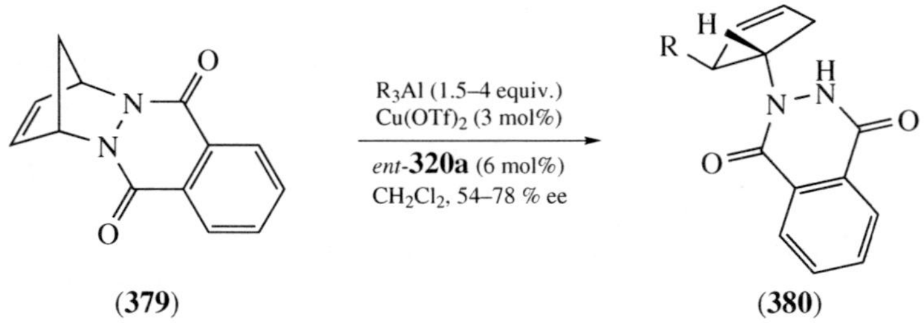

(63)

C. Dynamic Kinetic Asymmetric Transformation

Bäckvall and coworkers laid down the ground work for the possibility of performing dynamic kinetic asymmetric transformation (DYKAT) of racemic allylic acetates. Starting from enantiomerically pure allylic acetates **381** (X = OAc), they were able to slow the reductive coupling of the initially formed σ copper complex **384-A** by lowering the temperature, choosing the right solvent and copper catalyst, and adding DMAP or LiOAc (Scheme 41). An equilibrium between the two π-complexes **383-A** and **383-B** (presumably via the two σ-complexes **384-A** and **384-B**) was thus established and they collapsed to a racemic mixture of product (±)-**382**. An electron-deficient substrate (Ar = *p*-FC_6H_4) was more prone to lose chiral information this way than electron-rich substrates (Ar = *p*-$MeOC_6H_4$ or Ar = Me). Demonstrating the loss of chiral information from substrates like **381** is an essential step before a complete DYKAT with a chiral ligand can be performed[173]. DYKATs are more useful than kinetic resolutions to prepare enantiomerically pure compounds because, in principle, 100% of the racemic starting material can be transformed into the desired product.

As an interesting alternative, Bäckvall and coworkers performed an enzyme-catalyzed DYKAT of racemic allylic alcohols ((±)-**381**, X = OH) to non-racemic allylic acetates

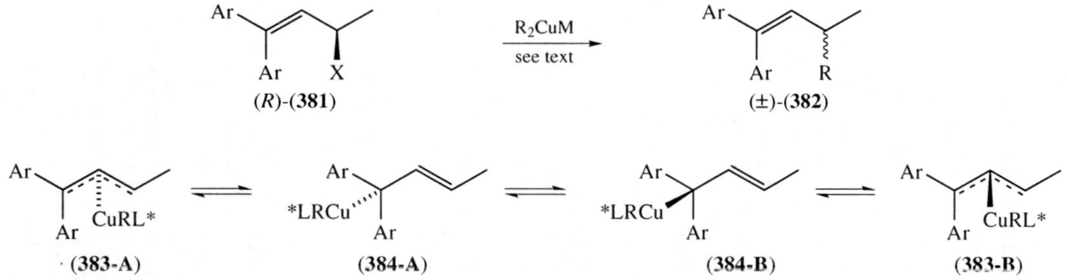
Ar
Ar
X
(R)-(381)
R2CuM
see text
Ar
Ar
R
(±)-(382)
Ar
Ar
CuRL*
(383-A)
Ar
*LRCu
Ar
(384-A)
Ar
*LRCu
Ar
(384-B)
Ar
Ar
CuRL*
(383-B)

SCHEME 41

(+)-**381** or (−)-**381** (X = OAc). Once the allylic acetates were obtained, a copper-catalyzed allylic substitution afforded the desired enantioenriched product **382** under reaction conditions that did not favor equilibration of the σ- and π-complexes[174].

V. CONCLUSIONS AND OUTLOOK

The asymmetric allylic alkylation is a valuable transformation in synthetic chemistry because it combines a good degree of flexibility in the number of different structures obtainable by it (S_N2, S_N2', E, Z etc.) with a high degree of predictability of its regio- and stereochemical outcome. As a result of the discovery of ever more selective organocopper reagents and reaction conditions, its appeal has increased dramatically in the last decade. Moreover, the alkylation with hard nucleophiles is not a trivial undertaking as many side reactions are possible and frequently occur, such as elimination reactions and competing additions to other reactive sites. Yet, the copper-mediated allylic alkylation is remarkably chemoselective and there are many examples of its use on fairly complex substrates.

The diastereospecific allylic alkylation is better understood and continues to be used elegantly in total syntheses. The survey of other leaving groups, copper reagents and additives may yet lead to even higher regio- and stereocontrol of the reaction. The development of reaction conditions allowing for the preparation of enantiomerically pure quaternary stereocenters is a more recent and worthwhile development. While several reactions have been used in tandem with an allylic alkylation, we have not seen as yet the use of two or more allylic alkylations from a single substrate.

Large strides have been made in the realm of enantioselective allylic alkylation in the last decade. Nonetheless, the substrate scope still needs to be widened and more catalytic systems need to be developed that combine high enantioselectivity with high regioselectivity on a wide range of substrates. To achieve this, a clearer picture of the reaction mechanism (or mechanisms) is required, especially as they relate to the existence and interrelation between σ- and π-complexes. What's more, not enough information is known on the chiral ligands themselves and their mode of action. Nonetheless, the copper-catalyzed asymmetric allylic alkylation is well on its way to becoming a leading reaction for the stereoselective formation of carbon–carbon bonds.

VI. REFERENCES AND NOTES

1. (a) B. Breit and P. Demel, in *Modern Organocopper Chemistry* (Ed. N. Krause), Wiley-VCH, Weinheim, 2002, pp. 210–223.
   (b) A. S. E. Karlström and J.-E. Bäckvall, in *Modern Organocopper Chemistry* (Ed. N. Krause), Wiley-VCH, Weinheim, 2002, pp. 261–288.
   (c) N. Krause and A. Gerold, *Angew. Chem., Int. Ed. Engl.*, **36**, 186 (1997).
   (d) B. H. Lipshutz and S. Sengupta, in *Organic Reactions* (Ed. L. Paquette), John Wiley & Sons, Inc., New York, 1992, p. 135.
   (e) T. Ibuka and Y. Yamamoto, in *Organocopper Reagents* (Ed. R. J. K. Taylor), Oxford University Press, London, 1994, pp. 143–158.
   (f) Y. Yamamoto, *Angew. Chem., Int. Ed. Engl.*, **25**, 947 (1986).
   (g) E. Nakamura, *Synlett*, 539 (1991).
   (h) P. Wipf, *Synthesis*, 537 (1993).
   (i) Y. Yamamoto, *Houben-Weyl, Methods for Organic Synthesis*, Vol. E21b, 1995, pp. 2011–2040.
   (j) R. M. Magid, *Tetrahedron*, **36**, 1901 (1980).
2. For recent reviews that include the subject of copper-catalyzed allylic alkylation see:
   (a) H. Yorimitsu and K. Oshima, *Angew. Chem., Int. Ed.*, **44**, 4435 (2005).
   (b) K. Geurts, S. P. Fletcher, A. W. van Zijl, A. J. Minnaard and B. L. Feringa, *Pure Appl. Chem.*, **80**, 1025 (2008).

(c) A. Alexakis, C. Malan, L. Lea, K. Tissot-Croset, D. Polet and C. Falciola, *Chimia*, **60**, 124 (2006).
(d) A. Kar and N. P. Argade, *Synthesis*, 2995 (2005).
(e) S. Woodward, *Angew. Chem., Int. Ed.*, **44**, 5560 (2005).
(f) M. Pineschi, *New J. Chem.*, **28**, 657 (2004).
(g) J. Christoffers and A. Mann, *Angew. Chem., Int. Ed.*, **40**, 4591 (2001).
(h) J. Christoffers and A. Baro, *Adv. Synth. Catal.*, **347**, 1473 (2005).

3. (a) P. Rona and P. Crabbé, *J. Am. Chem. Soc.*, **90**, 4733 (1968).
(b) P. Rona and P. Crabbé, *J. Am. Chem. Soc.*, **91**, 3289 (1969).
4. (a) P. Rona, L. Tökes, J. Tremble and P. Crabbé, *J. Chem. Soc., Chem. Commun.*, 43 (1969).
(b) P. Crabbé, J.-M. Dollat, J. Gallina, J.-L. Luche, E. Velarde, M. L. Maddox and L. Tökès, *J. Chem. Soc., Perkin Trans. 1*, 730 (1978).
5. R. J. Anderson, *J. Am. Chem. Soc.*, **92**, 4978 (1970).
6. For selected examples, see: (a) Reference 4b.
(b) S. D. Burke, A. D. Piscopio and M. E. Kort, *Tetrahedron Lett.*, **32**, 855 (1991).
(c) C. Spino and G. Liu, *J. Org. Chem.*, **58**, 817 (1993).
(d) S. Perrone and P. Knochel, *Org. Lett.*, **9**, 1041 (2007).
7. J. Staroscik and B. Rickborn, *J. Am. Chem. Soc.*, **93**, 3046 (1971).
8. H. L. Goering and V. D. Singleton, Jr., *J. Am. Chem. Soc.*, **98**, 7854 (1976).
9. (a) T. L. Underiner, S. D. Paisley, J. Schmitter, L. Lesheski and H. L. Goering, *J. Org. Chem.*, **54**, 2369 (1989).
(b) T. L. Underiner and H. L. Goering, *J. Org. Chem.*, **53**, 1140 (1988).
(c) H. L. Goering, S. S. Kantner and E. P. Seitz, Jr., *J. Org. Chem.*, **50**, 5495 (1985).
(d) H. L. Goering and S. S. Kantner, *J. Org. Chem.*, **49**, 422 (1984).
(e) H. L. Goering and C. C. Tseng, *J. Org. Chem.*, **48**, 3986 (1983).
(f) H. L. Goering and V. D. Singleton, Jr., *J. Org. Chem.*, **48**, 1531 (1983).
(g) H. L. Goering and S. S. Kantner, *J. Org. Chem.*, **48**, 721 (1983).
(h) H. L. Goering and S. S. Kantner, *J. Org. Chem.*, **46**, 2144 (1981).
(i) C. C. Tseng, S. J. Yen and H. L. Goering, *J. Org. Chem.*, **51**, 2892 (1986).
10. (a) A. S. E. Karlström and J.-E. Bäckvall, *Chem. Eur. J.*, **7**, 1981 (2001).
(b) E. S. M. Persson, M. van Klaveren, D. M. Grove, J.-E. Bäckvall and G. van Koten, *Chem. Eur. J.*, **1**, 351 (1995).
(c) J.-E. Bäckvall, E. S. M. Persson and A. Bombrun, *J. Org. Chem.*, **59**, 4126 (1994).
(d) J.-E. Bäckvall, M. Sellén and B. Grant, *J. Am. Chem. Soc.*, **112**, 6515 (1990).
11. (a) E. Nakamura and S. Mori, *Angew. Chem., Int. Ed.*, **39**, 3750 (2000).
(b) M. Yamanaka, S. Kato and E. Nakamura, *J. Am. Chem. Soc.*, **126**, 6287 (2004).
(c) J. Norinder, J.-E. Bäckvall, N. Yoshikai and E. Nakamura, *Organometallics*, **25**, 2129 (2006).
(d) N. Yoshikai, S.-L. Zhang and E. Nakamura, *J. Am. Chem. Soc.*, **130**, 12862 (2008).
12. (a) R. A. J. Smith and A. S. Vellekoop, in *Advances in Detailed Reaction Mechanisms* (Ed. J. M. Coxon), JAI, Greenville, 1994, pp. 79–130.
(b) M. F. Mechelke and D. F. Wiemer, *J. Org. Chem.*, **64**, 4821 (1999).
(c) A. E. Dorigo, J. Wanner and P. v. R. Schleyer, *Angew. Chem., Int. Ed. Engl.*, **34**, 476 (1995).
13. E. Nakamura, K. Sekiya, M. Arai and S. Aoki, *J. Am. Chem. Soc.*, **111**, 3091 (1989).
14. M. van Klaveren, E. S. M. Persson, A. del Villar, D. M. Grove, J.-E. Bäckvall and G. van Koten, *Tetrahedron Lett.*, **36**, 3059 (1995).
15. For examples that each include at least one enumerated functional group, see.
(a) Ref. 19 (ketones); Ref. 27a (ketones and nitriles).
(b) Ref. 60a (esters, lactones).
(c) Ref. 31e (ethers, acetals).
(d) Ref. 107 (alcohol).
(e) Ref. 132 (amide).
(f) Ref. 102a (nitrile).
(g) Ref. 157 (nitro).
(h) Ref. 117 (azides, acetals).
(i) S. Perreault and C. Spino, *Org. Lett.*, **8**, 4385 (2006) (alcohol, dithianes).

16. (a) T. Ibuka, T. Nakao, S. Nishii and Y. Yamamoto, *J. Am. Chem. Soc.*, **108**, 7420 (1986).
(b) T. Ibuka and Y. Yamamoto, *Synlett*, 769 (1992) and references cited therein.
17. (a) T. Ibuka, N. Akimoto, M. Tanaka, S. Nishii and Y. Yamamoto, *J. Org. Chem.*, **54**, 4055 (1989).
(b) T. Ibuka, M. Tanaka, S. Nishii and Y. Yamamoto, *J. Am. Chem. Soc.*, **111**, 4864 (1989).
18. M. Arai, B. H. Lipshutz and E. Nakamura, *Tetrahedron*, **48**, 5709 (1992).
19. See:
(a) K. Sekiya and E. Nakamura, *Tetrahedron Lett.*, **29**, 5155 (1988).
(b) A. Yanagisawa, Y. Noritake, N. Nomura and H. Yamamoto, *Synlett*, 251 (1991).
20. V. Caló, C. De Nitti, L. Lopez and A. Scilimati, *Tetrahedron*, **48**, 6051 (1992).
21. (a) J. Levisalles, M. Rudler-Chauvin and H. Rudler, *J. Organomet. Chem.*, **136**, 103 (1977).
(b) H. L. Goering, E. P. Seitz, Jr. and C. C. Tseng, *J. Org. Chem.*, **46**, 5304 (1981).
(c) T. L. Underiner and H. L. Goering, *J. Org. Chem.*, **56**, 2563 (1991).
22. (a) J. K. Kochi, *Organometallic Mechanisms and Catalysis,* Academic Press, New York, 1978, p. 381.
(b) M. Tamura and J. Kochi, *Synthesis*, 303 (1971).
23. (a) K. Maruyama and Y. Yamamoto, *J. Am. Chem. Soc.*, **99**, 8068 (1977).
(b) Y. Yamamoto, S. Yamamoto, H. Yatagai and K. Maruyama, *J. Am. Chem. Soc.*, **102**, 2318 (1980).
24. For a review on the use of Lewis acids with organocopper reagents see: Y. Yamamoto, *Angew. Chem., Int. Ed. Engl.*, **25**, 947 (1986).
25. (a) J.-P. Gorlier, L. Hamon, J. Levisalles and J. Wagnon, *J. Chem. Soc., Chem. Commun.*, 88 (1973).
(b) B. H. Lipshutz, R. S. Wilhelm and D. M. Floyd, *J. Am. Chem. Soc.*, **103**, 7672 (1981).
26. (a) B. M. Trost and T. P. Klun, *J. Org. Chem.*, **45**, 4256 (1980).
(b) C. C. Tseng, S. D. Paisley and H. L. Goering, *J. Org. Chem.*, **51**, 2884 (1986).
(c) M. Yus and J. Gomis, *Eur. J. Org. Chem.*, 2043 (2003).
27. (a) Y. Yamamoto, Y. Chounan, M. Tanaka and T. Ibuka, *J. Org. Chem.*, **57**, 1024 (1992).
(b) N. Fujii, K. Nakai, H. Habashita, H. Yoshizawa, T. Ibuka, F. Garrido, A. Mann, Y. Chounan and Y. Yamamoto, *Tetrahedron Lett.*, **34**, 4227 (1993).
28. E. Nakamura, K. Sekiya, M. Arai and S. Aoki, *J. Am. Chem. Soc.*, **111**, 3091 (1989).
29. G. L. van Mourik and H. J. J. Pabon, *Tetrahedron Lett.*, 2705 (1978).
30. G. Fouquet and M. Schlosser, *Angew. Chem., Int. Ed. Engl.*, **13**, 82 (1974).
31. (a) P. Barsanti, V. Caló, L. Lopez, G. Marchese, F. Naso and G. Pesce, *J. Chem. Soc., Chem. Commun.*, 1085 (1978).
(b) V. Caló, L. Lopez, G. Pesce and A. Calianno, *J. Org. Chem.*, **47**, 4482 (1982).
(c) V. Caló, L. Lopez and W. F. Carlucci, *J. Chem. Soc., Perkin Trans. 1*, 2953 (1983).
(d) S. Valverde, M. Bernabé, S. Garcia-Ochoa and A. M. Gómez, *J. Org. Chem.*, **55**, 2294 (1990).
(e) S. Valverde, M. Bernabé, A. M. Gómez and P. Puebla, *J. Org. Chem.*, **57**, 4546 (1992).
32. (a) J.-E. Bäckvall, M. Sellén and B. Grant, *J. Am. Chem. Soc.*, **112**, 6615 (1990).
(b) M. van Klaveren, E. S. M. Persson, D. M. Grove, J.-E. Bäckvall and G. van Koten, *Tetrahedron Lett.*, **35**, 5931 (1994).
(c) J.-E. Bäckvall and M. Sellén, *J. Chem. Soc., Chem. Commun.*, 827 (1987).
(d) E. S. M. Persson and J.-E. Bäckvall, *Acta Chem. Scand.*, **49**, 899 (1995).
33. (a) E. J. Corey and N. W. Boaz, *Tetrahedron Lett.*, **25**, 3063 (1984).
(b) L. Hamon and J. Levisalles, *Tetrahedron*, **45**, 489 (1989).
34. For discussions of alkene geometry in the products of allylic alkylation, see References 4b, 9a, 9d, 9e, 9g, 9h, 22a and
(a) H. L. Goering and C. C. Tseng, *J. Org. Chem.*, **50**, 1597 (1985).
(b) T. L. Underiner and H. L. Goering, *J. Org. Chem.*, **55**, 2757 (1990).
35. (a) J. H. Smitrovich and K. A. Woerpel, *J. Am. Chem. Soc.*, **120**, 12998 (1998).
(b) J. H. Smitrovich and K. A. Woerpel, *J. Org. Chem.*, **65**, 1601 (2000).
(c) Z.-H. Peng and K. A. Woerpel, *Org. Lett.*, **3**, 675 (2001).
(d) Z.-H. Peng and K. A. Woerpel, *Org. Lett.*, **2**, 1379 (2000).
36. (a) C. Gallina and P. G. Ciattini, *J. Am. Chem. Soc.*, **101**, 1035 (1979).
(b) C. Gallina, *Tetrahedron Lett.*, **23**, 3093 (1982).

37. See, for example, I. Fleming, D. Higgins, N. J. Lawrence and A. P. Thomas, *J. Chem. Soc., Perkin Trans. 1*, 3331 (1992).
38. (a) P. Demel, M. Keller and B. Breit, *Chem. Eur. J.*, **12**, 6669 (2006).
(b) B. Breit, P. Demel and C. Studte, *Angew. Chem., Int. Ed.*, **43**, 3785 (2004).
(c) B. Breit and C. Herber, *Angew. Chem., Int. Ed.*, **43**, 3790 (2004).
(d) B. Breit and P. Demel, *Adv. Synth. Catal.*, **343**, 429 (2001).
39. B. M. Trost and T. P. Klun, *J. Org. Chem.*, **45**, 4256 (1980).
40. (a) T. Ibuka, T. Nakao, S. Nishii and Y. Yamamoto, *J. Am. Chem. Soc.*, **108**, 7420 (1986).
(b) T. Ibuka, T. Taga, H. Habashita, K. Nakai, H. Tamamura, N. Fujii, Y. Chounan, H. Nemoto and Y. Yamamoto, *J. Org. Chem.*, **58**, 1207 (1993).
41. (a) T. Ibuka, T. Aoyagi and Y. Yamamoto, *Chem. Pharm. Bull.*, **34**, 2417 (1986).
(b) T. Ibuka, T. Aoyagi, K. Kitada, F. Yoneda and Y. Yamamoto, *J. Organomet. Chem.*, **287**, C18 (1985).
(c) T. Ibuka, T. Aoyagi and F. Yoneda, *J. Chem. Soc., Chem. Commun.*, 1452 (1985).
(d) C. Sahlberg and A. Claesson, *J. Org. Chem.*, **49**, 4120 (1984).
42. (a) T. Ibuka, H. Yoshizawa, H. Habashita, N. Fujii, Y. Chounan, M. Tanaka and Y. Yamamoto, *Tetrahedron Lett.*, **33**, 3783 (1992).
(b) A. Otaka, F. Katagiri, T. Kinoshita, Y. Odagaki, S. Oishi, H. Tamamura, N. Hamanaka and N. Fujii, *J. Org. Chem.*, **67**, 6152 (2002).
43. H. Yang, X. C. Sheng, E. M. Harrington, K. Ackermann, A. M. Garcia and M. D. Lewis, *J. Org. Chem.*, **64**, 242 (1999).
44. J. L. Belelie and J. M. Chong, *J. Org. Chem.*, **66**, 5552 (2001).
45. (a) R. M. Hanson and K. B. Sharpless, *J. Org. Chem.*, **51**, 1922 (1986).
(b) Y. Gao, J. M. Klunder, R. M. Hanson, H. Masamune, S. Y. Ko and K. B. Sharpless, *J. Am. Chem. Soc.*, **109**, 5765 (1987).
(c) Z.-M. Wang and W.-S. Zhou, *Tetrahedron*, **43**, 2935 (1987).
(d) B. E. Rossiter, in *Chiral Catalysis, Asymmetric Synthesis* (Ed. J. D. Morrison), Vol. 5, Academic Press, Orlando, 1985, pp. 193–246.
(e) R. A. Johnson and K. B. Sharpless, in *Catalytic Asymmetric Synthesis* (Ed. I. Ojima), VCH Publishers, New York, 1993, p 103.
46. (a) N. Harrington-Frost, H. Leuser, M. I. Calaza, F. F. Kneisel and P. Knochel, *Org. Lett.*, **5**, 2111 (2003).
(b) H. Leuser, S. Perrone, F. Liron, F. F. Kneisel and P. Knochel, *Angew. Chem., Int. Ed.*, **44**, 4627 (2005).
47. Y. Kiyotsuka, H. P. Acharya, Y. Katayama, T. Hyodo and Y. Kobayashi, *Org. Lett.*, **10**, 1719 (2008).
48. M. Kimura, T. Yamazaki, T. Kitazume and T. Kubota, *Org. Lett.*, **6**, 4651 (2004).
49. (a) J. P. Marino, A. Viso, R. Fernandez de la Pradilla and P. Fernandez, *J. Org. Chem.*, **56**, 1349 (1991).
(b) J. P. Marino, A. Viso, J.-D. Lee, R. Fernandez de la Pradilla, P. Fernandez and M. B. Rubio, *J. Org. Chem.*, **62**, 645 (1997).
50. (a) C. Spino, M.-C. Granger and M.-C. Tremblay, *Org. Lett.*, **4**, 4735 (2002).
(b) C. Spino, M.-C. Granger, L. Boisvert and C. Beaulieu, *Tetrahedron Lett.*, **43**, 4183 (2002).
51. (a) C. Spino and C. Beaulieu, *J. Am. Chem. Soc.*, **120**, 11832 (1998).
(b) C. Spino, C. Beaulieu and J. Lafrenière, *J. Org. Chem.*, **65**, 7091 (2000).
(c) C. Beaulieu and C. Spino, *Tetrahedron Lett.*, **40**, 1637 (1999).
(d) C. Spino and M. Allan, *Can. J. Chem.*, **82**, 177 (2004).
52. (a) C. Spino and C. Beaulieu, *Angew. Chem., Int. Ed.*, **39**, 1930 (2000).
(b) C. Spino, C. Godbout, C. Beaulieu, M. Harter, T. M. Mwene-Mbeja and L. Boisvert, *J. Am. Chem. Soc.*, **126**, 13312 (2004).
53. (a) C. Spino and C. Godbout, *J. Am. Chem. Soc.*, **125**, 12106 (2003).
(b) C. Spino, M.-C. Tremblay and C. Godbout, *Org. Lett.*, **6**, 2801 (2004).
54. (a) L. Boisvert, F. Beaumier and C. Spino, *Can. J. Chem.*, **84**, 1290 (2006).
(b) C. Spino, L. Boisvert, J. Douville, S. Roy, S. Lauzon, J. Minville, D. Gagnon, F. Beaumier and C. Chabot, *J. Organomet. Chem.*, **691**, 5336 (2006).
55. C. Spino, V. G. Gund and C. Nadeau, *J. Comb. Chem.*, **7**, 345 (2005).
56. C. Agami, F. Couty, G. Evano and H. Mathieu, *Tetrahedron*, **56**, 367 (2000).
57. (a) C. Herber and B. Breit, *Chem. Eur. J.*, **12**, 6684 (2006).

(b) C. Herber and B. Breit, *Angew. Chem., Int. Ed.*, **44**, 5267 (2005).
58. M. Torneiro, Y. Fall, L. Castedo and A. Mouriño, *J. Org. Chem.*, **62**, 6344 (1997).
59. (a) J. S. Bindra and R. Bindra, *Prostaglandin Synthesis*, Academic Press, London, 1977, pp 536.
(b) A. Mitra, *The Synthesis of Prostaglandins*, John Wiley & Sons, Inc., New York, 1977, pp 448.
(c) S. M. Roberts and R. F. Newton, *Prostaglandins and Thromboxanes*, Butterworths, London, 1982, pp 164.
(d) S. M. Roberts and F. Scheinman, *New Synthetic Routes to Prostaglandins and Thromboxanes*, Academic Press, London, 1982, pp 265.
60. (a) I. Fleming and S. B. D. Winter, *J. Chem. Soc., Perkin Trans. 1*, 2687 (1998).
(b) I. Fleming and S. B. D. Winter, *Tetrahedron Lett.*, **36**, 1733 (1995).
(c) See also: H.-F. Chow and I. Fleming, *J. Chem. Soc., Perkin Trans. 1*, 2651 (1998).
61. J. M. Hutchison, H. A. Lindsay, S. S. Dormi, G. D. Jones, D. A. Vicic and M. C. McIntosh, *Org. Lett.*, **8**, 3663 (2006).
62. (a) R. W. Herr and C. R. Johnson, *J. Am. Chem. Soc.*, **92**, 4979 (1970).
(b) See also: D. M. Wieland and C. R. Johnson, *J. Am. Chem. Soc.*, **93**, 3047 (1971).
63. For a review see: J. A. Marshall, *Chem. Rev.*, **89**, 1503 (1989).
64. (a) J. A. Marshall and J. D. Trometer, *Tetrahedron Lett.*, **28**, 4985 (1987).
(b) J. A. Marshall, J. D. Trometer, B. E. Blough and T. D. Crute, *Tetrahedron Lett.*, **29**, 913 (1988).
(c) T. Ibuka, M. Tanaka, H. Nemoto and Y. Yamamoto, *Tetrahedron*, **45**, 435 (1989).
65. (a) J. A. Marshall, J. D. Trometer, B. E. Blough and T. D. Crute, *J. Org. Chem.*, **53**, 4274 (1988).
(b) J. A. Marshall, J. D. Trometer and D. G. Cleary, *Tetrahedron*, **45**, 391 (1989).
66. B. H. Lipshutz, K. Woo, T. Gross, D. J. Buzard and R. Tirado, *Synlett*, 477 (1997).
67. (a) J. A. Marshall and B. E. Blough, *J. Org. Chem.*, **55**, 1540 (1990).
(b) J. A. Marshall and B. E. Blough, *J. Org. Chem.*, **56**, 2225 (1991).
(c) J. A. Marshall, T. D. Crute III and J. D. Hsi, *J. Org. Chem.*, **57**, 115 (1992).
68. (a) T. Ibuka, K. Nakai, H. Habashita, Y. Hotta, N. Fujii, N. Mimura, Y. Miwa, T. Taga and Y. Yamamoto, *Angew. Chem., Int. Ed. Engl.*, **33**, 652 (1994).
(b) N. Fujii, K. Nakai, H. Tamamura, A. Otaka, N. Mimura, Y. Miwa, T. Taga, Y. Yamamoto and T. Ibuka, *J. Chem. Soc., Perkin Trans. 1*, 1359 (1995).
(c) T. Ibuka, N. Mimura, H. Ohno, K. Nakai, M. Akaji, H. Habashita, H. Tamamura, Y. Miwa, T. Taga, N. Fujii and Y. Yamamoto *J. Org. Chem.*, **62**, 2982 (1997).
(d) T. Ibuka, M. Tanaka, S. Nishii and Y. Yamamoto, *J. Chem. Soc., Chem. Commun.*, 1596 (1987).
69. P. Wipf and P. C. Fritch, *J. Org. Chem.*, **59**, 4875 (1994).
70. (a) J. P. Marino, L. J. Anna, R. Fernandez de la Pradilla, M. V. Martínez, C. Montero and A. Viso, *Tetrahedron Lett.*, **37**, 8031 (1996).
(b) J. P. Marino, L. J. Anna, R. Fernandez de la Pradilla, M. V. Martínez, C. Montero and A. Viso, *J. Org. Chem.*, **65**, 6462 (2000).
71. For a review see: F. Chemla and F. Ferreira, *Curr. Org. Chem.*, **6**, 539 (2002).
72. For early references on the stereospecific S_N2' displacement of propargylic systems see:
(a) P. Vermeer, H. Westmijze, H. Kleijn and L. A. van Dijck, *Recl. Trav. Chim. Pays-Bas*, **97**, 56 (1978).
(b) J. Mattay, M. Conrads and J. Runsink, *Synthesis*, 595 (1988).
(c) S. W. Djuric, M. Miyano, M. Clare and R. M. Rydzewski, *Tetrahedron Lett.*, **28**, 299 (1987).
(d) T. L. Macdonald, D. R. Reagan and R. S. Brinkmeyer, *J. Org. Chem.*, **45**, 4740 (1980).
(e) C. J. Elsevier and P. Vermeer, *J. Org. Chem.*, **54**, 3726 (1989).
(f) K. Ruitenberg, H. Westmijze, J. Meijer, C. J. Elsevier and P. Vermeer, *J. Organomet. Chem.*, **241**, 417 (1983).
(g) K. Ruitenberg, H. Westmijze, H. Kleijn and P. Vermeer, *J. Organomet. Chem.*, **277**, 227 (1984).
(h) J. A. Marshall and X.-J. Wang, *J. Org. Chem.*, **55**, 6246 (1990).
(i) G. Uccello-Barretta, F. Balzano, A. M. Caporusso, A. Iodice and P. Salvadori, *J. Org. Chem.*, **60**, 2227 (1995).

(j) M. J. C. Buckle and I. Fleming, *Tetrahedron Lett.*, **34**, 2383 (1993).
73. (a) For a recent review, see: N. Krause and A. Hoffmann-Röder, *Tetrahedron*, **60**, 11671 (2004).
(b) See also: A. Jansen and N. Krause, *Inorg. Chem. Acta*, **359**, 1761 (2006).
74. (a) G. M. R. Tombo, E. Didier and B. Loubinoux, *Synlett*, 547 (1990).
(b) S. Niwa and K. Soai, *J. Chem. Soc., Perkin Trans. 1*, 937 (1990).
75. (a) L. Pu and H. B. Yu, *Chem. Rev.*, **101**, 757 (2001).
(b) D. E. Frantz, R. Fässler, C. S. Tomoka and E. M. Carreira, *Acc. Chem. Res.*, **33**, 373 (2000).
(c) N. K. Anand and E. M. Carreira, *J. Am. Chem. Soc.*, **123**, 9687 (2001).
(d) D. E. Frantz, R. Fässler and E. M. Carreira, *J. Am. Chem. Soc.*, **122**, 1806 (2000).
76. (a) R. V. Kolakowski and L. J. Williams, *Tetrahedron Lett.*, **48**, 4761 (2007).
(b) S. D. Lotesta, S. Kiren, R. R. Sauers and L. J. Williams, *Angew. Chem., Int. Ed.*, **46**, 7108 (2007).
(c) Z. Wang, N. Shangguan, J. R. Cusick and L. J. Williams, *Synlett*, 213 (2008).
(d) P. Ghosh, S. D. Lotesta and L. J. Williams, *J. Am. Chem. Soc.*, **129**, 2438 (2007).
(e) S. D. Lotesta, Y. Hou and L. J. Williams, *Org. Lett.*, **9**, 869 (2007).
(f) N. Shangguan, S. Kiren and L. J. Williams, *Org. Lett.*, **9**, 1093 (2007).
77. (a) A. Hoffmann-Röder and N. Krause, *Org. Lett.*, **3**, 2537 (2001).
(b) N. Krause, A. Hoffmann-Röder and J. Canisius, *Synthesis*, 1759 (2002).
(c) A. S. K. Hashmi, *Chem Rev.*, **107**, 3180 (2007).
78. J. A. Marshall and K. G. Pinney, *J. Org. Chem.*, **58**, 7180 (1993).
79. (a) Y. Tu, Z.-X. Wang and Y. Shi, *J. Am. Chem. Soc.*, **118**, 9806 (1996).
(b) Y. Tu, Z.-X. Wang, M. Frohn, M. He, H. Yu, Y. Thang and Y. Shi, *J. Org. Chem.*, **63**, 8475 (1998).
(c) G.-A. Cao, Z.-X. Wang, Y. Tu and Y. Shi, *Tetrahedron Lett.*, **39**, 4425 (1998).
(d) Z.-X. Wang, G.-A. Cao and Y. Shi, *J. Org. Chem.*, **64**, 7646 (1999).
80. C. Spino and S. Fréchette, *Tetrahedron Lett.*, **41**, 8033 (2000).
81. A. C. Oehlschlager and E. Czyzewska, *Tetrahedron Lett.*, **24**, 5587 (1983).
82. F. Volz and N. Krause, *Org. Biomol. Chem.*, **5**, 1519 (2007).
83. S.-K. Kang, S.-G. Kim and D.-G. Cho, *Tetrahedron: Asymmetry*, **3**, 1509 (1992).
84. (a) A. Claesson and L.-I. Olsson, *J. Chem. Soc., Chem. Commun.*, 524 (1979).
(b) J. H. B. Chenier, J. A. Howard and B. Mile, *J. Am. Chem. Soc.*, **107**, 4190 (1985).
85. (a) A. Alexakis, *Pure Appl. Chem.*, **64**, 387 (1992).
(b) A. Alexakis, P. Mangeney, A. Ghribi, I. Marek, R. Sedrani, C. Guir and J. Normant, *Pure Appl. Chem.*, **60**, 49 (1988).
(c) A. Alexakis, I. Marek, P. Mangeney and J. F. Normant, *J. Am. Chem. Soc.*, **112**, 8042 (1990).
86. C. Deutsch, A. Hoffmann-Röder, A. Domke and N. Krause, *Synlett*, 737 (2007).
87. C. Deutsch, B. H. Lipshutz and N. Krause, *Angew. Chem., Int. Ed.*, **46**, 1650 (2007).
88. (a) H. Ohno, A. Toda, N. Fujii, Y. Takemoto, T. Tanaka and T. Ibuka, *Tetrahedron*, **56**, 2811 (2000).
(b) H. Ohno, A. Toda, Y. Miwa, T. Taga, N. Fujii and T. Ibuka, *Tetrahedron Lett.*, **40**, 349 (1999).
89. See, for example, R. K. Dieter and H. Yu, *Org. Lett.*, **3**, 3855 (2001) and references cited therein.
90. H. Ohno, M. Anzai, A. Toda, S. Ohishi, N. Fujii, T. Tanaka, Y. Takemoto and T. Ibuka, *J. Org. Chem.*, **66**, 4904 (2001).
91. A. Horváth, J. Benner and J.-E. Bäckvall, *Eur. J. Org. Chem.*, 3240 (2004).
92. (a) R. K. Dieter, N. Chen and V. K. Gore, *J. Org. Chem.*, **71**, 8755 (2006).
(b) R. K. Dieter, N. Chen, H. Yu, L. E. Nice and V. K. Gore, *J. Org. Chem.*, **70**, 2109 (2005).
(c) R. K. Dieter and L. E. Nice, *Tetrahedron Lett.*, **40**, 4293 (1999).
93. E. J. Corey and N. W. Boaz, *Tetrahedron Lett.*, **25**, 3059 (1984).
94. (a) F. D'Aniello, A. Mann, M. Taddei and C.-G. Wermuth, *Tetrahedron Lett.*, **35**, 7775 (1994).
(b) A. Maria Caporusso, C. Polizzi and L. Lardicci, *J. Org. Chem.*, **52**, 3920 (1987).
(c) F. D'Aniello, A. Mann, A. Schoenfelder and M. Taddei, *Tetrahedron*, **53**, 1447 (1997).
(d) F. D'Aniello, A. Mann and M. Taddei, *J. Org. Chem.*, **61**, 4870 (1996).

95. Propargylic epoxides can be opened with α-selectivity by Grignard reagent in the presence of a Lewis acid. See:
(a) F. Chemla, N. Bernard and J. Normant, *Eur. J. Org. Chem.*, 2067 (1999).
(b) F. Ferreira, A. Denichoux, F. Chemla and J. Bejjani, *Synlett*, 2051 (2004).
96. F. D'Aniello, A. Schoenfelder, A. Mann and M. Taddei, *J. Org. Chem.*, **61**, 9631 (1996).
97. A. Krief, *Tetrahedron Lett.*, **18**, 1035 (1977).
98. (a) J. P. Marino and D. M. Floyd, *Tetrahedron Lett.*, **8**, 675 (1979).
For other contributions including cyclohexadiene monoepoxides, see:
(b) J. P. Marino and H. Abe, *J. Org. Chem.*, **46**, 5379 (1981).
(c) J. P. Marino and N. Hatanaka, *J. Org. Chem.*, **44**, 4467 (1979).
(d) J. P. Marino and J. S. Farina, *J. Org. Chem.*, **41**, 3213 (1976).
99. K. Tamao, A. Kawachi, Y. Tanaka, H. Ohtani and Y. Ito, *Tetrahedron*, **52**, 5765 (1996).
100. J. P. Marino and J. C. Jaén, *J. Am. Chem. Soc.*, **104**, 3165 (1982).
101. T. Sato, Y. Gotoh, M. Watanabe and T. Fujisawa, *Chem. Lett.*, 1533 (1983).
102. (a) G. Teutsch and A. Belanger, *Tetrahedron Lett.*, 2051 (1979).
(b) G. Teutsch, *Tetrahedron Lett.*, **23**, 4697 (1982).
103. (a) M. I. Calaza, E. Hupe and P. Knochel, *Org. Lett.*, **5**, 1059 (2003).
(b) D. Soorukram and P. Knochel, *Org. Lett.*, **6**, 2409 (2004).
(c) S. Demay, K. Harms and P. Knochel, *Tetrahedron Lett.*, **40**, 4981 (1999).
104. D. Soorukram and P. Knochel, *Org. Lett.*, **9**, 1021 (2007).
105. (a) A. Niida, H. Tanigaki, E. Inokuchi, Y. Sasaki, S. Oishi, H. Ohno, H. Tamamura, Z. Wang, S. C. Peiper, K. Kitaura, A. Otaka and N. Fujii, *J. Org. Chem.*, **71**, 3942 (2006).
(b) A. Niida, S. Oishi, Y. Sasaki, M. Mizumoto, H. Tamamura, N. Fujii and A. Otaka, *Tetrahedron Lett.*, **46**, 4183 (2005).
106. (a) T. Sugai and K. Mori, *Synthesis*, 19 (1988).
(b) K. Laumen and M. P. Schneider, *J. Chem. Soc., Chem. Commun.*, 1298 (1986).
(c) K. Laumen and M. Schneider, *Tetrahedron Lett.*, **25**, 5875 (1984).
(d) Y.-F. Wang, C.-S. Chen, G. Girdaukas and C. J. Sih, *J. Am. Chem. Soc.*, **106**, 3695 (1984).
107. (a) T. Ainai, M. Matsuumi and Y. Kobayashi, *J. Org. Chem.*, **68**, 7825 (2003).
(b) T. Ainai, M. Ito and Y. Kobayashi, *Tetrahedron Lett.*, **44**, 3983 (2003).
(c) Y. Kobayashi, K. Nakata and T. Ainai, *Org. Lett.*, **7**, 183 (2005).
(d) K. Nakata and Y. Kobayashi, *Org. Lett.*, **7**, 1319 (2005).
(e) M. Ito, M. Matsuumi, M. G. Murugesh and Y. Kobayashi, *J. Org. Chem.*, **66**, 5881 (2001).
(f) K. Yagi, H. Nonaka, H. P. Acharya, K. Furukawa, T. Ainai and Y. Kobayashi, *Tetrahedron*, **62**, 4933 (2006).
(g) K. Nakata, Y. Kiyotsuka, T. Kitazume and Y. Kobayashi, *Org. Lett.*, **10**, 1345 (2008).
108. (a) J. P. Marino, R. Fernandez de la Pradilla and E. Laborde, *J. Org. Chem.*, **52**, 4898 (1987).
(b) J. P. Marino and M. G. Kelly, *J. Org. Chem.*, **46**, 4389 (1981).
(c) J. P. Marino, R. Fernãndez de la Pradilla and E. Laborde, *J. Org. Chem.*, **49**, 5279 (1984).
109. (a) D. P. Curran, M.-H. Chen, D. Leszczweski, R. L. Elliott and D. M. Rakiewicz, *J. Org. Chem.*, **51**, 1612 (1986).
(b) D. P. Curran and D. M. Rakiewicz, *Tetrahedron*, **41**, 3943 (1985).
(c) T. L. Fevig, R. L. Elliott and D. P. Curran, *J. Am. Chem. Soc.*, **110**, 5064 (1988).
(d) D. P. Curran and M.-H. Chen, *Tetrahedron Lett.*, **26**, 4991 (1985).
(e) D. P. Curran and D. M. Rakiewicz, *J. Am. Chem. Soc.*, **107**, 1448 (1985).
110. For other examples of opening of cyclic or bicyclic lactones and derivatives, see:
(a) S. Borthwick, W. Dohle, P. R. Hirst and K. I. Booker-Milburn, *Tetrahedron Lett.*, **47**, 7205 (2006).
(b) G. Balme and D. Bouyssi, *Tetrahedron*, **50**, 403 (1994).
(c) K. Weinges, H. Reichert, U. Huber-Patz and H. Irngartinger, *Liebigs Ann. Chem.*, 403 (1993).
111. (a) E. Montenegro, B. Gabler, G. Paradies, M. Seemann and G. Helmchen, *Angew. Chem., Int. Ed.*, **42**, 2419 (2003).
(b) S. Kudis and G. Helmchen, *Angew. Chem., Int. Ed.*, **37**, 3047 (1998).
(c) S. Kudis and G. Helmchen, *Tetrahedron*, **54**, 10449 (1998).
112. R. W. Rickards and H. Rönneberg, *J. Org. Chem.*, **49**, 572 (1984).
113. A. E. Greene, F. Coelho, J.-P. Després and T. J. Brocksom, *Tetrahedron Lett.*, **29**, 5661 (1988).

114. J. P. Marino and H. Abe, *J. Am. Chem. Soc.*, **103**, 2907 (1981).
115. M. de los Angeles Rey, J. A. Martínez-Pérez, A. Fernández-Gacio, K. Halkes, Y. Fall, J. Granja and A. Mouriño, *J. Org. Chem.*, **64**, 3196 (1999).
116. (a) T. Hudlicky, X. Tian, K. Königsberger, R. Maurya, J. Rouden and B. Fan, *J. Am. Chem. Soc.*, **118**, 10752 (1996).
(b) X. Tian, T. Hudlicky and K. Königsberger, *J. Am. Chem. Soc.*, **117**, 3643 (1995).
(c) T. Hudlicky, X. Tian, K. Königsberger and J. Rouden, *J. Org. Chem.*, **59**, 4037 (1994).
(d) T. Hudlicky, U. Rinner, D. Gonzalez, H. Akgun, S. Schilling, P. Siengalewicz, T. A. Martinot and G. R. Pettit, *J. Org. Chem.*, **67**, 8726 (2002).
(e) X. Tian, R. Maurya, K. Königsberger and T. Hudlicky, *Synlett*, 1125 (1995).
117. B. M. Trost and S. R. Pulley, *J. Am. Chem. Soc.*, **117**, 10143 (1995).
118. (a) M. Ernst and G. Helmchen, *Angew. Chem., Int. Ed.*, **41**, 4054 (2002).
(b) M. Ernst and G. Helmchen, *Synthesis*, 1953 (2002).
119. M. Purpura and N. Krause, *Eur. J. Org. Chem.*, 267 (1999).
120. N. Krause and M. Purpura, *Angew. Chem., Int. Ed.*, **39**, 4355 (2000).
121. C. Zelder and N. Krause, *Eur. J. Org. Chem.*, 3968 (2004).
122. M. Arai, T. Kawasuji and E. Nakamura, *J. Org. Chem.*, **58**, 5121 (1993).
123. A. Mengel and O. Reiser, *Chem. Rev.*, **99**, 1191 (1999).
124. M. Arai, E. Nakamura and B. H. Lipshutz, *J. Org. Chem.*, **56**, 5489 (1991).
125. Y. Wu, P. De Clercq, M. Vandewalle, R. Bouillon and A. Verstuyf, *Bioorg. Med. Chem. Lett.*, **12,** 1633 (2002).
126. M. A. Kacprzynski and A. H. Hoveyda, *J. Am. Chem. Soc.*, **126**, 10676 (2004).
127. (a) S. Flemming, J. Kabbara, K. Nickisch, J. Westermann and J. Mohr, *Synlett*, 183 (1995).
(b) H.-J. Liu and L.-K. Ho, *Can. J. Chem.*, **61**, 632 (1983).
128. J. Bund, H.-J. Gais and I. Erdelmeier, *J. Am. Chem. Soc.*, **113**, 1442 (1991).
129. J. L. Belelie and J. M. Chong, *J. Org. Chem.*, **67**, 3000 (2002).
130. A. Sato, H. Ito and T. Taguchi, *J. Org. Chem.*, **70**, 709 (2005).
131. T. Spangenberg, A. Schoenfelder, B. Breit and A. Mann, *Org. Lett.*, **9**, 3881 (2007).
132. G. Girard, G. Mandville and R. Bloch, *Tetrahedron: Asymmetry*, **4**, 613 (1993).
133. Y. Nakamura, M. Okada, A. Sato, H. Horikawa, M. Koura, A. Saito and T. Taguchi, *Tetrahedron*, **61**, 5741 (2005).
134. (a) A. Ghribi, A. Alexakis and J. F. Normant, *Tetrahedron Lett.*, **25**, 3079 (1984).
(b) A. Ghribi, A. Alexakis and J. F. Normant, *Tetrahedron Lett.*, **25**, 3083 (1984).
(c) A. Alexakis, P. Mangeney and J. F. Normant, *Tetrahedron Lett.*, **28**, 2363 (1987).
(d) P. Mangeney, A. Alexakis and J. F. Normant, *Tetrahedron Lett.*, **27**, 3143 (1986).
(e) A. Alexakis, P. Mangeney, A. Ghribi, I. Marek, R. Sedrani, C. Guir and J. Normant, *Pure Appl. Chem.*, **60**, 49 (1988).
(f) A. Alexakis and P. Mangeney, *Tetrahedron: Asymmetry*, **1**, 477 (1990).
135. H. Rakotoarisoa, R. Guttierez Perez, P. Mangeney and A. Alexakis, *Organometallics*, **15**, 1957 (1996).
136. (a) A. Alexakis, P. Mangeney and J. F. Normant, *Tetrahedron Lett.*, **26**, 4197 (1985).
(b) I. Marek, P. Mangeney, A. Alexakis and J. F. Normant, *Tetrahedron Lett.*, **27**, 5499 (1986).
137. S. E. Denmark and L. K. Marble, *J. Org. Chem.*, **55**, 1984 (1990).
138. (a) V. Caló, V. Fiandanese, A. Nacci and A. Scilimati, *Tetrahedron*, **50**, 7283 (1994).
(b) V. Caló, A. Nacci and V. Fiandanese, *Tetrahedron*, **52**, 10799 (1996).
139. B. Breit and D. Breuninger, *Synthesis*, 147 (2005).
140. H.-J. Gais, H. Müller, J. Bund, M. Scommoda, J. Brandt and G. Raabe, *J. Am. Chem. Soc.*, **117**, 2453 (1995).
141. (a) R. Tamura, K.-i. Watabe, N. Ono and Y. Yamamoto, *J. Org. Chem.*, **57**, 4895 (1992).
(b) R. Tamura, K.-i. Watabe, H. Katayama, H. Suzuki and Y. Yamamoto, *J. Org. Chem.*, **55**, 408 (1990).
142. (a) D. M. Knotter, M. D. Janssen, D. M. Grove, W. J. J. Smeets, E. Horn, A. L. Spek and G. van Koten, *Inorg. Chem.*, **30**, 4361 (1991).
(b) D. M. Knotter, H. L. van Maanen, D. M. Grove, A. L. Spek and G. van Koten, *Inorg. Chem.*, **30**, 3309 (1991).

143. (a) G. J. Meuzelaar, A. S. E. Karlström, M. van Klaveren, E. S. M. Persson, A. del Villar, G. van Koten and J.-E. Bäckvall, *Tetrahedron*, **56**, 2895 (2000).
(b) M. van Klaveren, E. S. M. Persson, D. M. Grove, J.-E. Bäckvall and G. van Koten, *Tetrahedron Lett.*, **35**, 5931 (1994).
144. (a) A. S. E. Karlström, F. F. Huerta, G. J. Meuzelaar and J.-E. Bäckvall, *Synlett*, 923 (2001).
(b) H. K. Cotton, J. Norinder and J.-E. Bäckvall, *Tetrahedron*, **62**, 5632 (2006).
145. (a) F. Dübner and P. Knochel, *Angew. Chem., Int. Ed.*, **38**, 379 (1999).
(b) F. Dübner and P. Knochel, *Tetrahedron Lett.*, **41**, 9233 (2000).
146. H. Malda, A. W. van Zijl, L. A. Arnold and B. L. Feringa, *Org. Lett.*, **3**, 1169 (2001).
147. S. Ongeri, U. Piarulli, M. Roux, C. Monti and C. Gennari, *Helv. Chim. Acta*, **85**, 3388 (2002).
148. (a) A. Alexakis, C. Malan, L. Lea, C. Benhaim and X. Fournioux, *Synlett*, 927 (2001).
(b) A. Alexakis and K. Croset, *Org. Lett.*, **4**, 4147 (2002).
149. (a) K. Tissot-Croset, D. Polet and A. Alexakis, *Angew. Chem., Int. Ed.*, **43**, 2426 (2004).
(b) K. Tissot-Croset and A. Alexakis, *Tetrahedron Lett.*, **45**, 7375 (2004).
(c) K. Tissot-Croset, D. Polet, S. Gille, C. Hawner and A. Alexakis, *Synthesis*, 2586 (2004).
150. W.-J. Shi, L.-X. Wang, Y. Fu, S.-F. Zhu and Q.-L. Zhou, *Tetrahedron: Asymmetry*, **14**, 3867 (2003).
151. (a) F. López, A. W. van Zijl, A. J. Minnaard and B. L. Feringa, *Chem. Commun.*, 409 (2006).
(b) A. W. van Zijl, F. López, A. J. Minnaard and B. L. Feringa, *J. Org. Chem.*, **72**, 2558 (2007).
152. S. R. Harutyunyan, F. López, W. R. Browne, A. Correa, D. Peña, R. Badorrey, A. Meetsma, A. J. Minnaard and B. L. Feringa, *J. Am. Chem. Soc.*, **128**, 9103 (2006).
153. C. A. Luchaco-Cullis, H. Mizutani, K. E. Murphy and A. H. Hoveyda, *Angew. Chem., Int. Ed.*, **40**, 1456 (2001).
154. K. E. Murphy and A. H. Hoveyda, *J. Am. Chem. Soc.*, **125**, 4690 (2003).
155. A. O. Larsen, W. Leu, C. N. Oberhuber, J. E. Campbell and A. H. Hoveyda, *J. Am. Chem. Soc.*, **126**, 11130 (2004).
156. S. Tominaga, Y. Oi, T. Kato, D. K. An and S. Okamoto, *Tetrahedron Lett.*, **45**, 5585 (2004).
157. J. J. Van Veldhuizen, J. E. Campbell, R. E. Giudici and A. H. Hoveyda, *J. Am. Chem. Soc.*, **127**, 6877 (2005).
158. K. Geurts, S. P. Fletcher and B. L. Feringa., *J. Am. Chem. Soc.*, **128**, 15572 (2006).
159. J. K. Cha and N.-S. Kim, *Chem. Rev.*, **95**, 1761 (1995).
160. C. A. Falciola, K. Tissot-Croset and A. Alexakis, *Angew. Chem., Int. Ed.*, **45**, 5995 (2006).
161. Y. Lee, K. Akiyama, D. G. Gillingham, M. K. Brown and A. H. Hoveyda, *J. Am. Chem. Soc.*, **130**, 446 (2008).
162. (a) P. J. Goldsmith, S. J. Teat and S. Woodward, *Angew. Chem., Int. Ed.*, **44**, 2235 (2005).
(b) C. Börner, J. Gimeno, S. Gladiali, P. J. Goldsmith, D. Ramazzotti and S. Woodward, *Chem. Commun.*, 2433 (2000).
163. C. A. Falciola and A. Alexakis, *Angew. Chem., Int. Ed.*, **46**, 2619 (2007).
164. (a) F. Badalassi, P. Crotti, F. Macchia, M. Pineschi, A. Arnold and B. L. Feringa, *Tetrahedron Lett.*, **39**, 7795 (1998).
(b) F. Bertozzi, P. Crotti, B. L. Feringa, F. Macchia and M. Pineschi, *Synthesis*, 483 (2001).
165. O. Equey and A. Alexakis, *Tetrahedron: Asymmetry*, **15**, 1531 (2004). See Reference 172c concerning the reaction of phosphoramidate ligands with trialkylaluminum.
166. R. Millet and A. Alexakis, *Synlett*, 435 (2007).
167. F. Bertozzi, P. Crotti, F. Macchia, M. Pineschi, A. Arnold and B. L. Feringa, *Org. Lett.*, **2**, 933 (2000).
168. (a) F. Bertozzi, P. Crotti, F. Macchia, M. Pineschi and B. L. Feringa, *Angew. Chem., Int. Ed.*, **40**, 930 (2001).
(b) M. Pineschi, F. Del Moro, P. Crotti, V. Di Bussolo and F. Macchia, *J. Org. Chem.*, **69**, 2099 (2004).
169. F. Bertozzi, M. Pineschi, F. Macchia, L. A. Arnold, A. J. Minnaard and B. L. Feringa, *Org. Lett.*, **4**, 2703 (2002).
170. U. Piarulli, P. Daubos, C. Claverie, M. Roux and C. Gennari, *Angew. Chem., Int. Ed.*, **42**, 234 (2003).
171. U. Piarulli, C. Claverie, P. Daubos C. Gennari, A. J. Minnaard and B. L. Feringa, *Org. Lett.*, **5**, 4493 (2003).

172. (a) M. Pineschi, F. Del Moro, P. Crotti and F. Macchia, *Org. Lett.*, **7**, 3605 (2005).
(b) M. Pineschi, F. Del Moro, P. Crotti and F. Macchia, *Pure Appl. Chem.*, **78**, 463 (2006).
(c) C. Bournaud, C. Falciola, T. Lecourt, S. Rosset, A. Alexakis and L. Micouin, *Org. Lett.*, **8**, 3581 (2006).
173. J. Norinder and J.-E. Bäckvall, *Chem. Eur. J.*, **13**, 4094 (2007).
174. J. Norinder, K. Bogár, L. Kanupp and J.-E. Bäckvall, *Org. Lett.*, **9**, 5095 (2007).

CHAPTER **14**

Copper-catalyzed enantioselective conjugate addition

DAMIEN POLET and ALEXANDRE ALEXAKIS

University of Geneva, Department of Organic Chemistry, 30, quai Ernest Ansermet 1211 Genève 4, Switzerland
Fax: +41-22-3793215; e-mail: alexandre.alexakis@unige.ch

I. INTRODUCTION 693
II. NEW SUBSTRATES AND NUCLEOPHILES 695
III. CONSTRUCTION OF QUATERNARY STEREOCENTERS 700
IV. TANDEM REACTIONS 707
V. APPLICATIONS IN SYNTHESIS 714
VI. CONCLUSIONS AND PERSPECTIVE 728
VII. REFERENCES 728

I. INTRODUCTION

The search for new synthetic methods for constructing chiral centers is driven by the needs in exciting fields, especially those on the borderline of chemistry. Indeed, the development of new active pharmaceutical ingredients, the diversity-oriented synthesis of the library of molecules and the discovery of new materials place synthetic methodology as a central discipline in science. Great emphasis has been put on the development of catalytic tools, and the last 30 years of research have witnessed a tremendous amount of effort from organic chemistry groups around the world. From this perspective, copper-catalyzed asymmetric conjugate addition[1] (Scheme 1) has evolved to a mature synthetic tool, allowing high yields and enantioselectivities on a general scope of substrates and a broad range of organometallic nucleophiles. Highly efficient and easy-to-use catalytic systems have been developed, so that the organic chemist can now select a suitable ligand for its substrate in the toolbox. Besides, it has also found numerous applications in tandem and cascade reactions, and was successfully used in the total synthesis of natural or non-natural products.

The chemistry of organocopper compounds
Edited by Z. Rappoport and I. Marek

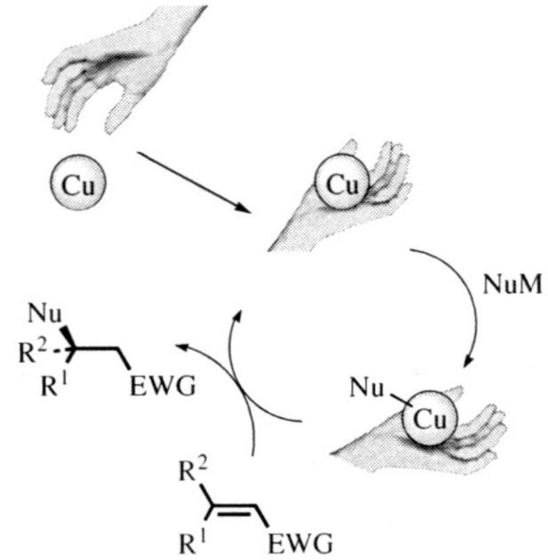

SCHEME 1. General scheme of the copper-catalyzed conjugate addition (the hand represents a chiral ligand for copper, M = metal such as MgX, EWG = electron-withdrawing)

In this chapter, we will focus our attention on the most recent developments of the copper-catalyzed conjugate addition of alkyl and aryl nucleophiles and its application in synthesis.

Copper-catalyzed conjugate addition traces back to its roots, in 1941, when Kharasch and Tawney observed that the addition of MeMgBr to isophorone preferentially occurs in a 1,4-manner in the presence of a catalytic amount of copper(I) salt[2]. Since then and for a long time, all the developments towards an enantioselective version were done using Grignard reagents, organolithium being much more reactive towards the carbonyl group.

The introduction of diorganozincs as primary organometallics by our laboratory in 1993 (Scheme 2) represents a major breakthrough in the field of asymmetric conjugate addition (ACA)[3]. Thanks to **L1**, a promising enantioselectivity of 32% was obtained in **2**, the addition product of diethylzinc to cyclohexenone **1**. The intrinsic tolerance of organozinc compounds towards a broad spectrum of synthetic groups opened the way to a multitude of possibilities in terms of substrate types or even functionalized nucleophiles. Associated to a rate-accelerating phosphorus ligand[4], a copper salt serves as an efficient catalyst for the transfer of organozinc nucleophiles.

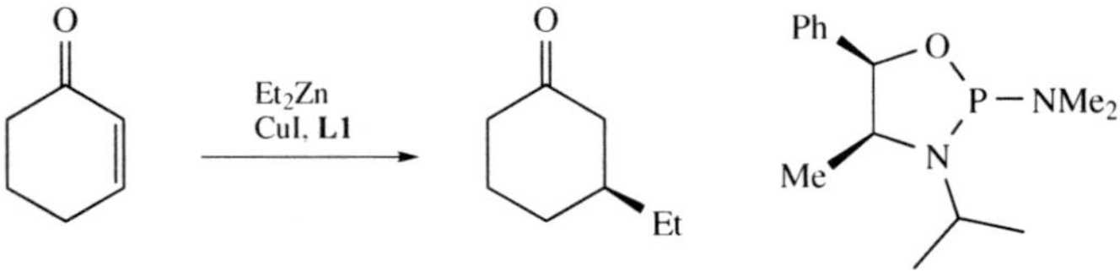

(**1**) (**2**), 32% ee (**L1**)

SCHEME 2. First example of copper-catalyzed ACA of organozinc

In 1997, Feringa and coworkers described the synthesis and use of the new phosphoramidite ligand **L2** that was particularly efficient (98% ee) in the conjugate addition of diorganozinc compounds to a wide range of substrates (Scheme 3)[5]. A few years later, our

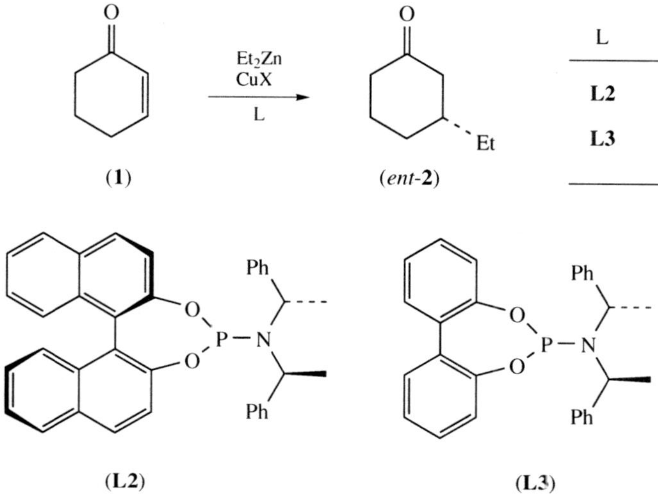

L	ee
L2	98%
L3	96%

SCHEME 3. Feringa and Alexakis ligand systems

group proposed **L3**, a simplified version of **L2**, in which the atropoisomerically flexible biphenyl replaces the binaphthyl moiety as in **L2**[6]. Providing almost as good enantioselectivity as with **L2** (96% vs 98%)[7], **L3** would later give birth to a large family of modular ligands, whose members are suitably adapted to any given substrate.

There now exist a huge number of chiral ligands and among them the trivalent phosphorus type is over-represented. They provide very high enantiomeric excesses for the addition of ZnR_2 to all types of α,β-unsaturated compounds including ketones, lactones, lactams, nitroalkenes, amides, malonates or imines. In Scheme 4 the most used ligands (**L2–L7**) are represented as well as classical types of substrates in copper-catalyzed ACA[8].

A large number of ligands provide high enantioselectivites in the 1,4-addition of Et_2Zn, as extensively reviewed elsewhere[1]. The present chapter intentionally avoids duplicating the content of existing reviews and focuses on recent (after 2002) and synthetically useful developments in the field. Particular emphasis will be placed on new substrates and organometallic reagents, on the construction of all-carbon quaternary centers, on trapping and cascade reactions as well as on applications in total synthesis of natural and non-natural products.

II. NEW SUBSTRATES AND NUCLEOPHILES

N-Containing Michael acceptors represent an interesting new class of substrates (Scheme 5), as their adducts can provide useful homochiral building blocks for the synthesis of alkaloids. α,β-Unsaturated lactams (**3**) were found to give high ee values with diethylzinc and dibutylzinc but were not alkylated by the less reactive Me_2Zn[9]. The use of trimethylaluminum fully converted **3** to the desired adduct **4** (R = Me), but a much lower ee was obtained (68%).

N-Substituted-2,3-dehydro-4-piperidones **5** were slightly more reactive than lactam **3**, as even dimethylzinc gave rise to the adduct (**6**), although in lower yield[10].

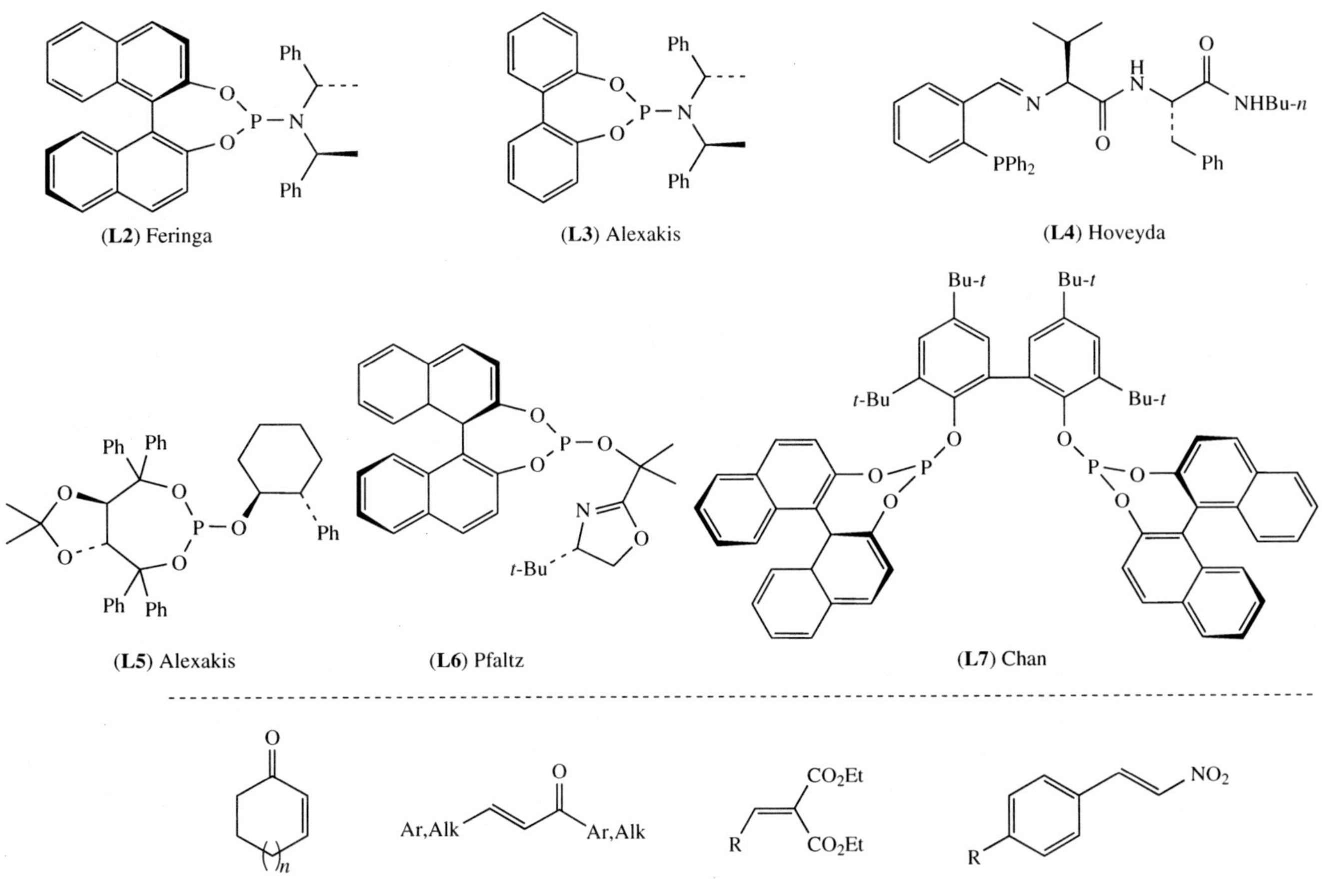

SCHEME 4. Classical ligands and substrates

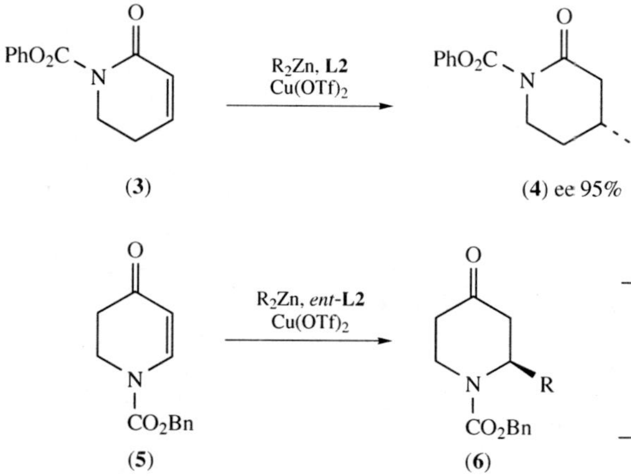

R	ee
Et	95%
n-Bu	>90%

R	ee
Me	96%
Et	94%
i-Pr	94%

SCHEME 5. Development of ACA on *N*-containing Michael acceptors

The group of Carretero designed a new type of electrophile: sulfonyl imines **7** derived from chalcones (Scheme 6)[11]. Interestingly, an *ortho*-pyridyl group was attached to the sulfonyl moiety and was found to be essential to the reactivity, probably through metal coordination. Resulting product **8** was obtained with high E:Z ratio (95:5) and good enantiomeric excess. Hydrolysis of **8** under acidic conditions yielded the corresponding chalcone, while an aldehyde was obtained by ozone cleavage.

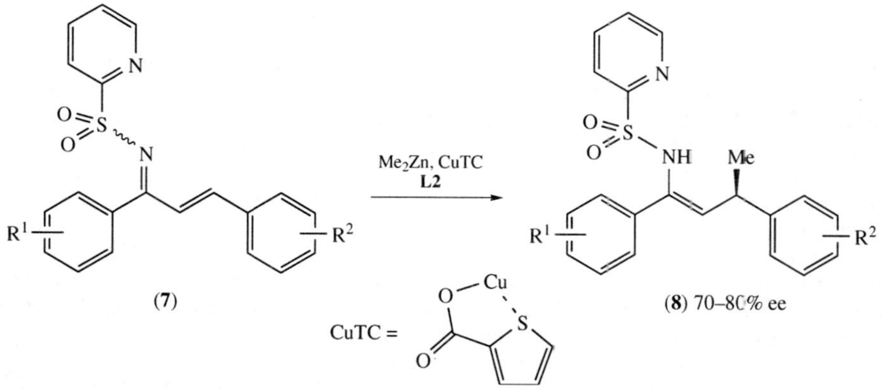

SCHEME 6. ACA of dimethylzinc to α,β-unsaturated ketimines

During the development of the asymmetric conjugate addition of diethylzinc to α-haloenones **9**, Li and Alexakis unraveled a parasitic radical pathway that could be suppressed by the use of styrene as a stoichiometric additive. Thanks to **L8**, high enantioselectivities could then be achieved as in **10** (Scheme 7)[12], together with modest to good diastereomeric ratios (up to 3:1 in favor of the *trans* adduct for X = Cl and Br, as opposed to 98:2 in favor of the *cis* adduct for X = I).

A large amount of recent literature has focused on the use of diorganozinc compounds as nucleophiles. However, other organometallic species represent interesting alternatives, either for their enhanced reactivity or their ready availability. After an early example

(9) (10)

n	X	ee
1	Cl	98%
1	Br	97%
1	I	95%
0	Br	79%

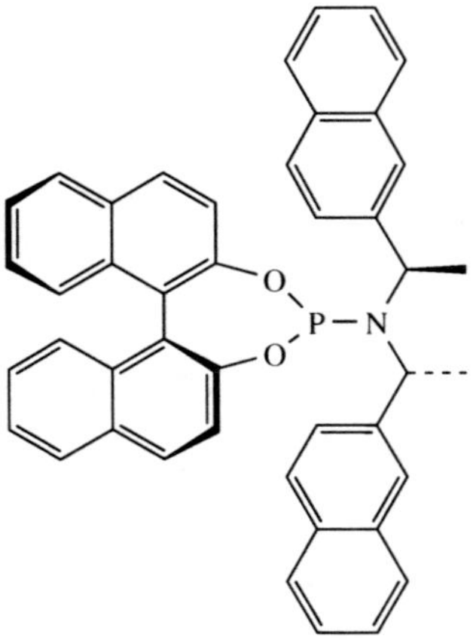

(L8)

SCHEME 7. α-haloenones as substrates

from Iwata and coworkers[13] as well as other pioneering work from several groups[14], there is now a general system that provides enantioenriched adducts (Scheme 8) based on ligand **L2**[15]. More than 90% ee is obtained for cyclohexen-1-one **1** and cyclohepten-1-one **11**, either with Me_3Al or Et_3Al as the nucleophile (formation of **2** and **12**). Although considered as more challenging substrates, acyclic enones **13** work just as well: up to 96% ee was reached as in **14**. Of particular interest is the use of either Me or Et as transferable groups without any reactivity difference.

Great efforts have culminated in the development of a general and reliable catalytic system for the ACA of Grignard reagents on disubstituted Michael acceptors. The first example of such a reaction implied cyclic enones, but consecutive reports made use of acyclic enones, α,β-unsaturated esters and thioesters.

Combination of copper(I) chloride and Taniaphos **L9** was the key for achieving high regio- (1,4- versus 1,2-addition) and enantioselectivity in the reaction of **1** (Scheme 9)[16]. The great advantage of using organomagnesium over diorganozinc compounds is the possibility of selecting the desired alkyl nucleophile among the enormous variety of commercially available Grignard reagents. However, the ee drops significantly when branched organometallics are transferred. The ligand has to be changed when dealing with more challenging substrates like cyclopentenone. In this case, Josiphos **L10** (*vide infra*) gives 92% ee.

Using Josiphos ligand **L10**, acyclic enones **13** give excellent results with a broad spectrum of Grignard reagents (Scheme 10)[17]. Both aliphatic and aromatic enones react well, allowing different alkyl chains to be transferred with high enantioselectivities.

Copper complex **16** containing Josiphos **L10** was prepared and used in the ACA of Grignard reagents to α,β-unsaturated esters (Scheme 11)[18]. Under these conditions, a

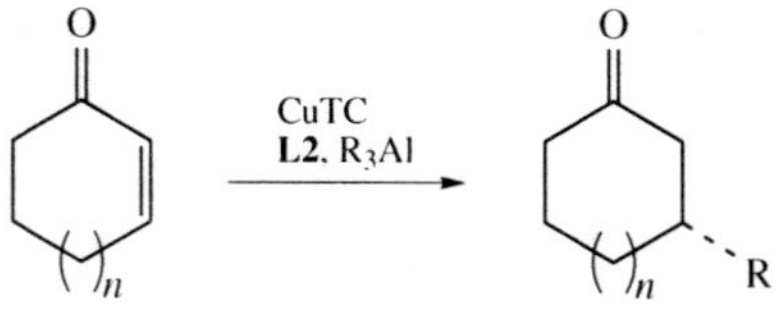

$n = 1$ (**1**) $n = 2$ (**11**) → $n = 1$ (**2**) $n = 2$ (**12**)

R	n	ee
Et	1	90%
Et	2	98%
Me	1	95%
Me	2	92%
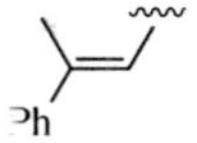	1	77%
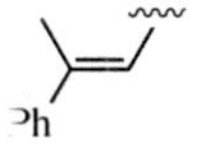	2	72%

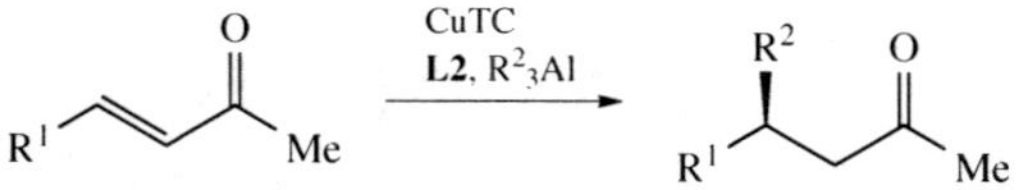

(**13**) → (**14**)

R^1	R^2	ee
n-Pen	Et	89%
i-Pr	Et	90%
n-Pen	Me	72%
i-Pr	Me	96%
c-Hex	Me	96%

SCHEME 8. General system for the ACA of alanes to enones

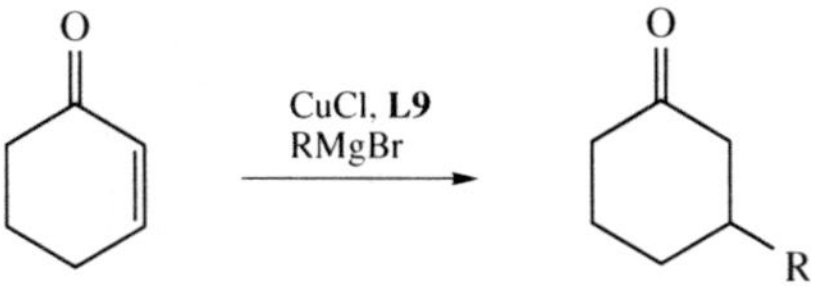

(**1**) → (**2**)

R	ee
Et	96%
Me	90%
n-Pr	94%
n-Bu	96%

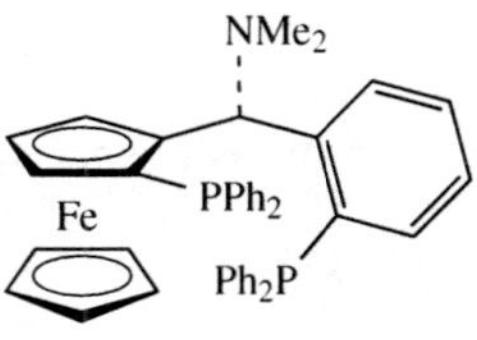

(**L9**)

SCHEME 9. General methodology for the conjugate addition of Grignard reagents

broad range of substrates and nucleophiles can be used, giving **17** with high ee values. However, substrates of type **15** have to deal with low conversion when MeMgBr is used. The Feringa group has then developed a system in which the substrate is more reactive. α,β-Unsaturated thioesters **18** allow high yields and ee values of **19** and a wider range of substrates and nucleophiles. In this case, the catalyst is formed *in situ*[9].

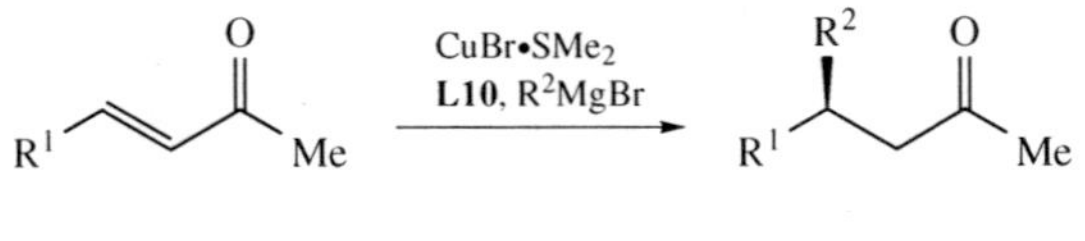

(13) (14)

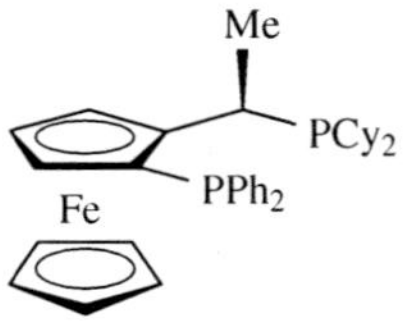

(L10)

R^1	R^2	ee
n-Bu	Et	90%
n-Bu	Me	98%
n-Pen	*n*-Pr	90%
n-Bu	*i*-Bu	86%
n-Pr	Ph	76%
i-Pr	Me	94%
Ph	Me	97%

SCHEME 10. 1,4-Addition of Grignard reagents to α,β-unsaturated acyclic ketones

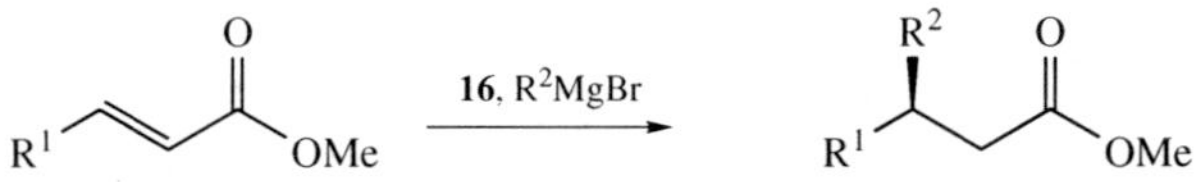

(15) (17)

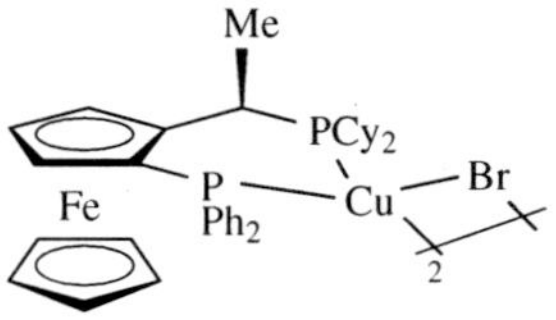

(16)

R^1	R^2	ee
n-Pr	Me	95%
Me	*n*-Bu	98%
n-Pr	Et	93%
$BnOCH_2$	Et	86%
i-Bu	*n*-Bu	92%

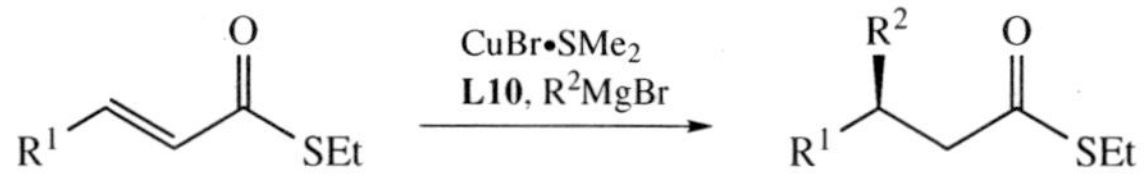

(18) (19)

R^1	R^2	ee
n-Pen	Me	96%
n-Pen	Et	96%
n-Pr	Et	96%
$BnO(CH_2)_3$	Et	95%
Ph	Et	95%

SCHEME 11. ACA of Grignard reagents to α,β-unsaturated esters and thioesters

An alternative system to the methodology described above makes use of Tol-BINAP ligand **L11** and copper(I) iodide (Scheme 12)[20]. Here again a wide range of substrates and nucleophiles are allowed, although MeMgBr suffers from low yields, due to the formation of β-methyl-substituted methylketone. The authors later managed to improve these conditions, successfully suppressing this competitive reaction[21].

III. CONSTRUCTION OF QUATERNARY STEREOCENTERS

Although diorganozinc reagents gained an incredible success and are now the most used nucleophiles, the construction of quaternary centers with these organometallic reagents represented an enormous challenge and was not possible until very recently.

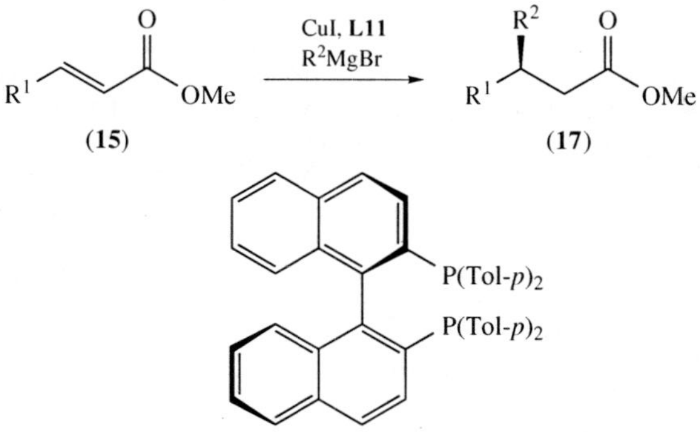

(15) (17) (L11)

R^1	R^2	ee
$Ph(CH_2)_2$	Me	>98%
$Ph(CH_2)_2$	Et	93%
$Ph(CH_2)_2$	*i*-Pr	91%
Ph	Et	93%
Et	Ph	74%
i-Pr	Et	95%

SCHEME 12. Alternative system for the ACA of Grignard compounds to α,β-unsaturated esters

When used in the presence of β-disubstituted enones, the reactivity of diorganozincs in copper-catalyzed ACA drops dramatically. This phenomenon forced researchers to find alternative methods for the elaboration of optically active all-carbon quaternary stereocenters. Enhancing the reactivity of the Michael acceptor represents an early successful solution. Focusing on more reactive nitroalkenes **20**, Hoveyda and coworkers described the formation of quaternary carbons by ACA using **L12** (Scheme 13)[22]. The generality of this method is noteworthy, as it allows the use of several diorganozinc reagents, even the less reactive Me_2Zn. However, the latter needs to be used under forcing conditions and, as a consequence, the ee drops significantly.

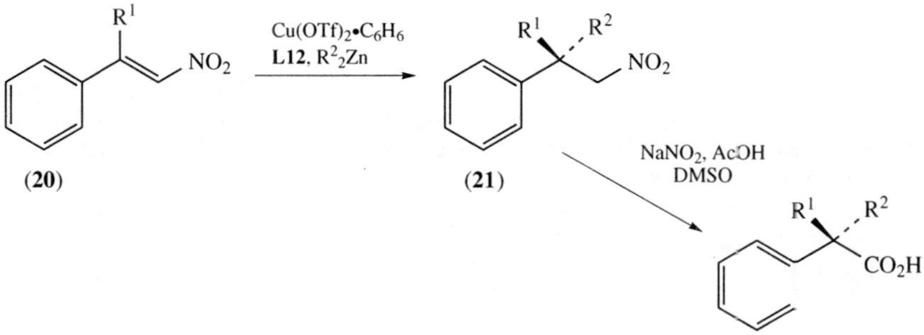

(20) (21) (22)

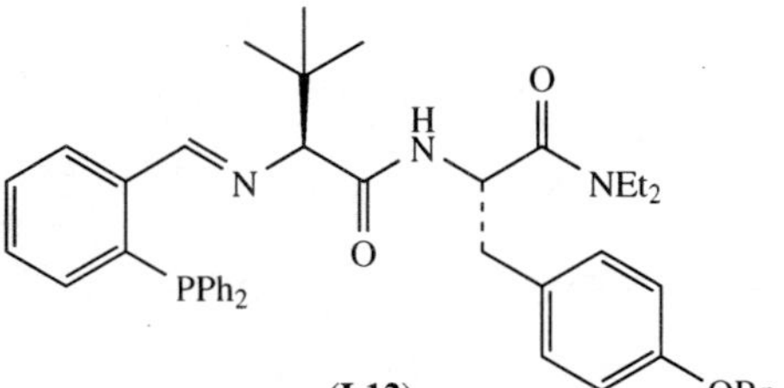

(L12)

$R^2{}_2Zn$	R^1	ee
n-Bu_2Zn	Me	93%
Et_2Zn	Me	94%
Me_2Zn	*i*-Pr	73%

SCHEME 13. Nitroalkenes as efficient substrates for the construction of quaternary centers

A subsequent Nef reaction allows the transformation of **21** into α-quaternary carboxylic acids **22**, which are not trivial to obtain otherwise.

Another way to increase the substrate reactivity is the introduction of an additional electron-withdrawing group. β-Ketoesters **23** were designed for this purpose. The authors found specific conditions for this particular process, and the required non-phosphorus-based ligand **L13** was successfully discovered after a screening of around 90 congeners (Scheme 14)[23]. The ester functionality as in **24** was easily removed by decarboxylation under thermal conditions to give **25**.

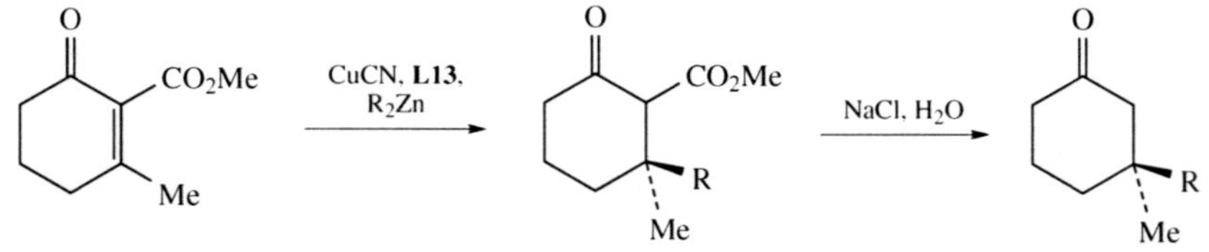

(**23**) (**24**) (**25**)

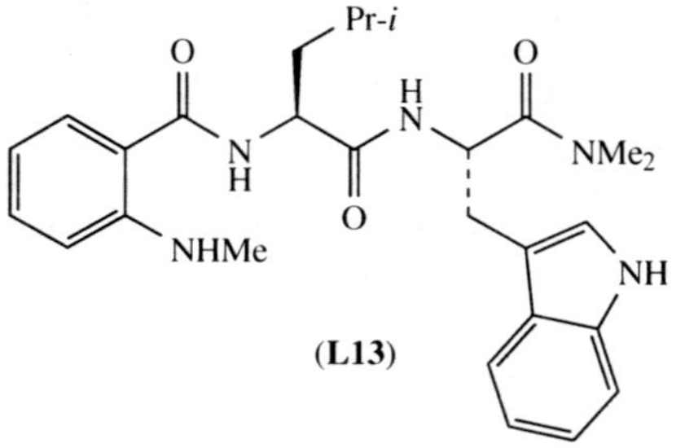

(**L13**)

R_2Zn	ee
Et_2Zn	82%
n-Bu_2Zn	92%
i-Pr_2Zn	86%

SCHEME 14. ACA on ketoesters **23**

This reactivity-enhancement strategy was similarly used by Fillion and Wilsily (Scheme 15)[24]. This research group showed that the combination of the general conditions and *ent*-**L2** as ligand could be applied to substrates of type **26**, affording adducts **27** with impressive ee values. But once again, dimethylzinc could not be used, as it gave rise to a much lower conversion.

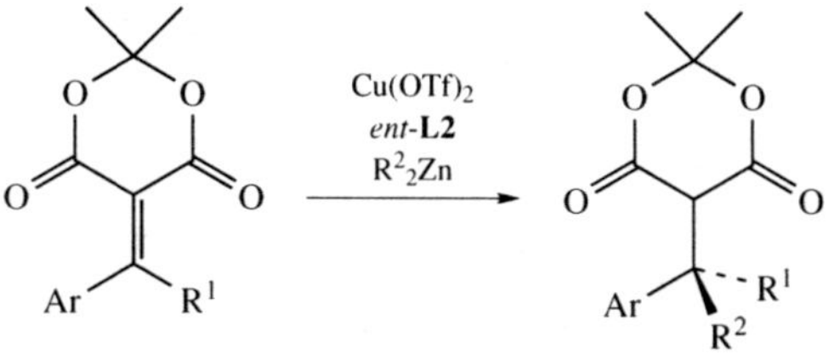

(**26**) (**27**)

Ar	R^2_2Zn	R^1	ee
Ph	Et_2Zn	Me	84%
p-ClC_6H_4	Et_2Zn	Me	95%
p-ClC_6H_4	Et_2Zn	*n*-Bu	94%
p-ClC_6H_4	*n*-Bu_2Zn	Me	87%

SCHEME 15. Meldrum's acid derivatives **26** as substrates in ACA

Another communication by Wilsily and Fillion demonstrated the value of Meldrum's acid derivatives as substrates in ACA (Scheme 16)[25]. Using the extremely activated tetra-substituted alkenes **28**, this group was able to achieve respectable enantioselectivities on

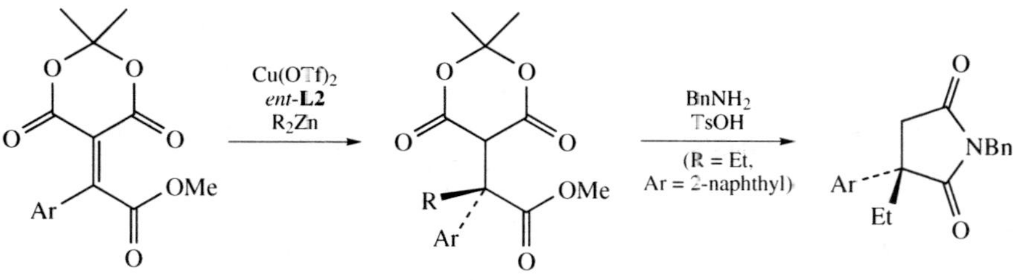

Ar	R_2Zn	ee
Ph	Et_2Zn	88%
p-ClC_6H_4	Et_2Zn	94%
p-$MeOC_6H_4$	Et_2Zn	92%
o-ClC_6H_4	Et_2Zn	80%
p-$MeOC_6H_4$	n-Bu_2Zn	90%
p-$MeOC_6H_4$	i-Pr_2Zn	84%
2-naphthyl	Me_2Zn	62%

SCHEME 16. Other Meldrum's acid derivatives, and their use in ACA

a broad range of substrates and nucleophiles. As an application, chiral succinimides **30** bearing the constructed stereocenter were prepared by treating **29** with benzylamine under acid catalysis.

Playing with the electronic and steric nature of the chiral ligand was also successfully explored. Indeed, chiral *N*-heterocyclic carbene (NHC) as in complex **32** enhances the feasibility to achieve higher reactivities in ACA (Scheme 17)[26]. By premixing **32** and the copper salt, the Hoveyda group was able to deliver alkyl or aryl groups to **31**. Reaction of dimethylzinc with successful outcome (91% ee measured on **33**) illustrates this ligand-triggered shift in reactivity.

Thanks to this family of NHC ligands such as in **35**, it is now possible to employ non-activated β-disubstituted enones **34** (Scheme 18)[27]. Alkyl- and arylzinc reagents react to give excellent yields of the ketones **36** bearing an all-carbon quaternary center. Of particular interest is the trapping of the zinc enolate intermediate, which can give the enol triflate or enol silane, as shown previously[28]. Although dimethylzinc does not give acceptable conversion, diphenylzinc is efficiently transferred with high enantioface selectivity. This protocol represents therefore a significant advantage over the rhodium-catalyzed conjugate addition of boronic acids or esters[29].

Along with the enhancement of the reactivity of either the substrate or the catalyst resides the use of more Lewis-acidic organometallic sources. Noteworthy is the use of triorganoaluminum, which can transfer alkyl, alkenyl or aryl nucleophiles in ACA. Our

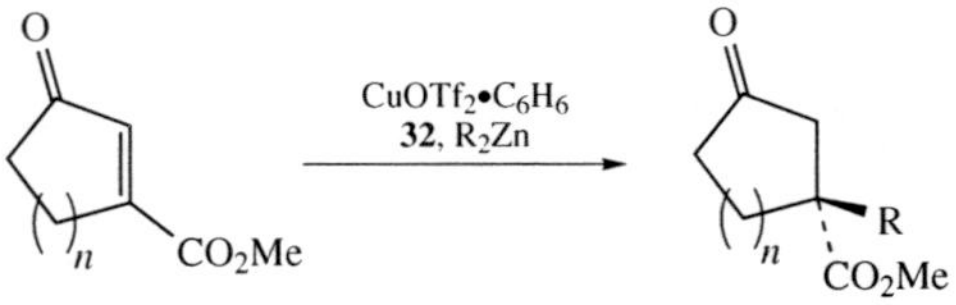

(31) (33)

n	R_2Zn	ee
1	Me_2Zn	91%
1	Et_2Zn	88%
1	Ph_2Zn	81%
2	Me_2Zn	89%
2	Ph_2Zn	93%

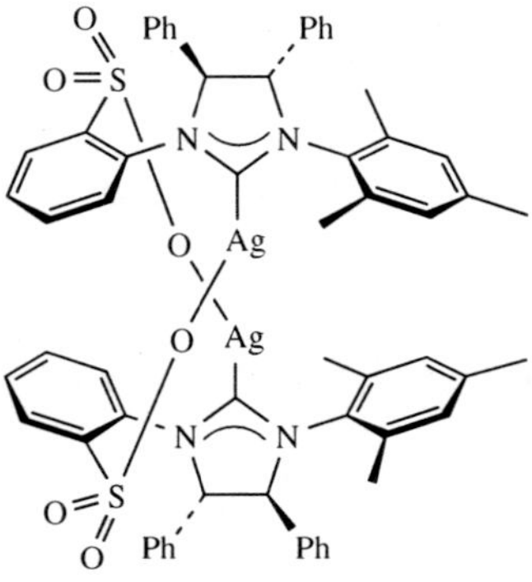

(32)

SCHEME 17. Carbene ligand-based catalyst and ketoesters as substrates

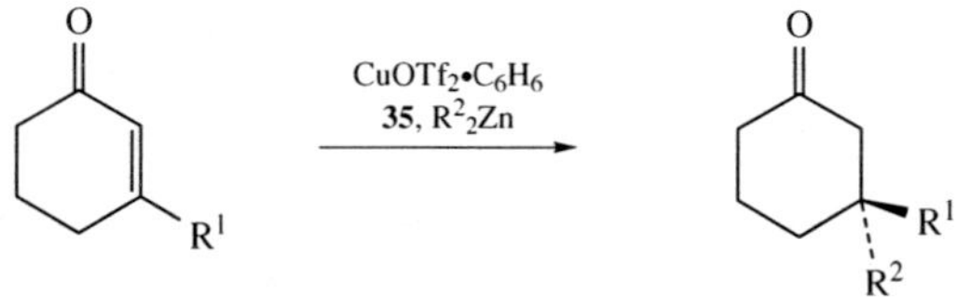

(34) (36)

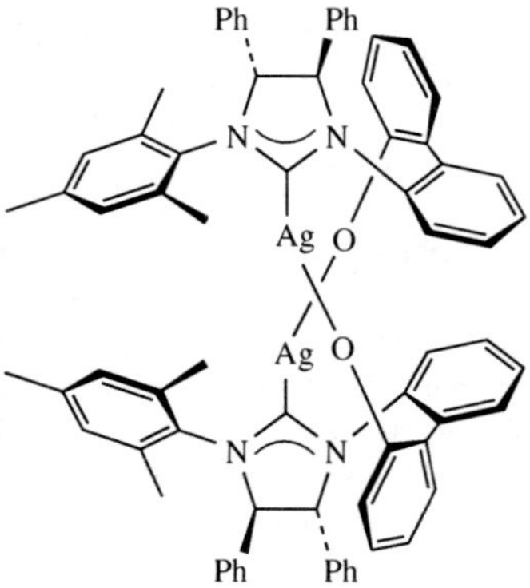

(35)

R^1	$R^2{}_2Zn$	ee
Me	Et_2Zn	93%
Ph	Et_2Zn	90%
	Et_2Zn	84%
Me	Ph_2Zn	97%

SCHEME 18. Non-activated β-substituted cyclohexen-1-ones as substrates

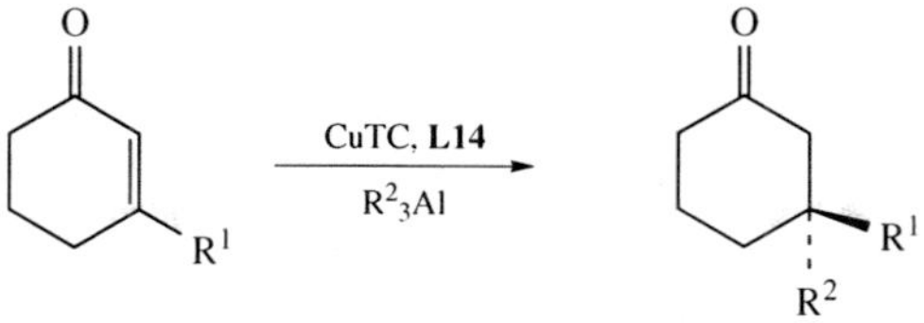

(34) **(36)**

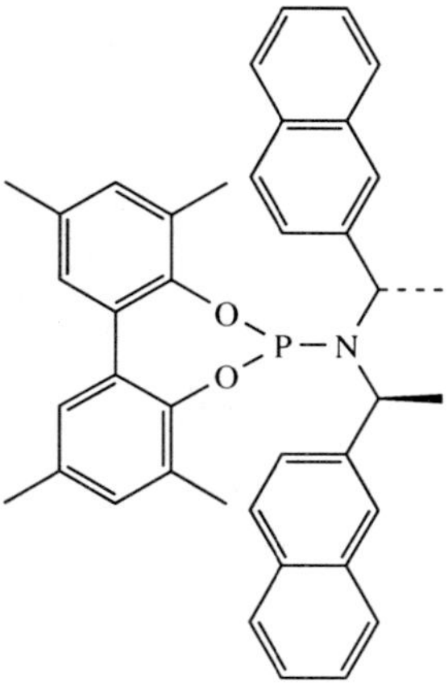

(L14)

R^2_3Al	R^1	ee
Et_3Al	Me	97%
Me_3Al	Et	96%
Me_3Al	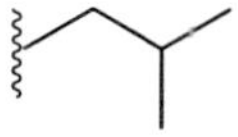	93%
Me_3Al	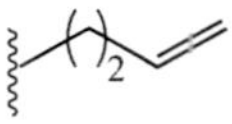	95%
Me_3Al	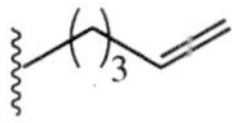	95%
Me_3Al	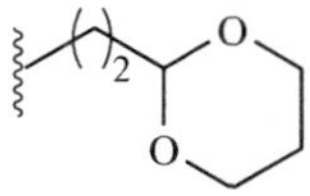	95%

SCHEME 19. Successful combination of substrates **34** and organoaluminum reagents

laboratory recognized the potential of such reagents and developed a very general and reliable method for the construction of quaternary stereocenters (Scheme 19)[30].

Easily accessible substrates like **34** react smoothly to give the desired ketone adduct **36** in high yields and enantioselectivities. It is therefore possible to transfer a methyl group onto a non-activated enone. An interesting application is the straightforward access to quaternary carbon-containing natural products (*vide infra*) starting from functionalized substrates.

One should note that the methodology can be extended to more challenging substrates and different ring sizes (Scheme 20), such as **37** ($n = 1, 3$). The results achieved on **38** are comparable to the stereoselectivity obtained with 3-methylcyclohexen-2-one **34** (R^1 = Me)[31]. Alternatively, the aluminum enolate **39** can be trapped by various electrophiles, generating an enol acetate, a silyl enol ether or an enol carbonate[31a]. Enol acetate as in **40** can be used to generate the versatile lithium enolate.

Aryl aluminum reagents, which can be prepared *in situ* through a halogen/Li exchange–Li/Al transmetalation sequence, are now used under the same conditions to yield adducts **36** (Scheme 21)[32]. Either electron-donating or electron-withdrawing groups on the aryl nucleophile generate full conversion and very good ee. Of particular synthetic interest, vinyl alanes (generated through hydroalumination or a I/Li exchange reaction) can be used successfully as well.

Similar results were obtained by Hoveyda and coworkers[33].

Grignard reagents have been studied and used in conjugate addition, but until recently no general method was described using this organometallic reagent in catalytic ACA.

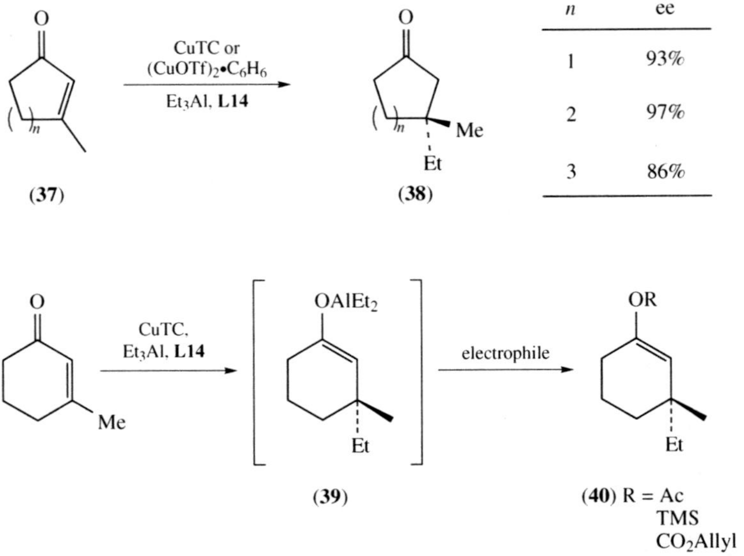

n	ee
1	93%
2	97%
3	86%

SCHEME 20. Variation of the ring size of **37** and trapping reactions

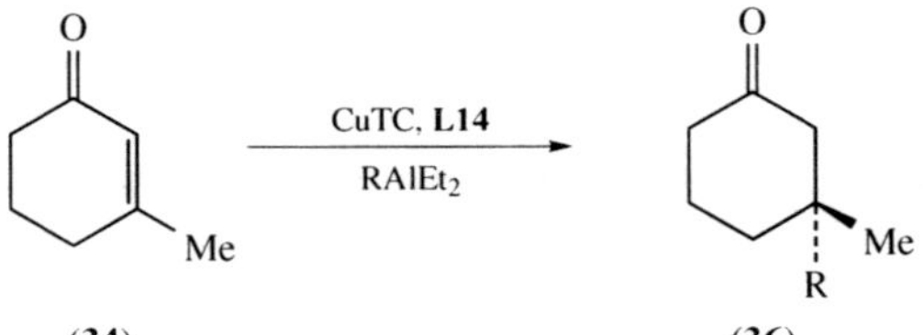

R	ee
Ph	97%
p-MeC_6H_4	96%
p-$MeOC_6H_4$	95%
p-$CF_3C_6H_4$	98%
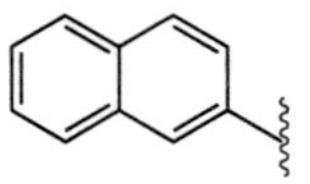	99%
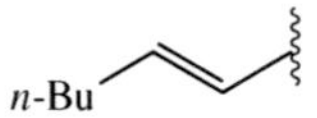	84%

SCHEME 21. Aryl and vinyl alanes as nucleophiles in the ACA

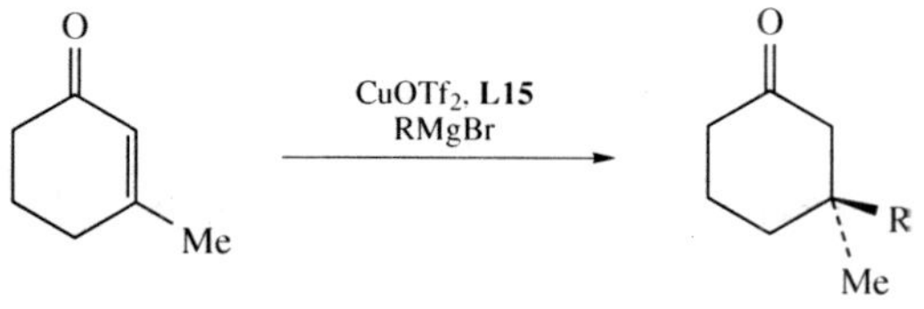

(**34**) (**36**)

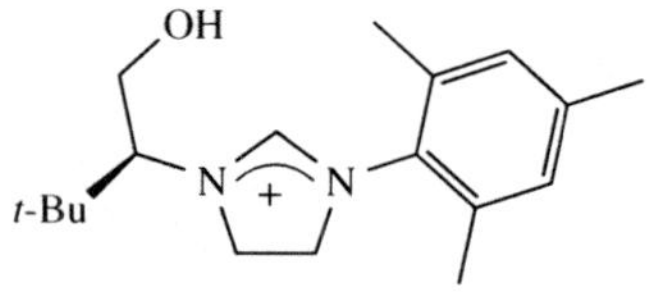

(**L15**)

R	ee
Et	80%
n-Bu	77%
	90%
i-Bu	96%
i-Pr	77%
c-Pen	85%
Ph	66%

SCHEME 22. Enantioselective construction of quaternary centers by ACA of RMgBr

Diaminocarbenes emerged as amazing ligands for copper in this reaction, giving impressive regioselectivity (1,2 versus 1,4 addition) and high enantioselectivity using enone **34** (Scheme 22)[34]. This represents a powerful implementation of the ACA reaction, as Grignards are one of the most nucleophilic, versatile and easy-to-prepare organometallics.

An extensive screening led to **L15** as the most optimal ligand, yielding **36** with up to 96% ee.

Another similar system using NHC ligands has been described by Tomioka and coworkers[35].

IV. TANDEM REACTIONS

Using only a catalytic amount of copper represents a great advantage over selecting stoichiometric organocopper or organocuprate reagents. For instance, one can employ the reaction intermediate—the zinc enolate—as shown in Scheme 23. In addition to saving steps and time, tandem reactions contribute to rapidly build complexity[36].

Despite their low reactivity, zinc enolates **41** (or **42**, as they should be seen as dimers[37]) have been reported to react with various electrophiles such as aldehydes to give **43**[38,39] or acetals (using $BF_3{\bullet}Et_2O$ as additive)[40] to yield **44** and ultimately **45**. Alkyl electrophiles were also used, such as allylic acetates (with Pd catalysis)[38,41], homopropargylic iodide[42], methyl iodide[43] or benzyl iodide[44] (**46** would then be obtained using a tenfold excess of $PhCH_2I$ and 10 equivalents HMPA as additive). Yus and coworkers recently disclosed

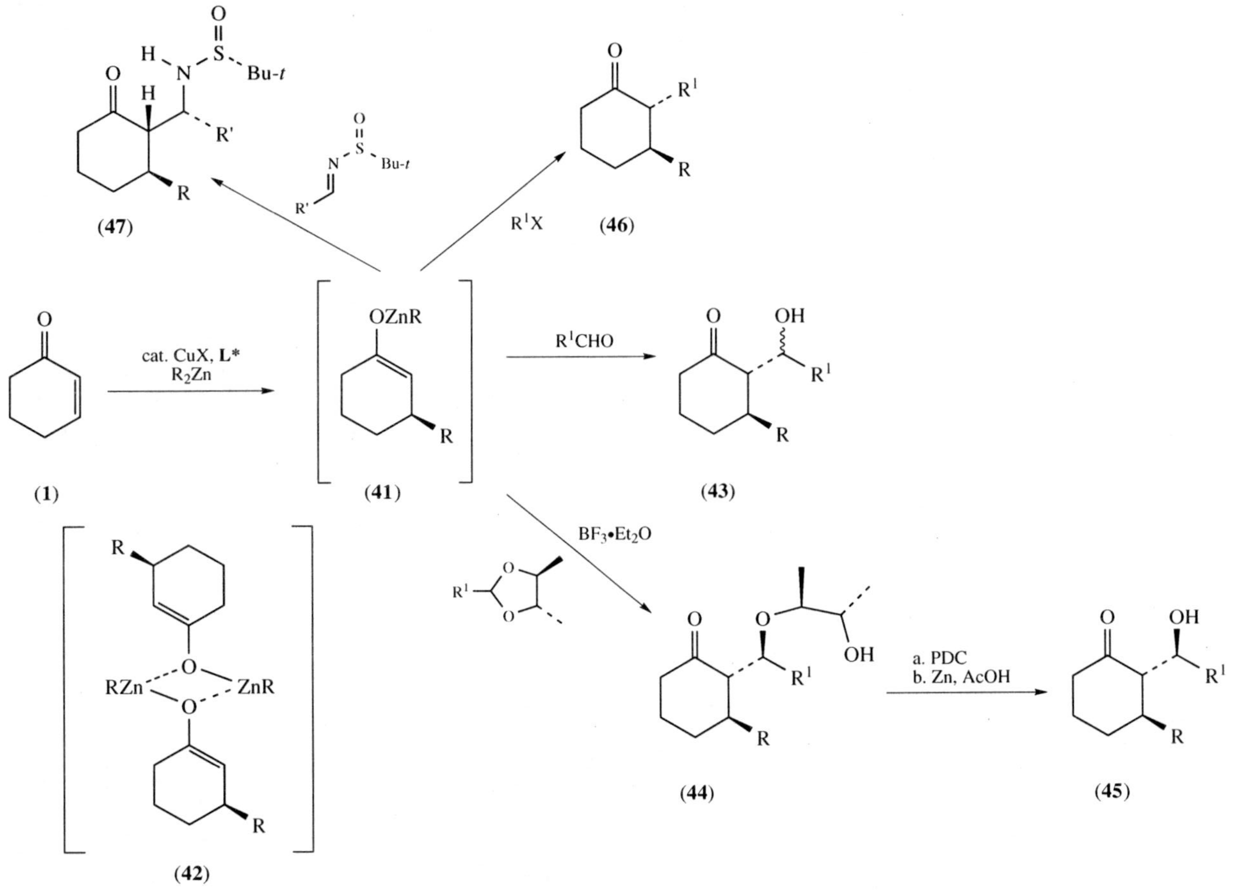

SCHEME 23. Trapping reactions

access to enantiopure aminoketones: as in the case of acetals, trapping with chiral *tert*-butylsulfonimines creates three contiguous stereocenters as in **47**, each of them perfectly controlled[45].

One may also use oxophilic electrophiles, whose products lead to enol acetates[46], enol triflates[27] or silyl enol ether **48**[28,47]. The latter—arising from the zinc enolate obtained with the aid of **L16**—give rise to useful synthetic intermediates such as the ones shown in Scheme 24. It should be noted that enol acetates, silylenol ethers and enol carbonates were also obtained from the trapping of aluminum enolates[31a,46].

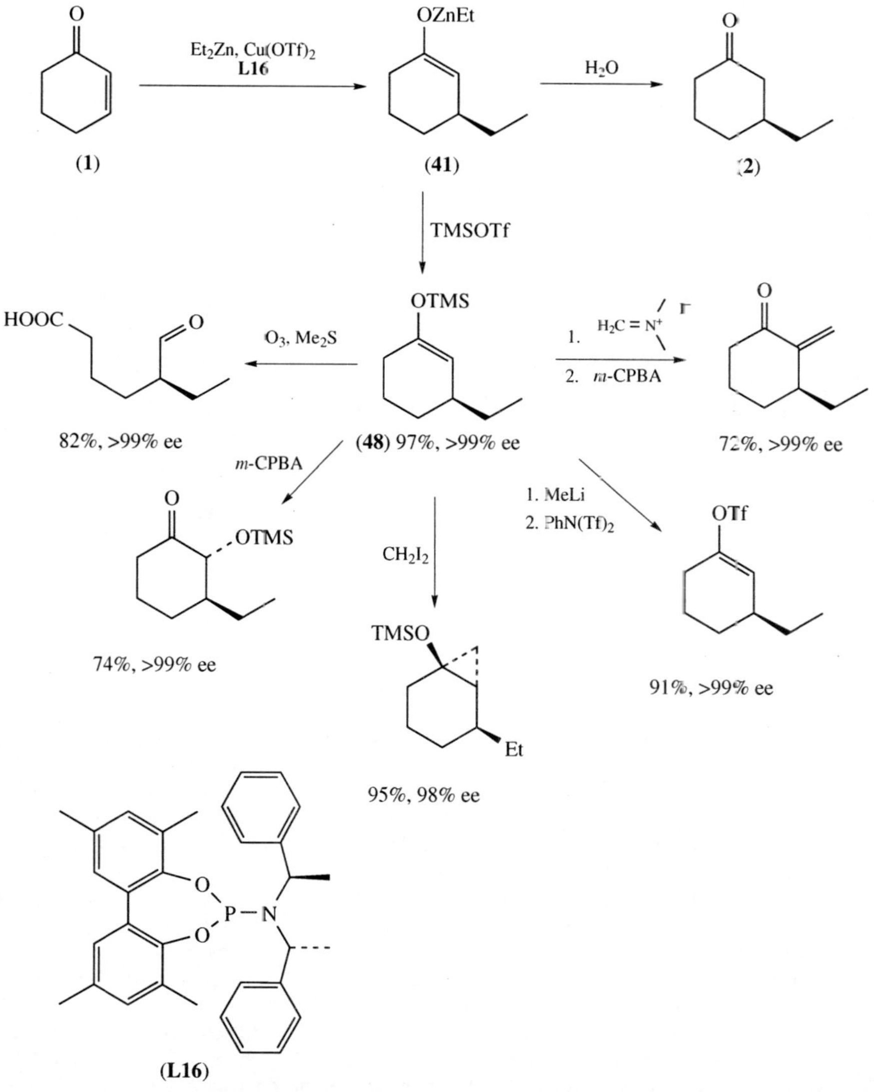

SCHEME 24. Use of the TMS enolether intermediate

One pot enol triflate formation was used by Hoveyda and coworkers on silane-containing enones **49** (simple peptide-based ligand **L17** as chiral information) in order to prepare synthetically useful building blocks (Scheme 25)[48]. Coupling of the triflate product **50** led to allylic silanes **51**, which could then be used in known processes to construct a diversity of skeletons **52–54**.

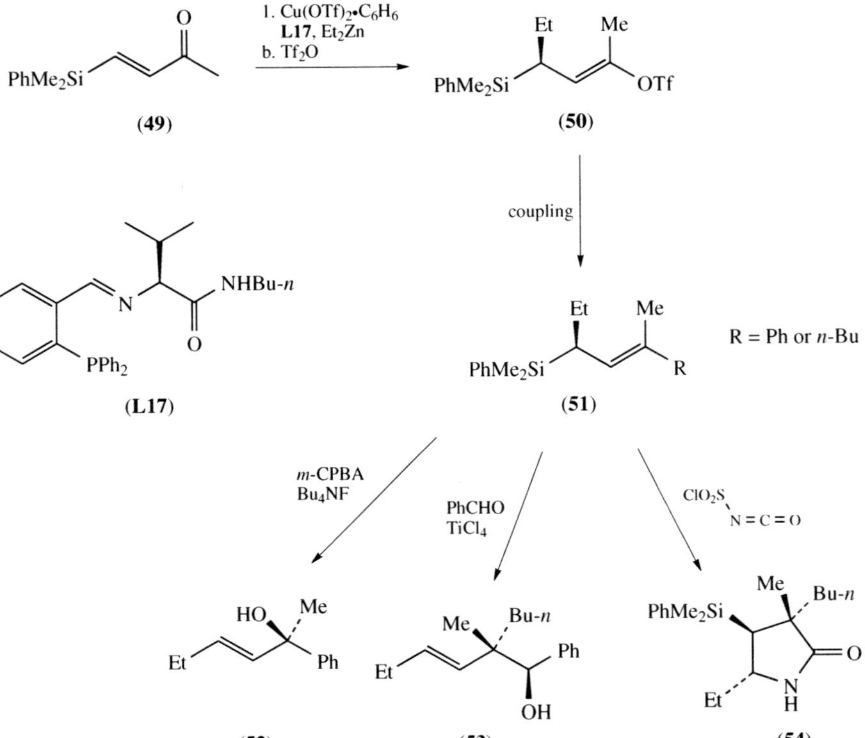

SCHEME 25. Silane-containing enones leading to useful chiral allylic silanes **51**

A three-component system, which represents an extremely efficient and general protocol for the ACA/aldehyde trapping reaction, has been disclosed (Scheme 26)[49]. Indeed, the presence of an aldehyde avoids any undesired intermolecular reaction from the zinc enolate intermediate. This system works on several diorganozinc reagents and different types of substrates **55** or **58**. After oxidation of product **56**, the ee of diketone **57** was found to be more than 98%.

Another example of intermolecular trapping makes use of several types of halogen donors as electrophiles (Scheme 27)[50]. Hence, the zinc enolate intermediate is trapped as ketone **10** with up to 74:26 of diastereoisomeric ratio. Similarly, halogenation of the zinc enolate arising from enone **59** provides **60**, which can then cyclize in a radical manner to give product **61** as a mixture of diastereoisomers (Scheme 28).

Following the same trend, a catalytic tandem 1,4-addition/*N*-nitroso-aldol reaction of organozinc reagents and acyclic enones **62** was described (Scheme 29)[51]. Although the diastereocontrol was poor, desired α-hydroxyamino ketone **63** was obtained with a high ee, thanks to ligand **L19**.

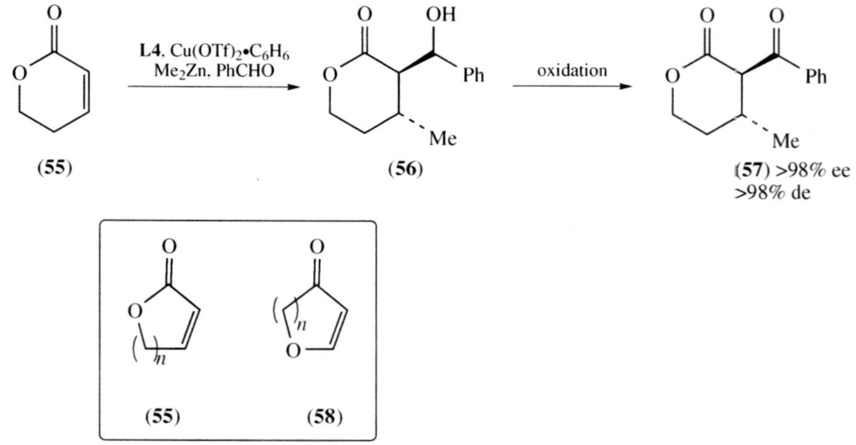

SCHEME 26. Three-component ACA/aldehyde trapping reaction

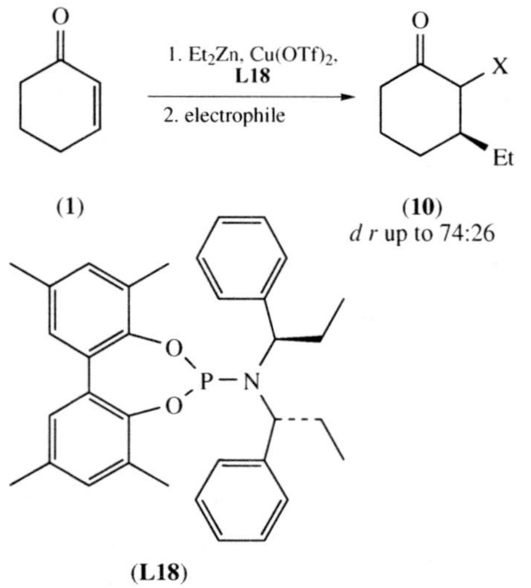

Electrophile	X	ee
NBS	Br	98%
NCS	Cl	97%
I_2	I	98%

SCHEME 27. ACA/halogenation sequence

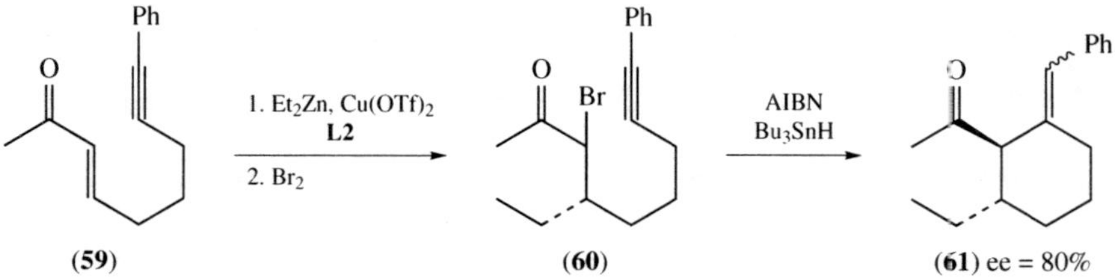

SCHEME 28. Halogenation of the zinc enolate intermediate and subsequent radical cyclization

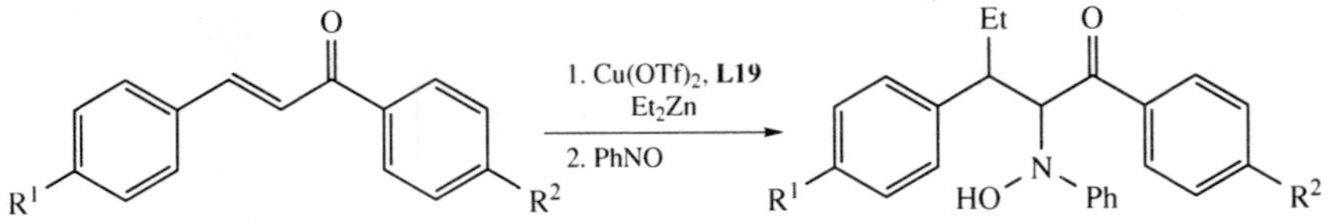

(**62**) (**63**) up to 92% ee up to 3:2 dr

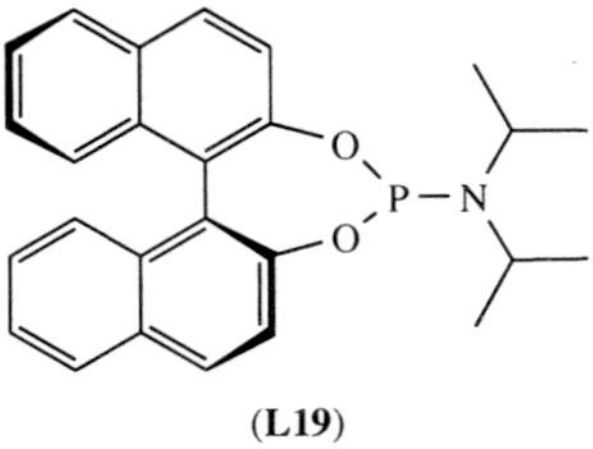

(**L19**)

SCHEME 29. Tandem 1,4-addition/*N*-nitroso-aldol reaction

As seen above, some electrophiles such as allylic acetate need the use of palladium catalysis. On the other hand, activated allylic electrophiles **64** were shown to react by themselves (Scheme 30)[52]. Using **L20** as ligand, high *trans* selectivity and excellent enantioselectivities could be attained as in **65**. In addition to introducing an alkene moiety, the functionalized nature of the electrophiles represents a great potential for further elaboration.

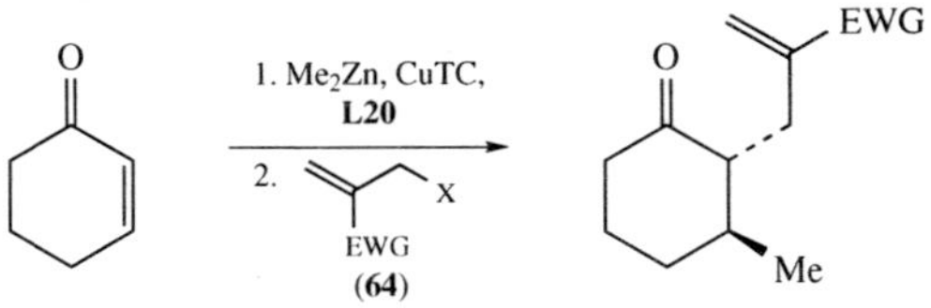

(**1**) (**65**)

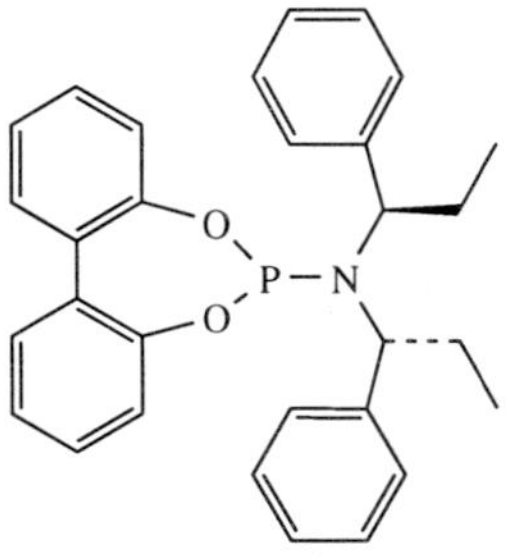

EWG	X	dr	ee
NO_2	OPiv	90:10	98.5%
CO_2Et	Br	94:6	98.9%
SO_2Tol-*p*	Br	89:11	99.1%
$P(O)(OEt)_2$	Br	n.d.	>98%

(**L20**)

SCHEME 30. Use of activated allylic electrophiles in the trapping process

Intramolecular trapping of the zinc enolate intermediate is also possible. An early example makes use of sulfonate enones **66**, which were prepared by a cross-metathesis reaction (Scheme 31)[44]. Ligand **L21** was especially designed for general ACA onto acyclic enones with high enantioselectivities (up to 95% ee measured on **67**). Copper-catalyzed conjugate addition of the alkyl group was followed by the desired cyclization in a very selective *trans* fashion (>98% de).

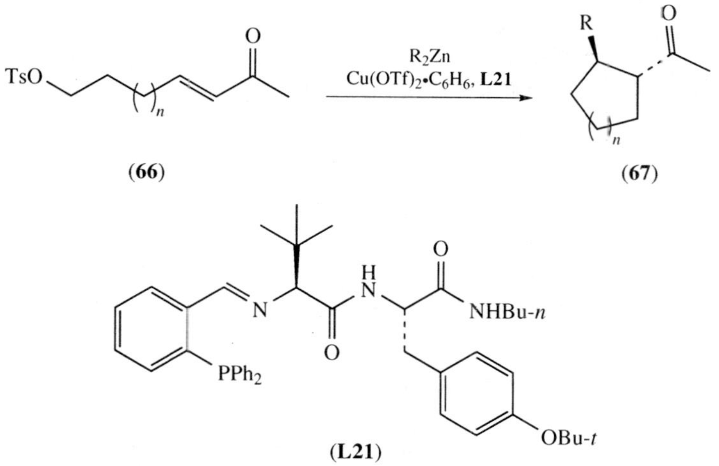

(66) **(67)**

(L21)

SCHEME 31. Intramolecular trapping of the zinc enolate intermediate

Tandem ACA/intramolecular aldol has been performed on substrates like **68** (Scheme 32)[53], thus forming four stereocenters (as in **69** and **70**) in a single step. Good to high enantioselectivities were obtained using *ent*-**L2** as the chiral information, although with modest diastereoselectivities.

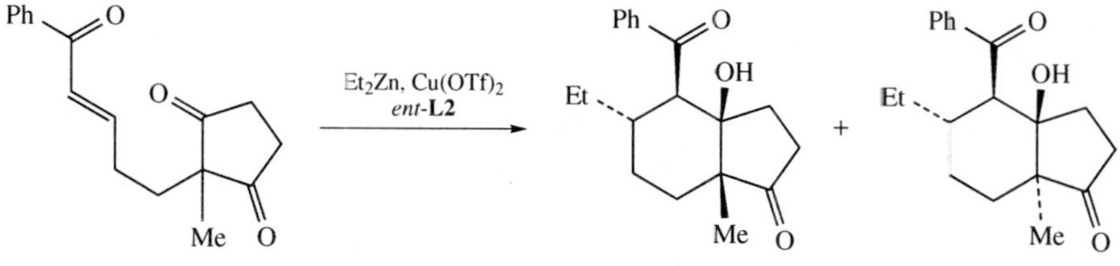

(68) **(69)** 80% ee **(70)** 98% ee

(2.3:1 dr)

SCHEME 32. Tandem ACA/intramolecular aldol

Substrate type **71** was designed to undergo a tandem ACA/intramolecular conjugate addition reaction (Scheme 33)[54]. No double ACA could be observed, therefore **72** was obtained in high yield and high enantio- and diastereopurity.

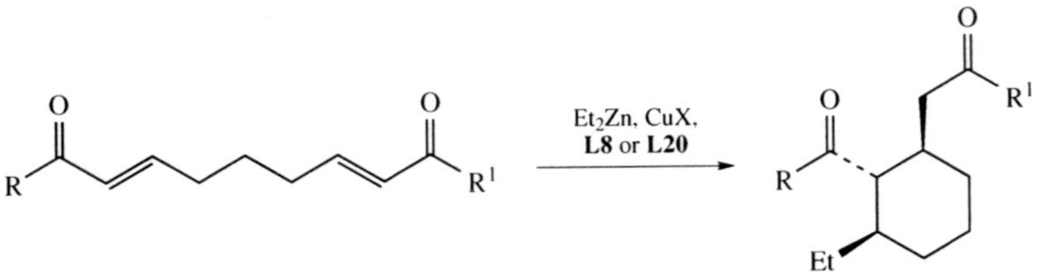

(**71**) (**72**)

Substituents	ee	dr
R = R^1 = Me	79%	80:20
R = R^1 = Ph	88%	>99:<1
R = Me, R^1 = OMe	81%	81:19
R = Ph, R^1 = OMe	92%	93:7

SCHEME 33. ACA/intramolecular conjugate addition

V. APPLICATIONS IN SYNTHESIS

In the last few years, an impressive amount of the ACA literature has been focused on the total synthesis of natural products. An early synthetic example of 1,4-addition/tandem alkylation reaction is the short total synthesis of clavularin B (**74**, Scheme 34)[42]. Using peptide-based phosphane **L4**, Hoveyda and coworkers were able to add dimethylzinc to **11** with high stereocontrol. Trapping the zinc enolate intermediate with the required iodoalkene gave *trans*-**73** as the major product. Only two additional steps were necessary to render the anticancer compound **74**.

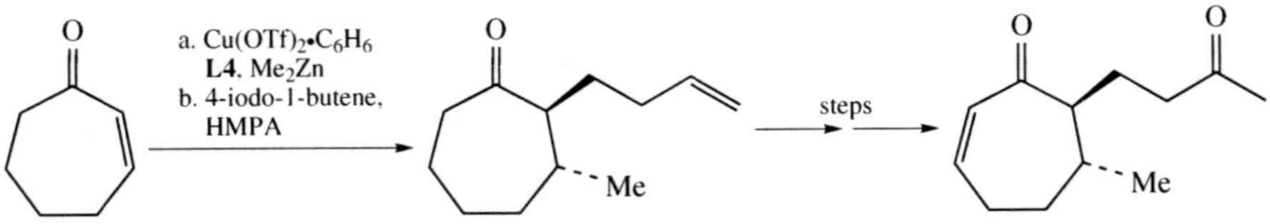

(**11**) (**73**) >15:< 1 *trans/cis* 97% *ee* (**74**) clavularin B

SCHEME 34. Total synthesis of clavularin B

The total synthesis of (−)-Prostaglandin E$_1$ methyl ester **78** (Scheme 35)[55] represents one of the most impressive applications of ACA. By designing the rightly functionalized diorganozinc compound and an extremely elaborated electrophile (**76**), the Feringa group was able to assemble three components (including **75**) in the same pot into **77** as the major diastereomer. An ee of 94% was measured at a later stage of the synthesis.

As seen above, conservation of the regiochemistry of the enolate allows the creation of a second stereocenter in a stereocontrolled fashion. One could also trap the zinc enolate **79** and isolate the TMS enol ether as in **80** in order to incorporate an extra carbon through a cyclopropanation/opening sequence **80** → **81** → **82** (Scheme 36)[47]. Preparation of homologated **82** was key in the first catalytic and asymmetric synthesis of

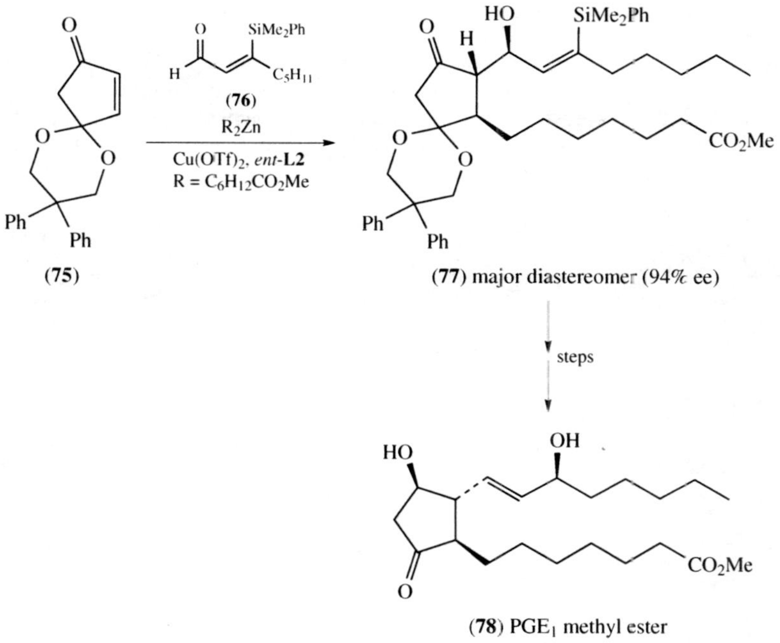

SCHEME 35. Trapping strategy developed for the total synthesis of (−)-Prostaglandin E$_1$ methyl ester

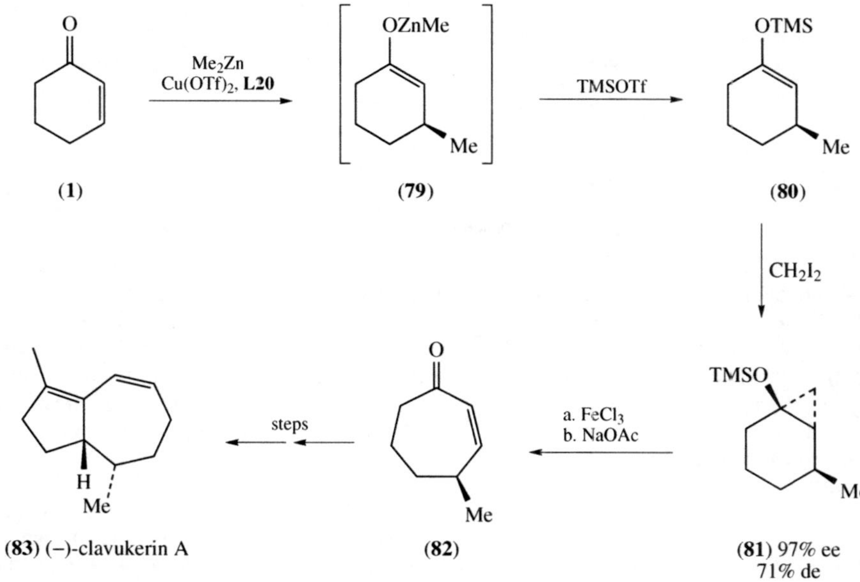

SCHEME 36. Formal synthesis of (−)-clavukerin A

(−)-clavukerin A **83**, a sesquiterpene isolated from the Okinawan soft coral *Clavularia koellikeri*.

A dienone strategy, starting from **84** and **85**, was employed in order to introduce two stereocenters through consecutive ACA (Scheme 37). The second zinc enolate was trapped as a TMS enol ether (**86**) and isolated with a high *trans*-stereochemistry. This compound was then reacted further to provide apple leafminer pheromones **88** and **89**[56] from a common intermediate **87**.

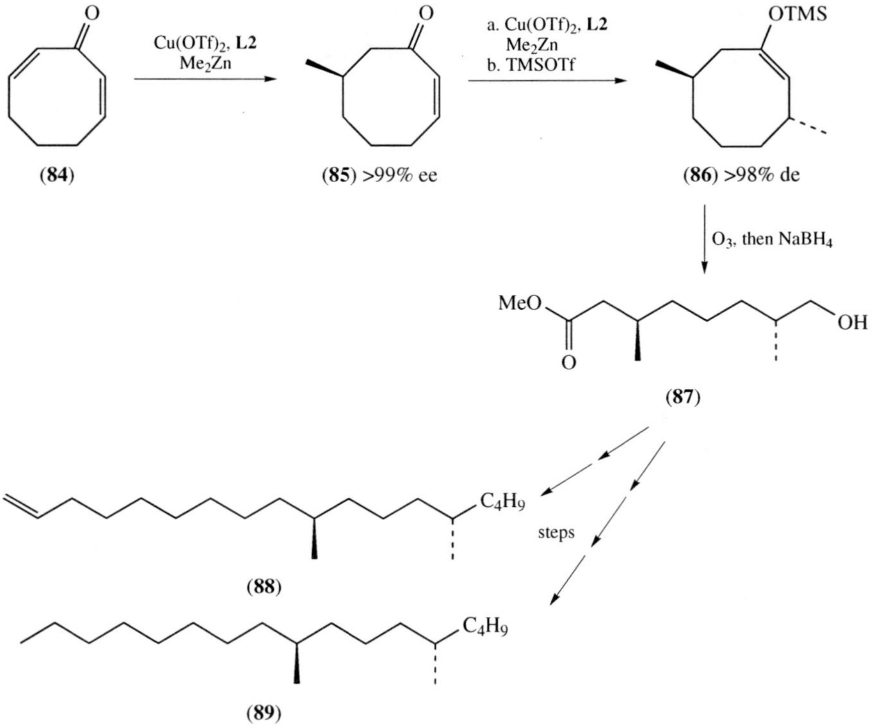

SCHEME 37. Total syntheses of two apple leafminer pheromones

Although there already exist a number of preparations of the musk odorant (*R*)-muscone **92** by conjugate addition, the Pfaltz group took advantage of such an above-mentioned dienone strategy as in **90** (Scheme 38)[57]. A conformational rigidity as well as a greater reactivity of this substrate allowed a much higher enantioselectivity than the corresponding mono-unsaturated ketone. Using Hoveyda's very modulable peptide-based phosphine ligand **L22**, they were able to obtain enantiopure **91**, which yielded (*R*)-muscone **92** after a hydrogenation step.

Starting from **93**, a short synthesis of erogorgiaene, an inhibitor of *Mycobacterium tuberculosis*, takes advantage of two asymmetric conjugate additions onto acyclic enones (Scheme 39)[58]. The first ACA afforded ketone **94**, which could be elaborated further to **95**. One should note that the transformation of dienoate derivative **95** into **96** with the aid of another ligand (**L23**) provides not only an excellent diastereocontrol, but also an impressive regiochemistry.

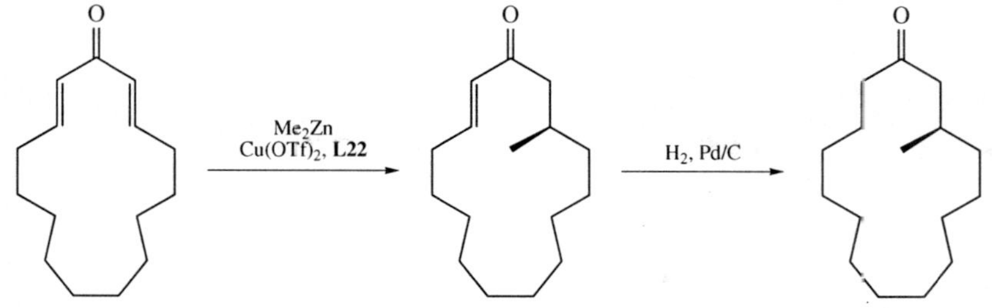

(**90**) (**91**) 98% ee (**92**) (*R*)-muscone

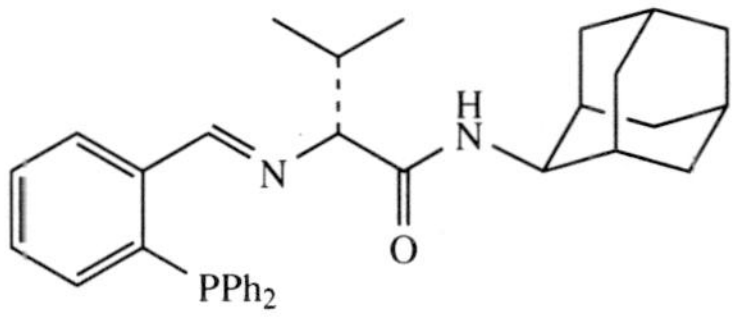

(**L22**)

SCHEME 38. Total synthesis of (*R*)-muscone by Pfaltz

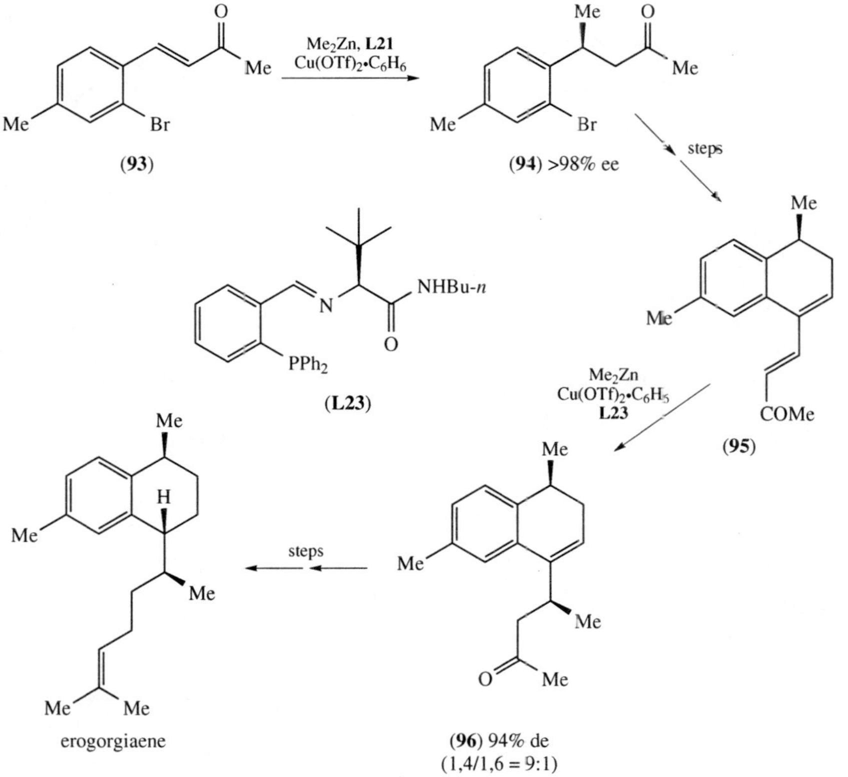

(**93**) (**94**) >98% ee (**L23**) (**95**)

erogorgiaene

(**96**) 94% de
(1,4/1,6 = 9:1)

SCHEME 39. Total synthesis of erogorgiaene

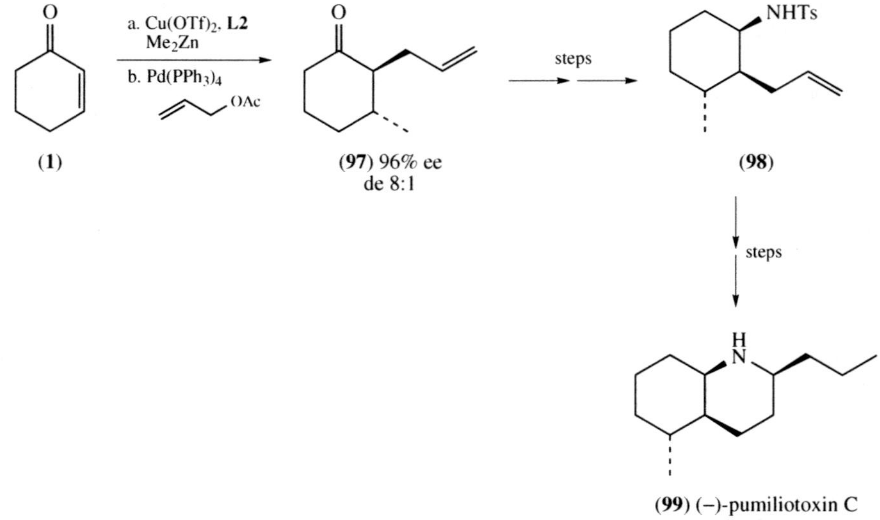

SCHEME 40. Total synthesis of (−)-pumiliotoxin C

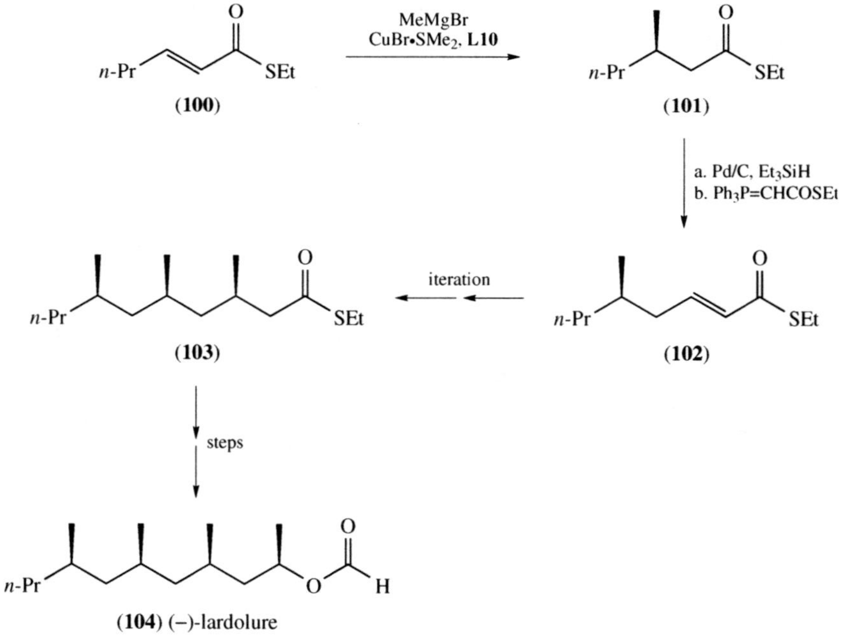

SCHEME 41. Total synthesis of (−)-lardolure following an iteration strategy

The zinc enolate intermediate formed by conjugate addition to **1** was treated with allyl acetate under palladium catalysis (Scheme 40)[59]. The resulting α-allylated product **97** was transformed into **98** and used further in the straightforward synthesis of (−)-pumiliotoxin C (**99**), a potent neurotoxin isolated from *Dendrobates pumilio* (poison dart frogs) that acts as a non-competitive blocker for acetylcholine receptor channels.

Using (*R*,*S*)-Josiphos, the Feringa group achieved the development of an efficient method for the conjugate addition of methyl Grignard reagents to α,β-unsaturated thioesters such as **100** (Scheme 41)[60]. Adduct **101** was submitted to a practical reduction/olefination procedure, which gave the elaborated Michael acceptor **102**. Iteration of this three-step procedure allowed the construction of the desired skeleton as in **103**, which ultimately led to **104**.

This highly efficient and enantioselective iterative catalytic protocol was used for the synthesis of other deoxypropionate skeletons, such as Phthioceranic acid **109** (Scheme 42)[61]. The desired skeleton **108** was prepared using as low as 1% of the complex (*R*,*S*)-Josiphos•CuBr and a 7-time sequence of conjugate addition/reduction/olefination as shown in **105** → **106** → **107**.

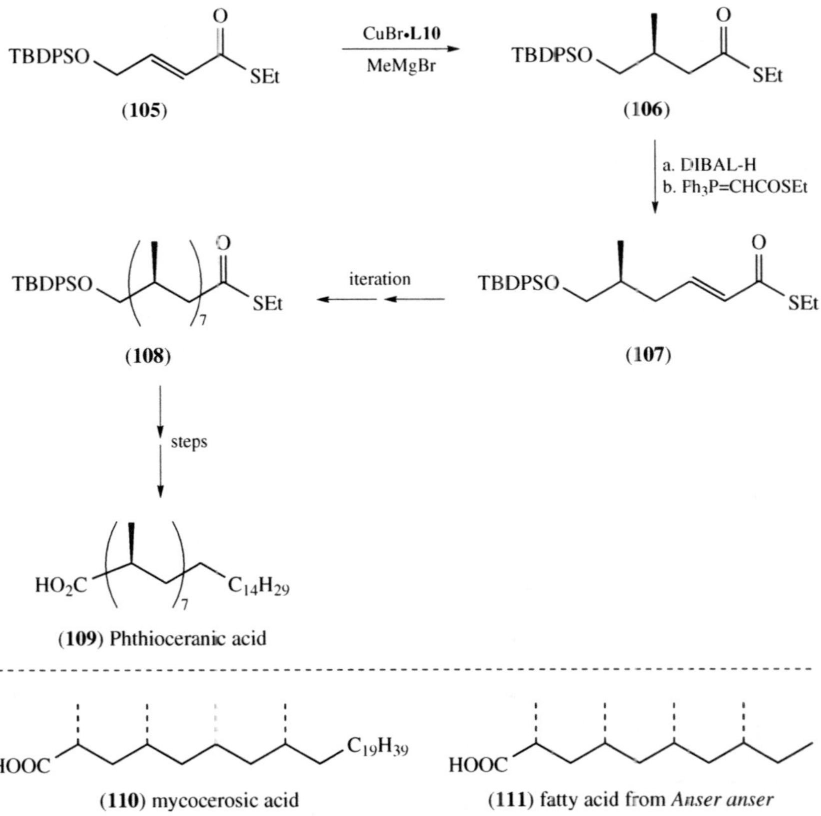

SCHEME 42. First total synthesis of Phthioceranic acid and total syntheses of mycocerosic acid and a fatty acid from *Anser anser*

Two fatty acids (**110** and **111**, Scheme 42), namely mycocerosic acid and a preen-gland wax of the graylag goose *Anser anser*, were prepared following this strategy[62].

Combining the methodologies developed for the conjugate addition of dimethylzinc[63] and methyl Grignard reagents[64], it was possible to achieve the first asymmetric total synthesis of phthiocerol dimycocerosate A, PDIM A (**115**, Scheme 43)[65]. The latter is a constituent of the cell wall of *Mycobacterium tuberculosis* and contains two units of the above tetramethyl-substituted saturated acid (mycocerosic acid[42] **110**) esterified with phthiocerol (**114**). The synthesis owes its efficiency to the expedient construction of two stereocenters in the same pot: conjugate addition of the methyl group is followed by ethylation to give **112** with a remarkable diastereoselectivity. Synthesis of **114** requires the introduction of an oxygen atom, which is done by a Baeyer–Villiger reaction to give intermediate **113**.

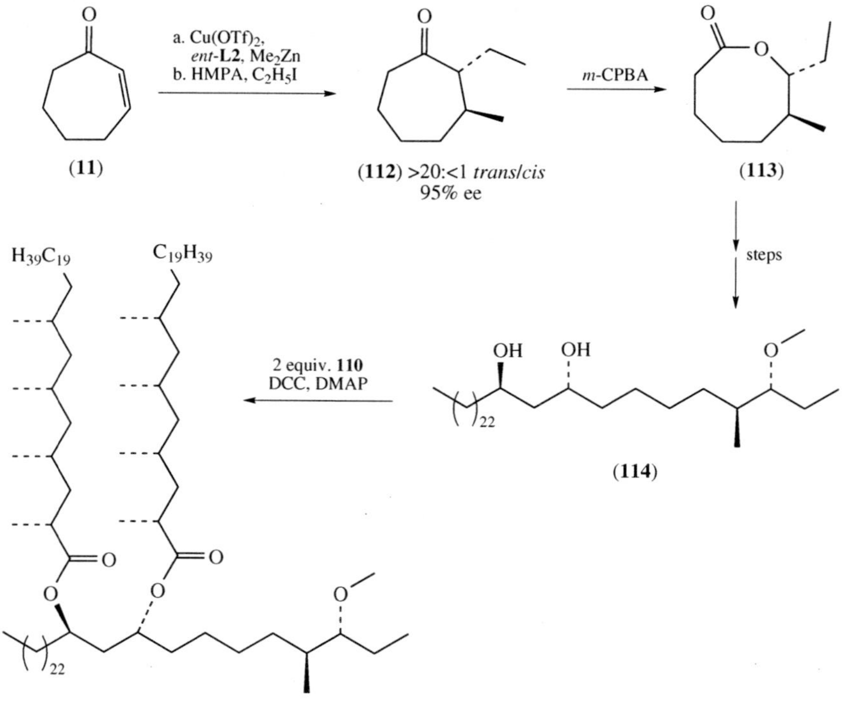

SCHEME 43. Total synthesis of PDIM A

Another utilization of the α,β-unsaturated thioesters like **116** was described by the Feringa group (Scheme 44)[66]. These authors developed a tandem 1,4-addition-aldol reaction using aromatic or aliphatic aldehydes to trap magnesium enolate **117**. One should note the high *syn*-aldol selectivity as well as the formation of the three contiguous stereocenters as in **119** in high diastereo- and enantioselectivity. Transition state **118** was

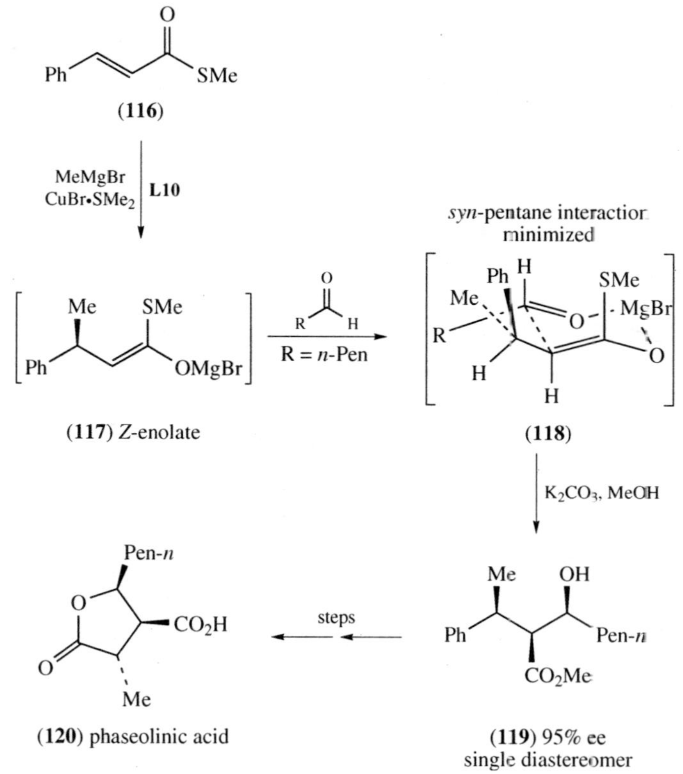

SCHEME 44. Concise route to phaseolinic acid

postulated to explain the stereochemical origin, probably triggered by the minimization of the *syn*-pentane interaction. This methodology was demonstrated by the total synthesis of phaseolinic acid **120** with a 54% overall yield.

By combining the asymmetric addition of diorganozinc[63] and Grignard reagents[64], the Feringa group was able to achieve the total synthesis of a β-D-Mannosyl phosphomycoketides **131** isolated from Mycobacterium tuberculosis (Scheme 45 and Scheme 46), thereby solving the stereochemistry of the lipid part[67].

Intermediate **85** was prepared using the well-established 1,4-addition of dimethylzinc onto dienone **84** with **L2** and $Cu(OTf)_2$ as catalyst[56]. The second stereocenter was created with the aid of *ent*-**L2**. Trapping with a silylating agent allowed for the preservation of the chiral information as in **121**. Further elaboration afforded **122**, which was then used in a divergent manner for the preparation of both **123** and **124**. Coupling of these partners led to precursor **125**, and ultimately **126**.

The last stereocenter of the lipid moiety **130** was installed by the 1,4-addition of MeMgBr on **127**. The reaction was directed by $CuBr{\bullet}SMe_2$ and (*R*,*S*)-Josiphos **L10** as a catalyst and allowed to isolate **128** with a 93% ee. Assembly of the two fragments **126**

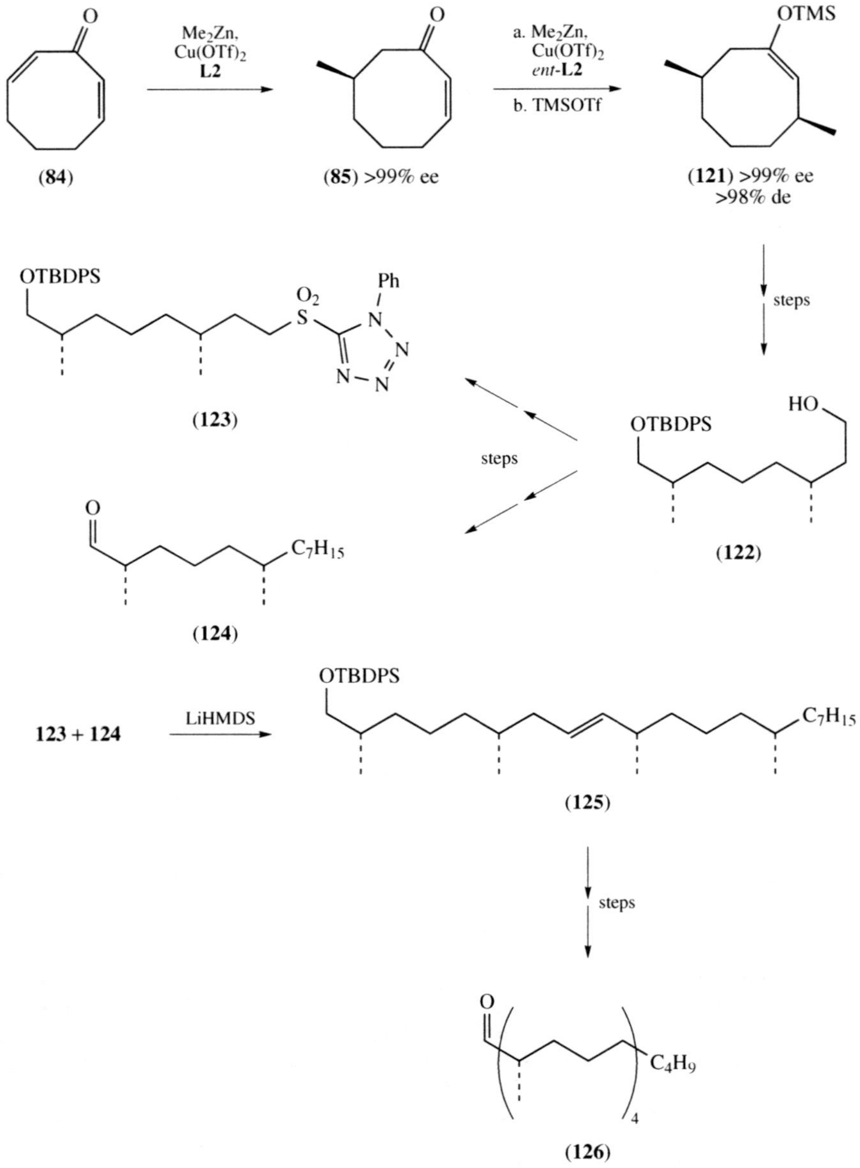

SCHEME 45. Towards the total synthesis of **131**: preparation of intermediate **126**

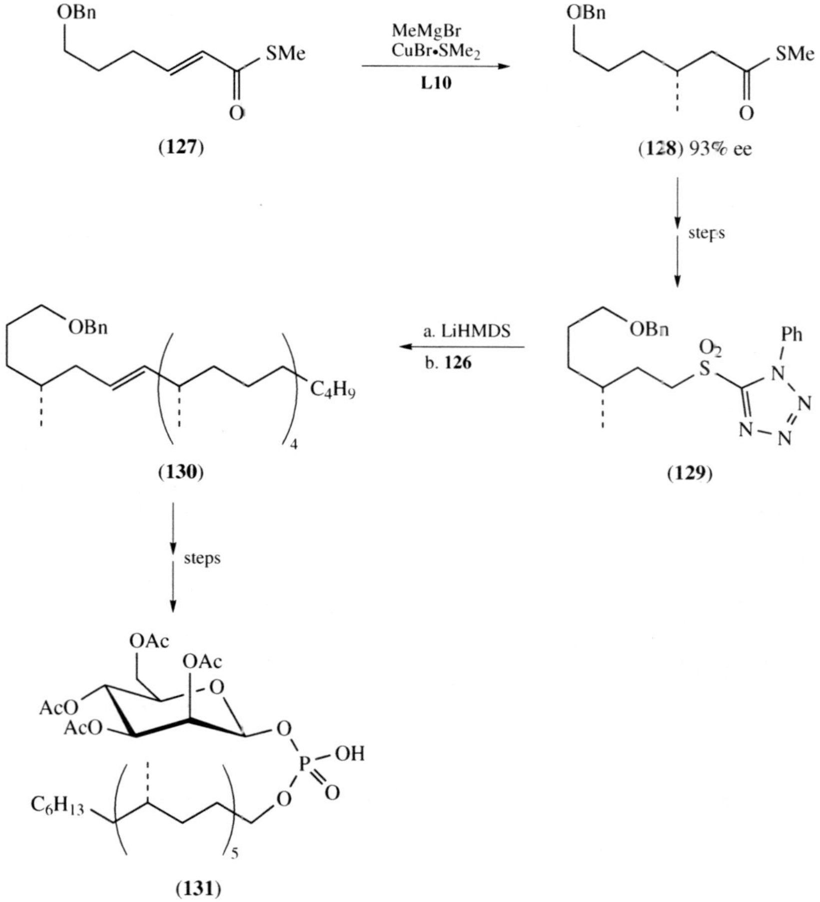

SCHEME 46. Total synthesis of β-D-Mannosyl Phosphomycoketides from Mycobacterium tuberculosis **131**

and **129** led to **130**, and ultimately to **131** after further elaboration and attachment to the β-mannopyranosyl phosphate moiety.

A mating hormone of plant pathogen *Phytophthora* has 16 possible stereoisomers, two of which were synthesized according to Scheme 47 (only the two synthesized diastereoisomers are drawn)[68]. Conjugate addition of MeMgBr on α,β-unsaturated thioester derivatives catalyzed by **L10**•CuBr (or *ent*-**L10** in the case of C7) was used to selectively control three of the stereocenters as in **132**.

A highly regio- and enantioselective 1,6-conjugate addition of Grignard reagent **134** to diene ester **133** has been developed. A very high degree of regioselectivity was observed in favor of 1,6-adduct **135**. This allowed the total synthesis of a sulfated alkene (**136**) isolated from the Echinus *Temnopleurus hardwickii* (Scheme 48)[69]. One should note the use of reversed Josiphos **L24** for this particular application.

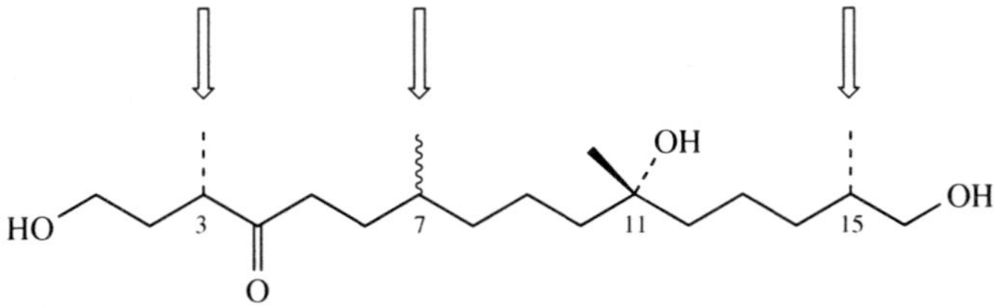

(**132**)

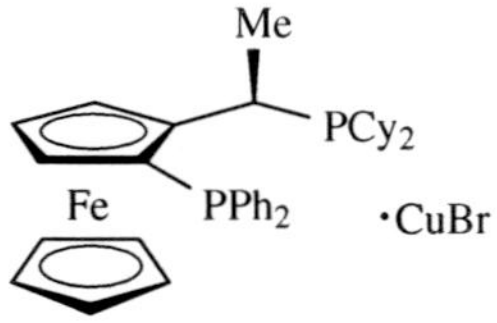

(**L10**·CuBr)

SCHEME 47. Mating hormone **132** (MH-α1) of a species of *Phytophthora*. Arrows designate the stereocenters created by the aid of conjugate addition

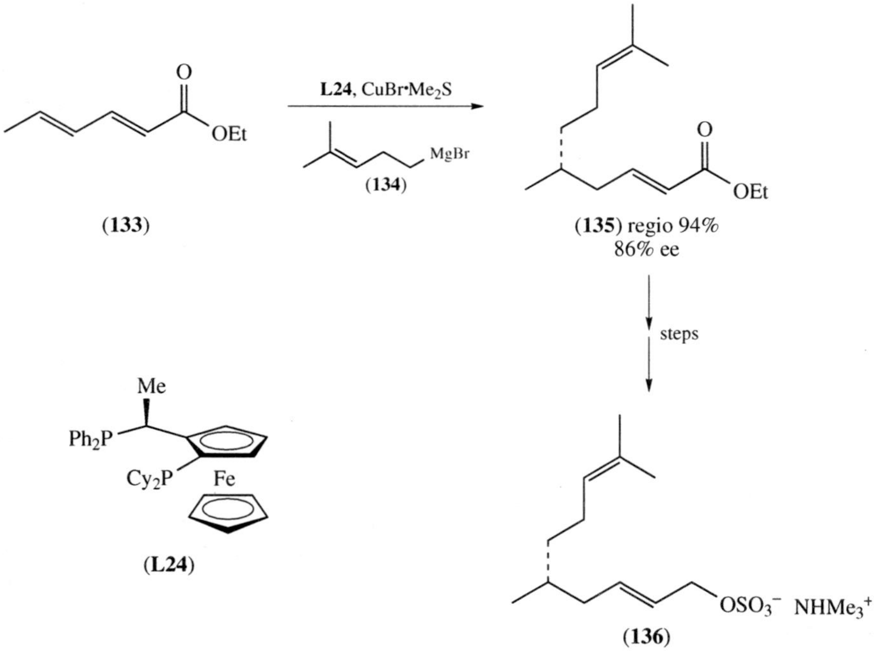

SCHEME 48. Total synthesis of sulfated alkene **136**

Loh's group developed its own iterative strategy, which was exemplified by the total syntheses of two marine natural products (Scheme 49)[70]. Starting this time from α,β-unsaturated esters such as **137**, the authors used the conjugate addition of methyl Grignard catalyzed by the combination of CuI and (S)-Tol-BINAP (*ent*-**L11**). Very efficiently, they were able to control the stereochemistry of **138** (96% ee) and transform it into the homologated Michael acceptor **139** in a reduction/olefination process. Repetition of this whole procedure gave ester **140**, which was then used in a Michael addition/trapping reaction step to yield **141**. The latter was subsequently transformed into both siphonarienal **142** and siphonarienone **143**.

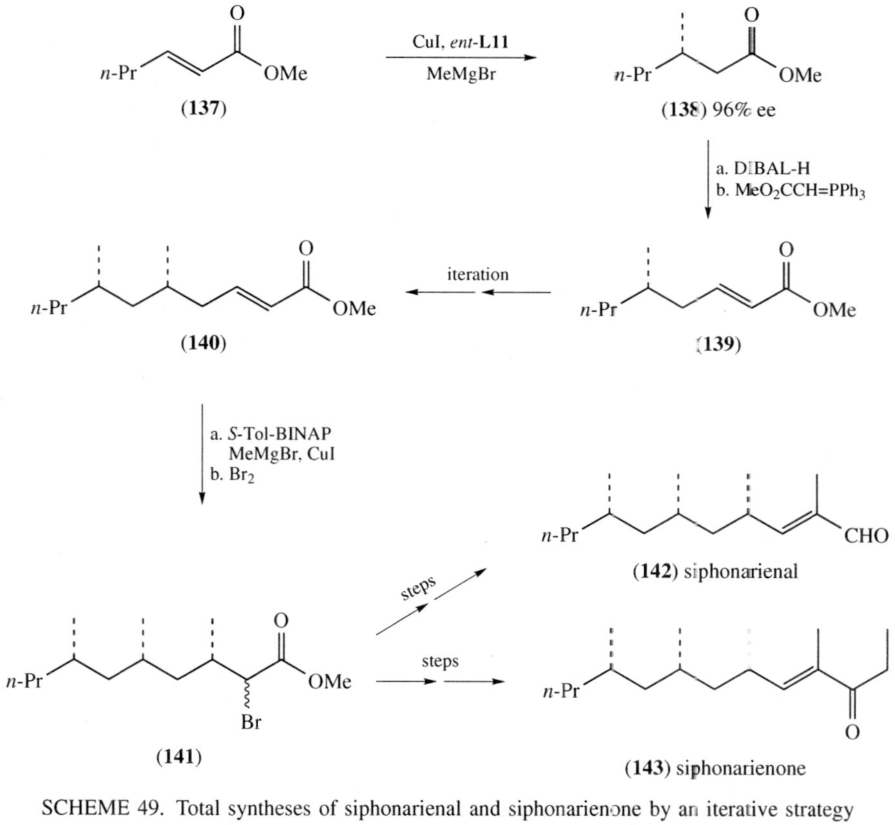

SCHEME 49. Total syntheses of siphonarienal and siphonarienone by an iterative strategy

The construction of all-carbon quaternary stereocenters with organoaluminum found an illustration in the preparation of **145**—a precursor of the Axane derivative **146**—via the formation of enantioenriched **144** (Scheme 50). The Axane family of natural products was isolated from the marine sponge *Axinella cannabia*[31a].

Our group developed a specific ligand for the 1,4-addition of trimethylaluminum to nitroalkenes (Scheme 51)[71]. The application to the total synthesis of the antiinflammatory agent (+)-ibuprofen illustrates this appealing methodology. Starting from compound **147**, adduct **148** was obtained in 82% ee thanks to **L25**. Transformation of the nitromethylene group to the carboxylic acid went smoothly using a modified Kornblum reaction. This process afforded ibuprofen **149** without any racemization.

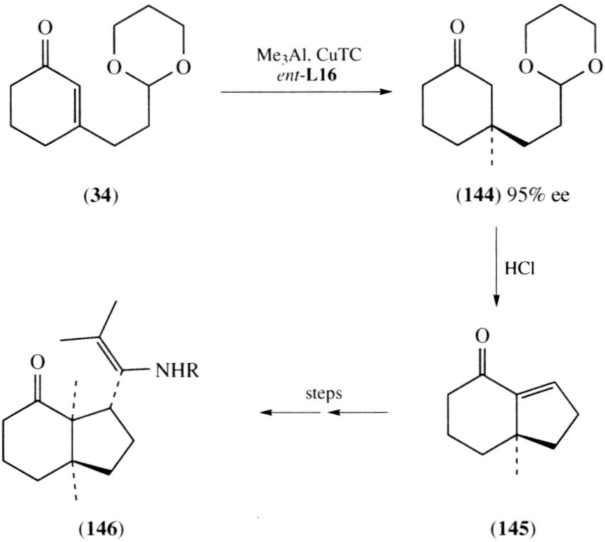

SCHEME 50. Enantioselective construction of an Axane precursor (**145**)

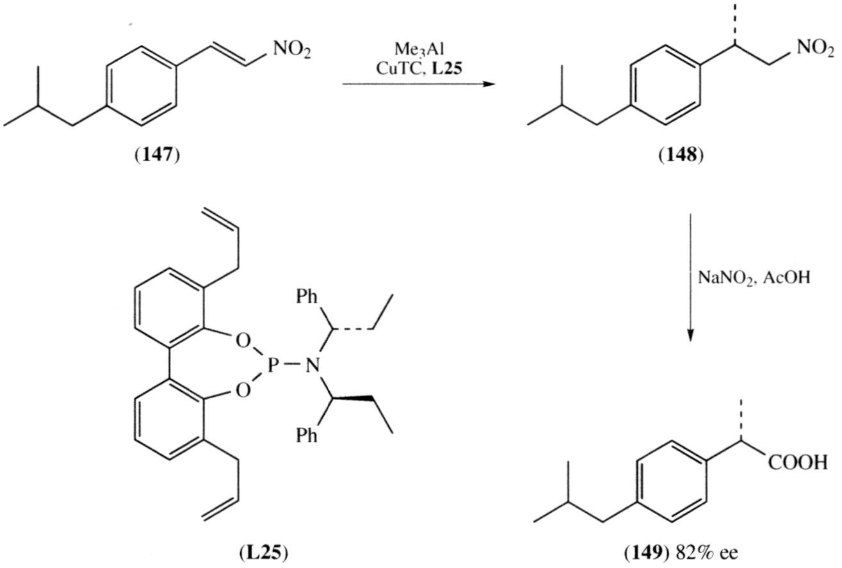

SCHEME 51. Total synthesis of (+)-ibuprofen **149**

Brown and Hoveyda designed an efficient and convergent strategy for the first enantioselective total synthesis of clavirolide C **156** (Scheme 52)[72]. This tricyclic natural product from the dolabellane family of diterpenes was isolated from the soft coral *Clavularia viridis*. In order to elaborate aldehyde **151**, the authors relied on the three-component

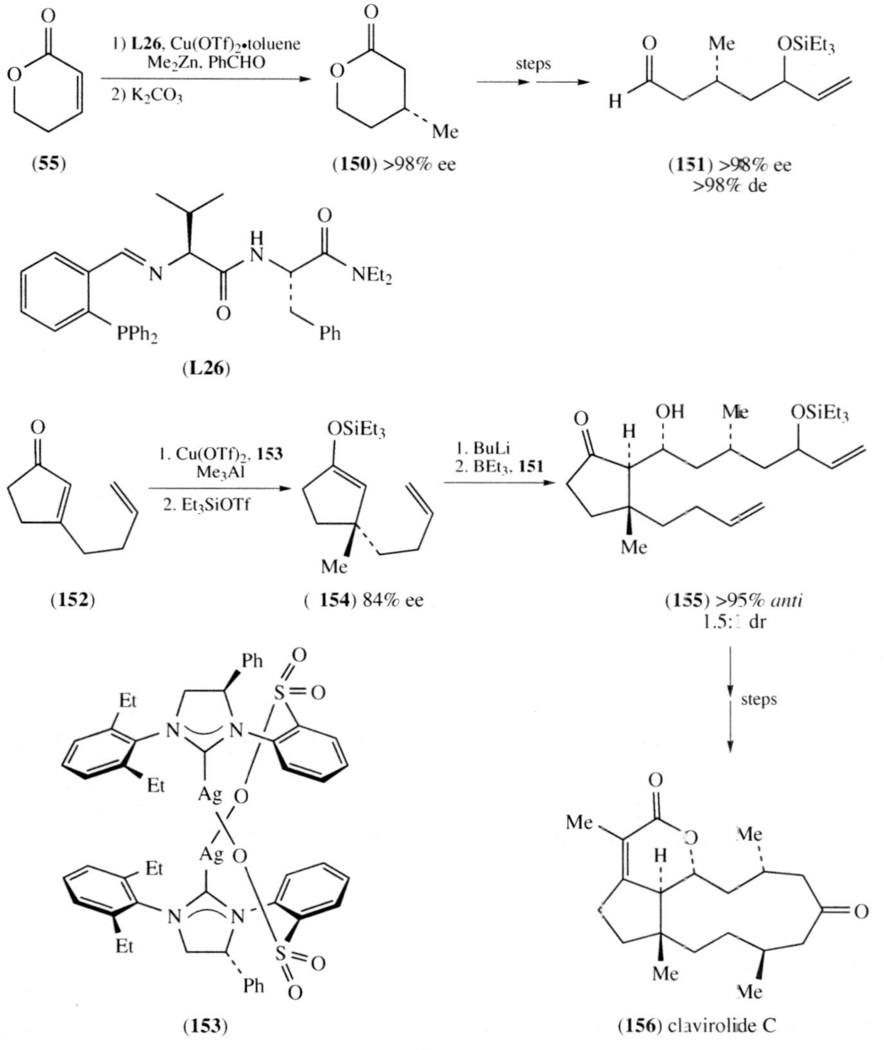

SCHEME 52. Total synthesis of clavirolide C

reaction described in Scheme 26, which generates keto alcohol **56** as the product. Desired lactone **150** was obtained after a retro-aldol step.

NHC-sulfonate ligand contained in **153** and a copper source efficiently catalyzed the conjugate addition to trimethylaluminum to β-alkyl-substituted cyclopentenone **152**. The adduct was trapped as a silyl enol ether (**154**), and an ee of 84% was obtained, which represents a great achievement for this extremely challenging substrate. The overall strategy owes its convergence to the aldol step, which uses aldehyde **151** as coupling partner. The *anti*-aldol reaction between **151** and **154** provided **155**, although the stereochemistry of the α-center was not controlled. Ring-closing metathesis and further elaboration finished the synthesis of clavirolide C.

Finally, asymmetric copper-catalyzed conjugate addition of organozinc reagents allowed an original entry to the hetisine familiy of alkaloids (Scheme 53)[73]. The first asymmetric synthesis of nominine **158** was developed by Peese and Gin using Hoveyda's NHC-sulfonate catalyst for the installation of the all-carbon quaternary stereocenter as in **157** (84% ee).

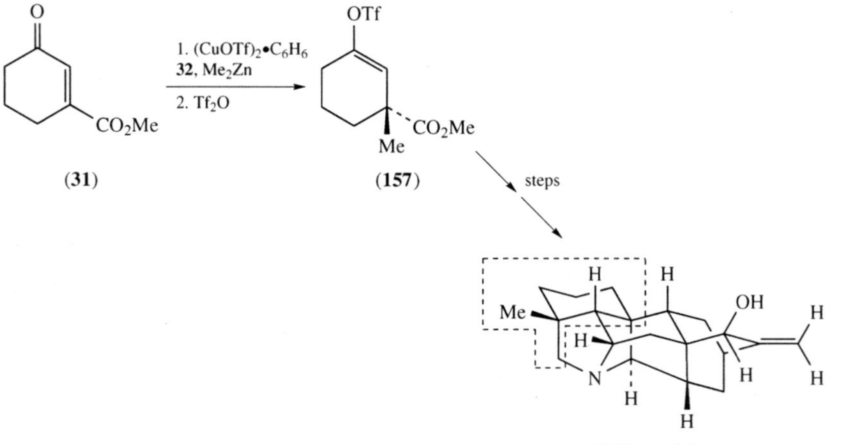

SCHEME 53. Total synthesis of nominine by Peese and Gin. Highlighted is the carbon structure constructed from **157**

VI. CONCLUSIONS AND PERSPECTIVE

The last few years have been a rich and exciting period in the field of copper-catalyzed asymmetric conjugate addition. From the over-used organozinc reagents, the interest has progressively moved to organoaluminum and Grignard reagents. Particularly, the development of the enantioselective addition of MeMgBr on acyclic substrates has opened the way to numerous syntheses of deoxypropionate skeletons.

The most impressive achievement was probably the conjugate addition to β-disubstituted substrates, which allowed the construction of all-carbon quaternary stereocenters using a catalytic amount of chiral information[74].

However, there remain some challenges and limitations. Among them, the use of an organolithium is still not possible, due to its reactivity. Other limitations are also due to the nature of the organocopper intermediate. For instance, alkyne groups are difficult to transfer by copper and have been mostly used as spectator ligands for this metal[75].

More than a viable methodology, the copper-catalyzed ACA is now becoming an essential tool one might use when designing the synthesis of a new molecule, efficiently constructing building blocks for active pharmaceutical ingredients or improving an existing synthetic route.

A large number of challenges still exist and we are convinced that they will continue to fuel the excitement and the stimulation in this field.

VII. REFERENCES

1. (a) M. P. Sibi and S. Manyem, *Tetrahedron*, **56**, 8033 (2000).
(b) N. Krause and A. Hoffmann-Röder, *Synthesis*, 171 (2001).

(c) O. M. Berner, L. Tedeschi and D. Enders, *Eur. J. Org. Chem.*, 1877 (2002).
(d) B. L. Feringa, R. Naasz, R. Imbos and L. A. Arnold, in *Modern Organocopper Chemistry* (Ed. N. Krause), Wiley-VCH, Weinheim, 2002, p. 224.
(e) A. Alexakis and C. Benhaim, *Eur. J. Org. Chem.*, 3221 (2002).
(f) J. Christoffers, G. Koripelly, A. Rosiak and M. Rössle, *Synthesis*, 1279 (2007).
(g) A. Alexakis, J. E. Bäckvall, N. Krause, O. Pamies and M. Dieguez, *Chem. Rev.*, **108**, 2796 (2008).

2. M. S. Kharasch and P. O. Tawney, *J. Am. Chem. Soc.*, **63**, 2308 (1941).
3. A. Alexakis, J. Frutos and P. Mangeney, *Tetrahedron: Asymmetry*, **4**, 2427 (1993).
4. A. Alexakis, J. Vastra and P. Mangeney, *Tetrahedron Lett.*, **38**, 7745 (1997).
5. B. L. Feringa, M. Pineschi, L. A. Arnold, R. Imbos and A. H. M. De Vries, *Angew. Chem., Int. Ed. Engl.*, **36**, 2620 (1997).
6. A. Alexakis, S. Rosset, J. Allamand, S. March, F. Guillen and C. Benhaim, *Synlett*, 1375 (2001).
7. A. Alexakis, C. Benhaim, S. Rosset and M. Humam, *J. Am. Chem. Soc.*, **124**, 5262 (2002).
8. **L2**: Reference 5; **L3**: Reference 6; **L4**: A. H. Hoveyda, A. W. Hird and M. A. Kacprzynski, *Chem. Commun.*, 1779 (2004); **L5**: A. Alexakis, J. Burton, J. Vastra, C. Benhaim, X. Fournioux, A. Van den Heuvel, J.-M. Leveque, F. Maze and S. Rosset, *Eur. J. Org. Chem.*, 4011 (2000); **L6**: I. H. Escher and A. Pfaltz, *Tetrahedron*, **56**, 2879 (2000); **L7**: M. Yan, Z. Y. Zhou and A. S. C. Chan, *Chem. Commun.*, 115 (2000).
9. M. Pineschi, F. Del Moro, F. Gini, A. J. Minnaard and B. L. Feringa, *Chem. Commun.*, 1244 (2004).
10. R. Sebesta, M. G. Pizzuti, A. J. Boersma, A. J. Minnaard and B. L. Feringa, *Chem. Commun.*, 1711 (2005).
11. J. Esquivias, R. G. Arrayas and J. C. Carretero, *J. Org. Chem.*, **70**, 7451 (2005).
12. K. Y. Li and A. Alexakis, *Angew. Chem., Int. Ed.*, **45**, 7600 (2006).
13. Y. Takemoto, S. Kuraoka, N. Hamaue and C. Iwata, *Tetrahedron: Asymmetry*, **7**, 993 (1996).
14. (a) L. Liang and A. S. C. Chan, *Tetrahedron: Asymmetry*, **13**, 1393 (2002).
(b) S. Woodward and P. K. Fraser, *Chem. Eur. J.*, **9**, 776 (2003).
15. A. Alexakis, V. Albrow, K. Biswas, M. d'Augustin, O. Prieto and S. Woodward, *Chem. Commun.*, 2843 (2005).
16. B. L. Feringa, R. Badorrey, D. Pena, S. R. Harutyunyan and A. J. Minnaard, *Proc. Nat. Acad. Sci. U.S.A.*, **101**, 5834 (2004).
17. F. Lopez, S. R. Harutyunyan, A. J. Minnaard and B. L. Feringa, *J. Am. Chem. Soc.*, **126**, 12784 (2004).
18. F. Lopez, S. R. Harutyunyan, A. Meetsma, A. J. Minnaard and B. L. Feringa, *Angew. Chem., Int. Ed.*, **44**, 2752 (2005).
19. B. M. Ruiz, K. Geurts, M. A. Fernandez-Ibanez, B. ter Horst, A. J. Minnaard and B. L. Feringa, *Org. Lett.*, **9**, 5123 (2007).
20. S. Y. Wang, S. J. Ji and T. P. Loh, *J. Am. Chem. Soc.*, **129**, 276 (2007).
21. S. Y. Wang, T. K. Lum, S. J. Ji and T. P. Loh, *Adv. Synth. Catal.*, **350**, 673 (2008).
22. J. Wu, D. M. Mampreian and A. H. Hoveyda, *J. Am. Chem. Soc.*, **127**, 4584 (2005).
23. A. W. Hird and A. H. Hoveyda, *J. Am. Chem. Soc.*, **127**, 14988 (2005).
24. E. Fillion and A. Wilsily, *J. Am. Chem. Soc.*, **128**, 2774 (2006).
25. A. Wilsily and E. Fillion, *Org. Lett.*, **10**, 2801 (2008).
26. M. K. Brown, T. L. May, C. A. Baxter and A. H. Hoveyda, *Angew. Chem., Int. Ed.*, **46**, 1097 (2007).
27. K. S. Lee, M. K. Brown, A. W. Hird and A. H. Hoveyda, *J. Am. Chem. Soc.*, **128**, 7182 (2006).
28. A. Alexakis and O. Knopff, *Org. Lett.*, **4**, 3835 (2002).
29. T. Hayashi and K. Yamasaki, *Chem. Rev.*, **103**, 2829 (2003).
30. M. d'Augustin, L. Palais and A. Alexakis, *Angew. Chem., Int. Ed.*, **44**, 1376 (2005).
31. (a) M. Vuagnoux-d'Augustin and A. Alexakis, *Chem. Eur. J.*, **13**, 9647 (2007).
(b) M. Vuagnoux-d'Augustin, S. Kehrli and A. Alexakis, *Synlett*, 2057 (2007).
32. C. Hawner, K. Li, V. Cirriez and A. Alexakis, *Angew. Chem., Int. Ed.*, **47**, 8211 (2008).
33. T. L. May, M. K. Brown and A. H. Hoveyda, *Angew. Chem., Int. Ed.*, **120**, 7468 (2008).
34. D. Martin, S. Kehrli, M. d'Augustin, H. Clavier, M. Mauduit and A. Alexakis, *J. Am. Chem. Soc.*, **128**, 8416 (2006).
35. Y. Matsumoto, K. I. Yamada and K. Tomioka, *J. Org. Chem.*, **73**, 4578 (2008).

36. H.-C. Guo and J.-A. Ma, *Angew. Chem., Int. Ed.*, **45**, 354 (2006).
37. M. Kitamura, T. Miki, K. Nakano and R. Noyori, *Bull. Chem. Soc. Jpn.*, **73**, 999 (2000).
38. M. Kitamura, T. Miki, K. Nakano and R. Noyori, *Tetrahedron Lett.*, **37**, 5141 (1996).
39. E. Keller, J. Maurer, R. Naasz, T. Schader, A. Meetsma and B. L. Feringa, *Tetrahedron: Asymmetry*, **9**, 2409 (1998).
40. A. Alexakis, G. P. Trevitt and G. Bernardinelli, *J. Am. Chem. Soc.*, **123**, 4358 (2001).
41. R. Naasz, L. A. Arnold, A. J. Minnaard and B. L. Feringa, *Chem. Commun.*, 735 (2001).
42. S. J. Degrado, H. Mizutani and A. H. Hoveyda, *J. Am. Chem. Soc.*, **123**, 755 (2001).
43. S. J. Degrado, H. Mizutani and A. H. Hoveyda, *J. Am. Chem. Soc.*, **124**, 13362 (2002).
44. H. Mizutani, S. J. Degrado and A. H. Hoveyda, *J. Am. Chem. Soc.*, **124**, 779 (2002).
45. J. C. Gonzalez-Gomez, F. Foubelo and M. Yus, *Tetrahedron Lett.*, **49**, 2343 (2008).
46. M. Vuagnoux-d'Augustin and A. Alexakis, *Tetrahedron Lett.*, **48**, 7408 (2007).
47. A. Alexakis and S. March, *J. Org. Chem.*, **67**, 8753 (2002).
48. M. A. Kacprzynski, S. A. Kazane, T. L. May and A. H. Hoveyda, *Org. Lett.*, **9**, 3187 (2007).
49. M. K. Brown, S. J. Degrado and A. H. Hoveyda, *Angew. Chem., Int. Ed.*, **44**, 5306 (2005).
50. K. Y. Li and A. Alexakis, *Tetrahedron Lett.*, **46**, 5823 (2005).
51. Y. J. Xu, Q. Z. Liu and L. Dong, *Synlett*, 273 (2007).
52. X. Rathgeb, S. March and A. Alexakis, *J. Org. Chem.*, **71**, 5737 (2006).
53. K. Agapiou, D. F. Cauble and M. J. Krische, *J. Am. Chem. Soc.*, **126**, 4528 (2004).
54. K. Y. Li and A. Alexakis, *Chem. Eur. J.*, **13**, 3765 (2007).
55. L. A. Arnold, R. Naasz, A. J. Minnaard and B. L. Feringa, *J. Org. Chem.*, **67**, 7244 (2002).
56. R. P. van Summeren, S. J. W. Reijmer, B. L. Feringa and A. J. Minnaard, *Chem. Commun.*, 1387 (2005).
57. B. Bulic, U. Lücking and A. Pfaltz, *Synlett*, 1031 (2006).
58. R. R. Cesati, J. de Armas and A. H. Hoveyda, *J. Am. Chem. Soc.*, **126**, 96 (2004).
59. E. W. Dijk, L. Panella, P. Pinho, R. Naasz, A. Meetsma, A. J. Minnaard and B. L. Feringa, *Tetrahedron*, **60**, 9687 (2004).
60. R. D. Mazery, M. Pullez, F. Lopez, S. R. Harutyunyan, A. J. Minnaard and B. L. Feringa, *J. Am. Chem. Soc.*, **127**, 9966 (2005).
61. B. ter Horst, B. L. Feringa and A. J. Minnaard, *Org. Lett.*, **9**, 3013 (2007).
62. B. ter Horst, B. L. Feringa and A. J. Minnaard, *Chem. Commun.*, 489 (2007).
63. B. L. Feringa, *Acc. Chem. Res.*, **33**, 346 (2000).
64. F. Lopez, A. J. Minnaard and B. L. Feringa, *Acc. Chem. Res.*, **40**, 179 (2007).
65. E. Casas-Arce, B. ter Horst, B. L. Feringa and A. J. Minnaard, *Chem. Eur. J.*, **14**, 4157 (2008).
66. G. P. Howell, S. P. Fletcher, K. Geurts, B. ter Horst and B. L. Feringa, *J. Am. Chem. Soc.*, **128**, 14977 (2006).
67. R. P. van Summeren, D. B. Moody, B. L. Feringa and A. J. Minnaard, *J. Am. Chem. Soc.*, **128**, 4546 (2006).
68. S. R. Harutyunyan, Z. Zhao, T. den Hartog, K. Bouwmeester, A. J. Minnaard, B. L. Feringa and F. Govers, *Proc. Natl. Acad. Sci. U.S.A.*, **105**, 8507 (2008).
69. T. den Hartog, S. R. Harutyunyan, D. Font, A. J. Minnaard and B. L. Feringa, *Angew. Chem., Int. Ed.*, **47**, 398 (2008).
70. T. K. Lum, S. Y. Wang and T. P. Loh, *Org. Lett.*, **10**, 761 (2008).
71. D. Polet and A. Alexakis, *Tetrahedron Lett.*, **46**, 1529 (2005).
72. M. K. Brown and A. H. Hoveyda, *J. Am. Chem. Soc.*, **130**, 12904 (2008).
73. K. M. Peese and D. Y. Gin, *Chem. Eur. J.*, **14**, 1654 (2008).
74. B. M. Trost and C. Jiang, *Synthesis*, 369 (2006).
75. H. O. House and W. F. Fischer, *J. Org. Chem.*, **34**, 3615 (1969). For successful enantioselective conjugate additions of the alkyne group using other metals, see: S. Fujimori, T. F. Knöpfel, P. Zarotti, T. Ichikawa, D. Boyall and E. M. Carreira, *Bull. Chem. Soc. Jpn.*, **80**, 1635 (2007).

CHAPTER **15**

Copper(I) hydride reagents and catalysts

OLIVIER RIANT

Unité de chimie organique et médicinale, departement de chimie, Place Louis Pasteur 1, Université catholique de Louvain, 1348 Louvain la Neuve, Belgium
Fax: +32010474168; email: olivier.riant@uclouvain.be

I. INTRODUCTION 731
II. PREPARATION AND CHARACTERIZATION OF COPPER HYDRIDE COMPLEXES AND CATALYSTS 732
III. STRYKER'S REAGENT AS A CHEMOSELECTIVE REDUCING AGENT 735
A. Pioneer Work of Stryker with Copper(I) Hydride Cluster 2 735
B. Other Stoichiometric Reagents as an Alternative to Stryker's Reagent . 738
IV. CATALYTIC HYDROSILYLATION OF α,β-UNSATURATED SUBSTRATES 738
A. Catalytic Conjugated Hydrosilylation with Non-chiral Ligands 738
B. Catalytic Enantioselective Conjugated Hydrosilylation with Chiral Ligands 741
V. CATALYTIC HYDROSILYLATION OF KETONES AND IMINES 749
A. Catalytic Hydrosilylation with Non-chiral Ligands 749
B. Catalytic Hydrosilylation with Chiral Ligands 752
VI. CATALYTIC DOMINO REACTIONS 757
A. Catalytic Domino Reactions with Non-chiral Ligands 757
B. Catalytic Domino Reactions with Chiral Ligands 761
VII. HYDROCUPRATION OF ALKYNES 768
VIII. REFERENCES 771

I. INTRODUCTION

Since the first report on the synthesis and characterization of a copper(I) hydride complex in 1971 by the group of Churchill and Osborn, the application of such soft hydride for the reduction of electrophilic double bonds has known a growing success due to the work of Stryker. It was also rapidly shown that such complexes could also be used in catalytic

The chemistry of organocopper compounds
Edited by Z. Rappoport and I. Marek

reduction reactions using various sources of hydrides such as molecular hydrogen, silanes and boranes. Despite its soft character, the copper(I) hydride complex has also exhibited an excellent ability to reduce harder electrophiles such as aldehydes and ketones, and gave rise to some of the most active families of reduction catalysts for these functional groups. More recent development in this field showed that the addition of a copper hydride complex to an electrophilic double bond leads to the formation of a copper enolate that can be used in domino transformation and leads to the formation of carbon–carbon bonds. The aim of this review[1] is to illustrate the principle for the preparation of such reagents and catalytic systems and their use in reduction and domino processes.

II. PREPARATION AND CHARACTERIZATION OF COPPER HYDRIDE COMPLEXES AND CATALYSTS

The first example of a fully characterized phosphine copper(I) hydride complex was reported in 1971 by Churchill, Osborn and coworkers[2]. The hydrogenation of the tetrameric copper(I) *tert*-butoxide **1** in the presence of triphenylphosphine gave an orange crystalline solid identified as the hexameric triphenylphosphine copper(I) hydride **2** (equation 1). This crystalline complex is air-sensitive, but it can be easily stored under inert atmosphere and was later commercialized as a chemoselective reducing agent.

$$\underset{(\mathbf{1})}{1/4\,[\mathrm{CuOBu}\text{-}t]_4} + \mathrm{PPh_3} + \mathrm{H_2} \xrightarrow{\mathrm{THF}} \underset{(\mathbf{2})}{1/6\,[\mathrm{CuH(PPh_3)}]_6} + t\text{-BuOH} \qquad (1)$$

The structure of this new complex was proved by X-ray crystallography, which shows the six copper atoms at each position of a regular octahedron with an average Cu–Cu distance of 2.599 Å and one phosphine linked to each copper atom. The structure of this type of complex was later confirmed by Bau and coworkers[3], who analyzed the $(p\text{-Tol})_3P$ analogue by neutron diffraction and confirmed the facial disposition of the hydride ligands in the cluster. The ^1H NMR spectra of cluster **2** in C_6D_6 shows a broad singlet at 3.5 ppm, this singlet becoming sharp when the spectra is recorded with phosphorus decoupling. Apart from a simple triarylphosphine cluster with structures analogous to **2**, there are still very few examples of fully characterized copper(I) hydride complexes described in the literature (Chart 1).

There are, indeed, to date only two examples of copper hydride–diphosphine complexes which have been prepared and characterized. Those examples show that the structure and stoichiometry P/Cu is dependent on the structure of the ligand used. Caulton and coworkers used dppp (1,3-bis(diphenylphosphino)propane) and showed that the corresponding cluster **3** incorporated two copper atoms per diphosphine ligand[4]. However, when they used triphos (1,1,1-tris(diphenylphosphinomethyl)ethane) as ligand, they isolated a dimeric hydride **4** in which each copper atom is chelated by two phosphorus atoms from the triphos ligand[5]. Two other copper hydride complexes have been recently reported and fully characterized. Sadighi and coworkers used a bulky NHC ligand on copper and isolated the unstable hydride dimer **5** which could be characterized by X-ray crystallography[6]. Che and coworkers used the restricted ligand bis(cyclohexylphosphino)methane and isolated the air-stable cationic μ_3-H hydride **6**[7]. However, the reactivity of the hydride was not reported for this complex.

As will be detailed later, the hexameric cluster **2** (now known as Stryker's reagent) was used from 1988 by the group of J. Stryker as an efficient chemoselective reducing agent of electrophilic double bonds and later as a precatalyst for various hydrogenation

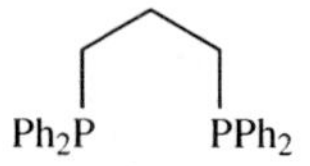

(dppp)

$H_8Cu_8(dppp)_4$

(**3**)

K. G. Caulton (1985)

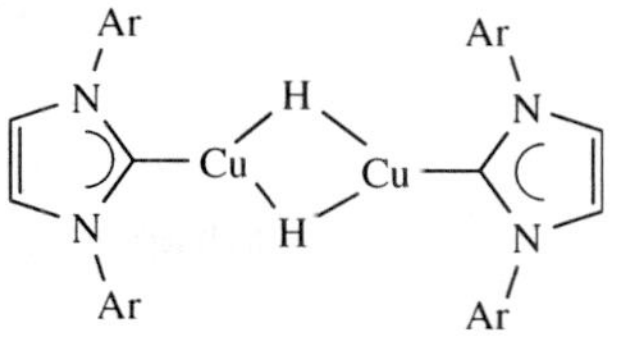

(**5**)

J. P. Sadighi (2004)

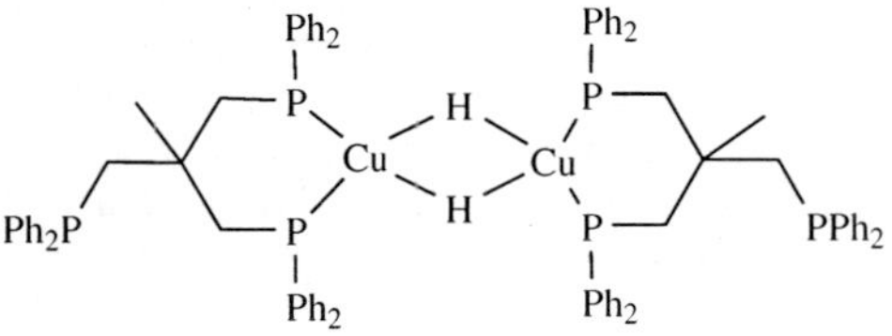

(**4**)

K. G. Caulton (1986)

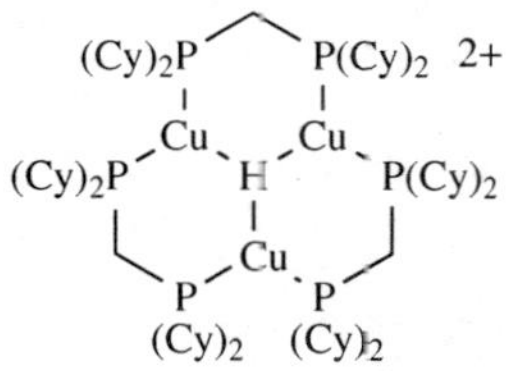

(**6**)

C.-M. Che (2005)

CHART 1

and hydrosilylation reactions. As the initial procedure involved the use of the strongly air-sensitive copper(I) *tert*-butoxide and the manipulation of hydrogen, Stryker first reported a more convenient procedure for the multi-gram scale preparation of **2** (equation 2)[8].

$$\text{CuCl} + \text{PPh}_3 + \text{NaOBu-}t \xrightarrow[\text{toluene}\ C_6H_6]{H_2} \tfrac{1}{6}\,[\text{CuH(PPh}_3)]_6 + t\text{-BuOH} + \text{NaCl} \quad (2)$$

(**2**)

The copper alkoxide can thus be generated *in situ* from copper(I) chloride and potassium *tert*-butoxide and the reduction to the copper hydride occurs under a positive pressure (1 atm) of hydrogen. This procedure allows the preparation of the complex **2** using standard Schlenck tube procedures and **2** is isolated in good yields after crystallization from benzene/acetonitrile. This procedure was later ameliorated by Chiu and coworkers, who used dimethylphenylsilane as reducing agent, thus avoiding the manipulation of molecular hydrogen (equation 3)[9].

$$\text{CuCl} + \text{PPh}_3 + \text{KOBu-}t \xrightarrow[C_6H_6]{\text{PhMe}_2\text{SiH}} \tfrac{1}{6}\,[\text{CuH(PPh}_3)]_6 + t\text{-BuOH} + \text{KCl} \quad (3)$$

(**2**)

Yun and Lee also showed that copper(II) salts could be directly used for the preparation of Stryker's reagent, thus avoiding the preparation of the sensitive copper(I) alkoxide[10]. Copper(I) acetate can be directly reduced by diphenylsilane in the presence of triphenyl phosphine and delivers the cluster **2** in a 82% yield after crystallization.

It was also shown that this complex could be used as a useful precursor for the *in situ* generation of diphosphine copper hydride catalysts for hydrogenation and hydrosilylation reactions. A chelating ligand such as **7** is expected to depolymerize the cluster **2** and to displace the triphenyl phosphine ligand on the copper atom to give diphosphine copper(I) hydride oligomers **8** (equation 4).

(**7**) (**8**) (4)

This procedure was useful in some cases, but the characterization of the species in solution was far from easy as the equilibrium was not completely shifted toward the formation of the chelate[11]. During the examination of a catalytic system optimized for asymmetric hydrosilylation, Lipshutz and coworkers examined this equilibrium through the disappearance of the hydride signal of **2** by ^{1}H NMR (when increased amounts of chiral diphosphine were added to a solution of **2** in C_6D_6)[11]. They observed that more than one equivalent of the diphosphine is required for the full disappearance of the hydride shift of **2** with the concomitant appearance of a new hydride signal, and that the signal of **2** reappears when triphenylphosphine is added to the NMR tube, thus suggesting a rapid equilibrium between the PPh_3 and diphosphine ligated Cu–H species.

Although there are still very few studies on the observation of those catalytic species in solution as well as the corresponding equilibrium involved, they show that the protocol used for the generation might have a strong influence on the catalytic activity as well as the selectivity targeted for such catalysts.

Various experimental protocols have thus been devised for the generation of ligated copper hydride catalysts in the literature. The main goal of those protocols was to offer a straightforward and reliable preparation of the active copper hydride species which would avoid the prior isolation of the sensitive cluster **2**, and that would also give an easy access to ligand screening for optimization of the catalytic systems. The main protocols used in the literature are summarized in equations 5–7.

$$\text{Cu(I)Cl} + \text{L} \xrightarrow[\textit{complexation}]{} \underset{(\mathbf{9})}{[\text{LCu(I)}]\text{-Cl}} \xrightarrow[\textit{ligand exchange}]{\text{RO}^-} \underset{(\mathbf{10})}{[\text{LCu(I)}]\text{-OR}} \xrightarrow[\textit{activation}]{\text{Si-H}} \underset{(\mathbf{11})}{[\text{LCu(I)}]\text{-H}} \qquad (5)$$

$$\text{Cu(II)F}_2 + \text{L} \xrightarrow[\textit{reduction + complexation}]{\text{Si-H}} \underset{(\mathbf{12})}{[\text{LCu(I)}]\text{-F}} \xrightarrow[\textit{activation}]{\text{Si-H}} \underset{(\mathbf{11})}{[\text{LCu(I)}]\text{-H}} \qquad (6)$$

$$\underset{(\mathbf{13})}{\text{Cu(I)F(PPh}_3)_3} + \text{L} \xrightarrow[\textit{ligand exchange}]{} \underset{(\mathbf{12})}{[\text{LCu(I)}]\text{-F}} \xrightarrow[\textit{activation}]{\text{Si-H}} \underset{(\mathbf{11})}{[\text{LCu(I)}]\text{-H}} \qquad (7)$$

These equations describe a general guideline for the *in situ* preparation of the ligated copper hydride catalysts without taking into account the exact structure of those complex species which often exist in various aggregation states. The most widely used method (equation 5) requires the use of copper(I) chloride and a chelating diphosphine which first generated the diphosphine copper(I) chloride complex **9**. As the Cu−Cl bond cannot be efficiently cleaved by a hydride source (such as a silane), it is necessary to activate the complex by ligand exchange with a hard ligand such as an alkoxide. The sensitive copper(I) alkoxide **10** is thus generated and can be directly activated by the silane to give the desired ligated copper(I) hydride catalyst **11**. Copper(II) sources bearing hard ligands such as fluoride can also be used as direct precursors (equation 6). However, as the diphosphine cannot complex the copper(II) salt, *in situ* reduction of the copper(II) salt to a copper(I) species still bearing one fluoride ligand should occur by stoichiometric reducing agent prior to the complexation by the diphosphine ligand. The intermediate copper(I) fluoride complex **12** is thus directly transformed into the copper(I) hydride catalyst **11** by the silane. This procedure was also used with copper(II) acetate in some hydrosilylation catalytic procedures. As both procedures require either an exchange with a hard ligand for activation or an *in situ* reduction of a copper(II) salt, a third procedure which uses a precursor that already bears a hard ligand on a copper(I) precursor complex was also devised and widely used for the generation of copper(I) hydride catalyst (equation 7). Copper(I) fluoride tris(triphenylphosphine) **13** is an air-stable copper(I) fluoride source that can be easily prepared by reaction of commercially available copper(II) fluoride and triphenylphosphine. It is isolated as a methanol or ethanol adduct as a stable crystalline solid and can be manipulated and kept in the presence of air without any degradation. Mixing this complex with a chelating diphosphine will lead to ligand exchange, and the fluoride will be directly eliminated by the reducing agent (silane) to yield the catalytic active copper(I) hydride **11**.

Those procedures as well as the cluster **2** have been used for various chemical transformations using either catalytic or stoichiometric amounts of copper hydride species.

III. STRYKER'S REAGENT AS A CHEMOSELECTIVE REDUCING AGENT

A. Pioneer Work of Stryker with Copper(I) Hydride Cluster 2

The generation of copper(I) hydride ate complexes for the chemoselective reduction of electrophilic double bond was extensively studied from the early seventies and various reagents such as copper(I) hydride alkynyl or alkyl ate-complexes[12], copper(I) halide aluminum hydride combinations[13] and methylcopper/DIBAH[14] were successfully described for this type of transformation. However, Stryker and coworkers described for the first time in 1988 the use of the triphenylphosphine copper(I) hydride cluster **2** as a mild and efficient reducing agent for various types of α,β-unsaturated carbonyl compounds[15]. Two representative examples are described in Scheme 1.

This seminal paper showed that full reduction of electrophilic double bonds activated by ketone and ester could be carried out with the reagent **2** and that all six hydrides of the cluster could be delivered to the substrate. They also showed an excellent chemoselectivity toward the electrophilic double bond, as this reagent did not reduce unactivated double bonds and did not carry out 1,2-reduction on ketones (such as in substrate **14**). Complex substrates such as **15** also showed that this reagent displayed an excellent chemical compatibility with various functional groups and could thus be used as a very mild and chemoselective reducing agent. This reagent was later used with α,β-unsaturated aldehydes and unactivated alkynes as in the examples described in Scheme 2.

When unsaturated aldehydes were used, a silyl chloride electrophile was usually added to form the silyl enol ether resulting from the conjugated addition and trapping of the copper enolate[16]. This procedure allowed working with base-sensitive substrates. The use of water as a protonating agent for the generated enolate was efficient for non-sensitive substrates and the silyl chloride reagents were not necessary in such cases. Stryker and coworkers showed that the reagent **2** could also reduce unactivated triple bonds[17]. However, such substrates proved to be much less reactive, and heating in benzene with water

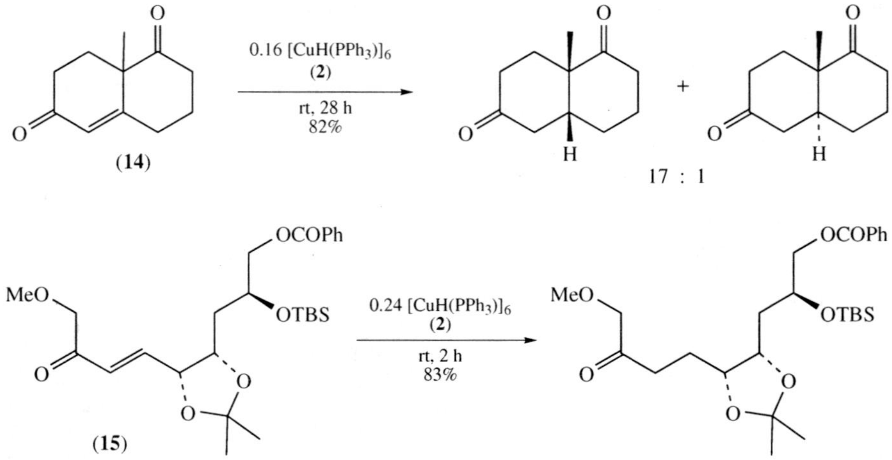

SCHEME 1

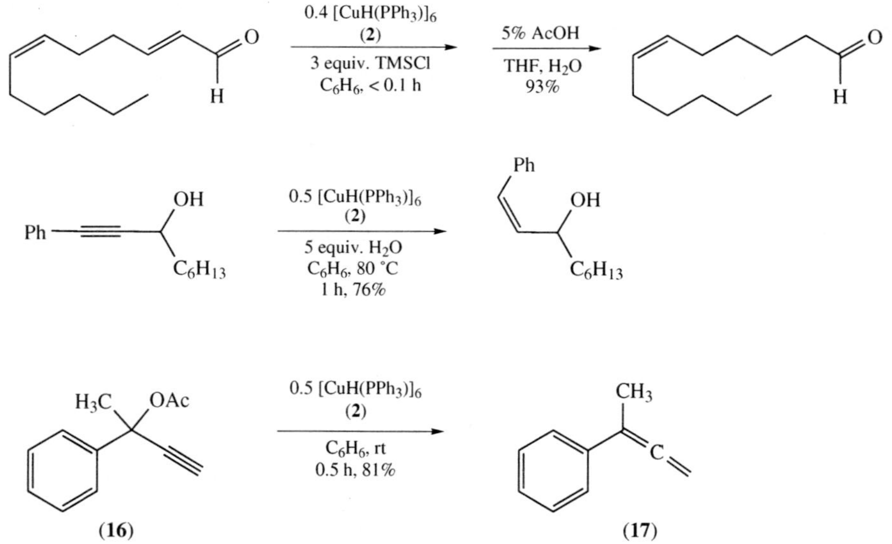

SCHEME 2

as a trapping agent was necessary to achieve good conversion for most substrates. One interesting example also showed that when a leaving group (acetate) was present next to the triple bond of the substrate, such as in **16**, the conjugated reductive elimination occurs at room temperature to yield the allene **17** in 81% yield.

Soon after reporting the use of cluster **2** as a chemoselective reagent, Stryker and coworkers showed that it could also be used as a catalyst in hydrogenation reactions, and those seminal publications set the bases for the use of copper hydride catalysts in reduction of various families of substrates (Scheme 3)[18].

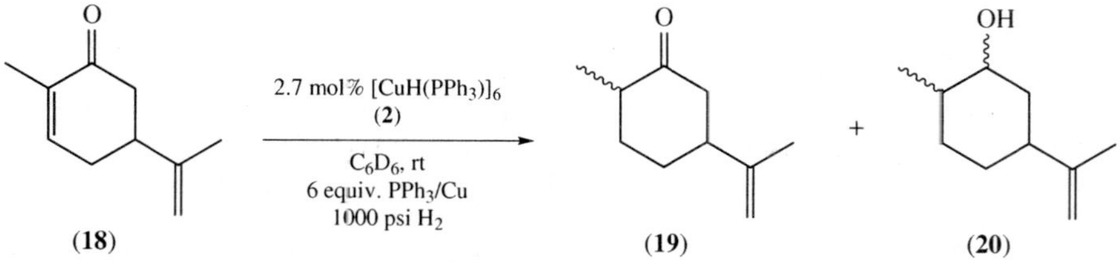

	19:20
25 h,	78:22
25 h (THF)	100:0
75 h	0:100

SCHEME 3

They first showed that electrophilic double bonds could be hydrogenated at room temperature with a catalytic amount of the cluster **2** (Scheme 3). The catalytic system usually required an excess of triphenylphosphine per copper atom (6–12 equivalents) and it was also observed that some reduction of the carbonyl double bond could also be obtained, depending on the reaction conditions. The group showed that the reduction occurred in homogeneous conditions on substrates such as **18** with a preference for the activated electron-deficient double bond over the non-activated one, yielding the corresponding saturated ketone **19**. When the reaction time was increased, reduction of the ketone occurred and led to the fully reduced alcohol **20**.

Later reports from the same group focused on the use of dimethylphenylphosphine for the chemoselective reduction of aldehydes and ketones and 1,2-selective hydrogenation of α,β-unsaturated substrates[19] (Scheme 4)

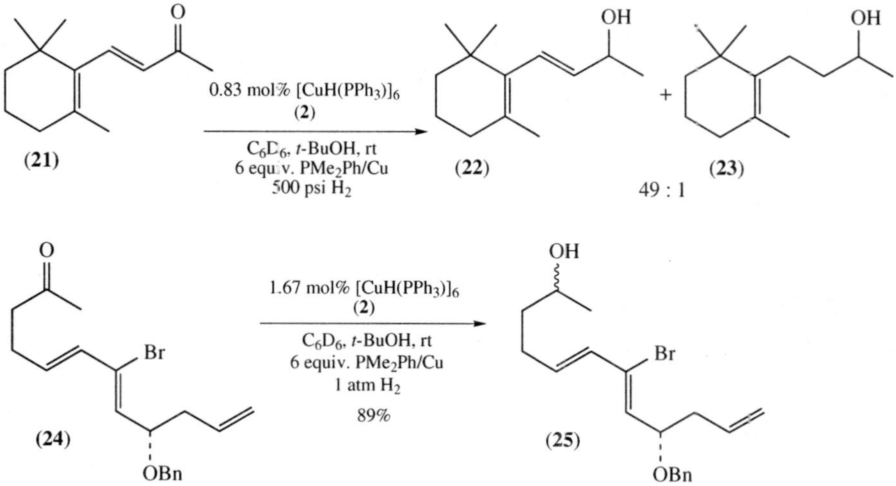

SCHEME 4

Although it was not possible to isolate or identify any copper(I) hydride species when Me_2PhP was used as a ligand, this more basic phosphine proved to be well suited for the chemoselective reduction of ketones, even when an electron-deficient double bond is conjugated to the ketone (such as **21**, which yields the reduction alcohol **22** and a trace amount of the reduced alcohol **23**). The catalytic system can then be generated by ligand exchange between the hydride complex **2** and the ligand, or by generation from CuCl/*t*-BuONa/$PPhMe_2$ (1:1:6) in anhydrous benzene under H_2 atmosphere. Several examples of polyfunctionalized substrates (such as **24**, which gives the corresponding alcohol **25** without any reduction of either double bonds or carbon–bromine bond) were reported, and showed that only ketone and aldehyde were reduced to the corresponding alcohols, and various functional groups (such as unactivated double bond, vinylic bromide, aliphatic tosylate, epoxide, alkyne and ester) were left unchanged, thus leading to a mild and very selective method for the reduction of ketones. The use of *tert*-butanol as an additive (10–20 equivalents/Cu) allowed the use of lower hydrogen pressure and gave prolonged catalyst lifetimes.

B. Other Stoichiometric Reagents as an Alternative to Stryker's Reagent

An early report from Ryu, Sonoda and coworkers showed that a stereoselective reduction of acetylenic sulfones could be mediated by copper(II) salts and diethylmethylsilane in isopropanol[20]. The corresponding *Z*-ethylenic sulfones were isolated in good yield with 2 equivalents of copper(II) tetrafluoroborates and a cationic copper(II) hydride was postulated as the active reducing species.

Hosomi and coworkers in 1997 reported an interesting alternative to Stryker's reagent without stabilizing phosphine ligands[21]. The use of very polar solvents such as DMI (1,3-dimethylimidazolidinone) or DMF allowed them to generate active and phosphine-less copper hydride species from copper(I) chloride and phenyldimethylsilane. This reagent was used for the chemoselective reduction of the double bond of α,β-unsaturated ketones and esters in good yield when two equivalents of copper(I) chloride were used. The formation of the Cu–H species could not occur when less polar solvents such as dichloromethane and THF were tested, thus showing the stabilizing effect of the polar solvent on the copper hydride species. This reagent was later used by the same group for the reduction of ketones and aromatic conjugated olefins[22].

Mori and coworkers used the stable fluoride complex $CuF(PPh_3)_3{\bullet}2EtOH$ to generate copper(I) hydride species after activation with various silanes for the chemoselective reduction of electrophilic double bonds with good to excellent yields[23]. They also showed that the active species could be generated with a CuCl/PPh_3/Bu_4NF combination. With this latter system, the amount of copper salt could be decreased to 20 mol% for the most activated substrates, thus showing one of the first examples of catalytic application of copper(I) hydride in reduction reactions.

IV. CATALYTIC HYDROSILYLATION OF α,β-UNSATURATED SUBSTRATES

A. Catalytic Conjugated Hydrosilylation with Non-chiral Ligands

A significant breakthrough in the field of copper(I) hydride catalyzed reductions occurred when the group of Lipshutz proposed the use of silanes as a stoichiometric reducing agent in conjugated 1,4-reduction of electrophilic double bonds[24]. They first reported that molecular hydrogen could be replaced by either tributyltin hydride or phenylsilane when the Stryker catalyst **2** was used for the chemoselective reduction of double bonds activated with aldehydes, ketones and esters (Scheme 5).

Both hydride sources proceed with similar efficiencies under very mild conditions to afford the reduced products with good to excellent yields. The authors also showed that the inexpensive silyl hydride source PMHS (polymethylhydrosiloxane) could be used and gave very high efficiencies for some substrates such as **26**, allowing catalyst loading to be as low as 0.2 mol% for the conjugated reduction to the saturated ketone **27**. The primary

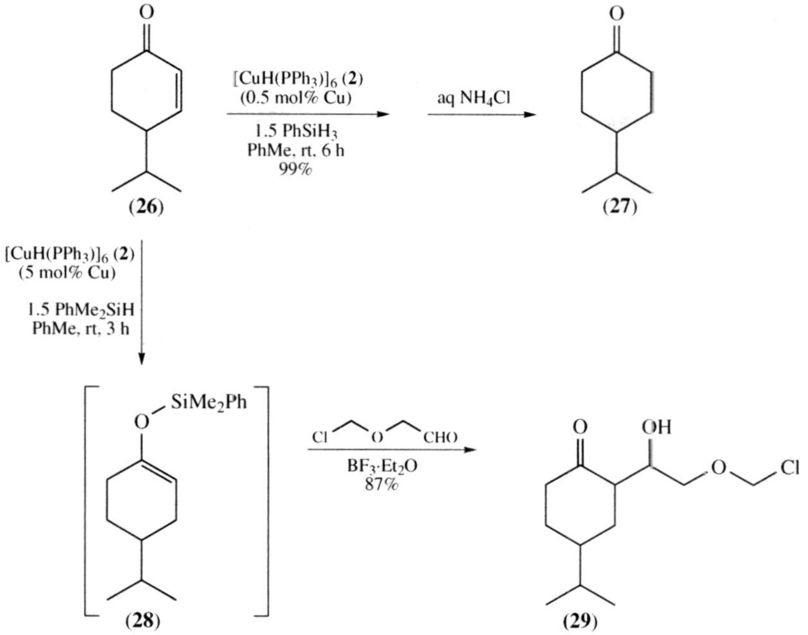

SCHEME 5

product of the conjugated reduction is the silyl enol ether **28** and the *in situ* reaction of this silyl enol ether with aldehyde and a Lewis acid catalyst gave the corresponding aldol adduct **29** in good yield. Chandler and Phillips later applied this method for the conjugated reduction of a functionalized cyclopentenone during the course of the total synthesis of *rac-trans*-Kumausyne[25].

The main problem with tandem methodology is that the aldol adducts were obtained as mixture of diastereoisomers. Replacing the silane for a borane as reducing agent allowed Lipshutz and Papa to offer an elegant alternative inspired by the known positive effect of boron enolates on the stereochemical outcome of aldol reactions (Scheme 6)[26].

Cyclic enone **30** (but also acyclic enones) were then efficiently reduced by a catalytic amount of Stryker's reagent **2** with diethyl borane as the hydride source. The 1,4-addition gave a boron enolate intermediate **31** that could be trapped with an aldehyde to give the corresponding aldol adduct **32** in high yield and complete diastereoselectivity (*syn*-aldol adduct for acyclic enone and *anti*-aldol adduct for cyclic enone). The example of carvone **30** also shows that the use of enone as precursor of the enolate nucleophile is an interesting equivalent methodology for the regioselective enolate generation.

Recently, Lipshutz and coworkers reported the preparation of a stable catalytic solution of a diphosphine copper hydride catalyst **33** from copper acetate and the ligand DPB (1,2-bis(diphenylphosphino)benzene) in toluene[27] (Scheme 7).

While the copper loading was kept at 1 mol%, the substrate-to-ligand ratio was usually modulated from 250:1 to 1000:1. This catalytic solution could be kept for several months at room temperature under inert atmosphere and could be used directly for the conjugated reduction of functionalized electrophilic double bonds bearing electron-withdrawing groups such as ketone, ester and cyano group.

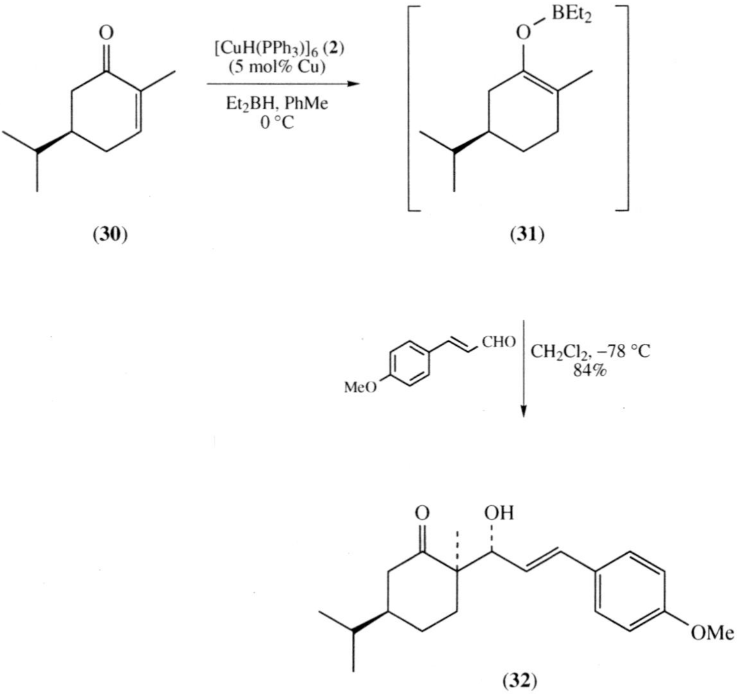

SCHEME 6

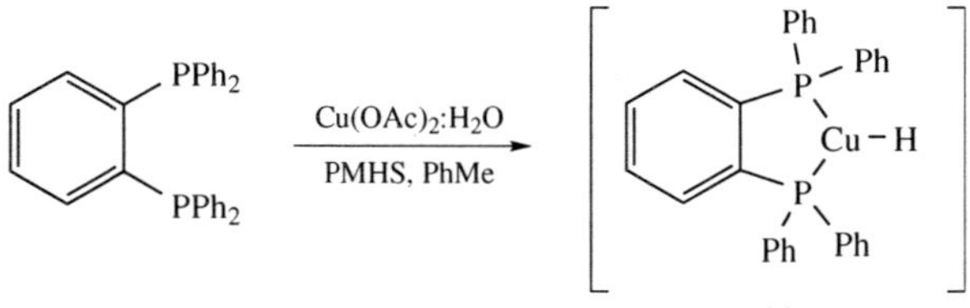

SCHEME 7

Buchwald, Sadighi and coworkers reported for the first time in 2003 that *N*-heterocyclic carbene ligands could be used for the preparation of highly efficient copper catalysts for the conjugated hydrosilylation of various types of electrophilic double bonds[28] (Scheme 8).

The NHC copper(I) halide complex **34** is readily prepared by reaction of the corresponding imidazolium salt with copper chloride and sodium *tert*-butoxide and proved to be an air-stable and easy-to-handle crystalline solid. The generation of the NHC–copper(I) hydride active catalyst requires the use of a metal alkoxide as cocatalyst. Sodium chloride elimination occurs *in situ* to generate the NHC–copper *tert*-butoxide that gives the copper hydride through activation with the silane source. The copper hydride catalyst provided impressive activity, considering that catalyst loading as low as 0.3–0.1 mol% allowed the reduction of trisubstituted α,β-unsaturated ester **35** and cyclic enone **36** into the corresponding saturated carbonyl compounds in short reactions times. Fast reactions require

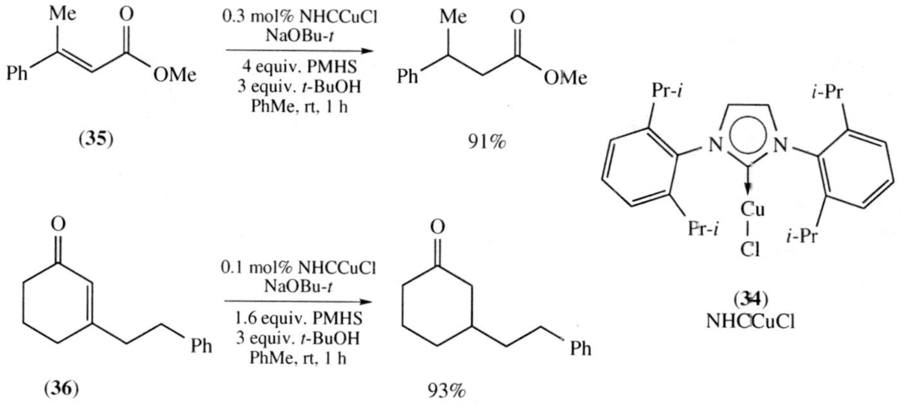

SCHEME 8

the use of a protonation agent such as *tert*-butanol. It was suggested that this alcohol protonates the copper enolate generated after the conjugated addition of the copper hydride catalyst on the electrophilic double bond, generating the saturated ester or ketone and a copper alkoxide that can then re-enter the catalytic cycle through silane activation. This method was later used by Madec, Poli and coworkers in one of the late steps of the total synthesis of kainic acid[29].

B. Catalytic Enantioselective Conjugated Hydrosilylation with Chiral Ligands

Following the pioneer work of Stryker in the field of copper hydride catalysts and Lipshutz in the field of conjugated hydrosilylation, another major breakthrough occurred in 1999 with the first report of an asymmetric catalytic conjugated hydrosilylation of α,β-unsaturated esters by Buchwald and coworkers[30] (Scheme 9).

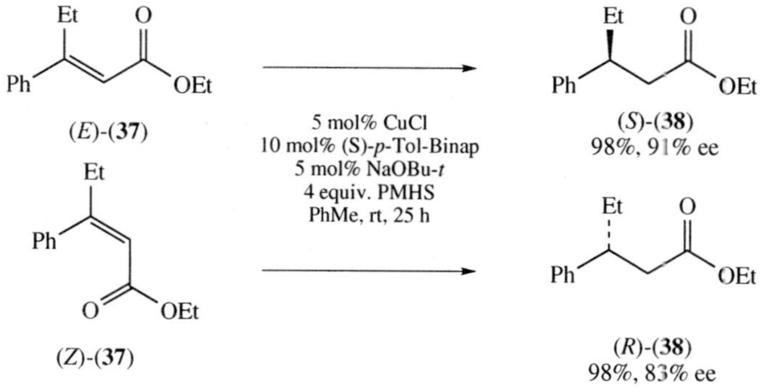

SCHEME 9

An efficient method for the *in situ* generation of chiral diphosphine copper hydride catalyst was reported by mixing chiral diphosphine, copper(I) chloride and sodium *tert*-butoxide as cocatalyst. The generated diphosphine copper(I) alkoxide is then reduced to the corresponding hydride by a silane (PMHS) and efficiently reduced various β,β-disubstituted α,β-unsaturated esters with high yields and excellent enantioselectivities (up to 92% with *p*-Tol-Binap as chiral ligand). It was also shown that the enantioselection was dependent on the stereochemistry of the double bond of the substrate as (*Z*)-**37** and (*E*)-**37** isomers gave the corresponding reduction products (*R*) and (*S*) **38** with similar enantioselectivities, albeit with opposite absolute configurations. The relationship between the ee's of the products and the ee's of the chiral ligand reveals a linear relationship, and it was therefore suggested that a monomeric 1:1 ligand:metal complex was involved in the catalytic cycle.

This first method was later applied to various families of substrates such as cyclic enones[31] and unsaturated lactones and lactams[32], yielding the corresponding saturated products **39–44** with good to high enantioselectivities. Representative examples of the conjugated reduction adducts are represented in Chart 2 with ee's >90% when simple chiral ligands such as *p*-Tol-Binap were used with PMHS as the reducing agent.

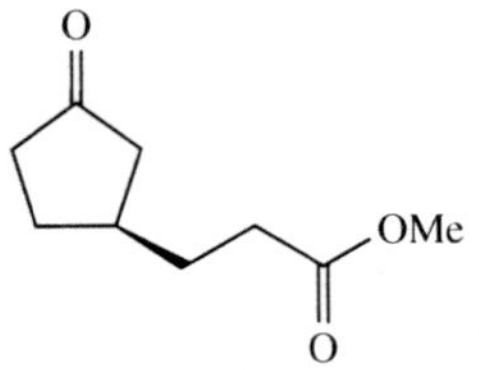

(**39**)
92% ee

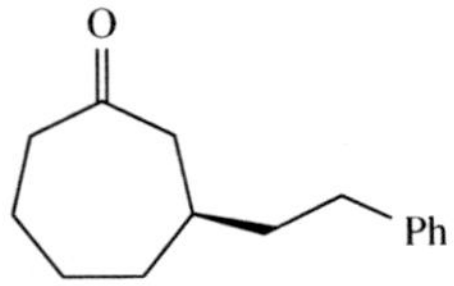

(**40**)
96% ee

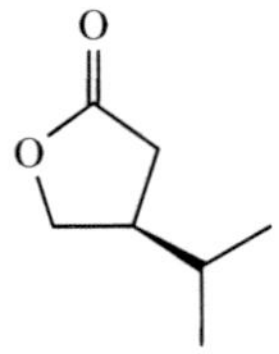

(**41**)
92% ee

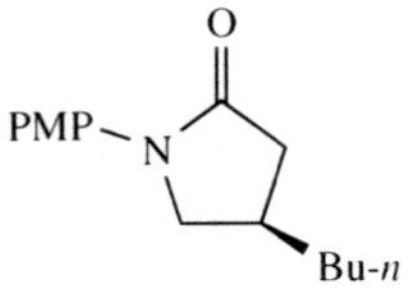

(**42**)
94% ee

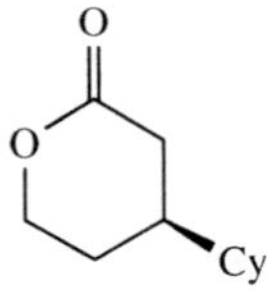

(**43**)
94% ee

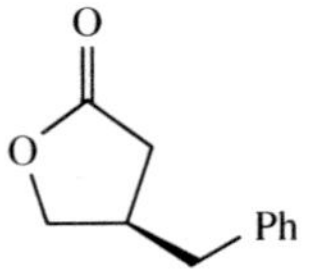

(**44**)
92% ee

CHART 2

N-Protected unsaturated lactams also gave excellent results, and this methodology was applied to the enantioselective synthesis of the antidepressant (−)-paroxetine **45** (Chart 3).

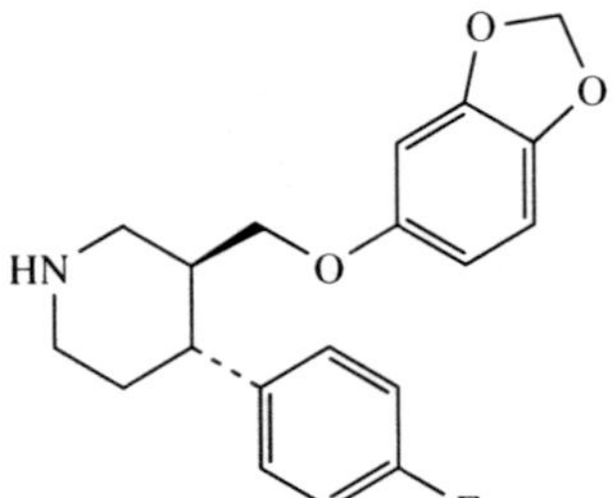

(**45**)
(−)-paroxetine

CHART 3

More recently, α,β-unsaturated esters bearing azaheterocycles at the β-position were also tested[33] (Chart 4).

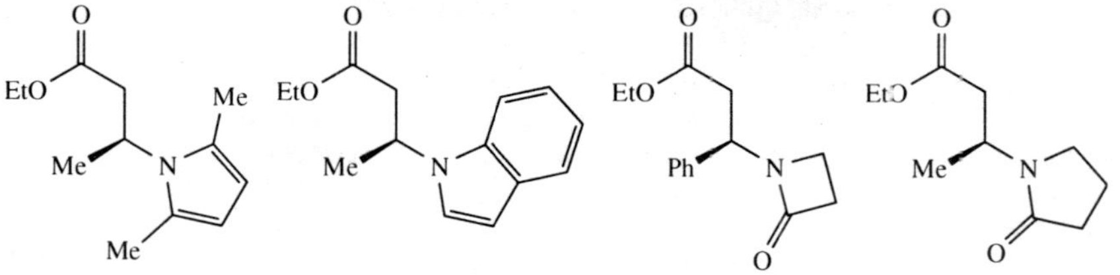

(**46**) 86% ee | (**47**) 90% ee | (**48**) 99% ee | (**49**) 82% ee

CHART 4

In order to avoid deactivation of the electrophilic double bond by introduction of a nitrogen atom at the β-position, the heterocyclic groups such as lactam and pyrrole were selected and the corresponding conjugated addition adducts were isolated with excellent yield and ee's in the range of 81–99%. Measurement of the relative rates for the conjugated addition for various nitrogen-based substituents revealed that pyrrole-based derivatives (leading to adducts **46** and **47**) gave the most active substrates over γ- and β-lactams (leading to adducts **48** and **49**), and confirms that the lone pair of the nitrogen atom should be involved in maximum conjugation in order to avoid the deactivation of the double bond.

An elegant application of this method was reported by Movassaghi and Ondrus as a key step for the total synthesis of tricyclic myrmicarin alkaloids **52**[34] (Scheme 10).

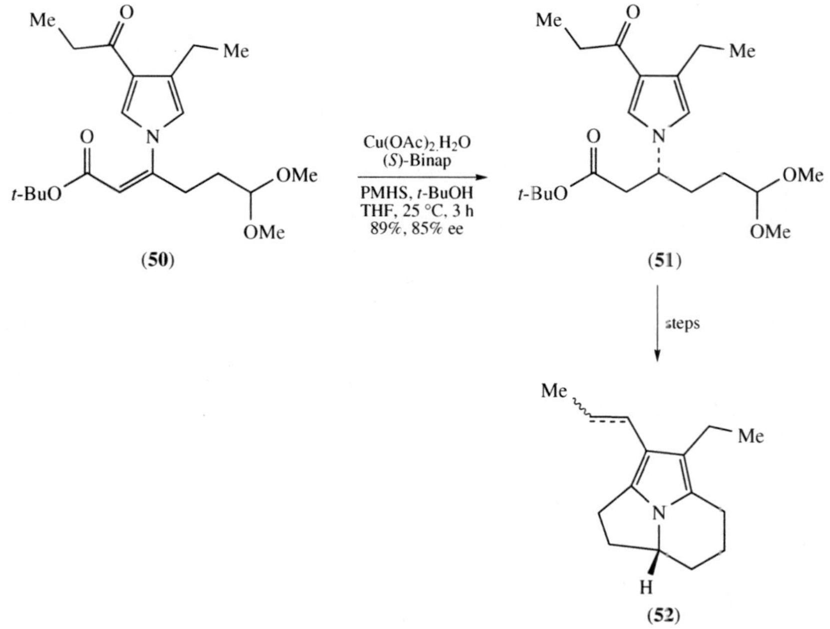

SCHEME 10

The conjugated reduction of trisubstituted activated ester **50** with the copper(II) acetate/(*S*)-Binap catalyst and PMHS proceeds efficiently at room temperature and gives the key intermediate **51** in 89% yield and 85% enantiomeric excess on a 2 g scale. The pyrrole ring can then be used for further cyclizations to give the central core of a series of myrmicarin alkaloids **52**.

Yun and Buchwald also showed that the methodology could lead to the formation of silyl enol ethers arising from the conjugated addition of the copper hydride catalyst followed by the σ-bond metathesis of the copper enolates with diphenylsilane[35] (Scheme 11).

This reaction occurred when the protonating agent was omitted and the corresponding silyl ether could then be trapped with activated alkylating agents. TBAT (tetrabutylammonium difluorotriphenylsilicate) was then used to activate the enol ether **53** and the corresponding *trans*-2,3-disubstituted cyclopentanone **54** was isolated in good yield. The *cis/trans* diastereoselectivity ranged from 76:24 to 94:6 for the crude adduct **54**, but could be improved in all cases by equilibration with a catalytic amount of sodium methoxide. The same authors later reported that the generated silyl enol ether could be directly used in a palladium-catalyzed arylation with aryl bromides, using cesium fluoride as activating agent for the silyl enol ether[36]. In that case, keeping a 5 mol% of the palladium salt, the copper/diphosphine ratio should be reduced to 1 mol% in order to avoid the deactivation of the palladium by the bidentate phosphine ligand (Binap) used in the first step.

Further opportunities for the preparation of functionalized cyclopentanones were reported by Jurkauskas and Buchwald and involved an elegant example of dynamic kinetic resolution (DKR) via asymmetric conjugated hydrosilylation of racemic cyclopentenones[37] (Scheme 12).

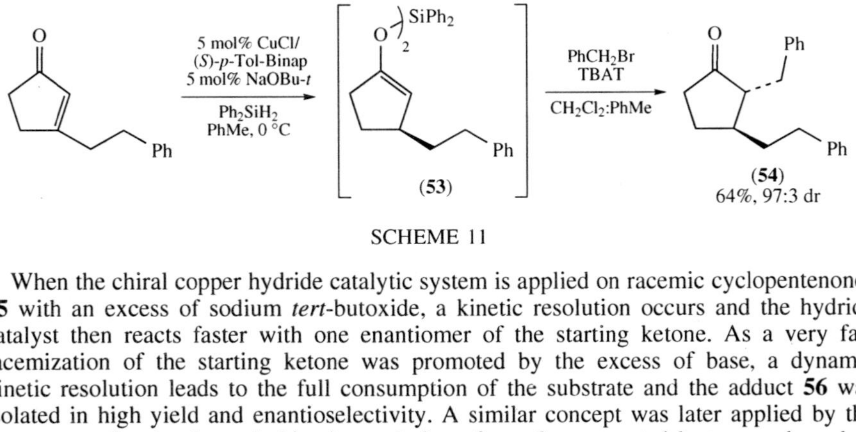

SCHEME 11

When the chiral copper hydride catalytic system is applied on racemic cyclopentenones **55** with an excess of sodium *tert*-butoxide, a kinetic resolution occurs and the hydride catalyst then reacts faster with one enantiomer of the starting ketone. As a very fast racemization of the starting ketone was promoted by the excess of base, a dynamic kinetic resolution leads to the full consumption of the substrate and the adduct **56** was isolated in high yield and enantioselectivity. A similar concept was later applied by the same group for the dynamic kinetic resolution of a α,β-unsaturated lactone and used as the key enantioselective step for the total synthesis of eupomatilone 3[38].

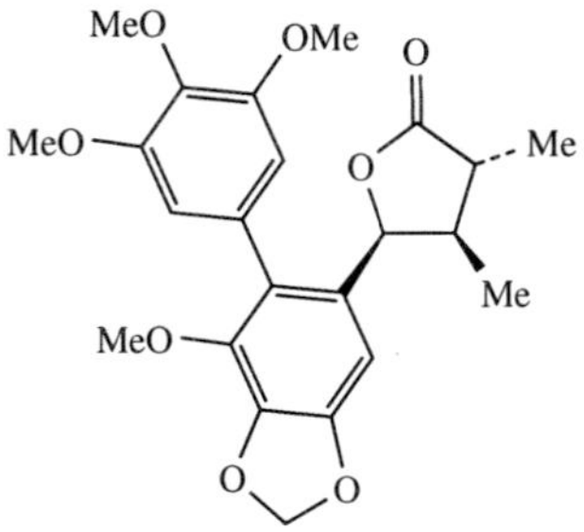

eupomatilone 3

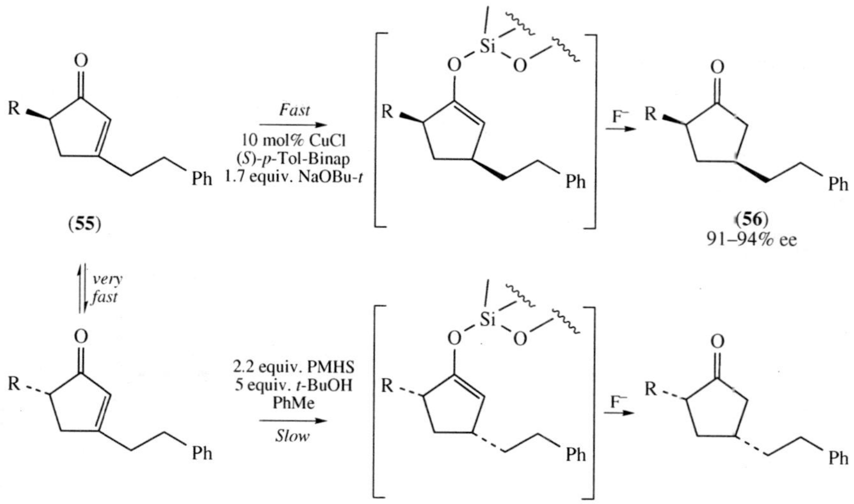

SCHEME 12

Following the first developments of the group of Buchwald on the enantioselective conjugated reduction of various families of acyclic and cyclic conjugated substrates, Lipshutz and coworkers reported the use of different chiral diphosphine ligands that allowed one to reach high enantioselectivities for a wide range of substrates. The first account was reported in 2003 and focused on the conjugated reduction of acyclic enones[39] (Scheme 13).

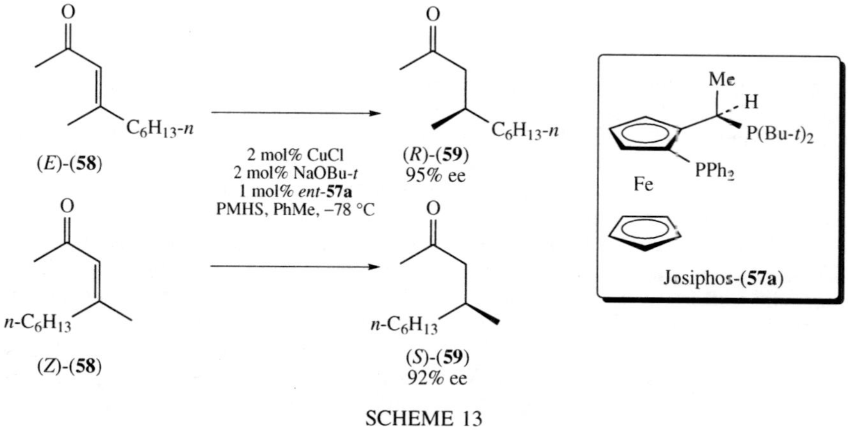

SCHEME 13

A survey of various chiral ligands showed that high enantioselectivities could be reached with Togni's Josiphos ligand **57a** (commercialized by Solvias) and that most enones could be successfully reduced with PMHS and ligand loading as low as 1–2 mol%. Enantioselectivities over 90% were reached in all reported cases and the influence of the stereochemistry of the double bond of the starting enone on the enantiodiscrimination was also studied. Thus, starting from enone *E*-**58**, the reduction product **59** was obtained with a 95% ee with *R* stereochemistry, while starting with the enone *Z*-**58**, a similar enantioinduction was observed, albeit with the opposite configuration.

The use of Solvia's Josiphos ligand as well as Tagasago's Segphos ligand was later successfully applied to the conjugated reduction of another family of substrates by the same group, including α,β-unsaturated esters and lactones[40], cycloalkenones[40], and β-silyl-substituted α,β-unsaturated esters[41]. Representative examples are listed in Chart 5.

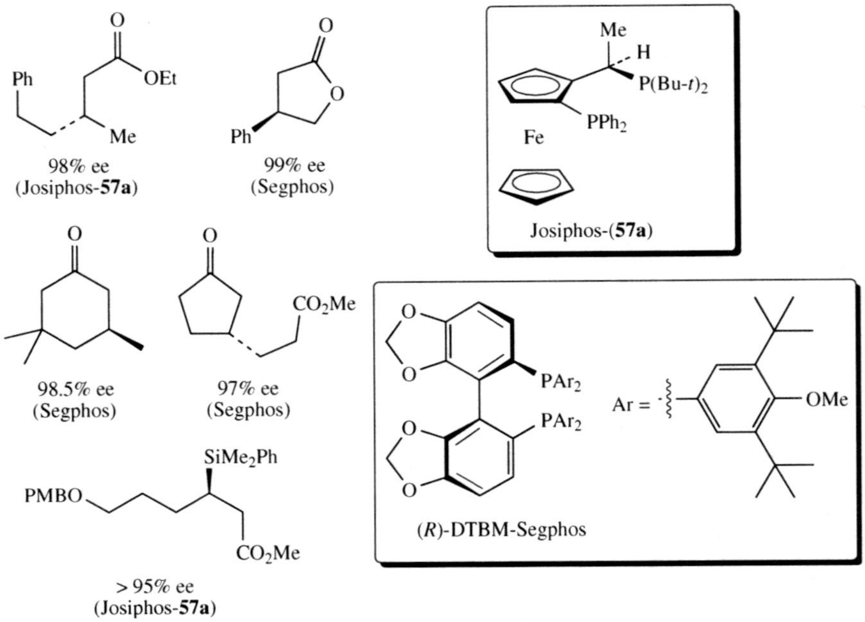

CHART 5

While Josiphos ligand **57a** gave the highest enantioselectivities for acyclic enones and esters, the chiral atropo-diphosphine DTBM-Segphos gave the best enantioselection for cyclic substrates. The authors also showed that the copper hydride catalysts gave impressive reactivity because substrate-to-ligand ratios as high as 275,000:1 could be used for this transformation, without any decrease in the enantioselectivity of the reaction. This was exemplified by one experiment conduced on 64.76 g of isophorone with a substrate:ligand ratio of 275,000:1 and two equivalents of PMHS. The use of 1 mol% CuCl/NaOBu-*t* required only 2 mg of the chiral ligand and gave 88% of the saturated ketone with a 98.5% ee. Since in most cases the loading of the copper salt is always higher than the chiral ligand, those experiments show a strong ligand acceleration effect for those conjugated reduction reactions. This strategy was recently applied to the total synthesis of the C-9 epimer of amphidinoketide I **60**[42] (Chart 6). The three stereogenic centers were introduced through enantioselective Josiphos **57a**/copper(II) acetate conjugated reductions of enones with high ee's.

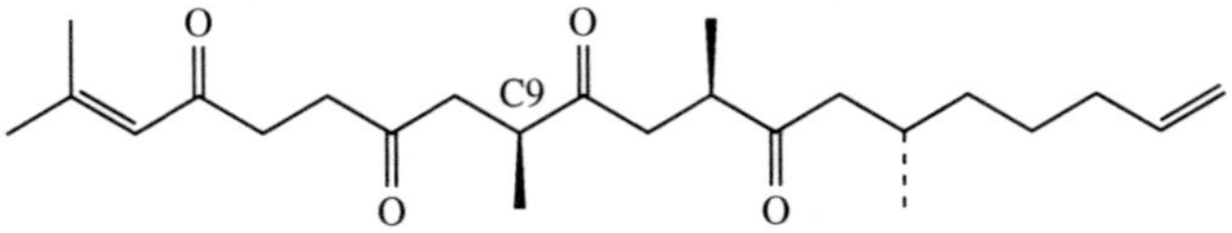

amphidinoketide I (C-9 epimer) (**60**)

CHART 6

The pioneer work of Buchwald in asymmetric conjugated addition of copper hydride catalyst led other groups to investigate this methodology on various types of electrophilic double bonds bearing electron-withdrawing groups such as nitro, cyano and sulfonyl groups.

Carreira and Czekelius investigated the reactivity of nitroolefins and showed that high enantiomeric excesses could be easily reached for the conjugated reduction into enantiomerically enriched β-substituted nitroalkanes[43–45] (Scheme 14).

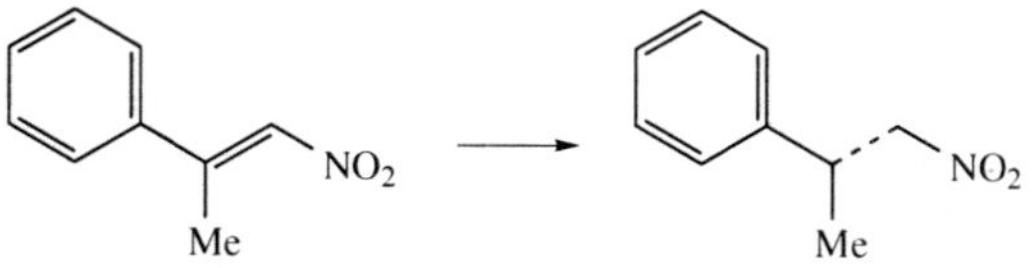

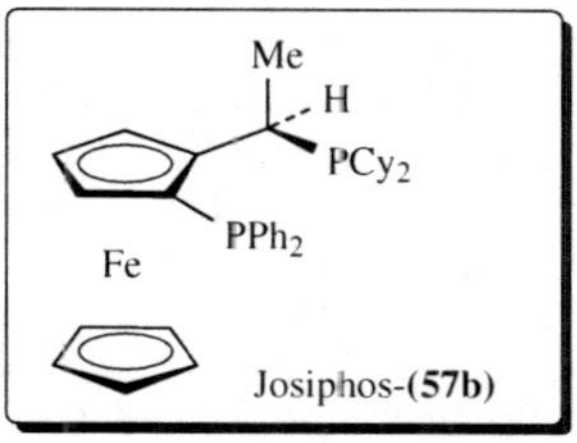

initial conditions : 0.1 mol% CuOBu-*t*/Josiphos **57b**
0.1 equiv. PMHS, 1.2 equiv. $PhSiH_3$
1.2 equiv. water, PhMe, rt, 24 h
=> 77%, 88% ee

optimized conditions with CuF_2:
1 mol% CuF_2/Josiphos **57b**
0.1 equiv. PMHS, 1.5 equiv. $PhSiH_3$
10 mol% CH_3NO_2
1 equiv. water, PhMe, rt, 16 h
=> 76%, 92% ee

SCHEME 14

The authors initially chose to work with copper *tert*-butoxide as the precursor of the copper hydride catalyst[43]. A survey of chiral ligands showed that Josiphos disphosphine **57b** is the most efficient in terms of enantioselectivity; 66–92% ee's could be obtained with catalyst loading of 0.1–1 mol%. The addition of 1.2 equivalent of water to the reaction mixture avoids over-reduction of the nitro group to the corresponding oximes of aldehydes. It was postulated that the addition of water allowed the hydrolysis of the primary product of the reaction, a silyl nitronate, which gave the corresponding nitroalkane without further reduction to the oxime by-product. A limitation is the use of an air-sensitive catalyst precursor, since the use of purified copper *tert*-butoxide was necessary for a good catalytic reaction. It was noted that the *in situ* formation of copper alkoxide from CuCl and NaOBu-*t* led to the formation of a much less active catalyst with concomitant formation of by-products. Furthermore, it was shown that various inorganic salts (such as NaCl) led to diminished reaction rates. The authors initially noted that isomerization of the nitroalkanes occurred and that diminished yield could result from this phenomenon. The use of a catalytic amount of tetra-*n*-butyl ammonium hydroxide as an additive allowed the *in situ* equilibration of the nitroalkenes mixture and to drive the reaction to completion[44]. As copper(I) *tert*-butoxide is known to be strongly air- and moisture-sensitive, the same authors later proposed an optimized protocol that used air-stable and easy-to-handle copper(II) fluoride as metal source for the catalyst preparation. A catalyst loading of 1 mol% including Josiphos **57b** as chiral ligand gave high enantioselectivities (82–96%) for the corresponding nitroalkanes[45]. The addition of a catalytic amount of nitromethane also allowed the reduction of unreactive substrates bearing electron-rich aryl and heteroaryl groups.

Yun and coworkers focused on the enantioselective reduction of unsaturated nitriles with copper/Josiphos catalysts with high ee's[46] (Scheme 15).

In a previous report, Yun and coworkers showed that copper(II) acetate could be used as a copper(I) hydride precursor for the conjugated hydrosilylation of α,β-unsaturated nitriles such as **61**, when rigid diphosphines such as DPEphos and Xantphos were used as

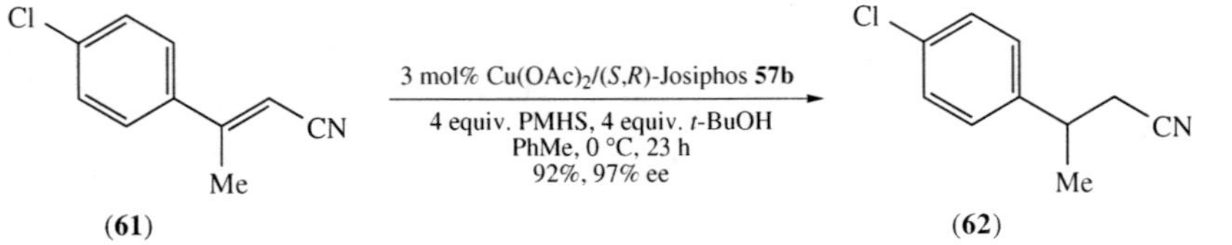

SCHEME 15

ligands and PMHS as reducing agent[47]. It was postulated that a diphosphine Cu(I) hydride catalyst was formed *in situ* by reduction of the copper(II) acetate by the silane and complexation by the diphosphine ligand. They then used Josiphos **57b** as chiral ligand to carry out the enantioselective reduction of β-disubstituted α,β-unsaturated nitriles with the same catalytic system and observed excellent reactivity and high enantioselectivities for the resulting saturated β-substituted nitriles **62** (absolute configuration not determined) (Scheme 15). They later reported that 3,3-diarylacrylonitrile bearing two different aryl groups on the electrophilic double bond could also be used as substrate and that the corresponding reduction adduct could be obtained with high enantioselectivities[48] (up to 94%). Xu and coworkers reported at the same time a similar catalytic system for the enantioselective reduction of substituted dicyanovinylene double bond with ee's up to 93% when Binap was used as chiral ligand and phenylsilane as reducing agent[48].

Finally, the enantioselective conjugated reduction of vinylic sulfones was independently reported in 2007 by Carretero, Charette and coworkers (Scheme 16)[49,50].

Carretero and coworkers

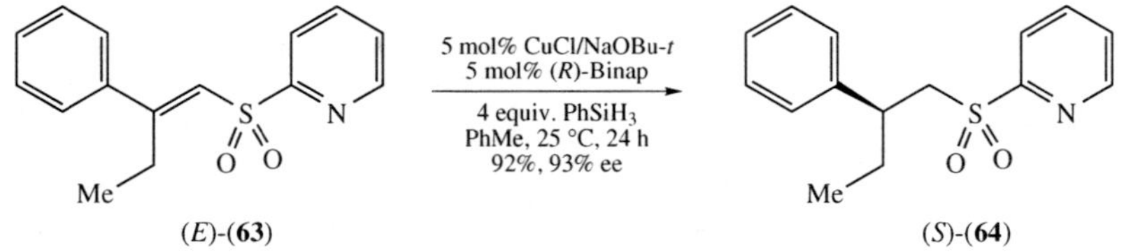

Desrosiers and Charette

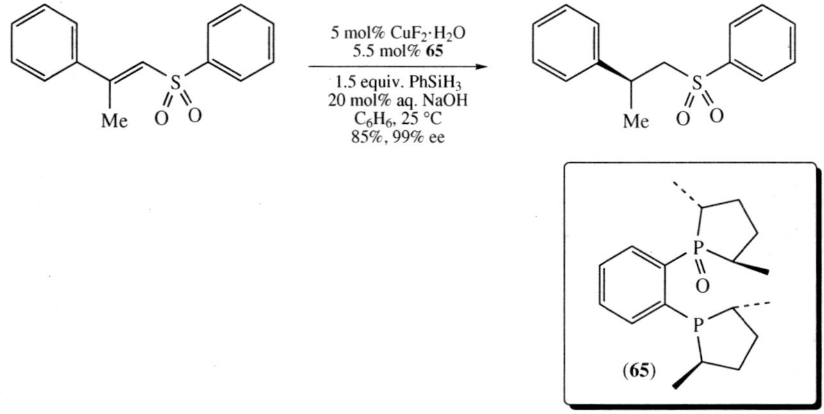

SCHEME 16

Carretero and coworkers showed that the use of a 2-pyridylsulfonyl group on the double bond leads to a dramatic increase in the reactivity of the conjugated reduction with the *in situ* generated Binap copper(I) hydride and phenylsilane (no reactivity was observed with a phenylsulfonyl group). This phenomenon was ascribed to a possible coordinating effect of the 2-pyridyl group on the copper atom which could increase the reaction rate for the hydride addition to the electrophilic double bond. High enantioselectivities were obtained for a broad range of disubstituted vinylic sulfones **63** and the saturated adduct **64** were also transformed using standard chemical group transformations (such as Julia–Kocienski olefination) into various enantioenriched synthons. Desrosiers and Charette optimized an original catalytic system for the enantioselective conjugated reduction of β,β-disubstituted vinyl phenyl sulfones by using methyl-DuPhos monoxide ligand **65** to activate the copper hydride catalyst. Copper(II) fluoride was used as the metal source and was believed to be reduced and activated by the silane to give the chiral copper(I) hydride catalyst. Most substrates tested gave high yields and enantioselectivities over 90%, and the utility of the new adducts was also demonstrated though a transformation using the Julia olefination.

V. CATALYTIC HYDROSILYLATION OF KETONES AND IMINES

A. Catalytic Hydrosilylation with Non-chiral Ligands

Although the use of copper(I) hydride catalysts for the conjugated reduction of an electrophilic double bond was logical due to the soft character of the hydride, their application for the reduction of harder carbonyl electrophiles such as aldehydes and ketones was less obvious and the first known example was published early 2001 by Lipshutz and coworkers[51]. They initially used a catalytic amount of Stryker's reagent **2** to catalyze the hydrosilylation of aliphatic and aromatic aldehydes with various silanes, to yield the reduction adducts as the corresponding silyl ethers (Scheme 17).

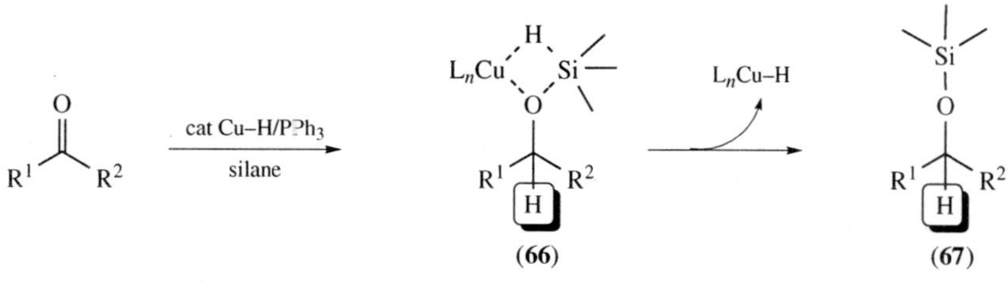

SCHEME 17

The mechanism should occur through a copper hydride addition to the carbonyl that led to the formation of the corresponding copper alkoxide. Regeneration of the catalytic hydride can then proceed through a σ-bond metathesis of the copper alkoxide and the silane via a 4-centered transition state **66** that releases the silyl ether **67** and the copper hydride L_nCuH. Later on, the same group improved the reaction that could be extended to poorly activated silanes such as triethylsilane and *tert*-butyldimethylsilane[52] (Scheme 18).

The diphosphine copper hydride catalyst is generated *in situ* first by formation of the chelated copper *tert*-butoxide complex **10**. The σ-bond metathesis with the silane **68** releases the silyl alkoxide **69** and the desired copper hydride catalyst **11**. When triethylsilane (5 equivalents) was used as the stoichiometric reducing agent with DM-Segphos as the diphosphine ligand, the reduction of dialkyl ketones occurs at room temperature with 3 mol% copper and 0.05 mol% ligand loading. The corresponding reduced triethylsilyl

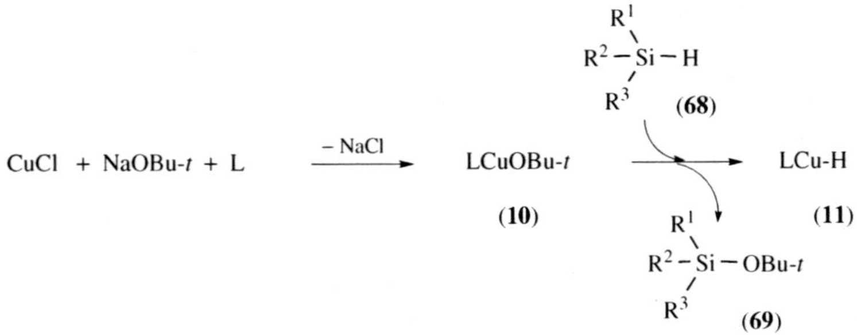

SCHEME 18

ethers are isolated in excellent yields. When the more hindered *tert*-butyldimethylsilane is used, a very low reactivity is observed and the authors attributed this disappointing reactivity to the steric hindrance of the silane which cannot efficiently activate the copper(I) *tert*-butoxide precatalyst. However, switching from sodium *tert*-butoxide to sodium methoxide gave a less hindered copper alkoxide precatalyst which could then be efficiently activated by the more bulky silane.

N-Heterocyclic carbene displays strong σ-donor properties to the metal center that can be compared to tertiary alkyl phosphine. It has been shown that such ligands give well-defined and air-stable monomeric copper halide complexes. It could be expected that the corresponding copper hydride complex might display a harder character, which might in turn prove more favorable for the nucleophilic addition on a ketone. Therefore, it may lead to an increased reactivity for the hydrosilylation of ketones. The reactivity of a wide range of NHC ligands on copper in the hydrosilylation reaction of cyclohexanone was examined by Nolan and coworkers, and optimization of the ligand structure led to the selection of the IPrCuCl complex **70** as the catalytic precursor for the hydrosilylation of various aryl alkyl and dialkyl ketones **71** (Scheme 19)[53].

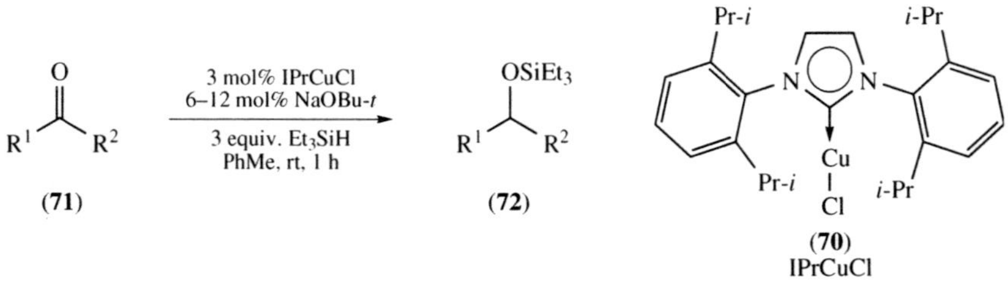

SCHEME 19

The catalytic system requires the use of an excess of sodium *tert*-butoxide as a cocatalyst. It is expected that the ligand exchange leads to the *in situ* formation of the sensitive NHC copper alkoxide that can be then activated by the silane to give the NHC copper hydride catalyst. This procedure was applied with triethylsilane, an inexpensive silane that led to its triethysilyl ether **72**. Further optimizations of the catalytic system lead to an efficient procedure for the hydrosilylation of hindered ketones with triethylsilane[54] (Scheme 20).

A survey of various carbene salts on the reduction of dicyclohexyl ketone showed that the *N*,*N*′-dicyclohexyl-substituted carbene ligand gave the best reactivities. A protocol

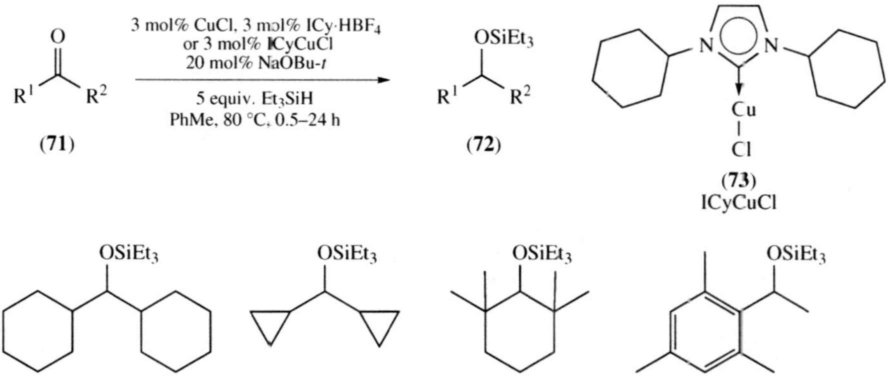

(71) (72) (73) (74) (75) (76) (77)

SCHEME 20

using a 3 mol% of ICyCuCl **73**, and sodium *tert*-butoxide as cocatalysts and triethylsilane at 80 °C reduces a wide range of aliphatic and aromatic hindered ketones into the corresponding triethylsilyl ethers (**74–77**) in excellent yields. It was also shown that the NHC copper complex could be prepared *in situ* from copper(I) chloride and ICy•HBF_4, albeit with a reduced reactivity. Various hindered and electron-rich monophosphines were also tested for comparison with the NHC ligands and displayed lower reactivities.

The same group later reported that air-stable cationic copper complexes bearing two NHC ligands per copper atom, such as $[IPr_2Cu]BF_4$, could be used as hydrosilylation catalysts after activation with sodium *tert*-butoxide[55]. Simple aliphatic and aromatic ketones can be fully reduced with 3 mol% of the cationic precatalyst and only two equivalents of triethylsilane when the reaction is carried out at room temperature in THF. More hindered ketones can also be reduced with excellent yield in the same solvent when the reaction is carried out at 55 °C. Mechanistic investigations by ^1H and ^{13}C NMR allowed the determination of the true catalyst precursor, as the reaction of NaOBu-*t* with $[IPr_2Cu]BF_4$ releases the IPrCuOBu-*t* complex, the direct precursor of the catalytic species IPrCuH[56]. The formation of the copper alkoxide and the release of the free carbene IPr was unambiguously proved by ^1H NMR in $[D_8]$THF and the positive effect of the tetrafluoroborate counteranion was also investigated.

Two other groups reported the use of NHC copper carbene complexes with their application in hydrosilylation of aldehydes, ketones and electrophilic double bonds.

Buchmeiser and coworkers designed the copper(I) complexes **78** and **79** (Chart 7) which were tested for their performance in carbonyl cyanosilylation, carbonyl hydrosilylation and atom transfer radical polymerization of methyl methacrylate[57]. Both precatalysts, when activated with NaOBu-*t*, gave very high activities with triethylsilane when the reactions were carried out in THF at 65 °C with a turnover up to 100,000 could be achieved with some activated substrates (such as benzaldehyde and 4-bromoacetophenone). Yun and coworkers reported a rare example of a NHC copper(II) complex **80** (Chart 7) which was prepared by direct complexation of the IPr carbene ligand and copper(II) acetate[58]. This air-stable copper complex was characterized by X-ray crystallography and was successfully used as a precatalyst for the hydrosilylation of propiophenone and electrophilic double bonds with PMHS or TMDS (tetramethyldisiloxane). Only 1 mol% of the precatalyst is required and it was postulated that the NHC copper(II) acetate **80** is reduced *in situ* by the silane to give the NHC copper(I) hydride catalytic species.

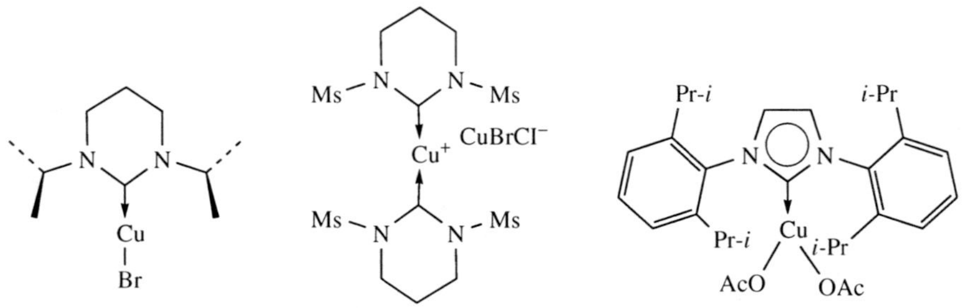

(78) (79) (80)

CHART 7

B. Catalytic Hydrosilylation with Chiral Ligands

Although often forgotten in the scientific literature of the field, the true pioneer work in the field of enantioselective catalysis with chiral copper(I) hydride complex was reported in 1984 by Brunner and Miehling[59]. They reported the first generation of chiral diphosphine copper(I) hydride catalyst and its application in the asymmetric hydrosilylation of acetophenone (Scheme 21).

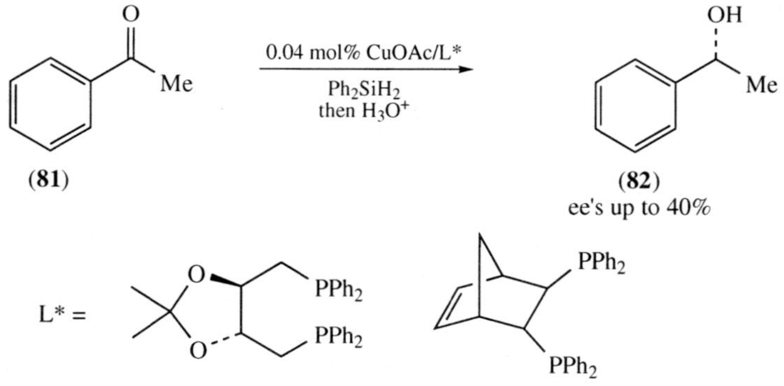

SCHEME 21

Copper(I) acetate and benzoate were used as metallic precursors and combined with chiral diphosphines such as Diop and Norphos. The hydrosilylation of acetophenone **81** gave quantitative yield of the corresponding carbinol **82** with catalyst loading as low as 0.04 mol%, thus showing the excellent reactivity of such hydride catalyst for the reduction of ketones. Although the enantioselectivity remains modest with the chiral ligand available at that time, the pioneer work of Brunner and Miehling was reported four years before the first publication of Stryker on the use of the $[CuH(PPh_3)]_6$ complex for the chemoselective reduction of electrophilic double bonds.

It took another 17 years after this publication before significant ameliorations were reported in this field with simultaneous reports in 2001 by the groups of Lipshutz and Riant on the asymmetric hydrosilylation of prochiral ketones with chiral copper catalysts[60,61] (Scheme 22).

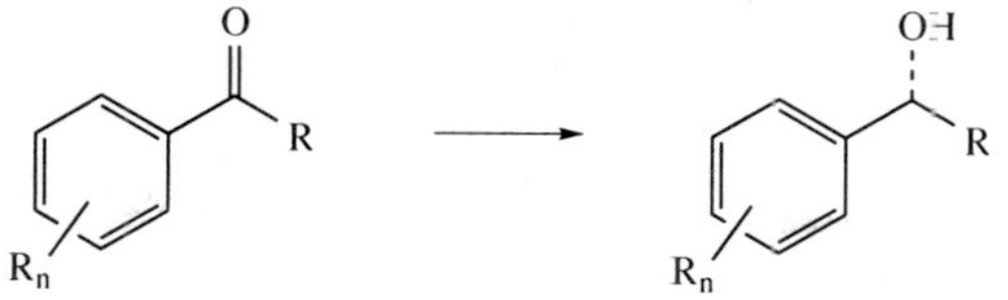

Lipshutz and coworkers:
3 mol% CuCl, 3 mol% NaOBu-*t*
3 mol (*R*) –3,5-xyl-MeOBiphep
0.34 PMHS, PhMe, –50 or –78 °C
=> ee's up to 97%
ee = 75% for acetophenone with Binap

Ligand acceleration effect
substrate/ligand up to 20,000

Riant and coworkers:
1 mol% CuF_2, 1 mol% Binap
1.2 equiv. $PhSiH_3$
PhMe, rt
=> ee's up to 92%
ee = 76% for acetophenone with Binap

air acceleration effect

SCHEME 22

Lipshutz and coworkers optimized a catalytic procedure in which the chiral catalyst is generated *in situ* by combining copper(I) chloride, chiral diphosphines and sodium *tert*-butoxide with silane activation[60]. A survey of various chiral diphosphines showed that a member of the MeOBiphep family (Biphep = 2,2′-bis(diphenylphosphine)-3,3-diphenylbiphenyl) bearing hindered xylyl groups on the phosphorus atom allowed one to reach enantioselectivities over 90% for a range of aryl alkyl ketones. Furthermore, in the presence of 3 mol% of CuCl, they also discovered a strong ligand acceleration effect as the amount of the chiral ligand could be decreased to 0.005 mol% without any decrease in the ee's of the chiral alcohol, thus showing a substrate-to-ligand ratio of 20,000. Riant and coworkers used copper(II) fluoride as metal precursor and Binap as chiral ligand and observed enantioselectivities in the range of 64–92% for a range of aryl alkyl ketones[61]. They also observed that the catalytic activity decreased strongly for dialkyl ketones with low enantioselection. The most striking feature of this catalytic system resides in the observation that the system is not only compatible with the presence of air, but that it is actually strongly accelerated by oxygen. This allowed the authors to build up a practical procedure that used undried standard glassware and undistilled solvents for this catalytic reaction. It should be noted that both catalytic systems reported by the two groups involve active species which are probably of the same nature, as both systems give the same enantioselectivity for acetophenone (75–76%) when Binap is used as ligand.

Both research groups later optimized their initial catalytic system and studied both the ligand acceleration effect and the effect of molecular oxygen on the reactivity of the catalytic system. Riant and coworkers used $CuF(PPh_3)_3{\cdot}2MeOH$ with a chiral diphosphine ligand and confirmed their initial observations of the unexpected acceleration effect of the hydrosilylation reaction by molecular oxygen[62]. They carried out a survey of chiral diphosphine ligand and found that 4-MeO-3,5-di-*t*-Bu-MeOBiphep bearing bulky groups on the phosphorus atom gave the highest enantioselectivity. An optimized protocol using 0.05 mol% of the chiral copper catalyst was then applied to various aryl alkyl ketones with high enantioselectivities when the reaction was carried out in air. Lipshutz and coworkers reported further optimization of their initial catalytic system with chiral ligands from the Biphep and Seghos family[63]. They found that shorter reaction times and higher enantioselectivities could be reached with 4-MeO-3,5-di-*t*-Bu-Segphos (DTBM-Segphos), and they showed that the strong ligand acceleration effect allowed one to work with a substrate/ligand ratio up to 100,000:1 when the copper loading was kept at 3 mol%.

Other research groups reported some variations on the ligand or on the copper sources in the enantioselective hydrosilylation of aryl alkyl ketones. Yun and Lee used copper(II) acetate and Binap with phenylsilane (with a 3 mol% catalyst loading) and obtained enantioselectivities in the range of 79–89%[64]. Dagorne, Bellemin-Laponnaz and coworkers used phenyldimethylsilane as the hydride source with a CuCl/NaOBu-*t*/Binap catalyst (5 mol% loading) and observed enantioselectivities over 90% when the reaction was carried out in toluene at −78 °C[65]. Finally, Chan and coworkers used Xylyl-P-Phos with copper(II) fluoride (1.2–3 mol% loading) and phenylsilane as the hydride source and observed enantioselectivities in the range of 72–96% for a wide range of aryl methyl ketones when the reaction was carried out in air[66]. They could obtain a substrate-to-ligand ratio up to 100,000 for the most activated substrates (such as *p*-nitroacetophenone), and they also studied their catalytic system for the enantioselective hydrosilylation of dissymmetric benzophenones **83** (Scheme 23).

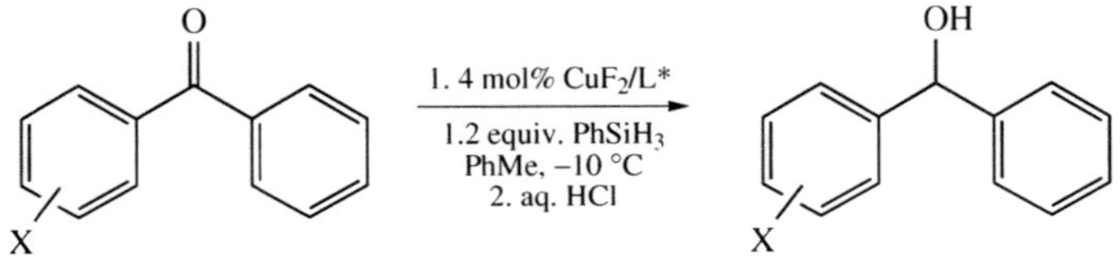

(**83**) (**84**)

X = 2, 3 and 4-Me, Cl, F, CF_3

Substituents at 2 position, 63–98% ee
Substituents at 3 and 4 position,6–43% ee

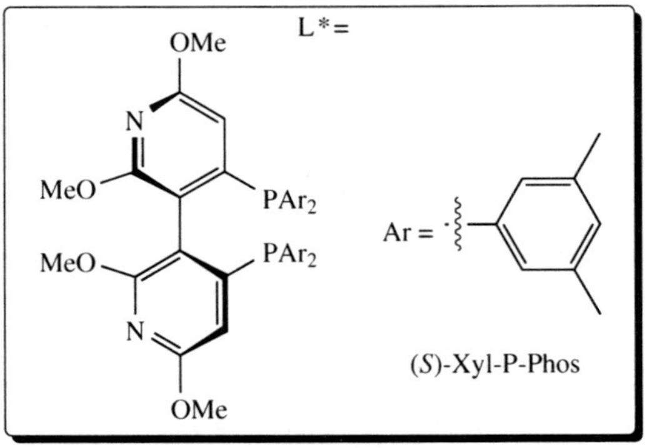

SCHEME 23

As expected, high enantiomeric excesses could be reached when a good differentiation was made between the two aryl groups of the starting benzophenones **83**. Good to high enantioselectivities were then obtained for 2-substituted benzophenones with an optimum of 98% for the 2-CF_3 substituent. Low to modest enantioselectivities were also observed with substitution at 3- and 4-position of the aromatic and the inversion of configuration of the *p*-substituted benzhydrols **84** was obtained with the ligand bearing the same configuration compared to the corresponding *o*-substituted benzhydrols.

The study of the asymmetric hydrosilylation of diaryl (and heteroaryl) ketones was later generalized by Lee and Lipshutz, who used the DTBM-Segphos/CuH catalyst for the enantioselective reduction of a wide variety of benzophenones[67]. They showed that high enantioselectivities could be reached with substituents on both aryl groups of the starting ketones, and they studied the effect of the electronic and steric properties of the substituents on the two aryl moieties as well as the influence of some heterocycles such as pyridines, thiophene and furan.

Early studies on the asymmetric hydrosilylation of prochiral ketones showed that high reactivities and enantioselectivities were limited to the case of aromatic ketones. Lipshutz and coworkers focused on families of substrates that have the potential to be produced on a large scale as key intermediates for pharmaceutical industries[68,69]. Some representative examples of enantiomerically enriched alcohols are presented in Chart 8.

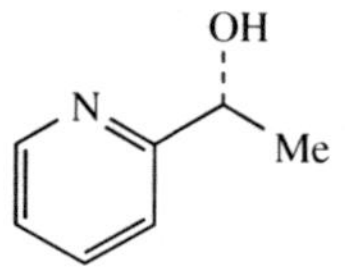

(85)
90% ee

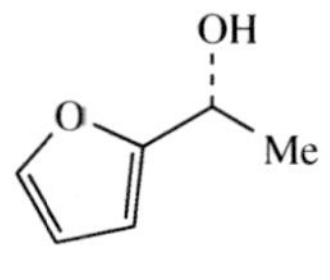

(86)
90% ee

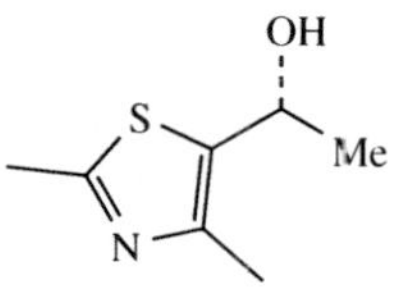

(87)
99% ee

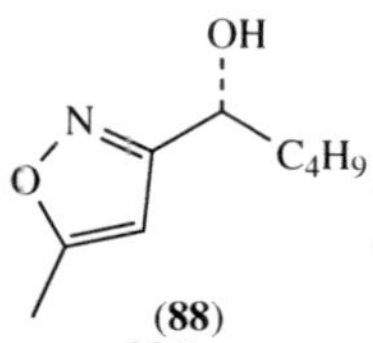

(88)
83% ee

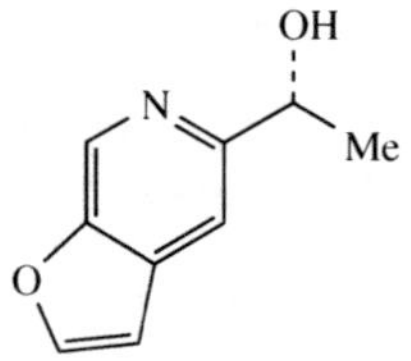

(89)
97.1% ee

HIV-1 non-nucleosidic reverse transcriptase inhibitor precursor

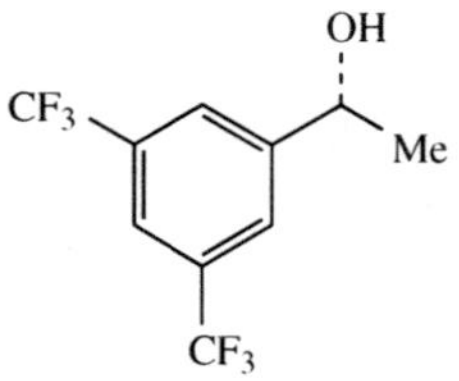

(90)
95% ee

NK-1 antagonist precursor

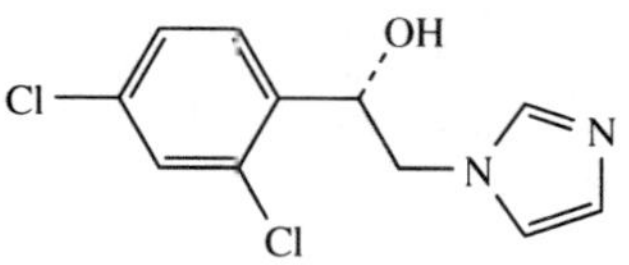

(91)
86.1% ee

antifungal miconazole type precursor

CHART 8

They first studied various heteroaryl alkyl ketones for which corresponding alcohols have found wide applications in medicinal chemistry. The reduction of all the reported susbstrates occurred with good to high yield with 1–3 mol% loading of CuCl/NaOBu-*t* and 0.05 mol% of DTBM-Segphos with PMHS as the hydride source[68]. They observed that good to high enantioselectivities for the formation of **85–88** could be obtained regardless of the electron-deficient (such as pyridine) or electron-rich (such as furan) nature of the heterocycle. They later focused on the synthesis of key intermediates of known physiologically active compounds using this methodology[69]. The copper hydride precursors were either CuCl/NaOBu-*t* (1–2 mol%) and DMTB-Segphos or 3,5-Xylyl-MeOBiphep as chiral ligand. PMHS (1 equiv.) was used as the hydride source with the addition of 1 equivalent of *tert*-butanol for optimum reactivity of the catalytic system. High enantioselectivities (three representative examples **89–91** are described in Chart 8) ranging from 86.1 to 99.4% were reported, thus showing the high potential of such methodology in the production of enantiomerically enriched key intermediates for the pharmaceutical industry.

While there are many examples reported for the enantioselective hydrosilylation of prochiral ketones with copper(I) hydride catalyst, there is only one example in the application of such methodology for more challenging imine substrates[70] (Scheme 24).

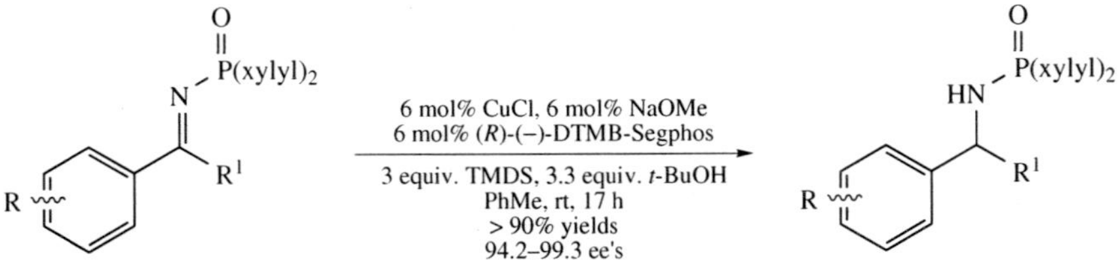

(**92**)

SCHEME 24

This system uses a copper(I) hydride catalyst generated from CuCl/NaOMe and TMDS (tetramethyldisiloxane) as the hydride source with DTBM-Segphos as the chiral ligand. Imine **92** activated by a bisarylphosphinyl group was also required for a good reactivity and steric hindrance was increased on the phosphorus atom by introducing xylyl groups. Three equivalents of TMDS (6 equivalents of Si−H per imino bond) and *tert*-butanol were also required for optimum reactivity. Although this system still requires a fairly high catalyst loading (compared to the catalyst loading usually required for simple ketones and electrophilic olefins), it remains one of the most enantioselective reported to date as all the enantiomeric excesses reported for various aryl alkyl ketimines were in the range of 94.2−99.3%. The same authors also later reported that the heterogeneous precatalyst copper on charcoal (Cu/C) could be used as the metal source for this transformation with equal efficiency when the reaction was carried out under ultra-sonication[71]. This heterogeneous catalytic system was also investigated for the enantioselective hydrosilylation of ketones and electrophilic double bonds and compared with the corresponding homogeneous catalytic system.

The catalytic use of Stryker's reagent in reduction processes was initially used with molecular hydrogen as the hydride source and there is only one example of the development of an enantioselective version of this method for the reduction of prochiral ketones to date. Shimizu and coworkers recently reported a new catalytic system, which generated a chiral copper(I) hydride catalyst for the enantioselective hydrogenation of aryl alkyl ketones[72] (Scheme 25).

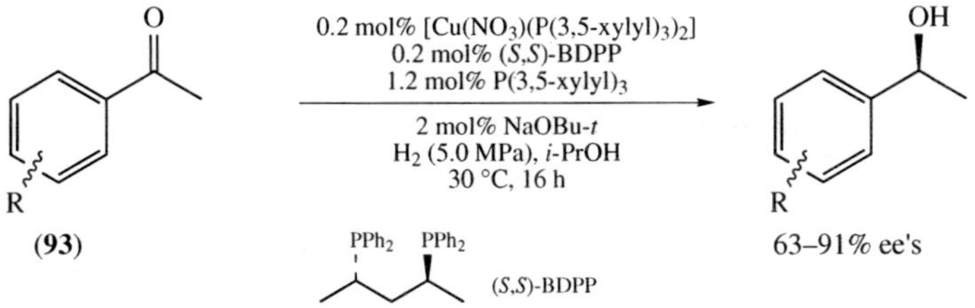

(**93**)

63–91% ee's

SCHEME 25

They first found that copper(I) nitro complex $[Cu(NO_3)(PPh_3)_2]$ gave a better reactivity for the hydrogenation of acetophenone when it was activated with an excess of sodium *tert*-butoxide. A survey of various chiral ligands with a 500−3000 substrate to catalyst ratio showed that (*S*,*S*)-BDPP gave the best enantioselectivity (48%). They also studied the influence of the monophosphine coligand and found that the enantioselectivity could be increased from 48% to 56% when triphenylphosphine was replaced by the more bulky tris(3,5-xylyl) phosphine. Using this optimized catalytic system and a catalyst loading of 0.2 mol%, they carried out the hydrogenation of various aryl methyl ketones **93** and observed that optimum enantioselectivities were reached when the aryl group bears an *ortho* substituent (81–91% ee).

VI. CATALYTIC DOMINO REACTIONS

A. Catalytic Domino Reactions with Non-chiral Ligands

The conjugated addition of a copper(I) hydride complex on an electrophilic double bond bearing an electron-withdrawing group leads to the formation of a nucleophilic copper enolate, which can then be used for further transformations according to the reaction conditions (Scheme 26).

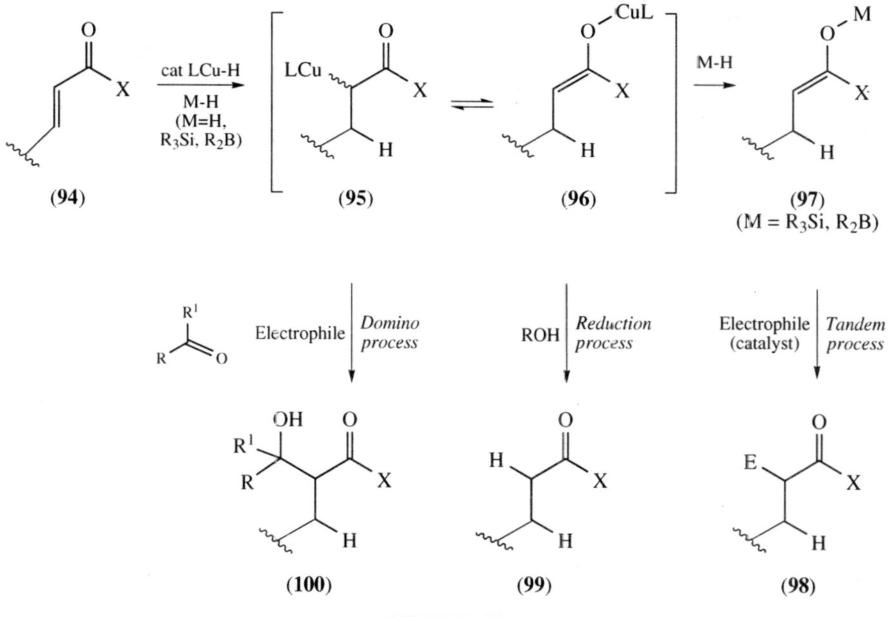

SCHEME 26

As shown before, when no electrophile is added, the copper enolate **96** can then be trapped by the hydride source (silane or borane) to give the corresponding silyl (or boryl) enol ethers **97**. Those pronucleophiles can react with various electrophiles (aryl and alkyl halides for arylation and alkylation, aldehydes for aldolization) after a proper activation (palladium, fluorides, Lewis acids in the case of silyl enol ethers) to yield the corresponding adducts **98** (E = Ar, Alk, RCH(OH)). For this method, two steps are required, although it is not necessary to isolate **97**. When the catalytic addition leading to **96** (the real structure of the copper enolate intermediates remains still to be established (**95** versus **96**) as well as its mode of reaction with electrophiles) is carried out and treated with a proton source such as *tert*-butanol, the enolate should be protonated to give the reduction adduct **99** and a copper alkoxide which will be easily transformed into the copper(I) hydride catalyst.

This shows that the nucleophilic copper hydride reacts faster with the electrophilic double bond of **94** than with the proton source (the choice of *tert*-butanol as a hindered and less basic proton source is often crucial for good reactivity). This observation led to the idea that when an electrophilic double bond and a harder electrophile (such as an aldehyde or a ketone) are subjected together to the action of a copper(I) hydride catalyst, a domino process could take place with a preference of the copper hydride for the softer electrophilic double bond. The resulting copper enolate **96** could then in turn react with the aldehyde or ketone, to give at the end of the catalytic process an aldol adduct **100** in a *one pot* domino process.

Those reductive aldol domino processes have been well described with cobalt, rhodium, iridium and palladium catalysts and have been now developed for the past ten years with copper(I) hydride catalysts[73].

Although the early publications in this field do not truly rely on catalytic systems, it should be noted that the pioneer work was reported in 1998 by Chiu and coworkers, who reported over several years elegant variations of intramolecular versions of reductive aldol reactions using stoichiometric and catalytic amounts of Stryker's reagent **2**. Chiu and coworkers reported the first example of this reaction for the construction of a 5,7-bicycle as model compound in the total synthesis of pseudolaric acid A[74] (Scheme 27).

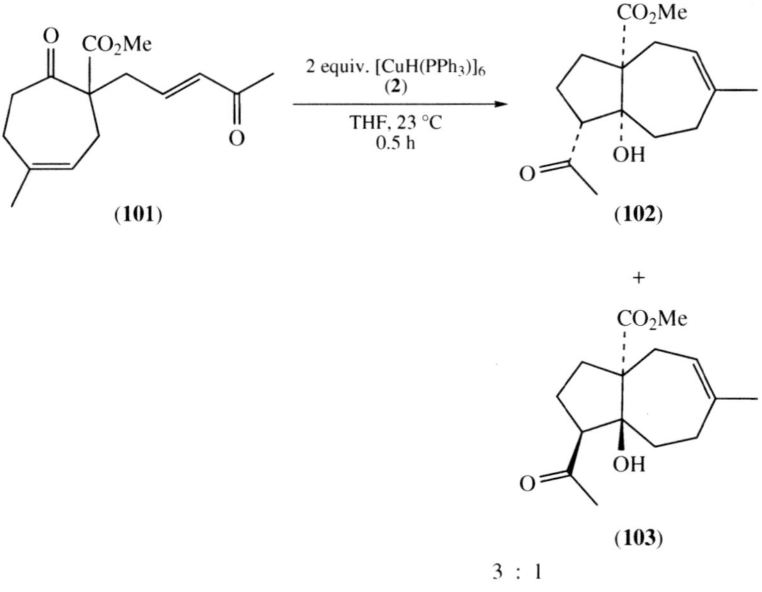

SCHEME 27

They initially noted that the intramolecular base-catalyzed condensation of the saturated diketone analogue of **101** gave a mixture of products. The cyclization precursor **101** was then subjected to the action of an excess of Stryker's reagent **2** to yield, after a few minutes, the corresponding stereoisomeric adducts **102** and **103** in 66% and 16% isolated yields.

This pioneer idea led Chiu and coworkers to study the scope of this methodology for the formation of mono- and bicycle derivatives of various sizes and to investigate the nature of the Michael acceptor part.

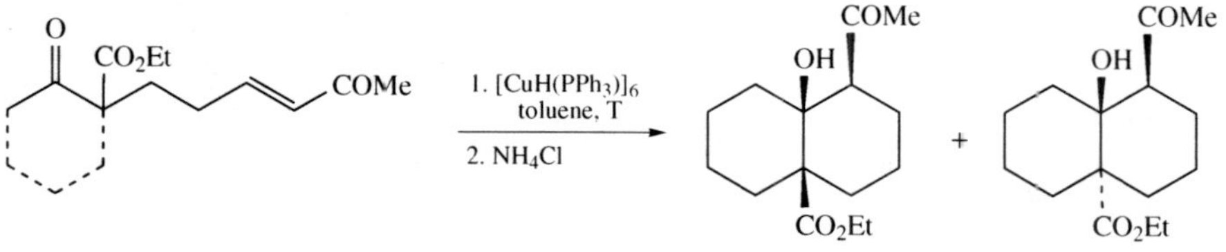

(**104**) (**105**) (**106**)

	(105)	(106)	
−40 °C (1 h)	93%	0%	(isolated yield)
−25 °C (0.5 h)	79%	3%	
−10 °C (0.25 h)	12%	58%	

SCHEME 28

They first reported a meticulous study of the reaction parameters and substrate variations using a stoichiometric amount of Stryker's reagent **2** to carry out the domino process[75] (Scheme 28).

The influence of the temperature on the reductive cyclization of diketo ester **104** has a strong influence on the stereochemistry of the bicycle junction of the corresponding adducts. Indeed, the kinetic *cis*-adduct **105** was obtained as the sole product when the reaction was carried out at −40 °C. However, when the reaction temperature was raised to −10 °C, the retro-aldol reaction gives rise to the formation of the thermodynamic *trans*-adduct **106** as the major product.

Keeping the reaction temperature at −40 °C, they then studied the scope of this reaction on various mono- and bicycle precursors. Representative examples are shown in Scheme 29.

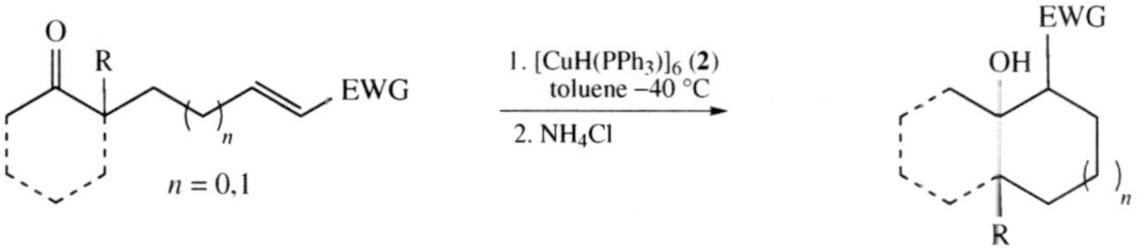

(**107**)

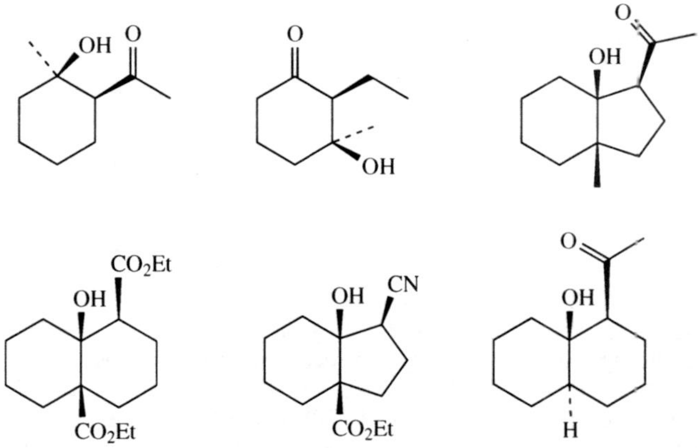

SCHEME 29

Most of the reported reactions proceed under the optimized reaction condition in good to high yields for the corresponding reductive aldol adducts (41–93%). All adducts **107** were isolated as single diastereoisomers when conditions for kinetic control (−40 °C) were used for the reaction and the cyclization gives 5- and 6-membered rings with little or no formation of the simple reduction of the double bond. When precursors of 4- and 7-membered rings were tested, the cyclization did not occur and the product arising from the reduction of the Michael acceptor was the only isolated product. The reaction is tolerant to various types of electron-withdrawing groups such as ketones, esters and cyano groups and the stereochemistry of the adduct is also independent of the stereochemistry of the Michael acceptor, although it was shown that Z-isomers were much less reactive than the E-electrophilic olefins. This methodology was later used for a short synthesis of lucinone from (+)-dihydrocarvone[76].

The adaptation of this procedure to the case of nitroalkene groups as Michael acceptors gave the corresponding reductive Henry reaction and was also investigated by Chiu and Chung[77]. Although the reaction proved to be less stereoselective in most cases, several precursors could be cyclized and the corresponding adducts were usually isolated in good yields along with variable amounts of the product arising from the reduction of the nitroolefin.

Finally, Chiu and Leung investigated the possibility of using electrophilic triple bonds as Michael acceptors for the reductive aldol reaction of a series of keto-alkyne precursors using either stoichiometric or catalytic amounts (using PMHS as the hydride source) of Stryker's reagent **2**[78] (Scheme 30).

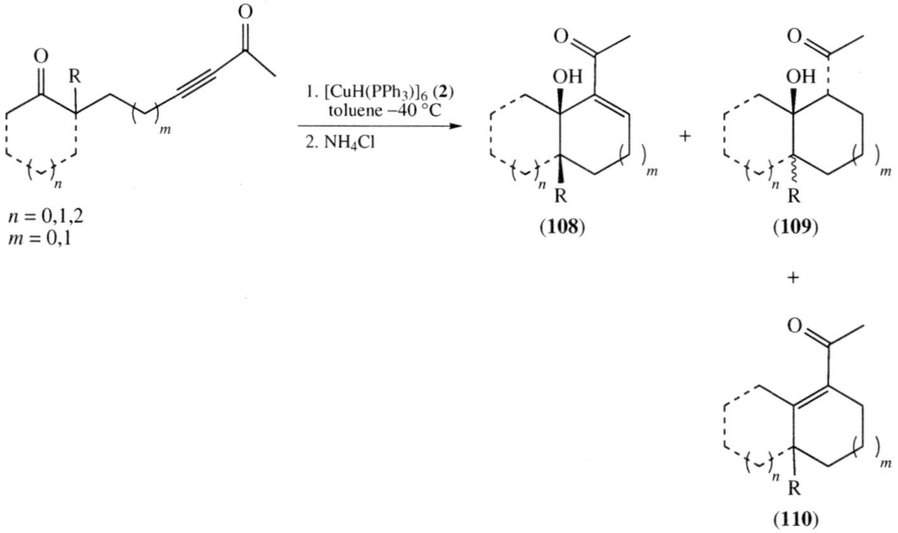

SCHEME 30

The main interest for using such substrates is that they give the equivalent of Baylis–Hillman adducts through reductive cyclization reactions. Although the desired adduct **108** was isolated in fair yield (50–60% typical isolated yield), the formation of by-products was also noticed in many cases. The adducts **108** being themselves Michael acceptors, the formation of the reduction products **109** was also observed as well as of the enones **110**, which probably originates from water elimination from the hydroxy ketones **109**.

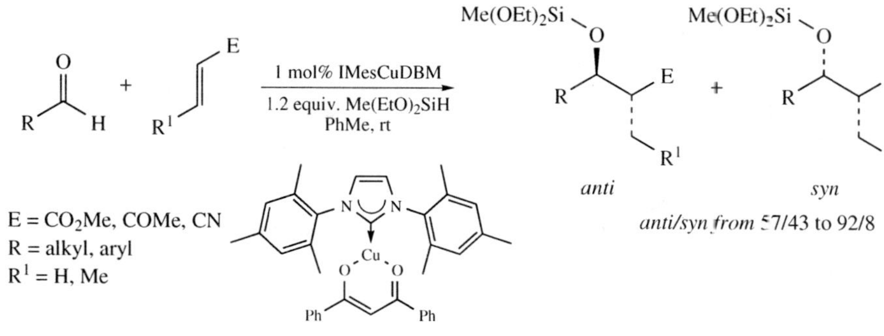

(**111**)
IMesCuDBM

SCHEME 31

There is to date only one report of an intermolecular version of the reductive aldol reaction between Michael acceptors and aldehydes[79] (Scheme 31).

It was shown earlier that NHC copper(I) chloride complexes could be used as precatalyst for the hydrosilylation reaction of various types of soft electrophiles such as conjugated double bonds as well as harder electrophiles such as aldehydes and ketones. The formation of the copper(I) hydride active species required preactivation with an excess of a metal alkoxide, and an alternative to the use of the cocatalyst was investigated by Nolan, Riant and coworkers. They found that dibenzoylmethanato ligand stabilized the copper(I) oxidation state and could be directly exchanged with a hydride ligand by reaction with a silane. The corresponding NHC copperDBM catalysts such as **111** proved to be extremely reactive in the intermolecular condensation of methyl acrylate and aldehydes as a TOF > 15,000 h^{-1} could be evaluated with a catalyst loading of 0.1 mol% (reaction complete in less than 4 min). The *anti/syn* selectivity proved to be quite insensitive to the structure of the diaminocarbene ligand, and low to fair diastereoselectivities were observed for various Michael acceptors and aromatic and aliphatic aldehydes.

B. Catalytic Domino Reactions with Chiral Ligands

The first enantioselective version of an intermolecular reductive aldol reaction between methyl acrylate and aryl alkyl ketones as electrophiles was reported simultaneously by the groups of Shibasaki, Kanai and coworkers and Riant and coworkers in 2006[80,81] (Scheme 32). They examined the asymmetric reductive aldol reaction between methyl acrylate and acetophenone catalyzed by a catalytic amount of $CuF(PPh_3)_3{\cdot}2EtOH$ associated to 3 different chiral diphosphine ligands[80]. The best combination in terms of yield, diastereo- and enantioselectivity was obtained with Tol-Binap, with either triethoxysilane or pinacolborane as the hydride source. This observation suggests that the actual nucleophile was a copper enolate intermediate rather than a silyl enolate.

Riant and coworkers investigated the use of chiral diphosphine-modified copper fluoride as catalyst for the stereoselective reductive aldol reaction between methyl acrylate and aromatic ketones with phenylsilane as reducing agent[81]. After optimization of the parameters of the reaction, a survey of the chiral diphosphine ligands showed that Solvia's Taniaphos ligand **112a** leads to higher diastereoselectivities in favor of the *erythro* adducts and with enantioselectivities up to 95%. A variety of aromatic and heteroaromatic ketones participate successfully in the tandem reaction with methyl acrylate. For the range of substrates considered, the diastereo- and enantioselectivity of the reaction were moderate to high

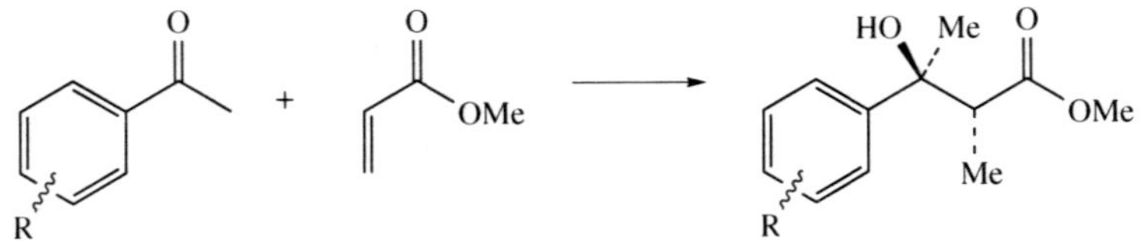

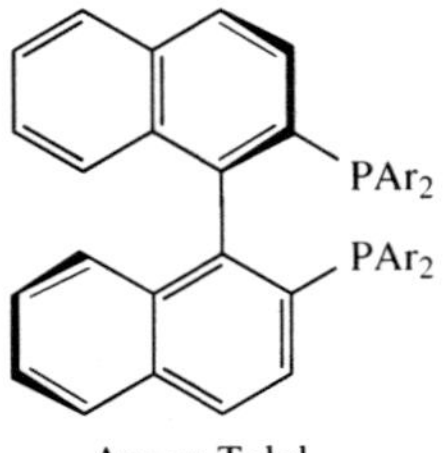

Ar = *p*-Tolyl

Shibasaki, Kanai and coworkers
2.5 mol% $CuF(PPh_3)_3{\cdot}2EtOH$, 5 mol% Tol-Binap
$(EtO)_3SiH$ (1.6 eq.), THF, 0 °C, 12 h
100% yield, *erythro:threo* 55:45
29% ee *erythro*; 1% ee *threo*

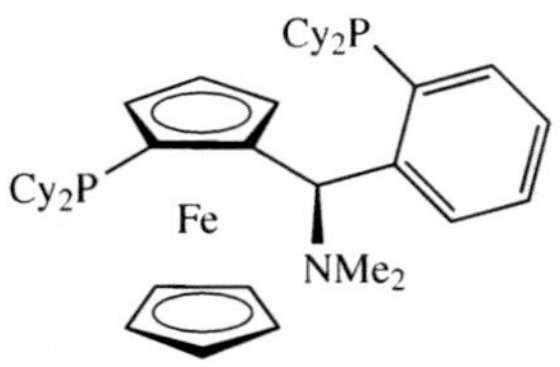

Taniaphos-(**112a**)

Riant and coworkers
1 mol% $CuF(PPh_3)_3{\cdot}2MeOH$, 1 mol% **112a**
$PhSiH_3$ (1.4 eq.), toluene, −50 °C, 1 h
98% yield, *erythro:threo* 92:8
95% ee *erythro*; 72% ee *threo*

aryl methyl and heteroaromatic ketones
31–98% yield, dr 80:20–96:4, 82–95% ee

SCHEME 32

with dr's and ee's for the *erythro* compound ranging from 80:20 to 96:4 and 82 to 95%, respectively.

Despite the somewhat limited potential application of such adducts, the results reported by both groups remain interesting since they show the high nucleophilic character of copper enolates as there are still few examples of true catalytic reactions of metal enolates on ketones.

Riant and coworkers also investigated the enantioselective catalytic aldol reaction of methyl acrylate and aldehydes[82]. They observed that their catalytic system ($CuF(PPh_3)_3$/diphosphine and phenylsilane) was extremely reactive with aldehydes as TOF up to 40,000 h^{-1} could be observed with benzaldehyde at room temperature. The competitive reduction of the aldehydes could be avoided when the reaction was carried out at low temperature and the selectivities of the reaction could be optimized to a 88:12 *syn*-to-*anti* ratio with a 96% ee with Taniaphos chiral diphosphine ligand and cyclohexane carboxaldehyde. While the enantioselectivities remain quite high for a large range of aromatic and aliphatic aldehydes, the *syn:anti* diastereoselectivities decreased when aromatic aldehydes were employed for the aldol condensation.

Recently Shibasaki, Kanai and coworkers investigated the reductive Mannich reaction with *N*-diphenylphosphinoyl substituted ketimines[83] (Scheme 33). The authors first optimized a catalytic procedure without chiral ligand to investigate the influence of the reaction parameters on the yield and diastereoselectivity of the new adducts. They found that the best catalytic system was obtained starting from copper(I) acetate and triphenylphosphine as ligand and pinacolborane as the hydride source. No asymmetric induction was, however, observed when chiral diphosphine ligands were used and some investigations led the authors to suspect the fast formation of a boron enolate (which would occur from a rapid transmetallation of the copper enolate with the borane), which would directly condense with the ketimines to deliver the racemic adducts. As silanes react much more slowly with copper enolates, pinacolborane was replaced by triethoxysilane. Although the reaction proved to be slower, high yield and enantioselectivities were reached with this silane when DiFluorPhos was used as chiral ligand.

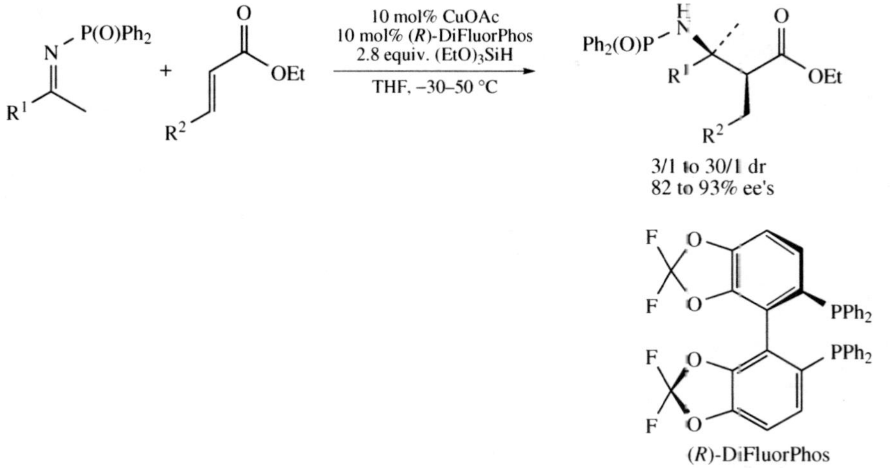

SCHEME 33

An elegant variation of the Michael acceptor was also studied by Shibasaki, Kanai and coworkers, who investigated the reductive aldol domino reaction between allenoates and aryl methyl ketones[84] (Scheme 34).

In a preliminary account of this work, the authors investigated the reaction between ethyl allenoate **113**, R = Et and acetophenone **114** with a chiral copper(I) hydride catalyst[80]. Despite moderate γ:α regioselectivity, excellent enantioselectivity and acceptable diastereoselectivity were, however, obtained for the γ and α isomers, respectively, using $CuF(PPh_3)_3{\bullet}2EtOH$/DTMB-Segphos as catalyst. In a later report, the regioselectivity was improved for the reductive aldol reaction between allenic ester (R = Et) and acetophenone by employing CuOAc as copper source and adding achiral basic phosphine PCy_3[84]. Indeed, a γ:α regioselectivity of 25:1 was obtained for the predominant *cis* isomer **115**, isolated in 96% yield with an enantiomeric excess of 99%. Similar trends were observed for a variety of aryl, alkenyl and alkyl methyl ketones. During the optimization stage, a change in the regioselectivity in favor of the α isomer **116** was noted when Taniaphos-type ligand **112b** was employed. This ligand in combination with $CuF(PPh_3)_3{\bullet}2EtOH$ catalyzed the tandem reaction between acetophenone and allenic ester (R = Me) in high yield (90%), a good ee value (84%) and moderate diastereoselectivity (dr = 10:1). This change of regioselectivity, controlled by the ligand, was also observed with other aryl and cinnamyl methyl ketones. Under this catalytic condition, these substrates gave the corresponding adducts with excellent stereoinduction.

The first example of a catalytic enantioselective intramolecular reductive aldol reaction was reported in 2005 by Lam and Joensuu[85] (Scheme 35). They started from easily available substrates **117** containing an α,β-unsaturated ester linked to a ketone through an ester linkage. The reductive cyclization was first studied with achiral ligands and an optimized protocol using a 5 mol% loading of copper(II) acetate, and dppf gave good to

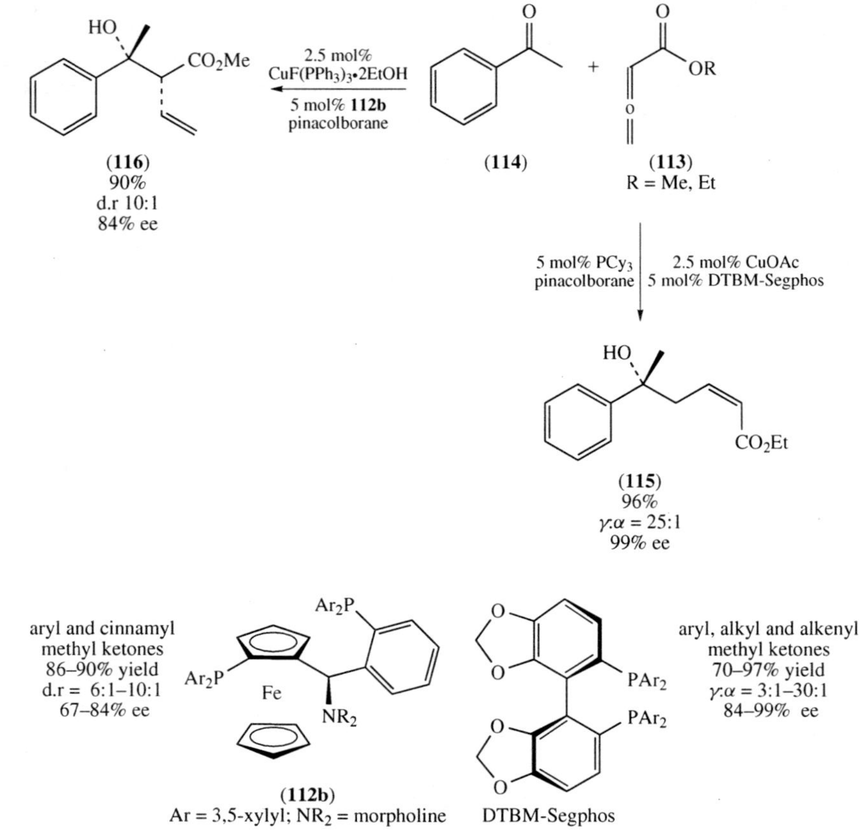

SCHEME 34

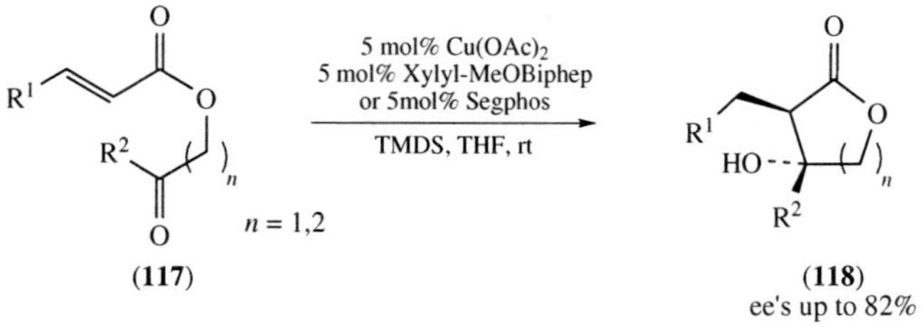

SCHEME 35

excellent yields of the corresponding 5- and 6-membered ring hydroxyl lactones **118**. A survey of chiral ligands showed that atropophosphines, such as members of the Biphep and Segphos families, gave enantiomeric excesses up to 82%. A catalytic cycle was proposed and involves a cyclic transition state in which the oxygen atom of the carbonyl was complexed to the copper atom of the enolate. This work was soon followed by the investigation of new substrates in which the oxygen atom linker was replaced by a protected nitrogen atom[86] (Chart 9).

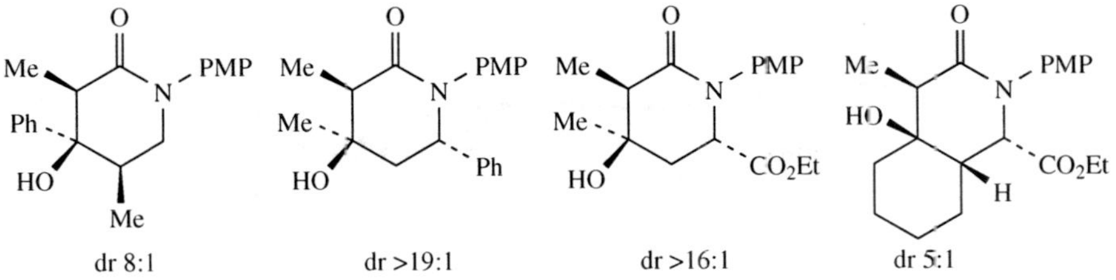

CHART 9

The chiral β-aminoketones required for the preparation of the cyclization precursors were easily prepared by an L-proline catalyzed asymmetric Mannich reaction. The cyclization of the corresponding acrylamide with the copper(II) acetate/dppf catalyst and TMDS as the hydride source gave the corresponding 4-hydroxypiperidinones in good to high diastereoselectivities.

Recently, Lipshutz and coworkers reported the enantioselective reductive cyclization of (E)- and (Z)-β-disubstituted enones **119**[87] (Scheme 36).

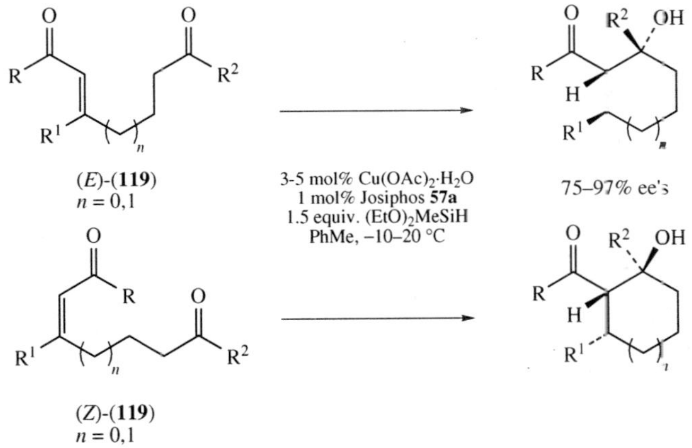

SCHEME 36

The reductive aldol cyclization occurred in good yields and high enantioselectivities with both (E)-**119** and (Z)-**119** enones with copper acetate/Josiphos-**57a** and diethoxymethylsilane as the hydride source at −20 to −10 °C. As the enantioselection occurs during the addition of the chiral copper(I) hydride catalyst on the substituted double bond, the sense of the stereoinduction occurs similarly to previously studied systems for the enantioselective reduction of related electrophilic olefins. This method allows the generation of three contiguous asymmetric stereocenters in a single step, the sense of induction for the aldol step being dependent on the Z- or E-stereochemistry of the starting olefin.

Finally, Deschamp and Riant recently reported an efficient method for the preparation of enantiomerically enriched bicyclic derivatives through an enantioselective copper(I) hydride catalyzed reductive aldol cyclization[88] (Scheme 37).

The reductive cyclization of monocyclic prochiral precursor **120** was optimized with Taniaphos-type chiral ligand **112c**. High diastereoselectivities and enantioselectivities were

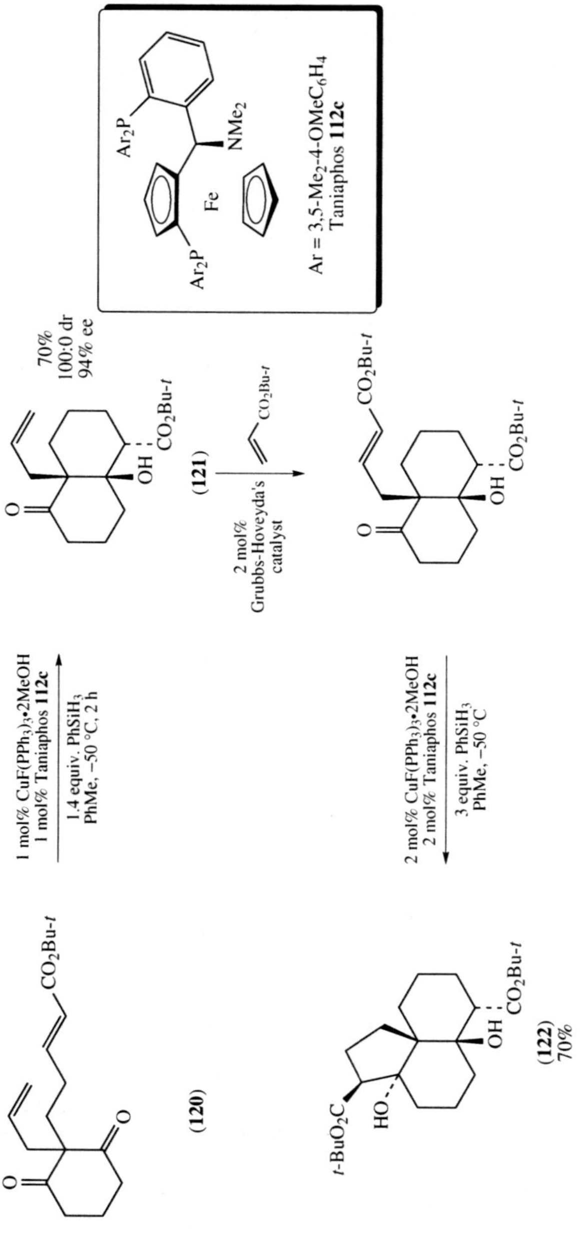

1 mol% CuF(PPh3)3•2MeOH
1 mol% Taniaphos 112c
1.4 equiv. PhSiH3
PhMe, −50 °C, 2 h
(120)
(121)
70%
100:0 dr
94% ee
2 mol%
Grubbs-Hoveyda's
catalyst
Ar = 3,5-Me2-4-OMeC6H4
Taniaphos 112c
2 mol% CuF(PPh3)3•2MeOH
2 mol% Taniaphos 112c
3 equiv. PhSiH3
PhMe, −50 °C
(122)
70%

SCHEME 37

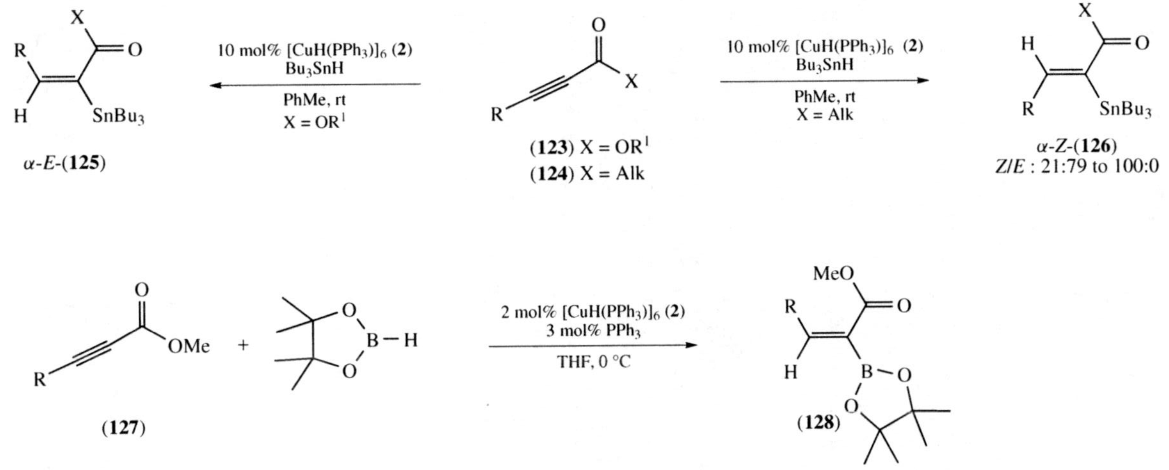
α-E-(125)
10 mol% [CuH(PPh3)]6 (2)
Bu3SnH
PhMe, rt
X = OR1
(123) X = OR1
(124) X = Alk
10 mol% [CuH(PPh3)]6 (2)
Bu3SnH
PhMe, rt
X = Alk
α-Z-(126)
Z/E : 21:79 to 100:0
(127)
2 mol% [CuH(PPh3)]6 (2)
3 mol% PPh3
THF, 0 °C
(128)

SCHEME 38

reached when *tert*-butyl ester was introduced on the Michael acceptor. The formation of cyclohexane ring occurred with enantioselectivities >94% and only one diastereoisomer was formed in most cases. The aldol adduct **121** bearing an allyl group on the bicycle junction could be further transformed and a second Michael acceptor was introduced through cross-metathesis with *tert*-butyl acrylate and a catalytic amount of the Grubbs–Hoveyda catalyst. The second reductive cyclization occurs smoothly at −50 °C with 2 mol% of chiral hydride catalyst and the angular tricyclic adduct **122** bearing 5 contiguous asymmetric centers was isolated in good yield and as a single diastereoisomer.

VII. HYDROCUPRATION OF ALKYNES

Soon after reporting the first examples of the use of copper(I) hydride cluster **2** for the chemoselective reduction of electrophilic double bonds, Stryker and coworkers investigated the reactivity of this reagent on non-activated triple bonds[17] (Scheme 2). Lower reactivity required higher temperature for the hydrocupration reaction to occur, and no further investigations were reported until Chiu and coworkers investigated the hydrocupration reaction of various alkynes activated by a carbonyl group in the presence of tributyltin hydride[89] (Scheme 38).

The reaction of various activated alkynes **123**, **124** with a catalytic amount of Stryker's reagent **2** and tributyltin hydride in toluene afforded the corresponding α-stannylated alkenes **125** and **126** in good yields. The proposed mechanism occurs through a hydrocupration reaction on the triple bond followed by transmetallation with the tin hydride and reductive elimination. When alkynoates **123** were used as substrates, the α-E-**125** stannylated adducts were isolated, showing that the transmetallation–elimination occurs without isomerization after the *cis*-hydrocupration step. On the other hand, when alkynones **124** were used, the intermediate isomerizes (presumably through a copper allenoate intermediate) prior to the transmetallation step and α-Z-**126** was isolated as the major adduct.

A similar strategy was recently reported by Lipshutz, Aue and coworkers who studied the reaction on alkynoates **127** with pinacolborane as the hydride source[90] (Scheme 38). The treatment of various alkynoates **127** with pinacolborane and a catalytic amount of Stryker's reagent **2** and triphenylphosphine afforded the Z-adducts **128** in good to high

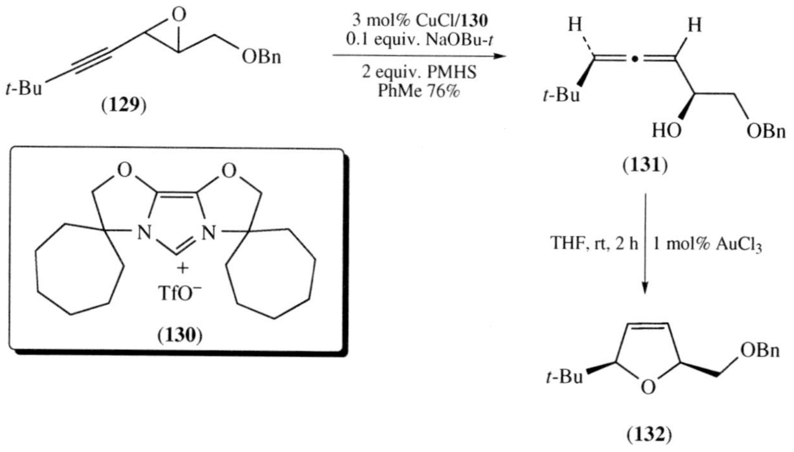

SCHEME 39

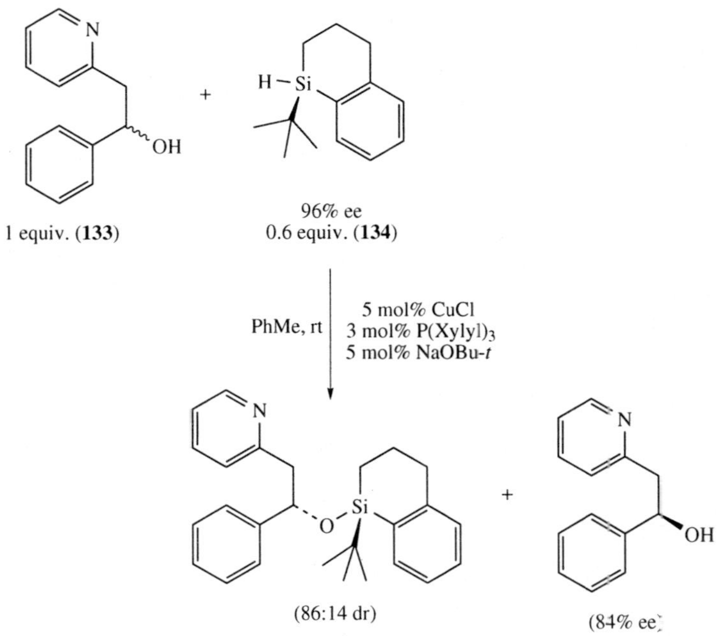

SCHEME 40

yields. The authors showed the utility of the resulting adducts through a series of transformations involving the boronic ester group.

The hydrocupration reaction of alkynes **129** bearing an epoxide group at the α-position of the alkyne was also recently reported by Krause and coworkers for the preparation of α-(hydroxy-substituted) allene **131**[91] (Scheme 39). The copper hydride catalyst was generated *in situ* by reaction of copper(I) chloride, sodium *tert*-butoxide and the imidazolium salt **130**. The influence of the structure of the NHC carbene ligand that stabilizes the copper(I) hydride was shown to be crucial not only for the S_N2' addition of the hydride on the triple bond with opening of the oxirane, but also for the chirality transfer that afforded the corresponding α-(hydroxy-substituted) allene **132** with high diastereoselectivity (93:7). The utility of the resulting α-(hydroxy-substituted) allene was also shown, as they can easily cyclize into the corresponding disubstituted dihydrofuran **132** by treatment with a catalytic amount of gold(III) chloride.

Finally, a conceptually elegant method for the kinetic resolution of racemic alcohols through copper(I)-catalyzed kinetic resolution with silicon stereogenic silanes was reported in 2008 by Oestreich and coworkers[92] (Scheme 40). Phosphine-stabilized copper(I) hydride complexes catalyze the dehydrogenative Si−O coupling between alcohol and silane, and the authors used this catalytic reaction to optimize the kinetic resolution of racemic alcohol **133** with chiral silane **134**. A careful examination of the reaction parameters allowed them to reach selectivity factors >15 for the kinetic resolution with enantioselectivities >85% for the coupling adducts and remaining alcohol.

This report shows that new reactions involving copper(I) hydride catalyst have yet to be discovered, and that its high efficiency will hopefully lead to their use in the chemical industry as an alternative to existing methodologies for the chemical and stereoselective reduction of various functional groups. Still, it remains to investigate the structures and dynamic behavior in solution of such copper(I) hydride catalytic species.

The chiral ligands cited in this review are presented in Chart 10.

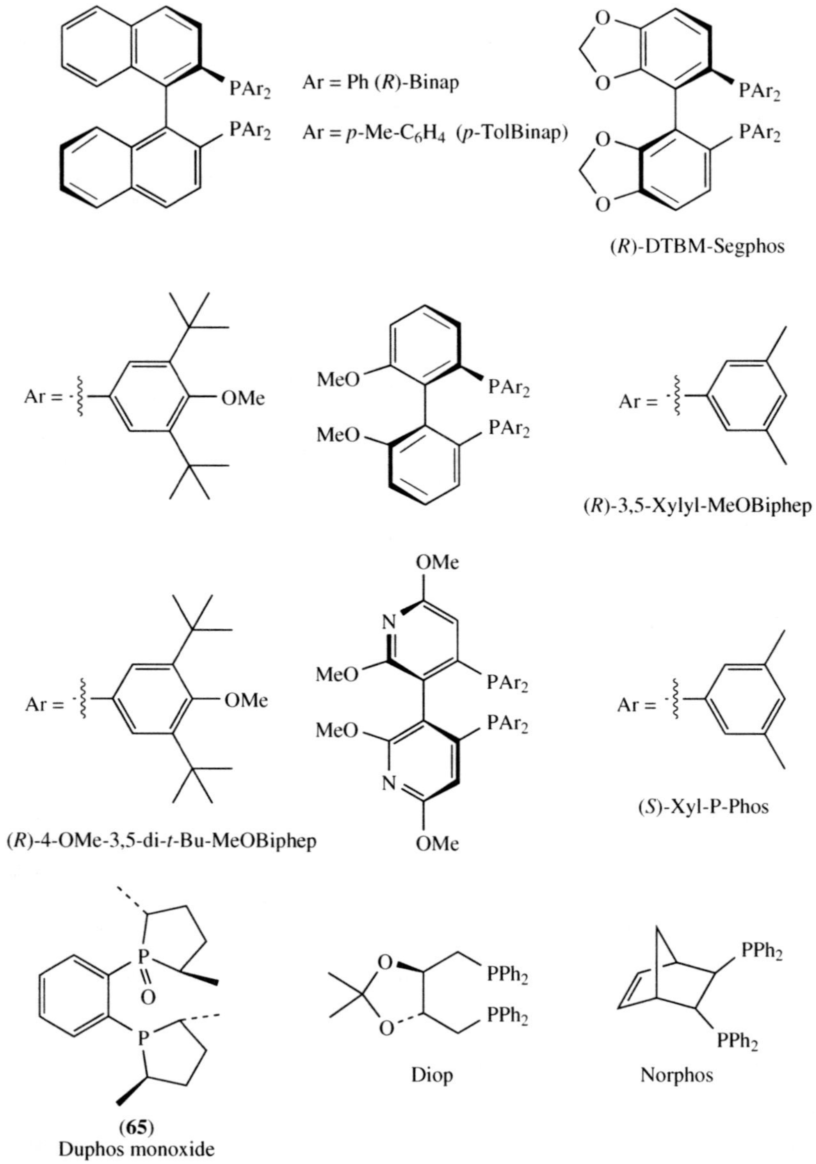

CHART 10

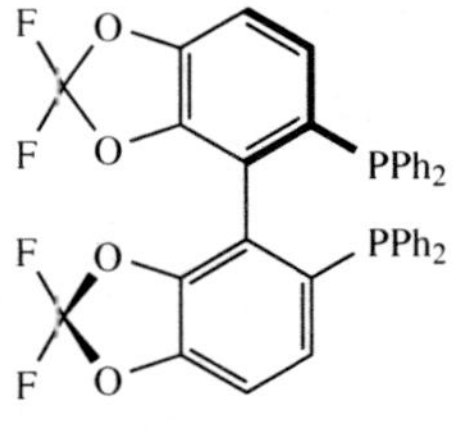

(*R*)-DiFluorPhos

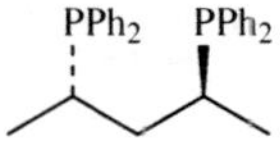

(*S*,*S*)-BDPP

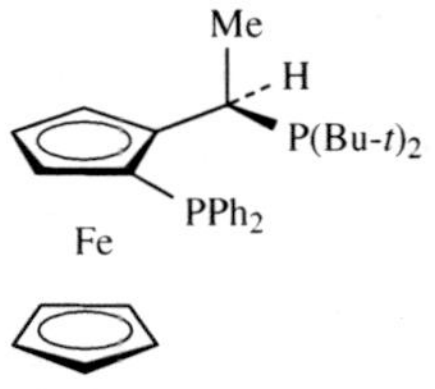

Josiphos-(**57a**)

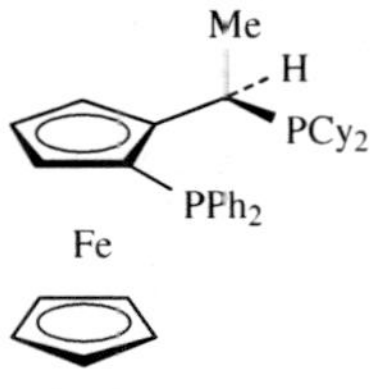

Josiphos-(**57b**)

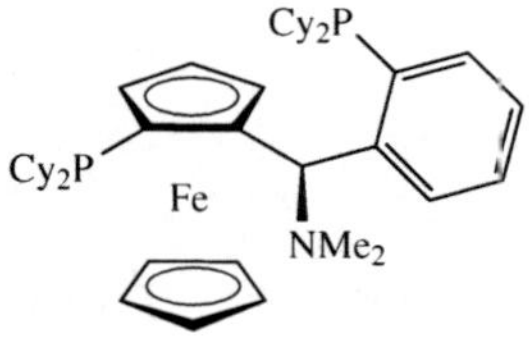

Taniaphos-(**112a**)

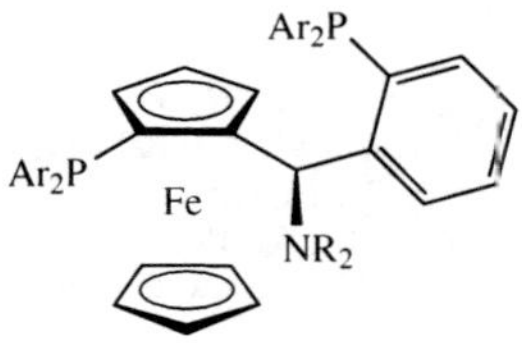

Ar = 3,5-xylyl; NR_2 = morpholine
Taniaphos-(**112b**)

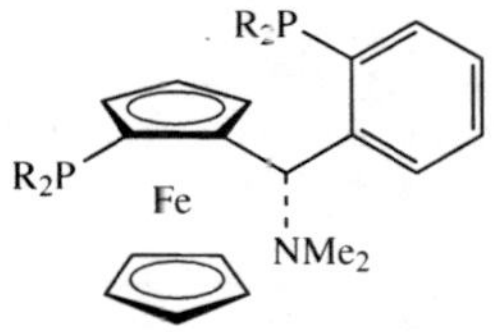

R = $(3,5\text{-Me}_2\text{-4-OMe})C_6H_4$
TaniaPhos-(**112c**)

CHART 10. (*continued*)

VIII. REFERENCES

1. (a) B. H. Lipshutz, in *Modern Organocopper Chemistry* (Ed. N. Krause), Wiley-VCH, Weinheim, 2002, p. 167.
   (b) A. de Fátima, *Synlett*, 1805 (2005).
   (c) O. Riant, N. Mostafaï and J. Courmarcel, *Synthesis*, 2943 (2004).
   (d) S. Rendler and M. Oestreich, *Angew. Chem., Int. Ed.*, **46**, 498 (2007).
   (e) M. Shibasaki and M. Kanai, *Chem. Rev.*, **108**, 2853 (2008).
   (f) C. Deutsch, N. Krause and B. H. Lipshutz, *Chem. Rev.*, **108**, 2916 (2008).
2. (a) S. A. Bezman, M. R. Churchill, J. A. Osborn and J. Wormald, *J. Am. Chem. Soc.*, **93**, 2063 (1971).
   (b) M. R. Churchill, S. A. Bezman, J. A. Osborn and J. Wormald, *Inorg. Chem.*, **11**, 1818 (1972).
3. R. C. Stevens, M. R. McLean and R. Bau, *J. Am. Chem. Soc.*, **111**, 3472 (1989).
4. T. H. Lemmen, K. Folting, J. C. Huffman and K. G. Caulton, *J. Am. Chem. Soc.*, **107**, 7774 (1985).
5. G. V. Goeden, J. C. Huffman and K. G. Caulton, *Inorg. Chem.*, **25**, 2484 (1986).
6. N. P. Mankad, D. S. Laitar and J. P. Sadighi, *Organometallics*, **23**, 3369 (2004).
7. Z. Mao, J.-S. Huang, C.-M. Che, N. Zhu, S.-Y. Leung and Z. Y. Zhou, *J. Am. Chem. Soc.*, **127**, 4562 (2005).
8. D. M. Brestensky, D. E. Huseland, C. McGettigan and J. M. Stryker, *Tetrahedron Lett.*, **29**, 3749 (1988).

9. P. Chiu, Z. Li and K. C. M. Fung, *Tetrahedron Lett.*, **44**, 455 (2003).
10. D. W. Lee and J. Yun, *Tetrahedron Lett.*, **46**, 2037 (2005).
11. B. H. Lipshutz and B. A. Frieman, *Angew. Chem., Int. Ed.*, **44**, 6345 (2005).
12. (a) R. K. Boekman Jr. and R. Michalak, *J. Am. Chem. Soc.*, **96**, 1623 (1974).
    (b) S. Masamune, G. S. Bates and P. E. Georghious, *J. Am. Chem. Soc.*, **96**, 3686 (1974).
13. (a) M. F. Semmelhack, R. D. Stauffer and A. Yamashita, *J. Org. Chem.*, **42**, 3180 (1977).
    (b) T. Tsuda, T. Fujii, K. Kawasaki and T. Saegusa, *J. Chem. Soc., Chem. Commun.*, 1013 (1980).
14. T. Tsuda, H. Satomi, T. Hayashi and T. Saegusa, *J. Org. Chem.*, **52**, 439 (1987).
15. W. S. Mahoney, D. M. Brestensky and J. M. Stryker, *J. Am. Chem. Soc.*, **110**, 291 (1988).
16. D. M. Brestensky and J. M. Stryker, *Tetrahedron Lett.*, **30**, 5677 (1989).
17. J. F. Daeuble, C. McGettigan and J. M. Stryker, *Tetrahedron Lett.*, **31**, 2397 (1990).
18. W. S. Mahoney and J. M. Stryker, *J. Am. Chem. Soc.*, **111**, 8818 (1989).
19. (a) J. X. Chen, J. F. Daeuble, D. M. Brestensky and J. M. Stryker, *Tetrahedron*, **56**, 2153 (2000).
    (b) J. X. Chen, J. F. Daeuble and J. M. Stryker, *Tetrahedron*, **56**, 2789 (2000).
20. I. Ryu, N. Kusumoto, A. Ogawa, N. Kambe and N. Sonoda, *Organometallics*, **8**, 2279 (1989).
21. H. Ito, I. Ishizuka, K. Arimoto, K. Miura and A. Hosomi, *Tetrahedron Lett.*, **38**, 8887 (1997).
22. H. Ito, H. Yamanaka, T. Ishizuka, T. Tomoko, J.-I. Tateiwa and A. Hosomi, *Synlett*, 479 (2000).
23. (a) A. Mori, A. Fujita, Y. Nishihara and T. Hiyama, *Chem. Commun.*, 2159 (1997).
    (b) A. Mori, A. Fujita, H. Kajiro, Y. Nishihara and T. Hiyama, *Tetrahedron*, **55**, 4573 (1999).
24. (a) B. H. Lipshutz, J. Keith, P. Papa and R. Vivian, *Tetrahedron Lett.*, **39**, 4627 (1998).
    (b) B. H. Lipshutz, W. Chrisman, K. Noson, P. Papa, J. A. Sclafani, R. W. Vivian and J. M. Keith, *Tetrahedron*, **56**, 2779 (2000).
25. C. L. Chandler and A. J. Phillips, *Org. Lett.*, **7**, 3493 (2005).
26. B. H. Lipshutz and P. Papa, *Angew. Chem., Int. Ed.*, **41**, 4580 (2002).
27. B. A. Baker, Z. V. Boskovic and B. H. Lipshutz, *Org. Lett.*, **10**, 289 (2008).
28. J. Jurkauskas, J. P. Sadighi and S. L. Buchwald, *Org. Lett.*, **5**, 2417 (2003).
29. M. Bui The Thuong, S. Sottocornola, G. Prestat, G. Broggini, D. Madec and G. Poli, *Synlett*, 1521 (2007).
30. D. H. Appella, Y. Moritani, R. Shintani, E. M. Ferreira and S. L. Buchwald, *J. Am. Chem. Soc.*, **121**, 9473 (1999).
31. Y. Moritani, D. H. Appella, V. Jurkauskas and S. L. Buchwald, *J. Am. Chem. Soc.*, **122**, 6797 (2000).
32. G. Hughes, M. Kimura and S. L. Buchwald, *J. Am. Chem. Soc.*, **125**, 11253 (2003).
33. M. P. Rainka, Y. Aye and S. L. Buchwald, *Proc. Natl. Acad. Sci. U.S.A.*, **101**, 5821 (2004).
34. M. Movassaghi and A. E. Ondrus, *Org. Lett.*, **7**, 4423 (2005).
35. J. Yun and S. L. Buchwald, *Org. Lett.*, **3**, 1129 (2001).
36. J. Chae, J. Yun and S. L. Buchwald, *Org. Lett.*, **6**, 4809 (2004).
37. V. Jurkauskas and S. L. Buchwald, *J. Am. Chem. Soc.*, **124**, 2892 (2002).
38. M. P. Rainka, J. E. Milne and S. L. Buchwald, *Angew. Chem., Int. Ed.*, **44**, 6177 (2005).
39. B. H. Lipshutz and J. M. Servesko, *Angew. Chem., Int. Ed.*, **42**, 4789 (2003).
40. B. H. Lipshutz, J. M. Servesko and B. R. Taft, *J. Am. Chem. Soc.*, **126**, 8353 (2004).
41. (a) B. H. Lipshutz, N. Tanaka, B. R. Taft and C.-T. Lee, *Org. Lett.*, **8**, 1963 (2006).
    (b) B. H. Lipshutz, C.-T. Lee and B. R. Taft, *Synthesis*, 3257 (2007).
42. B. H. Lipshutz, C.-T. Lee and J. M. Servesko, *Org. Lett.*, **9**, 4713 (2007).
43. C. Czekelius and E. M. Carreira, *Angew. Chem., Int. Ed.*, **42**, 4793 (2003).
44. C. Czekelius and E. M. Carreira, *Org. Proc. Res. & Dev.*, **11**, 633 (2007).
45. (a) C. Czekelius and E. M. Carreira, *Org. Lett.*, **6**, 4575 (2004).
    (b) D. Lee, D. Kim and J. Yun, *Angew. Chem., Int. Ed.*, **45**, 2785 (2006).
    (c) D. Lee, Y. Yand and J. Yun, *Synthesis*, 2233 (2007).
46. D. Kim, B.-M. Park and J. Yun, *Chem. Commun.*, 1755 (2005).
47. D. Lee, Y. Yang and J. Yun, *Org. Lett.*, **9**, 2749 (2007).
48. Y. Ren, X. Xu, K. Sun and J. Xu, *Tetrahedron: Asymmetry*, **16**, 4010 (2005).
49. T. Lamas, R. Gòmez Arrayàs and J. C. Carretero, *Angew. Chem., Int. Ed.*, **46**, 3329 (2007).
50. J.-N. Desrosiers and A. B. Charette, *Angew. Chem., Int. Ed.*, **46**, 5955 (2007).
51. B. H. Lipshutz, W. Chrisman and K. Noson, *J. Organomet. Chem.*, **624**, 367 (2001).

52. B. H. Lipshutz, C. C. Caires, P. Kuipers and W. Chrisman, *Org. Lett.*, **5**, 3085 (2003).
53. H. Kaur, F. Kauer Zinn, E. D. Stevens and S. P. Nolan, *Organometallics*, **23**, 1157 (2004). Reviews (a) S. Díez-González and S. P. Nolan, *Synlett*, 2158 (2007).
(b) S. Díez-González and S. P. Nolan, *Acc. Chem. Res.*, **41**, 349 (2008).
54. S. Díez-González, H. Kaur, F. Kauer Zinn, E. D. Stevens and S. P. Nolan, *J. Org. Chem.*, **70**, 4784 (2005).
55. S. Díez-González, N. M. Scott and S. P. Nolan, *Organometallics*, **25**, 2355 (2006).
56. S. Díez-González, E. D. Stevens, N. M. Scott, J. L. Petersen and S. P. Nolan, *Chem. Eur. J.*, **14**, 158 (2008).
57. B. Bantu, D. Wang, K. Wurst and M. R. Buchmeiser, *Tetrahedron*, **61**, 12154 (2005).
58. J. Yun, D. Kim and H. Yun, *Chem. Commun.*, 5181 (2005).
59. H. Brunner and W. Miehling, *J. Organomet. Chem.*, **275**, C17 (1984).
60. B. H. Lipshutz, K. Noson and W. Chrisman, *J. Am. Chem. Soc.*, **123**, 12917 (2001).
61. (a) S. Sirol, J. Courmarcel, N. Mostefaï and O. Riant, *Org. Lett.*, **3**, 4111 (2001).
(b) J. Courmarcel, N. Mostefaï, S. Sirol, S. Choppin and O. Riant, *Isr. J. Chem.*, **41**, 231 (2001).
62. N. Mostefaï, S. Sirol, J. Courmarcel and O. Riant, *Synthesis*, 1265 (2007).
63. B. H. Lipshutz, K. Noson, W. Chrisman and A. Lower, *J. Am. Chem. Soc.*, **125**, 8779 (2003).
64. D.-W. Lee and J. Yun, *Tetrahedron Lett.*, **45**, 5415 (2004).
65. J. T. Issenhuth, S. Dagorne and S. Bellemin-Laponnaz, *Adv. Synth. Catal.*, **348**, 1991 (2006).
66. J. Wu, J.-X. Ji and A. S. C. Chan, *Proc. Natl. Acad. Sci. U.S.A.*, **102**, 3570 (2005).
67. C.-T. Lee and B. Lipshutz, *Org. Lett.*, **10**, 4187 (2008).
68. B. H. Lipshutz, A. Lower and K. Noson, *Org. Lett.*, **4**, 4045 (2002).
69. B. H. Lipshutz, A. Lower, R. J. Kucejko and K. Noson, *Org. Lett.*, **8**, 2969 (2006).
70. B. H. Lipshutz and H. Shimizu, *Angew. Chem., Int. Ed.*, **43**, 2228 (2004).
71. B. H. Lipshutz, B. A. Frieman and A. E. Tomaso Jr., *Angew. Chem., Int. Ed.*, **45**, 1259 (2005).
72. H. Shimizu, D. Igarashi, W. Kuriyama, Y. Yusa, N. Sayo and T. Saito, *Org. Lett.*, **9**, 1655 (2007).
73. (a) H. Iida and M. J. Krische, *Top. Curr. Chem.*, **279**, 77 (2007).
(b) H. Nishiyama and T. Shiomi, *Top. Curr. Chem.*, **279**, 105 (2007).
74. P. Chiu, B. Chen and K. F. Cheng, *Tetrahedron Lett.*, **39**, 9229 (1998).
75. (a) P. Chiu, C.-P. Szeto, Z. Geng and K.-F. Cheng, *Org. Lett.*, **3**, 1901 (2001).
(b) P. Chiu, *Synthesis*, 2210 (2004).
76. P. Chiu, C. P. Szeto, Z. Geng and K. F. Cheng, *Tetrahedron Lett.*, **42**, 4091 (2001).
77. W. K. Chung and P. Chiu, *Synlett*, 55 (2005).
78. P. Chiu and S. K. Leung, *Chem. Commun.*, 2308 (2004).
79. A. Welle, S. Díez-Gonzàlez, B. Tinant, S. P. Nolan and O. Riant, *Org. Lett.*, **8**, 6059 (2006).
80. D. Zhao, K. Oisaki, M. Kanai and M. Shibasaki, *Tetrahedron Lett.*, **47**, 1403 (2006).
81. J. Deschamp, O. Chuzel, J. Hannedouche and O. Riant, *Angew. Chem., Int. Ed.*, **45**, 1292 (2006).
82. O. Chuzel, J. Deschamp, C. Chausteur and O. Riant, *Org. Lett.*, **8**, 5943 (2006).
83. Y. Du, L.-W Xu, Y. Shimizu, K. Oisaki, M. Kanai and M. Shibasaki, *J. Am. Chem. Soc.*, **130**, 16146 (2008).
84. D. Zhao, K. Oisaki, M. Kanai and M. Shibasaki, *J. Am. Chem. Soc.*, **128**, 14440 (2006).
85. H. W. Lam and P. M. Joensuu, *Org. Lett.*, **7**, 4225 (2005).
86. H. W. Lam, G. J. Murray and J. D. Firth, *Org. Lett.*, **7**, 5743 (2005).
87. B. H. Lipshutz, B. Amorelli and J. B. Inger, *J. Am. Chem. Soc.*, **130**, 14378 (2008).
88. J. Deschamp and O. Riant, *Org. Lett.*, **11**, 1217 (2009).
89. (a) L. T. Leung, S. K. Leung and P. Chiu, *Org. Lett.*, **7**, 5249 (2005).
(b) L. T. Leung and P. Chiu, *Pure Appl. Chem.*, **78**, 281 (2006).
(c) R. Miao, S. Li and P. Chiu, *Tetrahedron*, **63**, 6737 (2007).
90. B. H. Lipshutz, Z. V. Boskovic and D. H. Aue, *Angew. Chem., Int. Ed.*, **47**, 10183 (2008).
91. C. Deutsch, B. H. Lipshutz and N. Krause, *Angew. Chem., Int. Ed.*, **46**, 1650 (2007).
92. S. Rendler, O. Plefka, B. Karatas, G. Auer, R. Fröhlich, C. Mück-Lichtenfeld, S. Grimme and M. Oestreich, *Chem. Eur. J.*, **14**, 11512 (2008).

CHAPTER **16**

Silyl and stannyl derivatives of organocopper compounds

FRANCISCO J. PULIDO and ASUNCIÓN BARBERO

Departamento de Química Orgánica, Facultad de Ciencias, Universidad de Valladolid, 47011 Valladolid, Spain
Fax: 34-983-423013; e-mail: pulido@qo.uva.es

I. INTRODUCTION 776
II. SILYLCUPRATES 777
A. Preparation of Silylcuprates 777
1. Introduction 777
2. Homosilylcuprates 777
3. Cyanosilylcuprates 778
4. Mixed silylcuprates 779
5. Other silylcuprates 780
B. The C−Si to C−OH Conversion 781
C. The Structure of Silylcuprates 788
D. Reactions of Silylcuprates 790
1. Conjugate additions to unsaturated carbonyl compounds 790
2. Substitution reactions with allylic acetates, acid chlorides, epoxides and related systems 797
3. Reaction of silylcuprates with unactivated acetylenes and allenes . . . 803
III. STANNYLCUPRATES 817
A. Preparation of Stannylcuprates 817
1. Introduction 817
2. Methods of preparation 818
B. Reactions of Stannylcuprates 821
1. Introduction 821
2. Conjugate additions to unsaturated carbonyl compounds 822
3. Substitution reactions with allylic and propargylic electrophiles, epoxides, vinyl triflates and β-haloenones 831
4. Reaction of stannylcuprates with acetylenes and allenes 834

The chemistry of organocopper compounds
Edited by Z. Rappoport and I. Marek

C. Mechanisms 848
IV. REFERENCES 850

I. INTRODUCTION

In the mid-1960s organocuprate chemistry emerged as a fundamental topic in organic chemistry being, over the last decades and till today, a cornerstone of organic synthesis and one of the most popular methods for C−C bond formation. Gilman and coworkers' paper from 1952, describing the preparation of lithium dimethylcuprate[1], is often used to mark the beginning of cuprate chemistry. This was the first literature report of what we know as organocuprate reagents. Two posterior fundamental reactions: the conjugate addition reaction[2] reported by House and coworkers in 1966 and Corey's substitution reaction[3] in 1967, consecrate definitively organocopper chemistry as a powerful tool in synthesis. Half a century later, cuprate chemistry is still one of the leaders in the organometallic field, and convincingly demonstrates the value of stoichiometric organocopper reagents in the stereoselective preparation of complex structures.

Nowadays, rapid development is seen in their innumerable uses in total synthesis of natural products. This assessment is well illustrated by a 1992 *Organic Reaction* compilation[4] which lists 522 natural products synthesized by employing organocopper reagents.

Pioneering studies employed mainly lithium dialkylcuprates or mixed dialkyl versions. The fact that only the less electronegative of the two ligands is selectively transferred to the electrophile provides a good strategy to develop lithium heteroatom(alkyl)cuprates with a non-transferable ligand, usually a heteroatom of Group VA or VIA, and a transferable alkyl group[5]. Chiral heteroatom(alkyl)cuprates (RCuX*Li; X = heteroatomic ligand) prepared from non-tranferable heteroatom ligands induce asymmetric transformations, enabling the transfer of the alkyl group to the electrophile with high enantiomeric excesses[6].

Although conventional organocuprate reagents of type R_2CuLi have been used for over half a century, the analogous silyl and stannyl reagents have been reported only recently. A great deal of the current knowledge in silyl and stannylcuprate chemistry is due to the effort and brightness of two excellent chemists (I. Fleming and E. Piers) whose reports[7] opened a new field in metallocuprate chemistry which have proved to be extremely useful in synthetic organic chemistry.

Heteroatomcuprates containing Group IVA ligands (Si, Sn, Ge) readily transfer these ligands to a wide variety of electrophiles with subsequent formation of C−Si and C−Sn bonds. The synthetic value of this reaction lay in the posterior transformation of the resulting C−Si and C−Sn bonds. Thus, compounds carrying silyl or stannyl substituents may undergo multiple transformations in which these substituents are replaced by other functional groups or by carbon substituents. This fact has been exploited in the construction of new C−C bonds. Several reviews covering the late advances in the chemistry and synthetic application of organosilyl and organostannylcuprates have recently appeared[8].

Silyl- and stannylcuprates seem to be thermodynamically more stable than their carbon-based counterparts, allowing them to be treated with substrates at temperatures near 0 °C and above for several hours[8a]. At the same time, they show a higher kinetic reactivity than their carbon analogues. For instance, reaction of these reagents with unactivated allenes[9], styrenes[10] and 1,3-dienes[11] has no counterpart in alkylcuprate chemistry. In particular, the rich chemisty of the silylcupration and stannylcupration of allenes offers a high potential for the development of novel synthetic strategies[12].

Other families of metallocuprates bearing metals different from Si and Sn have also been an object of attention, for example, germylcuprates (with Ge−Cu bond)[13], zinc-cuprates (with Zn−Cu bond)[14] etc. Catalytic metallo-metalations mediated by Cu(I) are also closely related processes[15]. We do not intend to cover these developments in this chapter, which is devoted solely to the chemistry of silicon and tin cuprates and their application to organic synthesis.

II. SILYLCUPRATES

A. Preparation of Silylcuprates

1. Introduction

As their carbocuprate analogues, a silylcuprate reagent can be prepared from the corresponding silyllithium reagent by mixing it with an appropriate amount of a copper(I) salt, in THF solution[8a]. They possess enhanced nucleophilicity and stability compared to silylcopper reagents which often lead to more selective and reproducible reactions. Although other ethereal solutions have been used, THF is the most popularly employed solvent. The nature of copper(I) salt used, substitution pattern of silyl group and the ligand/copper ratio results in formation of a wide variety of cuprates with different stoichiometries. Some of the more usual reagents are collected in Table 1 (below in Section II.C). Their reactivity depends on the nature of the silyl group, stoichiometry and the presence of non-transferable ligands. As expected from the lower electronegativity of silicon, silylcuprates transfer selectively the silyl group in mixed alkyl(silyl)cuprates of type $R_3SiCu(R')Li$. Singer and Oehlschlager have attributed this preferential transfer to HOMO/LUMO orbital interactions between copper ligands and the electrophile, the interaction between the Cu−Si HOMO and the electrophile LUMO being energetically more favorable than the intereraction with the Cu−C HOMO[16a]. The factors which influence the relative migratory aptitudes of ligands attached to copper in organocuprates have been reviewed[16b].

2. Homosilylcuprates

The most simple reagent, bis(trimethylsilyl)cuprate[17], is not the most extensively used. The disadvantages of the tedious preparation of trimethylsilyllithium clearly reduce their use in synthesis. Trimethylsilyllithium[18] needs to be prepared in neat HMPA by treating hexamethyldisilane with methyllithium at 0 °C for half an hour (Scheme 1). HMPA is not only an undesirable reagent but also causes interferences in many reactions and separation problems after the work-up of the mixture. Other reported preparation methods do not improve this procedure[19].

Phenyldimethylsilyllithium[20] is much easier to prepare than trimethylsilyllithium, and is simply obtained by mixing chlorodimethylphenylsilane and lithium shots in THF, and stirring the mixture at 0 °C overnight (Scheme 1). Because of its immediacy, phenyldimethylsilylcuprates[21] are the most commonly used silylcuprates. The only disadvantage of using a phenyldimethylsilyl group is that the silicon by-products generated in the desilylative steps or in the silylcupration reaction are relatively involatile. Different reagents bearing the phenyldimethylsilyl group, e.g. $(PhMe_2Si)_2CuLi{\cdot}LiX$, have been prepared using copper(I) iodide, copper(I) bromide−dimethyl sulfide complex or even copper(I) chloride. All of them are considered formally lower-order cuprates and their reactivity is quite similar. The phenyldimethylsilyl group imparts very similar reactivity to that of trimethylsilyl group: both are replaced by a proton or an electrophilic group in allyl- and vinylsilane derivatives and undergo desilylative reactions in similar conditions, showing equivalent reaction rates. However, the phenyldimethylsilyl group has the advantage of being converted into a hydroxyl group, a feature associated with silyl groups bearing at least one phenyl substituent (see Section II.B).

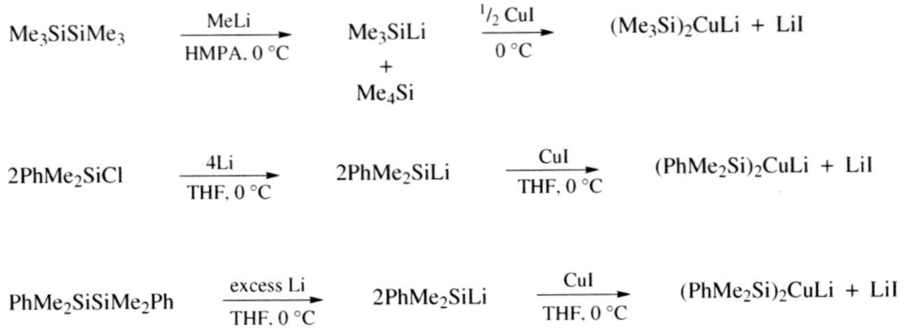

SCHEME 1

An alternative preparation of phenyldimethylsilyllithium, which is free of halide ion, is the cleavage of diphenyltetramethyldisilane with lithium powder (Scheme 1)[22]. This method was used for NMR characterization of several copper-containing species[22]. Cuprates made from disilanes behave as usual cuprates when reacting with electrophiles.

Other homosilylcuprates based on the triphenylsilyl, *t*-butyldimethylsilyl or *t*-butyldiphenylsilyl groups were eventually prepared for specific applications, but their use is not so extended as that of TMS and $PhMe_2Si$ groups[8a].

3. *Cyanosilylcuprates*

Cyanosilylcuprates are extremely versatile synthetic reagents and combine the stability of heterocuprates and the reactivity of homocuprates. Phenyldimethylsilylcyanocuprates **1**, **2** and **3** made from phenyldimethylsilyllithium, copper(I) cyanide and methyllithium exhibit often higher reactivity and better selectivity than the ordinary halocuprates[23]. The use of CuCN is generally more reliable than that of bromo or iodo derivatives[24]. CuCN is inexpensive, non-hygroscopic, light insensitive and very stable to change in oxidation level. Moreover, the cyanide-based reagent is also advantageous because it is easier to make and it keeps dry, which prevent the problems of diminished yield when using halocuprates.

Although cyanocuprate (**1**) and cuprates of type $(R_3Si)_2Cu(CN)Li_2$ and $R_3Si(R')Cu(CN)Li_2$ (Scheme 2) are described as higher-order cuprates, and the structures of many of them have been confirmed by NMR experiments[22,25], the true role of the cyano group in the reactions of higher-order cyanocuprates still remains obscure. In fact, it is known that the triply coordinated species $[Cu(CN)R_2]^{2-}$ is not a stable structure in ethereal solvents[26], however, the Lipshutz reagent[27] is still formulated as $R_2Cu(CN)Li_2$ and still remains the best choice in many synthetic transformations. NMR characterization of several cyanide-derived silylcopper species clearly shows that different stoichiometric ratios of copper-to-silicon lead to different species. Thus, the 1:1 reagent (**2**) and the 2:1 reagent (**1**) are distinct organocopper species, and the cyanide-based reagent (**1**) is also different from the iodide-based $(PhMe_2Si)_2CuLi{\bullet}LiI$[22,25]. Examination of ^{13}C ($SiCH_3$) chemical shifts (in ppm) show variations up to 10 ppm, while ^{29}Si signals appear in an interval of around 8 ppm and the ^{1}H proton NMR spectrum gives differences in value of approximately 0.3 ppm[16,22,25]. Dynamic NMR experiments on samples of $R_3Si(R')Cu(CN)Li_2$ carried out at low temperatures are consistent with a rapid ligand exchange[16]. This fact might lead to controversy concerning the composition and formulation of these silylcopper species and consequently in the term 'higher order' which could be open to revision[28].

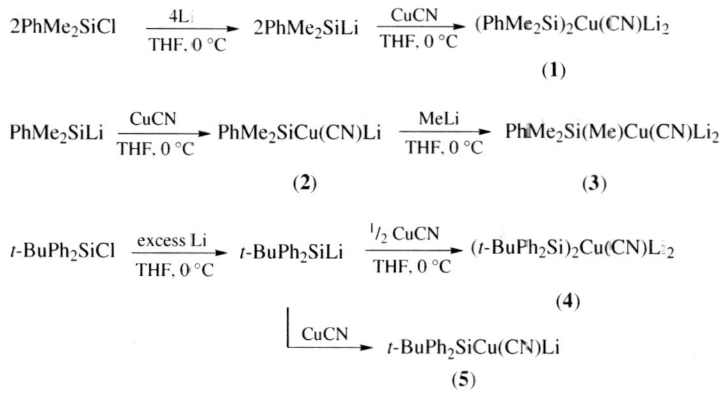

SCHEME 2

(t-Butyldiphenylsilyl)cuprates **4** and **5** are also distinct organocopper species, and show a reactivity pattern quite similar to the cyanocuprates **1** and **2**[9].

4. Mixed silylcuprates

Mixed alkyl(silyl)cuprates are readily synthesized by mixing a 1:1:1 equimolar amount of the alkyllithium compound, the silyllithium reagent and the copper(I) salt in THF as the preferred solvent[29]. The order is ambiguous but more preferably the silyllithium is added to the copper salt with a short stirring, and this is followed by adding one equivalent of the alkyllithium. Equilibration is usually fast between −40 °C and 0 °C, yielding the mixed cuprate as a homogeneous colored mixture ready to use. $PhMe_2Si(Me)Cu(CN)Li_2$[22,29a] (**3**) and $PhMe_2Si(n\text{-}Bu)Cu(CN)Li_2$[29] (**6**) are prepared by this method (Scheme 3), both are equally effective reagents and their reactivity is similar to that of the bis(silyl)cuprate. Mixed cuprates have some advantages. Since only one of the silyl groups of homosilylcuprates is used for reaction, mixed alkyl(silyl)cuprates greatly improve the reaction economy, especially when using expensive[30] or difficult to prepare silyl groups. This is especially true in total synthesis of natural product where convergent key steps that need the addition of a complex cuprate must proceed with high yield and ligand economy. In 1972 Corey and coworkers[31] introduced mixed cuprates bearing an alkynyl residual ligand and a transferable ligand, having demonstrated previously that alkynyl ligands are transferred much more slowly than other organic groups. Corey's conjugated addition–alkylation route to prostaglandins exemplifies well this strategy, where a mixed cyanoalkylcuprate transfers a rather complex alkyl chain to the conjugated position of the pentacarbocycle, whereas the cyano behaves as a residual ligand. Mixed alkynyl(silyl)cuprates, prepared by stirring an equimolar mixture of hexynyl lithium, phenyldimethylsilyllithium and copper(I) cyanide in THF at 0 °C for 0.5 h, have proved to show quite the same reactiviy as $(Ph_2MeSi)_2Cu(CN)Li_2$ when reacting with allenes, giving only vinylsilanes resulting from selective silicon transfer[29b].

Moreover, mixed alkyl(silyl)cuprates highly minimize silicon-containing by-products (such as R_3SiSiR_3, R_3SiH, R_3SiOH, $R_3SiOSiR_3$) formed with homosilylcuprates because the only by-product derived from the silylcupration step is methane or butane. On the other hand, the markedly higher electronegativity of carbon atom assures the selective transfer of the silyl group, as mentioned before. In order to get the best yield, mixed cuprates should be added in slight excess to the electrophile, when the reaction stoichiometry is 1:1.

An alternative preparation of mixed alkyl(silyl)cuprates that may occasionally be useful is the cleavage of silylstannanes like $R^2{}_3SnSiR^1{}_3$ with lithium dialkylcuprates of type $R^3{}_2CuLi$[32]. The resulting mixed cuprate $R^1{}_3Si(R^3)CuLi$ is formed by ligand exchange and an equivalent of tetraalkylstannane $R^2{}_3SnR^3$ is subsequently formed (Scheme 3). This procedure is restricted to the preparation of hindered mixed silyl cuprates like *t*-$BuMe_2Si(n\text{-}Bu)Cu(CN)Li_2$[32] (**7**) or the corresponding thexyldimethylsilyl derivative (**8**). Obviously, the mixed silylstannane can afford either silyl or stannylcuprates; the smaller the silyl group, the higher will be the opportunity for obtaining the stannylcuprate[33]. Furthermore, the ligand exchange methodology avoids the preparation of silyllithium salts, which are sometimes difficult to manipulate in the laboratory.

$$PhMe_2SiLi + BuLi \xrightarrow[\text{THF, 0 °C}]{\text{CuCN}} PhMe_2Si(n\text{-}Bu)Cu(CN)Li_2 \quad \mathbf{(6)}$$

$$t\text{-}BuMe_2SiSnMe_3 + n\text{-}Bu_2Cu(CN)Li_2 \xrightarrow[-78 \text{ to } 0\text{ °C}]{\text{THF}} t\text{-}BuMe_2Si(n\text{-}Bu)Cu(CN)Li_2 + n\text{-}BuSnMe_3 \quad \mathbf{(7)}$$

$$\text{thexyl}Me_2SiSnMe_3 + Me_2Cu(CN)Li_2 \xrightarrow[-78 \text{ to } 0\text{ °C}]{\text{THF}} \text{thexyl}Me_2Si(Me)Cu(CN)Li_2 + Me_4Sn \quad \mathbf{(8)}$$

SCHEME 3

Mixed hetero(silyl)cuprates of the type $R_3SiCu(Z)Li$, Z = heteroatom from Group VA and VIA (excluding CN), have been little used[8a].

5. *Other silylcuprates*

Apart from TMS-cuprates and $PhMe_2Si$-cuprates, other silylcuprates that have been used with some assiduity are triphenylsilyl(cyano)cuprates[34], *t*-butyldiphenylsilylcuprate **4**[30a] and **5**[30b], *t*-butyldimethylsilyl(butyl)cuprate (**7**)[32], bis(2-methylbut-2-en-1-yldiphenylsilyl)(cyano)cuprate[35] (**9**), bis[tris(trimethylsilyl)silyl]cuprate[36] and diethylaminodiphenylsilyl(cyano)cuprate[37] (**10**) prepared as in Scheme 4.

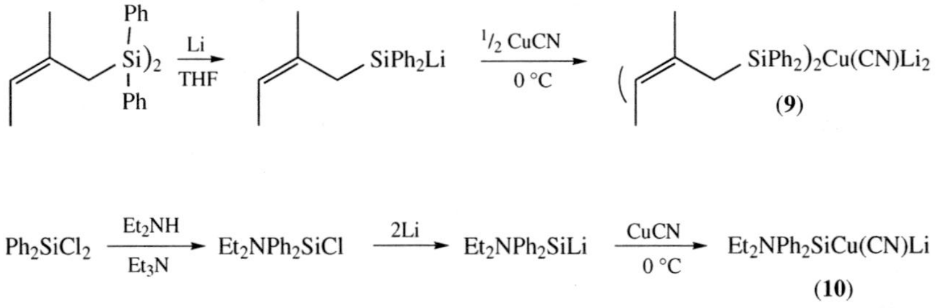

SCHEME 4

Except for the steric implications that the TBDMS-cuprate can impart, the TBDMS (*t*-$BuMe_2Si$) group is not always appropriate for carrying out the useful standard silicon chemistry. It cannot be converted into an OH (see Section II.B) group, nor is it suitable for electrophilic substitutions because the TBDMS group is not as good a leaving group as the TMS or $PhMe_2Si$ groups. The bis(2-methylbut-2-en-1-yldiphenylsilyl)(cyano)cuprate[35] (**9**)

(Scheme 4), although a rather complex reagent, was developed to facilitate the transformation of C−Si to C−OH. This conversion is particularly risky in case of allylsilanes (see Section II.B); however, the use of the 2-methylbut-2-enyldiphenylsilyl group permits this conversion to take place under very mild acidic conditions which are resisted by the allylic residue of the substrate, giving allyl alcohols in good yield[35]. The bulky (*t*-butyldiphenylsilyl)cyanocuprates **4** and **5** (Scheme 2) were first prepared by Pulido and coworkers[30] and their reactivity was extensively studied by this group (see Section II.D.3).[38] The TBDPS (t-$BuPh_2Si$) group imparts a regio- and stereochemistry different from the more usual TMS and $PhMe_2Si$ groups and it can also be transformed into a hydroxyl group[38a]. Moreover, the TBDPS group is significantly more electronegative than standard silyl groups. This enhances the acidity of the α-protons and facilitates their base-catalyzed abstraction to give α-silyl carbanions containing the bulky *t*-butyldiphenylsilyl group[39]. The diethylaminodiphenylsilyl(cyano)cuprate[37] (**10**) (Scheme 4) has gained special relevance because of its ability to introduce a silyl group which is extremely easy to be converted into a hydroxyl group in the presence of many functional groups (see Section II.B)[37,40].

Some procedures, catalytic in copper, have been recently developed for the preparation of silicon and tincuprates[41] and a procedure using disilanes and $(CuOTf)_2{\bullet}C_6H_6$ affords silylcuprates, avoiding the preparation of silyllithium reagents[42].

B. The C−Si to C−OH Conversion

The interest of silylcuprates lies in their high reactivity toward electrophiles and the synthetic versatility of the resulting silanes. These are useful synthons for C−C bond formation[43] and undergo a large number of highly stereocontrolled reactions of much interest in organic synthesis[44]. Moreover, the use of appropriate silylcuprates permits the conversion of the C−Si bond formed into a C−OH bond. While TMS and $PhMe_2Si$ groups show the same reactivity in allyl- and vinylsilane chemistry, only the latter can be converted to an OH group. The presence of a phenyl group is responsible for this powerful transformation, where the $PhMe_2Si$ group can be formally considered synthetically equivalent to a hydroxyl group[45]. Furthermore, the reaction proceeds with retention of configuration at the migrating carbon. Three main methods[45–47] have been developed: (a) protodesilylation, to remove the phenyl group from the silicon atom, followed by oxidation of the resulting functionalized silicon atom using peracids or H_2O_2; (b) removal of the phenyl group with mercuric acetate (or a mixture of palladium and mercuric salts) followed by the oxidative step, which can be done in one pot by addition of peracetic acid; (c) use of bromine in conjunction with peracetic acid in a one pot reaction. Bromine removes the phenyl group by electrophilic attack and the peracid completes the oxidation. Br_2 can be added by syringe or generated *in situ* from potassium bromide, which is oxidized by the peracetic acid. Variations or combinations of these methods have also been reported to improve the final outcome. Some of the more typical recipes[45] are summarized in Scheme 5.

The scope and limitations of the reaction have been tested with a wide range of structures carrying many of the most common functional groups (Scheme 6)[45]. The benzene ring, alcohols, ethers, esters, amides and nitriles are compatible with the reported procedures. Ketones do not undergo Baeyer−Villiger oxidations under any of the conditions used; however, they may be brominated when using recipe (c). Amines (Scheme 6) are oxidized and consequently interfere. When the $PhMe_2Si$ group is in the presence of neighboring nucleofugal groups, such as acetoxy, the use of acidic conditions is very risky. In these cases, the acid-free protocol (c) can be very valuable. The mechanistic pathway of this useful transformation is depicted in Scheme 7.

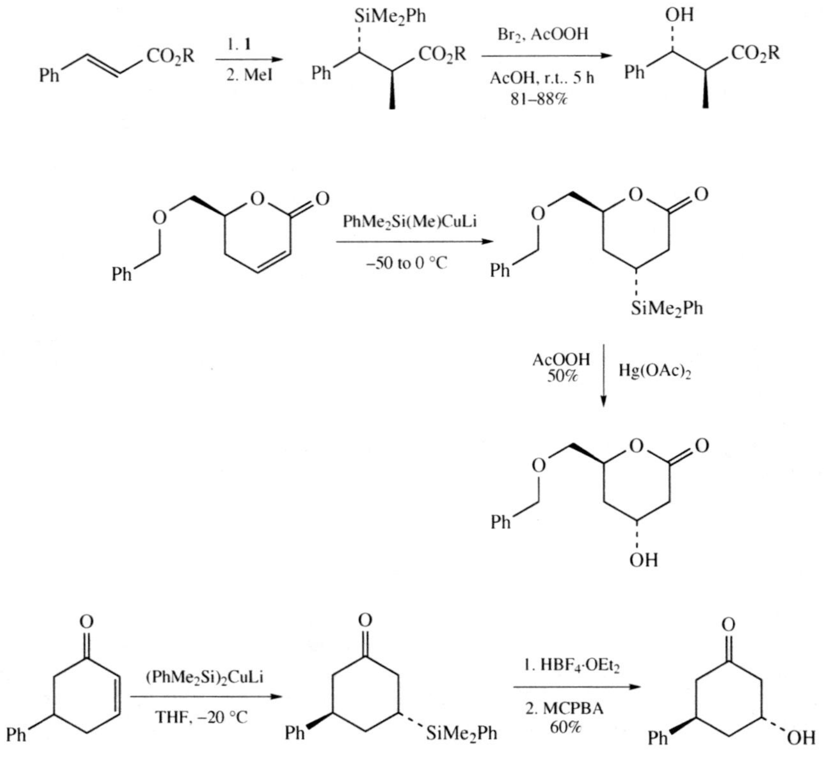

SCHEME 5

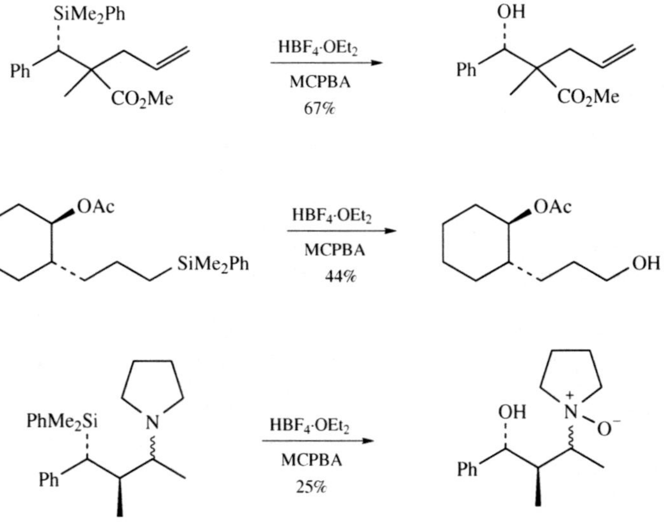

SCHEME 6

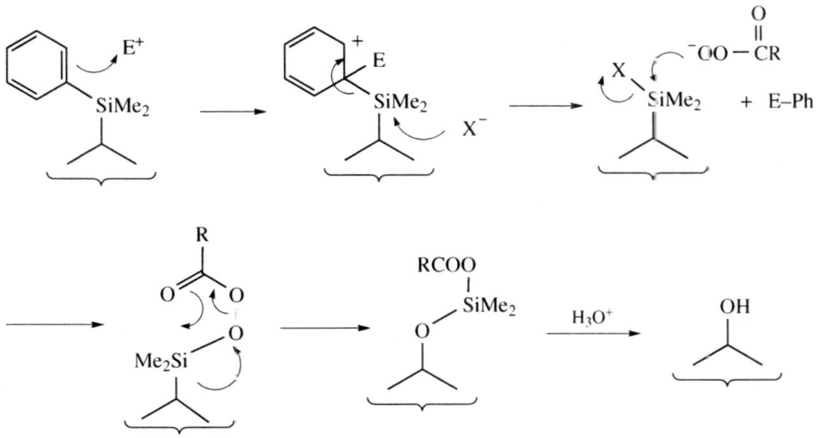

SCHEME 7

The reaction proceeds by attack of an electrophile (proton, bromine, mercury(II) cation) on the phenyl group. Electrophilic attack takes place at the carbon attached to silicon, to take advantage of the formation of a highly stabilized carbocation β to silicon (the so-called β effect)[43]. Removal of the phenyl group takes place simultaneously with the attachment of the nucleofugal group X on the Si atom. This step is followed by treatment with peracid or H_2O_2 and a base. The subsequent rearrangement resembles the Baeyer–Villiger oxidation. Finally, hydrolysis renders the alcohol.

Carbon–carbon double bonds are incompatible with the method reported for the $PhMe_2Si$ group because whichever electrophile is used, attack on the double bond occurs faster than phenyl removal. This problem has been partially overcome using a significantly more reactive silyl group. The (*Z*)-2-methylbut-2-en-1-yl(diphenyl)silyl group[35] (Scheme 8) is considerably more reactive, with respect to the silyl-to-hydroxy conversion, than the dimethylphenylsilyl group. The C–Si to C–OH conversion is particularly difficult with allylsilanes, but the development of the lithium bis(2-methylbut-2-en-1-yldiphenylsilyl)cuprate[35] (**9**) (Scheme 4) for this purpose makes suitable this powerful transformation. Thus, by using this silyl group, allylsilanes **11** and **12** can be transformed into allyl alcohols **13** and **14** in acceptable to good yield (Scheme 8). The removal of the (*Z*)-2-methylbut-2-en-1-yl(diphenyl)silyl group by protonation (or epoxide formation) and the easy loss of the silyl group leads to a functionalized diphenylsilyl group which undergoes the usual silyl-to-hydroxy transformation, thus allowing the (*Z*)-2-methylbut-2-en-1-yl(diphenyl)silyl group to be converted into the hydroxyl group without the need to remove the phenyl group.

The diethylaminodiphenylsilyl group[40,48] is even more reactive, toward the silyl-to-hydroxy conversion, than phenyldimethylsilyl and 2-methylbut-2-en-1-yl(diphenyl)silyl groups. It can be converted into the OH group even in the presence of a trisubstituted double bond[48]. This group is introduced by reaction with $Et_2NPh_2SiCu(CN)Li$[37] (**10**) (Scheme 4). The cuprate transfers a heteroatom-substituted silyl group that even in the presence of an allylic double bond can be converted into an alcohol functionality (Scheme 9). The silyl group already carries a nucleofugal group, which allows it to be converted directly into a hydroxyl group without the need to remove a phenyl group by electrophilic substitution. Conditions are very mild; they normally involve the use of basic ($KHCO_3$) hydrogen peroxide in THF/MeOH for several hours[40]. The aminosilane group

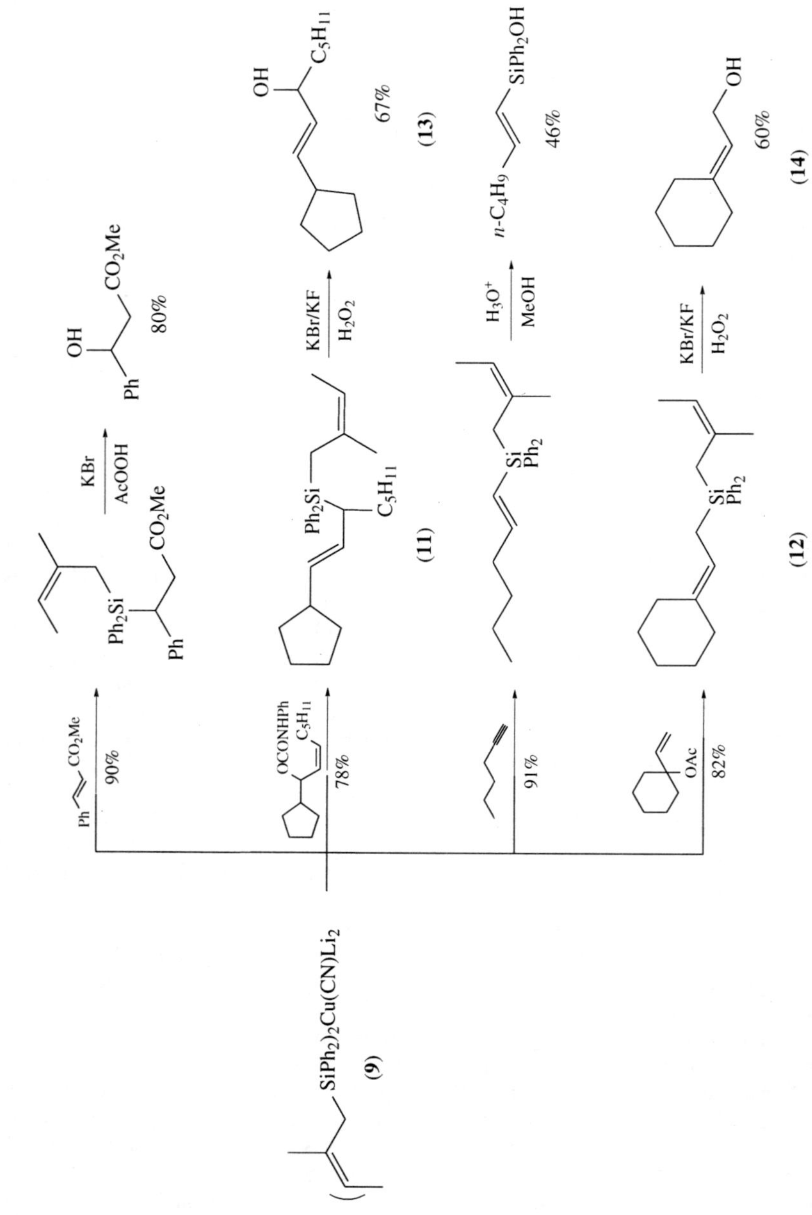

SCHEME 8

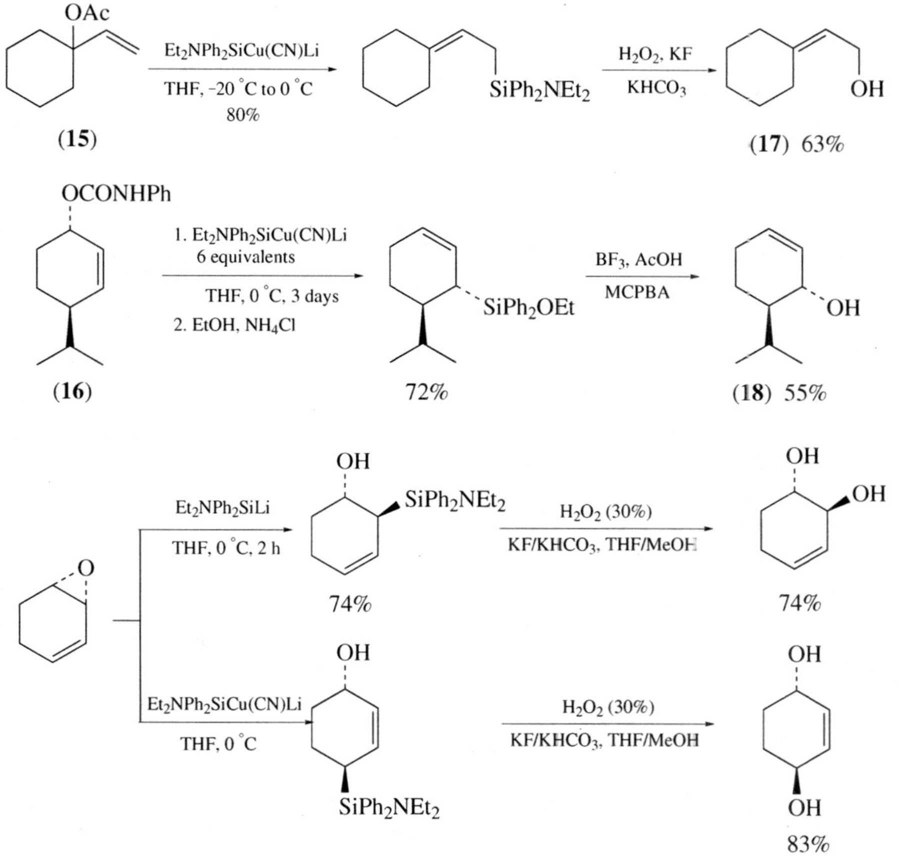

SCHEME 9

is unstable to chromatography and must be handled with care; however, direct oxidation of the crude in the conditions reported before affords the hydroxy derivatives in good yield[48]. In Scheme 9, some examples are collected that illustrate the ability of this group for converting C—Si to C—OH in the presence of allyllic double bonds. In addition, transformation of allylic acetate **15** and allylic carbamate **16** to allyl alcohols **17** and **18** allows overall allylic inversion.

As discussed in Schemes 5–9, the reactivity of silyl groups with respect to protodesilylation or silyl-to-hydroxy oxidation depends strongly on the substituents carried on the silicon atom. We have already shown some silyl groups that are more reactive than the $PhMe_2Si$ group, but in some cases there is a need for one that could be less reactive or even unreactive to protodesilylation or C—Si oxidation. This is the case of the bulky *t*-butyldiphenylsilyl group, a little reactive silyl group which is frequently introduced by reaction with the *t*-butyldiphenylsilylcuprates[30] **4** and **5** or the respective organolithium precursor (Scheme 2). The disilane **20**, prepared from the readily available silyl alcohol **19**, smoothly undergoes silyl-to-hydroxy conversion to the *t*-butyldiphenylsilyl monoalcohol **21** (Scheme 10); the *t*-butyl group evidently hinders the electrophilic attack on the phenyl ring, since both phenyl groups on the *t*-butyldiphenylsilyl group were intact[38a].

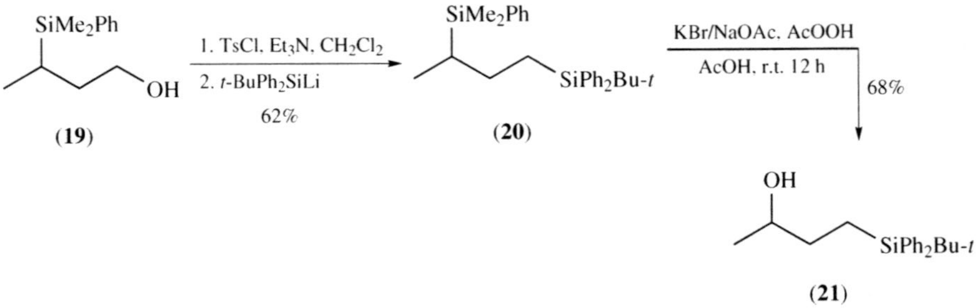

SCHEME 10

Due to the lower reactivity of the *t*-butyldiphenylsilyl group, this may or may not be the electrofugal group following electrophilic attack. It has been reported that it can function as an electrofugal group in the epoxidation of an allylsilane or in the formation of an α,β-unsaturated ketone from a β-silylketone (Scheme 11) and with some limitations in the oxidation to alcohols[30a]. Elimination of silicon in the β-silylketone **22** restores the initial double bond.

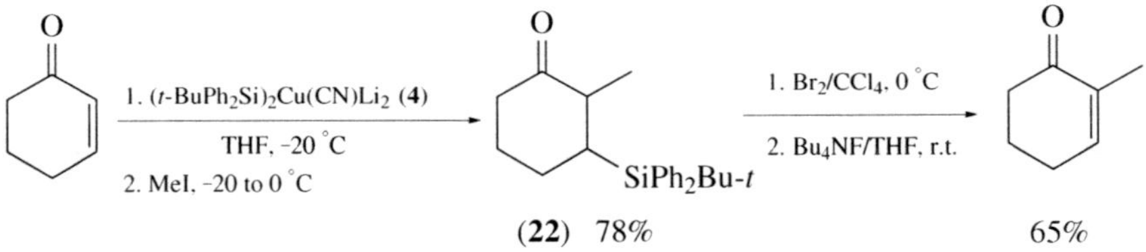

SCHEME 11

This bulky group can also function as a leaving group in the electrophilic substitutions of vinylsilanes, where it cleanly undergoes normal desilylative acetylation (Scheme 12). However, the desilylative acetylation of the vinyldisilane **23** is completely regioselective, where the trimethylsilyl group is exclusively removed to give **24**, in which the remaining *t*-butyldiphenylsilyl group is ready for an ulterior electrophilic substitution[38a]. The desilylative acetylation of divinylsilane **25** is not very selective, thus showing how similar are trimethylsilyl and phenyldimethylsilyl. Scheme 12 exemplifies well the different reactivity of three common silyl groups: trimethylsilyl, phenyldimethylsilyl and *t*-butyldiphenylsilyl, which can be chemoselectively used or removed according to the different conditions used[38a]. All three silyl groups can be adequately introduced into a desired molecule by attack of the corresponding cuprates (Schemes 1–3) on an appropriate electrophilic center.

A larger silyl group might also have some advantages in the control of the regio- and stereochemistry of the addition to a carbonyl group in β-silylketones. Comparison of the reaction of ketones **26** and **27** with complex metal hydrides is very illustrative (Scheme 13). Whereas ketone **26** unremarkably gives the alcohols **28** in a ratio of 73:27 on reduction with sodium borohydride, the β-silyl group in ketone **27** sufficiently hinders this reagent so that no reaction takes place, even under reflux in ethanol for several hours. This ketone is, however, reduced by lithium aluminum hydride to give only the equatorial alcohol **29**. Both ketones undergo methylation exclusively at C-6[38a].

Although a relatively large number of silicon groups, readily convertible to a hydroxyl function, have been introduced in the last few years[49], it is difficult in certain cases to

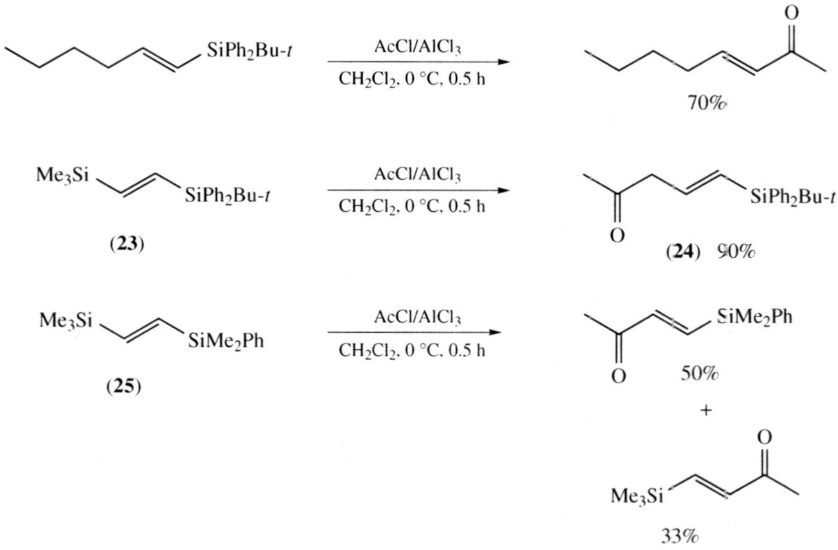

SCHEME 12

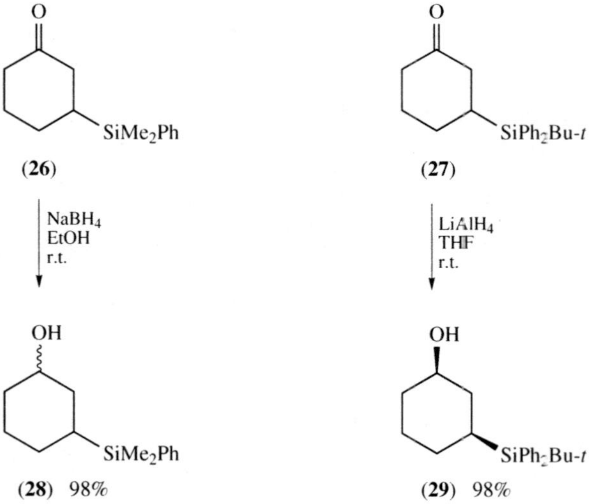

SCHEME 13

combine both the stability of the silicon group and its ability to be oxidized in mild conditions.

Recently, Landais and coworkers[50,51] have developed some new silyl groups which can be used as latent OH groups. A silyl group carrying a thienyl fragment attached to the silicon atom has been devised and has been shown to be oxidized under both electrophilic and nucleophilic conditions[50]. They found that thiophene derivatives might be particularly useful in this context since they are easy to introduce, relatively stable and should be easy

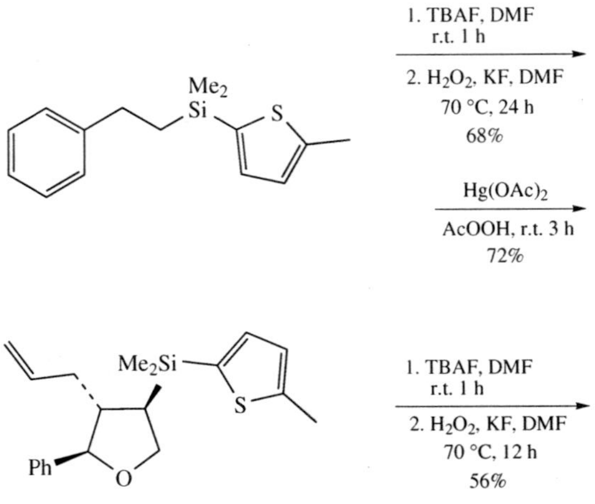

SCHEME 14

to displace using both electrophilic ($Hg(OAc)_2$ and AcOOH) and nucleophilic (TBAF and H_2O_2) conditions (Scheme 14). Electrophilic conditions may be troublesome when one has to deal with sensitive functions such as double bonds.

The thienylsilyl group depicted in Scheme 14 is a useful masked OH group that is more stable to electrophiles than double bonds. It is not as good a leaving group as the allyl group upon treatment with TBAF but is quite more reactive than a phenyl group. The mild conditions used for removing the thienyl group are compatible with numerous functionalities.

The (phenylthio)cyclopropylsilyl group was also used as a masked hydroxyl group[51]. Several procedures have been devised to allow oxidation of this group to an OH group, in the presence of various functionalities. A general behavior involves first oxidation of the sulfide to give a sulfoxide or sulfone, followed by a sila-Pummerer rearrangement[52] to provide the alkoxysilane required for the Tamao–Kumada oxidation[40] to OH (Scheme 15). Some examples of oxidation of the (phenylthio)cyclopropylsilyl group in the presence of sensitive groups such as allylic double bonds and hydroxyl groups are summarized in Scheme 15. As usual, introduction of OH occurs with retention of configuration.

C. The Structure of Silylcuprates

The exact structure of higher-order silylcuprates is not well known. There are not many reports on structural work carried out with cuprates containing silyl groups. Most of the current knowledge on silylcuprate structure is inferred by comparison with known data on the corresponding carbocuprates. Organocopper species containing at least three anionic groups $[R_3SiCu(R')_2]^{-2}$ attached to the central copper core are termed higher-order cuprates[8,27,53]. For purposes of differentiation, conventional cuprate(I) species of type $[R_3SiCuR']^-$ may be referred to as lower-order cuprates. Cyanosilylcuprates of stoichiometry $[R_3SiCu(CN)]^-$ are formally considered lower-order cuprates, although some discussion on this subject ascribed it to the formation of clusters in ethereal solutions[54]. Bicoordinate species $R_3Si(R')CuLi$ exist as higher aggregates where the copper bicoordinate C–Cu–Si fragment is linear with angles always close to 180°[55].

The structure of higher-order cyanosilylcuprates $R_3Si(R')Cu(CN)Li_2$ bearing a cyanide anion and two anionic residues has also been the subject of controversy (Scheme 16).

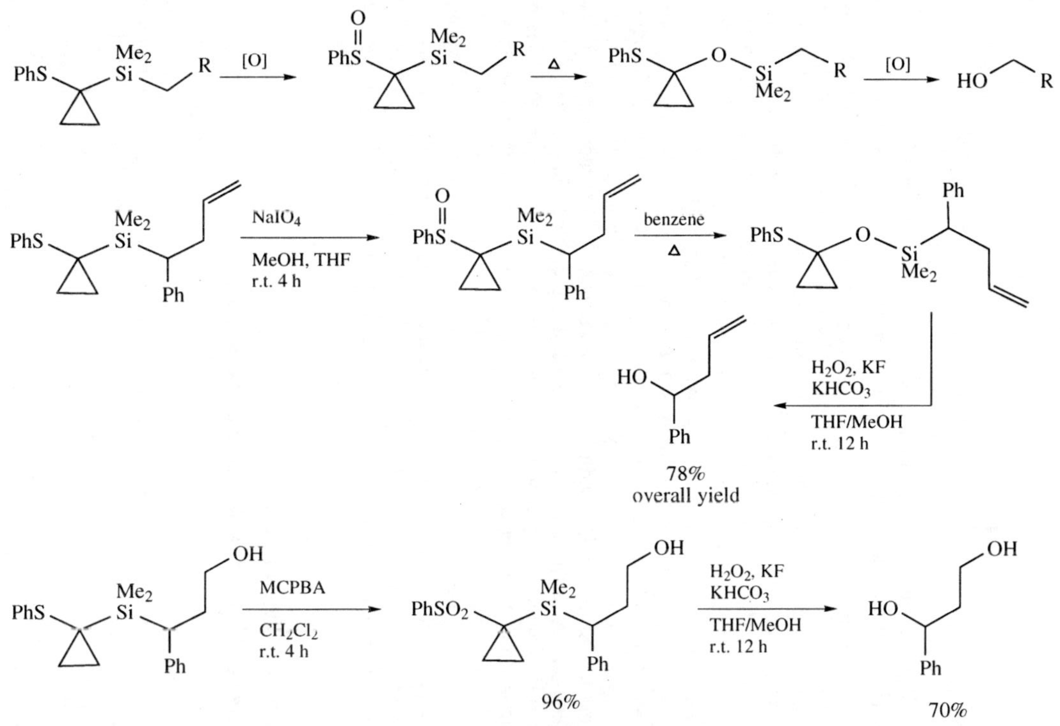
PhS
Me2
Si
R
[O]
O
PhS
Me2
Si
R
Δ
PhS
O
Si
Me2
R
[O]
HO
R
PhS
Me2
Si
Ph
NaIO4
MeOH, THF
r.t. 4 h
O
PhS
Me2
Si
Ph
benzene
Δ
Ph
PhS
O
Si
Me2
H2O2, KF
KHCO3
THF/MeOH
r.t. 12 h
HO
Ph
78%
overall yield
PhS
Me2
Si
OH
Ph
MCPBA
CH2Cl2
r.t. 4 h
PhSO2
Me2
Si
OH
Ph
96%
H2O2, KF
KHCO3
THF/MeOH
r.t. 12 h
HO
OH
Ph
70%

SCHEME 15

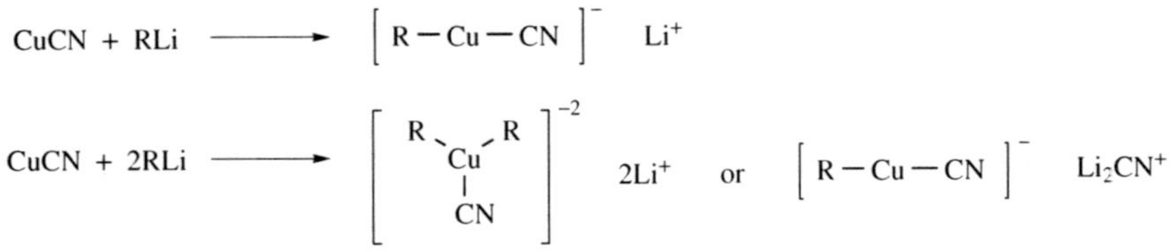

(**30**)

SCHEME 16

It has been pointed out that they don't exist as stable species. This controversy has been the origin of numerous mechanistic and structural studies. Physical measurements[56], NMR data[22,26,57] and theoretical studies[58] also contributed to this discussion. Moreover, crystallographic data[59] for higher-order cyanosilylcuprates indicated that the cyanide anion is coordinated to lithium rather than to copper. In conclusion, there is a general consensus among theoretical chemists that the triply coordinated species **30** (Scheme 16) is not stable enough in solution. However, the presence of a triply coordinated cuprate(I) dianion was claimed by Cabezas and Oehlschlager[60] in a recent NMR study of the ^{13}C$-^{13}$CN carbon coupling in cyanostannylcuprates using THF/HMPA as solvent.

In summary, silylcuprates prepared by similar methodologies to those developed for the more common carbocuprates are considerably more reactive than the latter. The cyanosilycuprates particularly show[8a] high regio- and stereoselectivity when reacting with electrophiles (see Section II.D) leading to silicon-containing products in high yield. The cuprate methodology, consequently, is one of the preferred alternatives for introducing silicon into a molecule, provided that the molecule carries a reactive electrophilic center. Silicon-containing synthons are useful intermediates in organic synthesis, particularly for the total synthesis of natural products[61] where they impart a high level of stereoselectivity[44] in many of their intervening reactions. Thus, the chemistry of allyl and vinylsilanes, two of the more largely and successfully used functions, has been used extensively because both functions undergo a great variety of silicon-assisted transformations and are excellent building blocks for C$-$C bond formation[12a]. In Table 1 are presented some of the more frequently reported and extensively employed silyl cuprates.

D. Reactions of Silylcuprates

1. Conjugate additions to unsaturated carbonyl compounds

Silylcuprate reagents readily add to the β-position of α,β-enones and α,β-unsaturated esters. In this respect, their behavior doesn't seem to differ from that observed for common carbocuprates. Surprisingly, the silyllithium precursors also add to α,β-unsaturated ketones exclusively at the conjugated position even at low temperatures ($-78\,^{\circ}$C) (Scheme 17), thus showing that silicon anions are considerably softer than carbanions. Hindered enones at the β-position, such as **31**, do not react usually with silyllithium compounds; however, addition of the respective cuprate takes place smoothly[7a].

Other functions such as unsaturated aldehydes, esters, amides and nitriles do not undergo conjugate addition by reaction with silyllithiums and consequently the use of silylcuprates is necessary (Scheme 18)[21,62,63].

The stereochemistry[21] of the resulting β-silyl ketone (see conversion of **32** to **33** in Scheme 18) is that expected for axial attack by the silylcuprate reagent followed by transmethylation by methyl iodide, showing that the silylcuprate is similar in this respect to carbocuprates[64].

TABLE 1. Some frequently used silylcuprate reagents

Silylcuprate reagent	Reference	Silylcuprate reagent	Reference
$(Me_3Si)_2CuLi$	21	$(t\text{-}BuPh_2Si)_2CuLi$	30a
$(Me_3Si)_2Cu(CN)Li_2$	29a	$t\text{-}BuPh_2SiCu(CN)Li$	30b
$(PhMe_2Si)_2CuLi$	21	$(t\text{-}BuPh_2Si)_2Cu(CN)Li_2$	30a
$PhMe_2SiCu(CN)Li$	22	$[(MeCH{=}CMeCH_2)Ph_2Si]_2Cu(CN)Li_2$	35b
$(PhMe_2Si)_2Cu(CN)Li_2$	23	$Et_2NPh_2SiCu(CN)Li$	37
$PhMe_2Si(Me)Cu(CN)Li_2$	29a	Ph_3SiCu	34
$t\text{-}BuMe_2Si(n\text{-}Bu)Cu(CN)Li_2$	32		

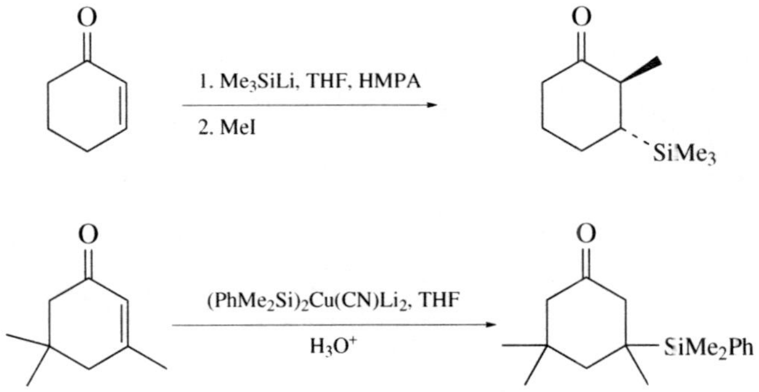

(**31**)

SCHEME 17

Yields are usually high; when low yields are observed, they are attributed to the high reactivity of the intermediate enolate, which induces oligomerization[65]. The use of trimethylchlorosilane (TMSCl) suppresses this side reaction and it is recommended for improving the final outcome. TMSCl presumably works as an enolate trapper, forming silyl enol ether intermediates that stabilize the reaction; it has also the effect of accelerating the reaction facilitating the coordination of the enone to the cuprate cluster[66].

β-Silyl ketones obtained in these processes can be converted into the starting enone by elimination of the silyl group with $CuBr_2$. This silylcuprate conjugate addition has been used for the protection of an enone double bond that is further retained by bromination–desilylation (Scheme 19). The process is illustrated with the conversion of ester **33** to carvone (**34**) (Scheme 19)[21].

The enolates resulting from conjugate addition of silylcuprate to the enone can be captured by a wide variety of electrophiles, thus providing good routes for asymmetric induction. For instance, conjugate addition to α,β-unsaturated carbonyl compounds having an α-alkyl group as **35** usually leads, after final work-up, to an excess *syn* diastereomer **37**, whereas the corresponding unsubstituted enone **36** affords almost exclusively the

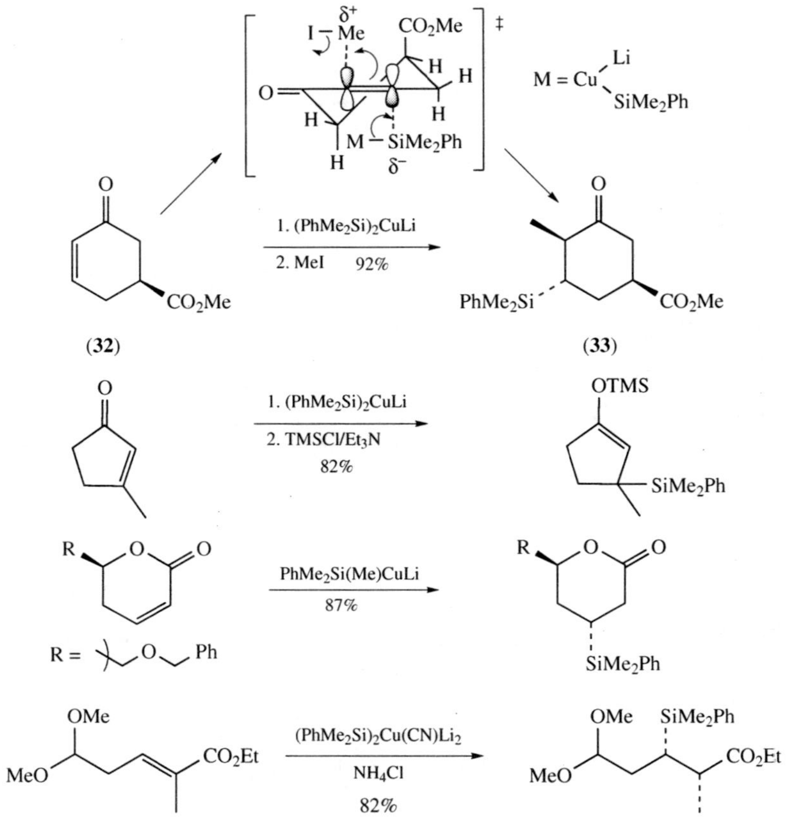

SCHEME 18

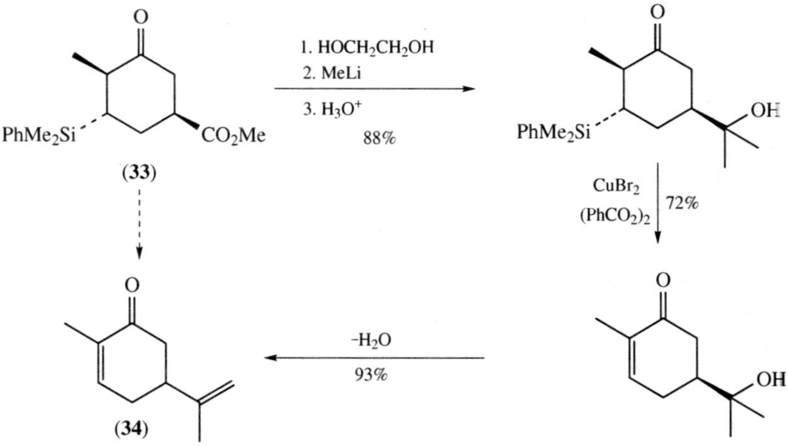

SCHEME 19

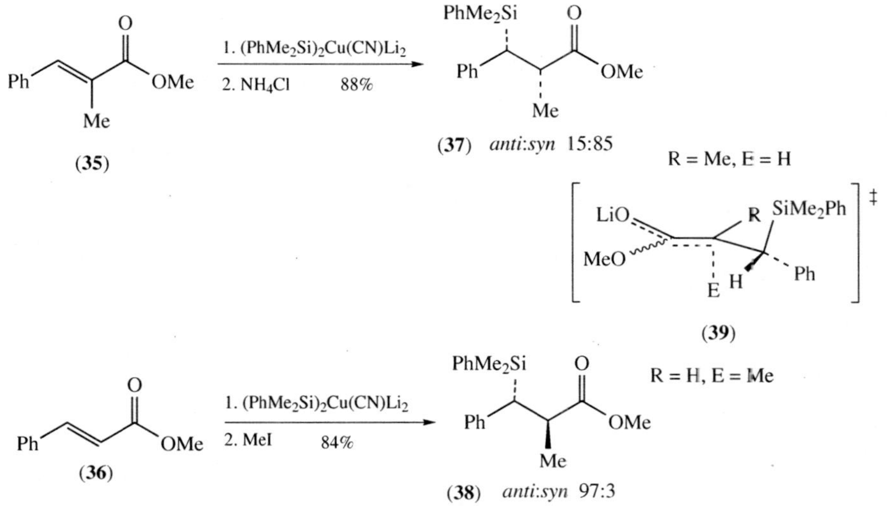

SCHEME 20

anti isomer **38** by alkylation of the intermediate enolate (Scheme 20)[67]. Carbocuprates, however, do not easily react with α-substituted enones.

The stereochemistry observed in Scheme 20 is in good agreement with the formation of a favored transition state **39** (Scheme 20) in which the silyl group occupies an *anti*-periplanar position with respect to the enolate moiety, for better orbital interactions. The larger of the two remaining groups on the stereocenter (Ph in the Scheme) moves away from the enolate group to avoid steric interactions and then the electrophile approaches axially from the less hindered side *anti* to the silyl group[67]. The lower stereoselectivities observed with aldehydes and nitriles can be associated to the geometry of the enolate, where the small size of H (in aldehydes) or the linear structure of nitriles results in formation of mixtures of diastereomers[67]. The silylcuprate conjugate addition strategy combined with the silylcuprate substitution reaction on the allylic acetate **40** has been recently used in a synthesis of the Prelog–Djerassi lactone (Scheme 21)[63].

When the enone is linked to an appropriate chiral auxiliary, the silylcuprate adds with high level of diastereoselectivity to create a stereogenic center bearing the silyl group. Oppolzer's camphorsultam[68] has been used for this purpose, leading to homochiral versions of β-silyl ketones and esters after removing the chiral auxiliary.

Enantiopure β-silyl ketones have been prepared by conjugate addition of TMS-cuprates to the appropriate enone enantiomer. For example, the enantiomerically pure trimethylsilylcyclohexenones **41**[69] have been synthesized as described in Scheme 22 and the regio- and stereocontrol imparted by the silyl group used in synthesis of natural products[44]. This methodology intends to exploit the *anti*-directing effect of the silyl group in ulterior conjugate additions.

Hindered β-silyl ketones and esters carrying bulky silyl substituents have been prepared from mixed cyanosilylcuprates by conjugate addition to enones (Scheme 23). In spite of their large size, *t*-butyldimethylsilylcuprate, thexyldimethylsilylcuprate and *t*-butyldiphenylsilylcuprate add to α,β-unsaturated carbonyl compounds in high yield and with a remarkable regio- and stereoselectivity even in the case of β-substituted enones. Bulky β-silyl ketones can lead to highly selective carbonyl additions due to the stereoelectronic effect of the silyl group[32,38a].

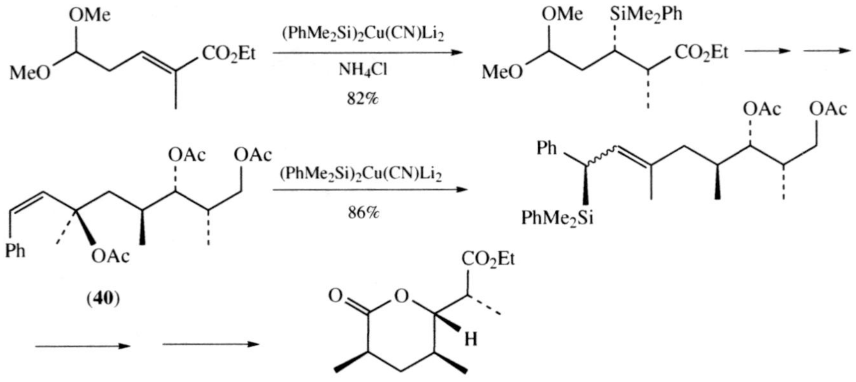

SCHEME 21

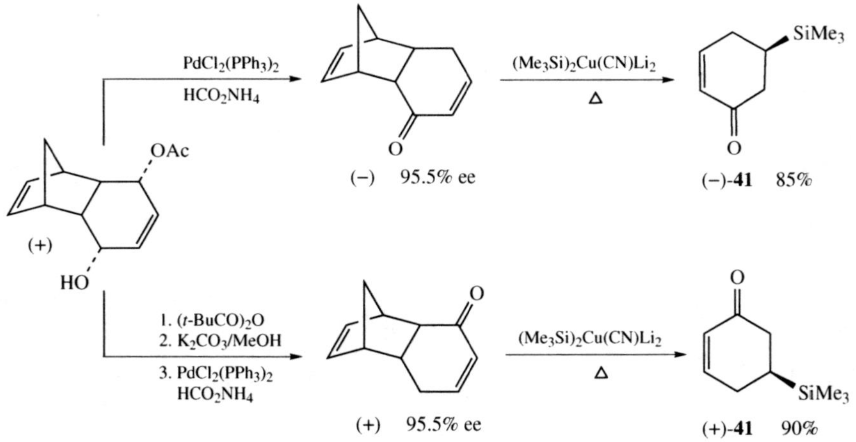

SCHEME 22

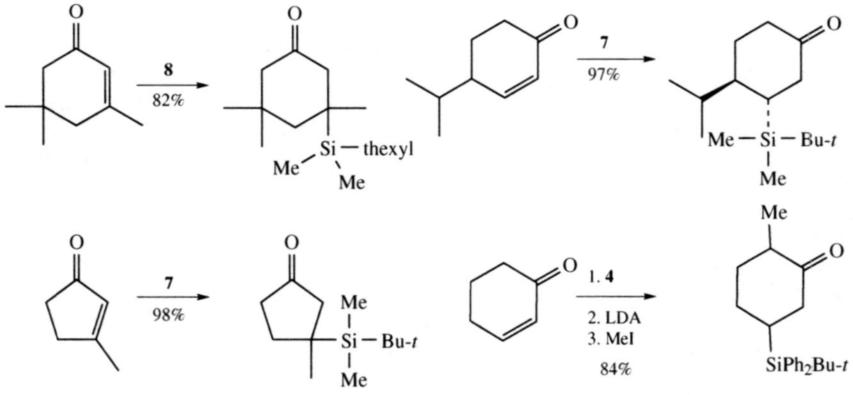

SCHEME 23

Conjugate addition of silylcuprates to unsaturated esters gives β-silyl esters, which can be reduced with lithium aluminum hydride followed by a Grieco dehydration to give allylsilanes. The procedure is a powerful strategy for the synthesis of allylsilanes with different structures and substitution pattern, as illustrated in Scheme 24 and Table 2.[70]

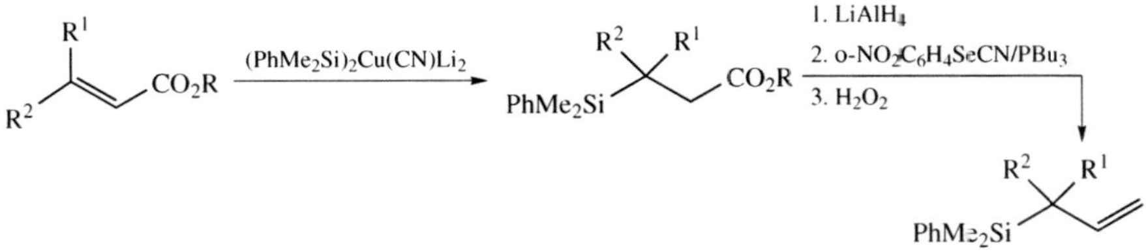

SCHEME 24

The selectivity for *anti*-attack on β-silyl enolates (Scheme 20) extends also to cyclic enolates. Thus, β-silyl enolates prepared *in situ* from lactones **42** and **43** are similarly selective in the *anti* sense with complete selectivity in the conjugate addition–alkylation reaction, giving only the lactones **44** and **45** (Scheme 25)[71,72]. The corresponding β-silyl enolate generated from carvone (**34**) undergoes *anti* protonation.

The enolate anions resulting from silylcuprate conjugate addition to α,β-unsaturated carbonyl compounds can also be trapped with aldehydes (Scheme 26). Results are similar to that of the above-mentioned alkylation process, with *syn:anti* diastereoselectivity ratios again favoring the *anti* product. Thus, silylcuprate addition to enoates (**46**) affords solely the *E*-enolate (**47**) by direct attack of cuprate on the β-position, whereas protonation, recovery of the silyl ester and regeneration of enolate by reaction with lithium diisopropylamide (LDA) leads to the *Z*-enolate (**48**). The direct formation of the *E*-enolate indicates that conjugate addition takes place selectively from the s-*cis* enolate

TABLE 2. Synthesis of allylsilanes from α,β-unsaturated esters

Substrate	1,4-Addition product	Allylsilane	Yield (%)[a]
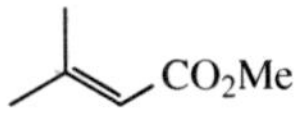			54
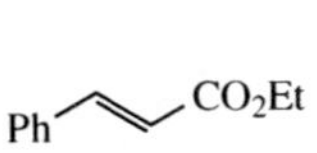	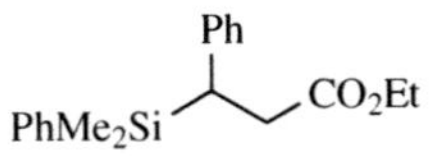	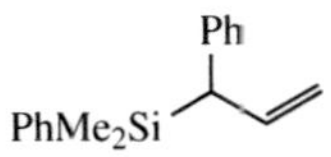	34
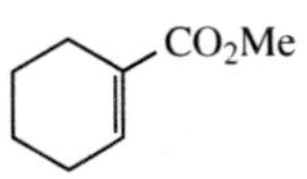	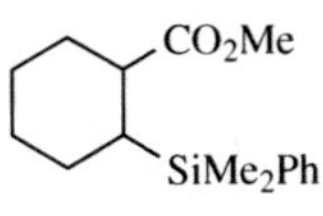	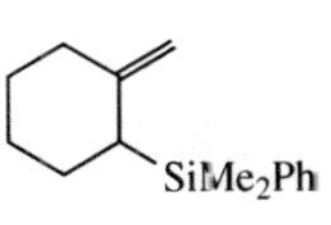	21
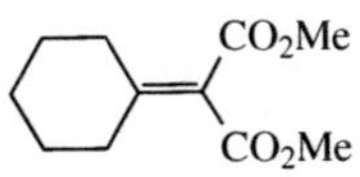	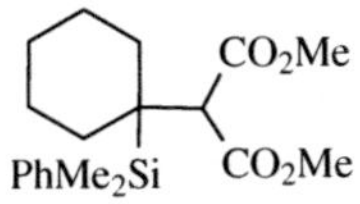	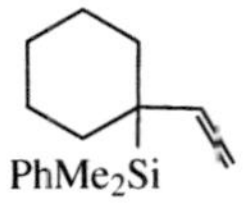	45

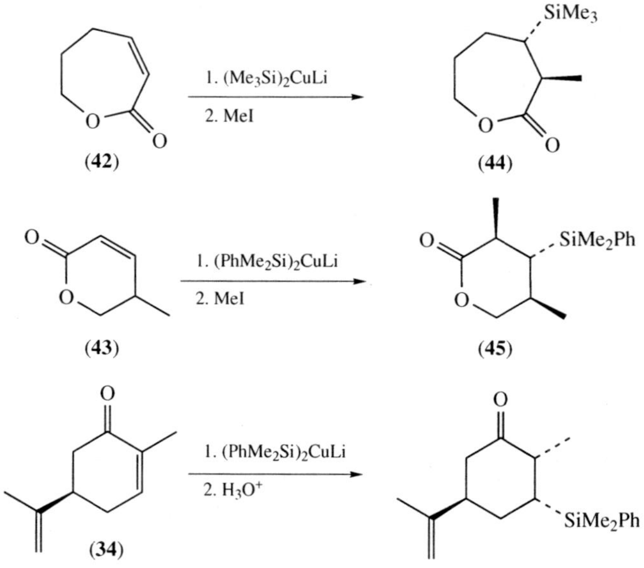

SCHEME 25

conformer, which can be explained by cuprate coordination to the carbonyl group and intramolecular transfer of the silyl group. The diastereoselectivity observed for both *Z*- and *E*-enolates in the aldol reaction with aldehydes is unexceptional, and can be well explained in terms of the Zimmeman–Traxler model. The selective formation of aldols **49** and/or **50** has the additional interest to proceed with control of three stereogenic centers at once (Scheme 26)[73].

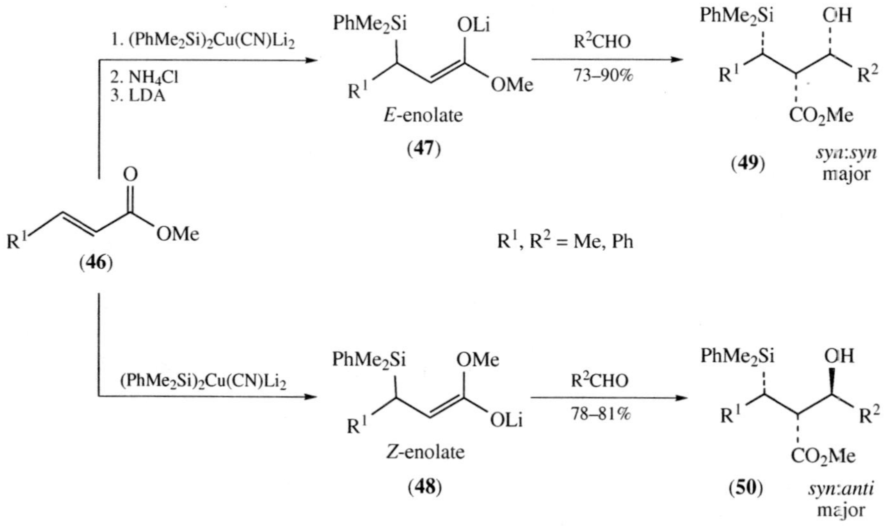

SCHEME 26

Saturated ketones are known to react slowly with carbocuprates, making the reaction of little synthetic interest. Silylcuprates react faster giving α-hydroxysilanes that can be used in Brook rearrangements[30a], but this alternative has not been extensively used in synthesis.

In summary, β-silyl carbonyl compounds obtained from silylcuprates and enones or enoates are powerful synthetic intermediates that can be used in different ways for further synthetic transformations: (a) the double bond can be recovered by desilylation–debromination; (b) the silyl group can be converted to a hydroxyl group and (c) they can be transformed into allylsilanes that are extremely useful synthons for C–C bond formation.

2. *Substitution reactions with allylic acetates, acid chlorides, epoxides and related systems*

Silylcuprates react with allylic acetates to give directly allylsilanes in good yields. The reaction is not always selective and problems of regio- and stereocontrol have frequently been reported. Nucleophilic displacement may occur with or without allyllic inversion. Depending on the reaction pathway (S_N2 or S_N2') the geometry of the double bond of the allylsilane may be affected (Scheme 27). Secondary allylic acetates such as **51** give both regioisomers **52** and **53** with low selectivity; however, some improvement of the regioisomeric ratio can be attained using the *cis*-alkene **54**, which encourages allylic rearrangement giving **53** and **55** in a 82:18 ratio (Scheme 27)[74].

It has been reported that by using mixed cuprates and allylic carbamates **56** a remarkable control of regioselectivity is observed, usually leading exclusively to **53** with allylic inversion. Interestingly, the carbamate and acetate reactions depicted in Scheme 27 are stereochemically complementary, since the former reaction occurs stereospecifically *syn* and the latter is stereospecifically *anti*. This feature has been used to prepare homochiral allylsilanes with high levels of stereospecificity. Good regioselectivities have also been found when the allylic system is more substituted at one end than at the other, when the steric hindrance at one end is very high (neopentyl) or because the silyl group is becoming attached to the less sterically hindered end of the allylic system[75]. The carbamate methodology has also been used by Tamao and coworkers[48] for a high regio- and stereoselective synthesis of allylsilanes (Scheme 9). Cyclic secondary allylic acetates do not show a great bias toward S_N2 or S_N2', the outcome depending on many different factors such as the cyclic size, the substitution pattern on the ring, the endo- or exocyclic position of the double bond, etc. Again, the use of carbamates highly increases the allylic inversion due to intramolecular delivery of the silyl group by way of an amido(silyl)cuprate reagent generated *in situ* (Scheme 28)[75].

Tertiary allylic acetates are much more predictable in their behavior, leading selectively to allylsilanes with the silyl group placed at the less substituted end of the allylic system (Scheme 29)[75,76]. An S_N2' process is almost exclusively observed.

The reaction proceeds *anti* stereospecifically. This stereospecificity has been proved in the reaction of silylcuprates with allyl and propargyl acetates prepared from cyclohexanones[77]. Thus, the reaction of acetates **57–59** and **60** with halide-based phenyldimethylsilylcuprates was straightforward and gave a single allylsilane (**61** or **62**) in each case with no cross-contamination of other substitution products (Scheme 30). The relative configurations of the two allylsilanes were confirmed by a series of reactions of known stereospecificity (osmilation, Peterson elimination and related fluoride-induced eliminations) that gave back the starting allyl alcohol. By analogous reactions of propargyl acetates **63** and **64** with the same silylcuprate, Fleming prepared allenylsilanes **65** and **66** differing in their *cis/trans* geometry. X-ray crystal-structure determination of **65** clearly

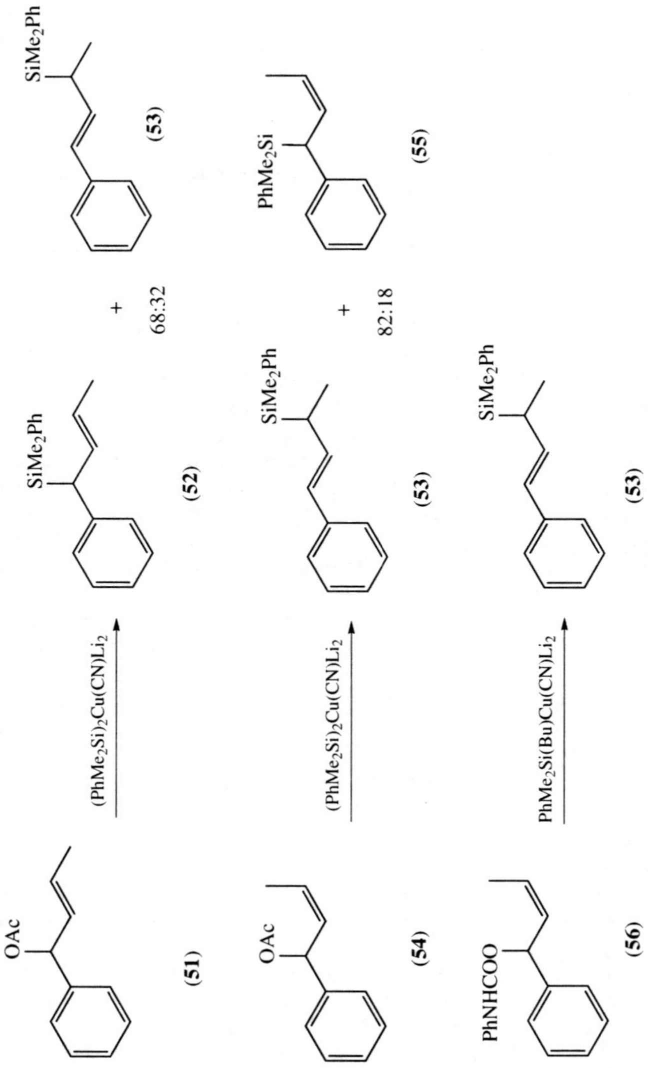
OAc
(51)
$(PhMe_2Si)_2Cu(CN)Li_2$
SiMe$_2$Ph
(52)
+
68:32
SiMe$_2$Ph
(53)
OAc
(54)
$(PhMe_2Si)_2Cu(CN)Li_2$
SiMe$_2$Ph
(53)
+
82:18
PhMe$_2$Si
(55)
PhNHCOO
(56)
$PhMe_2Si(Bu)Cu(CN)Li_2$
SiMe$_2$Ph
(53)

SCHEME 27

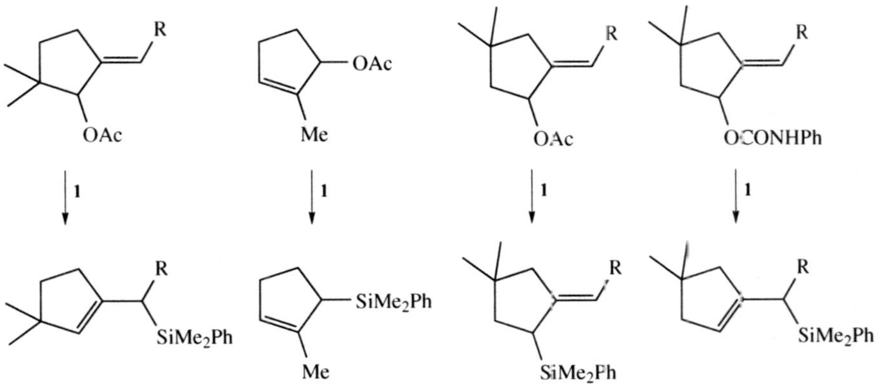

Major products of reactions: R = Ph, Me, *i*-Pr; Yields: 50–80%

SCHEME 28

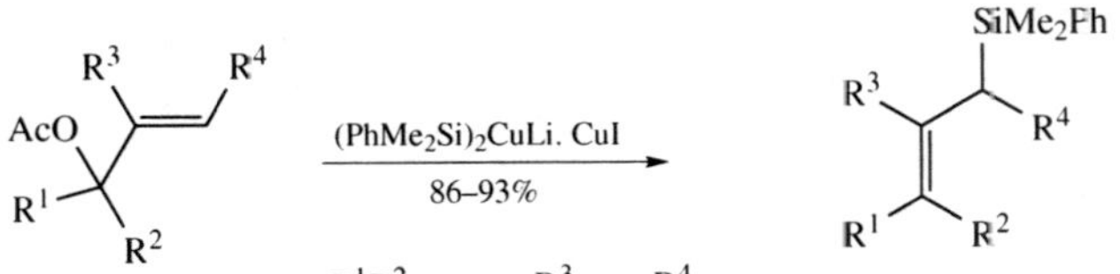

R^1R^2	R^3	R^4
$-(CH_2)_5-$	H	H
$-(CH_2)_4-$	H	H
$-(CH_2)_4-$	Me	H
$-(CH_2)_4-$	H	Me

SCHEME 29

shows that the reaction is stereospecifically *anti*. Silylated allenes[78] are useful intermediates in stereoselective synthesis with a remarkable reactivity toward addition and a silicon atom directing the regio- and stereochemistry of the reaction.

A consequence of the *anti* S_N2' mechanism of the silylcuprate substitution is that a mixture of allylic substrates differing in the configuration of both the olefin and the stereogenic center linked to acetate (for example **57** and **58**, or **59** and **60**) will afford selectively the same diastereomeric allylsilane. This feature has been exploited in a clever synthesis of racemic dihydronepetalactone (Scheme 31)[79]. Similarly, the high regio- and stereocontrol achieved in rigid bicyclic systems such as fused bicyclopentanes was used in a synthesis of carbacyclin analogues (Scheme 31)[80].

Many other natural product syntheses have taken advantage of the silylcuprate displacement reaction of allylic esters to generate intermediate allylsilanes that were used in a later transformation[81–83].

Silylcuprates also participate in substitution of allylic halides and sulfonates[84]. Labeling studies of allylic chlorides found that the pathway is predominantly *anti* S_N2'[84b].

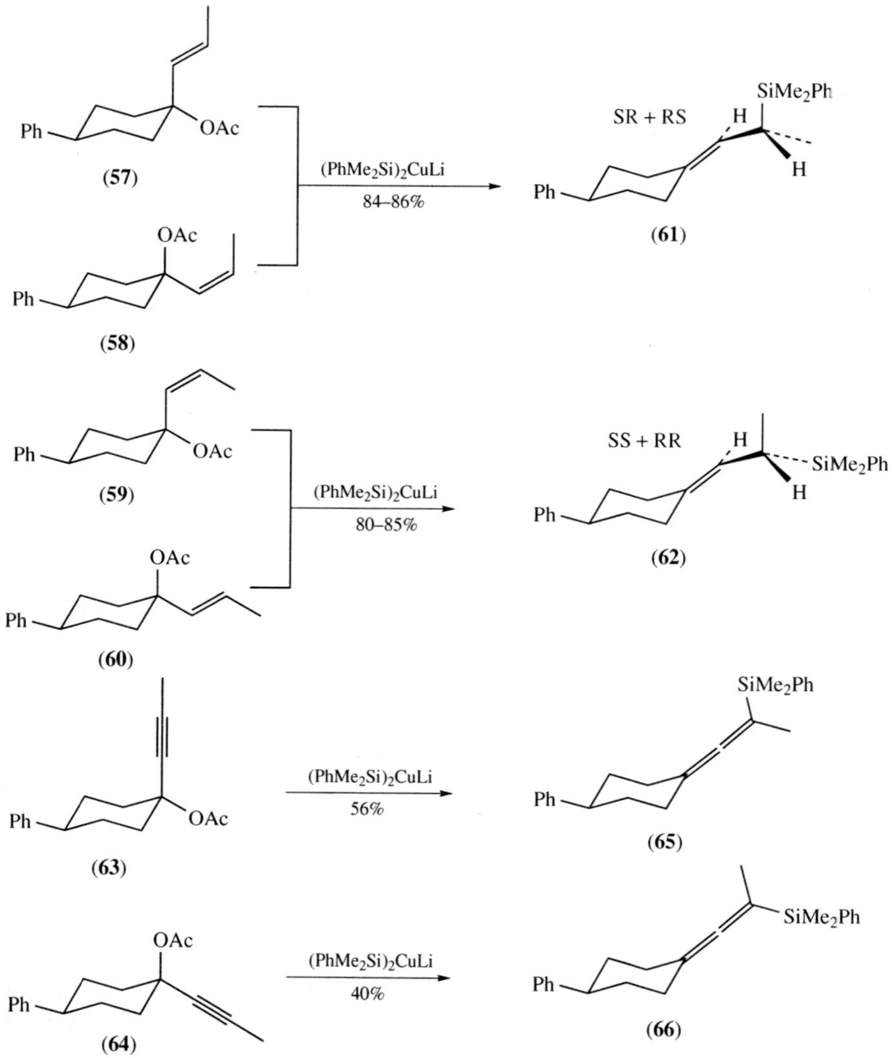

SCHEME 30

Better regiocontrol is usually achieved with halides than with sulfonate esters, whereas the regioselectivity decreases when increasing solvent polarity. 4-Bromo-2-enoates react with silylcuprates giving preferentially allylic substitution[85].

Although less frequently, alkyl halides have also been used in reactions with silylcuprates[16a]. This methodology has been exploited in a synthesis of silicon-containing alanines with moderate yields (Scheme 32)[86].

Silylcuprates also give substitution with acid chlorides[30a–87] or acyl imidazoles[88]. The reaction is fast and high yielding, although the high reactivity of the acyl chloride makes necessary the use of lower-order cuprates. By using zinc silylcuprates instead

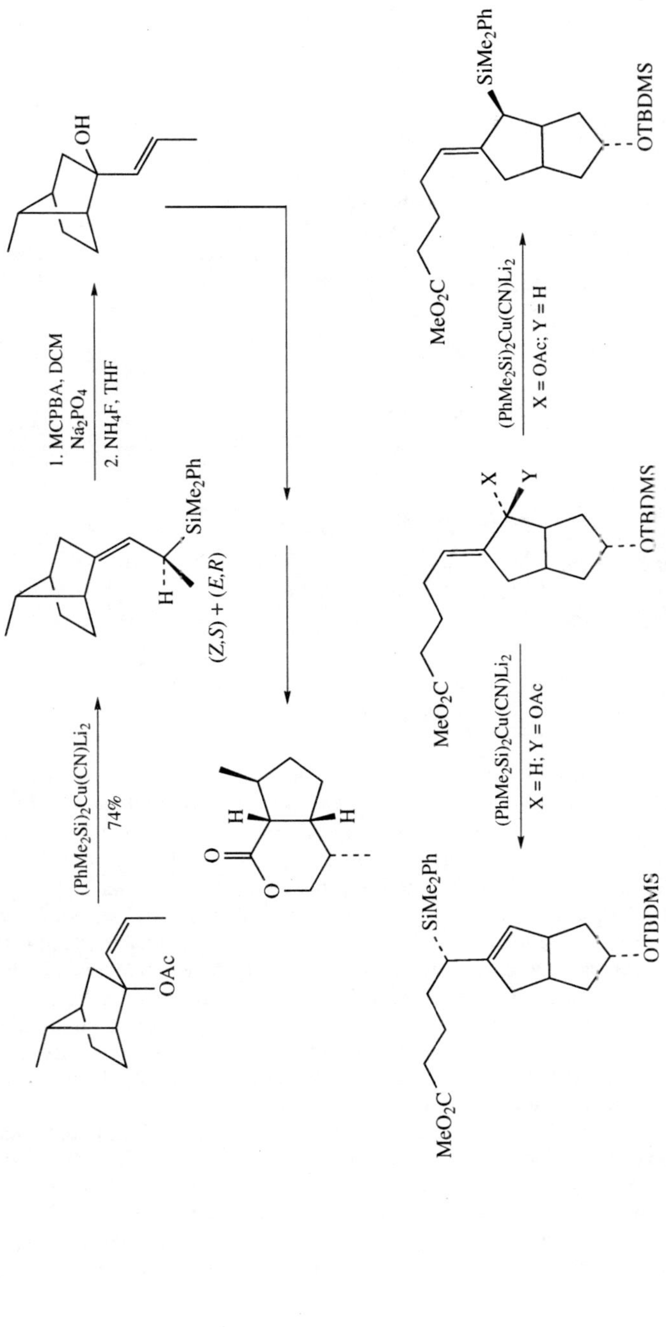
OAc
(PhMe2Si)2Cu(CN)Li2
74%
H
SiMe2Ph
(Z,S) + (E,R)
1. MCPBA, DCM
Na2PO4
2. NH4F, THF
OH
O
O
H
H
MeO2C
SiMe2Ph
OTBDMS
(PhMe2Si)2Cu(CN)Li2
X = H; Y = OAc
MeO2C
X
Y
OTBDMS
(PhMe2Si)2Cu(CN)Li2
X = OAc; Y = H
MeO2C
SiMe2Ph
OTBDMS

SCHEME 31

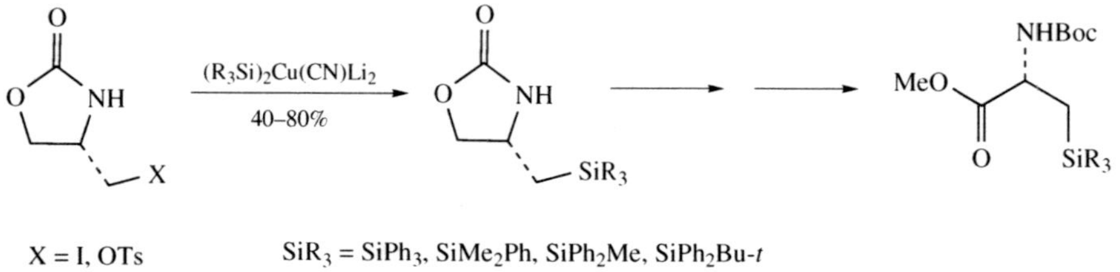

SCHEME 32

of lithium derivatives, highly functionalized acyl chlorides can be selectively transformed into acylsilanes[87]. Treatment of silylcuprates with α-amino acid chlorides or acyl imidazoles leads to α-aminoacylsilanes, which have been employed in synthesis of enantiopure β-amino alcohols (Scheme 33)[87,88]. Silylcuprates also react with 1,2-azole derivatives giving substitution and cleavage products[89].

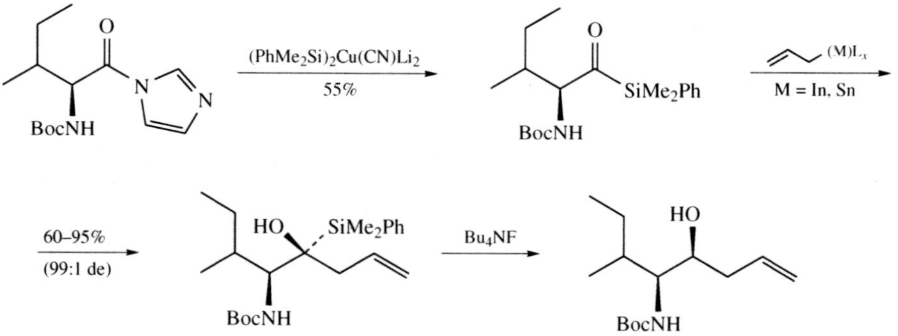

SCHEME 33

Epoxides are useful substrates to be used in substitution reactions with silylcuprates. They are highly reactive toward nucleophilic attack, undergoing cleavage of the oxirane ring to give β-hydroxysilanes that are the required precursors for Peterson olefination[43]. The development of selective alternatives for the conversion of β-hydroxysilanes into E- or Z-alkenes increases the utility of the silylcuprate-epoxide reaction.

Lipshutz and coworkers[32] and Oehlschlager and coworkers[16a,90] reported early examples of the reaction of epoxides with cyanosilylcuprates. Similarly, Fleming and coworkers[23] and Pulido and coworkers[38b] also showed some examples of cleavage of epoxides, carried out in connection with the silylcuprate-acetylene reaction. As expected, cuprate attacks the less hindered end of the epoxide moiety. Use of bulky silylcuprates reinforces this trend. Cleavage of vinyl epoxides is accompanied by allylic shift in a process that resembles that of allylic acetates. By using mixed alkylsilylcuprates, some transfer of the carbon ligand is usually observed (Scheme 34).

In a recent example Tamao and coworkers[48] and others[91] found that treatment of allylic cyclohexene-epoxides, such as those described in Scheme 9, with silylcuprate **10** carrying the Et_2NPh_2Si group proceeds with allylic rearrangement, leading to cyclohex-2-en-1,4-diols, while the corresponding silyllithium derivative (Et_2NPh_2SiLi) reacts by direct substitution, leading to cyclohex-3-en-1,2-diols. This route provides complementary procedures for the preparation of regioisomeric cycloalkenediols (Scheme 9).

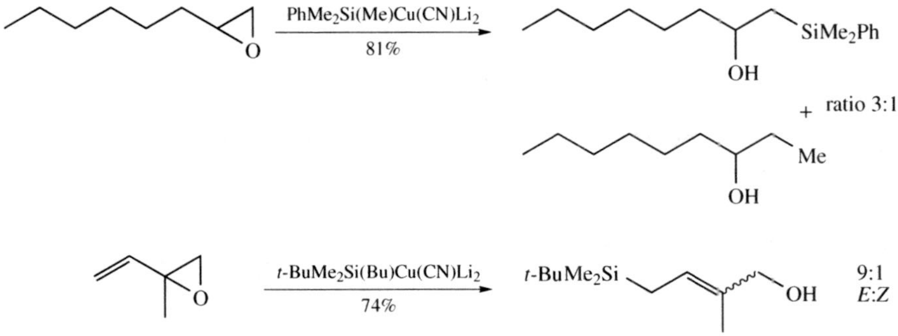

SCHEME 34

3. Reaction of silylcuprates with unactivated acetylenes and allenes

The silylcupration of unactivated acetylenes and allenes has been recently reviewed[12a] on the basis of the work of Fleming and Pulido. Silylcuprates react with acetylenes by *syn* stereospecific silyl-copper addition to the triple bond giving vinylsilanes in high yields[23, 92]. The intermediate vinylcuprate resulting from the initial addition to unsymmetrical acetylenes might consist of a mixture of regioisomers if the preference for the orientation of the Si−Cu pair is not favored in any sense. With terminal alkynes, the silyl group becomes attached to the terminal carbon atom, with a high level of regioselectivity. This result is the opposite of that which was expected by analogy with the corresponding carbocupration reaction[64], which shows a preference for transferring the carbo-ligand to the C-2 position of the acetylene. Silylcupration of a propargylsilane was used as a test for the reversibility of the reaction[92c].

The intermediate vinylcuprate **67** reacts with a wide variety of electrophiles, such as deuteron, iodine, acyl and alkyl halides, allyl halides, enones and epoxides to give vinylsilanes **68–73** (Scheme 35)[23]. Overall *syn*-addition of a silyl group and an electrophile to the acetylene is always observed. Protonation gives cleanly *E*-vinylsilanes. Cyanocuprate **1** gives better yield and selectivity than halide-based silylcuprates. Lower cyanosilylcuprates such as $PhMe_2Cu(CN)Li$ (**2**) are reported to give modest regioselectivity[25].

An interesting feature is that the silylcuprate reagents do not remove the proton from the alkyne as alkylcuprates do; this fact shows that silylcopper species are considerably less basic than carbanions[64]. When the acetylenic proton is deliberately removed from the hexyne by treatment with BuLi, the hexynyllithium obtained still reacts well with the silylcuprate reagent **1**, although not quite as regioselectively (10% of the regioisomeric 2-dimethylphenylsilylhex-1-ene is also formed). Quenching the intermediate cuprate **74** with D_2O resulted in deuterium incorporation in a 1,2-*trans* relationship, as is observed in compound **75**. This result accounts for the intervention of a trimetallated species vinyl−copper−lithium−silane (**74**) as shown in Scheme 36.

It is interesting to note that cuprate **67**, formulated as is shown in Scheme 35, has the stoichiometry of a mixed silicon−carbon cuprate and consequently we should have expected selective transfer of silicon to electrophiles as mixed silylalkylcuprates do. Instead, it transfers the carbon-based group, i.e. the vinylsilane moiety, to all the electrophiles. Since intermediate cuprates **67** were not characterized, it is very possible that the exact nature of these cuprates is not as simple as represented in this picture. The authors[9] were unable to define precisely the structure of the cuprate produced in the silylcupration and consequently one cannot be dogmatic about the exact constitution or stoichiometry

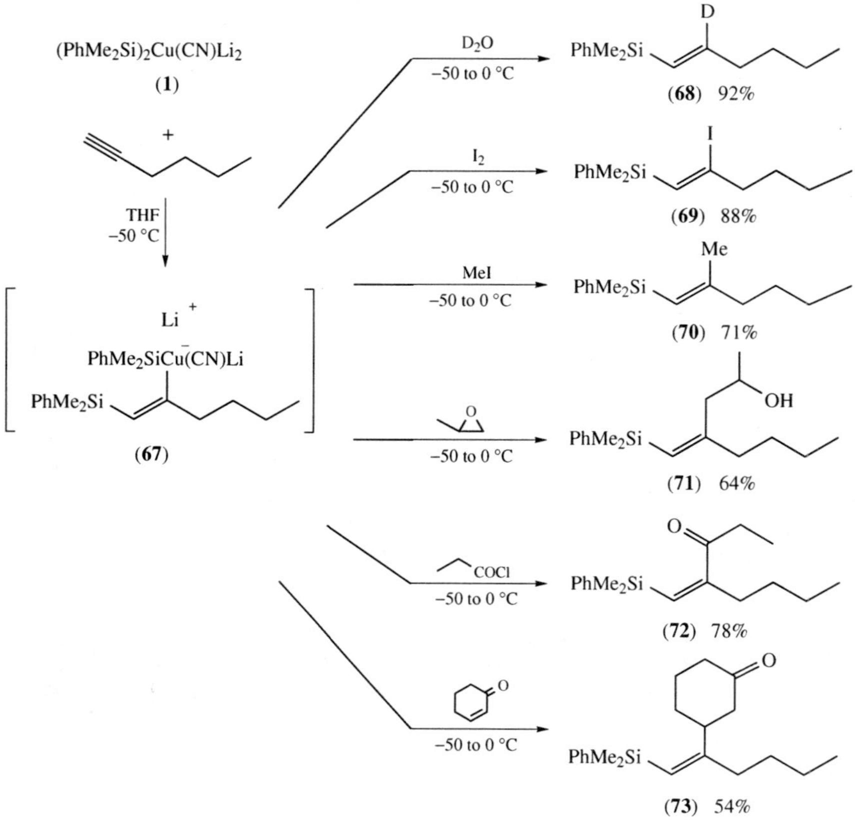

SCHEME 35

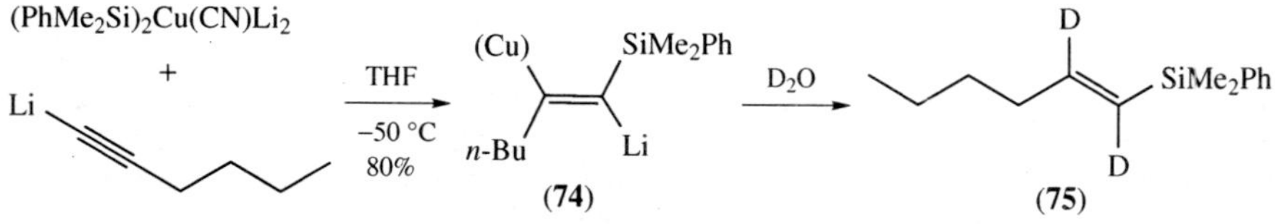

SCHEME 36

of the intermediates. In any case, the reaction is high yielding, remarkably selective and very reproducible and therefore there is no doubt about its synthetic interest.

Disubstituted alkynes also react with silylcuprates. The regioselectivity depends on how different in size are the substituents. With a methyl on one side and a branched group on the other, the regiocontrol is still highly in favor of the silyl group approaching the less hindered end. Reaction of alkynes with trimethylsilylcuprates was also used in a minor extension with similar results[92].

Silylcupration of acetylenes with the bulky lithium *t*-butyldiphenylsilylcyanocuprate (**4**) was studied by Pulido and coworkers in order to prepare vinylsilanes **77–82** carrying the hindered and relatively unreactive *t*-butyldiphenylsilyl group (Scheme 37)[38a]. Formation of intermediate vinylcuprate **76** takes place again by *syn*-addition of the cuprate to the triple bond, as can be deduced from the stereochemistry of the resulting vinylsilanes. Thus, addition to monosubstituted alkynes gives, after protonation, *E*-vinylsilanes (e.g. **79**), whereas addition to acetylene followed by reaction with electrophiles affords the *Z*-vinylsilanes **77** and **78**. The intermediate vinylcuprate **76** shows a normal reactivity toward simple electrophiles, leading to a wide variety of functionalized vinylsilanes with different substitution patterns (Scheme 37).

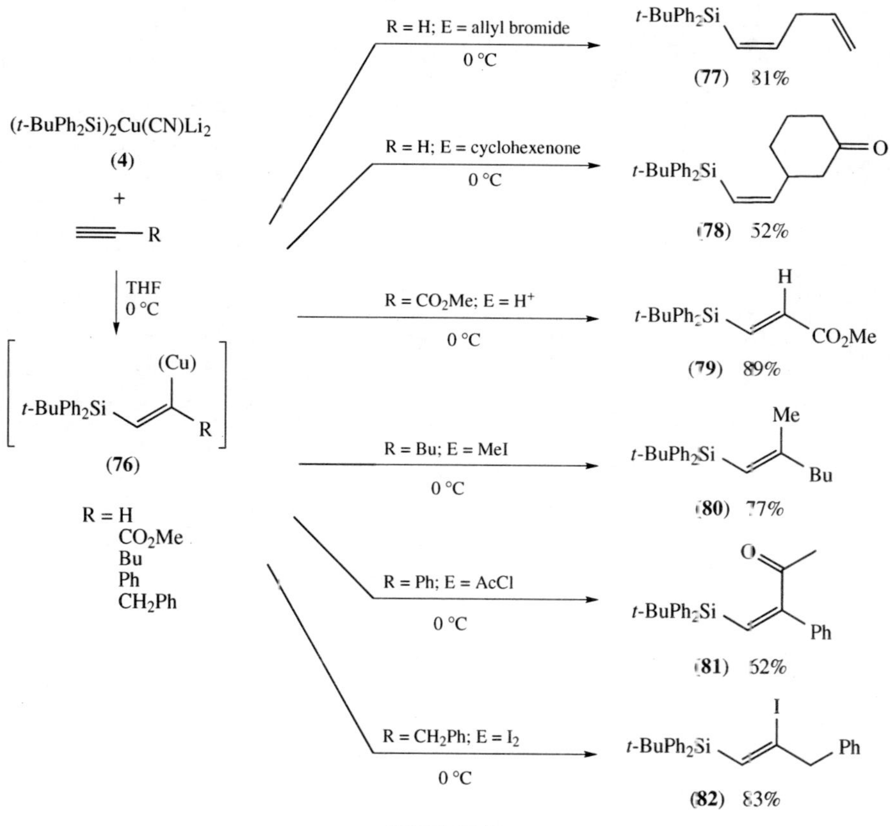

SCHEME 37

Using this strategy[38a] it has been possible to prepare polysilylated and stannylated alkenes **83–91** of known configuration (Scheme 38). Since trimethylsilyl or dimethylphenylsilyl groups may be introduced as the electrophile (Me_3SiCl or $PhMe_2SiCl$) or may be the resident group on acetylene, it is possible to make either stereoisomer of differentially disilylated 1,2-disilylethylenes (**83**, **86**) and (**88**, **91**). It is also easy to make the differentially metallated alkenes **85**, **87**, **89** and **90** having one silyl and one stannyl group.

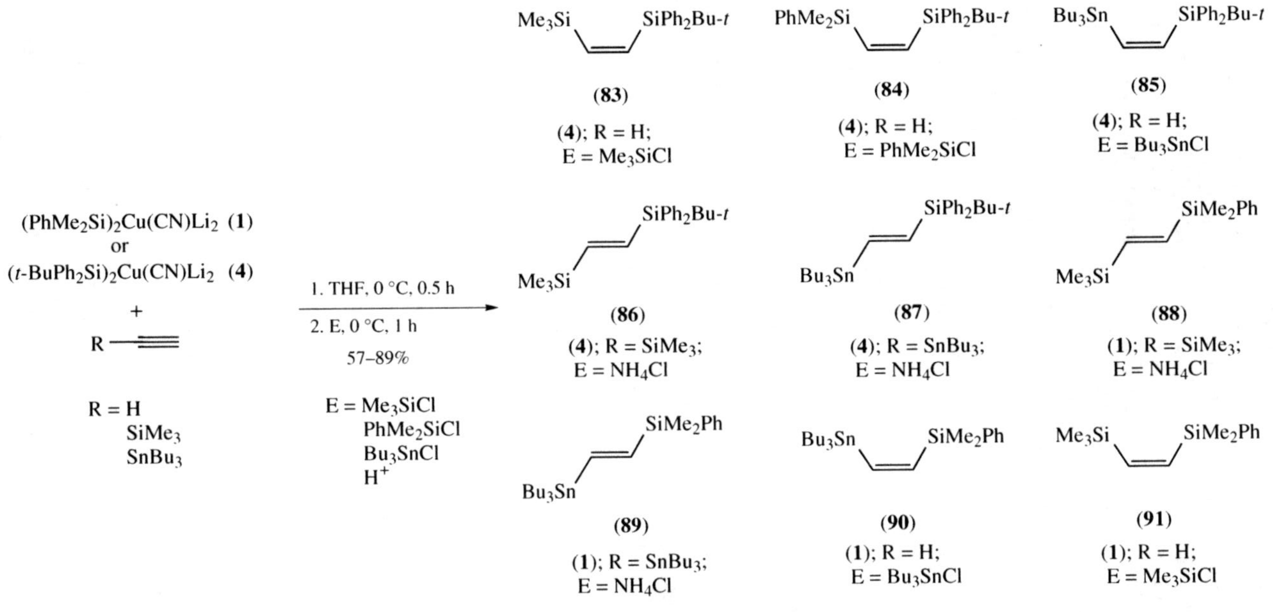
(PhMe2Si)2Cu(CN)Li2 (1)
or
(t-BuPh2Si)2Cu(CN)Li2 (4)
+
R
R = H
SiMe3
SnBu3
1. THF, 0 °C, 0.5 h
2. E, 0 °C, 1 h
57–89%
E = Me3SiCl
PhMe2SiCl
Bu3SnCl
H+
Me3Si
SiPh2Bu-t
(83)
(4); R = H;
E = Me3SiCl
PhMe2Si
SiPh2Bu-t
(84)
(4); R = H;
E = PhMe2SiCl
Bu3Sn
SiPh2Bu-t
(85)
(4); R = H;
E = Bu3SnCl
Me3Si
SiPh2Bu-t
(86)
(4); R = SiMe3;
E = NH4Cl
Bu3Sn
SiPh2Bu-t
(87)
(4); R = SnBu3;
E = NH4Cl
Me3Si
SiMe2Ph
(88)
(1); R = SiMe3;
E = NH4Cl
Bu3Sn
SiMe2Ph
(89)
(1); R = SnBu3;
E = NH4Cl
Bu3Sn
SiMe2Ph
(90)
(1); R = H;
E = Bu3SnCl
Me3Si
SiMe2Ph
(91)
(1); R = H;
E = Me3SiCl

SCHEME 38

Polymetallated alkene chemistry is a promising area of research due to the differential reactivity and chemoselectivity that the groups attached to the alkene can impart[38c].

In comparative tests, the *t*-butyldiphenylsilyl group shows some properties that are similar to those of the $PhMe_2Si$ group (protonation, bromine-desilylative elimination, allyl- and vinylsilane electrophilic substitution) and others that are usefully different from those of relatively unhindered silyl groups Me_3Si and $PhMe_2Si$. Since *t*-butyldiphenylsilyl group does leave too easily, it has found some applications in the transformation of silyl epoxides. Epoxides **92** derived from vinylsilanes, prepared as described in Scheme 37, undergo acid-catalyzed rearrangement by treatment with a protic acid giving aldehydes **93**, which retain the α-silyl group (Scheme 39)[93]. Contrary to other reported examples for common silyl groups (Me_3Si, $PhMe_2Si$), where the Stork–Colvin reaction occurs predominantly[94], it should be noted that in this case the α-silyl group is not lost, probably due to the bulky size and low electrofugal properties of the *t*-butyldiphenylsilyl group. The silyl aldehydes thus obtained arise from mechanisms involving cleavage of the α C–O bond followed by alkyl migration. They have been used for a highly stereoselective alkene synthesis by way of an *erythro*-β-silyl alcohol **94** and Peterson elimination, where the nucleophilic attack on the aldehyde is Cram-selective and the *t*-butyldiphenylsilyl group is evidently still capable of taking part in the *syn*-stereospecific elimination step to give **95** (Scheme 39).

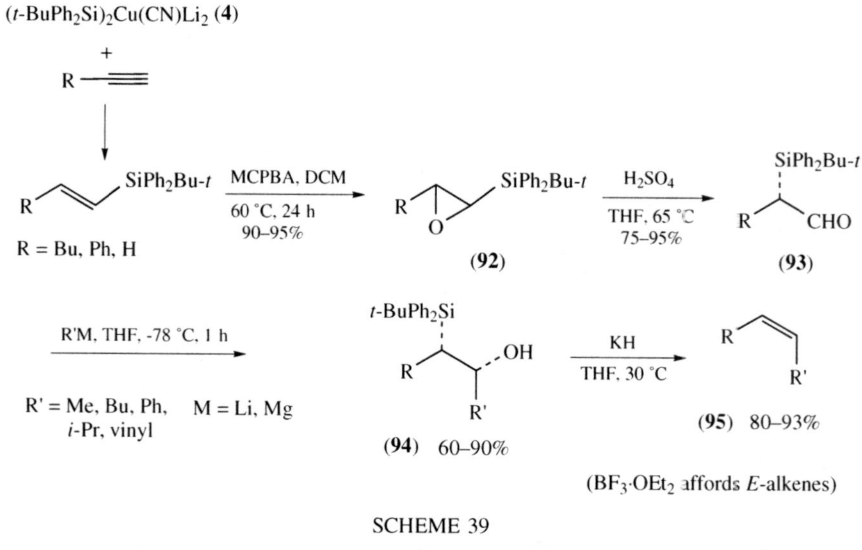

SCHEME 39

Vinylsilanes prepared according to the procedures above have many uses as carbon nucleophiles in synthesis[12a]. An easy transformation is desilylation with bromine, which takes place with inversion of configuration of the double bond[43]. This method was used to convert a long-chain C_{13} *E*-vinylsilane, made from 1-tridecyne, into the corresponding *Z*-vinyl bromide, which was transformed into a *cis*-vinylcuprate needed in a synthesis of tetrahydrolipstatin[95a] (Scheme 40). Vinylsilanes have also been used in a synthesis of ebelactone A[95b].

Silylcupration has also been carried out using 1-aminoalkynes[96a], 1-phosphinoalkynes[96b], propargyl sulfides[78a] and propargyl ethers[97]—where it was used in a synthesis of (+)-crotanecine—and propargyl amines[17,98], in this case used for a synthesis of γ-silyl-α-amino acids **96** (Scheme 41). Most of the work in this area has been reported by

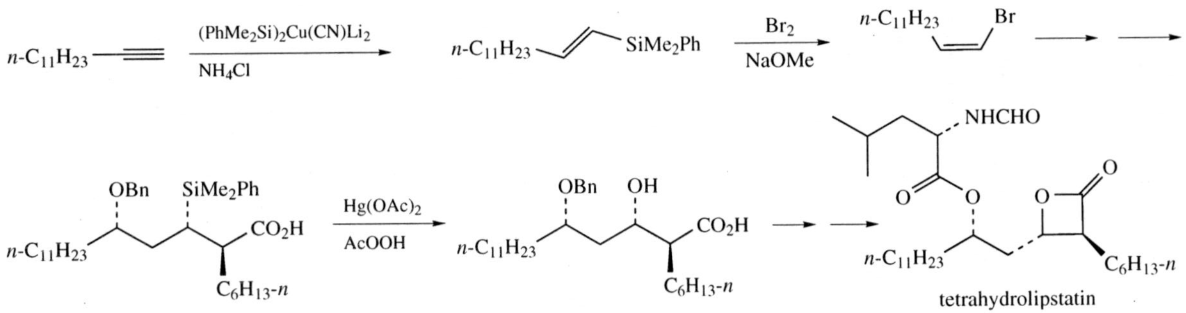
n-C11H23
(PhMe2Si)2Cu(CN)Li2
NH4Cl
SiMe2Ph
Br2
NaOMe
Br
OBn
SiMe2Ph
CO2H
C6H13-n
Hg(OAc)2
AcOOH
OH
NHCHO
O
tetrahydrolipstatin

SCHEME 40

Ricci and coworkers, who have taken advantage of the silylcupration of highly functionalized alkynes for the synthesis of amino acids and related compounds[99]. The intermediate *cis*-trimethylsilylalkenylcuprate obtained from propargyl amines has been trapped with a wide range of electrophiles such as vinyl, allyl and propargyl halides, 2-halothiophenes, carbon dioxide, methyl chloroformate and iodine, with yields ranging from 60 to 90%[17]. The vinylsilanes obtained in these reactions can also participate in Heck reactions, leading to stereodefined aryl-substituted olefins[100].

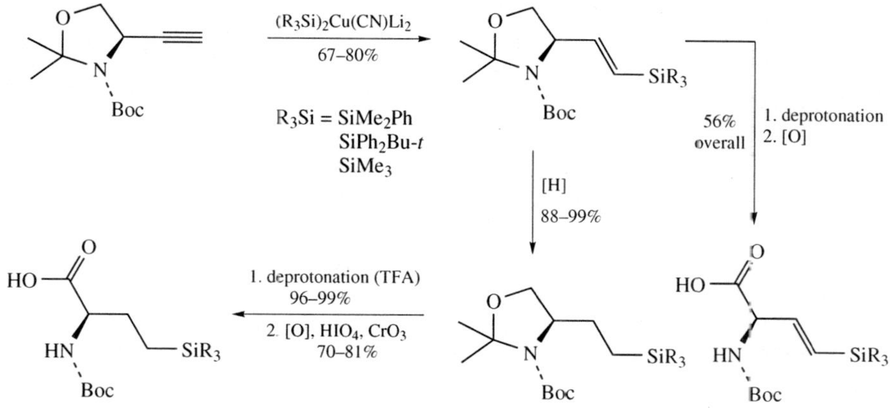

SCHEME 41

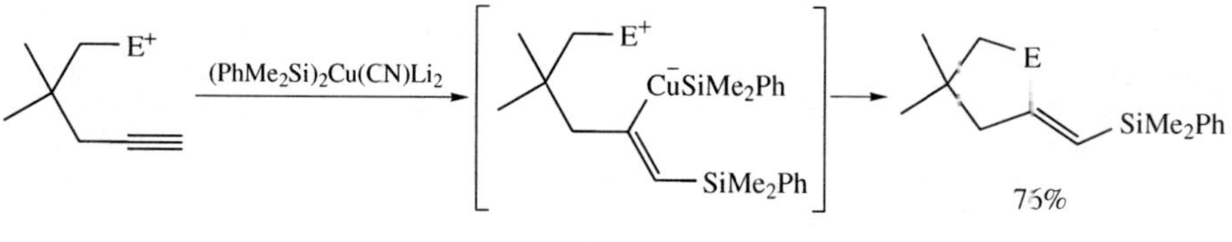

SCHEME 42

Intramolecular reaction of intermediate vinylcuprates with electrophilic units contained within the same molecule provides good chances for ring formation (Scheme 42).

The success of the cyclization depends upon the relative reactivity of the silylcuprate, the electrophilic function used and the nature of the alkyne. Terminal ω-alkynes are more reactive and very convenient to get control on the regiochemistry. The silylcupration-cyclization reaction has been carried out with alkynyl tosylates and mesylates, epoxides and ketones (Schemes 43 and 44)[101,102].

Yields go from acceptable to low, but even if modest yields are obtained, the reaction is worthwhile to attain exocyclic alkenylsilanes, which are useful synthons not easily available. Reaction of alkynyl epoxides and ketones with lithium cyanosilylcuprates is illustrated in Scheme 43[101] and the reaction of alkynyl sulfonates with silyl Grignard reagents in the presence of CuI is depicted in Scheme 44[102].

Although the silyl Grignard-copper-catalyzed reaction has been described as involving attachment of silylmagnesium species to the triple bond, due to the presence of CuI it is feasible that the actual process is an organocuprate addition. Oshima and coworkers

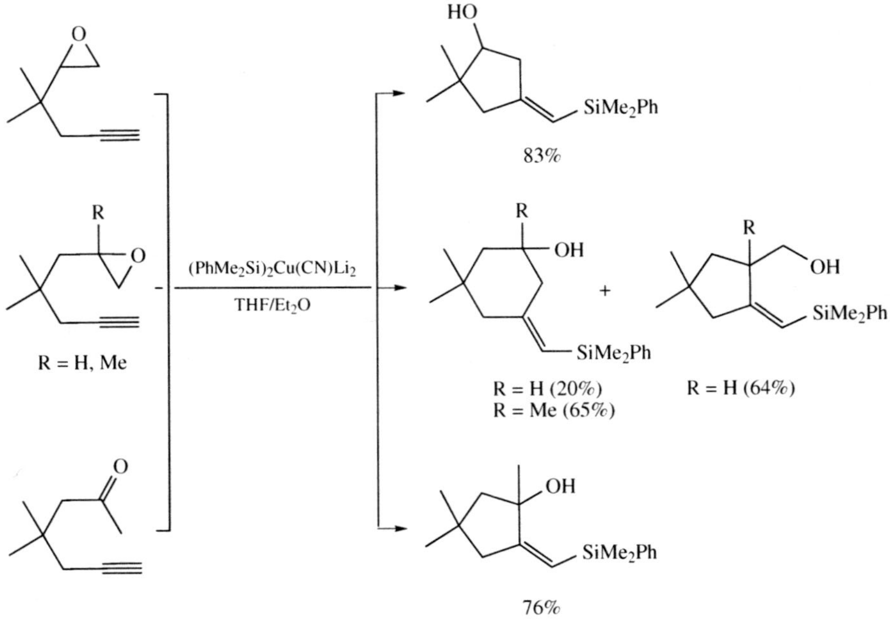

SCHEME 43

have reported more examples of addition of silicon–magnesium, silicon–aluminum and silicon–zinc bonds to acetylenes, catalyzed by copper(I)[103]. The species involved in the catalytic cycle are not clear.

Addition of the cuprate to the acetylene moiety is sometimes a competing reaction leading to endo- or exocyclic alkenylsilanes; this situation can be seen in the first example of Scheme 44, where the relative formation of 3- versus 4-membered rings seems to be balanced by both the ring strain and the so-called Thorpe–Ingold effect, the latter favoring formation of the smaller ring. Competitive processes leading to alkenylsilane-containing rings of different size are also observed in the intramolecular cyclization of alkynyl epoxides. Attack of the intermediate vinylcuprate can take place at both electrophilic centers of the epoxide, as seen in the second example shown in Scheme 43, giving mixtures of 5- and 6-membered rings in a ratio that depends mainly on the substitution pattern of the starting epoxide. An increase of alkyl substitution on the epoxide enhances formation of the larger ring by intramolecular attack of cuprate to the less hindered end of the epoxide; on the other hand, the presence of a *gem*-dimethyl group in the backbone facilitates formation of the smaller ring due to the Thorpe–Ingold effect. Some transfer of the silicon ligand instead of the vinylsilane carbon ligand has been sometimes observed in these reactions. It can be minimized using a less reactive mixed cuprate such as $PhMe_2Si(Me)Cu(CN)Li_2$.

Bäckvall has reported that silylcupration of acetylenes with the lower-order cuprate $PhMe_2SiCu(CN)Li$ followed by addition of allyl phosphates gives silylated 1,4-dienes[104].

In conclusion, the silylcupration of alkynes is generally faster than the silylcupration of sulfonates, ketones and epoxides and can be used successfully in cyclization reactions where the electrophilic partner is one of the former functions. Comparable rates have been observed for the 1,4-addition to enones and therefore in this case their use in cyclization is limited to some particular examples. Finally, silylcupration of alkynes is slower than reaction of the cuprate with allylic esters with which cyclization reaction fails.

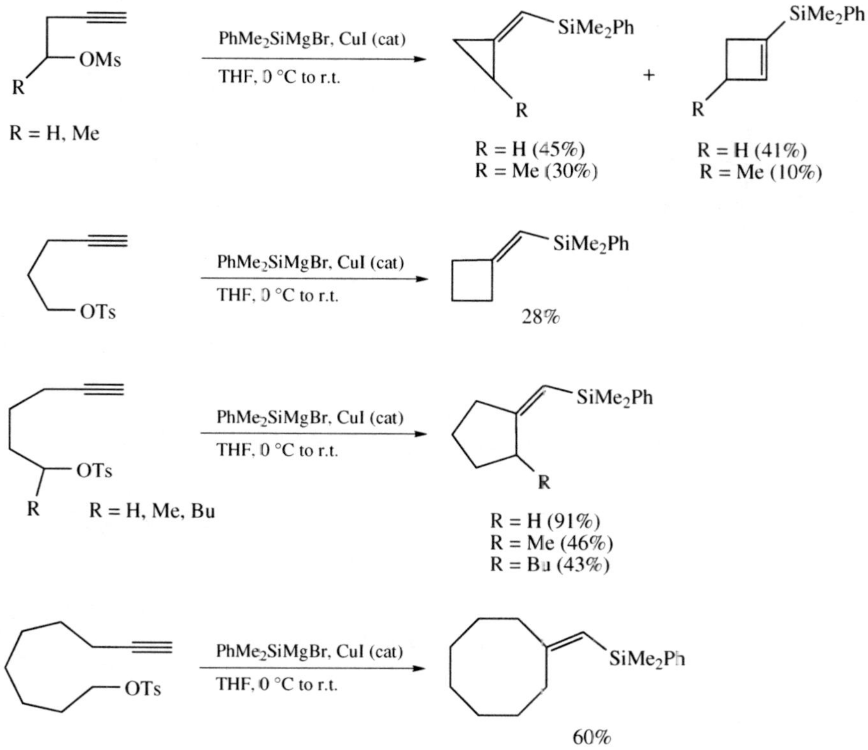

SCHEME 44

Silylcupration of allenes was first reported[9,105] by Fleming's group and later developed by Pulido and coworkers[30b,38b,106–108]. Unactivated allenes react with silylcuprates even at low temperatures, in contrast with the corresponding carbocupration of allenes which is unknown, except when activating substituents, either carbonyl[109] or alkoxy[110] are present. With allene itself and the higher-order cyanocuprate **1**, the overall regiochemistry places the silyl group at the central carbon and copper at the end carbon (Scheme 45).

The intermediate allylcuprate, indefinitely formulated as **97**, with the stoichiometry shown in Scheme 45, is easily captured by a wide variety of electrophiles giving differently substituted and functionalized vinylsilanes[9]. Enones give 1,2-addition rather than conjugate addition. There is a striking exception: iodine, which affords the allylsilane-vinyl iodide **99**, thus suggesting the possibility that regioisomeric intermediates **97** and **98** are in equilibrium. The most obvious explanation for this surprising change in regioselectivity is that silylcupration is a rapid and reversible reaction, with allylcopper-vinylsilane species **97** being more abundant or reactive than vinylcopper-allylsilane species **98**. In any case, whatever the explanation, the overall regiochemistry in the silylcupration of allene is easily controlled in either sense because lithiation (by BuLi) of the vinyl iodide followed by reaction with electrophiles leads to the regioisomeric allylsilanes (Scheme 46).

Substituted allenes give very selective silylcuprate-addition reactions. In fact, the degree of regio- and stereoselectivity of every kind seen in these reactions is remarkable, in view of the opportunity the reaction presents for the formation of mixtures of isomers.

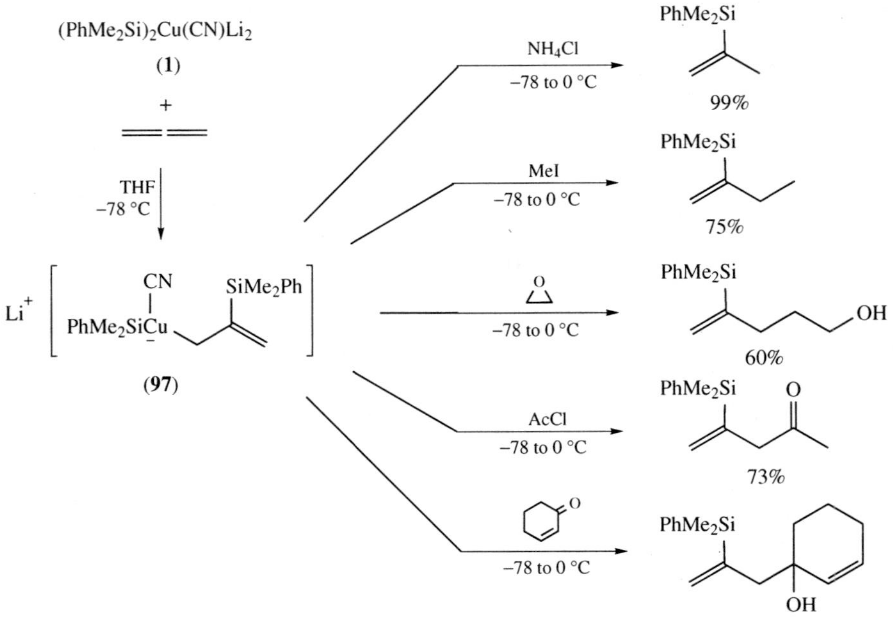

SCHEME 45

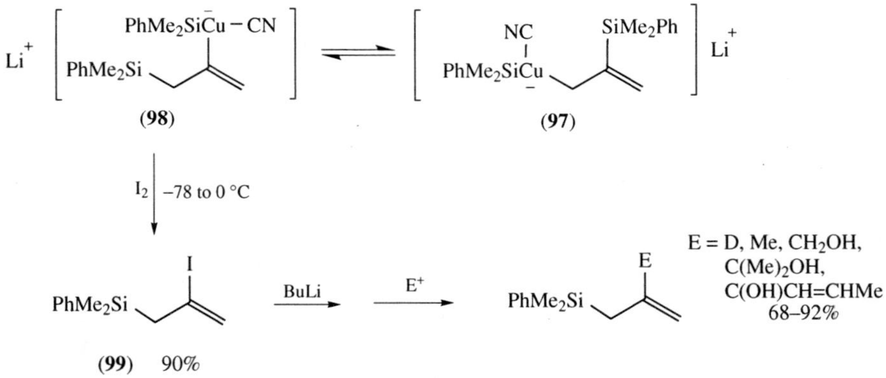

SCHEME 46

Substituted allenes fall into two groups, one giving vinylsilanes (and allylcopper intermediates) and the other giving allylsilanes (and vinylcopper intermediates). Phenyl-substituted allenes give mainly or exclusively vinylsilanes through a phenyl-stabilized allyl-cuprate, whereas alkyl-substituted allenes afford selectively allylsilanes because alkyl groups destabilize the allyl-cuprate regioisomer[9]. With unsymmetrically substituted allenes, addition takes place selectively by attack on the silylcuprate at the less substituted end of the allene, approaching from the less hindered face (*anti* to the phenyl group in Scheme 47). Silylcupration of allenes was also achieved with trimethylsilylcuprates leading to comparable results, with the inconvenience that HMPA must be present in the reaction mixture[9,105].

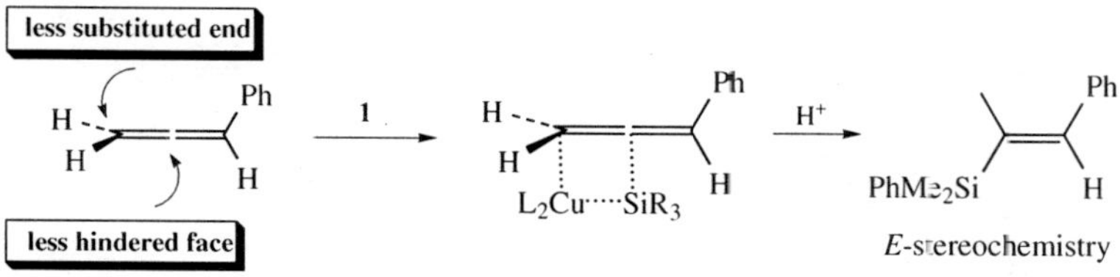

SCHEME 47

The metallocupration step is stereoespecifically *syn*[111]. This fact has been unequivocally proven by reaction of stereoisomeric cyclohexanyl(methyl)allenes, such as **100** and **101**, with silylcuprate **1** and identification of the stereochemistry of the resulting allylsilanes **102** and **103**. Both examples proceed with a high level of stereoespecifity and high yield (Scheme 48).

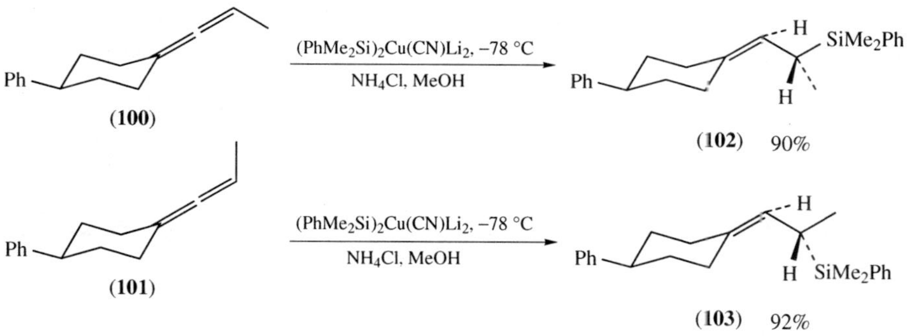

SCHEME 48

The reaction of *t*-butyldiphenylsilylcyanocuprate (**4**) with allene shows a different regioselectivity from that of $(PhMe_2Si)_2Cu(CN)Li_2$ (**1**), giving allylsilanes with all the electrophiles tested (Scheme 49). It seems likely that the change of regiochemistry may be a consequence of the steric preference of the bulky *t*-butyldiphenylsilyl group for the end of the allenic system. Again, alkyl-substituted allenes react unexceptionally giving allylsilanes. This route is an easy entry to the synthesis of diversely substituted allylsilanes carrying the voluminous *t*-butyldiphenylsilyl group[38b].

The intermediate allylsilane-vinylcuprate species **104** is not regiochemically stable. The results collected in Scheme 49 are those obtained when the reaction is carried out under kinetic control (−78 °C). When the reaction of **4** with allene was performed at CO_2-acetone temperature, and then the mixture was warmed to 0 °C before quenching, the only isolated product was vinylsilane **106**, in contrast with the regioisomer allylsilane **105** obtained by protonation of the low-temperature cuprate (Scheme 50). Presumably, the addition of the silylcuprate **4** is reversible. Thus the overall regiochemistry for the reaction of allene with **4** is easily controlled in either sense by the temperature. Unfortunately, the high-temperature cuprate (presumably an allylcopper intermediate regioisomeric of **104**) produced at 0 °C does not react with electrophiles apart from proton. As reported above for phenyldimethylsilylcuprates, the nature of the copper-containing intermediates is a matter of uncertainty and the copper intermediates cannot be as simple as they are represented here[38d].

Lower-order cyanosilylcuprates (**2**) and *t*-$BuPh_2SiCu(CN)Li$ (**5**), prepared by mixing one equivalent each of silyllithium and copper cyanide, also react with allenes, showing

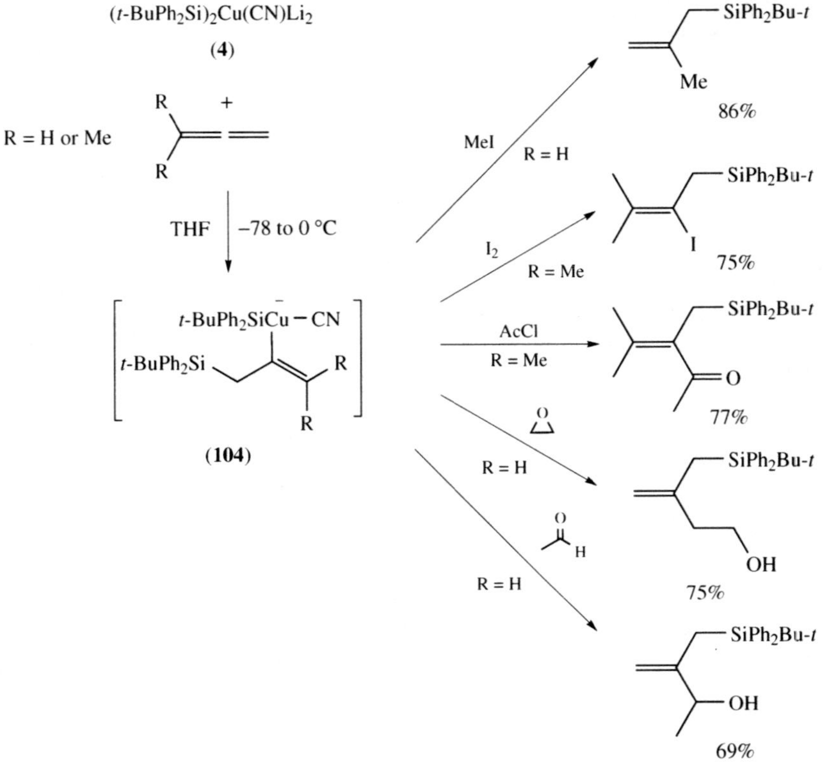

SCHEME 49

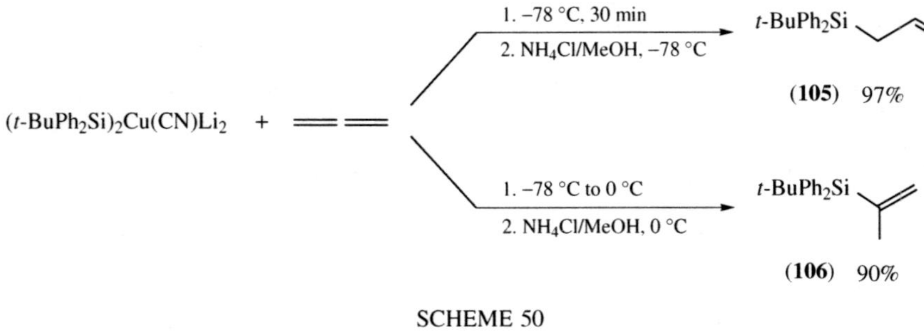

SCHEME 50

some interesting singularities. The allylsilane-vinylcuprate species **107** generated at low temperature by reaction of allene with **2** (or **5**) react with most of the electrophiles tested: proton, alkyl and allyl halides, halogens, acid chlorides, epoxides and α,β-unsaturated aldehydes and ketones, giving good yield of allysilanes differently functionalized[30b]. In particular, 2-enones undergo selectively 1,4-addition when treated with silylcuprates **2** and **5** (Scheme 51). This contrasts with the result obtained with higher-order cuprates **1** and **4**, where products of 1,2-addition are frequently observed (Scheme 45). It should be noted that the regiochemistry observed with cuprate **2** (Scheme 51) is opposite to that

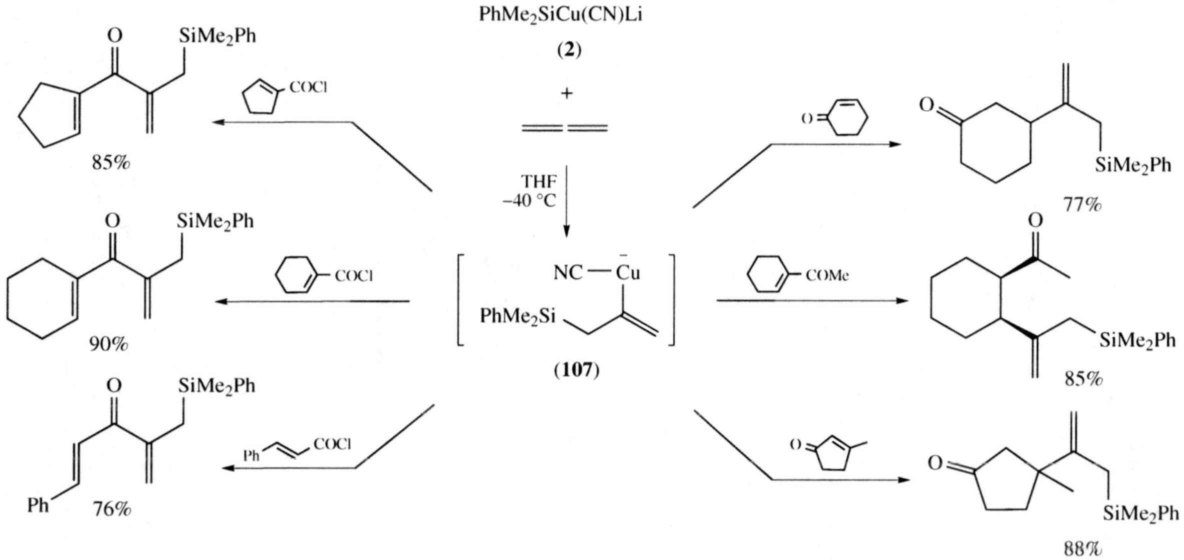
PhMe$_2$SiCu(CN)Li
(2)
+
THF
−40 °C
NC — Cu
PhMe$_2$Si
(107)
COCl
COCl
Ph
COCl
COMe
O
SiMe$_2$Ph
85%
90%
76%
77%
85%
88%

SCHEME 51

found for cuprate **1** (Scheme 45). Alternatively, use of the cyanosilylcuprate reagent **2** under thermodynamic control (warming-up the reaction to 0 °C before quenching) affords the allylcuprate-vinylsilane intermediate, which, once more, cannot be trapped with electrophiles other than proton[30b].

The reactions between intermediate **107**, formed by reaction of allene itself and (**2**), and acid chlorides, enals and enones provide excellent routes for silicon-directed Nazarov cyclizations or Lewis-acid-promoted intramolecular cyclizations, thus allowing the design and development of new strategies for cyclopentane annulations. Moreover, these silicon-assisted cyclizations occur with a high degree of stereocontrol. A simple protocol for triquinane systems based on the consecutive application of (a) silylcupration of allene, (b) slicon-directed Nazarov reaction and (c) allylsilane-terminated cyclization has been outlined for the synthesis of a precursor of β-capnellenol (Scheme 52).

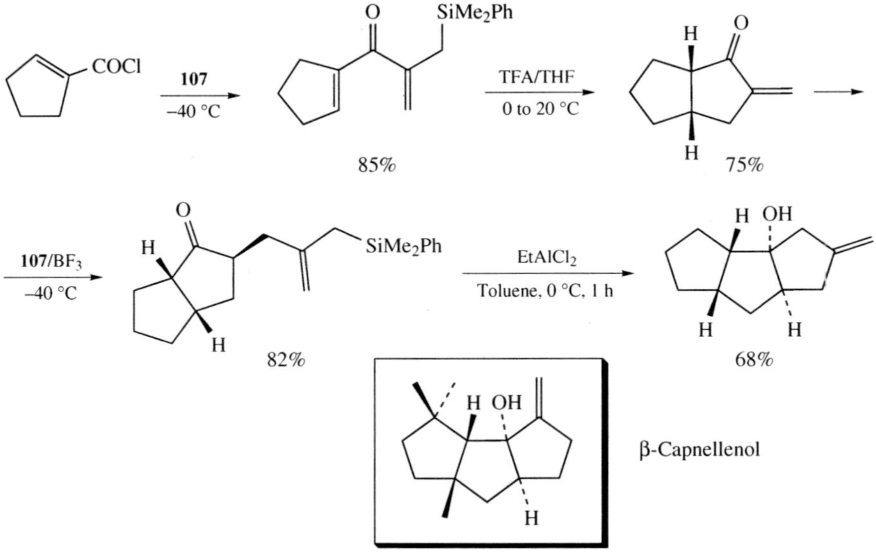

SCHEME 52

In recent years, many stereoselective routes for the synthesis of three- to seven-membered rings from allylsilanes, based on the silylcuprate addition to allene, have been developed[112–118].

Silylmetallation of cyclic allenes was reported by Oshima and coworkers some time ago[119]. The addition of silylmagnesium, silylaluminum and silylzinc reagents to unactivated cyclic allenes catalyzed by copper(I) salts seems to proceed smoothly, giving vinylsilanes. Although the authors claim that the reaction takes place with addition of the Si−Mg, Si−Al and Si−Zn species across the double bond, the presence of the copper salts seems to be more consistent with the intervention of a silylcuprate. Silylmetallation of cyclic allenes bearing a good leaving group provides a simple route to bicyclic compounds having a vinylsilane moiety that can be subsequently transformed (Scheme 53).

More recently Bäckvall and coworkers[120] found that silylcupration of terminal allenes with the cuprate **2** followed by treatment of the intermedite allylsilane-vinylcuprate with allyl phosphonates provides a facile synthesis of silylated 1,4-dienes containing an allylsilane unit. Reaction is highly regioselective (in a S_N2' manner) and isomerization to

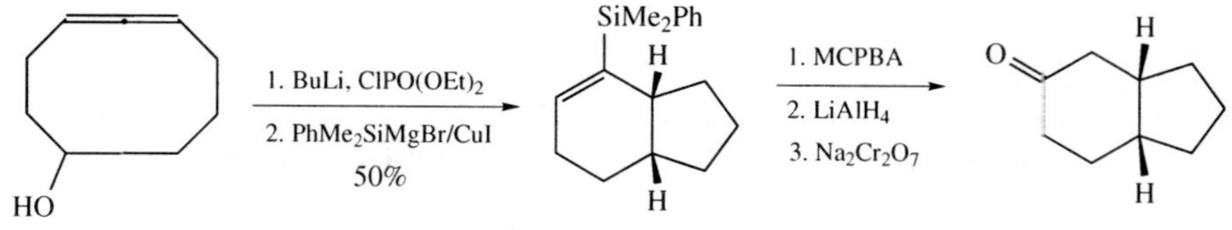

SCHEME 53

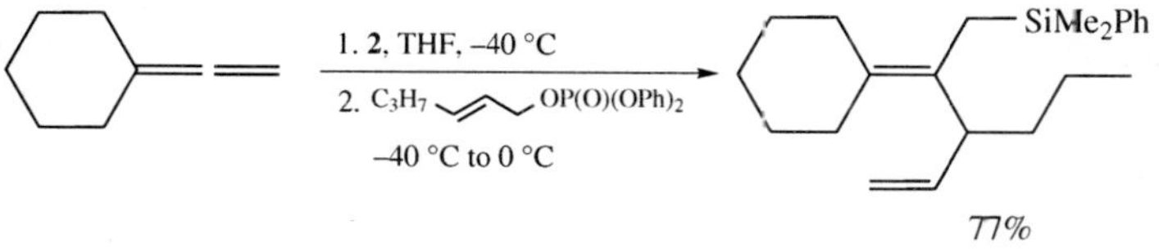

SCHEME 54

1,3-dienes was not observed. Nevertheless, there is some *Z* to *E* double-bond equilibration in some of the studied examples (Scheme 54).

The same group has discovered that the scope of silylcupration of multiple bonds is not solely restricted to acetylenes and allenes but can be also extended to simple double bonds such as styrenes[10] and 1,3-dienes[11].

III. STANNYLCUPRATES

A. Preparation of Stannylcuprates

1. Introduction

The demand for organotin compounds in organic chemistry has increased greatly in the last two decades, due to the large number of transformations that the C–Sn bond can undergo[121]. The development of new methodologies that allow the synthesis of these synthons has attracted the attention of the scientific community. One of the most interesting ways of forming C–Sn bonds is by stannylmetallation of an electrophilic function. In particular, stannyllithium derivatives and mainly stannylcopper derivatives have been used for this purpose. This methodology provides efficient procedures for introducing tin into organic molecules, giving rise to a wide variety of tin-containing carbon skeletons[12b].

Alternatively, the development of new heteroatom copper or cuprate reagents has significantly increased the utility of copper(I) chemistry. Among them, metallocuprates from the group IVA, and in particular stannylcuprates, have gained considerable importance in synthesis. The availability of cuprate reactivity patterns for the introduction of a stannyl substituent into an organic substrate has dramatically enhanced the strategic approaches to these compounds, the rich chemistry of which can be exploited in C–C bond formation and in chemocontrol, regiocontrol and stereocontrol.

Stannyl anions are even better nucleophilic species than silyl anions, in fact, stannyllithium compounds readily react with carbonyl derivatives giving conjugate addition to enones[122] and enoate[123] systems. Therefore, stannylcuprates are rarely used for this purpose. While stannyllithium reagents have been largely employed in reactions with saturated and unsaturated carbonyl compounds[121], stannylcuprates are most often used with acetylenic substrates[12b]. Pioneering work in this area was accomplished by Piers and Morton[7b]. Piers has been particularly active in showing how powerful the stannylcupration of alkynes can be.

The substituents on copper most often used are the trimethylstannyl group and the tributylstannyl group. Trimethylstannyl derivatives are easier to manipulate and lead to simple spectra, but they are highly toxic and therefore not suitable for scaling a regular lab synthesis into an applied commercial procedure. Thus, even though trimethylstannylorganocuprates frequently give better yields, the most widely used reagents are tributylstannylcuprates. Triphenylstannylcuprates[124] have been less frequently used. Halide-based cuprates or cyanocuprates have been employed indistinctly. The copper(I) bromide–dimethyl sulfide complex is a good choice because complexation on copper usually results in cleaner reactions. Compounds with different copper-to-tin stoichiometries leading to lower- or higher-order cuprates can be prepared (Section III.A.2). In general, cyanostannylcuprates and higher-order stannylcuprates show a greater reactivity and sometimes they are the only possible choice to get reaction. This is especially true with acetylenes, where the nature of the stannylcuprate determines the outcome of the process (Section III.B.4). Since one stannyl group is not used, reaction with bis(stannyl)cuprates produces, after quenching the mixture, involatile tin by-products (H−Sn and HO−Sn) which are difficult to eliminate and need further chromatographic purification. Mixed stannylalkylcuprates, although less reactive, diminish undesired by-products, usually giving better overall results. Tributyl- and trimethylstannylcuprates are sometimes noticeably different in their reactivities.

Structurally different species have been characterized by NMR spectroscopy for various lower- and higher-order cuprate reagents[22,25]. Methyl ^{13}C signals at −4.5 ppm of the lower-order trimethylstannylcyanocuprate $Me_3SnCu(CN)Li$ are shifted upfield with respect to the corresponding methyl ^{13}C signals of the higher-order bis(trimethylstannyl) cyanocuprate $(Me_3Sn)_2Cu(CN)Li_2$ at −0.04 ppm. The respective mixed stannyl(alkyl) cyanocuprate $Me_3Sn(Me)Cu(CN)Li_2$ is also different, appearing at −2.0 ppm. Spectroscopic studies[22] on mixed cyanostannylcuprates reveal a rapid and dynamic ligand exchange with the stoichiometric composition $R_3Sn(R')Cu(CN)Li_2$ as the thermodynamic stabilized species.

2. *Methods of preparation*

Like silylcuprates, stannylcuprates are usually prepared from the respective stannyllithium compounds (Scheme 55). Trimethyl- and tributylstannyl lithium can easily be made by reaction of trialkylstannyl halides with lithium shots[125], by cleavage of hexaalkylditin derivatives with methyllithium or butyllithium[122,123,126], by cleavage of the same ditin compounds with lithium metal[127] or by reaction of trialkyltin hydride with lithium diisopropylamide[126a–128] (this method is less convenient and has some limitations due to *in situ* formation of i-Pr_2NH). Reaction of the so-formed stannyllithiums with a copper(I) salt, using 1:1 or 2:1 stoichiometries, leads to mono- or bis(stannyl) cuprates[22,125b]. Homocuprates are not always advantageous; when needed, mixed stannyl (alkyl)cuprates[129] can be easily made by mixing one equivalent of the stannyllithium reagent with another equivalent of a different organolithium and one equivalent of the copper(I) salt (Scheme 55). These mixed stannyl(alkyl)cuprates transfer selectively the stannyl group[129], as expected from their relative electronegativity. They usually give cleaner reactions and one expensive tin group is economized. As reported before for silylcuprates, cyanostannylcuprates are more reactive than halide-based stannylcuprates, the former showing frequently a different reactivity pattern[125b].

On the other hand, Lipshutz and coworkers have developed methods for preparing mixed stannyl(alkyl)cuprates without using stannyllithiums reagents, by treating either tin hydrides[130,131] or silylstannanes[32,33] with dialkylcuprates (Schemes 56 and 57). When silylstannanes are used, it is required for the effectiveness of the reaction that the silyl group is much less hindered than the stannyl group.

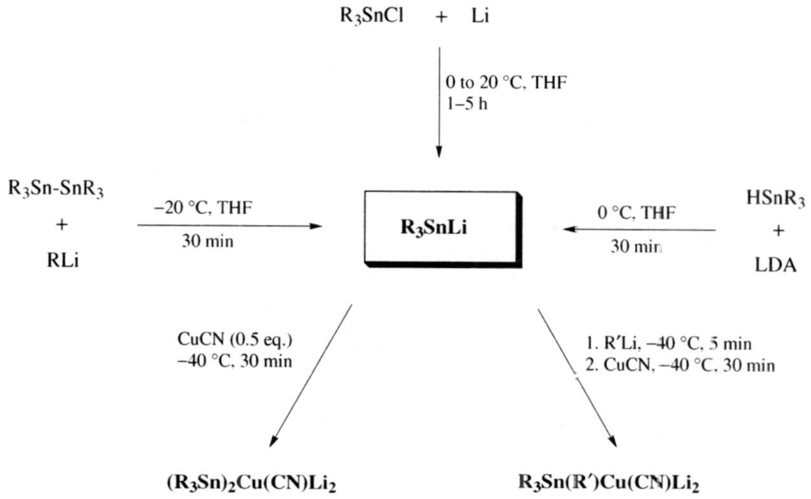

SCHEME 55

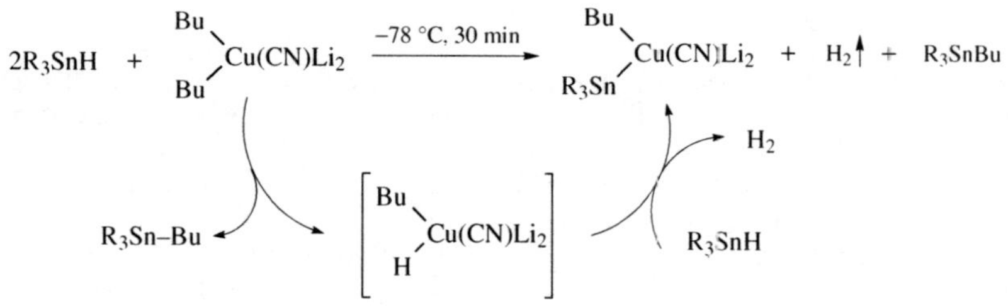

SCHEME 56

Oehlschlager and coworkers[132] have reported a similar procedure using hexaalkylditin derivatives (Scheme 57), while similar mixed alkylstannylcuprates have been also prepared by ligand exchange between homoalkylcuprates and homostannylcuprates. The ability of dialkylcuprate reagents to undergo transmetallation with organotin compounds provides a wide scope of alternatives for the preparation of stannylcuprates (Scheme 57)[133].

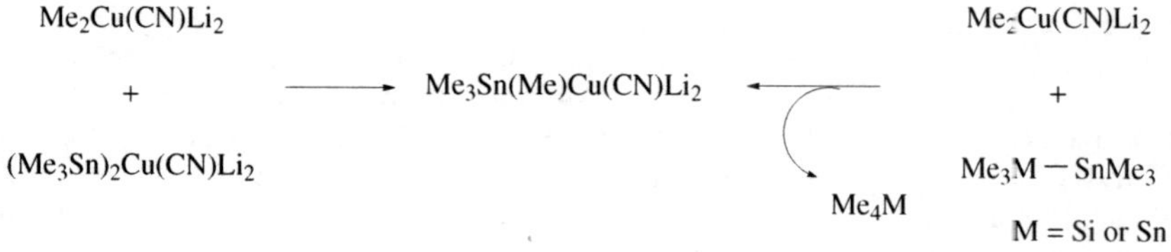

SCHEME 57

Mixed cuprates having a silyl and a stannyl group have been recently prepared[134]. In particular, phenyldimethylsilyl(tributylstannyl)cuprate (**108**) made by mixing 1 mol equivalent each of phenyldimethylsilyllithium, tributylstannyllithium and copper(I) cyanide

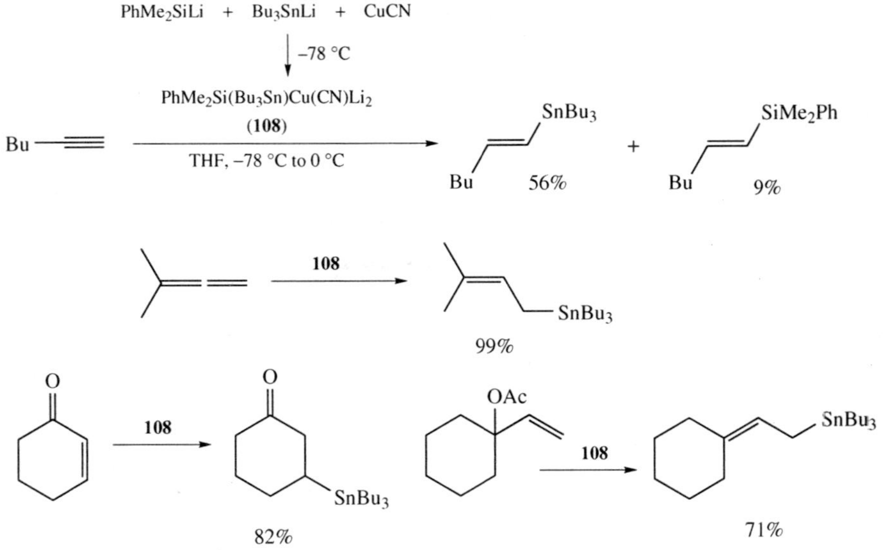

SCHEME 58

reacts with electrophiles (enones, allylic acetates, allenes and acetylenes), transferring selectively the stannyl group (Scheme 58). Except for the addition to 1-hexyne, where 9% of the vinylsilane is formed, the product mixtures show by GC-MS no trace of the corresponding products of silyl transfer. This is what should be expected according to the higher metallic character of the tin atom.

Although in this account formulations such as $(R_3Sn)_2Cu(CN)Li_2$, or the corresponding mixed cuprate $R_3Sn(R')Cu(CN)Li_2$, are considered higher-order compositions and formulations $R_3SnCu(CN)Li$ and $(R_3Sn)_2CuLi$ lower-order structures, earlier NMR studies[22] reveal multiple species in solution, depending upon the $R_3Sn/R'Li/CuCN$ stoichiometry and the nature of the copper salt (halide or cyanide). Thus, alternative complexation arrays are possible and ligand exchange, sometimes faster than the NMR timescale, doesn't help to clarify structural proposals[16a]. Therefore, formulas used in this survey reflect more the stoichiometry used in the original literature than the actual higher- or lower-order nature of stannylcuprates.

Structures of stannylcuprates are not well known, although probably they reflect those that were proposed for carbocuprates. A few structures of silyl- and stannylcuprates have been confirmed by X-ray analysis[135], but probably more work is needed before we can determine more precisely the nature of stannylcuprates in ethereal solution. However, stannyllithiums are better understood. Me_3SnLi is monomeric in ether. If HMPA is added to a THF solution of Bu_3SnLi, NMR spectra show that the first species formed is $Bu_3SnLi(HMPA)(THF)_2$[136]. This is an indication that the solvent probably plays a more important role than we suspected before. The ^{13}C NMR spectrum of Ph_3SnLi shows that the negative charge is built up mainly on the *para* position, suggesting an inductive polarization of the π-cloud, in contrast to the conjugative polarization in $Ph_3C{-}Metal$ which places the negative charge particularly on the *ortho* and *para* positions[137]. Many organostannyllithiums don't contain Sn−Li bonds. Tri(2-furyl)tinlithium has a lithium cation which is octahedrally coordinated by six oxygen atoms of two pyramidal trifurylstannyl anions, with a second separate and solvated lithium cation[138]. The first stannyllithium compound with a covalent Sn–Li bond to be identified was the pentamethyldiethylenetriamine complex of Ph_3SnLi. This stucture is preserved in solution[139].

As mentioned before, the synthetic power of these stannylcuprates lies in the utility of the resulting organostannanes for carbon–carbon bond formation and also in the relatively facile transmetallation of stannyl groups[12b].

Stannylcuprates undergo similar transformations to silylcuprates, independently of the method chosen for their preparation. While allylstannanes are used as allylic nucleophiles[121b, 140], vinyl- and arylstannanes are frequently employed in the Pd(0) catalyzed Stille coupling with vinyl, aryl and alkynyl halides and sulfonates[141]. A survey of representative transformations of stannylcuprates, which will be extensively discussed in the next section, is collected in Scheme 59[33, 130].

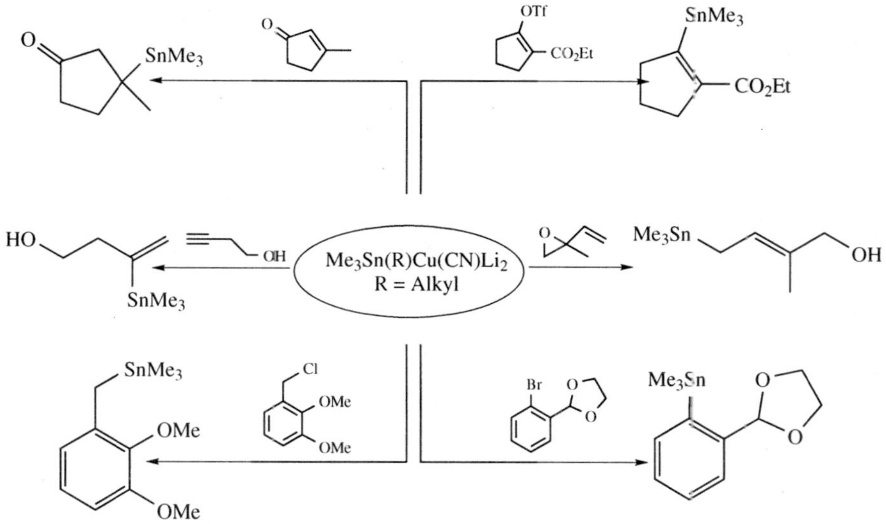

SCHEME 59

In conclusion, organotin compounds in which tin is bonded to copper provide an important source of nucleophilic tin in synthesis. These tin-nucleophiles compete favorably in chemo-, regio- and stereoselectivity with the more reactive alkali organotin derivatives, from which they are readily prepared. The different ways to prepare stannnylcuprates via stannyllithiums, or by transmetallation with alkylcuprates, are described in Schemes 55–57. In addition, Table 3 lists a compendium of the most frequently used stannylcuprates.

B. Reactions of Stannylcuprates

1. Introduction

Stannyl cuprate chemistry has emerged as a cornerstone of organic synthesis because of the wide range of copper-assisted transformations on offer, and particularly because this chemistry is often complementary to palladium chemistry and groups IA and IIA organometallic chemistry. Most of the early examples of application of stannylcuprates are stoichiometric reactions; however, the rapid development of new procedures catalytic in copper should greatly enhance the synthetic utility of this reaction. The presence of

TABLE 3. Representative stannylcuprate reagents

R	R′	$R_3Sn(R')Cu(CN)Li_2$ and $R_3Sn(R')CuLi$	Reference
Bu	Bu_3Sn	$(Bu_3Sn)_2Cu(CN)Li_2$	125b
Me	Me_3Sn	$(Me_3Sn)_2Cu(CN)Li_2$	142
Bu	Bu_3Sn	$(Bu_3Sn)_2CuLi$	9
Me	Me_3Sn	$(Me_3Sn)_2CuLi$	143
Bu	Bu	$Bu_3Sn(Bu)Cu(CN)Li_2$	130
Bu	Me	$Bu_3Sn(Me)Cu(CN)Li_2$	144
Me	Me	$Me_3Sn(Me)Cu(CN)Li_2$	132
Me	Bu	$Me_3Sn(Bu)Cu(CN)Li_2$	131
Me	CN	$Me_3SnCu(CN)Li$	22
Me	SPh	$Me_3SnCu(SPh)Li$	145
Ph	Ph_3Sn	$(Ph_3Sn)_2Cu(CN)Li_2$	124
Ph	Bu	$Ph_3Sn(Bu)Cu(CN)Li_2$	128
Ph	Ph_3Sn	$(Ph_3Sn)_2CuLi$	146
Me		$Me_3SnCu.SMe_2$	147
Bu		$Bu_3SnCu.SMe_2$	148

residual non-transferable ligands is a convenient alternative when looking for outcome efficiency, moderation of the reactivity or cuprate stability. Non-transferable ligands are also used in the preparation of chiral cuprates for asymmetric synthesis. On the other hand, homo bis(stannyl)cuprates show a high reactivity pattern where other cuprates fail to react. The increased availability of highly functionalized stannylcuprate reagents has allowed the design of new strategies in synthesis which would not have been possible a few years ago.

2. Conjugate additions to unsaturated carbonyl compounds

Stannylcuprate reagents are good nucleophilic reagents. They are less basic than their stannyllithium precursors and may sometimes work better, leading to more selective processes. The conjugate additions of stannylcopper and stannylcuprate reagents to enones **109–114** seems to proceed with better yield and chemoselectivity than the stannyllithium reagent, giving β-stannyl ketones **115–119** and β-stannyl ester **120** (Scheme 60)[33, 123, 131, 149, 150]. Double addition of the trimethylstannyl group, as well as aldol condensation of intermediate enolates, is observed when using the lithium reagent. Moreover, stannylcuprates give better stereochemical purity if diastereomers can be

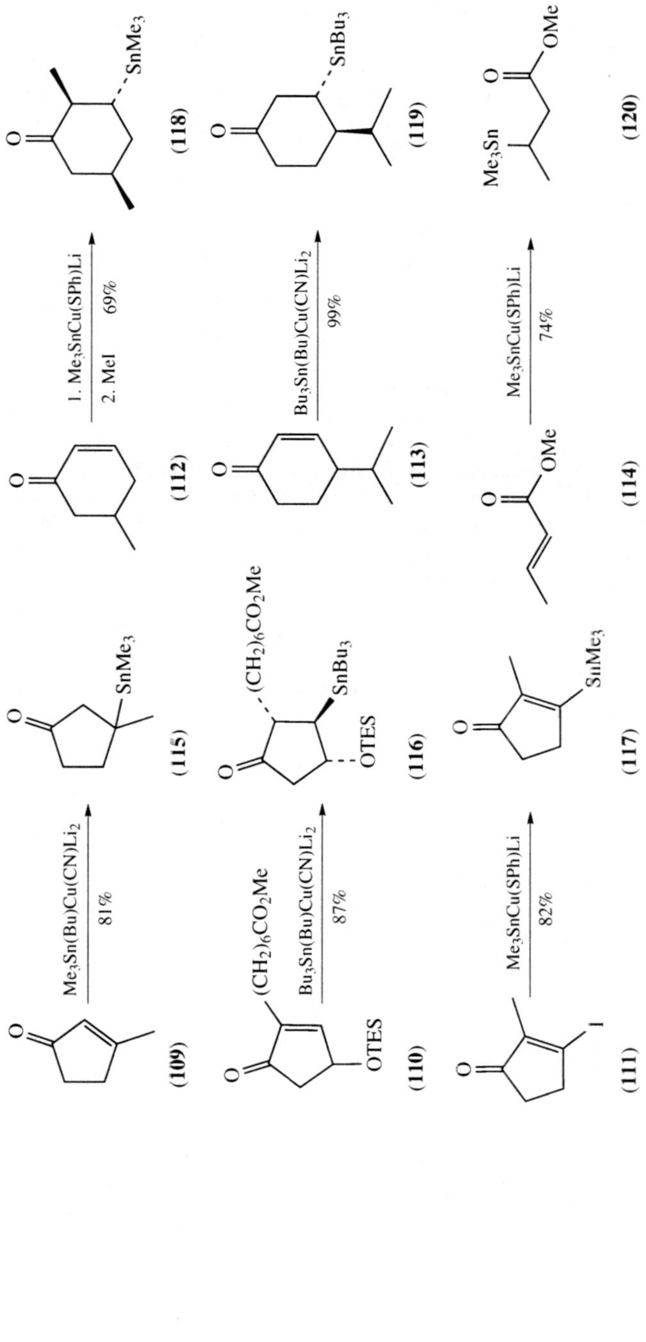
(109)
Me3Sn(Bu)Cu(CN)Li2
81%
(115)
SnMe3
(110)
(CH2)6CO2Me
OTES
Bu3Sn(Bu)Cu(CN)Li2
87%
(116)
SnBu3
(111)
Me3SnCu(SPh)Li
82%
(117)
(112)
1. Me3SnCu(SPh)Li
2. MeI 69%
(118)
(113)
Bu3SnBu)Cu(CN)Li2
99%
(119)
(114)
OMe
Me3SnCu(SPh)Li
74%
(120)
Me3Sn

SCHEME 60

formed[130,145]. β-Iodoenones such as **111** lead to β-stannylenones such as **117**, in a process that appears to take place through a conjugate addition–elimination pathway (see Section III.B.3)[123].

A common problem associated with the addition of stannylcuprates to 2-enoates is the low reactivity showed by β-dialkyl-substituted 2-enoates, which dramatically diminish the final yield. In these cases, the stannyllithium reagent successfully undergo 1,4-addition providing good yield of 3,3-dialkyl-substituted β-stannylketones[123].

Chiral 2-enoates carrying a 4-heteroatomic substituent provide good opportunities for diastereoselection, resulting from the conjugate addition of a stannyl group which generates a new chiral center, and a simultaneous stereoinduction by the adjacent stereogenic center[151]. The relative reactivity of stannyllithium, stannylcuprate and stannylzincate reagents has been compared by reaction of the former reagents with enantiopure 4-heteroatom-substituted 2-enoates (Scheme 61). *Z*-enoates afford excellent diastereoselectivities with all the reagents; the cuprate and zincate derivatives show the same diastereoselection but opposite to that found for the stannyllithium reagent. *E*-enoates, however, give poor diastereocontrol, except for the stannylzincate which still shows a high stereoselectivity in the same sense as for the *Z*-isomer. From the results observed in Scheme 61, it is clear that the reaction with stannyllithium reagents follows a somewhat different pathway from that of stannylcopper and zincate derivatives, but a predictive model of behavior has not been outlined[151].

β-Stannylketones are useful intermediates of much potential in organic synthesis (Scheme 62). The carbonyl group can be converted into a tertiary hydroxyl group that has been used to synthesize cyclopropanes, as in the conversion of **121** to **122**[152]. The Sn−C bond can be also oxidized to alcohol or ketone (**123** to **124**), a transformation that formally involves a 1,3-shift of the keto group[122]. Alternatively, the stannyl group can be eliminated, after having served to control stereochemistry, by reduction with dissolving metals (**125** to **126**)[145]. It is also used as a leaving group in radical-based cyclizations[153]. Halodestannylation and metallation are frequently used reactions[121b]. In addition, enantiopure β-stannylketones analogous to **41** but stannylated (Scheme 22) have been obtained as described therein[69].

Conjugate addition reactions of stannylcopper and cuprate reagents are most often employed with ynones and ynoates[154–156]. Acetylenic esters have proved to be so far the most versatile acetylenic substrates for this type of reaction. The stannyl group adds to the β-position of the α,β-acetylenic esters and amides to give the product **127**, resulting from *syn*-addition of the Sn−Cu pair. Protonation under kinetic control (−100 °C) occurs with retention of initial configuration, leading to the *E*-vinylstannane **128**. If the stannylcupration is carried out at −78 °C and then warmed up to −48 °C before quenching, the product obtained is the *Z*-vinylstannane **130** resulting now from protonation *anti* to the stannyl group of the thermodynamically more stable allenic intermediate **129** (Scheme 63)[154].

Lower-order cuprates $Me_3SnCu(SPh)Li$ and $Me_3SnCu(CN)Li$ work well to give both the kinetic and the thermodynamic vinylstannane, while other stannylcuprates give only the *E*-isomer. A serious limitation of this reaction when lower-order cuprates are used is the difficulty of persuading the intermediate cuprate to react with electrophiles other than proton, which remarkably limits the synthetic scope of the reaction. In these cases, the use of a more reactive higher-order cuprate such as $Bu_3Sn(Me)Cu(CN)Li_2$ solves the problem, producing high yields of substituted vinylstannanes (see Section III.B.4).

α,β-Acetylenic *N*,*N*-dimethylamides (**131**) have been tested with organocopper $Me_3SnCu{\bullet}SMe_2$ (**132**) and cuprate $Me_3SnCu(SPh)Li$ (**133**). A study of the reaction shows (a) that the overall process can be controlled so as to produce either (*Z*)-**134** or (*E*)-3-trimethylstannyl-2-alkenamides (**135**), (b) that the initially formed intermediate cuprate derived from addition of **132** or **133** to ynamides is more stable than that obtained with

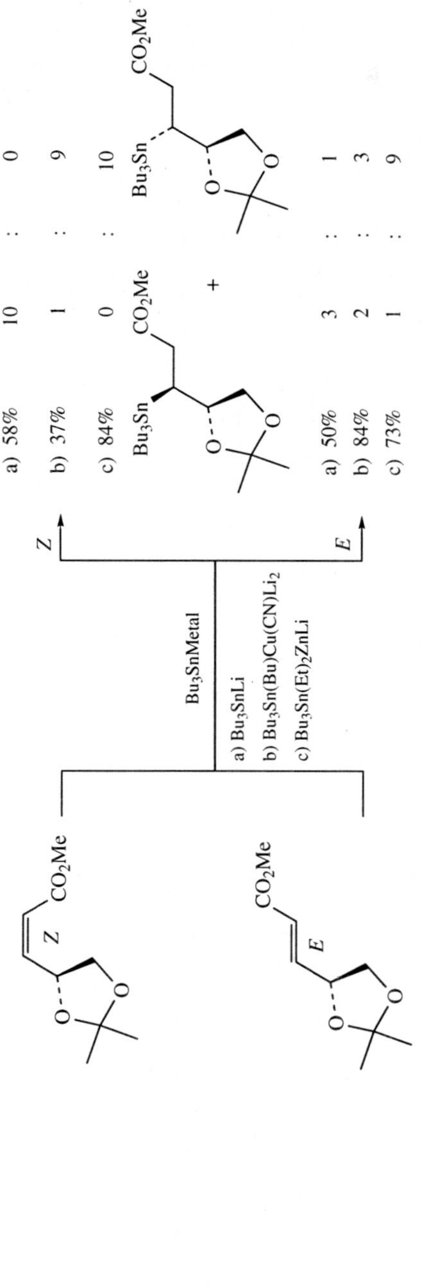
CO2Me
Z
E
Bu3SnMetal
a) Bu3SnLi
b) Bu3Sn(Bu)Cu(CN)Li2
c) Bu3Sn(Et)2ZnLi
Bu3Sn
a) 58% 10 : 0
b) 37% 1 : 9
c) 84% 0 : 10
+
a) 50% 3 : 1
b) 84% 2 : 3
c) 73% 1 : 9

SCHEME 61

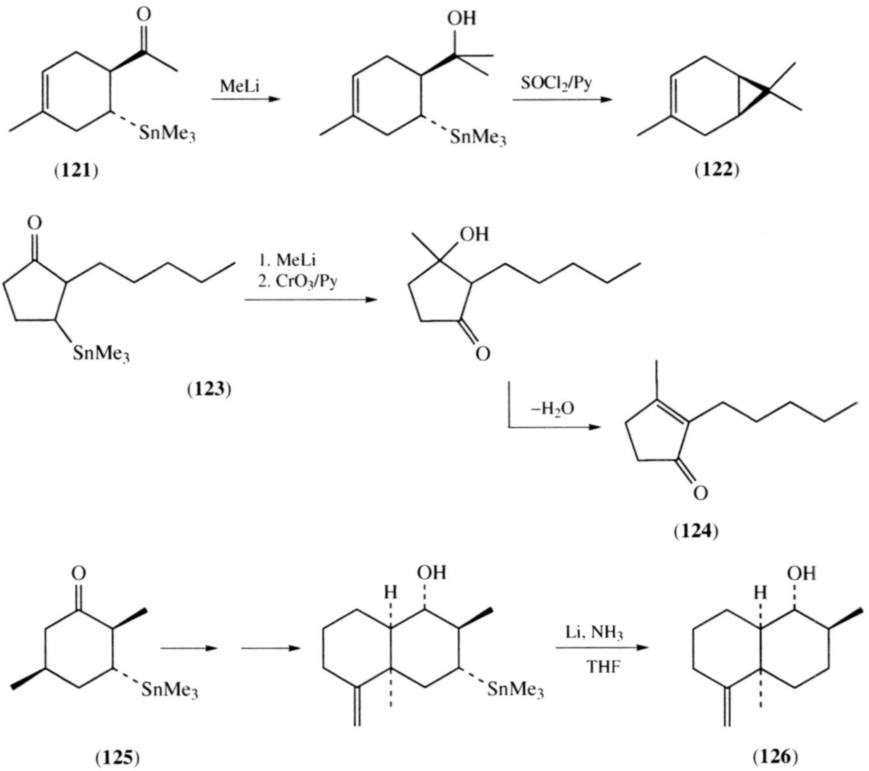

SCHEME 62

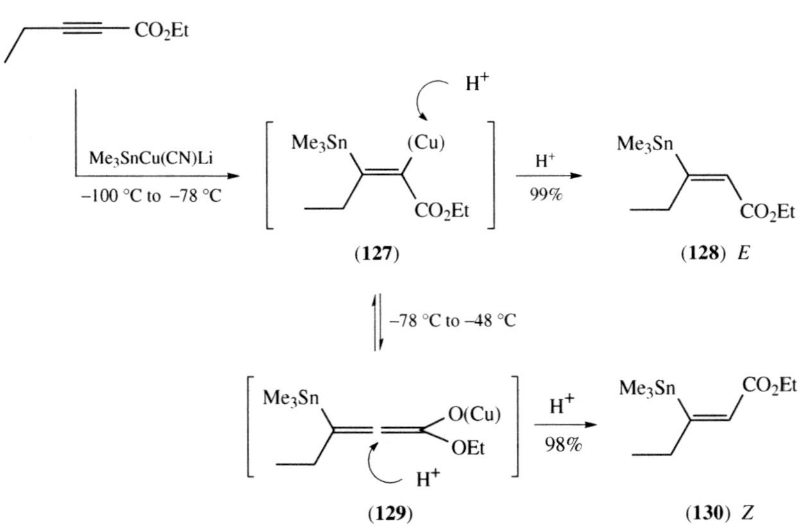

SCHEME 63

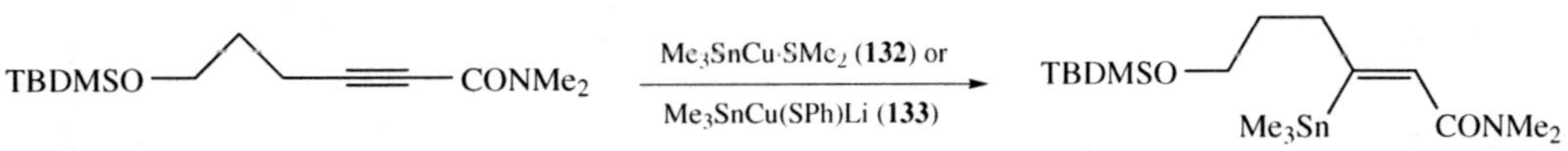

(**131**) (**134**)

+

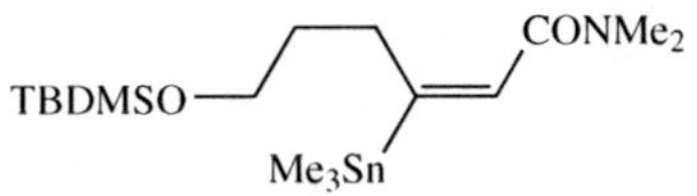

(**135**)

Reagent	Conditions	Ratio **134** : **135**	Yield (%)
132	A	1 : 99	77
133	B	95 : 5	76

A: THF, −78 °C, 3 h
B: THF, −48 °C, 1 h, add 2 volumes of Et_2O and then −48 °C to −20 °C to 0 °C, 4 h

SCHEME 64

ynoates and (c) that the intermediate cuprate produced in the reaction can be trapped with a few electrophiles other than proton (e.g. methyl iodide, allyl bromide) (Scheme 64)[157].

Z/*E* selectivities have been also achieved using chiral ynoates. Thus, 4-amino-2-ynoates as **136** react with the higher-order cyanocuprate $Bu_3Sn(Bu)Cu(CN)Li_2$ giving, after protonation with methanol at low temperature, vinylstannanes **137** and **138**, largely favoring the *E*-isomer. The latter can be converted into 4-tributylstannylpyrrolin-2-ones (Scheme 65). The resulting lactams are not stable enough and need to be protected for isolation, usually as *t*-butoxycarbonyl derivatives. The so-formed vinylstannanes undergo a palladium-catalyzed coupling reaction with vinyl halides[158].

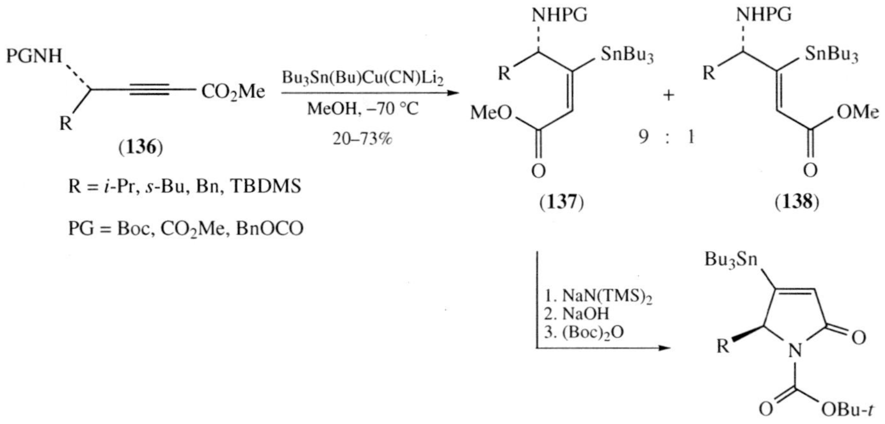

SCHEME 65

Diynoates can undergo intramolecular cyclization in two steps. The first involves attack of the stannylcuprate on one of the acetylenic ester units with stereoselective formation of an *E*-vinylstannane. Transmetallation of the vinylstannane to a vinylcuprate by treatment with a copper(I) salt in DMF and subsequent intramolecular addition to the remaining ynoate provides a versatile ring-forming procedure (Scheme 66)[159].

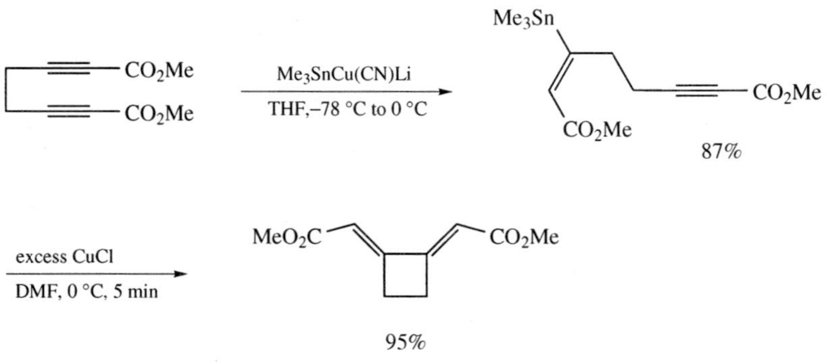

SCHEME 66

As briefly commented before, the stereoselectivity observed in the stannylcupration of alkynoates has been interpreted in terms of an α-cuprio ester (**127**, Scheme 63) generated by *syn*-addition of Sn−Cu at low temperature and trapped by proton quenching to afford the *E*-adduct **128**[154]. Stereoselective formation of the *Z*-diastereomer **130** at higher temperatures requires either direct isomerization of the *Z* α-cuprio ester to the *E* α-cuprio ester followed by selective protonation, or equilibration of **127** with the allenyl enolate **129** and stereoselective protonation from the less hindered side of the allenic system, that is *anti* to the bulky stannyl group. Mechanistic studies[160] involving alkylcuprates and alkynoates have found that isomerization between *E* and *Z* α-cuprio esters takes place through the allenyl enolate, with the resulting adduct *E*:*Z* diastereomeric ratio reflecting the alkenyl cuprate *E*:*Z* equilibrium ratio. Extrapolation of this argument to the stannylcupration of ynoates requires the *E* α-cuprio esters to be thermodynamically more stable than *Z* α-cuprio esters. This difference in stability should be sufficiently high so as to account for the high stereoselectivity observed. Stannylcuprate additions, as other metallo-metallation reactions, have been shown to be reversible and can lead sometimes to the 2-stannyl regioisomer[154]. The initial vinylcuprate, formed by conjugate addition of stannylcuprates to alkynyl esters, cannot generally be trapped with electrophiles other than proton, although some exceptions have been reported depending on the cuprate used and the nature of the alkynyl ester used[161]. As discussed above, 2-alkynyl amides are also exceptional in the sense that they give good yields of trapping products[157]. Intramolecular trapping has been also achieved to give a β-trimethylstannylcyclopentane carboxylate[154].

The formation of either 2- or 3-stannyl regioisomer can be attained with a judicious election of reagent and conditions. For example, addition of $Bu_3Sn(Bu)Cu(CN)Li_2$ to acetylenic acids affords 3-stannylenoic acids which can be trapped with different electrophiles, while treatment with a tributylstannyl organoaluminum compound in the presence of copper(I) cyanide as catalyst gives the opposite regioselectivity, leading to 2-stannylenoic acids (Scheme 67)[162].

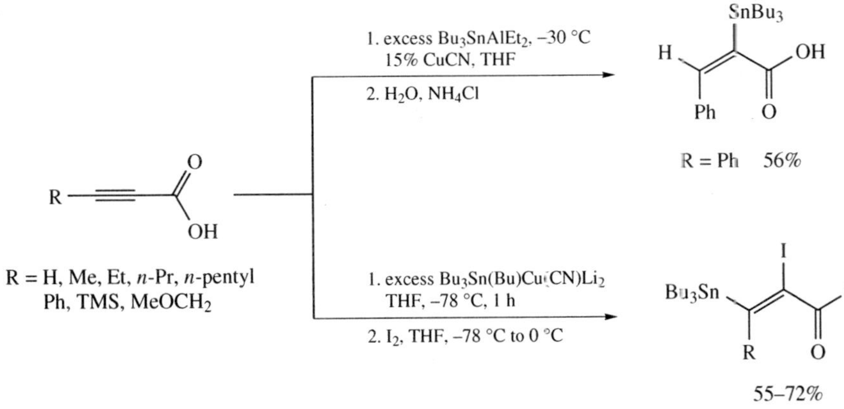

SCHEME 67

The *E*- and *Z*-products resulting from reaction of ynoates with stannylcuprates have been exploited by Piers and coworkers to prepare the chlorides **139** and **140**, and other related compounds, by deconjugative double-bond migration (Scheme 68). They are valuable synthons in natural product synthesis because they can be used as bifunctional donor–acceptor reagents for annelation. This sequence of reactions has the added advantage that exocyclic double bonds can be readily prepared with complete stereocontrol, due to the remarkable stereoselectivity of the deprotonation with LDA-reprotonation with AcOH step (Scheme 68). Chlorides **139** and **140**, and related compounds, have been

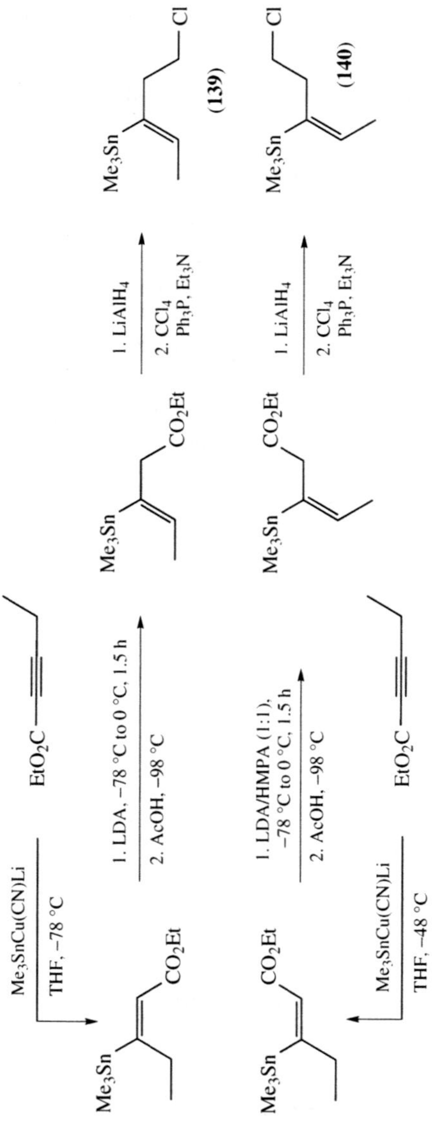

Me3SnCu(CN)Li
THF, −78 °C
EtO2C
Me3Sn
CO2Et
1. LDA, −78 °C to 0 °C, 1.5 h
2. AcOH, −98 °C
1. LiAlH4
2. CCl4
Ph3P, Et3N
Cl
(139)
Me3SnCu(CN)Li
THF, −48 °C
1. LDA/HMPA (1:1),
−78 °C to 0 °C, 1.5 h
2. AcOH, −98 °C
(140)

SCHEME 68

employed in synthetic routes to dolastane diterpenoids, (±)-amijitrienol and marine sesterpenoid (±)-palouolide[163].

3. *Substitution reactions with allylic and propargylic electrophiles, epoxides, vinyl triflates and β-haloenones*

The reaction of stannylcuprates with alkyl halides and tosylates has been reported by Lipshutz and coworkers[130, 131], but in spite of the high yield obtained (Scheme 69) its use in synthesis is rather limited. Trimethylstannyl groups, as well as tributylstannyl groups, have been equally efficiently introduced by means of a mixed tincuprate generated *in situ* by transmetallation of trialkyltin hydride with $Bu_2Cu(CN)Li_2$.

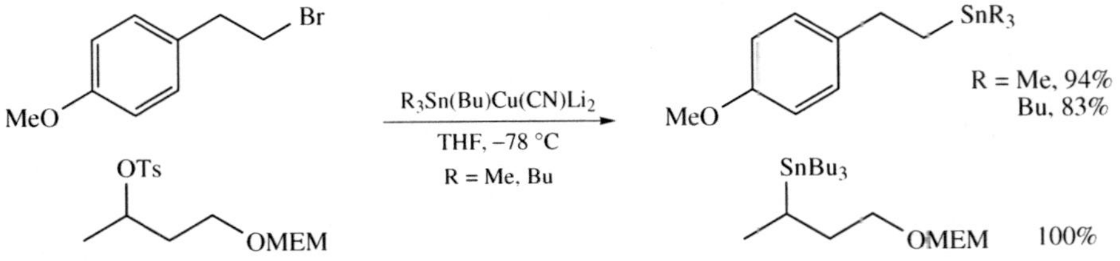

SCHEME 69

The reaction of stannylcuprates with allylic and propargylic halides and acetates is better known and has found application in synthesis of natural products. Thus, although the allyl bromide **141** cleanly reacts with stannyllithium reagents giving only the allylstannane **143** (resulting from direct S_N2 substitution), the corresponding cuprate gives mixtures of regioisomers favoring the allylstannane **142** arising from a S_N2' allylic inversion (Scheme 70)[84a, 164].

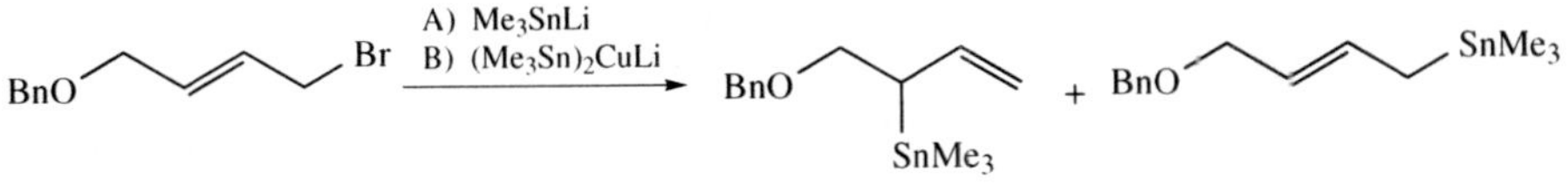

(**141**) (**142**) (**143**)

A) 0 : 100
B) 67 : 33

SCHEME 70

Allylstannanes prepared by reaction of allylic acetates with stannylcuprates have been employed in the strategy of synthesis of (+)-10-*epi*-elemol (Scheme 71).

The mixed dilithium tributylstannyl(butyl)cyanocuprate reacts chemoselectively with the tertiary allylic acetate **144**, affording an intermediate allystannane **145** that was later cyclized, by a palladium-catalyzed intramolecular cross-coupling reaction with a new unit (now a primary one) of allyl acetate, to give the 1,2-divinylcyclohexane **146** (Scheme 71). This was successfully transformed onto the 10-*epi*-elemol. The remarkable chemoselectivity of the stannylcuprate for the attack on the tertiary allyl acetate and the regioselectivity of the cross-coupling step which gives the 6-membered ring should be noted[165].

Propargylic electrophiles are even more reactive toward stannylcuprates than the respective allylic derivative. Propargylic mesylates usually give high yields of the S_N2' rearranged product leading to allenylstannanes, but the regioselectivity is very sensitive to the

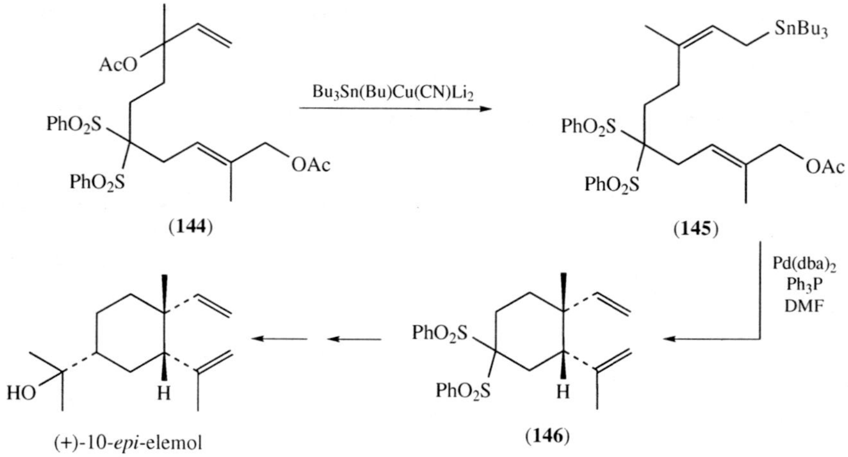

SCHEME 71

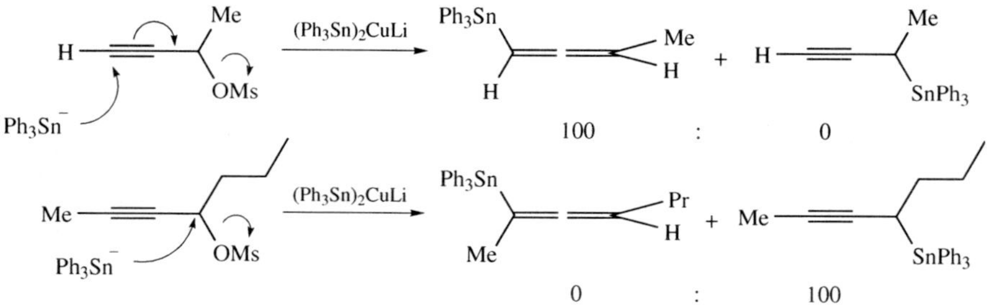

SCHEME 72

substitution pattern (Scheme 72)[146]. Ricci and coworkers[78] found that the regiochemistry of the addition of stannylcuprates to propargylic sulfides depends highly on the reaction conditions as well as the nature of the cuprate. By choosing these conditions adequately, in combination with the possibility of a metal-mediated desulfuration, a versatile and flexible method for the synthesis of a wide range of allenyl- and vinylstannanes has been developed. The method is also valid in the silicon series[78]. Allylstannanes and allenylstannanes are much used as carbon nucleophiles in organic synthesis[166].

Stannylcuprates participate in substitution reactions with 3-halo- and 3-sulfonyl-substituted 2-enones (Scheme 73)[167]. 2-Enoates having good leaving groups at the β-position (such as Cl, I, PhS) also undergo substitution reactions with stannylcuprates (Scheme 74).

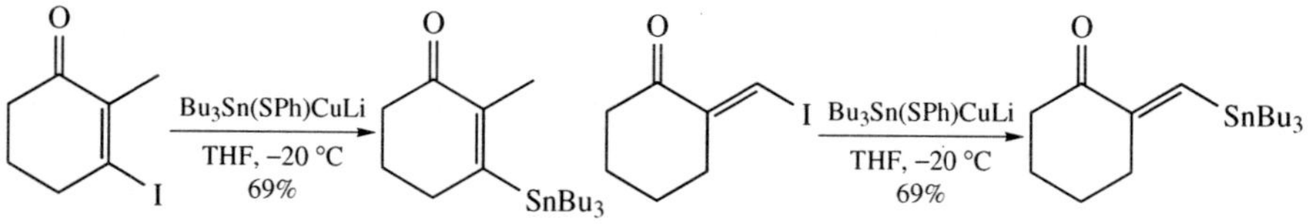

SCHEME 73

These substitution reactions may follow a conjugate addition–elimination reaction sequence or may proceed through a mechanism involving direct displacement of the leaving group. Examples of both pathways have been observed. β-Haloacrylates, β-(phenylthio)acrylates and β-tosylacrylates give dirty reactions and poor yields when reacting with higher-order cuprates; on the contrary, stannyllithium derivatives, although more reactive, afford distannyl adducts. The optimal conditions for a good yield and stereocontrol are attained using tributylstannylcopper at low temperature, which gives 3-tributylstannylacrylates with retention of the acrylate double-bond configuration (Scheme 73)[167a]. β-(Phenylthio)acrylates and β-tosylacrylates react slowly at $-78\,^{\circ}$C, and require higher temperatures to get acceptable yields. Warming-up the reaction of methyl (E)-β-chloroacrylate from $-78\,^{\circ}$C to $0\,^{\circ}$C causes an almost complete inversion of configuration (E/Z ratio of β-stannylacrylates 1:4). The isomerization observed for the (E)-β-(phenylthio)acrylates (Scheme 74) is apparently the result of competing equilibration

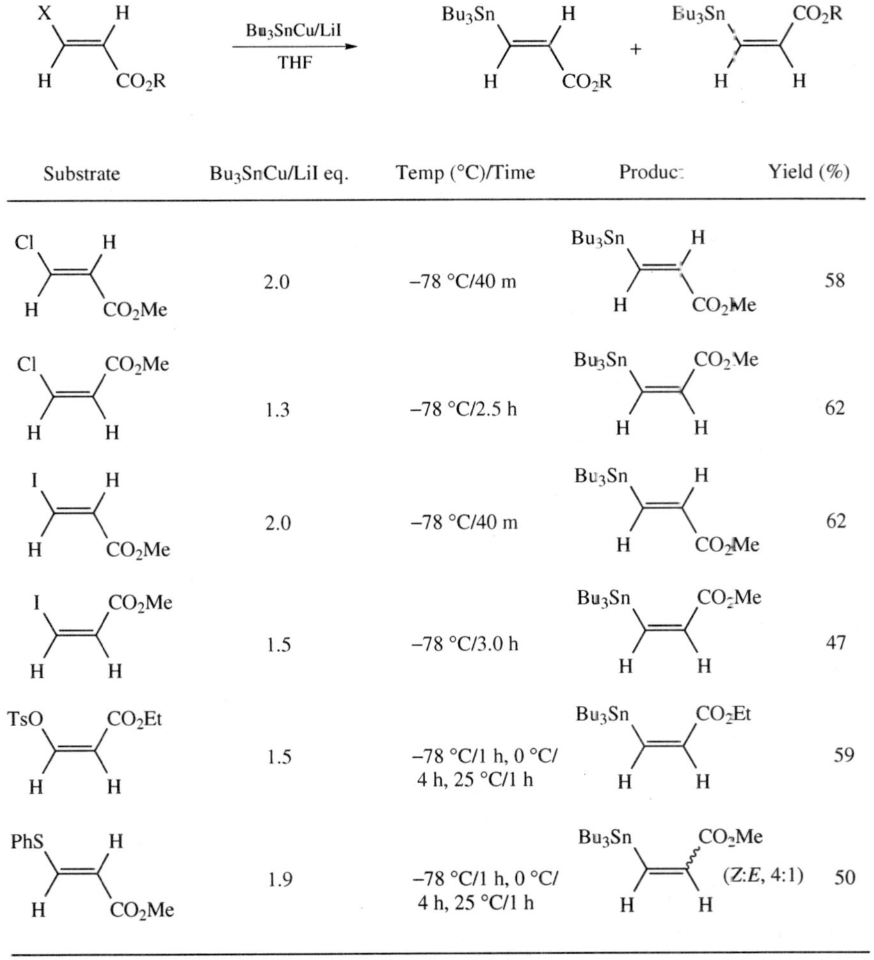

SCHEME 74

of the copper intermediate in the absence of a suitable proton donor that quenches the reaction. A similar substitution reaction has been carried out with tributylstannyl(2-thienyl)cyanocuprates and 3-sulfonyl-2-enoates[167b].

The regioselective cleavage of epoxides has been reported by Lipshutz, Oehlschlager and coworkers[130, 132]. Attack of the stannylcuprate on the less hindered end is usually observed (Scheme 75). Although stannyllithium and stannylcuprate derivatives show a similar behavior, the regioselectivity of the latter is frequently better and more selective processes are obtained. Vinyl epoxides are cleaved with allylic rearrangement. Contrary to silylcuprates where some alkyl group is transferred, no transfer of alkyl group is observed when a mixed stannyl(alkyl)cuprate is used.

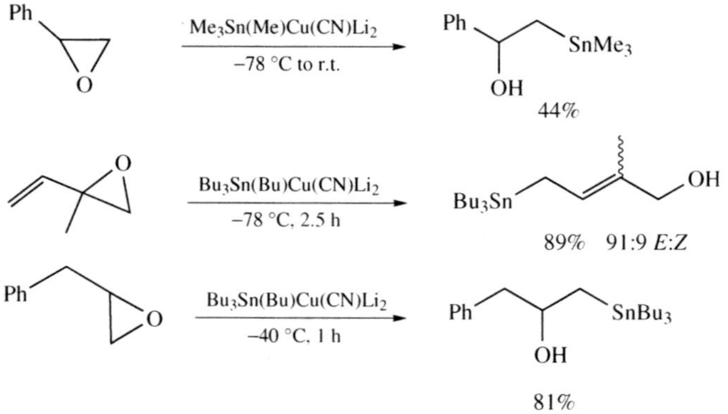

SCHEME 75

Higher-order tributylstannyl cuprates react with vinyl triflates, prepared from ketones, leading to vinylstannanes[33, 159, 168–170]. The reaction is carried out at low temperature to avoid formation of mixtures. Work-up of the mixture is usually accompanied by formation of much hexabutyldistannane. This is a very involatile by-product, but it can be mostly removed using silver acetate. This route to vinylstannanes is especially useful for preparing either regioisomers of an unsymmetrical ketone, which are readily available via the kinetic or thermodynamic enolate (Scheme 76).

Enol triflates have been used in the key step of an annulation method involving cyclic β-trifluoromethanesulfonyloxy α,β-unsaturated esters such as **147** and $Me_3Sn(PhS)CuLi$[168a]. Alkylation of the resulting vinylstannane **148** with α,ω-dihaloalkanes, followed by transmetallation-cyclization, leads to bicyclo[4.3.0]non-1-enes **149** with excellent yield. The 5,6-membered fused bicyclic system is a structural unit appearing in many natural products; 5,7-membered ring fusions were also prepared less efficiently (Scheme 77).

Annulation methods based on tincuprate chemistry and enol triflates derived from cyclic β-keto esters continue to be of great importance in the area of natural product synthesis. This methodology has been exploited in a cyclization strategy leading to (±)-chiloscyphone (Scheme 78)[169a].

4. Reaction of stannylcuprates with acetylenes and allenes

The addition of stannylcopper and stannylcuprate species to simple alkenes was first reported by Piers and Chong[171] and Westmijze and coworkers[124] and widely developed in later years[22, 33, 130–132, 147, 172–176]. Stannylcopper and stannylcuprate reagents react with

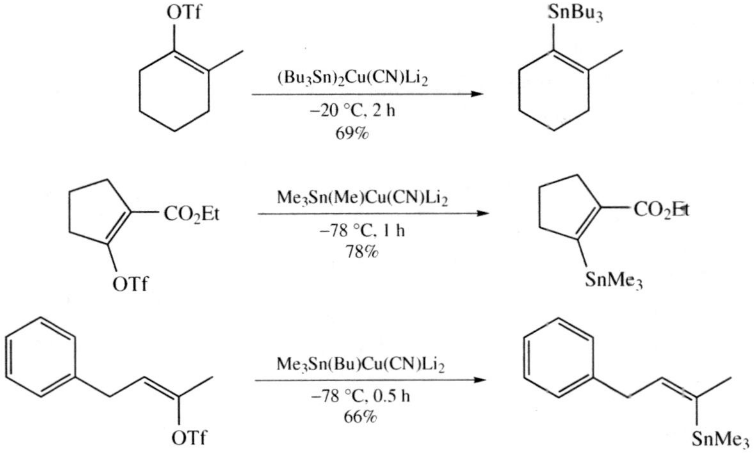

SCHEME 76

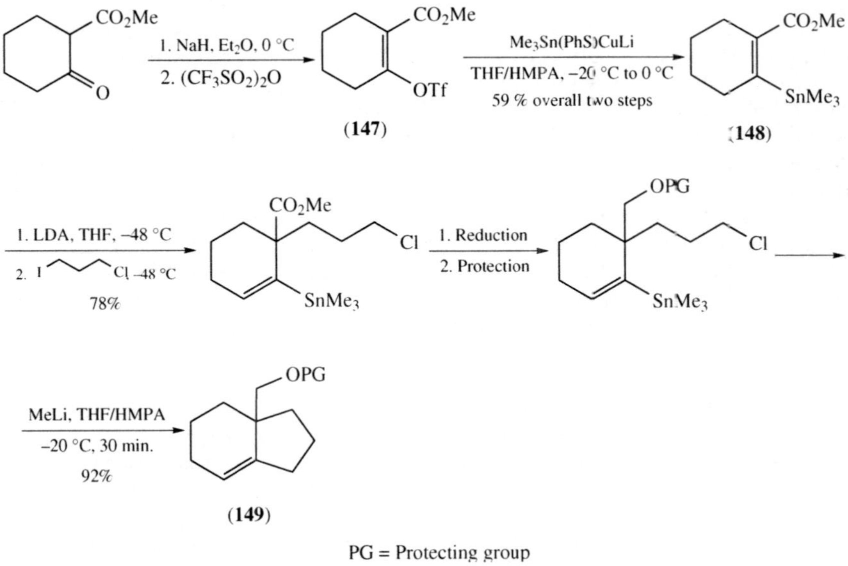

PG = Protecting group

SCHEME 77

1-alkynes **150** with high but not complete regioselectivity, introducing the stannyl group predominantly at C-2 and the copper unit at C-3. Nevertheless, the ratio **151:152** is highly affected by the nature of the cuprate. This reaction is easily reversible, as shown in Scheme 79, and a major limitation of these otherwise powerful processes is that the addition of an electrophile (E^+) does not always give high yields of products **154**. Frequently, it is difficult to persuade the intermediate vinylcopper species to react with anything more interesting than a proton. This is particularly true when using stannylcopper species and less drastic when real stannylcuprate species are present. The problem appears to be

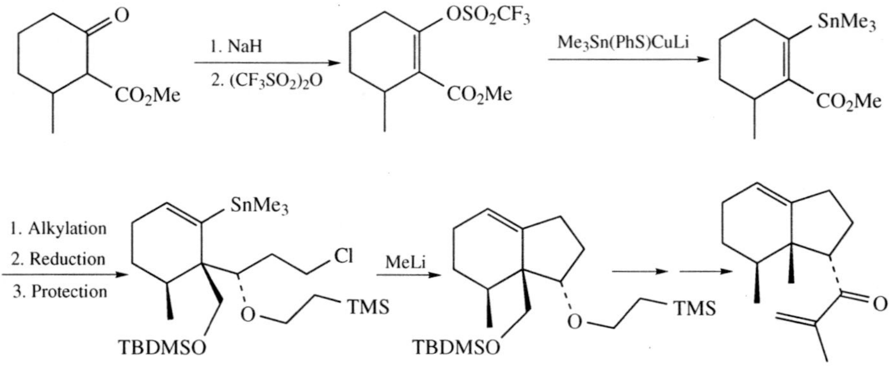

(±)-chiloscyphone

SCHEME 78

that the stannylcupration step, although well shifted to the right at equilibrium, is easily reversible, and the stannylcopper reagent itself is more reactive than the vinylcopper intermediate **151** toward most electrophiles, leading to the alternative products **153** instead of the expected compounds **154**[147, 173]. Stannylcopper and cuprate reagents are evidently not very basic, with the result that a proton, usually delivered from methanol included in the reaction mixture, is relatively selective for the vinylcopper intermediate, making the formation of the addition product **154** (E = H) high yielding.

In the early studies, the vinylcopper species could only be trapped with a proton[171]. Later, Marino and coworkers[177] using $Bu_3SnCu(CN)Li$ and acetylene, and Fleming and coworkers[144] using $Bu_3Sn(Me)Cu(CN)Li_2$ and substituted acetylenes, successfully achieved the trapping of the resulting intermediate vinylcuprates with a wide variety of electrophiles (alkyl and allyl halides, halogens, epoxides, acyl chlorides, enones and enoates). Reaction takes place stereospecifically by *syn* addition of the tin–copper pair to the triple bond of the acetylene (Scheme 79). Structural and mechanistic studies accomplished by Oehlschlager and coworkers[173] definitively established the reversibility of the stannylcupration of alkynes and shed some light on the nature of the intermediate vinylcuprate species involved.

The procedure for the stannylcupration of acetylene, mono-substituted and disubstituted acetylenes followed by protonation or reaction with electrophiles has been generalized by Pulido and coworkers showing that this methodology is synthetically more powerful than was earlier expected[144]. Thus, cuprates $Bu_3Sn(Me)Cu(CN)Li_2$ (**155**) and $(Me_3Sn)_2Cu(CN)Li_2$ (**172**) react with acetylenes **156–163** and **173** to give regioselectively the vinylstannanes **164–171** and **174** without contamination of regioisomers (Scheme 80).

The only exception to this general behavior is 1-hexyne (**163**), which gives **171** (78%) along with 18% of the regioisomeric 1-stannylated hex-1-ene. Interestingly, stannylcupration of the corresponding lithium acetylide **175** produces exclusively vinylstannane **171**, with reaction presumably taking place via the triply differentially metallated alkene **176**. Such poly-metallated alkenes are intermediates of much potential in synthesis due to the different reactivity pattern that each metal can impart. Furthermore, a surprising tetrametallated alkene intermediate having copper, tin, silicon and lithium linked to a double bond could be devised by stannylcupration of the readily available lithium acetylide of trimethylsilylacetylene.

More significantly, C–C bond formation can be easily achieved by successive addition of a stannylcuprate to unactivated acetylenes and further attack on copper of a carbon

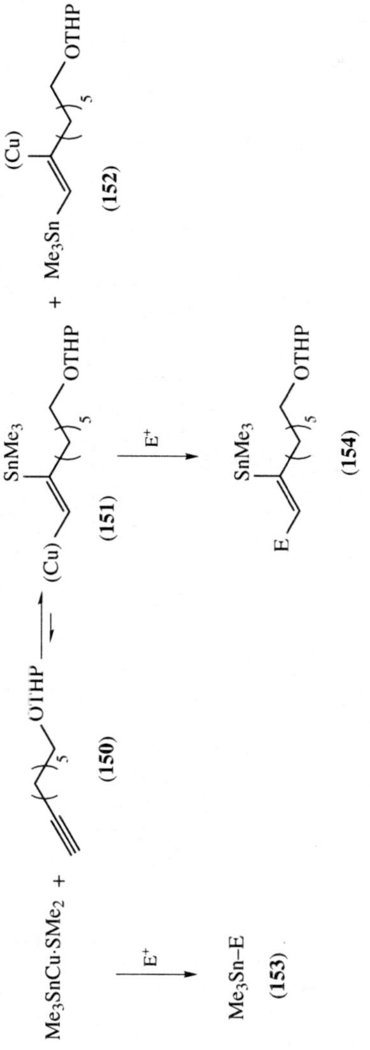

SCHEME 79

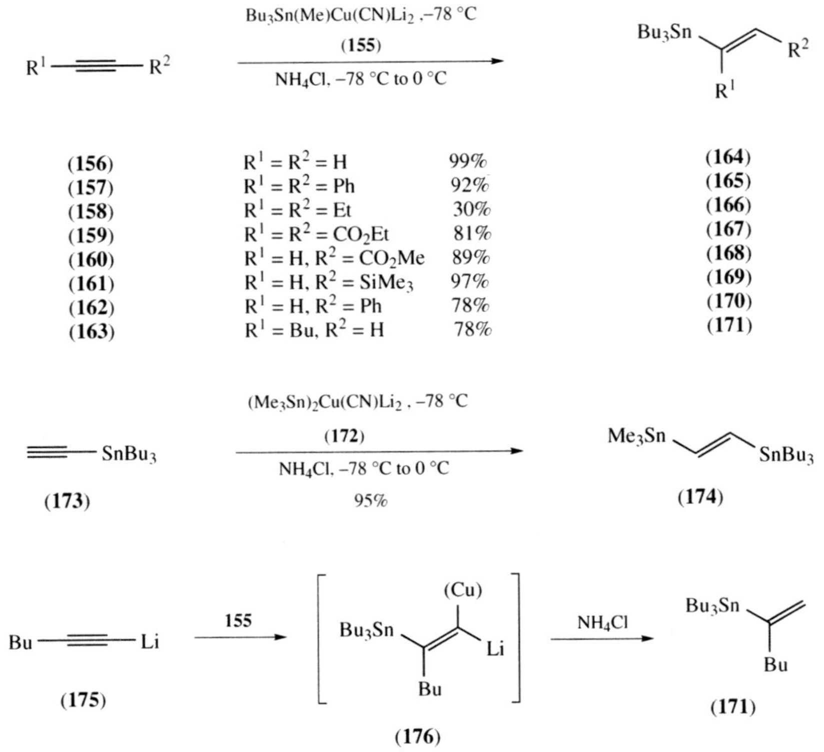

SCHEME 80

electrophile. The intermediate vinylcuprate **177**, generated by *syn* addition of cuprate **155** to acetylene, reacts with methyl iodide, ethylene oxide, cyclohexenone, methyl crotonate, acetyl chloride, bromine, iodine, trimethylsilyl chloride and tributyltin chloride, giving cleanly the vinylstannanes **178–186** (Scheme 81). The reaction is not merely limited to acetylene; likewise, substituted acetylenes also give high yields of vinylstannanes by reaction with cuprate **155** and subsequent trapping with electrophiles. Starting from acetylene itself, a Z-stereochemistry is cleanly obtained for all the vinylstannanes isolated. It should be noted that this strategy permits one to design easy methods for preparing both (E)- and (Z)-2-silylvinylstannanes **169** and **178** and (E)- and (Z)-vinylbisstannanes **174** and **179**[144].

It is unclear why cuprate **155** overcomes the limitations reported with Piers' cuprate, but most probably it is due to the fact that **155** is a higher-order cuprate and the earlier work had most often been carried out using stannylcopper reagents. Other reports point in the same direction[124, 177b, 178, 179]. Whatever the explanation, this route makes available a wide number of potentially useful small tin-synthons of much interest in synthesis (Scheme 82).

For instance, Corey's oxidation[180] of vinylstannanes **185** and **186** with lead tetraacetate makes the overall reaction a conjugate addition of acetylene to an α,β-unsaturated ketone or ester. Stille reactions on vinylstannane **186** gave 1,3- and 1,4-dienes. (Z)-β-Stannylvinyl ketones and tosylates such as **185** and **187** undergo tin–lithium exchange when treated with butyllithium followed by intramolecular addition to carbonyl or displacement of

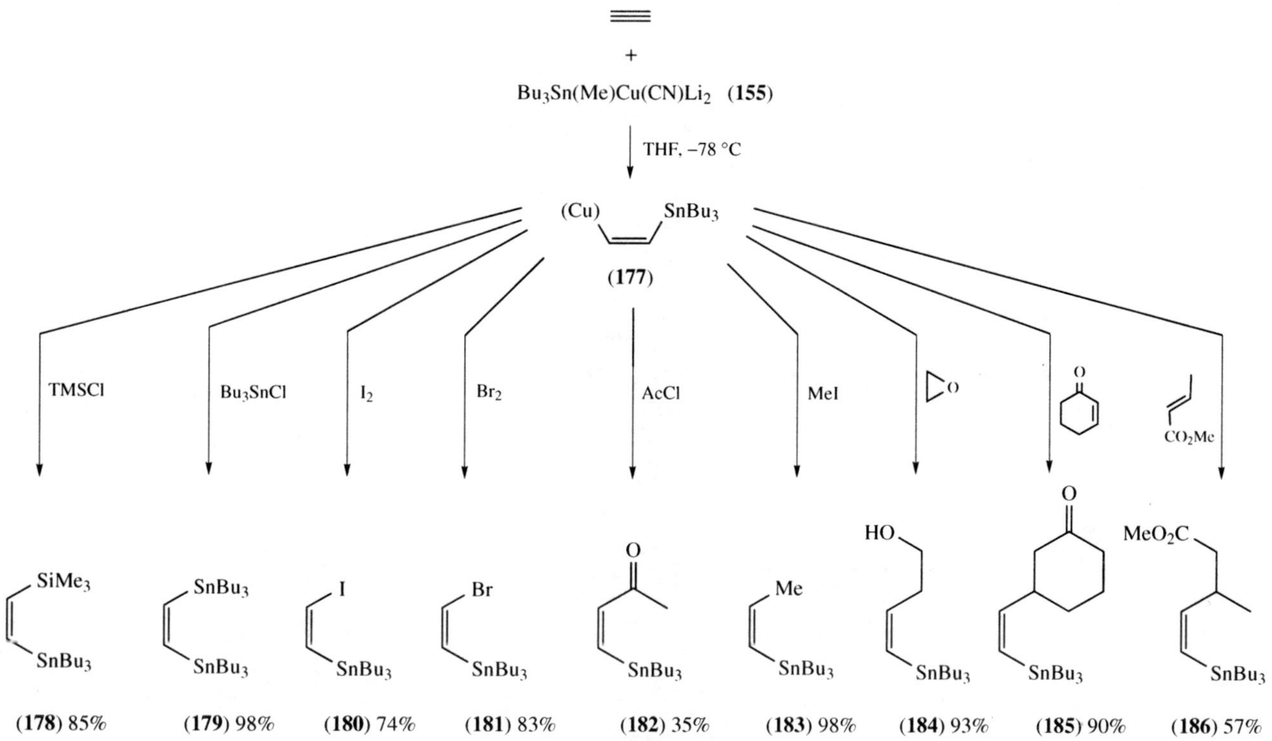
Bu3Sn(Me)Cu(CN)Li2 (155)
THF, −78 °C
(Cu)
SnBu3
(177)
TMSCl
Bu3SnCl
I2
Br2
AcCl
MeI
O
CO2Me
SiMe3
SnBu3
I
Br
Me
HO
MeO2C
(178) 85%
(179) 98%
(180) 74%
(181) 83%
(182) 35%
(183) 98%
(184) 93%
(185) 90%
(186) 57%

SCHEME 81

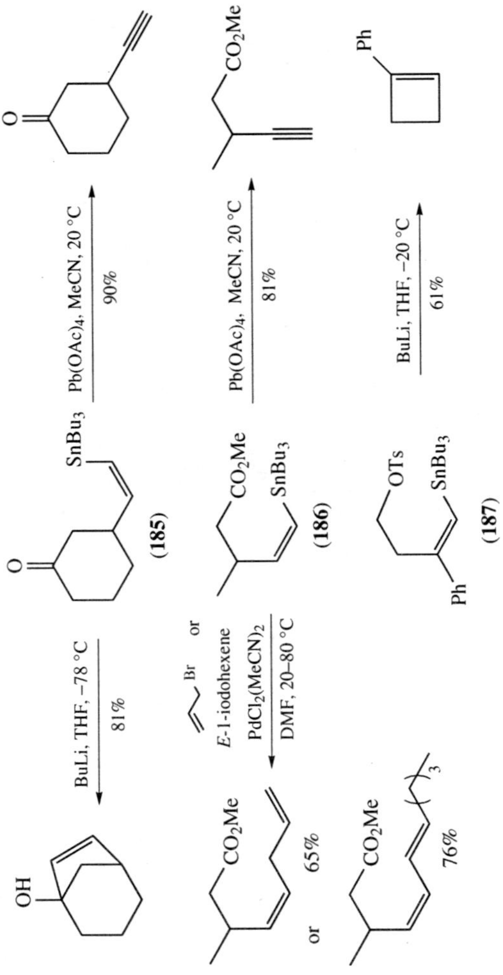
OH
BuLi, THF, –78 °C
81%
O
SnBu$_3$
(185)
Pb(OAc)$_4$, MeCN, 20 °C
90%
O
CO$_2$Me
SnBu$_3$
(186)
Pb(OAc)$_4$, MeCN, 20 °C
81%
CO$_2$Me
Br
or
E-1-iodohexene
PdCl$_2$(MeCN)$_2$
DMF, 20–80 °C
CO$_2$Me
or
65%
CO$_2$Me
3
76%
OTs
SnBu$_3$
Ph
(187)
BuLi, THF, –20 °C
61%
Ph

SCHEME 82

tosyl group, thus providing good methodologies for making five- and four-membered rings, respectively (Scheme 82)[144, 181].

Although the regioselectivity of the stannylcupration of alkynes is generally excellent, it proved to be quite sensitive to conditions, cuprate reagent and substrate structure. In general, by choosing judiciously an appropriate combination of the former factors, a good regiocontrol of the resulting products can be attained.

Some efforts made to control the regiochemistry of the addition of stannylcuprates to 3-butynoic acid derivatives clearly show how sensitive to the mentioned factors the process is. Thus, the usual stannylcupration affords 4-stannyl-3-enolates, which are readily protonated or trapped with MeI. Ulterior iodo-destannylation gives the vinyl iodide **188** (Scheme 83). However, the regiochemistry is easily reversed using a cuprate reagent made from a stannylaluminum reagent, leading in this case to vinyl iodide **189**[162].

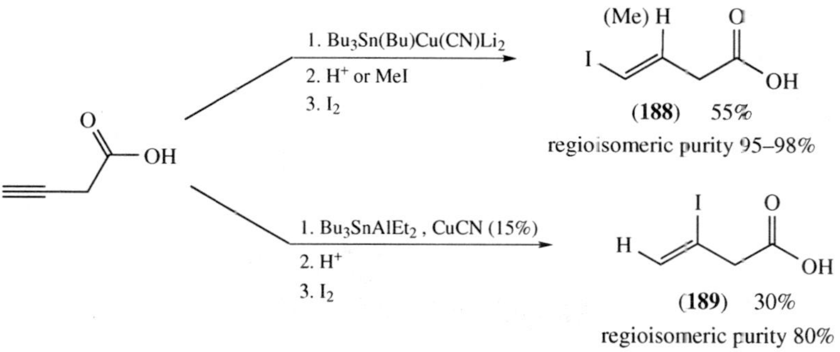

SCHEME 83

Copper-catalyzed reaction of terminal alkynes with stannylmagnesium reagents provides excellent regioselectivities of 1-stannylalkenes[170].

The regiochemistry of the addition to ethynyl ethers[182] depends on the presence or not of HMPA in the mixture. The stannyl group can be attached to either end of the acetylene. Under kinetic conditions (THF, $-78\,^{\circ}$C) 1-stannyl-1-alkoxyalkenes are obtained, whereas at 0 °C in the presence of HMPA (*E*)-2-stannylvinyl ethers are formed (Scheme 84). Complexation effects are responsible for the observed results. The initial (*E*)-2-alkoxyvinylcuprate intermediate, generated in kinetic conditions, must be trapped with proton, otherwise it decomposes to give stannylacetylenes. However, the presence of HMPA highly stabilizes the initial intermediate, thus allowing its isomerization, at higher temperatures, into the thermodynamic 1-alkoxyvinylcuprate, which by quenching gives (*E*)-2-stannylvinyl ethers. The greater stability of the thermodynamic 1-alkoxyvinylcuprate is due to intramolecular oxygen–copper complexation.

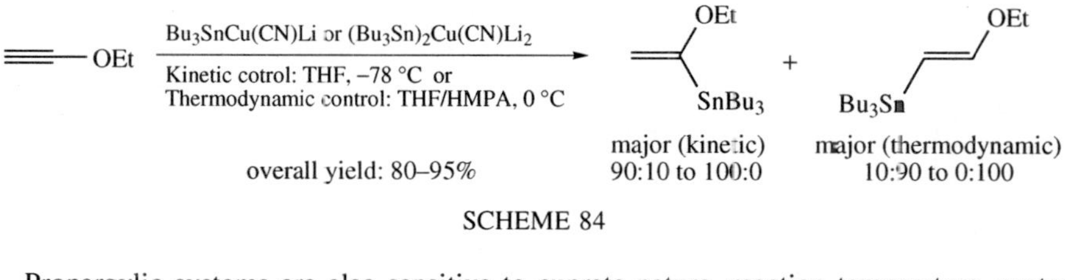

SCHEME 84

Propargylic systems are also sensitive to cuprate nature, reaction temperature, proton source and steric factors, but in general excellent regioselectivities are found. Care should

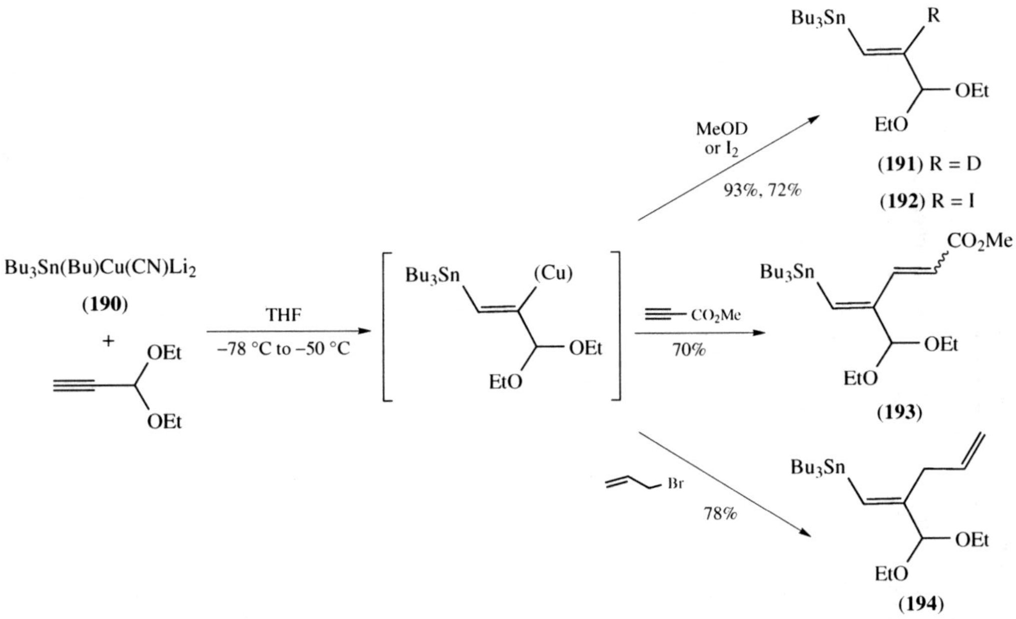

Bu3Sn(Bu)Cu(CN)Li2
(190)
+
OEt
OEt
THF
−78 °C to −50 °C
Bu3Sn
(Cu)
OEt
EtO
MeOD
or I2
93%, 72%
Bu3Sn
R
OEt
EtO
(191) R = D
(192) R = I
CO2Me
70%
Bu3Sn
CO2Me
OEt
EtO
(193)
Br
78%
Bu3Sn
OEt
EtO
(194)

SCHEME 85

be taken when strong electron-withdrawing groups are attached to the propargylic system due to eventual deprotonation of the alkyne by the stannylcuprate. Use of a less basic tin-cuprate or a stannylcopper reagent is required in these cases. Propargyl acetals have been employed by Quintard[178], Normant[179], Carreira[183] and their coworkers. Stannylcupration of 3,3-diethoxy-1-propyne using the higher-order cuprate $BuSn(Bu)Cu(CN)Li_2$[130] (**190**) occurs with complete stereo- and regiocontrol by *syn* addition of the tin–copper species, leading stereoselectively to 2-substituted *trans*-1-tributylstannyl-3,3-diethoxy-1-propenes **191–194** after quenching the mixture with deuteriomethanol, iodine, allyl bromide and methyl propiolate. Stannylated 1,3- and 1,4-dienes are easily made in this way (Scheme 85)[179].

Similar results have also been obtained with propargyl amines[184]. Stannylcupration of chiral propargylamines has been used in a peptide synthesis; thus, (*R*)-2,2-dimethyl-3-(*t*-butoxycarbonyl)-4-ethynyl oxazolidine (**195**) reacts with $Bu_3Sn(Bu)Cu(CN)Li_2$ to give an intermediate vinylstannane-oxazolidine **196**, which is efficiently converted into β,γ-unsaturated-α-amino acids **197** by palladium-mediated coupling with vinyl iodides (Scheme 86)[185]. The stannylcupration of propargyl ethers has been employed for the synthesis of the C(14)–C(26) chain segment of macrolide antitumor agent rhizoxin[186] and also in the synthesis of a tetrahydrofuran moiety of elfamycin antibiotic aurodox[187].

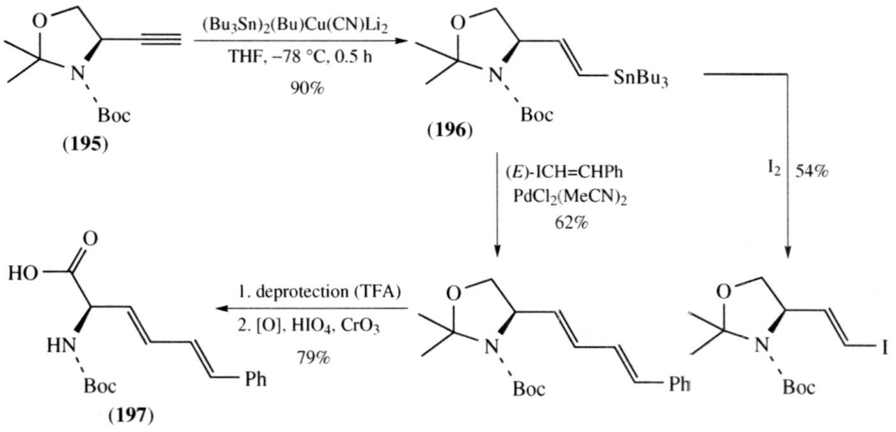

SCHEME 86

Propargyl alcohols were also reacted with stannylcuprates. Studies by Oehlschlager[175,176], Lipshutz[130], Pancrazi[188] and their coworkers show that the reaction can be a rather complex process with regio- and stereoselectivity arising from many different factors including the nature and order of the stannylcuprate, the kinetic or thermodynamic conditions used, the substitution pattern on the alkyne and conjugative effects of substituents. A summary of the most common behaviors of these substrates is given in Scheme 87.

Stannylcupration of long-chain alkynes carrying a good leaving group (or an electrophilic group such as carbonyl, epoxide etc.) at an appropriate distance from the triple bond leads to intramolecular trapping of the intermediate vinylcuprate, resulting in ring formation (Scheme 88)[189]. Stannylcupration has also been used to add a tributylstannyl group to the triple bond of butynediol; the resulting product is easily converted into 3-tributylstannylfuran (Scheme 88)[148].

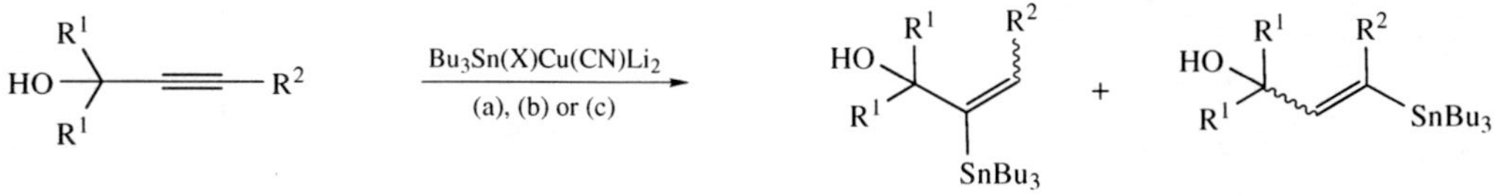

R^1	R^2	X	Conditions	(*E*):(*Z*), Yield (%)	(*E*):(*Z*), Yield (%)
H	Me	Bu	a	0 : 100, 42	
H	Me	Bu	b		100 : 0, 71
H	Me	Me	b		100 : 0, 65
H	Me	Bu_3Sn	a	*E* + Z, 57	*E* + Z, 25
H	Me	Bu_3Sn	b		100 : 0, 70
Me	H	Bu	c	87	

Conditions: (a) 2 eq. cuprate, THF, −10 °C, 12 h; (b) 2 eq. cuprate, MeOH, THF, −10 °C, 12 h; (c) 1.1 eq. cuprate, THF, −78 °C, 5 min.

SCHEME 87

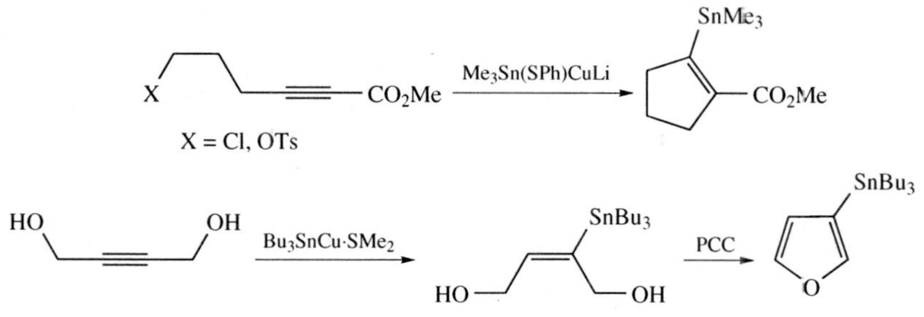

SCHEME 88

Stannylcuprates have also been used with diynes[190]. The addition leads to stannyl group attachment to both atoms of acetylene. Magriotis and coworkers[174] have used the reaction in a synthetic approach to enediynes. From a strategic point of view, the stannylcupration of acetylenes becomes a powerful tool in organic synthesis. It has been used in the total synthesis of a number of natural products such as certain mollusk pheromones[191] and retinol derivatives[192, 193]. Stannylcupration of acetylene and reaction with cyclic enones affords (Z)-β-stannylvinyl ketones, which undergo selective *syn*-to-tin addition of organometallic compounds to the carbonyl group. This remarkable remote stereocontrol, promoted by the vinyltin group, is a consequence of the anchoring of the organometallic reagent by the tin and the carbonyl groups[194].

In summary, the products of stannylcupration of acetylenes are very useful because the vinylstannanes produced can be converted by transmetallation with MeLi or BuLi into vinyllithium derivatives; these can be used directly or transformed into vinylcuprates again. This particular reactivity shown by vinylstannanes, together with their ability for intervening in Stille coupling reactions, convert these compounds to powerful synthons for annelation strategies, as shown in Scheme 89[195, 196].

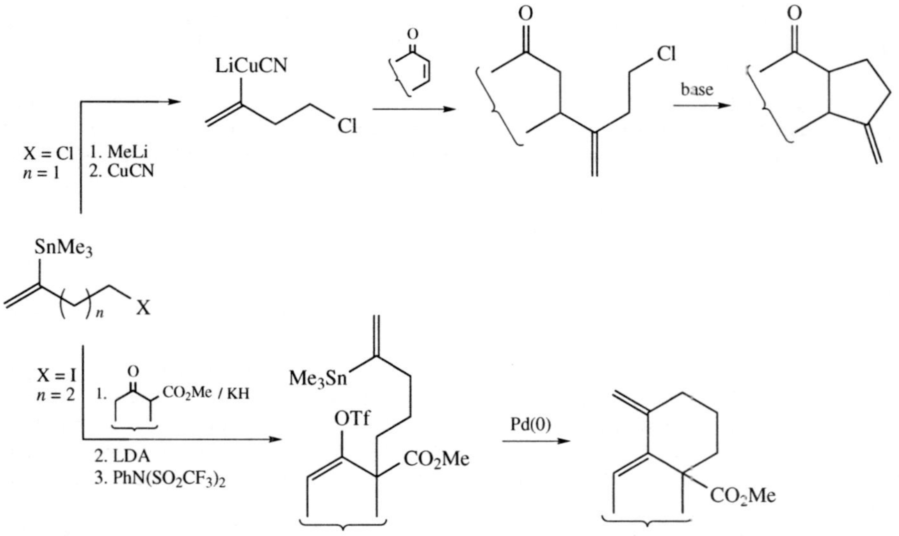

SCHEME 89

The stannylcupration of allenes was reported in 1989 by Fleming, Pulido and co-workers[9]. In the early work, a lower-order cuprate $(Bu_3Sn)_2CuLi$ (**198**), prepared from two equivalents of Bu_3SnLi and one equivalent of the copper(I) bromide–dimethyl sulfide complex, was used for the reactions. Allene affords vinylstannanes **200–202** after quenching the mixture with carbon electrophiles. Addition of the stannyl group and copper atom takes place respectively on the C-2 and C-1 of the allene to give **199**. 1,1-Dimethylallene shows the opposite regiochemistry leading selectively to allylstannanes resulting from addition of the copper–tin pair to the unsubstituted double bond with tin occupying the end of the former allenic system (Scheme 90)[9].

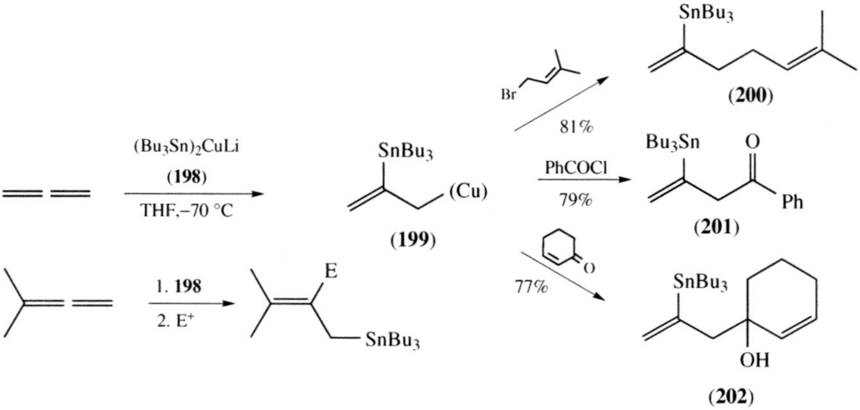

SCHEME 90

The higher-order stannylcuprate $Bu_3Sn(Bu)Cu(CN)Li_2$ (**190**) reacts with allene itself at $-100\,^\circ C$ under kinetic control, to give the intermediate cuprate **203**. If the reaction mixture is warmed to $-78\,^\circ C$, **203** rearranges to the thermodynamically more stable vinylstannane-allylcuprate **204**. Both intermediates react with electrophiles (E^+) affording allylstannanes **205** and **206** and vinylstannanes **207–213** (Scheme 91)[125b, 129]. As shown in Scheme 91, allylcuprate **204** can be trapped with a wide variety of electrophiles whereas the vinylcuprate **203** reacts only with strong electrophiles which are reactive enough to capture **203** before it isomerizes to the more stable **204**. This seriously limits the range of allylstannane products that can be prepared. Equilibration between vinyl- and allylcuprate species takes place at $-78\,^\circ C$. In effect, quenching of the reaction at any temperature between $-78\,^\circ C$ and $0\,^\circ C$ reveals the presence of only vinylstannanes. Attack of **204** on enones takes place selectively on the carbonyl group, giving the 1,2-addition product **213**. A similar result was observed in the reaction of **198** and cyclohexenone, leading to the formation of **202** (Scheme 90). These results are not surprising since allylcuprates are considerably harder nucleophiles than vinylcuprates, showing a great preference for the carbonyl addition[129].

There are two exceptions to this general behavior. Methyl iodide gives mixtures of allyl- and vinylstannanes whatever conditions are used, and methyl propiolate always gives the allylstannane product (a conjugated diene) even under thermodynamic conditions. The stannylcuprate reagent $(Bu_3Sn)_2Cu(CN)Li_2$ (**214**) displays the same kinetic and thermodynamic selectivity with allene leading to results comparable to that of the mixed cuprate **190**.

Substituted allenes also react, giving allyl- or vinylstannanes (**215–220**) depending upon the substitution pattern of the allene[90, 129]. In general, the cuprate adds to the less

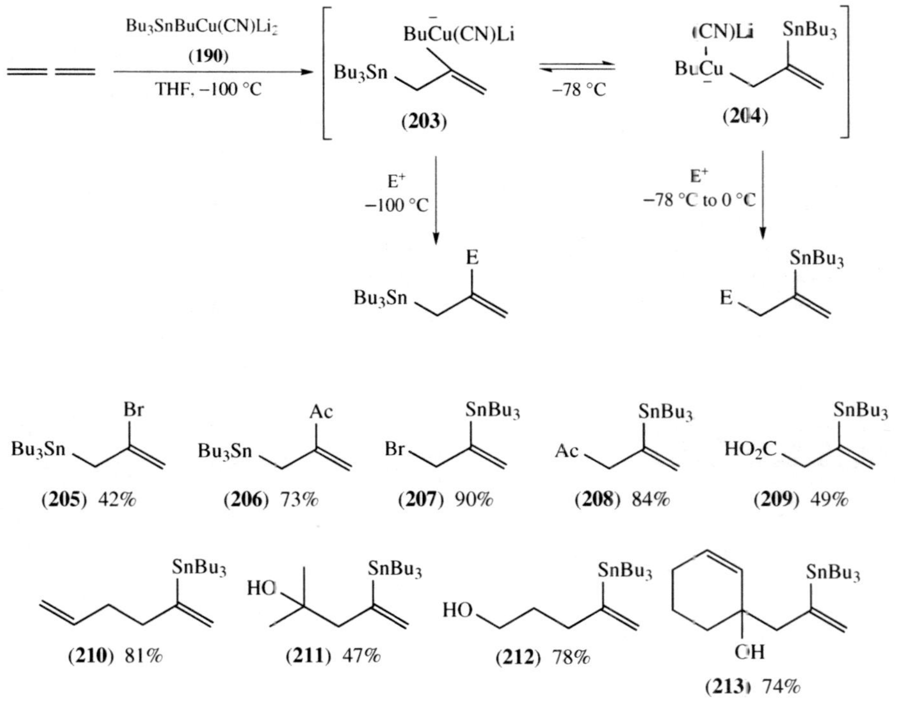

SCHEME 91

substituted double bond of the allene, approaching the system from the less hindered face of the allenic functionality. As a consequence, the stereochemistry of the resulting vinylstannane **216** is cleanly *E*, and that of the allylstannane **220** *Z*, which supports the suggested *syn*-addition mechanism. Phenylallenes give mixtures of regioisomers favoring largely the vinylstannane. On the other hand, alkyl-substituted allenes afford exclusively allystannanes with the tin atom attached to the less substituted carbon atom of the allene (Scheme 92)[129].

Allyl- and vinylstannanes are powerful synthons possessing the usual ability provided by the stannyl group for transmetallation and transition-metal-catalyzed coupling reactions. A combination of these capacities is shown in Scheme 93, which illustrates the remarkably rapid lithiation step, thus allowing ring formation without addition of BuLi to the carbonyl group[181a].

The regiochemistry of the stannylcupration of allenes using the lower-order cyanocuprate $(Bu_3Sn)Cu(CN)Li$ (**221**) was recently studied[197]. This cuprate, prepared by mixing one equivalent of tributylstannyllithium and one equivalent of copper(I) cyanide, reacts with allene at $-40\,^{\circ}C$, showing a regiochemistry opposite to that previously reported for higher-order cuprates **190** and **214** and Piers' reagent **198**. Capture of the intermediate allystannane-vinylcuprate species **222** with different electrophiles allows the selective formation of allylstannanes with different substitution pattern (Scheme 94). In this manner the overall regiochemistry in the stannylcupration of allene can be easily controlled in either sense just managing adequately the nature of the cuprate reagent. This method provides an easy entry to functionalized allylstannanes, which are masked allylic nucleophilic species of wide application in synthesis.The reaction has been extended to acetylenes, but

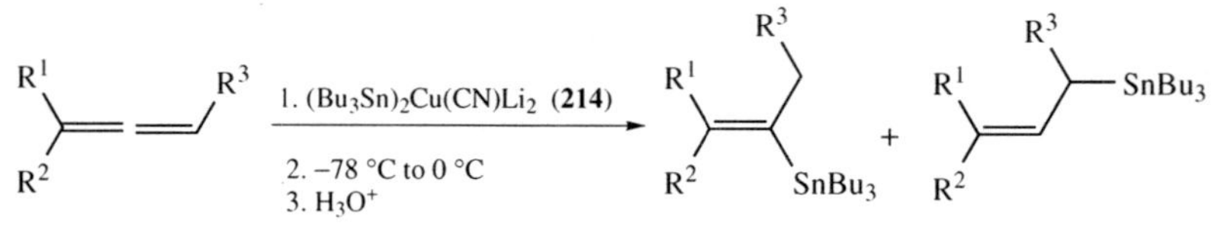

$R^1 = R^2 = R^3 = H$	(**215**) 92%	
$R^1 = Ph, R^2 = R^3 = H$	(**216**) 91%	(**217**) 9%
$R^1 = Me, R^2 = H, R^3 = Ph$	(**218**) 89%	
$R^1 = R^2 = Me, R^3 = H$		(**219**) 99%
$R^1 = Bu, R^2 = R^3 = H$		(**220**) 80%

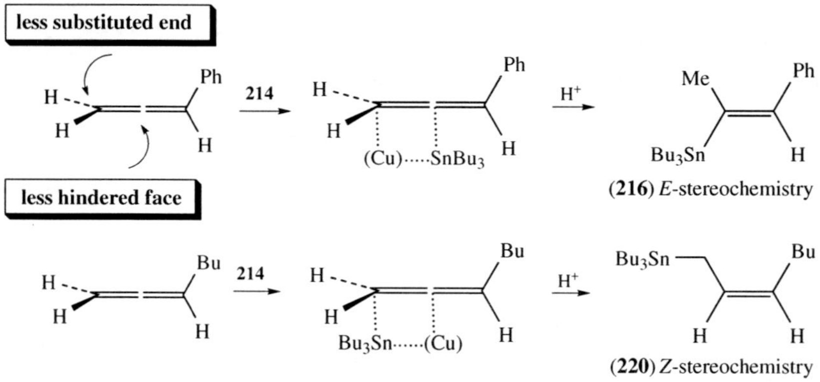

SCHEME 92

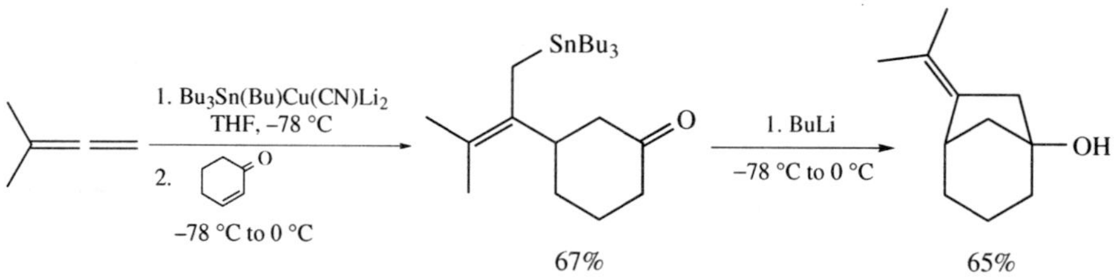

SCHEME 93

in this case the behavior of cuprate **221** is essentially the same as that observed for higher-order cuprates.

C. Mechanisms

In recent years, numerous investigations have been conducted in order to shed light on the *syn*-addition mechanism of cuprates to multiple bonds. Although the four-center

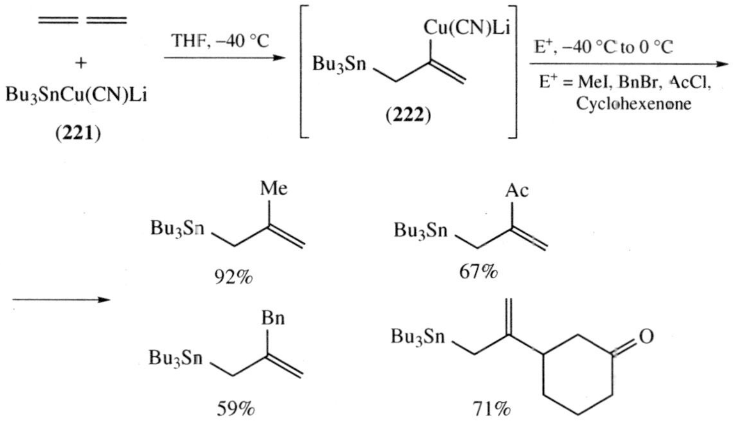

SCHEME 94

mechanism has been widely accepted so far, Nakamura and Mori[198], based on a detailed mechanistic study, have proposed that copper(III) species are probably intervening in the process of carbocupration of acetylenes. Bäckvall and coworkers[11, 104] have also supported this proposal for the silylcupration of dienes and acetylenes. According to the Nakamura hypothesis, the silyl- and stannylcupration of allene might proceed with an initial oxidative addition of metallocuprate to allene to give an intermediate metallocycle species **223** containing a copper(III) atom. Intermediate **223** evolves, through the equilibrium species **224** and **225**, toward the kinetic or thermodynamic cuprates **226** and **227**, after the final reductive elimination step. In this way, working below −40 °C under kinetic control, selective formation of cuprate **226** is achieved, whereas increasing the temperature allows the rapid equilibration of species **224** into **225**, leading to the formation of the thermodynamically controlled cuprate **227** (Scheme 95).

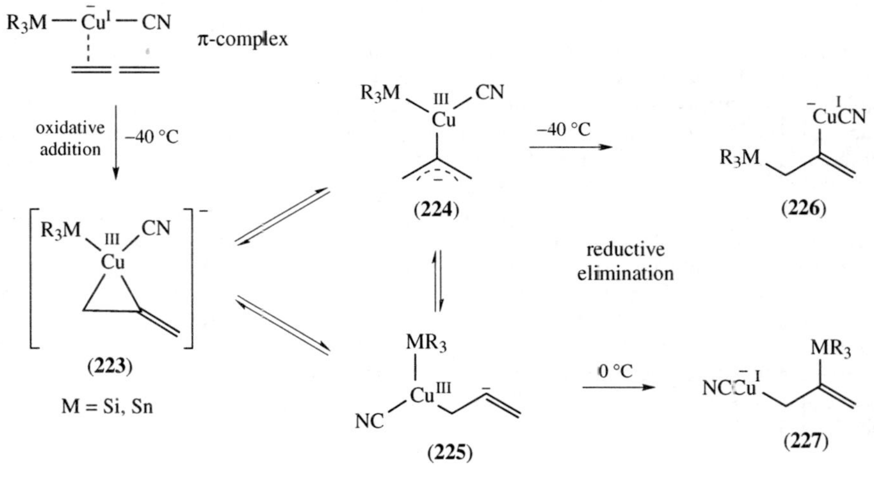

SCHEME 95

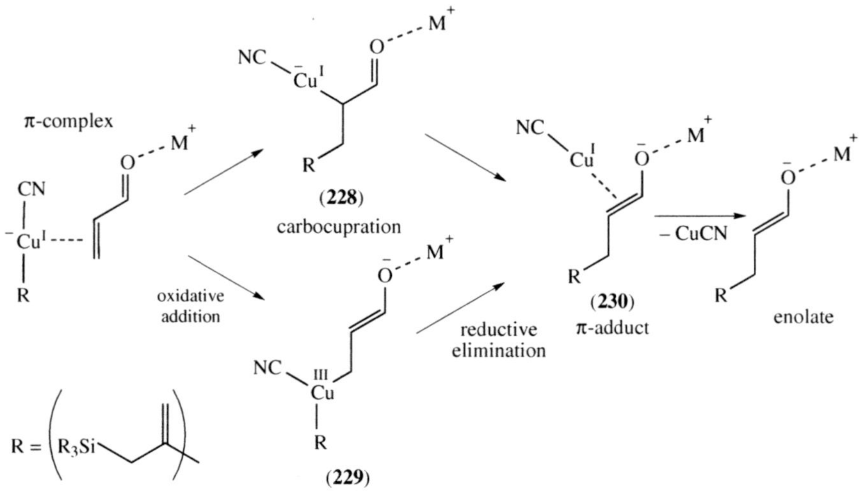

SCHEME 96

Lately, other authors have claimed the intervention of copper(III) species in organocuprate reactions. Thus, Woodward[199] suggests two possible pathways for the addition of organocuprates to enones. The first, based on the sole participation of copper(I) intermediates **228**, involves carbocupration of the enone and formation of a π-adduct **230** which, upon elimination of copper cyanide, affords the final addition product. Alternatively, conjugate addition of the cuprate to the enone (formally an oxidative addition step) affords the intermediate copper(III) species **229**, which by reductive elimination leads to the π-adduct and then to the final product (Scheme 96). Analogous silyl- and stannylcupration processes should proceed similarly.

More recently, an experimental and theoretical study of the mechanism of stannylcupration of ynones and ynoates has been reported and compared to the corresponding carbocupration reaction, with particular emphasis on stereoselectivity[200].

In conclusion, silicon and tincuprates are a particular class of heterocuprates that not only exhibit the extremely powerful possibilities of organocopper compounds, but also add the rich chemistry of silicon and tin elements to be exploited in synthesis. Their chemistry is of wide applicability, very efficient and easy to perform. Frequently, the problem is the choice of the most convenient reagent to be used. A huge amount of total synthesis involving the use of organocopper compounds has been published and many have used silicon and tincuprate reagents. In spite of the great progress achieved in the area, there is still much lack of knowledge of the mechanistic insights. Organosilicon- and organotincopper and cuprate compounds are not always stable enough to be easily trapped. No reactive intermediates have been captured and therefore their reaction pathways are considered by analogy with other transition metals. Although sometimes we do not know all the details of the reaction, there is no doubt about the extraordinary usefulness and the reproducibility of the cuprate chemistry, which has emerged at present as a cornerstone of organic synthesis. Ongoing developments in this area will bring new advances in the near future.

IV. REFERENCES

1. H. Gilman, R. G. Jones and L. A. Woods, *J. Org. Chem.*, **17**, 1630 (1952).

75. I. Fleming, D. Higgins, N. J. Lawrence and A. P. Thomas, *J. Chem. Soc., Perkin Trans. 1*, 3331 (1992).
76. I. Fleming and D. Marchi, *Synthesis*, 560 (1981).
77. I. Fleming and N. Terret, *J. Organomet. Chem.*,**264**, 99 (1984).
78. (a) A. Casarini, B. Jousseaume, D. Lazzari, E. Porciatti, G. Reginato, A. Ricci and G. Seconi, *Synlett.*, 981 (1992).
(b) J. Pornet, in *Science of Synthesis* (Ed. I. Fleming), Vol. 4, Thieme-Verlag, Stuttgart, 2002, pp. 669–683.
79. I. Fleming and N. Terret, *J. Chem. Soc., Perkin Trans. 1*, 2645 (1998).
80. I. Fleming and D. Higgins, *J. Chem. Soc., Perkin Trans. 1*, 2673 (1998).
81. T. K. Sarkar, in *Science of Synthesis* (Ed. I. Fleming), Vol. 4, Thieme-Verlag, Stuttgart, 2002, pp. 837–925.
82. (a) G. Majetich, J. S. Song, C. Ringold, G. A. Nemeth and M. G. Newton, *J. Org. Chem.*, **56**, 3973 (1991).
(b) C. Y. Hong, N. Kado and L. E. Overman, *J. Am. Chem. Soc.*, **115**, 11028 (1993).
83. E. J. Corey, J. M. Lee and D. R. Liu, *Tetrahedron Lett.*, **35**, 9149 (1994).
84. (a) J. G. Smith, S. E. Drozda, S. P. Petraglia, N. R. Quinn, E. M. Rice, B. S. Taylor and M. Viswanathan, *J. Org. Chem.*, **49**, 4112 (1984).
(b) B. Laycock, W. Kitching and G. Wickham, *Tetrahedron Lett.*, **24**, 5785 (1983).
85. P. H. Dussault, C. T. Eary, R. J. Lee and U. R. Zope, *J. Chem. Soc., Perkin Trans. 1*, 2189 (1999).
86. M. P. Sibi, B. J. Harris, J. J. Shay and S. Hajra, *Tetrahedron*, **54**, 7221 (1998).
87. (a) A. Degl'Innocenti, C. Faggi and A. Ricci, *J. Org. Chem.*, **53**, 3612 (1988).
(b) B. F. Bonini, M. Comes-Franchini, M. Fochi, J. Gawronski, G. Mazzanti, A. Ricci and G. Varchi, *Eur. J. Org. Chem.*, **2**, 437 (1999).
(c) B. F. Bonini, M. Comes-Franchini, M. Fochi, G. Mazzanti and A. Ricci, *J. Organomet. Chem.*, **567**, 181 (1998).
88. (a) B. F. Bonini, M. Comes-Franchini, M. Fochi, F. Laboroi, G. Mazzanti, A. Ricci and G. Varchi, *J. Org. Chem.*, **64**, 8008 (1999).
(b) B. F. Bonini, M. Comes-Franchini, M. Fochi, G. Mazzanti, A. Ricci and G. Varchi, *Polyhedron*, **19**, 529 (2000).
89. (a) M. Calle, P. Cuadrado, A. M. González and R. Valero, *Synthesis*, 1949 (2001).
(b) M. A. González, M. Calle and P. Cuadrado, *Eur. J. Org. Chem.*, **36**, 6089 (2007).
90. S. M. Singh and A. C. Oehlschlager, *Can. J. Chem.*, **69**, 1872 (1991).
91. D. L. J. Clive, C. Zhang, Y. Zhou and Y. Tao, *J. Organomet. Chem.*, **489**, C35 (1995).
92. (a) I. Fleming and F. Roessler, *J. Chem. Soc., Chem. Commun.*, 276 (1980).
(b) I. Fleming and N. J. Lawrence, *J. Chem. Soc., Perkin Trans. 1*, 2679 (1998).
(c) M. A. Cubillo, I. Fleming, W. Friedhoff and P. D. W. Woode, *J. Organomet. Chem.*, **624**, 69 (2001).
93. A. Barbero, Y. Blanco, C. García and F. J. Pulido, *Synthesis*, 1223 (2000).
94. G. Stork and E. Colvin, *J. Am. Chem. Soc.*, **93**, 2080 (1971).
95. (a) I. Fleming and N. J. Lawrence, *Tetrahedron Lett.*, **31**, 3645 (1990).
(b) S. C. Archibald, D. J. Barden, J. F. Y. Bazin, I. Fleming, C. F. Foster, Ajay K. Mandal, Amit K. Mandal, D. Parker, K. Takaki, A. C. Ware, A. R. B. Williams and A. B. Zwicky, *Org. Biomol. Chem.*, **2**, 1051 (2004). See also Amit K. Mandal, *Org. Lett.*, **4**, 2043 (2002).
96. (a) L. Capella, A. Capperucci, D. Lazzari, P. Dembech, G. Reginato and A. Ricci, *Tetrahedron Lett.*, **34**, 3311 (1993).
(b) X. Huang and L. Xu, *Synthesis*, 231 (2006).
97. S. E. Denmark and A. Thorarensen, *J. Am. Chem. Soc.*, **119**, 125 (1997).
98. G. Reginato, A. Mordini, M. Valachi and E. Grandini, *J. Org. Chem.*, **64**, 9211 (1999).
99. (a) A. Ricci, E. Blart, M. Comes-Franchini, G. Reginato and P. Zani, *Pure Appl. Chem.*, **68**, 679 (1996).
(b) G. Reginato, P. Meffre and F. Gaggini, *Amino-Acids*, **29**, 81 (2005).
100. D. Alvisi, E. Blart, B. F. Bonini, G. Mazzanti, A. Ricci and P. Zani, *J. Org. Chem.*, **61**, 7139 (1996).
101. I. Fleming and E. M. de Marigorta, *J. Chem. Soc., Perkin Trans. 1*, 889 (1999).
102. Y. Okuda, Y. Morizawa, K. Oshima and H. Nozaki, *Tetrahedron Lett.*, **25**, 2483 (1984).

103. T. Nonaka, Y. Okuda, W. Tuckmantel, K. Wakamatsu, K. Oshima, K. Utimoto and H. Nozaki, *Tetrahedron*, **42**, 4427 (1986).
104. V. Liepins, A. S. E. Karlstrom and J. E. Bäckvall, *J. Org. Chem.*, **67**, 2136 (2002).
105. I. Fleming and F. J. Pulido, *J. Chem. Soc., Chem. Commun.*, 1010 (1986).
106. P. Cuadrado, A. M. González, F. J. Pulido and I. Fleming, *Tetrahedron Lett.*, **29**, 1825 (1988).
107. F. J. Blanco, P. Cuadrado, A. M. González and F. J. Pulido, *Tetrahedron Lett.*, **35**, 8881 (1994).
108. A. Barbero, C. García and F. J. Pulido, *Tetrahedron Lett.*, **40**, 6649 (1999).
109. (a) M. Bertrand, G. Cul and J. Viala, *Tetrahedron Lett.*, 1785 (1977).
(b) S. Ma, *Aldrichimica Acta*, **40**, 91 (2007).
110. A. Alexakis and J. F. Normant, *Israel J. Chem.*, **24**, 113 (1984).
111. I. Fleming, Y. Landais and P. Raithby, *J. Chem. Soc., Perkin Trans. 1*, 715 (1991).
112. A. Barbero, P. Castreño, C. García and F. J. Pulido, *J. Org. Chem.*, **66**, 7723 (2001).
113. A. Barbero, P. Castreño and F. J. Pulido, *Org. Lett.*, **5**, 4045 (2003).
114. A. Barbero and F. J. Pulido, *Synthesis*, 779 (2004).
115. A. Barbero, Y. Blanco and F. J. Pulido, *J. Org. Chem.*, **70**, 6876 (2005).
116. A. Barbero, P. Castreño and F. J. Pulido, *J. Am. Chem. Soc.*, **127**, 8022 (2005).
117. A. Barbero, P. Castreño, G. Fernández and F. J. Pulido, *J. Org. Chem.*, **70**, 10747 (2005).
118. A. Barbero, F. J. Pulido and M. C. Sañudo, *Beilstein J. Org. Chem.*, **3**, 16 (2007).
119. Y. Morizawa, H. Oda, K. Oshima and H. Nozaki, *Tetrahedron Lett.*, **25**, 1163 (1984).
120. V. Liepins, A. S. E. Karlstrom and J. E. Bäckvall, *Org. Lett.*, **2**, 1237 (2000).
121. (a) M. Pereyre, J. P. Quintard and A. Rahm, *Tin in Organic Synthesis*, Butterworth, London, 1987.
(b) A. G. Davies, *Organotin Chemistry*, VCH, Weinheim, 2004.
122. W. C. Still, *J. Am. Chem. Soc.*, **99**, 4836 (1977).
123. E. Piers, H. E. Morton and J. Chong, *Can. J. Chem.*, **65**, 78 (1987).
124. H. Westmijze, K. Ruitenberg, J. Meijer and P. Vermeer, *Tetrahedron Lett.*, **23**, 2797 (1982).
125. (a) C. Tamborskii, F. E. Ford and E. J. Soloski, *J. Org. Chem.*, **28**, 237 (1963).
(b) A. Barbero, P. Cuadrado, I. Fleming, A. M. González and F. J. Pulido, *J. Chem. Soc., Chem. Commun.*, 1030 (1990).
126. (a) W. C. Still, *J. Am. Chem. Soc.*, **100**, 1481 (1978).
(b) J. San Filippo Jr. and J. Silberman, *J. Am. Chem. Soc.*, **104**, 2831 (1982).
127. B. L. Chenard and C. M. Van Zyl, *J. Org. Chem.*, **51**, 3561 (1986).
128. J. A. Rincon, Ph.D thesis, Univesidad de Valladolid, Valladolid, Spain, 1999.
129. A. Barbero, P. Cuadrado, I. Fleming, A. M. González and F. J. Pulido, *J. Chem. Soc., Perkin Trans. 1*, 327 (1992).
130. B. H. Lipshutz, E. L. Ellsworth, S. H. Dimock and D. C. Reuter, *Tetrahedron Lett.*, **30**, 2065 (1989).
131. B. H. Lipshutz and D. C. Reuter, *Tetrahedron Lett.*, **30**, 4617 (1989).
132. A. C. Oehlschlager, M. W. Hutzinger, R. Aksela, S. Sharma and S. M. Sing, *Tetrahedron Lett.*, **31**, 165 (1990).
133. J. R. Behling, K. A. Babiak, J. S. Ng, A. L. Campbell, R. Moretti, M. Koerner and B. H. Lipshutz, *J. Am. Chem. Soc.*, **110**, 2641 (1988).
134. A. Barbero, P. Cuadrado, I. Fleming, A. M. González and F. J. Pulido, *J. Chem. Res. (S)*, 291 (1990).
135. P. P. Power, *Prog. Inorg. Chem.*, **39**, 75 (1991).
136. R. R. Dykstra, J. P. Borst and H. J. Reich, *Abs. ACS Natl. Meeting*, **206**, no.ORGN229 (1993).
137. E. Buncel, R. D. Gordon and T. K. Venkatachalam, *J. Organomet. Chem.*, **507**, 81 (1996).
138. M. Veith, C. Ruloff, V. Huch and F. Tollner, *Angew. Chem., Int. Ed. Engl.*, **27**, 1381 (1988).
139. D. Reed, D. Stalke and D. S. Wright, *Angew. Chem., Int. Ed. Engl.*, **30**, 1459 (1991).
140. H. Nozaki, in *Organometallics in Synthesis: A Manual* (Ed. M. Schlosser), Chap. 8, John Wiley & Sons, Ltd, Chichester, 1994.
141. L. S. Hegedus, in *Organometallics in Synthesis: A Manual* (Ed. M. Schlosser), Chap. 5, John Wiley & Sons, Ltd, Chichester, 1994.
142. A. Barbero, P. Cuadrado, I. Fleming, A. M. González and F. J. Pulido, *J. Chem. Soc., Chem. Commun.*, 352 (1992).
143. E. Piers, J. M. Chong and H. E. Morton, *Tetrahedron Lett.*, **22**, 4905 (1981).

144. A. Barbero, P. Cuadrado, I. Fleming, A. M. González, F. J. Pulido and R. Rubio, *J. Chem. Soc., Perkin Trans. 1*, 1657 (1993).
145. E. Piers and J. Y. Roberge, *Tetrahedron Lett.*, **32**, 5219 (1991).
146. K. Ruitenberg, H. Westmijze, J. Meijer, C. J. Elsevier and P. Vermeer, *J. Organomet. Chem.*, **241**, 417 (1983).
147. E. Piers and J. M. Chong, *Can. J. Chem.*, **66**, 1425 (1988).
148. I. Fleming and M. Taddei, *Synthesis*, 898 (1985).
149. B. H. Lipshutz, *Synlett*, **3**, 119 (1990).
150. S. Kusuda, Y. Ueno, T. Hagiwara and T. Toru, *J. Chem. Soc., Perkin Trans. 1*, 1981 (1993).
151. A. Krief, L. Provins and W. Dumont, *Angew. Chem., Int. Ed.*, **38**, 1946 (1999).
152. C. R. Johnson and J. F. Kadow, *J. Org. Chem.*, **52**, 1493 (1987).
153. J. E. Baldwin, R. M. Adlington and J. Robertson, *Tetrahedron*, **47**, 6795 (1991).
154. (a) E. Piers, J. M. Chong and H. E. Morton, *Tetrahedron*, **45**, 363 (1989).
(b) E. Piers, T. Wong and K. A. Ellis, *Can. J. Chem.*, **70**, 2058 (1992).
155. D. S. Dodd, H. D. Pierce and A. C. Oehlschlager, *J. Org. Chem.*, **57**, 5250 (1992).
156. T. E. Nielsen, M. A. Cubillo and D. Tanner, *J. Org. Chem.*, **67**, 7309 (2002).
157. E. Piers, J. M. Chong and B. A. Keay, *Tetrahedron. Lett.*, **26**, 6265 (1985).
158. (a) G. Reginato, A. Mordini, A. Degl'Innocenti, S. Manganiello, A. Cappperucci and G. Poli, *Tetrahedron*, **54**, 10227 (1998).
(b) G. Reginato, A. Cappperucci, A. Degl'Innocenti, A. Mordini and S. Pecchi, *Tetrahedron*, **51**, 2129 (1995).
159. E. Piers, E. M. Boehringer and J. G. K. Yee, *J. Org. Chem.*, **63**, 8642 (1998).
160. K. Nilsson, T. Andersson, C. Ullenius, A. Gerold and N. Krause, *Chem. Eur. J.*, **4**, 2051 (1998).
161. E. Piers and R. D. Tillyer, *J. Org. Chem.*, **53**, 5366 (1988).
162. (a) J. Thibonnet, V. Launay, M. Abarbri, A. Duchene and J. L. Parrain, *Tetrahedron. Lett.*, **39**, 4277 (1998).
(b) J. Thibonnet, M. Abarbri, A. Duchene and J. L. Parrain, *Synlett*, 141 (1999).
(c) J. Thibonnet, M. Abarbri, J. L. Parrain and A. Duchene, *Tetrahedron*, **59**, 4433 (2003).
163. (a) E. Piers and A. V. Gavai, *J. Org. Chem.*, **55**, 2374 (1990).
(b) E. Piers and R. W. Friesen, *Can. J. Chem.*, **70**, 1204 (1992).
(c) E. Piers and J. S. M. Wai, *Can. J. Chem.*, **72**, 146 (1994).
164. (a) Y. Naruta and K. Maruyama, *J. Chem. Soc., Chem. Commun.*, 1264 (1983).
(b) T. Takeda, S. Ogawa, N. Ohta and T. Fujiwara, *Chem. Lett.*, 1967 (1987).
165. J. M. Cuerva, E. Gómez-Bengoa, M. Méndez and A. M. Echavarren, *J. Org. Chem.*, **62**, 7540 (1997).
166. B. M. Trost and I. Fleming (Eds.), *Comprehensive Organic Synthesis*, Vol. 2, Chap. 2.2, 3.1 and 4.3, Pergamon, Oxford, 1991.
167. (a) D. E. Seitz and S. H. Lee, *Tetrahedron Lett.*, **22**, 4909 (1981).
(b) H. Imanieh, D. MacLeod, P. Quayle, Y. Zhao and G. M. Davies, *Tetrahedron Lett.*, **33**, 405 (1992).
168. (a) E. Piers and H. L. A. Tse, *Tetrahedron Lett.*, **25**, 3155 (1984).
(b) S. R. Gilbertson, C. A. Challener, M. E. Boss and W. D. Wulff, *Tetrahedron Lett.*, **29**, 4975 (1988).
169. (a) E. Piers and H. L. A. Tse, *Can. J. Chem.*, **71**, 983 (1993).
(b) E. Piers, E. J. McEachern, M. A. Romero and P. L. Gladston, *Can. J. Chem.*, **75**, 694 (1997).
(c) I. N. Houpis, L. DiMichele and A. Molina, *Synlett*, 365 (1993).
170. (a) S. Matsubara, J. I. Hibino, Y. Morizawa, K. Oshima and H. Nozaki, *J. Organomet. Chem.*, **285**, 163 (1985).
(b) J. I. Hibino, S. Matsubara, Y. Morizawa, K. Oshima and H. Nozaki, *Tetrahedron Lett.*, **25**, 2151 (1984).
171. E. Piers and J. M. Chong, *J. Chem. Soc., Chem. Commun.*, 934 (1983).
172. S. D. Cox and F. Wudl, *Organometallics*, **2**, 184 (1983).
173. M. W. Hutzinger, R. D. Singer and A. C. Oehlschlager, *J. Am. Chem. Soc.*, **112**, 9397 (1990).
174. P. A. Magriotis, M. E. Scott and K. D. Kim, *Tetrahedron Lett.*, **32**, 6085 (1991).
175. R. Aksela and A. C. Oehlschlager, *Tetrahedron*, **47**, 1163 (1991).
176. R. D. Singer, M. W. Hutzinger and A. C. Oehlschlager, *J. Org. Chem.*, **56**, 4933 (1991).

177. (a) J. P. Marino and J. K. Long, *J. Am. Chem. Soc.,* **110**, 7916 (1988).
(b) J. P. Marino, M. V. M. Emonds, P. J. Stengel, A. R. M. Oliveira, J. Simonelli and J. T. B. Ferreira, *Tetrahedron Lett.*, **33**, 49 (1992).
178. I. Beaudet, J. L. Parrain and J. P. Quintard, *Tetrahedron Lett.*, **32**, 6333 (1991).
179. I. Marek, A. Alexakis and J. F. Normant, *Tetrahedron Lett.*, **32**, 6337 (1991).
180. E. J. Corey and R. H. Wollenberg, *J. Am. Chem. Soc.,* **96**, 5581 (1974).
181. (a) A. Barbero, P. Cuadrado, I. Fleming, A. M. González, F. J. Pulido and R. Rubio, *Tetrahedron Lett.*, **33**, 5841 (1992).
(b) A. Barbero, P. Cuadrado, C. García, J. A. Rincón and F. J. Pulido, *J. Org. Chem.*, **63**, 7531 (1998).
182. J. A. Cabezas and A. C. Oehlschlager, *Synthesis,* 432 (1994).
183. Y. Kim, R. A. Singer and E. M. Carreira, *Angew. Chem., Int. Ed.*, **37**, 1261 (1998).
184. (a) L. Capella, A. Degl'Innocenti, A. Mordini, G. Reginato, A. Ricci and G. Seconi, *Synthesis,* 1201 (1991).
(b) G. Reginato, A. Mordini, F. Messina and A. Degl'Innocenti, *Tetrahedron,* **52**, 10985 (1996).
(c) G. Reginato, A. Mordini, M. Valacchi and R. Piccardi, *Tetrahedron: Asymmetry,* **13**, 595 (2002).
185. G. Reginato, A. Mordini and M. Caracciolo, *J. Org. Chem.*, **62**, 6187 (1997).
186. J. D. White, M. A. Holoboski and N. J. Green, *Tetrahedron Lett.,* **38**, 7333 (1997).
187. D. Craig, A. H. Payne and P. Warner, *Synlett*, 1264 (1998).
188. (a) J. F. Betzer, F. Delaloge, B. Muller, A. Pancrazi and J. Prunet, *J. Org. Chem.*, **62**, 7768 (1997).
(b) J. F. Betzer and A. Pancrazi, *Synlett*, 1129 (1998).
(c) J. F. Betzer and A. Pancrazi, *Synthesis*, 629 (1999).
189. P. J. Parson, K. Booker-Milburn, S. H. Brooks, S. Martel and M. Stefinovic, *Synth. Commun.*, **24**, 2159 (1994).
190. G. Zweifel and W. Leong, *J. Am. Chem. Soc.,* **109**, 6409 (1987).
191. R. Alvarez, M. Herrero, S. López and A. R. de Lera, *Tetrahedron,* **54**, 6793 (1998).
192. R. Alvarez, B. Iglesias, S. López and A. R. de Lera, *Tetrahedron Lett.*, **39**, 5659 (1998).
193. J. Thibonnet, G. Prie, M. Abarbri, A. Duchêne and J. L. Parrain, *Tetrahedron Lett.,* **40**, 3151 (1999).
194. (a) A. Barbero, F. J. Pulido, J. A. Rincón, P. Cuadrado, D. Galisteo and H. Martinez-García, *Angew. Chem., Int. Ed.*, **40**, 1261 (2001).
(b) A. Barbero, F. J. Pulido and J. A. Rincón, *J. Am. Chem. Soc.,* **125**, 12049 (2003).
195. E. Piers and J. Renaud, *Synthesis,* 74 (1992).
196. E. Piers, R. W. Friesen and B. A. Keay, *Tetrahedron,* **47**, 4555 (1991).
197. A. Barbero and F. J. Pulido, *Tetrahedron Lett.,* **45**, 3765 (2004).
198. E. Nakamura and S. Mori, *Angew. Chem., Int. Ed.*, **39**, 3750 (2000).
199. S. Woodward, *Chem. Soc. Rev.*, **29**, 393 (2000).
200. M. Ahlquist, T. E. Nielsen, S. Le Quement, D. Tanner and P.-O. Norrby, *Chem. Eur. J.,* **12**, 2866 (2006).

CHAPTER **17**

Copper-mediated and copper-catalyzed addition and substitution reactions of extended multiple bond systems

NORBERT KRAUSE and ÖZGE AKSIN-ARTOK

Dortmund University of Technology, Organic Chemistry II, Otto-Hahn-Strasse 6, D-44227 Dortmund, Germany
Fax: +49-231-7553884; e-mail: norbert.krause@tu-dortmund.de

I. INTRODUCTION 857
II. COPPER-MEDIATED AND COPPER-CATALYZED ADDITION REACTIONS TO EXTENDED MICHAEL ACCEPTORS 858
  A. Acceptor-substituted Dienes 858
  B. Acceptor-substituted Enynes 863
  C. Acceptor-substituted Polyenynes 871
III. COPPER-MEDIATED AND COPPER-CATALYZED SUBSTITUTION REACTIONS OF EXTENDED SUBSTRATES 872
IV. CONCLUSION 875
V. REFERENCES 875

I. INTRODUCTION

Since the pioneering work of Gilman and coworkers, who carried out the first investigations on organocopper compounds RCu[1] and lithium diorganocuprates R_2CuLi[2], the latter reagents (referred to as Gilman cuprates) have become highly popular for carbon–carbon bond formation. Nowadays, organocuprates are the reagents of choice not only for substitution reactions of many saturated (haloalkanes, acid chlorides, oxiranes) and unsaturated (allylic and propargylic derivatives) electrophiles, but also for 1,4-addition reactions of α,β-unsaturated carbonyl compounds and for carbocuprations of non-activated alkynes[3,4]. In these processes, the unique reactivity of organocuprates R_2CuLi relies on the

The chemistry of organocopper compounds
Edited by Z. Rappoport and I. Marek

interplay of the 'soft', nucleophilic copper and the 'hard', electrophilic lithium, offering control over reactivity and selectivity through 'fine-tuning' of the reagent. The tremendous achievements in various fields of organocopper chemistry over the last decades include the elucidation of the structure of organocopper compounds[5,6] and the mechanism of their transformations[7], new copper-mediated and copper-catalyzed processes[8], diastereoselective reactions[9], as well as highly enantioselective substitution[10,11] and conjugate addition reactions[10,12]. The high standards attained in these fields are documented in numerous applications of copper-promoted transformations in target-oriented synthesis[13].

While 1,4-addition and 1,3-substitution (S_N2') reactions of simple unsaturated substrates have predominated so far, analogous transformations of ambident substrates with extended multiple bond systems (i.e. with two or more reactive positions) have gained importance in the last two decades since they usually proceed with high regio- and stereoselectivity, in particular when the substrate contains a triple bond besides one or more conjugated double bonds. These new reaction types not only open up unprecedented routes to important target molecules, but also provide insights into the mechanism of copper-mediated carbon–carbon bond formation[14].

II. COPPER-MEDIATED AND COPPER-CATALYZED ADDITION REACTIONS TO EXTENDED MICHAEL ACCEPTORS

A. Acceptor-substituted Dienes

Thanks to their ambident character, acceptor-substituted dienes can provide several isomeric products in copper-mediated or copper-catalyzed Michael additions, therefore making it particularly important to control the regio- and stereoselectivity of these transformations (Scheme 1).

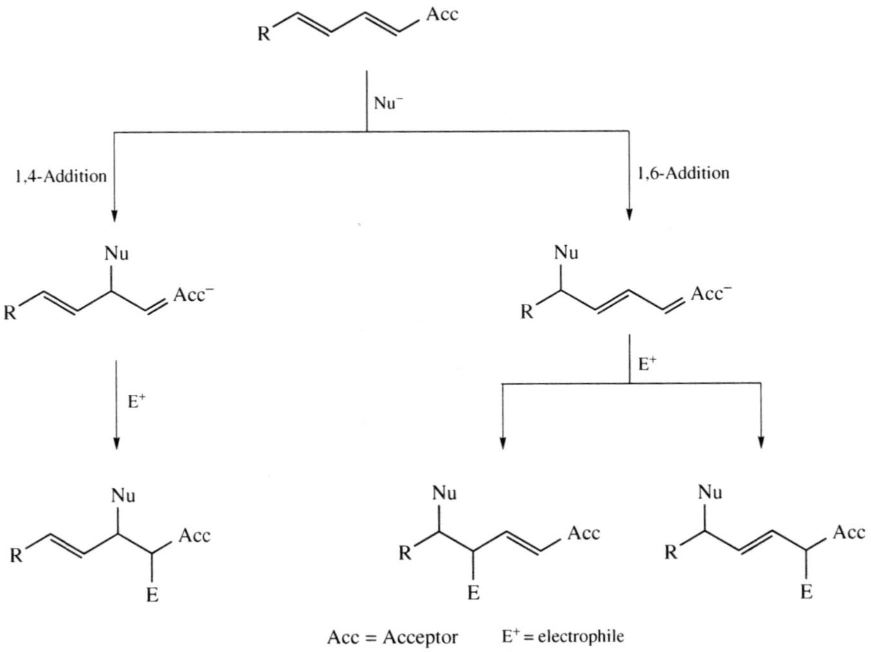

SCHEME 1. Regioselectivity in conjugate addition reactions to acceptor-substituted dienes

Besides direct nucleophilic attack at the acceptor group, an activated diene may undergo a 1,4- or 1,6-addition; in the latter case, capture of the ambident enolate with a soft electrophile can take place at two different positions. Hence, the nucleophilic addition can result in the formation of three regioisomeric alkenes, which may in addition be formed as *E*/*Z* isomers. Moreover, depending on the nature of the nucleophile and electrophile, the addition products may contain one or two stereogenic centers and, as a further complication, basic conditions may give rise to the isomerization of the initially formed β,γ-unsaturated carbonyl compounds (and other acceptor-substituted alkenes of this type) to the thermodynamically more stable conjugated isomer (equation 1).

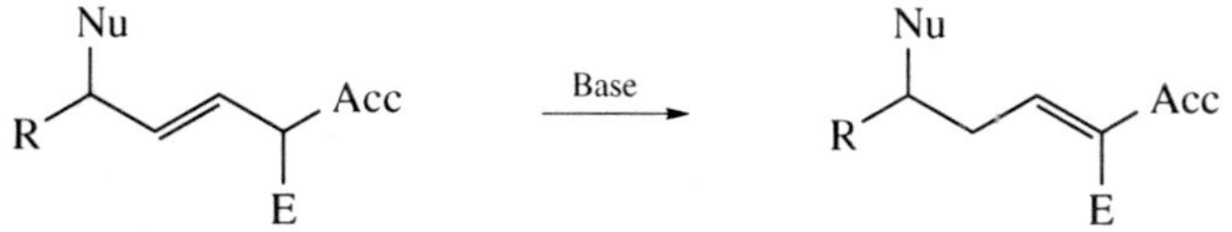

(1)

The first example of a cuprate addition to an acceptor-substituted diene was reported by Näf and coworkers[15], who used lithium di-(*Z*)-1-heptenylcuprate in a Michael addition to dienoate **1** (equation 2). The reaction proceeds with high regioselectivity, furnishing a 1:1 mixture of the two isomeric 1,6-adducts **2**, which were converted into the Bartlett pear constituent ethyl (2*E*,6Z)-2,6-dodecadienoate (**3**) by basic isomerization.

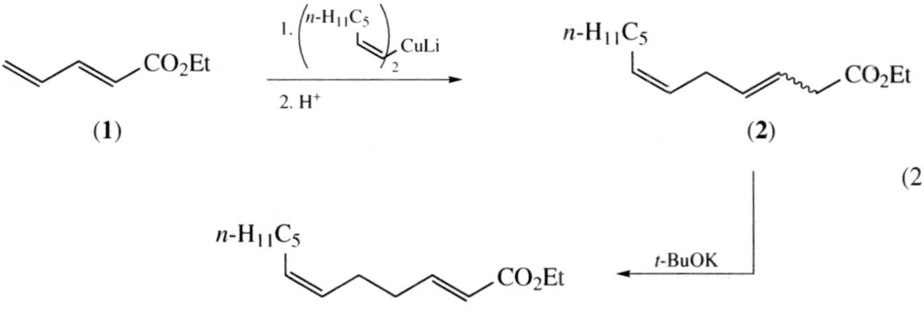

(**1**) (**2**) (**3**)

(2)

In analogous reactions, several other groups reported the exclusive formation of 1,6-addition products, suggesting that the regioselectivity of the transformation is not affected by the choice of the organocopper reagent[16]. Whereas the use of organocopper compounds RCu predominantly results in the formation of adducts with *E* configuration, the corresponding Gilman cuprates R_2CuLi usually afford 1:1 mixtures of the *E* and *Z* isomer[16b]. Ultimately, Yamamoto and coworkers[4f, 17] demonstrated that also 1,4-additions of organocopper reagents to activated dienes are feasible: while the reaction between methyl sorbate (**4**) and the reagent formed from n-butylcopper and boron trifluoride mainly gave the 1,4-adduct **5**, the corresponding Gilman cuprate n-Bu_2CuLi provided the 1,6-addition product **6** exclusively (equation 3). The synthetically very useful organocopper compounds $RCu{\bullet}BF_3$ are commonly referred to as Yamamoto reagents[4f].

Michael additions of organocopper reagents to acceptor-substituted dienes have found widespread application in target-oriented stereoselective synthesis[18]. For example, the chiral cuprate **8**, containing a Schöllkopf bislactim ether moiety, was used in the total synthesis of the antimycotic dipeptide chlorotetaine (**10**; equation 4)[18d]. Although the nucleophilic addition to dienone **7** took place only with moderate regioselectivity to furnish a 63:37 mixture of the 1,6- and 1,4-adduct, the major product **9** was successfully converted over several steps into diastereomerically and enantiomerically pure chlorotetaine (**10**).

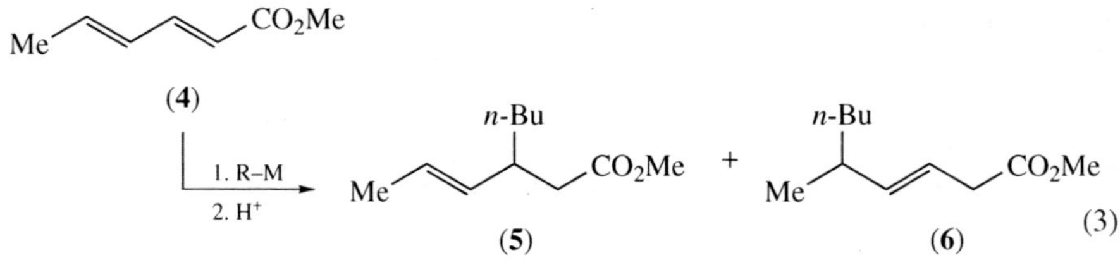

(3)

	(5) : (6)
n-BuCu·BF_3	93 : 7
n-Bu_2CuLi:	< 1 : 99

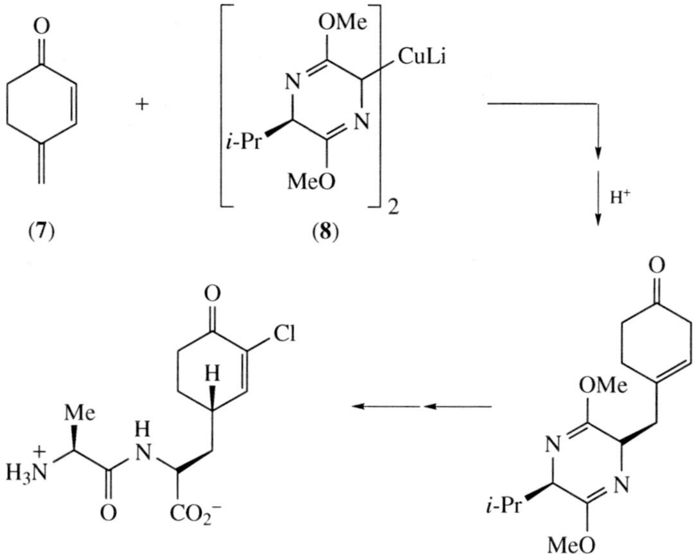

(4)

(**10**)
Chlorotetaine

(**9**)

While copper-catalyzed Michael additions to acceptor-substituted dienes using Grignard reagents as nucleophiles were reported even earlier than the corresponding additions of stoichiometric organocuprates, the former transformations have largely been restricted to the diastereoselective synthesis of steroid hormones. Besides tetrahydro-3*H*-naphthalen-2-ones, which are used as model substrates for doubly unsaturated steroids[19,20], estradiol derivatives bearing an alkyl chain in the 7α-position are especially interesting target molecules, due to their high affinity for and specificity towards estrogen receptors which renders these unsaturated steroids particularly useful for the treatment of mammary tumors (breast cancer)[21,22]. Besides the regioselectivity of the nucleophilic addition to doubly unsaturated $\Delta^{4,6}$-steroids[19,20,21b,22], the diastereoselectivity is particularly important since only the 7α-isomers are effective enzyme inhibitors[21b]. Although the diastereoselectivity of the copper-catalyzed 1,6-addition of methyl Grignard reagents to $\Delta^{4,6}$-steroids may depend on the substitution pattern of the substrate[19a], a general preference for attack from the α-side has been observed frequently[19]. Wieland and Auner[19e], for example, reported an α:β ratio of 90:10 in the copper-catalyzed 1,6-addition of MeMgBr to dienone **11** (equation 5). The product **12** was converted into 7α-methylestrone (**13**), a precursor of several highly active steroidal hormones.

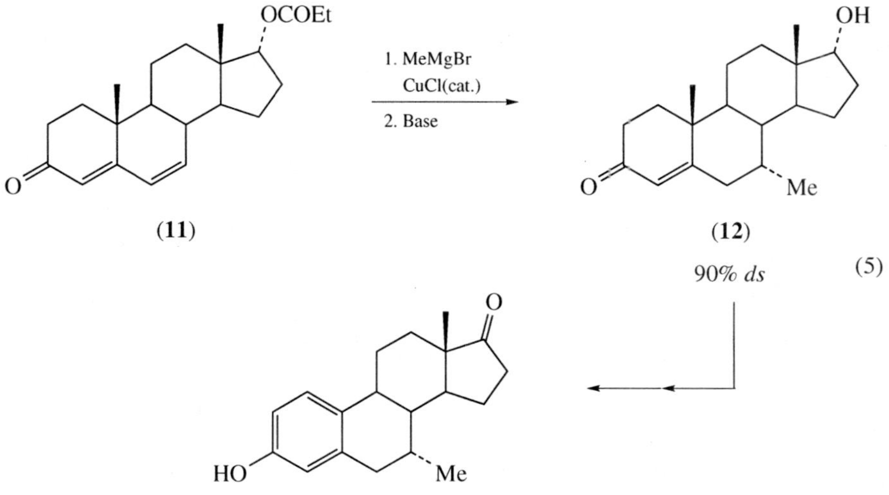

(5)

In contrast to this, the introduction of longer alkyl chains by copper-promoted 1,6-addition reactions of $\Delta^{4,6}$-steroids normally proceeds with unsatisfactory α:β ratios[21b,22]. In some cases, however, improvement of the diastereoselectivity by 'fine-tuning' of the reaction conditions has been possible. For example, the ratio of the epimeric products **15** and **16** in the copper-catalyzed 1,6-addition of 4-pentenylmagnesium bromide to dienone **14** was increased from 58:42 to 82:18 by variation of the solvent and the excess of nucleophile (equation 6)[22f].

A different behavior was observed for tetrahydro-3*H*-naphthalen-2-ones (e.g. **17**) as Michael acceptors: the 1,6-addition of cyano-Gilman cuprates or Grignard reagents (catalyzed by copper arene thiolate **18**) proceeds with high *trans*-selectivity to afford adducts **19**, irrespective of the transferred group (equation 7)[23]. NMR spectroscopic investigations have revealed that formation of π-complexes at the double bond adjacent to the carbonyl group, similar to those observed in 1,6-cuprate additions to acceptor-substituted enynes (Section II.C), are involved in these transformations.

More recently, enantioselective conjugate addition reactions to acceptor-substituted dienes have also been reported. The first example was the 1,6-addition of dialkylzinc reagents to Meldrum's acid derivatives **20** in the presence of catalytic amounts of $Cu(OTf)_2$ and Feringa's phosphoramidite **21** which affords the products **22** with 70–83% *ee* (equation 8)[24].

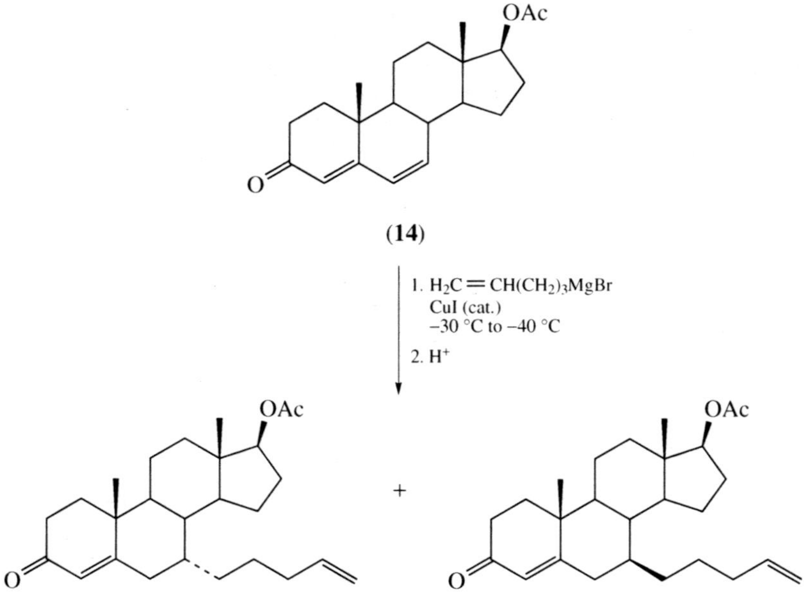

(14)

(15) (16)

(6)

Eq. RMgBr	Ratio THF / diethyl ether	**15 : 16**
12	1 : 9	58 : 42
12	1 : 4	60 : 40
12	1 : 1	78 : 22
4	1 : 1	82 : 18

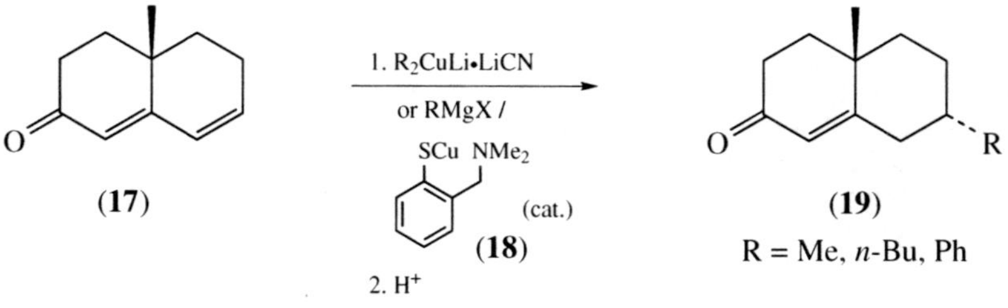

(17) (18) (19)

R = Me, *n*-Bu, Ph (7)

Even better results were obtained by Feringa and coworkers[25] with acyclic dienoates like ethyl sorbate (**23**) and Grignard reagents as the nucleophile. In this case, the JOSIPHOS-type ligand **24** provided both high regioselectivities in favor of the 1,6-addition products **25** and excellent enantioselectivities of 92–97% *ee* if a primary group R is transferred (equation 9). In contrast to this, the reaction of **23** with isopropylmagnesium bromide gave the 1,6-addition product with only 72% *ee*.

An intriguing case of ligand control of both the regio- and enantioselectivity in copper-catalyzed conjugate addition reactions to acceptor-substituted dienes was recently reported by Alexakis and coworkers[26]. Whereas the reaction of dienone **26** with organozinc or organoaluminum reagents in the presence of catalytic amounts of copper(II) triflate and phosphoramidites such as **27** affords the expected 1,6-addition product **28** with moderate to high enantioselectivity, the regioselectivty is reversed in favor of the 1,4-adduct **30** when the copper–NHC complex formed *in situ* from $Cu(OTf)_2$ and **29** is used as catalyst (equation 10). With unbranched groups R, the products **30** are obtained with >95% regioselectivity and excellent enantioselectivities. The presence of the OH group in the *N*-heterocyclic carbene is crucial for regiocontrol since simple Arduengo-type carbenes afford the 1,6-adduct exclusively.

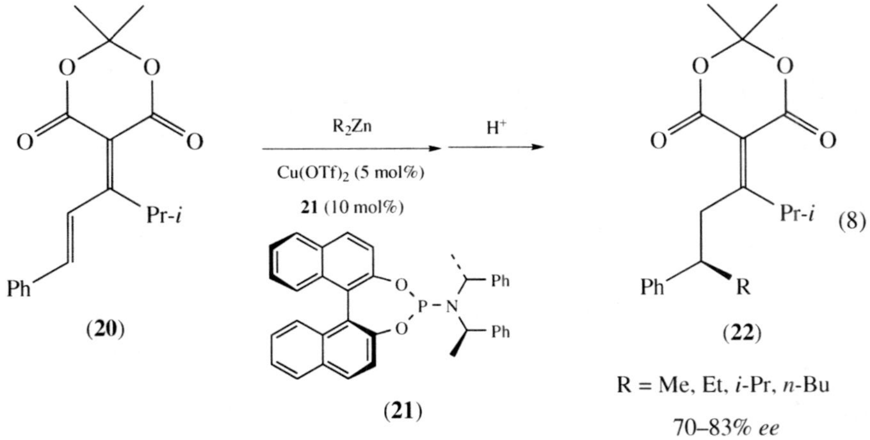

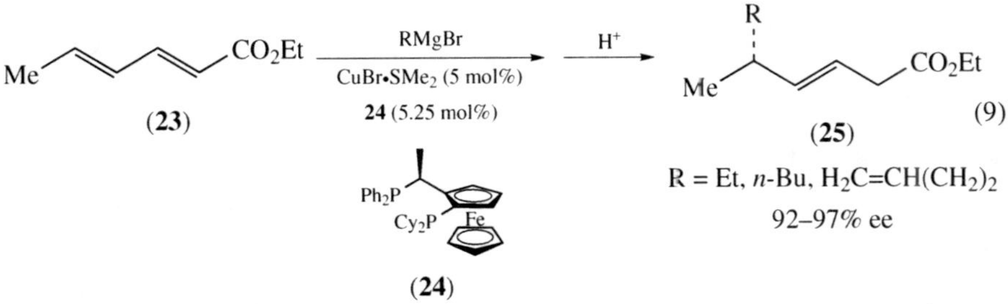

B. Acceptor-substituted Enynes

As for conjugate addition reactions of carbon nucleophiles to activated dienes, organocopper compounds are the reagents of choice for regio- and stereoselective Michael additions to acceptor-substituted enynes. While the reaction of substrates bearing an acceptor-substituted triple bond in conjugation with one or more double bonds (such as **31**) with organocuprates affords the 1,4-addition products **32** exclusively (equation 11)[27], the corresponding transformation of enynes bearing an acceptor substituent at the double bond can result in the formation of several regioisomeric products[14].

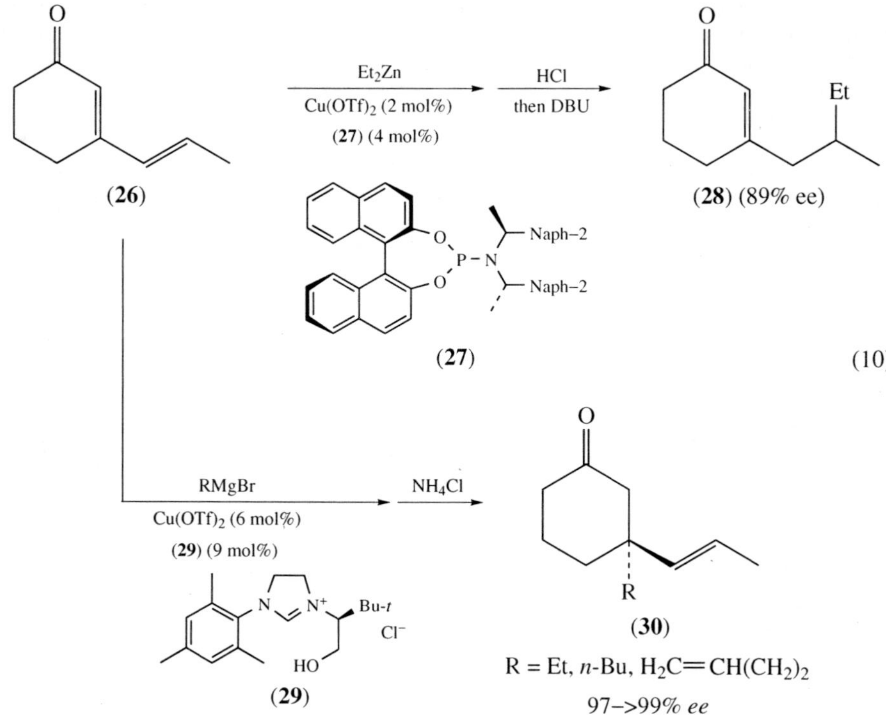

(10)

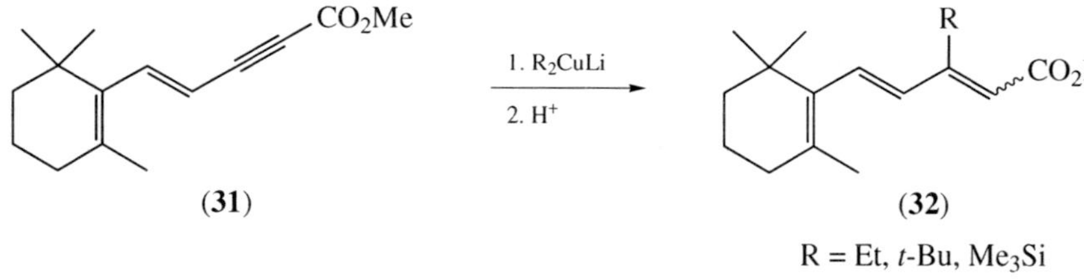

(11)

Analogously to acceptor-substituted dienes (Scheme 1), the outcome of the reaction depends on the regioselectivity of both the nucleophilic attack of the copper reagent (1,4- or 1,6-addition) and of the electrophilic trapping of the enolate formed (Scheme 2). Since the allenyl enolate formed by 1,6-addition can furnish either an allene or a conjugated diene upon reaction with a soft electrophile and hence offers the possibility of creating axial chirality, this transformation is of special interest both from the preparative and mechanistic point of view. Gratifyingly, the regio- and stereoselectivity of both steps can be controlled by the choice of the reactants, in particular by 'fine-tuning' of the organocopper reagent and the electrophile.

The first copper-mediated addition reactions to enynes with an acceptor group at the triple bond were reported by Hulce and coworkers[28] who found that 3-alkynylcycloalk-2-enones **33** react with cuprates regioselectively at the triple bond, i.e. in a 1,6-addition (equation 12). The allenyl enolates thus formed are protonated at C-4 to provide conjugated dienones **34** as a mixture of E/Z isomers. Interestingly, substrates of this type

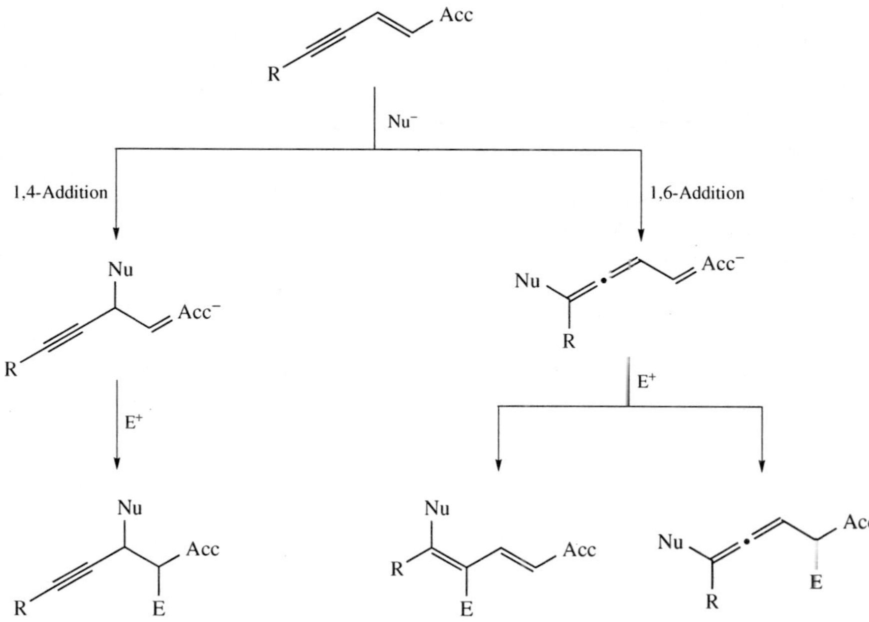

SCHEME 2. Regioselectivity in conjugate addition reactions to acceptor-substituted enynes

can also undergo tandem 1,6- and 5,6-additions, indicating that the allenyl enolate is suffciently reactive to undergo a carbometalation of the allenic double bond distal to the electron-releasing enolate moiety (equation 13)[28b]. Hence, it is possible to obtain products of the type **36** bearing two different groups, either by successive reaction of Michael acceptor **35** with two organocopper reagents or by employing a mixed cuprate.

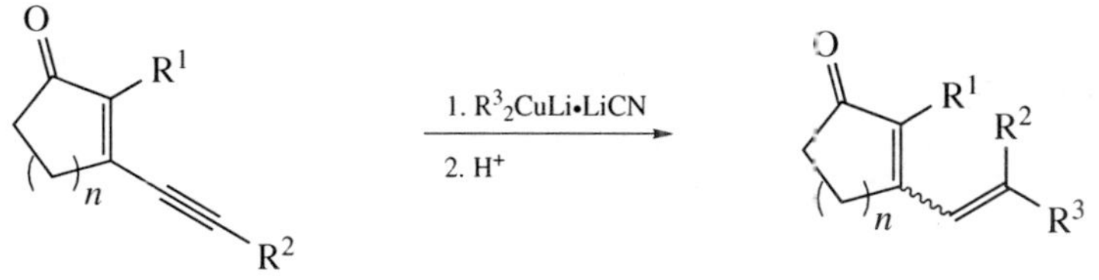

(**33**) (**34**) (12)

n = 1, 2; R^1 = H, Me; R^2 = H, Ph, Me_3Si; R^3 = Me, Et

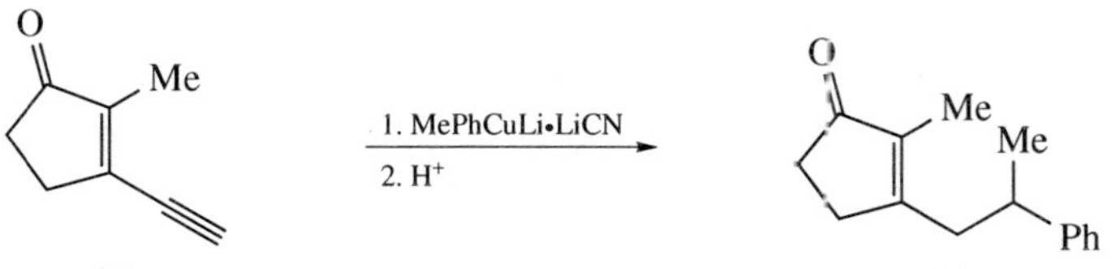

(**35**) (**36**) (13)

With regard to preparative applications, however, shifting the regioselectivity of the electrophilic quenching reaction towards the formation of allenes is more interesting, since the scope of synthetic methods for the preparation of functionalized allenes is limited[29]. Moreover, a stereoselective reaction of this type would open up a route to these axially chiral compounds in enantiomerically enriched or even pure form. Indeed, the Gilman cuprate $Me_2CuLi{\bullet}LiI$ and cyano-Gilman reagents $R_2CuLi{\bullet}LiCN$ ($R \neq Me$) undergo a completely regioselective 1,6-addition with various substituted 2-en-4-ynoates **37** in diethyl ether, and protonation of the allenyl enolate with dilute sulfuric acid affords the β-allenic esters **38** bearing alkyl, alkenyl, aryl and silyl substituents with good chemical yield (equation 14)[30]. Gratifyingly, these 1,6-addition reactions can also be carried out with catalytic amounts of the cuprate or copper arenethiolate **18**[31] by simultaneous addition of the substrate and the organolithium reagent to the copper source.

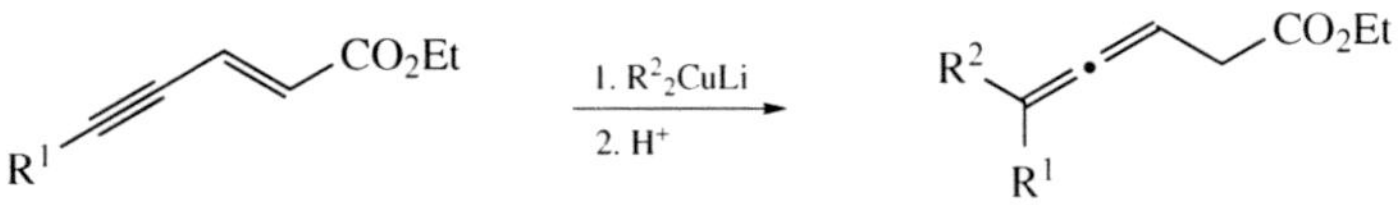

(**37**) (**38**)

(14)

R^1 = Ph, R^2 = Me:	79%
R^1 = Ph, R^2 = *t*-Bu:	81%
R^1 = *n*-Bu, R^2 = Ph:	62%
R^1 = Me_3Si, R^2 = Me:	57%

The nature of the acceptor substituent has almost no influence on the regioselectivity of the cuprate addition to acceptor-substituted enynes. Hence, enynes **39** comprising thioester, lactone, dioxanone as well as keto, sulfonyl, sulfinyl, cyano or oxazolidino groups undergo a 1,6-addition with Gilman or cyano-Gilman cuprates to furnish functionalized allenes **40** (equation 15). In contrast, 1-nitro-l-en-3-ynes are attacked at the C−C double bond with formation of the corresponding 1,4-adducts[30c]. The reactivity differences can be described qualitatively by the following order: Acceptor (Acc) = NO_2 > COR, CO_2R, COSR > CN, SO_3R, oxazolidino > SO_2R > SOR ≫ $CONR_2$. In order to achieve acceptable chemical yields with less reactive Michael acceptors, e.g. sulfones and sulfoxides, it is often necessary to use more reactive organocopper reagents (e.g. Me_3CuLi_2 instead of Me_2CuLi) or to activate the substrate by Lewis acid catalysis. Here, mild Lewis acids like Me_3SiI or Me_3SiOTf proved to be effective[30c]. Only enyne amides completely fail to form 1,6-adducts even under these conditions.

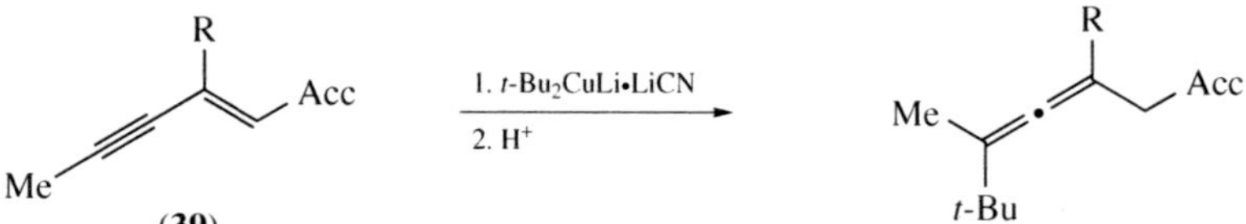

(**39**) (**40**) (15)

Acc = CN, R = Me:	79%
Acc = SO_2Ph, R = H:	91%
Acc = SO_3Et, R = H:	49%

Remarkably, the regioselectivity of the cuprate addition to acceptor-substituted enynes is insensitive to the steric properties of the substrate as well. Thus, enynes with *t*-butyl substituents at the triple bond (e.g. **41**) undergo a 1,6-addition even if the cuprate itself is also sterically demanding (equation 16)[30b]. This fact renders the method highly useful for the preparation of sterically encumbered allenes of type **42**.

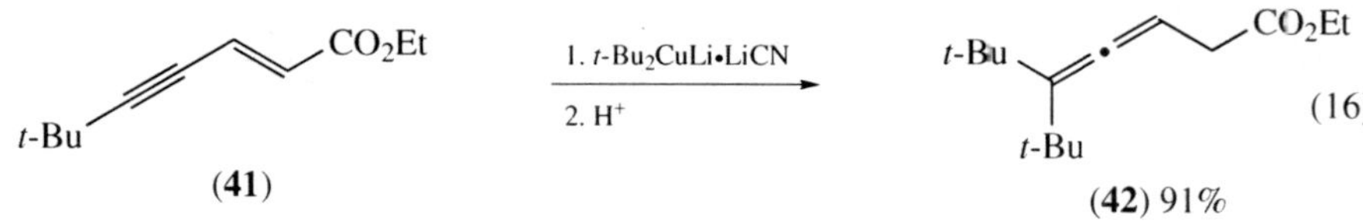

(16)

Unlike the substrate, the type of organocuprate used for the addition to acceptor-substituted enynes has a pronounced influence on the regiochemical course. While the Gilman cuprate $Me_2CuLi{\bullet}LiI$ or cyano-Gilman reagents $R_2CuLi{\bullet}LiCN$ ($R \neq Me$) readily undergo 1,6-additions, the Yamamoto reagent $RCu{\bullet}BF_3$, as well as silylcuprates[32] and organocopper compounds RCu activated by Me_3SiI[33] afford the 1,4-addition products[14a]. Interestingly, the use of lithium di-*s*-butylcyanocuprate in diethyl ether predominantly leads to the formation of reduced allenes as the major product[34], this may be attributed to the hydrolysis of a rather stable copper intermediate. Lower-order cyanocuprates RCu(CN)Li again show a different behavior; whereas these usually do not react with acceptor-substituted enynes, the cyanocuprate *t*-BuCu(CN)Li affords *anti*-Michael additions of the type **43** with 2-en-4-ynoates and nitriles (e.g. **41**; equation 17)[35]. Unfortunately, an adequate interpretation of the abnormal behavior of this particular cuprate is lacking.

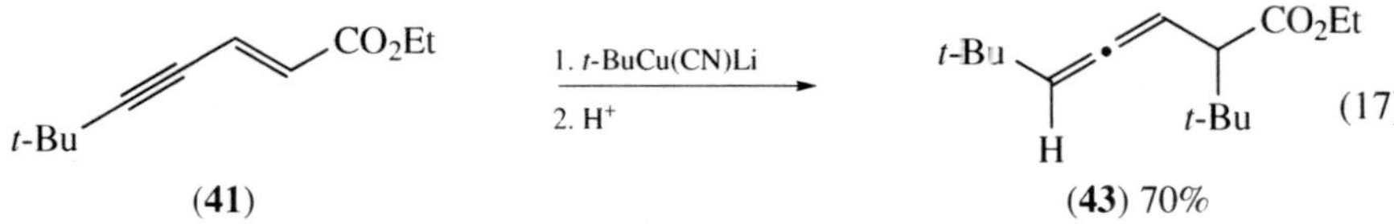

(17)

As already mentioned at the beginning of this section, allenes can only be obtained by 1,6-addition to acceptor-substituted enynes if the intermediate allenyl enolate reacts regioselectively with an electrophile at C-2 (or at the enolate oxygen atom to give an allenyl ketene acetal; see Scheme 2). Interestingly, the regioselectivity of the simplest trapping reaction, the protonation, depends on the steric and electronic properties of both the substrate and the proton source. Whereas the allenyl enolates obtained from alkynyl enones **33** provide conjugated dienones **34** by protonation at C-4 (possibly via allenyl enols; see equation 12)[28], the corresponding ester enolates are usually protonated at C-2 (equation 14), especially if sterically demanding groups at C-5 block the attack of a proton at C-4 (equation 16)[14,30]. In the presence of a substituent at C-2 of the enolate (e.g. substrate **44**), however, mixtures of allene **45** and conjugated diene **46** are formed for steric reasons (equation 18). Fortunately, this problem can be solved by using weak organic acids as the proton source. In particular, pivalic acid (2,2-dimethylpropionic acid) at low temperatures gives rise to the exclusive formation of allenes[30a].

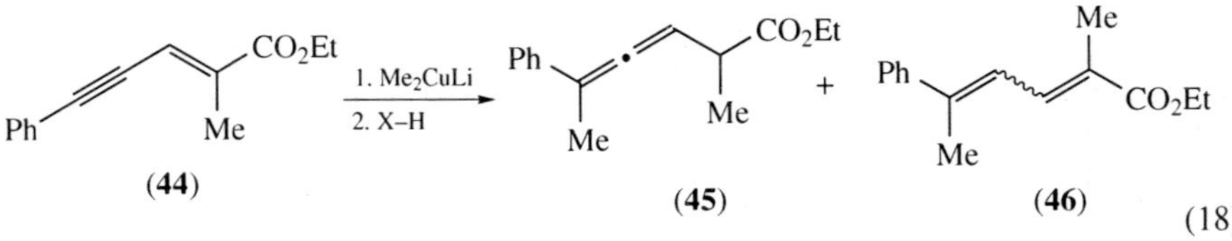

(18)

X–H = 2N H_2SO_4, 25 °C: 50 : 50
AcOH, 25 °C: 64 : 36
t-BuCO_2H, 25 °C: 82 : 18
t-BuCO_2H, −80 °C: > 99 : 1

In contrast to protonation, the regioselectivity of the reaction of other electrophiles with allenyl enolates derived from 2-en-4-ynoates is independent of the steric and electronic properties of the reaction partners (Scheme 3)[14,36]. As expected according to the HSAB principle, hard electrophiles such as silyl halides and triflates react at the enolate oxygen atom to form allenyl ketene acetals, while soft electrophiles such as carbonyl compounds attack at C-2. Only allylic and propargylic halides afford conjugated dienes bearing an unsaturated substituent at C-4. Again, cyclic allenyl enolates obtained by 1,6-cuprate addition to 3-alkynylcycloalk-2-enones **33** show a deviating behavior; treatment with iodomethane gives product mixtures derived from attack of the electrophile at C-2 and C-4, whereas the reaction with aldehydes and silyl halides takes place at C-4 exclusively[37].

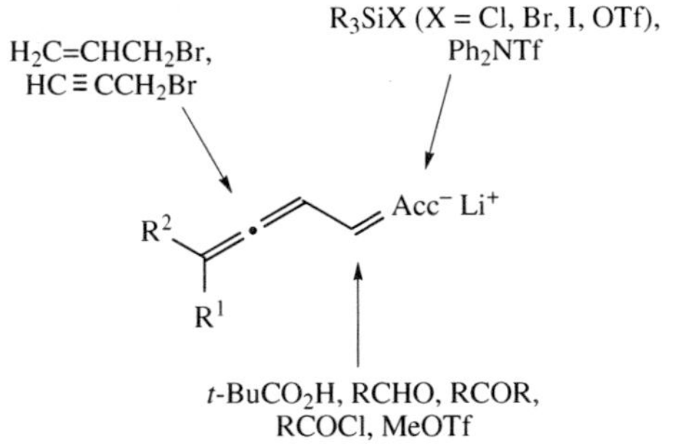

SCHEME 3. Regioselectivity of trapping reactions of acyclic allenyl enolates with different electrophiles

In order to control the configuration of the chirality axis of the resulting allenes, the 1,6-addition has to proceed diastereo- or enantioselectively. Among many different chiral substrates examined, 5-alkynylidene-1,3-dioxan-4-ones of type **47** have proven to be valuable synthetic precursors, since these Michael acceptors adopt a very rigid conformation. Due to the equatorial position of the t-butyl group, the trifluoromethyl residue shields the top face of the enyne moiety, exposing the underside of the molecule to be preferably attacked by the nucleophile (equation 19)[38]. Therefore, reaction with lithium dimethylcuprate and pivalic acid affords the allene **48** with a diastereoselectivity of 98%, and the stereochemical information generated in this step remains intact during the subsequent conversion into the chiral vinylallene **49**. In contrast to this work, all attempts to establish an enantioselective 1,6-addition by treatment of acceptor-substituted enynes with various chirally modified organocopper reagents failed, due to the low reactivity of the latter compounds towards the Michael acceptors. This problem was recently solved by Hayashi and coworkers[39] who employed aryltitanates as nucleophiles in rhodium-catalyzed enantioselective 1,6-addition reactions to 2-en-4-ynones.

The 1,6-cuprate addition to acceptor-substituted enynes has found several preparative applications. In natural product synthesis, the method has been used to generate the insect pheromone methyl 2,4,5-tetradecatrienoate[14a,40] as well as the precursor **51** of the fungal metabolite (±)-sterpurene (**53**) and its oxygenated metabolites (equation 20)[41]. In the latter case, the reaction sequence started with the 1,6-addition of lithium dimethylcuprate to enynoate **50** and subsequent regioselective enolate trapping with methyl triflate. The vinyllallene **51** thus formed underwent an intramolecular [4 + 2]-cycloaddition to the tricyclic product **52**, which was converted into the target molecule **53**.

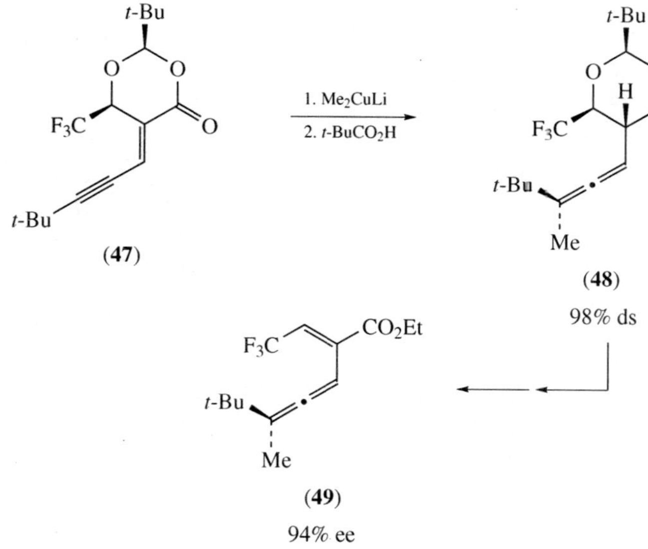

(19)

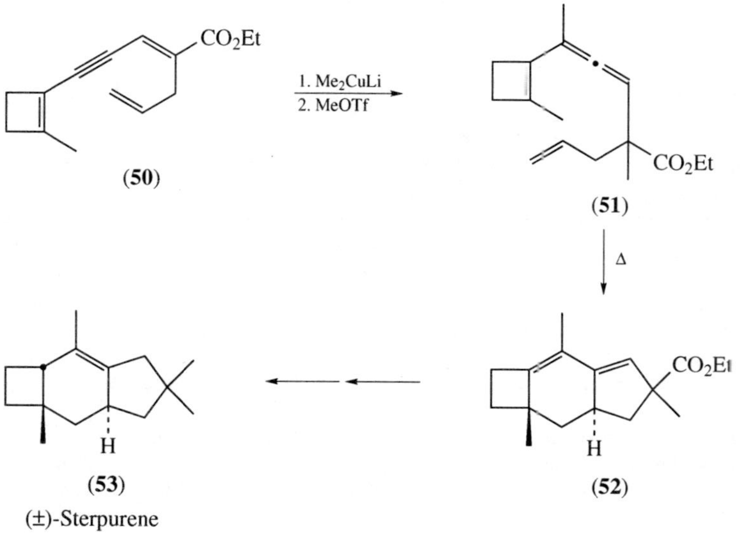

(20)

This Diels–Alder reaction is a typical example for the utilization of axially chiral allenes, accessible through 1,6-addition, to selectively generate new stereogenic centers. The chirality transfer is also possible by means of intermolecular Diels–Alder reactions of vinylallenes[42], aldol reactions of allenyl enolates[43] and Ireland–Claisen rearrangements of silyl allenylketene acetals[44]. Furthermore, it has been utilized recently in the diastereoselective oxidation of titanium or zirconium allenyl enolates (formed by deprotonation of β-allenylcarboxylates of type **54** and transmetalation with Cp_2TiCl_2 or Cp_2ZrCl_2) with dimethyl dioxirane (DMDO) which proceeds regioselectively at C-2 to afford 2-hydroxy-3,4-dienoates (e.g. **55**) with up to 90% *ds* (equation 21)[45]. α-Hydroxyallenes of this type are synthetically valuable precursors for 2,5-dihydrofurans, found not only in several natural products but also in biologically active compounds. Thus, the cyclization of allene **55** to heterocycle **56** took place with complete axis-to-center chirality transfer, being easily achieved by treatment with HCl gas in chloroform, acidic ion exchange resins such as Amberlyst 15 or with catalytic amounts of gold(I) or gold(III) salts[45,46]. The latter method is particularly useful for substrates containing acid-sensitive groups and has already found application in natural product synthesis[46].

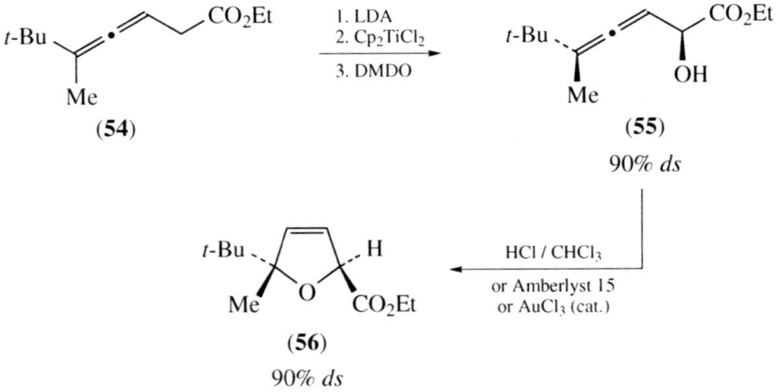

(21)

In contrast to copper-mediated addition reactions of activated dienes (Section II.A), the mechanism of 1,6-cuprate additions to acceptor-substituted enynes is quite well understood, largely thanks to kinetic and NMR spectroscopic investigations[14]. ^{13}C NMR spectroscopic studies have revealed that these transformations proceed through π-complexes, with an interaction between the π-system of the C–C double bond and the nucleophilic copper atom (a soft–soft interaction in terms of the HSAB principle), together with a second interaction between the hard lithium ion of the cuprate and the hard carbonyl oxygen atom (Scheme 4)[47]. In particular, the use of ^{13}C-labeled substrates has shed light on the structure of the metal-containing part of these π-complexes, indicating, for example, that the cuprate does not interact with the triple bond[47b,c]. Recently determined ^{13}C kinetic isotope effects prove that bond formation between C-5 of the acceptor-substituted enyne and the cuprate occurs in the rate-determining step[48]. Moreover, with the aid of kinetic measurements, activation parameters for these transformations have been determined experimentally[49]. All these experimental results are in accordance with a mechanistic model that comprises the formation of a σ-copper(III) species which might be in equilibrium with an allenic copper(III) intermediate (Scheme 4). Both intermediates can undergo reductive elimination of an alkylcopper compound to produce the 1,4- and 1,6-adduct, respectively. The experimentally observed exclusive formation of the 1,6-addition product, however, may indicate that the latter undergoes a much faster reductive elimination than the first intermediate.

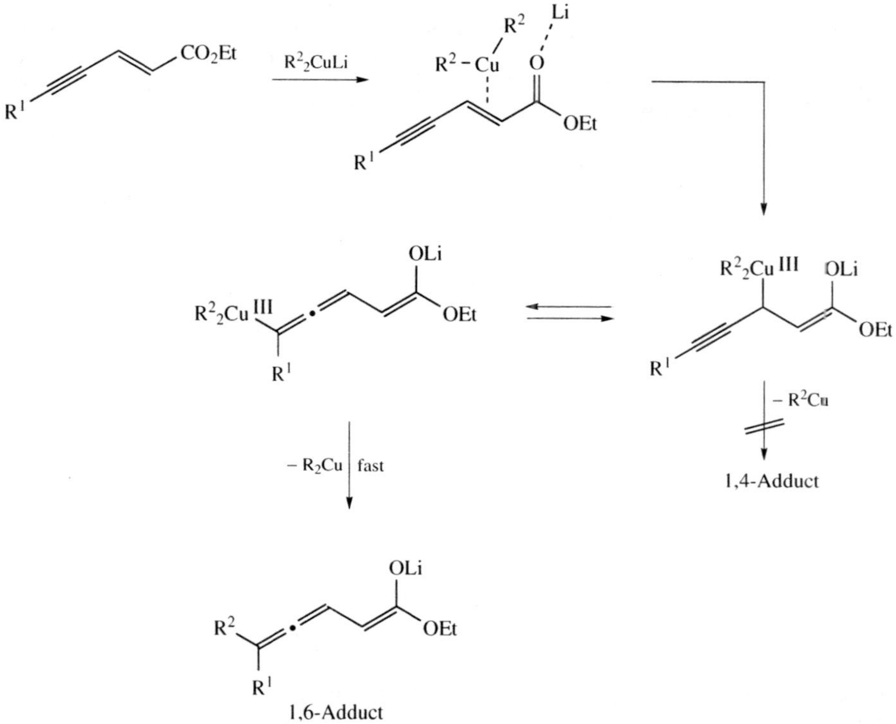

SCHEME 4. Proposed mechanism for the 1,6-addition of organocuprates to acceptor-substituted enynes

C. Acceptor-substituted Polyenynes

In view of the high value of the 1,6-cuprate addition to acceptor-substituted enynes for the synthesis of functionalized allenes, it seemed interesting to explore whether extended Michael acceptors with further C−C double bonds between the acceptor group and the triple bond can also be utilized for the formation of extended unsaturated allenic systems. Of course, the number of possible regioisomeric products rises with increasing length of the Michael acceptor. The 2,4-dien-6-ynoate **57**, for example, can be attacked by an organocopper reagent at C-3, C-5 or C-7, and the latter possibility leads to a vinylogous allenyl enolate possessing four reactive positions for the subsequent trapping reaction (enolate oxygen, C-2, C-4, C-6). Therefore, the high regioselectivity observed experimentally for the reaction of **57** with lithium dimethylcuprate is striking, the cuprate attacking the triple bond exclusively, and protonation with pivalic acid occurring selectively at C-2 of the enolate to afford the vinylallene **58** (a 1,8-addition product) as the only detectable regioisomer with 90% yield (equation 22)[42].

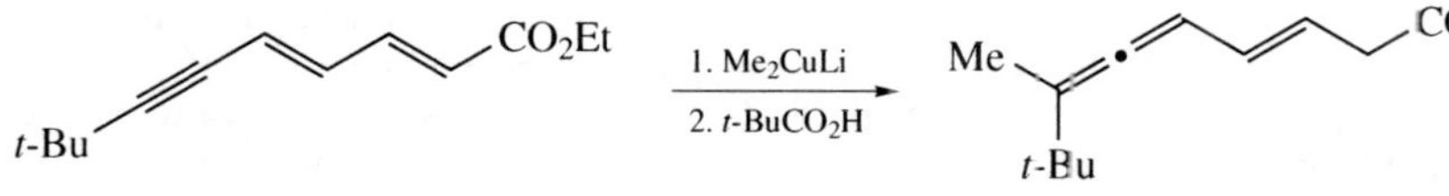

(**57**) (**58**) 90% (22)

Analogously, the trienynoate **59** reacts in a 1,10-addition to give the 3,5,7,8-tetraenoate **60**, and the even higher unsaturated allene **62** is obtained from the Michael acceptor **61** containing four double bonds between the triple bond and the acceptor substituent (equation 23)[42]. The latter case, however, seems to be the limit of the method since the yield of **62** was only 26%.

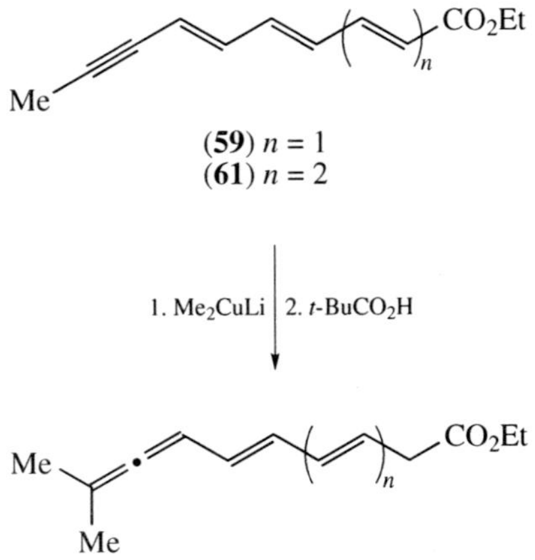

(**59**) $n = 1$
(**61**) $n = 2$

(**60**) $n = 1$: 68%
(**62**) $n = 2$: 26%

(23)

These transformations and those summarized in the previous section indicate that Michael acceptors containing any combination of double and triple bonds undergo highly regioselective copper-mediated addition reactions. The following rule holds: *Michael acceptors with any given arrangement of conjugated double and triple bonds react regioselectively with organocuprates at the triple bond closest to the acceptor substituent.* Like the 1,6-cuprate addition to acceptor-substituted enynes (Scheme 4), these reactions start with the formation of a cuprate π-complex at the double bond adjacent to the acceptor group[47]. Subsequently, an equilibrating mixture of σ-copper(III) intermediates is probably formed, and the regioselectivity of the reaction may be controlled by different rates of the reductive elimination of these intermediates.

III. COPPER-MEDIATED AND COPPER-CATALYZED SUBSTITUTION REACTIONS OF EXTENDED SUBSTRATES

In contrast to the addition reactions discussed so far, the number of examples for copper-mediated or copper-catalyzed substitutions of extended electrophiles is rather limited. Investigations of substitution reactions of various dienylic carboxylates with organocuprates (or Grignard reagents in the presence of catalytic amounts of copper salts) indicate that the ratio of the three possible regioisomers (α-, γ- and ε-alkylated products) depends strongly on the substrate and reaction conditions[50]. For example, treatment of dienyl acetate **63** with n-BuMgBr and stoichiometric quantities of CuI mainly affords the S_N2'-(1,3)-substitution product **64** (equation 24), whereas the conjugated diene **65** formed by S_N2''-(1,5)-substitution is obtained exclusively when catalytic quantities of CuI and THF as solvent are used[51]. This result gives credence to the postulate that different organocopper species are responsible for the formation of the regioisomeric products. With equimolar amounts of Grignard reagent and copper salt, the active species is probably the alkylcopper compound n–BuCu•MgBrI, which produces **64**, whereas an excess of the Grignard reagent should produce the magnesium cuprate n-$Bu_2CuMgBr$ as the reactive nucleophile, providing the 1,5-substitution product **65**.

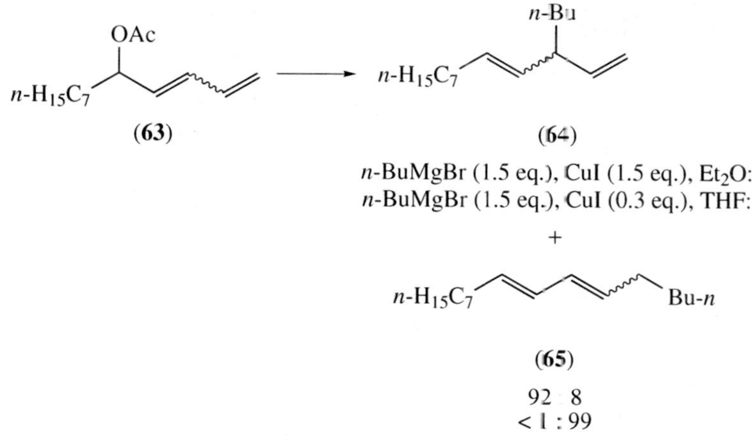

(24)

Stereoselective substitution reactions of chiral dienyl electrophiles have also been carried out. In analogy to the copper-promoted S_N2'-reactions of simple allylic electrophiles[3,4], the corresponding S_N2'-(1,3)-substitutions of dienyl carbonates have been reported to proceed with high *anti*-selectivity[52]. Interestingly, treatment of the chiral dienyl acetal **66** with the Yamamoto reagent PhCu•BF_3 gives rise to the formation of a mixture of the *anti*-S_N2'-substitution product **67** and the *syn*-S_N2''-(1,5)-substitution product **68** (equation 25)[53]. A mechanistic explanation of this puzzling result has yet to be put forward.

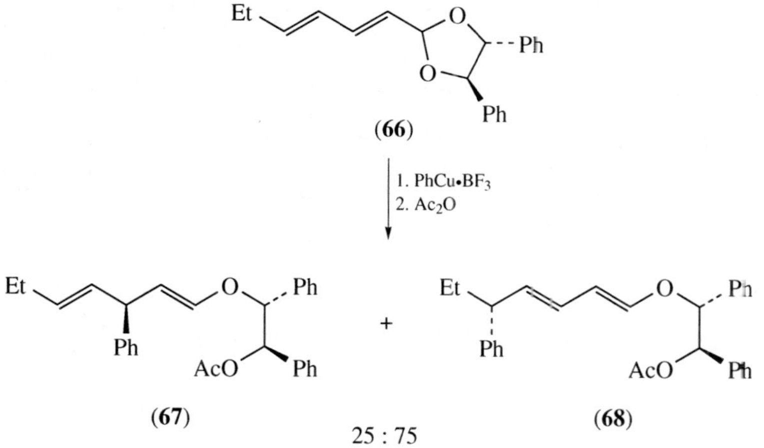

(25)

The corresponding copper-mediated S_N2''-(1,5)-substitution reactions of conjugated enyne acetates **69** also take place with high regioselectivity to afford vinylallenes **70** with variable substitution patterns (equation 26)[54]. Although the substitution products are usually obtained as a mixture of E/Z isomers, complete stereoselection with regard to the olefinic double bond of the vinylallene is achieved in some cases. Besides enyne acetates, the corresponding oxiranes (e.g. **71**) also participate in the 1,5-substitution and are transformed into hydroxy-substituted vinylallenes of the type **72** (equation 27)[54]. Moreover, these transformations can also be conducted under catalytic conditions by simultaneous addition of the organolithium compound and the substrate to catalytic amounts of the cuprate.

(**69**) (**70**)

(26)

$R^1 = R^2 = Me, R^3 = n$-Bu: 90% ($E : Z = 67 : 33$)

$R^1 = R^2 = Me, R^3 = t$-Bu: 93% ($E : Z = 25 : 75$)

$R^1 = Me_3Si, R^2 = Me, R^3 = t$-Bu: 96% ($E : Z = 40 : 60$)

$R^1 = n$-Bu, $R^2 = t$-Bu, R^3 = Me: 92% ($E : Z > 99 : 1$)

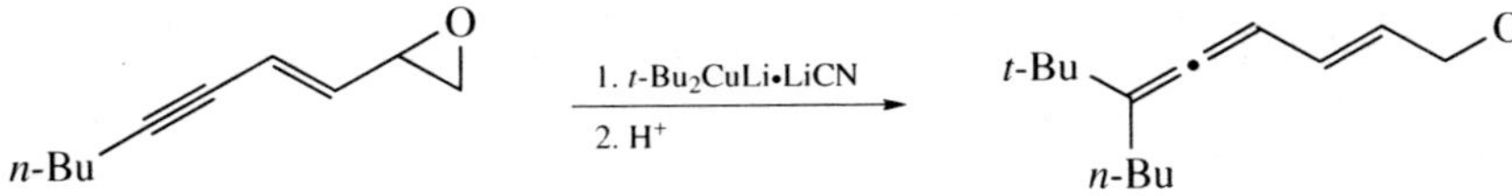

(27)

(**71**) (**72**) 72%

Highly enantioselective 1,5-substitution reactions of enyne acetates are also possible under carefully controlled conditions. For example, treatment of enantiomerically pure substrate **73** with the cyano-Gilman reagent t-Bu$_2$CuLi•LiCN at $-90\,^\circ$C provides vinylallene **74** as a 1:3 mixture of E/Z isomers with 20% and 74% *ee*, respectively (equation 28)[55]. This unsatisfactory stereoselection may be attributed to a racemization of the allene by the cuprate or other reactive copper species formed in the reaction mixture. Fortunately, the use of phosphines or phosphites as additive can effectively prevent such racemizations (which probably occur by one-electron transfer steps)[56]. Thus, vinylallene **74** was obtained with 92% *ee* for the E isomer and 93% *ee* for the Z isomer if the substitution was performed at $-80\,^\circ$C in the presence of 4 equivalents of n-Bu$_3$P. This method enables various substituted vinylallenes to be prepared with >90% *ee*[55].

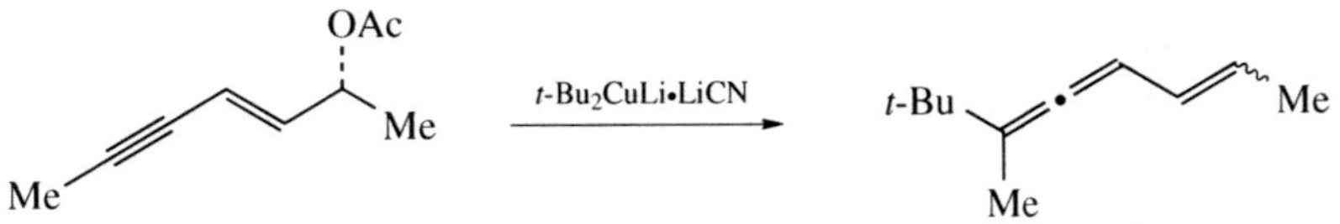

(**73**) (**74**) (28)

$E : Z = 25 : 75$

Without Additive: 20% *ee* (E) / 74% *ee* (Z)

With 4 equivalents of n-Bu$_3$P: 92% *ee* (E) / 93% *ee* (Z)

IV. CONCLUSION

Over the last four decades, organocopper reagents have been utilized with great success in organic synthesis. The results presented in this chapter highlight the excellent performance of these organometallic compounds in regio- and stereoselective transformations of substrates with extended π-systems, in particular in 1,6-addition or 1,5-substitution reactions of acetylenic electrophiles which profit from the high affinity of organocopper reagents towards sp-hybridized carbon atoms. In many cases, protocols involving stoichiometric amounts of organocopper reagents have been complemented by catalytic procedures with similar efficiency, selectivity and reliability. These transformations not only provide insights into the mechanism of copper-mediated carbon–carbon bond formation, but they also open up new opportunities in target-oriented synthesis.

V. REFERENCES

1. H. Gilman and J. M. Straley, *Recl. Trav. Chim. Pays-Bas*, **55**, 821 (1936).
2. H. Gilman, R. G. Jones and L. A. Woods, *J. Org. Chem.*, **17**, 1630 (1952).
3. N. Krause, in *Modern Organocopper Chemistry* (Ed. N. Krause), Wiley-VCH, Weinheim, 2002.
4. Reviews:
   (a) G. H. Posner, *Org. React.*, **19**, 1–113 (1972).
   (b) G. H. Posner, *Org. React.*, **22**, 253–400 (1975).
   (c) J. F. Normant and A. Alexakis, *Synthesis*, 841–870 (1981).
   (d) E. Erdik, *Tetrahedron*, **40**, 641–657 (1984).
   (e) R. J. K. Taylor, *Synthesis*, 364–392 (1985).
   (f) Y. Yamamoto, *Angew. Chem.*, **98**, 945–957 (1986); *Angew. Chem., Int. Ed. Engl.*, **25**, 947–959 (1986).
   (g) B. H. Lipshutz, *Synthesis*, 325–341 (1987).
   (h) B. H. Lipshutz, *Synlett*, 119–128 (1990).
   (i) E. Nakamura, *Synlett*, 539–547 (1991).
   (j) B. H. Lipshutz and S. Sengupta, *Org. React.*, **41**, 135–631 (1992).
   (k) *Conjugate Addition Reactions in Organic Synthesis* (Ed. P. Perlmutter), Pergamon Press, Oxford, 1992.
   (l) *Organocopper Reagents* (Ed. R. J. K. Taylor), Oxford University Press, Oxford, 1994.
   (m) B. H. Lipshutz, in *Organometallics in Synthesis* (Ed. M. Schlosser), John Wiley & Sons, Ltd, Chichester, 1994, pp. 283–382.
   (n) B. H. Lipshutz, in *Advances in Metal-Organic Chemistry* (Ed. L. S. E. Liebeskind), Vol. 4, JAI Press, Greenwich, 1995, pp. 1–64.
   (o) N. Krause and N. Morita, in *Comprehensive Organometallic Chemistry III* (Eds R. H. Crabtree and D. M. P. Mingos), Vol. 9, Elsevier, Oxford, 2007, pp. 501–586.
5. Solid state structures:
   (a) G. van Koten, *J. Organomet. Chem.*, **400**, 283 (1990).
   (b) P. P. Power, *Prog. Inorg. Chem.*, **39**, 75 (1991).
   (c) G. van Koten, S. L. James and J. T. B. H. Jastrzebski, in *Comprehensive Organometallic Chemistry II* (Eds. E. W. Abel, F. G. A. Stone and G. Wilkinson), Vol. 3, Pergamon/Elsevier, Oxford, 1995, pp. 57–133.
   (d) N. Krause, *Angew. Chem.*, **111**, 83 (1999); *Angew. Chem., Int. Ed.*, **38**, 79 (1999).
   Solid-state structures of cyanocuprates:
   (e) C. M. P. Kronenburg, J. T. B. H. Jastrzebski, A. L. Spek and G. van Koten, *J. Am. Chem. Soc.*, **120**, 9688 (1998).
   (f) G. Boche, F. Bosold, M. Marsch and K. Harms, *Angew. Chem.*, **110**, 1779 (1998); *Angew. Chem., Int. Ed.*, **37**, 1684 (1998).
   (g) C.-S. Hwang and P. P. Power, *J. Am. Chem. Soc.*, **120**, 6409 (1998).
   (h) J. T. B. H. Jastrzebski and G. van Koten, in *Modern Organocopper Chemistry* (Ed. N. Krause), Wiley-VCH, Weinheim, 2002, pp. 1–44.
6. Solution structures:
   (a) R. M. Gschwind, *Chem. Rev.*, **108**, 3029–3053 (2008) (Review).

(b) R. M. Gschwind, P. R. Rajamohanan, M. John and G. Boche, *Organometallics*, **19**, 2868 (2000).
(c) M. John, C. Auel, C. Behrens, M. Marsch, K. Harms, F. Bosold, R. M. Gschwind, P. Rajamohanan and G. Boche, *Chem. Eur. J.*, **6**, 3060 (2000).
(d) R. M. Gschwind, X. Xie, P. R. Rajamohanan, C. Auel and G. Boche, *J. Am. Chem. Soc.*, **123**, 7299 (2001).
(e) W. Henze, A. Vyaters, N. Krause and R. M. Gschwind, *J. Am. Chem. Soc.*, **127**, 17335 (2005).
(f) W. Henze, T. Gärtner and R. M. Gschwind, *J. Am. Chem. Soc.*, **130**, 13718 (2008).

7. (a) R. A. J. Smith and A. S. Vellekoop, in *Advances in Detailed Reaction Mechanisms* (Ed. J. M. Coxon), Vol. 3, JAI Press, Greenwich, 1994, pp. 79–130.
(b) S. Woodward, *Chem. Soc. Rev.*, **29**, 393–401 (2000).
(c) E. Nakamura and S. Mori, *Angew. Chem.*, **112**, 3902 (2000); *Angew. Chem., Int. Ed.*, **39**, 3750 (2000).
(d) S. Mori and E. Nakamura, in *Modern Organocopper Chemistry* (Ed. N. Krause), Wiley-VCH, Weinheim, 2002, pp. 315–346.
(e) S. H. Bertz, C. M. Carlin, D. A. Deadwyler, M. D. Murphy, G. A. Ogle and P. H. Seagle, *J. Am. Chem. Soc.*, **124**, 13650 (2002).
(f) H. Zhang and R. M. Gschwind, *Angew. Chem.*, **118**, 6540 (2006); *Angew. Chem., Int. Ed.*, **45**, 6391 (2006).
(g) S. H. Bertz, S. Cope, D. Dorton, M. Murphy and G. A. Ogle, *Angew. Chem.*, **119**, 7212 (2007); *Angew. Chem., Int. Ed.*, **46**, 7082 (2007).
(h) T. Gärtner, W. Henze and R. M. Gschwind, *J. Am. Chem. Soc.*, **129**, 11362 (2007).
(i) E. R. Bartholomew, S. H. Bertz, S. Cope, M. Murphy and C. A. Ogle, *J. Am. Chem. Soc.*, **130**, 11244 (2008).

8. (a) P. Knochel, M. J. Rozema, C. E. Tucker, C. Retherford, M. Furlong and S. A. Rao, *Pure Appl. Chem.*, **64**, 361 (1992).
(b) P. Knochel and R. D. Singer, *Chem. Rev.*, **93**, 2117–2188 (1993).
(c) P. Knochel, *Synlett*, 393 (1995).
(d) P. Knochel, J. J. Almena Perea and P. Jones, *Tetrahedron*, **54**, 8275 (1998).
(e) A. Boudier, L. O. Bromm, M. Lotz and P. Knochel, *Angew. Chem.*, **112**, 4584 (2000); *Angew. Chem., Int. Ed.*, **39**, 4414 (2000).
(f) P. Wipf, *Synthesis*, 537 (1993).
(g) B. H. Lipshutz, A. Bhandari, C. Lindsey, R. Keil and M. R. Wood, *Pure Appl. Chem.*, **66**, 1493 (1994).
(h) B. H. Lipshutz, *Acc. Chem. Res.*, **30**, 277 (1997).
(i) P. Knochel and B. Betzemeier, in *Modern Organocopper Chemistry* (Ed. N. Krause), Wiley-VCH, Weinheim, 2002, pp. 45–78.
(j) R. K. Dieter, in *Modern Organocopper Chemistry* (Ed. N. Krause), Wiley-VCH, Weinheim, 2002, pp. 79–144.
(k) B. H. Lipshutz, in *Modern Organocopper Chemistry* (Ed. N. Krause), Wiley-VCH, Weinheim, 2002, pp. 167–187.
(l) S. Rendler and M. Oestreich, *Angew. Chem.*, **119**, 504 (2007); *Angew. Chem., Int. Ed.*, **46**, 498 (2007).
(m) C. Deutsch, N. Krause and B. H. Lipshutz, *Chem. Rev.*, **108**, 2916–2927 (2008).

9. (a) B. Breit and P. Demel, *Tetrahedron*, **56**, 2833 (2000).
(b) B. Breit and P. Demel, in *Modern Organocopper Chemistry* (Ed. N. Krause), Wiley-VCH, Weinheim, 2002, pp. 188–223.
(c) B. Breit and Y. Schmidt, *Chem. Rev.*, **108**, 2928–2951 (2008).

10. Reviews:
(a) A. Alexakis, J. E. Bäckvall, N. Krause, O. Pamies and M. Dieguez, *Chem. Rev.*, **108**, 2796–2823 (2008).
(b) S. R. Harutyunyan, T. Den Hartog, K. Geurts, A. J. Minnaard and B. L. Feringa, *Chem. Rev.*, **108**, 2824–2852 (2008).

11. (a) M. van Klaveren, E. S. M. Persson, A. Del Villar, D. M. Grove, J.-E. Bäckvall and G. van Koten, *Tetrahedron Lett.*, **36**, 3059 (1995).
(b) G. J. Meuzelaar, A. S. E. Karlström, M. van Klaveren, E. S. M. Persson, A. Del Villar, G. van Koten and J.-E. Bäckvall, *Tetrahedron*, **56**, 2895 (2000).

(c) F. Dübner and P. Knochel, *Angew. Chem.*, **111**, 391 (1999); *Angew. Chem., Int. Ed.*, **38**, 379 (1999).
(d) F. Dübner and P. Knochel, *Tetrahedron Lett.*, **41**, 9233 (2000).
(e) A. Alexakis, C. Malan, L. Lea, C. Benhaim and X. Fournioux, *Synlett*, 927 (2001).
(f) A. S. E. Karlström and J.-E. Bäckvall, in *Modern Organocopper Chemistry* (Ed. N. Krause), Wiley-VCH, Weinheim, 2002, pp. 259–288.
(g) A. Alexakis and K. Croset, *Org. Lett.*, **4**, 4147 (2002).
(h) K. Tissot-Croset, D. Polet and A. Alexakis, *Angew. Chem.*. **116**, 2480 (2004); *Angew. Chem., Int. Ed.*, **43**, 2426 (2004).
(i) K. Tissot-Croset, D. Polet, S. Gille, C. Hawner and A. Alexakis, *Synthesis*, 2586 (2004).
(j) C. A. Falciola, K. Tissot-Croset and A. Alexakis, *Angew. Chem.*, **118**, 6141 (2006); *Angew. Chem., Int. Ed.*, **45**, 5995 (2006).
(k) C. A. Falciola and A. Alexakis, *Eur. J. Org. Chem.*, 3765 (2008).

12. Reviews:
(a) B. E. Rossiter and N. M. Swingle, *Chem. Rev.*, **92**, 771–806 (1992).
(b) N. Krause, *Angew. Chem.*, **110**, 295–297 (1998); *Angew. Chem., Int. Ed.*, **37**, 283–285 (1998).
(c) J. Leonard, E. Diez-Barra and S. Merino, *Eur. J. Org. Chem.*, 2051–2061 (1998).
(d) B. L. Feringa, *Acc. Chem. Res.*, **33**, 346–353 (2000).
(e) N. Krause and A. Hoffmann-Röder, *Synthesis*, 171–196 (2001).
(f) B. L. Feringa R. Naasz, R. Imbos and L. A. Arnold, in *Modern Organocopper Chemistry* (Ed. N. Krause), Wiley-VCH, Weinheim, 2002, pp. 224–258.
(g) A. Alexakis and C. Benhaim, *Eur. J. Org. Chem.*, 3221–3235 (2002).
(h) J. Christoffers, G. Koripelly, A. Rosiak and M. Rössle, *Synthesis*, 1279–1300 (2007).

13. (a) Y. Chounan and Y. Yamamoto, in *Modern Organocopper Chemistry* (Ed. N. Krause), Wiley-VCH, Weinheim, 2002, pp. 289–314.
(b) G. Helmchen, M. Ernst and G. Paradies, *Pure Appl. Chem*., **76**, 495 (2004).

14. Reviews:
(a) N. Krause and A. Gerold, *Angew. Chem.*, **109**, 194–213 (1997); *Angew. Chem., Int. Ed. Engl.*, **36**, 186–204 (1997).
(b) M. A. Fredrick and M. Hulce, *Tetrahedron*, **53**, 10197–10227 (1997).
(c) N. Krause and S. Thorand, *Inorg. Chim. Acta*, **296**, 1–11 (1999).
(d) N. Krause and C. Zelder, in *The Chemistry of Dienes and Polyenes* (Ed. Z. Rappoport), Vol. 2, Wiley, New York, 2000, pp. 645–691.
(e) N. Krause and A. Hoffmann-Röder, in *Modern Organocopper Chemistry* (Ed. N. Krause), Wiley-VCH, Weinheim, 2002, pp. 145–166.

15. F. Näf, P. Degen and G. Ohloff, *Helv. Chim. Acta*, **55**, 82 (1972).

16. (a) E. J. Corey, C. U. Kim, R. H. K. Chen and M. Takeda, *J. Am. Chem. Soc.*, **94**, 4395 (1972).
(b) E. J. Corey and R. H. K. Chen, *Tetrahedron Lett.*, **14**, 1611 (1973).
(c) E. J. Corey and N. W. Boaz, *Tetrahedron Lett.*, **26**, 6019 (1985).
(d) B. Ganem, *Tetrahedron Lett.*, 4467 (1974).
(e) S. F. Martin and P. J. Garrison, *Synthesis*, 394 (1982).
(f) F. Barbot, A. Kadib-Elban and P. Miginiac, *J. Organomet. Chem.*, **255**, 1 (1983).
(g) J. Bigorra, J. Font, C. Jaime, R. M. Ortuno and F. Sanchez-Ferrando, *Tetrahedron*, **41**, 5577 (1985).
(h) J. Bigorra, J. Font, C. Jaime, R. M. Ortuno, F. Sanchez-Ferrando, F. Florencio, S. Martinez Carrera and S. Garcia-Blanco, *Tetrahedron*, **41**, 5589 (1985).
(i) H. Liu, L. M. Gayo, R. W. Sullivan, A. Y. H. Choi and H. W. Moore, *J. Org. Chem.*, **59**, 3284 (1994).

17. (a) Y. Yamamoto, S. Yamamoto, H. Yatagai, Y. Ishihara and K. Maruyama, *J. Org. Chem.*, **47**, 119 (1982).
(b) See also: R. R. Cesati III, J. de Armas and A. H. Hoveyda, *J. Am. Chem. Soc.*, **126**, 96 (2004).
(c) M. Bella, M. Cianflone, G. Montemurro, P. Passacantilli and G. Piancatelli, *Tetrahedron*, **60**, 4821 (2004).

18. (a) L. Novak, J. Rohaly, P. Kolonits, J. Fekete, L. Varjas and C. Szantay, *Liebigs Ann. Chem.*, 1173 (1982).
(b) F. Näf, R. Decorzant and S. D. Escher, *Tetrahedron Lett.*, **23**, 5043 (1982).

(c) U. Schöllkopf, D. Pettig, E. Schulze, M. Klinge, E. Egert, B. Benecke and M. Noltemeyer, *Angew. Chem.*, **100**, 1238 (1988); *Angew. Chem., Int. Ed. Engl.*, **27**, 1194 (1988).
(d) H. Wild and L. Born, *Angew. Chem.*, **103**, 1729 (1991); *Angew. Chem., Int. Ed. Engl.*, **30**, 1685 (1991).
(e) K. Sabbe, C. D'Hallewyn, P. de Clercq, M. Vanderwalle, R. Bouillon and A. Verstuyf, *Bioorg. Med. Chem. Lett.*, **6**, 1697 (1996).

19. (a) J. A. Campbell and J. C. Babcock, *J. Am. Chem. Soc.*, **81**, 4069 (1959).
(b) N. W. Atwater, R. H. Bible, E. A. Brown, R. R. Burtner, J. S. Mihina, L. N. Nysted and P. B. Sollman, *J. Org. Chem.*, **26**, 3077 (1961).
(c) R. Wiechert, U. Kerb and K. Kieslich, *Chem. Ber.*, **96**, 2765 (1963).
(d) U. Kerb and R. Wiechert, *Chem. Ber.*, **96**, 2772 (1963).
(e) P. Wieland and G. Auner, *Helv. Chim. Acta*, **50**, 289 (1967).
(f) J.-C. Jacquesy, R. Jacquesy and C. Narbonne, *Bull. Soc. Chim. Fr.*, 1240 (1976).
20. (a) J. A. Marshall and H. Roebke, *J. Org. Chem.*, **31**, 3109 (1966).
(b) J. A. Marshall, R. A. Ruden, L. K. Hirsch and M. Philippe, *Tetrahedron Lett.*, **12**, 3795 (1971).
(c) J. A. Marshall and R. E. Conrow, *J. Am. Chem. Soc.*, **105**, 5679 (1983).
(d) J. A. Marshall, J. E. Audia and B. G. Shearer, *J. Org. Chem.*, **51**, 1730 (1986).
21. (a) R. Bucourt, M. Vignau, V. Torrelli, H. Richard-Foy, C. Geynet, C. Secco-Millet, G. Redeuilh and E.-E. Baulieu, *J. Biol. Chem.*, **253**, 8221 (1978).
(b) J. M. O'Reilly, N. Li, W. L. Duax and R. W. Brueggemeier, *J. Med. Chem.*, **38**, 2842 (1995).
22. (a) J. F. Grunwell, H. D. Benson, J. O. Johnston and V. Petrow, *Steroids*, **27**, 759 (1976).
(b) A. J. Solo, C. Caroli, M. V. Darby, T. McKay, W. D. Slaunwhite and P. Hebborn, *Steroids*, **40**, 603 (1982).
(c) B. Mühlenbruch, F. Kirmeier and H. J. Roth, *Arch. Pharm. (Weinheim)*, **319**, 177 (1986).
(d) J. Bowler, T. J. Lilley, J. D. Pittam and A. E. Wakeling, *Steroids*, **54**, 71 (1989).
(e) S. P. Modi, J. O. Gardner, A. Milowsky, M. Wierzba, L. Forgione, P. Mazur, A. J. Solo, W. L. Duax, Z. Galdecki, P. Grochulski and Z. Wawrzak, *J. Org. Chem.*, **54**, 2317 (1989).
(f) A. N. French, S. R. Wilson, M. J. Welch and J. A. Katzenellenbogen, *Steroids*, **58**, 157 (1993).
23. M. Uerdingen and N. Krause, *Tetrahedron*, **56**, 2799 (2000).
24. E. Fillion, A. Wilsily and E.-T. Liao, *Tetrahedron: Asymmetry*, **17**, 2957 (2006).
25. T. Den Hartog, S. R. Harutyunyan, D. Font, A. J. Minnaard and B. L. Feringa, *Angew. Chem.*, **120**, 404 (2008); *Angew. Chem., Int. Ed.*, **47**, 398 (2008).
26. H. Henon, M. Mauduit and A. Alexakis, *Angew. Chem.*, **120**, 9262 (2008); *Angew. Chem., Int. Ed.*, **47**, 9122 (2008).
27. (a) L. Ernst, H. Hopf and N. Krause, *J. Org. Chem.*, **52**, 398 (1987).
(b) H. Hopf and N. Krause, in *Chemistry and Biology of Synthetic Retinoids* (Eds M. I. Dawson and W. H. Okamura), CRC Press, Boca Raton, 1990, pp. 177–199.
(c) Y. L. Bennani, *J. Org. Chem.*, **61**, 3542 (1996).
28. (a) M. Hulce, *Tetrahedron Lett.*, **29**, 5851 (1988).
(b) S.-H. Lee and M. Hulce, *Tetrahedron Lett.*, **31**, 311 (1990).
(c) M. Cheng and M. Hulce, *J. Org. Chem.*, **55**, 964 (1990).
29. (a) *The Chemistry of Ketenes, Allenes and Related Compounds* (Ed. S. Patai), Wiley, New York, 1980.
(b) *The Chemistry of the Allenes* (Ed. S. R. Landor), Academic Press, London, 1982.
(c) H. F. Schuster and G. M. Coppola, *Allenes in Organic Synthesis*, John Wiley & Sons, Inc., New York, 1984.
(d) N. Krause and A. S. K. Hashmi, in *Modern Allene Chemistry* (Eds. N. Krause and A. S. K. Hashmi), Wiley-VCH, Weinheim, 2004.
(e) N. Krause and A. Hoffmann-Röder, *Tetrahedron*, **60**, 11671 (2004).
(f) H. H. A. M. Hassan, *Curr. Org. Synth.*, **4**, 413 (2007).
(g) K. M. Brummond and J. E. DeForest, *Synthesis*, 795 (2007).
(h) *Cumulenes and Allenes* (Ed. N. Krause), Vol. 44, *Science of Synthesis*, Thieme, Stuttgart, 2007.
30. (a) N. Krause, *Chem. Ber.*, **123**, 2173 (1990).
(b) N. Krause, *Chem. Ber.*, **124**, 2633 (1991).

(c) M. Hohmann and N. Krause, *Chem. Ber.*, **128**, 851 (1995).
31. A. Haubrich, M. van Klaveren, G. van Koten, G. Handke and N. Krause, *J. Org. Chem.*, **58**, 5849 (1993).
32. J. Dambacher and M. Bergdahl, *J. Org. Chem.*, **70**, 580 (2005).
33. M. Bergdahl, M. Eriksson, M. Nilsson and T. Olsson, *J. Org. Chem.*, **58**, 7238 (1993).
34. N. Krause and G. Handke, *Tetrahedron Lett.*, **32**, 7229 (1991).
35. A. Gerold and N. Krause, *Chem. Ber.*, **127**, 1547 (1994).
36. (a) S. Arndt, G. Handke and N. Krause, *Chem. Ber.*, **126**, 251 (1993).
(b) N. Krause and S. Arndt, *Chem. Ber.*, **126**, 261 (1993).
37. (a) S.-H. Lee, M. Shih and M. Hulce, *Tetrahedron Lett.*, **33**, 185 (1992).
(b) S.-H. Lee and M. Hulce, *Synlett*, 485 (1992).
38. (a) G. Handke and N. Krause, *Tetrahedron Lett.*, **34**, 6037 (1993).
(b) N. Krause, G. Handke and U. Wecker, in *Stereoselective Reactions of Metal-Activated Molecules* (Eds. H. Werner and J. Sundermeyer), Vieweg, Braunschweig, 1995, pp. 153–155.
39. (a) T. Hayashi, N. Tokunaga and K. Inoue, *Org. Lett.*, **6**, 305 (2004).
40. For a review on allenic natural products, see: A. Hoffmann-Röder and N. Krause, *Angew. Chem.*, **116**, 1216–1236 (2004); *Angew. Chem., Int. Ed.*, **43**, 1196–1216 (2004).
41. N. Krause, *Liebigs Ann. Chem.*, 521 (1993).
42. U. Koop, G. Handke and N. Krause, *Liebigs Ann.*, 1487 (1996).
43. M. Laux, N. Krause and U. Koop, *Synlett*, 87 (1996).
44. M. Becker and N. Krause, *Liebigs Ann./Recueil*, 725 (1997).
45. (a) N. Krause, M. Laux and A. Hoffmann-Röder, *Tetrahedron Lett.*, **41**, 9613 (2000).
(b) A. Hoffmann-Röder and N. Krause, *Synthesis*, 2143 (2006).
46. (a) A. Hoffmann-Röder and N. Krause, *Org. Lett.*, **3**, 2537 (2001).
(b) N. Krause, A. Hoffmann-Röder and J. Canisius, *Synthesis*, 1759 (2002).
(c) C. Deutsch, B. Gockel, A. Hoffmann-Röder and N. Krause, *Synlett*, 1790 (2007).
(d) N. Krause, V. Belting, C. Deutsch, J. Erdsack, H.-T. Fan, B. Gockel, A. Hoffmann-Röder, N. Morita and F. Volz, *Pure Appl. Chem.*, **80**, 1063 (2008).
47. (a) N. Krause, *J. Org. Chem.*, **57**, 3509 (1992).
(b) N. Krause, R. Wagner and A. Gerold, *J. Am. Chem. Soc.*, **116**, 381 (1994).
(c) J. Canisius, T. A. Mobley, S. Berger and N. Krause, *Chem. Eur. J.*, **7**, 2671 (2001).
48. (a) S. Mori, M. Uerdingen, N. Krause and K. Morokuma, *Angew. Chem.*, **117**, 4795 (2005); *Angew. Chem., Int. Ed.*, **44**, 4715 (2005).
(b) N. Yoshikai, T. Yamashita and E. Nakamura, *Angew. Chem.*, **117**, 4799 (2005); *Angew. Chem., Int. Ed.*, **44**, 4721 (2005).
(c) N. Yoshikai, T. Yamashita and E. Nakamura, *Chem. Asian J.*, **1**, 322 (2006).
49. J. Canisius, A. Gerold and N. Krause, *Angew. Chem.*, **111**, 1727 (1999); *Angew. Chem., Int. Ed.*, **38**, 1644 (1999).
50. (a) T. L. Underiner and H. L. Goering, *J. Org. Chem.*, **52**, 897 (1987).
(b) T. L. Underiner and H. L. Goering, *J. Org. Chem.*, **53**, 1140 (1988).
(c) T. L. Underiner, S. D. Paisley, J. Schmitter, L. Leheski and H. L. Goering, *J. Org. Chem.*, **54**, 2369 (1989).
(d) T. L. Underiner and H. L. Goering, *J. Org. Chem.*, **55**, 2757 (1990).
(e) T. L. Underiner and H. L. Goering, *J. Org. Chem.*, **56**, 2563 (1991).
51. N. Nakanishi, S. Matsubara, K. Utimoto, S. Kozima and R. Yamaguchi, *J. Org. Chem.*, **56**, 3278 (1991).
52. S.-K. Kang, D.-G. Cho, J.-U. Chung and D.-Y. Kim, *Tetrahedron: Asymmetry*, **5**, 21 (1994).
53. H. Rakotoarisoa, R. G. Perez, P. Mangeney and A. Alexakis, *Organometallics*, **8**, 1957 (1996).
54. M. Purpura and N. Krause, *Eur. J. Org. Chem.*, 267 (1999).
55. N. Krause and M. Purpura, *Angew. Chem.*, **112**, 4512 (2000); *Angew. Chem., Int. Ed.*, **39**, 4355 (2000).
56. (a) A. Alexakis, P. Mangeney, A. Ghribi, I. Marek, R. Sedrani, C. Guir and J. F. Normant, *Pure Appl. Chem.*, **60**, 49 (1988).
(b) A. Alexakis, I. Marek, P. Mangeney and J. F. Normant, *J. Am. Chem. Soc.*, **112**, 8042 (1990).

CHAPTER **18**

Copper-mediated cross-coupling reactions

LIZA PENN and DMITRI GELMAN

Institute of Chemistry, The Hebrew University of Jerusalem, Edmund Safra Campus, Givat Ram, 91904 Jerusalem, Israel
Fax: (+) 972-2-6585279; e-mail: elizabet.penn@mail.huji.ac.il; dgelman@chem.ch.huji.ac.il

I. GENERAL INTRODUCTION . . . 882
II. CARBON–HETEROATOM BOND-FORMING REACTIONS . . . 883
A. Classical Ullmann Chemistry . . . 883
1. C–O, C–S and C–Se bond-forming reactions . . . 883
a. Diaryl ethers . . . 883
b. Alkyl aryl ethers . . . 885
c. Sulfides and selenides . . . 887
2. C–N and C–P bond-forming reactions . . . 887
a. Amination of aryl halides . . . 887
b. Amidation of aryl and vinyl halides . . . 890
c. Early mechanistic considerations . . . 891
d. Coupling of aryl and vinyl halides with *P*-nucleophiles . . . 892
B. Modern Ullmann Chemistry . . . 893
1. C–O, C–S and C–Se bond-forming reactions . . . 893
a. Diaryl ethers . . . 893
b. Alkyl aryl ethers . . . 897
c. Vinyl ethers . . . 899
d. Miscellaneous *O*-nucleophiles . . . 902
e. Sulfides . . . 902
f. Selenides . . . 904
g. Copper-catalyzed coupling in syntheses of *O*- and *S*-containing heterocycles . . . 906
2. C–N and C–P bond-forming reactions . . . 909
a. Coupling of aliphatic amines with aryl halides . . . 909
b. Coupling of aromatic amines with aryl halides . . . 919
c. Coupling of NH heterocycles with aryl halides . . . 921

The chemistry of organocopper compounds
Edited by Z. Rappoport and I. Marek

d. Coupling of amides and related compounds with aryl halides 922
e. Coupling of amides and related compounds with vinyl and alkynyl halides 927
f. Miscellaneous *N*-nucleophiles 937
g. Coupling of P-nucleophiles with organic halides 938
3. Copper-catalyzed halogen exchange reactions 940
4. Mechanistic considerations 942
C. Non-Ullmann Substrates in Copper-mediated Carbon–Heteroatom Coupling Reactions 943
1. Iodonium salts 943
a. Coupling of O-nucleophiles with iodonium salts 943
b. Coupling of N-nucleophiles with iodonium salts 943
c. Coupling of P-nucleophiles with iodonium salts 945
2. Organobismuth compounds 946
a. Coupling of O-nucleophiles with organobismuth compounds 946
b. Coupling of N-nucleophiles with organobismuth compounds 947
3. Organolead compounds 949
a. Coupling of O-nucleophiles with organolead compounds 949
b. Coupling of N-nucleophiles with organolead compounds 949
4. Organoboron compounds 950
a. Coupling of O- and S-nucleophiles with organoboron compounds 950
b. Coupling of N-nucleophiles with organoboron compounds 954
5. Organotin compounds 959
a. Coupling of O-nucleophiles with organotin compounds 959
b. Coupling of N-nucleophiles with organotin compounds 960
6. Organosilicon compounds 961
7. Mechanistic considerations 961
8. Cross-dehydrogenative coupling to form C–N bonds 963
III. CARBON–CARBON BOND FORMING REACTIONS 965
A. Copper-catalyzed Cross-coupling of *C*-Nucleophiles with Organic Halides 965
1. Sonogashira reaction 965
2. Rosenmund–von Braun reaction 967
3. Hurley and related reactions 968
4. Cross-dehydrogenative coupling 973
B. Copper-catalyzed Cross-coupling of Organometallic Reagents with Organic Halides 973
1. Organoboron reagents 973
2. Organotin compounds 975
3. Organosilicon compounds 977
4. Organomagnesium compounds 978
IV. REFERENCES 980

I. GENERAL INTRODUCTION

Copper-assisted coupling chemistry has very old roots growing from the Ullmann reaction reported in 1901. For many years, these rather harsh transformations (200–250 °C heating in polar high-boiling solvents or even fusion of the starting materials with stoichiometric copper bronze or substoichiometric copper salts) have played an exceptionally important role in organic synthesis, being a complementary method to the nucleophilic aromatic substitution at highly activated aromatic rings.

In the mid-1980s, copper was displaced from the leading position by palladium that was capable of mediating carbon–heteroatom and carbon–carbon bond-forming reactions in catalytic fashion and under milder reaction conditions. However, paradoxically, the development of more efficient and general cross-coupling methodologies did not bury copper but lead to a revival of interest in its chemistry. Moreover, the knowledge gained from the recent studies on palladium-catalyzed cross-coupling reactions promoted very much the driving of old Ullmann chemistry into true catalytic rails. Since then, a number of excellent protocols and new reaction schemes utilizing catalytic copper compounds have been developed, offering a wide range of synthetic opportunities—alternative or orthogonal to those offered by palladium catalysis. Even more important, organocopper chemistry rapidly extends its scope, and the quest for better and new reactivity is not over.

II. CARBON–HETEROATOM BOND-FORMING REACTIONS

A. Classical Ullmann Chemistry

1. C–O, C–S and C–Se bond-forming reactions

a. Diaryl ethers. Ullmann cross-coupling reaction between aryl halides and phenols was originally described as a transformation *catalytic* in copper leading to the formation of diaryl ethers (equation 1)[1–4].

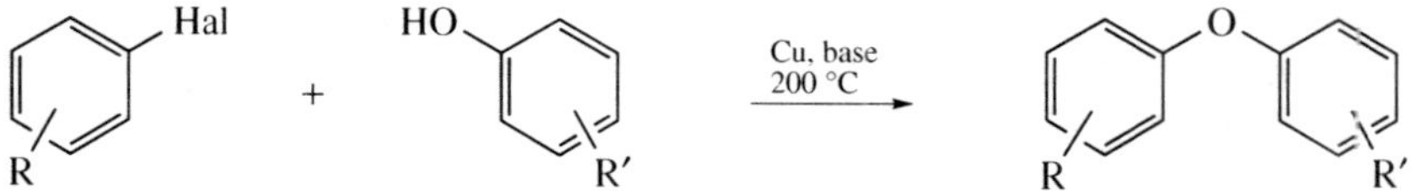

(1)

As reported, the reaction comprises heating aryl halides with phenols in the presence of a base (traditionally, potassium or sodium hydroxide) and substoichiometric copper powder, cupric oxide or cuprous halides at above 200 °C in polar solvents such as DMA, DMF, DMSO, HMPT or without a solvent. Normally, the reaction of unfunctionalized substrates results in the formation of diaryl ethers in good yield. However, the yields drop when substituents at either the phenol or the aryl halide coupling partners are present. Indeed, while simple electron-releasing groups such as methyl or methoxy slightly slow the reaction, the presence of hydroxyl, amino, nitro or carboxyl substituents generally leads to only poor yields of products at the expense of the competitive reductive dehalogenation or decarboxylation of the starting materials under the harsh reaction conditions[5].

Early mechanistic studies performed by Weingarten led to the conclusion that Ullmann diaryl ether synthesis apparently proceeds through a nucleophilic aromatic substitution pathway with a rate-limiting carbon–halogen bond cleavage[6,7]. These assumptions were made based on kinetic measurements of the reaction between potassium phenolate and substituted aryl halides that revealed a slightly positive Hammett ρ value of $+0.61$[8], but an unusual reactivity order $I \sim Br > Cl \gg F$ for halogen leaving groups[7]. Additionally, by comparing rates of diaryl ether formation promoted by cuprous and cupric salts under similar reaction conditions, Weingarten found essentially equal reactivity of the precatalysts in different oxidation states[6]. This important observation suggested that the reactions have a common intermediate, apparently a copper(I) species. The latter conclusion relies on the fact that different reaction rates were observed under air and air-free conditions.

Summarizing these findings, Weingarten suggested a plausible mechanism for the Ullmann coupling of phenols and aryl halides that was widely accepted at that time (Scheme 1). According to his hypothesis, the reaction starts with a formal ligand exchange between the Cu(I) precatalyst and phenoxide ions (if Cu(II) compounds are employed, a single electron reduction is required prior to the ligand exchange step) yielding catalytically active Cu(I) cuprate-like species. At latter stages, the reaction follows the S_NAr route, but proceeds through the formation of a cuprous–arene π complex, which can explain both the carbon–halogen bond cleavage as the rate-limiting step, and the *catalytic* influence of copper.

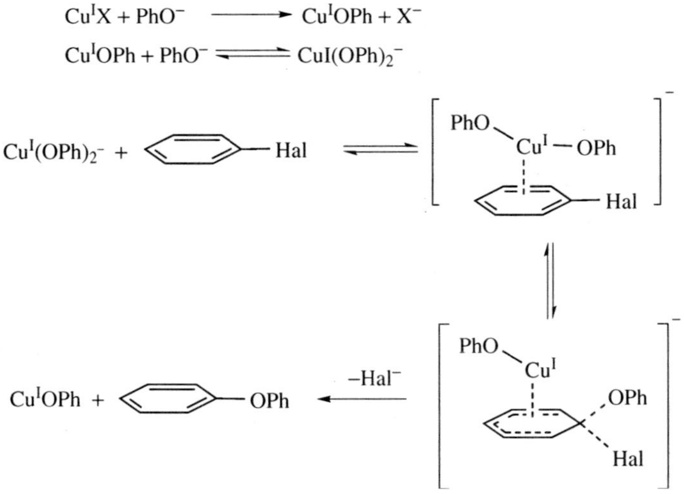

SCHEME 1

Even more spectacularly, during the same studies, a ligand acceleration effect in Ullmann diaryl ether synthesis was detected for the first time[6,7]. Careful work by Weingarten clearly demonstrated that impurities present in the diglyme reaction medium imparted enhanced catalytic activity. This activity disappeared by purification of the solvent by $LiAlH_4$ treatment. Subsequently, the impurity was identified as 2-(formyloxy)ethyl 2-methoxyacetate. A mechanistic rationale for this phenomenon was not clarified then, but increased catalyst solubility was suspected, and catalyst competence appeared to be related to the ligating ability of the diester. Indeed, it was found that other diesters such as 1,2-diacetoxycyclohexane, ethylene carbonate, propylene glycol mono- and diacetate deliberately added to the reaction mixture also accelerate the Ullmann cross-coupling.

Remarkably, understanding these mechanistic details greatly promoted the development of improved reaction protocols. Only three years after the fortunately discovered ligand acceleration effect, Williams and coworkers reported the CuCl-catalyzed *solvent*-assisted Ullmann coupling[9]. They found that 1,3-diphenoxybenzene, which was obtained with difficulty under classical Ullmann conditions, was prepared in 70% yield at only 90 °C when pyridine was chosen as a reaction medium. The role of pyridine (and similar coordinating solvents such as collidine or quinoline) in solubilizing the catalytic species was again proposed. However, addition of stronger complexing agents such as triarylphosphines or 2,2′-bipyridine was found deleterious.

Under such milder conditions, copper-catalyzed arylation of phenols with more sophisticated aromatic halides became possible. Thus, ferrocenyl aryl ethers[10] and heteroaryl aryl ethers[11] could be prepared in acceptable yields (equations 2 and 3).

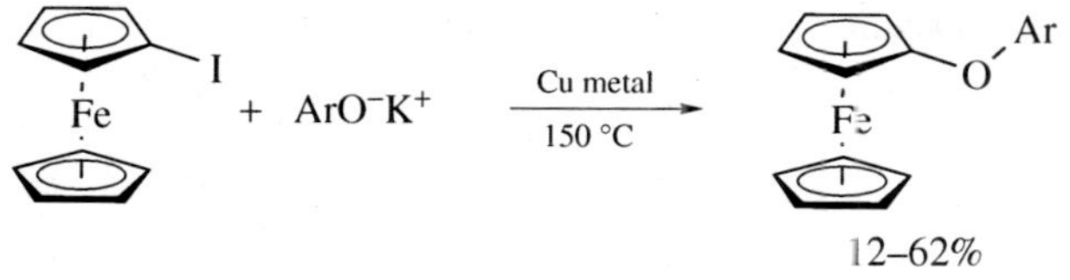

(2)

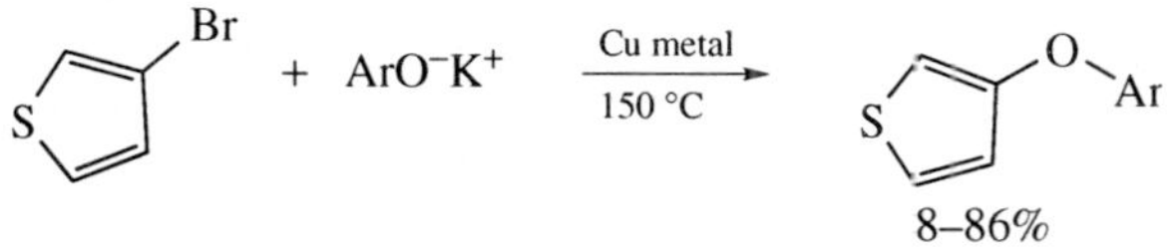

(3)

These improvements made the Ullmann ether synthesis a central tool for the synthesis of complex organic molecules[12–15]. For example, *rac*-dauricine[16,17] and *rac*-methylsterigmatocystin[18] were prepared using Ullmann chemistry as a key step (Chart 1).

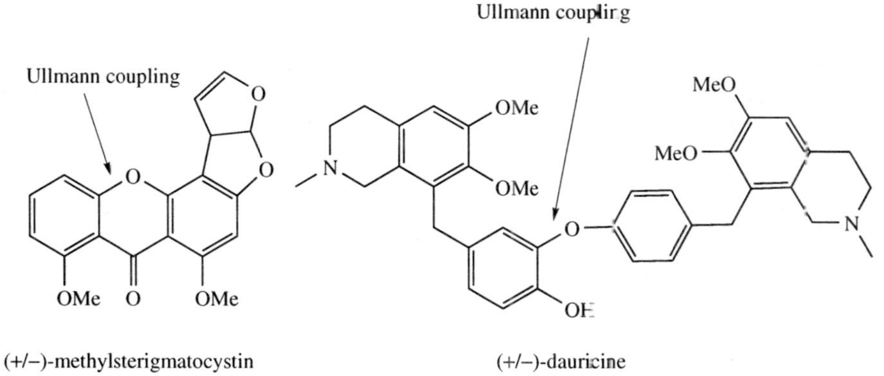

CHART 1

b. Alkyl aryl ethers. Over the years, the scope of the original Ullmann protocol was expanded to include aliphatic alcohols (equation 4)[19].

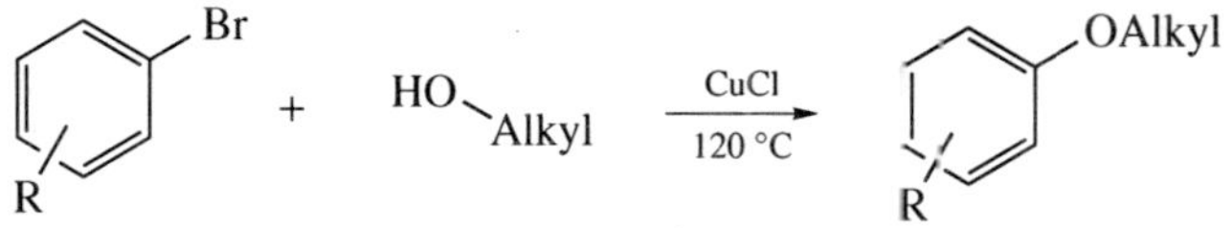

(4)

Unlike copper-catalyzed diaryl ether synthesis, the coupling of aryl halides and aliphatic alcohols in the presence of a base (or, alternatively, using presynthesized alkali metal alkoxides) proceeds under relatively mild conditions (100–150 °C) perhaps due to a higher basicity of the latter. Unfortunately, the reaction also suffers from multiple limitations. First, this reaction is strongly catalyst-dependent—satisfactory results can only be obtained in the presence of soluble cuprous salts, while reductive dehalogenation of the starting halides dominates under the heterogeneous cupric oxide catalysis[20]. Second, the reaction is generally high-yielding (80–100%) only when primary alkoxides are employed as the nucleophiles. Here again, reductive dehalogenation becomes predominant with secondary and tertiary alcohols[20]. Third, the extent of the competitive dehalogenation is greater with *ortho*-substituted aryl halides[21]. Finally, the coupling is very sensitive to the presence of functional groups—the yields remain high with a COOH substituent at the halide, but the presence of nitro or carbonyl groups is deleterious, presumably owing to greater vulnerability of these groups to the basic conditions and to their propensity to redox processes[19].

Since copper alkoxides are possible intermediates in the Ullmann ether synthesis, information concerning their properties and reactivity is pertinent.

Characterizable copper(I) alkoxides may be prepared either by direct reaction between CuCl and alkali metal alkoxides (equation 5)[22] or by reaction between alcohols and *in situ* formed alkyl copper(I) organometallics (equation 6)[23].

$$\text{ROLi} + \text{CuCl} \longrightarrow \text{ROCu} + \text{LiCl} \qquad (5)$$

$$\text{CH}_3\text{Li} + (\text{Bu}_2\text{S})_2\text{CuI} \longrightarrow [\text{CH}_3\text{Cu}] \xrightarrow[0\,^{\circ}\text{C}]{\text{ROH}} \text{ROCu} + \text{CH}_4 \qquad (6)$$

The isolated cuprous alkoxides were found active in stoichiometric reactions demonstrating comparable reactivity with that observed in catalyzed ones, indicating their participation in the catalytic turnover[23].

Copper(I) alkoxides are generally stable at ambient conditions but decompose at temperatures relevant to Ullmann reaction conditions. Therefore, thermal decomposition appears as a likely cause for the occurrence of certain side-reactions, such as reductive dehalogenation, decarboxylation and reduction of nitro and carbonyl functional groups.

The major products from the thermal decomposition of primary, secondary and tertiary copper(I) alkoxides are copper metal, alcohol and the corresponding ketone or aldehyde (equation 7).

$$\text{RCH}_2\text{OCu} \longrightarrow \text{RCH}_2\text{OH} + \text{RCHO} + \text{Cu(0)} \qquad (7)$$

Two different decomposition mechanisms can take place and none of them can be completely ruled out. The first involves disproportionation reaction via vicinal elimination of copper hydride from one equivalent of the starting alkoxide and subsequent reduction of the second equivalent of copper alkoxide by the latter. The second mechanism comprises a homolytic cleavage of the copper–oxygen bond leading to the formation of alkoxy radicals (Scheme 2)[23].

$$\text{RCH}_2\text{OCu} \longrightarrow \text{RCHO} + \text{CuH}$$

$$\text{RCH}_2\text{OCu} + \text{CuH} \longrightarrow \text{Cu(0)} + \text{RCH}_2\text{OH}$$

$$\text{RCH}_2\text{OCu} \longrightarrow \text{RCH}_2\text{O}^{\bullet} + \text{Cu(0)}$$

$$2\,\text{RCH}_2\text{O}^{\bullet} \longrightarrow \text{RCH}_2\text{OH} + \text{RCHO}$$

SCHEME 2

Both mechanisms have pros and cons; however, the radical pathway represents a more general case. At least for primary, some secondary and tertiary alkoxides the formation of radical species seems predominant. For example, the comparison between product distribution obtained from decomposition of copper(I) alkoxides in different solvents revealed that the alcohol/carbonyl ratio is solvent-sensitive and higher ratios were to be obtained in solvents with greater hydrogen atom donor ability. For example, thermal decomposition of copper(I) *n*-heptoxide leads to alcohol/aldehyde ratios of 2.9 in pentane and of 10 in THF, indicating that the reaction proceeds through the formation of alkoxy radicals. A similar behavior was observed for some secondary alkoxides such as copper(I) cyclohexoxide and cyclopentoxide. Interestingly, the decomposition of tertiary alkoxides was solvent-insensitive, for which the β-hydride elimination pathway obviously does not exist.

In addition, copper(I) alkoxides and phenoxides are prone to redox processes. For example, the exposure of Cu(I) alkoxides to molecular oxygen led to the formation of Cu(II) alkoxide species, whereas the reaction of copper(I) *n*-butoxide with nitrobenzene yielded *n*-butanal[23].

c. Sulfides and selenides. Similar copper-mediated coupling reactions between unactivated aryl halides and alkyl or aryl thiolates were reported[24–30]. Using these protocols, synthesis of diaryl sulfides in high yield was possible starting from aryl iodides or bromides and alkali metal arylthiolates under relatively mild conditions (120–150 °C in HMPT, DMF or pyridine) (equations 8 and 9). It should be noticed that these reactions require at least a stoichiometric amount of either copper(I) salts or corresponding Cu(I) thiolates.

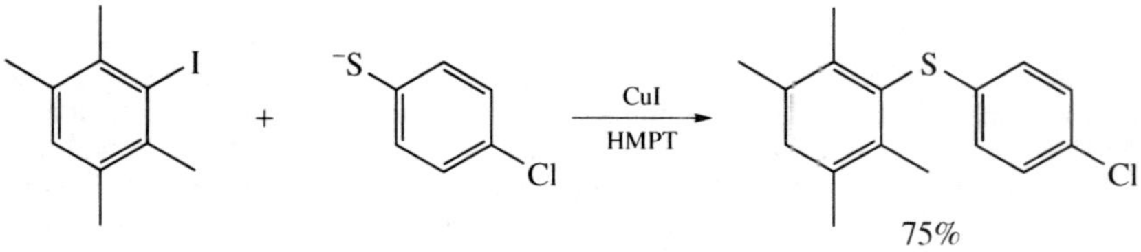

(8)

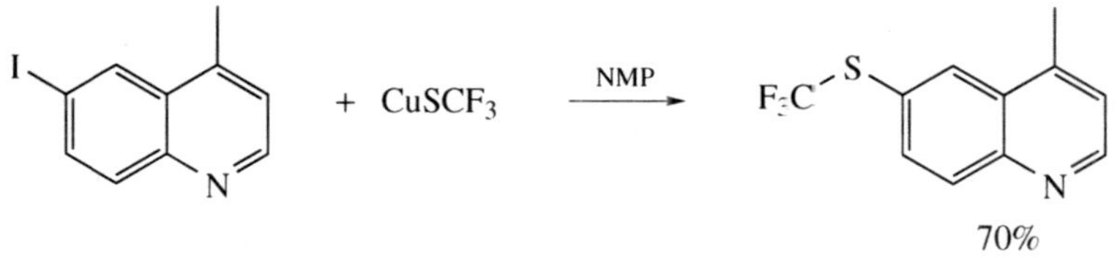

(9)

Synthesis of diaryl selenides involves heating aryl iodides, alkali metal selenoates and CuI in HMPT (equation 10). Yields of 40–75% may be obtained. Here again stoichiometric use of CuI was reported[31,32].

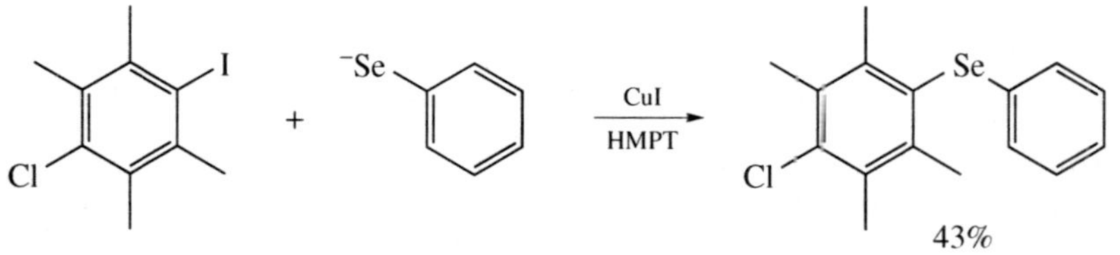

(10)

2. C–N and C–P bond-forming reactions

a. Amination of aryl halides. The Ullmann condensation of aryl halides with amines was discovered coincidentally in 1903 during an attempted synthesis of diphenic acid from 2-chlorobenzoic acid in the presence of stoichiometric copper powder in aniline as the solvent, which unexpectedly led to the formation of *N*-phenylanthranilic acid (equation 11)[33].

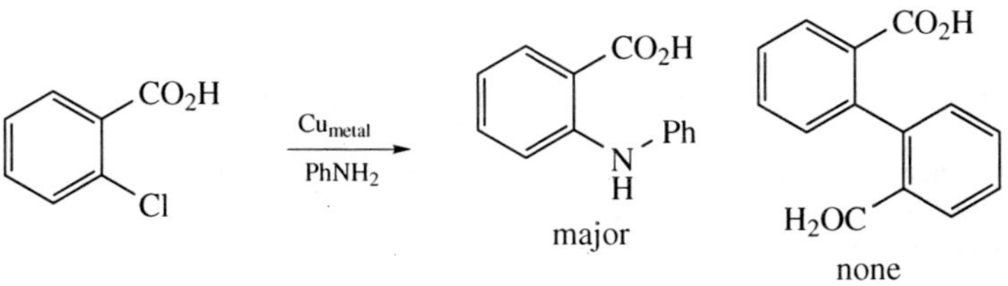

(11)

Under optimized conditions, this reaction is carried out in the presence of a substoichiometric amount of copper powder (10–30 mol%), sodium carbonate or sodium hydroxide as a base, in polar solvents such as nitrobenzene, DMSO, DMF, high-boiling alcohols or neat at 150–200 °C.

Further scope and limitation studies demonstrated that the functional group compatibility of the method is not high, but substrates bearing NO_2, OH, OMe, CN or NMe_2 groups can, in principle, be employed[34,35]. Unfortunately, yields are hardly predictable in all cases owing to very harsh reactions conditions and may vary at the expense of the competitive reductive dehalogenation or hydrolysis of the starting halides[36,37].

The Ullmann amination is more efficient with electron poor aryl halides[36,38] such as 2-halobenzoic acids; however, further activation by means of an additional electron-withdrawing group in the *para* position does not further facilitate the reaction. For example, the reactivities of 2-chlorobenzoic acid and 5-nitro-2-chlorobenzoic acid toward anilines are only slightly different, indicating that neighboring group assistance is more important than the inductive activation[36]. The reactions of 2,4-dichlorobenzoic acid with 2-pyridylamine leading to exclusive formation of the *ortho*-substituted isomer is even a better example (equation 12)[39].

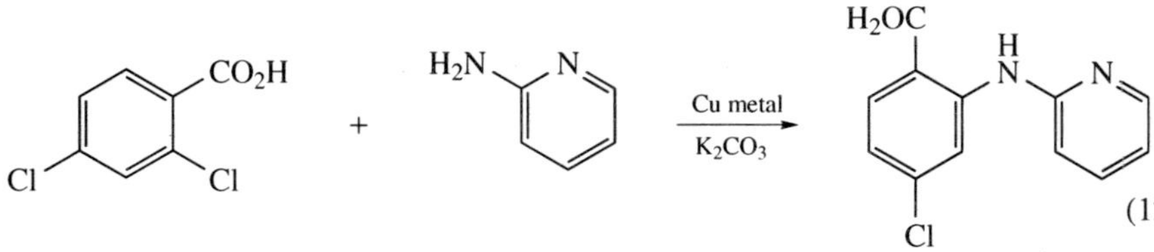

(12)

On the other hand, it was qualitatively established that the yield of Ullmann condensation of this type increases with an increase in the basicity of the amines[40].

The reaction was routinely applied to the synthesis of nonsymmetrical diaryl amines (after decarboxylation)[41] and acridone derivatives (after successive acid-promoted cyclization) (Scheme 3)[38,39,41–44].

Amination of substituted 1-haloanthraquinones with aromatic and aliphatic amines was accomplished using either copper(I) catalysts or Cu(II) catalysts[37,45–51]. It is noteworthy that amines with a hydroxyl group at the β- or γ-positions were more reactive than simple alkyl amines (e.g. butylamine)[47]. The selectivity toward the *N*-arylated product in such cases is achieved only in the presence of a large excess (100-fold) of the amino alcohol (equation 13).

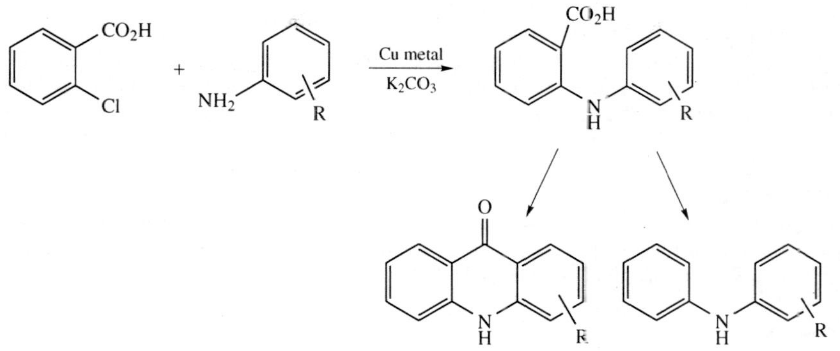

SCHEME 3

N-arylated heterocyclic compounds can also be prepared using this transformation. For example, syntheses of 10-arylphenoxazines (equation 14)[52] and 9-arylcarbazoles (equation 15)[53,54] were carried out by heating the neat N−H heterocycles with aryl bromides or iodides in the presence of a catalytic copper bronze and anhydrous potassium carbonate. As previously described, bromoarenes are less reactive than the analogous iodoarenes. The use of solvents for these reactions was found to be deleterious.

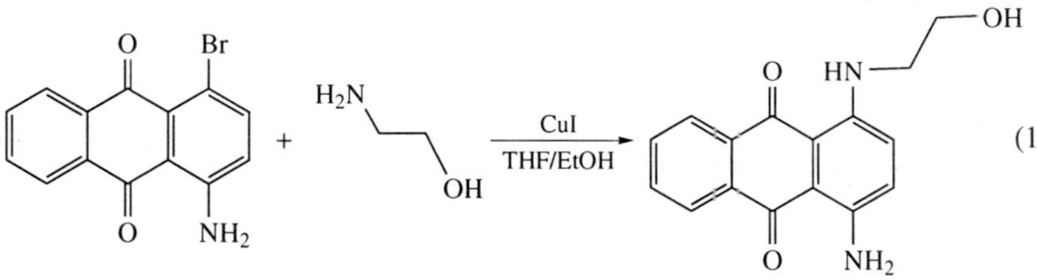

(13)

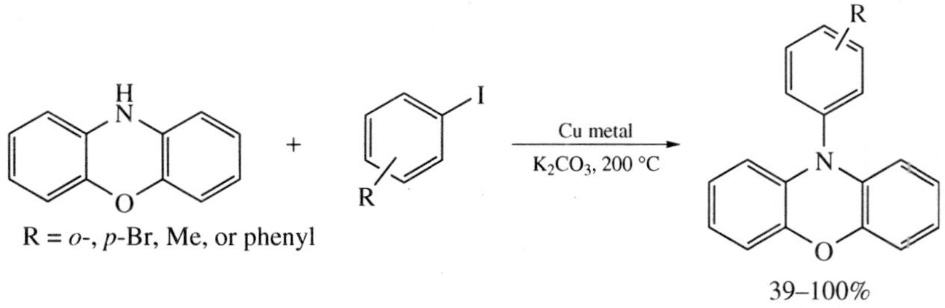

(14)

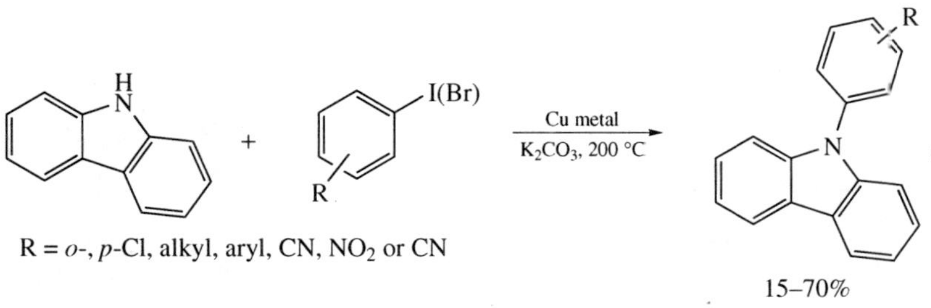

(15)

Reports on copper-catalyzed ammonolysis of aryl halides appeared in the literature only several years after the fortunate discovery of the Ullmann amination. A variety of copper catalysts in different oxidation states (copper powder, copper oxides, cuprous and cupric halides) were found effective in conversion of aryl and heteroaryl chlorides into anilines. The reaction is traditionally performed in aqueous ammonia in the presence of stoichiometric or substoichiometric copper at high temperatures. Participation of a base is not required in these reactions because ammonia is used in large excess[55–63].

Most primary and secondary amines displace halogen more readily than does ammonia, and therefore formation of secondary and tertiary amines often accompany the ammonolysis reaction. In addition, reductive dehalogenation and hydrolysis of the starting aryl halides is a serious problem (equation 16)[64].

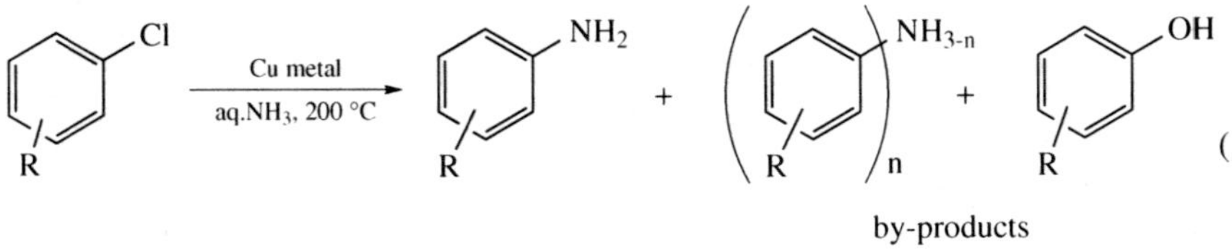

(16)

R = *o*-, *p*-Cl, NO_2, SO_3H, CF_3 or NH_2

Ammonolysis of hydrolytically unstable compounds under nonaqueous conditions was also reported[65].

Later, Cohen, Kondratov and their coworkers found that the use of well-soluble copper complexes such as CuOTf, $Cu(OAc)_2$ or Cu(II) complexes of 2,2′-bipyridine, 8-quinolinol or ethylene diamine allow more efficient ammonolysis of electron-deficient aryl halides under very mild reaction conditions[64, 66–68].

b. Amidation of aryl and vinyl halides. Copper-mediated coupling of aryl halides with acetamide or benzamide is recognized as the Goldberg reaction[69–71]. The reaction proceeds under reaction conditions quite similar to those developed by Ullmann for amine–aryl halide coupling: usually, good yields of *N*-arylated amides are obtained if the coupling partners are heated together in polar solvents such as nitrobenzene in the presence of catalytic copper metal or, better, cuprous iodide and anhydrous potassium carbonate (equation 17). After subsequent hydrolysis of the products, the reaction gives an access to differently substituted primary and secondary amines.

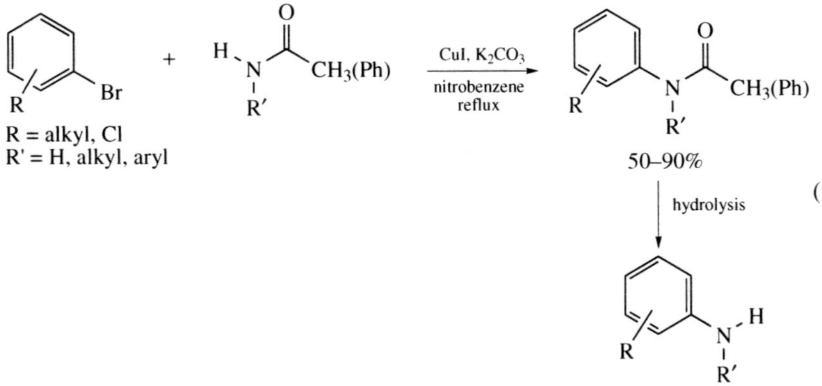

(17)

Although the Goldberg reaction was somehow overshadowed by the direct copper-mediated amine synthesis, *N*-arylation of amides has obvious advantages over the parent

Ullmann condensation. First, the Goldberg reaction is chemoselective giving no rise to distribution of polyarylated products. This is particularly important when applied to the synthesis of primary amines as an alternative to copper-mediated ammonolysis of aryl halides. Second, the reaction is less sensitive to the electronic and steric properties of the coupling partners[72,73]. The following example is illustrative: even *o,o'*-disubstituted compounds can be coupled in the presence of 10 mol% CuI in refluxing nitrobenzene (equation 18)[73].

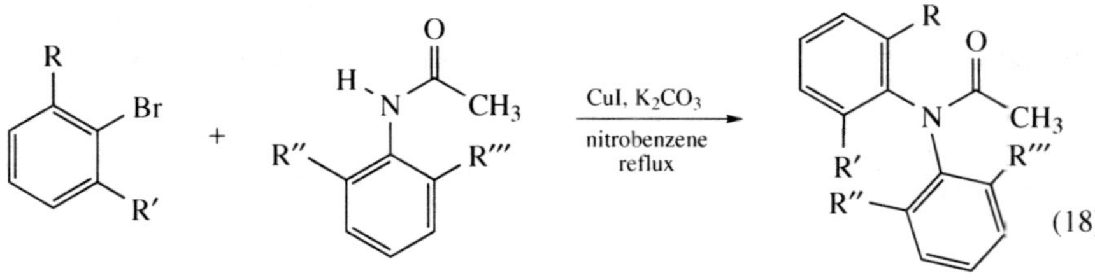

(18)

$R = CH_3$; $R' = R'' = R''' = H$	69 %
$R = R' = CH_3$; $R'' = R''' = H$	88 %
$R = R' = R'' = CH_3$; $R''' = H$	90 %
$R = R'' = R''' = CH_3$; $R' = H$	84 %
$R = R' = R'' = R''' = CH_3$	34 %
	18 %

The Goldberg reaction between aryl and vinyl halides and phthalimide, succinimide, benzenesulfonamide, maleimide[74,75] and pyridin-2-one[76] under similar conditions was also reported.

c. Early mechanistic considerations. Several interesting studies were performed in order to clarify the mechanistic details of the Ullmann and Goldberg condensations.

As was established experimentally, copper in various forms (copper metal prepared via different ways, cuprous oxide, cuprous halides, cupric salts or even their combination[38,42,44]) successfully catalyzes the reaction. Based on the fact that equally good results can be achieved under heterogeneous and homogeneous conditions, Weston and Adkins suggested in 1928 that the reaction takes place on soluble copper species[77]. A very simple experiment supported the hypothesis: metallic copper was fused with *N*-acetyl-*p*-toluidine, dissolved in alcohol, filtered and evaporated. The resulting alcohol soluble substance was found to be an active catalyst for the Goldberg arylation of *N*-acetyl-*p*-toluidine with bromobenzene in the presence of anhydrous K_2CO_3. More recent studies led to the same conclusion concerning Ullmann amine condensation[37,47,48,50,51].

The most comprehensive study on triarylamine formation from aryl halides and lithium diphenylamide in the presence of various copper catalysts was reported by Paine in 1987[78]. After a series of experiments, he concluded that a single catalytic species is formed from three different oxidation states of copper under Ullmann conditions. Moreover, he established that the actual catalytic species in these reactions are cuprous ions.

As was shown, the reduction of cupric bromide takes place readily in the presence of lithium diphenylamide, resulting in the formation of a large amount of 1,1,2,2-tetraphenylhydrazine and a mixture of crystalline cuprous and cupric oxides in a molar ratio of 3.3 (equation 19).

$$Ph_2NLi \quad + \quad CuBr_2 \longrightarrow Ph_2N\text{-}NPh_2 \quad + \quad Cu_2O/CuO \tag{19}$$

The formation of soluble cuprous species in heterogeneously catalyzed reactions apparently originates from the oxide layer always covering the metal surface. Indeed, quantitative XRPD measurements indicated that the usual amounts of copper catalysts employed in reactions of this type contain enough cuprous oxide to promote the transformation[78]. Of course, Cu_2O is insoluble in organic media[79], but the dissolution may be amine-assisted. The fact that *crystalline* Cu_2O was isolated after the workup of runs employing metallic copper is also illustrative[78].

In addition, the kinetic measurements revealed the same trends as have been previously observed in Ullmann diaryl ether synthesis: an I > Br > Cl ≫ F reactivity order in the aryl halide and a small substituent effect compared to uncatalyzed S_NAr reactions.

Based on these findings, Paine proposed a plausible mechanism for Ullmann amine condensation which is very similar to the mechanism suggested by Weingarten (Scheme 4)[6].

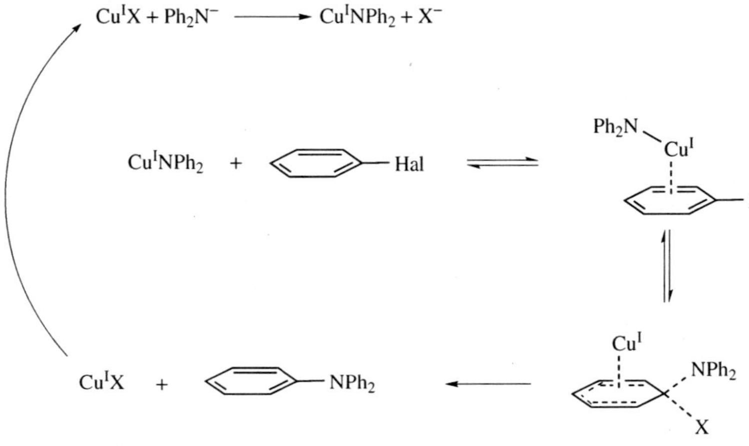

SCHEME 4

d. Coupling of aryl and vinyl halides with P-nucleophiles. Examples of copper-assisted carbon–phosphorus bond-forming coupling reactions of aryl halides were known, but limited. They mainly refer to the reaction of activated (electron-deficient) and unactivated aryl iodides with tertiary alkyl phosphites in the presence of at least stoichiometric copper bronze or copper complexes in polar solvents such as DMSO, DMF, NMP, HMPT or neat at above 150 °C (equation 20)[80,81].

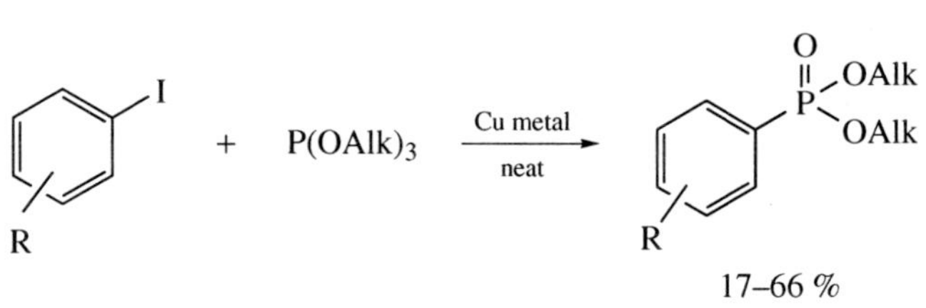

(20)

Yields in these reactions vary as a function of the steric properties of the starting materials but rarely exceed 80%[82].

It was reported that the presence of chelating groups in *ortho* position to a leaving halide facilitates the reaction (which is typical for the classical Ullmann type chemistry)[83–86].

Axelrad and coworkers extended the methodology to include vinyl halides as coupling partners[87–90]. Stoichiometric amounts of copper(I) chloride complexes of the corresponding trialkyl phosphites were employed for this purpose (equation 21).

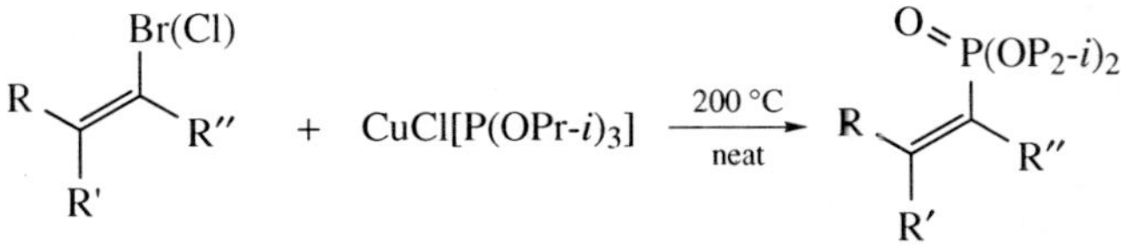

(21)

R, R′, R″ = H, alkyl, Ar
R‴ = Et, *i*-Pr, Ph

Several points are noteworthy. First, the reaction appears to be sensitive to the degree of substitution at the vinyl halide. Usually, α-halostyrenes react more sluggishly than the corresponding β-isomers. Second, vinyl bromides are generally more reactive than the corresponding chloro derivatives (giving moderate versus poor yields). Third, the reaction appears to be partially stereoselective. A pure *E* vinyl halide forms only *E* product; with an *E*/*Z* mixture there appears to be a preference for *E* product formation, although it is not overwhelming.

Yields were not large in these transformation, from 30% (for sterically demanding vinyl halides) through 95% (for β-bromostyrene), at the expense of a competitive copper-promoted halide exchange reaction taking place under Axelrad's conditions. In reactions with vinyl bromides as starting materials, the formation of the corresponding vinyl chlorides was always observed along with the phosphinated product (equation 22).

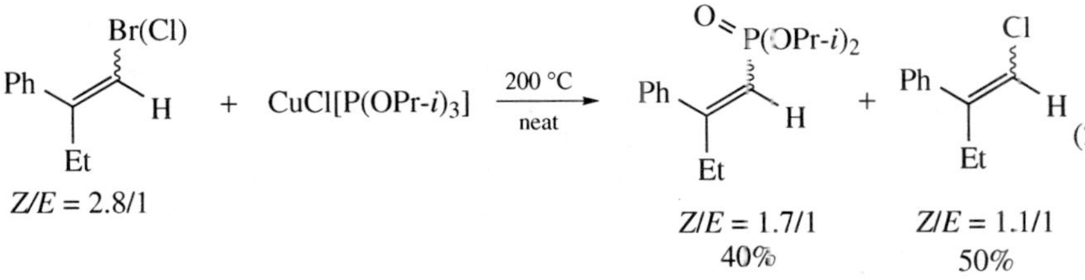

(22)

Quaternary phosphonium salts can be synthesized by the reaction between unactivated aryl bromides and tertiary phosphines in the presence of stoichiometric copper bronze or copper(II) acetate (equation 23)[80,81,91]. The reaction proceeds under much milder reaction conditions than its copper-free version and is facilitated by the presence of *ortho*-chelating groups.

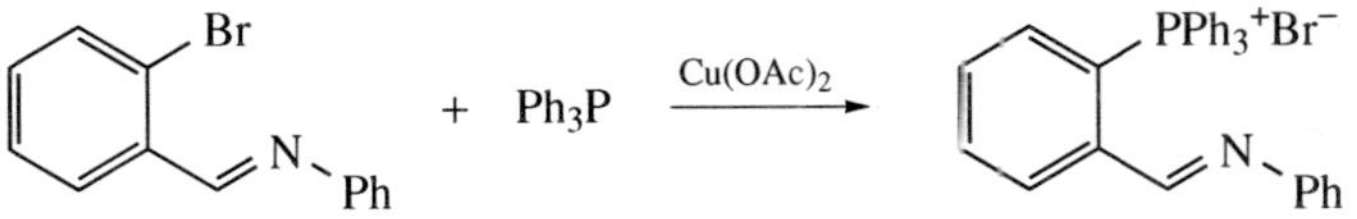

(23)

B. Modern Ullmann Chemistry

1. C–O, C–S and C–Se bond-forming reactions

a. Diaryl ethers. The major breakthrough in Ullmann carbon–oxygen cross-coupling was achieved in 1997 by Buchwald and coworkers, who offered the first really mild and efficient protocol for diaryl ether synthesis[92]. The most striking feature of Buchwald's modification was the utilization of Cs_2CO_3 as a base.

Remarkably, the superiority of cesium over other alkali metal bases has previously been mentioned in the context of Williamson ether synthesis[93,94]. The better performance was attributed to a higher solubility and stability of cesium alkoxides. In addition, cesium has a low charge/surface area ratio (0.03 Z/A^2) and therefore *in situ* prepared cesium alkoxides may be regarded as more nucleophilic 'naked anions' even in solvents with low dielectric constants[95]. Taking advantage of this 'cesium effect', Buchwald discovered new reaction conditions which comprise mild heating (100–110 °C) of functionalized *unactivated* aryl bromides or iodides with substituted phenols in *nonpolar* solvents (e.g. toluene) under air-free atmosphere in the presence of copper catalyst (0.5–2.5 mol%) and ethyl acetate as an additive (5 mol%) to obtain good to excellent yields of diaryl ethers (Scheme 5). The role of the additive appears to be related to the solubility of copper intermediates as was suggested in the early works[6]. It is worth noting that the nature of the copper precatalyst was not critical—CuCl, CuBr, CuI, $CuBr_2$, $CuSO_4$ exhibited similar reactivity, although a perfectly soluble $Cu(OTf)_2{\bullet}(C_6H_6)$ was somewhat superior in terms of reaction rate. The beneficial effect of cesium bases was also well-pronounced in more polar solvents and the reaction can be successfully performed in pyridine, DMF etc.[92,96].

These reaction conditions are compatible with a variety of functional groups such as ethers, esters, carboxylic acids, ketones, nitriles, nitro or dialkylamines, whereas aryl halides possessing amide functions appear to be less reactive. The reaction is relatively insensitive to steric demand of the starting materials, although the coupling of overcrowded substrates (e.g. 2,6-dimethylphenol) is inefficient.

Interestingly, in some cases the reactivity could be further enhanced upon addition of carboxylic acids (e.g. 1-naphthoic acid) as a co-additive. The authors explained the beneficial effect of the co-additive based on the mechanism proposed by Weingarten (Scheme 1). According to their hypothesis the solubility of cuprate-like species ($[Cu(I)(OR)_2]^- \ Cs^+$) increases due to the presence of the carboxylate ligand.

The method was successfully used in the preparation of complex compounds. For instance, the synthesis of verbenachalcone, a representative of the neurotrophine family, was accomplished using the Buchwald protocol at the key step (Scheme 6)[96].

Several groups reported slightly modified procedures using different copper precursors under ligandless conditions. For example, Snieckus and coworkers found that very soluble and air-stable $CuPF_6(CH_3CN)_4$ without any additives may be used as an efficient catalyst in phenoxylation of *ortho*-halogenated benzamides and sulfonamides in refluxing xylene in the presence of cesium carbonate. Taking into account that the starting halides can be readily prepared by the directed *ortho*-lithiation, the Snieckus modification (Scheme 7) is very valuable[97]. Primary and secondary carboxamides or sulfonamides can be used as *ortho*-directing groups.

Remarkably, under the Snieckus conditions, the usually unreactive aryl chlorides demonstrated acceptable conversions. However, their reactivity is most likely ascribed to the neighboring-group assistance rather than to the uniqueness of $CuPF_6(CH_3CN)_4$ catalyst (Scheme 7).

In addition, copper containing perovskites[98] or nanosized copper[99] were reported as efficient catalysts in Ullmann coupling of aryl bromides and iodides with phenols in the presence of cesium carbonate.

Despite earlier reports on the deleterious effect of strongly complexing agents (phosphines or 2,2′-bipyridine) on the rate of Ullmann-type reactions[9], further variants of the Buchwald procedure mainly focused on ligated copper compounds. Amazingly, this fact reemphasizes the importance of the 'cesium carbonate effect' discovered by Buchwald.

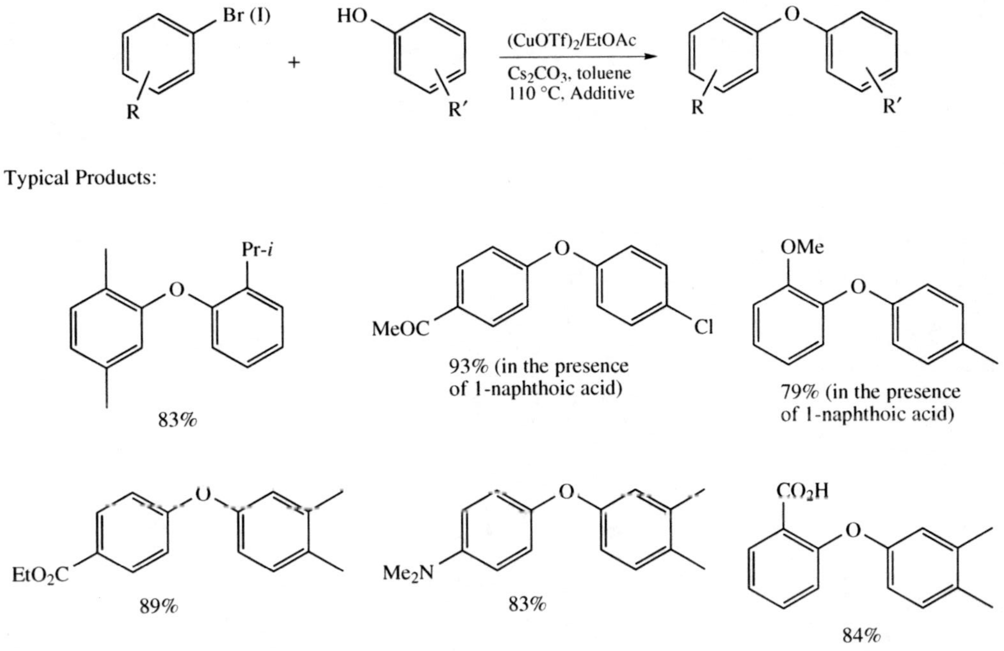
Br (I)
+
HO
R
R′
(CuOTf)$_2$/EtOAc
Cs$_2$CO$_3$, toluene
110 °C, Additive
O
R
R′
Typical Products:
Pr-*i*
O
83%
MeOC
O
Cl
93% (in the presence
of 1-naphthoic acid)
OMe
O
79% (in the presence
of 1-naphthoic acid)
EtO$_2$C
O
89%
Me$_2$N
O
83%
CO$_2$H
O
84%

SCHEME 5

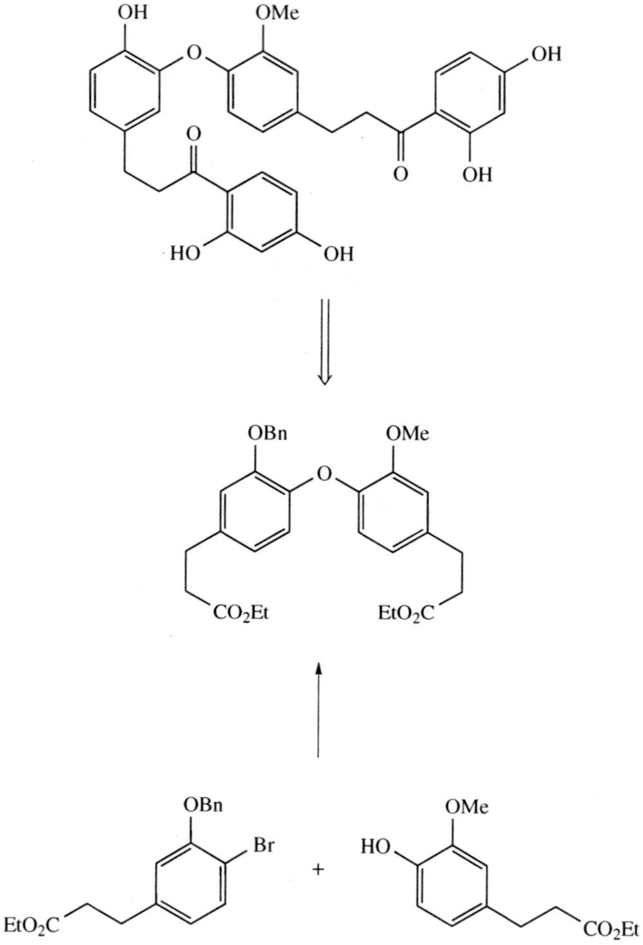

SCHEME 6

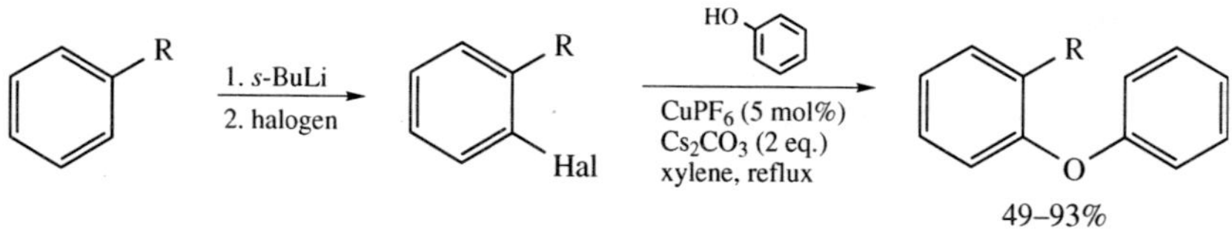

R = CONHR, $CONR_2$, SO_2NHR or SO_2NR_2; Hal = Cl, Br, I

SCHEME 7

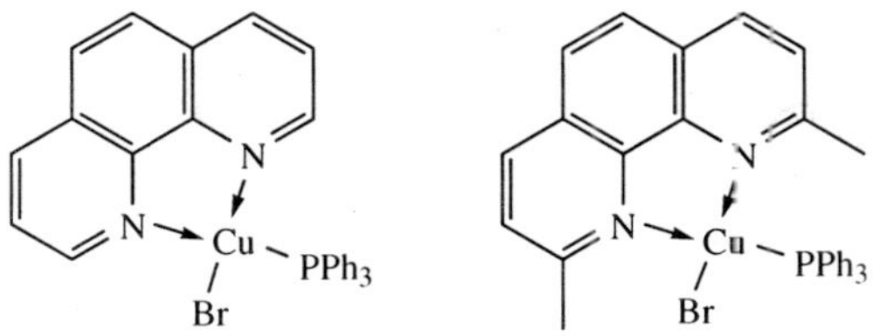

CHART 2

Naturally, amines, nitrogen-containing heterocycles and dicarbonyl compounds became the ligands of choice in copper-catalyzed C–O bond-forming reactions[6,9].

Soluble and structurally well-defined copper(I) complexes bearing 1,10-phenanthroline and neocuproine (Chart 2) were found to serve as efficient promoters for diaryl ether synthesis from aryl bromides and phenols in the presence of cesium carbonate at 110 °C in toluene[100]. The scope shown in this study is relatively limited; however, general trends are clear—the method is compatible with the presence of some functional groups (e.g. nitro and carbonyl), but is very sensitive to the steric bulk of the starting materials, particularly of the aryl halides (yield range 31–99%). Curiously, a simple combination of CuBr and 1,10-phenanthroline in 1:1 ratio was ineffective as reported[101].

Ma and Cai employed a very simple but very efficient catalyst derived from a 1:3 ratio of CuI and *N*,*N*-dimethylglycine for the same transformation[102]. The catalyst operates under milder than usual reaction conditions (80–90 °C in dioxane and in the presence of Cs_2CO_3) and was originally developed for arylation of L-tyrosine derivatives (equation 24)[103]. However, it did not remain limited to them: the scope of the reaction is very impressive and includes substrates possessing a variety of functional groups. Remarkably, no racemization of chiral amino acids takes place under the described reactions conditions, making the catalyst applicable even to peptide synthesis. The method was successfully engaged for the synthesis of compound K-13[104].

Another ligand-assisted protocol comprises heating of aryl bromides or iodides with phenols in the presence of Cs_2CO_3 and CuCl/2,2,6,6-tetramethylheptane-3,5-dione (TMHD which is, arguably, a structural descendant of 2-(formyloxy)ethyl 2-methoxyacetate[6]) in NMP at 130 °C (equation 25)[105]. The method is generally efficient; however, reductive elimination is a major side reaction that becomes more pronounced when electron-poor phenols are employed.

Likewise, other ligands were reported in the context of the copper-catalyzed diaryl ether synthesis (Chart 3)[106–110]. However, three structural motifs mentioned previously (phenanthroline, α-amino acid and β-diketone) represent the three major families of chelates used for this purpose.

b. Alkyl aryl ethers. Modern copper-catalyzed C–O bond-forming coupling is not limited to aromatic alcohols. Buchwald and coworkers found that the synthesis of alkyl aryl ethers can be efficiently accomplished using 10 mol% CuI in combination with 20 mol% of differently substituted 1,10-phenanthrolines as a catalyst in the presence of Cs_2CO_3 in toluene at 80–110 °C[111,112]. Although even in the presence of the catalyst, reductive dehalogenation was still a serious competitive process, the scope of the method became much more impressive in comparison with the classical Ullmann reaction. Thus, under these reaction conditions even deactivated (e.g. *o*-, *m*- and *p*-MeO substituted) and functionalized (NH_2, CN and heterocyclic) aryl iodides and bromides couple with primary and sometimes secondary (including allylic and benzylic) alcohols in very good to excellent yield. Remarkably, the general reactivity trend in alcohols is benzyl > 1° > 2°cyclic > 2° acyclic. This difference in reactivity can be exploited in

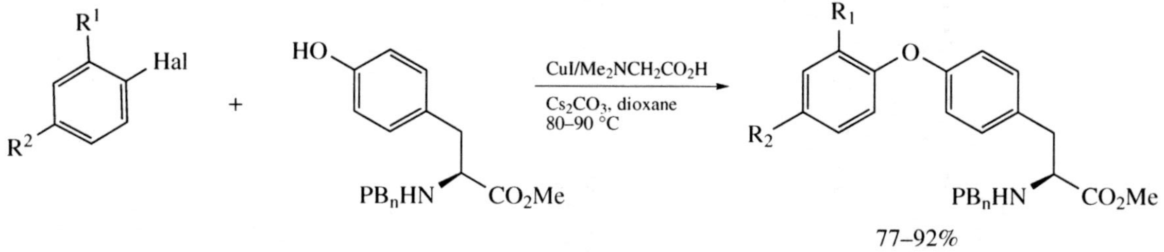

R^1 = $NHCOCF_3$, $NHSO_2Me$
R^2 = CO_2R, CN, OMe, NR_2, COR
Hal = Br, I

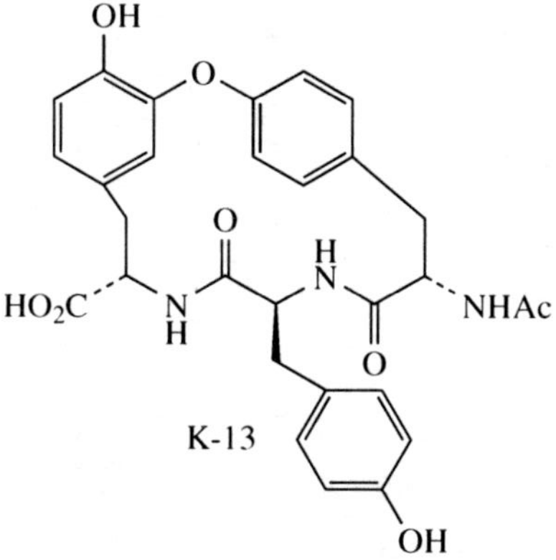

K-13

(24)

selective cross-coupling of primary in the presence of secondary alcohols. The catalyst was also remarkably chemoselective even in the presence of other potential nucleophiles such as aromatic or aliphatic amines (Scheme 8)[113]. Advantageously, inexpensive alcohols themselves can be used as reaction media. However, the reaction can also be carried out in toluene, if desired.

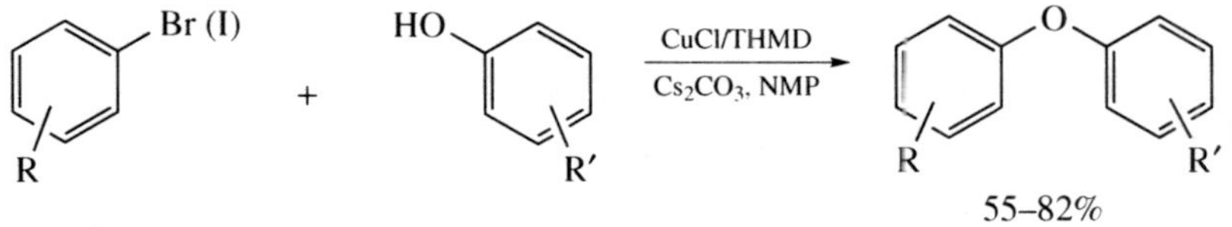

(25)

R, R′ = F, Me, OMe, CN, CO_2Pr or NMe_2

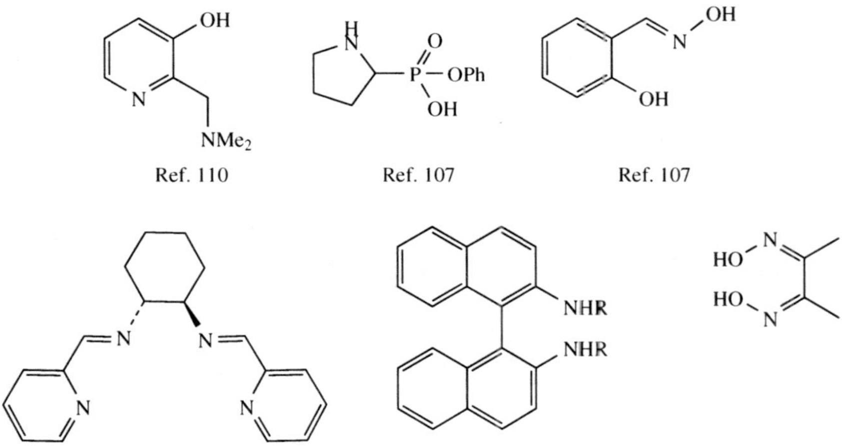

Ref. 110 Ref. 107 Ref. 107

Ref. 108 Ref. 106 Ref. 110

CHART 3

Remarkably, when chiral enantiopure 1-phenylethanol was used as coupling partner, no racemization and complete retention of configuration was observed. This result is particularly important because analogous palladium-catalyzed reactions normally fail[114].

The method was successfully applied as a key step (equation 26) to the total synthesis of Paliurine F[115]. Alternatively, *N*,*N*-dimethylglycine may be used for this purpose[116].

c. Vinyl ethers. Enol ethers are very important structural motifs in organic intermediates and polymers. These compounds were traditionally synthesized via *O*-alkylation of enolates or base-catalyzed addition of alcohols to alkynes.

In 1992, the copper-catalyzed cross-coupling was first adopted to the synthesis of these valuable compounds. The original procedure employing vinyl bromides and sodium methoxide in NMP/methanol at 80–120 °C along with 10 mol% of CuBr was described by Keegstra (equation 27)[117]. The reaction proceeded with complete retention of configuration at the double bonds and yielded 40–90% of products.

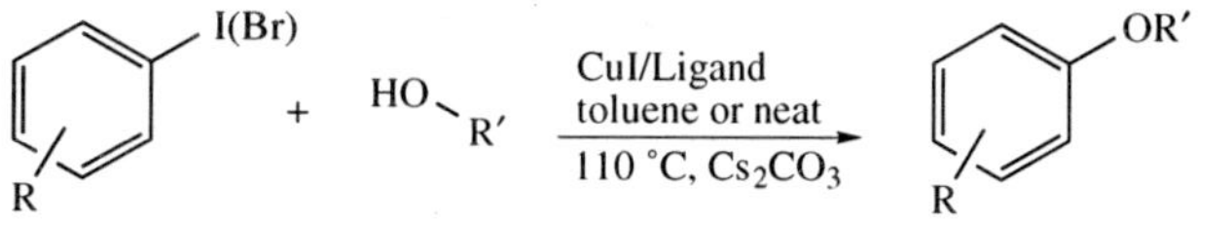

Ligand = 1,10-phenanthroline or 3,4,7,8-tetramethyl-1,10-phenanthroline

Typical Products:

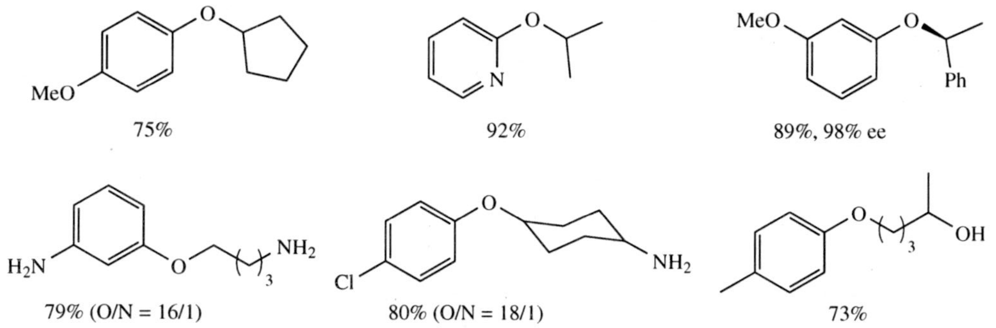

SCHEME 8

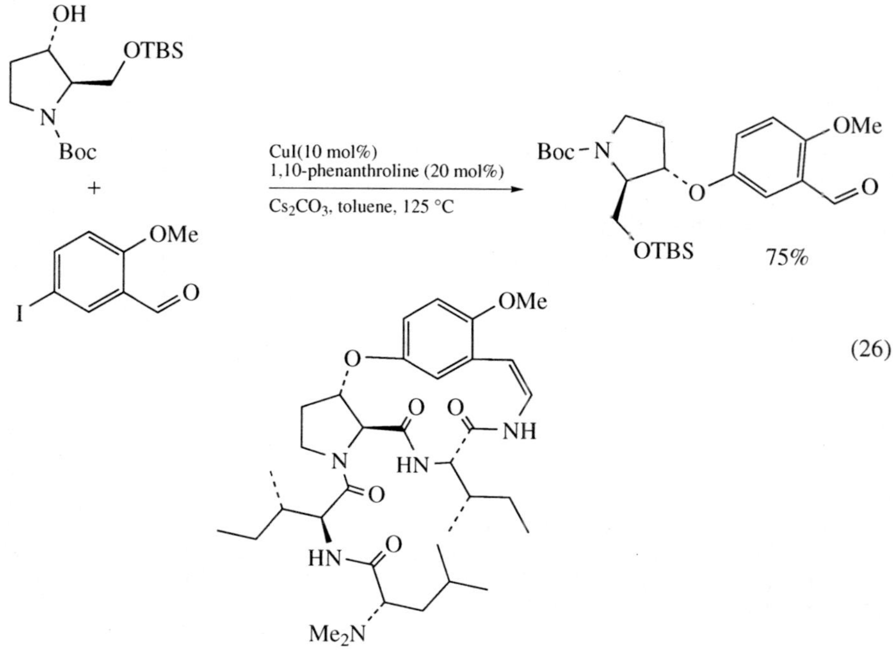

Paliurine F

(26)

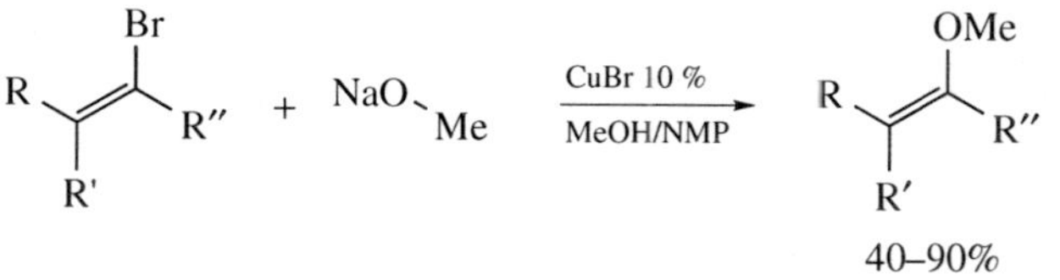

(27)

Following this precedent and taking advantage of the progress made in copper-catalyzed diaryl ether synthesis, several groups reported the phenoxylation of vinyl bromides and iodides using substituted phenols/Cs_2CO_3 combination as a nucleophile (equation 28). Wan and coworkers, for example, performed the reaction in the presence of 10 mol% of CuCl/*N*-(2-(4-methoxyphenoxy)ethyl)morpholine catalyst in refluxing toluene (though simpler ligands such as *N*-methylmorpholine or dipivaloylmethane are also applicable)[118]. Ma and coworkers engaged the previously described CuI/*N*,*N*-dimethylglycine system in dioxane for the preparation of a wide range of functionalized aryl vinyl ethers. In none of the cases was double-bond isomerization detected[119].

Buchwald and coworkers demonstrated a very interesting extension of this methodology. They used 10 mol% of CuI and 20 mol% of 1,10-phenanthroline in the presence of Cs_2CO_3 (1.5 equivalents) in toluene at 80 °C to prepare allyl vinyl ethers[111,120]. They found, however, that the corresponding Claisen rearrangement product was formed in *ca* 7% yield along with the expected ether (Scheme 9). The result was improved by switching the ligand to 3,4,7,8-tetramethyl-1,10-phenanthroline and using harsher reaction conditions.

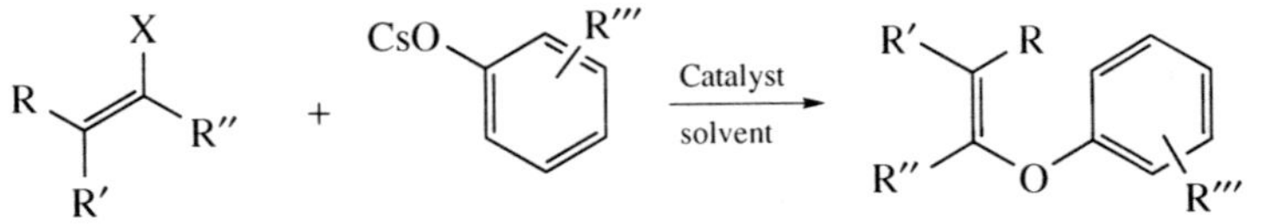

X = I or Br

R, R′, R″ = Alkyl, aryl, CO_2R

R‴ = halogen, alkyl, OMe, NH_2, CN, keto

41–95%

(28)

Wan and coworkers
Catalyst in toluene

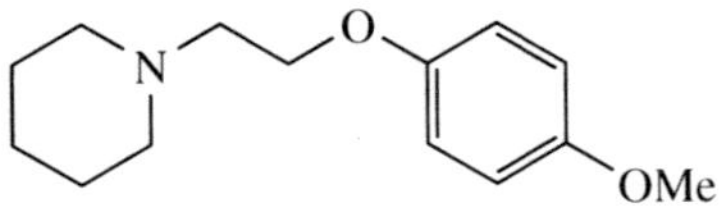

CuCl

Ma and coworkers
Catalyst in dioxane

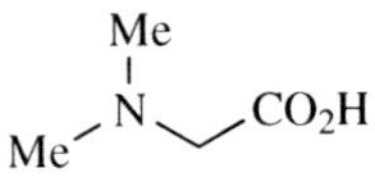

CuI

This elegant domino copper-catalyzed C–O coupling/Claisen rearrangement process proceeds with a high degree of stereospecificity (no double-bond isomerization takes place during the coupling step and *dr* of products > 88 : 12 was observed after the rearrangement step) and allows for the formation of a collection of γ-enones[120].

d. Miscellaneous O-nucleophiles. Very recently, Maitra and coworkers reported that *O*-aryloximes can be efficiently prepared via direct CuI/1,10-phenanthroline-catalyzed coupling of aryl halides with oximes in toluene in the presence of Cs_2CO_3 (equation 29)[121]. The procedure is operationally simple but moderately efficient. The functional group compatibility is not clear so far.

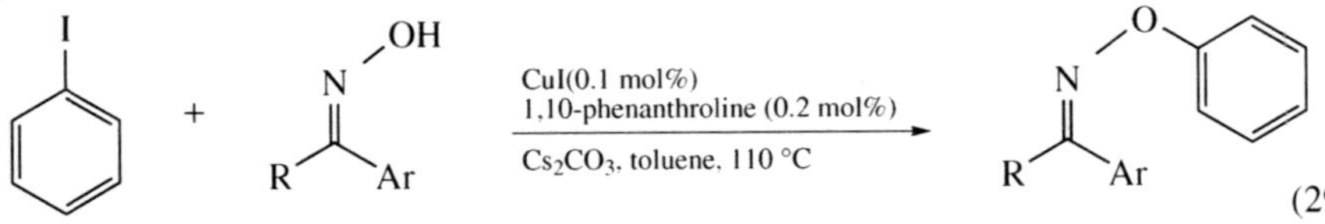

(29)

e. Sulfides. There is no much information available regarding the thermal and chemical stability of structurally well-defined copper(I) thiolates; however, they are believed to be relatively robust due to the efficient soft acid–soft base (Cu–S) interactions. On the other hand, these species are widely regarded as polymeric or, more precisely, cluster compounds[122, 123]. It is thus reasonable to suggest that the rate of Ullmann-type C–S coupling reactions should be greatly dependent on the cluster (inactive form)–monomer (active form) equilibrium and that the presence of a complexing agent is essential to keep the concentration of active species high over the course of a catalytic cycle[124, 125]. Indeed, most known successful protocols for mild copper-catalyzed cross-coupling of aryl halides with thiols are ligand-assisted.

The first report appeared in the literature in 2000. Palomo and coworkers described the synthesis of biaryl thioethers from aryl iodides and thiophenols in the presence of 5 mol% of CuBr and phosphazene bases in toluene at reflux temperature (Scheme 10)[126]. The authors found that the use of the phosphazene bases is essential and suspected that the

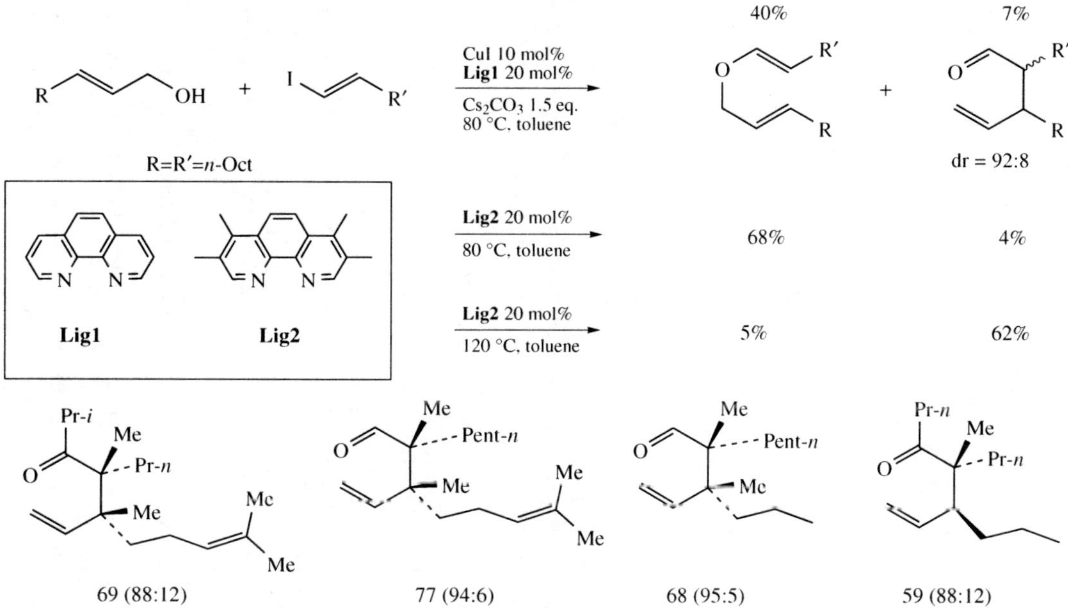

R
OH
I
R′
R=R′=n-Oct
Lig1
Lig2
CuI 10 mol%
Lig1 20 mol%
Cs2CO3 1.5 eq.
80 °C, toluene
Lig2 20 mol%
80 °C, toluene
Lig2 20 mol%
120 °C, toluene
40%
7%
dr = 92:8
68%
4%
5%
62%
69 (88:12)
77 (94:6)
68 (95:5)
59 (88:12)

SCHEME 9

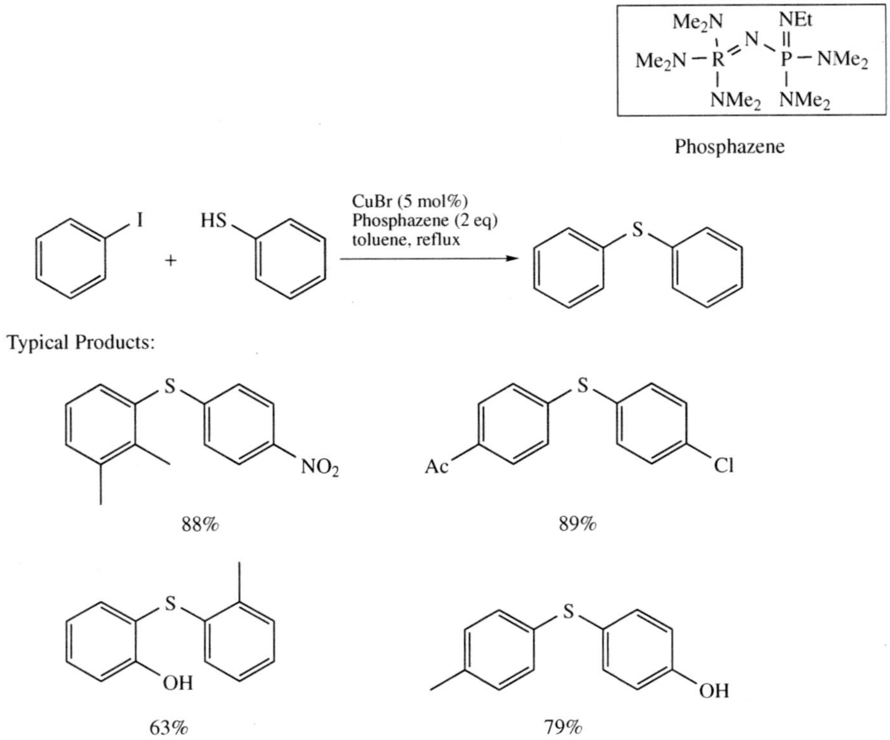

SCHEME 10

formation of 'naked' anions greatly facilitate the reaction. However, the binding ability of phosphazene should not be neglected[127–129].

At any rate, the protocol is efficient and demonstrates good functional group compatibility as well as a remarkable selectivity toward coupling of thiophenols even in the presence of unprotected hydroxyl groups.

Kwong and Buchwald suggested operationally simpler reaction conditions[130]. They found that a wide spectrum of diaryl thioethers and alkyl aryl sulfides can be prepared selectively using a catalyst derived from CuI (5 mol%) and ethylene glycol (2 equivalents) in the presence of a user-friendly K_2CO_3 in isopropanol at 80 °C (Scheme 11). The yields reported are generally very high (70–99%) and functional group compatibility is spectacular.

Venkataraman and coworkers were able to extend the methodology to the synthesis of vinyl sulfides using the soluble $[Cu(phen)(PPh_3)_2]NO_3$ catalyst[131]. The desired vinyl sulfides were obtained in good to excellent yield with complete retention of stereochemistry. It is also remarkable that the method tolerates a variety of functional groups and heterocycles (Scheme 12).

Additional catalytic systems are also available[132–139].

f. Selenides. Copper-catalyzed synthesis of diaryl selenides appeared as a natural extension of the C−S bond-forming methods. These compounds can now be prepared from aryl iodides and aryl selenols using 10 mol% CuI/neocuproine catalyst and NaOBu-*t* or

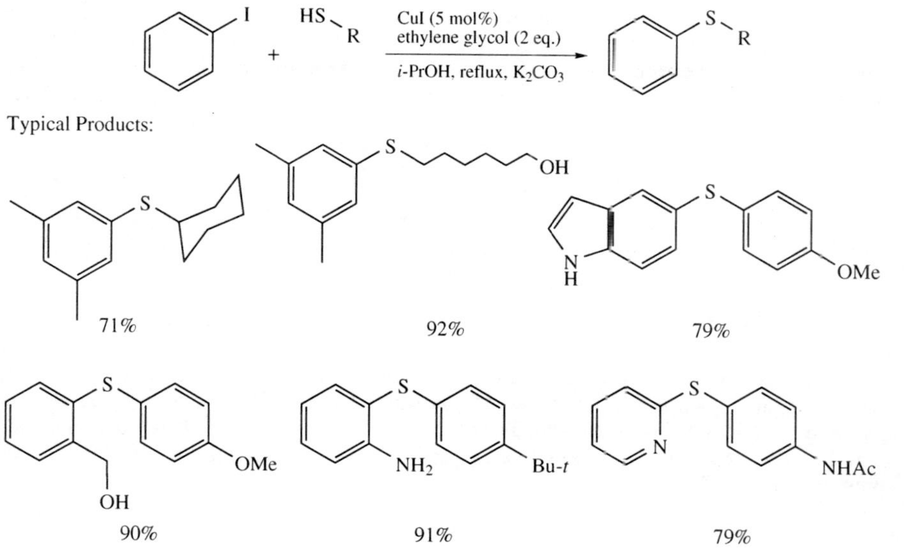

SCHEME 11

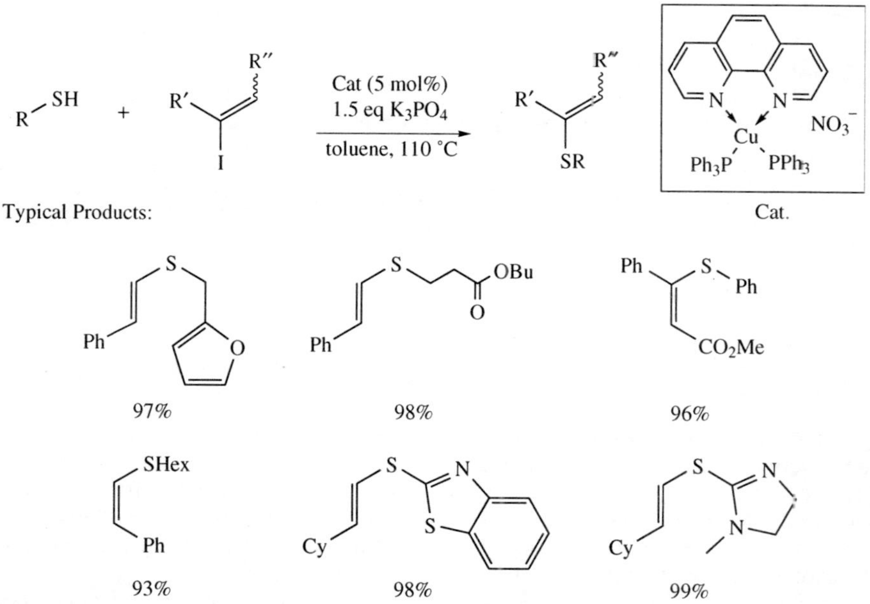

SCHEME 12

K_2CO_3 in refluxing toluene (equation 30)[140]. Using this protocol, diaryl selenides were obtained in 76–90% yield starting from electron-rich, neutral or electron-poor substrates. The reaction tolerates the presence of some functional groups such as NH_2, NO_2, CO_2R and COR; however, less basic K_2CO_3 was recommended for electron-deficient substrates.

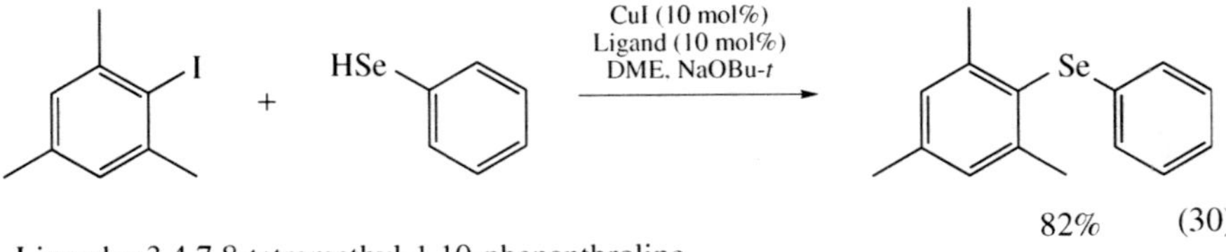

(30)

Ligand = 3,4,7,8-tetramethyl-1,10-phenanthroline

Selenols are prone to oxidation in air yielding the corresponding diselenides[141,142]. The latter are stable and easy-to-handle compounds and, fortunately, can be engaged as coupling partners in the above-mentioned transformation. Taniguchi and Onami demonstrated that the ArSe–SeAr bond could be catalytically cleaved by CuI/bpy or Cu_2O/bpy in the presence of Mg, Zn, Al or Sm reductants (equation 31)[143].

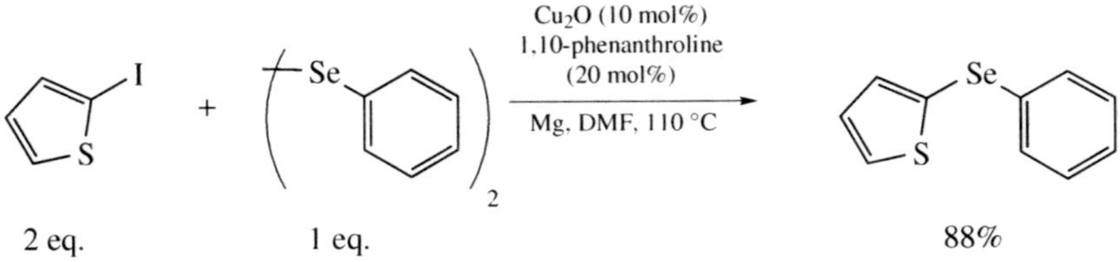

(31)

They proposed the cleavage mechanism shown in Scheme 13.

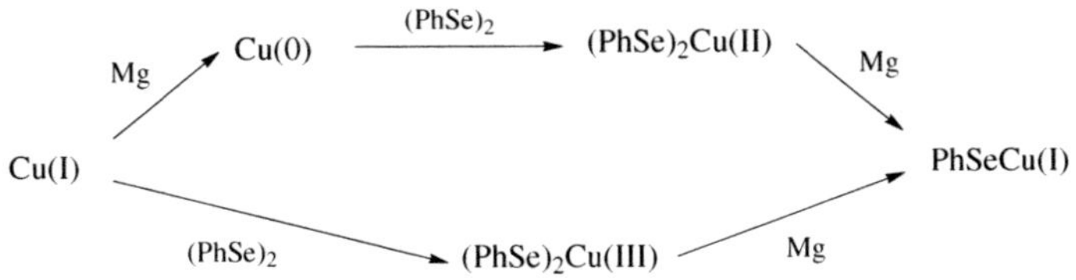

SCHEME 13

According to this hypothesis, either Cu(I) or prereduced Cu(0) species add oxidatively to the starting diselenide, yielding appropriate Cu(III)$(SePh)_2$ or Cu(II)$(SePh)_2$ complexes. The latter form catalytically active Cu(I)SePh after successive metal-assisted reduction. The same work demonstrated that the stoichiometric reaction between Cu(I)SePh[144] and aryl iodide provides the corresponding diaryl selenides in comparable yield, indicating the participation of these intermediates in the catalytic turnover.

Similar procedures were employed in stereoselective synthesis of vinyl selenides under conventional[145] or microwave heating[146].

g. Copper-catalyzed coupling in syntheses of O- and S-containing heterocycles. Copper-catalyzed carbon–oxygen and carbon–sulfur bond-forming reactions represent a practical approach toward the assembly of various heterocyclic compounds (Scheme 14).

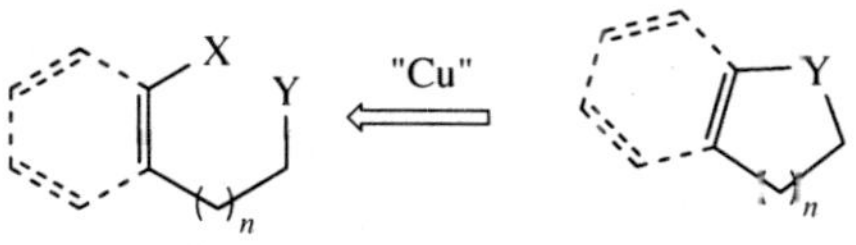

X = halogen
Y = O or S

SCHEME 14

Fang and Li successfully applied this strategy to the synthesis of small and medium-sized heterocycles[147]. They found that intramolecular 4-*exo* ring closure by means of ligand-assisted copper-catalyzed *O*-vinylation of γ-bromohomoallylic alcohols leads to the formation of 2-methyleneoxetanes (equation 32). The reaction is most efficient in the presence of CuI/1,10-phenathroline catalyst and Cs_2CO_3 as a base in refluxing acetonitrile. Under either ligandless conditions or in the presence of another ligand/base combination, undesired dehydrobromination product forms predominantly.

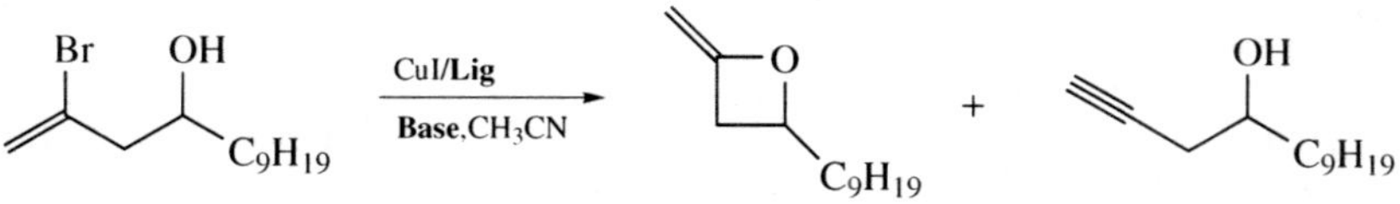

(32)

Lig	**Base**		
No ligand	Cs_2CO_3	0%	85%
1,10-phenanthroline	Cs_2CO_3	98%	0%
1,10-phenanthroline	NaOBu-*t*	0%	98%

Scope and limitations studies revealed that primary, secondary and tertiary alcohols are equally reactive, although aliphatic alcohols are more reactive than allylic ones. On the other hand, the substituents at the double bond retard the cyclization and higher reaction temperatures might be required to drive it to completion (equations 33–38).

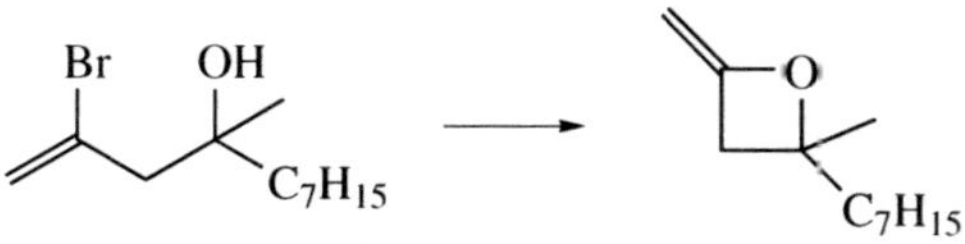

74%

(33)

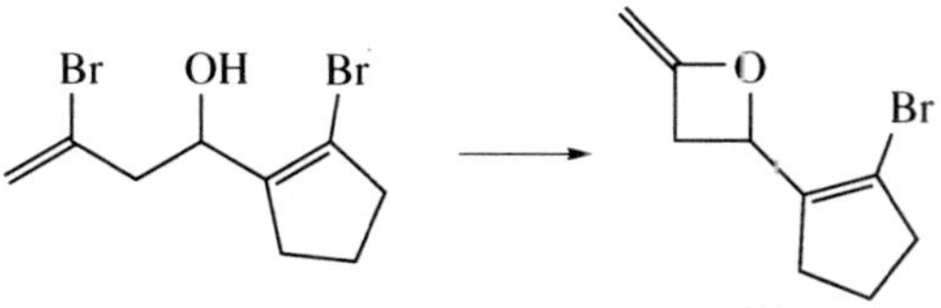

66%

(34)

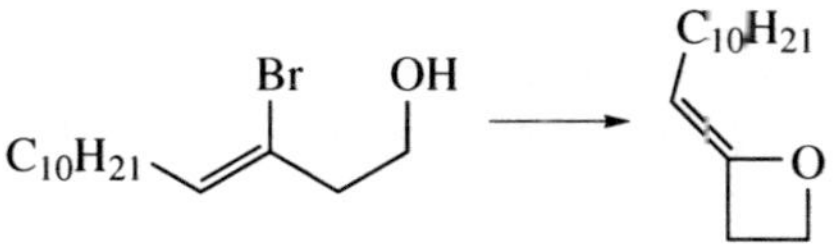

84%

(35)

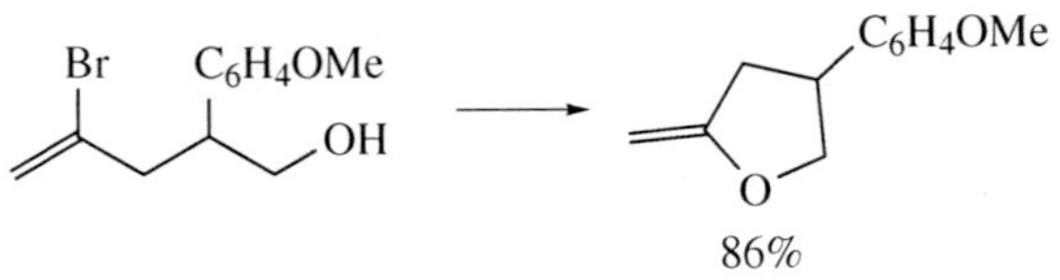

(36)

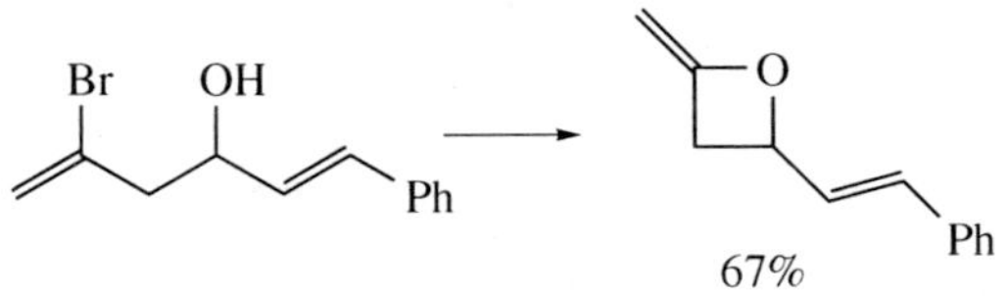

(37)

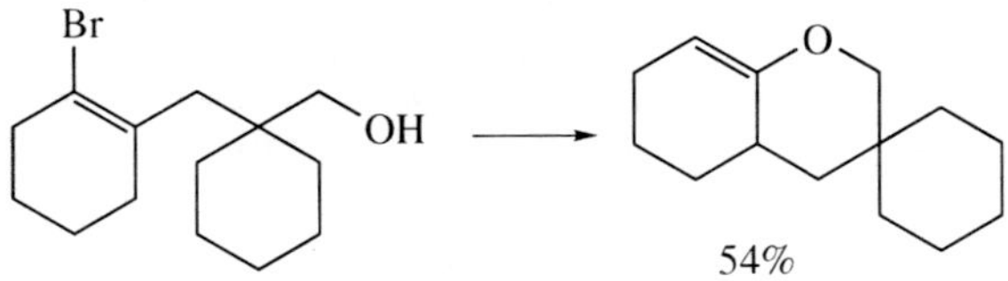

(38)

The method can be extended to the synthesis of 5- and 6-membered cycles. However, when possible, the 4-*exo* ring closure is favored over other cyclization modes (equation 39). The exact reason for this preference is not known.

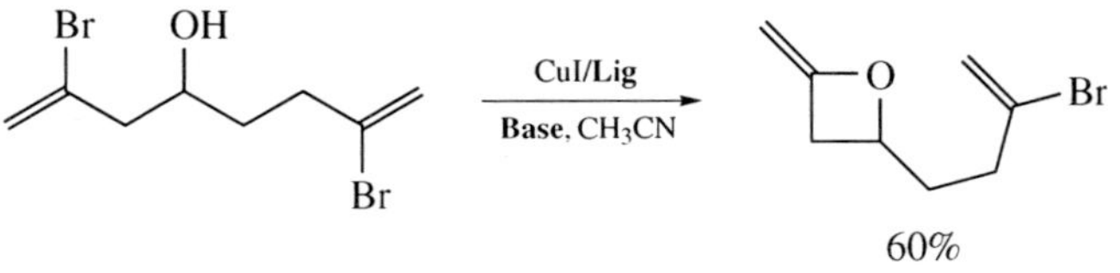

(39)

A concise synthesis of dibenzofurans and fused benzofuro heterocycles was demonstrated (Scheme 15)[148–152].

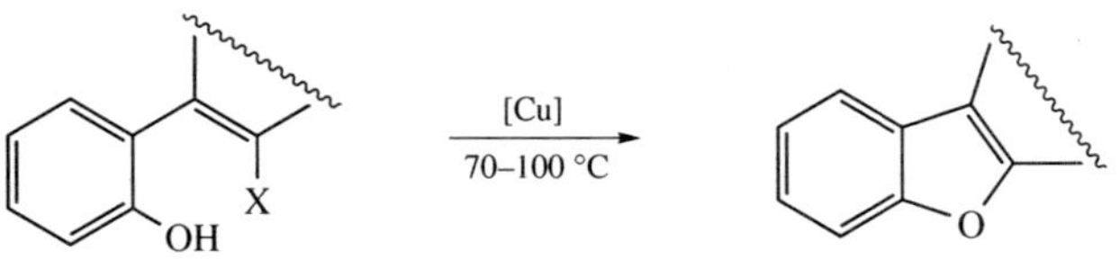

SCHEME 15

The mildest protocol taking advantage of Liebeskind's catalyst, copper(I) thiophene-2-carboxylate (CuTC), was developed by Liu and coworkers[150]. This protocol advantageously obviates the need in basic conditions and, therefore, is compatible with a variety of functional groups and heteroaromatic starting materials (Chart 4). For example, cyclization of pyridyl phenols to form the corresponding benzo[4,5]furo heterocycles proceeds smoothly, giving significantly higher yields than in the presence of the usual catalysts.

The strategy was also applied to the synthesis of benzoxazoles, benzothiazoles and their derivatives. The method involves an intramolecular C—O or C—S cross-coupling of easily accessible *ortho*-halophenyl(thio)amides[153,154] or *ortho*-halophenyl(thio)ureas (equations 40 and 41)[155,156].

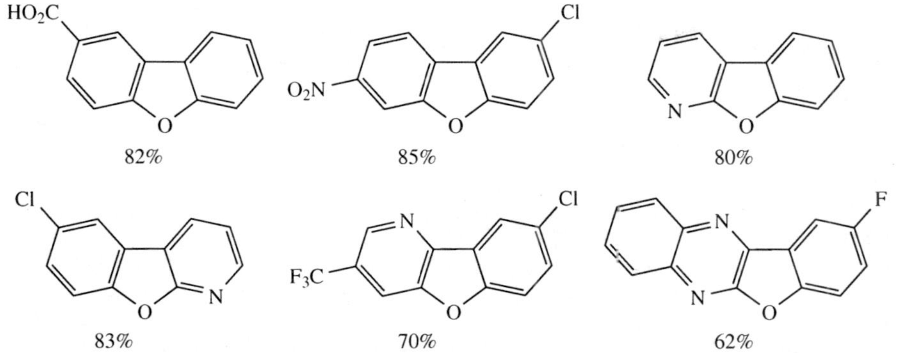

Reaction conditions: CuTC (1 mol%), DMA, 120–140 °C

CHART 4

The protocols are efficient and operationally simple, since they employ standard copper(I) iodide/1,10-phenanthroline catalyst and cesium carbonate in DME as a base.

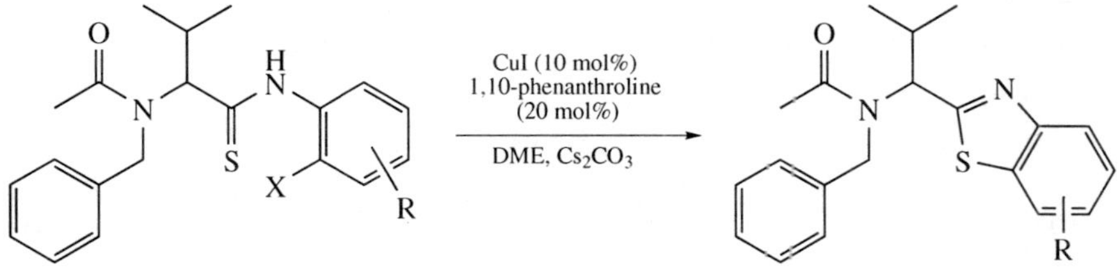

$X = Cl, Br, F$
$R = H, 4\text{-}F, 6\text{-}Br, 3\text{-}CF_3, CO_2Me$

55–94% (40)

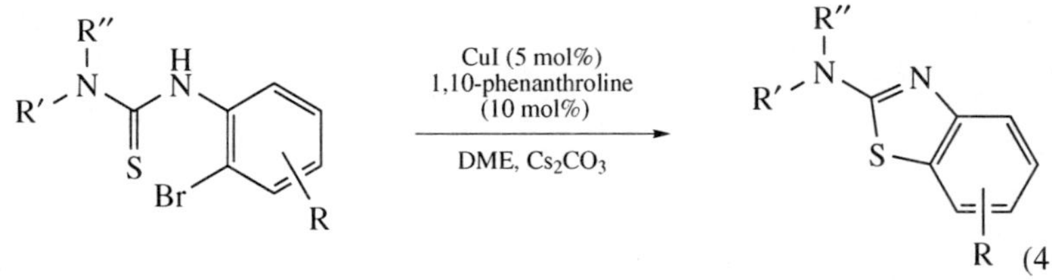

(41)

R′, R″= H, alkyl, aryl
$R = H, 4\text{-}F, 6\text{-}Br, 5\text{-}CF_3, 4\text{-}Me$

69–100%

2. C–N and C–P bond-forming reactions

a. Coupling of aliphatic amines with aryl halides. In 1996, Ma and Yao communicated the development of a copper-assisted modification of the palladium-catalyzed Buchwald–Hartwig amination of aryl halides with chiral enantiopure amines that proceeded without racemization under mild heating (equation 42)[157]. More careful investigation, however, revealed that copper alone was responsible for the unusual catalytic activity. Thus, the same reaction repeated in the presence of CuI (10 mol%) and K_2CO_3 in DMA at 90 °C yielded 80% of the arylated product[158].

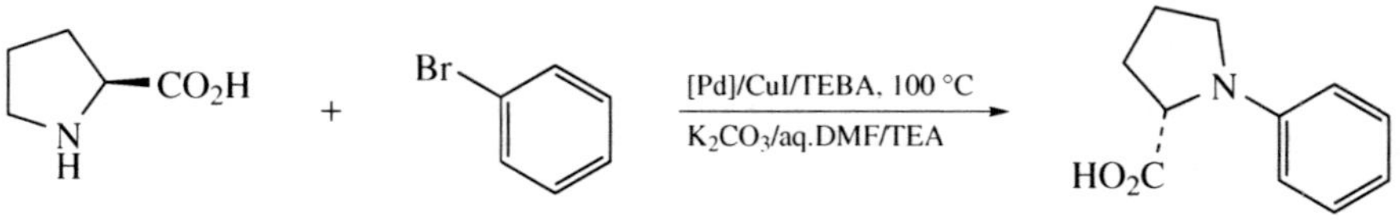

95%

(42)

Under optimized conditions, both electron-rich (e.g. *o*-, *m*- and *p*-bromoanisole or *o*-bromophenol) and electron-deficient aryl bromides or iodides (e.g. *o*-, *m*- and *p*-bromobenzoic acid, *o*-iodobenzoic acid or *p*-bromochlorobenzene) were equally reactive (64–85%) in the cross-coupling with a model L-valine. Aryl chlorides were generally inactive with the exception of the chelating 2-chlorobenzoic acid that could be aminated in comparable yield. In addition, steric hindrance at aryl bromides strongly disfavors the reaction (for example, 2,6-dimethylbromobenzene and L-valine gave the corresponding arylated product in only 14%).

Interestingly, the reaction was found to be influenced by the type of α-amino acid partner (Scheme 16). For instance, amino acids with larger hydrophobic groups (e.g. valine, proline, phenylalanine or methionine) couple more efficiently than those possessing smaller hydrophobic groups (e.g. alanine) or hydrophilic groups (e.g. serine or glycine do not react at all). In all the cases, complete retention of the configuration at the chiral centers was observed (Scheme 16).

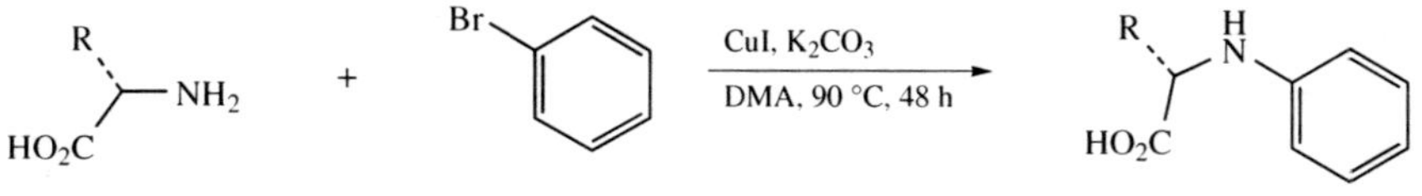

L- or D-valine	75–81%
L-proline	80%
L-phenylalanine	92%
L-alanine	65%
L-methionine	74%
L-tyrosine	60%
L-tryptophan	73%
glycine	0%

SCHEME 16

In many aspects, the observed reactivity (halide displacement order and well-pronounced neighboring group effect) and reaction conditions (the need for a polar reaction medium) resembled a classical Ullmann case. However, the mild heating required in order to achieve complete conversion of the starting materials implied an accelerating effect induced by amino acids.

Ma and Yao hypothesized that Cu(I)-α-amino acid chelates are the true catalytic species which can form a π-complex with the corresponding aryl halide as was suggested earlier

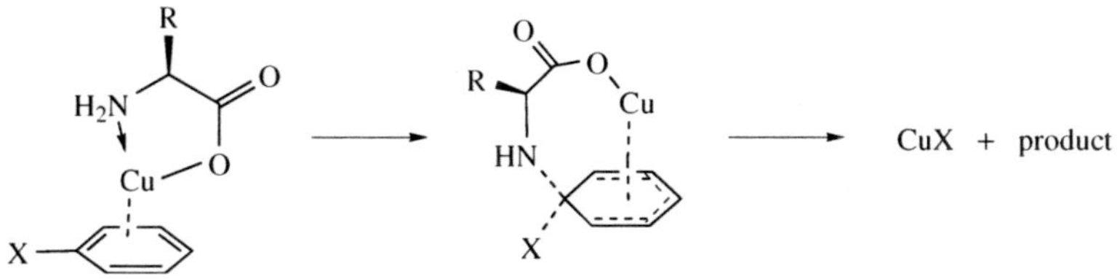

SCHEME 17

(Scheme 17). In this case, intramolecular nucleophilic substitution of the halide by the amine site should be energetically favorable.

This mild and successful methodology was employed in multistep syntheses of various complex molecules[159,160]. For example, Ma and coworkers demonstrated that *N*-(2′-hydroxymethylphenyl)-L-valine, a key precursor toward benzolactam-V8 (proteine kinase C activator)[161], can be synthesized in 47% yield (equation 43)[158].

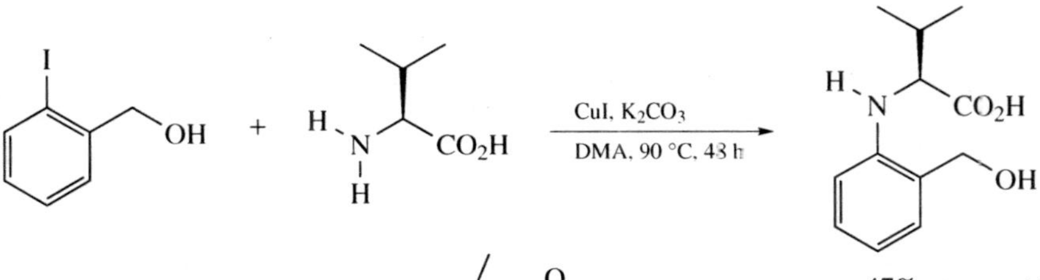

(43)

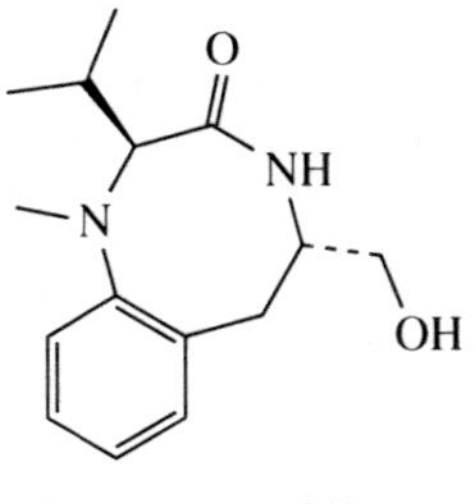

Benzolactam-V8

Later, Ma and Xia found that β-amino acids exert a similar accelerating effect[162]. For example, 3-aminobutanoic acid couples with bromobenzene in the presence of CuI, K_2CO_3 in DMF at 100 °C. The reaction is significantly more sluggish compared with α-amino acids and more sensitive to the electronic character of the starting aryl halides, i.e. under optimized conditions, electron-rich substrates show only diminished or no conversion (4-bromoanisole—0% or 4-iodoanisol—31%). However, functional group compatibility found for this method is very good—nitro, amino, amido and ester groups had no pronounced deleterious effect on conversion.

An intramolecular version of this reaction was used as the key step in the synthesis of SB-214857, a GPIIb/IIIa receptor antagonist (Chart 5)[163]. The step proceeded in acceptable yield and without notable racemization[162].

Although other chelating (e.g. α-amino-acid-derived aminoalcohols or 4-aminobutanoic acid) and nonchelating (e.g. benzylamine) amino ligands were found to be ineffective[158,162], the observation of the spectacular chelation assistance described here along with significant progress achieved by that time in Ullmann ether synthesis initiated a new phase of interest in copper-catalyzed amination chemistry.

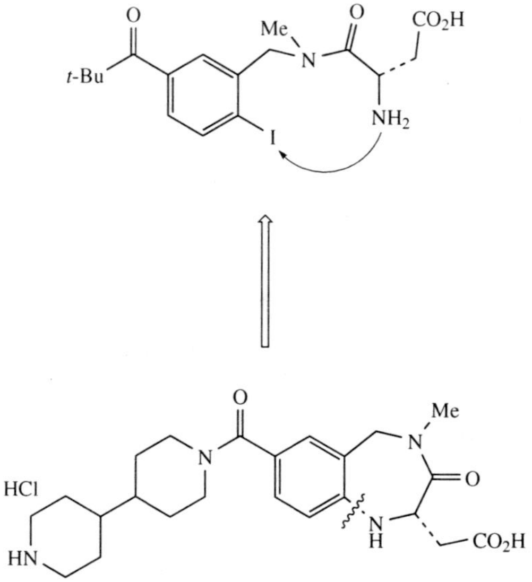

SB-214857

CHART 5

In 2001, Lang and coworkers reported a facile Cu_2O-catalyzed ammonolysis of electron-poor heteroaromatic bromides that proceeded at very mild heating of 80 °C with anhydrous ammonia in ethylene glycol (Scheme 18)[164]. A variety of heterocycles have been subjected to the reaction. The transformation was not completely chemoselective and the desired product was often contaminated with an *O*-alkylated by-product, which originated from the competitive Ullmann C−O coupling between the starting bromide and ethylene glycol solvent. However, the ultimate goal of this study, i.e. the multigram preparation of *N*-(1-((6-aminopyridin-2-yl)methyl)piperidin-4-yl)acetamide, was achieved.

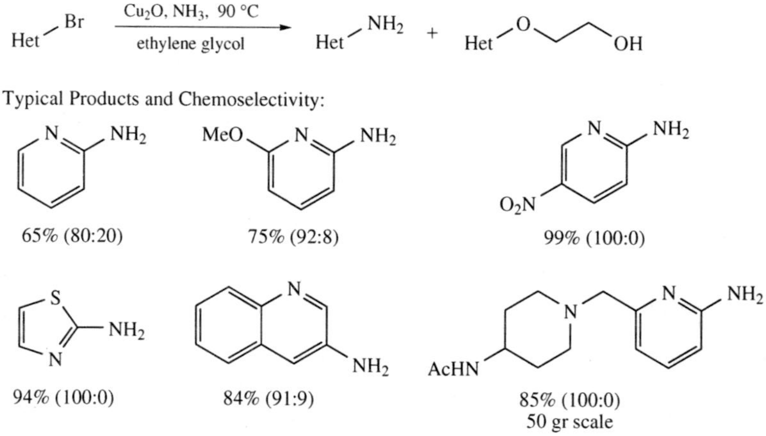

SCHEME 18

Buchwald and coworkers extended the method to include aryl iodides and primary and secondary aliphatic amines. In addition, they found that the use of ethylene glycol as the solvent was not necessary and the reaction could be carried out in nonchelating 2-propanol in the presence of 5 mol% of CuI, 200 mol% of ethylene glycol as an additive and 200 mol% of K_3PO_4 as a base. Interestingly, other diols (propylene glycol, butylene glycol, cyclohexene-1,2-diol etc.) were proved inefficient[165]. The reaction conditions are very mild and functional groups tolerant (Scheme 19). Under these mild conditions, a variety of functionalized aryl iodides may be coupled with primary and secondary alkyl amines in 58–95% yield. No significant electronic effects were observed for *para*- and *meta*-substituted aryl iodides and the yields were comparable. In contrast, *ortho*-substituted substrates show somewhat lower yields, but higher catalyst loading (up to 10 mol%) may improve the situation. Notably, the coupling of enantiopure (*R*)-α-methylbenzylamine proceeds without racemization. This is important because a significant enantiopurity loss often occurs during analogous Pd-catalyzed reactions. It is also noteworthy that the reaction is chemoselective with K_3PO_4 but the chemoselectivity drops if other bases are employed. For example, up to 10% of arylated ethylene glycol was observed in a Cs_2CO_3-assisted run.

Likewise, several groups demonstrated that ethylene glycol monoalkyl ether[138] and *N,N*-dimethylaminoethanol[166–168] in combination with CuI or Cu metal efficiently promote similar coupling reactions even under solvent-free conditions or in aqueous media. For instance, libraries of differently substituted aminothiophenes have been synthesized including in multigram scale using these methods (equation 44)[166].

The catalytic systems described above are not suitable for the amination of less reactive aryl bromides. However, moderate to excellent conversions were observed when ethylene glycol was replaced with bulkier phenol ligands (2-hydroxybiphenyl or 2,6-dimethylphenol). After screening different phenol derivatives, Kwong and Buchwald were

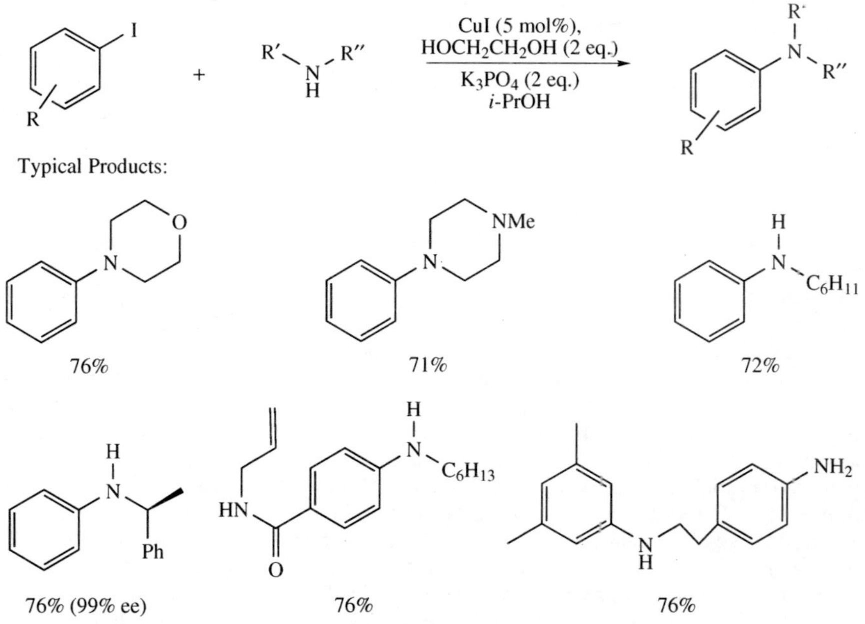

SCHEME 19

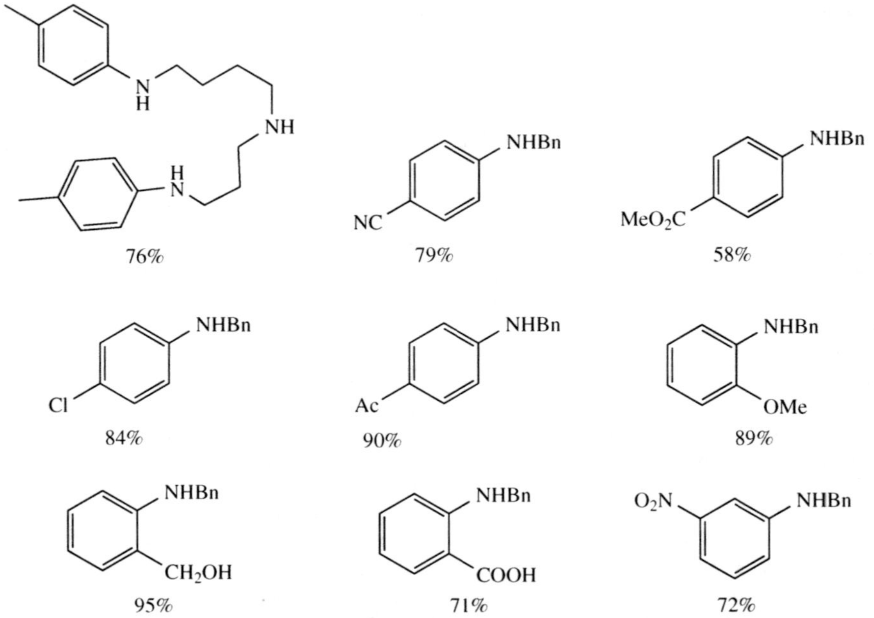

SCHEME 19. (*continued*)

able to identify structurally simple *N,N*-diethylsalicylamide as an efficient ligand in CuI-catalyzed arylation of primary alkyl amines[169].

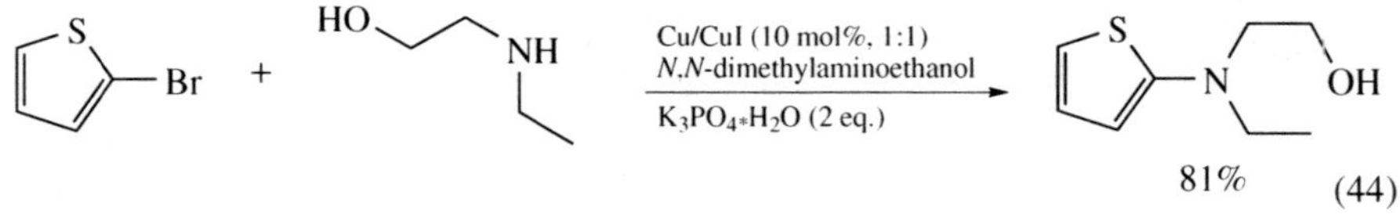

(44)

The optimized reaction conditions involve the heating of aryl or heteroaryl bromides and an appropriate amine in DMF in the presence of 5 mol% of CuI, 5–20 mol% of *N,N*-diethylsalicylamide ligand and 2 equivalents of K_3PO_4 at 90 °C. It was established that a variety of functional groups are compatible with this protocol: thioether-, hydroxyl-, cyano-, keto- and nitro-substituted examples are shown in Scheme 20.

Enolizable β-diketones (mimicking the binding site of the previously described *N,N*-diethylsalicylamide) were also found as efficient ligands for copper-catalyzed C–N coupling reactions[113,170,171]. Within a series of tested catalysts, CuI/2-isobutyrylcyclohexanone was chosen as the most efficient combination which catalyzes *room temperature* arylation of primary and some secondary aliphatic amines with aryl and heteroaryl iodides (employment of aryl bromides requires mild heating of 90 °C) in the presence of Cs_2CO_3 in DMF or NMP. The catalyst is highly chemoselective and tolerates a variety of potentially reactive functional groups such as Boc-protected aliphatic amines, ketones, carboxyl etc. Remarkably, anilines, phenols as well as aliphatic alcohols are not reactive under these reaction conditions and the coupling can be carried out even with unprotected bifunctional substrates (Scheme 21)[113].

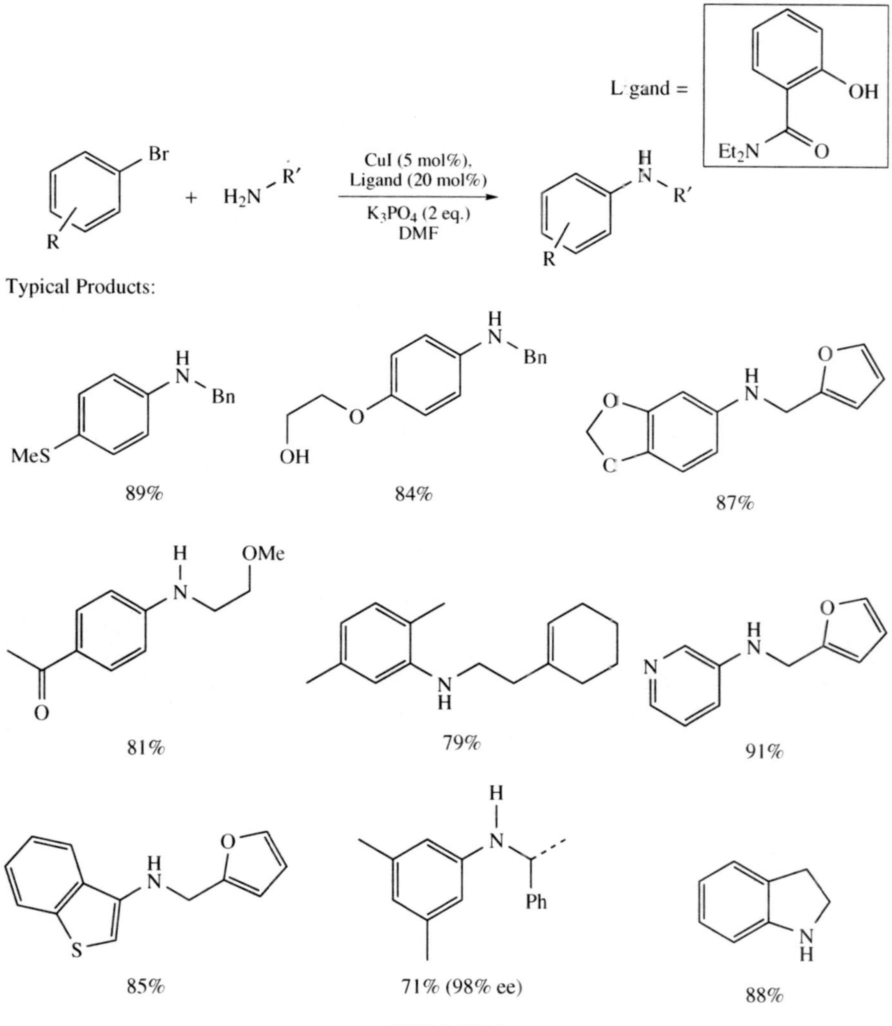

SCHEME 20

Considering the fact that most catalytic systems described so far in this and the previous section are capable of promoting both C−N and C−O bond-forming processes (although under different conditions), high chemoselectivity in such complex cases is rare[164, 166, 172].

Ma and coworkers suggested another operationally simple and general catalytic system for cross-amination of aryl bromides, iodides and sometimes chlorides. They took advantage of a previously disclosed accelerating effect induced by amino acids. In a series of papers, they reported spectacularly mild and efficient protocols employing CuI/amino acid catalysts operating in polar solvents such as DMSO at only 40–100 °C[173–176]. In order to avoid catalyst deactivation via possible ligand arylation[158], *N*-alkylated amino acids and their derivatives were tested. Although the screening study revealed that most α- and β-amino acids accelerate the reactions to a different extent[173, 177, 178], two ‘privileged’

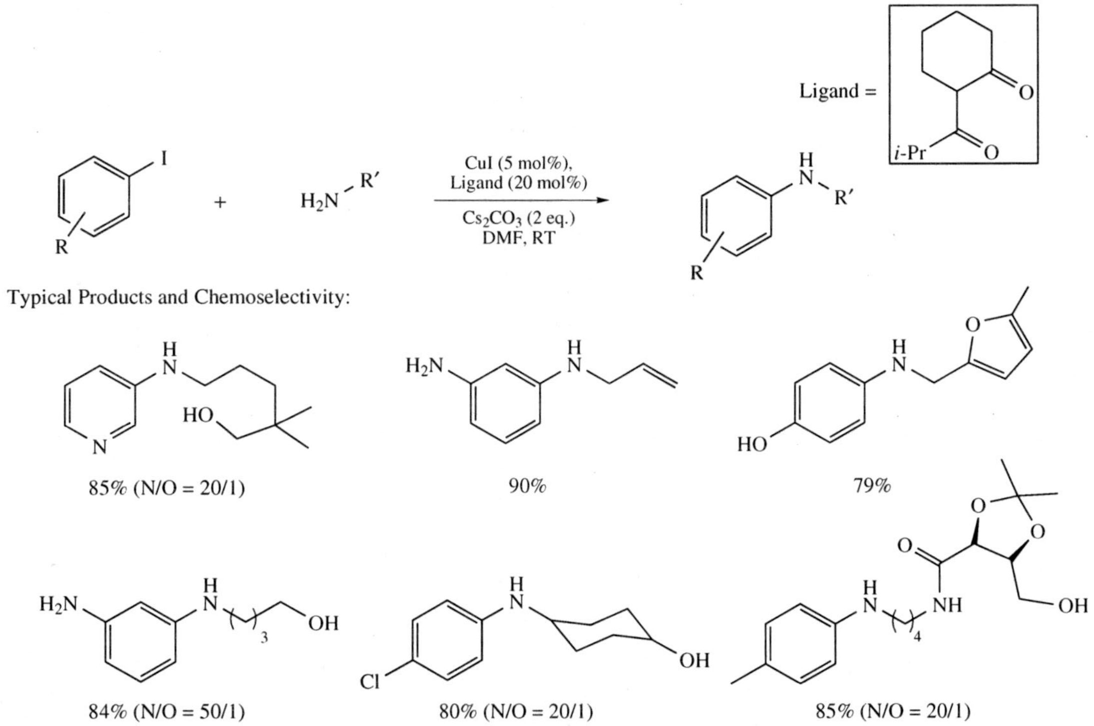
Ligand =
i-Pr
O
O
I
R
+
H_2N R′
CuI (5 mol%), Ligand (20 mol%)
Cs_2CO_3 (2 eq.) DMF, RT
H N R′
R
Typical Products and Chemoselectivity:
85% (N/O = 20/1)
90%
79%
84% (N/O = 50/1)
80% (N/O = 20/1)
85% (N/O = 20/1)

SCHEME 21

structural motifs were identified: proline (L-proline and 4-hydro-L-proline) and glycine (*N,N*-dimethylglycine and *N*-methylglycine).

Initial scope and limitation studies showed that *N*-methylglycine is somewhat more reactive than *N,N*-dimethylglycine and L-proline in coupling of aryl iodides with the model benzylamine, and good conversions were achieved *under* 60 °C with electron-deficient substrates. However, incomplete conversion was detected when electron-rich starting materials were employed. On the other hand, elevated temperatures did not provide a satisfactory solution to this problem due to the deactivation of the catalyst mentioned above[158]. Thus, L-proline and *N,N*-dimethylglycine were found superior in more difficult cases.

The general reactivity trends of these catalytic systems are as follows: more sterically hindered secondary amines are less reactive than the primary ones and therefore require elevated reaction temperatures (90 °C) to reach full conversion. Similarly, the difference in reactivity between aryl bromides and iodides is reflected in the reaction temperature only (<60 °C for aryl iodides and >60 °C for aryl bromides), but yields are comparable. In addition, functional group compatibility is excellent, which allows the employment of Ma's catalyst in syntheses of heterocyclic and highly functionalized amines (Scheme 22)[174, 178].

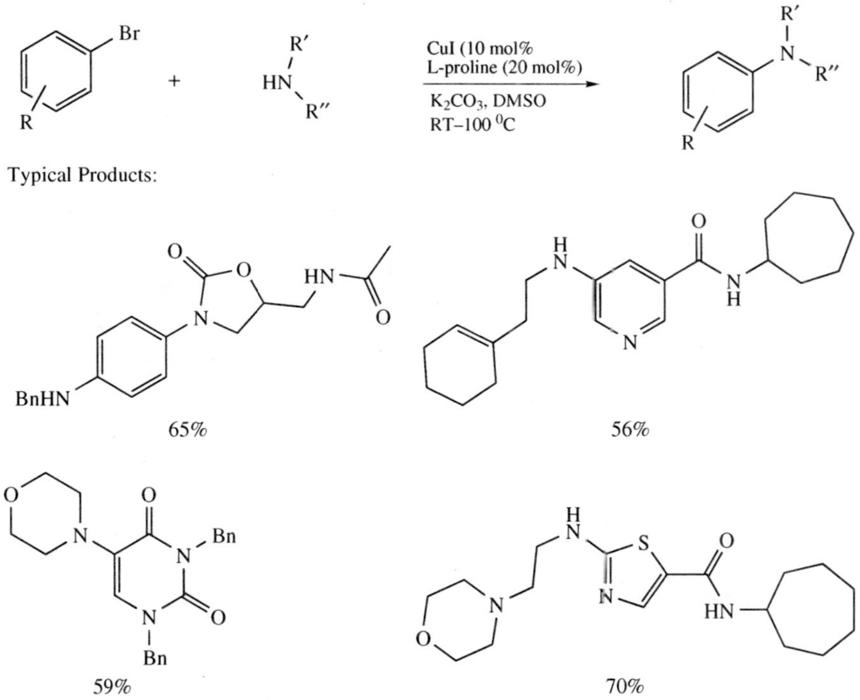

SCHEME 22

The same catalyst was applied to the mild (RT–50 °C) two-step one-pot synthesis of 1,2-substituted benzimidazoles (Scheme 23)[175].

It is interesting to note that although the catalysts are spectacularly active, the neighboring group assistance remains important and such a mild cross-coupling step is limited to aryl halides possessing an amide group at the *ortho* position. The final cyclization

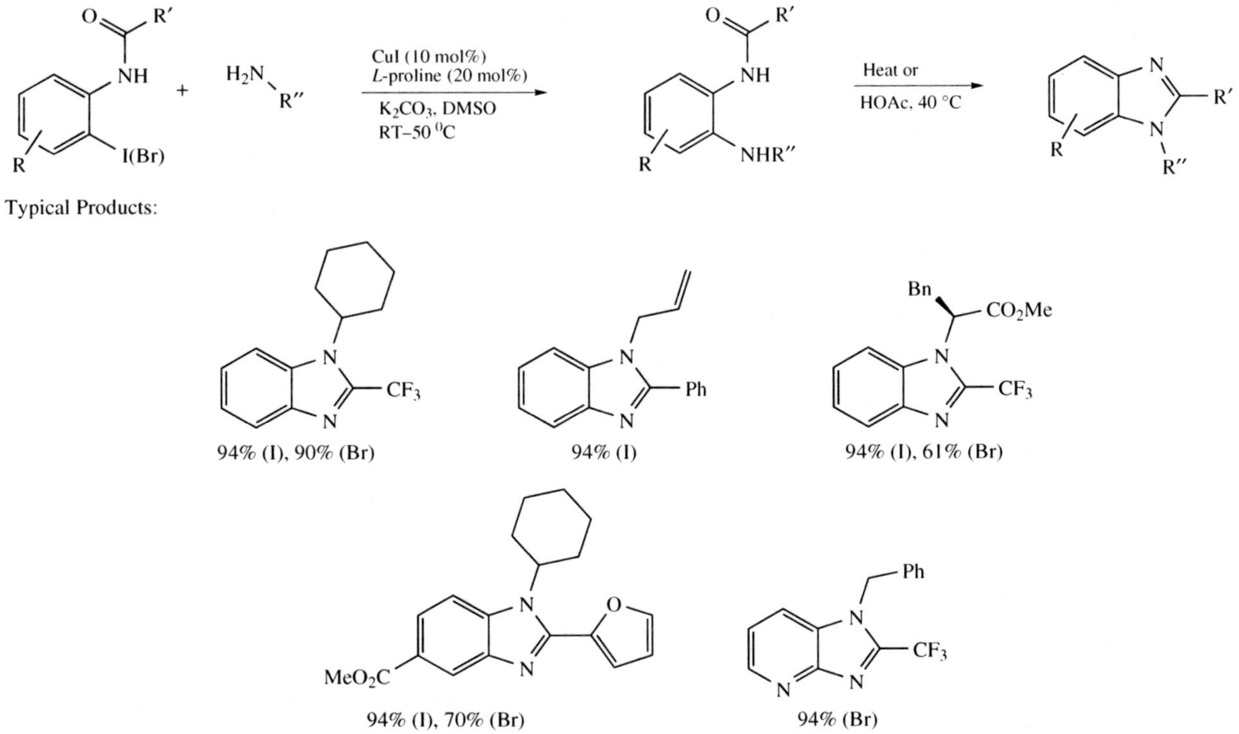
O
R′
NH
I(Br)
R
+
H_2N
R″
CuI (10 mol%)
L-proline (20 mol%)
K_2CO_3, DMSO
RT–50 °C
NHR″
Heat or
HOAc, 40 °C
N
R″
Typical Products:
CF_3
94% (I), 90% (Br)
Ph
94% (I)
Bn
CO_2Me
94% (I), 61% (Br)
MeO_2C
94% (I), 70% (Br)
94% (Br)

SCHEME 23

step proceeds spontaneously or may be acid-promoted. For nonchelating cases (e.g. 2-halophenylcarbamates), structurally similar *trans*-4-hydroxy-L-proline as a ligand demonstrated superior reactivity[176]. For example, coupling between functionalized amines (e.g. amino acids, amino amides, allylamine etc.) and 2-bromophenylcarbamates proceeds at 70 °C in the presence of CuI (20 mol%), *trans*-4-hydroxy-L-proline (20 mol%) and K_3PO_4 in DMSO. The successive heating of the reaction mixture to cyclization temperature (130 °C) without intermediate product isolation represents a straightforward two-stage one-pot synthesis of 1,3-dihydrobenzimidazol-2-ones (Scheme 24).

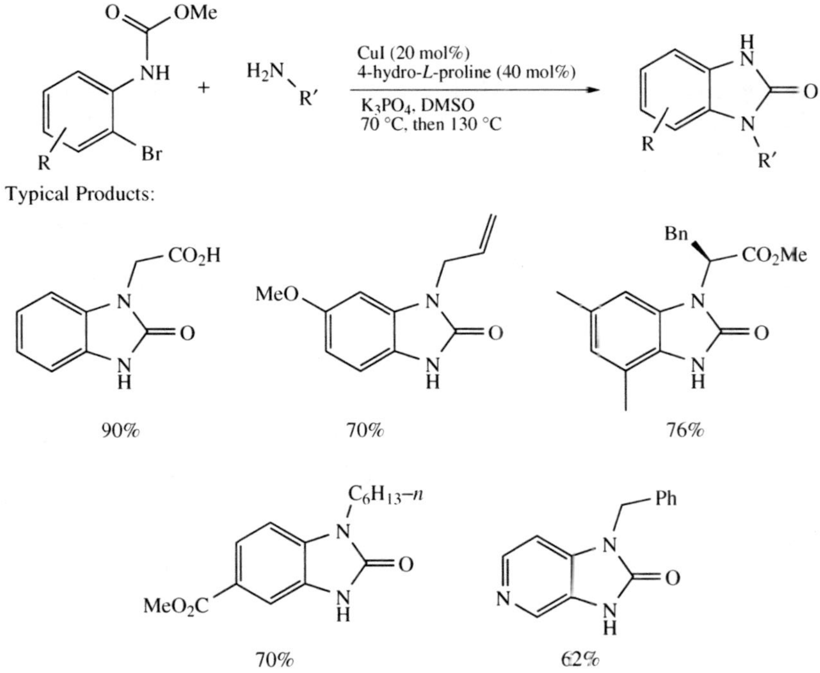

SCHEME 24

Likewise, several ligandless catalytic systems[179–182] and catalytic systems based on high molecular weight ligands have been offered[107, 181, 183–187].

b. Coupling of aromatic amines with aryl halides. Although synthesis of aromatic amines was possible via classical Ullmann coupling, the development of ligand-assisted protocols significantly improved the scope of the method.

In 1999, Goodbrand and Hu reported the results of a limited screening study demonstrating a significant rate acceleration of the coupling between aryl iodides and primary or secondary aromatic amines catalyzed by 3.5 mol% of CuI in refluxing toluene in the presence of 1,10-phenanthroline, pyridine, 2,2′-bipyridine, 8-hydroxyquinoline, TMEDA, 1,8-bis(dimethylamino)naphthalene and racemic sparteine as ligands and KOH as a base[188]. Although in all cases a rate acceleration was observed, 1,10-phenanthroline showed unambiguous superiority in terms of efficiency and selectivity (equation 45).

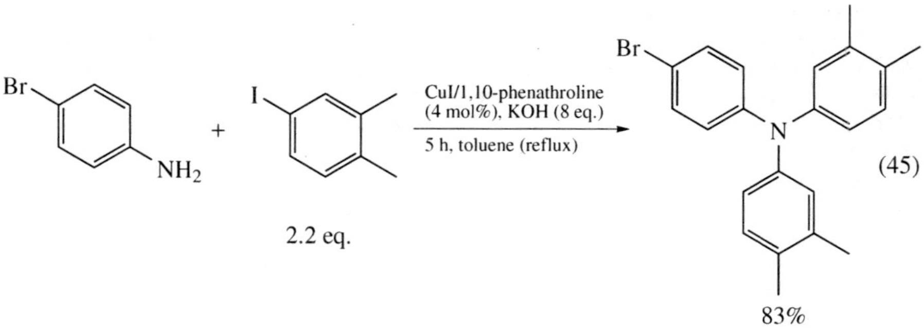

(45)

Comparable reactivity was observed using both well-defined copper complexes bearing 1,10-phenanthroline or neocuproine ligands and catalysts prepared *in situ*[100, 189, 190]. It was also found that the use of harsh bases such as KOH is not necessary, and the latter can be replaced with more conventional and mild K_2CO_3, Cs_2CO_3, K_3PO_4 etc. The solvent choice appears not to be critical as well, and DMSO, DMF or NMP were satisfactory.

Very simple ligands such as DMEDA[191], 2-ethoxyethanol[192, 193] or L-proline[194] and more complex ones[183, 195] can be successfully employed in performing this coupling. Generally, good efficiency and functional group tolerance are observed in reactions of aromatic and heteroaromatic amines with aryl bromides and iodides (e.g. Scheme 25)[196–201].

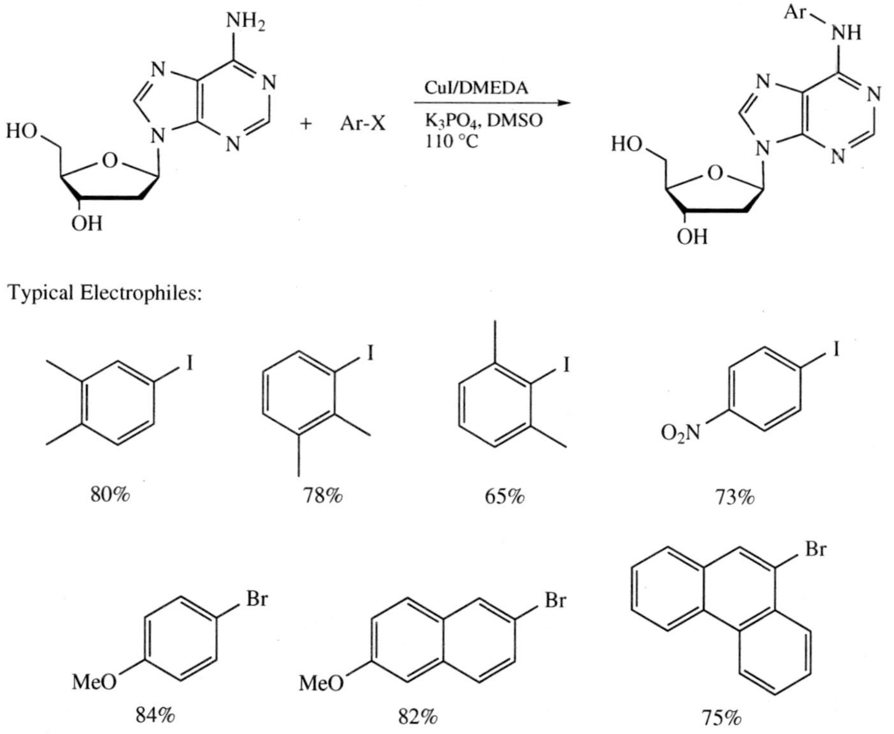

SCHEME 25

c. Coupling of NH heterocycles with aryl halides. Arylation of the N−H unsaturated heterocyclic compounds became possible using essentially similar catalytic systems.

The first ligand-assisted copper-catalyzed arylation of imidazoles by aryl halides was suggested by Buchwald and coworkers in 1999[202]. To reach the goal, they engaged the catalyst derived from a very soluble $Cu(OTf)_2$•benzene complex (5 mol%) and 1,10-phenanthroline (100 mol%), *trans,trans*-dibenzylideneacetone (dba) (5 mol%) as an additive and Cs_2CO_3 as a base, in xylenes at 100−125 °C (equation 46). Under these conditions, the reactions proceed smoothly with little, if any, formation of dehalogenated or homocoupled by-products[202,203]. Sterically nondemanding aryl iodides can be quantitatively coupled with unsubstituted imidazoles independently of their electronic nature. Lower yields, however, were observed when *ortho*-substituents were present at either of the coupling partners.

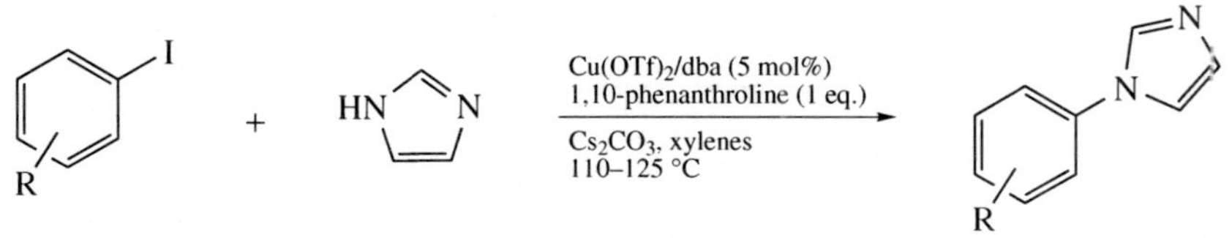

(46)

R = *o*-, *m*-, *p-alkyl*, *p*-MeO, *m*-CF_3—10 examples 62–97%

Further studies showed that arylation of imidazoles[186,187,204–213] and analogous benzimidazoles[209–211], pyrroles[210,214,215], indoles[210,216–218], pyrazoles[214,215,217,219,220], indazoles[214], triazoles[214,217] and tetrazoles[217] can be carried out in the presence of more user-friendly copper precursors (CuI, CuBr, Cu_2O etc.) and in the presence of an array of substoichiometric ligands such as those shown (Chart 6).

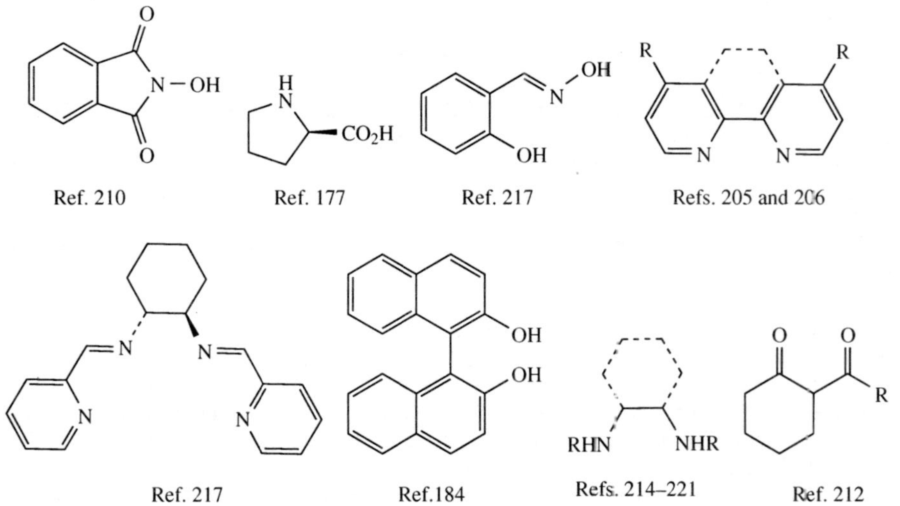

CHART 6

The reaction conditions are relatively typical and comprise heating of the coupling partners together with the catalyst in polar (DMF, dioxane, NMP etc.) or nonpolar (toluene) solvents in the presence of Cs_2CO_3, K_2CO_3 or K_3PO_4. The reaction time and temperature depend, however, on the type of aryl halides involved. For example, aryl iodides are normally more reactive than the corresponding bromides but both react at 80–110 °C. Coupling of aryl chlorides is limited to activated (electron-poor) substrates and normally requires higher reaction temperatures of 125–150 °C to achieve acceptable conversions[208,210,211]. In addition, the yields are generally high with sterically nondemanding coupling partners, but drop when substituents are present near the halogen atom of the aryl halide or the NH nucleophile[204].

CuI/*N*,*N*′-dimethyl-1,2-cyclohexanediamine (DMCyDA)[214,216] or Cu_2O/4,7-dimethoxy-1,10-phenanthroline[205,206] were found to be more effective in coupling of sterically demanding substrates (naturally, the use of more reactive aryl iodides is advised in these cases). It is interesting to note that poly(ethylene glycol) (PEG) phase-transfer catalyst further accelerates the reaction promoted by less soluble cuprous oxide (Scheme 26).

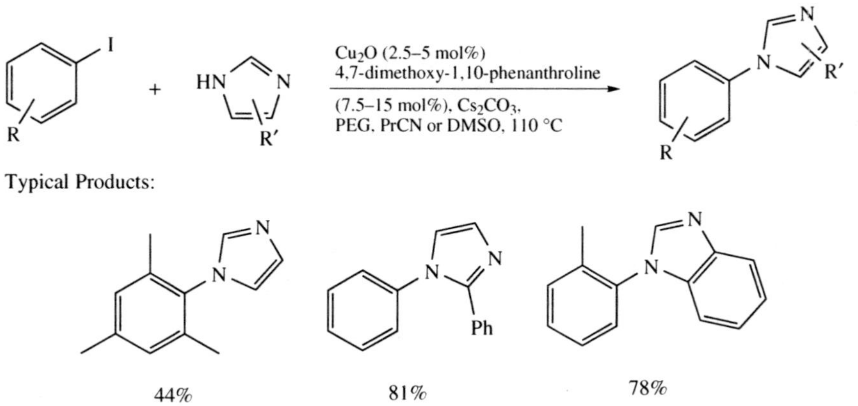

SCHEME 26

Copper-catalyzed arylation of asymmetrically substituted azoles may lead to the formation of several regioisomeric products (Scheme 27).

The degree of selectivity in such cases depends mainly on the steric properties of the starting materials and predominantly leads to the formation of the less substituted isomer (Scheme 28).

The functional group compatibility of this method is very good and a variety of organic substituents at both the aryl halides and the heterocycles are tolerated.

d. Coupling of amides and related compounds with aryl halides. A general and efficient catalyst for Goldberg-type coupling relying on the combination of CuI and racemic *trans*-1,2-cyclohexanediamine (CyDA) has been identified after extensive screening of various diamine ligands. The catalytic system operates in a variety of solvents (e.g. toluene, dioxane or DMF) in the presence of simple inorganic bases such as K_3PO_4 or K_2CO_3 and is applicable to a wide range of substrates[221].

As shown, lactams, primary and secondary amides can be coupled with functionalized aryl and heteroaryl iodides at 110 °C in the presence of only 1 mol% of CuI and 10 mol% of the ligand. Bromides are less reactive but can be successfully employed as coupling partners under higher catalyst loading (up to 10 mol%) delivering high yields of products in a chemoselective manner even in the presence of unprotected aromatic amines (Scheme 29).

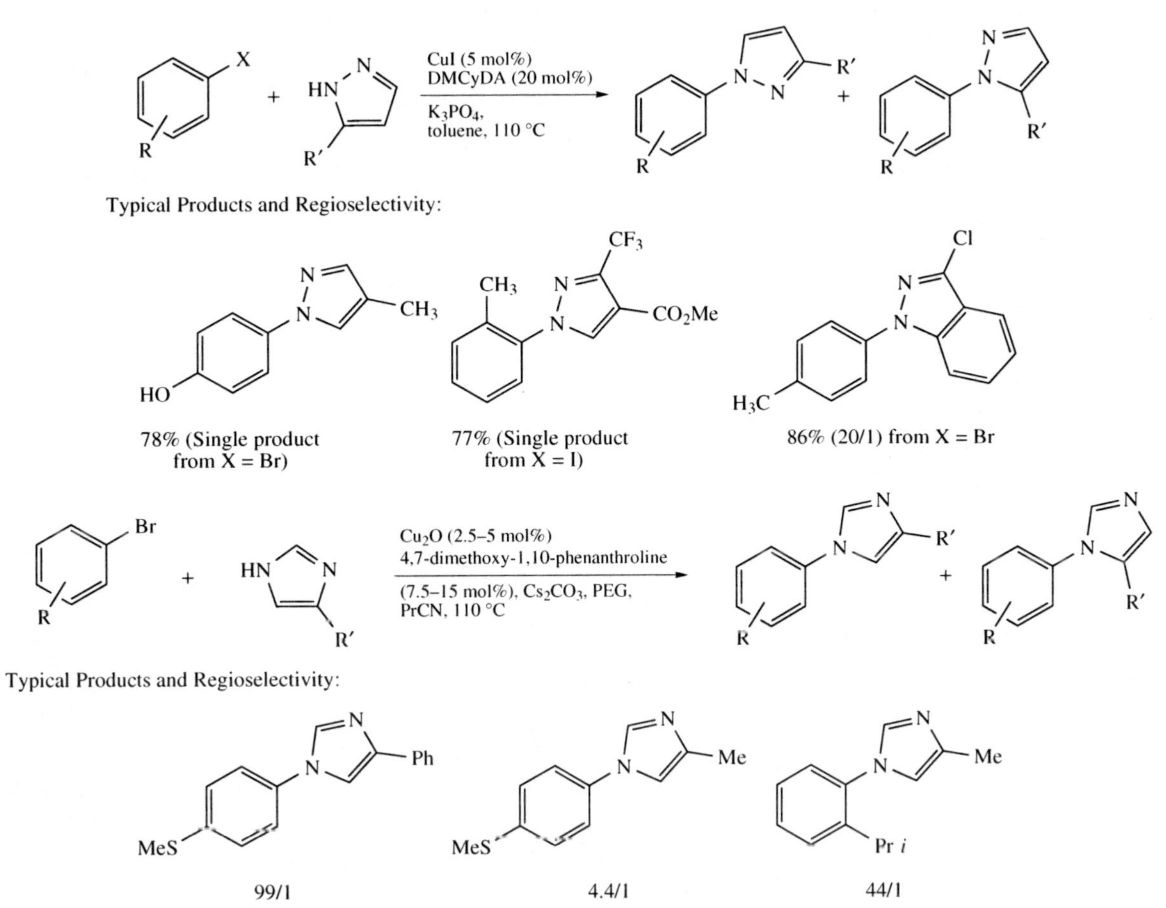

X
HN
N
R
R′
CuI (5 mol%)
DMCyDA (20 mol%)
K3PO4,
toluene, 110 °C
+
Typical Products and Regioselectivity:
HO
CH3
CF3
CO2Me
Cl
H3C
78% (Single product from X = Br)
77% (Single product from X = I)
86% (20/1) from X = Br
Br
Cu2O (2.5–5 mol%)
4,7-dimethoxy-1,10-phenanthroline
(7.5–15 mol%), Cs2CO3, PEG,
PrCN, 110 °C
Typical Products and Regioselectivity:
MeS
Ph
Me
Pr i
99/1
4.4/1
44/1

SCHEME 27

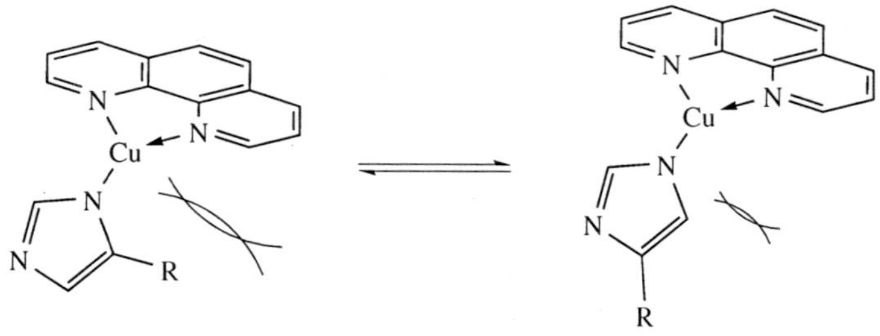

SCHEME 28

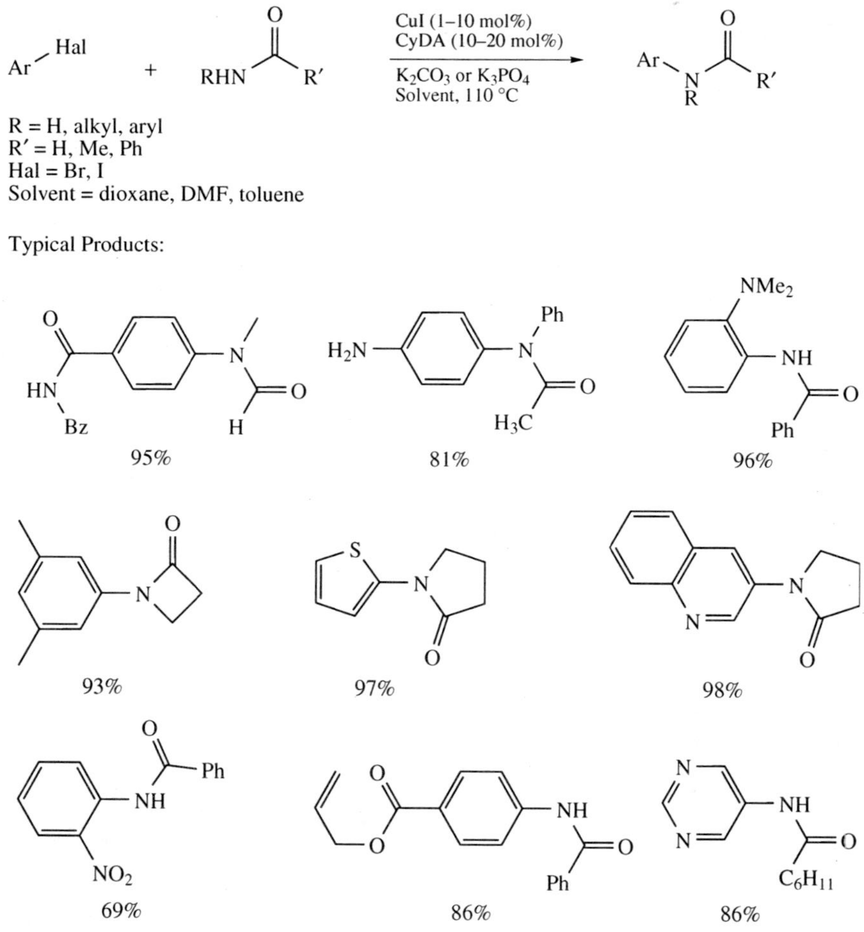

SCHEME 29

Intermolecular arylation of more challenging aryl chlorides having no chelating *ortho* groups became possible, but required the employment of the second-generation vicinal diamine ligands such as *N*,*N*′-dimethylcyclohexane-1,2-diamine (DMCyDA) (Scheme 30)[221]. A better performance by DMCyDA is likely due to a slower catalyst deactivation via self-arylation of the ligand under the harsher reaction conditions required to activate the less reactive aryl chlorides (110–130 °C, 23 h, neat).

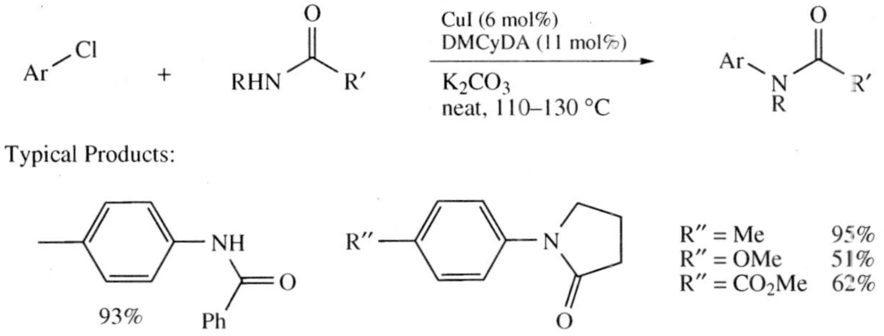

SCHEME 30

Nowadays, this second-generation ligand as well as its structurally simpler descendant, *N*,*N*-dimethylethylene diamine (DMEDA), are widely used in the Goldberg reaction to ensure good efficiency in all cases[222–227]. The general features of the reaction remained similar for all these catalysts: i) 1–10 mol% CuI as a convenient air- and moisture-stable copper precursor (although other available copper compounds can be used instead); ii) 10–20 mol% of the corresponding ligand; iii) 90–130 °C heating (or lower for electron-poor or -neutral aryl iodides and primary amides) in toluene, dioxane, NMP or DMF; iv) K_2CO_3, K_3PO_4, Cs_2CO_3 or rarely NaOBu-*t* as a base. The latter point is worth noting: the use of strong bases is normally avoided in order to prevent the formation of inactive diligated copper(I) amidate species (Scheme 31): optimally, the pK_{AH} value of the base employed should be below the pK_{AH} value of the amide. The use of a large excess of the chelating ligand is, apparently, beneficial for the same reason[227,228].

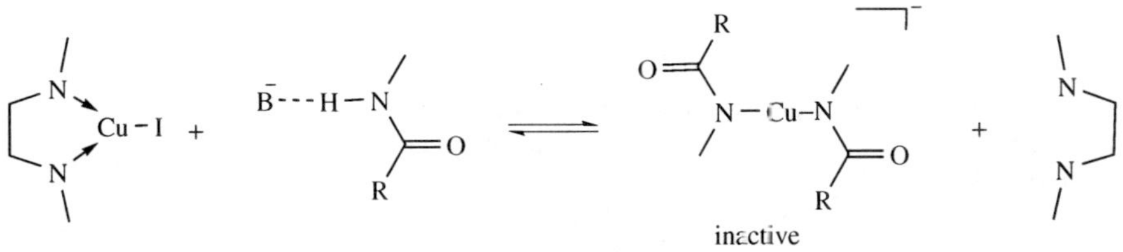

SCHEME 31

In principle, other ligands can also be employed for intermolecular arylation of amides. For example, ethylene diamine[229], 8-hydroxyquinoline[230], L-proline[231], bipyridine[190,232], polyoles[233], *β*-diketo[109,234] and Schiff base compounds[217,218] were reported to operate under the standard conditions.

Besides carboxylic acid amides, other amide-like compounds can be arylated using copper catalysis. For example, differently substituted oxazolidones efficiently couple with aryl bromides using CuI/*trans*-1,2-cyclohexanediamine catalyst (K_2CO_3 in dioxane)[235,236]. The method displays a very good functional group compatibility and was used in the synthesis of enantiopure κ-opioid receptor antagonist CJ-15161 precursor (50 gr scale, 97% yield with no racemization at the coupling step) (equation 47).

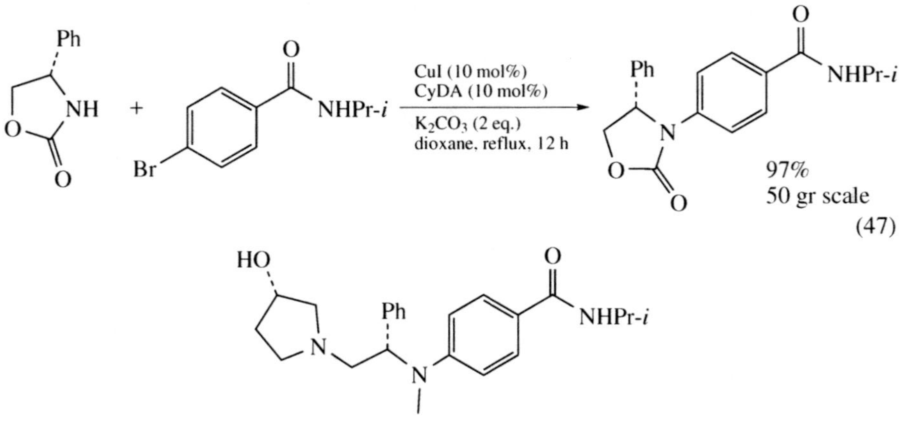

(47)

CJ-15161

Similarly, Boc-protected amines, phosphoramides[237], sulfonamides[238], sulfoximines[239–241] and sulfonimidamides[242] became useful coupling partners in intra- and intermolecular reactions using CuI in combination with L-proline or *N*-methylglycine catalyst (equation 48).

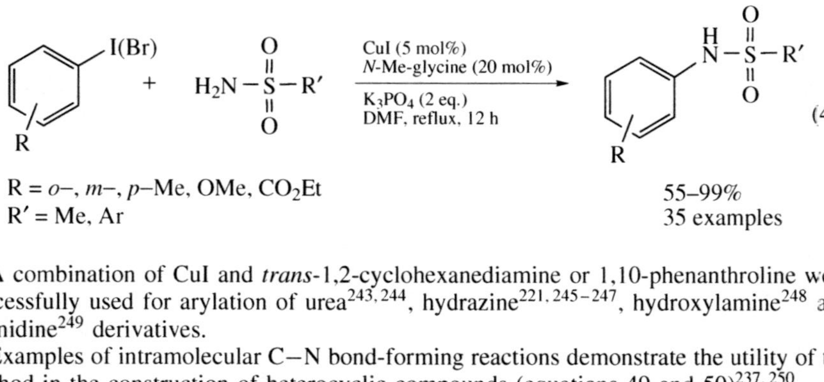

(48)

R = *o*–, *m*–, *p*–Me, OMe, CO_2Et
R′ = Me, Ar

55–99%
35 examples

A combination of CuI and *trans*-1,2-cyclohexanediamine or 1,10-phenanthroline were successfully used for arylation of urea[243,244], hydrazine[221,245–247], hydroxylamine[248] and guanidine[249] derivatives.

Examples of intramolecular C−N bond-forming reactions demonstrate the utility of the method in the construction of heterocyclic compounds (equations 49 and 50)[237,250].

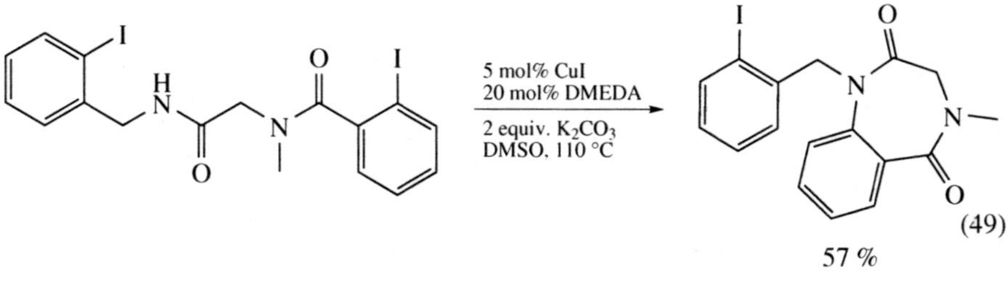

(49)

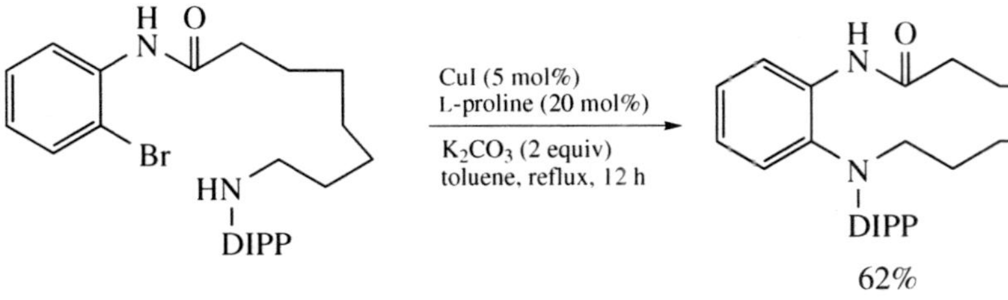
(50)

DIPP = diisopropylphosphoamine

Domino reactions comprising copper-catalyzed arylation of amides as one of the steps became a popular strategy toward construction of various heterocyclic compounds. The following example represents an expeditious route to benzoxazoles via one-pot sequential C−N and C−O bond-forming reactions[251]: CuI/DMEDA catalyst efficiently promotes arylation of aromatic or aliphatic amides as well as intramolecular *O*-arylation of the intermediate (Scheme 32). Unsubstituted or symmetrically substituted benzoxazoles can be synthesized starting from the corresponding 1,2-dibromo- and diiodobenzenes (dichlorobenzenes are inactive under these reaction conditions). The use of 1-bromo-2-chlorobenzenes is necessary in order to prepare asymmetrically substituted benzoxazoles in a regioselective manner.

A copper-catalyzed Goldberg reaction/condensation sequence was applied to the synthesis of *N*-alkylbenzimidazoles from *N*-substituted *o*-bromoanilines[252]. In this case, however, the *N*-aryl amide intermediate does not cyclize spontaneously and a different set of conditions was required in order to accomplish the assembly (Scheme 33).

Similarly, 2-aryl-4-quinolones can be prepared using an amidation/Camps cyclization sequence (Scheme 34)[253].

A useful strategy toward the construction of medium-sized nitrogen heterocycles via sequential β-lactam arylation and intramolecular ring expansion with a neighboring nitrogen nucleophile was suggested based on the same catalytic system (Scheme 35)[254]. The first step takes place cleanly in a chemoselective manner under the standard amidation conditions (CuI/DMEDA catalyst in the presence of K_2CO_3 in toluene). The sequential ring-expansion proceeds spontaneously with *o*-bromobenzylamine derivatives possessing unprotected or sterically nondemanding monoalkylated amine groups, leading to the formation of the corresponding 8-membered heterocycles. However, mixtures of the *N*-arylated β-lactam intermediate and the desired transamidated product were observed with bulky *N*-alkyl substituents. The same result was obtained when the protocol was applied to the synthesis of 7-membered rings. In these cases, acetic acid or Ti(OPr-*i*)$_4$ were suggested as efficient promoters for the ring-expansion at a separate step.

The method was also applied to the synthesis of 9- and 10-membered heterocycles (equation 51). However, in these experiments a facile intramolecular amination reaction appears as a competitive process in the first step. In addition, the sequential ring-expansion is slow and acidic catalysis is required to drive the reaction to completion.

e. Coupling of amides and related compounds with vinyl and alkynyl halides. Even early reports on the copper-catalyzed Ullmann and Goldberg reactions of vinyl halides pointed out a complete retention of the double-bond configuration[74]. Therefore, the development of a ligand-assisted Goldberg amidation allowing facile and stereocontrolled synthesis of enamides was much desired.

Shen, Porco and coworkers demonstrated that Liebeskind's copper(I) thiophenecarboxylate (CuTC) in combination with Cs_2CO_3 or Rb_2CO_3 in NMP or DMA at only 90 °C

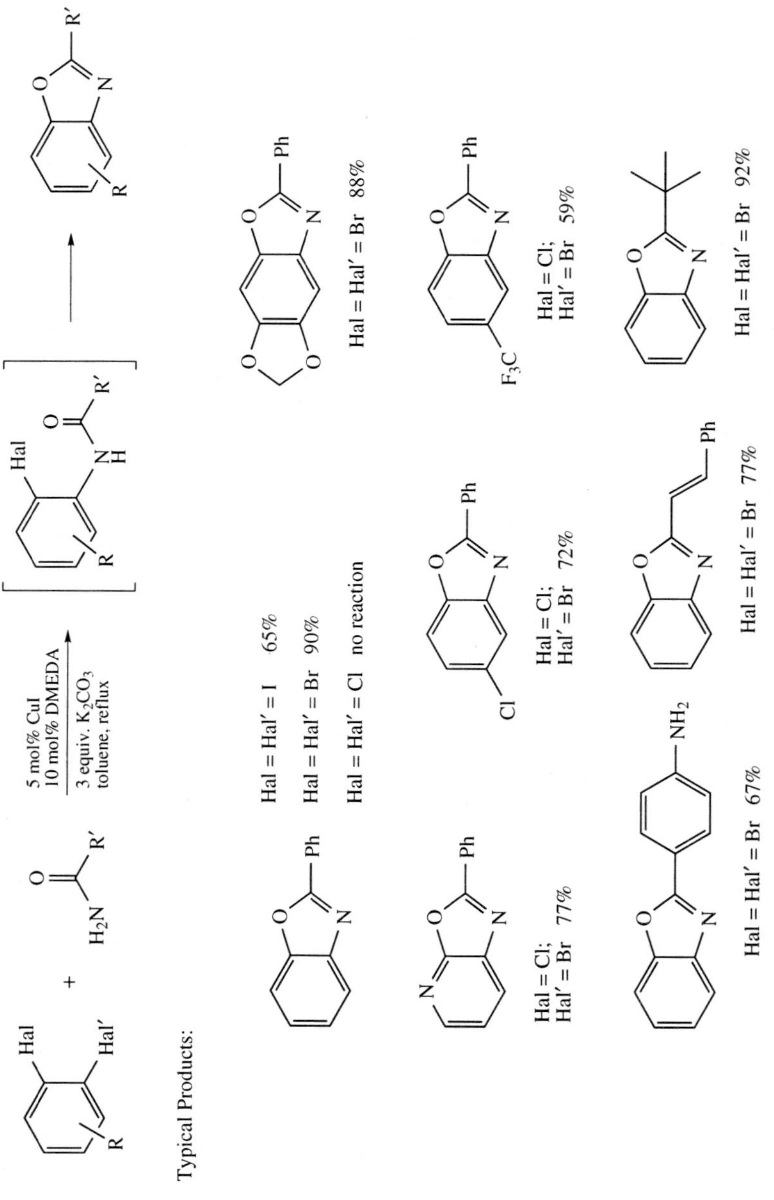
5 mol% CuI
10 mol% DMEDA
3 equiv. K2CO3
toluene, reflux
Typical Products:
Hal = Hal′ = I 65%
Hal = Hal′ = Br 90%
Hal = Hal′ = Cl no reaction
Hal = Hal′ = Br 88%
Hal = Cl;
Hal′ = Br 77%
Hal = Cl;
Hal′ = Br 72%
Hal = Cl;
Hal′ = Br 59%
Hal = Hal′ = Br 67%
Hal = Hal′ = Br 77%
Hal = Hal′ = Br 92%

SCHEME 32

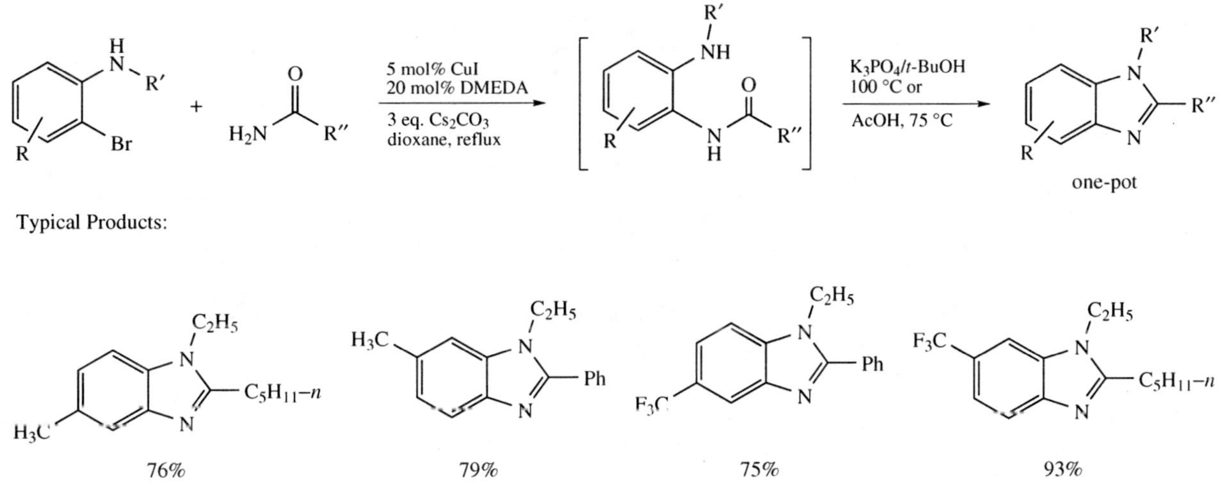
5 mol% CuI
20 mol% DMEDA
3 eq. Cs2CO3
dioxane, reflux
K3PO4/t-BuOH
100 °C or
AcOH, 75 °C
one-pot
Typical Products:
76%
79%
75%
93%

SCHEME 33

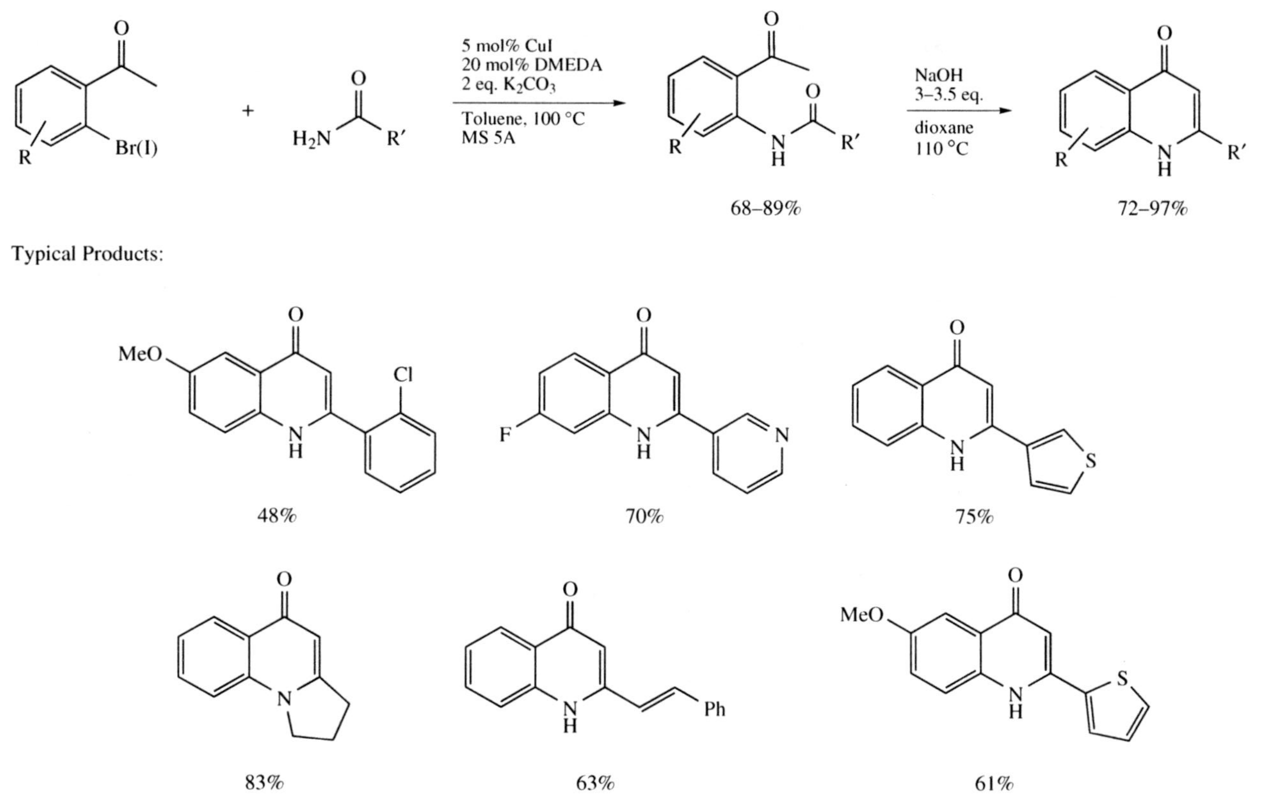
O
Br(I)
R
+
H$_2$N
R′
5 mol% CuI
20 mol% DMEDA
2 eq. K$_2$CO$_3$
Toluene, 100 °C
MS 5A
N
H
68–89%
NaOH
3–3.5 eq.
dioxane
110 °C
72–97%
Typical Products:
MeO
Cl
48%
F
70%
S
75%
83%
Ph
63%
61%

SCHEME 34

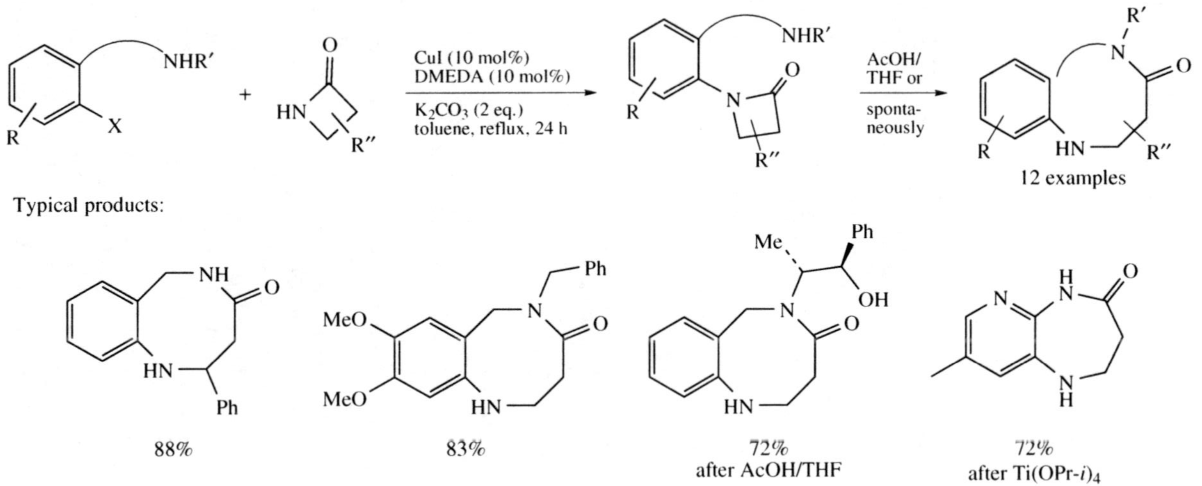
NHR′
X
R
HN
O
R″
CuI (10 mol%)
DMEDA (10 mol%)
K2CO3 (2 eq.)
toluene, reflux, 24 h
AcOH/
THF or
sponta-
neously
R′
N
HN
12 examples
Typical products:
NH
Ph
MeO
MeO
Me
OH
N
H
88%
83%
72%
after AcOH/THF
72%
after Ti(OPr-i)4

SCHEME 35

catalyzes cross-coupling of vinyl iodides with primary amides. Although the reaction conditions comprised high-catalyst loading (30 mol% of CuTC) the result was satisfactory in terms of both the yield and selectivity (equation 52)[255,256].

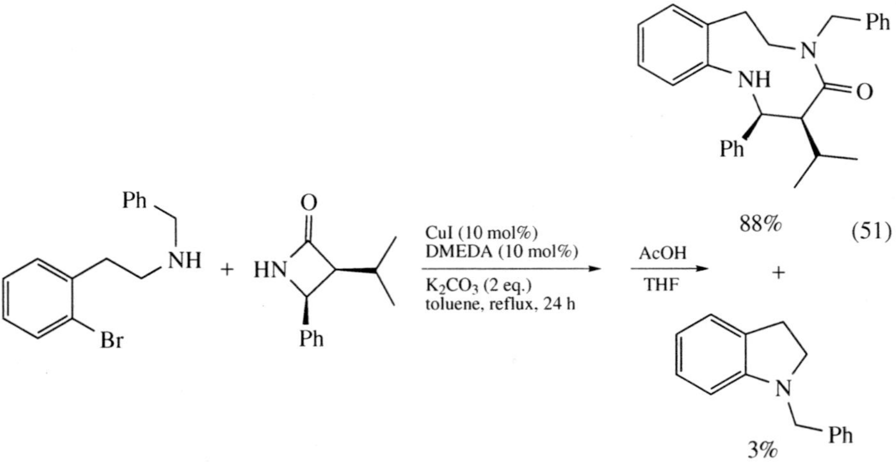

(51)

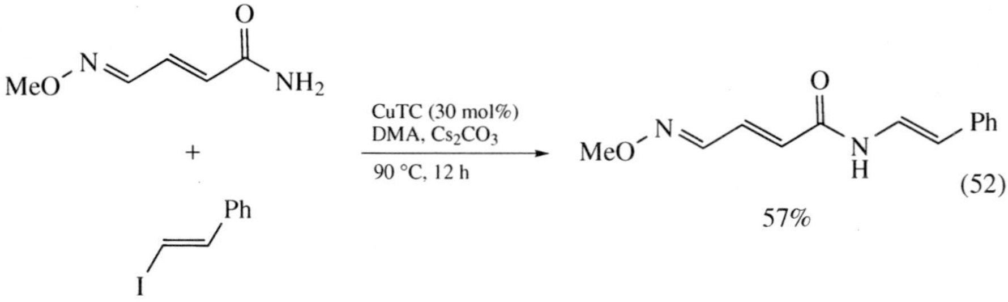

(52)

Very soon, more economic catalytic systems based on DMEDA[257,258], *N*,*N*-dimethylglycine[259] or phenanthrolines[260] became the usual choice for both intermolecular and intramolecular[261,262] transformations of this type. These catalytic systems operate under very much similar reaction conditions to those developed for Goldberg coupling of aryl halides, i.e. 5–10 mol% of the copper precursor (usually CuI), 2–3 equivalents of Cs_2CO_3, K_3PO_4 or K_2CO_3 in dioxane, DMA or THF medium at RT–110 °C.

The most comprehensive scope and limitation studies[258,259] showed that vinyl iodides are generally more reactive than the corresponding bromides and usually require lower reaction temperatures to reach complete conversions. Similarly, the reaction depends on the steric properties of both coupling partners. Usually, a longer reaction time and a higher reaction temperature are required when highly substituted aryl halides are employed and lower yields are obtained with sterically demanding amides (e.g. acyclic secondary amides are bad coupling partners). On the other hand, primary cyclic and acyclic amides, 4–6-membered lactams as well as substituted carbamates all show comparable reactivity (equations 53 and 54).

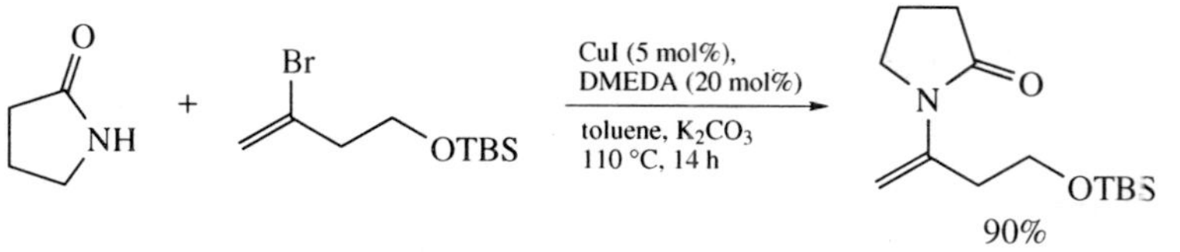
(53)

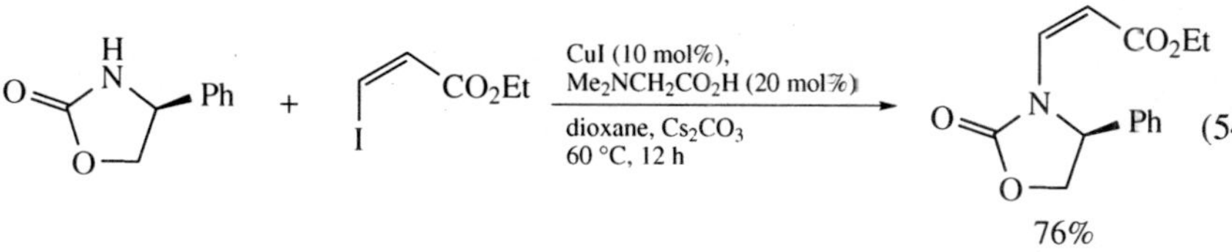
(54)

Hsung and coworkers demonstrated that the method is applicable to the direct synthesis of chiral ynamides (Scheme 36)[263]. In their report, CuCN/DMEDA was used as a catalyst (whereas more reproducible results were obtained using this particular copper precursor, more convenient copper sources can be used instead)[264,265], in the presence of K_3PO_4 in toluene at 110 °C. A variety of chiral oxazolidones as well as achiral acyclic carbamates efficiently couple with differently substituted (1°-, 2°-, 3°-alkyl, aryl or $SiAlk_3$) alkynyl bromides delivering the corresponding ynamides in moderate to good yield.

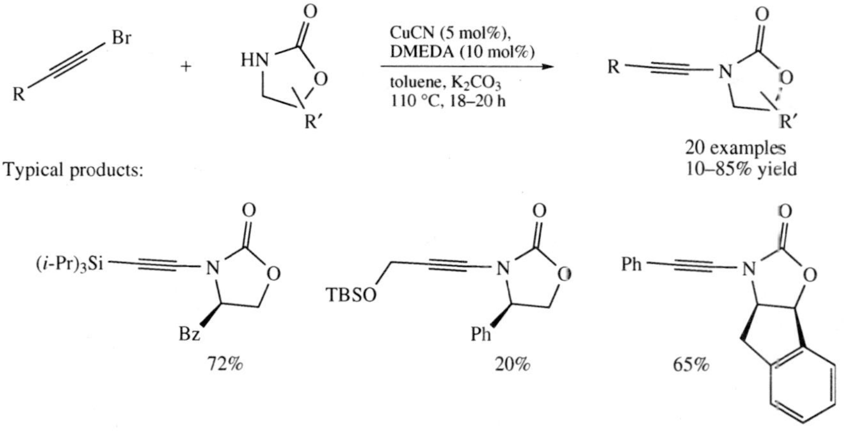

SCHEME 36

The same catalyst[266], or alternatively, a catalyst based on the CuTC/N,N'-dimethylcyclohexane-1,2-diamine system[267], was applied to the synthesis of chiral allenamides from allenyl iodides (Scheme 37). The catalysts offer a unique opportunity to access these valuable building blocks in a stereospecific fashion and under spectacularly mild conditions that might be applied even on highly functionalized chiral starting materials with only minor loss (if any) of enantiopurity. Amides, carbamates and ureas are subjects for successful coupling.

Intramolecular versions of the copper-catalyzed vinylation and alkynylation of amides represent a straightforward route toward the synthesis of lactams and other N-heterocyclic

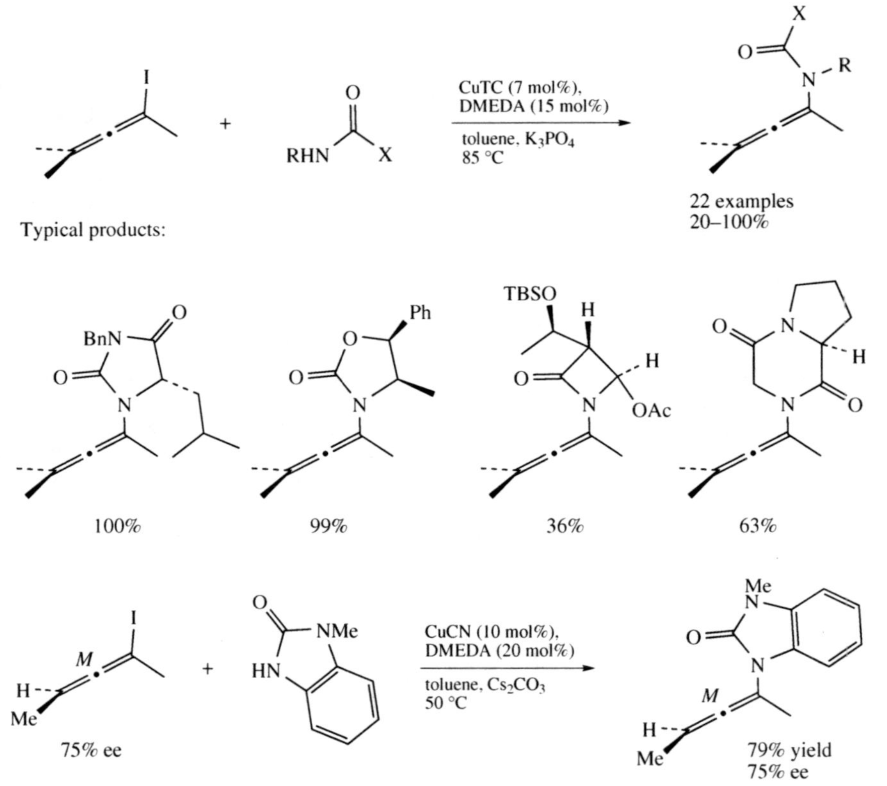

SCHEME 37

compounds (equations 55 and 56 and Scheme 38)[261, 262, 264]. Remarkably, even vinyl chlorides can be used as starting materials in these cases.

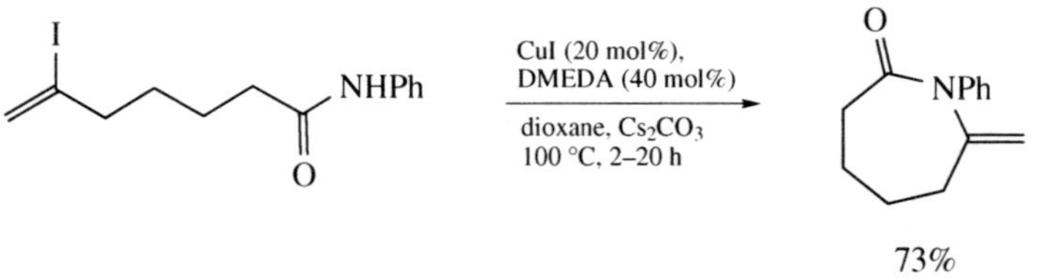

(55)

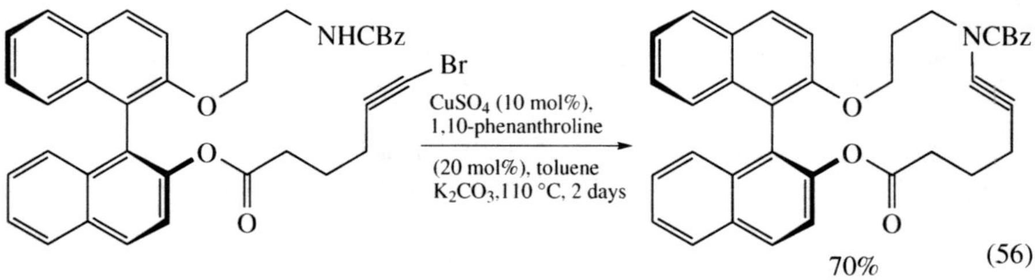

(56)

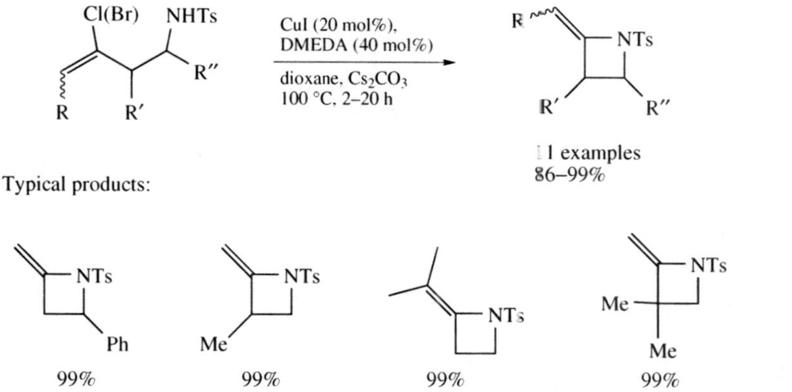

SCHEME 38

Extension of these fascinating intramolecular transformations to domino reactions makes the modified Goldberg reactions an extremely useful tool in synthesis of heterocycles. For example, sequential inter-/intramolecular amidation of vinyl halides was suggested as an expeditious route toward differently substituted pyrroles. This route can start from either carboxylic acid amides[268] or *t*-butyl carbamate[269] and can be performed using standard conditions (CuI/DMEDA or CuI/DMCyDA catalysts in THF, dioxane or toluene in the presence of Cs_2CO_3 or K_2CO_3 as the base). This transformation allows the synthesis of di-, tri- and tetrasubstituted pyrroles in 30–99% yield and tolerates a number of functional groups, including esters, ethers, alkyl halides, alkenes, heterocycles and silyl groups (Scheme 39).

Another interesting synthetic approach toward highly substituted pyrroles comprises the stepwise double alkenylation of Boc-hydrazides with vinyl iodides followed by a thermal [3,3]-rearrangement and acid-catalyzed aromatization (Scheme 40)[270]. Both amidation steps are catalyzed by CuI/1,10-phenanthroline in the presence of Cs_2CO_3 in DMF.

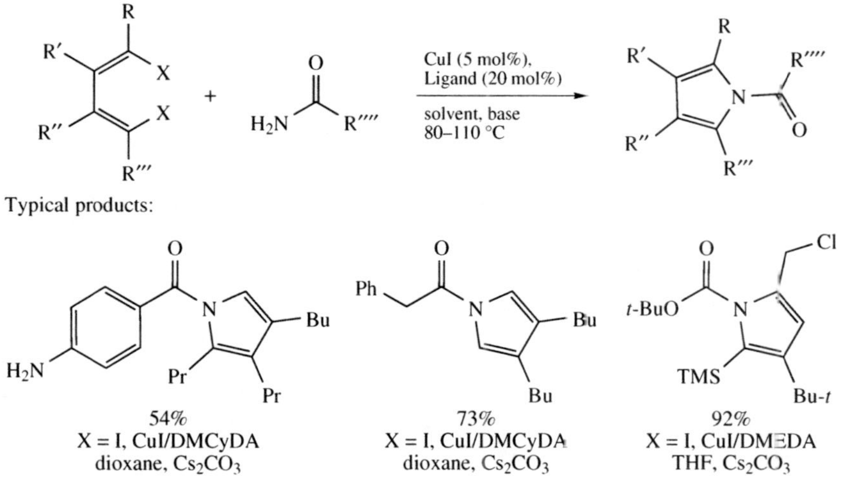

SCHEME 39

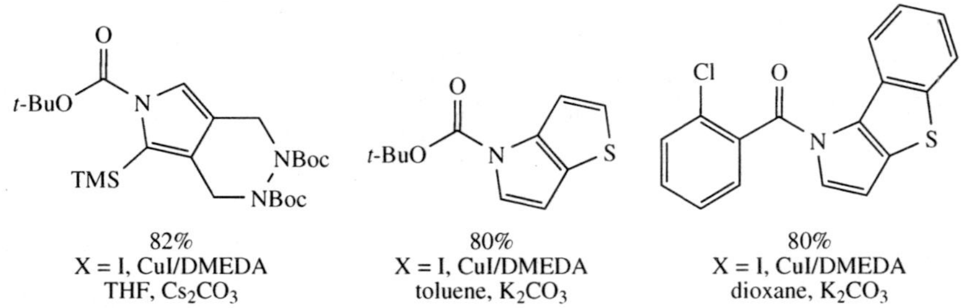

SCHEME 39. (*continued*)

The second amidation step generally requires longer reaction times and higher catalyst loading, but delivers the desired pyrroles in acceptable yields (sometimes as a mixture of Boc-protected and NH products). It was demonstrated that in a limited number of cases the whole sequence can be carried out in one pot without purification of the intermediates.

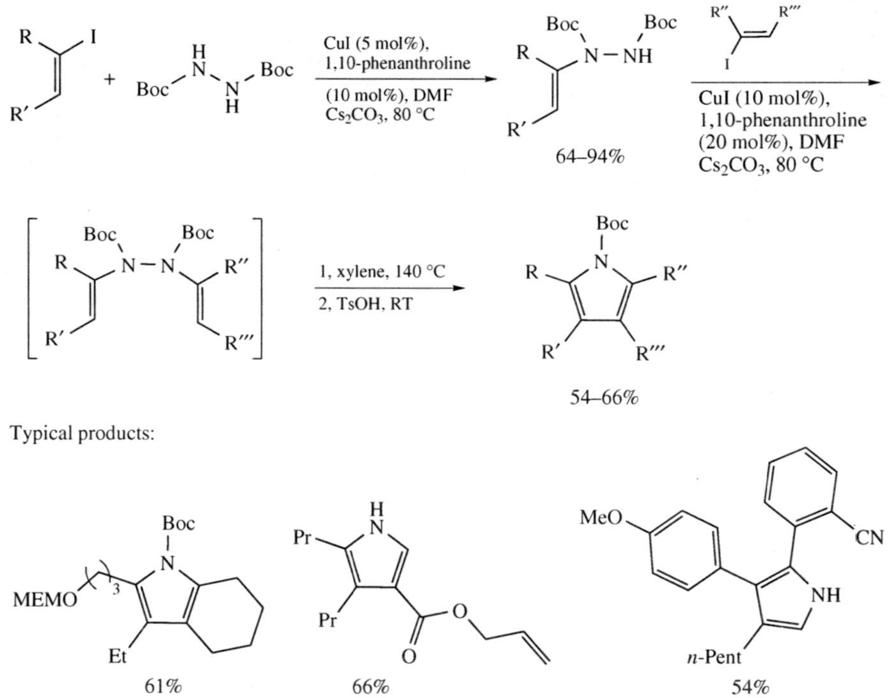

SCHEME 40

Alternatively, pyrroles can be prepared in good yield using a one-pot copper-catalyzed C−N coupling/hydroamidation sequence starting from bromo- or iodoenynes and *t*-butyl carbamate (equation 57)[271]. Following the same strategy, coupling of iodoenynes with Boc-hydrazides selectively delivers the corresponding pyrazoles in good to excellent yield (equation 58).

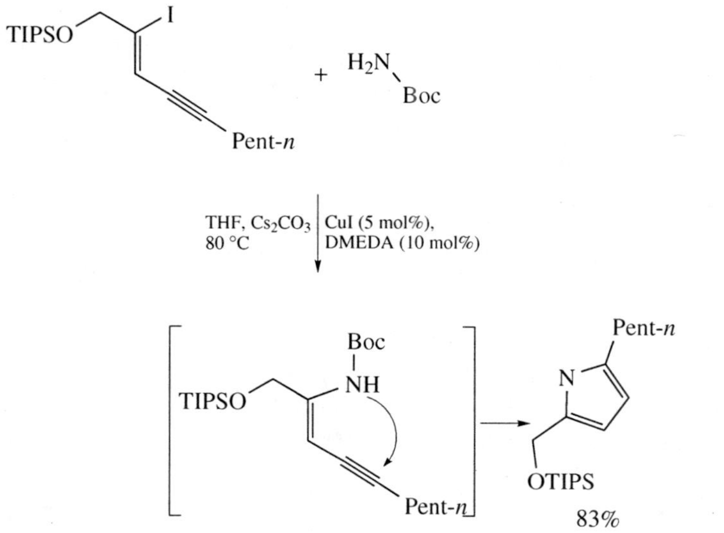

(57)

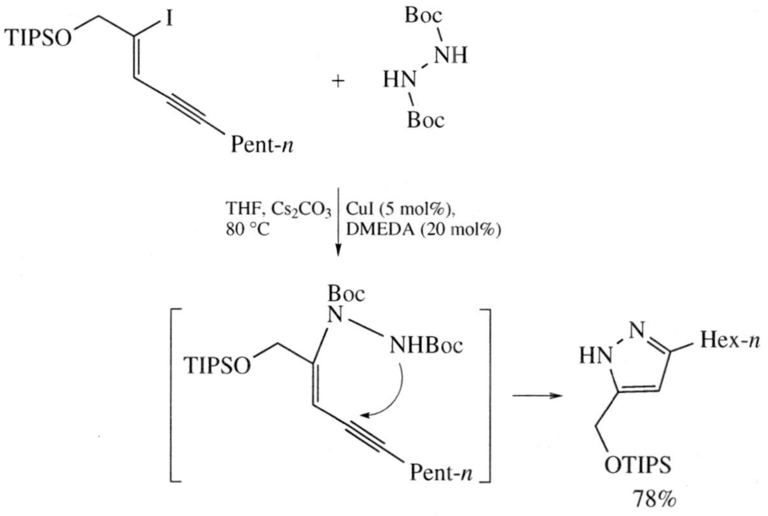

(58)

Lautens and coworkers used the CuI/CyDA catalyst for an intramolecular sequence starting from *gem*-dibromovinyl compounds to prepare differently substituted imidazindolones (Scheme 41)[272]. Although the double cyclization proceeds efficiently with a variety of functionalized and functionless natural amino acid derivatives, a significant loss of enantiopurity via facile epimerization of the chiral center may take place, apparently, at the product-forming step.

Syntheses of substituted oxazoles[273, 274] and tetrahydropyrazines[275] using domino reactions with Goldberg-type amidation in the first step were also reported.

f. Miscellaneous N-nucleophiles. Very recently, synthesis of aryl azides from aryl iodides or bromides was described (equation 59)[276–278]. This transformation is catalyzed

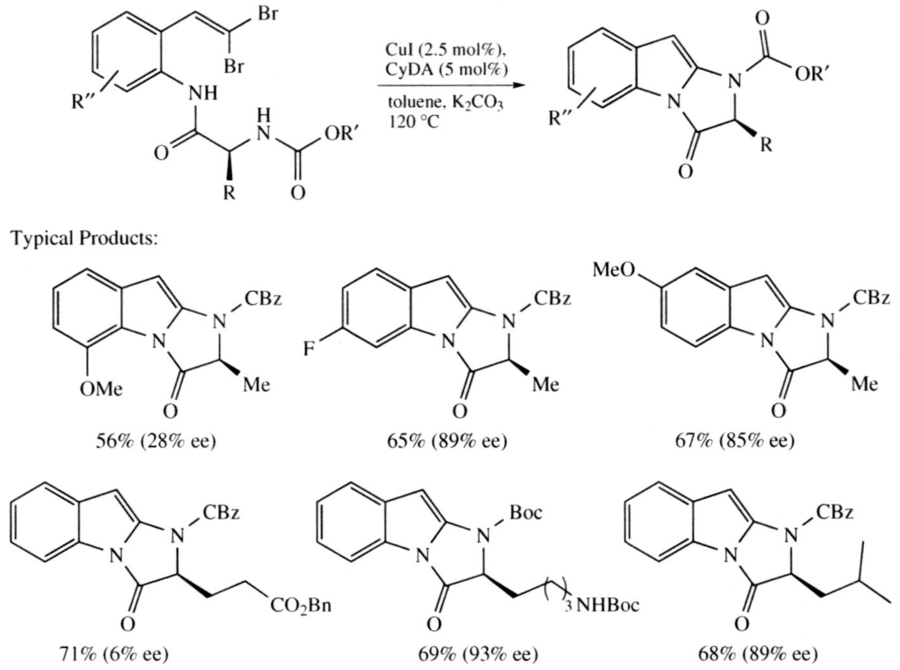

SCHEME 41

by the CuI/DMCyDA or CuI/L-proline combinations and proceeds in aqueous DMSO or EtOH in the presence or absence of a base at mild temperatures.

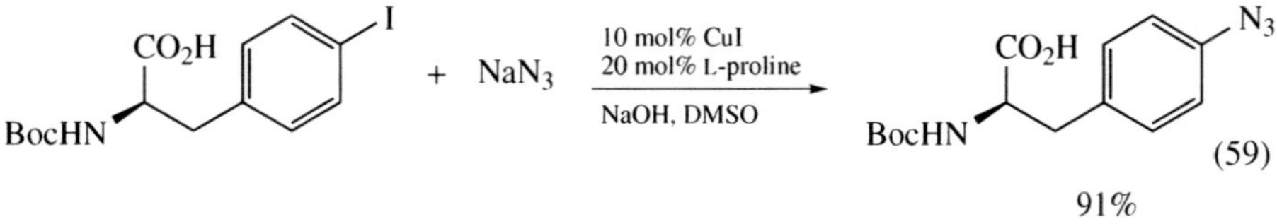

(59)

Remarkably, stereodefined vinyl azides can also be prepared using this mild method without notable isomerization[278].

g. Coupling of P-nucleophiles with organic halides. Taking advantage of the progress made over recent years in Ullmann chemistry, two groups independently suggested a variant of a CuI-catalyzed carbon–phosphorus bond-forming reaction using aryl or vinyl iodides and secondary phosphines[279, 280].

Optimization studies led to the conclusion that 5–10% of CuI can promote the coupling between functionalized aryl iodides and a slight excess of aromatic secondary phosphines in toluene under 'ligand-less' conditions. Apparently, the starting secondary and the resulting tertiary phosphines can stabilize and solubilize catalytically active copper species. Note that although other bases such as K_3PO_4 and K_2CO_3 can be applied in the phosphine synthesis, the selected cesium carbonate showed better performance and reproducibility, apparently, due to a greater solubility of cesium phosphides[281] in toluene (Scheme 42).

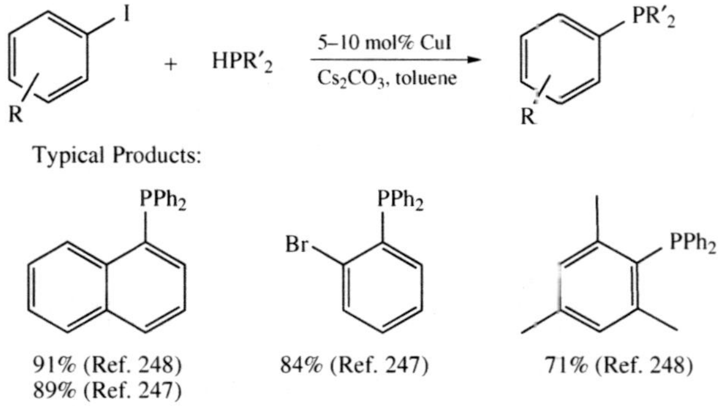

SCHEME 42

Although the 'ligand-less' protocol provides a satisfactory solution to the synthesis of asymmetrically substituted aromatic tertiary phosphines, coupling of aryl iodides with aliphatic secondary phosphines is ineffective at the expense of deiodination of the starting materials. If the target is dialkylaryl phosphines, the use of excessive DMEDA was proven beneficial (Scheme 43)[279]. Under the ligand-assisted conditions, the reaction is more tolerant to the steric bulk of the starting aryl iodides and to the presence of certain functional groups (cyano, ester, amino, methoxy etc.), but is still sensitive to the electronic and steric properties of the starting phosphines, i.e. diarylphosphines are more reactive than their dialkyl analogues. Similarly, bulkier aliphatic coupling partners (e.g. t-Bu_2PH) react slowly, if at all.

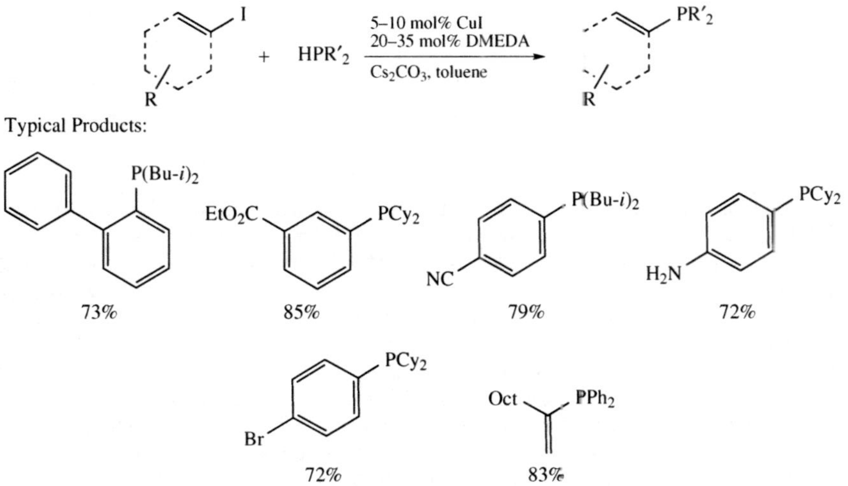

SCHEME 43

Although aryl bromides are significantly less reactive in these reactions, a slightly modified protocol was applied for the synthesis of a limited library of chiral phosphinoxazoline (PHOX) ligands starting from (2-bromophenyl)oxazolines and aromatic or aliphatic secondary phosphines (Scheme 44)[282,283]. The products were obtained in 53–93% yield, including the large-scale reactions.

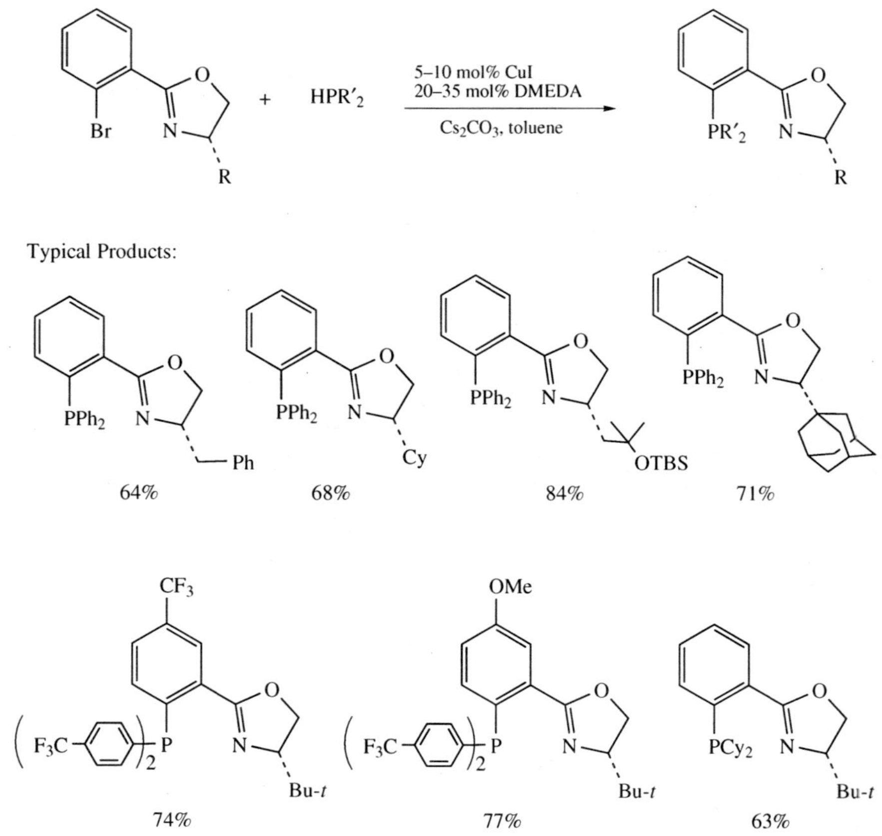

SCHEME 44

The method is also suitable for the synthesis of aryl and alkenyl phosphonates. These valuable substances may be synthesized from hydrolytically stable dibutyl phosphite and the corresponding aryl or vinyl iodides (vinyl bromides are equally reactive in this case) using the earlier described CuI/DMEDA catalyst[279]. It is noteworthy that the reaction proceeds with complete retention of the double-bond geometry of the precursors (Scheme 45).

Alternatively, construction of a C−P bond may be accomplished using CuI/pyrrolidine-2-phosphonic acid[107] or CuI/L-proline[284] combinations. Both catalysts operate under mild reaction conditions and give products in good yields. Reported examples include coupling of aryl iodides and bromides with aromatic and aliphatic phosphites, secondary phosphine oxides, monoalkylated phosphinic acids and even hypophosphites. For the last two cases, a milder base such as DMAP was employed instead of Cs_2CO_3 (Scheme 46).

3. Copper-catalyzed halogen exchange reactions

Halogen exchange was observed as a side reaction in several copper-mediated transformations[87–89]. It was found later that bromine–iodine exchange at aromatic[285–287],

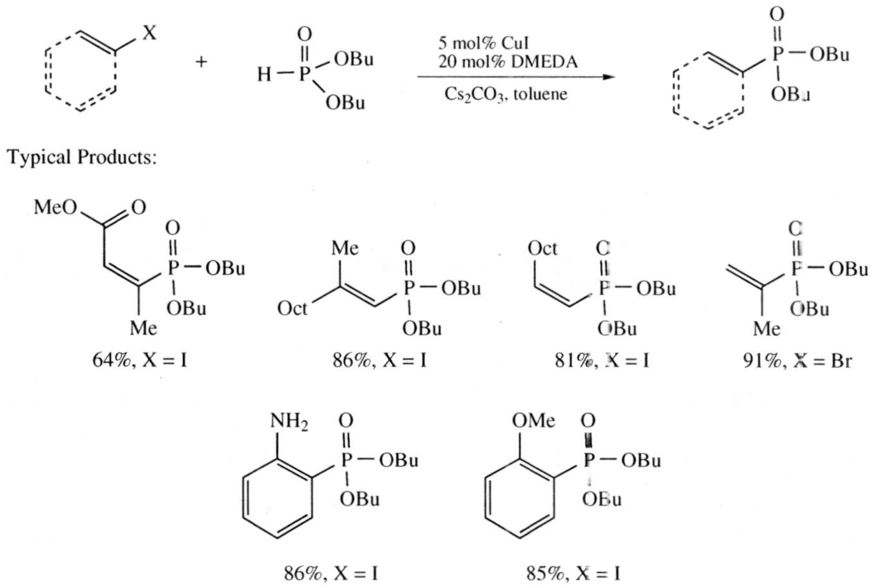

SCHEME 45

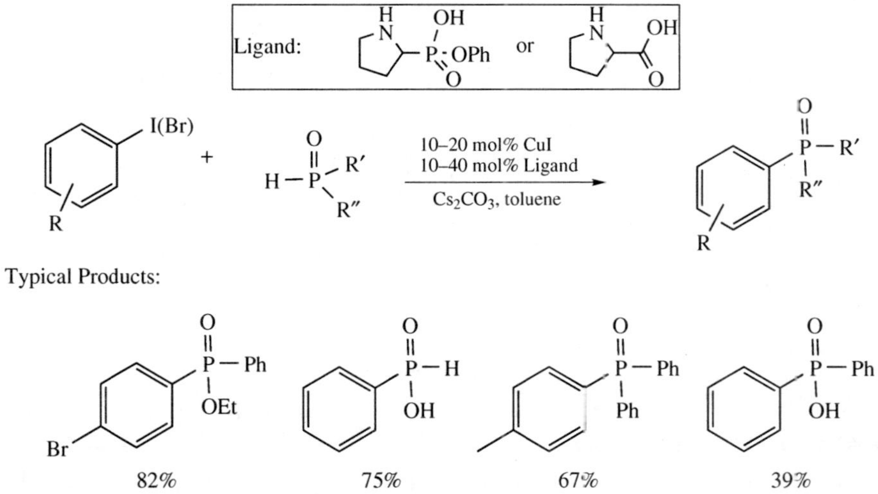

SCHEME 46

olefinic[286] or acetylenic[288] carbons can be carried out in the presence of stoichiometric (or higher) copper(I) iodide in HMPT or neat at 120 °C (equations 60 and 61). The transformation proceeds with complete retention of configuration. However, the yields are hardly predictable and vary at the expense of the competitive debromination reaction[286].

(60)

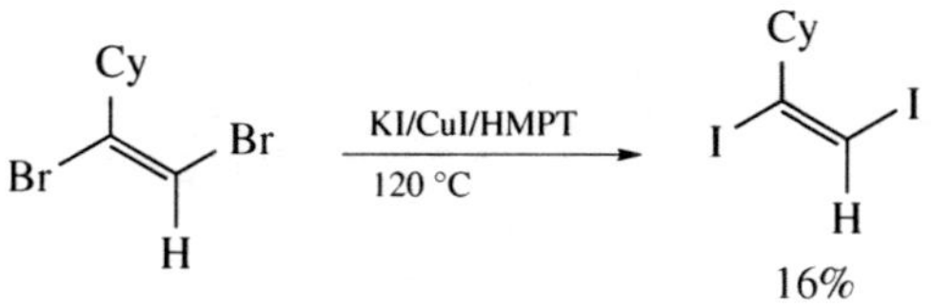

(61)

This valuable transformation was significantly improved with the disclosure of ligand-assisted protocols[289]. It was demonstrated that in the presence of DMCyDA, the reaction takes place in the presence of only 5 mol% of CuI in dioxane at 110 °C. The scope includes aryl, heteroaryl and vinyl bromides and the functional group compatibility is spectacular—heterocycles, esters, amines, carboxylic acids, hydroxyl groups etc. are perfectly tolerated (Scheme 47).

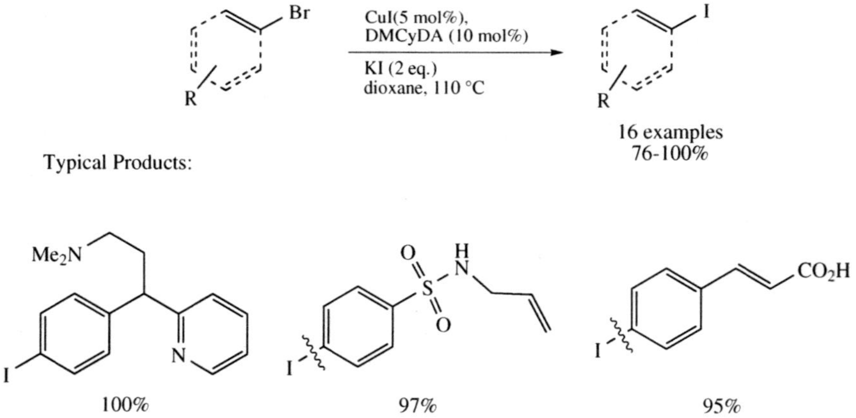

SCHEME 47

4. Mechanistic considerations

The exact mechanism of ligand-assisted copper-catalyzed cross-coupling reactions has not been established so far. However, the very similar reactivity trends, such as the reversed uncatalyzed S_NAr reactivity order of aryl halides (Ar-I > Ar-Br $\gg$ Ar-Cl), and the slightly positive effect of electron-withdrawing substituents on the reaction rates found in all carbon–heteroatom bond-forming transformations, imply a common mechanistic sequence.

Previous works suggested a copper-assisted S_NAr-type mechanism (Schemes 1 and 4)[6,78,92,158] to explain the observed reactivity and the catalytic effect of copper. However, the Ullmann reactions described in this section can also proceed via an oxidative addition/reductive elimination mechanism driven by Cu(I)/Cu(III) interconversion.

The involvement of Cu(I) and Cu(III) intermediates in the Ullmann chemistry was first proposed by Cohen and coworkers[290,291] and has been acknowledged[217,228,229,231,250,292] after detailed mechanistic studies on cross-coupling reactions catalyzed by palladium(0), which is isoelectronic to Cu(I), were obtained.

The key steps, according to this mechanism are: i) transmetallation of the Cu(I) precursor with the nucleophile (ligand exchange); ii) oxidative addition of the catalytically active Cu(I) species to aryl halide (carbon–halogen bond cleavage); and iii) reductive elimination (carbon–heteroatom bond formation) (Scheme 48).

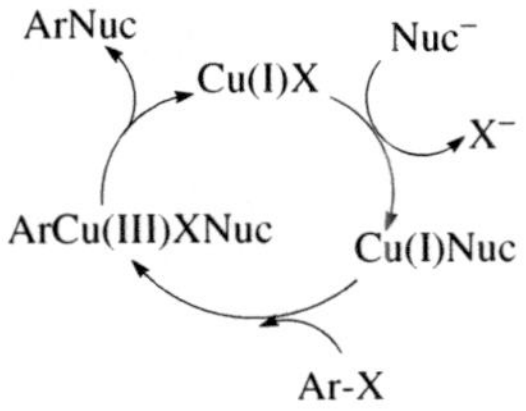

SCHEME 48

Whereas structurally characterized Cu(III) complexes are not very common[293–295], and thus raise questions concerning their participation in the catalytic cycle, the reactivity order in the aryl halides series as well as the higher reaction rates of electron-deficient substrates are in prefect agreement with a rate-limiting carbon–halogen cleavage (oxidative addition) step[296].

According to this mechanism, chelating amine ligands can play an additional pivotal role besides trivial solubilization of the catalytic species. It is reasonable that hard amine donors stabilize the hard Cu(III) intermediate and thus facilitate the rate limiting oxidative addition[297,298].

However, since no direct evidence has been reported so far, none of the mechanisms can be ruled out.

C. Non-Ullmann Substrates in Copper-mediated Carbon–Heteroatom Coupling Reactions

1. Iodonium salts

a. Coupling of O-nucleophiles with iodonium salts. Diaryliodonium salts have been known for a long time to participate in S_NAr-type reactions[299]. Among others, alcoholysis of diaryliodonium salts is facilitated by the presence of catalytic copper bronze[300,301] or even better by cupric acetate[302]. The nucleophilic displacement usually proceeds at relatively mild reaction conditions (i.e. milder than the classical Ullmann condensations but comparable to the ligand-assisted reactions) and produce diaryl or aryl alkyl ethers in modest yield (equation 62). The need to synthesize the diaryliodonium compounds obviously limits a wide utilization of the method[303].

b. Coupling of N-nucleophiles with iodonium salts. Similarly, diaryliodonium salts react with *N*-nucleophiles. For example, $Cu(OAc)_2$-catalyzed reactions of diphenyliodonium-

2-carboxylate with aniline or sulfonamide result in the formation of aminated or amidated products in moderate yield (equations 63 and 64)[302].

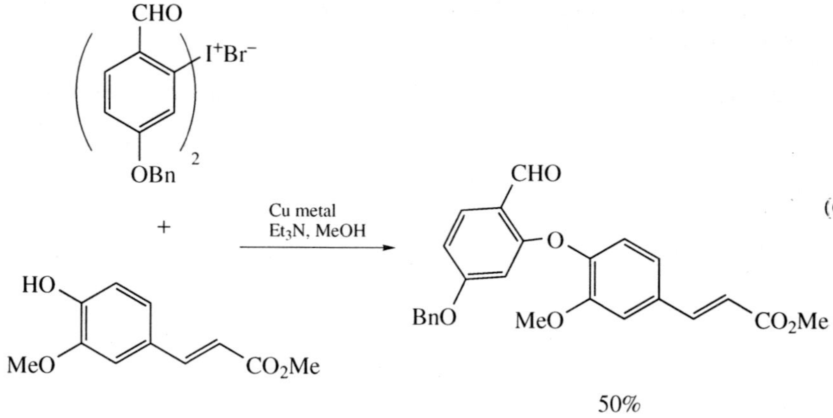

(62)

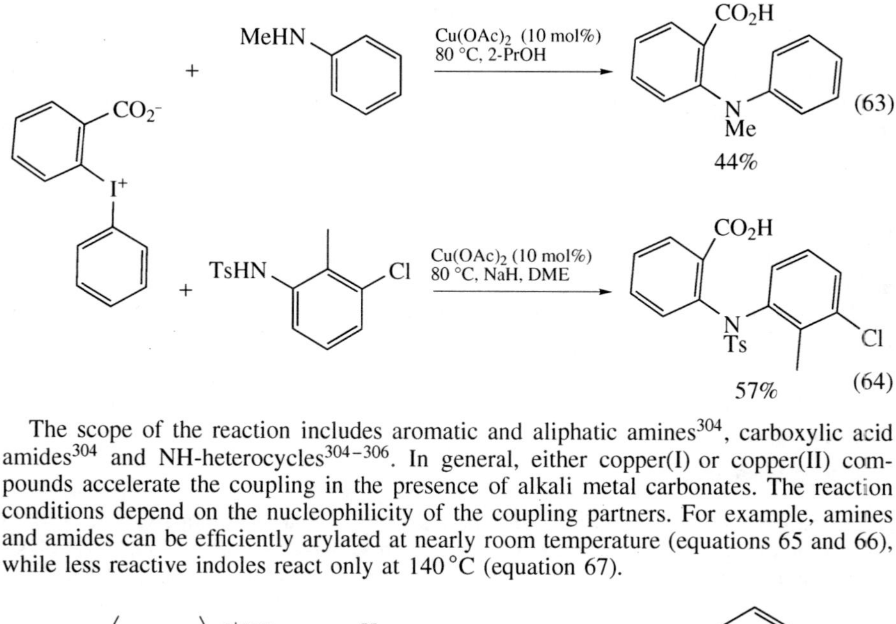

(63)

(64)

The scope of the reaction includes aromatic and aliphatic amines[304], carboxylic acid amides[304] and NH-heterocycles[304–306]. In general, either copper(I) or copper(II) compounds accelerate the coupling in the presence of alkali metal carbonates. The reaction conditions depend on the nucleophilicity of the coupling partners. For example, amines and amides can be efficiently arylated at nearly room temperature (equations 65 and 66), while less reactive indoles react only at 140 °C (equation 67).

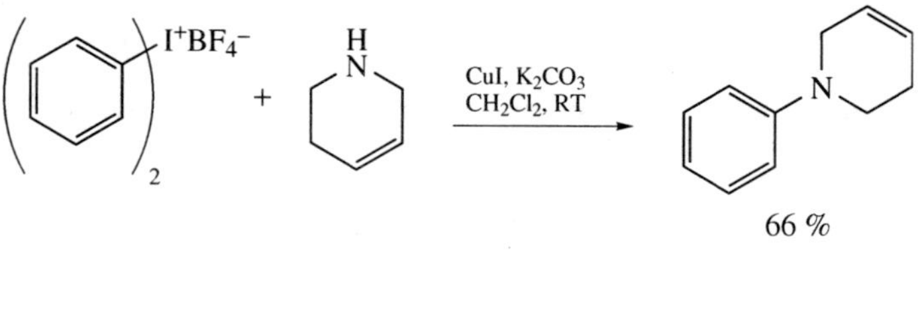

(65)

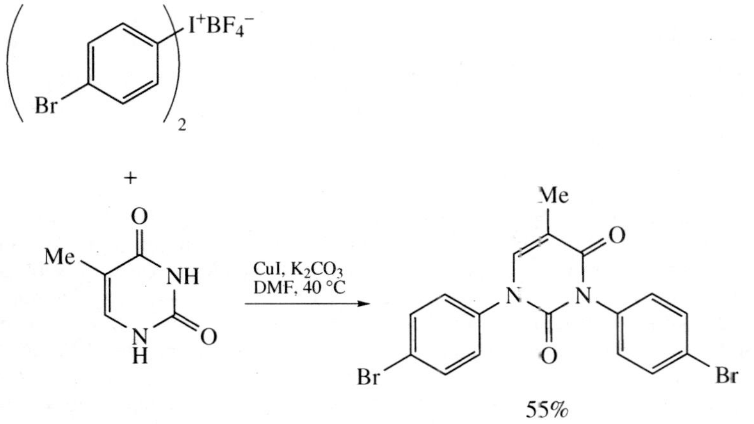

(66)

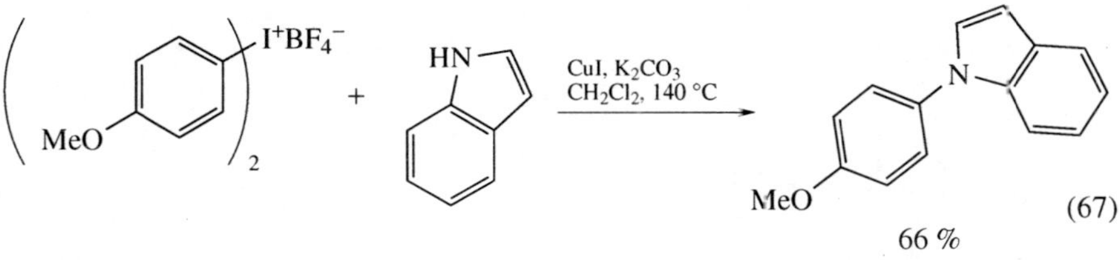

(67)

c. Coupling of P-nucleophiles with iodonium salts. An example of CuI/TMEDA-catalyzed coupling of *H*-phosphonates with vinyliodonium salts was reported (Scheme 49)[307]. The method allows the synthesis of differently substituted vinylphosphonates in

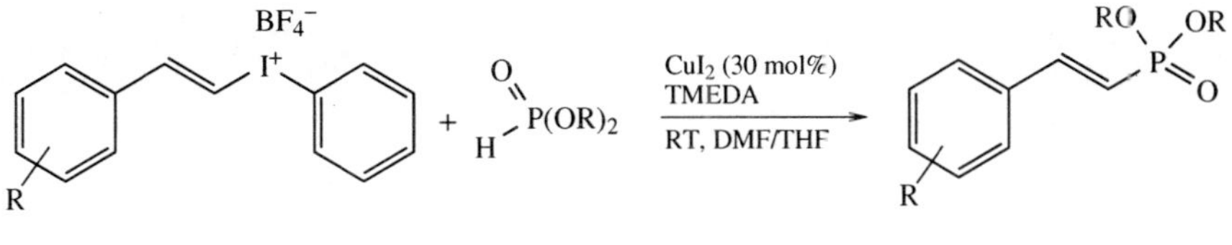

Typical Products:

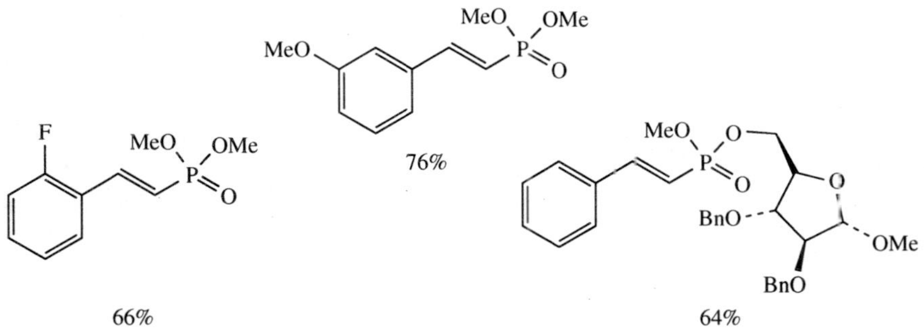

SCHEME 49

good yields under very mild conditions (30 mol% of CuI_2 in THF/DMF mixture at room temperature). Here again, the mild nonbasic reaction conditions compensate for the required laborious synthesis of the starting materials.

2. Organobismuth compounds

a. Coupling of O-nucleophiles with organobismuth compounds. *O*-phenylation of alcohols can be accomplished by reaction with pentavalent triphenylbismuth diacetate in the presence of simple copper salts[308–313]. Usually, the reaction proceeds under very mild reaction conditions (ambient temperature or very mild heating) in a variety of common organic solvents (toluene, CH_2Cl_2, THF etc.) and can be promoted by different copper salts, although soluble $Cu(OAc)_2$ was found to be superior in many cases. For example, addition of copper diacetate or metallic copper in a catalytic amount (10 mol%) to the mixture of a phenol and triarylbismuth diacetate results in a room-temperature formation of the corresponding *O*-phenylated product in poor to good yield depending on the steric demand of the starting materials (equations 68–70)[314–317].

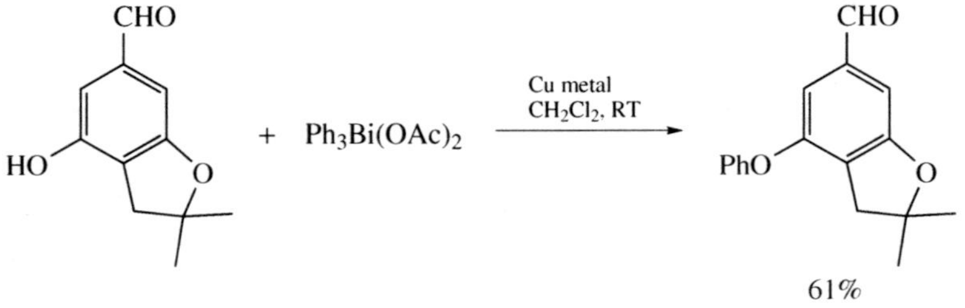

(68)

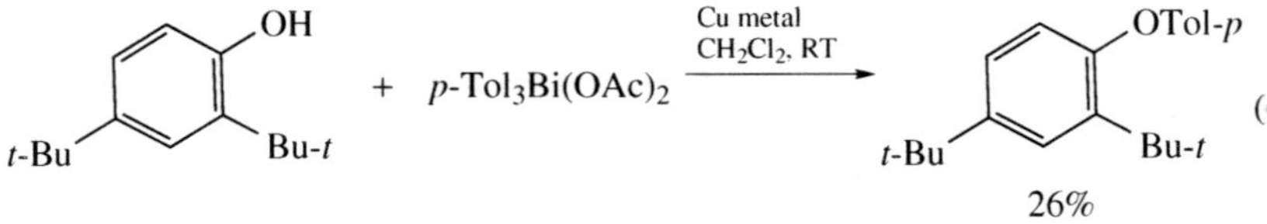

(69)

Arylation of aliphatic alcohols can also be achieved using arylbismuth reagents under copper-catalyzed conditions. The yields are strongly dependent on the steric properties of the starting materials. They are 60–90% for secondary and primary, but negligible for tertiary alcohols[308, 309].

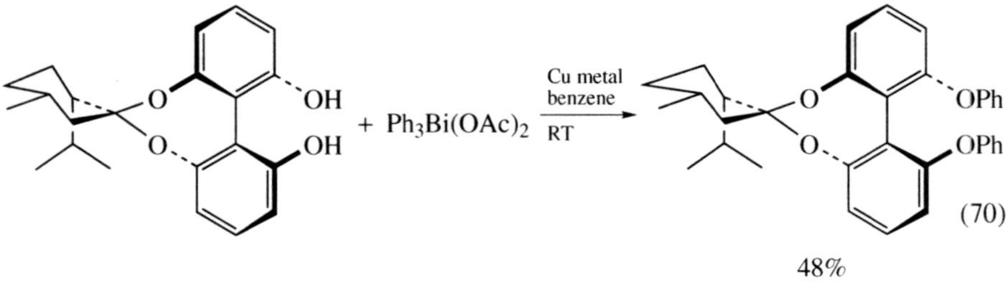

(70)

Brunner and coworkers developed an asymmetric version of this method[318,319]. Moderate ee values (up to 50%) were obtained in attempted desymmetrization of *meso*-diols using triphenylbismuth diacetate in the presence of $Cu(OAc)_2$ and chiral pyridineoxazoline ligand (equation 71). It is noteworthy that the chemical yield in the reaction was significantly lower due to the ligand participation.

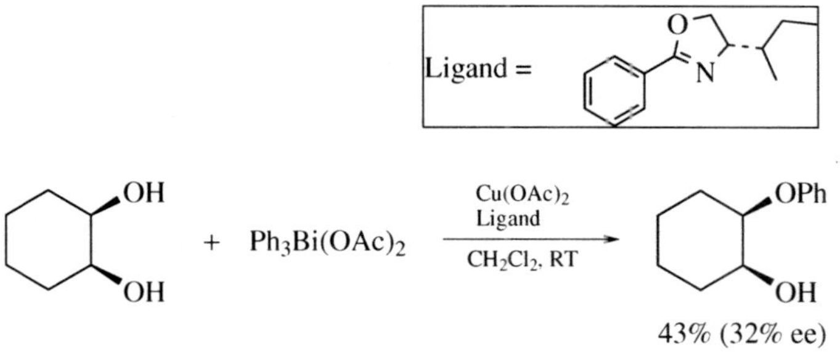

(71)

In principle, the method is not limited to phenyl-containing organobismuth reagents. The use of more sophisticated reagents is exemplified in the large-scale synthesis of an ascomycin derivative (Scheme 50)[320].

Enols were also found as suitable coupling partners under these conditions[314,321].

b. Coupling of N-nucleophiles with organobismuth compounds. Similarly, pentavalent triphenylbismuth reagents can be used as effective coupling partners in *N*-arylation reactions in the presence of copper catalysts[309,322]. A variety of aromatic[323–326], heteroaromatic (equation 72)[327,328], aliphatic cyclic (equation 73)[329] and acyclic amines[322,325,326] as well as azole derivatives (equation 74)[330] can be arylated in the presence of catalytic (less than 10 mol%) $Cu(OAc)_2$, $Cu(OPiv)_2$ or metallic copper in CH_2Cl_2 at room temperature and in a very short time. The yields are generally good and reaction times are relatively short with all sorts of nucleophiles but, naturally, depend on the steric properties of the starting materials[331].

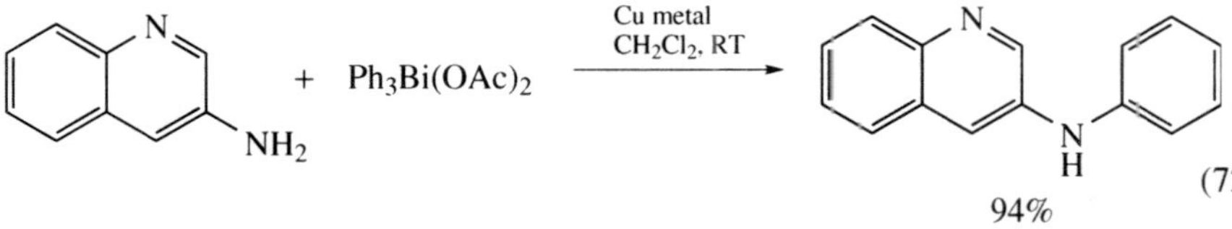

(72)

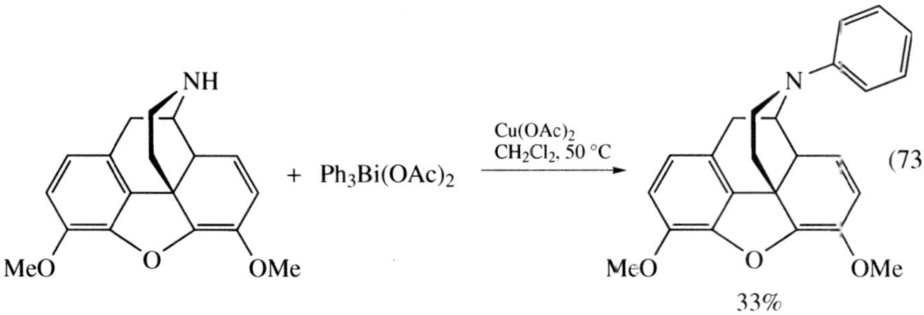

(73)

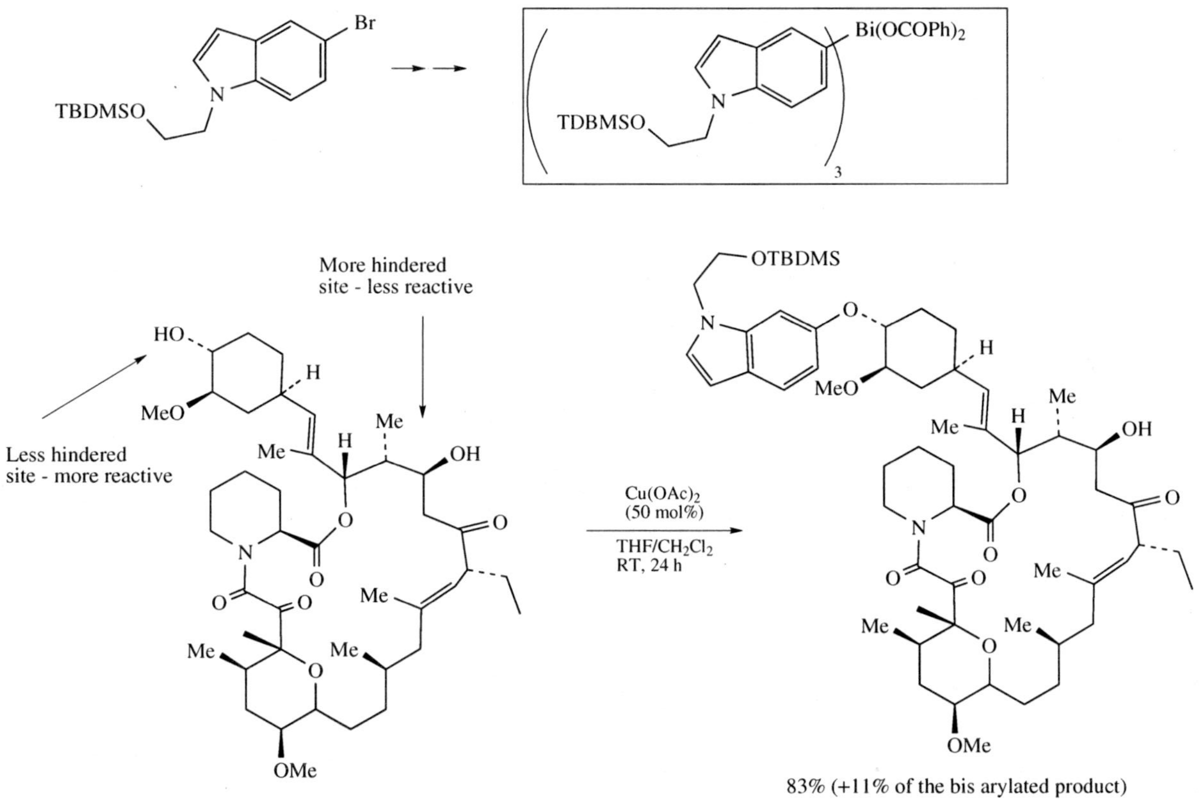
Br
TBDMSO
Bi(OCOPh)₂
TDBMSO
3
More hindered
site - less reactive
Less hindered
site - more reactive
HO
MeO
H
Me
OH
O
N
OMe
Cu(OAc)₂
(50 mol%)
THF/CH₂Cl₂
RT, 24 h
OTBDMS
83% (+11% of the bis arylated product)

SCHEME 50

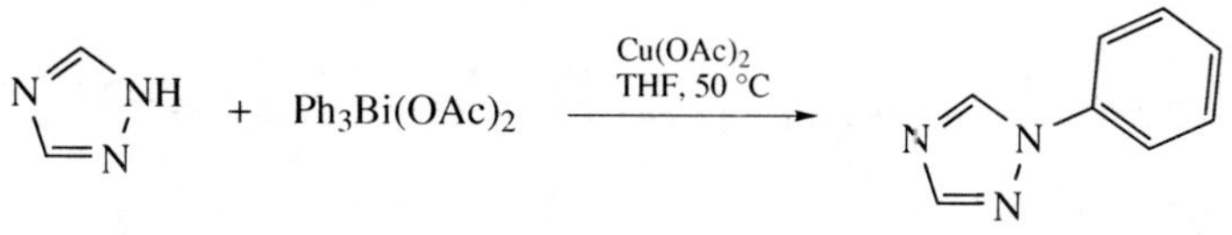

60% (74)

α-Amino acids do not participate directly in arylation reactions with triarylbismuth reagents, but can be arylated under the standard reaction conditions after being converted to esters[332–334]. Good selectivity toward monoarylated product is usually observed due to the sensitivity of the reaction to steric effects (equation 75).

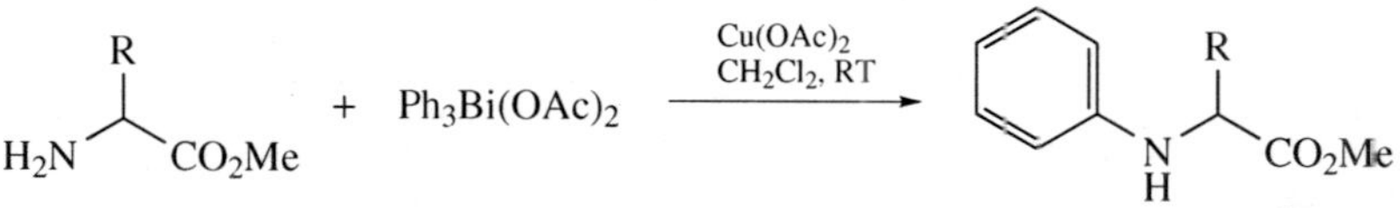

(75)

R = (*R*)-Ph, 92%
R = (*S*)-*t*-Bu, 86%

N-nucleophiles such as imines, oximes, amides, guanidines and semicarbazones are generally difficult coupling partners and normally do not react under these reaction conditions[322].

3. Organolead compounds

a. Coupling of O-nucleophiles with organolead compounds. Tetraphenyllead[335] and diphenyllead dichloride[336] were reported as moderately efficient aryl transfer reagents for aliphatic alcohols (equation 76) or phenols in the presence of catalytic $Cu(OAc)_2$. When alcohol is used as the solvent, only one phenyl group is usually transferred. The overall phenylating activity of the diphenyllead dichloride is significantly lower than that of tetraphenyllead and, apparently, this explains the low yields of the ethers per mole of phenylating agent.

$$Ph_4Pb + t\text{-BuOH} \xrightarrow[\text{neat, RT}]{Cu(OAc)_2} PhOBu\text{-}t \qquad (76)$$

b. Coupling of N-nucleophiles with organolead compounds. Neither tetraphenyllead nor diphenyllead dichloride are suitable arylating agents for *N*-nucleophiles. However, it was found that phenyllead triacetate reacts smoothly with anilines and simple aliphatic amines in the presence of 10 mol% of $Cu(OAc)_2$ in CH_2Cl_2 at room temperature producing the corresponding arylated products[337–339]. Good yields were obtained with various substituted anilines. The best yields were observed for anilines substituted with electron-donating groups, whereas electron-withdrawing substituents such as nitro or ester strongly inhibited the reaction. The steric hindrance had only moderate influence on the reaction rate and yield. For example, *o,o'*-substituted arylamines could be synthesized in acceptable yields using this method (equations 77 and 78)[338, 339]. The steric effect is more pronounced in reactions of less reactive aliphatic amines (1° > 2° and 3° alkylamines are essentially inactive) (equation 79)[337, 338]. Although the method is not limited to phenyllead compounds, the arsenal of functionalized reagents is relatively poor and contains only simple aryl derivatives[340, 341].

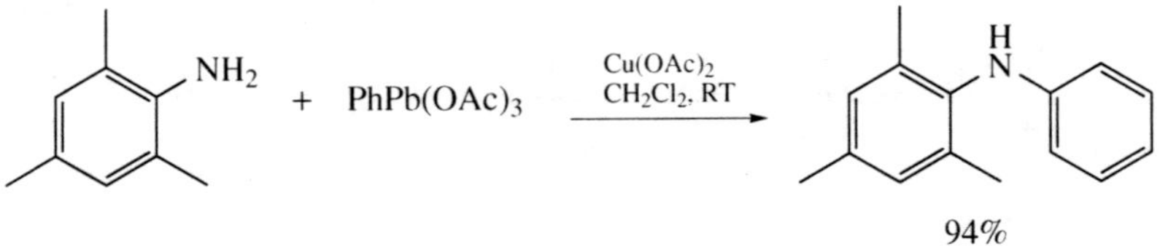

(77)

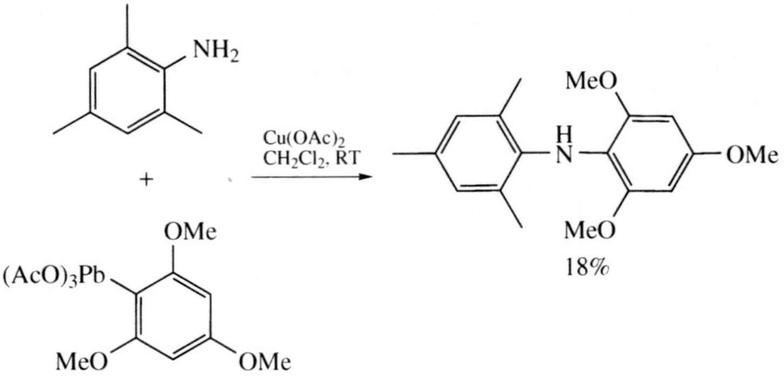

(78)

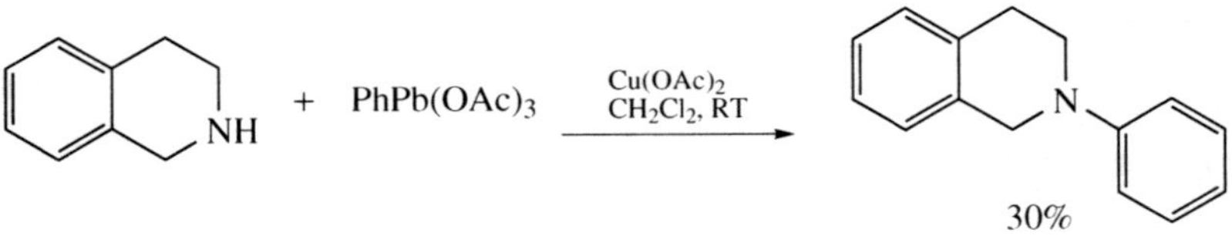

(79)

In addition, hydrazones[337], carboxylic acid amides[342,343], imides[343] and sulfonamides[342,343] are suitable candidates for the $Cu(OAc)_2$-catalyzed arylation with aryllead triacetates. Base assistance (usually, sodium hydride) and mild heating were often required to achieve the desired conversion (Scheme 51).

N-Arylation of azoles (imidazoles, benzimidazoles, triazoles, benzotriazoles, indazoles, pyrazoles, indoles) can be accomplished at room temperature in good yield and selectivity[344–347]. Interestingly, arylation of asymmetrically substituted imidazoles proceeds in excellent regioselectivity toward the formation of the *N*-1 isomer (equation 80).

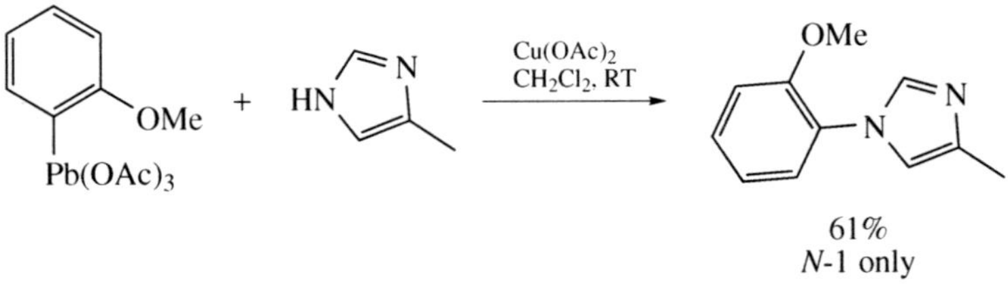

(80)

4. Organoboron compounds

a. Coupling of O- and S-nucleophiles with organoboron compounds. The use of phenylboronic acid as the aryl donor in copper-mediated arylation of amines and phenols started with independent reports by Chan[348] and Evans[349] and their coworkers in 1998.

Chan and coworkers[348] demonstrated that the method is applicable to a wide range of nucleophilic reaction partners, including phenols under very mild conditions, i.e. room

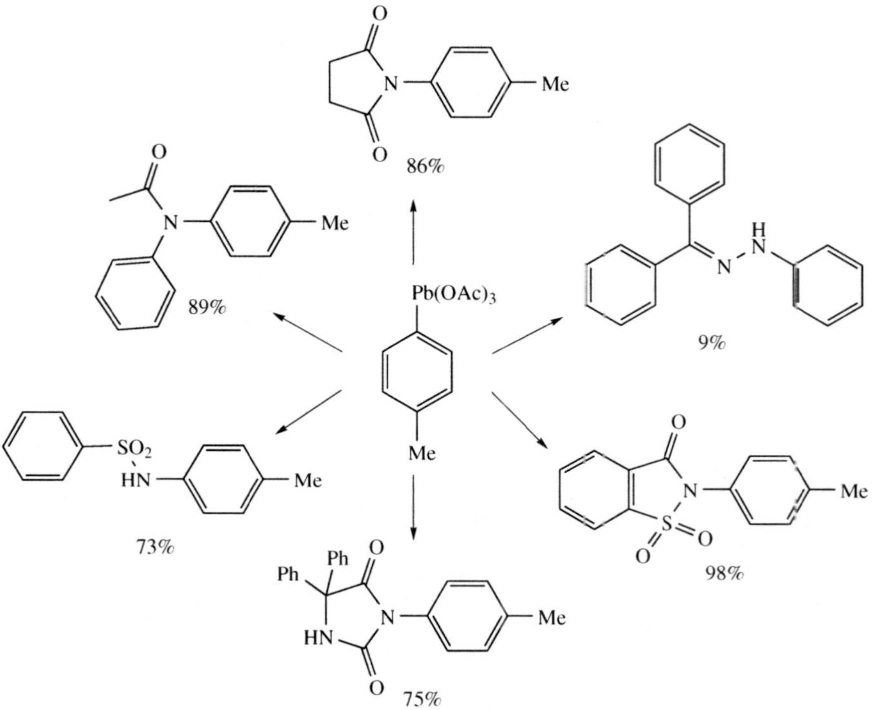

Reaction conditions: $Cu(OAc)_2$ (10 mol%), CH_2Cl_2-DMF, NaH (>1 eq.), 50–90 °C.

SCHEME 51

temperature, amine as a base and a stoichiometric amount of $Cu(OAc)_2$ (equation 81). However, the yields of *O*-arylated products were moderate at the expense of hydrolysis of the arylboronic acids, as well as by competition of the by-produced phenol for *O*-arylation.

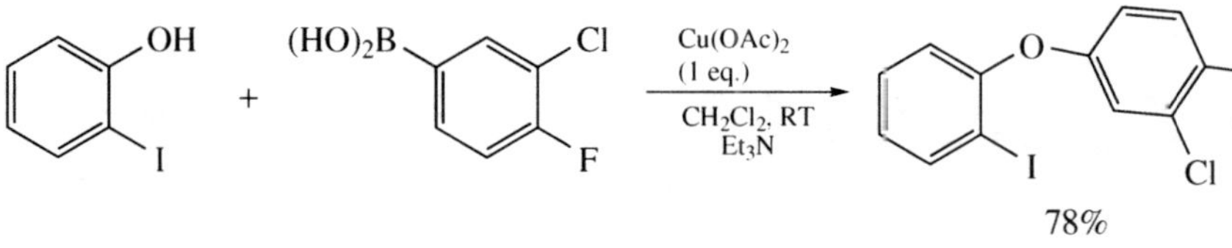

(81)

Further studies performed by Evans and coworkers[349] revealed that side reactions took place even under anhydrous conditions, indicating that water was generated from the phenylboronic acid upon triarylboroxine formation. Therefore, optimized reaction conditions suggested addition of powdered 4 Å molecular sieves in order to trap the water and to suppress the undesired processes[349]. Additionally, it was discovered that exposure of the reaction to atmospheric oxygen leads to better results (equation 82). Finally, while methylene chloride proved to be the optimal solvent, the reaction is tolerant of a wide range of solvents from acetonitrile to toluene, affording acceptable yields of the desired diaryl ether.

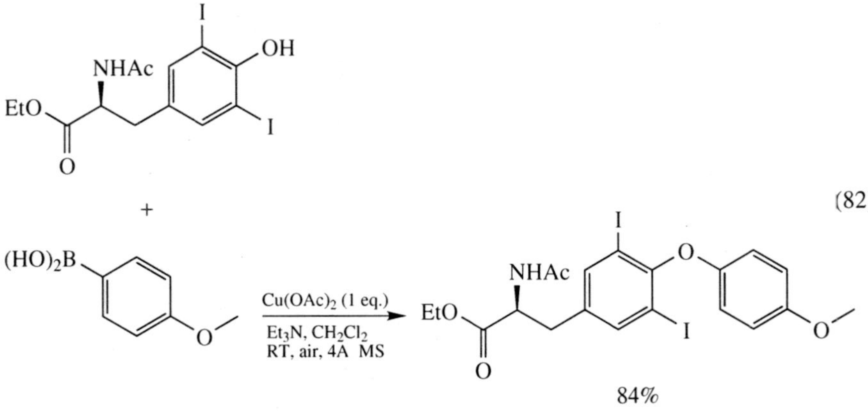

(82)

During scope and limitation studies, a number of structurally and electronically diverse substrates were evaluated[349–353]. While yields were generally good, electron-rich phenols underwent arylation more efficiently. The reaction is tolerant of proximal substitution, as *ortho*-substituted phenols were arylated in good yield. With several exceptions, the reaction is also general with respect to the arylboronic acid component, although yields are poor with *ortho*-heteroatom-substituted boronic acids. In the more sluggish reactions, a second equivalent of boronic acid was required to achieve reasonable yields. Vinylboronic acids were also reported as suitable coupling partners[351,354].

Formation of symmetrical diaryl ethers was achieved via *in situ* generation of phenols and their one-pot *O*-arylation with aryl boronic acids under the standard conditions[355,356]. For example, Prakash, Olah and coworkers reported on the regioselective reaction of arylboronic acids (1 equivalent) with hydrogen peroxide (0.25 equivalent) to form the corresponding phenols, followed by reaction with the remaining arylboronic acid in the presence of $Cu(OAc)_2$ and Et_3N in methylene chloride (equation 83)[356].

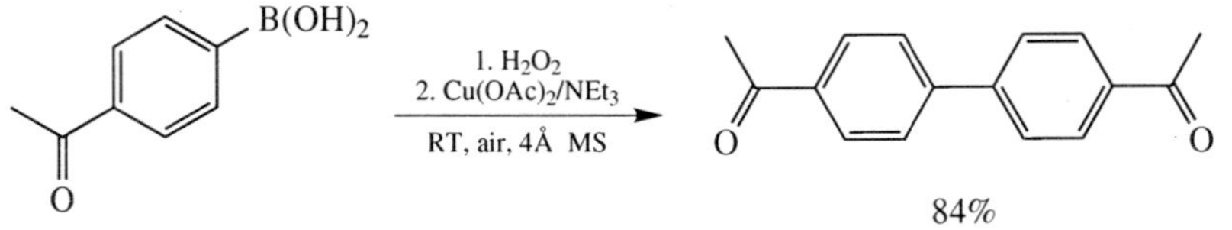

(83)

A novel route to *N*-aryloxyphthalimides via the copper-mediated cross-coupling of *N*-hydroxyphthalimide and phenylboronic acids was reported by Sharpless and coworkers[357]. It was shown that in the presence of pyridine and molecular sieves, different copper sources ($Cu(OAc)_2$, $Cu(OTf)_2$, CuCl and $CuBr{\bullet}SMe_2$) catalyze the reactions under ambient air atmosphere in 1,2-dichloroethane (equation 84).

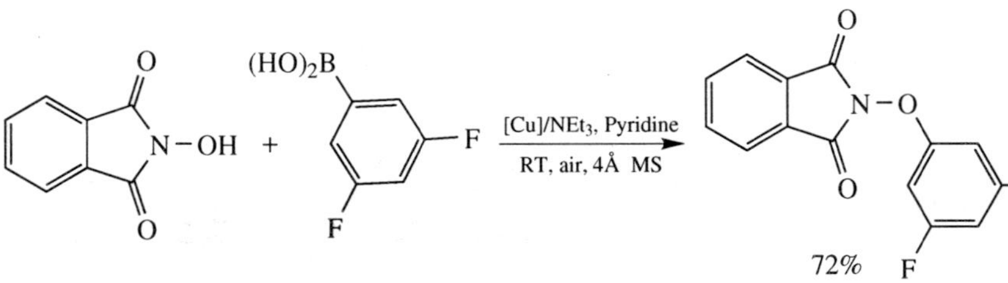

(84)

The inter-[358,359] and intramolecular[360,361] stoichiometric C—O bond-forming reactions were adopted for the syntheses of naturally occurring compounds. For example, total synthesis of (*S*,*S*)-isodityrosine (equation 85)[358] and macrocyclic metalloproteinase inhibitors (equation 86)[360] were recently reported.

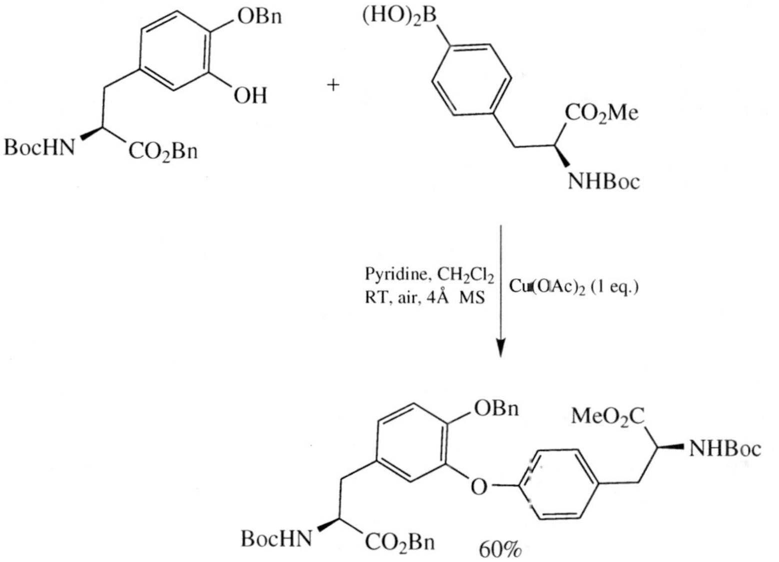

(85)

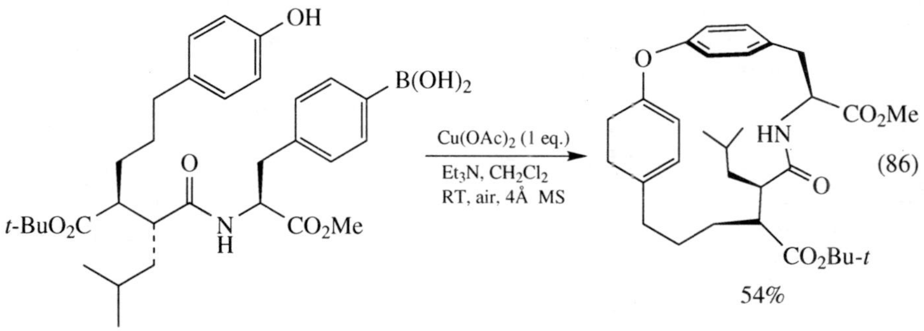

(86)

More recent publications demonstrated that reactions described in this section can be performed in a catalytic fashion[351,362,363]. Under catalytic conditions, the reaction utilizes only 10–40 mol% of $Cu(OAc)_2$, atmosphere of oxygen or additional co-oxidant such as TEMPO or pyridinium *N*-oxide to deliver products in modest yield.

A carbon–sulfur bond-forming version of this transformation was reported by Guy and coworkers[364]. They demonstrated that alkyl and aryl thiols couple with differently substituted arylboronic acids in the presence of an overstoichiometric amount of $Cu(OAc)_2$, 4 Å MS and pyridine as a base in DMF at reflux temperature. Despite the harsh reaction conditions, the functional group compatibility is acceptable and products are usually obtained in moderate to good yield (Scheme 52).

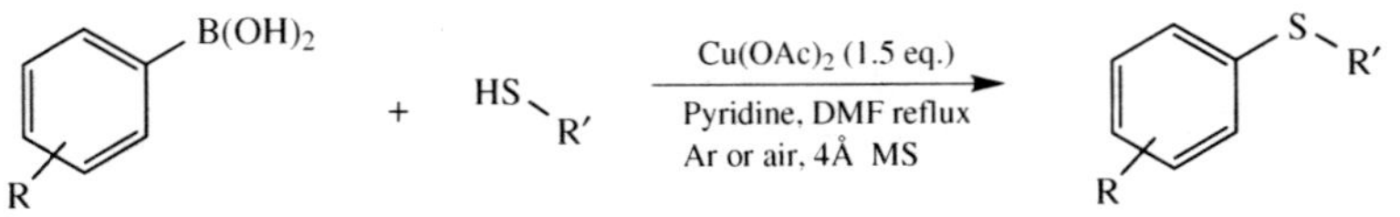

Typical Products:

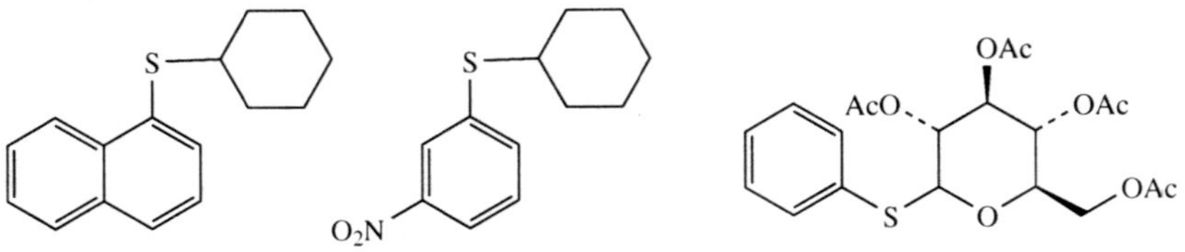

41% 65% 51%

SCHEME 52

b. Coupling of N-nucleophiles with organoboron compounds. The first examples of copper(II)-mediated *N*-arylation by arylboronic acids were published by Chan and coworkers[348] and by Lam and coworkers[365]. They demonstrated a robust and convenient method to arylate a wide range of NH-containing nucleophiles (e.g. aliphatic amines, anilines, amides, imides, ureas, sulfonamides, carbamates and heterocycles) by excessive arylboronic acids in the presence of stoichiometric $Cu(OAc)_2$ and a base (pyridine or triethylamine) at ambient temperature (Scheme 53). The choice of the amine base plays a critical role in determining the yield of the reaction, but no clear substrate–base–yield relationship has emerged from the results.

As a wide variety of electronically diverse boronic acids are available, Cundy and Forsyth explored the generality of Chan's procedure by examining the efficiency of the *N*-arylation by substrates of varied NH basicity/nucleophilicity[366]. The results did not reveal any obvious trends for either electron-rich or electron-poor boronic acids in their reactivity with the *NH* substrates.

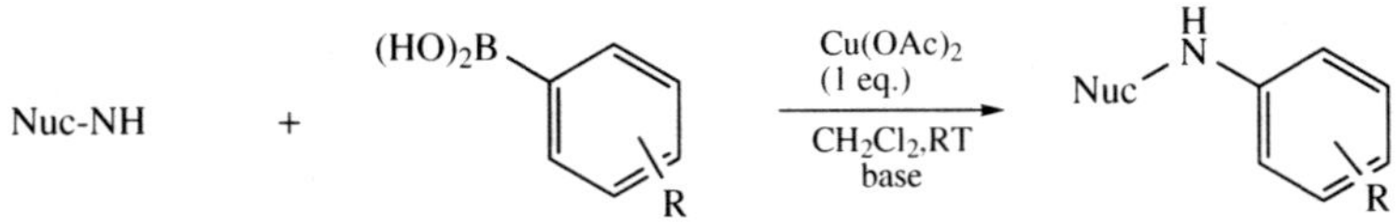

Typical Products:

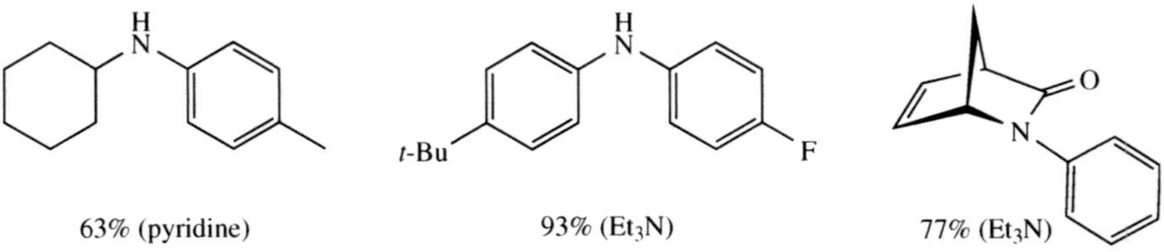

63% (pyridine) 93% (Et_3N) 77% (Et_3N)

SCHEME 53

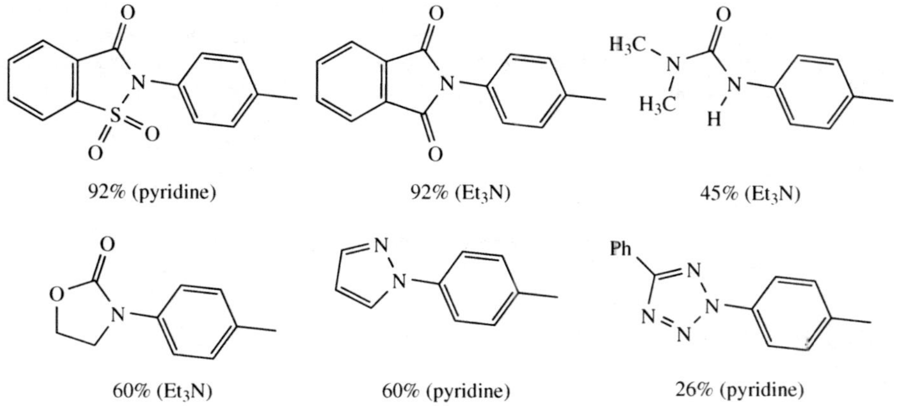

SCHEME 53. (*continued*)

Major improvements, however, were introduced after the discovery by Evans and coworkers that molecular sieves have a pronounced beneficial effect on the copper-promoted coupling of arylboronic acids[349]. The new reaction conditions were essentially identical to those described originally, but suggested addition of 4 Å MS to improve yields and reproducibility of the reactions. Subsequently, the improved Lam–Chan–Evans conditions were applied to the synthesis of various *N*-arylated compounds[367,368]. For example, Mederski and coworkers showed that a variety of 2-pyridones and some 3-pyridazinones can be coupled with different arylboronic acids in good yields (equation 87)[369].

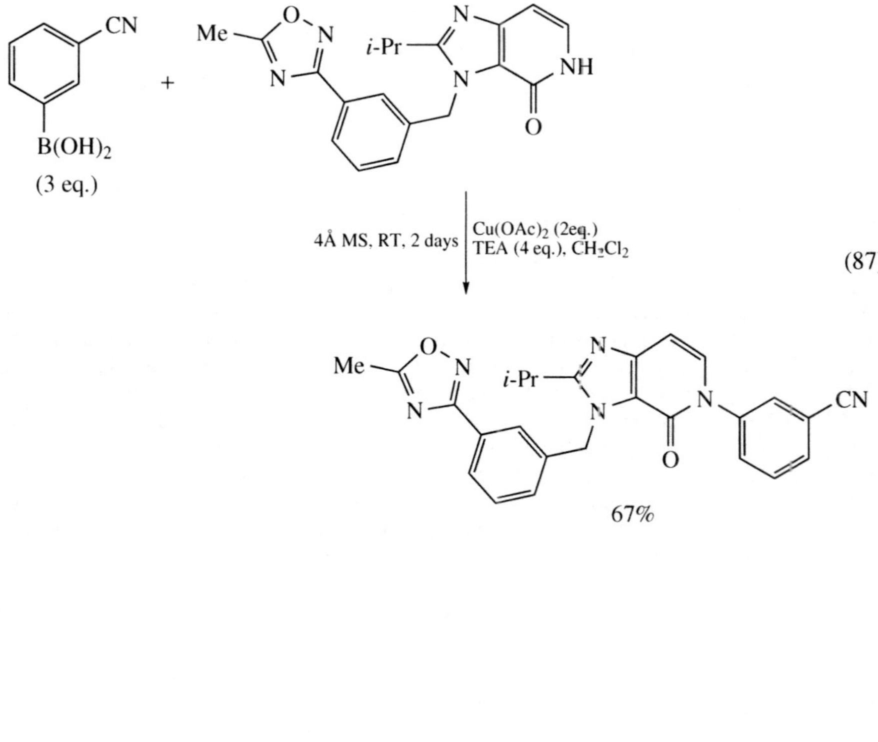

(87)

Srirangam and coworkers prepared *N*-arylpyrroles under Chan–Lam conditions from arylboronic acids and electron-deficient pyrroles[370]. This coupling reaction showed very good functional group compatibility (functional groups such as nitro, ether, ester, ketone, aldehyde, amide, bromo and iodo survived the reaction conditions), but was somewhat sensitive to the steric bulk of the starting materials. However, a necessary precursor toward AG3433, a MMP inhibitor, was synthesized in 93% yield using this reaction at the key step (equation 88).

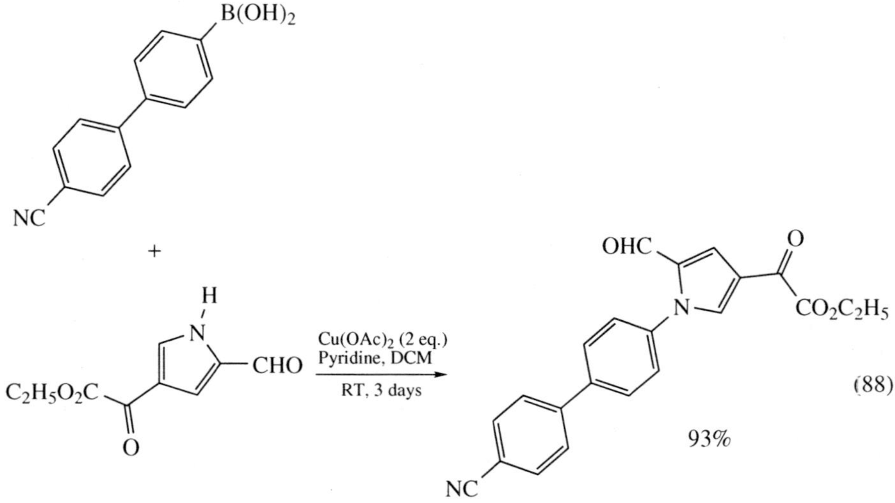

(88)

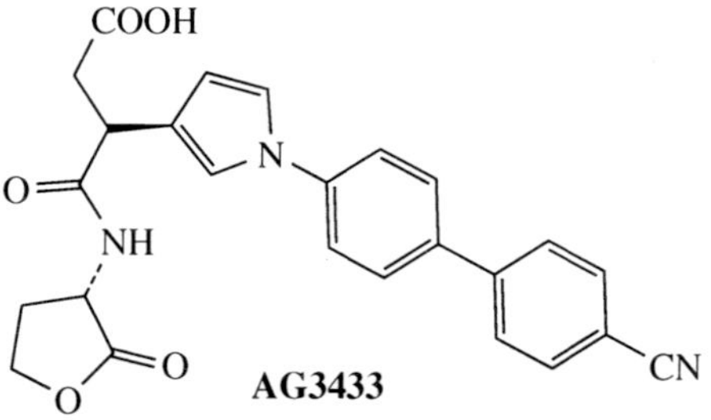

AG3433

α-Amino acids and their derivatives are important chemical building blocks in organic and biological chemistry. A reaction involving these building blocks should be mild enough not to cause epimerization of the chiral centers. Several groups demonstrated that *N*-arylation of enantiomerically pure α-amino acid derivatives can be achieved in 17–67% yield using Lam–Chan–Evans conditions, with little or no racemization (equations 89 and 90)[371,372]. The reaction may also be applied to solid supported substrates[371].

Gothelf and coworkers examined the ability of protected derivatives of nucleobases (thymine, uracil, cytosine, adenine and guanine) to undergo copper-mediated *N*-arylation and *N*-alkenylation with various boronic acids[373]. The reaction tolerated various substitution patterns on the arylboronic acids including *ortho*-substitution and electron-donating groups and formed products in good to high yields (Scheme 54). Vinylboronic acids were found to be acceptable coupling partners in these cases. *N*-arylation of unprotected nucleobases mediated by simple copper salts was also demonstrated[374].

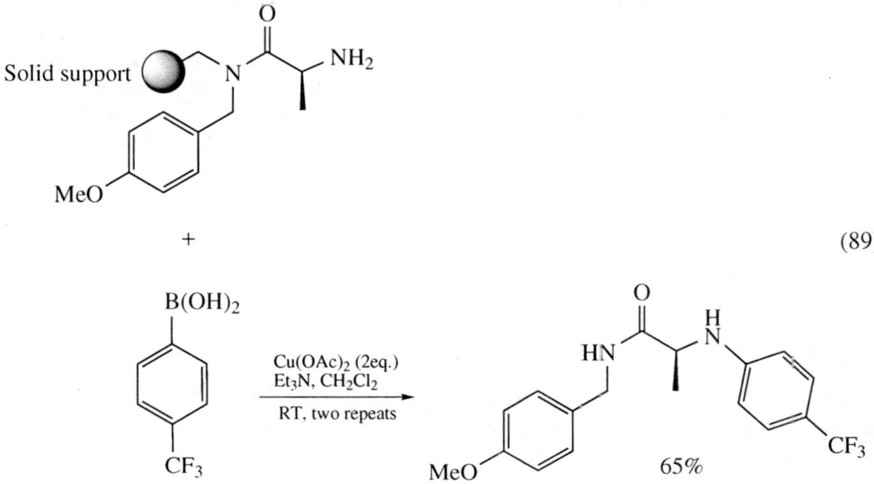

(89)

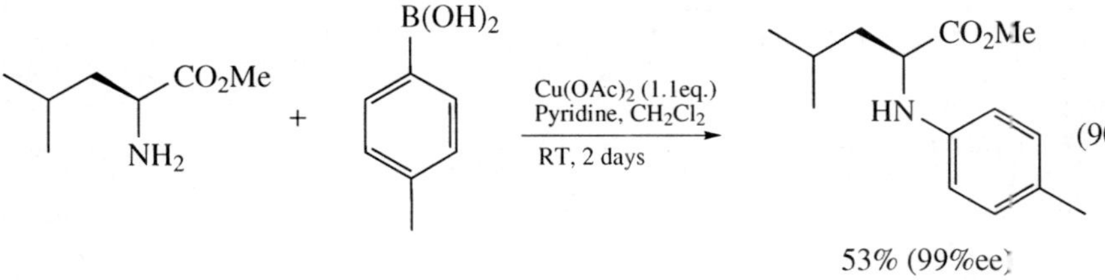

(90)

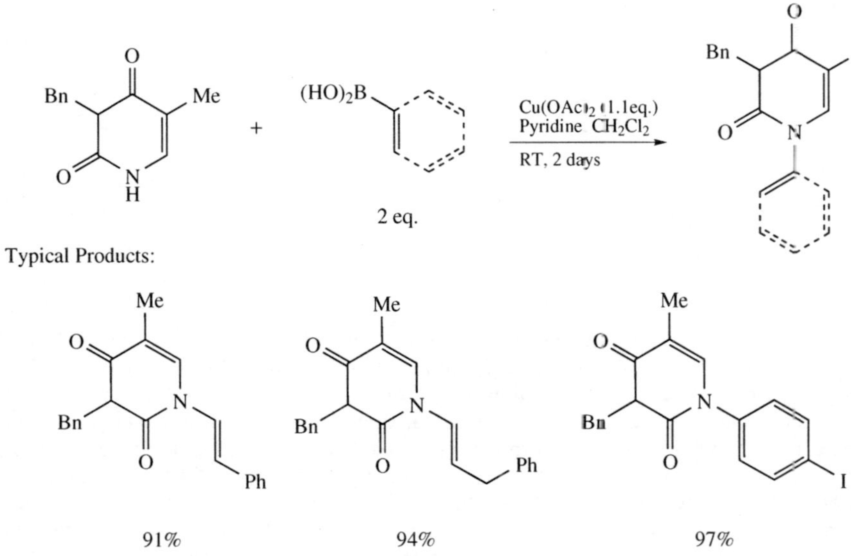

SCHEME 54

Coupling of arylboronic acids with imidazoles in the presence of $[Cu(OH)TMEDA]_2Cl_2$ complex was the first catalytic process described in the literature by Collman and coworkers[375–377]. They found that under oxygen atmosphere a mixture of 2 mmol of phenylboronic acid and 1 mmol of imidazoles react in dry dichloromethane and in the presence of only 0.1 mmol of $[Cu(OH)TMEDA]_2Cl_2$ to give phenylimidazole in 71% yield (equation 91). The presence of air or oxygen is necessary to obtain satisfactory yields of products under catalytic reaction conditions.

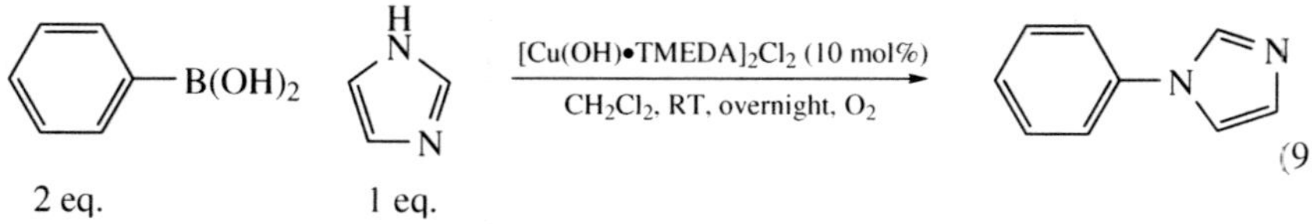

(91)

Buchwald and Antilla discovered and reported in 2001 that addition of a catalytic amount of myristic acid to the mixture of 1.0 equivalent of aniline, 1.5 equivalent of *p*-tolylboronic acid and 1.0 equivalent of 2,6-lutidine as a base in the presence of 5 mol% of $Cu(OAc)_2$ at ambient temperature was essential to achieve a complete conversion to *N*-arylated product[378]. This additive may operate by coordination to the copper center, thereby increasing the solubility of the catalyst. Both electron-poor and electron-rich anilines were found equally reactive under the reaction conditions. Sterically demanding anilines required higher catalyst loading (10 mol%) but were also efficiently arylated.

Differently substituted arylboronic acids were also tested and the results showed that substitution of the *para* position does not affect the reaction yields, but *ortho* substituents dramatically slow down the reaction rate. For example, relatively unhindered 2-methylphenylboronic acid coupled in only 50% yield.

The arylation reaction was also explored with alkylamines. Such reactions resulted in the formation of the desired products in moderate 50–64% yields (Scheme 55).

Yudin and coworkers applied Buchwald's protocol to the synthesis of *N*-arylated aziridines (equation 92)[379].

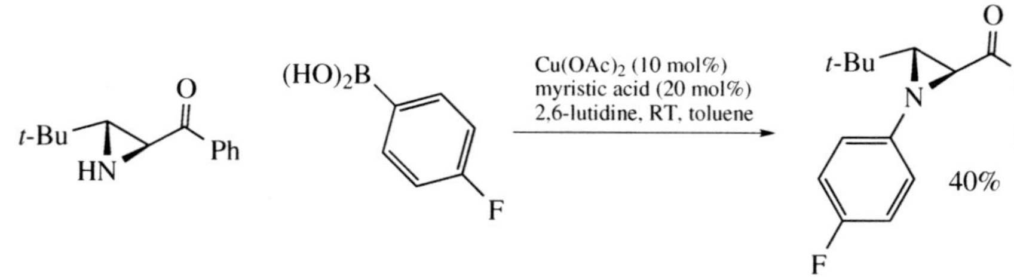

(92)

It was also established that the method is not limited to arylboronic acids. Several groups reported on the successful utilization of vinyl[380] and alkyl[381] boronic acid as coupling partners under similar reaction conditions.

Moessner and Bolm described the *N*-arylation reaction of sulfoximines with arylboronic acid[382]. Optimization experiments showed that the best results were achieved with 10 mol% of $Cu(OAc)_2$ and 2.3:1 equivalents of arylboronic acid to sulfoximine in methanol. To evaluate the substrate scope of this approach toward *N*-arylated sulfoximines, coupling of various arylboronic acids and different sulfoximines was investigated. Good to excellent (71–93%) yields were achieved with *para*-substituted boronic acids

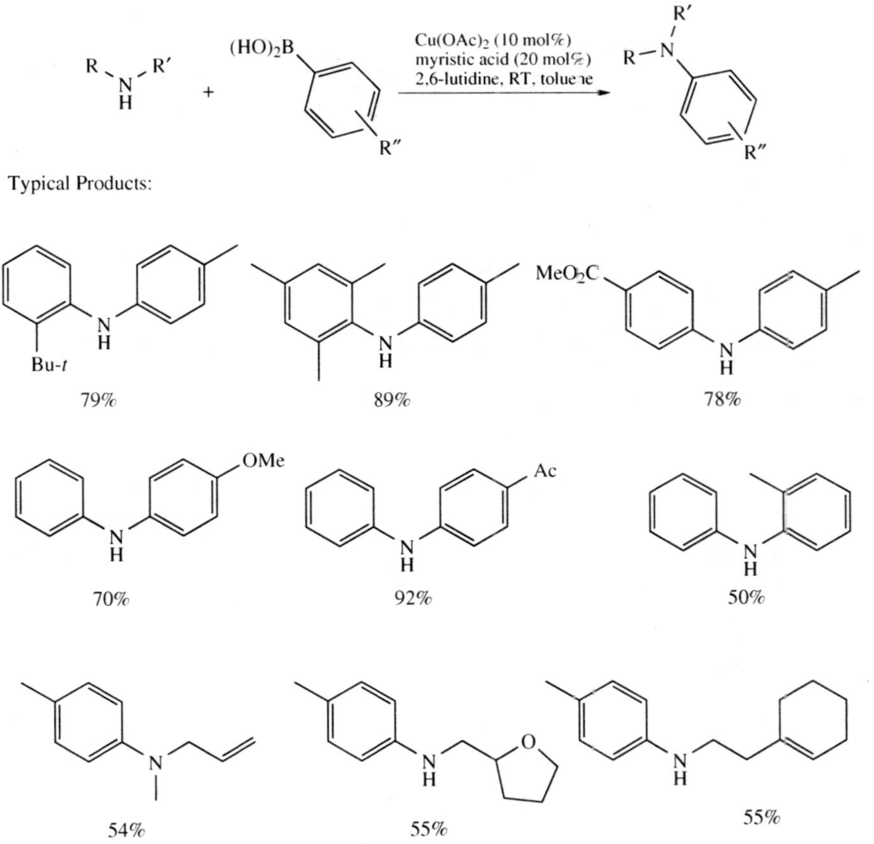

SCHEME 55

irrespectively of their electronic nature. The same conclusion is reached on reactions with *ortho*-substituted boronic acids, although in these cases the yields (62–91%) were slightly lower. Lower yields in most couplings of *ortho*-substituted boronic acids indicated the importance of steric effects and, indeed, whereas boronic acid with one *ortho* substituent were suitable substrates, overcrowded 2,4,6-trimethylphenylboronic acid failed to react (Scheme 56).

5. Organotin compounds

a. Coupling of O-nucleophiles with organotin compounds. Several very interesting and useful protocols for $Cu(OAc)_2$-promoted coupling of readily available stannanes with phenols have been described recently. For example, the reaction of tetravinyltin with differently substituted phenols takes place in acetonitrile under air at RT or with very mild heating (60 °C) in the presence of stoichiometric $Cu(OAc)_2$ and represents a general route toward aryl vinyl ethers (equation 93)[383]. Although the transformation requires the use of stoichiometric copper, the method is useful due to the excellent functional group

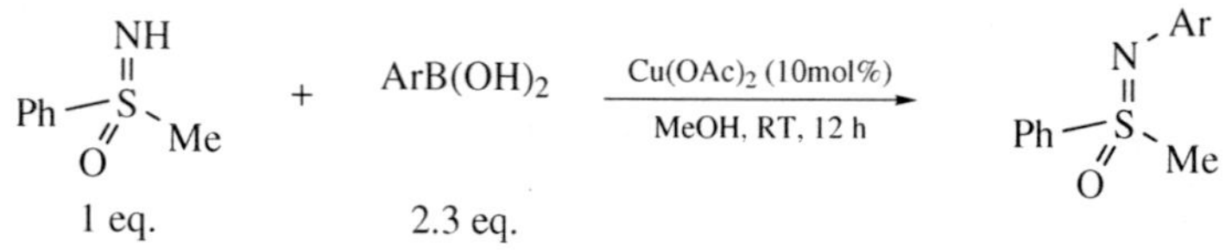

Typical Products:

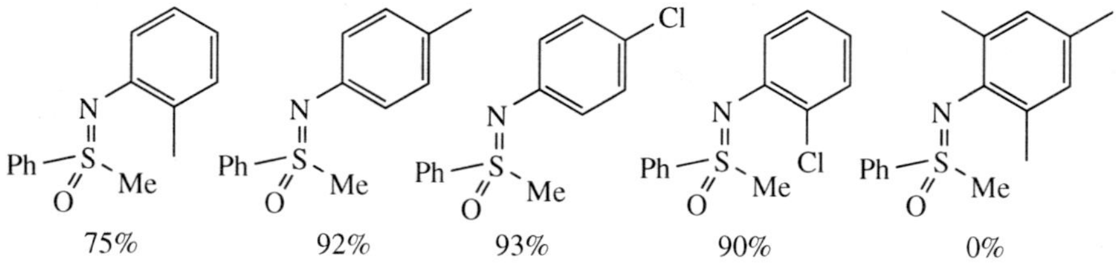

SCHEME 56

compatibility, mild conditions and generally high yields. Remarkably, tributyl(vinyl)tin showed incomparably slower reaction under the conditions described.

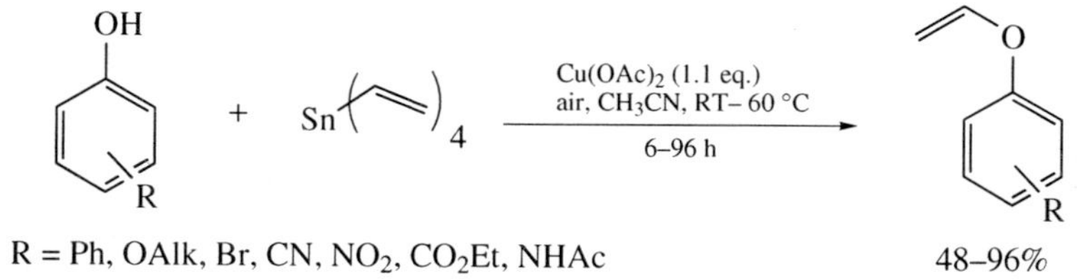

(93)

In situ prepared functionalized arylstannanes couple with differently substituted phenols and benzyl alcohols under essentially similar conditions (equation 94)[384]. However, this transformation is capricious and requires the use of excessive phenol or, alternatively, phenylstannane, in order to obtain reasonable conversions. Otherwise, the results are unsatisfactory because of the reduction of the arylstannanes to arenes. The use of DMAP as a base and molecular sieves somewhat improves the situation.

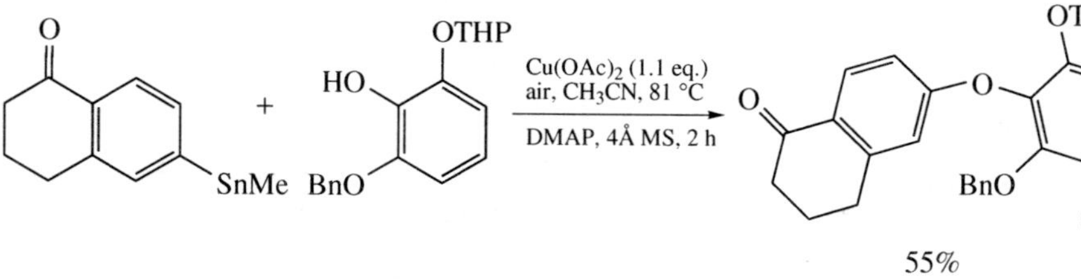

(94)

b. Coupling of N-nucleophiles with organotin compounds. C−N coupling is also possible using phenylstannanes as the aryl-transfer agents (equation 95)[385, 386]. Arylation of anilines, carboxylic acid amides, sulfonamides and benzimidazolinones was exemplified in the presence of stoichiometric $Cu(OAc)_2$ and TBAF as an additive. The presence of the fluoride ion apparently accelerates the tin–copper transmetallation step.

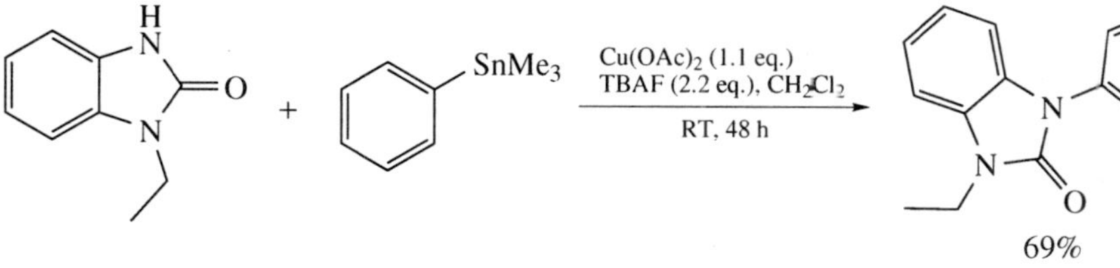

(95)

6. Organosilicon compounds

Lam and coworkers discovered that stoichiometric cupric acetate accelerates the coupling of hypervalent aryl and vinyl siloxanes with various *N*-nucleophiles[387, 388]. They demonstrated that anilines, aliphatic amines, amides and benzimidazoles can be arylated at room temperature in CH_2Cl_2 or DMF to the *N*-arylated products in acceptable yield (equations 96 and 97).

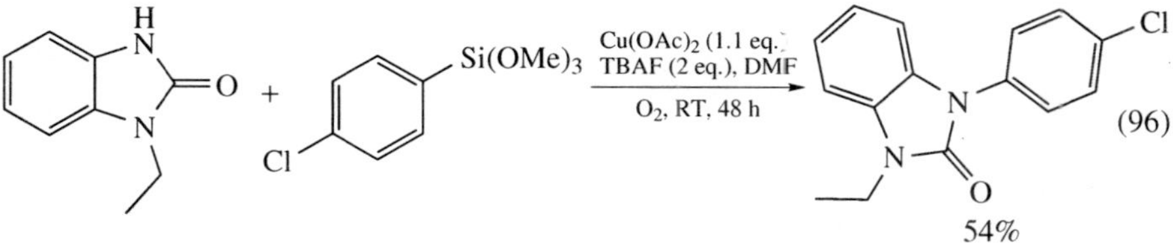

(96)

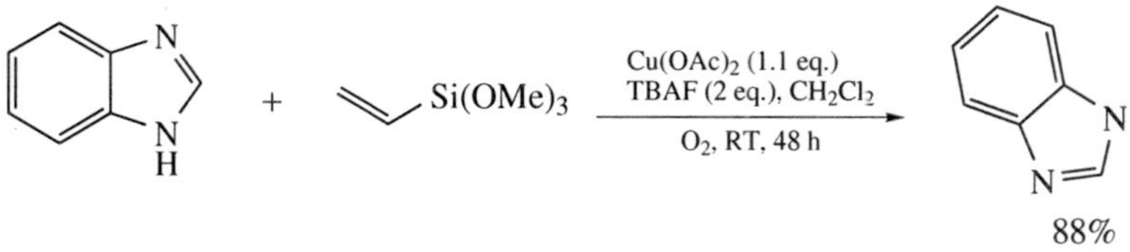

(97)

7. Mechanistic considerations

Currently, available information on possible mechanisms that can explain reactivity of the non-Ullmann substrates described in this section is very fragmental.

In one of the early studies, Lockhart suggested that transformations involving hypervalent diaryliodonium salts (see Section II.C.1) apparently follow the same mechanistic scheme as previously described for Ullmann condensation[339]. On the basis of comparison studies of the reactions promoted by copper compounds in different oxidation states, he concluded that Cu(I) species were actually responsible for the transformation. He hypothesized that further steps include oxidative addition of the active Cu(I) species to the hypervalent diaryliodonium compounds resulting in the formation of Cu(III) intermediate, and successive reductive elimination of the products (Scheme 57).

A similar mechanism has been invoked to explain the reactivity of hypervalent organobismuth (see Section II.C.2)[390] and organolead (see Section II.C.3)[339].

A different mechanism was postulated for C–O and C–N bond-forming reactions of other organometallic reagents (B, Si and Sn) under stoichiometric copper conditions. These

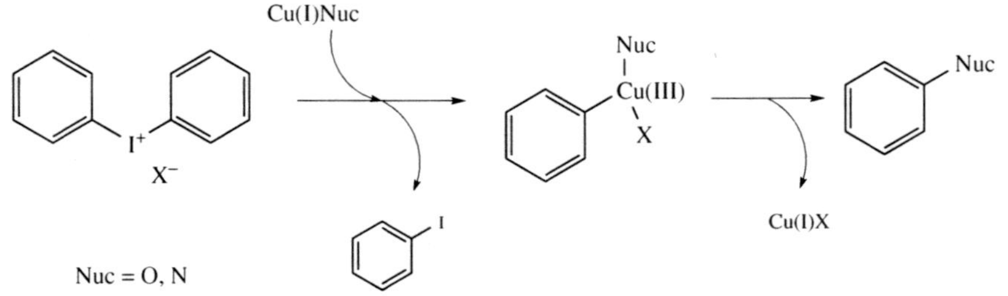

SCHEME 57

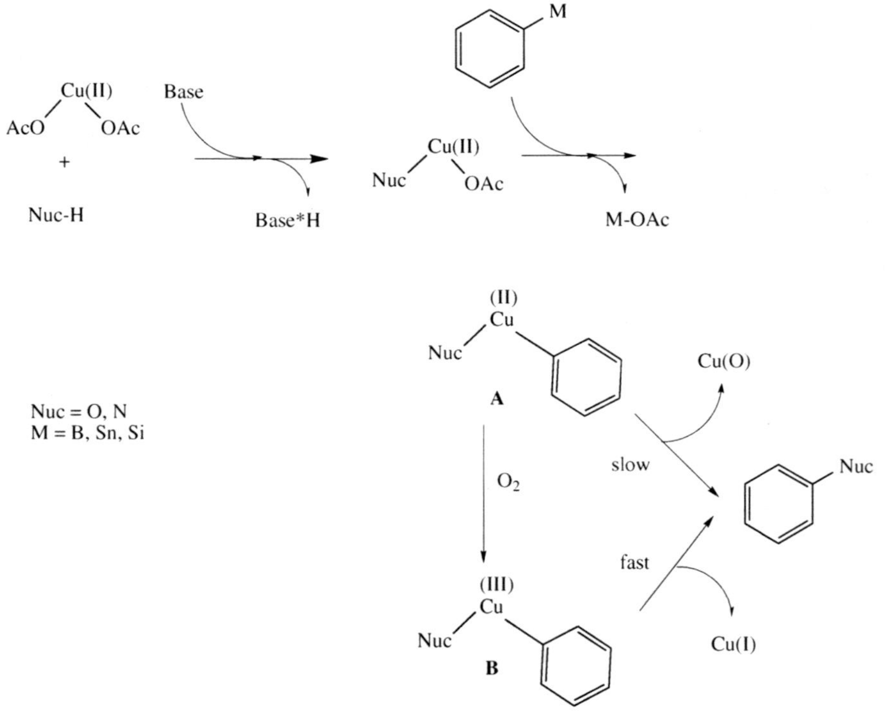

SCHEME 58

reactions apparently benefit from the '+2' oxidation state of the copper promoter. In addition, the presence of atmospheric oxygen, although not mandatory, improves the reaction rates. Generalizing these findings, several groups proposed the sequence of elementary steps shown in Scheme 58[349, 351, 386, 387].

According to this hypothesis, sequential transmetallation (ligand exchange) reactions of $Cu(OAc)_2$ with a deprotonated nucleophile (i.e. alcohol, amine, amide etc.) and an organoboron, organotin or organosilicon reagent leads to the formation of Cu(II) intermediate **A** (Scheme 58) capable of slow reductive elimination of the product and elemental copper. Alternatively, **A** may be oxidized to Cu(III) intermediate **B** that will eliminate the product while being reduced to Cu(I).

To perform such transformations in catalytic fashion, the presence of oxygen (or other oxidants) is obviously necessary in order to regenerate the catalytically active Cu(II) species after the reductive elimination step. Therefore, the third mechanism that involves the reoxidation step was proposed by Collman and coworkers to explain the catalytic C−N bond-forming reactions of the organoboron reactants (Scheme 59)[377].

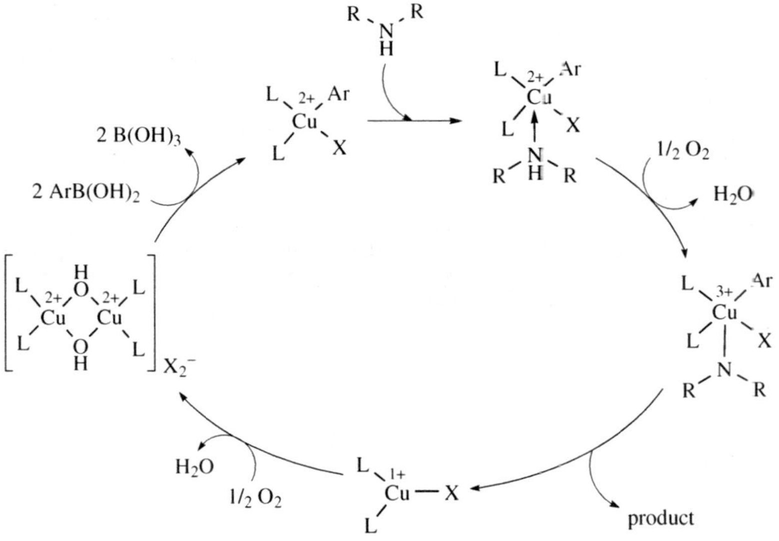

SCHEME 59

However, no strong evidence for these hypotheses has been found so far.

8. *Cross-dehydrogenative coupling to form C−N bonds*

Selective activation and functionalization of C(sp^2)−H bonds appears as an attractive approach to the development of waste-free cross-coupling methodology.

In 2005, Buchwald and coworkers reported the palladium-catalyzed and copper(II) acetate-promoted synthesis of carbazoles via C−H activation/C−N coupling (equation 98)[391].

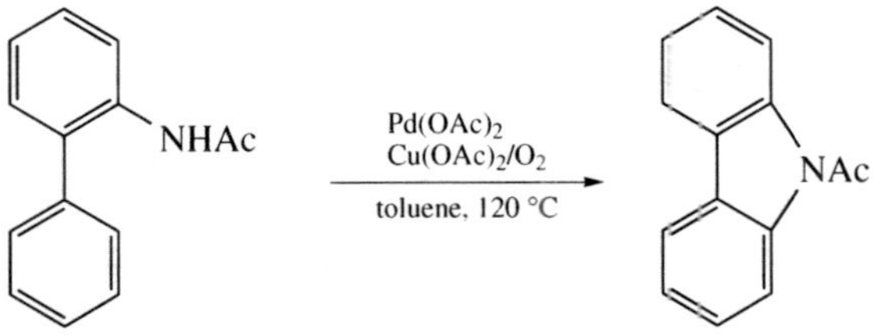

(98)

It was found, however, that copper alone can promote a similar transformation. For example, 2-phenylpyridines react intermolecularly with substituted and unsubstituted anilines in the presence of stoichiometric $Cu(OAc)_2$ in hot mesitylene forming the corresponding amine derivatives in moderate yield (equation 99)[392].

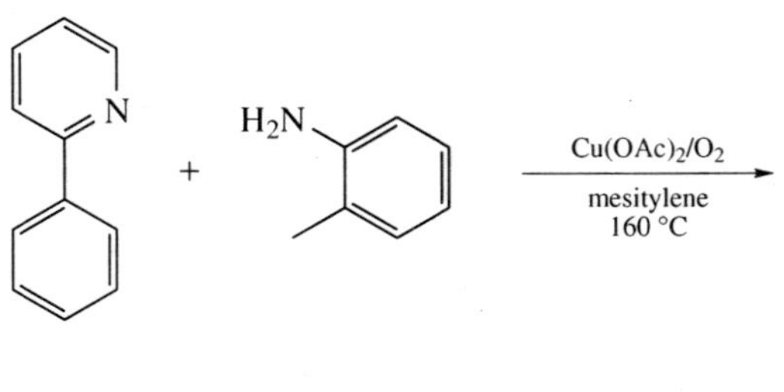

(99)

Very recently, a mild and really *catalytic* synthesis of benzimidazoles from amidines was suggested[393]. The intramolecular transformation proceeds under oxygen atmosphere in the presence of only 15 mol% of $Cu(OAc)_2$ and 2–5 equivalents of acetic acid additive in DMSO at 100 °C (Scheme 60).

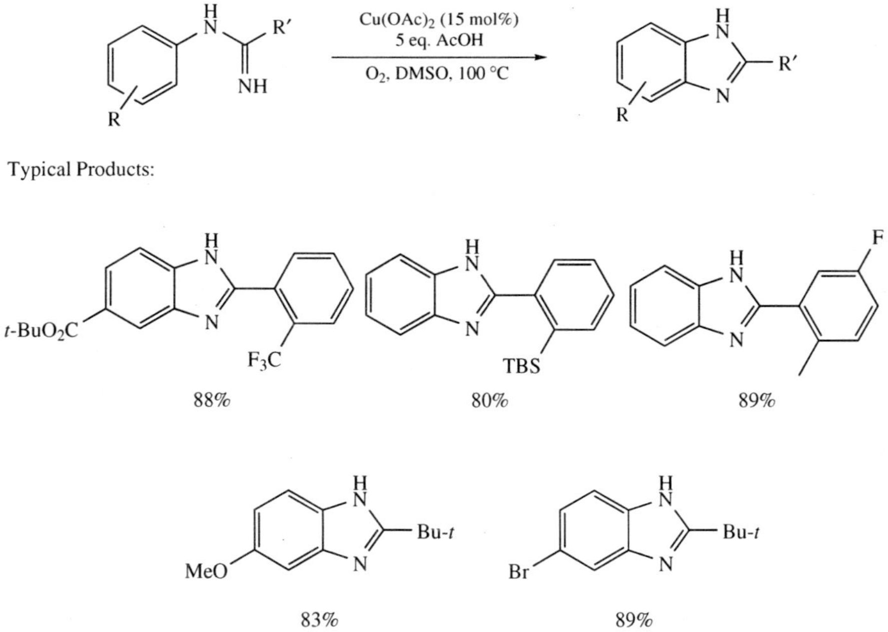

SCHEME 60

The mechanism of the reaction is not clear but may include the formation of a cyclometalated Cu(III) species[394] that reductively eliminates copper(I) which, upon reoxidation with oxygen, starts a new turnover (Scheme 61).

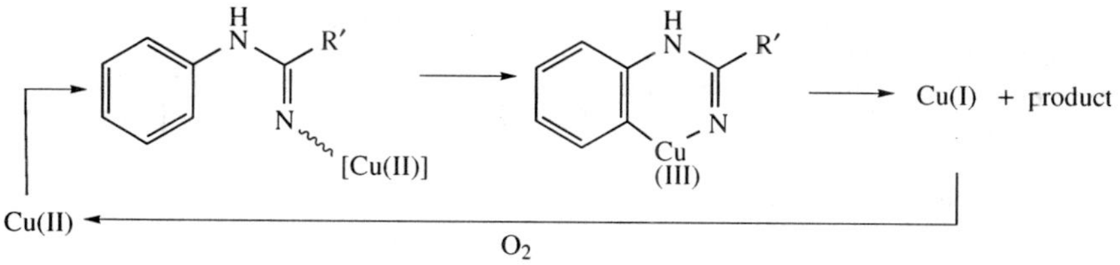

SCHEME 61

III. CARBON–CARBON BOND FORMING REACTIONS

A. Copper-catalyzed Cross-coupling of *C*-Nucleophiles with Organic Halides

1. Sonogashira reaction

A traditional method for the synthesis of internal alkynes via stoichiometric reaction between cuprous acetylides and aryl iodides is known as the Stephens–Castro substitution reaction (equation 100)[395, 396].

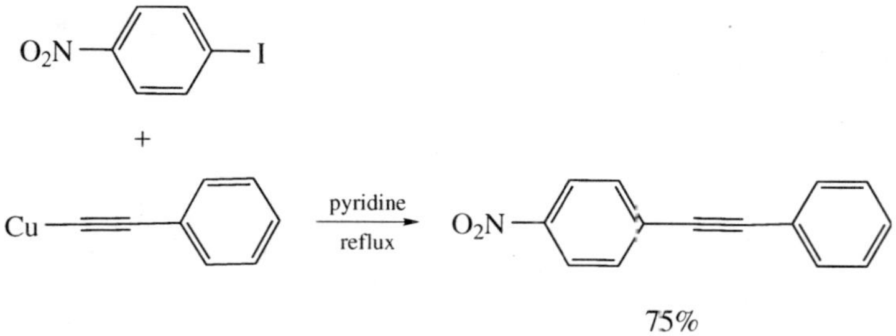

(100)

In the late 1980s, this rather harsh and synthetically limited transformation was replaced with a universal palladium-catalyzed and copper-co-catalyzed cross-coupling of aryl (or vinyl) halide with terminal acetylenes (Sonogashira–Hagihara reaction) which became a leading methodology toward the synthesis of substituted alkynes[397]. However, more recent studies demonstrated that this transformation can be performed using copper catalysts only. For example, CuI/PPh_3 was found capable of promoting coupling of terminal acetylenes and vinyl or aryl iodides in an intermolecular (equation 101)[398, 399] and intramolecular (equation 102)[400] fashion in the presence of K_2CO_3 in DMF or DMSO at 80–120 °C.

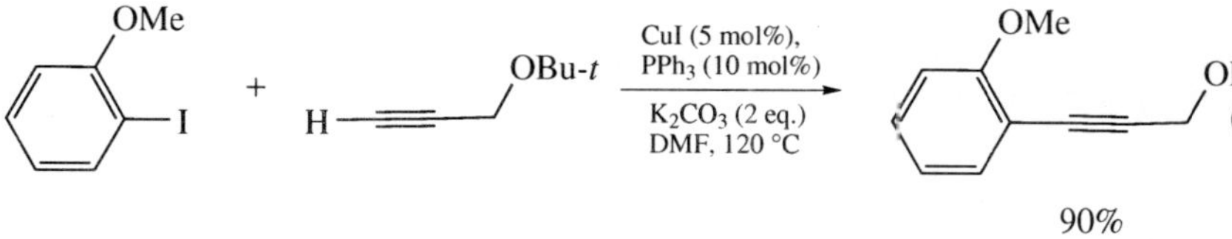

(101)

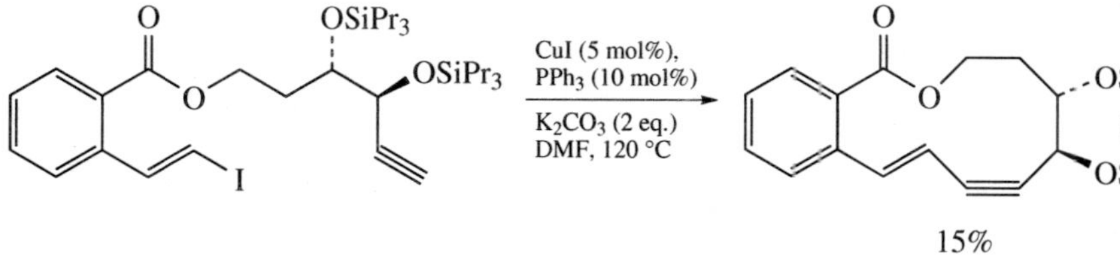

(102)

Although some ligandless protocols were also suggested[401,402], real improvement in the scope of the transformation came together with the progress made in the field of carbon–heteroatom bond-forming chemistry. For example, the use of Venkataraman's catalyst (Cu(phen)(PPh_3)Br or Cu(phen)$(PPh_3)_2NO_2$) allowed the synthesis of substituted alkynes from aryl iodides in a less polar medium (toluene) under relatively mild conditions[100] and led to the development of a straightforward synthetic route toward 2-substituted indoles (equation 103)[403].

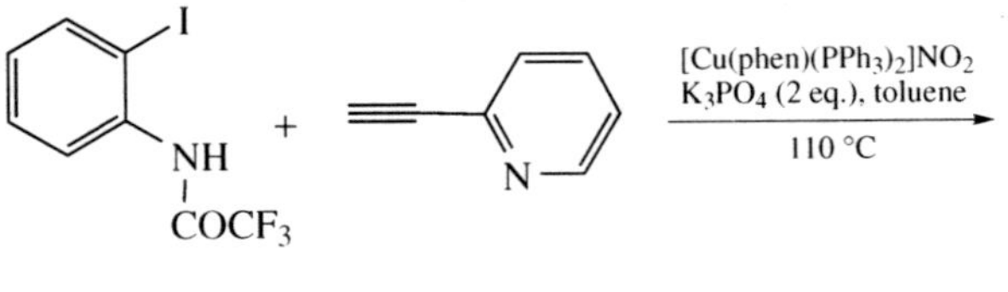

(103)

Coupling of less reactive aryl bromides became possible in the presence of CuI/*N,N*-dimethylglycine[404] or CuI/DABCO[405] catalysts. The catalysts operate in DMF in the presence of K_2CO_3 or Cs_2CO_3 at 80–120 °C and show clean reactions even in the presence of vulnerable functional groups such as carboxyl, methoxy, nitro, cyano, heteroaryl and other groups. The same conditions can be employed for coupling of stereodefined vinyl halides. No double isomerization is taking place.

If performed under the atmosphere of carbon monoxide, the transformation leads to the formation of aryl alkynones in good yield and selectivity (equation 104)[398,406].

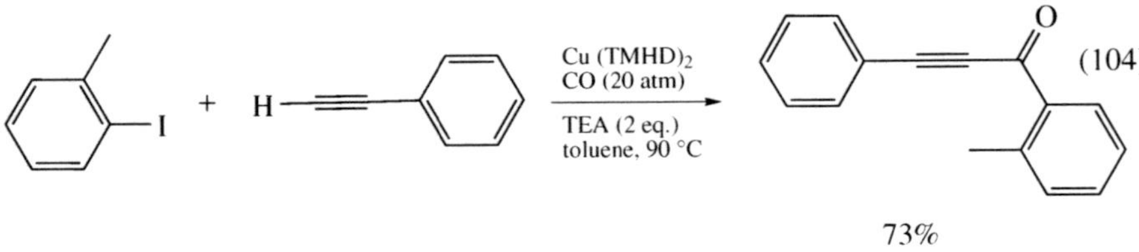

(104)

Although no detailed mechanistic studies have been performed, this transformation indicates that copper-catalyzed Sonogashira-like reactions proceed via the Cu(I)–Cu(III) catalytic cycle (i.e. Scheme 48) since formation of the carbonylated products via nucleophilic substitution pathway is unlikely.

Another approach toward aromatic or aliphatic alkynones is the copper-catalyzed reaction of terminal alkynes with carboxylic acid chlorides, which proceeds in the presence of *ca* 5 mol% of CuI under ligandless conditions in the presence of triethylamine in polar solvents such as THF at room temperature (equation 105)[407,408]. However, the cross-coupling nature of these reactions is not proven.

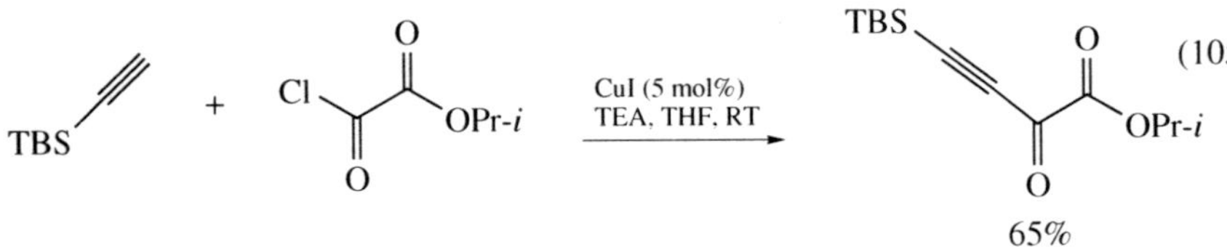

(105)

The utilization of aryliodonium salts as electrophiles was also reported (equation 106)[409]. As usual for this type of substrate, the transformation proceeds under very mild conditions.

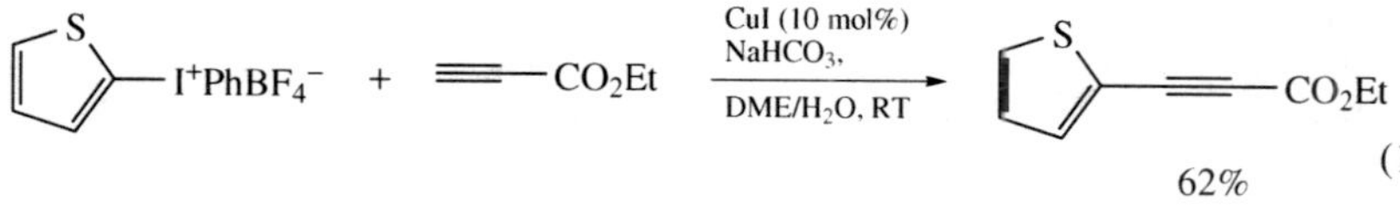

(106)

2. Rosenmund–von Braun reaction

Reaction of the stoichiometric copper(I) cyanide with aryl halides under extreme reaction conditions (150–250 °C), also known as Rosenmund–von Braun reaction, was a classical method for the preparation of aromatic nitriles[410]. Recently, this reaction was performed in catalytic fashion (equation 107). For example, 5 mol% of cuprous salts (CuI, CuCN, CuBr or CuCl) successfully catalyze the halide–cyanide exchange of aryl iodides in the presence of excess of NaCN in alkyl imidazolinium halide ionic liquids (bmil) as the reaction medium[411]. Aryl bromides showed unsatisfactory conversions under these conditions.

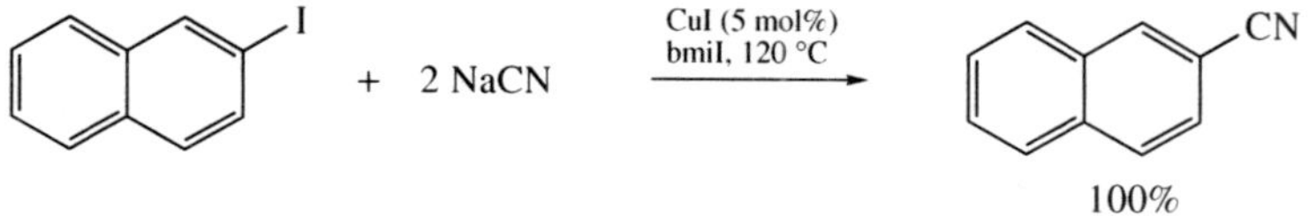

(107)

To overcome the lack of reactivity of more challenging aryl bromides, an alternative elegant approach relying on the copper-catalyzed aromatic Finkelstein reaction (Scheme 47) was developed. The straightforward cyanation of aryl bromides is accomplished via single copper species-catalyzed domino halide exchange-catalytic Rosenmund–von Braun reaction[412,413]. It was found that 10 mol% of CuI in combination with 20 mol% of DMEDA in the presence of 20 mol% of KI and 120 mol% of NaCN in toluene allows complete conversion of aryl and heteroaryl bromides into the corresponding nitriles (Scheme 62). Alternatively, the reaction can be carried out in DMF in the presence of CuI/1,10-phenanthroline catalyst and acetone cyanohydrin as a cyanide source.

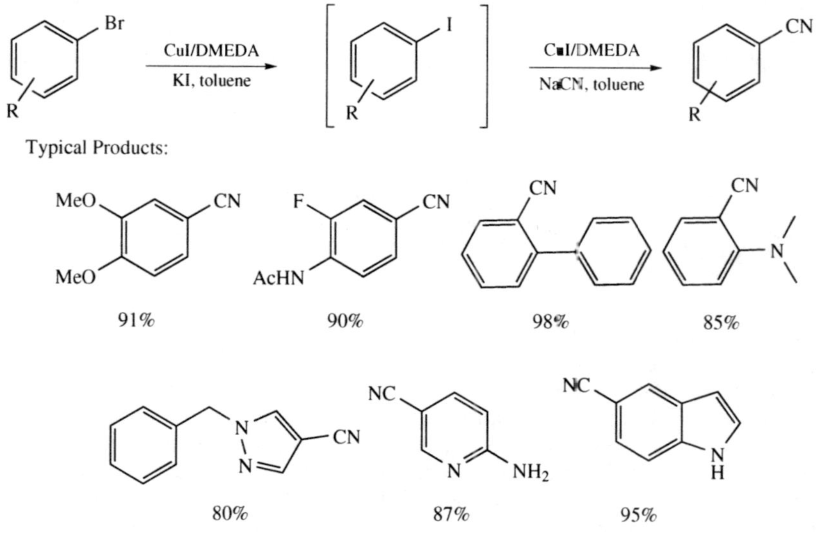

SCHEME 62

Due to the spectacular functional group compatibility, this method provides an excellent opportunity for the preparation of functionalized nitriles.

A more user-friendly version of this transformation was developed by Beller and coworkers, who utilized nontoxic potassium hexacyanoferrate as a cyanide source[414–416]. The coupling takes place in the presence of 10 mol% of CuI and 200 mol% of 1-butyl imidazoles as a ligand in refluxing toluene. Under these conditions, aryl and heteroaryl bromides couple smoothly almost independently of their steric and electronic properties: a variety of functional groups are tolerated and even highly substituted aryl bromides show decent conversion (equations 108 and 109).

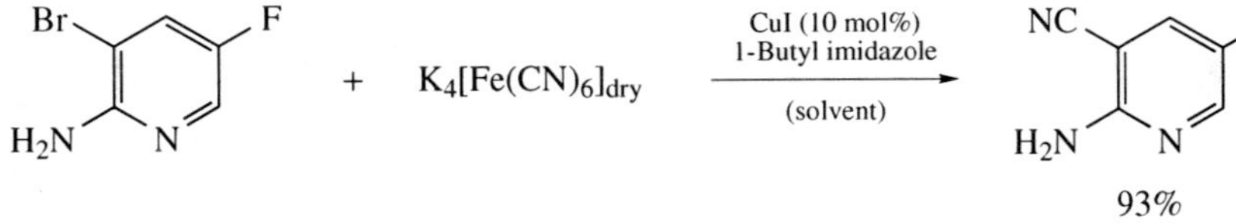

(108)

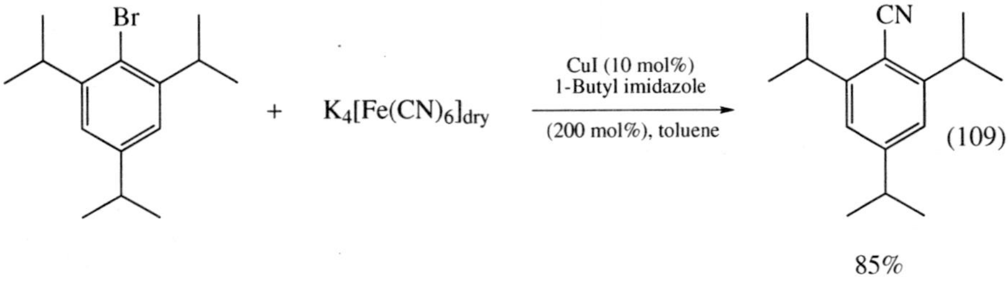

(109)

3. Hurley and related reactions

The original report by Hurtley in 1929 described a substitution of halogen in 2-bromobenzoic acid with deprotonated acetylacetone, ethyl malonate and ethyl acetoacetate in the presence of catalytic copper bronze[417]. The conditions were very mild and the yield was acceptable; however, the reaction was found to be limited to chelating 2-halobenzoic acids (equation 110).

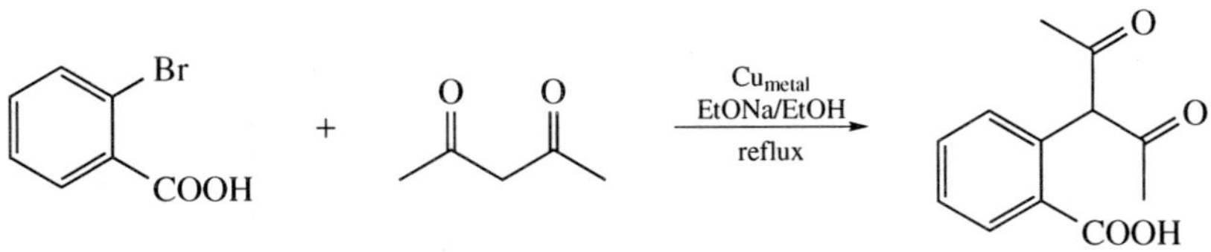

(110)

Modern studies led to the development of better protocols utilizing homogenous copper catalysts such as $Cu(OAc)_2$, CuI or CuBr instead of insoluble copper bronze, and NaH instead of NaOEt or NaOH[418–424]. Under these conditions the scope of the active hydrogen compounds was greatly improved to include even sterically hindered enolates. However, attempts to use aryl halides other than *o*-halobenzoic acid met with only limited success[425–427]. Substrates lacking an *o*-assisting group normally required the utilization of hard-to-manipulate solvents (DMSO or HMPT), stoichiometric copper in combination with the presynthesized sodium enolate (or isolated copper enolates) and harsher reaction conditions to form products in modest to high yield (equations 111–114).

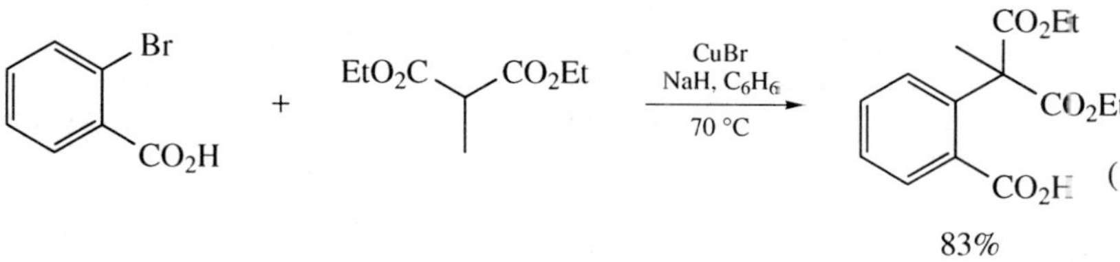

(111)

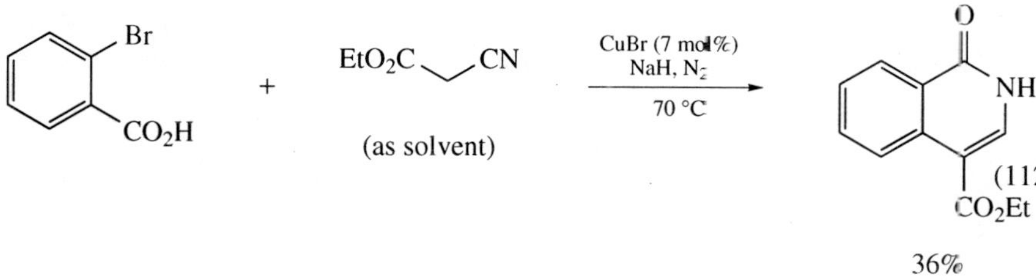

(112)

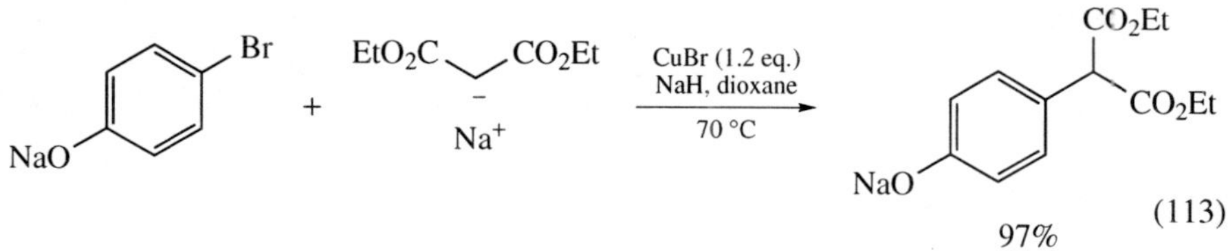

(113)

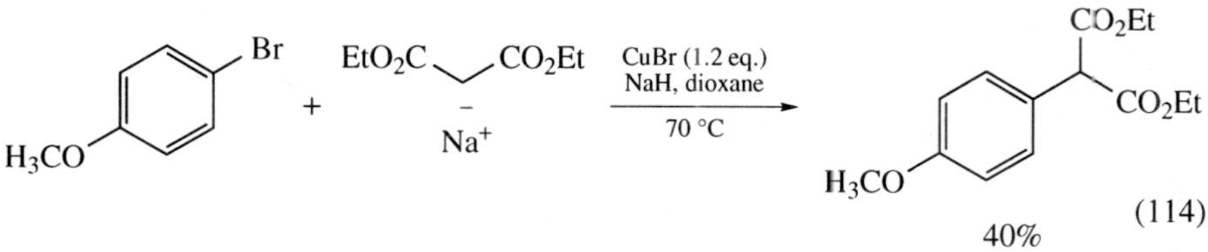

(114)

In 1993, Miura and coworkers reported the CuI-catalyzed Hurtley reaction of unactivated (lacking an *ortho*-assisting group) aryl iodides and bromides with active methylene compounds (malononitrile, ethyl cyanoacetate and ethyl malonate) that proceeded in the presence of K_2CO_3 in DMSO at 120 °C and resulted in the formation of arylated products in acceptable (up to 80%) yield (equation 115)[428].

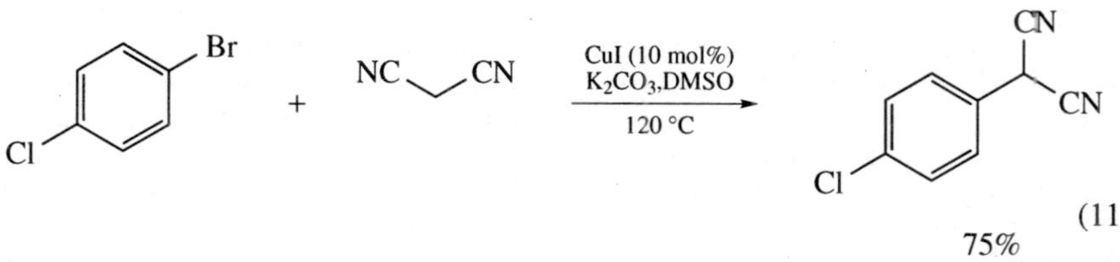

(115)

Although the procedure was more user-friendly due to the utilization of a simple potassium carbonate, the harsh reaction conditions required to obtain reasonable conversions limited the potential applications of the reaction in organic synthesis.

Under ligand assistance, the reaction proceeds at lower temperature and in a low boiling solvent such as THF[429]. For example, Hennessy and Buchwald demonstrated that

5 mol% of CuI and 10 mol% of 2-phenylphenol catalyze the Hurtley reaction between aryl iodides and diethyl malonate in the presence of Cs_2CO_3 as a base. High yields of the products were achieved with aryl iodides of different electronic properties even in the presence of sterically demanding *ortho*-substituents. Good functional group compatibility was observed (Scheme 63).

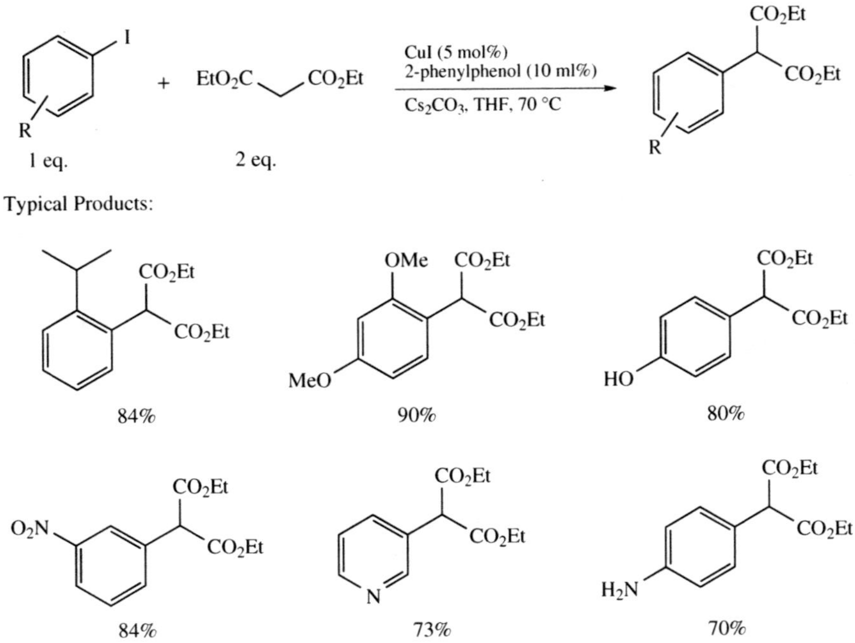

SCHEME 63

Unfortunately, this reaction was limited to dialkyl malonates, and other active methylene compounds such as 1,3-cyclopentandienone, 1,3-cyclohexanedione or Meldrum's acid were inactive.

Other catalytic systems were also found suitable for this transformation. For instance, CuI/DMEDA was used for the intramolecular coupling of aryl bromides with 1,3-dicarbonyl compounds (Scheme 64)[430].

The six-membered ring closure of α-(2-bromobenzyl)-β-keto esters proceeds smoothly in the presence of a substituent in the aromatic ring or in the β-keto ester chain, but it changes the cyclization mode when δ-(2-bromophenyl)-β-keto esters are employed as starting materials. In these cases the *O*-arylation mode is preferred and benzopyranes are formed in good yield (equation 116).

A cascade arylation–condensation process was suggested as a straightforward route toward polysubstituted indoles[431, 432]. The reaction is catalyzed by a CuI/L-proline couple and proceeds at very mild heating (RT–50 °C) due to the presence of an *ortho*-activator. The condensation step is acid-promoted and does not require the prior isolation of the arylated intermediate (Scheme 65).

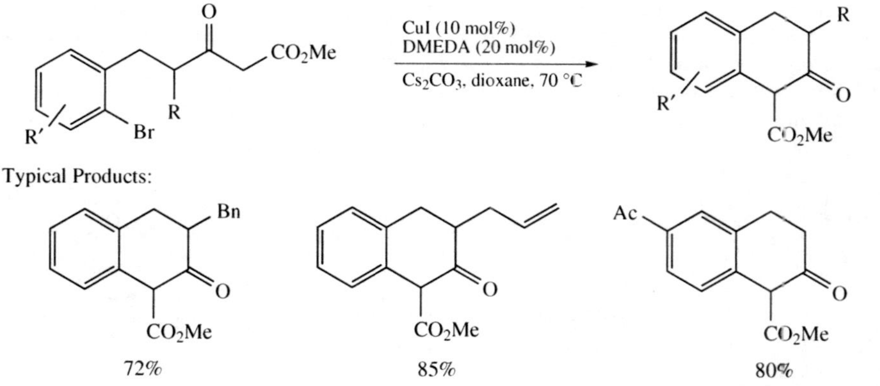

SCHEME 64

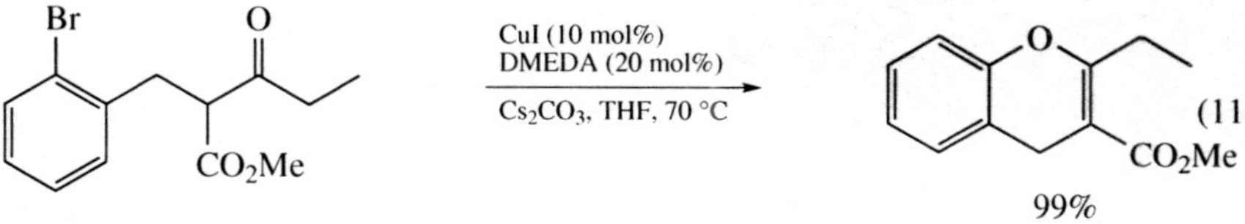

(116)

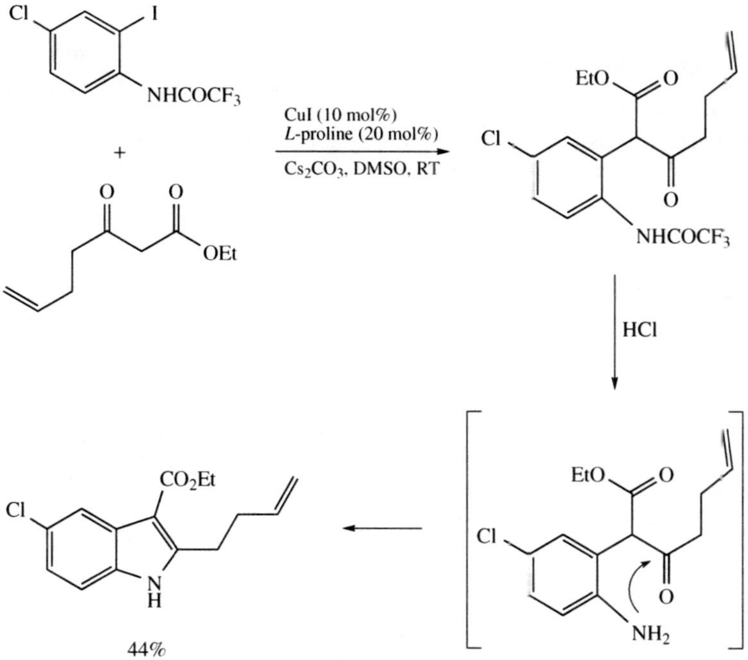

SCHEME 65

The reaction conditions applicable to aryl bromides require 30–50 °C heating to achieve maximal conversion, differently substituted β-keto esters and β-keto amides, and tolerate a variety of functional groups at the aromatic ring.

Similarly, 2,3-disubstituted benzofurans were synthesized from 1-bromo-2-iodobenzenes and β-keto esters via sequential C–C/C–O bond-forming reactions (Scheme 66)[433].

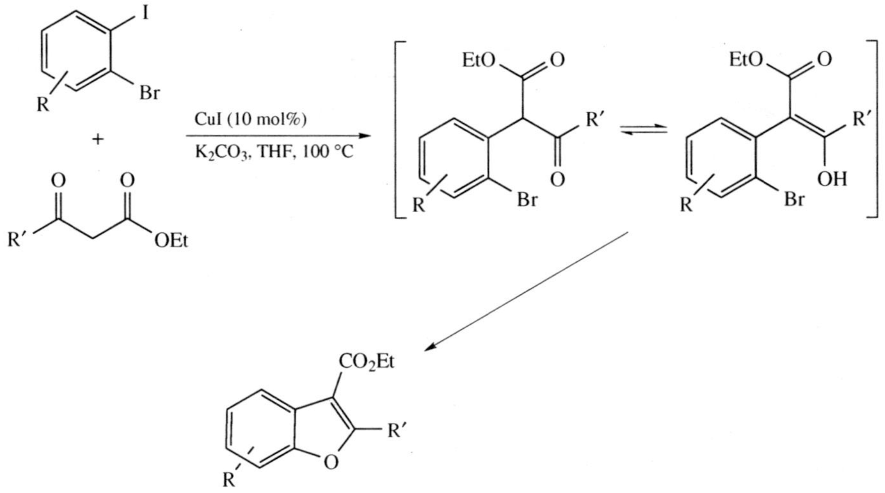

SCHEME 66

Finally, an enantioselective version of this transformation was suggested by Ma and coworkers[434]. The reaction of alkyl 2-methylacetoactates with 2-iodotrifluoroacetanilides proceeds even at very low temperature (−45 °C) when catalyzed by CuI/(2*S*,4*R*)-4-hydroxyproline catalyst in aqueous DMF in the presence of NaOH (Scheme 67). It is worth noting that other ligands showed no activity at such a low temperature.

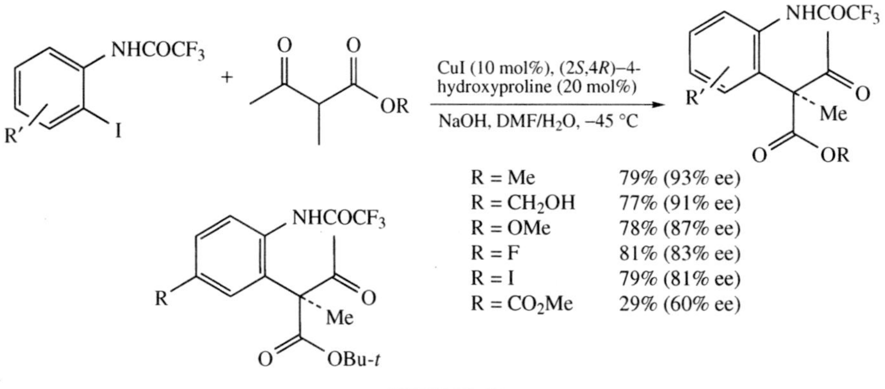

R = Me	79% (93% ee)
R = CH_2OH	77% (91% ee)
R = OMe	78% (87% ee)
R = F	81% (83% ee)
R = I	79% (81% ee)
R = CO_2Me	29% (60% ee)

SCHEME 67

The quaternary chiral centers form in moderate to excellent enantioselectivity depending on several parameters: i) the size of the ester moiety, i.e. the presence of bulkier alkyl groups dramatically improves the enantioselectivity; ii) the electronic nature of the aryl bromide, i.e. electron-rich substrates generally show better enantioselectivity.

4. Cross-dehydrogenative coupling

Several examples of the copper-mediated cross-dehydrogenative coupling processes appeared in the literature very recently. For example, active organocopper(I) intermediates can be generated by interaction of simple copper salts with *in situ* formed sp^2-carbon nucleophiles. The latter can, in principle, participate in cross-coupling reactions with aryl or vinyl halides following the usual Cu(I)–Cu(III) catalytic cycle. This strategy was applied to the arylation and vinylation of heterocycles such as oxazole, benzoxazole, imidazoles and triazole derivatives that can be easily deprotonated with relatively strong bases such as LiOBu-*t* or KOBu–*t*[435,436]. Unfortunately, the reaction conditions are rather harsh and require 100–140 °C heating in order to achieve acceptable conversions (Scheme 68).

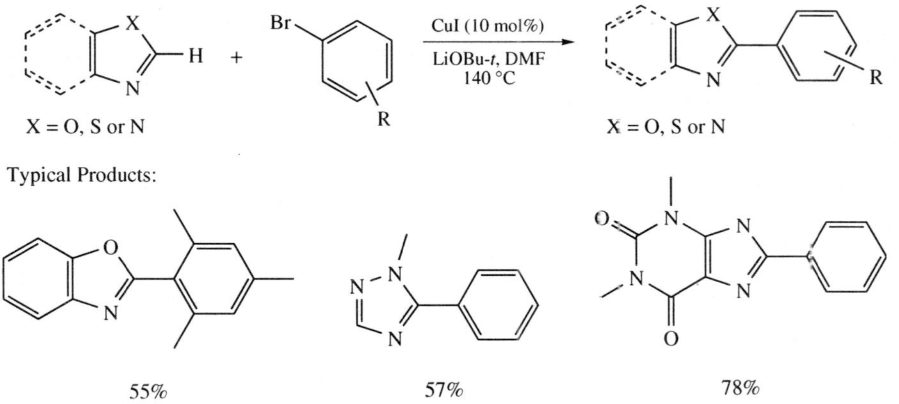

SCHEME 68

The acidity of the aromatic hydrogens in perfluorinated arenes is high enough to allow a facile deprotonation even with K_3PO_4 in DMF at 120–140 °C[437]. Therefore, such compounds can be arylated with aryl iodides and bromides in the presence of CuI/1,10-phenanthroline catalyst (Scheme 69).

B. Copper-catalyzed Cross-coupling of Organometallic Reagents with Organic Halides

1. Organoboron reagents

A very reliable palladium-catalyzed Suzuki–Miyaura reaction became a leading method for the construction of C–C bonds from aryl or vinyl halides and organoboron compounds. Although the scope of the method is remarkable and limitations are negligible, several copper-catalyzed palladium-free versions of this reaction were developed.

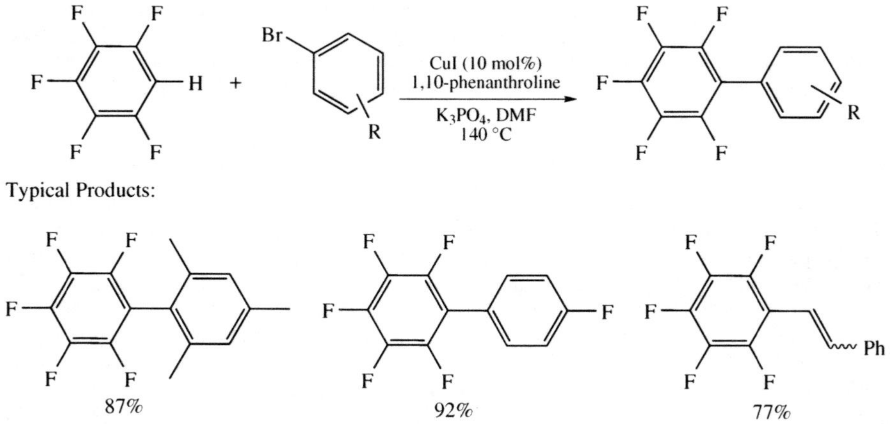

SCHEME 69

One of the first examples of the copper-catalyzed Suzuki-type chemistry refers to the reaction of alkynyl bromides and *in situ* prepared organoboron compounds which proceeds in the presence of 5 mol% of $Cu(acac)_2$ and LiOH as a base at very low temperature (−15 °C) (equation 117)[438–440].

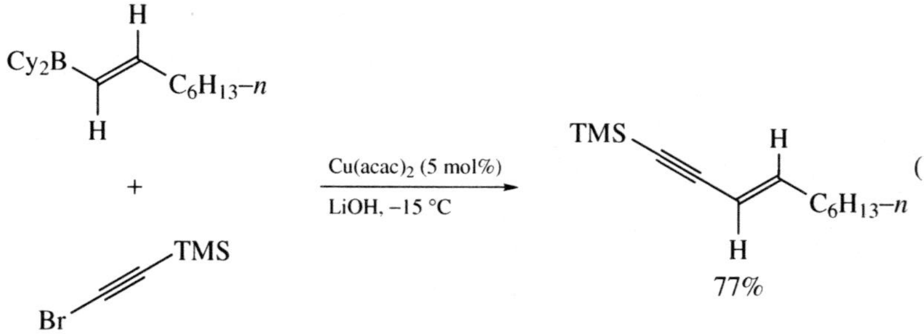

(117)

Later, Rothenberg and coworkers demonstrated that reactions of aryl iodides (bromides were found inactive under these reaction conditions) and phenylboronic acid take place in the presence of palladium-free colloidal copper[441, 442]. The same method was extended to include unactivated aryl bromides and even some activated (electron-poor) aryl chlorides when the reaction was performed in coordinating PEG-400 as a solvent and in the presence of 20 mol% of molecular iodine (whose role remained unclear) at 110 °C (equation 118)[443].

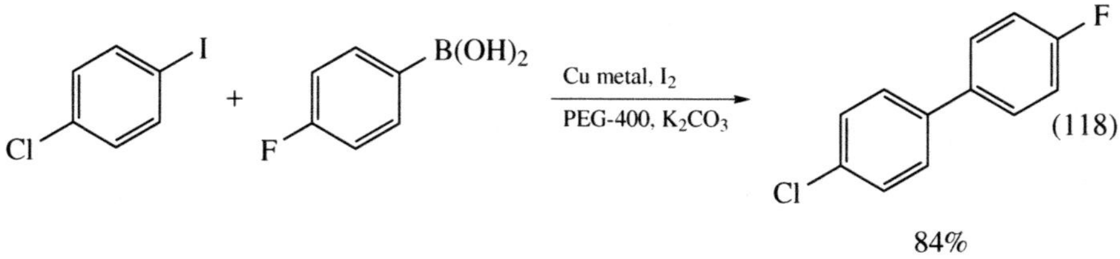

(118)

CuI/DABCO[405,444] and CuI/TBAB[445] can be employed for the same transformation. The reactions are carried out in DMSO or DMF in the presence of Cs_2CO_3 at high temperature.

The utilization of aryl iodonium salts as electrophiles was also reported (equation 119)[446].

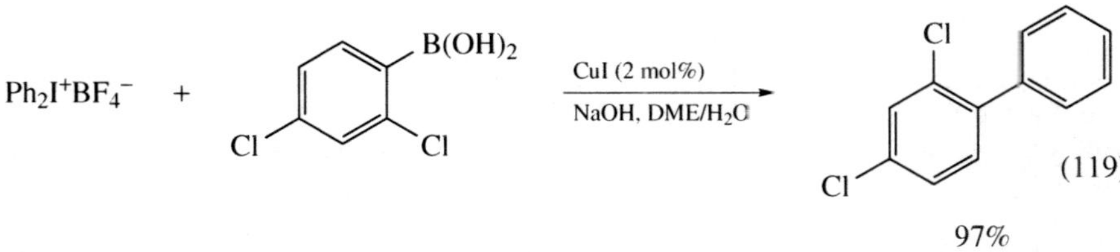

(119)

It is worth noting that copper-mediated reactions are not superior to the traditional palladium-catalyzed ones either in scope or in reaction conditions.

2. Organotin compounds

In 1990, the beneficial effect of copper cocatalyst on palladium-catalyzed Stille cross-coupling was identified[447]. Almost immediately, it was found that overstoichiometric copper chloride is capable of promoting intramolecular Stille reaction under palladium-free conditions (equation 120)[448,449]. The reported transformations took place within minutes and were generally cleaner than their palladium-catalyzed counterparts.

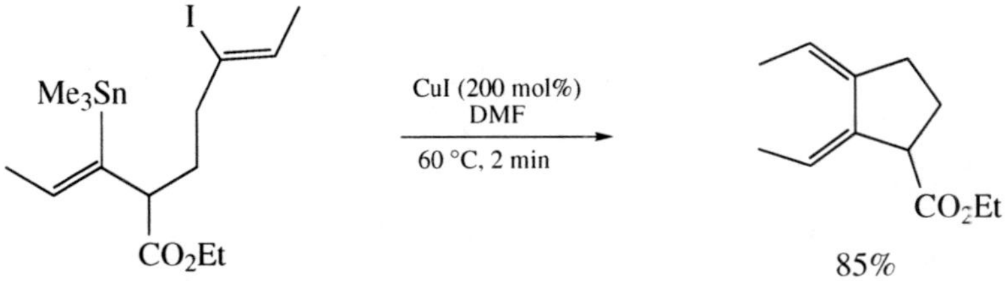

(120)

In 1996, Allred and Liebeskind developed a mild coupling of arylstannanes with vinyl iodides promoted by stoichiometric (or higher) copper(I) thiophenecarboxylate (CuTC) in NMP (equation 121)[450]. The coupling is very useful for thermally unstable substrates since it proceeds at 0 °C to RT[451–453]. They also found that the transformation can be effected with catalytic copper but in the presence of stoichiometric CsF additive. Unfortunately, the catalytic version appeared to be less general (equation 122).

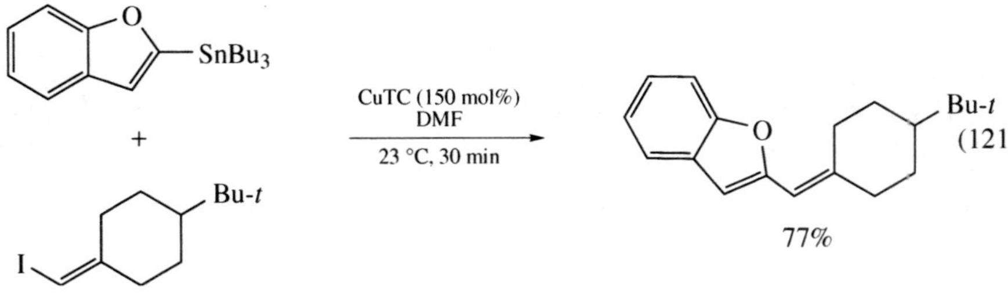

(121)

(122)

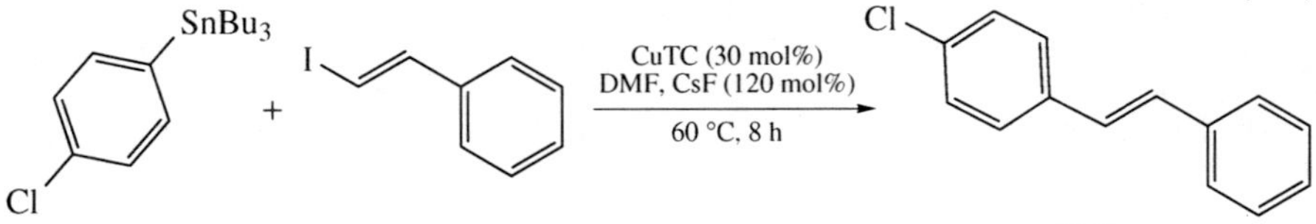

91%

After this precedent, several groups reported the development of CuI-catalyzed protocols for the Stille cross-coupling of aryl[454,455], vinyl[454], allyl[456] and alkynyl iodides[457] with organostanannes. The scope of organostanannes includes aryl, heteroaryl, alkenyl, alkynyl and allyl derivatives (equations 123–125).

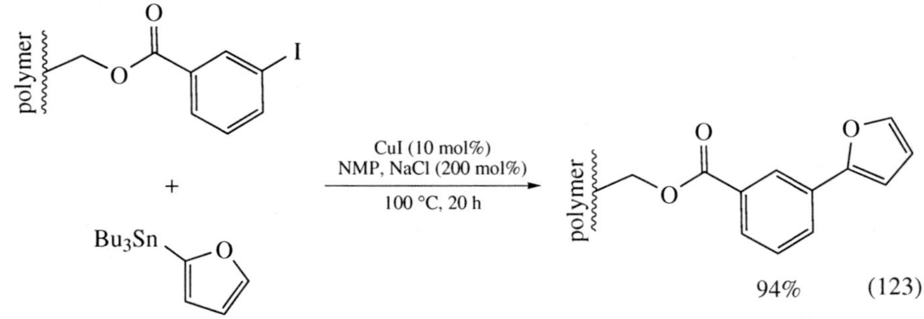

94% (123)

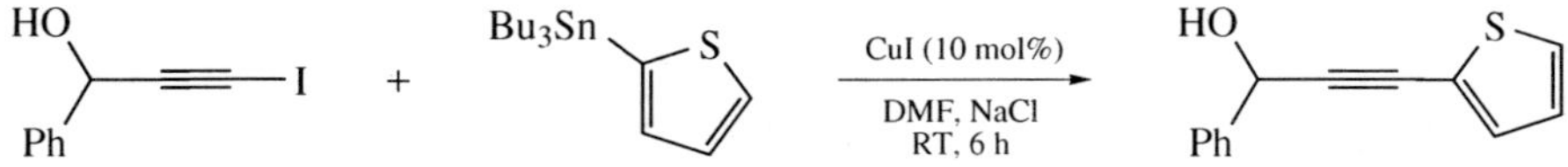

76% (124)

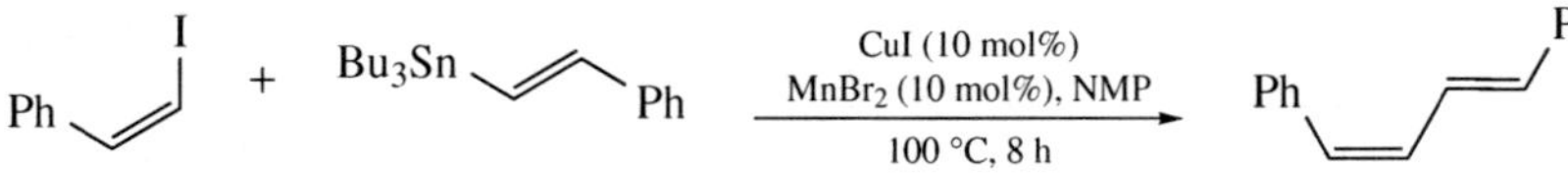

(125)

53%

Unfortunately, all the suggested catalytic reactions are slow in comparison with the stoichiometric ones and often require elevated temperatures in order to reach complete conversions. The possible explanation for this difference is the reversibility of the transmetallation of $RSnBu_3$ with the copper species (equation 126)[449,450].

$$RSnBu_3 + CuI \rightleftharpoons Cu\text{-}R + I\text{-}SnBu_3 \qquad (126)$$

Although the presence of stoichiometric alkali metal chlorides or fluorides can improve the situation by converting the Bu_3SnI into the corresponding chloride or fluoride, which do not participate in the back-reaction, the performance of these catalytic systems is relatively weak.

The use of aryl iodonium electrophiles in very mild carbonylative Stille-type coupling was also demonstrated (equation 127)[446,458].

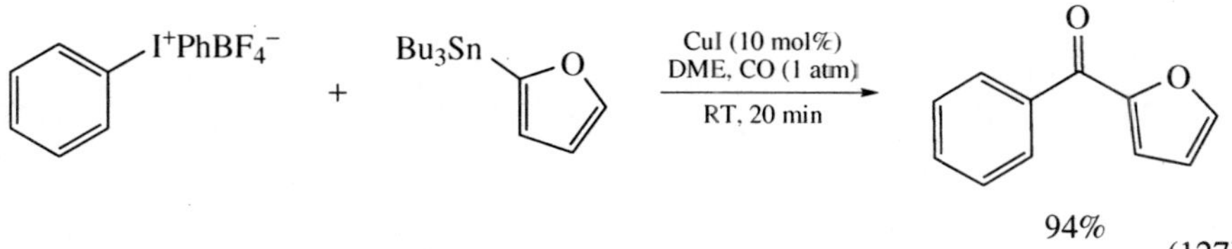

(127)

3. Organosilicon compounds

Transmetallation from silicon to copper to form active organocopper(I) compounds is a process that was observed directly or suggested without obvious evidence (Scheme 70)[459–463]. This fact along with indications that copper(I) species are capable of oxidative addition/reductive elimination reactions make the copper-catalyzed Hiyama-type cross-coupling theoretically possible.

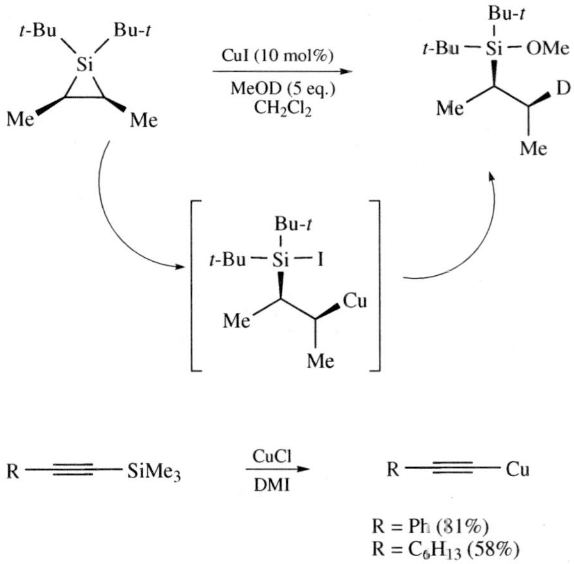

DMI = 1,3-dimethyl-2-imidazolidinone

SCHEME 70

Despite this possibility, examples of this transformation are very rare. For example, heteroarylsilanes[464] (e.g. 2-trimethylsilylthiazole) or alkynylsilanes[465] react with aryl or vinyl iodides in the presence of overstoichiometric cuprous halides in 1,3-dimethyl-2-imidazolidinone (DMI) at 120–130 °C (equations 128 and 129). The reactions of arylsilanes under the same conditions are more satisfactory.

Stoichiometric copper-promoted synthesis of trifluoromethyl-substituted arenes[466] and heteroarenes[467] was reported. The method is limited to activated aryl iodides and electron-deficient heterocycles but delivers products in acceptable yields (equations 130 and 131).

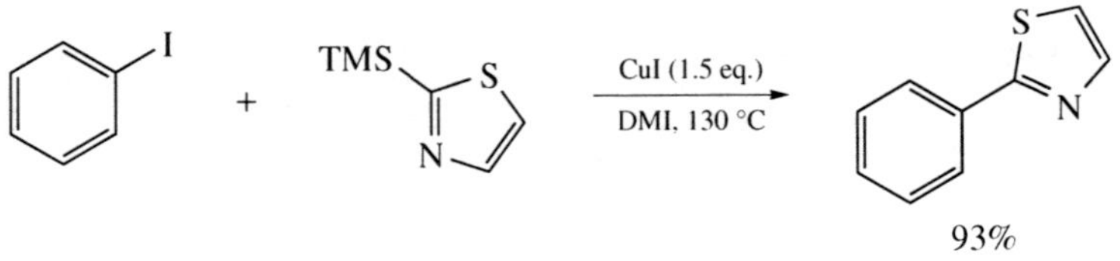

(128)

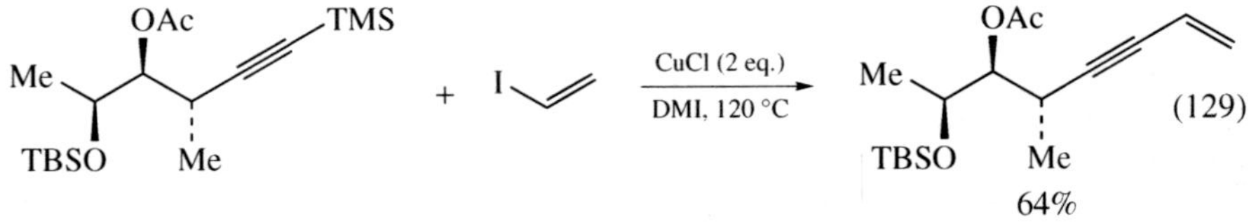

(129)

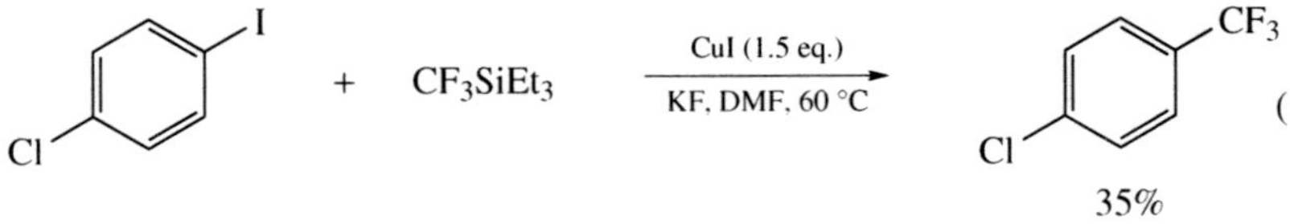

(130)

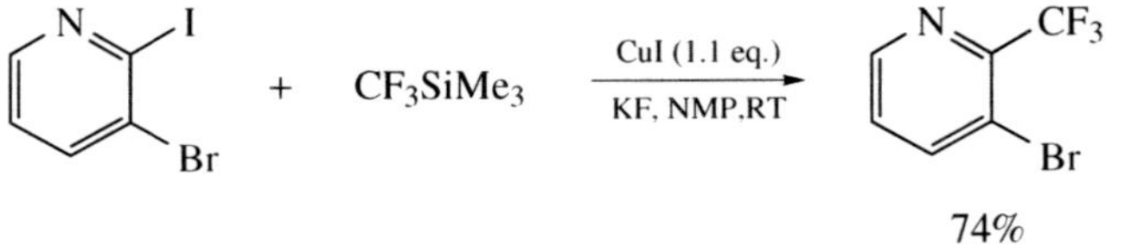

(131)

The only copper-catalyzed cross-coupling of alkynylsilanes with 1-chloroalkynes was suggested by Mori and coworkers[468]. They found that unsymmetrical diynes can be prepared in moderate yields using only 10 mol% of copper(I) chloride (equation 132).

(132)

4. Organomagnesium compounds

Back in 1936, Linn and Noller reported that if cuprous bromide is allowed to react with an excess of ethylmagnesium bromide in diethyl ether, a copper-containing solution forms that is capable of catalyzing the reaction of ethylmagnesium bromide and ethyl bromide to give a mixture of ethane and ethylene[469]. Later studies, however, showed that not only a disproportionation pathway but also catalytic cross-coupling can be achieved if the reaction is performed in tetrahydrofuran solution (equation 133)[470–472].

$$\text{Et-MgBr} + \text{EtBr} \xrightarrow[\text{THF}]{\text{CuBr}} \text{Et-Et} + \text{MgBr}_2 \qquad (133)$$

The method became synthetically useful after suitable conditions for cross-coupling of alkyl[473], vinyl[474–476], allyl[475,477] and aryl[475,478,479] Grignard reagents with organic halides were found. The reactions are usually catalyzed by 1–10 mol% of either cuprous halides (CuCl, CuBr or CuI) or more soluble copper complexes such as $CuCl_2$/TMEDA[480], CuBr/HMPA[478], Li_2CuCl_4[479] or CuBr/LiBr/PhSLi[481] in THF and proceed at ambient temperatures (equation 134).

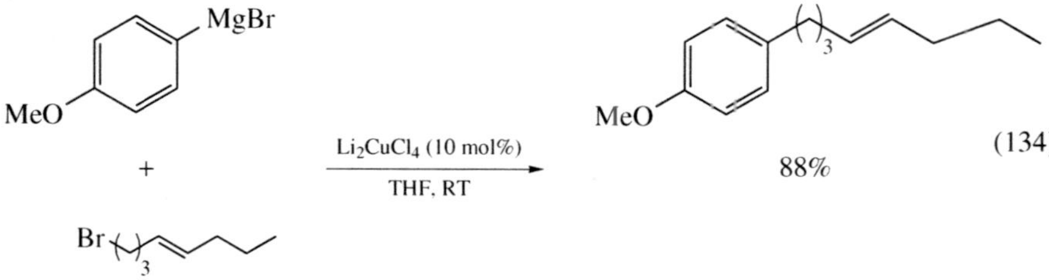

(134)

Obvious limitation of the organomagnesium chemistry is a low functional group compatibility of the highly reactive organometallics. However, at mild reaction temperature the catalyzed reactions are often faster than possible side reactions associated with the nucleophilic attack by the Grignard reagent, which allows, in some cases, the synthesis of functionalized targets (equations 135 and 136).

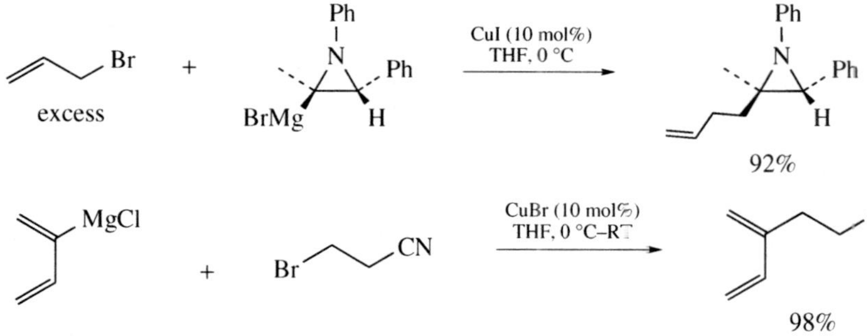

(135)

(136)

These reactions are usually limited to organic bromides or iodides, but alkyl chlorides were recently reported as suitable coupling partners in the presence of $CuCl_2$ (2 mol%)/1-phenylpropyne (10 mol%) catalyst[482]. It was suggested that 1-phenylpropyne is responsible for the stabilization of thermally unstable alkyl copper intermediates which allows one to carry out the reaction at the elevated temperatures necessary for the activation of less active alkyl chlorides (equation 137).

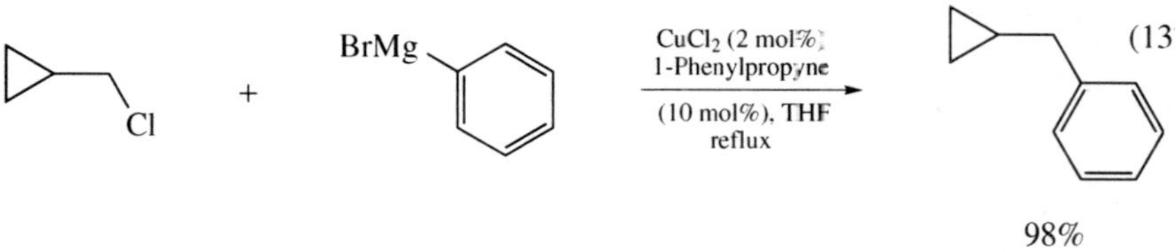

(137)

Apart from organic halides, alkyl triflates[483–486], tosylates[431,487–492] and phosphates[493] couple with Grignard reagents in the presence of copper catalysts. These readily accessible starting materials greatly extend the scope of the method (equations. 138–140).

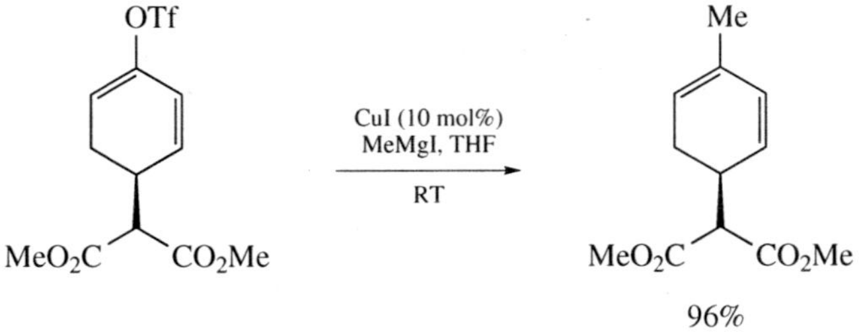

(138)

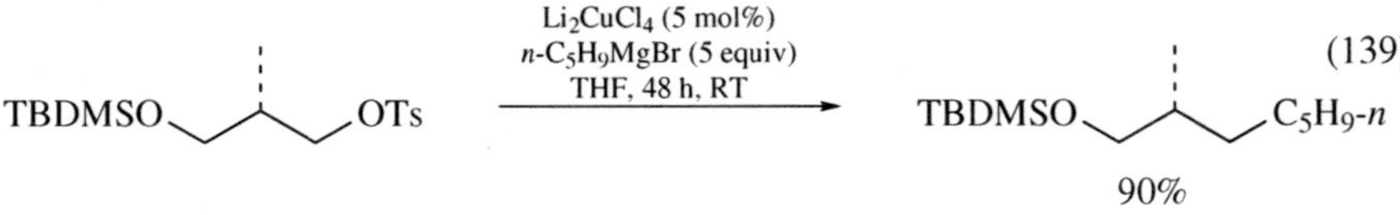

(139)

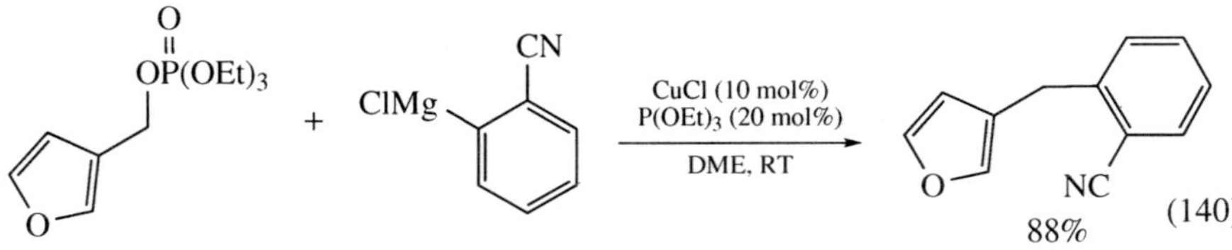

(140)

IV. REFERENCES

1. F. Ullmann and A. Lehner, *Ber. Dtsch. Chem. Ges.*, **37**, 853 (1904).
2. F. Ullmann and W. Minajeff, *Ber. Dtsch. Chem. Ges.*, **45**, 687 (1912).
3. F. Ullmann and P. Sponagel, *Justus Liebigs Ann. Chem.*, **350**, 83 (1907).
4. F. Ullmann and J. Tscherniack, *Ber. Dtsch. Chem. Ges.*, **38**, 4110 (1905).
5. R. G. R. Bacon and O. J. Stewart, *J. Chem. Soc.*, 4953 (1965).
6. H. Weingarten, *J. Org. Chem.*, **29**, 3624 (1964).
7. H. Weingarten, *J. Org. Chem.*, **29**, 977 (1964).
8. V. V. Litvak and S. M. Shein, *Zh. Org. Khim.*, **10**, 550 (1974); *Chem. Abstr.*, **80**, 132619 (1974).
9. A. L. Williams, R. E. Kinney and R. F. Bridger, *J. Org. Chem.*, **32**, 2501 (1967).
10. M. D. Rausch, *J. Org. Chem.*, **26**, 1802 (1961).
11. J. Sice, *J. Am. Chem. Soc.*, **75**, 3697 (1953).
12. E. Morishita and S. Shibata, *Chem. Pharm. Bull.*, **15**, 1765 (1967).
13. M. A. Ragan, *Can. J. Chem.*, **63**, 291 (1985).
14. R. R. Renshaw and R. C. Conn, *J. Am. Chem. Soc.*, **59**, 297 (1937).
15. J. W. H. Watthey and M. Desai, *J. Org. Chem.*, **47**, 1755 (1982).
16. T. Kametani and K. Fukumoto, *Tetrahedron Lett.*, 2771 (1964).
17. M. Tomita, K. Ito and H. Yamaguchi, *Chem. Pharm. Bull.*, **3**, 449 (1955).
18. M. J. Rance and J. C. Roberts, *Tetrahedron Lett.*, 2799 (1970).
19. R. G. R. Bacon and S. C. Rennison, *J. Chem. Soc. C*, 312 (1969).
20. R. G. R. Bacon and S. C. Rennison, *J. Chem. Soc. C*, 308 (1969).
21. R. G. R. Bacon and J. R. Wright, *J. Chem. Soc. C*, 1978 (1969).
22. T. Tsuda, T. Hashimoto and T. Saegusa, *J. Am. Chem. Soc.*, **94**, 958 (1972).
23. G. M. Whitesides, J. S. Sadowski and J. Lilburn, *J. Am. Chem. Soc.*, **96**, 2829 (1974).
24. L. J. Belf, M. W. Buxton and G. Fuller, *J. Chem. Soc.*, 3372 (1965).
25. L. M. Yagupol'skii, N. V. Kondratenko and V. P. Sambur, *Synthesis*, 721 (1975).
26. K. Fujiki, *Bull. Chem. Soc. Jpn.*, **50**, 3065 (1977).

27. W. J. Frazee and M. E. Peach, *J. Fluorine Chem.*, **13**, 223 (1979).
28. H. Suzuki, H. Abe and A. Osuka, *Chem. Lett.*, 1363 (1980).
29. M. E. Peach and D. J. Sutherland, *J. Fluorine Chem.*, **17**, 225 (1981).
30. A. Osuka, N. Ohmasa, Y. Uno and H. Suzuki, *Synthesis*, 68 (1983).
31. H. Suzuki, H. Abe and A. Osuka, *Chem. Lett.*, 151 (1981).
32. A. Osuka, N. Ohmasa and H. Suzuki, *Synthesis*, 857 (1982).
33. F. Ullmann, *Ber. Dtsch. Chem. Ges.*, **36**, 2382 (1903).
34. R. G. R. Bacon and S. D. Hamilton, *J. Chem. Soc., Perkin Trans. 1*, 2391 (1972).
35. R. G. R. Bacon and S. G. Pande, *J. Chem. Soc. C*, 1967 (1970).
36. N. Tuttle, *J. Am. Chem. Soc.*, **45**, 1906 (1923).
37. T. Tran Dinh and M. Hida, *Bull. Chem. Soc. Jpn.*, **43**, 1763 (1970).
38. A. A. Goldberg and W. Kelly, *J. Chem. Soc.*, 595 (1947).
39. D. M. Besly and A. A. Goldberg, *J. Chem. Soc.*, 2448 (1954).
40. W. G. Dauben, *J. Am. Chem. Soc.*, **70**, 2420 (1948).
41. F. Ullmann, *Justus Liebigs Ann. Chem.*, **355**, 312 (1908).
42. A. A. Goldberg and W. Kelly, *J. Chem. Soc.*, 102 (1946).
43. A. A. Goldberg, W. Kelly and F. Haynes, *J. Chem. Soc.*, 637 (1947).
44. D. M. Besly and A. A. Goldberg, *J. Chem. Soc.*, 5085 (1957).
45. S. Arai, Y. Hashimoto, T. Yamagishi and M. Hida, *Bull. Chem. Soc. Jpn.*, **62**, 3143 (1989).
46. S. Arai, M. Hida and T. Yamagishi, *Bull. Chem. Soc. Jpn.*, **51**, 277 (1978).
47. S. Arai, M. Hida, T. Yamagishi and S. Ototake, *Bull. Chem. Soc. Jpn.*, **50**, 2982 (1977).
48. S. Arai, A. Tanaka, M. Hida and T. Yamagishi, *Bull. Chem. Soc. Jpn.*, **52**, 1731 (1979).
49. T. Tran Dinh and M. Hida, *Bull. Chem. Soc. Jpn.*, **44**, 765 (1971).
50. T. Tran Dinh and M. Hida, *J. Chem. Soc., Perkin Trans. 2*, 675 (1974).
51. D. Bethell, I. L. Jenkins and P. M. Quan, *J. Chem. Soc., Perkin Trans. 2*, 1789 (1985).
52. H. Gilman and L. O. Moore, *J. Am. Chem. Soc.*, **79**, 3485 (1957).
53. H. G. Dunlop and S. H. Tucker, *J. Chem. Soc.*, 1945 (1939).
54. D. Hellwinkel and H. Seifert, *Chem. Commun.*, 1683 (1968).
55. A. J. Quick, *J. Am. Chem. Soc.*, **42**, 1033 (1920).
56. N. N. Vorozhtsov and Jr., V. A. Kobelev, *Zh. Obshch. Khim.*, **8**, 1330 (1938); *Chem. Abstr.*, **33**, 29121 (1939).
57. N. N. Vorozhtsov Jr. and V. A. Kobelev, *Zh. Obshch. Khim.*, **8**, 1106 (1938); *Chem. Abstr.*, **33**, 26671 (1939).
58. N. N. Vorozhtsov Jr. and V. A. Kobelev, *Zh. Obshch. Khim.*, **9**, 1569 (1939); *Chem. Abstr.*, **34**, 17481 (1940).
59. N. N. Vorozhtsov Jr. and V. A. Kobelev, *Zh. Obshch. Khim.*, **9**, 1515 (1939); *Chem. Abstr.*, **34**, 17480 (1940).
60. N. N. Vorozhtsov Jr. and V. A. Kobelev, *Zh. Obshch. Khim.*, **9**, 1465 (1939); *Chem. Abstr.*, **34**, 17479 (1940).
61. N. N. Vorozhtsov Jr. and V. A. Kobelev, *Zh. Obshch. Khim.*, **9**, 1043 (1939); *Chem. Abstr.*, **33**, 59696 (1939).
62. F. W. Bergstrom, *Chem. Rev.*, **35**, 77 (1944).
63. H. J. den Hertog, J. C. M. Schogt, J. de Bruyn and A. de Klerk, *Recl. Trav. Chim. Pays-Bas*, **69**, 673 (1950).
64. S. A. Kondratov and S. M. Shein, *Kinet. Katal.*, **21**, 1232 (1980); *Chem. Abstr.*, **94**, 64770 (1981).
65. L. P. Seiwell, *J. Org. Chem.*, **44**, 4731 (1979).
66. T. Cohen and J. G. Tirpak, *Tetrahedron Lett.*, 143 (1975).
67. S. A. Kondratov and S. M. Shein, *Zh. Org. Khim.*, **15**, 2387 (1979); *Chem. Abstr.*, **92**, 163296 (1980).
68. S. A. Kondratov and S. M. Shein, *Ukr. Khim. Zh.*, **52**, 858 (1986); *Chem. Abstr.*, **107**, 77050 (1987).
69. I. Goldberg, *Ber. Dtsch. Chem. Ges.*, **39**, 1691 (1906).
70. I. Goldberg, *Ber. Dtsch. Chem. Ges.*, **40**, 4541 (1907).
71. I. Goldberg, *Ber. Dtsch. Chem. Ges.*, **41**, 4541 (1908).
72. R. G. R. Bacon and A. Karim, *J. Chem. Soc., Perkin Trans. 1*, 272 (1973).
73. H. S. Freeman, J. R. Butler and L. D. Freedman, *J. Org. Chem.*, **43**, 4975 (1978).
74. R. G. R. Bacon and A. Karim, *J. Chem. Soc., Perkin Trans. 1*, 278 (1973).

75. T. Yamamoto, Y. Ehara, M. Kubota and A. Yamamoto, *Bull. Chem. Soc. Jpn.*, **53**, 1299 (1980).
76. T. Yamamoto and Y. Kurata, *Chem. Ind.*, 737 (1981).
77. P. E. Weston and H. Adkins, *J. Am. Chem. Soc.*, **50**, 859 (1928).
78. A. J. Paine, *J. Am. Chem. Soc.*, **109**, 1496 (1987).
79. W. F. Linke and A. Seidell (Eds), *Solubilities of Inorganic and Metal Organic Compounds*, American Chemical Society, Washington, D.C., 1966, Vol. II.
80. P. Tavs and F. Korte, *Tetrahedron*, **23**, 4677 (1967).
81. P. Tavs, *Chem. Ber.*, **103**, 2428 (1970).
82. A. Osuka, N. Ohmasa, Y. Yoshida and H. Suzuki, *Synthesis*, 69 (1983).
83. N. Hall and R. Price, *J. Chem. Soc., Perkin Trans. 1*, 2873 (1979).
84. N. Hall and R. Price, *J. Chem. Soc., Perkin Trans. 1*, 2634 (1979).
85. R. L. Beddoes, J. A. Connor, D. Dubowski, A. C. Jones, O. S. Mills and R. Price, *J. Chem. Soc., Dalton Trans.*, 2119 (1981).
86. J. A. Connor, D. Dubowski, A. C. Jones and R. Price, *J. Chem. Soc., Perkin Trans. 1*, 1143 (1982).
87. G. Axelrad, S. Laosooksathit and R. Engel, *J. Org. Chem.*, **46**, 5200 (1981).
88. G. Axelrad, S. Laosooksathit and R. Engel, *Synth. Commun.*, **10**, 933 (1980).
89. G. Axelrad, S. Laosooksathit and R. Engel, *Synth. Commun.*, **11**, 405 (1981).
90. S. Banerjee, R. Engel and G. Axelrad, *Phosphorus Sulfur*, **15**, 15 (1983).
91. D. W. Allen, I. W. Nowell, L. A. March and B. F. Taylor, *Tetrahedron Lett.*, **23**, 5417 (1982).
92. J.-F. Marcoux, S. Doye and S. L. Buchwald, *J. Am. Chem. Soc.*, **119**, 10539 (1997).
93. D. N. Reinhoudt, F. De Jong and H. P. M. Tomassen, *Tetrahedron Lett.*, 2067 (1979).
94. W. H. Kruizinga and R. M. Kellogg, *J. Am. Chem. Soc.*, **103**, 5183 (1981).
95. C. L. Liotta, E. E. Grisdale and H. P. Hopkins, Jr., *Tetrahedron Lett.*, 4205 (1975).
96. X. Xing, D. Padmanaban, L.-A. Yeh and G. D. Cuny, *Tetrahedron*, **58**, 7903 (2002).
97. A. V. Kalinin, J. F. Bower, P. Riebel and V. Snieckus, *J. Org. Chem.*, **64**, 2986 (1999).
98. S. Lohmann, S. P. Andrews, B. J. Burke, M. D. Smith, J. P. Attfield, H. Tanaka, K. Kaneko and S. V. Ley, *Synlett*, 1291 (2005).
99. M. Kidwai, N. K. Mishra, V. Bansal, A. Kumar and S. Mozumdar, *Tetrahedron Lett.*, **48**, 8883 (2007).
100. R. K. Gujadhur, C. G. Bates and D. Venkataraman, *Org. Lett.*, **3**, 4315 (2001).
101. B. H. Lipshutz, J. B. Unger and B. R. Taft, *Org. Lett.*, **9**, 1089 (2007).
102. D. Ma and Q. Cai, *Org. Lett.*, **5**, 3799 (2003).
103. Q. Cai, G. He and D. Ma, *J. Org. Chem.*, **71**, 5268 (2006).
104. Q. Cai, B. Zou and D. Ma, *Angew. Chem., Int. Ed.*, **45**, 1276 (2006).
105. E. Buck, Z. J. Song, D. Tschaen, P. G. Dormer, R. P. Volante and P. J. Reider, *Org. Lett.*, **4**, 1623 (2002).
106. G. Sekar, A. B. Naidu, O. R. Raghunath and D. J. C. Prasad, *Tetrahedron Lett.*, **49**, 1057 (2008).
107. H. Rao, Y. Jin, H. Fu, Y. Jiang and Y. Zhao, *Chem.-Eur. J.*, **12**, 3636 (2006).
108. A. Ouali, J.-F. Spindler, H.-J. Cristau and M. Taillefer, *Adv. Synth. Catal.*, **348**, 499 (2006).
109. X. Lv and W. Bao, *J. Org. Chem.*, **72**, 3863 (2007).
110. H.-J. Cristau, P. P. Cellier, S. Hamada, J.-F. Spindler and M. Taillefer, *Org. Lett.*, **6**, 913 (2004).
111. M. Wolter, G. Nordmann, G. E. Job and S. L. Buchwald, *Org. Lett.*, **4**, 973 (2002).
112. R. A. Altman, A. Shafir, A. Choi, P. A. Lichtor and S. L. Buchwald, *J. Org. Chem.*, **73**, 284 (2008).
113. A. Shafir, P. A. Lichtor and S. L. Buchwald, *J. Am. Chem. Soc.*, **129**, 3490 (2007).
114. M. Palucki, J. P. Wolfe and S. L. Buchwald, *J. Am. Chem. Soc.*, **119**, 3395 (1997).
115. M. Toumi, F. Couty and G. Evano, *Angew. Chem., Int. Ed.*, **46**, 572 (2007).
116. H. Zhang, D. Ma and W. Cao, *Synlett*, 243 (2007).
117. M. A. Keegstra. *Tetrahedron*, **48**, 2681 (1992).
118. Z. Wan, C. D. Jones, T. M. Koenig, Y. J. Pu and D. Mitchell, *Tetrahedron Lett.*, **44**, 8257 (2003).
119. D. Ma, Q. Cai and X. Xie, *Synlett*, 1767 (2005).
120. G. Nordmann and S. L. Buchwald. *J. Am. Chem. Soc.*, **125**, 4978 (2003).
121. P. D. Nonappa, K. Pandurangan, U. Maitra and S. Wailes, *Org. Lett.*, **9**, 2767 (2007).

122. W. E. Duncan, E. Ott and E. E. Reid, *J. Ind. Eng. Chem.*, **23**, 381 (1931).
123. P. G. Eller and G. J. Kubas, *J. Am. Chem. Soc.*, **99**, 4346 (1977).
124. S. Schneider, J. A. S. Roberts, M. R. Salata and T. J. Marks *Angew. Chem., Int. Ed.*, **45**, 1733 (2006).
125. L. M. Nguyen, M. E. Dellinger, J. T. Lee, R. A. Quinlan, A. L. Rheingold and R. D. Pike, *Inorg. Chim. Acta*, **358**, 1331 (2005).
126. C. Palomo, M. Oiarbide, R. Lopez and E. Gomez-Bengoa, *Tetrahedron Lett.*, **41**, 1283 (2000).
127. H. R. Allcock, K. D. Lavin, N. M. Tollefson and T. L. Evans, *Organometallics*, **2**, 267 (1983).
128. P. Oulie, C. Freund, N. Saffon, B. Martin-Vaca, L. Maron and D. Bourissou, *Organometallics*, **26**, 6793 (2007).
129. M. Witt and H. W. Roesky, *Polyhedron*, **8**, 1736 (1989).
130. F. Y. Kwong and S. L. Buchwald, *Org. Lett.*, **4**, 3517 (2002).
131. C. G. Bates, P. Saejueng, M. Q. Doherty and D. Venkataraman, *Org. Lett.*, **6**, 5005 (2004).
132. D. Zhu, L. Xu, F. Wu and B. Wan, *Tetrahedron Lett.*, **47**, 5781 (2006).
133. Y. Zheng, X. Du and W. Bao, *Tetrahedron Lett.*, **47**, 1217 (2006).
134. H. Zhang, W. Cao and D. Ma, *Synth. Commun.*, **37**, 25 (2007).
135. G. Zeni, *Tetrahedron Lett.*, **46**, 2647 (2005).
136. M. Carril, R. SanMartin, E. Dominguez and I. Tellitu, *Chem.-Eur. J.*, **13**, 5100 (2007).
137. P. Buranaprasertsuk, J. W. W. Chang, W. Chavasiri and P. W. H. Chan, *Tetrahedron Lett.*, **49**, 2023 (2008).
138. S. Liu, J. P. C. Pestano and C. Wolf, *Synthesis*, 3519 (2007).
139. B. C. Ranu, A. Saha and R. Jana, *Adv. Synth. Catal.*, **349**, 2690 (2007).
140. R. K. Gujadhur and D. Venkataraman, *Tetrahedron Lett.*, **44**, 81 (2003).
141. M. Clarembeau, A. Cravador, W. Dumont, L. Hevesi, A. Krief, J. Lucchetti and D. Van Ende, *Tetrahedron*, **41**, 4793 (1985).
142. S. Oae and H. Togo, *Bull. Chem. Soc. Jpn.*, **57**, 232 (1984).
143. N. Taniguchi and T. Onami, *J. Org. Chem.*, **69**, 915 (2004).
144. T. G. Back, S. Collins, M. V. Krishna and K. W. Law, *J. Org. Chem.*, **52**, 4258 (1987).
145. D. Z. Chang and W. L. Bao, *Synlett*, 1786 (2006).
146. S. Kumar and L. Engman, *J. Org. Chem.*, **71**, 5400 (2006).
147. Y. Fang and C. Li, *J. Am. Chem. Soc.*, **129**, 8092 (2007).
148. A. Martinez, M. Fernandez, J. C. Estevez, R. J. Estevez and L. Castedo, *Tetrahedron*, **61**, 1353 (2005).
149. A. Martinez, M. Fernandez, J. C. Estevez, R. J. Estevez and L. Castedo, *Tetrahedron*, **61**, 485 (2005).
150. J. Liu, A. E. Fitzgerald and N. S. Mani, *J. Org. Chem.*, **73**, 2951 (2008).
151. K. Kawaguchi, K. Nakano and K. Nozaki, *J. Org. Chem.*, **72**, 5119 (2007).
152. S. Y. Wen and J. L. Jie, *Org. Lett.*, **4**, 2201 (2002).
153. J. H. Spatz, T. Bach, M. Umkehrer, J. Bardin, G. Ross, C. Burdack and J. Kolb, *Tetrahedron Lett.*, **48**, 9030 (2007).
154. R. D. Viirre, G. Evindar and R. A. Batey, *J. Org. Chem.*, **73**, 3452 (2008).
155. J. K. Wang, F. Peng, J. L. Jiang, Z. J. Lu, L. Y. Wang, J. F. Bai and Y. Pan, *Tetrahedron Lett.*, **49**, 467 (2008).
156. L. L. Joyce, G. Evindar and R. A. Batey, *Chem. Commun.*, 446 (2004).
157. D. Ma and J. Yao, *Tetrahedron: Asymmetry*, **7**, 3075 (1996).
158. D. W. Ma, Y. D. Zhang, J. C. Yao, S. H. Wu and F. G. Tao, *J. Am. Chem. Soc.*, **120**, 12459 (1998).
159. D. J. Reynolds and S. A. Hermitage, *Tetrahedron*, **57**, 7765 (2001).
160. S. Roettger, P. J. R. Sjoeberg and M. Larhed, *J. Comb. Chem.*, **9**, 204 (2007)
161. Y. Endo, M. Ohno, M. Hirano, A. Itai and K. Shudo, *J. Am. Chem. Soc.*, **118**, 1841 (1996).
162. D. Ma and C. Xia, *Org. Lett.*, **3**, 2583 (2001).
163. R. M. Scarborough and D. D. Gretler, *J. Med. Chem.*, **43**, 3453 (2000).
164. F. Lang, D. Zewge, I. N. Houpis and R. P. Volante, *Tetrahedron Lett.*, **42**, 3251 (2001).
165. F. Y. Kwong, A. Klapars and S. L. Buchwald, *Org. Lett.*, **4**, 581 (2002).
166. Z. Lu and R. J. Twieg, *Tetrahedron*, **61**, 903 (2005).
167. Z. Lu and R. J. Twieg, *Tetrahedron Lett.*, **46**, 2997 (2005).
168. Z. Lu, R. J. Twieg and S. D. Huang, *Tetrahedron Lett.*, **44**, 6289 (2003).

169. F. Y. Kwong and S. L. Buchwald, *Org. Lett.*, **5**, 793 (2003).
170. B. de Lange, M. H. Lambers-Verstappen, L. Schmieder-van de Vondervoort, N. Sereinig, R. de Rijk, A. H. M. de Vries and J. G. de Vries, *Synlett*, 3105 (2006).
171. A. Shafir and S. L. Buchwald, *J. Am. Chem. Soc.*, **128**, 8742 (2006).
172. G. E. Job and S. L. Buchwald, *Org. Lett.*, **4**, 3703 (2002).
173. D. Ma, Q. Cai and H. Zhang, *Org. Lett.*, **5**, 2453 (2003).
174. H. Zhang, Q. Cai and D. Ma, *J. Org. Chem.*, **70**, 5164 (2005).
175. B. Zou, Q. Yuan and D. Ma, *Angew. Chem., Int. Ed.*, **46**, 2598 (2007).
176. B. L. Zou, O. L. Yuan and D. W. Ma, *Org. Lett.*, **9**, 4291 (2007).
177. X. Guo, H. Rao, H. Fu, Y. Jiang and Y. Zhao, *Adv. Synth. Catal.*, **348**, 2197 (2006).
178. V. S. C. Yeh and P. E. Wiedeman, *Tetrahedron Lett.*, **47**, 6011 (2006).
179. L. Ackermann, *Org. Lett.*, **7**, 439 (2005).
180. K. Okano, H. Tokuyama and T. Fukuyama, *Org. Lett.*, **5**, 4987 (2003).
181. Q. Jiang, D. Jiang, Y. Jiang, H. Fu and Y. Zhao, *Synlett*, 1836 (2007).
182. M. Sarkar and A. Samanta, *Synthesis*, 3425 (2006).
183. A. S. Gajare, K. Toyota, M. Yoshifuji and F. Ozawa, *Chem. Commun.*, 1994 (2004).
184. D. S. Jiang, H. Fu, Y. Y. Jiang and Y. F. Zhao, *J. Org. Chem.*, **72**, 672 (2007).
185. M. Yang and F. Liu, *J. Org. Chem.*, **72**, 8969 (2007).
186. Z. Zhang, J. Mao, D. Zhu, F. Wu, H. Chen and B. Wan, *Tetrahedron*, **62**, 4435 (2006).
187. D. Zhu, R. Wang, J. Mao, L. Xu, F. Wu and B. Wan, *J. Mol. Catal. A*, **256**, 256 (2006).
188. B. H. Goodbrand and N.-X. Hu, *J. Org. Chem.*, **64**, 670 (1999).
189. A. A. Kelkar, N. M. Patil and R. V. Chaudhari, *Tetrahedron Lett.*, **43**, 7143 (2002).
190. K. Moriwaki, K. Satoh, M. Takada, Y. Ishino and T. Ohno, *Tetrahedron Lett.*, **46**, 7559 (2005).
191. C. Z. Ran, Q. Dai and R. G. Harvey, *J. Org. Chem.*, **70**, 3724 (2005).
192. X. Mei, A. T. August and C. Wolf, *J. Org. Chem.*, **71**, 142 (2006).
193. X. F. Mei, R. M. Martin and C. Wolf, *J. Org. Chem.*, **71**, 2854 (2006).
194. V. Ramalingam, N. Bhagirath and R. S. Muthyala, *J. Org. Chem.*, **72**, 3976 (2007).
195. H. Rao, H. Fu, Y. Jiang and Y. Zhao, *J. Org. Chem.*, **70**, 8107 (2005).
196. Y. Liu, Y. Bai, J. Zhang, Y. Li, J. Jiao and X. Qi, *Eur. J. Org. Chem.*, 6084 (2007).
197. Y. H. Liu, C. Chen and L. M. Yang, *Tetrahedron Lett.*, **47**, 9275 (2006).
198. M. B. Maradolla, M. Amaravathi, V. N. Kumar and G. V. P. C. Mouli, *J. Mol. Catal. A*, **266**, 47 (2007).
199. N. S. Nandurkar, M. J. Bhanushali, M. D. Bhor and B. M. Bhanage. *Tetrahedron Lett.*, **48**, 6573 (2007).
200. D. Vina, E. del Olmo, J. L. Lopez-Perez and A. S. Feliciano, *Org. Lett.*, **9**, 525 (2007).
201. Y. H. Zhao, Y. S. Wang, H. W. Sun, L. Li and H. B. Zhang, *Chem. Commun.*, 3186 (2007).
202. A. Kiyomori, J.-F. Marcoux and S. L. Buchwald, *Tetrahedron Lett.*, **40**, 2657 (1999).
203. J. B. Arterburn, M. Pannala and A. M. Gonzalez, *Tetrahedron Lett.*, **42**, 1475 (2001).
204. E. Alcalde, I. Dinares, S. Rodriguez and C. G. de Miguel, *Eur. J. Org. Chem.*, 1637 (2005).
205. R. A. Altman and S. L. Buchwald, *Org. Lett.*, **8**, 2779 (2006).
206. R. A. Altman, E. D. Koval and S. L. Buchwald, *J. Org. Chem.*, **72**, 6190 (2007).
207. B. M. Choudary, C. Sridhar, M. L. Kantam, G. T. Venkanna and B. Sreedhar, *J. Am. Chem. Soc.*, **127**, 9948 (2005).
208. Y.-Z. Huang, J. Gao, H. Ma, H. Miao and J. Xu, *Tetrahedron Lett.*, **49**, 948 (2008).
209. L. Liu, M. Frohn, N. Xi, C. Dominguez, R. Hungate and P. J. Reider, *J. Org. Chem.*, **70**, 10135 (2005).
210. H.-C. Ma and X.-Z. Jiang, *J. Org. Chem.*, **72**, 8943 (2007).
211. H. Maheswaran, G. G. Krishna, K. L. Prasanth, V. Srinivas, G. K. Chaitanya and K. Bhanuprakash, *Tetrahedron*, **64**, 2471 (2008).
212. Z. Xi, F. Liu, Y. Zhou and W. Chen, *Tetrahedron*, **64**, 4254 (2008).
213. L. Zhu, L. Cheng, Y. Zhang, R. Xie and J. You, *J. Org. Chem.*, **72**, 2737 (2007).
214. J. C. Antilla, J. M. Baskin, T. E. Barder and S. L. Buchwald, *J. Org. Chem.*, **69**, 5578 (2004).
215. C. Enguehard, H. Allouchi, A. Gueiffier and S. L. Buchwald, *J. Org. Chem.*, **68**, 5614 (2003).
216. J. C. Antilla, A. Klapars and S. L. Buchwald, *J. Am. Chem. Soc.*, **124**, 11684 (2002).
217. H.-J. Cristau, P. P. Cellier, J.-F. Spindler and M. Taillefer, *Chem.-Eur. J.*, **10**, 5607 (2004).
218. T. Mino, Y. Harada, H. Shindo, M. Sakamoto and T. Fujita, *Synlett*, 614 (2008).
219. H. J. Cristau, P. P. Cellier, J. F. Spindler and M. Taillefer, *Eur. J. Org. Chem.*, 695 (2004).

220. A. Ouali, R. Laurent, A.-M. Caminade, J.-P. Majoral and M. Taillefer, *J. Am. Chem. Soc.*, **128**, 15990 (2006).
221. A. Klapars, J. C. Antilla, X. Huang and S. L. Buchwald, *J. Am. Chem. Soc.*, **123**, 7727 (2001).
222. C. Enguehard-Gueiffier, I. Thery, A. Gueiffier and S. L. Buchwald, *Tetrahedron*, **62**, 6042 (2006).
223. A. Padwa, K. R. Crawford, P. Rashatasakhon and M. Rose, *J. Org. Chem.*, **68**, 2609 (2003).
224. D. P. Phillips, A. R. Hudson, B. Nguyen, T. L. Lau, M. H. McNeill, J. E. Dalgard, J.-H. Chen, R. J. Penuliar, T. A. Miller and L. Zhi, *Tetrahedron Lett.*, **47**, 7137 (2006).
225. C. Z. Tao, J. Li, Y. Fu, L. Liu and Q. X. Guo, *Tetrahedron Lett.*, **49**, 70 (2008).
226. P. S. Wang, C. K. Liang and M. K. Leung, *Tetrahedron*, **61**, 2931 (2005).
227. A. Klapars, X. Huang and S. L. Buchwald, *J. Am. Chem. Soc.*, **124**, 7421 (2002).
228. E. R. Strieter, D. G. Blackmond and S. L. Buchwald, *J. Am. Chem. Soc.*, **127**, 4120 (2005).
229. O. S. D. Barros, C. W. Nogueira, E. C. Stangherlin, P. H. Menezes and G. Zeni. *J, Org. Chem.*, **71**, 1552 (2006).
230. K. J. Filipski, J. T. Kohrt, A. Casimiro-Garcia, C. A. Van Huis, D. A. Dudley, W. L. Cody, C. F. Bigge, S. Desiraju, S. Y. Sun, S. N. Maiti, M. R. Jaber and J. J. Edmunds, *Tetrahedron Lett.*, **47**, 7677 (2006).
231. G. Feng, J. Wu and W.-M. Dai, *Tetrahedron Lett.*, **48**, 401 (2007).
232. R. Hosseinzadeh, M. Tajbakhsh, M. Mohadjerani and H. Mehdinejad, *Synlett*, 1517 (2004).
233. Y. J. Chen and H. H. Chen, *Org. Lett.*, **8**, 5609 (2006).
234. A. van den Hoogenband, J. H. M. Lange, J. A. J. den Hartog, R. Henzen and J. W. Terpstra, *Tetrahedron Lett.*, **48**, 4461 (2007).
235. A. Ghosh, J. E. Sieser, S. Caron, M. Couturier, K. Dupont-Gaudet and M. Girardin, *J. Org. Chem.*, **71**, 1258 (2006).
236. B. Mallesham, B. M. Rajesh, P. R. Reddy, D. Srinivas and S. Trehan, *Org. Lett.*, **5**, 963 (2003).
237. T. Yang, C. Lin, H. Fu, Y. Jiang and Y. Zhao, *Org. Lett.*, **7**, 4781 (2005).
238. W. Deng, L. Liu, C. Zhang, M. Liu and Q. X. Guo, *Tetrahedron Lett.*, **46**, 7295 (2005).
239. G. Y. Cho, P. Remy, J. Jansson, C. Moessner and C. Bolm, *Org. Lett.*, **6**, 3293 (2004).
240. A. Correa and C. Bolm, *Adv. Synth. Catal.*, **349**, 2673 (2007).
241. J. Sedelmeier and C. Bolm, *J. Org. Chem.*, **70**, 6904 (2005).
242. C. Worch and C. Bolm, *Synthesis*, 739 (2008).
243. M. V. Nandakumar, *Tetrahedron Lett.*, **45**, 1989 (2004).
244. T. Hafner and D. Kunz, *Synthesis*, 1403 (2007).
245. J. E. Golden, S. D. Sanders, K. M. Muller and R. W. Buerli, *Tetrahedron Lett.*, **49**, 794 (2008).
246. K.-Y. Kim, J.-T. Shin, K.-S. Lee and C.-G. Cho, *Tetrahedron Lett.*, **45**, 117 (2004).
247. M. Wolter, A. Klapars and S. L. Buchwald, *Org. Lett.*, **3**, 3803 (2001).
248. K. L. Jones, A. Porzelle, A. Hall, M. D. Woodrow and N. C. O. Tomkinson, *Org. Lett.*, **10**, 797 (2008).
249. G. Evindar and R. A. Batey, *Org. Lett.*, **5**, 133 (2003).
250. G. Cuny, M. Bois-Choussy and J. Zhu, *J. Am. Chem. Soc.*, **126**, 14475 (2004).
251. G. Altenhoff and F. Glorius, *Adv. Synth. Catal.*, **346**, 1661 (2004).
252. N. Zheng and S. L. Buchwald, *Org. Lett.*, **9**, 4749 (2007).
253. C. P. Jones, K. W. Anderson and S. L. Buchwald, *J. Org. Chem.*, **72**, 7968 (2007).
254. A. Klapars, S. Parris, K. W. Anderson and S. L. Buchwald, *J. Am. Chem. Soc.*, **126**, 3529 (2004).
255. R. Shen and J. A. Porco, Jr., *Org. Lett.*, **2**, 1333 (2000).
256. R. Shen, C. T. Lin, E. J. Bowman, B. J. Bowman and J. A. Porco, Jr., *J. Am. Chem. Soc.*, **125**, 7889 (2003).
257. R. S. Coleman and P. H. Liu, *Org. Lett.*, **6**, 577 (2004).
258. L. Jiang, G. E. Job, A. Klapars and S. L. Buchwald, *Org. Lett.*, **5**, 3667 (2003).
259. X. Pan, Q. Cai and D. Ma, *Org. Lett.*, **6**, 1809 (2004).
260. C. Han, R. C. Shen, S. Su and J. A. Porco, Jr., *Org. Lett.*, **6**, 27 (2004).
261. T. S. Hu and C. Z. Li, *Org. Lett.*, **7**, 2035 (2005).
262. H. J. Lu and C. Z. Li, *Org. Lett.*, **8**, 5365 (2006).
263. M. O. Frederick, J. A. Mulder, M. R. Tracey, R. P. Hsung, J. Huang, K. C. M. Kurtz, L. C. Shen and C. J. Douglas, *J. Am. Chem. Soc.*, **125**, 2368 (2003).

264. X. Zhang, Y. Zhang, J. Huang, R. P. Hsung, K. C. M. Kurtz, J. Oppenheimer, M. E. Petersen, I. K. Sagamanova, L. Shen and M. R. Tracey, *J. Org. Chem.*, **71**, 4170 (2006).
265. Y. Zhang, R. P. Hsung, M. R. Tracey, K. C. M. Kurtz and E. L. Vera, *Org. Lett.*, **6**, 1151 (2004).
266. L. C. Shen, R. P. Hsung, Y. S. Zhang, J. E. Antoline and X. J. Zhang, *Org. Lett.*, **7**, 3081 (2005).
267. B. M. Trost and D. T. Stiles, *Org. Lett.*, **7**, 2117 (2005).
268. X. Yuan, X. Xu, X. Zhou, J. Yuan, L. Mai and Y. Li, *J. Org. Chem.*, **72**, 1510 (2007).
269. R. Martin, C. H. Larsen, A. Cuenca and S. L. Buchwald, *Org. Lett.*, **9**, 3379 (2007).
270. M. R. Rivero and S. L. Buchwald, *Org. Lett.*, **9**, 973 (2007).
271. R. Martin, M. R. Rivero and S. L. Buchwald, *Angew. Chem., Int. Ed.*, **45**, 7079 (2006).
272. J. Yuen, Y.-Q. Fang and M. Lautens, *Org. Lett.*, **8**, 653 (2006).
273. R. Martin, A. Cuenca and S. L. Buchwald, *Org. Lett.*, **9**, 5521 (2007).
274. K. Schuh and F. Glorius, *Synthesis*, 2297 (2007).
275. Y. Fukudome, H. Naito, T. Hata and H. Urabe, *J. Am. Chem. Soc.*, **130**, 1820 (2008).
276. J. Andersen, U. Madsen, F. Bjorkling and X. Liang, *Synlett*, 2209 (2005).
277. Q. Cai, W. Zhu, H. Zhang, Y. Zhang and D. Ma, *Synthesis*, 496 (2005).
278. W. Zhu and D. Ma, *Chem. Commun.*, 888 (2004).
279. D. Gelman, L. Jiang and S. L. Buchwald, *Org. Lett.*, **5**, 2315 (2003).
280. D. Van Allen and D. Venkataraman, *J. Org. Chem.*, **68**, 4590 (2003).
281. M. T. Honaker, B. J. Sandefur, J. L. Hargett, A. L. McDaniel and R. N. Salvatore, *Tetrahedron Lett.*, **44**, 8373 (2003).
282. D. C. Behenna and B. M. Stoltz, *J. Am. Chem. Soc.*, **126**, 15044 (2004).
283. K. Tani, D. C. Behenna, R. M. McFadden and B. M. Stoltz, *Org. Lett.*, **9**, 2529 (2007).
284. C. Huang, X. Tang, H. Fu, Y. Jiang and Y. Zhao, *J. Org. Chem.*, **71**, 5020 (2006).
285. H. Suzuki, A. Kondo, M. Inouye and T. Ogawa, *Synthesis*, 121 (1986).
286. H. Suzuki, M. Aihara, H. Yamamoto, Y. Takamoto and T. Ogawa, *Synthesis*, 236 (1988).
287. J. H. Clark, C. W. Jones, C. V. A. Duke and J. M. Miller, *J. Chem. Res., Synop.*, 238 (1989).
288. H. Abe and H. Suzuki, *Bull. Chem. Soc. Jpn.*, **72**, 787 (1999).
289. A. Klapars and S. L. Buchwald, *J. Am. Chem. Soc.*, **124**, 14844 (2002).
290. T. Cohen, J. Wood and A. G. Dietz, Jr., *Tetrahedron Lett.*, 3555 (1974).
291. A. H. Lewin and T. Cohen, *Tetrahedron Lett.*, 4531 (1965).
292. K. Yamada, T. Kubo, H. Tokuyama and T. Fukuyama, *Synlett*, 231 (2002).
293. G. Barone, G. La Manna and D. Duca, *Eur. J. Inorg. Chem.*, 410 (2005).
294. J. L. DuBois, P. Mukherjee, T. D. P. Stack, B. Hedman, E. I. Solomon and K. O. Hodgson, *J. Am. Chem. Soc.*, **122**, 5775 (2000).
295. J. P. Fox, B. Ramdhanie, A. A. Zareba, R. S. Czernuszewicz and D. P. Goldberg, *Inorg. Chem.*, **43**, 6600 (2004).
296. C. Amatore and A. Jutand, in *Handbook of Organopalladium Chemistry for Organic Synthesis* (Ed. E.-i. Negishi), John Wiley & Sons, Inc., Hoboken, N.J., 2002, p. 943.
297. S. L. Zhang, L. Liu, Y. Fu and Q. X. Guo, *Organometallics*, **26**, 4546 (2007).
298. M. A. Z. Eltayeb and Y. Sulfab, *Polyhedron*, **26**, 1 (2007).
299. F. M. Beringer, A. Brierley, M. Drexler, E. M. Gindler and C. C. Lumpkin, *J. Am. Chem. Soc.*, **75**, 2708 (1953).
300. P. Block, Jr. and D. H. Coy, *J. Chem. Soc., Perkin Trans. 1*, 633 (1972).
301. M. J. Humora, J. Quick and D. E. Seitz, *Tetrahedron Lett.*, **21**, 3971 (1980).
302. R. A. Scherrer and H. R. Beatty, *J. Org. Chem.*, **45**, 2127 (1980).
303. F. M. Beringer, M. Drexler, E. M. Gindler and C. C. Lumpkin, *J. Am. Chem. Soc.*, **75**, 2705 (1953).
304. S.-K. Kang, S.-H. Lee and D. Lee, *Synlett*, 102 (2000).
305. T. Zhou and Z.-C. Chen, *Synth. Commun.*, **32**, 903 (2002).
306. T. Zhou, T.-C. Li and Z.-C. Chen, *Helv. Chim. Acta*, **88**, 290 (2005).
307. S. Thielges, P. Bisseret and J. Eustache, *Org. Lett.*, **7**, 681 (2005).
308. V. A. Dodonov, A. V. Gushchin and T. G. Brilkina, *Zh. Obshch. Khim.*, **54**, 2157 (1984); *Chem. Abstr.*, **103**, 45543 (1985).
309. V. A. Dodonov, A. V. Gushchin and T. G. Brilkina, *Zh. Obshch. Khim.*, **55**, 2514 (1985); *Chem. Abstr.*, **106**, 5162 (1987).

310. V. A. Dodonov, A. V. Gushchin, T. G. Brilkina and L. V. Muratova, *Zh. Obshch. Khim.*, **56**, 2714 (1986); *Chem. Abstr.*, **107**, 197657 (1987).
311. V. A. Dodonov, T. I. Starostina, Y. L. Kuznetsova and A. V. Gushchin, *Izv. Akad. Nauk, Ser. Khim.*, 156 (1995); *Chem. Abstr.*, **123**, 198346 (1995).
312. D. H. R. Barton and J. P. Finet, *Pure Appl. Chem.*, **59**, 937 (1987).
313. D. H. R. Barton, J. P. Finet and C. Pichon, *J. Chem. Soc., Chem. Commun.*, 65 (1986).
314. D. H. R. Barton, J. P. Finet, J. Khamsi and C. Pichon, *Tetrahedron Lett.*, **27**, 3619 (1986).
315. A. Y. Fedorov and J.-P. Finet, *J. Chem. Soc., Perkin Trans. 1*, 3775 (2000).
316. T. Harada, S. Ueda, T. M. T. Tuyet, A. Inoue, K. Fujita, M. Takeuchi, N. Ogawa, A. Oku and M. Shiro, *Tetrahedron*, **53**, 16663 (1997).
317. T. Harada, S. Ueda, T. Yoshida, A. Inoue, M. Takeuchi, N. Ogawa, A. Oku and M. Shiro, *J. Org. Chem.*, **59**, 7575 (1994).
318. H. Brunner, U. Obermann and P. Wimmer, *J. Organomet. Chem.*, **316**, C1 (1986).
319. H. Brunner, U. Obermann and P. Wimmer, *Organometallics*, **8**, 821 (1989).
320. K. M. J. Brands, U.-H. Dolling, R. B. Jobson, G. Marchesini, R. A. Reamer and J. M. Williams, *J. Org. Chem.*, **63**, 6721 (1998).
321. S. Combes and J.-P. Finet, *Tetrahedron*, **55**, 3377 (1999).
322. D. H. R. Barton, J. P. Finet and J. Khamsi, *Tetrahedron Lett.*, **27**, 3615 (1986).
323. M. A. Esteves, N. Narender, M. J. Marcelo-Curto and B. Gigante, *J. Nat. Prod.*, **64**, 761 (2001).
324. S. Morel, F. Chatel, G. Boyer and J.-P. Galy, *J. Chem. Res., Synop.*, 4 (1998).
325. S. Combes and J.-P. Finet, *Tetrahedron*, **54**, 4313 (1998).
326. T. Arnauld, D. H. R. Barton and E. Doris, *Tetrahedron*, **53**, 4137 (1997).
327. S. Y. Ablordeppey, P. Fan, A. M. Clark and A. Nimrod, *Bioorg. Med. Chem.*, **7**, 343 (1999).
328. D. H. R. Barton, J. P. Finet and J. Khamsi, *Tetrahedron Lett.*, **29**, 1115 (1988).
329. S. K. Moiseev, I. V. Bakhanova, H. Schmidhammer and V. N. Kalinin, *Russ. Chem. Bull.*, **48**, 589 (1999).
330. A. Y. Fedorov and J.-P. Finet, *Tetrahedron Lett.*, **40**, 2747 (1999).
331. A. Y. Fedorov, S. Combes and J.-P. Finet, *Tetrahedron*, **55**, 1341 (1999).
332. D. H. R. Barton, J. P. Finet and J. Khamsi, *Tetrahedron Lett.*, **30**, 937 (1989).
333. J. C. Anderson and M. Harding, *Chem. Commun.*, 393 (1998).
334. J. C. Anderson, R. Cubbon, M. Harding and D. S. James, *Tetrahedron: Asymmetry*, **9**, 3461 (1998).
335. V. A. Dodonov, T. I. Starostina, A. V. Gushchin and T. A. Egorova, *Metalloorg. Khim.*, **2**, 682 (1989); *Chem. Abstr.*, **111**, 232200 (1989).
336. V. A. Dodonov, T. I. Starostina and V. V. Balukova, *Zh. Obshch. Khim.*, **62**, 621 (1992); *Chem. Abstr.*, **117**, 233191 (1992).
337. D. H. R. Barton, N. Yadav-Bhatnagar, J. P. Finet and J. Khamsi, *Tetrahedron Lett.*, **28**, 3111 (1987).
338. D. H. R. Barton, D. M. X. Donnelly, J. P. Finet and P. J. Guiry, *Tetrahedron Lett.*, **30**, 1377 (1989).
339. D. H. R. Barton, D. M. X. Donnelly, J. P. Finet and P. J. Guiry, *J. Chem. Soc., Perkin Trans. 1*, 2095 (1991).
340. R. P. Kozyrod, J. Morgan and J. T. Pinhey, *Aust. J. Chem.*, **38**, 1147 (1985).
341. R. P. Kozyrod and J. T. Pinhey, *Org. Synth.*, **62**, 24 (1984).
342. P. Lopez-Alvarado, C. Avendano and J. C. Menendez, *Tetrahedron Lett.*, **33**, 6875 (1992).
343. P. Lopez-Alvarado, C. Avendano and J. C. Menendez, *J. Org. Chem.*, **61**, 5865 (1996).
344. R. A. Abramovitch, D. H. R. Barton and J. P. Finet, *Tetrahedron*, **44**, 3039 (1988).
345. G. I. Elliott and J. P. Konopelski, *Org. Lett.*, **2**, 3055 (2000).
346. P. Lopez-Alvarado, C. Avendano and J. C. Menendez, *Tetrahedron Lett.*, **33**, 659 (1992).
347. P. Lopez-Alvarado, C. Avendano and J. C. Menendez, *J. Org. Chem.*, **60**, 5678 (1995).
348. D. M. T. Chan, K. L. Monaco, R.-P. Wang and M. P. Winters, *Tetrahedron Lett.*, **39**, 2933 (1998).
349. D. A. Evans, J. L. Katz and T. R. West, *Tetrahedron Lett.*, **39**, 2937 (1998).
350. D. M. T. Chan, K. L. Monaco, R. Li, D. Bonne, C. G. Clark and P. Y. S. Lam, *Tetrahedron Lett.*, **44**, 3863 (2003).
351. P. Y. S. Lam, G. Vincent, D. Bonne and C. G. Clark, *Tetrahedron Lett.*, **44**, 4927 (2003).
352. G. C. H. Chiang and T. Olsson, *Org. Lett.*, **6**, 3079 (2004).

353. N. F. McKinley and D. F. O'Shea, *J. Org. Chem.*, **69**, 5087 (2004).
354. P. Y. S. Lam, G. Vincent, C. G. Clark, S. Deudon and P. K. Jadhav, *Tetrahedron Lett.*, **42**, 3415 (2001).
355. A. D. Sagar, R. H. Tale and R. N. Adude, *Tetrahedron Lett.*, **44**, 7061 (2003).
356. J. Simon, S. Salzbrunn, G. K. S. Prakash, N. A. Petasis and G. A. Olah, *J. Org. Chem.*, **66**, 633 (2001).
357. H. M. Petrassi, K. B. Sharpless and J. W. Kelly, *Org. Lett.*, **3**, 139 (2001).
358. M. E. Jung and T. I. Lazarova, *J. Org. Chem.*, **64**, 2976 (1999).
359. D. A. Evans, J. L. Katz, G. S. Peterson and T. Hintermann, *J. Am. Chem. Soc.*, **123**, 12411 (2001).
360. C. P. Decicco, Y. Song and D. A. Evans, *Org. Lett.*, **3**, 1029 (2001).
361. Y. Hitotsuyanagi, H. Ishikawa, S. Naito and K. Takeya, *Tetrahedron Lett.*, **44**, 5901 (2003).
362. A. Biffis, M. Gardan and B. Corain, *J. Mol. Catal. A*, **250**, 1 (2006).
363. T. D. Quach and R. A. Batey, *Org. Lett.*, **5**, 1381 (2003).
364. P. S. Herradura, K. A. Pendola and R. K. Guy, *Org. Lett.*, **2**, 2019 (2000).
365. P. Y. S. Lam, C. G. Clark, S. Saubern, J. Adams, M. P. Winters, D. M. T. Chan and A. Combs, *Tetrahedron Lett.*, **39**, 2941 (1998).
366. D. J. Cundy and S. A. Forsyth, *Tetrahedron Lett.*, **39**, 7979 (1998).
367. E. T. Chernick, M. J. Ahrens, K. A. Scheidt and M. R. Wasielewski, *J. Org. Chem.*, **70**, 1486 (2005).
368. Y. Hari, Y. Shoji and T. Aoyama, *Tetrahedron Lett.*, **46**, 3771 (2005).
369. W. W. K. R. Mederski, M. Lefort, M. Germann and D. Kux, *Tetrahedron*, **55**, 12757 (1999).
370. S. Yu, J. Saenz and J. K. Srirangam, *J. Org. Chem.*, **67**, 1699 (2002).
371. A. P. Combs, S. Tadesse, M. Rafalski, T. S. Haque and P. Y. S. Lam, *J. Comb. Chem.*, **4**, 179 (2002).
372. P. Y. S. Lam, D. Bonne, G. Vincent, C. G. Clark and A. P. Combs, *Tetrahedron Lett.*, **44**, 1691 (2003).
373. M. F. Jacobsen, M. M. Knudsen and K. V. Gothelf, *J. Org. Chem.*, **71**, 9183 (2006).
374. Y. Yue, Z.-G. Zheng, B. Wu, C.-Q. Xia and X.-Q. Yu, *Eur. J. Org. Chem.*, 5154 (2005).
375. J. P. Collman, M. Zhong, L. Zeng and S. Costanzo, *J. Org. Chem.*, **66**, 1528 (2001).
376. J. P. Collman, M. Zhong, C. Zhang and S. Costanzo, *J. Org. Chem.*, **66**, 7892 (2001).
377. J. P. Collman and M. Zhong, *Org. Lett.*, **2**, 1233 (2000).
378. J. C. Antilla and S. L. Buchwald, *Org. Lett.*, **3**, 2077 (2001).
379. M. Sasaki, S. Dalili and A. K. Yudin, *J. Org. Chem.*, **68**, 2045 (2003).
380. A. Deagostino, C. Prandi, C. Zavattaro and P. Venturello, *Eur. J. Org. Chem.*, 1318 (2007).
381. T. Tsuritani, N. A. Strotman, Y. Yamamoto, M. Kawasaki, N. Yasuda and T. Maset, *Org. Lett.*, **10**, 1653 (2008).
382. C. Moessner and C. Bolm. *Org. Lett.*, **7**, 2667 (2005).
383. M. Blouin and R. Frenette, *J. Org. Chem.*, **66**, 9043 (2001).
384. A. Vakalopoulos, X. Kavazoudi and J. Schoof, *Tetrahedron Lett.*, **47**, 8607 (2006).
385. P. Y. S. Lam, G. Vincent, D. Bonne and C. G. Clark, *Tetrahedron Lett.*, **43**, 3091 (2002).
386. P. Y. S. Lam, C. G. Clark, S. Saubern, J. Adams, K. M. Averill, D. M. T. Chan and A. Combs, *Synlett*, **674**, (2000).
387. P. Y. S. Lam, S. Deudon, K. M. Averill, R. Li, M. Y. He, P. DeShong and C. G. Clark, *J. Am. Chem. Soc.*, **122**, 7600 (2000).
388. P. Y. S. Lam, S. Deudon, E. Hauptman and C. G. Clark, *Tetrahedron Lett.*, **42**, 2427 (2001).
389. T. P. Lockhart, *J. Am. Chem. Soc.*, **105**, 1940 (1983).
390. D. H. R. Barton, J. P. Finet and J. Khamsi, *Tetrahedron Lett.*, **28**, 887 (1987).
391. W. C. P. Tsang, N. Zheng and S. L. Buchwald, *J. Am. Chem. Soc.*, **127**, 14560 (2005).
392. T. Uemura, S. Imoto and N. Chatani, *Chem. Lett.*, **35**, 842 (2006).
393. G. Brasche and S. L. Buchwald, *Angew. Chem., Int. Ed.*, **47**, 1932 (2008).
394. X. Ribas, D. A. Jackson, B. Donnadieu, J. Mahia, T. Parella, R. Xifra, B. Hedman, K. O. Hodgson, A. Llobet and T. D. P. Stack, *Angew. Chem., Int. Ed.*, **41**, 2991 (2002).
395. C. E. Castro and R. D. Stephens, *J. Org. Chem.*, **28**, 2163 (1963).
396. R. D. Stephens and C. E. Castro, *J. Org. Chem.*, **28**, 3313 (1963).
397. R. Chinchilla and C. Najera, *Chem. Rev.*, **107**, 874 (2007).
398. K. Okuro, M. Furuune, M. Enna, M. Miura and M. Nomura, *J. Org. Chem.*, **58**, 4716 (1993).
399. K. Okuro, M. Furuune, M. Miura and M. Nomura, *Tetrahedron Lett.*, **33**, 5363 (1992).

400. R. S. Coleman and R. Garg, *Org. Lett.*, **3**, 3487 (2001).
401. A. Biffis, E. Scattolin, N. Ravasio and F. Zaccheria, *Tetrahedron Lett.*, **48**, 8761 (2007).
402. M. B. Thathagar, J. Beckers and G. Rothenberg, *Green Chemistry*, **6**, 215 (2004).
403. S. Cacchi, G. Fabrizi and L. M. Parisi, *Org. Lett.*, **5**, 3843 (2003).
404. D. Ma and F. Liu, *Chem. Commun.*, 1934 (2004).
405. J.-H. Li, J.-L. Li, D.-P. Wang, S.-F. Pi, Y.-X. Xie, M.-B. Zhang and X.-C. Hu, *J. Org. Chem.*, **72**, 2053 (2007).
406. P. J. Tambade, Y. P. Patil, N. S. Nandurkar and B. M. Bhanage, *Synlett*, 886 (2008).
407. C. Chowdhury and N. G. Kundu, *Tetrahedron*, **55**, 7011 (1999).
408. M. Guo, D. Li and Z. Zhang, *J. Org. Chem.*, **68**, 10172 (2003).
409. S.-K. Kang, S.-K. Yoon and Y.-M. Kim, *Org. Lett.*, **3**, 2697 (2001).
410. K. W. Rosenmund and E. Struck, *Ber. Dtsch. Chem. Ges. B*, **52B**, 1749 (1919).
411. J. X. Wu, B. Beck and R. X. Ren, *Tetrahedron Lett.*, **43**, 387 (2002).
412. J. Zanon, A. Klapars and S. L. Buchwald, *J. Am. Chem. Soc.*, **125**, 2890 (2003).
413. H.-J. Cristau, A. Ouali, J.-F. Spindler and M. Taillefer, *Chem.-Eur. J.*, **11**, 2483 (2005).
414. T. Schareina, A. Zapf and M. Beller, *Tetrahedron Lett.*, **46**, 2585 (2005).
415. T. Schareina, A. Zapf, W. Maegerlein, N. Mueller and M. Beller, *Synlett*, 555 (2007).
416. T. Schareina, A. Zapf, W. Maegerlein, N. Mueller and M. Beller, *Chem.-Eur. J.*, **13**, 6249 (2007).
417. W. R. H. Hurtley, *J. Chem. Soc.*, 1870 (1929).
418. R. Adams, B. R. Baker and R. B. Wearn, *J. Am. Chem. Soc.*, **62**, 2204 (1940).
419. R. Adams, C. K. Cain and B. R. Baker, *J. Am. Chem. Soc.*, **62**, 2201 (1940).
420. A. McKillop and D. P. Rao, *Synthesis*, 760 (1977).
421. A. McKillop and D. P. Rao, *Synthesis*, 759 (1977).
422. S. Pivsa-Art, Y. Fukui, M. Miura and M. Nomura, *Bull. Chem. Soc. Jpn.*, **69**, 2039 (1996).
423. J. Setsune, T. Ueda, K. Shikata, K. Matsukawa, T. Iida and T. Kitao, *Tetrahedron*, **42**, 2647 (1986).
424. A. Bruggink and A. McKillop, *Tetrahedron*, **31**, 2607 (1975).
425. A. Osuka, T. Kobayashi and H. Suzuki, *Synthesis*, 67 (1983).
426. J. Setsune, K. Matsukawa and T. Kitao, *Tetrahedron Lett.*, **23**, 663 (1982).
427. J. Setsune, K. Matsukawa, H. Wakemoto and T. Kitao, *Chem. Lett.*, 367 (1981).
428. K. Okuro, M. Furuune, M. Miura and M. Nomura, *J. Org. Chem.*, **58**, 7606 (1993).
429. E. J. Hennessy and S. L. Buchwald, *Org. Lett.*, **4**, 269 (2002).
430. Y. Fang and C. Li. *J. Org, Chem.*, **71**, 6427 (2006).
431. Y. Chen, X. Xie and D. Ma, *J. Org. Chem.*, **72**, 9329 (2007).
432. S. Tanimori, H. Ura and M. Kirihatali, *Eur. J. Org. Chem.*, 3977 (2007).
433. B. Lu, B. Wang, Y. Zhang and D. Ma, *J. Org. Chem.*, **72**, 5337 (2007).
434. X. Xie, Y. Chen and D. Ma, *J. Am. Chem. Soc.*, **128**, 16050 (2006).
435. F. Besselievre, S. Piguel, F. Mahuteau-Betzer and D. S. Grierson, *Org. Lett.*, **10**, 4029 (2008).
436. H.-Q. Do and O. Daugulis, *J. Am. Chem. Soc.*, **129**, 12404 (2007).
437. H.-Q. Do and O. Daugulis, *J. Am. Chem. Soc.*, **130**, 1128 (2008).
438. M. Hoshi, N. Kawamura and K. Shirakawa, *Synthesis*, 1961 (2006).
439. M. Hoshi, Y. Masuda and A. Arase, *Bull. Chem. Soc. Jpn.*, **56**, 2855 (1983).
440. M. Hoshi, K. Shirakawa and K. Takeda, *Synlett*, 403 (2001).
441. M. B. Thathagar, J. Beckers and G. Rothenberg, *J. Am. Chem. Soc.*, **124**, 11858 (2002).
442. M. B. Thathagar, J. Beckers and G. Rothenberg, *Adv. Synth. Catal.*, **345**, 979 (2003).
443. J. Mao, J. Guo, F. Fang and S.-J. Ji, *Tetrahedron*, **64**, 3905 (2008).
444. J.-H. Li and D.-P. Wang, *Eur. J. Org. Chem.*, 2063 (2006).
445. J.-H. Li, J.-L. Li and Y.-X. Xie, *Synthesis*, 984 (2007).
446. S.-K. Kang, T. Yamaguchi, T.-H. Kim and P.-S. Ho, *J. Org. Chem.*, **61**, 9082 (1996).
447. L. S. Liebeskind and R. W. Fengl, *J. Org. Chem.*, **55**, 5359 (1990).
448. E. Piers and T. Wong, *J. Org. Chem.*, **58**, 3609 (1993).
449. J. R. Falck, R. K. Bhatt and J. Ye, *J. Am. Chem. Soc.*, **117**, 5973 (1995).
450. G. D. Allred and L. S. Liebeskind, *J. Am. Chem. Soc.*, **118**, 2748 (1996).
451. J. R. Falck, P. K. Patel and A. Bandyopadhyay, *J. Am. Chem. Soc.*, **129**, 790 (2007).
452. J. Schuppan, H. Wehlan, S. Keiper and U. Koert, *Angew. Chem., Int. Ed.*, **40**, 2063 (2001).
453. J. S. Bradshaw, J. A. South and R. H. Hales, *J. Org. Chem.*, **37**, 2381 (1972).
454. S.-K. Kang, J.-S. Kim and S.-C. Choi, *J. Org. Chem.*, **62**, 4208 (1997).

455. S.-K. Kang, J.-S. Kim, S.-K. Yoon, K.-H. Lim and S. S. Yoon, *Tetrahedron Lett.*, **39**, 3011 (1998).
456. N. S. Nudelman and C. Carro, *Synlett*, 1942 (1999).
457. S.-K. Kang, W.-Y. Kim and X. Jiao, *Synthesis*, 1252 (1998).
458. C.-M. Yu, J.-H. Kweon, P.-S. Ho, S.-C. Kang and G. Y. Lee, *Synlett*, 2631 (2005).
459. A. K. Franz and K. A. Woerpel, *J. Am. Chem. Soc.*, **121**, 949 (1999).
460. H. Ito, K. Arimoto, H.-o. Sensui and A. Hosomi, *Tetrahedron Lett.*, **38**, 3977 (1997).
461. S.-K. Kang, T.-H. Kim and S.-J. Pyun, *J. Chem. Soc., Perkin Trans. 1*, 797 (1997).
462. H. Taguchi, K. Ghoroku, M. Tadaki, A. Tsubouchi and T. Takeda, *J. Org. Chem.*, **67**, 8450 (2002).
463. Y. Nishihara, M. Takemura, A. Mori and K. Osakada, *J. Organomet. Chem.*, **620**, 282 (2001).
464. H. Ito, H.-o. Sensui, K. Arimoto, K. Miura and A. Hosomi, *Chem. Lett.*, 639 (1997).
465. J. A. Marshall, H. R. Chobanian and M. M. Yanik, *Org. Lett.*, **3**, 4107 (2001).
466. H. Urata and T. Fuchikami, *Tetrahedron Lett.*, **32**, 91 (1991).
467. F. Cottet, M. Marull, O. Lefebvre and M. Schlosser, *Eur. J. Org. Chem.*, 1559 (2003).
468. Y. Nishihara, K. Ikegashira, A. Mori and T. Hiyama, *Tetrahedron Lett.*, **39**, 4075 (1998).
469. C. B. Linn and C. R. Noller, *J. Am. Chem. Soc.*, **58**, 816 (1936).
470. J. K. Kochi and M. Tamura, *J. Am. Chem. Soc.*, **93**, 1485 (1971).
471. M. Tamura and J. K. Kochi, *Synthesis*, 303 (1971).
472. M. Tamura and J. K. Kochi, *J. Organomet. Chem.*, **42**, 205 (1972).
473. T. Satoh, R. Matsue, T. Fujii and S. Morikawa, *Tetrahedron*, **57**, 3891 (2001).
474. S. Nunomoto, Y. Kawakami and Y. Yamashita, *Bull. Chem. Soc. Jpn.*, **54**, 2831 (1981).
475. R. Shimizu, E. Yoneda and T. Fuchikami, *Tetrahedron Lett.*, **37**, 5557 (1996).
476. F. F. Fleming and T. Jiang, *J. Org. Chem.*, **62**, 7890 (1997).
477. D. K. Johnson, J. P. Ciavarri, F. T. Ishmael, K. J. Schillinger, T. A. P. van Geel and S. M. Stratton, *Tetrahedron Lett.*, **36**, 8565 (1995).
478. J. Nishimura, N. Yamada, Y. Horiuchi, E. Ueda, A. Ohbayashi and A. Oku, *Bull. Chem. Soc. Jpn.*, **59**, 2035 (1986).
479. J. R. Vyvyan, C. L. Holst, A. J. Johnson and C. M. Schwenk, *J. Org. Chem.*, **67**, 2263 (2002).
480. K. Onuma and H. Hashimoto, *Bull. Chem. Soc. Jpn.*, **45**, 2582 (1972).
481. D. H. Burns, J. D. Miller, H.-K. Chan and M. O. Delaney, *J. Am. Chem. Soc.*, **119**, 2125 (1997).
482. J. Terao, H. Todo, S. A. Begum, H. Kuniyasu and N. Kambe, *Angew. Chem., Int. Ed.*, **46**, 2086 (2007).
483. H. Kotsuki, I. Kadota and M. Ochi, *Tetrahedron Lett.*, **30**, 1281 (1989).
484. A. S. E. Karlstroem, M. Roenn, A. Thorarensen and J.-E. Bäckvall, *J. Org. Chem.*, **63**, 2517 (1998).
485. A. Nivlet, L. Dechoux, J.-P. Martel, G. Proess, D. Mannes, L. Alcaraz, J. J. Harnett, T. Le Gall and C. Mioskowski, *Eur. J. Org. Chem.*, 3241 (1999).
486. C. Jonasson, M. Roenn and J.-E. Bäckvall, *J. Org. Chem.*, **65**, 2122 (2000).
487. H. Tamagawa, H. Takikawa and K. Mori, *Eur. J. Org. Chem.*, 973 (1999).
488. D. H. Burns, H.-K. Chan, J. D. Miller, C. L. Jayne and D. M. Eichhorn, *J. Org. Chem.*, **65**, 5185 (2000).
489. M. Seki and K. Mori, *Eur. J. Org. Chem.*, 3797 (2001).
490. P. T. Baraldi, P. H. G. Zarbin, P. C. Vieira and A. G. Correa, *Tetrahedron: Asymmetry*, **13**, 621 (2002).
491. J. A. Moreira and A. G. Correa, *Tetrahedron: Asymmetry*, **14**, 3787 (2003).
492. M. G. Organ and J. Wang, *J. Org. Chem.*, **68**, 5568 (2003).
493. C. C. Kofink and P. Knochel, *Org. Lett.*, **8**, 4121 (2006).

CHAPTER **19**

Fluorinated organocopper reagents

CHARLES R. DAVIS

Sigma-Aldrich, 6000 N. Teutonia Avenue, Milwaukee, WI 53209, USA
Fax: +1-414-438-4224; e-mail: Charles.Davis@sial.com

and

DONALD J. BURTON

Department of Chemistry, University of Iowa, Iowa City, IA 52242, USA
Fax: +1-319-335-1270; e-mail: donald-burton@uiowa.edu

I. INTRODUCTION 992
II. FLUOROALKENYLCOPPER REAGENTS 992
A. Introduction: Symmetrical Coupling of Fluoroalkenyl Halides 992
B. Pre-generation of Perfluoroalkenylcopper Reagents 994
1. Via metathesis reactions 994
a. From perfluoroalkenylsilver reagents 994
b. From fluoroalkenylzinc and cadmium reagents 994
c. From fluoroalkenyltin reagents 996
C. Formation of Fluorinated Cumulenes via Decomposition of α-Halo-(X = I, Br, Cl)-substituted Fluoroalkenylcopper Reagents 997
D. FUNCTIONALIZATION REACTIONS OF FLUOROALKENYLCOPPER REAGENTS 998
1. Protonation 998
2. Allylation 998
3. Acylation 998
4. Benzylation 999
5. Alkylation 999
6. Alkenylation 999
7. Arylation 999
8. Silylation 999

The chemistry of organocopper compounds
Edited by Z. Rappoport and I. Marek

9. Addition to perfluoroalkynes 1000
10. Insertion of CF_2 units 1002
III. DIFLUOROVINYLCOPPER REAGENTS 1003
IV. PERFLUOROALLYLCOPPER REAGENTS 1004
V. PERFLUOROALKYNYLCOPPER REAGENTS 1005
VI. CARBOALKOXYDIFLUOROMETHYLENECOPPER REAGENTS 1007
VII. DIALKOXYPHOSPHINYLFLUOROMETHYLCOPPER REAGENT 1008
VIII. DIALKOXYPHOSPHINYLDIFLUOROMETHYLCOPPER REAGENTS . 1008
IX. PERFLUOROBENZYLCOPPER REAGENT 1012
X. PERFLUOROALKYLCOPPER REAGENTS 1012
A. Perfluoroalkylation 1013
B. Trifluoromethylation 1016
C. Difluoromethylation 1022
XI. PENTAFLUOROPHENYLCOPPER REAGENT 1023
XII. 2,3,5,6-TETRAFLUOROPYRIDYLCOPPER REAGENTS 1027
XIII. REFERENCES 1028

I. INTRODUCTION

Fluorinated copper reagents have been utilized as replacements for many of the unstable fluorinated lithium or magnesium reagents. This chapter details methodology for the preparation of a wide variety of fluorinated copper reagents and specific applications of these reagents in functionalization processes. The coverage is designed as a general overview of each type of fluorinated copper reagent and is not intended to be a comprehensive review of every application of a fluorinated copper reagent in the chemical literature. Our goal is to acquaint the reader with the general types of fluorinated copper reagents available, their preparation and specific examples of their functionalization, so as to inform the reader of the current state of knowledge of this class of organometallic reagents and how these types of organometallic reagents can be utilized in their own research. Our emphasis is on fluorinated copper reagents that have been observed or detected or for which the experimental evidence is in agreement with the formation of a fluorinated copper reagent. There are numerous examples of copper mediated reactions in the literature, which most likely proceed via a fluorinated radical intermediate. These types of copper mediated processes are not covered in this chapter.

II. FLUOROALKENYLCOPPER REAGENTS

A. Introduction: Symmetrical Coupling of Fluoroalkenyl Halides

Attempts at the preparation of fluoroalkenylcopper reagents directly from the corresponding fluoroalkenyl iodides or bromides with various forms of activated copper powder have not been successful. The corresponding symmetrical diene was obtained as the major product of this type of reaction, as illustrated in equations 1 and 2[1–6]. Similar diene formation with (*E*)- or (*Z*)-substituted fluoroalkenyl iodides were reported by Yagupol'skii and coworkers[7] and Burton and coworkers[8] (equations 3 and 4). Riess and coworkers reported similar couplings of a series of bis(1,2-(perfluoroalkyl)-iodoethenes to afford 1,2,3,4-tetrakis(perfluoroalkyl)-1,3-butadienes (equation 5)[9]. Burton, MacNeil and coworkers utilized similar methodology to stereospecifically prepare (2*E*,4*E*,6*E*,8*E*)-perfluoro-2,4,6,8-decatetraene (equation 6)[8]. Cyclic analogs, such as 1-iodo-2-chloroperfluorocyclobutene, cyclopentene and cyclohexene, yielded the corresponding 2,2′-dichloroperfluoro(bis-1-cycloalken-1-yl) derivatives on coupling with copper powder (equation 7)[10,11]. Coupling of 1,2-diiodoperfluorocyclobutene and cyclopentene under similar conditions gave good yields of cyclic trimers and tetramers[10–12]. When only 0.5% wt of DMF was used at 130 °C, 50% of the cyclic trimer and 34% of perfluorocyclooctatetraene (tetramer) was formed from perfluoro-1,2-diiodocyclobutene. X-ray crystal structures of these compounds showed that the tetramer was planar[13], and that the central ring of the trimer[14] had bond lengths and angles essentially identical with those of benzene itself. Haszeldine and coworkers reported a similar copper mediated self-coupling of hexafluoro-2-iodo-3-trifluoromethylbut-2-ene. However, it was not possible to assign unequivocally the highly hindered structure of the final product as 1,6-di(trifluoromethyl)-2,3,4,5-tetra(trifluoromethyl)-hexa-2,4-diene or its cyclobutene isomer, formed at the high temperature of the reaction (equation 8)[15].

$$2\ \mathrm{FClC{=}CFI} + \mathrm{Cu} \xrightarrow[145\,^{\circ}\mathrm{C}]{24\ \mathrm{h}} \mathrm{FClC{=}CF{-}CF{=}CFCl} \qquad (1)$$

mixture of isomers — mixture of 3 isomers

$$2\ \mathrm{F_2C{=}CFI} + \mathrm{Cu} \xrightarrow{\mathrm{DMF}} \mathrm{F_2C{=}CF{-}CF{=}CF_2} \qquad (2)$$

63–67%

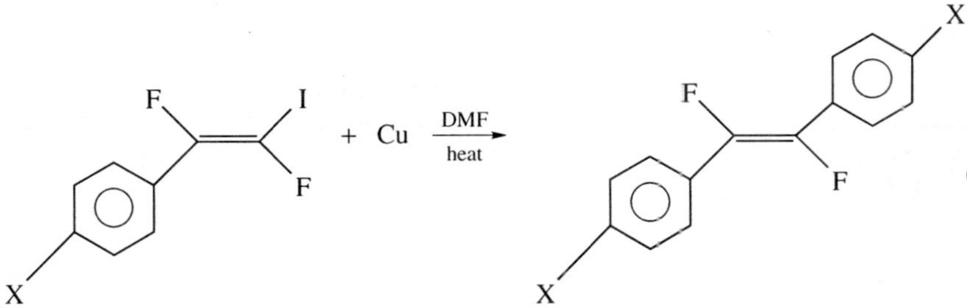
(3)

X = NO$_2$, 95%

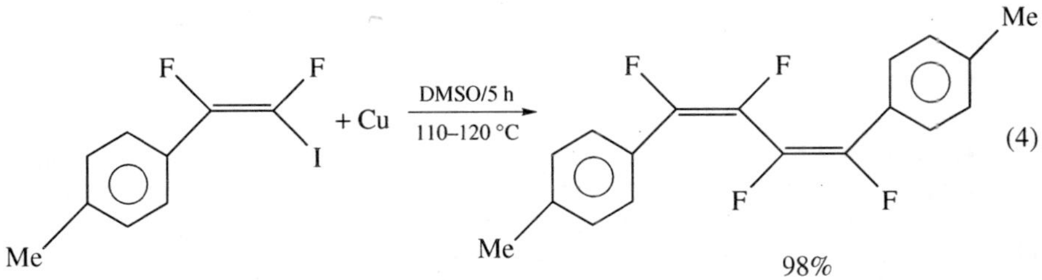
(4)

98%

$$\mathrm{R_fCH{=}CIR_f'} + \mathrm{Cu} \xrightarrow[220\,^{\circ}\mathrm{C}]{24\ \mathrm{h}} \mathrm{R_fCH{=}C(R_f'){-}C(R_f'){=}CHR_f} \qquad (5)$$

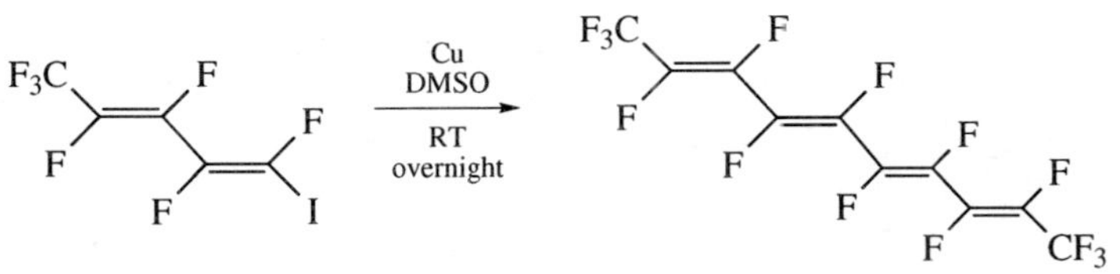
(6)

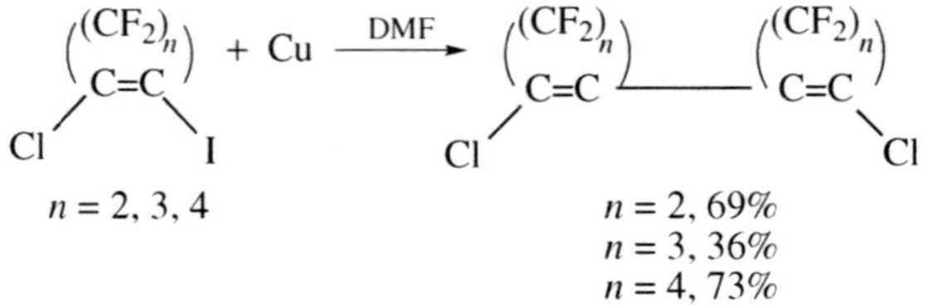

(7)

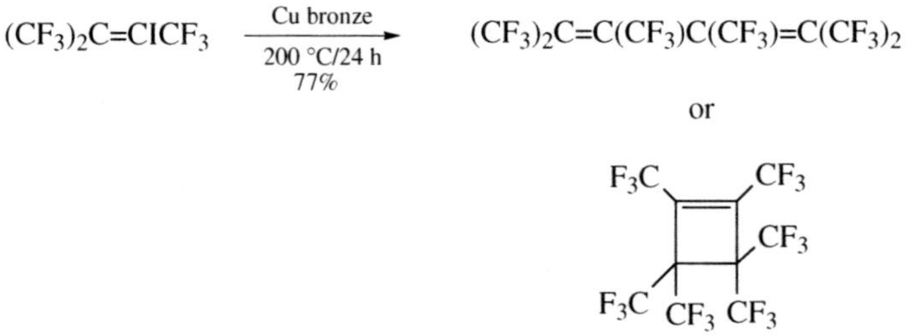

(8)

B. Pre-generation of Perfluoroalkenylcopper Reagents

1. Via metathesis reactions

a. From perfluoroalkenylsilver reagents. The first example of a pre-generated fluoroalkenylcopper reagent was reported by Miller and coworkers, who reacted (E)-perfluoro-2-buten-2-ylsilver with copper bronze and $(CH_3)_2S$ to afford the corresponding perfluoroalkenylcopper reagent (equation 9)[16]. Similar results were obtained when AgF and Cu were reacted with perfluoro-2-butyne in dimethyl sulfide at 30 °C for three hours. Since it is not possible to prepare a variety of perfluorovinylsilver precursors from unsymmetrical perfluoroalkynes in a stereo- and regio-controlled manner, this method is not a general route to fluoroalkenylcopper reagents.

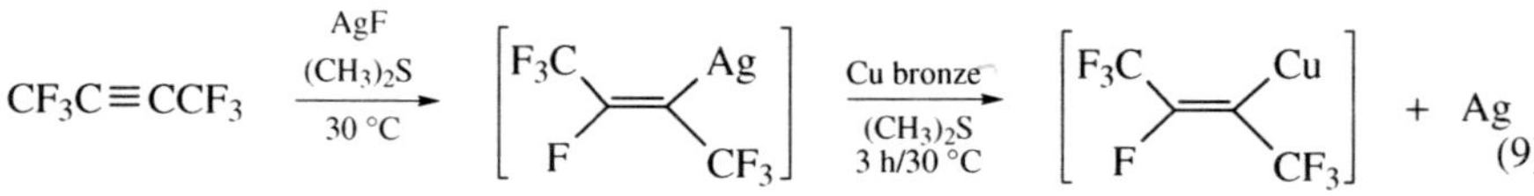

(9)

b. From fluoroalkenylzinc and cadmium reagents. As noted in Section II.A, no direct observation of a stable fluoroalkenylcopper reagent was observed when fluoroalkenyl iodides were reacted with Cu(0). Hansen, Burton and coworkers rationalized that the diene formation most likely occurred in two steps[2,3,8]: (1) slow formation of the fluoroalkenylcopper reagent (equation 10) followed by: (2) rapid reaction of the fluoroalkenylcopper reagent with a second equivalent of the fluoroalkenyl iodide precursor (equation 11). The overall reaction is shown in equation 12.

$$F_2C{=}CFI + 2Cu \xrightarrow{\text{slow}} [F_2C{=}CFCu] + CuI \qquad (10)$$

$$[F_2C{=}CFCu] + F_2C{=}CFI \xrightarrow{\text{fast}} F_2C{=}CF\text{-}CF{=}CF_2 + CuI \qquad (11)$$

$$2\,F_2C{=}CFI + 2\,Cu \longrightarrow F_2C{=}CF\text{-}CF{=}CF_2 + 2\,CuI \qquad (12)$$

In later work, Hansen, Burton and coworkers demonstrated that pre-generated trifluorovinylcopper reagent reacted readily with iodotrifluoroethene to afford perfluoro-1,3-butadiene in 63% yield[8]. Thus, if the initially formed fluoroalkenylcopper reagent is to survive, this reagent must be generated in the *absence* of any alkenyl iodide precursor, thus preventing the rapid second reaction outlined in equation 11 (similar to the exchange process employed by Miller and coworkers, equation 9). We had previously reported that fluoroalkenyl iodides and bromides could be stereospecifically converted into the corresponding fluoroalkenylzinc or cadmium reagents via direct reaction of the fluoroalkenyl halides with Zn(0) or Cd(0) (equation 13)[17,18]. With these metals, **no** diene formation occurred under the reaction conditions employed.

$$F_2C{=}CFX + M \xrightarrow[\text{RT to 60 °C}]{\text{DMF}} [F_2C{=}CFMX] + [F_2C{=}CF]_2M + MX_2 \qquad (13)$$

$X = I, Br \qquad M = Zn, Cd$

We utilized this lack of diene formation to affect the formation of the fluoroalkenylcopper reagent via metathesis of the corresponding fluoroalkenylzinc or cadmium reagent with copper (I) halides (equation 14)[19]. Since no fluoroalkenyl halide is present in the fluoroalkenylzinc or cadmium solution, the fluoroalkenylcopper reagent survives and is readily detected by ^{19}F NMR. The fluoroalkenylcopper reagents exhibit excellent stability at room temperature in the absence of oxygen and/or moisture. At higher temperatures (>50 °C) they undergo rapid decomposition. With substituted fluoroalkenylzinc or cadmium reagents, similar rapid exchange occurs with Cu(I) halides to stereospecifically afford the corresponding fluoroalkenylcopper reagents in good yield, as summarized in Table 1. Thus, if the requisite fluoroalkenyl iodide or bromide is available or can be prepared, the fluoroalkenylzinc or cadmium can be prepared stereospecifically and exchanged with Cu(I) halides to give the corresponding fluoroalkenylcopper reagent. Both regiochemistry and stereochemistry can be controlled, thus providing a general entry to this class of reagents. Subsequent to our initial report[19], Choi and coworkers extended this strategy to the corresponding fluorocycloalkenyl analogs[20–22]. In a series of papers, they converted 1-iodo-2-chloroperfluorocycloalkenes into the corresponding 2-chloroperfluorocycloalkenylzinc reagents. Subsequent exchange with Cu(I)Br afforded the corresponding stable 2-chloroperfluorocycloalkenylcopper reagents (equation 15).

$$[F_2C{=}CFMX] + Cu(I)Y \xrightarrow[\text{RT}]{\text{DMF}} [F_2C{=}CFCu] \qquad (14)$$

$M = Zn, Cd \qquad Y = I, Br$

$X = \text{halide or } F_2C{=}CF$

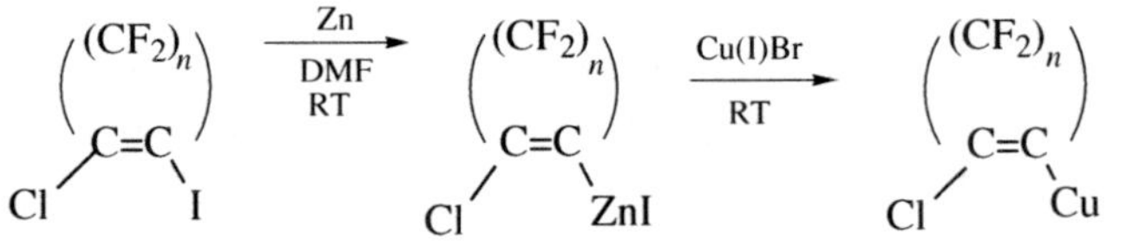

(15)

$n = 2, 3, 4$ 92–98%

TABLE 1. Preparation of fluoroalkenylcopper reagents

$$R_fC(Y){=}CFX + \text{Metal} \xrightarrow[\text{RT to 60 °C; 2. Cu(I)Br}]{\text{1. DMF}} [R_fC(Y){=}CFCu]$$

R_f	Y	X	M	% $R_f(Y){=}CFCu$ [a]
(Z)-CF_3	F	I	Cd	92
(Z)-CF_3	F	I	Zn	76
(E)-CF_3	F	I	Cd	83
(Z)-$CF_3(CF_2)_4$	F	I	Cd	87
(Z)-CF_3	Cl	I	Cd	78
(E/Z)-CF_3	Ph	Br	Zn	84 (E/Z)

[a] Yields determined by ^{19}F NMR.

Internal fluoroalkenyl halides also readily form fluoroalkenylzinc reagents, which readily exchange with Cu(I) halides to afford the internal fluoroalkenylcopper reagents (equation 16)[8,23].

$$CF_3C(Br){=}CF_2 \xrightarrow[\text{DMF}]{\text{Zn}} [CF_3C(ZnBr){=}CF_2] \xrightarrow{\text{Cu(I)Br}} [CF_3C(Cu){=}CF_2] \qquad (16)$$

The internal fluoroalkenylzinc or cadmium reagents can also be prepared from the corresponding 2-hydroperfluoroalkene via lithiation followed by reaction of the lithium reagent with zinc or cadmium halides, as illustrated in equation 17[23]. Functionalized fluoroalkenylzinc reagents also readily exchange with Cu(I)Br to afford the corresponding functionalized fluoroalkenylcopper reagents. The configuration is preserved (as expected) in the exchange process (equation 18)[24].

$$CF_3CH{=}CF_2 \xrightarrow[-78\,°C]{\text{LDA}} [CF_3C(Li){=}CF_2] \xrightarrow[\text{or } CdI_2]{ZnI_2} [CF_3C(MI){=}CF_2] \quad M = Zn, Cd \qquad (17)$$

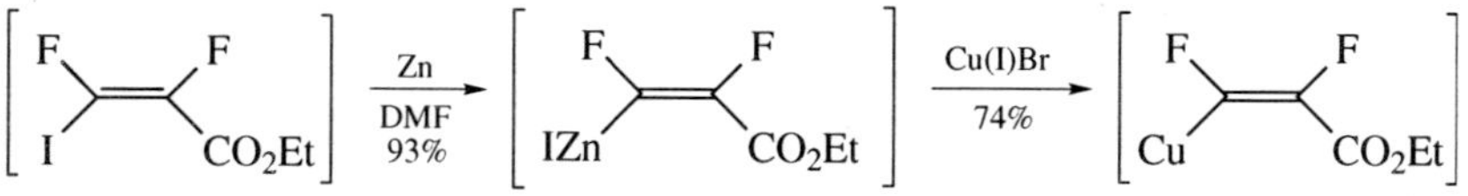

(18)

c. From fluoroalkenyltin reagents. Organotin precursors can also be utilized in exchange processes with Cu(I) halides to produce the corresponding fluoroalkenylcopper reagents (equation 19)[24]. Fluoroalkenylcopper species were also detected as intermediates in the copper (II) mediated homo-coupling of 1,2-difluorovinylstannanes (equation 20)[25].

(19)

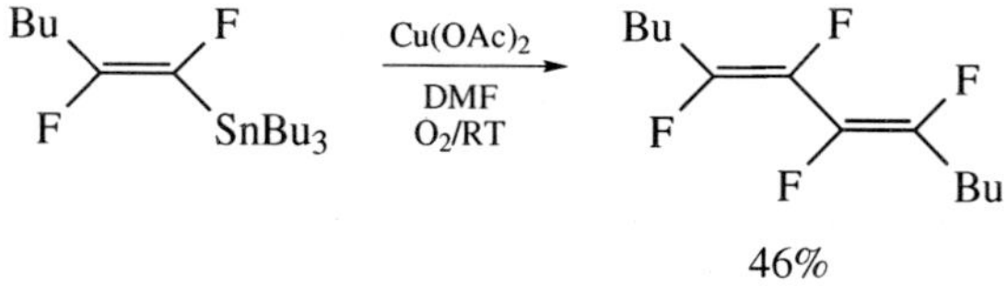

(20)

C. Formation of Fluorinated Cumulenes via Decomposition of α-Halo-(*X* = *I*, Br, Cl)-substituted Fluoroalkenylcopper Reagents

In the fluoroalkenylcopper reagent methodology outlined in Section II.A above, the copper reagents all contained an α-fluorine or an α-perfluoroalkyl substituent. They were all stable to *ca* 50 °C. When the preparation of a fluoroalkenylcopper reagent that contained an α-halogen other than fluorine was attempted (by a similar strategy to that obtained in the previous section), **no stable** fluoroalkenylcopper reagent was detected at room temperature. Thus, when 1,1-dibromo-2-phenyl-3,3,3-trifluoropropene was reacted with zinc, a stable monoorganozinc reagent was formed. However, upon exchange of the α-bromovinylzinc reagent with Cu(I)Br, only the (*E*)- and (*Z*)-butatriene isomers were detected (equation 21)[26,27]. The *E*/*Z* cumulenes were readily separated, and pure multigram quantities of (*E*)- and (*Z*)-cumulenes could be isolated by chromatography. Other *E*/*Z* mixtures of cumulenes prepared by this methodology[27] in the yields given are: $CF_3CF_2C(Ph)C{=}C{=}C{=}C(Ph)CF_2CF_3$, 65%; $CF_3CF_2CF_2(Ph)C{=}C{=}C{=}C(Ph)CF_2CF_2CF_3$, 70% and $CF_3(C_6F_5)C{=}C{=}C{=}C(C_6F_5)CF_3$, 68%. X-ray crystallographic analysis of one of each of the *E*/*Z* isomeric pairs noted above unequivocally confirmed the structural assignments[26,28]. Thermal isomerization of the isolated butatrienes was achieved at 110 °C (6 h)[27]. Note that a β-halo-substituent in the fluoroalkenylcopper reagent does not destabilize the copper reagent (cf. Table 1 and References 20–22). Mechanistic work with the *E*/*Z* isomers of $CF_3C(Ph){=}CClCu$ indicated that the α-chlorovinylcopper reagent could be detected at −45 °C by ^{19}F NMR. On warming to 0 °C, the ^{19}F NMR signals assigned to the α-chlorovinylcopper reagent disappeared and the integration of the ^{19}F NMR signals for the dienyl isomers accounted for 99% of the fluorines. Quenching the reaction mixture with HCl afforded the *E*/*Z* isomers of the dienylcopper intermediate. When the dienylcopper intermediate was warmed to room temperature, **only** the *E*/*Z*-butatrienes $CF_3(Ph)C{=}C{=}C{=}C(Ph)CF_3$ were observed by ^{19}F NMR analysis of the reaction mixture. The mechanism shown in Scheme 1 for cumulene formation was proposed[27].

$$CF_3C(Ph){=}CBr_2 \xrightarrow[\text{DMF, RT}]{\text{Zn}} \underset{90\%}{[CF_3C(Ph){=}C(Br)ZnBr]} \xrightarrow[\text{Cu(I)Br}]{\text{DMF, −40 °C to RT, 72\%}} \underset{E/Z}{CF_3(Ph)C{=}C{=}C{=}C(Ph)CF_3} \qquad (21)$$

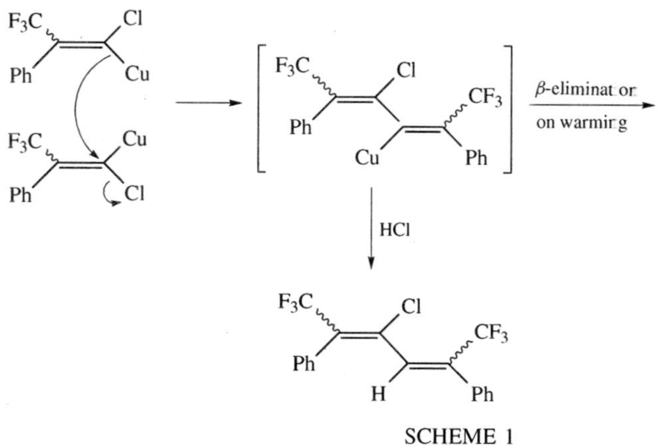

SCHEME 1

Defluorination of the cumulenes can be easily accomplished under mild conditions and affords a convenient route to divinylacetylene derivatives (equation 22)[29,30]. Both the (*E*)- and (*Z*)-cumulenes gave >95% of the (*E*,*E*)-divinylacetylene product. X-ray analysis of (*E*,*E*)-$CF_3C(F){=}C(Ph)\text{-}C{\equiv}C{-}C(Ph){=}CFCF_3$ confirmed the structural assignment[30]. Japanese workers utilized these cumulenes to prepare tetraaryltetrakis(trifluoromethyl)[4]-radialenes[31–33].

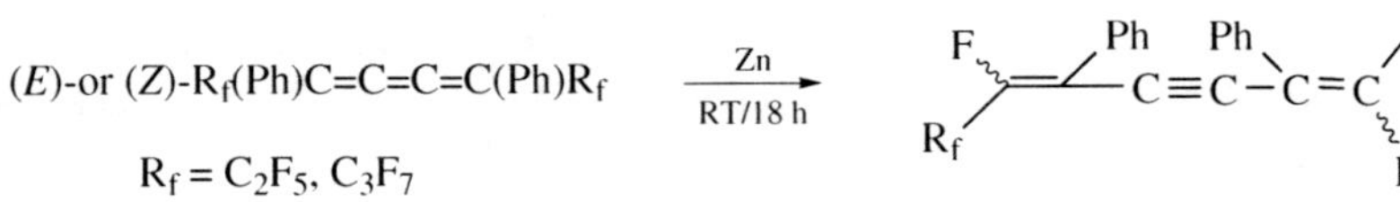

(22)

D. FUNCTIONALIZATION REACTIONS OF FLUOROALKENYLCOPPER REAGENTS

In this section, representative reactions of fluoroalkenylcopper reagents are summarized.

1. *Protonation* (equation 23)

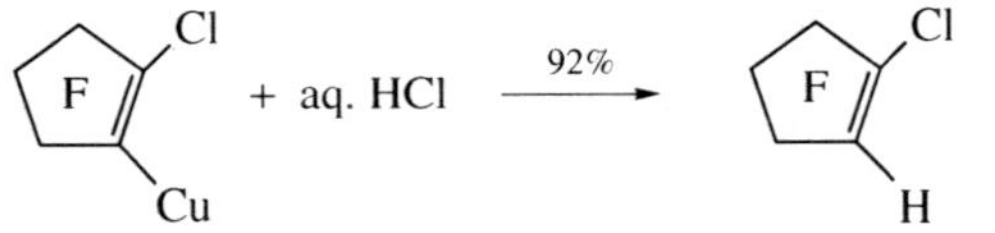

20

(23)

2. *Allylation* (equation 24)

19

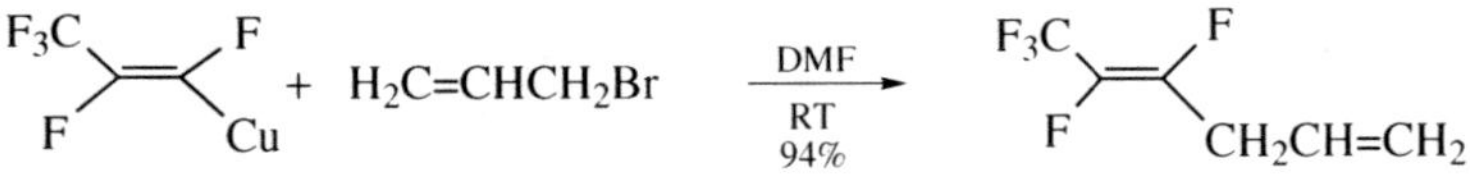

(24)

3. *Acylation* (equation 25)

19

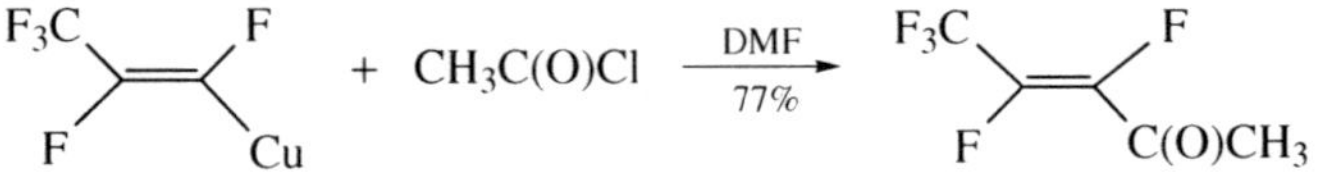

(25)

The fluoroalkenylcopper reagent need not be pre-formed to successfully accomplish acylation. Spawn and Burton developed a useful Cu(I) mediated acylation of the trifluorovinylzinc reagent as a route to trifluorovinyl ketones (equation 26)[34].

$$F_2C{=}CFI \xrightarrow[\text{triglyme or tetraglyme}]{Zn} [F_2C{=}CFZnI]\ (85\text{–}95\%) \xrightarrow[\text{Cu(I)Br (cat.), RT}]{RC(O)Cl} F_2C{=}CFC(O)R\ (50\text{–}87\%) \qquad (26)$$

$R = CH_3, CH_3CH_2, CH_3CH_2CH_2, H_3C(CH_2)_4, (CH_3)_2CH, (CH_3)_3C, C(O)OCH_2CH_3$

4. Benzylation (equation 27)

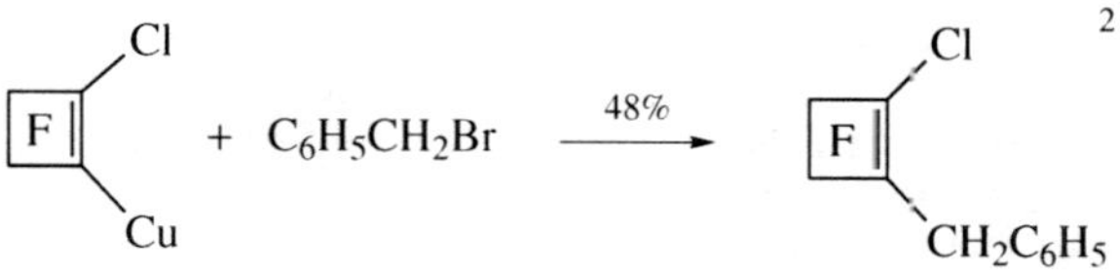

[21] (27)

5. Alkylation (equation 28)

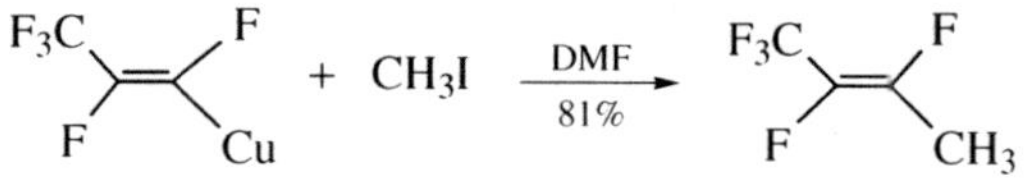

[19] (28)

6. Alkenylation (equation 29)

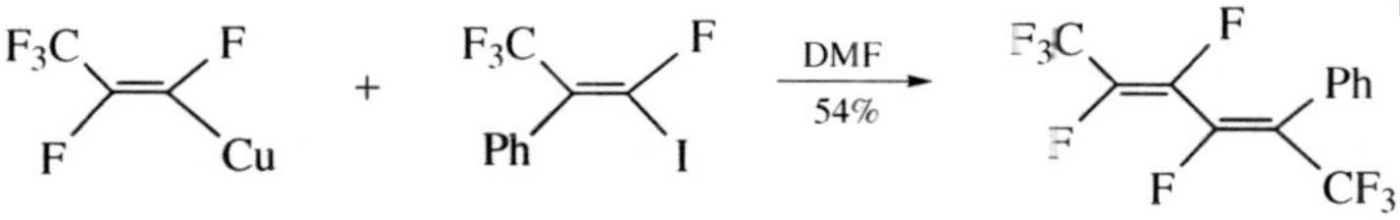

[19] (29)

7. Arylation (equation 30)

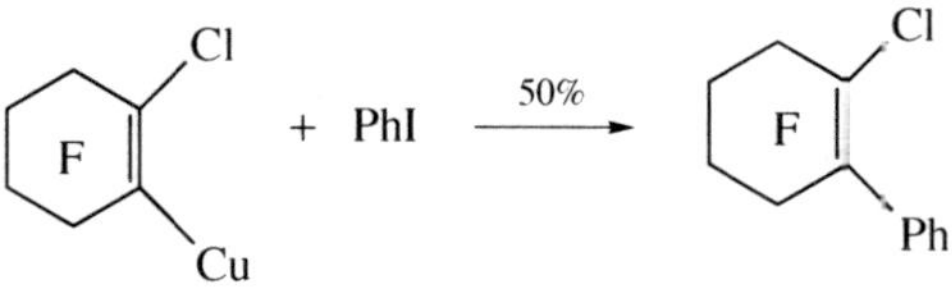

[22] (30)

8. Silylation

The preparation of trifluorovinylsilanes has been known for many years. When trifluorovinyllithium or its corresponding Grignard reagent is reacted at low temperature with chlorotrimethylsilane, 1,2,2-trifluoro-2-trimethylsilane is formed in reasonable yields. However, the use of $(CH_3)_3SiCl$ presented serious isolation problems and silanes, such as $(CH_3CH_2)_3SiCl$ and $ClSi(CH_3)_2Ph$, were utilized to permit easy separation of the trifluorovinylsilane from solvent and $RLi/F_2C{=}CFX$ exchange product[35]. Recently, Jairaj and Burton have reported a useful, large scale preparation of $F_2C{=}CFSiMe_3$ from trifluorovinylcopper (equation 31)[36]. This methodology avoids the previously described problems associated with lithium or magnesium reagents, is easily scaled up and avoids any low temperature reactions and unstable intermediates.

$$F_2C{=}CFBr \xrightarrow[\text{DMF, RT}]{\text{Zn}} [F_2C{=}CFZnBr] \xrightarrow[\text{DMF, Cu(I)Br (cat.)}]{\text{Me}_3\text{SiCl}} \underset{65\%}{F_2C{=}CFSiMe_3} \qquad (31)$$

9. Addition to perfluoroalkynes

Hansen, Burton and coworkers[2,5] found that trifluorovinylcopper added to perfluoro-2-butyne stereospecifically to afford the *syn*-addition product. Quenching the addition product with iodine gave only the *syn*-product. **No** *anti*-addition product was detected. The coupling constant between the trifluoromethyl groups provided evidence for the assignment of the *syn*-addition mode (equation 32)[37]. Reaction of the *syn*-addition product with iodobenzene afforded the corresponding arylated derivative. Reaction of (*Z*)-1,1,2,5,5,5-hexafluoro-4-phenyl-3-trifluoromethyl-1,3-pentadiene with pyridine resulted in the formation of the diastereomeric 4-quinolizone derivatives (after hydrolysis) (equation 33)[37]. The structure of the two diasteromeric products was determined by X-ray crystallography.

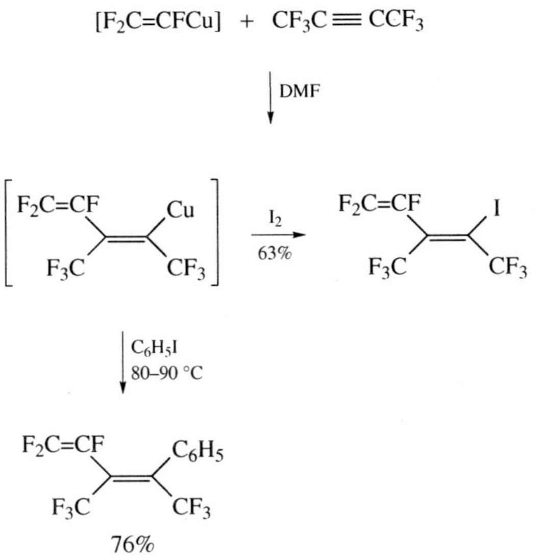

(32)

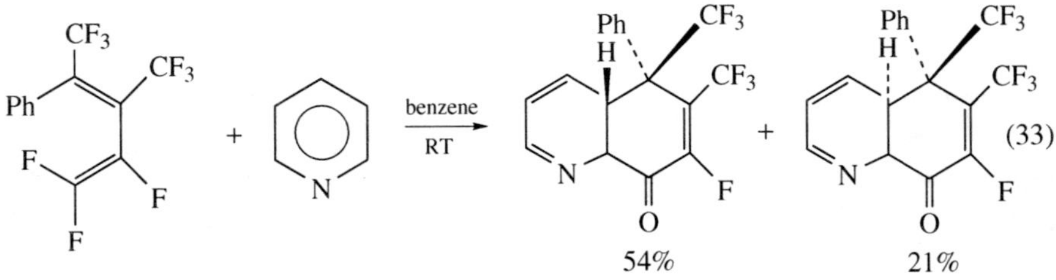

(33)

When the trifluorovinylcopper addition product to perfluoro-2-butyne reacted with 1,4-diiodobenzene, and the resultant product photolyzed at 254 nm, two isomers of $C_{18}H_2F_{16}$ were isolated (after purification) (equation 34)[38,39]. The structure of these two isomers were determined by X-ray crystallography to be 3,4,5,6-tetrafluoro-1,2,7,8-tetrakis(trifluoromethyl)phenanthrene and 3,4,7,8-tetrafluoro-1,2,5,6-tetrakis(trifluoromethyl)anthracene.

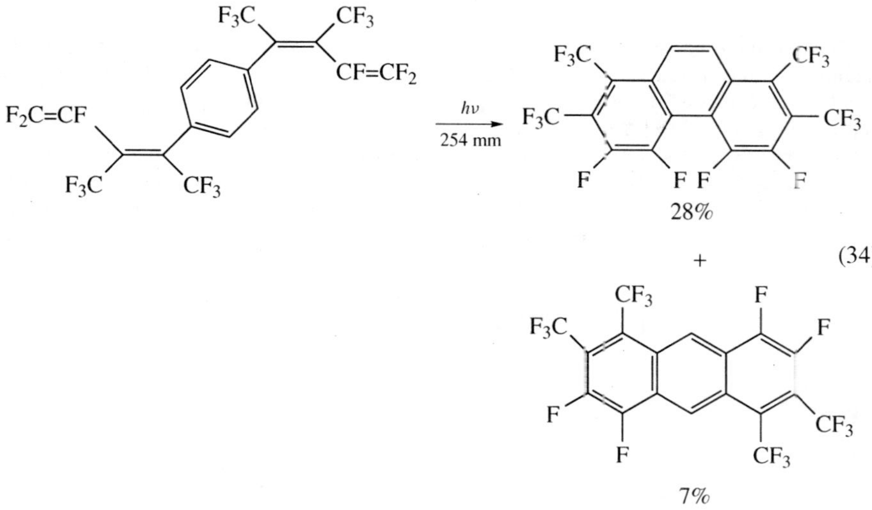

(34)

The trifluorovinylcopper addition product to perfluoro-2-butyne readily undergoes the expected acylation reaction. However, the acylated product spontaneously undergoes an electrocyclization reaction to afford the perfluoropyran derivative (equation 35)[24]. If the perfluorovinylcopper reagent contains bulky groups at the β-position, cyclization of the dienyl ketone is not spontaneous, and the dienyl ketone can be isolated (equation 36). Heating the hindered ketone promotes formation of the pyran product.

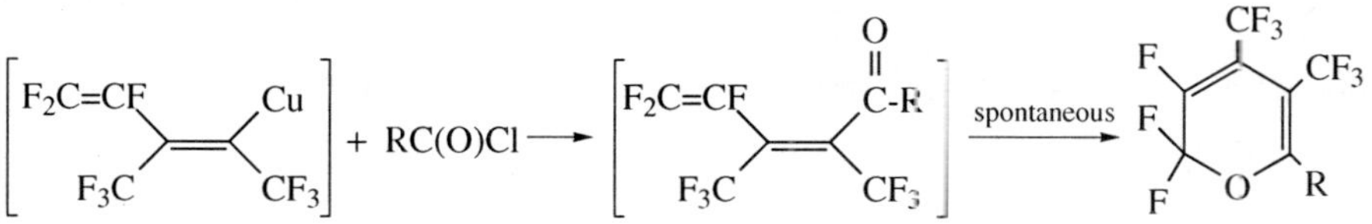

(35)

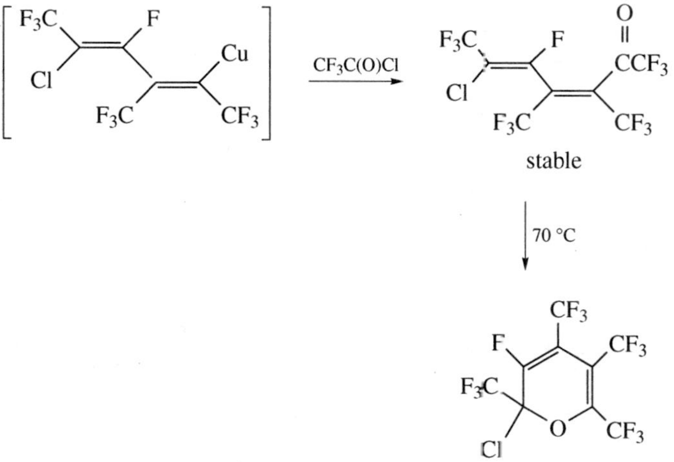

(36)

10. Insertion of CF_2 units

Trifluoromethylcopper is much less thermally stable than its higher homologs. On standing at room temperature (in DMF), CF_3Cu slowly converts to CF_3CF_2Cu. Thus, CF_3Cu can act as a CF_2 unit transfer agent. When (*Z*)-pentafluoropropenylcopper reacts with CF_3Cu in DMF at $-30\,^{\circ}C$, and the mixture warmed to room temperature, ^{19}F NMR analysis of the reaction mixture indicated that a double insertion of CF_2 units had occurred to form a new homoallylic copper reagent with retention of stereochemistry (equation 37)[40,41]. The assigned structure of the new homoallylic copper reagent was confirmed by functionalization with allyl bromide.

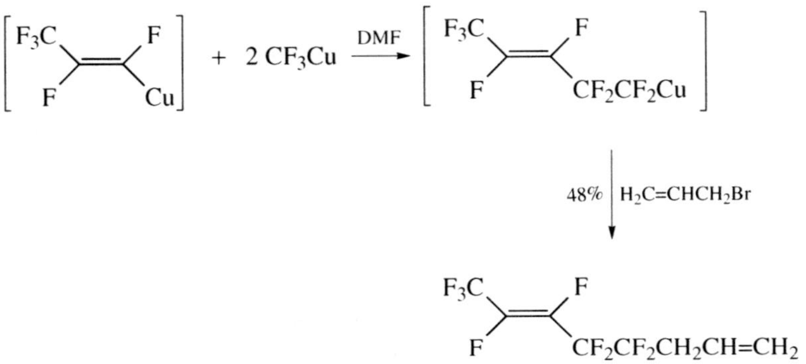

(37)

Trifluorovinylcopper also reacts with CF_3Cu in DMF; the only identifiable product was (*E*)-$CF_3CF{=}CFCF_2CF_2Cu$, insertion of three CF_2 units. Although the mechanism of formation of the triple insertion product is not clear, it was proposed that double insertion occurs to provide perfluorobutenylcopper, which is isomerized by fluoride ion (formed in small amounts in the formation of $F_2C{=}CFCu$) to the perfluoroallylcopper reagent, which inserts another CF_2 unit to give the observed product (equation 38)[41].

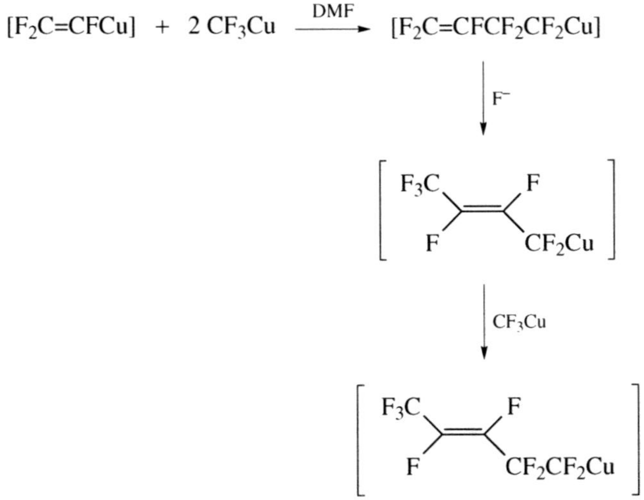

(38)

III. DIFLUOROVINYLCOPPER REAGENTS

Ichikawa and coworkers have developed a facile synthesis of 1,1-difluoro-1-alkenes via double transmetallation of 2,2-difluorovinylboranes (equation 39)[42,43]. Although the difluorovinylcopper reagent was not directly observed, the results are consistent with its formation.

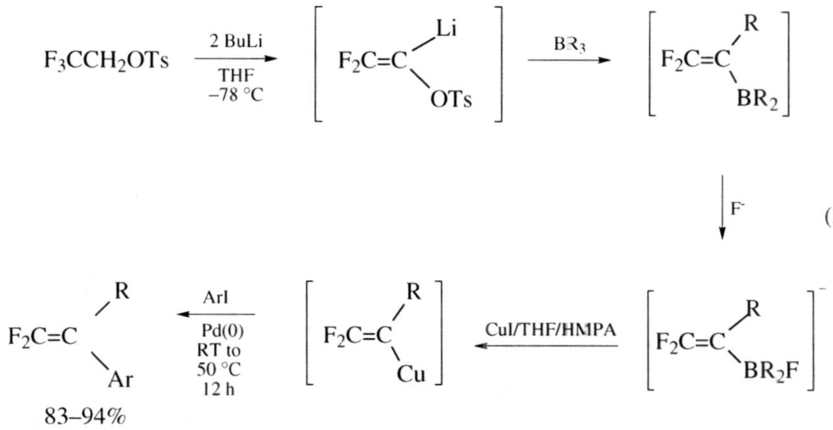

(39)

Similar methodology utilizing acyl chlorides in place of aryl iodides provided a one-flask synthesis of 2,2-difluorovinyl carbonyl compounds (equation 40)[44]. A chloroformate analog gave the 3,3-difluoroacrylate in 55% yield.

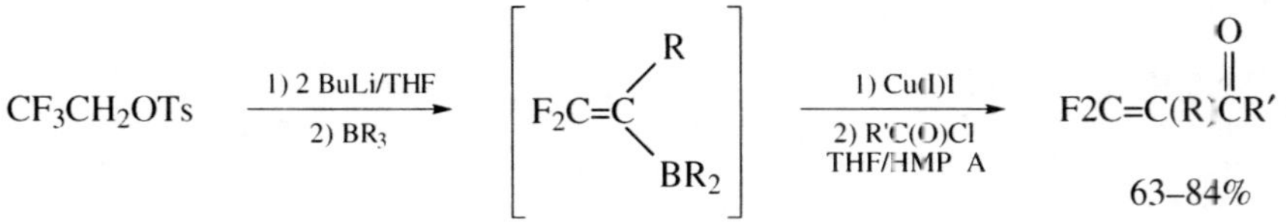

(40)

Extension of this methodology to coupling of 1-haloalkynes afforded a one-flask preparation of conjugated fluoroenynes (equation 41)[45]. Coupling with iodoethynylsilane, followed by cleavage of the eneyne with tetrabutylammonium fluoride at low temperature, gave the enyne with a terminal alkyne group (equation 42)[45].

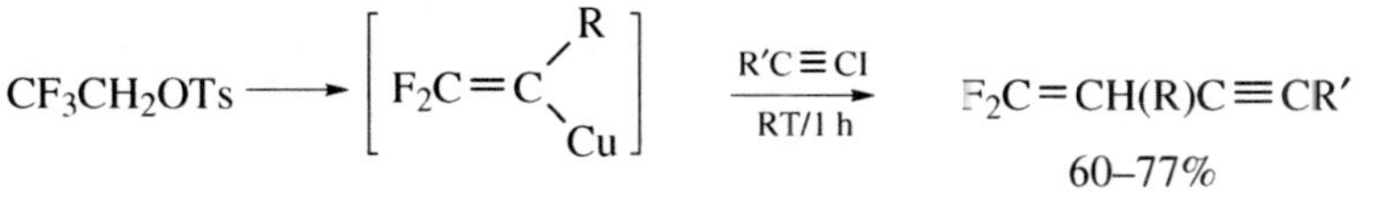

(41)

$$F_2C{=}C(CH_2Ph)C{\equiv}CSiMe_3 \xrightarrow[-78\,^\circ C/10\ min]{Bu_4NF/THF} F_2C{=}C(CH_2Ph)C{\equiv}CH \quad (42)$$

94%

Percy and coworkers employed similar methodology to accomplish transmetallation of an α-lithio-β,β-difluoroenol carbamate with CuX_3Li_2 to produce a difluorovinylcopper reagent, which reacted with activated haloalkanes and acid chlorides (equation 43)[46].

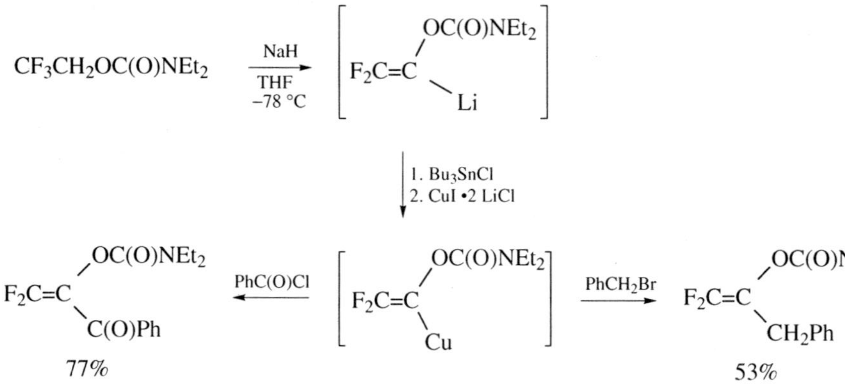

(43)

IV. PERFLUOROALLYLCOPPER REAGENTS

Perfluoroallyl iodide reacts readily with acid washed cadmium powder in DMF to give the perfluoroallylcadmium reagent. Metathesis of the perfluoroallylcadmium reagent with Cu(I)Br at $-35\,^{\circ}C$ in DMF affords the perfluoroallylcopper reagent (equation 44)[47]. The perfluoroallylcopper reagent was significantly less stable than the corresponding perfluoroallylcadmium reagent. At $-30\,^{\circ}C$ a mixture of decomposition products was detected, and decomposition was rapid at temperatures $> -20\,^{\circ}C$. Thus, this reagent must be formed and utilized at temperatures $< -30\,^{\circ}C$. The perfluoroallylcopper reagent could be detected by ^{19}F NMR analysis of the reaction mixture and captured with iodine and allyl bromide (equations 45 and 46)[47]. The perfluoroallylzinc reagent could not be utilized here. When perfluoroallyl iodide reacted with zinc, the main product formed was perfluoro-1,5-hexadiene. Only 7–14% of the perfluoroallylzinc reagent was detected under a variety of reaction conditions[47].

$$F_2C{=}CFCF_2I + Cd \xrightarrow[\text{2. RT/1 h}\;\;71\%]{\text{1. DMF/0 °C/1 h}} [F_2C{=}CFCF_2CdI] + [(F_2C{=}CFCF_2)_2Cd]$$

$$[F_2C{=}CFCF_2CdI] + [(F_2C{=}CFCF_2)_2Cd] \xrightarrow[-35\,^{\circ}C]{Cu(I)Br} [(F_2C{=}CFCF_2Cu] \quad 61\% \tag{44}$$

$$[F_2C{=}CFCF_2Cu] + I_2 \xrightarrow[-40\,^{\circ}C]{DMF} F_2C{=}CFCF_2I \quad 40\% \tag{45}$$

$$[F_2C{=}CFCF_2Cu] + H_2C{=}CHCH_2Br \xrightarrow[-40\,^{\circ}C\text{ to RT}]{DMF} F_2C{=}CFCF_2CH_2CH{=}CH_2 \quad 86\% \tag{46}$$

V. PERFLUOROALKYNYLCOPPER REAGENTS

The first perfluoroalkynylcopper reagent was reported by Haszeldine[48] as illustrated in equation 47. Direct preparation of a perfluoroalkynylcopper reagent via insertion of copper into the carbon–iodine bond of 1-iodoperfluorohexyne was unsuccessful. The diyne was the major product (equation 48)[49]. Similar to the preparation of triflurovinylcopper reagents (see Section II.A), metathesis of perfluoroalkynylzinc reagents with Cu(I) halides was found by Spawn to be a convenient, practical and general route to the corresponding perfluoroalkynylcopper reagent (equation 49)[5,49,50]. Burton and Spawn prepared the precursor to the perfluoroalkynylzinc reagents from the commercially available 1,1,2-trihydroperfluoro-1-alkenes, illustrated in equation 50[50,51].

$$CF_3C{\equiv}CH + CuCl \xrightarrow{NH_4OH} [CF_3C{\equiv}CCu] \qquad (47)$$

$$C_4F_9C{\equiv}CI + Cu \xrightarrow{DMF} \underset{52\%}{C_4F_9C{\equiv}C\text{-}C{\equiv}CC_4F_9} \qquad (48)$$

$$[R_fC{\equiv}CZnX] + Cu(I)X \longrightarrow [R_fC{\equiv}CCu] + ZnX_2 \qquad (49)$$

$$R_fCH{=}CH_2 + \text{excess } Cl_2 \xrightarrow[72\text{–}94\%]{h\nu} R_fCCl_2CCl_3 \xrightarrow{3\,Zn} [R_fC{\equiv}CZnCl] \qquad (50)$$

$R_f = C_3F_7, C_4F_9, C_6F_{13}, C_8F_{17}, C_{10}F_{21}$

This route was based on the earlier report of Finnegan and Norris[52], who first prepared trifluoropropynylzinc from 1,1,2-trichloro-3,3-trifluoropropene and excess zinc (equation 51)[52].

$$CF_3CCl{=}CCl_2 + \text{excess Zn} \longrightarrow [CF_3C{\equiv}CZnCl] + [(CF_3C{\equiv}C)_2Zn] \qquad (51)$$

Metallation of perfluoroalkynyllithium reagents with zinc chloride in THF has also been utilized as a route to perfluoroalkynylzinc reagents[53]. Spawn demonstrated that direct metallation of 1-iodo-1-perfluoroalkynes also provided a route to the requisite perfluoroalkynylzinc reagents (equation 52)[49].

$$R_fC{\equiv}CI + Zn \xrightarrow[\text{or DMF}]{\text{triglyme}} [(R_fC{\equiv}C)_2Zn] \qquad (52)$$

The perfluoroalkynylcopper reagents were generated from the corresponding perfluoroalkynylzinc reagents via exchange with Cu(I)Br in DMF (equation 49). The copper acetylides are soluble in DMF and exhibit excellent thermal stability. The $F_3CC{\equiv}CCu$ began to decompose slowly at 100 °C, whereas the longer chain analogs, such as $C_4F_9C{\equiv}CCu$, showed only 15% decomposition after 19 h at 100 °C[49]. The copper acetylides were not affected by water. Addition of water to samples of perfluoroalkynylcopper reagents caused no change in the ^{19}F NMR spectrum of these compounds, and the water treated reagents coupled normally with allyl chloride[49]. The copper acetylides reacted with aryl iodides at higher temperatures to give good yields of the corresponding perfluoroalkylalkynes, as illustrated in equation 53[49].

$$[R_fC{\equiv}CZnX] \xrightarrow[\text{2. ArI, 130–150 °C}]{\text{1. Cu(I)Br/DMF}} R_fC{\equiv}CAr \qquad (53)$$

TABLE 2.

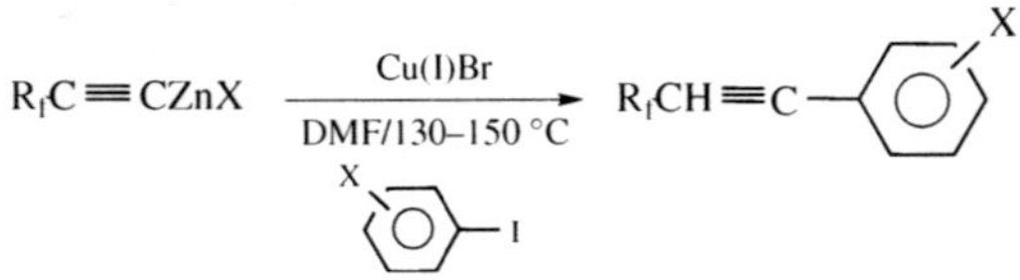

R_f	X	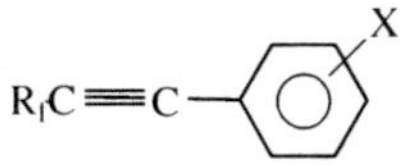 , yield (%)[a]
C_4F_9	H	68
C_6F_{13}	H	65
C_8F_{17}	H	46
C_4F_9	*m*-NO_2	76
C_4F_9	*m*-CF_3	73
C_6F_{13}	*p*-OMe	64
C_6F_{13}	*m*-NO_2	77
C_8F_{17}	*p*-OMe	55
C_8F_{17}	*m*-CF_3	60

[a] Isolated yield.

Representative examples are tabulated in Table 2[49]. A variety of functionalities is tolerated in the aryl iodide, and this preparation of arylperfluoroalkylalkynes is accomplished in three steps from the corresponding perfluoroalkylethenes.

In situ reaction of the perfluoroalkynylcopper reagent with 1-iodoalkynes affords the corresponding diynes (equation 54)[49].

$$R_fC{\equiv}CI + Zn \xrightarrow{\text{DMF}} [R_fC{\equiv}CZnI] \xrightarrow[\text{2. RC}\equiv\text{CI}]{\text{1. Cu(I)Br}} R_fC{\equiv}C{-}C{\equiv}CR \qquad (54)$$

$R_f = C_4F_9$; $R = C_4F_9$, 56%
$R_f = C_6F_{13}$; $R = C_6F_{13}$, 67%
$R_f = C_8F_{17}$; $R = C_3F_7$, 70%

Coupling of the perfluoroalkynylcopper reagent with vinyl iodides stereospecifically affords the corresponding eneyne derivative (equation 55)[49]. Allylation of the perfluoroalkynylcopper reagents is a facile reaction as illustrated in equations 56 and 57[49].

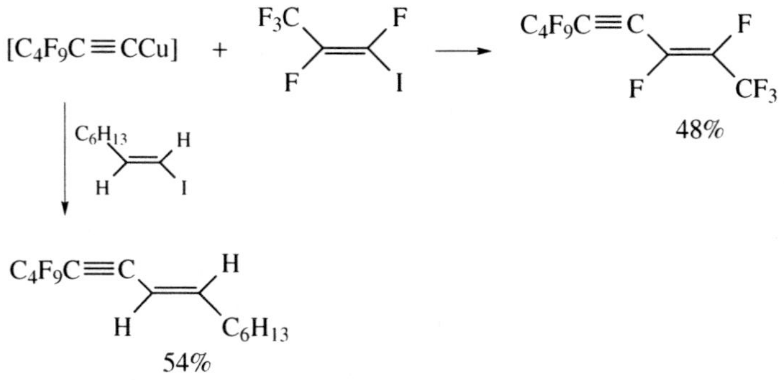

(55)

$$CF_3C{\equiv}CCu + H_2C{=}CHCH_2Cl \xrightarrow{72\%} CF_3C{\equiv}CCH_2CH{=}CH_2 \quad (56)$$

$$C_4F_9C{\equiv}CCu + ClCH_2CH{=}C(CH_3)_2 \xrightarrow{63\%} C_4F_9C{\equiv}CCH_2CH{=}C(CH_3)_2 \quad (57)$$

VI. CARBOALKOXYDIFLUOROMETHYLENECOPPER REAGENTS

Kobayashi and coworkers reacted methyl iododifluoroacetate with copper in DMSO to form $CuCF_2CO_2CH_3$ (equation 58)[54,55]. Subsequent addition of (*E*)-1-iodoheptene afforded the coupled product. HMPA and DMF were also found to be suitable solvents. Methyl bromodifluoroacetate was much less useable than the iodoester. The bromoester did not react with copper at room temperature in either HMPA or DMSO. At 80 °C/20 h, the bromoester did not give a stable copper reagent. The coupling reaction was applicable to a variety of vinyl iodides, acetylenic halides and alkyl halides, as illustrated in Table 3. In subsequent work, Kobayashi and coworkers investigated the ^{19}F NMR spectrum of $CuCF_2CO_2CH_3$[55]. The number of signals (and their chemical shifts) was significantly solvent-dependent. HMPA was found to be a specific solvent in the coupling reaction with various halides. Also, the $CuCF_2CO_2CH_3$ reagent in HMPA exhibited greater stability at RT than in DMSO or DMF. No specific structural assignments were made for the

TABLE 3. Coupling reactions of $CuCF_2CO_2CH_3$

RX	RT	Time	$RCF_2CO_2CH_3$	Yield (%)
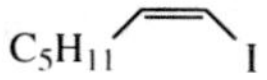	RT	30 min	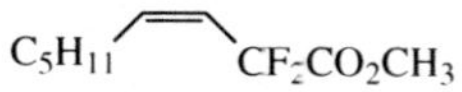	84
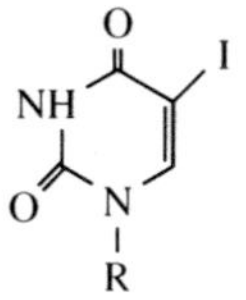	RT	30 min	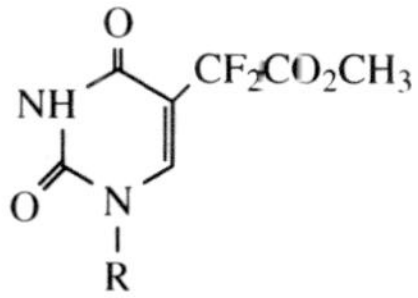	88
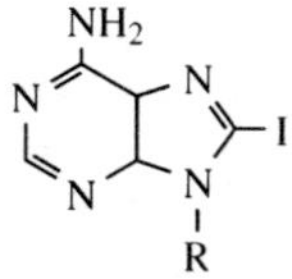	RT	25 min	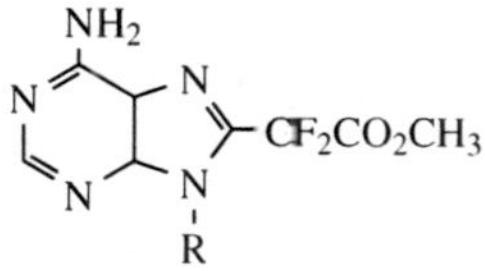	60
$BnOCH_2C{\equiv}CI$	RT	10 min	$BnOCH_2C{\equiv}CCF_2CO_2CH_3$	21
BnBr	RT	20 min	$BnCF_2CO_2CH_3$	79
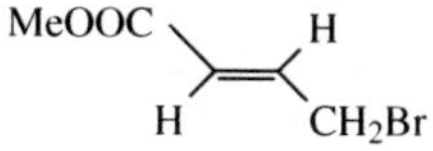	RT	30 min	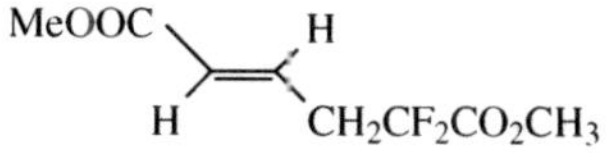	84
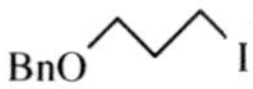	70 °C	3 h		38

three different ^{19}F NMR signals detected for this reagent[55].

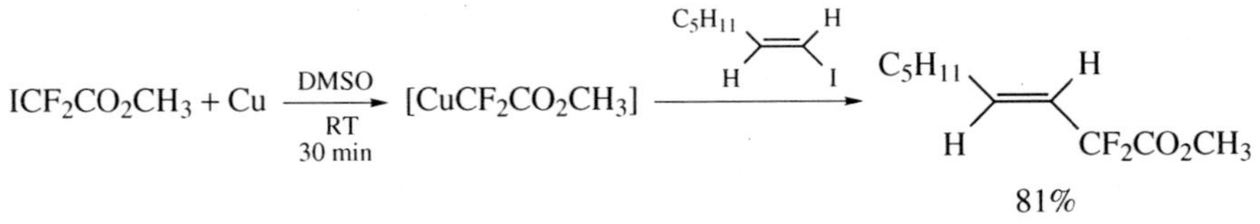

(58)

In later work, Kumadaki and coworkers developed a procedure for the synthesis of alkenyl and aryldifluoroacetates using a 'copper complex' from ethyl bromodifluoroacetate[56]. This work complements the Kobayashi work. However, since all the reagents are mixed immediately, it is not clear whether this reaction involves $CuCF_2CO_2Et$ or is a copper mediated free radical process (cf. References 57 and 58 for related copper mediated work with ethyl bromodifluoroacetate and methyl iododifluoroacetate).

VII. DIALKOXYPHOSPHINYLFLUOROMETHYLCOPPER REAGENT

Only one report has appeared in the chemical literature for the preparation of $(EtO)_2P(O)CFHCu$. This organometallic reagent was prepared by metathesis of $(EtO)_2P(O)CFHZnBr$ with Cu(I)Br, similar to the strategy employed for the preparation of trifluorovinylcopper (cf. Section II.A), as illustrated in equation 59[59]. The diethoxyphosphinylfluoromethylcopper reagent is less stable than the corresponding zinc reagent. It undergoes total decomposition in 48 h at room temperature. However, this stability limit gave no significant problem in cross-coupling processes when 1.2 to 1.5 equivalent of $(EtO)_2P(O)CFHCu$ was employed. Typical cross-coupling reactions are illustrated in Table 4.

$$[(EtO)_2P(O)CFHZn] \xrightarrow[0\,°\text{ to }5\,°C]{\text{Cu(I)Br, THF/DMF}} [(EtO)_2P(O)CFHCu] \qquad (59)$$

VIII. DIALKOXYPHOSPHINYLDIFLUOROMETHYLCOPPER REAGENTS

Dialkoxyphosphinyldifluoromethylcopper has not been prepared via direct insertion of copper into a carbon–halogen bond. However, it can be readily prepared by transmetallation of dialkoxyphosphinyldifluoromethylzinc and dialkoxyphosphinyldifluoromethylcadmium with Cu(I) halides. When acid washed zinc dust or powder reacts with dialkylbromodifluoromethyl phosphonates in ether solvents, such as THF, monoglyme, triglyme or

TABLE 4. Coupling reactions of $(EtO)_2P(O)CFHCu$

$$(EtO)_2P(O)CFHCu + R{-}X \longrightarrow (EtO)_2P(O)CFHR$$

R–X	Product	Yield (%)[a]
$PhC{\equiv}CBr$	$(EtO)_2P(O)CFHC{\equiv}CPh$	74
$C_6H_{13}C{\equiv}CBr$	$(EtO)_2P(O)CFHC{\equiv}CC_6H_{13}$	71
(*E*)-$PhCH{=}CHI$	(*E*)-$(EtO)_2P(O)CFHCH{=}CHPh$	69
(*E*)-$C_6H_{13}CH{=}CHI$	(*E*)-$(EtO)_2P(O)CFHCH{=}CHC_6H_{13}$	62
C_6H_5I	$(EtO)_2P(O)CFHC_6H_5$	34
p-$NO_2C_6H_4I$	$(EtO)_2P(O)CFHC_6H_4NO_2$-*p*	63

[a] Isolated yields.

dioxane, at room temperature to 60 °C, the **stable** dialkoxyphosphinyldifluoromethylzinc reagent is formed in good yields (equation 60)[60]. The dialkoxyphosphinyldifluoromethylzinc reagent was readily acylated to afford the corresponding dialkyl-2-oxo-1,1-difluoroalkylphosphonates (equation 61)[60].

$$(RO)_2P(O)CF_2Br \ + \ Zn \xrightarrow[\text{RT to 60 °C}]{\text{1 to 6 h, DMF}} [(RO)_2P(O)CF_2ZnBr] \qquad (60)$$

$$[n\text{-}BuO)_2P(O)CF_2ZnBr] + RC(O)X \xrightarrow[\text{20–24 h}]{\text{RT}} (n\text{-}BuO)_2P(O)CF_2C(O)R \qquad (61)$$

$X = Cl$
$R = CH_3$, 77%; $R = CH_2Cl$, 69%;
$R = (CH_3)_2CH$, 72%; $R = Ph$, 62%;
$R = CF_3$, 77%; $R = CH_3OCOCH_2CH_2$, 67%

The corresponding stable dialkoxyphosphinyldifluoromethylcadmium reagent was similarly prepared by Burton and coworkers in solvents such as triglyme, dioxane, DMF and HMPA (equation 62)[61]. The cadmium reagent readily participates in halogenation, acylation, silylation and allylation and adds to benzaldehyde in the presence of NaI to afford $PhCH{=}CF_2$[61].

$$(EtO)_2P(O)CF_2Br + Cd \xrightarrow[\text{RT/2 h}]{\text{DMF}} \underbrace{[(EtO)_2P(O)CF_2CdBr] + [(EtO)_2P(O)CF_2]_2Cd}_{80\%} \qquad (62)$$

Burton, Sprague and coworkers found that the addition of Cu(I)Br to the dialkoxyphosphinyldifluoromethylzinc reagent significantly enhanced the reactivity of the zinc reagent, which does not react with ethyl chloroformate. However, addition of Cu(I)Br to the zinc reagent affords the dialkoxyphosphinyldifluoromethylcopper reagent, which readily reacts with ethyl chloroformate, ethyl oxalyl chloride and diethyl carbamoyl chloride (equation 63)[62,63]. The product from the ethyl chloroformate reaction was subsequently converted to difluorophosphonoacetic acid and its derivatives.

$$(EtO)_2P(O)CF_2ZnBr \ + \ ClC(O)R \xrightarrow[\text{1 h/RT, monoglyme}]{\text{Cu(I)Br}} (EtO)_2P(O)CF_2C(O)R \qquad (63)$$

$R = OEt$ 50%
$R = C(O)OEt$ 62%
$R = NEt_2$ 38%

Subsequent work by Burton and Sprague developed a convenient allylation route to 1,1-difluoro-3-alkene phosphonates (equation 64)[64]. Table 5 summarizes some representative results of this reaction.

$$(EtO)_2P(O)CF_2ZnBr \ + \ H_2C{=}CHCH_2Cl \xrightarrow[\text{monoglyme, 87\%}]{\text{Cu(I)Br}} (EtO)_2P(O)CF_2CH_2CH{=}CH_2 \qquad (64)$$

Mechanistic studies indicated an S_N2 or S_N2' type mechanism, rather than involvement of a symmetrical (α-allyl)Cu(III) intermediate and an oxidative addition/reductive elimination type mechanism.

TABLE 5. Reaction of $(EtO)_2P(O)CF_2Cu$ with allylic halides

Allylic halide	Product	Yield (%)
$ClCH_2CH{=}CH_2$	$(EtO)_2P(O)CF_2CH_2CH{=}CH_2$	87
$BrCH_2CBr{=}CH_2$	$(EtO)_2P(O)CF_2CH_2CBr{=}CH_2$	52
$BrCH_2CH{=}CHCH_3$	$(EtO)_2P(O)CF_2CH_2CH{=}CHCH_3$	49
$H_2C{=}CHCHClCH_3$	$(EtO)_2P(O)CF_2CH_2CH{=}CHCH_3$	76
(E)-$ClCH_2CH{=}CHPh$	(E)-$(EtO)_2P(O)CF_2CH_2CH{=}CHPh$	55
$BrCF_2CH{=}CF_2$	$(EtO)_2P(O)CF_2CH_2CH{=}CF_2$	55

When $(EtO)_2P(O)CF_2Cu$ was generated in glyme solvents via metathesis of the corresponding zinc reagent with a catalytic amount of Cu(I)Br, the reagent spontaneously decomposed to yield a mixture of both geometrical isomers of 1,2-difluoroethenediylbisphosphonates (equation 65)[65].

$$(EtO)_2P(O)CF_2ZnBr \xrightarrow[\text{Cu(I)Br, glyme}]{\text{catalytic}} (EtO)_2P(O)CF{=}CFP(O)(OEt)_2 \quad 50\%,\ E/Z = 7/3 \tag{65}$$

Treatment of diethoxyphosphinyldifluoromethylcadmium with Cu(I)Br at 0 °C in DMF generated a stable $(EtO)_2P(O)CF_2Cu$ reagent (equation 66)[66].

$$(EtO)_2P(O)CF_2CdX \xrightarrow[\text{0 °C, DMF}]{\text{Cu(I)Br}} [(EtO)_2P(O)CF_2Cu] \quad 77\% \tag{66}$$

This reagent undergoes facile addition to perfluoro-2-butyne to give the *syn*-alkenylcopper reagent illustrated in equation 67[66]. This *syn*-alkenylcopper reagent undergoes the expected alkylation, arylation and coupling reactions (equation 68)[66].

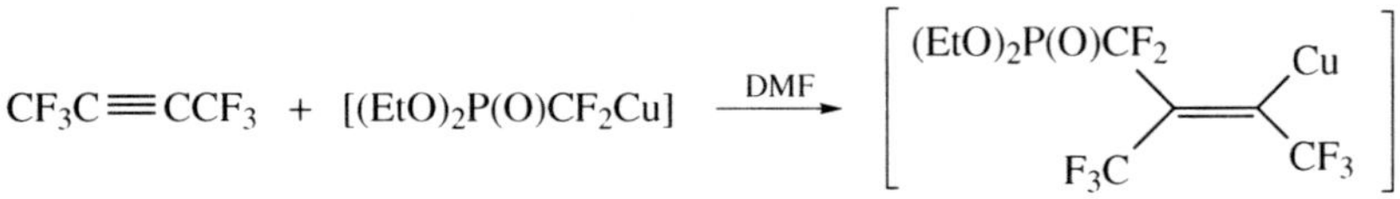

(67)

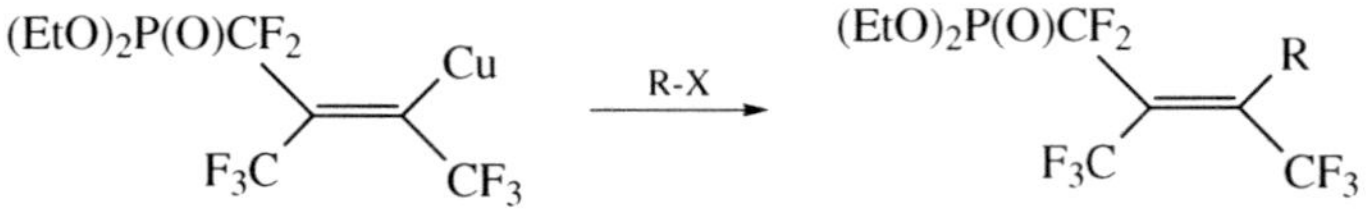

(68)

R = Me 54%
R = allyl 40%
R = p-$NO_2C_6H_4$ 58%
R = (Z)-$CF_3CF{=}CF_2$ 15%

Qiu and Burton developed a mild, useful preparation of α,α-difluorobenzylic phosphonates via the Cu(I)Cl promoted coupling reaction of the diethoxyphosphinyldifluoromethylcadmium reagent with aryl iodides (equation 69)[67]. A variety of functional groups, such as nitro, ether, ester and halides, were tolerated in this reaction. They also noted that in the presence of Cu(I)Cl, the diethoxyphosphinyldifluoromethylzinc reagent also coupled with aryl halides to afford the α,α-difluorobenzylic phosphonates[67]. This methodology has been utilized by Park and Standaert in the preparation of a series of novel, photoregulated phosphoamino acid analogs, based on an azobenzene bearing an α,α-difluoromethylphosphonate as a hydrolytically stable phosphate isostere[68].

$$[(EtO)_2P(O)CF_2CdBr] \xrightarrow[\text{ArI/RT, DMF}]{\text{Cu(I)Br}} (EtO)_2P(O)CF_2Ar \quad 65\text{–}88\% \qquad (69)$$

The dialkoxyphosphinyldifluoromethylcopper reagent also undergoes insertion of a CF_2 unit (similar to trifluorovinylcopper) to give the next higher homologous copper reagent (equation 70)[69].

$$[(EtO)_2P(O)CF_2Cu] + [Ph_3\overset{+}{P}CF_2Br]Br^- \xrightarrow[\text{DMF}]{Na_2CO_3} [(EtO)_2P(O)CF_2CF_2Cu] \xrightarrow{X_2} (EtO_2P(O)CF_2CF_2X \quad X = I, Br \quad 25\text{–}30\% \qquad (70)$$

Shibuya and coworkers further developed the Cu(I)Br/dialkoxyphosphinyldifluoromethylzinc reagent in a series of publications. For example, they studied the synthesis of α,α-difluoroallyl phosphonates from alkenyl halides and acetylenes (equation 71)[70]. This methodology was utilized in the preparation of optically active *trans*-1-(diethoxyphosphinyl)difluoromethyl-2-hydroxymethylcyclopropanes[71,72]. Subsequent extension of this methodology afforded 1,1-difluoro-5-(1*H*-9-purinyl)-2-pentenylphosphonic acids and related analogs[73].

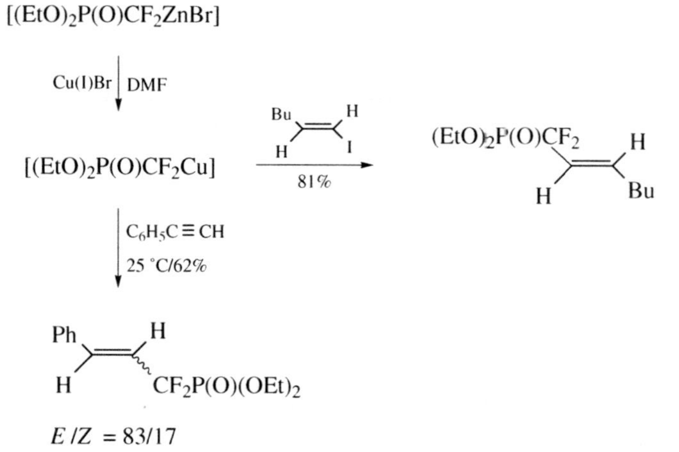

(71)

Similar to the report of Qiu and Burton[67], Shibuya and coworkers reported Cu(I)Br mediated cross-coupling reactions of diethoxyphosphinyldifluoromethylzinc with aryl iodides (equations 72 and 73)[74].

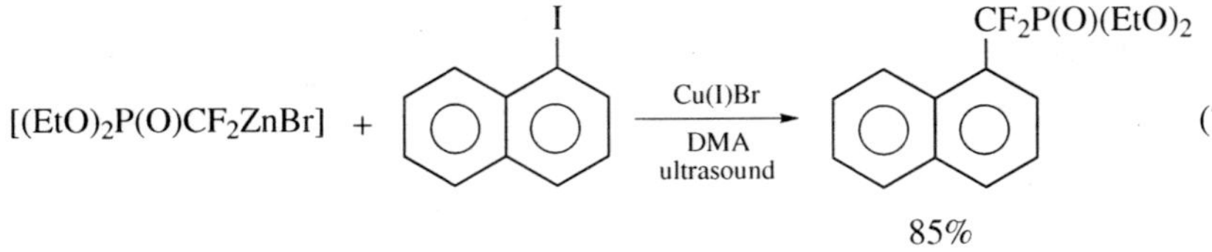

(72)

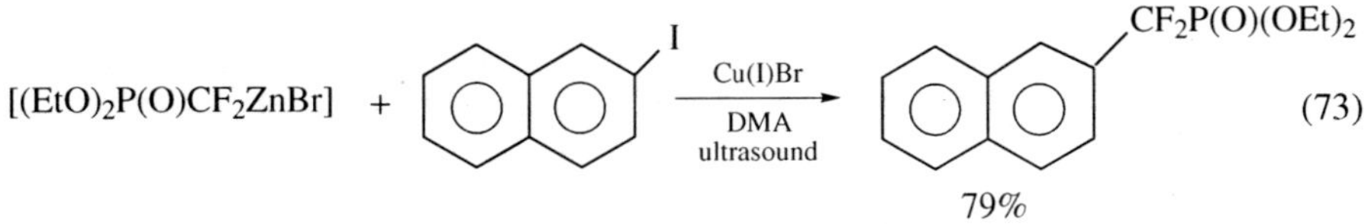

(73)

Extension of this aryl coupling methodology afforded (phosphonodifluoromethyl)phenylalanine (F_2PmP), hydrolytically stable phosphorylated-Tyr analogs from 2-benzyl-1,3-propanediols through chemoenzymatic sequences[75]. The synthesis of benzylphosphonic acid derivatives as small molecule inhibitors of protein-Tyrosine Phosphatase 1B was effected by a Stille coupling reaction with halogenated α,α-difluorobenzylphosphonates (equation 74)[76].

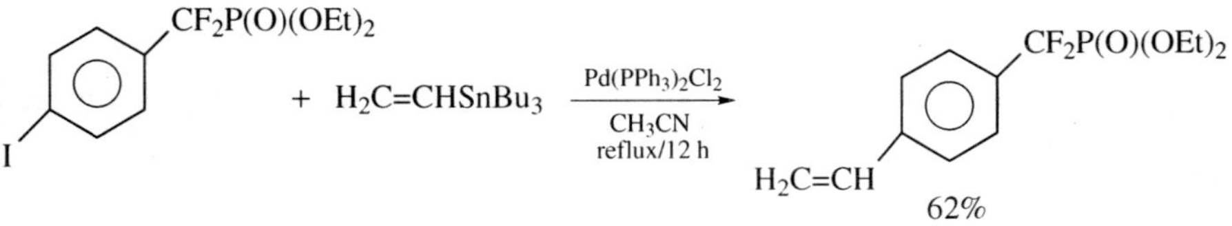

(74)

IX. PERFLUOROBENZYLCOPPER REAGENT

Only one report of the perfluorobenzylcopper has appeared in the chemical literature. Burton and coworkers prepared this reagent via metathesis of the corresponding cadmium reagent (equation 75)[77]. Attempts to prepare the perfluorobenzylcopper reagent at room temperature failed. Only decomposition products were observed. The copper reagent could be trapped at $-35\,^{\circ}C$ with allyl bromide to afford $C_6F_5CF_2CH_2CH{=}CH_2$.

$$C_6F_5CF_2Br + Cd \xrightarrow[\text{RT}]{\text{DMF}} \underset{85\text{–}90\%}{[C_6F_5CF_2CdBr]} \xrightarrow[-40\,^{\circ}C \text{ to } -35\,^{\circ}C]{\text{Cu(I)Br}} [C_6F_5CF_2Cu] \qquad (75)$$

X. PERFLUOROALKYLCOPPER REAGENTS

Perfluoroalkylcopper reagents are among the most studied perfluoroorganometallic reagents due to their combination of chemical reactivity and thermal stability. They are prepared by the following primary methods:

(1) Copper metal insertion reactions with perfluoroalkyl halides in a coordinating solvent at elevated temperatures.
(2) Decarboxylation of perfluoroalkanoic acid salts in the presence of copper(I) halides.

(3) Metathesis of perfluoroalkylorganometallic reagents with copper metal or copper(I) salts.
(4) Decomposition of methyl fluorosulfonyldifluoroacetate analogs in presence of copper(I) halide.
(5) Reaction of trialkyl(perfluoroalkyl)silanes with fluoride ion and copper(I) iodide.

These methods will be summarized in the context of general perfluoroalkylation and separately as they relate to the industrially important process of trifluoromethylation.

A. Perfluoroalkylation

The pioneering work of McLoughlin and Thrower documented the first preparation of perfluoroalkylcopper reagents by reaction of perfluoroalkyl iodides with excess copper metal in a coordinating solvent at 110–120 °C to give copper reagents in good yields (equation 76)[78].

$$R_fI + 2\,Cu \xrightarrow[110\text{–}120\,^\circ C]{DMSO} R_fCu + CuI \qquad (76)$$

These reactions were typically carried out *in situ* with aromatic halides to give fluoroalkyl arenes in good yields. A range of perfluoroalkyl iodides could be utilized and one example of trifluoromethylation was reported. Typical examples are outlined below (equations 77 and 78). α,ω-Diiodoperfluoroalkanes underwent disubstitution with aryl and heteroaryl iodide substrates.

$$C_7F_{15}I + C_6H_5I \xrightarrow[DMSO,\ 110\,^\circ C]{Cu} \underset{70\%}{C_7F_{15}C_6H_5} \qquad (77)$$

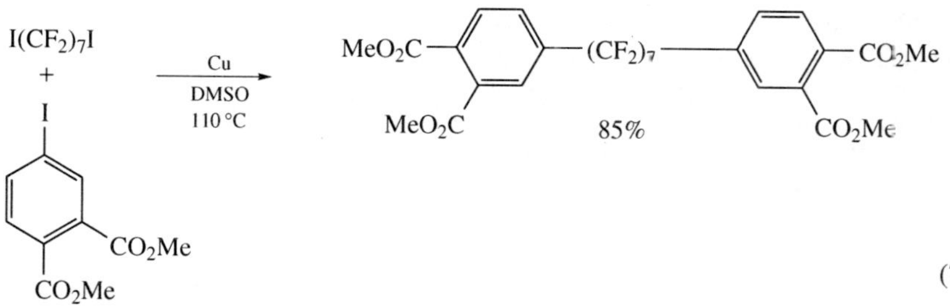

(78)

Preparation of perfluoroalkylated benzenes of interest as precursors of polymeric materials is a representative application of this method (equation 79)[79].

(79)

$R_f = C_3F_7, C_6F_{13}, C_7F_{15}, C_8F_{17}, (CF_3)_2CF$

DMSO is the most common solvent in the *in situ* generation of perfluoroalkylcopper but Me_2S, HMPA, DMF and pyridine have been used. Chen and coworkers have demonstrated the dependence of the mechanism of copper reagent formation on the donor number (DN) of the solvent[80] in studies carried out with 4-chlorooctafluorobutyl iodide. Initially, the generalized mechanism (Scheme 2) is proposed to involve SET to form an ion pair within a solvent cage. In solvents of low DN (MeCN, hexane, dioxane), diffusional control of a caged ion pair leads to a perfluoroalkyl radical which could be trapped selectively by 1-heptene in competition experiments with iodobenzene. Solvents of larger DN, however, lead to collapse of the ion pair to give a radical in a solvent cage, adsorbed on the surface of copper metal, which undergoes electron transfer to form R_fCu.

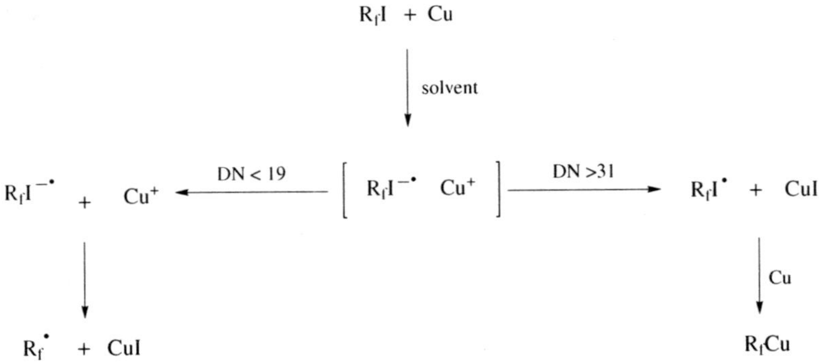

SCHEME 2

Aryl iodides undergo coupling more readily than aryl bromides. However, Chen and Tamborski demonstrated that aryl bromides[81] could be competent substrates in some cases and bromoheterocycles[82] were utilized in the preparation of perfluoroalkylated pyridines, pyrimidines, furans and thiophenes. Perfluoroalkylation of halothiophene substrates has also been reported to give isomeric mixtures due to rearrangement (equation 80)[83].

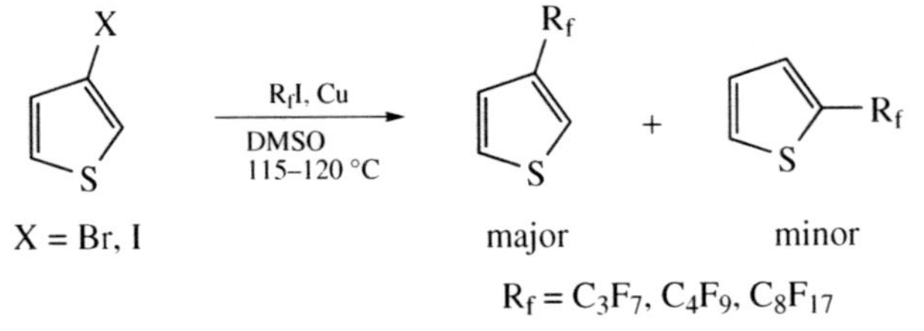

(80)

Although coupling of aryl halides has been the most extensively studied reaction of perfluoroalkylcopper reagents, these reagents couple readily with a range of substrates. The couplings can be carried out either with pre-generated or *in situ* generated R_fCu[84]. Vinyl bromides are capable substrates (equation 81)[84] and disubstitution occurs in high yield when α,ω-perfluoroalkyl diiodides are used (equation 82)[85]. Mono or bis coupling could be selectively achieved when 1,2-diiodoethylene was used[85–87].

$$PhCH{=}CHBr + C_3F_7I \xrightarrow[\text{Py, 100 °C}]{\text{Cu}} PhCH{=}CHC_3F_7 \quad (82\%) \qquad (81)$$

$$I(CF_2)_3I + ClCH{=}CHI \xrightarrow[\text{Py, 100 °C}]{\text{Cu}} \underset{96\%}{ClHC{=}CH(CF_2)_3CH{=}CHCl} \quad (82)$$

Similarly, R_fCu reagents couple with allyl halides (equation 83)[88] or thiocyanates[89]. An α-diketone was reported via capture of R_fCu with α-ketoaryl acyl halide (equation 84)[90]. The pre-generated copper reagent route typically gives the best yields with active halides such as allyl, propargyl and acyl halides.

$$C_7F_{15}Cu + H_2C{=}CHCH_2Br \longrightarrow H_2C{=}CHCH_2C_7F_{15} \quad (83)$$

$$C_3F_7Cu + C_6H_5C(O)COCl \xrightarrow{\text{MeCN}} \underset{49\%}{C_6H_5C(O)COC_3F_7} \quad (84)$$

Propargyl bromide was initially reported by Coe and Milner to undergo a highly exothermic reaction with perfluoroheptylcopper to give low yields of allene (equation 85)[91].

$$C_7F_{15}Cu + HC{\equiv}CCH_2Br \longrightarrow \underset{<10\%}{C_7F_{15}CH{=}C{=}CH_2} \quad (85)$$

The allene synthesis was later achieved in reasonable yields by Burton and coworkers[92] utilizing propargyl chlorides or tosylates in DMF or DMSO (equation 86). Related work by Hung demonstrated preparation of perfluorodiallenes by reacting perfluoroalkylene dicopper reagents with propargyl bromides (equation 87)[93].

$$R_fCu + HC{\equiv}CCR^1R^2X \xrightarrow[\text{or DMSO, 0 °C–RT}]{\text{DMF}} \underset{30\text{–}73\%}{R_fCH{=}C{=}CR^1R^2} \quad (86)$$

$R_f = CF_3, C_3F_7, C_6F_{13}, C_8F_{17}$ $\quad R^1, R^2 = H, CH_3; R^1R^2 = -(CH_2)_5-$

$X = Cl, OTs$

$$\underset{n = 6, 8}{Cu(CF_2)_nCu} + HC{\equiv}CCH_2Br \xrightarrow[\text{15–20 °C}]{\text{DMSO}} \underset{27\text{–}32\%}{H_2C{=}C{=}CH(CF_2)_nCH{=}C{=}CH_2} \quad (87)$$

Coupling of 1-iodo-2-phenylacetylene with pre-generated R_fCu gave the perfluoroalkyl phenylalkyne (equation 88)[49]. The procedure required low temperatures to suppress formation of 1,4-diphenylbutadiyne. Similar coupling with a 1-iodoperfluoroalkylacetylene substrate, however, resulted only in perfluorodiyne.

$$C_6F_{13}Cu + IC{\equiv}CPh \longrightarrow C_6F_{13}C{\equiv}CPh \quad (88)$$

R_fCu reagents undergo additions to olefins. Both pre-generation[88] (equation 89) and *in situ* conditions[94] (equation 90) have been employed. The additions are presumed to occur via perfluoroalkyl radicals formed from decomposition of the R_fCu reagent. Reactions with methyl acrylate, allyl alcohol or 1,2-dihydropyran gave polymerization or complex mixtures. Perfluoroalkylethylenes gave polyfluorinated dienes[94] resulting from addition followed by a double elimination process.

$$C_7F_{15}Cu + H_2C{=}CHC_5H_{11} \xrightarrow[110\,^{\circ}C]{DMSO} C_7F_{15}CH_2CH_2C_5H_{11} + C_7F_{15}CH_2CH{=}CHC_4H_9 \quad (89)$$

1:2

$$R_f^1CF_2I + R_fCF_2CH{=}CH_2 + Cu \xrightarrow[150\,^{\circ}C]{DMF} R_fCF{=}CHCH{=}CFR_f^1 \quad (90)$$

50–73%

Chen and Yang utilized copper metal to initiate addition of perfluoroalkyl iodides to olefins (equation 91)[95].

$$R_fI + R^1CH{=}CHR^2 \xrightarrow[\text{Ac}_2\text{O or diglyme, } 100\,^{\circ}C]{Cu} R^1CHICHR^2R_f \quad (91)$$

38–70%

B. Trifluoromethylation

Trifluoromethylation remains the most significant perfluoroalkylation reaction as incorporation of this moiety into pharmaceutical and agricultural chemicals often leads to enhanced bioactivity. As a result considerable studies aimed at developing cost-effective, efficient methods for introduction of the trifluoromethyl group into a variety of substrates via generation and coupling of trifluoromethylcopper have been undertaken. It is recognized that methods outside the scope of organocopper chemistry have appeared in the literature including nucleophilic[96a,b], electrophilic[97a] and radical trifluoromethylation[97b]. Classical methods include halogen exchange of trichloromethyl groups on exposure to hydrogen fluoride or antimony trifluoride and the conversion of carboxylic acids with sulfur tetrafluoride in pressure vessels[98–100]. These classical methods, however, are less practical and require special equipment.

The pioneering studies for *in situ* generation of trifluoromethylcopper were carried out by Kobayashi and coworkers[101,102]. These workers reacted CF_3I or CF_3Br with allyl, vinyl, aryl and heterocyclic halides in presence of copper powder in aprotic solvents at elevated temperatures or utilized HMPA solutions to facilitate milder conditions. Typical examples are shown in equations 92 and 93.

$$CF_3I + Cu + PhCH_2Br \xrightarrow[RT,\ 12\ h]{HMPA} PhCH_2CF_3 \quad (92)$$

65%

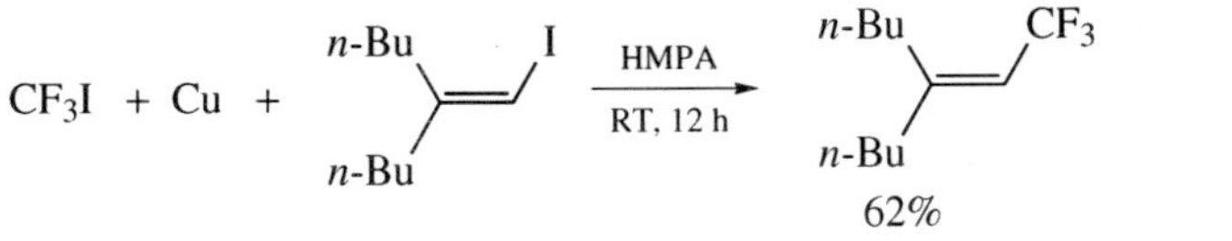

(93)

The method has been applied in the preparation of pyrimidine and purine nucleosides[103,104]. In some cases, heterocyclic substrates such as benzofurans[105] resulted in complex mixtures due to rearrangement and formation of pentafluoroethyl substituted products arising from decomposition of CF_3Cu.

The expense of CF_3I and the requirement for high temperatures have lead to investigations utilizing cheaper sources of the trifluoromethyl group. Trifluoromethylalkanoic acid salts undergo decarboxylation in presence of copper(I) halides. Early work by Matsui and coworkers showed that aromatic halides with electron-withdrawing and electron-releasing groups could be successfully trifluoromethylated using sodium trifluoroacetate (equation 94)[106].

$$CF_3CO_2Na + CuI + ArI \xrightarrow[140\text{–}160\ °C]{NMP} ArCF_3 \quad (40\text{–}78\%) \qquad (94)$$

Chambers and coworkers successfully extended this method to the pentafluoroethyl system (equation 95)[107]. Sodium heptafluoropropylbutyrate, however, gave only low yields of heptafluoropropyl aryl product under similar conditions.

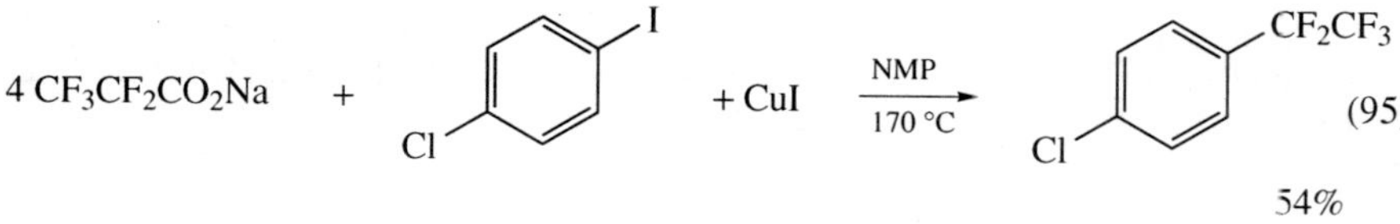

(95)

The CF_3CO_2Na/CuI method does not avoid high temperatures and the cost advantage of using CF_3CO_2Na is somewhat offset by the requirement of up to a fourfold excess.

In an improved protocol requiring only 2 equivalents of CF_3CO_2Na, Frekos employed a toluene/DMF solvent to azeotropically remove water prior to decarboxylation (equation 96)[108].

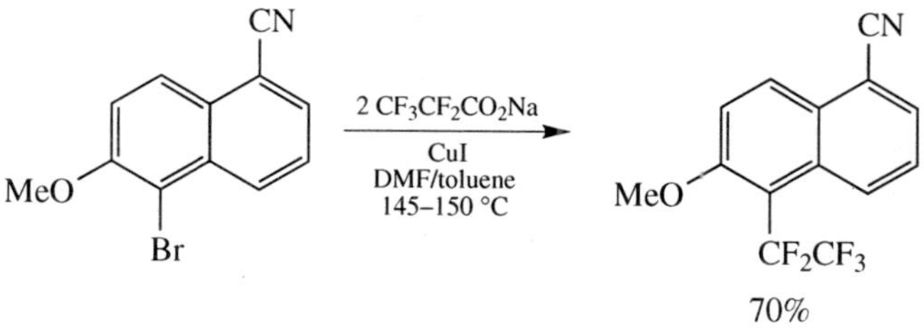

(96)

Recently, this method was applied as a key transformation in the synthesis of bioactive 2-alkoxymethyl-3-trifluoromethyl-1,4-napthoquinones (equation 97)[109].

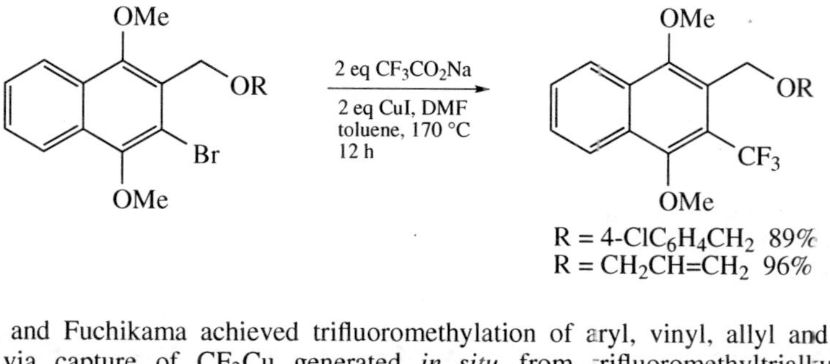

(97)

Urata and Fuchikama achieved trifluoromethylation of aryl, vinyl, allyl and benzyl halides via capture of CF_3Cu generated *in situ* from trifluoromethyltrialkylsilanes and CuI/KF[110]. In some cases, pentafluoroethyl-substituted side products were formed (equation 98). The scope of this method was extended to pentafluoroethylation utilizing $CF_3CF_2SiMe_3$ and one example of perfluoropropylation was reported.

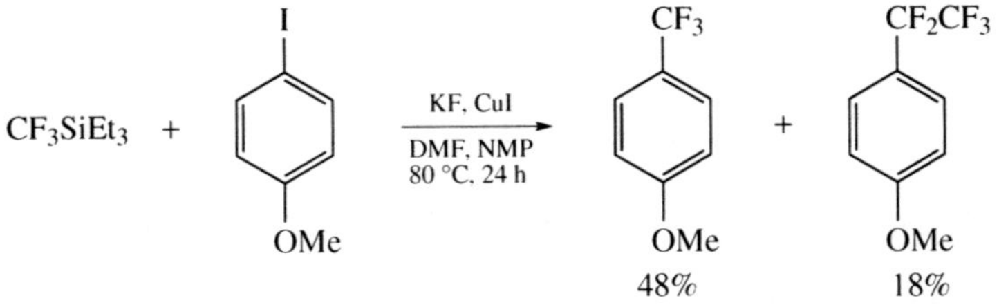

(98)

Cottet and Schlosser recently extended this method to the conversion of 2-iodopyridines to 2-(trifluoromethyl)pyridines at room temperature in good yields (Scheme 3)[111]. The presence of multiple non-exchangeable halogen atoms is well tolerated. However, 3- and 4-iodopyridyl substrates gave only modest yields.

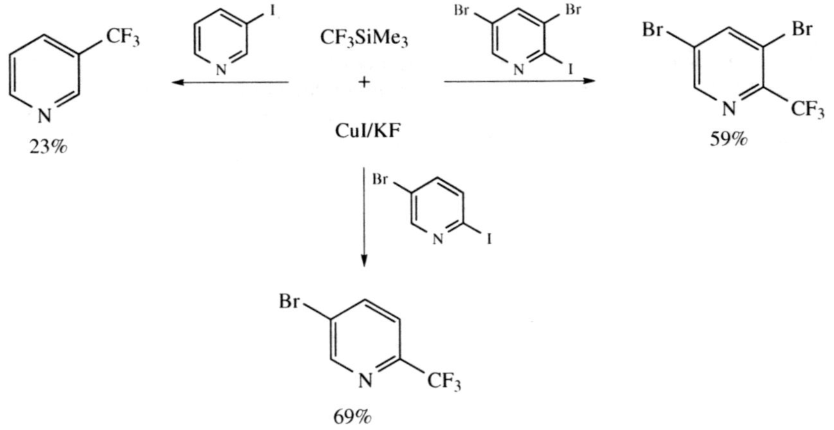

SCHEME 3

This procedure was similarly employed in a recent synthesis of bis-(aryl)thiourea inhibitors of human cytomegalovirus (Scheme 4)[112].

Although simple trifluoromethyl(trialkyl)silanes are generally commercially available, Fuchikama's approach using higher analogs is dependent on availability or synthesis of R_fSiR_3 via the Ruppert reaction[113].

Several pre-generative routes to CF_3Cu have been reported. Umemoto and Ando demonstrated a useful and mild route to a CF_3Cu complex via reaction of *N*-trifluoromethyl-*N*-nitrosotrifluoromethanesulfonamide (TNS-Tf) with copper in polar aprotic solvents (equation 99)[114]. The CF_3Cu complex coupled with aryl iodides to give reasonable yields of trifluoromethyl aromatics. One drawback of this method, however, is the multistep preparation and thermal lability of TNS-Tf.

$$\underset{\text{TMS-Tf}}{F_3C{-}\overset{\overset{\displaystyle N{=}O}{|}}{N}{-}SO_2CF_3} \xrightarrow[\text{solvent, RT}]{\text{Cu}} \left[CF_3Cu\cdot\text{solvent}\right] \xrightarrow[100\ ^\circ C]{\text{ArI}} \underset{\sim 60\%}{ArCF_3} \qquad (99)$$

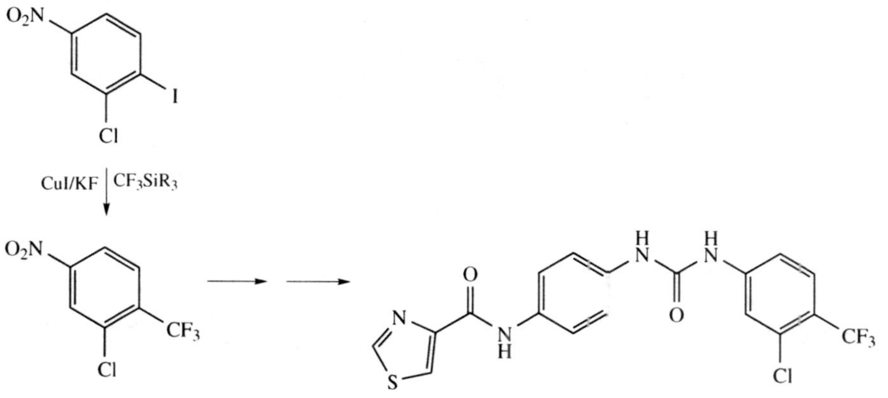

SCHEME 4

Generation of CF_3Cu from reaction of bis(trifluoromethy)mercury and copper has been reported by Yagupol'skii and coworkers[115]. Although the CF_3Cu solution could be successfully utilized for trifluoromethylation of aryl halides (equation 100), the toxicity of bis(trifluoromethyl)mercury has limited application of this method.

$$Hg(CF_3)_2 + Cu \longrightarrow CF_3Cu + Hg \xrightarrow{ArI} \underset{60\text{–}90\%}{ArCF_3} \qquad (100)$$

Burton and Weimers reported a unique and practical metathesis route to CF_3Cu extending from the discovery that dihalodifluoromethanes react with cadmium or zinc in DMF at room temperature to give zinc and cadmium reagent solutions of high thermal stability (equation 101)[116].

$$\begin{array}{llll} CF_2XY & + M & \xrightarrow[RT]{DMF} & CF_3MX + (CF_3)_2M \\ X = Br, Cl & & M = Cd, Zn & \quad 80\text{–}95\% \\ Y = Br, Cl & & & \end{array} \qquad (101)$$

While the zinc reagent undergoes metathesis only slowly at room temperature, the cadmium reagent exchanges readily at −40 °C, leading to a highly useful pregenerative route to CF_3Cu (equation 102). Two copper species, $CF_3Cu{\bullet}L$(L = metal halide) and $CdI^+[(CF_3)_2Cu]^-$, were assigned depending on CuX and stoichiometry of the metathesis reaction.

$$CF_3CdX + (CF_3)_2Cd \xrightarrow[Y = I, Br, Cl, CN]{CuY} \underset{90\text{–}100\%}{[CF_3Cu]} \qquad (102)$$

The CF_3Cu solution generated from metathesis of trifluoromethylcadmium undergoes slow decomposition to CF_3CF_2Cu at room temperature. Addition of HMPA to the CF_3Cu/DMF solution, however, inhibits formation of CF_3CF_2Cu and facilitates utility of CF_3Cu even at high temperatures[117,118]. The general scope of the trifluoromethylation is illustrated in typical examples as shown in Scheme 5[117,119–121]. Notably, vinyl, allyl and propargyl

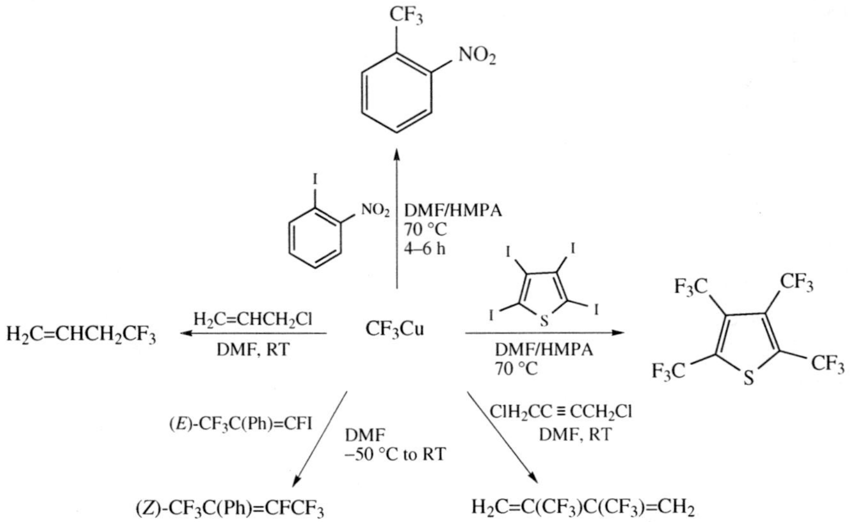

SCHEME 5

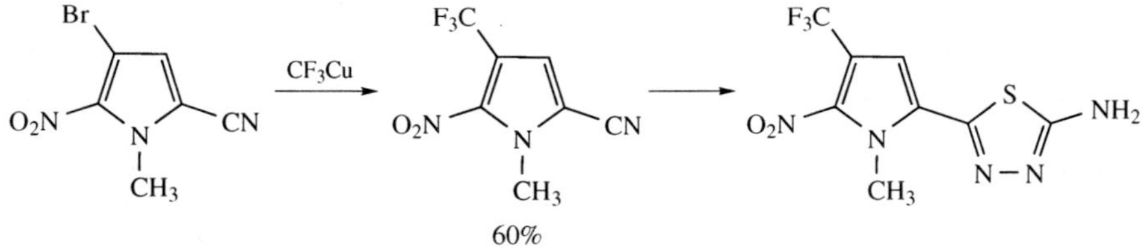

SCHEME 6

halide substrates do not require stabilization of the CF_3Cu solution with HMPA and the trifluoromethylation occurs readily at or below room temperature.

Recently, Chauviere and coworkers employed this methodology in the trifluoromethylation of a nitroimidazole intermediate enroute to megazol analogs studied as antiparasitic agents (Scheme 6)[122].

CF_3Cu has been generated directly from dihalodifluoromethanes at 85–95 °C in DMF. However, under these conditions oligomerization competes with trifluoromethylation to give an oligomeric product distribution (equation 103)[123]. The oligomerization could be effectively suppressed by addition of KF to the reaction mixture[123]. Clark and coworkers also reported that addition of charcoal to a CF_2Br_2/Cu/DMAC solution, or use of CuI-charcoal supported reagent, successfully suppressed oligomerization allowing trifluoromethylation of aryl chlorides[124].

$$CF_2BrCl + PhI \xrightarrow[\text{DMF, 85–95 °C}]{\text{Cu}} Ph(CF_2)_nCF_3 \quad n = 1\text{–}14 \tag{103}$$

Formation of CF_3M (M = Zn, Cd, Cu) directly from dihalodifluoromethanes occurs via a unique mechanistic pathway. Remarkably, DMF functions both as solvent and reagent. Inhibition and trapping experiments support a SET mechanism producing difluorocarbene which, on further reaction with fluoride, formed from reaction of difluorocarbene with DMF, leads to formation of trifluoromethide. The trifluoromethide reacts with *in situ* formed metal halide to give the trifluoromethyl organometallic reagent (Scheme 7)[116,123].

$$CF_2XY + M(Zn,Cd,Cu) \longrightarrow MXY - [:CF_2]$$

$$[:CF_2] + Me_2NCHO \longrightarrow Me_2NCF_2H + CO$$

$$Me_2NCF_2H \rightleftharpoons [Me_2\overset{+}{N}{=}CHF]F^-$$

$$[:CF_2] + F^- \rightleftharpoons CF_3^-$$

$$CF_3^- + MXY \longrightarrow CF_3MX + (CF_3)_2M$$

SCHEME 7

In a process mechanistically related to Scheme 7, Chen and coworkers have used fluorosulfonyldifluoroacetate, fluorosulfonyldifluoromethyl iodide or methyl perfluoro [2-(fluorosulfonyl)ethoxy]acetate as difluorocarbene precursors to achieve a similar overall transformation, forming and capturing trifluoromethylcopper *in situ*[125–127]. Representative examples are outlined in equations 104 and 105.

(104)

$$FSO_2CF_2I + PhCH_2Br \xrightarrow[\text{DMF, 80 °C}]{\text{Cu}} PhCH_2CF_3 \quad 83\% \qquad (105)$$

In a recent application, olefinic bromides were converted to a class of trifluoromethylated 5-α-reductase inhibitors (equation 106)[128]. Similarly, Qing and Zhang recently utilized this method in synthesis of (E)-α-trifluoromethyl-α,β-unsaturated esters (equation 107)[129]. The transformation was generally stereoselective although minor E/Z isomerization occurred in some cases.

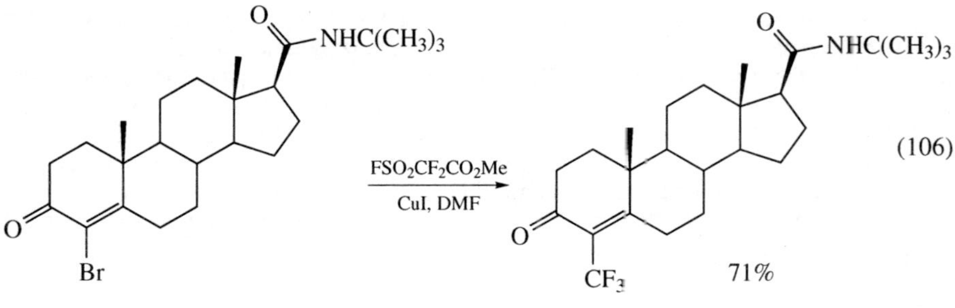

(106)

(107)

A recent modification of Chen's method was reported by Qing and coworkers in the trifluoromethylation of 1,1-dibromoalkenes in the presence of palladium catalyst[130]. 2-Aryl substrates gave bis trifluoromethylated products whereas 2-alkyl substrates gave exclusively monotrifluoromethylated products as *E*/*Z* mixtures (equations 108 and 109). The bis trifluoromethylation method is an alternative to reaction of aldehydes with 2,2-dichlorohexafluoropropane[131] or tetrakis(trifluoromethyl)-1,3-dithietane[132] in the presence of PPh_3.

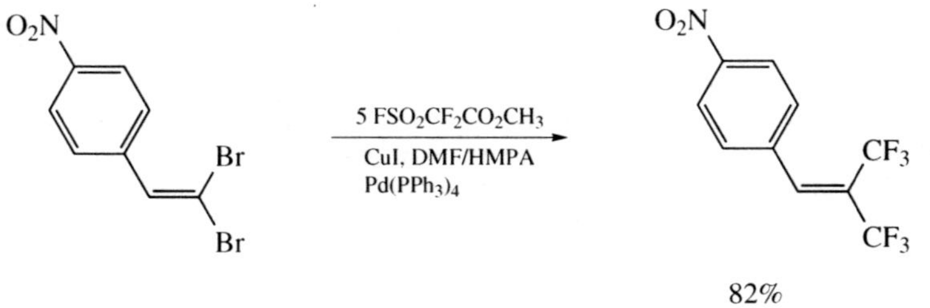

(108)

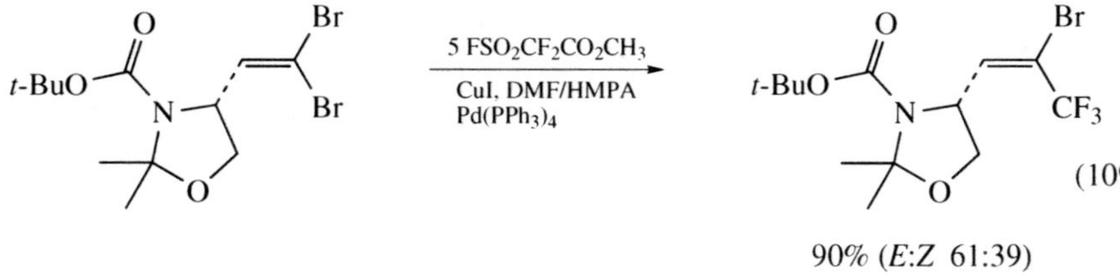

(109)

One drawback of utilizing these CF_3Cu precursors stems from the hazards associated with the use of tetrafluoroethylene and sulfur trioxide to prepare FSO_2CF_2COF, which is subsequently converted to both $FSO_2CF_2CO_2Me$ and FSO_2CF_2I.

C. Difluoromethylation

In a process analogous to formation of CF_3Cu via metathesis of the corresponding cadmium reagent, difluoromethylcadmium, prepared from insertion reaction of cadmium with difluoroiodomethane under mild conditions, undergoes rapid methathesis with Cu(I) halide to give difluoromethylcopper in good yield (equation 110)[133, 134]. The reagent exists in two forms in DMF as a function of temperature. A later report by Eujen and coworkers[135] also demonstrated the rapid exchange of difluoromethylcadmium in diglyme at low temperature to form two copper reagents denoted as $Cu(CF_2H)$ and $Cu(CF_2H)_2^-$.

Although unstable at temperatures greater than $-30\,^\circ C$, HF_2CCu is a useful difluoromethyl transfer agent which underwent coupling with allyl halides to give α- and γ-substituted products (equations 111 and 112). The difluoromethylcopper reagent is presumably aggregated in solution and in some cases exhibits complete selectivity for the least hindered site.

$$HF_2CI + 2\,Cd \xrightarrow[\text{RT, 1 h}]{\text{DMF}} HF_2CCdX + (HF_2C)_2Cd \xrightarrow[-50\text{–}60\,^\circ C]{CuX\ (X = Br, Cl)} HF_2CCu \qquad (110)$$

$$HF_2CCu + H_2C{=}CHCH(Cl)CH_3 \xrightarrow[-50\,^\circ C \text{ to RT}]{\text{DMF}} \underset{\gamma\text{-substitution, } 74\%}{HF_2CCH_2CH{=}CHCH_3} \qquad (111)$$

$$HF_2CCu + (CH_3)_2C{=}CHCH_2Cl \xrightarrow[-50\,^\circ C \text{ to RT}]{\text{DMF}} \underset{\alpha\text{-substitution, } 80\%}{(CH_3)_2C{=}CHCH_2CF_2H} \qquad (112)$$

With simple propargyl chlorides, complete regioselectivity for the γ-substituted product was observed in most cases (equation 113) except for 1-bromo-2-butyne, which gave 13% α-substituted product. However, yields were modest (7–58%) with the exception of 1-tosyl-2-butyne (78%). The low reactivity of HF_2CCu with propargyl derivatives could be explained by complexation of HF_2CCu with the triple bond of the alkyne.

$$HF_2CCu \xrightarrow[\text{DMF},\ -55\,^\circ C \text{ to RT}]{RC{\equiv}CCR^1R^2X} \underset{\alpha\text{-substitution}}{RC{\equiv}CCR^1R^2CF_2H} + \underset{\gamma\text{-substitution}}{HCF_2C(R){=}C{=}CR^1R^2} \qquad (113)$$

It should be noted that the HF_2CCdX precursor reacted with simple propargyl halides and tosylates in higher yield and also with high regioselectivity for the allene product. Thus, HF_2CCdX may be the reagent of choice in these cases whereas HF_2CCu would be the reagent of choice for more reactive propargyl substrates.

HF_2CCu coupled readily with iodoalkynes to give good yields of difluoromethylalkynes. When 1-iodoperfluoropropyne was used, only a trace of the difluoromethylalkyne was observed. However, when longer chain perfluoroalkynes were used, good yields of difluoromethylalkyne were obtained (Scheme 8).

$$\underset{44\text{–}75\%}{HF_2CC{\equiv}CR} \xleftarrow[\text{DMF, } -55\,^\circ C]{IC{\equiv}CR} HF_2CCu \xrightarrow[\text{DMF, } -55\,^\circ C]{IC{\equiv}CR_f} \underset{52\text{–}66\%}{HF_2CC{\equiv}CR_f}$$

SCHEME 8

XI. PENTAFLUOROPHENYLCOPPER REAGENT

Pentafluorophenylcopper has been prepared by the following methods:

1) Metathesis of pentafluorophenyl magnesium halides[136,137] or pentafluorophenyllithium[138,139] with copper(I) halides.
2) Reaction of $LiCu(Me)_2$ with C_6F_5H or C_6F_5I[138,139].
3) Metathesis of C_6F_5CdX with copper(I) halide in DMF[40,140].

4) Decarboxylation of cuprous pentafluorobenzoate in quinoline at 60 °C[141].
5) Direct reaction of C_6F_5I with activated copper[142–144].

Each method represents a viable entry to C_6F_5Cu (Scheme 9). However, exchange of magnesium and lithium reagents requires low temperatures and pre-generation of these thermally unstable reagents[136,137]. Similarly, exchange with $LiCu(Me)_2$ is carried out at low temperatures[138,139]. Rieke's method allows for the generation of C_6F_5Cu directly from C_6F_5I, and the mild conditions avoid Ullmann coupling normally expected for reaction of pentafluorophenyl halide with copper. However, activation of the copper by reduction of cuprous iodide with potassium in 10% naphthalene is required[142–144]. In the method of Burton and coworkers, the thermally stable C_6F_5CdX reagent is conveniently prepared from C_6F_5Br and Cd powder in DMF at room temperature; subsequent metathesis of the cadmium reagent with copper(I) halide proceeds without the need for solvent exchange at room temperature in DMF in near-quantitative yield[40,140].

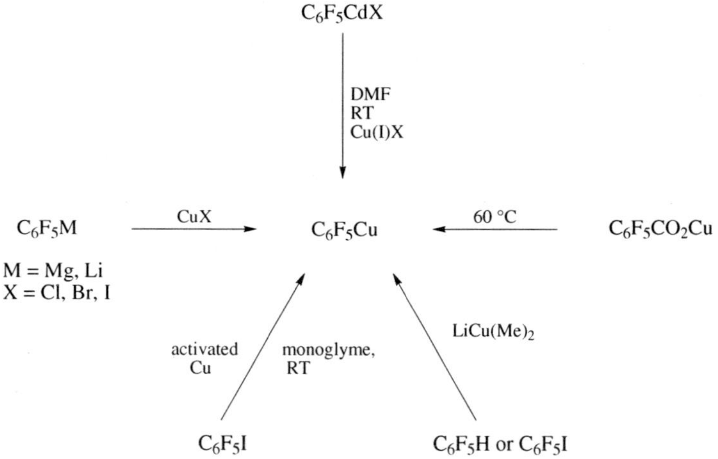

SCHEME 9

C_6F_5Cu is soluble in many organic solvents and exhibits excellent thermal stability. The copper reagent has been isolated as dioxane complexes[136], decomposes above 200 °C to form decafluorobiphenyl, and undergoes slow hydrolysis and oxidation in moist air. C_6F_5Cu exhibits excellent reactivity and couples with aryl iodides[136,137], fluorinated vinyl iodides[137,145,146], alkynyl halides[147,148], allyl halides and methyl iodide[136]. Longer chain alkyl halides give only low yields[139]. Reactions with *p*-nitrobenzoyl chloride[136] and CO_2[137] gave the corresponding ketone and carboxylic acid products, respectively. Representative examples are shown in Scheme 10. C_6F_5Cu generated by Rieke's method exhibited similar reactivity[142–144].

Coupling of C_6F_5Cu was recently applied by Suzuki and coworkers in the synthesis of perfluorinated oligo(*p*-phenylene)s studied as semiconductors for organic light-emitting diodes (equation 114)[149]. Nonafluoronaphthylcopper reagents were also utilized in this study.

Recently, Do and Daugulis reported a useful, direct copper-catalyzed arylation of polyfluorinated arenes[150]. Pentafluorobenzene was arylated in good yield with aryl and heteroaryl bromides or iodides (equation 115). A variety of functional groups is well tolerated and an example of vinyl bromide coupling was included. A mechanism involving

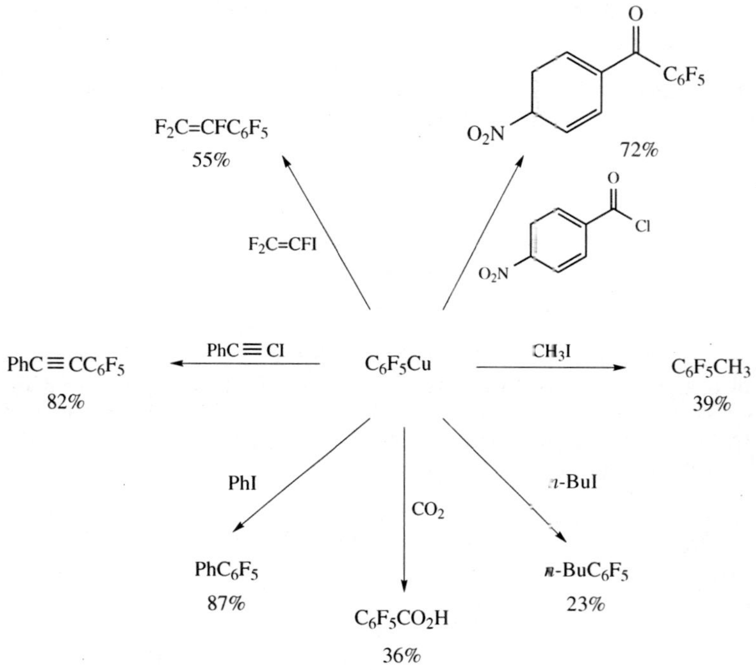

SCHEME 10

base-promoted formation of arylcopper reagent followed by coupling with aryl halide is proposed. Under similar conditions, several isomeric tetrafluorobenzenes as well as 1,3,5-trifluorobenzene underwent arylation. 1,3-Difluorobenzene underwent arylation but required the stronger base lithium *t*-butoxide to achieve deprotonation.

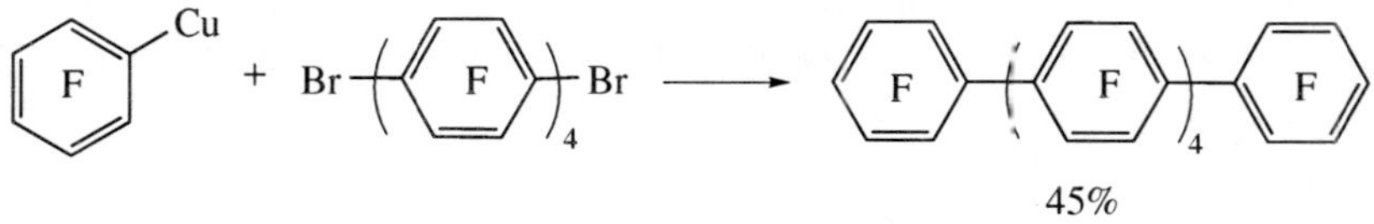

(114)

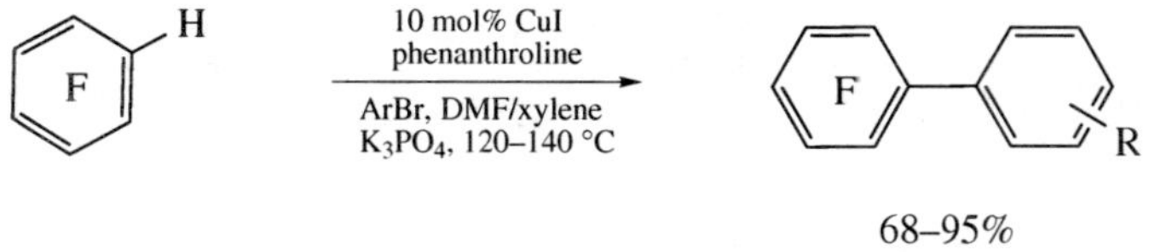

(115)

C_6F_5Cu undergoes *syn*-addition to hexafluorobutyne to form a vinylcopper reagent solution which could be quenched with substrates such as benzoyl chloride and methyl iodide to give coupled products (equation 116)[140, 151].

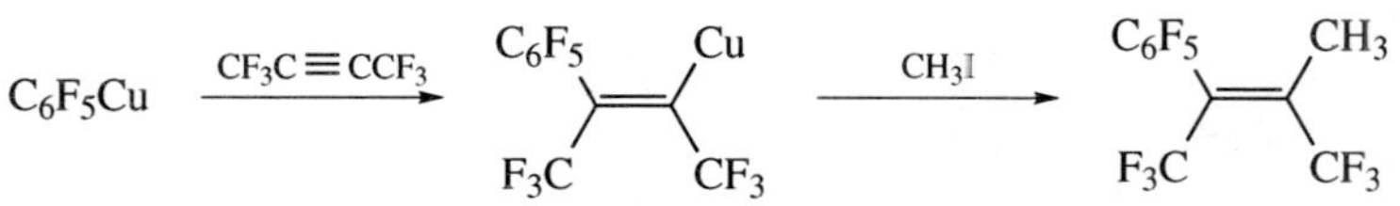

(116)

In a remarkable transformation, C_6F_5Cu, generated from C_6F_5CdX and CuBr, reacted with CF_3Cu to give the double insertion product $C_6F_5CF_2CF_2Cu$ (Scheme 11)[41]. In this case, CF_3Cu was prepared by metathesis of CF_3CdX with CuBr. Since CF_3CdX and C_6F_5CdX exchange rapidly with CuBr, the procedure could be operationally simplified to *in situ* generation of the copper reagents.

$$C_6F_5Cu \xrightarrow[-40\,^\circ C \text{ to RT}]{2\,CF_3Cu} C_6F_5CF_2CF_2Cu \xleftarrow[RT]{CuBr} C_6F_5CdX + 2\,CF_3MX$$

70–80% 70–80% M = Cd, Zn

SCHEME 11

The mechanism is proposed to involve a copper difluorocarbenoid complex intermediate, which inserts into the C−Cu bond of C_6F_5Cu to give a reactive benzylcopper intermediate $C_6F_5CF_2Cu$ which undergoes a second insertion to give $C_6F_5CF_2CF_2Cu$. The insertion process stops cleanly once a primary alkyl copper reagent is formed. $C_6F_5CF_2CF_2Cu$ exhibited good thermal stability and underwent halogenation and coupling with allyl, vinyl and aryl halides (Scheme 12)[41].

$$\underset{55\%}{C_6F_5CF_2CF_2Br} \xleftarrow{Br_2} C_6F_5CF_2CF_2Cu \xrightarrow{YC_6H_4I} \underset{35\text{–}72\%}{C_6F_5CF_2CF_2C_6H_4Y}$$

$$C_6F_5CF_2CF_2Cu \xrightarrow{PhCH=CHBr} \underset{58\%}{C_6F_5CF_2CF_2CH=CHPh}$$

SCHEME 12

Metathesis of cadmium reagents with copper halide has been extended to preparation of 1,4-dicuprotetrafluorobenzene[152]. The exchange occurs readily at room temperature in DMF and the copper reagent undergoes allylation and acylation (equation 117). This copper reagent had been previously reported from the corresponding 1,4-dilithium reagent at $-70\,^\circ C$[153].

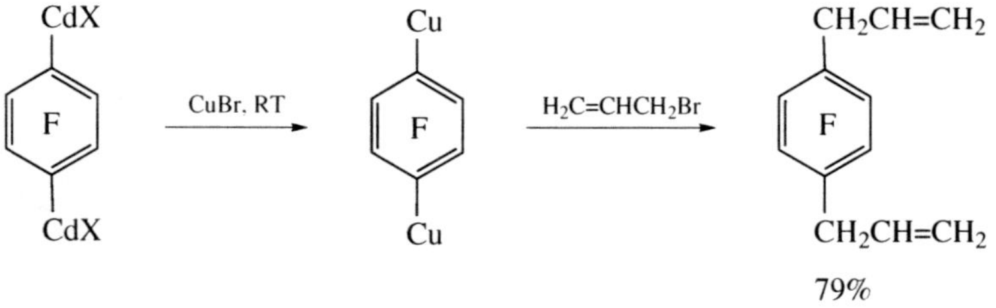

(117)

Similarly, this approach was used to prepare substituted (4-tetrafluoroaryl)copper reagents[154]. The copper reagents coupled with 2,2-difluoro-1-iodoethene to give 2,2-difluorostyrenes in good yield (equation 118).

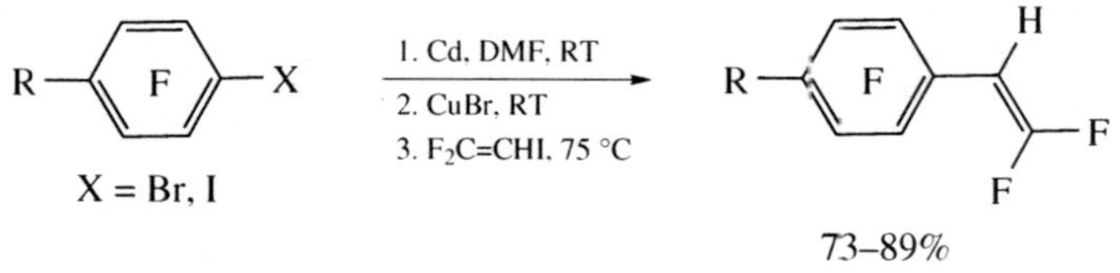

$R = F, CF_3, (CH_3)_2N, Br,$ (118)

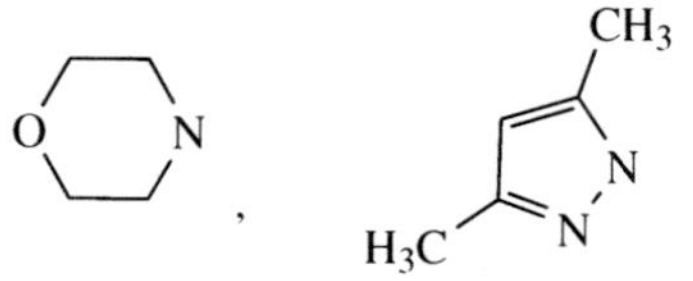

XII. 2,3,5,6-TETRAFLUOROPYRIDYLCOPPER REAGENTS

Tamborski and coworkers reported preparation of 2,3,5,6-tetrafluoropyridylcopper via metathesis of 4-tetrafluoropyridylmagnesium bromide with CuBr (equation 119)[146].

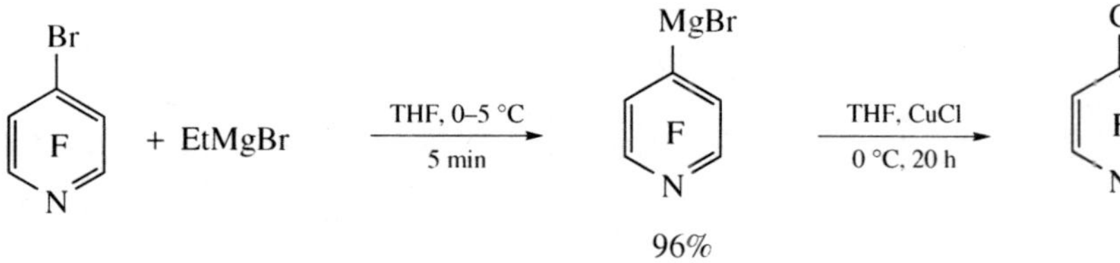

(119)

More recently, 2,3,5,6-tetrafluoropyridylcopper has been prepared by Nguyen and Burton by metathesis of the corresponding cadmium reagent with CuBr[155]. The thermally stable cadmium reagent is prepared from 4-iodotetrafluoropyridine, and both the cadmium and copper reagents were obtained in quantitative yields under mild conditions (equation 120). The copper reagent exhibits excellent thermal stability in a degassed sealed tube at room temperature for at least 48 hours, and no significant decomposition was observed by ^{19}F NMR after 3 hours at 90 °C.

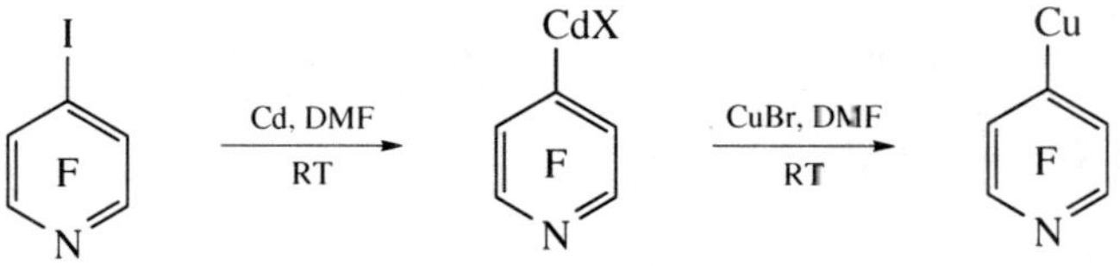

(120)

$X = I, C_5F_4N$

The tetrafluoropyridylcopper reagent is highly reactive with allyl halides, propargyl tosylate, vinyl iodides, acetylenic iodides and acyl halides to give the coupled products in good yields. Primary and secondary allyl halides give exclusively α-substitution whereas propargyl tosylate gave only the allene resulting from γ-substitution. 2-Bromobenzoyl chloride gave the ketone in good yield, indicating that coupling with acyl halides is faster than with aryl bromides. Typical examples are shown in Scheme 13[155].

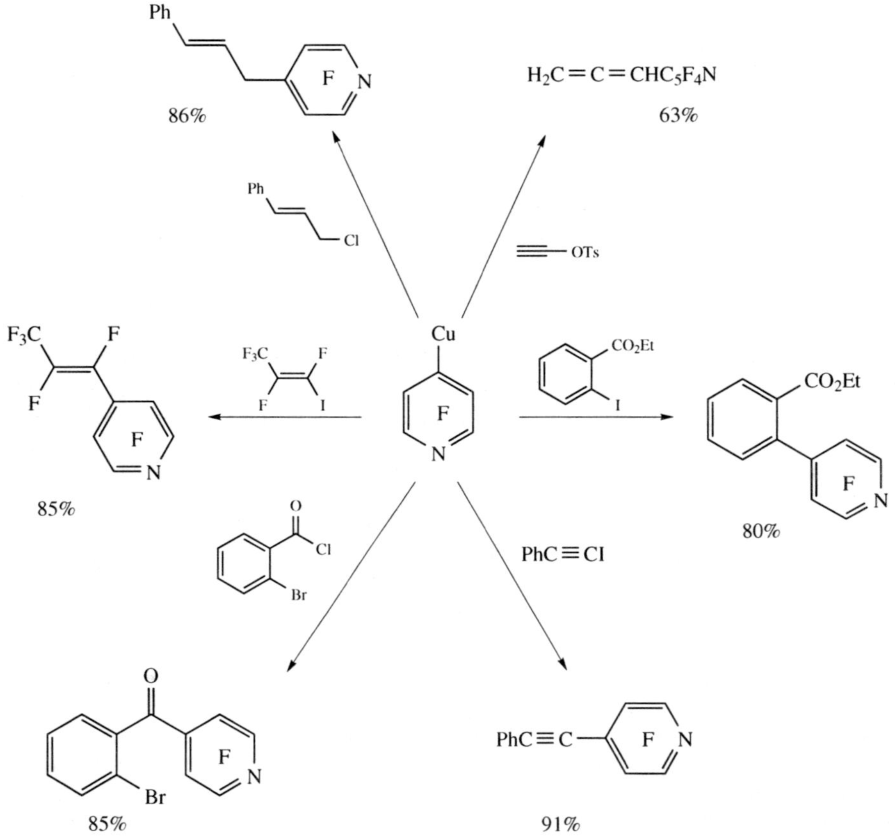

SCHEME 13

Direct copper-catalyzed arylation of 2,3,5,6-tetrafluoropyridine with 4-iodotoluene (equation 121) has been recently described by Do and Daugulis as an extension of the methodology described above (see equation 115)[150].

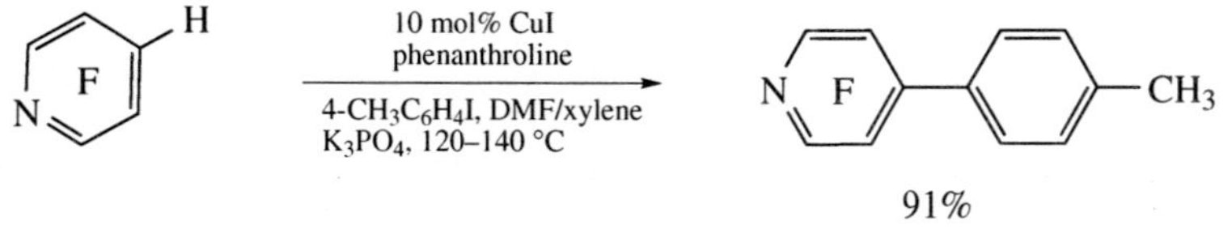

(121)

XIII. REFERENCES

1. G. Camaggi, S. F. Campbell, D. R. A. Perry, R. Stephens and J. C. Tatlow, *Tetrahedron*, **22**, 1755 (1966).
2. S. W. Hansen, PhD. Thesis, University of Iowa, 1984.

3. D. J. Burton, in *Synthetic Fluorine Chemistry* (Eds G. A. Olah, R. D. Chambers and G. K. S. Prakash), Chap. 9, Wiley-Interscience, New York, 1992, p. 205.
4. D. J. Burton, in *ACS Symposium Series 555* (Eds J. S. Thrasher and S. J. Strauss), Chap. 18, American Chemical Society, Washington, DC, 1994, p. 297.
5. D. J. Burton, Z. Y. Yang and P. A. Morken, *Tetrahedron*, **50**, 2993 (1994).
6. P. Veeraraghavan Ramachandran and G. Venkat Reddy, *J. Fluorine Chem.*, **129**, 443 (2008).
7. A. P. Sevast'yan, Y. A. Fialkov, V. A. Khromovskii and L. M. Yagupol'skii, *Zh. Org. Khim.*, **14**, 204 (1978); *Chem. Abstr.*, **88**, 169692 (1978).
8. D. J. Burton, S. W. Hansen, P. A. Morken, K. J. MacNeil, C. R. Davis and L. Xue, *J. Fluorine Chem.*, **129**, 435 (2008).
9. F. Jeanneaux, G. Santini, M. Le Blanc, A. Cambon and J. G. Riess, *Tetrahedron*, **30**, 4197 (1974).
10. A. W. Wu, S. K. Choi, J. D. Park and R. L. Soulen, *J. Fluorine Chem.*, **13**, 379 (1979).
11. R. L. Soulen, S. K. Choi and J. D. Park, *J. Fluorine Chem.*, **3**, 141 (1973/74).
12. G. Camaggi, *J. Chem. Soc. (C)*, 2382 (1971).
13. F. W. B. Einstein, A. C. Willis, W. R. Cullen and R. L. Soulen, *J. Chem. Soc., Chem. Commun.*, 526 (1981).
14. R. P. Thummel, J. D. Korp, I. Bernal, R. L. Harlow and R. L. Soulen, *J. Am. Chem. Soc.*, **99**, 6916 (1977).
15. H. H. Evans, R. Fields, R. N. Haszeldine and M. Illingworth, *J. Chem. Soc., Perkin Trans. 1*, 649 (1973).
16. W. T. Miller, T. R. Opie and H. S. Tsao, The 9th International Symposium on Fluorine Chemistry, Avignon, France, paper 027 (1979).
17. S. W. Hansen, T. D. Spawn and D. J. Burton, *J. Fluorine Chem.*, **35**, 415 (1987).
18. D. J. Burton and S. W. Hansen, *J. Fluorine Chem.*, **31**, 461 (1986).
19. D. J. Burton and S. W. Hansen, *J. Am. Chem. Soc.*, **108**, 4229 (1986).
20. S. K. Choi and Y. T. Jeong, *J. Chem. Soc., Chem. Commun.*, 1478 (1988).
21. S. K. Shin and S. K. Choi, *J. Fluorine Chem.*, **43**, 439 (1989).
22. Y. T. Jeong, J. H. Jung, S. K. Shin, Y. G. Kim, I. H. Jeong and S. K. Choi, *J. Chem. Soc., Perkin Trans. 1*, 1601 (1991).
23. P. A. Morken, H. Lu, A. Nakamura and D. J. Burton, *Tetrahedron Lett.*, **32**, 4271 (1991).
24. Y. Wang and D. J. Burton, *Org. Lett.*, **8**, 1109 (2006).
25. E. J. Blumenthal and D. J. Burton, *Israel J. Chem.*, **39**, 109 (1999).
26. P. A. Morken, N. C. Baenziger, D. J. Burton, P. C. Bachand, C. R. Davis, S. D. Pedersen and S. W. Hansen, *J. Chem. Soc., Chem. Commun.*, 566 (1991).
27. P. A. Morken, P. C. Bachand, D. C. Swenson and D. J. Burton, *J. Am. Chem. Soc.*, **115**, 5430 (1993).
28. D. C. Swenson, P. A. Morken and D. J. Burton, *Acta Crystallogr., Sect, C*, **C53**, 946 (1997).
29. P. A. Morken, D. J. Burton and D. C. Swenson, *J. Org. Chem.*, **59**, 2119 (1994).
30. D. C. Swenson, P. A. Morken and D. J. Burton, *Acta Crystallogr., Sect.C*, **C52**, 2349 (1996).
31. H. Uno, N. Nibu, Y. Yamaoka and N. Mizobe, *Chem. Lett.*, 105 (1998).
32. H. Uno, N. Nibu and N. Misobe, *Bull. Chem. Soc. Jpn.*, **72**, 1365 (1999).
33. H. Uno, K-i. Kasahara, N. Nibu, S-i. Nagaoka and N. Ono, *J. Org. Chem.*, **65**, 1615 (2000).
34. T. D. Spawn and D. J. Burton, *Bull. Soc. Chim. France*, 1 (1986).
35. P. Martinet, R. Sauvetre and J. F. Normant, *J. Organomet. Chem.*, **367**, 1 (1989) and references cited therein.
36. V. Jairaj and D. J. Burton, *J. Fluorine Chem.*, **121**, 75 (2003)
37. M. Yamamoto, D. J. Burton and D. C. Swenson, *J. Fluorine Chem.*, **72**, 49 (1995).
38. M. Yamamoto, University of Iowa, unpublished results.
39. D. C. Swenson, M. Yamamoto and D. J. Burton, *Acta Crystallogr., Sect.C*, **C54**, 846 (1998).
40. Z. Y. Yang, D. M. Wiemers and D. J. Burton, *J. Am. Chem. Soc.*, **114**, 4402 (1992).
41. Z. Y. Yang and D. J. Burton, *J. Fluorine Chem.*, **102**, 89 (2000).
42. J. Ichikawa, T. Moriya, T. Sonoda and H. Kobayashi, *Chem. Lett.*, 961 (1991).
43. J. Ichikawa, T. Minami, T. Sonoda and H. Kobayashi, *Tetrahedron Lett.*, **33**, 3779 (1992).
44. J. Ichikawa, S. Hamada, T. Sonoda and H. Kobayashi, *Tetrahedron Lett.*, **33**, 337 (1992).
45. J. Ichikawa, C. Ikeura and T. Minami, *J. Fluorine Chem.*, **63**, 281 (1993).
46. P. J. Crowley, J. A. Howarth, W. M. Owton, J. M. Percy and K. Stansfield, *Tetrahedron Lett.*, **37**, 5975 (1996).

47. D. J. Burton, Y. Tarumi and P. L. Heinze, *J. Fluorine Chem.*, **50**, 257 (1990).
48. R. N. Haszeldine, *J. Chem. Soc.*, 588 (1951).
49. T. D. Spawn, PhD. Thesis, University of Iowa, 1987.
50. D. J. Burton and T. D. Spawn, The 10th Winter Fluorine Conference, St. Petersburg Beach, FL, January 1991, Abstract #16.
51. D. J. Burton and T. D. Spawn, *J. Fluorine Chem.*, **38**, 119 (1988).
52. W. G. Finnegan and W. P. Norris, *J. Org. Chem.*, **28**, 1139 (1963).
53. J. E. Bunch and C. L. Bumgardner, *J. Fluorine Chem.*, **36**, 313 (1987).
54. T. Taguchi, O. Kitagawa, T. Morikawa, T. Nishiwaki, H. Uehara, H. Endo and Y. Kobayashi, *Tetrahedron Lett.*, **27**, 6103 (1986).
55. O. Kitagawa, T. Taguchi and Y. Kobayashi, *Chem. Lett.*, 389 (1989).
56. K. Sato, R. Kawata, F. Ama, M. Omote, A. Ando and I. Kumadaki, *Chem. Pharm. Bull.*, **47**, 1013 (1999).
57. K. Sato, Y. Ogawa, M. Tamura, M. Harada, T. Ohara, M. Omote, A. Ando and I. Kumadaki, *Collect. Czech. Chem. Commun.*, **67**, 1285 (2002).
58. O. Kitagawa, A. Miura, Y. Kobayashi and T. Taguchi, *Chem. Lett.*, 1011 (1990)
59. X. Zhang, W. Qiu and D. J. Burton, *Tetrahedron Lett.*, **40**, 2681 (1999).
60. D. J. Burton, T. Ishihara and M. Maruta, *Chem. Lett.*, 755 (1982).
61. D. J. Burton, R. Takei and S. Shin-Ya, *J. Fluorine Chem.*, **18**, 197 (1981).
62. D. J. Burton, L. G. Sprague, D. J. Pietrzyk and S. H. Edelmuth, *J. Org. Chem.*, **49**, 3437 (1984).
63. D. J. Burton and L. G. Sprague, *J. Org. Chem.*, **53**, 1523 (1988).
64. D. J. Burton and L. G. Sprague, *J. Org. Chem.*, **54**, 613 (1989).
65. L. G. Sprague, D. J. Burton, R. D. Guneratne and W. E. Bennett, *J. Fluorine Chem.*, **49**, 75 (1990).
66. R. D. Guneratne and D. J. Burton, *J. Fluorine Chem.*, **98**, 11 (1999).
67. W. Qiu and D. J. Burton, *Tetrahedron Lett.*, **37**, 2745 (1996).
68. S. B. Park and R. F. Standaert, *Tetrahedron Lett.*, **40**, 6557 (1999).
69. H. K. Nair and D. J. Burton, *J. Am. Chem. Soc.*, **119**, 9137 (1997).
70. T. Yokomatsu, K. Suemune, T. Murano and S. Shibuya, *J. Org. Chem.*, **61**, 7207 (1996).
71. T. Yokomatsu, M. Sato, H. Abe, K. Suemune, K. Matsumoto, T. Kihara, S. Soeda, H. Shimeno and S. Shibuya, *Tetrahedron*, **53**, 11297 (1997).
72. T. Yokomatsu, H. Abe, T. Yamagishi, K. Suemune and S. Shibuya, *J. Org. Chem.*, **64**, 8413 (1999).
73. T. Yokomatsu, H. Abe, M. Sato, K. Suemune, T. Kihara, S. Soeda, H. Shimeno and S. Shibuya, *Bioorg. Med. Chem.*, **6**, 2495 (1998).
74. T. Yokomatsu, T. Murano, K. Suemune and S. Shibuya, *Tetrahedron*, **53**, 815 (1997).
75. T. Yokomatsu, T. Minowa, T. Murano and S. Shibuya, *Tetrahedron*, **54**, 9341 (1998).
76. T. Yokomatsu, T. Murano, I. Umesue, S. Soeda, H. Shimeno and S. Shibuya, *Bioorg. Med. Chem. Lett.*, **9**, 529 (1999).
77. D. J. Burton, Z. Y. Yang and V. Platonov, *J. Fluorine Chem.*, **66**, 23 (1994).
78. V. C. R. McLoughlin and J. Thrower, *Tetrahedron*, **25**, 5921 (1969).
79. J. R. Griffith and J. G. O'Rear, *Synthesis,* 493 (1974).
80. Q-Y. Chen, Z. Y. Yang and Y-B. He, *J. Fluorine Chem.*, **37,** 171 (1987).
81. G. J. Chen and C. Tamborski, *J. Fluorine Chem.*, **43**, 207 (1989).
82. G. J. Chen and C. Tamborski, *J. Fluorine Chem.*, **46**, 137 (1990).
83. J. LeRoy, M. Rubinstein and C. Wakselman, *J. Fluorine Chem.,* **27**, 291 (1985).
84. J. Burdon, P. L. Coe, C. R. Marsh and J. C. Tatlow, *J. Chem. Soc., Chem. Commun.,* 1259 (1967).
85. P. L. Coe, N. E. Milner and J. A. Smith, *J. Chem. Soc., Perkin Trans. 1,* 654 (1975).
86. A. E. Pedler, R. C. Smith and J. C. Tatlow, *J. Fluorine Chem.*, **1**, 337 (1971/72).
87. J. Burdon, P. L. Coe, C. R. Marsh and J. C. Tatlow, *J. Chem. Soc., Perkin Trans. 1,* 639 (1972).
88. P. L. Coe and N. E. Milner, *J. Organomet. Chem.,* **39**, 395 (1972).
89. T. Nguyen, M. Rubinstein and C. Wakselman, *J. Org. Chem.*, **46**, 1938 (1981).
90. M. Hudlicky, *J. Fluorine Chem.,* **18**, 383 (1981).
91. P. L. Coe and N. E. Milner, *J. Organomet. Chem.,* **70**, 147 (1974).
92. D. J. Burton, G. A. Hartgraves and J. Hsu, *Tetrahedron Lett.*, **31**, 3699 (1990).

93. M-H. Hung, *Tetrahedron Lett.*, **31**, 3703 (1990).
94. M. LeBlanc, G. Santini, J. Guion and J. G. Riess *Tetrahedron*, **29**, 3195 (1973).
95. Q-Y. Chen and Z-Y. Yang, *J. Fluorine Chem.*, **28**, 399 (1985).
96. (a) G. K. S. Prakash and A. Y. Yudin, *Chem. Rev.*, **97**, 757 (1997).
(b) G. K. S. Prakash and G. A. Olah, *Org. Lett.*, **5**, 3253 (2003).
97. (a) T. Umemoto, *Chem. Rev.*, **96**, 1757 (1996).
(b) W. R. Dolbier, Jr., *Chem. Rev.*, **96**, 1557 (1996).
98. M. Hudlicky, *Chemistry of Organic Fluorine Compounds I*, 2nd edn, Ellis Horwood, New York, 1992, pp. 96–106 and 154–158.
99. M. Hudlicky and A. E. Pavlath (Eds), *Chemistry of Organic Fluorine Compounds II*, American Chemical Society, Washington, DC, 1995, pp. 178–184 and 243–249.
100. B. Baasner, H. Hagemann and J. C. Tatlow (Eds), *Houben-Weyl: Methods of Organic Chemistry*, Vol. *E10a*, Thieme, Stuttgart, 1999, pp. 133–141, 348–370 and 509–525.
101. Y. Kobayashi and I. Kumadaki, *Tetrahedron Lett.*, 4095 (1969).
102. Y. Kobayashi, K. Yamamoto and I. Kumadaki, *Tetrahedron Lett.*, 4071 (1979).
103. Y. Kobayashi, I. Kumadaki and K. Yamamoto, *J. Chem. Soc., Chem. Commun.*, 536 (1977).
104. Y. Kobayashi, K. Yamamoto, T. Asai, M. Nakano and I. Kumadaki, *J. Chem. Soc., Perkin Trans. 1*, 2755 (1980).
105. Y. Kobayashi and I. Kumadaki, *J. Chem. Soc., Perkin Trans. 1*, 661 (1980).
106. K. Matsui, E. Tobita, M. Ando and K. Kondo, *Chem. Lett.*, 1719 (1981).
107. G. E. Carr, R. D. Chambers and T. F. Holmes, *J. Chem. Soc., Perkin Trans. 1*, 921 (1988).
108. J. N. Frekos, *Synth. Commun.*, **18**, 965 (1988).
109. N. Van Tuyen, B. Kesteleyn and N. De Kimpe, *Tetrahedron*, **58**, 121 (2002).
110. H. Urata and T. Fuchikama, *Tetrahedron Lett.*, **32**, 91 (1991).
111. F. Cottet and M. Schlosser, *Eur. J. Org. Chem.*, 327 (2002).
112. J. D. Bloom, M. J. DiGrandi, R. G. Dushin, K. J. Curran, A. A. Ross, E. B. Norton, E. Terefenko, T. R. Jones, B. Feld and S. A. Lang, *Bioorg. & Med. Chem. Lett.*, **13**, 2929 (2003).
113. I. Ruppert, K. Schlich and W. Volbach, *Tetrahedron Lett.*, **25**, 2195 (1984).
114. T. Umemoto and A. Ando, *Bull. Chem. Soc. Jpn.*, **59**, 447 (1986).
115. N. V. Kondratenko, E. P. Verchirko and L. M. Yagupol'skii, *Synthesis*, 932 (1980).
116. D. J. Burton and D. M. Wiemers, *J. Am. Chem. Soc.*, **107**, 5014 (1985).
117. D. M. Wiemers and D. J. Burton, *J. Am. Chem. Soc.*, **108**, 832 (1986).
118. V. Nair and G. S. Buenger, *J. Am. Chem. Soc.*, **111**, 8502 (1989).
119. D. J. Burton, G. A. Hartgraves and J. Hsu, *Tetrahedron Lett.*, **31**, 3699 (1990).
120. D. M. Wiemers, PhD. Thesis, University of Iowa, 1987.
121. T. S. Chou, PhD. Thesis, University of Iowa, 1987.
122. G. Chauviere, B. Bouteille, B. Enanga, C. de Alburquerque, S. Croft, M. Dumas and J. Perie, *J. Med. Chem.*, **46**, 427 (2003).
123. J. C. Easdon, PhD. Thesis, University of Iowa, 1987.
124. J. H. Clark, M. A. McClinton, C. W. Jones, P. Landon, D. Bishop and R. J. Blade, *Tetrahedron Lett.*, **30**, 2133 (1989).
125. Q-Y. Chen and S-W. Wu, *J. Chem. Soc., Chem. Commun.*, 705 (1989).
126. Q-Y. Chen and S-W. Wu, *J. Chem. Soc., Perkin Trans. 1*, 2385 (1989).
127. Q-Y. Chen and J-X. Duan, *J. Chem. Soc. Chem. Commun.*, 1389 (1993).
128. X-S. Fei, W-S. Tian and Q-Y. Chen, *Bioorg. & Med. Chem. Lett.*, **7**, 3113 (1997).
129. F-L. Qing and X. Zhang, *Tetrahedron Lett.*, **42**, 5929 (2001).
130. F-L. Qing, X. Zhang and Y. Peng, *J. Fluorine Chem.*, **111**, 185 (2001).
131. M. Hanack and C. Korhummel, *Synthesis*, 944 (1987).
132. D. J. Burton and Y. Inouye, *Tetrahedron Lett.*, 3397 (1976).
133. Reported initially, G. A. Hartgraves and D. J. Burton, 3rd Chemical Congress of North America, Toronto, Canada, June 1988, Abstract FLUO #30.
134. D. J. Burton and G. A. Hartgraves, *J. Fluorine Chem.*, **128**, 1198 (2007).
135. R. Eujen, B. Hoge and D. J. Brauer, *J. Organomet. Chem.*, **519**, 7 (1996).
136. A. Cairncross and W. A. Sheppard, *J. Am. Chem. Soc.*, **90**, 2186 (1968).
137. R. J. DePasquale and C. Tamborski, *J. Org. Chem.*, **34**, 1736 (1969).
138. A. E. Jukes, S. S. Dua and H. J. Gilman, *J. Organomet. Chem.*, **21**, 241 (1970).
139. A. E. Jukes, S. S. Dua and H. J. Gilman, *J. Organomet. Chem.*, **21**, 791 (1970).

140. K. J. MacNeil and D. J. Burton, *J. Org. Chem.*, **58**, 4411 (1993).
141. A. Cairncross, J. R. Roland, R. M. Henderson and W. A. Sheppard, *J. Am. Chem. Soc.*, **92**, 3187 (1970).
142. R. D. Rieke and L. D. Rhyne, *J. Org. Chem.*, **44**, 3445 (1979).
143. G. W. Ebert and R. D. Rieke, *J. Org. Chem.*, **49**, 5280 (1984).
144. G. W. Ebert and R. D. Rieke, *J. Org. Chem.*, **53**, 4482 (1988).
145. C. Tamborski, E. J. Soloski and R. J. DePasquale, *J. Organomet. Chem.*, **15**, 494 (1968).
146. E. J. Soloski, W. E. Ward and C. Tamborski, *J. Fluorine Chem.*, **2**, 361 (1972/73).
147. R. G. Gastinger, E. F. Tokas and M. D. Rausch, *J. Org. Chem.*, **43**, 159 (1978).
148. F. Waugh and D. R. M. Walton, *J. Organomet. Chem.*, **39**, 275 (1972).
149. S. B. Heidenhain, Y. Sakamoto, T. Suzuki, A. Miura, H. Fujikawa, T. Mori, S. Tokito and Y. Taga, *J. Am. Chem. Soc.*, **122**, 10240 (2000).
150. H-Q. Do and O. Daugulis, *J. Am. Chem. Soc.*, **130**, 1129 (2008).
151. K. J. MacNeil and D. J. Burton, *J. Org. Chem.*, **60**, 4085 (1995).
152. Z-Y. Yang, K. MacNeil and D. J. Burton, *J. Fluorine Chem.*, **52**, 251 (1991).
153. G. M. Brooke and S. D. Mawson, *J. Fluorine Chem.*, **50**, 101 (1990).
154. B. V. Nguyen and D. J. Burton, *J. Org. Chem.*, **62**, 7758 (1997).
155. B. V. Nguyen and D. J. Burton, *J. Fluorine Chem.*, **67**, 205 (1994).

Author Index

This author index is designed to enable the reader to locate an author's name and work with the aid of the reference numbers appearing in the text. The page numbers are printed in normal type in ascending numerical order, followed by the reference numbers in parentheses. The numbers in *italics* refer to the pages on which the references are actually listed.

Aalten, H. L. 119 (119), 120 (120), *138*
Aamouche, A. 198 (198), *214*
Abarbri, M. 404 (125, 126), *439*, 841 (162), 845 (193), *855*, *856*
Abboud, J. L. 310 (170), 337 (169), *344*
Abe, H. 640 (98), 647 (114), *688*, *689*, 887 (28, 31), 941 (288), *981*, *986*, 1011 (71–73), *1030*
Abe, T. 228 (60), 230 (77), *242*, *243*
Abel, E. W. 2 (14), *19*, 27 (40), 34 (34), *136*, 858 (5), *875*
Ablordeppey, S. Y. 947 (327), *987*
Abolmaali, B. 349 (2), *361*
Abramczyk, H. 228 (64), 241 (143), *243*, *245*
Abramovitch, R. A. 950 (344), *987*
Abu-Salah, O. M. 76 (168, 169), *139*
Accorsi, G. 220 (10), *241*
Acharya, H. P. 616 (47), 644 (107), *685*, *688*
Achenbach, C. A. 114 (288), *142*
Achyutha Rao, S. 573 (12), *581*
Ackermann, K. 616 (43), *685*
Ackermann, L. 919 (179), *984*
Adam, W. 540 (59), *582*
Adamo, C. 283 (44), 289 (72), *341*, *342*
Adams, C. J. 258 (42), *277*
Adams, J. 324 (197), *345*, 954 (365), 962 (386), *988*
Adams, K. J. 70 (150), *139*
Adams, R. 968 (418, 419), *989*
Adkins, H. 891 (77), *982*
Adlington, R. M. 386 (63), *437*, 551 (94), *583*, 824 (153), *855*
Adlkofer, J. 100 (229), *141*
Adman, E. T. 349 (24), *361*
Adolf, P. K. 354 (46), *362*
Adude, R. N. 952 (355), *988*
Afonso, C. 326 (203, 208), *345*
Agami, C. 622 (56), *685*
Agapiou, K. 713 (53), *730*
Agarkov, A. 418 (145), *440*
Ager, D. J. 791 (21), 817 (7), *851*
Aggerholm, T. 326 (207), *345*
Aguirre, A. 80 (177), 82 (188), 83 (195), 85 (198), *139*, *140*
Ahlberg, P. 800 (54), *852*
Ahlquist, M. 850 (200), *856*
Ahlrichs, R. 157 (126), *162*
Åhman, A. B. 29 (51), *136*
Ahmar, M. 382 (54), *437*
Ahn, K. H. 430 (22), *436*
Ahrens, M. J. 955 (367), *988*
Ai, J. Y. 358 (69), *363*
Aihara, M. 941 (286), *986*
Aiken, L. 158 (135), *162*
Ainai, T. 644 (107), *688*
Aitken, G. 289 (66), 305 (64), *342*
Akaji, M. 629 (68), *686*
Akgun, H. 647 (116), *689*
Akhtar, F. 153 (67), *160*
Akimoto, N. 614 (17), *684*
Akiyama, K. 509 (228), *524*, 674 (161), *690*
Akman, S. A. 337 (247), *346*
Aksela, R. 834 (132), 842 (175), *854*, *855*
Al-Laham, M. A. 283 (44), *342*
Alam, T. M. 173 (164), *213*
Alami, M. 449 (35, 36), 450 (38–41), *520*
Alberti, A. 378 (45), *436*, 599 (35), *602*
Albright, T. A. 3 (41), *19*
Albrow, V. 506 (223), *524*, 698 (15), *729*
Alcalde, E. 922 (204), *984*

The chemistry of organocopper compounds
Edited by Z. Rappoport and I. Marek

Alcamí, M. 281 (15), 283 (48), 284 (32), 285 (50), 293 (86, 87), 295 (85), 311 (171, 172), 318 (150, 177), 333 (235), 337 (84, 125, 151, 169), *341–346*
Alcaraz, L. 979 (485), *990*
Aldrich-Wright, J. R. 163 (9), *210*
Alexakis, A. 2 (22, 23), 10 (64), *19*, *20*, 24 (17), *135*, 201 (223, 225), 202 (23, 227, 229), 203 (230), *210*, *215*, 401 (104, 106), 430 (12), *435*, *438*, *439*, 450 (37), 480 (149, 155), 503 (223), 505 (225–227), *520*, *523–525*, 506(223, 225, 226), 508(227), *522*, *524*, 529 (15), 530 (11, 20–24), 532 (13, 14, 26), 535 (42, 45), 536 (47), 537 (46), 539 (52–54), 540 (8), 545 (70, 72–74), 547 (71), 550 (66, 80), 561 (3), 569 (107), *581–583*, 663 (85, 134–136), 670 (148, 149), 674 (160), 675 (2, 163), 676 (165, 166), 679 (172), *683*, *687*, *689–691*, 694 (3, 4), 695 (1, 6–8), 697 (12), 698 (15), 705 (30, 32), 707 (34, 40), 709 (28, 46), 710 (50), 712 (52), 713 (54), 714 (47), 725 (31, 71), *729*, *730*, 811 (110), 842 (179), *854*, *856*, 858 (10–12), 863 (26), 873 (4, 53), 874 (56), *875–879*
Alexander, C. W. 386 (72, 73), 389 (74), 390 (84), *437*, *438*
Alexandrova, A. N. 156 (101), *161*
Allamand, J. 201 (223), *215*, 695 (6), *729*
Allan, M. 620 (51), *685*
Allcock, H. R. 904 (127), *983*
Allen, D. W. 893 (91), *982*
Allen, N. S. 231 (89), 232 (92), *243*
Allinger, N. L. 309 (147), *344*
Allison, J. 310 (159), *344*
Allouchi, H. 921 (215), *984*
Allred, G. D. 976 (450), *989*
Almena Perea, J. J. 858 (8), *876*
Alouane, N. 399 (97), *438*
Altenhoff, G. 927 (251), *985*
Alterman, M. A. 361 (80), *363*
Altman, R. A. 897 (112), 922 (205, 206), *982*, *984*
Alvarez, K. 800 (58), *852*
Alvarez, M. L. 358 (69), *363*
Alvarez, R. 845 (191, 192), *856*
Alvarez-Rúa, C. 451 (44), *520*
Alveraz, K. 175 (154), *213*
Alvhäll, J. 452 (53), *520*
Alvisi, D. 809 (100), *853*
Ama, F. 1008 (56), *1030*
Amaravathi, M. 920 (198), *984*
Amatore, C. 943 (296), *986*
Amekraz, B. 293 (29), 311 (171), 315 (121, 174), 318 (177), 330 (227), 334 (223), *341*, *343–346*
Amijs, C. H. M. 101 (232), *141*, 175 (156), *213*
Amorelli, B. 765 (87), *773*
Amsallem, D. 565 (122), *583*
Amunugama, R. 295 (56, 98), 300 (88), *342*, *343*
Amzel, L. M. 280 (2), *341*, 350 (30), *362*
An, D. K. 401 (107), *439*, 672 (156), *690*
Anand, N. K. 629 (75), *687*
Andersen, J. 937 (276), *986*
Anderson, J. C. 949 (333, 334), *987*
Anderson, K. W. 927 (253, 254), *985*
Anderson, M. A. 234 (97), *243*
Anderson, O. P. 73 (158), *139*
Anderson, Q. T. 64 (135), *138*
Anderson, R. J. 13 (81), *20*, 625 (5), *683*
Andersson, T. 8 (58), *20*, 191 (191), *214*, 388 (80), *438*, 534 (34, 35), *582*, 829 (160), *855*
Andert, J. 324 (201), *345*
Ando, A. 1008 (56, 57), 1018 (114), *1030*, *1031*
Ando, M. 1017 (106), *1031*
Andrews, D. 149 (29), *159*
Andrews, L. 157 (105), 158 (138), *161*, *162*
Andrews, S. P. 894 (98), *982*
Andrews, T. D. 68 (143), *138*
Andrey, O. 800 (50), *852*
Andrianov, V. G. 33 (62), *136*
Angelaud, R. 800 (51), *852*
Angeles Rey, M. 380 (49), *437*
Anna, L. J. 629 (70), *686*
Annby, U. 452 (53), *520*
Antholine, W. A. 352 (35), *362*
Anthonsen, H. W. 168 (101), *212*
Anthony, N. J. 502 (215), *524*
Antilla, J. C. 922 (214, 216), 926 (221), 958 (378), *984*, *985*, *988*
Antipas, A. 220 (29), *242*
Antoline, J. E. 933 (266), *986*
Antoniadis, A. 234 (99), *243*
Anzai, M. 635 (90), *687*
Aoki, S. 416 (138), *440*, 609 (28), 658 (13), *683*, *684*
Aoyagi, K. 222 (39), *242*
Aoyagi, T. 614 (41), *685*
Aoyama, H. 352 (38), *362*
Aoyama, T. 955 (368), *988*
Apanasevich, P. A. 220 (20), *241*
Appella, D. H. 741 (30), 742 (31), *772*
Arai, M. 15 (84, 85), *20*, 444 (10), 456 (11), *519*, 609 (28), 658 (13), 659 (18, 122, 124), *683*, *684*, *689*
Arai, S. 888 (45, 46), 891 (47, 48), *981*
Arai, T. 230 (86, 87), *243*
Araki, K. 240 (139), *244*, 257 (41), *277*
Araki, Y. 221 (37), *242*
Arase, A. 974 (439), *989*
Arbour, C. 229 (72), *243*
Archibald, S. C. 382 (55), *437*, 807 (95), *853*

Arduengo, A. J. 120 (309, 310), 123 (321), *142*, *143*
Arends, I. W. C. E. 353 (43), *362*
Argade, N. P. 675 (2), *683*
Arimoto, K. 513 (240), *524*, 738 (21), *772*, 977 (460, 464), *990*
Arita, T. 285 (13), *341*
Aritome, I. 117 (296), *142*
Armaroli, N. 220 (10), 221 (36), *241*, *242*
Armentrout, P. B. 150 (34), 151 (50), 152 (1), 157 (109, 117), *158–161*, 293 (82, 89), 294 (92, 97), 295 (99), 299 (100), 310 (160, 161, 164, 167, 168), *342–344*
Armstrong, A. 502 (215), *524*
Armstrong, D. R. 177 (161), *213*
Armstrong, F. A. 356 (57), *363*
Armstrong, J. D. 373 (33), *436*
Armstrong, N. R. 226 (53–57), *242*
Arnauld, T. 947 (326), *987*
Arndt, S. 868 (36), *879*
Arnold, A. 676 (164), 677 (167), *690*
Arnold, A. E. 201 (224), *215*
Arnold, L. A. 203 (228), *215*, 401 (105), *439*, 474 (133), *522*, 670 (146), 679 (169), *690*, 694 (5), 695 (1), 707 (41), 714 (55), *729*, *730*, 858 (12), *877*
Arnold, P. L. 119 (305), 127 (328), *142*, *143*, 402 (110), *439*
Arrayás, R. G. 492 (175), *523*, 697 (11), *729*
Arseniev, A. 330 (231), *346*
Arslan, I. 233 (95), *243*
Arterburn, J. B. 921 (203), *984*
Aruoma, O. I. 337 (246), *346*
Arvidsson, P. I. 166 (83), 178 (109), *212*, 369 (9), *435*, 800 (54), *852*
Asai, T. 1016 (104), *1031*
Asamoto, B. 294 (95, 96), *343*
Asano-Someda, M. 222 (39–41), *242*
Ashby, E. C. 169 (106), 173 (125), *212*
Ashida, M. 228 (66), *243*
Astruc, D. 248 (6), *276*
Asunskis, D. 361 (80), *363*
Attfield, J. P. 894 (98), *982*
Atwater, N. W. 861 (19), *878*
Aube, J. 430 (171), *441*
Aubrecht, K. B. 177 (162), *213*
Audergon, L. 201 (220), *215*
Audia, J. E. 860 (20), *878*
Audrain, H. 551 (85), *583*
Aue, D. H. 768 (90), *773*
Auel, C. 2 (29), *19*, 133 (222), *140*, 175 (129), 179 (87), 181 (169), *212*, *213*, 858 (6), *876*
Auer, G. 382 (52), *437*, 769 (92), *773*
Aufauvre, L. 565 (121), *583*
Augugliaro, V. 236 (107), *244*
August, A. T. 920 (192), *984*
Augustine, A. J. 348 (16), *361*
Aumann, R. 451 (43), *520*
Auner, G. 861 (19), *878*
Austin, J. 283 (44), *342*
Avendano, C. 950 (342, 343, 346, 347), *987*
Avent, A. G. 166 (75), *211*
Averill, K. M. 962 (386, 387), *988*
Avram, L. 168 (96), *212*
Axelrad, G. 893 (90), 940 (87–89), *982*
Ayala, P. Y. 283 (44), *342*
Aye, Y. 742 (33), *772*
Azab, H. A. 354 (48), *362*
Azhakar, R. 238 (134), *244*
Aziz, E. F. 157 (113), *161*

Baasner, B. 1016 (100), *1031*
Babcock, J. C. 861 (19), *878*
Babiak, K. A. 453 (61), 509 (231), *520*, *524*, 819 (133), *854*
Babko, A. K. 155 (98), *161*
Baboul, A. G. 283 (44), *342*
Bach, T. 908 (153), *983*
Bacha, J. D. 154 (78), *160*
Bachand, P. C. 997 (26, 27), *1029*
Back, S. 78 (171, 172), *139*, 259 (44), 271 (63, 65, 66), 275 (69), 276 (72), *277*, *278*
Back, T. G. 906 (144), *983*
Backen, W. 158 (128), *162*
Bäckvall, J.-E. 2 (23), 11 (71), *19*, *20*, 24 (17), *135*, 198 (173), *214*, 202 (23), *210*, 386 (65), *437*, 480 (153), *522*, 606 (11), 607 (10), 609 (32), 635 (91), 669 (14, 143, 144), 672 (1), 680 (173), 682 (174), *682–684*, *687*, *690*, *691*, 695 (1), *729*, 816 (120), 817 (10), 849 (11), *851*, 849 (104), *854*, 858 (10, 11), *876*, *877*, 979 (484, 486), *990*
Bacon, R. G. R. 883 (5), 886 (19–21), 888 (34, 35), 891 (72), 927 (74), *980*, *981*
Badalassi, F. 676 (164), *690*
Badiei, Y. M. 130 (333), *143*
Badorrey, R. 18 (118), *21*, 200 (216), *214*, 493 (177), *523*, 670 (152), *690*, 698 (16), *729*
Baenziger, N. C. 113 (31), *136*, 155 (87), *161*, 459 (86), *521*, 997 (26), *1029*
Baer, M. R. 157 (126), *162*
Baerends, E. J. 58 (58), *136*
Bagheri-Majdi, E. 329 (216), *346*
Bai, J. F. 908 (155), *983*
Bai, Y. 920 (196), *984*
Bailey, P. J. 87 (205–207), *140*
Bailey, S. M. 155 (3), *158*
Bailey, T. R. 386 (67), 421 (70), *437*
Baizer, M. M. 247 (2, 3), *276*
Baker, B. A. 739 (27), *772*
Baker, B. R. 968 (418, 419), *989*
Baker, E. A. 223 (43), *242*
Baker, E. N. 349 (27), *362*
Baker, P. K. 158 (135), *162*
Baker, R. 502 (214), *523*

Bakhanova, I. V. 947 (329), *987*
Bakker, W. I. I. 399 (96), *438*
Balakrishnan, G. 360 (79), *363*
Balamurugan, B. 155 (93), *161*
Balasubramanian, K. 153 (69), *160*
Balciodlu, I. A. 233 (95), *243*
Baldwin, J. E. 386 (63), *437*, 551 (94), *583*, 824 (153), *855*
Baldwin, M. A. 332 (225), *346*
Baliarsingh, N. 237 (113), *244*
Balkany, A. D. 238 (121), *244*
Ball, H. L. 332 (225), *346*
Balme, G. 574 (135, 136, 138), 576 (137), 579 (141–144), *584*, 646 (110), *688*
Balukova, V. V. 949 (336), *987*
Balzano, F. 629 (72), *686*
Bamford, C. H. 231 (90), *243*
Banci, L. 354 (48), *362*
Bandyopadhyay, A. 975 (451), *989*
Banerjee, S. 893 (90), *982*
Bansal, V. 894 (99), *982*
Bantu, B. 129 (331), *143*, 751 (57), *773*
Bao, W. 904 (133), 925 (109), *982*, *983*
Bao, W. L. 906 (145), *983*
Baptista, M. S. 240 (139), *244*
Baraldi, P. T. 979 (490), *990*
Barbe, G. 518 (263), *525*
Barbero, A. 800 (44), 807 (93), 811 (108), 813 (38), 816 (30, 112–118), 819 (134), 821 (12, 142), 841 (144), 845 (194), 846 (125), 847 (129, 181, 197), *851–856*
Barbot, F. 859 (16), *877*
Barden, D. J. 382 (55), *437*, 807 (95), *853*
Barder, T. E. 922 (214), *984*
Bardin, J. 908 (153), *983*
Bare, S. R. 148 (15), 150 (43), *159*, *160*
Barkigia, K. M. 220 (24), *242*
Barluenga, J. 451 (44, 45), *520*
Barnes, K. 247 (1), *276*
Barnhard, T. M. 800 (58), *852*
Barnhart, T. M. 108 (258, 259), *141*, 175 (154, 160), *213*, 800 (56), *852*
Baro, A. 675 (2), *683*
Baron, O. 424 (123), *439*
Barone, G. 943 (293), *986*
Barone, V. 283 (44), *341*
Barr, J. M. 320 (185), *345*
Barran, P. E. 288 (62), 310 (158), *342*, *344*
Barreira, T. 158 (134), *162*
Barreiro, E. 430 (172), *441*
Barrio, P. 451 (45), *520*
Barros, O. S. D. 943 (229), *985*
Barsanti, P. 613 (31), *684*
Barthel, A.-M. 536 (47), *582*
Bartholomew, E. R. 3 (39), *19*, 112 (280), *142*, 154 (83), *160*, 198 (64), 200 (57), *211*, 534 (39, 40), *582*, 858 (7), *876*
Bartolotti, L. J. 386 (71), *437*
Barton, D. H. R. 946 (312, 313), 947 (314, 326, 328), 949 (322, 332, 338), 950 (337, 344), 961 (339, 390), *987*, *988*
Basha, A. 373 (32), *436*
Bashe, R. W. 2 (19), *19*, 24 (9), 112 (275), *135*, *142*, 173 (140), *213*, 589 (15), *602*
Baskin, J. M. 922 (214), *984*
Basu, A. 386 (69), *437*
Bates, C. G. 904 (131), 966 (100), *982*, *983*
Bates, G. S. 735 (12), *772*
Batey, R. A. 908 (154, 156), 926 (249), 953 (363), *983*, *985*, *988*
Bau, R. 106 (227, 247), *141*, 152 (62, 63), *160*, 732 (3), *771*
Baudoin, O. 238 (128), *244*
Bauer, W. 166 (76–80), 178 (165, 166), *211*, *213*
Baulieu, E.-E. 861 (21), *878*
Baumeister, U. 254 (34), *277*
Bauschlicher, C. W. 282 (25), 283 (47), 294 (91), 296 (107), 302 (122–124), *341–343*
Bax, A. 189 (192, 193), *214*
Bax, B. 349 (29), *362*
Baxter, C. A. 402 (109), *439*, 480 (146), *522*, 703 (26), *729*
Baxter, C. W. 87 (206), *140*
Bayless, W. R. 114 (288), *142*
Bayse, C. A. 324 (200), *345*
Bazin, J. F. Y. 382 (55), *437*, 807 (95), *853*
Beak, P. 373 (32), 386 (69), 391 (89), *436–438*, 493 (180), *523*
Beames, D. J. 16 (97), *21*, 800 (31), *851*
Bearinger, J. P. 225 (48), *242*
Beatty, H. R. 944 (302), *986*
Beauchamp, J. L. 310 (160), *344*
Beaudet, I. 842 (178), *856*
Beaulieu, C. 620 (50–52), *685*
Beaumier, F. 622 (54), *685*
Beck, B. 967 (411), *989*
Beck, K. 324 (201), *345*
Becke, A. D. 282 (22), *341*
Becker, M. 870 (44), *879*
Beckers, J. 966 (402), 974 (441, 442), *989*
Beddoes, R. L. 892 (85), *982*
Beem, K. M. 354 (31), *362*
Begum, S. A. 979 (482), *990*
Behenna, D. C. 939 (282, 283), *986*
Behling, J. R. 108 (261), *141*, 175 (159), *213*, 453 (61), 509 (231), 517 (233), *520*, *524*, 819 (133), *854*
Behrens, C. 2 (29), *19*, 133 (222), *140*, 175 (129), *212*, 858 (6), *876*
Behrens, U. 32 (61), 82 (178), *136*, *139*
Bejjani, J. 638 (95), *688*
Belanger, A. 640 (102), *688*
Belanzoni, P. 58 (58), *136*
Belcastro, M. 303 (78), *342*
Belelie, J. L. 616 (44), 659 (129), *685*, *689*

Belf, L. J. 887 (24), *980*
Bell, A. P. 452 (56), *520*
Bell, R. 399 (98), *438*
Bella, M. 859 (17), *877*
Belle, C. 356 (54), *362*
Bellemin-Laponnaz, S. 120 (307), 127 (327), *142*, *143*, 754 (65), *773*
Bellemin-Loponnaz, S. 400 (101), *438*
Beller, M. 968 (414–416), *989*
Bellier, B. 325 (202), 326 (203, 208), *345*
Belting, V. 870 (46), *879*
Bemis, A. 154 (77), *160*
Bender, B. R. 290 (79), *342*
Benecke, B. 386 (60), *437*, 859 (18), *878*
Benet-Buchholz, J. 114 (291), *142*
Benhaim, C. 201 (223), 202 (227), 203 (230), *215*, 670 (148), *690*, 695 (1, 6–8), *729*, 858 (11, 12), *877*
Bennani, Y. L. 863 (27), *878*
Bennasar, M.-L. 510 (239), *524*
Benner, J. 386 (65), *437*, 635 (91), *687*
Bennett, R. A. 149 (27), *159*
Bennett, W. E. 1010 (65), *1030*
Benson, H. D. 861 (22), *878*
Bent, B. E. 148 (12–15, 17, 18, 20, 21), 150 (19, 41, 43), 153 (37), *159*, *160*
Bento, I. 356 (51), *362*
Bera, J. 303 (127), 309 (93), *343*
Bercaw, J. E. 452 (54), *521*
Bérces, A. 305 (133), *344*
Bergander, K. 451 (43), *520*
Bergbreiter, D. E. 493 (182), *523*
Bergdahl, M. 459 (90), 518 (89), *521*, 867 (32, 33), *879*
Berger, S. 108 (256), *141*, 166 (81), 167 (90), 168 (99), 172 (137), 187 (62), *211–213*, 416 (128), *439*, 872 (47), *879*
Bergstrom, F. W. 890 (62), *981*
Berillon, L. 404 (125), *439*
Beringer, F. M. 943 (299, 303), *986*
Berkheij, M. 399 (96), *438*
Berkowitz, W. F. 471 (113, 114), *521*
Berks, B. C. 358 (62), *363*
Berlan, J. 430 (172), *441*, 450 (37), *520*
Berman, A. M. 375 (37), *436*, 493 (178, 179), *524*
Bermel, W. 189 (195), *214*
Bernabé, M. 613 (31), *684*
Bernal, I. 993 (14), *1029*
Bernard, N. 638 (95), *688*
Bernardi, P. 374 (34), *436*
Bernardinelli, G. 201 (225), *215*, 707 (40), *730*
Bernaud, F. 399 (97), *438*
Berner, O. M. 695 (1), *729*
Berners-Price, S. J. 163 (1), *210*, 165 (40), *211*
Bernstein, N. 148 (17, 18), *159*
Berriel, J. N. 514 (244), *524*
Berthelot, M. P. E. 157 (122), *162*
Bertini, I. 348 (6), *361*
Bertozzi, F. 676 (164), 677 (167, 168), 679 (169), *690*
Bertrán, J. 303 (126), 309 (81), *342*, 310 (156, 157), *343*, *344*
Bertrand, G. 76 (76), *137*
Bertrand, M. 811 (109), *854*
Bertz, S. H. 3 (35, 37, 39), 5 (48), 16 (94, 100, 101), 17 (105), *19–21*, 105 (238–241), 108 (253–255, 262), 112 (279, 280, 282), *141*, *142*, 152 (55, 66), 154 (81–83), *160*, 169 (119), 173 (123, 136, 138), 174 (143), 175 (158), 179 (128, 152), 184 (180), 193 (133, 184), 195 (58), 198 (64, 65), 200 (57), *211–214*, 366 (4), 367 (6), 380 (3), 430 (16, 20), *435*, *436*, 534 (36–40), *582*, 587 (10), *602*, 778 (24), 790 (26), *851*, 858 (7), *876*
Besace, Y. 450 (37), *520*
Besly, D. M. 888 (39), 891 (44), *981*
Besselievre, F. 973 (435), *989*
Bessho, Y. 18 (111), *21*
Besson, S. 358 (67), *363*
Bethell, D. 891 (51), *981*
Betzemeier, B. 471 (5), *519*, 858 (8), *876*
Betzer, J. F. 842 (188), *856*
Beugelmans, R. 386 (62), *437*
Bezman, S. A. 732 (2), *771*
Bhagirath, N. 920 (194), *984*
Bhanage, B. M. 920 (199), 966 (406), *984*, *989*
Bhandari, A. 858 (8), *876*
Bhanuprakash, K. 922 (211), *984*
Bhanushali, M. J. 920 (199), *984*
Bhargava, S. K. 37 (74), *137*
Bhatt, R. K. 976 (449), *989*
Bhinderwala, N. S. 390 (84), *438*
Bhor, M. D. 920 (199), *984*
Bible, R. H. 861 (19), *878*
Bierbaum, V. M. 148 (16), *159*
Biffis, A. 953 (362), 966 (401), *988*, *989*
Bigge, C. F. 925 (230), *985*
Bigorra, J. 859 (16), *877*
Bigot, A. 386 (62), *437*
Bilello, L. 18 (116), *21*
Billotte, S. 410 (129), *439*
Billups, W. E. 150 (36), *159*
Bilodeau, R. C. 149 (28), *159*
Bilyarska, L. 237 (112), *244*
Bindra, J. S. 622 (59), *686*
Bindra, R. 622 (59), *686*
Biran, C. 800 (34), *851*
Bischoff, L. 373 (33), *436*
Bishop, D. 1020 (124), *1031*
Bishop, M. B. 285 (55), *342*
Bisseret, P. 945 (307), *986*
Biswas, D. 499 (202), *523*
Biswas, K. 506 (223), *524*, 698 (15), *729*
Bjorkling, F. 937 (276), *986*
Black, J. R. 165 (38), *211*

Blackmond, D. G. 373 (29), *436*, 943 (228), *985*
Blackwell, D. T. 595 (21), *602*
Blackwell, H. E. 593 (25), *602*
Blade, R. J. 1020 (124), *1031*
Blake, A. J. 127 (328), *143*, 402 (110), *439*
Blanchard, N. 164 (20), *210*
Blanco, F. J. 800 (39), 811 (107), *852*, *854*
Blanco, Y. 807 (93), 816 (115), *853*, *854*
Blart, E. 809 (99, 100), *853*
Blazy, C. M. 547 (75), *582*
Blitz, M. A. 157 (102), *161*
Bloch, R. 660 (132), *689*
Block, P. 943 (300), *986*
Bloom, J. D. 1018 (112), *1031*
Bloomfield, V. A. 181 (172), *214*
Blough, B. E. 625 (64, 65, 67), *686*
Blouin, M. 959 (383), *988*
Blue, E. D. 121 (313, 314), *143*, 403 (118), *439*
Bluemel, J. 200 (215), *214*
Bluestein, A. 224 (44), *242*
Bluhm, B. K. 324 (194, 200), *345*
Blumenthal, E. J. 996 (25), *1029*
Boaz, N. W. 11 (65), 16 (90, 91), *20*, 610 (33), 638 (93), *684*, *687*, 859 (16), *877*
Boche, G. 2 (29), *19*, 110 (265, 268), 133 (222), *140*, *141*, 175 (129, 142), 179 (87), *212*, *213*, 800 (59), *852*, 858 (5, 6), *875*, *876*
Bochkarev, L. N. 90 (209), *140*
Bock, C. W. 147 (9), *159*
Boehme, M. 153 (68), *160*
Boehringer, E. M. 834 (159), *855*
Boekman, R. K. 735 (12), *772*
Boere, R. T. 28 (46), *136*
Boersma, A. J. 695 (10), *729*
Boersma, J. 101 (231), *141*, 175 (157), *213*
Boese, A. D. 149 (30), *159*
Boettger, R. 585 (1), *601*
Boezio, A. A. 489 (172, 173), *523*
Bogár, K. 682 (174), *691*
Bohandy, J. 220 (18), *241*
Bohm, V. P. W. 426 (165), *440*
Bohme, D. K. 304 (131), 339 (250), *343*, *346*
Böhme, M. 108 (259), *141*, 175 (160), 179 (153), *213*
Bois-Choussy, M. 386 (62), *437*, 943 (250), *985*
Boisvert, L. 620 (50, 52), 622 (54), *685*
Boldyrev, A. I. 156 (101), *161*
Bollinger, J. A. 359 (72), *363*
Bolm, C. 565 (126), *583*, 926 (239–242), 958 (382), *985*, *988*
Bolognesi, M. 352 (34), 355 (50), *362*
Bolte, M. 60 (60), *136*
Bombrun, A. 607 (10), *683*
Bomparola, R. 178 (168), *213*
Bonakdar, A. 551 (83), *583*
Bonaventura, C. 351 (33), *362*
Bonaventura, J. 351 (33), *362*
Bondi, A. 179 (170), *213*
Bonini, B. F. 802 (87, 88), 809 (100), *853*
Bonne, D. 952 (350), 956 (372), 960 (385), 962 (351), *987*, *988*
Bonnet, V. 421 (152), *440*
Bonnett, R. 238 (117), *244*
Booker-Milburn, K. 842 (189), *856*
Booker-Milburn, K. I. 646 (110), *688*
Borah, S. 238 (126), *244*
Borders, D. B. 238 (137), *244*
Bordo, D. 355 (50), *362*
Borge, J. 451 (44), *520*
Born, L. 386 (61), *437*, 859 (18), *878*
Börner, A. 201 (221), *215*
Börner, C. 675 (162), *690*
Borsari, M. 354 (48), *362*
Borst, J. P. 820 (136), *854*
Borthwick, S. 646 (110), *688*
Bos, H. J. T. 540 (57, 58), *582*
Bosch, J. 510 (239), *524*
Boskovic, Z. V. 739 (27), 768 (90), *772*, *773*
Bosold, F. 2 (29), *19*, 110 (265, 268), 133 (222), *140*, *141*, 175 (129), *212*, 800 (59), *852*, 858 (5, 6), *875*, *876*
Boss, M. E. 834 (168), *855*
Bossée, A. 325 (202), 326 (203, 208), *345*
Böttger, R. C. 23 (1), *135*
Bouab, W. 310 (170), 337 (169), *344*
Bouchareb, S. 153 (73), *160*
Boudet, N. 169 (116), *212*, 378 (46), *437*, 600 (40), *602*
Boudier, A. 418 (144), 421 (152), *440*, 858 (8), *876*
Boudon, C. 221 (36), *242*
Bouillon, R. 654 (125), *689*, 859 (18), *878*
Bourcier, S. 319 (183), *345*
Bourdelande, J. L. 240 (142), *245*
Bourgain-Commerçon, M. 550 (63), 559 (62, 64), *582*
Bourgain, M. 540 (9, 10), *581*
Bourissou, D. 904 (128), *983*
Bournaud, C. 679 (172), *691*
Bouteille, B. 1020 (122), *1031*
Boutreau, L. 332 (234), 333 (235), 334 (223), *346*
Bouwmeester, K. 723 (68), *730*
Bouyssi, D. 574 (135, 138), *584*, 646 (110), *688*
Bouyssi, D. D. 576 (137), *584*
Bower, J. F. 894 (97), *982*
Bowler, J. 861 (22), *878*
Bowman, B. J. 932 (256), *985*
Bowman, E. J. 932 (256), *985*
Boyall, D. 728 (75), *730*
Boyer, G. 947 (324), *987*

Boyle, P. D. 121 (313, 315), *143*, 403 (118), *439*
Boyle, T. J. 178 (164), *213*
Brade, K. 169 (118), *212*, 586 (7), *601*
Bradshaw, J. S. 975 (453), *989*
Brancaleoni, D. 378 (45), *436*, 599 (36), *602*
Brand, T. 166 (81), *212*
Brands, K. M. J. 947 (320), *987*
Brandsma, L. 166 (77), *211*
Brandt, J. 666 (140), *689*
Brandt, J. C. 238 (118), *244*
Brasca, M. G. 502 (215), *524*
Brasche, G. 964 (393), *988*
Brauer, D. J. 459 (85), *521*, 1022 (135), *1031*
Braun, W. 330 (231), *346*
Breit, B. 2 (26), 11 (67), *19*, *20*, 24 (18), *135*, 164 (22), *210*, 480 (152), 484 (159–161), *522*, 528 (7), *581*, 622 (38, 57), 659 (131), 665 (139), 672 (1), *682*, *685*, *686*, *689*, 858 (9), *876*
Brestensky, D. M. 733 (8), 735 (15, 16), 737 (19), *771*, *772*
Breuninger, D. 665 (139), *689*
Brevard, C. 94 (216), *140*, 165 (40), *211*
Brich, Z. 191 (102), *212*
Bridger, R. F. 897 (9), *980*
Brierley, A. 943 (299), *986*
Brilkina, T. G. 946 (308, 310), 947 (309), *986*, *987*
Brinkmeyer, R. S. 629 (72), *686*
Brischwein, M. 349 (20), *361*
Brocksom, T. J. 380 (49), *437*, 647 (113), *688*
Brodbelt, J. S. 335 (239), *346*
Broggini, G. 741 (29), *772*
Bromm, L. O. 858 (8), *876*
Bronstrup, M. 426 (165), *440*
Brooke, G. M. 1026 (153), *1032*
Brookhart, M. 296 (109), *343*
Brooks, E. E. 165 (42), *211*
Brooks, S. H. 842 (189), *856*
Brown, E. A. 861 (19), *878*
Brown, H. C. 496 (189, 196, 197), *523*
Brown, J. 310 (168), *344*
Brown, K. 352 (36), 358 (67, 70), *362*, *363*
Brown, M. K. 402 (109), 430 (27), *436*, *439*, 480 (146, 148), 508 (224, 228), 509 (262), *522–525*, 674 (161), *690*, 703 (26), 705 (33), 709 (27), 710 (49), 726 (72), *729*, *730*
Brownawell, M. L. 9 (61), *20*
Browne, W. R. 18 (118), *21*, 200 (216), *214*, 493 (177), *523*, 670 (152), *690*
Brożek-Płuska, B. 228 (64), *243*
Bruce, M. I. 79 (175), 83 (197), *139*, *140*
Bruckner, R. 13 (76), *20*
Brueggemeier, R. W. 861 (21), *878*
Bruggink, A. 968 (424), *989*
Brummond, K. M. 866 (29), *878*
Brunelle, P. 306 (142, 143), *344*
Brunner, H. 447 (28), *519*, 752 (59), *773*, 947 (318, 319), *987*
Bslashlaszczyk, T. 228 (64), *243*
Bu, Y. 309 (152), *344*
Bubacco, L. 230 (3), *341*
Buchmeiser, M. R. 129 (331), *143*, 751 (57), *773*
Büchner, M. 41 (96), *137*, 259 (44), *277*
Buchner, W. 100 (229), *141*
Buchwald, S. L. 373 (29), 403 (113), *436*, *439*, 459 (82, 83), *521*, 740 (28), 741 (30), 742 (31–33), 744 (35–38), *772*, 897 (112), 899 (114), 901 (111), 902 (120), 904 (130), 913 (165), 914 (113, 169, 171), 915 (172), 921 (202, 215), 922 (205, 206, 214, 216), 925 (222, 227), 926 (221, 247), 927 (252–254), 932 (258), 935 (269, 270), 936 (271), 937 (273), 940 (279), 942 (92, 289), 943 (228), 958 (378), 963 (391), 964 (393), 967 (412), 969 (429), *982–986*, *988*, *989*
Buck, E. 897 (105), *982*
Buck, P. 423 (150), *440*
Buckle, M. J. C. 629 (72), *687*
Buckton, G. 23 (2), *135*
Bucourt, R. 861 (21), *878*
Buendia, J. 447 (26), *519*
Buenger, G. S. 1019 (118), *1031*
Buerli, R. W. 926 (245), *985*
Bui The Thuong, M. 741 (29), *772*
Bulic, B. 716 (57), *730*
Bumgardner, C. L. 1005 (53), *1030*
Buncel, E. 800 (52), 820 (137), *852*, *854*
Bunch, J. E. 1005 (53), *1030*
Bunlaksananusorn, T. 421 (152), *440*
Buranaprasertsuk, P. 904 (137), *983*
Burant, J. C. 283 (44), *341*
Burda, J. V. 281 (17), 282 (18, 21), 298 (114), 309 (20, 115), *341*, *343*
Burdack, C. 908 (153), *983*
Burdon, J. 73 (160), *139*, 1014 (84, 87), *1030*
Burford, S. C. 537 (16, 25), *581*
Burke, B. J. 894 (98), *982*
Burke, S. D. 616 (6), *683*
Burkholder, L. 156 (99), *161*
Burlingame, A. L. 332 (225), *346*
Burns, D. H. 979 (481, 488), *990*
Burns, P. A. 510 (237, 238), *524*
Burrows, C. J. 337 (245), *346*
Burstyn, J. N. 238 (127), *244*
Burtner, R. R. 861 (19), *878*
Burton, D. J. 113 (31, 284), *136*, *142*, 155 (87), *161*, 459 (86), *521*, 548 (76, 77), *582*, 994 (3), 995 (17–19), 996 (8, 23–25), 997 (26–28), 998 (29, 30, 34), 999 (36), 1000 (37, 39), 1001 (4), 1004 (47), 1005 (5, 50, 51), 1008 (59), 1009 (60–64), 1010 (65, 66), 1011 (67, 69), 1012 (77), 1015 (92),

1019 (117, 119), 1021 (116), 1022 (132–134), 1024 (40), 1025 (140, 151), 1026 (41, 152, 154), 1028 (155), *1029–1032*
Burton, J. 695 (8), *729*
Bury, P. 551 (92), *583*
Butler, I. S. 163 (6), *210*
Butler, J. R. 891 (73), *981*
Buxton, M. W. 887 (24), *980*
Buzard, D. J. 179 (131), *213*, 625 (66), *686*, 800 (56), *852*
Byrne, L. 165 (49), *211*

Cabezas, J. A. 800 (60), 841 (182), *852*, *856*
Cabrita, E. J. 166 (81), 168 (99), *212*
Cabrito, I. 352 (36), *362*
Cacchi, S. 966 (403), *989*
Cachata, V. 158 (134), *162*
Caeners, B. 113 (286), *142*
Caggiano, L. 418 (149), *440*
Cahiez, G. 421 (151), *440*, 447 (25, 26, 31, 32, 34), 449 (33, 35, 36), 450 (38–41), *519–520*, 530 (11, 22), 532 (26), 535 (42, 45), 540 (10), 569 (107), *581–583*
Cai, Q. 897 (102–104), 901 (119), 915 (173), 917 (174), 932 (259), 937 (277), *982*, *984–986*
Cain, C. K. 968 (419), *989*
Caires, C. C. 749 (52), *773*
Cairncross, A. 27 (24, 25), 60 (121), *135*, *136*, *138*, 1024 (136, 141), *1031*, *1032*
Calabrese, J. C. 123 (321), *143*
Calas, R. 800 (34), *851*
Calaza, I. 421 (152), *440*
Calaza, M. I. 484 (156), *522*, 616 (46), 641 (103), *685*, *688*
Calderón, M. 231 (88), *243*
Calianno, A. 613 (31), *684*
Caliman, A. F. 234 (99), *243*
Callahan, J. H. 336 (242), *346*
Calle, M. 802 (89), *853*
Caló, V. 613 (31), 663 (20, 138), *684*, *689*
Calvo, V. 88 (208), *140*
Camaggi, G. 992 (1), 993 (12), *1028*, *1029*
Cambillau, C. 352 (36), 358 (67, 70), *362*, *363*
Cambon, A. 992 (9), *1029*
Cameron, D. 322 (189), *345*
Cameron, J. F. 423 (161), *440*
Caminade, A.-M. 921 (220), *985*
Cammi, R. 283 (44), *342*
Campbell, A. 509 (231), *524*
Campbell, A. L. 819 (133), *854*
Campbell, J. A. 861 (19), *878*
Campbell, J. B. 496 (189, 196), *523*
Campbell, J. E. 133 (337), *143*, 402 (108), *439*, 508 (150, 151), *522*, 672 (155, 157), *690*
Campbell, M. J. 375 (37), *436*, 493 (181), *523*
Campbell, S. F. 992 (1), *1028*
Camus, A. 2 (12), *19*, 27 (22, 23), 39 (83, 84), 71 (71), *135*, *137*
Canè, F. 378 (45), *436*, 599 (35, 36), *602*
Canisius, J. 5 (47), *20*, 187 (62), *211*, 635 (77), *687*, 870 (46, 49), 872 (47), *879*
Canonica, L. 356 (56), *362*
Canters, G. W. 350 (26), *362*
Cao, G.-A. 631 (79), *687*
Cao, W. 899 (116), 904 (134), *982*, *983*
Capella, L. 807 (96), 809 (17), 842 (184), *851*, *853*, *856*
Caporusso, A. M. 629 (72), *686*
Cappalletti, P. 356 (56), *362*
Capperucci, A. 807 (96), *853*
Cappperucci, A. 828 (158), *855*
Caracciolo, M. 842 (185), *856*
Caraiman, D. 304 (131), *343*
Card, G. 349 (29), *362*
Cardinali, F. 220 (10), *241*
Carlin, C. M. 193 (133), *213*, 534 (36), *582*, 858 (7), *876*
Carlson, K. D. 114 (288), *142*
Carlucci, W. F. 613 (31), *684*
Caroli, C. 861 (22), *878*
Caron, S. 925 (235), *985*
Carpenter, S. D. 223 (43), *242*
Carr, D. B. 457 (70), *520*
Carr, G. E. 1017 (107), *1031*
Carreira, E. M. 629 (75), *687*, 728 (75), *730*, 747 (43–45), *772*, 842 (183), *856*
Carretero, J. C. 492 (175), *523*, 697 (11), *729*, 748 (49), *772*
Carriedo, G. A. 66 (140, 141), *138*
Carril, M. 904 (136), *983*
Carrillo-Marquez, T. 418 (149), *440*
Carro, C. 976 (456), *990*
Carrondo, M. A. 356 (51), *362*
Casarini, A. 374 (34), *436*, 832 (78), *853*
Casas-Arce, E. 720 (65), *730*
Casella, L. 280 (3), *341*
Casey, C. P. 151 (47), *160*, 586 (8), *601*
Casimiro-Garcia, A. 925 (230), *985*
Casson, S. 528 (5), *581*
Castedo, L. 622 (58), *686*, 908 (148, 149), *983*
Castleman, A. W. 156 (100), 157 (116), *161*
Castracane, J. 229 (69), *243*
Castreño, P. 816 (112, 113, 116, 117), *854*
Castro, C. E. 73 (159, 161), *139*, 965 (395, 396), *988*
Castro, J. L. 502 (214), *523*
Castro-Rodriguez, I. 126 (324, 325), *143*
Cataldo, F. 155 (94, 95), *161*
Cauble, D. F. 713 (53), *730*
Caughey, W. S. 220 (16), *241*
Caulton, K. C. 123 (320), *143*
Caulton, K. G. 87 (87), *137*, 165 (48), *211*, 732 (4, 5), *771*

Cavaleiro, J. A. S. 238 (123), *244*
Cavicchioli, M. 579 (143), *584*
Cederbaum, F. E. 461 (99), *521*
Cellier, P. P. 897 (110), 921 (219), 943 (217), *982*, *984*
Cenini, S. 18 (116), *21*
Cerda, B. A. 290 (76), *342*
César, V. 120 (307), *142*, 127 (327), *143*, 400 (101), *438*
Cesati, R. R. 716 (58), *730*, 859 (17), *877*
Cha, J. K. 672 (159), *690*
Chabot, C. 622 (54), *685*
Chachisvilis, M. 287 (67), *342*
Chae, J. 549 (78, 79), *583*, 744 (36), *772*
Chai, M. 236 (109), *244*
Chaitanya, G. K. 922 (211), *984*
Chakrabarti, S. 71 (154), *139*
Chakravarthy, A. R. 238 (132), *244*
Chakravarty, A. R. 238 (130), *244*
Challacombe, M. 283 (44), *342*
Challener, C. A. 834 (168), *855*
Chambers, R. D. 994 (3), 1017 (107), *1029*, *1031*
Chambionnat, A. 155 (90), *161*
Chamot-Rooke, J. 315 (178, 179), *345*
Chan, A. S. C. 695 (8), 698 (14), *729*, 754 (66), *773*
Chan, C.-L. 81 (185, 186), *139*
Chan, D. M. T. 952 (350), 954 (348, 365), 962 (386), *987*, *988*
Chan, H. C. 240 (140), *245*
Chan, H.-K. 979 (481, 488), *990*
Chan, J. 329 (218), *346*
Chan, J. M. 359 (72), *363*
Chan, P. W. H. 904 (137), *983*
Chan, W.-H. 82 (187), *139*
Chandler, C. L. 739 (25), *772*
Chandra, T. 238 (135), *244*
Chandrasekhar, J. 309 (93), *343*
Chandrasekhar, V. 238 (134), *244*
Chandrashekar, T. K. 238 (134), *244*
Chandupatla, K. R. 395 (90), 399 (93), *438*
Chang, C. C. 238 (137), *244*
Chang, C. J. 223 (43), *242*
Chang, D. Z. 906 (145), *983*
Chang, J. W. W. 904 (137), *983*
Chang, M. C. 235 (100), *243*
Chang, S. C. 150 (36), *159*
Charette, A. 518 (263), *525*
Charette, A. B. 131 (335), *143*, 489 (172–174), *523*, 748 (50), *772*
Chatani, N. 964 (392), *988*
Chatel, F. 947 (324), *987*
Chatt, J. 297 (111), *343*
Chau, L. K. 226 (53), *242*
Chaudhari, R. V. 158 (130, 132), *162*, 920 (189), *984*
Chausteur, C. 762 (82), *773*
Chauviere, G. 1020 (122), *1031*
Chavant, P.-Y. 447 (34), 496 (127), *520*, *522*
Chavasiri, W. 904 (137), *983*
Che, C.-M. 77 (166), 82 (187), *139*, 732 (7), *771*
Chechik-Lankin, H. 452 (6), *519*, 560 (111, 112), 569 (127), *583*, *584*
Cheeseman, J. R. 283 (44), *341*
Chemla, F. 545 (67), *582*, 629 (71), 638 (95), *686*, *688*
Chemler, S. R. 373 (29), *436*
Chen, B. 432 (176), *441*, 758 (74), *773*
Chen, C. 595 (30, 31), *602*, 920 (197), *984*
Chen, C.-S. 643 (106), *688*
Chen, F. 236 (110), *244*
Chen, G. 149 (24), *159*
Chen, G. J. 1014 (81, 82), *1030*
Chen, H. 430 (25), *436*, 921 (186), *984*
Chen, H. H. 925 (233), *985*
Chen, H. M. 800 (36), *851*
Chen, H. Z. 230 (75), *243*
Chen, J.-H. 925 (224), *985*
Chen, J. X. 737 (19), *772*
Chen, L. 404 (122), *439*
Chen, M. 157 (119), *162*
Chen, M.-H. 646 (109), *688*
Chen, N. 390 (86), 393 (87), 397 (92), 399 (94), *438*, 635 (92), *687*
Chen, N. Y. 587 (11), *602*
Chen, P. 348 (11), 359 (63), 360 (52), *361–363*
Chen, Q-Y. 1014 (80), 1016 (95), 1021 (125–128), *1030*, *1031*
Chen, R. H. K. 859 (16), *877*
Chen, W. 283 (44), *342*, 921 (212), *984*
Chen, X. 309 (152), *344*, 358 (71), *363*
Chen, Y. 970 (431), 972 (434), *989*
Chen, Y. J. 925 (233), *985*
Chen, Y.-M. 157 (109), *161*
Chen, Z.-C. 944 (305, 306), *986*
Chen, Z.-N. 76 (167), *139*, 85 (202), 90 (210), *140*
Chenard, B. L. 818 (127), *854*
Cheng, E. C.-C. 81 (185), *139*
Cheng, K.-F. 758 (74), 759 (75), *773*, 760 (76), *773*
Cheng, L. 921 (213), *984*
Cheng, M. 318 (176), *345*, 867 (28), *878*
Cheng, P. 339 (250), *346*
Cheng, P. Y. 285 (55), *342*
Chenier, J. H. B. 157 (104), *161*, 632 (84), *687*
Chernick, E. T. 955 (367), *988*
Chernyh, I. N. 249 (8), 250 (4), *276*, *277*
Cheshnovsky, O. 152 (61), *160*
Cheung, K.-K. 78 (174), *139*, 81 (180, 182, 185, 186), 83 (189, 191, 193, 194), 85 (199–201), *139*, *140*, 262 (53, 57), 266 (55, 56), 267 (58), *278*

Chiang, C. M. 148 (20), 153 (37), *159*
Chiang, G. C. H. 952 (352), *987*
Chieffi, A. 514 (246), *524*
Chiesi-Villa, A. 30 (55), 40 (88), 48 (56), *136*, *137*
Chikira, M. 330 (228), *346*
Chinchilla, R. 965 (397), *988*
Chinkov, N. 456 (68), *520*
Chirik, P. J. 452 (54), *520*
Chirvony, V. S. 220 (20), *241*
Chiu, P. 733 (9), 758 (74), 759 (75), 760 (76–78), 768 (89), *772*, *773*
Cho, C.-G. 926 (246), *985*
Cho, D.-G. 632 (83), *687*, 873 (52), *879*
Cho, D. W. 220 (26), *242*
Cho, G. Y. 926 (239), *985*
Cho, S. 236 (108), *244*
Cho, W. 445 (12), *519*
Chobanian, H. R. 977 (465), *990*
Choe, Y. 240 (141), *245*
Choi, A. 897 (112), *982*
Choi, A. Y. H. 859 (16), *877*
Choi, H. 489 (169), *523*
Choi, S.-C. 976 (454), *989*
Choi, S. K. 993 (10, 11), 995 (20–22), *1029*
Choi, S. W.-K. 81 (185, 186), *139*
Choi, W. 236 (108), *244*
Chong, J. 824 (123), *854*
Chong, J. M. 616 (44), 659 (129), *685*, *689*, 821 (143), 829 (154, 157), 836 (147, 171), *854*, *855*
Choppin, S. 753 (61), *773*
Chopra, A. 105 (241), *141*, 173 (123), *212*, 778 (24), *851*
Chou, T. S. 383 (58), *437*, 1019 (121), *1031*
Choudary, B. M. 921 (207), *984*
Chounan, Y. 276 (71), *278*, 516 (247), *524*, 616 (27), 625 (40, 42), *684*, *685*, 858 (13), *877*
Chow, H.-F. 622 (60), *686*, 800 (63, 71), *852*
Chowdhury, C. 966 (407), *989*
Chrisman, W. 738 (24), 749 (51, 52), 753 (60, 63), *772*, *773*
Christenson, B. 184 (181), 186 (186), *214*
Christian, A. T. 225 (48), *242*
Christie, S. 380 (1), *435*
Christoffers, J. 474 (132), *522*, 675 (2), *683*, 695 (1), *729*, 858 (12), *877*
Christopher, J. A. 399 (98), *438*
Chu, C.-Y. 459 (92), *521*
Chu, I. K. 324 (196), 329 (212, 213, 216, 218–220), 331 (221), *345*, *346*
Chu, J. 391 (89), *438*
Chu, Y. 299 (118, 119), *343*
Chui, S. S. Y. 77 (166), *139*
Chuit, C. 540 (10), *581*
Chung, J.-U. 873 (52), *879*
Chung, W. K. 760 (77), *773*
Churchill, M. R. 732 (2), *771*
Churney, K. L. 155 (3), *158*
Chuzel, O. 761 (81), 762 (82), *773*
Ciancaleoni, G. 168 (93), *212*
Cianflone, M. 859 (17), *877*
Ciattini, P. G. 11 (66), *20*, 379 (47), *437*, 640 (36), *684*
Ciavarri, J. P. 979 (477), *990*
Ciavatta, L. 157 (129), *162*
Ciccarella, G. 235 (105), *243*
Cini, R. 339 (249), *346*
Cioslowski, J. 283 (44), *342*
Cirriez, V. 508 (227), *524*, 705 (32), *729*
Citineni, J. R. 424 (121), *439*
Ciunik, L. Z. 252 (21), *277*
Ciunik, Z. 252 (22), *277*
Claesson, A. 386 (64), *437*, 614 (41), 632 (84), *685*, *687*
Clardy, J. 173 (148), *213*
Clare, M. 629 (72), *686*
Clarembeau, M. 906 (141), *983*
Clark, A. M. 947 (327), *987*
Clark, C. G. 952 (350, 354), 954 (365), 956 (372), 960 (385), 961 (388), 962 (351, 386, 387), *987*, *988*
Clark, J. H. 940 (287), *986*, 1020 (124), *1031*
Clark, J. S. 560 (110), *583*
Clark, T. 166 (76), *211*, 309 (147), *344*
Clarke, T. 502 (215), *524*
Claverie, C. 679 (170, 171), *690*
Clavier, H. 401 (104, 106), *439*, 480 (149), *522*, 707 (34), *729*
Cleary, D. G. 625 (65), *686*
Clifford, S. 283 (44), *342*
Clique, B. 579 (141, 142), *584*
Clive, D. L. J. 416 (136), *440*, 802 (91), *853*
Cloke, G. L. 119 (305), *142*
Clososki, G. C. 429 (168), 430 (169), *440*
Coan, P. S. 87 (87), *137*
Cobranchi, D. P. 150 (35), *159*
Cobranchi, S. T. 147 (6), 150 (35), *159*
Cody, W. L. 925 (230), *985*
Coe, P. L. 73 (160), *139*, 1014 (84, 85, 87), 1015 (88, 91), *1030*
Coelho, F. 380 (49), *437*, 647 (113), *688*
Cohen, F. E. 332 (225), *346*
Cohen, H. 256 (37), *277*
Cohen, J. S. 194 (194), *214*
Cohen, T. 112 (274), *142*, 890 (66), 943 (290, 291), *981*, *986*
Cohen, Y. 168 (96), *212*
Coia, N. 574 (138), *584*
Coish, P. D. G. 452 (48), *520*
Cojocaru, C. 234 (99), *243*
Coleman, R. S. 588 (13), *602*, 932 (257), 965 (400), *985*, *989*
Collins, S. 906 (144), *983*

Collman, J. P. 3 (32), *19*, 958 (375, 376), 963 (377), *988*
Collonges, F. 550 (81), *583*
Collum, D. B. 166 (66, 70), 173 (148, 150, 151), 177 (162), *211*, *213*
Collyer, C. A. 350 (25), *361*
Colman, P. M. 352 (39), *362*
Coltart, D. M. 416 (136), *440*
Colvin, E. 807 (94), *853*
Comasseto, J. V. 514 (244–246), *524*
Combariza, M. Y. 301 (120), *343*
Combes, S. 947 (321, 325, 331), *987*
Combs, A. 954 (365), 962 (386), *988*
Combs, A. P. 956 (371, 372), *988*
Comes-Franchini, M. 802 (87, 88), 809 (99), *853*
Commerçon, A. 536 (27, 28), 545 (73), 547 (71), *581*, *582*
Compagnini, G. 155 (95), *161*
Concepción Gutierréz, M. 236 (107), *244*
Conn, R. C. 885 (14), *980*
Conn, R. S. E. 445 (22), *519*
Connolly, J. S. 221 (32), *242*
Connor, J. A. 165 (46), *211*, 892 (85, 86), *982*
Conrads, M. 629 (72), *686*
Conrow, R. E. 860 (20), *878*
Conry, R. R. 64 (135), *138*, 200 (209), *214*
Constantino, E. 309 (31), *341*
Cooks, R. G. 149 (24), *159*, 322 (189–192), 326 (206, 207), *345*
Cooper, I. L. 285 (50), *342*
Cope, S. 3 (36, 37, 39), *19*, 112 (279, 280, 282), *142*, 154 (81–83), *160*, 195 (58), 198 (64, 65), 200 (57), *211*, 534 (37–40), *582*, 858 (7), *876*
Coppola, G. M. 866 (29), *878*
Corain, B. 953 (362), *988*
Corey, E. J. 11 (65), 16 (90, 91, 97), 17 (17), 18 (18), *19–21*, 24 (10), *135*, 430 (21), 431 (170), *436*, *440*, 509 (229, 230), 516 (248), *524*, 610 (33), 638 (93), *684*, *687*, 780 (3), 800 (31, 83), 838 (180), *851*, *853*, *856*, 859 (16), *877*
Corfield, P. W. R. 81 (35), *136*
Coronado, J. M. 235 (101), *243*
Corongiu, G. 293 (33), *341*
Corral, I. 308 (41, 104, 105), 309 (40, 147), *341*, *343*, *344*
Correa, A. 18 (118), *21*, 200 (216), *214*, 493 (177), *523*, 670 (152), *690*, 926 (240), *985*
Correa, A. G. 979 (490, 491), *990*
Corruble, A. 166 (67), *211*
Corsten, M. A. 134 (63), *137*
Cossi, M. 283 (44), *341*
Costa, G. 2 (12), *19*, 27 (22), 39 (80, 83), *135*, *137*
Costa, L. L. 237 (115), *244*
Costanzo, S. 958 (375, 376), *988*
Costello, C. E. 336 (240), *346*
Côté, A. 489 (173), *523*
Cottet, F. 977 (467), *990*, 1018 (111), *1031*
Cotton, F. A. 63 (131), 133 (133), *138*
Cotton, H. K. 669 (144), *690*
Coulentianos, C. 545 (73), 547 (71), *582*
Courmarcel, J. 732 (1), 753 (61, 62), *771*, *773*
Courtney, B. H. 224 (45), *242*
Coutable, L. 401 (104), *439*
Couturier, M. 925 (235), *985*
Couty, F. 622 (56), *685*, 899 (115), *982*
Cowie, J. 70 (150), *139*
Cowley, A. H. 103 (236), *141*
Cox, H. 289 (66), 305 (64), *342*
Cox, J. D. 148 (8), *159*
Cox, S. D. 834 (172), *855*
Coxon, J. M. 605 (12), *683*, 858 (7), *876*
Coy, D. H. 943 (300), *986*
Coyle, C. L. 359 (60), *363*
Crabbé, P. 2 (20), *19*, 430 (172), *441*, 629 (4), 674 (3), *683*
Crabtree, R. H. 41 (41), 120 (306), *136*, *142*, 873 (4), *875*
Craig, D. 842 (137), *856*
Cramer, C. J. 333 (236), *346*
Cran, G. A. 430 (19), *436*
Crandall, J. K. 550 (81), *583*
Cravador, A. 906 (141), *983*
Crawford, K. R. 925 (223), *985*
Cremer, D. 30 (57), *136*, 283 (46), 308 (148), *342*, *344*
Cremer, U. 74 (164), *139*
Cresson, P. 430 (172), *441*
Cristau, H.-J. 897 (108, 110), 921 (219), *984*, 943 (217), 967 (413), *982*, *984*, *989*
Cristea, I. 112 (274), *142*
Croft, S. 1020 (122), *1031*
Cronje, S. 129 (332), *143*
Croset, K. 670 (148), *690*, 858 (11), *877*
Crotti, P. 676 (164), 677 (167, 168), 679 (172), *690*, *691*
Crow, R. 517 (233), *524*
Crowley, P. J. 1004 (46), *1029*
Crump, R. A. 800 (67), *852*
Crump, W. 113 (287), *142*, 154 (84), *161*, 459 (84), *521*
Crute, T. D. 625 (64, 65, 67), *686*
Crystall, B. 85 (199), *140*, 262 (53), *278*
Csizmadia, I. G. 324 (198, 199), *345*
Cuadrado, P. 800 (39), 802 (89), 811 (106, 107), 813 (38), 816 (30), 819 (134), 821 (142), 841 (144), 845 (194), 846 (9, 125), 847 (129, 181), *851–856*
Cubbon, R. 949 (334), *987*
Cubillo, M. A. 804 (92), 824 (156), *853*, *855*
Cuenca, A. 935 (269), 937 (273), *985*
Cuerva, J. M. 831 (165), *855*
Cui, Q. 283 (44), *342*, 800 (58), *852*

Cul, G. 811 (109), *854*
Cullen, W. R. 993 (13), *1029*
Culshaw, D. 502 (215), *524*
Cummins, J. M. 173 (151), *213*
Cundari, T. R. 121 (313, 314), *143*, 403 (118), *439*
Cundy, D. J. 954 (366), *988*
Cuny, G. 943 (250), *985*
Cuny, G. D. 894 (96), *982*
Curran, D. P. 471 (111), *521*, 646 (109), *688*
Curran, K. J. 1018 (112), *1031*
Curtiss, L. A. 282 (28), *341*
Cusick, J. R. 631 (76), *687*
Czajkowski, W. 228 (64), *243*
Czekelius, C. 747 (43–45), *772*
Czernecki, S. 557 (106), *583*
Czernuszewicz, R. S. 943 (295), *986*
Czyzewska, E. 545 (69), *582*, 631 (81), *687*

D'Aniello, F. 638 (94, 96), *687*, *688*
D'Augustin, M. 401 (106), *439*, 480 (149), 503 (223), 506 (225), *522*, *524*, 698 (15), 705 (30), 707 (34), *729*
D'Hallewyn, C. 859 (18), *878*
da Silva Fraústo, J. J. R. 348 (1), *361*
Dabbagh, G. 16 (100), *21*, 152 (55), *160*, 173 (138), 174 (143), *213*, 366 (4), 380 (3), 430 (16), *435*
Daeuble, J. F. 737 (19), 768 (17), *772*
Dagorne, S. 754 (65), *773*
Dai, Q. 920 (191), *984*
Dai, W.-M. 943 (231), *985*
Dai, X. 130 (318), *143*
Daitoh, T. 549 (78, 79), *583*
Dale, M. J. 220 (25), *242*
Dalgard, J. E. 925 (224), *985*
Dalili, S. 958 (379), *988*
Dalko, P. I. 517 (260, 261), *525*
Dambacher, J. 867 (32), *879*
Danieli, B. 356 (56), *362*
Daniels, A. D. 283 (44), *342*
Dannenberg, J. J. 283 (44), *342*
Danopoulos, A. A. 127 (326), *143*
Dapprich, S. 283 (44), *342*
Darby, M. V. 861 (22), *878*
Darcel, C. 418 (144), *440*
Dargel, T. K. 290 (79), 308 (39), *341*, *342*
Das, J. 237 (113), *244*
Daskapan, T. 377 (35, 40), *436*
Dauben, W. G. 888 (40), *981*
Daubos, P. 679 (170, 171), *690*
Daugulis, O. 973 (436, 437), *989*, 1028 (150), *1032*
Davico, G. E. 148 (16), *159*
Davidson, E. R. 283 (43), *341*
Davidson, F. 123 (321), *143*
Davidson, O. 166 (83), *212*
Davidsson, Ö. 108 (255), *141*, 178 (109), 179 (128), *212*, 184 (187), *212*, *214*, 369 (9), *435*
Davies, A. G. 824 (121), *854*
Davies, G. M. 834 (167), *855*
Davies, J. A. 164 (10), *210*
Davies, R. P. 98 (225), 102 (234), 111 (271), *140*, *141*, 178 (167, 168), *213*, 368 (10), *435*
Davis, A. V. 360 (79), *363*
Davis, B. D. 335 (239), *346*
Davis, C. R. 996 (8), 997 (26), *1029*
Davis, J. A. 310 (164), *344*
Davis, K. M. 402 (110), *439*
Davison, Ô. 790 (26), *851*
Davoust, D. 166 (67), *211*
Davydov, V. N. 252 (14, 24), *277*
Dawson, A. 87 (207), *140*
Dawson, B. T. 149 (32), *159*
Dawson, M. I. 863 (27), *878*
Day, M. W. 452 (54), *520*
de Alburquerque, C. 1020 (122), *1031*
de Armas, J. 716 (58), *730*, 859 (17), *877*
de Bruyn, J. 890 (63), *981*
De Castro, F. J. V. 233 (96), *243*
De Clercq, P. 654 (125), *689*, 859 (18), *878*
de, F. 429 (168), 430 (169), *440*
de Fátima, A. 732 (1), *771*
de Jersey, J. 354 (46), *362*
De Jong, F. 894 (93), *982*
De Kimpe, N. 1017 (109), *1031*
de Klerk, A. 890 (63), *981*
de Lange, B. 914 (170), *984*
de Lera, A. R. 845 (191, 192), *856*
de los Angeles Rey, M. 647 (115), *689*
de Marigorta, E. M. 809 (101), *853*
de Meijere, A. 13 (75), *20*
de Miguel, C. G. 922 (204), *984*
De Nitti, C. 663 (20), *684*
de Oliveira, A. R. M. 429 (168), 430 (169), *440*
de Rijk, R. 914 (170), *984*
de Vries, A. H. M. 200 (207), 201 (226), 203 (228), *214*, *215*, 694 (5), *729*, 914 (170), *984*
de Vries, J. G. 201 (226), *215*, 914 (170), *984*
Deadwyler, D. A. 193 (133), *213*, 534 (36), *582*, 858 (7), *876*
Deagostino, A. 958 (380), *988*
Deagusa, T. 60 (122), *138*
Deboves, H. J. C. 416 (140), 418 (142), *440*
Dechoux, L. 979 (485), *990*
Decicco, C. P. 953 (360), *988*
Decker, C. 232 (93), *243*
Decorzant, R. 859 (18), *877*
Dedieu, A. 41 (95), *137*, 276 (73), *278*
DeFelippis, M. R. 354 (47), *362*
DeForest, J. E. 866 (29), *878*
Degen, P. 859 (15), *877*

Degl'Innocenti, A. 802 (87), 809 (17), 828 (158), 842 (184), *851*, *853*, *855*, *856*
Degrado, S. J. 430 (27), *436*, 707 (43), 710 (49), 713 (44), 720 (42), *730*
Deguchi, K. 165 (47), *211*
Dehmel, F. 404 (125, 126), *439*
Dehner, A. 168 (92), *212*
Dehnicke, K. 100 (228), *141*
del Amo, V. 378 (46), *436*, 600 (37, 38), *602*
Del Moro, F. 474 (139), *522*, 677 (168), 679 (172), *690*, *691*, 695 (9), *729*
del Olmo, E. 920 (200), *984*
Del Sole, R. 235 (101, 102, 104), *243*
del Villar, A. 669 (14, 143), *683*, *690*, 858 (11), *876*
Delacroix, T. 447 (31), *520*
Delaloge, F. 842 (188), *856*
Delaney, M. O. 979 (481), *990*
Delbaere, L. T. J. 63 (132), *138*
Dellinger, M. E. 902 (125), *983*
Delp, S. A. 121 (313–315), *143*, 403 (118), *439*
Delpuech, J. J. 165 (51, 52), *211*
Demartin, F. 81 (179), *139*
Demay, S. 421 (152), *440*, 641 (103), *688*
Dembech, P. 373 (28), 374 (34), 378 (45), *436*, 599 (35, 36), *602*, 807 (96), *853*
Demel, P. 11 (67), *20*, 480 (152), 484 (159, 160), *522*, 622 (38), 672 (1), *682*, *685*, 858 (9), *876*
Dempsey, D. F. 105 (243), *141*, 149 (26), *159*
den Boer, D. J. 399 (96), *438*
den Hartog, J. A. J. 925 (234), *985*
den Hartog, T. 202 (18), *210*, 723 (68, 69), *730*, 858 (10), 862 (25), *876*, *878*
den Hertog, H. J. 890 (63), *981*
Deng, M.-Z. 551 (97), *583*
Deng, W. 926 (238), *985*
Deng, Y. 223 (43), *242*
Deng, Z. 236 (110), *244*
Denichoux, A. 638 (95), *688*
Denmark, S. E. 382 (53), *437*, 516 (256), *524*, 663 (137), *689*, 807 (97), *853*
Dennison, C. 348 (12), *361*
Deo, N. 430 (17), *435*
DeOliveira, E. 237 (115), *244*
DePasquale, R. J. 1024 (137, 145), *1031*, *1032*
Depres, J.-P. 380 (49), *437*
DePue, R. T. 173 (148), *213*
DePuy, C. H. 148 (16), *159*
Desai, M. 885 (15), *980*
Deschamp, J. 761 (81), 762 (82), 765 (88), *773*
Descoins, C. 539 (51), *582*
DeShong, P. 962 (387), *988*
Deshpande, S. D. 228 (61), *242*
Desiraju, S. 925 (230), *985*
Desjardins, S. 166 (67), *211*
Desmazières, B. 315 (178), *345*
Despotopoulou, C. 425 (163), *440*
Després, J.-P. 647 (113), *688*
Desrosiers, J.-N. 748 (50), *772*
Dessideri, M. 355 (50), *362*
Desvergnes-Breuil, V. 200 (210), *214*
Deudon, S. 952 (354), 961 (388), 962 (387), *988*
Deutsch, C. 2 (25), *19*, 200 (21), *210*, 632 (86, 87), *687*, 732 (1), 769 (91), *771*, *773*, 858 (8), 870 (46), *876*, *879*
Deutsch, E. 165 (42), *211*
Devasagayaraj, A. 496 (127), *522*
Devlin, F. J. 198 (198), *214*
Dewar, M. J. S. 231 (90), *243*, 297 (110), *343*
Dexter, C. S. 416 (139), 418 (150), *440*, 474 (129), *522*
Dhanak, D. 551 (92), *583*
Dhar, S. 238 (130), *244*
Di Bussolo, V. 677 (168), *690*
Dias, H. V. R. 123 (321), *143*
Diaz-Requejo, M. M. 41 (41), *136*
Dicelio, L. E. 240 (142), *245*
Didier, E. 629 (74), *687*
Diederich, F. 13 (75), *20*
Diéguez, M. 2 (23), *19*, 24 (17), *135*, 202 (23), *210*, 695 (1), *729*, 858 (10), *876*
Diehl, P. 197 (204), *214*
Dieter, J. W. 390 (84), 430 (17), *435*, *438*
Dieter, R. K. 178 (167), *213*, 367 (8), 386 (72, 73), 387 (75), 388 (85), 389 (74, 76), 390 (82–84, 86), 393 (77–79, 87), 395 (90), 396 (91), 397 (92), 399 (93, 94, 100), 430 (17), *435*, *437*, *438*, 476 (140), *522*, 587 (11), *602*, 635 (89, 92), *687*, 858 (8), *876*
Dietrich-Buchecker, C. 200 (210), *214*
Dietz, A. G. 943 (290), *986*
Diez-Barra, E. 858 (12), *877*
Díez-González, S. 119 (302), *142*, 120 (316), 121 (312), *143*, 400 (102), 403 (111, 114), *438*, *439*, 750 (53, 54), 751 (55, 56), *773*, 761 (79), *773*
Diez, J. 80 (177), 82 (188), 83 (195), *139*, *140*
DiGrandi, M. J. 1018 (112), *1031*
Dijk, E. W. 719 (59), *730*
DiMartino, M. J. 197 (205), *214*
DiMichele, L. 834 (169), *855*
Dimitrov, V. 143 (22), *159*
Dimock, S. H. 16 (93), *20*, 456 (63), 503 (217), 517 (233), *520*, *524*, *525*, 800 (66), 842 (130), *852*, *854*
Dinares, I. 922 (204), *984*
Ding, H. 229 (71), *243*
DiPasquale, A. G. 60 (60), *136*
Disch, S. 74 (164), *139*
DiSpirito, A. A. 361 (80), *363*
Dixon, D. A. 304 (113), *343*
Djinovic-Carugo, K. 352 (36), 358 (70), *362*, *363*

Djuric, S. W. 629 (72), *686*
Do, H-Q. 973 (436, 437), 1028 (150), *989*, *1032*
Do, Y. 70 (146), 71 (147, 153), *138*, *139*
Dobbek, H. 360 (37, 73, 74), *362*, *363*
Dobromil'skaya, I. M. 155 (89), *161*
Dodd, D. S. 824 (155), *855*
Dodonov, V. A. 946 (308, 310, 311), 947 (309), 949 (335, 336), *986*, *987*
Doel, C. L. 165 (41), *211*
Doelle, A. 165 (50), *211*
Doering, W. v. E. 151 (49), *160*
Doherty, M. Q. 904 (131), *983*
Dohle, W. 404 (124), 421 (151–153), 422 (156), 423 (159), 424 (157), 425 (154), *439*, *440*, 600 (19), *602*, 646 (110), *688*
Dolbier, W. R. 151 (49), *160*, 1016 (97), *1031*
Dolgopol'skii, I. M. 155 (89), *161*
Dollat, J.-M. 629 (4), *683*
Dolling, U.-H. 947 (320), *987*
Dolphin, D. 219 (5), *241*
Dominguez, C. 921 (209), *984*
Dominguez, E. 904 (136), *983*
Domke, A. 632 (86), *687*
Domon, B. 336 (240), *346*
Dong, L. 710 (51), *730*
Donkervoort, J. G. 87 (204), *140*
Donley, C. L. 226 (57), *242*
Donnadieu, B. 114 (290, 291), *142*, 964 (394), *988*
Donnelly, D. M. X. 949 (338), 961 (339), *987*
Donovan, T. A. 153 (75), *160*
Dooley, D. M. 358 (62, 69), 359 (72), *363*
Dorigo, A. E. 3 (33), *19*, 605 (12), *683*
Doris, E. 947 (326), *987*
Dormer, P. G. 897 (105), *982*
Dormi, S. S. 622 (61), *686*
Dörner, W. 473 (121), *522*
Dorton, D. 3 (37), *19*, 112 (279), *142*, 154 (82), *160*, 198 (65), *211*, 858 (7), *876*
Dorton, D. C. 112 (280), *142*, 154 (83), *160*, 198 (64), *211*, 534 (37, 40), *582*
dos Santos, A. A. 429 (168), 430 (169), *440*
Doster, M. 63 (126), *138*
Doty, M. J. 553 (95), *583*
Dougherty, T. J. 238 (119, 120), *244*
Douglas, C. J. 478 (145), *522*, 933 (263), *985*
Douville, J. 622 (54), *685*
Doye, S. 942 (92), *982*
Doyle, T. W. 238 (136), *244*
Drager, A. S. 226 (54, 56, 57), *242*
Dressaire, G. 105 (237), *141*, 173 (139), *213*
Drexler, M. 943 (299, 303), *986*
Driess, A. 41 (96), 78 (171), *137*, *139*
Drohat, A. C. 280 (8), *341*
Drouin, R. 337 (247), *346*
Drozda, S. E. 831 (84), *853*
Druzhkova, O. N. 90 (209), *140*
Du, X. 904 (133), *983*
Du, Y. 762 (83), *773*
Dua, S. S. 1024 (138, 139), *1031*
Duan, J-X. 1021 (127), *1031*
Duax, W. L. 861 (21, 22), *878*
Dubay, G. 238 (136), *244*
Dubbaka, S. R. 169 (116–118), *212*, 378 (46), *436*, *437*, 586 (7), 589 (16), 600 (37, 38, 40), *601*, *602*
Dube, H. 423 (162), *440*
Dubinia, G. G. 44 (99), *137*
Dübner, F. 670 (145), *690*, 858 (11), *877*
DuBois, G. C. 354 (46), *362*
DuBois, J. L. 943 (294), *986*
Dubois, L. H. 150 (41), *160*
Duboudin, J. G. 551 (83, 93), 553 (82), *583*
Dubowski, D. 892 (85, 86), *982*
Duca, D. 943 (293), *986*
Ducan, M. A. 285 (55), *342*
Duchêne, A. 841 (162), *855*, 845 (193), *856*
Ducry, L. 800 (50), *852*
Duddu, R. 416 (128), *439*
Dudek, J. D. 114 (288), *142*
Dudley, D. A. 925 (230), *985*
Duffaut, N. 800 (34), *851*
Duffy, K. J. 517 (260, 261), *525*
Dugan, L. 225 (48), *242*
Duisenburg, A. J. M. 166 (77), *211*
Duke, C. V. A. 940 (287), *986*
Dumas, M. 1020 (122), *1031*
Dumesic, J. A. 158 (136), *162*
Dumont, W. 824 (151), *855*, 906 (141), *983*
Dunbar, R. C. 294 (95, 96), *343*
Duncan, W. E. 902 (122), *983*
Duncanson, L. A. 297 (111), *343*
Dunford, J. A. 530 (19), *581*
Dunks, G. B. 68 (144), *138*
Dunlop, H. G. 889 (53), *981*
Dunn, M. J. 416 (133, 134), *440*
Dunne, T. S. 238 (137), *244*
Dunoguès, J. 800 (34), *851*
Duplais, C. 447 (26), *519*
Dupont-Gaudet, K. 925 (235), *985*
Duran, M. 114 (291), *142*
Dushin, R. G. 1018 (112), *1031*
Dussault, P. H. 800 (85), *853*
Dutra, G. A. 9 (59), *20*
Dutton, P. L. 358 (71), *363*
Duursma, A. D. 489 (167), *523*
Duyck, C. 382 (54), *437*
Dvorak, D. 407 (127), *439*
Dybowski, C. 163 (2), *210*
Dygos, J. H. 453 (61), *520*
Dykstra, R. R. 820 (136), *854*
Dyrda, G. 235 (104), 236 (106), *243*, *244*
Dyrkacz, G. R. 354 (46), *362*

Eaborn, C. 37 (75), 106 (245), 110 (269), 111 (266), *137*, *141*, 166 (75), *211*
Eady, R. 307 (146), *344*
Eary, C. T. 800 (85), *853*
Easdon, J. C. 1021 (123), *1031*
Eastham, G. 127 (326), *143*
Eastwood, D. 220 (17), *241*
Eberhardt, W. 157 (113), *161*
Ebert, G. W. 153 (75), *160*, 1024 (143, 144), *1032*
Echavarren, A. M. 831 (165), *855*
Echegoyen, L. 221 (35, 36), *242*
Eckert, J. F. 221 (36), *242*
Eckhardt, M. 416 (128), *439*
Ecorchard, P. 269 (62), 275 (70), *278*
Edelmuth, S. H. 1009 (62), *1030*
Edmunds, J. J. 925 (230), *985*
Edwards, A. J. 106 (176), *139*
Edwards, P. D. 386 (67), 421 (70), *437*
Edwards, P. G. 106 (227, 247), *141*, 152 (62, 63), *160*
Egert, E. 386 (60), *437*, 859 (18), *878*
Eggert, J. 155 (91), *161*
Egorova, T. A. 949 (335), *987*
Eguchi, M. 430 (18), *435*, *436*
Ehara, M. 283 (44), *341*
Ehara, Y. 891 (75), *982*
Ehlers, A. W. 293 (80), *342*
Eichhorn, D. M. 979 (488), *990*
Eicken, C. 356 (53), *362*
Eidam, H. S. 474 (136), *522*
Eidam, P. 474 (136), *522*
Einstein, F. W. B. 993 (13), *1029*
Eipper, B. A. 280 (2), *341*, 350 (30), *362*
Eisch, J. J. 445 (18), *519*
Eisenberg, D. 349 (28), *362*
Eisenstein, O. 61 (125), *138*
El-Batta, A. 459 (90), 518 (89), *521*
El Firdoussi, A. 334 (237), *346*
El-Hamruni, S. M. 110 (269), *141*
El-Kehli, M. N. A. 166 (75), *211*
El-Khouly, M. E. 221 (37), *242*
Elguero, E. 282 (26), *341*
Elguero, J. 282 (26), *341*
Eller, K. 310 (162), *344*
Eller, P. G. 902 (123), *983*
Ellestad, G. A. 238 (137), *244*
Elliott, G. I. 950 (345), *987*
Elliott, J. 474 (129), *522*
Elliott, R. L. 646 (109), *688*
Ellis, B. D. 166 (60), *211*
Ellis, K. A. 829 (154), *855*
Ellison, G. B. 148 (16), *159*
Ellisworth, E. L. 24 (15), *135*
Ellsworth, E. L. 456 (63), 459 (78), 510 (234), 517 (233), 518 (59), *520*, *524*, 818 (32), 842 (130), *851*, *854*
Elsevier, C. J. 163 (3), *210*, 629 (72), *686*, 832 (146), *855*
Eltayeb, M. A. Z. 943 (298), *986*
Emonds, M. V. M. 838 (177), *856*
Enanga, B. 1020 (122), *1031*
Enders, D. 431 (170), *440*, 695 (1), *729*
Endo, H. 1007 (54), *1030*
Endo, K. 165 (47), *211*
Endo, Y. 911 (161), *983*
Eng, J. 150 (43), *160*
Engel, R. 893 (90), 940 (87–89), *982*
Engelhard, M. 349 (20), *361*
Engelmann, F. M. 240 (139), *244*
Engman, L. 906 (146), *983*
Enguehard, C. 921 (215), *984*
Enguehard-Gueiffier, C. 925 (222), *985*
Enna, M. 966 (398), *988*
Enozawa, H. 595 (28), *602*
Equey, O. 676 (165), *690*
Erb, W. 777 (19), *851*
Erdelmeier, I. 566 (128), *689*
Erdik, E. 377 (35, 41), *436*, 473 (119), 474 (130), *521*, *522* 873 (4), *875*
Erdsack, J. 870 (46), *879*
Eriksson, H. 23 (44, 45), 29 (51), 30 (54), 58 (48), *136*
Eriksson, J. 178 (109), 184 (187), *212*, *214*, 369 (9), *435*
Eriksson, L. A. 281 (16), *341*
Eriksson, M. 16 (95, 101), *21*, 105 (239–241), *141*, 173 (123), *212*, 367 (6), *435*, 459 (91, 93), *521*, 778 (24), *851*, 867 (33), *879*
Erkamp, C. J. M. 56 (116), *138*
Erker, G. 445 (23), *519*
Erkizia, E. 64 (135), *138*
Ermolenko, M. S. 535 (44), *582*
Ernst, L. 863 (27), *878*
Ernst, M. 649 (118), *689*, 858 (13), *877*
Ernsting, J. M. 163 (3), *210*
Eroglu, F. 377 (35), *436*
Ervin, K. M. 152 (60), 157 (114), *160*, *161*
Escher, I. H. 695 (8), *729*
Escher, S. D. 859 (18), *877*
Esquivias, J. 492 (175), *523*, 697 (11), *729*
Esseffar, M. 310 (170), 337 (169), *344*
Esteves, M. A. 947 (323), *987*
Estevez, J. C. 908 (148, 149), *983*
Estevez, R. J. 908 (148, 149), *983*
Eujen, R. 459 (85), *521*, 1022 (135), *1031*
Euler, W. B. 229 (69), *243*
Eustache, J. 945 (307), *986*
Evano, G. 164 (20), *210*, 622 (56), *685*, 899 (115), *982*
Evans, D. A. 953 (359, 360), 962 (349), *987*, *988*
Evans, H. H. 993 (15), *1029*
Evans, T. L. 904 (127), *983*

Evans, W. H. 155 (3), *158*
Evans, W. J. 63 (134), *138*
Evindar, G. 908 (154, 156), 926 (249), *983*, *985*
Eylander, C. 564 (114), *583*

Fabbri, G. 374 (34), *436*
Fabrizi, G. 966 (403), *989*
Faggi, C. 802 (87), *853*
Fairhurst, S. 359 (64), *363*
Fajer, J. 220 (24), *242*
Fajii, Y. 238 (129), *244*
Falciola, C. 675 (2), 679 (172), *683*, *691*
Falciola, C. A. 674 (160), 675 (163), *690*, 858 (11), *877*
Falck, J. R. 535 (43), *582*, 975 (451), 976 (449), *989*
Falconi, M. 355 (50), *362*
Fall, Y. 380 (49), *437*, 622 (58), 647 (115), *686*, *689*
Fallis, A. G. 528 (4), 551 (84–91), *581*, *583*
Falvello, L. R. 80 (177), *139*
Fan, B. 647 (116), *689*
Fan, H.-T. 870 (46), *879*
Fan, P. 947 (327), *987*
Fang, F. 974 (443), *989*
Fang, G.-H. 551 (97), *583*
Fang, W. 151 (48), *160*
Fang, Y. 907 (147), 970 (430), *983*, *989*
Fang, Y.-Q. 937 (272), *986*
Faraggi, M. 354 (47), *362*
Farina, J. S. 640 (98), *688*
Farkas, O. 283 (44), *342*
Fässler, R. 629 (75), *687*
Fathallah, S. 83 (189), *140*, 267 (58), *278*
Fatheree, P. 366 (5), *435*
Febles, M. 574 (139), *584*
Fedorov, A. Y. 946 (315), 947 (330, 331), *987*
Fei, X-S. 1021 (128), *1031*
Fekete, J. 859 (18), *877*
Feld, B. 1018 (112), *1031*
Felder, D. 221 (36), *242*
Feliciano, A. S. 920 (200), *984*
Felisberti, M. I. 231 (91), *243*
Feller, D. 304 (113, 130), *343*
Fendrich, J. A. 114 (288), *142*
Feng, G. 943 (231), *985*
Fengl, R. W. 975 (447), *989*
Fenical, W. 196 (196), *214*
Fenske, D. 103 (235), *141*
Ferguson, R. B. 63 (132), *138*
Feringa, B. L. 2 (24), 18 (118), *19*, *21*, 200 (207, 216–218), 201 (226), 202 (18), 203 (228), *210*, *214*, *215*, 401 (105), 430 (12), *435*, *439*, 474 (131, 133, 134, 139), 478 (141), 489 (167), 492 (176), 493 (177), *522*, *523*, 670 (146, 151, 152), 672 (158), 675 (2), 676 (164), 677 (167, 168), 679 (169, 171), *682*, *690*, 694 (5), 695 (1, 9, 10), 698 (16–18), 699 (19), 707 (39, 41), 714 (55), 719 (59–61), 720 (62, 65, 66), 721 (56, 63, 64, 67), 723 (68, 69), *729*, *730*, 858 (10, 12), 862 (25), *876–878*
Fernández de la Pradilla, R. 618 (49), 629 (70), 646 (108), *685*, *686*, *688*
Fernández, G. 816 (117), *854*
Fernández-Gacio, A. 380 (49), *437*, 647 (115), *689*
Fernandez, I. 168 (86), *212*
Fernandez-Ibanez, M. A. 699 (19), *729*
Fernandez, M. 908 (148, 149), *983*
Fernandez, P. 618 (49), *685*
Ferraudi, G. 220 (12, 28), *241*, *242*
Ferreira, E. M. 741 (30), *772*
Ferreira, F. 373 (33), *436*, 545 (67), *582*, 629 (71), 638 (95), *686*, *688*
Ferreira, J. T. B. 838 (177), *856*
Ferrer, M. L. 236 (107), *244*
Fevig, T. L. 646 (109), *688*
Fialkov, Y. A. 992 (7), *1029*
Fiandanese, V. 563 (48), 564 (117), *582*, *583*, 663 (138), *689*
Fichou, M. P. T. 238 (128), *244*
Fields, R. 993 (15), *1029*
Fife, D. J. 165 (39), *211*
Filippo, J. S. 112 (275), *142*
Filipski, K. J. 925 (230), *985*
Fillion, E. 702 (24, 25), *729*, 861 (24), *878*
Finazzi-Agrò, A. 352 (34), *362*
Finet, J. P. 946 (312, 313, 315), 947 (314, 321, 325, 328, 330, 331), *987*, 949 (322, 332, 338), 950 (337, 344), 961 (339, 390), *987*, *988*
Finke, R. G. 3 (32), *19*
Finn, M. G. 105 (105), *137*
Finnegan, W. G. 1005 (52), *1030*
Fiorin, V. 157 (108), *161*
Firth, J. D. 764 (86), *773*
Firth, S. 305 (65, 132), *342*, *343*
Fischer, D. A. 150 (43), *160*
Fischer, R. 105 (244), *141*, 152 (54), *160*
Fischer, W. F. 2 (19), *19*, 112 (275), *142*, 173 (140), *213*, 589 (15), *602*, 728 (75), *730*
Fisher, E. R. 150 (34), 152 (1), *158*, *159*, 293 (89), *343*
Fisher, W. F. 24 (9), *135*
Fishpaugh, J. R. 103 (236), *141*
Fitzgerald, A. E. 908 (150), *983*
Flachsman, F. 418 (144), *440*
Fleming, F. F. 426 (164, 166), 427 (167), *440*, 979 (476), *990*
Fleming, I. 381 (50), 382 (54, 55), *437*, 613 (37), 622 (60), 629 (72), *685–687*, 779 (29), 790 (8), 791 (21), 800 (35, 44–47, 49, 63, 65, 67, 70–77, 79–81), 803 (23), 804 (92),

807 (95), 809 (101), 811 (106), 812 (105), 813 (38, 111), 817 (7), 819 (134), 821 (142), 832 (78, 166), 841 (144), 842 (148), 846 (9, 125), 847 (129, 181), *851–856*
Flemming, S. 15 (86), *20*, 503 (219–222), *524*, 654 (127), *689*
Flessner, T. 374 (36), *436*
Fletcher, S. P. 672 (158), 675 (2), *682*, *690*, 720 (66), *730*
Florencio, F. 859 (16), *877*
Floriani, C. 30 (55), 40 (88), 48 (56), 58 (58), *136*, *137*
Floyd, D. M. 17 (104), *21*, 108 (249), *141*, 169 (120), *212*, 609 (25), 640 (98), *684*, *688*, 788 (27), *851*
Fluck, E. 197 (204), *214*
Fluckiger, E. 430 (18), *436*
Flynn, A. B. 528 (6), *581*
Flynn, G. W. 150 (43), *160*
Foces-Foces, C. 282 (26), *341*, 574 (139), *584*
Fochi, M. 802 (87, 88), *853*
Focia, P. 360 (79), *363*
Fohlman, J. 323 (193), *345*
Foley, S. R. 41 (89), *137*
Folting, K. 87 (87), *137*, 165 (48), *211*, 732 (4), *771*
Font, D. 723 (69), *730*, 862 (25), *878*
Font, J. 859 (16), *877*
Ford, F. E. 846 (125), *854*
Fordham, P. J. 315 (179), *345*
Foresman, J. B. 283 (44), *342*
Forgione, L. 861 (22), *878*
Forgione, P. 528 (4), 551 (86–88, 91), *581*, *583*
Forsyth, S. A. 954 (366), *988*
Foster, C. F. 382 (55), *437*, 807 (95), *853*
Fouassier, J. P. 232 (93), *243*
Foubelo, F. 709 (45), *730*
Foulon, J. P. 550 (63), 559 (62, 64), *582*
Fouquet, G. 609 (30), *684*
Fournier, F. 325 (202), 326 (203, 208), *345*
Fournier, M. 238 (124), *244*
Fournioux, X. 670 (148), *690*, 695 (8), *729*, 858 (11), *877*
Fox, D. J. 283 (44), *342*, 593 (18, 20), 595 (21), *602*
Fox, J. P. 943 (295), *986*
Fox, K. R. 388 (81), *438*
Foxman, B. M. 553 (99), *583*
Foxton, M. W. 445 (18), *519*
Frampton, C. 165 (41), *211*
Franckevicius, V. 593 (20), *602*
Fränkel, A. M. 166 (68), *211*
Fränkel, G. 166 (68), *211*
Frankevich, V. 309 (149), *344*
Frantz, D. E. 5 (49), 16 (96), *20*, *21*, 371 (23), *436*, 629 (75), *687*
Franz, A. K. 977 (459), *990*
Franz, K. J. 360 (78), *363*
Fraser, P. K. 400 (103), *438*, 698 (14), *729*
Frazee, W. J. 887 (27), *981*
Freas, R. B. 310 (159), *344*
Frechet, J. M. J. 423 (161), *440*
Fréchette, S. 631 (80), *687*
Frederick, M. O. 933 (263), *985*
Fredrick, M. A. 870 (14), *877*
Freedman, L. D. 891 (73), *981*
Freeman, H. C. 350 (25), 352 (39, 41), *361*, *362*
Freeman, H. S. 891 (73), *981*
Freire de Lima, M. E. 380 (49), *437*
Freiser, B. S. 289 (74), 296 (107), 310 (167), *342–344*
Frekos, J. N. 1017 (108), *1031*
French, A. N. 861 (22), *878*
Frenette, R. 959 (383), *988*
Frenking, G. 108 (259), *141*, 153 (68), *160*, 175 (160), 179 (153), *213*
Fressigne, C. 166 (67), *211*
Freund, C. 904 (128), *983*
Friedhoff, W. 804 (92), *853*
Frieman, B. A. 734 (11), 756 (71), *772*, *773*
Friend, J. N. 157 (123), *162*
Friesen, R. W. 831 (163), 845 (196), *855*, *856*
Friess, S. D. 309 (149), *344*
Friestad, G. K. 489 (170), *523*
Frisch, M. J. 282 (27), 283 (44), *341*
Frish, L. 168 (96), *212*
Fritch, P. C. 484 (162), *522*, 629 (69), *686*
Fröhlich, R. 769 (92), *773*
Frohn, M. 631 (79), *687*, 921 (209), *984*
Fronczek, F. R. 224 (45), *242*
Frosch, W. 78 (171, 172), *139*, 259 (44), 275 (69), 276 (72), *277*, *278*
Frutos, J. 694 (3), *729*
Fu, H. 915 (177), 919 (181, 184), 920 (195), 926 (237), 940 (107, 284), *982*, *984–986*
Fu, Y. 206 (208), *214*, 670 (150), *690*, 925 (225), 943 (297), *985*, *986*
Fuan, Z. 237 (111), *244*
Fuchikama, T. 1017 (110), *1031*
Fuchikami, T. 977 (466), 979 (475), *990*
Fuenzalida, V. 83 (208), *140*
Fuerstenau, D. W. 296 (102), *343*
Fugier, C. 536 (47), *582*
Fujihira, M. 225 (50), *242*
Fujii, H. 165 (31), *210*
Fujii, N. 484 (165), *522*, 616 (27), 625 (40, 42), 629 (68), 635 (88, 90), 641 (105), *684–688*
Fujii, T. 735 (13), *772*, 979 (473), *990*
Fujikawa, H. 1024 (149), *1032*
Fujiki, K. 887 (26), *980*
Fujimori, S. 728 (75), *730*
Fujisawa, T. 457 (71), *520*, 640 (101), *688*
Fujishima, A. 230 (76), *243*

Fujita, A. 738 (23), *772*
Fujita, K. 946 (316), *987*
Fujita, T. 925 (218), *984*
Fujitsuka, M. 221 (37), *242*
Fujiwara, R. 230 (78), *243*
Fujiwara, T. 831 (164), *855*
Fukin, G. K. 90 (209), *140*
Fukuda, R. 283 (44), *341*
Fukudome, Y. 937 (275), *986*
Fukui, Y. 968 (422), *989*
Fukumoto, K. 885 (16), *980*
Fukumura, H. 228 (59), 230 (80), *242*, *243*
Fukushi, E. 167 (89), *212*
Fukuyama, T. 919 (180), 943 (292), *984*, *986*
Fuller, G. 887 (24), *980*
Fuller, J. 303 (128), *343*
Fuller, P. H. 373 (29), *436*
Fung, K. C. M. 733 (9), *772*
Fung, W. K.-M. 81 (182), 83 (192–194), 85 (200, 201), *139*, *140*, 262 (57), 264 (54), 266 (55, 56), *278*
Furber, M. 537 (16, 25), *581*
Furlong, M. 416 (128), *439*, 858 (8), *876*
Furlong, M. T. 416 (135), *440*
Fürstner, A. 447 (28), *519*
Furukawa, J. 474 (125), *522*
Furukawa, K. 644 (107), *688*
Furuta, H. 117 (296, 297), 118 (298, 299), 119 (300), *142*, 240 (139), *244*, 257 (41), *277*
Furutachi, H. 44 (99), *137*
Furuune, M. 965 (399), 966 (398), 969 (428), *988*, *989*
Furuyama, T. 5 (42), *19*
Fyfe, H. B. 248 (5), *276*

Gabler, B. 647 (111), *688*
Gabriel, J. 166 (73), *211*
Gade, L. H. 120 (307), 127 (327), *142*, *143*, 400 (101), *438*
Gaggini, F. 809 (99), *853*
Gagnon, D. 622 (54), *685*
Gais, H.-J. 666 (128, 140), *689*
Gajare, A. S. 920 (183), *984*
Galdecki, Z. 861 (22), *878*
Galeva, N. 361 (80), *363*
Galiano, L. 293 (86, 87), *343*
Galisteo, D. 845 (194), *856*
Gallagher, D. J. 386 (69), *437*
Gallardo, A. L. 320 (185), *345*
Gallina, C. 11 (66), *20*, 379 (47), *437*, 640 (36), *684*
Gallina, J. 629 (4), *683*
Gallo, E. 18 (116), *21*
Galy, J.-P. 947 (324), *987*
Gamasa, M. P. 80 (177), 81 (181), 82 (188), 83 (195), 85 (198), *139*, *140*
Gambarotta, S. 30 (55), 40 (88), 48 (56), *136*, *137*
Gämers, S. 163 (3), *210*
Gamez, P. 353 (43), 356 (54), *362*
Ganem, B. 859 (16), *877*
Ganzer, D. 200 (211), *214*
Gao, J. 922 (208), *984*
Gao, S. W. 337 (247), *346*
Gao, Y. 445 (17, 19), *519*, 632 (45), *685*
Garber, S. B. 132 (336), *143*
Garcia, A. M. 616 (43), *685*
Garcia-Blanco, S. 859 (16), *877*
García, C. 807 (93), 811 (108), 816 (30, 112), 847 (181), *851*, *853*, *854*, *856*
Garcia, E. 587 (12), *602*
Garcia, F. 39 (79), *137*
García-Granda, S. 80 (177), 82 (188), 83 (195), 85 (198), *139*, *140*, 451 (44), *520*
García-López, E. 235 (101, 102, 104, 105), 236 (106), *243*, *244*
Garcia-Ochoa, S. 613 (31), *684*
Garcia-Prieto, J. 147 (5), *159*
Garcia, V. S. 237 (115), *244*
Gard, G. L. 154 (85), *161*
Gardan, M. 953 (362), *988*
Gardette, M. 532 (13, 14), 537 (46), *581*, *582*
Gardner, J. O. 861 (22), *878*
Garg, R. 965 (400), *989*
Garrido, F. 616 (27), *684*
Garrison, P. J. 859 (16), *877*
Gärtner, T. 3 (38), *19*, 112 (281), *142*, 196 (59, 63), *211*, 858 (6, 7), *876*
Gasteiger, J. 309 (147), *344*
Gastinger, R. G. 1024 (147), *1032*
Gatlin, C. L. 327 (153, 154), 328 (61, 211), *342*, *344*, *345*
Gatti, G. 352 (34), *362*
Gatti, L. 2 (12), *19*, 27 (22), 39 (83), *135*, *137*
Gaudinski, C. M. 73 (158), *139*
Gavai, A. V. 831 (163), *855*
Gawley, R. E. 386 (71), 430 (171), *437*, *441*
Gawronski, J. 802 (87), *853*
Gayo, L. M. 859 (16), *877*
Gebreyes, Z. 337 (247), *346*
Geckle, M. J. 166 (68), *211*
Geerts, R. L. 165 (48), *211*
Gegnas, L. D. 517 (252), *524*
Geiser, U. 114 (288, 289), *142*, 154 (85, 86), *161*
Gellert, R. W. 106 (247), *141*, 152 (62), *160*
Gelman, D. 940 (279), *986*
Genet, J. P. 373 (33), *436*
Geng, Z. 759 (75), 760 (76), *773*
Gennari, C. 18 (116), *21*, 670 (147), 679 (170, 171), *690*
George, M. V. 777 (20), *851*
George, P. 147 (9), *159*
Georghious, P. E. 735 (12), *772*
Georgiadis, R. 152 (1), *158*
Georgieva, I. 296 (103), 309 (37), *341*, *343*

Gerdemann, C. 356 (53), *362*
Germann, M. 955 (369), *988*
Germon, C. 530 (20, 21), 539 (53, 54), *581*, *582*
Gerold, A. 5 (47), 6 (6), 8 (58), *19*, *20*, 173 (126), 185 (135), 191 (191), *212–214*, 388 (80), 430 (13), *435*, *438*, 534 (35), *582*, 672 (1), *682*, 829 (160), *855*, 867 (35), 870 (14, 49), 872 (47), *877*, *879*
Gerval, J. 800 (34), *851*
Getzoff, E. D. 354 (31, 49), *362*
Geurts, K. 2 (24), *19*, 202 (18), *210*, 672 (158), 675 (2), *682*, *690*, 699 (19), 720 (66), *729*, *730*, 858 (10), *876*
Geynet, C. 861 (21), *878*
Ghanty, T. K. 283 (43), *341*
Ghilardi, C. A. 106 (246), *141*
Gholap, R. V. 158 (130, 132), *162*
Ghoroku, K. 977 (462), *990*
Ghosh, A. 925 (235), *985*
Ghosh, P. 631 (76), *687*
Ghosh, S. 426 (166), *440*
Ghosh, S. A. 348 (16), *361*
Ghribi, A. 10 (64), *20*, 663 (85, 134), *687*, *689*, 874 (56), *879*
Gibson, A. M. 165 (41), *211*
Gibson, C. L. 430 (19), *436*
Gibson, C. P. 587 (10), *602*
Gibson, T. 456 (66), *520*
Gielen, M. 178 (168), *213*
Giessner-Prettre, C. 166 (67), *211*
Gigante, B. 947 (323), *987*
Gil-Luna, M. D. 236 (107), *244*
Gilbertson, S. R. 418 (143, 145), *440*, 834 (168), *855*
Gill, D. S. 165 (49–53), *211*
Gill, P. M. W. 283 (44), *342*
Gille, S. 670 (149), *690*, 858 (11), *877*
Gilles, T. 113 (286), *142*
Gillet, J. P. 474 (122), *520*
Gillingham, D. G. 509 (228), 516 (259), *524*, 674 (161), *690*
Gilman, H. 11 (11), *19*, 24 (4, 5), *135*, 169 (104), *212*, 585 (2), *601*, 776 (1), 777 (20), *850*, *851*, 857 (1, 2), *875*, 889 (52), *981*
Gilman, H. J. 1024 (138, 139), *1031*
Gimeno, J. 80 (177), 81 (181), 82 (188), 83 (195), 85 (198), *139*, *140*, 675 (162), *690*
Gin, D. Y. 728 (73), *730*
Ginah, F. O. 153 (75), *160*
Gindler, E. M. 943 (299, 303), *986*
Gini, F. 474 (139), *522*, 695 (9), *729*
Giolando, D. M. 103 (236), *141*
Giordano, M. 231 (91), *243*
Girard, G. 660 (132), *689*
Girardin, M. 925 (235), *985*
Girdaukas, G. 643 (106), *688*
Girolami, G. S. 105 (243), *141*, 149 (26), *159*
Gischig, S. 129 (329, 330), *143*
Gisselbrecht, J. P. 221 (36), *242*
Giudice, E. 315 (179), *345*
Giudici, R. E. 508 (151), *522*, 672 (157), *690*
Gladiali, S. 675 (162), *690*
Gladston, P. L. 834 (169), *855*
Gladys, M. J. 157 (108), *161*
Gland, J. L. 150 (43), *160*
Glaser, C. 585 (4), *601*
Gleason, J. L. 516 (252), *524*
Gleiter, R. 445 (23), *519*
Glendening, E. D. 296 (108), 304 (130), *343*
Glish, G. L. 324 (195), *345*
Glorius, F. 120 (308), *142*, 400 (101), *438*, 927 (251), 937 (274), *985*, *986*
Glovyak, T. 74 (163), *139*
Gobdout, C. 553 (96), *583*
Gockel, B. 870 (46), *879*
Godbout, C. 620 (52, 53), *685*
Goeden, G. V. 123 (320), *143*, 732 (5), *771*
Goel, A. B. 169 (106), *212*
Goering, H. L. 2 (21), 11 (68–70), *19*, *20*, 379 (48), *437*, 609 (26), 612 (34), 638 (8), 647 (9, 21), *683*, *684*, 872 (50), *879*
Gogoll, A. 493 (184), *523*
Goj, L. A. 121 (313–315), *143*, 403 (118), *439*
Gokhale, A. A. 158 (136), *162*
Goldberg, A. A. 888 (39, 43), 891 (38, 42, 44), *981*
Goldberg, D. P. 943 (295), *986*
Goldberg, I. 373 (31), *436*, 890 (69–71), *981*
Goldberg, N. L. 112 (272), *141*
Golden, J. E. 926 (245), *985*
Goldsmith, P. J. 675 (162), *690*
Golik, J. 238 (136), *244*
Gomer, C. J. 238 (119), *244*
Gómez, A. M. 613 (31), *684*
Gòmez Arrayàs, R. 748 (49), *772*
Gómez-Balderas, R. 306 (142, 143), *344*
Gómez-Bengoa, E. 831 (165), *855*, 902 (126), *983*
Gomez, F. A. 71 (154), *139*
Gómez-Gallego, M. 450 (42), *520*
Gomis, J. 609 (26), *684*
Gommermann, N. 404 (124), 423 (162), *439*, *440*, 444 (3), *519*
Gomperts, R. 283 (44), *341*
Gong, L. 471 (116), 480 (117), *521*
González, A. M. 800 (39), 802 (89), 811 (106, 107), 813 (38), 816 (30), 819 (134), 821 (142), 841 (144), 846 (9, 125), 847 (129, 181), *851–856*, 921 (203), *984*
González, B. 816 (30), *851*
Gonzalez, C. 283 (44), *342*
Gonzalez, D. 647 (116), *689*
Gonzalez-Gomez, J. C. 709 (45), *730*
González, M. A. 802 (89), *853*
Goodbrand, B. H. 919 (188), *984*

Goodfellow, R. 165 (56), *211*
Goodman, D. W. 157 (107), *161*
Goodwin, K. V. 220 (13), *241*
Gordon, P. F. 232 (94), *243*
Gordon, R. D. 820 (137), *854*
Gore, J. 574 (135, 136), *584*
Gore, V. 393 (87), *438*
Gore, V. K. 390 (82, 86), 393 (87), 399 (93), *438*, 635 (92), *687*
Gorelenko, Y. K. 252 (21), *277*
Gorelsky, S. I. 359 (63), *363*
Goren, A. 305 (64), *342*
Goreshnik, E. 252 (18, 28, 29), *277*
Goreshnik, E. A. 252 (10–17, 19–27, 30–32), 254 (33), *277*
Gorlier, J.-P. 16 (98), *21*, 609 (25), *684*
Görls, H. 105 (244), *141*, 152 (54), *160*
Görner, H. 238 (123), *244*
Gornitzka, H. 76 (76), *137*
Gosh, S. 359 (63), *363*
Gossage, R. A. 134 (338), *143*
Goswami, R. 396 (91), *438*
Gothelf, K. V. 956 (373), *988*
Gotoh, Y. 640 (101), *688*
Göttker-Schnetmann, I. 451 (43), *520*
Goubitz, K. 119 (119), 120 (120), *138*
Gouloumis, A. 221 (38), *242*
Gouteman, M. 220 (17), *241*
Gouterman, M. 219 (5), 220 (14–16, 29, 30), *241*, *242*
Govers, F. 723 (68), *730*
Grabowska, U. 414 (132), 416 (140), 418 (149), *440*
Grabowski, E. J. J. 173 (151), *213*
Graham, D. W. 361 (80), *363*
Granander, J. 166 (74), *211*
Granata, A. 280 (3), *341*
Grand, A. 301 (83), 305 (141), *342*, *344*
Grandini, E. 807 (98), *853*
Granger, M.-C. 620 (50), *685*
Granger, P. 165 (25), *210*
Granja, J. 380 (49), *437*, 647 (115), *689*
Grant, B. 607 (10), 609 (32), *683*, *684*
Gratzel, M. 229 (73), *243*
Grauel, H. 290 (79), *342*
Grauer, D. 484 (160), *522*
Gray, T. G. 121 (100), *137*, 402 (117), *439*
Grebel'skaya, M. I. 155 (98), *161*
Greck, C. 373 (33), *436*, 502 (215), *524*
Green, J. C. 119 (305), *142*
Green, M. L. H. 296 (109), *343*
Green, N. J. 842 (186), *856*
Greenberg, A. 310 (164), *344*
Greene, A. E. 380 (49), *437*, 647 (113), *688*
Greenfield, S. J. 418 (143, 145), *440*
Gregory, B. W. 147 (6), *159*
Gregory, C. D. 9 (60), *20*, 91 (213), *140*, 170 (127), 173 (141), *212*, *213*
Gregory, K. 166 (69), *211*
Gregory, P. 232 (94), *243*
Gremer, L. 360 (37, 73), *362*, *363*
Gretler, D. D. 911 (163), *983*
Grewell, C. L. 359 (72), *363*
Grice, P. 502 (215), *524*
Grierson, D. S. 973 (435), *989*
Griesinger, C. 196 (196), *214*
Griffith, J. R. 1013 (79), *1030*
Grimme, S. 769 (92), *773*
Grisdale, E. E. 894 (95), *982*
Grochulski, P. 861 (22), *878*
Grodkowski, J. 158 (137), *162*
Groen, M. B. 382 (54), *437*
Groenewold, G. 238 (136), *244*
Gronowitz, S. 452 (53), 456 (67), *520*
Gross, M. 221 (36), *242*
Gross, M. L. 154 (79), *160*, 326 (204, 205), *345*
Gross, T. 625 (66), *686*
Grove, D. M. 41 (95), 43 (98), 48 (107), 61 (13), 78 (173), 87 (203, 204), 94 (94), 117 (117), 134 (63), *135*, *137–140*, 276 (73), *278*, 607 (10), 609 (32), 669 (14, 142, 143), *683*, *684*, *689*, *690*, 858 (11), *876*
Grunwell, J. F. 861 (22), *878*
Gruver, J. M. 173 (150), *213*
Gschwind, R. M. 2 (28, 29), 3 (38), 18 (112–114), *19*, *21*, 112 (281), 133 (19, 222), *135*, *140*, *142*, 168 (91), 175 (129, 142), 179 (87), 181 (169), 183 (124), 196 (59, 63), 200 (19), 205 (231, 232), 208 (61), *210–213*, *215*, 858 (6, 7), *875*, *876*
Guan, J. 289 (66), *342*
Guascito, M. R. 235 (101), *243*
Guastini, C. 30 (55), 40 (88), 48 (56), *136*, *137*
Gueiffier, A. 921 (215), 925 (222), *984*, *985*
Güell, M. 280 (1), *340*
Guenther, H. 163 (4), *210*
Guighges, N. P. 228 (63), *242*
Guilard, R. 218 (2, 3), 219 (6, 7), *241*
Guillemin, J.-C. 401 (104), *439*
Guillen, F. 201 (223), *215*, 401 (104), *438*, 695 (6), *729*
Guion, J. 1015 (94), *1031*
Guir, C. 663 (85, 134), *687*, *689*, 874 (56), *879*
Guiry, P. J. 949 (338), 961 (339), *987*
Gujadhur, R. K. 906 (140), 966 (100), *982*, *983*
Guldi, D. M. 221 (35, 38), *242*
Gulevich, Y. 388 (81), *438*
Gund, V. G. 622 (55), *685*
Guneratne, R. D. 548 (77), *582*, 1010 (65, 66), *1030*
Gunnoe, T. B. 121 (313–315), *143*, 403 (118), *439*
Gunter, M. J. 165 (27), *210*
Günther, H. 166 (82), 178 (168), *212*, *213*

Guo, C.-Y. 9 (61), *20*
Guo, F. 476 (140), *522*
Guo, H.-C. 707 (36), *730*
Guo, J. 517 (260, 261), *525*, 974 (443), *989*
Guo, M. 966 (408), *989*
Guo, Q. X. 925 (225), 926 (238), 943 (297), *985*, *986*
Guo, T. 157 (115), *161*
Guo, X. 324 (196), *345*, 915 (177), *984*
Guo, Y. 330 (230), *346*
Guodong, Q, 237 (111), *244*
Gurevich, A. B. 148 (15), *159*
Gurrala, S. R. 588 (13), *602*
Gushchin, A. V. 946 (308, 310, 311), 947 (309), 949 (335), *986*, *987*
Guss, J. M. 34 (36), 134 (29), *136*, 350 (25), 352 (39, 41), *361*, *362*
Gust, D. 219 (7), *241*
Gut, S. A. 173 (148), *213*
Gutiérrez, M. I. 231 (88), *243*
Guttierez Perez, R. 663 (135), *689*
Guy, R. K. 953 (364), *988*

Håkansson, M. 157 (124, 125), *162*
Habashita, H. 616 (27), 625 (40, 42), 629 (68), *684–686*
Habu, H. 158 (133), *162*
Hackett, P. A. 153 (67), 157 (102), *160*, *161*
Hada, M. 283 (44), *341*
Hadziioannou, G. 221 (36), *242*
Hafner, T. 926 (244), *985*
Hage, T. R. 518 (89), *521*
Hagemann, H. 1016 (100), *1031*
Hagen, W. 366 (5), *435*
Hagiwara, T. 822 (150), *855*
Haifeng, C. 237 (111), *244*
Hajra, A. 18 (119), *21*, 478 (142), *522*
Hajra, S. 800 (86), *853*
Hakansson, M. J. 166 (83), *212*
Hakemian, A. S. 361 (81), *363*
Haldys, V. 315 (178), 332 (234), *345*, *346*
Hale, M. R. 381 (50), *437*
Hales, R. H. 975 (453), *989*
Halet, J.-F. 83 (189), *140*, 267 (58), *278*
Halkes, K. 380 (49), *437*, 647 (115), *689*
Hall, A. 926 (248), *985*
Hall, B. C. 83 (197), *140*
Hall, M. B. 324 (200), *345*
Hall, N. 892 (83, 84), *982*
Hallberg, A. 452 (53), 456 (67), *520*
Halliwell, B. 337 (246), *346*
Hallnemo, G. 173 (122), 184 (134), *212*, *213*
Halmann, M. 229 (73), *243*
Halow, I. 155 (3), *158*
Haltia, T. 352 (36), 358 (70), *362*, *363*
Hamada, S. 496 (193), *523*, 897 (110), *982*, 1003 (44), *1029*
Hamada, T. 445 (16), *519*
Hamada, Y. 800 (31), *851*
Hamanaka, N. 625 (42), *685*
Hamaue, N. 698 (13), *729*
Hamazawa, R. T. 117 (295), *142*
Hamelin, O. 380 (49), *437*
Hamilton, E. J. M. 71 (152), *139*
Hamilton, G. A. 354 (46), *362*
Hamilton, J. 225 (48), *242*
Hamilton, S. D. 888 (34), *981*
Hamon, L. 16 (98), *21*, 609 (25), 610 (33), *684*, 820 (16), *851*
Hampson, C. A. 157 (104), *161*
Hampton, J. A. 238 (121, 122), *244*
Han, C. 932 (260), *985*
Han, H. 63 (126), *138*
Han, K.-L. 152 (59), *160*
Han, X. 553 (95), *583*
Hanack, M. 1022 (131), *1031*
Handa, S. 430 (19), *436*
Handke, G. 866 (31), 867 (34), 868 (36, 38), 872 (42), *879*
Hanks, T. W. 178 (167), *213*, 367 (8), *435*
Hanna, J. V. 166 (60), *211*
Hannedouche, J. 761 (81), *773*
Hannon, F. J. 430 (21), *436*
Hansch, C. 11 (74), *20*, 266 (59), *278*
Hansen, R. C. 288 (70), *342*
Hansen, S. W. 995 (17–19), 996 (8), 997 (26), 1001 (2), *1028*, *1029*
Hanson, M. V. 447 (30), *519*
Hanson, R. M. 632 (45), *685*
Hanusa, T. P. 63 (134), *138*
Hanzawa, Y. 461 (94–97), *521*
Haque, T. S. 956 (371), *988*
Hara, R. 468 (105), *521*
Hara, S. 496 (198), *523*
Harada, K. 445 (19), *519*
Harada, M. 1008 (57), *1030*
Harada, T. 946 (316, 317), *987*
Harada, Y. 925 (218), *984*
Harder, S. 166 (77), *211*
Harding, M. 949 (333, 334), *987*
Hareau, G. 551 (92), *583*
Harford, C. 337 (244), *346*
Hargett, J. L. 938 (281), *986*
Hari, Y. 955 (368), *988*
Harima, Y. 229 (68), *243*
Harlow, R. L. 120 (309), *142*, 993 (14), *1029*
Harmon, H. J. 237 (116), *244*
Harms, A. C. 157 (116), *161*
Harms, K. 2 (29), 18 (115), *19*, *21*, 110 (265, 268), 133 (222), *140*, *141*, 175 (129), *212*, 641 (103), *688*, 800 (59), *852*, 858 (5, 6), *875*, *876*
Harnett, J. J. 979 (485), *990*
Harrimana, A. 230 (79), *243*
Harrington, E. M. 616 (43), *685*

Harrington-Frost, N. 484 (156), *522*, 616 (46), *685*
Harris, B. J. 800 (86), *853*
Harrison, A. G. 324 (198, 199), *345*
Harrison-Marchand, A. 166 (67), *211*
Hart, G. C. 386 (71), *437*
Hart, P. J. 348 (17), 349 (28), *361*, *362*
Harter, M. 620 (52), *685*
Hartgraves, G. A. 1015 (92), 1019 (119), 1022 (133, 134), *1030*, *1031*
Hartley, F. R. 164 (10), *210*
Hartmann, B. 380 (49), *437*, 579 (143), *584*
Hartog, T. D. 2 (24), *19*
Harutyunyan, S. R. 2 (24), 18 (118), *19*, *21*, 200 (216, 217), 202 (18), *210*, *214*, *215*, 493 (177), *523*, 670 (152), *690*, 698 (16–18), 719 (60), 723 (68, 69), *729*, *730*, 858 (10), 862 (25), *876*, *878*
Harvey, D. J. 336 (241), *346*
Harvey, R. G. 920 (191), *984*
Harwell, D. E. 73 (157), *139*
Harwig, C. 551 (91), *583*
Hasegawa, J. 283 (44), *341*
Hashimoto, H. 979 (480), *990*
Hashimoto, S. 800 (31), *851*
Hashimoto, T. 886 (22), *980*
Hashimoto, Y. 588 (14), *602*, 888 (45), *981*
Hashmi, A. S. K. 635 (77), *687*, 866 (29), *878*
Hasnain, S. 307 (146), *344*
Hassan, H. H. A. M. 866 (29), *878*
Hässig, R. 166 (73), *211*
Hassner, A. 445 (20), *519*
Haszeldine, R. N. 993 (15), 1005 (48), *1029*, *1030*
Hata, T. 445 (19), *519*, 937 (275), *986*
Hatanaka, N. 640 (98), *688*
Hathaway, B. J. 25 (32), *136*
Haubrich, A. 866 (31), *879*
Hauge, R. H. 150 (36), *159*
Haugen, H. K. 149 (28), *159*
Hauptman, E. 961 (388), *988*
Havlin, R. 73 (161), *139*
Hawner, C. 508 (227), *524*, 670 (149), *690*, 705 (32), *729*, 858 (11), *877*
Hawthorne, M. F. 68 (142, 143), 70 (146), 71 (147, 153–155), 73 (156, 157), *138*, *139*, 257 (38), *277*
Hayakashi, Y. 376 (39), *436*
Hayashi, M. 461 (94), *521*
Hayashi, T. 543 (65), *582*, 703 (29), *729*, 735 (14), *772*, 869 (39), *879*
Hayashi, Y. 271 (64), *278*
Hayes, R. N. 154 (79), *160*
Haynes, F. 888 (43), *981*
Hayward, M. M. 517 (260, 261), *525*
Hazes, B. 351 (33), *362*
He, G. 897 (103), *982*
He, M. 631 (79), *687*
He, M. Y. 962 (387), *988*
He, X. 49 (92), 50 (109), 118 (118), *137*, *138*, 152 (55, 65, 66), *160*, 174 (143), *213*
He, Y-B. 1014 (80), *1030*
He, Z. 283 (46), *342*
Heaney, H. 380 (1), *435*
Hebbe, V. 200 (210), *214*
Hebborn, P. 861 (22), *878*
Hecquet, E. 321 (155), *344*
Hedman, B. 114 (290), *142*, 943 (294), 964 (394), *986*, *988*
Hefter, G. 179 (171), *213*
Hegedus, L. S. 3 (32), *19*, 821 (141), *854*
Hegg, E. L. 238 (127), *244*
Heidenhain, S. B. 1024 (149), *1032*
Heijdenrijk, D. 35 (64), *137*
Heinecke, J. 328 (209), *345*
Heinemann, F. W. 66 (139), *138*
Heinis, D. 430 (18), *435*
Heinze, K. 220 (23), *242*
Heinze, P. L. 1004 (47), *1030*
Held, G. 157 (108), *161*
Hellwinkel, D. 889 (54), *981*
Helmchen, G. 647 (111), 649 (118), *688*, *689*, 858 (13), *877*
Helquist, P. 502 (216), *524*
Henderson, B. W. 238 (119), *244*
Henderson, K. W. 177 (161), *213*
Henderson, R. M. 60 (121), *138*, 1024 (141), *1032*
Henderson, S. G. D. 70 (150), *139*
Hennessy, E. J. 969 (429), *989*
Henning, R. 800 (45, 46), *852*
Henon, H. 863 (26), *878*
Henrick, C. A. 13 (81), *20*
Henze, W. 3 (38), *19*, 112 (281), *142*, 181 (169), 183 (124), 186 (132), 196 (59, 63), *211–213*, 858 (6, 7), *876*
Henzen, R. 925 (234), *985*
Herber, C. 622 (38, 57), *685*, *686*
Hermitage, S. A. 911 (159), *983*
Hernández, A. E. 430 (18), *435*, *436*
Hernandez-Alonso, M. D. 235 (101), *243*
Hernandez, R. M. 158 (135), *162*
Herold, M. 474 (136), *522*
Herr, R. W. 625 (62), *686*
Herradura, P. S. 953 (364), *988*
Herranz, M. A. 221 (35), *242*
Herrera-Urbina, R. 296 (102), *343*
Herrero, M. 845 (191), *856*
Herres, M. 41 (95), 78 (173), 94 (94), *137*, *139*, 276 (73), *278*
Herrmann, R. G. 349 (28), *362*
Hersan, M. C. 228 (63), *242*
Hertwig, R. H. 290 (79), 308 (39, 90), *341–343*
Hettich, R. L. 339 (251), *346*
Hevesi, L. 906 (141), *983*

Hibino, J. I. 841 (170), *855*
Hida, M. 888 (45, 46, 49), 891 (37, 47, 48, 50), *981*
Hiddessen, A. L. 225 (48), *242*
Hiemstra, H. 399 (96), 413 (130), 414 (132), *438–440*
Higashi, Y. 824 (69), *852*
Higgins, D. 381 (50), 382 (55), *437*, 613 (37), *685*, 800 (75, 80), *853*
Higgs, T. C. 87 (205–207), *140*
Highsmith, T. K. 516 (252), *524*
Higuchi, K. 457 (71), *521*
Hihara, T. 230 (85), *243*
Hilinski, E. F. 220 (21), *241*
Hill, D. T. 197 (205), *214*
Hill, J. H. M. 800 (67), *852*
Hill, M. S. 37 (75), 110 (269), 111 (266), *137*, *141*
Hill, Y. D. 296 (107), *343*
Hillier, I. 307 (146), *344*
Hilmersson, G. 166 (74, 83), *211*, *212*, 800 (54), *852*
Himeno, K. 230 (81), *243*
Hintermann, T. 953 (359), *988*
Hirahara, M. 157 (112), *161*
Hirai, A. 3 (31), *19*, 585 (6), *601*, 800 (55), *852*
Hiramatsu, K. 484 (165), *522*
Hirano, M. 285 (13), *341*, 911 (161), *983*
Hirayama, T. 60 (122), *138*
Hird, A. W. 478 (147), 480 (148), *522*, 695 (8), 702 (23), 709 (27), *729*
Hirose, M. 230 (83, 84), *243*
Hirotsu, M. 117 (295), *142*
Hirsch, L. K. 860 (20), *878*
Hirst, P. R. 646 (110), *688*
Hirva, P. 309 (101), *343*
Hisada, H. 238 (129), *244*
Hisaeda, Y. 117 (296), *142*
Hitchcock, P. B. 37 (75, 77), 102 (234), 106 (245), 110 (269), 111 (266), *137*, *141*, 178 (167), *213*, 261 (49, 50), *277*, 368 (10), *435*
Hitotsuyanagi, Y. 953 (361), *988*
Hiyama, T. 447 (27), *519*, 738 (23), *772*, 978 (468), *990*
Hjelmencrantz, A. 459 (93), *521*
Hjelmgaard, T. 413 (131), *439*, 474 (128), *522*
Ho, J. 152 (60), *160*
Ho, L.-K. 654 (127), *689*
Ho, P.-S. 976 (446, 458), *989*, *990*
Ho, W. 157 (106), *161*
Hoang, A. 557 (106), *583*
Hobza, P. 282 (21), 309 (20), *341*
Hodgson, K. O. 114 (290), *142*, 943 (294), 964 (394), *986*, *988*
Hodson, B. E. 70 (151), *139*
Hoffmann, R. 3 (41), *19*, 129 (129), *138*
Hoffmann, R. W. 426 (165), *440*
Hoffmann-Röder, A. 629 (73), 632 (86), 635 (77), *687*, 695 (1), *728*, 858 (12), 866 (29), 869 (40), 870 (14, 45, 46), *877–879*
Hofman, P. 310 (168), *344*
Hoge, B. 459 (85), *521*, 1022 (135), *1031*
Hohmann, M. 867 (30), *879*
Hol, W. G. 351 (33), *362*
Holbrook, M. T. 148 (15), 150 (43), *159*, *160*
Holm, R. H. 348 (8, 9), *361*
Holm, T. 153 (76), *160*
Holmes, T. F. 1017 (107), *1031*
Holmquist, G. P. 337 (247), *346*
Holoboski, M. A. 842 (186), *856*
Holst, C. L. 979 (479), *990*
Holten, D. 220 (14, 30), *241*, *242*
Holthausen, M. C. 152 (52), *160*, 309 (42), *341*
Holubova, J. 80 (177), *139*
Holze, R. 273 (46), 274 (45), *277*
Honaker, M. T. 938 (281), *986*
Honda, K. 230 (76), *243*
Honda, Y. 283 (44), *341*
Hong, C. Y. 800 (82), *853*
Hong, X. Y. 595 (30, 31), *602*
Honwad, V. K. 73 (161), *139*
Hope, H. 106 (242, 248), *141*, 152 (53, 64), *160*
Hopf, H. 863 (27), *878*
Hopkins, A. D. 39 (79), *137*
Hopkins, H. F. 894 (95), *982*
Hopkinson, A. C. 304 (131), 320 (77), 329 (212, 213, 216), *342*, *343*, *345*, *346*
Hoppilliard, Y. 319 (183), 320 (182), 321 (155), *344*, *345*
Horiguchi, S. 158 (133), *162*
Horiguchi, Y. 16 (92), *20*, 516 (249, 250), *524*
Horikawa, H. 661 (133), *689*
Horiuchi, Y. 979 (478), *990*
Horn, E. 669 (142), *689*
Hornauer, S. 98 (225), 102 (234), 111 (271), *140*, *141*, 178 (167, 168), *213*, 368 (10), *435*
Horváth, A. 386 (65), *437*, 635 (91), *687*
Hoshi, M. 974 (438–440), *989*
Hoshino, K. 225 (47), *242*
Hosomi, A. 496 (201), 513 (240, 242), *523*, *524*, 738 (21, 22), *772*, 800 (42), *852*, 977 (460, 464), *990*
Hosono, M. 238 (129), *244*
Hosseinzadeh, R. 925 (232), *985*
Hotop, H. 149 (27), *159*
Hotta, Y. 629 (68), *686*
Hou, Y. 631 (76), *687*
Hou, Z. 403 (116), *439*
Houde, D. 223 (42), 238 (124, 125), *242*, *244*
Hough, M. 307 (146), *344*
Houpis, I. N. 834 (169), *855*, 915 (164), *983*
House, H. O. 2 (19), 16 (16), *19*, 24 (8, 9), 91 (212), 112 (273, 275), *135*, *140*, *142*, 173

(140), *213*, 459 (92), *521*, 589 (15), *602*, 728 (75), *730*, 778 (2), *851*
Houte, H. 166 (67), *211*
Hoveyda, A. H. 132 (336), 133 (337), *143*, 381 (50), 402 (108, 109), 430 (27), *436*, *437*, *439*, 478 (147), 480 (146, 148), 484 (166), 508 (150, 151, 224), 509 (228), 513 (243), 516 (259), 517 (262), *522–525*, 670 (153), 671 (126), 672 (154, 155, 157), 674 (161), *689*, *690*, 695 (8), 701 (22), 702 (23), 703 (26), 705 (33), 707 (43), 709 (27), 710 (48, 49), 713 (44), 716 (58), 720 (42), 726 (72), *729*, *730*, 859 (17), *877*
Howard, J. A. 153 (74), 157 (104), *160*, *161*, 632 (84), *687*
Howard, J. A. K. 66 (140, 141), *138*
Howarth, J. A. 1004 (46), *1029*
Howe, D. V. 68 (143), *138*
Howell, G. P. 720 (66), *730*
Howthorne, M. F. 68 (144), *138*
Hoyau, S. 301 (30), *341*
Hrusak, J. 308 (90), *343*
Hsi, J. D. 625 (67), *686*
Hsu, J. 1015 (92), 1019 (119), *1030*, *1031*
Hsu, W. Y. 295 (52), *342*
Hsung, R. P. 933 (263, 265, 266), 934 (264), *985*, *986*
Hu, C. H. 295 (52), 318 (176), *342*, *345*
Hu, H. 112 (278), *142*, 196 (199), *214*
Hu, N.-X. 919 (188), *984*
Hu, P. 326 (204, 205), *345*
Hu, T. S. 934 (261), *985*
Hu, X. 126 (324, 325), *143*, 430 (25), *436*
Hu, X.-C. 975 (405), *989*
Hua, Z. 489 (169), *523*
Huang, C. 940 (284), *986*
Huang, H. 108 (258, 260), *141*, 175 (154, 155), *213*, 300 (88), *343*, 800 (54, 56, 58), *852*
Huang, J. 933 (263), 934 (264), *985*, *986*
Huang, J.-S. 732 (7), *771*
Huang, S. D. 913 (168), *983*
Huang, X. 564 (118), 571 (130), *583*, *584*, 807 (96), *853*, 925 (227), 926 (221), *985*
Huang, Y. 318 (175), *345*
Huang, Y. C. 295 (52), *342*
Huang, Y.-Z. 922 (208), *984*
Hubbell, J. A. 225 (48), *242*
Huber, H. 153 (70, 71), 157 (103), *160*, *161*
Huber-Patz, U. 646 (110), *688*
Huber, R. 348 (7), 350 (26), 352 (34, 41), 354 (45), 355 (50), 356 (53, 58), 360 (37, 73), *361–363*
Huch, V. 820 (138), *854*
Huche, M. 430 (172), *441*
Hudlicky, M. 1015 (90), 1016 (98, 99), *1030*, *1031*
Hudlicky, T. 647 (116), *689*
Hudson, A. R. 925 (224), *985*
Huerta, F. F. 669 (144), *690*
Huffman, J. C. 123 (320), *143*, 165 (48), 173 (151), *211*, *213*, 238 (135), *244*, 732 (4, 5), *771*
Huffman, L. M. 203 (203), *214*
Huffmann, J. C. 87 (87), *137*
Hughes, G. 742 (32), *772*
Hulce, M. 867 (28), 868 (37), 870 (14), *877–879*
Humam, M. 203 (230), *215*, 695 (7), *729*
Hummel, G. 418 (144), *440*
Humora, M. J. 943 (301), *986*
Humphrey, J. M. 386 (68), *437*
Hung, M-H. 1015 (93), *1031*
Hungate, R. 921 (209), *984*
Hunter, C. 417 (141), 418 (150), *440*, 474 (129), *522*
Hunter, E. P. 147 (9), *159*
Hupe, E. 421 (152), *440*, 641 (103), *688*
Hursthouse, M. B. 127 (326), *143*
Hurtley, W. R. H. 968 (417), *989*
Huseland, D. E. 733 (8), *771*
Hussain, M. S. 76 (168, 169), *139*
Hutchison, J. M. 622 (61), *686*
Hütten, D. 113 (286), *142*
Huttner, G. 41 (96), 78 (171), *137*, *139*, 259 (44), *277*
Hutzinger, M. W. 834 (132), 836 (173), 842 (176), *854*, *855*
Hwang, C.-S. 106 (233), 109 (267), 110 (270), 118 (118), *138*, *141*, 858 (5), *875*
Hyla-Kryspin, I. 445 (23), *519*
Hyodo, T. 616 (47), *685*
Håkansson, M. 28 (44, 45), 29 (51), 30 (54), 58 (48), 61 (125), *136*, *138*

Iamamoto, Y. 237 (115), *244*
Ibuka, T. 276 (71), *278*, 614 (17, 41), 616 (27), 625 (16, 40, 42, 64), 629 (68), 635 (88, 90), 672 (1), *682*, *684–687*
Ichikawa, J. 496 (193–195), *523*, 1003 (42–45), *1029*
Ichikawa, M. 228 (59), 230 (80), *242*, *243*
Ichikawa, T. 376 (39), *436*, 728 (75), *730*
Ichimura, A. 116 (293, 294), *142*
Ichino, T. 222 (41), *242*
Igarashi, D. 756 (72), *773*
Iglesias, B. 845 (192), *856*
Iida, H. 758 (73), *773*
Iida, R. 13 (82), *20*, 133 (21), *135*
Iida, T. 968 (423), *989*
IIjima, M. 430 (174), 433 (173), *441*
Iimura, K. 461 (94), *521*
Ikeda, S. 81 (81), *137*
Ikegashira, K. 978 (468), *990*
Ikeura, C. 1003 (45), *1029*

Ila, H. 424 (123), *439*
Iliefski, T. 459 (91), *521*
Iliev, V. 237 (112), *244*
Illingworth, M. 993 (15), *1029*
Imai, H. 386 (59), *437*
Imamoto, T. 430 (15, 172), *435*, *441*
Imanieh, H. 834 (167), *855*
Imbos, R. 203 (228), *215*, 401 (105), *439*, 474 (133), *522*, 694 (5), 695 (1), *729*, 858 (12), *877*
Imoto, S. 964 (392), *988*
Inagaki, A. 133 (277), *142*
Inayama, T. 225 (47), *242*
Inderwildi, O. R. 157 (108), *161*
Inger, J. B. 765 (87), *773*
Inokuchi, E. 641 (105), *688*
Inoue, A. 946 (316, 317), *987*
Inoue, K. 869 (39), *879*
Inoue, M. 457 (72), *520*
Inoue, S. 220 (31), *242*
Inouye, M. 940 (285), *986*
Inouye, Y. 1022 (132), *1031*
Iodice, A. 629 (72), *686*
Irangu, J. K. 165 (29), *210*
Ireland, R. E. 503 (212), 516 (251–253), *523*, *524*
Irngartinger, H. 646 (110), *688*
Isago, H. 218 (4), *241*
Iseki, Y. 220 (31), *242*
Ishida, M. 283 (44), *341*
Ishida, N. 800 (40), *852*
Ishida, T. 462 (104), *521*
Ishihara, T. 549 (78, 79), 571 (131), *583*, *584*, 1009 (60), *1030*
Ishihara, Y. 859 (17), *877*
Ishikawa, H. 953 (361), *988*
Ishikawa, Y. 117 (296, 297), 119 (300), *142*
Ishikura, M. 386 (66), *437*
Ishino, Y. 925 (190), *984*
Ishiyama, T. 496 (199, 200), *523*
Ishizuka, I. 738 (21), *772*
Ishizuka, T. 119 (300), *142*, 513 (242), *523*, 738 (22), *772*, 800 (42), *852*
Ishmael, F. T. 979 (477), *990*
Isobe, K. 116 (292), *142*
Isobe, M. 330 (226), *346*, 418 (146), *440*
Israel, G. 232 (93), *243*
Issenhuth, J. T. 754 (65), *773*
Itai, A. 911 (161), *983*
Itami, K. 572 (133, 134), *584*
Ito, H. 461 (100, 101), 496 (201), 499 (204–207), 513 (240, 242), *521*, *523*, *524*, 659 (130), *689*, 738 (21, 22), *772*, 800 (42), *852*, 977 (460, 464), *990*
Ito, K. 885 (17), *980*
Ito, M. 644 (107), *688*
Ito, N. 351 (32), 354 (44), *362*
Ito, O. 221 (37), *242*
Ito, S. 499 (205), *523*
Ito, Y. 150 (42), *160*, 382 (54), 386 (59), *437*, 640 (99), *688*, 800 (37), 802 (48), *851*, *852*
Itoh, M. 496 (185–187), *523*
Itoh, T. 238 (129), *244*
Iuliano, M. 157 (129), *162*
Ivanov, S. V. 73 (158), *139*
Iwao, M. 378 (44), *436*, 599 (34), *602*
Iwata, C. 698 (13), *729*
Iyengar, S. S. 283 (44), *341*
Iyod, J. 157 (111), *161*
Iyoda, M. 595 (22, 28), *602*

Jaber, M. R. 925 (230), *985*
Jabri, N. 530 (24), 537 (46), *581*, *582*
Jackson, D. A. 114 (290), *142*, 964 (394), *988*
Jackson, R. F. W. 416 (133, 134, 139, 140), 417 (141), 418 (142, 148–150), *440*, 474 (129), *522*
Jacobsen, M. F. 956 (373), *988*
Jacquesy, J.-C. 861 (19), *878*
Jacquesy, R. 861 (19), *878*
Jadhav, P. K. 952 (354), *988*
Jaén, J. C. 640 (100), *688*
Jagdmann, G. E. 386 (67), *437*
Jagner, S. 28 (44, 45), 29 (51), 61 (125), *136*, *138*, 157 (124, 125), *162*
Jahn, H. 452 (47, 48), *520*
Jaime, C. 859 (16), *877*
Jairaj, V. 999 (36), *1029*
Jäkle, F. 30 (52), 53 (53), 60 (60), 93 (93), *136*, *137*
Jakob, A. 275 (70), *278*
Jakob, W. 359 (60), *363*
James, B. 16 (93), *20*, 800 (66), *852*
James, D. S. 949 (334), *987*
James, P. F. 149 (25), *159*
James, S. L. 2 (14), *19*, 34 (34), *136*, 272 (43), *277*, 858 (5), *875*
Jana, R. 904 (139), *983*
Jansen, A. 629 (73), *687*
Janssen, M. D. 41 (95), 43 (98), 78 (173), 87 (204), 94 (94), 134 (63), *137*, *139*, *140*, 276 (73), *278*, 669 (142), *689*
Jansson, J. 926 (239), *985*
Janza, B. 18 (115), *21*
Jaramillo, J. 283 (44), *341*
Jarvis, J. A. J. 27 (42), *136*
Jarvis, L. M. 294 (92), *343*
Jäschke, G. 201 (220), *215*
Jastrzebski, J. T. B. 800 (59), *852*
Jastrzebski, J. T. B. H. 2 (14, 15), *19*, 34 (34), 37 (78), 39 (39), 52 (43), 56 (116), 93 (50), 94 (216), 95 (217, 220), 97 (224), 101 (231, 232), 108 (264), 113 (113), 114 (114), 134 (49, 338), *136–138*, *140*, *141*, *143*, 173 (126), 175 (156, 157), *212*, *213*, 858 (5), *875*

Jay-Gerin, J. P. 223 (42), *242*
Jayne, C. L. 979 (488), *990*
Jeanneaux, F. 992 (9), *1029*
Jenkins, I. L. 891 (51), *981*
Jenks, C. J. 148 (13, 14, 17, 18, 20), *159*
Jensen, A. E. 421 (151), 422 (156), 425 (154), *440*, 600 (19), *602*
Jensen, L. T. 330 (224), *346*
Jeong, I. H. 995 (22), *1029*
Jeong, N. 383 (38), *436*
Jeong, Y. T. 995 (20, 22), *1029*
Jeoung, J.-H. 360 (74), *363*
Jeoung, S. C. 220 (26), *242*
Jerphagnon, T. 474 (131), *522*
Jerris, P. J. 15 (87), *20*
Jerschow, A. 204 (100), *212*
Jesser, F. 430 (15), *435*
Ji, J.-X. 754 (66), *773*
Ji, S.-J. 700 (20, 21), *729*, 974 (443), *989*
Jiang, C. 728 (74), *730*
Jiang, D. 919 (181), *984*
Jiang, D. S. 919 (184), *984*
Jiang, J. 418 (144), *440*
Jiang, J. L. 908 (155), *983*
Jiang, L. 932 (258), 940 (279), *985*, *986*
Jiang, Q. 919 (181), *984*
Jiang, T. 979 (476), *990*
Jiang, X.-Z. 922 (210), *984*
Jiang, Y. 294 (58), *342*, 915 (177), 919 (181), 920 (195), 926 (237), 940 (107, 284), *982*, *984–986*
Jiang, Y. Y. 919 (184), *984*
Jiao, J. 920 (196), *984*
Jiao, L. 224 (45), *242*
Jiao, X. 976 (457), *990*
Jie, J. L. 908 (152), *983*
Jin, A. J. 330 (229), *346*
Jin, C. 157 (115), *161*
Jin, Y. 940 (107), *982*
Job, G. E. 901 (111), 915 (172), 932 (258), *982*, *984*, *985*
Jobran, L. 418 (144), *440*
Jobson, R. B. 947 (320), *987*
Joensuu, P. M. 763 (85), *773*
Johansson, A. 16 (95), *21*
John, M. 2 (29), *19*, 133 (222), *140*, 175 (129, 142), *212*, *213*, 858 (6), *876*
Johnels, D. 163 (4), *210*
Johnson, A. A. 63 (126), *138*
Johnson, A. J. 979 (479), *990*
Johnson, B. 283 (44), *342*
Johnson, C. R. 9 (59), *20*, 625 (62), *686*, 824 (152), *855*
Johnson, D. K. 979 (477), *990*
Johnson, J. S. 375 (37), *436*, 493 (178, 179, 181), *523*, *524*
Johnson, R. A. 632 (45), *685*
Johnson, S. E. 71 (154), *139*
Johnson, T. A. 386 (69), *437*
Johnston, J. O. 861 (22), *878*
Joly, H. A. 153 (74), *160*
Jonasson, C. 979 (486), *990*
Jones, A. B. 502 (215), *524*
Jones, A. C. 87 (206, 207), *140*, 220 (25), *242*, 892 (85, 86), *982*
Jones, C. D. 901 (118), *982*
Jones, C. P. 927 (253), *985*
Jones, C. W. 940 (287), *986*, 1020 (124), *1031*
Jones, G. C. 147 (6), *159*
Jones, G. D. 622 (61), *686*
Jones, K. L. 926 (248), *985*
Jones, P. 858 (8), *876*
Jones, R. A. 103 (236), *141*
Jones, R. G. 11 (11), *19*, 24 (4), *135*, 585 (2), *601*, 776 (1), *850*, 857 (2), *875*
Jones, T. R. 1018 (112), *1031*
Jong, W. A. d. 304 (130), *343*
Jordan, R. B. 165 (29), *210*
Jordan, R. F. 41 (89), *137*
Jørgensen, K. A. 200 (14), *210*
Jori, G. 238 (119), *244*
Joullie, M. M. 418 (144), *440*
Jousseaume, B. 551 (83, 93), 553 (82), *583*, 832 (78), *853*
Joyce, L. L. 908 (156), *983*
Juan, C. 510 (239), *524*
Jubert, C. 383 (58), *437*
Judai, K. 285 (13), *341*
Jukes, A. E. 1024 (138, 139), *1031*
Jun, L. 235 (104), *243*
Jung, J. H. 995 (22), *1029*
Jung, M. E. 953 (358), *988*
Jurkauskas, J. 740 (28), *772*
Jurkauskas, V. 403 (113), *439*, 742 (31), 744 (37), *772*
Jutand, A. 200 (213), *214*, 943 (296), *986*
Jutzi, P. 66 (138), *138*

Kabbara, J. 15 (86), *20*, 503 (219–222), *524*, 654 (127), *689*
Kacprzynski, M. A. 402 (108), *439*, 484 (166), *523*, 671 (126), *689*, 695 (8), 710 (48), *729*, *730*
Kadib-Elban, A. 859 (16), *877*
Kadish, K. M. 218 (2, 3), 219 (6, 7), *241*
Kado, N. 800 (82), *853*
Kadota, I. 979 (483), *990*
Kadow, J. F. 824 (152), *855*
Kafafi, Z. H. 150 (36), *159*
Kaffashan, A. 334 (238), *346*
Kagan, H. B. 374 (36), *436*
Kagi, J. H. R. 330 (231), *346*
Kahlal, S. 83 (189), *140*, 267 (58), *278*
Kahne, D. 173 (148), *213*
Kahya, D. 377 (35), *436*

Kaim, W. 348 (3), *361*
Kaito, R. 228 (65), *243*
Kaizu, Y. 222 (40, 41), *242*
Kajiro, H. 738 (23), *772*
Kalaji, M. 158 (135), *162*
Kalinin, A. V. 894 (97), *982*
Kalinin, V. N. 947 (329), *987*
Kalish, V. J. 453 (61), *520*
Kallmerten, J. 445 (14), *519*
Kamada, K. 595 (28), *602*
Kambe, N. 556 (105), *583*, 738 (20), *772*, 979 (482), *990*
Kamei, T. 572 (133, 134), *584*
Kametani, T. 885 (16), *980*
Kamikubo, T. 824 (69), *852*
Kamp, U. 165 (50), *211*
Kamte, M. F. 254 (34–36), *277*
Kanai, M. 200 (13), *210*, 732 (1), 762 (83), 763 (80, 84), *771*, *773*
Kanazawa, A. M. 380 (49), *437*
Kaneko, K. 894 (98), *982*
Kaneko, Y. 230 (86, 87), *243*
Kanemura, S. 569 (128), *584*
Kang, H. 196 (196), *214*
Kang, H. C. 70 (146), 71 (147, 153), *138*, *139*
Kang, J. 445 (12), *519*
Kang, S.-C. 976 (458), *990*
Kang, S.-K. 632 (83), *687*, 873 (52), *879*, 944 (304), 966 (409), 976 (446, 454, 455, 457), 977 (461), *986*, *989*, *990*
Kano, K. 221 (34), *242*
Kanoh, M. 434 (175), *441*
Kańska, I. 119 (301), *142*
Kantam, M. L. 921 (207), *984*
Kanters, J. A. 166 (77), *211*
Kantner, S. 2 (21), *19*
Kantner, S. S. 11 (70), *20*, 379 (48), *437*, 647 (9), *683*
Kanupp, L. 682 (174), *691*
Kapteijn, G. M. 117 (117), *138*
Kar, A. 675 (2), *683*
Karakatsani, S. 157 (108), *161*
Karatas, B. 769 (92), *773*
Kargin, Yu. M. 249 (8), 250 (4), *276*, *277*
Karim, A. 891 (72), 927 (74), *981*
Karlin, K. D. 94 (216), *140*
Karlsson, S. 452 (53), 456 (67), *520*
Karlstroem, A. S. E. 979 (484), *990*
Karlström, A. S. E. 11 (71), *20*, 480 (153), *522*, 607 (10), 669 (143, 144), 672 (1), *682*, *683*, *690*, 816 (120), 849 (104), *854*, 858 (11), *876*, *877*
Karplus, M. 173 (147), *213*
Karpova, Yu. G. 158 (131), *162*
Karpyak, V. 252 (29), *277*
Karras, M. 445 (22), *519*
Karstens, W. F. J. 413 (130), 414 (132), *439*, *440*
Karzazi, M. 240 (142), *245*
Kasahara, K-i. 998 (33), *1029*
Kasai, K. 462 (103), *521*
Kasai, P. H. 153 (72), *160*
Kase, K. 401 (107), *439*
Kash, P. W. 150 (43), *160*
Katagiri, F. 625 (42), *685*
Katayama, H. 666 (141), *689*
Katayama, Y. 616 (47), *685*
Kato, K. 457 (74), *520*
Kato, N. 559 (109), *583*
Kato, S. 11 (73), *20*, 198 (174), *214*, 459 (88), *521*, 606 (11), *683*
Kato, T. 230 (83, 84), *243*, 401 (107), *439*, 672 (156), *690*
Katz, J. L. 953 (359), 962 (349), *987*, *988*
Katzenellenbogen, J. A. 861 (22), *878*
Kauer Zinn, F. 750 (53, 54), *773*
Kauffman, T. 585 (3), *601*
Kaufman, M. J. 173 (149), *213*
Kaur, H. 121 (311, 312), *142*, *143*, 403 (111), *439*, 750 (53, 54), *773*
Kavazoudi, X. 960 (384), *988*
Kawabata, J. 167 (89), *212*
Kawabata, N. 474 (125), *522*
Kawachi, A. 332 (54), *437*, 640 (99), *688*, 800 (37), 802 (43), *851*, *852*
Kawaguchi, H. 238 (136), *244*
Kawaguchi, K. 908 (151), *983*
Kawai, H. 309 (116), *343*
Kawakami, C. 499 (204), *523*
Kawakami, Y. 979 (474), *990*
Kawamura, N. 974 (438), *989*
Kawasaki, K. 735 (13), *772*
Kawasaki, M. 958 (381), *988*
Kawashima, T. 288 (71), *342*
Kawasuji, T. 15 (84), *20*, 659 (122), *689*
Kawata, R. 1008 (56), *1030*
Kaya, K. 285 (13), *341*
Kayser, F. 587 (12), 592 (23, 24), *602*
Kazane, S. A. 402 (108), *439*, 484 (166), *523*, 710 (48), *730*
Ke, Y. 329 (216), *346*
Kealy, T. J. 63 (127), *138*
Keay, B. A. 829 (157), 845 (196), *855*, *856*
Keck, R. W. 238 (121), *244*
Keegstra, M. A. 899 (117), *982*
Keen, A. L. 63 (126), *138*
Keen, K. P. 351 (32), *362*
Keenan, G. A. 220 (25), *242*
Kehrli, S. 401 (106), *439*, 480 (149), *522*, 707 (34), 725 (31), *729*
Keil, R. 456 (62), 459 (78), *520*, 858 (8), *876*
Keiper, S. 975 (452), *989*
Keith, J. 179 (131), *213*, 738 (24), *772*, 800 (56), *852*
Keith, J. M. 738 (24), *772*
Keith, T. 283 (44), *342*

Kelkar, A. A. 920 (189), *984*
Keller, E. 707 (39), *730*
Keller, M. 622 (38), *685*
Kellogg, R. M. 894 (94), *982*
Kelly, J. W. 952 (357), *988*
Kelly, M. G. 646 (108), *688*
Kelly, W. 888 (43), 891 (38, 42), *981*
Kendall, C. 131 (335), *143*, 452 (51), *520*
Kende, A. S. 553 (100), *583*
Kennedy, A. R. 177 (161), *213*, 430 (19), *436*
Kennedy, R. J. 165 (46), *211*
Kennepohl, P. 348 (8), *361*
Kerb, U. 861 (19), *878*
Kerpel, J. O. D. 310 (10), *341*
Kerr, W. J. 177 (161), *213*
Kerrick, S. T. 391 (89), *438*
Kerssebaum, R. 189 (195), 196 (196), *214*
Kessel, D. 238 (119, 120), *244*
Kessler, H. 168 (92), *212*
Kesteleyn, B. 1017 (109), *1031*
Keszei, E. 223 (42), *242*
Kettle, S. F. A. 157 (125), *162*
Khairallah, G. N. 152 (23), *159*
Khamsi, J. 947 (314, 328), 949 (322, 332), 950 (337), 961 (390), *987*, *988*
Khan, F. A. 151 (50), *160*, 294 (97), *343*
Khan, S. I. 106 (227), *141*, 152 (63), *160*
Kharasch, M. S. 2 (10), *19*, 24 (6), *135*, 694 (2), *729*
Khorshev, S. Ya. 90 (209), *140*
Khouw, C. 324 (198), *345*
Khromovskii, V. A. 992 (7), *1029*
Kickel, B. L. 310 (167), *344*
Kidwai, M. 894 (99), *982*
Kiefersauer, R. 360 (37), *362*
Kienle, M. 169 (117, 118), *212*, 378 (46), *436*, 586 (7), 589 (16), 600 (38), *601*, *602*
Kieslich, K. 861 (19), *878*
Kihara, T. 1011 (71, 73), *1030*
Kikuchi, O. 177 (163), *213*
Kilburn, J. D. 800 (73), *852*
Kim, B.-E. 348 (18), *361*
Kim, B. F. 220 (18), *241*
Kim, C. U. 859 (16), *877*
Kim, D. 131 (334), *143*, 220 (14, 26), *241*, *242*, 403 (111), *439*, 747 (45, 46), 751 (58), *772*, *773*
Kim, D. D. 386 (69), *437*
Kim, D.-Y. 873 (52), *879*
Kim, H. J. 361 (80), *363*
Kim, J.-S. 976 (454, 455), *989*, *990*
Kim, K. D. 845 (174), *855*
Kim, K.-Y. 926 (246), *985*
Kim, N.-S. 672 (159), *690*
Kim, S.-G. 632 (83), *687*
Kim, S.-H. 447 (29, 30), *519*
Kim, T. 240 (141), *245*
Kim, T.-H. 976 (446), 977 (461), *989*, *990*
Kim, W. 240 (141), *245*
Kim, W.-Y. 976 (457), *990*
Kim, Y. 842 (183), *856*
Kim, Y. G. 995 (22), *1029*
Kim, Y.-M. 966 (409), *989*
Kimber, M. C. 560 (110), *583*
Kimura, K. 116 (292), *142*
Kimura, M. 616 (48), *685*, 742 (32), *772*
Kimura, T. 543 (65), *582*
King, S. M. 459 (82, 83), *521*
Kingsbury, J. S. 132 (336), *143*
Kini, A. M. 154 (85), *161*
Kinney, R. E. 897 (9), *980*
Kinoshita, I. 116 (292–294), 117 (295), *142*
Kinoshita, T. 625 (42), *685*
Kirchhoff, J. R. 220 (13), *241*
Kiren, S. 631 (76), *687*
Kirihatali, M. 970 (432), *989*
Kirkwood, D. A. 288 (62), *342*
Kirmeier, F. 861 (22), *878*
Kirova-Snover, M. 516 (255), *524*
Kiselgof, J. Y. 553 (99), *583*
Kishi, Y. 517 (260, 261), *525*
Kitada, K. 614 (41), *685*
Kitagawa, O. 1007 (54), 1008 (55, 58), *1030*
Kitagawa, S. 165 (33, 54), *210*, *211*
Kitagawa, T. 222 (39), *242*
Kitamura, M. 18 (110, 111), *21*, 430 (26), *436*, 707 (37, 38), *730*
Kitao, O. 283 (44), *341*
Kitao, T. 968 (423, 426, 427), *989*
Kitaura, K. 641 (105), *688*
Kitazume, T. 616 (48), 644 (107), *685*, *688*
Kitching, W. 831 (84), *853*
Kiyomori, A. 921 (202), *984*
Kiyotsuka, Y. 616 (47), 644 (107), *685*, *688*
Klapars, A. 913 (165), 922 (216), 925 (227), 926 (221, 247), 927 (254), 932 (258), 942 (289), 967 (412), *983–986*, *989*
Klapper, M. H. 354 (47), *362*
Klassen, R. B. 430 (22), *436*
Klecka, M. 407 (127), *439*
Kleemann, L. 18 (115), *21*
Kleijn, H. 540 (55), 555 (104), 571 (132), *582–584*, 629 (72), *686*, 800 (53), *852*
Klein, A. 261 (50), *277*
Klein, W. R. 149 (31), *159*
Kleinhenz, S. 127 (326), *143*
Kline, M. 120 (309), *142*
Klinge, M. 386 (60), *437*, 859 (18), *878*
Klinkhammer, K. W. 800 (59), *852*
Klippenstein, S. J. 294 (36, 94), *341*, *343*
Klun, T. P. 609 (26), 614 (39), *684*, *685*
Klunder, J. M. 632 (45), *685*
Klusener, P. A. A. 166 (77), *211*
Kneisel, F. F. 404 (124), *439*, 471 (116), 480 (117), 484 (156, 157), *521*, *522*, 616 (46), *685*

Knight, L. B. 147 (6), 150 (35), *159*
Knobler, C. B. 70 (146), 71 (147, 153, 154), 73 (157), *138*, *139*
Knochel, P. 108 (259), *141*, 169 (107, 116–118), 175 (160), *212*, *213*, 378 (46), 383 (38, 58), 404 (122, 124–126), 416 (128, 135), 418 (144, 147), 421 (151–153, 155), 422 (156), 423 (158, 159, 161, 162), 424 (121, 123, 157), 425 (154, 163), 426 (164), *436*, *437*, *439*, *440*, 444 (3), 447 (25), 452 (6), 471 (5, 116), 473 (118, 120, 121), 474 (123, 124), 480 (117), 484 (156–158), 496 (126, 127, 190), 506 (135), 516 (254), *519*, *521–524*, 565 (120, 121, 124), 573 (12), *581*, *583*, 586 (7), 589 (16), 600 (17, 19, 37–40), *601*, *602*, 616 (6, 46), 641 (103, 104), 670 (145), *683*, *685*, *688*, *690*, 776 (14), *851*, 858 (8, 11), *876*, *877*, 979 (493), *990*
Knochenmuss, R. 309 (149), *344*
Knoess, H. P. 416 (128, 135), *439*, *440*
Knöpfel, T. F. 728 (75), *730*
Knopff, O. 709 (28), *729*
Knotter, D. M. 56 (116), 61 (13), 87 (12, 203), 108 (108), *135*, *138*, *140*, 669 (142), *689*
Knotter, M. 35 (64), *137*
Knowles, P. F. 351 (32), 354 (44), *362*
Knudsen, M. M. 956 (373), *988*
Ko, S. Y. 632 (45), *685*
Kobayashi, H. 496 (193, 194), *523*, 1003 (42–44), *1029*
Kobayashi, S. 502 (210), *523*
Kobayashi, T. 386 (56), *437*, 968 (425), *989*
Kobayashi, Y. 616 (47), 644 (107), *685*, *688*, 1007 (54), 1008 (55, 58), 1016 (101–105), *1030*, *1031*
Kobelev, V. A. 890 (56–61), *981*
Koch, K. A. 348 (10), *361*
Koch, W. 290 (79), 308 (39, 90), *341–343*
Kochel, P. 108 (257), *141*
Kochi, J. 260 (48), *277*, 609 (22), *684*
Kochi, J. K. 3 (41), *19*, 154 (77, 78), *160*, 609 (22), *684*, 978 (470–472), *990*
Kocienski, P. 528 (5), 551 (92), *581*, *583*
Kocienski, P. J. 399 (98), *438*
Köck, M. 196 (196), *214*
Kočovský, P. 493 (183, 184), *523*
Kodama, H. 502 (73), *520*
Kodweiss, J. 165 (30), *210*
Koebrich, G. 423 (160), *440*
Koehn, J. 155 (96), *161*
Koenig, T. M. 901 (118), *982*
Koerner, M. 509 (231), *524*, 819 (133), *854*
Koert, U. 975 (452), *989*
Kofink, C. 421 (155), *440*
Kofink, C. C. 425 (163), *440*, 979 (493), *990*
Koga, N. 572 (30), *581*
Köhler, J. 39 (82), *137*
Köhler, K. 41 (95, 96), 78 (170–172), *137*, *139*, 276 (72, 73), *278*
Kohrt, J. T. 925 (230), *985*
Koizumi, H. 295 (99), 299 (100), *343*
Koizumi, T. 177 (163), *213*
Kojima, M. 384 (57), 386 (56), *437*
Kokado, H. 225 (47), *242*
Kokko, B. 373 (32), *436*
Kolakowski, R. V. 631 (76), *687*
Kolb, J. 908 (153), *983*
Kolesov, V. P. 155 (97), *161*
Kolhekar, A. S. 280 (2), *341*
Kollman, P. A. 309 (147), *344*
Kollmann, P. A. 354 (49), *362*
Kolodney, G. 565 (124), *583*
Kolonits, P. 859 (18), *877*
Komaromi, I. 283 (44), *342*
Komarov, I. V. 201 (221), *215*
Komatsu, M. 16 (92), *20*
Komiya, S. 3 (41), *19*
Kondo, A. 940 (285), *986*
Kondo, K. 1017 (106), *1031*
Kondoh, A. 569 (128), *584*
Kondratenko, N. V. 887 (25), *980*, 1019 (115), *1031*
Kondratov, S. A. 890 (64, 67, 68), *981*
Kone, O. 334 (237), *346*
Kong, Q. 157 (119), *162*
Königsberger, K. 647 (116), *689*
Konishi, M. 238 (136), *244*
Konishi, Y. 800 (31), *851*
Konno, T. 549 (78, 79), 571 (131), *583*, *584*
Konopelski, J. P. 950 (345), *987*
Koop, U. 870 (43), 872 (42), *879*
Kopf, J. 74 (165), *139*
Kopp, F. 404 (124), 421 (152), 422 (156), *439*, *440*
Korbelik, M. 238 (119), *244*
Korbutov, S. V. 158 (131), *162*
Korhummel, C. 1022 (131), *1031*
Koripelly, G. 474 (132), *522*, 695 (1), *729*, 858 (12), *877*
Korn, T. 404 (124), *439*
Körner, H. 358 (65), 359 (60), *363*
Korp, J. D. 993 (14), *1029*
Kort, M. E. 616 (6), *683*
Korte, F. 893 (80), *982*
Korth, O. 220 (27), *242*
Kosaka, Y. 499 (206), *523*
Kosfeld, R. 197 (204), *214*
Koshigoe, T. 330 (228), *346*
Koslowski, J. A. 24 (14), *135*
Kosman, D. J. 348 (17), 356 (55), *361*, *362*
Koster, C. G. d. 293 (80), *342*
Kosugi, H. 453 (58), *520*
Kotchetkov, N. K. 535 (44), *582*
Kotiaho, T. 322 (191), *345*

Kotora, M. 462 (103, 104), 468 (105, 106), 469 (108, 109), 470 (110), *521*, 573 (61), *582*
Kotsuki, H. 979 (483), *990*
Koura, M. 661 (133), *689*
Koval, E. D. 922 (206), *984*
Koval, I. A. 356 (54), *362*
Kowenicki, R. A. 39 (79), *137*
Kozima, S. 872 (51), *879*
Kozlowski, J. A. 169 (121), *212*
Kozyrod, R. P. 949 (340, 341), *987*
Kraft, B. J. 238 (135), *244*
Kraka, E. 308 (148), *344*
Kralik, J. 259 (44), *277*
Kramer, J. R. 304 (129), *343*
Kramer, S. W. 453 (61), *520*
Krane, J. 168 (101), *212*
Krasnikov, V. 221 (36), *242*
Krasovskaya, V. 600 (17), *602*
Krasovskiy, A. 378 (46), *436*, 600 (17, 37, 39), *602*
Krause, N. 1 (7), 2 (15, 23, 25), 5 (47), 6 (6, 52, 53), 8 (57, 58), 9 (9), 17 (106), *19–21*, 24 (17), 39 (39), 108 (263), *135*, *136*, *141*, 169 (113), 173 (110, 126), 183 (124), 184 (178, 182, 188), 185 (135), 187 (62), 191 (191), 200 (21, 200), 202 (23, 108), *210–214*, 380 (1), 388 (80), 401 (105), 430 (13), *435*, *438*, *439*, 444 (2), 471 (5), 474 (133), 480 (152, 153), 516 (247), *519*, *522*, *524*, *525*, 534 (35), *582*, 629 (73), 632 (82, 86, 87), 635 (77), 649 (119), 651 (120), 653 (121), 672 (1), *682*, *687*, *689*, 695 (1), *728*, *729*, 732 (1), 769 (91), *771*, *773*, 778 (28), 790 (8), 800 (57), 829 (160), *851*, *852*, *855*, 858 (5–13), 861 (23), 863 (27), 866 (29, 31), 867 (30, 34, 35), 868 (36, 38), 869 (40, 41), 870 (14, 43–46, 48, 49), 872 (42, 47), 873 (3, 4), 874 (54, 55), *875–879*
Kravchuk, T. 156 (99), *161*
Krebs, B. 356 (53), *362*
Krebs, O. 537 (17), *581*
Kretchmer, R. A. 430 (14), *435*
Kretzschmar, I. 157 (117), *161*
Krief, A. 640 (97), *688*, 824 (151), *855*, 906 (141), *983*
Krieger, J. K. 151 (47), *160*
Krische, M. J. 713 (53), *730*, 758 (73), *773*
Krishna, G. G. 922 (211), *984*
Krishna, M. V. 906 (144), *983*
Krishnan, B. 238 (136), *244*
Krishnan, S. 593 (25, 26), *602*
Krishnan, V. 221 (33), *242*
Kroneck, P. 165 (30, 34–36), *210*
Kroneck, P. M. H. 352 (35), 356 (57), 358 (65), 359 (60, 61), 360 (59, 76), *362*, *363*
Kronenburg, C. M. P. 97 (224), 101 (231, 232), 108 (264), *140*, *141*, 173 (126), 175 (156, 157), *212*, *213*, 800 (59), *852*, 858 (5), *875*
Krüeger, C. 445 (23), *519*
Kruglik, S. G. 220 (20), *241*
Kruizinga, W. H. 894 (94), *982*
Krzyczmonik, P. 228 (64), *243*
Kubas, G. J. 902 (123), *983*
Kubo, S. 116 (292), *142*
Kubo, T. 943 (292), *986*
Kubota, M. 891 (75), *982*
Kubota, T. 616 (48), *685*
Kucejko, R. J. 755 (69), *773*
Kucera, G. L. 238 (131), *244*
Kucharek, R. 165 (43–45), *211*
Kudin, K. N. 283 (44), *341*
Kudis, S. 647 (111), *688*
Kuendig, E. P. 157 (103), *161*
Kuenzel, S. 324 (201), *345*
Kühnert, J. 275 (68), *278*
Kuipers, P. 749 (52), *773*
Kujime, M. 165 (31), *210*
Kulagina, T. G. 147 (7), *159*
Kumada, P. 800 (40), *852*
Kumadaki, I. 1008 (56, 57), 1016 (101–105), *1030*, *1031*
Kumar, A. 894 (99), *982*
Kumar, A. P. G. 168 (86), *212*
Kumar, M. R. 238 (134), *244*
Kumar, P. G. A. 163 (9), *210*
Kumar, S. 906 (146), *983*
Kumar, V. N. 920 (198), *984*
Kundu, N. G. 966 (407), *989*
Kuniyasu, H. 556 (105), *583*, 979 (482), *990*
Kunz, D. 926 (244), *985*
Kunz, K. 200 (211), *214*
Kunz, R. W. 197 (204), *214*
Kurahashi, T. 165 (31), *210*, 330 (226), *346*
Kuraoka, S. 698 (13), *729*
Kurasako, K. 225 (47), *242*
Kurata, Y. 891 (76), *982*
Kurczewska, M. 228 (64), *243*
Kurczewski, K. 228 (64), *243*
Kurikawa, T. 285 (13), *341*
Kurinovich, M. A. 280 (6), *341*
Kuriyama, M. 430 (27), *436*
Kuriyama, N. 157 (112), *161*
Kuriyama, W. 756 (72), *773*
Kuroda, K. 228 (65), *243*
Kuroda, R. 197 (205), *214*
Kurtz, K. C. M. 933 (263, 265), 934 (264), *985*, *986*
Kusano, S. 484 (165), *522*
Kusuda, K. 230 (78), *243*
Kusuda, S. 822 (150), *855*
Kusumoto, N. 738 (20), *772*
Kuwajima, I. 15 (88), 16 (92), *20*, 416 (138), *440*, 516 (249, 250), *524*
Kuwata, K. 225 (49), *242*

Kuwatani, Y. 595 (22), *602*
Kux, D. 955 (369), *988*
Kuznetsova, Y. L. 946 (311), *987*
Kvach, V. V. 220 (20), *241*
Kweon, J.-H. 976 (458), *990*
Kwok, W.-K. 114 (288), *142*
Kwong, F. Y. 904 (130), 913 (165), 914 (169), *983*, *984*

La Manna, G. 943 (293), *986*
Labinger, J. A. 452 (54), 459 (52), *520*
Laborde, E. 646 (108), *688*
Laboroi, F. 802 (88), *853*
Laboue, B. 447 (32), 449 (33), *520*
Lacerda, P. S. S. 238 (123), *244*
Lacour, J. 200 (210), *214*
Ladenstein, R. 352 (34), *362*
Lafitte, M. 334 (237), *346*
Lafrenière, J. 620 (51), *685*
Lagarrigue, M. 326 (208), *345*
Lagu, B. 178 (167), *213*, 367 (8), 430 (17), *435*
Lai, C. B. 595 (30), *602*
Lai, T.-F. 83 (196), *140*, 262 (51), *277*
Laine, R. M. 248 (5), *276*
Laitar, D. S. 44 (102), 46 (103), 121 (100), 122 (319), 123 (101), *137*, *143*, 151 (44), *160*, 402 (112, 117, 119, 120), *439*, 732 (6), *771*
Lalancette, R. A. 30 (52), *136*
Lam, C. 329 (218, 220), *346*
Lam, C.-H. 81 (184), 83 (189, 191, 192), *139*, *140*, 267 (58), *278*
Lam, C. N. W. 329 (219), 331 (221), *346*
Lam, H. W. 763 (85), 764 (86), *773*
Lam, P. Y. S. 952 (350, 354), 954 (365), 956 (371, 372), 960 (385), 961 (388), 962 (351, 386, 387), *987*, *988*
Lam, W. H. 81 (185), *139*
Lamač, M. 275 (68), *278*
LaMaire, S. J. 459 (82, 83), *521*
Lamas, T. 748 (49), *772*
Lambers-Verstappen, M. H. 914 (170), *984*
Lammertsma, K. 293 (80), *342*
Lamsabhi, A. M. 303 (49), 310 (170), 318 (150), 337 (84, 125, 151, 169), *342–344*
Landais, Y. 800 (50, 51), 813 (111), *852*, *854*
Landon, P. 1020 (124), *1031*
Landor, S. R. 866 (29), *878*
Lang, F. 915 (164), *983*
Lang, H. 41 (95–97), 43 (98), 78 (170–173), 94 (94), *137*, *139*, 259 (44), 269 (62), 271 (63, 65, 66), 273 (46, 67), 274 (45), 275 (61, 68–70), 276 (72, 73), *277*, *278*
Lang, S. A. 1018 (112), *1031*
Lange, J. H. M. 925 (234), *985*
Langer, F. 496 (126, 127), *522*
Langer, W. 430 (172), *441*
Langford, C. H. 229 (72), *243*
Langhoff, S. R. 294 (91), 302 (124), *343*
Langkoft, E. 800 (61), *852*
Langlois, Y. 105 (237), *141*, 173 (139), *213*
Langridge-Smith, P. R. R. 220 (25), *242*
Laosooksathit, S. 940 (87–89), *982*
Lappert, M. F. 27 (26, 42), 37 (77), *136*, *137*, 261 (49, 50), *277*
Lardicci, L. 638 (94), *687*
Largo, A. 281 (16), *341*
Larhed, M. 911 (160), *983*
Larive, C. K. 361 (80), *363*
Larock, R. C. 553 (95), 576 (140), *583*, *584*
Larsen, A. O. 133 (337), *143*, 402 (108), *439*, 508 (150), *522*, 672 (155), *690*
Larsen, C. H. 935 (269), *986*
Lasne, M.-C. 166 (67), *211*
Lastra, E. 81 (181), 83 (195), 85 (198), *139*, *140*
Laszlo, P. 129 (129), *138*
Latos-Grażyński, L 119 (301), *142*
Lau, T. 329 (212), *345*
Lau, T. C. 324 (196), *345*
Lau, T. L. 925 (224), *985*
Lau, V. C.-Y. 83 (192, 193), *140*, 266 (55), *278*
Laucher, D. 430 (15), *435*
Laumen, K. 643 (106), *688*
Launay, V. 841 (162), *855*
Laurent, A. 551 (89), *583*
Laurent, R. 921 (220), *985*
Lautens, M. 937 (272), *986*
Laux, M. 870 (43, 45), *879*
Lauzon, C. 489 (174), *523*
Lauzon, S. 622 (54), *685*
Lavanant, H. 320 (182), 321 (155), *344*, *345*
Lavin, K. D. 904 (127), *983*
Law, K. W. 906 (144), *983*
Lawrence, N. J. 381 (50), *437*, 613 (37), *685*, 800 (75), 804 (92), 807 (95), *853*
Laycock, B. 831 (84), *853*
Layh, M. 37 (77), *137*, 261 (49, 50), *277*
Lazarova, T. I. 953 (358), *988*
Lazzari, D. 374 (34), 378 (45), *436*, 599 (35), *602*, 807 (96), 832 (78), *853*
Le Blanc, M. 992 (9), *1029*
Le Clainche, L. 330 (227), *346*
Le Gall, T. 979 (485), *990*
Le Guennic, B. 83 (189), *140*, 267 (58), *278*
Le Quement, S. 850 (200), *856*
Le, T.-H. 157 (114), *161*
Lea, L. 670 (148), 675 (2), *683*, *690*, 858 (11), *877*
Leal, J. P. 158 (134), *162*
Leary, J. A. 334 (238), *346*
Lebedev, B. V. 147 (7), *159*
Lebioda, A. 241 (143), *245*

LeBlanc, F. 157 (120), *162*
LeBlanc, M. 1015 (94), *1031*
Leclerc, E. 315 (178), *345*
Lecourt, T. 679 (172), *691*
Ledon, H. 220 (22), *241*
Lee, C. 282 (23), *341*
Lee, C. C. 800 (52), *852*
Lee, C.-T. 746 (41, 42), 755 (67), *772*, *773*
Lee, D. 747 (45), 748 (47), *772*, 800 (65), *852*, 944 (304), *986*
Lee, D.-W. 733 (10), *772*, 754 (64), *773*
Lee, G. Y. 976 (458), *990*
Lee, H. F. 295 (52), *342*
Lee, H. J. 157 (106), *161*
Lee, H.-K. 78 (174), *139*
Lee, J.-D. 618 (49), *685*
Lee, J. I. 509 (232), *524*
Lee, J. K. 280 (6, 7, 9, 11), *341*
Lee, J. M. 800 (83), *853*
Lee, J. T. 902 (125), *983*
Lee, K.-S. 480 (148), 513 (243), *522*, *524*, 709 (27), *729*, 926 (246), *985*
Lee, M. D. 238 (137), *244*
Lee, R. J. 800 (85), *853*
Lee, S.-H. 834 (167), *855*, 867 (28), 868 (37), *878*, *879*, 944 (304), *986*
Lee, S.-K. 360 (52), *362*
Lee, W.-K. 78 (174), 81 (180), 83 (196), 85 (199), *139*, *140*, 261 (52), 262 (51, 53), *277*, *278*, 445 (12), *519*
Lee, Y. 509 (228), *524*, 674 (161), *690*
Lefebvre, O. 977 (467), *990*
Lefort, M. 955 (369), *988*
Legault, C. Y. 131 (335), *143*
Legoupy, S. 551 (91), *583*
Leheski, L. 872 (50), *879*
Lehmann, R. 456 (62), 461 (77), *520*
Lehmkuhl, H. 247 (2), *276*
Lehn, J. M. 238 (128), *244*
Lehner, A. 883 (1), *980*
Leibfritz, D. 189 (195), *214*
Lelj, F. 289 (73), *342*
Lemaire, J. 152 (52), *160*, 309 (42), *341*
Lemieu, R. M. 820 (13), *851*
Lemmen, T. H. 165 (48), *211*, 732 (4), *771*
Lemr, K. 326 (207), *345*
Lenders, B. 48 (107), *138*
Lennartz, H. W. 151 (49), *160*
Lentz, R. 473 (121), *521*
Leo, A. 11 (74), *20*, 266 (59), *278*
Leon, E. 334 (223), *346*
Leonard, J. 858 (12), *877*
Leong, W. 845 (190), *856*
Leoni, P. 106 (246), *141*
Leroux, M. 536 (47), *582*
LeRoy, J. 1014 (83), *1030*
Lesheski, L. 647 (9), *683*
Leszczweski, D. 646 (109), *688*
Leszczynski, J. 282 (21), 298 (114), 309 (115), *341*, *343*
Leu, W. 133 (337), *143*, 402 (108), *439*, 508 (150), *522*, 672 (155), *690*
Leuchtner, R. E. 157 (116), *161*
Leung, L. T. 768 (89), *773*
Leung, M. K. 925 (226), *985*
Leung, S. K. 760 (78), 768 (89), *773*
Leung, S.-Y. 732 (7), *771*
Leung, W.-P. 78 (174), *139*
Leuser, H. 471 (116), 480 (117), 484 (156, 157), *521*, *522*, 616 (46), *685*
Leusink, A. J. 34 (27, 28), 51 (30), 52 (111), 91 (215), *136*, *138*, *140*
Leussing, D. L. 288 (70), *342*
Levason, W. 165 (38), *211*
Leveque, J.-M. 695 (8), *729*
Lever, A. B. P. 288 (69), *342*
Levin, A. 456 (68), *520*
Levisalles, J. 16 (98), *21*, 609 (25), 610 (33), 647 (21), *684*, 820 (16), *851*
Levitsky, I. A. 229 (69), *243*
Levy, A. B. 530 (19), *581*
Lewin, A. H. 112 (272), *141*, 943 (291), *986*
Lewis, J. 272 (43), *277*
Lewis, M. D. 616 (43), *685*
Lewis, R. 423 (162), *440*
Ley, S. V. 200 (212), *214*, 380 (1), *435*, 502 (215), *524*, 894 (98), *982*
Leyendecker, F. 430 (15), *435*
Li, C. 907 (147), 970 (430), *983*, *989*
Li, C. Z. 934 (261, 262), *985*
Li, D. 966 (408), *989*
Li, H. 330 (232), *346*
Li, H. Y. 230 (75), *243*
Li, J. 235 (103), *243*, 925 (225), *985*
Li, J.-H. 975 (405, 444, 445), *989*
Li, J.-L. 975 (405, 445), *989*
Li, K. 202 (229), *215*, 221 (35), *242*, 508 (227), *524*, 705 (32), *729*
Li, K. Y. 697 (12), 710 (50), 713 (54), *729*, *730*
Li, L. 920 (201), *984*
Li, N. 861 (21), *878*
Li, Q.-S. 123 (322), *143*
Li, R. 952 (350), 962 (387), *987*, *988*
Li, S. 303 (128), *343*, 390 (84), 399 (100), *438*, 768 (89), *773*
Li, S. J. 587 (11), *602*
Li, T.-C. 944 (306), *986*
Li, X. 236 (110), *244*
Li, Y. 462 (102), *521*, 920 (196), 935 (268), *984*, *986*
Li, Z. 733 (9), *772*
Lian, L. 153 (67), *160*
Liang, B. 158 (138), *162*
Liang, C. H. 108 (260), *141*, 175 (155), *213*, 800 (54), *852*

Liang, C. K. 925 (226), *985*
Liang, L. 698 (14), *729*
Liang, X. 937 (276), *986*
Liao, E.-T. 861 (24), *878*
Lias, S. G. 147 (9), *159*
Liashenko, A. 283 (44), *342*
Libby, R. D. 354 (46), *362*
Lichtor, P. A. 897 (112), 914 (113), *982*
Liebeskind, L. S. 108 (252), *141*, 377 (42), *436*, 975 (447), 976 (450), *989*
Liebeskind, L. S. E. 873 (4), *875*
Liebman, J. F. 147 (4, 9), 149 (33), 151 (46), 153 (76), *159*, *160*, 310 (164), *344*
Liedl, K. R. 305 (134), 322 (188), *344*, *345*
Liepins, V. 816 (120), 817 (10), 849 (11, 104), *851*, *854*
Light, M. E. 127 (326), *143*
Lilburn, J. 887 (23), *980*
Lilley, T. J. 861 (22), *878*
Lim, K.-H. 976 (455), *990*
Lim, S. H. 386 (69), *437*
Lin, C. 926 (237), *985*
Lin, C. T. 932 (256), *985*
Lin, F. 295 (51), *342*
Lin, J. L. 150 (19), *159*
Lin, M. C. 282 (24), *341*
Lin, S.-Y. 386 (73), 430 (17), *435*, *437*
Lin, T. 324 (195), *345*
Lin, W. 404 (122), 425 (163), *439*, *440*
Linderman, R. J. 474 (137), *522*
Lindley, P. 349 (29), *362*
Lindley, P. F. 356 (51), *362*
Lindquist, N. 238 (126, 131), *244*
Lindsay, D. M. 421 (152, 153, 155), 422 (156), 425 (154), *440*, 600 (19), *602*
Lindsay, H. A. 622 (61), *686*
Lindsey, C. 858 (8), *876*
Lindstedt, E. L. 184 (183), *214*
Lineberger, W. C. 149 (27, 29), 152 (60), *159*, *160*
Ling, Y. 330 (230), *346*
Linke, W. F. 892 (79), *982*
Linn, C. B. 978 (469), *990*
Linstrom, P. J. 303 (53), *342*
Liotta, C. L. 894 (95), *982*
Liou, L. R. 173 (150), *213*
Lippard, S. J. 430 (22), *436*
Lippert, B. 337 (243), *346*
Lipscomb, W. N. 68 (145), *138*
Lipshutz, B. 755 (67), *773*
Lipshutz, B. H. 1 (4), 2 (25), 15 (85), 16 (93), 17 (104), *19–21*, 24 (14, 15), 108 (16, 249–252), *135*, *141*, 164 (17), 169 (114, 120, 121), 179 (131), 200 (21), *210*, *212*, *213*, 366 (5), 380 (1), *435*, 444 (1, 10), 452 (50), 456 (11, 62, 63), 457 (74, 75), 459 (78), 474 (138), 503 (217), 509 (231, 232), 510 (234, 235), 517 (233), 518 (59), *519*, *520*, *522–524*, 571 (2), *581*, 587 (12), 592 (23, 24), *602*, 609 (25), 625 (66), 632 (87), 659 (18, 124), 672 (1), *682*, *684*, *686*, *687*, *689*, 732 (1), 734 (11), 738 (24), 739 (26, 27), 745 (39), 746 (40–42), 749 (51, 52), 753 (60, 63), 755 (68, 69), 756 (70, 71), 765 (87), 768 (90), 769 (91), *771–773*, 776 (4), 788 (27), 800 (41, 56, 66), 818 (32), 819 (133), 822 (149), 834 (33, 131), 842 (130), *851*, *852*, *854*, *855*, 858 (8), 873 (4), *875*, *876*, 897 (101), *982*
Lipskier, J. F. 223 (42), *242*
Liron, F. 484 (157), *522*, 616 (46), *685*
Listorti, A. 220 (10), *241*
Litter, M. I. 240 (142), *245*
Litvak, V. V. 883 (8), *980*
Liu, D. R. 800 (33), *853*
Liu, F. 919 (185), 921 (212), 966 (404), *984*, *989*
Liu, G. 283 (44), *342*, 616 (6), *683*
Liu, H. 152 (58), *160*, 859 (16), *877*
Liu, H.-J. 654 (127), *689*
Liu, J. 908 (150), *983*
Liu, L. 516 (253), *524*, 564 (119), *583*, 921 (209), 925 (225), 926 (238), 943 (297), *984–986*
Liu, M. 926 (238), *985*
Liu, P. H. 932 (257), *985*
Liu, Q. 175 (154), *213*
Liu, Q. Z. 710 (51), *730*
Liu, S. 225 (50), *242*, 377 (42), *436*, 913 (138), *983*
Liu, S. G. 221 (36), *242*
Liu, W. 426 (164, 166), 427 (167), *440*
Liu, X. 163 (5), *210*
Liu, X.-J. 152 (59), *160*
Liu, Y. 468 (106), 469 (108, 109), *521*, 573 (61), *582*, 920 (196), *984*
Liu, Y. H. 920 (197), *984*
Liu, Y. Y. 595 (31), *602*
Liu, Z. 576 (140), *584*
Liu, Z. P. 592 (24), *602*
Livshin, S. 569 (127), *584*
Llamas-Saiz, A. L. 282 (26), *341*
Llobet, A. 114 (290, 291), *142*, 964 (394), *988*
Lo, W.-Y. 83 (189, 192), *140*, 267 (58), *278*
Löber, O. 451 (44), *520*
Lockhart, T. P. 961 (389), *988*
Loh, T. P. 700 (20, 21), 725 (70), *729*, *730*
Loh, Z. H. 223 (43), *242*
Lohmann, S. 894 (98), *982*
Lohr, L. L. 68 (145), *138*
Long, J. K. 838 (177), *856*
Loo, D. 373 (32), *436*
Loosli, H. R. 191 (102), *212*
Loots, M. J. 453 (58), 459 (57), *520*
Lopez-Alvarado, P. 950 (342, 343, 346, 347), *987*

López, F. 18 (118), *21*, 200 (216–218), *214*, *215*, 474 (134), 493 (177), *522*, *523*, 670 (151, 152), *690*, 698 (17, 18), 719 (60), 721 (64), *729*, *730*
López, L. 451 (44, 45), *520*, *521*, 613 (31), 663 (20), *684*
Lopez-Perez, J. L. 920 (200), *984*
Lopez, R. 902 (126), *983*
López, S. 845 (191, 192), *856*
Lopez, X. 281 (16), *341*
Lorenzen, N. P. 95 (221), *140*
Lotesta, S. D. 631 (76), *687*
Lotz, M. 858 (8), *876*
Loubinoux, B. 629 (74), *687*
Lower, A. 753 (63), 755 (68, 69), *773*
Lu, B. 972 (433), *989*
Lu, H. 996 (23), *1029*
Lu, H. J. 934 (262), *985*
Lu, K. 389 (76), 393 (78, 79), 395 (90), *438*
Lu, Z. 555 (102, 103), *583*, 913 (167, 168), 915 (166), *983*
Lu, Z. J. 908 (155), *983*
Lucchetti, J. 906 (141), *983*
Luchaco-Cullis, C. A. 670 (153), *690*
Luche, J.-L. 430 (172), *441*, 629 (4), *683*
Luchinat, C. 354 (48), *362*
Lucht, B. L. 166 (70), 177 (162), *211*, *213*
Lücking, U. 716 (57), *730*
Lum, T. K. 700 (21), 725 (70), *729*, *730*
Lumpkin, C. C. 943 (299, 303), *986*
Luna, A. 283 (48), 284 (32), 293 (29), 295 (85), 311 (171, 172), 315 (121, 174, 178), 318 (177), 333 (235), *341–346*
Lund, H. 247 (3), *276*
Lusch, M. J. 564 (113, 115), *583*
Lutz, M. 97 (224), 101 (231, 232), *140*, *141*, 175 (156, 157), *213*
Lutz, O. 165 (30, 34–36), *210*
Lv, X. 925 (109), *982*
Lyding, J. W. 228 (63), *242*
Lygo, B. 502 (215), *524*
Lynch, B. J. 283 (45), *342*
Lynd, R. A. 502 (209), *523*

Ma, D. 897 (102–104), 899 (116), 901 (119), 904 (134), 909 (157), 911 (162), 915 (173), 917 (174, 175), 932 (259), 937 (277), 938 (278), 966 (404), 970 (431), 972 (433, 434), *982–986*, *989*
Ma, D. W. 919 (176), 942 (158), *983*, *984*
Ma, H. 922 (208), *984*
Ma, H.-C. 922 (210), *984*
Ma, J.-A. 707 (36), *730*
Ma, S. 555 (102, 103), *583*, 811 (109), *854*
Maas, G. 416 (137), *440*
Macchia, F. 676 (164), 677 (167, 168), 679 (169, 172), *690*, *691*
Macchioni, A. 168 (84, 93, 98), *212*
Macdonald, C. L. B. 166 (60), *211*
MacDonald, J. G. 238 (123), *244*
Macdonald, S. J. F. 593 (18, 20), *602*
Macdonald, T. L. 629 (72), *686*
Mack, J. 219 (6), *241*
MacLeod, D. 834 (167), *855*
MacMillan, D. K. 154 (79, 80), *160*
MacNeil, K. 1026 (152), *1032*
MacNeil, K. J. 548 (76), *582*, 996 (8), 1025 (140, 151), *1029*, *1032*
Maddaluno, J. 166 (67), *211*
Maddox, M. L. 629 (4), *683*
Madec, D. 741 (29), *772*
Madhavaiah, C. 238 (133, 134), *244*
Madin, A. 502 (215), *524*
Madsen, U. 937 (276), *986*
Maeda, H. 117 (296, 297), 118 (298, 299), *142*, 240 (139), *244*, 257 (41), *277*
Maeda, M. 384 (57), 386 (56), *437*
Maegerlein, W. 968 (415, 416), *989*
Magid, R. M. 672 (1), *682*
Magnus, K. A. 351 (33), *362*
Magriotis, P. A. 845 (174), *855*
Maheswaran, H. 922 (211), *984*
Mahia, J. 114 (290, 291), *142*, 964 (394), *988*
Mahoney, W. S. 735 (15), 736 (18), *772*
Mahuteau-Betzer, F. 447 (25), *519*, 973 (435), *989*
Mai, L. 935 (268), *986*
Mains, R. E. 280 (2), *341*, 350 (30), *362*
Mair, F. S. 177 (161), *213*
Mais, D. E. 382 (54), *437*
Maiti, S. N. 925 (230), *985*
Maitra, U. 902 (121), *982*
Maiya, G. B. 221 (33), *242*
Majetich, G. 800 (82), *853*
Majid, T. N. 423 (161), *440*, 474 (123), *522*
Majoral, J.-P. 921 (220), *985*
Mak, A. N. S. 35 (64), *137*
Mak, T. C. W. 82 (187), *139*
Malan, C. 670 (148), 675 (2), *683*, *690*, 858 (11), *877*
Malda, H. 670 (146), *690*
Malick, D. K. 283 (44), *342*
Malitesta, C. 235 (101), *243*
Malito, J. 166 (26), *210*
Malkin, R. 351 (23), *361*
Mallard, W. G. 303 (53), *342*
Mallart, S. 373 (33), *436*
Mallesham, B. 925 (236), *985*
Malmström, B. G. 351 (23), 352 (40), *361*, *362*
Malte, A. 73 (161), *139*
Mampreian, D. M. 701 (22), *729*
Manahan, S. E. 255 (7), *277*
Manassero, M. 81 (179), *139*
Mancheño, M. J. 450 (42), *520*
Manchot, W. 157 (123), *162*

Mandal, A. K. 382 (55), *437*
Mandal, Ajay K. 807 (95), *853*
Mandal, Amit K. 807 (95), *853*
Manderville, R. A. 238 (126, 131), *244*
Mandeville, W. H. 587 (9), *601*
Mandoli, A. 203 (228), *215*
Mandville, G. 660 (132), *689*
Manganiello, S. 828 (158), *855*
Mangeney, P. 399 (97), 401 (104), *438*, 545 (70, 72, 74), *582*, 663 (85, 134–136), *687*, *689*, 694 (3, 4), *729*, 873 (53), 874 (56), *879*
Mani, N. S. 908 (150), *983*
Mankad, N. P. 121 (100), 123 (101), *137*, 151 (44), *160*, 402 (112, 117), *439*, 732 (6), *771*
Mann, A. 616 (27), 638 (94, 96), 659 (131), 675 (2), *683*, *684*, *687–689*
Mann, C. K. 247 (1), *276*
Mannes, D. 979 (485), *990*
Mäntele, W. 349 (20), *361*
Manyem, S. 695 (1), *728*
Mao, J. 921 (186, 187), 974 (443), *984*, *989*
Mao, Z. 732 (7), *771*
Maradolla, M. B. 920 (198), *984*
Marat, X. 579 (143), *584*
Marble, L. K. 382 (53), *437*, 663 (137), *689*
Marcelo-Curto, M. J. 947 (323), *987*
March, L. A. 893 (91), *982*
March, S. 201 (223), *215*, 695 (6), 712 (52), 714 (47), *729*, *730*
Marchese, G. 563 (48), 564 (117), *582*, *583*, 613 (31), *684*
Marchesini, A. 352 (34), 356 (56, 57), *362*, *363*
Marchesini, G. 947 (320), *987*
Marchi, D. 800 (76), *853*
Marci, G. 235 (101), *243*
Marcin, L. R. 516 (256), *524*
Marconi, G. 221 (36), *242*
Marcoux, J.-F. 921 (202), 942 (92), *982*, *984*
Marcus, R. 349 (22), *361*
Marcus, Y. 179 (171), *213*
Marder, T. B. 248 (5), *276*
Marek, I. 153 (76), *160*, 163 (4), *210*, 219 (8, 9), *241*, 444 (8), 452 (6, 50), 456 (68), 461 (95), 462 (102), 480 (117), *519–521*, 545 (70, 72), 560 (111, 112), 565 (120–126), 569 (127), *582–584*, 663 (85, 134, 136), *687*, *689*, 842 (179), *856*, 874 (56), *879*
Marfat, A. 502 (216), *524*
Margl, P. 305 (133), *344*
Margrave, J. L. 150 (36), *159*
Maria Caporusso, A. 638 (94), *687*
Marini, E. 374 (34), *436*
Marini, G. W. 305 (134), *344*
Marino, J. P. 514 (245, 246), *524*, 618 (49), 629 (70), 640 (98, 100), 646 (108), 647 (114), *685*, *686*, *688*, *689*, 838 (177), *856*
Marino, T. 290 (34), 302 (35), 303 (78), 305 (141), *341*, *342*, *344*
Mark, G. 238 (123), *244*
Marker, A. 165 (27), *210*
Marks, M. W. 106 (247), *141*, 152 (62), *160*
Marks, T. J. 63 (131), *138*, 902 (124), *983*
Maron, L. 904 (128), *983*
Marquais, S. 449 (36), *520*
Marques, A. 429 (168), 430 (169), *440*
Marquess, D. G. 551 (94), *583*
Marr, J. 430 (18), *435*
Marrot, J. 399 (97), *438*
Marsch, M. 2 (29), *19*, 110 (265, 268), 133 (222), *140*, *141*, 175 (129), *212*, 800 (59), *852*, 858 (5, 6), *875*, *876*
Marsch, R. E. 113 (285), *142*
Marschke, K. B. 382 (54), *437*
Marsh, C. R. 73 (160), *139*, 1014 (84, 87), *1030*
Marshall, J. A. 474 (136), *522*, 545 (68), *582*, 625 (64, 65, 67), 629 (72), 631 (78), 647 (63), *686*, *687*, 860 (20), *878*, 977 (465), *990*
Marsich, N. 2 (12), *19*, 27 (22, 23), 39 (83, 84), 71 (71), *135*, *137*
Marsman, J. W. 91 (215), *140*
Martel, J.-P. 979 (485), *990*
Martel, R. 150 (38–40), *159*
Martel, S. 842 (189), *856*
Martin, A. 447 (31), *520*
Martin, D. 401 (106), *439*, 480 (149), *522*, 707 (34), *729*
Martin Hernando, J. I. 553 (100), *583*
Martín, J. D. 574 (139), *584*
Martin, R. 935 (269), 936 (271), 937 (273), *986*
Martin, R. L. 283 (44), *342*
Martin, R. M. 920 (193), *984*
Martin, S. F. 103 (236), *141*, 386 (68), *437*, 859 (16), *877*
Martin-Vaca, B. 904 (128), *983*
Martinet, P. 999 (35), *1029*
Martinez, A. 908 (148, 149), *983*
Martinez Carrera, S. 859 (16), *877*
Martinez-Garcia, H. 845 (194), *856*
Martínez, M. V. 629 (70), *686*
Martínez-Pérez, J. A. 380 (49), *437*, 647 (115), *689*
Martinho Simões, J. A. 147 (4), *159*
Martinot, T. A. 647 (116), *689*
Marull, M. 977 (467), *990*
Maruoka, K. 377 (43), *436*, 598 (33), *602*
Maruta, M. 1009 (60), *1030*
Maruyama, K. 15 (83), *20*, 456 (64, 65), 496 (192), *520*, *523*, 613 (23), *684*, 831 (164), *855*, 859 (17), *877*
Masamune, H. 632 (45), *685*
Masamune, S. 735 (12), *772*

Masarawa, A. 256 (37), *277*
Maset, T. 958 (381), *988*
Mashima, H. 13 (79), *20*, 478 (144), *522*
Mason, J. 165 (24, 56), *210, 211*
Mason, R. 34 (36), 134 (29), *136*
Masuda, J. D. 28 (46), *136*
Masuda, Y. 974 (439), *989*
Masuhara, H. 228 (59), 230 (80), *242, 243*
Mathies, A. K. 489 (170), *523*
Mathies, R. A. 220 (16), *241*
Mathieu, H. 622 (56), *685*
Matsubara, S. 841 (170), *855*, 872 (51), *879*
Matsubara, T. 358 (66), *363*
Matsubayashi, T. 229 (68), *243*
Matsuda, H. 13 (78), *20*
Matsuda, T. 117 (297), *142*
Matsue, R. 979 (473), *990*
Matsui, K. 1017 (106), *1031*
Matsukawa, K. 968 (423, 426, 427), *989*
Matsumoto, A. 595 (28), *602*
Matsumoto, K. 430 (24), *436*, 1011 (71), *1030*
Matsumoto, Y. 401 (106), *439*, 707 (35), *729*
Matsunaga, D. 230 (86), *243*
Matsuo, T. 225 (49), *242*
Matsuo, Y. 2 (27), *19*, 164 (16), *210*
Matsushita, K. 165 (47), *211*
Matsushita, T. 116 (293), *142*
Matsuumi, M. 644 (107), *688*
Matsuura, K. 499 (205), *523*
Matsuura, T. 386 (59), *437*
Mattay, J. 629 (72), *686*
Mattila, K. 358 (70), *363*
Matxain, J. M. 281 (16), *341*
Mauduit, M. 401 (104, 106), *439*, 480 (149), *522*, 707 (34), *729*, 863 (26), *878*
Maullin, N. 592 (23), *602*
Maurer, J. 707 (39), *730*
Maurya, R. 647 (116), *689*
Mavrikakis, M. 158 (136), *162*
Mawson, S. D. 1026 (153), *1032*
May, T. L. 402 (108, 109), *439*, 480 (146), 484 (166), 508 (224), *522–524*, 703 (26), 705 (33), 710 (48), *729, 730*
Mayer, I. 240 (139), *244*
Mayer, P. 47 (104), *137*, 184 (190), *214*
Mayr, H. 169 (117), *212*, 589 (16), *602*
Mazal, C. 598 (32), *602*
Maze, F. 695 (8), *729*
Mazery, R. D. 719 (60), *730*
Mazid, M. A. 197 (205), *214*
Mazur, P. 861 (22), *878*
Mazzanti, G. 802 (87, 88), 809 (100), *853*
McBreen, P. H. 150 (38–40), *159*
McBride, D. W. 63 (132), *138*
McClinton, M. A. 1020 (124), *1031*
McCormick, G. J. 70 (150), *139*
McDaniel, A. L. 938 (281), *986*
McEachern, E. J. 510 (237, 238), *524*, 834 (169), *855*
McErlean, C. S. P. 560 (110), *583*
McFadden, R. M. 939 (283), *986*
McFadyen, W. D. 329 (214, 215, 217), 339 (248), *346*
McGahren, W. J. 238 (137), *244*
McGarrity, J. F. 191 (102, 103), *212*
McGettigan, C. 733 (8), 768 (17), *771, 772*
McGlashan, D. W. 296 (102), *343*
McGrath, T. D. 70 (151), *139*
McGuirk, P. R. 502 (216), *524*
McIntosh, D. 153 (70, 71), *160*
McIntosh, M. C. 622 (61), *686*
McKay, T. 861 (22), *878*
McKenzie, J. R. 474 (137), *522*
McKie, R. 124 (323), *143*
McKillop, A. 968 (420, 421, 424), *989*
McKinley, N. F. 952 (353), *988*
McLachlan, F. 87 (207), *140*
McLaughlin, K. T. 453 (61), *520*
McLean, M. R. 732 (3), *771*
McLeod, D. 153 (72), *160*
McLoughlin, V. C. R. 113 (283), *142*, 1013 (78), *1030*
McLuckey, S. A. 322 (189, 191), *345*
McMillan, J. 73 (157), *139*
McMillin, D. R. 220 (13), *241*
McNeil, A. J. 173 (150), *213*
McNeill, M. H. 925 (224), *985*
McPartlin, M. 39 (79), *137*
McPherson, J. N. 351 (32), *362*
McPherson, M. J. 354 (45), *362*
McWilliams, J. C. 373 (33), *436*
Md, A. B. 403 (115), *439*
Mebel, A. M. 282 (24), *341*
Mebs, S. 238 (118), *244*
Mechelke, M. F. 605 (12), *683*
Mederski, W. W. K. R. 955 (369), *988*
Medforth, C. J. 220 (24), *242*
Medich, J. 430 (18), *435*
Meetsma, A. 18 (118), *21*, 200 (207, 216, 217), *214, 215*, 493 (177), *523*, 670 (152), *690*, 698 (18), 707 (39), 719 (59), *729, 730*
Meffre, P. 809 (99), *853*
Mehdinejad, H. 925 (232), *985*
Mehrabian, Z. B. 349 (28), *362*
Mehta, B. R. 155 (93), *161*
Mei, X. 920 (192), *984*
Mei, X. F. 920 (193), *984*
Meier, R. J. 293 (80), *342*
Meijer, J. 530 (18), 540 (57, 58), 564 (114, 116), *581–583*, 629 (72), *686*, 800 (53), 832 (146), 838 (124), *852, 854, 855*
Meinwald, J. 547 (75), *582*
Meldal, M. 200 (15), *210*
Mele, G. 235 (101–105), 236 (106), *243, 244*
Melvin, M. S. 238 (126, 131), *244*

Méndez, M. 831 (165), *855*
Menendez, J. C. 950 (342, 343, 346, 347), *987*
Menezes, P. H. 943 (229), *985*
Mengel, A. 659 (123), *689*
Mennucci, B. 283 (44), *341*
Mercero, J. M. 281 (16), *341*
Merino, S. 858 (12), *877*
Merkle, H. 356 (57), *363*
Merle-Mejean, T. 153 (73), *160*
Merzouk, M. 401 (104), *439*
Meshitsuka, S. 228 (58), *242*
Messaoudi, S. 83 (189), *140*, 267 (58), *278*
Messerschmidt, A. 348 (7), 350 (26), 352 (34, 41), 354 (45), 355 (50), 356 (53, 58), *361–363*
Messina, F. 842 (184), *856*
Metais, E. 447 (34), *520*
Metz, M. 360 (52), *362*
Meuzelaar, G. J. 669 (143, 144), *690*, 858 (11), *876*
Meyer, E. M. 48 (56), *136*
Meyer, F. 151 (50), 157 (109), *160*, *161*, 294 (97), *343*
Meyer, K. 41 (41), 126 (324, 325), *136*, *143*
Meyer, M. 360 (73), *363*
Meyer, O. 360 (37), *362*
Meyers, A. I. 386 (67), 421 (70), *437*
Meyerstein, D. 256 (37), *277*
Miao, G. 105 (238–240), *141*, 367 (6), 430 (18), *435*, *436*
Miao, G. B. 16 (94, 101), *21*
Miao, H. 922 (208), *984*
Miao, R. 768 (89), *773*
Michalak, R. 735 (12), *772*
Michl, J. 598 (32), *602*
Micouin, L. 418 (144), *440*, 679 (172), *691*
Miehling, W. 752 (59), *773*
Miginiac, P. 859 (16), *877*
Mihina, J. S. 861 (19), *878*
Mikhel, I. S. 201 (225), *215*
Miki, T. 18 (110), *21*, 430 (26), *436*, 707 (37, 38), *730*
Milbank, J. B. J. 553 (100), *583*
Mile, B. 153 (74), 157 (104), *160*, *161*, 632 (84), *687*
Milko, P. 152 (52), *160*, 309 (42), *341*
Millam, J. M. 283 (44), *341*
Millar, J. G. 537 (49), 553 (98), *582*, *583*
Miller, J. A. 459 (80), *521*
Miller, J. D. 979 (481, 488), *990*
Miller, J. M. 940 (287), *986*
Miller, R. L. 502 (208), *523*
Miller, S. A. 63 (128), *138*
Miller, S. E. 223 (43), *242*
Miller, T. A. 925 (224), *985*
Miller, W. T. 994 (16), *1029*
Millet, R. 676 (166), *690*
Mills, O. S. 892 (85), *982*
Mills, R. J. 800 (68), *852*
Milne, J. E. 744 (38), *772*
Milner, N. E. 1014 (85), 1015 (88, 91), *1030*
Milowsky, A. 861 (22), *878*
Mimura, N. 629 (68), *686*
Minajeff, W. 963 (2), *980*
Minami, T. 495 (194), *523*, 1003 (43, 45), *1029*
Minas da Piedade, M. E. 158 (134), *162*
Minch, B. A. 225 (57), *242*
Mingos, D. M. P. 41 (41), *136*, 873 (4), *875*
Minnaard, A. 200 (216), *214*
Minnaard, A. J. 2 (24), 18 (118), *19*, *21*, 200 (217, 218), 201 (226), 202 (18), *210*, *215*, 474 (131, 134, 139), 478 (141), 489 (167), 492 (176), 493 (177), *522*, *523*, 670 (151, 152), 675 (2), 679 (169, 171), *682*, *690*, 695 (9, 10), 698 (16–18), 699 (19), 707 (41), 714 (55), 719 (59–61), 720 (62, 65), 721 (56, 64, 67), 723 (68, 69), *729*, *730*, 858 (10), 862 (25), *876*, *878*
Mino, T. 925 (218), *984*
Minowa, T. 1012 (75), *1030*
Minville, J. 622 (54), *685*
Mioskowski, C. 535 (43), *582*, 979 (485), *990*
Mir, Y. 238 (125), *244*
Mishra, N. K. 894 (99), *982*
Misobe, N. 998 (32), *1029*
Mitchell, D. 901 (118), *982*
Mitchell, M. B. 386 (63), *437*
Mitchell, S. A. 147 (5), 157 (102), *159*, *161*
Mitra, A. 622 (59), *686*
Mitra, S. 588 (13), *602*
Miura, A. 226 (52), *242*, 1008 (58), 1024 (149), *1030*, *1032*
Miura, K. 738 (21), *772*, 977 (464), *990*
Miura, M. 965 (399), 966 (398), 968 (422), 969 (428), *988*, *989*
Miwa, M. 386 (56), *437*
Miwa, Y. 629 (68), 635 (88), *686*, *687*
Miyabe, T. 571 (131), *584*
Miyachi, Y. 572 (30), *581*
Miyake, Y. 595 (22), *602*
Miyakushi, S. 228 (60), *242*
Miyamoto, N. 228 (65), *243*
Miyamoto, R. 116 (292–294), 117 (295), *142*
Miyano, M. 629 (72), *686*
Miyashita, A. 81 (81), *137*
Miyaura, N. 496 (185–187, 199, 200), *523*, 559 (109), *583*
Miyawaki, J. 286 (14, 59), *341*, *342*
Miyazaki, A. 330 (226), *346*
Mizobe, N. 998 (31), *1029*
Mizumoto, A. 228 (67), *243*
Mizumoto, M. 641 (105), *688*
Mizutani, H. 670 (153), *690*, 707 (43), 713 (44), 720 (42), *730*
Mizutani, J. 167 (89), *212*

Mlekuz, M. 248 (5), *276*
Mo, H. 166 (85), *212*
Mo, J. 571 (129), *584*
Mó, O. 281 (15), 282 (26), 283 (48), 284 (32), 285 (50), 293 (86, 87), 295 (85), 303 (49), 308 (41, 104, 105), 309 (40, 147), 311 (171, 172), 315 (121, 174, 178), 318 (150, 177), 333 (235), 337 (84, 125, 151, 169), *341–346*
Moan, J. 238 (119), *244*
Mobley, T. A. 108 (256), *141*, 172 (137), 187 (62), *211*, *213*, 872 (47), *879*
Modi, S. P. 861 (22), *878*
Moessner, C. 926 (239), 958 (382), *985*, *988*
Mohadjerani, M. 925 (232), *985*
Mohamadi, F. 173 (148), *213*
Mohamed, M. M. 237 (114), *244*
Mohamed, R. M. 237 (114), *244*
Mohr, B. 165 (42), *211*
Mohr, F. 37 (74), *137*
Mohr, J. 15 (86), *20*, 654 (127), *689*
Moir, J. H. 177 (161), *213*
Moiseev, S. K. 947 (329), *987*
Moiseeva, K. S. 155 (89), *161*
Moje, S. 73 (161), *139*
Molander, G. A. 496 (197), *523*
Molina, A. 834 (169), *855*
Molla, M. E. 166 (75), *211*
Monaco, K. L. 952 (350), 954 (348), *987*
Monerris, M. 511 (239), *524*
Montalbetti, C. A. G. N. 418 (142), *440*
Monteiro, N. 574 (136), 576 (137), 579 (141–144), *584*
Montemurro, G. 859 (17), *877*
Montenegro, E. 647 (111), *688*
Montero, C. 629 (70), *686*
Montgomery, J. A. 282 (27), *341*
Montgomery, J. J. A. 283 (44), *341*
Monti, C. 670 (147), *690*
Monzani, E. 280 (3), *341*
Moody, D. B. 721 (67), *730*
Moolenaar, M. J. 414 (132), *440*
Moon, C. 503 (213), *523*
Moonen, C. T. W. 194 (194), *214*
Moore, E. B. 68 (145), *138*
Moore, H. W. 859 (16), *877*
Moore, L. O. 889 (52), *981*
Moore, T. 401 (104), *439*
Moore, T. A. 219 (7), *241*
Moore, W. M. 165 (39), *211*
Moran, P. H. 177 (161), *213*
Mordini, A. 807 (98), 828 (158), 842 (184, 185), *853*, *855*, *856*
Moreira, J. A. 979 (491), *990*
Morel, S. 947 (324), *987*
Moretti, R. 509 (231), *524*, 819 (133), *854*
Morey, J. V. 588 (14), *602*
Morgan, A. R. 238 (121), *244*
Morgan, J. 949 (340), *987*
Mori, A. 513 (241), *524*, 738 (23), *772*, 977 (463), 978 (468), *990*
Mori, K. 643 (106), *688*, 979 (487, 489), *990*
Mori, S. 3 (30, 31, 35), 5 (44, 45), 6 (52), 7 (54, 55), 8 (8, 56), 9 (9, 62, 63), 17 (108, 109), *19–21*, 133 (20), *135*, 173 (144–146), 184 (177, 178), 198 (111), 200 (200), 201 (201), *212–214*, 459 (87), 478 (143), *521*, *522*, 534 (29, 31–33), *581*, *582*, 585 (6), *601*, 606 (11), *683*, 800 (55), 849 (198), *852*, *856*, 858 (7), 870 (48), *876*, *879*
Mori, T. 457 (71), *520*, 1024 (149), *1032*
Morigaki, A. 571 (131), *584*
Morihashi, K. 177 (163), *213*
Morikawa, S. 979 (473), *990*
Morikawa, T. 1007 (54), *1030*
Morishita, E. 885 (12), *980*
Morita, N. 870 (46), 873 (4), *875*, *879*
Morita, Z. 230 (81, 83–85), *243*
Moritani, Y. 741 (30), 742 (31), *772*
Moriwaki, K. 925 (190), *984*
Moriya, T. 1003 (42), *1029*
Morizawa, Y. 777 (15), 809 (102), 816 (119), 841 (170), *851*, *853–855*
Morizur, J. 315 (174), *345*
Morizur, J. P. 311 (171, 172), 315 (121, 178, 179), *343–345*
Morken, P. A. 996 (8, 23), 997 (26–28), 998 (29, 30), 1005 (5), *1029*
Morokuma, K. 5 (42, 44), 6 (52), 7 (54), 8 (56), 9 (62, 63), *19*, *20*, 173 (145, 146), 184 (177, 178), 201 (201), *213*, *214*, 282 (24), 283 (44), *341*, *342*, 459 (87), *521*, 534 (31), 572 (30), *581*, 870 (48), *879*
Morrison, J. D. 632 (45), *685*
Morse, K. W. 165 (39), *211*
Morton, G. O. 238 (137), *244*
Morton, H. E. 817 (7), 821 (143), 824 (123), 829 (154), *851*, *854*, *855*
Moser, C. C. 358 (71), *363*
Moshkov, K. 349 (29), *362*
Moskau, D. 166 (82), *212*
Moskovits, M. 157 (103), *161*
Mostafaï, N. 732 (1), *771*
Mostefaï, N. 753 (61, 62), *773*
Motiya, M. 824 (69), *852*
Motomura, H. 230 (83, 84), *243*
Mouli, G. V. P. C. 920 (198), *984*
Moulin, C. 330 (227), 334 (223), *346*
Moura, A. O. 237 (115), *244*
Moura, I. 352 (36), 358 (67), *362*, *363*
Moura, J. J. 358 (67), *363*
Moura, J. J. G. 352 (36), *362*
Mouriño, A. 380 (49), *437*, 622 (58), 647 (115), *686*, *689*
Moutiez, M. 330 (227), *346*
Movassaghi, M. 432 (176), *441*, 743 (34), *772*

Mozumdar, S. 894 (99), *982*
Mu, J. 229 (74), *243*
Mück-Lichtenfeld, C. 769 (92), *773*
Mueller, N. 968 (415, 416), *989*
Mühlenbruch, B. 861 (22), *878*
Mukaiyama, T. 430 (15, 172), *435*, *441*
Mukherjee, P. 943 (294), *986*
Mulder, J. A. 933 (263), *985*
Müller, A. 100 (228), *141*
Muller, B. 842 (188), *856*
Müller, F. 108 (256), *141*, 134 (49), *136*, 172 (137), *213*
Müller, G. 65 (137), *138*, 166 (78), *211*
Müller, H. 78 (171), *139*, 666 (140), *689*
Muller, J. G. 337 (245), *346*
Muller, K. M. 926 (245), *985*
Muller, M. 426 (165), *440*
Müller, N. 204 (100), *212*
Muller, P. 122 (319), *143*
Mulvey, R. E. 166 (71), *211*
Munakata, M. 165 (33, 54), *210*, *211*
Munch-Petersen, J. 24 (7), *135*
Munro-Leighton, C. 121 (315), *143*, 403 (118), *439*
Murahashi, S.-I. 496 (191, 192), *523*
Murano, T. 1011 (70, 74), 1012 (75, 76), *1030*
Murata, M. 352 (39), *362*
Muratova, L. V. 946 (310), *987*
Murphy, J. A. 124 (323), *143*
Murphy, K. E. 670 (153), 672 (154), *690*
Murphy, M. 3 (36, 37, 39), *19*, 112 (279, 280, 282), *142*, 154 (81–83), *160*, 195 (58), 198 (64, 65), 200 (57), *211*, 858 (7), *876*
Murphy, M. D. 193 (133, 184), *213*, *214*, 534 (36–40), *582*, 858 (7), *876*
Murray, G. J. 764 (86), *773*
Murray, P. J. 418 (150), *440*
Murrell, J. N. 310 (158), *344*
Murthy, C. P. 354 (47), *362*
Murugesh, M. G. 644 (107), *688*
Muthyala, R. S. 920 (194), *984*
Mwene-Mbeja, T. M. 620 (52), *685*
Mykhalichko, B. M. 74 (163), *139*, 249 (9), *277*
Mys'kiv, M. 252 (18, 28, 29), *277*
Mys'kiv, M. G. 74 (163), *139*, 249 (9), 252 (10, 12–17, 19–27, 30–32), 254 (33), *277*

Naasz, R. 203 (228), *215*, 401 (105), *439*, 474 (133), *522*, 695 (1), 707 (39, 41), 714 (55), 719 (59), *729*, *730*, 858 (12), *877*
Nacci, A. 663 (138), *689*
Nadeau, C. 622 (55), *685*
Naef, R. 430 (21), *436*
Näf, F. 859 (15, 18), *877*
Nagai, K. 228 (60), 230 (77), *242*, *243*
Nagane, R. 330 (228), *346*
Nagao, S. 285 (13), *341*
Nagaoka, S-i. 998 (33), *1029*
Nagarajan, L. 516 (257), *524*
Nagendran, S. 238 (134), *244*
Naidu, A. B. 897 (106), *982*
Nair, H. K. 1011 (69), *1030*
Nair, V. 1019 (118), *1031*
Naito, H. 937 (275), *986*
Naito, K. 226 (52), *242*
Naito, S. 953 (351), *988*
Najera, C. 965 (397), *988*
Nakada, M. 502 (210), *523*
Nakagawa, Y. 430 (24), *436*
Nakai, K. 616 (27), 625 (40), 629 (68), *684–686*
Nakai, T. 391 (88), *438*
Nakajima, A. 285 (13), *341*
Nakajima, K. 462 (103), 468 (106), 469 (108), 470 (110), *521*
Nakajima, T. 283 (44), *341*
Nakamura, A. 447 (27), *519*, 996 (23), *1029*
Nakamura, E. 2 (27), 3 (3, 30, 31, 35, 40), 5 (42–46), 6 (50, 51), 7 (54, 55), 8 (8, 56), 9 (9, 62, 63), 11 (72, 73), 13 (77–79, 82), 15 (84, 85, 88), 16 (89, 102, 103), 17 (107–109), 18 (119), *19–21*, 133 (20, 21, 277), *135*, *142*, 164 (16), 173 (144–146), 184 (112, 177, 179), 194 (175, 176), 198 (111, 173, 174, 206), 200 (200), 201 (201), *210*, *212–214*, 367 (7), 416 (138), *435*, *440*, 444 (10), 456 (11), 459 (87, 88), 471 (115), 478 (142–144), 516 (249, 250), *519–523*, *525*, 534 (29, 31–33), 572 (30), *581*, *582*, 585 (6), *601*, 606 (11), 609 (19, 28), 658 (13), 659 (18, 122, 124), 672 (1), *682–684*, *689*, 800 (55), 849 (198), *852*, *856*, 858 (7), 870 (48), 873 (4), *875*, *876*, *879*
Nakamura, M. 3 (31), 5 (43), 7 (54), *19*, *20*, 173 (146), *213*, 459 (87), *521*, 534 (31), *581*, 585 (6), *601*, 800 (55), *852*
Nakamura, S. 5 (42, 43), *19*, *20*
Nakamura, S.-i. 502 (210), *523*
Nakamura, Y. 461 (101), *521*, 661 (133), *689*
Nakanishi, N. 872 (51), *879*
Nakanishi, W. 3 (40), *19*
Nakano, K. 18 (110, 111), *21*, 430 (26), *436*, 707 (37, 38), *730*, 908 (151), *983*
Nakano, M. 1016 (104), *1031*
Nakao, T. 625 (16, 40), *684*, *685*
Nakashima, H. 484 (165), *522*
Nakashima, R. 352 (38), *362*
Nakata, K. 644 (107), *688*
Nakatsuji, H. 283 (44), *341*
Nakatsuka, T. 50 (122), *138*
Nakayama, K. 150 (42), *160*
Nalbandyan, R. M. 349 (28), *362*
Naldini, L. 81 (179), *139*
Nanayakkara, A. 283 (44), *342*

Nandakumar, M. V. 926 (243), *985*
Nandurkar, N. S. 920 (199), 966 (406), *984*, *989*
Nankina, F. B. 155 (89), *161*
Nar, H. 350 (26), *362*
Narasaka, K. 376 (39), *436*
Narbonne, C. 861 (19), *878*
Nardin, G. 71 (71), 101 (230), *137*, *141*
Narender, N. 947 (323), *987*
Narita, K. 461 (96, 97), *521*
Naruta, Y. 831 (164), *855*
Naso, F. 563 (48), 564 (117), *582*, *583*, 613 (31), *684*
Natarajan, G. S. 148 (10), *159*
Naumann, D. 113 (286, 287), 114 (288, 289), *142*, 154 (84–86), *161*, 459 (84), *521*
Nava, P. 293 (33), *341*
Navon, N. 256 (37), *277*
Nebesny, K. 226 (55, 57), *242*
Neeb, M. 157 (113), *161*
Neese, F. 358 (65), *363*
Negishi, E. 452 (46), 459 (79, 80), 461 (98, 99), *520*, *521*
Negishi, E.-i. 943 (296), *986*
Neh, H. 503 (219–222), *524*
Neipp, C. E. 386 (68), *437*
Nemeth, G. A. 800 (82), *853*
Nemoto, H. 625 (40, 64), *685*, *686*
Neri, C. R. 237 (115), *244*
Nersissian, A. M. 349 (28), *362*
Nesmeyanov, A. N. 33 (62), *136*
Neta, P. 158 (137), *162*
Nethaji, M. 238 (132), *244*
Netzel, T. 220 (22), *241*
Neue, G. 163 (2), *210*
Neuhaus, D. 176 (88), *212*
Neumann, B. 66 (138), *138*
Neumüller, B. 100 (228), *141*
Neuville, L. 386 (62), *437*
Neves, M. G. P. M. S. 238 (123), *244*
Nevitt, T. 348 (18), *361*
Newton, M. G. 800 (82), *853*
Newton, R. F. 622 (59), *686*
Newton, T. W. 779 (29), 803 (23), *851*
Ng, J. S. 453 (61), 509 (231), *520*, *524*, 819 (133), *854*
Ng, M. F. Y. 77 (166), *139*
Nguyen, B. 925 (224), *985*
Nguyen, B. V. 1026 (154), 1028 (155), *1032*
Nguyen, L. M. 902 (125), *983*
Nguyen, T. 459 (79), *520*, 1015 (89), *1030*
Ni, B. 304 (129), *343*
Nibu, N. 998 (31–33), *1029*
Nice, L. E. 387 (75), 388 (85), 389 (74), 393 (87), *438*, 635 (92), *687*
Nickisch, K. 15 (86), *20*, 503 (218–222), *524*, 654 (127), *689*
Nicolaou, K. C. 238 (138), *244*
Nicolini, C. 229 (71), *243*
Nicoud, J. F. 221 (36), *242*
Nielsen, R. B. 459 (82, 83), *521*
Nielsen, T. E. 824 (156), 850 (200), *855*, *856*
Niemann, N. C. 93 (50), *136*
Niemeyer, M. 31 (59), 44 (90), 60 (123, 124), *136–138*, 152 (57), *160*
Nierengarten, J. F. 221 (36), *242*
Niida, A. 641 (105), *688*
Niimura, K. 800 (31), *851*
Nikolaev, E. N. 322 (190), *345*
Nikolaitchik, A. V. 220 (27), *242*
Nilsson, K. 8 (57, 58), *20*, 108 (255), *141*, 179 (128), 184 (188), 191 (191), *212*, *214*, 388 (80), *438*, 534 (34, 35), *582*, 790 (26), 829 (160), *851*, *855*
Nilsson, K. B. 165 (55), *211*
Nilsson, M. 16 (95), *21*, 184 (183), *214*, 459 (91, 93), *521*, 867 (33), *879*
Nimrod, A. 947 (327), *987*
Ninomiya, K. 571 (131), *584*
Nishihara, Y. 468 (105), 513 (241), *521*, *524*, 738 (23), *772*, 977 (463), 978 (468), *990*
Nishii, S. 614 (17), 625 (16, 40), 629 (68), *684–686*
Nishikawa, T. 418 (146), *440*
Nishimura, J. 474 (125), *522*, 979 (478), *990*
Nishimura, Y. 157 (112), *161*, 230 (86), *243*
Nishinaga, T. 595 (28), *602*
Nishioka, T. 116 (292–294), 117 (295), *142*
Nishiura, M. 403 (116), *439*
Nishiwaki, T. 1007 (54), *1030*
Nishiyama, H. 758 (73), *773*
Nitzan, A. 228 (62), *242*
Nivlet, A. 979 (485), *990*
Niwa, S. 629 (74), *687*
Nixon, P. G. 154 (85), *161*
Nobel, D. 28 (47), *136*
Nocera, G. 223 (43), *242*
Noda, Y. 459 (79), *520*
Nogueira, C. W. 943 (229), *985*
Noguera, M. 303 (126), 310 (156, 157), *343*, *344*
Nohr, R. S. 238 (123), *244*
Noiri, A. 549 (78, 79), *583*
Nolan, S. P. 119 (302), 120 (316), 121 (311, 312), *142*, *143*, 400 (101, 102), 403 (111, 114), *438*, *439*, 750 (53, 54), 751 (55, 56), 761 (79), *773*
Nolle, A. 165 (30, 34), *210*
Noller, C. R. 978 (469), *990*
Nolte, C. 47 (104), *137*, 184 (190), *214*
Noltemeyer, M. 386 (60), *437*, 859 (18), *878*
Noltes, J. G. 25 (33), 27 (40), 34 (27, 28, 36, 37), 36 (67, 69), 39 (85), 40 (86), 51 (30), 52 (43, 110, 111), 53 (65), 66 (66), 68 (68), 70 (70), 91 (214, 215), 94 (218, 219), 95

(220), 112 (112), 113 (113), 134 (29), *136–138*, *140*
Noma, S. 384 (57), *437*
Nomura, M. 965 (399), 966 (398), 968 (422), 969 (428), *988*, *989*
Nomura, N. 609 (19), *684*
Nonaka, H. 644 (107), *688*
Nonaka, T. 810 (103), *854*
Nonappa, P. D. 902 (121), *982*
Nonoyama, K. 499 (206), *523*
Noort, M. 220 (19), *241*
Nordmann, G. 901 (111), 902 (120), *982*
Norimatsu, T. 228 (60), *242*
Norinder, J. 198 (173), *214*, 606 (11), 669 (144), 680 (173), 682 (174), *683*, *690*, *691*
Noritake, Y. 609 (19), *684*
Normant, J. 2 (22), *19*, 638 (95), 663 (85, 134), *687–689*
Normant, J. F. 1 (1), 10 (64), *19*, *20*, 169 (105), *212*, 474 (122), *521*, 529 (15), 530 (11, 20–22, 24), 532 (13, 14, 26), 535 (42, 45), 536 (27, 28, 47), 537 (46), 539 (52–54), 540 (8–10), 543 (1), 545 (70, 72–74), 547 (71), 550 (63, 66, 80), 559 (62, 64), 561 (3), 569 (107), *581–583*, 663 (85, 134, 136), *687*, *689*, 811 (110), 842 (179), *854*, *856*, 873 (4), 874 (56), *875*, *879*, 999 (35), *1029*
Norrby, P.-O. 850 (200), *856*
Norris, V. A. 352 (39), *362*
Norris, W. P. 1005 (52), *1030*
Norton, E. B. 1018 (112), *1031*
Norton, J. R. 3 (32), *19*
Noson, K. 452 (50), *520*, 738 (24), 749 (51), 753 (60, 63), 755 (68, 69), *772*, *773*
Notario, R. 310 (170), *344*
Novak, L. 859 (18), *877*
Nowell, I. W. 893 (91), *982*
Noyes, W. A. 157 (121), *162*
Noyori, R. 18 (110), *21*, 430 (26), *436*, 707 (37, 38), *730*
Nozaki, H. 502 (211), *523*, 777 (15), 809 (102), 810 (103), 816 (119), 821 (140), 841 (170), *851*, *853–855*
Nozaki, K. 908 (151), *983*
Nudelman, N. S. 976 (456), *990*
Nukada, T. 305 (133), *344*
Numata, T. 800 (52), *852*
Nunes, P. 158 (134), *162*
Nunes, R. L. 452 (55), *520*
Nunn, C. M. 103 (236), *141*
Nunomoto, S. 979 (474), *990*
Nuttall, R. L. 155 (3), *158*
Nuzzo, R. G. 150 (41), *160*
Nygren, M. A. 157 (115), *161*
Nyokong, T. 218 (4), *241*
Nysted, L. N. 861 (19), *878*
Nyzam, V. 424 (121), *439*
O Peña, M. M. 348 (10), *361*
O'Brien, D. F. 226 (53–57), *242*
O'Brien, P. 399 (95), *438*
O'Hair, R. A. J. 149 (25), 152 (23), *159*, 329 (214, 215, 217), 339 (248), *346*
O'Halloran, T. V. 360 (77, 79), *363*
O'Rear, J. G. 1013 (79), *1030*
O'Reilly, J. M. 861 (21), *878*
O'Shea, D. F. 952 (353), *988*
Oae, S. 800 (52), *852*, 906 (142), *983*
Oba, G. 395 (90), *438*
Obata, M. 116 (294), *142*
Obayashi, M. 447 (27), *519*
Oberhuber, C. N. 133 (337), *143*, 402 (108), *439*, 508 (150), *522*, 672 (155), *690*
Obermann, U. 947 (318, 319), *987*
Ochi, M. 979 (433), *990*
Ochsenbein, U. 165 (32), *210*
Ochterski, J. 282 (27), 283 (44), *341*, *342*
Oconnor, T. R. 337 (247), *346*
Oda, H. 777 (15), 816 (119), *851*, *854*
Odagaki, Y. 625 (42), *685*
Odinokov, V. N. 534 (41), *582*
Oehler, H. 165 (34–36), *210*
Oehlschläger, A. C. 184 (185), *214*, 545 (69), *582*, 631 (81), *687*, 800 (60), 818 (25), 820 (16), 824 (155), 834 (22, 132), 836 (173), 841 (182), 842 (175, 176), 846 (90), *851–856*
Oestreich, M. 382 (52), 418 (144), *437*, *440*, 732 (1), 769 (92), *771*, *773*, 858 (8), *876*
Öffle, K. 119 (304), *142*
Oganesyan, V. S. 359 (64), *363*
Ogasawara, K. 824 (69), *852*
Ogawa, A. 738 (20), *772*
Ogawa, M. 228 (65), *243*
Ogawa, N. 946 (316, 317), *987*
Ogawa, S. 831 (164), *855*
Ogawa, T. 221 (34), *242*, 940 (285), 941 (286), *986*
Ogawa, Y. 1008 (57), *1030*
Ogel, Z. B. 351 (32), *362*
Ogilvie, W. W. 528 (6), *581*
Ogimura, Y. 430 (174), 433 (173), *441*
Ogle, C. 112 (279), *142*
Ogle, C. A. 3 (36, 37, 39), *19*, 105 (241), 112 (280, 282), *141*, *142*, 154 (81–83), *160*, 169 (119), 173 (123), 191 (102), 193 (133, 184), 195 (58), 198 (64, 65), 200 (57), *211–214*, 430 (20), *436*, 534 (36–40), *582*, 778 (24), *851*, 858 (7), *876*
Ogle, G. A. 858 (7), *876*
Oguro, K. 457 (72), *520*
Oh, D. Y. 571 (129), *584*
Ohanessian, G. 301 (30), 319 (183), *341*, *345*
Ohara, T. 1008 (57), *1030*
Ohbayashi, A. 979 (478), *990*
Ohishi, S. 635 (90), *687*

Ohishi, T. 403 (116), *439*
Ohkuma, H. 238 (136), *244*
Ohloff, G. 859 (15), *877*
Ohmasa, N. 887 (30, 32), 892 (82), *981*, *982*
Ohno, H. 629 (68), 635 (88, 90), 641 (105), *686–688*
Ohno, M. 502 (210), *523*, 911 (161), *983*
Ohno, T. 925 (190), *984*
Ohta, N. 831 (164), *855*
Ohtani, H. 382 (54), *437*, 640 (99), *688*, 802 (48), *852*
Ohwada, T. 5 (43), *20*
Oi, Y. 401 (107), *439*, 672 (156), *690*
Oiarbide, M. 902 (126), *983*
Oisaki, K. 762 (83), 763 (80, 84), *773*
Oishi, S. 625 (42), 641 (105), *685*, *688*
Ojima, I. 489 (169), *523*, 632 (45), *685*
Okada, M. 461 (101), *521*, 661 (133), *689*
Okada, Y. 230 (81–85), *243*
Okamoto, S. 401 (107), *439*, 444 (9), *519*, 672 (156), *690*
Okamura, W. H. 863 (27), *878*
Okano, K. 919 (180), *984*
Oku, A. 946 (316, 317), 979 (478), *987*, *990*
Okuda, Y. 809 (102), 810 (103), *853*, *854*
Okuro, K. 965 (399), 966 (398), 969 (428), *988*, *989*
Olah, G. A. 952 (356), *988*, 994 (3), 1016 (96), *1029*, *1031*
Olbrich, F. 32 (61), 74 (165), 82 (178), *136*, *139*
Oliveira, A. R. M. 838 (177), *856*
Oliver, J. P. 800 (36), *851*
Olivier, P. J. 129 (332), *143*
Olmstead, M. M. 49 (92), 95 (106, 223), 99 (226), 106 (242), 118 (118), *137*, *138*, *140*, *141*, 152 (53, 56, 65), *160*, 179 (130), *213*
Olsen, K. 126 (325), *143*
Olsson, L.-I. 632 (84), *687*
Olsson, M. H. M. 307 (145), 310 (10, 144), *341*, *344*
Olsson, T. 16 (95), *21*, 184 (134, 181, 183), *213*, *214*, 459 (91, 93), *521*, 867 (33), *879*, 952 (352), *987*
Omote, M. 1008 (56, 57), *1030*
Omur, O. 377 (41), *436*
Omura, H. 27 (24), *135*
Onami, T. 906 (143), *983*
Ondrus, A. E. 743 (34), *772*
Ongeri, S. 670 (147), *690*
Ono, N. 666 (141), *689*, 998 (33), *1029*
Onuma, K. 979 (480), *990*
Oostveen, J. M. 558 (108), *583*
Opie, T. R. 994 (16), *1029*
Oppenheimer, J. 934 (264), *986*
Oppolzer, W. 800 (68), *852*
Oram, D. 106 (248), *141*, 152 (64), *160*
Organ, M. G. 979 (492), *990*
Orlova, G. 329 (216, 220), *346*
Orlovich, V. A. 220 (20), *241*
Örtendahl, M. 58 (48), *136*
Ortiz, J. V. 283 (44), *342*
Ortuno, R. M. 859 (16), *877*
Osada, Y. 228 (67), *243*
Osakada, K. 81 (183), *139*, 513 (241), *524*, 977 (463), *990*
Osawa, M. 271 (64), *278*
Osborn, J. A. 732 (2), *771*
Osburn, E. J. 226 (53), *242*
Oshima, K. 447 (24), 480 (154), 502 (211), *519*, *522*, *523*, 569 (128), *584*, 675 (2), *682*, 777 (15), 809 (102), 810 (103), 816 (119), 841 (170), *851*, *853–855*
Oshino, H. 416 (138), *440*
Osugi, M. E. 233 (96), 234 (97), *243*
Osuka, A. 117 (296, 297), 118 (298, 299), 119 (300), *142*, 240 (139), *244*, 257 (41), *277*, 887 (28, 30–32), 892 (82), 968 (425), *981*, *982*, *989*
Otaka, A. 484 (165), *522*, 625 (42), 629 (68), 641 (105), *685*, *686*, *688*
Ototake, S. 891 (47), *981*
Ott, E. 902 (122), *983*
Otting, G. 197 (197), *214*
Otto, B. 399 (98), *438*
Otvos, J. D. 330 (232), *346*
Ouali, A. 200 (213), *214*, 897 (108), 921 (220), 967 (413), *982*, *985*, *989*
Ouali, L. 221 (36), *242*
Ouannes, C. 173 (139), *213*
Ouellet, R. 238 (124), *244*
Oulie, P. 904 (128), *983*
Oulyadi, H. 166 (67), *211*
Ousset, J. B. 535 (43), *582*
Overman, L. E. 478 (145), *522*, 800 (82), *853*
Owen, D. A. 71 (155), 73 (156), *139*
Owton, W. M. 1004 (46), *1029*
Ozawa, F. 920 (183), *984*
Ozin, G. A. 147 (5), 153 (70, 71), 157 (103), *159–161*
Ozutsumi, K. 288 (71), *342*

Pabon, H. J. J. 609 (29), *684*
Pac, C. 230 (87), *243*
Pachinger, W. 800 (68), *852*
Packheiser, R. 269 (62), 275 (61, 70), *278*
Padmanaban, D. 894 (96), *982*
Padwa, A. 925 (223), *985*
Page, C. C. 358 (71), *363*
Pagelot, A. 165 (40), *211*
Pagliaro, M. 236 (107), *244*
Paine, A. J. 942 (78), *982*
Paisley, S. D. 609 (26), 647 (9), *683*, *684*, 872 (50), *879*
Palais, L. 506 (225), *524*, 705 (30), *729*

Palmer, A. E. 360 (52), *362*
Palmisano, G. 236 (107), *244*
Palmisano, L. 235 (101–105), 236 (106), *243*, *244*
Palomo, C. 902 (126), *983*
Palucki, M. 899 (114), *982*
Pàmies, O. 2 (23), *19*, 24 (17), *135*, 202 (23), *210*, 695 (1), *729*, 858 (10), *876*
Pan, S. 280 (9, 11), *341*
Pan, X. 932 (259), *985*
Pan, Y. 908 (155), *983*
Pancrazi, A. 842 (188), *856*
Pande, S. G. 888 (35), *981*
Pandurangan, K. 902 (121), *982*
Panek, E. J. 586 (8), *601*
Panella, L. 719 (59), *730*
Pannala, M. 921 (203), *984*
Papa, P. 738 (24), 739 (26), *772*
Papadopoulos, S. 37 (74), *137*
Papasergio, R. I. 37 (72), 73 (73), *137*, 261 (47), *277*
Paquette, B. 238 (125), *244*
Paquette, L. 672 (1), *682*
Paradies, G. 647 (111), *688*, 858 (13), *877*
Paraskevopoulos, K. 307 (146), *344*
Paraskos, A. J. 598 (32), *602*
Parella, T. 114 (290, 291), *142*, 964 (394), *988*
Parida, K. M. 237 (113), *244*
Parisi, L. M. 966 (403), *989*
Park, B.-M. 747 (46), *772*
Park, J. D. 993 (10, 11), *1029*
Park, S. B. 1011 (68), *1030*
Park, S. R. 124 (323), *143*
Park, Y. S. 386 (69), *437*
Parker, D. 382 (55), *437*, 800 (67), 807 (95), *852*, *853*
Parker, D. C. 800 (45), *852*
Parker, K. A. 554 (101), *583*
Parker, V. B. 155 (3), *158*
Parkin, A. 87 (207), *140*
Parkinson, A. 232 (92), *243*
Parkinson, B. A. 226 (53), *242*
Parnis, J. M. 147 (5), *159*
Parr, R. G. 282 (23), *341*
Parrain, J. L. 841 (162), 842 (178), 845 (193), *855*, *856*
Parris, S. 927 (254), *985*
Parson, P. J. 842 (189), *856*
Parsons, M. R. 354 (45), *362*
Parsons, S. 87 (205–207), *140*
Partridge, H. 294 (91), 302 (124), *343*
Pasquali, M. 106 (246), *141*
Passacantilli, P. 859 (17), *877*
Patai, S. 147 (4), *159*, 164 (10), *210*, 866 (29), *878*
Patel, P. K. 975 (451), *989*
Patel, S. K. 791 (21), *851*
Patil, N. M. 920 (189), *984*
Patil, Y. P. 966 (406), *989*
Patra, B. S. 237 (113), *244*
Patrick, J. S. 322 (191), *345*
Patterson, R. G. 430 (18), *435*, *436*
Pauson, P. L. 63 (127), *138*
Pavelka, M. 281 (17), 282 (18), 309 (115), *341*, *343*
Paver, M. A. 106 (176), *139*
Pavl'uk, O. V. 252 (15), *277*
Pavlath, A. E. 1016 (99), *1031*
Pavlyuk, A. V. 252 (13, 14), *277*
Pavlyuk, O. V. 252 (22), *277*
Pawlicki, M. 119 (301), *142*
Payne, A. D. 595 (29), *602*
Payne, A. H. 324 (195), *345*, 842 (187), *856*
Pazderski, L. 155 (45), *211*
Peach, M. E. 837 (27, 29), *981*
Peake, D. A. 154 (79), *160*
Pearce, R. 27 (26, 42), *136*
Pearson, R. G. 9 (60), *20*, 91 (213), *140*, 170 (127), 173 (141), *212*, *213*, 348 (19), *361*
Pecchi, S. 828 (158), *855*
Pedersen, S. D. 997 (26), *1029*
Pedler, A. E. 1014 (86), *1030*
Pedram, B. 382 (54), *437*
Pedrotty, D. M. 178 (164), *213*
Peese, K. M. 728 (73), *730*
Peiper, S. C. 484 (165), *522*, 641 (105), *688*
Peisach, J. 349 (28), *362*
Peisong, T. 237 (111), *244*
Pekel, Ö. Ö. 474 (130), *522*
Pellizer, G. 39 (80), *137*
Peltier, J. M. 330 (224), *346*
Peña, D. 18 (118), *21*, 200 (216), *214*, 478 (141), 493 (177), *522–523*, 670 (152), *690*, 698 (16), *729*
Pendola, K. A. 953 (364), *988*
Peng, C. Y. 283 (44), *342*
Peng, F. 908 (155), *983*
Peng, Q. 238 (119), *244*
Peng, Y. 1022 (130), *1031*
Peng, Z.-H. 382 (54), *437*, 612 (35), *684*
Penner, G. H. 163 (5), *210*
Penner-Hahn, J. E. 108 (257–260), *141*, 175 (154, 155, 160), *213*, 360 (79), *363*, 800 (54, 56, 58), *852*
Penuliar, R. J. 925 (224), *985*
Penwell, A. J. 551 (90), *583*
Pepe, C. 326 (203), *345*
Pépin, C. 223 (42), *242*, 238 (124), *244*
Percy, J. M. 1004 (46), *1029*
Pereira, A. S. 358 (67), *363*
Perera, B. A. 320 (185), *345*
Pereyre, M. 824 (121), *854*
Pérez, C. 574 (139), *584*
Pérez-Hernández, N. 574 (139), *584*
Pérez, P. J. 41 (41), *136*
Pérez, R. 574 (139), *584*

Perez, R. G. 873 (53), *879*
Periasamy, M. 151 (45), *160*, 540 (56), *582*
Perie, J. 1020 (122), *1031*
Peris, E. 120 (306), *142*
Perlmutter, P. 873 (4), *875*
Perreault, S. 607 (15), *683*
Perrone, S. 471 (116), 480 (117), 484 (157, 158), *521*, *522*, 565 (124), *583*, 616 (6, 46), *683*, *685*
Perry, D. R. A. 992 (1), *1028*
Persson, E. S. M. 607 (10), 609 (32), 669 (14, 143), *683*, *684*, *690*, 858 (11), *876*
Persson, I. 165 (55), *211*
Pesce, A. 355 (50), *362*
Pesce, G. 613 (31), *684*
Pestano, J. P. C. 913 (138), *983*
Petasis, N. A. 952 (356), *988*
Petersen, J. L. 121 (313–315), *143*, 403 (118), *439*, 751 (56), *773*
Petersen, M. E. 934 (264), *986*
Peterson, D. J. 777 (20), *851*
Peterson, G. S. 953 (359), *988*
Peterson, M. R. 324 (198, 199), *345*
Peterson, R. 226 (55), *242*
Petersson, G. A. 282 (27), 283 (44), *341*
Petit, L. 289 (72), *342*
Petraglia, S. P. 831 (84), *853*
Petrassi, H. M. 952 (357), *988*
Petrov, L. 237 (112), *244*
Petrow, V. 861 (22), *878*
Petruzelli, R. 352 (34), *362*
Pettiette-Hall, C. L. 152 (61), *160*
Pettig, D. 386 (60), *437*, 859 (18), *878*
Pettit, G. R. 647 (116), *689*
Petty, J. 150 (35), *159*
Pfaltz, A. 695 (8), 716 (57), *729*, *730*
Pfeiffer, S. S. 452 (50), *520*
Pfennig, D. R. 153 (75), *160*
Pfretzschner, T. 18 (115), *21*
Philippe, M. 860 (20), *878*
Phillips, A. J. 739 (25), *772*
Phillips, D. 85 (199), *140*, 262 (53), *278*
Phillips, D. P. 925 (224), *985*
Phillips, S. E. V. 351 (32), 354 (44), *362*
Pi, S.-F. 975 (405), *989*
Piancatelli, G. 859 (17), *877*
Pianet, I. 200 (214), *214*
Piarulli, U. 18 (116), *21*, 670 (147), 679 (170, 171), *690*
Piccardi, R. 842 (184), *856*
Pichon, C. 946 (313), 947 (314), *987*
Pichota, A. 18 (117), *21*, 201 (220), 202 (219), *215*
Pierce, H. D. 824 (155), *855*
Pierloot, K. 310 (10, 144), *341*, *344*
Pierpont, A. W. 121 (313), *143*, 403 (118), *439*
Piers, E. 510 (236–238), *524*, 817 (7), 820 (13), 821 (143), 824 (123, 145), 829 (154, 157, 161), 831 (163), 834 (159, 168, 169), 836 (147, 171), 845 (195, 196), *851*, *854–856*, 975 (448), *989*
Pietruszka, J. 416 (134), *440*
Pietrzyk, D. J. 1009 (62), *1030*
Pifferi, C. 339 (249), *346*
Piguel, S. 973 (435), *989*
Pike, R. D. 902 (125), *983*
Pilcher, G. 148 (8), *159*
Pilling, R. L. 68 (143), *138*
Pineschi, M. 474 (139), *522*, 675 (2), 676 (164), 677 (167, 168), 679 (169, 172), *683*, *690*, *691*, 694 (5), 695 (9), *729*
Pinhey, J. T. 949 (340, 341), *987*
Pinho, P. 719 (59), *730*
Pinney, K. G. 545 (68), *582*, 631 (78), *687*
Piper, T. S. 130 (130), *138*
Piscopio, A. D. 616 (6), *683*
Piskorz, P. 283 (44), *342*
Pistorio, B. J. 223 (43), *242*
Pitt, A. R. 551 (94), *583*
Pittam, J. D. 861 (22), *878*
Pitts, A. D. 68 (143), *138*
Piveteau, E. 373 (33), *436*
Pivsa-Art, S. 968 (422), *989*
Pizzuti, M. G. 474 (131), 492 (176), *522*, *523*, 695 (10), *729*
Planchenault, D. 800 (50), *852*
Platonov, V. 1012 (77), *1030*
Plaut, H. E. 800 (45, 46), *852*
Plefka, O. 769 (92), *773*
Plitt, H. S. 157 (126), *162*
Plotkin, S. 518 (89), *521*
Poater, A. 114 (291), *142*
Poater, J. 284 (38), *341*
Pochapsky, S. 189 (192), *214*
Pochapsky, T. C. 166 (85), *212*
Polet, D. 480 (155), *522*, 670 (149), 675 (2), *683*, *690*, 725 (71), *730*, 858 (11), *877*
Poli, G. 741 (29), *772*, 828 (158), *855*
Polizzi, C. 638 (94), *687*
Polyakov, O. G. 73 (158), *139*
Pomelli, C. 283 (44), *342*
Pople, J. A. 30 (57), *136*, 173 (147), *213*, 282 (28), 283 (44), *341*, *342*
Porciatti, E. 832 (78), *853*
Porco, J. A. 932 (255, 256, 260), *985*
Porento, M. 309 (101), *343*
Pornet, J. 832 (78), *853*
Porter, M. J. 551 (94), *583*
Porzelle, A. 926 (248), *985*
Posey, L. A. 286 (60), *342*
Posner, G. H. 13 (80), 16 (99), 17 (17), 18 (18), *19–21*, 24 (10, 11), *135*, 366 (2), *435*, 776 (5), 780 (3), 803 (64), *851*, *852*, 873 (4), *875*
Posset, T. 200 (215), *214*
Poulios, I. 234 (99), *243*

Poulos, T. 348 (7), 352 (41), 354 (45), 355 (50), 356 (53, 58), *361–363*
Poulsen, T. B. 200 (14), *210*
Pour, M. 493 (184), *523*
Pourcelot, G. 430 (172), *441*
Power, J. M. 103 (236), *141*
Power, P. P. 13 (13), *19*, 41 (91), 49 (92), 50 (109), 95 (106, 223), 99 (226), 106 (233, 242, 248), 109 (267), 110 (270), 118 (118), *137*, *138*, *140*, *141*, 152 (53, 55, 56, 64–66), *160*, 174 (143), 179 (130), *213*, 820 (135), *854*, 858 (5), *875*
Prabhakar, R. 280 (5), *341*
Prado, A. G. S. 237 (115), *244*
Prahov, L. 237 (112), *244*
Prakash, G. K. S. 952 (356), *988*, 994 (3), 1016 (96), *1029*, *1031*
Prandi, C. 958 (380), *988*
Pranowo, H. D. 305 (135, 139), *344*
Prasad, A. S. B. 424 (121), *439*
Prasad, D. J. C. 897 (106), *982*
Prasad, D. R. 220 (28), *242*
Prasanth, K. L. 922 (211), *984*
Pratihar, D. 499 (202, 203), *523*
Pregosin, P. S. 18 (117), *21*, 168 (86, 94), 197 (204), 202 (219), *212*, *214*, *215*
Prestat, G. 741 (29), *772*
Pretzel, D. 324 (201), *345*
Price, R. 892 (83–86), *982*
Price, S. J. B. 197 (205), *214*
Price, W. S. 163 (9), *210*
Prie, G. 845 (193), *856*
Prieto, O. 506 (223), *524*, 698 (15), *729*
Prigge, S. T. 280 (2), *341*, 350 (30), *362*
Priqueler, J. R. L. 163 (6), *210*
Pritzkow, H. 78 (170, 172), *139*, 271 (65), 276 (72), *278*
Priyakumar, U. D. 238 (134), *244*
Prodolliet, J. 191 (103), *212*
Proess, G. 979 (485), *990*
Provins, L. 824 (151), *855*
Prudêncio, M. 358 (67), *363*
Prunet, J. 842 (188), *856*
Prusiner, S. B. 332 (225), *346*
Pu, L. 629 (75), *687*
Pu, Y. J. 901 (118), *982*
Puebla, P. 613 (31), *684*
Puig, S. 348 (4), *361*
Pulido, F. J. 779 (29), 800 (39), 807 (93), 811 (106–108), 812 (105), 813 (38), 816 (30, 112–118), 819 (134), 821 (12, 142), 841 (144), 845 (194), 846 (9, 125), 847 (129, 181, 197), *851–856*
Pulkkinen, S. 303 (126), *343*
Pulley, S. R. 649 (117), *689*
Pullez, M. 719 (60), *730*
Purpura, M. 649 (119), 651 (120), *689*, 874 (54, 55), *879*
Puskar, L. 286 (63), 288 (54, 62), 305 (64), *342*
Py, S. 551 (91), *583*
Pytkowicz, J. 401 (104), *438*, 489 (173), *523*
Pyun, S.-J. 977 (461), *990*
Pyykkö, P. 90 (211), *140*

Qi, X. 920 (196), *984*
Qin, O. 157 (127), *162*
Qing, F-L. 1021 (129), 1022 (130), *1031*
Qiu, W. 1008 (59), 1011 (67), *1030*
Quach, T. D. 953 (363), *988*
Quan, P. M. 891 (51), *981*
Quannes, C. 105 (237), *141*
Quartarolo, A. D. 289 (73), *342*
Quayle, P. 834 (167), *855*
Que, D. L. 225 (51), *242*
Queguiner, G. 421 (152), *440*
Quick, A. J. 890 (55), *981*
Quick, J. 943 (301), *986*
Quickenden, T. I. 165 (49, 53), *211*
Quinlan, R. A. 902 (125), *983*
Quinn, N. R. 831 (84), *853*
Quintanar, L. 348 (17), *361*
Quintard, J. P. 824 (121), 842 (178), *854*, *856*

Raabe, G. 666 (140), *689*
Rabani, J. 221 (32), *242*
Rabuck, A. D. 283 (44), *342*
Rack, J. J. 157 (110), *161*
Radhakrishnan, S. 228 (61), *242*
Rae, A. 418 (149), *440*
Rafalski, M. 956 (371), *988*
Raffa, D. F. 306 (142, 143), 309 (4), *341*, *344*
Ragaini, F. 18 (116), *21*
Ragan, M. A. 885 (13), *980*
Raghavachari, K. 282 (28), 283 (44), *341*, *342*
Raghunath, O. R. 897 (106), *982*
Rahm, A. 824 (121), *854*
Rahman, M. J. 595 (22, 28), *602*
Rainka, M. P. 742 (33), 744 (38), *772*
Raithby, P. 813 (111), *854*
Raithby, P. R. 106 (176), *139*, 258 (42), 272 (43), *277*
Rajagopal, D. 383 (58), *437*
Rajamohanan, P. 858 (6), *876*
Rajamohanan, P. R. 2 (29), *19*, 133 (222), *140*, 168 (91), 175 (129, 142), 179 (87), *212*, *213*, 858 (6), *876*
Rajesh, B. M. 925 (236), *985*
Rakiewicz, D. M. 646 (109), *688*
Rakotoarisoa, H. 663 (135), *689*, 873 (53), *879*
Rall, J. 348 (3), *361*
Ralph, A. 349 (29), *362*
Ram, M. K. 229 (71), *243*
Ramachandran, P. V. 499 (202, 203), *523*
Ramalingam, V. 920 (194), *984*

Ramaprasad, S. 163 (7), *210*
Ramazzotti, D. 675 (162), *690*
Ramdhanie, B. 943 (295), *986*
Rami, H. K. 417 (141), *440*
Ramiandrasoa, F. 539 (50, 51), *582*
Raminhos, H. 158 (134), *162*
Ramirez, A. 173 (150), *213*
Ramirez, M. A. 121 (315), *143*, 403 (118), *439*
Ramshaw, J. A. M. 352 (39), *362*
Ran, C. Z. 920 (191), *984*
Rance, M. J. 885 (18), *980*
Randaccio, L. 71 (71), 101 (230), *137*, *141*
Rannulu, N. S. 299 (118), 300 (57), 304 (106), *342*, *343*
Ranu, B. C. 904 (139), *983*
Rao, C. P. 430 (22), *436*
Rao, D. P. 968 (420, 421), *989*
Rao, H. 915 (177), 920 (195), 940 (107), *982*, *984*
Rao, R. D. 328 (211), *345*
Rao, S. A. 151 (45), *160*, 858 (8), *876*
Rao, S. A. A. 540 (56), *582*
Rappoli, B. J. 336 (242), *346*
Rappoport, Z. 147 (4), 151 (46), 153 (76), *159*, *160*, 163 (4), *210*, 219 (8, 9), *241*, 480 (117), *521*, 870 (14), *877*
Rashatasakhon, P. 925 (223), *985*
Rasmussen, T. 358 (62), 359 (64), *363*
Rassu, G. 81 (179), *139*
Rastogi, A. 169 (119), *212*, 430 (20), *436*
Raston, C. L. 37 (72, 74), 73 (73), *137*, 261 (47), *277*
Rath, N. 165 (42), *211*
Rathbone, G. J. 149 (29), *159*
Rathgeb, X. 712 (52), *730*
Ratner, M. A. 228 (62), *242*
Raubenheimer, H. G. 129 (332), *143*
Rauk, A. 306 (142, 143), 308 (112), 309 (4), *341*, *343*, *344*
Rausch, M. D. 885 (10), *980*, 1024 (147), *1032*
Ravasio, N. 966 (401), *989*
Ray, K. 238 (134), *244*
Rayner, D. M. 153 (67), *160*
Ready, J. M. 542 (60), *582*
Reagan, D. R. 629 (72), *686*
Reamer, R. A. 173 (151), *213*, 947 (320), *987*
Reddy, K. V. 430 (18), *436*
Reddy, M. V. R. 499 (202), *523*
Reddy, N. L. 800 (67, 72), *852*
Reddy, P. A. N. 238 (132), *244*
Reddy, P. R. 925 (236), *985*
Redeuilh, G. 861 (21), *878*
Redfern, P. C. 282 (28), *341*
Reed, A. E. 296 (108), *343*
Reed, D. 821 (139), *854*
Reed, J. N. 378 (44), *436*, 599 (34), *602*
Reedijk, J. 356 (54), *362*
Reetz, M. T. 179 (153), *213*, 444 (7), *519*
Rega, N. 283 (44), *341*
Reginato, G. 374 (34), *436*, 807 (96, 98), 809 (17, 99), 828 (158), 832 (78), 842 (184, 185), *851*, *853*, *855*, *856*
Reich, H. J. 820 (136), *854*
Reich, M. R. 125 (3), *135*
Reichert, H. 646 (110), *688*
Reichoux, M. C. 230 (79), *243*
Reid, E. E. 902 (122), *983*
Reid, G. 165 (41), *211*
Reider, P. J. 173 (151), *213*, 897 (105), 921 (209), *982*, *984*
Reif, B. 196 (196), *214*
Reijmer, S. J. W. 721 (56), *730*
Reinhammar, B. 349 (21), *361*
Reinhart, A. 220 (23), *242*
Reinhoudt, D. N. 894 (93), *982*
Reintjes, M. 68 (143), *138*
Reis, A. M. M. V. 155 (88), *161*
Reiser, O. 659 (123), *689*
Reiss, P. 103 (235), *141*
Reisser, M. 416 (137), *440*
Reissmann, S. 324 (201), *345*
Remy, P. 926 (239), *985*
Ren, B. 294 (58), *342*
Ren, H. 65 (136), *138*
Ren, R. X. 967 (411), *989*
Ren, Y. 748 (48), *772*
Renaud, J. 845 (195), *856*
Rendler, S. 732 (1), 769 (92), *771*, *773*, 858 (8), *876*
Renner, M. W. 220 (24), *242*
Rennie, M.-A. 106 (176), *139*
Rennison, S. C. 886 (19, 20), *980*
Renshaw, R. R. 885 (14), *980*
Rentzepis, P. M. 220 (21), *241*
Respess, W. L. 16 (16), *19*, 24 (8), *135*, 778 (2), *851*
Retherford, C. 858 (8), *876*
Reuter, D. C. 510 (234, 235), *524*, 818 (32), 834 (33, 131), 842 (130), *851*, *854*
Reynolds, D. J. 911 (159), *983*
Rheingold, A. L. 30 (52), 60 (60), 93 (93), *136*, *137*, 902 (125), *983*
Rheinwald, G. 41 (96), 78 (172), *137*, *139*, 259 (44), 271 (63, 66), 276 (72), *277*, *278*
Rhyne, L. D. 1024 (142), *1032*
Riant, O. 120 (316), *143*, 374 (36), 403 (114), *436*, *439*, 732 (1), 753 (61, 62), 761 (79, 81), 762 (82), 765 (88), *771*, *773*
Ribas, X. 114 (290, 291), *142*, 964 (394), *988*
Ribeiro, A. O. 237 (115), *244*
Ribeiro da Silva, M. A. V. 155 (88), *161*
Ricca, A. 282 (25), 283 (47), *341*, *342*
Ricci, A. 373 (28), 374 (34), 378 (45), 421 (151, 155), *436*, *440*, 599 (35, 36), *602*, 802 (87, 88), 807 (96), 809 (17, 99, 100), 832 (78), 842 (184), *851*, *853*, *856*

Rice, E. M. 831 (84), *853*
Richardson, D. C. 354 (49), *362*
Richard-Foy, H. 861 (21), *878*
Richardson, D. C. 354 (31), *362*
Richardson, J. S. 354 (31, 49), *362*
Rickard, G. A. 306 (142, 143), 309 (4), *341*, *344*
Rickards, R. W. 445 (13), *519*, 647 (112), *688*
Rickborn, B. 638 (7), *683*
Ridge, D. P. 310 (159), *344*
Riebel, P. 894 (97), *982*
Rieke, R. D. 149 (31, 32), *159*, 447 (29, 30), *519*, 1024 (142–144), *1032*
Rieker, W. F. 421 (70), *437*
Riess, J. G. 992 (9), 1015 (94), *1029*, *1031*
Riester, J. 352 (35), 359 (61), *362*, *363*
Rietveld, T. 551 (91), *583*
Rigby, J. H. 516 (255), *524*
Rijs, N. 152 (23), *159*
Rilatt, I. 418 (150), *440*
Rimkus, A. 487 (168), *523*
Rimola, A. 284 (38), 309 (31), 339 (184), *341*, *345*
Rincón, J. A. 821 (128), *854*, 845 (194), 847 (181), *856*
Ringold, C. 800 (82), *853*
Rinner, U. 647 (116), *689*
Rivero, M. R. 935 (270), 936 (271), *986*
Rizzo, A. 416 (140), *440*
Robb, M. A. 283 (44), *341*
Roberge, J. Y. 824 (145), *855*
Roberts, J. A. S. 902 (124), *983*
Roberts, J. C. 885 (18), *980*
Roberts, S. M. 622 (59), *686*
Robertson, J. 560 (110), *583*, 824 (153), *855*
Roca, T. 510 (239), *524*
Rocca, M. 156 (99), *161*
Rochefort, A. 150 (39, 40), *159*
Rochon, F. D. 163 (6), *210*
Rockwell, J. J. 73 (158), *139*
Rodden, M. 402 (110), *439*
Rode, B. M. 305 (19, 134–140), 322 (187, 188), *341*, *344*, *345*
Rodehüser, L. 165 (51, 52), *211*
Rodgers, M. A. J. 220 (27), *242*
Rodgers, M. T. 152 (51), *160*, 293 (82), 295 (56, 75, 98), 299 (118, 119), 300 (57, 88), 304 (106), *342*, *343*
Rodriguez, C. F. 329 (213), *345*
Rodriguez, F. 329 (213), *345*
Rodriguez, H. 337 (247), *346*
Rodriguez, J. A. 157 (107), *161*
Rodríguez, M. L. 574 (139), *584*
Rodriguez, S. 922 (204), *984*
Rodríguez-Santiago, L. 284 (38), 303 (126), 309 (31, 37, 81), 339 (184), *341–343*, *345*
Rodriquez, C. 320 (77), 329 (212, 218), *342*, *345*, *346*
Roebke, H. 860 (20), *878*
Roenn, M. 979 (484, 486), *990*
Roepstorff, P. 323 (193), *345*
Roesky, H. W. 904 (129), *983*
Roessler, F. 803 (23), 804 (92), *851*, *853*
Roettger, S. 911 (160), *983*
Rogers, D. W. 151 (48), *160*
Rogers, M. C. 39 (79), *137*
Rohaly, J. 859 (18), *877*
Roithová, J. 152 (52), *160*, 309 (42), *341*
Roland, J. R. 60 (121), *138*, 1024 (141), *1032*
Roland, S. 401 (104), *438*
Romero, M. A. 834 (169), *855*
Rona, P. 2 (20), *19*, 629 (4), 674 (3), *683*
Ronconi, L. 163 (1), *210*
Rönneberg, H. 445 (13), *519*, 647 (112), *688*
Ronzini, L. 563 (48), *582*
Rooney, C. M. 147 (9), *159*
Roos, B. O. 310 (10, 144), *341*, *344*
Roper, T. D. 516 (251, 253), *524*
Rosch, L. 777 (19), *851*
Rosche, K. 232 (93), *243*
Rose, M. 925 (223), *985*
Rosenblum, L. D. 13 (81), *20*
Rosenmund, K. W. 967 (410), *989*
Rosenzweig, A. C. 356 (13), 358 (68), 360 (77), 361 (81, 82), *361*, *363*
Rosi, M. 58 (58), *136*, 302 (122, 123), *343*
Rosiak, A. 474 (132), *522*, 695 (1), *729*, 858 (12), *877*
Rößler, K. 273 (46), *277*
Ross, A. A. 1018 (112), *1031*
Ross, G. 908 (153), *983*
Rosset, S. 201 (223), 203 (230), *215*, 679 (172), *691*, 695 (6–8), *729*
Rossi, A. 352 (34), *362*
Rossiter, B. E. 15 (94), *21*, 105 (238), 108 (261), *141*, 175 (159), *213*, 430 (11, 18), *435*, *436*, 632 (45), *685*, 776 (6), *851*, 858 (12), *877*
Rössle, M. 474 (132), *522*, 695 (1), *729*, 858 (12), *877*
Roszak, S. 153 (69), *160*
Roth, H. J. 861 (22), *878*
Roth, R. M. 517 (260, 261), *525*
Roth, W. R. 151 (49), *160*
Rothenberg, G. 966 (402), 974 (441, 442), *989*
Rotteveel, M. A. 37 (78), *137*
Rottlander, M. 404 (125), *439*
Rotunno, D. 563 (48), *582*
Rouden, J. 647 (116), *689*
Roux, M. 670 (147), 679 (170), *690*
Rowley, D. A. 336 (242), *346*
Rowley, M. 846 (9), *851*
Roy, S. 622 (54), *685*
Roy, S. C. 516 (257), *524*
Roy, T. 113 (286, 287), 114 (288, 289), *142*, 154 (84–86), *161*, 459 (84), *521*

Rozema, M. J. 383 (38), 416 (135), *436*, *440*, 474 (124), *522*, 858 (8), *876*
Ruan, C. 152 (51), *160*
Ruan, C. H. 295 (75), *342*
Rubessa, F. 39 (80), *137*
Rubini, P. 165 (52), *211*
Rubinstein, M. 1014 (83), 1015 (89), *1030*
Rubio, M. B. 618 (49), *685*
Rubio, R. 841 (144), 847 (181), *855*, *856*
Ruden, R. A. 860 (20), *878*
Rudler-Chauvin, M. 647 (21), *684*
Rudler, H. 647 (21), *684*
Rudolf, M. 360 (76), *363*
Rue, C. 157 (117), *161*
Rüffer, T. 275 (68), *278*
Ruhlandt-Senge, K. 50 (109), *138*, 152 (66), *160*
Ruitenberg, K. 629 (72), *686*, 832 (146), 838 (124), *854*, *855*
Ruiz, B. M. 699 (19), *729*
Ruiz-Cabello, J. 194 (194), *214*
Rulisek, L. 299 (117), *343*
Ruloff, C. 820 (138), *854*
Runsink, J. 629 (72), *686*
Ruppert, I. 1018 (113), *1031*
Ruschewitz, U. 74 (164), *139*
Russell, A. T. 551 (94), *583*
Russell, C. A. 106 (176), *139*
Russell, D. H. 324 (194, 200), *345*
Russo, N. 289 (72, 73), 290 (34), 301 (83), 302 (35), 303 (78), 305 (141), *341*, *342*, *344*
Rutjes, F. P. J. T. 414 (132), *440*
Rutjes, P. J. T. 413 (130), *439*
Ryan, W. 390 (82), *438*
Rychnovsky, S. D. 399 (99), *438*
Ryde, U. 307 (145), 310 (10, 144), *341*, *344*
Rydzewski, R. M. 629 (72), *686*
Ryu, I. 738 (20), *772*

Sabbe, K. 859 (18), *878*
Sadhukhan, N. 303 (127), *343*
Sadighi, J. P. 44 (102), 46 (103), 121 (100), 122 (319), 123 (101), *137*, *143*, 151 (44), *160*, 402 (112, 117, 119, 120), 403 (113), *439*, 732 (6), 740 (28), *771*, *772*
Sadler, P. J. 163 (1), 165 (40), 197 (205), *210*, *211*, *214*
Sadowski, J. S. 887 (23), *980*
Saegusa, T. 150 (42), 158 (133), *160*, *162*, 386 (56, 59), *437*, 735 (13, 14), *772*, 886 (22), *980*
Saejueng, P. 904 (131), *983*
Saenz, J. 956 (370), *988*
Saffon, N. 904 (128), *983*
Sagamanova, I. K. 934 (264), *986*
Sagar, A. D. 952 (355), *988*
Saha, A. 904 (139), *983*
Sahin, C. 540 (59), *582*
Sahlberg, C. 386 (64), *437*, 614 (41), *685*
Saillard, J.-Y. 88 (208), *140*
Saino, N. 401 (107), *439*
Saito, A. 461 (94), *521*, 661 (133), *689*
Saito, I. 309 (116), *343*
Saito, T. 756 (72), *773*
Saitoh, K. 238 (136), *244*
Saji, T. 225 (46, 50), *242*
Sakamoto, M. 925 (218), *984*
Sakamoto, Y. 1024 (149), *1032*
Sakata, A. 116 (292), *142*
Sakuragi, H. 230 (86, 87), *243*
Salata, M. R. 902 (124), *983*
Salomon, R. G. 516 (257), *524*
Salpin, J.-Y. 332 (234), 334 (222, 223, 237), *346*, 337 (151), *344*, *346*
Saluta, G. R. 238 (131), *244*
Salvador, P. 283 (44), *342*
Salvadori, P. 629 (72), *686*
Salvatore, R. N. 938 (281), *986*
Salzbrunn, S. 952 (356), *988*
Samanta, A. 919 (182), *984*
Sambur, V. P. 887 (25), *980*
Samuel, O. 374 (36), *436*
Samuelson, A. 303 (127), 309 (93), *343*
San Filippo, J. 2 (19), 9 (61), *19*, *20*, 24 (9), *135*, 586 (8), 589 (15), *601*, *602*, 818 (126), *854*
San Román, E. 240 (142), *245*
Sánchez, A. 813 (38), *851*, *852*
Sánchez, E. I. 231 (88), *243*
Sanchez-Ferrando, F. 859 (16), *877*
Sandefur, B. J. 938 (281), *986*
Sandell, J. 106 (242), *141*, 152 (53), *160*
Sanders-Loehr, J. 358 (62, 69), *363*
Sanders, S. D. 926 (245), *985*
Sanderson, A. J. 399 (95), *438*
Sanderson, P. E. J. 800 (45, 47), *852*
SanFilippo, J. 173 (140), *213*
Sanford, T. 149 (29), *159*
SanMartin, R. 904 (136), *983*
Sano, H. 157 (111), *161*
Sansoni, M. 81 (179), *139*
Santa, L. E. 169 (107), *212*
Santini, G. 992 (9), 1015 (94), *1029*, *1031*
Santo, R. 116 (293, 294), *142*
Sañudo, M. C. 816 (118), *854*
Sapountzis, I. 404 (124), 421 (152), 423 (158, 162), 425 (154, 163), *439*, *440*, 600 (19), *602*
Sapse, A.-M. 178 (166), *213*
Sarangi, R. 348 (16), *361*
Saraste, M. 352 (36), 358 (70), *362*, *363*
Sarkar, A. K. 800 (70), *852*
Sarkar, B. 337 (244), *346*
Sarkar, M. 919 (182), *984*
Sarkar, T. K. 800 (81), *853*

Sarmentero, M. A. 813 (38), *852*
Saron, C. 231 (91), *243*
Sasaki, M. 165 (33), *210*, 958 (379), *988*
Sasaki, N. 496 (186, 187), *523*
Sasaki, Y. 499 (205–207), *523*, 641 (105), *688*
Saspe, A.-M. 166 (72), *211*
Sastry, G. N. 238 (134), *244*
Sata, T. 221 (34), *242*
Sato, A. 461 (100, 101), *521*, 659 (130), 661 (133), *689*
Sato, F. 444 (8, 9), 445 (15–17, 19, 21), 457 (72), 502 (73), *519*, *520*
Sato, K. 116 (294), *142*, 1008 (56, 57), *1030*
Sato, M. 403 (115), *439*, 457 (72), 502 (73), *520*, 1011 (71, 73), *1030*
Sato, S. I. 222 (39), *242*
Sato, T. 309 (116), *343*, 457 (71), *520*, 640 (101), *688*
Satoh, K. 925 (190), *984*
Satoh, T. 979 (473), *990*
Satoh, Y. 496 (198), *523*
Satomi, H. 735 (14), *772*
Saubern, S. 954 (365), 962 (386), *988*
Sauers, R. R. 631 (76), *687*
Sauvage, J.-P. 200 (210), *214*
Sauvêtre, R. 474 (122), *521*, 999 (35), *1029*
Sawamura, M. 499 (204–207), *523*
Sayo, N. 756 (72), *773*
Sazinsky, M. H. 356 (13), *361*
Sazonova, V. A. 33 (62), *136*
Scalmani, G. 283 (44), *341*
Scandurra, A. 155 (95), *161*
Scarborough, R. M. 911 (163), *983*
Scarisbrick, A. C. 127 (328), *143*, 402 (110), *439*
Scattolin, E. 966 (401), *989*
Schabbert, S. 800 (62), *852*
Schader, T. 707 (39), *730*
Schaefer, H. F. 309 (147), *344*
Schaefer, W. 254 (35), *277*
Schäfer, W. 254 (34), *277*
Schaper, F. 41 (89), *137*
Schareina, T. 968 (414–416), *989*
Scharf, B. 349 (20), *361*
Schauman, E. 800 (62), *852*
Sche, P. P. 154 (85), *161*
Scheer, M. 149 (28), *159*
Scheidt, K. A. 955 (367), *988*
Scheinman, F. 622 (59), *686*
Scherrer, R. A. 944 (302), *986*
Schiavello, M. 235 (102, 105), *243*
Schiemenz, B. 41 (91), 78 (171), *137*, *139*
Schilling, S. 647 (116), *689*
Schillinger, K. J. 979 (477), *990*
Schimeczek, M. 426 (165), *440*
Schinzer, D. 800 (61), *852*
Schläpfer, C. W. 165 (32), *210*
Schlegel, H. B. 283 (44), *341*
Schleltwein, D. 228 (66), *243*
Schlemper, E. O. 76 (168), *139*
Schleyer, P. v. R. 3 (33), *19*, 166 (69, 72, 76–80), 178 (166), *211*, *213*, 309 (147), *344*, 605 (12), *683*
Schlich, K. 1018 (113), *1031*
Schloss, F. 166 (68), *211*
Schlosser, M. 108 (251), *141*, 169 (114), *212*, 444 (1, 7), 452 (46), 471 (115), *519–521*, 609 (30), *684*, 821 (140, 141), *854*, 873 (4), *875*, 977 (467), *990*, 1018 (111), *1031*
Schlueter, J. A. 114 (288, 289), *142*, 154 (85, 86), *161*
Schmalz, D. 166 (82), *212*
Schmidbaur, H. 100 (229), *141*
Schmidhammer, H. 947 (329), *987*
Schmidhauser, J. C. 151 (49), *160*
Schmidt, Y. 2 (26), *19*, 24 (18), *135*, 164 (22), *210*, 484 (161), *522*, 528 (7), *581*, 858 (9), *876*
Schmieder-van de Vondervoort, L. 914 (170), *984*
Schmitter, J. 647 (9), *683*, 872 (50), *879*
Schmuff, N. R. 380 (49), *437*
Schneider, H. 149 (30), *159*
Schneider, M. 540 (59), *582*, 643 (106), *688*
Schneider, M. P. 643 (106), *688*
Schneider, N. 127 (327), *143*
Schneider, S. 902 (124), *983*
Schnoeckel, H. 157 (126), *162*
Schober, K. 18 (114), *21*, 208 (61), *211*
Schoenfelder, A. 638 (94, 96), 659 (131), *687–689*
Schogt, J. C. M. 890 (63), *981*
Scholl, H. 228 (64), *243*
Schöllkopf, U. 386 (60), *437*, 859 (18), *878*
Schollmeier, D. 252 (18), *277*
Schollmeyer, D. 252 (13, 15–17, 23, 26–28), 254 (33), *277*
Scholz, U. 200 (211), *214*
Schoof, J. 960 (384), *988*
Schoone, J. C. 66 (66), *137*
Schrader, T. 18 (115), *21*
Schreiber, S. L. 593 (25, 26), *602*
Schreiner, P. R. 309 (147), *344*
Schrirber, J. E. 114 (288), *142*
Schröder, D. 152 (52), *160*, 290 (79), 308 (90), *343*, 311 (173), *342*, *345*
Schröder, D. S. 309 (42), *341*, 318 (180), *345*
Schroeder, D. 157 (117), *161*
Schuchmann, H. P. 238 (123), *244*
Schudde, E. P. 201 (226), *215*
Schuh, K. 937 (274), *986*
Schulte, P. 32 (61), *136*
Schultze, P. 330 (231), *346*
Schulze, E. 386 (60), *437*, 859 (18), *878*
Schulze, M. L. C. 154 (80), *160*
Schulze, V. 425 (165), *440*

Schumacher, K. K. 418 (144), *440*
Schumm, R. H. 155 (3), *158*
Schuppan, J. 975 (452), *989*
Schurko, R. W. 166 (60), *211*
Schuster, D. I. 221 (35), *242*
Schuster, H. F. 866 (29), *878*
Schuster, T. M. 181 (172), *214*
Schwartz, J. 453 (58), 457 (70), 459 (52, 57), *520*
Schwarz, G. 360 (75), *363*
Schwarz, H. 152 (52), 154 (80), 157 (117), *160*, *161*, 290 (79), 308 (90), 309 (42), 310 (162), 311 (173), 318 (180), *341–345*
Schwenk, C. F. 305 (19, 136–138, 140), *341*, *344*
Schwenk, C. M. 979 (479), *990*
Schwerdtfeger, P. 308 (90), *343*, 426 (165), *440*
Schwink, L. 496 (127), *522*
Scilimati, A. 663 (20, 138), *684*, *689*
Sclafani, J. A. 738 (24), *772*, 800 (41), *852*
Scommoda, M. 666 (140), *689*
Scott, M. E. 845 (174), *855*
Scott, N. M. 403 (111), *439*, 751 (55, 56), *773*
Scur, R. G. 119 (305), *142*
Scuseria, G. E. 283 (44), *341*
Seagle, P. 105 (241), *141*, 173 (123), 193 (133), *212*, *213*, 778 (24), *851*
Seagle, P. H. 534 (36), *582*, 858 (7), *876*
Sebesta, R. 695 (10), *729*
Secco-Millet, C. 861 (21), *878*
Seconi, G. 373 (28), 374 (34), 378 (45), *436*, 599 (35, 36), *602*, 809 (17), 832 (78), 842 (184), *851*, *853*, *856*
Sedelmeier, J. 926 (241), *985*
Sedova, N. N. 33 (62), *136*
Sedrani, R. 663 (85, 134), *687*, *689*, 874 (56), *879*
Seebach, D. 18 (117), *21*, 166 (73), 201 (220), 202 (219), *211*, *215*, 430 (172), *441*
Seemann, M. 647 (111), *688*
Seib, P. A. 356 (57), *363*
Seidel, N. 421 (152), *440*
Seidell, A. 892 (79), *982*
Seifert, H. 889 (54), *981*
Seitz, D. E. 834 (167), *855*, 943 (301), *986*
Seitz, E. P. 647 (9, 21), *683*, *684*
Seiwell, L. P. 890 (65), *981*
Sekar, G. 897 (106), *982*
Seki, M. 979 (489), *990*
Sekiya, K. 416 (138), *440*, 609 (19, 28), 658 (13), *683*, *684*
Seland, J. G. 168 (101), *212*
Selim, K. 430 (27), *436*
Sellén, M. 607 (10), 609 (32), *683*, *684*
Selling, G. W. 373 (32), *436*
Selman, S. H. 238 (121, 122), *244*
Selmeczib, K. 356 (54), *362*
Semmelhack, M. F. 735 (13), *772*
Senge, M. O. 218 (1), 219 (8, 9), 238 (118), *241*, *244*
Sengupta, S. 1 (4), *19*, 108 (16), *135*, 380 (1), *435*, 571 (2), *581*, 672 (1), *682*, 776 (4), *851*, 873 (4), *875*
Sensui, H.-o. 513 (240), *524*, 977 (460, 464), *990*
Seota, T. 430 (27), *436*
Sereinig, N. 914 (170), *984*
Sergeeva, N. N. 219 (8, 9), *241*
Serpone, N. 220 (22), *241*
Serra, O. A. 237 (115), *244*
Servesko, J. M. 745 (39), 746 (40, 42), *772*
Setiaji, A. H. B. 305 (135), *344*
Seto, C. 321 (186), *345*
Setsune, J. 968 (423, 426, 427), *989*
Sevast'yan, A. P. 992 (7), *1029*
Sevens, E. D. 403 (111), *439*
Sewald, N. 487 (168), *523*
Sewing, C. 399 (96), *438*
Seymour, J. 328 (209–211), *345*
Sgamellotti, A. 58 (58), *136*
Shabli, A. 565 (122), *583*
Shafir, A. 897 (112), 914 (113, 171), *982*, *984*
Shahgholi, M. 336 (242), *346*
Shaker, Y. M. 238 (118), *244*
Shamim, K. 516 (258), *524*
Shang, J. 236 (109), *244*
Shangguan, N. 631 (76), *687*
Shao, L. 157 (127), *162*
Sharma, D. K. 229 (72), *243*
Sharma, R. R. 390 (82, 83), *438*
Sharma, S. 24 (15), *135*, 184 (185), *214*, 510 (235), *524*, 818 (25), 834 (22, 33, 132), *851*, *854*
Sharpless, K. B. 105 (105), *137*, 632 (45), *685*, 952 (357), *988*
Shattuck, J. C. 547 (75), *582*
Shay, J. J. 800 (86), *853*
Shearer, B. G. 860 (20), *878*
Shearer, H. M. M. 81 (35), *136*
Shein, S. M. 883 (8), 890 (64, 67, 68), *980*, *981*
Sheldon, R. A. 353 (43), *362*
Shen, B. 469 (109), *521*, 573 (61), *582*
Shen, L. 934 (264), *986*
Shen, L. C. 933 (263, 266), *985*, *986*
Shen, R. 932 (255, 256), *985*
Shen, R. C. 932 (260), *985*
Sheng, X. C. 616 (43), *685*
Shepherd, J. W. 157 (121), *162*
Sheppard, R. N. 502 (215), *524*
Sheppard, W. A. 27 (24, 25), 60 (121), *135*, *136*, *138*, 1024 (136, 141), *1031*, *1032*
Sherburn, M. S. 595 (29), *602*
Sherwood, P. M. A. 361 (80), *363*
Shi, L.-X. 85 (202), *140*

Shi, W.-J. 206 (208), *214*, 670 (150), *690*
Shi, Y. 631 (79), *687*
Shibasaki, M. 200 (13), *210*, 732 (1), 762 (83), 763 (80, 84), *771*, *773*
Shibata, S. 885 (12), *980*
Shibuya, S. 1011 (70–74), 1012 (75, 76), *1030*
Shields, S. J. 324 (194, 200), *345*
Shih, M. 868 (37), *879*
Shikata, K. 968 (423), *989*
Shimeno, H. 1011 (71, 73), 1012 (76), *1030*
Shimizu, H. 756 (70, 72), *773*
Shimizu, R. 979 (475), *990*
Shimizu, Y. 762 (83), *773*
Shimoda, K. 401 (107), *439*
Shin, J.-T. 926 (246), *985*
Shin, S. K. 995 (21, 22), *1029*
Shin-Ya, S. 1009 (61), *1030*
Shindo, H. 925 (218), *984*
Shintani, R. 741 (30), *772*
Shinzawa-Itoh, K. 352 (38), *362*
Shiomi, T. 758 (73), *773*
Shirakawa, E. 543 (65), *582*
Shirakawa, K. 974 (438, 440), *989*
Shiro, M. 946 (316, 317), *987*
Shivaprasad, S. M. 155 (93), *161*
Shoeib, T. 304 (131), 320 (77), *342*, *343*
Shoji, Y. 955 (368), *988*
Shu, H. Y. 235 (100), *243*
Shudo, K. 911 (161), *983*
Shukla, M. K. 298 (114), 309 (115), *343*
Sibi, M. P. 200 (12), *210*, 695 (1), *728*, 800 (86), *853*
Sice, J. 885 (11), *980*
Sicilia, E. 289 (73), *342*
Sidduri, A. R. 474 (124), *522*
Siegbahn, P. E. M. 157 (115), *161*, 280 (1, 5), *340*, *341*
Siegel, M. M. 238 (137), *244*
Siegmann, K. 587 (12), *602*
Siengalewicz, P. 647 (116), *689*
Sierra, M. A. 450 (42), *520*
Sieser, J. E. 925 (235), *985*
Sievers, M. R. 294 (92), *343*
Sigalas, M. P. 61 (125), *138*
Sigel, A. 348 (6), *361*
Sigel, H. 348 (6), *361*
Sih, C. J. 643 (106), *688*
Silberman, J. 818 (126), *854*
Simanek, M. 282 (18), *341*
Simon, J. 952 (356), *988*
Simonelli, F. 429 (168), 430 (169), *440*
Simonelli, J. 838 (177), *856*
Sing, S. M. 834 (132), *854*
Singer, R. A. 842 (183), *856*
Singer, R. D. 473 (118), *521*, 776 (14), 820 (16), 836 (173), 842 (176), *851*, *855*, 858 (8), *876*
Singh, J. 165 (53), *211*
Singh, R. 165 (53), *211*
Singh, S. M. 846 (90), *853*
Singleton, D. A. 5 (49), 16 (96), *20*, *21*, 371 (23), *436*
Singleton, V. D. 11 (68, 69), *20*, 638 (8), 647 (9), *683*
Sinz, A. 330 (229), *346*
Sirol, S. 753 (61, 62), *773*
Siu, K. M. 304 (131), *343*
Siu, K. W. M. 320 (77), 324 (196), 329 (212, 213, 216), 330 (230), *342*, *345*, *346*
Siu, S. 329 (218, 220), *346*
Siu, S. O. 329 (219), *346*
Sivasankar, C. 303 (127), *343*
Sjoeberg, P. J. R. 911 (160), *983*
Skalkos, D. 238 (121, 122), *244*
Skelton, B. W. 37 (74), 79 (175), 83 (197), *137*, *139*, *140*
Sklute, G. 565 (122–126), *583*
Skuratov, S. M. 155 (97), *161*
Slaunwhite, W. D. 861 (22), *878*
Slawin, A. M. Z. 502 (215), *524*
Slayden, S. W. 147 (4), 153 (76), *159*, *160*
Słota, R. 235 (104), 236 (106), *243*, *244*
Smalley, R. E. 152 (61), 157 (115), *160*, *161*
Smeets, W. J. J. 41 (95), 43 (98), 48 (107), 53 (115), 61 (13), 108 (108), 117 (117), *135*, *137*, *138*, 276 (73), *278*, 669 (142), *689*
Smith, A. B. 15 (87), *20*
Smith, A. L. 238 (138), *244*
Smith, B. E. 220 (15, 16), *241*
Smith, G. 334 (238), *346*
Smith, J. A. 1014 (85), *1030*
Smith, J. D. 37 (75), 106 (245), 110 (269), 111 (266), *137*, *141*, 166 (75), *211*
Smith, J. G. 831 (84), *853*
Smith, K. M. 218 (2, 3), 219 (6, 7), 220 (24), 224 (45), *241*, *242*
Smith, M. D. 894 (98), *982*
Smith, M. E. 83 (197), *140*
Smith, R. A. 184 (180), 189 (189), *214*
Smith, R. A. J. 5 (48), *20*, 179 (152), *213*, 517 (233), *524*, 605 (12), *683*, 858 (7), *876*
Smith, R. C. 1014 (86), *1030*
Smith, R. S. 169 (106), *212*
Smitrovich, J. H. 382 (51), *437*, 457 (60), 461 (77), 503 (213), *520*, *523*, 612 (35), *684*
Smolenyak, P. 226 (55), *242*
Smolenyak, P. E. 226 (56), *242*
Snaith, R. 166 (69), 177 (161), *211*, *213*
Snider, B. B. 445 (22), *519*, 553 (99), *583*
Snieckus, V. 378 (44), *436*, 599 (34), *602*, 894 (97), *982*
Snyder, J. P. 3 (34), 5 (49), 16 (94, 101), *19–21*, 105 (238, 240), 108 (255, 261, 262), 112 (276, 278), *141*, *142*, 175 (154, 158, 159), 179 (128, 152), 196 (199), 202 (202),

212–214, 367 (6), 399 (93), *435*, *438*, 790 (26), 800 (58), *851*, *852*
Soai, K. 629 (74), *687*
Soderbaum, H. G. 73 (162), *139*
Soderquist, J. A. 445 (20), *519*
Sodupe, M. 284 (38), 303 (126), 309 (31, 37, 81), 310 (156, 157), 339 (184), *341–345*
Soeda, S. 1011 (71, 73), 1012 (76), *1030*
Soeta, T. 430 (27), *436*
Sohnlein, B. 303 (128), *343*
Solà, M. 114 (291), *142*, 284 (38), *341*, 354 (48), *362*
Solans, X. 81 (181), *139*
Sollman, P. B. 861 (19), *878*
Solntsev, K. A. 73 (158), *139*
Solo, A. J. 861 (22), *878*
Solomon, E. I. 348 (8, 9, 11, 16, 17), 352 (14), 359 (63), 360 (52), *361–363*, 943 (294), *986*
Solooki, D. 294 (95, 96), *343*
Soloski, E. J. 846 (125), *854*, 1024 (145), 1027 (146), *1032*
Somers, N. 79 (175), *139*
Somorjai, G. A. 310 (166), *344*
Song, H. 65 (136), *138*
Song, H.-B. 123 (322), *143*
Song, J. S. 800 (82), *853*
Song, Y. 953 (360), *988*
Song, Z. J. 897 (105), *982*
Sonoda, A. 496 (191, 192), *523*
Sonoda, M. 513 (242), *524*, 800 (42), *852*
Sonoda, N. 738 (20), *772*
Sonoda, T. 496 (193, 194), *523*, 1003 (42–44), *1029*
Soorukram, D. 641 (103, 104), *688*
Sørland, G. H. 168 (101), *212*
Søtofte, I. 34 (36), *136*
Sott, R. 166 (74), *211*
Sottocornola, S. 741 (29), *772*
Soulen, R. L. 993 (10, 11, 13, 14), *1029*
Souma, Y. 157 (111, 112), *161*
South, J. A. 975 (453), *989*
Souza, F. E. S. 551 (90), *583*
Spandl, R. J. 595 (27), *602*
Spangenberg, T. 659 (131), *689*
Spangler, D. P. 108 (261), *141*, 175 (159), *213*
Spasov, V. A. 157 (114), *161*
Spatz, J. H. 908 (153), *983*
Spawn, T. D. 995 (17), 998 (34), 1005 (50, 51), 1015 (49), *1029*, *1030*
Speckamp, W. M. 414 (132), *440*
Spek, A. L. 28 (47), 36 (69), 40 (86), 41 (95), 43 (98), 48 (107), 53 (115), 61 (13), 66 (66), 78 (173), 87 (12, 203, 204), 94 (94), 97 (224), 101 (231, 232), 108 (108, 264), 117 (117), 134 (63), *135–141*, 175 (156, 157), *213*, 276 (73), *278*, 669 (142), *689*, 800 (59), *852*, 858 (5), *875*
Spelten, H. 35 (64), *137*
Spence, T. G. 286 (60), *342*
Spicer, M. D. 124 (323), *143*, 165 (38), *211*
Spindler, J.-F. 200 (213), *214*, 897 (108, 110), 921 (219), *984*, 943 (217), 967 (413), *982*, *984*, *989*
Spino, C. 553 (96), *583*, 607 (15), 616 (6), 620 (50–53), 622 (54, 55), 631 (80), *683*, *685*, *687*
Spiro, T. G. 360 (79), *363*
Spodine, E. 88 (208), *140*
Sponagel, P. 883 (3), *980*
Sponer, J. 282 (21), 299 (117), 309 (20), *341*, *343*
Spooner, R. K. 354 (45), *362*
Sprague, L. G. 1009 (62–64), 1010 (65), *1030*
Spring, D. R. 373 (30), *436*, 586 (5), 593 (18, 20, 25), 595 (21, 27), *601*, *602*
Sreedhar, B. 921 (207), *984*
Sridhar, C. 921 (207), *984*
Srinivas, D. 925 (236), *985*
Srinivas, V. 922 (211), *984*
Srinivasan, A. 238 (134), *244*
Srirangam, J. K. 956 (370), *988*
Srivastava, A. 499 (202), *523*
Srivatsan, S. G. 238 (133), *244*
Šrogl, J. 493 (183, 184), *523*
Stace, A. J. 286 (63), 288 (54, 62, 68), 289 (66), 305 (64, 65, 132), 310 (158), *342–344*
Stack, D. E. 149 (31, 32), *159*
Stack, T. D. P. 114 (290, 291), *142*, 943 (294), 964 (394), *986*, *988*
Stadtmüller, H. 473 (121), 516 (254), *521*, *524*
Staehlin, B. M. 360 (79), *363*
Stahl, S. S. 203 (203), *214*
Stalke, D. 76 (76), *137*, 821 (139), *854*
Stam, C. H. 35 (64), 37 (78), 56 (116), 93 (50), 94 (216), 119 (119), 120 (120), 134 (49), *136–138*, *140*
Stammler, H.-G. 66 (138), *138*
Standaert, R. F. 1011 (68), *1030*
Stangherlin, E. C. 943 (229), *985*
Stanley, J. R. 295 (56), *342*
Stanley, L. M. 200 (12), *210*
Stansfield, K. 1004 (46), *1029*
Stanzel, J. 157 (113), *161*
Staroscik, J. 638 (7), *683*
Starostina, T. I. 946 (311), 949 (335, 336), *987*
Stasser, J. P. 360 (79), *363*
Staubitz, A. 423 (159), 424 (157), *440*
Stauffer, R. D. 735 (13), *772*
Stead, D. 399 (95), *438*
Stefanov, B. B. 283 (44), *342*
Stefinovic, M. 842 (189), *856*
Stein, T. 41 (97), *137*, 259 (44), 271 (66), 274 (45), *277*, *278*
Stein, Th. 273 (67), *278*
Stejskal, E. O. 168 (97), *212*
Stemmler, T. 108 (257), *141*

Stemmler, T. L. 108 (259), *141*, 175 (160), *213*
Stengel, P. J. 838 (177), *856*
Stephens, P. J. 198 (198), *214*
Stephens, R. 992 (1), *1028*
Stephens, R. D. 73 (159), *139*, 965 (395, 396), *988*
Štěpnička, P. 275 (68), *278*
Sterling, J. J. 16 (99), *21*, 24 (11), *135*, 366 (2), *435*, 776 (5), *851*
Stevens, C. 351 (32), *362*
Stevens, E. D. 121 (311, 312), *142*, *143*, 403 (111), *439*, 750 (53, 54), 751 (56), *773*
Stevens, K. L. 366 (5), *435*, 517 (260, 261), *525*
Stevens, R. C. 732 (3), *771*
Stevenson, T. 800 (68), *852*
Stevenson, T. M. 424 (121), *439*
Steward, O. W. 426 (166), *440*
Stewart, O. J. 883 (5), *980*
Stilbs, P. 168 (95), *212*
Stiles, D. T. 933 (267), *986*
Still, B. M. 163 (9), *210*
Still, W. C. 777 (18), 818 (126), 824 (122), *851*, *854*
Stillman, M. J. 219 (6), *241*, 330 (233), *346*
Stinn, D. E. 149 (32), *159*
Stivers, J. T. 280 (8), *341*
Stöcklin, R. 330 (227), *346*
Stockton, C. 225 (48), *242*
Stoj, C. 348 (17), *361*
Stol, M. 413 (130), *439*
Stoltz, B. M. 939 (282, 283), *986*
Stone, F. G. A. 2 (14), *19*, 27 (40), 34 (34), 66 (140, 141), 70 (151), *136*, *138*, *139*, 858 (5), *875*
Stone, G. 225 (48), *242*
Stone, J. A. 318 (181), 321 (186), *345*
Stone, R. L. 453 (61), *520*
Stork, G. 807 (94), *853*
Strahle, J. 41 (96), *137*
Strain, M. C. 283 (44), *342*
Straley, J. M. 24 (5), *135*, 169 (104), *212*, 857 (1), *875*
Stratmann, R. E. 283 (44), *341*
Stratton, S. M. 979 (477), *990*
Straub, B. F. 47 (104), *137*, 184 (190), *214*
Straub, K. D. 220 (21), *241*
Strauss, S. H. 73 (158), *139*, 157 (110), *161*
Strauss, S. J. 1001 (4), *1029*
Strazzulla, G. 155 (95), *161*
Streitwieser, A. 173 (149), *213*
Strekowski, L. 388 (81), *438*
Striejewske, W. S. 200 (209), *214*
Strieter, E. R. 373 (29), *436*, 943 (228), *985*
Stringer, G. 248 (5), *276*
Strologo, S. 40 (88), *137*
Stroppolo, M. E. 355 (50), *362*
Strotman, N. A. 958 (381), *988*
Struchkov, Y. T. 90 (209), *140*
Struchkov, Yu. T. 33 (62), *136*
Struck, E. 967 (410), *989*
Stryker, J. M. 733 (8), 735 (15, 16), 736 (18), 737 (19), 768 (17), *771*, *772*
Stüdemann, T. 473 (121), *521*
Studte, C. 484 (159, 160), *522*, 622 (38), *685*
Su, P. 295 (51), 318 (175), *342*, *345*
Su, S. 932 (260), *985*
Su, T. 318 (181), *345*
Su, X. B. 593 (20), 595 (21, 27), *602*
Suárez, R. M. 478 (141), *522*
Suchan, S. D. 153 (75), *160*
Suemune, K. 1011 (70–74), *1030*
Sugai, T. 643 (106), *688*
Sugawara, K. 286 (14, 59), *341*, *342*
Sugihara, J. M. 224 (44), *242*
Sukachev, A. N. 155 (92), *161*
Sulfab, Y. 943 (298), *986*
Sullivan, A. C. 106 (245), *141*, 166 (75), *211*
Sullivan, R. W. 859 (16), *877*
Sumiya, T. 238 (129), *244*
Summers, M. F. 189 (193), *214*
Sun, A. 234 (98), *243*
Sun, H. W. 920 (201), *984*
Sun, K. 748 (48), *772*
Sun, S.-T. 152 (59), *160*
Sun, S. Y. 925 (230), *985*
Sun, W.-H. 468 (106, 107), *521*
Sun, X. J. 280 (9, 11), *341*
Sundararajan, G. 430 (16), *435*
Sundararajan, M. 307 (146), *344*
Sundararaman, A. 30 (52), 93 (93), *136*, *137*
Sundermeyer, J. 868 (38), *879*
Surendran, R. 307 (146), *344*
Surry, D. S. 373 (30), *436*, 586 (5), 593 (18, 20), 595 (27), *601*, *602*
Sutcliffe, M. J. 348 (5), *361*
Sutherland, D. J. 887 (29), *981*
Sutin, N. 349 (22), *361*
Suwan, S. 330 (226), *346*
Suzuki, A. 496 (185–187, 198), *523*
Suzuki, H. 666 (141), *689*, 887 (28, 30–32), 892 (82), 940 (285), 941 (286, 283), 968 (425), *981*, *982*, *986*, *989*
Suzuki, N. 462 (103), *521*
Suzuki, T. 1024 (149), *1032*
Suzuki, Y. 403 (115), *439*
Svatos, A. 547 (75), *582*
Svenson, R. 452 (53), *520*
Sviridov, A. F. 535 (44), *582*
Swanson, D. R. 459 (79), *520*
Swenson, D. C. 997 (27, 28), 998 (29, 30), 1000 (37, 39), *1029*
Swingle, N. M. 430 (11, 18), *435*, *436*, 776 (6), *851*, 858 (12), *877*
Sykora, J. 220 (11), *241*

Sylvester-Hvid, K. O. 229 (70), *243*
Szantay, C. 859 (18), *877*
Szeto, C.-P. 759 (75), *773*, 760 (76), *773*
Szlyk, E. 165 (37, 43–45), *210*, *211*
Szymanska, I. 165 (28, 37, 43–45), *210*, *211*
Szymczyk, I. 228 (64), 241 (143), *243*, *245*

Tabet, J. 325 (202), 326 (203, 208), *345*
Tadaki, M. 977 (462), *990*
Taddei, M. 638 (94, 96), *687*, *688*, 809 (17), 842 (148), *851*, *855*
Tadesse, S. 956 (371), *988*
Taft, B. R. 746 (40, 41), *772*, 897 (101), *982*
Taft, R. W. 11 (74), *20*, 266 (59), *278*
Taga, T. 625 (40), 629 (68), 635 (88), *685–687*
Taga, Y. 1024 (149), *1032*
Taguchi, H. 977 (462), *990*
Taguchi, T. 461 (96, 97, 100, 101), *521*, 659 (130), 661 (133), *689*, 1007 (54), 1008 (55, 58), *1030*
Taher, D. 273 (46), *277*
Taillefer, M. 200 (213), *214*, 897 (108, 110), 921 (219, 220), 943 (217), 967 (413), *982*, *984*, *985*, *989*
Tainer, J. A. 354 (31, 49), *362*
Tait, C. D. 220 (30), *242*
Tajbakhsh, M. 925 (232), *985*
Takada, M. 925 (190), *984*
Takahashi, H. 452 (48, 49), 458 (76), *520*
Takahashi, K. 496 (199, 200), *523*
Takahashi, T. 461 (98, 99), 462 (102–104), 468 (105–107), 469 (108, 109), 470 (110), *521*, 573 (61), *582*
Takai, K. 502 (211), *523*
Takaki, K. 382 (55), *437*, 800 (72), 807 (95), *852*, *853*
Takamoto, Y. 941 (286), *986*
Takanami, T. 800 (41), *852*
Takano, S. 824 (69), *852*
Takats, J. 133 (133), *138*
Takeda, H. 285 (13), *341*
Takeda, K. 974 (440), *989*
Takeda, M. 859 (16), *877*
Takeda, T. 831 (164), *855*, 977 (462), *990*
Takei, R. 1009 (61), *1030*
Takemoto, Y. 635 (88, 90), *687*, 698 (13), *729*
Takemura, M. 513 (241), *524*, 977 (463), *990*
Takeuchi, M. 946 (316, 317), *987*
Takeya, K. 953 (361), *988*
Takikawa, H. 979 (487), *990*
Takizawa, T. 81 (183), *139*
Takui, T. 116 (294), *142*
Talakin, O. G. 155 (97), *161*
Talaty, E. R. 320 (185), *345*
Tale, R. H. 952 (355), *988*
Talley, P. 530 (19), *581*
Tamagawa, H. 979 (487), *990*
Tamamura, H. 484 (165), *522*, 625 (40, 42), 629 (68), 641 (105), *685*, *686*, *688*
Tamao, K. 382 (54), *437*, 640 (99), *688*, 800 (37, 40), 802 (48), *851*, *852*
Tamaru, K. 228 (58), *242*
Tambade, P. J. 966 (406), *989*
Tamborski, C. 1014 (81, 82), 1024 (137, 145), 1027 (146), *1030–1032*
Tamborskii, C. 846 (125), *854*
Tamura, M. 260 (48), *277*, 609 (22), *684*, 978 (470–472), *990*, 1008 (57), *1030*
Tamura, R. 666 (141), *689*
Tan, X. 309 (152), *344*
Tanaka, A. 891 (48), *981*
Tanaka, H. 894 (98), *982*
Tanaka, K. 595 (21), *602*
Tanaka, M. 614 (17), 616 (27), 625 (42, 64), 629 (68), *684–686*
Tanaka, N. 746 (41), *772*
Tanaka, P. 800 (40), *852*
Tanaka, R. 116 (294), *142*
Tanaka, T. 635 (88, 90), *687*
Tanaka, Y. 382 (54), *437*, 640 (99), *688*, 802 (48), *852*
Tang, H. Y. 240 (140), *245*
Tang, J. A. 166 (60), *211*
Tang, W. 201 (222), *215*
Tang, X. 940 (284), *986*
Tang, Z. 152 (58), *160*
Tang, Z.-C. 152 (59), *160*
Tani, K. 939 (283), *986*
Tanigaki, H. 641 (105), *688*
Taniguchi, N. 906 (143), *983*
Tanihata, S. 157 (112), *161*
Tanimori, S. 970 (432), *989*
Tanner, D. 413 (131), *439*, 474 (128), *522*, 824 (156), 850 (200), *855*, *856*
Tanner, J. E. 168 (97), *212*
Tao, C. Z. 925 (225), *985*
Tao, F. G. 942 (158), *983*
Tao, W. A. 322 (190, 192), 326 (206), *345*
Tao, Y. 802 (91), *853*
Tarumi, Y. 1004 (47), *1030*
Tasker, P. A. 87 (205–207), *140*
Tasseau, O. 325 (202), 326 (203), *345*
Tateiwa, J. 800 (42), *852*
Tateiwa, J.-i. 496 (201), 513 (242), *523*, *524*, 738 (22), *772*
Tatlow, J. C. 73 (160), *139*, 992 (1), 1014 (84, 86, 87), 1016 (100), *1028*, *1030*, *1031*
Taudien, S. 374 (36), *436*
Tauler, R. 322 (187), *345*
Tavs, P. 893 (80, 81), *982*
Tawney, P. O. 2 (10), *19*, 24 (6), *135*, 694 (2), *729*
Taylor, A. B. 348 (17), *361*
Taylor, B. F. 893 (91), *982*

Taylor, B. J. 3 (36), *19*, 112 (282), *142*, 154 (81), *160*, 195 (58), *211*, 534 (38), *582*
Taylor, B. S. 831 (84), *853*
Taylor, H. V. 349 (2), *361*
Taylor, K. J. 152 (61), *160*
Taylor, N. J. 248 (5), *276*
Taylor, P. M. 238 (122), *244*
Taylor, R. J. K. 5 (5), 16 (89), *19*, *20*, 169 (115), *212*, 537 (16, 17, 25), 543 (1), *581*, 672 (1), *682*, 790 (8), *851*, 873 (4), *875*
Teat, S. J. 675 (162), *690*
Tebbe, K.-F. 113 (286, 287), *142*, 154 (84), *161*, 459 (84), *521*
Tebboth, J. A. 63 (128), *138*
Tedeschi, L. 695 (1), *729*
Teesch, L. M. 324 (197), *345*
Tegoni, M. 352 (36), 358 (67, 70), *362*, *363*
Teki, Y. 116 (293), *142*
Tellier, F. 539 (50), *582*
Tellitu, I. 904 (136), *983*
ten Hoedt, R. W. M. 36 (69), 53 (65), 66 (66), 68 (68), 70 (70), *137*
Teplyakov, A. V. 148 (15, 21), *159*
ter Horst, B. 699 (19), 719 (61), 720 (62, 65, 66), *729*, *730*
Terakubo, S. 484 (165), *522*
Terao, J. 556 (105), *583*, 979 (482), *990*
Terashima, M. 386 (66), *437*
Terefenko, E. 1018 (112), *1031*
Termine, E. J. 430 (171), *441*
Terpstra, J. W. 399 (96), *438*, 925 (234), *985*
Terret, N. 800 (77, 79), *853*
Terzulli, A. J. 356 (55), *362*
Tessier, C. A. 294 (95, 96), *343*
Tessier, P. J. 551 (90), *583*
Teutsch, G. 640 (102), *688*
Thaler, T. 506 (135), *522*
Thang, Y. 631 (79), *687*
Thathagar, M. B. 966 (402), 974 (441, 442), *989*
Thayumanavan, S. 386 (69), *437*
Thery, I. 925 (222), *985*
Thibonnet, J. 404 (125), *439*, 841 (162), 845 (193), *855*, *856*
Thiele, D. J. 348 (4, 10, 18), *361*
Thiele, K.-H. 39 (82), *137*, 148 (22), *159*
Thielges, S. 945 (307), *986*
Thomas, A. P. 381 (50), *437*, 613 (37), *685*, 800 (74, 75), *852*, *853*
Thomas, A. W. 200 (212), *214*
Thomas, K. M. 134 (29), *136*
Thomas, P. D. 322 (192), *345*
Thomsen, J. 157 (118), *161*
Thomson, A. J. 358 (62), 359 (64), *363*
Thomson, B. A. 330 (230), *346*
Thorand, S. 6 (53), *20*, 870 (14), *877*
Thorarensen, A. 807 (97), *853*, 979 (484), *990*
Thrasher, J. S. 1001 (4), *1029*
Thrower, J. 113 (283), *142*, 1013 (78), *1030*
Thummel, R. P. 993 (14), *1029*
Tian, W-S. 1021 (128), *1031*
Tian, X. 647 (116), *689*
Tian, Z. 294 (58), *342*
Tiederman, R. 800 (62), *852*
Tillyer, R. 173 (151), *213*
Tillyer, R. D. 829 (161), *855*
Tinant, B. 120 (316), *143*, 403 (114), *439*, 761 (79), *773*
Tirado, R. 474 (138), *522*, 625 (66), *686*
Tirpak, J. G. 890 (66), *981*
Tissot-Croset, K. 480 (155), *522*, 670 (149), 674 (160), 675 (2), *683*, *690*, 858 (11), *877*
Tjepkema, M. W. 551 (85), *583*
Tobita, E. 1017 (106), *1031*
Tobrman, T. 407 (127), *439*
Toda, A. 635 (88, 90), *687*
Todo, H. 556 (105), *583*, 979 (482), *990*
Togni, A. 129 (329, 330), *143*
Togo, H. 906 (142), *983*
Tokas, E. F. 1024 (147), *1032*
Tökès, L. 2 (20), *19*, 629 (4), *683*
Tokito, S. 1024 (149), *1032*
Tokles, M. 430 (17), *435*
Tokranova, N. 229 (69), *243*
Tokumaru, K. 230 (86, 87), *243*
Tokunaga, N. 859 (39), *879*
Tokuyama, H. 919 (180), 943 (292), *984*, *986*
Tolbert, B. M. 356 (57), *363*
Tollari, S. 355 (56), *362*
Tollefson, N. M. 904 (127), *983*
Tollner, F. 820 (138), *854*
Tolman, W. B. 121 (317), *143*, 359 (15), *361*
Toma, H. E. 240 (139), *244*, 257 (41), *277*
Tomás, M. 451 (44, 45), *521*
Tomasi, J. 283 (44), *341*
Tomaso, A. E. 756 (71), *773*
Tomassen, H. P. M. 894 (93), *982*
Tombo, G. M. R. 629 (74), *687*
Tomé, A. C. 238 (123), *244*
Tomilov, A. P. 249 (8), 250 (4), *276*, *277*
Tominaga, S. 401 (107), *439*, 672 (156), *690*
Tomioka, K. 200 (11), *210*, 401 (105), 430 (24, 27), *436*, *439*, 489 (171), *523*, 707 (35), *729*
Tomioka, T. 452 (50), *520*
Tomita, M. 885 (17), *980*
Tomizaki, T. 352 (38), *362*
Tomkinson, N. C. O. 926 (248), *985*
Tomlinson, J. T. 238 (131), *244*
Tomoka, C. S. 629 (75), *687*
Tomoko, T. 738 (22), *772*
Tomooka, K. 391 (88), *438*
Tomova, D. 237 (112), *244*
Tomoyasu, T. 391 (88), *438*
Tomura, M. 165 (31), *210*
Tomuro, Y. 502 (73), *520*

Tonthat, H. 351 (33), *362*
Toombes, G. E. S. 173 (150), *213*
Topping, C. M. 387 (75), 395 (90), *438*
Törker, M. 226 (55), *242*
Torneiro, M. 622 (58), *686*
Tornøe, C. W. 200 (15), *210*
Torrelli, V. 861 (21), *878*
Torres, T. 221 (38), *242*
Tortajada, J. 283 (48), 293 (29), 311 (171, 172), 315 (121, 174, 178, 179), 318 (150, 177), 332 (234), 333 (235), 334 (222, 223, 237), 337 (125, 151), 339 (184), *341–346*
Toru, T. 822 (150), *855*
Toscano, M. 290 (34), 301 (83), 302 (35), 303 (78), 305 (141), *341*, *342*, *344*
Totleben, M. J. 471 (111), *521*
Toumi, M. 164 (20), *210*, 899 (115), *982*
Toupet, L. 401 (104), *439*
Toyota, K. 116 (294), *142*, 283 (44), *341*, 920 (183), *984*
Tozer, M. J. 418 (149), *440*
Tracey, M. R. 933 (263, 265), 934 (264), *985*, *986*
Trachtman, M. 147 (9), *159*
Tramontano, A. 516 (248), *524*
Tran Dinh, T. 888 (49), 891 (37, 50), *981*
Tran, P. 28 (46), *136*
Tran-Thi, T. H. 223 (42), *242*
Tranquille, M. 153 (73), *160*
Trehan, S. 925 (236), *985*
Tremaine, J. F. 63 (128), *138*
Tremblay, M.-C. 620 (50, 53), *685*
Tremble, J. 2 (20), *19*, 629 (4), *683*
Trendafilova, N. 296 (103), 309 (37), *341*, *343*
Trent, J. O. 484 (165), *522*
Trevitt, G. P. 707 (40), *730*
Trometer, J. D. 625 (64, 65), *686*
Trost, B. M. 380 (49), *437*, 609 (26), 614 (39), 649 (117), *684*, *685*, *689*, 728 (74), *730*, 832 (166), *855*, 933 (267), *986*
Trotter, B. T. 286 (60), *342*
Truce, W. E. 564 (113, 115), *583*
Trucks, G. W. 283 (44), *341*
Truhlar, D. G. 283 (45), 333 (236), *342*, *346*
Trujillo, C. 303 (49), *342*
Truong, C. M. 157 (107), *161*
Tsang, W. C. P. 963 (391), *988*
Tsao, H. S. 994 (16), *1029*
Tschaen, D. 897 (105), *982*
Tscherniack, J. 883 (4), *980*
Tse, H. L. A. 834 (168, 169), *855*
Tseng, C. C. 379 (48), *437*, 609 (26), 612 (34), 647 (9, 21), *683*, *684*
Tsubota, S. 157 (112), *161*
Tsubouchi, A. 977 (462), *990*
Tsuda, T. 60 (122), *138*, 158 (133), *162*, 386 (56), *437*, 735 (13, 14), *772*, 886 (22), *980*
Tsui, E. Y. 44 (102), 46 (103), *137*, 402 (119, 120), *439*
Tsuji, J. 430 (174), 434 (175), *441*
Tsukatani, K. 228 (66), *243*
Tsukihara, T. 352 (38), *362*
Tsumori, N. 157 (112), *161*
Tsuritani, T. 958 (381), *988*
Tsutsui, H. 376 (39), *436*
Tu, Y. 631 (79), *687*
Tucci, F. 514 (245), *524*
Tucci, F. C. 514 (246), *524*
Tucker, C. E. 108 (259), *141*, 175 (160), *213*, 423 (161), *440*, 473 (121), 474 (123), *521–522*, 858 (8), *876*
Tucker, S. H. 889 (53), *981*
Tuckmantel, W. 810 (103), *854*
Tulloch, A. A. D. 127 (326), *143*
Tura, G. M. 240 (142), *245*
Tureček, F. 327 (153, 154), 328 (61, 209–211), 334 (12), *341*, *342*, *344*, *345*
Turner, D. 416 (134), *440*
Tuttle, N. 888 (36), *981*
Tuyet, T. M. T. 946 (316), *987*
Twieg, R. J. 913 (167, 168), 915 (166), *983*
Tysoe, W. T. 156 (99), *161*
Tyurina, L. S. 158 (131), *162*

Ubayama, H. 468 (107), *521*
Uccello-Barretta, G. 629 (72), *686*
Uchiyama, M. 5 (42, 43), *19*, *20*, 588 (14), *602*
Uechi, G. T. 294 (96), *343*
Ueda, E. 979 (478), *990*
Ueda, S. 484 (165), *522*, 946 (316, 317), *987*
Ueda, T. 968 (423), *989*
Uehara, H. 1007 (54), *1030*
Uemura, T. 964 (392), *988*
Ueno, Y. 822 (150), *855*
Uerdingen, M. 6 (52), *20*, 184 (178), *214*, 861 (23), 870 (48), *878*, *879*
Ugalde, J. M. 281 (16), *341*
Ulibarri, T. A. 63 (134), *138*
Ullenius, C. 8 (57, 58), *20*, 173 (122), 184 (134, 181, 188), 186 (186), 191 (191), *212–214*, 388 (80), *438*, 534 (34, 35), *582*, 829 (160), *855*
Ullmann, F. 883 (1, 3, 4), 888 (33, 41), 963 (2), *980*, *981*
Umbuzeiro, G. A. 233 (96), 234 (97), *243*
Umeda, C. 462 (104), *521*
Umemoto, T. 1016 (97), 1018 (114), *1031*
Umen, M. J. 91 (212), *140*, 459 (92), *521*
Umesue, I. 1012 (76), *1030*
Umkehrer, M. 908 (153), *983*
Underiner, T. L. 612 (34), 647 (9, 21), *683*, *684*, 872 (50), *879*
Unger, J. B. 897 (101), *982*
Uno, H. 998 (31–33), *1029*

Uno, Y. 887 (30), *981*
Ura, H. 970 (432), *989*
Urabe, H. 444 (8, 9), 445 (15, 16, 19), *519*, 937 (275), *986*
Urata, H. 977 (466), *990*, 1017 (110), *1031*
Ürtendahl, M. 61 (125), *138*
Urvoas, A. 330 (227), *346*
Usui, S. 588 (14), *602*
Usui, Y. 238 (129), *244*
Utimoto, K. 810 (103), *854*, 872 (51), *879*
Uwatoko, Y. 119 (300), *142*

Vachet, R. W. 301 (120), *343*
Vaisar, T. 327 (153, 154), 328 (61, 209, 211), *342*, *344*, *345*
Vakalopoulos, A. 960 (384), *988*
Valacchi, M. 842 (184), *856*
Valachi, M. 807 (98), *853*
Valentine, J. S. 349 (28), *362*
Valentini, M. 18 (117), *21*, 202 (219), *215*
Valero, R. 802 (89), *853*
Valéry, J.-M. 557 (106), *583*
Valina, A. B. 300 (88), *343*
Valnot, J.-Y. 166 (67), *211*
Valverde, S. 613 (31), *684*
Van Aken, K. 388 (81), *438*
Van Allen, D. 938 (280), *986*
van Berlekom, S. B. 87 (204), *140*
Van de Kamp, M. 350 (26), *362*
van den Ancker, T. R. 37 (74), *137*
van den Berg, M. 201 (226), *215*
Van den Heuvel, A. 695 (8), *729*
van den Hoogenband, A. 399 (96), *438*, 925 (234), *985*
van der Sluis, L. S. 399 (96), *438*
van der Sluis, P. 48 (107), *138*
van der Waals, J. H. 220 (19), *241*
van Dijck, L. A. 629 (72), *686*
Van Dijk, N. 220 (19), *241*
Van Duyne, G. 173 (148), *213*
Van Ende, D. 906 (141), *983*
van Esch, J. 201 (226), *215*
van Geel, T. A. P. 979 (477), *990*
van Gelderen, P. 194 (194), *214*
Van Huis, C. A. 925 (230), *985*
van Klaveren, M. 607 (10), 609 (32), 669 (14, 143), *683*, *684*, *690*, 858 (11), 866 (31), *876*, *879*
van Koten, G. 2 (14, 15), *19*, 27 (40), 28 (47), 34 (27, 28, 34, 36, 37), 35 (64), 36 (67, 69), 37 (78), 39 (39, 85), 40 (86), 41 (95), 43 (98), 48 (107), 51 (30), 52 (43, 110, 111), 53 (65), 56 (116), 61 (13), 66 (66), 68 (68), 70 (70), 78 (173), 87 (12, 203, 204), 91 (214, 215), 93 (50), 94 (94, 216, 218, 219), 95 (217, 220), 97 (224), 101 (231, 232), 108 (108, 264), 112 (112), 113 (113), 114 (114), 117 (117), 119 (119), 120 (120), 134 (29, 49, 63, 338), *135–141*, *143*, 173 (126), 175 (156, 157), *212*, *213*, 276 (73), *278*, 607 (10), 609 (32), 669 (14, 142, 143), *683*, *684*, *689*, *690*, 800 (59), *852*, 858 (5, 11), 866 (31), *875*, *876*, *879*
van Lier, J. E. 238 (124, 125), *244*
van Maanen, H. L. 669 (142), *689*
van Maarseveen, J. H. 399 (96), *438*
van Mourik, G. L. 609 (29), *684*
van Oeveren, A. 382 (54), *437*
van Spanning, R. J. M. 358 (65), *363*
Van Stipdonk, M. J. 320 (185), *345*
van Summeren, R. P. 721 (56, 67), *730*
Van Tuyen, N. 1017 (109), *1031*
Van Veldhuizen, J. J. 132 (336), *143*, 508 (151), *522*, 672 (157), *690*
van Zijl, A. W. 670 (146, 151), 675 (2), *682*, *690*
van Zijl, P. C. M. 194 (194), *214*
Van Zyl, C. M. 818 (127), *854*
Vanderwalle, M. 859 (18), *878*
Vandewalle, M. 654 (125), *689*
VanZile, M. L. 154 (85), *161*
Varadarajan, A. 71 (154), *139*
Varchi, G. 421 (151, 155), *440*, 802 (87, 88), *853*
Varghese, J. P. 565 (120–122), *583*
Varjas, L. 859 (18), *877*
Vasak, M. 330 (231), *346*
Vasapollo, G. 235 (101–105), *243*
Vassiliou, S. 579 (142), *584*
Vastra, J. 694 (4), 695 (8), *729*
Vattuone, L. 156 (99), *161*
Vaupel, A. 473 (120), *521*
Vázquez, P. 221 (38), *242*
Veeraraghavan Ramachandran, P. 992 (6), *1029*
Vega, A. 88 (208), *140*
Veith, M. 820 (138), *854*
Velarde, E. 629 (4), *683*
Vellekoop, A. S. 179 (152), 189 (189), *213*, *214*, 605 (12), *683*, 858 (7), *876*
Velu, S. E. 388 (85), 389 (76), *438*
Venanzi, L. M. 456 (62), 461 (77), *520*
Venkanna, G. T. 921 (207), *984*
Venkat Reddy, G. 992 (6), *1029*
Venkatachalam, K. A. 148 (10), *159*
Venkatachalam, T. K. 820 (137), *854*
Venkatappa, M. P. 352 (39), *362*
Venkataraman, D. 904 (131), 906 (140), 938 (280), 966 (100), *982*, *983*, *986*
Venkatasubbaiah, K. 60 (60), *136*
Venkatraman, S. 471 (112), *521*
Venturello, P. 958 (380), *988*
Vera, E. L. 933 (265), *986*
Verbost, P. M. 382 (54), *437*
Verchirko, E. P. 1019 (115), *1031*

Verma, S. 238 (133, 134), *244*
Vermeer, P. 530 (18), 540 (55, 57, 58), 555 (104), 558 (108), 564 (114, 116), 571 (132), *581–584*, 629 (72), *686*, 800 (53), 832 (146), 838 (124), *852*, *854*, *855*
Verstuyf, A. 654 (125), *689*, 859 (18), *878*
Vestin, R. 158 (128), *162*
Viala, J. 811 (109), *854*
Vicente, R. 451 (45), *520*
Vicic, D. A. 44 (99), *137*, 622 (61), *686*
Vicinelli, V. 221 (36), *242*
Vickers, D. 430 (18), *435*, *436*
Vieira, P. C. 979 (490), *990*
Viezzoli, M. S. 354 (48), *362*
Vignau, M. 861 (21), *878*
Vigneron, J. P. 238 (128), *244*
Viirre, R. D. 908 (154), *983*
Villacorta, G. M. 16 (100), *21*, 380 (3), 430 (22), *435*, *436*
Villava-Servin, N. P. 551 (89), *583*
Villiéras, J. 2 (22), *19*, 536 (27, 28), 540 (8, 10), 550 (66, 80), *581–583*
Vina, D. 920 (200), *984*
Vincent, G. 952 (354), 956 (372), 960 (385), 962 (351), *987*, *988*
Vincent, J.-M. 200 (214), *214*
Vinent, V. 240 (142), *245*
Viso, A. 618 (49), 629 (70), *685*, *686*
Viswanathan, M. 831 (84), *853*
Vitale, G. 300 (88), *343*
Vittal, J. J. 83 (190), *140*, 268 (60), *278*
Vivian, R. 738 (24), *772*
Vivian, R. W. 738 (24), *772*
Volante, R. P. 373 (33), *436*, 897 (105), 915 (164), *982*, *983*
Volbach, W. 1018 (113), *1031*
Volgin, Yu. V. 33 (62), *136*
Volz, F. 632 (82), *687*, 870 (46), *879*
von Sonntag, C. 238 (123), *244*
Vorozhtsov, N. N. 890 (56–61), *981*
Voth, G. A. 283 (44), *342*
Vrancken, E. 399 (97), *438*
Vreven, T. 283 (44), *341*
Vu, V. A. 404 (124), 421 (152), 425 (154), *439*, *440*, 600 (19), *602*
Vuagnoux-d'Augustin, M. 506 (226), *524*, 530 (23), *581*, 709 (46), 725 (31), *729*, *730*
Vuister, G. W. 194 (194), *214*
Vukomanovic, D. 318 (181), *345*
Vyater, A. 183 (124), *212*
Vyaters, A. 858 (6), *876*
Vyvyan, J. R. 979 (479), *990*

Waas, J. 496 (126), *522*
Wada, K. 260 (48), *277*, 418 (146), *440*
Wagener, T. 153 (68), *160*
Wagman, D. D. 155 (3), *158*
Wagner, A. J. 424 (123), *439*
Wagner, C. 254 (35), *277*
Wagner, G. 330 (231), *346*
Wagner, R. 167 (90), 185 (135), *212*, *213*, 872 (47), *879*
Wagnon, J. 16 (98), *21*, 609 (25), *684*
Wai, J. S. M. 831 (163), *855*
Wailes, P. C. 452 (56), 459 (81), *520–521*
Wailes, S. 902 (121), *982*
Wakamatsu, K. 810 (103), *854*
Wakatsuki, Y. 271 (64), *278*
Wakeling, A. E. 861 (22), *878*
Wakemoto, H. 968 (427), *989*
Wakita, Y. 384 (57), 386 (56), *437*
Wakselman, C. 1014 (83), 1015 (89), *1030*
Walfort, B. 269 (62), 273 (46), 275 (61, 68, 70), *277*, *278*
Waliszewska, G. 241 (143), *245*
Walker, N. R. 305 (65, 132), 310 (158), *342–344*
Walter, D. 800 (44), *852*
Walton, D. R. M. 1024 (148), *1032*
Wan, B. 904 (132), 921 (186, 187), *983*, *984*
Wan, X.-Jian 123 (322), *143*
Wan, Z. 901 (118), *982*
Wanat, R. A. 173 (148), *213*
Wang, B. 65 (136), *138*, 972 (433), *989*
Wang, C. 235 (103), *243*
Wang, D. 129 (331), *143*, 229 (74), *243*, 751 (57), *773*
Wang, D.-P. 975 (405, 444), *989*
Wang, F. 322 (192), *345*
Wang, H. H. 114 (288), *142*, 154 (85, 86), *161*
Wang, J. 564 (119), *583*, 979 (492), *990*
Wang, J. K. 908 (155), *983*
Wang, L. S. 156 (101), *161*
Wang, L.-X. 206 (208), *214*, 670 (150), *690*
Wang, L. Y. 908 (155), *983*
Wang, M. 225 (51), 230 (75), *242*, *243*
Wang, P. S. 925 (226), *985*
Wang, R. 921 (187), *984*
Wang, R. J. 595 (30), *602*
Wang, R.-P. 954 (348), *987*
Wang, S. 564 (119), *583*
Wang, S. Y. 700 (20, 21), 725 (70), *729*, *730*
Wang, X. 303 (128), *343*
Wang, X.-J. 629 (72), *686*
Wang, Y. 230 (75), *243*, 380 (49), *437*, 996 (24), *1029*
Wang, Y.-F. 643 (106), *688*
Wang, Y. S. 920 (201), *984*
Wang, Z. 484 (165), *522*, 631 (76), 641 (105), *687*, *688*
Wang, Z.-M. 632 (45), *685*
Wang, Z.-X. 631 (79), *687*
Wanner, J. 3 (33), *19*, 605 (12), *683*
Wanzlick, H.-W. 119 (303), *142*
Ward, W. E. 1027 (146), *1032*

Ware, A. C. 382 (55), *437*, 800 (72), 807 (95), *852*, *853*
Warner, P. 842 (187), *856*
Warren, L. F. 68 (143), *138*, 257 (38), *277*
Warren, T. H. 130 (318, 333), *143*, 166 (60), *211*
Wasielewski, M. R. 955 (367), *988*
Watabe, K.-i. 666 (141), *689*
Watanabe, H. 556 (105), *583*
Watanabe, M. 640 (101), *688*
Watanabe, T. 153 (72), *160*
Waters, T. 152 (23), *159*
Waterson, D. J. 800 (67, 70), *852*
Watkins, J. J. 173 (125), *212*
Watson, B. T. 459 (82, 83), *521*
Watson, P. L. 310 (164), *344*
Watson, R. 397 (92), *438*
Watson, R. T. 395 (90), 396 (91), 397 (92), 399 (93), *438*
Watthey, J. W. H. 885 (15), *980*
Waugh, F. 1024 (148), *1032*
Wawrzak, Z. 861 (22), *878*
Wearn, R. B. 968 (418), *989*
Webb, J. D. 157 (110), *161*
Weber, J. M. 149 (30, 33), *159*
Weber, V. 800 (50), *852*
Weber, W. 807 (43), *852*
Webster, M. 165 (38), *211*
Wecker, U. 868 (38), *879*
Wee, S. 329 (214, 215, 217), 339 (248), *346*
Wegner, P. A. 68 (142, 143), *138*
Wehlan, H. 975 (452), *989*
Wehman, E. 35 (64), 37 (78), 56 (116), 114 (114), *137*, *138*
Wehman-Ooyevaar, I. C. M. 117 (117), *138*
Wei, Q.-H. 76 (167), *139*, 85 (202), 90 (210), *140*
Wei, Y. 427 (167), *440*
Weigelt, J. 197 (197), *214*
Weigold, H. 452 (56), 459 (81), *520*, *521*
Weil, A. 146 (2), *158*
Weiner, P. K. 354 (49), *362*
Weingarten, H. 884 (7), 942 (6), *980*
Weinges, K. 646 (110), *688*
Weinhold, F. 296 (108), *343*
Weiping, M. 237 (111), *244*
Weisblum, B. 484 (164), *524*
Weiske, T. 311 (173), *345*
Weiss, E. 74 (165), 82 (178), 95 (221), *139*, *140*
Weisshaar, J. C. 310 (163, 165), *344*
Welch, A. J. 70 (150), 71 (152), *139*
Welch, M. J. 861 (22), *878*
Welle, A. 120 (316), *143*, 403 (114), *439*, 761 (79), *773*
Wen, S. Y. 908 (152), *983*
Wentzlaff, T. H. 153 (37), *159*
Wermuth, C.-G. 638 (94), *687*
Werner, H. 868 (38), *879*
Werstiuk, N. H. 304 (129), *343*
Wesdemiotis, C. 290 (76), 318 (180), *342*, *345*
Wesendrup, R. 290 (79), *342*
Weser, U. 349 (2), *361*
West, T. R. 962 (349), *987*
Westerhausen, M. 105 (244), *141*, 152 (54), *160*
Westermann, J. 15 (86), *20*, 503 (218–222), *524*, 654 (127), *689*
Westmije, H. 540 (55), *582*
Westmijze, H. 530 (18), 540 (57, 58), 558 (108), 571 (132), *581–584*, 629 (72), *686*, 800 (53), 832 (146), 838 (124), *852*, *854*, *855*
Weston, P. E. 891 (77), *982*
Wheatley, A. E. H. 588 (14), *602*
White, A. H. 37 (72, 74), 73 (73), 79 (175), 83 (197), *137*, *139*, *140*, 261 (47), *277*
White, A. J. P. 98 (225), *140*, 178 (168), *213*
White, D. A. 247 (3), *276*
White, J. D. 842 (186), *856*
Whitesides, G. M. 2 (19), 16 (16), *19*, 24 (8, 9), 112 (275), *135*, *142*, 151 (47), *160*, 173 (140), *213*, 493 (182), *523*, 586 (8), 587 (9), 589 (15), *601*, *602*, 778 (2), *851*, 887 (23), *980*
Whittaker, J. W. 353 (42), *362*
Whittal, R. M. 332 (225), *346*
Whitten, C. E. 13 (80), 16 (99), *20*, *21*, 24 (11), *135*, 356 (2), *435*, 776 (5), *851*
Wickham, G. 831 (84), *853*
Wiechert, R. 861 (19), *878*
Wiedeman, P. E. 917 (178), *984*
Wieghardt, K. 348 (7), 352 (41), 354 (45), 355 (50), 356 (53, 58), *361–363*
Wiehe, A. 238 (118), *244*
Wieland, D. M. 625 (62), *686*
Wieland, P. 861 (19), *878*
Wieland, W. 66 (138), *138*
Wiemer, D. F. 605 (12), *683*
Wiemers, D. M. 113 (284), *142*, 1019 (117, 120), 1021 (116), 1024 (40), *1029*, *1031*
Wierzba, M. 861 (22), *878*
Wild, H. 386 (61), *437*, 859 (18), *878*
Wilhelm, R. S. 17 (104), *21*, 24 (14), 108 (249), *135*, *141*, 169 (120, 121), *212*, 609 (25), *684*, 788 (27), *851*
Wilkins, C. L. 146 (2), *158*
Wilkins, J. M. 459 (92), *521*
Wilkinson, G. 2 (14), *19*, 25 (32), 27 (40), 130 (130), *136*, *138*, 858 (5), *875*
Willem, R. 178 (168), *213*
Willert-Porada, M. A. 113 (31), *136*, 155 (87), *161*, 459 (86), *521*
Willey, K. F. 285 (55), *342*
Williams, A. L. 897 (9), *980*

Williams, A. R. B. 382 (55), *437*, 807 (95), *853*
Williams, D. J. 502 (215), *524*
Williams, D. R. 516 (258), *524*
Williams, J. M. 114 (288, 289), *142*, 154 (85, 86), *161*, 947 (320), *987*
Williams, L. J. 631 (76), *687*
Williams, N. 401 (104), *439*
Williams, R. J. P. 348 (1), *361*
Williamson, G. 34 (34), *136*
Williamson, M. 176 (88), *212*
Willis, A. C. 595 (29), *602*, 993 (13), *1029*
Willker, W. 189 (195), *214*
Wilmot, C. M. 354 (45), *362*
Wilsily, A. 702 (24, 25), *729*, 861 (24), *878*
Wilson, C. 127 (328), *143*, 402 (110), *439*, 560 (110), *583*
Wilson, D. W. 388 (81), *438*
Wilson, I. 418 (148), *440*
Wilson, P. D. 551 (85, 87, 88), *583*
Wilson, S. R. 861 (22), *878*
Wimmer, P. 947 (318, 319), *987*
Wing, R. M. 68 (148, 149), *138*, 256 (39, 40), *277*
Winge, D. R. 330 (224), *346*
Wingerter, S. 76 (76), *137*
Winn, C. L. 401 (104), *438*
Winnischofer, H. 257 (41), *277*
Winstead, J. L. 147 (9), *159*
Winter, R. W. 154 (85), *161*
Winter, S. B. D. 382 (54), *437*, 622 (60), *686*, 800 (35), *851*
Winters, M. P. 954 (348, 365), *987*, *988*
Wipf, P. 452 (47–49, 51, 55), 457 (60, 69), 458 (76), 461 (77), 471 (111, 112), 484 (162–164), 503 (212), 503 (213), 509 (4), 516 (251), *519–524*, 629 (69), 672 (1), *682*, *686*, 858 (8), *876*
Witt, M. 904 (129), *983*
Woerpel, K. A. 382 (51, 54), *437*, 612 (35), *684*, 977 (459), *990*
Woertink, J. S. 348 (16), *361*
Wöhrle, D. 228 (66), *243*
Wolckenhauer, S. A. 399 (99), *438*
Wolf, C. 913 (138), 920 (192, 193), *983*, *984*
Wolfe, J. P. 899 (114), *982*
Wollenberg, R. H. 509 (229, 230), *524*, 838 (180), *856*
Wolter, M. 901 (111), 926 (247), *982*, *985*
Wong, K. M.-C. 83 (192, 193), *140*, 266 (55), *278*
Wong, K.-Y. 83 (190), *140*, 268 (60), *278*
Wong, M. T. 264 (54), *278*
Wong, M. W. 283 (44), *342*
Wong, T. 510 (236), *524*, 551 (85), *583*, 829 (154), *855*, 975 (448), *989*
Woo, K. 625 (66), *686*
Woo, S. 551 (91), *583*
Wood, J. 943 (290), *986*
Wood, M. R. 457 (75), 474 (138), *520*, *522*, 858 (8), *876*
Woode, P. D. W. 804 (92), *853*
Woodrow, M. D. 926 (248), *985*
Woods, A. L. 24 (4), *135*
Woods, L. A. 11 (11), *19*, 585 (2), *601*, 776 (1), *850*, 857 (2), *875*
Woods, M. 579 (144), *584*
Woodward, S. 400 (103), *438*, 506 (223), *524*, 675 (2, 162), *683*, *690*, 698 (14, 15), *729*, 850 (199), *856*, 858 (7), *876*
Worch, C. 926 (242), *985*
Worgotter, E. 330 (231), *346*
Worle, M. 18 (117), *21*, 202 (219), *215*
Wormald, J. 732 (2), *771*
Wrackmeyer, B. 163 (8), 178 (168), *210*, *213*
Wright, D. S. 39 (79), 106 (176), *137*, *139*, 821 (139), *854*
Wright, J. R. 886 (21), *980*
Wright, L. J. 116 (292), 117 (295), *142*
Wright, R. R. 288 (62), 305 (65), 310 (158), *342*, *344*
Wu, A. W. 993 (10), *1029*
Wu, B. 956 (374), *988*
Wu, D. 294 (58), *342*
Wu, F. 904 (132), 921 (186, 187), *983*, *984*
Wu, G. 289 (66), *342*
Wu, I-Y. 259 (44), *277*
Wu, J. 83 (190), *140*, 268 (60), *278*, 318 (180), *345*, 701 (22), *729*, 754 (66), *773*, 943 (231), *985*
Wu, J. X. 967 (411), *989*
Wu, K. J. J. 225 (48), *242*
Wu, L. 225 (48), *242*, 326 (206, 207), *345*
Wu, M. 595 (22), *602*
Wu, S. 391 (89), *438*
Wu, S. H. 942 (158), *983*
Wu, S-W. 1021 (125, 126), *1031*
Wu, Y. 471 (113, 114), *521*, 654 (125), *689*
Wu, Z. 571 (130), *584*
Wudl, F. 834 (172), *855*
Wulff, W. D. 834 (168), *855*
Wunsch, P. 358 (65), *363*
Wurst, K. 129 (331), *143*, 751 (57), *773*
Wuthrich, K. 330 (231), *346*

Xi, C. 468 (107), 470 (110), *521*
Xi, C. J. 595 (30, 31), *602*
Xi, M. 148 (14), *159*
Xi, N. 921 (209), *984*
Xi, Z. 468 (107), 469 (108), *521*, 921 (212), *984*
Xia, C. 911 (162), *983*
Xia, C.-Q. 956 (374), *988*
Xia, W. 226 (57), *242*
Xianping, F. 237 (111), *244*

Xiao, J. 484 (163, 164), *522*
Xie, M. 564 (119), *583*
Xie, Q. 554 (101), *583*
Xie, R. 921 (213), *984*
Xie, X. 168 (91), 179 (87), 181 (169), *212*, *213*, 858 (6), *876*, 901 (119), 970 (431), 972 (434), *982*, *989*
Xie, Y. 240 (140), *245*
Xie, Y.-X. 975 (405, 445), *989*
Xifra, R. 114 (290, 291), *142*, 964 (394), *988*
Xing, D. 309 (152), *344*
Xing, X. 152 (58), *160*, 894 (96), *982*
Xing, X.-P. 152 (59), *160*
Xu, B. 229 (69), *243*
Xu, F. 173 (151), *213*
Xu, F.-B. 123 (322), *143*
Xu, J. 748 (48), *772*, 922 (208), *984*
Xu, J. B. 225 (51), *242*
Xu, L. 807 (96), *853*, 904 (132), 921 (187), *983*, *984*
Xu, L.-W 762 (83), *773*
Xu, M. S. 225 (51), *242*
Xu, Q. 157 (112, 119), *161*, *162*, 564 (118), *583*
Xu, S. 65 (136), *138*
Xu, W. 452 (48), 457 (69), 461 (77), *520*
Xu, X. 106 (242), *141*, 152 (53), *160*, 294 (58), *342*, 748 (48), *772*, 935 (268), *986*
Xu, Y. 234 (98), *243*
Xu, Y. J. 710 (51), *730*
Xue, L. 996 (8), *1029*
Xue, Y. 360 (79), *363*

Yabe, M. 461 (97), *521*
Yadav-Bhatnagar, N. 950 (337), *987*
Yadav, K. K. S. 351 (32), *362*
Yadav, K. S. S. 354 (44), *362*
Yagi, K. 644 (107), *688*
Yagupol'skii, L. M. 887 (25), *980*, 992 (7), 1019 (115), *1029*, *1031*
Yalcin, T. 324 (198, 199), *345*
Yam, V. W. 83 (196), *140*
Yam, V. W.-W. 78 (174), 81 (180, 182, 184–186), 83 (189, 191–194), 85 (199–201), *139*, *140*, 261 (52), 262 (51, 53, 57), 264 (54), 266 (55, 56), 267 (58), *277*, *278*
Yamada, K. 401 (106), 430 (27), *436*, *439*, 943 (292), *986*
Yamada, K.-i. 200 (11), *210*, 489 (171), *523*, 707 (35), *729*
Yamada, N. 979 (478), *990*
Yamada, S. 221 (34), 225 (49), *242*
Yamada, Y. 156 (100), *161*
Yamagami, T. 543 (65), *582*
Yamagishi, T. 888 (45, 46), 891 (47, 48), *981*, 1011 (72), *1030*
Yamaguchi, H. 228 (66), *243*, 352 (38), *362*, 885 (17), *980*
Yamaguchi, K. 309 (116), *343*
Yamaguchi, R. 872 (51), *879*
Yamaguchi, S. 543 (65), *582*
Yamaguchi, T. 976 (446), *989*
Yamakawa, J. 595 (28), *602*
Yamamoto, A. 81 (81), *137*, 891 (75), *982*
Yamamoto, H. 377 (43), *436*, 598 (33), *602*, 609 (19), *684*, 941 (286), *986*
Yamamoto, K. 165 (47), *211*, 430 (174), 433 (173), 434 (175), *441*, 1016 (102–104), *1031*
Yamamoto, M. 1000 (37–39), *1029*
Yamamoto, N. 434 (175), *441*, 484 (165), *522*
Yamamoto, S. 456 (65), *520*, 613 (23), *684*, 859 (17), *877*
Yamamoto, T. 81 (81, 183), *137*, *139*, 891 (75, 76), *982*
Yamamoto, Y. 2 (2), 15 (83), *19*, *20*, 164 (17), *210*, 276 (71), *278*, 456 (64, 65), 496 (191, 192), 515 (247), *520*, *523*, *524*, 609 (24), 613 (23), 614 (17, 41), 616 (27), 625 (16, 40, 42, 64), 629 (68), 666 (141), 672 (1), *682*, *684–686*, *689*, 858 (13), 859 (17), 873 (4), *875*, *877*, 958 (381), *988*
Yamanaka, H. 496 (201), *523*, 549 (73, 79), *583*, 738 (22), *772*
Yamanaka, M. 3 (35, 40), 5 (46), 11 (73), 16 (102, 103), 17 (108, 109), *19–21*, 133 (277), *142*, 184 (179), 194 (175), 198 (174), *214*, 367 (7), *435*, 459 (88), *521*, 606 (11), *683*
Yamanouchi, H. 225 (46), *242*
Yamaoka, Y. 998 (31), *1029*
Yamasaki, K. 703 (29), *729*
Yamashita, A. 735 (13), *772*
Yamashita, E. 352 (38), *362*
Yamashita, K. 229 (68), *243*
Yamashita, T. 6 (50, 51), *20*, 194 (176), *214*, 870 (48), *879*
Yamashita, Y. 979 (474), *990*
Yamazaki, A. 469 (108), *523*
Yamazaki, T. 616 (48), *685*
Yan, M. 695 (8), *729*
Yan, Z.-J. 551 (97), *583*
Yanagi, H. 228 (66), *243*
Yanagisawa, A. 609 (19), *684*
Yand, Y. 747 (45), *772*
Yáñez, M. 281 (15), 282 (26), 283 (48), 284 (32), 285 (50), 293 (86, 87), 295 (85), 303 (49), 308 (41, 104, 105), 309 (40, 147), 310 (170), 311 (171, 172), 315 (121, 174, 178), 318 (150, 177), 333 (235), 337 (84, 125, 151, 169), *341–346*
Yang, C. N. 294 (36, 94), *341*, *343*
Yang, D. 303 (128), *343*
Yang, G.-M. 235 (103), *243*
Yang, H. 616 (43), *685*

Yang, J. 551 (97), *583*
Yang, L. M. 920 (197), *984*
Yang, M. 919 (185), *984*
Yang, M. X. 148 (14, 15), 150 (43), *159*, *160*
Yang, T. 926 (237), *985*
Yang, W. 282 (23), *341*
Yang, X. 444 (3), *519*
Yang, Y. 748 (47), *772*
Yang, Y.-L. 535 (43), *582*
Yang, Y. S. 295 (52), *342*
Yang, Z. 152 (51), *160*, 299 (118, 119), *343*
Yang, Z. B. 295 (75), *342*
Yang, Z-Y. 1005 (5), 1012 (77), 1014 (80), 1016 (95), 1024 (40), 1026 (41, 152), *1029*, *1030*, *1031*, *1032*
Yanik, M. M. 977 (465), *990*
Yano, S. 116 (294), *142*
Yano, T. 116 (292), *142*
Yanovsky, A. I. 90 (209), *140*
Yao, J. 909 (157), *983*
Yao, J. C. 942 (158), *983*
Yaono, R. 352 (38), *362*
Yap, G. P. A. 551 (88), *583*
Yarnaguchi, H. 230 (78), *243*
Yashunsky, D. V. 535 (44), *582*
Yasuda, N. 958 (381), *988*
Yatagai, H. 456 (65), 496 (188, 191, 192), *520*, *523*, 613 (23), *684*, 859 (17), *877*
Yazyev, O. 283 (44), *342*
Ye, J. 976 (449), *989*
Yee, J. G. K. 834 (159), *855*
Yeh, C. S. 295 (51, 52), 318 (175), *342*, *345*
Yeh, L.-A. 894 (96), *982*
Yeh, M. C. 418 (147), *440*
Yeh, M. C. P. 169 (107), *212*, 383 (38), *436*
Yeh, V. S. C. 917 (178), *984*
Yen, S. J. 647 (9), *683*
Yeung, K.-W. 83 (190), *140*, 268 (60), *278*
Yeung, P. K. Y. 261 (52), *277*
Yin, G.-Q. 76 (167), *139*, 90 (210), *140*
Yip, J. H. K. 83 (190), *140*, 268 (60), *278*
Yokomatsu, T. 1011 (70–74), 1012 (75, 76), *1030*
Yoneda, E. 979 (475), *990*
Yoneda, F. 614 (41), *685*
Yonemura, H. 225 (49), *242*
Yonezawa, K. 150 (42), *160*
Yoon, J. 348 (16), *361*
Yoon, S.-K. 966 (409), 976 (455), *989*, *990*
Yoon, S. S. 976 (455), *990*
Yorimitsu, H. 480 (154), *522*, 569 (128), *584*, 675 (2), *682*
York, D. M. 281 (16), *341*
York, J. T. 121 (317), *143*
Yoshida, J.-I. 572 (133, 134), *584*
Yoshida, T. 459 (80), *521*, 946 (317), *987*
Yoshida, Y. 892 (82), *982*
Yoshifuji, M. 459 (57), *520*, 920 (183), *984*
Yoshikai, N. 6 (50, 51), 11 (72), 13 (77–79, 82), 17 (107, 108), 18 (119), *20*, *21*, 133 (21), *135*, 184 (112), 194 (176), 198 (173, 206), *212*, *214*, 478 (142, 144), *522*, 606 (11), *683*, 870 (48), *879*
Yoshikawa, S. 352 (38), *362*
Yoshioka, Y. 309 (116), *343*
Yoshizawa, H. 616 (27), 625 (42), *684*, *685*
You, J. 921 (213), *984*
Young, D. C. 68 (142, 143), *138*
Young, V. G. 121 (317), *143*
Youngs, W. 294 (96), *343*
Youngs, W. J. 294 (95), *343*
Younus, M. 272 (43), *277*
Yu, C.-M. 976 (458), *990*
Yu, H. 390 (82), 393 (77, 87), *438*, 631 (79), 635 (89, 92), *687*
Yu, H. B. 629 (75), *687*
Yu, J. C. 240 (140), *245*
Yu, S. 956 (370), *988*
Yu, X.-Q. 956 (374), *988*
Yu, Y. 377 (42), *436*
Yuan, H. S. H. 106 (227), *141*, 152 (63), *160*
Yuan, J. 935 (268), *986*
Yuan, O. L. 919 (176), *984*
Yuan, Q. 917 (175), *984*
Yuan, X. 935 (268), *986*
Yudin, A. K. 958 (379), *988*
Yudin, A. Y. 1016 (96), *1031*
Yue, Y. 956 (374), *988*
Yuen, J. 937 (272), *986*
Yum, E. K. 576 (140), *584*
Yun, H. 131 (334), *143*, 403 (111), *439*, 751 (58), *773*
Yun, J. 131 (334), *143*, 403 (111), *439*, 733 (10), 744 (35, 36), 747 (45, 46), 748 (47), 751 (58), 754 (64), *772*, *773*
Yurdakal, S. 236 (107), *244*
Yus, M. 609 (26), *684*, 709 (45), *730*
Yusa, Y. 756 (72), *773*

Zaccheria, F. 966 (401), *989*
Zaera, F. 148 (11, 13, 17, 18), *159*
Zaitsev, A. 349 (29), *362*
Zaitseva, I. 349 (29), *362*
Zaitseva, N. N. 79 (175), *139*
Zakharov, L. N. 30 (52), 90 (209), 93 (93), *136*, *137*, *140*
Zakrzewski, V. G. 283 (44), *341*, *342*
Zaleski, J. M. 238 (135), *244*
Zamir, T. 165 (53), *211*
Zangmeister, R. A. P. 226 (54, 56, 57), *242*
Zangrando, E. 101 (230), *141*
Zani, P. 809 (99, 100), *853*
Zanon, J. 967 (412), *989*
Zanoni, M. B. 233 (96), *243*
Zanoni, M. V. B. 234 (97), *243*

Zapf, A. 968 (414–416), *989*
Zarate, A. 88 (208), *140*
Zarbin, P. H. G. 429 (168), 430 (169), *440*, 979 (490), *990*
Zareba, A. A. 943 (295), *986*
Zarotti, P. 728 (75), *730*
Zavattaro, C. 958 (380), *988*
Zegarski, B. R. 150 (41), *160*
Zeidler, M. D. 165 (50), *211*
Zelder, C. 653 (121), *689*, 870 (14), *877*
Zeller, M. 66 (139), *138*
Zeng, A. 157 (119), *162*
Zeng, L. 958 (375), *988*
Zeni, G. 904 (135), 943 (229), *983*, *985*
Zenneck, U. 66 (139), *138*
Zenobi, R. 309 (149), *344*
Zepf, D. 165 (30), *210*
Zewail, A. H. 287 (67), *342*
Zewge, D. 915 (164), *983*
Zhai, H.-J. 156 (101), *161*
Zhang, B. 229 (74), *243*
Zhang, C. 802 (91), *853*, 926 (238), 958 (376), *985*, *988*
Zhang, D. 322 (190, 192), *345*, 542 (60), 576 (140), *582*, *584*
Zhang, F. X. 235 (103), *243*
Zhang, G. 234 (98), *243*
Zhang, H. 18 (112–114), *21*, 205 (231, 232), 208 (61), *211*, *215*, 858 (7), *876*, 899 (116), 904 (134), 915 (173), 917 (174), 937 (277), *982–984*, *986*
Zhang, H. B. 920 (201), *984*
Zhang, J. 236 (110), *244*, 309 (149), *344*, 920 (196), *984*
Zhang, L. 157 (127), *162*, 240 (140), *245*
Zhang, L.-Y. 76 (167), *139*, 85 (202), 90 (210), *140*
Zhang, M.-B. 975 (405), *989*
Zhang, S.-L. 11 (72), *20*, 198 (206), *214*, 606 (11), *683*, 943 (297), *986*
Zhang, X. 152 (59), *160*, 201 (222), *215*, 229 (71), *243*, 299 (100), *343*, 430 (25), *436*, 934 (264), *986*, 1008 (59), 1021 (129), 1022 (130), *1030*, *1031*
Zhang, X. J. 933 (266), *986*
Zhang, Y. 220 (24), *242*, 921 (213), 933 (265), 934 (264), 937 (277), 972 (433), *984*, *986*, *989*
Zhang, Y. D. 942 (158), *983*
Zhang, Y. S. 933 (266), *986*
Zhang, Z. 426 (164), 427 (167), *440*, 921 (186), 966 (408), *984*, *989*
Zhang, Z.-Z. 82 (187), 123 (322), *139*, *143*
Zhao, D. 763 (80, 84), *773*
Zhao, J. 236 (110), 240 (140), *244*, *245*
Zhao, X. 65 (136), *138*
Zhao, Y. 834 (167), *855*, 915 (177), 919 (181), 920 (195), 926 (237), 940 (107, 284), *982*, *984–985*
Zhao, Y. F. 919 (184), *984*
Zhao, Y. H. 920 (201), *984*
Zhao, Z. 723 (68), *730*
Zheng, N. 373 (33), *436*, 927 (252), 963 (391), *985*, *988*
Zheng, Y. 904 (133), *983*
Zheng, Z.-G. 956 (374), *988*
Zhi, L. 382 (54), *437*, 925 (224), *985*
Zhiltsov, S. F. 90 (209), *140*
Zhiyu, W. 237 (111), *244*
Zhong, M. 958 (375, 376), 963 (377), *988*
Zhou, M. 157 (119, 127), 158 (138), *162*
Zhou, M. F. 157 (105), *161*
Zhou, Q.-L. 206 (208), *214*, 670 (150), *690*
Zhou, S.-Z. 124 (323), *143*
Zhou, T. 944 (305, 306), *986*
Zhou, W.-S. 632 (45), *685*
Zhou, X. 935 (268), *986*
Zhou, Y. 416 (136), *440*, 802 (91), *853*, 921 (212), *984*
Zhou, Z. Y. 695 (8), *729*, 732 (7), *771*
Zhu, D. 904 (132), 921 (186, 187), *983*, *984*
Zhu, J. 386 (62), *437*, 943 (250), *985*
Zhu, L. 921 (213), *984*
Zhu, N. 81 (184, 185), 83 (189, 192), *139*, *140*, 267 (58), *278*, 732 (7), *771*
Zhu, S.-F. 206 (208), *214*, 670 (150), *690*
Zhu, W. 937 (277), 938 (278), *986*
Zhu, Y. 236 (109), *244*
Zhuang, N. 452 (49), *520*
Ziegler, T. 305 (133), 308 (112), *343*, *344*
Zilbermann, I. 221 (38), *242*
Zinn, F. K. 121 (311, 312), *142*, *143*, 403 (111), *439*
Zope, U. R. 800 (85), *853*
Zoroddu, M. A. 81 (179), *139*
Zou, B. 897 (104), 917 (175), *982*, *984*
Zou, B. L. 919 (176), *984*
Zouev, I. 565 (121), *583*
Zschornig, O. 330 (229), *346*
Zsolnai, L. 41 (96), 78 (173), *137*, *139*
Zubieta, J. 94 (216), *140*
Zuccaccia, C. 168 (93), *212*
Zuccaccia, D. 168 (93, 98), *212*
Zulli, F. 231 (91), *243*
Zumft, W. G. 352 (35), 358 (62, 65, 66, 69), 359 (60, 61), 360 (59), *362*, *363*
Zweifel, G. 502 (208, 209), *523*, 845 (190), *856*
Zweig, J. S. 430 (172), *441*
Zwettler, R. 445 (23), *519*
Zwicky, A. B. 382 (55), *437*, 807 (95), *853*
Zybill, C. 65 (137), *138*

Subject Index

Entries are in letter-by-letter alphabetical order ignoring spaces and punctuation marks. Page numbers in *italic* refer to Figures and Tables not included in the relevant page ranges.

ab initio molecular orbital theory, gas-phase chemistry, 281, 282
Acetone, Cu complexes, 285, 287–288
Acetone cyanohydrin, Rosenmund–von Braun reaction, 967
Acetone *O*-(2,4,6-trimethylphenylsulfonyl)-oxime, *N*-arylated imines, 377
Acetylation, desilylative, 786, 787
Acetylene, carbocupration, 6–9, 529–539
Acetylene dicopper complex, theoretical, 153
Acetylenes
  stannylcupration, 834–845
  unactivated silylcupration, 803–811
α,β-Acetylenic *N*,*N*-dimethylamides, 824, 827
Acetylenic esters, stannylcuprate conjugate additions, 824, 826, 828
Acetylenides, thermochemistry, 155
Acetylides, thermochemistry, 155
Acid chlorides
  alkyl iodide cross-coupling, 474, 475
  β-cuprioamide reactions, 411–413
  β-cuprio-α-aminocarboxylate reactions, 416, 418
  eliminative oxidative addition, 13, 14
  organomanganese coupling, 447, 449
  organozirconium transmetalation, 457, 459
  silylcuprate substitution, 800–802
  Sonagashira reaction, 966
Acidity, uracil, 280
Acrylate esters, NHC–Cu complex reactions, 402
Activated copper, polycuprioalkanes, 153
Activation energies
  alkyl radicals on Cu surface, 148
  copper halide carbonyls, 158
Acyclic allylic systems, diastereospecific alkylation, 613–638, 639, 654, 655–657
Acyclic amines, organobismuth compound coupling, 947, 949
Acyclic dienoates, Michael addition, 862, 863
Acyclic enones, phosphoramidite complexes, 202
Acylation
  fluoroalkenylcoppers, 998
  mechanisms, 12–13, 14
Acyl halides
  α-(*N*-carbamoyl)alkylcuprate substitution, 388, 390
  β-nitroarylcuprate reactions, 423
Acyl imidazoles, silylcuprate substitution, 800, 802
N-Acyl oxazolidinones, organozirconium transmetalation, 458–459, 462
Acylsilanes, silylcuprate substitution, 802
Acylzirconocene, hydrozirconation preparation, 461, 464
Addition reactions
  1,2-addition, 461–462, 465
  1,4-addition, 24, 607, 863–864
    NHC–Cu complexes, 402
    NMR spectroscopy, 169–170, 189–191, 200–201
  1,6-addition, 861–863, 865–871
  1,10-addition, 872
  Cu-mediated/catalyzed, 200–201, 858–872
    acceptor-substituted dienes, 858–863, 864
    acceptor-substituted enynes, 861, 863–871
    acceptor-substituted polyenynes, 871–872
  Felkin–Ahn model, 654, *656*
  fluoroalkenylcoppers, 1000–1001
  perfluoroalkylcoppers, 1015–1016
  syn-addition
    mechanisms, 848–850

The chemistry of organocopper compounds
Edited by Z. Rappoport and I. Marek

Addition reactions (*continued*)
pentafluorophenylcopper, 1025
silyl–copper, 803
stannyl–copper, 836, 837, 838, 839, 846, 847
see also Conjugate addition; Oxidative addition
Adenine, organoboron compound coupling, 956, 957
Adrenosterone, transmetalation synthesis, 516
Adsorption, Cu surfaces, 150, 157
Aggregates
cluster of clusters, 88, *89*
2-cuprioterphenyl, 152
deaggregation, 40
heteroleptic, 32–33, 35–39
interaggregate exchange reaction, 60–61
structure, 25–27
copper(I) acetylides, 87–88, *89*
mixed copper–silver aggregate, 76, 78
organocopper copper iodide, 106, *107*
organocopper(I)copper(I)-monoanionic group aggregates, 51–63
supramolecular, 168, 170, 178–183
see also Clusters
Agostic interactions
Cu^+ binding energies, 296–298
methylcopper, 146
Alanes, conjugate addition, 503–508
Alcian Blue 8 GX, structure, *233*
Alcohols
C–O bond formation, 885–887
NHC–Cu complex reactions, 402
O-phenylation, 946
Alcoholysis, diaryliodonium salts, 943
Aldehydes
alkenyl cuprate reactions, 530, 531
arylsulfonyl allylic alcohol preparation, 564
tandem carbocupration–oxygenation, 540, 542
α,β-unsaturated, 450, 451, 459, 461
Aldimines, enantioselective organozinc addition, 489
Aldol adducts, domino reactions, 758–766, 768
Aliphatic amines
Cu-mediated cross-coupling
aryl halides, 909–919
iodonium salts, 944
organobismuth compounds, 947, 949
organoboron compounds, 954
organolead compounds, 949–950
organosilicon compounds, 961
Alkanoyl chlorides, β-cuprioamide reactions, 413, 414
(*E*)-Alkene dipeptide isosteres (EADIs), synthesis, 484, 488
Alkenes
hydrozirconation, 452–453, 456, 461, 464
poly-metalated, 805–807, 836, 837
Alkenoyl chlorides, β-cuprioamide reactions, 413, 414
Alkenylation
fluoroalkenylcoppers, 999
mechanisms, 10, 12–13, 14
Alkenyl coppers
dimerization, 595, 598
thermochemistry, 151
Alkenyl halides
coupling reactions, 13, 532–534
nucleophilic substitution, 12–13
Alkenyl organozirconium compounds, transmetalation, 452–462, 463–465
Alkenyl phosphonates, organic halide coupling, 940, 941
Alkenylsilanes
acetylene silylcupration, 809–811
transmetalation preparation, 445, 447, 484, 486
Alkenylstannanes, transmetalation preparation, 445, 447, 448
Alkenyltitanium, alkylidene malonic ester addition, 445, 447
Alkenyl-trifluorosilanes, conjugate addition to enones, 513, 514
Alkenylzirconocene, alkyne hydrozirconation, 453
Alkoxyamines, amine synthesis, 373
Alkoxy copper carbene complex, structure, 128–129
Alkoxy groups
diastereoselective allylic alkylation, 658–659
inside alkoxy effect, 658, 659
Alkylalanes, conjugate addition, 503, 505, 506, 508
Alkyl(amido)cuprates, amine synthesis, 377
Alkyl amines
arylation, 914, 915
aryl iodide coupling, 913, 914
Alkyl aryl ethers, C–O bond formation, 885–887, 897, 899, 900–901
Alkyl aryl sulfides, C–S bond formation, 904, 905
Alkylation
allylic, 444, 445, 603–691
fluoroalkenylcoppers, 999
NHC–Cu complexes, 401
ring-opening, 10
N-Alkylation
amines, 373
hydrazines, 378
N-Alkylbenzimidazoles, aryl halide amidation, 927, 929

Alkylcoppers
  allylic alkylation, 609, 613, 640, 661, 666
  dimerization, 595, 598
  organozirconium transmetalation, 456
  structure, 27–51
Alkylcyanocuprates, allylic alkylation, 607, 609, 620
Alkyl halides, S_N2 alkylation, 5, 9–10
Alkylidene malonic esters, transmetalation, 445, 447, 450
4-Alkylidene pyrrolidinones, α-(N-carbamoyl)-alkylcuprate reactions, 391, 393
Alkyl iodide, acid chloride cross-coupling, 474, 475
Alkyl organozirconium compounds, transmetalation, 452–462, 463–465
Alkyl phosphinous acids, β-cuprio-α-aminocarboxylate reactions, 418
Alkyl phosphites, C–P bond formation, 892–893
Alkyl radicals, Cu surface, 148
Alkyl(silyl)cuprates, preparation, 779–780
N-Alkyl(silyloxy)amines, amidocuprate formation, 374
Alkyl-substituted alkynes
  carbocupration, 539–550
    functionalized, 543–550
    non-functionalized, 539–543
Alkyl-substituted allenes
  silylcupration, 812–813
  stannylcupration, 847, 849
Alkylzinc halides, Cu-catalyzed addition, 492
Alkylzirconocene, hydrozirconation, 456, 457
Alkynes
  carboalumination, 502–503
  carbocupration, 502, 527–584
    alkyl-substituted alkynes, 539–550
    heteroatom-substituted alkynes, 557–572
    intramolecular reactions, 572–579
  hydrocupration, 767, 768–771
  hydrozirconation, 452–454, 461, 464
  organocuprate oxidation, 589
  organozirconium transmetalation, 468–470
  silylcupration, 803
  stannylcupration, 835–836, 837, 841, 842, 845
  terminal, 502–503, 803
Alkynoates, stannylcupration, 828–829
Alkynones, Sonagashira reaction, 966
Alkynyl acetals, carbocupration, 545, 547, 548
Alkynyl amides, carbocupration, 567–569
Alkynyl amines, carbocupration, 567
Alkynyl bromides, amide coupling, 933
N-Alkynyl carbamates
  carbocupration, 567–569
  carbomagnesiation, 567–569
O-Alkynyl carbamates
  carbocupration, 560, 562
  carbomagnesiation, 560, 562
Alkynyl complexes, mixed-metal, 257–258
Alkynyl epoxides
  carbocupration, 545, 546
  cyanosilylcupration, 809–811
α,β-Alkynyl esters
  β-cuprioamide reactions, 411–413
  β-cuprio-N-carbamoyl reactions, 410
  β-N-heteroarylcuprate reactions, 424
Alkynyl ethers, carbocupration, 557–560, 561
Alkynyl halides, amide coupling, 927, 932–937, 938
α,β-Alkynyl iminium triflates, β-cuprio-α-aminocarboxylate reactions, 416
Alkynyl iodides
  β-cuprio-α-aminocarboxylate reactions, 418, 419
  Stille cross-coupling, 976
Alkynyliodonium salts, transmetalation, 386
Alkynyl ketones, cyanosilylcupration, 809–811
Alkynyl ligands, mixed silylcuprates, 779
Alkynyl orthoesters, carbocupration, 547, 548
Alkynyl phosphine oxides, carbocupration, 569–571
Alkynyl phosphines, carbocupration, 569
Alkynyl phosphonates, carbocupration, 569–571
Alkynyl silanes
  carbocupration, 571–572
  carbomagnesiation, 572
  Hiyama-type cross-coupling, 977–978
Alkynyl(silyl)cuprates, preparation, 779
Alkynyl sulfonates, silyl Grignard reactions, 809–811
Alkynyl sulfones
  carbocupration, 563–567
  carbomagnesiation, 564–567
Alkynyl sulfoxides, carbocupration, 563–567
Alkynyl sulfoximines, carbocupration, 563–567
Alkynyl telluride, *syn*-reduction, 514
Alkynyl thioethers, carbocupration, 560–561, 563
All-carbon quaternary stereocenters
  enantioselective conjugate addition, 701, 725, 728
    transmetalation, 478, 480, 484, 485, 508
Allenamides, allenyl iodide coupling, 933, 934
Allenes
  allylic alkylation, 629–631, 638, 639, 649, 651, *653*
  α-cuprionitrile reactions, 428
  Michael addition, 864
  perfluoroalkylcopper reactions, 1015
  racemization, 632

Allenes (*continued*)
stannylcupration, 846–848, 849
terminal, 816–817
unactivated silylcupration, 811–817
α-Allenic amines, S_N2'-allylic substitution, 385
Allenic bromide, propargylation, 637, 638, 639
β-Allenic esters, Michael addition, 866
Allenyl enolates
Michael addition, 864
stannylcuprate conjugate additions, 829
α-Allenyl esters, α-(*N*-carbamoyl)alkylcuprate addition, 388, 389
Allenyl iodides, amide coupling, 933, 934
Allenylsilanes, silylcuprate substitution, 797, 799, 800
Allenylstannanes, substitution, 831–832
3-Allylated pyridine, β-*N*-heteroarylcuprate reactions, 425
Allylation
fluoroalkenylcoppers, 998
mechanisms, 10–12
S_N2 allylation, 169, *170*, 385
titanacycle mono-allylation, 445, 446
π-Allylcopper(III)
^{13}C NMR spectroscopy, 199–200
reductive elimination, 11, 12
Allylcopper–vinylsilane intermediate, 811, 812, 816
Allylcuprates, intermediate compounds, 811, 812, 846, 847
Allylic acetates
homocuprate substitution, 11–12
silylcuprate substitution, 797–800, 801, 802
stannylcuprate substitution, 831, 832
Allylic alcohols
diastereomeric compounds, 614–622, 623–628
enantioselective reduction, 672, 674, 680
Allylic alkylation
Cu-mediated asymmetric alkylation, 603–691
α-allylic, 604–605, 608
diastereoselective methods, 606, 653–663
diastereospecific methods, 610–653
enantioselective methods, 444, 606, 663–682
γ-allylic, 604–605, 608
organotitanium transmetalation, 444, 445
see also Allylic substitution
Allylic aziridines, stereospecific opening, 622, 629, 635
Allylic carbamates
N-carbamoyl(R_T)cuprates, 379–382
silylcuprate substitution, 381–382, 797, 798, 799
Allylic chlorides
allylic alkylation
diastereoselective, 654, 655, 658
enantioselective, 670
organotitanium transmetalation, 444, 445
^{13}C NMR spectroscopy, 199
Allylic compounds
acyclic, 613–638, 639
α-(*N*-carbamoyl)alkylcuprate substitution, 388, 390
cyclic, 638, 640–649, 650, 651
π-complexes, 11, 12, 199–200, 249–253, 388
Allylic *ortho*-diphenylphosphanylbenzoate (*o*-DPPB), quaternary stereocenter generation, 484, 486
Allylic epoxides
allylic alkylation, 640, 642, 676–678
organotellurium transmetalation, 516
stereospecific opening, 610, 622, 625, 628
Allylic fluorides, trialkylalane displacement, 660–661
Allylic halides
α-(*N*-carbamoyl)alkylcuprate substitution, 388, 390
dialkoxyphosphinyldifluoromethylcopper reactions, 1009–1010
β-nitroarylcuprate reactions, 423
organozirconium transmetalation, 456, *457*
silylcuprate substitution, 799–800
stannylcuprate substitution, 831
Allylic iodides, Stille cross-coupling, 976
Allylic organocuprates, synthesis, 509–510
Allylic pentafluorobenzoates, S_N2' addition of diorganozincs, 484, 485, 486
Allylic phosphates
allylic alkylation
diastereospecific, 616, 641, 644, 654, 658, 659
enantioselective, 670–671, 672, 674–675
organotitanium transmetalation, 444, 445
endocyclic, 418–419
S_N2' additions, 461–462, 465, 508–509
Allylic phosphonates, organozirconium transmetalation, 456
Allylic strain
allylic alkylation, 610–612, 620, 629, 654, 672
S_N2' addition of diorganozincs, 484, 485
Allylic substitution
asymmetric, 480, 483
Gilman cuprates, 169, *170*, 605, 609
α-*N*-heteroarylcuprates, 404, 406
inversion, 797, 798, 799
rearrangement, 444, 797, 798, 802
see also Allylic alkylation
Allylic sulfonates, silylcuprate substitution, 799–800
Allylpurines, α-*N*-heteroarylcuprate reactions, 404, 407

Allylsilanes
  conjugate addition, 795
  NHC–Cu complex reactions, 402
  preparation, 382, 484, 487, 489
  silylcupration, 811–816
  substitution, 380, 797, 798, 799
Allylsilane–vinylcuprate intermediates, 813, 814–816
Allylstannanes
  allylstannane–vinylcuprate intermediate, 847–848
  functionalized, 847–848
  nucleophiles, 821
  regiochemistry, 846
  substitution, 831, 832
Altohyrtin A, transmetalation synthesis, 516, 517
Aluminum, electronegativity, 502
Aluminum compounds *see* Organoaluminum reagents
Amidation
  aryl halides, 890–891, 922, 925–927, 928–932
  transferable *N*-ligands, 373
  vinyl halides, 891, 927, 933–937, 938
Amides
  allenyl iodide coupling, 933, 934
  aryl iodide coupling, 922, 924
  N-arylation, 890–891
  O-arylation, 927, 928
  from carbamoyl cuprates, 383, 384
  iodonium salt coupling, 944–945
  organoboron compound coupling, 954
  organosilicon compound coupling, 961
Amidines, cross-dehydrogenative coupling, 964
Amidocuprates
  ortho-cupration, 588–590
  N-functionalized
    non-transferable ligands
      chiral, 367–371
      cuprate structure, 366–367
    transferable ligands, 372, 373–379
  oxidation, 586, 598, 599
  synthesis, 169
Amido(silyl)cuprates, allylic carbamate reactions, 381–382
Amination
  aryl halides, 887–890, 909–919, 919–920
  Cu-mediated, 372, 373
  electrophilic, 493, 494, 586
Amines
  bicyclic, 413, 414
  Boc-protected, 926
  halogen–metal exchange, 378–379
  iodonium salt coupling, 944–945
  NHC–Cu complex reactions, 402
  organobismuth compound coupling, 947, 949
  organoboron coupling, 954
  organocuprate oxidation, 598–599, 600
  organolead compound coupling, 949–950
  organosilicon compound coupling, 961
  transferable *N*-ligands, 373
  unsaturated, 202
Amino acids
  abbreviations, *319*
  alkyne silylcupration, 809
  chiral recognition, 322–323
  β-cuprio-α-aminocarboxylate preparation, 415, 416
  gas-phase Cu(I)/Cu(II) reactivity, 318–332
  peptide synthesis catalyst, 897
  protected, 415, 416, 418
α-Amino acids
  alkyl halide amination, 910–911
  amidocuprate formation, 373, 374
  Cu^+ complexes, 282, 290–291
  enantiopure, 956, 957
  organobismuth compound coupling, 949
  organoboron compound coupling, 956, 957
  transmetalation, 383, 385
β-Amino acids
  alkyl halide amination, 911–912
  dialkylzinc addition to nitro-olefins, 487, 490
α-Aminoacylsilanes, silylcuprate substitution, 802
1-Aminoalkynes, silylcupration, 807–808
α-Amino allenes
  allylic alkylation, 635–636
  transmetalation, 385
2-Amino-2′-hydroxy-1,1′-binaphthyl derivatives, amidocuprate formation, 371
Aminohydroxyphosphine ligand, conjugate additions, 478, 479
cis-Aminoindanolyl chiral auxiliaries, amidocuprate formation, 373, 374
β-Aminothioethers, amidocuprate formation, 369
4-Amino-2-ynoates, stannylcuprate conjugate additions, 828
Ammonolysis, aryl halides, 890, 912
Anilines
  arylation, 954, 960–961
  β-*N*-arylcuprate formation, 419
Anionic copper
  Cu–C cluster ions, 156
  phenylcopper formation, 152
  thermochemistry, 149
Anionic groups, organocopper(I) copper(I) aggregates, 51–63
Anions, stannyl, 817
2-Anisylcopper, structure, 36, *37*
Annulation, stannylcuprates, 834, 835

Anthracene derivatives, organozirconium transmetalation, 467, 468
Anti-Bredt olefin (1-norbornene), Cu interaction, 148
anti-Michael additions
acceptor-substituted enynes, 867
π-complex intermediates, 191
anti-S_N2' substitution
N-carbamoyl(R_T)cuprates, 379, 382
NHC–Cu complexes, 401
silylcuprates, 799–800
anti-stereospecificity, allylic alkylation, 605
Arduengo-type carbenes, Michael addition, 863, 864
Arenes, polyfluorinated, 1024–1025
Arenethiolato complexes, copper(I) acetylides, 87
Aromatic amines
aryl halide coupling, 887–890, 919–920
iodonium salt coupling, 944
Aromatic compounds, theoretical models, 282–283
Aromatic Finkelstein reaction, Cu-catalyzed, 967
Aromatic nitriles, Rosenmund–von Braun reaction, 967–968
Aromatic tertiary phosphine, aryl iodide coupling, 939
Aromatization, hydrazides, 935, 936
Aroyl chlorides, β-cuprioamide reactions, 413, 414
Aryl acetylenes, carbocupration, 557
Arylalanes, conjugate addition, 506, 507, 508
Aryl alkyl ketones, organoboron transmetalation, 496, 498
Aryl alkynones, Sonagashira reaction, 966
Aryl(amido)cuprates, amine synthesis, 377–378
N-Arylanthranilamides, synthesis, 378
Arylation
β-lactam, 927, 931
N–H heterocycles, 921
polyfluorinated arenes, 1024–1025
2,3,5,6-tetrafluoropyridylcoppers, 1028
N-Arylation
amides, 890–891
amines, 373
hydrazines, 378
imines, 377
organobismuth compounds, 947, 949
organoboron compounds, 954–959, 960
organolead compounds, 950
organosilicon compounds, 961
O-Arylation, amides, 927, 928
Aryl azides, aryl halide coupling, 937, 938
Arylbismuth reagents, aliphatic alcohol arylation, 946
Arylboronic acids, *N*-arylation, 954, 958–959, 960
Aryl bromides
amine coupling, 913–914, 915, 917, 920
perfluoroalkylation, 1014
Rosenmund–von Braun reaction, 968
Sonogashira reaction, 966
Ullmann coupling with phenols, 894
9-Arylcarbazoles, aryl halide amination, 889
Aryl chlorides
aliphatic amine coupling, 915
intermolecular arylation, 925
N–H heterocycle coupling, 922
Arylcoppers
amidocuprate formation, 373, 374
structure, 27–51
arenthiolate aggregates, 60–61
heteroleptic aggregates, 32–33, 35–39
thermochemistry, 151–153
Arylcuprates, β-*N*-substituted, 419, 420, 421–423
Aryl halides
amide coupling, 890–891, 922, 925–927, 928–932
amine coupling
aliphatic, 909–919
aromatic, 887–890, 919–920
ammonolysis, 890, 912
C–C bond formation
C-nucleophiles, 965–973, 974
organometallic reagents, 973–980
N–H heterocycle coupling, 921–922, 923–924
P-nucleophile cross-coupling, 892
Aryl iodides
amide coupling, 922, 924
amine coupling, 913, 914, 915, 917, 920
Hiyama-type cross-coupling, 977–978
N–H heterocycle coupling, 921
P-nucleophile coupling, 938
Sonogashira reaction, 965
Stille cross-coupling, 976–977
Ullmann coupling, 894, 902, 904
Aryliodonium salts
C–C bond formation, 975, 976–977
O-nucleophile coupling, 943
Sonagashira reaction, 966–967
Aryllithiums, intramolecularly chelated, 97–98
O-Aryloximes, C–O bond formation, 902
N-Aryloxyphthalimides, organoboron compound coupling, 952
10-Arylphenoxazines, aryl halide amination, 889
Aryl phosphinous acids, β-cuprio-α-amino-carboxylate reactions, 418
N-Arylpyrroles, arylboronic acid coupling, 956
2-Aryl-4-quinolones, aryl halide amidation, 927, 930

Arylstannanes
functionalization, 960
Stille coupling, 821, 976
vinyl iodide coupling, 975
Aryl-trifluorosilanes, conjugate addition to enones, 513, 514
Aryl vinyl ethers, organotin compound coupling, 959–960
Ascorbate oxidase
multinuclear Cu sites, 356–357
type 1 Cu centers, 349, 356, *357*
A-strain
allylic alkylation
allylic alcohols, 616, 618, *620*
β-substituents, 672
diastereospecific, 611–612, 622, 627, 651
leaving group effects, 613, 614
N-carbamoyl(R_T)cuprates, 380–381
Asymmetric allylic alkylation
Cu-mediated, 603–691
diastereoselective methods, 606, 653–663
diastereospecific methods, 610–653
enantioselective methods, 480, 483, 606, 663–682
Asymmetric conjugate addition, alanes, 506–508
Asymmetric induction, silylcuprate conjugate addition, 793, 873
Aurates, structure, 91–93
Aza allylcuprates, transmetalation, 386
1-Azaallyl ligand, Cu(I) complexes, 260
Azetidines, allylic alkylation, 635
Aziridines
allylic alkylation, 622, 629, 635, 649, 650, 651
S_N2' addition of diorganozincs, 484, 488
α-Azoalkylcuprates, reactions, 386
Azoles, N-arylation, 950
Azurin, type 1 Cu centers, 349, *350*
Azurite, combustion of copper acetylides, 147–148

β effect, C–Si to C–OH conversion, 783
Back-donation, binding energies, 296
Baconipyrone C, transmetalation synthesis, 516
Baeyer–Villiger rearrangement, S_N2' addition of diorganozincs, 484, 485
Baylis–Hillman product, organoboron transmetalation, 499, 500
Bent structure, Cu(I) organocuprate intermediates, 186, 189
Benzene, theoretical models, 282–283
Benzenesulfonamide, Goldberg reaction, 891
Benzhydrylidene(methyl)amine, allylic alkylation, 635
Benzimidazoles
arylation, 921, 950, 961
aryl halide coupling, 917, 918
cross-dehydrogenative coupling, 964
enantioselective allylic alkylation, 663–665
Benzimidazolinones, arylation, 960–961
Benzofurans, Hurtley reaction, 972
Benzo[4,5]furo heterocycles, Cu-catalyzed coupling, 908
Benzophenones, enantioselective hydrosilylation, 754
Benzothiazoles
allylic alkylation, 608
Cu-catalyzed coupling, 908, 909
Benzotriazoles, N-arylation, 950
Benzoxazoles, Cu-catalyzed coupling, 908, 909
Benzylation, fluoroalkenylcoppers, 999
Benzylic organocopper compounds, borylcopper NHC complex, 44
Biaryls, organocuprate oxidation, 588, 590–595, 596
Bicyclic amines, β-cuprioamide preparation, 413, 414
Bidentate ligands, Cu^+ interaction, 295, 301
Binaphthol (BINOL), phosphoramidite ligands, 201–202, 203, 474, 476, 477, 492–493
Binary copper–carbon species, thermochemistry, 155–156
Binary mixtures, gas-phase Cu(I)/Cu(II) bioreactivity, 319–326
Binding energies
back-donation, 296
computational approach, 281
Cu^+ and Cu^{2+}, 289–300
agostic-type interactions, 296–298
solvated ions, 298–300
incremental, 298–299
sequential, 282, 299–300
Binding enthalpies
copper halide carbonyls, 158
solvated Cu ions, 298–300
BINOL (binaphthol), 201–202, 203, 474, 476, 477, 492–493
Binuclear compounds, electrochemical stability, 257–261
Biochemistry, 347–363
biological Cu centers, 348–352
novel centers, 352, 357–359
type 1, 349, 352, 356, *357*
type 2, 349–350, 356
type 3, 350–352, 355, 356
Cu proteins
blue Cu proteins, 280, 307–308
mononuclear Cu sites, 352–355
multinuclear Cu sites, 355–360
gas-phase chemistry, 279–346
Cu(I)/Cu(II) ion reactivity, 318–340
photobiochemistry, 237–241

Biphenol, phosphoramidite ligands, 201–202, 203
Bipyridine
amide intermolecular arylation, 925
unsubstituted, 266
2,2′-Bipyridine, aryl halide coupling, 919–920
Bis(alkynyl) titanocene complexes, electrochemical stability, 271–273
1,1-Bis(diphenylphosphino)methane (dppm), copper(I) acetylide complexes, 83–86
Bis(2-methylbut-2-en-1-yldiphenylsilyl)cuprate, 780, 783, 784
Bismuth compounds *see* Organobismuth compounds
Bis(silyl)cuprate, reactivity, 779
Bis(stannyl)cuprates
preparation, 818
reactivity, 818, 822
4,4′-Bis(trifluoromethyl)benzophenone-*o*-methylsulfonyloxime, amine synthesis, 375–376
Bis(trimethylsilyl)cuprate, homosilylcuprate preparation, 777
N,*O*-Bis(trimethylsilyl)hydroxylamine, amidocuprate formation, 374
Bis[(trimethylsilyl)methyl]cuprates, supramolecular aggregation, 179, *180*
Bis(trimethylstannyl)cyanocuprate, ^{13}C NMR spectroscopy, 818
Bis[tris(trimethylsilyl)silyl]cuprate, 779
Bis(vinylcopper) species, organozirconium transmetalation, 468–469
Blue copper proteins
biological Cu centers, 349
coordination, 307–308
reduction potentials, 280
N-Boc-azetidines, allylation, 410
Boc-protected amines, intermolecular arylation, 926
Bond critical point, nucleic acid Cu(I)/Cu(II) reactivity, 338
Bond energies
alkyl radicals on Cu surface, 148
copper carbonyls, 157, 158
Cu–methylene, 149–150
methylcopper, 146, 149
phenylcoppers, 151–152
Bonding
bridge bonding, 29
computational approach, 281
copper carborane compounds, 68–73
Cu^+ complexes, 308–309
Cu^{2+} complexes, 309–310
Cu(III) intermediate NMR, 198
energy density contour maps, 308–309
Bonds
syn-addition to triple bond, 836, 837
bond critical point, 338
C–H activation, 114, 115
C–heteroatom
C–P cleavage, 39–40
C–Si to C–OH conversion, 781–788
Ullmann bond formation, 883–965
C–N and C–P, 887–893, 909–940
C–O, C–S and C–Se, 883–887, 893–909
dative bonds, 308
deconjugative double-bond migration, 829–831
electron-deficient two-electron three-center bond, 26
ethylene bond-activation, 293
multiple bond systems
extended, 857–879
silylcupration, 817
Bongkrekic acid, transmetalation synthesis, 516
α-Boraalkylcopper, NHC complex, 45, *46*
η^3-Boroallylic bonding, copper carboranes, 71
Boron, electronegativity, 496
Boron ate complexes, formation, 496–499
Boron compounds *see* Organoboron compounds
Boronic acids, *N*-imination, 377
Boron trifluoride, conjugate addition effects, 15
Boron–zinc exchange, diorganozinc preparation, 474
Boroxyalkylcopper, NHC complex, 47
Borylation, Michael acceptors, 496, 500
Borylcopper, NHC complexes, 44–47
(*R*,*R*)-BozPHOS, *N*-diphenylphosphinoyl-imine addition, 489, 491
Bridge bonding, pentamethylphenylcopper, 29
Bromination–desilylation, silylcuprates, 791, 792
Bromine, Br–I exchange, 940–942
1-Bromoalkynes, β-cuprio-α-aminocarboxylate reactions, 416
Bromoallene, propargylation, 637, 638, 639
o-Bromoanilines, *N*-substituted, 927, 929
Bromoimidazoles, α-*N*-heteroarylcuprates, 404, 406
β-Bromopyridines, β-*N*-heteroarylcuprate reactions, 425
Buflavine, zinc organocuprate oxidation, 595
Build-up curves, NOE/HOE determinations, 166–167
Butterfly structure
2-[(dimethylamino)methyl]-5-(methyl)-phenylcopper, 34
organocopper copper bromide aggregate, 56
pentafluorophenylcopper, 30
t-Butyl carbamate, vinyl halide coupling, 935
Butylcycloheptanones, Cu(I)-catalyzed formation, 200–201

t-Butyldimethylsilyl(butyl)cuprate, 780
t-Butyldimethylsilylcuprate, conjugate additions, 793
t-(Butyldiphenylsilyl)cuprates, 779, 780–781, 785–786, 793
t-Butyldiphenylsilylcyanocuprates, silylcupration, 805, 813, 814
t-Butyldiphenylsilyl group, reactions, 807
1-Butyl imidazoles, Rosenmund–von Braun reaction, 968
3-Butynoic acid derivatives, stannylcuprate addition, 841

CAD *see* Collision-activated dissociation
Cadmium compounds
 dialkoxyphosphinyldifluoromethylcopper preparation, 1008–1012
 fluoroalkenylcadmium reagents, 994–996
 2,3,5,6-tetrafluoropyridylcopper metathesis, 1027
 trifluoromethyl reagents, 1019–1021
Calorimetric measurements, copper acetylides, 147
Camps cyclization, aryl halide amidation, 927, 930
β-Capnellenol, synthesis, 816
Carbamates
 allenyl iodide coupling, 933, 934
 from carbamoyl cuprates, 383, 384
 organoboron compound coupling, 954
 vinyl iodide coupling, 932–933
α-(*N*-Carbamoyl)alkylcuprates, transmetalation, 387–400
Carbamoylcuprates, reactions, 383, 384
N-Carbamoyl(R_T)cuprates, N–Cu bonds, 379–382
Carbazoles, cross-dehydrogenative coupling, 963
Carbenes
 Arduengo-type, 863, 864
 copper hydride complex structure, 122–123
 Fischer chromium carbene transmetalation, 450–452
 see also *N*-Heterocyclic carbenes
Carboalkoxydifluoromethylenecopper reagents, 1007–1008
Carboalumination, terminal alkynes, 502–503
Carbocupration
 acetylene, 6–9, 529–539
 alkynes, 502, 527–584
  alkyl-substituted, 539–550
  enynes/arylacetylenes, 555–557
  heteroatom-substituted, 557–572
  intramolecular reactions, 572–579
  tandem oxygenation, 540, 542
Carbocycles
 intramolecular carbocupration, 572–576
  non-stabilized organocoppers, 572–574, 575
  stabilized organocoppers, 574, 576
Carbohydrates, gas-phase Cu(I)/Cu(II) reactivity, 332–337
Carbomagnesiation
 N-alkynyl carbamates, 567–569
 O-alkynyl carbamates, 560
 alkynyl silanes, 572
 alkynyl sulfones, 564–567
 propargyl alcohols, 528, 550–555, 556
Carbometalation, alkynes, 469–470
Carbon–carbon bond formation
 Cu-mediated cross-coupling, 965–980
  C-nucleophiles, 965–973, 974
  organocuprate oxidation, 586–590, 592, 593, 594
  organometallic reagents, 973–980
 organocuprate oxidation, 586–598
  biaryl formation, 588, 590–595, 596
  cross-coupling, 586–590, 592, 593, 594
  dimerizations, 595–598
  intramolecular bond formation, 592–595, 596
  macrocycle formation, 595–598
 organostannanes, 821, 836, 837
Carbon dioxide, Cu complexes, 158
Carbon–heteroatom bond formation
 Cu-mediated cross-coupling, 883–965
  classical Ullmann chemistry, 883–893
  modern Ullmann chemistry, 893–943
  non-Ullmann substrates, 943–965
Carbon–hydrogen bond activation, 114, 115
Carbon monoxide
 acylzirconocene preparation, 461, 464
 Cu complexes, 156–158
  Cu(I) vs Cu(II), 158
Carbon monoxide dehydrogenase, multinuclear Cu sites, 359–360
Carbon–nitrogen bond formation
 Cu-mediated cross-coupling, 887–892, 909–938
  cross-dehydrogenative coupling, 963–965, 973, 974
  hydroamidation, 936–937
  N–H heterocycles, 921–922, 923–924, 944
 nitrogen enolate cuprates, 426
 oxidation, 598–601
Carbon–oxygen bond formation
 Cu-mediated cross-coupling, 883–887, 893–902
  Claisen rearrangement, 902
  heterocycles, 906–908
Carbon–phosphorus bond formation, Cu-mediated cross-coupling, 892–893, 938–940, 941

Carbon–selenium bond formation,
Cu-mediated cross-coupling, 887, 904, 906
Carbon–sulfur bond formation
Cu-mediated cross-coupling, 887, 902, 904, 905
heterocycles, 908–909
β-(N-Carbonyl)alkylcuprates, 406, 408–414
β-cuprioamide reagents, 411–414
β-cuprio-N-carbamoyl reagents, 408–409, 410–411
Carbonyl compounds
silylcuprate conjugate additions, 790–797
stannylcuprate conjugate additions, 822–831
β-Carbonyl siloxanes, preparation, 380
Carborane complexes, structure, 68–73
Carboxamides, diaryl ether synthesis, 894
O-Carboxylates, N-imination, 377
Carboxylic acid amides
Cu-mediated cross-coupling
iodonium salts, 944
organolead compounds, 950, 951
organotin compounds, 960–961
vinyl halides, 935
Carboxylic acid derivatives, α,β-unsaturated, 388, 389
Cascade reactions, enantioselective conjugate addition, 693, 695
Catalysts
biochemistry, 347–363
Fe, 425–426
Hoveyda–Grubbs, 484, 487, 489
Ni, 473, 478, 479
Pd, 473, 474, 478, 479, 559, 561
see also Copper-catalyzed reactions; Photocatalysts
Catechol oxidase, multinuclear Cu sites, 355–356
Cations
copper carbene complex structure, 123–126
Cu^+/Cu^{2+} reactions
biological molecules, 318–340
small organic bases, 310–318
CCSD(T) (coupled cluster approaches), 281, 282, 283, 284
β-Cedrene, transmetalation synthesis, 516
Cellulose, polysaccharide Cu(I)/Cu(II) reactivity, 336–337
Cesium carbonate
C–O bond formation, 894–899, 901, 909
C–P bond formation, 938–939
Cesium phosphides, aryl iodide coupling, 938–939
Charge transfer, dissociative, 287–288
Chelates
aryllithiums, 97–98
Cu^+ coordination, 301
Chemoselectivity
allylic alkylation, 606–608
stannylcuprate substitution, 831, 832
Stryker's reagent, 735–738
Chiral alkenylsilanes, α,β-unsaturated ketone synthesis, 484, 486
Chiral amido(alkyl)cuprate dimer, structure, 369
Chiral auxiliaries
amidocuprate formation, 367, 373, 374
diastereospecific allylic alkylation, 616, 618–622, 623, 660
Chiral centers
diastereoselective allyl alkylation
with heteroatom, 658–660
without heteratom, 654–658
Chiral cuprates
heteroatom(alkyl)cuprates, 776
Michael addition, 859–860
Chiral cyclic enones, diastereoselectivity, 189–190
Chiral 2-enoates, stannylcuprate conjugate additions, 824
Chiral N-heterocyclic carbenes, enantioselective conjugate addition, 513, 514
Chirality, S_N2' addition of diorganozincs, 484, 485
Chiral ligands
Cu(I) hydride catalysis, 770–771
domino reactions, 761–766, 768
hydrosilylation, 741–749, 752–757
ligand acceleration effect, 753
monodentate, 201–202
non-transferable, 367–371
see also Non-chiral ligands
Chiral phosphoramidites
asymmetric allylic substitution, 480, 483
dialkylzinc addition to nitro-olefins, 487, 489, 490
Chiral propargylamines, stannylcupration, 842
Chiral recognition, amino acids, 322–323
Chiral sulfoxides, allylic alcohol formation, 618, 619
Chiral sulfoximines, allylic alkylation, 654, 666
Chiral vinylallene, intramolecular [4 + 2]-cycloaddition, 868–869
Chiral ynoates, stannylcuprate conjugate additions, 828
Chitin/chitosan, polysaccharide Cu(I)/Cu(II) reactivity, 336–337
1-Chloroalkynes, alkynylsilane coupling, 978
Chlorodiphenylphosphine, β-cuprio-α-amino-carboxylate reactions, 418
Chlorotrimethylsilane
conjugate addition effects, 15–16
α-cuprioxazolidine reactions, 428

Chromium Fischer carbene complexes, transmetalation, 450–452
CID *see* Collision-induced dissociation
C.I. Direct Blue 199, structure, 232
Cinnamic ester, conjugate addition, 184
CIPs *see* Contact ion pairs
Claisen rearrangement, vinyl ether synthesis, 901–902, 903
Classical Ullmann chemistry, C–heteroatom bond formation, 883–893
Clavirolide C, transmetalation synthesis, 517
Cleavage
  C–P bond cleavage, 39–40
  diphenyltetramethyldisilane, 778
  DNA cleavage inhibition, 238
  epoxide regioselective cleavage, 834
  saccharide cross-ring cleavage, 334–335
  silylstannanes, 780
Click reaction, copper triazolide–NHC complex, 47
Clusters
  cluster of clusters, 88, *89*
  copper–carbon cluster ions, 152, 156
  copper(I) hydride reagents, 732–735
  coupled cluster approaches, 281, 282, 283, 284
  lithium dialkylcuprates, 3
  see also Aggregates
^{13}C NMR spectroscopy, *164*
  ethyl 2-en-4-ynoate, 185
  ^{1}H,X heteronuclear reactions, 166, 183, 196–200
  stannylcuprates, 818, 820
Collision-activated dissociation (CAD)
  amino acid reactivity, 322
  gas-phase photodissociation, 286, *287*
Collision-induced dissociation (CID)
  binding energies, 286
  Cu(III) species, 154
Combustion *see* Enthalpies of combustion
Computational studies
  amido ligands, 367
  gas-phase chemistry, 281
Configuration, C–Si to C–OH conversion, 781
Conformation, *N*-carbamoyl(R_T)cuprates, 380–381
Conjugate addition
  alanes, 503–508
  boron trifluoride effect, 15
  chlorotrimethylsilane effect, 15–16
  conjugate addition–elimination, 833–834
  Cu(I) organocuprate intermediates, 184
  enantioselective, 693–730
    NMR spectroscopy, 200–201
    transmetalation, 444
      organosilicon compounds, 513, 514
      organozinc compounds, 474, 478, 480, 481–482, 517–518
  Michael addition, 444, 474, 476, 478, 479, 864
  NHC–Cu complex reactions, 402
  organozinc reagents, 474–478
  Pd(0)-catalyzed, 474
  remote conjugate addition, 6, *7*
  and S_N2 alkylation, 5–10
  α,β-unsaturated carbonyl compounds, 5, 444
    esters, 493, 495
    lactams, 474, 476, 477, 510, 512
    nitriles, 402
    silylcuprates, 790–797
    stannylcuprates, 822–831
  vinylstannanes, 510, 512
Conjugated fluoroenynes, preparation, 1003
Conjugated hydrosilylation, α,β-unsaturated substrates, 738–749
Conjugate reduction, NHC–Cu complex reactions, 402, 403
Contact ion pairs (CIPs)
  cyanocuprates, 109
  diorganocuprate NMR, 173–178
  Gilman reagents, 2
  organocuprates, 104–105
Conventional π-complex, theoretical models, 282–283
Coordination
  Cu oxidation states, 281, 300–308
    Cu(II) in blue proteins, 307–308
    more than one ligand, 303–306
    one ligand, 301–303
  4,4*a*,5,6,7,8-hexahydro-4*a*-methyl-naphthalen-2(3*H*)-one, 187–191
Copper
  activated, 153
  alkyl radicals on surface, 148
  anionic, 149, 152
  CO complexes, 156–158
  CO_2 complexes, 158
  cyclopropane reactions, 150
  electronegativity, 493, 496, 509
  hydrocarbyl derivatives, 146–155
  methylene bonding, 149–150
  oxidation states, 146, 153–155, 280
  surfaces, 148, 150, 156, 157
Copper acetylides
  dimeric complex, 80–81
  electronic interactions, 267–268
  mixed copper–platinum acetylide complexes, 90
  oxidation potentials, 261–266
  perfluoroalkylalkyne synthesis, 1005
  Sonogashira reaction, 965
  structure, 73–90
    aggregates, 87–88, *89*
    arenethiolato groups, 87
    dppm complexes, 83–86
    heterocubanes, 81

Copper acetylides (*continued*)
triphos complexes, 79–80
thermochemistry, 147–148, 155
Copper anions *see* Anionic copper
Copper arylthiolate, enantioselective allylic substitution, 666, 668–669
Copper(I) 1-azaallyl complexes, electrochemical stability, 260–261
Copper(I) benzoate, Hurtley reaction, 58–60
Copper(I) bis-carbene species, catalyst regeneration, 451
Copper boryl complexes, structure, 122
Copper(I) bromide
allenic bromide displacement, 637, 638, 639
allylic alkylation, 609
acyclic systems, 616, 618, 622, 624, 657
cyclic systems, *646*, 647
enantioselective, 665–666, *669*, 670, 674
dialkoxyphosphinylfluoromethylcopper preparation, 1008
dimethyl sulfide complex, 777, 846
2,3,5,6-tetrafluoropyridylcopper metathesis, 1027
Copper(I) *tert*-butoxide, Cu(I) hydride complex formation, 732
Copper *tert*-butylacetylides, structure, 74, *75*, 82, *83*
Copper carbene complexes
structure, 119–133
cationic, 123–126
Cu(II) complexes, 131–132
see also *N*-Heterocyclic carbenes
Copper–carbon binary species, thermochemistry, 155–156
Copper carborane, structure, 68–73
Copper-catalyzed reactions
acylation of trifluorovinylzinc reagent, 998
aromatic Finkelstein reaction, 967
arylation
polyfluorinated arenes, 1024–1025
2,3,5,6-tetrafluoropyridine, 1028
carbomagnesiation
N-alkynyl carbamates, 567–569
O-alkynyl carbamates, 560
alkynyl silanes, 572
alkynyl sulfones, 564–567
propargyl alcohols, 528, 550–555, 556
Cu(I) hydride reagents, 731–773
domino reactions, 757–766, 768
hydrosilylation, 738–757
enantioselective conjugate addition, 200–201, 444, 693–730
natural product syntheses, 693, 695, 714–728
new substrates, 695–700, 701
quaternary stereocenters, 700–707, 725, 728
all-carbon, 701, 725, 728
transmetalation, 478, 480, 484, 485, 508
tandem reactions, 693, 707–714
extended multiple bond systems
addition to Michael acceptors, 858–872
substitution of substrates, 872–874
halogen exchange, 940–942, 967
NMR structure, 200–209
active species, 203–204
precatalytic complexes, 200
see also Copper-mediated reactions
Copper complexes
copper(I)
binding energies, 289–300
bioreactivity, 318–340
bonding, 308–309
coordination, 300–306
Cu(I) hydride complexes, 732–735
organic base reactions, 310–318
π-complex, 605–606
photodissociation, 285–286
theoretical models, 282–283, *284*
copper(II)
binding energies, 289–300
bioreactivity, 318–340
bonding, 309–310
carbene complexes, 131–132
coordination, 300–308
correlation effects, 282, 284–285
gas-phase preparation, 286
organic base reactions, 310–318
ternary complexes, 281, 326–330
copper(III)
σ-complex, 605–606
thermochemistry, 153–155
η-complexes, η_3 π-allyl complex, 199
N-heterocyclic carbenes, 400–402, 403, 404
π-complexes
allyl complexes, 11, 12, 199–200, 249–253, 605–606
α-(*N*-carbamoyl)alkylcuprate reactions, 388, 391
electrochemical synthesis, 248–255
η_3 π-allyl complex, 199
intermediates, 170, 184–189, 199–200
theoretical models, 282
σ-complexes, 194–196, 199–200
Copper(I) cyanide
allylic alkylation
acyclic systems, 609, 614–616, 617, 621, 659, 660
allenes, 633, 634, 635–636
allylic aziridines, 625, 629
cyclic systems, 640, 641, 649, 651, 654, 658
enantioselective, *669*, 671, 672
Copper(II) dithiocarbamates, thermochemistry, 155

Copper(I) diynyl compounds, oxidation potentials, 266–267
Copper enolates, organozirconium transmetalation, 458, 461
Copper enzymes *see* Enzymes
Copper halide carbonyls, thermochemistry, 157–158
Copper(I) halides
 dialkoxyphosphinyldifluoromethylcopper preparation, 1008–1009
 perfluoroalkynylzinc reagent metathesis, 1005
Copper(I) hydride reagents, 731–773
 alkyne hydrocupration, 767, 768–771
 catalytic domino reactions, 757–766, 768
 catalytic hydrosilylation
  ketones and imines, 749–757
  α,β-unsaturated substrates, 738–749
 copper carbene complex formation, 119, 122–123
 NHC complex, 45, *46*
 polymethylhydridosiloxane reaction, 632, 634
 preparation and characterization, 732–735
 Stryker's reagent, 735–738
Copper(I) intermediates, organocuprate reactions, 850
Copper(I) iodide
 allylic alkylation, 609, 612, 657, *669*, 675
 organocopper copper iodide aggregate, 106, *107*
 phenyldimethylsilylcuprate preparation, 777
Copper-mediated reactions
 asymmetric allylic alkylation, 603–691
  diastereoselective methods, 606, 653–663
  diastereospecific methods, 610–653
  enantioselective methods, 606, 663–682
 carbometalation, 469–470
 cross-coupling, 444, 881–990
  C–C bond formation, 965–980
  C–heteroatom bond formation, 883–965
  intramolecular, 510, 512
  mechanistic considerations, 942–943, 961–963
 extended multiple bond systems
  addition to Michael acceptors, 858–872
  substitution of substrates, 872–874
 see also Copper-catalyzed reactions
Copper phenylacetylide, structure, 76, *77*, 82, *83*
Copper phthalocyanines
 photochemical transformations, 217–245
  electron transfer systems, 219–232
Copper porphyrins
 photochemical transformations, 217–245
  electron transfer systems, 219–232
 structure, 117
Copper proteins *see* Proteins
Copper tetrapyrroles
 photophysics, 219–220
 uses, 218
Copper(I) thiophene-2-carboxylate (CuTC)
 amide coupling, 927, 932
 O-containing heterocycles, 908, 909
 vinyl iodide coupling, 975
Copper triazolide, NHC complex, 47
Copper trifluoromethanesulfonate (CuOTf), aggregate formation, 52
Copper/zinc superoxide dismutase
 mononuclear Cu sites, 354–355
 type 2 Cu centers, 350, *351*
Corey's oxidation, vinylstannanes, 838, 840
Correlation, gas-phase chemistry, 281–282
Correlation spectroscopy (COSY), phosphoramidite complexes, 206
Correlation times, NOE/HOE determinations, 166
Cortisone, transmetalation synthesis, 516
COSY (correlation spectroscopy), 206
Coupled cluster (CCSD(T)) approaches, gas-phase chemistry, 281, 282, 283, 284
Coupling constants, NMR spectroscopy, 166, 171–173, 177–178, 195–198, 206
Coupling reactions
 alkenyl coppers, 542, 545, 555, 556, 557, 558
 fluoroalkenyl halides, 992–994
 Glaser alkyne coupling, 585
 heterocycle syntheses, 579, 580–581
 pentafluorophenylcopper, 1024–1025
 reductive coupling
  acyclic systems, 638, 639
  cyclic systems, 640, 642
  enantioselective, 677–678, 680
  regioselective, 606, 608
 Sonogashira coupling, 447, 448, 589, 965–967
 Stille coupling, 821, 838, 840, 841, 976–977
 2,3,5,6-tetrafluoropyridylcoppers, 1027–1028
 three-component coupling protocol, 457–458, 461
 vinyl coppers, 532–534
 see also Cross-coupling reactions
Cross-coupling reactions
 alkenyl coppers, 572, 574, 576
 alkenyl halides, 13, 532–534
 Cu(III) intermediate formation, 198
 Cu-mediated, 444, 881–990
  C–C bond formation, 965–980
  C–heteroatom bond formation, 883–965
  cross-dehydrogenative coupling, 963–965, 973, 974
  intramolecular, 510, 512

Cross-coupling reactions (*continued*)
mechanistic considerations, 942–943, 961–963
heteroleptic organocopper aggregates, 36
Hiyama reaction, 977
Ni-catalyzed, 478, 479
organocuprate oxidation, 586–590, 592, 593, 594
Pd-catalyzed, 478, 479, 559, 561
transmetalation
organoaluminum compounds, 502
organoboron compounds, 496, 498
organozinc compounds, 473, 474, 475
organozirconium compounds, 469–470, 471
see also Coupling reactions
Cross-dehydrogenative coupling, C–N bond formation, 963–965, 973, 974
Cross-over experiments, amine synthesis, 373, 375
Cross-ring cleavage, saccharide Cu(I)/Cu(II) reactivity, 334–335
Crotyl chloride, organoboron transmetalation, 496, 497
Crown ethers
diorganocuprate NMR, 173
organocuprate [12]crown-4, 104–105
Cryogenic matrices, homoleptic copper carbonyls, 157
Cryoscopy
cyanocuprates, 109
diarylcuprates, 91
Cumulenes, fluorinated, 997–998
$^{63/65}$Cu NMR spectroscopy, 164–166
CuOTf (copper trifluoromethanesulfonate), 52
Cuprates *see* Organocuprates
ortho-Cupration, cross-coupling, 588–590
Cupredoxin fold, type 1 Cu proteins, 349
Cuprioalkanes, thermochemistry, 153
β-Cuprioamide reagents, 406, 411–414
β-Cuprio-α-aminocarboxylates, 415, 416–419
α-Cuprioazaenolates, 428, 430–434
β-Cuprio-*N*-carbamoyl reagents, 406, 408–409, 410–411
α-Cuprioimides, amidocuprate formation, 373, 374
α-Cuprionitriles, 426–428
α-Cuprioxazolidines, 428, 429, *431*
α-Cupriophosphonic esters, amidocuprate formation, 373, 374
2-Cuprioterphenyl
aggregation number, 152
structure, 31, *32*
Curtius rearrangement, S_N2' addition of diorganozincs, 484, 485
CuTC *see* Copper(I) thiophene-2-carboxylate
Cyanocuprates
higher-order
allylic alkylation, 616
structure, 108–109, *110*, 778, 788, 790
lower-order
Michael addition, 867
structure, 108, 109–110, *111*, 788
NMR spectroscopy, 171–173
organic synthesis, 24
reactivity, 17–18, 169
structure, 108–111
Cyanoformates, α-cuprionitrile reactions, 426–428
Cyano-Gilman cuprates
addition to extended Michael acceptors, 866, 867
kinetic reactivity, 17–18
structure, 108, *109*
substitution of extended substrates, 874
Cyanosilylcuprates
lower-order, 813–816
preparation, 778–779
structure, 788, 790
Cyanostannylcuprates, reactivity, 818
Cyclic allenes, silylmetalation, 816–817
Cyclic allylic systems, diastereospecific alkylation, 638, 640–649, 650, 651, 654, 658
Cyclic amines, organobismuth compound coupling, 947, 949
Cyclic enones
diastereoselectivity, 189–190
π-complex intermediates, 186–194
phosphoramidite complexes, 202
stannylcupration, 845
Cyclization
aryl halide coupling, 917, 919, 927, 930
Camps cyclization, 927, 930
Cu(I) hydride catalysis, 765–766, 768
intramolecular
dynoates, 828
Lewis-acid-promoted, 816
Nazarov cyclizations, 452, 816
Cycloaddition, intramolecular, 868–869
Cyclobutane-containing steroid derivative, organomercury transmetalation, 493, 495
Cyclobutanol, preparation, 493, 495
Cycloheptenones, organozinc transmetalation, 478
Cyclohexadiene oxide, allylic alkylation, 640, 641
1,2-Cyclohexanediamine (CyDA), aryl halide coupling, 922, 925–926
Cyclohexenone
Cu(III) σ-complex, 194
Michael addition, 478, 479
organomanganese addition, 449
Cyclopentadiene monoepoxides, diastereospecific ring-opening, 644–646
Cyclopentadiene oxide

kinetic resolution, 676
regioselective ring-opening, 645, 646
Cyclopentadienylcopper(I) compounds
structure, 63–68
NHC complexes, 65
Cyclopenta[*b*]furanones, diastereospecific ring-opening, 646–647
Cyclopentenediol derivatives, diastereospecific ring-opening, 643–644
Cyclophanes, organocuprate oxidation, 592, 595, 597
Cyclopropanes
Cu reactions, 150
organoboron synthesis, 499, 501
Cyclopropanol, formation, 499, 501
CyDA (1,2-cyclohexanediamine), 922, 925–926
Cytosine, organoboron compound coupling, 956, 957

Dative bonds, Cu^+ complexes, 308
Deaggregation, donor–acceptor complexes, 40
Decarboxylation
perfluoroalkanoic acid salts, 1012
ternary Cu(II) amino acid complexes, 328
trifluoromethylalkanoic acid salts, 1016–1017
Decomposition
methyl fluorosulfonyldifluoroacetate, 1013
substituted fluoroalkenylcoppers, 997–998
trifluoromethylcadmium, 1019–1020
Deconjugative double-bond migration, ynoate stannylcupration, 829–831
Defluorination, cumulenes, 998
Dehydrogenation, cross-dehydrogenative coupling, 963–965, 973, 974
[4]Dendralene, alkenyl cuprate oxidation, 595, 598
Density functional calculations *see* DFT calculations
Deoxypropionate skeletons, enantioselective conjugate addition, 719, 728
Deprotonation, Cu^{2+} complexes, 309–310
Desilylation
bromination–desilylation, 791, 792
vinylsilane acetylations, 786, 787
Destannylation, iodo-destannylation, 841
Desymmetrization
meso-diols, 947
enantioselective allylic alkylation, 676–680
DFT calculations
gas-phase chemistry, 281, 283–284
mesitylcopper structure, 30–31
organocopper(II) compounds, 116
reaction mechanisms, 2–3
Dialkoxy allylic zirconium species, Negeshi reagent treatment, 461–462, 465
Dialkoxyphosphinyldifluoromethylcopper reagents, 1008–1012
Dialkoxyphosphinylfluoromethylcopper reagents, 1008
Dialkylarylalanes, conjugate addition, 506, 507, 508
N-Dialkylbenzoyloxy amines, amine synthesis, 374–375
Dialkylcuprates
structure, 95–96
synthesis, 169–170
thermochemistry, 149
Dialkylcyanocuprates, allylic substitution, 607
β,β-Dialkyl-substituted enones, α-amino acid synthesis, 383, 385
Dialkylzinc reagents
Cu-mediated/catalyzed addition, 202, 203, 861, 863
enantioselective conjugate additions, 480–491
Diallylcuprate, substitution, 509–510
Diamagnetic compounds, NMR spectroscopy, 165–166
Diaminocarbenes, enantioselective conjugate addition, 707
Diaryl amines, Ullmann amination, 888, 889
Diarylargentate, structure, 91, *92*
Diarylcuprates, structure, 91–95
Diaryl ethers, C–O bond formation, 883–885, 893–897, 898, 899
Diaryliodonium salts, O-nucleophile coupling, 943
Diaryl selenides, synthesis, 887, 904, 906
Diaryl sulfides, synthesis, 887
Diastereomeric compounds
acyclic allylic alcohols, 614–622, 623–628
4,4*a*,5,6,7,8-hexahydro-4*a*-methyl-naphthalen-2(3*H*)-one complexes, 187
Diastereoselectivity
allylic alkylation, 606, 653–663
chirality at homoallylic position, 660–661
chirality δ to leaving group, 654–660
chirality on leaving group, 661–663
chiral cyclic enones, 189–191
chiral 2-enolate stannylcupration, 824
β-silyl enolates, 795
Diastereospecific methods
asymmetric allylic alkylation, 610–653
acyclic allylic systems, 613–638, 639
cyclic allylic systems, 638, 640–649, 650, 651
leaving group effects, 612–613, 614, 622, 629–636
substrates, 610–612
Dibenzofurans, Cu-catalyzed coupling, 908
1,1-Dibromoalkenes, trifluoromethylation, 1022

(3)-1,2-Dicarbollide ligand, Cu sandwich complexes, 256–257
1,4-Dicarbonyl compounds, organozirconium transmetalation, 461, 464
Dicuprioalkanes, thermochemistry, 153
Dienes
Michael addition, 858–863, 864
activated dienes, 859–860
conjugated dienes, 864
organozirconium transmetalation, 457, 460
silylcupration
1,3-dienes, 817
1,4-dienes, 810, 816–817
skipped diene transmetalation, 457, 460, 496, 499
Dienones, Michael addition, 859–860, 861, 863, 864
Dienyl carboxylates, Cu-mediated/catalyzed substitution, 872–873
Dienyl cuprates, carbocupration reactions, 529–530, 532, 548–549, 555, 557
Dienyne, organozirconium transmetalation, 467, 468
Diethylaminodiphenylsilyl(cyano)cuprate, 780–781
Diethylaminodiphenylsilyl group, C–Si to C–OH conversion, 783, 785
N,*N*-Diethylsalicylamide, alkyl amine arylation, 914, 915
Diffusion-ordered spectroscopy (DOSY), 168, 179–181, 204–205
3,3-Difluoroacrylate, synthesis, 1003
1,1-Difluoro-3-alkene phosphonates, allylation route, 1009–1010
α,α-Difluoroallyl phosphonates, synthesis, 1011
α,α-Difluorobenzylic phosphonates, formation, 1010–1011
Difluorocarbene precursors, fluorosulfonyl compounds, 1021
1,2-Difluoroethenediylbisphosphonates, isomers, 1010
Difluoromethylcopper reagents, 1022–1023
Difluorophosphonoacetic acid, synthesis, 1009
2,2′-Difluorovinyl carbonyl compounds, synthesis, 1003
Difluorovinylcopper reagents, 1003–1004
Dihalodifluoromethanes, trifluoromethylmetallic reagent formation, 1020–1021
trans-Dihydrindandione, α-cuprioazaenolate reactions, 430
1,3-Dihydrobenzimidazol-2-ones, aryl halide coupling, 919
Dihydrofurans
allene cyclization, 629–632, 633
Michael addition, 870
3,4-Dihydroisoquinoline, organozirconium transmetalation, 461, 464
3,4-Dihydro-2*H*-pyran, hydrozirconation, 456–457, 458
Dihydropyrrole, allylic alkylation, 635–636
Diimine copper complexes, NMR structure, 200
1,2-Diiodoethylene, perfluoroalkylation, 1014–1015
β-Diketiminates, copper carbene complexes, 120, 121, *122*, 130–131
β-Diketonates, NHC–copper complexes, 120–121, *122*
Diketopiperazine, allylic alkylation, 641, 644
Dilithium tetrachlorocuprate, allylic alkylation, 608, 609
Dimerization
Fischer chromium carbene complexes, 451–452
heteroaromatics, alkenyl and alkyl groups, 595–598
organocuprates, 586
Dimers
chiral amido(alkyl)cuprate, 369
copper acetylide complex structure, 80–81
diorganocuprate NMR, 173, 175–176
heterodimer electron transfer, 222–224
1,2-Dimetalla alkenyl compounds, structure, 43
Dimethoxyethane (DXE), Cu^+ affinity, *292*, 295
4,7-Dimethoxy-1,10-phenanthroline, aryl halide coupling, 922
1,1-Dimethylallene, regiochemistry, 846
N,*N*-Dimethylaminoethanol, aryl halide coupling, 913, 914
2-[(Dimethylamino)methyl]ferrocenylcopper, structure, 33
2-[(Dimethylamino)methyl]-5-(methyl)-phenylcopper, structure, 34
2-[(Dimethylamino)methyl]phenylcopper, structure, 34
8-(Dimethylamino)-1-naphthylcopper, structure, *34*, 35
Dimethylcuprates *see* Gilman reagents
N,*N*′-Dimethyl-1,2-cyclohexane-1,2-diamine (DMCyDA)
allenyl iodide coupling, 933, 934
aryl halide coupling, 922, 925, 938
Br–I exchange, 942
vinyl halide coupling, 935
Dimethyl ether (DME), Cu^+ affinity, *292*, 295
Dimethylethylenediamine (DMEDA)
aryl halide coupling, 920, 925, 927, 928, 931, 939
Hurtley reaction, 970, 971
Rosenmund–von Braun reaction, 967
vinyl halide coupling, 932, 933, 935, 936, 940, 941

N,N-Dimethylglycine
aryl halide coupling, 897, 899, 901, 917, 966
vinyl halide coupling, 932
N,N-Dimethylhydrazones, α-cuprioazaenolate preparation, 428, 430–434
2-(4,4-Dimethyl-2-oxazolinyl)phenylcopper, structure, 37, *39*
Dimethyl sulfide
Cu(I) bromide complex, 777, 846
diastereoselective allylic alkylation, 663
meso-Diols, desymmetrization, 947
Diorganocuprates
molecular orbitals, 3–5
NMR spectroscopy
dimer structure, 173, 175–176
monomer structure, 171–173
SSIPs vs CIPs, 173–178
structure, 100–101
Diorganozinc reagents
conjugate addition to enals, 474
imine additions, 492–493
preparation, 471, 473, 474, 476
S_N2' addition, 484–489
Diphenylcarbene complex, structure, 130–131
Diphenylcopperlithium, structure, 95, *96*
Diphenylcuprates
anion thermochemistry, 152
NMR spectroscopy, 173, *174*
2,6-Diphenylphenylcopper (2-cuprioterphenyl)
aggregation number, 152
structure, 31, *32*
ortho-Diphenylphosphanylbenzoate (*o*-DPPB), quaternary stereocenter generation, 484, 486
(Diphenylphosphanyl)ferrocene carboxylate, enantioselective allylic alkylation, 665
N-Diphenylphosphinoylimines, diorganozinc addition, 489, 491
Diphenyltetramethyldisilane, cleavage, 778
Diphosphine copper complexes, NMR structure, 200
Direct Blue 199, structure, 232
Directed ortho-metalation (DOM)
amine synthesis, 378–379
β-N-aryl/heteroarylcuprates, 419
Diselenides, selenol oxidation, 906
Disphenyllead dichloride, O-nucleophile coupling, 949
Dissociation energies
Cu(III) species, 154
methylcopper, 146
Dissociative charge transfer, photofragmentation, 287–288
Disubstitution
acetylene stannylcupration, 836, 838
alkyne silylcupration, 804
enones, 478, 480
perfluoroalkylcoppers, 1014–1015
Diversity-oriented synthesis, 693
Divinylacetylene derivatives, cumulene defluorination, 998
Diynes
perfluoroalkynylcopper synthesis, 1006
stannylcupration, 845
titanacycle mono-allylation, 445, 446
Diynoates, intramolecular cyclization, 828
Diynyl compounds, electrochemical stability, 266–267
DMCyDA *see* N,N'-Dimethyl-1,2-cyclohexane-1,2-diamine
DME (dimethyl ether), *292*, 295
DMEDA *see* Dimethylethylenediamine
DNA
cleavage inhibition, 238
Cu(I) interactions, 282
nucleic acid Cu(I)/Cu(II) reactivity, 337–340
DOM *see* Directed ortho-metalation
Domino reactions, Cu(I) hydride complexes, 757–766, 768
Donor–acceptor compounds
complex structures, 39–51
electron transfer, 220–222
Donor-base-free compounds, structure, 27–34
DOSY (diffusion-ordered spectroscopy), 168, 179–181, 204–205
Double bonds, deconjugative migration, 829–831
Double insertion
CF_2 units, 1002
pentafluorophenylcopper, 1026
DPPB (*ortho*-diphenylphosphanylbenzoate), 484, 486
dppm (1,1-bis(diphenylphosphino)methane), 83–86
Dummy ligand effect, mixed organocuprates, 16–17
DXE (dimethoxyethane), *292*, 295
Dye industry
copper porphyrins/phthalocyanines, 230–237
green chemistry, 232–237
photocatalysts, 235–237
Dynamic kinetic asymmetric transformation (DYKAT), allylic substitution, 680–682
Dynamic kinetic resolution, Cu(I) halide catalysis, 744, 745, 769
Dynamic NMR simulations, phosphoramidite complexes, 206–207

EADIs ((E)-alkene dipeptide isosteres), 484, 488
Ebelactone A, synthesis from vinylsilanes, 807, 808

Electric field gradient (EFG), NMR spectroscopy, 165
Electrochemical measurements, copper carbonyls, 158
Electrochemistry, 247–278
  stability, 255–276
    binuclear compounds, 257–261
      $\{Cu(\mu\text{-}L)\}_2$ compounds, 259–261
      Cu/Pt/Ti mixed compounds, 257–259
    mononuclear compounds, 255–257
    polynuclear compounds, 261–276
  synthesis, 248–255
Electron affinities, homoleptic copper carbonyls, 157
Electron correlation, gas-phase chemistry, 281–282
Electron-deficient two-electron three-center bonds, 26
Electronegativity
  ligand transfer, 444
  transmetalation, 471, 493, 496, 502, 509, 514
Electronic interactions
  Cu(I) acetylide–ferrocene, 267–268
  Ti(IV)–Cu(I) compounds, 271, 273
Electron paramagnetic resonance (EPR), *N*-functionalization, 378
Electron transfer systems
  allene racemization, 632
  copper porphyrins/phthalocyanines, 219–232
    donor–acceptor compounds, 220–222
    dye industry, 230–237
    heteroligand systems, 222–224
    hydrogen production, 230
    nanomaterials, 225–228
    solar energy, 228–230
  see also Photoinduced electron transfer
Electrophilic amination
  amidocuprate oxidation, 586
  organozinc transmetalation, 493, 494
Electrophilic substitution
  stannylcuprates, 831–834
  vinylsilanes, 786, 787
Elimination reactions
  acid chloride oxidative addition, 13, 14
  conjugate addition–elimination, 833–834
  β-hydride elimination, 367
  hydrozirconation–β-elimination–hydrozirconation, 456–457, 458
  Peterson elimination, 797, 807
  β-silylketone, 786
  see also Reductive elimination
Enals
  α-amino acid synthesis, 383, 385
  diorganozinc conjugate additions, 474
α,β-Enals, α-(*N*-carbamoyl)alkylcuprate addition, 388, 389
Enantiomeric complexes, achiral enones, 187, *188*
Enantiopure compounds
  α-amino acids, 956, 957
  β-silyl ketones, 793, 794
Enantioselective conjugate addition, 444, 693–730
  alanes, 503–508
  dialkylzincs, 480, 481–491
  Grignard reagents, 480, 481–482
  hydrogenation, 756–757
  hydrosilylation, 741–749, 752–757
  Michael acceptors, 474, 476, 477, 478, 479
  natural product syntheses, 693, 695, 714–728
  new substrates, 695–700, 701
  NMR structure, 200–201
  organosilicon transmetalation, 513, 514
  phosphoramidites, 476, 478, 492–493
  quaternary stereocenters, 700–707
    all-carbon, 701, 725, 728
      transmetalation, 478, 480, 484, 485, 508
  β-silylketones, 793, 794
  β-stannylketones, 824
  tandem reactions, 693, 707–714
Enantioselectivity
  allylic alkylation, 444, 606, 663–682
    chirality on leaving group, 663–666, 667
    chirality ligands on Cu, 666, 668–680
    dynamic kinetic asymmetric transformation, 680–682
  Hurtley reaction, 972–973
  NHC–Cu complex alkylation, 401
Endocyclic allylic phosphates, preparation, 418–419
Endocyclic restriction test, amine synthesis, 373, 375, 493
Energy density contour maps, Cu^+ complexes, 308–309
Eneyne derivatives, perfluoroalkynylcopper synthesis, 1006
Enoates, NHC–Cu complex reactions, 402, 403
2-Enoates
  conjugate additions, 824
  substitution, 832–833
Enolates
  Cu(III) organocuprate intermediates, 194
  domino reaction intermediate, 757
  E/Z isomers, 795–796
  nitrogen enolate cuprates, 426–434
Enol ethers, C–O bond formation, 899, 901
Enones
  β,β-dialkyl-substituted, 383, 385
  diastereoselectivity, 189–190
  enantioselective additions, 474, 476–478, 480, 481, 513, 514

NHC–Cu complex reactions, 402
organomanganese addition, 449
π-complex intermediates, 186–194
phosphoramidite complexes, 202
stannylcuprate conjugate additions, 510, 512, 822–824
α,β-Enones
α-(*N*-carbamoyl)alkylcuprate addition, 388, 389
cuprate–enone d–π* complex, 391
β-cuprio-α-aminocarboxylate reactions, 416
α-cuprioazaenolate reactions, 432
NHC–Cu complex reactions, 402, 403
2-Enones, 3-halo and 3-sulfonyl-substituted, 832–834
Enthalpies of combustion, copper acetylides, 147–148
Enthalpies of formation, 146
copper acetylides, 155
copper carbonyls, 158
methylcopper, 147
Enthalpies of reaction, 146
Enthalpy of dissociation, methylcopper, 146
Enthalpy of hydrogenation, copper acetylides, 147
Enyne acetates, Cu-mediated/catalyzed substitution, 874
Enynes
carbocupration, 555–557
Michael addition, 861, 863–871
organoboron transmetalation, 496, 499
titanacycle mono-allylation, 445, 446
Enzymes
biological Cu centers, 348–352
multi-copper enzymes, 349
mononuclear Cu sites, 352–354
multinuclear Cu sites, 355–360
(−)-Ephedrine, amidocuprate formation, 367–370
Epoxides
carbocupration, 530, 531, 535, 545, 546
organozirconium transmetalation, 457, 460
regioselective cleavage, 834
ring-opening
alkylation, 10, 610, 622, 625, 628
transmetalation, 444, 509–510, 514–516
silylcuprate substitution, 802–803, 807
stannylcuprate substitution, 834, 835
EPR (electron paramagnetic resonance), 378
Esters
α,β-unsaturated
α-(*N*-carbamoyl)alkylcuprate addition, 388, 389
α-cuprioazaenolate preparation, 428
transmetalation, 449, 478, 480, 493, 495, 496, 500, 517, 1021–1022
(*E*)-α-trifluoromethyl, 1021–1022
Ethanol, Cu(II) coordination, 305
Ethyl cyanoacetate, Hurtley reaction, 969
Ethylene, Cu^+ affinity, 293–294
Ethyl 2-en-4-ynoate, ^{13}C NMR spectroscopy, 185
Ethyl malonate, Hurtley reaction, 969
Ethynediides, thermochemistry, 155
Ethynyl ethers, stannycupration, 841
Eupomatilones 4 and 6, cross-coupling synthesis, 588
EXAFS (extended x-ray absorption fine structure), 108
Exchange equilibria, π-complex intermediates, 192–193
Exchange rates, phosphoramidite complexes, 206
Exchange reactions
interaggregate, 60–61
Sn–Li, 838, 840, 841
see also Halogen exchange
Exchange spectroscopy (EXSY), 167–168
4-*exo*-ring closure, Cu-catalyzed coupling, 908
Extended multiple bond systems, 857–879
addition to Michael acceptors, 858–872
substitution of extended substrates, 872–874
Extended x-ray absorption fine structure (EXAFS), cyanocuprates, 108
E/*Z* isomers
allylic carbamates, 379–382
α-halo-substituted fluoroalkenylcoppers, 997–998
β-silylenolates, 795–796
stannylcupration, 824–831, 838, 839

α- and β-Face coordination, 4,4*a*,5,6,7,8-hexahydro-4*a*-methylnaphthalen-2(3*H*)-one, 187–191
Felkin–Ahn model, addition, 654, *656*
Ferrocene
Cu(I) acetylide electronic interactions, 267–268
cyclopentadienylcopper(I) compounds, 63
diphosphine copper complexes, 200
Finkelstein reaction, Cu-catalyzed, 967
Fischer carbene complexes, transmetalation, 450–452
Five-membered rings, stannylcupration, 840, 841
FK-506
transmetalation synthesis, 516
carboalumination, 503
Flavonoids, gas-phase Cu(I)/Cu(II) reactivity, 335
Fleming–Tamao oxidation
allylic silane preparation, 487, 489
organoboron transmetalation, 499, 501
9-Fluorenylmethoxycarbonyl (Fmoc)-protected amino acids, 418

Fluoride-induced transmetalation, organoboron compounds, 496, 498
Fluorinated cumulenes, fluoroalkenylcopper decomposition, 997–998
Fluorinated organocopper reagents, 991–1032
Fluoroalkenylcadmium reagents, perfluoroalkenylcopper pre-generation, 994–996
Fluoroalkenylcopper reagents
  CF_2 insertion, 1002
  fluorinated cumulene formation, 997–998
  functionalization reactions, 996, 998–1002
  halide symmetrical coupling, 992–994
  perfluoroalkenylcopper pre-generation, 994–996
    internal fluoroalkenylcopper reagents, 996
Fluoroalkenyl halides, symmetrical coupling, 992–994
Fluoroalkenyltin reagents, perfluoroalkenylcopper pre-generation, 996
Fluoroalkenylzinc reagents
  acylation, 998
  perfluoroalkenylcopper pre-generation, 994–996
Fluoroalkylated alkynes, carbocupration, 548–550
Fluoroenynes, conjugated, 1003
Fluorosulfonyldifluoroacetate, difluorocarbene precursors, 1021
Fluorosulfonyldifluoromethyl iodide, difluorocarbene precursors, 1021
Fmoc (9-fluorenylmethoxycarbonyl), 418
Formamides
  from carbamoyl cuprates, 383, 384
  Cu^+/Cu^{2+} reactions with small organic bases, 310–318
Formamidine
  β-N-arylcuprate formation, 419, 420, 422
  β-N-heteroarylcuprate formation, 425
α-Formamidinylcuprates, reactions, 386
Formation, *see also* Enthalpies of formation
Four-center mechanism, *syn*-addition, 848–849
Four-membered rings, stannylcupration, 840, 841
Free-energy differences, alkyl radicals on Cu surface, 148
Fujimycin *see* FK-506
Fullerenes, photoinduced electron transfer, 221–222
Functionalization
  allylstannanes, 847–848
  fluoroalkenylcoppers, 996, 998–1002
  N-functionalized organocuprates, 365–441
    N–Cu bonds, 366–382
    α-N-substituents, 383–406, 407
    β-N-substituents, 406, 408–432, 433, 434
Furans, perfluoroalkylation, 1014
Galactose oxidase, mononuclear Cu sites, 352–354
Gallina's carbamate, allylic alkylation, 609
Gas-phase chemistry, 279–346
  binding energies, 289–300
  bioreactivity of Cu(I)/Cu(II) ions, 318–340
  bonding, 308–310
  coordination, 281, 300–308
  Cu^+/Cu^{2+} reactions, 310–318
    Cu^+/alkane reactions, 146
  ligand-field spectroscopy, 285–289
  theoretical models, 281–285
Gattermann synthesis, CO reactions, 156
Germylcuprates, 776–777
Gilman reagents
  addition to extended Michael acceptors, 859, 866, 867
  allylic substitution, 169, *170*, 605, 609, 654
  cuprate synthesis, 24
  intermediates, 112
  NMR spectroscopy, 169–170, 197–198
  oxidation, 585
  reaction mechanisms, 2
  structure, 90–91
  see also Cyano-Gilman cuprates
Glaser alkyne coupling, 585
D-Glucose, gas-phase Cu(I)/Cu(II) reactivity, 332–334
Glycine
  aryl halide coupling, 917
  Cu complexes, 282, 301, 302
  oligomers, 290–291
α-C-Glycosylamino acids, preparation, 418, 419
Goldberg reactions, aryl/vinyl chloride amidation, 373, 890–891, 922, 927, 929, 932, 935
Gold compounds, aurate structure, 91–93
Gold(I) heteromultimetallic compounds, electrochemical stability, 273–276
Green chemistry, copper porphyrins/phthalocyanine dyes, 232–237
Grignard reagents
  acid chloride addition, 457
  C–C bond formation, 979–980
  enantioselective conjugate addition, 480, 481–482
  Michael addition, 860–863
  2-vinylphenyl, 58
Group IIB
  transmetalation, 471, 473–496
    organomercury compounds, 444, 493, 495–496
    organozinc compounds, 444, 458, 461, 471–493, 494, 517
Group IIIA
  transmetalation, 496–508

organoaluminum compounds, 444, 502–508, 517
organoboron compounds, 444, 496–501
Group IVA
transmetalation, 509–514
organosilicon compounds, 444, 513–514
organotin compounds, 444, 447, 509–512
stannylcuprates, 819, 821
vinylstannanes, 447, 509–510, 512, 828, 838, 839, 845–848
Group VIA
transmetalation, 514–516
organotellurium compounds, 444, 514–516
Guanine
binding energies, 282
organoboron compound coupling, 956, 957

Halide ions, thermochemistry, 149
2-Halobenzoic acids, Ullmann amination, 888
β-Haloenones, stannylcuprate substitution, 831–834
Halogen exchange
Cu-catalyzed, 940–942, 967
halogen–metal exchange, 378–379, 419
halogen–zinc exchange, 473, 474
Li–halogen exchange, 502, 590
see also Iodine exchange; Magnesium–halogen exchange
ortho-Halophenyl(thio)amides, Cu-catalyzed coupling, 908, 909
ortho-Halophenyl(thio)ureas, Cu-catalyzed coupling, 908, 909
3-Halo-substituted 2-enones, stannylcuprate substitution, 832–834
α-Halo-substituted fluoroalkenylcoppers, decomposition, 997–998
Halothiophene substrates, perfluoroalkylation, 1014
Hammett equation, substituent constants, 261, 263–264
Hammett plots, amidocuprate formation, 374, 377
Hard-sphere increments, phosphoramidite complex NMR, 205
Hemocyanins, CO binding by Cu, 156
Heteroaromatics, dimerization, 595, 597
Heteroaryl bromides
aliphatic amine coupling, 914, 915
Br–I exchange, 942
Heteroaryl compounds, transmetalation, 386
Heteroarylcuprates, β-*N*-substituted, 419, 423–426
α-*N*-Heteroarylcuprates, reactions, 402–406, 407
Heteroaryl iodides
aliphatic amine coupling, 914, 916
amide coupling, 922, 924
Heteroarylsilanes, Hiyama-type cross-coupling, 977–978
N-(Heteroarylsulfonyl)aldimines, alkylzinc halide addition, 492
Heteroatom cuprates
ligand transfer, 776–777
structure, 49, 101–104
Heteroatoms
C–heteroatom bond formation, 883–965
classical Ullmann chemistry, 883–893
modern Ullmann chemistry, 893–943
non-Ullmann substrates, 943–965
diastereoselective allylic alkylation
chiral centers with heteroatoms, 658–660
chiral centers without heteroatoms, 654–658
substituted alkyne carbocupration, 557–572
Heterobimetallic compounds, electrochemical stability, 258, 268, 270–272
Heterocubanes
copper(I) acetylide complexes, 81
potassium cyanocuprate, 111
Heterocycles
C–O bond formation, 906–908
C–S bond formation, 908–909
intramolecular carbocupration, 576–579, 580–581
N–H heterocycle coupling, 921–922, 923–924, 944
organoboron compound coupling, 954
N-Heterocyclic carbenes (NHC)
Cu complexes, 400–402, 403, 404, 435
Cu(I) halide complexes, 740–741, 750–752
cyclopentadienylcopper complexes, 65
donor–acceptor complexes, 43–47
enantioselective conjugate addition, 480, 481–482, 513, 514, 517–518, 703–704
Michael addition, 863, 864
silver(I) complex, 480, 481–482, 508
S_N2' additions
allylic phosphates, 508–509
diorganozincs, 484, 489
structure, 119–121, *122*, 123–124, 127–128, 131–133
Heterodimers
diorganocuprate NMR, 175–176
electron transfer, 222–224
Heteroleptic organocopper aggregates, structure, 32–33, 35–39, 87–88
Heteroleptic organocuprates
NMR spectroscopy, 169, 171
structure, 48–49, 97–98
thermochemistry, 149
Heteroligand systems, photoinduced electron transfer, 222–224

Heteromultimetallic complexes, electrochemical stability, 268–276
Heteronuclear multiple bond correlation (HMBC), 189
Heteronuclear multiple quantum coherence (HMQC)
$^{1}H,^{13}C$ HMQC, *188*
HMQC–ROESY pulse sequences, 167
Heteronuclear Overhauser effect (HOE), 166–167, 173, 174–175, 181–183
Heteronuclear Overhauser effect spectroscopy (HOESY), 173, 174–178, 181–183
Heteronuclear single quantum coherence (HSQC), HSQC–NOESY pulse sequences, 167, *168*, 176, *177*
Heterotetranuclear iron(II)–silver(I)–copper(I)–titanium(IV) complexes, electrochemical stability, 273–276
Heterovalent copper (I,II) π,σ-complexes, electrochemical synthesis, 252–255
Hexaalkylditin derivatives, stannyllithium preparation, 818, 819
Hexabutyldistannane, substitution, 834, 835
Hexafluorobutyne, pentafluorophenylcopper *syn*-addition, 1025
4,4*a*,5,6,7,8-Hexahydro-4*a*-methylnaphthalen-2(3*H*)-one, π-complex intermediate, 187–191
Hexamethylphosphoramide (HMPA)
amidocuprate formation, 371
α-(*N*-carbamoyl)alkylcuprate reactions, 391
α-cuprioxazolidine reactions, 428
diorganocuprate NMR, 173
1-Hexene, hydrozirconation, 456, 458–459, 462
Higher-order cuprates
allene silylcupration, 811, 812, 814–816
allylic alkylation, 609
cross-coupling, 587
cyanocuprates, 108–109, *110*, 616, 778
silylcuprates, 778, 788, 790
stannylcuprates, 818, 820, 828, 834, 846, 847, 848
structure, 99, *100*
Highest occupied molecular orbital (HOMO), diorganocuprate(I) species, 3–5
Hindered compounds
silylcuprate preparation, 780
β-silyl ketones and esters, 793, 794
Histidines, ligand exchange, 306
Hiyama cross-coupling, C–C bond formation, 977
HMBC (heteronuclear multiple bond correlation), 189
HMPA *see* Hexamethylphosphoramide
HMQC *see* Heteronuclear multiple quantum coherence
^{1}H NMR spectroscopy, *164*
$^{1}H,^{1}H$ homonuclear reactions, 166, 181–183
^{1}H,X heteronuclear reactions, 166, 181–183, 196–200
HOE (heteronuclear Overhauser effect), 166–167, 173, 174–175, 181–183
HOESY (heteronuclear Overhauser effect spectroscopy), 173, 174–178, 181–183
HOMO (highest occupied molecular orbital), 3–5
Homoallylic amines, transmetalation, 386
Homoallylic position, diastereoselective allylic alkylation, 660–661
Homoatomic copper anion clusters, phenylcopper formation, 152
Homochiral allylsilanes, silylcuprate substitution, 797
Homodimers, diorganocuprate NMR, 175–176
Homoleptic copper carbonyls, thermochemistry, 156–157
Homoleptic organocuprates
synthesis, 169
thermochemistry, 149
Homologous series, copper acetylides, 147
Homosilylcuprates, preparation, 777–778
Homostannylcuprates, ligand exchange, 819
Horner–Emmons olefination, organomercury transmetalation, 493, 495
Hoveyda–Grubbs catalyst, allylic silane–ester coupling, 484, 487, 489
HSQC (heteronuclear single quantum coherence), 167, *168*, 176, *177*
Hurtley reaction
C–C bond formation, 968–973
copper(I) benzoate synthesis, 58–60
Hydration, Cu^{2+} ions, 299
Hydrazides, vinyl iodide coupling, 935, 936
Hydrazines, *N*-functionalization, 378
Hydrazones
nitrogen enolate cuprate formation, 426
organolead compound coupling, 950, 951
β-Hydride elimination, amidocuprate formation, 367
Hydroalumination, vinylalane preparation, 506
Hydroamidation, C–N coupling, 936–937
Hydroboration, allylic silane preparation, 487, 489
Hydrocarbyl derivatives of copper, thermochemistry, 146–155
Hydrocupration
alkynes, 767, 768–771
hexynes, 151
NHC–Cu complex reactions, 402, 403
Hydrodynamic radius, self-diffusion coefficients, 168
Hydrogen, C–H bond activation, 114, 115
Hydrogenation
Cu(I) hydride catalysts, 736–737, 756–757

see also Enthalpy of hydrogenation
Hydrogen bonding, blue Cu proteins, 280
Hydrogen production, photochemical processes, 230
4-Hydro-L-proline, aryl halide coupling, 917
Hydrosilylation
  ketones and imines, 749–757
  NHC–Cu complex reactions, 402, 403
  α,β-unsaturated substrates
    chiral ligands, 741–749
    non-chiral ligands, 738–741
Hydrotitanation, transmetalation, 445–447, 448
α-Hydroxyallenes
  allylic alkylation, 631, 632, 634, 635
  Michael addition, 870
Hydroxyl groups
  C–Si to C–OH conversion, 781–791
    masked hydroxyl group, 788
Hydroxyphthalimide, organoboron compound coupling, 952
trans-4-Hydroxy-L-proline, aryl halide coupling, 919, 972–973
8-Hydroxyquinoline, amide intermolecular arylation, 925
β-Hydroxysilanes, epoxide substitution, 802
Hydroxy-substituted vinylallenes, formation, 874
Hydrozirconation
  alkynes, 518
  β-elimination–hydrozirconation, 456–457, 458
  organozirconium reagent preparation, 452–462, 463–465

Imidazindolones, vinyl halide coupling, 937, 938
Imidazole, Cu^+ affinity, *292*, 295
Imidazoles
  arylation, 921, 950
  α-N-heteroarylcuprates, 404, 406
  organoboron compound coupling, 958
Imides
  α-(N-carbamoyl)alkylcuprate addition, 388, 389
  organoboron compound coupling, 954
  organolead compound coupling, 950, 951
α-Imidocoppers, reactions, 383
Imidoyl compounds, transferable N-ligands, 373
Imines
  1,2-addition, 461–462, 465
  N-arylation, 377
  catalytic hydrosilylation, 749–757
  diorganozinc addition, 492–493
  nitrogen enolate cuprate formation, 426
  unsaturated, 202
Iminium chlorides, α-cuprioazaenolate reactions, 432
α-Iminomethylcuprates, reactions, 385
INADEQUATE (incredible natural abundance double quantum transfer experiment), 166, 189–190
Indazoles, arylation, 921, 950
Indoles
  arylation, 921, 950
  β-N-arylcuprate reactions, 419
  α-N-heteroarylcuprate reactions, 402
  β-N-heteroarylcuprate reactions, 425
  Hurtley reaction, 970, 971
  iodonium salt coupling, 944, 945
  organotin transmetalation, 510, 512
  Sonogashira reaction, 966
Indolizidine alkaloids, α-(N-carbamoyl)-alkylcuprate reactions, 394, 397
Indolizidines, α-(N-carbamoyl)-alkylcuprate reactions, 391, 392
INEPT (insensitive nuclei enhanced by polarization transfer), 189–190
Initial build-up rates, NOE/HOE determinations, 167
Insertion reactions
  fluoroalkenylcoppers, 1002
  pentafluorophenylcopper double insertion, 1026
  perfluoroalkyl halides, 1012
Inside alkoxy effect, allylic chlorides, 658, 659
Interaggregate exchange reaction, aryl copper arenethiolate copper aggregates, 60–61
Interconversion
  temperature-dependence, 208–209
  time-dependence, 199–200
Intermediate compounds
  allylcuprates, 811, 812, 816
  allylstannane–vinylcuprate, 847–848
  copper enolates, 458, 461
  Cu(I) compounds, 184–194, 850
  dialkoxy allylic copper, 462, 465
  domino reactions, 757
  Gilman reagents, 112, 197–198
  metallocycle species, 849
  organocuprate complex NMR, 170, 184–200
    Cu(I) organocuprates, 184–194
    Cu(III) organocuprates, 194–200
  oxidative addition, 3, 5–6, *7*
  vinylcuprates
    silylcupration, 803, 804, 805, 809, 813–816
    stannylcupration, 836, 837, 838, 839
    vinylstannane–oxazolidine, 842
Intermolecular reactions
  C–O bond formation, 953
  vinyl halide amidation, 935

Internal fluoroalkenylcopper reagents, pre-generation, 996
Intramolecularly coordinating substituted compounds, structure, 34–39, 97–98
Intramolecular reactions
alkyne carbocupration, 572–579
non-stabilized organocoppers, 572–574, 575
stabilized organocoppers, 574, 576
C–C bond formation, 592–595, 596
C–O bond formation, 953
conjugate addition, 493, 495, 510, 512
cross-coupling, 510, 512
[4 + 2]-cycloaddition, 868–869
diynoate cyclization, 828
intermediate vinylcuprates, 809
Lewis-acid-promoted cyclizations, 816
vinyl halide amidation, 935
Iodine exchange
Br–I exchange, 940–942
I–Zn exchange, 474
Mg–I exchange, 593, 594, 595, 596
see also Halogen exchange
1-Iodoalkynes, β-cuprio-α-aminocarboxylate reactions, 416
3-Iodo-*N*-boc-azetidine, allylation, 410
Iodocuprates, reactivity, 169
3-Iodocycloenones, zirconacycle transmetalation, 470
Iodo-destannylation, 841
3-Iodo-1,3-dimethyl-1,3,5,6-tetrahydropyrimidine-2,4-dione, 424
Iodoimidazoles, α-*N*-heteroarylcuprates, 402, 404, 406
Iodoindoles
α-*N*-heteroarylcuprate reactions, 404, 406
β-*N*-heteroarylcuprate reactions, 425
Iodolysis, alkyne carbocupration, 529–530, 545, 548, 553–555, 556
2-Iodo-1-methylimidazole, organozinc reagent preparation, 402, 405
Iodonitrobenzenes, β-nitroarylcuprate formation, 423
Iodonium salts
C–C bond formation, 966–967, 975
non-Ullmann Cu-mediated coupling
N-nucleophiles, 943–945
O-nucleophiles, 943, 944
P-nucleophiles, 945–946
2-Iodoquinoline, organozinc reagent preparation, 402, 405
3-Iodoquinoline, β-*N*-heteroarylcuprates, 424
2-Iodo-1,3-thiazole, organozinc reagent preparation, 402, 405
Ion affinities, Cu^+, 282, 289–295, 296–298
Ionic structures
Cu(III) intermediate NMR, 198
organocuprates, 104–108
Ion pairs
Gilman reagents, 2
organocuprates, 104–105
Iron-catalyzed reactions, β-*N*-heteroarylcuprates, 425–426
Iron(II) heteromultimetallic compounds, electrochemical stability, 273–276
2-Isobutyrylcyclohexanone, aliphatic amine coupling, 914, 916
Isocyanates, S_N2' addition of diorganozincs, 484, 485
Iso[7]-levuglandin D_2, transmetalation synthesis, 516
Isomeric differentiation, peptides, 325–326, 330, *331*
Isomerization
Cu(III) intermediate NMR, 198
α-halo-substituted fluoroalkenylcoppers, 997–998
β-silyl enolates, 793, 795–796
Isonitrile, S_N2' addition of diorganozincs, 484, 485
Isotopes
NMR spectroscopy, 164–165
organocuprate intermediates, 184

Jahn–Teller distortion, Cu(II) cations, 281, 282

Kendomycin, transmetalation synthesis, 516
Ketimines
domino reactions, 762–763
enantioselective reactions
hydrosilylation, 756
organozinc addition, 489
Ketones
catalytic hydrosilylation, 749–757
organoboron transmetalation, 496, 498
α,β-unsaturated
asymmetric conjugate addition, 506
N-functionalized organocuprate formation, 416, 428, *430*, 433–434
organotin transmetalation, 509
organozinc transmetalation, 484, 486
organozirconium transmetalation, 452–453, 459, 461, 464
KIEs *see* Kinetic isotope effects
Kinetic control
amino acid chiral recognition, 322–323
silylcupration, 813, 814
stannylcupration, 846, 847
Kinetic isotope effects (KIEs)
amidocuprates, 371
reaction mechanisms, 3
Kinetic mixed organocuprates, cross-coupling, 587–588
Kinetic resolution
allylic alkylation, 616, 676–680

  Cu(I) halide catalysis, 744, 745, 769
Koch carbonylation, CO reactions, 156
Kochi's salt, allylic alkylation, 609

β-Lactam, arylation, 927, 931
Lactams
  aryl iodide coupling, 922, 924
  β-cuprioamide preparation, 413, 414
  enantioselective conjugated hydrosilylation, 742–743
  α,β-unsaturated, 474, 476, 477, 510, 512
  vinyl/alkynyl halide coupling, 932, 933–935
Lactones
  domino reactions, 764–765
  enantioselective conjugated hydrosilylation, 742, 744, 746
Lanthanides, transmetalation, 471, 472, 473
Laser excitation, Cu formation, 147
Lead, electronegativity, 509
Lead compounds *see* Organolead compounds
Leaving group effects
  allylic alkylation, 612–614, 622, 629–636, 654–667
  quaternary stereocenter generation, 484, 487
(+)-Lepadin B, transmetalation synthesis, 518
Lewis acids/bases
  boron trifluoride effect, 15
  cyano–Gilman cuprates, 17–18
  dummy ligand effect, 16–17
  intramolecular cyclizations, 816
  reaction mechanisms, 15–18
  transmetalation complexes, 445, 456
Ligand-field spectroscopy, gas-phase, 285–289
Ligands
  accelerated catalysis, 200, 753
  alkynyl, 779
  bidentate Cu^{+} interaction, 295, 296
  coordination effects
    Cu(II) in blue proteins
    more than one ligand, 303–306
    one ligand, 301–303
  copper(I) acetylide complexes
    arenethiolato, 87
    dppm, 83–86
    triphos, 79–80
  dummy ligand effect, 16–17
  exchange, 306, 457, 460, 819
  heteroligand systems, 222–224
  isocyanide, 201, *202*
  metal–ligand repulsion, 284–285
  mixed silylcuprate preparation, 779–780
  monodentate
    binding energies, 295
    thiol–TADDOL, 201–202
  non-transferable
    chiral, 367–371
    cuprate structure, 366–367
    organoaluminum transmetalation, 502
    organozirconium transmetalation, 457, 460
    stannylcuprates, 822
  pinwheel array, 71, *72*
  polydentate Cu complexes, 301
  privileged structures, 202, 203
  residual ligands, 779
  thiol–TADDOL, 200–201, *202*
  transferable
    electronegativity, 444
    Fischer chromium carbene complexes, 451–452
    N-functionalized amidocuprates, 372, 373–379
  transmetalation, 201, *202*, 205, 819
  tweezer-type, 41–43, 78, *79*
  see also Chiral ligands; Non-chiral ligands
Line widths, NMR spectroscopy, 25–26
^{6}Li NMR spectroscopy, *164*
^{7}Li NMR spectroscopy, *164*
  ^{1}H,X heteronuclear reactions, 166, 181–183
Lipshutz cuprates
  oxidation, 587–589, 592, 595
  synthetic transformations, 778
ortho-Lithiation
  biaryl formation, 590–591, 593, 594
  C–N bond formation, 599
Lithium, Sn–Li exchange, 838, 840, 841
Lithium bislactim ethers, 383, 385
Lithium cyanocuprates, NMR spectroscopy, 171–173
Lithium dialkylcuprates, 776
  cluster reaction mechanisms, 3
  structure, 95–96
t-Lithium dibutylcuprate(I), π-complex intermediate, 184–186
Lithium dimethylcuprate
  addition to extended Michael acceptors, 868–869, 871–872
  allylic alkylation, 605, 610, 613
    acyclic systems, 622, 623
    cyclic systems, 638, 640
  propargyl acetate reaction, 629
Lithium–halogen exchange
  biaryl formation, 590
  organoaluminum transmetalation, 502
Lithium hexamethyldisilazido(*n*-Bu)cuprate, reactivity, 366–367
Lithium organocuprates
  structure, 95–99
  see also Gilman reagents
Lithium trimethylcyanocuprate(III), square–planar structure, *196*, 197
Lower-order cuprates
  acetylene silylcupration, 810
  allylic alkylation, 609
  cross-coupling, 587

Lower-order cuprates (*continued*)
cyanocuprates
cyanosilylcuprates, 813–816
cyanostannylcuprates, 847, 848
Michael addition, 867
structure, 108, 109–110, *111*, 788
homosilylcuprates, 777–778
stannylcuprates, 818, 820, 846, 847, 848
Low-temperature NMR
Cu(I) organocuprate intermediates, 184
diffusion-ordered spectroscopy, 168
phosphoramidite complexes, 205–206, 207–209

Macrocycles
allyl alkylation, 653
organocuprate oxidation, 595–598
Magic numbers, binary Cu–C species stability, 156
Magnesium compounds *see* Organomagnesium compounds
Magnesium–halogen exchange
β-*N*-arylcuprates, 419, 421–423
α-*N*-heteroarylcuprates, 404, 406
β-*N*-heteroarylcuprates, 425
organocuprates with N–Cu bonds, 593, 594, 595, 596
Malachite, combustion of copper acetylides, 147–148
Maleimide, Goldberg reaction, 891
Malonates, phosphoramidite complexes, 202
Malononitrile, Hurtley reaction, 969
Manganese compounds, transmetalation, 444, 447, 449–450, 451
Masked hydroxyl groups, (phenylthio)-cyclopropylsilyl group, 788
Matrix environment, methylcopper formation, 147
Medium ring synthesis, biaryls, 590, 593
p-Menthane-3-carboxaldehyde, chiral auxiliary, 618, 620
Menthone, allylic alkylation, 618, 620
Mercury, electronegativity, 493
Mercury compounds, transmetalation, 444, 493, 495–496
(−)-Mesembrine, transmetalation synthesis, 516
Mesitylamidocuprate, structure, 102–103
Mesitylcopper
organocopper(I) copper(I) aggregates, 58–63
organocuprates, 98–99, 102–103
structure, 30–31, 48–49
γ-Mesyloxy-*α*,*β*-enolate, allylic substitution, 607, 610–611
α-Metalated oxazolidines, synthons, 428
Metalation *see* Transmetalation
Metal carborane complexes, structure, 68–73
closo-Metallacarborane structure, copper carboranes, 69
Metal–ligand repulsion, theoretical models, 284–285
α-Metallo aza enolates, *α*-amino acid synthesis, 383, 385
Metallocycle intermediate, silyl/stannylcupration, 849
α-Metallo imines, *α*-amino acid synthesis, 383, 385
Metallothioneins (MTs), gas-phase Cu(I)/Cu(II) reactivity, 330
Metathesis
allylic silanes, 484, 487, 489
dialkoxyphosphinylfluoromethylcopper, 1008
perfluoroalkenylcopper pre-generation, 994–996
perfluoroalkyl metallic reagents, 1013, 1019
perfluoroalkynylzincs, 1005
2,3,5,6-tetrafluoropyridylcoppers, 1027
Methanides, thermochemistry, 155
Methanol, Cu(II) coordination, 305
Methionine, Cu complex coordination, 306
(*Z*)-2-Methylbut-2-en-1-yl(diphenyl)silyl group, C–Si to C–OH conversion, 783, 784
Methylcopper
formation, 24
synthesis reactions, 39
thermochemistry, 146–147
N-Methyldiphenylmethanimine, transmetalation, 385
Methylene, Cu bonding, 149–150
2-Methyleneoxetanes, Cu-catalyzed coupling, 907
Methyl epijasmonate, transmetalation synthesis, 516
Methyl fluorosulfonyldifluoroacetate, decomposition, 1013
N-Methylglycine, aryl halide coupling, 917, 926
O-Methylhydroxylamine, amidocuprate formation, 374
O-Methylnorephedrine, amidocuprate formation, 369, 371
β-Methyloctalone, *cis*-selective 1,4-addition, 189–191
Methyl perfluoro[2-(fluorosulfonyl)ethoxy]-acetate, difluorocarbene precursors, 1021
Methylphosphidocuprate, structure, 103–104
Michael acceptors
borylation, 496, 500
Cu-mediated/catalyzed addition reactions, 858–872
acceptor-substituted dienes, 858–863, 864

acceptor-substituted enynes, 861, 863–871
acceptor-substituted polyenynes, 871–872
anti-Michael additions, 191, 867
transmetalation, 444, 510, 512, 513
enantioselective 1,4-conjugate additions, 474, 476, 477, 478, 479
protective group-free Michael addition, 457
Microwave titration, methylcopper reactions, 39
Mixed copper–platinum acetylide complexes, photoluminescence, 90
Mixed copper–silver aggregate, structure, 76, 78
Mixed-metal alkynyl complexes, electrochemical stability, 257–258
Mixed organocuprates, dummy ligand effect, 16–17
Mixed silylcuprates, preparation, 779–780
Mixed stannyl(alkyl)cuprates, preparation, 818–820
Modern Ullmann chemistry
C–heteroatom bond formation, 893–943
halogen exchange reactions, 940–942
mechanistic considerations, 942–943
Molecular conductance junctions, copper phthalocyanines, 228
Molecular electronic devices, photoinduced electron transfer, 225–228
Molecular orbitals
ab initio molecular orbital theory, 281
conjugate addition vs S_N2 alkylation, 5–6, 7
diorganocuprate(I) species, 3–5
N-Monoalkylbenzoyloxy amines, amine synthesis, 374–375
Monoalkylcopper reagents
allylic alkylation, 609, 613, 640, 661, 666
cyanocuprates, 620
Mono-allylation, titanacycles, 445, 446
Monoanionic groups, organocopper(I) copper(I) aggregates, 51–63
Monodentate ligands
binding energies, 295
NMR structure, 201–202
Monomer structure, diorganocuprate NMR, 171–173
Mononuclear compounds, electrochemical stability, 255–257
Mononuclear copper sites, Cu proteins, 352–355
Mono(stannyl)cuprates, preparation, 818
Mono-substituted acetylenes, stannylcupration, 836, 838
MP2 method, gas-phase chemistry, 282, 283, 284
MTs (metallothioneins), 330
Multi-component reactions, conjugate addition to lactams, 474, 476, 477
Multi-copper enzymes, type 1 Cu centers, 349
Multinuclear copper sites, Cu proteins, 355–360
Multiple bond systems
extended, 857–879
silylcupration, 817

Nanomaterials, photoinduced electron transfer, 225–228
Naphthalene derivatives, organozirconium transmetalation, 467, 468
Naphthalene radical anions, Cu(I) complex reduction 149
Natural product synthesis
enantioselective conjugate addition, 693, 695, 714–728
transmetalation, 516–518
Nazarov cyclizations
Fischer chromium carbene dimerization, 452
Si-directed, 816
Negeshi reagent
organozinc transmetalation, 478, 479
organozirconium transmetalation, 461–462, 465, 466, 468
Neocuproine complexes
aryl halide coupling, 920
diaryl ether synthesis, 897
diaryl selenide synthesis, 904, 905
Neutral organocuprates
structure, 90–101
heterocuprates, 101–104
NHC *see* *N*-Heterocyclic carbenes
Nickel(0)-catalyzed reactions
cross-coupling, 478, 479
halogen–zinc exchange, 473
Nitriles
enantioselective conjugated hydrosilylation, 747–748
nitrogen enolate cuprate formation, 426
Rosenmund–von Braun reaction, 967–968
α,β-unsaturated, 402
Nitroalkenes, α-cupriooxazolidine reactions, 428, 429
β-Nitroarylcuprates, 423
Nitrogen
C–N bond formation
Cu-mediated cross-coupling, 887–892, 909–938
cross-dehydrogenative coupling, 963–965, 973, 974
hydroamidation, 936–937
N–H heterocycles, 921–922, 923–924, 944
nitrogen enolate cuprates, 426
oxidation, 598–601

Nitrogen–copper bonds, *N*-functionalized organocuprates, 366–382
Nitrogen enolates, cuprate formation, 426–434
Nitro-olefins
  dialkylzinc additions, 487, 489, 490
  enantioselective conjugated hydrosilylation, 747
  unsaturated, 202
Nitrous oxide reductase, multinuclear Cu sites, 357–359
NMR spectroscopy, 163–215
  coupling constants, 166, 171–173, 177–178, 195–198, 206
  dynamic simulations, 206–207
  isotopes, 164–165
  line widths, 25–26
  low-temperature, 168, 184
  quadrupole moments, 164–165
  stannylcuprates, 820
  structure determination
    Cu(I)-catalyzed reactions, 200–209
    cyanosilylcuprates, 778, 790
    intermediate complexes, 170, 184–200
    organocoppers, 169–183
  techniques, 166–168
^{15}N NMR spectroscopy, *164*
  ^{1}H,X heteronuclear reactions, 166
NOE (nuclear Overhauser effect), 166–168, 176, *177*, 181–183
NOESY (nuclear Overhauser effect spectroscopy), 167–168, 176, *177*, 181–183
Non-chiral ligands
  domino reactions, 757–761
  hydrosilylation, 738–741, 749–752
  see also Chiral ligands
Non-conventional complexes, Si- and Ge-containing, 297–298, *299*
Non-linear effect, butylcycloheptanone formation, 200–201
Non-transferable groups, cuprate reagents, 24
Non-transferable ligands *see* Ligands
Non-Ullmann substrates
  C–heteroatom bond formation, 943–965
    cross-dehydrogenative coupling, 963–965
    mechanistic considerations, 961–963
1-Norbornene, Cu interaction, 148
Norephedrine, amidocuprate formation, 369, 371
NQR (nuclear quadrupole resonance), 165
Nuclear magnetic resonance *see* NMR spectroscopy
Nuclear Overhauser effect (NOE), 166–168, 176, *177*, 181–183
Nuclear Overhauser effect spectroscopy (NOESY), 167–168, 176, *177*, 181–183
Nuclear quadrupole resonance (NQR), NMR spectroscopy, 165
Nucleic acid building blocks
  gas-phase Cu(I)/Cu(II) reactivity, 337–340
    nucleobases/nucleosides/nucleotides, 339–340
  Nucleobases
    gas-phase Cu(I)/Cu(II) reactivity, 339–340
    organoboron compound coupling, 956, 957
Nucleophiles
  C-nucleophile coupling, C–C bond formation, 965–973, 974
  enantioselective conjugate addition, 695–700, 701
  N-nucleophile coupling
    aryl/vinyl azide synthesis, 937, 938
    iodonium salts, 944–945
    organobismuth compounds, 947, 949, 951
    organoboron compounds, 954–959, 960
    organolead compounds, 949–950
    organotin compounds, 960–961
  O-nucleophile coupling
    aryl halides, 902
    iodonium salts, 943, 944
    organobismuth compounds, 946–947, 948
    organoboron compounds, 950–953
    organolead compounds, 949
    organotin compounds, 959–960
  P-nucleophile coupling
    iodonium salts, 945–946
    organic halides, 892–893, 938–940, 941
  S-nucleophile coupling, organoboron compounds, 953–954
  stannylcuprates, 821, 822, 959–961
Nucleophilic substitution, alkenyl halides, 12–13

Odd-electron peptide radical cations, dissociation, 330, *331*
Olefination
  Horner–Emmons olefination, 493, 495
  Peterson olefination, 802
Olefins
  Cu(I) π-complex electrochemical synthesis, 248–249
  perfluoroalkylcopper additions, 1015–1016
Oligomers
  glycine, 290–291
  supramolecular aggregation, 179–180
Oligonucleotides, gas-phase Cu(I)/Cu(II) reactivity, 339–340
One-pot reactions, transmetalation, 459, 462, 496, 498
Organic bases, Cu^{+}/Cu^{2+} reactions, 310–318
Organic electronic devices, photoinduced electron transfer, 226
Organoaluminum reagents

Michael addition, 863
transmetalation, 444, 502–508, 517
Organobismuth compounds
non-Ullmann Cu-mediated coupling
N-nucleophiles, 947, 949
O-nucleophiles, 946–947, 948
Organoboron compounds
C–C bond formation, 973–975
non-Ullmann Cu-mediated coupling
N-nucleophiles, 954–959, 960
O-nucleophiles, 950–953
S-nucleophiles, 953–954
transmetalation, 444, 496–501
Organochromium compounds, transmetalation, 444, 450–452
Organocopper aggregates, structure, 25–27, 53–58, 106, *107*
Organocopper(I) compounds
intermediate complex NMR, 184–194
monoanionic group aggregates, 51–63
π-complexes, 170, 184–189
perfluoroalkyl compounds, 113
stoichiometric, 169–171
Organocopper(II) compounds
copper carbene complexes, 131–132
structure, 112, 115–117, 119
Organocopper(III) compounds
Gilman reaction, 112
intermediate complex NMR, 194–200
organocuprate reactions, 849, 850
oxidative addition, 3, 5–6, *7*
reductive elimination, 3–4, 6, *7*, 10, 11, 12
structure, 112–114, 112–119, 116, 117–118
superconductors, 114
thermochemistry, 153–155
Organocopper copper bromide aggregate, structure, 53–58
Organocopper copper iodide aggregate, structure, 106, *107*
Organocopper–titanium Lewis acid complex, 445
Organocuprates
aggregates, 58
supramolecular aggregation, 168, 170, 178–183
alkyl cuprate generation, 456
allylic, 509–510
dummy ligand effect, 16–17
N-functionalized, 365–441
N–Cu bonds, 366–382
α-*N*-substituents, 383–406, 407
β-*N*-substituents, 406, 408–432, 433, 434
kinetic mixed organocuprates, 587–588
oligomers, 179–180
oxidation, 585–602
C–N bond formation, 598–601
cross-coupling, 586–590, 592, 593, 594
dimerization and macrocycle formation, 595–598
higher-order cuprates, 587
lower-order cuprates, 587
structure, 90–111
cyanocuprates, 108–111
higher-order, 108–109, *110*, 778, 788, 790
lower-order, 108, 109–110, *111*, 788
ionic structures, 104–108
neutral organocuprates, 90–101
heterocuprates, 101–104
higher-order, 99, *100*
NMR determination, 169–171
silylcuprates, 788–790
theoretical aspects, 1–21
see also Chiral cuprates
Organogermanium compounds, Cu^+ complex, 297–298
Organolead compounds
non-Ullmann Cu-mediated coupling
N-nucleophiles, 949–950, 951
O-nucleophiles, 949
Organomagnesium compounds
C–C bond formation, 978–980
Cu-catalyzed carbomagnesiation
N-alkynyl carbamates, 567–569
O-alkynyl carbamates, 560
alkynyl silanes, 572
alkynyl sulfones, 564–567
propargyl alcohols, 528, 550–555, 556
oxidation states, 153
see also Magnesium–halogen exchange
Organomanganese compounds, transmetalation, 444, 447, 449–450, 451
Organomercury compounds, transmetalation, 444, 493, 495–496
Organosamarium compounds, transmetalation, 444, 471, 472, 473
Organosilicon compounds
C–C bond formation, 977–978
Cu^+ complex, 297–298
non-Ullmann Cu-mediated coupling, 961
transmetalation, 444, 513–514
Organostannanes
C–C bond formation, 821
N-imination, 377
Organotellurium compounds, transmetalation, 444, 514–516
Organotin compounds
C–C bond formation, 975–977
fluoroalkenyltin reagents, 996
non-Ullmann Cu-mediated coupling
N-nucleophiles, 960–961
O-nucleophiles, 959–960
synthons, 838, 839
transmetalation, 444, 447, 509–512
see also Stannylcuprates

Organotitanium compounds
mixed bi/polynuclear compounds, 258–259, 268–276
transmetalation, 444, 444–447, 448
tweezer-type ligands, 41–43, 78, *79*
Organozinc halides, preparation, 471, 473–474
Organozinc reagents
acyl chloride substitution, 800, 802
Cu-mediated/catalyzed addition, 202, 203, 861, 863
dialkoxyphosphinylfluoromethylcopper preparation, 1008–1012
dialkylzinc reagents, 202, 203, 480–491, 861, 863
fluoroalkenylzinc, 994–996, 998
α-*N*-heteroarylcuprate reactions, 402, 405
oxidation, 592, 595
perfluoroalkynylzinc metathesis, 1005
stannylzincate, 824, 825
transmetalation, 444, 458, 461, 471–493, 494, 517
trifluoromethyl, 1020–1021
Organozirconium compounds
transmetalation, 444
alkyl/alkenyl compounds, 452–462, 463–465
(+)-lepadin B synthesis, 518
zirconacycles, 462, 466–470
Oxabicyclo compounds, desymmetrization, 679
Oxamic esters, from carbamoyl cuprates, 383, 384
Oxazoles, amide coupling, 937
Oxazolidines
α-metalated, 428
nitrogen enolate cuprate formation, 426
vinylstannane–oxazolidine intermediate, 842
Oxazolidones
alkynyl bromide coupling, 933
aryl bromide coupling, 925–926
Oxidation, 585–602
C–C bond formation, 586–598
C–N bond formation, 598–601
Corey's oxidation, 838, 840
Cu(II) coordination in blue proteins, 307–308
Fleming–Tamao oxidation, 487, 489, 499, 501
Glaser alkyne coupling, 585
hydroboration–oxidation/Fleming–Tamao oxidation, 487, 489
single electron oxidation, 73
Tamao–Kumada oxidation, 788, 789
zinc organocuprates, 592, 595
see also Photooxidation
Oxidation potentials
electrochemical stability, 258–267, 275
substituent effects, 261–266
Oxidation states
Cu, 146, 153–155, 280–281
coordination, 281, 300–308
organomagnesium compounds, 153
Oxidative addition
acid chlorides, 13, 14
allene silyl/stannylcupration, 849
Cu(III) intermediate, 3, 5–6, *7*, 850
trans effect, 11–12, 16–17
Oxime ethers, α-cuprioazaenolate reactions, 430, 432
Oximes
aryl halide coupling, 902
nitrogen enolate cuprate formation, 426
Oxiranes, Cu-mediated/catalyzed substitution, 874
α-Oxoamides, from carbamoyl cuprates, 383
Oxygen
C–O bond formation, 883–887, 893–902
heterocycles, 906–908
Oxygenation, tandem alkyne carbocupration, 540, 542

Palladium(0)-catalyzed reactions
conjugate additions, 474
cross-coupling, 478, 479, 559, 561
halogen–zinc exchange, 473, 474
Paramagnetic compounds, NMR spectroscopy, 165–166
Pauli repulsion, Cu complexes, 299–300
PDT (photodynamic therapy), 237–241
Pentafluorobenzoates, allylic alkylation, 616, 617
Pentafluoroethylation, perfluoroalkylcoppers, 1018
Pentafluorophenylcopper
reactions, 1024–1027
structure, 30
synthesis, 24, 1023–1024
Pentametallic titanium(IV)–copper(I)–iron(II)–copper(I)–titanium(IV) compounds, electrochemical stability, 273, 274, *275*
Pentamethylphenylcopper, structure, 28–29
(Pentaphenylcyclopentadienyl)(triphenylphosphine)copper, structure, 63–64
Pentavalent triphenylbismuth diacetate, alcohol *O*-phenylation, 946
Peptides
gas-phase Cu(I)/Cu(II) reactivity, 318, 323–332
isomeric differentiation, 325–326, 330, *331*
primary sequence information, 323–324
radical cations, 328–330, *331*
odd-electron cations, 330, *331*
Peptidic amides, amidocuprate formation, 371
Perfluoroalkanoic acid salts, decarboxylation, 1012

Perfluoroalkenylcoppers, pre-generation, 994–996
Perfluoroalkenylsilver reagents, perfluoroalkenylcopper pre-generation, 994
Perfluoroalkylalkynes, synthesis from copper acetylides, 1005
Perfluoroalkylation, perfluoroalkylcoppers, 1013–1016
Perfluoroalkylcopper reagents, 1012–1023
  difluoromethylation, 1022–1023
  pentafluoroethylation, 1018
  perfluoroalkylation, 1013–1016
  preparation, 1012–1013, 1014
  reactivity, 113
  trifluoromethylation, 1016–1022
Perfluoroalkyl halides, Cu insertion, 1012
Perfluoroalkynes, fluoroalkenylcopper addition, 1000–1001
Perfluoroalkynylcopper reagents, 1004–1007
Perfluoroalkynylzinc reagents, metathesis with Cu(I) halides, 1005
Perfluoroallylcopper reagents, 1004
Perfluorobenzylcopper reagent, 1012
Perfluoropyran derivative, trifluorovinylcopper addition, 1001
(+)-*O*-Permethyltellimagrandin II, intramolecular bond formation, 592–593
Perovskites, catalysts in Ullmann cross-coupling, 894
PES (potential energy surfaces), 283
PET *see* Photoinduced electron transfer
Peterson elimination
  allylsilane configuration, 797, 800
  silyl aldehydes, 807
Peterson olefination, β-hydroxysilane precursors, 802
PFG–SE (pulsed-field-gradient spin-echo), 168
1,10-Phenanthroline complexes
  amide coupling, 935, 936
  aryl halide coupling, 919–920, 921, 926, 967
  C–O bond formation
    alkyl aryl ethers, 897, 899, 900, 901
    O-aryloximes, 902
    heterocycles, 907, 909
Phenanthrolines, amide coupling, 932
Phenols, aryl bromide/iodide Ullmann coupling, 894
Phenylalanine, gas-phase Cu(I)/Cu(II) reactivity, 320–321
Phenylallenes, stannylcupration, 848, 849
N-Phenylanthranilic acid, aryl halide amination, 888
O-Phenylation, organobismuth compounds, 946
Phenylboronic acid, O-nucleophile coupling, 950
Phenylcopper
  donor–acceptor complexes, 40–41, *42*, 47–48
  isolation, 24
  structure, 26–27
  thermochemistry, 151–153
Phenylcopper acetylides, thermochemistry, 147
Phenyldimethylsilylcuprates, 777–778
  propargyl acetate reactions, 797, 799, 800
Phenyldimethylsilylcyanocuprates, preparation, 778–779
Phenyldimethylsilyllithium
  homosilylcuprate preparation, 777–778
  phenyldimethylsilyl(tributylstannyl)cuprate preparation, 819–820
Phenyldimethylsilyl(tributylstannyl)cuprate, preparation, 819–820
Phenylglycine, amidocuprate formation, 368–371
Phenyllead triacetate, N-nucleophile coupling, 949–950
Phenyl-substituted allenes, silylcupration, 812–813
(Phenylthio)cyclopropylsilyl group, C–Si to C–OH conversion, 788
Pheromones
  acetylene carbocupration, 534–539
  alkynyl thioether carbocupration, 563
Phosphazene bases, C–S bond formation, 902, 904
Phosphidocuprates, stability, 366
Phosphines, C–P bond formation, 938
Phosphinoxazoline (PHOX), aryl bromide coupling, 939–940
Phosphonates, vinyliodonium salt coupling, 945–946
Phosphoramides, aryl halide coupling, 926
Phosphoramidite ligands
  allylic alkylation, 480, 483, 670, 676, 679
  conjugate addition
    asymmetric, 506
    enantioselective, 474, 476–478, 487, 490, 492–493, 694–695
  NMR structure, 200, 201–209, 694–695
Phosphorus, C–P bond formation, 892–893, 938–940, 941
Photoactivation, titanium dioxide, 235–236
Photobiochemistry
  copper porphyrins/pthalocyanines, 237–241
  see also Biochemistry
Photocatalysts, copper porphyrins/ phthalocyanine dyes, 235–237
Photochemistry
  copper porphyrins/pthalocyanines, 217–245
    applied photochemistry, 225–232
    electron transfer systems, 219–232
    photobiochemistry, 237–241
    photochemical reactions, 224–225

Photodegradation, copper porphyrins/ phthalocyanine dyes, 230–235
Photodissociation, gas-phase, 285–289
Photodynamic therapy (PDT), copper porphyrins/pthalocyanines, 237–241
Photoelectrodes, molecular-based, 228
Photofragmentation processes, gas-phase, 281, 285–289
Photoinduced electron transfer (PET)
applied photochemistry, 225–232
heteroligand systems, 222–224
see also Electron transfer systems
Photoionization
Cu oxidation states, 153
phenylcoppers, 152
Photoionization spectroscopy, gas-phase, 285
Photoluminescence, mixed copper–platinum acetylide complexes, 90
Photooxidation, copper porphyrins/ phthalocyanine dyes, 230–232
Photoreduction, copper porphyrins/ phthalocyanine dyes, 230–232
Photovoltaic cells, copper porphyrins/ phthalocyanines, 228–230
PHOX (phosphinoxazoline), 939–940
Phthalimide, Goldberg reaction, 891
Phthalocyanine, structure, 218–219
Phthalocyanine Blue BN, structure, 232
Phthalocyanine Green G, structure, *233*
Physical measurements, silylcuprate structure, 790
Pinacolborane
domino reactions, 761–763
hydrocupration, 767, 768–769
NHC–Cu complexes, 402, 404
Pinwheel ligand array, copper carboranes, 71, *72*
Piperidine rings, amidocuprate formation, 369
Piperidines, α-(N-carbamoyl)alkylcuprate reactions, 391, 392
Piperidones, unsaturated, 202
Piperizines, α-(N-carbamoyl)alkylcuprate reactions, 399
Plastocyanin
mononuclear Cu sites, 352, *353*
type 1 Cu centers, 349, *350*
Platinum compounds
mixed bi/polynuclear compounds, 257–258, 268, 269
mixed copper–platinum acetylide complexes, 90
PMHS (polymethylhydridosiloxane), 632, 634
^{31}P NMR spectroscopy, *164*, 203, 206–209
Polarization, gas-phase chemistry, 281–282
Polycarbonate, copper porphyrins/phthalocyanine dyes, 231–232
Polycopperalkanes, thermochemistry, 153
Polycuprioalkanes (polycopperalkanes), thermochemistry, 153
Polydentate ligands, Cu^+ interaction, 295, 301
Polyenynes, Michael addition, 871–872
Polyfluorinated arenes, arylation, 1024–1025
Polymers, copper porphyrins/phthalocyanine dyes, 230–232
Polymer-supported synthesis, biaryls, 593–594
Polymetal anions, enthalpy of formation, 152
Polymetalated alkenes
silylcupration, 805–807
stannylcupration, 836, 837
Polymethylhydridosiloxane (PMHS), Cu(I) hydride formation, 632, 634
Polynuclear compounds
electrochemical stability, 261–276
structure, 25
Polysaccharides, gas-phase Cu(I)/Cu(II) reactivity, 335–337
Porphyrin
N-confused, 117
structure, 218–219
Potassium cyanocuprate, structure, 111
Potassium hexacyanoferrate, Rosenmund–von Braun reaction, 968
Potential energy surfaces (PES), theoretical models, 283
Pre-generation
perfluoroalkenylcoppers, 994–996
perfluoroalkylcoppers, 1015, 1018, 1019
Primary alcohols, arylbismuth arylation, 946
Primary alkyl amines, arylation, 914, 915
Primary sequence information, peptides, 323–324
Prion protein (PrP), gas-phase Cu(I)/Cu(II) reactivity, 330, 332
Privileged ligand structures, phosphoramidites, 202, 203
Prochiral substrates
allylic alkylation, 669–675
β-substituents, 672, 674–675
di- and trisubstituted, 669–672, 673
L-Proline
amide intermolecular arylation, 925
aryl halide coupling, 917, 920, 938, 970, 971
C–P bond formation, 940, 941
Propargyl acetates
α-hydroxyallene formation, 632, 634
lithium dimethylcuprate reactions, 629
macrocyclic, 653
silylcuprate substitution, 797, 799, 800
Propargyl alcohols
carbomagnesiation, 528, 550–555, 556
enantioenriched, 629
stannylcupration, 842, 844
Propargyl amines
silylcupration, 807–808

stannylcupration, 842
Propargyl aziridines, stereospecific opening, 635
Propargyl carbonates, organoboron transmetalation, 499, 501
Propargyl compounds
α-(N-carbamoyl)alkylcuprate substitution, 388, 390
carbocupration, 543–545
leaving group displacement, 629–636
Propargyl epoxides
enantioselective allylic alkylation, 676–678
α-hydroxyallene formation, 631
Propargyl ethers, stannylcupration, 842
Propargyl halides
α-(N-carbamoyl)alkylcuprate substitution, 388, 390
β-cuprio-α-aminocarboxylate reactions, 416, *417*
stannylcuprate substitution, 831
Propargyl mesylates, β-cuprioamide reactions, 413
Propargyl phosphates, S_N2' additions, 461–462, 465
Propargylsilane, silylcupration, 803
Propargyl sulfides
silylcupration, 807–808
stannylcuprate addition, 832
Propargyl tosylates, β-cuprioamide reactions, 413
Prostaglandin misoprostol, large-scale synthesis, 453, 455
Prostaglandins
carboalumination synthesis, 503, 505
diastereospecific allylic alkylation, 622, 626, 645
organozirconium transmetalation, 458, 461
Protected amino acids
β-cuprio-α-aminocarboxylate preparation, 415, 416
9-fluorenylmethoxycarbonyl, 418
Protective group-free Michael addition, 457
Proteins
biological Cu centers, 348–352
Cu proteins
blue Cu proteins, 280, 307–308, 349
mononuclear Cu sites, 352–355
multinuclear Cu sites, 355–360
gas-phase Cu(I)/Cu(II) reactivity, 330–332
Protonation
deprotonation, 309–310
fluoroalkenylcoppers, 998
Protonolysis, mesitylcopper, 58–59, 63
Proton transfer, Cu^{2+} complexes, 303
PrP (Prion protein), 330, 332
Pseudoephedrine, amidocuprate formation, 368–370
Pulsed-field-gradient spin-echo (PFG–SE), 168
Purvalanol A, synthesis, 600–601
Pyrazoles
amide coupling, 936, 937
arylation, 921, 950
3-Pyridazinones, arylboronic acid coupling, 955
Pyridine
Cu complex dissociation, 285, 286–289
Cu(III) intermediate stabilization, 198
Pyridines
α-N-heteroarylcuprate reactions, 402
β-N-heteroarylcuprate reactions, 425
perfluoroalkylation, 1014
Pyridin-2-one, Goldberg reaction, 891
Pyridones
alkylzinc halide addition, 492
organozinc conjugate addition, 476, 478
2-Pyridones, arylboronic acid coupling, 955
2-Pyridylcopper, structure, 39
N-Pyridylsulfonyl imines, alkylzinc halide addition, 492
(2-Pyridyl)(trimethylsilyl)methylcopper, structure, 37, *38*
Pyrimidine, Cu^+ affinity, *292*, 294
Pyrimidines, perfluoroalkylation, 1014
Pyrroles
allylic alkylation, 635, 636
arylation, 921
α-(N-carbamoyl)alkylcuprate reactions, 391, 393
vinyl halide coupling, 935–937
Pyrrolidine-2-phosphonic acid, C–P bond formation, 940, 941
3-Pyrrolidin-2-ones, α-(N-carbamoyl)alkylcuprate reactions, 391, 393
Pyrrolines, allylic alkylation, 635
3-Pyrrolines, α-(N-carbamoyl)alkylcuprate reactions, 391, 393
Pyrrolizidine, α-(N-carbamoyl)alkylcuprate reactions, 394, 397
Pyrrolizidinones, α-(N-carbamoyl)alkylcuprate reactions, 391, 393

Quadrupole moments, Cu NMR spectroscopy, 164–165
Quaternary phosphonium salts, C–P bond formation, 893
Quaternary stereocenters
directed-non-directing leaving group, 484, 487
enantioselective conjugate addition, 700–707
all-carbon, 701, 725, 728
transmetalation, 478, 480, 484, 485, 508
NHC–Cu complex reactions, 401–402

Quinazolinones, β-*N*-arylcuprate reactions, 419
Quinolines
  β-*N*-arylcuprate reactions, 419
  α-*N*-heteroarylcuprate reactions, 402
Quinolizidines, α-(*N*-carbamoyl)alkylcuprate reactions, 391, 392
Quinones, photoinduced electron transfer, 220–221

Radiation chemistry, Cu–CO_2 complexes, 158
Radical cations, peptides, 328–330, *331*
Rapid injection NMR (RI–NMR), 168, 191–192, 193–194
Rate constants, π-complex intermediate formation, 192–193
Reaction pathways, silylcuprate substitution, 797
Reaction rates, intermediate complex formation, 192–193, 198
Reactivity
  C–H bond activation, 114, 115
  C–Si to C–OH conversion, 785
  cyanocuprates, 17–18, 169
  gas-phase Cu(I)/Cu(II) bioreactivity, 318–340
    binary mixtures, 319–326
    protein interactions, 330–332
    ternary complexes, 326–330
  iodocuprates, 169
  mixed silylcuprates, 779
  oxidation states, 280
  perfluoroalkylcoppers, 113
  theoretical aspects, 1–21
  see also Enthalpies of reaction
Rearrangements
  3,3-rearrangement, 935, 936
  Baeyer–Villiger rearrangements, 484, 485
  Claisen rearrangements, 901–902, 903
  Curtius rearrangements, 484, 485
Reduction
  alkynyltelluride, 514
  chemoselective Stryker's reagent, 735–738
  conjugate reductions, 402, 403
  Cu complex coordination, 306
  zirconocene dichloride, 459, 463
Reduction potentials, electrochemical stability, 258–260, *264*, 268, *269*, *272*, 273, *275*
Reductive aldol reactions, domino reactions, 758–766, 768
Reductive coupling
  allylic alkylation
    acyclic systems, 638, 639
    cyclic systems, 640, 642
    enantioselective, 677–678, 680
    regioselective, 606, 608
Reductive elimination
  π-allylcopper(III), 11, 12
  amidocuprate formation, 371, 377
  silyl/stannylcupration, 849, 850
  trialkylcopper(III) species, 3–4, 6, *7*, 10
Regiochemistry
  allenes
    silylcupration, 811, 812, 813, 814–816
    stannylcupration, 846–848, 849
  1,1-dimethylallene, 846
  β-silylketone addition, 786, 787
Regioselectivity
  1-alkyne stannylcupration, 835–836, 837, 841
  allylic alkylation, 608–610
  N-carbamoyl(R_T)cuprates, 380–382
  Cu(III) intermediates, 198
  epoxide cleavage, 834
Remazol Turquoise Blue 15, 233–234
Remote conjugate addition, organocopper(III) species, 6, *7*
Residual ligands, mixed silylcuprate preparation, 779
Resolution *see* Kinetic resolution
Reversible reactions, alkyne stannylcupration, 836
Rieke zinc
  organozinc halide preparation, 473
  zinc organocuprate oxidation, 592, 595
Ring formation
  4-*exo*-ring closure, 908
  intermediate vinylcuprates, 809
  stannylcupration, 840, 841
Ring-opening alkylation
  aziridines, 622, 629
  cyclopentadiene monoepoxides, 644–646
  cyclopenta[*b*]furanones, 646–647
  cyclopentenediol derivatives, 643–644
  epoxides, 10, 610, 622, 625, 628
    transmetalation, 444, 509–510, 514–516
RI–NMR (rapid injection NMR), 168, 191–192, 193–194
ROESY (rotating-frame Overhauser effect spectroscopy), 167
Rosenmund–von Braun reaction, aromatic nitrile preparation, 967–968
Rotating-frame Overhauser effect spectroscopy (ROESY), 167
Ruthenium compounds, α-(*N*-carbamoyl)alkylcuprate reactions, 391, 393

Saccharides
  gas-phase Cu(I)/Cu(II) reactivity, 332–337
    cross-ring cleavage, 334–335
    doubly-charged species, 334
    polysaccharides, 335–337
Samarium compounds, transmetalation, 444, 471, 472, 473

Samarium(II) iodide, organosamarium compound preparation, 471, 472
Sandwich structures
binary Cu–C species, 156
(3)-1,2-dicarbollide ligand, 256–257
slipped-sandwich structures, 65
see also Structure
Sanguiin H-5, intramolecular bond formation, 593, 594, 595, 596
Schwartz reagent, hydrozirconation, 452–453, 456–457, 459
Secondary alcohols
arylbismuth arylation, 946
C–O bond formation, 886–887
Secondary allylic acetates, silylcuprate substitution, 797, 799
Secondary phosphines, C–P bond formation, 938
Second essential intermediates, Cu(III) organocuprates, 194
(R)-Segphos complex, organoboron transmetalation, 499, 501
cis-Selective 1,4-addition, β-methyloctalone, 189–191
Selective transfer, alkyl(silyl)cuprate preparation, 779–780
Selenium, C–S bond formation, 887, 904, 906
Selenols, oxidation, 906
Self-diffusion coefficients, diffusion-ordered spectroscopy, 168, 181
Sequential binding energies, theoretical models, 282
Serine, β-cuprio-α-aminocarboxylate preparation, 415, 416
Shape correction factors, supramolecular aggregation, 179–180
Silicon, electronegativity, 509
Silicon compounds *see* Organosilicon compounds
Silver acetylides
mixed copper–silver aggregate, 76, 78
thermochemistry, 155
Silver compounds
diarylargentate structure, 91, *92*
NHC complexes, 480, 481–482, 508
perfluoroalkenylsilver reagents, 994
erythro-β-Silyl alcohol, alkene synthesis, 807
Silyl aldehydes, *t*-butyldiphenylsilyl reactions, 807
Silylated allenes, silylcuprate substitution, 799
Silylated 1,4-dienes, silylcupration, 810, 816–817
Silylation, fluoroalkenylcoppers, 999
β-Silyl-carbonyl compounds, 790–797
transmetalation, 513
Silylcuprates, 776, *791*
C–Si to C–OH conversion, 781–788
hindered, 780
mixed, 779–780
preparation, 777–781
reactions, 381–382, 790–817
structure, 788–790
synthons, 790, *791*
β-Silyl enolates, *anti*-attack, 793, 795–796
Silyl enol ethers
conjugate additions, 791, 792
Cu(III) organocuprate intermediates, 194
transmetalation, 449–450, 451, 471, 473, 474, 476, 517–518
β-Silyl esters, conjugate additions, 793, 794, 795
β-Silyl ketones
enantiopure, 793, 794
reactions, 786, 787, 791, 792, 793, 794
stereochemistry, 790, 792
transmetalation, 510, 511
N-Silyl-*O*-methylnorephredrine, amidocuprate formation, 369, 371
Silylstannanes
cleavage, 780
transmetalation, 510, 511
2-Silylvinylstannanes, preparation, 838, 839
Single electron oxidation, copper carboranes, 73
Single electron transfer, allene racemization, 632
Singlet oxygen production, copper porphyrins/pthalocyanines, 237–241
Skipped dienes, transmetalation, 457, 460, 496, 499
Slipped-sandwich structures, cyclopentadienylcopper(I) compounds, 65
S_N2' addition
allylic phosphates, 461–462, 465, 508–509
diorganozincs, 484, 485
S_N2 alkylation
alkyl halides, 5, 9–10
asymmetric allylic alkylation, 604–605, 606–607, 608
acyclic systems, 629, 644, 655, 659, 662, 663
cyclic systems, 643–644, 645–646, 647–649, 662, 663
enantioselective, 676, 677
vs conjugate addition, 5–10
S_N2 allylation
α-allenic amines, 385
Gilman cuprates, 169, *170*
S_N2 substitution
amine synthesis, 373, 374–375
anti-selective, 379, 382, 401, 799–800
N-carbamoyl(R_T)cuprates, 379–382
β-cuprio-α-aminocarboxylate preparation, 415, 416
α-cuprionitriles, 428
Gilman cuprates, 169, *170*

S_N2 substitution (*continued*)
α-*N*-heteroarylcuprates, 404, 407
β-*N*-heteroarylcuprates, 424
β-nitroarylcuprates, 423
organozirconium transmetalation, 461–462, 465
silylcuprates, 797, 799–800
S_N2'-(1,3)-substitution, 872–873
S_N2''-(1,5)-substitution, 872–874
syn-selective, 379, 401
Solar energy, copper porphyrins/ phthalocyanines, 228–230
Solomon equations diorganocuprate NMR, 176
Solvation
binding energies, 298–300
Cu^+ Jahn–Teller effects, 282
Solvation indices, diffusion-ordered spectroscopy, 180–181
Solvent effects, allylic alkylation, 609
Solvent-separated ion pairs (SSIPs)
cyanocuprates, 109, *110*
diorganocuprate NMR, 173–178
Gilman reagents, 2
organocuprates, 104–105
Sonogashira coupling, alkyne preparation, 447, 448, 589, 965–967
(−)-Sparteine, amidocuprate formation, 367
Spectroscopic measurements, copper carbonyls, 158
Sphos, organoboron transmetalation, 499, 501
Spirocycles, organozirconium transmetalation, 470
Spiro diepoxide, allene oxidation, 631
Square–planar structure, Cu(III) σ-complex, 196–197
SSIPs *see* Solvent-separated ion pairs
Stabilization
Cu(III) intermediate NMR, 198
dialkoxyphosphinyldifluoromethyl reagents, 1009, 1010
electrochemistry stability, 255–276
intramolecular carbocupration, 572–576
intramolecular coordination, 34–39
1-Stannyl-1-alkoxyalkenes, formation, 841
Stannyl(alkyl)cuprates, preparation, 818–820
Stannylcopper derivatives
conjugate additions, 824
stannylcupration, 817, 835–836, 837
Stannylcuprates, 776, *822*
preparation, 817–821
methods, 818–821
reactions, 821–848
limitations, 835–836, 837
mechanisms, 848–850
structure, 820–821
transmetalation, 819, 821, 828, 838, 839, 845–847, 848
see also Organotin compounds
Stannylenoic acids, formation, 829
β-Stannylenones, stannylcuprate conjugate additions, 823, 824
β-Stannylesters, stannylcuprate conjugate additions, 822–824
β-Stannylketones
stannylcuprate conjugate additions, 822–824
transmetalation, 510, 511
Stannyllithium derivatives
reactivity, 824, 825
stannylcuprate preparation, 818, 819
stannylmetalation, 817
Stannylmetalation, C–Sn bond formation, 817, 821
(*E*)-1-Stannylvinyl ethers, formation, 841
(*Z*)-β-Stannylvinyl ketones
formation, 845
Sn–Li exchange, 838, 840, 841
(*Z*)-β-Stannylvinyl tosylates, Sn–Li exchange, 838, 840, 841
Stannylzincate reagents, reactivity, 824, 825
Stereochemistry
allylic alkylation, 605, 610
silylcuprate conjugate additions, 793, 794
β-silylketone additions, 786, 787, 790, 792
stannylcupration, 822, 824, 828–829, 845, 847
Stereoselectivity
allene silylcupration, 811–813
allylic alkylation, 605–606, 610–612
conjugate addition
Cu-catalyzed, 700–707, 725, 728
to lactams, 474, 476, 477
Cu(III) intermediates, 198
syn-addition, 803, 847, 848
Steric hindrance
N-carbamoyl(R_T)cuprates, 381
endocyclic restriction test, 373, 375
organozirconium transmetalation, 456, 457
Steroids, cyclobutane-containing, 493, 495
Stille coupling
arylstannanes, 821, 976–977
vinylstannanes, 821, 838, 840, 841
Stoichiometry
C–O bond formation, 953
organocopper(I) compounds, 169–171
stannylcuprates, 820, 821
Stork–Colvin reaction, *t*-butyldiphenylsilyl group, 807
Structure, 23–143
alkyl- and arylcopper(I) compounds, 27–51
donor–acceptor complexes, 39–51
donor-base-free compounds, 27–34
intramolecularly coordinating substituents, 34–39
copper(I) acetylides, 73–90
copper carbene complexes, 119–133
copper carborane compounds, 68–73

cyanocuprates, 108–111
cyclopentadienylcopper(I) compounds, 63–68
N-functionalized amidocuprates, 366–367
organocopper(I)copper(I)-monoanionic group aggregates, 51–63
organocopper(II)/organocopper(III) compounds, 112–119
organocuprates, 90–111
ionic structures, 104–108
neutral heterocuprates, 101–104
neutral organocuprates, 90–101
NMR determination, 169–183
silylcuprates, 788–790
stannylcuprates, 820–821
vinylcuprate intermediate, 803
see also Sandwich structures
Structure–activity relationships, organocuprates, 105
Stryker's reagent
Cu(I) hydride reagents, 735–738
NHC–Cu complexes, 402, 403
Styrenes, silylcupration, 817
Substituents
intramolecularly coordinating, 34–39
N-functionalized organocuprates
α-*N*-substituents, 383–406, 407
β-*N*-substituents, 406, 408–434
oxidation potentials, 261–266
Substitution
acetylene stannylcupration, 836, 838
alkyl-substituted alkyne carbocupration, 539–550
allenes
silylcupration, 811–813, 812–813
stannylcupration, 846–847, 848
Cu-mediated/catalyzed, 872–874
β,β-dialkyl-substituted enones, 383, 385
disubstitution
alkyne silylcupration, 804
enones, 383, 385, 478, 480
perfluoroalkylcoppers, 1014–1015
γ-substitution, 379–380, 382
α-halo-substituted fluoroalkenylcoppers, 997–998
3-halo and 3-sulfonyl-substituted 2-enones, 832–834
α-*N*-heteroarylcuprates, 404, 406
heteroatom-substituted alkyne carbocupration, 557–572
hydroxy-substituted vinylallenes, 874
N-substitution
arylcuprates, 419, 420, 421–423
o-bromoanilines, 927, 929
heteroarylcuprates, 419, 423–426
silylcuprates, 797–803
stannylcuprates, 831–834
substrates for allylic alkylation, 669–675
see also Allylic substitution; Electrophilic substitution; Nucleophilic substitution; S_N2 substitution
Substrates
allylic alkylation
diastereospecific methods, 610–612, 614–616, 617, 618
prochiral substrates, 669–675
regioselective, 608
enantioselective conjugate addition, 695–700, 701
non-Ullmann C–heteroatom coupling, 943–955
substitution reactions, 872–874
Succinimide, Goldberg reaction, 891
Sulfonamides
amidocuprate formation, 371
aryl halide coupling, 926
diaryl ether synthesis, 894
organoboron compound coupling, 954
organolead compound coupling, 950, 951
organotin compound coupling, 960–961
Sulfonimidamides, aryl halide coupling, 926
3-Sulfonyl-substituted 2-enones, stannylcuprate substitution, 832–834
Sulfoximines
N-arylation, 958–959
aryl halide coupling, 926
Sulfur
C–S bond formation, 887, 902, 904, 905
heterocycles, 908–909
Sultam, chiral auxiliary, 202
Superconductors, organocopper(III) compounds, 114
Superoxide dismutase, Cu,Zn, 350, *351*, 354–355
Supramolecular aggregation
dialkylcuprates, 170, 178–183
diffusion-ordered spectroscopy, 168, 179–180
Surfaces of metallic copper
adsorption, 150, 157
thermochemistry, 148, 150
undercoordinated (410) surface, 156
Suzuki coupling, S_N2' addition of diorganozincs, 484, 486
Suzuki–Miyaura reaction
C–C bond formation, 973–974
organoboron transmetalation, 499, 501
Symmetrical coupling, fluoroalkenyl halides, 992–994
syn-addition
mechanisms, 848–850
pentafluorophenylcopper, 1025
silyl–copper, 803
stannyl–copper, 836, 837, 838, 839, 846, 847
syn-reduction, alkynyl telluride, 514

syn-S_N2' substitution
  N-carbamoyl(R_T)cuprates, 379
  NHC–Cu complexes, 401
Synthesis
  diastereospecific allylic alkylation, 622, 624–627
  diversity-oriented, 693
  intramolecular carbocupration
    carbocycles, 572–576
    heterocycles, 576–579, 580–581
  natural products
    enantioselective conjugate addition, 693, 695, 714–728
    transmetalation, 516–518
  prostaglandin misoprostol, 453, 455
  Si-synthons, 790, *791*
  Sn-synthons, 838, 839
Synthons
  organotin compounds, 838, 839
  silylcuprates, 790, *791*

T_2 filters, diffusion-ordered spectroscopy, 204
Tacrolimus *see* FK-506
TADDOL (tetraaryl-1,3-dioxolane-4,5-dimethanol), 200–201, *202*
Tamao–Kumada oxidation, C–Si to C–OH conversion, 788, 789
Tandem reactions
  alkyne carbocupration–oxygenation, 540, 542
  enantioselective conjugate addition, 693, 707–714
  hydrozirconation–β-elimination–hydrozirconation, 456–457, 458
Tellurium ate compounds, transmetalation, 444, 514–516
Temperature-dependent interconversion, phosphoramidite complexes, 208–209
Terminal alkynes
  carboalumination, 502–503
  silylcupration, 803
Terminal allenes, silylcupration, 816–817
Ternary complexes, Cu(II), 281, 326–330
Tertiary alcohols
  arylbismuth arylation, 946
  C–O bond formation, 886–887
Tertiary alkyl phosphites, C–P bond formation, 892
Tertiary allylic acetates, silylcuprate substitution, 797, 799
Tertiary phosphines, C–P bond formation, 938–939
Tertiary stereocenters, NHC–Cu complex reactions, 402
Tetraalkylcopper(III) species, NMR spectroscopy, 198
Tetraaryl-1,3-dioxolane-4,5-dimethanol (TADDOL), thiol–TADDOL complexes, 200–201, *202*
1,1,2,2-Tetracuprioethane, theoretical, 153
2,3,5,6-Tetrafluoropyridylcopper reagents, 1027–1028
3,4,7,8-Tetrafluoro-1,2,5,6-tetrakis(trifluoromethyl)anthracene, structure, 1000–1001
3,4,5,6-Tetrafluoro-1,2,7,8-tetrakis(trifluoromethyl)phenanthrene, structure, 1000–1001
Tetrahydrolipstatin, synthesis from vinylsilanes, 807, 808
Tetrahydropyrazines, amide coupling, 937
Tetrametalated alkene, stannylcupration, 836, 837
Tetrametallic bis(alkynyl) titanocene complexes, electrochemical stability, 271–273
Tetramethylethylenediamine (TMEDA)
  aryl halide coupling, 919–920, 979
  iodonium salt coupling, 945–946
2,2,6,6-Tetramethylheptane-3,5-dione (TMHD), diaryl ether synthesis, 897, 899
3,4,7,8-Tetramethyl-1,10-phenanthroline, vinyl ether synthesis, 901
Tetranuclear iron(II)–silver(I)–copper(I)–titanium(IV) complexes, electrochemical stability, 273–276
Tetraphenyllead, O-nucleophile coupling, 949
Tetravinyltin, O-nucleophile coupling, 959–960
Tetrazoles, arylation, 921
Theoretical aspects
  organocopper reactions, 1–21
  silylcuprate structure, 790
Thermal stability
  2,3,5,6-tetrafluoropyridylcoppers, 1027
  trihydrocarbyl copper species, 154
Thermochemistry, 145–162
  binary Cu–C species, 155–156
  Cu complexes
    CO_2, 158
    CO, 156–158
  hydrocarbyl derivatives, 146–155
Thermodynamic control
  silylcupration, 816
  stannylcupration, 846
Thexyldimethylsilylcuprate, conjugate additions, 793
Thiazoles, α-*N*-heteroarylcuprate reactions, 402
2-Thienylcopper, structure, 29
Thienylsilyl group, C–Si to C–OH conversion, 788
Thioesters, substitution of acyl electrophile, 13
Thiolesters, α-(*N*-carbamoyl)alkylcuprate addition, 388, 389

Thiols, NHC–Cu complex reactions, 402
Thiol–TADDOL ligands, NMR structure, 200–201, *202*
Thiophenes, perfluoroalkylation, 1014
Thiophenols, C–S bond formation, 902, 904
Thiouracil, binding energy, 291, *292*
Thiourea, Cu complex coordination, 303
Thorpe–Ingold effect, alkenylsilanes, 810
Three-component coupling protocol, organozirconium transmetalation, 457–459, 461
L-Threonine, β-cuprio-α-aminocarboxylate reactions, 418
Thymine, organoboron compound coupling, 956, 957
Time-dependent interconversion, Cu(III) intermediate complexes, 199–200
Tin, electronegativity, 509
Tin compounds *see* Organotin compounds
Tin–lithium exchange, (*Z*)-β-stannylvinylketones/tosylates, 838, 840, 841
Titanacycles, mono-allylation, 445, 446
Titanacyclopentene, transmetalation preparation, 445, 446
Titanium compounds *see* Organotitanium compounds
Titanium(IV)–copper(I) compounds, electrochemical stability, 258–259, 271–276
Titanium dioxide, copper porphyrins/phthalocyanine photocatalysts, 235–237
TMEDA *see* Tetramethylethylenediamine
TMHD (2,2,6,6-tetramethylheptane-3,5-dione), 897, 899
TMSCl *see* Trimethylsilyl chloride
TMSCN (trimethylsilyl cyanide), 194
TNS–Tf (*N*-trifluoromethyl-*N*-nitrosotrifluoromethanesulfonamide), 1018–1019
o-Tolylcopper, structure, 48, *49*
N-Tosylimines, NHC–Cu complex reactions, 402
trans effect, oxidative addition, 11–12, 16–17
Transferable ligands *see* Ligands
Transition metals
  transmetalation
    first series, 444–452
    second series, 452–470, 518
Transmetalation, 443–525
  catalytic complexes, 201, *202*, 205, 819
  α-cuprioazaenolates, 430, 433
  dialkoxyphosphinyldifluoromethylcopper preparation, 1008–1009
  directed ortho-metalation, 378–379
  N-functionalized organocuprates, 383–400
    β-*N*-arylcuprates, 423
    α-(*N*-carbamoyl)alkylcuprates, 394, *396*
    β-cuprio-α-aminocarboxylates, 417
    β-cuprio-*N*-carbamoyl reagents, 410
    α-*N*-heteroarylcuprates, 404, 406, 407
  group IIB, 471, 473–496
    organomercury compounds, 444, 493, 495–496
    organozinc compounds, 444, 458, 461, 471–493, 494, 517
  group IIIA, 496–508
    organoaluminum compounds, 444, 502–508, 517
    organoboron compounds, 444, 496–501
  group IVA, 509–514
    organosilicon compounds, 444, 513–514
    organotin compounds, 444, 447, 509–512
    stannylcuprates, 819, 821
    vinylstannanes, 447, 509–510, 512, 828, 838, 839, 845–848
  group VIA, 514–516
    organotellurium compounds, 444, 514–516
  lanthanide series, 471, 472, 473
  natural product synthesis, 516–518
  transition metals
    first series, 444–452
    second series, 452–470, 518
Trapping, intermediate vinylcuprates, 836
Trialkylalanes
  allylic fluoride displacement, 660–661
  conjugate addition, 503, 505, 506, 508
Trialkylcopper(III) species
  Gilman reaction, 112
  NMR spectroscopy, 198
  reductive elimination, 3–4, 6, 7, 10
Trialkyl(perfluoroalkyl)silanes, perfluoroalkylcopper preparation, 1013
Trialkyl phosphites, C–P bond formation, 893
Triarylboroxine, O-nucleophile coupling, 951–952
Triazoles, arylation, 921, 950
Tributylphosphine
  Cu(III) intermediate stabilization, 198
  diastereoselective allylic alkylation, 663
3-Tributylstannylacrylates, substitution, 833
Tributylstannyl(butyl)cyanocuprate, allylstannane preparation, 831, 832
Tributylstannylcuprates, reactions, 818, 834
Tributylstannyllithium
  cyanostannylcuprate preparation, 847, 848
  phenyldimethylsilyl(tributylstannyl)cuprate preparation, 819–820
  preparation, 818
4-Tributylstannylpyrrolin-2-ones, stannylcuprate conjugate additions, 828
Tributyl(vinyl)tin, O-nucleophile coupling, 960
Trienynoates, Michael addition, 872
Trifluoromethylalkanoic acid salts, decarboxylation, 1016–1017

Trifluoromethylation
1,1-dibromoalkenes, 1022
perfluoroalkylcoppers, 1016–1022
Trifluoromethylcopper
in situ generation, 1016
metathesis, 1019–1020
Trifluoromethylmetallic reagents, from dihalodifluoromethanes, 1020–1021
N-Trifluoromethyl-*N*-nitrosotrifluoromethanesulfonamide (TNS–Tf), 1018–1019
2-(Trifluoromethyl)pyridines, preparation, 1018
Trifluoromethyltrialkylsilanes, trifluoromethylation, 1017–1018
(*E*)-α-Trifluoromethyl-α,β-unsaturated esters, synthesis, 1021–1022
Trifluorovinylcopper
addition to perfluoro-2-butyne, 1000–1001
silylation, 999
Trifluorovinyl ketones, fluoroalkenylcopper acylation, 998
Trifluorovinylzinc reagent, Cu(I)-mediated acylation, 998
Trihydrocarbyl copper species, thermal decomposition, 154
Trimetallic titanium(IV)–copper(I)–iron(II) compounds, electrochemical stability, 273, 274, *275*
Trimethylalane, conjugate addition, 506
Trimethylchlorosilane *see* Trimethylsilyl chloride
Trimethylsilyl chloride (TMSCl)
amidocuprate formation, 371
α-(*N*-carbamoyl)alkylcuprate reactions, 391
conjugate additions, 791, 792
Cu(III) organocuprate intermediates, 194
transmetalation
organomanganese reagents, 449–450, 451
organosamarium compounds, 471, 473
organotitanium compounds, 444–445
organozinc compounds, 474, 476, 478
Trimethylsilyl cyanide (TMSCN), Cu(III) organocuprate intermediates, 194
Trimethylsilyllithium, homosilylcuprate preparation, 777
(Trimethylsilyl)methylcuprates
structure, 24, 27–28
supramolecular aggregation, 179, *180*
N(Trimethylsilyl)triethylphosphinimine, diorganocuprate formation, 100–101
3-Trimethylstannyl-2-alkenamides, 824, 827
Trimethylstannylcuprates, stannylcupration, 818
Trimethylstannylcyanocuprate, ^{13}C NMR spectroscopy, 818
Trimethylstannyl lithium, preparation, 818
Trinuclear copper(I) acetylides
electronic interactions, 267–268
oxidation potentials, 261–266
Trinuclear copper(I) diynyl compounds, oxidation potentials, 266–267
Trinuclear copper sites, ascorbate oxidase, 356, *357*
Triorganocopper(III) complexes, Gilman reaction, 112
Triorganozincates, preparation, 471, 473–474
Triphenylbismuth diacetate, O-nucleophile coupling, 946–947
2,4,6-Triphenylphenylcopper, structure, 49–50
Triphenylsilylcuprates, preparation, 780–781
Triphenylstannylcuprates, stannylcupration, 818
triphos (1,1,1-tris(diphenylphosphinomethyl)-ethane), 79–80
Triple bonds, stannyl–copper *syn* addition, 836, 837
Triple insertion, CF_2 units, 1002
1,1,1-Tris(diphenylphosphinomethyl)ethane (triphos), copper(I) acetylides, 79–80
Tris(triphenylphosphine)methylcopper, structure, 40, *41*
Tryptophan, gas-phase Cu(I)/Cu(II) reactivity, 320
Tunable leaving groups, diastereoselective allylic alkylation, 622, 659, 660
Tweezer-type ligands
copper(I) acetylides, 78, *79*
donor–acceptor complexes, 41–43
Tyrosine, gas-phase Cu(I)/Cu(II) reactivity, 320
L-Tyrosine derivatives, arylation, 897, 898

Ullmann chemistry
classical, 883–893
mechanistic considerations, 891–892, 942–943
modern, 893–943
non-Ullmann substrates, 943–965
Unactivated acetylenes/allenes, silylcuprate reactions, 803–817
Unsaturated compounds
nitrile hydrosilylation, 747–748
phosphoramidite complexes, 202
α,β-Unsaturated compounds
aldehydes, 450, 451, 459, 461
carboxylic acid derivatives, 388, 389
catalytic hydrosilylation, 738–749
chiral ligands, 741–749
non-chiral ligands, 738–741
conjugate addition
alkylidene malonic esters, 450
esters, 493, 495
Michael acceptors, 444
nitriles, 402
silylcuprates, 790–797

stannylcuprates, 822–831
vs S_N2 alkylation, 5
esters
α-cuprioazaenolate preparation, 428
transmetalation, 449, 478, 480, 493, 495, 517
(*E*)-α-trifluoromethyl, 1021–1022
ketones
asymmetric conjugate addition, 506
N-functionalized organocuprate formation, 416, 428, *430*, 433–434
organotin transmetalation, 509
organozinc transmetalation, 484, 486
organozirconium transmetalation, 452–453, 459, 461, 464
lactam conjugate additions, 474, 476, 477, 510, 512
(*E*)-α-trifluoromethyl-α,β-unsaturated esters, 1021–1022
Uracil
acidity, 280
binding energy, 291, *292*
organoboron compound coupling, 956, 957
Ureas
allenyl iodide coupling, 933, 934
Cu complex coordination, 301
organoboron compound coupling, 954

Vinylalanes, conjugate addition, 503, 504, 505, 506, 508–509
Vinylallenes
Michael addition, 868–869, 871
substitution, 874
Vinylbisstannanes, preparation, 838, 839
Vinylboronic acids, nucleobase *N*-arylation, 956
Vinyl bromides
amide coupling, 932–933
Br–I exchange, 942
perfluoroalkylation, 1014
Vinyl chlorides, amide coupling, 934
Vinylcopper acetylides, thermochemistry, 147
Vinylcopper–allylsilane species, 811, 812
Vinylcoppers
allylic alkylation, 622, 645, 662, 663
organocopper(I) copper(I) aggregates, 53–54, 58
stannylcupration intermediate, 836, 837
thermochemistry, 150–151
Vinylcuprates
intermediate compounds
silylcupration, 803, 804, 805, 809, 813–816
stannylcupration, 836, 837, 838, 839, 845, 846, 847–848
organotellurium transmetalation, 514–516
Vinyl ethers, C–O bond formation, 899, 901–902
Vinyl halides
amide coupling, 891, 927, 932–937, 938
inter/intramolecular amidation, 935
P-nucleophile cross-coupling, 893
Vinyl iodides
amide coupling, 932–933
arylstannane coupling, 975
α-(*N*-carbamoyl)alkylcuprate substitution, 388, 390
Hiyama-type cross-coupling, 977–978
P-nucleophile coupling, 938
Stille cross-coupling, 976
Vinyliodonium salts, phosphonate coupling, 945–946
Vinyllithium derivatives, stannylcupration, 845
Vinylmanganese chloride, conjugate addition, 450
Vinylogous carbonates, Fischer chromium carbene coupling, 451
Vinylogous esters, organomanganese addition, 449–450
2-Vinylphenyl Grignard reagent, organocopper(I) copper(I) aggregates, 58
Vinylphosphonates, vinyliodonium salt coupling, 945–946
Vinylsilanes
Cu^+ complex, 297, *298*, *299*
electrophilic substitution, 786, 787
epoxide rearrangement, 807
silylcupration, 803–809, 811, 812
E-Vinylsilanes, protonation, 803, 805
Z-Vinylsilanes, addition, 805
Vinylstannane–allylcuprate, intermediate, 846, 847
Vinylstannane–oxazolidine, intermediate, 842
Vinylstannanes
conjugate additions, 510, 512, 824, 826, 828
Corey's oxidation, 838, 840
regioselectivity, 836, 837
Stille reactions, 821, 838, 840, 841
substitution, 834, 835
transmetalation, 447, 509–510, 512, 828, 838, 839, 845–848
Vinyl sulfides, C–S bond formation, 904, 905
Vinyl sulfones, enantioselective conjugated hydrosilylation, 748
Vinyl sulfoxide, allylic epoxide formation, 629
Vinyltellurides, transmetalation, 514–516
Vinyl triflates
α-(*N*-carbamoyl)alkylcuprate substitution, 388, 390
stannylcuprate substitution, 834, 835, 836

XANES (x-ray absorption near edge spectra), 108

Xantphos ligand, organoboron transmetalation, 499, 500
X-ray absorption near edge spectra (XANES), cyanocuprates, 108
X-ray analysis, stannylcuprates, 820

Yamamoto reagents
  addition to extended Michael acceptors, 859, 867
  diastereospecific allylic alkylation, 622, 625
  substitution of extended substrates, 873
  transmetalation, 445, 456
Ynamides, vinyl/alkynyl halide coupling, 933
Ynoates, stannylcupration, 829
α,β-Ynoates, α-(N-carbamoyl)alkylcuprate addition, 388, 389
α,β-Ynones, α-(N-carbamoyl)alkylcuprate addition, 388, 389

Zimmerman–Traxler model
  organozirconium transmetalation, 459, 462
  β-silyl enolate diastereoselectivity, 796
Zinc
  Cu,Zn superoxide dismutase
    type 2 Cu centers, 350, *351*
    mononuclear Cu sites, 354–355
  Rieke zinc, 473, 592, 595
Zinc compounds *see* Organozinc reagents
Zinc enolates
  clavirolide C synthesis, 517–518
  conjugate additions, 474, 476, 477, 478, 479
  organozirconium transmetalation, 458, 461
Zinc organocuprates, oxidation, 592, 595
Zinc silylcuprates, acyl chloride substitution, 800, 802
Zirconacycles, organozirconium transmetalation, 462, 466–470
Zirconacyclopentadienes, transmetalation, 462, 466–469, 470
Zirconacyclopentanes
  alkyne carbocupration, 542, 557, 558, 573–574, 575
  transmetalation, 469–470
Zirconacyclopentene, spirocycle formation, 470
Zirconium compounds, transmetalation, 444, 452–470
Zirconocene dichloride, reduction, 459, 463
Zwitterions, glycine complex coordination, 302–303